CONVERSION FACTORS

LENGTH

1 in. = 2.54 cm
1 ft = 0.3048 m
1 mi = 5280 ft = 1.609 km
1 m = 3.281 ft
1 km = 0.6214 mi
1 angstrom (Å) = 10^{-10} m

MASS

1 slug = 14.59 kg
1 kg = 1000 grams = 6.852×10^{-2} slug
1 atomic mass unit (u) = 1.6605×10^{-27} kg
(1 kg has a weight of 2.205 lb where the
   acceleration due to gravity is 32.174 ft/s²)

TIME

1 day = 24 h = 1.44×10^3 min = 8.64×10^4 s
1 yr = 365.24 days = 3.156×10^7 s

SPEED

1 mi/h = 1.609 km/h = 1.467 ft/s = 0.4470 m/s
1 km/h = 0.6214 mi/h = 0.2778 m/s = 0.9113 ft/s

FORCE

1 lb = 4.448 N
1 N = 10^5 dynes = 0.2248 lb

WORK AND ENERGY

1 J = 0.7376 ft · lb = 10^7 ergs
1 kcal = 4186 J
1 Btu = 1055 J
1 kWh = 3.600×10^6 J
1 eV = 1.602×10^{-19} J

POWER

1 hp = 550 ft · lb/s = 745.7 W
1 W = 0.7376 ft · lb/s

PRESSURE

1 Pa = 1 N/m² = 1.450×10^{-4} lb/in.²
1 lb/in.² = 6.895×10^3 Pa
1 atm = 1.013×10^5 Pa = 1.013 bar =
   14.70 lb/in.² = 760 torr

VOLUME

1 liter = 10^{-3} m³ = 1000 cm³ = 0.03531 ft³
1 ft³ = 0.02832 m³ = 7.481 U.S. gallons
1 U.S. gallon = 3.785×10^{-3} m³ = 0.1337 ft³

ANGLE

1 radian = 57.30°
1° = 0.01745 radian

STANDARD PREFIXES USED TO DENOTE MULTIPLES OF TEN

Prefix	Symbol	Factor
Tera	T	10^{12}
Giga	G	10^9
Mega	M	10^6
Kilo	k	10^3
Hecto	h	10^2
Deka	da	10^1
Deci	d	10^{-1}
Centi	c	10^{-2}
Milli	m	10^{-3}
Micro	μ	10^{-6}
Nano	n	10^{-9}
Pico	p	10^{-12}
Femto	f	10^{-15}

BASIC MATHEMATICAL FORMULAE

Area of a circle = πr^2

Circumference of a circle = $2\pi r$

Surface area of a sphere = $4\pi r^2$

Volume of a sphere = $\frac{4}{3}\pi r^3$

Pythagorean theorem: $h^2 = h_o^2 + h_a^2$

Sine of an angle: $\sin\theta = h_o/h$

Cosine of an angle: $\cos\theta = h_a/h$

Tangent of an angle: $\tan\theta = h_o/h_a$

Law of cosines: $c^2 = a^2 + b^2 - 2ab\cos\gamma$

Law of sines: $a/\sin\alpha = b/\sin\beta = c/\sin\gamma$

Quadratic formula:
   If $ax^2 + bx + c = 0$, then, $x = (-b \pm \sqrt{b^2 - 4ac})/(2a)$

PHYSICS

JOHN D. CUTNELL & KENNETH W. JOHNSON

SOUTHERN ILLINOIS UNIVERSITY AT CARBONDALE

PHYSICS

FIFTH EDITION

JOHN WILEY & SONS, INC.

New York / Chichester / Weinheim / Brisbane / Singapore / Toronto

Acquisitions Editor   *Stuart Johnson*

Developmental Editor   *Suzanne Thibodeau*

Marketing Manager   *Sue Lyons*

Production Editor   *Barbara Russiello*

Cover and Text Designer   *Madelyn Lesure*

Photo Department Manager   *Hilary Newman*

Photo Researcher   *Elyse Rieder*

Illustration Editor   *Sigmund Malinowski*

Electronic Illustrations   *Precision Graphics*

Cover Photo   © *Karl Weatherly/Tony Stone Images, New York.*

This book was set in 10/12 Times Roman by Progressive Information Technologies and printed and bound by Von Hoffmann Press. The cover was printed by Lehigh Press.

Recognizing the importance of preserving what has been written, it is a policy of John Wiley & Sons, Inc. to have books of enduring value published in the United States printed on acid-free paper, and we exert our best efforts to that end.

This book is printed on acid free paper.

Main text 0471 32146-X (hardcover)
Volume 1 0471 38717-7 (paperback)
Volume 2 0471 38718-5 (paperback)
Set: Vol. 1/2 paperbacks 0471 39377-0
Wiley International Edition 0471 39219-7

Printed in the United States of America

10 9 8 7 6 5 4 3 2 1

To my wife Joan Cutnell,
my patient friend and my support
throughout this project.

To my wife Anne Johnson
and family, Rob, Rick and Cash, Lauri and Bryan.
One of life's great pleasures is being with warm
and caring people.

BRIEF CONTENTS

CONTENTS

PREFACE

We have written this text for students and teachers who are partners in a one-year algebra-based physics course. The positive feedback that we have received from many users of the fourth edition has encouraged us to make this new edition even more attuned to the needs of students and teachers. In taking a fresh look at the text, we have added new features and material and refined areas in need of improvement. In addition, we have expanded our Web sites considerably by including many new offerings for both students and teachers.

GOALS

Our first goal is to help students develop a conceptual understanding of physics principles. One of the challenges for teachers is to dispel the notion that physics is merely a collection of equations that can be used blindly to solve problems. Developing a good problem-solving technique starts with a firm grasp of the concepts and how they fit together to provide a coherent description of the physical world. The features in this edition that work toward this goal are *Concepts & Calculations* sections (a new feature), *Conceptual Examples,* and *Concepts-at-a-Glance* charts that illustrate diagrammatically the interrelationships between concepts.

Physicists know that the ability to reason in an organized manner is essential to solving problems, and helping students to improve their reasoning ability is our second goal. To this end, we have included *Interactive LearningWare* (a new on-line tutorial that steps students through 60 different problems), *Explicit Reasoning Steps* in all examples, and *Reasoning Strategies* that summarize the logical steps involved in solving certain classes of problems. A strong reasoning ability, combined with a good conceptual understanding, is the legacy that we wish to leave to our students.

Finally, we want to show students that physics principles come into play over and over again in their lives. It is always easier to learn something new if it is directly related to our lives. Direct applications of physics principles are identified in the margin by the label *The Physics of* . . . . Many of these applications are biomedical in nature and deal with human physiology. Others, such as CDs and DVDs, digital photography, and laser printers, deal with modern technology. We have also incorporated real-world situations into many of the worked-out examples and the homework material at the end of each chapter.

ORGANIZATION AND COVERAGE

The text consists of 32 chapters and is organized in a fairly standard fashion according to the following sequence: *Mechanics, Thermal Physics, Wave Motion, Electricity and Magnetism, Light and Optics,* and *Modern Physics.* Within each chapter, material that is a likely candidate for omission is typically located in a subsection at the end of a main section or in a separate section near the end of the chapter. Sections marked with an asterisk can be omitted with little impact on the overall development of the material.

Based on feedback from users and reviewers, we have added new material, made judicial deletions, and carefully tightened other material. The new material consists of the 60 *Interactive-LearningWare* problems and the *Concepts-&-Calcu-lations* sections (and associated homework problems) found at the end of each chapter. We have rearranged Chapter 10 by placing simple harmonic motion ahead of elasticity, reflecting the relative importance of these subjects. The discussion of the mole and molecular mass in Chapter 14 has been rewritten to improve clarity. We have updated our applications of physics principles and added a number of new ones. Many of these applications deal with cutting-edge technologies, such as charge-coupled devices and digital cameras, electronic ink, and digital movie projectors.

For instructors who wish to cover surface tension, we have included a module on our Web site (**http://www.wiley.com/college/cutnell**). This module discusses the nature of surface tension, capillary action, as well as the pressure inside a soap bubble and a liquid drop. Homework problems are also included for assignment.

In addition to the one-volume hardbound edition, the book is available as a two-volume paperback version. Volume I consists of Chapters 1–17 (*Mechanics, Thermal Physics,* and *Wave Motion*), while Volume 2 consists of Chapters 18–32 (*Electricity and Magnetism, Light and Optics,* and *Modern Physics*).

NEW FEATURES OF THE FIFTH EDITION

Interactive LearningWare   This new feature consists of 60 interactive calculational examples presented on our Web site (**http://www.wiley.com/college/cutnell**). References to the *LearningWare* are contained in nearly every chapter and include the problem statement, along with a short explanation of the purpose of each example. Each example is designed in an interactive format that consists of five steps: (1) Visualization, (2) Reasoning, (3) Knowns and Unknowns, (4) Modeling the Problem, and (5) Numerical Solution. We believe that these steps are essential to the development of problem-solving skills and that students will benefit from seeing them explicitly identified in each example. The Visualization step uses animations and drawings to help students understand the general nature of the situation being analyzed. The Reasoning step establishes the conceptual basis for the solution with the aid of multiple-choice questions and answers. The Knowns-and-Unknowns step deals with the important process of identifying all the variables and the symbols that represent them. This information is not simply given to the student but, rather, is interactively developed in the form of tables that evolve as the example proceeds. In the Modeling step, the student, using the results of the previous steps, models the problem with equations developed in the text. Finally, in the Solution step, the solution to the problem is obtained. Throughout this process, the student can identify exactly what step is being executed by referring to the highlighted button in the navigational tool that is always on-screen. These Interactive LearningWare examples include extensive audio

INTERACTIVE LEARNINGWARE #13

A person is standing on the edge of a cliff that is 8.10 m above a lake. Using a rope tied to a tree limb, he swings downward, lets go of the rope, and subsequently splashes into the water. He can let go of the rope either as it is swinging downward or a little later as it is swinging upward. Ignore air resistance and, for each case, determine the speed of the person just before entering the water.

This problem has two features that make it interesting: (1) It demonstrates that the conservation of mechanical energy can be used to find the speed of the person just before entering the water, even though a nonconservative force (the tension in the rope) is present. (2) It also shows that if the total mechanical energy is conserved, it is the same everywhere along a path. It does not matter which particular path the person follows into the water.

Go to our Web site at

http://www.wiley.com/college/cutnell

tracks, and, for convenience, they use the drag-and-drop technique to manipulate data and symbols in tables and drawings. The LearningWare performs intermediate modeling steps in a way that keeps students from getting lost in the details. For instance, when it is necessary to substitute another expression for part of the starting equation, an identifiable sub-screen appears in which the analysis is carried out. The main screen remains visible, however, and is used again only when the intermediate work is completed. In summary, the hallmark of these Interactive-LearningWare examples is a user-friendly clarity that allows the student to see how conceptual understanding and quantitative analysis work together in a problem-solving environment.

Concepts & Calculations   Conceptual understanding is an essential part of a good problem solving technique. To emphasize the role of conceptual understanding, we have included in every chapter a new section, entitled *Concepts & Calculations*. These sections are organized around a special type of example, each of which begins with several conceptual questions that are answered before the quantitative problem is worked out. The purpose of the questions is to put into sharp focus the concepts with which the problem deals. We hope that these examples will serve as models for students as they do their homework. These examples also provide mini-reviews of material studied earlier in the chapter and in previous chapters. There are 64 of these special examples in the text, in addition to the 82 purely conceptual examples and the 276 standard quantitative examples.

Concepts & Calculations — Group Learning Problems   Many students learn more quickly when working in small groups. These groups, often consisting of 3–5 students, can meet out of class or as part of an in-class lecture/discussion strategy. A group usually focuses on a well-defined topic, and members are free to put their ideas "on the table." Such peer-instruction works well when a particular problem is being solved, especially when the "plug and chug" impulse is avoided by having the group discuss the concepts before attempting a numerical solution.

To provide the stimulus for small group discussions, a new type of problem set, called *Concepts & Calculations—Group Learning Problems,* has been added at the end of the Problems section in each chapter. The problems in this set have been chosen to cover the main ideas in the chapter and to provide a useful review of the material. Each problem is organized in a two-part format that begins with a series of conceptual questions and ends with a related quantitative problem. The questions are designed to help the group identify and focus on the relevant concepts. The quantitative problem, often in the form of multiple parts, provides a numerical check on the conclusions drawn by the group in discussing the conceptual questions.

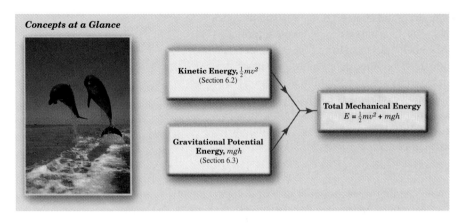

Concepts at a Glance

OTHER FEATURES OF THE FIFTH EDITION

Concepts at a Glance   To provide a coherent picture of how new concepts are built upon previous ones and to reinforce fundamental unifying ideas, the text contains 66 *Concepts-at-a-Glance* flowcharts. These charts show in a visual way the conceptual development of physics principles. Within a chart, new concepts are placed in gold panels, while previously introduced and related concepts are placed in light blue panels. Each chart is discussed in a separate and highlighted paragraph that helps students connect the new concept with the real world.

Conceptual Examples   A good problem-solving technique begins with a foundation of conceptual understanding. Therefore, the text includes 82 examples that are entirely conceptual in nature. These examples are found in every chapter and worked out in a rigorous, but qualitative fashion, with no (or very few) equations. Our intent is to provide students with explicit models of how to "think through" a problem before attempting to solve it numerically. The *Conceptual Examples* deal with a wide range of topics and a large number of issues that often confuse students. Wherever possible, we have focused on real-world situations and have structured the examples so that they lead naturally to homework material found at the ends of the chapters. The related homework material is explicitly identified, and that material contains a cross reference that encourages students to review the pertinent *Conceptual Example.*

CONCEPTUAL EXAMPLE 7
Deceleration Versus Negative Acceleration

A car is traveling along a straight road and is decelerating. Does the car's acceleration a necessarily have a negative value?

Reasoning and Solution   We begin with the meaning of the term "decelerating," which has nothing to do with whether the acceleration a is positive or negative. The term means only that the acceleration vector points opposite to the velocity vector and indicates that the moving object is slowing down. When a moving object slows down, its instantaneous speed (the magnitude of the instantaneous velocity) decreases. One possibility is that the velocity vector of the car points to the right, in the positive direction, as Figure 2.13*a* shows. The term "decelerating" implies that the acceleration vector points opposite, or to the left, which is the negative direction. Here, the value of the acceleration a would indeed be negative. However, there is another possibility. The car could be traveling to the left, as in Figure 2.13*b*. Now, since the velocity vector points to the left, the acceleration vector would point opposite or to the right, according to the meaning of the term "decelerating." But right is the positive direction, so the acceleration a would have a positive value in Figure 2.13*b*. We see, then, that *a decelerating object does not necessarily have a negative acceleration.*

Related Homework Material: *Problems 20 and 38*

Explicit Reasoning Steps   Since reasoning is the cornerstone of problem solving, we believe that students will benefit from seeing the reasoning stated explicitly. Therefore, the format in which examples are worked out includes an explicit reasoning step. In this step we explain what motivates our procedure for solving the problem before any algebraic or numerical work is done. In the new *Concepts-&-Calculations* examples, the reasoning is included in the form of conceptual questions and answers.

Reasoning Strategies
A number of the examples in the text deal with well-defined strategies for solving certain types of problems. In such cases, we have included summaries of the steps involved. These summaries, which are titled *Reasoning Strategies,* encourage frequent review of the techniques used and help students focus on the related concepts.

> **REASONING STRATEGY**
>
> **Applying the Principle of Conservation of Linear Momentum**
>
> **1.** Decide which objects are included in the system.
>
> **2.** Relative to the system that you have chosen, identify the internal forces and the external forces.
>
> **3.** Verify that the system is isolated. In other words, verify that the sum of the external forces applied to the system is zero. Only if this sum is zero can the conservation principle be applied. If the sum of the external forces is not zero, consider a different system for analysis.
>
> **4.** Set the total final momentum of the isolated system equal to the total initial momentum. Remember that linear momentum is a vector. If necessary, apply the conservation principle separately to the various vector components.

The Physics of . . .
This edition contains over 210 real-world applications that reflect our commitment to show students how prevalent physics is in their lives. Each application is identified in the margin with the label *The Physics of . . . ,* and those that deal with biomedical material are further marked with an icon in the shape of a microscope ⚕. New applications in a variety of areas have been added, such as a pyroelectric ear thermometer, charge-coupled devices and digital cameras, electronic ink, digital movie projectors and micromirrors, and bursting water pipes. The range of applications is wide, including, for example, the Global Positioning System, Next Generation Weather Radar or NEXRAD, magnetohydrodynamic propulsion, and automobile features. Many biomedical applications are discussed, such as Gamma Knife radiosurgery, cochlear implants, endoscopy, photorefractive keratectomy, and others. A complete list of the applications can be found on pages xxii–xxiv.

THE PHYSICS OF...

a pyroelectric ear thermometer.

Problem Solving Insights
To reinforce the problem-solving techniques illustrated in the worked-out examples, we have included short statements in the margins, identified by the label *Problem Solving Insight.* These statements help students to develop good problem-solving skills by providing the kind of advice that a teacher gives when explaining a calculation in detail.

> ● **PROBLEM SOLVING INSIGHT**
> The electrostatic force is a vector and has a direction as well as a magnitude. When adding electrostatic forces, take into account the directions of all forces, using vector components as needed.

Homework Problems and Conceptual Questions
The fifth edition contains over 2400 problems and nearly 600 conceptual questions for assignment as homework. About 20% of the problems are new or modified. In providing so many problems and questions, we have used a wide variety of real-world situations with realistic data. The problems are ranked according to difficulty, with the most difficult problems marked with a double asterisk (**) and those of intermediate difficulty marked with a single asterisk (*). The easiest problems are unmarked. Those whose solutions appear in the *Student Solutions Manual* are identified with the label **ssm**. Those whose solutions are available on the World Wide Web (**http://www.wiley.com/college/cutnell**) are marked with the label **www**. Problems and questions that are biomedical in nature are identified with an icon in the shape of a microscope ⚕. The problems are organized in several groupings. Most are arranged by section. However, teachers sometimes want to assign homework without identifying a particular section. Therefore, under the heading *Additional Problems,* we have provided a group of problems without references to sections. A new grouping of problems has been added to the fifth edition

under the heading *Concepts & Calculations—Group Learning Problems.* Each problem in this category begins with conceptual questions that lead into a quantitative problem. The questions are designed to help students identify the concepts that are used in solving the problem and provide a focus for group discussions.

•••

In spite of our best efforts to produce an error-free book, there are no doubt errors that still remain. They are solely our responsibility, and we would appreciate hearing of any that you find. We hope that this text makes learning and teaching physics easier and more enjoyable and look forward to hearing about your experiences with it. Please feel free to write us care of Physics Editor, College Division, John Wiley & Sons, Inc., 605 Third Avenue, New York, NY 10158-0012 or contact us at **http://www.wiley.com/college/cutnell.**

SUPPLEMENTS

An extensive package of supplements to accompany *Physics,* fifth edition, is available to assist both the teacher and the student.

INSTRUCTOR'S SUPPLEMENTS

Instructor's Resource Guide   by David T. Marx, Southern Illinois University at Carbondale. This guide contains an extensive listing of Web-based physics education resources. It also includes teaching ideas, lecture notes, demonstration suggestions, alternative syllabi for courses of different lengths and emphasis, as well as conversion notes that allow instructors to use their class notes from other texts. For instructors who have homework problems from the fourth edition that they especially like to assign, there is a Problem Locator Guide. This locator guide provides an easy way to correlate fourth edition problem numbers with the corresponding fifth edition numbers.

Instructor's Solutions Manual   by John D. Cutnell and Kenneth W. Johnson. This manual provides worked-out solutions for all end-of-chapter Conceptual Questions and Problems. Volume 1 contains the solutions for Chapters 1–17. Volume 2 contains the solutions for Chapters 18–32.

Test Bank   by David T. Marx, Southern Illinois University at Carbondale. This manual includes more than 2200 multiple-choice questions. These items are also available in the *Computerized Test Bank* (see below).

Instructor's Resource CD-ROM.   This CD contains:

* The entire *Instructor's Solutions Manual* in both Microsoft Word © (IBM and Macintosh) and PDF files.

* A *Computerized Test Bank,* for use with both IBM compatible and Macintosh PC's with full editing features to help you customize tests.

* Multiple-choice format for 1,200 of the end-of-chapter problems in the text. This material allows instructors to assign homework in a machine-gradable format.

* All text illustrations, suitable for both classroom projection, printing, and web posting.

* Simulations for classroom demonstrations.

Transparencies.   More than 200 four-color illustrations from the text are provided in a form suitable for projection in the classroom.

On-line Homework and Quizzing.   *Physics,* fifth edition, supports WebAssign, CAPA, and eGrade, which are programs that give instructors the ability to deliver and grade homework and quizzes over the Internet.

WebCT.   A variety of course management materials has been prepared in a WebCT-ready format. WebCT is a powerful course management program that allows instructors to set up a complete on-line course with chat rooms, bulletin boards, quizzing, as well as built in student tracking tools, etc.

STUDENT'S SUPPLEMENTS

Student Web Site.   This Web site, **http://www.wiley.com/college/cutnell,** was developed specifically for *Physics,* fifth edition, and is designed to further assist students in the study of physics. At this site, students can access the *Interactive LearningWare* tutorials that are described earlier in this Preface. The site also includes solutions to selected end-of-chapter problems, which are identified with a **www** icon in the text.

Student Study Guide   by John D. Cutnell, Kenneth W. Johnson, and the late Mark J. Comella. This student study guide consists of traditional print materials and an accompanying web site that together provide a rich, interactive environment for review and study. The web site includes self-quizzes, simulation exercises, additional *Interactive LearningWare* tutorials, review material for the MCAT exam, and links to other Web sites that offer physics tutorial help.

Student Solutions Manual   by John D. Cutnell and Kenneth W. Johnson. This manual provides students with complete worked-out solutions for approximately 600 of the odd-numbered problems found at the end of each chapter within the text. These problems are indicated in the text with an **ssm** icon.

Cutnell and Johnson Multimedia 2.0.   This CD-ROM-based version of *Physics,* fifth edition, contains the complete text, the *Student Study Guide,* the *Student Solutions Manual,* the *Interactive LearningWare,* hints for solving end-of-chapter problems, and numerous simulations all connected with extensive hyperlinking.

Take Note!   This art notebook lets students take notes directly onto 8.5 × 11, black-and-white versions of textbook illustrations. All of the illustrations from the transparency set are included. This supplement substantially reduces time spent in class copying illustrations.

ACKNOWLEDGMENTS

In putting together this text we have had the help of many individuals. It has been an exhilarating experience to work with them and witness their talents, dedication, and professionalism. It is a pleasure to acknowledge their contributions.

Special thanks go to our editor, Stuart Johnson, for his expert and on-target advice at all stages of the project. We are fortunate to have had his help and encouragement. His keen understanding of college publishing is an asset to be treasured.

Suzanne Thibodeau, our developmental editor, is a joy to work with. Her sharp eye for details, her aesthetic judgment, and her articulate common sense contributed immeasurably to the final product. Thanks, Suzanne.

We are especially indebted to Barbara Russiello, production editor, for coordinating the all-electronic production of the book. She was always there for us with answers to our questions, help, and encouragement. Her tireless efforts made it possible to produce a book that meets the highest standards of production.

In modifying and adding to the illustrations for this edition, we worked with Sigmund Malinowski of the illustrations department. He was always ready and willing to help with advice and suggestions. We are grateful for his efforts and enjoy working with him immensely.

The design of the text and its cover is the work of Maddy Lesure, design director. The clear and pleasing presentation of the text's many features and elements is due to her immense talent in coordinating color, design, and fonts. It has been a pleasure, Maddy.

We are grateful to Hilary Newman, photo department manager, and Elyse Rieder, photo researcher, for the photos in the text. They were able to find great photos that express just the right physics. We think that you will like the photos as much as we do.

Outstanding proofreaders are essential in producing an error-free book. Our thanks go to Betty Pessagno for catching all types of errors from awkwardly worded sentences to wrong fonts.

Our sincere gratitude goes to Jennifer Cerciello for coordinating the extensive supplements package and to Virginia Dunn for copyediting the manuscript.

The sales representatives of John Wiley & Sons, Inc. are a very special group. We deeply appreciate their efforts on our behalf as they are in constant contact with physics departments throughout the country. In particular, we wish to acknowledge Jane Doub, senior sales representative and "rep" for our area. She regularly stops by to give us user feedback, and we try out many new ideas on her. We are fortunate to be able to take advantage of her considerable expertise.

Many physicists have generously shared their ideas with us about good pedagogy and helped us by pointing out errors in our presentation. For all of their suggestions we are grateful. They have helped us to write more clearly and accurately and have influenced markedly the evolution of this text. To the reviewers of this and previous editions, we especially owe a large debt of gratitude. Specifically, we thank the reviewers who provided us guidance in preparing the fifth edition:

Zaven Altounian, *McGill University*
Doyle Davis, *New Hampshire Community Technical College*
Greg Francis, *Montana State University*
William Gregg, *Louisiana State University*

Rod Nave, *Georgia State University*

Barry C. Robertson, *Queens University (Ontario)*

Larry Rowan, *University of North Carolina*

Roy S. Rubin, *University of Texas, Arlington*

Patrick Tam, *Humboldt State University*

Finally, we gratefully acknowledge the students who participated in focus groups during the testing stages of the Interactive LearningWare; their comments were insightful and enlightening: Issam Alfalah, Robert D. Eugenio, Lawrence Ettarh, Christian Gobo, Kristine Mecca, Shoheen Owhady, and Saif Salim of Bergen County Community College, and Sushil Battarai of Connecticut College.

THE PHYSICS OF . . .

THE PHYSICS OF...

🔬 applications of physics principles.

To show students that physics has a widespread impact on their lives, we have included a large number of applications of physics principles. Many of these applications are not found in other texts. The most important ones are listed below along with the page number locating the corresponding discussion. They are identified in the margin of the page on which they occur with the label "**THE PHYSICS OF . . .**" Biomedical applications are marked with an icon in the shape of a microscope 🔬. The discussions are integrated into the text, so that they occur as a natural part of the physics being presented. It should be noted that the list is not a complete list of all the applications of physics principles to be found in the text. There are many additional applications that are discussed only briefly or occur in the homework questions and problems.

In the movie *Star Wars Episode I — The Phantom Menace*, these battle droids were created using animation techniques that rely heavily on computers and mathematical concepts such as trigonometry and vectors. Trigonometry and vectors will be useful throughout this book in dealing with the laws of physics.

CHAPTER 1

INTRODUCTION AND MATHEMATICAL CONCEPTS

1.1 THE NATURE OF PHYSICS

The science of physics has developed out of the efforts of men and women to explain our physical environment. These efforts have been so successful that the laws of physics now encompass a remarkable variety of phenomena, including planetary orbits, radio and TV waves, magnetism, and lasers, to name just a few.

The laws of physics are equally remarkable for their scope. They describe the behavior of particles many times smaller than an atom and objects many times larger than our sun. The same laws apply to the heat generated by a burning match and the heat generated by a rocket engine. The same laws guide an astronomer in using the light from a distant star to determine how fast the star is moving and a police officer in using radar to catch a speeder. Physics can be applied fruitfully to objects as different as subatomic particles, distant stars, or speeding automobiles because it focuses on issues that are truly basic to the way nature works.

The strength of physics derives from the fact that its laws are based on experiment. This is not to say that intuition and educated guesses are unimportant. The great creative geniuses in science, as in art and music, work in leaps and bounds that no one can fully understand. In physics, however, a flash of insight never becomes accepted law unless its implications can be verified by experiment. This insistence on experimental verification has enabled physicists to build a rational and coherent understanding of nature.

The exciting feature of physics is its capacity for predicting how nature will behave in one situation on the basis of experimental data obtained in another situation. Such predictions place physics at the heart of modern technology and, therefore, can have a tremendous impact on our lives. Rocketry and the development of space travel have their roots firmly planted in the physical laws of Galileo Galilei (1564–1642) and Isaac Newton (1642–1727). The transportation industry relies heavily on physics in the development of engines and the design of aerodynamic vehicles. Entire electronics and computer industries owe their existence to the invention of the transistor, which grew directly out of the laws of physics that describe the electrical behavior of solids. The telecommunications industry depends extensively on electromagnetic waves, whose existence was predicted by James Clerk Maxwell (1831–1879) in his theory of electricity and magnetism. The medical profession uses X-ray, ultrasonic, and magnetic resonance methods for obtaining images of the interior of the human body, and physics lies at the core of all these. Perhaps the most widespread impact in modern technology is that due to the laser. Fields ranging from space exploration to medicine benefit from this incredible device, which is a direct application of the principles of atomic physics.

Because physics is so fundamental, it is a required course for students in a wide range of major areas. We welcome you to the study of this fascinating topic. You will learn how to see the world through the "eyes" of physics and to reason as a physicist does. In the process, you will learn how to apply physics principles to a wide range of problems. We hope that you will come to recognize that physics has important things to say about your environment.

1.2 UNITS

Physics experiments involve the measurement of a variety of quantities, and a great deal of effort goes into making these measurements as accurate and reproducible as possible. The first step toward ensuring accuracy and reproducibility is defining the units in which the measurements are made.

TABLE 1.1   Units of Measurement

	System		
	SI	CGS	BE
Length	meter (m)	centimeter (cm)	foot (ft)
Mass	kilogram (kg)	gram (g)	slug (sl)
Time	second (s)	second (s)	second (s)

In this text, we emphasize the system of units known as **SI units,** which stands for the French phrase "Le **S**ystème **I**nternational d'Unités." By international agreement, this system employs the **meter** (m) as the unit of length, the **kilogram** (kg) as the unit of mass, and the **second** (s) as the unit of time. Two other systems of units are also in use today, however. The CGS system utilizes the centimeter (cm), the gram (g), and the second for length, mass, and time, respectively, and the BE or British Engineering system (the gravitational version) uses the foot (ft), the slug (sl), and the second. Table 1.1 summarizes the units used for length, mass, and time in the three systems.

Originally, the meter was defined in terms of the distance measured along the earth's surface between the north pole and the equator. Eventually, a more accurate measurement standard was needed, and by international agreement the meter became the distance between two marks on a bar of platinum–iridium alloy (see Figure 1.1) kept at a temperature of 0 °C. Today, to meet further demands for increased accuracy, the meter is defined as the distance that light travels in a vacuum in a time of 1/299 792 458 second. This definition arises because the speed of light is a universal constant that is defined to be 299 792 458 m/s.

The definition of a kilogram as a unit of mass has also undergone changes over the years. As Chapter 4 discusses, the mass of an object indicates the tendency of the object to continue in motion with a constant velocity. Originally, the kilogram was expressed in terms of a specific amount of water. Today, one kilogram is defined to be the mass of a standard cylinder of platinum–iridium alloy, like the one in Figure 1.2.

As with the units for length and mass, the present definition of the second as a unit of time is different from the original definition. Originally, the second was defined according to the average time for the earth to rotate once about its axis, one day being set equal to 86 400 seconds. The earth's rotational motion was chosen because it is naturally repetitive, occurring over and over again. Today, we still use a naturally occurring repetitive phenomenon to define the second, but of a very different kind. We use the electromagnetic waves emitted by cesium-133 atoms in an atomic clock like that in Figure 1.3. One second is defined as the time needed for 9 192 631 770 wave cycles to occur.*

The units for length, mass, and time, along with a few other units that will arise later, are regarded as **base** SI units. The word "base" refers to the fact that these units are used along with various laws to define additional units for other important physical quantities, such as force and energy. The units for such other physical quantities are referred to as **derived** units, since they are combinations of the base units. Derived units will be introduced from time to time, as they arise naturally along with the related physical laws.

---

* See Chapter 16 for a discussion of waves in general and Chapter 24 for a discussion of electromagnetic waves in particular.

Figure 1.1 The standard platinum–iridium meter bar.

Figure 1.2 The standard platinum–iridium kilogram is kept at the International Bureau of Weights and Measures in Sèvres, France.

Figure 1.3 A cesium atomic clock.

The value of a quantity in terms of base or derived units is sometimes a very large or very small number. In such cases, it is convenient to introduce larger or smaller units that are related to the normal units by multiples of ten. Table 1.2 summarizes the prefixes that are used to denote multiples of ten. For example, 1000 or 10^3 meters are referred to as 1 kilometer (km), and 0.001 or 10^{-3} meter is called 1 millimeter (mm). Similarly, 1000 grams and 0.001 gram are referred to as 1 kilogram (kg) and 1 milligram (mg), respectively. Appendix A contains a discussion of scientific notation and powers of ten, such as 10^3 and 10^{-3}.

1.3 THE ROLE OF UNITS IN PROBLEM SOLVING

The Conversion of Units

Since any quantity, such as length, can be measured in several different units, it is important to know how to convert from one unit to another. For instance, the foot can be used to express the distance between the two marks on the standard platinum–iridium meter bar. There are 3.281 feet in one meter, and this number can be used to convert from meters to feet, as the following example demonstrates.

EXAMPLE 1   The World's Highest Waterfall

The highest waterfall in the world is Angel Falls in Venezuela, with a total drop of 979.0 m (see Figure 1.4). Express this drop in feet.

Reasoning When converting between units, we write down the units explicitly in the calculations and treat them like any algebraic quantity. In particular, we will take advantage of the following algebraic fact: Multiplying or dividing an equation by a factor of 1 does not alter an equation.

Solution   Since 3.281 feet = 1 meter, it follows that (3.281 feet)/(1 meter) = 1. Using this factor of 1 to multiply the equation "Length = 979.0 meters," we find that

$$\text{Length} = (979.0 \text{ meters})(1) = (979.0 \text{ meters})\left(\frac{3.281 \text{ feet}}{1 \text{ meter}}\right) = \boxed{3212 \text{ feet}}$$

The colored lines emphasize that the units of meters behave like any algebraic quantity and cancel when the multiplication is performed, leaving only the desired unit of feet to describe the answer. In this regard, note that 3.281 feet = 1 meter also implies that (1 meter)/(3.281 feet) = 1. However, we chose not to multiply by a factor of 1 in this form, because the units of meters would not have canceled.

A calculator gives the answer as 3212.099 feet. Standard procedures for significant figures, however, indicate that the answer should be rounded off to four significant figures, since the value of 979.0 meters is accurate to only four significant figures. In this regard, the "1 meter" in the denominator does not limit the significant figures of the answer, because this number is precisely one meter by definition of the conversion factor. Appendix B contains a review of significant figures.

In any conversion, if the units do not combine algebraically to give the desired result, the conversion has not been carried out properly. With this in mind, the next example stresses the importance of writing down the units and illustrates a typical situation in which several conversions are required.

TABLE 1.2   Standard Prefixes Used to Denote Multiples of Ten

Prefix	Symbol	Factor[a]
Tera	T	10^{12}
Giga[b]	G	10^{9}
Mega	M	10^{6}
Kilo	k	10^{3}
Hecto	h	10^{2}
Deka	da	10^{1}
Deci	d	10^{-1}
Centi	c	10^{-2}
Milli	m	10^{-3}
Micro	μ	10^{-6}
Nano	n	10^{-9}
Pico	p	10^{-12}
Femto	f	10^{-15}

[a] Appendix A contains a discussion of powers of ten and scientific notation.
[b] Pronounced jig′a.

EXAMPLE 2   Interstate Speed Limit

Express the speed limit of 65 miles/hour in terms of meters/second.

Reasoning As in Example 1, it is important to write down the units explicitly in the calculations and treat them like any algebraic quantity. Here, two well-known relationships come into play, namely, 5280 feet = 1 mile and 3600 seconds = 1 hour. As a result, (5280 feet)/(1 mile) = 1 and (3600 seconds)/(1 hour) = 1. In our solution we will use the fact that multiplying and dividing by these factors of unity does not alter an equation.

Solution Multiplying and dividing by factors of unity, we find the speed limit in feet per second as shown below:

$$\text{Speed} = \left(65\,\frac{\text{miles}}{\text{hour}}\right)(1)(1) = \left(65\,\frac{\cancel{\text{miles}}}{\cancel{\text{hour}}}\right)\left(\frac{5280\text{ feet}}{1\text{ \cancel{mile}}}\right)\left(\frac{1\text{ \cancel{hour}}}{3600\text{ s}}\right) = 95\,\frac{\text{feet}}{\text{second}}$$

To convert feet into meters, we use the fact that (1 meter)/(3.281 feet) = 1:

$$\text{Speed} = \left(95\,\frac{\text{feet}}{\text{second}}\right)(1) = \left(95\,\frac{\cancel{\text{feet}}}{\text{second}}\right)\left(\frac{1\text{ meter}}{3.281\text{ \cancel{feet}}}\right) = \boxed{29\,\frac{\text{meters}}{\text{second}}}$$

Figure 1.4 Angel Falls in Venezuela is the highest waterfall in the world.

In addition to their role in guiding the use of conversion factors, units serve a useful purpose in solving problems. They can provide an internal check to eliminate errors, if they are carried along during each step of a calculation and treated like any algebraic factor. In particular, remember that ***only quantities with the same units can be added or subtracted.*** Thus, at one point in a calculation, if you find yourself adding 12 miles to 32 kilometers, stop and reconsider. Either miles must be converted into kilometers or kilometers must be converted into miles before the addition can be carried out.

A collection of useful conversion factors is given on the page facing the inside of the front cover. The reasoning strategy that we have followed in Examples 1 and 2 for converting between units is outlined as follows:

REASONING STRATEGY

Converting Between Units

1. In all calculations, write down the units explicitly.

2. Treat all units as algebraic quantities. In particular, when identical units are divided, they are eliminated algebraically.

3. Use the conversion factors located on the page facing the inside of the front cover. In your calculations, be guided by the fact that multiplying or dividing an equation by a factor of 1 does not alter the equation. For instance, the conversion factor of 3.281 feet = 1 meter might be applied in the form (3.281 feet)/(1 meter) = 1. This factor of 1 would be used to multiply an equation such as "Length = 5.00 meters" in order to convert meters to feet.

4. Check to see that your calculations are correct by verifying that the units combine algebraically to give the desired unit for the answer. Only quantities with the same units can be added or subtracted.

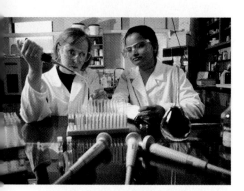

These scientists are using an automatic pipette to deliver a fixed volume of liquid into a test tube.

Dimensional Analysis

We have seen that many quantities are denoted by specifying both a number and a unit. For example, the distance to the nearest telephone may be 8 meters, or the speed of a car might be 25 meters/second. Each quantity, according to its physical nature, requires a certain *type* of unit. Distance must be measured in a length unit such as meters, feet, or miles, and a time unit will not do. Likewise, the speed of an object must be specified as a length unit divided by a time unit. In physics, the term *dimension* is used to refer to the physical nature of a quantity and the type of unit used to specify it. Distance has the dimension of length, which is symbolized as [L], while speed has the dimensions of length [L] divided by time [T], or [L/T]. Many physical quantities can be expressed in terms of a combination of fundamental dimensions such as length [L], time [T], and mass [M]. Later on, we will encounter certain other quantities, such as temperature, which are also fundamental. A fundamental quantity like temperature cannot be expressed as a combination of the dimensions of length, time, mass, or any other fundamental dimension.

Dimensional analysis is used to check mathematical relations for the consistency of their dimensions. As an illustration, consider a car that starts from rest and accelerates to a speed v in a time t. Suppose we wish to calculate the distance x traveled by the car but are not sure whether the correct relation is $x = \frac{1}{2}vt^2$ or $x = \frac{1}{2}vt$. We can decide by checking the quantities on both sides of the equals sign to see if they have the same dimensions. If the dimensions are not the same, the relation is incorrect. For $x = \frac{1}{2}vt^2$, we use the dimensions for distance [L], time [T], and speed [L/T] in the following way:

$$x = \tfrac{1}{2}vt^2$$

Dimensions
$$[L] \overset{?}{=} \left[\frac{L}{\cancel{T}}\right][T]^{\cancel{2}} = [L][T]$$

Dimensions cancel just like algebraic quantities, and pure numerical factors like $\frac{1}{2}$ have no dimensions, so they can be ignored. The dimension on the left of the equals sign does not match those on the right, so the relation $x = \frac{1}{2}vt^2$ cannot be correct. On the other hand, applying dimensional analysis to $x = \frac{1}{2}vt$, we find that

$$x = \tfrac{1}{2}vt$$

Dimensions
$$[L] \overset{?}{=} \left[\frac{L}{\cancel{T}}\right][\cancel{T}] = [L]$$

The dimension on the left of the equals sign matches that on the right, so this relation is dimensionally correct. If we know that one of our two choices is the right one, then $x = \frac{1}{2}vt$ is it. In the absence of such knowledge, however, dimensional analysis cannot identify the correct relation. It can only identify which choices *may be* correct, since it does not account for numerical factors like $\frac{1}{2}$ or for the manner in which an equation was derived from physics principles.

1.4 TRIGONOMETRY

Scientists use mathematics to help them describe how the physical universe works, and trigonometry is an important branch of mathematics. Three trigonometric functions are utilized throughout this text. They are the sine, the cosine, and the tangent of the angle θ (Greek theta), abbreviated as sin θ, cos θ, and tan θ, respectively. These functions are defined below in terms of the symbols given along with the right triangle in Figure 1.5.

Figure 1.5 A right triangle.

● **DEFINITION OF SIN θ, COS θ, AND TAN θ**

$$\sin \theta = \frac{h_{o}}{h} \qquad (1.1)$$

$$\cos \theta = \frac{h_{a}}{h} \qquad (1.2)$$

$$\tan \theta = \frac{h_{o}}{h_{a}} \qquad (1.3)$$

h = length of the **hypotenuse** of a right triangle
h_{o} = length of the side **opposite** the angle θ
h_{a} = length of the side **adjacent** to the angle θ

The sine, cosine, and tangent of an angle are numbers without units, because each is the ratio of the lengths of two sides of a right triangle. Example 3 illustrates a typical application of Equation 1.3.

EXAMPLE 3   Using Trigonometric Functions

On a sunny day, a tall building casts a shadow that is 67.2 m long. The angle between the sun's rays and the ground is $\theta = 50.0°$, as Figure 1.6 shows. Determine the height of the building.

Reasoning We want to find the height of the building. Therefore, we begin with the colored right triangle in Figure 1.6 and identify the height as the length h_{o} of the side opposite the angle θ. The length of the shadow is the length h_{a} of the side that is adjacent to the angle θ. The ratio of the length of the opposite side to the length of the adjacent side is the tangent of the angle θ, which can be used to find the height of the building.

Solution We use the tangent function in the following way, with $\theta = 50.0°$ and $h_{a} = 67.2$ m:

$$\tan \theta = \frac{h_{o}}{h_{a}} \qquad (1.3)$$

$$h_{o} = h_{a} \tan \theta = (67.2 \text{ m})(\tan 50.0°) = (67.2 \text{ m})(1.19) = \boxed{80.0 \text{ m}}$$

The value of tan 50.0° is found by using a calculator.

Figure 1.6 From a value for the angle θ and the length h_{a} of the shadow, the height h_{o} of the building can be found using trigonometry.

The sine, cosine, or tangent may be used in calculations such as that in Example 3, depending on which side of the triangle has a known value and which side is asked for. However, *the choice of which side of the triangle to label h_o (opposite) and which to label h_a (adjacent) can be made only after the angle θ is identified.*

Often the values for two sides of the right triangle in Figure 1.5 are available, and the value of the angle θ is unknown. The concept of *inverse trigonometric functions* plays an important role in such situations. Equations 1.4–1.6 give the inverse sine, inverse cosine, and inverse tangent in terms of the symbols used in the drawing. For instance, Equation 1.4 is read as "θ equals the angle whose sine is h_o/h."

$$\theta = \sin^{-1}\left(\frac{h_o}{h}\right) \tag{1.4}$$

$$\theta = \cos^{-1}\left(\frac{h_a}{h}\right) \tag{1.5}$$

$$\theta = \tan^{-1}\left(\frac{h_o}{h_a}\right) \tag{1.6}$$

The use of "-1" as an exponent in Equations 1.4–1.6 *does not mean* "take the reciprocal." For instance, $\tan^{-1}(h_o/h_a)$ does not equal $1/\tan(h_o/h_a)$. Another way to express the inverse trigonometric functions is to use arc sin, arc cos, and arc tan instead of $\sin^{-1}$, $\cos^{-1}$, and $\tan^{-1}$. Example 4 illustrates the use of an inverse trigonometric function.

EXAMPLE 4   Using Inverse Trigonometric Functions

A lakefront drops off gradually at an angle θ, as Figure 1.7 indicates. For safety reasons, it is necessary to know how deep the lake is at various distances from the shore. To provide some information about the depth, a lifeguard rows straight out from the shore a distance of 14.0 m and drops a weighted fishing line. By measuring the length of the line, the lifeguard determines the depth to be 2.25 m. (a) What is the value of θ? (b) What would be the depth d of the lake at a distance of 22.0 m from the shore?

Reasoning   Near the shore, the lengths of the opposite and adjacent sides of the right triangle in Figure 1.7 are $h_o = 2.25$ m and $h_a = 14.0$ m, relative to the angle θ. Having made this identification, we can use the inverse tangent to find the angle in part (a). For part (b) the opposite and adjacent sides farther from the shore become $h_o = d$ and $h_a = 22.0$ m. With the value for θ obtained in part (a), the tangent function can be used to find the unknown depth. Considering the way in which the lake bottom drops off in Figure 1.7, we expect the unknown depth to be greater than the value of 2.25 m that applies nearer the shore.

Solution
(a) Using the inverse tangent given in Equation 1.6, we find that

$$\theta = \tan^{-1}\left(\frac{h_o}{h_a}\right) = \tan^{-1}\left(\frac{2.25\text{ m}}{14.0\text{ m}}\right) = \boxed{9.13^\circ}$$

(b) With $\theta = 9.13^\circ$, the tangent function given in Equation 1.3 can be used to find the unknown depth farther from the shore, where $h_o = d$ and $h_a = 22.0$ m. Since $\tan \theta = h_o/h_a$, it follows that

Figure 1.7 If the distance from the shore and the depth of the water at any one point are known, the angle θ can be found with the aid of trigonometry. Knowing the value of θ is useful, because then the depth d at another point can be determined.

$$h_{\mathrm{o}} = h_{\mathrm{a}} \tan \theta$$
$$d = (22.0 \text{ m})(\tan 9.13°) = \boxed{3.54 \text{ m}}$$

which is greater than 2.25 m, as expected.

The right triangle in Figure 1.5 provides the basis for defining the various trigonometric functions according to Equations 1.1–1.3. These functions always involve an angle and two sides of the triangle. There is also a relationship among the lengths of the three sides of a right triangle. This relationship is known as the **Pythagorean theorem** and is used often in this text.

● **PYTHAGOREAN THEOREM**

The square of the length of the hypotenuse of a right triangle is equal to the sum of the squares of the lengths of the other two sides:

$$h^2 = h_{\mathrm{o}}^{\,2} + h_{\mathrm{a}}^{\,2} \qquad\qquad (1.7)$$

1.5 THE NATURE OF PHYSICAL QUANTITIES: SCALARS AND VECTORS

The volume of water in a swimming pool might be 50 cubic meters, or the winning time of a race could be 11.3 seconds. In cases like these, only the size of the numbers matters. In other words, *how much* volume or time is there? The "50" specifies the amount of water in units of cubic meters, while the "11.3" specifies the amount of time in seconds. Volume and time are examples of scalar quantities. A *scalar quantity* is one that can be described by a single number (including any units) giving its size or magnitude. Some other common scalars are temperature (e.g., 20 °C) and mass (e.g., 85 kg).

While many quantities in physics are scalars, there are also many that are not scalars, quantities for which magnitude tells only part of the story. Consider Figure 1.8, which depicts a car that has moved 2 km along a straight line from start to finish. When describing the motion, it is incomplete to say that "the car moved a distance of 2 km." This statement would indicate only that the car ends up somewhere on a circle whose center is at the starting point and whose radius is 2 km. A complete description must include the direction along with the distance, as in the statement "the car moved a distance of 2 km in a direction 30° north of east." A quantity that deals inherently with both magnitude and direction is called a *vector quantity.* Because direction is an important characteristic of vectors, arrows are used to represent them; *the direction of the arrow gives the direction of the vector.* The colored arrow in Figure 1.8, for example, is called the displacement vector, because it shows how the car is displaced from its starting point. Chapter 2 discusses this particular vector.

The length of the arrow in Figure 1.8 represents the magnitude of the displacement vector. If the car had moved 4 km instead of 2 km from the starting point, the arrow would have been drawn twice as long. *By convention, the length of a vector arrow is proportional to the magnitude of the vector.*

In physics there are many important kinds of vectors, and the practice of using the length of an arrow to represent the magnitude of a vector applies to each of them. All forces, for instance, are vectors. In common usage a force is a push or a

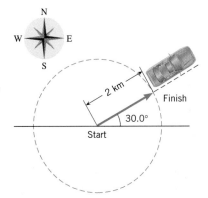

Figure 1.8 A vector quantity has a magnitude and a direction. The arrow in this drawing represents a displacement vector.

The velocity of this skier is another example of a vector quantity.

pull, and the direction in which a force acts is just as important as the strength or magnitude of the force. The magnitude of a force is measured in SI units called newtons (N). An arrow representing a force of 20 newtons is drawn twice as long as one representing a force of 10 newtons.

The fundamental distinction between scalars and vectors is the characteristic of direction. Vectors have it, and scalars do not. Conceptual Example 5 helps to clarify this distinction and explains what is meant by the "direction" of a vector.

CONCEPTUAL EXAMPLE 5
Vectors, Scalars, and the Role of Plus and Minus Signs

There are places where the temperature is +20 °C at one time of the year and −20 °C at another time. Do the plus and minus signs that signify positive and negative temperatures imply that temperature is a vector quantity?

Reasoning and Solution   A vector has a physical direction associated with it, due east or due west, for example. The question, then, is whether such a direction is associated with temperature. In particular, do the plus and minus signs that go along with temperature imply this kind of direction? On a thermometer, the algebraic signs simply mean that the temperature is a number less than or greater than zero on the scale and have nothing to do with east, west, or any other physical direction. Temperature, then, is not a vector. It is a scalar, and scalars can sometimes be negative. *The fact that a quantity is positive or negative does not necessarily mean that the quantity is a scalar or a vector.*

Often, for the sake of convenience, quantities such as volume, time, displacement, and force are represented by symbols. This text follows the usual practice of writing vectors in boldface symbols* **(this is boldface)** and writing scalars in italic symbols (*this is italic*). Thus, a displacement vector is written as "**A** = 750 m, due east," where the **A** is a boldface symbol. By itself, however, separated from the direction, the magnitude of this vector is a scalar quantity. Therefore, the magnitude is written as "*A* = 750 m," where the *A* is an italic symbol.

1.6   VECTOR ADDITION AND SUBTRACTION

Addition

Often it is necessary to add one vector to another, and the process of addition must take into account both the magnitude and the direction of the vectors. The simplest situation occurs when the vectors point along the same direction, that is, when they are colinear, as in Figure 1.9. Here, a car first moves along a straight line, with a displacement vector **A** of 275 m, due east. Then, the car moves again in the same direction, with a displacement vector **B** of 125 m, due east. These two vectors add to give the total displacement vector **R**, which would apply if the car had moved from start to finish in one step. The symbol **R** is used because the total vector is often called the *resultant vector.* With the tail of the second arrow located at the head of the first arrow, the two lengths simply add to give the length of the total displacement. This kind of vector addition is identical to the familiar addition of two scalar

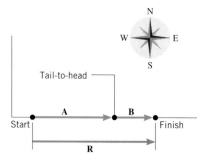

Figure 1.9 Two colinear displacement vectors **A** and **B** add to give the resultant displacement vector **R**.

---

* A vector quantity can also be represented without boldface symbols, by including an arrow above the symbol, e.g., $\vec{A}$.

numbers (2 + 3 = 5) *and can be carried out here only because the vectors point along the same direction.* In such cases we add the individual magnitudes to get the magnitude of the total, knowing in advance what the direction must be. Formally, the addition is written as follows:

$$\mathbf{R} = \mathbf{A} + \mathbf{B}$$

$$\mathbf{R} = 275 \text{ m, due east} + 125 \text{ m, due east} = 400 \text{ m, due east}$$

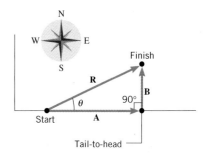

Figure 1.10 The addition of two perpendicular displacement vectors **A** and **B** gives the resultant vector **R**.

Perpendicular vectors are frequently encountered, and Figure 1.10 indicates how they can be added. This figure applies to a car that first travels with a displacement vector **A** of 275 m, due east, and then with a displacement vector **B** of 125 m, due north. The two vectors add to give a resultant displacement vector **R**. Once again, the vectors to be added are arranged in a tail-to-head fashion, and the resultant vector points from the tail of the first to the head of the last vector added. The resultant displacement is given by the vector equation

$$\mathbf{R} = \mathbf{A} + \mathbf{B}$$

The addition in this equation cannot be carried out by writing R = 275 m + 125 m, because the vectors have different directions. Instead, we take advantage of the fact that the triangle in Figure 1.10 is a right triangle and use the Pythagorean theorem (Equation 1.7). According to this theorem, the magnitude of **R** is

$$R = \sqrt{(275 \text{ m}} \sqrt{(275 \text{ m})^2 + (125 \text{ m})^2} = 302 \text{ m}$$

The angle θ in Figure 1.10 gives the direction of the resultant vector. Since the lengths of all three sides of the right triangle are now known, either sin θ, cos θ, or tan θ can be used to determine θ. Noting that tan $\theta = B/A$ and using the inverse trigonometric function, we find that :

$$\theta = \tan^{-1}\left(\frac{B}{A}\right) = \tan^{-1}\left(\frac{125 \text{ m}}{275 \text{ m}}\right) = 24.4°$$

Thus, the resultant displacement of the car has a magnitude of 302 m and points north of east at an angle of 24.4°. This displacement would bring the car from the start to the finish in Figure 1.10 in a single straight-line step.

When two vectors to be added are not perpendicular, the tail-to-head arrangement does not lead to a right triangle, and the Pythagorean theorem cannot be used. Figure 1.11a illustrates such a case for a car that moves with a displacement **A** of 275 m, due east and then with a displacement **B** of 125 m in a direction 55.0° north of west. As usual, the resultant displacement vector **R** is directed from the tail of the first to the head of the last vector added. The vector addition is still given according to

$$\mathbf{R} = \mathbf{A} + \mathbf{B}$$

However, the magnitude of **R** is not $R = \sqrt{A^2 + B^2}$, because the vectors **A** and **B** are not perpendicular and the Pythagorean theorem does not apply. Some other means must be used to find the magnitude and direction of the resultant vector.

One approach uses a graphical technique. In this method, a diagram is constructed in which the arrows are drawn tail to head. The lengths of the vector arrows are drawn to scale, and the angles are drawn accurately (with a protractor, perhaps). Then, the length of the arrow representing the resultant vector is measured with a ruler. This length is converted into the magnitude of the resultant vector by

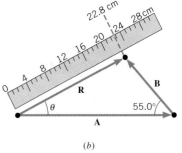

Figure 1.11 (*a*) The two displacement vectors **A** and **B** are neither colinear nor perpendicular but add to give the resultant vector **R**. (*b*) In one method for adding them together, a graphical technique is used.

Figure 1.12 (*a*) The displacement vector for a woman climbing 1.2 m up a ladder is **D**. (*b*) The displacement vector for a woman climbing 1.2 m down a ladder is −**D**.

using the scale factor with which the drawing is constructed. In Figure 1.11*b*, for example, a scale of one centimeter of arrow length for each 10.0 m of displacement is used, and it can be seen that the length of the arrow representing **R** is 22.8 cm. Since each centimeter corresponds to 10.0 m of displacement, the magnitude of **R** is 228 m. The angle θ, which gives the direction of **R**, can be measured with a protractor to be $\theta = 26.7°$ north of east.

Subtraction

The subtraction of one vector from another is carried out in a way that depends on the following fact. *When a vector is multiplied by −1, the magnitude of the vector remains the same, but the direction of the vector is reversed.* Conceptual Example 6 illustrates the meaning of this statement.

CONCEPTUAL EXAMPLE 6   Multiplying a Vector by −1

Consider the two vectors described below:

1. A woman climbs 1.2 m up a ladder, so that her displacement vector **D** is 1.2 m, upward along the ladder, as in Figure 1.12*a*.

2. A man is pushing with 450 N of force on his stalled car, trying to move it eastward. The force vector **F** that he applies to the car is 450 N, due east, as in Figure 1.13*a*.

What are the physical meanings of the vectors −**D** and −**F**?

Reasoning and Solution A displacement vector of −**D** is (−1)**D** and has the same magnitude as the vector **D**, but is opposite in direction. Thus, −**D** would represent the displacement of a woman climbing 1.2 m down the ladder, as in Figure 1.12*b*. Similarly, a force vector of −**F** has the same magnitude as the vector **F** but has the opposite direction. As a result, −**F** would represent a force of 450 N applied to the car in a direction of due west, as in Figure 1.13*b*.

Related Homework Material: *Question 15, Problem 61*

Figure 1.13 (*a*) The force vector for a man pushing on a car with 450 N of force in a direction due east is **F**. (*b*) The force vector for a man pushing on a car with 450 N of force in a direction due west is −**F**.

In practice, vector subtraction is carried out exactly as vector addition, except that one of the vectors added is multiplied by a scalar factor of −1. To see why, look in Figure 1.14*a* at the two vectors **A** and **B**. These vectors add together to give a third vector **C**, according to **C** = **A** + **B**. Therefore, we can calculate vector **A** as **A** = **C** − **B**, which is an example of vector subtraction. However, we can also write this result as **A** = **C** + (−**B**) and treat it as vector addition. Figure 1.14*b* shows how to calculate vector **A** by adding the vectors **C** and −**B**. Notice that vectors **C** and −**B** are arranged tail to head and that any suitable method of vector addition can be employed to determine **A**.

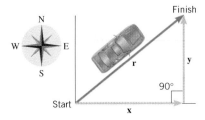

Figure 1.14 (*a*) Vector addition according to **C** = **A** + **B**. (*b*) Vector subtraction according to **A** = **C** − **B** = **C** + (−**B**).

1.7 THE COMPONENTS OF A VECTOR

Vector Components

Suppose a car moves along a straight line from start to finish in Figure 1.15, the corresponding displacement vector being **r**. The magnitude and direction of the vector **r** give the distance and direction traveled along the straight line. However, the car could also arrive at the finish point by first moving due east, turning through 90°, and then moving due north. This alternative path is shown in the drawing and is associated with the two displacement vectors **x** and **y**. The vectors **x** and **y** are called the *x* vector component and the *y* vector component of **r**.

Vector components are very important in physics and have two basic features that are apparent in Figure 1.15. One is that the components add together to equal the original vector:

$$\mathbf{r} = \mathbf{x} + \mathbf{y}$$

The components **x** and **y**, when added vectorially, convey exactly the same meaning as does the original vector **r**: they indicate how the finish point is displaced relative to the starting point. In general, ***the components of any vector can be used in place of the vector itself in any calculation where it is convenient to do so.*** The other feature of vector components that is apparent in Figure 1.15 is that **x** and **y** are not just any two vectors that add together to give the original vector **r**: they are perpendicular vectors.* This perpendicularity is a valuable characteristic, as we will soon see.

Any type of vector may be expressed in terms of its components, in a way similar to that illustrated for the displacement vector in Figure 1.15. Figure 1.16 shows an arbitrary vector **A** and its vector components **A**$_x$ and **A**$_y$. The components are drawn parallel to convenient *x* and *y* axes and are perpendicular. They add vectorially to equal the original vector **A**:

$$\mathbf{A} = \mathbf{A}_x + \mathbf{A}_y$$

There are times when a drawing such as Figure 1.16 is not the most convenient way to represent vector components, and Figure 1.17 presents an alternative method. The disadvantage of this alternative is that the tail-to-head arrangement of **A**$_x$ and **A**$_y$ is missing, an arrangement that is a nice reminder that **A**$_x$ and **A**$_y$ add together to equal **A**.

The definition given below summarizes the meaning of vector components:

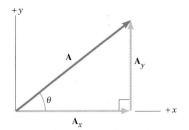

Figure 1.15 The displacement vector **r** and its vector components *x* and *y*.

Figure 1.16 An arbitrary vector **A** and its vector components **A**$_x$ and **A**$_y$.

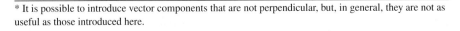

* It is possible to introduce vector components that are not perpendicular, but, in general, they are not as useful as those introduced here.

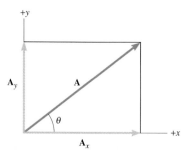

Figure 1.17 This alternative way of drawing the vector **A** and its vector components is completely equivalent to that shown in Figure 1.16.

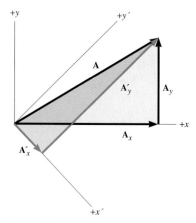

Figure 1.18 The vector components of the vector depend on the orientation of the axes used as a reference.

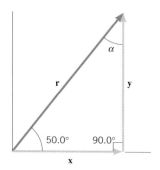

Figure 1.19 The x and y components of the displacement vector **r** can be found using trigonometry.

● **DEFINITION OF VECTOR COMPONENTS**

In two dimensions, the vector components of a vector **A** are two perpendicular vectors $\mathbf{A}_x$ and $\mathbf{A}_y$ that are parallel to the x and y axes, respectively, and add together vectorially so that $\mathbf{A} = \mathbf{A}_x + \mathbf{A}_y$.

The values calculated for vector components depend on the orientation of the vector relative to the axes used as a reference. Figure 1.18 illustrates this fact for a vector **A** by showing two sets of axes, one set being rotated clockwise relative to the other. With respect to the black axes, vector **A** has perpendicular vector components $\mathbf{A}_x$ and $\mathbf{A}_y$; with respect to the colored rotated axes, vector **A** has different vector components $\mathbf{A}_x{'}$ and $\mathbf{A}_y{'}$. The choice of which set of components to use is purely a matter of convenience.

Scalar Components

It is often easier to work with the *scalar components,* A_x and A_y (note the italic symbols), rather than the vector components $\mathbf{A}_x$ and $\mathbf{A}_y$. Scalar components are positive or negative numbers (with units) that are defined as follows. The component A_x has a magnitude that is equal to that of $\mathbf{A}_x$ and is given a positive sign if $\mathbf{A}_x$ points along the $+x$ axis and a negative sign if $\mathbf{A}_x$ points along the $-x$ axis. The component A_y is defined in a similar manner. The following table shows an example of vector and scalar components:

Vector Components	Scalar Components
$\mathbf{A}_x = 8$ meters, directed along the $+x$ axis	$A_x = +8$ meters
$\mathbf{A}_y = 10$ meters, directed along the $-y$ axis	$A_y = -10$ meters

In this text, when we use the term "component," we will be referring to a scalar component, unless otherwise indicated.

Resolving a Vector into Its Components

If the magnitude and direction of a vector are known, it is possible to find the components of the vector. The process of finding the components is called "resolving the vector into its components." As Example 7 illustrates, this process can be carried out with the aid of trigonometry, because the two perpendicular vector components and the original vector form a right triangle.

EXAMPLE 7   Finding the Components of a Vector

A displacement vector **r** has a magnitude of $r = 175$ m and points at an angle of $50.0°$ relative to the x axis in Figure 1.19. Find the x and y components of this vector.

Reasoning We will base our solution on the fact that the triangle formed in Figure 1.19 by the vector **r** and its components **x** and **y** is a right triangle. This fact enables us to use the trigonometric sine and cosine functions, as defined in Equations 1.1 and 1.2.

Solution 1 The y component can be obtained using the $50.0°$ angle and Equation 1.1, $\sin \theta = y/r$:

$$y = r \sin \theta = (175 \text{ m})(\sin 50.0°) = \boxed{134 \text{ m}}$$

In a similar fashion, the x component can be obtained using the 50.0° angle and Equation 1.2, $\cos \theta = x/r$:

$$x = r \cos \theta = (175 \text{ m})(\cos 50.0°) = \boxed{112 \text{ m}}$$

Solution 2   The angle α in Figure 1.19 can also be used to find the components. Since $\alpha + 50.0° = 90.0°$, it follows that $\alpha = 40.0°$. The solution using α yields the same answers as in Solution 1:

$$\cos \alpha = \frac{y}{r}$$

$$y = r \cos \alpha = (175 \text{ m})(\cos 40.0°) = \boxed{134 \text{ m}}$$

$$\sin \alpha = \frac{x}{r}$$

$$x = r \sin \alpha = (175 \text{ m})(\sin 40.0°) = \boxed{112 \text{ m}}$$

• **PROBLEM SOLVING INSIGHT**
Either acute angle of a right triangle can be used to determine the components of a vector. The choice of angle is a matter of convenience.

Since the vector components and the original vector form a right triangle, the Pythagorean theorem can be applied to check the validity of calculations such as those in Example 7. Thus, with the components obtained in Example 7, the theorem can be used to verify that the magnitude of the original vector is indeed 175 m, as given initially:

$$r = \sqrt{(112 \text{ m})^2 + (134 \text{ m})^2} = 175 \text{ m}$$

• **PROBLEM SOLVING INSIGHT**
When a vector is resolved into components, one can check to see if they are correct; substitute the components into the Pythagorean theorem and verify that the result is the magnitude of the original vector.

It is possible for one of the components of a vector to be zero. This does not mean that the vector itself is zero, however. ***For a vector to be zero, every vector component must individually be zero.*** Thus, in two dimensions, saying that $\mathbf{A} = 0$ is equivalent to saying that $\mathbf{A}_x = 0$ and $\mathbf{A}_y = 0$. Or, stated in terms of scalar components, if $\mathbf{A} = 0$, then $A_x = 0$ and $A_y = 0$.

Two vectors are equal if, and only if, they have the same magnitude and direction. Thus, if one displacement vector points east and another points north, they are *not* equal, even if each has the same magnitude of 480 m. In terms of vector components, two vectors, $\mathbf{A}$ and $\mathbf{B}$, are equal if, and only if, each vector component of one is equal to the corresponding vector component of the other. In two dimensions, if $\mathbf{A} = \mathbf{B}$, then $\mathbf{A}_x = \mathbf{B}_x$ and $\mathbf{A}_y = \mathbf{B}_y$. Alternatively, using scalar components, we write that $A_x = B_x$ and $A_y = B_y$.

1.8   ADDITION OF VECTORS BY MEANS OF COMPONENTS

The components of a vector provide the most convenient and accurate way of adding (or subtracting) any number of vectors. For example, suppose that vector $\mathbf{A}$ is added to vector $\mathbf{B}$. The resultant vector is $\mathbf{C}$, where $\mathbf{C} = \mathbf{A} + \mathbf{B}$. Figure 1.20a illustrates this vector addition, along with the x and y vector components of $\mathbf{A}$ and $\mathbf{B}$. In part *b* of the drawing, the vectors $\mathbf{A}$ and $\mathbf{B}$ have been removed, because we can use the vector components of these vectors in place of them. The vector component $\mathbf{B}_x$ has been shifted downward and arranged tail-to-head with the vector component $\mathbf{A}_x$. Similarly, the vector component $\mathbf{A}_y$ has been shifted to the right and arranged tail-to-head with the vector component $\mathbf{B}_y$. The x components are colinear and add together to give the x component of the resultant vector $\mathbf{C}$. In like fashion, the y components are colinear and add together to give the y component of $\mathbf{C}$. In terms of scalar components, we can write

$$C_x = A_x + B_x \quad \text{and} \quad C_y = A_y + B_y$$

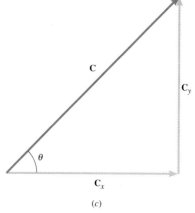

(a)

(b)

(c)

Figure 1.20 (a) The vectors **A** and **B** add together to give the resultant vector **C**. The x and y vector components of **A** and **B** are also shown. (b) The drawing illustrates that $C_x = A_x + B_x$ and $C_y = A_y + B_y$. (c) Vector **C** and its components form a right triangle.

(a)

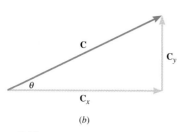

(b)

Figure 1.21 (a) The vectors **A** and **B** add together to give the resultant vector **C**. The vector components of **A** and **B** are also shown. (b) The resultant vector **C** can be obtained once its components have been found.

The vector components C_x and C_y of the resultant vector form the sides of the right triangle shown in Figure 1.20c. Thus, we can find the magnitude of **C** by using the Pythagorean theorem:

$$C = \sqrt{C_x^{\,2} + C_y^{\,2}}$$

The angle θ that **C** makes with the x axis is given by $\theta = \tan^{-1}(C_y/C_x)$. Example 8 illustrates how to add several vectors using the component method.

EXAMPLE 8   The Component Method of Vector Addition

A jogger runs 145 m in a direction 20.0° east of north (displacement vector **A**) and then 105 m in a direction 35.0° south of east (displacement vector **B**). Determine the magnitude and direction of the resultant vector **C** for these two displacements.

Reasoning Figure 1.21a shows the vectors **A** and **B**, assuming that the y axis corresponds to the direction due north. Since the vectors are not given in component form, we will begin by using the given magnitudes and directions to find the components. Then, the components of **A** and **B** can be used to find the components of the resultant **C**. Finally, with the aid of the Pythagorean theorem and trigonometry, the components of **C** can be used to find its magnitude and direction.

Solution The first two rows of the following table give the x and y components of the vectors **A** and **B**. Note that the component B_y is negative, because B_y points downward, in the negative y direction in the drawing.

Vector	x Component	y Component
A	$A_x = (145 \text{ m}) \sin 20.0° = 49.6 \text{ m}$	$A_y = (145 \text{ m}) \cos 20.0° = 136 \text{ m}$
B	$B_x = (105 \text{ m}) \cos 35.0° = 86.0 \text{ m}$	$B_y = -(105 \text{ m}) \sin 35.0° = -60.2 \text{ m}$
C	$C_x = A_x + B_x = 135.6 \text{ m}$	$C_y = A_y + B_y = 76 \text{ m}$

The third row in the table gives the x and y components of the resultant vector **C**: $C_x = A_x + B_x$ and $C_y = A_y + B_y$. Part b of the drawing shows **C** and its vector components. The magnitude of **C** is given by the Pythagorean theorem as

$$C = \sqrt{C_x^{\,2} + C_y^{\,2}} = \sqrt{(135.6 \text{ m})^2 + (76 \text{ m})^2} = \boxed{155 \text{ m}}$$

The angle θ that **C** makes with the x axis is

$$\theta = \tan^{-1}\left(\frac{C_y}{C_x}\right) = \tan^{-1}\left(\frac{76 \text{ m}}{135.6 \text{ m}}\right) = \boxed{29°}$$

In later chapters we will often use the component method for vector addition. For future reference, the main features of the reasoning strategy used in this technique are summarized below.

REASONING STRATEGY

The Component Method of Vector Addition

1. For each vector to be added, determine the x and y components relative to a conveniently chosen x, y coordinate system. Be sure to take into account the directions of the components by using plus and minus signs to denote whether the components point along the positive or negative axes.

2. Find the algebraic sum of the x components, which is the x component of the resultant vector. Similarly, find the algebraic sum of the y components, which is the y component of the resultant vector.

3. Use the x and y components of the resultant vector and the Pythagorean theorem to determine the magnitude of the resultant vector.

4. Use either the inverse sine, inverse cosine, or inverse tangent function to find the angle that specifies the direction of the resultant vector.

1.9 CONCEPTS & CALCULATIONS

This chapter has presented an introduction to the mathematics of trigonometry and vectors, which will be used throughout this text. Therefore, in this last section we consider several examples in order to review some of the important features of this mathematics. The three-part format of these examples stresses the role of conceptual understanding in problem solving. First, the problem statement is given. Then, there is a concept question-and-answer section, which is followed by the solution section. The purpose of the concept question-and-answer section is to provide help in understanding the solution and to illustrate how a review of the concepts can help in anticipating some of the characteristics of the numerical answers.

CONCEPTS & CALCULATIONS EXAMPLE 9   Equal Vectors

Figure 1.22 shows two displacement vectors **A** and **B**. Vector **A** points at an angle of 22.0° above the x axis but has an unknown magnitude. Vector **B** has an x component of $B_x = 35.0$ m but has an unknown y component B_y. These two vectors are equal. Find the magnitude of **A** and the value of B_y.

Concept Questions and Answers What does the fact that vector **A** equals vector **B** imply about the magnitudes and directions of the vectors?

Answer When two vectors are equal, each has the same magnitude and each has the same direction.

What does the fact that vector **A** equals vector **B** imply about the x and y components of the vectors?

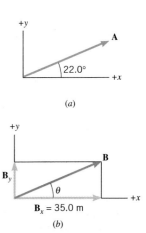

Figure 1.22 The two displacement vectors **A** and **B** are equal. Example 9 discusses what this equality means.

Answer   When two vectors are equal, the x component of vector **A** equals the x component of vector **B** ($A_x = B_x$) and the y component of vector **A** equals the y component of vector **B** ($A_y = B_y$).

Solution 1   This solution focuses on the fact that the magnitudes and directions of the vectors are the same, which allows us to write that

$$\underbrace{A}_{\substack{\text{Magnitude} \\ \text{of vector } \mathbf{A}}} = \underbrace{\sqrt{B_x^2 + B_y^2}}_{\substack{\text{Magnitude} \\ \text{of vector } \mathbf{B}}} = \sqrt{(35.0 \text{ m})^2 + B_y^2} \qquad (1.8)$$

$$\underbrace{\tan 22.0°}_{\substack{\text{Direction} \\ \text{of vector } \mathbf{A}}} = \underbrace{\tan \theta}_{\substack{\text{Direction} \\ \text{of vector } \mathbf{B}}} = \frac{B_y}{B_x} = \frac{B_y}{35.0 \text{ m}} \qquad (1.9)$$

In Equation 1.8, we have taken advantage of the Pythagorean theorem, since the components of vector **B** are perpendicular. From Equation 1.9, we find that

$$B_y = (35.0 \text{ m}) \tan 22.0° = \boxed{14.1 \text{ m}}$$

Substituting this value into Equation 1.8 gives

$$A = \sqrt{(35.0 \text{ m})^2 + (14.1 \text{ m})^2} = \boxed{37.7 \text{ m}}$$

Solution 2   This solution focuses on the fact that the x components of the vectors are the same and the y components of the vectors are the same. This allows us to write that

$$\underbrace{A \cos 22.0°}_{\substack{\text{Component } A_x \\ \text{of vector } \mathbf{A}}} = \underbrace{35.0 \text{ m}}_{\substack{\text{Component } B_x \\ \text{of vector } \mathbf{B}}} \qquad (1.10)$$

$$\underbrace{A \sin 22.0°}_{\substack{\text{Component } A_y \\ \text{of vector } \mathbf{A}}} = \underbrace{B_y}_{\substack{\text{Component } B_y \\ \text{of vector } \mathbf{B}}} \qquad (1.11)$$

Dividing Equation 1.11 by Equation 1.10 shows that

$$\frac{A \sin 22.0°}{A \cos 22.0°} = \frac{B_y}{35.0 \text{ m}}$$

$$B_y = (35.0 \text{ m}) \frac{\sin 22.0°}{\cos 22.0°} = (35.0 \text{ m}) \tan 22.0° = \boxed{14.1 \text{ m}}$$

Solving Equation 1.10 directly for A gives

$$A = \frac{35.0 \text{ m}}{\cos 22.0°} = \boxed{37.7 \text{ m}}$$

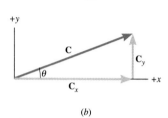

(a)

(b)

Figure 1.23 (a) Two displacement vectors **A** and **B** add together to give a resultant vector **C**. (b) The components of **C** are C_x and C_y.

CONCEPTS & CALCULATIONS EXAMPLE 10   Adding Vectors

Figure 1.23a shows two displacement vectors **A** and **B**, which add together to give a resultant displacement **C**. Find the magnitude and direction of **C**.

Concept Questions and Answers   Does the Pythagorean theorem apply directly to this problem, so that the magnitude of vector **C** is given by $C = \sqrt{A^2 + B^2}$?

Answer   The magnitude of the vector **C** is not given by the Pythagorean theorem in the form $C = \sqrt{A^2 + B^2}$, because the vectors **A** and **B** are not perpendicular (see Figure 1.23*a*). It is only when the two vectors are perpendicular that this equation for the magnitude applies.

Does the component method for vector addition apply to this problem, even though the vectors are not perpendicular?

Answer   Yes. The component method applies whether or not the vectors are perpendicular, and it applies to any number of vectors. We will use it in our solution.

Solution The components of the vectors **A** and **B** can be determined from the data in Figure 1.23*a* and are listed in the first two rows of the following table. Note that the vector **A** points along the *x* axis and, therefore, has no *y* component.

Vector	*x* Component	*y* Component
A	$A_x = 47.0$ m	$A_y = 0$ m
B	$B_x = (62.0$ m$) \cos 35.0° = 50.8$ m	$B_y = (62.0$ m$) \sin 35.0° = 35.6$ m
C	$C_x = A_x + B_x = 97.8$ m	$C_y = A_y + B_y = 35.6$ m

Figure 1.23*b* shows the components of the resultant vector, and the third row in the table gives their values. The components C_x and C_y are perpendicular, so that we can use the Pythagorean theorem to find the magnitude of **C**:

$$C = \sqrt{C_x{}^2 + C_y{}^2} = \sqrt{(97.8 \text{ m})^2 + (35.6 \text{ m})^2} = \boxed{104 \text{ m}}$$

The value of the directional angle θ is

$$\theta = \tan^{-1}\left(\frac{C_y}{C_x}\right) = \tan^{-1}\left(\frac{35.6 \text{ m}}{97.8 \text{ m}}\right) = \boxed{20.0°}$$

At the end of the problem set for this chapter, you will find homework problems that contain both conceptual and quantitative parts. These problems are grouped under the heading *Concepts & Calculations, Group Learning Problems*. They are designed for use by students working alone or in small learning groups. The conceptual part of each problem provides a convenient focus for goup discussions.

SUMMARY

Physics is an experimental science that uses precisely defined **units of measurement.** This text emphasizes SI (Système International) units, a system that includes the meter (m), the kilogram (kg), and the second (s) as base units for length, mass, and time, respectively. Units play an important role in solving problems, because the units on the left side of an equation must match the units on the right side. If the units on both sides do not match, either the equation is written incorrectly or the variables and constants in the equation are not expressed in a consistent set of units.

Trigonometry is used throughout physics. Particularly important are the sine, cosine, and tangent functions of an angle θ. They are defined in terms of a right triangle that contains θ. The side of the triangle opposite θ is h_o, the side adjacent to θ is h_a, and the hypotenuse is h. In terms of these quantities we have

$$\sin \theta = \frac{h_o}{h} \qquad \cos \theta = \frac{h_a}{h} \qquad \tan \theta = \frac{h_o}{h_a}$$

If the value of the sine, cosine, or tangent is known, the angle can be obtained using inverse trigonometric functions. The Pythagorean theorem also applies to the right triangle:

$$h^2 = h_o{}^2 + h_a{}^2$$

A **scalar quantity** is described completely by its size, which is also called its magnitude. For a **vector quantity,** however, both magnitude and direction must be specified. Vectors are often represented by arrows, the length of the arrow being proportional to the magnitude of the vector and the direction of the arrow indicating the direction of the vector. The **addition of vectors** to give a resultant vector must account for both magnitude and direction. When the vectors are all colinear, the addition proceeds in the same way as the simple addition of scalar quantities. When the vectors are not colinear, one procedure for addition utilizes a graphical technique, in which the vectors to be added are arranged in a tail-to-head fashion. The subtraction of a vector is treated as the addition of a vector that has been multiplied by a scalar factor of -1. Multiplying a vector by -1 reverses the direction of the vector.

In two dimensions, the **vector components** of a vector $\mathbf{A}$ are two perpendicular vectors $\mathbf{A}_x$ and $\mathbf{A}_y$ that are parallel to the x and y axes, respectively, and add together vectorially so that $\mathbf{A} = \mathbf{A}_x + \mathbf{A}_y$. The **scalar component** A_x has a magni-

tude that is equal to that of $\mathbf{A}_x$ and is given a positive sign if $\mathbf{A}_x$ points along the $+x$ axis and a negative sign if $\mathbf{A}_x$ points along the $-x$ axis. The scalar component A_y is defined in a similar manner. Components provide the best way of adding any number of vectors. A vector is zero if, and only if, each of its vector components is zero. Two vectors are equal in two dimensions if, and only if, the x vector components of each are equal and the y vector components of each are equal.

CONCEPTUAL QUESTIONS

1. The table below lists four variables along with their units:

Variable	Units
x	meters (m)
v	meters per second (m/s)
t	seconds (s)
a	meters per second squared (m/s^2)

These variables appear in the following equations, along with a few numbers that have no units. In which of the equations are the units on the left side of the equals sign consistent with the units on the right side?

(a) $x = vt$

(b) $x = vt + \frac{1}{2}at^2$

(c) $v = at$

(d) $v = at + \frac{1}{2}at^3$

(e) $v^3 = 2ax^2$

(f) $t = \sqrt{\dfrac{2x}{a}}$

2. The variables x and v have the units shown in the table that accompanies question 1. Is it possible for x and v to be related to an angle θ according to $\tan \theta = x/v$? Account for your answer.

3. You can always add two numbers that have the same units. However, you cannot always add two numbers that have the same dimensions. Explain why not, and include an example in your explanation.

4. (a) Is it possible for two quantities to have the same dimensions but different units? (b) Is it possible for two quantities to have the same units but different dimensions? In each case, support your answer with an example and an explanation.

5. In the equation $y = c^n at^2$ you wish to determine the integer value (1, 2, etc.) of the exponent n. The dimensions of y, a, and t are known. It is also known that c has no dimensions. Can dimensional analysis be used to determine n? Account for your answer.

6. Using your calculator, verify that $\sin \theta$ divided by $\cos \theta$ is equal to $\tan \theta$, for an angle θ. Try 30°, for example. Prove that this result is true in general by using the definitions for $\sin \theta$, $\cos \theta$, and $\tan \theta$ given in Equations 1.1–1.3.

7. $\sin \theta$ and $\cos \theta$ are called sinusoidal functions of the angle θ. The way in which these functions change as θ changes leads to a characteristic pattern when they are graphed. This pattern arises many times in physics. (a) To familiarize yourself with the sinus-

oidal pattern, use a calculator and construct a graph, with $\sin \theta$ plotted on the vertical axis and θ on the horizontal axis. Use 15° increments for θ between 0° and 720°. (b) Repeat for $\cos \theta$.

8. Which of the following quantities (if any) can be considered a vector: (a) the number of people attending a football game, (b) the number of days in a month, and (c) the number of pages in a book? Explain your reasoning.

9. Which of the following displacement vectors (if any) are equal? Explain your reasoning.

Vector	Magnitude	Direction
A	100 m	30° north of east
B	100 m	30° south of west
C	50 m	30° south of west
D	100 m	60° east of north

10. Are two vectors with the same magnitude necessarily equal? Give your reasoning.

11. A cube has six faces and twelve edges. You start at one corner, are allowed to move only along the edges, and may not retrace your path along any edge. Consistent with these rules, there are a number of ways to arrive back at your starting point. For instance, you could move around the four edges that make up one of the square faces. The four corresponding displacement vectors would add to zero. How many ways are there to arrive back at your starting point that involve *eight* displacement vectors that add to zero? Describe each possibility, using drawings for clarity.

12. (a) Is it possible for one component of a vector to be zero, while the vector itself is not zero? (b) Is it possible for a vector to be zero, while one component of the vector is not zero? Explain.

13. Can two nonzero perpendicular vectors be added together so their sum is zero? Explain.

14. Can three or more vectors with unequal magnitudes be added together so their sum is zero? If so, show by means of a tail-to-head arrangement of the vectors how this could occur.

15. In preparation for this question, review Conceptual Example 6. Vectors $\mathbf{A}$ and $\mathbf{B}$ satisfy the vector equation $\mathbf{A} + \mathbf{B} = 0$. (a) How does the magnitude of $\mathbf{B}$ compare with the magnitude of $\mathbf{A}$? (b) How does the direction of $\mathbf{B}$ compare with the direction of $\mathbf{A}$? Give your reasoning.

16. Vectors **A**, **B**, and **C** satisfy the vector equation **A** + **B** = **C**, and their magnitudes are related by the scalar equation $A^2 + B^2 = C^2$. How is vector **A** oriented with respect to vector **B**? Account for your answer.

17. Vectors **A**, **B**, and **C** satisfy the vector equation **A** + **B** = **C**, and their magnitudes are related by the scalar equation $A + B = C$. How is vector **A** oriented with respect to vector **B**? Explain your reasoning.

18. The magnitude of a vector has doubled, its direction remaining the same. Can you conclude that the magnitude of each component of the vector has doubled? Explain your answer.

19. The tail of a vector is fixed to the origin of an x, y axis system. Originally the vector points along the $+x$ axis. As time passes, the vector rotates counterclockwise. Describe how the sizes of the x and y components of the vector compare to the size of the original vector for rotational angles of (a) 90°, (b) 180°, (c) 270°, and (d) 360°.

20. A vector has a component of zero along the x axis of a certain axes system. Does this vector necessarily have a component of zero along the x axis of another (rotated) axes system? Use a drawing to justify your answer.

PROBLEMS

Problems that are not marked with a star are considered the easiest to solve. Problems that are marked with a single star (*) are more difficult, while those marked with a double star (**) are the most difficult.

ssm Solution is in the Student Solutions Manual. **www** Solution is available on the World Wide Web at http://www.wiley.com/college/cutnell
⚕ This icon represents a biomedical application.

Section 1.2 Units,
Section 1.3 The Role of Units in Problem Solving

1. ssm The largest diamond ever found had a size of 3106 carats. One carat is equivalent to a mass of 0.200 g. Use the fact that 1 kg (1000 g) has a weight of 2.205 lb under certain conditions, and determine the weight of this diamond in pounds.

2. Vesna Vulovic survived the longest fall on record without a parachute when her plane exploded and she fell 6 miles, 551 yards. What is this distance in meters?

3. How many seconds are there in (a) one hour and thirty-five minutes and (b) one day?

4. A speed of 515.2 kilometers per hour is the highest recorded train speed on any national rail system. It was achieved in France. Express this speed in miles per hour.

5. ssm The mass of the parasitic wasp *Caraphractus cintus* can be as small as 5×10^{-6} kg. What is this mass in (a) grams (g), (b) milligrams (mg), and (c) micrograms (μg)?

6. A bottle of wine known as a magnum contains a volume of 1.5 liters. A bottle known as a jeroboam contains 0.792 U.S. gallons. How many magnums are there in one jeroboam?

7. The following are dimensions of various physical parameters that will be discussed later on in the text. Here [L], [T], and [M] denote, respectively, dimensions of length, time, and mass.

Dimension		Dimension	
Distance (x)	[L]	Acceleration (a)	$[L]/[T]^2$
Time (t)	[T]	Force (F)	$[M][L]/[T]^2$
Mass (m)	[M]	Energy (E)	$[M][L]^2/[T]^2$
Speed (v)	[L]/[T]		

Which of the following equations are dimensionally correct?

(a) $F = ma$

(b) $x = \frac{1}{2}at^3$

(c) $E = \frac{1}{2}mv$

(d) $E = max$

(e) $v = \sqrt{Fx/m}$

8. Consider the equation $v = \frac{1}{3}zxt^2$. The dimensions of the variables x, v, and t are [L], [L]/[T], and [T], respectively. What must be the dimensions of the variable z, such that both sides of the equation have the same dimensions? Show how you determined your answer.

*9. ssm The depth of the ocean is sometimes measured in fathoms (1 fathom = 6 feet). Distance on the surface of the ocean is sometimes measured in nautical miles (1 nautical mile = 6076 feet). The water beneath a surface rectangle 1.20 nautical miles by 2.60 nautical miles has a depth of 16.0 fathoms. Find the volume of water (in cubic meters) beneath this rectangle.

*10. The CGS unit for measuring the viscosity of a liquid is the poise [P]: 1 P = 1 g/(s·cm). The SI unit is the kg/(s·m). The viscosity of water at 0 °C is 1.78×10^{-3} kg/(s·m). Express this viscosity in poise.

Section 1.4 Trigonometry

11. ssm www A highway is to be built between two towns, one of which lies 35.0 km south and 72.0 km west of the other. What is the shortest length of highway that can be built between the two towns, and at what angle would this highway be directed with respect to due west?

12. A hill that has a 12.0% grade is one that rises 12.0 m vertically for every 100.0 m of distance in the horizontal direction. At what angle is such a hill inclined above the horizontal?

13. The gondola ski lift at Keystone, Colorado, is 2830 m long. On average, the ski lift rises 14.6° above the horizontal. How high is the top of the ski lift relative to the base?

14. You are driving into St. Louis, Missouri, and in the distance you see the famous Gateway-to-the-West arch. This monument rises to a height of 192 m. You estimate your line of sight with the top of the arch to be 2.0° above the horizontal. Approximately how far (in kilometers) are you from the base of the arch?

15. The silhouette of a Christmas tree is an isosceles triangle. The angle at the top of the triangle is 30.0°, and the base measures 2.00 m across. How tall is the tree?

16. The corners of an equilateral triangle lie on a circle that has a radius of 0.25 m. What is the length of a side of the triangle?

*17. The drawing shows sodium and chlorine ions positioned at the corners of a cube that is part of the crystal structure of sodium chloride (common table salt). The edge of the cube is 0.281 nm (1 nm = 1 nanometer = 10^{-9} m) in length. Find the distance (in nanometers) between the sodium ion located at one corner of the cube and the chlorine ion located on the diagonal at the opposite corner.

Sodium ion
Chlorine ion
θ
0.281 nanometers

*18. What is the value of the angle θ in the drawing that accompanies problem 17?

*19. ssm What is the value of each of the angles of a triangle whose sides are 95, 150, and 190 cm in length? (*Hint: Consider using the law of cosines given in Appendix E.*)

**20. A regular tetrahedron is a three-dimensional object that has four faces, each of which is an equilateral triangle. Each of the edges of such an object has a length L. The height H of a regular tetrahedron is the perpendicular distance from one corner to the center of the opposite triangular face. Show that the ratio between H and L is $H/L = \sqrt{2/3}$.

Section 1.6 Vector Addition and Subtraction

21. A force vector $\mathbf{F}_1$ points due east and has a magnitude of 200 newtons. A second force $\mathbf{F}_2$ is added to $\mathbf{F}_1$. The resultant of the two vectors has a magnitude of 400 newtons and points along the east/west line. Find the magnitude and direction of $\mathbf{F}_2$. Note that there are two answers.

22. A chimpanzee sitting against his favorite tree gets up and walks 51 m due east and 39 m due south to reach a termite mound, where he eats lunch. (a) What is the shortest distance between the tree and the termite mound? (b) What angle does the shortest distance make with respect to due east?

23. ssm www One displacement vector $\mathbf{A}$ has a magnitude of 2.43 km and points due north. A second displacement vector $\mathbf{B}$ has a magnitude of 7.74 km and also points due north. (a) Find the magnitude and direction of $\mathbf{A} - \mathbf{B}$. (b) Find the magnitude and direction of $\mathbf{B} - \mathbf{A}$.

24. The drawing shows a triple jump on a checkerboard, starting at the center of square A and ending on the center of square B. Each side of a square measures 4.0 cm. What is the magnitude of the displacement of the colored checker during the triple jump?

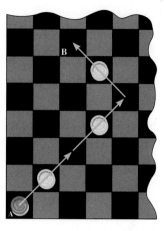
B

A

25. ssm Displacement vector $\mathbf{A}$ points due east and has a magnitude of 2.00 km. Displacement vector $\mathbf{B}$ points due north and has a magnitude of 3.75 km. Displacement vector $\mathbf{C}$ points due west and has a magnitude of 2.50 km. Displacement vector $\mathbf{D}$ points due south and has a magnitude of 3.00 km. Find the magnitude and direction (relative to due west) of the resultant vector $\mathbf{A} + \mathbf{B} + \mathbf{C} + \mathbf{D}$.

26. A circus performer begins his act by walking out along a nearly horizontal high wire. He slips and falls to the safety net, 25.0 ft below. The magnitude of his displacement from the beginning of the walk to the net is 26.7 ft. (a) How far out along the high wire did he walk? (b) Find the angle that his displacement vector makes below the horizontal.

*27. ssm www A car is being pulled out of the mud by two forces that are applied by the two ropes shown in

2900 newtons
30.0°
2900 newtons

the drawing. The dashed line in the drawing bisects the 30.0° angle. The magnitude of the force applied by each rope is 2900 newtons. Arrange the force vectors tail to head and use the graphical technique to answer the following questions. (a) How much force would a single rope need to apply to accomplish the same effect as the two forces added together? (b) How would the single rope be directed relative to the dashed line?

*28. A basketball player runs a pattern consisting of three segments. The corresponding three displacement vectors $\mathbf{A}$, $\mathbf{B}$, and $\mathbf{C}$ have equal magnitudes of 7.0 m. Displacement $\mathbf{A}$ is directed forward and parallel to one side of the court, $\mathbf{B}$ is directed forward at a 45° angle with respect to the side of the court, and $\mathbf{C}$ is directed forward and parallel to the side of the court. With a scale drawing, use the graphical technique to find the magnitude and direction of the displacement vector for a straight-line dash between the starting and finishing points.

*29. In wandering, a grizzly bear makes a displacement of 1563 m due west, followed by a displacement of 3348 m in a direction 32.0° north of west. What are (a) the magnitude and (b) the direction of the displacement needed for the bear to *return to its starting point?* Specify the direction relative to due east.

*30. A jogger travels a route that has two parts. The first is a displacement $\mathbf{A}$ of 2.50 km due south, and the second involves a displacement **B** that points due east. (a) The resultant displacement **A** + **B** has a magnitude of 3.75 km. What is the magnitude of **B**, and what is the direction of **A** + **B** relative to due south? (b) Suppose that **A** − **B** had a magnitude of 3.75 km. What then would be the magnitude of **B**, and what is the direction of **A** − **B** relative to due south?

Section 1.7 The Components of a Vector

31. ssm Vector **A** points along the +y axis and has a magnitude of 100.0 units. Vector **B** points at an angle of 60.0° above the +x axis and has a magnitude of 200.0 units. Vector **C** points along the +x axis and has a magnitude of 150.0 units. Which vector has (a) the largest x component and (b) the largest y component?

32. Your friend has slipped and fallen. To help her up, you pull with a force **F**, as the drawing shows. The vertical component of this force is 130 newtons, while the horizontal component is 150 newtons. Find (a) the magnitude of **F** and (b) the angle θ.

33. An ocean liner leaves New York City and travels 18.0° north of east for 155 km. How far east and how far north has it gone? In other words, what are the magnitudes of the components of the ship's displacement vector in the directions (a) due east and (b) due north?

34. A bicyclist is headed due east. A 5.00-m/s wind is blowing partially into the rider's face and is coming from a direction that is 35.0° south of east. The speed and direction of the wind constitute a vector quantity known as the velocity. In effect, then, the rider must "pump" against a component of the wind's velocity vector. What is the magnitude of this component?

35. ssm The x vector component of a displacement vector **r** has a magnitude of 125 m and points along the negative x axis. The y vector component has a magnitude of 184 m and points along the negative y axis. Find the magnitude and direction of **r**. Specify the direction with respect to the negative x axis.

36. On takeoff, an airplane climbs with a speed of 180 m/s at an angle of 34° above the horizontal. The speed and direction of the airplane constitute a vector quantity known as the velocity. The sun is shining directly overhead. How fast is the shadow of the plane moving along the ground? (That is, what is the magnitude of the horizontal component of the plane's velocity?)

*__37.__ The vector **A** in the drawing has a magnitude of 750 units. Determine the magnitude and direction of the x and y components of the vector **A**, relative to (a) the black axes and (b) the colored axes.

*__38.__ A force vector points at an angle of 52° above the

+x axis. It has a y component of +290 newtons. Find (a) the magnitude and (b) the x component of the force vector.

**__39. ssm www__ The drawing shows a force vector that has a magnitude of 475 newtons. Find the (a) x, (b) y, and (c) z components of the vector.

Section 1.8 Addition of Vectors by Means of Components

40. You are on a treasure hunt and your map says "Walk due west for 52 paces, then walk 30.0° north of west for 42 paces, and finally walk due north for 25 paces." What is the magnitude of the component of your displacement in the direction (a) due north and (b) due west?

41. ssm A golfer, putting on a green, requires three strokes to "hole the ball." During the first putt, the ball rolls 5.0 m due east. For the second putt, the ball travels 2.1 m at an angle of 20.0° north of east. The third putt is 0.50 m due north. What displacement (magnitude and direction relative to due east) would have been needed to "hole the ball" on the very first putt?

42. Two forces are applied to a tree stump to pull it out of the ground. Force **F_A** has a magnitude of 2240 newtons and points 34.0° south of east, while force **F_B** has a magnitude of 3160 newtons and points due south. Using the component method, find the magnitude and direction of the resultant force **F_A** + **F_B** that is applied to the stump. Specify the direction with respect to due east.

43. A football player runs the pattern given in the drawing by the three displacement vectors **A**, **B**, and **C**. The magnitudes of these vectors are A = 5.00 m, B = 15.0 m, and C = 18.0 m. Using the component method, find the magnitude and direction θ of the resultant vector **A** + **B** + **C**.

44. On a safari, a team of naturalists sets out toward a research station located 4.8 km away in a direction 42° north of east. After traveling in a straight line for 2.4 km, they stop and discover that they have been traveling 22° north of east, because their guide misread his compass. What are (a) the magnitude and (b) the direction (relative to due east) of the displacement vector now required to bring the team to the research station?

45. A baby elephant is stuck in a mud hole. To help pull it out, game keepers use a rope to apply force **F_A**, as part *a* of the drawing shows. By itself, however, force **F_A** is insufficient.

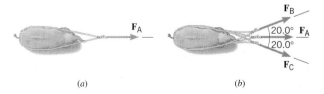

(a)   (b)

Therefore, two additional forces $\mathbf{F_B}$ and $\mathbf{F_C}$ are applied, as in part *b* of the drawing. Each of these additional forces has the same magnitude *F*. The magnitude of the resultant force acting on the elephant in part *b* of the drawing is twice that in part *a*. Find the ratio F/F_A.

*46. Two forces act on an object. One has a magnitude of 166 newtons and points at an angle of 60.0° above the +*x* axis. The second has a magnitude of 284 newtons and points at an angle of 30.0° above the +*x* axis. A third force is applied and balances to zero the effects of the other two. What are the magnitude and direction of this third force? Specify the direction relative to the negative *x* axis.

*47. ssm Vector **A** has a magnitude of 6.00 units and points due east. Vector **B** points due north. (a) What is the magnitude of **B**, if the vector **A** + **B** points 60.0° north of east? (b) Find the magnitude of **A** + **B**.

*48. The route followed by a hiker consists of three displacement vectors **A**, **B**, and **C**. Vector **A** is along a measured trail and is 1550 m in a direction 25.0° north of east. Vector **B** is not along a measured trail, but the hiker uses a compass and knows that the direction is 41.0° east of south. Similarly, the direction of vector **C** is 35.0° north of west. The hiker ends up back where she started, so the resultant displacement is zero, or **A** + **B** + **C** = 0. Find the magnitudes of (a) vector **B** and (b) vector **C**.

*49. ssm A sailboat race course consists of four legs, defined by the displacement vectors **A**, **B**, **C**, and **D**, as the drawing indicates. The magnitudes of the first three vectors are *A* = 3.20 km, *B* = 5.10 km, and *C* = 4.80 km. The finish line of the course coincides with the starting line. Using the data in the drawing, find the distance of the fourth leg and the angle θ.

*50. A grasshopper makes four jumps. The displacement vectors are (1) 27.0 cm, due west; (2) 23.0 cm, 35.0° south of west; (3) 28.0 cm, 55.0° south of east; and (4) 35.0 cm, 63.0° north of east. Find the magnitude and direction of the resultant displacement. Express the direction with respect to due west.

ADDITIONAL PROBLEMS

51. An observer, whose eyes are 1.83 m above the ground, is standing 32.0 m away from a tree. The ground is level, and the tree is growing perpendicular to it. The observer's line of sight with the treetop makes an angle of 20.0° above the horizontal. How tall is the tree?

52. (a) Two workers are trying to move a heavy crate. One pushes on the crate with a force **A**, which has a magnitude of 445 newtons and is directed due west. The other pushes with a force **B**, which has a magnitude of 325 newtons and is directed due north. What are the magnitude and direction of the resultant force **A** + **B** applied to the crate? (b) Suppose that the second worker applies a force −**B** instead of **B**. What then are the magnitude and direction of the resultant force **A** − **B** applied to the crate? In both cases express the direction relative to due west.

53. ssm The speed of an object and the direction in which it moves constitute a vector quantity known as the velocity. An ostrich is running at a speed of 17.0 m/s in a direction of 68.0° north of west. What is the magnitude of the ostrich's velocity component that is directed (a) due north and (b) due west?

54. The variables *x*, *v*, and *a* have the dimensions of [L], [L]/[T], and [L]/[T]², respectively. These variables are related by an equation that has the form $v^n = 2ax$, where *n* is an integer constant (1, 2, 3, etc.) without dimensions. What must be the value of *n*, so that both sides of the equation have the same dimensions? Explain your reasoning.

55. Find the resultant of the three displacement vectors in the drawing by means of the component method. The magnitudes of the vectors are *A* = 5.00 m, *B* = 5.00 m, and *C* = 4.00 m.

56. A frog hops four times: twice forward, once to the right, and once forward again. Each hop covers a distance of 28 cm. What is the magnitude of the frog's displacement?

57. ssm Two ropes are attached to a heavy box to pull it along the floor. One rope applies a force of 475 newtons in a direction due west; the other applies a force of 315 newtons in a direction due south. As we will see later in the text, force is a vector quantity. (a) How much force should be applied by a single rope, and (b) in what direction (relative to due west), if it is to accomplish the same effect as the two forces added together?

*58. A displacement vector **A** has a magnitude of 636 m and points 40.0° above the −*x* axis. Another displacement vector **B** is added to **A**. The resultant has the same magnitude as **A**, but the opposite direction. Find (a) the *x* component and (b) the *y* component of **B**.

*59. Consider the two vectors **A** and **B** in the drawing for problem 55. Determine (a) the vector sum **A** + **B** and (b) the vector difference **A** − **B** using the method of components. In each case, specify both magnitude and direction (relative to the negative *x* axis).

*60. At a picnic, there is a contest in which hoses are used to shoot water at a beach ball from three directions. As a result, three forces act on the ball, $\mathbf{F_1}$, $\mathbf{F_2}$, and $\mathbf{F_3}$ (see the drawing). The magnitudes of $\mathbf{F_1}$ and $\mathbf{F_2}$ are F_1 = 50.0 newtons and F_2 = 90.0 newtons. Using a scale drawing and the graphical technique,

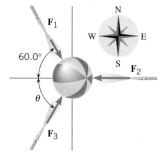

determine (a) the magnitude of $\mathbf{F_3}$ and (b) the angle θ such that the resultant force acting on the ball is zero.

*61. Before starting this problem, review Conceptual Example 6. The force vector $\mathbf{F_A}$ has a magnitude of 90.0 newtons and points due east. The force vector $\mathbf{F_B}$ has a magnitude of 135 newtons and points 75° north of east. Use the graphical method and find the magnitude and direction of (a) $\mathbf{F_A} - \mathbf{F_B}$ (give the direction with respect to due east) and (b) $\mathbf{F_B} - \mathbf{F_A}$ (give the direction with respect to due west).

*62. Three deer, A, B, and C, are grazing in a field. Deer B is located 62 m from deer A at an angle of 51° north of west. Deer C

is located 77° north of east relative to deer A. The distance between deer B and C is 95 m. What is the distance between deer A and C? (*Hint: Consider the law of cosines given in Appendix E.*)

**63. What are the x and y components of the vector that must be added to the following three vectors, so that the sum of the four vectors is zero? Due east is the $+x$ direction, and due north is the $+y$ direction.

$\mathbf{A} = 113$ units, 60.0° south of west
$\mathbf{B} = 222$ units, 35.0° south of east
$\mathbf{C} = 177$ units, 23.0° north of east

CONCEPTS & CALCULATIONS
GROUP LEARNING PROBLEMS

Note: Each of these problems consists of Concept Questions followed by a related quantitative Problem. They are designed for use by students working alone or in small learning groups. The Concept Questions involve little or no mathematics and are intended to stimulate group discussions. They focus on the concepts with which the problems deal. Recognizing the concepts is the essential initial step in any problem-solving technique.

64. Concept Questions (a) Considering the fact that 3.28 ft = 1 m, which is the larger unit for measuring area, 1 ft² or 1 m²? (b) Consider a 1330-ft² apartment. With your answer to part (a) in mind and without doing any calculations, decide whether this apartment has an area that is greater than or less than 1330 m².

Problem In a 1330-ft² apartment, how many square meters of area are there? Be sure that your answer is consistent with your answers to the Concept Questions.

65. Concept Question The corners of a square lie on a circle diameter D. Each side of the square has a length L. Is L smaller or larger than D? Explain your reasoning using the Pythagorean theorem.

Problem The diameter D of the circle is 0.35 m. Each side of the square has a length L. Find L. Be sure that your answer is consistent with your answer to the Concept Question.

66. Concept Question Can the x or y component of a vector ever have a greater magnitude than the vector itself has? Give your reasoning.

Problem A force vector has a magnitude of 575 newtons and points at an angle of 36.0° below the positive x axis. What are (a) the x scalar component and (b) the y scalar component of the vector? Verify that your answers are consistent with your answer to the Concept Question.

67. Concept Questions The components of vector $\mathbf{A}$ are A_x and A_y, and the angle that it makes with respect to the positive x axis is θ. (a) Does increasing the component A_x (while holding A_y constant)

increase or decrease the angle θ? (b) Does increasing the component A_y (while holding A_x constant) increase or decrease the angle θ? Account for your answers.

Problem (a) The components of displacement vector $\mathbf{A}$ are $A_x = 12$ m and $A_y = 12$ m. Find θ. (b) The components of displacement vector $\mathbf{A}$ are $A_x = 17$ m and $A_y = 12$ m. Find θ. (c) The components of displacement vector $\mathbf{A}$ are $A_x = 12$ m and $A_y = 17$ m. Find θ. Be sure that your answers are consistent with your answers to the Concept Questions.

68. Concept Questions Vector $\mathbf{A}$ points due west, while vector $\mathbf{B}$ points due south. (a) Does the direction $\mathbf{A} + \mathbf{B}$ point north or south of due west? (b) Does the direction of $\mathbf{A} - \mathbf{B}$ point north or south of due west? Give your reasoning in each case.

Problem Vector $\mathbf{A}$ has a magnitude of 63 units and points due west, while vector $\mathbf{B}$ has the same magnitude and points due south. Find the magnitude and direction of (a) $\mathbf{A} + \mathbf{B}$ and (b) $\mathbf{A} - \mathbf{B}$. Specify the directions relative to due west. Verify that your answers agree with your answers to the Concept Questions.

69. Concept Questions A pilot flies her route in two straight-line segments. The displacement vector $\mathbf{A}$ for the first segment has a magnitude of 244 km and a direction 30.0° north of east. The displacement vector $\mathbf{B}$ for the second segment has a magnitude of 175 km and a direction due west. The resultant displacement vector is $\mathbf{R} = \mathbf{A} + \mathbf{B}$ and makes an angle θ with the direction due east. Make a drawing to scale showing the vectors $\mathbf{A}$ and $\mathbf{B}$ placed tail-to-head and the resultant vector $\mathbf{R}$. Without doing any calculations decide if (a) the magnitude of $\mathbf{R}$ is greater or smaller than the magnitude of $\mathbf{A}$, (b) the magnitude of $\mathbf{R}$ is greater or smaller than the magnitude of $\mathbf{B}$, and (c) the angle θ is greater than, smaller than, or equal to 30.0°.

Problem Using the component method, find the magnitude of $\mathbf{R}$ and the directional angle θ. Check to see that your answers are consistent with your answers to the Concept Questions.

The USAF Thunderbird flight team. The motion of each aircraft entails three concepts that are discussed in this chapter: displacement, velocity, and acceleration.

CHAPTER 2

KINEMATICS IN ONE DIMENSION

2.1 DISPLACEMENT

There are two aspects to any motion. In a purely descriptive sense, there is the movement itself. Is it rapid or slow, for instance? Then, there is the issue of what causes the motion or what changes it, which requires that forces be considered. *Mechanics* is the branch of physics that focuses on the motion of objects and the forces that cause the motion to change. There are two parts to mechanics: kinematics and dynamics. *Kinematics* deals with the concepts that are needed to describe motion, without any reference to forces. The present chapter discusses these concepts as they apply to motion in one dimension, while the next chapter treats two-dimensional motion. *Dynamics* deals with the effect that forces have on motion, a topic that is considered in Chapter 4. We turn now to the first of the kinematics concepts to be discussed, which is displacement.

To describe the motion of an object, we must be able to specify the location of the object at all times, and Figure 2.1 shows how to do this for one-dimensional motion. In this drawing, the initial position of a car is indicated by the vector labeled $\mathbf{x_0}$. The length of $\mathbf{x_0}$ is the distance of the car from an arbitrarily chosen origin. At a later time the car has moved to a new position, which is indicated by the vector $\mathbf{x}$. The *displacement* of the car $\Delta\mathbf{x}$ (read as "delta x" or "the change in x") is a vector drawn from the initial position to the final position. Displacement is a vector quantity in the sense discussed in Section 1.5, for it conveys both a magnitude (the distance between the initial and final positions) and a direction. The displacement can be related to $\mathbf{x_0}$ and $\mathbf{x}$ by noting from the drawing that

$$\mathbf{x_0} + \Delta\mathbf{x} = \mathbf{x} \quad \text{or} \quad \Delta\mathbf{x} = \mathbf{x} - \mathbf{x_0}$$

Thus, the displacement $\Delta\mathbf{x}$ is the difference between $\mathbf{x}$ and $\mathbf{x_0}$, and the Greek letter delta (Δ) is used to denote this difference. It is important to note that the change in any variable is always the final value minus the initial value.

Figure 2.1 The displacement $\Delta\mathbf{x}$ is a vector that points from the initial position $\mathbf{x_0}$ to the final position $\mathbf{x}$.

● **DEFINITION OF DISPLACEMENT**

The displacement is a vector that points from an object's initial position toward its final position and has a magnitude that equals the shortest distance between the two positions.

SI Unit of Displacement: meter (m)

The SI unit for displacement is the meter (m), but there are other units as well, such as the centimeter and the inch. When converting between centimeters (cm) and inches (in.), remember that 2.54 cm = 1 in.

Often, we will deal with motion along a straight line. In such a case, a displacement in one direction along the line is assigned a positive value, and a displacement in the opposite direction is assigned a negative value. For instance, assume that a car is moving along an east/west direction and that a positive (+) sign is used to denote a direction due east. Then, $\Delta\mathbf{x} = +500$ m represents a displacement that points to the east and has a magnitude of 500 meters. Conversely, $\Delta\mathbf{x} = -500$ m is a displacement that has the same magnitude but points in the opposite direction, due west.

2.2 SPEED AND VELOCITY

Average Speed

One of the most obvious features of an object in motion is how fast it is moving. If a car travels 200 meters in 10 seconds, we say its average speed is 20 meters per second, the **average speed** being the distance traveled divided by the time required to cover the distance:

$$\text{Average speed} = \frac{\text{Distance}}{\text{Elapsed time}} \tag{2.1}$$

Equation 2.1 indicates that the unit for average speed is the unit for distance divided by the unit for time, or meters per second (m/s) in SI units. Example 1 illustrates how the idea of average speed is used.

EXAMPLE 1   Distance Run by a Jogger

How far does a jogger run in 1.5 hours (5400 s) if his average speed is 2.22 m/s?

Reasoning The average speed of the jogger is the average distance per second that he travels. Thus, the distance covered by the jogger is equal to the average distance per second (his average speed) multiplied by the number of seconds (the elapsed time) that he runs.

Solution To find the distance run, we rewrite Equation 2.1 as

$$\text{Distance} = (\text{Average speed})(\text{Elapsed time}) = (2.22 \text{ m/s})(5400 \text{ s}) = \boxed{12\ 000 \text{ m}}$$

Average Velocity

Speed is a useful idea, because it indicates how fast an object is moving. However, speed does not reveal anything about the direction of the motion. To describe both how fast an object moves and the direction of its motion, we need the vector concept of velocity.

● *Concepts at a Glance*   To define the velocity of an object, we will use two concepts that we have already encountered, displacement and time. The building of new concepts from more basic ones is a common theme throughout all of physics. The great strength of physics as a science lies in the fact that it builds a coherent understanding of nature through the development of interrelated concepts. This development is so important that it is emphasized throughout this text by Concepts-at-a-Glance charts, such as that in Figure 2.2. These charts illustrate diagrammatically how useful ideas (e.g., velocity) emerge from more basic ones (e.g., displacement and time). ●

Suppose that the initial position of the car in Figure 2.1 is x_0 when the time is t_0. A little later the car arrives at the final position x at the time t. The difference between these times is the time required for the car to travel between the two positions. We denote this difference by the shorthand notation Δt (read as "delta t"), where Δt represents the final time t minus the initial time t_0:

$$\underbrace{\Delta t = t - t_0}_{\text{Elapsed time}}$$

Note that Δt is defined in a manner analogous to Δx, which is the final position minus the initial position ($\Delta x = x - x_0$).

Concepts at a Glance

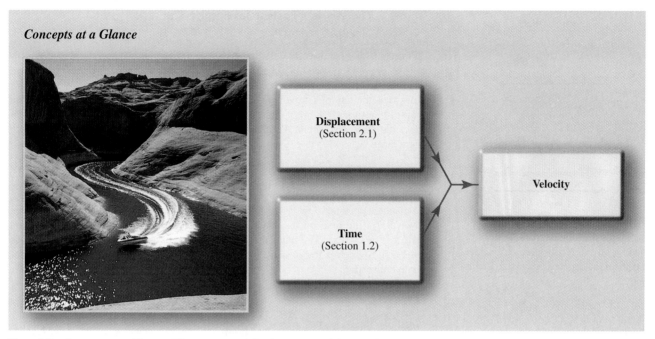

Figure 2.2 *Concepts at a Glance* The concepts of displacement and time are brought together to formulate the concept of velocity. The greater the velocity of this power boat, the greater is its displacement per unit of time.

Dividing the displacement $\Delta\mathbf{x}$ of the car by the elapsed time Δt gives the *average velocity* of the car. It is customary to denote the average value of a quantity by placing a horizontal bar above the symbol representing the quantity. The average velocity, then, is written as $\overline{\mathbf{v}}$, as specified in Equation 2.2:

● **DEFINITION OF AVERAGE VELOCITY**

$$\text{Average velocity} = \frac{\text{Displacement}}{\text{Elapsed time}}$$

$$\overline{\mathbf{v}} = \frac{\mathbf{x} - \mathbf{x_0}}{t - t_0} = \frac{\Delta\mathbf{x}}{\Delta t} \tag{2.2}$$

SI Unit of Average Velocity: meter per second (m/s)

Equation 2.2 indicates that the unit for average velocity is the unit for length divided by the unit for time, or meters per second (m/s) in SI units. Velocity can also be expressed in other units, such as kilometers per hour (km/h) or miles per hour (mi/h).

Average velocity is a vector that points in the same direction as the displacement in Equation 2.2. Figure 2.3 illustrates that the velocity of a car confined to move along a line can point either in one direction or in the opposite direction. As with displacement, we will use plus and minus signs to indicate the two possible directions. If the displacement points in the positive direction, the average velocity is positive. Conversely, if the displacement points in the negative direction, the average velocity is negative. Example 2 illustrates these features of average velocity.

Figure 2.3 In this time-lapse photo of traffic on the Golden Gate bridge, the velocity of the cars in the left lanes (white headlights) is opposite to that of the cars in the right lanes (red taillights).

Figure 2.4   The arrows in the box at the top of the drawing indicate the positive and negative directions for the displacements of the car, as explained in Example 2.

EXAMPLE 2   The World's Fastest Jet-Engine Car

Andy Green in the car *ThrustSSC* set a world record in 1997 of 341.1 m/s (763 mi/h). The car was powered by two jet engines, and it was the first one officially to exceed the speed of sound. To establish such a record, the driver makes two runs through the course, one in each direction, to nullify wind effects. Figure 2.4*a* shows that the car first travels from left to right and covers a distance of 1609 m (1 mile) in a time of 4.740 s. Figure 2.4*b* shows that in the reverse direction, the car covers the same distance in 4.695 s. From these data, determine the average velocity for each run.

Reasoning Average velocity is defined as the displacement divided by the elapsed time. In using this definition we recognize that the displacement is not the same as the distance traveled. Displacement takes the direction of the motion into account, and distance does not. During both runs, the car covers the same distance of 1609 m. However, for the first run the displacement is $\Delta x = +1609$ m, while for the second it is $\Delta x = -1609$ m. The plus and minus signs are essential, because the first run is to the right, which is the positive direction, and the second run is in the opposite or negative direction.

Solution According to Equation 2.2, the average velocities are

Run 1
$$\overline{\mathbf{v}} = \frac{\Delta \mathbf{x}}{\Delta t} = \frac{+1609 \text{ m}}{4.740 \text{ s}} = \boxed{+339.5 \text{ m/s}}$$

Run 2
$$\overline{\mathbf{v}} = \frac{\Delta \mathbf{x}}{\Delta t} = \frac{-1609 \text{ m}}{4.695 \text{ s}} = \boxed{-342.7 \text{ m/s}}$$

In these answers the algebraic signs convey the directions of the velocity vectors. In particular, for run 2 the minus sign indicates that the average velocity, like the displacement, points to the left in Figure 2.4*b*. The magnitudes of the velocities are 339.5 and 342.7 m/s. The average of these numbers is 341.1 m/s and is recorded in the record book.

Instantaneous Velocity

Suppose the magnitude of your average velocity for a long trip was 20 m/s. This value, being an average, does not convey any information about how fast you were moving at any instant during the trip. Surely there were times when your car traveled faster than 20 m/s and times when it traveled more slowly. The ***instantaneous velocity*** **v** of the car indicates how fast the car moves and the direction of the motion at each instant of time. The magnitude of the instantaneous velocity is called the ***instantaneous speed,*** and it is the number (with units) indicated by the speedometer.

The instantaneous velocity at any point during a trip can be obtained by measuring the time interval Δt for the car to travel a *very small* displacement $\Delta \mathbf{x}$. We can then compute the average velocity over this interval. If the time Δt is small enough, the instantaneous velocity does not change much during the measurement. Then, the instantaneous velocity **v** at the point of interest is approximately equal to ($\approx$) the average velocity $\overline{\mathbf{v}}$ computed over the interval, or $\mathbf{v} \approx \overline{\mathbf{v}} = \Delta \mathbf{x}/\Delta t$ (for sufficiently small Δt). In fact, in the limit that Δt becomes infinitesimally small, the instantaneous velocity and the average velocity become equal, so that

$$\mathbf{v} = \lim_{\Delta t \to 0} \frac{\Delta \mathbf{x}}{\Delta t} \qquad (2.3)$$

The notation $\lim_{\Delta t \to 0} (\Delta \mathbf{x}/\Delta t)$ means that the ratio $\Delta \mathbf{x}/\Delta t$ is defined by a limiting process in which smaller and smaller values of Δt are used, so small that they approach zero. As smaller values of Δt are used, $\Delta \mathbf{x}$ also becomes smaller. However, the ratio $\Delta \mathbf{x}/\Delta t$ does *not* become zero but, rather, approaches the value of the instantaneous velocity. For brevity, we will use the word *velocity* to mean "instantaneous velocity" and *speed* to mean "instantaneous speed."

In a wide range of motions, the velocity changes from moment to moment. To describe the manner in which it changes, the concept of acceleration is needed, as the next section discusses.

2.3 ACCELERATION

The velocity of a moving object may change in a number of ways. For example, it may increase, as it does when the driver of a car steps on the gas pedal to pass the car ahead. Or, it may decrease, as it does when the driver applies the brakes to stop at a red light. In either case, the change in velocity may occur over a short or a long time interval.

⬤ *Concepts at a Glance*   To describe the fact that the velocity of an object often changes during a given time interval, we now introduce the new idea of acceleration. As the Concepts-at-a-Glance chart in Figure 2.5 illustrates, this idea depends on two concepts that we have previously encountered, velocity and time. Specifically, the notion of acceleration emerges when the *change* in the velocity is combined with the time during which the change occurs. ⬤

The meaning of **average acceleration** can be illustrated by considering a plane during takeoff. Figure 2.6 focuses attention on how the plane's velocity changes along the runway. During an elapsed time interval $\Delta t = t - t_0$, the velocity changes from an initial value of $\mathbf{v_0}$ to a final value of $\mathbf{v}$. The change $\Delta \mathbf{v}$ in the plane's velocity is its final velocity minus its initial velocity, so that $\Delta \mathbf{v} = \mathbf{v} - \mathbf{v_0}$. The average acceleration $\overline{\mathbf{a}}$ is defined in the following manner, to provide a measure of how much the velocity changes per unit of elapsed time.

Figure 2.5 *Concepts at a Glance* To formulate the concept of acceleration, the change in velocity is combined with the time required for the change to occur. As this runner explodes out of the starting block, he is accelerating because his velocity is changing.

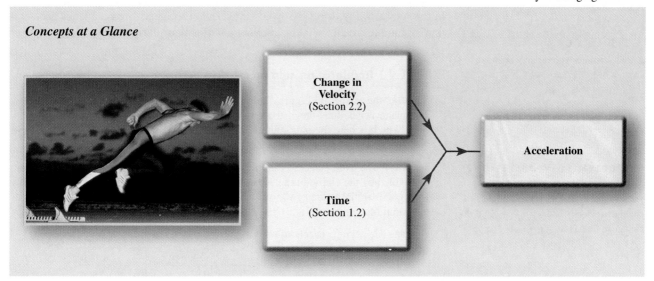

Concepts at a Glance

Change in Velocity (Section 2.2)

Time (Section 1.2)

Acceleration

Figure 2.6   During takeoff, the plane accelerates from an initial velocity $\mathbf{v_0}$ to a final velocity $\mathbf{v}$ during the time interval $\Delta t = t - t_0$.

● **DEFINITION OF AVERAGE ACCELERATION**

$$\text{Average acceleration} = \frac{\text{Change in velocity}}{\text{Elapsed time}}$$

$$\bar{\mathbf{a}} = \frac{\mathbf{v} - \mathbf{v_0}}{t - t_0} = \frac{\Delta \mathbf{v}}{\Delta t} \qquad (2.4)$$

SI Unit of Average Acceleration: meter per second squared (m/s^2)

The average acceleration $\bar{\mathbf{a}}$ is a vector that points in the same direction as $\Delta\mathbf{v}$, the change in the velocity. Following the usual custom, plus and minus signs indicate the two possible directions for the acceleration vector when the motion is along a straight line.

We are often interested in an object's acceleration at a particular instant of time. The *instantaneous acceleration* $\mathbf{a}$ can be defined by analogy with the procedure used in Section 2.2 for instantaneous velocity:

$$\mathbf{a} = \lim_{\Delta t \to 0} \frac{\Delta \mathbf{v}}{\Delta t} \qquad (2.5)$$

Equation 2.5 indicates that the instantaneous acceleration is a limiting case of the average acceleration. When the time interval Δt for measuring the acceleration becomes extremely small (approaching zero in the limit), the average acceleration and the instantaneous acceleration become equal. Moreover, in many situations the acceleration is constant, so the acceleration has the same value at any instant of time. In the future, we will use the word *acceleration* to mean "instantaneous acceleration." Example 3 deals with the acceleration of a plane during takeoff.

A long-distance runner accelerates down a hill.

EXAMPLE 3   Acceleration and Increasing Velocity

Suppose the plane in Figure 2.6 starts from rest ($\mathbf{v_0} = 0$ m/s) when $t_0 = 0$ s. The plane accelerates down the runway and at $t = 29$ s attains a velocity of $\mathbf{v} = +260$ km/h, where the plus sign indicates the velocity points to the right. Determine the average acceleration of the plane.

Reasoning   The average acceleration of the plane is defined as the change in its velocity divided by the elapsed time. The change in the plane's velocity is its final velocity $\mathbf{v}$ minus its initial velocity $\mathbf{v_0}$, or $\mathbf{v} - \mathbf{v_0}$. The elapsed time is the final time t minus the initial time t_0, or $t - t_0$.

Solution   The average acceleration is expressed by Equation 2.4 as

$$\bar{\mathbf{a}} = \frac{\mathbf{v} - \mathbf{v_0}}{t - t_0} = \frac{260 \text{ km/h} - 0 \text{ km/h}}{29 \text{ s} - 0 \text{ s}} = \boxed{+9.0 \; \frac{\text{km/h}}{\text{s}}}$$

● **PROBLEM SOLVING INSIGHT**

The change in any variable is the final value minus the initial value: e.g., the change in velocity is $\Delta\mathbf{v} = \mathbf{v} - \mathbf{v_0}$, and the change in time is $\Delta t = t - t_0$.

Figure 2.7 An acceleration of $+9.0 \dfrac{\text{km/h}}{\text{s}}$ means that the velocity of the plane changes by +9.0 km/h during each second of the motion. The "+" direction for **a** and **v** is to the right.

The average acceleration calculated in Example 3 is read as "nine kilometers per hour per second." Assuming the acceleration of the plane is constant, a value of $9.0 \dfrac{\text{km/h}}{\text{s}}$ means the velocity changes by 9.0 km/h during each second of the motion. During the first second, the velocity increases from 0 to 9.0 km/h; during the next second, the velocity increases by another 9.0 km/h to 18 km/h, and so on. Figure 2.7 illustrates how the velocity changes during the first two seconds. By the end of the 29th second, the velocity is 260 km/h.

It is customary to express the units for acceleration solely in terms of SI units. One way to obtain SI units for the acceleration in Example 3 is to convert the velocity units from km/h to m/s:

$$\left(260 \; \frac{\text{km}}{\text{h}}\right)\left(\frac{1000 \text{ m}}{1 \text{ km}}\right)\left(\frac{1 \text{ h}}{3600 \text{ s}}\right) = 72 \; \frac{\text{m}}{\text{s}}$$

The average acceleration then becomes

$$\bar{\mathbf{a}} = \frac{72 \text{ m/s} - 0 \text{ m/s}}{29 \text{ s} - 0 \text{ s}} = +2.5 \text{ m/s}^2$$

where we have used $2.5 \; \dfrac{\text{m/s}}{\text{s}} = 2.5 \; \dfrac{\text{m}}{\text{s} \cdot \text{s}} = 2.5 \; \dfrac{\text{m}}{\text{s}^2}$. An acceleration of $2.5 \; \dfrac{\text{m}}{\text{s}^2}$ is read as "2.5 meters per second per second" (or "2.5 meters per second squared") and means that the velocity changes by 2.5 m/s during each second of the motion.

Example 4 deals with a case where the motion becomes slower as time passes.

EXAMPLE 4   Acceleration and Decreasing Velocity

A drag racer crosses the finish line, and the driver deploys a parachute and applies the brakes to slow down, as Figure 2.8 illustrates. The driver begins slowing down when $t_0 = 9.0$ s and the car's velocity is $\mathbf{v_0} = +28$ m/s. When $t = 12.0$ s, the velocity has been reduced to $\mathbf{v} = +13$ m/s. What is the average acceleration of the dragster?

Reasoning The average acceleration of an object is always specified as its change in velocity, $\mathbf{v} - \mathbf{v_0}$, divided by the elapsed time, $t - t_0$. This is true whether the final velocity is less than the initial velocity or greater than the initial velocity.

Solution The average acceleration is, according to Equation 2.4,

$$\bar{\mathbf{a}} = \frac{\mathbf{v} - \mathbf{v_0}}{t - t_0} = \frac{13 \text{ m/s} - 28 \text{ m/s}}{12.0 \text{ s} - 9.0 \text{ s}} = \boxed{-5.0 \text{ m/s}^2}$$

(a)

Figure 2.8 (*a*) To slow down, a drag racer deploys a parachute and applies the brakes. (*b*) The velocity of the car is decreasing, giving rise to an average acceleration $\bar{\mathbf{a}}$ that points opposite to the velocity.

(b)

Figure 2.9 shows how the velocity of the dragster changes during the braking, assuming that the acceleration is constant throughout the motion. The acceleration calculated in Example 4 is negative, indicating that the acceleration points to the left in the drawing. As a result, the acceleration and the velocity point in *opposite* directions. ***Whenever the acceleration and velocity vectors have opposite directions, the object slows down and is said to be "decelerating."*** In contrast, the acceleration and velocity vectors in Figure 2.7 point in the *same* direction, and the object speeds up.

2.4 EQUATIONS OF KINEMATICS FOR CONSTANT ACCELERATION

● *Concepts at a Glance* It is now possible to describe the motion of an object traveling with a constant acceleration along a straight line. To do so, we will use a set of equations known as the equations of kinematics for constant acceleration. These equations entail no new concepts, because they will be obtained by combining the familiar ideas of displacement, velocity, and acceleration, as the Concepts-at-a-Glance chart in Figure 2.10 shows. However, they will provide a very convenient way to determine certain aspects of the motion, such as the final position and velocity of a moving object. ●

Figure 2.9 Here, an acceleration of -5.0 m/s^2 means the velocity decreases by 5.0 m/s during each second of elapsed time.

Concepts at a Glance

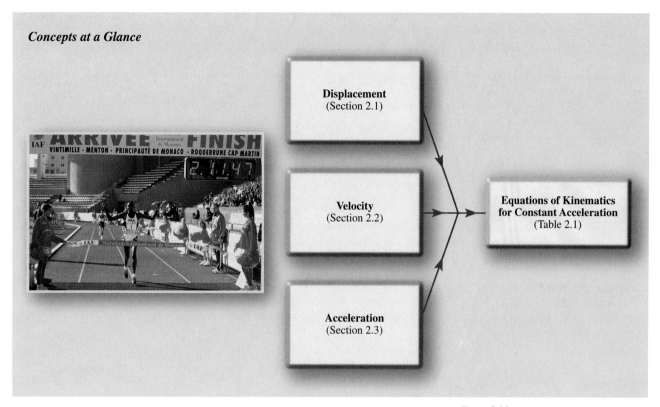

Figure 2.10 *Concepts at a Glance*
The equations of kinematics for constant acceleration are obtained by combining the concepts of displacement, velocity, and acceleration. Whenever Ezekiel Bitok's acceleration remained constant during this marathon, these equations could be used to calculate his displacement and velocity as a function of time.

In discussing the equations of kinematics, it will be convenient to assume that the object is located at the origin $x_0 = 0$ m when $t_0 = 0$ s. With this assumption, the displacement $\Delta x = x - x_0$ becomes $\Delta x = x$. Furthermore, it is customary to dispense with the use of boldface symbols for the displacement, velocity, and acceleration vectors in the equations that follow. We will, however, continue to convey the directions of these vectors with plus or minus signs.

Consider an object that has an initial velocity of v_0 at time $t_0 = 0$ s and moves for a time t with a constant acceleration a. For a complete description of the motion, it is also necessary to know the final velocity and displacement at time t. The final velocity v can be obtained directly from Equation 2.4:

$$\bar{a} = a = \frac{v - v_0}{t} \quad \text{or} \quad v = v_0 + at \tag{2.4}$$

The displacement x at time t can be obtained from Equation 2.2, if a value for the average velocity $\bar{v}$ can be obtained. Considering the assumption that $x_0 = 0$ m at $t_0 = 0$ s, we have

$$\bar{v} = \frac{x - x_0}{t - t_0} = \frac{x}{t} \quad \text{or} \quad x = \bar{v}t \tag{2.2}$$

Because the acceleration is constant, the velocity increases at a constant rate. Thus, the average velocity $\bar{v}$ is midway between the initial and final velocities:

$$\bar{v} = \tfrac{1}{2}(v_0 + v) \quad \text{(constant acceleration)} \tag{2.6}$$

Equation 2.6 applies only if the acceleration is constant and cannot be used when the acceleration is changing. The displacement at time t can now be determined as

$$x = \bar{v}t = \tfrac{1}{2}(v_0 + v)t \qquad \text{(constant acceleration)} \qquad (2.7)$$

Notice in Equations 2.4 ($v = v_0 + at$) and 2.7 [$x = \tfrac{1}{2}(v_0 + v)t$] that there are five kinematic variables:

1. x = displacement
2. $a = \bar{a}$ = acceleration (constant)
3. v = final velocity at time t
4. v_0 = initial velocity at time $t_0 = 0$ s
5. t = time elapsed since $t_0 = 0$ s

Each of the two equations contains four of these variables, so if three of them are known, the fourth variable can always be found. Example 5 illustrates how Equations 2.4 and 2.7 are used to describe the motion of an object.

EXAMPLE 5   The Displacement of a Speedboat

The speedboat in Figure 2.11 has a constant acceleration of $+2.0$ m/s^2. If the initial velocity of the boat is $+6.0$ m/s, find its displacement after 8.0 seconds.

Reasoning Numerical values for the three known variables are listed in the data table below. We wish to determine the displacement x of the speedboat, so it is an unknown variable. Therefore, we have placed a question mark in the displacement column of the data table.

Speedboat Data				
x	a	v	v_0	t
?	$+2.0$ m/s^2		$+6.0$ m/s	8.0 s

We can use $x = \tfrac{1}{2}(v_0 + v)t$ to find the displacement of the boat if a value for the final velocity v can be found. To find the final velocity, it is necessary to use the value given for the acceleration, because it tells us how the velocity changes, according to $v = v_0 + at$.

Solution The final velocity is

$$v = v_0 + at = 6.0 \text{ m/s} + (2.0 \text{ m/s}^2)(8.0 \text{ s}) = +22 \text{ m/s} \qquad (2.4)$$

The displacement of the boat can now be obtained:

$$x = \tfrac{1}{2}(v_0 + v)t = \tfrac{1}{2}(6.0 \text{ m/s} + 22 \text{ m/s})(8.0 \text{ s}) = \boxed{+110 \text{ m}} \qquad (2.7)$$

Figure 2.11 (a) An accelerating speed boat. (b) The boat's displacement x can be determined if the boat's acceleration, initial velocity, and time of travel are known.

(a)

(b)

A calculator would give the answer as 112 m, but this number must be rounded to 110 m, since the data are accurate to only two significant figures.

The solution to Example 5 involved two steps: finding the final velocity v and then calculating the displacement x. It would be helpful if we could find an equation that allows us to determine the displacement in a single step. Using Example 5 as a guide, we can obtain such an equation by substituting the final velocity v from Equation 2.4 ($v = v_0 + at$) into Equation 2.7 [$x = \frac{1}{2}(v_0 + v)t$]:

$$x = \frac{1}{2}(v_0 + v)t = \frac{1}{2}(v_0 + \boxed{v_0 + at})t = \frac{1}{2}(2v_0 t + at^2)$$

$$x = v_0 t + \frac{1}{2}at^2 \quad \text{(constant acceleration)} \quad (2.8)$$

You can verify that Equation 2.8 gives the displacement of the speedboat directly without the intermediate step of determining the final velocity. The first term ($v_0 t$) on the right side of this equation represents the displacement that would result if the acceleration were zero, and the velocity remained constant at its initial value of v_0. The second term ($\frac{1}{2}at^2$) gives the additional displacement that arises because the velocity changes (a is not zero) to values that are different from its initial value. We now turn to another example of accelerated motion.

EXAMPLE 6   Catapulting a Jet

A jet is taking off from the deck of an aircraft carrier, as Figure 2.12 shows. Starting from rest, the jet is catapulted with a constant acceleration of $+31$ m/s^2 along a straight line and reaches a velocity of $+62$ m/s. Find the displacement of the jet.

Reasoning The data are as follows:

Jet Data				
x	a	v	v_0	t
?	$+31$ m/s^2	$+62$ m/s	0 m/s	

The initial velocity v_0 is zero, since the jet starts from rest. The displacement x of the aircraft can be obtained from $x = \frac{1}{2}(v_0 + v)t$, if we can determine the time t during which the plane is being accelerated. But t is controlled by the value of the acceleration. With larger accelerations, the jet reaches its final velocity in shorter times, as can be seen by solving Equation 2.4 ($v = v_0 + at$) for t.

THE PHYSICS OF...

catapulting a jet from an aircraft carrier.

● PROBLEM SOLVING INSIGHT

"Implied data" are important. For instance, in Example 6 the phrase "starting from rest" means that the initial velocity is zero ($v_0 = 0$ m/s).

Figure 2.12 (*a*) A plane is being launched from an aircraft carrier. (*b*) During the launch, a catapult accelerates the jet down the flight deck.

(a)

(b)

TABLE 2.1 Equations of Kinematics for Constant Acceleration

Equation Number	Equation	Variables				
		x	a	v	v_0	t
(2.4)	$v = v_0 + at$	—	✓	✓	✓	✓
(2.7)	$x = \frac{1}{2}(v_0 + v)t$	✓	—	✓	✓	✓
(2.8)	$x = v_0 t + \frac{1}{2}at^2$	✓	✓	—	✓	✓
(2.9)	$v^2 = v_0^2 + 2ax$	✓	✓	✓	✓	—

Solution Solving Equation 2.4 for t, we find

$$t = \frac{v - v_0}{a} = \frac{62 \text{ m/s} - 0 \text{ m/s}}{31 \text{ m/s}^2} = 2.0 \text{ s}$$

Since the time is now known, the displacement can be found by using Equation 2.7:

$$x = \frac{1}{2}(v_0 + v)t = \frac{1}{2}(0 \text{ m/s} + 62 \text{ m/s})(2.0 \text{ s}) = \boxed{+62 \text{ m}} \tag{2.7}$$

When a, v, and v_0 are known, but the time t is not known, as in Example 6, it is possible to calculate the displacement x in a single step. Solving Equation 2.4 for the time $[t = (v - v_0)/a]$, and then substituting into Equation 2.7 $[x = \frac{1}{2}(v_0 + v)t]$ reveals that

$$x = \frac{1}{2}(v_0 + v)t = \frac{1}{2}(v_0 + v)\boxed{\frac{v - v_0}{a}} = \frac{v^2 - v_0^2}{2a}$$

Solving for v^2 shows that

$$v^2 = v_0^2 + 2ax \qquad \text{(constant acceleration)} \tag{2.9}$$

It is a straightforward exercise to verify that Equation 2.9 can be used to find the displacement of the jet in Example 6 without having to solve first for the time. Table 2.1 presents a summary of the equations that we have been considering. These equations are called the **equations of kinematics.** Each equation contains four variables, as indicated by the check marks (✓) in the table. The next section shows how to apply the equations of kinematics.

2.5 APPLICATIONS OF THE EQUATIONS OF KINEMATICS

The equations of kinematics can be used for any moving object, as long as the acceleration of the object is constant. However, to avoid errors when using these equations, it helps to follow a few sensible guidelines and to be alert for a few situations that can arise during your calculations.

Decide at the start which directions are to be called positive (+) and negative (−) relative to a conveniently chosen coordinate origin. This decision is arbitrary, but important, for displacement, velocity, and acceleration are vectors, and their directions must always be taken into account. In the examples that follow, the positive and negative directions will be shown in the drawings that accompany the problems. It does not matter which direction is chosen to be positive. However, once the choice has been made, it should not be changed during the course of the calculation.

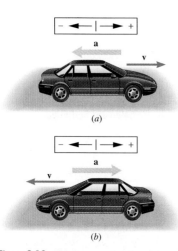

(a)

(b)

Figure 2.13 When a car decelerates along a straight road, the direction of the acceleration vector depends on the direction in which the car is traveling, as Conceptual Example 7 discusses.

As you reason through a problem before attempting to solve it, be sure to interpret the terms "decelerating" or "deceleration" correctly, should they occur in the problem statement. These terms are the source of frequent confusion, and Conceptual Example 7 offers help in understanding them.

CONCEPTUAL EXAMPLE 7
Deceleration Versus Negative Acceleration

A car is traveling along a straight road and is decelerating. Does the car's acceleration a necessarily have a negative value?

Reasoning and Solution We begin with the meaning of the term "decelerating," which has nothing to do with whether the acceleration a is positive or negative. The term means only that the acceleration vector points opposite to the velocity vector and indicates that the moving object is slowing down. When a moving object slows down, its instantaneous speed (the magnitude of the instantaneous velocity) decreases. One possibility is that the velocity vector of the car points to the right, in the positive direction, as Figure 2.13a shows. The term "decelerating" implies that the acceleration vector points opposite, or to the left, which is the negative direction. Here, the value of the acceleration a would indeed be negative. However, there is another possibility. The car could be traveling to the left, as in Figure 2.13b. Now, since the velocity vector points to the left, the acceleration vector would point opposite or to the right, according to the meaning of the term "decelerating." But right is the positive direction, so the acceleration a would have a positive value in Figure 2.13b. We see, then, that *a decelerating object does not necessarily have a negative acceleration.*

Related Homework Material: *Problems 20 and 38*

INTERACTIVE LEARNINGWARE #1

START

Visualization

Reasoning

Knowns and Unknowns

Modeling the Problem

Numerical Solution

FINISH

The important issue of choosing which directions are to be called positive (+) and negative (−) is discussed further in this Interactive LearningWare example, which can be found on our Web site at the address given below. In this example we revisit Example 5, in which the positive direction was assumed originally to point toward the right (see Figure 2.11). Now, however, we assume that the positive direction points to the left and will show how to deal with such a choice and how to interpret the results. To help you develop good problem-solving skills, this Interactive LearningWare example, and all others to be introduced later, employ the five steps outlined at the left: visualization, reasoning, knowns and unknowns, modeling the problem, and numerical solution.

Go to our Web site at

http://www.wiley.com/college/cutnell

Sometimes there are two possible answers to a kinematics problem, each answer corresponding to a different situation. Example 8 discusses one such case.

EXAMPLE 8   An Accelerating Spacecraft

The spacecraft shown in Figure 2.14a is traveling with a velocity of $+3250$ m/s. Suddenly the retrorockets are fired, and the spacecraft begins to slow down with an acceleration whose magnitude is 10.0 m/s². What is the velocity of the spacecraft when the displacement of the craft is $+215$ km, relative to the point where the retrorockets began firing?

Reasoning Since the spacecraft is slowing down at this stage of the motion, the acceleration must be opposite to the velocity. The velocity points to the right in the drawing, so the acceleration must point to the left, which is the negative direction; thus, $a = -10.0$ m/s². The three known variables are listed as follows:

Spacecraft Data				
x	a	v	v_0	t
$+215\,000$ m	-10.0 m/s²	?	$+3250$ m/s	

THE PHYSICS OF...

the acceleration caused by a retrorocket.

● **PROBLEM SOLVING INSIGHT**
It is only when values are available for at least three of the five kinematic variables (x, a, v, v_0, and t) that the equations in Table 2.1 can be used to determine the fourth and fifth variables.

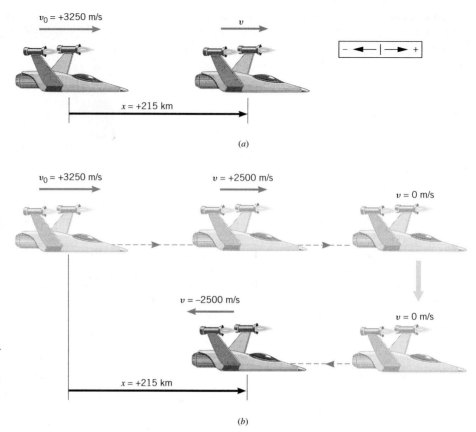

Figure 2.14 (a) Because of an acceleration of -10.0 m/s^2, the spacecraft changes its velocity from v_0 to v. (b) Continued firing of the retrorockets changes the direction of the craft's motion.

As this barn owl comes in for a landing, its velocity vector points in the same direction as its motion. However, since the owl is slowing down, the acceleration vector points opposite to the velocity vector.

The final velocity v of the spacecraft can be calculated using Equation 2.9, since it contains the four pertinent variables.

Solution From Equation 2.9 ($v^2 = v_0^2 + 2ax$), we find that

$$v = \pm\sqrt{v_0^2 + 2ax}$$

$$v = \pm\sqrt{(3250 \text{ m/s})^2 + 2(-10.0 \text{ m/s}^2)(215\ 000 \text{ m})}$$

$$= \boxed{+2500 \text{ m/s}} \text{ and } \boxed{-2500 \text{ m/s}}$$

Both of these answers correspond to the *same* displacement ($x = +215$ km), but each arises in a different part of the motion. The answer $v = +2500$ m/s corresponds to the situation in Figure 2.14a, where the spacecraft has slowed to a speed of 2500 m/s, but is still traveling to the right. The answer $v = -2500$ m/s arises because the retrorockets eventually bring the spacecraft to a momentary halt and cause it to reverse its direction. Then it moves to the left, and its speed increases due to the continually firing rockets. After a time, the velocity of the craft becomes $v = -2500$ m/s, giving rise to the situation in Figure 2.14b. In both parts of the drawing the spacecraft has the same displacement, but a greater travel time is required in part b compared to part a.

The motion of two objects may be interrelated, so they share a common variable. The fact that the motions are interrelated is an important piece of information. In such cases, data for only two variables need be specified for each object.

Often the motion of an object is divided into segments, each with a different acceleration. When solving such problems, it is important to realize that the final velocity for one segment is the initial velocity for the next segment, as Example 9 illustrates.

EXAMPLE 9 A Motorcycle Ride

A motorcycle, starting from rest, has an acceleration of $+2.6$ m/s^2. After the motorcycle has traveled a distance of 120 m, it slows down with an acceleration of -1.5 m/s^2 until its velocity is $+12$ m/s (see Figure 2.15). What is the total displacement of the motorcycle?

Reasoning The total displacement is the sum of the displacements for the first ("speeding up") and second ("slowing down") segments. The displacement for the first segment is $+120$ m. The displacement for the second segment can be found if the initial velocity for this segment can be determined, since values for two other variables are already known ($a = -1.5$ m/s^2 and $v = +12$ m/s). The initial velocity for the second segment can be determined, since it is the final velocity of the first segment.

Solution Recognizing that the motorcycle starts from rest ($v_0 = 0$ m/s), we can determine the final velocity v of the first segment from the given data:

Segment 1 Data				
x	a	v	v_0	t
$+120$ m	$+2.6$ m/s^2	?	0 m/s	

From Equation 2.9 ($v^2 = v_0^2 + 2ax$), it follows that

$$v = \sqrt{v_0^2 + 2ax} = \sqrt{(0 \text{ m/s})^2 + 2(2.6 \text{ m/s}^2)(120 \text{ m})} = +25 \text{ m/s}$$

Now we can use $+25$ m/s as the initial velocity for the second segment, along with the remaining data listed below:

Segment 2 Data				
x	a	v	v_0	t
?	-1.5 m/s^2	$+12$ m/s	$+25$ m/s	

Figure 2.15 This motorcycle ride consists of two segments, each with a different acceleration.

The displacement for segment 2 can be obtained by solving $v^2 = v_0^2 + 2ax$ for x:

$$x = \frac{v^2 - v_0^2}{2a} = \frac{(12 \text{ m/s})^2 - (25 \text{ m/s})^2}{2(-1.5 \text{ m/s}^2)} = +160 \text{ m}$$

The total displacement of the motorcycle is 120 m + 160 m = $\boxed{280 \text{ m}}$.

Now that we have seen how the equations of kinematics are applied to various situations, it's a good idea to summarize the reasoning strategy that has been used. This strategy, which is outlined below, will also be used when we consider freely falling bodies in Section 2.6 and two-dimensional motion in Chapter 3.

REASONING STRATEGY

Applying the Equations of Kinematics

1. Make a drawing to represent the situation being studied. A drawing helps us to "see" what's happening.

2. Decide which directions are to be called positive (+) and negative (−) relative to a conveniently chosen coordinate origin. Do not change your decision during the course of a calculation.

3. In an organized way, write down the values (with appropriate plus and minus signs) that are given for any of the five kinematic variables (x, a, v, v_0, and t). Be on the alert for "implied data," such as the phrase "starts from rest," which means that the value of the initial velocity is $v_0 = 0$ m/s. The data summary boxes used in the examples in the text are a good way of keeping track of this information. In addition, identify the variables that you are being asked to determine.

4. Before attempting to solve a problem, verify that the given information contains values for at least three of the five kinematic variables. Once the three known variables are identified along with the desired unknown variable, the appropriate relation from Table 2.1 can be selected. Remember that the motion of two objects may be interrelated, so they may share a common variable. The fact that the motions are interrelated is an important piece of information. In such cases, data for only two variables need be specified for each object.

5. When the motion of an object is divided into segments, as in Example 9, remember that the final velocity of one segment is the initial velocity for the next segment.

6. Keep in mind that there may be two possible answers to a kinematics problem as, for instance, in Example 8. Try to visualize the different physical situations to which the answers correspond.

2.6   FREELY FALLING BODIES

Everyone has observed the effect of gravity as it causes objects to fall downward. In the absence of air resistance, it is found that all bodies at the same location above the earth fall vertically with the same acceleration. Furthermore, if the distance of the fall is small compared to the radius of the earth, the acceleration remains essentially constant throughout the fall. This idealized motion, in which air resistance is neglected and the acceleration is nearly constant, is known as *free-fall.* Since the acceleration is constant in free-fall, the equations of kinematics can be used.

The acceleration of a freely falling body is called the **acceleration due to gravity,** and its magnitude (without any algebraic sign) is denoted by the symbol g. The acceleration due to gravity is directed downward, toward the center of the earth. Near the earth's surface g is approximately

$$g = 9.80 \text{ m/s}^2 \quad \text{or} \quad 32.2 \text{ ft/s}^2$$

Unless circumstances warrant otherwise, we will use either of these values for g in subsequent calculations. In reality, however, g decreases with increasing altitude and varies slightly with latitude.

Figure 2.16a shows the well-known phenomenon of a rock falling faster than a sheet of paper. The effect of air resistance is responsible for the slower fall of the paper, for when air is removed from the tube, as in Figure 2.16b, the rock and the paper have exactly the same acceleration due to gravity. In the absence of air, the rock and the paper both exhibit free-fall motion. Free-fall is closely approximated for objects falling near the surface of the moon, where there is no air to retard the motion. A nice demonstration of lunar free-fall was performed by astronaut David Scott who dropped a hammer and a feather simultaneously from the same height. Both experienced the same acceleration due to lunar gravity and consequently hit the ground at the same time. The acceleration due to gravity near the surface of the moon is approximately one-sixth as large as that on the earth.

When the equations of kinematics are applied to free-fall motion, it is natural to use the symbol y for the displacement that occurs, since the motion occurs in the vertical or y direction. Thus, when using the equations in Table 2.1 for free-fall motion, we will simply replace x with y. There is no significance to this change. The equations have the same algebraic form for either the horizontal or vertical direction, provided that the acceleration remains constant during the motion. We now turn our attention to several examples that illustrate how the equations of kinematics are applied to freely falling bodies.

Figure 2.16 (a) In the presence of air resistance, the acceleration of the rock is greater than that of the paper. (b) In the absence of air resistance, both the rock and the paper have the same acceleration.

EXAMPLE 10   A Falling Stone

A stone is dropped from rest from the top of a tall building, as Figure 2.17 indicates. After 3.00 s of free-fall, what is the displacement y of the stone?

Reasoning The upward direction is chosen as the positive direction. The three known variables are shown in the box below. The initial velocity v_0 of the stone is zero, because the stone is dropped from rest. The acceleration due to gravity is negative, since it points downward in the negative direction.

Stone Data				
y	a	v	v_0	t
?	-9.80 m/s^2		0 m/s	3.00 s

Equation 2.8 contains the appropriate variables and offers a direct solution to the problem. Since the stone moves downward and upward is the positive direction, we expect the displacement y to have a negative value.

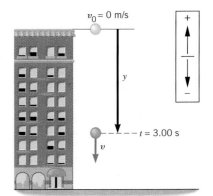

Figure 2.17 The stone, starting with zero velocity at the top of the building, is accelerated downward by gravity.

Solution Using Equation 2.8, we find that

$$y = v_0 t + \tfrac{1}{2}at^2 = (0 \text{ m/s})(3.00 \text{ s}) + \tfrac{1}{2}(-9.80 \text{ m/s}^2)(3.00 \text{ s})^2 = \boxed{-44.1 \text{ m}}$$

The answer for y is negative, as expected.

EXAMPLE 11   The Velocity of a Falling Stone

After 3.00 s of free-fall, what is the velocity v of the stone in Figure 2.17?

Reasoning Because of the acceleration due to gravity, the magnitude of the stone's downward velocity increases by 9.80 m/s during each second of free-fall. The data for the stone are the same as in Example 10, and Equation 2.4 offers a direct solution for the final velocity. Since the stone is moving downward in the negative direction, the value determined for v should be negative.

Solution Using Equation 2.4, we obtain

$$v = v_0 + at = 0 \text{ m/s} + (-9.80 \text{ m/s}^2)(3.00 \text{ s}) = \boxed{-29.4 \text{ m/s}}$$

The velocity is negative, as expected.

The acceleration due to gravity is always a downward-pointing vector. It describes how the speed increases for an object that is falling freely downward. This same acceleration also describes how the speed decreases for an object moving upward under the influence of gravity alone, in which case the object eventually comes to a momentary halt and then falls back to earth. Examples 12 and 13 show how the equations of kinematics are applied to an object that is moving upward under the influence of gravity.

Figure 2.18 At the start of a football game, a referee tosses a coin upward with an initial velocity of $v_0 = +6.00$ m/s. The velocity of the coin is momentarily zero when the coin reaches its maximum height.

• **PROBLEM SOLVING INSIGHT**
"Implied data" is important. In Example 12, for instance, the phrase "how high does the coin go" refers to the maximum height, which occurs when the final velocity v in the vertical direction is $v = 0$ m/s.

EXAMPLE 12   How High Does It Go?

A football game customarily begins with a coin toss to determine who kicks off. The referee tosses the coin up with an initial speed of 6.00 m/s. In the absence of air resistance, how high does the coin go above its point of release?

Reasoning The coin is given an upward initial velocity, as in Figure 2.18. But the acceleration due to gravity points downward. Since the velocity and acceleration point in opposite directions, the coin slows down as it moves upward. Eventually, the velocity of the coin becomes $v = 0$ m/s at the highest point. Assuming that the upward direction is positive, the data can be summarized as shown below:

Coin Data				
y	a	v	v_0	t
?	-9.80 m/s^2	0 m/s	$+6.00 \text{ m/s}$	

With these data, we can use Equation 2.9 ($v^2 = v_0^2 + 2ay$) to find the maximum height y.

Solution Rearranging Equation 2.9, we find that the maximum height of the coin above its release point is

$$y = \frac{v^2 - v_0^2}{2a} = \frac{(0 \text{ m/s})^2 - (6.00 \text{ m/s})^2}{2(-9.80 \text{ m/s}^2)} = \boxed{+1.84 \text{ m}}$$

EXAMPLE 13   How Long Is It in the Air?

In Figure 2.18, what is the total time the coin is in the air before returning to its release point?

Reasoning During the time the coin travels upward, gravity causes its speed to decrease to zero. On the way down, however, gravity causes the coin to regain the lost
speed. Thus, the time for the coin to go up is equal to the time for it to come down.
In other words, the total travel time is twice the time for the upward motion. The data
for the coin during the upward trip are the same as in Example 12. With these data,
we can use Equation 2.4 ($v = v_0 + at$) to find the upward travel time.

Solution Rearranging Equation 2.4, we find that

$$t = \frac{v - v_0}{a} = \frac{0 \text{ m/s} - 6.00 \text{ m/s}}{-9.80 \text{ m/s}^2} = 0.612 \text{ s}$$

The total up-and-down time is twice this value, or $\boxed{1.22 \text{ s}}$.

It is possible to determine the total time by another method. When the coin is
tossed upward and returns to its release point, the displacement for the *entire trip* is
$y = 0$ m. With this value for the displacement, Equation 2.8 ($y = v_0 t + \frac{1}{2}at^2$) can
be used to find the time for the entire trip directly.

Examples 12 and 13 illustrate that the expression "freely falling" does not necessarily mean an object is falling down. A freely falling object is any object moving
either upward or downward under the influence of gravity alone. In either case, the
object always experiences the same *downward acceleration* due to gravity, a fact
that is the focus of the next example.

CONCEPTUAL EXAMPLE 14   Acceleration Versus Velocity

There are three parts to the motion of the coin in Figure 2.18. On the way up, the
coin has a velocity vector that is directed upward and has a decreasing magnitude. At
the top of its path, the coin momentarily has a zero velocity. On the way down, the
coin has a downward-pointing velocity vector with an increasing magnitude. In the
absence of air resistance, does the acceleration of the coin, like the velocity, change
from one part of the motion to another?

Reasoning and Solution Since air resistance is absent, the coin is in free-fall motion.
Therefore, the acceleration vector is that due to gravity and has the same magnitude
and the same direction at all times. It has a magnitude of 9.80 m/s^2 and points downward during both the upward and downward portions of the motion. Furthermore,
just because the coin's instantaneous velocity is zero at the top of the motional path,
don't think that the acceleration vector is also zero there. Acceleration is the rate at
which velocity changes, and the velocity at the top is changing, even though at one
instant it is zero. In fact, the acceleration at the top has the same magnitude of
9.80 m/s^2 and the same downward direction as during the rest of the motion. Thus,
***the coin's velocity vector changes from moment to moment, but its acceleration
vector does not change.***

The motion of an object that is thrown upward and eventually returns to earth
contains a symmetry that is useful to keep in mind from the point of view of problem solving. The calculations just completed indicate that a time symmetry exists in
free-fall motion, in the sense that the time required for the object to reach maximum height equals the time for it to return to its starting point.

A type of symmetry involving the speed also exists. Figure 2.19 shows the coin
considered in Examples 12 and 13. At any displacement y above the point of re-

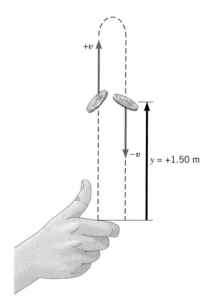

Figure 2.19 For a given displacement
along the motional path, the upward
speed of the coin is equal to its downward speed, but the two velocities point
in opposite directions.

(a)

(b)

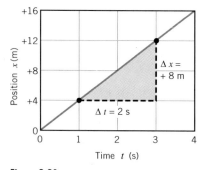

(c)

Figure 2.20 (a) From the edge of a cliff, a pellet is fired straight upward from a gun. The pellet's initial speed is 30 m/s. (b) The pellet is fired straight downward with an initial speed of 30 m/s. (c) In Conceptual Example 15 this drawing plays the central role in reasoning that is based on symmetry.

lease, the coin's speed during the upward trip equals the speed at the same point during the downward trip. For instance, when $y = +1.50$ m, Equation 2.9 gives two possible values for the final velocity v, assuming that the initial velocity is $v_0 = +6.00$ m/s:

$$v^2 = v_0{}^2 + 2ay = (6.00 \text{ m/s})^2 + 2(-9.80 \text{ m/s}^2)(1.50 \text{ m}) = 6.60 \text{ m}^2/\text{s}^2$$
$$v = \pm 2.57 \text{ m/s}$$

The value $v = +2.57$ m/s is the velocity of the coin on the upward trip, while $v = -2.57$ m/s is the velocity on the downward trip. The speed in both cases is identical and equals 2.57 m/s. Likewise, the speed just as the coin returns to its point of release is 6.00 m/s, which equals the initial speed. This symmetry involving the speed arises because the coin loses 9.80 m/s in speed each second on the way up and gains back the same amount each second on the way down. In Conceptual Example 15, we use just this kind of symmetry to guide our reasoning as we analyze the motion of a pellet shot from a gun.

CONCEPTUAL EXAMPLE 15   *Taking Advantage of Symmetry*

Figure 2.20a shows a pellet, having been fired from a gun, moving straight upward from the edge of a cliff. The initial speed of the pellet is 30 m/s. It goes up and then falls back down, eventually hitting the ground beneath the cliff. In Figure 2.20b the pellet has been fired straight downward at the same initial speed. In the absence of air resistance, does the pellet in part b strike the ground beneath the cliff with a smaller, a greater, or the same speed as the pellet in part a?

Reasoning and Solution Because air resistance is absent, the motion is that of free-fall, and the symmetry inherent in free-fall motion offers an immediate answer to the question. Figure 2.20c shows why. This part of the drawing shows the pellet after it has been fired upward and then fallen back down to its starting point. Symmetry indicates that the speed in part c is the same as in part a, namely, 30 m/s. Thus, part c is just like part b, where the pellet is actually fired downward with a speed of 30 m/s. Consequently, whether the pellet is fired as in part a or part b, it starts to move downward from the cliff edge at a speed of 30 m/s. In either case, there is the same acceleration due to gravity and the same displacement from the cliff edge to the ground below. Under these conditions, ***the pellet reaches the ground with the same speed no matter in which vertical direction it is fired initially.***

Related Homework Material: *Problems 44 and 46*

2.7   GRAPHICAL ANALYSIS OF VELOCITY AND ACCELERATION

Graphical techniques are helpful in understanding the concepts of velocity and acceleration. Suppose a bicyclist is riding with a constant velocity of $v = +4$ m/s. The position x of the bicycle can be plotted along the vertical axis of a graph, while the time t is plotted along the horizontal axis. Since the position of the bike increases by 4 m every second, the graph of x versus t is a straight line. Furthermore, if the bike is assumed to be at $x = 0$ m when $t = 0$ s, the straight line passes through the origin, as Figure 2.21 shows. Each point on this line gives the position of the bike at a particular time. For instance, at $t = 1$ s the position is 4 m, while at $t = 3$ s the position is 12 m.

In constructing the graph in Figure 2.21, we used the fact that the velocity was +4 m/s. Suppose, however, that we were given this graph, but did not have prior

Figure 2.21 A graph of position vs. time for an object moving with a constant velocity of $v = \Delta x/\Delta t = +4$ m/s.

knowledge of the velocity. The velocity could be determined by considering what happens to the bike between the times of 1 and 3 s, for instance. The change in time is $\Delta t = 2$ s. During this time interval, the position of the bike changes from $+4$ to $+12$ m, and the change in position is $\Delta x = +8$ m. The ratio $\Delta x/\Delta t$ is called the **slope** of the straight line.

$$\text{Slope} = \frac{\Delta x}{\Delta t} = \frac{+8 \text{ m}}{2 \text{ s}} = +4 \text{ m/s}$$

Notice that the slope is equal to the velocity of the bike. This result is no accident, because $\Delta x/\Delta t$ is the definition of average velocity (see Equation 2.2). Thus, for an object moving with a constant velocity, the slope of the straight line in a position–time graph gives the value of the velocity. Since the position–time graph is a straight line, any time interval Δt can be chosen to calculate the velocity. Choosing a different Δt will yield a different Δx, but the velocity $\Delta x/\Delta t$ will not change. In the real world, objects rarely move with a constant velocity at all times, as the next example illustrates.

EXAMPLE 16   A Bicycle Trip

A bicyclist maintains a constant velocity on the outgoing leg of a journey, zero velocity while stopped for lunch, and another constant velocity on the way back. Figure 2.22 shows the position–time graph for such a trip. Using the time and position intervals indicated in the drawing, obtain the velocities for each segment of the trip.

Reasoning The average velocity $\bar{v}$ is equal to the displacement Δx divided by the elapsed time Δt, $\bar{v} = \Delta x/\Delta t$. The displacement is the final position minus the initial position, which is a positive number for segment 1 and a negative number for segment 3. Note for segment 2 that $\Delta x = 0$ m, since the bicycle is at rest. The drawing shows values for Δx and Δt for each of the three segments.

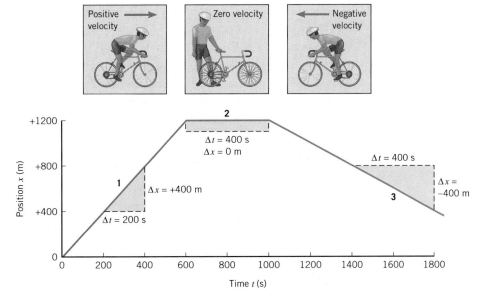

Figure 2.22 This position vs. time graph consists of three straight-line segments, each corresponding to a different constant velocity.

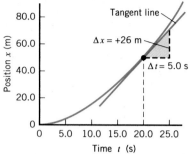

Figure 2.23 When the velocity is changing, the position vs. time graph is a curved line. The slope $\Delta x/\Delta t$ of the tangent line drawn to the curve at a given time is the instantaneous velocity at that time.

Solution The average velocities for the three segments are

Segment 1 $\overline{v} = \dfrac{\Delta x}{\Delta t} = \dfrac{800 \text{ m} - 400 \text{ m}}{400 \text{ s} - 200 \text{ s}} = \dfrac{+400 \text{ m}}{200 \text{ s}} = \boxed{+2 \text{ m/s}}$

Segment 2 $\overline{v} = \dfrac{\Delta x}{\Delta t} = \dfrac{1200 \text{ m} - 1200 \text{ m}}{1000 \text{ s} - 600 \text{ s}} = \dfrac{0 \text{ m}}{400 \text{ s}} = \boxed{0 \text{ m/s}}$

Segment 3 $\overline{v} = \dfrac{\Delta x}{\Delta t} = \dfrac{400 \text{ m} - 800 \text{ m}}{1800 \text{ s} - 1400 \text{ s}} = \dfrac{-400 \text{ m}}{400 \text{ s}} = \boxed{-1 \text{ m/s}}$

In the second segment of the journey the velocity is zero, reflecting the fact that the bike is stationary. Since the position of the bike does not change, segment 2 is a horizontal line that has a zero slope. In the third part of the motion the velocity is negative, because the position of the bike decreases from $x = +800$ m to $x = +400$ m during the 400-s interval shown in the graph. As a result, segment 3 has a negative slope, and the velocity is negative.

If the object is accelerating, its velocity is changing. When the velocity is changing, the x versus t graph is not a straight line, but is a curve, perhaps like that in Figure 2.23. This curve was drawn using Equation 2.8 ($x = v_0 t + \frac{1}{2}at^2$), assuming an acceleration of $a = 0.26$ m/s^2 and an initial velocity of $v_0 = 0$ m/s. The velocity at any instant of time can be determined by measuring the slope of the curve at that instant. The slope at any point along the curve is defined to be the slope of the tangent line drawn to the curve at that point. For instance, in Figure 2.23 a tangent line is drawn at $t = 20.0$ s. To determine the slope of the tangent line, a triangle is constructed using an arbitrarily chosen time interval of $\Delta t = 5.0$ s. The change in x associated with this time interval can be read from the tangent line as $\Delta x = +26$ m. Therefore,

$$\text{Slope of tangent line} = \frac{\Delta x}{\Delta t} = \frac{+26 \text{ m}}{5.0 \text{ s}} = +5.2 \text{ m/s}$$

The slope of the tangent line is the instantaneous velocity, which in this case is $v = +5.2$ m/s. This graphical result can be verified by using Equation 2.4 with $v_0 = 0$ m/s: $v = at = (+0.26 \text{ m/s}^2)(20.0 \text{ s}) = +5.2$ m/s.

Insight into the meaning of acceleration can also be gained with the aid of a graphical representation. Consider an object moving with a constant acceleration of $a = +6$ m/s^2. If the object has an initial velocity of $v_0 = +5$ m/s, its velocity at any time is represented by Equation 2.4 as

$$v = v_0 + at = 5 \text{ m/s} + (6 \text{ m/s}^2)t$$

This relation is plotted as the velocity versus time graph in Figure 2.24. The graph of v versus t is a straight line that intercepts the vertical axis at $v_0 = 5$ m/s. The slope of this straight line can be calculated from the data shown in the drawing:

$$\text{Slope} = \frac{\Delta v}{\Delta t} = \frac{+12 \text{ m/s}}{2 \text{ s}} = +6 \text{ m/s}^2$$

The ratio $\Delta v/\Delta t$ is, by definition, equal to the average acceleration (Equation 2.4), so the slope of the straight line in a velocity–time graph is the average acceleration.

Figure 2.24 A velocity vs. time graph that applies to an object with an acceleration of $\Delta v/\Delta t = +6$ m/s^2. The initial velocity is $v_0 = +5$ m/s when $t = 0$ s.

2.8 CONCEPTS & CALCULATIONS

In this chapter we have studied the displacement, velocity, and acceleration vectors. We conclude now by presenting several examples that review some of the important features of these concepts. The three-part format of these examples stresses the role of conceptual understanding in problem solving. First, the problem statement is given. Then, there is a concept question-and-answer section, followed by the solution section. The purpose of the concept question-and-answer section is to provide help in understanding the solution and to illustrate how a review of the concepts can help in anticipating some of the characteristics of the numerical answers.

CONCEPTS & CALCULATIONS EXAMPLE 17   The Acceleration of a Skydiver

A skydiver is falling straight down, along the negative y direction. (a) During the initial part of the fall, her speed increases from 16 to 28 m/s in 1.5 s. (b) Later, her parachute opens, and her speed decreases from 48 to 26 m/s in 11 s. In both instances, determine the magnitude and direction of her average acceleration.

Concept Questions and Answers Is her average acceleration positive or negative when her speed is increasing?

Answer Since her speed is increasing, the acceleration vector must point in the same direction as the velocity vector, which points in the negative y direction. Thus, the acceleration is negative.

Is her average acceleration positive or negative when her speed is decreasing?

Answer Since her speed is decreasing, the acceleration vector must point opposite to the velocity vector. Since the velocity vector points in the negative y direction, the acceleration must point in the positive y direction. Thus, the acceleration is positive.

Solution

(a) Since the skydiver is moving in the negative y direction, her initial velocity is $v_0 = -16$ m/s and her final velocity is $v = -28$ m/s. Her average acceleration $\bar{a}$ is the change in the velocity divided by the elapsed time:

$$\bar{a} = \frac{v - v_0}{t} = \frac{-28 \text{ m/s} - (-16 \text{ m/s})}{1.5 \text{ s}} = \boxed{-8.0 \text{ m/s}^2} \qquad (2.4)$$

As expected, her average acceleration is negative. Note that her acceleration is not that due to gravity (-9.8 m/s^2), because wind resistance is slowing her down.

(b) Now the skydiver is slowing down, but still falling along the negative y direction. Her initial and final velocities are $v_0 = -48$ m/s and $v = -26$ m/s, respectively. The average acceleration for this phase of the motion is

$$\bar{a} = \frac{v - v_0}{t} = \frac{-26 \text{ m/s} - (-48 \text{ m/s})}{11 \text{ s}} = \boxed{+2.0 \text{ m/s}^2} \qquad (2.4)$$

Now, as anticipated, her average acceleration is positive.

CONCEPTS & CALCULATIONS EXAMPLE 18   A Top-Fuel Dragster

A top-fuel dragster starts from rest and has a constant acceleration of 40.0 m/s^2. What are the (a) final velocities and (b) displacements of the dragster at the end of 2.0 s and at the end of twice this time, or 4.0 s?

Concept Questions and Answers At a time t the dragster has a certain velocity. When the time doubles to $2t$, does the velocity also double?

Answer Because the dragster has an acceleration of 40.0 m/s^2, its velocity changes by 40.0 m/s during each second of travel. That is, the velocity is 40.0 m/s at the end of the 1st second, 2 × 40.0 m/s at the end of the 2nd second, 3 ×

40.0 m/s at the end of the 3rd second, and so on. Thus, when the time doubles, the velocity also doubles.

When the time doubles to $2t$, does the displacement of the dragster also double?

Answer The displacement of the dragster is equal to its average velocity multiplied by the elapsed time. The average velocity $\overline{v}$ is just one-half the sum of the initial and final velocities, or $\overline{v} = \frac{1}{2}(v_0 + v)$. Since the initial velocity is zero, $v_0 = 0$ m/s and the average velocity is just one-half the final velocity, or $\overline{v} = \frac{1}{2}v$. However, as we have seen, the final velocity is proportional to the elapsed time, since when the time doubles, the final velocity also doubles. Therefore, the displacement, being the product of the average velocity and the time, is proportional to the time squared, or t^2. Consequently, as the time doubles, the displacement does not double, but increases by a factor of four.

Solution

(a) According to Equation 2.4, the final velocity v, the initial velocity v_0, the acceleration a, and the elapsed time t are related by $v = v_0 + at$. The final velocities at the two times are

$t = 2.0\ s$ $v = v_0 + at = 0$ m/s $+ (40.0$ m/s$^2)(2.0$ s$) = \boxed{80\ \text{m/s}}$

$t = 4.0\ s$ $v = v_0 + at = 0$ m/s $+ (40.0$ m/s$^2)(4.0$ s$) = \boxed{160\ \text{m/s}}$

We see that the velocity doubles when the time doubles, as expected.

(b) The displacement x of the dragster is equal to its average velocity multiplied by the time, so

$$x = \underbrace{\tfrac{1}{2}(v_0 + v)}_{\text{Average velocity}}t = \tfrac{1}{2}vt$$

where we have used the fact that $v_0 = 0$ m/s. According to Equation 2.4, the final velocity is related to the acceleration by $v = v_0 + at$, or $v = at$, since $v_0 = 0$ m/s. Therefore, the displacement can be written as $x = \frac{1}{2}vt = \frac{1}{2}(at)t = \frac{1}{2}at^2$. The displacements at the two times are then

$t = 2.0\ s$ $x = \frac{1}{2}at^2 = \frac{1}{2}(40.0$ m/s$^2)(2.0$ s$)^2 = \boxed{80\ \text{m}}$

$t = 4.0\ s$ $x = \frac{1}{2}at^2 = \frac{1}{2}(40.0$ m/s$^2)(4.0$ s$)^2 = \boxed{320\ \text{m}}$

As predicted, the displacement at $t = 4.0$ s is four times the displacement at $t = 2.0$ s.

At the end of the problem set for this chapter, you will find homework problems that contain both conceptual and quantitative parts. These problems are grouped under the heading *Concepts & Calculations, Group Learning Problems*. They are designed for use by students working alone or in small learning groups. The conceptual part of each problem provides a convenient focus for group discussions.

SUMMARY

Displacement is a vector that points from an object's initial position toward its final position. The magnitude of the displacement is the shortest distance between the two positions.

The **average speed** of an object is the distance traveled by the object divided by the time required to cover the distance: Average speed = (Distance)/(Elapsed time).

The **average velocity** $\overline{\mathbf{v}}$ of an object is the object's displacement $\Delta\mathbf{x}$ divided by the elapsed time Δt: $\overline{\mathbf{v}} = \Delta\mathbf{x}/\Delta t$. Average velocity is a vector that has the same direction as the displacement. When the elapsed time Δt is infinitesimally small, the average velocity becomes equal to the **instantaneous velocity v,** the velocity at an instant of time.

The **average acceleration** $\overline{\mathbf{a}}$ is a vector. It equals the change in the velocity $\Delta\mathbf{v}$ divided by the elapsed time Δt, the "change in velocity" being the final minus the initial velocity: $\bar{a} = \Delta\mathbf{v}/\Delta t$. When Δt is infinitesimally small, the average acceleration becomes equal to the **instantaneous acceleration a.** Acceleration is the rate at which the velocity is changing.

When an object moves with a constant acceleration along a straight line, its displacement $x - x_0$, final velocity v, initial velocity v_0, acceleration a, and the elapsed time t are related by the following equations, assuming that $x_0 = 0$ m at $t = 0$ s:

$$x = \bar{v}t = \tfrac{1}{2}(v_0 + v)t \quad \text{and} \quad v = v_0 + at$$

These equations can be combined algebraically to give two additional equations. See Table 2.1. The equations in this table are known as the **equations of kinematics.**

In **free-fall** motion, an object experiences negligible air resistance and a constant acceleration due to gravity. All objects at the same location above the earth have the same acceleration due to gravity. The acceleration due to gravity is directed toward the center of the earth and has a magnitude of approximately 9.80 m/s^2 near the earth's surface.

CONCEPTUAL QUESTIONS

1. A honeybee leaves the hive and travels 2 km before returning. Is the displacement for the trip the same as the distance traveled? If not, why not?

2. Two buses depart from Chicago, one going to New York and one to San Francisco. Each bus travels at a speed of 30 m/s. Do they have equal velocities? Explain.

3. Is the average speed of a vehicle a vector or a scalar quantity? Provide a reason for your answer.

4. Often, traffic lights are timed so that if you travel at a certain constant speed, you can avoid all red lights. Discuss how the timing of the lights is determined, considering that the distance between them varies from one light to the next.

5. One of the following statements is incorrect. (a) The car traveled around the track at a constant velocity. (b) The car traveled around the track at a constant speed. Which statement is incorrect and why?

6. Give an example from your own experience in which the velocity of an object is zero for just an instant of time, but its acceleration is not zero.

7. At a given instant of time, a car and a truck are traveling side by side in adjacent lanes of a highway. The car has a greater velocity than the truck. Does the car necessarily have a greater acceleration? Explain.

8. The average velocity for a trip has a positive value. Is it possible for the instantaneous velocity at any point during the trip to have a negative value? Justify your answer.

9. A runner runs half the remaining distance to the finish line every ten seconds. She runs in a straight line and does not ever reverse her direction. Does her acceleration have a constant magnitude? Give a reason for your answer.

10. An object moving with a constant acceleration can certainly slow down. But can an object ever come to a permanent halt if its acceleration truly remains constant? Explain.

11. An experimental vehicle slows down and comes to a halt with an acceleration whose magnitude is 9.80 m/s^2. After reversing direction in a negligible amount of time, the vehicle speeds up with an acceleration of 9.80 m/s^2. Other than being horizontal, how is this motion different, if at all, from the motion of a ball that is thrown straight upward, comes to a halt, and falls back to earth?

12. A ball is dropped from rest from the top of a building and strikes the ground with a speed v_f. From ground level, a second ball is thrown straight upward at the same instant that the first ball is dropped. The initial speed of the second ball is $v_0 = v_f$, the same speed with which the first ball will eventually strike the ground. Ignoring air resistance, decide whether the balls cross paths at half the height of the building, above the halfway point, or below the halfway point. Give your reasoning.

13. Two objects are thrown vertically upward, first one, and then, a bit later, the other. Is it possible that both reach the same maximum height at the same instant? Account for your answer.

14. The muzzle velocity of a gun is the velocity of the bullet when it leaves the barrel. The muzzle velocity of one rifle with a short barrel is greater than the muzzle velocity of another rifle that has a longer barrel. In which rifle is the acceleration of the bullet larger? Explain your reasoning.

PROBLEMS

ssm Solution is in the Student Solutions Manual. **www** Solution is available on the World Wide Web at http://www.wiley.com/college/cutnell
⚕ **This icon represents a biomedical application.**

Section 2.1 Displacement, Section 2.2 Speed and Velocity

1. ssm A whale swims due east for a distance of 6.9 km, turns around and goes due west for 1.8 km, and finally turns around again and heads 3.7 km due east. (a) What is the total distance traveled by the whale? (b) What are the magnitude and direction of the displacement of the whale?

2. One afternoon, a couple walks three-fourths of the way around a circular lake, the radius of which is 1.50 km. They start at the west side of the lake and head due south at the beginning of their walk. (a) What is the distance they travel? (b) What are the magnitude and direction (relative to due east) of the couple's displacement?

3. ssm A plane is sitting on a runway, awaiting takeoff. On an adjacent parallel runway, another plane lands and passes the stationary plane at a speed of 45 m/s. The arriving plane has a length of 36 m. By looking out of a window (very narrow), a passenger on the stationary plane can see the moving plane. For how long a time is the moving plane visible?

4. Due to continental drift, the North American and European continents are drifting apart at an average speed of about 3 cm per year. At this speed, how long (in years) will it take for them to drift apart by another 1500 m (a little less than a mile)?

5. The three-toed sloth is the slowest moving land mammal. On the ground, the sloth moves at an average speed of 0.037 m/s, considerably slower than the giant tortoise, which walks at 0.076 m/s. After 12 minutes of walking, how much further would the tortoise have gone relative to the sloth?

6. As the earth rotates through one revolution, a person standing on the equator traces out a circular path whose radius is equal to the radius of the earth (6.38×10^6 m). What is the average speed of this person in (a) meters per second and (b) miles per hour?

7. A tourist being chased by an angry bear is running in a straight line toward his car at a speed of 4.0 m/s. The car is a distance d away. The bear is 26 m behind the tourist and running at 6.0 m/s. The tourist reaches the car safely. What is the maximum possible value for d?

*8. A sky diver, with parachute unopened, falls 625 m in 15.0 s. Then she opens her parachute and falls another 356 m in 142 s. What is her average velocity (both magnitude and direction) for the entire fall?

*9. ssm www A woman and her dog are out for a morning run to the river, which is located 4.0 km away. The woman runs at 2.5 m/s in a straight line. The dog is unleashed and runs back and forth at 4.5 m/s between his owner and the river, until she reaches the river. What is the total distance run by the dog?

*10. A car makes a 60.0-km trip with an average velocity of 40.0 km/h in a direction due north. The trip consists of three parts. The car moves with a constant velocity of 25 km/h due north for the first 15 km and 62 km/h due north for the next 32 km. With what constant velocity (magnitude and direction) does the car travel for the last 13-km segment of the trip?

**11. You are on a train that is traveling at 3.0 m/s along a level straight track. Very near and parallel to the track is a wall that slopes upward at a 12° angle with the horizontal. As you face the window (0.90 m high, 2.0 m wide) in your compartment, the train is moving to the left, as the drawing indicates. The top edge of the wall first appears at window corner A and eventually disappears at window corner B. How much time passes be-tween appearance and disappearance of the upper edge of the wall?

Section 2.3 Acceleration

12. For a standard production car, the highest road-tested acceleration ever reported occurred in 1993, when a Ford RS200 Evolution went from zero to 26.8 m/s (60 mi/h) in 3.275 s. Find the magnitude of the car's acceleration.

13. ssm www A jogger accelerates from rest to 3.0 m/s in 2.0 s. A car accelerates from 38.0 to 41.0 m/s also in 2.0 s. (a) Find the acceleration (magnitude only) of the jogger. (b) Determine the acceleration (magnitude only) of the car. (c) Does the car travel farther than the jogger during the 2.0 s? If so, how much farther?

14. NASA has developed Deep-Space 1 (DS-1), a spacecraft that is scheduled to rendezvous with the asteroid named 1992 KD (which orbits the sun millions of miles from the earth). The propulsion system of DS-1 works by ejecting high-speed argon ions out the rear of the engine. The engine slowly increases the velocity of DS-1 by about +9.0 m/s per day. (a) How much time (in days) will it take to increase the velocity of DS-1 by +2700 m/s? (b) What is the acceleration of DS-1 (in m/s²)?

15. ssm A runner accelerates to a velocity of 5.36 m/s due west in 3.00 s. His average acceleration is 0.640 m/s², also directed due west. What was his velocity when he began accelerating?

16. Starting from rest, a speedboat reaches a speed of 3.2 m/s in 2.0 s. What is the boat's speed after an additional 3.0 s has elapsed, assuming the boat's acceleration remains the same?

*17. A car is traveling along a straight road at a velocity of +31.0 m/s when its engine cuts out. For the next ten seconds, the car slows down, and its average acceleration is $\bar{a}_1$. For the next five seconds, the car slows down further, and its average acceleration is $\bar{a}_2$. The velocity of the car at the end of the fifteen-second period is +24.5 m/s. The ratio of the average acceleration values is $\bar{a}_1/\bar{a}_2 = 1.67$. Find the velocity of the car at the end of the initial ten-second interval.

**18. Two motorcycles are traveling due east with different velocities. However, four seconds later, they have the same velocity. During this four-second interval, motorcycle A has an average acceleration of 2.0 m/s² due east, while motorcycle B has an average acceleration of 4.0 m/s² due east. By how much did the speeds *differ* at the beginning of the four-second interval, and which motorcycle was moving faster?

Section 2.4 Equations of Kinematics for Constant Acceleration, Section 2.5 Applications of the Equations of Kinematics

19. ssm The 1999 VW Beetle goes from 0 to 60.0 mi/h with an acceleration of $+2.35$ m/s^2. (a) How much time does it take for the Beetle to reach this speed? (b) A top-fuel dragster can go from 0 to 60.0 mi/h in 0.600 s. What is the acceleration (in m/s^2) of the dragster?

20. Review Conceptual Example 7 as background for this problem. A car is traveling to the left, which is the negative direction. The direction of travel remains the same throughout this problem. The car's initial speed is 27.0 m/s, and during a 5.0-s interval, it changes to a final speed of (a) 29.0 m/s and (b) 23.0 m/s. In each case, find the acceleration (magnitude and algebraic sign) and state whether or not the car is decelerating.

21. In getting ready to slam-dunk the ball, a basketball player starts from rest and sprints to a speed of 6.0 m/s in 1.5 s. Assuming that the player accelerates uniformly, determine the distance he runs.

22. (a) What is the magnitude of the average acceleration of a skier who, starting from rest, reaches a speed of 8.0 m/s when going down a slope for 5.0 s? (b) How far does the skier travel in this time?

23. ssm A jetliner, traveling northward, is landing with a speed of 69 m/s. Once the jet touches down, it has 750 m of runway in which to reduce its speed to 6.1 m/s. Compute the average acceleration (magnitude and direction) of the plane during landing.

24. 🖢 The left ventricle of the heart accelerates blood from rest to a velocity of $+26$ cm/s. (a) If the displacement of the blood during the acceleration is $+2.0$ cm, determine its acceleration (in cm/s^2). (b) How much time does it take for the blood to reach its final velocity?

25. The length of the barrel of a primitive blowgun is 1.2 m. Upon leaving the barrel, a dart has a speed of 14 m/s. Assuming that the dart is uniformly accelerated, how long does it take for the dart to travel the length of the barrel?

26. A speed trap is set up with two pressure-activated strips placed across a highway, 110 m apart. A car is speeding along at 33 m/s, while the speed limit is 21 m/s. At the instant the car activates the first strip, the driver begins slowing down. What minimum deceleration is needed in order that the average speed is within the limit by the time the car crosses the second marker?

27. ssm www A speed ramp at an airport is basically a large conveyor belt on which you can stand and be moved along. The belt of one speed ramp moves at a constant speed such that a person who stands still on it leaves the ramp 64 s after getting on. Clifford is in a real hurry, however, and skips the speed ramp. Starting from rest with an acceleration of 0.37 m/s^2, he covers the same distance as the ramp does, but in one-fourth the time. What is the speed at which the belt of the ramp is moving?

*28. A drag racer, starting from rest, speeds up for 402 m with an acceleration of $+17.0$ m/s^2. A parachute then opens, slowing the car down with an acceleration of -6.10 m/s^2. How fast is the racer moving 3.50×10^2 m after the parachute opens?

*29. A car is traveling on a dry road with a velocity of $+32.0$ m/s. The driver slams on the brakes and skids to a halt with an acceleration of -8.00 m/s^2. On an icy road, the car would have skidded to a halt with an acceleration of -3.00 m/s^2. How much farther would the car have skidded on the icy road compared to the dry road?

*30. A speedboat starts from rest and accelerates at $+2.01$ m/s^2 for 7.00 s. At the end of this time, the boat continues for an additional 6.00 s with an acceleration of $+0.518$ m/s^2. Following this, the boat accelerates at -1.49 m/s^2 for 8.00 s. (a) What is the velocity of the boat at $t = 21.0$ s? (b) Find the total displacement of the boat.

*31. A cab driver picks up a customer and delivers her 2.00 km away, driving a straight route. The driver accelerates to the speed limit and, upon reaching it, begins to decelerate immediately. The magnitude of the deceleration is three times the magnitude of the acceleration. Find the lengths of the acceleration and deceleration phases of the trip.

*32. A car is traveling at a constant speed of 27 m/s on a highway. At the instant this car passes an entrance ramp, a second car enters the highway from the ramp. The second car starts from rest and has a constant acceleration. What acceleration must it maintain, so that the two cars meet for the first time at the next exit, which is 1.8 km away?

*33. Two soccer players start from rest, 48 m apart. They run directly toward each other, both players accelerating. The first player has an acceleration whose magnitude is 0.50 m/s^2. The second player's acceleration has a magnitude of 0.30 m/s^2. (a) How much time passes before they collide? (b) At the instant they collide, how far has the first player run?

**34. A Boeing 747 "Jumbo Jet" has a length of 59.7 m. The runway on which the plane lands intersects another runway. The width of the intersection is 25.0 m. The plane decelerates through the intersection at a rate of 5.70 m/s^2 and clears it with a final speed of 45.0 m/s. How much time is needed for the plane to clear the intersection?

**35. ssm A locomotive is accelerating at 1.6 m/s^2. It passes through a 20.0-m-wide crossing in a time of 2.4 s. After the locomotive leaves the crossing, how much time is required until its speed reaches 32 m/s?

**36. In the one-hundred-meter dash a sprinter accelerates from rest to a top speed with an acceleration whose magnitude is 2.68 m/s^2. After achieving top speed, he runs the remainder of the race without speeding up or slowing down. If the total race is run in 12.0 s, how far does he run during the acceleration phase?

Section 2.6 Freely Falling Bodies

37. ssm The greatest height reported for a jump into an airbag is 99.4 m by stuntman Dan Koko. In 1948 he jumped from rest from the top of the Vegas World Hotel and Casino. He struck the airbag at a speed of 39 m/s (88 mi/h). To assess the effects of air resistance, determine how fast he would have been traveling on impact had air resistance been absent.

38. In preparation for this problem, review Conceptual Example 7. From the top of a cliff, a person uses a slingshot to fire a pebble straight downward, which is the negative direction. The initial speed of the pebble is 9.0 m/s. (a) What is the acceleration (magnitude and direction) of the pebble during the downward motion? Is the pebble decelerating? Explain. (b) After 0.50 s, how far beneath the cliff top is the pebble?

39. The drawing shows a device that you can make with a piece of cardboard, which can be used to measure a person's reaction time. Hold the card at the top and suddenly drop it. Ask a friend to try to catch the card between his or her thumb and index finger. Initially, your friend's fingers must be level with the asterisks at the bottom. By noting where your friend catches the card, you can determine his or her reaction time in milliseconds (ms). Calculate the distances d_1, d_2, and d_3.

40. An astronaut on a distant planet wants to determine its acceleration due to gravity. The astronaut throws a rock straight up with a velocity of +15 m/s and measures a time of 20.0 s before the rock returns to his hand. What is the acceleration (magnitude and direction) due to gravity on this planet?

41. ssm From her bedroom window a girl drops a water-filled balloon to the ground, 6.0 m below. If the balloon is released from rest, how long is it in the air?

42. A hot-air balloon is rising upward with a constant speed of 2.50 m/s. When the balloon is 3.00 m above the ground, the balloonist accidentally drops a compass over the side of the balloon. How much time elapses before the compass hits the ground?

43. A diver springs upward with an initial speed of 1.8 m/s from a 3.0-m board. (a) Find the velocity with which he strikes the water. *[Hint: When the diver reaches the water, his displacement is $y = -3.0$ m (measured from the board), assuming that the downward direction is chosen as the negative direction.]* (b) What is the highest point he reaches above the water?

44. Review Conceptual Example 15 before attempting this problem. Two identical pellet guns are fired simultaneously from the edge of a cliff. These guns impart an initial speed of 30.0 m/s to

each pellet. Gun A is fired straight upward, with the pellet going up and then falling back down, eventually hitting the ground beneath the cliff. Gun B is fired straight downward. In the absence of air resistance, how long after pellet B hits the ground does pellet A hit the ground?

45. ssm A wrecking ball is hanging at rest from a crane when suddenly the cable breaks. The time it takes for the ball to fall halfway to the ground is 1.2 s. Find the time it takes for the ball to fall from rest all the way to the ground.

46. Before working this problem, review Conceptual Example 15. A pellet gun is fired straight downward from the edge of a cliff that is 15 m above the ground. The pellet strikes the ground with a speed of 27 m/s. How far above the cliff edge would the pellet have gone had the gun been fired straight upward?

47. A ball is thrown straight upward and rises to a maximum height of 16 m above its launch point. At what height above its launch point has the speed of the ball decreased to one-half of its initial value?

* **48.** A woman on a bridge 90.0 m high sees a raft floating at a constant speed on the river below. She drops a stone from rest in an attempt to hit the raft. The stone is released when the raft has 6.00 m more to travel before passing under the bridge. The stone hits the water 2.00 m in front of the raft. Find the speed of the raft.

* **49.** Two arrows are shot vertically upward. The second arrow is shot after the first one, but while the first is still on its way up. The initial speeds are such that both arrows reach their maximum heights at the same instant, although these heights are different. Suppose that the initial speed of the first arrow is 25.0 m/s and that the second arrow is fired 1.20 s after the first. Determine the initial speed of the second arrow.

* **50.** Two students, Anne and Joan, are bouncing straight up and down on a trampoline. Anne bounces twice as high as Joan. Assuming both are in free-fall, find the ratio of the time Anne spends between bounces to the time Joan spends.

* **51. ssm** A cement block accidentally falls from rest from the ledge of a 53.0-m-high building. When the block is 14.0 m above the ground, a man, 2.00 m tall, looks up and notices that the block is directly above him. How much time, at most, does the man have to get out of the way?

* **52.** (a) Just for fun, a person jumps from rest from the top of a tall cliff overlooking a lake. In falling through a distance H, she acquires a certain speed v. Assuming free-fall conditions, how much farther must she fall in order to acquire a speed of $2v$? Express your answer in terms of H. (b) Would the answer to part (a) be different, if this event were to occur on another planet where the acceleration due to gravity had a value other than 9.80 m/s^2? Explain.

* **53. ssm www** A spelunker (cave explorer) drops a stone from rest into a hole. The speed of sound is 343 m/s in air, and the sound of the stone striking the bottom is heard 1.50 s after the stone is dropped. How deep is the hole?

* **54.** A ball is thrown upward from the top of a 25.0-m-tall building. The ball's initial speed is 12.0 m/s. At the same instant, a person is running on the ground at a distance of 31.0 m from the

building. What must be the average speed of the person if he is to catch the ball at the bottom of the building?

**55. A hot air balloon is ascending straight up at a constant speed of 7.0 m/s. When the balloon is 12.0 m above the ground, a gun fires a pellet straight up from ground level with an initial speed of 30.0 m/s. Along the paths of the balloon and the pellet, there are two places where each of them has the same altitude at the same time. How far above ground level are these places?

**56. A roof tile falls from rest from the top of a building. An observer inside the building notices that it takes 0.20 s for the tile to pass her window, whose height is 1.6 m. How far above the top of this window is the roof?

Section 2.7 Graphical Analysis of Velocity and Acceleration

57. ssm For the first 10.0 km of a marathon, a runner averages a velocity that has a magnitude of 15.0 km/h. For the next 15.0 km, he averages 10.0 km/h, while for the last 15.0 km, he averages 5.0 km/h. Construct, to scale, the position–time graph for the runner.

58. A snowmobile moves according to the velocity–time graph shown in the drawing. What is the snowmobile's average acceleration during each of the segments A, B, and C?

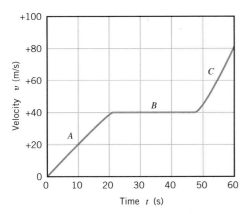

59. A bus makes a trip according to the position–time graph shown in the drawing. What is the average velocity (magnitude and direction) of the bus during each of the segments labeled A, B, and C? Express your answers in km/h.

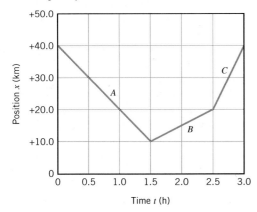

60. A person who walks for exercise produces the position–time graph given with this problem. (a) Without doing any calculations, decide which segments of the graph (A, B, C, or D) indicate positive, negative, and zero average velocities. (b) Calculate the average velocity for each segment to verify your answers to part (a).

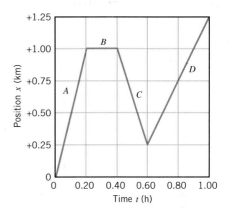

*61. ssm A bus makes a trip according to the position–time graph shown in the illustration. What is the average acceleration (in km/h^2) of the bus for the entire 3.5-h period shown in the graph?

*62. A runner is at the position $x = 0$ m when time $t = 0$ s. One hundred meters away is the finish line. Every ten seconds, this runner runs half the remaining distance to the finish line. During each ten-second segment, the runner has a constant velocity. For the first forty seconds of the motion, construct (a) the position–time graph and (b) the velocity–time graph.

**63. Two runners start one hundred meters apart and run toward each other. Each runs ten meters during the first second. During each second thereafter, each runner runs ninety percent of the distance he ran in the previous second. Thus, the velocity of each person changes from second to second. However, during any one second, the velocity remains constant. Make a position–time graph for one of the runners. From this graph, determine (a) how much time passes before the runners collide and (b) the speed with which each is running at the moment of collision.

ADDITIONAL PROBLEMS

64. The Concorde jetliner achieves a lift-off speed of 112 m/s in 20.0 s, starting from rest and traveling due east. What are the magnitude and direction of its average acceleration?

65. ssm A penny is dropped from rest from the top of the Sears Tower in Chicago. Considering the height of the building is 427 m and ignoring air resistance, find the speed with which the penny strikes the ground.

66. Two runners in a 1609-m race (one mile) finish with times of 3:52.46 (3 minutes and 52.46 seconds) and 3:52.72. Assuming that both run at their average speeds during the entire time, what distance separates them at the end of the race?

67. ssm A motorcycle has a constant acceleration of 2.5 m/s². Both the velocity and acceleration of the motorcycle point in the same direction. How much time is required for the motorcycle to change its speed from (a) 21 to 31 m/s, and (b) 51 to 61 m/s?

68. At the beginning of a basketball game, a referee tosses the ball straight up with a speed of 4.6 m/s. A player cannot touch the ball until after it reaches its maximum height and begins to fall down. What is the minimum time that a player must wait before touching the ball?

69. The Space Shuttle travels at a speed of about 7.6×10^3 m/s. The blink of an astronaut's eye lasts about 110 ms. How many football fields (length = 91.4 m) does the Space Shuttle cover in the blink of an eye?

70. An automobile starts from rest and accelerates to a final velocity in two stages along a straight road. Each stage occupies the same amount of time. In stage 1, the magnitude of the car's acceleration is 3.0 m/s². The magnitude of the car's velocity at the end of stage 2 is 2.5 times greater than it is at the end of stage 1. Find the magnitude of the acceleration in stage 2.

71. An arrow is shot straight up with an initial speed of 50.0 m/s. After reaching its maximum height, the arrow starts down. On the descent, a slight breeze blows the arrow laterally, and it strikes a tree limb that is 30.0 m above the ground. Determine the velocity of the arrow just before the arrow strikes the limb.

*72. Along a straight road through town, there are three speed-limit signs. They occur in the following order: 55, 35, and 25 mi/h, with the 35-mi/h sign being midway between the other two. Obeying these speed limits, the smallest possible time t_A that a driver can spend on this part of the road is to travel between the first and second signs at 55 mi/h and between the second and third signs at 35 mi/h. More realistically, a driver could slow down from 55 to 35 mi/h with a constant deceleration and then do a similar thing from 35 to 25 mi/h. This alternative requires a time t_B. Find the ratio t_B/t_A.

*73. ssm A bicyclist makes a trip that consists of three parts, each in the same direction (due north) along a straight road. During the first part, she rides for 22 minutes at an average speed of 7.2 m/s. During the second part, she rides for 36 minutes at an average speed of 5.1 m/s. Finally, during the third part, she rides for 8.0 minutes at an average speed of 13 m/s. (a) How far has the bicyclist traveled during the entire trip? (b) What is the average velocity of the bicyclist for the trip?

*74. In reaching her destination, a backpacker walks with an average velocity of 1.34 m/s, due west. This average velocity results, because she hikes for 6.44 km with an average velocity of 2.68 m/s, due west, turns around, and hikes with an average velocity of 0.447 m/s, due east. How far east did she walk?

*75. ssm A log is floating on swiftly moving water. A stone is dropped from rest from a 75-m-high bridge and lands on the log as it passes under the bridge. If the log moves with a constant speed of 5.0 m/s, what is the horizontal distance between the log and the bridge when the stone is released?

*76. In a bicycle race, the leader is traveling with a constant velocity of +11.1 m/s and is 10.0 m ahead of the second-place cyclist. The second-place cyclist has a velocity of +9.50 m/s and an acceleration of +1.20 m/s². How much time elapses before the second-place cyclist catches the leader?

*77. A race driver has made a pit stop to refuel. After refueling, he leaves the pit area with an acceleration whose magnitude is 6.0 m/s², and after 4.0 s he enters the main speedway. At the same instant, another race car that is on the speedway and traveling at a constant speed of 70.0 m/s overtakes and passes the entering car. If the entering car maintains its acceleration, how much time is required for it to catch the other car?

**78. A ball is dropped from rest from the top of a cliff that is 24 m high. From ground level, a second ball is thrown straight upward at the same instant that the first ball is dropped. The initial speed of the second ball is exactly the same as that with which the first ball eventually hits the ground. In the absence of air resistance, the motions of the balls are just the reverse of each other. Determine how far below the top of the cliff the balls cross paths.

**79. ssm A train has a length of 92 m and starts from rest with a constant acceleration at time $t = 0$ s. At this instant, a car just reaches the end of the train. The car is moving with a constant velocity. At a time $t = 14$ s, the car just reaches the front of the train. Ultimately, however, the train pulls ahead of the car, and at time $t = 28$ s, the car is again at the rear of the train. Find the magnitudes of (a) the car's velocity and (b) the train's acceleration.

**80. A football player, starting from rest at the line of scrimmage, accelerates along a straight line for a time of 3.0 s. Then, during a negligible amount of time, he changes the magnitude of his acceleration to a value of 1.1 m/s². With this acceleration, he continues in the same direction for another 2.0 s, until he reaches a speed of 6.4 m/s. What is the value of his acceleration (assumed to be constant) during the initial 3.0-s period?

CONCEPTS & CALCULATIONS
GROUP LEARNING PROBLEMS

Note: Each of these problems consists of Concept Questions followed by a related quantitative Problem. They are designed for use by students working alone or in small learning groups. The Concept Questions involve little or no mathematics and are intended to stimulate group discussions. They focus on the concepts with which the problems deal. Recognizing the concepts is the essential initial step in any problem-solving technique.

81. Concept Question Listed below are three pairs of initial and final positions that lie along the x axis. Using the concept of displacement discussed in Section 2.1, decide which pairs give a positive displacement and which give a negative displacement. Explain.

	Initial position x_0	Final position x
(a)	+2 m	+6 m
(b)	+6 m	+2 m
(c)	−3 m	+7 m

Problem For each of the three pairs of positions listed above, determine the magnitude and direction (positive or negative) of the displacement. Verify that your answers are consistent with those determined in the Concept Question.

82. Concept Question Suppose that each pair of positions in Problem 81 represents the initial and final positions of a moving car. In each case, is the direction of the car's average velocity the same as its initial position x_0, its final position x, or the difference $x - x_0$ between its final and initial positions? Decide which pairs give a positive average velocity and which give a negative average velocity. Provide reasons for your answers.

Problem The elapsed time for each pair of positions is 0.5 s. Review the concept of average velocity in Section 2.2 and then determine the average velocity (magnitude and direction) for each of the three pairs. Check to see that the directions of the velocities agree with the directions found in the Concept Question.

83. Concept Questions (a) In general, does the average acceleration of an object have the same direction as its initial velocity v_0, its final velocity v, or the difference $v - v_0$ between its final and initial velocities? Provide a reason for your answer. (b) The following table lists four pairs of initial and final velocities for a boat traveling along the x axis. Use the concept of acceleration presented in Section 2.3 to determine the direction (positive or negative) of the average acceleration.

Problem The elapsed time for each of the four pairs of velocities is 2.0 s. Find the average acceleration (magnitude and direction) for each of the four pairs. Be sure that your directions agree with those found in the Concept Questions.

	Initial velocity v_0	Final velocity v
(a)	+2.0 m/s	+5.0 m/s
(b)	+5.0 m/s	+2.0 m/s
(c)	−6.0 m/s	−3.0 m/s
(d)	+4.0 m/s	−4.0 m/s

84. Concept Questions The initial velocity v_0 and acceleration a of four moving objects at a given instant in time are given in the table:

	Initial velocity v_0	Acceleration a
(a)	+12 m/s	+3.0 m/s^2
(b)	+12 m/s	−3.0 m/s^2
(c)	−12 m/s	+3.0 m/s^2
(d)	−12 m/s	−3.0 m/s^2

Draw vectors for v_0 and a and, in each case, state whether the *speed* of the object is increasing or decreasing in time. Account for your answers.

Problem For each of the four pairs in the table above, determine the final *speed* of the object if the elapsed time is 2.0 s. Compare your final speeds with the initial speeds and make sure that your answers are consistent with your answers to the Concept Questions.

85. Concept Questions (a) The acceleration of a NASCAR race car is zero. Does this necessarily mean that the velocity of the car is also zero? (b) If the speed of the car is constant as it goes around the track, is its average acceleration necessarily zero? Justify your answers.

Problem (a) Suppose that the race car is moving to the right with a constant velocity of +82 m/s. What is the acceleration of the car? (b) Twelve seconds later, the car is halfway around the track and traveling in the opposite direction with the same speed. What is the average acceleration of the car? Verify that your answers agree with your answers to the Concept Questions.

86. Concept Questions A ball is thrown vertically upward, which is the positive direction. A little later it returns to its point of release. (a) Does the acceleration of the ball reverse direction when the ball starts its downward trip? (b) What is the displacement of the ball when it returns to its point of release? Explain your answers.

Problem If the ball is in the air for a total time of 8.0 s, what is its initial velocity?

Once this bottle-nosed dolphin launches itself into the air, it follows a familiar arc-shaped path that depends on the launch velocity and the acceleration due to gravity.

CHAPTER 3

KINEMATICS
IN TWO DIMENSIONS

3.1 DISPLACEMENT, VELOCITY, AND ACCELERATION

In Chapter 2 the concepts of displacement, velocity, and acceleration are used to describe an object moving in one dimension. There are also situations in which the motion is along a curved path that lies in a plane. Such two-dimensional motion can also be described using the same concepts. In Grand Prix racing, for example, the course follows a curved road, and Figure 3.1 shows a race car at two different positions along it. These positions are identified by the vectors $\mathbf{r}$ and $\mathbf{r_0}$, which are drawn from an arbitrary coordinate origin. The *displacement* $\Delta\mathbf{r}$ of the car is the vector drawn from the initial position $\mathbf{r_0}$ at time t_0 to the final position $\mathbf{r}$ at time t. The magnitude of $\Delta\mathbf{r}$ is the shortest distance between the two positions. In the drawing, the vectors $\mathbf{r_0}$ and $\Delta\mathbf{r}$ are drawn tail-to-head, so it is evident that $\mathbf{r}$ is the vector sum of $\mathbf{r_0}$ and $\Delta\mathbf{r}$. This means that $\mathbf{r} = \mathbf{r_0} + \Delta\mathbf{r}$, or

$$\text{Displacement} = \Delta\mathbf{r} = \mathbf{r} - \mathbf{r_0}$$

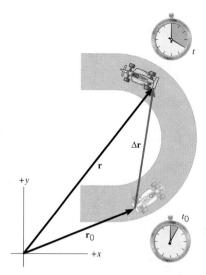

Figure 3.1 The displacement $\Delta\mathbf{r}$ of the car is a vector that points from the initial position of the car at time t_0 to the final position at time t. The magnitude of $\Delta\mathbf{r}$ is the shortest distance between the two positions.

The displacement here is defined as it is in Chapter 2. Now, however, the displacement vector can lie anywhere in a plane, rather than just along a straight line.

The average velocity $\overline{\mathbf{v}}$ of the car between two positions is defined in a manner similar to that in Equation 2.2, as the displacement, $\Delta\mathbf{r} = \mathbf{r} - \mathbf{r_0}$, divided by the elapsed time $\Delta t = t - t_0$:

$$\overline{\mathbf{v}} = \frac{\mathbf{r} - \mathbf{r_0}}{t - t_0} = \frac{\Delta\mathbf{r}}{\Delta t} \tag{3.1}$$

Since both sides of Equation 3.1 must agree in direction, the average velocity vector has the same direction as the displacement. The velocity of the car at an instant of time is its *instantaneous velocity* $\mathbf{v}$. The average velocity becomes equal to the instantaneous velocity $\mathbf{v}$ in the limit that Δt becomes infinitesimally small:

$$\mathbf{v} = \lim_{\Delta t \to 0} \frac{\Delta\mathbf{r}}{\Delta t}$$

Figure 3.2 illustrates that the instantaneous velocity $\mathbf{v}$ is tangent to the path of the car. The drawing also shows the vector components $\mathbf{v}_x$ and $\mathbf{v}_y$ of the velocity, which are parallel to the x and y axes, respectively. Using the components of a vector is advantageous when describing two-dimensional motion.

The *average acceleration* $\overline{\mathbf{a}}$ is defined just as it is for one-dimensional motion, namely, as the change in velocity, $\Delta\mathbf{v} = \mathbf{v} - \mathbf{v_0}$, divided by the elapsed time Δt:

$$\overline{\mathbf{a}} = \frac{\mathbf{v} - \mathbf{v_0}}{t - t_0} = \frac{\Delta\mathbf{v}}{\Delta t} \tag{3.2}$$

The average acceleration vector has the same direction as the change in velocity. In the limit that the elapsed time becomes infinitesimally small, the average acceleration becomes equal to the *instantaneous acceleration* $\mathbf{a}$:

$$\mathbf{a} = \lim_{\Delta t \to 0} \frac{\Delta\mathbf{v}}{\Delta t}$$

The acceleration has a vector component $\mathbf{a}_x$ along the x direction and a vector component $\mathbf{a}_y$ along the y direction.

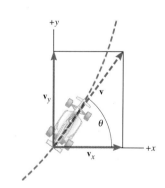

Figure 3.2 The instantaneous velocity $\mathbf{v}$ and its two vector components $\mathbf{v}_x$ and $\mathbf{v}_y$.

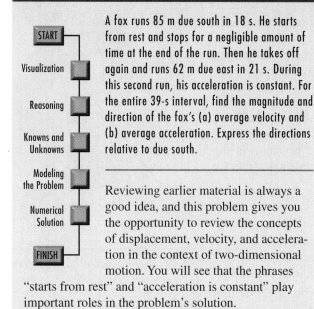

A fox runs 85 m due south in 18 s. He starts from rest and stops for a negligible amount of time at the end of the run. Then he takes off again and runs 62 m due east in 21 s. During this second run, his acceleration is constant. For the entire 39-s interval, find the magnitude and direction of the fox's (a) average velocity and (b) average acceleration. Express the directions relative to due south.

Reviewing earlier material is always a good idea, and this problem gives you the opportunity to review the concepts of displacement, velocity, and acceleration in the context of two-dimensional motion. You will see that the phrases "starts from rest" and "acceleration is constant" play important roles in the problem's solution.

Go to our Web site at

http://www.wiley.com/college/cutnell

3.2 EQUATIONS OF KINEMATICS IN TWO DIMENSIONS

To understand how displacement, velocity, and acceleration are applied to two-dimensional motion, consider a spacecraft equipped with two engines that are mounted perpendicular to each other. These engines produce the only forces that the craft experiences, and it is assumed to be at the coordinate origin when $t_0 = 0$ s, so that $\mathbf{r_0} = 0$ m. At a later time t, its displacement is $\Delta \mathbf{r} = \mathbf{r} - \mathbf{r_0} = \mathbf{r}$. Relative to the x and y axes, the displacement $\mathbf{r}$ has vector components of $\mathbf{x}$ and $\mathbf{y}$, respectively.

In Figure 3.3 only the engine oriented along the x direction is firing, and the vehicle accelerates along this direction. It is assumed that the velocity in the y direction is zero, and it remains zero, since the y engine is turned off. The motion of the spacecraft along the x direction is described by the five kinematic variables x, a_x, v_x, v_{0x}, and t. Here the symbol "x" reminds us that we are dealing with the x components of the displacement, velocity, and acceleration vectors. The variables x, a_x, v_x, and v_{0x} are scalar components (or "components," for short). As Section 1.7 discusses, these components are positive or negative numbers (with units), depending on whether the associated vector components point along the $+x$ or the $-x$ axis. If the spacecraft has a constant acceleration along the x direction, the motion is exactly like that described in Chapter 2, and the equations of kinematics can be used. For convenience, these equations are written in the left column of Table 3.1.

Figure 3.4 is analogous to Figure 3.3, except that now only the y engine is firing, and the spacecraft accelerates along the y direction. Such a motion can be described in terms of the kinematic variables y, a_y, v_y, v_{0y}, and t. And if the acceleration along the y direction is constant, these variables are related by the equations of kinematics, as written in the right column of Table 3.1. Like their counterparts in the x direction, the components, y, a_y, v_y, and v_{0y}, may be positive ($+$) or negative ($-$) numbers (with units).

If both engines of the spacecraft are firing *at the same time,* the resulting motion takes place in part along the x axis and in part along the y axis, as Figure 3.5 illus-

Figure 3.3 The spacecraft is moving with a constant acceleration a_x parallel to the x axis. There is no motion in the y direction, and the y engine is turned off.

TABLE 3.1 Equations of Kinematics for Constant Acceleration in Two-Dimensional Motion

x Component		Variable	y Component	
x		Displacement	y	
a_x		Acceleration	a_y	
v_x		Final velocity	v_y	
v_{0x}		Initial velocity	v_{0y}	
t		Elapsed time	t	
$v_x = v_{0x} + a_x t$	(3.3a)		$v_y = v_{0y} + a_y t$	(3.3b)
$x = \frac{1}{2}(v_{0x} + v_x)t$	(3.4a)		$y = \frac{1}{2}(v_{0y} + v_y)t$	(3.4b)
$x = v_{0x}t + \frac{1}{2}a_x t^2$	(3.5a)		$y = v_{0y}t + \frac{1}{2}a_y t^2$	(3.5b)
$v_x^2 = v_{0x}^2 + 2a_x x$	(3.6a)		$v_y^2 = v_{0y}^2 + 2a_y y$	(3.6b)

Figure 3.4 The spacecraft is moving with a constant acceleration a_y parallel to the y axis. There is no motion in the x direction, and the x engine is turned off.

trates. The thrust of each engine gives the vehicle a corresponding acceleration component. The x engine accelerates the ship in the x direction and causes a change in the x component of the velocity. Likewise, the y engine causes a change in the y component of the velocity. ***It is important to realize that the x part of the motion occurs exactly as it would if the y part did not occur at all. Similarly, the y part of the motion occurs exactly as it would if the x part of the motion did not exist.*** In other words, the x and y motions are independent of each other.

● *Concepts at a Glance* The independence of the x and y motions lies at the heart of two-dimensional kinematics. It allows us to treat two-dimensional motion as two distinct one-dimensional motions, one for the x direction and one for the y direction. As the Concepts-at-a-Glance chart in Figure 3.6 illustrates, everything that we have learned in Chapter 2 about kinematics in one dimension will now be applied separately to each of the two directions. In so doing, we will be able to describe the x and y variables separately and then bring these descriptions together to understand the two-dimensional picture. Example 1 takes this approach in dealing with a moving spacecraft. ●

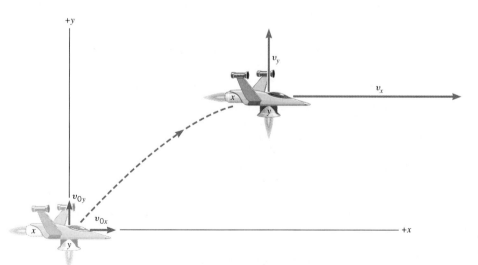

Figure 3.5 The two-dimensional motion of the spacecraft can be viewed as the combination of the separate x and y motions.

Concepts at a Glance

Figure 3.6 *Concepts at a Glance* In two dimensions, motion along the *x* direction and motion along the *y* direction are independent of each other. As a result, each can be analyzed separately according to the procedures for one-dimensional kinematics discussed in Chapter 2. On the space shuttle *Challenger*, motion in perpendicular directions is controlled by thrusters. The photographs show the *Challenger* in orbit with different thrusters activated.

EXAMPLE 1   A Moving Spacecraft

In the *x* direction, the spacecraft in Figure 3.5 has an initial velocity component of $v_{0x} = +22$ m/s and an acceleration component of $a_x = +24$ m/s². In the *y* direction, the analogous quantities are $v_{0y} = +14$ m/s and $a_y = +12$ m/s². The directions to the right and upward have been chosen as the positive directions. After a time of 7.0 s, find (a) *x* and v_x, (b) *y* and v_y, and (c) the final velocity (magnitude and direction) of the spacecraft.

Reasoning The motion in the *x* direction and the motion in the *y* direction can be treated separately, each as a one-dimensional motion. We will follow this approach in parts (a) and (b) to obtain the location and velocity components of the spacecraft. Then, in part (c) the velocity components will be combined to give the final velocity.

Solution

(a) The data for the motion in the *x* direction are listed below.

x-Direction Data				
x	a_x	v_x	v_{0x}	t
?	+24 m/s²	?	+22 m/s	7.0 s

The *x* component of the craft's displacement can be found by using Equation 3.5a.

$$x = v_{0x}t + \tfrac{1}{2}a_x t^2 = (22 \text{ m/s})(7.0 \text{ s}) + \tfrac{1}{2}(24 \text{ m/s}^2)(7.0 \text{ s})^2 = \boxed{+740 \text{ m}}$$

The velocity component v_x can be calculated with the aid of Equation 3.3a:

$$v_x = v_{0x} + a_x t = (22 \text{ m/s}) + (24 \text{ m/s}^2)(7.0 \text{ s}) = \boxed{+190 \text{ m/s}}$$

(b) The data for the motion in the y direction are listed below.

y-Direction Data

y	a_y	v_y	v_{0y}	t
?	$+12$ m/s^2	?	$+14$ m/s	7.0 s

Proceeding in the same manner as in part (a), we find that

$$\boxed{y = +390 \text{ m}} \quad \text{and} \quad \boxed{v_y = +98 \text{ m/s}}$$

(c) Figure 3.7 shows the velocity of the vehicle and its components v_x and v_y. The magnitude v of the velocity can be found by using the Pythagorean theorem:

$$v = \sqrt{v_x^2 + v_y^2} = \sqrt{(190 \text{ m/s})^2 + (98 \text{ m/s})^2} = \boxed{210 \text{ m/s}}$$

The direction of the velocity vector is given by the angle θ in the drawing:

$$\tan \theta = \frac{v_y}{v_x} \quad \text{or} \quad \theta = \tan^{-1}\left(\frac{v_y}{v_x}\right) = \tan^{-1}\left(\frac{98 \text{ m/s}}{190 \text{ m/s}}\right) = \boxed{27°}$$

After 7.0 s, the spacecraft has a velocity of 210 m/s in a direction of 27° above the positive x axis. At this time the craft is located at a point that is 740 m to the right and 390 m above the origin, as in Figure 3.5.

The following Reasoning Strategy gives an overview of how the equations of kinematics are applied to describe motion in two dimensions, such as that in Example 1. We will use this strategy for solving other types of kinematics problems in the rest of this chapter.

• **PROBLEM SOLVING INSIGHT**
When the motion is two-dimensional, the time variable *t* has the same value for both the x and y directions.

Figure 3.7 The magnitude of the velocity vector gives the speed of the spacecraft, and the angle θ gives the direction of travel relative to the positive x direction.

REASONING STRATEGY

Applying the Equations of Kinematics in Two Dimensions

1. Make a drawing to represent the situation being studied.

2. Decide which directions are to be called positive ($+$) and negative ($-$) relative to a conveniently chosen coordinate origin. Do not change your decision during the course of a calculation. There are positive and negative directions for both the x axis and the y axis.

3. Remember that the time variable *t* has the same value for the part of the motion along the x axis and the part along the y axis.

4. In an organized way, write down the values (with appropriate $+$ and $-$ signs) that are given for any of the five kinematic variables associated with the x direction and the y direction. Be on the alert for "implied data," such as the phrase "starts from rest," which means that the values of the initial velocity components are zero: $v_{0x} = 0$ m/s and $v_{0y} = 0$ m/s. The data summary boxes used in the examples are a good way of keeping track of this information. In addition, identify the variables that you are being asked to determine.

5. Before attempting to solve a problem, verify that the given information contains values for at least three of the kinematic variables. Do this for the x and the y direction of the motion. Once the three known variables are identified along with the desired unknown variable, the appropriate relations from Table 3.1 can be selected.

6. When the motion of an object is divided into "segments," remember that the final velocity for one segment becomes the initial velocity for the next segment.

7. Keep in mind that there may be two possible answers to a kinematics problem. Try to visualize the different physical situations to which the answers correspond.

3.3 PROJECTILE MOTION

The biggest thrill in baseball is a home run. The motion of the ball on its curving path into the stands is a common type of two-dimensional motion called "projectile motion." A good description of projectile motion can often be obtained with the assumption that air resistance is absent.

● *Concepts at a Glance*   Following the approach outlined in Figure 3.6, we consider the horizontal and vertical parts of the motion separately. In the horizontal or x direction, the moving object (the projectile) does not slow down in the absence of air resistance. Thus, the x component of the velocity remains constant at its initial value or $v_x = v_{0x}$, and the x component of the acceleration is $a_x = 0$ m/s^2. In the vertical or y direction, however, the projectile experiences the effect of gravity. As a result, the y component of the velocity v_y is not constant, but changes. The y component of the acceleration a_y is the downward acceleration due to gravity. If the path or trajectory of the projectile is near the earth's surface, a_y has a magnitude of 9.80 m/s^2. In this text, then, the phrase "projectile motion" means that $a_x = 0$ m/s^2 and a_y equals the acceleration due to gravity, as the Concept-at-a-Glance chart in Figure 3.8 summarizes. Example 2 and other examples in this section illustrate how the equations of kinematics are applied to projectile motion. ●

Figure 3.8 *Concepts at a Glance* In projectile motion, the horizontal or x component of the acceleration is zero, and the vertical or y component of the acceleration is the acceleration due to gravity. In this time-lapse photograph, a long jumper exhibits projectile motion after he leaves the ground, assuming that the effects of air resistance can be ignored.

Concepts at a Glance

Kinematics in One Dimension (Chapter 2)
x Direction

Variables: x, a_x, v_x, v_{0x}, t

Kinematics in One Dimension (Chapter 2)
y Direction

Variables: y, a_y, v_y, v_{0y}, t

Kinematics in Two Dimensions
Projectile Motion

x **Variables:** x, a_x, v_x, v_{0x}, t
  1. $a_x = 0$
  2. $v_x = v_{0x} = constant$

y **Variables:** y, a_y, v_y, v_{0y}, t
  1. $a_y = acceleration$
     $due\ to\ gravity$
  2. $v_y\ is\ not\ constant$

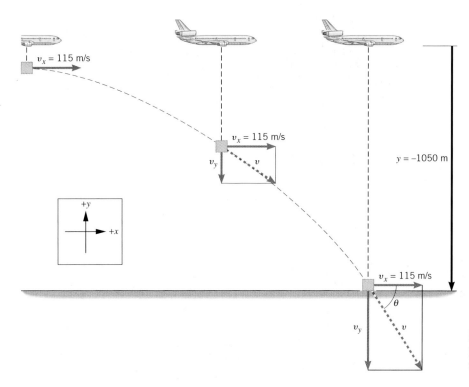

Figure 3.9 The package falling from the plane is an example of projectile motion, as Examples 2 and 3 discuss.

EXAMPLE 2   A Falling Care Package

Figure 3.9 shows an airplane moving horizontally with a constant velocity of $+115$ m/s at an altitude of 1050 m. The directions to the right and upward have been chosen as the positive directions. The plane releases a "care package" that falls to the ground along a curved trajectory. Ignoring air resistance, determine the time required for the package to hit the ground.

Reasoning The time required for the package to hit the ground is the time it takes for the package to fall through a vertical distance of 1050 m. In falling, it moves to the right, as well as downward, but these two parts of the motion occur independently. Therefore, we can focus solely on the vertical part. We note that the package is not moving in the y direction at the instant of release, so that $v_{0y} = 0$ m/s. (The package is moving initially in the horizontal or x direction, not in the y direction.) Furthermore, when the package hits the ground, the y component of its displacement is $y = -1050$ m, as the drawing shows. The acceleration is that due to gravity, so $a_y = -9.80$ m/s^2. These data are summarized as follows:

y-Direction Data				
y	a_y	v_y	v_{0y}	t
-1050 m	-9.80 m/s^2		0 m/s	?

With these data, Equation 3.5b ($y = v_{0y}t + \frac{1}{2}a_yt^2$) can be used to find the fall time.

Solution Since $v_{0y} = 0$ m/s, it follows from Equation 3.5b that $y = \frac{1}{2}a_yt^2$ and

$$t = \sqrt{\frac{2y}{a_y}} = \sqrt{\frac{2(-1050 \text{ m})}{-9.80 \text{ m/s}^2}} = \boxed{14.6 \text{ s}}$$

● **PROBLEM SOLVING INSIGHT**
The variables y, a_y, v_y, and v_{0y} are scalar components. Therefore, an algebraic sign (+ or −) must be included with each one to denote direction.

Figure 3.10 Package A and package B are released simultaneously at the same height and strike the ground at the same time because their y variables (y, a_y, and v_{0y}) are the same.

The freely falling package in Example 2 picks up vertical speed on the way down. The horizontal component of the velocity, however, retains its initial value of $v_{0x} = +115$ m/s throughout the entire descent. Since the plane also travels at a constant horizontal velocity of $+115$ m/s, it remains directly above the falling package. The pilot always sees the package directly beneath the plane, as the dashed vertical lines in Figure 3.9 show. This result is a direct consequence of the fact that the package has no acceleration in the horizontal direction. In reality, air resistance would slow down the package, and it would not remain directly beneath the plane during the descent. Figure 3.10 further clarifies this point by illustrating what happens to two packages that are released simultaneously from the same height. Package B is given an initial velocity component of $v_{0x} = +115$ m/s in the horizontal direction, as in Example 2, and the package follows the curved path shown in the figure. Package A, on the other hand, is dropped from a stationary balloon and falls straight down toward the ground, since $v_{0x} = 0$ m/s. Both packages hit the ground at the same time.

Not only do the packages in Figure 3.10 reach the ground at the same time, but the y components of their velocities are also equal at all points on the way down. However, package B does hit the ground with a greater speed than package A. Remember, speed is the magnitude of the velocity vector, and the velocity of B has an x component, whereas the velocity of A does not. The magnitude and direction of the velocity vector for package B at the instant just before the package hits the ground is computed in Example 3.

EXAMPLE 3   The Velocity of the Care Package

For the situation shown in Figure 3.9, find the speed of the package and the direction of the velocity vector just before the package hits the ground.

Reasoning Since the speed v of the package is given by $v = \sqrt{v_x{}^2 + v_y{}^2}$, it is necessary to know values for v_x and v_y at the instant before impact. The component v_x is constant and has a value of 115 m/s. The component v_y can be determined by using Equation 3.3b and the data from Example 2 ($a_y = -9.80$ m/s^2, $v_{0y} = 0$ m/s, $t = 14.6$ s). Thus, we expect that the final speed v of the package will be greater than 115 m/s.

Solution From Equation 3.3b, it follows that

$$v_y = v_{0y} + a_y t = 0 \text{ m/s} + (-9.80 \text{ m/s}^2)(14.6 \text{ s}) = -143 \text{ m/s}$$

The speed of the package at the instant before impact is

$$v = \sqrt{(115 \text{ m/s})^2 + (-143 \text{ m/s})^2} = \boxed{184 \text{ m/s}}$$

which is greater than 115 m/s, as expected. The velocity vector makes an angle θ with the horizontal, as Figure 3.9 indicates:

$$\cos\theta = \frac{v_x}{v} \quad \text{or} \quad \theta = \cos^{-1}\left(\frac{v_x}{v}\right) = \cos^{-1}\left(\frac{115 \text{ m/s}}{184 \text{ m/s}}\right) = \boxed{51.3°}$$

• **PROBLEM SOLVING INSIGHT**
The speed of a projectile at any location along its path is the magnitude v of its velocity at that location: $v = \sqrt{v_x{}^2 + v_y{}^2}$. Both the horizontal and vertical velocity components contribute to the speed.

An important feature of projectile motion is that there is no acceleration in the horizontal or x direction. Conceptual Example 4 discusses an interesting implication of this feature.

CONCEPTUAL EXAMPLE 4   I Shot a Bullet into the Air . . . .

Suppose you are driving in a convertible with the top down. The car is moving to the right at a constant velocity. As Figure 3.11 illustrates, you point a rifle straight upward and fire it. In the absence of air resistance, where would the bullet land—behind you, ahead of you, or in the barrel of the rifle?

Reasoning and Solution If air resistance were present, it would slow down the bullet and cause it to land behind you, toward the rear of the car. However, air resistance is absent, so we must consider the bullet's motion more carefully. Before the rifle is fired, the bullet, rifle, and car are moving together, so the bullet and rifle have the same horizontal velocity as the car. When the rifle is fired, the bullet is given an additional velocity component in the vertical direction; the bullet retains its initial horizontal velocity component since the rifle is pointed straight up. Because there is no air resistance to slow it down, the bullet experiences no acceleration in the horizontal direction. Thus, the bullet's horizontal velocity component does not change. It retains its initial value, and remains matched to that of the rifle and the car. As a result, *the bullet remains directly above the rifle at all times during the flight and would fall directly back into the barrel of the rifle,* as the drawing indicates. This situation is analogous to that in Figure 3.9, where the care package, as it falls, remains directly below the plane.

Related Homework Material: *Question 12 and Problem 34*

Figure 3.11 The car is moving with a constant velocity to the right, and the rifle is pointed straight up. In the absence of air resistance, a bullet fired from the rifle has no acceleration in the horizontal direction. As a result, the bullet would land back in the barrel of the rifle.

Often projectiles, like footballs and baseballs, are sent into the air at an angle with respect to the ground. From a knowledge of the projectile's initial velocity, a wealth of information can be obtained about the motion. For instance, Example 5 demonstrates how to calculate the maximum height reached by the projectile.

EXAMPLE 5   The Height of a Kickoff

A placekicker kicks a football at an angle of $\theta = 40.0°$ above the horizontal axis, as Figure 3.12 shows. The initial speed of the ball is $v_0 = 22$ m/s. Ignore air resistance and find the maximum height H that the ball attains.

Reasoning The maximum height is a characteristic of the vertical part of the motion. The vertical part of the motion can be treated separately from the horizontal part. In preparation for making use of this fact, we calculate the vertical component of the initial velocity:

$$v_{0y} = v_0 \sin\theta = (22 \text{ m/s}) \sin 40.0° = +14 \text{ m/s}$$

Figure 3.12 A football is kicked with an initial speed of v_0 at an angle of θ above the ground. The ball attains a maximum height H and a range R.

● **PROBLEM SOLVING INSIGHT**
When a projectile reaches maximum height, the vertical component of its velocity is momentarily zero ($v_y = 0$ m/s). However, the horizontal component of its velocity is not zero.

The maximum height H can be determined by noting that the y component of the velocity v_y decreases as the ball moves upward. Eventually, $v_y = 0$ m/s at the top of the trajectory. The data below can be used in Equation 3.6b ($v_y^2 = v_{0y}^2 + 2a_y y$) to find the maximum height:

y-Direction Data				
y	a_y	v_y	v_{0y}	t
$H = ?$	-9.80 m/s^2	0 m/s	$+14$ m/s	

Solution   From Equation 3.6b, we find that

$$y = H = \frac{v_y^2 - v_{0y}^2}{2a_y} = \frac{(0 \text{ m/s})^2 - (14 \text{ m/s})^2}{2(-9.80 \text{ m/s}^2)} = \boxed{+10 \text{ m}}$$

The height H depends only on the y variables; the same height would have been reached had the ball been thrown *straight up* with an initial velocity of $v_{0y} = +14$ m/s.

It is also possible to find the total time or "hang time" during which the football in Figure 3.12 is in the air. Example 6 shows how to determine this time.

THE PHYSICS OF...

the "hang time" of a football.

EXAMPLE 6   The Time of Flight of a Kickoff

For the motion illustrated in Figure 3.12, ignore air resistance and use the data from Example 5 to determine the time of flight between kickoff and landing.

Reasoning Given the initial velocity, it is the acceleration due to gravity that determines how long the ball stays in the air. Thus, to find the time of flight we deal with the vertical part of the motion. Since the ball starts at ground level and returns to ground level, the displacement in the y direction is zero. The initial velocity component in the y direction is the same as that in Example 5, i.e., $v_{0y} = +14$ m/s. Therefore, we have

y-Direction Data				
y	a_y	v_y	v_{0y}	t
0 m	-9.80 m/s^2		$+14$ m/s	?

The time of flight can be determined from Equation 3.5b ($y = v_{0y}t + \frac{1}{2}a_y t^2$).

Solution   Using Equation 3.5b, we find

$$0 \text{ m} = (14 \text{ m/s})t + \tfrac{1}{2}(-9.80 \text{ m/s}^2)t^2 = [(14 \text{ m/s}) + \tfrac{1}{2}(-9.80 \text{ m/s}^2)t]t$$

There are two solutions to this equation. One is given by

$$(14 \text{ m/s}) + \tfrac{1}{2}(-9.80 \text{ m/s}^2)t = 0 \quad \text{or} \quad t = 2.9 \text{ s}$$

The other is given by $t = 0$ s. The solution we seek is $\boxed{t = 2.9 \text{ s}}$, because $t = 0$ s corresponds to the initial kickoff.

Another important feature of projectile motion is called the "range." The range, as Figure 3.12 shows, is the horizontal distance traveled between launching and landing, assuming the projectile returns to the *same vertical level* at which it was fired. Example 7 shows how to obtain the range.

EXAMPLE 7   The Range of a Kickoff

For the motion shown in Figure 3.12 and discussed in Examples 5 and 6, ignore air resistance and calculate the range R of the projectile.

Reasoning The range is a characteristic of the horizontal part of the motion. Thus, our starting point is to determine the horizontal component of the initial velocity:

$$v_{0x} = v_0 \cos \theta = (22 \text{ m/s}) \cos 40.0° = +17 \text{ m/s}$$

Recall from Example 6 that the time of flight is $t = 2.9$ s. Since there is no acceleration in the x direction, v_x remains constant, and the range is simply the product of $v_x = v_{0x}$ and the time.

Solution The range is

$$x = R = v_{0x}t = (17 \text{ m/s})(2.9 \text{ s}) = \boxed{+49 \text{ m}}$$

The range in the previous example depends on the angle θ at which the projectile is fired above the horizontal. When air resistance is absent, the maximum range results when $\theta = 45°$.

The examples considered thus far have used information about the initial location and velocity of a projectile to determine the final location and velocity. Example 8 deals with the opposite situation and illustrates how the final parameters can be used with the equations of kinematics to determine the initial parameters.

EXAMPLE 8   A Home Run

A baseball player hits a home run, and the ball lands in the left-field seats, 7.5 m above the point at which the ball was hit. The ball lands with a velocity of 36 m/s at an angle of 28° below the horizontal (see Figure 3.13). Ignoring air resistance, find the initial velocity with which the ball leaves the bat.

Reasoning To find the initial velocity, we must determine its magnitude (the initial speed v_0) and its direction (the angle θ in the drawing). These quantities are related

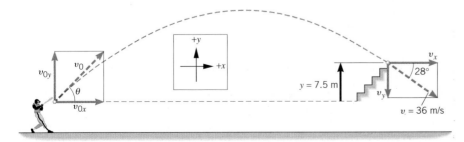

Figure 3.13 The velocity and location of the baseball upon landing can be used to determine its initial velocity, as Example 8 illustrates.

to the horizontal and vertical components of the initial velocity (v_{0x} and v_{0y}) by the relations

$$v_0 = \sqrt{v_{0x}^2 + v_{0y}^2} \quad \text{and} \quad \tan\theta = \frac{v_{0y}}{v_{0x}}$$

Therefore, it is necessary to find v_{0x} and v_{0y}, which we will do with the equations of kinematics.

Solution Since air resistance is being ignored, the horizontal component of the velocity v_x remains constant throughout the motion. Thus,

$$v_{0x} = v_x = (36\text{ m/s})\cos 28° = 32\text{ m/s}$$

The value for v_{0y} can be obtained from Equation 3.6b and the data displayed below (see Figure 3.13 for the positive and negative directions):

y-Direction Data				
y	a_y	v_y	v_{0y}	t
$+7.5$ m	-9.80 m/s^2	$(-36\sin 28°)$ m/s	?	

$$v_y^2 = v_{0y}^2 + 2a_y y \quad \text{or} \quad v_{0y} = +\sqrt{v_y^2 - 2a_y y}$$

$$v_{0y} = +\sqrt{[(-36\sin 28°)\text{ m/s}]^2 - 2(-9.80\text{ m/s}^2)(7.5\text{m})} = +21\text{ m/s}$$

In determining v_{0y} we choose the plus sign for the square root, because the vertical component of the initial velocity points upward in Figure 3.13, which is the positive direction. The initial speed v_0 and angle θ of the baseball are

$$v_0 = \sqrt{v_{0x}^2 + v_{0y}^2} = \sqrt{(32\text{ m/s})^2 + (21\text{ m/s})^2} = \boxed{38\text{ m/s}}$$

$$\theta = \tan^{-1}\left(\frac{v_{0y}}{v_{0x}}\right) = \tan^{-1}\left(\frac{21\text{ m/s}}{32\text{ m/s}}\right) = \boxed{33°}$$

Figure 3.14 The speed v of a projectile at a given height above the ground is the same on the upward and downward parts of the trajectory. The velocities are different, however, since they point in different directions.

In projectile motion, the magnitude of the acceleration due to gravity affects the trajectory in a significant way. For example, a baseball or a golf ball would travel much farther and higher on the moon than on the earth, when launched with the same initial velocity. The reason is that the moon's gravity is only about one-sixth as strong as the earth's.

Section 2.6 points out that certain types of symmetry with respect to time and speed are present for freely falling bodies. These symmetries are also found in projectile motion, since projectiles are falling freely in the vertical direction. In particular, the time required for a projectile to reach its maximum height H is equal to the time spent returning to the ground. In addition, Figure 3.14 shows that the speed v of the object at any height above the ground on the upward part of the trajectory is equal to the speed v at the same height on the downward part. Although the two speeds are the same, the velocities are different, because they point in different directions. Conceptual Example 9 shows how to use this type of symmetry in your reasoning.

Figure 3.15 Two stones are thrown off the cliff with identical initial speeds v_0, but at equal angles θ that are below and above the horizontal. Conceptual Example 9 compares the velocities with which the stones hit the water beneath the cliff.

CONCEPTUAL EXAMPLE 9   Two Ways to Throw a Stone

From the top of a cliff overlooking a lake, a person throws two stones. The stones have identical initial speeds v_0, but stone 1 is thrown downward at an angle θ below the horizontal, while stone 2 is thrown upward at the same angle above the horizontal, as Figure 3.15 shows. Neglect air resistance and decide which stone, if either, strikes the water with the greater velocity.

Reasoning and Solution We might guess that stone 1, being hurled downward, would strike the water with the greater velocity. To show that this is not true, let's follow the path of stone 2 as it rises to its maximum height and falls back to earth. Notice point *P* in the drawing, where stone 2 returns to its initial height; here the speed of stone 2 is v_0, but its velocity is directed at an angle θ below the horizontal. This is exactly the type of projectile symmetry illustrated in Figure 3.14. At this point, then, stone 2 has a velocity that is identical to the velocity with which stone 1 is thrown downward from the top of the cliff. From this point on, the velocity of stone 2 changes in exactly the same way as that for stone 1, so **both stones strike the water with the same velocity.**

Related Homework Material: *Problems 25 and 33*

In all the examples in this section, the projectiles follow a curved trajectory. In general, if the only acceleration is that due to gravity, the shape of the path can be shown to be a *parabola.*

* **3.4 RELATIVE VELOCITY**

To someone hitchhiking along a highway, two cars speeding by in adjacent lanes seem like a blur. But if the cars have the same velocity, each driver sees the other remaining in place, one lane away. The hitchhiker observes a velocity of perhaps 30 m/s. But each driver observes the other's velocity to be zero. Clearly, the velocity of an object is relative to the observer who is making the measurement.

Figure 3.16 illustrates the concept of relative velocity by showing a passenger walking toward the front of a moving train. The people sitting on the train see the passenger walking with a velocity of +2.0 m/s, where the plus sign denotes a direction to the right. Suppose the train is moving with a velocity of +9.0 m/s relative to an observer standing on the ground. Then the ground-based observer would see the passenger moving with a velocity of +11 m/s, due in part to the walking motion and in part to the train's motion. As an aid in describing relative velocity, let us define the following symbols:

$$\mathbf{v}_{\boxed{\text{PT}}} = \text{velocity of the } \boxed{\text{Passenger}} \text{ relative to the } \boxed{\text{Train}} = +2.0 \text{ m/s}$$

$$\mathbf{v}_{\boxed{\text{TG}}} = \text{velocity of the } \boxed{\text{Train}} \text{ relative to the } \boxed{\text{Ground}} = +9.0 \text{ m/s}$$

$$\mathbf{v}_{\boxed{\text{PG}}} = \text{velocity of the } \boxed{\text{Passenger}} \text{ relative to the } \boxed{\text{Ground}} = +11 \text{ m/s}$$

Ground-based observer

Figure 3.16 The velocity of the passenger relative to the ground-based observer is $\mathbf{v}_{PG}$. It is the vector sum of the velocity $\mathbf{v}_{PT}$ of the passenger relative to the train and the velocity $\mathbf{v}_{TG}$ of the train relative to the ground: $\mathbf{v}_{PG} = \mathbf{v}_{PT} + \mathbf{v}_{TG}$.

Mid-air refueling of military aircraft offers an interesting example of relative velocity. To refuel, the F-14 aircraft (lower plane) matches its velocity to that of the tanker (upper plane) and couples to the tanker's delivery tube. During the refueling maneuver, the velocity of the F-14 relative to the tanker is zero.

In terms of these symbols, the situation in Figure 3.16 can be summarized as follows:

$$\mathbf{v}_{PG} = \mathbf{v}_{PT} + \mathbf{v}_{TG} \tag{3.7}$$

or

$$\mathbf{v}_{PG} = (2.0 \text{ m/s}) + (9.0 \text{ m/s}) = +11 \text{ m/s}$$

According to Equation 3.7, $\mathbf{v}_{PG}$ is the vector sum of $\mathbf{v}_{PT}$ and $\mathbf{v}_{TG}$, and this sum is shown in the drawing. Had the passenger been walking toward the rear of the train, rather than the front, the velocity relative to the ground-based observer would have been $\mathbf{v}_{PG} = (-2.0 \text{ m/s}) + (9.0 \text{ m/s}) = +7.0 \text{ m/s}$.

Each velocity symbol in Equation 3.7 contains a two-letter subscript. The first letter in the subscript refers to the body that is moving, while the second letter indicates the object relative to which the velocity is measured. For example, $\mathbf{v}_{TG}$ and $\mathbf{v}_{PG}$ are the velocities of the Train and Passenger measured relative to the Ground. Similarly, $\mathbf{v}_{PT}$ is the velocity of the Passenger measured by an observer sitting on the Train.

The ordering of the subscript symbols in Equation 3.7 follows a definite pattern. The first subscript (P) on the left side of the equation is also the first subscript on the right side of the equation. Likewise, the last subscript (G) on the left side of the equation is also the last subscript on the right side of the equation. The third subscript (T) appears only on the right side of the equation as the two "inner" subscripts. The colored boxes below emphasize the pattern of the symbols in the subscripts:

$$\mathbf{v}_{\boxed{PG}} = \mathbf{v}_{\boxed{P}T} + \mathbf{v}_{T\boxed{G}}$$

In other situations, the subscripts will not necessarily be P, G, and T, but will be compatible with the names of the objects involved in the motion.

Equation 3.7 has been presented in connection with one-dimensional motion, but the result is also valid for two-dimensional motion. Figure 3.17 depicts a common situation that deals with relative velocity in two dimensions. Part a of the drawing shows a boat being carried downstream by a river; the engine of the boat is turned off. In part b, the engine has been turned on, and now the boat moves across the river in a diagonal fashion because of the combined motion produced by the current and the engine. The list below gives the velocities for this type of motion and the objects relative to which they are measured:

$$\mathbf{v}_{\boxed{BW}} = \text{velocity of the } \boxed{\text{Boat}} \text{ relative to the } \boxed{\text{Water}}$$

$$\mathbf{v}_{\boxed{WS}} = \text{velocity of the } \boxed{\text{Water}} \text{ relative to the } \boxed{\text{Shore}}$$

$$\mathbf{v}_{\boxed{BS}} = \text{velocity of the } \boxed{\text{Boat}} \text{ relative to the } \boxed{\text{Shore}}$$

The velocity $\mathbf{v}_{BW}$ of the boat relative to the water is the velocity measured by an observer who, for instance, is floating on an inner tube and drifting downstream with the current. When the engine is turned off, the boat also drifts downstream with the current, and $\mathbf{v}_{BW}$ is zero. When the engine is turned on, however, the boat can move relative to the water, and $\mathbf{v}_{BW}$ is no longer zero. The velocity $\mathbf{v}_{WS}$ of the water relative to the shore is the velocity of the current measured by an observer on the shore. The velocity $\mathbf{v}_{BS}$ of the boat relative to the shore is due to the combined motion of the boat relative to the water and the motion of the water relative to the shore. In symbols,

$$\mathbf{v}_{\boxed{BS}} = \mathbf{v}_{\boxed{B}W} + \mathbf{v}_{W\boxed{S}}$$

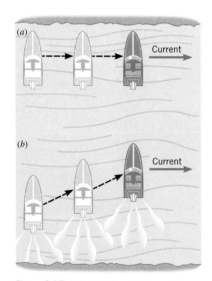

Figure 3.17 (a) A boat with its engine turned off is carried along by the current. (b) With the engine turned on, the boat moves across the river in a diagonal fashion.

The ordering of the subscripts in this equation is identical to that in Equation 3.7, although the letters have been changed to reflect a different physical situation. Example 10 illustrates the concept of relative velocity in two dimensions.

EXAMPLE 10   Crossing a River

The engine of a boat drives it across a river that is 1800 m wide. The velocity $\mathbf{v}_{BW}$ of the boat relative to the water is 4.0 m/s, directed perpendicular to the current, as in Figure 3.18. The velocity $\mathbf{v}_{WS}$ of the water relative to the shore is 2.0 m/s. (a) What is the velocity $\mathbf{v}_{BS}$ of the boat relative to the shore? (b) How long does it take for the boat to cross the river?

Reasoning

(a) The velocity of the boat relative to the shore is $\mathbf{v}_{BS}$. It is the vector sum of the velocity $\mathbf{v}_{BW}$ of the boat relative to the water and the velocity $\mathbf{v}_{WS}$ of the water relative to the shore: $\mathbf{v}_{BS} = \mathbf{v}_{BW} + \mathbf{v}_{WS}$. Since $\mathbf{v}_{BW}$ and $\mathbf{v}_{WS}$ are both known, we can use this relation among the relative velocities, with the aid of trigonometry, to find the magnitude and directional angle of $\mathbf{v}_{BS}$.

(b) The component of $\mathbf{v}_{BS}$ that is parallel to the width of the river (see Figure 3.18) determines how fast the boat moves across the river; this parallel component is $v_{BS} \sin \theta = v_{BW} = 4.0$ m/s. The time for the boat to cross the river is equal to the width of the river divided by the magnitude of this velocity component.

Solution

(a) Since the vectors $\mathbf{v}_{BW}$ and $\mathbf{v}_{WS}$ are perpendicular (see Figure 3.18), the magnitude of $\mathbf{v}_{BS}$ can be determined from the Pythagorean theorem:

$$v_{BS} = \sqrt{v_{BW}^2 + v_{WS}^2} = \sqrt{(4.0 \text{ m/s})^2 + (2.0 \text{ m/s})^2} = \boxed{4.5 \text{ m/s}}$$

Thus, the boat moves at a speed of 4.5 m/s with respect to an observer on shore. The direction of the boat relative to the shore is given by the angle θ in the drawing:

$$\tan \theta = \frac{v_{BW}}{v_{WS}} \quad \text{or} \quad \theta = \tan^{-1}\left(\frac{v_{BW}}{v_{WS}}\right) = \tan^{-1}\left(\frac{4.0 \text{ m/s}}{2.0 \text{ m/s}}\right) = \boxed{63^\circ}$$

(b) The time t for the boat to cross the river is

$$t = \frac{\text{Width}}{v_{BS} \sin \theta} = \frac{1800 \text{ m}}{4.0 \text{ m/s}} = \boxed{450 \text{ s}}$$

Figure 3.18 The velocity of the boat relative to the shore is $\mathbf{v}_{BS}$. It is the vector sum of the velocity $\mathbf{v}_{BW}$ of the boat relative to the water and the velocity $\mathbf{v}_{WS}$ of the water relative to the shore:
$$\mathbf{v}_{BS} = \mathbf{v}_{BW} + \mathbf{v}_{WS}.$$

Occasionally, situations arise when two vehicles are in relative motion, and it is useful to know the relative velocity of one with respect to the other. Example 11 considers this type of relative motion.

EXAMPLE 11   Approaching an Intersection

Figure 3.19a shows two cars approaching an intersection along perpendicular roads. The cars have the following velocities:

$\mathbf{v}_{\boxed{AG}}$ = velocity of $\boxed{\text{car A}}$ relative to the $\boxed{\text{Ground}}$ = 25.0 m/s, eastward

$\mathbf{v}_{\boxed{BG}}$ = velocity of $\boxed{\text{car B}}$ relative to the $\boxed{\text{Ground}}$ = 15.8 m/s, northward

Find the magnitude and direction of $\mathbf{v}_{AB}$, where

$\mathbf{v}_{\boxed{AB}}$ = velocity of $\boxed{\text{car A}}$ as measured by a passenger in $\boxed{\text{car B}}$

Figure 3.19 Two cars are approaching an intersection along perpendicular roads.

Reasoning To find $\mathbf{v}_{AB}$, we use an equation whose subscripts follow the order outlined earlier. Thus,

$$\mathbf{v}\boxed{AB} = \mathbf{v}\boxed{A}G + \mathbf{v}_G\boxed{B}$$

In this equation, the term $\mathbf{v}_{GB}$ is the velocity of the ground relative to a passenger in car B, rather than $\mathbf{v}_{BG}$, which is given as 15.8 m/s, northward. In other words, the subscripts are reversed. However, $\mathbf{v}_{GB}$ is related to $\mathbf{v}_{BG}$ according to

$$\mathbf{v}_{GB} = -\mathbf{v}_{BG}$$

This relationship reflects the fact that a passenger in car B, moving northward relative to the ground, looks out the car window and sees the ground moving southward, in the opposite direction. Therefore, the equation $\mathbf{v}_{AB} = \mathbf{v}_{AG} + \mathbf{v}_{GB}$ may be used to find $\mathbf{v}_{AB}$, provided we recognize $\mathbf{v}_{GB}$ as a vector that points opposite to the given velocity $\mathbf{v}_{BG}$. With this in mind, Figure 3.19b illustrates how $\mathbf{v}_{AG}$ and $\mathbf{v}_{GB}$ are added vectorially to give $\mathbf{v}_{AB}$.

Solution From the vector triangle shown in Figure 3.19b, the magnitude and direction of $\mathbf{v}_{AB}$ can be calculated as

$$v_{AB} = \sqrt{v_{AG}^2 + v_{GB}^2} = \sqrt{(25.0 \text{ m/s})^2 + (-15.8 \text{ m/s})^2} = \boxed{29.6 \text{ m/s}}$$

and

$$\cos\theta = \frac{v_{AG}}{v_{AB}} \quad \text{or} \quad \theta = \cos^{-1}\left(\frac{v_{AG}}{v_{AB}}\right) = \cos^{-1}\left(\frac{25.0 \text{ m/s}}{29.6 \text{ m/s}}\right) = \boxed{32.4°}$$

Thus, passengers in car B measure the velocity of car A to be 29.6 m/s in a direction 32.4° south of east.

THE PHYSICS OF...

raindrops falling on car windows.

While driving a car, have you ever noticed that the rear window sometimes remains dry, even though rain is falling? This phenomenon is a consequence of relative velocity, as Figure 3.20 helps to explain. Part *a* shows a car traveling horizontally with a velocity of $\mathbf{v}_{CG}$ and a raindrop falling vertically with a velocity of $\mathbf{v}_{RG}$. Both velocities are measured relative to the ground. To determine whether the raindrop hits the window, however, we need to consider the velocity of the raindrop relative to the car, not the ground. This velocity is $\mathbf{v}_{RC}$, and we know that

$$\mathbf{v}_{RC} = \mathbf{v}_{RG} + \mathbf{v}_{GC} = \mathbf{v}_{RG} - \mathbf{v}_{CG}$$

Here, we have used the fact that $\mathbf{v}_{GC'} = -\mathbf{v}_{CG}$. Part *b* of the drawing shows the tail-to-head arrangement corresponding to this vector addition and indicates that the direction of $\mathbf{v}_{RC}$ is given by the angle θ_R. In comparison, the rear window is inclined at

Figure 3.20 (*a*) With respect to the ground, a car is traveling at a velocity of $\mathbf{v}_{CG}$ and a raindrop is falling at a velocity of $\mathbf{v}_{RG}$. The rear window of the car is inclined at an angle θ_W with respect to the vertical. (*b*) This tail-to-head arrangement of vectors corresponds to the equation $\mathbf{v}_{RC} = \mathbf{v}_{RG} - \mathbf{v}_{CG}$.

an angle θ_W with respect to the vertical (see the blowup in part *a*). Only if θ_R is less than θ_W will the raindrop hit the window. However, θ_R is determined by the speed v_{RG} of the raindrop and the speed v_{CG} of the car, according to $\tan \theta_R = v_{CG}/v_{RG}$. At higher car speeds, the angle θ_R becomes too large for the drop to hit the window. At a high enough speed, then, the car simply drives out from under each falling drop!

3.5 CONCEPTS & CALCULATIONS

A primary focus of this chapter has been projectile motion. This section presents two additional examples that serve as a review of the basic features of this type of motion. Example 12 deals with the fact that projectile motion consists of a horizontal and a vertical part, which occur independently of one another. Example 13 stresses the fact that the time variable has the same value for both the horizontal and vertical parts of the motion.

CONCEPTS & CALCULATIONS EXAMPLE 12   Projectile Motion

In a circus act, Burpy the clown is fired from a cannon at an initial velocity $\mathbf{v_0}$ directed at an angle θ above the horizontal, as Figure 3.21 shows. Simultaneously, two other clowns are also launched. Bingo is fired straight upward at a speed of 10.0 m/s and reaches the same maximum height at the same instant as Burpy. Bongo, however, is launched horizontally on roller skates at a speed of 4.6 m/s. He rolls along the ground while Burpy flies through the air. When Burpy returns to the ground, he lands side by side with his roller-skating friend, who is gliding by just at the instant of landing. Ignore air resistance, and assume that the roller skates are unimpeded by friction. Find the speed v_0 and the angle θ for Burpy.

Concept Questions and Answers Is Bongo's left-to-right motion the same as or different from the horizontal part of Burpy's motion along his trajectory?

Answer The horizontal and vertical parts of projectile motion occur independently of one another, so Bongo's left-to-right motion is identical to the horizontal part of Burpy's motion along his trajectory.

Is Burpy's motion in the horizontal direction determined by the initial velocity $\mathbf{v_0}$, just its horizontal component v_{0x}, or just its vertical component v_{0y}?

Answer Just the horizontal component v_{0x} determines Burpy's motion in the horizontal direction.

Is Bingo's up-and-down motion the same as or different than the vertical part of Burpy's motion along his trajectory?

Answer The horizontal and vertical parts of projectile motion occur independently of one another, so Bingo's up-and-down motion is identical to the vertical part of Burpy's motion along his trajectory.

Is Burpy's initial motion in the vertical direction determined by the initial velocity $\mathbf{v_0}$, just its horizontal component v_{0x}, or just its vertical component v_{0y}?

Answer Just the vertical component v_{0y} determines Burpy's motion in the vertical direction.

Solution Based on the Concept Questions and Answers, we can identify the *x* and *y* components of Burpy's initial velocity as follows:

$$v_{0x} = 4.6 \text{ m/s} \quad \text{and} \quad v_{0y} = 10.0 \text{ m/s}$$

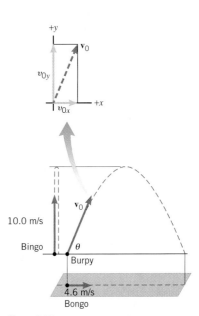

Figure 3.21 Example 12 discusses the projectile motion shown here, in which three circus clowns, Burpy, Bongo, and Bingo, are launched simultaneously.

The initial speed v_0 is the magnitude of the initial velocity, and it, along with the directional angle θ, can be determined from the components:

$$v_0 = \sqrt{v_{0x}^2 + v_{0y}^2} = \sqrt{(4.6 \text{ m/s})^2 + (10.0 \text{ m/s})^2} = \boxed{11 \text{ m/s}}$$

$$\theta = \tan^{-1}\left(\frac{v_{0y}}{v_{0x}}\right) = \tan^{-1}\left(\frac{10.0 \text{ m/s}}{4.6 \text{ m/s}}\right) = \boxed{65°}$$

CONCEPTS & CALCULATIONS EXAMPLE 13   Time and Projectile Motion

A projectile is launched from and returns to ground level. Air resistance is absent. The horizontal range of the projectile is $R = 175$ m, and the horizontal component of the launch velocity is $v_{0x} = 25$ m/s. Find the vertical component v_{0y} of the launch velocity. (See Figure 3.12 for a drawing that approximates this situation.)

Concept Questions and Answers What is the final value of the horizontal component v_x of the projectile's velocity?

Answer   The final value v_x of the horizontal component of the projectile's velocity is the same as the initial value in the absence of air resistance. In other words, the horizontal motion occurs at a constant velocity of 25 m/s.

Can the time be determined for the horizontal part of the motion?

Answer   Yes. In constant-velocity motion, the time is just the horizontal distance (the range) divided by the magnitude of the horizontal component of the projectile's velocity.

Is the time for the horizontal part of the motion the same as the time for the vertical part of the motion?

Answer   Yes. The value for the time calculated for the horizontal part of the motion can be used to analyze the vertical part of the motion.

For the vertical part of the motion, what is the displacement of the projectile?

Answer   Since the projectile is launched from and returns to ground level, the vertical displacement is zero.

Solution From the constant-velocity horizontal motion, we find that the time is

$$t = \frac{R}{v_{0x}} = \frac{175 \text{ m}}{25 \text{ m/s}} = 7.0 \text{ s}$$

For the vertical part of the motion, we also know that the displacement is zero and that the acceleration due to gravity is -9.80 m/s^2, assuming that upward is the positive direction. Therefore, we can use Equation 3.5b to find the initial y component of the velocity:

$$y = v_{0y}t + \tfrac{1}{2}a_y t^2$$

$$0 \text{ m} = v_{0y}t + \tfrac{1}{2}a_y t^2$$

$$v_{0y} = -\tfrac{1}{2}a_y t = -\tfrac{1}{2}(-9.80 \text{ m/s}^2)(7.0 \text{ s}) = \boxed{34 \text{ m/s}}$$

At the end of the problem set for this chapter, you will find homework problems that contain both conceptual and quantitative parts. These problems are grouped under the heading *Concepts & Calculations, Group Learning Problems*. They are designed for use by students working alone or in small learning groups. The conceptual part of each problem provides a convenient focus for group discussions.

SUMMARY

Motion in two dimensions can be described in terms of the time t and the x and y components of four vectors: the displacement, the acceleration, the final velocity, and the initial velocity. The motion can be analyzed by treating the x and y components of the vectors separately and realizing that the time t is the same for each. The directions of these components are conveyed by assigning a plus ($+$) or minus ($-$) sign to each one. When the acceleration vector is constant, the x components of these vectors (x, a_x, v_x, and v_{0x}) are related by the equations of kinematics, as are the y components (y, a_y,

v_y, and v_{0y}). Table 3.1 summarizes the equations of kinematics for two-dimensional motion.

Projectile motion is a kind of two-dimensional motion that occurs when the moving object (the projectile) experiences only the acceleration due to gravity, which acts in the vertical direction. The acceleration of the projectile has no horizontal component ($a_x = 0$), the effects of air resistance being negligible. The vertical component of the acceleration is a_y, and it equals the acceleration due to gravity.

CONCEPTUAL QUESTIONS

1. Suppose an object could move in three dimensions. What additions to the equations of kinematics in Table 3.1 would be necessary to describe three-dimensional motion?

2. An object is thrown upward at an angle θ above the ground, eventually returning to earth. (a) Is there any place along the trajectory where the velocity and acceleration are perpendicular? If so, where? (b) Is there any place where the velocity and acceleration are parallel? If so, where? In each case, explain.

3. Is the acceleration of a projectile equal to zero when it reaches the top of its trajectory? If not, why not?

4. In baseball, the pitcher's mound is raised to compensate for the fact that the ball falls downward as it travels from the pitcher toward the batter. If baseball were played on the moon, would the pitcher's mound have to be higher than, lower than, or the same height as it is on earth? Give your reasoning.

5. A tennis ball is hit upward into the air and moves along an arc. Neglecting air resistance, where along the arc is the speed of the ball (a) a minimum and (b) a maximum? Justify your answers.

6. Suppose there is a wind blowing parallel to the ground and toward the kicker in Figure 3.12. Then the acceleration component in the horizontal direction would not be zero. How would you expect the time of flight of the football to be affected, if at all? Explain.

7. A wrench is accidentally dropped from the top of the mast on a sailboat. Will the wrench hit at the same place on the deck whether the sailboat is at rest or moving with a constant velocity? Justify your answer.

8. A rifle, at a height H above the ground, fires a bullet parallel to the ground. At the same instant and at the same height, a second bullet is dropped from rest. In the absence of air resistance, which bullet strikes the ground first? Explain.

9. Two projectiles are launched from ground level at the same angle above the horizontal, and both return to ground level. Projectile A has a launch speed that is twice that of projectile B. Assuming that air resistance is absent, sketch the trajectories of both projectiles. If your drawings are to be accurate, what should be the ratio of the maximum heights in your drawings and what should be the ratio of the ranges? Justify your answers.

10. A stone is thrown horizontally from the top of a cliff and eventually hits the ground below. A second stone is dropped from rest from the same cliff, falls through the same height, and also hits the ground below. Ignore air resistance. Discuss whether each of the following quantities is different or the same in the two cases; if there is a difference, describe the difference: (a) displacement, (b) speed just before impact with the ground, and (c) time of flight.

11. A leopard springs upward at a 45° angle and then falls back to the ground. Does the leopard, at any point on its trajectory, ever have a speed that is one-half its initial value? Give your reasoning.

12. As background for this question, review Conceptual Example 4. A football quarterback throws a pass on the run and then keeps running without changing his velocity. Can he throw the pass and then catch it himself? Give your reasoning.

13. On a riverboat cruise, a plastic bottle is accidentally dropped overboard. A passenger on the boat estimates that the boat pulls ahead of the bottle by 5 meters each second. Is it possible to conclude that the boat is moving at 5 m/s with respect to the shore? Account for your answer.

14. A plane takes off at St. Louis, flies straight to Denver, and then returns the same way. The plane flies at the same speed with respect to the ground during the entire flight, and there are no head winds or tail winds. Since the earth revolves around its axis once a day, you might expect that the times for the outbound trip and the return trip differ, depending on whether the plane flies against the earth's rotation or with it. However, under the conditions given, the two flight times are identical. Explain why.

15. A child is playing on the floor of a recreational vehicle (RV) as it moves along the highway at a constant velocity. He has a toy cannon, which shoots a marble at a fixed angle and speed with respect to the floor. The cannon can be aimed toward the front or the rear of the RV. Is the range toward the front the same as, less than, or greater than the range toward the rear? Answer this question (a) from the child's point of view and (b) from the point of view of an observer standing still on the ground. Justify your answers.

16. Three swimmers can swim equally fast relative to the water. They have a race to see who can swim across a river in the least time. Swimmer A swims perpendicular to the current and lands on the far shore downstream, because the current has swept him in that direction. Swimmer B swims upstream at an angle to the current and lands on the far shore directly opposite the starting point. Swimmer C swims downstream at an angle to the current in an attempt to take advantage of the current. Who crosses the river in the least time? Account for your answer.

PROBLEMS

Section 3.1 Displacement, Velocity, and Acceleration

1. ssm A jetliner is moving at a speed of 245 m/s. The vertical component of the plane's velocity is 40.6 m/s. Determine the magnitude of the horizontal component of the plane's velocity.

2. A radar antenna is tracking a satellite orbiting the earth. At a certain time, the radar screen shows the satellite to be 162 km away. The radar antenna is pointing upward at an angle of 62.3° from the ground. Find the x and y components (in km) of the position of the satellite.

3. A mountain-climbing expedition establishes two intermediate camps, labeled A and B in the drawing, above the base camp. What is the magnitude Δr of the displacement between camp A and camp B?

4. The altitude of a hang glider is increasing at a rate of 6.80 m/s. At the same time, the shadow of the glider moves along the ground at a speed of 15.5 m/s when the sun is directly overhead. Find the magnitude of the glider's velocity.

5. ssm In diving to a depth of 750 m, an elephant seal also moves 460 m due east of his starting point. What is the magnitude of the seal's displacement?.

6. A dolphin leaps out of the water at an angle of 35° above the horizontal. The horizontal component of the dolphin's velocity is 7.7 m/s. Find the magnitude of the vertical component of the velocity.

7. A wild horse starts from rest and runs in a straight line 29° north of west. After 36 s of running in this direction, the horse has a speed of 12 m/s. (a) What is the magnitude of the horse's average acceleration? Assuming that north and east are the positive directions, find the component of the horse's acceleration

that points along (b) the north–south line and (c) the east–west line.

8. In a football game a kicker attempts a field goal. The ball remains in contact with the kicker's foot for 0.050 s, during which time it experiences an acceleration of 340 m/s². The ball is launched at an angle of 51° above the ground. Determine the horizontal and vertical components of the launch velocity.

9. ssm www In a mall, a shopper rides up an escalator between floors. At the top of the escalator, the shopper turns right and walks 9.00 m to a store. The magnitude of the shopper's displacement from the bottom of the escalator is 16.0 m. The vertical distance between the floors is 6.00 m. At what angle is the escalator inclined above the horizontal?

* **10.** A bird watcher meanders through the woods, walking 0.50 km due east, 0.75 km due south, and 2.15 km in a direction 35.0° north of west. The time required for this trip is 2.50 h. Determine the magnitude and direction (relative to due west) of the bird watcher's (a) displacement and (b) average velocity. Use kilometers and hours for distance and time, respectively.

* **11.** The earth moves around the sun in a nearly circular orbit of radius 1.50×10^{11} m. During the three summer months (an elapsed time of 7.89×10^{6} s), the earth moves one-fourth of the distance around the sun. (a) What is the average speed of the earth? (b) What is the magnitude of the average velocity of the earth during this period?

Section 3.2 Equations of Kinematics in Two Dimensions, Section 3.3 Projectile Motion

12. On a spacecraft two engines fire for a time of 565 s. One gives the craft an acceleration in the x direction of $a_x = 5.10$ m/s², while the other produces an acceleration in the y direction of $a_y = 7.30$ m/s². At the end of the firing period, the craft has velocity components of $v_x = 3775$ m/s and $v_y = 4816$ m/s. Find the magnitude and direction of the initial velocity. Express the direction as an angle with respect to the $+x$ axis.

13. The punter on a football team tries to kick a football so that it stays in the air for a long "hang time." If the ball is kicked with an initial velocity of 25.0 m/s at an angle of 60.0° above the ground, what is the "hang time"?

14. A volleyball is spiked so that it has an initial velocity of 15 m/s directed downward at an angle of 55° below the horizontal. What is the horizontal component of the ball's velocity when the opposing player fields the ball?

15. ssm Suppose that the plane in Example 2 is traveling with twice the horizontal velocity, that is, with a velocity of +230 m/s. If all other factors remain the same, determine the time required for the package to hit the ground.

16. A rock climber throws a small first aid kit to another climber who is higher up the mountain. The initial velocity of the kit is 11 m/s at an angle of 65° above the horizontal. At the instant when the kit is caught, it is traveling horizontally, so its vertical speed is zero. What is the vertical height between the two climbers?

17. ssm www A diver runs horizontally with a speed of 1.20 m/s off a platform that is 10.0 m above the water. What is his speed just before striking the water?

18. The acceleration due to gravity on the moon has a magnitude of 1.62 m/s². Examples 5–7 deal with a placekicker kicking a football. Assume that the ball is kicked on the moon instead of on the earth. Find (a) the maximum height H and (b) the range that the ball would attain on the moon.

19. When chasing a hare along a flat stretch of ground, a greyhound leaps into the air at a speed of 10.0 m/s, at an angle of 31° above the horizontal. (a) What is the range of his leap and (b) for how much time is he in the air?

20. During a baseball game, a fly ball is hit to center field and is caught 115 m from home plate. Just when the ball is caught, a runner on third base takes off for home, and the center fielder throws the ball to the catcher standing on home plate. The runner takes 3.50 s to reach home, while the baseball is thrown with a velocity whose *horizontal* component is 41 m/s. Which reaches home first, the runner or the ball, and by how much time?

21. ssm A fire hose ejects a stream of water at an angle of 35.0° above the horizontal. The water leaves the nozzle with a speed of 25.0 m/s. Assuming that the water behaves like a projectile, how far from a building should the fire hose be located to hit the highest possible fire?

22. The highest barrier that a projectile can clear is 13.5 m, when the projectile is launched at an angle of 15.0° above the horizontal. What is the projectile's launch speed?

23. A car drives straight off the edge of a cliff that is 54 m high. The police at the scene of the accident note that the point of impact is 130 m from the base of the cliff. How fast was the car traveling when it went over the cliff?

24. The 1994 Winter Olympics included the aerials competition in skiing. In this event skiers speed down a ramp that slopes sharply upward at the end. The sharp upward slope launches them into the air, where they perform acrobatic maneuvers. In the women's competition, the end of a typical launch ramp is directed 63° above the horizontal. With this launch angle, a skier attains a height of 13 m above the end of the ramp. What is the skier's launch speed?

25. ssm As preparation for this problem, review Conceptual Example 9. The drawing shows an empty fuel tank being dropped by two different planes. At the moment of release each plane has the same speed of 135 m/s, and each tank is at the same height of 2.00 km above the ground. While the speeds are the same, the velocities are different at the instant of release, because one plane is flying at an angle of 15.0° above the horizontal and the other is flying at an angle of 15.0° below the horizontal. Find the magnitude and direction of the velocity with which the fuel tank hits the ground if it is from (a) plane A and (b) plane B. In each part, give the directional angles with respect to the horizontal.

Fuel tank

Plane A

Plane B

26. Michael Jordan, formerly of the Chicago Bulls basketball team, has some fanatic fans. They claim that he is able to jump and remain in the air for two full seconds from launch to landing. Evaluate this claim by calculating the maximum height that such a jump would attain. For comparison, Jordan's maximum jump height has been estimated at about one meter.

27. A motorcycle daredevil is attempting to jump across as many buses as possible (see the drawing). The takeoff ramp makes an angle of 18.0° above the horizontal, and the landing ramp is identical to the takeoff ramp. The buses are parked side by side, and each bus is 2.74 m wide. The cyclist leaves the ramp with a speed of 33.5 m/s. What is the maximum number of buses over which the cyclist can jump?

33.5 m/s

18.0°   18.0°

2.74 m →

28. A horizontal rifle is fired at a bull's-eye. The muzzle speed of the bullet is 670 m/s. The barrel is pointed directly at the center of the bull's-eye, but the bullet strikes the target 0.025 m below the center. What is the horizontal distance between the end of the rifle and the bull's-eye?

29. ssm A hot-air balloon is rising straight up with a speed of 3.0 m/s. A ballast bag is released from rest relative to the balloon when it is 9.5 m above the ground. How much time elapses before the ballast bag hits the ground?

30. A rocket is fired at a speed of 75.0 m/s from ground level, at an angle of 60.0° above the horizontal. The rocket is fired toward an 11.0-m high wall, which is located 27.0 m away. By how much does the rocket clear the top of the wall?

31. Suppose the water at the top of Niagara Falls has a horizontal speed of 2.7 m/s just before it cascades over the edge of the falls. At what vertical distance below the edge does the velocity vector of the water point downward at a 75° angle below the horizontal?

*32. A cliff has a height of 12 m. From the top of this cliff stone A is thrown upward at an angle (less than 90°). It rises on its trajectory and eventually hits the ground below. At the same instant that stone A is thrown, stone B is thrown upward at an angle (less than 90°) from the base of the cliff, directly below stone A. Stone B rises on its trajectory and eventually returns to the same level from which it was launched. The launch angles and speeds of the stones are different, but 3.0 s after launch, they both strike the ground at the same place. (a) Find the difference $v_{0xB} - v_{0xA}$ between the x components of the launch velocities. (b) Find the difference $v_{0yB} - v_{0yA}$ between the y components of the launch velocities.

*33. ssm www As preparation for this problem, review Conceptual Example 9. The two stones described there have identical initial speeds of $v_0 = 13.0$ m/s and are thrown at an angle $\theta = 30.0°$, one below the horizontal and one above the horizontal. What is the distance *between* the points where the stones strike the ground?

*34. Review Conceptual Example 4 before beginning this problem. You are traveling in a convertible with the top down. The car is moving at a constant velocity of 25 m/s, due east along flat ground. You throw a tomato straight upward at a speed of 11 m/s. How far has the car moved when you get a chance to catch the tomato?

*35. A soccer player kicks the ball toward a goal that is 29.0 m in front of him. The ball leaves his foot at a speed of 19.0 m/s and an angle of 32.0° above the ground. Find the speed of the ball when the goalie catches it in front of the net. (*Note: The answer is not 19.0 m/s.*)

*36. A diver springs upward from a board that is three meters above the water. At the instant she contacts the water her speed is 8.90 m/s and her body makes an angle of 75.0° with respect to the horizontal surface of the water. Determine her initial velocity, both magnitude and direction.

*37. ssm An Olympic long jumper leaves the ground at an angle of 23° and travels through the air for a horizontal distance of 8.7 m before landing. What is the takeoff speed of the jumper?

*38. After leaving the end of a ski ramp, a ski jumper lands downhill at a point that is displaced 55 m horizontally from the end of the ramp. His velocity, just before landing, is 25 m/s and points in a direction 38° below the horizontal. Neglecting air resistance and any lift that he experiences while airborne, find his initial velocity (magnitude and direction) when he left the end of the ramp.

*39. Stones are thrown horizontally with the same velocity from the tops of two different buildings. One stone lands twice as far from the base of the building from which it was thrown as does the other stone. Find the ratio of the height of the taller building to the height of the shorter building.

*40. The lob in tennis is an effective tactic when your opponent is near the net. It consists of lofting the ball over his head, forcing him to move quickly away from the net (see the drawing). Suppose that you loft the ball with an initial speed of 15.0 m/s, at an angle of 50.0° above the horizontal. At this instant your opponent is 10.0 m away from the ball. He begins moving away from you 0.30 s later, hoping to reach the ball and hit it back at the moment

that it is 2.10 m above its launch point. With what minimum average speed must he move? (Ignore the fact that he can stretch, so that his racket can reach the ball before he does.)

*41. ssm An airplane is flying with a velocity of 240 m/s at an angle of 30.0° with the horizontal, as the drawing shows. When the altitude of the plane is 2.4 km, a flare is released from the plane. The flare hits the target on the ground. What is the angle θ?

**42. Water from a garden hose that is pointed 25° above the horizontal lands directly on a sunbather lying on the ground 4.4 m away in the horizontal direction. If the hose is held 1.4 m above the ground, at what speed does the water leave the nozzle?

**43. ssm From the top of a tall building, a gun is fired. The bullet leaves the gun at a speed of 340 m/s, parallel to the ground. As the drawing shows, the bullet puts a hole in a window of another building and hits the wall that faces the window. Using the data in the drawing, determine the distances D and H, which locate the point where the gun was fired. Assume that the bullet does not slow down as it passes through the window.

**44. A projectile is launched from ground level at an angle of 12.0° above the horizontal. It returns to ground level. To what value should the launch angle be adjusted, without changing the launch speed, so that the range doubles?

**45. Two cannons are mounted as shown in the drawing and rigged to fire simultaneously. They are used in a circus act in which two clowns serve as human cannon balls. The clowns are fired toward each other and collide at a height of 1.00 m above the muzzles of the cannons. Clown A is launched at a 75.0° angle, with a speed of 9.00 m/s. The horizontal separation between the clowns as they leave the cannons is 6.00 m. Find the launch speed v_{0B} and the launch angle θ_B ($>45.0°$) for clown B.

**46. A small can is hanging from the ceiling. A rifle is aimed directly at the can, as the figure illustrates. At the instant the gun is fired, the can is released. Ignore air resistance and show that the bullet will always strike the can, regardless of the initial speed of the bullet. Assume that the bullet strikes the can before the can reaches the ground.

Section 3.4 Relative Velocity

47. ssm Two passenger trains are passing each other on adjacent tracks. Train A is moving east with a speed of 13 m/s, and train B is traveling west with a speed of 28 m/s. (a) What is the velocity (magnitude and direction) of train A as seen by the passengers in train B? (b) What is the velocity (magnitude and direction) of train B as seen by the passengers in train A?

48. In a marathon race Chad is out in front, running due north at a speed of 4.00 m/s. John is 95 m behind him, running due north at a speed of 4.50 m/s. How long does it take for John to pass Chad?

49. At some airports there are speedramps to help passengers get from one place to another. A speedramp is a moving conveyor belt that you can either stand or walk on. Suppose a speedramp has a length of 105 m and is moving at a speed of 2.0 m/s relative to the ground. In addition, suppose you can cover this distance in 75 s when walking on the ground. If you walk at the same rate with respect to the speedramp that you walk with respect to the ground, how long does it take for you to travel the 105 m using the speedramp?

50. On a pleasure cruise a boat is traveling relative to the water at a speed of 5.0 m/s due south. Relative to the boat, a passenger walks toward the back of the boat at a speed of 1.5 m/s. (a) What is the magnitude and direction of the passenger's velocity relative to the water? (b) How long does it take for the passenger to walk a distance of 27 m on the boat? (c) How long does it take for the passenger to cover a distance of 27 m on the water?

51. ssm A swimmer, capable of swimming at a speed of 1.4 m/s in still water (i.e., the swimmer can swim with a speed of 1.4 m/s relative to the water), starts to swim directly across a 2.8-km-wide river. However, the current is 0.91 m/s, and it carries the swimmer downstream. (a) How long does it take the swimmer to cross the river? (b) How far downstream will the swimmer be upon reaching the other side of the river?

52. The escalator that leads down into a subway station has a length of 30.0 m and a speed of 1.8 m/s relative to the ground. A student is coming out of the station by running in the wrong direction on this escalator. The local record time for this trick is 11 s. Relative to the escalator, what speed must the student exceed in order to beat the record?

53. You are in a hot-air balloon that, relative to the ground, has a velocity of 6.0 m/s in a direction due east. You see a hawk moving directly away from the balloon in a direction due north. The speed of the hawk relative to you is 2.0 m/s. What are the magnitude and direction of the hawk's velocity relative to the ground? Express the directional angle relative to due east.

54. A remote-controlled model airplane is flying due east in still air. The airplane travels with a speed of 22.6 m/s relative to the air. A wind suddenly begins to blow from the north toward the south with a speed of 8.70 m/s. Find the velocity (magnitude and direction) of the airplane as seen by the controller standing on the ground. Determine the directional angle relative to due east.

55. A person looking out the window of a stationary train notices that raindrops are falling vertically down at a speed of 5.0 m/s relative to the ground. When the train moves at a constant velocity, the raindrops make an angle of 25° when they move past the window, as the drawing shows. How fast is the train moving?

*56. A ferry boat is traveling in a direction 35.1° north of east with a speed of 5.12 m/s relative to the water. A passenger is walking with a velocity of 2.71 m/s due east relative to the boat. What is the velocity (magnitude and direction) of the passenger with respect to the water? Determine the directional angle relative to due east.

*57. ssm Mario, a hockey player, is skating due south at a speed of 7.0 m/s relative to the ice. A teammate passes the puck to him. The puck has a speed of 11.0 m/s and is moving in a direction of 22° west of south, relative to the ice. What are the magnitude and direction (relative to due south) of the puck's velocity, as observed by Mario?

*58. Two twins set out to row separately across a swiftly moving river. They have identical canoes and can row at the same speed relative to the water. Twin A heads straight across the river but, due to the current, is carried downstream before reaching the opposite bank. Twin B heads upstream at an angle of 55° from the bank so as to arrive at the opposite bank at a point that is directly across from the starting point. Determine the ratio of the time it takes twin A to cross the river to the time it takes twin B to cross.

59. ssm www A Coast Guard ship is traveling at a constant velocity of 4.20 m/s, due east, relative to the water. On his radar

screen the navigator detects an object that is moving at a constant velocity. The object is located at a distance of 2310 m with respect to the ship, in a direction 32.0° south of east. Six minutes later, he notes that the object's position relative to the ship has changed to 1120 m, 57.0° south of west. What are the magnitude and direction of the velocity of the object relative to the water? Express the direction as an angle with respect to due west.

ADDITIONAL PROBLEMS

60. A baseball player hits a triple and ends up on third base. A baseball "diamond" is a square, each side of length 27.4 m, with home plate and the three bases on the four corners. What is the magnitude of his displacement?

61. A tennis ball is struck such that it leaves the racket horizontally with a speed of 28.0 m/s. The ball hits the court at a horizontal distance of 19.6 m from the racket. What is the height of the tennis ball when it leaves the racket?

62. A bullet is fired from a rifle that is held 1.6 m above the ground in a horizontal position. The initial speed of the bullet is 1100 m/s. Find (a) the time it takes for the bullet to strike the ground and (b) the horizontal distance traveled by the bullet.

63. ssm A golf ball rolls off a horizontal cliff with an initial speed of 11.4 m/s. The ball falls a vertical distance of 15.5 m into a lake below. (a) How much time does the ball spend in the air? (b) What is the speed v of the ball just before it strikes the water?

64. A spacecraft is traveling with a velocity of $v_{0x} = 5480$ m/s along the $+x$ direction. Two engines are turned on for a time of 842 s. One engine gives the spacecraft an acceleration in the $+x$ direction of $a_x = 1.20$ m/s², while the other gives it an acceleration in the $+y$ direction of $a_y = 8.40$ m/s². At the end of the firing, find (a) v_x and (b) v_y.

65. ssm An eagle is flying horizontally at 6.0 m/s with a fish in its claws. It accidentally drops the fish. (a) How much time passes before the fish's speed doubles? (b) How much additional time would be required for the fish's speed to double again?

66. On a distant planet, golf is just as popular as it is on earth. A golfer tees off and drives the ball 3.5 times as far as he would have on earth, given the same initial velocities on both planets. The ball is launched at a speed of 45 m/s at an angle of 29° above the horizontal. When the ball lands, it is at the same level as the tee. On the distant planet, what is (a) the maximum height and (b) the range of the ball?

*67. A soccer ball is kicked and launched at an angle of 11.0° above the ground. All other factors remaining the same, at what angle should the ball be launched to double the maximum height that it attains?

*68. In the javelin throw at a track and field event, the javelin is launched at a speed of 29 m/s at an angle of 36° above the horizontal. As the javelin travels upward, its velocity points above the horizontal at an angle that decreases as time passes. How much time is required for the angle to be reduced from 36° at launch to 18°?

*69. **ssm** A small aircraft is headed due south with a speed of 57.8 m/s with respect to still air. Then, for 9.00×10^2 s a wind blows the plane so that it moves in a direction 45.0° west of south, even though the plane continues to point due south. The plane travels 81.0 km with respect to the ground in this time. Determine the velocity (magnitude and direction) of the wind with respect to the ground. Determine the directional angle relative to due south.

*70. An oceanliner is heading due north with a speed of 8.5 m/s relative to the water. A small sailboat is heading 45° east of north with a speed of 1.0 m/s relative to the water. Find the relative velocity (magnitude and direction) of the sailboat as observed by the passengers on the oceanliner. Determine the directional angle relative to due north.

71. ssm The drawing shows an exaggerated view of a rifle that has been "sighted in" for a 91.4-meter target. If the muzzle speed of the bullet is $v_0 = 427$ m/s, what are the two possible angles θ_1 and θ_2 between the rifle barrel and the horizontal such that the bullet will hit the target? One of these angles is so large that it is never used in target shooting. (*Hint: The following trigonometric identity may be useful:* $2 \sin \theta \cos \theta = \sin 2\theta$.)

72. A jetliner can fly 6.00 hours on a full load of fuel. Without any wind it flies at a speed of 2.40×10^2 m/s. The plane is to make a round-trip by heading due west for a certain distance, turning around, and then heading due east for the return trip. During the entire flight, however, the plane encounters a 57.8 m/s wind from the jet stream, which blows from west to east. What is the maximum distance that the plane can travel due west and just be able to return home?

73. A baseball is hit into the air at an initial speed of 36.6 m/s and an angle of 50.0° above the horizontal. At the same time, the center fielder starts running away from the batter and catches the ball 0.914 m above the level at which it was hit. If the center fielder is initially 1.10×10^2 m from home plate, what must be his average speed?

CONCEPTS & CALCULATIONS

GROUP LEARNING PROBLEMS

Note: Each of these problems consists of Concept Questions followed by a related quantitative Problem. They are designed for use by students working alone or in small learning groups. The Concept Questions involve little or no mathematics and are intended to stimulate group discussions. They focus on the concepts with which the problems deal. Recognizing the concepts is the essential initial step in any problem-solving technique.

74. Concept Questions A puck is moving on an air hockey table. Relative to an x, y coordinate system at time $t = 0$ s, the x components of the puck's initial velocity and acceleration are $v_{0x} = +1.0$ m/s and $a_x = +2.0$ m/s^2. The y components of the initial velocity and acceleration are $v_{0y} = +2.0$ m/s and $a_y = -2.0$ m/s^2. (a) Is the magnitude of the x component of the velocity increasing or decreasing in time? (b) Is the magnitude of the y component of the velocity increasing or decreasing in time?

Problem Find the magnitude and direction of the puck's velocity at a time of $t = 0.50$ s. Specify the direction relative to the $+x$ axis. Be sure that your calculations are consistent with your answers to the Concept Questions.

75. Concept Questions (a) A projectile is launched at a speed v_0 and at an angle above the horizontal; its initial velocity components are v_{0x} and v_{0y}. In the absence of air resistance, what is the speed of the projectile at the peak of its trajectory? (b) What is its speed just before it lands at the same vertical level from which it was launched? (c) Consider its speed at a point that is at a vertical level between that in (a) and (b). How does the speed at this point compare with the speeds identified in (a) and (b)? In each case, give your reasoning.

Problem A golfer hits a shot to a green that is elevated 3.0 m above the point where the ball is struck. The ball leaves the club at a speed of 14.0 m/s at an angle of 40.0° above the horizontal. It rises to its maximum height and then falls down to the green. Ignoring air resistance, find the speed of the ball just before it lands. Check to see that your answer is consistent with your answer to part (c) of the Concept Questions.

76. Concept Questions (a) When a projectile is launched horizontally from a rooftop at a speed v_{0x}, does its horizontal velocity component ever change in the absence of air resistance? (b) Can you calculate the horizontal distance D traveled after launch simply as $D = v_{0x}t$, where t is the fall time of the projectile? (c) In calculating the fall time, is the vertical part of the motion just like that of a ball dropped from rest? Explain each of your answers.

Problem A criminal is escaping across a rooftop and runs off the roof horizontally at a speed of 5.3 m/s, hoping to land on the roof of an adjacent building. The horizontal distance between the two buildings is D, and the roof of the adjacent building is 2.0 m below the jumping-off point. Find the maximum value for D.

77. Concept Questions (a) When a projectile is thrown at a speed v_0 at an angle above the horizontal, which component of the launch velocity determines the maximum height reached by the projectile? (b) Do the horizontal and vertical components of the launch velocity have larger or smaller magnitudes than v_0? (c) If the projectile is thrown straight upward at a speed v_0, does it attain a greater or lesser height than that in part (a)? In each case, account for your answer.

Problem A ball is thrown upward at a speed v_0 at an angle of 52° above the horizontal. It reaches a maximum height of 7.5 m. How high would this ball go if it were thrown straight upward at speed v_0? Verify that your answer is consistent with your answer to part (c) of the Concept Questions.

78. Concept Questions (a) For a projectile that follows a trajectory like that in Figure 3.12 the range is given by $R = v_{0x}t$, where v_{0x} is the horizontal component of the launch velocity and t is the time of flight. Is v_{0x} proportional to the launch speed v_0? (b) Is the time of flight t proportional to the launch speed v_0? (c) Is the range R proportional to v_0 or v_0^2? Explain each answer.

Problem In the absence of air resistance, a projectile is launched from and returns to ground level. It follows a trajectory similar to that in Figure 3.12 and has a range of 23 m. Suppose the launch speed is doubled, and the projectile is fired at the same angle above the ground. What is the new range?

79. Concept Questions Two friends, Barbara and Neil, are out rollerblading. With respect to the ground, Barbara is skating due south. Neil is in front of her and to her left. With respect to the ground, he is skating due west. (a) Does Barbara see him moving toward the east or toward the west? (b) Does Barbara see him moving toward the north or toward the south? (c) Considering your answers to parts (a) and (b), how does Barbara see Neil moving relative to herself, toward the east and north, toward the east and south, toward the west and north, or toward the west and south? Justify your answers in each case.

Problem With respect to the ground, Barbara is skating due south at a speed of 4.0 m/s. With respect to the ground, Neil is skating due west at a speed of 3.2 m/s. Find Neil's velocity (magnitude and direction relative to due west) as seen by Barbara. Make sure that your answer agrees with your answer to part (c) of the Concept Question.

The acceleration of this windsurfer depends on the forces that act on it, as well as its mass. The relationship between acceleration, forces, and mass is known as Newton's second law, and it is one of the central ideas of this chapter.

CHAPTER 4

FORCES AND NEWTON'S LAWS OF MOTION

4.1 THE CONCEPTS OF FORCE AND MASS

In common usage, a *force* is a push or a pull, as the examples in Figure 4.1 illustrate. In basketball, a player launches a shot by pushing on the ball. The tow bar attached to a speeding boat pulls a water skier. Forces such as those that launch the basketball or pull the skier are called *contact forces*, because they arise from the physical contact between two objects. There are circumstances, however, in which two objects exert forces on one another even though they are not touching. Such forces are referred to as *noncontact forces* or *action-at-a-distance forces.* One example of such a noncontact force occurs when a skydiver is pulled toward the earth because of the force of gravity. The earth exerts this force even when it is not in direct contact with the skydiver. In Figure 4.1, arrows are used to represent the forces. It is appropriate to use arrows, because a force is a vector quantity and has both a magnitude and a direction. The direction of the arrow gives the direction of the force, and the length is proportional to its strength or magnitude.

The word *mass* is just as familiar as the word force. A massive supertanker, for instance, is one that contains an enormous amount of mass. As we will see in the next section, such massive objects are difficult to get moving and are hard to stop once they are in motion. In comparison, a penny does not contain much mass. The emphasis here is on the amount of mass, and the idea of direction is of no concern. Therefore, mass is a scalar quantity.

During the seventeenth century, Isaac Newton, starting with the work of Galileo, developed three important laws that deal with force and mass. Collectively they are called "Newton's laws of motion" and provide the basis for understanding the effect that forces have on an object. Because of the importance of these laws, a separate section will be devoted to each one.

4.2 NEWTON'S FIRST LAW OF MOTION

The First Law

To gain some insight into Newton's first law, think about the game of ice hockey (Figure 4.2). If a player does not hit a stationary puck, it will remain at rest on the ice. After the puck is struck, however, it coasts on its own across the ice, slowing down only slightly because of friction. Since ice is very slippery, there is only a relatively small amount of friction to slow down the puck. In fact, if it were possible to remove all friction and wind resistance, and if the rink were infinitely large, the puck would coast forever in a straight line at a constant speed. Left on its own, the puck would lose none of the velocity imparted to it at the time it was struck. This is the essence of Newton's first law of motion:

> **● NEWTON'S FIRST LAW OF MOTION**
>
> An object continues in a state of rest or in a state of motion at a constant speed along a straight line, unless compelled to change that state by a net force.

The first law uses the phrase "net force." Often, several forces act simultaneously on a body, and *the net force is the vector sum of all of them*. Individual forces matter only to the extent that they contribute to the total. For instance, if friction and other opposing forces were absent, a car could travel forever at 30 m/s in a

(a)

(b)

(c)

Figure 4.1 The arrow labeled **F** represents the force that acts on the basketball, the water skier, and the skydiver.

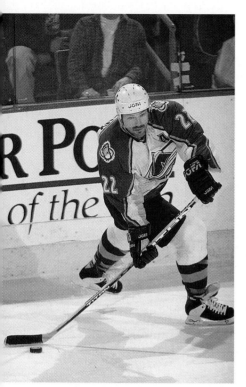

Figure 4.2 The game of ice hockey can give some insight into Newton's laws of motion.

straight line, without using any gas after it has come up to speed. In reality gas is needed, but only so that the engine can produce the necessary force to cancel opposing forces such as friction. This cancellation ensures that there is no net force to change the state of motion of the car.

When an object moves at a constant speed along a straight line, its velocity is constant. Newton's first law indicates that a state of rest (zero velocity) and a state of constant velocity are completely equivalent, in the sense that neither one requires the application of a net force to sustain it. *The purpose served when a net force acts on an object is not to sustain the object's velocity, but, rather, to change it.*

Inertia and Mass

A greater net force is required to change the velocity of some objects than of others. For instance, a net force that is just enough to cause a bicycle to pick up speed will cause an imperceptible change in the motion of a freight train. In comparison to the bicycle, the train has a much greater tendency to remain at rest. Accordingly, we say that the train has more ***inertia*** than the bicycle. Quantitatively, the inertia of an object is measured by its ***mass***. The following definition of inertia and mass indicates why Newton's first law is sometimes called the law of inertia:

● DEFINITION OF INERTIA AND MASS

Inertia is the natural tendency of an object to remain at rest or in motion at a constant speed along a straight line. The mass of an object is a quantitative measure of inertia.

SI Unit of Inertia and Mass: kilogram (kg)

The SI unit for mass is the kilogram (kg), whereas the units in the CGS system and the BE system are the gram (g) and the slug (sl), respectively. Conversion factors between these units are given on the page facing the inside of the front cover. Figure 4.3 gives the masses of various objects, ranging from a penny to a supertanker. The larger the mass, the greater is the inertia. Often the words "mass" and "weight" are used interchangeably, but this is incorrect. Mass and weight are different concepts, and Section 4.7 will discuss the distinction between them.

Figure 4.4 shows a useful application of inertia. Automobile seat belts unwind freely when pulled gently, so they can be buckled. But in an accident, they hold you safely in place. One seat-belt mechanism consists of a ratchet wheel, a locking bar, and a pendulum. The belt is wound around a spool mounted on the ratchet wheel. While the car is at rest or moving at a constant velocity, the pendulum hangs straight down, and the locking bar rests horizontally, as the gray part of the drawing shows. Consequently, nothing prevents the ratchet wheel from turning, and the seat belt can be pulled out easily. When the car suddenly slows down in an accident, however, the relatively massive, lower part of the pendulum keeps moving forward because of its inertia. The pendulum swings on its pivot into the position shown in color and causes the locking bar to block the rotation of the ratchet wheel, thus preventing the seat belt from unwinding.

THE PHYSICS OF...

seat belts.

An Inertial Reference Frame

Newton's first law (and also the second law) can appear to be invalid to certain observers. Suppose, for instance, that you are a passenger riding in a friend's car.

While the car moves at a constant speed along a straight line, you do not feel the seat pushing against your back to any unusual extent. This experience is consistent with the first law, which indicates that in the absence of a net force you should move with a constant velocity. Suddenly the driver floors the gas pedal. Immediately you feel the seat pressing against your back as the car accelerates. Therefore, you sense that a force is being applied to you. The first law leads you to believe that your motion should change, and, relative to the ground outside, your motion does change. But *relative to the car*, you can see that your motion does *not* change, because you remain stationary with respect to the car. Clearly, Newton's first law does not hold for observers who use the accelerating car as a frame of reference. As a result, such a reference frame is said to be noninertial. All accelerating reference frames are noninertial. In contrast, observers for whom the law of inertia is valid are said to be using **inertial reference frames** for their observations, as defined below:

● **DEFINITION OF AN INERTIAL REFERENCE FRAME**

An inertial reference frame is one in which Newton's law of inertia is valid.

The acceleration of an inertial reference frame is zero, so it moves with a constant velocity. All of Newton's laws of motion are valid in inertial reference frames, and when we apply these laws, we will be assuming such a reference frame. In particular, the earth itself is a good approximation of an inertial reference frame.

4.3 NEWTON'S SECOND LAW OF MOTION

Newton's first law indicates that if no net force acts on an object, then the velocity of the object remains unchanged. The second law deals with what happens when a net force does act. Consider a hockey puck once again. When a player strikes a stationary puck, he causes the velocity of the puck to change. In other words, he makes the puck accelerate. The cause of the acceleration is the force that the hockey stick applies. As long as this force acts, the velocity increases, and the puck accelerates. Now, suppose another player strikes the puck and applies twice as much force as the first player does. The greater force produces a greater acceleration. In fact, if the friction between the puck and the ice is negligible, and if there is no wind resistance, the acceleration of the puck is directly proportional to the force. Twice the force produces twice the acceleration. Moreover, the acceleration is a vector quantity, just as the force is, and points in the same direction as the force.

Often, several forces act on an object simultaneously. Friction and wind resistance, for instance, do have some effect on a hockey puck. In such cases, it is the net force, or the vector sum of all the forces acting, that is important. Mathematically, the net force is written as $\Sigma \mathbf{F}$, where the Greek capital letter Σ (sigma) denotes the vector sum. Newton's second law states that the acceleration is proportional to the net force acting on the object.

In Newton's second law, the net force is only one of two factors that determine the acceleration. The other is the inertia or mass of the object. After all, the same net force that imparts an appreciable acceleration to a hockey puck (small mass) will impart very little acceleration to a semitrailer truck (large mass). Newton's second law states that for a given net force, the magnitude of the acceleration is inversely proportional to the mass. Twice the mass means one-half the acceleration, if

Penny
(0.003 kg)

Book
(2 kg)

Bicycle
(15 kg)

Car
(2000 kg)

Jetliner
(1.2×10^5 kg)

Supertanker
(1.5×10^8 kg)

Figure 4.3 The masses of various objects.

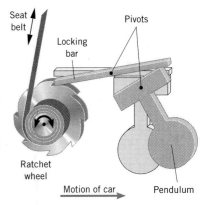

Figure 4.4 Inertia plays a central role in one seat belt mechanism. The gray part of the drawing applies when the car is at rest or moving at a constant velocity. The colored parts show what happens when the car suddenly slows down, as in an accident.

the same net force acts on both objects. Thus, the second law shows how the acceleration depends on both the net force and the mass, as given in Equation 4.1.

● **NEWTON'S SECOND LAW OF MOTION**

When a net external force $\Sigma\mathbf{F}$ acts on an object of mass m, the acceleration $\mathbf{a}$ that results is directly proportional to the net force and has a magnitude that is inversely proportional to the mass. The direction of the acceleration is the same as the direction of the net force.

$$\mathbf{a} = \frac{\Sigma\mathbf{F}}{m} \quad \text{or} \quad \Sigma\mathbf{F} = m\mathbf{a} \tag{4.1}$$

SI Unit of Force: $\text{kg}\cdot\text{m/s}^2 = \text{newton (N)}$

Note that the net force in Equation 4.1 includes only the forces that the environment exerts on the object of interest. Such forces are called **external forces.** In contrast, **internal forces** are forces that one part of an object exerts on another part of the object and are not included in Equation 4.1.

According to Equation 4.1, the SI unit for force is the unit for mass (kg) times the unit for acceleration (m/s^2), or

$$\text{SI unit for force} = (\text{kg})\left(\frac{\text{m}}{\text{s}^2}\right) = \frac{\text{kg}\cdot\text{m}}{\text{s}^2}$$

The combination of $\text{kg}\cdot\text{m/s}^2$ is called a *newton* (N) and is a derived SI unit, not a base unit; $1 \text{ newton} = 1 \text{ N} = 1 \text{ kg}\cdot\text{m/s}^2$.

In the CGS system, the procedure for establishing the unit of force is the same as with SI units, except that mass is expressed in grams (g) and acceleration in cm/s^2. The resulting unit for force is the *dyne*; $1 \text{ dyne} = 1 \text{ g}\cdot\text{cm/s}^2$.

In the BE system, the unit for force is defined to be the pound (lb),* while the unit for acceleration is ft/s^2. With this procedure, Newton's second law can then be used to obtain the unit for mass:

$$\text{BE unit for force} = \text{lb} = (\text{unit for mass})\left(\frac{\text{ft}}{\text{s}^2}\right)$$

$$\text{Unit for mass} = \frac{\text{lb}\cdot\text{s}^2}{\text{ft}}$$

The combination of $\text{lb}\cdot\text{s}^2/\text{ft}$ is the unit for mass in the BE system and is called the *slug* (sl); $1 \text{ slug} = 1 \text{ sl} = 1 \text{ lb}\cdot\text{s}^2/\text{ft}$.

TABLE 4.1   Units for Mass, Acceleration, and Force

System	Mass	Acceleration	Force
SI	kilogram (kg)	meter/second^2 (m/s^2)	newton (N)
CGS	gram (g)	centimeter/second^2 (cm/s^2)	dyne (dyn)
BE	slug (sl)	foot/second^2 (ft/s^2)	pound (lb)

---

* We refer here to the gravitational version of the BE system, in which a force of one pound is defined to be the pull of the earth on a certain standard body at a location where the acceleration due to gravity is 32.174 ft/s^2.

Figure 4.5 (*a*) Two people push a stalled car, in opposition to a force created by friction and the pavement. (*b*) A free-body diagram that shows the horizontal forces acting on the car.

Table 4.1 summarizes the various units for mass, acceleration, and force. Conversion factors between force units from different systems are provided on the page facing the inside of the front cover.

When using the second law to calculate the acceleration, it is necessary to determine the net force that acts on the object. In this determination a *free-body diagram* helps enormously. A free-body diagram is a diagram that represents the object and the forces that act on it. Only the forces that *act on the object* appear in a free-body diagram. Forces that the object exerts on its environment are not included. Example 1 illustrates the use of a free-body diagram.

● **PROBLEM SOLVING INSIGHT**
A free-body diagram is very helpful when applying Newton's second law. Always start a problem by drawing the free-body diagram.

EXAMPLE 1   Pushing a Stalled Car

Two people are pushing a stalled car, as Figure 4.5*a* indicates. The mass of the car is 1850 kg. One person applies a force of 275 N to the car, while the other applies a force of 395 N. Both forces act in the same direction. A third force of 560 N also acts on the car, but in a direction opposite to that in which the people are pushing. This force arises because of friction and the extent to which the pavement opposes the motion of the tires. Find the acceleration of the car.

Reasoning According to Newton's second law, the acceleration is the net force divided by the mass of the car. To determine the net force, we use the free-body diagram in Figure 4.5*b*. In this diagram, the car is represented as a dot, and its motion is along the $+x$ axis. The diagram makes it clear that the forces all act along one direction. Therefore, they can be added as co-linear vectors to obtain the net force.

Solution The net force is

$$\Sigma F = +275 \text{ N} + 395 \text{ N} - 560 \text{ N} = +110 \text{ N}$$

The acceleration can now be obtained:

$$a = \frac{\Sigma F}{m} = \frac{+110 \text{ N}}{1850 \text{ kg}} = \boxed{+0.059 \text{ m/s}^2} \qquad (4.1)$$

The plus sign indicates that the acceleration points along the $+x$ axis, in the same direction as the net force.

● **PROBLEM SOLVING INSIGHT**
The direction of the acceleration is always the same as the direction of the net force.

4.4 THE VECTOR NATURE OF NEWTON'S SECOND LAW OF MOTION

When a football player throws a pass, the direction of the force he applies to the ball is important. Both the force and the resulting acceleration of the ball are vector quantities, as are all forces and accelerations. The directions of these vectors can be taken into account in two dimensions by using x and y components. The net force $\Sigma\mathbf{F}$ in Newton's second law has components ΣF_x and ΣF_y, while the acceleration $\mathbf{a}$ has components a_x and a_y. Consequently, Newton's second law, as expressed in Equation 4.1, can be written in an equivalent form as two equations, one for the x components and one for the y components:

$$\Sigma F_x = ma_x \tag{4.2a}$$

$$\Sigma F_y = ma_y \tag{4.2b}$$

This procedure is similar to that employed in Chapter 3 for the equations of two-dimensional kinematics (see Table 3.1). The components in Equations 4.2a and 4.2b are scalar components and will be either positive or negative numbers, depending on whether they point along the positive or negative x or y axis. The remainder of this section deals with examples that show how these equations are used.

EXAMPLE 2   Applying Newton's Second Law Using Components

A man is stranded on a raft (mass of man and raft = 1300 kg), as shown in Figure 4.6a. By paddling, he causes an average force $\mathbf{P}$ of 17 N to be applied to the raft in a direction due east (the $+x$ direction). The wind also exerts a force $\mathbf{A}$ on the raft. This force has a magnitude of 15 N and points 67° north of east. Ignoring any resistance from the water, find the x and y components of the raft's acceleration.

Reasoning Since the mass of the man and the raft is known, Newton's second law can be used to determine the acceleration components from the given forces. According

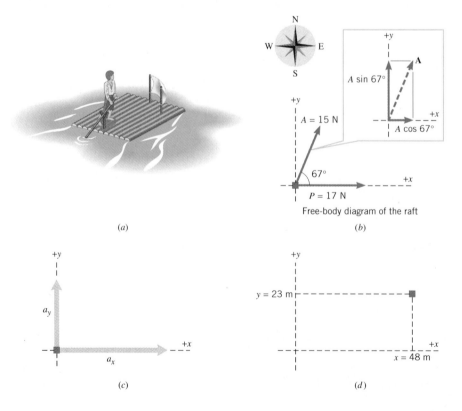

Figure 4.6 (a) A man is paddling a raft, as in Example 2. (b) The free-body diagram shows the forces $\mathbf{P}$ and $\mathbf{A}$ that act on the raft. Forces acting on the raft in a direction perpendicular to the surface of the water play no role in the example and are omitted for clarity. (c) The raft's acceleration components a_x and a_y. (d) In 65 s, the components of the raft's displacement are $x = 48$ m and $y = 23$ m.

to the form of the second law in Equations 4.2a and 4.2b, the acceleration component in a given direction is the component of the net force in that direction divided by the mass. As an aid in determining the components ΣF_x and ΣF_y of the net force, we use the free-body diagram in Figure 4.6b. In this diagram, the direction due east is the $+x$ direction.

Solution Figure 4.6b shows the force components:

Force	x Component	y Component
P	$+17$ N	0 N
A	$+(15$ N$)\cos 67°$	$+(15$ N$)\sin 67°$
	$= +6$ N	$= +14$ N
	$\Sigma F_x = +17$ N $+ 6$ N $= +23$ N	$\Sigma F_y = +14$ N

The plus signs indicate that ΣF_x points in the direction of the $+x$ axis and ΣF_y points in the direction of the $+y$ axis. The x and y components of the acceleration point in the directions of ΣF_x and ΣF_y, respectively, and can now be calculated:

$$a_x = \frac{\Sigma F_x}{m} = \frac{+23 \text{ N}}{1300 \text{ kg}} = \boxed{+0.018 \text{ m/s}^2} \quad (4.2a)$$

$$a_y = \frac{\Sigma F_y}{m} = \frac{+14 \text{ N}}{1300 \text{ kg}} = \boxed{+0.011 \text{ m/s}^2} \quad (4.2b)$$

These acceleration components are shown in Figure 4.6c.

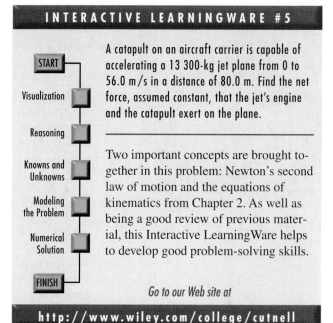

INTERACTIVE LEARNINGWARE #5

START
Visualization
Reasoning
Knowns and Unknowns
Modeling the Problem
Numerical Solution
FINISH

A catapult on an aircraft carrier is capable of accelerating a 13 300-kg jet plane from 0 to 56.0 m/s in a distance of 80.0 m. Find the net force, assumed constant, that the jet's engine and the catapult exert on the plane.

Two important concepts are brought together in this problem: Newton's second law of motion and the equations of kinematics from Chapter 2. As well as being a good review of previous material, this Interactive LearningWare helps to develop good problem-solving skills.

Go to our Web site at

http://www.wiley.com/college/cutnell

• **PROBLEM SOLVING INSIGHT**
Applications of Newton's second law always involve the net external force, which is the vector sum of all the external forces that act on an object.

EXAMPLE 3   The Displacement of a Raft

At the moment the forces **P** and **A** begin acting on the raft in Example 2, the velocity of the raft is 0.15 m/s, in a direction due east (the $+x$ direction). Assuming that the forces are maintained for 65 s, find the x and y components of the raft's displacement during this time interval.

Reasoning Once the net force acting on an object and the object's mass have been used in Newton's second law to determine the acceleration, it becomes possible to use the equations of kinematics to describe the resulting motion. We know from Example 2 that the acceleration components are $a_x = +0.018$ m/s^2 and $a_y = +0.011$ m/s^2, and it is given here that the initial velocity components are $v_{0x} = +0.15$ m/s and $v_{0y} = 0$ m/s. Thus, Equation 3.5a ($x = v_{0x}t + \frac{1}{2}a_x t^2$) and Equation 3.5b ($y = v_{0y}t + \frac{1}{2}a_y t^2$) can be used with $t = 65$ s to determine the x and y components of the raft's displacement.

Solution According to Equations 3.5a and 3.5b, the x and y components of the displacement are

$$x = v_{0x}t + \frac{1}{2}a_x t^2 = (0.15 \text{ m/s})(65 \text{ s}) + \frac{1}{2}(0.018 \text{ m/s}^2)(65 \text{ s})^2 = \boxed{48 \text{ m}}$$

$$y = v_{0y}t + \frac{1}{2}a_y t^2 = (0 \text{ m/s})(65 \text{ s}) + \frac{1}{2}(0.011 \text{ m/s}^2)(65 \text{ s})^2 = \boxed{23 \text{ m}}$$

Figure 4.6d shows the final location of the raft.

The two elephants exert action and reaction forces on each other.

4.5 NEWTON'S THIRD LAW OF MOTION

Imagine you are in a football game. You line up facing your opponent, the ball is snapped, and the two of you crash together. No doubt, you feel a force. But think about your opponent. He too feels something, for while he is applying a force to you, you are applying a force to him. In other words, there isn't just one force on the line of scrimmage; there is a pair of forces. Newton was the first to realize that all forces occur in pairs and there is no such thing as an isolated force, existing all by itself. His third law of motion deals with this fundamental characteristic of forces.

● NEWTON'S THIRD LAW OF MOTION

Whenever one body exerts a force on a second body, the second body exerts an oppositely directed force of equal magnitude on the first body.

The third law is often called the "action–reaction" law, for it is sometimes quoted as follows: "For every action (force) there is an equal, but opposite, reaction."

Figure 4.7 illustrates how the third law applies to an astronaut who is drifting just outside a spacecraft and who pushes on the spacecraft with a force **P**. According to the third law, the spacecraft pushes back on the astronaut with a force **−P** that is equal in magnitude but opposite in direction. In Example 4, we examine the accelerations produced by each of these forces.

Figure 4.7 The astronaut pushes on the spacecraft with a force **+P**. According to Newton's third law, the spacecraft simultaneously pushes back on the astronaut with a force **−P**.

● **PROBLEM SOLVING INSIGHT**
Even though the magnitudes of the action and reaction forces are always equal, these forces do not necessarily produce accelerations that have equal magnitudes, since each force acts on a different object that may have a different mass.

EXAMPLE 4   The Accelerations Produced by Action and Reaction Forces

Suppose that the mass of the spacecraft in Figure 4.7 is $m_S = 11\,000$ kg and that the mass of the astronaut is $m_A = 92$ kg. In addition, assume that the astronaut exerts a force of **P** = +36 N on the spacecraft. Find the accelerations of the spacecraft and the astronaut.

Reasoning According to Newton's third law, when the astronaut applies the force **P** = +36 N to the spacecraft, the spacecraft applies a reaction force **−P** = −36 N to the astronaut. As a result, the spacecraft and the astronaut accelerate in opposite directions. Although the action and reaction forces have the same magnitude, they do not create accelerations of the same magnitude, because the spacecraft and the astronaut have different masses. According to Newton's second law, the astronaut, having a much smaller mass, will experience a much larger acceleration. In applying the second law, we note that the net force acting on the spacecraft is $\Sigma \mathbf{F} = \mathbf{P}$, while the net force acting on the astronaut is $\Sigma \mathbf{F} = -\mathbf{P}$.

Solution The acceleration of the spacecraft is

$$\mathbf{a_S} = \frac{\mathbf{P}}{m_S} = \frac{+36\,\text{N}}{11\,000\,\text{kg}} = \boxed{+0.0033\,\text{m/s}^2}$$

The acceleration of the astronaut is

$$\mathbf{a_A} = \frac{-\mathbf{P}}{m_A} = \frac{-36\,\text{N}}{92\,\text{kg}} = -0.39\,\text{m/s}^2$$

THE PHYSICS OF...

automatic trailer brakes.

There is a clever application of Newton's third law in some rental trailers. As Figure 4.8 illustrates, the tow bar connecting the trailer to the rear bumper of a car contains a mechanism that can automatically actuate brakes on the trailer wheels.

Mechanism for actuating
trailer brakes

Figure 4.8 Some rental trailers include an automatic brake-actuating mechanism.

This mechanism works without the need for electrical connections between the car and the trailer. When the driver applies the car brakes, the car slows down. Because of inertia, however, the trailer continues to roll forward and begins pushing against the bumper. In reaction, the bumper pushes back on the tow bar. The reaction force is used by the mechanism in the tow bar to "push the brake pedal" for the trailer.

4.6 TYPES OF FORCES: AN OVERVIEW

⬤ *Concepts at a Glance* Newton's three laws of motion make it clear that forces play a central role in determining the motion of an object. In the next four sections some common forces will be discussed: the gravitational force (Section 4.7), the normal force (Section 4.8), frictional forces (Section 4.9), and the tension force (Section 4.10). In later chapters, we will encounter still others, such as electric and magnetic forces. The importance of the force concept is that Newton's second law is always valid, regardless of which of these forces may act on an object. One does not have a different law for every type of common force. Thus, we need only to determine what forces are acting on an object, add them together to form the net force, and then use Newton's second law to determine the object's acceleration. The Concepts-at-a-Glance chart in Figure 4.9 illustrates this important idea. ⬤

In nature there are two general types of forces, fundamental and nonfundamental. Fundamental forces are the ones that are truly unique, in the sense that all other

Figure 4.9 *Concepts at a Glance* When any of the external forces listed here act on an object, they are included as part of the net force $\Sigma\mathbf{F}$ in any application of Newton's second law. In this tug-of-war, each of the four external forces acts on the team members.

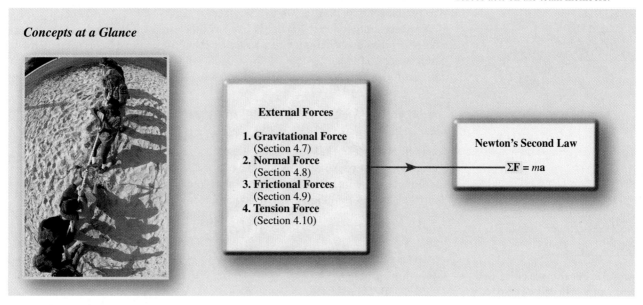

Concepts at a Glance

External Forces

1. **Gravitational Force** (Section 4.7)
2. **Normal Force** (Section 4.8)
3. **Frictional Forces** (Section 4.9)
4. **Tension Force** (Section 4.10)

Newton's Second Law

$$\Sigma\mathbf{F} = m\mathbf{a}$$

forces can be explained in terms of them. Only three fundamental forces have been discovered:

1. Gravitational force
2. Strong nuclear force
3. Electroweak force

The gravitational force is discussed in the next section. The strong nuclear force plays a primary role in the stability of the nucleus of the atom (see Section 31.2). The electroweak force is a single force that manifests itself in two ways (see Section 32.6). One manifestation is the electromagnetic force that electrically charged particles exert on one another (see Sections 18.5, 21.2, and 21.8). The other manifestation is the so-called weak nuclear force that plays a role in the radioactive disintegration of certain nuclei (see Section 31.5).

Except for the gravitational force, all of the forces discussed in this chapter are nonfundamental, because they are related to the electromagnetic force. They arise from the interactions between the electrically charged particles that comprise atoms and molecules. Our understanding of which forces are fundamental, however, is continually evolving. For instance, in the 1860s and 1870s James Clerk Maxwell showed that the electric force and the magnetic force could be explained as manifestations of a single electromagnetic force. Then, in the 1970s, Sheldon Glashow (1932–), Abdus Salam (1926–), and Steven Weinberg (1933–) presented the theory that explains how the electromagnetic force and the weak nuclear force are related to the electroweak force. They received a Nobel prize in 1979 for their achievement. Today, efforts continue that have the goal of reducing further the number of fundamental forces.

4.7 THE GRAVITATIONAL FORCE

Newton's Law of Universal Gravitation

Objects fall downward because of gravity, and Chapters 2 and 3 discuss how to describe the effects of gravity by using a value of $g = 9.80$ m/s^2 for the downward acceleration it causes. However, nothing has been said about why g is 9.80 m/s^2. The reason is fascinating, as we will now see.

The acceleration due to gravity is like any other acceleration, and Newton's second law indicates that it must be caused by a net force. In addition to his famous three laws of motion, Newton also provided a coherent understanding of the **gravitational force**. His "law of universal gravitation" is stated as follows:

● **NEWTON'S LAW OF UNIVERSAL GRAVITATION**

Every particle in the universe exerts an attractive force on every other particle. A particle is a piece of matter, small enough in size to be regarded as a mathematical point. For two particles, which have masses m_1 and m_2 and are separated by a distance r, the force that each exerts on the other is directed along the line joining the particles (see Figure 4.10) and has a magnitude given by

$$F = G \frac{m_1 m_2}{r^2} \qquad (4.3)$$

The symbol G denotes the universal gravitational constant, whose value is found experimentally to be

$$G = 6.672\ 59 \times 10^{-11}\ \mathrm{N \cdot m^2/kg^2}$$

Figure 4.10 The two particles, whose masses are m_1 and m_2, are attracted by gravitational forces $+\mathbf{F}$ and $-\mathbf{F}$.

The constant G that appears in Equation 4.3 is called the ***universal gravitational constant***, because it has the same value for all pairs of particles anywhere in the universe, no matter what their separation. The value for G was first measured in an experiment by the English scientist Henry Cavendish (1731–1810), more than a century after Newton proposed his law of universal gravitation.

To see the main features of Newton's law of universal gravitation, look at the two particles in Figure 4.10. They have masses m_1 and m_2 and are separated by a distance r. In the picture, it is assumed that a force pointing to the right is positive. The gravitational forces point along the line joining the particles and are

$+\mathbf{F}$, the gravitational force exerted on particle 1 by particle 2
$-\mathbf{F}$, the gravitational force exerted on particle 2 by particle 1

These two forces have equal magnitudes and opposite directions. They act on different bodies, causing them to be mutually attracted. In fact, these forces are an action–reaction pair, as required by Newton's third law. Example 5 shows that the magnitude of the gravitational force is extremely small for ordinary values of the masses and the distance between them.

EXAMPLE 5   Gravitational Attraction

What is the magnitude of the gravitational force that acts on each particle in Figure 4.10, assuming $m_1 = 12$ kg (approximately the mass of a bicycle), $m_2 = 25$ kg, and $r = 1.2$ m?

Reasoning and Solution The magnitude of the gravitational force can be found using Equation 4.3:

$$F = G \frac{m_1 m_2}{r^2} = (6.67 \times 10^{-11} \text{ N} \cdot \text{m}^2/\text{kg}^2) \frac{(12 \text{ kg})(25 \text{ kg})}{(1.2 \text{ m})^2} = \boxed{1.4 \times 10^{-8} \text{ N}}$$

For comparison, you exert a force of about 1 N when pushing a doorbell, so that the gravitational force is exceedingly small in circumstances such as those here. This result is due to the fact that G itself is very small. However, if one of the bodies has a large mass, like that of the earth (5.98×10^{24} kg), the gravitational force can be large.

As expressed by Equation 4.3, Newton's law of gravitation applies only to particles. However, most familiar objects are too large to be considered particles. Nevertheless, the law of universal gravitation can be applied to such objects with the aid of calculus. Newton was able to prove that an object of finite size can be considered to be a particle for purposes of using the gravitation law, provided the mass of the object is distributed with spherical symmetry about its center. Thus, Equation 4.3 can be applied when each object is a sphere whose mass is spread uniformly over its entire volume. Figure 4.11 shows this kind of application, assuming that the earth and the moon are such uniform spheres of matter. In this case, r is the distance *between the centers of the spheres* and not the distance between the outer surfaces. The gravitational forces that the spheres exert on each other are the same as if the entire mass of each were concentrated at its center. Even if the objects are not uniform spheres, Equation 4.3 can be used to a good degree of approximation if the sizes of the objects are small relative to the distance of separation r.

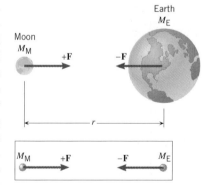

Figure 4.11 The gravitational force that each uniform sphere of matter exerts on the other is the same as if each sphere were a particle with its mass concentrated at its center. The earth (mass M_E) and the moon (mass M_M) approximate such uniform spheres.

Weight

The weight of an object arises because of the gravitational pull of the earth.

Object of mass m

Mass of earth = M_E

Figure 4.12 On or near the earth, the weight **W** of an object is the gravitational force exerted on the object by the earth.

> ● **DEFINITION OF WEIGHT**
>
> The weight of an object on the earth is the gravitational force that the earth exerts on the object. The weight always acts downward, toward the center of the earth. On another astronomical body, the weight is the gravitational force exerted on the object by that body.
>
> **SI Unit of Weight:** newton (N)

Using W for the magnitude of the weight,* m for the mass of the object, and M_E for the mass of the earth, it follows from Equation 4.3 that

$$W = G \frac{M_E m}{r^2} \tag{4.4}$$

Equation 4.4 and Figure 4.12 both emphasize that an object has weight whether or not it is resting on the earth's surface, because the gravitational force is acting even when the distance r is not equal to the radius R_E of the earth. However, the gravitational force becomes weaker as r increases, since r is in the denominator of Equation 4.4. Figure 4.13, for example, shows how the weight of the Hubble Space Telescope becomes smaller as the distance r from the center of the earth increases. In Example 6 the telescope's weight is determined when it is on earth and in orbit.

EXAMPLE 6   The Hubble Space Telescope

The mass of the Hubble Space Telescope is 11 600 kg. Determine the weight of the telescope (a) when it was resting on the earth and (b) as it is in its orbit 598 km above the earth's surface.

Reasoning The weight of the Hubble Space Telescope is the gravitational force exerted on it by the earth. According to Equation 4.4, the weight varies inversely as the square of the radial distance r. Thus, we expect the telescope's weight on the earth's surface (r smaller) to be greater than its weight in orbit (r larger).

Solution

(a) On the earth's surface, the weight is given by Equation 4.4 with $r = 6.38 \times 10^6$ m (the earth's radius):

$$W = G \frac{M_E m}{r^2} = \frac{(6.67 \times 10^{-11} \text{ N·m}^2/\text{kg}^2)(5.98 \times 10^{24} \text{ kg})(11\ 600 \text{ kg})}{(6.38 \times 10^6 \text{ m})^2}$$

$$\boxed{W = 1.14 \times 10^5 \text{ N}}$$

(b) When the telescope is 598 km above the surface, its distance from the center of the earth is

$$r = 6.38 \times 10^6 \text{ m} + 598 \times 10^3 \text{ m} = 6.98 \times 10^6 \text{ m}$$

The weight now can be calculated as in part (a), except the new value of r must be used: $\boxed{W = 0.950 \times 10^5 \text{ N}}$. As expected, the weight is less in orbit.

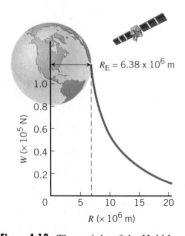

$R_E = 6.38 \times 10^6$ m

Figure 4.13 The weight of the Hubble Space Telescope decreases as it gets farther from the earth. The distance from the center of the earth to the telescope is r.

---

* Often, the word "weight" and the phrase "magnitude of the weight" are used interchangeably, even though weight is a vector. Generally, the context makes it clear when the direction of the weight vector must be taken into account.

The space age has forced us to broaden our ideas about weight. For instance, an astronaut weighs only about one-sixth as much on the moon as on the earth. To obtain his weight on the moon from Equation 4.4, it is only necessary to replace M_E by M_M (the mass of the moon) and let $r = R_M$ (the radius of the moon).

Relation Between Mass and Weight

Although massive objects weigh a lot on the earth, mass and weight are not the same quantity. As Section 4.2 discusses, mass is a quantitative measure of inertia. As such, mass is an intrinsic property of matter and does not change as an object is moved from one location to another. Weight, in contrast, is the gravitational force acting on the object and can vary, depending on how far the object is above the earth's surface or whether it is located near another body such as the moon.

The relation between weight W and mass m can be written in one of two ways:

$$W = \boxed{G\,\frac{M_E}{r^2}}\,m \tag{4.4}$$

$$W = m\,\boxed{g} \tag{4.5}$$

• **PROBLEM SOLVING INSIGHT**
Mass and weight are different quantities. They cannot be interchanged when solving problems.

Equation 4.4 is Newton's law of universal gravitation, while Equation 4.5 is Newton's second law (net force equals mass times acceleration) incorporating the acceleration g due to gravity. These expressions make the distinction between mass and weight stand out. The weight of an object whose mass is m depends on the values for the universal gravitation constant G, the mass M_E of the earth, and the distance r. These three parameters together determine the acceleration g due to gravity. The specific value of $g = 9.80 \text{ m/s}^2$ applies only when r equals the radius R_E of the earth. For larger values of r, as would be the case on top of a mountain, the effective value of g is less than 9.80 m/s^2. The fact that g decreases as the distance r increases means that the weight likewise decreases. The mass of the object, however, does not depend on these effects and does not change. Conceptual Example 7 further explores the difference between mass and weight.

CONCEPTUAL EXAMPLE 7   Mass Versus Weight

A vehicle is being designed for use in exploring the moon's surface and is being tested on earth, where it weighs roughly six times more than it will on the moon. In one test, the acceleration of the vehicle along the ground is measured. To achieve the same acceleration on the moon, will the net force acting on the vehicle be greater than, less than, or the same as that required on earth?

Reasoning and Solution The net force ΣF required to accelerate the vehicle is specified by Newton's second law as $\Sigma \mathbf{F} = m\mathbf{a}$, where m is the vehicle's mass and $\mathbf{a}$ is the acceleration along the ground. For a given acceleration, the net force depends only on the mass. But the mass is an intrinsic property of the vehicle and is the same on the moon as it is on the earth. Therefore, the same net force would be required for a given acceleration on the moon as on the earth. Do not be misled by the fact that the vehicle weighs more on earth. The greater weight occurs only because the earth's mass and radius are different than the moon's. In any event, *in Newton's second law, the net force is proportional to the vehicle's mass, not its weight.*

Related Homework Material: *Problems 19 and 98*

The lunar exploration vehicle that this astronaut is driving on the moon has the same mass that it has on the earth. However, its weight is different on the moon than on the earth, as Conceptual Example 7 discusses.

Figure 4.14 Two forces act on the block, its weight **W** and the normal force **F$_N$** exerted by the surface of the table.

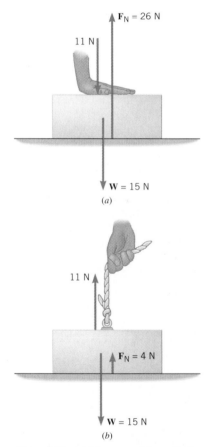

Figure 4.15 (*a*) The normal force **F$_N$** is greater than the weight of the box, because the box is being pressed downward with an 11-N force. (*b*) The normal force is smaller than the weight, because the rope supplies an upward force of 11 N that partially supports the box.

4.8   THE NORMAL FORCE

The Normal Force and Newton's Third Law

In many situations, an object is in contact with a surface, such as a tabletop. Because of the contact, there is a force acting on the object. The present section discusses only one component of this force, the component that acts perpendicular to the surface. The next section discusses the component that acts parallel to the surface. The perpendicular component is called the ***normal force.***

> ● **DEFINITION OF THE NORMAL FORCE**
>
> The normal force **F$_N$** is one component of the force that a surface exerts on an object with which it is in contact, namely, the component that is perpendicular to the surface.

Figure 4.14 shows a block resting on a horizontal table and identifies the two forces that act on the block, the weight **W** and the normal force **F$_N$**. To understand how an inanimate object, such as a tabletop, can exert a normal force, think about what happens when you sit on a mattress. Your weight causes the springs in the mattress to compress. As a result, the compressed springs exert an upward force (the normal force) on you. In a similar manner, the weight of the block causes invisible "atomic springs" in the surface of the table to compress, thus producing a normal force on the block.

Newton's third law plays an important role in connection with the normal force. In Figure 4.14, for instance, the block exerts a force on the table by pressing down on it. Consistent with the third law, the table exerts an oppositely directed force of equal magnitude on the block. This reaction force is the normal force. The magnitude of the normal force indicates how hard the two objects press against each other.

If an object is resting on a horizontal surface and there are no vertically acting forces except the object's weight and the normal force, the magnitudes of these two forces are equal, i.e., $F_N = W$. This is the situation in Figure 4.14. The weight must be balanced by the normal force for the object to remain at rest on the table. If the magnitudes of these forces were not equal, there would be a net force acting on the block, and the block would accelerate either upward or downward, in accord with Newton's second law.

If other forces in addition to **W** and **F$_N$** act in the vertical direction, the magnitudes of the normal force and the weight are no longer equal. In Figure 4.15*a*, for instance, a box whose weight is 15 N is being pushed downward against a table. The pushing force has a magnitude of 11 N. Thus, the total downward force exerted on the box is 26 N, and this must be balanced by the upward-acting normal force if the box is to remain at rest. In this situation, then, the normal force is 26 N, which is considerably larger than the weight of the box.

Figure 4.15*b* illustrates a different situation. Here, the box is being pulled upward by a rope that applies a force of 11 N. The net force acting on the box due to its weight and the rope is only 4 N, downward. To balance this force, the normal force needs to be only 4 N. It is not hard to imagine what would happen if the force applied by the rope were increased to 15 N—exactly equal to the weight of the box. In this situation, the normal force would become zero. In fact, the table could be removed, since the block would be supported entirely by the rope. The situations in Figure 4.15 are consistent with the idea that the magnitude of the normal force indicates how hard two objects press against each other. Clearly, the box and the table press against each other harder in part *a* of the picture than in part *b*.

Like the box and the table in Figure 4.15, various parts of the human body press against one another and exert normal forces. Example 8 illustrates the remarkable ability of the human skeleton to withstand a wide range of normal forces.

EXAMPLE 8   A Balancing Act

In a circus balancing act, a woman performs a headstand on top of a man's head, as Figure 4.16*a* illustrates. The woman weighs 490 N, and the man's head and neck weigh 50 N. It is primarily the seventh cervical vertebra in the spine that supports all the weight above the shoulders. What is the normal force that this vertebra exerts on the neck and head of the man (a) before the act and (b) during the act?

THE PHYSICS OF...

a circus balancing act.

Reasoning To begin, we draw a free-body diagram for the neck and head of the man. Before the act, there are only two forces, the weight of the man's head and neck, and the normal force. During the act, an additional force is present due to the woman's weight. In both cases, the upward and downward forces must balance for the head and neck to remain at rest. This condition of balance will lead us to values for the normal force.

Solution

(a) Figure 4.16*b* shows the free-body diagram for the man's head and neck before the act. The only forces acting are the normal force $\mathbf{F_N}$ and the 50-N weight. These two forces must balance for the man's head and neck to remain at rest. Therefore, the seventh cervical vertebra exerts a normal force of $\boxed{F_N = 50 \text{ N}}$.

(b) Figure 4.16*c* shows the free-body diagram that applies during the act. Now, the total downward force exerted on the man's head and neck is 50 N + 490 N = 540 N, which must be balanced by the upward normal force, so that $\boxed{F_N = 540 \text{ N}}$.

Figure 4.16 (*a*) A balancing act and free-body diagrams for the man's body above the shoulders (*b*) before the act and (*c*) during the act. For convenience, the scales used for the vectors in parts *b* and *c* are different.

(*a*)

(*b*)

(*c*)

(a) No acceleration (**v** = constant)   (b) Upward acceleration   (c) Downward acceleration   (d) Free-fall

Figure 4.17 (a) When the elevator is not accelerating, the scale registers the true weight (W = 700 N) of the person. (b) When the elevator accelerates upward, the apparent weight (1000 N) exceeds the true weight. (c) When the elevator accelerates downward, the apparent weight (400 N) is less than the true weight. (d) The apparent weight is zero if the elevator falls freely, that is, if it falls with the acceleration due to gravity.

In summary, the normal force does not necessarily have the same magnitude as the weight of the object. The value of the normal force depends on what other forces are present. It also depends on whether the objects in contact are accelerating. In one situation that involves accelerating objects, the magnitude of the normal force can be regarded as a kind of "apparent weight," as we will now see.

Apparent Weight

Usually, the weight of an object can be determined with the aid of a scale. However, even though a scale is working properly, there are situations in which it does not give the correct weight. In such situations, the reading on the scale gives only the "apparent" weight, rather than the gravitational force or "true" weight. The apparent weight is the force that the object exerts on the scale with which it is in contact.

To see the discrepancies that can arise between true weight and apparent weight, consider the scale in the elevator in Figure 4.17. The reasons for the discrepancies will be explained shortly. A person whose true weight is 700 N steps on the scale. If the elevator is at rest or moving with a constant velocity (either upward or downward), the scale registers the true weight, as Figure 4.17a illustrates.

If the elevator is accelerating, the apparent weight and the true weight are not equal. When the elevator accelerates upward, the apparent weight is greater than the true weight, as Figure 4.17b shows. Conversely, if the elevator accelerates downward, as in part c, the apparent weight is less than the true weight. In fact, if the elevator falls freely, so its acceleration is equal to the acceleration due to gravity, the apparent weight becomes zero, as part d indicates. In a situation such as this, where the apparent weight is zero, the person is said to be "weightless." The apparent weight, then, does not equal the true weight if the scale and the person on it are accelerating.

The discrepancies between true weight and apparent weight can be understood with the aid of Newton's second law. Figure 4.18 shows a free-body diagram of the person in the elevator. The two forces that act on him are the true weight $\mathbf{W} = m\mathbf{g}$ and the normal force $\mathbf{F_N}$ exerted by the platform of the scale. Applying Newton's second law in the vertical direction gives

Figure 4.18 A free-body diagram showing the forces acting on the person riding in the elevator of Figure 4.17. **W** is the true weight, and $\mathbf{F_N}$ is the normal force exerted on the person by the platform of the scale.

$$\Sigma F_y = +F_N - mg = ma$$

where a is the acceleration of the elevator and person. In this result, the symbol g stands for the magnitude of the acceleration due to gravity and can never be a negative quantity. However, the acceleration a may be either positive or negative, depending on whether the elevator is accelerating upward $(+)$ or downward $(-)$. Solving for the normal force F_N shows that

$$\underbrace{F_N}_{\substack{\text{Apparent}\\\text{weight}}} = \underbrace{mg}_{\substack{\text{True}\\\text{weight}}} + ma \tag{4.6}$$

In Equation 4.6, F_N is the magnitude of the normal force exerted on the person by the scale. But in accord with Newton's third law, F_N is also the magnitude of the downward force that the person exerts on the scale, namely, the apparent weight.

Equation 4.6 contains all the features shown in Figure 4.17. If the elevator is not accelerating, $a = 0$ m/s^2, and the apparent weight equals the true weight. If the elevator accelerates upward, a is positive, and the equation shows that the apparent weight is greater than the true weight. If the elevator accelerates downward, a is negative, and the apparent weight is less than the true weight. If the elevator falls freely, $a = -g$, and the apparent weight is zero. The apparent weight is zero, because when both the person and the scale fall freely, they cannot push against one another. In this text, when the weight is given, it is assumed to be the true weight, unless stated otherwise.

Figure 4.19 This photo, shot from underneath a transparent surface, shows two tires rolling under wet conditions. The channels in the tires collect and divert the water away from the regions where the tires contact the surface, thus providing better traction.

4.9 STATIC AND KINETIC FRICTIONAL FORCES

When an object is in contact with a surface, there is a force acting on the object. The previous section discusses the component of this force that is perpendicular to the surface, which is called the normal force. When the object moves or attempts to move along the surface, there is also a component of the force that is parallel to the surface. This parallel force component is called the ***frictional force,*** or simply ***friction.***

In many situations considerable engineering effort is expended trying to reduce friction. For example, oil is used to reduce the friction that causes wear and tear in the pistons and cylinder walls of an automobile engine. Sometimes, however, friction is absolutely necessary. Without friction, car tires could not provide the traction needed to move the car. In fact, the raised tread on a tire is designed to maintain friction. On a wet road, the spaces in the tread pattern (see Figure 4.19) provide channels for the water to collect and be diverted away. Thus, these channels largely prevent the water from coming between the tire surface and the road surface, where it would reduce friction and allow the tire to skid.

Surfaces that appear to be highly polished can actually look quite rough when examined under a microscope. Such an examination reveals that two surfaces in contact touch only at relatively few spots, as Figure 4.20 illustrates. The microscopic area of contact for these spots is substantially less than the apparent macroscopic area of contact between the surfaces—perhaps thousands of times less. At these contact points the molecules of the different bodies are close enough together to exert strong attractive intermolecular forces on one another, leading to what are known as "cold welds." Frictional forces are associated with these welded spots, but the exact details of how frictional forces arise are not well understood. However, some empirical relations have been developed that make it possible to account for the effects of friction.

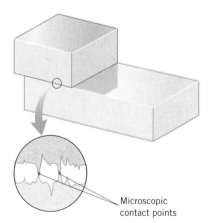

Microscopic contact points

Figure 4.20 Even when two highly polished surfaces are in contact, they touch only at a relatively few points.

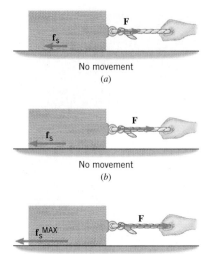

Figure 4.21 Applying a small force **F** to the block, as in parts *a* and *b*, produces no movement, because the static frictional force **f**$_s$ exactly balances the applied force. (*c*) The block just begins to move when the applied force is slightly greater than the maximum static frictional force **f**$_s^{MAX}$.

Figure 4.21 helps to explain the main features of the type of friction known as *static friction*. The block in this drawing is initially at rest on a table, and as long as there is no attempt to move the block, there is no static frictional force. Then, a horizontal force **F** is applied to the block by means of a rope. If **F** is small, as in part *a,* experience tells us that the block still does not move. Why? It does not move because the static frictional force **f**$_s$ exactly cancels the effect of the applied force. The direction of **f**$_s$ is opposite to that of **F**, and the magnitude of **f**$_s$ equals the magnitude of the applied force, $f_s = F$. Increasing the applied force in Figure 4.21 by a small amount still does not cause the block to move. There is no movement because the static frictional force also increases by an amount that cancels out the increase in the applied force (see part *b* of the drawing). If the applied force continues to increase, however, there comes a point when the block finally "breaks away" and begins to slide. The force just before breakaway represents the *maximum static frictional force* **f**$_s^{MAX}$ that the table can exert on the block (see part *c* of the drawing). Any applied force that is greater than **f**$_s^{MAX}$ cannot be balanced by static friction, and the resulting net force accelerates the block to the right.

Experimental evidence shows that, to a good degree of approximation, the maximum static frictional force between a pair of dry, unlubricated surfaces has two main characteristics. It is independent of the apparent macroscopic area of contact between the objects, provided that the surfaces are hard or nondeformable. For instance, in Figure 4.22 the maximum static frictional force that the surface of the table can exert on a block is the same, whether the block is resting on its largest or its smallest side. The other main characteristic of **f**$_s^{MAX}$ is that its magnitude is proportional to the magnitude of the normal force **F**$_N$. As Section 4.8 points out, the magnitude of the normal force indicates how hard two surfaces are being pressed together. The harder they are pressed, the larger is f_s^{MAX}, presumably because the number of "cold-welded," microscopic contact points is increased. Equation 4.7 expresses the proportionality between f_s^{MAX} and F_N with the aid of a proportionality constant μ_s, which is called the *coefficient of static friction.*

● **STATIC FRICTIONAL FORCE**

The magnitude f_s of the static frictional force can have any value from zero up to a maximum value of f_s^{MAX}, depending on the applied force. In other words, $f_s \leq f_s^{MAX}$, where the symbol "$\leq$" is read as "less than or equal to." The equality holds only when f_s attains its maximum value, which is

$$f_s^{MAX} = \mu_s F_N \tag{4.7}$$

In Equation 4.7, μ_s is the coefficient of static friction, and F_N is the magnitude of the normal force.

It should be emphasized that Equation 4.7 relates only the magnitudes of **f**$_s^{MAX}$ and **F**$_N$, *not the vectors themselves*. This equation does not imply that the directions of the vectors are the same. In fact, **f**$_s^{MAX}$ is parallel to the surface, while **F**$_N$ is perpendicular to it.

The coefficient of static friction, being the ratio of the magnitudes of two forces ($\mu_s = f_s^{MAX}/F_N$), is a unitless number. It depends on the type of material from which each surface is made (steel on wood, rubber on concrete, etc.), the condition of the surfaces (polished, rough, lubricated, etc.), and other variables such as temperature. Typical values for μ_s range from about 0.01 for smooth surfaces to about

Figure 4.22 The maximum static frictional force **f**$_s^{MAX}$ would be the same, no matter which side of the block is in contact with the table.

1.5 for rough surfaces. Example 9 illustrates the use of Equation 4.7 for the maximum static frictional force.

EXAMPLE 9   The Force Needed to Start a Sled Moving

A sled is resting on a horizontal patch of snow, and the coefficient of static friction is $\mu_s = 0.350$. The sled and its rider have a total mass of 38.0 kg. What is the magnitude of the maximum horizontal force that can be applied to the sled before it just begins to move?

Reasoning The maximum horizontal force occurs when its magnitude equals that of the maximum force of static friction, as indicated in Figure 4.21c. Equation 4.7 ($f_s^{MAX} = \mu_s F_N$) specifies how the magnitude f_s^{MAX} of the maximum static frictional force is related to the magnitude F_N of the normal force. We can determine F_N by noting that the sled does not accelerate in the vertical direction. Thus, the net force acting vertically on the sled must be zero. Consequently, the normal force and the weight of the sled and its rider must balance, so that $F_N = mg$.

Solution The magnitude of the maximum horizontal force is

$$f_s^{MAX} = \mu_s F_N = \mu_s mg = (0.350)(38.0 \text{ kg})(9.80 \text{ m/s}^2) = \boxed{130 \text{ N}}$$

Once two surfaces begin sliding over one another, the static frictional force is no longer of any concern. Instead, a type of friction known as ***kinetic* friction*** comes into play. The kinetic frictional force opposes the relative sliding motion. If you have ever pushed an object across a floor, you may have noticed that it takes less force to keep the object sliding than it takes to get it going in the first place. In other words, the kinetic frictional force is usually less than the static frictional force.

Experimental evidence indicates that the kinetic frictional force $\mathbf{f}_k$ has three main characteristics, to a good degree of approximation. It is independent of the apparent area of contact between the surfaces (see Figure 4.22). It is independent of the speed of the sliding motion, if the speed is small. And lastly, the magnitude of the kinetic frictional force is proportional to the magnitude of the normal force. Equation 4.8 expresses this proportionality with the aid of a proportionality constant μ_k, which is called the ***coefficient of kinetic friction.***

● KINETIC FRICTIONAL FORCE

The magnitude f_k of the kinetic frictional force is given by

$$f_k = \mu_k F_N \qquad (4.8)$$

In Equation 4.8, μ_k is the coefficient of kinetic friction, and F_N is the magnitude of the normal force.

Equation 4.8, like Equation 4.7, is a relationship between only the magnitudes of the frictional and normal forces. The directions of these forces are perpendicular to each other. Moreover, like the coefficient of static friction, the coefficient of kinetic friction is a unitless number and depends on the type and condition of the two surfaces that are in contact. Values for μ_k are typically less than those for μ_s, reflecting the fact that kinetic friction is generally less than static friction. The next example illustrates the effect of kinetic friction.

* The word "kinetic" is derived from the Greek word *kinetikos*, meaning "of motion."

To maneuver his way up Devil's Tower in Wyoming, this rock climber uses the static frictional force between his hands and feet and the vertical rock walls.

Good grooming depends on the kinetic frictional force.

Figure 4.23 (*a*) The moving sled decelerates because of the kinetic frictional force. (*b*) Three forces act on the moving sled, its weight **W**, the normal force **F**$_N$, and the kinetic frictional force, **f**$_K$. The free-body diagram for the sled shows these forces.

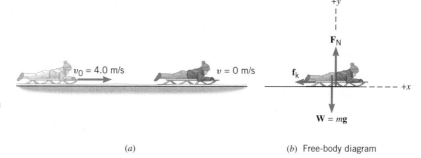

(*a*)   (*b*) Free-body diagram

EXAMPLE 10   Sled Riding

A sled is traveling at 4.00 m/s along a horizontal stretch of snow, as Figure 4.23*a* illustrates. The coefficient of kinetic friction is $\mu_k = 0.0500$. How far does the sled go before stopping?

Reasoning The sled comes to a halt because the kinetic frictional force opposes the motion and causes the sled to slow down. Therefore, we will determine the kinetic frictional force and use it in Newton's second law to find the acceleration of the sled. Knowing the acceleration, we can determine the stopping distance by employing the appropriate equation of kinematics, as discussed in Chapter 3.

Solution To determine the magnitude f_k of the kinetic frictional force, it is necessary to know the magnitude F_N of the normal force, since $f_k = \mu_k F_N$. Part *b* of Figure 4.23 shows the free-body diagram for the sled. Since the sled does not accelerate in the vertical direction, there can be no net force acting vertically on the sled. As a result, the normal force and the weight **W** must balance, so the magnitude of the normal force is $F_N = mg$. The magnitude of the kinetic frictional force is

$$f_k = \mu_k F_N = \mu_k mg \tag{4.8}$$

The kinetic frictional force is the only force acting on the sled in the *x* direction, so it is the net force. Newton's second law then gives the acceleration as

$$a_x = \frac{-f_k}{m} = \frac{-\mu_k mg}{m} = -\mu_k g \tag{4.2a}$$

The minus sign before f_k arises because the kinetic frictional force opposes the sliding motion of the sled and is directed to the left, or along the $-x$ axis, in Figure 4.23*b*. Therefore, the acceleration also points along the $-x$ axis. Notice that the acceleration does not depend on the mass of the sled and rider, since it cancels algebraically. The stopping distance *x* can be obtained with the aid of Equation 3.6a from the equations of kinematics ($v_x^2 = v_{0x}^2 + 2a_x x$) as

$$x = \frac{v_x^2 - v_{0x}^2}{2a_x}$$

where v_x and v_{0x} are the final and initial velocities, respectively. Using the fact that $a_x = -\mu_k g$, we find

$$x = \frac{v_x^2 - v_{0x}^2}{-2\mu_k g} = \frac{(0 \text{ m/s})^2 - (4.00 \text{ m/s})^2}{-2(0.0500)(9.80 \text{ m/s}^2)} = \boxed{16.3 \text{ m}}$$

Static friction opposes the impending relative motion between two objects, while kinetic friction opposes the relative sliding motion that actually does occur. In either case, *relative motion* is opposed. However, this opposition to relative motion does not mean that friction prevents or works against the motion of *all* objects. For instance, the foot of a person walking exerts a force on the earth, and the earth exerts a reaction force on the foot. This reaction force is a static frictional force, and it opposes the impending backward motion of the foot, propelling the person forward in the process. Kinetic friction can also cause an object to move, all the while opposing relative motion, as it does in Example 10. In this example the kinetic frictional force acts on the sled and opposes the relative motion of the sled and the earth. Newton's third law indicates, however, that since the earth exerts the kinetic frictional force on the sled, the sled must exert a reaction force on the earth. In response, the earth accelerates, but because of the earth's huge mass, the motion is too slight to be noticed.

4.10 THE TENSION FORCE

Forces are often applied by means of cables or ropes that are used to pull an object. For instance, Figure 4.24*a* shows a force **T** being applied to the right end of a rope attached to a box. Each particle in the rope, in turn, applies a force to its neighbor. As a result, the force is applied to the box, as part *b* of the drawing shows.

In situations such as that in Figure 4.24, we say that "the force **T** is applied to the box because of the tension in the rope," meaning that the tension and the force applied to the box have the same magnitude. However, the word "tension" is commonly used to mean the tendency of the rope to be pulled apart. To see the relationship between these two uses of the word "tension," consider the left end of the rope, which applies the force **T** to the box. In accordance with Newton's third law, the box applies a reaction force to the rope. The reaction force has the same magnitude as **T** but is oppositely directed. In other words, a force − **T** acts on the left end of the rope. Thus, forces of equal magnitude act on opposite ends of the rope, as in Figure 4.24*c*, and tend to pull it apart.

In the previous discussion, we have used the concept of a "massless" rope (*m* = 0 kg) without saying so. In reality, a massless rope does not exist, but it is useful as an idealization when applying Newton's second law. According to the second law, a net force is required to accelerate an object that has mass. In contrast, no net force is needed to accelerate a massless rope, since Σ**F** = *m***a** and *m* = 0 kg. Thus, when a force **T** is applied to one end of a massless rope, none of the force is needed to accelerate the rope. As a result, the force **T** is also applied undiminished to the object attached at the other end, as we assumed in Figure 4.24.* If the rope had mass, however, some of the force **T** would have to be used to accelerate the rope. The force applied to the box would then be less than **T**, and the tension would be different at different locations along the rope. In this text we will assume that a rope connecting one object to another is massless, unless stated otherwise. The ability of a massless rope to transmit tension undiminished from one end to the other is not

---

* If a rope is not accelerating, **a** is zero in the second law, and Σ**F** = *m***a** = 0, regardless of the mass of the rope. Then, the rope can be ignored, no matter what mass it has.

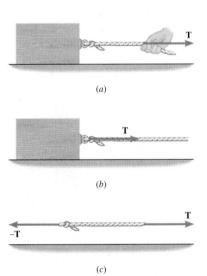

Figure 4.24 (*a*) A force **T** is being applied to the right end of a rope. (*b*) The force is transmitted to the box. (*c*) Forces are applied to both ends of the rope. These forces have equal magnitudes and opposite directions.

Figure 4.25 The force **T** applied at one end of a massless rope is transmitted undiminished to the other end, even when the rope bends around a pulley, provided the pulley is also massless and there is no friction.

Figure 4.26 *Concepts at a Glance* Both equilibrium and nonequilibrium problems can be solved with the aid of Newton's second law. For equilibrium situations, such as the tightrope walker in the left photograph, the acceleration is zero (**a** = 0 m/s²), while for nonequilibrium situations, such as the freely falling performer in the right photograph, it is not zero (**a** ≠ 0 m/s²).

affected when the rope passes around objects such as the pulley in Figure 4.25 (provided the pulley itself is massless and frictionless).

4.11 EQUILIBRIUM APPLICATIONS OF NEWTON'S LAWS OF MOTION

Have you ever been so upset that it took days to recover your "equilibrium?" In this context, the word "equilibrium" refers to a balanced state of mind, one that is not changing wildly. In physics, the word "equilibrium" also refers to a lack of change, but in the sense that the velocity of an object isn't changing. If its velocity doesn't change, an object is not accelerating. Our definition of equilibrium, then, is as follows:

● **DEFINITION OF EQUILIBRIUM**

An object is in equilibrium when it has zero acceleration.

● *Concepts at a Glance* The concept of equilibrium arises directly from Newton's second law. The Concepts-at-a-Glance chart in Figure 4.26, which is an enhanced version of the chart in Figure 4.9, illustrates this important point. When the acceleration of an object is zero (**a** = 0 m/s²), the object is in equilibrium, as the upper-right part of the drawing indicates. This section presents several examples involving equilibrium situations. On the other hand, when the acceleration is not zero (**a** ≠ 0 m/s²), we have a nonequilibrium situation, as the lower-right part of the chart suggests. Section 4.12 deals with nonequilibrium applications of Newton's second law. ●

Since the acceleration is zero for an object in equilibrium, all of the acceleration components are also zero. In two dimensions, this means that $a_x = 0$ m/s² and $a_y = 0$ m/s². Substituting these values into the second law ($\Sigma F_x = ma_x$ and $\Sigma F_x = ma_y$)

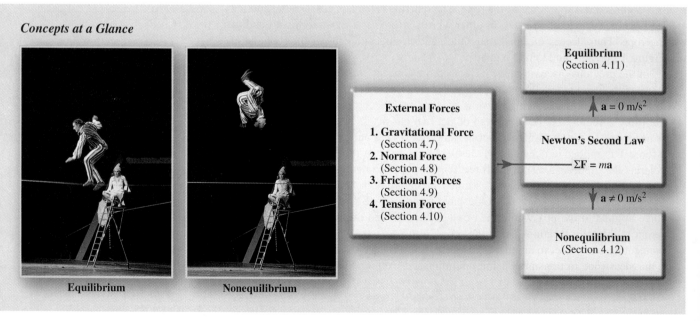

Concepts at a Glance

Equilibrium

Nonequilibrium

External Forces

1. **Gravitational Force** (Section 4.7)
2. **Normal Force** (Section 4.8)
3. **Frictional Forces** (Section 4.9)
4. **Tension Force** (Section 4.10)

Equilibrium (Section 4.11)

a = 0 m/s²

Newton's Second Law

$\Sigma\mathbf{F} = m\mathbf{a}$

a ≠ 0 m/s²

Nonequilibrium (Section 4.12)

shows that the x component and the y component of the net force must each be zero. Thus, in two dimensions, the equilibrium condition is expressed by two equations:

$$\Sigma F_x = 0 \qquad (4.9a)$$
$$\Sigma F_y = 0 \qquad (4.9b)$$

In other words, the forces acting on an object in equilibrium must balance.

In using Equations 4.9a and 4.9b to solve equilibrium problems, we will use the following five-step reasoning strategy:

REASONING STRATEGY

Analyzing Equilibrium Situations

1. Select the object (often called the "system") to which Equations 4.9a and 4.9b are to be applied. Generally, this will be the object about which the most information is known. It may be that two or more objects are connected by means of a rope or a cable. Then, it may be necessary to treat each object separately according to the following steps.

2. Draw a free-body diagram for each object chosen above. As Section 4.3 discusses, a free-body diagram is a drawing that represents the object and shows the forces that act on it, each force with its proper direction. Be sure to include only forces that act on the object. *Do not include forces that the object exerts on its environment.*

3. Choose a set of x, y axes for each object and resolve all forces in the free-body diagram into components that point along these axes. Select the axes so that as many forces as possible point directly along the x axis or the y axis. Such a choice minimizes the number of calculations needed to determine the force components.

4. Apply Equations 4.9a and 4.9b by setting the sum of the x components of the forces equal to zero and the sum of the y components of the forces equal to zero.

5. Solve the two equations obtained in Step 4 for the desired unknown quantities, remembering that two equations can yield answers for only two unknowns at most.

Example 11 illustrates how these steps are followed. It deals with a traction device in which three forces act together to bring about the equilibrium.

EXAMPLE 11   Traction for the Foot

Figure 4.27a shows a traction device used with a foot injury. The weight of the 2.2-kg object creates a tension in the rope that passes around the pulleys. Therefore, tension forces T_1 and T_2 are applied to the pulley on the foot. It may seem surprising that the rope applies a force to either side of the foot pulley. A similar effect occurs when you place a finger inside a rubber band and push downward. You can feel each side of the rubber band pulling upward on the finger. The foot pulley is kept in equilibrium, because the foot also applies a force F to it. This force arises in reaction (Newton's third law) to the pulling effect of the forces T_1 and T_2. Ignoring the weight of the foot, find the magnitude of F.

THE PHYSICS OF...

traction for a foot injury.

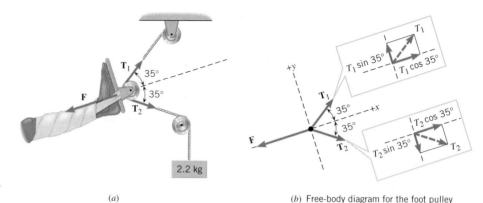

Figure 4.27 (*a*) A traction device for the foot. (*b*) The free-body diagram for the pulley on the foot.

(*a*)

(*b*) Free-body diagram for the foot pulley

Reasoning The forces **T₁**, **T₂**, and **F** keep the pulley on the foot at rest. The pulley, therefore, has no acceleration and is in equilibrium. As a result, the sum of the *x* components and the sum of the *y* components of the three forces must each be zero. Figure 4.27*b* shows the free-body diagram of the pulley on the foot. The *x* axis is chosen to be along the direction of force **F**, and the components of the forces are indicated in the drawing. (See Section 1.7 for a review of vector components.)

Solution Since the sum of the *y* components of the forces is zero, it follows that

$$\Sigma F_y = +T_1 \sin 35° - T_2 \sin 35° = 0 \tag{4.9b}$$

or $T_1 = T_2$. In other words, the magnitudes of the tension forces are equal. In addition, the sum of the *x* components of the forces is zero, so we have that

$$\Sigma F_x = +T_1 \cos 35° + T_2 \cos 35° - F = 0 \tag{4.9a}$$

Solving for *F* and letting $T_1 = T_2 = T$, we find that $F = 2T \cos 35°$. However, the tension *T* in the rope is determined by the weight of the 2.2-kg object: $T = mg$, where *m* is its mass and *g* is the acceleration due to gravity. Therefore, the magnitude of **F** is

$$F = 2T \cos 35° = 2mg \cos 35° = 2(2.2 \text{ kg})(9.80 \text{ m/s}^2) \cos 35° = \boxed{35 \text{ N}}$$

• **PROBLEM SOLVING INSIGHT**
Choose the orientation of the *x*, *y* axes for convenience. Here, the axes have been rotated so the force F points along the *x* axis. Since F does not have a component along the *y* axis, the analysis is simplified.

Example 12 presents another situation in which three forces are responsible for the equilibrium of an object. However, in this example all the forces have different magnitudes.

EXAMPLE 12   Replacing an Engine

An automobile engine has a weight **W**, whose magnitude is $W = 3150$ N. This engine is being positioned above an engine compartment, as Figure 4.28*a* illustrates. To position the engine, a worker is using a rope. Find the tension **T₁** in the supporting cable and the tension **T₂** in the positioning rope.

Reasoning Under the influence of the forces **W**, **T₁**, and **T₂** the ring is at rest and, therefore, in equilibrium. Consequently, the sum of the *x* components and the sum of the *y* components of these forces must each be zero; $\Sigma F_x = 0$ and $\Sigma F_y = 0$. By using these relations, we can find T_1 and T_2. Figure 4.28*b* shows the free-body diagram of the ring and the force components for a suitable *x*, *y* axes system.

Solution The free-body diagram shows the components for each of the three forces, and the components are listed in the following table:

(a)

(b) Free-body diagram for the ring

Figure 4.28 (a) The ring is in equilibrium because of the three forces $\mathbf{T}_1$ (the tension force in the supporting cable), $\mathbf{T}_2$ (the tension force in the positioning rope), and $\mathbf{W}$ (the weight of the engine). (b) The free-body diagram for the ring.

Force	x Component	y Component
$\mathbf{T}_1$	$-T_1 \sin 10.0°$	$+T_1 \cos 10.0°$
$\mathbf{T}_2$	$+T_2 \sin 80.0°$	$-T_2 \cos 80.0°$
$\mathbf{W}$	0	$-W$

The plus signs in the table denote components that point along the positive axes, while the minus signs denote components that point along the negative axes. Setting the sum of the x components and the sum of the y components equal to zero leads to the following two equations:

$$\Sigma F_x = -T_1 \sin 10.0° + T_2 \sin 80.0° = 0 \qquad (4.9a)$$

$$\Sigma F_y = +T_1 \cos 10.0° - T_2 \cos 80.0° - W = 0 \qquad (4.9b)$$

Solving the first of these equations for T_1 shows that

$$T_1 = \left(\frac{\sin 80.0°}{\sin 10.0°} \right) T_2$$

Substituting this expression for T_1 into the second equation gives

$$\left(\frac{\sin 80.0°}{\sin 10.0°} \right) T_2 \cos 10.0° - T_2 \cos 80.0° - W = 0$$

$$T_2 = \frac{W}{\left(\dfrac{\sin 80.0°}{\sin 10.0°} \right) \cos 10.0° - \cos 80.0°}$$

Setting $W = 3150$ N in this result yields $\boxed{T_2 = 582 \text{ N}}$.

Since $T_1 = \left(\dfrac{\sin 80.0°}{\sin 10.0°} \right) T_2$ and $T_2 = 582$ N, it follows that $\boxed{T_1 = 3.30 \times 10^3 \text{ N}}$.

● **PROBLEM SOLVING INSIGHT**
When an object is in equilibrium, as here in Example 12, the net force is zero, $\Sigma F = 0$. This does not mean that each individual force is zero. It means that the vector sum of all the forces is zero.

An object can be moving and still be in equilibrium, provided there is no acceleration. Example 13 illustrates such a case, and the solution is again obtained using the five-step reasoning strategy summarized at the beginning of the section.

EXAMPLE 13   Equilibrium at Constant Velocity

A jet plane is flying with a constant speed along a straight line, at an angle of 30.0° above the horizontal, as Figure 4.29a indicates. The plane has a weight **W** whose magnitude is $W = 86\,500$ N, and its engines provide a forward thrust **T** of magnitude $T = 103\,000$ N. In addition, the lift force **L** (directed perpendicular to the wings) and the force **R** of air resistance (directed opposite to the motion) act on the plane. Find **L** and **R**.

Reasoning Figure 4.29b shows the free-body diagram of the plane, including the forces **W**, **L**, **T**, and **R**. Since the plane is not accelerating, it is in equilibrium, and the sum of the x components and the sum of the y components of these forces must be zero. The lift force **L** and the force **R** of air resistance can be obtained from these equilibrium conditions. To calculate the components, we have chosen axes in the free-body diagram that are rotated by 30.0° from their usual horizontal-vertical positions. This has been done purely for convenience, since the weight **W** is then the only force that does not lie along either axis.

Solution When determining the components of the weight, it is necessary to realize that the angle β in Figure 4.29a is 30.0°. Part c of the drawing focuses attention on the geometry that is responsible for this fact. There it can be seen that $\alpha + \beta = 90°$ and $\alpha + 30° = 90°$, with the result that $\beta = 30°$. The table below lists the components of the forces that act on the jet.

Force	x Component	y Component
W	$-W \sin 30.0°$	$-W \cos 30.0°$
L	0	$+L$
T	$+T$	0
R	$-R$	0

Setting the sum of the x component of the forces to zero gives
$$\Sigma F_x = -W \sin 30.0° + T - R = 0 \qquad (4.9a)$$
$$R = T - W \sin 30.0° = 103\,000 \text{ N} - (86\,500 \text{ N}) \sin 30.0° = \boxed{59\,800 \text{ N}}$$

Setting the sum of the y component of the forces to zero gives
$$\Sigma F_y = -W \cos 30.0° + L = 0 \qquad (4.9b)$$
$$L = W \cos 30.0° = (86\,500 \text{ N}) \cos 30.0° = \boxed{74\,900 \text{ N}}$$

Figure 4.29 (a) A plane moves with a constant velocity at an angle of 30.0° above the horizontal due to the action of four forces, the weight **W**, the lift **L**, the engine thrust **T**, and the air resistance **R**. (b) The free-body diagram for the plane. (c) This geometry occurs often in physics.

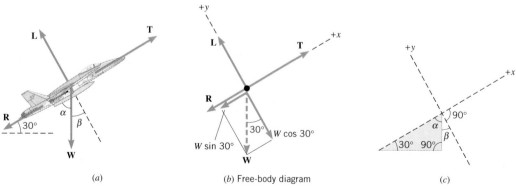

(a)   (b) Free-body diagram   (c)

4.12 NONEQUILIBRIUM APPLICATIONS OF NEWTON'S LAWS OF MOTION

When an object is accelerating, it is not in equilibrium, as indicated in Figure 4.26. The forces acting on it are not balanced, so the net force is not zero in Newton's second law. However, with one exception, the reasoning strategy followed in solving nonequilibrium problems is identical to that used in equilibrium situations. The exception occurs in step 4 of the five steps outlined at the beginning of the last section. Since the object is now accelerating, the representation of Newton's second law in Equations 4.2a and 4.2b applies instead of Equations 9.9a and 9.9b:

$$\Sigma F_x = ma_x \quad (4.2a) \quad \text{and} \quad \Sigma F_y = ma_y \quad (4.2b)$$

Example 14 uses these equations in a situation where the forces are applied in directions similar to those in Example 11, except that now an acceleration is present.

EXAMPLE 14   Towing a Supertanker

A supertanker of mass $m = 1.50 \times 10^8$ kg is being towed by two tugboats, as in Figure 4.30a. The tensions in the towing cables apply the forces $\mathbf{T}_1$ and $\mathbf{T}_2$ at equal angles of 30.0° with respect to the tanker's axis. In addition, the tanker's engines produce a forward drive force $\mathbf{D}$, whose magnitude is $D = 75.0 \times 10^3$ N. Moreover, the water applies an opposing force $\mathbf{R}$, whose magnitude is $R = 40.0 \times 10^3$ N. The tanker moves forward with an acceleration that points along the tanker's axis and has a magnitude of 2.00×10^{-3} m/s². Find the magnitudes of the tensions $\mathbf{T}_1$ and $\mathbf{T}_2$.

Reasoning The unknown forces $\mathbf{T}_1$ and $\mathbf{T}_2$ contribute to the net force that accelerates the tanker. To determine $\mathbf{T}_1$ and $\mathbf{T}_2$, therefore, we analyze the net force, which we will do using components. The various force components can be found by referring to the free-body diagram for the tanker in Figure 4.30b, where the ship's axis is cho-

Figure 4.30 (a) Four forces act on a supertanker: $\mathbf{T}_1$ and $\mathbf{T}_2$ are the tension forces due to the towing cables, $\mathbf{D}$ is the forward drive force produced by the tanker's engines, and $\mathbf{R}$ is the force with which the water opposes the tanker's motion. (b) The free-body diagram for the tanker.

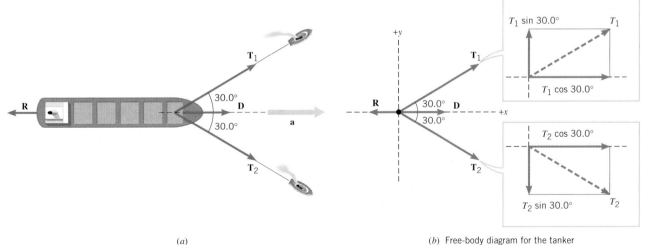

(a)   (b) Free-body diagram for the tanker

sen as the x axis. We will then use Newton's second law in its component form, $\Sigma F_x = ma_x$ and $\Sigma F_y = ma_y$, to obtain the magnitudes of $\mathbf{T_1}$ and $\mathbf{T_2}$.

Solution The individual force components are summarized as follows:

Force	x Component	y Component
$\mathbf{T_1}$	$+T_1 \cos 30.0°$	$+T_1 \sin 30.0°$
$\mathbf{T_2}$	$+T_2 \cos 30.0°$	$-T_2 \sin 30.0°$
$\mathbf{D}$	$+D$	0
$\mathbf{R}$	$-R$	0

Since the acceleration points along the x axis, there is no y component of the acceleration ($a_y = 0$ m/s²). Consequently, the sum of the y components of the forces must be zero:

$$\Sigma F_y = +T_1 \sin 30.0° - T_2 \sin 30.0° = 0$$

This result shows that the magnitudes of the tensions in the cables are equal, $T_1 = T_2$. Since the ship accelerates along the x direction, the sum of the x components of the forces is not zero. The second law indicates that

$$\Sigma F_x = T_1 \cos 30.0° + T_2 \cos 30.0° + D - R = ma_x$$

Since $T_1 = T_2$, we can replace the two separate tension symbols by a single symbol T, the magnitude of the tension. Solving for T gives

$$T = \frac{ma_x + R - D}{2 \cos 30.0°}$$

$$= \frac{(1.50 \times 10^8 \text{ kg})(2.00 \times 10^{-3} \text{ m/s}^2) + 40.0 \times 10^3 \text{ N} - 75.0 \times 10^3 \text{ N}}{2 \cos 30.0°}$$

$$= \boxed{1.53 \times 10^5 \text{ N}}$$

It often happens that two objects are connected somehow, perhaps by a drawbar like that used when a truck pulls a trailer. If the tension in the connecting device is of no interest, the objects can be treated as a single composite object when applying Newton's second law. However, if it is necessary to find the tension, as in the next example, then the second law must be applied separately to at least one of the objects.

EXAMPLE 15   Hauling a Trailer

A truck is hauling a trailer along a level road, as Figure 4.31a illustrates. The mass of the truck is $m_1 = 8500$ kg and that of the trailer is $m_2 = 27\ 000$ kg. The two move along the x axis with an acceleration of $a_x = 0.78$ m/s². Ignoring the retarding forces of friction and air resistance, determine (a) the magnitude of the tension in the horizontal drawbar between the trailer and the truck and (b) the force $\mathbf{D}$ that propels the truck forward.

Reasoning Since the truck and the trailer accelerate along the horizontal direction and friction is being ignored, only forces that have components in the horizontal direction are of interest. Therefore, Figure 4.31 omits the weight and the normal force, which act vertically. To determine the tension force $\mathbf{T}$ in the drawbar, we draw the free-body diagram for the trailer and apply Newton's second law, $\Sigma F_x = ma_x$. Similarly, we can determine the force $\mathbf{D}$ that propels the truck forward by drawing the free-body diagram for the truck and applying Newton's second law to relate the net force to the acceleration.

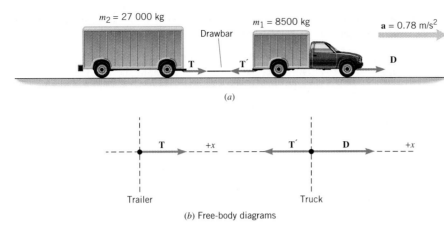

(a)

(b) Free-body diagrams

Figure 4.31 (a) The force **D** acts on the truck and propels it forward. The drawbar exerts the tension force **T′** on the truck and the tension force **T** on the trailer. (b) The free-body diagrams for the truck and the trailer, ignoring the vertical forces.

Solution

(a) The free-body diagram for the trailer is shown in Figure 4.31b. There is only one horizontal force acting on the trailer, the tension force **T** due to the drawbar. Therefore, it is straightforward to obtain the tension from $\Sigma F_x = m_2 a_x$, since the mass of the trailer and the acceleration are known:

$$\Sigma F_x = T = m_2 a_x = (27\ 000\ \text{kg})(0.78\ \text{m/s}^2) = \boxed{21\ 000\ \text{N}}$$

(b) Two horizontal forces act on the truck, as the free-body diagram in Figure 4.31b shows. One is the desired force **D**. The other is the force **T′**. According to Newton's third law, **T′** is the force with which the trailer pulls back on the truck, in reaction to the truck pulling forward. If the drawbar has negligible mass, the magnitude of **T′** is equal to the magnitude of **T**, namely, 21 000 N. Since the magnitude of **T′**, the mass of the truck, and the acceleration are known, $\Sigma F_x = m_1 a_x$ can be used to determine the drive force:

$$\Sigma F_x = +D - T' = m_1 a_x$$

$$D = m_1 a_x + T' = (8500\ \text{kg})(0.78\ \text{m/s}^2) + 21\ 000\ \text{N} = \boxed{28\ 000\ \text{N}}$$

In Section 4.11 we examined situations where the net force acting on an object is zero, and in this section we have considered two examples where the net force is not zero. Conceptual Example 16 illustrates a common situation where the net force is zero at certain times but is not zero at other times.

CONCEPTUAL EXAMPLE 16   The Motion of a Water Skier

Figure 4.32 shows a water skier at four different moments:

(a) The skier is floating motionless in the water.

(b) The skier is being pulled out of the water and up onto the skis.

(c) The skier is moving at a constant speed along a straight line.

(d) The skier has let go of the tow rope and is slowing down.

Figure 4.32 A water skier (a) floating in water, (b) being pulled up by the boat, (c) moving at a constant velocity, and (d) slowing down.

(a)   (b)   (c)   (d)

INTERACTIVE LEARNINGWARE #8

START

Visualization

Reasoning

Knowns and Unknowns

Modeling the Problem

Numerical Solution

FINISH

A passenger is pulling on the strap of a 15.0-kg suitcase with a force of 70.0 N. The strap makes an angle of 35.0° above the horizontal. A 37.8-N friction force opposes the horizontal motion of the suitcase. Determine the acceleration of the suitcase.

The pulling and friction forces combine to give a net force in the horizontal direction, so the suitcase experiences an acceleration. This problem shows further how Newton's second law is applied in a nonequilibrium situation, emphasizes the importance of the free-body diagram, and reviews how a force is resolved into its components.

Go to our Web site at

http://www.wiley.com/college/cutnell

For each moment, explain whether the net force acting on the skier is zero.

Reasoning and Solution According to Newton's second law, if an object has zero acceleration, the net force acting on it is zero. In such a case, the object is in equilibrium. In contrast, if the object has an acceleration, the net force acting on it is not zero. Such an object is not in equilibrium. We will apply this criterion to each of the four phases of the motion to decide if the net force is zero.

(a) The skier is floating motionless in the water, so her velocity and acceleration are both zero. Therefore, the net force acting on her is zero, and she is in equilibrium.

(b) As the skier is being pulled up and out of the water, her velocity is increasing. Thus, she is accelerating, and the net force acting on her is not zero. The skier is not in equilibrium. The direction of the net force is shown in Figure 4.32b.

(c) The skier is now moving at a constant speed along a straight line, so her velocity is constant. Since her velocity is constant, her acceleration is zero. Thus, the net force acting on her is zero, and she is again in equilibrium, even though she is moving.

(d) After the skier lets go of the tow rope, her speed decreases, so she is decelerating. Thus, the net force acting on her is not zero, and she is not in equilibrium. The direction of the net force is shown in Figure 4.32d, and it is opposite to that in (b).

Related Homework Material: *Problem 66*

The force of gravity is often present among the forces that affect the acceleration of an object. Examples 17–19 deal with typical situations.

EXAMPLE 17   Hauling a Crate

A flatbed truck is carrying a crate up a 10.0° hill, as Figure 4.33a illustrates. The coefficient of static friction between the truck bed and the crate is $\mu_s = 0.350$. Find the maximum acceleration that the truck can attain before the crate begins to slip backward relative to the truck.

Figure 4.33 (*a*) A crate on a truck is kept from slipping by the static frictional force f_s^{MAX}. The other forces that act on the crate are its weight **W** and the normal force F_N. (*b*) The free-body diagram of the crate.

(a)

(b) Free-body diagram of the crate

Reasoning The crate will not slip as long as it has the same acceleration as the truck. Therefore, a net force must act on the crate to accelerate it, and the static frictional force $\mathbf{f_s}$ contributes in a major way to this net force. Since the crate tends to slip backward, the static frictional force must be directed forward, up the hill. As the acceleration of the truck increases, $\mathbf{f_s}$ must also increase to produce a corresponding increase in the acceleration of the crate. However, the static frictional force can increase only until its maximum magnitude $f_s^{\text{MAX}} = \mu_s F_N$ is reached, at which point the crate and the truck have the maximum acceleration $\mathbf{a}^{\text{MAX}}$. If the acceleration increases even more, the crate will slip. To find $\mathbf{a}^{\text{MAX}}$, we focus our attention on the crate, and part b of the drawing shows its free-body diagram. The three forces acting on the crate at the instant slipping begins are its weight $\mathbf{W} = m\mathbf{g}$, the normal force $\mathbf{F_N}$ exerted by the truck bed, and the maximum static frictional force $\mathbf{f_s}^{\text{MAX}}$.

Solution Using the x components of the forces (see the free-body diagram for the magnitudes of the components) and Newton's second law ($\Sigma F_x = ma_x$), we have

$$\Sigma F_x = -mg \sin 10.0° + \mu_s F_N = ma^{\text{MAX}}$$

$$a^{\text{MAX}} = \frac{-mg \sin 10.0° + \mu_s F_N}{m}$$

Before this equation can be used to calculate a^{MAX}, however, a value is needed for F_N. This can be obtained by considering the force components along the y axis. Since the crate does not accelerate along this axis, the sum of the y components must be zero according to Newton's second law ($\Sigma F_y = ma_y = 0$):

$$\Sigma F_y = -mg \cos 10.0° + F_N = 0 \quad \text{or} \quad F_N = mg \cos 10.0°$$

Substituting this expression for F_N into the equation for a^{MAX} and algebraically eliminating the mass m reveals that

$$a^{\text{MAX}} = -g \sin 10.0° + \mu_s g \cos 10.0°$$
$$= -(9.80 \text{ m/s}^2) \sin 10.0° + (0.350)(9.80 \text{ m/s}^2) \cos 10.0° = \boxed{1.68 \text{ m/s}^2}$$

- **PROBLEM SOLVING INSIGHT**
The magnitude F_N of the normal force is not necessarily equal to the weight of the object.

EXAMPLE 18 Accelerating Blocks

Block 1 (mass $m_1 = 8.00$ kg) is moving on a frictionless 30.0° incline. This block is connected to block 2 (mass $m_2 = 22.0$ kg) by a cord that passes over a massless and frictionless pulley (see Figure 4.34a). Find the acceleration of each block and the tension in the cord.

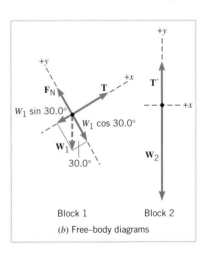

Figure 4.34 (a) Three forces act on block 1: its weight $\mathbf{W_1}$, the normal force $\mathbf{F_N}$, and the force $\mathbf{T}$ due to the tension in the cord. Two forces act on block 2: its weight $\mathbf{W_2}$ and the force $\mathbf{T}'$ due to the tension. The acceleration is labeled a. (b) The free-body diagrams for the two blocks.

(a)

(b) Free–body diagrams

Reasoning Since both blocks accelerate, there must be a net force acting on each one. The key to this problem is to realize that Newton's second law can be used separately for each block to relate the net force and the acceleration. Note also that both blocks have accelerations of the same magnitude a, since they move as a unit.

Solution We assume that block 1 accelerates up the incline and choose this direction to be the $+x$ axis. If block 1 in reality accelerates down the incline, then the value obtained for the acceleration will be a negative number. Three forces act on block 1: (1) $\mathbf{W_1}$ is its weight [$W_1 = m_1g = (8.00 \text{ kg})(9.80 \text{ m/s}^2) = 78.4 \text{ N}$], (2) $\mathbf{T}$ is the force applied because of the tension in the cord, and (3) $\mathbf{F_N}$ is the normal force that the incline exerts. Figure 4.34b shows the free-body diagram for block 1. The weight is the only force that does not point along the x, y axes, and its x and y components are given in the diagram. Applying Newton's second law ($\Sigma F_x = m_1a_x$) to block 1 shows that

$$\Sigma F_x = -W_1 \sin 30.0° + T = m_1a$$

where we have set $a_x = a$. This equation cannot be solved as it stands, since both T and a are unknown quantities. To complete the solution, we next consider block 2.

Two forces act on block 2, as the free-body diagram in Figure 4.34b indicates: (1) $\mathbf{W_2}$ is its weight [$W_2 = m_2g = (22.0 \text{ kg})(9.80 \text{ m/s}^2) = 216 \text{ N}$] and (2) $\mathbf{T'}$ is exerted as a result of block 1 pulling back on the connecting cord. If the masses of the cord and the frictionless pulley are negligible, the magnitudes of $\mathbf{T'}$ and $\mathbf{T}$ are the same: $T' = T$. Applying Newton's second law ($\Sigma F_y = m_2a_y$) to block 2 reveals that

$$\Sigma F_y = T - W_2 = m_2(-a)$$

The acceleration a_y has been set equal to $-a$ since block 2 moves downward along the $-y$ axis in the free-body diagram, consistent with the assumption that block 1 moves up the incline. Now there are two equations in two unknowns, and they may be solved simultaneously (see Appendix C) to give T and a:

$$\boxed{T = 86.3 \text{ N}} \quad \text{and} \quad \boxed{a = 5.89 \text{ m/s}^2}$$

EXAMPLE 19   Hoisting a Scaffold

A window washer on a scaffold is hoisting the scaffold up the side of a building by pulling downward on a rope, as in Figure 4.35a. The magnitude of the pulling force is 540 N, and the combined mass of the worker and the scaffold is 155 kg. Find the upward acceleration of the unit.

Reasoning The worker and the scaffold form a single unit, on which the rope exerts a force in three places. The left end of the rope exerts an upward force $\mathbf{T}$ on the worker's hands. This force arises because he pulls downward with a 540-N force, and the rope exerts an oppositely directed force of equal magnitude on him, in accord with Newton's third law. Thus, the magnitude T of the upward force is $T = 540$ N and is the magnitude of the tension in the rope. If the masses of the rope and each pulley are negligible, and if the pulleys are friction-free, the tension is transmitted undiminished along the rope. Then, a 540-N tension force $\mathbf{T}$ acts upward on the left side of the scaffold pulley (see part a of the drawing). A tension force is also applied to the point P, where the rope attaches to the roof. The roof pulls back on the rope in accord with the third law, and this pull leads to the 540-N tension force $\mathbf{T}$ that acts on the right side of the scaffold pulley. In addition to the three upward forces, the weight of the unit must be taken into account [$W = mg = (155 \text{ kg}) \times (9.80 \text{ m/s}^2) = 1520 \text{ N}$]. Part b of the drawing shows the free-body diagram.

(a)

(b) Free-body diagram of the unit

Figure 4.35 (a) A window washer pulls down on the rope to hoist the scaffold up the side of a building. The force $\mathbf{T}$ results from the effort of the window washer and acts on him and the scaffold in three places, as discussed in Example 19. (b) The free-body diagram of the unit comprising the man and the scaffold.

Solution Newton's second law ($\Sigma F_y = ma_y$) can be applied to calculate the acceleration a_y:

$$\Sigma F_y = +T + T + T - W = ma_y$$

$$a_y = \frac{3T - W}{m} = \frac{3(540 \text{ N}) - 1520 \text{ N}}{155 \text{ kg}} = \boxed{0.65 \text{ m/s}^2}$$

4.13 CONCEPTS CALCULATIONS

Newton's three laws of motion provide the basis for understanding the effect of forces on the motion of an object, as we have seen. The second law is especially important, because it provides the quantitative relationship between force and acceleration. The examples in this section serve as a review of the essential features of this relationship.

CONCEPTS & CALCULATIONS EXAMPLE 20
Velocity, Acceleration, and Newton's Second Law of Motion

Figure 4.36 shows two forces, $\mathbf{F_1} = +3000$ N and $\mathbf{F_2} = +5000$ N acting on a space-craft, where the plus signs indicate that the forces are directed along the $+x$ axis. A third force $\mathbf{F_3}$ also acts on the spacecraft but is not shown in the drawing. The craft is moving with a constant velocity of $+850$ m/s. Find the magnitude and direction of $\mathbf{F_3}$.

Concept Questions and Answers Suppose the spacecraft were stationary. What would be the direction of $\mathbf{F_3}$?

Answer If the spacecraft is stationary, its acceleration is zero. According to Newton's second law, the acceleration of an object is proportional to the net force acting on it. Thus, the net force must also be zero. But the net force is the vector sum of the three forces in this case. Therefore, the force $\mathbf{F_3}$ must have a direction such that it balances to zero the forces $\mathbf{F_1}$ and $\mathbf{F_2}$. Since $\mathbf{F_1}$ and $\mathbf{F_2}$ point to the right in Figure 4.36, $\mathbf{F_3}$ must then point to the left.

When the spacecraft is moving at a constant velocity of $+850$ m/s, what is the direction of $\mathbf{F_3}$?

Answer Since the velocity is constant, the acceleration is still zero. As a result, everything we said in the stationary case applies again here. The net force is zero, and the force $\mathbf{F_3}$ must point to the left in Figure 4.36.

Solution Since the velocity is constant, the acceleration is zero. The net force must also be zero, so that

$$\Sigma F_x = F_1 + F_2 + F_3 = 0$$

Solving for F_3 yields

$$F_3 = -(F_1 + F_2) = -(3000 \text{ N} + 5000 \text{ N}) = \boxed{-8000 \text{ N}}$$

The minus sign in the answer means that $\mathbf{F_3}$ points opposite to the sum of $\mathbf{F_1}$ and $\mathbf{F_2}$, or along the $-x$ axis, which is to the left in Figure 4.36. The force $\mathbf{F_3}$ has a magnitude of 8000 N, which is the magnitude of the sum of the forces $\mathbf{F_1}$ and $\mathbf{F_2}$. The answer is independent of the velocity of the spacecraft, as long as that velocity remains constant.

Figure 4.36 Two horizontal forces, $\mathbf{F_1}$ and $\mathbf{F_2}$, act on the spacecraft. A third force $\mathbf{F_3}$ also acts but is not shown.

a_x

T

(a)

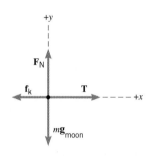

$+y$

F_N

f_k T $+x$

$m\mathbf{g}_{moon}$

(b) Free-body diagram for the block

Figure 4.37 (a) A block is sliding on a horizontal surface on the moon. The tension in the rope is **T**. (b) The free-body diagram for the block, including a kinetic frictional force $\mathbf{f}_k$.

CONCEPTS & CALCULATIONS EXAMPLE 21   The Importance of Mass

On earth a block has a weight of 88 N. This block is sliding on a horizontal surface on the moon, where the acceleration due to gravity is 1.60 m/s². As Figure 4.37a shows, the block is being pulled by a horizontal rope in which the tension is $T = 24$ N. The coefficient of kinetic friction between the block and the surface is $\mu_k = 0.20$. Determine the acceleration of the block.

Concept Questions and Answers Which of Newton's laws of motion provides a way to determine the acceleration of the block?

Answer Newton's second law allows us to calculate the acceleration as $a_x = \Sigma F_x/m$, where ΣF_x is the net force acting in the horizontal direction and m is the mass of the block.

This problem deals with a situation on the moon, but the block's mass on the moon is not given. Instead, the block's earth weight is given. Why can the earth weight be used to obtain a value for the block's mass that applies on the moon?

Answer Since the block's earth weight W_{earth} is related to the block's mass according to $W_{earth} = mg_{earth}$, we can use $W_{earth} = 88$ N and $g_{earth} = 9.80$ m/s² to obtain m. But mass is an intrinsic property of the block and does not depend on whether it is on the earth or on the moon. Therefore, the value obtained for m applies on the moon as well as on the earth.

Does the net force ΣF_x equal the tension T?

Answer No. The net force ΣF_x is the vector sum of all the external forces acting in the horizontal direction. It includes the kinetic frictional force f_k as well as the tension T.

Solution Figure 4.37b shows the free-body diagram for the block. The net force along the x axis is $\Sigma F_x = +T - f_k$, where T is the magnitude of the tension in the rope and f_k is the magnitude of the kinetic frictional force. According to Equation 4.8, f_k is related to the magnitude F_N of the normal force by $f_k = \mu_k F_N$, where μ_k is the coefficient of kinetic friction. The acceleration a_x of the block is given by Newton's second law as

$$a_x = \frac{\Sigma F_x}{m} = \frac{+T - \mu_k F_N}{m}$$

We can obtain an expression for F_N by noting that the block does not move in the y direction, so $a_y = 0$ m/s². Therefore, the net force ΣF_y along the y direction must also be zero. An examination of the free-body diagram in Figure 4.37b reveals that $\Sigma F_y = +F_N - mg_{moon} = 0$, so that $F_N = mg_{moon}$. The acceleration in the x direction becomes

$$a_x = \frac{+T - \mu_k mg_{moon}}{m}$$

Using the earth weight of the block to determine its mass, we find

$$W_{earth} = mg_{earth} \quad \text{or} \quad m = \frac{W_{earth}}{g_{earth}} = \frac{88 \text{ N}}{9.80 \text{ m/s}^2} = 9.0 \text{ kg}$$

The acceleration of the block is, then,

$$a_x = \frac{+T - \mu_k mg_{moon}}{m} = \frac{24 \text{ N} - (0.20)(9.0 \text{ kg})(1.60 \text{ m/s}^2)}{9.0 \text{ kg}} = \boxed{+2.3 \text{ m/s}^2}$$

At the end of the problem set for this chapter, you will find homework problems that contain both conceptual and quantitative parts. These problems are grouped under the heading *Concepts & Calculations, Group Learning Problems*. They are designed for use by students working alone or in small learning groups. The conceptual part of each problem provides a convenient focus for group discussions.

SUMMARY

Newton's first law of motion or **law of inertia** states that an object continues in a state of rest or in a state of motion at a constant speed along a straight line unless compelled to change that state by a net force. **Inertia** is the natural tendency of an object to remain at rest or in motion at a constant speed along a straight line. The **mass** of a body is a quantitative measure of inertia and is measured in an SI unit called the **kilogram** (kg). An **inertial reference frame** is one in which Newton's law of inertia is valid.

Newton's second law of motion states that when a net force $\Sigma\mathbf{F}$ acts on an object of mass m, the acceleration $\mathbf{a}$ of the object is given by $\Sigma\mathbf{F} = m\mathbf{a}$. The SI unit of force is the **newton** (N). When determining the net force, a **free-body diagram** is helpful. A free-body diagram is a diagram that represents the object and the forces acting on it.

Newton's third law of motion, often called the **action–reaction law**, states that whenever one object exerts a force on a second object, the second object exerts an oppositely directed force of equal magnitude on the first object.

Newton's law of universal gravitation states that every particle in the universe exerts an attractive force on every other particle. For two particles that are separated by a distance r and have masses m_1 and m_2, the law states that the magnitude of this attractive force is $F = Gm_1m_2/r^2$, while its direction lies along the line between the particles. The constant G has a value of $G = 6.673 \times 10^{-11}\,\text{N}\cdot\text{m}^2/\text{kg}^2$ and is called the universal gravitational constant. The gravitational force is one of nature's three fundamental forces. The **weight** of an object on earth is the gravitational force that the earth exerts on the object.

The **apparent weight** is the force that an object exerts on the platform of a scale and may be larger or smaller than the true weight, if the object and the scale are accelerating.

A surface exerts a force on an object with which it is in contact. The component of the force perpendicular to the surface is called the **normal force**. The component parallel to the surface is called friction. The **force of static friction** between two surfaces opposes any impending relative motion of the surfaces. The magnitude of the static friction force depends on the magnitude of the applied force and can assume any value up to a maximum of $f_s^{\text{MAX}} = \mu_s F_N$, where μ_s is the **coefficient of static friction** and F_N is the magnitude of the normal force. The **force of kinetic friction** between two surfaces sliding against one another opposes the relative motion of the surfaces. This force has a magnitude given by $f_k = \mu_k F_N$, where μ_k is the **coefficient of kinetic friction**.

The word **tension** is commonly used to mean the tendency of a rope to be pulled apart due to forces that are applied at each end. Because of tension, a rope transmits a force from one end to the other. When a rope is accelerating, the force is transmitted undiminished only if the rope is massless. Neither the tension force, the normal force, nor the friction force is one of nature's fundamental forces.

An object is in **equilibrium** when the object moves at a constant velocity (which may be zero), or, in other words, when it is not accelerating. The sum of the forces that act on an object in equilibrium is zero. Under equilibrium conditions in two dimensions, the separate sums of the force components in the x direction and in the y direction must each be zero: $\Sigma F_x = 0$ and $\Sigma F_y = 0$. If an object is not in equilibrium, then Newton's second law must be used to account for the acceleration: $\Sigma F_x = ma_x$ and $\Sigma F_y = ma_y$. Sections 4.11 and 4.12 discuss a five-step reasoning strategy that facilitates the use of Newton's laws of motion.

CONCEPTUAL QUESTIONS

1. Why do you lunge forward when your car suddenly comes to a halt? Why are you pressed backward against the seat when your car rapidly accelerates? In your explanation, refer to the most appropriate one of Newton's three laws of motion.

2. A bird feeder of large mass is hung from a tree limb, as the drawing shows. A cord attached to the bottom of the feeder has been left dangling free. Curiosity gets the best of a child, who pulls on the dangling cord in an attempt to see what's in the feeder. The dangling cord is cut from the same source as the cord attached to the limb. Is the cord between the feeder and the limb more likely to snap with a slow continuous pull or a sudden downward pull? Give your reasoning.

3. The net external force acting on an object is zero. Is it possible for the object to be traveling with a velocity that is *not* zero? If your answer is yes, state whether any conditions must be placed on the magnitude and direction of the velocity. If your answer is no, provide a reason for your answer.

4. Is a net force being applied to an object when the object is moving downward (a) with a constant acceleration of $9.80\ \text{m/s}^2$ and (b) with a constant velocity of $9.80\ \text{m/s}$? Explain.

5. Newton's second law indicates that when a net force acts on an object, it must accelerate. Does this mean that when two or more forces are applied to an object simultaneously, it must accelerate? Explain.

6. A father and his seven-year-old daughter are facing each other on ice skates. With their hands, they push off against one another. (a) Compare the magnitudes of the pushing forces that they experience. (b) Which one, if either, experiences the larger acceleration? Account for your answers.

7. A gymnast is bouncing on a trampoline. After a high bounce the gymnast comes down and hits the elastic surface of the trampoline. In so doing the gymnast applies a force to the trampoline. (a) Describe the effect this force has on the elastic surface. (b) The surface applies a reaction force to the gymnast. Describe the effect that this reaction force has on the gymnast.

8. According to Newton's third law, when you push on an object, the object pushes back on you with an oppositely directed force of equal magnitude. If the object is a massive crate resting on the floor, it will probably not move. Some people think that the reason the crate does not move is that the two oppositely directed pushing forces cancel. Explain why this logic is faulty and why the crate does not move.

9. Three particles have identical masses. Each particle experiences only the gravitational forces due to the other two particles. How should the particles be arranged so each one experiences a net gravitational force that has the same magnitude? Give your reasoning.

10. When a body is moved from sea level to the top of a mountain, what changes—the body's mass, its weight, or both? Explain.

11. The force of air resistance acts to oppose the motion of an object moving through the air. A ball is thrown upward and eventually returns to the ground. (a) As the ball moves upward, is the net force that acts on the ball greater than, less than, or equal to its weight? Justify your answer. (b) Repeat part (a) for the downward motion of the ball.

12. Object A weighs twice as much as object B at the same spot on the earth. Would the same be true at a given spot on Mars? Account for your answer.

13. Does the acceleration of a freely falling object depend to any extent on the location, i.e., whether the object is on top of Mt. Everest or in Death Valley, California? Explain.

14. A "bottle rocket" is a type of fireworks that has a long thin tail that you insert into an empty bottle, to provide a launch platform. One of these rockets is fired with the bottle pointing vertically upward. An identical rocket is fired with the bottle lying on its side, pointing horizontally. In which case does the rocket leave the bottle with the greater acceleration? Explain, ignoring air resistance and friction.

15. A 10-kg suitcase is placed on a scale that is in an elevator. Is the elevator accelerating up or down when the scale reads (a) 75 N and (b) 120 N? Justify your answers.

16. A stack of books whose true weight is 165 N is placed on a scale in an elevator. The scale reads 165 N. Can you tell from this information whether the elevator is moving with a constant velocity of 2 m/s upward or 2 m/s downward or whether the elevator is at rest? Explain.

17. Suppose you are in an elevator that is moving upward with a constant velocity. A scale inside the elevator shows your weight to be 600 N. (a) Does the scale register a value that is greater than, less than, or equal to 600 N during the time when the elevator slows down as it comes to a stop? (b) What is the reading when the elevator is stopped? (c) How does the value registered on the scale compare to 600 N during the time when the elevator picks up speed again on its way back down? Give your reasoning in each case.

18. A person has a choice of either pushing or pulling a sled at a constant velocity, as the drawing illustrates. Friction is present. If the angle θ is the same in both cases, does it require less force to push or to pull? Account for your answer.

19. Suppose that the coefficients of static and kinetic friction have values such that $\mu_s = 2.0\mu_k$ for a crate in contact with a cement floor. Does this mean that the magnitude of the static frictional force acting on the crate at rest would always be twice the magnitude of the kinetic frictional force acting on the moving crate? Give your reasoning.

20. A box rests on the floor of an elevator. Because of static friction, a force is required to start the box sliding across the floor when the elevator is (a) stationary, (b) accelerating upward, and (c) accelerating downward. Rank the forces required in these three situations in ascending order, i.e., smallest first. Explain.

21. A rope is used in a tug-of-war between two teams of five people each. Both teams are equally strong, so neither team wins. An identical rope is tied to a tree, and the same ten people pull just as hard on the loose end as they did in the contest. In both cases, the people pull steadily with no jerking. Which rope, if either, is more likely to break? Justify your answer.

22. A stone is thrown from the top of a cliff. As the stone falls, is it in equilibrium? Explain, ignoring air resistance.

23. Can an object ever be in equilibrium if the object is acted on by only (a) a single nonzero force, (b) two forces that point in mutually perpendicular directions, and (c) two forces that point in directions that are not perpendicular? Account for your answers.

24. A circus performer hangs stationary from a rope. She then begins to climb upward by pulling herself up, hand-over-hand. When she starts climbing, is the tension in the rope less than, equal to, or greater than it is when she hangs stationary? Explain.

25. During the final stages of descent, a sky diver with an open parachute approaches the ground with a constant velocity. The wind does not blow him from side to side. Is the sky diver in equilibrium and, if so, what forces are responsible for the equilibrium?

26. A weight hangs from a ring at the middle of a rope, as the drawing illustrates. Can the person who is pulling on the right end of the rope ever make the rope perfectly horizontal? Explain your answer in terms of the forces that act on the ring.

27. A freight train is accelerating on a level track. Other things being equal, would the tension in the coupling between the engine and the first car change if some of the cargo in the last car were transferred to any one of the other cars? Account for your answer.

PROBLEMS

Section 4.3 Newton's Second Law of Motion

1. An airplane has a mass of 3.1×10^4 kg and takes off under the influence of a constant net force of 3.7×10^4 N. What is the net force that acts on the plane's 78-kg pilot?

2. A bicycle has a mass of 13.1 kg, and its rider has a mass of 81.7 kg. The rider is pumping hard, so that a horizontal net force of 9.78 N accelerates them. What is the acceleration?

3. ssm A person with a black belt in karate has a fist that has a mass of 0.70 kg. Starting from rest, this fist attains a velocity of 8.0 m/s in 0.15 s. What is the magnitude of the average net force applied to the fist to achieve this level of performance?

4. Scientists are experimenting with a kind of gun that may eventually be used to fire payloads directly into orbit. In one test, this gun accelerates a 5.0-kg projectile from rest to a speed of 4.0×10^3 m/s. The net force accelerating the projectile is 4.9×10^5 N. How much time is required for the projectile to come up to speed?

5. ssm When a 58-g tennis ball is served, it accelerates from rest to a speed of 45 m/s. The impact with the racket gives the ball a constant acceleration over a distance of 44 cm. What is the magnitude of the net force acting on the ball?

6. During a circus performance, a 72-kg human cannonball is shot out of an 18-m-long cannon. If the human cannonball spends 0.95 s in the cannon, determine the average net force exerted on him in the barrel of the cannon.

7. In the amusement park ride known as Magic Mountain Superman, powerful magnets accelerate a car and its riders from rest to 45 m/s (about 100 mi/h) in a time of 7.0 s. The mass of the car and riders is 5.5×10^3 kg. Find the average net force exerted on the car and riders by the magnets.

*__8.__ An arrow, starting from rest, leaves the bow with a speed of 25.0 m/s. If the average force exerted on the arrow by the bow were doubled, all else remaining the same, with what speed would the arrow leave the bow?

*__9. ssm www__ Two forces $\mathbf{F_A}$ and $\mathbf{F_B}$ are applied to an object whose mass is 8.0 kg. The larger force is $\mathbf{F_A}$. When both forces point due east, the object's acceleration has a magnitude of 0.50 m/s². However, when $\mathbf{F_A}$ points due east and $\mathbf{F_B}$ points due west, the acceleration is 0.40 m/s², due east. Find (a) the magnitude of $\mathbf{F_A}$ and (b) the magnitude of $\mathbf{F_B}$.

Section 4.4 The Vector Nature of Newton's Second Law of Motion, Section 4.5 Newton's Third Law of Motion

10. A force vector has a magnitude of 720 N and a direction of 38° north of east. Determine the magnitude and direction of the components of the force that point along the north–south line and along the east–west line.

11. Only two forces act on an object (mass = 4.00 kg), as in the drawing. Find the magnitude and direction (relative to the x axis) of the acceleration of the object.

12. Two forces, $\mathbf{F_1}$ and $\mathbf{F_2}$, act on the 5.00-kg block shown in the drawing. The magnitudes of the forces are $F_1 = 45.0$ N and $F_2 = 25.0$ N. What is the horizontal acceleration (magnitude and direction) of the block?

13. ssm Only two forces act on an object (mass = 3.00 kg), as in the drawing. Find the magnitude and direction (relative to the x axis) of the acceleration of the object.

14. A duck has a mass of 2.5 kg. As the duck paddles, a force of 0.10 N acts on it in a direction due east. In addition, the current of the water exerts a force of 0.20 N in a direction of 52° south of east. When these forces begin to act, the velocity of the duck is 0.11 m/s in a direction due east. Find the magnitude and direction (relative to due east) of the displacement that the duck undergoes in 3.0 s while the forces are acting.

*__15.__ A billiard ball strikes the cushion of a pool table perpendicularly. The mass of the ball is 0.38 kg. The ball approaches the cushion with a velocity of $+2.1$ m/s and rebounds with a velocity of -2.0 m/s. The ball remains in contact with the cushion for a time of 3.3×10^{-3} s. What is the average force (magnitude and direction) exerted on the ball by the cushion?

**16. At a time when mining asteroids has become feasible, astronauts have connected a line between their 3500-kg space tug and a 6200-kg asteroid. Using their ship's engine, they pull on the asteroid with a force of 490 N. Initially the tug and the asteroid are at rest, 450 m apart. How much time does it take for the ship and the asteroid to meet?

**17. ssm www A 325-kg boat is sailing 15.0° north of east at a speed of 2.00 m/s. Thirty seconds later, it is sailing 35.0° north of east at a speed of 4.00 m/s. During this time, three forces act on the boat: a 31.0-N force directed 15.0° north of east (due to an auxiliary engine), a 23.0-N force directed 15.0° south of west (resistance due to the water), and $\mathbf{F_W}$ (due to the wind). Find the magnitude and direction of the force $\mathbf{F_W}$. Express the direction as an angle with respect to due east.

Section 4.7 The Gravitational Force

18. The mass of one small ball is 1.50 g, and the mass of another is 870.0 g. If the center-to-center distance between these two balls is 10.0 cm, find the magnitude of the gravitational force that each exerts on the other.

19. ssm In preparation for this problem, review Conceptual Example 7. A space traveler whose mass is 115 kg leaves earth. What are his weight and mass (a) on earth and (b) in interplanetary space where there are no nearby planetary objects?

20. A rock of mass 45 kg accidentally breaks loose from the edge of a cliff and falls straight down. The magnitude of the air resistance that opposes its downward motion is 250 N. What is the magnitude of the acceleration of the rock?

21. On earth, two parts of a space probe weigh 11 000 N and 3400 N. These parts are separated by a center-to-center distance of 12 m and may be treated as uniform spherical objects. Find the magnitude of the gravitational force that each part exerts on the other out in space, far from any other objects.

22. The drawing shows three particles far away from any other objects and located on a straight line. The masses of these particles are $m_A = 363$ kg, $m_B = 517$ kg, and $m_C = 154$ kg. Find the magnitude and direction of the net gravitational force acting on (a) particle A, (b) particle B, and (c) particle C.

23. ssm Saturn has an equatorial radius of 6.00×10^7 m and a mass of 5.67×10^{26} kg. (a) Compute the acceleration of gravity at the equator of Saturn. (b) What is the ratio of a person's weight on Saturn to that on earth?

24. The drawing (not to scale) shows one alignment of the sun, earth, and moon. The gravitational force $\mathbf{F_{SM}}$ that the sun exerts on the moon is perpendicular to the force $\mathbf{F_{EM}}$ that the earth exerts on the moon. The masses are: mass of sun $= 1.99 \times 10^{30}$ kg, mass of earth $= 5.98 \times 10^{24}$ kg, mass of moon $= 7.35 \times 10^{22}$ kg. The distances shown in the drawing are $r_{SM} = 1.50 \times 10^{11}$ m and $r_{EM} = 3.85 \times 10^8$ m. Determine the magnitude of the net gravitational force on the moon.

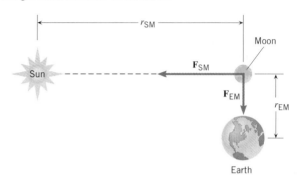

25. The mass of a robot is 5450 kg. This robot weighs 3620 N more on planet A than it does on planet B. Both planets have the same radius of 1.33×10^7 m. What is the difference $M_A - M_B$ in the masses of these planets?

26. A planet has a radius that is 0.50 times that of the earth and a mass that is 0.10 times that of the earth. What is the acceleration due to gravity on the surface of this planet?

27. ssm Synchronous communications satellites are placed in a circular orbit that is 3.59×10^7 m above the surface of the earth. What is the magnitude of the acceleration due to gravity at this distance?

*28. Three uniform spheres are located at the corners of an equilateral triangle. Each side of the triangle has a length of 1.20 m. Two of the spheres have a mass of 2.80 kg each. The third sphere (mass unknown) is released from rest. Considering only the gravitational forces that the spheres exert on each other, what is the magnitude of the initial acceleration of the third sphere?

*29. At a distance H above the surface of a planet, the true weight of a remote probe is one percent less than its true weight on the surface. The radius of the planet is R. Find the ratio H/R.

*30. The sun is more massive than the moon, but the sun is farther from the earth. Which one exerts a greater gravitational force on a person standing on the earth? Give your answer by determining the ratio F_{sun}/F_{moon} of the magnitudes of the gravitational forces. Use the data on the inside of the front cover.

*31. ssm www Several people are riding in a hot-air balloon. The combined mass of the people and balloon is 310 kg. The balloon is motionless in the air, because the downward-acting weight of the people and balloon is balanced by an upward-acting "buoyant" force. If the buoyant force remains constant, how much mass should be dropped overboard so the balloon acquires an upward acceleration of 0.15 m/s²?

*32. A neutron star has a mass of 2.0×10^{30} kg (about the mass of our sun) and a radius of 5.0×10^3 m (about the height of a good-sized mountain). Suppose an object falls from rest near the surface of such a star. How fast would it be moving after it had fallen a distance of 0.010 m? (Assume that the gravitational force is constant over the distance of the fall, and that the star is not rotating.)

**33. Two particles are located on the x axis. Particle 1 has a mass m and is at the origin. Particle 2 has a mass 2m and is at $x = +L$. A third particle is placed between particles 1 and 2. Where on the x axis should the third particle be located so that the magnitude of the gravitational force on *both* particle 1 and particle 2 doubles? Express your answer in terms of L.

Section 4.8 The Normal Force, Section 4.9 Static and Kinetic Frictional Forces

34. A 35-kg crate rests on a horizontal floor, and a 65-kg person is standing on the crate. Determine the magnitude of the normal force that (a) the floor exerts on the crate and (b) the crate exerts on the person.

35. ssm A rocket blasts off from rest and attains a speed of 45 m/s in 15 s. An astronaut has a mass of 57 kg. What is the astronaut's apparent weight during takeoff?

36. A 95.0-kg person stands on a scale in an elevator. What is the apparent weight when the elevator is (a) accelerating upward with an acceleration of 1.80 m/s², (b) moving upward at a constant speed, and (c) accelerating downward with an acceleration of 1.30 m/s²?

37. A block whose weight is 45.0 N rests on a horizontal table. A horizontal force of 36.0 N is applied to the block. The coefficients of static and kinetic friction are 0.650 and 0.420, respectively. Will the block move under the influence of the force, and, if so, what will be the block's acceleration? Explain your reasoning.

38. A cup of coffee is sitting on a table in an airplane that is flying at a constant altitude and a constant velocity. The coefficient of static friction between the cup and the table is 0.30. Suddenly, the plane accelerates, its altitude remaining constant. What is the maximum acceleration that the plane can have without the cup sliding backward on the table?

39. ssm A 60.0-kg crate rests on a level floor at a shipping dock. The coefficients of static and kinetic friction are 0.760 and 0.410, respectively. What horizontal pushing force is required to (a) just start the crate moving and (b) slide the crate across the dock at a constant speed?

40. A 92-kg baseball player slides into second base. The coefficient of kinetic friction between the player and the ground is $\mu_k = 0.61$. (a) What is the magnitude of the frictional force? (b) If the player comes to rest after 1.2 s, what is his initial speed?

41. A 6.00-kg box is sliding across the horizontal floor of an elevator. The coefficient of kinetic friction between the box and the floor is 0.360. Determine the kinetic frictional force that acts on the box when the elevator is (a) stationary, (b) accelerating upward with an acceleration whose magnitude is 1.20 m/s², and (c) accelerating downward with an acceleration whose magnitude is 1.20 m/s².

*42. Many new cars are equipped with antilock brakes that prevent tires from sliding on slippery roads during sudden stops. A Honda Accord is traveling at a speed of 17.9 m/s (about 40 mi/h) on a snow-packed road when the driver applies the brakes. The effective coefficient of static friction with and without antilock brakes is 0.764 and 0.615, respectively. How much *less* is the stopping distance with antilock brakes than without antilock brakes?

*43. ssm A skater with an initial speed of 7.60 m/s is gliding across the ice. Air resistance is negligible. (a) The coefficient of kinetic friction between the ice and the skate blades is 0.100. Find the deceleration caused by kinetic friction. (b) How far will the skater travel before coming to rest?

*44. The drawing shows a large cube (mass = 25 kg) being accelerated across a horizontal frictionless surface by a horizontal force **P**. A small cube (mass = 4.0 kg) is in contact with the front surface of the large cube and will slide downward unless **P** is sufficiently large. The coefficient of static friction between the cubes is 0.71. What is the smallest magnitude that **P** can have in order to keep the small cube from sliding downward?

**45. While moving in, a new homeowner is pushing a box across the floor at a constant velocity. The coefficient of kinetic friction between the box and the floor is 0.41. The pushing force is directed downward at an angle θ below the horizontal. When θ is greater than a certain value, it is not possible to move the box, no matter how large the pushing force is. Find that value of θ.

Section 4.10 The Tension Force, Section 4.11 Equilibrium Applications of Newton's Laws of Motion

46. A supertanker (mass = 1.70×10^8 kg) is moving with a constant velocity. Its engines generate a forward thrust of 7.40×10^5 N. Determine (a) the magnitude of the resistive force exerted on the tanker by the water and (b) the magnitude of the upward buoyant force exerted on the tanker by the water.

47. ssm A 12.0-kg lantern is suspended from the ceiling by two vertical wires. What is the tension in each wire?

48. As preparation for this problem, review Example 13. Suppose that the pilot suddenly jettisons 2800 N of fuel. If the plane is to continue moving with the same velocity under the influence of the same air resistance **R**, by how much does the pilot have to reduce (a) the thrust and (b) the lift?

49. ssm A 1.40-kg bottle of vintage wine is lying horizontally in the rack shown in the drawing. The two surfaces on which the bottle rests are 90.0° apart, and the right surface makes an angle of 45.0° with respect to the ground. Each surface exerts a force on the bottle that is perpendicular to the surface. What is the magnitude of each of these forces?

50. The steel I-beam in the drawing has a weight of 8.00 kN and is being lifted at a constant velocity. What is the tension in each cable attached to its ends?

70.0° 70.0°

I-beam

51. A student presses a book between his hands, as the drawing indicates. The forces that he exerts on the front and back covers of the book are perpendicular to the book and are horizontal. The book weighs 31 N. The coefficient of static friction between his hands and the book is 0.40. To keep the book from falling, what is the magnitude of the minimum pressing force that each hand must exert?

52. The helicopter in the drawing is moving horizontally to the right at a constant velocity. The weight of the helicopter is $W = 53\,800$ N. The lift force **L** generated by the rotating blade makes an angle of 21.0° with respect to the vertical. (a) What is the magnitude of the lift force? (b) Determine the magnitude of the air resistance **R** that opposes the motion.

53. ssm Three forces act on a moving object. One force has a magnitude of 80.0 N and is directed due north. Another has a magnitude of 60.0 N and is directed due west. What must be the magnitude and direction of the third force, such that the object continues to move with a constant velocity?

54. The drawing shows a circus clown who weighs 890 N. The coefficient of static friction between the clown's feet and the ground is 0.53. He pulls vertically downward on a rope that passes around three pulleys and is tied around his feet. What is the minimum pulling force that the clown must exert to yank his feet out from under himself?

*****55.** A bicyclist is coasting down an 8.0° hill at a constant velocity of +8.9 m/s. The mass of the bicycle and rider is 85 kg. A force due to air resistance is directed opposite to the motion and has a magnitude f_{air} that is proportional to the cyclist's speed v, so that $f_{air} = cv$, where c is a constant. Using the fact that the cyclist is coasting at a constant velocity, determine the numerical value (including units) for the constant c.

*****56.** A skier is pulled up a slope at a constant velocity by a tow bar. The slope is inclined at 25.0° with respect to the horizontal. The force applied to the skier by the tow bar is parallel to the slope. The skier's mass is 55.0 kg, and the coefficient of kinetic friction between the skis and the snow is 0.120. Find the magnitude of the force that the tow bar exerts on the skier.

*****57. ssm** A 44-kg chandelier is suspended 1.5 m below a ceiling by three wires, each of which has the same tension and the same length of 2.0 m (see the drawing). Find the tension in each wire.

1.5 m

2.0 m

*****58.** A 43.8-kg sign is suspended by two wires, as the drawing shows. Find the tension in wire 1 and in wire 2.

43.0°   55.0°

Wire 1   Wire 2

43.8 kg

*****59.** During a storm, a limb falls from a tree. It comes to rest across a barbed wire fence, one-fifth of the way between two fence posts that are 4.00 m apart. The limb exerts a downward force of 151 N on the wire, depressing it 0.200 m below the horizontal. Find the tension in the section of wire that is (a) shorter and (b) longer.

******60.** The weight of the block in the drawing is 88.9 N. The coefficient of static friction between the block and the vertical wall is 0.560. (a) What minimum force **F** is required to prevent the block from sliding down the wall? (Hint: The static frictional force exerted on the block is directed upward, parallel to the wall.) (b) What minimum force is required to start the block moving up the wall? (Hint: The static frictional force is now directed down the wall.)

Block

Vertical wall

F   40°

******61. ssm** A 225-kg crate rests on a surface that is inclined above the horizontal at an angle of 20.0°. A horizontal force (magnitude = 535 N and parallel to the ground, not the incline) is required to start the crate moving down the incline. What is the coefficient of static friction between the crate and the incline?

******62.** A damp washcloth is hung over the edge of a table to dry. Thus, part (mass $= m_{on}$) of the washcloth rests on the table and part (mass = m_{off}) does not. The coefficient of static friction between the table and the washcloth is 0.40. Determine the maximum fraction $[m_{\text{off}}/(m_{\text{on}} + m_{\text{off}})]$ that can hang over the edge without causing the whole washcloth to slide off the table.

Section 4.12 Nonequilibrium Applications of Newton's Laws of Motion

63. A 350-kg sailboat has an acceleration of 0.62 m/s² at an angle of 64° north of east. Find the magnitude and direction of the net force that acts on the sailboat.

64. The speed of a bobsled is increasing, because it has an acceleration of 2.4 m/s². At a given instant in time, the forces resisting the motion, including kinetic friction and air resistance, total 450 N. The mass of the bobsled and its riders is 270 kg. (a) What is the magnitude of the force propelling the bobsled forward? (b) What is the magnitude of the net force that acts on the bobsled?

65. ssm A 1380-kg car is moving due east with an initial speed of 27.0 m/s. After 8.00 s the car has slowed down to 17.0 m/s. Find the magnitude and direction of the net force that produces the deceleration.

66. Review Conceptual Example 16 as background for this problem. The water skier there has a mass of 73 kg. Find the magnitude of the net force acting on the skier when (a) she is accelerated from rest to a speed of 11 m/s in 8.0 s and (b) she lets go of the tow rope and glides to a halt in 21 s.

67. In the drawing, the weight of the block on the table is 422 N and that of the hanging block is 185 N. Ignoring all frictional effects and assuming the pulley to be massless, find (a) the acceleration of the two blocks and (b) the tension in the cord.

68. A rocket of mass 4.5×10^5 kg is accelerating straight up. Its engines produce an upward thrust of 7.5×10^6 N, and the air resistance produces a force that is opposite the motion and of magnitude 2.4×10^6 N. Find the acceleration of the rocket.

69. ssm www A student is skateboarding down a ramp that is 6.0 m long and inclined at 18° with respect to the horizontal. The initial speed of the skateboarder at the top of the ramp is 2.6 m/s. Neglect friction and find the speed at the bottom of the ramp.

70. A cable is lifting a construction worker and a crate, as the drawing shows. The weights of the worker and crate are 965 and 1510 N, respectively. The acceleration of the cable is 0.620 m/s², upward. What is the tension in the cable (a) below the worker and (b) above the worker?

$a = 0.620$ m/s²

71. ssm A 15-g bullet is fired from a rifle. It takes 2.50×10^{-3} s for the bullet to travel the length of the barrel, and it exits the barrel with a speed of 715 m/s. Assum-

ing that the acceleration of the bullet is constant, find the average net force exerted on the bullet.

72. A passenger is pulling on the strap of a 15.0-kg suitcase with a force of 70.0 N. The strap makes an angle of 35.0° above the horizontal. A 37.8-N friction force opposes the motion (horizontal) of the suitcase. Determine the acceleration of the suitcase.

73. A 1.14×10^4-kg lunar landing craft is about to touch down on the surface of the moon, where the acceleration due to gravity is 1.60 m/s². At an altitude of 165 m the craft's downward velocity is 18.0 m/s. To slow down the craft, a retrorocket is firing to provide an upward thrust. Assuming the descent is vertical, find the magnitude of the thrust needed to reduce the velocity to zero at the instant when the craft touches the lunar surface.

*__74.__ The drawing shows Robin Hood (mass = 82 kg) about to escape from a dangerous situation. With one hand, he is gripping the rope that holds up a chandelier (mass = 220 kg). When he cuts the rope where it is tied to the floor, the chandelier will fall, and he will be pulled up toward a balcony above. Ignore the friction between the rope and the beams over which it slides, and find (a) the acceleration with which Robin is pulled upward and (b) the tension in the rope while Robin escapes.

*__75. ssm__ A person whose weight is 5.20×10^2 N is being pulled up vertically by a rope from the bottom of a cave that is 35.1 m deep. The maximum tension that the rope can withstand without breaking is 569 N. What is the shortest time, starting from rest, in which the person can be brought out of the cave?

*__76.__ A 205-kg log is pulled up a ramp by means of a rope that is parallel to the surface of the ramp. The ramp is inclined at 30.0° with respect to the horizontal. The coefficient of kinetic friction between the log and the ramp is 0.900, and the log has an acceleration of 0.800 m/s². Find the tension in the rope.

*__77.__ A crate is resting on the bed of a moving truck. The coefficient of static friction between the crate and the truck bed is 0.40. The driver hits the brakes. Assuming the truck is traveling on level ground, determine the maximum deceleration that the truck can have without the crate slipping forward relative to the truck.

*__78.__ A train consists of 50 cars, each of which has a mass of 6.8×10^3 kg. The train has an acceleration of $+8.0 \times 10^{-2}$ m/s². Ignore friction and determine the tension in the coupling (a) between the 30th and 31st cars and (b) between the 49th and 50th cars.

*__79. ssm__ A box is sliding up an incline that makes an angle of 15.0° with respect to the horizontal. The coefficient of kinetic friction between the box and the surface of the incline is 0.180. The initial speed of the box at the bottom of the incline is

1.50 m/s. How far does the box travel along the incline before coming to rest?

*80. A girl is sledding down a slope that is inclined at 30.0° with respect to the horizontal. A moderate wind is aiding the motion by providing a steady force of 105 N that is parallel to the motion of the sled. The combined mass of the girl and sled is 65.0 kg, and the coefficient of kinetic friction between the runners of the sled and the snow is 0.150. How much time is required for the sled to travel down a 175-m slope, starting from rest?

*81. A sports car is accelerating up a hill that rises 18° above the horizontal. The coefficient of static friction between the wheels and the road is $\mu_s = 0.88$. It is the static frictional force that propels the car forward. (a) What is the magnitude of the maximum acceleration that the car can have? (b) What is the magnitude of the maximum acceleration if the car is being driven down the hill?

*82. A truck is traveling at a speed of 25.0 m/s along a level road. A crate is resting on the bed of the truck, and the coefficient of static friction between the crate and the truck bed is 0.650. Determine the shortest distance in which the truck can come to a halt without causing the crate to slip forward relative to the truck.

**83. ssm A penguin slides at a constant velocity of 1.4 m/s down an icy incline. The incline slopes above the horizontal at an angle of 6.9°. At the bottom of the incline, the penguin slides onto a horizontal patch of ice. The coefficient of kinetic friction between the penguin and the ice is the same for the incline as for the horizontal patch. How much time is required for the penguin to slide to a halt after entering the horizontal patch of ice?

**84. As part *a* of the drawing shows, two blocks are connected by a rope that passes over a set of pulleys. One block has a weight of 412 N, and the other has a weight of 908 N. The rope and the pulleys are massless and there is no friction. (a) What is the acceleration of the lighter block? (b) Suppose that the heavier block is removed, and a downward force of 908 N is provided by someone pulling on the rope, as part *b* of the drawing shows. Find the

(a)   (b)

acceleration of the remaining block. (c) Explain why the answers in (a) and (b) are different.

**85. The drawing shows three objects. They are connected by strings that pass over massless and friction-free pulleys. The objects move, and the coefficient of kinetic friction between the middle object and the surface of the table is 0.100. (a) What is the acceleration of the three objects? (b) Find the tension in each of the two strings.

**86. A 5.00-kg block is placed on top of a 12.0-kg block that rests on a frictionless table. The coefficient of static friction between the two blocks is 0.600. What is the maximum horizontal force that can be applied before the 5.00-kg block begins to slip relative to the 12.0-kg block, if the force is applied to (a) the more massive block and (b) the less massive block?

ADDITIONAL PROBLEMS

87. ssm Mars has a mass of 6.46×10^{23} kg and a radius of 3.39×10^6 m. (a) What is the acceleration due to gravity on Mars? (b) How much would a 65-kg person weigh on this planet?

88. A 92-kg water skier is being pulled at a constant velocity. The horizontal pulling force is 350 N. Find (a) the magnitude of the total resistive force exerted on the skier by the water and air and (b) the magnitude of the upward force exerted on the skier by the water.

89. A stuntman is being pulled along a rough road at a constant velocity, by a cable attached to a moving truck. The cable is parallel to the ground. The mass of the stuntman is 109 kg, and the coefficient of kinetic friction between the road and him is 0.870. Find the tension in the cable.

90. (a) Calculate the magnitude of the gravitational force exerted on a 425-kg satellite that is a distance of two earth radii from the center of the earth. (b) What is the magnitude of the gravitational force exerted on the earth by the satellite? (c) Determine the magnitude of the satellite's acceleration. (d) What is the magnitude of the earth's acceleration?

91. ssm A 20.0-kg sled is being pulled across a horizontal surface at a constant velocity. The pulling force has a magnitude of 80.0 N and is directed at an angle of 30.0° above the horizontal. Determine the coefficient of kinetic friction.

92. A falling skydiver has a mass of 110 kg. (a) What is the magnitude of the skydiver's acceleration when the upward force of air resistance has a magnitude that is equal to one-third of his

weight? (b) After the parachute opens, the skydiver descends at a constant velocity. What is the force of air resistance (magnitude and direction) that acts on the skydiver?

93. A rescue helicopter is lifting a man (weight = 822 N) from a capsized boat by means of a cable and harness. (a) What is the tension in the cable when the man is given an initial upward acceleration of 1.10 m/s^2? (b) What is the tension during the remainder of the rescue when he is pulled upward at a constant velocity?

94. A woman stands on a scale in a moving elevator. Her mass is 60.0 kg, and the combined mass of the elevator and scale is an additional 815 kg. Starting from rest, the elevator accelerates upward. During the acceleration, there is a tension of 9410 N in the hoisting cable. What does the scale read during the acceleration?

95. ssm In a supermarket parking lot, an employee is pushing ten empty shopping carts, lined up in a straight line. The acceleration of the carts is 0.050 m/s^2. The ground is level, and each cart has a mass of 26 kg. (a) What is the net force acting on any one of the carts? (b) Assuming friction is negligible, what is the force exerted by the fifth cart on the sixth cart?

96. A 1580-kg car is traveling with a speed of 15.0 m/s. What is the magnitude of the horizontal net force that is required to bring the car to a halt in a distance of 50.0 m?

97. A spacecraft has a mass of 3.50×10^4 kg and is drifting along a straight line through deep space. Its speed is 1820 m/s. An engine is suddenly turned on and provides a thrust of 2240 N in the direction of the spacecraft's velocity. (a) Find the acceleration of the spacecraft. (b) What is the distance (in km) traveled while the spacecraft increases its speed to 2310 m/s?

98. Review Conceptual Example 7 in preparation for this problem. In tests on earth a lunar surface exploration vehicle (mass = 5.90×10^3 kg) achieves a forward acceleration of 0.220 m/s^2. To achieve this same acceleration on the moon, the vehicle's engines must produce a drive force of 1.43×10^3 N. What is the magnitude of the frictional force that acts on the vehicle on the moon?

99. ssm Two objects (45.0 and 21.0 kg) are connected by a massless string that passes over a massless, frictionless pulley. The pulley hangs from the ceiling. Find (a) the acceleration of the objects and (b) the tension in the string.

100. Traveling at a speed of 16.1 m/s, the driver of an automobile suddenly locks the wheels by slamming on the brakes. The coefficient of kinetic friction between the tires and the road is 0.720. How much time does it take for the car to come to a halt? Ignore the effects of air resistance.

101. A toboggan slides down a hill and has a constant velocity. The angle of the hill is 11.3° with respect to the horizontal. What is the coefficient of kinetic friction between the surface of the hill and the toboggan?

102. The drawing shows box 1 resting on a table, with box 2 resting on top of box 1. A massless rope passes over a massless, frictionless pulley. One end of the rope is connected to box 2 and the other end is connected to box 3. The weights of the three

boxes are $W_1 = 55$ N, $W_2 = 35$ N, and $W_3 = 28$ N. Determine the magnitude of the normal force that the table exerts on box 1.

103. ssm To hoist himself into a tree, a 72.0-kg man ties one end of a nylon rope around his waist and throws the other end over a branch of the tree. He then pulls downward on the free end of the rope with a force of 358 N. Neglect any friction between the rope and the branch, and determine the man's upward acceleration.

104. A space probe has two engines. Each generates the same amount of force when fired, and the directions of these forces can be independently adjusted. When the engines are fired simultaneously and each applies its force in the same direction, the probe, starting from rest, takes 28 s to travel a certain distance. How long does it take to travel the same distance, again starting from rest, if the engines are fired simultaneously and the forces that they apply to the probe are perpendicular?

105. At an airport, luggage is unloaded from a plane into the three cars of a luggage carrier, as the drawing shows. The acceleration of the carrier is 0.12 m/s^2, and friction is negligible. The coupling bars have negligible mass. By how much would the tension in *each* of the coupling bars A, B, and C change if 39 kg of luggage has been removed from car 2 and placed in (a) car 1 and (b) car 3? If the tension changes, specify whether it increases or decreases.

106. Jupiter is the largest planet in our solar system, having a mass and radius that are, respectively, 318 and 11.2 times that of earth. Suppose that an object falls from rest near the surface of each planet and that the acceleration due to gravity remains constant during the fall. Each object falls the same distance before striking the ground. Determine the ratio of the time of fall on Jupiter to that on earth.

107. ssm A person is trying to judge if a picture (mass = 1.10 kg) is properly positioned by temporarily pressing it against a wall. The pressing force is perpendicular to the wall. The coefficient of static friction between the picture and the wall is 0.660. What is the minimum amount of pressing force that must be used?

108. A small sphere is hung by a string from the ceiling of a van. When the van is stationary, the sphere hangs vertically. However, when the van accelerates, the sphere swings backward so that the string makes an angle of θ with respect to the vertical. (a) Derive an expression for the magnitude a of the acceleration of the van in terms of the angle θ and the magnitude g of the acceleration due to gravity. (b) Find the acceleration of the van when $\theta = 10.0°$. (c) What is the angle θ when the van moves with a constant velocity?

109. In the drawing, the rope and the pulleys are massless, and there is no friction. Find (a) the tension in the rope and (b) the acceleration of the 10.0-kg block. (*Hint: The larger mass moves twice as far as the smaller mass.*)

10.0 kg

3.00 kg

110. The drawing shows a top view of a 25.0-kg crate that is initially at rest. Note that the view is one looking down on the top of the crate. Two forces, F_1 and F_2, are applied to the crate, and it begins to move. The coefficient of kinetic friction between the crate and the floor is $\mu_k = 0.350$. Determine the magnitude and direction (relative to the x axis) of the acceleration of the crate.

Top view

111. ssm A bicyclist is coasting straight down a hill at a constant speed. The mass of the rider and bicycle is 80.0 kg, and the hill is inclined at 15.0° with respect to the horizontal. Air resistance opposes the motion of the cyclist. Later, the bicyclist climbs the same hill at the same constant speed. How much force (directed parallel to the hill) must be applied to the bicycle in order for the bicyclist to climb the hill?

CONCEPTS & CALCULATIONS

GROUP LEARNING PROBLEMS

Note: Each of these problems consists of Concept Questions followed by a related quantitative Problem. They are designed for use by students working alone or in small learning groups. The Concept Questions involve little or no mathematics and are intended to stimulate group discussions. They focus on the concepts with which the problems deal. Recognizing the concepts is the essential initial step in any problem-solving technique.

112. Concept Questions Two skaters, a man and a woman, are standing on ice. Neglect any friction between the skate blades and the ice. The woman pushes on the man with a certain force that is parallel to the ground. (a) Must the man accelerate under the action of this force? If so, what three factors determine the magnitude and direction of his acceleration? (b) Is there a corresponding force exerted on the woman? If so, where does it originate? Is this force related to the magnitude and direction of the force the woman exerts on the man? If so, how?

Problem The mass of the man is 82 kg and that of the woman is 48 kg. The woman pushes on the man with a force of 45 N due east. Determine the acceleration (magnitude and direction) of (a) the man and (b) the woman.

113. Concept Questions A person is attempting to push a refrigerator across a room. He exerts a horizontal force on the refrigerator, but it does not move. (a) What other horizontal force must be acting on the refrigerator? (b) How are the magnitude and direction of this force related to the force that the person exerts? (c) Suppose that the person applies the force such that it is the largest

possible force before the refrigerator begins to move. What factors determine the magnitude of this force?

Problem A person pushes on a 57-kg refrigerator with a horizontal force of -267 N; the minus sign indicates that the force is directed along the $-x$ direction. The coefficient of static friction is 0.65. (a) If the refrigerator does not move, what is the magnitude and direction of the static frictional force that the floor exerts on the refrigerator? (b) What is the magnitude of the largest pushing force that can be applied to the refrigerator before it just begins to move?

114. Concept Questions A large rock and a small pebble are held at the same height above the ground. (a) Is the gravitational force exerted on the rock greater than, less than, or equal to that exerted on the pebble? Justify your answer. (b) When released, is the downward acceleration of the rock greater than, less than, or equal to that of the pebble? Why?

Problem A 5.0-kg rock and a 3.0×10^{-4}-kg pebble are held near the surface of the earth. (a) Determine the magnitude of the gravitational force exerted on each by the earth. (b) Calculate the acceleration of each object when released. Verify that your answers are consistent with your answers to the Concept Questions.

115. Concept Question The earth exerts a gravitational force on a falling raindrop, pulling it down. Does the raindrop exert a gravitational force on the earth, pulling it up? If so, is this force greater than, less than, or equal to the force that the earth exerts on the raindrop? Provide a reason for your answer.

Problem A raindrop has a mass of 5.2×10^{-7}-kg, and is falling near the surface of the earth. Calculate the magnitude of the gravitational force exerted (a) on the raindrop by the earth and (b) on the earth by the raindrop.

116. Concept Questions Two horizontal forces, $\mathbf{F}_1$ and $\mathbf{F}_2$, are acting on a box, but only $\mathbf{F}_1$ is shown (see the drawing). $\mathbf{F}_2$ can

point either to the right or to the left. The box moves only along the x axis. There is no friction between the box and the surface. What is the direction of $\mathbf{F}_2$ and how does its magnitude compare to the magnitude of $\mathbf{F}_1$ when the acceleration of the box is (a) positive, (b) negative, and (c) zero?

Problem Suppose that $\mathbf{F}_1 = +9.0$ N and the mass of the box is 3.0 kg. Find the magnitude and direction of $\mathbf{F}_2$ when the accelera-

tion of the box is (a) $+5.0$ m/s^2, (b) -5.0 m/s^2, and (c) 0 m/s^2. Be sure that your answers are consistent with your answers to the Concept Questions.

117. Concept Questions A car is driving up a hill. (a) Is the magnitude of the normal force exerted on the car equal to the magnitude of its weight? Why or why not? (b) If the car drives up a steeper hill, does the normal force increase, decrease, or remain the same? Justify your answer. (c) Does the magnitude of the normal force depend on whether the car is traveling up the hill or down the hill? Give your reasoning.

Problem A car is traveling up a hill that is inclined at an angle of θ above the horizontal. Determine the ratio of the magnitude of the normal force to the weight of the car when (a) $\theta = 15°$ and (b) $\theta = 35°$. Check to see that your answers are consistent with your answers to the Concept Questions.

In traveling on a vertical circular arc, these roller coaster fans experience a net force and an acceleration that point toward the center of the circle.

CHAPTER 5

DYNAMICS OF UNIFORM CIRCULAR MOTION

5.1 UNIFORM CIRCULAR MOTION

There are many examples of motion on a circular path. Of the many possibilities, we want to single out those that satisfy the following definition:

● **DEFINITION OF UNIFORM CIRCULAR MOTION**

Uniform circular motion is the motion of an object traveling at a constant (uniform) speed on a circular path.

Figure 5.1 The motion of an airplane flying at a constant speed on a horizontal circular path is an example of uniform circular motion.

As an example of uniform circular motion, Figure 5.1 shows a model airplane on a guideline. The speed of the plane is the magnitude of the velocity vector **v**, and since the speed is constant, the vectors in the drawing have the same magnitude at all points on the circle.

Sometimes it is more convenient to describe uniform circular motion by specifying the period of the motion, rather than the speed. The **period T** is the time required to travel once around the circle, that is, to make one complete revolution. There is a relationship between period and speed, since speed v is the distance traveled (circumference of the circle $= 2\pi r$) divided by the time T:

$$v = \frac{2\pi r}{T} \qquad (5.1)$$

If the radius is known, as in Example 1, the speed can be calculated from the period or vice versa.

EXAMPLE 1   A Tire-Balancing Machine

The wheel of a car has a radius of 0.29 m and is being rotated at 830 revolutions per minute (rpm) on a tire-balancing machine. Determine the speed (in m/s) at which the outer edge of the wheel is moving.

Reasoning The speed v can be obtained directly from $v = 2\pi r/T$, but first the period T is needed. The period is the time for one revolution, and it must be expressed in seconds, because the problem asks for the speed in meters per second.

Solution Since the tire makes 830 revolutions in one minute, the number of minutes required for a single revolution is

$$\frac{1}{830 \text{ revolutions/min}} = 1.2 \times 10^{-3} \text{ min/revolution}$$

Therefore, the period is $T = 1.2 \times 10^{-3}$ min, which corresponds to 0.072 s. Equation 5.1 can now be used to find the speed:

$$v = \frac{2\pi r}{T} = \frac{2\pi(0.29 \text{ m})}{0.072 \text{ s}} = \boxed{25 \text{ m/s}}$$

The definition of uniform circular motion emphasizes that the speed or the magnitude of the velocity vector is constant. It is equally significant that the direction of the vector is *not constant*. In Figure 5.1, for instance, the velocity vector changes direction as the plane moves around the circle. Any change in the velocity vector, even if it is only a change in direction, means that an acceleration is occurring. This particular acceleration is called "centripetal acceleration," because it points toward the center of the circle, as the next section explains.

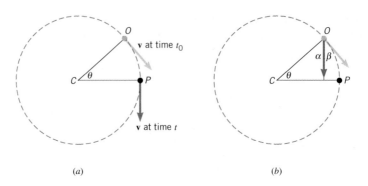

Figure 5.2 (*a*) For an object (●) in uniform circular motion, the velocity **v** has different directions at different places on the circle. (*b*) The velocity vector has been removed from point *P*, shifted parallel to itself, and redrawn with its tail at point *O*.

5.2 CENTRIPETAL ACCELERATION

In this section we determine how the magnitude a_c of the centripetal acceleration depends on the speed v of the object and the radius r of the circular path. We will see that $a_c = v^2/r$.

In Figure 5.2*a* an object (symbolized by a dot ●) is in uniform circular motion. At time t_0 the velocity is tangent to the circle at point *O*, while at a later time *t* the velocity is tangent at point *P*. As the object moves from *O* to *P*, the radius traces out the angle θ, and the velocity vector changes direction. To emphasize the change, part *b* of the picture shows the velocity vector removed from point *P*, shifted parallel to itself, and redrawn with its tail at point *O*. The angle β between the two vectors indicates the change in direction. Since the radii *CO* and *CP* are perpendicular to the tangents at points *O* and *P*, respectively, it follows that $\alpha + \beta = 90°$ and $\alpha + \theta = 90°$. Therefore, angle β and angle θ are equal.

As always, acceleration is the change $\Delta\mathbf{v}$ in velocity divided by the elapsed time Δt, or $\mathbf{a} = \Delta\mathbf{v}/\Delta t$. Figure 5.3*a* shows the two velocity vectors oriented at the angle θ with respect to one another, together with the vector $\Delta\mathbf{v}$ that represents the change in velocity. The change $\Delta\mathbf{v}$ is the increment that must be added to the velocity at time t_0, so that the resultant velocity has the new direction after an elapsed time $\Delta t = t - t_0$. Figure 5.3*b* shows the sector of the circle *COP*. In the limit that Δt is very small, the arc length *OP* is approximately a straight line whose length is the distance $v\,\Delta t$ traveled by the object. In this limit, *COP* is an isosceles triangle, as is the triangle in part *a* of the drawing. Since both triangles have equal apex angles θ, they are similar, so that

$$\frac{\Delta v}{v} = \frac{v\,\Delta t}{r}$$

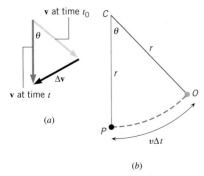

Figure 5.3 (*a*) The directions of the velocity vector at times *t* and t_0 differ by the angle θ. (*b*) When the object moves along the circle from *O* to *P*, the radius *r* traces out the same angle θ. Here, the sector *COP* has been rotated clockwise by 90° relative to its orientation in Figure 5.2.

This equation can be solved for $\Delta v/\Delta t$, to show that the magnitude a_c of the centripetal acceleration is given by $a_c = v^2/r$.

Centripetal acceleration is a vector quantity and, therefore, has a direction as well as a magnitude. The direction is toward the center of the circle, and Conceptual Example 2 helps us to set the stage for explaining this important fact.

CONCEPTUAL EXAMPLE 2   Which Way Will the Object Go?

In Figure 5.4 an object, such as a model airplane on a guideline, is in uniform circular motion. The object is symbolized by a dot (●), and at point *O* it is released suddenly from its circular path. For instance, the guideline for a model plane is cut suddenly. Does the object move along the straight tangent line between points *O* and *A* or along the circular arc between points *O* and *P*?

Reasoning and Solution Newton's first law of motion guides our reasoning. An object continues in a state of rest or in a state of motion at a constant speed along a straight line unless compelled to change that state by a net force. When the object is suddenly released from its circular path, there is no longer a net force being applied to the object. In the case of a model airplane, the guideline cannot apply a force, since it is cut. Gravity certainly acts on the plane, but the wings provide a lift force that balances the weight of the plane. In the absence of a net force, then, the plane or any object would continue to move at a constant speed along a straight line in the direction it had at the time of release. This speed and direction are given in Figure 5.4 by the velocity vector **v**. As a result, *the object would move along the straight line between points O and A, not on the circular arc between points O and P.*

Related Homework Material: *Problems 4 and 7*

As Example 2 discusses, the object in Figure 5.4 would travel on a tangent line if it were released from its circular path suddenly at point *O*. It would move in a straight line to point *A* in the time it would have taken to travel on the circle to point *P*. It is as if the object drops through the distance *AP* in the process of remaining on the circle, and *AP* is directed toward the center of the circle in the limit that the angle *θ* is small. Thus, the object accelerates toward the center of the circle at every moment. The acceleration is called *centripetal acceleration,* because the word "centripetal" means "center-seeking."

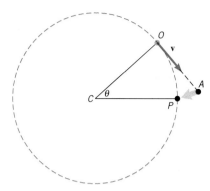

Figure 5.4 If an object (●) moving on a circular path were released from its path at point *O*, it would move along the straight tangent line *OA* in the absence of a net force.

● **CENTRIPETAL ACCELERATION**

Magnitude: The centripetal acceleration of an object moving with a speed v on a circular path of radius r has a magnitude a_c given by

$$a_c = \frac{v^2}{r} \qquad (5.2)$$

Direction: The centripetal acceleration vector always points toward the center of the circle and continually changes direction as the object moves.

The following example illustrates the effect of the radius r on the centripetal acceleration.

EXAMPLE 3 The Effect of Radius on Centripetal Acceleration

The bobsled track at the 1994 Olympics in Lillehammer, Norway, contained turns with radii of 33 m and 24 m, as Figure 5.5 illustrates. Find the centripetal acceleration at each turn for a speed of 34 m/s, a speed that was achieved in the 2-man event. Express the answers as multiples of $g = 9.8$ m/s^2.

Reasoning In each case, the magnitude of the centripetal acceleration can be obtained from $a_c = v^2/r$. Since the radius r is in the denominator on the right side of this expression, we expect the acceleration to be smaller when r is larger.

Solution From $a_c = v^2/r$ it follows that

Radius = 33 m $a_c = \dfrac{(34 \text{ m/s})^2}{33 \text{ m}} = 35 \text{ m/s}^2 = \boxed{3.6\ g}$

Radius = 24 m $a_c = \dfrac{(34 \text{ m/s})^2}{24 \text{ m}} = 48 \text{ m/s}^2 = \boxed{4.9\ g}$

Figure 5.5 This bobsled travels at the same speed around two curves with different radii. For the turn with the larger radius, the sled has a smaller centripetal acceleration.

THE PHYSICS OF...

a bobsled track.

Car A negotiates a curve at a speed of 32 m/s and experiences a centripetal acceleration of 6.4 m/s². Car B negotiates this same curve at a speed of 16 m/s. What centripetal acceleration does it experience?

Centripetal acceleration depends on both the radius of the circular path and the speed of the motion. Example 3 illustrates the effect of the radius. This Interactive LearningWare example explores the effect of the speed, which is particularly strong, since centripetal acceleration depends on the square of the speed.

Go to our Web site at

http://www.wiley.com/college/cutnell

The centripetal acceleration is indeed smaller when the radius is larger. In fact, with r in the denominator on the right of $a_c = v^2/r$, the acceleration approaches zero when the radius becomes very large. Uniform circular motion along the arc of an infinitely large circle entails no acceleration, because it is just like motion at a constant speed along a straight line.

In Section 4.11 we learned that an object is in equilibrium when it has zero acceleration. Conceptual Example 4 discusses whether an object undergoing uniform circular motion can ever be at equilibrium.

CONCEPTUAL EXAMPLE 4
Uniform Circular Motion and Equilibrium

A car moves at a constant speed, and there are three parts to the motion. It moves along a straight line toward a circular turn, goes around the turn, and then moves away along a straight line. In each of these parts, is the car in equilibrium?

Reasoning and Solution An object in equilibrium has no acceleration, according to the definition given in Section 4.11. As the car approaches the turn, both the speed and direction of the motion are constant. Thus, the velocity vector does not change, and there is no acceleration. The same is true as the car moves away from the turn. For these parts of the motion, then, the car is in equilibrium. As the car goes around the turn, however, the direction of travel changes, so the car has a centripetal acceleration that is characteristic of uniform circular motion. Because of this acceleration, the car is not in equilibrium during the turn. In general, *an object that is in uniform circular motion can never be in equilibrium.*

Related Homework Material: *Problem 11*

We have seen that going around tight turns (smaller r) and gentle turns (larger r) at the same speed entails different centripetal accelerations. And most drivers know that such turns "feel" different. This feeling is associated with the force that is present in uniform circular motion, and we now turn to this topic.

5.3 CENTRIPETAL FORCE

● *Concepts at a Glance*   Newton's second law indicates that whenever an object accelerates, there must be a net force to create the acceleration. Thus, in uniform circular motion there must be a net force to produce the centripetal acceleration. As the Concept-at-a-Glance chart in Figure 5.6 indicates, the second law gives this net force as the product of the object's mass m and its acceleration v^2/r. This chart is an expanded version of the chart shown previously in Figure 4.10. The net force causing the centripetal acceleration is called the *centripetal force* F_c and points in the same direction as the acceleration, that is, toward the center of the circle. ●

Concepts at a Glance

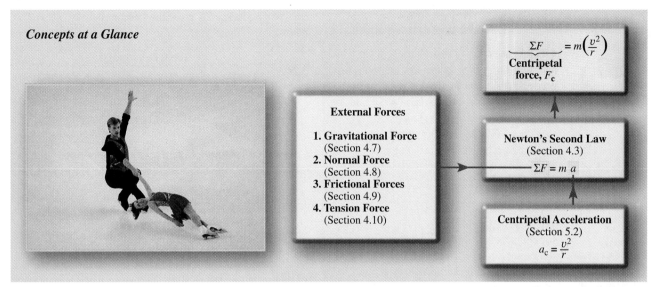

$$\frac{\Sigma F}{} = m\left(\frac{v^2}{r}\right)$$
Centripetal force, F_c

External Forces

1. **Gravitational Force** (Section 4.7)
2. **Normal Force** (Section 4.8)
3. **Frictional Forces** (Section 4.9)
4. **Tension Force** (Section 4.10)

Newton's Second Law (Section 4.3)
$$\Sigma F = m\,a$$

Centripetal Acceleration (Section 5.2)
$$a_c = \frac{v^2}{r}$$

Figure 5.6 *Concepts at a Glance* Uniform circular motion entails a centripetal acceleration $a_c = v^2/r$. The net force required to produce this acceleration is called the centripetal force and is given by Newton's second law of motion as $F_c = mv^2/r$. In the death spiral in pairs figure skating, the woman is spun around in a circle by her partner, and the centripetal force is largely provided by a component of the force applied to her by his hand.

● **CENTRIPETAL FORCE**

Magnitude: The centripetal force is the name given to the net force required to keep an object of mass m, moving at a speed v, on a circular path of radius r and has a magnitude of

$$F_c = \frac{mv^2}{r} \tag{5.3}$$

Direction: The centripetal force always points toward the center of the circle and continually changes direction as the object moves.

The phrase "centripetal force" does not denote a new and separate force created by nature. The phrase merely labels the net force pointing toward the center of the circular path, and this net force is the vector sum of all the force components that point along the radial direction.

In some cases, it is easy to identify the source of the centripetal force, as when a model airplane on a guideline flies in a horizontal circle. The only force pulling the plane inward is the tension in the line, so this force alone (or a component of it) is the centripetal force. Example 5 illustrates the fact that higher speeds require greater tensions.

EXAMPLE 5   The Effect of Speed on Centripetal Force

The model airplane in Figure 5.7 has a mass of 0.90 kg and moves at a constant speed on a circle that is parallel to the ground. Find the tension T in the guideline (length = 17 m) for speeds of 19 and 38 m/s.

Reasoning Because gravity pulls downward on the plane, the guideline slopes downward. As a result, the radius of the circle is slightly less than 17 m, and only the horizontal component of the tension points toward the center of the circle to provide the centripetal force. For simplicity, however, we assume that the guideline is parallel to the ground. With this assumption, the radius of the circle is 17 m, and the entire tension, rather than just its horizontal component, provides the centripetal force.

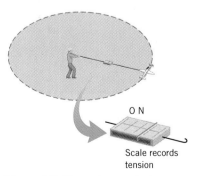

0 N

Scale records tension

Figure 5.7 The scale records the tension in the guideline. See Example 5.

INTERACTIVE LEARNINGWARE #10

START

Visualization

Reasoning

Knowns and
Unknowns

Modeling
the Problem

Numerical
Solution

FINISH

A 0.60-kg and a 1.2-kg model airplane fly in horizontal circles on guidelines that are each nearly parallel to the ground. The speeds of the planes are the same, and the same type of cord is used for each guideline. The smallest circle on which the 0.60-kg plane can fly without breaking its guideline has a radius of 3.5 m. What is the radius of the smallest circle on which the 1.2-kg plane can fly?

The centripetal force, like the centripetal acceleration, depends on both the radius and the speed of the motion. Example 5 examines the effect of the speed, and this Interactive Learning Ware example considers the effect of the radius and illustrates that the force is inversely proportional to the radius.

Go to our Web site at

http://www.wiley.com/college/cutnell

Solution Equation 5.3 gives the tension directly: $F_c = T = mv^2/r$.

$$\textit{Speed} = 19 \ m/s \quad T = \frac{(0.90 \ \text{kg})(19 \ \text{m/s})^2}{17 \ \text{m}} = \boxed{19 \ \text{N}}$$

$$\textit{Speed} = 38 \ m/s \quad T = \frac{(0.90 \ \text{kg})(38 \ \text{m/s})^2}{17 \ \text{m}} = \boxed{76 \ \text{N}}$$

Conceptual Example 6 deals with another case where it is easy to identify the source of the centripetal force.

CONCEPTUAL EXAMPLE 6   A Trapeze Act

In a circus, a man hangs upside down from a trapeze, legs bent over the bar and arms downward, holding his partner (see Figure 5.8). Is it harder for the man to hold his partner when the partner hangs straight downward and is stationary or when the partner is swinging through the straight-down position?

Reasoning and Solution When the man and his partner are stationary, the man's arms must support his partner's weight. When the two are swinging, however, the man's arms must do an additional job. Then, the partner is moving on a circular arc and has a centripetal acceleration. The man's arms must exert an additional pull so that there will be sufficient centripetal force to produce this acceleration. Because of the additional pull, *it is harder for the man to hold his partner while swinging than while stationary.*

Related Homework Material: *Problem 46*

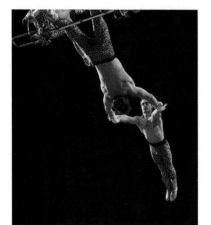

Figure 5.8 The man hanging upside down from the trapeze has a harder job holding his partner when the team is swinging through the straight-down position than when they are hanging straight down and stationary.

When a car moves at a steady speed around an unbanked curve, the centripetal force keeping the car on the curve comes from the static friction between the road and the tires, as Figure 5.9 indicates. It is static, rather than kinetic friction, because the tires are not slipping with respect to the radial direction. If the static frictional force is insufficient, given the speed and the radius of the turn, the car will skid off the road. Example 7 shows how an icy road can limit safe driving.

EXAMPLE 7   Centripetal Force and Safe Driving

Compare the maximum speeds at which a car can safely negotiate an unbanked turn (radius = 50.0 m) in dry weather (coefficient of static friction = 0.900) and icy weather (coefficient of static friction = 0.100).

Reasoning At the maximum speed, the maximum centripetal force acts on the tires, and static friction must provide it. The magnitude of the maximum force of static friction is specified by Equation 4.7 as $f_s^{\text{MAX}} = \mu_s F_N$, where μ_s is the coefficient of static friction and F_N is the magnitude of the normal force. Our strategy, then, is to find the normal force, substitute it in the expression for the maximum force of static friction, and then equate the result to mv^2/r. Experience indicates that the maximum speed should be greater for the dry road than for the icy road.

Solution Since the car does not accelerate in the vertical direction, the weight mg of the car is balanced by the normal force, so $F_N = mg$. From Equation 5.3 it follows, then, that

$$F_c = \mu_s F_N = \mu_s mg = \frac{mv^2}{r}$$

Consequently, $\mu_s g = v^2/r$, and

$$v = \sqrt{\mu_s gr}$$

The mass m of the car has been eliminated algebraically from this result. All cars, heavy or light, have the same maximum speed:

Dry road ($\mu_s = 0.900$) $v = \sqrt{(0.900)(9.80 \text{ m/s}^2)(50.0 \text{ m})} = \boxed{21.0 \text{ m/s}}$

Icy road ($\mu_s = 0.100$) $v = \sqrt{(0.100)(9.80 \text{ m/s}^2)(50.0 \text{ m})} = \boxed{7.00 \text{ m/s}}$

As expected, the dry road allows the greater maximum speed.

The passenger in Figure 5.9 must also experience a centripetal force to remain on the circular path. However, if the upholstery is very slippery, there may not be enough static friction to keep him in place as the car makes a tight turn at high speed. Then, when viewed from inside the car, he appears to be thrown toward the outside of the curve. What really happens is that the passenger slides off on a tangent to the circle, until he encounters a source of centripetal force to keep him in place while the car turns. This occurs when the passenger bumps into the side of the car, which pushes on him with the necessary force.

Sometimes the source of the centripetal force is not obvious. A pilot making a turn, for instance, banks or tilts the plane at an angle to create the centripetal force. As a plane flies, the air pushes upward on the wing surfaces with a net lifting force **L** that is perpendicular to the wing surfaces, as Figure 5.10a shows. Part b of the drawing illustrates that when the plane is banked at an angle θ, a component $L \sin \theta$ of the lifting force is directed toward the center of the turn. It is this component that provides the centripetal force. Greater speeds and/or tighter turns require greater centripetal forces. In such situations, the pilot must bank the plane at a larger angle, so that a larger component of the lift points toward the center of the turn. The technique of banking into a turn also has an application in the construction of high-speed roadways, where the road itself is banked to achieve a similar effect, as the next section discusses.

Figure 5.9 When the car moves without skidding around a curve, static friction between the road and the tires provides the centripetal force to keep the car on the road.

THE PHYSICS OF...

flying an airplane in a banked turn.

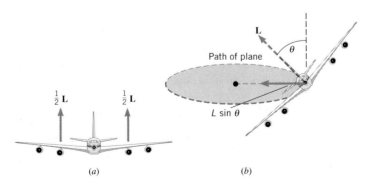

Figure 5.10 (a) The air exerts an upward lifting force $\frac{1}{2}$L on each wing. (b) When a plane executes a circular turn, the plane banks at an angle θ. The lift component $L \sin \theta$ is directed toward the center of the circle and provides the centripetal force.

(a)   (b)

(a)

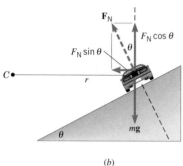

(b)

Figure 5.11 (*a*) A car travels on a circle of radius r on a frictionless banked road. The banking angle is θ, and the center of the circle is at C. (*b*) The forces acting on the car are its weight $m\mathbf{g}$ and the normal force $\mathbf{F_N}$. A component $F_N \sin \theta$ of the normal force provides the centripetal force.

5.4 BANKED CURVES

When a car travels without skidding around an unbanked curve, the static frictional force between the tires and the road provides the centripetal force. The reliance on friction can be eliminated completely for a given speed, however, if the curve is banked at an angle relative to the horizontal, much in the same way that a plane is banked while making a turn.

Figure 5.11*a* shows a car going around a friction-free banked curve. The radius of the curve is r, where r is measured parallel to the horizontal and not to the slanted surface. Part *b* shows the normal force $\mathbf{F_N}$ that the road applies to the car, the normal force being perpendicular to the road. Because the roadbed makes an angle θ with respect to the horizontal, the normal force has a component $F_N \sin \theta$ that points toward the center C of the circle and provides the centripetal force:

$$F_c = F_N \sin \theta = \frac{mv^2}{r}$$

The vertical component of the normal force is $F_N \cos \theta$ and, since the car does not accelerate in the vertical direction, must balance the weight mg of the car. Therefore, $F_N \cos \theta = mg$. Dividing this equation into the previous one shows that

$$\frac{F_N \sin \theta}{F_N \cos \theta} = \frac{mv^2/r}{mg}$$

$$\tan \theta = \frac{v^2}{rg} \qquad (5.4)$$

Equation 5.4 indicates that, for a given speed v, the centripetal force needed for a turn of radius r can be obtained from the normal force by banking the turn at an angle θ, independent of the mass of the vehicle. Greater speeds and smaller radii require more steeply banked curves, that is, larger values of θ. At a speed that is too small for a given θ, a car would slide down a frictionless banked curve; at a speed that is too large, a car would slide off the top. The next example deals with a famous banked curve.

EXAMPLE 8   The Daytona 500

THE PHYSICS OF...

the Daytona International Speedway.

The Daytona 500 is the major event of the NASCAR (National Association for Stock Car Auto Racing) season. It is held at the Daytona International Speedway in Daytona, Florida. The turns in this oval track have a maximum radius (at the top) of $r = 316$ m and are banked steeply, with $\theta = 31°$ (see Figure 5.11). Suppose these maximum-radius turns were frictionless. At what speed would the cars have to travel around them?

Reasoning No friction is available to provide the centripetal force. Therefore, the horizontal component of the normal force that the track exerts on the car must provide it, and the speed of the car is given by Equation 5.4.

Solution From Equation 5.4, it follows that

$$v = \sqrt{rg \tan \theta} = \sqrt{(316 \text{ m})(9.80 \text{ m/s}^2) \tan 31°} = \boxed{43 \text{ m/s (96 mph)}}$$

Drivers actually negotiate the turns at speeds up to 195 mph, however, which requires a greater centripetal force than that implied by Equation 5.4 for frictionless turns. Static friction provides the additional force.

5.5 SATELLITES IN CIRCULAR ORBITS

Today there are many satellites in orbit about the earth. The ones in circular orbits are examples of uniform circular motion. Like a model airplane on a guideline, each satellite is kept on its circular path by a centripetal force. The gravitational pull of the earth provides the centripetal force and acts like an invisible guideline for the satellite.

There is only one speed that a satellite can have if the satellite is to remain in an orbit with a fixed radius. To see how this fundamental characteristic arises, consider the gravitational force acting on the satellite of mass m in Figure 5.12. Since the gravitational force is the only force acting on the satellite in the radial direction, it alone provides the centripetal force. Therefore, using Newton's law of gravitation (Equation 4.3), we have

$$F_c = G \frac{mM_E}{r^2} = \frac{mv^2}{r}$$

where G is the universal gravitational constant, M_E is the mass of the earth, and r is the distance from the center of the earth to the satellite. Solving for the speed v of the satellite gives

$$v = \sqrt{\frac{GM_E}{r}} \qquad (5.5)$$

If the satellite is to remain in an orbit of radius r, the speed must have precisely this value. Note that the radius r of the orbit is in the denominator in Equation 5.5. This means that the closer the satellite is to the earth, the smaller is the value for r and the greater the orbital speed must be.

Many applications of satellite technology affect our lives. One of the most recent is a network of 24 satellites called the Global Positioning System (GPS), which can be used to determine the position of an object to within 15 m or less. Figure 5.13 illustrates how the system works. Each GPS satellite carries a highly accurate atomic clock, whose time is transmitted to the ground continually by means of radio waves. In the drawing, a car carries a computerized GPS receiver that can detect the waves and is synchronized to the satellite clock. The receiver can, therefore, determine the distance between the car and a satellite from a knowledge of the travel

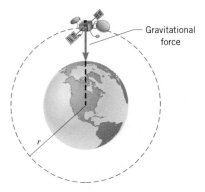

Figure 5.12 For a satellite in circular orbit around the earth, the gravitational force provides the centripetal force.

THE PHYSICS OF...

the Global Positioning System. (See below.)

Figure 5.13 The Navstar Global Positioning System (GPS) of satellites can be used with a GPS receiver to locate an object, such as a car, on the earth. (*a*) One satellite identifies the car as being somewhere on a circle. (*b*) A second places it on another circle, which identifies two possibilities for the exact spot. (*c*) A third provides the means for deciding where the car is.

(a)   (b)   (c)

Figure 5.14 The Hubble space telescope is pictured here in orbit in the cargo bay of the Space Shuttle Endeavour, where a variety of servicing tasks were performed.

time of the waves and the speed at which they move. This speed, as we will see in Chapter 24, is the speed of light and is known with great precision. A measurement using a single satellite locates the car somewhere on a circle, as Figure 5.13*a* shows, while a measurement using a second satellite locates the car on another circle. The intersection of the circles reveals two possible positions for the car, as in Figure 5.13*b*. With the aid of a third satellite, a third circle can be established, which intersects the other two and identifies the car's exact position, as in Figure 5.13*c*. The use of ground-based radio beacons to provide additional reference points leads to a system called Differential GPS, which can locate objects even more accurately than the satellite-based system alone. Navigational systems for automobiles and portable systems that tell hikers and people with visual impairments where they are located are two of the many uses for the GPS technique. Potential GPS applications are so numerous that they are expected to develop into a multibillion dollar industry in the next several years.

The mass *m* of the satellite does not appear in Equation 5.5, having been eliminated algebraically. ***Consequently, for a given orbit, a satellite with a large mass has exactly the same orbital speed as a satellite with a small mass.*** However, more effort is certainly required to lift the larger-mass satellite into orbit. The orbital speed of one famous artificial satellite is determined in the following example.

EXAMPLE 9   Orbital Speed of the Hubble Space Telescope

THE PHYSICS OF...

the Hubble space telescope.

Determine the speed of the Hubble space telescope (see Figure 5.14) orbiting at a height of 598 km above the earth's surface.

Reasoning Before Equation 5.5 can be applied, the orbital radius *r* must be determined *relative to the center of the earth*. Since the radius of the earth is approximately 6.38×10^6 m, and the height of the telescope above the earth's surface is 0.598×10^6 m, the orbital radius is $r = 6.98 \times 10^6$ m.

• PROBLEM SOLVING INSIGHT
The orbital radius *r* that appears in the relation $v = \sqrt{GM_E/r}$ is the distance from the satellite to the center of the earth (not to the surface of the earth).

Solution The orbital speed is

$$v = \sqrt{\frac{GM_E}{r}} = \sqrt{\frac{(6.67 \times 10^{-11}\,\text{N} \cdot \text{m}^2/\text{kg}^2)(5.98 \times 10^{24}\,\text{kg})}{6.98 \times 10^6\,\text{m}}}$$

$$\boxed{v = 7.56 \times 10^3\,\text{m/s}\ (16\ 900\ \text{mi/h})}$$

Equation 5.5 applies to man-made earth satellites or to the moon, which is a natural satellite of the earth. It also applies to circular orbits about any astronomical object, provided M_E is replaced by the mass of the object on which the orbit is centered. Example 10, for instance, shows how scientists have applied this equation to conclude that a supermassive black hole is probably located at the center of the galaxy known as M87. This galaxy is located at a distance of about 50 million light-years away from the earth. (One light-year is the distance that light travels in a year, or 9.5×10^{15} m.)

EXAMPLE 10   A Supermassive Black Hole

THE PHYSICS OF...

locating a black hole.

The Hubble telescope has detected the light being emitted from different regions of galaxy M87, which is shown in Figure 5.15. The black circle identifies the center of the galaxy. From the characteristics of this light, astronomers have determined an orbiting speed of 7.5×10^5 m/s for matter located at a distance of 5.7×10^{17} m from the center. Find the mass *M* of the object located at the galactic center.

Reasoning and Solution Replacing M_E in Equation 5.5 with M gives $v = \sqrt{GM/r}$, which can be solved to show that

$$M = \frac{v^2 r}{G} = \frac{(7.5 \times 10^5 \text{ m/s})^2(5.7 \times 10^{17} \text{ m})}{6.67 \times 10^{-11} \text{ N} \cdot \text{m}^2/\text{kg}^2} = \boxed{4.8 \times 10^{39} \text{ kg}}$$

The ratio of this incredibly large mass to the mass of our sun is $(4.8 \times 10^{39} \text{ kg})/(2.0 \times 10^{30} \text{ kg}) = 2.4 \times 10^9$. Thus, matter equivalent to 2.4 billion suns is located at the center of galaxy M87. Considering that the volume of space in which this matter is located contains relatively few visible stars, researchers have concluded that the data provide strong evidence for the existence of a supermassive black hole. The term "black hole" is used because the tremendous mass prevents even light from escaping. The light that forms the image in Figure 5.15 comes not from the black hole itself, but from matter that surrounds it.

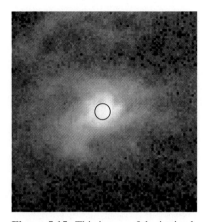

Figure 5.15 This image of the ionized gas (yellow) at the heart of galaxy M87 was obtained by the Hubble space telescope. The circle identifies the center of the galaxy, at which a black hole is thought to exist.

The period T of a satellite is the time required for one orbital revolution. As in any uniform circular motion, the period is related to the speed of the motion by $v = 2\pi r/T$. Substituting v from Equation 5.5 shows that

$$\sqrt{\frac{GM_E}{r}} = \frac{2\pi r}{T}$$

Solving this expression for the period T gives

$$T = \frac{2\pi r^{3/2}}{\sqrt{GM_E}} \tag{5.6}$$

Although derived for earth orbits, Equation 5.6 can also be used for calculating the periods of those planets in nearly circular orbits about the sun, if M_E is replaced by the mass M_S of the sun and r is interpreted as the distance between the center of the planet and the center of the sun. The fact that the period is proportional to the three-halves power of the orbital radius is known as Kepler's third law, for it is one of the laws discovered by Johannes Kepler (1571–1630) during his studies of planetary motion. Kepler's third law also holds for elliptical orbits, which will be discussed in Chapter 9.

An important application of Equation 5.6 occurs in the field of communications, where "synchronous satellites" are put into a circular orbit that is in the plane of the equator, as Figure 5.16 shows. The orbital period is chosen to be one day, which is also the time it takes for the earth to turn once about its axis. Therefore, these satellites move around their orbits in a way that is synchronized with the rotation of the earth. For earth-based observers, synchronous satellites have the useful characteristic of appearing in fixed positions in the sky and can serve as "stationary" relay stations for communication signals sent up from the earth's surface. This is exactly what is done in the digital satellite systems that are a popular alternative to cable TV. As the blowup in Figure 5.16 indicates, a small "dish" antenna on your house picks up the digital TV signals relayed to earth by the satellite. After being decoded, these signals are delivered to your TV set. All synchronous satellites are in orbit at the same height above the earth's surface, as Example 11 shows.

THE PHYSICS OF...

digital satellite system TV. (See below.)

Figure 5.16 A synchronous satellite orbits the earth once per day on a circular path that lies in the plane of the equator. Digital satellite system television uses such satellites as relay stations for TV signals that are sent up from the earth's surface and then rebroadcast down toward your own small dish antenna.

EXAMPLE 11 The Orbital Radius for Synchronous Satellites

What is the height above the earth's surface at which all synchronous satellites (regardless of mass) must be placed in orbit?

Reasoning The period T of a synchronous satellite is one day, so we can use Equation 5.6 to find the distance r from the center of the earth. However, the problem asks for the height H of the satellite above the earth's surface. To find H, we will have to take into account the fact that the earth itself has a radius of 6.38×10^6 m.

Solution A period of one day* corresponds to $T = 8.64 \times 10^4$ s. In using this value it is convenient to rearrange the equation $T = 2\pi r^{3/2}/\sqrt{GM_E}$ as follows:

$$r^{3/2} = \frac{T\sqrt{GM_E}}{2\pi} = \frac{(8.64 \times 10^4 \text{ s})\sqrt{(6.67 \times 10^{-11} \text{ N} \cdot \text{m}^2/\text{kg}^2)(5.98 \times 10^{24} \text{ kg})}}{2\pi}$$

By squaring and then taking the cube root, we find that $r = 4.23 \times 10^7$ m. Since the radius of the earth is approximately 6.38×10^6 m, the height of the satellite above the earth's surface is

$$H = 4.23 \times 10^7 \text{ m} - 0.64 \times 10^7 \text{ m} = \boxed{3.59 \times 10^7 \text{ m (22 300 mi)}}$$

5.6   APPARENT WEIGHTLESSNESS AND ARTIFICIAL GRAVITY

The idea of life on board an orbiting satellite conjures up visions of astronauts floating around in a state of "weightlessness," as in Figure 5.17. Actually, this state should be called "apparent weightlessness," because it is similar to the condition of zero apparent weight that occurs in an elevator during free-fall. Conceptual Example 12 explores this similarity.

CONCEPTUAL EXAMPLE 12   Apparent Weightlessness and Free-fall

Figure 5.18 shows a person on a scale in a freely falling elevator and in a satellite in a circular orbit. In each case, what apparent weight is recorded by the scale?

Reasoning and Solution As Section 4.8 discusses, apparent weight is the force that an object exerts on the platform of a scale. In the freely falling elevator in Figure 5.18a, the apparent weight is zero. The reason is that both the scale and the person on it fall together and, therefore, cannot push against one another. In the orbiting satellite in Figure 5.18b, both the person and the scale are in uniform circular motion. Objects in uniform circular motion continually accelerate or "fall" toward the center of the circle, in order to remain on the circular path. Consequently, the scale in Figure 5.18b and the person on it both "fall" with the same acceleration toward the center of the orbit and cannot push against one another. Thus, ***the apparent weight in the satellite is zero, just as it is in the freely falling elevator.*** The only difference between the satellite and the elevator is that the satellite moves in a circle, so that its "falling" does not bring it closer to the earth. In contrast to the apparent weight, the true weight is the gravitational force ($F = GmM_E/r^2$) that the earth exerts on an object and is not zero in a freely falling elevator or aboard an orbiting satellite.

The physiological effects of prolonged apparent weightlessness are only partially known. To minimize such effects, it is likely that artificial gravity will be provided in large space stations of the future. To help explain artificial gravity, Figure

Figure 5.17 As she orbits the earth, this NASA astronaut floats around in a state of apparent weightlessness.

THE PHYSICS OF...

apparent weightlessness.

---

* Successive appearances of the sun define the solar day of 24 h or 8.64×10^4 s. The sun moves against the background of the stars, however, and the time required for the earth to turn once on its axis relative to the fixed stars is 23 h 56 min, which is called the sidereal day. The sidereal day should be used in Example 11, but the neglect of this effect introduces an error of less than 0.4 percent in the answer.

Figure 5.18 (*a*) During free-fall, the elevator accelerates downward with the acceleration of gravity, and the apparent weight of the person is zero. (*b*) The orbiting space station is also in free-fall toward the center of the earth.

5.19 shows a space station rotating about an axis. Because of the rotational motion, any object located at a point *P* on the interior surface of the station experiences a centripetal force directed toward the axis. The surface of the station provides this force by pushing on the feet of an astronaut, for instance. The centripetal force can be adjusted to match the astronaut's earth weight by properly selecting the rotational speed of the space station, as Examples 13 and 14 illustrate.

EXAMPLE 13   Artificial Gravity

At what speed must the surface of the space station (*r* = 1700 m) move in Figure 5.19, so that the astronaut at point *P* experiences a push on his feet that equals his earth weight?

Reasoning The floor of the rotating space station exerts a normal force on the feet of the astronaut. This is the centripetal force ($F_c = mv^2/r$) that keeps the astronaut moving in a circular path. Since the magnitude of the normal force equals the astronaut's earth weight, we can determine the speed *v* of the space station.

Solution The earth weight of the astronaut (mass = *m*) is *mg*, and with this substitution, Equation 5.3 can be used to determine the required speed: $F_c = mg = mv^2/r$. Solving this equation for the speed, we find that

$$v = \sqrt{rg} = \sqrt{(1700 \text{ m})(9.80 \text{ m/s}^2)} = \boxed{130 \text{ m/s}}$$

Figure 5.19 The surface of the rotating space station pushes on an object with which it is in contact and thereby provides the centripetal force that keeps the object moving on a circular path.

THE PHYSICS OF...

artificial gravity.

EXAMPLE 14   A Rotating Space Laboratory

A space laboratory is rotating to create artificial gravity, as Figure 5.20 indicates. Its period of rotation is chosen so the outer ring (r_O = 2150 m) simulates the acceleration of gravity on earth (9.80 m/s²). What should be the radius r_I of the inner ring, so it simulates the acceleration of gravity on the surface of Mars (3.72 m/s²)?

Figure 5.20 The outer ring (radius = r_O) of this rotating space laboratory simulates gravity on earth, while the inner ring (radius = r_I) simulates gravity on Mars.

Reasoning It is important to realize that the value given for either acceleration corresponds to the centripetal acceleration $a_c = v^2/r$ in the corresponding ring. Moreover, the speed v and radius r are related according to $v = 2\pi r/T$ (Equation 5.1), where T is the period of the motion. Substituting this relation for the speed gives

$$a_c = \frac{v^2}{r} = \frac{(2\pi r/T)^2}{r} = \frac{4\pi^2 r}{T^2}$$

We will apply this expression to the outer and inner rings, both of which have the same period T. The periods of the two rings are the same, because the laboratory is rigid. All points on a rigid object require the same time to make one revolution.

Solution Applying the result that $a_c = 4\pi^2 r/T^2$, we find that

$$9.80 \text{ m/s}^2 = \underbrace{\frac{4\pi^2(2150 \text{ m})}{T^2}}_{\text{Outer ring}} \quad \text{and} \quad 3.72 \text{ m/s}^2 = \underbrace{\frac{4\pi^2 r_I}{T^2}}_{\text{Inner ring}}$$

Dividing the inner-ring expression by the outer-ring expression and eliminating the common algebraic factors of $4\pi^2$ and T^2 gives

$$\frac{3.72 \text{ m/s}^2}{9.80 \text{ m/s}^2} = \frac{r_I}{2150 \text{ m}} \quad \text{or} \quad r_I = \boxed{816 \text{ m}}$$

*5.7   VERTICAL CIRCULAR MOTION

Motorcycle stunt drivers often perform a feat in which they drive their cycles around a vertical circular track, as in Figure 5.21a. Usually, the speed varies in this stunt. When the speed of travel on a circular path changes from moment to moment, the motion is said to be "nonuniform." Nonetheless, we can use the concepts that apply to uniform circular motion to gain considerable insight into the motion that occurs on a vertical circle.

There are four points on a vertical circle where the centripetal force can be identified easily, as Figure 5.21b indicates. As you look at Figure 5.21b, keep in mind that the centripetal force is not a new and separate force of nature. Instead, at each point the centripetal force is the net sum of all the force components oriented along the radial direction and points toward the center of the circle. The drawing shows only the weight of the cycle plus rider (magnitude = mg) and the normal force pushing on the cycle (magnitude = F_N). The propulsion and braking forces are

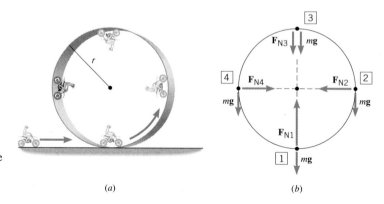

Figure 5.21 (a) A vertical loop-the-loop motorcycle stunt. (b) The normal force $\mathbf{F_N}$ and the weight mg of the cycle and the rider are shown here at four locations.

(a)   (b)

omitted for simplicity, because they do not act in the radial direction. The magnitude of the centripetal force at each of the four points is given as follows in terms of mg and F_N:

$$(1) \quad F_{c1} = \frac{mv_1^2}{r} = F_{N1} - mg \qquad (3) \quad F_{c3} = \frac{mv_3^2}{r} = F_{N3} + mg$$

$$(2) \quad F_{c2} = \frac{mv_2^2}{r} = F_{N2} \qquad\qquad (4) \quad F_{c4} = \frac{mv_4^2}{r} = F_{N4}$$

● PROBLEM SOLVING INSIGHT
Centripetal force is the name given to the net force that points toward the center of a circular path. As shown here, there may be several forces that contribute to this net force.

As the cycle goes around, the magnitude of the normal force changes. It changes because the speed changes and because the weight does not have the same effect at every point. At the bottom, the normal force and the weight oppose one another, giving a centripetal force of magnitude $F_{N1} - mg$. At the top, the normal force and the weight reinforce each other to provide a centripetal force whose magnitude is $F_{N3} + mg$. At points 2 and 4 on either side, only F_{N2} and F_{N4} provide the centripetal force. The weight is tangent to the circle at points 2 and 4 and has no component pointing toward the center. If the speed at each of the four places is known, along with the mass and the radius, the normal forces can be determined.

Riders who perform the loop-the-loop trick know that they must have at least a minimum speed at the top of the circle to remain on the track. This speed can be determined by considering the centripetal force at point 3. The speed v_3 in the equation $mv_3^2/r = F_{N3} + mg$ is a minimum when F_{N3} is zero. Then, the speed is given by $v_3 = \sqrt{rg}$. At this speed, the track does not exert a normal force to keep the cycle on the circle, because the weight mg provides all the centripetal force. Under these conditions, the rider experiences an apparent weightlessness like that discussed in Section 5.6, because for an instant the rider and the cycle are falling freely toward the center of the circle.

THE PHYSICS OF...

the loop-the-loop motorcycle stunt.

5.8 CONCEPTS & CALCULATIONS

In uniform circular motion the concepts of acceleration and force play central roles. Example 15 deals with acceleration, which we first discussed in Chapter 2, particularly in connection with the equations of kinematics. The example emphasizes the difference between the acceleration that is used in the equations of kinematics and the acceleration that arises in uniform circular motion.

CONCEPTS & CALCULATIONS EXAMPLE 15   Acceleration

At time $t = 0$ s automobile A is traveling at a speed of 18 m/s along a straight road and is picking up speed with an acceleration that has a magnitude of 3.5 m/s². At time $t = 0$ s automobile B is traveling at a speed of 18 m/s in uniform circular motion as it negotiates a turn. It has a centripetal acceleration whose magnitude is also 3.5 m/s². Determine the speed of each automobile when $t = 2.0$ s.

Concept Questions and Answers Which automobile has a constant acceleration?

Answer Acceleration is a vector, and for it to be constant, both its magnitude and direction must be constant. Automobile A has a constant acceleration, because its acceleration has a constant magnitude of 3.5 m/s² and its direction always points forward along the straight road. Automobile B has an acceleration with a constant magnitude of 3.5 m/s², but the acceleration does not have a constant direction. Automobile B is in uniform circular motion, so it has a centripetal acceleration, which points toward the center of the circle at every instant. There-fore, the direction of the acceleration vector continually changes, and the vector is not constant.

For which automobile do the equations of kinematics apply?

Answer The equations of kinematics apply for automobile A, because it has a constant acceleration, which must be the case when you use these equations. They do not apply to automobile B, because it does not have a constant accelera-tion.

Solution To determine the speed of automobile A at $t = 2.0$ s, we use Equation 2.4 from the equations of kinematics:

$$v = v_0 + at = 18 \text{ m/s} + (3.5 \text{ m/s}^2)(2.0 \text{ s}) = \boxed{25 \text{ m/s}}$$

To determine the speed of automobile B we note that this car is in uniform circular motion and, therefore, has a constant speed as it goes around the turn. At a time of $t = 2.0$ s its speed is the same as it was at $t = 0$ s, namely, $\boxed{v = 18 \text{ m/s}}$.

The next example deals with force, and it stresses that the centripetal force is the net force along the radial direction. As we will see, this means that all forces along the radial direction must be taken into account when identifying the centripetal force. In addition, the example is a good review of the tension force, which we first encountered in Chapter 4.

CONCEPTS & CALCULATIONS EXAMPLE 16   Centripetal Force

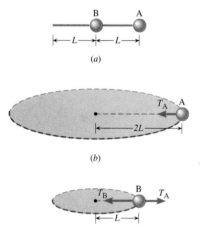

(a)

(b)

(c)

Figure 5.22 Two identical balls at-tached to a rigid massless rod are whirled around on horizontal circles, as Example 16 explains.

Ball A is attached to one end of a rigid massless rod, while an identical ball B is at-tached to the center of the rod, as Figure 5.22a illustrates. Each ball has a mass of $m = 0.50$ kg, and the length of each half of the rod is $L = 0.40$ m. This arrangement is held by the empty end and is whirled around in a horizontal circle at a constant rate, so each ball is in uniform circular motion. Ball A travels at a constant speed of $v_A = 5.0$ m/s. Find the tension in each half of the rod.

Concept Questions and Answers How many tension forces contribute to the centripetal force that acts on ball A?

Answer As Figure 5.22b illustrates, only a single tension force of magnitude T_A acts on ball A. It points to the left in the drawing and is due to the tension in the rod between the two balls. This force alone provides the centripetal force keeping ball A on its circular path of radius 2L.

How many tension forces contribute to the centripetal force that acts on ball B?

Answer As Figure 5.22c shows, two tension forces act on ball B. One has a magnitude T_B and points to the left in the drawing, which is the positive direction. It is due to the tension in the left half of the rod. The other has a magnitude T_A and points to the right. It is due to the tension in the right half of the rod. The cen-tripetal force acting on ball B points toward the center of the circle and is the vec-tor sum of these two forces, or $T_B - T_A$.

Is the speed of ball B the same as the speed of ball A?

Answer No, it is not. The reason is that ball A travels farther than ball B in the same time. Consider a time of one period, which is the same for either ball, since the arrangement is a rigid unit. It is the time for either ball to travel once around its circular path. In this time ball A travels a distance equal to the circumference of its path, which is $2\pi(2L)$. In contrast, the circumference for ball B is only $2\pi L$. Therefore, the speed of ball B is one-half the speed of ball A, or $v_B = 2.5$ m/s.

Solution Applying Equation 5.3 to each ball, we have

Ball A
$$\underbrace{T_A}_{\text{Centripetal force, } F_c} = \frac{m v_A{}^2}{2L}$$

Ball B
$$\underbrace{T_B - T_A}_{\text{Centripetal force, } F_c} = \frac{m v_B{}^2}{L}$$

The tension in the right half of the rod follows directly from the first of these equations:

$$T_A = \frac{m v_A{}^2}{2L} = \frac{(0.50 \text{ kg})(5.0 \text{ m/s})^2}{2(0.40 \text{ m})} = \boxed{16 \text{ N}}$$

Adding the equations for ball A and ball B gives the following result for the tension in the left half of the rod:

$$T_B = \frac{m v_B{}^2}{L} + \frac{m v_A{}^2}{2L} = \frac{(0.50 \text{ kg})(2.5 \text{ m/s})^2}{0.40 \text{ m}} + \frac{(0.50 \text{ kg})(5.0 \text{ m/s})^2}{2(0.40 \text{ m})} = \boxed{23 \text{ N}}$$

At the end of the problem set for this chapter, you will find homework problems that contain both conceptual and quantitative parts. These problems are grouped under the heading *Concepts & Calculations, Group Learning Problems*. They are designed for use by students working alone or in small learning groups. The conceptual part of each problem provides a convenient focus for group discussions.

SUMMARY

In **uniform circular motion,** an object of mass m travels at a constant speed v on a circular path of radius r. The **period** T of the motion is the time required for one revolution. The speed, the period, and the radius are related according to $v = 2\pi r/T$. The velocity vector in such motion is always changing direction, and, therefore, an acceleration exists. This acceleration is called **centripetal acceleration,** and its magnitude is $a_c = v^2/r$, while its direction is toward the center of the circle. To create this acceleration, a net force pointing toward the center of the circle is needed. This net force is called the **centripetal force,** and its magnitude is $F_c = mv^2/r$.

There are many examples of uniform circular motion, including the **banking of curves** and the **orbiting of satellites.** The angle θ at which a friction-free curve is banked depends on the radius r of the curve and the speed v at which the curve is to be negotiated, according to $\tan \theta = v^2/(rg)$. The speed and the period of a satellite in a circular orbit about the earth depend on the radius of the orbit, according to $v = \sqrt{GM_E/r}$ and $T = 2\pi r^{3/2}/\sqrt{GM_E}$, where G is the universal gravitational constant and M_E is the mass of the earth. **Apparent weightlessness** and **artificial gravity** in satellites can be explained in terms of uniform circular motion.

CONCEPTUAL QUESTIONS

1. The speedometer of your car shows that you are traveling at a constant speed of 35 m/s. Is it possible that your car is accelerating? If so, explain how this could happen.

2. Consider two people, one on the earth's surface at the equator and the other at the north pole. Which has the larger centripetal acceleration? Explain.

3. The equations of kinematics (Equations 3.3–3.6) describe the motion of an object that has a constant acceleration. These equations cannot be applied to uniform circular motion. Why not?

4. Is it possible for an object to have an acceleration when the velocity of the object is constant? When the speed of the object is constant? In each case, give your reasoning.

5. A car is traveling at a constant speed along the road *ABCDE* shown in the drawing. Sections *AB* and *DE* are straight. Rank the accelerations in each of the four sections according to magnitude, listing the smallest first. Justify your ranking.

6. Other things being equal, would it be easier to drive at high speed around an unbanked horizontal curve on the moon as compared to driving around the same curve on the earth? Explain.

7. A bug lands on a windshield wiper. Explain why the bug is more likely to be dislodged when the wipers are turned on at the high rather than the low setting.

8. What is the chance of a light car safely rounding an unbanked curve on an icy road as compared to that of a heavy car: worse, the same as, or better? Assume that both cars have the same speed and are equipped with identical tires. Account for your answer.

9. A container is filled with water and attached to a rope. Holding the free end of the rope, you whirl the container around in a horizontal circle at a constant speed. There is a hole in the container, so that as you whirl it, water continually leaks out. Explain what, if anything, you feel as the water leaks out.

10. A propeller, operating under test conditions, is being made to rotate at ever faster speeds. Explain what is likely to happen, and why, when the maximum rated speed of the propeller is exceeded.

11. A penny is placed on a rotating turntable. Where on the platter does the penny require the largest centripetal force to remain in place? Give your reasoning.

12. Explain why a real airplane must bank as it flies in a circle, while a model airplane on a guideline can fly in a circle without banking.

13. Would a change in the earth's mass affect (a) the banking of airplanes as they turn, (b) the banking of roadbeds, (c) the speeds with which satellites are put into circular orbits, and (d) the performance of the loop-the-loop motorcycle stunt? In each case, give your reasoning.

14. A stone is tied to a string and whirled around in a circle at a constant speed. Is the string more likely to break when the circle is horizontal or when it is vertical? Account for your answer, assuming the constant speed is the same in each case.

15. ⚕ A fighter plane is diving toward the earth. At the last instant, the pilot pulls out of the dive on a vertical circle and begins climbing upward. In such maneuvers, the pilot can black out if too much blood drains from his head. Why does blood drain from his head?

16. Two cars are identical, except for the type of tread design on their tires. The cars are driven at the same speed and enter the same unbanked horizontal turn. Car A cannot negotiate the turn, but car B can. Which tread design yields a larger coefficient of static friction between the tires and the road? Give your reasoning in terms of centripetal force.

PROBLEMS

ssm Solution is in the Student Solutions Manual. **www** Solution is available on the World Wide Web at http://www.wiley.com/college/cutnell
⚕ This icon represents a biomedical application.

**Section 5.1 Uniform Circular Motion,
Section 5.2 Centripetal Acceleration**

1. ssm How long does it take a plane, traveling at a speed of 110 m/s, to fly once around a circle whose radius is 2850 m?

2. A horse races once around a circular track in a time of 118 s, with a speed of 17 m/s. What is the radius of the track?

3. The tips of the blades in a food blender are moving with a speed of 21 m/s in a circle that has a radius of 0.053 m. How much time does it take for the blades to make one revolution?

4. Review Conceptual Example 2 in preparation for this problem. In Figure 5.4, an object, after being released from its circular path, travels the distance *OA* in the same time it would have moved from *O* to *P* on the circle. The speed of the object on and off the circle remains constant at the same value. Suppose that the radius of the circle in Figure 5.4 is 3.6 m and the angle θ is 25°. What is the distance *OA*?

5. ssm Speedboat A negotiates a curve whose radius is 120 m. Speedboat B negotiates a curve whose radius is 240 m. Each boat experiences the same centripetal acceleration. What is the ratio v_A/v_B of the speeds of the boats?

6. There is a clever kitchen gadget for drying lettuce leaves after you wash them. It consists of a cylindrical container mounted so that it can be rotated about its axis by turning a hand crank. The outer wall of the cylinder is perforated with small holes. You put the wet leaves in the container and turn the crank to spin off the water. The radius of the container is 12 cm. When the cylinder is rotating at 2.0 revolutions per second, what is the magnitude of the centripetal acceleration at the outer wall?

7. Review Conceptual Example 2 as background for this problem. One kind of slingshot consists of a pocket that holds a pebble and is whirled on a circle of radius *r*. The pebble is released from the circle at the angle θ, so that it will hit the target. The distance to the target from the center of the circle is d. (See the drawing, which is not to scale.) The circular path is parallel to the ground, and the target lies in the plane of the circle. The distance d is ten times the radius r. Ignore the effect of gravity in pulling the stone downward after it is released and find the angle θ.

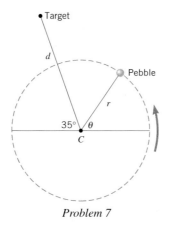

Problem 7

*8. The earth rotates once per day about an axis passing through the north and south poles, an axis that is perpendicular to the plane of the equator. Assuming the earth is a sphere with a radius of 6.38×10^6 m, determine the speed and centripetal acceleration of a person situated (a) at the equator and (b) at a latitude of 30.0° north of the equator.

*9. ssm The large blade of a helicopter is rotating in a horizontal circle. The length of the blade is 6.7 m, measured from its tip to the center of the circle. Find the ratio of the centripetal acceleration at the end of the blade to that which exists at a point located 3.0 m from the center of the circle.

*10. ☕ A centrifuge is a device in which a small container of material is rotated at a high speed on a circular path. Such a device is used in medical laboratories, for instance, to cause the more dense red blood cells to settle through the less dense blood serum and collect at the bottom of the container. Suppose the centripetal acceleration of the sample is 6.25×10^3 times as large as the acceleration due to gravity. How many revolutions per minute is the sample making, if it is located at a radius of 5.00 cm from the axis of rotation?

Section 5.3 Centripetal Force

11. ssm Review Example 3, which deals with the bobsled in Figure 5.5. Also review Conceptual Example 4. The mass of the sled and its two riders in Figure 5.5 is 350 kg. Find the magnitude of the centripetal force that acts on the sled during the turn with a radius of (a) 33 m and (b) 24 m.

12. A speed skater goes around a turn with a 31-m radius. The skater has a speed of 14 m/s and experiences a centripetal force of 460 N. What is the mass of the skater?

13. A car is safely negotiating an unbanked circular turn at a speed of 21 m/s. The maximum static frictional force acts on the tires. Suddenly a wet patch in the road reduces the maximum static frictional force by a factor of three. If the car is to continue safely around the curve, to what speed must the driver slow the car?

14. At an amusement park there is a ride in which cylindrically shaped chambers spin around a central axis. People sit in seats facing the axis, their backs against the outer wall. At one instant the outer wall moves at a speed of 3.2 m/s, and an 83-kg person feels a 560-N force pressing against his back. What is the radius of a chamber?

15. A child is twirling a 0.0120-kg ball on a string in a horizontal circle whose radius is 0.100 m. The ball travels once around the circle in 0.500 s. (a) Determine the centripetal force acting on the ball. (b) If the speed is doubled, does the centripetal force double? If not, by what factor does the centripetal force increase?

16. A stone has a mass of 6.0×10^{-3} kg and is wedged into the tread of an automobile tire, as the drawing shows. The coefficient of static friction between the stone and each side of the tread channel is 0.90. When the tire surface is rotating at 13 m/s, the stone flies out of the tread. The magnitude F_N of the normal force that each side of the tread channel exerts on the stone is 1.8 N. Assume that only static friction supplies the centripetal force, and determine the radius r of the tire.

*17. ssm www The hammer throw is a track and field event in which a 7.3-kg ball (the "hammer") is whirled around in a circle several times and released. It then moves upward on the familiar curving path of projectile motion and eventually returns to earth, some distance away. The world record for this distance is 86.75 m, achieved in 1986 by Yuriy Sedykh. Ignore air resistance and the fact that the ball is released above the ground rather than at ground level. Furthermore, assume that the ball is whirled on a circle that has a radius of 1.8 m and that its velocity at the instant of release is directed 41° above the horizontal. Find the magnitude of the centripetal force acting on the ball just prior to the moment of release.

*18. A "swing" ride at a carnival consists of chairs that are swung in a circle by 12.0-m cables attached to a vertical rotating pole, as the drawing shows. Suppose the total mass of a chair and its occupant is 220 kg. (a) Determine the tension in the cable attached to the chair. (b) Find the speed of the chair.

**19. A block is hung by a string from the inside roof of a van. When the van goes straight ahead at a speed of 28 m/s, the block hangs vertically down. But when the van maintains this same speed around an unbanked curve (radius = 150 m), the block swings toward the outside of the curve. Then the string makes an angle θ with the vertical. Find θ.

Section 5.4 Banked Curves

20. At what angle should a curve of radius 150 m be banked, so cars can travel safely at 25 m/s without relying on friction?

21. ssm A curve of radius 120 m is banked at an angle of 18°. At what speed can it be negotiated under icy conditions where friction is negligible?

22. A car can negotiate an unbanked curve safely at a certain maximum speed when the coefficient of static friction between the tires and the ground is 0.88. At what angle should the same curve be banked for the car to negotiate the curve safely at the same maximum speed without relying on friction?

23. On a banked race track, the smallest circular path on which cars can move has a radius of 112 m, while the largest has a radius of 165 m, as the drawing illustrates. The height of the outer wall is 18 m. Find (a) the smallest and (b) the largest speed at which cars can move on this track without relying on friction.

*24. There is a similarity between a plane banking into a turn and a car going around a banked curve. The lifting force **L** in Figure 5.10 plays the same role as the normal force $\mathbf{F_N}$ in Figure 5.11. (a) Derive an expression that relates the banking angle to the speed of the plane, the radius of the turn, and the acceleration due to gravity. (b) At what angle with respect to the horizontal should a plane be banked when traveling at 195 m/s around a turn whose radius is 8250 m?

*25. ssm www Refer to problem 24 before attempting to solve this problem. A jet ($m = 2.00 \times 10^5$ kg), flying at 123 m/s, banks to make a horizontal circular turn. The radius of the turn is 3810 m. Calculate the necessary lifting force.

**26. The drawing shows a baggage carousel at an airport. Your suitcase has not slid all the way down the slope and is going around at a constant speed on a circle ($r = 11.0$ m) as the carousel turns. The coefficient of static friction between the suitcase and the carousel is 0.760, and the angle θ in the drawing is 36.0°. How much time is required for your suitcase to go around once?

Section 5.5 Satellites in Circular Orbits,
Section 5.6 Apparent Weightlessness and Artificial Gravity

27. ssm www A satellite is in a circular orbit around an unknown planet. The satellite has a speed of 1.70×10^4 m/s, and the radius of the orbit is 5.25×10^6 m. A second satellite also has a circular orbit around this same planet. The orbit of this second satellite has a radius of 8.60×10^6 m. What is the orbital speed of the second satellite?

28. Suppose the surface (radius = r) of the space station in Figure 5.19 is rotating at 35.8 m/s. What must be the value of r for the astronauts to weigh one-half of their earth weight?

29. A satellite is placed in orbit 6.00×10^5 m above the surface of Jupiter. Jupiter has a mass of 1.90×10^{27} kg and a radius of 7.14×10^7 m. Find the orbital speed of the satellite.

30. A satellite circles the earth in an orbit whose radius is twice the earth's radius. The earth's mass is 5.98×10^{24} kg, and its radius is 6.38×10^6 m. What is the period of the satellite?

31. The earth travels around the sun once a year in an approximately circular orbit whose radius is 1.50×10^{11} m. From these data determine (a) the orbital speed of the earth and (b) the mass of the sun.

*32. The earth orbits the sun once a year at a distance of 1.50×10^{11} m. Venus orbits the sun at a distance of 1.08×10^{11} m. These distances are between the centers of the planets and the sun. How long (in earth days) does it take for Venus to make one orbit around the sun?

*33. ssm A satellite has a mass of 5850 kg and is in a circular orbit 4.1×10^5 m above the surface of a planet. The period of the orbit is two hours. The radius of the planet is 4.15×10^6 m. What is the true weight of the satellite when it is at rest on the planet's surface?

*34. A satellite is in a circular earth orbit that has a radius of 6.7×10^6 m. A model airplane is flying on a 15-m guideline in a horizontal circle. The guideline is nearly parallel to the ground. Find the speed of the plane such that the plane and the satellite have the same centripetal acceleration.

**35. To create artificial gravity, the space station shown in the drawing is rotating at a rate of 1.00 rpm. The radii of the cylindrically shaped chambers have the ratio $r_A/r_B = 4.00$. Each chamber A simulates an acceleration due to gravity of 10.0 m/s². Find values for (a) r_A, (b) r_B, and (c) the acceleration due to gravity that is simulated in chamber B.

Section 5.7 Vertical Circular Motion

36. A fighter pilot dives his plane toward the ground at 230 m/s. He pulls out of the dive on a vertical circle. What is the minimum radius of the circle, so that the normal force exerted on the pilot by his seat never exceeds three times his weight?

37. ssm The condition of apparent weightlessness for the passengers can be created for a brief instant when a plane flies over the top of a vertical circle. At a speed of 215 m/s, what is the radius of the vertical circle that the pilot must use?

38. For the normal force in Figure 5.21 to have the same magnitude at all points on the vertical track, the stunt driver must adjust the speed to be different at different points. Suppose, for example, that the track has a radius of 3.0 m and that the driver goes past point 1 at the bottom with a speed of 15 m/s. What speed must she have at point 3, so that the normal force at the top has the same magnitude as it did at the bottom?

39. ssm A motorcycle has a constant speed of 25.0 m/s as it passes over the top of a hill whose radius of curvature is 126 m. The mass of the motorcycle and driver is 342 kg. Find the magnitude of (a) the centripetal force and (b) the normal force that acts on the cycle.

*40. A stone is tied to a string (length = 0.950 m) and whirled in a circle at the same constant speed in two different ways. First, the circle is horizontal and the string is nearly parallel to the ground. Next, the circle is vertical. In the vertical case the maximum tension in the string is 10.0% larger than the tension that

exists when the circle is horizontal. Determine the speed of the stone.

*41. A roller-blader is traveling up one side of a hill and down the other side, trying to go fast enough to lose contact with the ground as she goes over the top. The crest is a circular arc with a radius of 20.0 m. Determine the minimum speed that she needs to achieve at the top of the hill.

**42. In an automatic clothes drier, a hollow cylinder moves the clothes on a vertical circle (radius $r = 0.32$ m), as the drawing shows. The appliance is designed so that the clothes tumble gently as they dry. This means that when a piece of clothing reaches an angle of θ above the horizontal, it loses contact with the wall of the cylinder

and falls onto the clothes below. How many revolutions per second should the cylinder make in order that the clothes lose contact with the wall when $\theta = 70.0°$?

ADDITIONAL PROBLEMS

43. ssm A 0.015-kg ball is shot from the plunger of a pinball machine. Because of a centripetal force of 0.028 N, the ball follows a circular arc whose radius is 0.25 m. What is the speed of the ball?

44. A rocket is used to place a synchronous satellite in orbit about the earth. What is the speed of the satellite in orbit?

45. A bicycle chain is wrapped around a rear sprocket ($r = 0.039$ m) and a front sprocket ($r = 0.10$ m). The chain moves with a speed of 1.4 m/s around the sprockets, while the bike moves at a constant velocity. Find the magnitude of the acceleration of a chain link that is in contact with (a) the rear sprocket, (b) neither sprocket, and (c) the front sprocket.

46. For background pertinent to this problem, review Conceptual Example 6. In Figure 5.8 the man hanging upside down is holding a partner who weighs 475 N. Assume that the partner moves on a circle that has a radius of 6.50 m. At a swinging speed of 4.00 m/s, what force must the man apply to his partner?

47. ssm Computer-controlled display screens provide drivers in the Indianapolis 500 with a variety of information about how their cars are performing. For instance, as a car is going through a turn, a speed of 221 mi/h (98.8 m/s) and a centripetal acceleration of 3.00g (three times the acceleration due to gravity) are displayed. Determine the radius of the turn (in meters).

48. A 9.5-kg monkey is hanging by one arm from a branch and swinging on a vertical circle. As an approximation, assume that the radial distance between the branch and the point where the monkey's mass is located to be 85 cm. As it swings through the lowest point on the circle, the monkey has a speed of 2.8 m/s. Find (a) the magnitude of the centripetal force acting on the monkey and (b) the magnitude of the tension in the monkey's arm.

49. ssm A 2100-kg demolition ball swings at the end of a 15-m cable on the arc of a vertical circle. At the lowest point of the swing, the ball is moving at a speed of 7.6 m/s. Determine the tension in the cable. (*Hint: The tension serves the same purpose as the normal force at point 1 in Figure 5.21.*)

50. The moon orbits the earth at a distance of 3.85×10^8 m. Assume that this distance is between the centers of the earth and the moon and that the mass of the earth is 5.98×10^{24} kg. Find the period for the moon's motion around the earth. Express the answer in days and compare it to the length of a month.

*51. A computer is reading data from a rotating CD-ROM. At a point that is 0.030 m from the center of the disc, the centripetal acceleration is 120 m/s². What is the centripetal acceleration at a point that is 0.050 m from the center of the disc?

*52. Each of the space shuttle's main engines is fed liquid hydrogen by a high-pressure pump. Turbine blades inside the pump rotate at 617 rev/s. A point on one of the blades traces out a circle with a radius of 0.020 m as the blade rotates. (a) What is the magnitude of the centripetal acceleration that the blade must sustain at this point? (b) Express this acceleration as a multiple of $g = 9.80$ m/s².

**53. ssm www Redo Example 5 without ignoring the effect of gravity. That is, the guideline now slopes downward due to the weight of the plane. For purposes of significant figures, use 0.900 kg for the mass of the plane, 17.0 m for the length of the guideline, and 19.0 and 38.0 m/s for the speeds.

**54. At amusement parks, there is a popular ride where the floor of a rotating cylindrical room falls away, leaving the backs of the riders "plastered" against the wall. Suppose the radius of the room is 3.30 m and the speed of the wall is 10.0 m/s when the floor falls away. (a) What is the source of the centripetal force acting on the riders? (b) How much centripetal force acts on a 55.0-kg rider? (c) What is the minimum coefficient of static friction that must exist between a rider's back and the wall, if the rider is to remain in place when the floor drops away?

**55. A rigid massless rod is rotated about one end in a horizontal circle. There is a mass m_1 attached to the center of the rod and a mass m_2 attached to the outer end of the rod. The inner section of the rod sustains three times as much tension as the outer section. Find the ratio m_2/m_1.

CONCEPTS CALCULATIONS

G R O U P   L E A R N I N G   P R O B L E M S

Note: Each of these problems consists of Concept Questions followed by a related quantitative Problem. They are designed for use by students working alone or in small learning groups. The Concept Questions involve little or no mathematics and are intended to stimulate group discussions. They focus on the concepts with which the problems deal. Recognizing the concepts is the essential initial step in any problem-solving technique.

56. Concept Questions The second hand and the minute hand on one type of clock are the same length. (a) What is the period T of the motion for the second hand and for the minute hand? (b) We have seen that the centripetal acceleration is given by $a_c = v^2/r$, where v is the speed and r is the radius. How is the centripetal acceleration related to the period? In each case, explain your reasoning.

Problem Find the ratio $(a_{c,\,second}/a_{c,\,minute})$ of the centripetal accelerations for the tips of the second hand and the minute hand.

57. Concept Questions The following table lists data for the speed and the radius in three examples of uniform circular motion.

	Radius	Speed
Example 1	0.50 m	12 m/s
Example 2	Infinitely large	35 m/s
Example 3	1.8 m	2.3 m/s

(a) Without doing any calculations, identify the example with the smallest centripetal acceleration. (b) Similarly identify the example with the greatest centripetal acceleration. In each case, justify your answer.

Problem Find the value for the centripetal acceleration for each example. Verify that your answers are consistent with your answers to the Concept Questions.

58. Concept Questions A penny is placed at the outer edge of a $33\frac{1}{3}$-rpm record (radius = 0.150 m). (a) What is producing the centripetal force that enables the penny to rotate along with the record—kinetic or static friction? (b) The centripetal force needed depends on the speed at which the penny is moving. How

can the speed be determined from the data given? Account for your answer in each case.

Problem Find the minimum coefficient of friction necessary to allow the penny to rotate along with the record.

59. Concept Questions Two cars are traveling at the same speed of 27 m/s on a curve that has a radius of 120 m. Car A has a mass of 1100 kg, while car B has a mass of 1600 kg. Without doing any calculations, decide (a) which car, if either, has the greatest centripetal acceleration and (b) which car, if either, experiences the greatest centripetal force. Justify your answers.

Problem Find the centripetal acceleration and the centripetal force for each car. Be sure that your answers are consistent with your answers to the Concept Questions.

60. Concept Questions Two satellites are in circular orbits around the earth. The orbit for satellite A is at a height of 360 km above the earth's surface, while that for satellite B is at a height of 720 km. (a) Which satellite has the greatest orbital speed? (b) The speed of a satellite in a circular orbit is given by Equation 5.5 ($v = \sqrt{GM_E/r}$). In this expression, do you substitute the heights of 360×10^3 m and 720×10^3 m for the term r? Explain your answers.

Problem Find the orbital speed for each satellite. Check to see that your answers are consistent with your answers to the Concept Questions.

61. Concept Questions A 0.20-kg ball on a string is whirled on a vertical circle at a constant speed. When the ball is at the three o'clock position, the tension is 16 N. (a) Is the centripetal force acting on the ball the same at each point on the circle? (b) When the ball is at the three o'clock position, what is providing the centripetal force—gravity, the tension in the string, or a combination of both? (c) Is the tension in the string greater when the ball is at the twelve o'clock position or when it is at the six o'clock position? Justify your answers.

Problem Find the tensions in the string when the ball is at the twelve o'clock and at the six o'clock positions. Verify that your answers are consistent with your answers to the Concept Questions.

Work and play do mix, for, as we shall see, work is being done by the gravitational force during the entire time these skydivers are descending.

CHAPTER 6
WORK AND ENERGY

Figure 6.1 Work is done when a force **F** pushes a car through a displacement **s**.

6.1 WORK DONE BY A CONSTANT FORCE

Work is a familiar concept. For example, it takes work to push a stalled car. In fact, more work is done when the pushing force is greater or when the displacement of the car is greater. Force and displacement, then, are the two essential elements of work, as Figure 6.1 illustrates. The drawing shows a constant pushing force **F** that points in the same direction as the resulting displacement **s**.* In such a case, the work W is defined as the magnitude F of the force times the magnitude s of the displacement: $W = Fs$. The work done to push a car is the same whether the car is moved north to south or east to west, provided that the amount of force used and the distance moved are the same. Since work does not convey directional information, it is a scalar quantity.

The equation $W = Fs$ indicates that the unit of work is the unit of force times the unit of distance, or the newton·meter in SI units. One newton·meter is referred to as a *joule* (J) (rhymes with "cool"), in honor of James Joule (1818–1889) and his research into the nature of work, energy, and heat. Table 6.1 summarizes the units for work in several systems of measurement.

The definition of work as $W = Fs$ does have one surprising feature: If the distance s is zero, the work is zero, even if a force is applied. Pushing on an immovable object, such as a brick wall, may tire your muscles, but there is no work done of the type we are discussing. In physics, the idea of work is intimately tied up with the idea of motion. If there is no movement of the object, the work done by the force acting on the object is zero.

Often, the force and displacement do not point in the same direction. For instance, Figure 6.2a shows a luggage carrier being pulled to the right by a force that is applied along the handle. The force is directed at an angle θ relative to the displacement. In such a case, only the component of the force along the displacement is used in defining work. As part b of the drawing illustrates, this component is $F \cos \theta$, and it appears in the general definition of work as follows:

Figure 6.2 (*a*) Work can be done by a force **F** that points at an angle θ relative to the displacement **s**. (*b*) The force component that points along the displacement is $F \cos \theta$.

● **DEFINITION OF WORK DONE BY A CONSTANT FORCE**

The work done on an object by a constant† force **F** is

$$W = (F \cos \theta)s \qquad (6.1)$$

where F is the magnitude of the force, s is the magnitude of the displacement, and θ is the angle between the force and the displacement.

SI Unit of Work: newton·meter = joule (J)

---

* When discussing work, it is customary to use the symbol **s** for the displacement, rather than **x** or **y**.
† Section 6.9 considers the work done by a variable force.

TABLE 6.1 Units of Measurement for Work

System	Force	×	Distance	=	Work
SI	newton (N)		meter (m)		joule (J)
CGS	dyne (dyn)		centimeter (cm)		erg
BE	pound (lb)		foot (ft)		foot · pound (ft · lb)

(a)

When the force points in the same direction as the displacement, then $\theta = 0°$, and Equation 6.1 reduces to $W = Fs$. The following example illustrates how Equation 6.1 is used to calculate work.

EXAMPLE 1   Pulling a Luggage Carrier

Find the work done by a 45.0-N force in pulling the luggage carrier in Figure 6.2a at an angle $\theta = 50.0°$ for a distance $s = 75.0$ m.

Reasoning The pulling force causes the luggage carrier to move a distance of 75.0 m and does work. However, the force makes an angle of 50.0° with the displacement, and we must take this angle into account by using the definition of work given by Equation 6.1.

Solution The work done by the 45.0-N force is

$$W = (F \cos \theta)s = [(45.0 \text{ N}) \cos 50.0°](75.0 \text{ m}) = \boxed{2170 \text{ J}}$$

The answer is expressed in newton · meters or joules (J).

(b)

The definition of work in Equation 6.1 takes into account only the component of the force in the direction of the displacement. The force component perpendicular to the displacement does no work. To do work, there must be a force *and* a displacement, and since there is no displacement in the perpendicular direction, there is no work done by the perpendicular component of the force.

Work can be either positive or negative, depending on whether a component of the force points in the same direction as the displacement or in the opposite direction. Example 2 illustrates how positive and negative work arise.

(c)

Figure 6.3 (a) In the bench press, work is done during both the lifting and lowering phases. (b) During the lifting phase, the force **F** does positive work. (c) During the lowering phase, the force does negative work.

EXAMPLE 2   Bench-pressing

The weight lifter in Figure 6.3a is bench-pressing a barbell whose weight is 710 N. In part *b* of the figure, he raises the barbell a distance of 0.65 m above his chest, and in part *c* he lowers it the same distance. The weight is raised and lowered at a constant velocity. Determine the work done on the barbell by the weight lifter during (a) the lifting phase and (b) the lowering phase.

Reasoning To calculate the work, it is necessary to know the force exerted by the weight lifter. The barbell is raised and lowered at a constant velocity and, therefore, is in equilibrium. Consequently, the force **F** exerted by the weight lifter must balance the weight of the barbell, so $F = 710$ N. During the lifting phase, the force **F** and displacement **s** are in the same direction, as Figure 6.3b shows. The angle between them is $\theta = 0°$. When the barbell is lowered, however, the force and displacement are in opposite directions, as in Figure 6.3c. The angle between the force and the displacement is now $\theta = 180°$. With these observations, we can find the work.

THE PHYSICS OF...

positive and negative "reps" in weight lifting.

INTERACTIVE LEARNINGWARE #11

START

Visualization

Reasoning

Knowns and Unknowns

Modeling the Problem

Numerical Solution

FINISH

A cyclist coasts down a hill, losing 20.0 m in height, and then coasts up an identical hill to the same initial vertical height. The bicycle and rider have a mass of 73.5 kg. Two forces act on the bicycle/rider system: the weight of the bicycle and rider and the normal force exerted on the tires by the ground. Determine the work done by each of these forces during (a) the downward part of the motion and (b) the upward part of the motion.

This problem illustrates two important aspects of work. First, it shows how to determine the work done in the common situation where the force is not parallel to the displacement. Second, the problem demonstrates that the work done by any particular force can be positive, negative, or zero, depending on how the force is oriented relative to the displacement of the moving object.

Go to our Web site at

http://www.wiley.com/college/cutnell

Solution

(a) During the lifting phase, the work done by the force **F** is given by Equation 6.1 as

$$W = (F \cos \theta)s = [(710 \text{ N}) \cos 0°](0.65 \text{ m}) = \boxed{460 \text{ J}}$$

(b) The work done during the lowering phase is

$$W = (F \cos \theta)s$$
$$= [(710 \text{ N}) \cos 180°](0.65 \text{ m}) = \boxed{-460 \text{ J}}$$

since $\cos 180° = -1$. The work is negative, because the force is opposite to the displacement. Weight lifters call each complete up-and-down movement of the barbell a repetition, or "rep." The lifting of the weight is referred to as the positive part of the rep, and the lowering is known as the negative part.

As Example 2 illustrates, work can be positive or negative. If the force has a component in the *same* direction as the displacement of the object, the work done by the force is *positive*. On the other hand, if a force component points in the direction *opposite* to the displacement, the work is *negative*. If the force is perpendicular to the displacement, the force has no component in the direction of the displacement, and the work is zero ($\theta = 90°$ in Equation 6.1). Example 3 deals with the work done by a static frictional force when it acts on a crate that is resting on the bed of an accelerating truck.

(a)

(b) Free-body diagram for the crate

Figure 6.4 (a) The truck and crate are accelerating to the right for a distance of $s = 65$ m. (b) The free-body diagram for the crate.

EXAMPLE 3   Accelerating a Crate

Figure 6.4a shows a 120-kg crate on the flatbed of a truck that is moving with an acceleration of $a = +1.5$ m/s² along the positive x axis. The crate does not slip with respect to the truck, as the truck undergoes a displacement whose magnitude is $s = 65$ m. What is the total work done on the crate by all of the forces acting on it?

Reasoning The free-body diagram in Figure 6.4b shows the forces that act on the crate: (1) the weight **W** of the crate, (2) the normal force **F$_N$** exerted by the flatbed, and (3) the static frictional force **f$_s$**, which is exerted by the flatbed in the forward direction and keeps the crate from slipping backward. The weight and the normal force are perpendicular to the displacement, so they do no work. Only the static frictional force does work, since it acts in the x direction. To determine the frictional force, we note that the crate does not slip and, therefore, must have the same acceleration of $a = +1.5$ m/s² as does the truck. The force creating this acceleration is the static frictional force, and knowing the mass of the crate and its acceleration, we can use Newton's second law to obtain its magnitude. Then, knowing the frictional force and the displacement, we can determine the total work done on the crate.

Solution From Newton's second law, we find that the magnitude f_s of the static frictional force is

$$f_s = ma = (120 \text{ kg})(1.5 \text{ m/s}²) = 180 \text{ N}$$

The total work is that done by the static frictional force and is

$$W = (f_s \cos \theta)s = (180 \text{ N})(\cos 0°)(65 \text{ m}) = \boxed{1.2 \times 10^4 \text{ J}} \qquad (6.1)$$

The work is positive, because the frictional force is in the same direction as the displacement ($\theta = 0°$).

6.2 THE WORK–ENERGY THEOREM AND KINETIC ENERGY

● *Concepts at a Glance* Most people expect that if you do work, you get something as a result. In physics, when a net force performs work on an object, there is always a result from the effort. The result is a change in the *kinetic energy* of the object. As we will now see, the relationship that relates work to the change in kinetic energy is known as the *work–energy theorem.* This theorem is obtained by bringing together three basic concepts that we've already learned about, as the Concepts-at-a-Glance chart in Figure 6.5 illustrates. First we'll apply Newton's second law of motion, $\Sigma F = ma$, which relates the net force ΣF to the acceleration a of an object. Then, we'll determine the work done by the net force when the object moves through a certain distance. Finally, we'll use Equation 2.9, one of the equations of kinematics, to relate the distance and acceleration to the initial and final speeds of the object. The result of this approach will be the work–energy theorem. ●

To gain some insight into the idea of kinetic energy and the work–energy theorem, look at Figure 6.6, where a constant net external force ΣF acts on an airplane of mass m. This net force is the vector sum of all the external forces acting on the plane, and, for simplicity, it is assumed to have the same direction as the displacement **s**. According to Newton's second law, the net force produces an acceleration a, given by $a = \Sigma F/m$. Consequently, the speed of the plane changes from an initial

Figure 6.5 *Concepts at a Glance* In this section, the familiar concepts of Newton's second law, work, and the equations of kinematics are integrated to produce the work–energy theorem. In head-butting a soccer ball, the player does work on it. This work causes the ball's kinetic energy to change in accordance with the work–energy theorem.

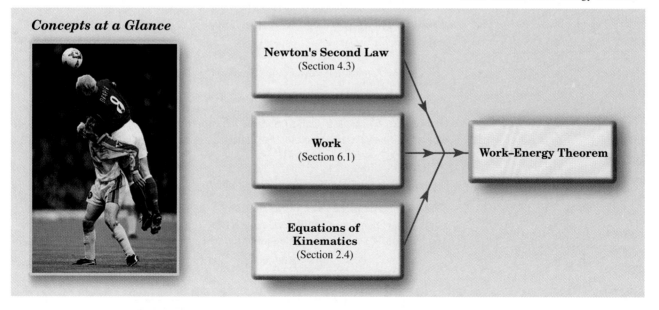

Concepts at a Glance

Newton's Second Law
(Section 4.3)

Work
(Section 6.1)

Equations of Kinematics
(Section 2.4)

Work–Energy Theorem

Figure 6.6 A constant net external force $\Sigma\mathbf{F}$ acts over a displacement $\mathbf{s}$ and does work on the plane. As a result of the work done, the plane's kinetic energy changes.

value of v_0 to a final value of v_f.* Multiplying both sides of $\Sigma F = ma$ by the distances s gives

$$\underbrace{(\Sigma F)s}_{\substack{\text{Work done} \\ \text{by net external} \\ \text{force}}} = mas$$

The left side of this equation is the work done by the net external force. The term as on the right side can be related to v_0 and v_f by using Equation 2.9 ($v_f^2 = v_0^2 + 2as$) from the equations of kinematics. Solving this equation to give $as = \frac{1}{2}(v_f^2 - v_0^2)$ and substituting into $(\Sigma F)s = mas$ shows that

$$\underbrace{(\Sigma F)s}_{\substack{\text{Work done by} \\ \text{net external} \\ \text{force}}} = \underbrace{\tfrac{1}{2}mv_f^2}_{\substack{\text{Final} \\ \text{kinetic} \\ \text{energy}}} - \underbrace{\tfrac{1}{2}mv_0^2}_{\substack{\text{Initial} \\ \text{kinetic} \\ \text{energy}}}$$

This expression is the work–energy theorem. Its left side is the work W done by the net external force, while its right side involves the difference between two terms, each of which has the form $\frac{1}{2}(\text{mass})(\text{speed})^2$. The quantity $\frac{1}{2}(\text{mass})(\text{speed})^2$ is called kinetic energy and plays a significant role in physics, as we will soon see.

● **DEFINITION OF KINETIC ENERGY**

The kinetic energy KE of an object with mass m and speed v is given by

$$\text{KE} = \tfrac{1}{2}mv^2 \tag{6.2}$$

SI Unit of Kinetic Energy: joule (J)

The SI unit of kinetic energy is the same as the unit for work, the joule. Kinetic energy, like work, is a scalar quantity. These are not surprising observations, for work and kinetic energy are closely related, as is clear from the following statement of the work–energy theorem.

● **THE WORK–ENERGY THEOREM**

When a net external force does work W on an object, the kinetic energy of the object changes from its initial value of KE_0 to a final value of KE_f, the difference between the two values being equal to the work:

$$W = \text{KE}_f - \text{KE}_0 = \tfrac{1}{2}mv_f^2 - \tfrac{1}{2}mv_0^2 \tag{6.3}$$

---

* For extra emphasis, the final speed is now represented by the symbol v_f, rather than v.

The work–energy theorem may be derived for any direction of the force relative to the displacement, not just the situation in Figure 6.6. In fact, the force may even vary from point to point along a path that is curved rather than straight, and the theorem remains valid. According to the work–energy theorem, a moving object has kinetic energy, because work was done to accelerate the object from rest to a speed v_f.* Conversely, an object with kinetic energy can perform work, if it is allowed to push or pull on another object. Example 4 illustrates the work–energy theorem and considers a single force that does work to change the kinetic energy of a space probe.

EXAMPLE 4   A Space Probe

A space probe of mass $m = 5.00 \times 10^4 \, \text{kg}$ is traveling at a speed of $v_0 = 1.10 \times 10^4 \, \text{m/s}$ through deep space. No forces act on the probe except that generated by its own engine. The engine exerts a constant external force $\mathbf{F}$ of magnitude $4.00 \times 10^5 \, \text{N}$, directed parallel to the displacement $\mathbf{s}$ of magnitude $2.50 \times 10^6 \, \text{m}$. (Figure 6.7). Determine the final speed of the probe.

Reasoning Since the force $\mathbf{F}$ is the only force acting on the probe, it is the net external force, and the work it does causes the kinetic energy of the spacecraft to change. The work can be calculated directly from Equation 6.1 and is positive, because $\mathbf{F}$ points in the same direction as the displacement $\mathbf{s}$. According to the work–energy theorem $(W = KE_f - KE_0)$, a positive value for W means that the kinetic energy increases. We can use the theorem to find the final kinetic energy and, hence, the final speed. Since the kinetic energy increases, we expect the final speed to be greater than the initial speed.

Solution From the definition of work (Equation 6.1), we have

$$W = (F \cos \theta)s = [(4.00 \times 10^5 \, \text{N}) \cos 0°](2.50 \times 10^6 \, \text{m}) = 1.00 \times 10^{12} \, \text{J}$$

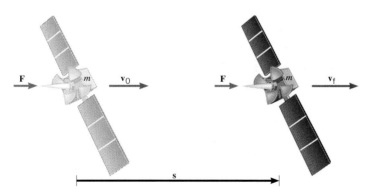

Figure 6.7 The engine of the space probe generates a single force $\mathbf{F}$ that points in the same direction as the displacement $\mathbf{s}$. The force performs positive work, causing the probe to gain kinetic energy.

---

* Strictly speaking, the work–energy theorem, as given by Equation 6.3, applies only to a single particle, which occupies a mathematical point in space. A macroscopic object, however, is a collection or system of particles and is spread out over a region of space. Therefore, when a force is applied to a macroscopic object, the point of application of the force may be anywhere on the object. To take into account this and other factors, a discussion of work and energy is required that is beyond the scope of this text. The interested reader may refer to A. B. Arons, *The Physics Teacher,* October 1989, p. 506.

where $\theta = 0°$, because the force and displacement are in the same direction (see the drawing). Since $W = KE_f - KE_0$ according to the work–energy theorem, the final kinetic energy of the probe is

$$KE_f = W + KE_0$$
$$= (1.00 \times 10^{12} \text{ J}) + \tfrac{1}{2}(5.00 \times 10^4 \text{ kg})(1.10 \times 10^4 \text{ m/s})^2 = 4.03 \times 10^{12} \text{ J}$$

The final kinetic energy is $KE_f = \tfrac{1}{2}mv_f^2$, so the final speed is

$$v_f = \sqrt{\frac{2(KE_f)}{m}} = \sqrt{\frac{2(4.03 \times 10^{12} \text{ J})}{(5.00 \times 10^4 \text{ kg})}} = \boxed{1.27 \times 10^4 \text{ m/s}}$$

As expected, the final speed is greater than the initial speed.

In Example 4 only the force of the engine does work. If several external forces act on an object, they must be added together vectorially to give the net force. The work done by the net force can then be related to the change in the object's kinetic energy by using the work–energy theorem, as in the next example.

EXAMPLE 5   Downhill Skiing

A 58-kg skier is coasting down a 25° slope, as Figure 6.8a shows. A kinetic frictional force of magnitude $f_k = 70$ N opposes her motion. Near the top of the slope, the skier's speed is $v_0 = 3.6$ m/s. Ignoring air resistance, determine the speed v_f at a point that is displaced 57 m downhill.

Reasoning As in Example 4, we will use the work–energy theorem to find the final speed. Note that the work done by the net external force is needed, not just the work done by any single force. The net force is the vector sum of the forces shown in the free-body diagram in Figure 6.8b. In this diagram, the normal force $\mathbf{F_N}$ is balanced by the component of the skier's weight ($mg \cos 25°$) perpendicular to the slope, because no acceleration occurs along that direction. Therefore, the net force points along the x axis.

Solution The net external force in Figure 6.8b points along the x axis and is

$$\Sigma F = mg \sin 25° - f_k = (58 \text{ kg})(9.80 \text{ m/s}^2) \sin 25° - 70 \text{ N} = +170 \text{ N}$$

The work done by the net force is

$$W = (\Sigma F \cos \theta)s = [(170 \text{ N}) \cos 0°](57 \text{ m}) = 9700 \text{ J} \qquad (6.1)$$

Figure 6.8 (a) A skier coasting downhill. (b) The free-body diagram for the skier.

(a)

(b) Free-body diagram for the skier

where $\theta = 0°$ because the net force and the displacement point in the same direction down the slope. From the work–energy theorem $(W = KE_f - KE_0)$, it follows that the final kinetic energy of the skier is

$$KE_f = W + KE_0$$
$$= 9700 \text{ J} + \tfrac{1}{2}(58 \text{ kg})(3.6 \text{ m/s})^2 = 10\ 100 \text{ J}$$

Since the final kinetic energy is $KE_f = \tfrac{1}{2}mv_f^2$, the final speed of the skier is

$$v_f = \sqrt{\frac{2(KE_f)}{m}} = \sqrt{\frac{2(10\ 100 \text{ J})}{58 \text{ kg}}} = \boxed{19 \text{ m/s}}$$

Example 5 emphasizes that *the work–energy theorem deals with the work done by the net external force. The work–energy theorem does not apply to the work done by an individual force,* unless that force happens to be the only one present, in which case it is the net force. If the work done by the net force is *positive,* as in Example 5, the kinetic energy of the object *increases.* If the work done is *negative,* the kinetic energy *decreases.* If the work is zero, the kinetic energy remains the same. Conceptual Example 6 explores these ideas further.

CONCEPTUAL EXAMPLE 6   **Work and Kinetic Energy**

Figure 6.9 illustrates a satellite moving about the earth in a circular orbit and in an elliptical orbit. The only external force that acts on the satellite is the gravitational force. For these two orbits, determine whether the kinetic energy of the satellite changes during the motion.

Reasoning and Solution For the circular orbit in Figure 6.9a the gravitational force **F** does no work on the satellite, since the force is perpendicular to the instantaneous displacement **s** at all times. The gravitational force is the only external force acting on the satellite, so it is the net force. Thus, the work done by the net force is zero, and according to the work–energy theorem (Equation 6.3), the kinetic energy of the satellite (and, hence, its speed) remains the same everywhere on the orbit.

In contrast, the gravitational force does do work on a satellite in an elliptical orbit. For example, as the satellite moves toward the earth in the top part of Figure 6.9b, there is a component of the gravitational force that points in the same direction as the displacement. Consequently, the gravitational force does positive work during this part of the orbit, and the kinetic energy of the satellite increases. When the satellite moves away from the earth, as in the lower part of Figure 6.9b, the gravitational force has a component that points opposite to the displacement. Now, the gravitational force does negative work, and the kinetic energy of the satellite decreases.

Can you identify the two places on an elliptical orbit where the gravitational force does no work?

Related Homework Material: *Problem 16*

(a)

(b)

Figure 6.9 (a) In a circular orbit, the gravitational force **F** is always perpendicular to the displacement **s** of the satellite and does no work. (b) In an elliptical orbit, there can be a component of the force along the displacement, and, consequently, work can be done.

6.3 GRAVITATIONAL POTENTIAL ENERGY

Work Done by the Force of Gravity

The gravitational force is a well-known force that can do positive or negative work, and Figure 6.10 helps to show how the work can be determined. This drawing depicts a basketball of mass m moving vertically downward, the force of gravity $m\mathbf{g}$ being the only force acting on the ball. The initial height of the ball is h_0, and the final height is h_f, both distances measured from the earth's surface. The displacement $\mathbf{s}$ is downward and has a magnitude of $s = h_0 - h_f$. To calculate the work W_{gravity} done on the ball by the force of gravity, we use $W = (F \cos \theta)s$ with $F = mg$ and $\theta = 0°$, since the force and displacement are in the same direction:

$$W_{\text{gravity}} = (mg \cos 0°)(h_0 - h_f) = mg(h_0 - h_f) \tag{6.4}$$

Equation 6.4 is valid for *any path* taken between the initial and final heights, and not just for the straight-down path shown in Figure 6.10. For example, the same expression can be derived for both paths shown in Figure 6.11. Thus, only the *difference in vertical distances* $(h_0 - h_f)$ need be considered when calculating the work done by gravity. Since the difference in the vertical distances is the same for each path in the drawing, the work done by gravity is the same in each case. We are assuming here that the difference in heights is small compared to the radius of the earth, so that the magnitude g of the acceleration due to gravity is the same at every height. Moreover, for positions close to the earth's surface, we can use the value of $g = 9.80 \text{ m/s}^2$.

Since only the difference between h_0 and h_f appears in Equation 6.4, the vertical distances themselves need not be measured from the earth. For instance, they could be measured relative to a zero level that is one meter above the ground, and $h_0 - h_f$ would still have the same value. Example 7 illustrates how the work done by gravity is used in conjunction with the work–energy theorem.

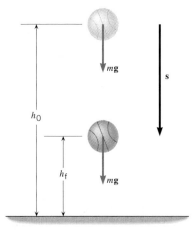

Figure 6.10 Gravity exerts a force $m\mathbf{g}$ on the basketball. Work is done by the gravitational force as the basketball falls from a height of h_0 to a height of h_f.

Figure 6.11 An object can move along different paths in going from an initial height of h_0 to a final height of h_f. In each case, the work done by the gravitational force is the same [$W_{\text{gravity}} = mg(h_0 - h_f)$], since the change in vertical distance $(h_0 - h_f)$ is the same.

(a)

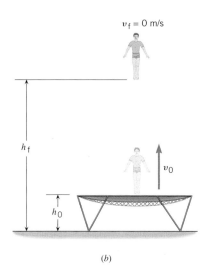

$v_f = 0$ m/s

h_f

v_0

h_0

(b)

Figure 6.12 (*a*) A gymnast bounces on a trampoline. (*b*) The gymnast moves upward with an initial speed v_0 and reaches maximum height with a final speed of zero.

EXAMPLE 7   A Gymnast on a Trampoline

A gymnast springs vertically upward from a trampoline as in Figure 6.12*a*. The gymnast leaves the trampoline at a height of 1.20 m and reaches a maximum height of 4.80 m before falling back down. All heights are measured with respect to the ground. Ignoring air resistance, determine the initial speed v_0 with which the gymnast leaves the trampoline.

Reasoning We can find the initial speed of the gymnast (mass = m) by using the work–energy theorem, provided the work done by the net external force can be determined. Since only the gravitational force acts on the gymnast in the air, it is the net force, and we can evaluate the work by using the relation $W_{\text{gravity}} = mg(h_0 - h_f)$.

Solution Figure 6.12*b* shows the gymnast moving upward. The initial and final heights are $h_0 = 1.20$ m and $h_f = 4.80$ m, respectively. The initial speed is v_0 and the final speed is $v_f = 0$ m/s, since the gymnast comes to a momentary halt at the highest point. Since $v_f = 0$ m/s, the final kinetic energy is $KE_f = 0$ J, and the work–energy theorem becomes $W = KE_f - KE_0 = -KE_0$.   The work W is that due to gravity, so this theorem reduces to $W_{\text{gravity}} = mg(h_0 - h_f) = -\frac{1}{2}mv_0^2$.   Solving for v_0 gives

$$v_0 = \sqrt{-2g(h_0 - h_f)} = \sqrt{-2(9.80 \text{ m/s}^2)(1.20 \text{ m} - 4.80 \text{ m})} = \boxed{8.40 \text{ m/s}}$$

Gravitational Potential Energy

We have seen that an object in motion has kinetic energy. Energy also occurs in other forms. For example, an object may possess energy by virtue of its position relative to the earth; such an object is said to have gravitational potential energy. A pile driver, for instance, is used by construction workers to pound "piles" or structural support beams into the ground. The pile driver contains a massive hammer that is raised to a height h and then dropped (see Figure 6.13). As a result, the hammer has the potential to do the work of driving the pile into the ground. The greater the height of the hammer, the greater is the potential for doing work, and the greater is the gravitational potential energy.

Hammer

mg

h

Pile

Figure 6.13 In a pile driver, the gravitational potential energy of the hammer relative to the ground is PE = mgh.

Now, let's obtain an expression for the gravitational potential energy. Our starting point is Equation 6.4 for the work done by the gravitational force as an object moves from an initial height h_0 to a final height h_f:

$$W_{\text{gravity}} = \underbrace{mgh_0}_{\substack{\text{Initial} \\ \text{gravitational} \\ \text{potential energy,} \\ \text{PE}_0}} - \underbrace{mgh_f}_{\substack{\text{Final} \\ \text{gravitational} \\ \text{potential energy} \\ \text{PE}_f}} \qquad (6.4)$$

This equation indicates that the work done by the gravitational force is equal to the difference between the initial and final values of the quantity mgh. The value of mgh is larger when the height is larger and smaller when the height is smaller. We are led, then, to identify the quantity mgh as the **gravitational potential energy.** The concept of potential energy is associated only with a type of force known as a "conservative" force. We will discuss this type of force in Section 6.4.

● DEFINITION OF GRAVITATIONAL POTENTIAL ENERGY

The gravitational potential energy PE is the energy that an object of mass m has by virtue of its position relative to the surface of the earth. That position is measured by the height h of the object relative to an arbitrary zero level:

$$\text{PE} = mgh \qquad (6.5)$$

SI Unit of Gravitational Potential Energy: joule (J)

   Gravitational potential energy, like work and kinetic energy, is a scalar quantity and has the same SI unit as they do, the joule. It is the *difference* between two potential energies that is related by Equation 6.4 to the work done by the force of gravity. Therefore, the zero level for the heights can be taken anywhere, as long as both h_0 and h_f are measured relative to the same zero level. The gravitational potential energy depends on both the object and the earth (m and g, respectively), as well as the height h. Therefore, the gravitational potential energy belongs to the object and the earth as a system, although one often speaks of the object alone as possessing the gravitational potential energy.

6.4 CONSERVATIVE FORCES, NONCONSERVATIVE FORCES, AND THE WORK–ENERGY THEOREM

The gravitational force has an interesting property that when an object is moved from one place to another, the work done by the gravitational force does not depend on the choice of path. In Figure 6.11, for instance, an object moves from an initial height h_0 to a final height h_f along two different paths. As Section 6.3 discusses, the work done by gravity depends only on the initial and final heights, and not on the path between these heights. For this reason, the gravitational force is called a **conservative force,** according to version 1 of the following definition:

● **DEFINITION OF A CONSERVATIVE FORCE**

Version 1 A force is conservative when the work it does on a moving object is independent of the path between the object's initial and final positions.

Version 2 A force is conservative when it does no net work on an object moving around a closed path, starting and finishing at the same point.

Figure 6.14 A roller coaster track is an example of a closed path.

The gravitational force is our first example of a conservative force. Later, we will encounter other examples, such as the elastic force of a spring and the electrical force of electrically charged particles. With each conservative force we will associate a potential energy, as we have already done in the gravitational case (see Equation 6.5). For other conservative forces, however, the algebraic form of the potential energy will differ from that in Equation 6.5.

Version 1 of the definition given above is not the only way to describe a conservative force. Figure 6.14 helps us to illustrate another way. The picture shows a roller coaster car racing through dips and double dips, ultimately returning to its starting point. This kind of path, which begins and ends at the same place, is called a *closed path*. Gravity provides the only force that does work on the car, assuming that there is no friction or air resistance. Of course, the track exerts a normal force, but this force is always directed perpendicular to the motion and does no work. On the downward parts of the trip, the gravitational force does positive work, increasing the car's kinetic energy. Conversely, on the upward parts of the motion, the gravitational force does negative work, decreasing the car's kinetic energy. Over the entire trip, the gravitational force does as much positive work as negative work, so the net work is zero, and the car returns to its starting point with the same kinetic energy it had at the start. Therefore, consistent with version 2 of the definition of a conservative force, $W_{\text{gravity}} = 0$ J for a closed path.

Not all forces are conservative forces. A force is nonconservative if the work it does on an object moving between two points depends on the path of the motion between the points. The kinetic frictional force is one example of a nonconservative force. When an object slides over a surface, the kinetic frictional force points opposite to the sliding motion and does negative work. Between any two points, greater amounts of work are done over longer paths between the points, so that the work depends on the choice of path. Thus, the kinetic frictional force is nonconservative. Air resistance is another nonconservative force. The concept of potential energy is not defined for a nonconservative force.

For a closed path, the total work done by a nonconservative force is not zero as it is for a conservative force. In Figure 6.14, for instance, a frictional force would oppose the motion and slow down the car. Unlike gravity, friction would do negative work on the car throughout the entire trip, on *both* the up and down parts of the motion. Assuming that the car makes it back to the starting point, the car would have *less* kinetic energy than it had originally. Table 6.2 gives some examples of conservative and nonconservative forces.

In normal situations, conservative forces (such as gravity) and nonconservative forces (such as friction and air resistance) act simultaneously on an object. Therefore, we write the work W done by the net external force as $W = W_{\text{c}} + W_{\text{nc}}$, where W_{c} is the work done by the conservative forces and W_{nc} is the work done by the nonconservative forces. According to the work–energy theorem, the work done by the net external force is equal to the change in the object's kinetic energy, or

TABLE 6.2 Some Conservative and Nonconservative Forces

Conservative Forces
Gravitational force (Ch. 4)
Elastic spring force (Ch. 10)
Electric force (Ch. 18, 19)

Nonconservative Forces
Static and kinetic frictional forces
Air resistance
Tension
Normal force
Propulsion force of a rocket

$W_c + W_{nc} = \frac{1}{2}mv_f^2 - \frac{1}{2}mv_0^2$. If the only conservative force acting is the gravitational force, then $W_c = W_{gravity} = mg(h_0 - h_f)$, and the work–energy theorem becomes

$$mg(h_0 - h_f) + W_{nc} = \frac{1}{2}mv_f^2 - \frac{1}{2}mv_0^2$$

The work done by the gravitational force can be moved to the right side of this equation, with the result that

$$W_{nc} = (\tfrac{1}{2}mv_f^2 - \tfrac{1}{2}mv_0^2) + (mgh_f - mgh_0) \qquad (6.6)$$

In terms of kinetic and potential energies, we find that

$$\underbrace{W_{nc}}_{\substack{\text{Net work done by} \\ \text{nonconservative} \\ \text{forces}}} = \underbrace{(KE_f - KE_0)}_{\substack{\text{Change in} \\ \text{kinetic energy}}} + \underbrace{(PE_f - PE_0)}_{\substack{\text{Change in} \\ \text{gravitational} \\ \text{potential energy}}} \qquad (6.7a)$$

● **PROBLEM SOLVING INSIGHT**
The change in both the kinetic and the potential energy is always the final value minus the initial value: $\Delta KE = KE_f - KE_0$ and $\Delta PE = PE_f - PE_0$.

Equation 6.7a states that the net work W_{nc} done by all the external nonconservative forces equals the change in the object's kinetic energy plus the change in its gravitational potential energy. It is customary to use the delta symbol (Δ) to denote such changes; thus, $\Delta KE = (KE_f - KE_0)$ and $\Delta PE = (PE_f - PE_0)$. With the delta notation, the work–energy theorem takes the form

$$W_{nc} = \Delta KE + \Delta PE \qquad (6.7b)$$

In the next two sections, we will show why the form of the work–energy theorem expressed by Equations 6.7a and 6.7b is useful.

6.5 THE CONSERVATION OF MECHANICAL ENERGY

◉ *Concepts at a Glance*   The concept of work and the work–energy theorem have led us to the conclusion that an object can possess two kinds of energy: kinetic energy, KE, and gravitational potential energy, PE. The sum of these two energies is called the *total mechanical energy* E, so that $E = KE + PE$. The Concepts-at-a-Glance chart in Figure 6.15 illustrates this summation. The concept of total mechanical energy will be extremely useful in describing the motion of objects in this and other chapters. Later on, in a number of places, we will be updating this chart to include other forms of energy. ◉

By simply rearranging the terms on the right side of Equation 6.7a, the work–energy theorem can be expressed in terms of the total mechanical energy:

$$W_{nc} = (KE_f - KE_0) + (PE_f - PE_0) \qquad (6.7a)$$
$$= \underbrace{(KE_f + PE_f)}_{E_f} - \underbrace{(KE_0 + PE_0)}_{E_0}$$

or

$$W_{nc} = E_f - E_0 \qquad (6.8)$$

Remember, Equation 6.8 is just another form of the work–energy theorem. It states that W_{nc}, the net work done by external nonconservative forces, changes the total mechanical energy from an initial value of E_0 to a final value of E_f.

The conciseness of the work–energy theorem in the form $W_{nc} = E_f - E_0$ allows an important basic principle of physics to stand out. This principle is known as the conservation of mechanical energy. Now, suppose that the net work W_{nc}

Concepts at a Glance

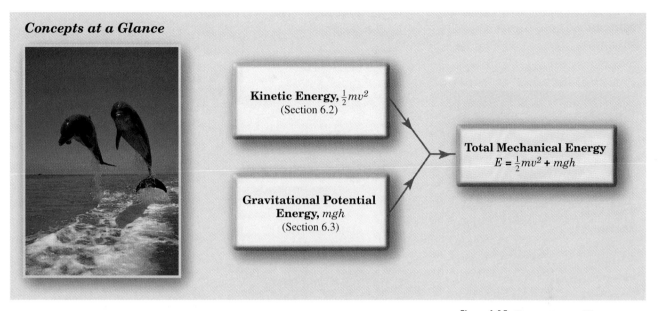

Kinetic Energy, $\frac{1}{2}mv^2$
(Section 6.2)

Gravitational Potential
Energy, mgh
(Section 6.3)

Total Mechanical Energy
$E = \frac{1}{2}mv^2 + mgh$

Figure 6.15 *Concepts at a Glance*
The total mechanical energy E is
formed by combining the concepts
of kinetic energy and gravitational
potential energy. These dolphins have
both types of energy as they sail
through the air.

done by external nonconservative forces is zero, so $W_{nc} = 0$ J. Then, Equation 6.8
reduces to

$$E_f = E_0 \tag{6.9a}$$

$$\underbrace{\tfrac{1}{2}mv_f^2 + mgh_f}_{E_f} = \underbrace{\tfrac{1}{2}mv_0^2 + mgh_0}_{E_0} \tag{6.9b}$$

In this last equation we have written E_f and E_0 explicitly as a sum of kinetic and
gravitational potential energies. Equation 6.9a indicates that the final mechanical
energy is equal to the initial mechanical energy. Consequently, the total mechanical
energy *remains constant all along the path between the initial and final points*,
never varying from the initial value of E_0. A quantity that remains constant throughout the motion is said to be "conserved." The fact that the total mechanical energy
is conserved when $W_{nc} = 0$ J is called the ***principle of conservation of mechanical
energy.***

● THE PRINCIPLE OF CONSERVATION OF MECHANICAL ENERGY

The total mechanical energy ($E = $ KE $+$ PE) of an object remains constant as
the object moves, provided that the net work done by external nonconservative
forces is zero, $W_{nc} = 0$ J.

⬤ *Concepts at a Glance*   The Concepts-at-a-Glance chart in Figure 6.16 outlines how the conservation of mechanical energy arises naturally from concepts that
we have encountered earlier. You might recognize that this chart is an expanded version of the one in Figure 6.5. In the middle are the three concepts that we used to
develop the work–energy theorem: Newton's second law, work, and the equations
of kinematics. On the right of the chart is the work–energy theorem, stated in the

Concepts at a Glance

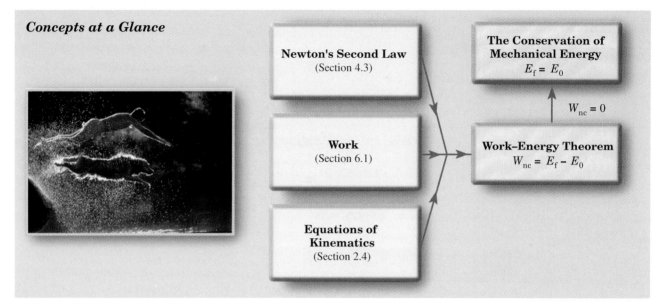

Figure 6.16 *Concepts at a Glance*
The work–energy theorem leads to the principle of conservation of mechanical energy under circumstances in which $W_{nc} = 0$ J, where W_{nc} is the net work done by external nonconservative forces. If air resistance, a nonconservative force, can be ignored, the total mechanical energy of this man and his best friend is conserved as they fall toward the water.

form $W_{nc} = E_f - E_0$. If, as the chart shows, $W_{nc} = 0$, then the final and initial total mechanical energies are equal. In other words, $E_f = E_0$, which is the mathematical statement of the conservation of mechanical energy.

The principle of conservation of mechanical energy offers keen insight into the way in which the physical universe operates. While the sum of the kinetic and potential energies at any point is conserved, the two forms may be interconverted or transformed into one another. Kinetic energy of motion is converted into potential energy of position, for instance, when a moving object coasts up a hill. Conversely, potential energy is converted into kinetic energy when an object is allowed to fall. Figure 6.17 illustrates such transformations of energy for a bobsled run, assuming that nonconservative forces, such as friction and wind resistance, can be ignored. The normal force, being directed perpendicular to the path, does no work. Only the force of gravity does work, so the total mechanical energy E remains constant at all points along the run. The conservation principle is well known for the ease with which it can be applied, as in the following examples.

Figure 6.17 If friction and wind resistance are ignored, a bobsled run illustrates how kinetic and potential energy can be interconverted, while the total mechanical energy remains constant. The total mechanical energy is 600 000 J, being all potential energy at the top and all kinetic energy at the bottom.

KE	PE	E = KE + PE	
0 J	600 000 J	600 000 J	$\mathbf{v}_0 = 0$ m/s
200 000 J	400 000 J	600 000 J	
400 000 J	200 000 J	600 000 J	
600 000 J	0 J	600 000 J	

EXAMPLE 8   A Daredevil Motorcyclist

A motorcyclist is trying to leap across the canyon shown in Figure 6.18 by driving horizontally off the cliff at a speed of 38.0 m/s. Ignoring air resistance, find the speed with which the cycle strikes the ground on the other side.

Reasoning Once the cycle leaves the cliff, no forces other than gravity act on the cycle, since air resistance is being ignored. Thus, the work done by external nonconservative forces is zero, $W_{nc} = 0$ J. Accordingly, the principle of conservation of mechanical energy holds, so the total mechanical energy is the same at the final and initial positions of the motorcycle. We will use this important observation to determine the final speed of the cyclist.

Figure 6.18 A daredevil jumping a canyon.

Solution The principle of conservation of mechanical energy is written as

$$\underbrace{\tfrac{1}{2}mv_f^2 + mgh_f}_{E_f} = \underbrace{\tfrac{1}{2}mv_0^2 + mgh_0}_{E_0} \tag{6.9b}$$

The mass m of the rider and cycle can be eliminated algebraically from this equation, since m appears as a factor in every term. Solving for v_f gives

$$v_f = \sqrt{v_0^2 + 2g(h_0 - h_f)}$$

$$v_f = \sqrt{(38.0 \text{ m/s})^2 + 2(9.80 \text{ m/s}^2)(70.0 \text{ m} - 35.0 \text{ m})} = \boxed{46.2 \text{ m/s}}$$

● **PROBLEM SOLVING INSIGHT**
Be on the alert for factors, such as the mass m here in Example 8, that sometimes can be eliminated algebraically when using the conservation of mechanical energy.

Examples 9 and 10 emphasize that the principle of conservation of mechanical energy can be applied even when forces act perpendicular to the path of a moving object.

CONCEPTUAL EXAMPLE 9   The Favorite Swimming Hole

A rope is tied to a tree limb and used by a swimmer to swing into the water below. The person starts from rest with the rope held in the horizontal position, as in Figure 6.19, swings downward, and then lets go of the rope. Three forces act on him: his weight, the tension in the rope, and the force due to air resistance. His initial height h_0 and final height h_f are known. Considering the nature of these forces, conservative versus nonconservative, can we use the principle of conservation of mechanical energy to find his speed v_f at the point where he lets go of the rope?

Reasoning and Solution The principle of conservation of mechanical energy can be used only if the net work W_{nc} done by nonconservative forces is zero, $W_{nc} = 0$ J.

● **PROBLEM SOLVING INSIGHT**
When nonconservative forces are perpendicular to the motion, we can still use the principle of conservation of mechanical energy, because such "perpendicular" forces do no work.

Figure 6.19 During the downward swing, the tension **T** in the rope acts perpendicular to the circular arc and, hence, does no work on the person.

INTERACTIVE LEARNINGWARE #13

- START
- Visualization
- Reasoning
- Knowns and Unknowns
- Modeling the Problem
- Numerical Solution
- FINISH

A person is standing on the edge of a cliff that is 8.10 m above a lake. Using a rope tied to a tree limb, he swings downward, lets go of the rope, and subsequently splashes into the water. He can let go of the rope either as it is swinging downward or a little later as it is swinging upward. Ignore air resistance and, for each case, determine the speed of the person just before entering the water.

This problem has two features that make it interesting: (1) It demonstrates that the conservation of mechanical energy can be used to find the speed of the person just before entering the water, even though a nonconservative force (the tension in the rope) is present. (2) It also shows that if the total mechanical energy is conserved, it is the same everywhere along a path. It does not matter which particular path the person follows into the water.

Go to our Web site at

http://www.wiley.com/college/cutnell

The tension and the force due to air resistance are nonconservative forces (see Table 6.2), so we need to inquire if the net work done by these forces is zero. The tension **T** is always perpendicular to the circular path of the motion, as shown in the drawing. Thus, the angle θ between the tension and the displacement is 90°. According to Equation 6.1, the work depends on the cosine of this angle, or cos 90°, which is zero. Thus, the work done by the tension is zero. On the other hand, the force due to air resistance is directed opposite to the motion of the swinging person. The angle θ between this force and the displacement is 180°. The work done by the air resistance is not zero, because the cosine of 180° is not zero. Since the net work done by the two nonconservative forces is not zero, $W_{net} \neq 0$ J, we cannot use the principle of conservation of mechanical energy to determine the final speed. On the other hand, if the force due to air resistance has a magnitude that is very small, then the work done by this force is negligible. In this case, the net work done by the nonconservative forces is effectively zero, and the principle of conservation of mechanical energy can be used to determine the final speed of the swimmer.

Related Homework Material: *Problems 40 and 43*

The next example illustrates how the conservation of mechanical energy is applied to the breathtaking drop of a roller coaster.

EXAMPLE 10   The Magnum XL-200

THE PHYSICS OF...

a giant roller coaster.

One of the fastest roller coasters in the world is the Magnum XL-200 at Cedar Point Park in Sandusky, Ohio (Figure 6.20). The ride includes a vertical drop of 59.4 m. Assume that the coaster has a speed of nearly zero as it crests the top of the hill. Neglect friction and find the speed of the riders at the bottom of the hill.

Reasoning Since we are neglecting friction, we may set the work done by the frictional force equal to zero. A normal force from the seat acts on each rider, but this force is perpendicular to the motion, so it does not do any work. Thus, the work done by external nonconservative forces is zero, $W_{nc} = 0$ J, and we may use the principle of conservation of mechanical energy to find the speed of the riders at the bottom of the hill.

Solution The principle of conservation of mechanical energy states that

$$\underbrace{\tfrac{1}{2}mv_f^2 + mgh_f}_{E_f} = \underbrace{\tfrac{1}{2}mv_0^2 + mgh_0}_{E_0} \qquad (6.9b)$$

The mass m of the rider appears as a factor in every term in this equation and can be eliminated algebraically. Solving for the final speed gives

$$v_f = \sqrt{v_0^2 + 2g(h_0 - h_f)}$$

$$v_f = \sqrt{(0 \text{ m/s})^2 + 2(9.80 \text{ m/s}^2)(59.4 \text{ m})} = \boxed{34.1 \text{ m/s (about 76 mi/h)}}$$

where the initial speed of the coaster is assumed to be zero, $v_0 = 0$ m/s, and the vertical height of the hill is $h_0 - h_f = 59.4$ m.

When applying the principle of conservation of mechanical energy to solving problems, we have been using the following reasoning strategy:

REASONING STRATEGY

Applying the Principle of Conservation of Mechanical Energy

1. Identify the external conservative and nonconservative forces that act on the object. For this principle to apply, nonconservative forces must do no work, $W_{nc} = 0$ J. A nonconservative force that is perpendicular to the displacement of the object does no work, for example.

2. Choose the location where the gravitational potential energy is taken to be zero. This location is arbitrary but must not be changed during the course of solving a problem.

3. Set the final total mechanical energy of the object equal to the initial total mechanical energy, as in Equations 6.9a and 6.9b. The total mechanical energy is the sum of the kinetic and potential energies.

6.6 NONCONSERVATIVE FORCES AND THE WORK–ENERGY THEOREM

Figure 6.20 The Magnum XL-200 roller coaster; it is one of the fastest roller coasters in the world and includes a vertical drop of 59.4 m.

Most moving objects experience nonconservative forces, such as friction, air resistance, and propulsive forces, and the work W_{nc} done by the net external nonconservative force is not zero. In these situations, the difference between the final and initial total mechanical energies is equal to W_{nc}, according to $W_{nc} = E_f - E_0$ (Equation 6.8). Consequently, the total mechanical energy is not conserved. The next two examples illustrate how Equation 6.8 is used when nonconservative forces are present and do work.

EXAMPLE 11 The Magnum XL-200, Revisited

In Example 10, we ignored nonconservative forces, such as friction. In reality, however, such forces are present when the roller coaster descends the hill. The actual speed of the riders at the bottom is 32.2 m/s, which is less than that determined in Example 10. Assuming again that the coaster has a speed of nearly zero at the top of the hill, find the work done by nonconservative forces on a 55.0-kg rider during the descent from a height h_0 to a height h_f, where $h_0 - h_f = 59.4$ m.

Reasoning Since the speed at the top of the hill is nearly zero and the final speed and the vertical drop are given, we can determine the initial and final total mechanical energies of the rider. The work–energy theorem, $W_{nc} = E_f - E_0$, can then be used to determine the work W_{nc} done by the nonconservative forces.

Solution The work–energy theorem is

$$W_{nc} = \underbrace{(\tfrac{1}{2}mv_f^2 + mgh_f)}_{E_f} - \underbrace{(\tfrac{1}{2}mv_0^2 + mgh_0)}_{E_0} \qquad (6.8)$$

Rearranging the terms on the right side of this equation gives

$$W_{nc} = \tfrac{1}{2}m(v_f^2 - v_0^2) - mg(h_0 - h_f)$$

$$W_{nc} = \tfrac{1}{2}(55.0 \text{ kg})[(32.2 \text{ m/s})^2 - (0 \text{ m/s})^2] - (55.0 \text{ kg})(9.80 \text{ m/s}^2)(59.4 \text{ m}) = \boxed{-3500 \text{ J}}$$

● **PROBLEM SOLVING INSIGHT**
As illustrated here and in Example 3, a nonconservative force such as friction can do negative or positive work. It does negative work when it has a component opposite to the displacement and slows down the object. It does positive work when it has a component in the direction of the displacement and speeds up the object.

Figure 6.21 A fireworks rocket, moving along an erratic flight path, reaches a point P that is 29 m above the launch point.

EXAMPLE 12 Fireworks

A 0.20-kg rocket in a fireworks display is launched from rest and follows an erratic flight path to reach the point P, as Figure 6.21 shows. Point P is 29 m above the starting point. In the process, 425 J of work is done on the rocket by the nonconservative force generated by the burning propellant. Ignoring air resistance and the mass lost due to the burning propellant, find the speed v_f of the rocket at the point P.

Reasoning The only nonconservative force acting on the rocket is the force generated by the burning propellant, and the work done by this force is $W_{nc} = 425$ J. Because work is done by a nonconservative force, we use the work–energy theorem in the form $W_{nc} = E_f - E_0$ to find the final speed v_f of the rocket.

Solution From the work–energy theorem we have

$$W_{nc} = (\tfrac{1}{2}mv_f^2 + mgh_f) - (\tfrac{1}{2}mv_0^2 + mgh_0) \tag{6.8}$$

Solving this expression for the final speed of the rocket, we get

$$v_f = \sqrt{\frac{2[W_{nc} + \tfrac{1}{2}mv_0^2 - mg(h_f - h_0)]}{m}}$$

$$v_f = \sqrt{\frac{2[425 \text{ J} + \tfrac{1}{2}(0.20 \text{ kg})(0 \text{ m/s})^2 - (0.20 \text{ kg})(9.80 \text{ m/s}^2)(29 \text{ m})]}{0.20 \text{ kg}}} = \boxed{61 \text{ m/s}}$$

6.7 POWER

In many situations, the time it takes to do work is just as important as the amount of work that is done. Consider two automobiles that are identical in all respects (e.g., same mass), except that one has a "souped-up" engine. The car with the "souped-up" engine can go from 0 to 27 m/s (60 mph) in 4 seconds, while the other car requires 8 seconds to achieve the same speed. Each engine does work in accelerating its car, but one does it more quickly. Where cars are concerned, we associate the quicker performance with an engine that has a larger horsepower rating. A large horsepower rating means that the engine can do a large amount of work in a short time. In physics, the horsepower rating is just one way to measure an engine's ability to generate power. The idea of *power* incorporates both the concepts of work and time, for power is work done per unit time.

● **DEFINITION OF AVERAGE POWER**

Average power $\overline{P}$ is the average rate at which work W is done, and it is obtained by dividing W by the time t required to perform the work:

$$\overline{P} = \frac{\text{Work}}{\text{Time}} = \frac{W}{t} \tag{6.10a}$$

SI Unit of Power: joule/s = watt (W)

The definition of average power presented in Equation 6.10a involves the work that is done by the net force. However, the work–energy theorem relates this work to the change in the energy of the object (see, for example, Equations 6.3 and 6.8). Therefore, we can also define average power as the rate at which the energy is

TABLE 6.3   Units of Measurement for Power

System	Work	÷	Time	=	Power
SI	joule (J)		second (s)		watt (W)
CGS	erg		second (s)		erg per second (erg/s)
BE	foot · pound (ft · lb)		second (s)		foot · pound per second (ft · lb/s)

changing, or as the change in energy divided by the time during which the change occurs:

$$\bar{P} = \frac{\text{Change in energy}}{\text{Time}} \tag{6.10b}$$

Since work, energy, and time are scalar quantities, power is also a scalar quantity. The unit in which power is expressed is that of work divided by time, or a joule per second in SI units. One joule per second is called a watt (W), in honor of James Watt (1736–1819), the developer of the steam engine. The unit of power in the BE system is the foot-pound per second (ft · lb/s), although the familiar horsepower (hp) unit is used frequently for specifying the power generated by electric motors and internal combustion engines:

$$1 \text{ horsepower} = 550 \text{ foot} \cdot \text{pounds/second} = 745.7 \text{ watts}$$

Table 6.3 summarizes the units for power in the various systems of measurement.

An alternative expression for power can be obtained from Equation 6.1, which indicates that the work W done when a constant net force of magnitude F points in the same direction as the displacement is $W = (F \cos 0°)s = Fs$. Dividing both sides of this equation by the time t it takes for the force to move the object through the distance s, we obtain

$$\frac{W}{t} = \frac{Fs}{t}$$

But W/t is the average power $\bar{P}$, and s/t is the average speed $\bar{v}$, so that

$$\bar{P} = F\bar{v} \tag{6.11}$$

The next example illustrates the use of Equation 6.11.

EXAMPLE 13   **The Power to Accelerate a Car**

A 1.10×10^3-kg car, starting from rest, accelerates for 5.00 s. The magnitude of the acceleration is $a = 4.60 \text{ m/s}^2$. Determine the average power generated by the net force that accelerates the vehicle.

Reasoning We can find the average power by using the relation $\bar{P} = F\bar{v}$, provided the magnitude F of the net force acting on the car and the average speed $\bar{v}$ of the car can be determined. The net force can be obtained from Newton's second law, and the average speed can be calculated from the equations of kinematics.

Solution According to Newton's second law, the magnitude of the net force is

$$F = ma = (1.10 \times 10^3 \text{ kg})(4.60 \text{ m/s}^2) = 5060 \text{ N}$$

Since the car starts from rest ($v_0 = 0$ m/s) and has a constant acceleration, the average speed $\bar{v}$ of the car is one-half of its final speed $\bar{v}_f$:

$$\bar{v} = \tfrac{1}{2}(v_0 + v_f) = \tfrac{1}{2}v_f \qquad (2.6)$$

Because the initial speed of the car is zero, the final speed of the car after 5.00 s is the product of its acceleration and time:

$$v_f = v_0 + at = (0 \text{ m/s}) + (4.60 \text{ m/s}^2)(5.00 \text{ s}) = 23.0 \text{ m/s} \qquad (2.4)$$

Thus, the average speed is $\bar{v} = 11.5$ m/s, and the average power is

$$\bar{P} = F\bar{v} = (5060 \text{ N})(11.5 \text{ m/s}) = \boxed{5.82 \times 10^4 \text{ W (78.0 hp)}}$$

6.8 OTHER FORMS OF ENERGY AND THE CONSERVATION OF ENERGY

Up to now, we have considered only two types of energy, kinetic energy and gravitational potential energy. There are many other types, however. Electrical energy is used to run electrical appliances. Energy in the form of heat is utilized in cooking food. Moreover, the work done by the kinetic frictional force often appears as heat, as you can experience by rubbing your hands back and forth. Chemical energy is the energy stored in the molecules of fuels and food. When gasoline is burned, some of the stored chemical energy is released and does the work of moving cars, airplanes, and boats. The chemical energy stored in food provides the energy needed for metabolic processes.

One of the most controversial forms of energy is nuclear energy. The research of many scientists, most notably Albert Einstein, led to the discovery that mass itself is one manifestation of energy. Einstein's famous equation, $E_0 = mc^2$, describes how mass m and energy E_0 are related, where c is the speed of light in a vacuum and has a value of 3.00×10^8 m/s. Because the speed of light is so large, this equation implies that very small masses are equivalent to large amounts of energy. The relationship between mass and energy will be discussed further in Chapter 28.

We have seen that kinetic energy can be converted into gravitational potential energy and vice versa. In general, energy of all types can be converted from one form to another. Part of the chemical energy stored in food is transformed into the kinetic energy of walking and into the thermal energy needed to keep our bodies at a temperature near 98.6 °F. Similarly, in a moving car the chemical energy of gasoline is converted into kinetic energy, as well as electrical energy (to operate the radio, headlights, and air conditioner), and heat (to warm the car during the winter). Whenever energy is transformed from one form to another, it is found that no energy is gained or lost in the process; the total of all the energies before the process is equal to the total of the energies after the process. This observation leads to the following important principle:

● THE PRINCIPLE OF CONSERVATION OF ENERGY

Energy can neither be created nor destroyed, but can only be converted from one form to another.

Learning how to convert energy from one form to another more efficiently is one of the main goals of modern science and technology.

6.9 WORK DONE BY A VARIABLE FORCE

The work W done by a constant force (constant in both magnitude and direction) is given by Equation 6.1 as $W = (F \cos \theta)s$. Quite often, situations arise in which the force is not constant but changes with the displacement of the object. For instance, Figure 6.22a shows an archer using a high-tech compound bow. This type of bow consists of a series of pulleys and strings that produce a force-versus-displacement graph like that in Figure 6.22b. One of the key features of the compound bow is that the force rises to a maximum as the string is drawn back, and then falls to 60% of this maximum value when the string is fully drawn. The reduced force at $s = 0.500$ m makes it much easier for the archer to hold the fully drawn bow while aiming the arrow.

When the force varies with the displacement, as in Figure 6.22b, we cannot use the relation $W = (F \cos \theta)s$, Equation 6.1, to find the work, because this equation is valid only when the force is constant. However, we can use a graphical method. In this method we divide the total displacement into very small segments, Δs_1, Δs_2, and so on (see Figure 6.23a). For each segment, the *average value* of the force component is indicated by a short horizontal line. For example, the short horizontal line for segment Δs_1 is labeled $(F \cos \theta)_1$ in Figure 6.23a. We can then use this average value as the constant force component in Equation 6.1 and determine an approximate value for the work ΔW_1 done during the first segment: $\Delta W_1 = (F \cos \theta)_1 \Delta s_1$. But this work is just the area of the colored rectangle in the drawing. The word "area" here refers to the area of a rectangle that has a width of Δs_1 and a height of $(F \cos \theta)_1$; it does not mean an area in square meters, such as the area of a parcel of land. In a like manner, we can calculate an approximate value for the work for each segment. Then we add the results for the segments to get, approximately, the work W done by the variable force:

$$W \approx (F \cos \theta)_1 \Delta s_1 + (F \cos \theta)_2 \Delta s_2 + \cdots$$

The symbol "$\approx$" means "approximately equal to." The right side of this equation is the sum of all the rectangular areas in Figure 6.23a and is an approximate value for the area shaded in color under the graph in Figure 6.23b. If the rectangles are made narrower and narrower by decreasing each Δs, the right side of this equation eventually becomes equal to the area under the graph. Thus, we define the work done by a variable force as follows: ***The work done by a variable force in moving an object is equal to the area under the graph of $F \cos \theta$ versus s.*** Example 14 illustrates how to use this graphical method to determine the approximate work done when a high-tech compound bow is drawn.

The Physics of section

THE PHYSICS OF...

the compound bow.

(a)

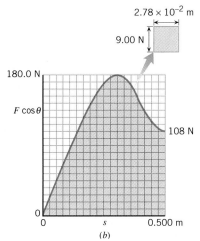

(b)

Figure 6.22 (a) A compound bow. (b) A plot of $F \cos \theta$ versus s as the bowstring is drawn back.

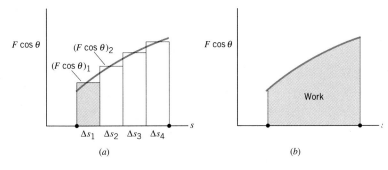

(a)   (b)

Figure 6.23 (a) The work done by the average force component $(F \cos \theta)_1$ during the small displacement Δs_1 is $(F \cos \theta)_1 \Delta s_1$, which is the area of the colored rectangle. (b) The work done by a variable force is equal to the colored area under the $F \cos \theta$-versus-s curve.

EXAMPLE 14 Work and the Compound Bow

Find the work that the archer must do in drawing back the string of the compound bow in Figure 6.22 from 0 to 0.500 m.

Reasoning The work is equal to the colored area under the curved line in Figure 6.22b. For convenience, this area is divided into a number of small squares, each having an area of $(9.00 \text{ N})(2.78 \times 10^{-2} \text{ m}) = 0.250 \text{ J}$. The area can be found by counting the number of squares under the curve and multiplying by the area per square.

Solution We estimate that there are 242 colored squares in the drawing. Since each square represents 0.250 J of work, the total work done is

$$W = (242 \text{ squares})\left(0.250 \, \frac{\text{J}}{\text{square}}\right) = \boxed{60.5 \text{ J}}$$

When the arrow is fired, part of this work is imparted to it as kinetic energy.

6.10 CONCEPTS *&* CALCULATIONS

This section contains examples that discuss one or more conceptual questions, followed by a related quantitative problem. Example 15 reviews the important concept of work and illustrates how forces can give rise to positive, negative, and zero work. Example 16 examines the all-important conservation of mechanical energy and the work–energy theorem.

CONCEPTS & CALCULATIONS EXAMPLE 15   Skateboarding and Work

The skateboarder in Figure 6.24a is coasting down a ramp, and there are three forces acting on her: her weight **W** (magnitude = 675 N), a frictional force **f** (magnitude = 125 N) that opposes her motion, and a normal force **F$_N$** (magnitude = 612 N). Determine the net work done by the three forces when she coasts for a distance of 9.2 m.

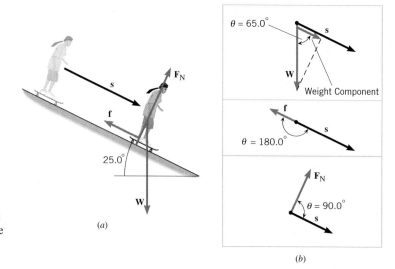

Figure 6.24 (a) Three forces act on the skateboarder: **W** = weight, **f** = frictional force, **F$_N$** = normal force. For clarity, the frictional force is not drawn to scale. (b) The orientation of the three forces relative to the displacement **s** of the skateboarder.

Concept Questions and Answers Figure 6.24*b* shows each force, along with the displacement **s** of the skateboarder. By examining these diagrams and without doing any numerical calculations, determine whether the work done by each force is positive, negative, or zero. Provide a reason for each answer.

Answer   The work done by a force is positive if the force has a component that points in the same direction as the displacement. The work is negative if there is a force component pointing opposite to the displacement. The work done by the weight **W** is positive, because the weight has a component that points in the same direction as the displacement. The top drawing in Figure 6.24*b* shows this situation. The work done by the frictional force **f** is negative, because it points opposite to the direction of the displacement ($\theta = 180.0°$ in the middle drawing). The work done by the normal force $\mathbf{F_N}$ is zero, because it is perpendicular to the displacement ($\theta = 90.0°$ in the bottom drawing). Consequently, the normal force does not have a component along the displacement.

Solution The work W done by a force is given by Equation 6.1 as $W = (F \cos \theta)s$, where F is the magnitude of the force, s is the magnitude of the displacement, and θ is the angle between the force and the displacement. Figure 6.24*b* shows this angle for each of the three forces. The work done by each force is computed in the table below. The net work is the algebraic sum of the three values.

Force	Magnitude F of the Force	Angle, θ	Magnitude s of the Displacement	Work Done by the Force $W = (F \cos \theta)\, s$
W	675 N	65.0°	9.2 m	$W = (675\ \text{N})(\cos 65.0°)$ $\times (9.2\ \text{m}) = +2620\ \text{J}$
f	125 N	180.0°	9.2 m	$W = (125\ \text{N})(\cos 180.0°)$ $\times (9.2\ \text{m}) = -1150\ \text{J}$
$\mathbf{F_N}$	612 N	90.0°	9.2 m	$W = (612\ \text{N})(\cos 90.0°)$ $\times (9.2\ \text{m}) = 0\ \text{J}$

The net work done by the three forces is

$$+2620\ \text{J} + (-1150\ \text{J}) + 0\ \text{J} = \boxed{+1470\ \text{J}}$$

CONCEPTS & CALCULATIONS EXAMPLE 16
Conservation of Mechanical Energy and the Work–Energy Theorem

Figure 6.25 shows a 0.41-kg block sliding from A to B along a frictionless surface. When the block reaches B, it continues to slide along the horizontal surface BC where a kinetic frictional force acts. As a result, the block slows down, coming to rest at C. The kinetic energy of the block at A is 37 J, and the heights of A and B are 12.0 and 7.0 m above the ground, respectively. (a) What is the kinetic energy of the block when it reaches B? (b) How much work does the kinetic frictional force do during the BC segment of the trip?

Figure 6.25 The block slides on a frictionless surface from A to B. From B to C, a kinetic frictional force slows down the block until it comes to rest at C.

Concept Questions and Answers Is the total mechanical energy of the block conserved as the block goes from A to B? Why or why not?

Answer The total mechanical energy of the block is conserved if we can show that the net work done by the nonconservative forces is zero, or $W_{nc} = 0$ J (see Section 6.5). There are only two forces acting on the block during its trip from A to B, its weight and the normal force. The block's weight, as we have seen in Section 6.4, is a conservative force. The normal force, on the other hand, is a nonconservative force. However, it is always perpendicular to the displacement of the block, so the normal force does no work. Thus, we conclude that $W_{nc} = 0$ J, with the result that the total mechanical energy is conserved during the AB part of the trip.

When the block reaches point B, has its kinetic energy increased, decreased, or remained the same relative to what it had at A? Provide a reason for your answer.

Answer As we have seen, the total mechanical energy is the sum of the kinetic and gravitational potential energies and remains constant during the trip. Therefore, as one type of energy decreases, the other must increase for the sum to remain constant. Since B is lower than A, the gravitational potential energy at B is less than that at A. As a result, the kinetic energy at B must be greater than that at A.

Is the total mechanical energy of the block conserved as the block goes from B to C? Justify your answer.

Answer During this part of the trip, a kinetic frictional force acts on the block. This force is nonconservative and it does work on the block, just as the frictional force does work on the skateboarder in Example 15. Consequently, the net work W_{nc} done by nonconservative forces is not zero ($W_{nc} \neq 0$ J), so the total mechanical energy is not conserved during this part of the trip.

Solution

(a) Since the total mechanical energy is conserved during the AB segment, we can set the total mechanical energy at B equal to that at A.

$$\underbrace{KE_B + mgh_B}_{\substack{\text{Total mechanical} \\ \text{energy at B}}} = \underbrace{KE_A + mgh_A}_{\substack{\text{Total mechanical} \\ \text{energy at A}}}$$

Solving this equation for the kinetic energy at B gives

$$KE_B = KE_A + mg(h_A - h_B)$$
$$= 37 \text{ J} + (0.41 \text{ kg})(9.80 \text{ m/s}^2)(12.0 \text{ m} - 7.0 \text{ m}) = \boxed{57 \text{ J}}$$

As expected, the kinetic energy at B is greater than that at A.

(b) During the BC part of the trip, the total mechanical energy is not conserved because a kinetic frictional force is present. The work W_{nc} done by this nonconservative force is given by the work–energy theorem (see Equation 6.8) as the final total mechanical energy minus the initial total mechanical energy:

$$W_{nc} = \underbrace{KE_C + mgh_C}_{\substack{\text{Total mechanical} \\ \text{energy at C}}} - \underbrace{(KE_B + mgh_B)}_{\substack{\text{Total mechanical} \\ \text{energy at B}}}$$

Rearranging this equation gives

$$W_{nc} = \underbrace{KE_C}_{= 0 \text{ J}} - KE_B + mg\underbrace{(h_C - h_B)}_{= 0 \text{ m}}$$

In the preceding equation we have noted that the kinetic energy KE_C at C is equal to zero, because the block comes to rest at this point. The term $(h_C - h_B)$ is also zero, since the two heights are the same. Thus, the work W_{nc} done by the kinetic frictional force during the BC part of the trip is

$$W_{nc} = -KE_B = \boxed{-57 \text{ J}}$$

The work done by the nonconservative frictional force is negative, because this force points opposite to the displacement of the block.

At the end of the problem set for this chapter, you will find homework problems that contain both conceptual and quantitative parts. These problems are grouped under the heading *Concepts & Calculations, Group Learning Problems*. They are designed for use by students working alone or in small learning groups. The conceptual part of each problem provides a convenient focus for group discussions.

SUMMARY

The **work** W done by a constant force acting on an object is $W = (F \cos \theta)s$, where F is the magnitude of the force, s is the magnitude of the displacement, and θ is the angle between the force and the displacement vectors. Work is a scalar quantity and can be positive or negative, depending on whether the force has a component that points, respectively, in the same direction as the displacement or in the opposite direction. The work is zero if the force is perpendicular ($\theta = 90°$) to the displacement.

The **kinetic energy** KE of an object of mass m and speed v is $KE = \frac{1}{2} mv^2$. The **work–energy theorem** states that the work W done by the net external force acting on an object equals the difference between the object's final kinetic energy KE_f and initial kinetic energy KE_0 : $W = KE_f - KE_0$. If the net force does positive work, the kinetic energy increases; if the net force does negative work, the kinetic energy decreases.

The **work done by the force of gravity** on an object of mass m is $W_{gravity} = mg(h_0 - h_f)$, where h_0 and h_f are the initial and final heights of the object, respectively.

Gravitational potential energy PE is the energy that an object has by virtue of its position. For an object near the surface of the earth, the gravitational potential energy is given by $PE = mgh$, where h is the height of the object relative to an arbitrary zero level.

A **conservative force** is one that in moving an object between two points does the same work, independent of the path taken between the points. Alternatively, a force is conservative if the work it does in moving an object around any closed path is zero. A force is a **nonconservative force** if the work it does on an object moving between two points depends on the path of the motion between the points.

The **total mechanical energy** E is the sum of the kinetic energy and the potential energy: $E = KE + PE$. The work–energy theorem can be expressed in an alternative form as $W_{nc} = E_f - E_0$, where W_{nc} is the net work done by the external nonconservative forces, and E_f and E_0 are the final and initial total mechanical energies, respectively.

The **principle of conservation of mechanical energy** states that the total mechanical energy E remains constant along the path of an object, provided that the net work done by external nonconservative forces is zero. While E is constant, however, KE and PE may be transformed into one another.

Average power $\overline{P}$ is the work done per unit time, $\overline{P} = $ Work/Time, or the rate at which work is done. It is also the rate at which energy changes.

The **principle of conservation of energy** states that energy cannot be created or destroyed but can only be transformed from one form to another.

CONCEPTUAL QUESTIONS

1. Two forces F_1 and F_2 are acting on the box shown in the drawing, causing the box to move across the floor. The two force vectors are drawn to scale. Which force does more work? Justify your answer.

Question 1

2. A box is being moved with a velocity v by a force P (parallel to v) along a level horizontal floor. The normal force is F_N, the kinetic frictional force is f_k, and the weight of the box is mg. Decide which forces

do positive, zero, or negative work. Provide a reason for each of your answers.

3. A force does positive work on a particle that has a displacement pointing in the $+x$ direction. This same force does negative work on a particle that has a displacement pointing in the $+y$ direction. In what quadrant does the force lie? Account for your answer.

4. A sailboat is moving at a constant velocity. (a) Is work being done by a net external force acting on the boat? Explain. (b) Recognizing that the wind propels the boat forward and the water resists the boat's motion, what does your answer in part (a) imply about the work done by the wind's force compared to the work done by the water's resistive force?

5. A ball has a speed of 15 m/s. Only one external force acts on the ball. After this force acts, the speed of the ball is 7 m/s. Has the force done positive or negative work? Explain.

6. A slow-moving car may have more kinetic energy than a fast-moving motorcycle. How is this possible?

7. A net external force acts on a particle. This net force is not zero. Is this sufficient information to conclude that (a) the velocity of the particle changes, (b) the kinetic energy of the particle changes, and (c) the speed of the particle changes? Give your reasoning in each case.

8. The speed of a particle doubles and then doubles again because a net external force acts on it. Does the net force do more work during the first or the second doubling? Justify your answer.

9. A shopping bag is hanging straight down from your hand as you walk across a horizontal floor at a constant velocity. (a) Does the force that your hand exerts on the bag's handle do any work? Explain. (b) Does this force do any work while you are riding up an escalator at a constant velocity? Give a reason for your answer.

10. In a simulation on earth, an astronaut in his space suit climbs up a vertical ladder. On the moon, the same astronaut makes the same climb. In which case does the gravitational potential energy of the astronaut change by a greater amount? Account for your answer.

11. A net external nonconservative force does positive work on a particle, and both its kinetic and potential energies change. What, if anything, can you conclude about (a) the change in the particle's total mechanical energy and (b) the individual changes in the kinetic and potential energies? Justify your answers.

12. Suppose the total mechanical energy of an object is conserved. (a) If the kinetic energy decreases, what must be true about the gravitational potential energy? (b) If the potential energy decreases, what must be true about the kinetic energy? (c) If the kinetic energy does not change, what must be true about the potential energy?

13. In Example 10 the Magnum XL-200 starts from rest at the top of the hill and attains a speed of 34.1 m/s when it reaches the bottom of the hill. If the roller coaster were to then start up an identical hill, would its speed be 0 m/s at the top of this hill? Assume that friction is negligible. Explain your answer in terms of energy concepts.

14. A person is riding on a Ferris wheel. When the wheel makes one complete turn, is the net work done by the gravitational force positive, negative, or zero? Justify your answer.

15. Consider the following two situations in which the retarding effects of friction and air resistance are negligible. Car A approaches a hill. The driver turns off the engine at the bottom of the hill, and the car coasts up the hill. Car B, its engine running, is driven up the hill at a constant speed. Which situation is an example of the principle of conservation of mechanical energy? Provide a reason for your answer.

16. A trapeze artist, starting from rest, swings downward on the bar, lets go at the bottom of the swing, and falls freely to the net. An assistant, standing on the same platform as the trapeze artist, jumps from rest straight downward. Friction and air resistance are negligible. (a) On which person, if either, does gravity do the greatest amount of work? Explain. (b) Who, if either, strikes the net with a greater speed? Why?

17. The drawing shows an empty fuel tank being released by three different jet planes. At the moment of release, each plane has the same speed and each tank is at the same height above the ground. However, the directions of travel are different. In the absence of air resistance, do the tanks have different speeds when they hit the ground? If so, which tank has the largest speed and which has the smallest speed? Explain.

Fuel tank   (a)   (b)   (c)

18. Is it correct to conclude that one engine is doing twice the work of another just because it is generating twice the power? Explain, neglecting friction and taking into account the time of operation of the engines.

PROBLEMS

Section 6.1 Work Done by a Constant Force

1. ssm During a tug-of-war, team A pulls on team B by applying a force of 1100 N to the rope between them. How much work does team A do if they pull team B toward them a distance of 2.0 m?

2. A student removes a 10.5-kg stereo amplifier from a shelf that is 1.82 m high. The amplifier is lowered at a constant speed to a height of 0.75 m. What is the work done by (a) the person and (b) the gravitational force that acts on the amplifier?

3. A person pulls a toboggan for a distance of 35.0 m along the snow with a rope directed 25.0° above the snow. The tension in the rope is 94.0 N. (a) How much work is done on the toboggan by the tension force? (b) How much work is done if the same tension is directed parallel to the snow?

4. The drawing shows a boat being pulled by two locomotives through a canal of length 2.00 km. The tension in each cable is

5.00 × 10³ N, and θ = 20.0°. What is the net work done on the boat by the two locomotives?

5. ssm Suppose in Figure 6.2 that +1.10 × 10³ J of work are done by the force **F** (magnitude = 30.0 N) in moving the luggage carrier a distance of 50.0 m. At what angle θ is the force oriented with respect to the ground?

6. The drawing shows a plane diving toward the ground and then climbing back upward. During each of these motions, the lift force **L** acts perpendicular to the displacement **s**, which has the same magnitude of 1.7 × 10³ m in each case. The engines of the plane exert a thrust **T**, which points in the direction of the displacement and has the same magnitude during the dive and the climb. The weight **W** of the plane has a magnitude of 5.9 × 10⁴ N. In both motions, net work is performed due to the combined action of the forces **L**, **T**, and **W**. (a) Is more net work done during the dive or the climb? Explain. (b) Find the difference between the net work done during the dive and the climb.

(a) Dive   75°

(b) Climb   115°

7. A person pushes a 16.0-kg shopping cart at a constant velocity for a distance of 22.0 m. She pushes in a direction 29.0° below the horizontal. A 48.0-N frictional force opposes the motion of the cart. (a) What is the magnitude of the force that the shopper exerts? Determine the work done by (b) the pushing force, (c) the frictional force, and (d) the gravitational force.

*8. A 2.40 × 10²-N force is pulling an 85.0-kg refrigerator across a horizontal surface. The force acts at an angle of 20.0° above the surface. The coefficient of kinetic friction is 0.200, and the refrigerator moves a distance of 8.00 m. Find (a) the work done by the pulling force, and (b) the work done by the kinetic frictional force.

*9. **ssm** A husband and wife take turns pulling their child in a wagon along a horizontal sidewalk. Each exerts a constant force and pulls the wagon through the same displacement. They do the same amount of work, but the husband's pulling force is directed 58° above the horizontal, while the wife's pulling force is directed 38° above the horizontal. The husband pulls with a force whose magnitude is 67 N. What is the magnitude of the pulling force exerted by his wife?

*10. A 1.00 × 10²-kg crate is being pulled across a horizontal floor by a force **P** that makes an angle of 30.0° above the horizontal. The coefficient of kinetic friction is 0.200. What should be the magnitude of **P**, so that the net work done by it and the kinetic frictional force is zero?

11. A 1200-kg car is being driven up a 5.0° hill. The frictional force is directed opposite to the motion of the car and has a magnitude of f = 524 N. A force **F is applied to the car by the road and propels the car forward. In addition to these two forces, two other forces act on the car: its weight **W** and the normal force **F_N** directed perpendicular to the road surface. The length of the road up the hill is 290 m. What should be the magnitude of **F**, so that the net work done by all the forces acting on the car is +150 kJ?

Section 6.2 The Work–Energy Theorem and Kinetic Energy

12. A 0.075-kg arrow is fired horizontally. The bowstring exerts an average force of 65 N on the arrow over a distance of 0.90 m. With what speed does the arrow leave the bow?

13. ssm The hammer throw is a track and field event in which a 7.3-kg ball (the "hammer"), starting from rest, is whirled around in a circle several times and released. It then moves upward on the familiar curving path of projectile motion. In one throw, the hammer is given a speed of 29 m/s. For comparison, a .22 caliber bullet has a mass of 2.6 g and, starting from rest, exits the barrel of a gun with a speed of 410 m/s. Determine the work done to launch the motion of (a) the hammer and (b) the bullet.

14. The aircraft carrier *Nimitz* has a fully loaded mass of 8.35 × 10⁷ kg. It travels a distance of 208 km in 3.5 hours. What is the average kinetic energy (in joules) of the *Nimitz?*

15. ssm www Two cars, A and B, are traveling with the same speed of 40.0 m/s, each having started from rest. Car A has a mass of 1.20 × 10³ kg, and car B has a mass of 2.00 × 10³ kg. Compared to the work required to bring car A up to speed, how much *additional* work is required to bring car B up to speed?

16. As background for this problem, review Conceptual Example 6. A 7420-kg satellite has an elliptical orbit, as in Figure 6.10b. The point on the orbit that is farthest from the earth is called the *apogee* and is at the far right side of the drawing. The point on the orbit that is closest to the earth is called the *perigee* and is at the far left side of the drawing. Suppose that the speed of the satellite is 2820 m/s at the apogee and 8450 m/s at the perigee. Find the work done by the gravitational force when the satellite moves from (a) the apogee to the perigee and (b) the perigee to the apogee.

17. When a 0.045-kg golf ball takes off after being hit, its speed is 41 m/s. (a) How much work is done on the ball by the club? (b) Assume that the force of the golf club acts parallel to the motion of the ball and that the club is in contact with the ball for a distance of 0.010 m. Ignore the weight of the ball and determine the average force applied to the ball by the club.

18. A 5.0 × 10⁴-kg space probe is traveling at a speed of 11 000 m/s through deep space. Retrorockets are fired along the line of motion to reduce the probe's speed. The retrorockets generate a force of 4.0 × 10⁵ N over a distance of 2500 km. What is the final speed of the probe?

*19. **ssm** A sled is being pulled across a horizontal patch of snow. Friction is negligible. The pulling force points in the same direction as the sled's displacement, which is along the $+x$ axis. As a result, the kinetic energy of the sled increases by 38%. By what percentage would the sled's kinetic energy have increased if this force had pointed 62° above the $+x$ axis?

*20. A rescue helicopter lifts a 79-kg person straight up by means of a cable. The person has an upward acceleration of 0.70 m/s^2 and is lifted from rest through a distance of 11 m. (a) What is the tension in the cable? How much work is done by (b) the tension in the cable and (c) the person's weight? (d) Use the work–energy theorem and find the final speed of the person.

*21. An extreme skier, starting from rest, coasts down a mountain that makes an angle of 25.0° with the horizontal. The coefficient of kinetic friction between her skis and the snow is 0.200. She coasts for a distance of 10.4 m before coming to the edge of a cliff. Without slowing down, she skis off the cliff and lands downhill at a point whose vertical distance is 3.50 m below the edge. How fast is she going just before she lands?

*22. A power boat of mass 480 kg is cruising at a constant speed of 8.9 m/s. The propeller provides a drive force of 760 N. The driver of the boat shuts off the engine, and the boat coasts to a halt. Assume–contrary to fact–that the resistive force due to the water is constant, independent of the boat's speed. (a) How far does the boat coast? (b) How much time does it take for the boat to come to rest after the engine is turned off?

*23. **ssm www** A 6200-kg satellite is in a circular earth orbit that has a radius of 3.3 × 10^7 m. A net external force must act on the satellite to make it change to a circular orbit that has a radius of 7.0 × 10^6 m. What work must the net external force do?

**24. The model airplane in Figure 5.7 is flying at a speed of 22 m/s on a horizontal circle of radius 16 m. The mass of the plane is 0.90 kg. The person holding the guideline pulls it in until the radius of the circle becomes 14 m. The plane speeds up, and the tension in the guideline becomes four times greater. What is the net work done on the plane?

Section 6.3 Gravitational Potential Energy,
Section 6.4 Conservative Forces, Nonconservative Forces, and the Work–Energy Theorem

25. Relative to the ground, what is the gravitational potential energy of a 55.0-kg person who is at the top of the Sears Tower, a height of 443 m above the ground?

26. A 0.60-kg basketball is dropped out of a window that is 6.1 m above the ground. The ball is caught by a person whose hands are 1.5 m above the ground. (a) How much work is done on the ball by its weight? What is the gravitational potential energy of the basketball, relative to the ground, when it is (b) released and (c) caught? (d) How is the change (PE$_f$ − PE$_0$) in the ball's gravitational potential energy related to the work done by its weight?

27. **ssm** A bicyclist rides 5.0 km due east, while the resistive force from the air has a magnitude of 3.0 N and points due west. The rider then turns around and rides 5.0 km due west, back to her starting point. The resistive force from the air on the return

trip has a magnitude of 3.0 N and points due east. (a) Find the work done by the resistive force during the round trip. (b) Based on your answer to part (a), is the resistive force a conservative force? Explain.

28. A shot-putter puts a shot (weight = 71.1 N) that leaves his hand at a distance of 1.52 m above the ground. (a) Find the work done by the gravitational force when the shot has risen to a height of 2.13 m above the ground. Include the correct sign for the work. (b) Determine the change (ΔPE = PE$_f$ − PE$_0$) in the gravitational potential energy of the shot.

29. A 75.0-kg skier rides a 2830-m-long lift to the top of a mountain. The lift makes an angle of 14.6° with the horizontal. What is the change in the skier's gravitational potential energy?

30. A pole-vaulter just clears the bar at 5.80 m and falls back to the ground. The change in the vaulter's potential energy is −3.70 × 10^3 J. What is his weight?

31. **ssm** "Rocket man" has a propulsion unit strapped to his back. He starts from rest on the ground, fires the unit, and is propelled straight upward. At a height of 16 m, his speed is 5.0 m/s. His mass, including the propulsion unit, has the approximately constant value of 136 kg. Find the work done by the force generated by the propulsion unit.

Section 6.5 The Conservation of Mechanical Energy

32. A gymnast is swinging on a high bar. The distance between his waist and the bar is 1.1 m, as the drawing shows. At the top of the swing his speed is momentarily zero. Ignoring friction and treating the gymnast as if all of his mass is located at his waist, find his speed at the bottom of the swing.

$r = 1.1$ m

33. The height of the upper falls at Yellowstone Falls is 33.2 m. When the water reaches the bottom of the falls, its speed is 25.8 m/s. Neglecting air resistance, what is the speed of the water at the top of the falls?

34. A water-skier lets go of the tow rope upon leaving the end of a jump ramp at a speed of 14.0 m/s. As the drawing indicates, the skier has a speed of 13.0 m/s at the highest point of the jump. Ignoring air resistance, determine the skier's height H above the *top of the ramp* at the highest point.

13.0 m/s

14.0 m/s

H

35. ssm A 2.00-kg rock is released from rest at a height of 20.0 m. Ignore air resistance and determine the kinetic energy, gravitational potential energy, and total mechanical energy at each of the following heights: 20.0, 10.0, and 0 m.

36. A 47.0-g golf ball is driven from the tee with an initial speed of 52.0 m/s and rises to a height of 24.6 m. (a) Neglect air resistance and determine the kinetic energy of the ball at its highest point. (b) What is its speed when it is 8.0 m below its highest point?

37. A cyclist approaches the bottom of a gradual hill at a speed of 11 m/s. The hill is 5.0 m high, and the cyclist estimates that she is going fast enough to coast up and over it without pedaling. Ignoring air resistance and friction, find the speed at which the cyclist crests the hill.

* **38.** A grappling hook, attached to a 1.5-m rope, is whirled in a circle that lies in the vertical plane. The lowest point on this circle is at ground level. The hook is whirled at a constant rate of three revolutions per second. In the absence of air resistance, to what maximum height can the hook be cast?

* **39. ssm www** Two pole-vaulters just clear the bar at the same height. The first lands at a speed of 8.90 m/s, while the second lands at a speed of 9.00 m/s. The first vaulter clears the bar at a speed of 1.00 m/s. Ignore air resistance and friction and determine the speed at which the second vaulter clears the bar.

* **40.** Before starting this problem, review Conceptual Example 9. A wrecking ball swings at the end of a 10.0-m cable on a vertical circular arc. The crane operator manages to give the ball a speed of 6.50 m/s as the ball passes through the lowest point of its swing and then gives the ball no further assistance. Friction and air resistance are negligible. What speed v_f does the ball have when the cable makes an angle of 32.0° with respect to the vertical?

* **41. ssm** A water slide is constructed so that swimmers, starting from rest at the top of the slide, leave the end of the slide traveling horizontally. As the drawing shows, one person hits the water 5.00 m from the end of the slide in a time of 0.500 s after leaving the slide. Ignoring friction and air resistance, find the height H in the drawing.

Water slide

H

|← 5.00 m →|

* **42.** A skier starts from rest at the top of a hill. The skier coasts down the hill and up a second hill, as the drawing illustrates. The crest of the second hill is circular, with a radius of $r = 36$ m. Ne-

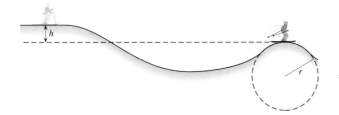

h

r

glect friction and air resistance. What must be the height h of the first hill so that the skier just loses contact with the snow at the crest of the second hill?

** **43.** Conceptual Example 9 provides background for this problem. A swing is made from a rope that will tolerate a maximum tension of 8.00×10^2 N without breaking. Initially, the swing hangs vertically. The swing is then pulled back at an angle of 60.0° with respect to the vertical and released from rest. What is the mass of the heaviest person who can ride the swing?

** **44.** A person starts from rest at the top of a large frictionless spherical surface, and slides into the water below (see the drawing). At what angle θ does the person leave the surface? (*Hint: When the person leaves the surface, the normal force is zero.*)

r

θ

Section 6.6 Nonconservative Forces and the Work–Energy Theorem

45. ssm A roller coaster (375 kg) moves from A (5.00 m above the ground) to B (20.0 m above the ground). Two nonconservative forces are present: friction does -2.00×10^4 J of work on the car, and a chain mechanism does $+3.00 \times 10^4$ J of work to help the car up a long climb. What is the change in the car's kinetic energy, $\Delta KE = KE_f - KE_0$, from A to B?

46. A basketball player makes a jump shot. The 0.600-kg ball is released at a height of 2.00 m above the floor with a speed of 7.20 m/s. The ball goes through the net 3.10 m above the floor at a speed of 4.20 m/s. What is the work done on the ball by air resistance, a nonconservative force?

47. A 55.0-kg skateboarder starts out with a speed of 1.80 m/s. He does $+80.0$ J of work on himself by pushing with his feet against the ground. In addition, friction does -265 J of work on him. In both cases, the forces doing the work are nonconservative. The final speed of the skateboarder is 6.00 m/s. (a) Calculate the change ($\Delta PE = PE_f - PE_0$) in the gravitational potential energy. (b) How much has the vertical height of the skater changed, and is the skater above or below the starting point?

48. A projectile of mass 0.750 kg is shot straight up with an initial speed of 18.0 m/s. (a) How high would it go if there were no air friction? (b) If the projectile rises to a maximum height of only 11.8 m, determine the magnitude of the average force due to air resistance.

49. A pitcher throws a 0.140-kg baseball, and it approaches the bat at a speed of 40.0 m/s. The bat does $W_{nc} = 70.0$ J of work on the ball in hitting it. Ignoring air resistance, determine the speed of the ball after the ball leaves the bat and is 25.0 m above the point of impact.

* **50.** A 63-kg skier coasts up a snow-covered hill that makes an angle of 25° with the horizontal. The initial speed of the skier is 6.6 m/s. After coasting a distance of 1.9 m up the slope, the speed

of the skier is 4.4 m/s. (a) Find the work done by the kinetic frictional force that acts on the skis. (b) What is the magnitude of the kinetic frictional force?

*51. At the end of the semester, you are loading your gear into a van to go home. In loading the van, you can either lift a 45.0-kg crate straight up (method 1) or push it up a 3.00-m-long ramp that is inclined at an angle θ above the horizontal (method 2). In both methods, the crate is moved at a constant speed, and the vertical distance from the ground to the floor of the van has the same value h (see the drawing). In method 1, no friction is present. In method 2, however, a frictional force does -485 J of work on the crate. In applying your nonconservative pushing force to the crate, you do twice as much work in method 2 as in method 1. Find the angle θ.

Method 1

Method 2

*52. A basketball of mass 0.60 kg is dropped from rest from a height of 1.22 m. It rebounds to a height of 0.69 m. (a) How much mechanical energy was lost during the collision with the floor? (b) A basketball player dribbles the ball from a height of 1.22 m by exerting a constant downward force on it for a distance of 0.13 m. In dribbling, the player compensates for the mechanical energy lost during each bounce. If the ball now returns to a height of 1.22 m, what is the magnitude of the force?

*53. ssm At a carnival, you can try to ring a bell by striking a target with a 9.00-kg hammer. In response, a 0.400-kg metal piece is sent upward toward the bell, which is 5.00 m above. Suppose that 25.0% of the hammer's kinetic energy is used to do the work of sending the metal piece upward. How fast must the hammer be moving when it strikes the target so that the bell just barely rings?

**54. A 3.00-kg model rocket is launched straight up. It reaches a maximum height of 1.00×10^2 m above where its engine cuts out, even though air resistance performs -8.00×10^2 J of work on the rocket. What would have been this height if there were no air resistance?

Section 6.7 Power

55. In an action-adventure movie, the hero lifts the 91-kg villain straight upward through a distance of 1.2 m in 0.51 s at a constant speed. What power does the hero produce while doing this?

56. A person is making homemade ice cream. She exerts a force of magnitude 22 N on the free end of the crank handle, and this end moves in a circular path of radius 0.28 m. The force is always applied parallel to the motion of the handle. If the handle is turned once every 1.3 s, what is the average power being expended?

57. ssm One kilowatt·hour (kWh) is the amount of work or energy generated when one kilowatt of power is supplied for a time of one hour. A kilowatt·hour is the unit of energy used by power companies when figuring your electric bill. Determine the number of joules of energy in one kilowatt·hour.

58. A 3.00×10^2-kg piano is being lifted at a steady speed from ground level straight up to an apartment 10.0 m above the ground. The crane that is doing the lifting produces a steady power of 4.00×10^2 W. How much time does it take to lift the piano?

*59. The cheetah is one of the fastest acclerating animals, for it can go from rest to 27 m/s (about 60 mi/h) in 4.0 s. If its mass is 110 kg, determine the average power developed by the cheetah during the acceleration phase of its motion. Express your answer in (a) watts and (b) horsepower.

*60. A motorcycle (mass of cycle plus rider $= 2.50 \times 10^2$ kg) is traveling at a steady speed of 20.0 m/s. The force of air resistance acting on the cycle and rider is 2.00×10^2 N. Find the power necessary to sustain this speed if (a) the road is level and (b) the road is sloped upward at 37.0° with respect to the horizontal.

**61. ssm The motor of a ski boat generates an average power of 7.50×10^4 W when the boat is moving at a constant speed of 12 m/s. When the boat is pulling a skier at the same speed, the engine must generate an average power of 8.30×10^4 W. What is the tension in the tow rope that is pulling the skier?

**62. A 1900-kg car experiences a combined force of air resistance and friction that has the same magnitude whether the car goes up or down a hill at 27 m/s. Going up a hill, the car's engine needs to produce 47 hp more power to sustain the constant velocity than it does going down the same hill. At what angle is the hill inclined above the horizontal?

Section *6.9 Work Done by a Variable Force

63. ssm The drawing shows the force-versus-displacement graph for two different bows. These graphs give the force that an archer must apply to draw the bowstring. (a) For which bow is more work required to draw the bow fully from $s = 0$ to $s = 0.50$ m? Give your reasoning. (b) Estimate the additional work required for the bow identified in part (a) compared to the other bow.

64. The force component along the displacement varies with the magnitude of the displacement, as shown in the graph. Find the work done by the force in the interval from (a) 0 to 1.0 m, (b) 1.0 to 2.0 m, and (c) 2.0 to 4.0 m. (*Note: In the last interval the force component is negative, so the work is negative.*)

65. The graph shows the net external force component $F \cos \theta$ along the displacement as a function of the magnitude s of the displacement. The graph applies to a 65-kg ice skater. How much work does the net force component do on the skater from (a) 0 to 3.0 m and (b) 3.0 m to 6.0 m? (c) If the initial speed of the skater is 1.5 m/s when $s = 0$ m, what is the speed when $s = 6.0$ m?

66. Review Example 14, in which the work done in drawing the bowstring in Figure 6.22 from $s = 0$ to $s = 0.500$ m is determined. In part b of the figure, the force comonent $F \cos \theta$ reaches a maximum at $s = 0.306$ m. Find the percentage of the total work that is done when the bowstring is moved (a) from $s = 0$ to 0.306 m and (b) from $s = 0.306$ to 0.500 m.

67. A net external force is applied to a 6.00-kg object that is initially at rest. The net force component along the displacement of the object varies with the magnitude of the displacement as shown in the drawing. (a) How much work is done by the net force? (b) What is the speed of the object at $s = 20.0$ m?

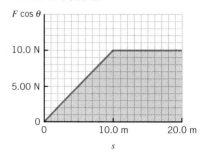

ADDITIONAL PROBLEMS

68. A cable lifts a 1200-kg elevator at a constant velocity for a distance of 35 m. What is the work done by (a) the tension in the cable and (b) the elevator's weight?

69. ssm A pole-vaulter approaches the takeoff point at a speed of 9.00 m/s. Assuming that only this speed determines the height to which he can rise, find the maximum height at which the vaulter can clear the bar.

70. A softball pitcher has a "windmill" windup in which a 0.25-kg ball moves on a vertical arc of radius $r = 0.51$ m. To accelerate the ball, she exerts a 28-N force parallel to the ball's motion along the circular arc. The speed of the ball is 12 m/s at the top of the arc. With what speed is the ball released one-half a revolution later at the bottom of the arc?

71. ssm The brakes of a truck cause it to slow down by applying a retarding force of 3.0×10^3 N to the truck over a distance of 850 m. What is the work done by this force on the truck? Is the work positive or negative? Why?

72. A 5.00×10^2-kg hot-air balloon takes off from rest at the surface of the earth. The nonconservative wind and lift forces take the balloon up, doing $+9.70 \times 10^4$ J of work on the balloon in the process. At what height above the surface of the earth does the balloon have a speed of 8.00 m/s?

*73. A 55-kg box is being pushed a distance of 7.0 m across the floor by a force **P** whose magnitude is 150 N. The force **P** is parallel to the displacement of the box. The coefficient of kinetic friction is 0.25. Determine the work done on the box by each of the *four* forces that act on the box. Be sure to include the proper plus or minus sign for the work done by each force.

*74. In 2.0 minutes, a ski lift raises four skiers at constant speed to a height of 140 m. The average mass of each skier is 65 kg. What is the average power provided by the tension in the cable pulling the lift?

*75. ssm The (nonconservative) force propelling a 1.50×10^3-kg car up a mountain road does 4.70×10^6 J of work on the car. The car starts from rest at sea level and has a speed of 27.0 m/s at an altitude of 2.00×10^2 m above sea level. Obtain the work done on the car by the combined forces of friction and air resistance, both of which are nonconservative forces.

*76. A 55.0-kg person jumps from rest off a 10.0-m-high tower straight down into the water. Neglect air resistance during the descent. She comes to rest 3.00 m under the surface of the water. Determine the average force that the water exerts on the diver. This force is nonconservative.

*77. A particle, starting from point A in the drawing, is projected down the curved runway. Upon leaving the runway at point B, the particle is traveling straight upward and reaches a height of 4.00 m above the floor before falling back down. Ignoring friction and air resistance, find the speed of the particle at point A.

78. The drawing shows a version of the loop-the-loop trick for a small car. If the car is given an initial speed of 4.0 m/s, what is the largest value that the radius r can have if the car is to remain in contact with the circular track at all times?

79. ssm www A truck is traveling at 11.1 m/s down a hill when the brakes on all four wheels lock. The hill makes an angle of 15.0° with respect to the horizontal. The coefficient of kinetic friction between the tires and the road is 0.750. How far does the truck skid before coming to a stop?

CONCEPTS & CALCULATIONS

GROUP LEARNING PROBLEMS

Note: Each of these problems consists of Concept Questions followed by a related quantitative Problem. They are designed for use by students working alone or in small learning groups. The Concept Questions involve little or no mathematics and are intended to stimulate group discussions. They focus on the concepts with which the problems deal. Recognizing the concepts is the essential initial step in any problem-solving technique.

80. Concept Questions You are moving into an apartment and take the elevator to the 6th floor. Does the force exerted on you by the elevator do positive or negative work when the elevator (a) goes up and (b) goes down? Explain your answers.

Problem Suppose your weight is 685 N and that of your belongings is 915 N. (a) Determine the work done by the elevator in lifting you and your belongings up to the 6th floor (15.2 m) at a constant velocity. (b) How much work does the elevator do on you alone (without belongings) on the downward trip, which is also made at a constant velocity? Check to see that your answers are consistent with your answers to the Concept Questions.

81. Concept Questions An asteroid is moving along a straight line. A force acts along the displacement of the asteroid and slows it down. (a) Is the direction of the force the same as or opposite to the direction of the displacement of the asteroid? Why? (b) Does the force do positive, negative, or zero work? Justify your answer. (c) What type of energy is changing as the object slows down? (d) What is the relationship between the work done by this force and the change in the object's energy?

Problem The asteroid has a mass of 4.5×10^4 kg, and the force causes its speed to change from 7100 to 5500 m/s. (a) What is the work done by the force? (b) If the asteroid slows down over a distance of 1.8×10^6 m, determine the magnitude of the force. Verify that your answers are consistent with the answers to the Concept Questions.

82. Concept Questions The drawing shows two boxes resting on frictionless ramps. One box is relatively light and sits on a steep ramp. The other box is heavier and rests on a ramp that is less steep. The boxes are released from rest at A and allowed to slide down the ramps. Which box, if either, has (a) the greatest speed and (b) the greatest kinetic energy at B? Provide a reason for each answer.

Problem The two boxes have masses of 11 and 44 kg. If A and B are 4.5 and 1.5 m, respectively, above the ground, determine the speed of (a) the lighter box and (b) the heavier box when each reaches B. (c) What is the ratio of the kinetic energy of the heavier box to that of the lighter box at B? Be sure that your answers are consistent with your answers to the Concept Questions.

83. Concept Questions A student, starting from rest, slides down a water slide. On the way down, a kinetic frictional force (a nonconservative force) acts on her. (a) Does the kinetic frictional force do positive, negative, or zero work? Provide a reason for your answer. (b) Does the total mechanical energy of the student increase, decrease, or remain the same as she descends the slide? Why? (c) If the kinetic frictional force does work, how is this work related to the change in the total mechanical energy of the student?

Problem The student has a mass of 83.0 kg and the height of the water slide is 11.8 m. If the kinetic frictional force does -6.50×10^3 J of work, how fast is the student going at the bottom of the slide?

84. Concept Questions A helicopter, starting from rest, accelerates straight up from the roof of a hospital. The lifting force does work in raising the helicopter. (a) What two types of energy are changing? Is each type increasing or decreasing? Why? (b) How are the two types of energy related to the work done by the lifting force? (c) If you want to determine the average power generated by the lifting force, what other variable besides the work must be known?

Problem An 810-kg helicopter rises from rest to a speed of 7.0 m/s in a time of 3.5 s. During this time it climbs to a height of 8.2 m. What is the average power generated by the lifting force?

85. Concept Questions A water-skier is being pulled by a towrope attached to a boat. As the driver pushes the throttle forward, the skier accelerates. (a) What type of energy is changing? (b) Is the work being done by the net external force acting on the skier positive or negative? Why? (c) Explain how this work is related to the change in the energy of the skier.

Problem A 70.3-kg water-skier has an initial speed of 6.10 m/s. Later, the speed increases to 11.3 m/s. Determine the work done by the net external force acting on the skier.

The concepts of impulse and momentum discussed in this chapter aid us in understanding collisions such as this one between a bowling ball and the ten pins.

CHAPTER 7

IMPULSE AND MOMENTUM

(a)

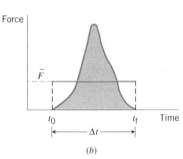

(b)

Figure 7.1   (*a*) Mark McGuire hitting his 70th home run. (*b*) When the bat strikes the ball, the magnitude of the force exerted on the ball rises to a maximum value and then returns to zero when the ball leaves the bat. The time interval during which the force acts is Δt, and the magnitude of the average force is $\overline{F}$.

7.1 THE IMPULSE — MOMENTUM THEOREM

There are many situations in which the force acting on an object is not constant, but varies with time. For instance, Figure 7.1*a* shows a baseball being hit, and part *b* of the figure illustrates approximately how the force applied to the ball by the bat changes during the time of contact. The magnitude of the force is zero at the instant t_0 just before the bat touches the ball. During contact, the force rises to a maximum and then returns to zero at the time t_f when the ball leaves the bat. The time interval $\Delta t = t_f - t_0$ during which the bat and ball arc in contact is quite short, being only a few-thousandths of a second, although the maximum force can be very large, often exceeding thousands of newtons. For comparison, the graph also shows the magnitude $\overline{F}$ of the average force exerted on the ball during the time of contact. Figure 7.2 depicts other situations in which a time-varying force is applied to a ball.

    ● *Concepts at a Glance*   To describe how a time-varying force affects the motion of an object, we will introduce two new ideas: the impulse of a force and the linear momentum of an object. As the Concepts-at-a-Glance chart in Figure 7.3 illustrates, these ideas will be used with Newton's second law of motion to produce an important result known as the impulse–momentum theorem. This theorem plays a central role in describing collisions, such as that between a ball and a bat. Later on, we will see also that the theorem leads in a natural way to one of the most fundamental laws in physics, the conservation of momentum. ●

    If a baseball is to be hit well, both the size of the force and the time of contact are important. When a large average force acts on the ball for a long enough time, the ball is hit solidly. To describe such situations, we bring together the average force and the time of contact, calling the product of the two the **_impulse_** of the force.

● **DEFINITION OF IMPULSE**

The impulse of a force is the product of the average force $\overline{\mathbf{F}}$ and the time interval Δt during which the force acts:

$$\text{Impulse} = \overline{\mathbf{F}}\,\Delta t \qquad (7.1)$$

Impulse is a vector quantity and has the same direction as the average force.

SI Unit of Impulse: newton·second (N·s)

When a ball is hit, it responds to the value of the impulse. A large impulse produces a large response; that is, the ball departs from the bat with a large velocity. However, we know from experience that the more massive the ball, the less velocity it has after leaving the bat. Both mass and velocity play a role in how an object responds to a given impulse. The effect of mass and velocity is included in the concept of **_linear momentum_**, which is defined as follows:

● **DEFINITION OF LINEAR MOMENTUM**

The linear momentum **p** of an object is the product of the object's mass m and velocity **v**:

$$\mathbf{p} = m\mathbf{v} \qquad (7.2)$$

Linear momentum is a vector quantity that points in the same direction as the velocity.

SI Unit of Momentum: kilogram·meter/second (kg·m/s)

Figure 7.2 In each of these situations the force applied to the ball varies with time. The time of contact is small, but the maximum force can be large.

Newton's second law of motion can now be used to reveal a relationship between impulse and momentum. Figure 7.4 shows a ball with an initial velocity of $\mathbf{v_0}$ approaching a bat, being struck by the bat, and then leaving with a final velocity of $\mathbf{v_f}$. When the velocity of an object changes from $\mathbf{v_0}$ to $\mathbf{v_f}$ during a time interval Δt, the average acceleration $\overline{\mathbf{a}}$ is given by Equation 2.4 as

$$\overline{\mathbf{a}} = \frac{\mathbf{v_f} - \mathbf{v_0}}{\Delta t}$$

Figure 7.3 *Concepts at a Glance* Two new concepts, impulse and momentum, will be integrated with Newton's second law of motion to produce the impulse–momentum theorem, a theorem used for describing collisions. When the golf club collides with the ball in this photograph, an impulse is imparted to the ball. In response, the ball's momentum changes, in accord with the impulse–momentum theorem.

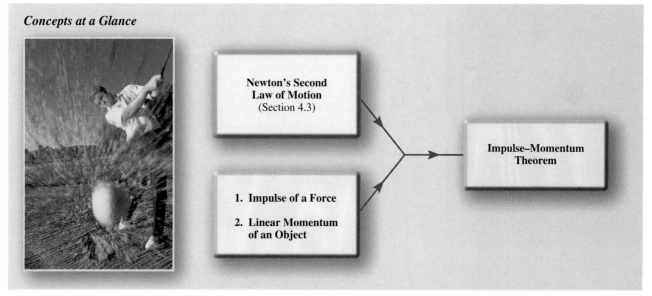

Concepts at a Glance

Newton's Second Law of Motion (Section 4.3)

1. Impulse of a Force
2. Linear Momentum of an Object

Impulse–Momentum Theorem

Figure 7.4 When a bat hits a ball, an average force $\overline{\mathbf{F}}$ is applied to the ball. As a result, the ball's velocity changes from an initial value of $\mathbf{v_0}$ (top drawing) to a final value of $\mathbf{v_f}$ (bottom drawing).

According to Newton's second law, the cause of the average acceleration is an average net force $\overline{\mathbf{F}} = m\overline{\mathbf{a}}$. Thus,

$$\overline{\mathbf{F}} = m\left(\frac{\mathbf{v_f} - \mathbf{v_0}}{\Delta t}\right) = \frac{m\mathbf{v_f} - m\mathbf{v_0}}{\Delta t} \tag{7.3}$$

In this result, the numerator on the far right is the final momentum minus the initial momentum, which is the change in momentum. Thus, the average net force is given by the change in momentum per unit of time.* Multiplying both sides of Equation 7.3 by Δt yields Equation 7.4, which is known as the *impulse–momentum theorem.*

● **IMPULSE–MOMENTUM THEOREM**

When a net force acts on an object, the impulse of the net force is equal to the change in momentum of the object:

$$\underbrace{\overline{\mathbf{F}}\,\Delta t}_{\text{Impulse}} = \underbrace{m\mathbf{v_f}}_{\substack{\text{Final} \\ \text{momentum}}} - \underbrace{m\mathbf{v_0}}_{\substack{\text{Initial} \\ \text{momentum}}} \tag{7.4}$$

Impulse = Change in momentum

During a collision, it is often difficult to measure the average net force $\overline{\mathbf{F}}$, so it is not easy to determine the impulse, $\overline{\mathbf{F}}\,\Delta t$, directly. On the other hand, it is usually straightforward to measure the mass and velocity of an object, so that its momentum just after the collision $m\mathbf{v_f}$ and just before it $m\mathbf{v_0}$ can be found. Thus, the impulse–momentum theorem allows us to gain information about the impulse indirectly by measuring the change in momentum that the impulse causes. Then, armed with a knowledge of the contact time Δt, we can evaluate the average force. Examples 1 and 2 illustrate how the theorem is used in this way.

EXAMPLE 1   A Well-Hit Ball

A baseball ($m = 0.14$ kg) has an initial velocity of $\mathbf{v_0} = -38$ m/s as it approaches a bat. We have chosen the direction of approach as the negative direction. The bat applies a force that is much larger than the weight of the ball, and the ball departs from the bat with a final velocity of $\mathbf{v_f} = +58$ m/s. (a) Determine the impulse applied to the ball by the bat. (b) Assuming that the time of contact is $\Delta t = 1.6 \times 10^{-3}$ s, find the average net force exerted on the ball by the bat.

Reasoning In hitting the ball, the bat imparts an impulse to it. We cannot use Equation 7.1 (**Impulse** $= \overline{\mathbf{F}}\,\Delta t$) to determine the impulse, since the average net force $\overline{\mathbf{F}}$ is not known. We can find the impulse, however, by turning to the impulse–momentum theorem, which states that the impulse is equal to the ball's final momentum minus its initial momentum. With values for the impulse and the time of contact, Equation 7.1 can be used to determine the average net force applied to the ball. In the present case, the net force is just the force applied by the bat, assuming that the ball's weight is negligible in comparison.

Solution

(a) According to the impulse–momentum theorem, the impulse applied to the ball is

This karate expert applies an impulse to the stack of tiles. As a result, the momentum of the stack changes as the tiles break.

---

* The equality between force and the change in momentum per unit time is the version of the second law of motion presented originally by Newton.

$$\text{Impulse} = m\mathbf{v_f} - m\mathbf{v_0}$$

$$= \underbrace{(0.14\ \text{kg})(+58\ \text{m/s})}_{\text{Final momentum}} - \underbrace{(0.14\ \text{kg})(-38\ \text{m/s})}_{\text{Initial momentum}} = \boxed{+13.4\ \text{kg} \cdot \text{m/s}}$$

(b) Now that the impulse is known, the contact time can be used in Equation 7.1 to find the average force:

$$\overline{\mathbf{F}} = \frac{\text{Impulse}}{\Delta t} = \frac{+13.4\ \text{kg} \cdot \text{m/s}}{1.6 \times 10^{-3}\ \text{s}} = \boxed{+8400\ \text{N}}$$

The force is positive, indicating that it points opposite to the velocity of the approaching ball. A force of 8400 N corresponds to 1900 lb, such a large value being necessary to change the ball's momentum during the brief contact time.

EXAMPLE 2   A Rain Storm

During a storm, rain comes straight down with a velocity of $\mathbf{v_0} = -15$ m/s and hits the roof of a car perpendicularly (see Figure 7.5). The mass of rain per second that strikes the car roof is 0.060 kg/s. Assuming that the rain comes to rest upon striking the car ($\mathbf{v_f} = 0$ m/s), find the average force exerted by the rain on the roof.

Reasoning This example differs from Example 1 in an important way. Example 1 gives information about the ball and asks for the force applied to the ball. In contrast, the present example gives information about the rain but asks for the force acting on the roof. However, the force exerted on the roof by the rain and the force exerted on the rain by the roof have equal magnitudes and opposite directions, according to Newton's law of action and reaction (see Section 4.5). Thus, we will find the force exerted on the rain and then apply the law of action and reaction to obtain the force on the roof. Assuming that the weight of a raindrop can be ignored, the average net force $\overline{\mathbf{F}}$, as calculated from the impulse–momentum theorem, is the force exerted by the roof on the rain.

Solution The average force $\overline{\mathbf{F}}$ needed to reduce the rain's velocity from $\mathbf{v_0} = -15$ m/s to $\mathbf{v_f} = 0$ m/s is given by Equation 7.3 as

$$\overline{\mathbf{F}} = \frac{m\mathbf{v_f} - m\mathbf{v_0}}{\Delta t} = -\left(\frac{m}{\Delta t}\right)\mathbf{v_0}$$

The term $m/\Delta t$ is the mass of rain per second that strikes the roof, so that $m/\Delta t = 0.060$ kg/s. Thus, the average force acting on the rain is

$$\overline{\mathbf{F}} = -(0.060\ \text{kg/s})(-15\ \text{m/s}) = +0.90\ \text{N}$$

This force is in the positive or upward direction, which is reasonable, since the roof exerts an upward force on each falling drop in order to bring it to rest. According to the action–reaction law, the force exerted on the roof by the rain also has a magnitude of 0.90 N but points downward: Force on roof = $\boxed{-0.90\ \text{N}}$

As you reason through problems such as those in Examples 1 and 2, take advantage of the impulse–momentum theorem. It is a powerful statement that can lead to significant insights. The following Conceptual Example further illustrates its use.

CONCEPTUAL EXAMPLE 3   Hailstones Versus Raindrops

In Example 2 rain is falling on the roof of a car and exerts a force on it. Instead of rain, suppose hail is falling. The hail comes straight down at a mass rate of $m/\Delta t = 0.060$ kg/s and an initial velocity of $\mathbf{v_0} = -15$ m/s and strikes the roof perpendicularly, just

• PROBLEM SOLVING INSIGHT

Momentum is a vector and, as such, has a magnitude and a direction. For motion in one dimension, be sure to indicate the direction by assigning a plus or minus sign to it, as in this example.

Figure 7.5   A raindrop falling on a car roof has an initial velocity of $\mathbf{v_0}$ just before striking the roof. The final velocity of the raindrop is $\mathbf{v_f} = 0$, because it comes to rest on the roof.

Hailstone

$\mathbf{v}_0$ $\mathbf{v}_f$

Figure 7.6 Hailstones have a downward velocity of $\mathbf{v}_0$ just before striking this car roof. They rebound with an upward velocity of $\mathbf{v}_f$.

as the rain does in Example 2. However, unlike rain, hail usually does not come to rest after striking a surface. Instead, the hailstones bounce off the roof of the car. If hail fell instead of rain, would the force on the roof be smaller than, equal to, or greater than that calculated in Example 2?

Reasoning and Solution The raindrops and the hailstones fall in exactly the same way. That is, they both fall with the same initial velocity and mass rate, and they both strike the roof perpendicularly. However, there is an important difference: the raindrops come to rest (see Figure 7.5), while the hailstones bounce upward after striking the roof (see Figure 7.6). According to the impulse–momentum theorem, the impulse that acts on an object is given by the change in the momentum of the object. This change is $m\mathbf{v}_f - m\mathbf{v}_0 = m\Delta\mathbf{v}$ and is proportional to the change in velocity $\Delta\mathbf{v}$. For a hailstone, the change in velocity is from $\mathbf{v}_0$ (downward) to $\mathbf{v}_f$ (upward). This is a larger change than for the raindrop, for which the change is only from $\mathbf{v}_0$ (downward) to zero. Therefore, the change in momentum is greater for the hailstone, and, correspondingly, a greater impulse must act on the hailstone. But an impulse is the product of the average force and the time interval Δt. Since the same amount of mass falls in the same time interval in either case, Δt is the same for hailstones as for raindrops. The greater impulse acting on the hailstones, then, means that the car roof must exert a greater force on the hailstones than on the raindrops. Conversely, according to Newton's action–reaction law, *the car roof experiences a greater force from the hailstones than from the raindrops.*

Related Homework Material: *Problems 6, 50*

In the next section we discuss a principle that is based on the impulse–momentum theorem, namely, the principle of conservation of linear momentum.

7.2 THE PRINCIPLE OF CONSERVATION OF LINEAR MOMENTUM

It is worthwhile comparing the impulse–momentum theorem to the work–energy theorem discussed in Chapter 6. The impulse–momentum theorem states that the impulse produced by a net force is equal to the change in the object's momentum, while the work–energy theorem states that the work done by a net force is equal to the change in the object's kinetic energy. The work–energy theorem leads directly to the principle of conservation of mechanical energy (see Figure 6.16), and, as we will see, the impulse–momentum theorem also leads to a conservation principle, known as the conservation of linear momentum.

We begin by applying the impulse–momentum theorem to a midair collision between two objects. The two objects (masses m_1 and m_2) are approaching each other with initial velocities $\mathbf{v}_{01}$ and $\mathbf{v}_{02}$, as Figure 7.7a shows. The collection of objects being studied is referred to as the "system." In this case, the system contains only the two objects. They interact during the collision in part b of the drawing and then depart with the final velocities $\mathbf{v}_{f1}$ and $\mathbf{v}_{f2}$ shown in part c. Because of the collision, the initial and final velocities are not the same.

Two types of forces act on the system:

1. **Internal forces**—Forces that the objects within the system exert on each other.

2. **External forces**—Forces exerted on the objects by agents that are external to the system.

During the collision in Figure 7.7b, $\mathbf{F}_{12}$ is the force exerted on object 1 by object 2, while $\mathbf{F}_{21}$ is the force exerted on object 2 by object 1. These forces are action–reaction forces that are equal in magnitude but opposite in direction, so $\mathbf{F}_{12} = -\mathbf{F}_{21}$. They are also internal forces, since they are forces that the two objects within the system exert on each other. The force of gravity also acts on the objects, their weights being $\mathbf{W}_1$ and $\mathbf{W}_2$. These weights, however, are external forces, because they are applied by the earth, which is outside the system. Friction and air resistance would also be considered external forces, although these forces are ignored here for the sake of simplicity. The impulse–momentum theorem, as applied to each object, gives the following results:

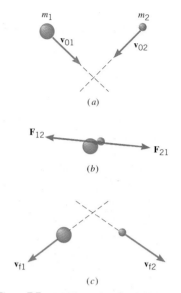

Figure 7.7 (a) The velocities of the two objects before the collision are $\mathbf{v}_{01}$ and $\mathbf{v}_{02}$. (b) During the collision, each object exerts a force on the other. These forces are $\mathbf{F}_{12}$ and $\mathbf{F}_{21}$. (c) The velocities after the collision are $\mathbf{v}_{f1}$ and $\mathbf{v}_{f2}$.

Object 1
$$(\underbrace{\mathbf{W}_1}_{\substack{\text{External} \\ \text{force}}} + \underbrace{\overline{\mathbf{F}}_{12}}_{\substack{\text{Internal} \\ \text{force}}}) \Delta t = m_1 \mathbf{v}_{f1} - m_1 \mathbf{v}_{01}$$

Object 2
$$(\underbrace{\mathbf{W}_2}_{\substack{\text{External} \\ \text{force}}} + \underbrace{\overline{\mathbf{F}}_{21}}_{\substack{\text{Internal} \\ \text{force}}}) \Delta t = m_2 \mathbf{v}_{f2} - m_2 \mathbf{v}_{02}$$

Adding these equations produces a single result for the system as a whole:

$$(\underbrace{\mathbf{W}_1 + \mathbf{W}_2}_{\substack{\text{External} \\ \text{forces}}} + \underbrace{\overline{\mathbf{F}}_{12} + \overline{\mathbf{F}}_{21}}_{\substack{\text{Internal} \\ \text{forces}}}) \Delta t = \underbrace{(m_1 \mathbf{v}_{f1} + m_2 \mathbf{v}_{f2})}_{\substack{\text{Total final} \\ \text{momentum } \mathbf{P}_f}} - \underbrace{(m_1 \mathbf{v}_{01} + m_2 \mathbf{v}_{02})}_{\substack{\text{Total initial} \\ \text{momentum } \mathbf{P}_0}}$$

On the right side of this equation, the quantity $m_1 \mathbf{v}_{f1} + m_2 \mathbf{v}_{f2}$ is the vector sum of the final momenta for each object, or the total final momentum $\mathbf{P}_f$ of the system. Likewise, $m_1 \mathbf{v}_{01} + m_2 \mathbf{v}_{02}$ is the total initial momentum $\mathbf{P}_0$. Therefore, the result above can be rewritten as

$$\left(\begin{array}{c} \textbf{Sum of average} \\ \textbf{external forces} \end{array} + \begin{array}{c} \textbf{Sum of average} \\ \textbf{internal forces} \end{array} \right) \Delta t = \mathbf{P}_f - \mathbf{P}_0 \qquad (7.5)$$

The advantage of the internal/external force classification is that the internal forces always add together to give zero, as a consequence of Newton's law of action–reaction; $\overline{\mathbf{F}}_{12} = -\overline{\mathbf{F}}_{21}$ so that $\overline{\mathbf{F}}_{12} + \overline{\mathbf{F}}_{21} = 0$. Cancellation of the internal forces occurs no matter how many parts there are to the system and allows us to ignore the internal forces, as Equation 7.6 indicates:

$$(\textbf{Sum of average external forces}) \Delta t = \mathbf{P}_f - \mathbf{P}_0 \qquad (7.6)$$

We developed this result with gravity as the only external force. But, in general, the sum of the external forces on the left includes *all* external forces.

With the aid of Equation 7.6, it is possible to see how the conservation of linear momentum arises. Suppose that the sum of the external forces is zero. A system for which this is true is called an **isolated system.** Then Equation 7.6 indicates that

$$0 = \mathbf{P}_f - \mathbf{P}_0 \quad \text{or} \quad \mathbf{P}_f = \mathbf{P}_0 \qquad (7.7)$$

Concepts at a Glance

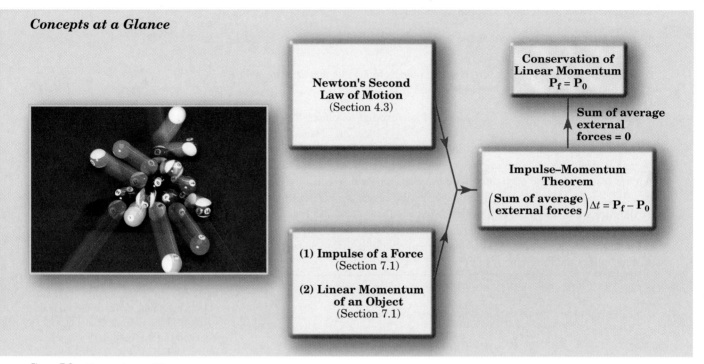

Figure 7.8 *Concepts at a Glance*
The impulse–momentum theorem
leads to the principle of conservation of
linear momentum when the sum of the
average external forces acting on a system is zero. In the collision in the photograph, the conservation of linear momentum means that the total linear
momentum of the cue ball and the other
balls before the collision is equal to
the total linear momentum after the
collision.

In other words, the final total momentum of the isolated system after the objects in
Figure 7.7 collide is the same as the initial total momentum*. This is an example of
a general principle known as the ***principle of conservation of linear momentum.***

● PRINCIPLE OF CONSERVATION OF LINEAR MOMENTUM

The total linear momentum of an isolated system remains constant (is conserved). An isolated system is one for which the vector sum of the external
forces acting on the system is zero.

● *Concepts at a Glance*   The Concepts-at-a-Glance chart in Figure 7.8 presents an overview of how concepts that we have studied earlier lead to the conservation of linear momentum. The chart is an extension of that shown in Figure 7.3.
On the left, it shows that the origin of the impulse–momentum theorem lies in
Newton's second law and the ideas of the impulse of a force and the linear momentum of an object. The theorem is stated in the middle of the chart in the form
(Sum of average external forces) $\Delta t = \mathbf{P_f} - \mathbf{P_0}$. If the sum of the average external
forces is zero, then the final and initial total momenta are equal, or $\mathbf{P_f} = \mathbf{P_0}$, which
is the mathematical statement of the conservation of linear momentum. ●
This principle applies to a system containing any number of objects, regardless
of the internal forces, provided the system is isolated. Whether the system is isolated depends on whether the vector sum of the external forces is zero. Judging
whether a force is internal or external depends on which objects are included in the
system, as Conceptual Example 4 illustrates.

---

*Technically, the initial and final momenta are equal when the impulse of the sum of the external forces
is zero, that is, when the left-hand side of Equation 7.6 is zero. Sometimes, however, the initial and final
momenta are very nearly equal even when the sum of the average external forces is not zero. This occurs
when the time Δt during which the forces act is so short that it is effectively zero. Then, the left-hand
side of Equation 7.6 is approximately zero.

CONCEPTUAL EXAMPLE 4   Is the Total Momentum Conserved?

Imagine two balls on a billiard table that is friction-free. Use the momentum conservation principle in answering the following questions. (a) Just before the balls collide, suppose that a hole opens in the table beneath them. In this hypothetical situation, is the total momentum of the two-ball system the same before and after the collision? (b) In a normal situation, when there is no hole in the table, is the total momentum of the two-ball system the same before and after the collision? (c) Answer part (b) for a system that contains only one of the balls.

Reasoning and Solution

(a) Figure 7.9a shows this situation. The dashed outline around the two balls identifies them as the system. The only external forces acting on this system are W_1 and W_2, which are the weights of the balls. The forces that one ball exerts on the other are not included in the drawing, because they are internal forces and cannot cause the total momentum of the system to change. Clearly, the vector sum of the external forces is not zero and points downward. This external force causes the balls to accelerate downward, leading to a downward-pointing momentum that increases during the collision. Thus, the total momentum of the two-ball system is not the same before and after the collision. ***The total momentum is not conserved.***

(b) Figure 7.9b shows the balls colliding on a normal table. Again, the dashed outline emphasizes that both balls are part of the system. As in the first part of this example, the collision forces are not shown, because they are internal forces. The external forces include the weights W_1 and W_2 of the balls. There are now two additional external forces, however: the upward-pointing normal force that the table exerts on each ball. These normal forces are F_{N1} and F_{N2}. Since the balls do not accelerate in the vertical direction, the normal forces must balance the weights, so that the vector sum of the four external forces in Figure 7.9b is zero. Furthermore, the table is friction-free. Thus, there is no net external force to change the total momentum of the two-ball system, and it is the same before and after the collision. ***The total momentum of this two-ball system is conserved.***

(c) In Figure 7.9c only one ball is included in the system, as the dashed outline indicates. The forces acting on this system are all external and include the weight W_1 of the ball and the normal force F_{N1}. As in part b of the drawing, these two forces balance. However, there is a third external force to consider. Ball 2 is outside the system, so the force F_{12} that it applies to the system during the collision is now an external force. As a result, the vector sum of the three external forces is not zero, and the net external force causes the total momentum of the one-ball system to be different after the collision than it is before the collision. ***The total momentum of this one-ball system is not conserved.***

(a)

(b)

(c)

Figure 7.9 (a) In this hypothetical situation, a hole opens up beneath two colliding billiard balls, which have weights W_1 and W_2. The dashed box emphasizes that both balls are included in the system. (b) The balls collide on a normal billiard table, where F_{N1} and F_{N2} are the normal forces that the table applies to the balls. Both balls are again included in the system. (c) The balls collide on a normal billiard table, but now only ball 1 is included in the system.

Next, we apply the principle of conservation of linear momentum to the problem of assembling a freight train.

EXAMPLE 5   Assembling a Freight Train

A freight train is being assembled in a switching yard, and Figure 7.10 shows two boxcars. Car 1 has a mass of $m_1 = 65 \times 10^3$ kg and moves at a velocity of $v_{01} = +0.80$ m/s. Car 2, with a mass of $m_2 = 92 \times 10^3$ kg and a velocity of $v_{02} = +1.3$ m/s, overtakes car 1 and couples to it. Neglecting friction, find the common velocity v_f of the cars after they become coupled.

Reasoning The two boxcars constitute the system. The sum of the external forces acting on the system is zero, because the weight of each car is balanced by a corre-

Figure 7.10 (a) The boxcar on the left eventually catches up with the other boxcar and (b) couples to it. The coupled cars move together with a common velocity after the collision.

(a) Before   (b) After

sponding normal force, and friction is being neglected. Thus, the system is isolated, and the principle of conservation of linear momentum applies. The coupling forces that each car exerts on the other are internal forces and do not affect the applicability of this principle.

Solution Momentum conservation indicates that

$$\underbrace{(m_1 + m_2)v_f}_{\substack{\text{Total momentum} \\ \text{after collision}}} = \underbrace{m_1 v_{01} + m_2 v_{02}}_{\substack{\text{Total momentum} \\ \text{before collision}}}$$

This equation can be solved for v_f, the common velocity of the two cars after the collision:

$$v_f = \frac{m_1 v_{01} + m_2 v_{02}}{m_1 + m_2}$$

$$= \frac{(65 \times 10^3 \text{ kg})(0.80 \text{ m/s}) + (92 \times 10^3 \text{ kg})(1.3 \text{ m/s})}{(65 \times 10^3 \text{ kg} + 92 \times 10^3 \text{ kg})} = \boxed{+1.1 \text{ m/s}}$$

In the previous example it can be seen that the velocity of car 1 increases, while the velocity of car 2 decreases as a result of the collision. The acceleration and deceleration arise at the moment the cars become coupled, because the cars exert internal forces on each other. The powerful feature of the momentum conservation principle is that it allows us to determine the changes in velocity without knowing what the internal forces are. Example 6 further illustrates this feature.

(a)

(b)

Figure 7.11 (a) In the absence of friction, two skaters pushing on each other constitute an isolated system. (b) As the skaters move away, the total linear momentum of the system remains zero, which is what it was initially.

EXAMPLE 6   Ice Skaters

Starting from rest, two skaters "push off" against each other on smooth level ice, where friction is negligible. As Figure 7.11a shows, one is a woman ($m_1 = 54$ kg), and one is a man ($m_2 = 88$ kg). Part b of the drawing shows that the woman moves away with a velocity of $v_{f1} = +2.5$ m/s. Find the "recoil" velocity v_{f2} of the man.

Reasoning For a system consisting of the two skaters on level ice, the sum of the external forces is zero. This is because the weight of each skater is balanced by a corresponding normal force and friction is negligible. The skaters, then, constitute an isolated system, and the principle of conservation of linear momentum applies. We expect the man to have a smaller recoil speed for the following reason. The internal forces that the man and woman exert on each other during pushoff have equal magnitudes, but opposite directions, according to Newton's action–reaction law. The man, having the larger mass, experiences a smaller acceleration and, hence, a smaller recoil speed.

Solution The total momentum of the skaters before they push on each other is zero, since they are at rest. Momentum conservation requires that the total momentum remains zero after the skaters have separated, as in part b of the drawing:

$$\underbrace{m_1 v_{f1} + m_2 v_{f2}}_{\substack{\text{Total momentum} \\ \text{after pushing}}} = \underbrace{0}_{\substack{\text{Total momentum} \\ \text{before pushing}}}$$

Solving for the recoil velocity of the man gives

$$v_{f2} = \frac{-m_1 v_{f1}}{m_2} = \frac{-(54 \text{ kg})(2.5 \text{ m/s})}{88 \text{ kg}} = \boxed{-1.5 \text{ m/s}}$$

The minus sign indicates that the man moves to the left in the drawing. After the skaters separate, the total momentum of the system remains zero, because momentum is a vector quantity, and the momenta of the man and the woman have equal magnitudes but opposite directions.

It is important to realize that the total linear momentum may be conserved even when the kinetic energies of the individual parts of a system change. In Example 6, for instance, the initial kinetic energy is zero since the skaters are stationary. But after they push off, the skaters are moving, so each has kinetic energy. The kinetic energy changes, because work is done by the internal force that each skater exerts on the other. However, internal forces cannot change the total linear momentum of a system, since the total linear momentum of an isolated system is conserved in the presence of such forces.

When applying the principle of conservation of linear momentum, we have been following a definite reasoning strategy that is summarized below.

INTERACTIVE LEARNINGWARE #15

START
Visualization
Reasoning
Knowns and Unknowns
Modeling the Problem
Numerical Solution
FINISH

An astronaut is motionless in outer space. Upon command, the propulsion unit strapped to his back ejects some gas with a velocity of $+32$ m/s, and the astronaut recoils with a velocity of -0.30 m/s. After the gas is ejected, the mass of the astronaut is 160 kg. What is the mass of the ejected gas?

The principle of conservation of linear momentum can be applied in a wide variety of situations. This Interactive LearningWare example deals with a practical matter that designers of space suits must consider. It illustrates that the masses of the various parts of the system play critical roles in applications of the conservation principle.

Go to our Web site at

http://www.wiley.com/college/cutnell

REASONING STRATEGY

Applying the Principle of Conservation of Linear Momentum

1. Decide which objects are included in the system.

2. Relative to the system that you have chosen, identify the internal forces and the external forces.

3. Verify that the system is isolated. In other words, verify that the sum of the external forces applied to the system is zero. Only if this sum is zero can the conservation principle be applied. If the sum of the external forces is not zero, consider a different system for analysis.

4. Set the total final momentum of the isolated system equal to the total initial momentum. Remember that linear momentum is a vector. If necessary, apply the conservation principle separately to the various vector components.

7.3 COLLISIONS IN ONE DIMENSION

As discussed in the last section, the total linear momentum is conserved when two objects collide, provided they constitute an isolated system. When the objects are atoms or subatomic particles, the total kinetic energy of the system is often conserved also. In other words, the total kinetic energy of the particles before the collision equals the total kinetic energy of the particles after the collision, so that kinetic energy gained by one particle is lost by another.

In contrast, when two macroscopic objects collide, such as two cars, the total kinetic energy after the collision is generally less than that before the collision. Kinetic energy is lost mainly in two ways: (1) it can be converted into heat because of friction, and (2) it is spent in creating permanent distortion or damage, as in an automobile collision. With very hard objects, such as a solid steel ball and a marble floor, the permanent distortion suffered upon collision is much smaller than with softer objects and, consequently, less kinetic energy is lost.

Collisions are often classified according to whether the total kinetic energy changes during the collision:

1. *Elastic collision*—One in which the total kinetic energy of the system after the collision is equal to the total kinetic energy before the collision.

2. *Inelastic collision*—One in which the total kinetic energy of the system is *not* the same before and after the collision; if the objects stick together after colliding, the collision is said to be completely inelastic.

The boxcars coupling together in Figure 7.10 is an example of a completely inelastic collision. When a collision is completely inelastic, the greatest amount of kinetic energy is lost. Example 7 shows how one particular elastic collision is described using the conservation of linear momentum and the fact that no kinetic energy is lost.

EXAMPLE 7   A Collision in One Dimension

- **PROBLEM SOLVING INSIGHT**
As long as the net external force is zero, the conservation of linear momentum applies to any type of collision. This is true whether the collision is elastic or inelastic.

As in Figure 7.12, a ball of mass $m_1 = 0.250$ kg and velocity $v_{01} = +5.00$ m/s collides head-on with a ball of mass $m_2 = 0.800$ kg that is initially at rest ($v_{02} = 0$ m/s). No external forces act on the balls. If the collision is elastic, what are the velocities of the balls after the collision?

Reasoning The total linear momentum of the two-ball system is conserved, because no external forces act on the system. Momentum conservation applies whether or not the collision is elastic:

$$\underbrace{m_1 v_{f1} + m_2 v_{f2}}_{\substack{\text{Total momentum} \\ \text{after collision}}} = \underbrace{m_1 v_{01} + 0}_{\substack{\text{Total momentum} \\ \text{before collision}}}$$

Before and after an elastic collision, the total kinetic energy is the same.

$$\underbrace{\tfrac{1}{2} m_1 v_{f1}^2 + \tfrac{1}{2} m_2 v_{f2}^2}_{\substack{\text{Total kinetic energy} \\ \text{after collision}}} = \underbrace{\tfrac{1}{2} m_1 v_{01}^2 + 0}_{\substack{\text{Total kinetic energy} \\ \text{before collision}}}$$

We expect that ball 1, having the smaller mass, will rebound to the left after striking ball 2, which is more massive. Ball 2 will be driven to the right in the process. One solution to the equations above, then, should give a value for v_{f1} that is negative (the left direction in Figure 7.12) and a value for v_{f2} that is positive.

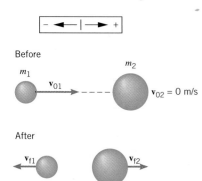

Before

After

Figure 7.12 A 0.250-kg ball, traveling with an initial velocity of $v_{01} = +5.00$ m/s, undergoes an elastic collision with a 0.800-kg ball that is initially at rest.

Solution The equations above are simultaneous equations containing the two unknown quantities v_{f1} and v_{f2}. To solve them, we begin by rearranging the equation expressing momentum conservation to show that $v_{f2} = m_1(v_{01} - v_{f1})/m_2$. Substituting this result into the equation expressing the conservation of kinetic energy leads to the following equation for v_{f1}:

$$v_{f1} = \left(\frac{m_1 - m_2}{m_1 + m_2}\right) v_{01} \qquad (7.8a)$$

The expression for v_{f1} in Equation 7.8a can be substituted into either of the equations obtained in the Reasoning to show that

$$v_{f2} = \left(\frac{2m_1}{m_1 + m_2}\right) v_{01} \qquad (7.8b)$$

With the given values for m_1, m_2, and v_{01}, Equations 7.8 yield the following values for v_{f1} and v_{f2}:

$$\boxed{v_{f1} = -2.62 \text{ m/s}} \quad \text{and} \quad \boxed{v_{f2} = +2.38 \text{ m/s}}$$

The negative value for v_{f1} indicates that ball 1 rebounds to the left after the collision in Figure 7.12, while the positive value for v_{f2} indicates that ball 2 moves to the right, as expected.

We can get a feel for an elastic collision by dropping a steel ball onto a hard surface, such as a marble floor. If the collision is elastic, the ball will rebound to its original height, as Figure 7.13a illustrates. In contrast, a partially deflated basketball exhibits little rebound from a relatively soft asphalt surface, as in part b, indicating that a fraction of the ball's kinetic energy is dissipated during the inelastic collision. The completely deflated basketball in part c has no bounce at all, and a maximum amount of kinetic energy is lost during the completely inelastic collision.

The next example illustrates a completely inelastic collision in a device called a "ballistic pendulum." This device can be used to measure the speed of a bullet.

EXAMPLE 8   A Ballistic Pendulum

A ballistic pendulum is sometimes used in laboratories to measure the speed of a projectile, such as a bullet. The ballistic pendulum shown in Figure 7.14a consists of a block of wood (mass $m_2 = 2.50$ kg) suspended by a wire of negligible mass. A bullet (mass $m_1 = 0.0100$ kg) is fired with a speed v_{01}. Just after the bullet collides with it, the block (with the bullet in it) has a speed v_f and then swings to a maximum height of 0.650 m above the initial position (see part b of the drawing). Find the speed v_{01} of the bullet, assuming that air resistance is negligible.

Reasoning The physics of the ballistic pendulum can be divided into two parts. The first is the completely inelastic collision between the bullet and the block. The second is the resulting motion of the block and bullet as they swing upward. The total momentum of the system (block plus bullet) is conserved during the collision, because the suspension wire supports the system's weight, which means that the sum of the external forces acting on the system is nearly zero. Furthermore, as the system swings upward, the principle of conservation of mechanical energy applies, because nonconservative forces do no work. The tension force in the wire does no work because it acts perpendicular to the motion. Since air resistance is negligible, we can ignore the work it does.

Solution Applying the momentum conservation principle gives

$$\underbrace{(m_1 + m_2)v_f}_{\substack{\text{Total momentum} \\ \text{after collision}}} = \underbrace{m_1 v_{01}}_{\substack{\text{Total momentum} \\ \text{before collision}}}$$

This equation can be solved for the initial speed v_{01} of the bullet:

$$v_{01} = \frac{m_1 + m_2}{m_1} v_f$$

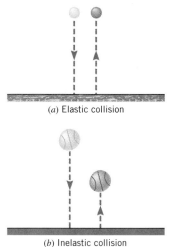

(a) Elastic collision

(b) Inelastic collision

(c) Completely inelastic collision

Figure 7.13 (a) A hard steel ball would rebound to its original height after striking a hard marble surface if the collision were elastic. (b) A partially deflated basketball has little bounce on a soft asphalt surface. (c) A deflated basketball has no bounce at all.

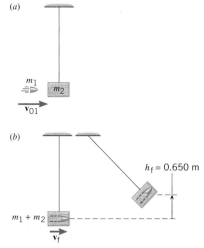

Figure 7.14 (a) A bullet approaches a ballistic pendulum. (b) The block and bullet swing upward after the collision.

THE PHYSICS OF...

measuring the speed of a bullet.

A value is now needed for the speed v_f immediately after the collision. It can be obtained from the maximum height to which the system swings, by using the principle of conservation of mechanical energy:

$$\underbrace{(m_1 + m_2)gh_f}_{\substack{\text{Total mechanical energy} \\ \text{at the top of the swing,} \\ \text{all potential}}} = \underbrace{\tfrac{1}{2}(m_1 + m_2)v_f^2}_{\substack{\text{Total mechanical energy} \\ \text{at the bottom of the} \\ \text{swing, all kinetic}}}$$

Solving this expression for v_f gives $v_f = \sqrt{2gh_f}$. Substituting this result into the equation for v_{01}, we find that

$$v_{01} = \left(\frac{m_1 + m_2}{m_1}\right)\sqrt{2gh_f} = \left(\frac{0.0100 \text{ kg} + 2.50 \text{ kg}}{0.0100 \text{ kg}}\right)\sqrt{2(9.80 \text{ m/s}^2)(0.650 \text{ m})}$$

$$= \boxed{+896 \text{ m/s}}$$

7.4 COLLISIONS IN TWO DIMENSIONS

The collisions discussed so far have been "head-on" or one-dimensional, for the velocities of the objects all point along a single line before and after contact. Collisions often occur, however, in two or three dimensions. Figure 7.15 shows a two-dimensional case in which two balls collide on a horizontal frictionless table.

For the system consisting of the two balls, the external forces include the weights of the balls and the corresponding normal forces produced by the table. Since each weight is balanced by a normal force, the sum of the external forces is zero, and the total momentum of the system is conserved, as Equation 7.7 indicates ($\mathbf{P_f} = \mathbf{P_0}$). Momentum is a vector quantity, however, and in two dimensions the x and y components of the total momentum are conserved separately. In other words, Equation 7.7 is equivalent to the following two equations:

x Component

$$\underbrace{m_1 v_{f1x} + m_2 v_{f2x}}_{P_{fx}} = \underbrace{m_1 v_{01x} + m_2 v_{02x}}_{P_{0x}} \qquad (7.9\text{a})$$

y Component

$$\underbrace{m_1 v_{f1y} + m_2 v_{f2y}}_{P_{fy}} = \underbrace{m_1 v_{01y} + m_2 v_{02y}}_{P_{0y}} \qquad (7.9\text{b})$$

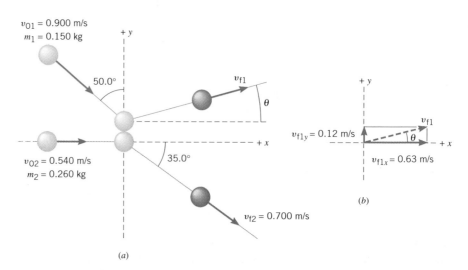

Figure 7.15   (*a*) Top view of two balls colliding on a horizontal frictionless table. (*b*) This part of the drawing shows the x and y components of the velocity of ball 1 after the collision.

These equations are written for a system that contains two objects. If a system contains more than two objects, a mass-times-velocity term must be included for each additional object on either side of Equations 7.9a and 7.9b. Example 9 shows how to deal with a two-dimensional collision when the total linear momentum of the system is conserved.

EXAMPLE 9   A Collision in Two Dimensions

For the data given in Figure 7.15a, use momentum conservation to determine the magnitude and direction of the velocity of ball 1 after the collision.

Reasoning The magnitude and direction of the velocity of ball 1 can be obtained once the components v_{f1x} and v_{f1y} of this velocity are known. We will begin by determining these components with the aid of the momentum conservation principle.

Solution Applying momentum conservation (Equation 7.9a), we find that

x Component

$$\underbrace{(0.150 \text{ kg})(v_{f1x})}_{\text{Ball 1, after}} + \underbrace{(0.260 \text{ kg})(0.700 \text{ m/s})(\cos 35.0°)}_{\text{Ball 2, after}}$$

$$= \underbrace{(0.150 \text{ kg})(0.900 \text{ m/s})(\sin 50.0°)}_{\text{Ball 1, before}} + \underbrace{(0.260 \text{ kg})(0.540 \text{ m/s})}_{\text{Ball 2, before}}$$

This equation can be solved to show that $v_{f1x} = +0.63$ m/s. Applying momentum conservation (Equation 7.9b), we find that

y Component

$$\underbrace{(0.150 \text{ kg})(v_{f1y})}_{\text{Ball 1, after}} + \underbrace{(0.260 \text{ kg})[-(0.700 \text{ m/s})(\sin 35.0°)]}_{\text{Ball 2, after}}$$

$$= \underbrace{(0.150 \text{ kg})[-(0.900 \text{ m/s})(\cos 50.0°)]}_{\text{Ball 1, before}} + \underbrace{0}_{\text{Ball 2, before}}$$

The solution to this equation reveals that $v_{f1y} = +0.12$ m/s.

Figure 7.15b shows the x and y components of the final velocity of ball 1. The magnitude of the velocity is

$$v_{f1} = \sqrt{(0.63 \text{ m/s})^2 + (0.12 \text{ m/s})^2} = \boxed{0.64 \text{ m/s}}$$

The direction of the velocity is given by the angle θ:

$$\theta = \tan^{-1}\left(\frac{0.12 \text{ m/s}}{0.63 \text{ m/s}}\right) = \boxed{11°}$$

INTERACTIVE LEARNINGWARE #16

START
Visualization
Reasoning
Knowns and Unknowns
Modeling the Problem
Numerical Solution
FINISH

A 1080-kg car moves with a speed of 28.0 m/s and is headed 30.0° north of east. A second car has a mass of 1630 kg and is headed due south. A third car has a mass of 1350 kg and is headed due west. What are the speeds of the second and third cars if the total linear momentum of the three-car system is zero?

The total linear momentum of a system is the sum of the individual momentum vectors of the objects that comprise the system. When adding vectors, it is often convenient to use vector components, which we first studied in Chapter 1. This Interactive LearningWare example provides an opportunity to review the use of components in the context of linear momentum.

Go to our Web site at

http://www.wiley.com/college/cutnell

● **PROBLEM SOLVING INSIGHT**
Momentum, being a vector quantity, has a magnitude and a direction. In two dimensions, take into account the direction by using vector components and assigning a plus or minus sign to each component, as illustrated in this example.

7.5 CENTER OF MASS

In previous sections, we have encountered situations in which objects interact with one another, such as the two skaters pushing off in Example 6. In these situations, the mass of the system is located in several places, and the various contributions to the total move about before, after, and even during the interaction. It is possible, however, to speak of a kind of average location for the total mass by introducing a concept known

Figure 7.16 The center of mass "cm" of the two particles is located on a line between them and lies closer to the more massive one.

as the **center of mass** (abbreviated as "cm"). With the aid of this concept, we will be able to gain additional insight into the principle of conservation of linear momentum.

The center of mass is a point that represents the average location for the total mass of a system. Figure 7.16, for example, shows two particles of mass m_1 and m_2 that are located on the x axis at the positions x_1 and x_2, respectively. The position x_{cm} of the center-of-mass point from the origin is defined to be

Center of mass
$$x_{cm} = \frac{m_1 x_1 + m_2 x_2}{m_1 + m_2}$$
(7.10)

Each term in the numerator of this equation is the product of a particle's mass and position, while the denominator is the total mass of the system. If the two masses are equal, we expect the average location of the total mass to be midway between the particles. With $m_1 = m_2 = m$, Equation 7.10 becomes $x_{cm} = (mx_1 + mx_2)/(m + m) = \frac{1}{2}(x_1 + x_2)$, which indeed corresponds to the point midway between the particles. Alternatively, suppose that $m_1 = 5.0$ kg and $x_1 = 2.0$ m, while $m_2 = 12$ kg and $x_2 = 6.0$ m. Then we expect the average location of the total mass to be located closer to particle 2, since it is more massive. Equation 7.10 is also consistent with this expectation, for it gives

$$x_{cm} = \frac{(5.0 \text{ kg})(2.0 \text{ m}) + (12 \text{ kg})(6.0 \text{ m})}{5.0 \text{ kg} + 12 \text{ kg}} = 4.8 \text{ m}$$

If a system contains more than two particles, the center-of-mass point can be determined by generalizing Equation 7.10. For three particles, for instance, the numerator would contain a third term $m_3 x_3$, and the total mass in the denominator would be $m_1 + m_2 + m_3$. For a macroscopic object, which contains many, many particles, the center-of-mass point is located at the geometric center of the object, provided that the mass is distributed symmetrically about the center. Such would be the case for a billiard ball. For objects such as a golf club, the mass is not distributed symmetrically and the center-of-mass point is not located at the geometric center of the club. The driver used to launch a golf ball from the tee, for instance, has more mass in the club head than in the handle, so the center-of-mass point is closer to the head than to the handle.

To see how the center-of-mass concept is related to momentum conservation, suppose that the two particles in a system are moving, as they would be during a collision. With the aid of Equation 7.10, we can determine the velocity v_{cm} of the center-of-mass point. During a time interval Δt, the particles experience displacements of Δx_1 and Δx_2, as Figure 7.17 shows. They have different displacements during this time, because they have different velocities. Equation 7.10 can be used to find the displacement Δx_{cm} of the center of mass by replacing x_{cm} by Δx_{cm}, x_1 by Δx_1, and x_2 by Δx_2:

$$\Delta x_{cm} = \frac{m_1 \Delta x_1 + m_2 \Delta x_2}{m_1 + m_2}$$

Now we divide both sides of this equation by the time interval Δt. In the limit as Δt becomes infinitesimally small, the ratio $\Delta x_{cm}/\Delta t$ becomes equal to the instantaneous velocity v_{cm} of the center of mass. (See Section 2.2 for a review of instantaneous velocity.) Likewise, the ratios $\Delta x_1/\Delta t$ and $\Delta x_2/\Delta t$ become equal to the instantaneous velocities v_1 and v_2, respectively. Thus, we have

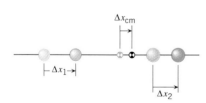

Figure 7.17 During a time interval Δt, the displacements of the particles are Δx_1 and Δx_2, while the displacement of the center of mass is Δx_{cm}.

Velocity of center of mass
$$v_{cm} = \frac{m_1 v_1 + m_2 v_2}{m_1 + m_2}$$
(7.11)

The numerator $(m_1v_1 + m_2v_2)$ on the right-hand side in Equation 7.11 is the momentum of particle 1 (m_1v_1) plus the momentum of particle 2 (m_2v_2), which is the total linear momentum of the system. In an isolated system, the total linear momentum does not change because of an interaction such as a collision. Therefore, Equation 7.11 indicates that the velocity v_{cm} of the center of mass does not change either. To emphasize this important point, consider the collision discussed in Example 7. With the data from that example, we can apply Equation 7.11 to determine the velocity of the center of mass before and after the collision:

Before collision
$$v_{cm} = \frac{(0.250 \text{ kg})(+5.00 \text{ m/s}) + (0.800 \text{ kg})(0 \text{ m/s})}{0.250 \text{ kg} + 0.800 \text{ kg}} = +1.19 \text{ m/s}$$

After collision
$$v_{cm} = \frac{(0.250 \text{ kg})(-2.62 \text{ m/s}) + (0.800 \text{ kg})(+2.38 \text{ m/s})}{0.250 \text{ kg} + 0.800 \text{ kg}} = +1.19 \text{ m/s}$$

Thus, the velocity of the center of mass is the same before and after the objects interact during a collision in which the total linear momentum is conserved.

7.6 CONCEPTS & CALCULATIONS

Momentum is one of the most fundamental concepts in physics, and energy is another. As we have seen in this chapter, momentum is a vector and, like all vectors, has a magnitude and a direction. In contrast, energy is a scalar quantity, as Chapter 6 discusses, and does not have direction associated with it. Example 10 provides the opportunity to review how the vector nature of momentum and the scalar nature of kinetic energy influence calculations using these quantities.

CONCEPTS & CALCULATIONS EXAMPLE 10   A Scalar and a Vector

Two joggers, Jim and Tom, are both running at a speed of 4.00 m/s. Jim has a mass of 90.0 kg, while Tom has a mass of 55.0 kg. Find the kinetic energy and momentum of the two-jogger system when (a) Jim and Tom are both running due north and (b) Jim is running due north and Tom is running due south.

Concept Questions and Answers   Does the total kinetic energy have a smaller value in case (a) or (b) or neither?

Answer   Everything is the same in both cases, except that both joggers are running due north in (a), but one is running due north and one is running due south in (b). Kinetic energy is a scalar quantity and does not depend on directional differences such as these. Therefore, the kinetic energy of the two-jogger system is the same in both cases.

Does the total momentum have a smaller magnitude in case (a) or (b) or neither?

Answer   Momentum is a vector. As a result, when we add Jim's momentum to Tom's momentum to get the total for the two-jogger system, we must take the directions into account. For example, we can use positive to denote north and negative to denote south. In case (a), two positive momentum vectors are added, while in case (b) one positive and one negative vector are added. Due to the partial cancellation in case (b), the resulting total momentum will have a smaller magnitude than in case (a).

Solution

(a) The kinetic energy of the two-jogger system can be obtained by applying the definition of kinetic energy as $\frac{1}{2}mv^2$ (see Equation 6.2) to each jogger:

$$KE_{system} = \tfrac{1}{2}m_{Jim}v_{Jim}^2 + \tfrac{1}{2}m_{Tom}v_{Tom}^2$$
$$= \tfrac{1}{2}(90.0 \text{ kg})(4.00 \text{ m/s})^2 + \tfrac{1}{2}(55.0 \text{ kg})(4.00 \text{ m/s})^2 = \boxed{1160 \text{ J}}$$

The momentum of each jogger is his mass times his velocity, according to the definition given in Equation 7.2. Using P_{system} to denote the momentum of the two-jogger system and denoting due north as the positive direction, we have

$$P_{system} = m_{Jim}v_{Jim} + m_{Tom}v_{Tom}$$
$$= (90.0 \text{ kg})(+4.00 \text{ m/s}) + (55.0 \text{ kg})(+4.00 \text{ m/s}) = \boxed{580 \text{ kg} \cdot \text{m/s}}$$

(b) Proceeding as in part (a), we find
$$KE_{system} = \tfrac{1}{2}m_{Jim}v_{Jim}^2 + \tfrac{1}{2}m_{Tom}v_{Tom}^2$$
$$= \tfrac{1}{2}(90.0 \text{ kg})(4.00 \text{ m/s})^2 + \tfrac{1}{2}(55.0 \text{ kg})(-4.00 \text{ m/s})^2 = \boxed{1160 \text{ J}}$$
$$P_{system} = m_{Jim}v_{Jim} + m_{Tom}v_{Tom}$$
$$= (90.0 \text{ kg})(+4.00 \text{ m/s}) + (55.0 \text{ kg})(-4.00 \text{ m/s}) = \boxed{140 \text{ kg} \cdot \text{m/s}}$$

In these calculations, we note that Tom's velocity is in the due south direction, so it is -4.00 m/s, not $+4.00$ m/s.

Momentum and kinetic energy are not the same concepts, and the next example explores further some of the differences between them.

CONCEPTS & CALCULATIONS EXAMPLE 11   Momentum and Kinetic Energy

The following table gives mass and speed data for two objects. Find the magnitude of the momentum and the kinetic energy for each object.

	Mass	Speed
Object A	2.0 kg	6.0 m/s
Object B	6.0 kg	2.0 m/s

Concept Questions and Answers   Is it possible for two objects to have different speeds and yet each has a momentum that has the same magnitude?

Answer   Yes. The magnitude of an object's momentum is the product of its mass and its speed, according to Equation 7.2. Both mass and speed play a role. The speed can be reduced, for instance, but if the mass is increased proportionally, the magnitude of the momentum will be the same.

If two objects have the same momentum, do they necessarily have the same kinetic energy?

Answer   No. Kinetic energy is $KE = \tfrac{1}{2}mv^2$. Momentum is $p = mv$, which means that $v = p/m$. Substituting into the kinetic energy expression shows that $KE = \tfrac{1}{2}m(p/m)^2 = p^2/(2m)$. Thus, two objects can have the same momentum p, but if each has different mass m, their kinetic energies will be different.

Solution   Using Equation 7.2 for the momentum and Equation 6.2 for the kinetic energy, we find that

Momenta

$$p_A = m_A v_A = (2.0 \text{ kg})(6.0 \text{ m/s}) = \boxed{12 \text{ kg} \cdot \text{m/s}}$$
$$p_B = m_B v_B = (6.0 \text{ kg})(2.0 \text{ m/s}) = \boxed{12 \text{ kg} \cdot \text{m/s}}$$

Kinetic energy

$$KE_A = \tfrac{1}{2}m_A v_A^2 = \tfrac{1}{2}(2.0 \text{ kg})(6.0 \text{ m/s})^2 = \boxed{36 \text{ J}}$$
$$KE_B = \tfrac{1}{2}m_B v_B^2 = \tfrac{1}{2}(6.0 \text{ kg})(2.0 \text{ m/s})^2 = \boxed{12 \text{ J}}$$

At the end of the problem set for this chapter, you will find homework problems that contain both conceptual and quantitative parts. These problems are grouped under the heading *Concepts & Calculations, Group Learning Problems*. They are designed for use by students working alone or in small learning groups. The conceptual part of each problem provides a convenient focus for group discussions.

SUMMARY

The **impulse** of a force is the product of the average force $\overline{\mathbf{F}}$ and the time interval Δt during which the force acts: **Impulse** $= \overline{\mathbf{F}} \Delta t$. Impulse is a vector that points in the same direction as the average force.

The **linear momentum p** of an object is the product of the object's mass m and velocity $\mathbf{v}$, or $\mathbf{p} = m\mathbf{v}$. Linear momentum is a vector that points in the direction of the velocity. The total momentum of a system of objects is the vector sum of the momenta of the individual objects.

The **impulse–momentum theorem** states that an impulse produces a change in an object's momentum, according to $\overline{\mathbf{F}} \Delta t = m\mathbf{v_f} - m\mathbf{v_0}$, where $m\mathbf{v_f}$ is the final momentum and $m\mathbf{v_0}$ is the initial momentum.

The **principle of conservation of linear momentum** states that the total linear momentum of an isolated system remains constant. An isolated system is one for which the sum of the external forces acting on the system is zero. External forces are those that agents external to the system exert on objects within the system. In contrast, internal forces are those that the objects within the system exert on each other.

An **elastic collision** is one in which the total kinetic energy is the same before and after the collision. An **inelastic collision** is one in which the total kinetic energy is not the same before and after the collision. If the objects stick together after colliding, the collision is said to be **completely inelastic.**

The location x_{cm} of the **center of mass** of two objects lying on the x axis is given by $x_{cm} = \dfrac{m_1 x_1 + m_2 x_2}{m_1 + m_2}$, where m_1 and m_2 are the masses of the objects, and x_1 and x_2 are their positions relative to the coordinate origin. If the objects move with velocities v_1 and v_2, the velocity v_{cm} of the center of mass is $v_{cm} = \dfrac{m_1 v_1 + m_2 v_2}{m_1 + m_2}$. If the total linear momentum of a system of objects remains constant during an interaction such as a collision, the velocity of the center of mass also remains constant.

CONCEPTUAL QUESTIONS

1. Two identical automobiles have the same speed, one traveling east and one traveling west. Do these cars have the same momentum? Explain.

2. In Times Square in New York City, people celebrate on New Year's Eve. Some just stand around, but many move about randomly. Consider a system comprised of all of these people. Approximately, what is the total linear momentum of this system at any given instant? Justify your answer.

3. Two objects have the same momentum. Do the velocities of these objects necessarily have (a) the same directions and (b) the same magnitudes? Give your reasoning in each case.

4. (a) Can a single object have kinetic energy but no momentum? (b) Can a system of two or more objects have a total kinetic energy that is not zero but a total momentum that is zero? Account for your answers.

5. An airplane is flying horizontally with a constant momentum during a time interval Δt. (a) With the aid of Equation 7.4, decide whether a net impulse is acting on the plane during this time interval. (b) In the horizontal direction, both the thrust generated by the engines and air resistance act on the plane. What does the answer in part (a) imply about the impulse of the thrust compared to the impulse of the resistive force?

6. You have a choice. You may get hit head-on either by an adult moving slowly on a bicycle or by a child that is moving twice as fast on a bicycle. The mass of the child and bicycle is one-half that of the adult and bicycle. Considering only the issues of mass and velocity, which collision do you prefer? Or doesn't it matter? Account for your answer.

7. An object slides along the surface of the earth and slows down because of kinetic friction. If the object itself is considered as the system, the kinetic frictional force must be identified as an external force that, according to Equation 7.4, decreases the momentum of the system. (a) If *both* the object and the earth are considered to be part of the system, is the force of kinetic friction still an external force? (b) Can the friction force change the total linear momentum of the two-body system? Give your reasoning for both answers.

8. When driving a golf ball, a good "follow-through" helps to increase the distance of the drive. A good follow-through means that the club head is kept in contact with the ball as long as possible. Using the impulse–momentum theorem, explain why this technique allows you to hit the ball farther.

9. The drawing shows a garden sprinkler that whirls around a vertical axis. From each of the four arms of the sprinkler, water

exits through a tapered nozzle. Because of this nozzle, the water leaves
each nozzle with a speed that is
greater than the speed inside the arm.
(a) Apply the impulse–momentum
theorem to deduce the direction of
the force applied to the water. (b)
Then, with the aid of Newton's third
law, explain how the water causes the
whirling motion.

10. In movies, Superman hovers in midair, grabs a villain by the
neck, and throws him forward. Superman, however, remains stationary. Using the conservation of linear momentum, explain
what is wrong with this scene.

11. A satellite explodes in outer space, far from any other body,
sending thousands of pieces in all directions. How does the linear
momentum of the satellite before the explosion compare with the
total linear momentum of all the pieces after the explosion? Account for your answer.

12. You are a passenger on a jetliner that is flying at a constant
velocity. You get up from your seat and walk toward the front of
the plane. Because of this action, your forward momentum increases. What, if anything, happens to the forward momentum of
the plane? Give your reasoning.

13. An ice boat is coasting along on a frozen lake. Friction between the ice and the boat is negligible, and so is air resistance.
Nothing is propelling the boat. From a bridge someone jumps
straight down and lands in the boat, which continues to coast
straight ahead. (a) Does the horizontal momentum of the boat

change? (b) Does the speed of the boat increase, decrease, or remain the same? Explain your answers.

14. On a distant asteroid, a large catapult is used to "throw"
chunks of stone into space. Could such a device be used as a propulsion system to move the asteroid closer to the earth? Explain.

15. A collision occurs between three moving billiard balls such
that no net external force acts on the three-ball system. Is the momentum of *each* ball conserved during the collision? If so, explain why. If not, what quantity is conserved?

16. In an elastic collision, is the kinetic energy of *each* object the
same before and after the collision? Explain.

17. Review Example 7. Now, suppose both objects have the
same mass, $m_1 = m_2$. Describe what happens to the velocities of
both objects as a result of the collision, using Equations 7.8a and
7.8b to justify your answers.

18. Where would you expect the center of mass of a doughnut to
be located? Why?

19. Would you expect the center of mass of a baseball bat to
be located halfway between the ends of the bat, nearer the
lighter end or nearer the heavier end? Provide a reason for your
answer.

20. A sunbather is lying on a floating raft that is stationary. She
then gets up and walks to one end of the raft. Consider the sunbather and raft as an isolated system. (a) What is the velocity of
the center of mass of this system while she is walking? Why? (b)
Does the raft itself move while she is walking? If so, what is the
direction of the raft's velocity relative to that of the sunbather?
Provide a reason for your answer.

PROBLEMS

Section 7.1 The Impulse–Momentum Theorem

1. ssm A soccer player kicks a ball with an average force of
$+1400$ N, and her foot remains in contact with the ball for a time
of 7.9×10^{-3} s. What is the impulse (magnitude and direction)
of this force?

2. Two men pushing a stalled car generate a net force of $+680$ N
for 7.2 s. What is the final momentum of the car?

3. A 62.0-kg person, standing on a diving board, dives straight
down into the water. Just before striking the water, her speed
is 5.50 m/s. At a time of 1.65 s after entering the water, her
speed is reduced to 1.10 m/s. What is the average net force
(magnitude and direction) that acts on her when she is in the
water?

4. Two arrows are fired horizontally with the same speed of
30.0 m/s. Each arrow has a mass of 0.100 kg. One is fired due
east and the other due south. Find the magnitude and direction of

the total momentum of this two-arrow system. Specify the direction with respect to due east.

5. ssm A volleyball is spiked so that its incoming velocity of
$+4.0$ m/s is changed to an outgoing velocity of -21 m/s. The
mass of the volleyball is 0.35 kg. What impulse does the player
apply to the ball?

6. Refer to Conceptual Example 3 as an aid in understanding this
problem. A hockey goalie is standing on ice. Another player fires
a puck ($m = 0.17$ kg) at the goalie with a velocity of $+65$ m/s.
(a) If the goalie catches the puck with his glove in a time of $5.0 \times
10^{-3}$ s, what is the average force (magnitude and direction) ex-
erted on the goalie by the puck? (b) Instead of catching the puck,
the goalie slaps it with his stick and returns the puck straight back
to the player with a velocity of -65 m/s. The puck and stick are
in contact for a time of 5.0×10^{-3} s. Now, what is the average
force exerted on the goalie by the puck? Verify that your answers

to parts (a) and (b) are consistent with the conclusion of Conceptual Example 3.

7. A 46-kg skater is standing still in front of a wall. By pushing against the wall she propels herself backward with a velocity of -1.2 m/s. Her hands are in contact with the wall for 0.80 s. Ignore friction and wind resistance. Find the magnitude and direction of the average force she exerts on the wall (which has the same magnitude, but opposite direction, as the force that the wall applies to her).

8. ⚄ When jumping straight down, you can be seriously injured if you land stiff-legged. One way to avoid injury is to bend your knees upon landing to reduce the force of the impact. A 75-kg man just before contact with the ground has a speed of 6.4 m/s. (a) In a stiff-legged landing he comes to a halt in 2.0 ms. Find the average net force that acts on him during this time. (b) When he bends his knees, he comes to a halt in 0.10 s. Find the average net force now. (c) During the landing, the force of the ground on the man points upward, while the force due to gravity points downward. The average net force acting on the man includes both of these forces. Taking into account the directions of the forces, find the force of the ground on the man in parts (a) and (b).

*9. ssm www A golf ball strikes a hard, smooth floor at an angle of 30.0° and, as the drawing shows, rebounds at the same angle. The mass of the ball is 0.047 kg, and its speed is 45 m/s just before and after striking the floor. What is the magnitude of the impulse applied to the golf

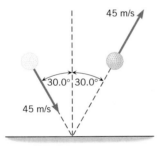

ball by the floor? *(Hint: Note that only the vertical component of the ball's momentum changes during impact with the floor, and ignore the weight of the ball.)*

*10. A basketball ($m = 0.60$ kg) is dropped from rest. Just before striking the floor, the magnitude of the basketball's momentum is 3.1 kg·m/s. At what height was the basketball dropped?

*11. A stream of water strikes a stationary turbine blade horizontally, as the drawing illustrates. The incident water stream has a velocity of $+18.0$ m/s, while the exiting water stream has a velocity of

-18.0 m/s. The mass of water per second that strikes the blade is 25.0 kg/s. Find the magnitude of the average force exerted on the water by the blade.

*12. An 85-kg jogger is heading due east at a speed of 2.0 m/s. A 55-kg jogger is heading 32° north of east at a speed of 3.0 m/s. Find the magnitude and direction of the sum of the momenta of the two joggers.

*13. A 0.500-kg ball is dropped from rest at a point 1.20 m above the floor. The ball rebounds straight upward to a height of 0.700 m. What are the magnitude and direction of the impulse of the net force applied to the ball during the collision with the floor?

**14. A dump truck is being filled with sand. The sand falls straight downward from rest from a height of 2.00 m above the truck bed, and the mass of sand that hits the truck per second is 55.0 kg/s. The truck is parked on the platform of a weight scale. By how much does the scale reading exceed the weight of the truck and sand?

Section 7.2 The Principle of Conservation of Linear Momentum

15. ⚄ For tests using a *ballistocardiograph,* a patient lies on a horizontal platform that is supported on jets of air. Because of the air jets, the friction impeding the horizontal motion of the platform is negligible. Each time the heart beats, blood is pushed out of the heart in a direction that is nearly parallel to the platform. Since momentum must be conserved, the body and the platform recoil, and this recoil can be detected to provide information about the heart. For each beat, suppose that 0.050 kg of blood is pushed out of the heart with a velocity of $+0.25$ m/s and that the mass of the patient and platform is 85 kg. Assuming that the patient does not slip with respect to the platform, and that the patient and platform start from rest, determine the recoil velocity.

16. With the engines off, a spaceship is coasting at a velocity of $+230$ m/s through outer space. The ship carries rockets that are mounted in firing tubes, the back ends of which are closed. It fires a rocket straight ahead at an enemy vessel. The mass of the rocket is 1300 kg, and the mass of the spaceship (not including the rocket) is 4.0×10^6 kg. The firing of the rocket brings the spaceship to a halt. What is the velocity of the rocket?

17. ssm Batman (mass = 91 kg) jumps straight down from a bridge into a boat (mass = 510 kg) in which a criminal is fleeing. The velocity of the boat is initially $+11$ m/s. What is the velocity of the boat after Batman lands in it?

18. Two friends, Al and Jo, have a combined mass of 168 kg. At an ice skating rink they stand close together on skates, at rest and facing each other, with a compressed spring between them. The spring is kept from pushing them apart, because they are holding each other. When they release their arms, Al moves off in one direction at a speed of 0.90 m/s, while Jo moves off in the opposite direction at a speed of 1.2 m/s. Assuming that friction is negligible, find Al's mass.

19. The lead female character in the movie *Diamonds Are Forever* is standing at the edge of an offshore oil rig. As she fires a gun, she is driven back over the edge and into the sea. Suppose the mass of a bullet is 0.010 kg and its velocity is $+720$ m/s. Her mass (including the gun) is 51 kg. (a) What recoil velocity does she acquire in response to a single shot from a stationary position, assuming that no external force keeps her in place? (b) Under the same assumption, what would be her recoil velocity if, instead, she shoots a blank cartridge that ejects a mass of 5.0×10^{-4} kg at a velocity of $+720$ m/s?

*20. Adolf and Ed are wearing harnesses and are hanging from the ceiling by means of ropes attached to them. They are face to face and push off against one another. Adolf has a mass of 120 kg, and Ed has a mass of 78 kg. Following the push, Adolf swings upward to a height of 0.65 m above his starting point. To what height above his starting point does Ed rise?

*21. ssm During July 1995 the comet Shoemaker–Levy 9 smashed into Jupiter in a spectacular fashion. The comet actually consisted of 21 distinct pieces, the largest of which had a mass of about 4.0×10^{12} kg and a speed of 6.0×10^4 m/s. Jupiter, the largest planet in the solar system, has a mass of 1.9×10^{27} kg and an orbital speed of 1.3×10^4 m/s. If this piece of the comet had hit Jupiter head on, what would have been the *change* (magnitude only) in Jupiter's orbital speed (not its final speed)?

*22. A fireworks rocket is moving at a speed of 50.0 m/s. The rocket suddenly breaks into two pieces of equal mass. They fly off with velocities $\mathbf{v_1}$ and $\mathbf{v_2}$, as shown in the drawing. What is the magnitude of (a) $\mathbf{v_1}$ and (b) $\mathbf{v_2}$?

Problem 22

**23. ssm Two people are standing on a 2.0-m-long platform, one at each end. The platform floats parallel to the ground on a cushion of air, like a hovercraft. One person throws a 6.0-kg ball to the other, who catches it. The ball travels nearly horizontally. Excluding the ball, the total mass of the platform and people is 118 kg. Because of the throw, this 118-kg mass recoils. How far does it move before coming to rest again?

**24. A wagon is coasting at a speed v_A along a straight and level road. When ten percent of the wagon's mass is thrown off the wagon, parallel to the ground and in the forward direction, the wagon is brought to a halt. If the direction in which this mass is thrown is exactly reversed, but the speed of this mass relative to the wagon remains the same, the wagon accelerates to a new speed v_B. Calculate the ratio v_B/v_A.

Section 7.3 and Section 7.4 Collisions in One and Two Dimensions

25. In a football game, a receiver is standing still, having just caught a pass. Before he can move, a tackler, running at a velocity of +4.5 m/s, grabs him. The tackler holds onto the receiver, and the two move off together with a velocity of +2.6 m/s. The mass of the tackler is 115 kg. Assuming that momentum is conserved, find the mass of the receiver.

26. Kevin has a mass of 87-kg and is skating with in-line skates. He sees his 22-kg younger brother up ahead standing on the sidewalk, with his back turned. Coming up from behind, he grabs his

brother and rolls off at a speed of 2.4 m/s. Ignoring friction, find Kevin's speed just before he grabbed his brother.

27. ssm A golf ball bounces down a flight of steel stairs, striking each stair once on the way down. The ball starts at the top step with a vertical velocity component of zero. If all the collisions with the stairs are elastic, and if the vertical height of the staircase is 3.00 m, determine the bounce height when the ball reaches the bottom of the stairs. Neglect air resistance.

28. A 3.00-kg block of wood rests on the muzzle opening of a vertically oriented rifle, the stock of the rifle being firmly planted on the ground. When the rifle is fired, an 8.00-g bullet (velocity = 8.00×10^2 m/s, straight upward) becomes completely embedded in the block. (a) Using the conservation of linear momentum, find the velocity of the block/bullet system immediately after the collision. (b) Ignoring air resistance, determine how high the block/bullet system rises above the muzzle opening of the rifle.

29. A cue ball (mass = 0.165 kg) is at rest on a frictionless pool table. The ball is hit dead center by a pool stick, which applies an impulse of +1.50 N·s to the ball. The ball then slides along the table and makes an elastic head-on collision with a second ball of equal mass that is initially at rest. Find the velocity of the second ball just after it is struck.

30. A 0.150-kg projectile is fired with a velocity of +715 m/s at a 2.00-kg wooden block that rests on a frictionless table. The velocity of the block, immediately after the projectile passes through it, is +40.0 m/s. Find the velocity with which the projectile exits from the block.

31. ssm www A 5.00-kg ball, moving to the right at a velocity of +2.00 m/s on a frictionless table, collides head-on with a stationary 7.50-kg ball. Find the final velocities of the balls if the collision is (a) elastic and (b) completely inelastic.

32. The drawing shows a collision between two pucks on an air-hockey table. Puck A has a mass of 0.025 kg and is moving along the x axis with a velocity of +5.5 m/s. It makes a collision with puck B, which has a mass of 0.050 kg and is initially at rest. The collision is not head-on. After the collision, the two pucks fly apart with the angles shown in the drawing. Find the final speed of (a) puck A and (b) puck B.

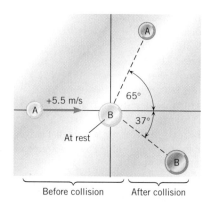

*33. A 0.010-kg bullet is fired straight up at a falling wooden block that has a mass of 2.0 kg. The bullet has a speed of

750 m/s when it strikes the block. The block originally was dropped from rest from the top of a building and had been falling for a time t when the collision with the bullet occurs. As a result of the collision, the block (with the bullet in it) reverses direction, rises, and comes to a momentary halt at the top of the building. Find the time t.

*34. Three guns are aimed at the center of a circle. They are mounted on the circle, 120.0° apart. They fire in a timed sequence, such that the three bullets collide at the center and mash into a *stationary* lump. Two of the bullets have identical masses of 4.50 g each and speeds of v_1 and v_2. The third bullet has a mass of 2.50 g and a speed of 575 m/s. Find the the unknown speeds.

*35. ssm A 50.0-kg skater is traveling due east at a speed of 3.00 m/s. A 70.0-kg skater is moving due south at a speed of 7.00 m/s. They collide and hold on to each other after the collision, managing to move off at an angle θ south of east, with a speed of v_f. Find (a) the angle θ and (b) the speed v_f, assuming that friction can be ignored.

*36. A mine car, whose mass is 440 kg, rolls at a speed of 0.50 m/s on a horizontal track, as the drawing shows. A 150-kg chunk of coal has a speed of 0.80 m/s when it leaves the chute. Determine the velocity of the car/coal system after the coal has come to rest in the car.

*37. A 60.0-kg person, running horizontally with a velocity of +3.80 m/s, jumps onto a 12.0-kg sled that is initially at rest. (a) Ignoring the effects of friction during the collision, find the velocity of the sled and person as they move away. (b) The sled and person coast 30.0 m on level snow before coming to rest. What is the coefficient of kinetic friction between the sled and the snow?

**38. A ball is dropped from rest at the top of a 6.10-m-tall building, falls straight downward, collides inelastically with the ground, and bounces back. The ball loses 10.0% of its kinetic energy every time it collides with the ground. How many bounces can the ball make and still reach a window sill that is 2.44 m above the ground?

**39. ssm Starting with an initial speed of 5.00 m/s at a height of 0.300 m, a 1.50-kg ball swings downward and strikes a 4.60-kg ball that is at rest, as the drawing shows. (a) Using the principle of conservation of mechanical energy, find the speed of the 1.50-kg ball just before impact. (b) Assuming that the collision is elastic, find the velocities (magnitude and direction) of both balls just after the collision. (c) How high does each ball swing after the collision, ignoring air resistance?

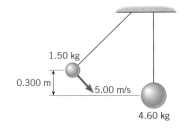

**40. Two identical balls are traveling toward each other with velocities of −4.0 and +7.0 m/s, and they experience an elastic head-on collision. Obtain the velocities (magnitude and direction) of each ball after the collision.

Section 7.5 Center of Mass

41. ssm The earth and moon are separated by a center-to-center distance of 3.85×10^8 m. The mass of the earth is 5.98×10^{24} kg and that of the moon is 7.35×10^{22} kg. How far does the center of mass lie from the center of the earth?

42. Two balls are approaching each other head-on. Their velocities are +9.70 and −11.8 m/s. Determine the velocity of the center of mass of the two balls if (a) they have the same mass and (b) if the mass of one ball ($v = 9.70$ m/s) is twice the mass of the other ball ($v = -11.8$ m/s).

43. The carbon monoxide molecule (CO) consists of a carbon atom and an oxygen atom separated by a distance of 1.13×10^{-10} m. The mass m_C of the carbon atom is 0.750 times the mass m_O of the oxygen atom: $m_C = 0.750 \, m_O$. Determine the location of the center of mass of this molecule relative to the carbon atom.

*44. Two particles have a total mass of 10.0 g. Particle A is located on the x axis at $x = 2.0$ cm, while particle B is located on the x axis at $x = 4.0$ cm. The center of mass of this two-particle system is located at $x = 3.4$ cm. Find the mass of each particle.

ADDITIONAL PROBLEMS

45. ssm www A lumberjack (mass = 98 kg) is standing at rest on one end of a floating log (mass = 230 kg) that is also at rest. The lumberjack runs to the other end of the log, attaining a velocity of +3.6 m/s relative to the shore, and then hops onto an identical floating log that is initially at rest. Neglect any friction and resistance between the logs and the water. (a) What is the velocity of the first log just before the lumberjack jumps off? (b) Determine the velocity of the second log if the lumberjack comes to rest on it.

46. A person is standing on a sheet of ice so slippery that friction may be ignored. This individual fires a gun parallel to the ground.

When a standard cartridge is used, a 14-g bullet is shot forward with a speed of 280 m/s, and the person recoils with a speed of v_C. When a blank cartridge is used, a mass of 0.10 g is shot forward with a speed of 55 m/s, and the recoil speed is v_B. Find the ratio v_B/v_C.

47. ssm In a science fiction novel two enemies, Bonzo and Ender, are fighting in outer space. From stationary positions they push against each other. Bonzo flies off with a velocity of +1.5 m/s, while Ender recoils with a velocity of −2.5 m/s. (a) Without doing any calculations, decide which person has the

greater mass. Give your reasoning. (b) Determine the ratio of the masses (m_{Bonzo}/m_{Ender}) of these two people.

48. (a) The earth travels with an approximate speed of 29.9 km/s (66 900 mi/h) in its journey around the sun. The mass of the earth is 5.98×10^{24} kg. Find the magnitude of its linear momentum. (b) Is the direction of the earth's linear momentum constant? If not, describe how it changes and specify the force that causes it to change.

49. A freight train moves due north with a speed of 1.4 m/s. The mass of the train is 4.5×10^5 kg. How fast would an 1800-kg automobile have to be moving due north to have the same momentum as the train?

50. Before starting this problem, review Conceptual Example 3. Suppose that the hail described there bounces off the roof of the car with a velocity of $+15$ m/s. Ignoring the weight of the hailstones, calculate the force exerted by the hail on the roof. Compare your answer to that obtained in Example 2 for the rain, and verify that your answer is consistent with the conclusion reached in Conceptual Example 3.

51. ssm A two-stage rocket moves in space at a constant velocity of 4900 m/s. The two stages are then separated by a small explosive charge placed between them. Immediately after the explosion the velocity of the 1200-kg upper stage is 5700 m/s in the same direction as before the explosion. What is the velocity (magnitude and direction) of the 2400-kg lower stage after the explosion?

52. Consider the two moving boxcars in Example 5. Determine the velocity of their center of mass (a) before and (b) after the collision. (c) Should your answer in part (b) be less than, greater than, or equal to the common velocity v_f of the two coupled cars after the collision? Justify your answer.

*53. ssm By accident, a large plate is dropped and breaks into three pieces. The pieces fly apart parallel to the floor. As the plate falls, its momentum has only a vertical component, and no component parallel to the floor. After the collision, the component of the total momentum parallel to the floor must remain zero, since the net external force acting on the plate has no component parallel to the floor. Using the data shown in the drawing, find the masses of pieces 1 and 2.

*54. A person stands in a stationary canoe and throws a 5.00-kg stone with a velocity of 8.00 m/s at an angle of 30.0° above the horizontal. The person and canoe have a combined mass of 105 kg. Ignoring air resistance and effects of the water, find the horizontal recoil velocity (magnitude and direction) of the canoe.

*55. An automobile has a mass of 2300 kg and a velocity of $+16$ m/s. It makes a rear-end collision with a stationary car whose mass is 1800 kg. The cars lock bumpers and skid off together with the wheels locked. (a) What is the velocity of the two cars just after the collision? (b) Find the impulse (magnitude and direction) that acts on the skidding cars from just after the collision until they come to a halt. (c) If the coefficient of kinetic friction between the wheels of the cars and the pavement is $\mu_k = 0.80$, determine how far the cars skid before coming to rest.

*56. The drawing shows the bond lengths and angles in the nitric acid (HNO_3) molecule, which is planar. The masses of the atoms are $m_H = 1.67 \times 10^{-27}$ kg, $m_N = 23.3 \times 10^{-27}$ kg, and $m_O = 26.6 \times 10^{-27}$ kg. Locate the center of mass of this molecule relative to the hydrogen atom. (*Hint: The oxygen atoms are located symmetrically on either side of the H−O−N line.*)

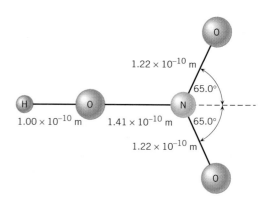

**57. ssm www A cannon of mass 5.80×10^3 kg is rigidly bolted to the earth so it can recoil only by a negligible amount. The cannon fires an 85.0-kg shell horizontally with an initial velocity of $+551$ m/s. Suppose the cannon is then unbolted from the earth, and no external force hinders its recoil. What would be the velocity of a shell fired by this loose cannon? (*Hint: In both cases assume that the burning gunpowder imparts the same kinetic energy to the system.*)

CONCEPTS   &   CALCULATIONS

GROUP LEARNING PROBLEMS

Note: Each of these problems consists of Concept Questions followed by a related quantitative Problem. They are designed for use by students working alone or in small learning groups. The Concept Questions involve little or no mathematics and are intended to stimulate group discussions. They focus on the concepts with which the problems deal. Recognizing the concepts is the essential initial step in any problem-solving technique.

58. Concept Questions In a performance test two cars take the same time to accelerate from rest up to the same speed. Car A has a mass of 1400 kg, while car B has a mass of 1900 kg. During the test, which car (a) has the greatest change in momentum, (b) experiences the greatest impulse, and (c) is acted upon by the greatest average net force? In each case, give your reasoning.

Problem Each car takes 9.0 s to accelerate from rest to 27 m/s. Find the average net force that acts on each car during the test. Verify that your answers are consistent with your answers to the Concept Questions.

59. Concept Questions As the drawing illustrates, two disks with masses m_1 and m_2 are moving horizontally to the right at a speed v_0. They are on an air-hockey table, which supports them with an essentially frictionless cushion of air. They move as a unit, with a compressed spring between them, which has a negligible mass. (a) When the spring is released and allowed to push outward, what are the directions of the forces that act on disk 1 and disk 2? (b) After the spring is released, is the speed of each disk larger, smaller, or the same as the speed v_0? Explain.

Problem Consider the situation where disk 1 comes to a momentary halt shortly after the spring is released. Assuming that $m_1 = 1.2$ kg, $m_2 = 2.4$ kg, and $v_0 = 5.0$ m/s, find the speed of disk 2 at that moment. Verify that your answer is consistent with your answers to the Concept Questions.

60. Concept Questions A wagon is rolling forward on level ground. Friction is negligible. The person sitting in the wagon throws a rock. Does the momentum of the wagon increase, decrease, or remain the same (a) when the rock is thrown directly forward and (b) when the rock is thrown directly backward? (c) In which case does the wagon have the greatest speed after the rock is thrown?

Problem The total mass of the wagon, rider, and the rock is 95.0 kg. The mass of the rock is 0.300 kg. Initially the wagon is rolling forward at a speed of 0.500 m/s. Then, the rock is thrown with a speed of 16.0 m/s. Both speeds are relative to the ground.

Find the speed of the wagon after the rock is thrown directly forward in one case and directly backward in another. Check to see that your answers are consistent with your answers to the Concept Questions.

61. Concept Questions One object is at rest and another is moving. Furthermore, one object is more massive than the other. The two collide in a one-dimensional completely inelastic collision. In other words, they stick together after the collision and move off with a common velocity. Momentum is conserved. (a) Consider the final momentum of the two-object system after the collision. Is it greater when the large-mass object is moving initially or the small-mass object is moving initially? (b) Is the final speed after the collision greater when the large-mass or the small-mass object is moving initially?

Problem The speed of the object that is moving initially is 25 m/s. The masses of the two objects are 3.0 and 8.0 kg. Determine the final speed of the two-object system after the collision for the case when the large-mass object is moving initially and the case when the small-mass object is moving initially. Be sure that your answers are consistent with your answers to the Concept Questions.

62. Concept Questions Object A is moving due east, while object B is moving due north. They collide and stick together in a completely inelastic collision. Momentum is conserved. (a) Is it possible that the two-object system has a final total momentum of zero after the collision? (b) Roughly, what is the direction of the final total momentum of the two-object system after the collision?

Problem Object A has a mass of $m_A = 17.0$ kg and an initial velocity of $\mathbf{v}_{0A} = 8.00$ m/s, due east. Object B has a mass of $m_B = 29.0$ kg and an initial velocity of $\mathbf{v}_{0B} = 5.00$ m/s, due north. Find the magnitude and direction of the total momentum of the two-object system after the collision. Make sure that your answers are consistent with your answers to the Concept Questions.

63. Concept Questions (a) Neil has a larger mass than Barbara. He is standing on the x axis at $x_N = +9.0$ m, while she is standing on the x axis at $x_B = +2.0$ m. Is their center-of-mass point closer to the 9.0-m point or the 2.0-m point? (b) They switch positions. Is their center-of-mass point now closer to the 9.0-m point or the 2.0-m point? (c) In which direction, toward or away from the origin, does their center of mass move as a result of the switch?

Problem Neil's mass is 86 kg, while Barbara's is 55 kg. How far and in which direction does their center of mass move as a result of the switch? Verify that your answer is consistent with your answers to the Concept Questions.

This rotating Ferris wheel illustrates the central theme of this chapter, rotational kinematics. Rotational motion is described by using the concepts of angular displacement, angular velocity, and angular acceleration, concepts that are analogous to those used in Chapters 2 and 3 for describing linear motion.

CHAPTER 8

ROTATIONAL KINEMATICS

8.1 ROTATIONAL MOTION AND ANGULAR DISPLACEMENT

In the simplest kind of rotation, points on a rigid object move on circular paths. In Figure 8.1, for example, we see the circular paths for points *A*, *B*, and *C* on a spinning skater. The centers of all such circular paths define a line, called the **axis of rotation.**

The angle through which a rigid object rotates about a fixed axis is called the **angular displacement.** Figure 8.2 shows how the angular displacement is measured for a rotating compact disc (CD). Here, the axis of rotation passes through the center of the disc and is perpendicular to its surface. On the surface of the CD we draw a radial line, which is a line that intersects the axis of rotation perpendicularly. As the CD turns, we observe the angle through which this line moves relative to a convenient reference line that does not rotate. The radial line moves from its initial orientation at angle θ_0 to a final orientation at angle θ (Greek letter theta). In the process, the line sweeps out the angle $\theta - \theta_0$. As with other differences that we have encountered ($\Delta x = x - x_0$, $\Delta v = v - v_0$, and $\Delta t = t - t_0$), it is customary to denote the difference between the final and initial angles by the notation $\Delta\theta$ (read as "delta theta"): $\Delta\theta = \theta - \theta_0$. The angle $\Delta\theta$ is the angular displacement. A rotating object may rotate either counterclockwise or clockwise, and standard convention calls a counterclockwise displacement positive and a clockwise displacement negative.

Figure 8.1 When an object rotates, points on the object, such as *A*, *B*, or *C*, move on circular paths. The centers of the circles form a line that is the axis of rotation.

● **DEFINITION OF ANGULAR DISPLACEMENT**

When a rigid body rotates about a fixed axis, the angular displacement is the angle $\Delta\theta$ swept out by a line passing through any point on the body and intersecting the axis of rotation perpendicularly. By convention, the angular displacement is positive if it is counterclockwise and negative if it is clockwise.

SI Unit of Angular Displacement: radian (rad)*

Angular displacement is often expressed in one of three units. The first is the familiar **degree,** and it is well known that there are 360 degrees in a circle. The second unit is the **revolution (rev),** one revolution representing one complete turn of 360°. The most useful unit from a scientific viewpoint is the SI unit called the **radian (rad).** Figure 8.3 shows how the radian is defined, again using a CD as an example. The picture focuses attention on a point *P* on the disc. This point starts out on the stationary reference line, so that $\theta_0 = 0$ rad, and the angular displacement is $\Delta\theta = \theta - \theta_0 = \theta$. As the disc rotates, the point traces out an arc of length *s*, which is measured along a circle of radius *r*. Equation 8.1 defines the angle θ in radians:

$$\theta \text{ (in radians)} = \frac{\text{Arc length}}{\text{Radius}} = \frac{s}{r} \qquad (8.1)$$

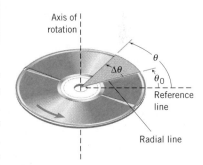

Figure 8.2 The angular displacement of a CD is the angle $\Delta\theta$ swept out by a radial line as the disc turns about its axis of rotation.

According to this definition, an angle in radians is the ratio of two lengths, for example, meters/meters. In calculations, therefore, the radian is treated as a unitless number and has no effect on other units that it multiplies or divides.

To convert between degrees and radians, it is only necessary to remember that the arc length of an entire circle of radius *r* is the circumference $2\pi r$. Therefore, ac-

Figure 8.3 In radian measure, the angle θ is defined to be the arc length *s* divided by the radius *r*.

---

* The radian is neither a base nor a derived SI unit. It is regarded as a supplementary SI unit.

cording to Equation 8.1, *the number of radians that corresponds to 360° or one revolution is*

$$\theta = \frac{2\pi r}{r} = 2\pi \text{ rad}$$

Since 2π rad corresponds to 360°, the number of degrees in one radian is

$$1 \text{ rad} = \frac{360°}{2\pi} = 57.3°$$

It is useful to express an angle θ in radians, because then the arc length s subtended at any radius r can be calculated by multiplying θ by r. Example 1 illustrates this point and also shows how to convert between degrees and radians.

THE PHYSICS OF...

communications satellites.

Figure 8.4 Two adjacent synchronous satellites have an angular separation of $\theta = 2.00°$. The distances and angles have been exaggerated for clarity.

THE PHYSICS OF...

a total solar eclipse.

EXAMPLE 1   Adjacent Synchronous Satellites

Synchronous or "stationary" communications satellites are put into an orbit whose radius is $r = 4.23 \times 10^7$ m, as Figure 8.4 illustrates. The orbit is in the plane of the equator, and two adjacent satellites have an angular separation of $\theta = 2.00°$. Find the arc length s (see the drawing) that separates the satellites.

Reasoning Since the radius r and the angle θ are known, we may find the arc length s by using the relation θ (in radians) $= s/r$. But first, the angle must be converted to radians from degrees.

Solution To convert 2.00° into radians, we use the fact that 2π radians is equivalent to 360°:

$$2.00° = (2.00 \text{ degrees})\left(\frac{2\pi \text{ radians}}{360 \text{ degrees}}\right) = 0.0349 \text{ radians}$$

From Equation 8.1, it follows that the arc length between the satellites is

$$s = r\theta = (4.23 \times 10^7 \text{ m})(0.0349 \text{ rad}) = \boxed{1.48 \times 10^6 \text{ m (920 miles)}}$$

The radian, being a unitless quantity, is dropped from the final result, leaving the answer expressed in meters.

Conceptual Example 2 takes advantage of the radian as a unit for measuring angles and explains the spectacular phenomenon of a total solar eclipse.

CONCEPTUAL EXAMPLE 2   A Total Eclipse of the Sun

The diameter of the sun is about 400 times greater than that of the moon. By coincidence, the sun is also about 400 times farther from the earth than is the moon. For an observer on earth, compare the angle subtended by the moon to the angle subtended by the sun, and show why this result leads to a total solar eclipse.

Reasoning and Solution Figure 8.5*a* shows a person on earth viewing the moon and the sun. The angle θ_{moon} subtended by the moon is given by Equation 8.1 as the arc length s_{moon} divided by the distance r_{moon} from the earth to the moon: $\theta_{\text{moon}} = s_{\text{moon}}/r_{\text{moon}}$. The earth and moon are sufficiently far apart that the arc length s_{moon} is very nearly equal to the moon's diameter. Similar reasoning applies for the sun, so that the angle subtended by the sun is $\theta_{\text{sun}} = s_{\text{sun}}/r_{\text{sun}}$. But since $s_{\text{sun}} \approx 400\, s_{\text{moon}}$ (where the symbol "$\approx$" means "approximately equal to") and $r_{\text{sun}} \approx 400\, r_{\text{moon}}$, the two angles are approximately equal: $\theta_{\text{moon}} \approx \theta_{\text{sun}}$. Figure 8.5*b* shows what happens when the moon comes between the sun and the earth. *Since the angle subtended by the moon is nearly equal to the angle subtended by the sun, the moon blocks most*

(a)

(b)

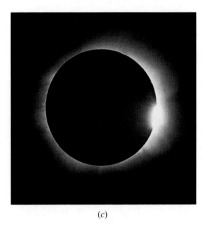

(c)

of the sun's light from reaching the observer's eyes, and a total solar eclipse like that in Figure 8.5c occurs.

Related Homework Material: *Problem 14*

Figure 8.5 (*a*) The angles subtended by the moon and sun at the eyes of the observer are θ_{moon} and θ_{sun}. (The distances and angles are exaggerated for the sake of clarity.) (*b*) Since the moon and sun subtend approximately the same angle, the moon blocks nearly all the sun's light from reaching the observer's eyes. (*c*) The result is a total solar eclipse.

8.2 ANGULAR VELOCITY AND ANGULAR ACCELERATION

Angular Velocity

In Section 2.2 we introduced the idea of linear velocity to describe how fast an object moves and the direction of its motion. A rotating object also exhibits motion. However, it is an angular type of motion that occurs about an axis of rotation. To describe the angular motion we now introduce the idea of angular velocity.

● *Concepts at a Glance*   To define the angular velocity of a rotating object, we use two concepts previously encountered. As the Concepts-at-a-Glance chart in Figure 8.6 illustrates, the angular velocity is obtained by combining the angular displacement and the time during which the displacement occurs. A comparison of this chart with that in Figure 2.2 shows that angular velocity is defined in a manner closely analogous to that used for linear velocity. Taking advantage of the similarity between the two types of velocities will help us understand rotational motion. ●

According to Equation 2.2 ($\overline{v} = \Delta x/\Delta t$), the average linear velocity is the linear displacement of the object divided by the time required for the displacement to occur. For rotational motion about a fixed axis, the *average angular velocity* $\overline{\omega}$ (Greek letter omega) is obtained in an analogous way, as the angular displacement divided by the elapsed time during which the displacement occurs.

● **DEFINITION OF AVERAGE ANGULAR VELOCITY**

$$\text{Average angular velocity} = \frac{\text{Angular displacement}}{\text{Elapsed time}}$$

$$\overline{\omega} = \frac{\theta - \theta_0}{t - t_0} = \frac{\Delta \theta}{\Delta t} \tag{8.2}$$

SI Unit of Angular Velocity: radian per second (rad/s)

Concepts at a Glance

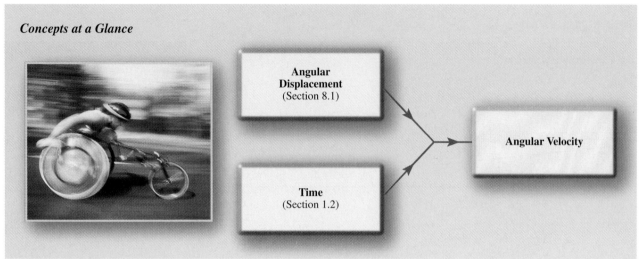

Figure 8.6 *Concepts at a Glance*
Angular displacement and time are combined to produce the concept of angular velocity. Each wheel on this racing cycle has an angular velocity that depends on how fast the wheel is spinning as well as the direction of its spin.

The SI unit for angular velocity is the radian per second (rad/s), although other units such as revolutions per minute (rev/min or rpm) are encountered. In agreement with the sign convention adopted for angular displacement, angular velocity is positive when the rotation is counterclockwise and negative when it is clockwise. Example 3 shows how the concept of average angular velocity is applied to a gymnast.

EXAMPLE 3   Gymnast on a High Bar

A gymnast on a high bar swings through two revolutions in a time of 1.90 s, as Figure 8.7 suggests. Find the average angular velocity (in rad/s) of the gymnast.

Reasoning The average angular velocity of the gymnast in rad/s is the angular displacement in radians divided by the elapsed time. However, the angular displacement is given as two revolutions, so we begin by converting this value into radian measure.

Solution The angular displacement (in radians) of the gymnast is

$$\Delta\theta = -2.00 \text{ revolutions} \left(\frac{2\pi \text{ radians}}{1 \text{ revolution}} \right) = -12.6 \text{ radians}$$

where the minus sign denotes that the gymnast rotates clockwise (see the drawing). The average angular velocity is

$$\overline{\omega} = \frac{\Delta\theta}{\Delta t} = \frac{-12.6 \text{ rad}}{1.90 \text{ s}} = \boxed{-6.63 \text{ rad/s}} \tag{8.2}$$

The *instantaneous angular velocity* ω is the angular velocity that exists at any given instant. To measure it, we follow the same procedure used in Chapter 2 for the instantaneous linear velocity. In this procedure, a small angular displacement $\Delta\theta$ occurs during a small time interval Δt. The time interval is so small that it approaches zero ($\Delta t \rightarrow 0$), and in this limit, the measured average angular velocity, $\overline{\omega} = \Delta\theta/\Delta t$, becomes the instantaneous angular velocity ω:

$$\omega = \lim_{\Delta t \to 0} \overline{\omega} = \lim_{\Delta t \to 0} \frac{\Delta\theta}{\Delta t} \tag{8.3}$$

Figure 8.7 Swinging on a high bar.

The magnitude of the instantaneous angular velocity, without reference to whether it is a positive or negative quantity, is called the **instantaneous angular speed.** If a rotating object has a constant angular velocity, the instantaneous value and the average value are the same.

Angular Acceleration

In linear motion, a changing velocity means that an acceleration is occurring. Such is also the case in rotational motion; a changing angular velocity means that an **angular acceleration** is occurring. There are many examples of angular acceleration. For instance, as a compact disc recording is played, the disc turns with an angular velocity that is continually decreasing. And when the push buttons of an electric blender are changed from a lower setting to a higher setting, the angular velocity of the blades increases.

● *Concepts at a Glance*   We now introduce the new idea of angular acceleration to describe how rapidly or slowly the angular velocity changes during a given time interval. The Concepts-at-a-Glance chart in Figure 8.8 illustrates that the notion of angular acceleration is formulated by bringing together the change in the angular velocity with the time required for the change to occur. A comparison of this chart with that in Figure 2.5 reveals the close analogy between angular acceleration and linear acceleration. This analogy, like that between angular velocity and linear velocity, will be useful in helping us to understand rotational motion. ●

When the linear velocity of an object changes, Equation 2.4 ($\overline{\mathbf{a}} = \Delta\mathbf{v}/\Delta t$) defines the average linear acceleration as the change in velocity per unit time. When the angular velocity changes from an initial value of ω_0 at time t_0 to a final value of

Figure 8.8 *Concepts at a Glance*
To define the concept of angular acceleration, we bring together the change in angular velocity and the time required for the change to occur. Each wheel on this mountain bike has an angular acceleration because the wheel's angular velocity changes as the bike accelerates down the hill.

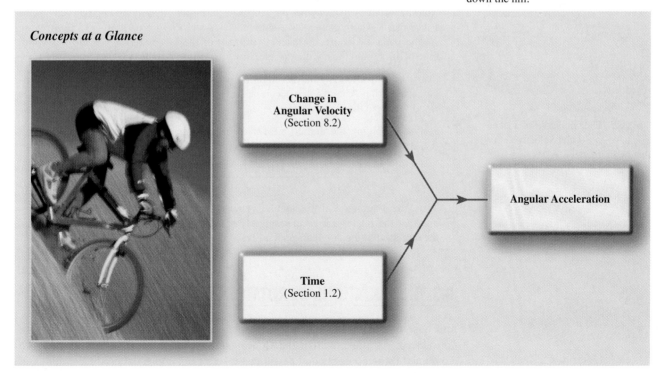

Concepts at a Glance

Change in
Angular Velocity
(Section 8.2)

Time
(Section 1.2)

Angular Acceleration

ω at time t, the average angular acceleration $\bar{\alpha}$ (Greek letter alpha) is defined similarly:

● **DEFINITION OF AVERAGE ANGULAR ACCELERATION**

$$\text{Average angular acceleration} = \frac{\text{Change in angular velocity}}{\text{Elapsed time}}$$

$$\bar{\alpha} = \frac{\omega - \omega_0}{t - t_0} = \frac{\Delta\omega}{\Delta t} \qquad (8.4)$$

SI Unit of Average Angular Acceleration: radian per second squared (rad/s^2)

The SI unit for average angular acceleration is the unit for angular velocity divided by the unit for time, or (rad/s)/s = rad/s^2. An angular acceleration of +5 rad/s^2, for example, means that the angular velocity of the rotating object increases by +5 radians per second during each second of acceleration.

The *instantaneous angular acceleration* α is the angular acceleration at a given instant. In discussing linear motion, we assumed a condition of constant acceleration, so that the average and instantaneous accelerations were identical ($\bar{\mathbf{a}} = \mathbf{a}$). Similarly, we assume here that the angular acceleration is constant, so that the instantaneous angular acceleration α and the average angular acceleration $\bar{\alpha}$ are the same ($\bar{\alpha} = \alpha$). The next example illustrates the concept of angular acceleration.

EXAMPLE 4   A Jet Revving Its Engines

A jet awaiting clearance for takeoff is momentarily stopped on the runway. As seen from the front of one engine, the fan blades are rotating with an angular velocity of −110 rad/s, where the negative sign indicates a clockwise rotation (see Figure 8.9). As the plane takes off, the angular velocity of the blades reaches −330 rad/s in a time of 14 s. Find the angular acceleration, assuming it to be constant.

Reasoning Since the angular acceleration is constant, it is equal to the average angular acceleration. The average acceleration is the change in the angular velocity, $\omega - \omega_0$, divided by the elapsed time, $t - t_0$.

Solution Applying the definition of average angular acceleration given in Equation 8.4, we find that

$$\bar{\alpha} = \frac{\omega - \omega_0}{t - t_0} = \frac{(-330 \text{ rad/s}) - (-110 \text{ rad/s})}{14 \text{ s}} = \boxed{-16 \text{ rad/s}^2}$$

Thus, the magnitude of the angular velocity increases by 16 rad/s during each second that the blades are accelerating. The negative sign in the answer indicates that the direction of the angular acceleration is also in the clockwise direction.

Figure 8.9 The fan blades of a jet engine have an angular velocity in a clockwise direction.

8.3 THE EQUATIONS OF ROTATIONAL KINEMATICS

In Chapters 2 and 3 the concepts of displacement, velocity, and acceleration were introduced. As illustrated in the Concepts-at-a-Glance chart in Figure 2.10, we combined these concepts and developed a set of equations called the equations of

Concepts at a Glance

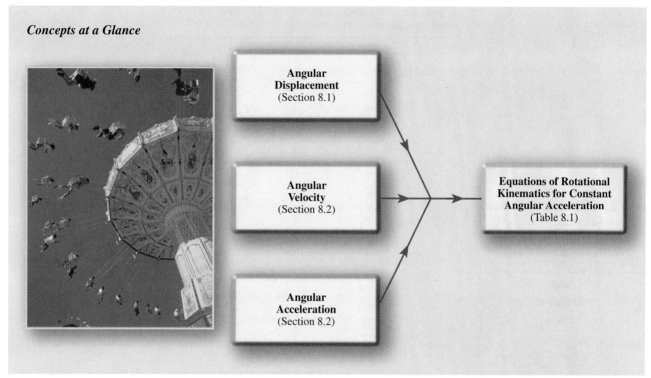

Figure 8.10 *Concepts at a Glance* The equations of rotational kinematics for constant angular acceleration are obtained by integrating the notions of angular displacement, angular velocity, and angular acceleration. The rotational motion of this swing ride can be described by using the equations of rotational kinematics, provided that the angular acceleration is constant.

kinematics for constant acceleration (see Tables 2.1 and 3.1). These equations were a great aid in solving problems involving linear motion in one and two dimensions.

● *Concepts at a Glance* We now take a similar approach for rotational motion. The Concepts-at-a-Glance chart in Figure 8.10, which is analogous to that in Figure 2.10, shows that we will bring together the ideas of angular displacement, angular velocity, and angular acceleration to produce a set of equations called the equations of kinematics for constant angular acceleration. These equations, like those developed in Chapters 2 and 3, will prove very useful in solving problems that involve rotational motion. ●

A complete description of rotational motion requires values for the angular displacement $\Delta\theta$, the angular acceleration α, the final angular velocity ω, the initial angular velocity ω_0, and the elapsed time Δt. In Example 4, for instance, only the angular displacement of the fan blades during the 14-s interval is missing. Such missing information can be calculated, however. For convenience in the calculations, we assume that the orientation of the rotating object is given by $\theta_0 = 0$ rad at time $t_0 = 0$ s. Then, the angular displacement becomes $\Delta\theta = \theta - \theta_0 = \theta$, and the time interval becomes $\Delta t = t - t_0 = t$.

In Example 4, the angular velocity of the fan blades changes at a constant rate from an initial value of $\omega_0 = -110$ rad/s to a final value of $\omega = -330$ rad/s. Therefore, the average angular velocity is midway between the initial and final values:

$$\overline{\omega} = \tfrac{1}{2}[(-110 \text{ rad/s}) + (-330 \text{ rad/s})] = -220 \text{ rad/s}$$

TABLE 8.1 The Equations of Kinematics for Rotational and Linear Motion

Rotational Motion (α = constant)		Linear Motion (a = constant)	
$\omega = \omega_0 + \alpha t$	(8.4)	$v = v_0 + at$	(2.4)
$\theta = \frac{1}{2}(\omega_0 + \omega)t$	(8.6)	$x = \frac{1}{2}(v_0 + v)t$	(2.7)
$\theta = \omega_0 t + \frac{1}{2}\alpha t^2$	(8.7)	$x = v_0 t + \frac{1}{2}at^2$	(2.8)
$\omega^2 = \omega_0^2 + 2\alpha\theta$	(8.8)	$v^2 = v_0^2 + 2ax$	(2.9)

TABLE 8.2 Symbols Used in Rotational and Linear Kinematics

Rotational Motion	Quantity	Linear Motion
θ	Displacement	x
ω_0	Initial velocity	v_0
ω	Final velocity	v
α	Acceleration	a
t	Time	t

In other words, when the angular acceleration is constant, the average angular velocity is given by

$$\overline{\omega} = \frac{1}{2}(\omega_0 + \omega) \tag{8.5}$$

With a value for the average angular velocity, Equation 8.2 can be used to obtain the angular displacement of the fan blades:

$$\theta = \overline{\omega}t = (-220 \text{ rad/s})(14 \text{ s}) = -3100 \text{ rad}$$

In general, when the angular acceleration is constant, the angular displacement can be obtained from

$$\theta = \overline{\omega}t = \frac{1}{2}(\omega_0 + \omega)t \tag{8.6}$$

This equation and Equation 8.4 provide a complete description of rotational motion under the condition of constant angular acceleration. Equation 8.4 (with $t_0 = 0$ s) and Equation 8.6 are compared with the analogous results for linear motion in the first two lines of Table 8.1. The purpose of this comparison is to emphasize that the mathematical forms of Equations 8.4 and 2.4 are identical, as are the forms of Equations 8.6 and 2.7. Of course, the symbols used for the rotational variables are different from those used for the linear variables, as Table 8.2 indicates.

In Chapter 2, Equations 2.4 and 2.7 are used to derive the remaining two equations of kinematics (Equations 2.8 and 2.9). These additional equations convey no new information but are convenient to have when solving problems. Similar derivations can be carried out here. The results are listed as Equations 8.7 and 8.8 below and in Table 8.1; they can be inferred directly from their counterparts in linear motion by making the substitution of symbols indicated in Table 8.2:

$$\theta = \omega_0 t + \frac{1}{2}\alpha t^2 \tag{8.7}$$

$$\omega^2 = \omega_0^2 + 2\alpha\theta \tag{8.8}$$

The four equations in the left column of Table 8.1 are called the **equations of rotational kinematics for constant angular acceleration.** The following example illustrates that they are used in the same fashion as the equations of linear kinematics.

Figure 8.11 The angular velocity of the blades in an electric blender changes each time a different push button is chosen.

EXAMPLE 5   Blending with a Blender

The blades of an electric blender are whirling with an angular velocity of $+375$ rad/s while the "puree" button is pushed in, as Figure 8.11 shows. When the "blend" button is pressed, the blades accelerate and reach a greater angular velocity after the blades have rotated through an angular displacement of $+44.0$ rad (seven revolutions). The angular acceleration has a constant value of $+1740$ rad/s^2. Find the final angular velocity of the blades.

Reasoning The three known variables are listed in the table below, along with a question mark indicating that a value for the final angular velocity ω is being sought.

θ	α	ω	ω_0	t
+44.0 rad	+1740 rad/s^2	?	+375 rad/s	

We can use Equation 8.8, because it relates the angular variables θ, α, ω, and ω_0.

Solution From Equation 8.8 we find that

$$\omega^2 = \omega_0^2 + 2\alpha\theta$$
$$= (375 \text{ rad/s})^2 + 2(1740 \text{ rad/s}^2)(44.0 \text{ rad})$$
$$= 2.94 \times 10^5 \text{ rad}^2/\text{s}^2$$

$$\boxed{\omega = +542 \text{ rad/s}}$$

The negative root is disregarded, since the blades do not reverse their direction of rotation.

The equations of rotational kinematics can be used with any self-consistent set of units for θ, α, ω, ω_0, and t. Radians are used in Example 5 only because data are given in terms of radians. Had the data for θ, α, and ω_0 been provided in rev, rev/s^2, and rev/s, respectively, then Equation 8.8 could have been used to determine the answer for ω directly in rev/s. In any case, the reasoning strategy for applying the kinematics equations is summarized below.

INTERACTIVE LEARNINGWARE #17

After 10.0 s, a spinning roulette wheel has slowed down to an angular velocity of +1.88 rad/s, where the plus sign indicates a counterclockwise rotation. During this time, the wheel rotates through an angular displacement of +44.0 rad. Determine the angular acceleration of the wheel.

This problem provides great practice in using the equations of kinematics for rotational motion, because they are used twice in arriving at the answer. They are employed to determine the initial angular velocity of the roulette wheel and then again to find its angular acceleration. Since the wheel is slowing down, we will see that the direction of the angular acceleration is opposite to that of the angular velocity.

Go to our Web site at

http://www.wiley.com/college/cutnell

REASONING STRATEGY

Applying the Equations of Rotational Kinematics

1. Make a drawing to represent the situation being studied, showing the direction of rotation.

2. Decide which direction of rotation is to be called positive (+) and which direction is to be called negative (−). In this text we choose the counterclockwise direction to be positive and the clockwise direction to be negative, but this is arbitrary. However, do not change your decision during the course of a calculation.

3. In an organized way, write down the values (with appropriate + and − signs) that are given for any of the five rotational kinematic variables (θ, α, ω, ω_0, and t). Be on the alert for implied data, such as the phrase "starts from rest," which means that the value of the initial angular velocity is $\omega_0 = 0$ rad/s. The data box in Example 5 is a good way to keep track of this information. In addition, identify the variable(s) that you are being asked to determine.

4. Before attempting to solve a problem, verify that the given information contains values for at least three of the five kinematic variables. Once the three variables are identified for which values are known, the appropriate relation from Table 8.1 can be selected.

• **PROBLEM SOLVING INSIGHT**
Each equation of rotational kinematics contains four of the five kinematic variables, θ, α, ω, ω_0, and t. Therefore, it is necessary to have values for three of these variables if one of the equations is to be used to determine a value for the unknown variable.

5. When the rotational motion is divided into segments, the final angular velocity of one segment becomes the initial angular velocity for the next segment.

6. Keep in mind that there may be two possible answers to a kinematics problem. Try to visualize the different physical situations to which the answers correspond.

8.4 ANGULAR VARIABLES AND TANGENTIAL VARIABLES

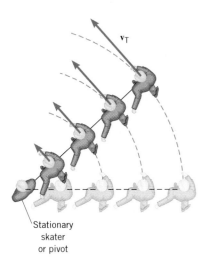

Figure 8.12 When doing a stunt known as "crack-the-whip," each skater along the radial line moves on a circular arc. The tangential velocity $\mathbf{v_T}$ of each skater is represented by an arrow that is tangent to each arc.

In the familiar ice-skating stunt known as "crack-the-whip," a number of skaters attempt to maintain a straight line as they skate around the one person (the pivot) who remains in place. Figure 8.12 shows each skater moving on a circular arc and includes the corresponding velocity vector at the instant portrayed in the picture. For every individual skater, the vector is drawn tangent to the appropriate circle and, therefore, is called the *tangential velocity* $\mathbf{v_T}$. The magnitude of the tangential velocity is referred to as the *tangential speed.*

Of all the skaters involved in the stunt, the one farthest from the pivot has the hardest job. Why? Because, in keeping the line straight, this skater covers more distance than anyone else. To accomplish this, he must skate faster than anyone else and, thus, must have the largest tangential speed. In fact, the line remains straight only if each person skates with the correct tangential speed. The skaters closer to the pivot must move with smaller tangential speeds than those farther out, as indicated by the magnitudes of the tangential velocity vectors in Figure 8.12.

With the aid of Figure 8.13, it is possible to show that the tangential speed of any skater is directly proportional to his distance r from the pivot, assuming a given angular speed for the rotating line. When the line rotates as a rigid unit for a time t, it sweeps out the angle θ shown in the drawing. The distance s through which a skater moves along a circular arc can be calculated from Equation 8.1, $s = r\theta$, provided θ is measured in radians. Dividing both sides of this equation by t gives $s/t = r(\theta/t)$. The term s/t is the tangential speed v_T (e.g., in meters/second) of the skater, while θ/t is the angular speed ω (in radians/second) of the line:

$$v_T = r\omega \qquad (\omega \text{ in rad/s}) \qquad (8.9)$$

Figure 8.13 During a time t, the line of skaters sweeps through an angle θ. An individual skater, located at a distance r from the stationary skater, moves through a distance s on a circular arc.

In this expression, the terms v_T and ω refer to the magnitudes of the tangential and angular velocities, respectively, and are numbers without algebraic signs.

It is important to emphasize that the angular speed ω in $v_T = r\omega$ must be expressed in radian measure (e.g., in rad/s); no other units, such as revolutions per second, are acceptable. This restriction arises because the equation was derived by using the definition of radian measure, $s = r\theta$.

The real challenge for the "crack-the-whip" skaters is to keep the line straight while making it pick up angular speed, that is, while giving it an angular acceleration. To make the angular speed of the line increase, each skater must increase his tangential speed, since the two speeds are related according to $v_T = r\omega$. Of course, the fact that a skater must skate faster and faster means that he must accelerate, and his tangential acceleration a_T can be related to the angular acceleration α of the line. If time is measured relative to $t_0 = 0$ s, the definition of linear acceleration is given by Equation 2.4 as $a_T = (v_T - v_{T0})/t$, where v_T and v_{T0} are the final and initial tangential speeds, respectively. Substituting $v_T = r\omega$ for the tangential speed shows that

$$a_T = \frac{v_T - v_{T0}}{t} = \frac{(r\omega) - (r\omega_0)}{t} = r\left(\frac{\omega - \omega_0}{t}\right)$$

Since $\alpha = (\omega - \omega_0)/t$ according to Equation 8.4, it follows that

$$a_T = r\alpha \quad (\alpha \text{ in rad/s}^2) \tag{8.10}$$

This result shows that, for a given value of α, the tangential acceleration a_T is proportional to the radius r, so the skater farthest from the pivot must have the largest tangential acceleration. In this expression, the terms a_T and α refer to the magnitudes of the numbers involved, without reference to any algebraic sign. Moreover, as is the case for ω in $v_T = r\omega$, only radian measure can be used for α in $a_T = r\alpha$.

There is an advantage to using the angular velocity ω and the angular acceleration α to describe the rotational motion of a rigid object. The advantage is that these angular quantities describe the motion of the *entire object*. In contrast, the tangential quantities v_T and a_T describe only the motion of a single point on the object, and Equations 8.9 and 8.10 indicate that different points located at different distances r have different tangential velocities and accelerations. Example 6 stresses this advantage.

EXAMPLE 6   A Helicopter Blade

A helicopter blade has an angular speed of $\omega = 6.50$ rev/s and an angular acceleration of $\alpha = 1.30$ rev/s^2. For points 1 and 2 on the blade in Figure 8.14, find the magnitudes of (a) the tangential speeds and (b) the tangential accelerations.

Reasoning Since the radius r of each point and the angular speed ω of the helicopter blade are known, we can find the tangential speed v_T of each point by using the relation $v_T = r\omega$. However, since this equation can be used only with radian measure, the angular speed ω must be converted to rad/s from rev/s. In a similar manner, the tangential acceleration a_T of points 1 and 2 can be found using $a_T = r\alpha$, provided the angular acceleration α is expressed in rad/s^2 rather than in rev/s^2.

Solution

(a) Converting the angular speed ω to rad/s from rev/s, we obtain

$$\omega = \left(6.50 \frac{\text{rev}}{\text{s}}\right)\left(\frac{2\pi \text{ rad}}{1 \text{ rev}}\right) = 40.8 \frac{\text{rad}}{\text{s}}$$

The tangential speed of each point is

Point 1 $\qquad v_T = r\omega = (3.00 \text{ m})(40.8 \text{ rad/s}) = \boxed{122 \text{ m/s (273 mph)}}$ $\qquad$ (8.9)

Point 2 $\qquad v_T = r\omega = (6.70 \text{ m})(40.8 \text{ rad/s}) = \boxed{273 \text{ m/s (611 mph)}}$ $\qquad$ (8.9)

The rad unit, being dimensionless, does not appear in the final answers.

(b) Converting the angular acceleration α to rad/s^2 from rev/s^2, we find

$$\alpha = \left(1.30 \frac{\text{rev}}{\text{s}^2}\right)\left(\frac{2\pi \text{ rad}}{1 \text{ rev}}\right) = 8.17 \frac{\text{rad}}{\text{s}^2}$$

The tangential accelerations can now be determined:

Point 1 $\qquad a_T = r\alpha = (3.00 \text{ m})(8.17 \text{ rad/s}^2) = \boxed{24.5 \text{ m/s}^2}$ $\qquad$ (8.10)

Point 2 $\qquad a_T = r\alpha = (6.70 \text{ m})(8.17 \text{ rad/s}^2) = \boxed{54.7 \text{ m/s}^2}$ $\qquad$ (8.10)

Figure 8.14 Points 1 and 2 on the rotating blade of the helicopter have the same angular speed and acceleration, but they have *different* tangential speeds and accelerations.

● **PROBLEM SOLVING INSIGHT**
When using Equations 8.9 and 8.10 to relate tangential and angular quantities, remember that the angular quantity is always expressed in radian measure. These equations are not valid if angles are expressed in degrees or revolutions.

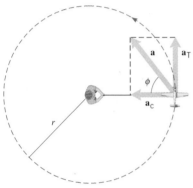

(a) Uniform circular motion

(b) Nonuniform circular motion

Figure 8.15 (a) If a model airplane flying on a guide wire has a constant tangential speed, the motion is uniform circular motion, and the plane experiences only a centripetal acceleration $\mathbf{a}_c$. (b) Nonuniform circular motion occurs when the tangential speed changes. Then there is a tangential acceleration $\mathbf{a}_T$ in addition to the centripetal acceleration.

8.5 CENTRIPETAL ACCELERATION AND TANGENTIAL ACCELERATION

When an object picks up speed as it moves around a circle, it has a tangential acceleration, as discussed in the last section. In addition, the object also has a centripetal acceleration, as emphasized in Chapter 5. That chapter deals with *uniform circular motion,* in which a particle moves at a constant tangential speed on a circular path. The tangential speed v_T is the magnitude of the tangential velocity vector. Even when the magnitude of the tangential velocity is constant, an acceleration is present, since the direction of the velocity changes continually. Because the resulting acceleration points toward the center of the circle, it is called the centripetal acceleration. Figure 8.15a shows the centripetal acceleration $\mathbf{a}_c$ for a model airplane flying in uniform circular motion on a guide wire. The magnitude of $\mathbf{a}_c$ is

$$a_c = \frac{v_T^2}{r} \tag{5.2}$$

The subscript "T" has now been included in this equation as a reminder that it is the tangential speed that appears in the numerator.

The centripetal acceleration can be expressed in terms of the angular speed ω by using $v_T = r\omega$ (Equation 8.9):

$$a_c = \frac{v_T^2}{r} = \frac{(r\omega)^2}{r} = r\omega^2 \qquad (\omega \text{ in rad/s}) \tag{8.11}$$

Only radian measure, such as rad/s, can be used for ω in this result, since the relation $v_T = r\omega$ presumes radian measure.

While considering uniform circular motion in Chapter 5, we ignored the details of how the motion is established in the first place. In Figure 8.15b, for instance, the engine of the plane produces a thrust in the tangential direction, and this force leads to a tangential acceleration. In response, the tangential speed of the plane increases from moment to moment, until the situation shown in the drawing results. While the tangential speed is changing, the motion is called *nonuniform circular motion.*

Figure 8.15b illustrates an important feature of nonuniform circular motion. Since the direction and the magnitude of the tangential velocity are both changing, the airplane experiences two acceleration components simultaneously. The changing direction means that there is a centripetal acceleration $\mathbf{a}_c$. The magnitude of $\mathbf{a}_c$ at any moment can be calculated using the value of the instantaneous angular speed and the radius: $a_c = r\omega^2$. The fact that the magnitude of the tangential velocity is changing means that there is also a tangential acceleration $\mathbf{a}_T$. The magnitude of $\mathbf{a}_T$ can be determined from the angular acceleration α according to $a_T = r\alpha$, as the previous section explains. If the magnitude F_T of the net tangential force and the mass m are known, a_T also can be calculated using Newton's second law, $F_T = ma_T$. Figure 8.15b shows the two acceleration components. The total acceleration is given by the vector sum of $\mathbf{a}_c$ and $\mathbf{a}_T$. Since $\mathbf{a}_c$ and $\mathbf{a}_T$ are perpendicular, the magnitude of the total acceleration $\mathbf{a}$ can be obtained from the Pythagorean theorem as $a = \sqrt{a_c^2 + a_T^2}$, while the angle ϕ in the drawing can be determined from $\tan \phi = a_T/a_c$. The next example applies these concepts to a discus thrower.

EXAMPLE 7   A Discus Thrower

Discus throwers often warm up by standing with both feet flat on the ground and throwing the discus with a twisting motion of their bodies. Figure 8.16a illustrates a top view of such a warm-up throw. Starting from rest, the thrower accelerates the discus to a final angular velocity of $+15.0$ rad/s in a time of 0.270 s before releasing it. During the acceleration, the discus moves on a circular arc of radius 0.810 m. Find (a) the magnitude a of the total acceleration of the discus just before it is released and (b) the angle ϕ that the total acceleration makes with the radius at this moment.

Reasoning Since the tangential speed of the discus is increasing, the discus simultaneously experiences a tangential acceleration $\mathbf{a}_T$ and a centripetal acceleration $\mathbf{a}_c$ that are oriented at right angles to each other. The magnitude of the total acceleration is $a = \sqrt{a_c^2 + a_T^2}$, where a_c and a_T are the magnitudes of the centripetal and tangential accelerations. The angle ϕ in Figure 8.16b is given by $\phi = \tan^{-1}(a_T/a_c)$. The magnitude of the centripetal acceleration can be evaluated from $a_c = r\omega^2$. The magnitude of the tangential acceleration follows from $a_T = r\alpha$, where α can be found from the definition of angular acceleration in Equation 8.4.

INTERACTIVE LEARNINGWARE #18

START
Visualization
Reasoning
Knowns and Unknowns
Modeling the Problem
Numerical Solution
FINISH

A thin rigid rod is rotating with a constant angular acceleration about an axis that passes perpendicularly through one of its ends. At one instant, the total acceleration vector (centripetal plus tangential) at the other end of the rod makes a 60.0° angle with respect to the rod and has a magnitude of 15.0 m/s². The rod has an angular speed of 2.00 rad/s at this instant. What is the rod's length?

This problem reviews how the total acceleration of a rotating object is comprised of components that are parallel and perpendicular to the rod. These components are, respectively, the centripetal and tangential accelerations.

Go to our Web site at

http://www.wiley.com/college/cutnell

Solution

(a) The magnitude of the centripetal acceleration is

$$a_c = r\omega^2 = (0.810 \text{ m})(15.0 \text{ rad/s})^2 = 182 \text{ m/s}^2 \qquad (8.11)$$

The magnitude of the tangential acceleration is $a_T = r\alpha$. But $\alpha = (\omega - \omega_0)/t$ according to Equation 8.4, so

$$a_T = r\alpha = r\left(\frac{\omega - \omega_0}{t}\right)$$

$$a_T = (0.810 \text{ m})\left(\frac{15.0 \text{ rad/s} - 0 \text{ rad/s}}{0.270 \text{ s}}\right) = 45.0 \text{ m/s}^2$$

Just before the moment of release, the magnitude of the total acceleration of the discus is

$$a = \sqrt{a_c^2 + a_T^2} = \sqrt{(182 \text{ m/s}^2)^2 + (45.0 \text{ m/s}^2)^2} = \boxed{187 \text{ m/s}^2}$$

(b) The angle ϕ in Figure 8.16b is given by

$$\phi = \tan^{-1}\left(\frac{a_T}{a_c}\right) = \tan^{-1}\left(\frac{45.0 \text{ m/s}^2}{182 \text{ m/s}^2}\right) = \boxed{13.9°}$$

(a)

(b)

8.6   ROLLING MOTION

Rolling motion is a familiar situation that involves rotation, as Figure 8.17 illustrates for the case of an automobile tire. The essence of rolling motion is that there is *no slipping* at the point of contact where the tire touches the ground. To a good approximation, the tires on a normally moving automobile roll and do not slip. On

Figure 8.16 (a) A discus thrower. (b) The total acceleration **a** of the discus just before the discus is released. The total acceleration is the vector sum of the centripetal acceleration $\mathbf{a}_c$ and the tangential acceleration $\mathbf{a}_T$.

Figure 8.17 (*a*) An automobile moves with a linear speed *v*. (*b*) If the tires roll and do not slip, the distance *d*, through which an axle moves, equals the circular arc length *s* along the outer edge of a tire.

the other hand, the squealing tires that accompany the start of a drag race are rotating, but they are not rolling as they rapidly spin and slip against the ground.

When the tires in Figure 8.17 roll, there is a relationship between the angular speed at which the tires rotate and the linear speed (assumed constant) at which the car moves forward. In part *b* of the drawing, consider the points labeled *A* and *B* on the left tire. Between these points we apply a coat of red paint to the tread of the tire; the length of this circular arc of paint is *s*. The tire then rolls to the right until point *B* comes in contact with the ground. As the tire rolls, all the paint comes off the tire and sticks to the ground, leaving behind the horizontal red line shown in the drawing. The axle of the wheel moves through a linear distance *d*, which is equal to the length of the horizontal strip of paint. Since the tire does not slip, the distance *d* must be equal to the circular arc length *s*, measured along the outer edge of the tire: *d* = *s*. Dividing both sides of this equation by the elapsed time *t* shows that *d*/*t* = *s*/*t*. The term *d*/*t* is the speed at which the axle moves parallel to the ground, namely, the linear speed *v* of the car. The term *s*/*t* is the tangential speed v_T at which a point on the outer edge of the tire moves relative to the axle. In addition, v_T is related to the angular speed ω about the axle according to $v_T = r\omega$ (Equation 8.9). Therefore, it follows that

$$\underbrace{v}_{\substack{\text{Linear} \\ \text{speed}}} = \underbrace{r\omega}_{\substack{\text{Tangential} \\ \text{speed, } v_T}} \qquad (\omega \text{ in rad/s}) \qquad (8.12)$$

If the car in Figure 8.17 has a linear acceleration **a** parallel to the ground, a point on the tire's outer edge experiences a tangential acceleration $\mathbf{a}_T$ relative to the axle. The same kind of reasoning used in the last paragraph reveals that the magnitudes of these accelerations are the same and that they are related to the angular acceleration α of the wheel relative to the axle:

$$\underbrace{a}_{\substack{\text{Linear} \\ \text{acceleration}}} = \underbrace{r\alpha}_{\substack{\text{Tangential} \\ \text{acceleration, } a_T}} \qquad (\alpha \text{ in rad/s}^2) \qquad (8.13)$$

Equations 8.12 and 8.13 may be applied to any rolling motion, because the object does not slip against the surface on which it is rolling. Example 8 illustrates the basic features of rolling motion.

EXAMPLE 8   An Accelerating Car

An automobile starts from rest and for 20.0 s has a constant linear acceleration of 0.800 m/s² to the right. During this period, the tires do not slip. The radius of the tires is 0.330 m. At the end of the 20.0-s interval what is the angle through which each wheel has rotated?

Reasoning As the car accelerates, the tires rotate faster and faster. Thus, each tire has an angular acceleration, which must be taken into account when we determine the angular displacement of each wheel. Since the tires roll and do not slip, the magnitude α of the angular acceleration is related to the magnitude a of the linear acceleration of the car by $a = r\alpha$ (Equation 8.13). Therefore, we find that

$$\alpha = \frac{a}{r} = \frac{0.800 \text{ m/s}^2}{0.330 \text{ m}} = 2.42 \text{ rad/s}^2$$

The angular data, then, are as follows:

θ	α	ω	ω_0	t
?	-2.42 rad/s²		0 rad/s	20.0 s

The angular acceleration is negative, because the tires rotate increasingly rapidly in the clockwise direction as the car moves to the right (see Figure 8.17). The angular displacement θ is given in terms of α, ω_0, and t by Equation 8.7.

Solution From Equation 8.7, we find that

$$\theta = \omega_0 t + \tfrac{1}{2}\alpha t^2 = (0 \text{ rad/s})(20.0 \text{ s}) + \tfrac{1}{2}(-2.42 \text{ rad/s}^2)(20.0 \text{ s})^2 = \boxed{-484 \text{ rad}}$$

The angular displacement θ is negative, because the wheels rotate in the clockwise direction.

*8.7 THE VECTOR NATURE OF ANGULAR VARIABLES

We have presented angular velocity and angular acceleration by taking advantage of the analogy between angular variables and linear variables. Like the linear velocity and the linear acceleration, the angular quantities are also vectors and have a direction as well as a magnitude. As yet, however, we have not discussed the directions of these vectors.

When a rigid object rotates about a fixed axis, it is the axis that identifies the motion, and the angular velocity vector points along this axis. Figure 8.18 shows how to determine the direction using a *right-hand rule:*

> ***Right-Hand Rule.*** Grasp the axis of rotation with your right hand, so that your fingers circle the axis in the same sense as the rotation. Your extended thumb points along the axis in the direction of the angular velocity vector.

Note that no part of the rotating object moves in the direction of the angular velocity vector.

Angular acceleration arises when the angular velocity changes, and the acceleration vector also points along the axis of rotation. The acceleration vector has the same direction as the *change* in the angular velocity. That is, when the angular velocity is increasing, the angular acceleration vector points in the same direction as the angular velocity. Conversely, when the angular velocity is decreasing, the angular acceleration vector points in the direction opposite to the angular velocity.

Right hand

Right hand

Figure 8.18 The angular velocity vector ω of a rotating object points along the axis of rotation. The direction along the axis depends on the sense of the rotation and can be determined with the aid of a right-hand rule (see text).

8.8 CONCEPTS & CALCULATIONS

In this chapter we have studied the concepts of angular displacement, angular velocity, and angular acceleration. We conclude with several examples that review some of the important features of these ideas. Example 9 illustrates that the angular acceleration can have the same or opposite direction as the angular velocity, depending on whether the angular speed is increasing or decreasing.

CONCEPTS & CALCULATIONS EXAMPLE 9   Riding a Mountain Bike

A rider on a mountain bike is traveling to the left in Figure 8.19. Each wheel has an angular velocity of $+21.7$ rad/s, where, as usual, the plus sign indicates that the wheel is rotating in the counterclockwise direction. (a) To pass another cyclist, the rider pumps harder, and the angular velocity of the wheels increases from $+21.7$ to $+28.5$ rad/s in a time of 3.50 s. (b) After passing the cyclist, the rider begins to coast, and the angular velocity of the wheels decreases from $+28.5$ to $+15.3$ rad/s in a time of 10.7 s. In both instances, determine the magnitude and direction of the angular acceleration (assumed constant) of the wheels.

(a) Angular speed increasing

(b) Angular speed decreasing

Figure 8.19 (a) When the angular speed of the wheel is increasing, the angular velocity ω and the angular acceleration α point in the same direction (counterclockwise in this drawing). (b) When the angular speed is decreasing, the angular velocity and the angular acceleration point in opposite directions.

Concept Questions and Answers Is the angular acceleration of the wheels positive or negative when the rider is passing the cyclist and the angular speed of the wheels is increasing?

Answer Since the angular speed is increasing, the angular acceleration has the same direction as the angular velocity, which is the counterclockwise, or positive, direction (see Figure 8.19*a*).

Is the angular acceleration positive or negative when the rider has passed the cyclist and the angular speed of the wheels is decreasing?

Answer Since the angular speed is decreasing during the coasting phase, the direction of the angular acceleration is opposite to that of the angular velocity. The angular velocity is in the counterclockwise, or positive direction, so the angular acceleration must be in the clockwise or negative direction (see Figure 8.19*b*).

Solution

(a) The angular acceleration α is the change in the angular velocity, $\omega - \omega_0$, divided by the elapsed time t:

$$a = \frac{\omega - \omega_0}{t} = \frac{+28.5 \text{ rad/s} - (+21.7 \text{ rad/s})}{3.50 \text{ s}} = \boxed{+1.9 \text{ rad/s}^2} \quad (8.4)$$

As expected, the angular acceleration is positive (counterclockwise).

(b) Now the wheels are slowing down, but still rotating in the positive (counterclockwise) direction. The angular acceleration for this part of the motion is

$$a = \frac{\omega - \omega_0}{t} = \frac{+15.3 \text{ rad/s} - (+28.5 \text{ rad/s})}{10.7 \text{ s}} = \boxed{-1.23 \text{ rad/s}^2} \quad (8.4)$$

Now, as anticipated, the angular acceleration is negative (clockwise).

Example 10 reviews the two different types of acceleration, centripetal and tangential, a car can have when it travels on a circular road.

CONCEPTS & CALCULATIONS EXAMPLE 10
A Circular Roadway and the Acceleration of Your Car

Suppose you are driving a car in a counterclockwise direction on a circular road whose radius is $r = 390$ m (see Figure 8.20). You look at the speedometer and it reads a steady 32 m/s (about 72 mi/h). (a) What is the angular speed of the car? (b) Determine the acceleration (magnitude and direction) of the car. (c) To avoid a rear-end collision with a vehicle ahead, you apply the brakes and reduce your angular speed to 4.9×10^{-2} rad/s in a time of 4.0 s. What is the tangential acceleration (magnitude and direction) of the car?

Concept Questions and Answers Does an object traveling at a constant tangential speed (for example, $v_T = 32$ m/s) along a circular path have an acceleration?

Answer Yes. Recall that an object has an acceleration if its velocity is changing in time. The velocity has two attributes, a magnitude (or speed) and a direction. In this instance the speed is not changing, since it is steady at 32 m/s. However, the direction of the velocity is changing continually as the car moves on the circular road. As Sections 5.2 and 8.5 discuss, this change in direction gives rise to an acceleration, called the centripetal acceleration $\mathbf{a_c}$. The centripetal acceleration is directed toward the center of the circle (see Figure 8.20*a*).

Is there a tangential acceleration $\mathbf{a_T}$ when the angular speed of an object changes (e.g., when the car's angular speed decreases to 4.9×10^{-2} rad/s)?

(*a*) Constant angular speed

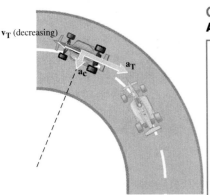

(*b*) Decreasing angular speed

Figure 8.20 (*a*) A car is traveling in a counterclockwise direction at a constant angular speed. The tangential velocity of the car is $\mathbf{v_T}$ and its centripetal acceleration is $\mathbf{a_c}$. (*b*) The car is slowing down, so the tangential acceleration $\mathbf{a_T}$ points opposite to the tangential velocity $\mathbf{v_T}$.

Answer   Yes. When the car's angular speed ω decreases, for example, its tangential speed v_T also decreases. This is because they are related by $v_T = r\omega$ (Equation 8.9), where r is the radius of the circular road. A decreasing tangential speed v_T, in turn, means that the car has a tangential acceleration $\mathbf{a}_T$. The direction of the tangential acceleration $\mathbf{a}_T$ must be *opposite* to that of the tangential velocity $\mathbf{v}_T$, because the tangential speed is decreasing. (If the two vectors were in the same direction, the tangential speed would increase.) Figure 8.20*b* shows these two vectors.

Solution

(a) The angular speed ω of the car is equal to its tangential speed v_T (the speed indicated by the speedometer) divided by the radius of the circular road:

$$\omega = \frac{v_T}{r} = \frac{32 \text{ m/s}}{390 \text{ m}} = \boxed{8.2 \times 10^{-2} \text{ rad/s}} \qquad (8.9)$$

(b) The centripetal acceleration arises because the tangential velocity is changing direction as the car travels around the circular path. The magnitude of the centripetal acceleration is

$$a_c = r\omega^2 = (390 \text{ m})(8.2 \times 10^{-2} \text{ rad/s})^2 = \boxed{2.6 \text{ m/s}^2} \qquad (8.11)$$

As always, the centripetal acceleration is directed toward the center of the circle.

(c) The tangential acceleration arises because the tangential speed of the car is changing in time. The magnitude of the tangential acceleration is $a_T = r\alpha$ (Equation 8.10), where α is the magnitude of the angular acceleration. The angular acceleration is given by Equation 8.4 as

$$\alpha = \frac{\omega - \omega_0}{t} = \frac{4.9 \times 10^{-2} \text{ rad/s} - 8.2 \times 10^{-2} \text{ rad/s}}{4.0 \text{ s}} = -8.3 \times 10^{-3} \text{ rad/s}^2$$

The magnitude of the angular acceleration is 8.3×10^{-3} rad/s². Therefore, the magnitude of the tangential acceleration is

$$a_T = r\alpha = (390 \text{ m})(8.3 \times 10^{-3} \text{ rad/s}^2) = \boxed{3.2 \text{ m/s}^2} \qquad (8.10)$$

Since the car is slowing down, the tangential acceleration is directed opposite to the direction of the tangential velocity, as shown in Figure 8.20*b*.

At the end of the problem set for this chapter, you will find homework problems that contain both conceptual and quantitative parts. These problems are grouped under the heading *Concepts & Calculations, Group Learning Problems*. They are designed for use by students working alone or in small learning groups. The conceptual part of each problem provides a convenient focus for group discussions.

S U M M A R Y

When a rigid body rotates about a fixed axis, the **angular displacement** is the angle swept out by a line passing through any point on the body and intersecting the axis of rotation perpendicularly. The **radian (rad)** is the SI unit of angular displacement. In radians, the angle θ is defined as the circular arc length s traveled by a point on the rotating body divided by the radial distance r of the point from the axis: $\theta = s/r$.

The **average angular velocity** $\overline{\omega}$ is the angular displacement $\Delta\theta$ divided by the elapsed time Δt: $\overline{\omega} = \Delta\theta/\Delta t$. As Δt approaches zero, the average angular velocity becomes equal to the **instantaneous angular velocity** ω. The magnitude of the instantaneous angular velocity is called the **instantaneous angular speed.**

The **average angular acceleration** $\overline{\alpha}$ is the change $\Delta\omega$ in the angular velocity divided by the elapsed time Δt: $\overline{\alpha} = \Delta\omega/\Delta t$. As Δt approaches zero, the average angular acceleration becomes equal to the **instantaneous angular acceleration** α.

When a rigid body rotates with constant angular acceleration about a fixed axis, the angular displacement θ, the final angular velocity ω, the initial angular velocity ω_0, the angular acceleration α, and the elapsed time t are related as follows, assuming that $\theta_0 = 0$ rad at $t_0 = 0$ s:

$$\omega = \omega_0 + \alpha t \quad \text{and} \quad \theta = \tfrac{1}{2}(\omega_0 + \omega)t$$

These two equations can be combined algebraically to give two additional equations, Equations 8.7 and 8.8:

$$\theta = \omega_0 t + \tfrac{1}{2}\alpha t^2 \quad \text{and} \quad \omega^2 = \omega_0^2 + 2\alpha\theta$$

Together, the four equations are known as the **equations of rotational kinematics for constant angular acceleration.** These equations may be used with any self-consistent set of units and are not restricted to radian measure.

When a rigid body rotates through an angle θ about a fixed axis, any point on the body moves on a circular arc of length s and radius r. Such a point has a tangential velocity $\mathbf{v_T}$ and, possibly, a tangential acceleration $\mathbf{a_T}$. The **angular and tangential variables are related** by the equations $s = r\theta$, $v_T = r\omega$, and $a_T = r\alpha$, where ω and α are, respectively, the magnitudes of the angular velocity and the angular acceleration. These equations refer to the magnitudes of the variables involved, without reference to positive or negative signs, and only radian measure can be used when applying them.

A point on an object rotating with **nonuniform circular motion** experiences a total acceleration that is the vector sum of two perpendicular acceleration components, the centripetal acceleration $\mathbf{a_c}$ and the tangential acceleration $\mathbf{a_T}$; $\mathbf{a} = \mathbf{a_c} + \mathbf{a_T}$.

The essence of **rolling motion** is that there is no slipping at the point where the object touches the surface upon which it is rolling. As a result, the tangential speed v_T of a point on the outer edge of a rolling object, measured relative to the axis through the center of the object, is equal to the linear speed v with which the object moves parallel to the surface. In other words, $v = v_T (= r\omega)$, where r is the radial distance of the point from the axis. The magnitudes of the tangential acceleration a_T and the linear acceleration a of a rolling object are similarly related: $a = a_T (= r\alpha)$.

CONCEPTUAL QUESTIONS

1. In the drawing, the flat triangular sheet ABC is lying in the plane of the paper. This sheet is going to rotate about an axis that also lies in the plane of the paper and passes through point A. Draw two such axes that are oriented so that points B and C will move on circular paths having the same radii.

2. A pair of scissors is being used to cut a string. Does each blade of the scissors have the same angular velocity (both magnitude and direction) at a given instant? Give your reasoning.

3. An electric clock is hanging on the wall in the living room. The clock is unplugged, and the second hand comes to a halt over a brief period of time. During this period, what is the direction of the angular acceleration of the second hand? Why?

4. The earth rotates once per day about its axis. Where on the earth's surface should you stand in order to have the smallest possible tangential speed? Justify your answer.

5. A thin rod rotates at a constant angular speed. Consider the tangential speed of each point on the rod for the case when the axis of rotation is perpendicular to the rod (a) at its center and (b) at one end. Explain for each case whether there are any points on the rod that have the same tangential speeds.

6. A car is up on a hydraulic lift at a garage. The wheels are free to rotate, and the drive wheels are rotating with a constant angular velocity. Does a point on the rim of a wheel have (a) a tangential acceleration and (b) a centripetal acceleration? In each case, give your reasoning.

7. Two points are located on a rigid wheel that is rotating with an increasing angular velocity about a fixed axis. The axis is perpendicular to the wheel at its center. Point 1 is located on the rim, and point 2 is halfway between the rim and the axis. At any given instant, which point (if either) has the greater (a) angular velocity, (b) angular acceleration, (c) tangential speed, (d) tangential acceleration, and (e) centripetal acceleration? Provide a reason for each of your answers.

8. A building is located on the earth's equator. Which has the greatest tangential speed due to the earth's rotation, the top floor, the bottom floor, or neither? Justify your answer.

9. Section 5.6 discusses how the uniform circular motion of a space station can be used to create "artificial" gravity for the astronauts. This can be done by adjusting the angular speed of the space station, so the centripetal acceleration at the astronaut's feet equals g, the magnitude of the acceleration due to gravity (see Figure 5.19). If such an adjustment is made, will the acceleration due to the "artificial" gravity be greater than, equal to, or less than g at the astronaut's head? Account for your answer.

10. It is possible, but not very practical, to build a clock in which the tips of the second hand, the minute hand, and the hour hand move with the same tangential speed. Explain why not.

11. Explain why a point on the rim of a tire has an acceleration when the tire is on a car that is moving at a constant linear velocity.

12. A bicycle is turned upside down, the front wheel is spinning (see the drawing), and there is an angular acceleration. At the instant shown, there are six points on the wheel that have arrows associated with them. Which of the following quantities could the arrows represent: (a) tangential velocity, (b) tangential acceleration, (c) centripetal acceleration? In each case, answer why the arrows do or do not represent the quantity.

13. Suppose that the speedometer of a truck is set to read the linear speed of the truck, but uses a device that actually measures the angular speed of the tires. If larger diameter tires are mounted on the truck, will the reading on the speedometer be correct? If not, will the reading be greater than or less than the true linear speed of the truck? Why?

14. The blades of a fan rotate more and more slowly after the fan is shut off. Eventually they stop rotating altogether. In such a situation, we sometimes assume that the angular acceleration of the blades is constant and apply the equations of rotational kinematics as an approximation. Explain why the angular acceleration can never really be constant in this kind of situation.

15. Rolling motion is one example that involves rotation about an axis that is not fixed. Give three other examples. In each case, identify the axis of rotation and explain why it is not fixed.

PROBLEMS

**Section 8.1 Rotational Motion and Angular Displacement,
Section 8.2 Angular Velocity and Angular Acceleration**

1. ssm In Europe, surveyors often measure angles in *grads*. There are 100 grads in one-quarter of a circle. How many grads are there in one radian?

2. A tachometer on the dashboard of a car indicates that the engine is "turning over" at an angular speed of 2500 rpm. (a) Express this angular speed in units of rad/s. (b) What is the magnitude of the angular displacement (in rad) of the engine when it has run for 5.1 minutes?

3. ssm A diver completes $3\frac{1}{2}$ somersaults in 1.7 s. What is the average angular speed (in rad/s) of the diver?

4. A pulsar is a rapidly rotating neutron star that continuously emits a beam of radio waves in a searchlight manner. Each time the pulsar makes one revolution, the rotating beam sweeps across the earth, and the earth receives a pulse of radio waves. For one particular pulsar, the time between two successive pulses is 0.033 s. Determine the average angular speed (in rad/s) of this pulsar.

5. A certain CD has a playing time of 74 minutes. When the music starts, the CD is rotating at an angular speed of 480 revolutions per minute (rpm). At the end of the music, the CD is rotating at 210 rpm. Find the magnitude of the average angular acceleration of the CD. Express your answer in rad/s².

6. A string trimmer is a grass and weed cutting tool that utilizes a 0.21-m length of nylon "string," rotating at 650 rad/s about an axis perpendicular to one end of the string. (a) What is the time needed for the string to sweep out an angle of 0.61 rad? (b) Assuming that the length of the string does not change, find the distance through which the tip of the string moves during this time.

7. ssm An electric circular saw is designed to reach its final angular speed, starting from rest, in 1.50 s. Its average angular acceleration is 328 rad/s². Obtain its final angular speed.

8. An automatic drier spins wet clothes at an angular speed of 5.2 rad/s. Starting from rest, the drier reaches its operating speed with an average angular acceleration of 4.0 rad/s². How long does it take the drier to come up to speed?

*9. A space station consists of two donut-shaped living chambers, A and B, that have the radii shown in the drawing. As the station rotates, an astronaut in chamber A is moved 2.40×10^2 m along a circular arc. How far along a circular arc is an astronaut in chamber B moved during the same time?

$r_A = 3.20 \times 10^2$ m

$r_B = 1.10 \times 10^3$ m

*10. The drawing shows a graph of the angular velocity of a rotating wheel as a function of time. Although not shown in the graph, the angular velocity continues to increase at the same rate until $t =$ 8.0 s. What is the angular displacement of the wheel from 0 to 8.0 s?

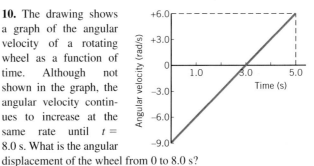

*11. ssm The drawing shows a device that can be used to measure the speed of a bullet. The device consists of two rotating disks, separated by a distance of $d = 0.850$ m, and rotating with an angular speed of 95.0 rad/s. The bullet first passes through the left disk and then through the right disk. It is found that the angular displacement between the two bullet holes is $\theta = 0.240$ rad. From these data, determine the speed of the bullet.

*12. Two people start at the same place and walk around a circular lake in opposite directions. One has an angular speed of 1.7×10^{-3} rad/s, while the other has an angular speed of 3.4×10^{-3} rad/s. How long will it be before they meet?

*13. A baton twirler throws a spinning baton directly upward. As it goes up and returns to the twirler's hand, the baton turns through four revolutions. Ignoring air resistance and assuming that the average angular speed of the baton is 1.80 rev/s, determine the height to which the center of the baton travels above the point of release.

**14. Review Conceptual Example 2 before attempting this problem. The moon has a diameter of 3.48×10^6 m and is a distance of 3.85×10^8 m from the earth. The sun has a diameter of 1.39×10^9 m and is 1.50×10^{11} m from the earth. (a) Determine (in radians) the angles subtended by the moon and the sun, as measured by a person standing on the earth. (b) Based on your answers to part (a), decide whether a total eclipse of the sun is really "total." Give your reasoning. (c) Determine the ratio, expressed as a percentage, of the apparent circular area of the moon to the apparent circular area of the sun.

**15. A quarterback throws a pass that is a perfect spiral. In other words, the football does not wobble, but spins smoothly about an axis passing through each end of the ball. Suppose the ball spins at 7.7 rev/s. In addition, the ball is thrown with a linear speed of 19 m/s at an angle of 55° with respect to the ground. If the ball is caught at the same height at which it left the quarterback's hand, how many revolutions has the ball made while in the air?

Section 8.3 The Equations of Rotational Kinematics

16. A figure skater is spinning with an angular velocity of $+15$ rad/s. She then comes to a stop over a brief period of time. During this time, her angular displacement is $+5.1$ rad. Determine (a) her average angular acceleration and (b) the time during which she comes to rest.

17. ssm The drill bit of a variable-speed electric drill has a constant angular acceleration of 2.50 rad/s². The initial angular speed of the bit is 5.00 rad/s. After 4.00 s, (a) what angle has the bit turned through and (b) what is the bit's angular speed?

18. A basketball player is balancing a spinning basketball on the tip of his finger. The angular velocity of the ball slows down from 18.5 to 14.1 rad/s. During the slow-down, the angular displacement is 85.1 rad. Determine the time it takes for the ball to slow down.

19. ssm www A flywheel has a constant angular deceleration of 2.0 rad/s². (a) Find the angle through which the flywheel turns as it comes to rest from an angular speed of 220 rad/s. (b) Find the time required for the flywheel to come to rest.

20. A 9.0-m-long string is wrapped around a pulley that is free to rotate about a fixed axle. The pulley has a radius of 6.0 cm and initially is at rest. The thickness of the string is negligible. Someone pulls on the free end of the string, thereby unwinding it and giving the pulley an angular acceleration of $+17$ rad/s². (a) Through what angle has the pulley rotated when the string is completely unwound? (b) What is the final angular velocity of the pulley?

21. The angular speed of the rotor in a centrifuge increases from 420 to 1420 rad/s in a time of 5.00 s. (a) Obtain the angle through which the rotor turns. (b) What is the magnitude of the angular acceleration?

*22. A motorcyclist is traveling along a road and accelerates for 4.50 s to pass another cyclist. The angular acceleration of each wheel is $+6.70$ rad/s², and, just after passing, the angular velocity of each is $+74.5$ rad/s, where the plus signs indicate counterclockwise directions. What is the angular displacement of each wheel during this time?

*23. ssm A spinning wheel on a fireworks display is initially rotating in a counterclockwise direction. The wheel has an angular acceleration of -4.00 rad/s². Because of this acceleration, the angular velocity of the wheel changes from its initial value to a final value of -25.0 rad/s. While this change occurs, the angular displacement of the wheel is zero. (Note the similarity to that of a ball being thrown vertically upward, coming to a momentary halt, and then falling downward to its initial position.) Find the time required for the change in the angular velocity to occur.

*24. Starting from rest at $t_0 = 0$ s, a wheel on a mountain bike has a constant angular acceleration. When $t = 5.00$ s, the angular velocity of the wheel is $+7.50$ rad/s. The angular acceleration continues until $t = 15.0$ s, after which time the angular velocity remains constant. What is the angular displacement (magnitude and direction) of the wheel in the time interval from $t = 0$ to 25.0 s?

*25. A dentist causes the bit of a high-speed drill to accelerate from an angular speed of 1.05×10^4 rad/s to an angular speed of 3.14×10^4 rad/s. In the process, the bit turns through 1.88×10^4 rad. Assuming a constant angular acceleration, how long would it take the bit to reach its maximum speed of 7.85×10^4 rad/s, starting from rest?

*26. At the local swimming hole, a favorite trick is to run horizontally off a cliff that is 8.3 m above the water. One diver runs off the edge of the cliff, tucks into a "ball," and rotates on the way down with an average angular speed of 1.6 rev/s. Ignore air resistance and determine the number of revolutions she makes while on the way down.

**27. ssm www A child, hunting for his favorite wooden horse, is running on the ground around the edge of a stationary merry-go-round. The angular speed of the child has a constant value of 0.250 rad/s. At the instant the child spots the horse, one-quarter of a turn away, the merry-go-round begins to move (in the direction the child is running) with a constant angular acceleration of 0.0100 rad/s². What is the shortest time it takes for the child to catch up with the horse?

Section 8.4 Angular Variables and Tangential Variables

28. Our sun rotates in a circular orbit about the center of the Milky Way galaxy. The radius of the orbit is 2.2×10^{20} m, and the angular speed of the sun is 1.2×10^{-15} rad/s. (a) What is the tangential speed of the sun? (b) How long (in years) does it take for the sun to make one revolution around the center?

29. An auto race is held on a circular track. A car completes one lap in a time of 18.9 s, with an average tangential speed of 42.6 m/s. Find (a) the average angular speed and (b) the radius of the track.

30. The tip of a tornado has a radius of 29 m, and the wind there has a tangential speed of 251 mph. Determine the angular speed of the tornado in revolutions per second.

31. ssm A disk (radius = 2.00 mm) is attached to a high-speed drill at a dentist's office and is turning at 7.85×10^4 rad/s. Determine the tangential speed of a point on the outer edge of this disk.

32. The earth has a radius of 6.38×10^6 m and turns on its axis once every 23.9 h. (a) What is the tangential speed (in m/s) of a person living in Ecuador, a country that lies on the equator? (b) At what latitude (i.e., the angle θ in the drawing) is the tangential speed one-third that of a person living in Ecuador?

Axis of rotation

θ

Equator

Problem 32

33. The take-up reel of a cassette tape has a radius of 2.5 cm. Find the length of tape that passes around the reel in 7.1 s when the reel rotates at an average angular speed of 1.9 rad/s.

*__34.__ A baseball pitcher throws a baseball horizontally at a linear speed of 42.5 m/s (about 95 mi/h). Before being caught, the baseball travels a horizontal distance of 16.5 m and rotates through an angle of 49.0 rad. The baseball has a radius of 3.67 cm and is rotating about an axis as it travels, much like the earth does. What is the tangential speed of a point on the "equator" of the baseball?

*__35. ssm__ A compact disc (CD) contains music on a spiral track. Music is put onto a CD with the assumption that, during playback, the music will be detected at a *constant tangential speed* at any point. Since $v_{\mathrm{T}} = r\omega$, a CD rotates at a smaller angular speed for music near the outer edge and a larger angular speed for music near the inner part of the disc. For music at the outer edge ($r = 0.0568$ m), the angular speed is 3.50 rev/s. Find (a) the constant tangential speed at which music is detected and (b) the angular speed (in rev/s) for music at a distance of 0.0249 m from the center of a CD.

*__36.__ A thin rod (length = 1.50 m) is oriented vertically, with its bottom end attached to the floor by means of a frictionless hinge. The

mass of the rod may be ignored, compared to the mass of the object fixed to the top of the rod. The rod, starting from rest, tips over and rotates downward. (a) What is the angular speed of the rod just before it strikes the floor? *(Hint: Consider using the principle of conservation of mechanical energy.)* (b) What is the magnitude of the angular acceleration of the rod just before it strikes the floor?

**__37.__ The drawing shows a golf ball passing through a windmill at a miniature golf course. The windmill has 8 blades and rotates at an angular speed of 1.25 rad/s. The opening between successive blades is equal to the width of a blade. A golf ball of diameter 4.50×10^{-2} m is just passing by one of the rotating blades. What must be the *minimum* speed of the ball so that it will not be hit by the next blade?

Golf ball

Section 8.5 Centripetal Acceleration and Tangential Acceleration

38. A ceiling fan has two different angular speed settings: $\omega_1 = 440$ rev/min and $\omega_2 = 110$ rev/min. What is the ratio a_1/a_2 of the centripetal accelerations of a given point on a fan blade?

39. ssm During a tennis serve, a racket is given an angular acceleration of magnitude 160 rad/s². At the top of the serve, the racket has an angular speed of 14 rad/s. If the distance between the top of the racket and the shoulder is 1.5 m, find the magnitude of the total acceleration of the top of the racket.

40. The earth orbits the sun once a year (3.16×10^7 s) in a nearly circular orbit of radius 1.50×10^{11} m. With respect to the sun, determine (a) the angular speed of the earth, (b) the tangential speed of the earth, and (c) the magnitude and direction of the earth's centripetal acceleration.

41. ssm A 220-kg speedboat is negotiating a circular turn (radius = 32 m) around a buoy. During the turn, the engine causes a net tangential force of magnitude 550 N to be applied to the boat. The initial tangential speed of the boat going into the turn is 5.0 m/s. (a) Find the tangential acceleration. (b) After the boat is 2.0 s into the turn, find the centripetal acceleration.

*__42.__ The front and rear sprockets on a bicycle have radii of 9.00 and 5.10 cm, respectively. The angular speed of the front sprocket is 9.40 rad/s. Determine (a) the linear speed (in cm/s) of the chain as it moves between the sprockets and (b) the centripetal acceleration (in cm/s²) of the chain as it passes around the rear sprocket.

*__43.__ A rectangular plate is rotating with a constant angular acceleration about an axis that passes perpendicularly through one corner, as the drawing shows. The tangential acceleration measured at corner A has twice the magnitude of that measured at corner B. What is the ratio L_1/L_2 of the lengths of the sides of the rectangle?

L_1

L_2

A

B

**44. The blades of a windmill start from rest and rotate with an angular acceleration of 22.0 rad/s². At any point on a blade, how much time passes before the magnitude of the tangential acceleration equals the magnitude of the centripetal acceleration?

**45. ssm An electric drill starts from rest and rotates with a constant angular acceleration. After the drill has rotated through a certain angle, the magnitude of the centripetal acceleration of a point on the drill is twice the magnitude of the tangential acceleration. What is the angle?

Section 8.6 Rolling Motion

Note: All problems in this section assume that there is no slipping of the surfaces in contact during the rolling motion.

46. Suppose you are riding a stationary exercise bicycle, and the electronic meter indicates that the wheel is rotating at 9.1 rad/s. The wheel has a radius of 0.45 m. If you ride the bike for 35 min, how far would you have gone if the bike could move?

47. ssm www A motorcycle accelerates uniformly from rest and reaches a linear speed of 22.0 m/s in a time of 9.00 s. The radius of each tire is 0.280 m. What is the magnitude of the angular acceleration of each tire?

48. An automobile tire has a radius of 0.330 m, and its center moves forward with a linear speed of $v = 15.0$ m/s. (a) Determine the angular speed of the wheel. (b) Relative to the axle, what is the tangential speed of a point located 0.175 m from the axle?

49. On an open-reel tape deck, the tape is being pulled past the playback head at a constant linear speed of 0.381 m/s. (a) Using the data in part *a* of the drawing, find the angular speed of the take-up reel. (b) After 2.40×10^3 s, the take-up reel is almost full, as part *b* of the drawing indicates. Find the average angular acceleration of the reel and specify whether the acceleration indicates an increasing or decreasing angular velocity.

0.0508 m

$v = 0.381$ m/s

(a)

0.114 m

$v = 0.381$ m/s

(b)

50. The warranty on a new tire says that an automobile can travel for a distance of 96 000 km before the tire wears out. The radius

of the tire is 0.31 m. How many revolutions does the tire make before wearing out?

*51. ssm A person lowers a bucket into a well by turning the hand crank, as the drawing illustrates. The crank handle moves with a constant tangential speed of 1.20 m/s on its circular path. Find the linear speed with which the bucket moves down the well.

0.100 m diameter

0.400 m diameter

Problem 51

*52. A ball of radius 0.200 m rolls along a horizontal table top with a constant linear speed of 3.60 m/s. The ball rolls off the edge and falls a vertical distance of 2.10 m before hitting the floor. What is the angular displacement of the ball while it is in the air?

*53. The two-gear combination shown in the drawing is being used to hoist the load L with a constant upward speed of 2.50 m/s. The rope attached to the load is being wound onto a cylinder behind the big gear. The depth of the teeth of the gears is negligible compared to the radii. Determine the angular velocity (magnitude and direction) of (a) the larger gear and (b) the smaller gear.

0.300 m

0.170 m

$v = 2.50$ m/s

L

**54. The chain on a bicycle connects the front sprocket (radius = 12.5 cm) to the rear sprocket (radius = 3.50 cm). The wheels have a radius of 40.0 cm. The rear wheel and the rear sprocket turn together as a single unit. The rider is peddling so the front sprocket makes 1.50 revolutions each second. What is the linear speed (in m/s) of the bicycle?

**55. A bicycle is rolling down a circular hill whose radius is 9.00 m. As the drawing illustrates, the angular displacement of the bicycle is 0.960 rad. The radius of each wheel is 0.400 m. What is the angle (in radians) through which each tire rotates?

0.960 rad

ADDITIONAL PROBLEMS

56. A commercial jetliner, starting from rest, accelerates down the runway and attains a linear speed of 72 m/s at liftoff. The radius of each tire is 0.42 m. (a) What is the angular speed of a tire at liftoff? (b) As the plane moves down the runway, the angular acceleration of each tire has a magnitude of 1.9 rad/s². Determine the magnitude of the angular displacement (in radians) of a tire from start to liftoff.

57. ssm A race car travels with a constant tangential speed of 75.0 m/s around a circular track of radius 625 m. Find (a) the magnitude of the car's total acceleration and (b) the direction of its total acceleration relative to the radial direction.

58. The drawing shows the blade of a chain saw. The rotating sprocket tip at the end of the guide bar has a radius of $3.5 \times$

10^{-2} m. The linear speed of a chain link at point A is 7.7 m/s. Find the angular speed of the sprocket tip in rev/s.

Guide bar

59. ssm An electric fan is running on HIGH. After the LOW button is pressed, the angular speed of the fan decreases to 83.8 rad/s in 1.75 s. The deceleration is 42.0 rad/s^2. Determine the initial angular speed of the fan.

60. A train is rounding a circular curve whose radius is 2.00×10^2 m. At one instant, the train has an angular acceleration of 1.50×10^{-3} rad/s^2 and an angular speed of 0.0500 rad/s. (a) Find the magnitude of the total acceleration (centripetal plus tangential) of the train. (b) Determine the angle of the total acceleration relative to the radial direction.

61. The shaft of a pump starts from rest and has an angular acceleration of 3.00 rad/s^2 for 18.0 s. At the end of this interval, what is (a) the shaft's angular speed and (b) the angle through which the shaft has turned?

* **62.** The sun has a mass of 1.99×10^{30} kg and is moving in a circular orbit about the center of our galaxy, the Milky Way. The radius of the orbit is 2.3×10^4 light-years (1 light-year = 9.5×10^{15} m), and the angular speed of the sun is 1.1×10^{-15} rad/s. (a) Determine the tangential speed of the sun. (b) What is the magnitude of the net force that acts on the sun to keep it moving around the center of the Milky Way?

* **63. ssm www** The drawing shows a view (from beneath) of the platter on a belt-drive turntable. The platter has an angular speed of 3.49 rad/s. The pulley on the motor shaft has a radius of 1.27 cm. Assuming that the belt does not slip, determine the angular speed of the motor shaft.

Belt   Turntable platter

$r = 0.102$ m

Motor

* **64.** A stroboscope is a light that flashes on and off at a constant rate. It can be used to illuminate a rotating object, and if the flashing rate is adjusted properly, the object can be made to appear stationary. (a) What is the shortest time between flashes of light that will make a three-bladed propeller appear stationary when it is rotating with an angular speed of 16.7 rev/s? (b) What is the next shortest time?

* **65.** A fan blade, whose angular acceleration is a constant 2.00 rad/s^2, rotates through an angle of 285 radians in 11.0 s. How long did it take the blade, starting from rest, to reach the *beginning* of the 11.0-s interval?

* **66.** A race car, starting from rest, travels around a circular turn of radius 23.5 m. At a certain instant, the car is still accelerating, and its angular speed is 0.571 rad/s. At this time, the total acceleration (centripetal plus tangential) makes an angle of 35.0° with respect to the radius. (The situation is similar to that in Figure 8.15b.) What is the magnitude of the total acceleration?

** **67. ssm** Take two quarters and lay them on a table. Press down on one quarter so it cannot move. Then, starting at the 12:00 position, roll the other quarter along the edge of the stationary quarter, as the drawing suggests. How many revolutions does the rolling quarter make when it travels once around the circumference of the stationary quarter? Surprisingly, the answer is *not* one revolution. (*Hint: Review the paragraph just before Equation 8.12 that discusses how the distance traveled by the axle of a wheel is related to the circular arc length along the outer edge of the wheel.*)

** **68.** One type of slingshot can be made from a length of rope and a leather pocket for holding the stone. The stone can be thrown by whirling it rapidly in a horizontal circle and releasing it at the right moment. Such a slingshot is used to throw a stone from the edge of a cliff, the point of release being 20.0 m above the base of the cliff. The stone lands on the ground below the cliff at a point X. The horizontal distance of point X from the base of the cliff (directly beneath the point of release) is thirty times the radius of the circle on which the stone is whirled. Determine the angular speed of the stone at the moment of release.

CONCEPTS & CALCULATIONS

GROUP LEARNING PROBLEMS

Note: Each of these problems consists of Concept Questions followed by a related quantitative Problem. They are designed for use by students working alone or in small learning groups. The Concept Questions involve little or no mathematics and are intended to stimulate group discussions. They focus on the concepts with which the problems deal. Recognizing the concepts is the essential initial step in any problem-solving technique.

69. Concept Question In general, is the direction of an object's average angular velocity the same as its initial angle θ_0, its final angle θ, or the difference $\theta - \theta_0$ between its final and initial angles? The table below lists four pairs of initial and final angles of a wheel on a moving car. Decide which pairs give a positive average angular velocity and which give a negative average angular velocity. Provide reasons for your answers.

	Initial angle θ_0	Final angle θ
(a)	0.45 rad	0.75 rad
(b)	0.94 rad	0.54 rad
(c)	5.4 rad	4.2 rad
(d)	3.0 rad	3.8 rad

Problem The elapsed time for each pair of angles is 2.0 s. Review the concept of average angular velocity in Section 8.2 and then determine the average angular velocity (magnitude and direction) for each of the four pairs of angles in the table. Check to see that the directions (positive or negative) of the angular velocities agree with the directions found in the Concept Question.

70. Concept Question In general, does the average angular acceleration of a rotating object have the same direction as its initial angular velocity ω_0, its final angular velocity ω, or the difference $\omega - \omega_0$ between its final and initial angular velocities? The table below lists four pairs of initial and final angular velocities for a rotating fan blade. Determine the direction (positive or negative) of the average angular acceleration for each pair. Provide reasons for your answers.

	Initial angular velocity ω_0	Final angular velocity ω
(a)	+2.0 rad/s	+5.0 rad/s
(b)	+5.0 rad/s	+2.0 rad/s
(c)	−7.0 rad/s	−3.0 rad/s
(d)	+4.0 rad/s	−4.0 rad/s

Problem The elapsed time for each of the four pairs of angular velocities is 4.0 s. Find the average angular acceleration (magnitude and direction) for each of the four pairs. Be sure that your directions agree with those found in the Concept Question.

71. Concept Question In the table are listed the initial angular velocity ω_0 and the angular acceleration α of four rotating objects at a given instant in time.

	Initial angular velocity ω_0	Angular acceleration α
(a)	+12 rad/s	+3.0 rad/s²
(b)	+12 rad/s	−3.0 rad/s²
(c)	−12 rad/s	+3.0 rad/s²
(d)	−12 rad/s	−3.0 rad/s²

In each case, state whether the angular *speed* of the object is increasing or decreasing in time. Account for your answers.

Problem For each of the four pairs in the table above, determine the final angular speed of the object if the elapsed time is 2.0 s. Compare your final angular speeds with the initial angular speeds and make sure that your answers are consistent with your answers to the Concept Question.

72. Concept Questions Does the tip of a rotating fan blade have a tangential acceleration when the blade is rotating (a) at a constant angular velocity and (b) at a constant angular acceleration? Provide a reason for your answers.

Problem A fan blade is rotating with a constant angular acceleration of $+12.0$ rad/s². At what point on the blade, measured from the axis of rotation, does the magnitude of the tangential acceleration equal that of the acceleration due to gravity?

73. Concept Questions During a time t, a wheel has a constant angular acceleration, so its angular velocity increases from an initial value of ω_0 to a final value of ω. (a) During this time, is the angular displacement less than, greater than, or equal to (a) $\omega_0 t$ or (b) ωt? In each case, justify your answer. (c) If you conclude that the angular displacement does not equal $\omega_0 t$ or ωt, then how does it depend on ω_0, ω, and t?

Problem A car is traveling along a road, and the engine is "turning over" with an angular velocity of $+220$ rad/s. The driver steps on the accelerator, and in a time of 10.0 s the angular velocity increases to $+280$ rad/s. (a) What would have been the angular displacement of the engine if its angular velocity had remained constant at the initial value of $+220$ rad/s during the 10.0 s? (b) What would have been the angular displacement if the angular velocity had been equal to its final value of $+280$ rad/s during the 10.0 s? (c) Determine the actual value of the angular displacement during this period. Check your answers to see that they are consistent with those in the Concept Questions.

74. Concept Questions Two identical dragsters, starting from rest, accelerate side-by-side along a straight track. The wheels on one of the cars roll without slipping, while the wheels on the other slip during part of the time. (a) For which car, the winner or the loser, do the wheels roll without slipping? Why? For the dragster whose wheels roll without slipping, is there (b) a relationship between its linear speed and the angular speed of its wheels, and (c) a relationship between the magnitude of its linear acceleration and the magnitude of the angular acceleration of its wheels? If a relationship exists in either case, what is it?

Problem A dragster starts from rest and accelerates down the track. Each tire has a radius of 0.320 m and rolls without slipping. At a distance of 384 m, the angular speed of the wheels is 288 rad/s. Determine, if possible, (a) the linear speed of the dragster and (b) the magnitude of the angular acceleration of its wheels.

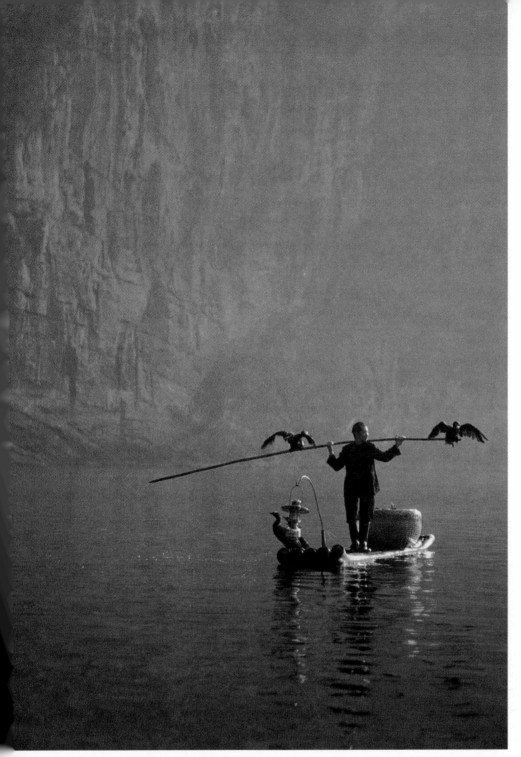

The two birds on the pole across the man's shoulders are cormorants, which he uses to catch fish. Because of its weight, each bird applies a force to the pole. The man's shoulders and hands also apply forces. The pole, however, remains balanced in equilibrium because of the locations of the forces. The location of a force and the force itself determine a quantity called *torque,* and we will see that the net torque acting on a rigid body in equilibrium must be zero.

CHAPTER 9

ROTATIONAL DYNAMICS

(a) Translation

(b) Combined translation and rotation

Figure 9.1 An example of (a) translational motion and (b) combined translational and rotational motion.

9.1 THE EFFECTS OF FORCES AND TORQUES ON THE MOTION OF RIGID OBJECTS

The mass of most rigid objects, such as a propeller or a wheel, is spread out and not concentrated at a single point. These objects can move in a number of ways. Figure 9.1a illustrates one possibility called translational motion, in which all points on the body travel on parallel paths (not necessarily straight lines). In pure translation there is no rotation of any line in the body. Because translational motion can occur along a curved line, it is often called curvilinear motion or linear motion. Another possibility is rotational motion, which may occur in combination with translational motion, as is the case for the somersaulting gymnast in Figure 9.1b.

We have seen many examples of how a net force affects linear motion by causing an object to accelerate. When there is no net force, there is no acceleration, and the object is in equilibrium. We now need to expand our concept of equilibrium to take into account the possibility that a rigid object can also rotate. A net external force causes linear motion to change, but what causes rotational motion to change? For example, something causes the rotational velocity of a speed boat's propeller to change when the boat accelerates. Is it simply the net force? As it turns out, it is not the net external force, but rather the net external torque that causes the rotational velocity to change. Just as greater net forces cause greater linear accelerations, greater net torques cause greater rotational or angular accelerations.

Figure 9.2 helps to explain the idea of torque. When you push on a door with a force **F** as in part a, the door opens more quickly when the force is larger. Other things being equal, a larger force generates a larger torque. However, the door does not open as quickly if you apply the same force at a point closer to the hinge, as in part b, because the force now produces less torque. Furthermore, if your push is directed nearly at the hinge, as in part c, you will have a hard time opening the door at all, because the torque is nearly zero. In summary, the torque depends on the magnitude of the force, on the point where the force is applied relative to the axis of rotation (the hinge in Figure 9.2), and on the direction of the force.

For simplicity, we deal with situations in which the force lies in a plane that is perpendicular to the axis of rotation. In Figure 9.3, for instance, the axis is perpendicular to the page and the force lies in the plane of the paper. The drawing shows the line of action and the lever arm of the force, two concepts that are important in the definition of torque. The **line of action** is an extended line drawn colinear with the force. The **lever arm** is the distance ℓ between the line of action and the axis of rotation, measured on a line that is perpendicular to both. The torque is represented by the symbol τ (Greek letter *tau*), and its magnitude is defined as the magnitude of the force times the lever arm:

Figure 9.2 It is easier to open a door with a force of a given magnitude by (a) pushing at the door's outer edge than by (b) pushing closer to the axis of rotation (the hinge). (c) Pushing nearly into the hinge makes it very difficult to open the door.

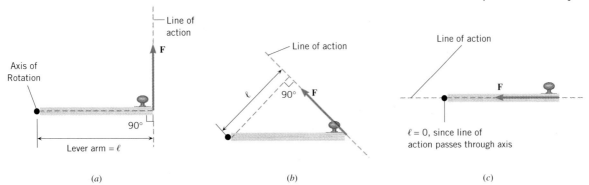

(a)   (b)   (c)

Figure 9.3 In this top view, the hinges of a door appear as a block dot (●) and define the axis of rotation. The line of action and lever arm ℓ are illustrated for a force applied to the door (a) perpendicularly and (b) at an angle. (c) The lever arm is zero because the line of action passes through the axis of rotation.

● **DEFINITION OF TORQUE**

Magnitude:

$$\text{Torque} = (\text{Magnitude of the force}) \times (\text{Lever arm})$$

$$\tau = F\ell \tag{9.1}$$

Direction: The torque is positive when the force tends to produce a counterclockwise rotation about the axis, and negative when the force tends to produce a clockwise rotation.

SI Unit of Torque: newton · meter (N · m)

Equation 9.1 indicates that forces of the same magnitude can produce *different* torques, depending on the value of the lever arm, and Example 1 illustrates this important feature.

EXAMPLE 1   Different Lever Arms, Different Torques

In Figure 9.3 a force whose magnitude is 55 N is applied to a door. However, the lever arms are different in the three parts of the drawing: (a) $\ell = 0.80$ m, (b) $\ell = 0.60$ m, and (c) $\ell = 0$. Find the magnitude of the torque in each case.

Reasoning In each case the lever arm is the perpendicular distance between the axis of rotation and the line of action of the force. In part *a* this perpendicular distance is equal to the width of the door. In parts *b* and *c*, however, the lever arm is less than the width. Because the lever arm is different in each case, the torque is different, even though the magnitude of the applied force is the same.

Solution Equation 9.1 gives the following values for the torques:

(a) $\tau = F\ell = (55 \text{ N})(0.80 \text{ m}) = \boxed{44 \text{ N} \cdot \text{m}}$

(b) $\tau = F\ell = (55 \text{ N})(0.60 \text{ m}) = \boxed{33 \text{ N} \cdot \text{m}}$

(c) $\tau = F\ell = (55 \text{ N})(0 \text{ m}) = \boxed{0 \text{ N} \cdot \text{m}}$

In part *c* the line of action of F passes through the axis of rotation (the hinge). Hence, the lever arm is zero, and the torque is zero.

INTERACTIVE LEARNINGWARE #19

START

Visualization

Reasoning

Knowns and Unknowns

Modeling the Problem

Numerical Solution

FINISH

A thin rod is mounted on an axle that passes perpendicularly through the center of the rod and allows the rod to rotate in the vertical plane. The weight of the rod is negligible. The rod is held at an angle of 33° above the horizontal by the action of two forces. These two forces produce a total torque of zero with respect to the axle. One force has a magnitude F and is applied perpendicularly to one end of the rod. The other force is due to a 49-N weight that hangs vertically from the other end of the rod. Find F.

When a number of forces act on an object, each can create a torque. The total torque in such a case is the sum of individual torques. This Interactive LearningWare example illustrates that the algebraic sign of each individual torque must be taken into account when calculating the total. As a result, the total may be zero, even though the individual torques are not zero.

Go to our Web site at

http://www.wiley.com/college/cutnell

In our bodies muscles and tendons produce torques about various joints. Example 2 illustrates how the Achilles tendon produces a torque about the ankle joint.

THE PHYSICS OF...

the Achilles tendon.

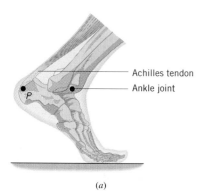

Achilles tendon

Ankle joint

P

(a)

F

Lever arm

55°

3.6×10^{-2} m

(b)

Figure 9.4 The force **F** generated by the Achilles tendon produces a clockwise (negative) torque about the ankle joint.

EXAMPLE 2   The Achilles Tendon

Figure 9.4a shows the ankle joint and the Achilles tendon attached to the heel at point P. The tendon exerts a force of magnitude $F = 720$ N, as Figure 9.4b indicates. Determine the torque (magnitude and direction) of this force about the ankle joint, which is located 3.6×10^{-2} m away from point P.

Reasoning To calculate the magnitude of the torque, it is necessary to have a value for the lever arm ℓ. However, the lever arm is not the given distance of 3.6×10^{-2} m. Instead, the lever arm is the perpendicular distance between the axis of rotation at the ankle joint and the line of action of the force **F**. In Figure 9.4b this distance is indicated by the dashed red line.

Solution From the drawing, it can be seen that the lever arm is $\ell = (3.6 \times 10^{-2}$ m$) \cos 55°$. The magnitude of the torque is

$$\tau = F\ell = (720 \text{ N})(3.6 \times 10^{-2} \text{ m})\cos 55°$$
$$= 15 \text{ N} \cdot \text{m} \tag{9.1}$$

The force **F** tends to produce a clockwise rotation about the ankle joint, so the torque is negative: $\boxed{\tau = -15 \text{ N} \cdot \text{m}}$

Now that we have introduced the idea of torque, the next step is to incorporate it into the concept of equilibrium. The following section shows that torques, as well as forces, must be balanced in order for a rigid object to be in equilibrium.

9.2   RIGID OBJECTS IN EQUILIBRIUM

● *Concepts at a Glance*   If a rigid body is in equilibrium, neither its linear motion nor its rotational motion changes. This lack of change leads to certain equations that apply for rigid-body equilibrium, as the Concepts-at-a-Glance chart in Figure 9.5 emphasizes. For instance, an object whose linear motion is not changing has no acceleration **a**. Therefore, the net force Σ**F** applied to the object must be zero, since Σ**F** $= m$**a** and **a** $= 0$. For two-dimensional motion the x and y components of the net force are separately zero: $\Sigma F_x = 0$ and $\Sigma F_y = 0$ (Equations 4.9a and 4.9b). In calculating the net force, we include only forces from external agents, or *external forces*.* In addition to linear motion, we must consider rotational motion, which also does not change under equilibrium conditions. This means that the net external torque acting on the object must be zero, because it is what causes rotational motion to change. Using the symbol $\Sigma\tau$ to represent the net external torque

---

* We ignore internal forces that one part of an object exerts on another part, because they occur in action–reaction pairs, each of which consists of oppositely directed forces of equal magnitude. The effect of one force cancels the effect of the other, as far as the acceleration of the entire object is concerned.

Concepts at a Glance

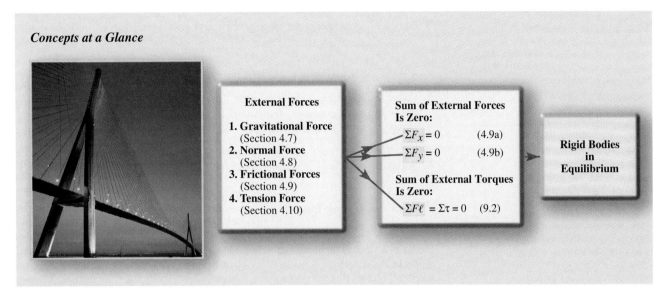

Figure 9.5 *Concepts at a Glance* A rigid body is in equilibrium when the sum of the external forces is zero and when the sum of the external torques is zero. This bridge in Normandy, France, is in equilibrium.

(the sum of all positive and negative torques), we have

$$\Sigma\tau = 0 \qquad (9.2)$$

We define rigid-body equilibrium, then, in the following way.

● EQUILIBRIUM OF A RIGID BODY

A rigid body is in equilibrium if it has zero translational acceleration and zero angular acceleration. In equilibrium, the sum of the externally applied forces is zero, and the sum of the externally applied torques is zero:

$$\Sigma F_x = 0 \quad \text{and} \quad \Sigma F_y = 0 \qquad (4.9a \text{ and } 4.9b)$$

$$\Sigma\tau = 0 \qquad (9.2)$$

The reasoning strategy for analyzing the forces and torques acting on a body in equilibrium is given below. The first four steps of the strategy are essentially the same as those outlined in Section 4.11, where only forces are considered. Steps 5 and 6 have been added to account for any external torques that may be present. Example 3 illustrates how this reasoning strategy is applied to a diving board.

REASONING STRATEGY

Applying the Conditions of Equilibrium to a Rigid Body

1. Select the object to which the equations for equilibrium are to be applied.
2. Draw a free-body diagram that shows all the external forces acting on the object, each force with its proper direction.

3. Choose a convenient set of x, y axes and resolve all forces into components that lie along these axes.

4. Apply the equations that specify the balance of forces at equilibrium: $\Sigma F_x = 0$ and $\Sigma F_y = 0$.

5. Select a convenient axis of rotation. Identify the point where each external force acts on the object, and calculate the torque produced by each force about the axis of rotation. Set the sum of the torques about this axis equal to zero: $\Sigma \tau = 0$.

6. Solve the equations in steps 4 and 5 for the desired unknown quantities.

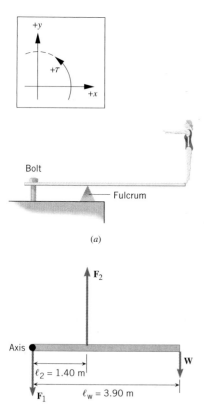

(a)

(b) Free-body diagram of the diving board

Figure 9.6 (a) A diver stands at the end of a diving board. (b) The free-body diagram for the diving board. The box at the upper left shows the positive x and y directions for the forces, as well as the positive (counterclockwise) direction for the torques.

EXAMPLE 3   A Diving Board

A woman whose weight is 530 N is poised at the right end of a diving board, whose length is 3.90 m. The board has negligible weight and is bolted down at the left end, while being supported 1.40 m away by a fulcrum, as Figure 9.6a shows. Find the forces $\mathbf{F_1}$ and $\mathbf{F_2}$ that the bolt and the fulcrum, respectively, exert on the board.

Reasoning Part b of the figure shows the free-body diagram of the diving board. Three forces act on the board: $\mathbf{F_1}$, $\mathbf{F_2}$, and the force due to the diver's weight $\mathbf{W}$. In choosing the directions of $\mathbf{F_1}$ and $\mathbf{F_2}$ we have used our intuition: $\mathbf{F_1}$ points downward, because the bolt must pull in that direction to counteract the tendency of the board to rotate clockwise about the fulcrum; $\mathbf{F_2}$ points upward because the board pushes downward against the fulcrum, which, in reaction, pushes upward on the board. Since the board is stationary, it is in equilibrium.

Solution Since the board is in equilibrium, the sum of the vertical forces must be zero:

$$\Sigma F_y = -F_1 + F_2 - 530 \text{ N} = 0 \qquad (4.9\text{b})$$

Similarly, the sum of the torques must be zero, $\Sigma \tau = 0$. For calculating torques, we select an axis that passes through the left end of the board and is perpendicular to the page. (We will see shortly that this choice is arbitrary.) The force $\mathbf{F_1}$ produces no torque since it passes through the axis and has a zero lever arm, while $\mathbf{F_2}$ creates a counterclockwise (positive) torque, and $\mathbf{W}$ produces a clockwise (negative) torque. The free-body diagram shows the lever arms for the torques:

$$\Sigma \tau = +F_2 \ell_2 - W\ell_w = +F_2(1.40 \text{ m}) - (530 \text{ N})(3.90 \text{ m}) = 0 \qquad (9.2)$$

$$\boxed{F_2 = 1480 \text{ N}}$$

This value for F_2 can be substituted into Equation 4.9b above to show that $\boxed{F_1 = 950 \text{ N}}$.

In Example 3 the sum of the external torques is calculated using an axis that passes through the left end of the diving board. *However, the choice of the axis is* ***completely arbitrary, because if an object is in equilibrium, it is in equilibrium*** ***with respect to any axis whatsoever.*** Thus, the sum of the external torques is zero, no matter where the axis is placed. One usually chooses the location so that the lines of action of one or more of the unknown forces pass through the axis. Such a choice simplifies the torque equation, because the torques produced by these forces are zero. For instance, in Example 3 the torque due to the force F_1 does not appear in Equation 9.2, because the lever arm of this force is zero.

In a calculation of torque, the lever arm of the force must be determined relative to the axis of rotation. In Example 3 the lever arms are obvious, but sometimes a little care is needed in determining them, as in the next example.

EXAMPLE 4 Fighting a Fire

In Figure 9.7a an 8.00-m ladder of weight $W_L = 355$ N leans against a smooth vertical wall. The term "smooth" means that the wall can exert only a normal force directed perpendicular to the wall and cannot exert a frictional force parallel to it. A firefighter, whose weight is $W_F = 875$ N, stands 6.30 m from the bottom of the ladder. Assume that the ladder's weight acts at the ladder's center and neglect the hose's weight. Find the forces that the wall and the ground exert on the ladder.

Reasoning Part *b* of the figure shows the free-body diagram of the ladder. The following forces act on the ladder:

1. Its weight $\mathbf{W_L}$

2. A force due to the weight $\mathbf{W_F}$ of the firefighter

3. The force $\mathbf{P}$ applied to the top of the ladder by the wall and directed perpendicular to the wall

4. The forces $\mathbf{G_x}$ and $\mathbf{G_y}$, which are the horizontal and vertical components of the force exerted by the ground on the bottom of the ladder.

The ground, unlike the wall, is not smooth, so that the force $\mathbf{G_x}$ is produced by static friction and prevents the ladder from slipping. The force $\mathbf{G_y}$ is the normal force applied to the ladder by the ground. The ladder is in equilibrium, so that the sum of these forces and the sum of the torques produced by them must be zero.

INTERACTIVE LEARNINGWARE #20

START
Visualization
Reasoning
Knowns and Unknowns
Modeling the Problem
Numerical Solution
FINISH

A woman whose weight is 530 N is poised at the right end of a diving board, whose length is 3.90 m. The board has negligible weight and is bolted down at the left end, while being supported 1.40 m away by a fulcrum, as Figure 9.6a shows. Find the forces $\mathbf{F_1}$ and $\mathbf{F_2}$ that the bolt and the fulcrum, respectively, exert on the board.

This Interactive LearningWare example is identical to Example 3. However, the solution is carried out differently, to emphasize that it does not matter where you choose the rotational axis. In Example 3, we used an axis at the left end of the diving board in Figure 9.6. Here, we use an axis at the right end of the board. As you will see, the answers obtained for $\mathbf{F_1}$ and $\mathbf{F_2}$ are the same as those in Example 3.

Go to our Web site at

http://www.wiley.com/college/cutnell

Solution Since the net force acting on the ladder is zero, we have

$$\Sigma F_x = G_x - P = 0 \tag{4.9a}$$
$$\Sigma F_y = G_y - W_L - W_F$$
$$= G_y - 355 \text{ N} - 875 \text{ N} = 0 \quad \text{or} \quad \boxed{G_y = 1230 \text{ N}} \tag{4.9b}$$

Equation 4.9a cannot be solved as it stands, because it contains two unknown variables. However, another equation can be obtained from the fact that the net torque acting on an object in equilibrium is zero. In calculating torques, it is convenient to

Figure 9.7 (*a*) A ladder leaning against a smooth wall. (*b*) The free-body diagram for the ladder. (*c*) Three of the forces that act on the ladder and their lever arms. The axis of rotation is at the lower end of the ladder and is perpendicular to the page.

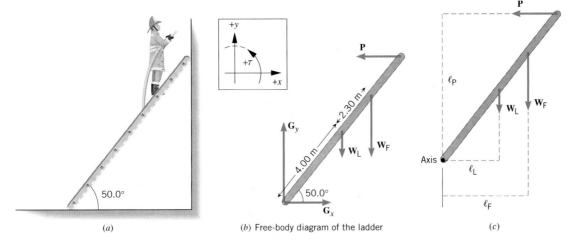

(a)

(b) Free-body diagram of the ladder

(c)

use an axis at the left end of the ladder, directed perpendicular to the page, as Figure 9.7c indicates. This axis is convenient, because $\mathbf{G}_x$ and $\mathbf{G}_y$ produce no torques about it, their lever arms being zero. Consequently, these forces will not appear in the equation representing the balance of torques. The lever arms for the remaining forces are shown in part c as red, dashed lines. The following list summarizes these forces, their lever arms, and the torques:

Force	Lever arm	Torque
$W_L = 355$ N	$\ell_L = (4.00 \text{ m}) \cos 50.0°$	$-W_L\ell_L$
$W_F = 875$ N	$\ell_F = (6.30 \text{ m}) \cos 50.0°$	$-W_F\ell_F$
P	$\ell_P = (8.00 \text{ m}) \sin 50.0°$	$+P\ell_P$

Setting the sum of the torques equal to zero gives

$$\Sigma\tau = -W_L\ell_L - W_F\ell_F + P\ell_P \qquad (9.2)$$
$$= -(355 \text{ N})(4.00 \text{ m})\cos 50.0° - (875 \text{ N})(6.30 \text{ m})\cos 50.0°$$
$$+ P(8.00 \text{ m})\sin 50.0° = 0$$

This equation can be solved to give $\boxed{P = 727 \text{ N}}$. Equation 4.9a indicates that $G_x = P$, so $\boxed{G_x = 727 \text{ N}}$.

To a large extent the directions of the forces acting on an object in equilibrium can be deduced using intuition. Sometimes, however, the direction of an unknown force is not obvious, and it is inadvertently drawn reversed in the free-body diagram. This kind of mistake causes no difficulty. ***Choosing the direction of an unknown force backward in the free-body diagram simply means that the value determined for the force will be a negative number,*** as the next example illustrates.

Figure 9.8 (a) The fully extended, horizontal arm of a bodybuilder supports a dumbbell. (b) The free-body diagram for the arm. (c) Three of the forces that act on the arm and their lever arms. The axis of rotation at the left end of the arm is perpendicular to the page. Force vectors are not to scale.

EXAMPLE 5   Bodybuilding

A bodybuilder holds a dumbbell of weight $\mathbf{W_d}$ as in Figure 9.8a. His arm is horizontal and weighs $W_a = 31.0$ N. The deltoid muscle is assumed to be the only muscle acting and is attached to the arm as shown. The maximum force $\mathbf{M}$ that the deltoid muscle can supply has a magnitude of 1840 N. Figure 9.8b shows the distances that

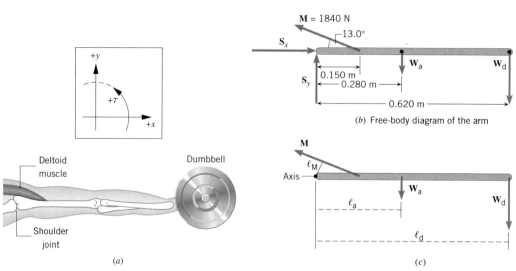

(b) Free-body diagram of the arm

(a)

(c)

locate where the various forces act on the arm. What is the weight of the heaviest dumbbell that can be held, and what are the horizontal and vertical force components, S_x and S_y, that the shoulder joint applies to the left end of the arm?

Reasoning Figure 9.8b is the free-body diagram for the arm. Note that S_x is directed to the right, because the deltoid muscle pulls the arm in toward the shoulder joint, and the joint pushes back in accordance with Newton's third law. The direction of the force S_y, however, is less obvious, and we are alert for the possibility that the direction chosen in the free-body diagram is backward. If so, the value obtained for S_y will be negative.

Solution The arm is in equilibrium, so the net force acting on it is zero:

$$\Sigma F_x = S_x - (1840 \text{ N}) \cos 13.0° = 0 \quad \text{or} \quad \boxed{S_x = 1790 \text{ N}} \tag{4.9a}$$

$$\Sigma F_y = S_y + (1840 \text{ N}) \sin 13.0° - 31.0 \text{ N} - W_d = 0 \tag{4.9b}$$

THE PHYSICS OF...

bodybuilding.

Equation 4.9b cannot be solved at this point, because it contains two unknowns. However, since the arm is in equilibrium, the torques acting on the arm must balance, and this fact provides another equation. To calculate torques, we choose an axis through the left end of the arm and perpendicular to the page. With this axis, the torques due to S_x and S_y are zero, because the line of action of each force passes through the axis and the lever arm of each force is zero. The list below summarizes the remaining forces, their lever arms (see Figure 9.8c), and the torques.

Force	Lever arm	Torque
$W_a = 31.0$ N	$\ell_a = 0.280$ m	$-W_a\ell_a$
W_d	$\ell_d = 0.620$ m	$-W_d\ell_d$
$M = 1840$ N	$\ell_M = (0.150 \text{ m}) \sin 13.0°$	$+M\ell_M$

The condition specifying a zero net torque is

$$\Sigma\tau = -W_a\ell_a - W_d\ell_d + M\ell_M \tag{9.2}$$
$$= -(31.0 \text{ N})(0.280 \text{ m}) - W_d(0.620 \text{ m}) + (1840 \text{ N})(0.150 \text{ m}) \sin 13.0° = 0$$

Equation 9.2 can be solved to show that the maximum dumbbell weight is $\boxed{W_d = 86.1 \text{ N}}$. This value for W_d can be substituted in Equation 4.9b to show that the value for S_y is $\boxed{S_y = -297 \text{ N}}$. The minus sign indicates that the choice of direction for S_y in the free-body diagram is wrong. In reality, S_y has a magnitude of 297 N but is directed downward, not upward.

● **PROBLEM SOLVING INSIGHT**
When a force is negative, such as $S_y = -297$ N in this example, it means that the direction of the force is opposite to that chosen originally.

9.3 CENTER OF GRAVITY

Often, it is important to know the torque produced by the weight of an *extended* body. In Examples 4 and 5, for instance, it is necessary to determine the torques caused by the weight of the ladder and the arm, respectively. In both cases the weight is considered to act at a definite point for the purpose of calculating the torque. This point is called the ***center of gravity*** (abbreviated "cg").

● **DEFINITION OF CENTER OF GRAVITY**
The center of gravity of a rigid body is the point at which its weight can be considered to act when calculating the torque due to the weight.

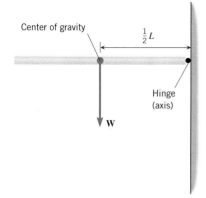

Figure 9.9 A thin, uniform, horizontal rod of length L is attached to a vertical wall by a hinge. The center of gravity of the rod is at its geometrical center.

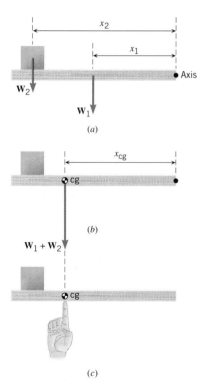

Figure 9.10 (*a*) A box rests near the left end of a horizontal board. (*b*) The total weight ($\mathbf{W_1} + \mathbf{W_2}$) acts at the center of gravity of the group. (*c*) The group can be balanced by applying an external force (due to the index finger) at the center of gravity.

When an object has a symmetric shape and its weight is distributed uniformly, the center of gravity lies at its geometrical center. For instance, Figure 9.9 shows a thin, uniform, horizontal rod of length L attached to a vertical wall by a hinge. The center of gravity of the rod is located at the geometrical center. The lever arm for the weight $\mathbf{W}$ is $L/2$, and the magnitude of the torque is $\tau = W(L/2)$. In a similar fashion, the center of gravity of any symmetrically shaped and uniform object, such as a sphere, disk, cube, or cylinder, is located at its geometrical center. However, this does not mean that the center of gravity must lie within the object itself. The center of gravity of a compact disc recording, for instance, lies at the center of the hole in the disc and is, therefore, "outside" the object.

Suppose we have a group of objects, with known weights and centers of gravity, and it is necessary to know the center of gravity for the group as a whole. As an example, Figure 9.10*a* shows a group composed of two parts: a horizontal uniform board (weight $\mathbf{W_1}$) and a uniform box (weight $\mathbf{W_2}$) near the left end of the board. The center of gravity can be determined by calculating the net torque created by the board and box about an axis that is picked arbitrarily to be at the right end of the board. Part *a* of the figure shows the weights $\mathbf{W_1}$ and $\mathbf{W_2}$ and their corresponding lever arms x_1 and x_2. The net torque is $\Sigma\tau = W_1x_1 + W_2x_2$. It is also possible to calculate the net torque by treating the total weight $\mathbf{W_1} + \mathbf{W_2}$ as if it were located at the center of gravity and had the lever arm x_{cg}, as part *b* of the drawing indicates: $\Sigma\tau = (W_1 + W_2)x_{cg}$. The two values for the net torque must be the same, so that

$$W_1x_1 + W_2x_2 = (W_1 + W_2)x_{cg}$$

This expression can be solved for x_{cg}, which locates the center of gravity relative to the axis:

Center of gravity $$x_{cg} = \frac{W_1x_1 + W_2x_2 + \cdots}{W_1 + W_2 + \cdots} \qquad (9.3)$$

The notation "$+ \cdots$" indicates that Equation 9.3 can be extended to account for any number of weights distributed along a horizontal line. Figure 9.10*c* illustrates that the group can be balanced by a single external force (due to the index finger), if the line of action of the force passes through the center of gravity, and if the force is equal in magnitude, but opposite in direction, to the weight of the group. Example 6 demonstrates how to calculate the center of gravity for the human arm.

EXAMPLE 6   The Center of Gravity of an Arm

The horizontal arm in Figure 9.11 is composed of three parts: the upper arm (weight $W_1 = 17$ N), the lower arm ($W_2 = 11$ N), and the hand ($W_3 = 4.2$ N). The drawing shows the center of gravity of each part, measured with respect to the shoulder joint. Find the center of gravity of the entire arm, relative to the shoulder joint.

Reasoning and Solution The coordinate x_{cg} of the center of gravity is given by

$$x_{cg} = \frac{W_1x_1 + W_2x_2 + W_3x_3}{W_1 + W_2 + W_3} \qquad (9.3)$$

$$= \frac{(17 \text{ N})(0.13 \text{ m}) + (11 \text{ N})(0.38 \text{ m}) + (4.2 \text{ N})(0.61 \text{ m})}{17 \text{ N} + 11 \text{ N} + 4.2 \text{ N}} = \boxed{0.28 \text{ m}}$$

The center of gravity plays an important role in determining whether a group of objects remains in equilibrium as the weight distribution within the group changes. A change in the weight distribution causes a change in the position of the center of gravity, and if the change is too great, the group will not remain in equilibrium. Conceptual Example 7 discusses a shift in the center of gravity that led to an embarrassing result.

CONCEPTUAL EXAMPLE 7 Overloading a Cargo Plane

Figure 9.12*a* shows a stationary cargo plane with its front landing gear 9 meters off the ground. This accident occurred because the plane was overloaded toward the rear. How did a shift in the center of gravity of the loaded plane cause the accident?

Reasoning and Solution Figure 9.12*b* shows a drawing of a correctly loaded plane, with the center of gravity located between the front and the main landing gears. The weight **W** of the plane and cargo acts downward at the center of gravity, and the normal forces $\mathbf{F}_{N1}$ and $\mathbf{F}_{N2}$ act upward at the front and at the main landing gear, respectively. With respect to an axis at the main landing gear, the counterclockwise torque due to $\mathbf{F}_{N1}$ balances the clockwise torque due to **W**, and the plane remains in equilibrium. Figure 9.12*c* shows the plane with too much cargo loaded toward the rear, just after the plane has begun to rotate counterclockwise. Because of the overloading, the center of gravity has shifted behind the main landing gear. The torque due to **W** is now counterclockwise and is not balanced by any clockwise torque. Due to the unbalanced counterclockwise torque, the plane rotates until its tail hits the ground, which applies an upward force to the tail. The clockwise torque due to this upward force balances the counterclockwise torque due to **W**, and the plane comes again into an equilibrium state, this time with the front landing gear 9 meters off the ground.

Related Homework Material: *Problems 14, 19*

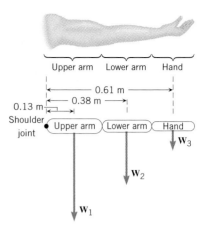

Figure 9.11 The three parts of a human arm, and the weight and center of gravity for each.

The center of gravity of an object with an irregular shape and a nonuniform weight distribution can be found by suspending the object from two different points P_1 and P_2, one at a time. Figure 9.13*a* shows the object at the moment of release,

(*a*)

Figure 9.12 (*a*) This stationary cargo plane is sitting on its tail at Los Angeles International Airport, after being overloaded toward the rear. (*b*) In a correctly loaded plane, the center of gravity is between the front and the main landing gear. (*c*) When the plane is overloaded toward the rear, the center of gravity shifts behind the main landing gear, and the accident in part *a* occurs.

(*b*)   (*c*)

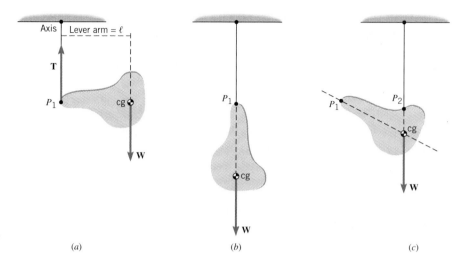

Figure 9.13 The center of gravity (cg) of an object can be located by suspending the object from two different points, P_1 and P_2, one at a time.

(a)   (b)   (c)

when its weight **W**, acting at the center of gravity, has a nonzero lever arm ℓ. At this instant the weight produces a torque about the axis. The tension force **T** applied to the object by the suspension cord produces no torque because its line of action passes through the axis. Hence, in part *a* there is a net torque applied to the object, and the object begins to rotate. Friction eventually brings the object to rest as in part *b*, where the center of gravity lies directly below the point of suspension. In such an orientation, the line of action of the weight passes through the axis, so there is no longer any net torque. In the absence of a net torque the object remains at rest. By suspending the object from a second point (see Figure 9.13*c*), a second line through the object can be established, along which the center of gravity must also lie. The center of gravity, then, must be at the intersection of the two lines.

The center of gravity is closely related to the center-of-mass concept discussed in Section 7.5. To see why they are related, let's replace each occurrence of the weight in Equation 9.3 by $W = mg$, where m is the mass of a given object and g is the acceleration due to gravity at the location of the object. Suppose that g has the same value everywhere that the objects are located. Then it can be algebraically canceled from each term on the right side of Equation 9.3. The resulting equation, which contains only masses and distances, is the same as Equation 7.10, which defines the location of the center of mass. Thus, the two points are identical. For ordinary-sized objects, like cars and boats, the center of gravity coincides with the center of mass.

9.4 NEWTON'S SECOND LAW FOR ROTATIONAL MOTION ABOUT A FIXED AXIS

Figure 9.14 A model airplane on a guideline has a mass m and is flying on a circle of radius r (top view). A net tangential force F_T acts on the plane.

The goal of this section is to put Newton's second law into a form suitable for describing the rotational motion of a rigid object about a fixed axis. We begin by considering a particle moving on a circular path. Figure 9.14 presents a good approximation of this situation by using a small model plane on a guideline of negligible mass. The plane's engine produces a net external tangential force F_T that gives the plane a tangential acceleration a_T. In accord with Newton's second law, it follows that $F_T = ma_T$. The torque τ produced by this force is $\tau = F_T r$, where the radius r of the circular path is also the lever arm. As a result, the torque is $\tau = ma_T r$. But the tangential acceleration is related to the angular acceleration α according to $a_T = r\alpha$ (Equation 8.10), where α must be expressed in rad/s^2. With this substitution for a_T, the torque becomes

$$\tau = \underbrace{(mr^2)}_{\substack{\text{Moment} \\ \text{of inertia } I}}\alpha \qquad (9.4)$$

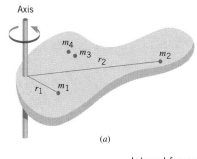

Equation 9.4 is the form of Newton's second law we have been seeking. It indicates that the net external torque τ is directly proportional to the angular acceleration α. The constant of proportionality is $I = mr^2$, which is called the **moment of inertia of the particle.** The SI unit for moment of inertia is kg · m^2.

If all objects were single particles, it would be just as convenient to use the second law in the form $F_T = ma_T$, as in the form $\tau = I\alpha$. The advantage in using $\tau = I\alpha$ is that it can be applied to any rigid body rotating about a fixed axis, and not just to a particle. To illustrate how this advantage arises, Figure 9.15a shows a flat sheet of material that rotates about an axis perpendicular to the sheet. The sheet is composed of a number of mass particles, $m_1, m_2, \ldots, m_N$, where N is very large. Only four particles are shown for the sake of clarity. Each particle behaves in the same way as the model airplane in Figure 9.14 and obeys the relation $\tau = (mr^2)\alpha$:

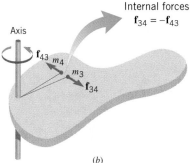

$$\tau_1 = (m_1 r_1^2)\alpha$$
$$\tau_2 = (m_2 r_2^2)\alpha$$
$$\vdots$$
$$\tau_N = (m_N r_N^2)\alpha$$

Figure 9.15 (a) A rigid body consists of a large number of particles, four of which are shown. (b) The internal forces that particles 3 and 4 exert on each other obey Newton's law of action and reaction.

In these equations each particle has the same angular acceleration α, since the rotating object is assumed to be rigid. Adding together the N equations and factoring out the common value of α, we find that

$$\underbrace{\Sigma\tau}_{\substack{\text{Net} \\ \text{external torque}}} = \underbrace{(\Sigma mr^2)}_{\substack{\text{Moment} \\ \text{of inertia}}}\alpha \qquad (9.5)$$

where $\Sigma\tau = \tau_1 + \tau_2 + \cdots + \tau_N$ is the sum of the torques, and $\Sigma mr^2 = m_1 r_1^2 + m_2 r_2^2 + \cdots + m_N r_N^2$ represents the sum of the individual moments of inertia. The latter quantity is the **moment of inertia of the body** I:

Moment of inertia of a body
$$I = \Sigma mr^2 \qquad (9.6)$$

In this equation, r is the perpendicular radial distance of each particle from the axis of rotation. Combining Equation 9.6 with Equation 9.5 gives the following result:

● **ROTATIONAL ANALOG OF NEWTON'S SECOND LAW FOR A RIGID BODY ROTATING ABOUT A FIXED AXIS**

Net external torque $= \left(\begin{array}{c}\text{Moment of} \\ \text{inertia}\end{array}\right) \times \left(\begin{array}{c}\text{Angular} \\ \text{acceleration}\end{array}\right)$

$$\Sigma\tau = I\alpha \qquad (9.7)$$

Requirement: α must be expressed in rad/s^2.

The form of the second law for rotational motion, $\Sigma\tau = I\alpha$, is similar to that for translational (linear) motion, $\Sigma F = ma$, and is valid only in an inertial frame. The moment of inertia I plays the same role for rotational motion that the mass m does for translational motion. Thus, I is a measure of the rotational inertia of a body. When using Equation 9.7, α must be expressed in rad/s^2, because the relation $a_T = r\alpha$ (which requires radian measure) was used in the derivation.

When calculating the sum of torques in Equation 9.7, it is necessary to include only the *external torques,* those applied by agents outside the body. The torques produced by internal forces need not be considered, because they always combine to produce a net torque of zero. Internal forces are those that one particle within the body exerts on another particle. They always occur in pairs of oppositely directed forces of equal magnitude, in accord with Newton's third law (see m_3 and m_4 in Figure 9.15b). The forces in such a pair have the same line of action, so they have identical lever arms and produce torques of equal magnitudes. One torque is counterclockwise, while the other is clockwise, the net torque from the pair being zero.

It can be seen from Equation 9.6 that the moment of inertia depends on both the mass of each particle and its distance from the axis of rotation. The farther a particle is from the axis, the greater is its contribution to the moment of inertia. Therefore, although a rigid object possesses a unique total mass, it does not have a unique moment of inertia, for ***the moment of inertia depends on the location and orientation of the axis relative to the particles that make up the object.*** Example 8 shows how the moment of inertia can change when the axis of rotation changes.

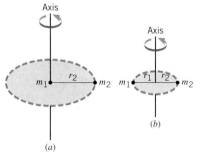

Figure 9.16 Two particles, masses m_1 and m_2, are attached to the ends of a massless rigid rod. The moment of inertia of this object is different, depending on whether the rod rotates about an axis through (a) the end or (b) the center of the rod.

● **PROBLEM SOLVING INSIGHT**
When solving problems involving the moment of inertia, be sure to identify the axis of rotation. The moment of inertia depends on where the axis is. If the location of the axis changes, the moment of inertia may also change.

EXAMPLE 8   The Moment of Inertia Depends on Where the Axis Is

Two particles each have a mass m and are fixed to the ends of a thin rigid rod, whose mass can be ignored. The length of the rod is L. Find the moment of inertia when this object rotates relative to an axis that is perpendicular to the rod at (a) one end and (b) the center. (See Figure 9.16.)

Reasoning When the axis of rotation changes, the distance r between the axis and each particle changes. In determining the moment of inertia using $I = \Sigma mr^2$, we must be careful to use the distances that apply for each axis.

Solution

(a) Particle 1 lies on the axis, as part a of the drawing shows, and has a zero radial distance: $r_1 = 0$. In contrast, particle 2 moves on a circle whose radius is $r_2 = L$. The moment of inertia is

$$I = \Sigma mr^2 = m_1 r_1^2 + m_2 r_2^2 = \boxed{mL^2} \qquad (9.6)$$

(b) Part b of the drawing shows that particle 1 no longer lies on the axis but now moves on a circle of radius $r_1 = L/2$. Particle 2 moves on a circle with the same radius, $r_2 = L/2$. Therefore,

$$I = \Sigma mr^2 = m_1 r_1^2 + m_2 r_2^2 = m(L/2)^2 + m(L/2)^2 = \boxed{\tfrac{1}{2}mL^2}$$

This value differs from that in part (a) because the axis of rotation is different.

The procedure illustrated in Example 8 can be extended using integral calculus to evaluate the moment of inertia of a rigid object with a continuous mass distribution, and Table 9.1 gives some typical results. These results depend on the total mass of the object, its shape, and the location and orientation of the axis.

● *Concepts at a Glance*   When forces act on a rigid object, they can affect its motion in two ways. They can produce a translational acceleration a (components

a_x and a_y). The forces can also produce torques, which can produce an angular acceleration α. In general, we can deal with the resulting combined motion by using Newton's second law. For the translational motion, we use the law in the form $\Sigma F = ma$. For the rotational motion, we use the law in the form $\Sigma \tau = I\alpha$. The Concepts-at-a-Glance chart in Figure 9.17 illustrates the essence of this joint usage of Newton's second law. When a (both components) and α are zero, there is no acceleration of any kind, and the object is in equilibrium. This is the situation already discussed in Section 9.2. If any component of a or α is nonzero, we have accelerated motion, and the object is not in equilibrium. Examples 9, 10, and 11 deal with this type of situation. ●

EXAMPLE 9   The Torque of an Electric Saw Motor

The motor in an electric saw brings the circular blade up to the rated angular speed of 80.0 rev/s in 240.0 rev. One type of blade has a moment of inertia of 1.41×10^{-3} kg·m². What net torque (assumed constant) must the motor apply to the blade?

Reasoning Newton's second law for rotational motion (Equation 9.7, $\Sigma \tau = I\alpha$) can be used to find the net torque $\Sigma \tau$, once the angular acceleration α is determined. The angular acceleration can be calculated from the data in the following table and the appropriate equation of rotational kinematics:

θ	α	ω	ω_0	t
1508 rad (240.0 rev)	?	503 rad/s (80.0 rev/s)	0 rad/s	

The data for the angular displacement θ and the angular velocity ω have been converted to radian measure because $\Sigma \tau = I\alpha$ requires that α be expressed in radian measure.

Solution From Equation 8.8 ($(\omega^2 = \omega_0^2 + 2\alpha\theta)$) it follows that

$$\alpha = \frac{\omega^2 - \omega_0^2}{2\theta}$$

Newton's second law for rotational motion (Equation 9.7) can now be used to obtain the net torque:

$$\Sigma \tau = I\alpha = I\left(\frac{\omega^2 - \omega_0^2}{2\theta}\right)$$

$$= (1.41 \times 10^{-3} \text{ kg·m}^2)\left[\frac{(503 \text{ rad/s})^2 - (0 \text{ rad/s})^2}{2(1508 \text{ rad})}\right] = \boxed{0.118 \text{ N·m}}$$

Example 9 shows how Newton's second law for rotational motion is used when design considerations demand an adequately large angular acceleration. There are also situations when it is desirable to have as little angular acceleration as possible, and Conceptual Example 10 deals with one of them.

CONCEPTUAL EXAMPLE 10   Modern Archery and Bow Stabilizers

Archers can shoot with amazing accuracy, especially using modern bows such as the one in Figure 9.18. Notice the bow stabilizer, a long, thin rod that extends from the front of the bow and has a relatively massive cylinder at the tip. Advertisements claim that the stabilizer helps to steady the archer's aim. Could there be any truth to this claim? Explain.

TABLE 9.1 Moments of Inertia for Various Rigid Objects of Mass M

Thin-walled hollow cylinder or hoop

$I = MR^2$

Solid cylinder or disk

$I = \frac{1}{2}MR^2$

Thin rod, axis perpendicular to rod and passing through center

$I = \frac{1}{12}ML^2$

Thin rod, axis perpendicular to rod and passing through one end

$I = \frac{1}{3}ML^2$

Solid sphere, axis through center

$I = \frac{2}{5}MR^2$

Solid sphere, axis tangent to surface

$I = \frac{7}{5}MR^2$

Thin-walled spherical shell, axis through center

$I = \frac{2}{3}MR^2$

Thin rectangular sheet, axis parallel to one edge and passing through center of other edge

$I = \frac{1}{12}ML^2$

Thin rectangular sheet, axis along one edge

$I = \frac{1}{3}ML^2$

Concepts at a Glance

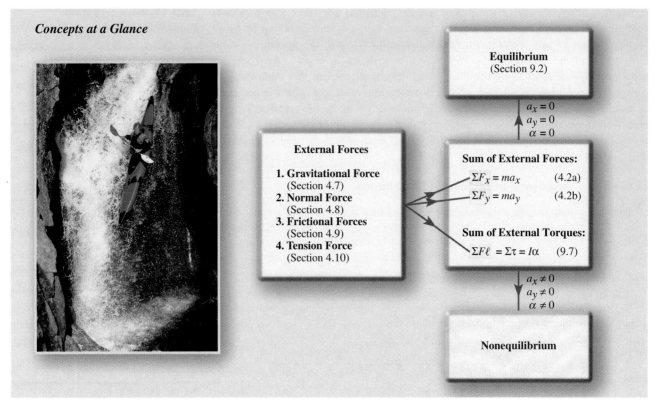

Equilibrium
(Section 9.2)

$a_x = 0$
$a_y = 0$
$\alpha = 0$

External Forces

1. **Gravitational Force**
   (Section 4.7)
2. **Normal Force**
   (Section 4.8)
3. **Frictional Forces**
   (Section 4.9)
4. **Tension Force**
   (Section 4.10)

Sum of External Forces:

$\Sigma F_x = ma_x$ (4.2a)
$\Sigma F_y = ma_y$ (4.2b)

Sum of External Torques:

$\Sigma F\ell = \Sigma \tau = I\alpha$ (9.7)

$a_x \neq 0$
$a_y \neq 0$
$\alpha \neq 0$

Nonequilibrium

Figure 9.17 *Concepts at a Glance* An object is in equilibrium when its translational acceleration components, a_x and a_y, and its angular acceleration α are zero. If a_x or a_y or α is not zero, the object has an acceleration and is not in equilibrium. As this man accelerates down the waterfall in his kayak, he is not in equilibrium.

Reasoning and Solution To help explain why the stabilizer works, we have added to the photograph an axis for rotation that passes through the archer's shoulder and is perpendicular to the plane of the paper. Any angular acceleration α about this axis will lead to a rotation of the bow that will degrade the archer's aim. The acceleration will be created by any unbalanced torques that occur while the archer's tensed muscles try to hold the drawn bow. Newton's second law for rotation indicates, however, that the angular acceleration is $\alpha = (\Sigma \tau)/I$. The moment of inertia I is in the denominator on the right side of this equation. Therefore, to the extent that I is larger, a given net torque $\Sigma \tau$ will

Figure 9.18 The stabilizer helps to steady the archer's aim, as Conceptual Example 10 discusses. Relative to an axis through the archer's shoulder, the moment of inertia of the bow is larger with the stabilizer than without it.

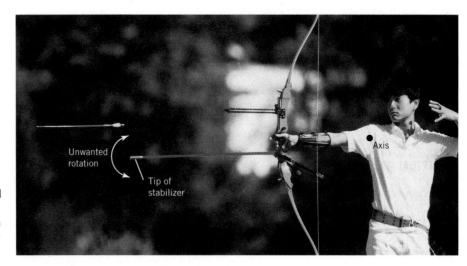

Unwanted rotation

Tip of stabilizer

Axis

create a smaller angular acceleration and less disturbance of the aim. It is to increase the moment of inertia of the bow that the stabilizer has been added. The relatively massive cylinder, which is placed far from the axis of rotation (a large value of r in the equation $I = \Sigma mr^2$), is particularly effective in increasing the moment of inertia.

Rotational motion and translation motion sometimes occur together. The next example deals with an interesting situation in which both angular acceleration and translational acceleration must be considered.

EXAMPLE 11   Hoisting a Crate

A crate that weighs 4420 N is being lifted by the mechanism shown in Figure 9.19a. The two cables are wrapped around their respective pulleys, which have radii of 0.600 and 0.200 m. The pulleys are fastened together to form a "dual" pulley and turn as a single unit about the center axle, relative to which the combined moment of inertia is $I = 50.0 \text{ kg} \cdot \text{m}^2$. A tension of magnitude $T_1 = 2150$ N is maintained in the cable attached to the motor. Find the angular acceleration of the "dual" pulley and the tension in the cable connected to the crate.

Reasoning To determine the angular acceleration of the dual pulley and the tension in the cable attached to the crate, we will apply Newton's second law to the pulley and the crate separately. Three external forces act on the dual pulley, as its free-body diagram shows (Figure 9.19b). These are (1) the tension $\mathbf{T}_1$ in the cable connected to the motor, (2) the tension $\mathbf{T}_2$ in the cable attached to the crate, and (3) the reaction force $\mathbf{P}$ exerted on the dual pulley by the axle. The force $\mathbf{P}$ arises because the two cables pull the pulley down and to the left into the axle, and the axle pushes back, thus keeping the pulley in place. The net torque that results from these forces obeys Newton's second law for rotational motion (Equation 9.7). Two external forces act on the crate, as its free-body diagram indicates (Figure 9.19c). These are (1) the cable tension $\mathbf{T}_2'$ and (2) the weight $\mathbf{W}$ of the crate. The net force that results from these forces obeys Newton's second law for translational motion (Equation 4.2b).

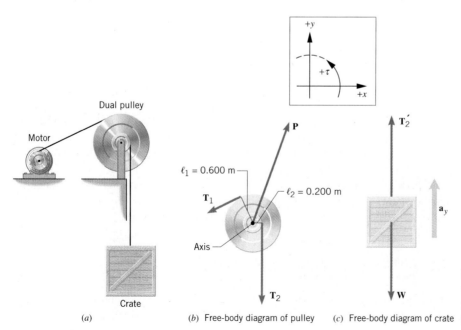

(a)

(b) Free-body diagram of pulley

(c) Free-body diagram of crate

Figure 9.19 (a) The crate is lifted upward by the motor and pulley arrangement. The free-body diagram for (b) the dual pulley and (c) the crate.

TABLE 9.2   Analogies Between Rotational and Translational Concepts

Physical Concept	Rotational	Translational
Displacement	θ	s
Velocity	ω	v
Acceleration	α	a
The cause of acceleration	Torque τ	Force F
Inertia	Moment of inertia I	Mass m
Newton's second law	$\Sigma \tau = I\alpha$	$\Sigma F = ma$
Work	$\tau\theta$	Fs
Kinetic energy	$\frac{1}{2}I\omega^2$	$\frac{1}{2}mv^2$
Momentum	$L = I\omega$	$p = mv$

Solution Using the lever arms ℓ_1 and ℓ_2 shown in part b of the figure, we can apply the second law to the rotational motion of the pulley. We note that the force **P** has a zero lever arm, since the line of action of **P** passes directly through the axle:

$$\Sigma \tau = T_1\ell_1 - T_2\ell_2 = I\alpha \tag{9.7}$$

$$(2150 \text{ N})(0.600 \text{ m}) - T_2(0.200 \text{ m}) = (50.0 \text{ kg} \cdot \text{m}^2)\alpha$$

This equation contains two unknown quantities, so a second equation is needed and may be obtained by applying Newton's second law to the upward translational motion of the crate. In doing so, we note that the magnitude of the tension in the cable attached to the crate is $T_2' = T_2$ and that the mass of the crate is $m = (4420 \text{ N})/(9.80 \text{ m/s}^2) = 451 \text{ kg}$:

$$\Sigma F_y = ma_y \tag{4.2b}$$

$$T_2 - (4420 \text{ N}) = (451 \text{ kg}) a_y$$

Because the cable attached to the crate rolls on the pulley without slipping, the linear acceleration a_y of the crate is related to the angular acceleration α of the pulley via Equation 8.13: $a_y = r\alpha = (0.200 \text{ m})\alpha$. With this substitution for a_y, Equation 4.2b becomes

$$T_2 - (4420 \text{ N}) = (451 \text{ kg})(0.200 \text{ m})\alpha$$

This result and Equation 9.7 can be solved simultaneously to yield

$$\boxed{T_2 = 4960 \text{ N}} \quad \text{and} \quad \boxed{\alpha = 6.0 \text{ rad/s}^2}$$

We have seen that Newton's second law for rotational motion, $\Sigma \tau = I\alpha$, has the same form as that for translational motion, $\Sigma F = ma$, so each rotational variable has a translational analog: torque τ and force F are analogous quantities, as are moment of inertia I and mass m, and angular acceleration α and linear acceleration a. The other physical concepts developed for studying translational motion, such as kinetic energy and momentum, also have rotational analogs. For future reference, Table 9.2 itemizes these concepts and their rotational analogs.

9.5 ROTATIONAL WORK AND ENERGY

Work and energy are among the most fundamental and useful concepts in physics. Chapter 6 discusses their application to translational motion. These concepts are equally useful for rotational motion, provided they are expressed in terms of angular variables.

The work W done by a constant force that points in the same direction as the displacement is $W = Fs$ (Equation 6.1), where F and s are the magnitudes of the force and displacement, respectively. To see how this expression can be rewritten using angular variables, consider Figure 9.20. Here a rope is wrapped around a wheel and is under a constant tension F. If the rope is pulled out a distance s, the wheel rotates through an angle $\theta = s/r$ (Equation 8.1), where r is the radius of the wheel and θ is in radians. Thus, $s = r\theta$, and the work done by the tension force in turning the wheel is $W = Fs = Fr\theta$. But Fr is the torque τ applied to the wheel by the tension, so the rotational work can be written as follows:

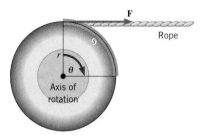

Figure 9.20 The force **F** does work in rotating the wheel through the angle θ.

● DEFINITION OF ROTATIONAL WORK

The rotational work W_R done by a constant torque τ in turning an object through an angle θ is

$$W_R = \tau\theta \qquad (9.8)$$

Requirement: θ must be expressed in radians.

SI Unit of Rotational Work: joule (J)

Section 6.2 discusses the work–energy theorem and kinetic energy. There we saw that the work done on an object by a net external force causes the translational kinetic energy ($\frac{1}{2}mv^2$) of the object to change. In an analogous manner, the rotational work done by a net external torque causes the rotational kinetic energy to change. A rotating body possesses kinetic energy, because its constituent particles are moving. If the body is rotating with an angular speed ω, the tangential speed v_T of a particle at a distance r from the axis is $v_T = r\omega$ (Equation 8.9). Figure 9.21 shows two such particles. If a particle's mass is m, its kinetic energy is $\frac{1}{2}mv_T^2 = \frac{1}{2}mr^2\omega^2$. The kinetic energy of the entire rotating body, then, is the sum of the kinetic energies of the particles:

Figure 9.21 The rotating wheel is composed of many particles, two of which are shown.

$$\text{Rotational KE} = \Sigma(\tfrac{1}{2}mr^2\omega^2) = \tfrac{1}{2}\underbrace{(\Sigma mr^2)}_{\substack{\text{Moment of}\\ \text{inertia, } I}}\omega^2$$

In this result, the angular speed ω is the same for all particles in a rigid body and, therefore, has been factored outside the summation. The term in parentheses is the moment of inertia, $I = \Sigma mr^2$, so the rotational kinetic energy takes the following form:

● DEFINITION OF ROTATIONAL KINETIC ENERGY

The rotational kinetic energy KE_R of a rigid object rotating with an angular speed ω about a fixed axis and having a moment of inertia I is

$$KE_R = \tfrac{1}{2}I\omega^2 \qquad (9.9)$$

Requirement: ω must be expressed in rad/s.

SI Unit of Rotational Kinetic Energy: joule (J)

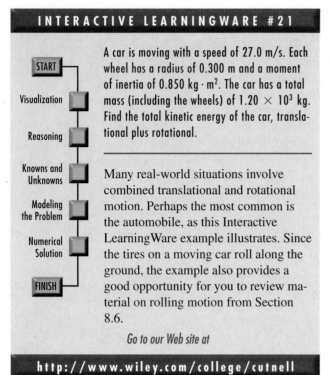

Go to our Web site at

http://www.wiley.com/college/cutnell

INTERACTIVE LEARNINGWARE #21

START

Visualization

Reasoning

Knowns and Unknowns

Modeling the Problem

Numerical Solution

FINISH

A car is moving with a speed of 27.0 m/s. Each wheel has a radius of 0.300 m and a moment of inertia of 0.850 kg·m². The car has a total mass (including the wheels) of 1.20×10^3 kg. Find the total kinetic energy of the car, translational plus rotational.

Many real-world situations involve combined translational and rotational motion. Perhaps the most common is the automobile, as this Interactive LearningWare example illustrates. Since the tires on a moving car roll along the ground, the example also provides a good opportunity for you to review material on rolling motion from Section 8.6.

● *Concepts at a Glance* Kinetic energy is one part of an object's total mechanical energy. The total mechanical energy is the sum of the kinetic and potential energies and obeys the principle of conservation of mechanical energy (see Section 6.5). The Concepts-at-a-Glance chart in Figure 9.22, which is an expanded version of that in Figure 6.15, shows how rotational kinetic energy is incorporated into this principle. Specifically, we need to remember that translational and rotational motion can occur simultaneously. When a bicycle coasts down a hill, for instance, its tires are both translating and rotating. For an object such as a rolling bicycle tire, the total kinetic energy is the sum of its translational and rotational kinetic energies:

$$\text{KE}_{\text{Total}} = \tfrac{1}{2}mv^2 + \tfrac{1}{2}I\omega^2$$

Here, m is the mass of the object, v is the translational speed of its center of mass, I is its moment of inertia about an axis through the center of mass, and ω is its angular speed. The total mechnical energy, then, is

$$\underbrace{E}_{\substack{\text{Total} \\ \text{mechanical} \\ \text{energy}}} = \underbrace{\tfrac{1}{2}mv^2}_{\substack{\text{Translational} \\ \text{kinetic energy}}} + \underbrace{\tfrac{1}{2}I\omega^2}_{\substack{\text{Rotational} \\ \text{kinetic energy}}} + \underbrace{mgh}_{\substack{\text{Gravitational} \\ \text{potential energy}}}$$

where h is the height of the object's center of mass relative to an arbitrary zero level. Mechanical energy is conserved if W_{nc}, the net work done by external nonconservative forces and torques, is zero. If the total mechanical energy is conserved as an object moves, its final total mechanical energy E_{f} equals its initial total mechanical energy E_0: $E_{\text{f}} = E_0$. Example 12 illustrates the effect of combined translational and rotational motion. ●

EXAMPLE 12   Rolling Cylinders

A thin-walled hollow cylinder (mass $= m_{\text{h}}$, radius $= r_{\text{h}}$) and a solid cylinder (mass $= m_{\text{s}}$, radius $= r_{\text{s}}$) start from rest at the top of an incline (Figure 9.23). Both cylinders start at the same vertical height h_0. All heights are measured relative to an arbitrarily chosen zero level that passes through the center of mass of a cylinder when it is at the bottom of the incline (see the drawing). Ignoring energy losses due to retarding forces, determine which cylinder has the greatest translational speed upon reaching the bottom.

Reasoning Only the conservative force of gravity does work on the cylinders, so the total mechanical energy is conserved as they roll down. The total mechanical energy E at any height h above the zero level is the sum of the translational kinetic energy ($\tfrac{1}{2}mv^2$), the rotational kinetic energy ($\tfrac{1}{2}I\omega^2$), and the gravitational potential energy (mgh):

$$E = \tfrac{1}{2}mv^2 + \tfrac{1}{2}I\omega^2 + mgh$$

As the cylinders roll down, potential energy is converted into kinetic energy. But the kinetic energy is shared between the translational form ($\tfrac{1}{2}mv^2$) and the rotational form ($\tfrac{1}{2}I\omega^2$). The object with more of its kinetic energy in the translational form will have the greater translational speed at the bottom of the incline. As we shall see, the solid cylinder has the greater translational speed, for more of its mass is located near the rotational axis and, thus, possesses less rotational kinetic energy.

Concepts at a Glance

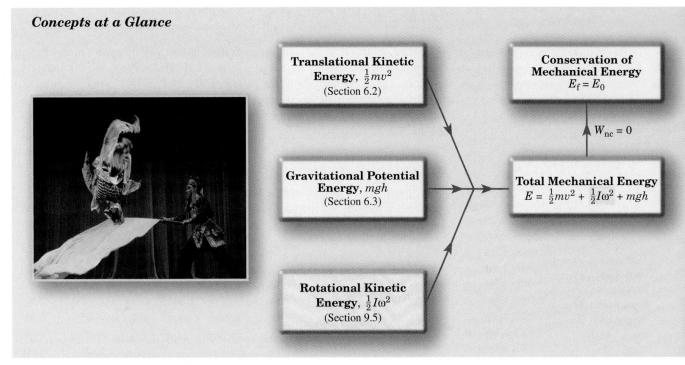

Figure 9.22 *Concepts at a Glance* The principle of conservation of mechanical energy can be applied to a rigid object that has both translational and rotational motion, provided the rotational kinetic energy is included in the total mechanical energy and provided that W_{nc}, the net work done by external nonconservative forces and torques, is zero. The principle can be applied to the translational and rotational motion of this tumbling acrobat, to the extent that his shape is constant and air resistance is negligible.

Solution The total mechanical energy E_f at the bottom ($h_f = 0$) is the same as the total mechanical energy E_0 at the top ($h = h_0$, $v_0 = 0$, $\omega_0 = 0$):

$$\tfrac{1}{2}mv_f^2 + \tfrac{1}{2}I\omega_f^2 + mgh_f = \tfrac{1}{2}mv_0^2 + \tfrac{1}{2}I\omega_0^2 + mgh_0$$

$$\tfrac{1}{2}mv_f^2 + \tfrac{1}{2}I\omega_f^2 = mgh_0$$

Since each cylinder rolls without slipping, the final rotational speed ω_f and the final translational speed v_f of its center of mass are related according to Equation 8.12, $\omega_f = v_f/r$, where r is the radius of the cylinder. Substituting this expression for ω_f into the conservation-of-mechanical-energy equation and solving for v_f yields

$$v_f = \sqrt{\frac{2mgh_0}{m + I/r^2}}$$

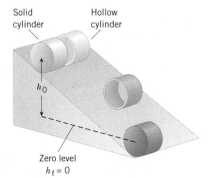

Setting $m = m_h$, $r = r_h$ and $I = m r_h^2$ for the hollow cylinder and then setting $m = m_s$, $r = r_s$ and $I = \tfrac{1}{2}m r_s^2$ for the solid cylinder (see Table 9.1), we find that the two cylinders have the following translational speeds at the bottom of the incline:

Hollow cylinder $v_f = \sqrt{gh_0}$

Solid cylinder $v_f = \sqrt{\dfrac{4gh_0}{3}} = 1.15\sqrt{gh_0}$

The solid cylinder, having the greater translational speed, arrives at the bottom first.

Figure 9.23 A hollow cylinder and a solid cylinder start from rest and roll down the incline plane. The conservation of mechanical energy can be used to show that the solid cylinder, having the greater translational speed, reaches the bottom first.

9.6 ANGULAR MOMENTUM

In Chapter 7 the linear momentum p of an object is defined as the product of its mass m and linear velocity v, that is, $p = mv$. For rotational motion the analogous concept is called the **angular momentum** L. The mathematical form of angular momentum is analogous to that of linear momentum, with the mass m and the linear velocity v being replaced with their rotational counterparts, the moment of inertia I and the angular velocity ω.

● **DEFINITION OF ANGULAR MOMENTUM**

The angular momentum L of a body rotating about a fixed axis is the product of the body's moment of inertia I and its angular velocity ω with respect to that axis:

$$L = I\omega \qquad (9.10)$$

Requirement: ω must be expressed in rad/s.

SI Unit of Angular Momentum: $\text{kg} \cdot \text{m}^2/\text{s}$

Figure 9.24 *Concepts at a Glance* The impulse–momentum theorem for rotational motion leads to the principle of conservation of angular momentum when the sum of the external torques is zero. This concept diagram is similar to that in Figure 7.8, which shows how the conservation of linear momentum arises from the impulse–momentum theorem. The angular momentum of the spinning diver is conserved while she is in the air, assuming that air resistance is negligible.

● *Concepts at a Glance* Linear momentum is an important concept in physics because the total linear momentum of a system is conserved when the sum of the average external forces acting on the system is zero. Then, the final total linear momentum P_f and the initial total linear momentum P_0 are the same: $P_f = P_0$. Figure 7.8 outlines the conceptual development of the conservation of linear momentum. The Concepts-at-a-Glance chart in Figure 9.24 outlines a similar development for

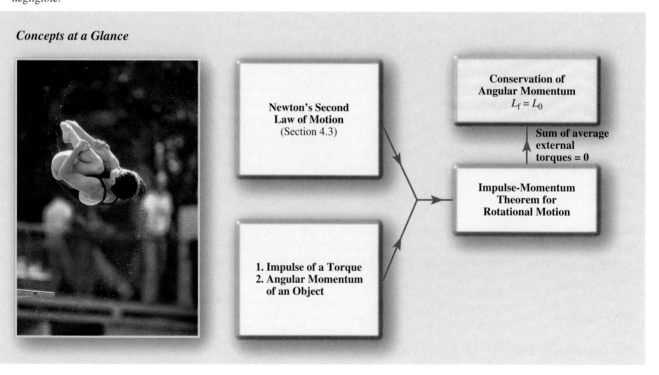

Concepts at a Glance

angular momentum. In constructing this chart, we recall that each translational variable is analogous to a corresponding rotational variable, as Table 9.2 indicates. Therefore, if we replace "force" by "torque" and "linear momentum" by "angular momentum" in Figure 7.8, we obtain Figure 9.24. This chart indicates that when the sum of the average external torques is zero, the final and initial angular momenta are the same: $L_f = L_0$, which is the ***principle of conservation of angular momentum.*** ●

● **PRINCIPLE OF CONSERVATION OF ANGULAR MOMENTUM**

The total angular momentum of a system remains constant (is conserved) if the net external torque acting on the system is zero.

Example 13 illustrates an interesting consequence of the conservation of angular momentum.

Figure 9.25 (*a*) A skater spins slowly on one skate, with both arms and one leg outstretched. (*b*) As she pulls her arms and leg in toward the rotational axis, her moment of inertia *I* decreases, and her angular velocity *ω* increases.

CONCEPTUAL EXAMPLE 13   A Spinning Skater

In Figure 9.25*a* an ice skater is spinning with both arms and a leg outstretched. In Figure 9.25*b* she pulls her arms and leg inward. As a result of this maneuver, her spinning motion changes dramatically. Using the principle of conservation of angular momentum, explain how and why it changes.

Reasoning and Solution Choosing the skater as the system, we can apply the conservation principle provided that the net external torque produced by air resistance and by friction between the skates and the ice is negligibly small. We assume that it is. Then the skater in Figure 9.25*a* would spin forever at the same angular velocity, since her angular momentum is conserved in the absence of a net external torque. In Figure 9.25*b* the inward movement of her arms and leg involves internal, not external, torques and, therefore, does not change her angular momentum. But angular momentum is the product of the moment of inertia *I* and angular velocity *ω*. By moving the mass of her arms and leg inward, the skater decreases the distance *r* of the mass from the axis of rotation and, consequently, decreases her moment of inertia *I* ($I = \Sigma mr^2$). If the product of *I* and *ω* is to remain constant, then *ω* must increase. Thus, the consequence of pulling her arms and leg inward is that ***she spins with a larger angular velocity.***

Related Homework Material: *Questions 20, 21, 22, and Problem 54*

THE PHYSICS OF...

a spinning ice skater.

The next example involves a satellite and illustrates another application of the principle of conservation of angular momentum.

EXAMPLE 14   A Satellite in an Elliptical Orbit

An artificial satellite is placed into an elliptical orbit about the earth, as in Figure 9.26. Telemetry data indicate that its point of closest approach (called the *perigee*) is $r_P = 8.37 \times 10^6$ m from the center of the earth, while its point of greatest distance (called the *apogee*) is $r_A = 25.1 \times 10^6$ m from the center of the earth. The speed of the satellite at the perigee is $v_P = 8450$ m/s. Find its speed v_A at the apogee.

Reasoning The only force of any significance that acts on the satellite is the gravitational force of the earth. However, at any instant, this force is directed toward the center of the earth and passes through the axis about which the satellite instantaneously rotates. Therefore, the gravitational force exerts *no torque* on the satellite

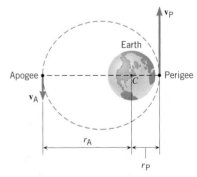

Figure 9.26 A satellite is moving in an elliptical orbit about the earth. The gravitational force exerts no torque on the satellite, so the angular momentum of the satellite is conserved.

THE PHYSICS OF...

a satellite in orbit about the earth.

(the lever arm is zero). Consequently, the angular momentum of the satellite remains constant at all times.

Solution Since the angular momentum is the same at the apogee (A) and the perigee (P), it follows that $I_A\omega_A = I_P\omega_P$. Furthermore, the orbiting satellite can be considered a point mass, so its moment of inertia is $I = mr^2$. In addition, the angular speed ω of the satellite is related to its tangential speed v_T by $\omega = v_T/r$ (Equation 8.9). If these relations are used at the apogee and perigee, the conservation of angular momentum gives the following result:

$$I_A\omega_A = I_P\omega_P$$

$$(mr_A^2)\left(\frac{v_A}{r_A}\right) = (mr_P^2)\left(\frac{v_P}{r_P}\right)$$

$$v_A = \frac{r_P v_P}{r_A} = \frac{(8.37 \times 10^6 \text{ m})(8450 \text{ m/s})}{25.1 \times 10^6 \text{ m}} = \boxed{2820 \text{ m/s}}$$

The answer is independent of the mass of the satellite. The satellite behaves just like the skater in Figure 9.25, because its speed is greater at the perigee, where the moment of inertia is smaller.

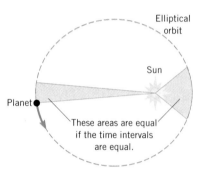

Figure 9.27 Kepler's second law of planetary motion states that a line joining a planet to the sun sweeps out equal areas in equal time intervals.

The result in Example 14 indicates that a satellite does not have a constant speed in an elliptical orbit. The speed changes from a maximum at the perigee to a minimum at the apogee; the closer the satellite comes to the earth, the faster it travels. Planets moving around the sun in elliptical orbits exhibit the same kind of behavior, and Johannes Kepler (1571–1630) formulated his famous second law based on observations of such characteristics of planetary motion. Kepler's second law states that, in a given amount of time, a line joining any planet to the sun sweeps out the same amount of area no matter where the planet is on its elliptical orbit, as Figure 9.27 illustrates. The conservation of angular momentum can be used to show why the law is valid, by means of a calculation similar to that in Example 14.

9.7 CONCEPTS & CALCULATIONS

In this chapter we have seen that a rotational or angular acceleration results when a net external torque acts on an object. In contrast, when a net external force acts on an object, it leads to a translational or linear acceleration, as Chapter 4 discusses. Torque and force, then, are fundamentally different concepts, and Example 15 focuses on this fact.

Figure 9.28 (a) A horizontal pushing force **P** is applied to a uniform crate, which has a square cross section and a weight **W**. The crate rests on the ground, up against a small obstruction. (b) Some of the forces acting on the crate and their lever arms.

CONCEPTS & CALCULATIONS EXAMPLE 15   Torque and Force

Figure 9.28a shows a crate resting on a horizontal surface. It has a square cross section and a weight of $W = 580$ N, which is uniformly distributed. At the bottom right edge is a small obstruction that prevents the crate from sliding when a horizontal pushing force **P** is applied to the left side. However, if this force is great enough, the crate will begin to tip or rotate over the obstruction. Determine the minimum pushing force that leads to tipping.

Concept Questions and Answers What causes tipping—the force **P** or the torque that it creates?

Answer Tipping is a rotational or angular motion. Since the crate starts from rest, an angular acceleration is needed, which can only be created by a net external torque. Thus, the torque created by the force **P** causes the tipping.

A given force can create a variety of torques, depending on the lever arm of the force with respect to the rotational axis. In this case, the rotational axis is located at the small obstruction in Figure 9.28a and is perpendicular to the page. For this axis, the lever arm of the force **P** is ℓ_P, as the drawing shows. Where should **P** be applied, so that a minimum force will give the necessary torque? In other words, should the lever arm be a minimum or a maximum?

Answer The magnitude of the torque is the product of the magnitude of the force and the lever arm. Thus, for a minimum force, the lever arm should be a maximum, and the force **P** should be applied at the upper left corner of the crate, as in Figure 9.28b.

Consider the crate just at the instant *before* it begins to rotate. At this instant, the crate is in equilibrium. What must be true about the sum of the external torques acting on the crate?

Answer Because the crate is in equilibrium, the sum of the external torques must be zero.

Solution Only the pushing force **P** and the weight **W** of the crate produce external torques with respect to a rotational axis through the lower right corner of the crate. The obstruction also applies a force to the crate, but it creates no torque, since its line of action passes through the axis. Since the sum of the external torques is zero, we refer to Figure 9.28b for the lever arms and write

$$\Sigma \tau = -P\ell_P + W\ell_W = 0$$

The lever arm for the force P is $\ell_P = L$, where L is the length of the side of the crate. The lever arm for the weight is $\ell_W = L/2$. This is because the crate is uniform, and the center of gravity is at the center of the crate. Substituting these lever arms in the torque equation, we obtain

$$-PL + W(\tfrac{1}{2}L) = 0 \quad \text{or} \quad P = \tfrac{1}{2}W = \tfrac{1}{2}(580 \text{ N}) = \boxed{290 \text{ N}}$$

For rotational motion, the moment of inertia plays a key role. We have seen in this chapter that the moment of inertia and the net external torque determine the angular acceleration of a rotating object, according to Newton's second law. The angular acceleration, in turn, can be used in the equations of rotational kinematics, provided that it remains constant, as Chapter 8 discusses. When applied together in this way, Newton's second law and the equations of rotational kinematics are particularly useful in accounting for a wide variety of rotational motion. The following example reviews this approach in a situation where a rotating object is slowing down.

CONCEPTS & CALCULATIONS EXAMPLE 16 Which Takes Longer to Stop?

Two spheres are each rotating at an angular speed of 24 rad/s about axes that pass through their centers. Each has a radius of 0.20 m and a mass of 1.5 kg. However, one is solid and the other is a thin-walled spherical shell. Suddenly, a net external torque due to friction (magnitude = 0.12 N·m) begins to act on each sphere and slows the motion down. How long does it take each to come to a halt?

Concept Questions and Answers Which sphere has the greater moment of inertia and why?

Answer Referring to Table 9.1, we see that the solid sphere has a moment of inertia of $\tfrac{2}{5}MR^2$, while the shell has a moment of inertia of $\tfrac{2}{3}MR^2$. Since the masses and radii of the spheres are the same, it follows that the shell has the greater moment of inertia. The reason is that more of the mass of the shell is located farther from the rotational axis than is the case for the solid sphere. In the solid sphere, some of the mass is located close to the axis and, therefore, does not contribute as much to the moment of inertia.

Which sphere has the angular acceleration (a deceleration) with the smaller magnitude?

Answer Newton's second law for rotation (Equation 9.7) specifies that the angular acceleration is $\alpha = (\Sigma\tau)/I$, where $\Sigma\tau$ is the net external torque and I is the moment of inertia. Since the moment of inertia is in the denominator, the angular acceleration is smaller when I is greater. Because it has the greater moment of inertia, the shell has the angular acceleration with the smaller magnitude.

Which sphere takes the longer time to come to a halt?

Answer Since the angular acceleration of the shell has the smaller magnitude, the shell requires a longer time for the deceleration to reduce its angular velocity to zero.

Solution According to Equation 8.4 from the equations of rotational kinematics, the time is given by

$$t = \frac{\omega - \omega_0}{\alpha}$$

where ω and ω_0 are, respectively, the final and initial angular velocities. From Newton's second law as given in Equation 9.7 we know that the angular acceleration is $\alpha = (\Sigma\tau)/I$. Substituting this into the expression for the time gives

$$t = \frac{\omega - \omega_0}{(\Sigma\tau)/I} = \frac{I(\omega - \omega_0)}{\Sigma\tau}$$

In applying this result, we arbitrarily choose the direction of the initial rotation to be positive. With this choice, the torque must be negative, since it causes a deceleration. Using the proper moments of inertia, we find the following times for the spheres to come to a halt:

Solid sphere $t = \dfrac{I(\omega - \omega_0)}{\Sigma\tau} = \dfrac{\frac{2}{5}MR^2(\omega - \omega_0)}{\Sigma\tau}$

$$t = \frac{\frac{2}{5}(1.5 \text{ kg})(0.20 \text{ m})^2[(0 \text{ rad/s}) - (24 \text{ rad/s})]}{-0.12 \text{ N} \cdot \text{m}} = \boxed{4.8 \text{ s}}$$

Spherical shell $t = \dfrac{I(\omega - \omega_0)}{\Sigma\tau} = \dfrac{\frac{2}{3}MR^2(\omega - \omega_0)}{\Sigma\tau}$

$$t = \frac{\frac{2}{3}(1.5 \text{ kg})(0.20 \text{ m})^2[(0 \text{ rad/s}) - (24 \text{ rad/s})]}{-0.12 \text{ N} \cdot \text{m}} = \boxed{8.0 \text{ s}}$$

As expected, the shell requires a longer time to come to a halt.

At the end of the problem set for this chapter, you will find homework problems that contain both conceptual and quantitative parts. These problems are grouped under the heading *Concepts & Calculations, Group Learning Problems.* They are designed for use by students working alone or in small learning groups. The conceptual part of each problem provides a convenient focus for group discussions.

SUMMARY

The **torque** τ of a force has a magnitude that is given by the magnitude F of the force times the lever arm ℓ: $\tau = F\ell$. The lever arm is the perpendicular distance between the line of action of the force and the axis of rotation.

A rigid body is in **equilibrium** if it has zero translational acceleration and zero angular acceleration, so the net external force and the net external torque acting on the body are zero. For forces acting only in the x, y plane, the conditions for equilibrium are $\Sigma F_x = 0$, $\Sigma F_y = 0$, and $\Sigma \tau = 0$.

The **center of gravity** of a rigid object is the point where its entire weight can be considered to act when calculating the torque due to the weight. For a symmetrical body with uniformly distributed weight, the center of gravity is at the geometrical center of the body. When a number of objects whose weights W_1, W_2, . . . are distributed along the x axis at locations x_1, x_2, . . . , the center of gravity is located at $x_{cg} = (W_1x_1 + W_2x_2 + \cdots)/(W_1 + W_2 + \cdots)$. The **center of mass** is identical to the center of gravity, provided the acceleration due to gravity does not vary over the physical extent of the objects.

For rigid bodies rotating about fixed axes, **Newton's second law for rotational motion** is $\Sigma \tau = I\alpha$. $\Sigma \tau$ is the net external torque applied to a body, I is the moment of inertia of the body, and α is the angular acceleration (in rad/s^2). The **moment of inertia** I of an object composed of N particles is $I = (m_1 r_1^2 + m_2 r_2^2 + \cdots + m_N r_N^2)$, where m_1, m_2, . . . , m_N are the masses of the particles and r_1, r_2, . . . , r_N are the perpendicular distances of the particles from the axis of rotation.

The **rotational work** W_R done by a constant torque τ in turning a rigid body through an angular displacement θ (in radians) is $W_R = \tau\theta$. The **rotational kinetic energy** of an object with angular speed ω (in rad/s) and moment of inertia I is $KE_R = \frac{1}{2}I\omega^2$. The **total mechanical energy** E of a rigid object is the sum of its translational kinetic energy, its rotational kinetic energy, and its potential energy. The **total mechanical energy is conserved,** provided that the net work done by external nonconservative forces and torques is zero.

The **angular momentum** L of a body rotating with angular velocity ω (in rad/s) about a fixed axis and having a moment of inertia I is $L = I\omega$. The **principle of conservation of angular momentum** states that the total angular momentum of a system remains constant if the net external torque acting on the system is zero.

CONCEPTUAL QUESTIONS

1. Sometimes, even with a wrench, one cannot loosen a nut that is frozen tightly to a bolt. It is often possible to loosen the nut by slipping a long pipe over the wrench handle. The purpose of the pipe is to extend the length of the handle, so that the applied force can be located farther away from the nut. Explain why this trick works.

2. Explain (a) how it is possible for a large force to produce only a small, or even zero, torque, and (b) how it is possible for a small force to produce a large torque.

3. A magnetic tape is being played on a cassette deck. The tension in the tape applies a torque to the supply reel. Assuming the tension remains constant during playback, discuss how this torque varies as the reel becomes empty.

4. A flat rectangular sheet of plywood can rotate about an axis perpendicular to the sheet through one corner. How should a force (acting in the plane of the sheet) be applied to the plywood so as to create the largest possible torque? Give your reasoning.

5. A torque is the product of a force and a distance (lever arm). Work is also the product of a force and a distance. Yet, torque and work *are different*. What is it about the distances that makes torque and work different?

6. Suppose you are standing on a train, both feet together, facing a window. The front of the train is to your left. The train starts moving forward. To keep from falling, you slide your right foot out toward the rear of the train. Explain in terms of torque how this action keeps you from falling over.

7. Starting in the spring, fruit begins to grow on the outer end of a branch on a pear tree. Explain how the center of gravity of the pear-growing branch shifts during the course of the summer.

8. The free-body diagram in the drawing shows the forces that act on a thin rod. The three forces are drawn to scale and lie in the plane of the paper. Are these forces sufficient to keep the rod in equilibrium, or are additional forces necessary? Explain.

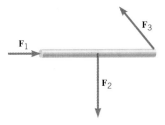

9. An A-shaped step ladder is standing on frictionless ground. The ladder consists of two sections joined at the top and kept from spreading apart by a horizontal crossbar. Draw a free-body diagram showing the forces that keep *one* section of the ladder in equilibrium.

10. The drawing shows a wine rack for a single bottle of wine that seems to defy common sense, as it balances on a table top. Treat the rack and wine bottle as a rigid body and draw the external forces that keep it in equilibrium. In

Wine rack

Rack is not attached to the base.

particular, where must the center of gravity of the rigid body be located? Give your reasoning. *(Hint: The physics here is the same as in Figure 9.10c.)*

11. A flat triangular sheet of uniform material is shown in the drawing. There are three possible axes of rotation, each perpendicular to the sheet and passing through one corner, A, B, or C.

For which axis is the greatest net external torque required to bring the triangle up to an angular speed of 10.0 rad/s in 10.0 s, starting from rest? Explain, assuming that the net torque is kept constant while it is being applied.

12. An object has an angular velocity. It also has an angular acceleration due to torques that are present. Therefore, the angular velocity is changing. What happens to the angular velocity if (a) additional torques are applied so as to make the net torque suddenly equal to zero and (b) all the torques are suddenly removed?

13. The space probe in the drawing is initially moving with a constant translational velocity and zero angular velocity. (a) When the two engines are fired, each generating a thrust of magnitude T, will the translational velocity increase, decrease, or remain the same? Why? (b) Explain what will happen to the angular velocity.

14. Sit-ups are more difficult to do with your hands placed behind your head instead of on your stomach. Why?

15. For purposes of computing the translational kinetic energy of a rigid body, its mass can be considered as concentrated at the center of mass. If one wishes to compute the body's moment of inertia, can the mass be considered as concentrated at the center of mass? If not, why not?

16. A thin sheet of plastic is uniform and has the shape of an equilateral triangle. Consider two axes for rotation. Both are perpendicular to the plane of the triangle, axis A passing through the center of the triangle and axis B passing through one corner. If the angular speed ω about each axis is the same, for which axis does the triangle have the greater rotational kinetic energy? Explain.

17. Bob and Bill have the same weight and wear identical shoes. Keeping both feet flat on the floor and the body straight, Bob can lean over farther than Bill can before falling. Whose center of gravity is closer to the ground? Account for your answer.

18. A hoop, a solid cylinder, a spherical shell, and a solid sphere are placed at rest at the top of an incline. All the objects have the same radius. They are then released at the same time. What is the order in which they reach the bottom? Justify your answer.

19. A woman is sitting on the spinning seat of a piano stool with her arms folded. What happens to her (a) angular velocity and (b) angular momentum when she extends her arms outward? Justify your answers.

20. Review Conceptual Example 13 as an aid in answering this question. Suppose the ice cap at the South Pole melted and the water was distributed uniformly over the earth's oceans. Would the earth's angular velocity increase, decrease, or remain the same? Explain.

21. The concept behind this question is discussed in Conceptual Example 13. Many rivers, like the Mississippi River, flow from north to south toward the equator. These rivers often carry a large amount of sediment that they deposit when entering the ocean. What effect does this redistribution of the earth's soil have on the angular velocity of the earth? Why?

22. Conceptual Example 13 provides background for this question. A cloud of interstellar gas is rotating. Because the gravitational force pulls the gas particles together, the cloud shrinks, and, under the right conditions, a star may ultimately be formed. Would the angular velocity of the star be less than, equal to, or greater than, the angular velocity of the rotating gas? Justify your answer.

23. A person is hanging motionless from a vertical rope over a swimming pool. She lets go of the rope and drops straight down. After letting go, is it possible for her to curl into a ball and start spinning? Justify your answer.

24. The photograph shows a workman struggling to keep a stack of boxes balanced on a dolly. The man's right foot is on the axle of the dolly. Assuming that the boxes are identical, which one creates the greatest torque with respect to the axle? Why?

PROBLEMS

ssm Solution is in the Student Solutions Manual. **www** Solution is available on the World Wide Web at http://www.wiley.com/college/cutnell
✍ This icon represents a biomedical application.

Section 9.1 The Effects of Forces and Torques on the Motion of Rigid Objects

1. ssm A post is driven perpendicularly into the ground and serves as the axis about which a gate rotates. A force of 12 N is applied perpendicular to the gate and acts parallel to the ground. How far from the post should the force be applied to produce a torque with a magnitude of 3.0 N·m?

2. You are installing a new spark plug in your car, and the manual specifies that it be tightened to a torque that has a magnitude of 45 N·m. Using the data in the drawing, determine the magnitude F of the force that you must exert on the wrench.

Problem 2

3. In San Francisco a very simple technique is used to turn around a cable car when it reaches the end of its route. The car rolls onto a turntable, which can rotate about a vertical axis through its center. Then, two people push perpendicularly on the car, one at each end, as in the drawing. The turntable is rotated one-half of a revolution to turn the car around. If the length of the car is 9.20 m and each person pushes with a 185-N force, what is the magnitude of the net torque applied to the car?

4. The drawing shows a jet engine suspended beneath the wing of an airplane. The weight **W** of the engine is 10 200 N and acts as shown in the drawing. In flight the engine produces a thrust **T** of 62 300 N that is parallel to the ground. The rotational axis in the drawing is perpendicular to the plane of the paper. With respect to this axis, find the magnitude of the torque due to (a) the weight and (b) the thrust.

5. ssm The steering wheel of a car has a radius of 0.19 m, while the steering wheel of a truck has a radius of 0.25 m. The same force is applied in the same direction to each. What is the ratio of the torque produced by this force in the truck to the torque produced in the car?

6. A pole has a length of 3.0 m and is attached to a building by a bracket mounted to the side of the building. The bracket holds the pole at an angle of 15° above the horizontal. A sign has a weight of $W = 110$ N. This sign is attached by a wire to the end of the pole and hangs straight downward. Calculate the magnitude of the torque due to the sign's weight with respect to an axis through the bracket.

7. Find the net torque (magnitude and direction) produced by the forces $\mathbf{F}_1$ and $\mathbf{F}_2$ about the rotational axis shown in the drawing. The forces are acting on a thin rigid rod, and the axis is perpendicular to the page.

*8. A pair of forces with equal magnitudes, opposite directions, and different lines of action is called a "couple." When a couple acts on a rigid object, the couple produces a torque that does *not* depend on the location of the axis. The drawing shows a couple acting on a tire wrench, each force being perpendicular to the wrench. Determine an expression for the torque produced by the couple when the axis is perpendicular to the tire and passes through (a) point A, (b) point B, and (c) point C. Express your answers in terms of the magnitude F of the force and the length L of the wrench.

*9. ssm One end of a meter stick is pinned to a table, so the stick can rotate freely in a plane parallel to the tabletop. Two forces, both parallel to the tabletop, are applied to the stick in such a way that the net torque is zero. One force has a magnitude of 2.00 N and is applied perpendicular to the length of the stick at the free end. The other force has a magnitude of 6.00 N and acts at a 30.0° angle with respect to the length of the stick. Where along the stick is the 6.00-N force applied? Express this distance with respect to the end that is pinned.

**10. A rotational axis is directed perpendicular to the plane of a square and is located as shown in the drawing. Two forces, $\mathbf{F}_1$ and $\mathbf{F}_2$, are applied to diagonally opposite corners, and act along the sides of the square, first as shown in part *a* and then as shown in part *b* of the drawing. In each case the net torque produced by the forces is zero. The square is one meter on a side, and the magnitude of $\mathbf{F}_2$ is three times that of $\mathbf{F}_1$. Find the distances *a* and *b* that locate the axis.

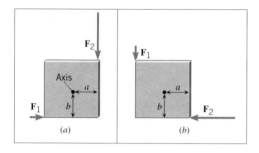

(a) (b)

Section 9.2 Rigid Objects in Equilibrium,
Section 9.3 Center of Gravity

11. The mass of an automobile is 1160 kg, and the distance between its front and rear axles is 2.54 m. The center of gravity of the car is between the front and rear tires, and the horizontal distance between the center of gravity and the front axle is 1.02 m. Determine the normal force that the ground applies to (a) *each* of the two front wheels and to (b) *each* of the two rear wheels.

12. The drawing shows a person whose weight is $W = 584$ N doing push-ups. Find the normal force exerted by the floor on *each* hand and *each* foot, assuming that the person holds this position.

0.840 m   0.410 m

13. ssm Review Example 4 before attempting this problem. What is the minimum value for the coefficient of static friction between the ladder and the ground, so that the ladder does not slip?

14. Review Conceptual Example 7 before starting this problem. A uniform plank of length 5.0 m and weight 225 N rests horizontally on two supports, with 1.1 m of the plank hanging over the right support (see the drawing). To what distance *x* can a person who weighs 450 N walk on the overhanging part of the plank before it just begins to tip?

15. ssm A uniform door (0.81 m wide and 2.1 m high) weighs 140 N and is hung on two hinges that fasten the long left side of the door to a vertical wall. The hinges are 2.1 m apart. Assume that the lower hinge bears all the weight of the door. Find the magnitude and direction of the horizontal component of the force

applied *to the door* by (a) the upper hinge and (b) the lower hinge. Determine the magnitude and direction of the force applied *by the door* to (c) the upper hinge and (d) the lower hinge.

16. A lunch tray is being held in one hand, as the drawing illustrates. The mass of the tray itself is 0.200 kg, and its center of gravity is located at its geometrical center. On the tray is a 1.00-kg plate of food and a 0.250-kg cup of coffee. Obtain the force $\mathbf{T}$ exerted by the thumb and the force $\mathbf{F}$ exerted by the four fingers. Both forces act perpendicular to the tray, which is being held parallel to the ground.

17. A person exerts a horizontal force of 170 N in the test apparatus shown in the drawing. Find the horizontal force $\mathbf{M}$ (magnitude and direction) that his flexor muscle exerts on his forearm.

Problem 17

18. A person is standing on a level floor. His head, upper torso, arms, and hands together weigh 438 N and have a center of gravity that is 1.28 m above the floor. His upper legs weigh 144 N and have a center of gravity that is 0.760 m above the floor. Finally, his lower legs and feet together weigh 87 N and have a center of gravity that is 0.250 m above the floor. Relative to the floor, find the location of the center of gravity for the entire body.

19. Review Conceptual Example 7 as background material for this problem. A jet transport has a weight of 1.00×10^6 N and is at rest on the runway. The two rear wheels are 15.0 m behind the front wheel, and the plane's center of gravity is 12.6 m behind the front wheel. Determine the normal force exerted by the ground on (a) the front wheel and on (b) *each* of the two rear wheels.

20. A uniform steel beam of length 5.00 m has a weight of 4.50×10^3 N. One end of the beam is bolted to a vertical wall. The beam is held in a horizontal position by a cable attached between the other end of the beam and a point on the wall. The cable makes an angle of 25.0° above the horizontal. A load whose weight is 12.0×10^3 N is hung from the beam at a point that is 3.50 m from the wall. Find (a) the magnitude of the tension in the supporting cable and (b) the magnitude of the force exerted on the end of the beam by the bolt that attaches the beam to the wall.

*21. **ssm www** A massless, rigid board is placed across two bathroom scales that are separated by a distance of 2.00 m. A person lies on the board. The scale under his head reads 425 N, and the scale under his feet reads 315 N. (a) Find the weight of the person. (b) Locate the center of gravity of the person relative to the scale beneath his head.

*22. ◢ A person is sitting with one leg outstretched, so that it makes an angle of 30.0° with the horizontal, as the drawing indicates. The weight of the leg below the knee is 44.5 N with the center of gravity located below the knee joint. The leg is being held in this position because of the force **M** applied by the quadriceps muscle, which is attached 0.100 m below the knee joint (see the drawing). Obtain the magnitude of **M**.

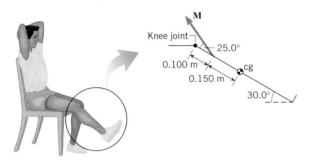

*23. **ssm** ◢ A man holds a 178-N ball in his hand, with the forearm horizontal (see the drawing). He can support the ball in this position because of the flexor muscle force **M**, which is applied perpendicular to the forearm. The forearm weighs 22.0 N and has a center of gravity as indicated. Find (a) the magnitude of **M** and (b) the magnitude and direction of the force applied by the upper arm bone to the forearm at the elbow joint.

*24. A wrecking ball (weight = 4800 N) is supported by a boom, which may be assumed to be uniform and has a weight of 3600 N. As the drawing shows, a support cable runs from the top of the boom to the tractor. The angle between the support cable and the horizontal is 32°, and the angle between the boom and the horizontal is 48°. Find (a) the tension in the support cable and

(b) the magnitude of the force exerted on the lower end of the boom by the hinge at point P.

**25. The drawing shows an inverted "A" that is suspended from the ceiling by two vertical ropes. Each leg of the "A" has a length of $2L$ and a weight of 120 N. The horizontal crossbar has a negligible weight. Find the force that the crossbar applies to each leg.

**26. Two vertical walls are separated by a distance of 1.5 m, as the drawing shows. Wall 1 is smooth, while wall 2 is not smooth. A uniform board is propped between them. The coefficient of static friction between the board and wall 2 is 0.98. What is the length of the longest board that can be propped between the walls?

**27. The drawing shows an A-shaped ladder. Both sides of the ladder are equal in length. This ladder is standing on a frictionless horizontal surface, and only the crossbar (which has a negligible mass) of the "A" keeps the ladder from collapsing. The ladder is uniform and has a mass of 20.0 kg. Determine the tension in the crossbar of the ladder.

Section 9.4 Newton's Second Law for Rotational Motion About a Fixed Axis

28. A solid circular disk has a mass of 1.2 kg and a radius of 0.16 m. Each of three identical thin rods has a mass of 0.15 kg. The rods are attached perpendicularly to the plane of the disk at its outer edge to form a three-legged stool (see the drawing). Find the moment of inertia of the stool with respect to an axis that is perpendicular to the plane of the disk at its center. (*Hint: When considering the moment of inertia of each rod, note that all of the mass of each rod is located at the same perpendicular distance from the axis.*)

29. ssm A clay vase on a potter's wheel experiences an angular acceleration of 8.00 rad/s² due to the application of a 10.0-N·m net torque. Find the total moment of inertia of the vase and potter's wheel.

30. The blades of a ceiling fan have a total moment of inertia of 0.16 kg·m² and an angular acceleration of 7.0 rad/s². What net torque is being applied to the blades?

31. A uniform solid disk with a mass of 24.3 kg and a radius of 0.314 m is free to rotate about a frictionless axle. Forces of 90.0 and 125 N are applied to the disk, as the drawing illustrates. What is (a) the net torque produced by the two forces and (b) the angular acceleration of the disk?

32. A rotating door is made from four rectangular glass panes, as shown in the drawing. The mass of each pane is 85 kg. A person pushes on the outer edge of one pane with a force of $F = 68$ N that is directed perpendicular to the pane. Determine the magnitude of the door's angular acceleration.

33. ssm A particle is located at each corner of an imaginary cube. Each edge of the cube is 0.25 m long, and each particle has a mass of 0.12 kg. What is the moment of inertia of these particles with respect to an axis that lies along one edge of the cube?

Problem 32

34. A cylinder is rotating about an axis that passes through the center of each circular end piece. The cylinder has a radius of 0.0750 m, an angular speed of 88.0 rad/s, and a moment of inertia of 0.850 kg·m². A brake shoe presses against the surface of the cylinder and applies a tangential frictional force to it. The frictional force reduces the angular speed of the cylinder by a factor of two during a time of 5.00 s. (a) Find the magnitude of the angular deceleration of the cylinder. (b) Find the magnitude of the force of friction applied by the brake shoe.

35. The circular blade on a radial arm saw is turning at 262 rad/s at the instant the motor is turned off. In 18.0 s the speed of the blade is reduced to 85 rad/s. Assume the blade to be a uniform solid disk of radius 0.130 m and mass 0.400 kg. Find the net torque applied to the blade.

***36.** As an approximation, suppose a baton consists of a thin uniform rod (length = 0.81 m, mass = 0.13 kg), to either end of which is attached a very small ball (mass = 0.13 kg). A baton twirler can bring this baton from rest up to an angular speed of 2.0 rev/s in 0.40 s by grasping the baton at its center and flicking her wrist. Effectively, this causes a force to be applied to the ba-

ton at a point 2.0 cm from the rotational axis through the center of the rod. Find the magnitude of the force.

***37. ssm** A thin, rigid, uniform rod has a mass of 2.00 kg and a length of 2.00 m. (a) Find the moment of inertia of the rod relative to an axis that is perpendicular to the rod at one end. (b) Suppose all the mass of the rod were located at a single point. Determine the perpendicular distance of this point from the axis in part (a), such that this point particle has the same moment of inertia as the rod. This distance is called the *radius of gyration* of the rod.

***38.** The drawing shows a model for the motion of the human forearm in throwing a dart. Because of the force **M** applied by the triceps muscle, the forearm can rotate about an axis at the elbow joint. Assume that the forearm has the dimensions shown in the drawing and a moment of inertia

of 0.065 kg·m² (including the effect of the dart) relative to the axis at the elbow. Assume also that the force **M** acts perpendicular to the forearm. Ignoring the effect of gravity and any frictional forces, determine the magnitude of the force **M** needed to give the dart a tangential speed of 5.0 m/s in 0.10 s, starting from rest.

***39. ssm** A stationary bicycle is raised off the ground, and its front wheel ($m = 1.3$ kg) is rotating at an angular velocity of 13.1 rad/s (see the drawing). The front brake is then applied for 3.0 s, and the wheel slows down to 3.7 rad/s. Assume that all the mass of the wheel is concentrated in the rim, the radius of which is 0.33 m. The coefficient of kinetic friction between each brake pad and the rim is $\mu_k = 0.85$. What is the magnitude of the normal force that *each* brake pad applies to the rim?

***40.** The *parallel axis theorem* provides a useful way to calculate the moment of inertia I about an arbitrary axis. The theorem states that $I = I_{cm} + Mh^2$, where I_{cm} is the moment of inertia of the object relative to an axis that passes through the center of mass and is parallel to the axis of interest, M is the total mass of the object, and h is the perpendicular distance between the two axes. Use this theorem and information to determine an expression for the moment of inertia of a solid cylinder of radius R relative to an axis that lies on the surface of the cylinder and is perpendicular to the circular ends.

****41.** The drawing shows the top view of two doors. The doors are uniform and identical. Door A rotates about an axis through its left edge, while door B rotates about an axis through the center. The same force **F** is applied perpendicular to each door at its right edge, and the force remains perpendicular as the door turns. Starting from rest, door A rotates through a certain angle in 3.00 s. How long does it take door B to rotate through the same angle?

Door A

Door B

42. The crane shown in the drawing is lifting a 180-kg crate upward with an acceleration of 1.2 m/s². The cable from the crate passes over a solid cylindrical pulley at the top of the boom. The pulley has a mass of 130 kg. The cable is then wound onto a hollow cylindrical drum that is mounted on the deck of the crane. The mass of the drum is 150 kg, and its radius is 0.76 m. The engine applies a counterclockwise torque to the drum in order to wind up the cable. What is the magnitude of this torque? Ignore the mass of the cable.

Pulley

Drum

Boom

Section 9.5 Rotational Work and Energy

43. ssm A flywheel is a solid disk that rotates about an axis that is perpendicular to the disk at its center. Rotating flywheels provide a means for storing energy in the form of rotational kinetic energy and are being considered as a possible alternative to batteries in electric cars. The gasoline burned in a 300-mile trip in a typical midsize car produces about 1.2×10^9 J of energy. How fast would a 13-kg flywheel with a radius of 0.30 m have to rotate to store this much energy? Give your answer in rev/min.

44. A solid sphere is rotating about an axis that passes through its center, and so is a thin-walled spherical shell. They have identical masses, radii, and kinetic energies. The angular speed of the shell is 7.0 rad/s. What is the angular speed of the solid sphere?

45. ssm Three objects lie in the x, y plane. Each rotates about the z axis with an angular speed of 6.00 rad/s. The mass m of each object and its perpendicular distance r from the z axis are as follows: (1) $m_1 = 6.00$ kg and $r_1 = 2.00$ m, (2) $m_2 = 4.00$ kg and $r_2 = 1.50$ m, (3) $m_3 = 3.00$ kg and $r_3 = 3.00$ m. (a) Find the tangential speed of each object. (b) Determine the total kinetic energy of this system using the expression KE $= \frac{1}{2}m_1 v_1^2 + \frac{1}{2}m_2 v_2^2 + \frac{1}{2}m_3 v_3^2$. (c) Obtain the moment of inertia of the system. (d) Find the rotational kinetic energy of the system using the relation $\frac{1}{2}I\omega^2$ to verify that the answer is the same as that in (b).

46. Calculate the kinetic energy that the earth has because of (a) its rotation about its own axis and (b) its motion around the sun. Assume that the earth is a uniform sphere and that its path around the sun is circular. For comparison, the total energy used in the United States in one year is about 9.3×10^{19} J.

*47. Review Example 12 before attempting this problem. A marble and a cube are placed at the top of a ramp. Starting from rest at the same height, the marble rolls and the cube slides (no kinetic friction) down the ramp. Determine the ratio of the center-of-mass speed of the cube to the center-of-mass speed of the marble at the bottom of the ramp.

*48. A thin-walled spherical shell is rolling on a surface. What fraction of its total kinetic energy is in the form of rotational kinetic energy about the center of mass?

*49. ssm Review Example 12 before attempting this problem. A solid cylinder and a thin-walled hollow cylinder (see Table 9.1) have the same mass and radius. They are rolling horizontally toward the bottom of an incline. The center of mass of each has the same translational speed. The cylinders roll up the incline and reach their highest points. Calculate the ratio of the distances $(s_{\text{solid}}/s_{\text{hollow}})$ along the incline through which each center of mass moves.

*50. Review Example 12 before attempting this problem. A bowling ball encounters a 0.760-m vertical rise on the way back to the ball rack, as the drawing illustrates. Ignore frictional losses and assume that the mass of the ball is distributed uniformly. The translational speed of the ball is 3.50 m/s at the bottom of the rise. Find the translational speed at the top.

0.760 m

**51. A tennis ball, starting from rest, rolls down the hill in the drawing. At the end of the hill the ball becomes airborne, leaving at an angle of 35° with respect to the ground. Treat the ball as a thin-walled spherical shell, and determine the range x.

1.8 m

35°

x

Section 9.6 Angular Momentum

52. For a certain satellite with an apogee distance of $r_A = 1.30 \times 10^7$ m, the ratio of the orbital speed at perigee to the orbital speed at apogee is 1.20. Find the perigee distance r_P.

53. ssm www Two disks are rotating about the same axis. Disk A has a moment of inertia of 3.4 kg·m² and an angular velocity of +7.2 rad/s. Disk B is rotating with an angular velocity of −9.8 rad/s. The two disks are then linked together without the aid of any external torques, so that they rotate as a single unit with an angular velocity of −2.4 rad/s. The axis of rotation for this unit is the same as that for the separate disks. What is the moment of inertia of disk B?

54. Before starting this problem, review Conceptual Example 13. A woman stands at the center of a platform. The woman and the platform rotate with an angular speed of 5.00 rad/s. Friction is negligible. Her arms are outstretched, and she is holding a dumb-bell in each hand. In this position the total moment of inertia of the rotating system (platform, woman, and dumbbells) is 5.40 kg·m². By pulling in her arms, the moment of inertia is reduced to 3.80 kg·m². Find her new angular speed.

55. A baggage carousel at an airport is rotating with an angular speed of 0.20 rad/s when the baggage begins to be loaded onto it. The moment of inertia of the carousel is 1500 kg·m². Ten pieces of baggage with an average mass of 15 kg each are dropped vertically onto the carousel and come to rest at a perpendicular distance of 2.0 m from the axis of rotation. (a) Assuming that no net external torque acts on the system of carousel and baggage, find the final angular speed. (b) In reality, the angular speed of a baggage carousel does not change. Therefore, what can you say qualitatively about the net external torque acting on the system?

56. A flat uniform circular disk (radius = 2.00 m, mass = 1.00×10^2 kg) is initially stationary. The disk is free to rotate in the horizontal plane about a frictionless axis perpendicular to the center of the disk. A 40.0-kg person, standing 1.25 m from the axis, begins to run on the disk in a circular path and has a tangential speed of 2.00 m/s relative to the ground. Find the resulting angular speed of the disk (in rad/s) and describe the direction of the rotation.

*****57. ssm** A cylindrically shaped space station is rotating about the axis of the cylinder to create artificial gravity. The radius of the cylinder is 82.5 m. The moment of inertia of the station without people is 3.00×10^9 kg·m². Suppose 500 people, with an average mass of 70.0 kg each, live on this station. As they move radially from the outer surface of the cylinder toward the axis, the angular speed of the station changes. What is the maximum possible percentage change in the station's angular speed due to the radial movement of the people?

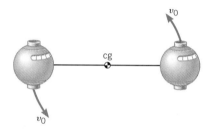

*****58.** In outer space two space modules are joined together by a massless cable. These probes are rotating about their center of mass, which is at the center of the wire, because the modules are identical (see the drawing). In each probe, the cable is connected to a motor, so that the modules can pull each other together. The initial tangential speed of each module is $v_0 = 17$ m/s. Then they pull together until the distance between them is reduced by a factor of two. Determine the final tangential speed, which is v_f for each module.

******59.** A small 0.500-kg object moves on a frictionless horizontal table in a circular path of radius 1.00 m. The angular speed is 6.28 rad/s. The object is attached to a string of negligible mass that passes through a small hole in the table at the center of the circle. Someone under the table begins to pull the string downward to make the circle smaller. If the string will tolerate a tension of no more than 105 N, what is the radius of the smallest possible circle on which the object can move?

******60.** A platform is rotating at an angular speed of 2.2 rad/s. A block is resting on this platform at a distance of 0.30 m from the axis. The coefficient of static friction between the block and the platform is 0.75. Without any external torque acting on the system, the block is moved toward the axis. Ignore the moment of inertia of the platform and determine the smallest distance from the axis at which the block can be relocated and still remain in place as the platform rotates.

ADDITIONAL PROBLEMS

61. ssm www A square, 0.40 m on a side, is mounted so that it can rotate about an axis that passes through the center of the square. The axis is perpendicular to the plane of the square. A force of 15 N lies in this plane and is applied to the square. What is the magnitude of the maximum torque that such a force could produce?

62. A turntable platter has a radius of 0.150 m and is rotating at 3.49 rad/s. When the power is shut off, the platter slows down and comes to rest in 15.0 s, due to a net retarding torque of magnitude 6.20×10^{-3} N·m. Assume the platter to be a uniform solid disk. Determine the mass of the platter.

63. A 71.0-kg boulder is placed on a 2.00-m-long board, at a point that is 1.40 m from one end. Cliff and Will support the board at each end so that it is horizontal. Cliff is nearest the boulder. If the weight of the board is negligible, what force (magnitude only) does (a) Cliff and (b) Will apply to the board?

64. A thin rod has a length of 0.20 m and rotates in a circle on a frictionless tabletop. The axis is perpendicular to the length of the rod at one of its ends. The rod has an angular velocity of 0.40 rad/s and a moment of inertia of 1.2×10^{-3} kg·m². A bug standing on the axis decides to crawl out to the other end of the rod. When the bug (mass = 5.0×10^{-3} kg) gets where it's going, what is the angular velocity of the rod?

65. ssm A bicycle wheel has a radius of 0.330 m and a rim whose mass is 1.20 kg. The wheel has 50 spokes, each with a mass of 0.010 kg. (a) Calculate the moment of inertia of the rim about the axle. (b) Determine the moment of inertia of *any one spoke*, assuming it to be a long, thin rod that can rotate about one end. (c) Find the *total* moment of inertia of the wheel, including the rim and all 50 spokes.

66. In an isometric exercise a person places a hand on a scale and pushes vertically downward, keeping the forearm horizontal. This is possible because the triceps muscle applies an upward force **M** perpendicular to the arm, as the drawing indicates. The forearm weighs 22.0 N and has a center of gravity as indicated. The scale registers 111 N. Determine the magnitude of **M**.

Problem 66

Problem 69

*67. ssm A uniform board is leaning against a smooth vertical wall. The board is at an angle θ above the horizontal ground. The coefficient of static friction between the ground and the lower end of the board is 0.650. Find the smallest value for the angle θ, such that the lower end of the board does not slide along the ground.

*68. A woman who weighs 5.00×10^2 N is leaning against a smooth vertical wall, as the drawing shows. Find (a) the force $\mathbf{F_N}$ (directed perpendicular to the wall) exerted on her shoulders by the wall and the (b) horizontal and (c) vertical components of the force exerted on her shoes by the ground.

Problem 68

*69. A thin uniform rod is rotating at an angular velocity of 7.0 rad/s about an axis that is perpendicular to the rod at its center. As the drawing indicates, the rod is hinged at two places, one-quarter of the length from each end. Without the aid of external torques, the rod suddenly assumes a "u" shape, with the arms of the "u" parallel to the rotation axis. What is the angular velocity of the rotating "u"?

*70. A thin uniform rod is initially positioned in the vertical direction, with its lower end attached to a frictionless axis that is mounted on the floor. The rod has a length of 2.00 m and is allowed to fall, starting from rest. Find the tangential speed of the free end of the rod, just before the rod hits the floor after rotating through 90°.

**71. ssm www An inverted "V" is made of uniform boards and weighs 356 N. Each side has the same length and makes a 30.0° angle with the vertical, as the drawing shows. Find the magnitude of the static frictional force that acts on the lower end of each leg of the "V."

**72. By means of a rope whose mass is negligible, two blocks are suspended over a pulley, as the drawing shows. The pulley can be treated as a uniform solid cylindrical disk. The downward acceleration of the 44.0-kg block is observed to be exactly one-half the acceleration due to gravity. Noting that the tension in the rope is not the same on each side of the pulley, find the mass of the pulley.

CONCEPTS & CALCULATIONS

GROUP LEARNING PROBLEMS

Note: Each of these problems consists of Concept Questions followed by a related quantitative Problem. They are designed for use by students working alone or in small learning groups. The Concept Questions involve little or no mathematics and are intended to stimulate group discussions. They focus on the concepts with which the problems deal. Recognizing the concepts is the essential initial step in any problem-solving technique.

73. Concept Questions The drawing shows a rectangular piece of wood. The forces applied to corners B and D have the same magnitude and are directed parallel to the long and short sides of the rectangle. An axis of rotation is shown perpendicular

to the plane of the rectangle at its center. (a) Relative to this axis, which force produces the torque with the greater magnitude? (b) A force is to be applied to corner A, directed along the short side of the rectangle. The net torque produced by the three forces is zero. How is the force at corner A directed—from A toward B or away from B? Give your reasoning.

Problem The magnitudes of the forces at corners B and D are each 12 N. The long side of the rectangle is twice as long as the short side. Find the magnitude and direction of the force applied to corner A.

74. Concept Questions The wheels, axle, and handles of a wheelbarrow weigh 60.0 N. The load chamber and its contents weigh 525 N. The drawing shows these two forces in two different wheelbarrow designs. To support the wheelbarrow in equilibrium,

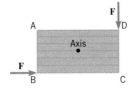

the man's hands apply a force **F** to the handles that is directed vertically upward. Consider a rotational axis at the point where the tire contacts the ground, directed perpendicular to the plane of the paper. (a) For which design is there a greater total torque from the 60.0- and 525-N forces? (b) For which design does the torque from the man's force need to be greater? (c) For which design does the man's force need to be greater? Account for your answers.

Problem Find the magnitude of the man's force for both designs. Be sure that your answers are consistent with your answers to the Concept Questions.

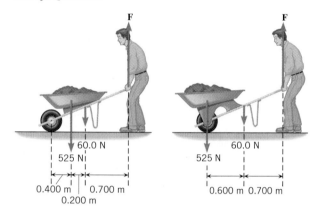

75. Concepts Questions
Part *a* of the drawing shows a uniform horizontal beam bolted to a vertical wall and supported from below at an angle θ by a brace that is attached to a pin. The forces that keep the beam in equilibrium are its weight **W** and the forces applied to the

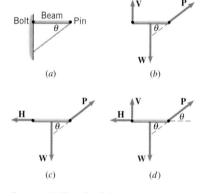

beam by the bolt and the brace. (a) Part *b* of the drawing shows the weight **W** and the force **P** from the brace, while assuming that the bolt applies a force **V** that is vertical. Can these three forces keep the beam in equilibrium? (b) Part *c* of the drawing is like part *b*, except that it assumes the bolt applies a force **H** that is horizontal rather than vertical. Can the three forces in part *c* keep the beam in equilibrium? (c) Part *c* of the drawing assumes that the bolt applies a force that has both a vertical component **V** and a horizontal component **H**. Can the forces in part *c* keep the beam in equilibrium? Justify your answers.

Problem The brace makes an angle $\theta = 39°$ with respect to the beam, which has a weight of 340 N. Find the magnitudes of the forces **V**, **H**, and **P**.

76. Concept Questions Two wheels have the same mass and radius. One has the shape of a hoop and the other the shape of a solid disk. Each wheel starts from rest and has a constant angular acceleration with respect to a rotational axis that is perpendicular to the plane of the wheel at its center. Each makes the same number of revolutions in the same time. (a) Which wheel, if either, has the greater angular acceleration? (b) Which, if either, has the greater moment of inertia? (c) To which wheel, if either, is a greater net external torque applied? Explain your answers.

Problem The mass and radius of each wheel are 4.0 kg and 0.35 m, respectively. Each wheel starts from rest and turns through an angle of 13 rad in 8.0 s. Find the net external torque that acts on each wheel. Check to see that your answers are consistent with your answers to the Concept Questions.

77. Concept Questions Two thin rods of length L are rotating with the same angular speed ω (in rad/s) about axes that pass perpendicularly through one end. Rod A is massless but has a particle of mass 0.66 kg attached to its free end. Rod B has a mass 0.66 kg, which is distributed uniformly along its length. (a) Which has the greater moment of inertia—rod A with its attached particle or rod B? (b) Which has the greater kinetic energy according to Equation 9.9 ($KE_R = \frac{1}{2}I\omega^2$)? Account for your answers.

Problem The length of each rod is 0.75 m, and the angular speed is 4.2 rad/s. Find the kinetic energies of rod A with its attached particle and rod B. Make sure answers are consistent with your answers to the Concept Questions.

78. Concept Questions As seen from above, a playground carousel is rotating counterclockwise about its center on frictionless bearings. A person standing still on the ground grabs onto one of the bars on the carousel very close to its outer edge and climbs aboard. Thus, this person begins with an angular speed of zero and ends up with a nonzero angular speed, which means that he underwent a counterclockwise angular acceleration. (a) What applies the force to the person to create the torque causing this acceleration? What is the direction of this force? (b) According to Newton's action–reaction law, what can you say about the direction of the force applied to the carousel by the person and about the nature (clockwise or counterclockwise) of the torque that it creates? (c) Does the torque identified in part (b) increase or decrease the angular speed of the carousel?

Problem The carousel has a radius of 1.50 m, an initial angular speed of 3.14 rad/s, and a moment of inertia of 125 kg·m². The mass of the person is 40 kg. Find the final angular speed of the carousel after the person climbs aboard. Verify that your answer is consistent with your answers to the Concept Questions.

The web spun by the orb weaver spider is both strong and elastic. As this chapter discusses, the elastic behavior of materials leads to an important kind of vibratory motion, known as simple harmonic motion.

CHAPTER 10

SIMPLE HARMONIC MOTION AND ELASTICITY

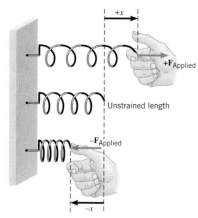

Figure 10.1 An ideal spring obeys the equation, $F_{\text{Applied}} = kx$, where $\mathbf{F}_{\text{Applied}}$ is the force applied to the spring, x is the displacement of the spring from its unstrained length, and k is the spring constant.

THE PHYSICS OF...

a tire pressure gauge.

Figure 10.2 In a tire pressure gauge, the pressurized air from the tire exerts a force $\mathbf{F}_{\text{Applied}}$ that compresses a spring.

10.1   THE IDEAL SPRING AND SIMPLE HARMONIC MOTION

Springs are familiar objects that have many applications, ranging from push-button switches on electronic components, to automobile suspension systems, to mattresses. They are useful because they can be stretched or compressed. For example, the top drawing in Figure 10.1 shows a spring being stretched. Here a hand applies a pulling force $+F_{\text{Applied}}$ to the spring, the $+$ sign indicating that the force is directed to the right. In response, the spring stretches and undergoes a displacement of $+x$ from its original, or "unstrained," length. The bottom drawing in Figure 10.1 illustrates the spring being compressed. Now the hand applies a pushing force $-F_{\text{Applied}}$ to the spring, and it undergoes a displacement of $-x$ from its unstrained length.

Experiment reveals that for relatively small displacements, the force F_{Applied} required to stretch or compress a spring is directly proportional to the displacement x, or $F_{\text{Applied}} \propto x$. As is customary, this proportionality may be converted into an equation by introducing a proportionality constant k:

$$F_{\text{Applied}} = kx \qquad (10.1)$$

The constant k is called the **spring constant,** and Equation 10.1 shows that it has the dimensions of force per unit length (N/m). A spring that behaves according to $F_{\text{Applied}} = kx$ is said to be an **ideal spring.** Example 1 illustrates one application of such a spring.

EXAMPLE 1   A Tire Pressure Gauge

In a tire pressure gauge, the air in the tire pushes against a plunger attached to a spring when the gauge is pressed against the tire valve, as in Figure 10.2. Suppose the spring constant of the spring is $k = 320$ N/m and the bar indicator of the gauge extends 2.0 cm when the gauge is pressed against the tire valve. What force does the air in the tire apply to the spring?

Reasoning We assume that the spring is an ideal spring, so that the relation $F_{\text{Applied}} = kx$ is obeyed. The spring constant k is known, as is the displacement x. Therefore, we can determine the force applied to the spring.

Solution The force needed to compress the spring is given by Equation 10.1 as

$$F_{\text{Applied}} = kx = (320 \text{ N/m})(0.020 \text{ m}) = \boxed{6.4 \text{ N}}$$

Thus, the exposed length of the bar indicator indicates the force that the air pressure in the tire exerts on the spring. We will see later that pressure is force per unit area. Since the area of the plunger surface is fixed, the bar indicator can be marked in units of pressure.

Sometimes the spring constant k is referred to as the **stiffness** of the spring, because a large value for k means the spring is "stiff," in the sense that a large force is required to stretch or compress it. Conceptual Example 2 examines what happens to the stiffness of a spring when the spring is cut into two shorter pieces.

CONCEPTUAL EXAMPLE 2   Are Shorter Springs Stiffer Springs?

Figure 10.3*a* shows a 10-coil spring that has a spring constant k. If this spring is cut in half, so there are two 5-coil springs, what is the spring constant of each of the smaller springs?

Reasoning and Solution It is tempting to say that each 5-coil spring has a spring constant of $\frac{1}{2}k$, or one-half the value for the 10-coil spring. However, the spring constant of each 5-coil spring is $2k$, twice that of the 10-coil spring (see part b of the drawing). Here's the logic. Suppose we apply a force that compresses the 10-coil spring by 1 cm, each coil being compressed by 0.1 cm. Let us also compress a 5-coil spring by 1 cm, each coil now being compressed by 0.2 cm. According to Equation 10.1 ($F_{\text{Applied}} = kx$), the necessary force is directly proportional to the displacement of the coil. The force needed to compress a single coil by 0.2 cm must be twice as large as the force needed to compress it by 0.1 cm. Thus, the spring constant of the 5-coil spring must be twice that of the 10-coil spring. In general, the spring constant is inversely proportional to the number of coils in the spring, so **shorter springs are stiffer springs,** all other things being equal.

Related Homework Material: *Problems 8 and 16*

Figure 10.3 (*a*) The 10-coil spring has a spring constant k. (*b*) Each 5-coil spring has a spring constant of $2k$, twice that of the 10-coil spring.

To stretch or compress a spring, a force must be applied to it. In accord with Newton's third law, the spring exerts an oppositely directed force of equal magnitude. This reaction force is applied by the spring to the agent that does the pulling or pushing. In other words, the reaction force is applied to the object attached to the spring. The reaction force is also called a "restoring force," for a reason that will be clarified shortly. The restoring force of an ideal spring is obtained from the relation $F_{\text{Applied}} = kx$ by including the minus sign required by Newton's action–reaction law, as indicated in Equation 10.2.

● HOOKE'S LAW* RESTORING FORCE OF AN IDEAL SPRING

The restoring force of an ideal spring is

$$F = -kx \tag{10.2}$$

where k is the spring constant and x is the displacement of the spring from its unstrained length. The minus sign indicates that the restoring force always points in a direction opposite to the displacement of the spring.

⬤ *Concepts at a Glance* In Chapter 4 we encountered four types of forces: the gravitational force, the normal force, the frictional force, and the tension force. The Concepts-at-a-Glance chart in Figure 4.9 illustrates that these forces can contribute to the net external force, which Newton's second law relates to the mass and acceleration of an object. The restoring force of a spring can also contribute to the net external force, as the Concepts-at-a-Glance chart in Figure 10.4 indicates. Except for this force, this chart is identical to that in Figure 4.9. Once again, we see the unifying theme of Newton's second law, in that individual forces contribute to the net force, which, in turn, is responsible for the acceleration. Newton's second law plays a central role in describing the motion of objects attached to springs. ⬤

Figure 10.5 helps to explain why the phrase "restoring force" is used. In the picture, an object of mass m is attached to a spring on a frictionless table. In part A, the spring has been stretched to the right, so it exerts the leftward-pointing force $-\mathbf{F}$. When the object is released, this force pulls it to the left, restoring it toward its equilibrium position. However, consistent with Newton's first law, the moving object has inertia and coasts beyond the equilibrium position, compressing the spring

---

* As we will see in Section 10.8, Equation 10.2 is similar to a relationship first discovered by Robert Hooke (1635–1703).

Concepts at a Glance

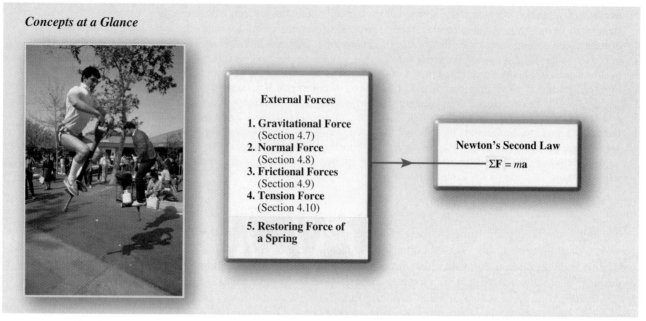

External Forces
1. Gravitational Force (Section 4.7)
2. Normal Force (Section 4.8)
3. Frictional Forces (Section 4.9)
4. Tension Force (Section 4.10)
5. Restoring Force of a Spring

Newton's Second Law

$$\Sigma \mathbf{F} = m\mathbf{a}$$

Figure 10.4 *Concepts at a Glance* The restoring force of a spring may contribute to the net external force $\Sigma\mathbf{F}$ that acts on an object. According to Newton's second law, the resulting acceleration **a** is directly proportional to the net force. The spring in a pogo stick, along with the weight of the person, provide the net force that causes the bouncing motion.

as in part *B*. The force exerted by the spring now points to the right and, after bringing the object to a momentary halt, acts to restore the object to its equilibrium position. But the object's inertia again carries it beyond the equilibrium position, this time stretching the spring and leading to the restoring force $-\mathbf{F}$ shown in part *C*. The back-and-forth motion illustrated in the drawing then repeats itself, continuing forever, since no friction acts on the object or the spring.

When the restoring force has the mathematical form given by $F = -kx$, the type of friction-free motion illustrated in Figure 10.5 is designated as "simple harmonic motion." By attaching a pen to the object and moving a strip of paper past it at a steady rate, we can record the position of the vibrating object as time passes. Figure 10.6 illustrates the resulting graphical record of simple harmonic motion. The maximum excursion from equilibrium is the *amplitude A* of the motion. The shape of this graph is characteristic of simple harmonic motion and is called "sinusoidal," because it has the shape of a trigonometric sine or cosine function.

Figure 10.5 The restoring force (see blue arrows) produced by an ideal spring always points opposite to the displacement (see black arrows) of the spring and leads to a back-and-forth motion of the object.

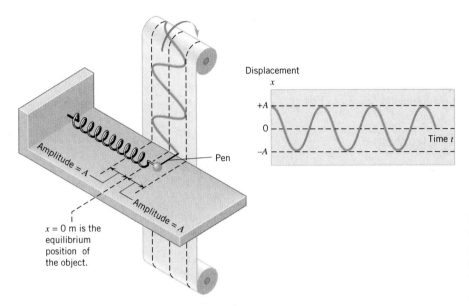

Figure 10.6 When an object moves in simple harmonic motion, a graph of its position as a function of time has a sinusoidal shape with an amplitude A. A pen attached to the object records the graph.

The restoring force also leads to simple harmonic motion when the object is attached to a vertical spring, just as it does when the spring is horizontal. When the spring is vertical, however, the weight of the object causes the spring to stretch, and the motion occurs with respect to the equilibrium position of the object on the stretched spring, as Figure 10.7 indicates. The amount of initial stretching d_0 due to the weight can be calculated by equating the weight to the magnitude of the restoring force that supports it; thus, $mg = kd_0$, which gives $d_0 = mg/k$.

10.2 SIMPLE HARMONIC MOTION AND THE REFERENCE CIRCLE

Simple harmonic motion, like any motion, can be described in terms of displacement, velocity, and acceleration, and the model in Figure 10.8 is helpful in explaining these

Figure 10.7 The weight of an object on a vertical spring stretches the spring by an amount d_0. Simple harmonic motion of amplitude A occurs with respect to the equilibrium position of the object on the stretched spring.

Figure 10.8 The ball mounted on the turntable moves in uniform circular motion, and its shadow, projected on a moving strip of film, executes simple harmonic motion.

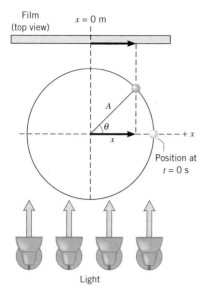

Figure 10.9 A top view of a ball on a turntable. The ball's shadow on the film has a displacement x that depends on the angle θ through which the ball has moved on the reference circle.

characteristics. The model consists of a small ball attached to the top of a rotating turntable. The ball is moving in uniform circular motion (see Section 5.1) on a path known as the *reference circle.* As the ball moves, its shadow falls on a strip of film, which is moving upward at a steady rate and records where the shadow is. A comparison of the film with the paper in Figure 10.6 reveals the same kind of patterns, suggesting that the shadow of the ball is a good model for simple harmonic motion.

Displacement

Figure 10.9 takes a closer look at the reference circle (radius $= A$) and indicates how to determine the displacement of the shadow on the film. The ball starts on the x axis at $x = +A$ and moves through the angle θ in a time t. Since the circular motion is uniform, the ball moves with a constant angular speed ω (in rad/s). Therefore, the angle has a value (in rad) of $\theta = \omega t$. The displacement x of the shadow is just the projection of the radius A onto the x axis:

$$x = A \cos \theta = A \cos \omega t \qquad (10.3)$$

Figure 10.10 shows a graph of this equation. As time passes, the shadow of the ball oscillates between the values of $x = +A$ and $x = -A$, corresponding to the limiting values of $+1$ and -1 for the cosine of an angle. The radius A of the reference circle, then, is the amplitude of the simple harmonic motion.

As the ball moves one revolution or cycle around the reference circle, its shadow executes one cycle of back-and-forth motion. For any object in simple harmonic motion, the time required to complete one cycle is the *period T,* as Figure 10.10 indicates. The value of T depends on the angular speed ω of the ball, because the greater the angular speed, the shorter the time it takes to complete one revolution. We can obtain the relationship between ω and T by recalling that $\omega = \Delta\theta/\Delta t$ (Equation 8.2), where $\Delta\theta$ is the angular displacement of the ball and Δt is the time. For one cycle, $\Delta\theta = 2\pi$ rad and $\Delta t = T$, so that

$$\omega = \frac{2\pi}{T} \qquad (10.4)$$

Often, instead of the period, it is more convenient to speak of the *frequency f* of the motion, the frequency being just the number of cycles of the motion per second. For example, if an object on a spring completes 10 cycles in one second, the frequency is $f = 10$ cycles/s. The period T, or the time for one cycle, would be $\frac{1}{10}$ s. Thus, frequency and period are related according to

$$f = \frac{1}{T} \qquad (10.5)$$

Usually one cycle per second is referred to as one hertz (Hz), the unit being named after Heinrich Hertz (1857–1894). One thousand cycles per second is called one kilohertz (kHz). Thus, five thousand cycles per second, for instance, can be written as 5 kHz.

Using the relationships $\omega = 2\pi/T$ and $f = 1/T$, we can relate the angular speed ω (in rad/s) to the frequency f (in cycles/s or Hz):

$$\omega = \frac{2\pi}{T} = 2\pi f \qquad (10.6)$$

Because ω is directly proportional to the frequency f, ω is often called the *angular frequency.*

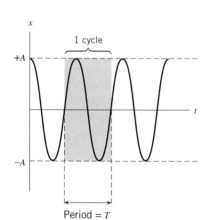

Figure 10.10 For simple harmonic motion, the graph of displacement x versus time t is a sinusoidal curve. The period T is the time required for one complete motional cycle.

Velocity

The reference circle model can also be used to determine the velocity of an object in simple harmonic motion. Figure 10.11 shows the tangential velocity $\mathbf{v_T}$ of the ball on the reference circle. The drawing indicates that the velocity $\mathbf{v}$ of the shadow is just the x component of the vector $\mathbf{v_T}$, that is, $v = -v_T \sin \theta$, where $\theta = \omega t$. The minus sign is necessary, since $\mathbf{v}$ points to the left, in the direction of the negative x axis. Since the tangential speed v_T is related to the angular speed ω by $v_T = r\omega$ (Equation 8.9) and since $r = A$, it follows that $v_T = A\omega$. Therefore, the velocity in simple harmonic motion is given by

$$v = -A\omega \sin \theta = -A\omega \sin \omega t \qquad (10.7)$$

This velocity is *not* constant, but varies between maximum and minimum values as time passes. When the shadow changes direction at either end of the oscillatory motion, the velocity is momentarily zero. When the shadow passes through the $x = 0$ m position, the velocity has a maximum magnitude of $A\omega$, since the sine of an angle is between $+1$ and -1:

$$v_{\max} = A\omega \qquad (\omega \text{ in rad/s}) \qquad (10.8)$$

Both the amplitude A and the angular frequency ω determine the maximum velocity, as Example 3 emphasizes.

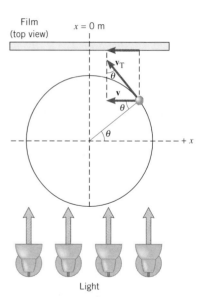

Figure 10.11 The velocity $\mathbf{v}$ of the ball's shadow is the x component of the tangential velocity $\mathbf{v_T}$ of the ball on the reference circle.

EXAMPLE 3   The Maximum Speed of a Loudspeaker Diaphragm

The diaphragm of a loudspeaker moves back and forth in simple harmonic motion to create sound, as in Figure 10.12. The frequency of the motion is $f = 1.0$ kHz and the amplitude is $A = 0.20$ mm. (a) What is the maximum speed of the diaphragm? (b) Where in the motion does this maximum speed occur?

Reasoning The maximum speed $v_{\max}$ of an object vibrating in simple harmonic motion is $v_{\max} = A\omega$, according to Equation 10.8. The angular frequency ω is related to the frequency f by $\omega = 2\pi f$, according to Equation 10.6.

Solution

(a) Using Equations 10.8 and 10.6, we find that the maximum speed of the vibrating diaphragm is

$$v_{\max} = A\omega = A(2\pi f) = (0.20 \times 10^{-3} \text{ m})(2\pi)(1.0 \times 10^3 \text{ Hz}) = \boxed{1.3 \text{ m/s}}$$

(b) The speed of the diaphragm is zero when the diaphragm momentarily comes to rest at the two ends of its motion: $x = +A$ and $x = -A$. Its maximum speed occurs midway between these two positions, or at $x = 0$ m.

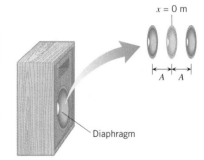

Figure 10.12 The diaphragm of a loudspeaker generates a 1.0-kHz sound by moving back and forth in simple harmonic motion.

Simple harmonic motion is not just any kind of vibratory motion. It is a very specific kind and, among other things, must have the velocity given by Equation 10.7. For instance, advertising signs often use a "moving light" display to grab your attention. Conceptual Example 4 examines the back-and-forth motion in one such display, to see if it is simple harmonic motion.

(a)

(b)

Figure 10.13 The motion of a lighted bulb is from (a) left to right and then from (b) right to left.

CONCEPTUAL EXAMPLE 4   Moving Lights

Over the entrance to a restaurant is mounted a strip of equally spaced light bulbs, as Figure 10.13a illustrates. Starting at the left end, each bulb turns on in sequence for one-half second. Thus, a lighted bulb appears to move from left to right. Once the

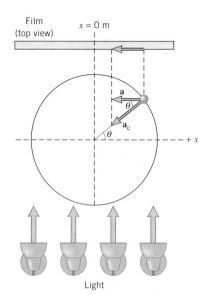

Figure 10.14 The acceleration **a** of the ball's shadow is the x component of the centripetal acceleration $\mathbf{a}_c$ of the ball on the reference circle.

apparent motion of a lighted bulb reaches the right side of the sign, the motion reverses. The lighted bulb then appears to move to the left, as part b of the drawing indicates. Thus, the lighted bulb appears to oscillate back and forth. Is the apparent motion simple harmonic motion?

Reasoning and Solution Since the bulbs are equally spaced and each bulb remains lit for the same amount of time, the apparent motion of a lighted bulb across the sign occurs at a constant speed. If the motion were simple harmonic motion, however, it would not have a constant speed (see Equation 10.7), for it would have zero speed at each end of the sign and increase to a maximum speed at the center of the sign. Therefore, although the apparent motion of a lighted bulb is oscillatory, ***it is not simple harmonic motion.***

Acceleration

In simple harmonic motion, the velocity is not constant; consequently, there must be an acceleration. This acceleration can also be determined with the aid of the reference-circle model. As Figure 10.14 shows, the ball on the reference circle moves in uniform circular motion, and, therefore, has a centripetal acceleration $\mathbf{a}_c$ that points toward the center of the circle. The acceleration **a** of the shadow is the x component of the centripetal acceleration; $a = -a_c \cos\theta$. The minus sign is needed because the acceleration of the shadow points to the left. Recalling that the centripetal acceleration is related to the angular speed ω by $a_c = r\omega^2$ (Equation 8.11) and using $r = A$, we find that $a_c = A\omega^2$. With this substitution, the acceleration in simple harmonic motion becomes

$$a = -A\omega^2 \cos\theta = -A\omega^2 \cos\omega t \qquad (10.9)$$

The acceleration, like the velocity, does *not* have a constant value as time passes. The maximum magnitude of the acceleration is

$$a_{max} = A\omega^2, \qquad (\omega \text{ in rad/s}) \qquad (10.10)$$

Although both the amplitude A and the angular frequency ω determine the maximum value, the frequency has a particularly strong effect, for it is squared. Example 5 shows that the acceleration can be remarkably large in a practical situation.

EXAMPLE 5   The Loudspeaker Revisited—The Maximum Acceleration

THE PHYSICS OF...

a loudspeaker diaphragm.

The loudspeaker diaphragm in Figure 10.12 is vibrating at a frequency of $f = 1.0$ kHz, and the amplitude of the motion is $A = 0.20$ mm. (a) What is the maximum acceleration of the diaphragm, and (b) where does this maximum acceleration occur?

Reasoning The maximum acceleration a_{max} of an object vibrating in simple harmonic motion is $a_{max} = A\omega^2$, according to Equation 10.10. Equation 10.6 shows that the angular frequency ω is related to the frequency f by $\omega = 2\pi f$.

Solution

(a) Using Equations 10.10 and 10.6, we find that the maximum acceleration of the vibrating diaphragm is

$$a_{max} = A\omega^2 = A(2\pi f)^2 = (0.20 \times 10^{-3} \text{ m})[2\pi(1.0 \times 10^3 \text{ Hz})]^2$$
$$= \boxed{7.9 \times 10^3 \text{ m/s}^2}$$

This is an incredible acceleration, being more than 800 times the acceleration due to gravity.

(b) The maximum acceleration occurs when the force acting on the diaphragm is a maximum. The maximum force arises when the diaphragm is at the ends of its path, where the displacement is greatest. Thus, the maximum acceleration occurs at $x = +A$ and $x = -A$ in Figure 10.12.

Frequency of Vibration

With the aid of Newton's second law ($\Sigma F = ma$), it is possible to determine the frequency at which an object of mass m vibrates on a spring. We assume that the mass of the spring itself is negligible and that the only force acting on the object in the horizontal direction is due to the spring, i.e., the Hooke's law restoring force. Thus, the net force is $\Sigma F = -kx$, and Newton's second law becomes $-kx = ma$, where a is the acceleration of the object. The displacement and acceleration of an oscillating spring are, respectively, $x = A \cos \omega t$ (Equation 10.3) and $a = -A\omega^2 \cos \omega t$ (Equation 10.9). Substituting these expressions for x and a into the relation $-kx = ma$, we find that

$$-k(A \cos \omega t) = m(-A\omega^2 \cos \omega t)$$

which yields

$$\omega = \sqrt{\frac{k}{m}} \qquad (\omega \text{ in rad/s}) \qquad (10.11)$$

In this expression, the angular frequency ω must be in radians per second. Larger spring constants k and smaller masses m result in larger frequencies. Example 6 illustrates the effect of the mass on the frequency.

INTERACTIVE LEARNINGWARE #22

START

Visualization

Reasoning

Knowns and Unknowns

Modeling the Problem

Numerical Solution

FINISH

Suppose that an object on a vertical spring oscillates up and down at a frequency of 5.00 Hz. By how much would this object, hanging from rest, stretch the spring?

This is a nice problem, because it reviews several important features of a spring: (1) how its frequency of oscillation depends on the spring constant and the mass of the object attached to it, and (2) how the restoring force of a spring depends on the displacement and the spring constant. This Interactive LearningWare problem combines these features with a review of Newton's second law as applied to an equilibrium situation. Interestingly, the mass of the object does not appear in the final answer.

Go to our Web site at

http://www.wiley.com/college/cutnell

EXAMPLE 6   A Vibrating Test Platform

Electronic equipment for high-performance aircraft like the Space Shuttle must be able to withstand large vibrations. Figure 10.15 shows a platform that can be used to test such equipment. The platform itself has a mass of 3.00 kg and is mounted on four springs, each of which has a spring constant of $k = 1150$ N/m. If electronic equipment is placed on the platform, find the frequency f at which the system vibrates when the electronic equipment has a mass of (a) 4.00 kg and (b) 7.00 kg.

Reasoning The vibrational frequency f is related to the angular frequency ω by $f = \omega/(2\pi)$ (Equation 10.6). The angular frequency of each spring is given by Equation 10.11 as $\omega = \sqrt{k/m}$, where m is the mass supported by each spring. Since there are four springs, each supports one-fourth of the total mass.

Solution

(a) The mass supported by each spring is $m = \frac{1}{4}(3.00 \text{ kg} + 4.00 \text{ kg}) = 1.75$ kg. The frequency of vibration is

$$f = \frac{1}{2\pi} \sqrt{\frac{k}{m}} = \frac{1}{2\pi} \sqrt{\frac{1150 \text{ N/m}}{1.75 \text{ kg}}} = \boxed{4.08 \text{ Hz}}$$

(b) The mass supported by each spring is now $m = \frac{1}{4}(3.00 \text{ kg} + 7.00 \text{ kg}) = 2.50$ kg, and the frequency is

$$f = \frac{1}{2\pi} \sqrt{\frac{k}{m}} = \frac{1}{2\pi} \sqrt{\frac{1150 \text{ N/m}}{2.50 \text{ kg}}} = \boxed{3.41 \text{ Hz}}$$

The frequency is lower here because of the greater inertia of the larger mass.

Platform

Figure 10.15 A piece of electronic equipment on a vibrating test platform.

● **PROBLEM SOLVING INSIGHT**
Do not confuse the vibrational frequency f with the angular frequency ω. The value for f is in hertz (cycles per second). The value for ω is in radians per second. The two are related by $\omega = 2\pi f$.

THE PHYSICS OF...

**detecting and measuring small
amounts of chemicals.**

Example 6 indicates that the mass of the vibrating object influences the frequency of simple harmonic motion. Electronic sensors are being developed that take advantage of this effect in detecting and measuring small amounts of chemicals. These sensors utilize tiny quartz crystals that vibrate when an electric current passes through them. If the crystal is coated with a substance that absorbs a particular chemical, then its mass increases as the chemical is absorbed and, according to the relation $f = \frac{1}{2\pi}\sqrt{k/m}$ discussed in Example 6, the frequency of the simple harmonic motion decreases. The change in frequency is detected electronically, and the sensor is calibrated to give the mass of the absorbed chemical.

10.3 ENERGY AND SIMPLE HARMONIC MOTION

THE PHYSICS OF...

a door-closing unit.

We saw in Chapter 6 that an object above the surface of the earth has gravitational potential energy. Therefore, when the object is allowed to fall, like the hammer of the pile driver in Figure 6.13, it can do work. A spring also has potential energy when the spring is stretched or compressed, which we refer to as *elastic potential energy.* Because of elastic potential energy, a stretched or compressed spring can do work on an object that is attached to the spring. For instance, Figure 10.16 shows a door-closing unit that is often found on screen doors. When the door is opened, a spring inside the unit is compressed and has elastic potential energy. When the door is released, the compressed spring expands and does the work of closing the door.

To find an expression for the elastic potential energy, we will determine the work done by the spring force on an object. Figure 10.17 shows an object attached to one end of a stretched spring. When the object is released, the spring contracts and pulls the object from its initial position x_0 to its final position x_f. The work W done by a constant force is given by Equation 6.1 as $W = (F \cos \theta)s$, where F is the magnitude of the force, s is the magnitude of the displacement ($s = x_0 - x_f$), and θ is the angle between the force and the displacement. The magnitude of the spring force is not constant, however. Equation 10.2 gives the spring force as $F = -kx$, and as the spring contracts, the magnitude of this force changes from kx_0 to kx_f. In using Equation 6.1 to determine the work, we can account for the changing magnitude by using an average magnitude $\overline{F}$ in place of the constant magnitude F. Because the dependence of the spring force on x is linear, the magnitude of the average force is just one-half the sum of the initial and final values, or $\overline{F} = \frac{1}{2}(kx_0 + kx_f)$. The work W_{elastic} done by the average spring force is, then,

Figure 10.16 A door-closing unit. The elastic potential energy stored in the compressed spring is used to close the door.

$$W_{\text{elastic}} = (\overline{F} \cos \theta)s = \tfrac{1}{2}(kx_0 + kx_f) \cos 0° \, (x_0 - x_f) \qquad (10.12)$$

$$W_{\text{elastic}} = \underbrace{\tfrac{1}{2}kx_0^2}_{\substack{\text{Initial elastic} \\ \text{potential energy}}} - \underbrace{\tfrac{1}{2}kx_f^2}_{\substack{\text{Final elastic} \\ \text{potential energy}}}$$

In the calculation above, θ is $0°$, since the spring force points to the left in Figure 10.17, which is the same direction as the displacement. Equation 10.12 indicates that the work done by the spring force is equal to the difference between the initial and final values of the quantity $\frac{1}{2}kx^2$. The quantity $\frac{1}{2}kx^2$ is analogous to the quantity mgh, which we identified in Section 6.3 as the gravitational potential energy. Here, we identify the quantity $\frac{1}{2}kx^2$ as the elastic potential energy. Equation 10.13 indicates that the elastic potential energy is a maximum for a fully stretched or compressed spring and zero for a spring that is neither stretched nor compressed ($x = 0$ m).

Position when spring is unstrained

x_f

x_0

Figure 10.17 When the object is released, its displacement changes from an initial value of x_0 to a final value of x_f.

● DEFINITION OF ELASTIC POTENTIAL ENERGY

The elastic potential energy $PE_{elastic}$ is the energy that a spring has by virtue of being stretched or compressed. For an ideal spring that has a spring constant k and is stretched or compressed by an amount x relative to its unstrained length, the elastic potential energy is

$$PE_{elastic} = \tfrac{1}{2}kx^2 \qquad (10.13)$$

SI Unit of Elastic Potential Energy: joule (J)

● *Concepts at a Glance* The total mechanical energy E is a familiar idea that was originally defined in Section 6.5 to be the sum of the translational kinetic energy and the gravitational potential energy, as illustrated by the Concepts-at-a-Glance chart in Figure 6.15. Then, in Section 9.5 we added the rotational kinetic energy to this list, and the chart in Figure 9.22 shows this addition. We now expand the total mechanical energy to include the elastic potential energy, as indicated by the Concepts-at-a-Glance chart in Figure 10.18. ●

In Equation 10.14 the elastic potential energy is included as part of the total mechanical energy:

$$E \;=\; \underbrace{\tfrac{1}{2}mv^2}_{\substack{\text{Translational}\\\text{kinetic}\\\text{energy}}} \;+\; \underbrace{\tfrac{1}{2}I\omega^2}_{\substack{\text{Rotational}\\\text{kinetic}\\\text{energy}}} \;+\; \underbrace{mgh}_{\substack{\text{Gravitational}\\\text{potential}\\\text{energy}}} \;+\; \underbrace{\tfrac{1}{2}kx^2}_{\substack{\text{Elastic}\\\text{potential}\\\text{energy}}} \qquad (10.14)$$

Total mechanical energy

Figure 10.18 *Concepts at a Glance* The elastic potential energy is added to other energies to give the total mechanical energy. The total mechanical energy is conserved if W_{nc}, the net work done by external nonconservative forces, is zero. As this Bungee jumper oscillates up and down at the end of the elastic cord, the total mechanical energy remains constant, to the extent that the nonconservative forces of friction and air resistance are negligible.

Concepts at a Glance

Figure 10.19 The total mechanical energy of this system is entirely elastic potential energy (A), partly elastic potential energy and partly kinetic energy (B), and entirely kinetic energy (C).

As Section 6.5 discusses, the total mechanical energy is conserved when external nonconservative forces (such as friction) do no net work; that is, when $W_{nc} = 0$ J. Then, the final and initial values of E are the same: $E_f = E_0$. The principle of conservation of total mechanical energy is the subject of the next example.

EXAMPLE 7   An Object on a Horizontal Spring

Figure 10.19 shows an object of mass $m = 0.200$ kg that is vibrating on a horizontal frictionless table. The spring has a spring constant $k = 545$ N/m. It is stretched initially to $x_0 = 4.50$ cm and then released from rest (see part A of the drawing). Determine the final translational speed v_f of the object when the final displacement of the spring is (a) $x_f = 2.25$ cm and (b) $x_f = 0$ cm.

Reasoning The conservation of mechanical energy indicates that, in the absence of friction (a nonconservative force), the final and initial total mechanical energies are the same (see Figure 10.18):

$$E_f = E_0$$
$$\tfrac{1}{2}mv_f^2 + \tfrac{1}{2}I\omega_f^2 + mgh_f + \tfrac{1}{2}kx_f^2 = \tfrac{1}{2}mv_0^2 + \tfrac{1}{2}I\omega_0^2 + mgh_0 + \tfrac{1}{2}kx_0^2$$

Since the object is moving on a horizontal table, the final and initial heights are the same: $h_f = h_0$. The object is not rotating, so its angular speed is zero: $\omega_f = \omega_0 = 0$ rad/s. And, as the problem states, the initial translational speed of the object is zero, $v_0 = 0$ m/s. With these substitutions, the equation above becomes

$$\tfrac{1}{2}mv_f^2 + \tfrac{1}{2}kx_f^2 = \tfrac{1}{2}kx_0^2$$

from which we can obtain v_f:

$$v_f = \sqrt{\frac{k}{m}(x_0^2 - x_f^2)}$$

Solution

(a) Since $x_0 = 0.0450$ m and $x_f = 0.0225$ m, the final translational speed is

$$v_f = \sqrt{\frac{545 \text{ N/m}}{0.200 \text{ kg}}[(0.0450 \text{ m})^2 - (0.0225 \text{ m})^2]} = \boxed{2.03 \text{ m/s}}$$

The total mechanical energy at this point is partly translational kinetic energy ($\tfrac{1}{2}mv_f^2 = 0.414$ J) and partly elastic potential energy ($\tfrac{1}{2}kx_f^2 = 0.138$ J). The total mechanical energy E is the sum of these two energies: $E = 0.414$ J $+ 0.138$ J $= 0.552$ J. Because the total mechanical energy remains constant during the motion, this value equals the initial total mechanical energy when the object is stationary and the energy is entirely elastic potential energy ($E_0 = \tfrac{1}{2}kx_0^2 = 0.552$ J).

(b) When $x_0 = 0.0450$ m and $x_f = 0$ m, we have

$$v_f = \sqrt{\frac{k}{m}(x_0^2 - x_f^2)} = \sqrt{\frac{545 \text{ N/m}}{0.200 \text{ kg}}[(0.0450 \text{ m})^2 - (0 \text{ m})^2]} = \boxed{2.35 \text{ m/s}}$$

Now the total mechanical energy is due entirely to the translational kinetic energy ($\tfrac{1}{2}mv_f^2 = 0.552$ J), since the elastic potential energy is zero. Note that the total mechanical energy is the same as it is in part (a). In the absence of friction, the simple harmonic motion of a spring converts the different types of energy between one form and another, the total always remaining the same.

Conceptual Example 8 takes advantage of energy conservation to illustrate what happens to the maximum speed, amplitude, and angular frequency of a simple harmonic oscillator when its mass is changed suddenly at a certain point in the motion.

CONCEPTUAL EXAMPLE 8
Changing the Mass of a Simple Harmonic Oscillator

(a)

Figure 10.20a shows a box of mass m attached to a spring that has a force constant k. The box rests on a horizontal, frictionless surface. The spring is initially stretched to $x = A$ and then released from rest. The box then executes simple harmonic motion that is characterized by a maximum speed v_{max}, an amplitude A, and an angular frequency ω. When the box is passing through the point where the spring is unstrained ($x = 0$ m), a second box of the same mass m and speed v_{max} is attached to it, as in part b of the drawing. Discuss what happens to (a) the maximum speed, (b) the amplitude, and (c) the angular frequency of the subsequent simple harmonic motion.

(b)

Figure 10.20 (a) A box of mass m, starting from rest at $x = A$, undergoes simple harmonic motion about $x = 0$ m. (b) When $x = 0$ m, a second box, with the same mass and speed, is attached.

Reasoning and Solution

(a) The maximum speed of an object in simple harmonic motion occurs when the object is passing through the point where the spring is unstrained ($x = 0$ m), as in Figure 10.20b. Since the second box is attached at this point with the same speed, **the maximum speed of the two-box system remains the same as that of the one-box system.**

(b) At the same speed, the maximum kinetic energy of the two boxes is twice that of a single box, since the mass is twice as much. Subsequently, when the two boxes move to the left and compress the spring, their kinetic energy is converted into elastic potential energy. Since the two boxes have twice as much kinetic energy as one box alone, the two will have twice as much elastic potential energy when they come to a halt at the extreme left. Here, we are using the principle of conservation of mechanical energy, which applies since friction is absent. But the elastic potential energy is proportional to the amplitude squared (A^2) of the motion, so **the amplitude of the two-box system is greater than that of the one-box system by a factor of $\sqrt{2}$.**

(c) The angular frequency ω of a simple harmonic oscillator is given by Equation 10.11 as $\omega = \sqrt{k/m}$. Since the mass of the two-box system is twice that of the one-box system, **the angular frequency of the two-box system is smaller than that of the one-box system by a factor of $\sqrt{2}$.**

Related Homework Material: *Question 6, Problem 32*

In the previous two examples, gravitational potential energy plays no role because the spring is horizontal. The next example illustrates that gravitational potential energy must be taken into account when a spring is oriented vertically.

EXAMPLE 9   A Falling Ball on a Vertical Spring

A 0.20-kg ball is attached to a vertical spring, as in Figure 10.21. The spring constant of the spring is 28 N/m. The ball, supported initially so that the spring is neither stretched nor compressed, is released from rest. In the absence of air resistance, how far does the ball fall before being brought to a momentary stop by the spring?

Reasoning Since air resistance is absent, only the conservative forces of gravity and the spring act on the ball. Therefore, the principle of conservation of mechanical energy applies:

$$E_f = E_0$$

$$\tfrac{1}{2}mv_f^2 + \tfrac{1}{2}I\omega_f^2 + mgh_f + \tfrac{1}{2}kx_f^2 = \tfrac{1}{2}mv_0^2 + \tfrac{1}{2}I\omega_0^2 + mgh_0 + \tfrac{1}{2}kx_0^2$$

The problem states that the final and initial translational speeds of the ball are zero; $v_f = v_0 = 0$ m/s. The ball and spring do not rotate, so the final and initial angular speeds are also zero: $\omega_f = \omega_0 = 0$ rad/s. As Figure 10.21 indicates, the initial height of the ball is h_0, while the final height is $h_f = 0$ m. In addition, the spring is un-

Figure 10.21 The ball is supported initially so that the spring is unstrained. After being released from rest, the ball falls through the distance h_0 before being momentarily stopped by the spring.

● **PROBLEM SOLVING INSIGHT**
When evaluating the total mechanical energy E, always include a potential energy term for every conservative force acting on the system. In Example 9 there are two such terms, gravitational and elastic.

strained ($x_0 = 0$ m) to begin with, and so it has no elastic potential energy initially. With these substitutions, the conservation of mechanical energy equation reduces to

$$\tfrac{1}{2}kx_f^2 = mgh_0$$

This result shows that the initial gravitational potential energy (mgh_0) is converted into elastic potential energy ($\tfrac{1}{2}kx_f^2$). When the ball falls to its lowest point, its displacement is $x_f = -h_0$, where the minus sign indicates that the displacement is downward. Substituting this result into the equation above and solving for h_0 yields $h_0 = 2mg/k$.

Solution The distance that the ball falls before coming to a momentary halt is

$$h_0 = \frac{2mg}{k} = \frac{2(0.20 \text{ kg})(9.8 \text{ m/s}^2)}{28 \text{ N/m}} = \boxed{0.14 \text{ m}}$$

10.4 THE PENDULUM

As Figure 10.22 shows, a **simple pendulum** consists of a particle of mass m, attached to a frictionless pivot P by a cable of length L and negligible mass. When the particle is pulled away from its equilibrium position by an angle θ and released, it swings back and forth. By attaching a pen to the bottom of the swinging particle and moving a strip of paper beneath it at a steady rate, we can record the position of the particle as time passes. The graphical record reveals a pattern that is similar (but not identical) to the sinusoidal pattern for simple harmonic motion.

The force of gravity is responsible for the back-and-forth rotation about the axis at P. The rotation speeds up as the particle approaches the lowest point on the arc and slows down on the upward part of the swing. Eventually the angular speed is reduced to zero, and the particle swings back. As Section 9.4 discusses, a net torque is required to change the angular speed. The gravitational force $m\mathbf{g}$ produces this torque. (The tension $\mathbf{T}$ in the cable creates no torque, because it points directly at the pivot P and, therefore, has a zero lever arm.) According to Equation 9.1, the magnitude of the torque τ is the product of the magnitude mg of the gravitational force and the lever arm ℓ, so that $\tau = -(mg)\ell$. The minus sign is included since the torque is a restoring torque: that is, it acts to reduce the angle θ (the angle θ is positive [counterclockwise], while the torque is negative [clockwise]). The lever arm ℓ is the perpendicular distance between the line of action of $m\mathbf{g}$ and the pivot P. From Figure 10.22 it can be seen that ℓ is very nearly equal to the arc length s of the circular path when the angle θ is small (about $10°$ or less). Furthermore, if θ is expressed in radians, the arc length and the radius L of the circular path are related, according to $s = L\theta$ (Equation 8.1). Under these conditions, it follows that $\ell \approx s = L\theta$, and the torque created by gravity is

$$\tau \approx \underbrace{-mgL}_{k'}\,\theta$$

In the equation above, the term mgL has a constant value k', independent of θ. *For small angles,* then, the torque that restores the pendulum to its vertical equilibrium position is proportional to the angular displacement θ. The expression $\tau = -k'\theta$ has the same form as the Hooke's law restoring force for an ideal spring, $F = -kx$. Therefore, we expect the frequency of the back-and-forth movement of the pendulum to be given by an equation analogous to Equation 10.11 ($\omega = 2\pi f = \sqrt{k/m}$). In place of the spring constant k, the constant $k' = mgL$ will appear, and, as usual in rotational motion, in place of the mass m, the moment of inertia I will appear:

$$\omega = 2\pi f = \sqrt{\frac{mgL}{I}} \qquad \text{(small angles only)} \qquad (10.15)$$

The moment of inertia of a particle of mass m, rotating at a radius $r = L$ about an axis, is given by $I = mL^2$ (Equation 9.6). Substituting this expression for I into Equation 10.15 reveals for a simple pendulum that

Simple pendulum $\qquad \omega = 2\pi f = \sqrt{\frac{g}{L}} \qquad \text{(small angles only)} \qquad (10.16)$

The mass of the particle has been eliminated algebraically from this expression, so only the length L and the acceleration g due to gravity determine the frequency of a simple pendulum. If the angle of oscillation is large, the pendulum does not exhibit simple harmonic motion, and Equation 10.16 does not apply. Equation 10.16 provides the basis for using a pendulum to keep time, as Example 10 demonstrates.

EXAMPLE 10 Keeping Time

Determine the length of a simple pendulum that will swing back and forth in simple harmonic motion with a period of 1.00 s.

Reasoning When a simple pendulum is swinging back and forth in simple harmonic motion, its frequency f is given by Equation 10.16 as $f = \frac{1}{2\pi}\sqrt{g/L}$, where g is the acceleration due to gravity and L is the length of the pendulum. We also know from Equation 10.5 that the frequency is the reciprocal of the period T, so $f = 1/T$. Thus, the equation above becomes $1/T = \frac{1}{2\pi}\sqrt{g/L}$. We can solve this equation for the length L of the pendulum.

Solution The length of the pendulum is

$$L = \frac{T^2 g}{4\pi^2} = \frac{(1.00 \text{ s})^2(9.80 \text{ m/s}^2)}{4\pi^2} = \boxed{0.248 \text{ m}}$$

Figure 10.23 shows a clock that uses a pendulum to keep time.

It is not necessary that the object in Figure 10.22 be a particle. It may be an extended object, in which case the pendulum is called a *physical pendulum*. For small oscillations, Equation 10.15 still applies, but the moment of inertia I is no longer mL^2. The proper value for the rigid object must be used. (See Section 9.4 for a discussion of moment of inertia.) In addition, the length L for a physical pendulum is the distance between the axis at P and the center of gravity of the object.

10.5 DAMPED HARMONIC MOTION

In simple harmonic motion, an object oscillates with a constant amplitude, because there is no mechanism for dissipating energy. In reality, however, friction or some other energy-dissipating mechanism is always present. In the presence of energy

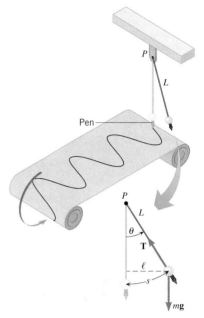

Figure 10.22 A simple pendulum swinging back and forth about the pivot P. If the angle θ is small, the swinging is approximately simple harmonic motion.

Figure 10.23 This pendulum clock keeps time as the pendulum swings back and forth.

dissipation, the amplitude of oscillation decreases as time passes, and the motion is no longer simple harmonic motion. Instead, it is referred to as ***damped harmonic motion,*** the decrease in amplitude being called "damping."

One widely used application of damped harmonic motion is in the suspension system of an automobile. Figure 10.24*a* shows a shock absorber attached to a main suspension spring of a car. A shock absorber is designed to introduce damping forces, which reduce the vibrations associated with a bumpy ride. As part *b* of the drawing shows, a shock absorber consists of a piston in a reservoir of oil. When the piston moves in response to a bump in the road, holes in the piston head permit the piston to pass through the oil. Viscous forces that arise during this movement cause the damping.

Figure 10.25 illustrates the different degrees of damping that can exist. As applied to the example of a car's suspension system, these graphs show the vertical position of the chassis after it has been pulled upward by an amount A_0 at time $t_0 = 0$ s and then released. Part *a* of the figure compares undamped or simple harmonic motion in curve 1 (red) to slightly damped motion in curve 2 (green). In damped harmonic motion, the chassis oscillates with decreasing amplitude and it eventually comes to rest. As the degree of damping is increased from curve 2 to curve 3 (gold), the car makes fewer oscillations before coming to a halt. Part *b* of the drawing shows that as the degree of damping is increased still further, there comes a point when the car does not oscillate at all after it is released but, rather, settles directly back to its equilibrium position, as in curve 4 (blue). The smallest degree of damping that completely eliminates the oscillations is termed "critical damping," and the motion is said to be *critically damped.*

Figure 10.25*b* also shows that the car takes the longest time to return to its equilibrium position in curve 5 (purple), where the degree of damping is above the value for critical damping. When the damping exceeds the critical value, the motion is said to be *overdamped.* In contrast, when the damping is less than the critical level, the motion is said to be *underdamped* (curves 2 and 3). Typical automobile shock absorbers are designed to produce underdamped motion somewhat like that in curve 3.

10.6 DRIVEN HARMONIC MOTION AND RESONANCE

In damped harmonic motion, a mechanism such as friction dissipates or reduces the energy of an oscillating system, with the result that the amplitude of the motion decreases in time. This section discusses the opposite effect, namely, the increase in amplitude that results when energy is continually added to an oscillating system.

THE PHYSICS OF...

a shock absorber.

Shock absorber

Suspension spring

(*a*)

Oil

Hole

Piston

(*b*)

Figure 10.24   (*a*) A shock absorber mounted in the suspension system of an automobile and (*b*) a simplified, cutaway view of the shock absorber.

Figure 10.25 Damped harmonic motion. The degree of damping increases from curve 1 to curve 5. Curve 1 represents undamped or simple harmonic motion. Curves 2 and 3 show underdamped motion. Curve 4 represents critically damped harmonic motion. Curve 5 illustrates overdamped motion.

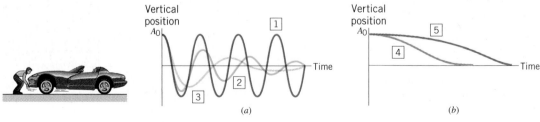

(*a*)

(*b*)

To set an object on an ideal spring into simple harmonic motion, some agent must apply a force that stretches or compresses the spring initially. Suppose that this force is applied at all times, not just for a brief initial moment. The force could be provided, for example, by a person who simply pushes and pulls the object back and forth. The resulting motion is known as ***driven harmonic motion,*** because the additional force drives or controls the behavior of the object to a large extent. The additional force is identified as the ***driving force.***

Figure 10.26 illustrates one particularly important example of driven harmonic motion. Here, the driving force has the same frequency as the spring system and always points in the direction of the object's velocity. The frequency of the spring system is $f = (1/2\pi)\sqrt{k/m}$ and is called a natural frequency, because it is the frequency at which the spring system naturally oscillates. Since the driving force and the velocity always have the same direction, positive work is done on the object at all times, and the total mechanical energy of the system increases. As a result, the amplitude of the vibration becomes larger and will increase without limit, if there is no damping force to dissipate the energy being added by the driving force. The situation depicted in Figure 10.26 is known as ***resonance.***

Figure 10.26 Resonance occurs when the frequency of the driving force matches a frequency at which the object naturally vibrates. The blue arrows represent the driving force, and the red arrows represent the velocity of the object.

● RESONANCE

Resonance is the condition in which a time-dependent force can transmit large amounts of energy to an oscillating object, leading to a large amplitude motion. In the absence of damping, resonance occurs when the frequency of the force matches a natural frequency at which the object will oscillate.

The role played by the frequency of a driving force is a critical one. The matching of this frequency with a natural frequency of vibration allows even a relatively weak force to produce a large amplitude vibration, because the effect of each push–pull cycle is cumulative.

Resonance can occur with any object that can oscillate, and springs need not be involved. The greatest tides in the world occur in the Bay of Fundy, which lies between the Canadian provinces of New Brunswick and Nova Scotia. Figure 10.27 shows the enormous difference between the water level at high and low tides, a difference that in some locations averages about 15 m. This phenomenon is partly due to resonance. The time, or period, that it takes for the tide to flow into and ebb out of a bay depends on the size of the bay, the topology of the bottom, and the config-

THE PHYSICS OF...

high tides at the Bay of Fundy.

Figure 10.27 The Bay of Fundy at (*a*) high tide and (*b*) low tide. In some places the water level changes by almost 15 m.

(*a*)

(*b*)

Figure 10.28 The forces between atoms act like springs. The atoms are represented by gold spheres, and the springs between some atoms have been omitted for clarity.

Figure 10.29 In this diagram, **F** denotes the stretching force, A the cross-sectional area, L_0 the original length of the rod, and ΔL the amount of stretch.

Figure 10.30 The entire weight of the balanced group is supported by the legs of the performer who is lying on his back. From the Ringling Brothers and Barnum & Bailey circus.

uration of the shoreline. The ebb and flow of the water in the Bay of Fundy has a period of 12.5 hours, which is very close to the lunar tidal period of 12.42 hours. The tide then "drives" water into and out of the Bay of Fundy at a frequency (once per 12.42 hours) that nearly matches the natural frequency of the bay (once per 12.5 hours). The result is the extraordinary high tide in the bay. (You can create a similar effect in a bathtub full of water by moving back and forth in synchronism with the waves you're causing.)

10.7 ELASTIC DEFORMATION

Stretching, Compression, and Young's Modulus

We have seen that a spring returns to its original shape when the force compressing or stretching it is removed. In fact, all materials become distorted in some way when they are squeezed or stretched, and many of them, such as rubber, return to their original shape when the squeezing or stretching is removed. Such materials are said to be "elastic." From an atomic viewpoint, elastic behavior has its origin in the forces that atoms exert on each other, and Figure 10.28 symbolizes these forces with the aid of springs. It is because of these atomic level "springs" a material tends to return to its initial shape once the forces that cause the deformation are removed.

The interatomic forces that hold the atoms of a solid together are particularly strong, so considerable force must be applied to stretch a solid object. Experiments have shown that the magnitude of the force can be expressed by the following relation, provided that the amount of stretching is small compared to the original length of the object:

$$F = Y\left(\frac{\Delta L}{L_0}\right) A \qquad (10.17)$$

As Figure 10.29 shows, F denotes the magnitude of the stretching force applied perpendicularly to the surface at the end, A is the cross-sectional area of the rod, ΔL is the increase in length, and L_0 is the original length. The term Y is a proportionality constant called **Young's modulus,** after Thomas Young (1773–1829), and its value depends on the nature of the material, as Table 10.1 reveals. Solving Equation 10.17 for Y shows that Young's modulus has units of force per unit area (N/m^2).

It should be noted that the magnitude of the force in Equation 10.17 is proportional to the fractional increase in length $\Delta L/L_0$, rather than the absolute increase ΔL. The magnitude of the force is also proportional to the cross-sectional area A, which need not be circular, but can have any shape (e.g., rectangular).

Forces that are applied as in Figure 10.29 and cause stretching are called "tensile" forces, because they create a tension in the material, much like the tension in a rope. Equation 10.17 also applies when the force compresses the material along its length. In this situation, the force is applied in a direction opposite to that shown in Figure 10.29, and ΔL stands for the amount by which the original length L_0 decreases.

Most solids have Young's moduli that are rather large, reflecting the fact that a large force is needed to change the length of a solid object by even a small amount, as Example 11 illustrates.

EXAMPLE 11   Bone Compression

In a circus act, a performer supports the combined weight (1640 N) of a number of colleagues (see Figure 10.30). Each thighbone (femur) of this performer has a length of 0.55 m and an effective cross-sectional area of 7.7×10^{-4} m^2. Determine the amount by which each thighbone compresses under the extra weight.

Reasoning The additional weight supported by each thighbone is $F = \frac{1}{2}(1640$ N$) = 820$ N and Table 10.1 indicates that Young's modulus for bone compression is 9.4×10^9 N/m^2. Since the length and cross-sectional area of the thighbone are also known, we may use Equation 10.17 to find the amount by which the additional weight compresses the thighbone.

Solution The amount of compression ΔL of each thighbone is

$$\Delta L = \frac{FL_0}{YA} = \frac{(820 \text{ N})(0.55 \text{ m})}{(9.4 \times 10^9 \text{ N/m}^2)(7.7 \times 10^{-4} \text{ m}^2)} = \boxed{6.2 \times 10^{-5} \text{ m}}$$

This is a very small change, the fractional decrease being $\Delta L/L_0 = 0.000\,11$.

Shear Deformation and the Shear Modulus

It is possible to deform a solid object in a way other than stretching or compressing it. For instance, place a book on a rough table and push on the top cover, as in Figure 10.31a. Notice that the top cover, and the pages below it, become shifted relative to the stationary bottom cover. The resulting deformation is called a **shear deformation** and occurs because of the combined effect of the force **F** applied (by the hand) to the top of the book and the force $-$**F** applied (by the table) to the bottom of the book. The directions of the forces are parallel to the covers of the book, each of which has an area A, as illustrated in part b of the drawing. These two forces have equal magnitudes, but opposite directions, and ensure that the book remains in equilibrium. Equation 10.18 gives the magnitude F of the force needed to produce an amount of shear ΔX for an object with thickness L_0:

$$F = S\left(\frac{\Delta X}{L_0}\right)A \qquad (10.18)$$

This equation is very similar to Equation 10.17. The constant of proportionality S is called the **shear modulus** and, like Young's modulus, has units of force per unit area (N/m^2). The value of S depends on the nature of the material, and Table 10.2 gives some representative values. Example 12 illustrates how to determine the shear modulus of a favorite dessert.

TABLE 10.1 Values for the Young's Modulus of Solid Materials

Material	Young's Modulus Y (N/m^2)
Aluminum	6.9×10^{10}
Bone	
Compression	9.4×10^9
Tension	1.6×10^{10}
Brass	9.0×10^{10}
Brick	1.4×10^{10}
Copper	1.1×10^{11}
Mohair	2.9×10^9
Nylon	3.7×10^9
Pyrex glass	6.2×10^{10}
Steel	2.0×10^{11}
Teflon	3.7×10^8
Tungsten	3.6×10^{11}

TABLE 10.2 Values for the Shear Modulus of Solid Materials

Material	Shear Modulus S (N/m^2)
Aluminum	2.4×10^{10}
Bone	8.0×10^{10}
Brass	3.5×10^{10}
Copper	4.2×10^{10}
Lead	5.4×10^9
Nickel	7.3×10^{10}
Steel	8.1×10^{10}
Tungsten	1.5×10^{11}

Figure 10.31 (a) An example of a shear deformation. The shearing forces **F** and $-$**F** are applied parallel to the top and bottom covers of the book. In general, shearing forces cause a solid object to change its shape. (b) The shear deformation is ΔX. The area of each cover is A, and the thickness of the book is L_0.

(a)   (b)

(a)

(b)

Figure 10.32 (a) A block of Jell-O and (b) a shearing force applied to it.

EXAMPLE 12 J-E-L-L-O

A block of Jell-O is resting on a plate. Figure 10.32a gives the dimensions of the block. You are bored, impatiently waiting for dinner, and push tangentially across the top surface with a force of $F = 0.45$ N, as in part b of the drawing. The top surface moves a distance $\Delta X = 6.0 \times 10^{-3}$ m relative to the bottom surface. Use this idle gesture to measure the shear modulus of Jell-O.

Reasoning The finger applies a force that is parallel to the top surface of the Jell-O block. The shape of the block changes, because the top surface moves a distance ΔX relative to the bottom surface. The magnitude of the force required to produce this change in shape is given by Equation 10.18 as $F = S(\Delta X/L_0)A$. We know the values for all the variables in this relation except S, which can now be determined.

Solution Solving Equation 10.18 for the shear modulus S, we find that $S = FL_0/(A\,\Delta X)$, where $A = (0.070$ m$)(0.070$ m$)$ is the area of the top surface, and $L_0 = 0.030$ m is the thickness of the block:

$$S = \frac{FL_0}{A\,\Delta X} = \frac{(0.45\text{ N})(0.030\text{ m})}{(0.070\text{ m})(0.070\text{ m})(6.0 \times 10^{-3}\text{ m})} = \boxed{460\text{ N/m}^2}$$

Jell-O can be deformed easily, so its shear modulus is significantly less than that of a more rigid material like steel (see Table 10.2).

Although Equations 10.17 and 10.18 are algebraically similar, they refer to different kinds of deformations. The tensile force in Figure 10.29 is perpendicular to the surface whose area is A, whereas the shearing force in Figure 10.31 is parallel to that surface. Furthermore, the ratio $\Delta L/L_0$ in Equation 10.17 is different from the ratio $\Delta X/L_0$ in Equation 10.18. The distances ΔL and L_0 are parallel, whereas ΔX and L_0 are perpendicular. Young's modulus refers to a *change in length* of one dimension of a solid object as a result of tensile or compressive forces. In contrast, the shear modulus refers to a *change in shape* of a solid object as a result of shearing forces.

Volume Deformation and the Bulk Modulus

When a compressive force is applied along one dimension of a solid, the length of that dimension decreases. It is also possible to apply compressive forces so that the size of every dimension (length, width, and depth) decreases, leading to a decrease in volume, as Figure 10.33 illustrates. This kind of overall compression occurs, for example, when an object is submerged in a liquid, and the liquid presses inward everywhere on the object. The forces acting in such situations are applied perpendicular to every surface, and it is more convenient to speak of the perpendicular force per unit area, rather than the amount of any one force in particular. The magnitude of the perpendicular force per unit area is called the ***pressure P.***

Figure 10.33 The arrows denote the forces that push perpendicularly on every surface of an object immersed in a liquid. The force per unit area is the pressure. When the pressure increases, the volume of the object decreases.

● **DEFINITION OF PRESSURE**

The pressure P is the magnitude F of the force acting perpendicular to a surface divided by the area A over which the force acts:

$$P = \frac{F}{A} \tag{10.19}$$

SI Unit of Pressure: N/m^2 = pascal (Pa)

Equation 10.19 indicates that the SI unit for pressure is the unit of force divided by the unit of area, or newton/meter^2 (N/m^2). This unit of pressure is often referred to as a *pascal* (Pa), named after the French scientist Blaise Pascal (1623–1662).

Suppose we change the pressure on an object by an amount ΔP, where, as usual, the "delta" notation ΔP represents the final pressure P minus the initial pressure P_0: $\Delta P = P - P_0$. Because of this change in pressure, the volume of the object changes by an amount $\Delta V = V - V_0$, where V and V_0 are the final and initial volumes, respectively. Such a pressure change occurs, for example, when a swimmer dives deeper into the water. Experiment reveals that the change ΔP in pressure needed to change the volume by an amount ΔV is directly proportional to the fractional change $\Delta V/V_0$ in the volume:

$$\Delta P = -B\left(\frac{\Delta V}{V_0}\right) \tag{10.20}$$

This relation is analogous to Equations 10.17 and 10.18, except that the area A in those equations does not appear here explicitly; the area is already taken into account by the concept of pressure (force per unit area). The proportionality constant B is known as the **bulk modulus.** The minus sign occurs because an increase in pressure (ΔP positive) always creates a decrease in volume (ΔV negative), and B is given as a positive quantity. Like Young's modulus and the shear modulus, the bulk modulus has units of force per unit area (N/m^2), and its value depends on the nature of the material. Table 10.3 gives representative values of the bulk modulus.

Example 13 illustrates that an extremely large change in pressure causes only a small change in the volume of a steel ball.

EXAMPLE 13   **The Mariana Trench**

The Mariana trench is located in the Pacific Ocean, and at one place it is nearly seven miles beneath the surface of the water. The water pressure at the bottom of the trench is enormous, being about $\Delta P = 1.1 \times 10^8$ Pa greater than the pressure at the surface of the ocean. A solid steel ball of volume $V_0 = 0.20$ m^3 is dropped into the ocean and falls to the bottom of the trench. What is the change ΔV in the volume of the ball when it reaches the bottom?

Reasoning As the ball falls, the water pressure increases, and the volume of the ball becomes smaller. The change in volume ΔV can be determined directly from Equation 10.20 as $\Delta V = -(V_0 \, \Delta P/B)$, where B is the bulk modulus of steel.

Solution From Equation 10.20 we have

$$\Delta V = \frac{-V_0 \, \Delta P}{B} = \frac{-(0.20 \text{ m}^3)(1.1 \times 10^8 \text{ Pa})}{1.4 \times 10^{11} \text{ N/m}^2} = \boxed{-1.6 \times 10^{-4} \text{ m}^3}$$

It is evident that the steel ball is compressed very little, even under the large pressure at the bottom of the Mariana trench.

10.8 STRESS, STRAIN, AND HOOKE'S LAW

Equations 10.17, 10.18, and 10.20 specify the amount of force needed for a given amount of elastic deformation, and they are repeated in Table 10.4 to emphasize their common features. The left side of each equation is the magnitude of the force per unit area required to cause an elastic deformation. In general, the ratio of the force to the area is called the **stress.** The right side of each equation involves the

TABLE 10.3 Values for the Bulk Modulus of Solid and Liquid Materials

Material	Bulk Modulus B (N/m^2)
Solids	
Aluminum	7.1×10^{10}
Brass	6.7×10^{10}
Copper	1.3×10^{11}
Lead	4.2×10^{10}
Nylon	6.1×10^{9}
Pyrex glass	2.6×10^{10}
Steel	1.4×10^{11}
Liquids	
Ethanol	8.9×10^{8}
Oil	1.7×10^{9}
Water	2.2×10^{9}

TABLE 10.4 Stress and Strain Relations for Elastic Behavior

$$\frac{F}{A} = Y\left(\frac{\Delta L}{L_0}\right) \tag{10.17}$$

$$\frac{F}{A} = S\left(\frac{\Delta X}{L_0}\right) \tag{10.18}$$

$$\Delta P = B\left(\frac{-\Delta V}{V_0}\right) \tag{10.20}$$

Stress   is proportional to   Strain

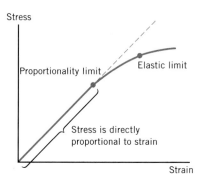

Figure 10.34 Hooke's law (stress is directly proportional to strain) is valid only up to the proportionality limit of a material. Beyond this limit, Hooke's law no longer applies. Beyond the elastic limit, the material remains deformed even when the stress is removed.

change in a quantity (ΔL, ΔX, or ΔV) divided by a quantity (L_0 or V_0) relative to which the change is compared. The terms $\Delta L/L_0$, $\Delta X/L_0$, and $\Delta V/V_0$ are unitless ratios, and each is referred to as the *strain* that results from the applied stress. In the case of stretch and compression, the strain is the fractional change in length, whereas in volume deformation it is the fractional change in volume. In shear deformation the strain refers to a change in shape of the object. Experiments show that these three equations, with constant values for Young's modulus, the shear modulus, and the bulk modulus, apply to a wide range of materials. Therefore, stress and strain are directly proportional to one another, a relationship first discovered by Robert Hooke (1635–1703) and now referred to as **Hooke's law.**

● HOOKE'S LAW FOR STRESS AND STRAIN

Stress is directly proportional to strain.

SI Unit of Stress: newton per square meter = pascal (Pa)

SI Unit of Strain: Strain is a unitless quantity.

In reality, materials obey Hooke's law only up to a certain limit, as Figure 10.34 shows. As long as stress remains proportional to strain, a plot of stress versus strain is a straight line. The point on the graph where the material begins to deviate from straight-line behavior is called the "proportionality limit." Beyond the proportionality limit stress and strain are no longer directly proportional. However, if the stress does not exceed the "elastic limit" of the material, the object will return to its original size and shape once the stress is removed. The "elastic limit" is the point beyond which the object no longer returns to its original size and shape when the stress is removed; the object remains permanently deformed.

10.9 CONCEPTS & CALCULATIONS

This chapter has examined an important kind of vibratory motion known as simple harmonic motion. We were able to determine the displacement, velocity, and acceleration, as well as the frequency, of such motion. In addition, we saw that the total mechanical energy is conserved if nonconservative forces, such as friction and air resistance, are absent. We conclude now by presenting some examples that review important features of simple harmonic motion.

CONCEPTS & CALCULATIONS EXAMPLE 14
A Diver Vibrating in Simple Harmonic Motion

Figure 10.35 A diver at the end of a diving board is bouncing up and down in simple harmonic motion.

A 75-kg diver is standing at the end of a diving board while it is vibrating up and down in simple harmonic motion, as indicated in Figure 10.35. The diving board has an effective spring constant of $k = 4100$ N/m, and the vertical distance between the highest and lowest points in the motion is 0.60 m. (a) What is the amplitude of the motion? (b) Starting when the diver is at the highest point, what is his speed one-quarter of a period later? (c) If the vertical distance between his highest and lowest points were doubled to 1.2 m, what would be the time required for the diver to make one complete motional cycle?

Concept Questions and Answers How is the amplitude A related to the vertical distance between the highest and lowest points of the diver's motion?

Answer The amplitude is the distance from the midpoint of the motion to either the highest point or to the lowest point. Thus, the amplitude is one-half the vertical distance between the highest and lowest points in the motion.

Starting from the top, where is the diver located one-quarter of a period later, and what can be said about his speed at this point?

Answer The time for the diver to complete one motional cycle is defined as the period. In one cycle, the diver moves downward from the highest point to the lowest point and then moves upward and returns to the highest point. In a time equal to one-quarter of a period, the diver completes one-quarter of this cycle and, therefore, is halfway between the highest and lowest points. Since his speed is momentarily zero at the highest and lowest points, his speed is a maximum at the halfway point.

If the amplitude of the motion were to double, would the period also double?

Answer No. The period is time to complete one cycle, and it is equal to the distance traveled during one cycle divided by the average speed. If the amplitude doubles, the distance also doubles. However, the average speed also doubles. We can verify this by examining Equation 10.7, which gives the diver's velocity as $v = -A\omega \sin \omega t$. The speed is the magnitude of this value, or $A\omega \sin \omega t$. Since the speed is proportional to the amplitude A, the speed at every point in the cycle also doubles when the amplitude doubles. Thus, the average speed doubles. However, the period, being the distance divided by the average speed, does not change.

Solution

(a) Since the amplitude A is one-half the vertical distance between the highest and lowest points in the motion, $A = \frac{1}{2}(0.60 \text{ m}) = \boxed{0.30 \text{ m}}$.

(b) When the diver is halfway between the highest and lowest points, his speed is a maximum. The maximum speed of an object vibrating in simple harmonic motion is given by Equation 10.8 as $v_{max} = A\omega$, where A is the amplitude of the motion and ω is the angular frequency. The angular frequency can be determined from Equation 10.11 as $\omega = \sqrt{k/m}$, where k is the effective spring constant of the diving board and m is the mass of the diver. The maximum speed is

$$v_{max} = A\omega = A\sqrt{\frac{k}{m}} = (0.30 \text{ m})\sqrt{\frac{4100 \text{ N/m}}{75 \text{ kg}}} = \boxed{2.2 \text{ m/s}}$$

(c) The period is the same, regardless of the amplitude of the motion. From Equation 10.4 we know that the period T and the angular speed ω are related by $T = 2\pi/\omega$, where $\omega = \sqrt{k/m}$. Thus, the period can be written as

$$T = \frac{2\pi}{\omega} = \frac{2\pi}{\sqrt{k/m}} = 2\pi\sqrt{\frac{m}{k}} = 2\pi\sqrt{\frac{75 \text{ kg}}{4100 \text{ N/m}}} = \boxed{0.85 \text{ s}}$$

As expected, the period does not depend on the amplitude of the motion.

CONCEPTS & CALCULATIONS EXAMPLE 15
Bungee Jumping and the Conservation of Mechanical Energy

A 68.0-kg bungee jumper is standing on a tall platform ($h_0 = 46.0$ m), as indicated in Figure 10.36. The bungee cord has an unstrained length of $L_0 = 9.00$ m, and when stretched, behaves like an ideal spring with a spring constant of $k = 66.0$ N/m. The jumper falls from rest, and the only forces acting on her are her weight and, for the latter part of the descent, the elastic force of the bungee cord. What is her speed (it is not zero) when she is at the following heights above the water (see the drawing): (a) $h_A = 37.0$ m and (b) $h_B = 15.0$ m?

Figure 10.36 A bungee jumper jumps from a height of $h_0 = 46.0$ m. The length of the unstrained bungee cord is $L_0 = 9.00$ m.

Concept Questions and Answers Can we use the conservation of mechanical energy to find her speed at any point during the descent?

Answer Yes. Her weight and the elastic force of the bungee cord are the only forces acting on her and are conservative forces. Therefore, the total mechanical energy remains constant (is conserved) during her descent.

What types of energy does she have when she is standing on the platform?

Answer Since she's at rest, she has neither translational or rotational kinetic energy. The bungee cord is not stretched, so there is no elastic potential energy. Relative to the water, however, she does have gravitational potential energy, since she is 46.0 m above it.

What types of energy does she have at point A?

Answer Since she's moving downward, she possesses translational kinetic energy. She is not rotating though, so her rotational kinetic energy is zero. Because the bungee cord is still not stretched at this point, there is no elastic potential energy. But she still has gravitational potential energy relative to the water, because she is 37.0 m above it.

What types of energy does she have at point B?

Answer She has translational kinetic energy, because she's still moving downward. She's not rotating, so her rotational kinetic energy is zero. The bungee cord is stretched at this point, so there is elastic potential energy. She has gravitational potential energy, because she is still 15.0 m above the water.

Solution

(a) The total mechanical energy is the sum of the kinetic and potential energies, as expressed in Equation 10.14:

$$E \;\; = \;\; \underbrace{\tfrac{1}{2}mv^2}_{\substack{\text{Translational} \\ \text{kinetic} \\ \text{energy}}} \;+\; \underbrace{\tfrac{1}{2}I\omega^2}_{\substack{\text{Rotational} \\ \text{kinetic} \\ \text{energy}}} \;+\; \underbrace{mgh}_{\substack{\text{Gravitational} \\ \text{potential} \\ \text{energy}}} \;+\; \underbrace{\tfrac{1}{2}kx^2}_{\substack{\text{Elastic} \\ \text{potential} \\ \text{energy}}}$$

$\underbrace{}_{\substack{\text{Total} \\ \text{mechanical} \\ \text{energy}}}$

The conservation of mechanical energy states that the total mechanical energy at point A is equal to that at the platform:

$$\underbrace{\tfrac{1}{2}mv_A^2 + \tfrac{1}{2}I\omega_A^2 + mgh_A + \tfrac{1}{2}kx_A^2}_{\text{Total mechanical energy at A}} = \underbrace{\tfrac{1}{2}mv_0^2 + \tfrac{1}{2}I\omega_0^2 + mgh_0 + \tfrac{1}{2}kx_0^2}_{\text{Total mechanical energy at the platform}}$$

While standing on the platform, the jumper is at rest, so $v_0 = 0$ m/s and $\omega_0 = 0$ rad/s. The bungee cord is not stretched, so $x_0 = 0$ m. At point A, the jumper is not rotating, $\omega_A = 0$ rad/s, and the bungee cord is still not stretched, $x_A = 0$ m. With these substitutions, the conservation of mechanical energy becomes

$$\tfrac{1}{2}mv_A^2 + mgh_A = mgh_0$$

Solving for the speed at point A yields

$$v_A = \sqrt{2g(h_0 - h_A)} = \sqrt{2(9.80 \text{ m/s}^2)(46.0 \text{ m} - 37.0 \text{ m})} = \boxed{13 \text{ m/s}}$$

(b) At point B the total mechanical energy is the same as it was on the platform, so

$$\underbrace{\tfrac{1}{2}mv_B^2 + \tfrac{1}{2}I\omega_B^2 + mgh_B + \tfrac{1}{2}kx_B^2}_{\text{Total mechanical energy at B}} = \underbrace{mgh_0}_{\substack{\text{Total mechanical} \\ \text{energy at} \\ \text{the platform}}}$$

We set $\omega_B = 0$ rad/s, since there is no rotational motion. Furthermore, the bungee cord stretches by an amount $x_B = h_0 - L_0 - h_B$ (see the drawing). Therefore, we have

$$v_B = \sqrt{2g(h_0 - h_B) - \left(\frac{k}{m}\right)(h_0 - L_0 - h_B)^2}$$

$$= \sqrt{2(9.80 \text{ m/s}^2)(46.0 \text{ m} - 15.0 \text{ m}) - \left(\frac{66.0 \text{ N/m}}{68.0 \text{ kg}}\right)(46.0 \text{ m} - 9.00 \text{ m} - 15.0 \text{ m})^2}$$

$$= \boxed{11.7 \text{ m/s}}$$

At the end of the problem set for this chapter, you will find homework problems that contain both conceptual and quantitative parts. These problems are grouped under the heading *Concepts & Calculations, Group Learning Problems*. They are designed for use by students working alone or in small learning groups. The conceptual part of each problem provides a convenient focus for group discussions.

SUMMARY

The force that must be applied to stretch or compress an **ideal spring** is $F_{Applied} = kx$, where k is the spring constant and x is the displacement of the spring from its unstrained length. A spring exerts a **restoring force** on an object attached to the spring. The restoring force produced by an ideal spring is $F = -kx$, where the minus sign indicates that the restoring force points opposite to the displacement of the spring.

Simple harmonic motion is the oscillatory motion that occurs when a restoring force of the form $F = -kx$ acts on an object. A graphical record of position versus time for an object in simple harmonic motion is sinusoidal. The **amplitude** A of the motion is the maximum distance that the object moves away from its equilibrium position. The **period** T is the time required to complete one cycle of the motion, while the **frequency** f is the number of cycles per second that occur. Frequency and period are related according to $f = 1/T$. The frequency f (in Hz) is related to the angular frequency ω (in rad/s) according to $\omega = 2\pi f$. For an object of mass m on a spring with spring constant k, the frequency is determined by $2\pi f = \sqrt{k/m}$. The velocity and the acceleration in simple harmonic motion are continually changing with time; the maximum speed is $v_{max} = A\omega$, and the maximum acceleration is $a_{max} = A\omega^2$.

The **elastic potential energy** of an object attached to an ideal spring is $PE_{elastic} = \frac{1}{2}kx^2$. The total mechanical energy E of such a system is the sum of its translational and rotational kinetic energies, gravitational potential energy, and elastic potential energy; $E = \frac{1}{2}mv^2 + \frac{1}{2}I\omega^2 + mgh + \frac{1}{2}kx^2$. If external nonconservative forces like friction do no net work, the total mechanical energy of the system is conserved, $E_f = E_0$.

A **simple pendulum** is a particle of mass m attached to a frictionless pivot by a cable whose length is L and whose mass is negligible. The small-angle ($\leq 10°$) back-and-forth swinging of a simple pendulum is simple harmonic motion, while large-angle motion is not. The frequency of small-angle motion is given by $2\pi f = \sqrt{g/L}$. A **physical pendulum** consists of a rigid object, with moment of inertia I and mass m, suspended from a frictionless pivot. For small-angle displacements, the frequency of simple harmonic motion for a physical pendulum is determined by $2\pi f = \sqrt{mgL/I}$, where L is the distance between the axis of rotation and the center of gravity of the rigid object.

Damped harmonic motion is motion in which the amplitude of oscillation decreases as time passes. **Critical damping** is the minimum degree of damping that eliminates any oscillations in the motion as the object returns to its equilibrium position.

Driven harmonic motion occurs when an additional driving force is applied to an object along with the restoring force. **Resonance** is the condition under which a driving force can transmit large amounts of energy to an oscillating object, leading to large amplitude motion. In the absence of damping, resonance occurs when the frequency of the driving force matches a natural frequency at which the object oscillates.

The possible kinds of **elastic deformation** include stretch and compression, shear deformation, and volume deformation. The forces required to create them are given by Equations 10.17, 10.18, and 10.20 and are characterized with the aid of proportionality constants called, respectively, **Young's modulus,** the **shear modulus,** and the **bulk modulus.**

The **pressure** P is the magnitude F of the force acting perpendicular to a surface divided by the area A over which the force acts: $P = F/A$. The SI unit of pressure is N/m², a unit known as a pascal (Pa); 1 Pa = 1 N/m².

Stress is the force per unit area applied to an object and causes **strain.** For a stretch/compression deformation, the strain is the resulting fractional change in length, $\Delta L / L_0$ (see Figure 10.29). For a shear deformation, the strain reflects the change in shape of an object and is given by $\Delta X / L_0$ (see Fig-

ure 10.31b). For a volume deformation, the strain is the negative of the fractional change in volume, $-\Delta V / V_0$, where ΔV is the change in volume and V_0 is the original volume. **Hooke's law** states that stress is directly proportional to strain, up to a limit called the **proportionality limit.**

CONCEPTUAL QUESTIONS

1. Two people pull on a horizontal spring that is attached to an immovable wall. Then, they detach it from the wall and pull on opposite ends of the horizontal spring. They pull just as hard in each case. In which situation, if either, does the spring stretch more? Account for your answer.

2. The drawing shows identical springs that are attached to a box in two different ways. Initially, the springs are unstrained. The box is then pulled to the right and released. In each case the initial displacement of the box is the same. At the moment of release, which box, if either, experiences the greater net force due to the springs? Provide a reason for your answer.

3. In Figures 10.11 and 10.14, the shadow moves in simple harmonic motion. Where on the shadow's path is (a) the velocity equal to zero and (b) the acceleration equal to zero?

4. A steel ball is dropped onto a concrete floor. Over and over again, it rebounds to its original height. Is the motion simple harmonic motion? Justify your answer.

5. Ignoring the damping introduced by the shock absorbers, explain why the number of passengers in a car affects the vibration frequency of the car's suspension system.

6. Review Conceptual Example 8 before answering this question. A block is attached to a horizontal spring and slides back and forth in simple harmonic motion on a frictionless horizontal surface. A second identical block is suddenly attached to the first block. The attachment is accomplished by joining the blocks at one extreme end of the oscillation cycle. The velocities of the blocks are exactly matched at the instant of joining. Explain how (a) the amplitude, (b) the frequency, and (c) the maximum speed of the oscillation change.

7. A particle is oscillating in simple harmonic motion. The time required for the particle to travel through one complete cycle is equal to the period of the motion, no matter what the amplitude is. But how can this be, since larger amplitudes mean that the particle travels farther? Explain.

8. An electric saber saw consists of a blade that is driven back and forth by a pin mounted on the circumference of a rotating circular disk. As the disk rotates at a constant angular speed, the pin engages a slot and forces the blade back and forth, in the sequence (a), (b), (c), and (d) shown in the drawing. Is the motion of the blade simple harmonic motion? Explain.

9. Is more elastic potential energy stored in a spring when the spring is compressed by one centimeter than when it is stretched by the same amount? Explain.

10. Suppose that a grandfather clock (a simple pendulum) is running slowly. That is, the time it takes to complete each cycle is longer than it should be. Should one shorten or lengthen the pendulum to make the clock keep the correct time? Why?

11. In principle, the motion of a simple pendulum and an object on an ideal spring can both be used to provide the basic time interval or period used in a clock. Which of the two kinds of clocks becomes more inaccurate when carried to the top of a high mountain? Justify your answer.

12. Suppose you were kidnapped and held prisoner by space invaders in a completely isolated room, with nothing but a watch and a pair of shoes (including shoelaces of known length). Explain how you might determine whether this room is on earth or on the moon.

13. Two people are sitting on playground swings. One person is pulled back 4° from the vertical and released, while the other is pulled back 8° from the vertical and released. If the two swings are started together, will they both come back to the starting points at the same time? Justify your answer.

14. A car travels over a road that contains a series of equally spaced bumps. Explain why a particularly jarring ride can result if the horizontal velocity of the car, the bump spacing, and the oscillation frequency of the car's suspension system are properly "matched."

15. The drawing shows two cylinders. They are identical in all respects, except one is hollow. In   a setup like that in Figure 10.29, identical forces are applied to the right end of each cylinder. Which cylinder, if either, stretches the most? Why?

16. Young's modulus for steel is thousands of times greater than that for rubber. All other factors being equal, does this mean that steel stretches much more easily than rubber? Explain.

17. A trash compactor crushes empty aluminum cans, thereby reducing the total volume of the cans by 75%. Can the value given in Table 10.3 for the bulk modulus of aluminum be used to calculate the change in pressure generated in the trash compactor? Give a reason for your answer.

18. Both sides of the relation $F = S(\Delta X/L_0)A$ (Equation 10.18) can be divided by the area A to give F/A on the left side. Why can't this F/A term be called a pressure, such as the pressure that appears in $\Delta P = -B(\Delta V/V_0)$ (Equation 10.20)?

19. The block in the drawing rests on the ground. Which face, A, B, or C, experiences (a) the largest stress and (b) the smallest stress when the block is resting on it? Explain.

20.0 cm

10.0 cm

30.0 cm

PROBLEMS

Note: Unless otherwise indicated, the values for Young's modulus Y, the Shear modulus S, and the bulk modulus B are given, respectively, in Table 10.1, Table 10.2, and Table 10.3.

ssm Solution is in the Student Solutions Manual. **www** Solution is available on the World Wide Web at http://www.wiley.com/college/cutnell
⚕ This icon represents a biomedical application.

Section 10.1 The Ideal Spring and Simple Harmonic Motion

1. ssm A hand exerciser utilizes a coiled spring. A force of 89.0 N is required to compress the spring by 0.0191 m. Determine the force needed to compress the spring by 0.0508 m.

2. A person who weighs 670 N steps onto a spring scale in the bathroom, and the spring compresses by 0.79 cm. (a) What is the spring constant? (b) What is the weight of another person who compresses the spring by 0.34 cm?

3. A spring has a spring constant of 248 N/m. Find the magnitude of the force needed (a) to stretch the spring by 3.00×10^{-2} m from its unstrained length and (b) to compress the spring by the same amount.

4. The graph shows the force F that an archer applies to the string of a long bow versus the string's displacement x. Drawing back this bow is analogous to stretching a spring. From the data in the graph determine the effective spring constant of the bow.

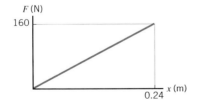

5. ssm A car is hauling a 92-kg trailer, to which it is connected by a spring. The spring constant is 2300 N/m. The car accelerates with an acceleration of 0.30 m/s². By how much does the spring stretch?

6. The drawing shows three identical springs hanging from the ceiling. Nothing is attached to the first spring, while a 4.50-N block hangs from the second spring. A block of unknown weight hangs from the third spring. From the drawing, determine (a) the spring constant (in N/m) and (b) the weight of the unknown block.

7. A spring (k = 830 N/m) is hanging from the ceiling of an elevator, and a 5.0-kg object is attached to the lower end. By how much does the spring stretch (relative to its unstrained length) when the elevator is accelerating upward at a = 0.60 m/s²?

* **8.** Review Conceptual Example 2 as an aid in solving this problem. An object is attached to the lower end of a 100-coil spring that is hanging from the ceiling. The spring stretches by 0.160 m. The spring is then cut into two identical springs of 50 coils each. As the drawing shows, each spring is attached between the ceiling and the object. By how much does each spring stretch?

50-coil spring

* **9.** A 10.1-kg uniform board is wedged into a corner and held by a spring at a 50.0° angle, as the drawing shows. The spring has a

spring constant of 176 N/m and is parallel to the floor. Find the amount by which the spring is stretched from its unstrained length.

Problem 9

* **10.** To measure the static friction coefficient between a 2.3-kg block and a vertical wall, the setup shown in the drawing is used. A spring (spring constant = 480 N/m) is attached to the block. Someone pushes on the end of the spring in a direction perpendicular to the wall until the block does not slip downward. If the spring in such a setup is compressed by 0.051 m, what is the coefficient of static friction?

** **11. ssm** A small ball is attached to one end of a spring that has an unstrained length of 0.200 m. The spring is held by the other end, and the ball is whirled around in a horizontal circle at a speed of 3.00 m/s. The spring remains nearly parallel to the ground during the motion and is observed to stretch by 0.010 m. By how much would the spring stretch if it were attached to the ceiling and the ball allowed to hang straight down, motionless?

** **12.** A 30.0-kg block is resting on a flat horizontal table. On top of this block is resting a 15.0-kg block, to which a horizontal spring is attached, as the drawing illustrates. The spring constant of

the spring is 325 N/m. The coefficient of kinetic friction between the lower block and the table is 0.600, while the coefficient of static friction between the two blocks is 0.900. A horizontal force **F** is applied to the lower block as shown. This force is increasing in such a way as to keep the blocks moving at a *constant speed.* At the point where the upper block begins to slip on the lower block determine (a) the amount by which the spring is compressed and (b) the magnitude of the force **F**.

** **13.** A 15.0-kg block rests on a horizontal table and is attached to one end of a massless, horizontal spring. By pulling horizontally on the other end of the spring, someone causes the block to accelerate uniformly and reach a speed of 5.00 m/s in 0.500 s. In the process, the spring is stretched by 0.200 m. The block is then pulled at a *constant speed* of 5.00 m/s, during which time the spring is stretched by only 0.0500 m. Find (a) the spring constant of the spring and (b) the coefficient of kinetic friction between the block and the table.

Section 10.2 Simple Harmonic Motion and the Reference Circle

14. The fan blades on a jet engine make one thousand revolutions in a time of 50.0 ms. Determine (a) the period (in seconds) and (b) the frequency (in Hz) of the rotational motion. (c) What is the angular frequency of the blades?

15. ssm The shock absorbers in the suspension system of a car are in such bad shape that they have no effect on the behavior of the springs attached to the axles. Each of the identical springs attached to the front axle supports 320 kg. A person pushes down on the middle of the front end of the car and notices that it vibrates through five cycles in 3.0 s. Find the spring constant of either spring.

16. Refer to Conceptual Example 2 as an aid in solving this problem. A 100-coil spring has a spring constant of 420 N/m. It is cut into four shorter springs, each of which has 25 coils. One end of a 25-coil spring is attached to a wall. An object of mass 46 kg is attached to the other end of the spring, and the system is set into horizontal oscillation. What is the angular frequency of the motion?

17. An 0.80-kg object is attached to one end of a spring, as in Figure 10.6, and the system is set into simple harmonic motion. The displacement x of the object as a func-

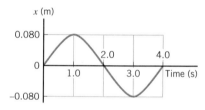

tion of time is shown in the drawing. With the aid of this data, determine (a) the amplitude A of the motion, (b) the angular frequency ω, (c) the spring constant k, (d) the speed of the object at $t = 1.0$ s, and (e) the magnitude of the object's acceleration at $t = 1.0$ s.

18. A person bounces up and down on a trampoline, while always staying in contact with it. The motion is simple harmonic motion, and it takes 1.90 s to complete one cycle. The height of each bounce above the equilibrium position is 45.0 cm. Determine (a) the amplitude and (b) the angular frequency of the motion. (c) What is the maximum speed attained by the person?

19. ssm www Atoms in a solid are not stationary, but vibrate about their equilibrium positions. Typically, the frequency of vibration is about $f = 2.0 \times 10^{12}$ Hz, and the amplitude is about 1.1×10^{-11} m. For a typical atom, what is its (a) maximum speed and (b) maximum acceleration?

* **20.** When an object of mass m_1 is hung on a vertical spring and set into vertical simple harmonic motion, its frequency is 12.0 Hz. When another object of mass m_2 is hung on the spring along with m_1, the frequency of the motion is 4.00 Hz. Find the ratio m_2/m_1 of the masses.

* **21. ssm** A spring stretches by 0.018 m when a 2.8-kg object is suspended from its end. How much mass should be attached to this spring so that its frequency of vibration is $f = 3.0$ Hz?

*22. In Figure 10.9, the radius of the reference circle is 0.500 m. Suppose the frequency of the simple harmonic motion of the shadow is 2.00 Hz. At time $t = 0.0500$ s calculate (a) the displacement x, (b) the magnitude of the velocity, and (c) the magnitude of the acceleration of the shadow.

**23. A tray is moved horizontally back and forth in simple harmonic motion at a frequency of $f = 2.00$ Hz. On this tray is an empty cup. Obtain the coefficient of static friction between the tray and the cup, given that the cup begins slipping when the amplitude of the motion is 5.00×10^{-2} m.

Section 10.3 Energy and Simple Harmonic Motion

24. An archer pulls the bowstring back for a distance of 0.470 m before releasing the arrow. The bow and string act like a spring whose spring constant is 425 N/m. (a) What is the elastic potential energy of the drawn bow? (b) The arrow has a mass of 0.0300 kg. How fast is it traveling when it leaves the bow?

25. ssm A 1.00×10^{-2}-kg block is resting on a horizontal frictionless surface and is attached to a horizontal spring whose spring constant is 124 N/m. The block is shoved parallel to the spring axis and is given an initial speed of 8.00 m/s, while the spring is initially unstrained. What is the amplitude of the resulting simple harmonic motion?

26. A spring is hung from the ceiling. A 0.450-kg block is then attached to the free end of the spring. When released from rest, the block drops 0.150 m before momentarily coming to rest. (a) What is the spring constant of the spring? (b) Find the angular frequency of the block's vibrations.

27. In preparation for shooting a ball in a pinball machine, a spring ($k = 675$ N/m) is compressed by 0.0650 m relative to its unstrained length. The ball ($m = 0.0585$ kg) is at rest against the spring at point A. When the spring is released, the ball slides (without rolling) to point B, which is 0.300 m higher than point A. How fast is the ball moving at B?

28. A 0.25-kg block is dropped straight downward onto a vertical spring. The spring constant of the spring is 58 N/m. The block sticks to the spring, and the spring compresses 0.15 m before coming to a momentary halt. What is the speed of the block just before it hits the spring?

29. ssm A 2.00-kg object is hanging from the end of a vertical spring. The spring constant is 50.0 N/m. The object is pulled 0.200 m downward and released from rest. Complete the table below by calculating the translational kinetic energy, the gravitational potential energy, the elastic potential energy, and the total mechanical energy E for each of the vertical positions indicated. The vertical positions h indicate distances above the point of release, where $h = 0$ m.

h (meters)	KE	PE (gravity)	PE (elastic)	E
0				
0.200				
0.400				

30. A heavy-duty stapling gun uses a 0.150-kg metal rod that rams against the staple to eject it. The rod is pushed by a stiff spring called a "ram spring" ($k = 34\ 000$ N/m). The mass of this spring may be ignored. Squeezing the handle of the gun first compresses the ram spring by 3.5×10^{-2} m from its unstrained length and then releases it. Assuming that the ram spring is oriented vertically and is still compressed by 1.0×10^{-2} m when the downward-moving ram hits the staple, find the speed of the ram at the instant of contact.

31. The drawing shows a block ($m = 1.7$ kg) and a spring ($k = 310$ N/m) on a frictionless incline. The spring is compressed by $x_0 = 0.31$ m relative to its unstrained position at $x = 0$ m and then released. What is the speed of the block when the spring is still compressed by $x_f = 0.14$ m?

*32. Review Conceptual Example 8 before starting this problem. A block is attached to a horizontal spring and oscillates back and forth on a frictionless horizontal surface at a frequency of 3.00 Hz. The amplitude of the motion is 5.08×10^{-2} m. At the point where the block has its maximum speed, it suddenly splits into two identical parts, only one part remaining attached to the spring. (a) What is the amplitude and the frequency of the simple harmonic motion that exists after the block splits? (b) Repeat part (a), assuming that the block splits when it is at one of its extreme positions.

*33. ssm A 1.1-kg object is suspended from a vertical spring whose spring constant is 120 N/m. (a) Find the amount by which the spring is stretched from its unstrained length. (b) The object is pulled straight down by an additional distance of 0.20 m and released from rest. Find the speed with which the object passes through its original position on the way up.

*34. A 14.6-kg block and a 29.2-kg block are resting on a horizontal frictionless surface. Between the two is squeezed a spring (spring constant = 1170 N/m). The spring is compressed by 0.152 m from its unstrained length and is not attached permanently to either block. With what speed does each block move away after the mechanism keeping the spring squeezed is released and the spring falls away?

*35. A spring is compressed by 0.0800 m and is used to launch an object horizontally with a speed of 2.40 m/s. If the object were attached to the spring, at what angular frequency (in rad/s) would it oscillate?

**36. A 0.200-m uniform bar has a mass of 0.750 kg and is released from rest in the vertical position, as the drawing indicates. The spring is initially unstrained and has a spring constant of $k = 25.0$ N/m. Find the tangential speed with which end A strikes the horizontal surface.

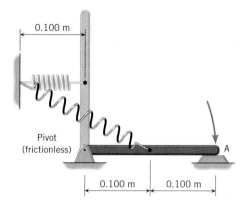

0.100 m

Pivot
(frictionless)

A

0.100 m   0.100 m

**37. ssm A 70.0-kg circus performer is fired from a cannon that is elevated at an angle of 40.0° above the horizontal. The cannon uses strong elastic bands to propel the performer, much in the same way that a slingshot fires a stone. Setting up for this stunt involves stretching the bands by 3.00 m from their unstrained length. At the point where the performer flies free of the bands, his height above the floor is the same as that of the net into which he is shot. He takes 4.00 s to travel the horizontal distance of 50.0 m between this point and the net. Ignore friction and air resistance and determine the effective spring constant of the firing mechanism.

**38. A spring is mounted vertically on the floor. The mass of the spring is negligible. A certain object is placed on the spring to compress it. When the object is pushed further down by just a bit and then released, one up/down oscillation cycle occurs in 0.250 s. However, when the object is pushed down by 5.00×10^{-2} m to point P and then released, the object flies entirely off the spring. To what height above point P does the object rise in the absence of air resistance?

Section 10.4 The Pendulum

39. A spiral staircase winds up to the top of a tower in an old castle. To measure the height of the tower, a rope is attached to the top of the tower and hung down the center of the staircase. However, nothing is available with which to measure the length of the rope. Therefore, at the bottom of the rope a small object is attached so as to form a simple pendulum that just clears the floor. The period of the pendulum is measured to be 9.2 s. What is the height of the tower?

40. A pendulum clock can be approximated as a simple pendulum of length 1.00 m and keeps accurate time at a location where $g = 9.83$ m/s². In a location where $g = 9.78$ m/s², what must be the new length of the pendulum, such that the clock continues to keep accurate time (that is, its period remains the same)?

41. ssm Astronauts on a distant planet set up a simple pendulum of length 1.2 m. The pendulum executes simple harmonic motion and makes 100 complete vibrations in 280 s. What is the acceleration due to gravity?

42. A simple pendulum is made from a 0.65-m-long string and a small ball attached to its free end. The ball is pulled to one side

through a small angle and then released from rest. After the ball is released, how much time elapses before it attains its greatest speed?

*43. ssm Pendulum A is a physical pendulum made from a thin, rigid, and uniform rod whose length is d. One end of this rod is attached to the ceiling by a frictionless hinge, so the rod is free to swing back and forth. Pendulum B is a simple pendulum whose length is also d. Obtain the ratio T_A/T_B of their periods for small-angle oscillations.

*44. A small object oscillates back and forth
at the bottom of a fric-
tionless hemispherical
bowl, as the drawing
illustrates. The radius
of the bowl is R, and
the angle θ is small
enough that the object oscillates in simple harmonic motion. Derive an expression for the angular frequency ω of the motion. Express your answer in terms of R and g, the acceleration due to gravity.

Hemispherical
bowl

R θ

**45. A point on the surface of a solid sphere (radius = R) is at-
tached directly to a pivot on the ceiling. The sphere swings back and forth as a physical pendulum with a small amplitude. What is the length of a simple pendulum that has the same period as this physical pendulum? Give your answer in terms of R.

Section 10.7 Elastic Deformation, Section 10.8 Stress, Strain, and Hooke's Law

46. A tow truck is pulling a car out of a ditch by means of a steel cable that is 9.1 m long and has a radius of 0.50 cm. When the car just begins to move, the tension in the cable is 890 N. How much has the cable stretched?

47. ssm www A 3500-kg statue is placed on top of a cylindrical concrete ($Y = 2.3 \times 10^{10}$ N/m²) stand. The stand has a cross-
sectional area of 7.3×10^{-2} m² and a height of 1.8 m. By how much does the statue compress the stand?

48. The drawing shows a 160-kg
crate hanging from the end of a
steel bar. The length of the bar is
0.10 m, and its cross-sectional
area is 3.2×10^{-4} m². Neglect
the weight of the bar itself and
determine (a) the shear stress on
the bar and (b) the vertical deflection ΔY of the right end of the
bar.

49. ssm A CD player is mounted on four cylindrical rubber blocks. Each cylinder has a height of 0.030 m and a cross-
sectional area of 1.2×10^{-3} m², and the shear modulus for rubber is 2.6×10^6 N/m². If a horizontal force of magnitude 32 N is applied to the CD player, how far will the unit move sideways? Assume that each block is subjected to one-fourth of the force.

50. 🔧 The femur is a bone in the leg whose minimum cross-sectional area is about 4.0×10^{-4} m^2. A compressional force in excess of 6.8×10^4 N will fracture this bone. (a) Find the maximum stress that this bone can withstand. (b) What is the strain that exists under a maximum-stress condition?

51. The pressure increases by 1.0×10^4 N/m^2 for every meter of depth beneath the surface of the ocean. At what depth does the volume of a Pyrex glass cube, 1.0×10^{-2} m on an edge at the ocean's surface, decrease by 1.0×10^{-10} m^3?

52. An 1800-kg car, being lifted at a steady speed by a crane, hangs at the end of a cable whose radius is 6.0×10^{-3} m. The cable is 15 m in length and stretches by 8.0×10^{-3} m because of the weight of the car. Determine (a) the stress, (b) the strain, and (c) Young's modulus for the cable.

53. ssm A piece of aluminum is surrounded by air at a pressure of 1.01×10^5 Pa. The aluminum is placed in a vacuum chamber where the pressure is reduced to zero. Determine the fractional change $\Delta V/V_0$ in the volume of the aluminum.

54. A copper cylinder and a brass cylinder are stacked end to end, as in the drawing. Each cylinder has a radius of 0.25 cm. A compressive force of $F = 6500$ N is applied to the right end of the brass cylinder. Find the amount by which the length of the stack decreases.

55. A die is designed to punch holes of radii 1.00×10^{-2} m in a metal sheet that is 3.0×10^{-3} m thick, as the drawing illustrates. To punch through the sheet, the die must exert a shearing stress of 3.5×10^8 Pa. What force **F** must be applied to the die? (*Hint: Consider carefully which area is used in your calculation.*)

56. A solid steel pipe is standing vertically on the floor. The length of the pipe is 3.6 m and its radius is 65 cm. When an object is placed on top of the pipe, the pipe compresses by an amount of 5.7×10^{-7} m. What is the weight of the object?

57. The shovel of a backhoe is controlled by hydraulic cylinders that are moved by oil under pressure. Determine the volume strain $\Delta V/V_0$ (including the algebraic sign) experienced by the oil, when the pressure increases from 1.8×10^5 Pa to 6.5×10^5 Pa while the shovel is digging a trench.

*58. A piece of mohair from an Angora goat has a radius of 31×10^{-6} m. What is the least number of identical pieces of mohair that should be used to suspend a 75-kg person, so the strain $\Delta L/L_0$ experienced by each piece is less than 0.010? Assume that the tension is the same in all the pieces.

*59. ssm A helicopter is lifting a 2100-kg jeep. The steel suspension cable is 48 m long and has a radius of 5.0×10^{-3} m. (a) Find the amount that the cable is stretched when the jeep is suspended motionless in the air. (b) What is the amount of cable stretch when the jeep is hoisted upward with an acceleration of 1.5 m/s^2?

*60. A block of copper is securely fastened to the floor. A force of 1800 N is applied to the top surface of the block, as the drawing shows. Find (a) the amount by which the height of the block is changed and (b) the shear deformation of the block.

Problem 60

*61. ssm www An 8.0-kg stone at the end of a steel wire is being whirled in a circle at a constant tangential speed of 12 m/s. The stone is moving on the surface of a frictionless horizontal table. The wire is 4.0 m long and has a radius of 1.0×10^{-3} m. Find the strain in the wire.

*62. Suppose that F newtons of force are required to stretch a wire (circular cross section) by an amount ΔL. This wire is melted down, and a new wire (circular cross section) is formed that has half the length of the first. How much force is needed to stretch this second wire by the same amount ΔL? Express your answer in terms of F.

*63. A 1.0×10^{-3}-kg spider is hanging vertically by a thread that has a Young's modulus of 4.5×10^9 N/m^2 and a radius of 13×10^{-6} m. Suppose that a 95-kg person is hanging vertically on an aluminum wire. What is the radius of the wire that would exhibit the same strain as the spider's thread, when the thread is stressed by the full weight of the spider?

*64. A square plate is 1.0×10^{-2} m thick, measures 3.0×10^{-2} m on a side, and has a mass of 7.2×10^{-2} kg. The shear modulus of the material is 2.0×10^{10} N/m^2. One of the square faces rests on a flat horizontal surface, and the coefficient of static friction between the plate and the surface is 0.90. A force is applied to the top of the plate, as in Figure 10.31a. Determine (a) the maximum possible amount of shear stress, (b) the maximum possible amount of shear strain, and (c) the maximum possible amount of shear deformation ΔX (see Figure 10.31b) that can be created by the applied force just before the plate begins to move.

**65. ssm www A solid brass sphere is subjected to a pressure of 1.0×10^5 Pa due to the earth's atmosphere. On Venus the pressure due to the atmosphere is 9.0×10^6 Pa. By what fraction $\Delta r / r_0$ (including the algebraic sign) does the radius of the sphere change when it is exposed to the Venusian atmosphere? Assume that the change in radius is very small relative to the initial radius.

**66. The drawing shows two crates that are connected by a steel wire that passes over a pulley. The unstretched length of the wire is 1.5 m, and its cross-sectional area is 1.3×10^{-5} m^2. The pulley is frictionless and massless. When the crates are accelerating, determine the change in length of the wire. Ignore the mass of the wire.

3.0 kg

5.0 kg

ADDITIONAL PROBLEMS

67. The drawing shows how a piston in an automobile engine is attached to the crankshaft, which is rotating with an angular speed of $\omega = 126$ rad/s. If the shadow of point P could be projected onto a screen, the shadow would move in simple harmonic motion. Find (a) the amplitude, (b) the period, and (c) the maximum speed of the shadow's motion.

Cylinder

Piston

Crankshaft

P

5.08×10^{-2} m

68. A block of mass $m = 0.750$ kg is fastened to an unstrained horizontal spring whose spring constant is $k = 82.0$ N/m. The block is given a displacement of $+0.120$ m, where the $+$ sign indicates that the displacement is along the $+x$ axis, and then released from rest. (a) What is the force (magnitude and direction) that the spring exerts on the block just before the block is released? (b) Find the angular frequency ω of the resulting oscillatory motion. (c) What is the maximum speed of the block? (d) Determine the magnitude of the maximum acceleration of the block.

69. If the period of a simple pendulum is to be 2.0 s, what should be its length?

70. A computer to be used in a satellite must be able to withstand accelerations of up to 25 times the acceleration due to gravity. In a test to see if it meets this specification, the computer is bolted to a frame that is vibrated back and forth in simple harmonic motion at a frequency of 9.5 Hz. What is the minimum amplitude of vibration that must be used in this test?

71. ssm Two metal beams are joined together by four rivets, as the drawing indicates. Each rivet has a radius of 5.0×10^{-3} m and is to be exposed to a shearing stress of no more than 5.0×10^8 Pa. What is the maximum tension **T** that can be applied to each beam, assuming that each rivet carries one-fourth of the total load?

T

Rivet

$-$T

72. When the rubber band in a slingshot is stretched, it obeys Hooke's law. Suppose that the "spring constant" for the rubber band is $k = 44$ N/m. When the rubber band is pulled back with a force of 9.6 N, how far does it stretch?

73. A copper cube, 0.30 m on a side, is subjected to two shearing forces, each of magnitude $F = 6.0 \times 10^6$ N (see the drawing). Find the angle θ (in degrees), which is one measure of how the shape of the block has been altered by shear deformation.

F

θ

$-$F

*74. A 3.0-kg block is placed between two horizontal springs. Neither spring is strained when the block is located at the position labeled $x = 0$ m in the drawing. The block is then displaced a distance of 0.070 m from the position where $x = 0$ m and released from rest. (a) What is the speed of the block when it passes back through the $x = 0$ m position? (b) Determine the angular frequency ω of this system.

$x = 0$ m

0.070 m

$k = 450$ N/m

$k = 650$ N/m

*75. ssm In 0.750 s, a 7.00-kg block is pulled through a distance of 4.00 m on a frictionless horizontal surface, starting from rest. The block has a constant acceleration and is pulled by means of a horizontal spring that is attached to the block. The spring constant of the spring is 415 N/m. By how much does the spring stretch?

*76. A gymnast does a one-arm handstand. The humerus, which is the upper arm bone between the elbow and the shoulder joint, may be approximated as a 0.30-m-long cylinder with an outer radius of 1.00×10^{-2} m and a hollow inner core with a radius of 4.0×10^{-3} m. Excluding the arm, the mass of the gymnast is 63 kg. (a) What is the compressional strain of the humerus? (b) By how much is the humerus compressed?

*77. The drawing shows one arm of a centrifuge. The arm is a solid cylinder of aluminum with a length of 0.10 m and a cross-sectional area of 1.3 × 10⁻⁴ m². At the end of the aluminum arm is a 0.58-kg specimen holder. By how much does the aluminum rod stretch when the centrifuge has an angular speed of 180 rad/s?

*78. The front spring of a car's suspension system has a spring constant of 1.50×10^6 N/m and supports a mass of 215 kg. The wheel has a radius of 0.400 m. The car is traveling on a bumpy road, on which the distance between the bumps is equal to the circumference of the wheel. Due to resonance, the wheel starts to vibrate strongly when the car is traveling at a certain minimum linear speed. What is this speed?

**79. ssm A steel wire is strung between two supports attached to a ceiling. Initially, there is no tension in the wire when it is horizontal. A 96-N picture is then hung from the center of the wire, as the drawing illustrates, so the ends of the wire make angles of 26° with respect to the horizontal. What is the radius of the wire?

**80. A 1.00×10^{-2}-kg bullet is fired horizontally into a 2.50-kg wooden block attached to one end of a massless, horizontal spring ($k = 845$ N/m). The other end of the spring is fixed in place, and the spring is unstrained initially. The block rests on a horizontal, frictionless surface. The bullet strikes the block perpendicularly and quickly comes to a halt within it. As a result of this completely inelastic collision, the spring is compressed along its axis and causes the block/bullet to oscillate with an amplitude of 0.200 m. What is the speed of the bullet?

**81. A copper rod (length = 2.0 m, radius = 3.0×10^{-3} m) hangs down from the ceiling. A 9.0-kg object is attached to the lower end of the rod. The rod acts as a "spring," and the object oscillates vertically with a small amplitude. Ignoring the rod's mass, find the frequency f of the simple harmonic motion.

**82. An 86.0-kg climber is scaling the vertical wall of a mountain. His safety rope is made of nylon that, when stretched, behaves like a spring with a spring constant of 1.20×10^3 N/m. He accidentally slips and falls freely for 0.750 m before the rope runs out of slack. How much is the rope stretched when it breaks his fall and momentarily brings him to rest?

**83. ssm The drawing shows a top view of a frictionless horizontal surface, where there are two springs with particles of mass m_1 and m_2 attached to them. Each spring has a spring constant of 120 N/m. The particles are pulled to the right and then released from the positions shown in the drawing. How much time passes before the particles are side by side for the first time at $x = 0$ m if (a) $m_1 = m_2 = 3.0$ kg and (b) $m_1 = 3.0$ kg and $m_2 = 27$ kg?

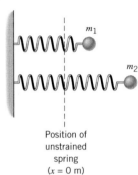

**84. A cylindrically shaped piece of collagen (a substance found in the body in connective tissue) is being stretched by a force that increases from 0 to 3.0×10^{-2} N. The length and radius of the collagen are, respectively, 2.5 and 0.091 cm, and Young's modulus is 3.1×10^6 N/m². (a) If the stretching obeys Hooke's law, what is the spring constant k for collagen? (b) How much work is done by the variable force that stretches the collagen? (See Section 6.9 for a discussion of the work done by a variable force.)

CONCEPTS CALCULATIONS

GROUP LEARNING PROBLEMS

Note: Each of these problems consists of Concept Questions followed by a related quantitative Problem. They are designed for use by students working alone or in small learning groups. The Concept Questions involve little or no mathematics and are intended to stimulate group discussions. They focus on the concepts with which the problems deal. Recognizing the concepts is the essential initial step in any problem-solving technique.

85. Concept Questions A spring lies on a horizontal table, and the left end is attached to a wall. The other end is connected to a box. The box is pulled to the right, stretching the spring. Static friction exists between the box and the table. When the box is released, it does not move. (a) Is the direction of the restoring force of the spring the same as or opposite to the direction of the static frictional force? Why? (b) How is the magnitude of the restoring force related to the magnitude of the static frictional force? (c)

What factors determine the *maximum* static frictional force that can be applied to the box?

Problem A 0.80-kg box is attached to a horizontal spring whose spring constant is 59 N/m. The coefficient of static friction between the box and the table on which it rests is $\mu_s = 0.74$. How far can the spring be stretched from its unstrained position without the box moving?

86. Concept Question The drawing shows three situations in which a block is attached to a spring. The position labeled "0 m" represents the unstrained position of the spring. The block is moved from an initial position x_0 to a final position x_f, the displacement being denoted by the symbol s. For each case, determine the direction of the restoring force as the block is moved. (Note that the direction of this force may change during the movement.) Then, without doing any calculations, decide if the total work

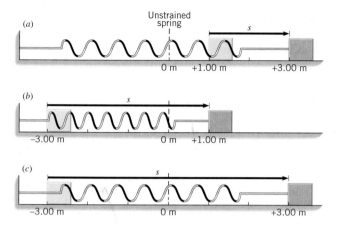

done by the restoring force of the spring is positive, negative, or zero. Give your reason for each answer in terms of the directions of the displacement and the restoring force.

Problem Suppose the spring constant is $k = 46.0$ N/m. Using the data provided in the drawing, determine the total work done by the restoring force of the spring for each situation. Verify that your answers are consistent with your answers to the Concept Question.

87. Concept Questions A spring is resting vertically on a table. A small box is dropped onto the top of the spring and compresses it. (a) Two forces are acting on the box: its weight and the restoring force of the spring. What law determines the relationship between these forces when the acceleration of the box is zero? (b) What is the speed of the box when the spring is fully compressed? (c) Many people say that the acceleration of the box is zero when spring is fully compressed. Is this correct? Explain why or why not.

Problem Suppose the spring has a spring constant of 450 N/m and the box has a mass of 1.5 kg. The speed of the box just before it makes contact with the spring is 0.49 m/s. (a) Determine the magnitude of the spring's displacement at the instant when the

acceleration of the box is zero. (b) What is the magnitude of the spring's displacement when the spring is fully compressed?

88. Concept Question The different types of energy possessed by a falling bungee jumper are discussed in Concepts & Calculations Example 15. When the jumper reaches her lowest point in the fall, which is above the water, what types of energy does she have? Provide a reason for each answer.

Problem Using the data given in Concepts & Calculations Example 15, determine how far she is from the water when she reaches the lowest point in the fall.

89. Concept Questions Two physical pendulums (not simple pendulums) are made from meter sticks that are suspended from the ceiling at one end. They are identical in all respects, except that one is made of wood and the other of metal. They are set into oscillation and execute simple harmonic motion. (a) How does the angular frequency of a physical pendulum depend on its mass? Take into account how the mass enters directly as well as indirectly through the moment of inertia of the pendulum. (b) Which pendulum, if either, has the greatest period? Why?

Problem The wood and metal pendulums have masses of 0.17 and 0.85 kg, respectively. Determine the periods of the two physical pendulums. Check to see that your answers are consistent with your answer to the Concept Questions.

90. Concept Questions One end of a piano wire is wrapped around a cylindrical tuning peg and the other end is fixed in place. The tuning peg is turned so as to stretch the wire. (a) For each turn of the tuning peg, by how much does the length of the wire change? Express your answer in terms of the radius of the tuning peg. (b) As the wire is stretched, do you expect the tension to increase, decrease, or remain the same? Why?

Problem The piano wire is made from steel $(Y = 2.0 \times 10^{11}$ N/m$^2)$. It has a radius of 0.80 mm, and an unstrained length of 0.76 m. The radius of the tuning peg is 1.8 mm. Initially, there is no tension in the wire. Find the tension in the wire when the tuning peg is turned through two revolutions.

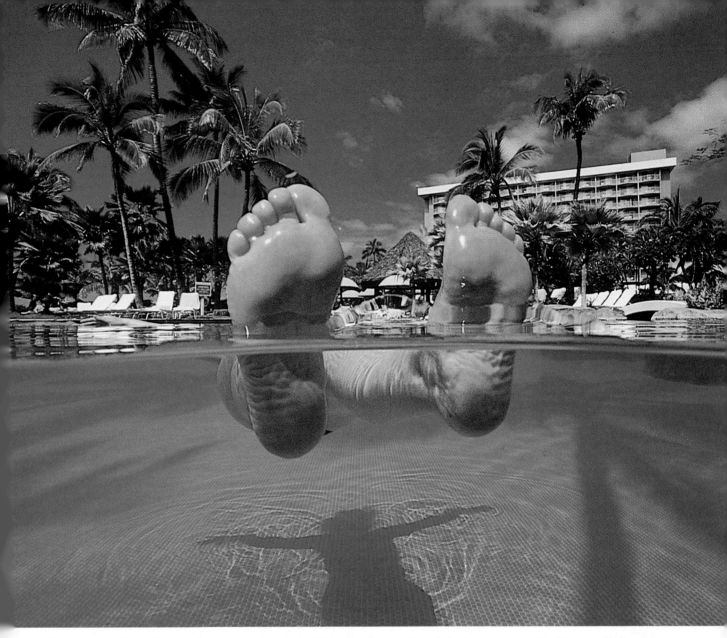

This swimmer is floating because the water exerts an upward force, the "buoyant force," that balances her weight. Water is a fluid, and this chapter examines the forces and pressures that fluids exert when they are at rest and when they are in motion.

CHAPTER 11

FLUIDS

11.1 MASS DENSITY

Fluids are materials that can flow, and they include both gases and liquids. Air is the most common gas, and moves from place to place as wind. Water is the most familiar liquid and has many uses, from generating hydroelectric power to white-water rafting. The *mass density* of a liquid or gas is an important factor that determines its behavior as a fluid. As indicated below, the mass density is the mass per unit volume and is denoted by the Greek letter rho (ρ).

TABLE 11.1   Mass Densities[a] of Common Substances

Substance	Mass Density ρ (kg/m^3)
Solids	
Aluminum	2 700
Brass	8 470
Concrete	2 200
Copper	8 890
Diamond	3 520
Gold	19 300
Ice	917
Iron (steel)	7 860
Lead	11 300
Quartz	2 660
Silver	10 500
Wood (yellow pine)	550
Liquids	
Blood (whole, 37 °C)	1 060
Ethyl alcohol	806
Mercury	13 600
Oil (hydraulic)	800
Water (4 °C)	1.000×10^3
Gases	
Air	1.29
Carbon dioxide	1.98
Helium	0.179
Hydrogen	0.0899
Nitrogen	1.25
Oxygen	1.43

[a] Unless otherwise noted, densities are given at 0 °C and 1 atm pressure.

● **DEFINITION OF MASS DENSITY**

The mass density ρ is the mass m of a substance divided by its volume V:

$$\rho = \frac{m}{V} \qquad (11.1)$$

SI Unit of Mass Density: kg/m^3

Equal volumes of different substances generally have different masses, so the density depends on the nature of the material as Table 11.1 indicates. Gases have the smallest densities because gas molecules are relatively far apart and a gas contains a large fraction of empty space. In contrast, the molecules are much more tightly packed in liquids and solids, and the tighter packing leads to larger densities. The densities of gases are very sensitive to changes in temperature and pressure. However, for the range of temperatures and pressures encountered in this text, the densities of liquids and solids do not differ much from the values in Table 11.1.

It is the mass of a substance, not its weight, that enters into the definition of density. In situations where weight is needed, it can be calculated from the mass density, the volume, and the acceleration of gravity, as Example 1 illustrates.

EXAMPLE 1   Blood as a Fraction of Body Weight

The body of a man whose weight is about 690 N contains about 5.2×10^{-3} m^3 (5.5 qt) of blood. (a) Find the blood's weight and (b) express it as a percentage of the body weight.

Reasoning To find the weight W of the blood, we need the mass m, since $W = mg$, where g is the magnitude of the acceleration due to gravity. According to Table 11.1, the density of blood is 1060 kg/m^3, so the mass of the blood can be found by using the given volume of 5.2×10^{-3} m^3 in Equation 11.1.

Solution

(a) The mass and the weight of the blood are

$$m = \rho V = (1060 \text{ kg/m}^3)(5.2 \times 10^{-3} \text{ m}^3) = 5.5 \text{ kg} \qquad (11.1)$$
$$W = mg = (5.5 \text{ kg})(9.80 \text{ m/s}^2) = \boxed{54 \text{ N}} \qquad (4.5)$$

(b) The percentage of body weight contributed by the blood is

$$\text{Percentage} = \frac{54 \text{ N}}{690 \text{ N}} \times 100 = \boxed{7.8\%}$$

A convenient way to compare densities is to use the concept of *specific gravity.* The specific gravity of a substance is its density divided by the density of a standard reference material, usually chosen to be water at 4 °C.

$$\text{Specific gravity} = \frac{\text{Density of substance}}{\text{Density of water at 4 °C}} = \frac{\text{Density of substance}}{1.000 \times 10^3 \text{ kg/m}^3} \quad (11.2)$$

Being the ratio of two densities, specific gravity has no units. For example, Table 11.1 reveals that diamond has a specific gravity of 3.52, since the density of diamond is 3.52 times greater than the density of water at 4 °C.

The next two sections deal with the important concept of pressure. We will see that the density of a fluid is one factor determining the pressure that a fluid exerts.

Figure 11.1 In colliding with the inner walls of the tire, the air molecules (blue dots) exert a force on every part of the wall surface. If a small cube were inserted inside the tire, the cube would experience forces acting perpendicular to each of its six faces.

11.2 PRESSURE

People who have fixed a flat tire know something about pressure. The final step in the job is to reinflate the tire to the proper pressure. The underinflated tire is soft because it contains an insufficient number of air molecules to push outward against the rubber and give the tire that solid feel. When air is added from a pump, the number of molecules and the collective force they exert are increased. The air molecules within a tire are free to wander throughout its entire volume, and in the course of their wandering they collide with one another and with the inner walls of the tire. The collisions with the walls allow the air to exert a force against every part of the wall surface, as Figure 11.1 shows. The pressure P exerted by a fluid is defined in Section 10.7 as the magnitude F of the force acting perpendicular to a surface divided by the area A over which the force acts:

$$P = \frac{F}{A} \quad (11.3)$$

The SI unit for pressure is a newton/meter^2 (N/m^2), a combination that is referred to as a pascal (Pa). A pressure of 1 Pa is a very small amount. Many common situations involve pressures of approximately 10^5 Pa, an amount referred to as one *bar* of pressure. Alternatively, force can be measured in pounds and area in square inches, so another unit for pressure is pounds per square inch (lb/in.^{2}), often abbreviated as "psi."

Because of its pressure, the air in a tire applies a force to any surface with which it is in contact. Suppose, for instance, that a small cube is inserted inside the tire. As Figure 11.1 shows, the air pressure causes a force to act perpendicularly on each face of the cube. In a similar fashion, a liquid such as water also exerts pressure. A swimmer, for example, feels the water pushing perpendicularly inward everywhere on her body, as Figure 11.2 illustrates. In general, a static fluid cannot produce a force parallel to a surface, for if it did, the surface would apply a reaction force to the fluid, consistent with Newton's action–reaction law. In response, the fluid would flow and would not then be static.

While fluid pressure can generate a force, pressure itself is not a vector quantity, as is the force. In the definition of pressure, $P = F/A$, the symbol F refers only to the magnitude of the force, so that pressure has no directional characteristic. The force generated by the pressure of a static fluid is always perpendicular to the surface that the fluid contacts, as Example 2 illustrates.

Figure 11.2 Water applies a force perpendicular to each surface within the water, including the walls and bottom of the swimming pool, and all parts of the swimmer's body.

Figure 11.3 Atmospheric pressure at sea level is 1.013×10^5 Pa, which is sufficient to crumple a can if the inside air is pumped out.

Area of base = 1.000 m²

Crumpled can

Pump

Force = 1.013×10^5 N

EXAMPLE 2   The Force on a Swimmer

Suppose the pressure acting on the back of a swimmer's hand is 1.2×10^5 Pa, a realistic value near the bottom of the diving end of a pool. The surface area of the back of the hand is 8.4×10^{-3} m². (a) Determine the magnitude of the force that acts on it. (b) Discuss the direction of the force.

Reasoning From the definition of pressure in Equation 11.3, we can see that the magnitude of the force is the pressure times the area. The direction of the force is always perpendicular to the surface that the water contacts.

Solution

(a) A pressure of 1.2×10^5 Pa is 1.2×10^5 N/m². From Equation 11.3, we find

$$F = PA = (1.2 \times 10^5 \text{ N/m}^2)(8.4 \times 10^{-3} \text{ m}^2) = \boxed{1.0 \times 10^3 \text{ N}}$$

This is a rather large force, about 230 lb.

(b) In Figure 11.2, the hand (palm downward) is oriented parallel to the bottom of the pool. Since the water pushes perpendicularly against the back of the hand, the force **F** is directed downward in the drawing. This downward-acting force is balanced by an upward-acting force on the palm, so that the hand is in equilibrium. If the hand were rotated by 90°, the direction of these forces would also be rotated by 90°, always being perpendicular to the hand.

A person need not be under water to experience the effects of pressure. Walking about on land, we are at the bottom of the earth's atmosphere, which is a fluid and pushes inward on our bodies just like the water in a swimming pool. As Figure 11.3 indicates, there is enough air above the surface of the earth to create the following pressure at sea level:

THE PHYSICS OF...

lynx paws. (See below.)

Atmospheric pressure at sea level 1.013×10^5 Pa = 1 atmosphere

This amount of pressure corresponds to 14.70 lb/in.² and is referred to as one *atmosphere (atm)*. One atmosphere of pressure is a significant amount. Look, for instance, in Figure 11.3 at the results of pumping out the air from within a gasoline can. With no internal air to push outward, the inward push of the external air is unbalanced and is strong enough to crumple the can.

In contrast to the situation in Figure 11.3, reducing the pressure is sometimes beneficial. Lynx, for example, are especially well suited for hunting on snow because of their oversize paws (see Figure 11.4). The large paws function as snowshoes that distribute the weight over a large area. Thus, they reduce the weight per unit area or pressure that the cat applies to the surface, which helps to keep it from sinking into the snow.

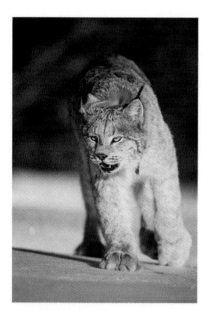

Figure 11.4 Lynx have large paws that act as natural snowshoes.

11.3   PRESSURE AND DEPTH IN A STATIC FLUID

● *Concepts at a Glance* The deeper an underwater swimmer goes, the more strongly the water pushes on his body and the greater is the pressure that he experiences. To determine the relation between pressure and depth, we turn to Newton's second law. The Concepts-at-a-Glance chart in Figure 11.5 outlines our approach. This chart is an expanded version of the charts in Figures 4.26 and 10.4. In using the second law, we will focus on two external forces that act on the fluid. One is the gravitational force, that is, the weight of the fluid. The other is the collisional force

Concepts at a Glance

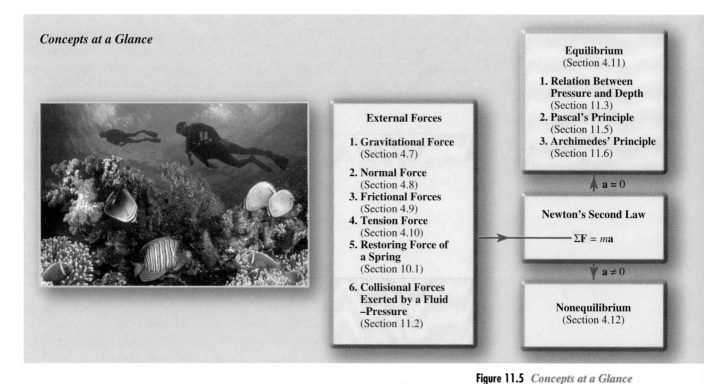

External Forces

1. **Gravitational Force**
   (Section 4.7)

2. **Normal Force**
   (Section 4.8)
3. **Frictional Forces**
   (Section 4.9)
4. **Tension Force**
   (Section 4.10)
5. **Restoring Force of
   a Spring**
   (Section 10.1)

6. **Collisional Forces
   Exerted by a Fluid
   –Pressure**
   (Section 11.2)

Equilibrium
(Section 4.11)

1. **Relation Between
   Pressure and Depth**
   (Section 11.3)
2. **Pascal's Principle**
   (Section 11.5)
3. **Archimedes' Principle**
   (Section 11.6)

$\mathbf{a} = 0$

Newton's Second Law

$\Sigma \mathbf{F} = m\mathbf{a}$

$\mathbf{a} \neq 0$

Nonequilibrium
(Section 4.12)

Figure 11.5 *Concepts at a Glance*
The gravitational force and the collisional forces exerted by a fluid are used
with Newton's second law to determine
the relation between pressure and depth
in a static ($\mathbf{a} = 0$) fluid. The deeper that
these scuba divers go, the greater will
be the water pressure on them.

that is responsible for fluid pressure, as the previous section discusses. Since the
fluid is at rest, its acceleration is zero ($\mathbf{a} = 0$), and it is in equilibrium. By applying
the second law in the form $\Sigma \mathbf{F} = 0$, we will derive a relation between pressure and
depth. This relation is especially important because it leads to Pascal's principle
(Section 11.5) and Archimedes' principle (Section 11.6), both of which are useful
in describing the properties of static fluids. ●

Figure 11.6 shows a container of fluid and focuses attention on one column of
the fluid. The free-body diagram in the figure shows all the vertical forces acting on
the column. On the top face (area = A), the fluid pressure P_1 generates a downward
force whose magnitude is P_1A. Similarly, on the bottom face, the pressure P_2 gener-

Pressure = P_1
Area = A

h

Pressure = P_2
Area = A

(a)

+y axis

P_1A

mg

P_2A

(b) Free-body diagram
of the column

Figure 11.6 (*a*) A container of fluid in
which one column of the fluid is outlined. The fluid is at rest. (*b*) The free-
body diagram, showing the vertical
forces acting on the column.

ates an upward force of magnitude P_2A. The pressure P_2 is greater than the pressure P_1 because the bottom face supports the weight of more fluid than the upper one does. In fact, the excess weight supported by the bottom face is exactly the weight of the fluid within the column. As the free-body diagram indicates, this weight is mg, where m is the mass of the fluid and g is the magnitude of the acceleration due to gravity. Since the column is in equilibrium, we can set the sum of the vertical forces equal to zero and find that

$$\Sigma F_y = P_2A - P_1A - mg = 0 \quad \text{or} \quad P_2A = P_1A + mg$$

The mass m is related to the density ρ and the volume V of the column by $m = \rho V$. Since the volume is the cross-sectional area A times the vertical dimension h, we have $m = \rho A h$. With this substitution, the condition for equilibrium becomes $P_2A = P_1A + \rho A h g$. The area A can be eliminated algebraically from this expression, with the result that

$$P_2 = P_1 + \rho g h \tag{11.4}$$

Equation 11.4 indicates that if the pressure P_1 is known at a higher level, the larger pressure P_2 at a deeper level can be calculated by adding the increment $\rho g h$. In determining the pressure increment $\rho g h$, we assumed that the density ρ is the same at any vertical distance h or, in other words, the fluid is incompressible. The assumption is reasonable for liquids, since the bottom layers can support the upper layers with little compression. In a gas, however, the lower layers are compressed markedly by the weight of the upper layers, with the result that the density varies with vertical distance. For example, the density of our atmosphere is larger near the earth's surface than it is at higher altitudes. When applied to gases, the relation $P_2 = P_1 + \rho g h$ can be used only when h is small enough that any variation in ρ is negligible.

A significant feature of Equation 11.4 is that the pressure increment $\rho g h$ is affected by the vertical distance h, but not by any horizontal distance within the fluid. Conceptual Example 3 helps to clarify this feature.

CONCEPTUAL EXAMPLE 3   The Hoover Dam

Lake Mead is the largest wholly artificial reservoir in the United States and was formed after the completion of the Hoover Dam in 1936. As Figure 11.7a suggests, the water in the reservoir backs up behind the dam for a considerable distance (about 200 km or 120 miles). Suppose that all the water were removed, except for a relatively narrow vertical column in contact with the dam. Figure 11.7b shows a side view of this hypothetical situation, in which the water against the dam has the same depth as in Figure 11.7a. Would the Hoover Dam still be needed to contain the water in this hypothetical reservoir, or could a much less massive structure do the job?

Reasoning and Solution Since our hypothetical reservoir contains much less water than Lake Mead, it is tempting to say that a much less massive structure than the Hoover Dam would be needed. Not so, however. Hoover Dam would still be needed to contain our hypothetical reservoir. To see why, imagine a small square on the inner face of the dam, located beneath the water. The magnitude of the force on this square is the product of its area and the pressure of the water. But the pressure depends on the depth, that is, on the vertical distance h in Equation 11.4. The horizontal distance of the water behind the dam does not appear in this equation and, therefore, has no effect on the pressure. Consequently, at a given depth the small square experiences a force that is the same in both parts of Figure 11.7. Certainly the dam experiences greater forces generated by greater water pressures at greater depths. But no matter where on the inner face of the dam the square is, the force that the water applies to it

(a)

(b)

Figure 11.7 (a) The Hoover Dam in Nevada and Lake Mead behind it. (b) This drawing shows a hypothetical reservoir formed by removing most of the water from Lake Mead. Conceptual Example 3 compares the dam needed for this hypothetical reservoir with the Hoover Dam.

Figure 11.8 The pressures at points A and B are the same, since both points are located at the same vertical distance of 5.50 m beneath the surface of the water.

depends only on the depth, not on the amount of water backed up behind the dam. ***Thus, the dam for our imaginary reservoir would sustain the same forces that the Hoover Dam sustains and would need to be equally large.***

The next example deals further with the relationship between pressure and depth given by Equation 11.4.

EXAMPLE 4   The Swimming Hole

Figure 11.8 shows the cross section of a swimming hole. Points A and B are both located at a distance of $h = 5.50$ m below the surface of the water. Find the pressure at each of these two points.

Reasoning The pressure at point B is the same as that at point A, since both are located at the *same vertical distance* beneath the surface and only the vertical distance h affects the pressure increment $\rho g h$ in Equation 11.4. To understand this important feature more clearly, consider the path $AA'B'B$ in Figure 11.8. The pressure decreases on the way up along the vertical segment AA' and increases by the same amount on the way back down along segment $B'B$. Since no change in pressure occurs along the horizontal segment $A'B'$, the pressure is the same at A and B.

Solution The pressure acting on the surface of the water is the atmospheric pressure of 1.01×10^5 Pa. Using this value as P_1 in Equation 11.4, we can determine a value for the pressure P_2 at either point A or B, both of which are located 5.50 m under the water. Table 11.1 gives the density of water as 1.000×10^3 kg/m³.

$$P_2 = P_1 + \rho g h$$
$$P_2 = 1.01 \times 10^5 \text{ Pa} + (1.000 \times 10^3 \text{ kg/m}^3)(9.80 \text{ m/s}^2)(5.50 \text{ m}) = \boxed{1.55 \times 10^5 \text{ Pa}}$$

• **PROBLEM SOLVING INSIGHT**
The pressure at any point in a fluid depends on the vertical distance h of the point beneath the surface. However, for a given vertical distance, the pressure is the same, no matter where the point is located horizontally in the fluid.

Figure 11.9 shows an irregularly shaped container of liquid. Reasoning similar to that used in Example 4 leads to the conclusion that the pressure is the same at points A, B, C, and D, since each is at the same vertical distance h beneath the surface. In effect, the arteries in our bodies constitute an irregularly shaped "container" for the blood. The next example examines the blood pressure at different places in this "container."

EXAMPLE 5   Blood Pressure

Blood in the arteries is flowing, but as a first approximation, the effects of this flow can be ignored and the blood can be treated as a static fluid. Estimate the amount by

Figure 11.9 Since points A, B, C, and D are at the same distance h beneath the liquid surface, the pressure at each of them is the same.

Blood pressure = P_1

1.35 m

Blood pressure = P_2

Figure 11.10 The blood pressure in the feet can exceed the blood pressure in the heart, depending on whether the body is (a) reclining horizontally or (b) standing erect.

THE PHYSICS OF...

pumping water from a well.

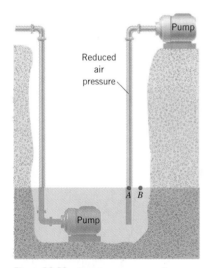

Figure 11.11 A water pump can be placed at the bottom of a well or at ground level. Conceptual Example 6 discusses how the depth of the well influences the choice of which method to use.

which the blood pressure P_2 in the anterior tibial artery at the foot exceeds the blood pressure P_1 in the aorta at the heart when the body is (a) reclining horizontally as in Figure 11.10a and (b) standing as in Figure 11.10b.

Reasoning and Solution

(a) When the body is horizontal, there is little or no vertical separation between the feet and the heart. Since $h = 0$ m,

$$P_2 - P_1 = \rho g h = \boxed{0 \text{ Pa}} \tag{11.4}$$

(b) When an adult is erect, the vertical separation between the feet and the heart is about 1.35 m, as Figure 11.10b indicates. Table 11.1 gives the density of blood as 1060 kg/m³, so that

$$P_2 - P_1 = \rho g h = (1060 \text{ kg/m}^3)(9.80 \text{ m/s}^2)(1.35 \text{ m}) = \boxed{1.40 \times 10^4 \text{ Pa}}$$

Sometimes fluid pressure places limits on how a job can be done. Conceptual Example 6 illustrates how fluid pressure restricts the height to which water can be pumped.

CONCEPTUAL EXAMPLE 6   Pumping Water

Figure 11.11 shows two methods for pumping water from a well. In one method, the pump is submerged in the water at the bottom of the well, while in the other, it is located at ground level. If the well is shallow, either technique can be used. However, if the well is very deep, only one of the methods works. Which is it?

Reasoning and Solution To answer this question, we need to examine the nature of the job done by each pump. The pump at the bottom of the well literally pushes water up the pipe. For a very deep well, the column of water becomes very tall, and the pressure at the bottom of the pipe becomes large, due to the pressure increment $\rho g h$ in Equation 11.4. However, as long as the pump can push with sufficient strength to overcome the large pressure, it can shove the next increment of water into the pipe, so the method can be used for very deep wells. In contrast, the pump at ground level does not push water at all. It removes air from the pipe. As the pump reduces the air pressure within the pipe (see point A), the greater air pressure outside (see point B) pushes water up the pipe. But even the strongest pump can only remove *all* of the air. Once the air is completely removed, an increase in pump strength does not increase the height to which the water is pushed by the external air pressure. ***Thus, a ground-level pump can only cause water to rise to a certain maximum height, so it cannot be used for very deep wells.*** (Incidentally, the pump at ground level works just the way you do when you drink through a straw. You draw the air out of the straw, and the external air pressure pushes the liquid up into it.)

Related Homework Material: *Problems 20 and 24*

11.4 PRESSURE GAUGES

One of the simplest pressure gauges is the mercury barometer used for measuring atmospheric pressure. This device is a tube sealed at one end, filled completely with mercury, and then inverted, so that the open end is under the surface of a pool of mercury (see Figure 11.12). Except for a negligible amount of mercury vapor, the space above the mercury in the tube is empty, and the pressure P_1 is nearly zero there. The pressure P_2 at point A at the bottom of the mercury column is the same as that at point B, namely, atmospheric pressure, for these two points are at the same level. With $P_1 = 0$ Pa and $P_2 = P_{atm}$, it follows from Equation 11.4 that $P_{atm} = 0$ Pa $+ \rho gh$. Thus, the atmospheric pressure can be determined from the height h of the mercury in the tube, the density ρ of mercury, and the acceleration due to gravity. Usually weather forecasters report the pressure in terms of the height h, expressing it in millimeters or inches of mercury. For instance, using $P_{atm} = 1.013 \times 10^5$ Pa and $\rho = 13.6 \times 10^3$ kg/m^3 for the density of mercury, we find that $h = P_{atm}/(\rho g) = 760$ mm (29.9 inches).* Slight variations from this value occur, depending on weather conditions and altitude.

Figure 11.13 shows another kind of pressure gauge, the open-tube manometer. The phrase "open-tube" refers to the fact that one side of the U-tube is open to atmospheric pressure. The tube contains a liquid, often mercury, and its other side is connected to the container whose pressure P_2 is to be measured. When the pressure in the container is equal to the atmospheric pressure, the liquid levels in both sides of the U-tube are the same. When the pressure in the container is greater than atmospheric pressure, as in Figure 11.13, the liquid in the tube is pushed downward on the left side and upward on the right side. The relation $P_2 = P_1 + \rho gh$ can be used to determine the container pressure. Atmospheric pressure exists at the top of the right column, so that $P_1 = P_{atm}$. The pressure P_2 is the same at points A and B, so we find that $P_2 = P_{atm} + \rho gh$, or

$$P_2 - P_{atm} = \rho gh$$

The height h is proportional to $P_2 - P_{atm}$, which is called the **gauge pressure.** The gauge pressure is the amount by which the container pressure differs from atmospheric pressure. The actual value for P_2 is called the **absolute pressure.**

The sphygmomanometer is a familiar device for measuring blood pressure. As Figure 11.14 illustrates, a squeeze bulb can be used to inflate the cuff with air, which cuts off the flow of blood through the artery below the cuff. When the release valve is opened, the cuff pressure drops. Blood begins to flow again when the pressure created by the heart at the peak of its beating cycle exceeds the cuff pressure. Using a stethoscope to listen for the initial flow, the operator can measure the corresponding cuff gauge pressure with an open-tube manometer. This cuff gauge pressure is called the *systolic* pressure. Eventually, there comes a point when even the pressure created by the heart at the low point of its beating cycle is sufficient to cause blood to flow. Identifying this point with the stethoscope, the operator can measure the corresponding cuff gauge pressure, which is referred to as the *diastolic* pressure. The systolic and diastolic pressures are reported in millimeters of mercury, and values of 120 and 80, respectively, are typical of a young, healthy heart.

11.5 PASCAL'S PRINCIPLE

As we have seen, the pressure in a fluid increases with depth, due to the weight of the fluid above the point of interest. A completely enclosed fluid may be subjected to an additional pressure by the application of an external force. For example, Figure 11.15a shows two interconnected cylindrical chambers. The chambers have different diameters and, together with the connecting tube, are completely filled with a liquid. The larger chamber is sealed at the top with a cap, while the smaller one is fitted with a movable piston. We now follow the approach outlined earlier in Figure 11.5

---

* A pressure of one millimeter of mercury is sometimes referred to as one *torr,* to honor the inventor of the barometer, Evangelista Torricelli (1608–1647). Thus, one atmosphere of pressure is 760 torr.

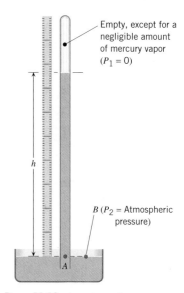

Figure 11.12   A mercury barometer.

● PROBLEM SOLVING INSIGHT
When solving problems that deal with pressure, be sure to note the distinction between gauge pressure and absolute pressure.

THE PHYSICS OF...

measuring blood pressure.

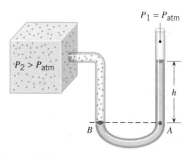

Figure 11.13   The U-shaped tube is called an open-tube manometer and can be used to measure the pressure P_2 in a container.

Cuff

Squeeze bulb

Release valve

Stethoscope

Open-tube manometer

Figure 11.14 A sphygmomanometer is used to measure blood pressure.

(a)

(b)

Figure 11.15 (a) An external force $\mathbf{F_1}$ is applied to the piston on the left. As a result, a force $\mathbf{F_2}$ is exerted on the cap on the chamber on the right. (b) The familiar hydraulic car lift.

THE PHYSICS OF...

a hydraulic car lift.

and begin by asking what determines the pressure P_1 at a point immediately beneath the piston. According to the definition of pressure, it is the magnitude F_1 of the external force divided by the area A_1 of the piston: $P_1 = F_1/A_1$. If it is necessary to know the pressure P_2 *at any other place in the liquid,* we just add to the value of P_1 the increment ρgh, which takes into account the depth h below the piston: $P_2 = P_1 + \rho gh$. The important feature here is this: The pressure P_1 adds to the pressure ρgh due to the depth of the liquid at any point, whether that point is in the smaller chamber, the connecting tube, or the larger chamber. Therefore, if the applied pressure P_1 is increased or decreased, the pressure at any other point within the confined liquid changes correspondingly. This behavior is described by *Pascal's principle.*

● **PASCAL'S PRINCIPLE**

Any change in the pressure applied to a completely enclosed fluid is transmitted undiminished to all parts of the fluid and the enclosing walls.

The usefulness of the arrangement in Figure 11.15a becomes apparent when we calculate the force F_2 applied by the liquid to the cap on the right side. The area of the cap is A_2 and the pressure there is P_2. As long as the tops of the left and right chambers are at the same level, the pressure increment ρgh is zero, so that the relation $P_2 = P_1 + \rho gh$ becomes $P_2 = P_1$. Consequently, $F_2/A_2 = F_1/A_1$, and

$$F_2 = F_1 \left(\frac{A_2}{A_1} \right) \qquad (11.5)$$

If area A_2 is larger than area A_1, a large force $\mathbf{F_2}$ can be applied to the cap on the right chamber, starting with a smaller force $\mathbf{F_1}$ on the left. Depending on the ratio of the areas A_2/A_1, the force $\mathbf{F_2}$ can be large indeed, as in the familiar hydraulic car lift shown in part b. In this device the force $\mathbf{F_2}$ is not applied to a cap that seals the larger chamber, but, rather, to a movable plunger that lifts a car. Example 7 deals with a hydraulic car lift.

EXAMPLE 7   A Car Lift

In a hydraulic car lift, the input piston has a radius of $r_1 = 0.0120$ m and a negligible weight. The output plunger has a radius of $r_2 = 0.150$ m. The combined weight of the car and the plunger is $F_2 = 20\,500$ N. The lift uses hydraulic oil that has a density of 8.00×10^2 kg/m³. What input force F_1 is needed to support the car and the output plunger when the bottom surfaces of the piston and plunger are at (a) the same level and (b) the levels shown in Figure 11.15b with $h = 1.10$ m?

Reasoning When the bottom surfaces of the piston and plunger are at the same levels, as in part (a), Equation 11.5 applies. However, this equation does not apply in part (b), where the bottom surface of the output plunger is $h = 1.10$ m below the input piston. Therefore, our solution in part (b) will take into account the pressure increment ρgh. In either case, we will see that the input force is less than the combined weight of the plunger and car.

Solution

(a) Using $A = \pi r^2$ for the circular areas of the piston and plunger, we find from Equation 11.5 that

$$F_1 = F_2 \left(\frac{A_1}{A_2} \right) = F_2 \left(\frac{\pi r_1^2}{\pi r_2^2} \right) = (20\,500 \text{ N}) \frac{(0.0120 \text{ m})^2}{(0.150 \text{ m})^2} = \boxed{131 \text{ N}}$$

(b) In Figure 11.15*b*, the bottom surface of the plunger at point *B* is at the same level as point *A*, which is at a depth *h* beneath the input piston. Therefore, we can apply $P_2 = P_1 + \rho gh$, with $P_2 = F_2/(\pi r_2^2)$ and $P_1 = F_1/(\pi r_1^2)$:

$$\frac{F_2}{\pi r_2^2} = \frac{F_1}{\pi r_1^2} + \rho gh$$

Solving for F_1 gives

$$F_1 = F_2\left(\frac{r_1^2}{r_2^2}\right) - \rho gh(\pi r_1^2)$$

$$= (20\ 500\ \text{N})\frac{(0.0120\ \text{m})^2}{(0.150\ \text{m})^2} - (8.00 \times 10^2\ \text{kg/m}^3)$$

$$\times (9.80\ \text{m/s}^2)(1.10\ \text{m})\ \pi\ (0.0120\ \text{m})^2 = \boxed{127\ \text{N}}$$

The answer here is less than in part (a) because the weight of the 1.10-m column of hydraulic oil provides some of the input force to support the car.

In a device such as a hydraulic car lift, the same amount of work is done by both the input and output forces in the absence of friction. The larger output force $\mathbf{F_2}$ moves through a smaller distance, while the smaller input force $\mathbf{F_1}$ moves through a larger distance. The work, being the product of the magnitude of the force and the distance, is the same in either case since mechanical energy is conserved.

An enormous variety of clever devices use hydraulic fluids, just as the car lift does. In a backhoe, for instance, the fluids multiply a small input force into the large output force required for digging (see Figure 11.16).

Figure 11.16 A backhoe uses a hydraulic fluid to generate a large output force, starting with a small input force.

11.6 ARCHIMEDES' PRINCIPLE

Anyone who has tried to push a beach ball under the water has felt how the water pushes back with a strong upward force. This upward force is called the **buoyant force,** and all fluids apply such a force to objects that are immersed in them. The buoyant force exists because fluid pressure is larger at greater depths.

In Figure 11.17 a cylinder of height *h* is being held under the surface of a liquid. The pressure P_1 on the top face generates the downward force P_1A, where *A* is the area of the face. Similarly, the pressure P_2 on the bottom face generates the upward force P_2A. Since the pressure is greater at greater depths, the upward force exceeds the downward force. Consequently, the liquid applies to the cylinder a net upward force, or buoyant force, whose magnitude F_B is

$$F_B = P_2A - P_1A = (P_2 - P_1)A$$

Substituting $P_2 - P_1 = \rho gh$ from Equation 11.4, we find that the buoyant force equals ρghA. In this result *hA* is the volume of liquid that the cylinder moves aside or displaces in being submerged, and ρ denotes the density of the liquid, not the density of the material from which the cylinder is made. Therefore, ρhA gives the mass *m* of the displaced fluid, so that the buoyant force equals *mg*, the weight of the displaced fluid. The phrase "weight of the displaced fluid" refers to the weight of the fluid that would spill out, if the container were filled to the brim before the cylinder is inserted into the liquid. The buoyant force is not a new type of force. It is just the name given to the net upward force exerted by the fluid on the object.

The shape of the object in Figure 11.17 is not important. No matter what the shape, the buoyant force arises in accord with ***Archimedes' principle.*** It was an im-

Figure 11.17 The fluid applies a downward force P_1A to the top face of the submerged cylinder and an upward force P_2A to the bottom face.

Figure 11.18 (*a*) An object of weight 100 N is being inserted into a liquid. The deeper the object is, the more liquid it displaces, and the stronger the buoyant force is. (*b*) The buoyant force matches the 100-N weight, so the object floats.

Figure 11.19 A raft floating with a distance *h* beneath the surface of the water.

---

pressive accomplishment that the Greek scientist Archimedes (ca. 287–212 B.C.) discovered the essence of this principle so long ago.

● ARCHIMEDES' PRINCIPLE

Any fluid applies a buoyant force to an object that is partially or completely immersed in it; the magnitude of the buoyant force equals the weight of the fluid that the object displaces:

$$\underbrace{F_B}_{\substack{\text{Magnitude of} \\ \text{buoyant force}}} = \underbrace{W_{\text{fluid}}}_{\substack{\text{Weight of} \\ \text{displaced fluid}}} \tag{11.6}$$

The effect that the buoyant force has depends on how strong it is compared with the other forces that are acting. For example, if the buoyant force is strong enough to balance the force of gravity, an object will float in a fluid. Figure 11.18 explores this possibility. In part *a*, a block that weighs 100 N displaces some liquid, and the liquid applies a buoyant force F_B to the block, according to Archimedes' principle. Nevertheless, if the block were released, it would fall further into the liquid because the buoyant force is not sufficiently strong to balance the weight. In part *b*, however, enough of the block is submerged to provide a buoyant force that can balance the 100-N weight, so the block is in equilibrium and floats when released. If the buoyant force were not sufficiently large to balance the weight, even with the block completely submerged, the block would sink. Even if an object sinks, there is still a buoyant force acting on it; it's just that the buoyant force is not large enough to balance the weight. Example 8 provides additional insight into what determines whether an object floats or sinks in a fluid.

EXAMPLE 8   A Swimming Raft

A solid, square, pinewood raft measures 4.0 m on a side and is 0.30 m thick. (a) Determine whether the raft floats in water, and (b) if so, how much of the raft is beneath the surface (see the distance *h* in Figure 11.19).

Reasoning To determine whether the raft floats, we will compare the weight of the raft to the maximum possible buoyant force and see if there could be enough buoyant force to balance the weight. If so, then the value of the distance *h* can be obtained by utilizing the fact that the floating raft is in equilibrium, with the magnitude of the buoyant force equaling the raft's weight.

Solution

(a) The weight of the raft can be calculated from the density $\rho_{\text{pine}} = 550$ kg/m³ (Table 11.1), the volume of the wood, and the acceleration due to gravity. The volume of the wood is $V_{\text{pine}} = 4.0$ m $\times$ 4.0 m $\times$ 0.30 m $= 4.8$ m³, so that

$$\begin{array}{l} \text{Weight} \\ \text{of raft} \end{array} = (\rho_{\text{pine}} V_{\text{pine}})g = (550 \text{ kg/m}^3)(4.8 \text{ m}^3)(9.80 \text{ m/s}^2) = 26\,000 \text{ N}$$

The maximum possible buoyant force occurs when the entire raft is under the surface, displacing 4.8 m³ of water. According to Archimedes' principle, the weight of this volume of water is the maximum buoyant force F_B^{max}. It can be obtained using the density of water:

$$F_B^{\text{max}} = \rho_{\text{water}} V_{\text{pine}} g = (1.000 \times 10^3 \text{ kg/m}^3)(4.8 \text{ m}^3)(9.80 \text{ m/s}^2) = 47\,000 \text{ N}$$

Since the maximum possible buoyant force exceeds the 26 000-N weight of the raft, the raft will float only partially submerged at a distance h beneath the water.

(b) We now find the value of h. The buoyant force balances the raft's weight, so $F_B = 26\ 000$ N. But according to Equation 11.6, the magnitude of the buoyant force is also the weight of the displaced fluid, so $F_B = W_{fluid} = 26\ 000$ N. Using the density of water, we can also express the weight of the displaced water as $W_{fluid} = \rho_{water} V_{water} g$, where the volume is $V_{water} = 4.0$ m $\times$ 4.0 m $\times h$. As a result,

$$W_{fluid} = \rho_{water}(4.0 \text{ m} \times 4.0 \text{ m} \times h)g = 26\ 000 \text{ N}$$

$$h = \frac{26\ 000 \text{ N}}{\rho_{water}(4.0 \text{ m} \times 4.0 \text{ m})g}$$

$$= \frac{26\ 000 \text{ N}}{(1.000 \times 10^3 \text{ kg/m}^3)(4.0 \text{ m} \times 4.0 \text{ m})(9.80 \text{ m/s}^2)} = \boxed{0.17 \text{ m}}$$

In deciding whether the raft floats in part (a) of Example 8, we compared the raft's weight $(\rho_{pine} V_{pine})g$ to the maximum possible buoyant force $(\rho_{water} V_{pine})g$. The comparison depends only on the densities ρ_{pine} and ρ_{water}. The take-home message is this: Any object that is *solid throughout* will float in a liquid if the density of the object is less than or equal to the density of the liquid. For instance, at 0 °C ice has a density of 917 kg/m^3, while water has a density of 1000 kg/m^3. Therefore, ice floats in water.

Although a solid piece of a high-density material like steel will sink in water, such materials can, nonetheless, be used to make floating objects. A supertanker, for example, floats because it is *not* solid metal. It contains enormous amounts of empty space and, because of its shape, displaces enough water to balance its own large weight. Conceptual Example 9 focuses on the reason ships float.

CONCEPTUAL EXAMPLE 9   How Much Water Is Needed to Float a Ship?

A ship floating in the ocean is a familiar sight. But is all that water really necessary? Can an ocean vessel float in the amount of water that a swimming pool contains, for instance?

Reasoning and Solution In principle, a ship can float in much less than the amount of water that a swimming pool contains; even a glass of water would do. To see why, look at Figure 11.20. Part *a* shows the ship floating in the ocean because it contains empty space within its hull and displaces enough water to balance its own weight. Part *b* shows the water that the ship displaces, which, according to Archimedes' principle, has a weight that equals the ship's weight. While this wedge-shaped portion of water represents the water *displaced* by the ship, it is not the amount that must be present to float the ship, as part *c* illustrates. This part of the drawing shows a canal, the cross section of which matches the shape in part *b*. **All that is needed, in principle, is a thin section of water that separates the hull of the floating ship from the sides of the canal.** This thin section of water could have a very small volume indeed and fit into a single glass!

A useful application of Archimedes' principle can be found in car batteries. To alert the owner that recharging is necessary, some batteries include a state-of-charge indicator, such as the one illustrated in Figure 11.21. The battery includes a viewing port that looks down through a plastic rod, which extends into the battery acid. Attached to the end of this rod is a "cage," containing a green ball. The cage has holes in it that allow the acid to enter. When the battery is charged, the density of the acid

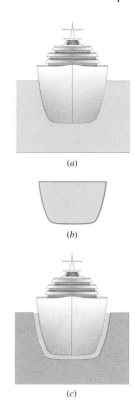

Figure 11.20 (*a*) A ship floating in the ocean. (*b*) This is the water that the ship displaces. (*c*) The ship floats here in a canal that has a cross section similar in shape to that in part *b*.

THE PHYSICS OF...

a state-of-charge battery indicator.

Port for viewing
state-of-charge
indicator

Green dot   Black dot

Plastic rod

Battery acid

Cage

Charged   Discharged

Figure 11.21 A state-of-charge indicator for a car battery.

is large enough so that its buoyant force makes the ball rise to the top of the cage, to just beneath the plastic rod. The viewing port shows a green dot. As the battery discharges, the density of the acid decreases. Since the buoyant force is the weight of the acid displaced by the ball, the buoyant force also decreases. As a result, the ball sinks into one of the chambers oriented at an angle. With the ball no longer visible, the viewing port shows a dark or black dot, warning that the battery charge is low.

Archimedes' principle has allowed us to determine how an object can float in a liquid. This principle also applies to gases, as the next example illustrates.

EXAMPLE 10   A Goodyear Airship

THE PHYSICS OF...

a Goodyear airship.

Normally, a Goodyear airship, such as that in Figure 11.22, contains about 5.40×10^3 m^3 of helium (He) whose density is 0.179 kg/m^3. Find the weight of the load W_L that the airship can carry in equilibrium at an altitude where the density of air is 1.20 kg/m^3.

Reasoning The airship and its load are in equilibrium. Thus, the buoyant force F_B applied to the airship by the surrounding air balances the weight W_{He} of the helium and the weight W_L of the load, including the solid parts of the airship. The free-body diagram in Figure 11.22b shows these forces.

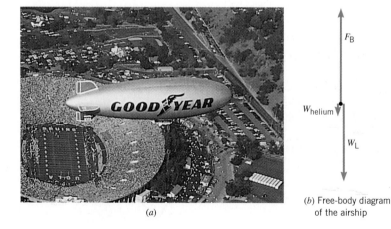

F_B

W_{helium}

W_L

Figure 11.22 (a) A helium-filled Goodyear airship. (b) The free-body diagram of the airship, including the load weight W_L.

(a)

(b) Free-body diagram of the airship

Solution Because the forces in the free-body diagram balance, we have

$$W_{He} + W_L = F_B \quad \text{or} \quad W_L = F_B - W_{He}$$

According to Archimedes' principle, the buoyant force is the weight of the displaced air: $F_B = W_{air} = \rho_{air} V_{ship} g$. The weight of the helium is $W_{He} = \rho_{He} V_{ship} g$, where we have assumed that the volume of the helium and the volume of the airship have nearly the same value of $V_{ship} = 5.40 \times 10^3 \text{ m}^3$. Thus, we find that

$$W_L = \rho_{air} V_{ship} g - \rho_{He} V_{ship} g = (\rho_{air} - \rho_{He}) V_{ship} g$$

$$W_L = (1.20 \text{ kg/m}^3 - 0.179 \text{ kg/m}^3)(5.40 \times 10^3 \text{ m}^3)$$

$$\times (9.80 \text{ m/s}^2) = \boxed{5.40 \times 10^4 \text{ N}}$$

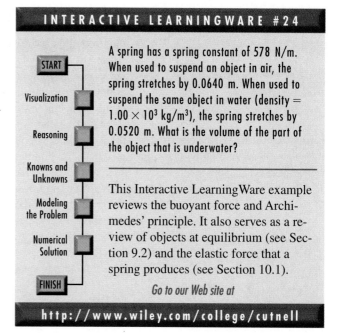

INTERACTIVE LEARNINGWARE #24

START
Visualization
Reasoning
Knowns and Unknowns
Modeling the Problem
Numerical Solution
FINISH

A spring has a spring constant of 578 N/m. When used to suspend an object in air, the spring stretches by 0.0640 m. When used to suspend the same object in water (density = $1.00 \times 10^3 \text{ kg/m}^3$), the spring stretches by 0.0520 m. What is the volume of the part of the object that is underwater?

This Interactive LearningWare example reviews the buoyant force and Archimedes' principle. It also serves as a review of objects at equilibrium (see Section 9.2) and the elastic force that a spring produces (see Section 10.1).

Go to our Web site at

http://www.wiley.com/college/cutnell

11.7 FLUIDS IN MOTION

Fluids can move or flow in many circumstances. Water may flow smoothly and slowly in a quiet stream or violently over a waterfall. The air may form a gentle breeze or a raging tornado. To deal with such diversity, it helps to identify some of the basic types of fluid flow.

Fluid flow can be steady or unsteady. In *steady flow* the velocity of the fluid particles at any point is constant as time passes. For instance, in Figure 11.23 a fluid particle flows with a velocity of $v_1 = +2$ m/s past point 1. In steady flow every particle passing through this point has this same velocity. At another location the velocity may be different, as in a river, which usually flows fastest near its center and slowest near its banks. Thus, at point 2, the fluid velocity is $v_2 = +0.5$ m/s, and if the flow is steady, all particles passing through this point have a velocity of $+0.5$ m/s.

Unsteady flow exists whenever the velocity at a point in the fluid changes as time passes. *Turbulent flow* is an extreme kind of unsteady flow and occurs when there are sharp obstacles or bends in the path of a fast-moving fluid, as in the rapids in Figure 11.24. In turbulent flow, the velocity at any particular point changes erratically from moment to moment, both in magnitude and direction.

Fluid flow can be compressible or incompressible. Most liquids are nearly incompressible; that is, the density of a liquid remains almost constant as the pressure changes. To a good approximation, then, liquids flow in an incompressible manner. In contrast, gases are highly compressible. However, there are situations in which the density of a flowing gas remains constant enough that the flow can be considered incompressible.

Fluid flow can be viscous or nonviscous. A viscous fluid, such as honey, does not flow readily and is said to have a large viscosity. In contrast, water is less viscous and flows more readily; water has a smaller viscosity than honey. The flow of a viscous fluid is an energy-dissipating process. The viscosity hinders neighboring layers of fluid from sliding freely past one another. A fluid with zero viscosity flows in an unhindered manner with no dissipation of energy. Although no real fluid has zero viscosity at normal temperatures, some fluids have negligibly small viscosities. An incompressible, nonviscous fluid is called an *ideal fluid*.

Figure 11.23 Two fluid particles in a stream. At different locations in the stream the particle velocities may be different, as indicated by v_1 and v_2.

Figure 11.24 The flow of water in a rapids is an example of turbulent flow.

Figure 11.25 At any point along a streamline, the velocity vector of the fluid particle at that point is tangent to the streamline.

When the flow is steady, *streamlines* are often used to represent the trajectories of the fluid particles. A streamline is a line drawn in the fluid such that a tangent to the streamline at any point is parallel to the fluid velocity at that point. Figure 11.25 shows the velocity vectors at three points along a streamline. The fluid velocity can vary (in both magnitude and direction) from point to point along a streamline, but at any given point, the velocity is constant in time, as required by the condition of steady flow. In fact, steady flow is often called *streamline flow.*

Figure 11.26a illustrates a method for making streamlines visible by using small tubes to release a colored dye into the moving liquid. The dye does not immediately mix with the liquid and is carried along a streamline. In the case of a flowing gas, such as that in a wind tunnel, streamlines are often revealed by smoke streamers, as part b of the figure shows.

In steady flow, the pattern of streamlines is steady in time, and, as Figure 11.26a indicates, no two streamlines cross one another. If they did, every particle arriving at the crossing point could go one way or the other. This would mean that the velocity at the crossing point would change from moment to moment, a condition that does not exist in steady flow.

(a)

(b)

Figure 11.26 (*a*) In the steady flow of a liquid, a colored dye reveals the streamlines. (*b*) A smoke streamer reveals a streamline pattern for the air flowing around this pursuit cyclist, as he tests his bike for wind resistance in a wind tunnel.

11.8 THE EQUATION OF CONTINUITY

Have you ever used your thumb to control the water flowing from the end of a hose, as in Figure 11.27? If so, you have seen that the water velocity increases when your thumb reduces the cross-sectional area of the hose opening. This kind of fluid behavior is described by the *equation of continuity.* This equation expresses the following simple idea: If a fluid enters one end of a pipe at a certain rate (e.g., 5 kilograms per second), then fluid must also leave at the same rate, assuming that there are no places between the entry and exit points to add or remove fluid. The mass of fluid per second (e.g., 5 kg/s) that flows through a tube is called the *mass flow rate.*

Figure 11.28 shows a small mass of fluid or fluid element moving along a tube. Upstream at position 2, where the tube has a cross-sectional area A_2, the fluid has a speed v_2 and a density ρ_2. Downstream at location 1, the corresponding quantities are A_1, v_1, and ρ_1. During a small time interval Δt, the fluid at point 2 moves a distance of $v_2\Delta t$, as the drawing shows. The volume of fluid that has flowed past this point is the cross-sectional area times this distance or $A_2v_2\Delta t$. The mass Δm of this fluid element is the product of the density and volume: $\Delta m = \rho_2 A_2 v_2 \Delta t$. Dividing Δm by Δt gives the mass flow rate (the mass per second):

$$\begin{matrix}\text{Mass flow rate} \\ \text{at position 2}\end{matrix} = \frac{\Delta m}{\Delta t} = \rho_2 A_2 v_2 \qquad (11.7a)$$

Similar reasoning leads to the mass flow rate at position 1:

$$\begin{matrix}\text{Mass flow rate} \\ \text{at position 1}\end{matrix} = \rho_1 A_1 v_1 \qquad (11.7b)$$

Since no fluid can cross the sidewalls of the tube, the mass flow rates at positions 1 and 2 must be equal. But these positions were selected arbitrarily, so the mass flow rate has the same value everywhere in the tube, an important result known as the *equation of continuity.* The equation of continuity is an expression of the fact that mass is conserved (i.e., neither created nor destroyed) as it flows along the tube.

● EQUATION OF CONTINUITY

The mass flow rate ($\rho A v$) has the same value at every position along a tube that has a single entry and a single exit point for fluid flow. For two positions along such a tube

$$\rho_1 A_1 v_1 = \rho_2 A_2 v_2 \qquad (11.8)$$

where ρ = fluid density (kg/m^3)
   A = cross-sectional area of tube (m^2)
   v = fluid speed (m/s)

SI Unit of Mass Flow Rate: kg/s

Figure 11.27 When the end of a hose is partially closed off, thus reducing its cross-sectional area, the fluid velocity increases.

The density of an incompressible fluid does not change during flow, so that $\rho_1 = \rho_2$, and the equation of continuity reduces to

Incompressible fluid $\qquad\qquad A_1 v_1 = A_2 v_2 \qquad (11.9)$

The quantity Av represents the volume of fluid per second that passes through the tube and is referred to as the ***volume flow rate Q:***

$$Q = \text{volume flow rate} = Av \qquad (11.10)$$

Equation 11.9 shows that where the tube area is large, the fluid speed is small, and, conversely, where the tube area is small, the speed is large. Example 11 explores this behavior in more detail for the hose in Figure 11.27.

EXAMPLE 11   A Garden Hose

A garden hose has an unobstructed opening with a cross-sectional area of 2.85×10^{-4} m^2, from which water fills a bucket in 30.0 s. The volume of the bucket is 8.00×10^{-3} m^3 (about two gallons). Find the speed of the water that leaves the hose through (a) the unobstructed opening and (b) an obstructed opening that has only half as much area.

Reasoning If we can determine the volume flow rate Q, the speed of the water can be obtained from Equation 11.10 as $v = Q/A$, since the area A is given. The volume flow rate can be found from the volume of the bucket and the time required to fill it.

Solution

(a) The volume flow rate Q is equal to the volume of the bucket divided by the fill time. Therefore, according to Equation 11.10, the speed of the water is

$$v = \frac{Q}{A} = \frac{(8.00 \times 10^{-3} \text{ m}^3)/(30.0 \text{ s})}{2.85 \times 10^{-4} \text{ m}^2} = \boxed{0.936 \text{ m/s}} \qquad (11.10)$$

Figure 11.28 In general, a fluid flowing in a tube that has different cross-sectional areas A_1 and A_2 at positions 1 and 2 also has different velocities $\mathbf{v_1}$ and $\mathbf{v_2}$ at these positions.

(b) Water can be considered incompressible, so the equation of continuity can be applied in the form $A_1v_1 = A_2v_2$. Since $A_2 = \frac{1}{2}A_1$, we find that

$$v_2 = \left(\frac{A_1}{A_2}\right)v_1 = \left(\frac{A_1}{\frac{1}{2}A_1}\right)(0.936 \text{ m/s}) = \boxed{1.87 \text{ m/s}} \tag{11.9}$$

The next example applies the equation of continuity to the flow of blood.

EXAMPLE 12   A Clogged Artery

In the condition known as atherosclerosis, a deposit or atheroma forms on the arterial wall and reduces the opening through which blood can flow. In the carotid artery in the neck, blood flows three times faster through a partially blocked region than it does through an unobstructed region. Determine the ratio of the effective radii of the artery at the two places.

Reasoning Blood, like most liquids, is incompressible, and the equation of continuity in the form of Equation 11.9 can be applied. In applying this equation, we use the fact that the area of a circle is πr^2.

Solution From Equation 11.9, it follows that

$$\underbrace{(\pi r_U^2)v_U}_{\substack{\text{Unobstructed} \\ \text{volume flow rate}}} = \underbrace{(\pi r_O^2)v_O}_{\substack{\text{Obstructed} \\ \text{volume flow rate}}}$$

The ratio of the radii is

$$\frac{r_U}{r_O} = \sqrt{\frac{v_O}{v_U}} = \sqrt{3} = \boxed{1.7}$$

11.9   BERNOULLI'S EQUATION

● *Concepts at a Glance*   For *steady flow,* the speed, pressure, and elevation of an *incompressible and nonviscous* fluid are related by an equation discovered by Daniel Bernoulli (1700–1782). To derive **Bernoulli's equation,** we will use the work–energy theorem, as outlined in the Concepts-at-a-Glance chart in Figure 11.29. This theorem, which is introduced in Chapter 6, states that the net work W_{nc} done on an object by external nonconservative forces is equal to the change in the total mechanical energy of the object (see Equation 6.8). As mentioned earlier, the pressure within a fluid is caused by collisional forces, which are nonconservative. Therefore, when a fluid is accelerated because of a difference in pressures, work is being done by nonconservative forces ($W_{nc} \neq 0$), and this work changes the total mechanical energy of the fluid from an initial value of E_0 to a final value of E_f. The total mechanical energy is not conserved. We will now see how the work–energy theorem leads directly to Bernoulli's equation. ●

To begin with, let us make two observations about a moving fluid. First, whenever a fluid is flowing in a horizontal pipe and encounters a region of reduced cross-sectional area, the pressure of the fluid drops, as Figure 11.30a indicates. The reason for this follows from Newton's second law. When moving from the wider region 2 to the narrower region 1, the fluid speeds up or accelerates, consistent with the conservation of mass (as expressed by the equation of continuity). According to the second law, the accelerating fluid must be subjected to an unbalanced force. But there can be an unbalanced force only if the pressure in region 2 exceeds the pres-

Concepts at a Glance

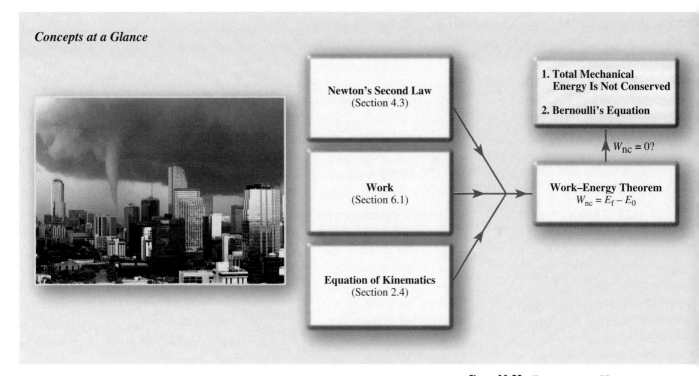

Figure 11.29 *Concepts at a Glance*
The work–energy theorem leads to
Bernoulli's equation when the net work
W_{nc} done by external nonconservative
forces is not zero. Bernoulli's equation
reveals that the pressure associated with
moving air, such as this tornado in Mi-
ami (May 12, 1997), is lower than that
of stationary air, such as that in the
buildings.

sure in region 1. We will see that the difference in pressures is given by Bernoulli's
equation. Our second observation is that if the fluid moves to a higher elevation, the
pressure at the lower level is greater than the pressure at the higher level, as in Figure 11.30*b*. The basis for this observation is our previous study of static fluids, and
Bernoulli's equation will confirm it, provided that the cross-sectional area of the
pipe does not change.

To derive Bernoulli's equation, consider Figure 11.31*a*. This drawing shows a
fluid element of mass *m*, upstream in region 2 of a pipe. Both the cross-sectional
area and the elevation are different at different places along the pipe. The speed,
pressure, and elevation in this region are v_2, P_2, and y_2, respectively. Downstream in region 1 these variables have the values v_1, P_1, and y_1. As Chapter 6
discusses, an object moving under the influence of gravity has a total mechanical
energy *E* that is the sum of the kinetic energy KE and the gravitational potential
energy PE: $E = KE + PE = \frac{1}{2}mv^2 + mgy$. When work W_{nc} is done on the fluid
element by external nonconservative forces, the total mechanical energy changes.

Figure 11.30 (*a*) In this horizontal pipe,
the pressure in region 2 is greater than
that in region 1. The difference in pressures leads to the net force that accelerates the fluid to the right. (*b*) When the
fluid changes elevation, the pressure at
the bottom is greater than that at the
top, assuming the cross-sectional area
of the pipe is constant.

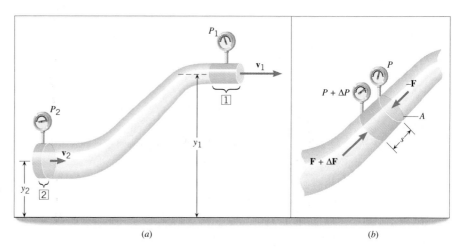

Figure 11.31 (*a*) A fluid element (dark blue) moving through a pipe whose cross-sectional area and elevation change. (*b*) The fluid element experiences a force $-\mathbf{F}$ on its top surface due to the fluid above it, and a force $\mathbf{F} + \Delta\mathbf{F}$ on its bottom surface due to the fluid below it.

According to the work–energy theorem, the work equals the change in the total mechanical energy:

$$W_{\text{nc}} = E_1 - E_2 = \underbrace{(\tfrac{1}{2}mv_1^2 + mgy_1)}_{\substack{\text{Total mechanical} \\ \text{energy in region 1}}} - \underbrace{(\tfrac{1}{2}mv_2^2 + mgy_2)}_{\substack{\text{Total mechanical} \\ \text{energy in region 2}}} \qquad (6.8)$$

Figure 11.31*b* helps us understand how the work W_{nc} arises. On the top surface of the fluid element, the surrounding fluid exerts a pressure P. This pressure gives rise to a force of magnitude $F = PA$, where A is the cross-sectional area. On the bottom surface, the surrounding fluid exerts a slightly greater pressure, $P + \Delta P$, where ΔP is the pressure difference between the ends of the element. As a result, the force on the bottom surface has a magnitude of $F + \Delta F = (P + \Delta P)A$. The magnitude of the *net* force pushing the fluid element up the tube is $\Delta F = (\Delta P)A$. When the fluid element moves through its own length s, the work done is the product of the magnitude of the net force and the distance, according to Equation 6.1: Work $= (\Delta F)s = (\Delta P)As$. The quantity As is the volume V of the element, so the work is $(\Delta P)V$. The total work done on the fluid element in moving it from region 2 to region 1 is the sum of the small increments of work $(\Delta P)V$ done as the element moves along the tube. This sum amounts to $W_{\text{nc}} = (P_2 - P_1)V$, where $P_2 - P_1$ is the pressure difference between the two regions. With this expression for W_{nc}, the work–energy theorem becomes

$$W_{\text{nc}} = (P_2 - P_1)V = (\tfrac{1}{2}mv_1^2 + mgy_1) - (\tfrac{1}{2}mv_2^2 + mgy_2)$$

By dividing both sides of this result by the volume V, recognizing that m/V is the density ρ of the fluid, and rearranging terms, we obtain Bernoulli's equation.

● **BERNOULLI'S EQUATION**

In the steady flow of a nonviscous, incompressible fluid of density ρ, the pressure P, the fluid speed v, and the elevation y at any two points (1 and 2) are related by

$$P_1 + \tfrac{1}{2}\rho v_1^2 + \rho g y_1 = P_2 + \tfrac{1}{2}\rho v_2^2 + \rho g y_2 \qquad (11.11)$$

Since the points 1 and 2 were selected arbitrarily, the term $P + \frac{1}{2}\rho v^2 + \rho gy$ has a constant value at all positions in the flow. For this reason, Bernoulli's equation is sometimes expressed as $P + \frac{1}{2}\rho v^2 + \rho gy = $ constant.

Equation 11.11 can be regarded as an extension of the earlier result that specifies how the pressure varies with depth in a static fluid ($P_2 = P_1 + \rho gh$), the terms $\frac{1}{2}\rho v_1^2$ and $\frac{1}{2}\rho v_2^2$ accounting for the effects of fluid speed. Bernoulli's equation reduces to the result for static fluids when the speed of the fluid is the same everywhere ($v_1 = v_2$), as it is when the cross-sectional area remains constant. Under such conditions, Bernoulli's equation is $P_1 + \rho gy_1 = P_2 + \rho gy_2$. After rearrangement, this result becomes

$$P_2 = P_1 + \rho g(y_1 - y_2) = P_1 + \rho gh$$

which is the result (Equation 11.4) for static fluids.

11.10 APPLICATIONS OF BERNOULLI'S EQUATION

When a moving fluid is contained in a horizontal pipe, all parts of it have the same elevation ($y_1 = y_2$), and Bernoulli's equation simplifies to

$$P_1 + \frac{1}{2}\rho v_1^2 = P_2 + \frac{1}{2}\rho v_2^2 \qquad (11.12)$$

Thus, the quantity $P + \frac{1}{2}\rho v^2$ remains constant throughout a horizontal pipe; if v increases, P decreases and vice versa. This is exactly the result that we deduced qualitatively from Newton's second law at the beginning of Section 11.9, and Conceptual Example 13 illustrates it.

CONCEPTUAL EXAMPLE 13   Tarpaulins and Bernoulli's Equation

A tarpaulin is a piece of canvas that is used to cover a cargo, like that pulled by the truck in Figure 11.32. When the truck is stationary the tarpaulin lies flat, but it bulges outward when the truck is speeding down the highway. Account for this behavior.

Reasoning and Solution The behavior is a direct consequence of the pressure changes that Bernoulli's equation describes for flowing fluids. When the truck is stationary, the air outside and inside the cargo area is stationary, so the air pressure is the same in both places. This pressure applies the same force to the outer and inner surfaces of the canvas, with the result that the tarpaulin lies flat. When the truck is moving, the outside air rushes over the top surface of the canvas. In accord with Bernoulli's equation (Equation 11.12), the moving air has a lower pressure than the stationary air within the cargo area. ***The greater inside pressure generates a greater force on the inner surface of the canvas, and the tarpaulin bulges outward.***

Related Homework Material: *Problems 57 and 58*

Figure 11.32 The tarpaulin that covers the cargo is flat when the truck is stationary but bulges outward when the truck is moving.

Example 14 applies Equation 11.12 to a dangerous physiological condition known as an aneurysm.

EXAMPLE 14   An Enlarged Blood Vessel

An aneurysm is an abnormal enlargement of a blood vessel such as the aorta. Suppose that, because of an aneurysm, the cross-sectional area A_1 of the aorta increases to a value $A_2 = 1.7A_1$. The speed of the blood ($\rho = 1060 \text{ kg/m}^3$) through a normal portion of the aorta is $v_1 = 0.40$ m/s. Assuming that the aorta is horizontal (the per-

THE PHYSICS OF...

an aneurysm.

son is lying down), determine the amount by which the pressure P_2 in the enlarged region exceeds the pressure P_1 in the normal region.

Reasoning According to the equation of continuity, the speed of the blood (incompressible) in the enlarged section of the artery is smaller than the speed in a healthy section. In turn, Bernoulli's equation for horizontal flow (Equation 11.12) indicates that the smaller speed leads to a greater pressure and may be used to find the pressure difference $P_2 - P_1$.

Solution From the equation of continuity we find the speed v_2 of the blood in the aneurysm to be

$$v_2 = \left(\frac{A_1}{A_2}\right) v_1 = \left(\frac{A_1}{1.7A_1}\right)(0.40 \text{ m/s}) = 0.24 \text{ m/s} \qquad (11.9)$$

From Bernoulli's equation for horizontal flow it follows that $P_1 + \frac{1}{2}\rho v_1^2 = P_2 + \frac{1}{2}\rho v_2^2$. Solving for $P_2 - P_1$, we find

$$P_2 - P_1 = \frac{1}{2}\rho(v_1^2 - v_2^2) = \frac{1}{2}(1060 \text{ kg/m}^3)[(0.40 \text{ m/s})^2 - (0.24 \text{ m/s})^2] = \boxed{54 \text{ Pa}}$$

This result is positive, indicating that P_2 is greater than P_1. The excess pressure puts added stress on the already weakened tissue of the arterial wall at the aneurysm.

THE PHYSICS OF...

household plumbing.

The impact of fluid flow on pressure is widespread. Figure 11.33, for instance, illustrates how household plumbing takes into account the implications of Bernoulli's equation. The U-shaped section of pipe beneath the sink is called a "trap," because it traps water, which serves as a barrier to prevent sewer gas from leaking into the house. Part *a* of the drawing shows poor plumbing. When water from the clotheswasher rushes through the sewer pipe, the high-speed flow causes the pressure at point *A* to drop. The pressure at point *B* in the sink, however, remains at the higher atmospheric pressure. As a result of this pressure difference, the water is pushed out of the trap and into the sewer line, leaving no protection against sewer gas. A correctly designed system is vented to the outside of the house, as in Figure 11.33*b*. The vent ensures that the pressure at *A* remains the same as that at *B* (atmospheric pressure), even when water from the clotheswasher is rushing through the pipe. Thus, the purpose of the vent is to prevent the trap from being emptied, not to provide an escape route for sewer gas.

One of the most spectacular examples of how fluid flow affects pressure is the dynamic lift on airplane wings. Figure 11.34*a* shows a wing (in cross section) moving to the right, with the air flowing past the wing to the left. Because of the shape

Figure 11.33 In a household plumbing system, a vent is necessary to equalize the pressures at points *A* and *B*, thus preventing the trap from being emptied. An empty trap allows sewer gas to enter the house.

(*a*) Without vent

(*b*) With vent

Figure 11.34 (*a*) Air flowing around an airplane wing. The wing is moving to the right. (*b*) The end of this wing has roughly the shape indicated in part *a*.

of the wing, the air travels faster over the curved upper surface than it does over the flatter lower surface. According to Bernoulli's equation, the pressure above the wing is lower (faster moving air), while the pressure below the wing is higher (slower moving air). Thus, the wing is lifted upward. Part *b* of the figure shows the wing of an airplane.

The curve ball, one of the main weapons in the arsenal of a baseball pitcher, is another illustration of the effects of fluid flow. Figure 11.35*a* shows a baseball moving to the right with no spin. The view is from above, looking down toward the ground. In this situation, air flows with the same speed around both sides of the ball, and the pressure is the same on both sides. No net force exists to make the ball curve to either side. However, when the ball is given a spin, the air close to its surface is dragged around with it; the air on one half of the ball is speeded up (lower pressure), while that on the other half is slowed down (higher pressure). Part *b* of the picture illustrates the effects of a counterclockwise spin. The baseball experiences a net deflection force and curves on its way from the pitcher's mound to the plate, as part *c* shows.

As a final application of Bernoulli's equation, Figure 11.36*a* shows a large tank from which water is emerging through a small pipe near the bottom. Bernoulli's equation can be used to determine the speed (called the efflux speed) at which the water leaves the pipe, as the next example shows.

THE PHYSICS OF...

airplane wings.

THE PHYSICS OF...

a curve ball.

Figure 11.35 These views of a baseball are from above, looking down toward the ground, with the ball moving to the right. (*a*) Without spin, the ball does not curve to either side. (*b*) A spinning ball curves in the direction of the deflection force. (*c*) The spin in part *b* causes the ball to curve as shown here.

EXAMPLE 15   **Efflux Speed**

The tank in Figure 11.36*a* is open to the atmosphere at the top. Find an expression for the speed of the liquid leaving the pipe at the bottom.

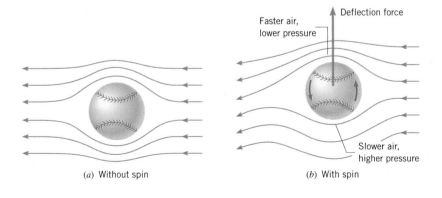

(*a*) Without spin   (*b*) With spin

(*c*)

(a)

(b)

Figure 11.36 (a) Bernoulli's equation
can be used to determine the speed of
the liquid leaving the small pipe. (b) An
ideal fluid (no viscosity) will rise to the
fluid level in the tank after leaving a
vertical outlet nozzle.

Reasoning We assume that the liquid behaves as an ideal
fluid. Therefore, we can apply Bernoulli's equation, and in
preparation for doing so, we locate two points in the liquid
in Figure 11.36a. Point 1 is just outside the efflux pipe, and
point 2 is at the top surface of the liquid. The pressure at
each of these points is equal to the atmospheric pressure, a
fact that will be used to simplify Bernoulli's equation.

Solution Since the pressures at points 1 and 2 are the same,
we have $P_1 = P_2$, and Bernoulli's equation becomes
$\frac{1}{2}\rho v_1^2 + \rho g y_1 = \frac{1}{2}\rho v_2^2 + \rho g y_2$. The density ρ can be eliminated algebraically from this result, which can then be
solved for the square of the efflux speed v_1:

$$v_1^2 = v_2^2 + 2g(y_2 - y_1) = v_2^2 + 2gh$$

We have substituted $h = y_2 - y_1$ for the height of the liquid
above the efflux tube. If the tank is very large, the liquid
level changes only slowly, and the speed at point 2 can be
set equal to zero, so that $\boxed{v_1 = \sqrt{2gh}}$.

In Example 15 the liquid is assumed to be an ideal fluid,
and the speed with which it leaves the pipe is the same as if
the liquid had freely fallen through a height h (see Equation
2.9 with $x = h$ and $a = g$). This result is known as
Torricelli's theorem. If the outlet pipe were pointed directly
upward, as in part b of the drawing, the liquid would rise to
a height h equal to the fluid level above the pipe. However,
if the liquid is not an ideal fluid, its viscosity cannot be neglected. Then, the efflux
speed would be less than that given by Bernoulli's equation, and the liquid would
rise to a height less than h.

*11.11 VISCOUS FLOW

In an ideal fluid there is no viscosity to hinder the fluid layers as they slide past one
another. Within a pipe of uniform cross section, every layer of an ideal fluid moves
with the same velocity, even the layer next to the wall, as Figure 11.37a shows.
When viscosity is present, the fluid layers have different velocities, as part b of the
drawing illustrates. The fluid at the center of the pipe has the greatest velocity. In
contrast, the fluid layer next to the wall surface does not move at all because it is
held tightly by intermolecular forces. So strong are these forces that if a solid surface moves, the adjacent fluid layer moves along with it and remains at rest *relative*
to the moving surface.

To help introduce viscosity in a quantitative fashion, Figure 11.38a shows a viscous fluid between two parallel plates. The top plate is free to move while the bottom one is stationary. If the top plate is to move with a velocity **v** relative to the bottom plate, a force **F** is required. For a highly viscous fluid, like thick honey, a large
force is needed; for a less viscous fluid, like water, a smaller one will do. As part b
of the drawing suggests, we may imagine the fluid to be composed of many thin
horizontal layers. When the top plate moves, the intermediate fluid layers slide over
each other. The velocity of each layer is different, changing uniformly from **v** at the

top plate to zero at the bottom plate. The resulting flow is called *laminar flow,* since a thin layer is often referred to as a lamina. As each layer moves, it is subjected to viscous forces from its neighbors. The purpose of the force **F** is to compensate for the effect of these forces, so that any layer can move with a constant velocity.

The amount of force required in Figure 11.38*a* depends on several factors. Larger areas *A*, being in contact with more fluid, require larger forces, so that the force is proportional to the contact area ($F \propto A$). For a given area, greater speeds require larger forces, with the result that the force is proportional to the speed ($F \propto v$). The force is also inversely proportional to the perpendicular distance *y* between the top and bottom plates ($F \propto 1/y$). The larger the distance *y*, the smaller is the force required to achieve a given speed with a given contact area. These three proportionalities can be expressed simultaneously in the following way: $F \propto Av/y$. Equation 11.13 expresses this relationship with the aid of a proportionality constant η (Greek letter *eta*), which is called the *coefficient of viscosity* or simply the *viscosity.*

Figure 11.37 (*a*) In ideal (nonviscous) fluid flow, all fluid particles across the pipe have the same velocity. (*b*) In viscous flow, the speed of the fluid is zero at the surface of the pipe and increases to a maximum along the center axis.

● **FORCE NEEDED TO MOVE A LAYER OF VISCOUS FLUID WITH A CONSTANT VELOCITY**

The tangential force **F** required to move a fluid layer at a constant speed *v*, when the layer has an area *A* and is located a perpendicular distance *y* from an immobile surface, is given by

$$F = \frac{\eta A v}{y} \qquad (11.13)$$

where η is the coefficient of viscosity.

SI Unit of Viscosity: Pa·s

Common Unit of Viscosity: poise (P)

By solving this equation for the viscosity, $\eta = Fy/(vA)$, it can be seen that the SI unit for viscosity is N·m/[(m/s)·m²] = Pa·s. Another common unit for viscosity is the *poise* (P), which is used in the cgs system of units and is named after the French physician Jean Poiseuille (1797–1869, pronounced, approximately, as Pwah-zoy′). The following relation exists between the two units:

1 poise (P) = 0.1 Pa·s

Values of viscosity depend on the nature of the fluid. Under ordinary conditions, the viscosities of liquids are significantly *larger* than those of gases. Moreover, the viscosities of either liquids or gases depend markedly on temperature. Usually, the viscosities of liquids decrease as the temperature is increased. Anyone who has heated honey or oil, for example, knows that these fluids flow much more freely at an elevated temperature. In contrast, the viscosities of gases increase as the temperature is raised. An ideal fluid has $\eta = 0$ P.

Viscous flow occurs in a wide variety of situations, such as oil moving through a pipeline or a liquid being forced through the needle of a hypodermic syringe. Figure 11.39 identifies the factors that determine the volume flow rate *Q* (in m³/s) of the fluid. First, a difference in pressures $P_2 - P_1$ must be maintained between any two locations along the pipe in order for the fluid to flow. In fact, *Q* is proportional to $P_2 - P_1$, a greater pressure difference leading to a larger flow rate. Second, a long pipe offers greater resistance to the flow than a short pipe does, and *Q* is in-

Figure 11.38 (*a*) A force **F** is applied to the top plate, which is in contact with a viscous fluid. (*b*) Because of the force **F**, the top plate and the adjacent layer of fluid move with a constant velocity **v**.

THE PHYSICS OF...

pipeline pumping stations.

Figure 11.39 For viscous flow, the difference in pressures $P_2 - P_1$, the radius R and length L of the tube, and the viscosity η of the fluid influence the volume flow rate.

Figure 11.40 As oil flows along the Alaskan pipeline, the pressure drops because oil is a viscous fluid. Pumping stations are located along the pipeline to compensate for the drop in pressure.

versely proportional to the length L. Because of this fact, long pipelines, such as the Alaskan pipeline, have pumping stations at various places along the line to compensate for a drop in pressure (see Figure 11.40). Third, high-viscosity fluids flow less readily than low-viscosity fluids do, and Q is inversely proportional to the viscosity η. Finally, the volume flow rate is larger in a pipe of larger radius, other things being equal. The dependence on the radius R is a surprising one, Q being proportional to the fourth power of the radius, or R^4. If, for instance, the pipe radius is reduced to one-half of its original value, the volume flow rate is reduced to one-sixteenth of its original value, assuming the other variables remain constant. The mathematical relation for Q in terms of these parameters was discovered by Poiseuille and is known as **Poiseuille's law.**

● POISEUILLE'S LAW

A fluid whose viscosity is η, flowing through a pipe of radius R and length L, has a volume flow rate Q given by

$$Q = \frac{\pi R^4 (P_2 - P_1)}{8 \eta L} \qquad (11.14)$$

where P_1 and P_2 are the pressures at the ends of the pipe.

Example 16 illustrates the use of Poiseuille's law.

EXAMPLE 16   Giving an Injection

THE PHYSICS OF...

a hypodermic syringe.

A hypodermic syringe is filled with a solution whose viscosity is 1.5×10^{-3} Pa·s. As Figure 11.41 shows, the plunger area of the syringe is 8.0×10^{-5} m^2, and the length of the needle is 0.025 m. The internal radius of the needle is 4.0×10^{-4} m. The gauge pressure in a vein is 1900 Pa (14 mm of mercury). What force must be applied to the plunger, so that 1.0×10^{-6} m^3 of solution can be injected in 3.0 s?

Reasoning The necessary force is the pressure applied to the plunger times the area of the plunger. Since viscous flow is occurring, the pressure is different at different points along the syringe. However, the barrel of the syringe is so wide that little pressure difference is required to sustain the flow up to point 2, where the fluid encounters the narrow needle. Consequently, the pressure applied to the plunger is nearly equal to the pressure P_2 at point 2. To find this pressure, we apply Poiseuille's law to the needle. This law indicates that $P_2 - P_1 = 8\eta L Q/(\pi R^4)$. We note that the pressure P_1 is given as a gauge pressure, which, in this case, is the amount of pressure in excess of atmospheric pressure. This causes no difficulty, because we need to find the amount of force in excess of that applied to the plunger by the atmosphere. The volume flow rate Q can be obtained from the time needed to inject the known volume of solution.

Solution The volume flow rate is $Q = (1.0 \times 10^{-6}$ m$^3)/(3.0$ s$) = 3.3 \times 10^{-7}$ m^3/s. According to Poiseuille's law (Equation 11.14), the required pressure difference is

$$P_2 - P_1 = \frac{8\eta L Q}{\pi R^4} = \frac{8(1.5 \times 10^{-3} \text{ Pa·s})(0.025 \text{ m})(3.3 \times 10^{-7} \text{ m}^3/\text{s})}{\pi (4.0 \times 10^{-4} \text{ m})^4} = 1200 \text{ Pa}$$

Since $P_1 = 1900$ Pa, the pressure P_2 must be $P_2 = 1200$ Pa $+ 1900$ Pa $= 3100$ Pa. The force that must be applied to the plunger is this pressure times the plunger area:

$$F = (3100 \text{ Pa})(8.0 \times 10^{-5} \text{ m}^2) = \boxed{0.25 \text{ N}}$$

11.12 CONCEPTS CALCULATIONS

In the behavior of fluids pressure plays an important role. As we have seen in this chapter, pressure is the magnitude of the force acting perpendicular to a surface divided by the area of the surface. Pressure should not be confused, however, with the force itself. The idea of a force is introduced in Chapter 4. The next example serves to emphasize that pressure and force are different concepts. In addition, it reviews the techniques that Chapters 4 and 9 discuss for analyzing objects that are in equilibrium.

Figure 11.41 The difference in pressure $P_2 - P_1$ required to sustain the fluid flow through a hypodermic needle can be found with the aid of Poiseuille's law.

CONCEPTS & CALCULATIONS EXAMPLE 17   Pressure and Force

Figure 11.42*a* shows a rear view of a loaded wheelbarrow on a horizontal surface. It has balloon tires and a weight of $W = 684$ N, which is uniformly distributed. The left tire has a contact area with the ground of $A_L = 6.6 \times 10^{-4}$ m^2, whereas the right tire is underinflated and has a contact area of $A_R = 9.9 \times 10^{-4}$ m^2. Find the force and the pressure that each tire applies to the ground.

Concept Questions and Answers Force is a vector. Therefore, both a direction and a magnitude are needed to specify it. Are both a direction and a magnitude needed to specify a pressure?

Answer No. Only a magnitude is needed to specify a pressure because pressure is a scalar quantity, not a vectory quantity. The definition of pressure is the magnitude of the force acting perpendicular to a surface divided by the area of the surface. This definition does not include directional information.

The problem asks for the force that each tire applies to the ground. How is this force related to the force that the ground applies to each tire?

Answer As Newton's law of action–reaction indicates, the force that a tire applies to the ground has the same magnitude and opposite direction as the force that the ground applies to the tire.

Do the left and right tires apply the same force to the ground?

Answer Since the weight is uniformly distributed, each tire supports half the load. Therefore, each tire applies the same force to the ground.

Do the left and right tires apply the same pressure to the ground?

Answer No. Pressure involves the contact area as well as the magnitude of the force. Since the force magnitudes are the same and the contact areas are different, the pressures are different, too.

Solution Since the wheelbarrow is at rest, it is in equilibrium and the net force acting on it must be zero, as Chapters 4 and 9 discuss. Referring to the free-body diagram in Figure 11.42*b* and taking upward as the positive direction, we have

$$\underbrace{\Sigma F_y}_{\text{Net force}} = F_L + F_R - W = 0$$

(a)   *(b)* Free-body diagram for the wheelbarrow

Figure 11.42 (*a*) Rear view of a wheelbarrow with balloon tires. (*b*) Free-body diagram of the wheelbarrow, showing its weight W and the forces F_L and F_R that the ground applies, respectively, to the left and right tires.

where F_L and F_R are, respectively, the forces that the ground applies to the left and right tires. At equilibrium the net torque acting on the wheelbarrow must also be zero, as Chapter 9 discusses. In the free-body diagram the weight acts at the center of gravity, which is at the center of the wheelbarrow. Using this point as the axis of rotation for torques, we can see that the forces F_L and F_R have the same lever arm ℓ, which is half the width of the wheelbarrow. For this axis, the weight W has a zero lever arm and, hence, a zero torque. Therefore, it does not contribute to the net torque. Remembering that counterclockwise torques are positive and setting the net torque equal to zero, we have

$$\underbrace{\Sigma\tau}_{\text{Net torque}} = F_R\ell - F_L\ell = 0 \quad \text{or} \quad F_R = F_L$$

Substituting this result into the force equation gives

$$F_L + F_L - W = 0$$

$$F_L = F_R = \tfrac{1}{2}W = \tfrac{1}{2}(684 \text{ N}) = 342 \text{ N}$$

The value of 342 N is the magnitude of the upward force that the ground applies to each tire. Using Newton's action–reaction law, we conclude that each tire applies to the ground a $\boxed{\text{downward force of 342 N}}$. Using this value and the given contact areas, we find that the left and right tires apply the following pressures P_L and P_R to the ground:

$$P_L = \frac{F}{A_L} = \frac{342 \text{ N}}{6.6 \times 10^{-4} \text{ m}^2} = \boxed{5.2 \times 10^5 \text{ Pa}}$$

$$P_R = \frac{F}{A_R} = \frac{342 \text{ N}}{9.9 \times 10^{-4} \text{ m}^2} = \boxed{3.5 \times 10^5 \text{ Pa}}$$

One manifestation of fluid pressure is the buoyant force that is applied to any object immersed in a fluid. Section 11.6 discusses this force, as well as Archimedes' principle, which specifies a convenient way to determine it. Example 18 focuses on the essence of this principle.

CONCEPTS & CALCULATIONS EXAMPLE 18   The Buoyant Force

A father (weight $W = 830$ N) and his daughter (weight $W = 340$ N) are spending the day at the lake. They each have beach balls, on which they are sitting, so that each ball is just submerged beneath the water. Ignoring the weight of the air within the balls and the parts of their legs that are underwater, find the radius of each ball.

Concept Questions and Answers Each person and his beach ball are in equilibrium, being stationary and having no acceleration. Thus, the net force acting on each is zero. What balances the downward-acting weight in each case?

Answer The downward-acting weight is balanced by the upward-acting buoyant force F_B that the water applies to the ball.

In which case is the buoyant force greater?

Answer The buoyant force acting on the father's beach ball is greater, since it must balance his greater weight.

In the situation described, what determines the magnitude of the buoyant force?

Answer According to Archimedes' principle, the magnitude of the buoyant force equals the weight of the fluid that the ball displaces. Since the ball is completely submerged, it displaces a volume of water that equals the ball's volume. The weight of this volume of water is the magnitude of the buoyant force.

Which beach ball has the larger radius?

Answer The father's ball has the larger volume and the larger radius. This follows because a bigger buoyant force acts on that ball. For the buoyant force to be bigger, that ball must displace a greater volume of water, according to Archimedes' principle. Therefore, the volume of that ball is larger, since the balls are completely submerged.

Solution For an object in equilibrium, the net force acting on the object must be zero. Therefore, taking upward to be the positive direction, we have

$$\underbrace{\Sigma F}_{\text{Net force}} = F_B - W = 0$$

Archimedes' principle specifies that the magnitude of the buoyant force is the weight of the water displaced by the ball. Using the definition of density given in Equation 11.1, the mass of the displaced water is $m = \rho V$, where $\rho = 1.00 \times 10^3 \text{ kg/m}^3$ is the density of water (see Table 11.1) and V is the volume displaced. Since all of the ball is submerged, $V = \frac{4}{3}\pi r^3$, assuming that the ball remains roughly spherical. The weight of the displaced water is $mg = \rho(\frac{4}{3}\pi r^3)g$. With this value for the buoyant force, the force equation becomes

$$F_B - W = \rho(\tfrac{4}{3}\pi r^3)g - W = 0$$

Solving for the radius we find that

Father $\quad r = \sqrt[3]{\dfrac{3W}{4\pi\rho g}} = \sqrt[3]{\dfrac{3(830 \text{ N})}{4\pi(1.00 \times 10^3 \text{ kg/m}^3)(9.80 \text{ m/s}^2)}} = \boxed{0.27 \text{ m}}$

Daughter $\quad r = \sqrt[3]{\dfrac{3W}{4\pi\rho g}} = \sqrt[3]{\dfrac{3(340 \text{ N})}{4\pi(1.00 \times 10^3 \text{ kg/m}^3)(9.80 \text{ m/s}^2)}} = \boxed{0.20 \text{ m}}$

At the end of the problem set for this chapter, you will find homework problems that contain both conceptual and quantitative parts. These problems are grouped under the heading *Concepts & Calculations, Group Learning Problems*. They are designed for use by students working alone or in small learning groups. The conceptual part of each problem provides a convenient focus for group discussions.

SUMMARY

Fluids are materials that can flow. The **mass density** ρ of any substance is its mass m divided by its volume V: $\rho = m/V$.

In the presence of gravity, the upper layers of a fluid push downward on the layers beneath, with the result that **fluid pressure is related to depth.** In an incompressible static fluid whose density is ρ, the relation is $P_2 = P_1 + \rho gh$, where P_1 is the pressure at one level and P_2 is the pressure at a level that is h meters deeper. The **gauge pressure** is the amount by which a pressure P differs from atmospheric pressure. The **absolute pressure** is the actual value for P.

According to **Pascal's principle,** any change in the pressure applied to a completely enclosed fluid is transmitted undiminished to all parts of the fluid and the enclosing walls.

The **buoyant force** is the net upward force that a fluid applies to any object that is immersed partially or completely in it. **Archimedes' principle** states that the magnitude of the buoyant force equals the weight of the fluid that the immersed object displaces.

The **mass flow rate** (in kg/s) of a fluid with a density ρ, flowing with a speed v in a pipe of cross-sectional area A, is the mass per second flowing past a point and is given by ρAv. The **equation of continuity** expresses the fact that mass is conserved; what flows into one end of a pipe flows out the other end, assuming there are no additional entry or exit points in between. In terms of the mass flow rate, the equation of continuity is $\rho_1 A_1 v_1 = \rho_2 A_2 v_2$, where the subscripts 1 and 2 denote two points along the pipe. If the fluid is incompressible, $\rho_1 = \rho_2$. The equation of continuity then becomes $A_1 v_1 = A_2 v_2$, where the product of the cross-sectional area and speed is called the **volume flow rate** Q; $Q = Av$.

Bernoulli's equation is a direct consequence of the work–energy theorem and describes the steady flow of an ideal fluid whose density is ρ. For any two points (1 and 2) in the fluid, this equation relates the pressure P, the speed v, and the elevation y: $P_1 + \frac{1}{2}\rho v_1^2 + \rho gy_1 = P_2 + \frac{1}{2}\rho v_2^2 + \rho gy_2$. When the flow is horizontal, Bernoulli's equation indicates that higher fluid speeds are associated with lower fluid pressures.

The **coefficient of viscosity** η is the proportionality constant that determines how much tangential force $\mathbf{F}$ is required to move a fluid layer at a constant speed v, when the layer has an area A and is located a perpendicular distance y from an immobile surface. The magnitude of the force is $F = \eta Av/y$. The SI unit of viscosity is Pa·s. A viscous fluid, flowing through a pipe of radius R and length L, has a volume flow rate Q given by $Q = \pi R^4 (P_2 - P_1)/(8\eta L)$, where P_1 and P_2 are the pressures at the ends of the pipe. This equation is known as **Poiseuille's law.**

CONCEPTUAL QUESTIONS

1. A pile of empty aluminum cans has a volume of 1.0 m³. The density of aluminum is 2700 kg/m³. Explain why the mass of the pile is *not* $\rho_{Al} V = (2700 \text{ kg/m}^3)(1.0 \text{ m}^3) = 2700$ kg.

2. A method for resealing a partially full bottle of wine under a vacuum uses a specially designed rubber stopper to close the bottle. A simple pump is attached to the stopper, and to remove air from the bottle, the plunger of the pump is pulled up and then released. After about 15 pull-and-release cycles the wine is under a partial vacuum. On the 15th pull-and-release cycle, it is harder to pull up the plunger than on the first cycle. Why?

3. A person could not balance her entire weight on the pointed end of a single nail, because it would penetrate her skin. However, she can lie safely on a "bed of nails." A "bed of nails" consists of many nails driven through a sheet of wood so that the pointed ends form a flat array. Why is the "bed of nails" trick safe?

4. ⚖ As you climb a mountain, your ears "pop" because of the changes in atmospheric pressure. In which direction does your eardrum move (a) as you climb up and (b) as you climb down? Give your reasoning.

5. A bottle of juice is sealed under partial vacuum, with a lid on which a red dot or "button" is painted. Around the button the following phrase is printed: "Button pops up when seal is broken." Explain why the button remains pushed in when the seal is intact.

6. A sealed container of water is full, except for a tube that is attached to it, as the drawing shows. Water is poured into this tube, and before the tube is full the container bursts. Explain why.

7. A closed tank is completely filled with water. A valve is then opened at the bottom of the tank and water begins to flow out. When the water stops flowing, will the tank be completely empty, or will there still be a noticeable amount of water in it? Give a reason for your answer.

8. Could you use a straw to sip a drink on the moon? Justify your answer.

9. Why does the cork fly out with a loud "pop"   *Question 6*
when a bottle of champagne is opened? To answer
this question on an exam a student says, "Because the gas pressure in the bottle is about 0.3×10^5 Pa." Explain whether the student is referring to absolute or gauge pressure.

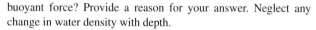

10. A scuba diver is below the surface of the water when a storm approaches, dropping the air pressure above the water. Would a sufficiently sensitive pressure gauge attached to his wrist register this drop in air pressure? Give your reasoning.

11. A steel beam is suspended completely under water by a cable that is attached to one end of the beam, so it hangs vertically. Another identical beam is also suspended completely under water, but by a cable that is attached to the center of the beam so it hangs horizontally. Which beam, if either, experiences the greater buoyant force? Provide a reason for your answer. Neglect any change in water density with depth.

12. A glass beaker, filled to the brim with water, is resting on a scale. A block is placed in the water, causing some of it to spill over. The water that spills is wiped away, and the beaker is still filled to the brim. How do the initial and final readings on the scale compare if the block is made from (a) wood and (b) iron? In both cases, explain your answer, using data from Table 11.1 and Archimedes' principle.

13. On a distant planet the acceleration due to gravity is less than it is on earth. Would you float more easily in water on this planet than on earth? Account for your answer.

14. Put an ice cube in a glass, and fill the glass to the brim with water. When the ice cube melts and the temperature of the water returns to its initial value, would the water level drop, remain the same, or rise (causing water to spill out)? Explain what you would observe in terms of Archimedes' principle.

15. As a person dives toward the bottom of a swimming pool, the pressure increases noticeably. Does the buoyant force also increase? Justify your answer. Neglect any change in water density with depth.

16. Suppose that an orbiting space station of the future had a swimming pool in it. If there is no artificial gravity, would a buoyant force be exerted on a swimmer? Explain.

17. The drawing shows a side view of part of a system of gates that is being installed to protect the city of Venice, Italy, against flooding due to very high tides. When not in use the hollow, triangular-shaped gate is filled with seawater and rests on the ocean floor. The gate is anchored to the ocean floor by a hinge mechanism, about which it can rotate. When a dangerously high tide is expected, water is pumped out of the gate, and the structure rotates into a position where it can protect against flooding. Explain why the gate rotates into position.

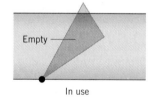

18. In steady flow, the velocity **v** of a fluid particle at any point is constant in time. On the other hand, a fluid accelerates when it moves into a region of smaller cross-sectional area. (a) Explain what causes the acceleration. (b) Explain why the condition of steady flow does not rule out such an acceleration.

19. The cross-sectional area of a stream of water becomes smaller as the water falls from a faucet. Account for this phenomenon in terms of the equation of continuity. What would you expect to happen to the cross-sectional area when the water is shot upward, as it is in a fountain?

20. Have you ever had a large truck pass you from the opposite direction on a narrow two-lane road? You probably noticed that your car was pulled toward the truck as it passed. What can you conclude about the speed of the air between your car and the truck compared to that on the opposite side of the car? Provide a reason for your answer.

21. Can Bernoulli's equation describe the flow of water that is cascading down a rock-strewn spillway? Explain.

22. Hold two sheets of paper by adjacent corners, so that they hang downward. They should be parallel and slightly separated, so that you can see the floor through the gap between them. Blow air strongly down through the gap. Do the sheets move further apart, or do they come closer together? Discuss what you observe in terms of Bernoulli's equation.

23. Which way would you have to spin a baseball, so that it curves upward on its way to the plate? In describing the spin, state how you are viewing the ball. Justify your answer.

24. You are traveling on a train with your window open. As the train approaches its rather high operating speed, your ears "pop."

Your eardrums respond to a decrease or increase in the air pressure by "popping" outward or inward, respectively. Do your ears "pop" outward or inward on the train? Give your reasoning in terms of Bernoulli's equation.

25. A passenger is smoking in the backseat of a moving car. To remove the smoke, the driver opens his window just a bit. Explain why the smoke is drawn to and out of the driver's window.

26. The airport in Phoenix, Arizona, has occasionally been closed to large planes because of weather conditions that do not entail storms or low visibility. Instead, the conditions combine to create an unusually low air density. Using Bernoulli's equation, explain why such an air density would make it difficult for a large, heavy plane to take off, especially if the runway were not exceptionally long.

27. To change the oil in a car, you remove a plug beneath the engine and let the old oil run out. Your car has been sitting in the garage on a cold day. Before changing the oil, it is advisable to run the engine for a while. Why?

PROBLEMS

ssm Solution is in the Student Solutions Manual. **www** Solution is available on the World Wide Web at http://www.wiley.com/college/cutnell
⚕ This icon represents a biomedical application.

Section 11.1 Mass Density

1. ssm The ice on a lake is 0.010 m thick. The lake is circular, with a radius of 480 m. Find the mass of the ice.

2. Neutron stars consist only of neutrons and have unbelievably high densities. A typical mass and radius for a neutron star might be 2.7×10^{28} kg and 1.2×10^3 m. (a) Find the density of such a star. (b) If a dime ($V = 2.0 \times 10^{-7}$ m^3) were made from this material, how much would it weigh (in pounds)?

3. A two-lane highway is 150 km in length and has a width of 7.0 m. The solid concrete roadbed has a depth of 0.30 m. Find the weight of the concrete in the highway. (For comparison, the weight of a typical car is about 1.3×10^5 N.)

4. One end of a wire is attached to a ceiling, and a solid brass ball is tied to the lower end. The tension in the wire is 120 N. What is the radius of the brass ball?

5. ssm Accomplished silver workers in India can pound silver into incredibly thin sheets, as thin as 3.00×10^{-7} m (about one hundredth of the thickness of this sheet of paper). Find the area of such a sheet that can be formed from 1.00 kg of silver.

***6.** A bar of gold measures 0.15 m × 0.050 m × 0.050 m. How many gallons of water have the same mass as this bar?

***7.** An irregularly shaped chunk of concrete has a hollow spherical cavity inside. The mass of the chunk is 33 kg, and the volume enclosed by the outside surface of the chunk is 0.025 m^3. What is the radius of the spherical cavity?

***8.** Planners of an experiment are evaluating the design of a helium-filled (0 °C, 1 atm pressure) sphere of radius R. Ultrathin silver foil of thickness T is used to make the sphere, and the designers claim that the mass of helium in the sphere equals the mass of silver used. Assuming that T is much less than R, calculate the ratio T/R for such a sphere.

****9. ssm www** An antifreeze solution is made by mixing ethylene glycol ($\rho = 1116$ kg/m^3) with water. Suppose the specific gravity of such a solution is 1.0730. Assuming that the total volume of the solution is the sum of its parts, determine the volume percentage of ethylene glycol in the solution.

Section 11.2 Pressure

10. An airtight box has a removable lid of area 1.3×10^{-2} m^2 and negligible weight. The box is taken up a mountain where the air pressure outside the box is 0.85×10^5 Pa. The inside of the box is completely evacuated. What is the magnitude of the force required to pull the lid off the box?

11. The circular top of a can of soda has a radius of 0.0320 m. The pull-tab has an area of 3.80×10^{-4} m^2. The absolute pressure of the carbon dioxide in the can is 1.40×10^5 Pa. Find the force that this gas generates (a) on the top of the can (including the pull-tab) and (b) on the pull-tab itself.

12. United States currency is printed using intaglio presses that generate a printing pressure of 8.0×10^4 lb/in.^{2} A twenty-dollar bill is 6.1 in. by 2.6 in. Calculate the magnitude of the force that the printing press applies to one side of the bill.

13. ssm High-heeled shoes can cause tremendous pressure to be applied to a floor. Suppose the radius of a heel is 6.00×10^{-3} m. At times during a normal walking motion, nearly the entire body weight acts perpendicular to the surface of such a heel. Find the pressure that is applied to the floor under the heel because of the weight of a 50.0-kg woman.

14. A storm approaches, and the air pressure outside your apartment suddenly drops to 0.960×10^5 Pa from 1.013×10^5 Pa. Before the pressure inside your apartment has had time to change, what is the magnitude of the net force exerted on a window that measures 2.0 m $\times$ 3.1 m?

15. A brick weighs 17.8 N and is resting on the ground. Its dimensions are 0.203 m $\times$ 0.0890 m $\times$ 0.0570 m. A number of the bricks are then stacked on top of this one. What is the smallest number of whole bricks (including the one on the ground) that could be used, so that their weight creates a pressure of at least one atmosphere on the ground beneath the first brick? (*Hint: First decide which face of the brick is in contact with the ground.*)

*16. A suitcase (mass $m = 16$ kg) is resting on the floor of an elevator. The part of the suitcase in contact with the floor measures 0.50 m by 0.15 m. The elevator is moving upward, the magnitude of its acceleration being 1.5 m/s^2. What pressure (in excess of atmospheric pressure) is applied to the floor beneath the suitcase?

*17. ssm A cylinder (with circular ends) and a hemisphere are solid throughout and made from the same material. They are resting on the ground, the cylinder on one of its ends and the hemisphere on its flat side. The weight of each causes the same pressure to act on the ground. The cylinder is 0.500 m high. What is the radius of the hemisphere?

**18. The drawing shows a cylinder fitted with a piston that has a mass m_1 of 0.500 kg and a radius of 2.50×10^{-2} m. The top of the piston is open to the atmosphere. The pressure beneath the piston is maintained at a reduced (but constant) value by means of the pump. As shown, a rope of negligible mass is attached to the piston and passes over two massless pulleys. The other end of the rope is attached to a block that has a mass of $m_2 = 9.50$ kg. The block falls from rest down through a distance of 1.25 m in 3.30 s. Ignoring friction, find the absolute pressure beneath the piston.

Section 11.3 Pressure and Depth in a Static Fluid, Section 11.4 Pressure Gauges

19. ssm Some researchers believe that the dinosaur Barosaurus held its head erect on a long neck, much as a giraffe does. If so, fossil remains indicate that its heart would have been about 12 m below its brain. Assume that the blood has the density of water, and calculate the amount by which the blood pressure in the heart would have exceeded that in the brain. Size estimates

for the single heart needed to withstand such a pressure range up to two tons. Alternatively, Barosaurus may have had a number of smaller hearts.

20. Review Conceptual Example 6 as an aid in understanding this problem. Consider the pump on the right side of Figure 11.11, which acts to reduce the air pressure in the pipe. The air pressure outside the pipe is one atmosphere. Find the maximum depth from which this pump can extract water from the well.

21. The Mariana trench is located in the Pacific Ocean at a depth of about 11 000 m below the surface of the water. The density of seawater is 1025 kg/m^3. (a) If an underwater vehicle were to explore such a depth, what force would the water exert on the vehicle's observation window (radius = 0.10 m)? (b) For comparison, determine the weight of a jetliner whose mass is 1.2×10^5 kg.

22. Measured along the surface of the water, a rectangular swimming pool has a length of 15 m. Along this length, the flat bottom of the pool slopes downward at an angle of 11° below the horizontal, from one end to the other. By how much does the pressure at the bottom of the deep end exceed the pressure at the bottom of the shallow end?

23. ssm The main water line enters a house on the first floor. The line has a gauge pressure of 1.90×10^5 Pa. (a) A faucet on the second floor, 6.50 m above the first floor, is turned off. What is the gauge pressure at this faucet? (b) How high could a faucet be before no water would flow from it, even if the faucet were open?

24. As background for this problem, review Conceptual Example 6. A submersible pump is put under the water at the bottom of a well and is used to push water up through a pipe. What minimum output gauge pressure must the pump generate to make the water reach the nozzle at ground level, 71 m above the pump?

25. The drawing shows an intravenous feeding. With the distance shown, nutrient solution ($\rho = 1030$ kg/m^3) can just barely enter the blood in the vein. What is the gauge pressure of the venous blood? Express your answer in millimeters of mercury.

Problem 25

*26. A mercury barometer reads 747.0 mm on the roof of a building and 760.0 mm on the ground. Assuming a constant value of 1.29 kg/m^3 for the density of air, determine the height of the building.

*27. ssm A tube is sealed at both ends and contains a 0.0100-m long portion of liquid. The length of the tube is large compared to 0.0100 m. There is no air in the tube, and the vapor in the space above the liquid may be ignored. The tube is whirled around in a horizontal circle at a constant angular speed. The axis of the rotation passes through one end of the tube, and during the motion, the liquid collects at the other end. The pressure experienced by the liquid is the same as it would experience at the bottom of the tube, if the tube were completely filled with liquid and allowed to hang vertically. Find the angular speed (in rad/s) of the tube.

*28. Figure 11.12 shows a mercury barometer. Consider two barometers, one using mercury and another using an unknown liquid. Suppose that the pressure above the liquid in each tube is maintained at the same value P, between zero and atmospheric pressure. The height of the unknown liquid is 16 times greater than the height of the mercury. Find the density of the unknown liquid.

*29. Two identical containers are open at the top and are connected at the bottom via a tube of negligible volume and a valve which is closed. Both containers are filled initially to the same height of 1.00 m, one with water, the other with mercury, as the drawing indicates. The valve is then opened. Water and mercury are immiscible. Determine the fluid level in the left container when equilibrium is reestablished.

**30. As the drawing illustrates, a pond has the shape of an inverted cone with the tip sliced off and has a depth of 5.00 m. The atmospheric pressure above the pond is 1.01×10^5 Pa. The circular top surface (radius = R_2) and circular bottom surface (radius = R_1) of the pond are both parallel to the ground. The magnitude of the force acting on the top surface is the same as the magnitude of the force acting on the bottom surface. Obtain (a) R_2 and (b) R_1.

Section 11.5 Pascal's Principle

31. ssm The atmospheric pressure above a swimming pool changes from 755 to 765 mm of mercury. The bottom of the pool is a 12 m × 24 m rectangle. By how much does the force on the bottom of the pool increase?

32. In a car lift the output plunger has a radius of 9.0 cm. The weight of the plunger and the car is 21 600 N. What is the gauge pressure of the hydraulic oil used in operating the lift?

33. In the hydraulic press used in a trash compactor, the radii of the input piston and the output plunger are 6.4×10^{-3} m and 5.1×10^{-2} m, respectively. The height difference between the input piston and the output plunger can be neglected. What force is applied to the trash when the input force is 330 N?

34. A dentist's chair with a patient in it weighs 2100 N. The output plunger of a hydraulic system begins to lift the chair, when the dentist's foot applies a force of 55 N to the input piston. Neglect any height difference between the plunger and the piston. What is the ratio of the radius of the plunger to the radius of the piston?

*35. ssm A dump truck uses a hydraulic cylinder, as the drawing illustrates. When activated by the operator, a pump injects hydraulic oil into the cylinder at an absolute pressure of 3.54×10^6 Pa and drives the output plunger, which has a radius of 0.150 m. Assuming the plunger remains perpendicular to the floor of the load bed, find the torque that the plunger creates about the axis identified in the drawing.

*36. The drawing shows a hydraulic chamber in which a spring (spring constant = 1600 N/m) is attached to the input piston, and a rock of mass 40.0 kg rests on the output plunger. The piston and plunger are nearly at the same height, and each has a negligible mass. By how much is the spring compressed from its unstrained position?

*37. The drawing shows a hydraulic system used with disc brakes. The force **F** is applied perpendicularly to the brake pedal. The pedal rotates about the axis shown in the drawing and causes a force to be applied perpendicularly to the input piston (radius = 9.50×10^{-3} m) in the master cylinder. The resulting pressure is transmitted by the brake fluid to the output plungers (radii = 1.90×10^{-2} m), which are covered with the brake linings. The linings are pressed against both sides of a disc attached to the rotating wheel. Suppose that the magnitude of **F** is 9.00 N. Assume that the input piston and the output plungers are at the same vertical level and find the force applied to each side of the rotating disc.

Section 11.6 Archimedes' Principle

38. A bodybuilder is holding a 29-kg steel barbell above her head. How much force would she have to exert if the barbell were lifted underwater?

39. ssm A 0.10-m × 0.20-m × 0.30-m block is suspended from a wire and is completely under water. What buoyant force acts on the block?

40. A solid object floats on ethyl alcohol, with 68.2% of the object's volume submerged. Using Table 11.1, identify the substance from which the object is made.

41. What is the total mass of swimmers that the raft in Example 8 can carry and float with its top surface at water level?

42. A paperweight, when weighed in air, has a weight of $W = 6.9$ N. When completely immersed in water, however, it has a weight of $W_{in\ water} = 4.3$ N. Find the volume of the paperweight.

43. ssm www A person can change the volume of his body by taking air into his lungs. The amount of change can be determined by weighing the person under water. Suppose that under water a person weighs 20.0 N with partially full lungs and 40.0 N with empty lungs. Find the change in body volume.

*44. What is the smallest number of whole logs ($\rho = 725$ kg/m^3, radius $= 0.0800$ m, length $= 3.00$ m) that can be used to build a raft that will carry four people, each of whom has a mass of 80.0 kg?

*45. A hollow cubical box is 0.30 m on an edge. This box is floating in a lake with one-third of its height beneath the surface. The walls of the box have a negligible thickness. Water is poured into the box. What is the depth of the water in the box at the instant the box begins to sink?

*46. To verify her suspicion that a rock specimen is hollow, a geologist weighs the specimen in air and in water. She finds that the specimen weighs twice as much in air as it does in water. The solid part of the specimen has a density of 5.0×10^3 kg/m^3. What fraction of the specimen's apparent volume is solid?

**47. ssm A solid cylinder (radius $= 0.150$ m, height $= 0.120$ m) has a mass of 7.00 kg. This cylinder is floating in water. Then oil ($\rho = 725$ kg/m^3) is poured on top of the water until the situation shown in the drawing results. How much of the height of the cylinder is in the oil?

Problem 47

**48. A lighter-than-air balloon and its load of passengers and ballast are floating stationary above the earth. Ballast is weight (of negligible volume) that can be dropped overboard to make the balloon rise. The radius of this balloon is 6.25 m. Assuming a constant value of 1.29 kg/m^3 for the density of air, determine how much weight must be dropped overboard to make the balloon rise 105 m in 15.0 s.

**49. A spring is attached to the bottom of an empty swimming pool, with the axis of the spring oriented vertically. An 8.00-kg block of wood ($\rho = 840$ kg/m^3) is fixed to the top of the spring and compresses it. Then the pool is filled with water, completely covering the block. The spring is now observed to be stretched twice as much as it had been compressed. Determine the percentage of the block's total volume that is hollow. Ignore any air in the hollow space.

Section 11.8 The Equation of Continuity

50. Oil is flowing with a speed of 1.22 m/s through a pipeline with a radius of 0.305 m. How many gallons of oil (1 gal $= 3.79 \times 10^{-3}$ m^3) flow in one day?

51. ssm ⚲ A patient recovering from surgery is being given fluid intravenously. The fluid has a density of 1030 kg/m^3, and

9.5×10^{-4} m^3 of it flows into the patient every six hours. Find the mass flow rate in kg/s.

52. A room has a volume of 120 m^3. An air-conditioning system is to replace the air in this room every twenty minutes, using ducts that have a square cross section. Assuming that air can be treated as an incompressible fluid, find the length of a side of the square if the air speed within the ducts is (a) 3.0 m/s and (b) 5.0 m/s.

53. ⚲ Suppose that blood flows through the aorta with a speed of 0.35 m/s. The cross-sectional area of the aorta is 2.0×10^{-4} m^2. (a) Find the volume flow rate of the blood. (b) The aorta branches into tens of thousands of capillaries whose total cross-sectional area is about 0.28 m^2. What is the average blood speed through them?

*54. In an adjustable nozzle for a garden hose, a cylindrical plug is aligned along the axis of the hose and can be inserted into the hose opening. The purpose of the plug is to change the speed of the water leaving the hose. The speed of the water passing around the plug is to be three times greater than the speed of the water before it encounters the plug. Find the ratio of the plug radius to the inside hose radius.

*55. ssm A water line with an internal radius of 6.5×10^{-3} m is connected to a shower head that has 12 holes. The speed of the water in the line is 1.2 m/s. (a) What is the volume flow rate in the line? (b) At what speed does the water leave one of the holes (effective hole radius $= 4.6 \times 10^{-4}$ m) in the head?

Section 11.9 Bernoulli's Equation,
Section 11.10 Applications of Bernoulli's Equation

56. ⚲ One way to administer an inoculation is with a "gun" that shoots the vaccine through a narrow opening. No needle is necessary, for the vaccine emerges with sufficient speed to pass directly into the tissue beneath the skin. The speed is high, because the vaccine ($\rho = 1100$ kg/m^3) is held in a reservoir where a high pressure pushes it out. The pressure on the surface of the vaccine in one gun is 4.1×10^6 Pa above the atmospheric pressure outside the narrow opening. The dosage is small enough that the vaccine's surface in the reservoir is nearly stationary during an inoculation. The vertical height between the vaccine's surface in the reservoir and the opening can be ignored. Find the speed at which the vaccine emerges.

57. ssm Review Conceptual Example 13 as an aid in understanding this problem. Suppose that a 15-m/s wind is blowing across the roof of your house. The density of air is 1.29 kg/m^3. (a) Determine the reduction in pressure (below atmospheric pressure of stationary air) that accompanies this wind. (b) Explain why some roofs are "blown outward" during high winds.

58. Review Conceptual Example 13 before attempting this problem. The truck in that example is traveling at 27 m/s. The density of air is 1.29 kg/m^3. By how much does the pressure inside the cargo area beneath the tarpaulin exceed the outside pressure?

59. ssm An airplane wing is designed so that the speed of the air across the top of the wing is 251 m/s when the speed of the air below the wing is 225 m/s. The density of the air is 1.29 kg/m^3. What is the lifting force on a wing of area 24.0 m^2?

60. The hold of a boat is the interior space beneath the deck. It is open to the atmosphere. On a lake, the hull of a boat has a hole in it, through which water leaks into the hold. The effective area of the hole is 8.0×10^{-3} m^2 and is located 2.0 m beneath the lake surface. At what volume flow rate (in gal/s) does water leak into the boat? Note that 1 gal $= 3.785 \times 10^{-3}$ m^3. (*Hint: This situation is like that shown in Figure 11.36a.*)

61. The water tower in the drawing is drained by a pipe that extends to the ground. The flow is nonviscous. (a) What is the absolute pressure at point 1 if the valve is *closed,* assuming that the top surface of the water at point 2 is at atmospheric pressure. (b) What is the absolute pressure at point 1 when the valve is opened and the water is flowing? Assume that the water speed at point 2 is negligible. (c) Assuming the effective cross-sectional area of the valve opening is 2.00×10^{-2} m^2, find the volume flow rate at point 1.

15.0 m

*62. Water is running out of a faucet, falling straight down, with an initial speed of 0.50 m/s. At what distance below the faucet is the radius of the stream reduced to one-half its value at the faucet?

*63. The construction of a flat rectangular roof (4.0 m $\times$ 5.5 m) allows it to withstand a maximum net outward force of 21 000 N. The density of the air is 1.29 kg/m^3. At what wind speed will this roof blow outward?

*64. An airplane has an effective wing surface area of 16 m^2 that is generating the lift force. In level flight the air speed over the top of the wings is 62.0 m/s, while the air speed beneath the wings is 54.0 m/s. What is the weight of the plane?

*65. ssm A Venturi meter is a device for measuring the speed of a fluid within a pipe. The drawing shows a gas flowing at speed v_2 through a horizontal section of pipe whose cross-sectional area is $A_2 = 0.0700$ m^2. The gas has a density of $\rho = 1.30$ kg/m^3. The Venturi meter has a cross-sectional area of $A_1 = 0.0500$ m^2 and has been substituted for a section of the larger pipe. The pressure difference between the two sections is $P_2 - P_1 = 120$ Pa. Find (a) the speed v_2 of the gas in the larger original pipe and (b) the volume flow rate Q of the gas.

Venturi meter

*66. A liquid is flowing through a horizontal pipe whose radius is 0.0200 m. The pipe bends straight upward through a height

of 10.0 m and joins another horizontal pipe whose radius is 0.0400 m. What volume flow rate will keep the pressures in the two horizontal pipes the same?

*67. A pump and its horizontal intake pipe are located 12 m beneath the surface of a reservoir. The speed of the water in the intake pipe causes the pressure there to decrease, in accord with Bernoulli's principle. Assuming nonviscous flow, what is the maximum speed with which water can flow through the intake pipe?

**68. Two circular holes, one larger than the other, are cut in the side of a large water tank whose top is open to the atmosphere. The center of one of these holes is located twice as far beneath the surface of the water as the other. The volume flow rate of the water coming out of the holes is the same. (a) Decide which hole is located nearest the surface of the water. (b) Calculate the ratio of the radius of the larger hole to the radius of the smaller hole.

**69. ssm A uniform rectangular plate is hanging vertically downward from a hinge that passes along its left edge. By blowing air at 11.0 m/s *over the top of the plate only,* it is possible to keep the plate in a horizontal position, as illustrated in part *a* of the drawing. To what value should the air speed be reduced so that the plate is kept at a 30.0° angle with respect to the vertical, as in part *b* of the drawing? (*Hint: Apply Bernoulli's equation in the form of Equation 11.12.*)

(a)   (b)

Section 11.11 Viscous Flow

70. A pipe is horizontal and carries oil that has a viscosity of 0.14 Pa·s. The volume flow rate of the oil is 5.3×10^{-5} m^3/s. The pipe has a length of 37 m and a radius of 0.60 cm. At the output end of the pipe the pressure is atmospheric pressure. What is the absolute pressure at the input end?

71. ssm www A 1.3-m length of horizontal pipe has a radius of 6.4×10^{-3} m. Water flows with a volume flow rate of 9.0×10^{-3} m^3/s out of the right end of the pipe and into the air. What is the pressure in the flowing water at the left end of the pipe if the water behaves as (a) an ideal fluid and (b) a viscous fluid ($\eta = 1.00 \times 10^{-3}$ Pa·s)?

72. A blood vessel is 0.10 m in length and has a radius of 1.5×10^{-3} m. Blood ($\eta = 4 \times 10^{-3}$ Pa·s) flows at a rate of 1.0×10^{-7} m^3/s. Determine the difference in pressure that must be maintained between the two ends of the vessel.

73. A cylindrical air duct in an air conditioning system has a length of 5.5 m and a radius of 7.2×10^{-2} m. A fan forces air

($\eta = 1.8 \times 10^{-5}$ Pa·s) through the duct, such that the air in a room (volume = 280 m^3) is replenished every ten minutes. Determine the difference in pressure between the ends of the air duct.

74. $\mathcal{S}$ A blood transfusion is being set up in an emergency room for an accident victim. Blood has a density of 1060 kg/m^3 and a viscosity of 4.0×10^{-3} Pa·s. The needle being used has a length of 3.0 cm and an inner radius of 0.25 mm. The doctor wishes to use a volume flow rate through the needle of 4.5×10^{-8} m^3/s. What is the distance h above the victim's arm where the level of the blood in the transfusion bottle should be located? As an approximation, assume that the level of the blood in the transfusion bottle and the point where the needle enters the vein in the arm have the same pressure of one atmosphere. (In reality, the pressure in the vein is slightly above atmospheric pressure.)

*75. ssm When an object moves through a fluid, the fluid exerts a viscous force **F** on the object that tends to slow it down. For a small sphere of radius R, moving slowly with a speed v, the magnitude of the viscous force is given by Stokes' law, $F = 6\pi\eta Rv$,

where η is the viscosity of the fluid. (a) What is the viscous force on a sphere of radius $R = 5.0 \times 10^{-4}$ m falling through water ($\eta = 1.00 \times 10^{-3}$ Pa·s) when the sphere has a speed of 3.0 m/s? (b) The speed of the falling sphere increases until the viscous force balances the weight of the sphere. Thereafter, no net force acts on the sphere, and it falls with a constant speed called the terminal speed. If the sphere has a mass of 1.0×10^{-5} kg, what is its terminal speed?

*76. Water ($\eta = 1.00 \times 10^{-3}$ Pa·s) is flowing through a horizontal pipe with a volume flow rate of 0.014 m^3/s. As the drawing shows, there are two vertical tubes that project from the pipe. From the data in the drawing, find the radius of the horizontal pipe.

ADDITIONAL PROBLEMS

77. At what depth beneath the surface of a lake is the absolute pressure three times the atmospheric pressure of 1.01×10^5 Pa that acts on the lake's surface?

78. Prairie dogs are burrowing rodents. They do not suffocate in their burrows, because the effect of air speed on pressure creates sufficient air circulation. The animals maintain a difference in the shapes of two entrances to the burrow, and because of this difference, the air ($\rho = 1.29$ kg/m^3) blows past the openings at different speeds, as the drawing indicates. Assuming that the openings are at the same vertical level, find the difference in air pressure between the openings and indicate which way the air circulates.

79. ssm What is the radius of a hydrogen-filled balloon that would carry a load of 5750 N (in addition to the weight of the hydrogen) when the density of air is 1.29 kg/m^3?

80. Suppose you are drinking a can of soda ($\rho = 1.0 \times 10^3$ kg/m^3) using a straw. When you suck on the straw the gauge pressure inside your mouth and lungs is 1200 Pa. How high is the soda drawn up into the straw?

81. A pirate in a movie is carrying a chest (0.30 m $\times$ 0.30 m $\times$ 0.20 m) that is supposed to be filled with gold. To see how ridicu-

lous this is, determine the weight (in newtons) of the gold. To judge how large this weight is, remember that 1 N = 0.225 lb.

82. The *karat* is a dimensionless unit that is used to indicate the proportion of gold in a gold-containing alloy. An alloy that is one karat gold contains a weight of pure gold that is one part in twenty-four. What is the volume of gold in a 14.0-karat gold necklace whose weight is 1.27 N?

83. ssm Water flows with a volume flow rate of 1.50 m^3/s in a pipe. Find the water speed where the pipe radius is 0.500 m.

84. $\mathcal{S}$ Poiseuille's law remains valid as long as the fluid flow is laminar. For sufficiently high speed, however, the flow becomes turbulent, even if the fluid is moving through a smooth pipe with no restrictions. It is found experimentally that the flow is laminar as long as the *Reynolds number* Re is less than about 2000: Re = $2\bar{v}\rho R/\eta$. Here $\bar{v}$, ρ, and η are, respectively, the average speed, density, and viscosity of the fluid, and R is the radius of the pipe. Calculate the highest average speed that blood ($\rho = 1060$ kg/m^3, $\eta = 4.0 \times 10^{-3}$ Pa·s) could have and still remain in laminar flow when it flows through the aorta ($R = 8.0 \times 10^{-3}$ m).

85. A cylinder is fitted with a piston, beneath which is a spring, as in the drawing. The cylinder is open at the top. Friction is absent. The spring constant of the spring is 3600 N/m. The piston has a negligible mass and a radius of 0.025 m. (a) When air beneath the piston is completely pumped out, how much does the atmospheric pressure cause the spring to compress? (b) How much work does the atmospheric pressure do in compressing the spring?

86. Only a small part of an iceberg protrudes above the water, while the bulk lies below the surface. The density of ice is

917 kg/m³ and that of seawater is 1025 kg/m³. Find the percentage of the iceberg's volume that lies below the surface.

87. ssm A water tower is a familiar sight in many towns. The purpose of such a tower is to provide storage capacity and to provide sufficient pressure in the pipes that deliver the water to customers. The drawing shows a spherical reservoir that contains 5.25×10^5 kg of water when full. The reservoir is vented to the atmosphere at the top. For a full reservoir, find the gauge pressure that the water has at the faucet in (a) house A and (b) house B. Ignore the diameter of the delivery pipes.

*88. Three fire hoses are connected to a fire hydrant. Each hose has a radius of 0.020 m. Water enters the hydrant through an underground pipe of radius 0.080 m. In this pipe the water has a speed of 3.0 m/s. (a) How many kilograms of water are poured onto a fire in one hour? (b) Find the water speed in each hose.

*89. ssm A 1967 Kennedy half-dollar has a mass of 1.150×10^{-2} kg. The coin is a mixture of silver and copper, and in water weighs 0.1011 N. Determine the mass of silver in the coin.

*90. A 1.00-m-tall container is filled to the brim, partway with mercury and the rest of the way with water. The container is open to the atmosphere. What must be the depth of the mercury, so the absolute pressure on the bottom of the container is twice the atmospheric pressure?

*91. ssm www Mercury is poured into a tall glass. Ethyl alcohol is then poured on top of the mercury until the height of the ethyl alcohol itself is 110 cm. The two fluids do not mix, and the air pressure at the top of the ethyl alcohol is one atmosphere. What is the absolute pressure at a point that is 7.10 cm below the ethyl alcohol–mercury interface?

*92. A full can of soda has a mass of 0.416 kg. It contains 3.54×10^{-4} m³ of liquid. Assuming that the soda has the same

density as water, find the volume of aluminum used to make the can.

*93. In a very large closed tank, the absolute pressure of the air above the water is 6.01×10^5 Pa. The water leaves the bottom of the tank through a nozzle that is directed straight upward. The opening of the nozzle is 4.00 m below the surface of the water. (a) Find the speed at which the water leaves the nozzle. (b) Ignoring air resistance and viscous effects, determine the height to which the water rises.

**94. One kilogram of glass ($\rho = 2.60 \times 10^3$ kg/m³) is shaped into a hollow spherical shell that just barely floats in water. What are the inner and outer radii of the shell? Do not assume the shell is thin.

**95. A house has a roof (colored gray) with the dimensions shown in the drawing. Determine the magnitude and direction of the net force that the atmosphere applies to the roof, when the outside pressure rises suddenly by 10.0 mm of mercury before the pressure in the attic can adjust.

**96. A siphon tube is useful for removing liquid from a tank. The siphon tube is first filled with liquid, and then one end is inserted into the tank. Liquid then drains out the other end, as the drawing illustrates. (a) Using reasoning similar to that employed in obtaining Torricelli's theorem, derive an expression for the speed v of the fluid emerging from the tube. This expression should give v in terms of the vertical height y and the acceleration due to gravity g. (Note that this speed does not depend on the depth d of the tube below the surface of the liquid.) (b) At what value of the vertical distance y will the siphon stop working? (c) Derive an expression for the absolute pressure at the highest point in the siphon (point A) in terms of the atmospheric pressure P_0, the fluid density ρ, g, and the heights h and y. (Note that the fluid speed at point A is the same as the speed of the fluid emerging from the tube, because the cross-sectional area of the tube is the same everywhere.)

CONCEPTS & CALCULATIONS

GROUP LEARNING PROBLEMS

Note: Each of these problems consists of Concept Questions followed by a related quantitative Problem. They are designed for

use by students working alone or in small learning groups. The Concept Questions involve little or no mathematics and are in-

tended to stimulate group discussions. They focus on the concepts with which the problems deal. Recognizing the concepts is the essential initial step in any problem-solving technique.

97. Concept Questions A gold prospector finds a solid rock that is composed solely of quartz and gold. (a) How is the total mass m_T of the rock related to the mass m_G of the gold and the mass m_Q of the quartz? (b) What is the relationship between the total volume V_T of the rock, the volume V_G of the gold, and the volume V_Q of the quartz? (c) How is the volume of a substance (gold or quartz) related to the mass of the substance and its density?

Problem The mass and volume of the rock are 12.0 kg and 4.00×10^{-3} m³. Find the mass of the gold in the rock.

98. Concept Questions A meat baster consists of a squeeze bulb attached to a plastic tube. When the bulb is squeezed and released, with the open end of the tube under the surface of the basting sauce, the sauce rises in the tube to a distance h, as the drawing shows. It can then be squirted over the meat. (a) Is the absolute pressure in the bulb in the drawing greater than or less than atmospheric pressure? (b) In a second trial, the distance h is somewhat less than it is in the drawing. Is the absolute pressure in the bulb in the second trial greater or smaller than in the case shown in the drawing? Explain your answers.

Problem Using 1.013×10^5 Pa for the atmospheric pressure and 1200 kg/m³ for the density of the sauce, find the absolute pressure in the bulb when the distance h is (a) 0.15 m and (b) 0.10 m. Verify that your answers are consistent with your answers to the Concept Questions.

h

99. Concept Questions A hydrometer is a device used to measure the density of a liquid. It is a cylindrical tube weighted at one end, so that it floats with the heavier end downward. It is contained inside a large "medicine dropper," into which the liquid is drawn using a squeeze bulb (see the drawing). For use with your car, marks are put on the tube so that the level at which it floats indicates whether the liquid is battery acid (more dense) or antifreeze (less dense). (a) Compared to the weight W of the tube, how much buoyant force is needed to make the tube float in either battery acid or antifreeze? (b) Is a greater volume of battery acid or a greater volume of antifreeze displaced by the hydrometer to provide the necessary buoyant force? (c) Which mark is farther up from the bottom of the tube. Justify your answers.

Problem The hydrometer has a weight of $W = 5.88 \times 10^{-2}$ N and a cross-sectional area of $A = 7.85 \times 10^{-5}$ m². How far from the bottom of the tube should the mark be put that denotes (a) battery acid ($\rho = 1280$ kg/m³) and (b) antifreeze ($\rho = 1073$ kg/m³)? Check to

Hydrometer

Marks for two types of liquids.

see that your answers are consistent with your answers to the Concept Questions.

100. Concept Questions Water flows straight down from an open faucet. The effects of air resistance and viscosity can be ignored. (a) After the water has fallen a bit below the faucet, is its speed less than, greater than, or the same as it was upon leaving the faucet? (b) Is the volume flow rate in cubic meters per second less than, greater than, or the same as it was when the water left the faucet? (c) Is the cross-sectional area of the water stream less than, greater than, or the same as it was when the water left the faucet? Give your reasoning.

Problem The cross-sectional area of the faucet is 1.8×10^{-4} m², and the speed of the water is 0.85 m/s as it leaves the faucet. Ignoring air resistance, find the cross-sectional area of the water stream at a point 0.10 m below the faucet. Make sure that your answer is consistent with your answers to the Concept Questions.

101. Concept Questions Water flowing out of a horizontal pipe emerges through a nozzle. The nozzle has a smaller radius than does the pipe. Treat the water as an ideal fluid. (a) What does the fact that the pipe is horizontal imply about the pipe's elevation y above the ground? (b) Is the speed at which the water flows in the pipe greater than, less than, or the same as the speed of the water emerging from the nozzle? (c) The emerging water is at atmospheric pressure. Is the absolute pressure of the water in the pipe greater than, less than, or the same as the atmospheric pressure? Explain your answers.

Problem The radius of the pipe is 1.9 cm, while the radius of the nozzle is 0.48 cm. The speed of the water in the pipe is 0.62 m/s. Determine the absolute pressure of the water in the pipe. Verify that your answer is consistent with your answers to the Concept Questions.

102. Concept Questions Two hoses are connected to the same outlet using a Y-connector, as the drawing shows. The hoses A and B have the same length, but hose B has the larger radius. Each is open to the atmosphere at the end where the water exits. Water flows through both hoses as a viscous fluid, and Poiseuille's law $[Q = \pi R^4(P_2 - P_1)/(8\eta L)]$ applies to each. In this law, P_2 is the pressure upstream, and P_1 is the pressure downstream, and Q is the volume flow rate. (a) For hoses A and B, is the value for the term $P_2 - P_1$ the same or different? (b) How is Q related to the radius of a hose and the speed of the water in the hose? Account for your answers.

Water from outlet

Hose A   Hose B

Problem The ratio of the radius of hose B to the radius of hose A is $R_B/R_A = 1.50$. Find the ratio of the speed of the water in hose B to that in hose A.

Temperature and heat, two subjects of this chapter, are evident in this fire-breathing performance. The chemical energy released during the burning process causes the gas to heat up, as evidenced by its high temperature.

CHAPTER 12

TEMPERATURE AND HEAT

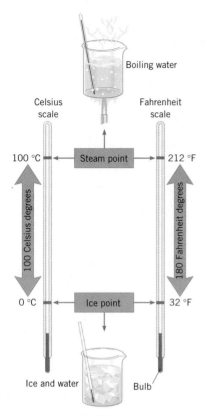

Figure 12.1 The Celsius and Fahrenheit temperature scales.

12.1 COMMON TEMPERATURE SCALES

To measure temperature we use a thermometer. Many thermometers make use of the fact that materials usually expand with increasing temperature. For example, Figure 12.1 shows the common mercury-in-glass thermometer, which consists of a mercury-filled glass bulb connected to a capillary tube. When the mercury is heated, it expands into the capillary tube, the amount of expansion being proportional to the change in temperature. The outside of the glass is marked with an appropriate scale for reading the temperature.

A number of different temperature scales have been devised, two popular choices being the *Celsius* (formerly, centigrade) and *Fahrenheit scales.* Figure 12.1 illustrates these scales. Historically,* both scales were defined by assigning two temperature points on the scale and then dividing the distance between them into a number of equally spaced intervals. One point was chosen to be the temperature at which ice melts under one atmosphere of pressure (the "ice point"), and the other was the temperature at which water boils under one atmosphere of pressure (the "steam point"). On the Celsius scale, an ice point of 0 °C (0 degrees Celsius) and a steam point of 100 °C were selected. On the Fahrenheit scale, an ice point of 32 °F (32 degrees Fahrenheit) and a steam point of 212 °F were chosen. The Celsius scale is used worldwide, while the Fahrenheit scale is used mostly in the United States, often in home medical thermometers.

There is a subtle difference in the way the temperature of an object is reported, as compared to a *change* in its temperature. For example, the temperature of the human body is about 37 °C, where the symbol °C stands for "degrees Celsius." However, the *change* between two temperatures is specified in "Celsius degrees" (C°)— not in "degrees Celsius." Thus, if the body temperature rises to 39 °C, the change in temperature is 2 Celsius degrees or 2 C°, not 2 °C.

As Figure 12.1 indicates, the separation between the ice and steam points on the Celsius scale is divided into 100 Celsius degrees, while on the Fahrenheit scale the separation is divided into 180 Fahrenheit degrees. Therefore, the size of the Celsius degree is larger than that of the Fahrenheit degree by a factor of $\frac{180}{100}$, or $\frac{9}{5}$. Examples 1 and 2 illustrate how to convert between the Celsius and Fahrenheit scales using this factor.

EXAMPLE 1   Converting from a Fahrenheit to a Celsius Temperature

A healthy person has an oral temperature of 98.6 °F. What would this reading be on the Celsius scale?

Reasoning and Solution A temperature of 98.6 °F is 66.6 Fahrenheit degrees above the ice point of 32.0 °F. Since 1 C° = $\frac{9}{5}$ F°, the difference of 66.6 F° is equivalent to

$$(66.6 \text{ F}°)\left(\frac{1 \text{ C}°}{\frac{9}{5} \text{ F}°}\right) = 37.0 \text{ C}°$$

Thus, the person's temperature is 37.0 Celsius degrees above the ice point. Adding 37.0 Celsius degrees to the ice point of 0 °C on the Celsius scale gives a Celsius temperature of $\boxed{37.0 \text{ °C}}$.

---

* Today, the Celsius and Fahrenheit scales are defined in terms of the Kelvin temperature scale; Section 12.2 discusses the Kelvin scale.

EXAMPLE 2 Converting from a Celsius to a Fahrenheit Temperature

A time and temperature sign on a bank indicates the outdoor temperature is −20.0 °C. Find the corresponding temperature on the Fahrenheit scale.

Reasoning and Solution The temperature of −20.0 °C is 20.0 Celsius degrees *below* the ice point of 0 °C. This number of Celsius degrees corresponds to

$$(20.0 \text{ C}°) \left(\frac{\frac{9}{5} \text{ F}°}{1 \text{ C}°} \right) = 36.0 \text{ F}°$$

The temperature, then, is 36.0 Fahrenheit degrees below the ice point. Subtracting 36.0 Fahrenheit degrees from the ice point of 32.0 °F on the Fahrenheit scale gives a Fahrenheit temperature of $\boxed{-4.0 \text{ °F}}$.

The reasoning strategy used in Examples 1 and 2 for converting between the Celsius and Fahrenheit scales is summarized below.

REASONING STRATEGY

Converting Between Different Temperature Scales

1. Determine the magnitude of the difference between the stated temperature and the ice point on the initial scale.

2. Convert this number of degrees from one scale to the other scale by using the fact that $1 \text{ C}° = \frac{9}{5}\text{F}°$.

3. Add or subtract the number of degrees on the new scale to or from the ice point on the new scale.

12.2 THE KELVIN TEMPERATURE SCALE

Although the Celsius and Fahrenheit scales are widely used, the ***Kelvin temperature scale*** has greater scientific significance. It was introduced by the Scottish physicist William Thompson (Lord Kelvin, 1824–1907), and in his honor each degree on the scale is called a kelvin (K). By international agreement, the symbol K is not written with a degree sign (°), nor is the word "degrees" used when quoting temperatures. For example, a temperature of 300 K (not 300 °K) is read as "three hundred kelvins," not "three hundred degrees kelvin." The kelvin is the SI base unit for temperature.

Figure 12.2 compares the Kelvin and Celsius scales. The size of one kelvin is identical to that of one Celsius degree, for there are one hundred divisions between the ice and steam points on both scales. As we will discuss shortly, experiments have shown that there exists a lowest possible temperature, below which no substance can be cooled. This lowest temperature is defined to be the zero point on the Kelvin scale and is referred to as absolute zero. Moreover, the ice point (0 °C) occurs at 273.15 K on the Kelvin scale. Thus, the Kelvin temperature T and the Celsius temperature T_c are related by

$$T = T_c + 273.15 \tag{12.1}$$

The number 273.15 in Equation 12.1 is an experimental result, obtained in studies that utilize a gas-based thermometer. When a gas is heated, its pressure in-

Figure 12.2 A comparison of the Kelvin and Celsius temperature scales.

Figure 12.3 A constant-volume gas thermometer.

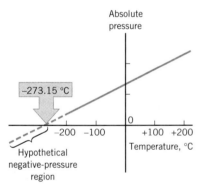

Figure 12.4 A plot of absolute pressure versus temperature for a low-density gas at constant volume. The graph is a straight line and, when extrapolated (dashed line), crosses the temperature axis at −273.15 °C.

creases, and when a gas is cooled, its pressure decreases, assuming the gas is confined to a fixed volume. For example, the air pressure in automobile tires can rise by as much as 20% after the car has been driven and the tires have become warm. The change in gas pressure with temperature is the basis for the ***constant-volume gas thermometer.***

A constant-volume gas thermometer consists of a gas-filled bulb to which a pressure gauge is attached, as in Figure 12.3. The gas is often hydrogen or helium at a low density, and the pressure gauge can be a U-tube manometer filled with mercury. The bulb is placed in thermal contact with the substance whose temperature is being measured. The volume of the gas is held constant by raising or lowering the *right* column of the U-tube manometer in order to keep the mercury level in the *left* column at the same reference level. The absolute pressure of the gas is proportional to the height h of the mercury on the right. As the temperature changes, the pressure changes and can be used to indicate the temperature, once the constant-volume gas thermometer has been calibrated.

Suppose the absolute pressure of the gas in Figure 12.3 is measured at different temperatures. If the results are plotted on a pressure versus temperature graph, a straight line is obtained, as in Figure 12.4. If the straight line is extended or extrapolated to lower and lower temperatures, the line crosses the temperature axis at −273.15 °C. In reality, no gas can be cooled to this temperature because all gases liquify before reaching it. However, helium and hydrogen liquify at such low temperatures that they are often used in the thermometer. This kind of graph can be obtained for different amounts and types of low-density gases. In all cases, it is found that the straight line extrapolates back to −273.15 °C on the temperature axis, which suggests that the value of −273.15 °C has fundamental significance. The significance of this number is that it is the ***absolute zero point*** for temperature measurement. The phrase "absolute zero" means that temperatures lower than −273.15 °C cannot be reached by continually cooling a gas or any other substance. If lower temperatures could be reached, then further extrapolation of the straight line in Figure 12.4 would suggest that negative absolute gas pressures could exist. Such a situation would be impossible, because a negative absolute gas pressure has no meaning. Thus, the Kelvin scale is chosen so that its zero temperature point is the lowest temperature attainable.

12.3 THERMOMETERS

All thermometers make use of the change in some physical property with temperature. A property that changes with temperature is called a ***thermometric property.*** For example, the thermometric property of the mercury thermometer is the length of the mercury column, while in the constant-volume gas thermometer it is the pressure of the gas. Several other thermometers and their thermometric properties will now be discussed.

The *thermocouple* is a thermometer used extensively in scientific laboratories. It consists of thin wires of different metals, welded together at the ends to form two junctions, as Figure 12.5 illustrates. Often the metals are copper and constantan (a copper–nickel alloy). One of the junctions, called the "hot" junction, is placed in thermal contact with the object whose temperature is being measured. The other junction, termed the "reference" junction, is kept at a known constant temperature (usually an ice–water mixture at 0 °C). The thermocouple generates a "voltage" that depends on the *difference in temperature* between the two junctions. This volt-

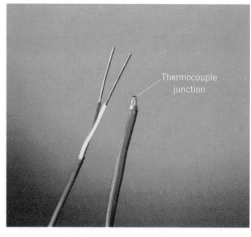

(a)

(b)

Figure 12.5 (a) A thermocouple is made from two different types of wires, copper and constantan in this case. (b) A thermocouple junction between two different wires.

age is the thermometric property and is measured by a voltmeter, as the drawing indicates. With the aid of calibration tables, the temperature of the hot junction can be obtained from the voltage. Thermocouples are used to measure temperatures as high as 2300 °C or as low as −270 °C.

Most substances offer resistance to the flow of electricity. Because this electrical resistance changes with temperature, electrical resistance is another thermometric property. *Electrical resistance thermometers* are often made from platinum wire, because platinum has excellent mechanical and electrical properties in the temperature range from −270 °C to +700 °C. The electrical resistance of platinum wire is known as a function of temperature. Thus, the temperature of a substance can be determined by placing the resistance thermometer in thermal contact with the substance and measuring the resistance of the platinum wire.

Radiation emitted by an object can also be used to indicate temperature. At low to moderate temperatures, the predominant type of radiation emitted is infrared radiation. As the temperature is raised, the intensity of the radiation increases substantially. In one interesting application, an infrared camera registers the intensity of the infrared radiation produced at different locations on the human body. The camera is connected to a color monitor that displays the different infrared intensities as different colors. This "thermal painting" of the body is called a *thermograph* or *thermogram*. Thermography is an important diagnostic tool in medicine. For example, breast cancer may be indicated in a thermograph by the elevated temperatures often associated with malignant tissue. Figure 12.6 shows a typical thermograph used to diagnose breast cancer. Figure 12.7 shows a thermogram of the head and neck in which the sympathetic nerves on the patient's right side have been blocked by a local anesthetic. With the nerves blocked, the temperature rises by several Celsius degrees, as is evident from the red color in the thermogram.

Oceanographers and meteorologists use thermographs extensively to map the temperature distribution on the surface of the earth. For example, Figure 12.8 shows a satellite image of the sea-surface temperature of the Pacific Ocean. The region depicted in red is the 1997/98 El Niño, a large area of the ocean, approximately twice the width of the United States, where temperatures reached abnormally high values. This El Niño caused major weather changes in certain regions of the earth.

Figure 12.6 The breast on the right in this thermograph has an invasive carcinoma (cancer) and registers colors from red to yellow/white, indicating markedly elevated temperatures.

THE PHYSICS OF...

thermography.

Figure 12.7 A thermogram of the face after the sympathetic nerves on the patient's right side (the left side of the thermogram) have been blocked by a local anesthetic. The temperature in the red area is an average of 35.75 °C, and in the dark blue area it is 33.25 °C.

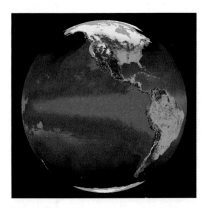

Figure 12.8 This thermograph vividly shows (in red) the 1997/98 El Niño, a large region of the Pacific Ocean that had abnormally high temperatures.

Figure 12.9 When the temperature of a rod is raised by an amount ΔT, the length of the rod increases by an amount ΔL.

12.4 LINEAR THERMAL EXPANSION

Normal Solids

Have you ever found the metal lid on a glass jar too tight to open? One solution is to run hot water over the lid, which loosens because the metal expands more than the glass does. To varying extents, most materials expand when heated and contract when cooled. The increase in any one dimension of a solid is called *linear expansion,* linear in the sense that the expansion occurs along a line. Figure 12.9 illustrates the linear expansion of a rod whose length is L_0 when the temperature is T_0. When the temperature increases to $T_0 + \Delta T$, the length becomes $L_0 + \Delta L$, where ΔT and ΔL are the magnitudes of the changes in temperature and length, respectively. Conversely, when the temperature decreases to $T_0 - \Delta T$, the length decreases to $L_0 - \Delta L$.

For modest temperature changes, experiments show that the change in length is directly proportional to the change in temperature ($\Delta L \propto \Delta T$). In addition, the change in length is proportional to the initial length of the rod, a fact that can be understood by considering Figure 12.10. Part *a* of the drawing shows two identical rods. Each rod has a length L_0 and expands by ΔL when the temperature increases by ΔT. Part *b* shows the two heated rods combined into a single rod, for which the total expansion is the sum of the expansions of each part, namely, $\Delta L + \Delta L = 2\Delta L$. Clearly, the amount of expansion doubles if the rod is twice as long to begin with. In other words, the change in length is directly proportional to the original length ($\Delta L \propto L_0$). Equation 12.2 expresses the fact that ΔL is proportional to both L_0 and ΔT ($\Delta L \propto L_0 \Delta T$) by using a proportionality constant α, which is called the *coefficient of linear expansion.*

● **LINEAR THERMAL EXPANSION OF A SOLID**

The length L_0 of an object changes by an amount ΔL when its temperature changes by an amount ΔT:

$$\Delta L = \alpha L_0 \Delta T \qquad (12.2)$$

where α is the coefficient of linear expansion.

Common Unit for the Coefficient of Linear Expansion: $\dfrac{1}{C^\circ} = (C^\circ)^{-1}$

Solving Equation 12.2 for α shows that $\alpha = \Delta L/(L_0 \Delta T)$. Since the length units of ΔL and L_0 algebraically cancel, the coefficient of linear expansion α has the unit of $(C^\circ)^{-1}$ when the temperature difference ΔT is expressed in Celsius degrees (C°). Different materials with the same initial length expand and contract by different amounts as the temperature changes, so the value of α depends on the nature of the material. Table 12.1 shows some typical values. Coefficients of linear expansion also vary somewhat depending on the range of temperatures involved, but the values in Table 12.1 are adequate approximations. Example 3 deals with a situation where a dramatic effect due to thermal expansion can be observed, even though the change in temperature is small.

EXAMPLE 3 Buckling of a Sidewalk

A concrete sidewalk is constructed between two buildings on a day when the temperature is 25 °C. The sidewalk consists of two slabs, each three meters in length and of negligible thickness (Figure 12.11*a*). As the temperature rises to 38 °C, the slabs expand, but no space is provided for thermal expansion. The buildings do not move, so the slabs buckle upward. Determine the vertical distance *y* in part *b* of the drawing.

Reasoning The expanded length of each slab is equal to its original length plus the change in length ΔL due to the rise in temperature. We know the original length, and Equation 12.2 can be used to find the change in length. Once the expanded length has been determined, the Pythagorean theorem can be employed to find the vertical distance *y* in Figure 12.11b.

Solution The change in temperature is $\Delta T = 38\ °C - 25\ °C = 13\ C°$, and the coefficient of linear expansion for concrete is given in Table 12.1. The change in length of each slab associated with this temperature change is

$$\Delta L = \alpha L_0 \Delta T = [12 \times 10^{-6}\ (C°)^{-1}](3.0\ \text{m})(13\ C°) = 0.000\ 47\ \text{m} \quad (12.2)$$

The expanded length of each slab is 3.000 47 m. The vertical distance *y* can be obtained by applying the Pythagorean theorem to the right triangle in Figure 12.11*b*:

$$y = \sqrt{(3.000\ 47\ \text{m})^2 - (3.000\ 00\ \text{m})^2} = \boxed{0.053\ \text{m}}$$

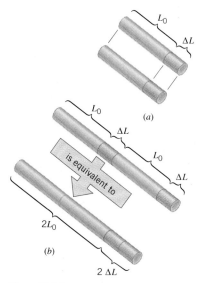

Figure 12.10 (*a*) Each of two identical rods expands by an amount ΔL when heated. (*b*) When the two rods are combined into a single rod of length $2L_0$, the "combined" rod expands by $2\,\Delta L$.

TABLE 12.1 Coefficients of Thermal Expansion for Solids and Liquids[a]

Substance	Coefficient of Thermal Expansion, $(C°)^{-1}$	
	Linear (α)	Volumetric (β)
Solids		
Aluminum	23×10^{-6}	69×10^{-6}
Brass	19×10^{-6}	57×10^{-6}
Concrete	12×10^{-6}	36×10^{-6}
Copper	17×10^{-6}	51×10^{-6}
Glass (common)	8.5×10^{-6}	26×10^{-6}
Glass (Pyrex)	3.3×10^{-6}	9.9×10^{-6}
Gold	14×10^{-6}	42×10^{-6}
Iron or steel	12×10^{-6}	36×10^{-6}
Lead	29×10^{-6}	87×10^{-6}
Nickel	13×10^{-6}	39×10^{-6}
Quartz (fused)	0.50×10^{-6}	1.5×10^{-6}
Silver	19×10^{-6}	57×10^{-6}
Liquids[b]		
Benzene	—	1240×10^{-6}
Carbon tetrachloride	—	1240×10^{-6}
Ethyl alcohol	—	1120×10^{-6}
Gasoline	—	950×10^{-6}
Mercury	—	182×10^{-6}
Methyl alcohol	—	1200×10^{-6}
Water	—	207×10^{-6}

[a] The values for α and β pertain to a temperature near 20 °C.
[b] Since liquids do not have fixed shapes, the coefficient of linear expansion is not defined for them.

Figure 12.11 (*a*) Two concrete slabs completely fill the space between the buildings. (*b*) When the temperature increases, each slab expands, causing the sidewalk to buckle.

The buckling of a sidewalk is one consequence of not providing sufficient room for thermal expansion. To eliminate such problems, engineers incorporate expansion joints or spaces at intervals along bridge roadbeds, as Figure 12.12 shows.

While Example 3 shows how thermal expansion can cause problems, there are also times when it can be useful. For instance, each year thousands of children are taken to emergency rooms suffering from burns caused by scalding tap water. Such accidents can be reduced with the aid of the antiscalding device shown in Figure 12.13. This device screws onto the end of a faucet and quickly shuts off the flow of water when it becomes too hot. As the water temperature rises, the actuator spring expands and pushes the plunger forward, shutting off the flow. When the water cools, the spring contracts and the water flow resumes.

THE PHYSICS OF...

an antiscalding device.

Thermal Stress

If the concrete slabs in Figure 12.11 had not buckled upward, they would have been subjected to immense forces from the buildings. The forces needed to prevent a solid object from expanding must be strong enough to counteract any change in length that would occur due to a change in temperature. Although the change in temperature may be small, the forces—and hence the stresses—can be enormous, as Example 4 illustrates.

EXAMPLE 4   The Stress on a Steel Beam

A steel beam is used in the roadbed of a bridge. The beam is mounted between two concrete supports when the temperature is 23 °C, with no room provided for thermal expansion (Figure 12.14). What compressional stress must the concrete supports apply to each end of the beam, if they are to keep the beam from expanding when the temperature rises to 42 °C?

Reasoning Recall from Section 10.8 that the stress (force per unit cross-sectional area or F/A) required to change the length L_0 of an object by an amount ΔL is

$$\text{Stress} = Y\frac{\Delta L}{L_0} \tag{10.17}$$

THE PHYSICS OF...

thermal stress.

where Y is Young's modulus. If the steel beam were free to expand because of the change in temperature, the length would change by $\Delta L = \alpha L_0 \Delta T$. Because the concrete supports do not permit any expansion, they must supply a stress to compress the beam by an amount ΔL. Thus,

$$\text{Stress} = Y\frac{\Delta L}{L_0} = Y\frac{\alpha L_0 \Delta T}{L_0} = Y\alpha\,\Delta T$$

Figure 12.12 An expansion joint in a bridge.

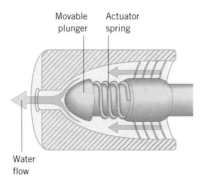

Movable plunger   Actuator spring

Water flow

Figure 12.13 An antiscalding device.

Beam

Concrete support   Concrete support

Figure 12.14 A steel beam is mounted between concrete supports with no room provided for thermal expansion of the beam.

Solution Young's modulus and the coefficient of linear expansion for steel are $Y = 2.0 \times 10^{11}$ N/m^2 (Table 10.1) and $\alpha = 12 \times 10^{-6}$ (C°)$^{-1}$ (Table 12.1), respectively. The change in temperature from 23 to 42 °C is $\Delta T = 19$ C°. The thermal stress is

$$\text{Stress} = Y\alpha\,\Delta T = (2.0 \times 10^{11} \text{ N/m}^2)[12 \times 10^{-6} \text{ (C°)}^{-1}](19 \text{ C°}) =$$

$$\boxed{4.6 \times 10^7 \text{ N/m}^2}$$

This stress is enormous. If the beam has a cross-sectional area of $A = 0.10$ m^2, the force applied to each end by a concrete support is $F = (\text{Stress})A = 4.6 \times 10^6$ N (over one million pounds).

The Bimetallic Strip

A *bimetallic strip* is made from two thin strips of metal that have *different* coefficients of linear expansion, as Figure 12.15a shows. Often brass [$\alpha = 19 \times 10^{-6}$ (C°)$^{-1}$] and steel [$\alpha = 12 \times 10^{-6}$ (C°)$^{-1}$] are selected. The two pieces are welded or riveted together. When the bimetallic strip is heated, the brass, having the larger value of α, expands more than the steel. Since the two metals are bonded together, the bimetallic strip bends into an arc as in part *b*, with the longer brass piece having a larger radius than the steel piece. When the strip is cooled, the bimetallic strip bends in the opposite direction, as in part *c*.

Bimetallic strips are frequently used as adjustable automatic switches in electrical appliances. Figure 12.16 shows an automatic coffee maker that turns off when the coffee is brewed to the selected strength. In part *a*, while the brewing cycle is on, electricity passes through the heating coil that heats the water. The electricity can flow because the contact mounted on the bimetallic strip touches the contact mounted on the "strength" adjustment knob, thus providing a continuous path for the electricity. When the bimetallic strip gets hot enough to bend away, as in part *b* of the drawing, the contacts separate. The electricity stops because it no longer has a continuous path along which to flow, and the brewing cycle is shut off. Turning the "strength" knob adjusts the brewing time by adjusting the distance through which the bimetallic strip must bend for the contact points to separate.

(a)   (b) Heated   (c) Cooled

Figure 12.15 (a) A bimetallic strip and how it behaves when (b) heated and (c) cooled.

THE PHYSICS OF...

an automatic coffee maker.

Figure 12.16 A bimetallic strip controls the brewing time on this automatic coffee maker. (a) When the bimetallic strip is cold, electricity flows through the heating coil. (b) When sufficiently hot, the bimetallic strip bends away from the adjustment knob contact, thus turning off the electricity.

Heating coil

Electricity

Heating coil "on"

Bimetallic switch (cold)
Contacts closed

Coffee "strength" adjustment knob

Bimetallic switch (hot)
Contacts separated

(a) Coffee pot "on"   (b) Coffee pot "off"

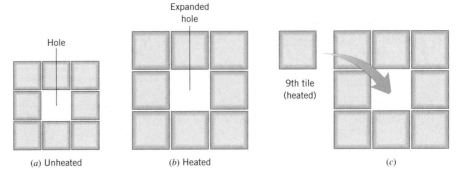

Figure 12.17 (*a*) The tiles are arranged to form a square pattern with a hole in the center. (*b*) When the tiles are heated, the pattern, as well as the hole in the center, expands. (*c*) The expanded hole is the same size as a heated tile.

The Expansion of Holes

An interesting example of linear expansion occurs when there is a hole in a piece of solid material. We know that the material itself expands when heated. But what about the hole? Does it expand, contract, or remain the same? Conceptual Example 5 provides some insight into the answer of this question.

CONCEPTUAL EXAMPLE 5
Do Holes Expand or Contract When the Temperature Increases?

Figure 12.17*a* shows eight square tiles that are arranged to form a square pattern with a hole in the center. If the tiles are heated, what happens to the size of the hole?

Reasoning and Solution We can analyze this problem by disassembling the pattern into separate tiles, heating them, and then reassembling the pattern. Because each tile expands upon heating, it is evident from Figure 12.17*b* that the heated pattern expands and so does the hole in the center. In fact, if we had a ninth tile that was identical to and also heated like the others, it would fit exactly into the center hole, as Figure 12.17*c* indicates. Thus, not only does the hole in the pattern expand, but it expands exactly as much as one of the tiles. Since the ninth tile is made of the same material as the others, we see that ***the hole expands just as if it were made of the material of the surrounding tiles.*** The thermal expansion of the hole and the surrounding material is analogous to a photographic enlargement; in both situations everything is enlarged, including holes.

Related Homework Material: *Problem 12*

Instead of the separate tiles in Example 5, we could have used a square plate with a square hole in the center. The hole in the plate would have expanded just like the hole in the pattern of tiles. Furthermore, the same conclusion applies to a hole of any shape. Thus, it follows that **a hole in a piece of solid material expands when heated and contracts when cooled, just as if it were filled with the material that surrounds it.** If the hole is circular, the equation $\Delta L = \alpha L_0 \Delta T$ can be used to find the change in any linear dimension of the hole, such as its radius or diameter. Example 6 illustrates this type of linear expansion.

EXAMPLE 6   A Heated Engagement Ring

A gold engagement ring has an inner diameter of 1.5×10^{-2} m and a temperature of 27 °C. The ring falls into a sink of hot water whose temperature is 49 °C. What is the change in the diameter of the hole in the ring?

Reasoning The hole expands as if it were filled with gold, so the change in the diameter is given by $\Delta L = \alpha L_0 \Delta T$, where $\alpha = 14 \times 10^{-6} (\text{C}°)^{-1}$ (Table 12.1) is the coefficient of linear expansion for gold, L_0 is the original diameter, and ΔT is the change in temperature.

Solution The change in the ring's diameter is

$$\Delta L = \alpha L_0 \Delta T = [14 \times 10^{-6} \, (\text{C}°)^{-1}](1.5 \times 10^{-2} \, \text{m})(49 \, °\text{C} - 27 \, °\text{C}) =$$

$$\boxed{4.6 \times 10^{-6} \, \text{m}}$$

The previous two examples illustrate that holes expand like the surrounding material when heated. Therefore, holes in materials with larger coefficients of linear expansion expand more than those in materials with smaller coefficients of linear expansion. Conceptual Example 7 explores this aspect of thermal expansion.

CONCEPTUAL EXAMPLE 7   Expanding Cylinders

Figure 12.18 shows a cross-sectional view of three cylinders, A, B, and C. Each is made from a different material; one is lead, one is brass, and one is steel. All three have the same temperature, and they barely fit inside each other. As the cylinders are heated to the same, but higher, temperature, cylinder C falls off, while cylinder A becomes tightly wedged to cylinder B. Which cylinder is made from which material?

Reasoning and Solution We need to consider how the outer and inner diameters of each cylinder change as the temperature is raised. With respect to the inner diameter, we will be guided by the fact that a hole expands as if it were filled with the surrounding material. According to Table 12.1, lead has the greatest coefficient of linear expansion, followed by brass, and then by steel. These data indicate that the outer and inner diameters of the lead cylinder change the most, while those of the steel cylinder change the least.

Since the steel cylinder expands the least, it cannot be the outer one, for if it were, the greater expansion of the middle cylinder would prevent the steel cylinder from falling off. The steel cylinder also cannot be the inner one, because then the greater expansion of the middle cylinder would allow the steel cylinder to fall out, contrary to what is observed. The only place left for the steel cylinder is in the middle, which leads to the two possibilities in Figure 12.18. In part *a*, lead is on the outside and will fall off as the temperature is raised, since lead expands more than steel. On the other hand, the inner brass cylinder expands more than the steel that surrounds it and becomes tightly wedged, as observed. Thus, one possibility is *A = brass, B = steel, and C = lead.*

In part *b* of the drawing, brass is on the outside. As the temperature is raised, brass expands more than steel, so the outer cylinder will again fall off. The inner lead cylinder has the greatest expansion and will be wedged against the middle steel cylinder. A second possible answer, then, is *A = lead, B = steel, and C = brass.*

Figure 12.18 Conceptual Example 7 discusses the arrangements of the three cylinders shown in these cutaway views in parts *a* and *b*.

12.5 VOLUME THERMAL EXPANSION

The volume of a normal material increases as the temperature increases. Most solids and liquids behave in this fashion. By analogy with linear thermal expansion, the change in volume ΔV is proportional to the change in temperature ΔT and to the initial volume V_0, provided the change in temperature is not too large. These two proportionalities can be converted into Equation 12.3 with the aid of a proportionality constant β, known as the ***coefficient of volume expansion.*** The algebraic form of this equation is similar to that for linear expansion, $\Delta L = \alpha L_0 \Delta T$.

> ● **VOLUME THERMAL EXPANSION**
>
> The volume V_0 of an object changes by an amount ΔV when its temperature changes by an amount ΔT:
>
> $$\Delta V = \beta V_0 \Delta T \qquad (12.3)$$
>
> where β is the coefficient of volume expansion.
>
> ***Common Unit for the Coefficient of Volume Expansion:*** $(C°)^{-1}$

The unit for β, like that for α, is $(C°)^{-1}$. Values for β depend on the nature of the material, and Table 12.1 lists some examples measured near 20 °C. The values of β for liquids are substantially larger than those for solids, because liquids typically expand more than solids, given the same initial volumes and temperature changes. Table 12.1 also shows that, for most solids, the coefficient of volume expansion is three times greater than the coefficient of linear expansion: $\beta = 3\alpha$.

If a cavity exists within a solid object, the volume of the cavity increases when the object expands, just as if the cavity were filled with the surrounding material. The expansion of the cavity is analogous to the expansion of a hole in a sheet of material. Accordingly, the change in volume of a cavity can be found using the relation $\Delta V = \beta V_0 \Delta T$, where β is the coefficient of volume expansion of the material that surrounds the cavity. Example 8 illustrates this point.

Coolant reservoir

Radiator

Figure 12.19 An automobile radiator and a coolant reservoir for catching the overflow from the radiator.

EXAMPLE 8   An Automobile Radiator

THE PHYSICS OF...

the overflow of an automobile radiator.

A small plastic container, called the coolant reservoir, catches the radiator fluid that overflows when an automobile engine becomes hot (see Figure 12.19). The radiator is made of copper, and the coolant has a coefficient of volume expansion of $\beta = 4.10 \times 10^{-4}\ (C°)^{-1}$. If the radiator is filled to its 15-quart capacity when the engine is "cold" (6.0 °C), how much overflow from the radiator will spill into the reservoir when the coolant reaches its operating temperature of 92 °C?

Reasoning When the temperature increases, both the coolant and radiator expand. If they were to expand by the same amount, there would be no overflow. However, the liquid coolant expands more than the radiator, and the amount of overflow is the amount of coolant expansion *minus* the expansion of the radiator cavity.

Solution When the temperature increases by 86 C°, the coolant expands by an amount

$$\Delta V = \beta V_0 \Delta T = [4.10 \times 10^{-4}\ (C°)^{-1}](15\ \text{quarts})(86\ C°) = 0.53\ \text{quarts} \quad (12.3)$$

The volume of the radiator cavity expands as if it were filled with copper [$\beta = 51 \times 10^{-6}\ (C°)^{-1}$, see Table 12.1]. The expansion of the radiator cavity is

$$\Delta V = \beta V_0 \Delta T = [51 \times 10^{-6}\ (C°)^{-1}](15\ \text{quarts})(86\ C°) = 0.066\ \text{quarts}$$

The amount of coolant overflow is 0.53 quarts − 0.066 quarts = $\boxed{0.46\ \text{quarts}}$.

● **PROBLEM SOLVING INSIGHT**
The way in which the level of a liquid in a container changes with temperature depends on the change in volume of both the liquid and the container.

While most substances expand when heated, a few do not. For instance, if water at 0 °C is heated, its volume *decreases* until the temperature reaches 4 °C. Above 4 °C water behaves normally, and its volume increases as the temperature increases.

Because a given mass of water has a minimum volume at 4 °C, the density (mass per unit volume) of water is greatest at 4 °C, as Figure 12.20 shows.

The fact that water has its greatest density at 4 °C, rather than at 0 °C, has important consequences for the way in which a lake freezes. When the air temperature drops, the surface layer of water is chilled. As the temperature of the surface layer drops toward 4 °C, this layer becomes more dense than the warmer water below. The denser water sinks and pushes up the deeper and warmer water, which in turn is chilled at the surface. This process continues until the temperature of the entire lake reaches 4 °C. Further cooling of the surface water below 4 °C makes it *less dense* than the deeper layers; consequently, the surface layer does not sink but stays on top. Continued cooling of the top layer to 0 °C leads to the formation of ice that floats on the water, because ice has a smaller density than water at any temperature. Below the ice, however, the water temperature remains above 0 °C. The sheet of ice acts as an insulator that reduces the loss of heat from the lake, especially if the ice is covered with a blanket of snow, which is also an insulator. As a result, lakes usually do not freeze solid, even during prolonged cold spells, so fish and other aquatic life can survive.

The fact that the density of ice is smaller than the density of water has an important consequence for home owners, who have to contend with the possibility of bursting water pipes during severe winters. Water often freezes in a section of pipe exposed to unusually cold temperatures. The ice can form an immovable plug that prevents the subsequent flow of water, as Figure 12.21 illustrates. When water (larger density) turns to ice (smaller density), its volume expands by 8.3 percent. Therefore, when more water freezes at the left side of the plug, the expanding ice pushes liquid back into the pipe leading to the street connection, and no damage is done. However, when ice forms on the right side of the plug, the expanding ice pushes liquid to the right. But it has nowhere to go if the faucet is closed. As ice continues to form and expand, the water pressure between the plug and faucet rises. Even a small increase in the amount of ice produces a large increase in the pressure. This situation is analogous to the thermal stress discussed in Example 4, where a small change in the length of the steel beam produces a large stress on the concrete supports. The entire section of pipe to the right of the blockage experiences the same elevated pressure, according to Pascal's principle (Section 11.5). Therefore, the pipe can burst at any point where it is structurally weak, even within the heated space of the building. If you should lose heat during the winter, there is a simple way to prevent pipes from bursting. Simply open the faucet, so it drips a little. The excessive pressure will be relieved.

THE PHYSICS OF...

bursting water pipes.

12.6 HEAT AND INTERNAL ENERGY

An object with a high temperature is said to be hot, and the word "hot" brings to mind the word "heat." *Heat* flows from a hotter object to a cooler object when the two are placed in contact. It is for this reason that a cup of hot coffee feels hot to the touch, while a glass of ice water feels cold. The temperature of hot coffee is higher than the

Figure 12.20 The density of water in the temperature range from 0 to 10 °C. Water has a maximum density of 999.973 kg/m³ at 4 °C. (This value for the density is equivalent to the often-quoted density of 1.000 00 grams per milliliter.)

Figure 12.21 As water freezes and expands, enormous pressure is applied to the liquid water between the ice and the faucet.

normal body temperature of 37 °C, while the temperature of ice water is lower than 37 °C. When the person in Figure 12.22*a* touches the coffee cup, heat flows from the hotter cup into the cooler hand. When the person touches the glass in part *b* of the drawing, heat again flows from hot to cold, in this case from the warmer hand into the colder glass. The response of the nerves in the hand to the arrival or departure of heat prompts the brain to identify the coffee cup as being hot and the glass as being cold.

But just what is heat? As the following definition indicates, heat is a form of energy, energy in transit from hot to cold.

> ● **DEFINITION OF HEAT**
>
> Heat is energy that flows from a higher-temperature object to a lower-temperature object because of the difference in temperatures.
>
> *SI Unit of Heat:* joule (J)

Being a kind of energy, heat is measured in the same units used for work, kinetic energy, and potential energy. Thus, the SI unit for heat is the joule.

The heat that flows from hot to cold in Figure 12.22 originates in the **internal energy** of the hot substance. The internal energy of a substance is the sum of the molecular kinetic energy (due to the random motion of the molecules), the molecular potential energy (due to forces that act between the atoms of a molecule and between molecules), and other kinds of molecular energy. When heat flows in circumstances where the work done is negligible, the internal energy of the hot substance decreases and the internal energy of the cold substance increases. While heat may originate in the internal energy supply of a substance, *it is not correct to say that a substance contains heat.* The substance has internal energy, not heat. The word "heat" is used only when referring to the energy actually in transit from hot to cold.

● *Concepts at a Glance* In the next two sections, we will consider some of the effects of heat. For instance, when preparing spaghetti for dinner, the first thing that the cook does is to place a pot of water on the stove. Heat from the stove causes the internal energy of the water to increase, as the right-hand side of the Concepts-at-a-Glance chart in Figure 12.23 suggests. Associated with this increase in internal energy is a rise in temperature. After a while, however, the temperature reaches 100 °C and the water begins to boil. During boiling, any heat added to the water goes into producing steam, a process in which water changes from a liquid phase to a vapor phase. The next section investigates how the addition (or removal) of heat causes the temperature of a substance to change (see the upper-right portion of the chart in Figure 12.23). Then, Section 12.8 discusses the relationship between heat and phase changes (see the lower-right part of Figure 12.23), such as that which occurs when water boils. ●

12.7 HEAT AND TEMPERATURE CHANGE: SPECIFIC HEAT CAPACITY

Solids and Liquids

Greater amounts of heat are needed to raise the temperature of solids or liquids to higher values. A greater amount of heat is also required to raise the temperature of a greater mass of material. Similar comments apply when the temperature is lowered, except that heat must be removed. For limited temperature ranges, experiment

Figure 12.22 Heat is energy in transit from hot to cold. (*a*) Heat flows from the hotter coffee cup to the colder hand. (*b*) Heat flows from the warmer hand to the colder glass of ice water.

Concepts at a Glance

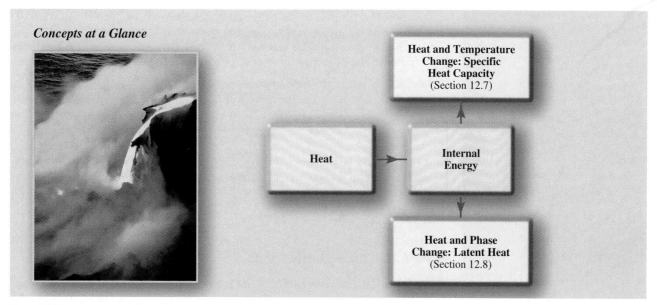

Figure 12.23 *Concepts at a Glance* When heat is added or removed from a substance, its internal energy can change. This change can cause a change in temperature (Section 12.7) or a change in phase (Section 12.8). This volcanic eruption in Hawaii is spewing yellow-hot lava into the ocean. The temperature is high because of the heat absorbed by the lava when it was deep within the earth.

shows that the amount of heat Q is directly proportional to the change in temperature ΔT and to the mass m. These two proportionalities are expressed below in Equation 12.4, with the help of a proportionality constant c that is referred to as the *specific heat capacity* of the material.

● **HEAT SUPPLIED OR REMOVED IN CHANGING THE TEMPERATURE OF A SUBSTANCE**

The heat Q that must be supplied or removed to change the temperature of a substance of mass m by an amount ΔT is

$$Q = cm\,\Delta T \qquad (12.4)$$

where c is the specific heat capacity of the substance.

Common Unit for Specific Heat Capacity: $\mathrm{J/(kg \cdot C°)}$

Solving Equation 12.4 for the specific heat capacity shows that $c = Q/(m\,\Delta T)$, so the unit for specific heat capacity is $\mathrm{J/(kg \cdot C°)}$. Table 12.2 reveals that the value of the specific heat capacity depends on the nature of the material. Examples 9 and 10 illustrate the use of Equation 12.4.

EXAMPLE 9 A Hot Jogger

In a half hour, a 65-kg jogger can generate 8.0×10^5 J of heat. This heat is removed from the jogger's body by a variety of means, including the body's own temperature regulating mechanisms. If the heat were not removed, how much would the body temperature increase?

TABLE 12.2 Specific Heat Capacities[a] of Some Solids and Liquids

Substance	Specific Heat Capacity, c $\mathrm{J/(kg \cdot C°)}$[b]
Solids	
Aluminum	9.00×10^2
Copper	387
Glass	840
Human body (37 °C, average)	3500
Ice (−15 °C)	2.00×10^3
Iron or steel	452
Lead	128
Silver	235
Liquids	
Benzene	1740
Ethyl alcohol	2450
Glycerin	2410
Mercury	139
Water (15 °C)	4186

[a] Except as noted, the values are for 25 °C and 1 atm of pressure.
[b] The values given are the same in units of $\mathrm{cal/(g \cdot C°)}$.

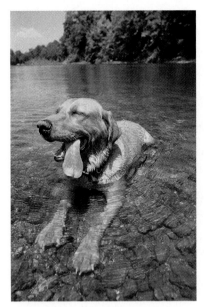

Figure 12.24 Dogs, such as this golden retriever, often pant to get rid of excess heat.

Reasoning The increase in the body temperature depends on the amount of heat Q generated by the jogger, her mass m, and the specific heat capacity c of the human body. Since numerical values are known for these three variables, we can determine the potential rise in temperature by using Equation 12.4.

Solution Table 12.2 gives the average specific heat capacity of the human body as 3500 J/(kg · C°). With this value, Equation 12.4 shows that

$$\Delta T = \frac{Q}{cm} = \frac{8.0 \times 10^5 \text{ J}}{[3500 \text{ J/(kg} \cdot \text{C°)}](65 \text{ kg})} = \boxed{3.5 \text{ C°}}$$

An increase in body temperature of 3.5 °C could be life-threatening. One way in which the jogger's body prevents it from occurring is to remove excess heat by perspiring. In contrast, dogs, such as the one in Figure 12.24, do not perspire but often pant to remove excess heat.

EXAMPLE 10   Taking a Hot Shower

Cold water at a temperature of 15 °C enters a heater, and the resulting hot water has a temperature of 61 °C. A person uses 120 kg of hot water in taking a shower. (a) Find the energy needed to heat the water. (b) Assuming that the utility company charges $0.10 per kilowatt · hour for electrical energy, determine the cost of heating the water.

Reasoning The amount Q of heat needed to raise the water temperature can be found from the relation $Q = cm \Delta T$, since the specific heat capacity, mass, and temperature change of the water are known. To determine the cost of this energy, we multiply the cost per unit of energy ($0.10 per kilowatt · hour) by the amount of energy used, expressed in energy units of kilowatt · hours.

Solution

(a) The amount of heat needed to heat the water is

$$Q = cm \, \Delta T = [4186 \text{ J/(kg} \cdot \text{C°)}](120 \text{ kg})(61 \text{ °C} - 15 \text{ °C}) = \boxed{2.3 \times 10^7 \text{ J}} \quad (12.4)$$

(b) The kilowatt · hour (kWh) is the unit of energy that utility companies use in your electric bill. To calculate the cost, we need to determine the number of joules in one kilowatt · hour. Recall that 1 kilowatt is 1000 watts (1 kW = 1000 W), one watt is one joule per second (1 W = 1 J/s, see Equation 6.10b), and one hour is equal to 3600 seconds (1 h = 3600 s). Thus,

$$1 \text{ kWh} = (1 \text{kWh}) \left(\frac{1000 \text{ W}}{1 \text{ kW}} \right) \left(\frac{1 \text{ J/s}}{1 \text{ W}} \right) \left(\frac{3600 \text{ s}}{1 \text{ h}} \right) = 3.60 \times 10^6 \text{ J}$$

The number of kilowatt · hours of energy used to heat the water is

$$(2.3 \times 10^7 \text{ J}) \left(\frac{1 \text{ kWh}}{3.60 \times 10^6 \text{ J}} \right) = 6.4 \text{ kWh}$$

At a cost of $0.10 per kWh, the bill for the heat is $\boxed{\$0.64}$ or 64 cents.

Gases

As we will see in Section 15.6, the value of the specific heat capacity depends on whether the pressure or volume is held constant while energy in the form of heat is added to or removed from a substance. The distinction between constant pressure and constant volume is usually not important for solids and liquids but is significant for gases. As we will see in Section 15.6, a greater value for the specific heat capacity is obtained for a gas at constant pressure than for a gas at constant volume.

Heat Units Other Than the Joule

There are three heat units other than the joule in common use. One kilocalorie (1 kcal) was defined historically as the amount of heat needed to raise the temperature of one kilogram of water by one Celsius degree.* With $Q = 1.00$ kcal, $m = 1.00$ kg, and $\Delta T = 1.00$ C°, the equation $Q = cm\,\Delta T$ shows that such a definition is equivalent to a specific heat capacity for water of $c = 1.00$ kcal/(kg·C°). Similarly, one calorie (1 cal) was defined as the amount of heat needed to raise the temperature of one gram of water by one Celsius degree, which yields a value of $c = 1.00$ cal/(g·C°). (Nutritionists use the word "Calorie," with a capital C, to specify the energy content of foods; this use is unfortunate, since 1 Calorie = 1000 calories = 1 kcal.) The British thermal unit (Btu) is the other commonly used heat unit and was defined historically as the amount of heat needed to raise the temperature of one pound of water by one Fahrenheit degree.

It was not until the time of James Joule (1818–1889) that the relationship between energy in the form of work (in units of joules) and energy in the form of heat (in units of kilocalories) was firmly established. Joule's experiments revealed that the performance of mechanical work, like rubbing your hands together, can make the temperature of a substance rise, just as the absorption of heat can. His experiments and those of later workers have shown that

$$1 \text{ kcal} = 4186 \text{ joules} \quad \text{or} \quad 1 \text{ cal} = 4.186 \text{ joules}$$

Because of its historical significance, this conversion factor is known as the *mechanical equivalent of heat.*

Calorimetry

In Section 6.8 we encountered the principle of conservation of energy, which states that energy can be neither created nor destroyed, but can only be converted from one form to another. There we dealt with kinetic and potential energies. In this chapter we have expanded our concept of energy to include heat, which is energy that flows from a higher-temperature object to a lower-temperature object because of the difference in temperature. No matter what its form, whether kinetic energy, potential energy, or heat, energy can be neither created nor destroyed. This fact governs the way objects at different temperatures come to an equilibrium temperature when they are placed in contact. If there is no heat loss to the external surroundings, the heat lost by the hotter objects equals the heat gained by the cooler ones, a process that is consistent with the conservation of energy. Just this kind of process occurs in a thermos. A perfect thermos would prevent any heat from leaking out or in. However, energy in the form of heat can flow *between* materials inside the thermos to the extent that they have different temperatures, for example, between ice cubes and warm tea. Such heat flow satisfies the conservation of energy, for the colder materials gain the energy that the hotter materials lose. The transfer of energy continues until a common temperature is reached at thermal equilibrium.

The kind of heat transfer that occurs in a thermos of iced tea also occurs in a calorimeter, which is the experimental apparatus used in a technique known as

---

* From 14.5 to 15.5 °C.

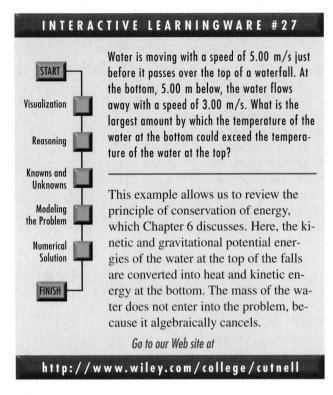

● **PROBLEM SOLVING INSIGHT**
In the equation "Heat gained = Heat lost," both sides must have the same algebraic sign. Therefore, when calculating heat contributions, always write any temperature changes as the higher minus the lower temperature.

Thermometer

Calorimeter cup

Insulating container

Unknown material

Figure 12.25 A calorimeter can be used to measure the specific heat capacity of an unknown material.

calorimetry. Figure 12.25 shows that, like a thermos, a calorimeter is essentially an insulated container. It can be used to determine the specific heat capacity of a substance, as the next example illustrates.

EXAMPLE 11　Measuring the Specific Heat Capacity

The calorimeter cup in Figure 12.25 is made from 0.15 kg of aluminum and contains 0.20 kg of water. Initially, the water and the cup have a common temperature of 18.0 °C. A 0.040-kg mass of unknown material is heated to a temperature of 97.0 °C and then added to the water. The temperature of the water, the cup, and the unknown material is 22.0 °C after thermal equilibrium is reestablished. Ignoring the small amount of heat gained by the thermometer, find the specific heat capacity of the unknown material.

Reasoning Since energy is conserved and there is negligible heat flow between the calorimeter and the outside surroundings, the heat gained by the cold water and the aluminum cup as they warm up is equal to the heat lost by the unknown material as it cools down. Each quantity of heat can be calculated using the relation $Q = cm \Delta T$, where we always write the change in temperature ΔT as the higher temperature minus the lower temperature. The equation "Heat gained = Heat lost" will then contain a single unknown quantity, the desired specific heat capacity.

Solution

$$\underbrace{(cm \, \Delta T)_{Al} + (cm \, \Delta T)_{water}}_{\substack{\text{Heat gained by} \\ \text{aluminum and water}}} = \underbrace{(cm \, \Delta T)_{unknown}}_{\substack{\text{Heat lost by} \\ \text{unknown material}}}$$

$$c_{unknown} = \frac{c_{Al}m_{Al} \, \Delta T_{Al} + c_{water}m_{water} \, \Delta T_{water}}{m_{unknown} \, \Delta T_{unknown}}$$

The changes in temperature for the three substances are $\Delta T_{Al} = \Delta T_{water} = 22.0 \text{ °C} - 18.0 \text{ °C} = 4.0 \text{ C°}$, and $\Delta T_{unknown} = 97.0 \text{ °C} - 22.0 \text{ °C} = 75.0 \text{ C°}$. Table 12.2 contains values for the specific heat capacities of aluminum and water. Substituting numbers into the equation above, we find that

$c_{unknown}$
$$= \frac{[9.00 \times 10^2 \text{ J/(kg·C°)}](0.15 \text{ kg})(4.0 \text{ C°}) + [4186 \text{ J/(kg·C°)}](0.20 \text{ kg})(4.0 \text{ C°})}{(0.040 \text{ kg})(75.0 \text{ C°})}$$

$$= \boxed{1300 \text{ J/(kg·C°)}}$$

12.8　HEAT AND PHASE CHANGE: LATENT HEAT

Surprisingly, there are situations in which the addition or removal of heat does not cause a temperature change. Consider a well-stirred glass of iced tea that has come to thermal equilibrium. Even though heat enters the glass from the warmer room, the temperature of the tea does not rise above 0 °C as long as ice cubes are present.

Apparently the heat is being used for some purpose other than raising the temperature. In fact, the heat is being used to melt the ice, and only when all of it is melted will the temperature of the liquid begin to rise.

An important point illustrated by the iced tea example is that there is more than one type or phase of matter. For instance, some of the water in the glass is in the solid phase (ice) and some in the liquid phase. The gas or vapor phase is the third familiar phase of matter. In the gas phase, water is referred to as water vapor or steam. All three phases of water are present in the scene depicted in Figure 12.26.

Matter can change from one phase to another, and heat plays a role in the change. Figure 12.27 summarizes the various possibilities. A solid can *melt* or *fuse* into a liquid if heat is added, while the liquid can *freeze* into a solid if heat is removed. Similarly, a liquid can *evaporate* into a gas if heat is supplied, while the gas can *condense* into a liquid if heat is taken away. Rapid evaporation, with the formation of vapor bubbles within the liquid, is called boiling. Finally, a solid can sometimes change directly into a gas if heat is provided. We say that the solid *sublimes* into a gas. Examples of sublimation are (1) solid carbon dioxide, CO_2 (dry ice), turning into gaseous CO_2 and (2) solid naphthalene (moth balls) turning into naphthalene fumes. Conversely, if heat is removed under the right conditions, the gas will condense directly into a solid.

Figure 12.28 displays a graph that indicates what typically happens when heat is added to a material that changes phase. The graph records temperature versus heat added and refers to water at the normal atmospheric pressure of 1.01×10^5 Pa. The water starts off as ice at the subfreezing temperature of -30 °C. As heat is added, the temperature of the ice increases, in accord with the specific heat capacity of ice [2000 J/(kg·C°)]. Not until the temperature reaches the normal melting/freezing point of 0 °C does the water begin to change phase. Then, when heat is added, the solid changes into the liquid, the temperature staying at 0 °C until *all the ice has melted.* Once all the material is in the liquid phase, additional heat causes the temperature to increase again, now in accord with the specific heat capacity of liquid water [4186 J/(kg·C°)]. When the temperature reaches the normal boiling/condensing point of 100 °C, the water begins to change from the liquid to the gas phase and continues to do so as long as heat is added. The temperature remains at 100 °C *until all liquid is gone.* When all of the material is in the gas phase, additional heat once again causes the temperature to rise, this time according to the specific heat capacity of water vapor at constant atmospheric pressure [2020 J/(kg·C°)]. Conceptual Example 12 applies the information in Figure 12.28 to a familiar situation.

Figure 12.26 The three phases of water: ice is floating in liquid water while (invisible) water vapor is present in the air.

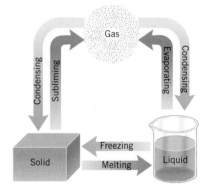

Figure 12.27 Three familiar phases of matter—solid, liquid, and gas—and the phase changes that can occur between any two of them.

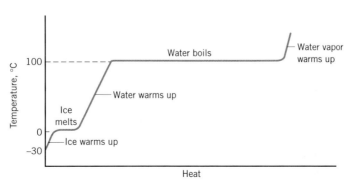

Figure 12.28 The graph shows the way the temperature of water changes as heat is added, starting with ice at -30 °C. The pressure is atmospheric pressure.

CONCEPTUAL EXAMPLE 12   Saving Energy

Suppose you are cooking spaghetti for dinner, and the instructions say to boil the pasta in water for ten minutes. To cook spaghetti in an open pot with the least amount of energy, should you turn up the burner to its fullest so the water vigorously boils, or should you turn down the burner so the water barely boils?

Reasoning and Solution The spaghetti needs to cook at a temperature of 100 °C for ten minutes. It doesn't matter whether the water is vigorously boiling or barely boiling, for its temperature is still 100 °C. Remember, as long as there is boiling water in the pot, and the pot is open to one atmosphere of pressure, no amount of additional heat will cause the water temperature to rise above 100 °C. Additional heat only vaporizes the water and produces steam, which is of no use in cooking the spaghetti. So, save your money and turn down the heat, for *the least amount of energy is expended when the water barely boils.*

When a substance changes from one phase to another, the amount of heat that must be added or removed depends on the type of material and the nature of the phase change. The heat per kilogram associated with a phase change is referred to as *latent heat:*

● HEAT SUPPLIED OR REMOVED IN CHANGING THE PHASE OF A SUBSTANCE

The heat Q that must be supplied or removed to change the phase of a mass m of a substance is

$$Q = mL \tag{12.5}$$

where L is the latent heat of the substance.

SI Unit of Latent Heat: J/kg

The *latent heat of fusion* L_f refers to the change between solid and liquid phases, the *latent heat of vaporization* L_v applies to the change between liquid and gas phases, and the *latent heat of sublimation* L_s refers to the change between solid and gas phases.

TABLE 12.3   Latent Heats[a] of Fusion and Vaporization

Substance	Melting Point (°C)	Latent Heat of Fusion, L_f (J/kg)	Boiling Point (°C)	Latent Heat of Vaporization, L_v (J/kg)
Ammonia	−77.8	33.2×10^4	−33.4	13.7×10^5
Benzene	5.5	12.6×10^4	80.1	3.94×10^5
Copper	1083	20.7×10^4	2566	47.3×10^5
Ethyl alcohol	−114.4	10.8×10^4	78.3	8.55×10^5
Gold	1063	6.28×10^4	2808	17.2×10^5
Lead	327.3	2.32×10^4	1750	8.59×10^5
Mercury	−38.9	1.14×10^4	356.6	2.96×10^5
Nitrogen	−210.0	2.57×10^4	−195.8	2.00×10^5
Oxygen	−218.8	1.39×10^4	−183.0	2.13×10^5
Water	0.0	33.5×10^4	100.0	22.6×10^5

[a] The values pertain to 1 atm pressure.

Table 12.3 gives some typical values of latent heats of fusion and vaporization. For instance, the latent heat of fusion for water is $L_f = 3.35 \times 10^5$ J/kg. Thus, 3.35×10^5 J of heat must be supplied to melt one kilogram of ice at 0 °C into liquid water at 0 °C; conversely, this amount of heat must be removed from one kilogram of liquid water at 0°C to freeze the liquid into ice at 0 °C. In comparison, the latent heat of vaporization for water has the much larger value of $L_v = 22.6 \times 10^5$ J/kg. When water boils at 100 °C, 22.6×10^5 J of heat must be supplied for each kilogram of liquid turned into steam. And when steam condenses at 100 °C, this amount of heat is released from each kilogram of steam that changes back into liquid. Liquid water at 100 °C is hot enough by itself to cause a bad burn, and the additional effect of the large latent heat can cause severe tissue damage if condensation occurs on the skin. Examples 13 and 14 illustrate how to take into account the effect of latent heat when using the conservation of energy principle.

THE PHYSICS OF...

steam burns.

EXAMPLE 13   Ice-cold Lemonade

Ice at 0 °C is placed in a Styrofoam cup containing 0.32 kg of lemonade at 27 °C. The specific heat capacity of lemonade is virtually the same as that of water, i.e., $c = 4186$ J/(kg·C°). After the ice and lemonade reach an equilibrium temperature, some ice still remains. The latent heat of fusion for water is $L_f = 3.35 \times 10^5$ J/kg. Assume that the mass of the cup is so small that it absorbs a negligible amount of heat, and ignore any heat lost to the surroundings. Determine the mass of ice that has melted.

Reasoning According to the principle of energy conservation, the heat gained by the melting ice equals the heat lost by the cooling lemonade. According to Equation 12.5, the heat gained by the melting ice is $Q = mL_f$, where m is the mass of the melted ice, and L_f is the latent heat of fusion for water. The heat lost by the lemonade is given by $Q = cm\Delta T$, where ΔT is the higher temperature of 27 °C minus the lower equilibrium temperature. The equilibrium temperature is 0 °C, because there is some ice remaining, and ice is in equilibrium with liquid water when the temperature is 0 °C.

Solution

$$\underbrace{(mL_f)_{ice}}_{\substack{\text{Heat gained}\\\text{by ice}}} = \underbrace{(cm\,\Delta T)_{lemonade}}_{\substack{\text{Heat lost}\\\text{by lemonade}}}$$

The mass m_{ice} of ice that has melted is

$$m_{ice} = \frac{(cm\,\Delta T)_{lemonade}}{L_f} = \frac{[4186\text{ J/(kg·C°)}](0.32\text{ kg})(27\text{ °C} - 0\text{ °C})}{3.35 \times 10^5\text{ J/kg}} = \boxed{0.11\text{ kg}}$$

EXAMPLE 14   Getting Ready for a Party

A 7.00-kg glass bowl [$c = 840$ J/(kg·C°)] contains 16.0 kg of punch at 25.0 °C. Two-and-a-half kilograms of ice [$c = 2.00 \times 10^3$ J/(kg·C°)] are added to the punch. The ice has an initial temperature of -20.0 °C, having been kept in a very cold freezer. The punch may be treated as if it were water [$c = 4186$ J/(kg·C°)], and it may be assumed that there is no heat flow between the punch bowl and the external environment. When thermal equilibrium is reached, all the ice has melted, and the final temperature of the mixture is above 0 °C. The latent heat of fusion for water is 3.35×10^5 J/kg. Determine this temperature.

Reasoning The final temperature can be found by using the conservation of energy, because the heat gained is equal to the heat lost. Heat is gained (a) by the ice in warming up to the melting point, (b) by the ice in changing phase from a solid to a

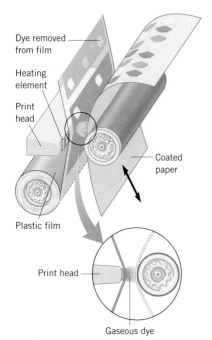

Figure 12.29 A dye sublimation printer. As the plastic film passes in front of the print head, the heat from a given heating element causes one of three pigments or dyes on the film to sublime from a solid to a gas. The gaseous dye is absorbed onto the coated paper as a dot of color. The size of the dots on the paper has been exaggerated for clarity.

THE PHYSICS OF...

a dye-sublimation color printer.

liquid, and (c) by the liquid that results from the ice warming up to the final temperature; heat is lost (d) by the punch and (e) by the bowl in cooling down. The heat gained or lost by each component in changing temperature can be determined from the relation $Q = cm\,\Delta T$. The heat gained when water changes phase from a solid to a liquid at 0 °C is $Q = mL_f$, where m is the mass of water and L_f is the latent heat of fusion.

Solution The heat gained or lost by each component is listed as follows:

(a) Heat gained when
ice warms to 0.0 °C
$= [2.00 \times 10^3 \text{ J/(kg} \cdot \text{C}°)](2.50 \text{ kg})[0.0 \text{ °C} - (-20.0 \text{ °C})]$

(b) Heat gained when
ice melts at 0.0 °C
$= (2.50 \text{ kg})(3.35 \times 10^5 \text{ J/kg})$

(c) Heat gained when
melted ice (liquid)
warms to
temperature T
$= [4186 \text{ J/(kg} \cdot \text{C}°)](2.50 \text{ kg})(T - 0.0 \text{ °C})$

(d) Heat lost when
punch cools to
temperature T
$= [4186 \text{ J/(kg} \cdot \text{C}°)](16.0 \text{ kg})(25.0 \text{ °C} - T)$

(e) Heat lost when
bowl cools to
temperature T
$= [840 \text{ J/(kg} \cdot \text{C}°)](7.00 \text{ kg})(25.0 \text{ °C} - T)$

Setting the heat gained equal to the heat lost gives:

$$\underbrace{(a) + (b) + (c)}_{\text{Heat gained}} = \underbrace{(d) + (e)}_{\text{Heat lost}}$$

This equation can be solved to show that $\boxed{T = 11 \text{ °C}}$.

An interesting application of the phase change between a solid and a gas is found in one kind of color printer used with computers. A dye-sublimation printer uses a thin plastic film coated with separate panels of cyan (blue), yellow, and magenta pigment or dye. A full spectrum of colors is produced by using combinations of tiny spots of these dyes. As Figure 12.29 shows, the coated film passes in front of a print head that extends across the width of the paper and contains 2400 heating elements. When a heating element is turned on, the dye in front of it absorbs heat and goes from a solid to a gas—it sublimes—with no liquid phase in between. A coating on the paper absorbs the gaseous dye on contact, producing a small spot of color. The intensity of the spot is controlled by the heating element, since each element can produce 256 different temperatures; the hotter the element, the greater the amount of dye transferred to the paper. The paper makes three separate passes across the print head, once for each of the dyes. The final result is an image that is of near-photographic quality.

*12.9 EQUILIBRIUM BETWEEN PHASES OF MATTER

Under specific conditions of temperature and pressure, a substance can exist in equilibrium in more than one phase at the same time. Consider Figure 12.30, which shows a container kept at a constant temperature by a large reservoir of heated sand. Initially the container is evacuated. Part *a* shows it just after it has been par-

Constant-temperature heated sand

(a)

Constant-temperature heated sand

(b)

Figure 12.30 (*a*) Some of the molecules begin entering the vapor phase in the evacuated space above the liquid. (*b*) Equilibrium is reached when the number of molecules entering the vapor phase equals the number returning to the liquid.

tially filled with a liquid and a few fast-moving molecules escape the liquid and form a vapor phase. These molecules pick up the required energy (the latent heat of vaporization) during collisions with neighboring molecules in the liquid. However, the reservoir of heated sand replenishes the energy carried away, thus maintaining the constant temperature. At first, the movement of molecules is predominantly from liquid to vapor, although some molecules in the vapor phase do reenter the liquid. As the molecules accumulate in the vapor, the number reentering the liquid eventually equals the number entering the vapor, and equilibrium becomes established, as in part *b*. From this point on, the concentration of molecules in the vapor phase does not change, and the vapor pressure remains constant. The pressure of the vapor that coexists in equilibrium with the liquid is called the ***equilibrium vapor pressure*** of the liquid.

The equilibrium vapor pressure does not depend on the volume of space above the liquid. If more space were provided, more liquid would vaporize, until equilibrium was reestablished at the same vapor pressure, assuming the same temperature is maintained. In fact, the equilibrium vapor pressure depends only on the temperature of the liquid; a higher temperature causes a higher pressure, as the graph in Figure 12.31 indicates for the specific case of water. Only when the temperature and vapor pressure correspond to a point on the curved line, which is called the ***vapor pressure curve*** or the ***vaporization curve,*** can liquid and vapor phases coexist in equilibrium.

To illustrate the use of a vaporization curve, let's see what happens when water boils in a pot that is *open to the air.* Assume that the air pressure acting on the water is 1.01×10^5 Pa (one atmosphere). When boiling occurs, bubbles of water vapor

It takes less heat to boil water high on a mountain because the boiling point becomes less than 100 °C as the air pressure decreases.

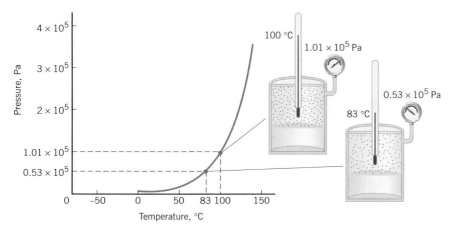

Figure 12.31 A plot of the equilibrium vapor pressure versus temperature is called the vapor pressure curve or the vaporization curve, the example shown being that for the liquid/vapor equilibrium of water.

(a) Water boiling

Cool water

Cork

(b) Water boiling again

Figure 12.32 (a) Water is boiling at a temperature of 100 °C and a pressure of one atmosphere. (b) The water boils at a temperature that is less than 100 °C, because the cool water reduces the pressure above the water in the flask.

THE PHYSICS OF...

spray cans.

form throughout the liquid, rise to the surface, and break. For these bubbles to form and rise, the pressure of the vapor inside them must at least equal the air pressure acting on the surface of the water. According to Figure 12.31, a value of 1.01×10^5 Pa corresponds to a temperature of 100 °C. Consequently, water boils at 100 °C at 1 atmosphere of pressure. In general, a *liquid boils at the temperature for which its vapor pressure equals the external pressure.* Water will not boil, then, at sea level if the temperature is only 83 °C, because at this temperature the vapor pressure of water is only 0.53×10^5 Pa (see Figure 12.31), a value less than the external pressure of 1.01×10^5 Pa. However, water does boil at 83 °C on a mountain at an altitude of just under five kilometers, because the atmospheric pressure there is 0.53×10^5 Pa.

The fact that water can boil at a temperature less than 100 °C leads to an interesting phenomenon that Conceptual Example 15 discusses.

CONCEPTUAL EXAMPLE 15   How to Boil Water That Is Cooling Down

Figure 12.32a shows water boiling in an open flask. Shortly after the flask is removed from the burner, the boiling stops. A cork is then placed in the neck of the flask to seal it. To restart the boiling, should you pour hot (but not boiling) water or cold water over the neck of the flask, as in part *b* of the drawing?

Reasoning and Solution When the open flask is removed from the burner, the water begins to cool. Boiling stops, because water cannot boil when its temperature is less than 100 °C and the external pressure is one atmosphere (1.01×10^5 Pa). Certainly, pouring hot water over the corked flask can rewarm the water within it to some extent. But since the water being poured is not boiling, its temperature must be less than 100 °C. Therefore, it cannot reheat the water within the flask to 100 °C and restart the boiling. However, when cold water is poured over the corked flask, it causes some of the water vapor inside to condense. Consequently, the pressure above the liquid in the flask drops. When the pressure drops to a certain level, boiling restarts. This occurs when the pressure in the flask becomes equal to the vapor pressure of the water at its current temperature (which is now less than 100 °C). *Thus, it is possible to restart the boiling by pouring cold water over the neck of the flask.*

The operation of spray cans is based on the equilibrium between a liquid and its vapor. Figure 12.33a shows that a spray can contains a liquid propellant that is mixed with the product (such as hair spray). Inside the can, propellant vapor forms over the liquid. A propellant is chosen that has an equilibrium vapor pressure that is greater than atmospheric pressure at room temperature. Consequently, when the nozzle of the can is pressed, as in part *b* of the drawing, the vapor pressure forces the liquid propellant and product up the tube in the can and out the nozzle as a spray. When the nozzle is released, the coiled spring reseals the can and the propellant vapor builds up once again to its equilibrium value.

As is the case for liquid/vapor equilibrium, a solid can be in equilibrium with its liquid phase only at specific conditions of temperature and pressure. For each temperature, there is a single pressure at which the two phases can coexist in equilibrium. A plot of the equilibrium pressure versus equilibrium temperature is referred to as the *fusion curve,* and Figure 12.34a shows a typical curve for a normal substance. A normal substance expands upon melting (e.g., carbon dioxide and sulfur). Since higher pressures make it more difficult for such materials to expand, a higher melting temperature is needed for a higher pressure, and the fusion curve slopes upward to the right. Part *b* of the picture illustrates the fusion curve for water, one of the few substances that contract when they melt. Higher pressures make it easier for

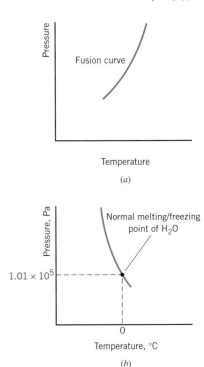

Figure 12.34 (*a*) The fusion curve for a normal substance that expands on melting. (*b*) The fusion curve for water, one of the few substances that contract on melting.

Figure 12.33 (*a*) A closed spray can containing liquid and vapor in equilibrium. (*b*) An open spray can.

such substances to melt. Consequently, a lower melting temperature is associated with a higher pressure, and the fusion curve slopes downward to the right.

It should be noted that just because two phases can coexist in equilibrium does not necessarily mean that they will. Other factors may prevent it. For example, water in an *open* bowl may never come into equilibrium with water vapor if air currents are present. Under such conditions the liquid, perhaps at a temperature of 25 °C, attempts to establish the corresponding equilibrium vapor pressure of 3.2×10^3 Pa. If air currents continually blow the water vapor away, however, equilibrium will never be established, and eventually the water will evaporate completely. Each kilogram of water that goes into the vapor phase takes along the latent heat of vaporization. Because of this heat loss, the remaining liquid would become cooler, except for the fact that the surroundings replenish the loss.

In the case of the human body, water is exuded by the sweat glands and evaporates from a much larger area than the surface of a typical bowl of water. The removal of heat along with the water vapor is called evaporative cooling and is one mechanism that the body uses to maintain its constant temperature. It's for the same reason that breath mints make your mouth feel cool. Mints contain sugar-alcohols—like, for instance, the menthol in peppermint. The sugar-alcohols evaporate when they come in contact with your warm mouth, and this evaporation is a cooling process.

THE PHYSICS OF...

evaporative cooling of the human body

*12.10 HUMIDITY

Air is a mixture of gases, including nitrogen, oxygen, and water vapor. The total pressure of the mixture is the sum of the partial pressures of the component gases. The partial pressure of a gas is the pressure it would exert if it alone occupied the entire volume at the same temperature as the mixture. The partial pressure of water vapor in air depends on weather conditions. It can be as low as zero or as high as the equilibrium vapor pressure of water at the given temperature.

THE PHYSICS OF...

relative humidity.

To provide an indication of how much water vapor is in the air, weather forecasters usually give the *relative humidity.* If the relative humidity is too low, the air contains such a small amount of water vapor that skin and mucous membranes tend to dry out. If the relative humidity is too high, especially on a hot day, we become very uncomfortable and our skin feels "sticky." Under such conditions, the air holds so much water vapor that the water exuded by sweat glands cannot evaporate efficiently. The relative humidity is defined as the ratio (expressed as a percentage) of the partial pressure of water vapor in the air to the equilibrium vapor pressure at a given temperature.

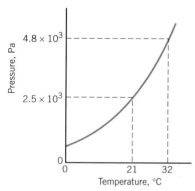

$$\begin{array}{c} \text{Percent} \\ \text{relative} \\ \text{humidity} \end{array} = \dfrac{\begin{array}{c}\text{Partial pressure}\\ \text{of water vapor}\end{array}}{\begin{array}{c}\text{Equilibrium vapor pressure of}\\ \text{water at the existing temperature}\end{array}} \times 100 \qquad (12.6)$$

The term in the denominator on the right of Equation 12.6 is given by the vaporization curve of water and is the pressure of the water vapor in equilibrium with the liquid. At a given temperature, the pressure of the water vapor in the air cannot exceed this value. If it did, the vapor would not be in equilibrium with the liquid and would condense as dew or rain to reestablish equilibrium.

Figure 12.35 The vaporization curve of water.

When the partial pressure of the water vapor equals the equilibrium vapor pressure of water at a given temperature, the relative humidity is 100%. In such a situation, the vapor is said to be *saturated* because it is present in the maximum amount, as it would be above a pool of liquid at equilibrium in a closed container. If the relative humidity is less than 100%, the water vapor is said to be *unsaturated.* Example 16 demonstrates how to find the relative humidity.

EXAMPLE 16   Relative Humidities

One day, the partial pressure of water vapor in the air is 2.0×10^3 Pa. Using the vaporization curve for water in Figure 12.35, determine the relative humidity if the temperature is (a) 32 °C and (b) 21 °C.

Reasoning and Solution

(a) According to Figure 12.35, the equilibrium vapor pressure of water at 32 °C is 4.8×10^3 Pa. Equation 12.6 reveals that the relative humidity is

$$\text{Relative humidity at 32 °C} = \frac{2.0 \times 10^3 \text{ Pa}}{4.8 \times 10^3 \text{ Pa}} \times 100 = \boxed{42\%}$$

(b) A similar calculation shows that

$$\text{Relative humidity at 21 °C} = \frac{2.0 \times 10^3 \text{ Pa}}{2.5 \times 10^3 \text{ Pa}} \times 100 = \boxed{80\%}$$

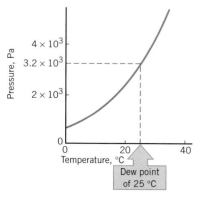

Figure 12.36 On the vaporization curve of water, the dew point is the temperature that corresponds to the actual partial pressure of water vapor in the air.

When air containing a given amount of water vapor is cooled, a temperature is reached in which the partial pressure of the vapor equals the equilibrium vapor pressure. This temperature is known as the *dew point.* For instance, Figure 12.36 shows that if the partial pressure of water vapor is 3.2×10^3 Pa, the dew point is 25 °C. This partial pressure would correspond to a relative humidity of 100%, if the ambient temperature were equal to the dew-point temperature. Hence, the dew point is the temperature below which water vapor in the air condenses in the form of liquid drops (dew or fog). The closer the actual temperature is to the dew point, the closer the relative humidity is to 100%. Thus, for fog to form, the air tempera-

THE PHYSICS OF...

fog formation.

ture must drop below the dew point. Similarly, water condenses on the outside of a cold glass when the temperature of the air next to the glass falls below the dew point. And the cold coils in a home dehumidifier (see Figure 12.37) function very much in the same way that the cold glass does. The coils are kept cold by a circulating refrigerant. When the air blown across them by the fan cools below the dew point, water vapor condenses in the form of droplets, which collect in a receptacle.

THE PHYSICS OF...

a home dehumidifier.

Figure 12.37 The cold coils of a dehumidifier cool the air blowing across them below the dew point, and water vapor condenses out of the air.

12.11 CONCEPTS & CALCULATIONS

This section contains examples that discuss one or more conceptual questions, followed by a related quantitative problem. Example 17 provides insight on the variables involved when the length and volume of an object change due to a temperature change. Example 18 discusses how different factors affect the temperature change of an object to which heat is being added.

CONCEPTS & CALCULATIONS EXAMPLE 17
Linear and Volume Thermal Expansion

Figure 12.38 shows three rectangular blocks made from the same material. The initial dimensions of each are expressed as a multiple of D, where $D = 2.00$ cm. They are heated and their temperatures increase by 35.0 C°. The coefficients of linear and volume expansion are $\alpha = 1.50 \times 10^{-5}$ (C°)$^{-1}$ and $\beta = 4.50 \times 10^{-5}$ (C°)$^{-1}$, respectively. Determine the change in their (a) vertical heights and (b) volumes.

Concept Questions and Answers Does the change in the vertical height of a block depend only on its height, or does it also depend on its width and depth? Without doing any calculations, rank the blocks according to their change in height, largest first.

Answer According to Equation 12.2, $\Delta L = \alpha L_0 \Delta T$, the change ΔL in the height of an object depends on its original height L_0, the change in temperature ΔT, and the coefficient of linear expansion, α. It does not depend on the width or depth of the object. Since blocks B and C have twice the height of A, their heights will increase by twice as much as that of A. The heights of B and C, however, will increase by the same amount, even though C is twice as wide.

Does the change in the volume of a block depend only on its height, or does it also depend on its width and depth? Without doing any calculations, rank the blocks according to their greatest change in volume, largest first.

Answer According to Equation 12.3, $\Delta V = \beta V_0 \Delta T$, the change ΔV in the volume of an object depends on its original volume V_0, the change in temperature ΔT, and the coefficient of volume expansion, β. Thus, the change in volume depends on height, width, and depth, because the original volume is the product of the three dimensions. The initial volumes of A, B, and C, are, respectively, $D \times 2D \times 2D = 4D^3$, $2D \times D \times D = 2D^3$, and $2D \times 2D \times D = 4D^3$. Blocks A and C have equal volumes, which are greater than that of B. Thus, we expect A and C to exhibit the greatest increase in volume, while B exhibits the smallest increase.

Solution

(a) The change in the height of each block is given by Equation 12.2 as

$$\Delta L_A = \alpha D \, \Delta T$$

$$= [1.50 \times 10^{-5} \,(\text{C}°)^{-1}](2.00 \text{ cm})(35.0 \text{ C}°) = \boxed{1.05 \times 10^{-3} \text{ cm}}$$

Figure 12.38 The temperatures of the three blocks are raised by the same amount. Which one(s) experience the greatest change in height and which the greatest change in volume?

$$\Delta L_B = \alpha(2D)\Delta T$$
$$= [1.50 \times 10^{-5}\ (\text{C}°)^{-1}](2 \times 2.00\ \text{cm})(35.0\ \text{C}°) = \boxed{2.10 \times 10^{-3}\ \text{cm}}$$
$$\Delta L_C = \alpha(2D)\Delta T$$
$$= [1.50 \times 10^{-5}\ (\text{C}°)^{-1}](2 \times 2.00\ \text{cm})(35.0\ \text{C}°) = \boxed{2.10 \times 10^{-3}\ \text{cm}}$$

As expected, the heights of B and C increase more than that of A.

(b) The change in the volume of each block is given by Equation 12.3 as

$$\Delta V_A = \beta(D \times 2D \times 2D)\Delta T$$
$$= [4.50 \times 10^{-5}\ (\text{C}°)^{-1}][4(2.00\ \text{cm})^3](35.0\ \text{C}°) = \boxed{5.04 \times 10^{-2}\ \text{cm}^3}$$
$$\Delta V_B = \beta(2D \times D \times D)\Delta T$$
$$= [4.50 \times 10^{-5}\ (\text{C}°)^{-1}][2(2.00\ \text{cm})^3](35.0\ \text{C}°) = \boxed{2.52 \times 10^{-2}\ \text{cm}^3}$$
$$\Delta V_C = \beta(2D \times 2D \times D)\Delta T$$
$$= [4.50 \times 10^{-5}\ (\text{C}°)^{-1}][4(2.00\ \text{cm})^3](35.0\ \text{C}°) = \boxed{5.04 \times 10^{-2}\ \text{cm}^3}$$

As discussed earlier, the greatest change in volume occurs with A and C followed by B.

CONCEPTS & CALCULATIONS EXAMPLE 18   Heat and Temperature Changes

Figure 12.39 Which block has the greatest change in temperature when the same heat is supplied to each?

Objects A and B in Figure 12.39 are made from copper, but the mass of B is three times that of A. Object C is made from glass and has the same mass as B. The *same amount of* heat Q is supplied to each one: $Q = 14$ J. Determine the rise in temperature for each.

Concept Questions and Answers Which object, A or B, experiences the greater rise in temperature?

Answer Consider an extreme example. Suppose a cup and a swimming pool are filled with water. For the same heat input, would you intuitively expect the temperature of the cup to rise more than that of the pool? Yes, because the cup has less mass. Another way to arrive at this conclusion is to solve Equation 12.4 for the change in temperature: $\Delta T = Q/cm$. Since the heat Q and the specific heat capacity c are the same for A and B, ΔT is inversely proportional to the mass m. So smaller masses give rise to larger temperature changes. Therefore, A has a greater temperature change than B.

Which object, B or C, experiences the greater rise in temperature?

Answer Objects B and C are made from different materials. To see how the rise in temperature depends on the type of material, let's again use Equation 12.4: $\Delta T = Q/cm$. Since the heat Q and the mass m are the same for B and C, ΔT is inversely proportional to the specific heat capacity c. So smaller specific heat capacities give rise to larger temperature changes. Table 12.2 indicates that the specific heat capacities of copper and glass are 387 and 840 J/(kg·C°), respectively. Since B is made from copper, which has the smaller specific heat capacity, it has the greater temperature change.

Solution According to Equation 12.4, the temperature change for each object is

$$\Delta T_A = \frac{Q}{c_A m_A} = \frac{14\ \text{J}}{[387\ \text{J/(kg·C}°)](2.0 \times 10^{-3}\ \text{kg})} = \boxed{18\ \text{C}°}$$

$$\Delta T_B = \frac{Q}{c_B m_B} = \frac{14\ \text{J}}{[387\ \text{J/(kg·C}°)](6.0 \times 10^{-3}\ \text{kg})} = \boxed{6.0\ \text{C}°}$$

$$\Delta T_C = \frac{Q}{c_C m_C} = \frac{14\ \text{J}}{[840\ \text{J/(kg·C}°)](6.0 \times 10^{-3}\ \text{kg})} = \boxed{2.8\ \text{C}°}$$

As anticipated, ΔT_A is greater than ΔT_B, and ΔT_B is greater than ΔT_C.

At the end of the problem set for this chapter, you will find homework problems that contain both conceptual and quantitative parts. These problems are grouped under the heading *Concepts & Calculations, Group Learning Problems.* They are designed for use by students working alone or in small learning groups. The conceptual part of each problem provides a convenient focus for group discussions.

SUMMARY

On the **Celsius temperature scale,** there are 100 equal divisions between the ice point (0 °C) and the steam point (100 °C). On the **Fahrenheit temperature scale,** there are 180 equal divisions between the ice point (32 °F) and the steam point (212 °F). For most scientific work, the **Kelvin temperature scale** is the scale of choice. One kelvin is equal in size to one Celsius degree; however, the temperature T on the Kelvin scale differs from the temperature T_c on the Celsius scale by an additive constant of 273.15: $T = T_c + 273.15$. The lower limit of temperature is called **absolute zero** and is designated as 0 K on the Kelvin scale. The operation of any thermometer is based on the change in some physical property with temperature; this physical property is called a **thermometric property.**

Most substances expand when heated. For **linear expansion,** an object of length L_0 experiences a change in length ΔL when the temperature changes by ΔT: $\Delta L = \alpha L_0 \Delta T$, where α is the **coefficient of linear expansion.** For **volume expansion,** the change in volume ΔV of an object of volume V_0 is given by $\Delta V = \beta V_0 \Delta T$, where β is the **coefficient of volume expansion.** When the temperature changes, a hole in a plate or a cavity in a piece of solid material expands or contracts as if the hole or cavity were filled with the surrounding material.

For an object held rigidly in place, a **thermal stress** can occur when the object attempts to expand or contract. The thermal stress can be extremely large, even when the temperature change is small.

The **internal energy** of a substance is the sum of the internal kinetic, potential, and other kinds of energy that the molecules of the substance have. **Heat** is energy that flows from a higher-temperature object to a lower-temperature object because of the difference in temperatures. The SI unit for heat is the joule.

The amount of heat Q that must be supplied or removed to change the temperature of a substance of mass m by an amount ΔT is $Q = cm\,\Delta T$, where c is the **specific heat capacity.** When materials are placed in thermal contact within a perfectly insulated container, the **principle of energy conservation** requires that heat lost by warmer materials equals heat gained by cooler materials.

Heat must be supplied or removed to make a material change from one phase to another. The heat Q that must be supplied or removed to change the phase of a mass m of a substance is $Q = mL$, where L is the latent heat of the substance. The **latent heats of fusion, vaporization,** and **sublimation** refer, respectively, to the solid/liquid, liquid/vapor, and solid/vapor phase changes.

The **equilibrium vapor pressure** of a substance is the pressure of the vapor phase that is in equilibrium with the liquid phase. Vapor pressure depends only on temperature. For a liquid, a plot of the equilibrium vapor pressure versus temperature is called the **vapor pressure** or **vaporization curve.** The **fusion curve** gives the combinations of temperature and pressure for equilibrium between solid and liquid phases.

The **relative humidity** is the ratio (expressed as a percentage) of the partial pressure of water vapor in the air to the equilibrium vapor pressure at the existing temperature. The **dew point** is the temperature below which the water vapor in the air condenses. On the vaporization curve of water, the dew point is the temperature that corresponds to the actual pressure of water vapor in the air.

CONCEPTUAL QUESTIONS

1. For the highest accuracy, would you choose an aluminum or a steel tape rule for year-round outdoor use? Why?

2. The first international standard of length was a metal bar kept at the International Bureau of Weights and Measures. One meter of length was defined to be the distance between two fine lines engraved near the ends of the bar. Why was it important that the bar be kept at a constant temperature?

3. A circular hole is cut through a flat aluminum plate. A spherical brass ball has a diameter that is slightly *smaller* than the diameter of the hole. The plate and the ball have the same temperature at all times. Should the plate and ball both be heated or both be cooled to *prevent* the ball from falling through the hole? Give your reasoning.

4. For added strength, many highways and buildings are constructed with reinforced concrete (concrete that is reinforced with embedded steel rods). Table 12.1 shows that the coefficient of linear expansion for concrete is the same as that for steel. Why is it important that these two coefficients be the same?

5. At a certain temperature, a rod is hung from an aluminum frame, as the drawing shows. A small gap exists between the rod and the floor. The frame and rod are heated uniformly. Explain whether the rod will ever touch the floor, assuming that the rod is made from (a) aluminum and (b) lead.

Problem 5

6. A simple pendulum is made using a long thin metal wire. When the temperature drops, does the period of the pendulum increase, decrease, or remain the same? Account for your answer.

7. Two bimetallic strips have identical shapes. One is made from aluminum and steel, while the other is made from copper and silver. Which is likely to bend more for a given change in temperature? Justify your answer.

8. A hot steel ring fits snugly over a cold brass cylinder. The temperatures of the ring and cylinder are, respectively, above and below room temperature. Account for the fact that it is nearly impossible to pull the ring off the cylinder once the assembly has reached room temperature.

9. For glass baking dishes, Pyrex glass is used instead of common glass. A cold Pyrex dish taken from the refrigerator can be put directly into a hot oven without cracking from thermal stress. A dish made from common glass would crack. With the aid of Table 12.1, explain why Pyrex is better in this respect than common glass.

10. Suppose liquid mercury and glass both had the same coefficient of volume expansion. Explain why such a mercury-in-glass thermometer would not work.

11. Is the buoyant force provided by warm water (above 4 °C) greater than, less than, or equal to that provided by cold water (also above 4 °C)? Explain your reasoning.

12. When the bulb of a mercury-in-glass thermometer is inserted into boiling water, the mercury column first drops slightly before it begins to rise. Account for this phenomenon. *(Hint: Consider what happens initially to the glass.)*

13. Two different objects are supplied with equal amounts of heat. Give the reason(s) why their temperature changes would not necessarily be the same.

14. Two objects are made from the same material. They have different masses and temperatures. If the two are placed in contact, which object will experience the greater temperature change? Explain.

15. Two identical mugs contain hot coffee from the same pot. One mug is full, while the other is only one-quarter full. Sitting on the kitchen table, which mug stays warmer longer? Explain.

16. Bicyclists often carry along a water bottle on hot days. When the bottle is wrapped in a wet sock, the water temperature stays near 75 °F for three or four hours, even when the air temperature is near 100 °F. Why does the water stay cool?

17. 🔊 To help lower the high temperature of a sick patient, an alcohol rub is sometimes used. Isopropyl alcohol is rubbed over the patient's back, arms, legs, etc., and allowed to evaporate. Why does the procedure work?

18. Your head feels colder under an air-conditioning vent when your hair is wet than when it is dry. Why?

19. 🔊 Suppose the latent heat of vaporization of H_2O were one-tenth its actual value. (a) Other things being equal, would it take the same time, a shorter time, or a longer time for a pot of water on a stove to boil away? (b) Would the evaporative cooling mechanism of the human body be as effective? Account for both answers.

20. Fruit blossoms are permanently damaged when the temperature drops below about −4 °C (a "hard freeze"). Orchard owners sometimes spray a film of water over the blossoms to protect them when a hard freeze is expected. From the point of view of phase changes, give a reason for the protection.

21. A camping stove is used to boil water on a mountain. Does it necessarily follow that the same stove can boil water at lower altitudes, such as at sea level? Provide a reason for your answer.

22. If a bowl of water is placed in a closed container and water vapor is pumped away rapidly enough, the remaining liquid will turn to ice. Explain why the ice appears.

23. 🔊 Medical instruments are sterilized under the hottest possible temperatures. Explain why they are sterilized in an autoclave, which is a device that is essentially a pressure cooker and heats the instruments in water under a pressure greater than one atmosphere.

24. A bottle of carbonated soda is left outside in subfreezing temperatures, although it remains in the liquid form. When the soda is brought inside and opened, it immediately freezes. Explain why this could happen.

25. A bowl of water is covered tightly and allowed to sit at a constant temperature of 23 °C for a long time. What is the relative humidity in the space between the surface of the water and the cover? Justify your answer.

26. Is it possible for dew to form on Tuesday night and not on Monday night, even though Monday night is the cooler night? Incorporate the idea of the dew point into your answer.

27. Two rooms in a house have the same temperature. One of the rooms contains an indoor swimming pool. On a cold day the windows of one of the rooms are "steamed up." Which room is it? Explain.

28. A jar is half filled with boiling water. The lid is then screwed on the jar. After the jar has cooled to room temperature, the lid is difficult to remove. Why?

PROBLEMS

Note: For problems in this set, use the values of α and β given in Table 12.1, and the values of c, L_f, and L_v given in Tables 12.2 and 12.3, unless stated otherwise.

ssm Solution is in the Student Solutions Manual. **www** Solution is available on the World Wide Web at http://www.wiley.com/college/cutnell

⚖ This icon represents a biomedical application.

Section 12.1 Common Temperature Scales, Section 12.2 The Kelvin Temperature Scale, Section 12.3 Thermometers

1. ssm What's your normal body temperature? It may not be 98.6 °F, the oft-quoted average that was determined in the nineteenth century. A recent study has reported an average temperature of 98.2 °F. What is the *difference* between these averages, expressed in Celsius degrees?

2. ⚖ Dermatologists often remove small precancerous skin lesions by freezing them quickly with liquid nitrogen, which has a temperature of 77 K. What is this temperature on the (a) Celsius and (b) Fahrenheit scales?

3. A personal computer is designed to operate over the temperature range from 50.0 to 104 °F. To what do these temperatures correspond (a) on the Celsius scale and (b) on the Kelvin scale?

4. Suppose the temperature of a normal body is 98.6 °F. Your doctor tells you that your temperature is 310.6 kelvins. Convert this temperature to the Fahrenheit scale.

5. ssm A temperature of absolute zero occurs at −273.15 °C. What is this temperature on the Fahrenheit scale?

*****6.** On the Rankine temperature scale, which is sometimes used in engineering applications, the ice point is at 491.67 °R and the steam point is at 671.67 °R. Determine a relationship (analogous to Equation 12.1) between the Rankine and Fahrenheit temperature scales.

*****7. ssm** A constant-volume gas thermometer (see Figures 12.3 and 12.4) has a pressure of 5.00×10^3 Pa when the gas temperature is 0.00 °C. What is the temperature (in °C) when the pressure is 2.00×10^3 Pa?

*****8.** A copper–constantan thermocouple generates a voltage of 4.75×10^{-3} volts when the temperature of the hot junction is 110.0 °C and the reference junction is kept at 0.0 °C. If the voltage is proportional to the difference in temperature between the junctions, what is the temperature of the hot junction when the voltage is 1.90×10^{-3} volts?

Section 12.4 Linear Thermal Expansion

9. A steel aircraft carrier is 370 m long when moving through the icy North Atlantic at a temperature of 2.0 °C. By how much does the carrier lengthen when it is traveling in the warm Mediterranean Sea at a temperature of 21 °C?

10. A steel section of the Alaskan pipeline had a length of 65 m and a temperature of 18 °C when it was installed. What is its change in length when the temperature drops to a frigid −45 °C?

11. ssm www Find the approximate length of the Golden Gate bridge if it is known that the steel in the roadbed expands by 0.53 m when the temperature changes from +2 to +32 °C.

12. Conceptual Example 5 provides background for this problem. A hole is drilled through a copper plate whose temperature is 11 °C. (a) When the temperature of the plate is increased, will the radius of the hole be larger or smaller than the radius at 11 °C? Why? (b) When the plate is heated to 110 °C, by what fraction $\Delta r / r_0$ will the radius of the hole change?

13. A steel beam is used in the construction of a skyscraper. By what fraction $\Delta L / L_0$ does the length of the beam increase when the temperature changes from that on a cold winter day (−15 °F) to that on a summer day (+105 °F)?

14. The Eiffel Tower is a steel structure whose height increases by 18 cm when the temperature changes from −9 to +41 °C. What is its height (in meters) at the lower temperature?

15. ssm A rod made from a particular alloy is heated from 25.0 °C to the boiling point of water. Its length increases by 8.47×10^{-4} m. The rod is then cooled from 25.0 °C to the freezing point of water. By how much does the rod shrink?

16. When a copper rod is stretched, it will rupture when a tensile stress of 2.3×10^7 N/m² is applied at each end. A copper rod is fastened securely at both ends to an immovable support. The rod is then cooled. What is the change in temperature of the rod when it ruptures?

*****17.** The brass bar and the aluminum bar in the drawing are each attached to an immovable wall. At 28 °C the air gap between the rods is 1.3×10^{-3} m. At what temperature will the gap be closed?

*****18.** A lead sphere has a diameter that is 0.050% larger than the inner diameter of a steel ring when each has a temperature of 70.0 °C. Thus, the ring will not slip over the sphere. At what common temperature will the ring just slip over the sphere?

*****19. ssm www** A simple pendulum consists of a ball connected to one end of a thin brass wire. The period of the pendulum is 2.0000 s. The temperature rises by 140 C°, and the length of the wire increases. Determine the period of the heated pendulum.

*****20.** A brass spring has an unstrained length of 0.18 m and a spring constant of 1.3×10^4 N/m. The temperature increases from 21 to 135 °C. What is the magnitude of the compressional force that must be applied to the heated spring to bring it back to its original length?

*****21.** The length of each edge of an aluminum cube is 0.050 m. The cube is heated to 145 °C from 25 °C. What is the increase in the surface area of the cube?

******22.** A steel ruler is calibrated to read true at 20.0 °C. A draftsman uses the ruler at 40.0 °C to draw a line on a 40.0 °C copper plate. As indicated on the warm ruler, the length of the line is 0.50 m. To what temperature should the plate be cooled, such that the length of the line truly becomes 0.50 m?

**23. ssm A wire is made by attaching two segments together, end to end. One segment is made of aluminum and the other is steel. The effective coefficient of linear expansion of the two-segment wire is 19×10^{-6} (C°)$^{-1}$. What fraction of the length is aluminum?

**24. An aluminum wire of radius 3.0×10^{-4} m is stretched between the ends of a concrete block, as the drawing illustrates. When the system (wire and concrete) is at 35 °C, the tension in the wire is 50.0 N. What is the tension in the wire when the system is heated to 185 °C?

Section 12.5 Volume Thermal Expansion

25. A copper kettle contains water at 24 °C. When the water is heated to its boiling point, the volume of the kettle expands by 1.2×10^{-5} m^3. Determine the volume of the kettle at 24 °C.

26. A swimming pool contains 110 m^3 of water. The sun heats the water from 17 to 27 °C. What is the change in the volume of the water?

27. ssm When heated at constant pressure, a given volume of gas expands much more than an equal volume of liquid. The coefficient of volume expansion for air is about 3.7×10^{-3} (C°)$^{-1}$ at room temperature and one atmosphere of pressure. Find the ratio of the change in volume of air to the change in volume of water, assuming the same initial volumes and temperature changes.

28. Inside an aluminum block is a hollow spherical cavity of radius 5.0 cm. The block is placed in a furnace and heated from 23 to 282 °C. (a) What is the change in volume (in cm^3) of the cavity? (b) Does the cavity become larger or smaller? Why?

29. A thin spherical shell of silver has an inner radius of 1.5×10^{-2} m when the temperature is 25 °C. The shell is heated to 135 °C. Find the change in the interior volume of the shell.

30. Many hot-water heating systems have a reservoir tank connected directly to the pipeline, so as to allow for expansion when the water becomes hot. The heating system of a house has 76 m of copper pipe whose inside radius is 9.5×10^{-3} m. When the water and pipe are heated from 24 to 78 °C, what must be the minimum volume of the reservoir tank to hold the overflow of water?

31. ssm Suppose you are selling apple cider for two dollars a gallon when the temperature is 4.0 °C. The coefficient of volume expansion of the cider is 280×10^{-6} (C°)$^{-1}$. If the expansion of the container is ignored, how much more money (in pennies) would you make per gallon by refilling the container on a day when the temperature is 26 °C?

32. During an all-night cram session, a student heats up a one-half liter (0.50×10^{-3} m^3) glass (Pyrex) beaker of cold coffee. Initially, the temperature is 18 °C, and the beaker is filled to the brim. A short time later when the student returns, the temperature

has risen to 92 °C. The coefficient of volume expansion of coffee is the same as that of water. How much coffee (in cubic meters) has spilled out of the beaker?

*33. ssm The bulk modulus of water is $B = 2.2 \times 10^9$ N/m^2. What change in pressure ΔP (in atmospheres) is required to keep water from expanding when it is heated from 15 to 25 °C?

*34. At the bottom of an old mercury-in-glass thermometer is a 45-mm^3 reservoir filled with mercury. When the thermometer was placed under your tongue, the warmed mercury would expand into a very narrow cylindrical channel, called a capillary, whose radius was 1.7×10^{-2} mm. Marks were placed along the capillary that indicated the temperature. Ignore the thermal expansion of the glass and determine how far (in mm) the mercury would expand into the capillary when the temperature changed by 1.0 C°.

*35. The density of mercury is 13 600 kg/m^3 at 0 °C. What would be its density at 166 °C?

*36. A solid aluminum sphere has a radius of 0.50 m and a temperature of 75 °C. The sphere is then completely immersed in a pool of water whose temperature is 25 °C. The sphere cools, while the water temperature remains nearly at 25 °C, because the pool is very large. The sphere is weighed in the water immediately after being submerged (before it begins to cool) and then again after cooling to 25 °C. (a) Which weight is larger? Why? (b) Use Archimedes' principle to find the magnitude of the *difference* between the weights.

**37. ssm www Two identical thermometers made of Pyrex glass contain, respectively, identical volumes of mercury and methyl alcohol. If the expansion of the glass is taken into account, how many times greater is the distance between the degree marks on the methyl alcohol thermometer than that on the mercury thermometer?

**38. The column of mercury in a barometer (see Figure 11.12) has a height of 0.760 m when the pressure is one atmosphere and the temperature is 0.0 °C. Ignoring any change in the glass containing the mercury, what will be the height of the mercury column for the same one atmosphere of pressure when the temperature rises to 38.0 °C on a hot day? *(Hint: The pressure in the barometer is given by* Pressure $= \rho g h$, *and the density ρ of the mercury changes when the temperature changes.)*

Section 12.6 Heat and Internal Energy, Section 12.7 Heat and Temperature Change: Specific Heat Capacity

39. A precious-stone dealer wishes to find the specific heat capacity of a 0.030-kg gemstone. The specimen is heated to 95.0 °C and then placed in a 0.15-kg copper vessel that contains 0.080 kg of water at equilibrium at 25.0 °C. The loss of heat to the external environment is negligible. When equilibrium is established, the temperature is 28.5 °C. What is the specific heat capacity of the specimen?

40. Blood can carry excess energy from the interior to the surface of the body, where the energy is dispersed in a number of ways. While a person is exercising, 0.6 kg of blood flows to the surface of the body and releases 2000 J of energy. The blood arriving at the surface has the temperature of the body interior, 37.0 °C. Assuming that blood has the same specific heat capacity as water, determine the temperature of the blood that leaves the surface and returns to the interior.

41. ssm At a fabrication plant, a hot metal forging has a mass of 75 kg and a specific heat capacity of 430 J/(kg·C°). To harden it, the forging is quenched by immersion in 710 kg of oil that has a temperature of 32 °C and a specific heat capacity of 2700 J/(kg·C°). The final temperature of the oil and forging at thermal equilibrium is 47 °C. Assuming that heat flows only between the forging and the oil, determine the initial temperature of the forging.

42. Ideally, when a thermometer is used to measure the temperature of an object, the temperature of the object itself should not change. However, if a significant amount of heat flows from the object to the thermometer, the temperature will change. A thermometer has a mass of 31.0 g, a specific heat capacity of $c = 815$ J/(kg·C°), and a temperature of 12.0 °C. It is immersed in 119 g of water, and the final temperature of the water and thermometer is 41.5 °C. What was the temperature of the water before the insertion of the thermometer?

43. ⚖ When resting, a person has a metabolic rate of about 4.2×10^5 joules per hour. The person is submerged neck-deep into a tub containing 1.0×10^3 kg of water at 27.00 °C. If the heat from the person goes only into the water, find the water temperature after half an hour.

44. When you take a bath, how many kilograms of hot water (49.0 °C) must you mix with cold water (13.0 °C) so that the temperature of the bath is 36.0 °C? The total mass of water (hot plus cold) is 191 kg. Ignore any heat flow between the water and its external surroundings.

45. ssm If the price of electrical energy is $0.10 per kilowatt·hour, what is the cost of using electrical energy to heat the water in a swimming pool (12.0 m × 9.00 m × 1.5 m) from 15 to 27 °C?

* **46.** In a passive solar house, the sun heats water stored in barrels to a temperature of 38 °C. The stored energy is then used to heat the house on cloudy days. Suppose that 2.4×10^8 J of heat are needed to maintain the inside of the house at 21 °C. How many barrels (1 barrel = 0.16 m³) of water are needed?

* **47.** The box of a well-known breakfast cereal states that one ounce of the cereal contains 110 Calories (1 food Calorie = 4186 J). If 2.0% of this energy could be converted by a weight lifter's body into work done in lifting a barbell, what is the heaviest barbell that could be lifted a distance of 2.1 m?

* **48.** The water in a swimming pool absorbs 2.00×10^9 J of heat from the sun. What is the change in volume of the water?

* **49. ssm** An electric hot water heater takes in cold water at 13.0 °C and delivers hot water. The hot water has a constant temperature of 45.0 °C, when the "hot" faucet is left open all the time and the volume flow rate is 5.0×10^{-6} m³/s. What is the minimum power rating of the hot water heater?

** **50.** A steel rod ($\rho = 7860$ kg/m³) has a length of 2.0 m. It is bolted at both ends between immobile supports. Initially there is no tension in the rod, because the rod just fits between the supports. Find the tension that develops when the rod loses 3300 J of heat.

Section 12.8 Heat and Phase Change: Latent Heat

51. ssm How much heat must be added to 0.45 kg of aluminum to change it from a solid at 130 °C to a liquid at 660 °C (its melting point)? The latent heat of fusion for aluminum is 4.0×10^5 J/kg.

52. A 10.0-kg block of ice has a temperature of −10.0 °C. The pressure is one atmosphere. The block absorbs 4.11×10^6 J of heat. What is the final temperature of the liquid water?

53. Suppose the amount of heat removed when 3.0 kg of water freezes at 0 °C were removed from ethyl alcohol at its freezing/melting point of −114 °C. How many kilograms of ethyl alcohol would freeze?

54. ⚖ A person eats a container of yogurt. The Nutritional Facts label states that it contains 240 Calories (1 Calorie = 4186 J). What mass of perspiration would one have to lose to get rid of this energy? At body temperature, the latent heat of vaporization of water is 2.42×10^6 J/kg.

55. ssm Liquid nitrogen boils at a chilly −195.8 °C when the pressure is one atmosphere. A silver coin of mass 1.5×10^{-2} kg and temperature 25 °C is dropped into the boiling liquid. What mass of nitrogen boils off as the coin cools to −195.8 °C?

56. ⚖ The latent heat of vaporization of H_2O at body temperature (37.0 °C) is 2.42×10^6 J/kg. To cool the body of a 75-kg jogger [average specific heat capacity = 3500 J/(kg·C°)] by 1.5 C°, how many kilograms of water in the form of sweat have to be evaporated?

57. A woman finds the front windshield of her car covered with ice at −12.0 °C. The ice has a thickness of 4.50×10^{-4} m, and the windshield has an area of 1.25 m². The density of ice is 917 kg/m³. How much heat is required to melt the ice?

58. At the beach a 2.4-kg bag of ice in a cooler completely melts in 8.0 h. The temperature of the ice and the resulting liquid water is 0 °C. What is the average power entering the cooler?

59. When it rains, water vapor in the air condenses into liquid water, and energy is released. (a) How much energy is released when 0.0254 m (one inch) of rain falls over an area of 2.59×10^6 m² (one square mile)? (b) If the average energy needed to heat one home for a year is 1.50×10^{11} J, how many homes can be heated with the energy determined in part (a)?

* **60.** To help keep his barn warm on cold days, a farmer stores 840 kg of solar-heated water ($L_f = 3.35 \times 10^5$ J/kg) in barrels. For how many hours would a 2.0-kW electric space heater have to operate to provide the same amount of heat as the water does, when it cools from 10.0 to 0.0 °C and completely freezes?

* **61. ssm** Ice at −10.0 °C and steam at 130 °C are brought together at atmospheric pressure in a perfectly insulated container.

After thermal equilibrium is reached, the liquid phase at 50.0 °C is present. Ignoring the container and the equilibrium vapor pressure of the liquid at 50.0 °C, find the ratio of the mass of steam to the mass of ice.

*62. Two grams of liquid water are at 0 °C, and another two grams are at 100 °C. Heat is removed from the water at 0 °C, completely freezing it at 0 °C. This heat is then used to vaporize some of the water at 100 °C. What is the mass (in grams) of the liquid water that remains?

*63. A 35-kg block of ice at 0 °C is sliding on a horizontal surface. The initial speed of the ice is 6.5 m/s and the final speed is 4.8 m/s. Assume that the part of the block that melts has a very small mass and that all the heat generated by kinetic friction goes into the block of ice, and determine the mass of ice that melts into water at 0 °C.

*64. Heat is extracted from a certain quantity of steam at 100.0 °C. As a result, the steam changes into ice at 0.0 °C. If this energy were used to accelerate the ice from rest, what would be the linear speed of the ice? For comparison, bullet speeds of about 700 m/s are common.

*65. **ssm** It is claimed that if a lead bullet goes fast enough, it can melt completely when it comes to a halt suddenly, and all its kinetic energy is converted into heat via friction. Find the minimum speed of a lead bullet (initial temperature = 30.0 °C) for such an event to happen.

**66. A locomotive wheel is 1.00 m in diameter. A 25.0-kg steel band has a temperature of 20.0 °C and a diameter that is 6.00×10^{-4} m less than that of the wheel. What is the smallest mass of water vapor at 100 °C that can be condensed on the steel band to heat it, so that it will fit onto the wheel? Do not ignore the water that results from the condensation.

Section 12.9 Equilibrium Between Phases of Matter

67. **ssm** Use the vapor pressure curve that accompanies this problem to determine the temperature at which liquid carbon dioxide exists in equilibrium with its vapor phase when the pressure is 3.5×10^6 Pa.

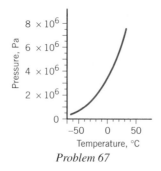

Problem 67

68. What pressure would be required for carbon dioxide to boil at a temperature of 30 °C? Use the vapor pressure curve that accompanies problem 67.

*69. A container is fitted with a movable piston of negligible mass and radius $r = 0.061$ m. Inside the container is liquid water in equilibrium with its vapor, as the drawing shows. The piston remains stationary with a 120-kg block on top of it. The air pressure acting on the top of the piston is one atmosphere. By using the vaporization curve for water in Figure 12.31, find the temperature of the water.

**70. A tall column of water is open to the atmosphere. At a depth of 10.3 m below the surface, the water is boiling. What is the temperature at this depth? Use the vaporization curve for water in Figure 12.31, as needed.

Section 12.10 Humidity

71. Using the vapor pressure curve for water that accompanies this problem, find the partial pressure of water vapor on a day when the weather forecast gives the relative humidity as 56.0% and the temperature as 30.0 °C.

72. On a warm spring day the relative humidity is 26% and the temperature is 30 °C. What will be the relative humidity that night when the temperature drops to 10 °C, assuming that the partial pressure of water vapor remains constant? Use the vapor pressure curve that accompanies problem 71.

73. **ssm** What is the relative humidity on a day when the temperature is 30 °C and the dew point is 10 °C? Use the vapor pressure curve that accompanies problem 71.

*74. The temperature of the air in a room is 36 °C. A person turns on a dehumidifier and notices that when the cooling coils reach 30 °C, water begins to condense on them. What is the relative humidity in the room? Use the vapor pressure curve that accompanies problem 71.

*75. **ssm** At a picnic, a glass contains 0.300 kg of tea at 30.0 °C, which is the air temperature. To make iced tea, someone adds 0.0670 kg of ice at 0.0 °C and stirs the mixture. When all the ice melts and the final temperature is reached, the glass begins to fog up, because water vapor condenses on the outer glass surface. Using the vapor pressure curve for water that accompanies problem 71, ignoring the specific heat capacity of the glass, and treating the tea as if it were water, estimate the relative humidity.

*76. A woman has been outdoors where the temperature is 10 °C. She walks into a 25 °C house, and her glasses "steam up." Using the vapor pressure curve for water that accompanies problem 71, find the smallest possible value for the relative humidity of the room.

ADDITIONAL PROBLEMS

77. The Concorde is 62 m long when its temperature is 23 °C. In flight, the outer skin of this supersonic aircraft can reach 105 °C due to air friction. The coefficient of linear expansion of the skin is 2.0×10^{-5} (C°)$^{-1}$. Find the amount by which the Concorde expands.

78. Assume that the pressure is one atmosphere and determine the heat required to produce 2.00 kg of water vapor at 100.0 °C, starting with (a) 2.00 kg of water at 100.0 °C and (b) 2.00 kg of liquid water at 0.0 °C.

79. ssm Find the mass of water that vaporizes when 2.10 kg of mercury at 205 °C is added to 0.110 kg of water at 80.0 °C.

80. A thermos contains 150 cm³ of coffee at 85 °C. To cool the coffee, you drop two 11-g ice cubes into the thermos. The ice cubes are initially at 0 °C and melt completely. What is the final temperature of the coffee? Treat the coffee as if it were water.

81. A commonly used method of fastening one part to another part is called "shrink fitting." A steel rod has a diameter of 2.0026 cm, and a flat plate contains a hole whose diameter is 2.0000 cm. The rod is cooled so that it just fits into the hole. When the rod warms up, the enormous thermal stress exerted by the plate holds the rod securely to the plate. By how many Celsius degrees should the rod be cooled?

82. When the temperature of a coin is raised by 75 C°, the coin's diameter increases by 2.3×10^{-5} m. If the original diameter is 1.8×10^{-2} m, find the coefficient of linear expansion.

83. ssm Suppose that the steel gas tank in your car is completely filled when the temperature is 17 °C. How many gallons will spill out of the twenty-gallon tank when the temperature rises to 35 °C?

84. An ice chest at a beach party contains 12 cans of soda at 5.0 °C. Each can of soda has a mass of 0.35 kg and a specific heat capacity of 3800 J/(kg·C°). Someone adds a 6.5-kg watermelon at 27 °C to the chest. The specific heat capacity of watermelon is nearly the same as that of water. Ignore the specific heat capacity of the chest and determine the final temperature T of the soda and watermelon.

85. The relative humidity is 35% when the temperature is 27 °C. Using the vapor pressure curve for water that accompanies problem 71, determine the dew point.

*86. Space invaders land on earth. On the invaders' temperature scale, the ice point is at 25 °I (I = invader), and the steam point is at 156 °I. The invaders' thermometer shows the temperature on earth to be 58 °I. Using logic similar to that in Example 1 in the text, what would this temperature be on the Celsius scale?

*87. ssm www An unknown material has a normal melting/freezing point of −25.0 °C, and the liquid phase has a specific heat capacity of 160 J/(kg·C°). One-tenth of a kilogram of the solid at −25.0 °C is put into a 0.150-kg aluminum calorimeter cup that contains 0.100 kg of glycerin. The temperature of the cup and the glycerin is initially 27.0 °C. All the unknown material melts, and the final temperature at equilibrium is 20.0 °C. The calorimeter loses no energy to the external environment. What is the latent heat of fusion of the unknown material?

*88. Occasionally, huge icebergs are found floating on the ocean's currents. Suppose one such iceberg is 120 km long, 35 km wide, and 230 m thick. (a) How much heat would be required to melt this iceberg (assumed to be at 0 °C) into liquid water at 0 °C? The density of ice is 917 kg/m³. (b) The annual energy consumption by the United States in 1994 was 9.3×10^{19} J. If this energy were delivered to the iceberg every year, how many years would it take before the ice melted?

*89. A 0.25-kg coffee mug is made from a material that has a specific heat capacity of 950 J/(kg·C°) and contains 0.30 kg of water. The cup and water are at 25 °C. To make a cup of coffee, a small electric heater is immersed in the water and brings it to a boil in two minutes. Assume that the cup and water always have the same temperature and determine the minimum power rating of this heater.

*90. A steel ruler is accurate when the temperature is 25 °C. When the temperature drops to −15 °C, the ruler no longer reads correctly, but it can be made to read correctly if a stress is applied to each end of the ruler. (a) Should the stress be a compression or a tension? Why? (b) What is the magnitude of the necessary stress?

*91. ssm A rock of mass 0.20 kg falls from rest from a height of 15 m into a pail containing 0.35 kg of water. The rock and water have the same initial temperature. The specific heat capacity of the rock is 1840 J/(kg·C°). Ignore the heat absorbed by the pail itself, and determine the rise in the temperature of the rock and water.

**92. A steel bicycle wheel (without the rubber tire) is rotating freely with an angular speed of 18.00 rad/s. The temperature of the wheel changes from −100.0 to +300.0 °C. No net external torque acts on the wheel, and the mass of the spokes is negligible. (a) Does the angular speed increase or decrease as the wheel heats up? Why? (b) What is the angular speed at the higher temperature?

**93. A 1.5-kg steel sphere will not fit through a circular hole in a 0.85-kg aluminum plate, because the radius of the sphere is 0.10% larger than the radius of the hole. If both the sphere and the plate are always kept at the same temperature, how much heat must be put into the two so the ball just passes through the hole?

**94. An 85.0-N backpack is hung from the middle of an aluminum wire, as the drawing shows. The temperature of the wire drops by 20.0 C°. Find the tension in the wire at the lower temperature. Assume that the distance between the supports does not change, and ignore any thermal stress.

CONCEPTS & CALCULATIONS

GROUP LEARNING PROBLEMS

Note: Each of these problems consists of Concept Questions followed by a related quantitative Problem. They are designed for use by students working alone or in small learning groups. The Concept Questions involve little or no mathematics and are intended to stimulate group discussions. They focus on the concepts with which the problems deal. Recognizing the concepts is the essential initial step in any problem-solving technique.

95. Concept Questions
The drawing shows two thermometers, A and B, whose temperatures are measured in °A and °B. The ice and boiling points of water are also indicated. (a) Is the size of one degree on the A scale larger or smaller than on the B scale? Why? (b) Is a

given temperature on the A scale (e.g., +20 °A) hotter or colder than the same reading on the B scale (e.g., +20 °B)? Provide a reason for your answer.

Problem (a) Using the data in the drawing, determine the number of B° on the B scale that correspond to 1 A° on the A scale. (b) If the temperature of a substance reads +40.0 °A on the A scale, what would that temperature read on the B scale?

96. Concept Questions (a) When an object, such as a rod, is heated, what factors determine how much its length increases? (b) Suppose two rods are made from different materials, but their lengths change by the same amount when their temperatures change by the same amount. Are their initial lengths the same or different? Explain.

Problem One rod is made from lead and another from quartz. They are heated and experience the same change in temperature. If the initial length of the lead rod is 0.10 m, what is the initial length of the quartz rod? Be sure that your answer is consistent with your answer to the Concept Questions.

97. Concept Questions An aluminum can is filled to the brim with a liquid. The can and the liquid are heated so their temperatures change by the same amount. (a) In general, what factors determine how the volume of an object changes when it is heated? (b) Aluminum has a smaller coefficient of volume expansion than the liquid does. Which will expand more, the can or the liquid? Pro-

vide a reason for your answer. (c) How is the volume of liquid that spills over related to the changes in the volume of the can and the liquid?

Problem The can's initial volume at 5 °C is 3.5×10^{-4} m^3. The coefficient of volume expansion for aluminum is 69×10^{-6} (C°)$^{-1}$, according to Table 12.1. When the can and the liquid are heated to 78 °C, 3.6×10^{-6} m^3 of liquid spills over. What is the coefficient of volume expansion of the liquid?

98. Concept Questions (a) When heat is added to an object, what factors determine its change in temperature? (b) Suppose the same amount of heat is applied to two bars. They have the same mass, but experience different changes in temperature. Are the specific heat capacities the same for each bar? Why or why not?

Problem Two bars of identical mass are at 25 °C. One is made from glass and the other from another substance in Table 12.2. When identical amounts of heat are supplied to each, the glass bar reaches a temperature of 88 °C, while the other reaches 250.0 °C. What is the other substance?

99. Concept Questions Suppose you have two solid objects, A and B, made from different materials. They have the same mass, and each solid is at its melting temperature. You then add heat to melt them. (a) It takes less heat to melt A than B. Which has the larger latent heat of fusion? Why? (b) The mass of each is doubled. Does it require twice as much heat to melt them? Justify your answer.

Problem (a) Objects A and B have the same mass of 3.0 kg. They melt when 3.0×10^4 J of heat is added to A and 9.0×10^4 J is added to B. Determine the latent heats of fusion. (b) Find the heat required to melt object A when its mass is 6.0 kg. Verify that your answers are consistent with your answers to the Concept Questions.

100. Concept Questions On a cool autumn morning, the relative humidity is 100%. (a) Does this mean that the partial pressure of water in the air equals atmospheric pressure? Why or why not? (b) Suppose the air warms up in the afternoon, but the partial pressure of water in the air does not change. Is the humidity still 100%? Provide a reason for your answer.

Problem The vapor pressure of water at 20 °C is 2500 Pa. (a) What percentage of atmospheric pressure is this? (b) What percentage of the total air pressure at 20 °C is due to water vapor if the relative humidity is 100%? (c) The vapor pressure of water at 35 °C is 5500 Pa. What is the relative humidity at this temperature if the partial pressure of water in the air has not changed from that at 20 °C?

These emperor penguins survive in the extreme cold of Antarctica. One reason for their survival is that their feathers trap air, which helps to reduce the loss of body heat. This chapter considers the processes by which heat is lost or gained.

CHAPTER 13
THE TRANSFER OF HEAT

13.1 CONVECTION

● *Concepts at a Glance*   When heat is transferred to or from a substance, the internal energy of the substance can change as we saw in Chapter 12. This change in internal energy is accompanied by a change in temperature or a change in phase, as suggested by the Concepts-at-a-Glance chart in Figure 12.23. The transfer of heat from one place to another is important to our own well-being. For instance, within our homes furnaces distribute heat on cold days, and air conditioners remove it on hot days. Our bodies constantly transfer heat in one direction or another, to prevent the adverse effects of hypo- and hyperthermia. And virtually all our energy originates in the sun and is transferred to us over a distance of 150 million kilometers through the void of space. Today's sunlight provides the energy to drive photosynthesis in the plants that provide our food and, hence, metabolic energy. Ancient sunlight nurtured the organic matter that became the fossil fuels of oil, natural gas, and coal. This chapter examines the three processes by which heat is transferred: convection, conduction, and radiation. The Concepts-at-a-Glance chart in Figure 13.1, which is a modification of that in Figure 12.23, illustrates how these three methods of heat transfer are related to our previous study of heat and internal energy. ●

When part of a fluid is warmed, such as the air above a fire, the volume of the fluid expands, and the density decreases. According to Archimedes' principle (see Section 11.6), the surrounding cooler and denser fluid exerts a buoyant force on the warmer fluid and pushes it upward. As warmer fluid rises, the surrounding cooler fluid replaces it. This cooler fluid, in turn, is warmed and pushed upward. Thus, a continuous flow is established, which carries along heat. Whenever heat is transferred by the bulk movement of a gas or a liquid, the heat is said to be transferred by *convection.* The fluid flow itself is called a *convection current.*

Figure 13.1 *Concepts at a Glance* Heat is transferred from one place to another by three methods: convection, conduction, and radiation. As discussed in Chapter 12, heat can change the internal energy of a substance, which results in a change in the temperature or phase of the substance. Igloos take advantage of the insulating ability of packed snow and ice to reduce the amount of heat lost by conduction through the walls.

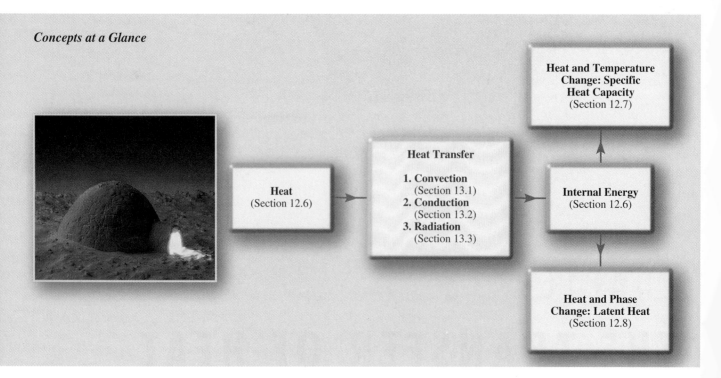

Concepts at a Glance

● CONVECTION

Convection is the process in which heat is carried from place to place by the bulk movement of a fluid.

The smoke rising above the fire in Figure 13.2 is one visible result of convection. Figure 13.3 shows an example of convection currents in a pan of water being heated on a gas burner. The currents distribute the heat from the burning gas to all parts of the water. Conceptual Example 1 deals with some of the important roles that convection plays in the home.

CONCEPTUAL EXAMPLE 1
Hot Water Baseboard Heating and Refrigerators

Hot water baseboard heating units are frequently used in homes, where they are mounted on the wall next to the floor, as in Figure 13.4a. In contrast, the cooling coil in a refrigerator is mounted near the top of the refrigerator, as in part b of the drawing. The locations for these heating and cooling devices are different, yet each location is designed to maximize the production of convection currents. Explain how.

Reasoning and Solution An important goal for a heating system is to distribute heat throughout a room. The analogous goal for the cooling coil is to remove heat from all of the space within a refrigerator. In each case, the heating or cooling device is positioned so that convection makes the goal achievable. The air above the baseboard unit is heated, like the air above a fire. Buoyant forces from the surrounding cooler air push the warm air upward. Cooler air near the ceiling is displaced downward and then warmed by the baseboard heating unit, leading to the convection current illustrated in Figure 13.4a. Had the heating unit been located near the ceiling, the warm air would have remained there, with very little convection to distribute the heat.

Within the refrigerator, as the air in contact with the top-mounted coil is cooled, its volume decreases and its density increases. The surrounding warmer and less

Figure 13.2 The smoke from this burning sugar cane field rises because of convection.

THE PHYSICS OF...

heating and cooling by convection.

Figure 13.3 Convection currents are set up when a pan of water is heated.

Convection current

Cooling coil

Hot water baseboard heating unit

(a)

(b)

Figure 13.4 (a) Air warmed by the baseboard heating unit is pushed to the top of the room by the cooler and denser air. (b) Air cooled by the cooling coil sinks to the bottom of the refrigerator. In both (a) and (b) a convection current is established.

dense air cannot provide sufficient buoyant force to support the colder air, which sinks downward. In the process, warmer air near the bottom is displaced upward and is then cooled by the coil, establishing the convection current shown in Figure 13.4*b*. Had the cooling coil been placed at the bottom of the refrigerator, stagnant, cool air would have collected there, with little convection to carry the heat from other parts of the refrigerator to the coil for removal.

THE PHYSICS OF...

"thermals."

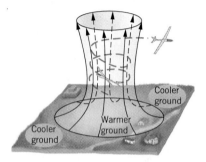

Figure 13.5 Updrafts, or thermals, are caused by the convective movement of air that the ground has warmed.

Another example of convection occurs when the ground, heated by the sun's rays, warms the neighboring air. Surrounding cooler and denser air pushes the heated air upward. The resulting updraft or "thermal" can be quite strong, depending on the amount of heat that the ground can supply. As Figure 13.5 illustrates, these thermals can be used by glider pilots to gain considerable altitude. Birds such as eagles utilize thermals in a similar fashion.

It is usual for air temperature to decrease with increasing altitude, and the resulting upward convection currents are important for dispersing pollutants from industrial sources and automobile exhaust systems. Sometimes, however, meteorological conditions cause a layer to form in the atmosphere where the temperature increases with increasing altitude. Such a layer is called an *inversion layer* because its temperature profile is inverted compared to the usual situation. An inversion layer arrests the normal upward convection currents, causing a stagnant-air condition in which the concentration of pollutants increases substantially. This leads to a smog layer that can often be seen hovering over large cities (Figure 13.6).

We have been discussing ***natural convection,*** in which a temperature difference causes the density at one place in a fluid to be different from that at another. Sometimes, natural convection is inadequate to transfer sufficient amounts of heat. In such cases ***forced convection*** is often used, and an external device such as a pump or a fan mixes the warmer and cooler portions of the fluid. Figure 13.7 shows an example of an automobile engine, where forced convection occurs in two ways. First, a pump circulates radiator fluid (water and antifreeze) through the engine to remove excess heat from the combustion process. Second, a radiator fan draws air through the radiator. Heat is transferred from the hotter radiator fluid to the cooler air, thereby cooling the fluid.

THE PHYSICS OF...

cooling by forced convection.

Figure 13.6 In the absence of upward convection currents in the air, pollutants accumulate and form a smog layer in Mexico City.

13.2 CONDUCTION

Anyone who has fried a hamburger in an all-metal skillet knows that the metal handle becomes hot. Somehow, heat is transferred from the burner to the handle. Clearly, heat is not being transferred by the bulk movement of the metal or the surrounding air, so convection can be ruled out. Instead, heat is transferred directly through the metal by a process called ***conduction.***

> ● **CONDUCTION**
>
> Conduction is the process whereby heat is transferred directly through a material, any bulk motion of the material playing no role in the transfer.

One mechanism for conduction occurs when the atoms or molecules in a hotter part of the material vibrate or move with greater energy than those in a cooler part. By means of collisions, the more energetic molecules pass on some of their energy to their less energetic neighbors. For example, imagine a gas filling the space between two walls that face each other and are maintained at different temperatures. Molecules strike the hotter wall, absorb energy from it, and rebound with a greater kinetic energy than when they arrived. As these more energetic molecules collide with their less energetic neighbors, they transfer some of their energy to them. Eventually, this energy is passed on until it reaches the molecules next to the cooler wall. These molecules, in turn, collide with the wall, giving up some of their energy to it in the process. Through such molecular collisions, heat is conducted from the hotter to the cooler wall.

A similar mechanism for the conduction of heat occurs in metals. Metals are different from most substances in having a pool of electrons that are more or less free to wander throughout the metal. These free electrons can transport energy and allow metals to transfer heat very well. The free electrons are also responsible for the excellent electrical conductivity that metals have.

Those materials that conduct heat well are called ***thermal conductors,*** while those that conduct heat poorly are known as ***thermal insulators.*** Most metals are excellent thermal conductors, while wood, glass, and most plastics are common thermal insulators. Thermal insulators have many important applications. Virtually all new housing construction incorporates thermal insulation in attics and walls to reduce heating and cooling costs. And the wooden or plastic handles on many pots and pans reduce the flow of heat to the cook's hand.

To illustrate the factors that influence the conduction of heat, Figure 13.8 displays a rectangular bar. The ends of the bar are in thermal contact with two bodies, one of which is kept at a constant higher temperature, while the other is kept at a constant lower temperature. Although not shown for the sake of clarity, the sides of the bar are insulated, so the heat lost through them is negligible. The amount of heat Q conducted through the bar from the warmer end to the cooler end depends on a number of factors:

1. Q is proportional to the length of time t during which conduction takes place ($Q \propto t$). More heat flows in longer time periods.

2. Q is proportional to the temperature difference ΔT between the ends of the bar ($Q \propto \Delta T$). A larger difference causes more heat to flow. No heat flows when both ends have the same temperature, so that $\Delta T = 0$.

3. Q is proportional to the cross-sectional area A of the bar ($Q \propto A$). Figure 13.9 helps to explain this fact by showing two identical bars (insulated sides not shown) placed between the warmer and cooler bodies. Clearly, twice as much heat flows through two bars as through one, since the cross-sectional area has been doubled.

4. Q is inversely proportional to the length L of the bar ($Q \propto 1/L$). Greater lengths of material conduct less heat. To experience this effect, put two insulated mittens (the kind that cooks keep around the stove) on the *same*

Figure 13.7 The forced convection generated by a pump circulates radiator fluid through an automobile engine to remove excessive heat.

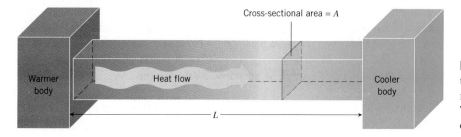

Figure 13.8 Heat is conducted through the bar when the ends of the bar are maintained at different temperatures. The heat flows from the warmer to the cooler end.

TABLE 13.1   Thermal Conductivities[a]
of Selected Materials

Substance	Thermal Conductivity, k [J/(s·m·C°)]
Metals	
Aluminum	240
Brass	110
Copper	390
Iron	79
Lead	35
Silver	420
Steel (stainless)	14
Gases	
Air	0.0256
Hydrogen (H_2)	0.180
Nitrogen (N_2)	0.0258
Oxygen (O_2)	0.0265
Other Materials	
Asbestos	0.090
Body fat	0.20
Concrete	1.1
Diamond	2450
Glass	0.80
Goose down	0.025
Ice (0 °C)	2.2
Styrofoam	0.010
Water	0.60
Wood (oak)	0.15
Wool	0.040

[a] Except as noted, the values pertain to temperatures near 20 °C.

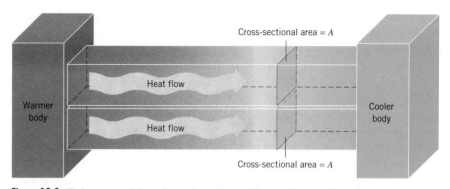

Figure 13.9 Twice as much heat flows through two identical bars as through one.

hand. Then, touch a hot pot and notice that it feels cooler than when you wear only one mitten, signifying that less heat passes through the greater thickness ("length") of material.

These proportionalities can be stated together as $Q \propto (A \, \Delta T)t/L$. Equation 13.1 expresses this result with the aid of a proportionality constant k, which is called the ***thermal conductivity.***

● **CONDUCTION OF HEAT THROUGH A MATERIAL**

The heat Q conducted during a time t through a bar of length L and cross-sectional area A is

$$Q = \frac{(kA \, \Delta T)t}{L} \qquad (13.1)$$

where ΔT is the temperature difference between the ends of the bar and k is the thermal conductivity of the material.

SI Unit of Thermal Conductivity: J/(s·m·C°)

Since $k = QL/(tA \, \Delta T)$, the SI unit for thermal conductivity is J·m/(s·m²·C°) or J/(s·m·C°). The SI unit of power is the joule per second (J/s) or watt (W), so the thermal conductivity is also given in units of W/(m·C°).

Different materials have different thermal conductivities, and Table 13.1 gives some representative values. Because metals are such good thermal conductors, they have large thermal conductivities. In comparison, liquids and gases generally have small thermal conductivities. In fact, in most fluids the heat transferred by conduction is negligible compared to that transferred by convection when there are strong convection currents. Air, for instance, with its small thermal conductivity, is an excellent thermal insulator when confined to small spaces where no appreciable convection currents can be established. Goose down, Styrofoam, and wool derive their fine insulating properties in part from the small dead-air spaces within them, as Figure 13.10 illustrates. We also take advantage of dead-air spaces when we dress "in layers" during very cold weather and put on several layers of relatively thin clothing rather than one thick layer. The air trapped between the layers acts as an excellent insulator.

THE PHYSICS OF...

dressing warm

Example 2 deals with the role that conduction through body fat plays in regulating body temperature.

EXAMPLE 2   Heat Transfer in the Human Body

When excessive heat is produced within the body, it must be transferred to the skin and dispersed if the temperature at the body interior is to be maintained at the normal value of 37.0 °C. One possible mechanism for transfer is conduction through body fat. Suppose that heat travels through 0.030 m of fat in reaching the skin, which has a total surface area of 1.7 m² and a temperature of 34.0 °C. Find the amount of heat that reaches the skin in half an hour (1800 s).

Reasoning and Solution In Table 13.1 the thermal conductivity of body fat is given as $k = 0.20$ J/(s·m·C°). According to Equation 13.1,

$$Q = \frac{(kA\,\Delta T)t}{L}$$

$$Q = \frac{[0.20 \text{ J/(s·m·C°)}](1.7 \text{ m}^2)(37.0 \text{ °C} - 34.0 \text{ °C})(1800 \text{ s})}{0.030 \text{ m}} = \boxed{6.1 \times 10^4 \text{ J}}.$$

For comparison, a jogger can generate over ten times this amount of heat in a half hour. Thus, conduction through body fat is not a particularly effective way of removing excess heat. Heat transfer via blood flow to the skin is more effective and has the added advantage that the body can vary the blood flow as needed (see problem 7).

Virtually all homes contain insulation in the walls to reduce heat loss. Example 3 illustrates how to determine this loss with and without insulation.

EXAMPLE 3   Layered Insulation

One wall of a house consists of 0.019-m-thick plywood backed by 0.076-m-thick insulation, as Figure 13.11 shows. The temperature at the inside surface is 25.0 °C, while the temperature at the outside surface is 4.0 °C, both being constant. The thermal conductivities of the insulation and plywood are, respectively, 0.030 and 0.080 J/(s·m·C°), and the area of the wall is 35 m². Find the heat conducted through the wall in one hour (a) with the insulation and (b) without the insulation.

Reasoning The temperature T at the insulation–plywood interface (see drawing) must be determined before the heat conducted through the wall can be obtained. In calculating this temperature, we use the fact that no heat is accumulating in the wall, for the inner and outer temperatures are constant. Therefore, the heat conducted through the insulation must equal the heat conducted through the plywood during the same time, that is, $Q_{\text{insulation}} = Q_{\text{plywood}}$. Each of the Q values can be expressed as $Q = (kA\,\Delta T)t/L$, according to Equation 13.1, leading to an expression that can be solved for the interface temperature. Once a value for T is available, Equation 13.1 can be used to obtain the heat conducted through the wall.

Solution

(a) Using the fact that $Q_{\text{insulation}} = Q_{\text{plywood}}$ and Equation 13.1, we find that

$$\left[\frac{(kA\,\Delta T)t}{L}\right]_{\text{insulation}} = \left[\frac{(kA\,\Delta T)t}{L}\right]_{\text{plywood}}$$

$$\frac{[0.030 \text{ J/(s·m·C°)}]A(25.0 \text{ °C} - T)t}{0.076 \text{ m}} = \frac{[0.080 \text{ J/(s·m·C°)}]A(T - 4.0 \text{ °C})t}{0.019 \text{ m}}$$

Eliminating the area A and time t algebraically and solving this equation for T reveals that the temperature at the insulation–plywood interface is $T = 5.8$ °C.

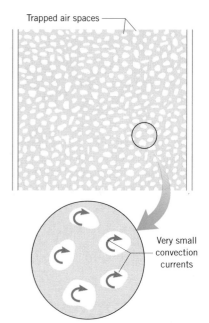

Trapped air spaces

Very small convection currents

Figure 13.10 Styrofoam is an excellent thermal insulator because it contains many small, dead-air spaces. These small spaces inhibit heat transfer by convection currents, and air itself has a very low thermal conductivity.

● **PROBLEM SOLVING INSIGHT**
When heat is conducted through a multiple-layer material, and the high and low temperatures are constant, the heat conducted through each layer is the same.

Plywood (0.019 m)

Insulation (0.076 m)

Interface temperature = T

Inside (25.0 °C)   Heat   Outside (4.0 °C)

Figure 13.11 Heat flows through the insulation and plywood from the warmer inside to the cooler outside. The temperature of the insulation–plywood interface is T.

THE PHYSICS OF...

protecting fruit plants from freezing.

The heat conducted through the wall is either $Q_{\text{insulation}}$ or Q_{plywood}, since the two quantities are equal. Choosing $Q_{\text{insulation}}$ and using $T = 5.8$ °C in Equation 13.1, we find that

$$Q_{\text{insulation}} = \frac{[0.030 \text{ J/(s} \cdot \text{m} \cdot \text{C°})](35 \text{ m}^2)(25.0 \text{ °C} - 5.8 \text{ °C})(3600 \text{ s})}{0.076 \text{ m}}$$

$$= \boxed{9.5 \times 10^5 \text{ J}}$$

(b) It is straightforward to use Equation 13.1 to calculate the amount of heat flowing through the plywood in one hour if the insulation were absent:

$$Q_{\text{plywood}} = \frac{[0.080 \text{ J/(s} \cdot \text{m} \cdot \text{C°})](35 \text{ m}^2)(25.0 \text{ °C} - 4.0 \text{ °C})(3600 \text{ s})}{0.019 \text{ m}}$$

$$= \boxed{110 \times 10^5 \text{ J}}$$

Without insulation, the heat loss is increased by a factor of about 12.

Fruit growers sometimes protect their crops by spraying them with water when overnight temperatures are expected to drop below freezing. Some fruit, like the strawberries in Figure 13.12, can withstand temperatures down to freezing (0 °C). However, as the temperature falls below freezing, the risk of damage rises significantly. When water is sprayed on the strawberries, it can freeze and form a covering of ice. When the water freezes it releases heat (see Section 12.8), some of which goes into warming the plant. In addition, both water and ice have relatively small thermal conductivities, as Table 13.1 indicates. Thus, they also protect the crop by acting as thermal insulators that reduce heat loss from the plants.

While a layer of ice may be beneficial to strawberry plants, it is not so desirable inside a refrigerator, as Conceptual Example 4 discusses.

CONCEPTUAL EXAMPLE 4   An Iced-up Refrigerator

In a refrigerator heat is removed by a cold refrigerant fluid that circulates within a tubular space embedded within a metal plate, as Figure 13.13 illustrates. A good refrigerator cools food as quickly as possible. Decide whether the plate should be made from aluminum or stainless steel and whether the arrangement works better or worse when it becomes coated with a layer of ice.

Reasoning and Solution Figure 13.13 (see blowups) shows the metal cooling plate with and without a layer of ice. Without ice, heat passes by conduction through the metal to the refrigerant fluid within. For a given temperature difference across the thickness of metal, heat is transferred more quickly through the metal with the largest thermal conductivity. Table 13.1 indicates that the thermal conductivity of aluminum is more than 17 times larger than that of stainless steel. Therefore, **the plate should be made from aluminum.** When the plate becomes coated with ice, any heat that is removed by the refrigerant fluid must first be transferred by conduction through the ice before it encounters the aluminum plate. But the conduction of heat through ice occurs much less readily than through aluminum, for Table 13.1 indicates that ice has a much smaller thermal conductivity. Moreover, Equation 13.1 indicates that the heat conducted per unit time (Q/t) is inversely proportional to the thickness L of the ice. Thus, as ice builds up, the heat removed per unit time by the cooling plate decreases. ***When covered with ice, the cooling plate works less well.***

Related Homework Material: *Problem 10*

13.3 RADIATION

Energy from the sun is brought to earth by large amounts of visible light waves, as well as by substantial amounts of infrared and ultraviolet waves. These waves belong to a class of waves known as electromagnetic waves, a class that also includes the microwaves used for cooking and the radio waves used for AM and FM broadcasts. The sunbather in Figure 13.14 feels hot because her body absorbs energy from the sun's electromagnetic waves. And anyone who has stood by a roaring fire or put a hand near an incandescent light bulb has experienced a similar effect. Thus, fires and light bulbs also emit electromagnetic waves, and when the energy of such waves is absorbed, it can have the same effect as heat.

The process of transferring energy via electromagnetic waves is called ***radiation,*** and, unlike convection or conduction, it does not require a material medium. Electromagnetic waves from the sun, for example, travel through the void of space during their journey to earth.

Figure 13.12 Fruit growers sometimes spray water to protect their crops against freezing. After a subzero night, these berries are visible in their insulating jackets of ice.

> ● **RADIATION**
> Radiation is the process in which energy is transferred by means of electromagnetic waves.

All bodies continuously radiate energy in the form of electromagnetic waves. Even an ice cube radiates energy, although so little of it is in the form of visible light that an ice cube cannot be seen in the dark. Likewise, the human body emits insufficient visible light to be seen in the dark. However, as Figures 12.6 and 12.7 illustrate, the infrared waves radiating from the body can be detected in the dark by electronic cameras. Generally, an object does not emit much visible light until the temperature of the object exceeds about 1000 K. Then a characteristic red glow appears, like that of a heating coil on an electric stove. When its temperature reaches about 1700 K, an object begins to glow white-hot, like the tungsten filament in an incandescent light bulb.

In the transfer of energy by radiation, the absorption of electromagnetic waves is just as important as the emission. The surface of an object plays a significant role in determining how much radiant energy the object will absorb or emit. The two blocks in sunlight in Figure 13.15, for example, are identical, except that one has a rough surface coated with lampblack (a fine black soot), while the other has a highly polished silver surface. As the thermometers indicate, the temperature of the black block rises at a much faster rate than that of the silvery block. This is because lampblack absorbs about 97% of the incident radiant energy, while the silvery surface absorbs only about 10%. The remaining part of the incident energy is reflected in each case. We see the lampblack as black in color because it reflects so little of the light falling on it, while the silvery surface looks like a mirror because it reflects so much light. Since the color black is associated with nearly complete absorption of visible light, the term ***perfect blackbody*** or, simply, ***blackbody*** is used when referring to an object that absorbs *all* the electromagnetic waves falling on it.

Tubular space containing refrigerant fluid

Aluminum plate

Refrigerant fluid

Heat   Heat

Aluminum   Ice

Figure 13.13 In a refrigerator, cooling is accomplished by a cold refrigerant fluid that circulates through a tubular space embedded within an aluminum plate. The arrangement works less well when coated with a layer of ice.

Figure 13.14 Suntans are produced by ultraviolet rays.

THE PHYSICS OF...

summer clothing.

THE PHYSICS OF...

a white sifaka lemur warming up.

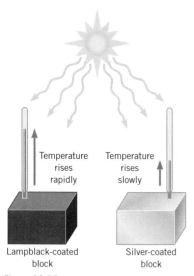

Figure 13.15 The temperature of the block coated with lampblack rises faster than the temperature of the block coated with silver because the black surface absorbs radiant energy from the sun at the greater rate.

All objects emit and absorb electromagnetic waves simultaneously. When a body has the same constant temperature as its surroundings, the amount of radiant energy being absorbed must balance the amount being emitted in a given interval of time. The block coated with lampblack absorbs and emits the same amount of radiant energy, and the silvery block does too. In either case, if absorption were greater than emission, the block would experience a net gain in energy. As a result, the temperature of the block would rise and not be constant. Similarly, if emission were greater than absorption, the temperature would fall. Since absorption and emission are balanced, *a material that is a good absorber, like lampblack, is also a good emitter, and a material that is a poor absorber, like polished silver, is also a poor emitter.* A perfect blackbody, being a perfect absorber, is also a perfect emitter.

The fact that a black surface is both a good absorber and a good emitter is the reason people are uncomfortable wearing dark clothes during the summer. Dark clothes absorb a large fraction of the sun's radiation and then reemit it in all directions. About one-half of the emitted radiation is directed inward toward the body and creates the sensation of warmth. Light-colored clothes, in contrast, are cooler to wear, since they absorb and reemit relatively little of the incident radiation.

The use of light colors for comfort also occurs in nature. Most lemurs, for instance, are nocturnal and have dark fur like the one shown in Figure 13.16*a*. Since they are active at night, the dark fur poses no disadvantage in absorbing excessive sunlight. Figure 13.16*b* shows a species of lemur called the white sifaka, which lives in semiarid regions where there is little shade. The white color of the fur may help in thermoregulation, by reflecting sunlight during the hot part of the day. However, during the cool mornings, reflection of sunlight would be a hindrance in warming up. It is interesting to note that these lemurs have black skin and only sparse fur on their bellies, and that to warm up in the morning, they turn their dark bellies toward the sun. The dark color enhances the absorption of sunlight.

The amount of radiant energy Q emitted by a perfect blackbody is proportional to the radiation time interval t ($Q \propto t$). The longer the time, the greater the amount of energy radiated. Experiment shows that Q is also proportional to the surface area A ($Q \propto A$). An object with a large surface area radiates more energy than one with a small surface area, other things being equal. Finally, experiment reveals that Q is proportional to the *fourth power of the Kelvin temperature T* ($Q \propto T^4$), so the emitted energy increases markedly with increasing temperature. If, for example, the Kelvin temperature of an object doubles, the object emits 2^4 or 16 times more energy. Combining these factors into a single proportionality, we see that $Q \propto T^4At$. This proportionality is converted into an equation by inserting a proportionality constant σ, known as the *Stefan–Boltzmann constant*. It has been found experimentally that $\sigma = 5.67 \times 10^{-8}$ J/(s·m^2·K^4):

$$Q = \sigma T^4 At$$

The relationship above holds only for a perfect emitter. Most objects are not perfect emitters, however. Suppose that an object radiates only about 80% of the visible light energy that a perfect emitter would radiate, so Q (for the object) = $(0.80)\sigma T^4 At$. The factor such as the 0.80 in this equation is called the *emissivity e* and is a dimensionless number between zero and one. The emissivity is the ratio of the energy an object actually radiates to the energy the object would radiate if it were a perfect emitter. For visible light, the value of e for the human body, for instance, varies between about 0.65 and 0.80, the smaller values pertaining to lighter skin colors. For infrared radiation, e is nearly one for all skin colors. For a perfect blackbody emitter, $e = 1$. Including the factor e on the right side of the expression $Q = \sigma T^4 At$ leads to the **Stefan–Boltzmann law of radiation.**

● THE STEFAN–BOLTZMANN LAW OF RADIATION

The radiant energy Q, emitted in a time t by an object that has a Kelvin temperature T, a surface area A, and an emissivity e, is given by

$$Q = e\sigma T^4 At \qquad (13.2)$$

where σ is the Stefan–Boltzmann constant and has a value of $5.67 \times 10^{-8} \text{ J/(s} \cdot \text{m}^2 \cdot \text{K}^4)$.

In Equation 13.2, the Stefan–Boltzmann constant σ is a universal constant in the sense that its value is the same for all bodies, regardless of the nature of their surfaces. The emissivity e, however, depends on the condition of the surface.

Example 5 shows how the Stefan–Boltzmann law can be used to determine the size of a star.

EXAMPLE 5   A Supergiant Star

The supergiant star Betelgeuse has a surface temperature of about 2900 K and emits a radiant power (in joules per second or watts) of approximately 4×10^{30} W. The temperature is about one-half and the power about 10 000 greater than that of our sun. Assuming that Betelgeuse is a perfect emitter (emissivity $e = 1$) and spherical, find its radius.

Reasoning According to the Stefan–Boltzmann law, the power emitted is $Q/t = e\sigma T^4 A$. A star with a relatively small temperature T can have a relatively large radiant power Q/t only if the area A is large. As we will see, Betelgeuse has a very large surface area, so its radius is enormous.

Solution Solving the Stefan–Boltzmann law for the area, we find

$$A = \frac{Q/t}{e\sigma T^4}$$

But the surface area of a sphere is $A = 4\pi r^2$, so $r = \sqrt{A/4\pi}$. Therefore, we have

$$r = \sqrt{\frac{Q/t}{4\pi e\sigma T^4}} = \sqrt{\frac{4 \times 10^{30} \text{ W}}{4\pi(1)[5.67 \times 10^{-8} \text{ J/(s} \cdot \text{m}^2 \cdot \text{K}^4)](2900 \text{ K})^4}}$$

$$= \boxed{3 \times 10^{11} \text{ m}}$$

For comparison, Mars orbits the sun at a distance of 2.28×10^{11} m. Betelgeuse is certainly a "supergiant."

(b)

Figure 13.16 (a) Most lemurs, like this one, are nocturnal and have dark fur. (b) The species of lemur called the white sifaka, however, is active during the day and has white fur.

● PROBLEM SOLVING INSIGHT
First solve an equation for the unknown in terms of the known variables. Then substitute numbers for the known variables, as this example shows.

The next example explains how to apply the Stefan–Boltzmann law when an object, such as a wood stove, simultaneously emits and absorbs radiant energy.

EXAMPLE 6   A Wood-Burning Stove

A wood-burning stove stands unused in a room where the temperature is 18 °C (291 K). A fire is started inside the stove. Eventually, the temperature of the stove surface reaches a constant 198 °C (471 K), and the room warms to a constant 29 °C (302 K). The stove has an emissivity of 0.900 and a surface area of 3.50 m². Determine the *net* radiant power generated by the stove when the stove (a) is unheated and has a temperature equal to room temperature and (b) has a temperature of 198 °C.

Reasoning The stove emits more radiant power when heated than when unheated. In both cases, however, the Stefan–Boltzmann law can be used to determine the amount of power emitted. Power is energy per unit time or Q/t. But in this problem we need to find the *net* power produced by the stove. The net power is the power the stove emits minus the power the stove absorbs. The power the stove absorbs comes from the walls, ceiling, and floor of the room, all of which emit radiation.

Solution

(a) Remembering that temperature must be expressed in kelvins when using the Stefan–Boltzmann law, we find that

$$\begin{matrix} \text{Power emitted} \\ \text{by unheated} \\ \text{stove at } 18°C \end{matrix} = \frac{Q}{t} = e\sigma T^4 A \tag{13.2}$$

$$= (0.900)[5.67 \times 10^{-8} \text{ J/(s} \cdot \text{m}^2 \cdot \text{K}^4)](291 \text{ K})^4(3.50 \text{ m}^2) = 1280 \text{ W}$$

The fact that the unheated stove emits 1280 W of power and yet maintains a constant temperature means that the stove also absorbs 1280 W of radiant power from its surroundings. Thus, the *net* power generated by the unheated stove is zero:

$$\begin{matrix} \text{Net power} \\ \text{generated by} \\ \text{stove at } 18 °C \end{matrix} = \underbrace{1280 \text{ W}}_{\substack{\text{Power emitted} \\ \text{by stove at} \\ 18 °C}} - \underbrace{1280 \text{ W}}_{\substack{\text{Power emitted by} \\ \text{room at } 18 °C \text{ and} \\ \text{absorbed by stove}}} = \boxed{0 \text{ W}}$$

(b) The hot stove (198 °C or 471 K) emits more radiant power than it absorbs from the cooler room. The radiant power the stove emits is

$$\begin{matrix} \text{Power emitted} \\ \text{by stove at} \\ 198 °C \end{matrix} = \frac{Q}{t} = e\sigma T^4 A$$

$$= (0.900)[5.67 \times 10^{-8} \text{ J/(s} \cdot \text{m}^2 \cdot \text{K}^4)](471 \text{ K})^4(3.50 \text{ m}^2) = 8790 \text{ W}$$

The radiant power the stove absorbs from the room is identical to the power that the stove would emit at the constant room temperature of 29 °C (302 K). The reasoning here is exactly like that in part (a):

$$\begin{matrix} \text{Power emitted by} \\ \text{room at } 29 °C \text{ and} \\ \text{absorbed by stove} \end{matrix} = \frac{Q}{t} = e\sigma T^4 A$$

$$= (0.900)[5.67 \times 10^{-8} \text{ J/(s} \cdot \text{m}^2 \cdot \text{K}^4)](302 \text{ K})^4(3.50 \text{ m}^2) = 1490 \text{ W}$$

The *net* radiant power the stove produces from the fuel it burns is

$$\begin{matrix} \text{Net power} \\ \text{generated by} \\ \text{stove at } 198 °C \end{matrix} = \underbrace{8790 \text{ W}}_{\substack{\text{Power emitted} \\ \text{by stove at} \\ 198 °C}} - \underbrace{1490 \text{ W}}_{\substack{\text{Power emitted by} \\ \text{room at } 29 °C \text{ and} \\ \text{absorbed by stove}}} = \boxed{7300 \text{ W}}$$

● **PROBLEM SOLVING INSIGHT**
In the Stefan–Boltzmann law of radiation, the temperature T must be expressed in kelvins, not in degrees Celsius or degrees Fahrenheit.

Example 6 illustrates that when an object has a higher temperature than its surroundings, the object emits a net radiant power $P_{\text{net}} = (Q/t)_{\text{net}}$. The net power is the power the object emits minus the power it absorbs. Applying the Stefan–Boltzmann law as in Example 6 leads to the following expression for P_{net} when the temperature of the object is T and the temperature of the environment is T_0:

$$P_{\text{net}} = e\sigma A(T^4 - T_0^4) \tag{13.3}$$

13.4 APPLICATIONS

To keep heating and air-conditioning bills to a minimum, it pays to use good thermal insulation in your home. Insulation inhibits convection between inner and outer walls and minimizes heat transfer by conduction. With respect to conduction, the logic behind home insulation comes directly from Equation 13.1. According to this equation, the heat per unit time Q/t flowing through a thickness of material is $Q/t = kA\,\Delta T/L$. Keeping the value for Q/t to a minimum means using materials that have small thermal conductivities k and large thicknesses L. Construction engineers, however, prefer to use Equation 13.1 in the slightly different form shown below:

$$\frac{Q}{t} = \frac{A\,\Delta T}{L/k}$$

The term L/k in the denominator is called the R value of the insulation. For a building material, it is convenient to talk about an R value because it expresses in a single number the combined effects of thermal conductivity and thickness. Larger R values reduce the heat per unit time flowing through the material and, therefore, mean better insulation. It is also convenient to use R values to describe layered slabs formed by sandwiching together a number of materials with different thermal conductivities and different thicknesses. The R values for the individual layers can be added to give a single R value for the entire slab (see problem 14). It should be noted, however, that R values are expressed using units of feet, hours, F°, and BTU for thickness, time, temperature, and heat, respectively.

When it is in the earth's shadow, an orbiting satellite is shielded from the intense electromagnetic waves emitted by the sun. But when a satellite moves out of the earth's shadow, the satellite experiences the full effect of these waves. As a result, the temperature within a satellite would decrease and increase sharply during an orbital period and sensitive electronic circuitry would suffer, unless precautions are taken. To minimize temperature fluctuations, satellites are often covered with a highly reflecting and, hence, poorly absorbing metal foil, as Figure 13.17 shows. By reflecting much of the sunlight, the foil minimizes temperature rises. Being a poor absorber, the foil is also a poor emitter and reduces radiant energy losses. Reducing these losses keeps the temperature from falling excessively when the satellite is in the earth's shadow.

A thermos bottle, sometimes referred to as a Dewar flask, reduces the rate at which hot liquids cool down or cold liquids warm up. A thermos usually consists of a double-walled glass vessel with silvered inner walls (see Figure 13.18) and accomplishes its job by minimizing heat transfer via convection, conduction, and radiation. The space between the walls is evacuated to minimize energy losses due to conduction and convection. The silvered surfaces reflect most of the radiant energy that would otherwise enter or leave the liquid in the thermos. Finally, little heat is lost through the glass or the rubberlike gaskets and stopper, since these materials have relatively small thermal conductivities.

Halogen cooktops are a relatively new idea in cooking, and they use radiant energy to heat pots and pans. A halogen cooktop uses several quartz–iodine lamps,

INTERACTIVE LEARNINGWARE #29

START
Visualization
Reasoning
Knowns and Unknowns
Modeling the Problem
Numerical Solution
FINISH

One day, the temperature of a radiator has to be 68 °C to keep the surrounding walls of a room at 24 °C. The next day, however, the temperature of the radiator needs to be only 49 °C to keep the walls at 24 °C because it is warmer outside. Assuming that the room is heated only by radiation, determine the ratio of the *net* power radiated by the unit on the colder day to that radiated on the warmer day.

This Interactive LearningWare example focuses on the essentials of using the Stefan–Boltzmann law of radiation. In so doing, it reviews the idea of power and the conversion between the Celsius and the Kelvin temperature scales. It also illustrates that when a ratio is asked for, sometimes fewer data are needed than you might think.

Go to our Web site at

http://www.wiley.com/college/cutnell

THE PHYSICS OF...

rating thermal insulation by R values.

THE PHYSICS OF...

regulating the temperature of an orbiting satellite.

THE PHYSICS OF...

a thermos bottle.

THE PHYSICS OF...

a halogen cooktop stove.

Figure 13.17 The Chandra X-ray telescope is shown here in the payload bay of the Space Shuttle *Columbia,* just before being released into orbit. The highly reflecting metal foil that covers the telescope minimizes the temperature changes that would otherwise occur as it travels around the earth.

Figure 13.18 A thermos bottle or Dewar flask minimizes energy transfer due to convection, conduction, and radiation.

like the ones used for ultrabright automobile headlights. These lamps are electrically powered and are mounted below a ceramic top. (See Figure 13.19.) They radiate a great deal of electromagnetic energy, which passes through the ceramic top and is absorbed directly by the bottom of the pot. Consequently, the pot heats up very quickly, rivaling the time of an open gas burner.

13.5 CONCEPTS & CALCULATIONS

Heat transfer by conduction is governed by Equation 13.1, as we have seen. Our first example illustrates a familiar application of this relation in the kitchen. It also gives us the opportunity to review the idea of latent heat of vaporization, which Section 12.8 discusses.

CONCEPTS & CALCULATIONS EXAMPLE 7   Boiling Water

Two pots are identical, except that in one case the flat bottom is aluminum and in the other it is copper. Each pot contains the same amount of boiling water and sits on a heating element that has a temperature of 155 °C. In the aluminum-bottom pot, the water boils away completely in 360 s. How long does it take for the water in the copper-bottom pot to boil away completely?

Concept Questions and Answers Is the heat needed to boil away the water in the aluminum-bottom pot less than, greater than, or the same as the heat needed in the copper-bottom pot?

Answer The heat Q needed is the same in each case. When water boils, it changes from the liquid phase to the vapor phase. The heat needed to make the water boil away is $Q = mL_v$, according to Equation 12.5, where m is the mass of the water and L_v is the latent heat of vaporization for water. Since the amount of water in each pot is the same, the mass of water is the same in each case. Moreover, the latent heat is a characteristic of water and, therefore, is also the same in each case.

One of the factors in Equation 13.1 that influences the amount of heat conducted through the bottom of each pot is the temperature difference ΔT between the upper and lower surfaces of the pot's bottom. Is this temperature difference for the aluminum-bottom pot less than, greater than, or the same as that for the copper-bottom pot?

Answer For each pot the temperature difference is the same. At the upper surface of each pot bottom the temperature is 100.0 °C, because water boils at 100.0 °C under normal conditions of atmospheric pressure. The temperature remains at 100.0 °C until all the water is gone. For each pot the temperature at the lower surface of the pot bottom is 155 °C, the temperature of the heating element. Therefore, $\Delta T = 155 \ °C - 100.0 \ °C = 55 \ C°$ for each pot.

Is the time required to boil away the water completely in the copper-bottom pot less than, greater than, or the same as that required for the aluminum-bottom pot?

Answer The time is less for the copper-bottom pot. The factors that influence the amount of heat conducted in a given time are the thermal conductivity, the area of the bottom, the temperature difference across the bottom, and the thickness of the bottom. All of these factors are the same for each pot except for the thermal conductivity, which is greater for copper (Cu) than for aluminum (Al) (see Table 13.1). The greater the thermal conductivity, the greater the heat that is conducted in a given time, other things being equal. Therefore, less time is required for the copper-bottom pot.

Solution Applying Equation 13.1 to the conduction of heat into both pots and using Equation 12.5 to express the heat needed to boil away the water, we have

$$Q_{Al} = \frac{(k_{Al}A\,\Delta T)t_{Al}}{L} = mL_v$$

$$Q_{Cu} = \frac{(k_{Cu}A\,\Delta T)t_{Cu}}{L} = mL_v$$

In these two equations the area A, the temperature difference ΔT, the thickness L of the pot bottom, the mass m of the water, and the latent heat of vaporization of water L_v have the same values. Therefore, we can set the two heats equal and obtain

$$\frac{(k_{Al}A\,\Delta T)t_{Al}}{L} = \frac{(k_{Cu}A\,\Delta T)t_{Cu}}{L} \quad \text{or} \quad k_{Al}t_{Al} = k_{Cu}t_{Cu}$$

Solving for t_{Cu} and taking values for the thermal conductivities from Table 13.1, we find

$$t_{Cu} = \frac{k_{Al}t_{Al}}{k_{Cu}} = \frac{[240\ \text{J/(s·m·C°)}](360\ \text{s})}{390\ \text{J/(s·m·C°)}} = \boxed{220\ \text{s}}$$

As expected, the boil-away time is less for the copper-bottom pot.

Figure 13.19 In a halogen cooktop, quartz–iodine lamps emit a large amount of electromagnetic energy that is absorbed directly by a pot or pan.

Heat transfer by conduction is only one way in which heat gets from place to place. Heat transfer by radiation is another way, and it is governed by the Stefan–Boltzmann law of radiation, as Section 13.3 discusses. Example 8 deals with a case in which heat loss by radiation leads to freezing of water. It stresses the importance of the area from which the radiation occurs and also provides a review of the idea of latent heat of fusion, which Section 12.8 discusses.

CONCEPTS & CALCULATIONS EXAMPLE 8   Freezing Water

One half of a kilogram of liquid water at 273 K (0 °C) is placed outside on a day when the temperature is 261 K (-12 °C). Assume that heat is lost from the water only by means of radiation and that the emissivity of the radiating surface is 0.60. How long does it take for the water to freeze into ice at 0 °C when the surface area from which the radiation occurs is (a) 0.035 m² (as it could be in a cup) and (b) 1.5 m² (as it would be if the water were spilled out to form a thin sheet)?

Concept Questions and Answers In case (a) is the heat that must be removed to freeze the water less than, greater than, or the same as in case (b)?

Answer The heat that must be removed is the same in both cases. When water freezes, it changes from the liquid phase to the solid phase. The heat that must be removed to make the water freeze is $Q = mL_f$, according to Equation 12.5, where m is the mass of the water and L_f is the latent heat of fusion for water. The mass is the same in both cases and so is L_f, since it is a characteristic of the water.

The loss of heat by radiation depends on the temperature of the radiating object. Does the temperature of the water change as the water freezes?

Answer No. The temperature of the water does not change as the freezing process takes place. The heat removed serves only to change the water from the liquid to the solid phase, as Section 12.8 discusses. Only after all the water has frozen does the temperature of the ice begin to fall below 0 °C.

The water both loses and gains heat by radiation. How, then, can heat transfer by radiation lead to freezing of the water?

Answer The water freezes because it loses more heat by radiation than it gains. The gain occurs because the environment radiates heat and the water absorbs it. But the temperature of the environment is less than the temperature of the water. As a result, the environmental radiation is insufficient to offset completely the loss of heat due to radiation from the water.

Will it take longer for the water to freeze in case (a) when the area is smaller or in case (b) when the area is larger?

Answer It will take longer when the area is smaller. This is because the amount of energy radiated in a given time is proportional to the area from which the radiation occurs. A smaller area means that less energy is radiated per second, so more time will be needed to remove the heat needed to make the water freeze.

Solution We use Equation 13.3 to take into account that the water both gains and loses heat via radiation. This expression gives the net power lost, the net power being the net heat divided by the time. Thus, we have

$$\frac{Q}{t} = e\sigma A(T^4 - T_0^4) \quad \text{or} \quad t = \frac{Q}{e\sigma A(T^4 - T_0^4)}$$

Using Equation 12.5 to express the heat Q as $Q = mL_f$ and taking the latent heat of fusion for water from Table 12.3 ($L_f = 33.5 \times 10^4$ J/kg), we find

(a) *Smaller area*

$$t = \frac{mL_f}{e\sigma A(T^4 - T_0^4)}$$

$$= \frac{(0.50 \text{ kg})(33.5 \times 10^4 \text{ J/kg})}{0.60[5.67 \times 10^{-8} \text{ J/(s} \cdot \text{m}^2 \cdot \text{K}^4)](0.035 \text{ m}^2)[(273 \text{ K})^4 - (261 \text{ K})^4]}$$

$$= \boxed{1.5 \times 10^5 \text{ s (42 h)}}$$

(b) *Larger area*

$$t = \frac{mL_f}{e\sigma A(T^4 - T_0^4)}$$

$$= \frac{(0.50 \text{ kg})(33.5 \times 10^4 \text{ J/kg})}{0.60[5.67 \times 10^{-8} \text{ J/(s} \cdot \text{m}^2 \cdot \text{K}^4)](1.5 \text{ m}^2)[(273 \text{ K})^4 - (261 \text{ K})^4]}$$

$$= \boxed{3.6 \times 10^3 \text{ s (1.0 h)}}$$

As expected, the freezing time is longer when the area is smaller.

At the end of the problem set for this chapter, you will find homework problems that contain both conceptual and quantitative parts. These problems are grouped under the heading *Concepts & Calculations, Group Learning Problems*. They are designed for use by students working alone or in small learning groups. The conceptual part of each problem provides a convenient focus for group discussions.

SUMMARY

Convection is the process in which heat is carried by the bulk movement of a fluid. During natural convection, the warmer, less dense part of a fluid is pushed upward by the buoyant force provided by the surrounding cooler and denser part. Forced convection occurs when an external device, such as a fan or a pump, causes the fluid to move.

Conduction is the process whereby heat is transferred directly through a material, any bulk motion of the material playing no role in the transfer. The heat Q conducted during a time t through a bar of length L and cross-sectional area A is expressed as $Q = (kA\,\Delta T)t/L$, where ΔT is the difference in temperature between the ends of the bar and k is the **thermal**

conductivity of the material. Materials that have large values of k, such as most metals, are known as thermal conductors. Materials with small values of k, such as Styrofoam, are referred to as thermal insulators.

Radiation is the process in which energy is transferred by electromagnetic waves. All objects, regardless of their temperatures, simultaneously absorb and emit electromagnetic waves. A body that is at the same constant temperature as its surroundings absorbs and emits equal amounts of radiant energy per unit time. **Objects that are good absorbers of radiant energy are also good emitters, and objects that are poor absorbers are also poor emitters.** An object that absorbs all the radiation incident upon it is called a **perfect**

blackbody. A perfect blackbody, being a perfect absorber, is also a perfect emitter.

The amount of radiant energy Q emitted during a time t by an object whose surface area is A and whose Kelvin temperature is T is given by the **Stefan–Boltzmann law,** $Q = e\sigma T^4 At$. In this equation, σ is the Stefan–Boltzmann constant [$\sigma = 5.67 \times 10^{-8}$ J/(s·m²·K⁴)] and e is the emissivity, a dimensionless number characterizing the surface of the object. The emissivity lies between 0 and 1, being zero for a nonemitting surface and one for a perfect blackbody. The net radiant power emitted by an object of temperature T located in an environment of temperature T_0 is $P_{net} = e\sigma A(T^4 - T_0^4)$.

CONCEPTUAL QUESTIONS

1. One often hears about heat transfer by convection in gases and liquids, but not in solids. Why?

2. A heavy drape, hung close to a cold window, reduces heat loss considerably by interfering primarily with one of the three processes of heat transfer. Explain which one.

3. The *windchill factor* is a term used by weather forecasters. Roughly speaking, it refers to the fact that you feel colder when the wind is blowing than when it is not, even though the air temperature is the same in either case. Which of the three processes for heat transfer plays the principal role in the windchill factor? Explain your reasoning.

4. Often the following warning sign is seen on bridges: "Caution—Bridge surface freezes before road surface." Account for the warning in terms of heat transfer processes. Note that, unlike the road, a bridge has both surfaces exposed to the air.

5. A piece of Styrofoam and a piece of wood are sandwiched together to form a layered slab. The two pieces have the same thickness and cross-sectional area. The exposed surfaces have constant temperatures. The temperature of the exposed Styrofoam surface is greater than the temperature of the exposed wood surface. Is the temperature of the Styrofoam–wood interface closer to the larger or smaller of the two temperatures? Give your reasoning, using Equation 13.1 and data from Table 13.1.

6. ✍ One way that heat is transferred from place to place inside the human body is by the flow of blood. Which one of the three heat transfer processes best describes this action of the blood? Justify your answer.

7. Some animals have hair, the strands of which are hollow, air-filled tubes. Other animals have hair that is composed of solid tubular strands. Which kind of hair would be more likely to give an animal an advantage for surviving in very cold climates? Why?

8. A poker used in a fireplace is held at one end, while the other end is in the fire. Why are pokers made of iron rather than copper? Ignore the fact that iron may be cheaper and stronger.

9. In Alaska, a lack of snow allowed the ground to freeze down to a depth of about one meter, causing buried water pipes to freeze and burst. Why did a lack of snow lead to this situation?

10. Concrete walls often contain steel reinforcement bars. Does the steel enhance or degrade the insulating value of the concrete? Explain.

11. Grandma says that it is quicker to bake a potato if you put a nail into it. In fact, she is right. Justify her baking technique in terms of one of the three processes of heat transfer.

12. Several days after a snowstorm, the roof on a house is uniformly covered with snow. On a neighboring house, however, the snow on the roof has completely melted. Which house is probably better insulated? Give your reasoning.

13. One car has a metal body, while another has a plastic body. On a cold winter day these cars are parked side by side. If you put a bare hand on each car, the metal body feels colder. Why?

14. A high-quality pot is designed so heat can enter readily and be distributed evenly, while the rate of energy loss from the pot is kept to a minimum. Many high-quality pots have copper bases and polished stainless steel sides. Based on conduction and radiation principles, explain why this design is better than all-copper or all-steel units.

15. Two objects have the same size and shape. Object A has an emissivity of 0.3, and object B has an emissivity of 0.6. Each radiates the same power. Is the Kelvin temperature of A twice that of B? Give your reasoning.

16. A concave mirror can be used to start a fire by directing sunlight onto a small spot on a piece of paper. Explain why the mirror does not get as hot as the paper.

17. Two strips of material, A and B, are identical, except they have emissivities of 0.4 and 0.7, respectively. The strips are heated to the same temperature and have a red glow. A brighter glow signifies that more energy per second is being radiated. Which strip has the brighter glow? Explain.

18. A pot of water is being heated on an electric stove. The diameter of the pot is smaller than the diameter of the heating element on which the pot rests. The exposed outer edges of the heating element are glowing cherry red. When you lift the pot, you see that the part of the heating element beneath it is not glowing cherry red, indicating that it is cooler than the outer edges. Why are the outer edges hotter?

19. To keep your hands as warm as possible during skiing, should you wear mittens or gloves? (Mittens, except for the thumb, do not have individual finger compartments.) Give a reason for your answer.

20. Two identical hot cups of cocoa are sitting on a table. One has a metal spoon in it and one does not. After five minutes, which cup is cooler? Explain in terms of heat transfer processes.

21. (a) Would a hot solid cube cool more rapidly if it were left intact or cut in half? Explain your answer in terms of one or more of the three heat transfer processes. (b) Using reasoning similar to that used in answering part (a), decide which cools

faster, one pound of wide and flat lasagna noodles or one pound of spaghetti noodles. Assume that both kinds of noodles are made from the same pasta and start out with the same temperature.

22. One day during the winter the sun has been shining all day. Toward sunset a light snow begins to fall. It collects without melting on a cement playground, but it melts immediately upon contact with a black asphalt road adjacent to the playground. Account for the fact that the snow collects in one place but not in the other.

23. If you were stranded in the mountains in cold weather, it would help to minimize energy losses from your body by curling up into the tightest ball possible. Which of the factors in Equation 13.2 are you using to the best advantage by curling into a ball? Why?

PROBLEMS

Note: For problems in this set, use the values for thermal conductivities given in Table 13.1 unless stated otherwise.

Section 13.2 Conduction

1. ssm One end of an iron poker is placed in a fire where the temperature is 502 °C, and the other end is kept at a temperature of 26 °C. The poker is 1.2 m long and has a radius of 5.0×10^{-3} m. Ignoring the heat lost along the length of the poker, find the amount of heat conducted from one end of the poker to the other in 5.0 s.

2. A refrigerator has a surface area of 5.3 m². It is lined with 0.075-m-thick insulation whose thermal conductivity is 0.030 J/(s·m·C°). The interior temperature is kept at 5 °C, while the temperature at the outside surface is 25 °C. How much heat per second is being removed from the unit?

3. A closed box is filled with dry ice at a temperature of -78.5 °C, while the outside temperature is 21.0 °C. The box is cubical, measuring 0.350 m on a side, and the thickness of the walls is 3.00×10^{-2} m. In one day, 3.10×10^6 J of heat is conducted through the six walls. Find the thermal conductivity of the material from which the box is made.

4. In an electrically heated home, the temperature of the ground in contact with a concrete basement wall is 12.8 °C. The temperature at the inside surface of the wall is 20.0 °C. The wall is 0.10 m thick and has an area of 9.0 m². Assume one kilowatt·hour of electrical energy costs $0.10. How many hours are required for one dollar's worth of energy to be conducted through the wall?

5. ssm Due to a temperature difference ΔT, heat is conducted through an aluminum plate that is 0.035 m thick. The plate is then replaced by a stainless steel plate that has the same temperature difference and cross-sectional area. How thick should the steel plate be, so that the same amount of heat per second is conducted through it?

6. A composite rod is made from stainless steel and iron and has a length of 0.50 m. The cross section of this composite rod is

shown in the drawing and consists of a square within a circle. The square cross section of the steel is 1.0 cm on a side. The temperature at one end of the rod is 78 °C, while it is 18 °C at the other end. Assuming that no heat exits through the cylindrical outer surface, find the total amount of heat conducted through the rod in two minutes.

7. ssm www ⚕ In the conduction equation $Q = (kA\,\Delta T)t/L$, the combination of factors kA/L is called the *conductance*. The human body has the ability to vary the conductance of the tissue beneath the skin by means of vasoconstriction and vasodilation, in which the flow of blood to the veins and capillaries underlying the skin is decreased and increased, respectively. The conductance can be adjusted over a range such that the tissue beneath the skin is equivalent to a thickness of 0.080 mm of Styrofoam or 3.5 mm of air. By what factor can the body adjust the conductance?

8. A skier wears a jacket filled with goose down that is 15 mm thick. Another skier wears a wool sweater that is 5.0 mm thick. Both have the same surface area. Assuming the temperature difference between the inner and outer surfaces of each garment is the same, calculate the ratio (wool/goose down) of the heat lost due to conduction during the same time interval.

*9. At a picnic, a Styrofoam cup contains lemonade and ice at 0 °C. The thickness of the cup is 2.0×10^{-3} m, and the area is 0.016 m². The temperature at the outside surface of the cup is 35 °C. The latent heat of fusion for ice is 3.35×10^5 J/kg. What mass of ice melts in one hour?

*10. Review Conceptual Example 4 before attempting this problem. To illustrate the effect of ice on the aluminum cooling plate, consider the drawing shown here and the data contained therein. Ignore any limitations due to significant figures. (a) Calculate the heat per second per square meter that is conducted through the ice–aluminum sandwich. (b) Calculate the heat per second per square meter that would be conducted through the aluminum if the ice were not present. Notice how much larger the answer is in (b) as compared to (a).

*11. ssm Two rods, one of aluminum and the other of copper, are joined end to end. The cross-sectional area of each is 4.0×10^{-4} m^2, and the length of each is 0.040 m. The free end of the aluminum rod is kept at 302 °C, while the free end of the copper rod is kept at 25 °C. The loss of heat through the sides of the rods may be ignored. (a) What is the temperature at the aluminum–copper interface? (b) How much heat is conducted through the unit in 2.0 s? (c) What is the temperature in the aluminum rod at a distance of 0.015 m from the hot end?

*12. Three building materials, plasterboard [$k = 0.30$ J/(s·m·C°)], brick [$k = 0.60$ J/(s·m·C°)], and wood [$k = 0.10$ J/(s·m·C°)], are sandwiched together as the drawing illustrates. The temperatures at the inside and outside surfaces are 27 °C and 0 °C, respectively. Each material has the same thickness and cross-sectional area. Find the temperature (a) at the plasterboard–brick interface and (b) at the brick–wood interface.

*13. In an aluminum pot, 0.20 kg of water at 100 °C boils away in five minutes. The bottom of the pot is 2.5×10^{-3} m thick and has a surface area of 0.020 m^2. To prevent the water from boiling too rapidly, a stainless steel plate has been placed between the pot and the heating element. The plate is 1.2×10^{-3} m thick, and its area matches that of the pot. Assuming that heat is conducted into the water only through the bottom of the pot, find the temperature at the steel surface in contact with the heating element.

**14. The drawing shows a layered slab formed by sandwiching together two pieces of different insulating materials that have thermal conductivities of k_1 and k_2. The thicknesses

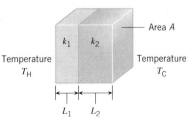

L_1 and L_2 are different, and the temperature T_H is greater than the temperature T_C. Show that the heat conducted through the slab in a time t is given by $Q = A(T_H - T_C)t/[(L_1/k_1) + (L_2/k_2)]$, so that the effective R value of the slab is $(L_1/k_1) + (L_2/k_2)$.

**15. ssm www The drawing shows a solid cylindrical rod made from a center cylinder of lead and an outer concentric jacket of copper. Except for its ends, the rod is insulated (not shown), so that the loss of heat from the curved surface is negligible. When a temperature difference is maintained between its ends, this rod conducts one-half the amount of heat that it would conduct if it were solid copper. Determine the ratio of the radii r_1/r_2.

Section 13.3 Radiation

16. The filament of a light bulb has a temperature of 3.0×10^3 °C and radiates sixty watts of power. The emissivity of the filament is 0.36. Find the surface area of the filament.

17. An object emits 30 W of radiant power. If it were a perfect blackbody, other things being equal, it would emit 90 W of radiant power. What is the emissivity of the object?

18. Assume that the earth has an average surface temperature of 22 °C and is a perfect emitter of radiation ($e = 1$). Find the energy per second per square meter that the earth radiates into space.

19. ssm www How many days does it take for a perfect blackbody cube (0.0100 m on a side, 30.0 °C) to radiate the same amount of energy that a one-hundred-watt light bulb uses in one hour?

20. By what factor should the Kelvin temperature of an object be increased to double the radiant energy per second emitted by the object?

21. ssm Review Example 6 before attempting this problem. Suppose the stove in that example had a surface area of only 2.00 m^2. What would its temperature (in kelvins) have to be so that it still generated a net power of 7300 W?

22. ⌀ A person's body is producing energy internally due to metabolic processes. If the body loses more energy than metabolic processes are generating, its temperature will drop. If the drop is severe, it can be life-threatening. Suppose a person is unclothed and energy is being lost via radiation from a body surface area of 1.40 m^2, which has a temperature of 34 °C and an emissivity of 0.700. Suppose that metabolic processes are producing energy at a rate of 115 J/s. What is the temperature of the coldest room in which this person could stand and not experience a drop in body temperature?

23. The concrete wall of a building is 0.10 m thick. The temperature inside the building is 20.0 °C, while the temperature outside is 0.0 °C. Heat is conducted through the wall. When the building is unheated, the inside temperature falls to 0.0 °C, and heat conduction ceases. However, the wall does emit radiant energy when its temperature is 0.0 °C. The radiant energy emitted per second per square meter is the same as the heat lost per second per square meter due to conduction. What is the emissivity of the wall?

*24. Liquid helium is stored at its boiling-point temperature of 4.2 K in a spherical container ($r = 0.30$ m). The container is a perfect blackbody radiator. The container is surrounded by a spherical shield whose temperature is 77 K. A vacuum exists in the space between the container and shield. The latent heat of vaporization for helium is 2.1×10^4 J/kg. What mass of liquid helium boils away through the venting valve in one hour?

**25. ssm A solid cylinder is radiating power. It has a length that is ten times its radius. It is cut into a number of smaller cylinders, each of which has the same length. Each small cylinder has the same temperature as the original cylinder. The total radiant power emitted by the pieces is twice that emitted by the original cylinder. How many smaller cylinders are there?

**26. A small sphere (emissivity = 0.90, radius = r_1) is located at the center of a spherical asbestos shell (thickness = 1.0 cm, outer radius = r_2). The thickness of the shell is small compared to the inner and outer radii of the shell. The temperature of the small sphere is 800.0 °C, while the temperature of the inner surface of the shell is 600.0 °C, both temperatures remaining constant. Assuming that $r_2/r_1 = 10.0$ and ignoring any air inside the shell, find the temperature of the outer surface of the shell.

ADDITIONAL PROBLEMS

27. ssm A person's body is covered with 1.6 m² of wool clothing. The thickness of the wool is 2.0×10^{-3} m. The temperature at the outside surface of the wool is 11 °C, and the skin temperature is 36 °C. How much heat per second does the person lose due to conduction?

28. The temperature in an electric oven is 160 °C. The temperature at the outer surface in the kitchen is 50 °C. The oven (surface area = 1.6 m²) is insulated with material that has a thickness of 0.020 m and a thermal conductivity of 0.045 J/(s·m·C°). (a) How much energy is used to operate the oven for six hours? (b) At a price of $0.10 per kilowatt·hour for electrical energy, what is the cost of operating the oven?

29. A solar collector is placed in direct sunlight where it absorbs energy at the rate of 880 J/s for each square meter of its surface. The emissivity of the solar collector is $e = 0.75$. What equilibrium temperature does the collector reach? Assume that the only energy loss is due to the emission of radiation.

30. ⚄ Suppose the skin temperature of a naked person is 34 °C when the person is standing inside a room whose temperature is 25 °C. The skin area of the individual is 1.5 m². (a) Assuming the emissivity is 0.80, find the net loss of radiant power from the body. (b) Determine the number of food Calories of energy (1 food Calorie = 4186 J) that is lost in one hour due to the net loss rate obtained in part (a). Metabolic conversion of food into energy replaces this loss.

31. ssm A car parked in the sun absorbs energy at a rate of 560 watts per square meter of surface area. The car reaches a temperature at which it radiates energy at this same rate. Treating the car as a perfect radiator ($e = 1$), find the temperature.

*32. One end of a brass bar is maintained at 295 °C, while the other end is kept at a constant but lower temperature. The cross-sectional area of the bar is 4.0×10^{-4} m². Because of insulation, there is negligible heat loss through the sides of the bar. Heat flows through the bar, however, at the rate of 2.7 J/s. What is the temperature of the bar at a point 0.20 m from the hot end?

*33. A solid sphere has a temperature of 773 K. The sphere is melted down and recast into a cube that has the same emissivity and emits the same radiant power as the sphere. What is the cube's temperature?

*34. In a house the temperature at the surface of a window is 25 °C. The temperature outside at the window surface is 5.0 °C. Heat is lost through the window via conduction, and the heat lost per second has a certain value. The temperature outside begins to fall, while the conditions inside the house remain the same. As a result, the heat lost per second increases. What is the temperature at the outside window surface when the heat lost per second doubles?

**35. ssm www Two cylindrical rods have the same mass. One is made of silver (density = 10 500 kg/m³), and one is made of iron (density = 7860 kg/m³). Both rods conduct the same amount of heat per second when the same temperature difference is maintained across their ends. What is the ratio (silver-to-iron) of (a) the lengths and (b) the radii of these rods?

**36. A 0.30-m-thick sheet of ice covers a lake. The air temperature at the ice surface is −15 °C. In five minutes, the ice thickens by a small amount. Assume that no heat flows from the ground below into the water and that the added ice is very thin compared to 0.30 m. Find the number of millimeters by which the ice thickens.

CONCEPTS & CALCULATIONS

GROUP LEARNING PROBLEMS

Note: Each of these problems consists of Concept Questions followed by a related quantitative Problem. They are designed for use by students working alone or in small learning groups. The Concept Questions involve little or no mathematics and are intended to stimulate group discussions. They focus on the concepts with which the problems deal. Recognizing the concepts is the essential initial step in any problem-solving technique.

37. Concept Questions A pot of water is boiling on a stove under one atmosphere of pressure. Assume that heat enters the pot only through its bottom, which is copper and rests on a heating element. In a certain time, a mass m of water boils away. (a) What is the temperature of the boiling water and does it change during this time? (b) What determines the amount of heat needed to boil the water? (c) Is the temperature of the heating element in contact with the pot greater than, smaller than, or equal to 100 °C? Explain.

Problem In two minutes, the mass of water boiled away is $m = 0.45$ kg. The radius of the pot bottom is $R = 6.5$ cm and the thickness is $L = 2.0$ mm. What is the temperature T_E of the heating element in contact with the pot? Verify that your answer is consistent with your answers to the Concept Questions.

38. Concept Questions A wall in a house contains a single window, the area and thickness of which are small relative to the area and thickness of the wall. The window consists of a single pane of glass. The wall may be treated as a slab of the insulating material Styrofoam. The temperature difference between the inside and outside is the same for the wall and the window. Heat is lost via conduction through the window and the wall. Decide, for a given time period, whether each of the following factors, by itself, causes the window to lose more than, less than, or the same amount of heat as the wall: (a) the area, other things being equal, (b) the thickness, other things being equal, and (c) the thermal conductivity (see Table 13.1), other things being equal. Give your reasoning.

Problem The glass window pane has an area of 0.16 m² and a thickness of 2.0 mm. The Styrofoam area and thickness are 18 m² and 0.10 m, respectively. Of the total heat lost by the wall and the window, what is the percentage lost by the window?

39. Concept Questions Two objects are maintained at constant temperatures, one hot and one cold. Two identical bars can be attached end to end, as in part a of the drawing, or one on top of the other, as in part b. When either of these arrangements is placed between the hot and the cold object for the same amount of time, heat Q flows. (a) Is the area through which the heat flows greater for arrangement a or arrangement b? (b) Is the thickness of the material through which the heat flows greater for arrangement a or arrangement b? (c) Is Q_a less than, greater than, or equal to Q_b? Account for your answers.

Problem Find the ratio Q_a/Q_b. Be sure that your answer is consistent with your answers to the Concept Questions.

(a)   (b)

40. Concept Questions Light bulb 1 operates with a higher filament temperature than light bulb 2, but both filaments have the same emissivity. (a) How is the power P expressed in terms of the energy Q radiated by a bulb and the time t during which energy is radiated? (b) Does a higher filament temperature generate more radiated power or less radiated power? (c) Does a smaller area for radiation promote more radiated power or less radiated power? (d) Suppose that both bulbs radiate the same power. Is the filament area of bulb 1 greater than, less than, or the same as the filament area of bulb 2? Give your reasoning.

Problem Bulb 1 has a filament temperature of 2700 K, whereas bulb 2 has a filament temperature of 2100 K. Both bulbs radiate the same power. Find the ratio A_1/A_2 of the filament areas of the bulbs. Verify that your answer is consistent with your answers to the Concept Questions.

41. Concept Questions Via radiation, an object emits more power than it absorbs from the room in which it is located. (a) The object has a temperature T (in kelvins). According to the Stefan–Boltzmann law, the power radiated by the object is $Q/t = e\sigma T^4 A$, where A is the area from which the radiation is emitted. What is the expression for the power absorbed from the room, which has a temperature T_0 (in kelvins)? (b) Is the temperature of the object greater than, less than, or equal to the temperature of the room? Explain, assuming the temperatures are constant.

Problem The temperature of the room is 293 K. The object emits three times as much power as it absorbs from the room. What is the temperature of the object? Check to see that your answer is consistent with your answers to the Concept Questions.

42. Concept Questions Sirius B is a white star that has a much greater surface temperature than our sun does. Assume that both Sirius B and our sun are spherical and have the same emissivity. (a) Other things being equal, would the greater surface temperature imply that the power radiated by Sirius B is greater than, less than, or equal to the power radiated by our sun? (b) The fact is that Sirius B radiates much less power than our sun does. Considering this fact, is the surface area of Sirius B greater than, less than, or equal to the surface area of our sun? (c) Is the radius of Sirius B greater than, less than, or equal to the radius of our sun? Explain your answers.

Problem Our sun has a radius of 6.96×10^8 m. Sirius B radiates only 0.04 times the power radiated by our sun. Its surface temperature is four times that of our sun. Find the radius of Sirius B. Be sure that your answer is consistent with your answers to the Concept Questions.

Frogs communicate by using a variety of calls. The males of many species, like the Barking Tree frog, possess vocal sacs that can be inflated with air. If the air within a sac behaves like an ideal gas, its pressure, volume, and temperature are related by the ideal gas law, the subject of this chapter.

CHAPTER 14

THE IDEAL GAS LAW AND KINETIC THEORY

14.1 THE MOLE, AVOGADRO'S NUMBER, AND MOLECULAR MASS

Often, we wish to compare the mass of one atom with another. To facilitate the comparison, a mass scale known as the ***atomic mass scale*** has been established. To set up this scale, a reference value (along with a unit) is chosen for one of the elements. The unit is called the ***atomic mass unit*** (symbol: u). By international agreement, the reference element is chosen to be the most abundant type or isotope* of carbon, called carbon-12, and its atomic mass† is defined to be exactly twelve atomic mass units, or 12 u. The relationship between the atomic mass unit and the kilogram is

$$1\,u = 1.6605 \times 10^{-27}\,kg$$

The atomic masses of all the elements are listed in the periodic table, part of which is shown in Figure 14.1. The complete periodic table is given on the inside of the back cover. In general, the masses listed are average values and take into account the various isotopes of an element that exist naturally. For brevity, the unit "u" is often omitted from the table. For example, a magnesium atom (Mg) has an average atomic mass of 24.305 u, while the corresponding average value for the lithium atom (Li) is 6.941 u; thus, atomic magnesium is more massive than atomic lithium by a factor of $(24.305\,u)/(6.941\,u) = 3.502$. In the periodic table, the atomic mass of carbon (C) is given as 12.011 u, rather than exactly 12 u, because a small amount (about 1%) of the naturally occurring material is an isotope called carbon-13. The value of 12.011 u is an average that reflects the small contribution of carbon-13.

The molecular mass of a molecule is the sum of the atomic masses of its atoms. For instance, hydrogen and oxygen have atomic masses of 1.007 94 u and 15.9994 u, respectively, so the molecular mass of a water molecule (H_2O) is $2(1.007\,94\,u) + 15.9994\,u = 18.0153\,u$.

Macroscopic amounts of materials contain large numbers of atoms or molecules. Even in a small volume of gas, 1 cm^3, for example, the number is enormous. It is convenient to express such large numbers in terms of a single unit, the ***gram-mole,*** or simply the ***mole*** (symbol: *mol*). ***One gram-mole of a substance contains as many particles (atoms or molecules) as there are atoms in 12 grams of the isotope carbon-12.*** Experiment shows that 12 grams of carbon-12 contain 6.022×10^{23} atoms. The number of atoms per mole is known as ***Avogadro's number N_A,*** after the Italian scientist Amedeo Avogadro (1776–1856):

$$N_A = 6.022 \times 10^{23}\,mol^{-1}$$

Thus, the number of moles n contained in any sample is the number of particles N in the sample divided by the number of particles per mole N_A (Avogadro's number):

$$n = \frac{N}{N_A}$$

Although defined in terms of carbon atoms, the concept of a mole can be applied to any collection of objects by noting that one mole contains Avogadro's number of objects. Thus, one mole of atomic sulfur contains 6.022×10^{23} sulfur atoms, one mole of water contains 6.022×10^{23} H_2O molecules, and one mole of golf balls

Figure 14.1 A portion of the periodic table showing the atomic number and atomic mass of each element. In the periodic table it is customary to omit the symbol "u" denoting the atomic mass unit.

---

* Isotopes are discussed in Section 31.1.

† In chemistry the expression "atomic weight" is frequently used in place of "atomic mass."

(a)

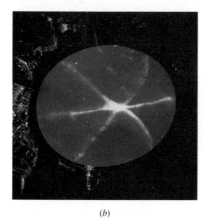

(b)

Figure 14.2 (a) The oval-shaped Hope diamond surrounded by 16 smaller diamonds. (b) The Rosser Reeves ruby. Both gems are on display at the Smithsonian Institution in Washington, D.C.

THE PHYSICS OF...

gemstones.

contains 6.022×10^{23} golf balls. Just as the meter is the SI base unit for length, the mole is the SI base unit for expressing "the amount of a substance."

The number of moles contained in a sample can also be found from its mass. Let's multiply and divide the right-hand side of the previous equation by the mass m_{particle} of a single particle, expressed in grams. Thus, the number n of moles can be written as

$$n = \frac{m_{\text{particle}}N}{m_{\text{particle}}N_{\text{A}}} = \frac{m}{\text{Mass per mole}}$$

The numerator $m_{\text{particle}}N$ is the mass of a particle times the number of particles in the sample, which is the mass m of the sample. The denominator $m_{\text{particle}}N_{\text{A}}$ is the mass of a particle times the number of particles per mole, which is the mass per mole, expressed in grams per mole. The mass per mole of carbon-12 is 12 g/mol, since, by definition, 12 grams of carbon-12 contain one mole of atoms. On the other hand, the mass per mole of sodium (Na) is 22.9898 g/mol for the following reason. As indicated in Figure 14.1, a sodium atom is more massive than a carbon-12 atom by the ratio of their atomic masses, $(22.9898 \text{ u})/(12 \text{ u}) = 1.915\,82$. Therefore, the mass per mole of sodium is 1.915 82 times greater than that of carbon-12, or $(1.915\,82)(12 \text{ g/mol}) = 22.9898$ g/mol. Note that the numerical value of the mass per mole of sodium (22.9898) is the same as the numerical value of its atomic mass. This is true in general, so ***the mass per mole (in g/mol) of a substance has the same numerical value as the atomic or molecular mass of the substance (in atomic mass units).***

Example 1 illustrates how to use the concepts of the mole, atomic mass, and Avogadro's number to determine the number of atoms and molecules present in two famous gemstones.

EXAMPLE 1   The Hope Diamond and the Rosser Reeves Ruby

Figure 14.2a shows the Hope diamond (44.5 carats), which is almost pure carbon. Figure 14.2b shows the Rosser Reeves ruby (138 carats), which is primarily aluminum oxide (Al_2O_3). One carat is equivalent to a mass of 0.200 g. Determine (a) the number of carbon atoms in the diamond and (b) the number of Al_2O_3 molecules in the ruby.

Reasoning The number N of atoms (or molecules) in a sample is the number of moles n times the number of atoms per mole N_{A} (Avogadro's number); $N = nN_{\text{A}}$. We can determine the number of moles by dividing the mass of the sample m by the mass per mole of the substance.

Solution

(a) The mass of the Hope diamond is $m = (44.5 \text{ carats})[(0.200 \text{ g})/(1 \text{ carat})] = 8.90$ g. Since the average atomic mass of naturally occurring carbon is 12.011 u, the mass per mole of this substance is 12.011 g/mol. The number of moles of carbon in the Hope diamond is

$$n = \frac{m}{\text{Mass per mole}} = \frac{8.90 \text{ g}}{12.011 \text{ g/mol}} = 0.741 \text{ mol}$$

The number of carbon atoms in the Hope diamond is

$$N = nN_{\text{A}} = (0.741 \text{ mol})(6.022 \times 10^{23} \text{ atoms/mol}) = \boxed{4.46 \times 10^{23} \text{ atoms}}$$

(b) The mass of the Rosser Reeves ruby is $m = (138 \text{ carats})[(0.200 \text{ g})/(1 \text{ carat})] = 27.6$ g. The molecular mass of an aluminum oxide molecule (Al_2O_3) is the sum of the atomic masses of its atoms, which are 26.9815 u for aluminum and 15.9994 u for oxygen:

$$\text{Molecular mass} = \underbrace{2(26.9815\ u)}_{\substack{\text{Mass of 2}\\ \text{aluminum}\\ \text{atoms}}} + \underbrace{3(15.9994\ u)}_{\substack{\text{Mass of 3}\\ \text{oxygen}\\ \text{atoms}}} = 101.9612\ u$$

Thus, the mass per mole of Al_2O_3 is 101.9612 g/mol. Calculations like those in part (a) reveal that the Rosser Reeves ruby contains 0.271 mol or 1.63×10^{23} molecules of Al_2O_3.

14.2 THE IDEAL GAS LAW

An ideal gas is an idealized model for real gases that have sufficiently low densities. The condition of low density means that the molecules of the gas are so far apart that they do not interact (except during collisions that are effectively elastic). The ideal gas law expresses the relationship between the absolute pressure, the Kelvin temperature, the volume, and the number of moles of the gas.

In discussing the constant volume gas thermometer, Section 12.2 has already explained the relationship between the absolute pressure and Kelvin temperature of a low-density gas. This thermometer utilizes a small amount of gas (e.g., hydrogen or helium) placed inside a bulb and kept at a constant volume. Since the density is low, the gas behaves as an ideal gas. Experiment reveals that a plot of gas pressure versus temperature is a straight line, as in Figure 12.4. This plot is redrawn in Figure 14.3, with the change that the temperature axis is now labeled in kelvins rather than in degrees Celsius. The graph indicates that the absolute pressure P is directly proportional to the Kelvin temperature T ($P \propto T$), for a fixed volume and a fixed number of molecules.

The relation between absolute pressure and the number of molecules of an ideal gas is simple. Experience indicates that it is possible to increase the pressure of a gas by adding more molecules; this is exactly what happens when a tire is pumped up. When the volume and temperature of a low-density gas are kept constant, doubling the number of molecules doubles the pressure. Thus, the absolute pressure of an ideal gas is proportional to the number of molecules or, equivalently, to the number of moles n of the gas ($P \propto n$).

To see how the absolute pressure of a gas depends on the volume of the gas, look at the partially filled balloon in Figure 14.4a. This balloon is "soft," because the pressure of the air is low. However, if all the air in the balloon is squeezed into a smaller "bubble," as in part b of the figure, the "bubble" has a very tight feel. This tightness indicates that the pressure in the smaller volume is high enough to stretch the rubber substantially. Thus, it is possible to increase the pressure of a gas by reducing its volume, and if the number of molecules and the temperature are kept constant, the absolute pressure of an ideal gas is inversely proportional to its volume V ($P \propto 1/V$).

The three relations just discussed for the absolute pressure of an ideal gas can be expressed as a single proportionality, $P \propto nT/V$. This proportionality can be written as an equation by inserting a proportionality constant R, called the **universal gas constant.** The value of R has been determined experimentally to be 8.31 J/(mol·K)

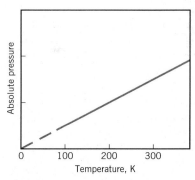

Figure 14.3 The pressure inside a constant-volume gas thermometer is directly proportional to the Kelvin temperature, which is characteristic of an ideal gas.

for any real gas whose density is sufficiently low to ensure ideal gas behavior. The resulting equation is known as the *ideal gas law.*

● IDEAL GAS LAW

The absolute pressure P of an ideal gas is directly proportional to the Kelvin temperature T and the number of moles n of the gas and is inversely proportional to the volume V of the gas: $P = R(nT/V)$. In other words,

$$PV = nRT \tag{14.1}$$

where R is the universal gas constant and has the value of 8.31 J/(mol·K).

Sometimes, it is convenient to express the ideal gas law in terms of the total number of particles N, instead of the number of moles n. To obtain such an expression, we multiply and divide the right side of the ideal gas law by Avogadro's number $N_A = 6.022 \times 10^{23}$ particles/mol* and recognize that the product nN_A is equal to the total number N of particles:

$$PV = nRT = nN_A \left(\frac{R}{N_A}\right) T = N \left(\frac{R}{N_A}\right) T$$

The constant term R/N_A is referred to as *Boltzmann's constant,* in honor of the Austrian physicist Ludwig Boltzmann (1844–1906), and is represented by the symbol k:

$$k = \frac{R}{N_A} = \frac{8.31 \text{ J/(mol·K)}}{6.022 \times 10^{23} \text{ mol}^{-1}} = 1.38 \times 10^{-23} \text{ J/K}$$

With this substitution, the ideal gas law becomes

$$PV = NkT \tag{14.2}$$

Example 2 presents an application of the ideal gas law.

EXAMPLE 2   Oxygen in the Lungs

In the lungs, the respiratory membrane separates tiny sacs of air (absolute pressure $= 1.00 \times 10^5$ Pa) from the blood in the capillaries. These sacs are called alveoli, and it is from them that oxygen enters the blood. The average radius of the alveoli is 0.125 mm, and the air inside contains 14% oxygen. Assuming that the air behaves as an ideal gas at body temperature (310 K), find the number of oxygen molecules in one of the sacs.

Reasoning The pressure and temperature of the air inside an alveolus are known, and its volume can be determined since we know the radius. Thus, the ideal gas law in the form $PV = NkT$ can be used directly to find the number N of air particles inside one of the sacs. The number of oxygen molecules is 14% of the number of air particles.

Solution The volume of a spherical sac is $V = \frac{4}{3}\pi r^3$. Solving Equation 14.2 for the number of air particles, we have

$$N = \frac{PV}{kT} = \frac{(1.00 \times 10^5 \text{ Pa})[\frac{4}{3}\pi(0.125 \times 10^{-3} \text{ m})^3]}{(1.38 \times 10^{-23} \text{ J/K})(310 \text{ K})} = 1.9 \times 10^{14}$$

The number of oxygen molecules is 14% of this value or $0.14N = \boxed{2.7 \times 10^{13}}$.

Figure 14.4 (*a*) The air pressure in the partially filled balloon can be increased by decreasing the volume of the balloon, as illustrated in (*b*).

THE PHYSICS OF...

oxygen in the lungs.

* "Particles" is not an SI unit and is often omitted. Then, particles/mol = 1/mol = mol^{-1}.

With the aid of the ideal gas law, it can be shown that one mole of an ideal gas occupies a volume of 22.4 liters at a temperature of 273 K (0 °C) and a pressure of one atmosphere (1.013×10^5 Pa). These conditions of temperature and pressure are known as *standard temperature and pressure (STP).* Conceptual Example 3 discusses another interesting application of the ideal gas law.

● PROBLEM SOLVING INSIGHT
In the ideal gas law, the temperature T must be expressed on the Kelvin scale. The Celsius and Fahrenheit scales cannot be used.

CONCEPTUAL EXAMPLE 3   Beer Bubbles on the Rise

The next time you get a chance, watch the bubbles rise in a glass of beer (see Figure 14.5). If you look carefully, you'll see them grow in size as they move upward, often doubling in volume by the time they reach the surface. Why does a bubble grow as it ascends?

Reasoning and Solution Beer bubbles contain mostly carbon dioxide (CO_2), a gas that is in the beer because of the fermentation process. The volume V of gas in a bubble is related to its temperature T, pressure P, and the number n of moles of CO_2 by the ideal gas law: $V = nRT/P$. Thus, one or more of these variables must be responsible for the growth of a bubble. Temperature can be eliminated immediately, since it is constant throughout the beer. What about the pressure? As a bubble rises, its depth decreases, and so does the fluid pressure. Since the volume is inversely proportional to pressure, part of the bubble growth is due to the decreasing pressure of the surrounding beer. However, some bubbles double in volume on the way up. To account for the doubling, there would have to be two atmospheres of pressure at the bottom of the glass, compared to the one atmosphere at the top. The pressure increment due to depth is $\rho g h$ according to Equation 11.4, so the extra pressure of one atmosphere at the bottom would mean 1.01×10^5 Pa $= \rho g h$. Solving for h with ρ equal to the density of water reveals that $h = 10.3$ m. Since most beer glasses are only about 0.2 m tall, we can rule out a change in pressure as the major cause of the change in volume. We are left with only one variable, the number of moles of CO_2 gas in the bubble. In fact, the number of moles does increase as the bubble rises. Each bubble acts as a nucleation site for CO_2 molecules, so as a bubble moves upward, it accumulates carbon dioxide from the surrounding beer and grows larger.

Related Homework Material: *Problem 22*

THE PHYSICS OF...

rising beer bubbles.

Figure 14.5 The bubbles in a glass of beer grow larger as they move upward.

Historically the work of several investigators led to the formulation of the ideal gas law. The Irish scientist Robert Boyle (1627–1691) discovered that at a constant temperature, the absolute pressure of a fixed mass (fixed number of moles) of a low-density gas is inversely proportional to its volume ($P \propto 1/V$). This fact is often called Boyle's law and can be derived from the ideal gas law by noting that $P = nRT/V = $ constant$/V$ when n and T are constants. Alternatively, if an ideal gas changes from an initial pressure and volume (P_i, V_i) to a final pressure and volume (P_f, V_f), it is possible to write $P_iV_i = nRT$ and $P_fV_f = nRT$. Since the right sides of these equations are equal, we may equate the left sides to give the following concise way of expressing *Boyle's law:*

Constant T, constant n $P_iV_i = P_fV_f$ (14.3)

Figure 14.6 illustrates how pressure and volume change according to Boyle's law for a fixed number of moles of an ideal gas at a constant temperature of 100 K. The gas begins with an initial pressure and volume of P_i and V_i and is compressed. The pressure increases as the volume decreases, according to $P = nRT/V$, until the final pressure and volume of P_f and V_f are reached. The curve that passes through the initial and final points is called an *isotherm,* meaning "same temperature." If the temperature had been 300 K, rather than 100 K, the compression would have oc-

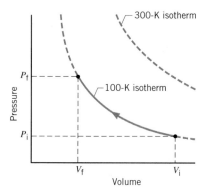

Figure 14.6 A pressure-versus-volume plot for an ideal gas at a constant temperature. Each isotherm is a plot of the equation $P = nRT/V = \text{constant}/V$, one with $T = 100$ K and the other with $T = 300$ K.

THE PHYSICS OF...

scuba diving.

● **PROBLEM SOLVING INSIGHT**
When using the ideal gas law, either directly or in the form of Boyle's law, remember that the pressure P must be the absolute pressure, not the gauge pressure.

Figure 14.7 The air pressure inside the body cavities of a scuba diver must be maintained at the same pressure as that of the surrounding water.

curred along the 300-K isotherm. Different isotherms do not intersect. Example 4 deals with an application of Boyle's law to scuba diving.

EXAMPLE 4   Scuba* Diving

In scuba diving, a greater water pressure acts on a diver at greater depths. The air pressure inside the body cavities (e.g., lungs, sinuses) must be maintained at the same pressure as that of the surrounding water; otherwise they would collapse. A special valve automatically adjusts the pressure of the air breathed from a scuba tank to ensure that the air pressure equals the water pressure at all times. The scuba gear in Figure 14.7 consists of a 0.0150-m³ tank filled with compressed air at an absolute pressure of 2.02×10^7 Pa. Assuming that air is consumed at a rate of 0.0300 m³ per minute and that the temperature is the same at all depths, determine how long the diver can stay under seawater at a depth of (a) 10.0 m and (b) 30.0 m.

Reasoning The time (in minutes) that a scuba diver can remain under water is equal to the volume of air that is available divided by the volume per minute consumed by the diver. The available volume is the volume of air at the pressure P_2 breathed by the diver. This pressure is determined by the depth h beneath the surface, according to $P_2 = P_1 + \rho gh$ (Equation 11.4), where $P_1 = 1.01 \times 10^5$ Pa is the atmospheric pressure at the surface. Since we know the pressure and volume of air in the scuba tank, and since the temperature is constant, we can use Boyle's law to find the volume of air available at the pressure P_2.

Solution

(a) Using $\rho = 1025$ kg/m³ for the density of seawater, we find that the absolute pressure P_2 at the depth of $h = 10.0$ m is

$$P_2 = P_1 + \rho gh = 1.01 \times 10^5 \text{ Pa} + (1025 \text{ kg/m}^3)(9.80 \text{ m/s}^2)(10.0 \text{ m})$$
$$= 2.01 \times 10^5 \text{ Pa}$$

The pressure and volume of the air in the tank are $P_i = 2.02 \times 10^7$ Pa and $V_i = 0.0150$ m³, respectively. According to Boyle's law, the volume of air V_f available at a pressure of $P_f = 2.01 \times 10^5$ Pa is

$$V_f = \frac{P_i V_i}{P_f} = \frac{(2.02 \times 10^7 \text{ Pa})(0.0150 \text{ m}^3)}{2.01 \times 10^5 \text{ Pa}} = 1.51 \text{ m}^3 \qquad (14.3)$$

Of this volume, only 1.51 m³ − 0.0150 m³ = 1.50 m³ is available for breathing, because 0.0150 m³ of air always remains in the tank. At a consumption rate of 0.0300 m³/min, the compressed air will last for

$$t = \frac{1.50 \text{ m}^3}{0.0300 \text{ m}^3/\text{min}} = \boxed{50.0 \text{ min}}$$

(b) The calculation here is like that in part (a). Equation 11.4 indicates that at a depth of 30.0 m, the absolute water pressure is 4.02×10^5 Pa. Because this pressure is twice that at the 10.0-m depth, Boyle's law reveals that the volume of air provided by the tank is now only $V_f = 0.754$ m³. The air available for use is 0.754 m³ − 0.0150 m³ = 0.739 m³. At a consumption rate of 0.0300 m³/min, the air will last for $\boxed{t = 24.6 \text{ min}}$, so the deeper dive must have a shorter duration.

Another investigator whose work contributed to the formulation of the ideal gas law was the Frenchman, Jacques Charles (1746–1823). He discovered that at a

---

* The word is an acronym for **s**elf-**c**ontained **u**nderwater **b**reathing **a**pparatus.

constant pressure, the volume of a fixed mass (fixed number of moles) of a low-density gas is directly proportional to the Kelvin temperature ($V \propto T$). This relationship is known as Charles' law and can be obtained from the ideal gas law by noting that $V = nRT/P = (\text{constant})T$, if n and P are constant. Equivalently, when an ideal gas changes from an initial volume and temperature (V_i, T_i) to a final volume and temperature (V_f, T_f), it is possible to write $V_i/T_i = nR/P$ and $V_f/T_f = nR/P$. Thus, one way of stating **Charles' law** is

Constant P, constant n
$$\frac{V_i}{T_i} = \frac{V_f}{T_f}$$
(14.4)

14.3 KINETIC THEORY OF GASES

● *Concepts at a Glance*   As useful as it is, the ideal gas law provides no insight as to how pressure and temperature are related to properties of the molecules themselves, such as their masses and speeds. To show how such microscopic properties are related to the pressure and temperature of an ideal gas, this section examines the dynamics of molecular motion. The pressure that a gas exerts on the walls of a container is due to the force exerted by the gas molecules when they collide with the walls. Therefore, as the Concepts-at-a-Glance chart in Figure 14.8 illustrates, we will begin by combining the notion of collisional forces exerted by a fluid (Section 11.2) with Newton's second and third laws of motion (Sections 4.3 and 4.5). These concepts will allow us to obtain an expression for the pressure in terms of microscopic properties. We will then combine this with the ideal gas law to show that the average translational kinetic energy $\overline{\text{KE}}$ of a particle in an ideal gas is $\overline{\text{KE}} = \frac{3}{2}kT$, where k is Boltzmann's constant and T is the Kelvin temperature. In the process, we will also see that the internal energy U of a monatomic ideal gas is $U = \frac{3}{2}nRT$, where n is the number of moles and R is the universal gas constant.   ●

Figure 14.8 *Concepts at a Glance* To formulate the kinetic theory of gases, Newton's second and third laws are brought together with the ideal gas law. One of the predictions of the theory is that the internal energy of an ideal gas is proportional to the Kelvin temperature of the gas.

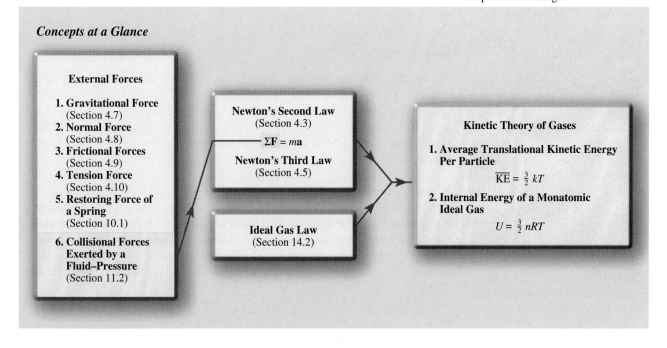

Concepts at a Glance

External Forces

1. Gravitational Force
   (Section 4.7)
2. Normal Force
   (Section 4.8)
3. Frictional Forces
   (Section 4.9)
4. Tension Force
   (Section 4.10)
5. Restoring Force of
   a Spring
   (Section 10.1)

6. Collisional Forces
   Exerted by a
   Fluid–Pressure
   (Section 11.2)

Newton's Second Law
(Section 4.3)

$\Sigma \mathbf{F} = m\mathbf{a}$

Newton's Third Law
(Section 4.5)

Ideal Gas Law
(Section 14.2)

Kinetic Theory of Gases

1. Average Translational Kinetic Energy
   Per Particle

   $\overline{\text{KE}} = \frac{3}{2}kT$

2. Internal Energy of a Monatomic
   Ideal Gas

   $U = \frac{3}{2}nRT$

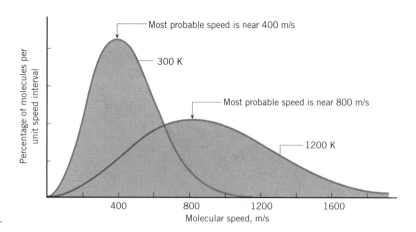

Figure 14.9 The Maxwell distribution curves for molecular speeds in oxygen gas at temperatures of 300 and 1200 K.

The Distribution of Molecular Speeds

A macroscopic container filled with a gas at standard temperature and pressure contains a large number of particles (atoms or molecules). These particles are in constant, random motion, colliding with each other and with the walls of the container. In the course of one second, a particle undergoes many collisions, and each one changes the particle's speed and direction of motion. As a result, the atoms or molecules have different speeds. It is possible, however, to speak about an average particle speed. At any given instant, some particles have speeds less than, some near, and some greater than the average. For conditions of low gas density, the distribution of speeds within a large collection of molecules at a constant temperature was calculated by the Scottish physicist James Clerk Maxwell (1831–1879). Figure 14.9 displays the Maxwell speed distribution curves for O_2 gas at two different temperatures. When the temperature is 300 K, the maximum in the curve indicates that the most probable speed is about 400 m/s. At a temperature of 1200 K, the distribution curve shifts to the right, and the most probable speed increases to about 800 m/s.

Figure 14.10 The pressure that a gas exerts is caused by the impact of its molecules on the walls of the container.

Kinetic Theory

If a ball is thrown against a wall, it exerts a force on the wall. As Figure 14.10 suggests, gas particles do the same thing, except that their masses are smaller and their speeds are greater. The number of particles is so great and they strike the wall so often that the effect of their individual impacts appears as a continuous force. Dividing the magnitude of this force by the area of the wall gives the pressure exerted by the gas.

To calculate the force, consider an ideal gas composed of N identical particles in a cubical container whose sides have length L. Except for elastic* collisions, these particles do not interact. Figure 14.11 focuses attention on one particle of mass m as it strikes the right wall perpendicularly and rebounds elastically. While approaching the wall, the particle has a velocity $+v$ and linear momentum $+mv$ (see Section 7.1 for a review of linear momentum). The particle rebounds with a velocity $-v$ and momentum $-mv$, travels to the left wall, rebounds again, and heads back toward

Figure 14.11 A gas particle is shown colliding elastically with the right wall of the container and rebounding from it.

---

* The term "elastic" is used here to mean that *on the average*, in a large number of particles, there is no gain or loss of translational kinetic energy because of collisions.

the right. The time t between collisions with the right wall is the round-trip distance $2L$ divided by the speed of the particle, that is, $t = 2L/v$. According to Newton's second law of motion, in the form of the impulse–momentum theorem, the average force exerted on the particle by the wall is given by the change in the particle's momentum per unit time:

$$\text{Average force} = \frac{\text{Final momentum} - \text{Initial momentum}}{\text{Time between successive collisions}} \qquad (7.4)$$

$$= \frac{(-mv) - (+mv)}{2L/v} = \frac{-mv^2}{L}$$

According to Newton's law of action–reaction, the force applied to the wall by the particle is equal in magnitude to this value, but oppositely directed (i.e., $+mv^2/L$). The magnitude F of the *total* force exerted on the right wall is equal to the number of particles that collide with the wall during the time t multiplied by the average force exerted by each particle. Since the N particles move randomly in three dimensions, one-third of them on the average strike the right wall during the time t. Therefore, the total force is

$$F = \left(\frac{N}{3}\right)\left(\frac{m\overline{v^2}}{L}\right)$$

In this result v^2 has been replaced by $\overline{v^2}$, *the average* value of the squared speed. The collection of particles possesses a Maxwell distribution of speeds, so an average value for v^2 must be used, rather than a value for any individual particle. The square root of the quantity $\overline{v^2}$ is called the ***root-mean-square speed,*** or, for short, the *rms speed;* $v_{rms} = \sqrt{\overline{v^2}}$. With this substitution, the total force becomes

$$F = \left(\frac{N}{3}\right)\left(\frac{mv_{rms}^2}{L}\right)$$

Pressure is force per unit area, so the pressure P acting on a wall of area L^2 is

$$P = \frac{F}{L^2} = \left(\frac{N}{3}\right)\left(\frac{mv_{rms}^2}{L^3}\right)$$

Since the volume of the box is $V = L^3$, the equation above can be written as

$$PV = \tfrac{2}{3}N(\tfrac{1}{2}mv_{rms}^2) \qquad (14.5)$$

Equation 14.5 relates the macroscopic properties of the gas, its pressure and volume, to the microscopic properties of the constituent particles, their mass and speed. Since the term $\tfrac{1}{2}mv_{rms}^2$ is the average translational kinetic energy $\overline{\text{KE}}$ of an individual particle, it follows that

$$PV = \tfrac{2}{3}N(\overline{\text{KE}})$$

This result is similar to the ideal gas law, $PV = NkT$. Both equations have identical terms on the left, so the terms on the right must be equal: $\tfrac{2}{3}N(\overline{\text{KE}}) = NkT$. Therefore,

$$\overline{\text{KE}} = \tfrac{1}{2}mv_{rms}^2 = \tfrac{3}{2}kT \qquad (14.6)$$

Equation 14.6 is significant, for it allows us to interpret temperature in terms of the motion of gas particles. This equation indicates that the Kelvin temperature is directly proportional to the average translational kinetic energy per particle in an ideal gas, no matter what the pressure and volume are. On the average, the particles have greater kinetic energies when the gas is hotter than when it is cooler. Conceptual

Example 5 discusses a common misconception about the relation between kinetic energy and temperature.

CONCEPTUAL EXAMPLE 5   Does a Single Particle Have a Temperature?

Each particle in a gas has kinetic energy. Furthermore, the equation $\frac{1}{2}mv_{rms}^2 = \frac{3}{2}kT$ establishes the relationship between the average kinetic energy per particle and temperature of an ideal gas. Is it valid, then, to conclude that a single particle has a temperature?

Reasoning and Solution We know that a gas contains an enormous number of particles that are traveling with a distribution of speeds, such as those indicated by the graphs in Figure 14.9. Therefore, the particles do not all have the same kinetic energy, but possess a distribution of kinetic energies ranging from very nearly zero to extremely large values. If each particle had a temperature that was associated with its kinetic energy, there would be a whole range of different temperatures within the gas. This is not so, for a gas at thermal equilibrium has only one temperature (see Section 15.2), a temperature that would be registered by a thermometer placed in the gas. Thus, temperature is a property that characterizes the gas as a whole, a fact that is inherent in the relation $\frac{1}{2}mv_{rms}^2 = \frac{3}{2}kT$. The term v_{rms} is a kind of *average particle speed*. Therefore, $\frac{1}{2}mv_{rms}^2$ is the *average kinetic energy* per particle and is characteristic of the gas as a whole. Since the Kelvin temperature is proportional to $\frac{1}{2}mv_{rms}^2$, it is also a characteristic of the gas as a whole and cannot be ascribed to each gas particle individually. Thus, *a single gas particle does not have a temperature.*

If two ideal gases have the same temperature, the relation $\frac{1}{2}mv_{rms}^2 = \frac{3}{2}kT$ indicates that the average kinetic energy of each kind of gas particle is the same. In general, however, the rms speeds of the different particles are not the same, for the masses may be different. The next example illustrates these facts and shows how rapidly gas particles move at normal temperatures.

EXAMPLE 6   The Speed of Molecules in Air

Air is primarily a mixture of nitrogen N_2 (molecular mass = 28.0 u) and oxygen O_2 (molecular mass = 32.0 u). Assume that each behaves as an ideal gas and determine the rms speed of the nitrogen and oxygen molecules when the temperature of the air is 293 K.

Reasoning The rms speed can be obtained from Equation 14.6 as $v_{rms} = \sqrt{2\overline{KE}/m_{particle}}$, where $m_{particle}$ is the mass of a single particle (nitrogen or oxygen molecule). In this expression, the average kinetic energy is $\overline{KE} = \frac{3}{2}kT$ and is the same for both types of molecules at the same temperature. However, the masses of the nitrogen and oxygen molecules are different. The mass of each particle is equal to the mass per mole divided by the number of particles per mole (Avogadro's number).

Solution The average kinetic energy per particle for both nitrogen and oxygen is

$$\overline{KE} = \tfrac{3}{2}kT = \tfrac{3}{2}(1.38 \times 10^{-23} \text{ J/K})(293 \text{ K}) = \boxed{6.07 \times 10^{-21} \text{ J}} \qquad (14.6)$$

Since the molecular mass of nitrogen is 28.0 u, its mass per mole is 28.0 g/mol. Therefore, the mass of a nitrogen molecule is

$$m_{particle} = \frac{\text{Mass per mole}}{\text{Number of particles per mole}}$$

$$= \frac{28.0 \text{ g/mol}}{6.022 \times 10^{23} \text{ mol}^{-1}} = 4.65 \times 10^{-23} \text{ g} = 4.65 \times 10^{-26} \text{ kg}$$

Similarly, we find the particle mass for oxygen to be 5.31×10^{-26} kg. The calculations of the rms speeds are shown below:

Nitrogen $v_{rms} = \sqrt{\dfrac{2(\overline{KE})}{m_{particle}}} = \sqrt{\dfrac{2(6.07 \times 10^{-21} \text{ J})}{4.65 \times 10^{-26} \text{ kg}}} = \boxed{511 \text{ m/s}}$

Oxygen $v_{rms} = \sqrt{\dfrac{2(\overline{KE})}{m_{particle}}} = \sqrt{\dfrac{2(6.07 \times 10^{-21} \text{ J})}{5.31 \times 10^{-26} \text{ kg}}} = \boxed{478 \text{ m/s}}$

For comparison, the speed of sound at a temperature of 293 K is 343 m/s (767 mi/h).

● PROBLEM SOLVING INSIGHT
The average translational kinetic energy is the same for all ideal-gas molecules at the same temperature, regardless of their masses. The rms translational speed of the molecules is not the same, however, but depends on the mass.

The equation $\overline{KE} = \frac{3}{2}kT$ has also been applied to particles much larger than atoms or molecules. The English botanist Robert Brown (1773–1858) observed through a microscope that pollen grains suspended in water move on very irregular, zigzag paths. This Brownian motion can also be observed with other particle suspensions, such as fine smoke particles in air. In 1905, Albert Einstein (1879–1955) showed that Brownian motion could be explained as a response of the large suspended particles to impacts from the moving molecules of the fluid medium (e.g., water or air). As a result of the impacts, the suspended particles have the same average translational kinetic energy as the fluid molecules, namely, $\overline{KE} = \frac{3}{2}kT$. But unlike the molecules, the particles are large enough to be seen through a microscope and, because of their relatively large mass, have a comparatively small average speed.

The Internal Energy of a Monatomic Ideal Gas

Chapter 15 deals with the science of thermodynamics, in which the concept of internal energy plays an important role. Using the results just developed for the average translational kinetic energy, we conclude this section by expressing the internal energy of a monatomic ideal gas in a form that is suitable for use later on.

The internal energy of a substance is the sum of the various kinds of energy that the atoms or molecules of the substance possess. A monatomic ideal gas is composed of single atoms. These atoms are assumed to be so small that the mass is concentrated at a point, with the result that the moment of inertia I about the center of mass is negligible. Thus, the rotational kinetic energy $\frac{1}{2}I\omega^2$ is also negligible. Vibrational kinetic and potential energies are absent, because the atoms are not connected by chemical bonds and, except for elastic collisions, do not interact. As a result, the internal energy U is the total translational kinetic energy of the N atoms that constitute the gas: $U = N(\frac{1}{2}mv_{rms}^2)$. Since $\frac{1}{2}mv_{rms}^2 = \frac{3}{2}kT$ according to Equation 14.6, the internal energy can be written in terms of the Kelvin temperature as

$$U = N(\tfrac{3}{2}kT)$$

Usually, U is expressed in terms of the number of moles n, rather than the number of atoms N. Using the fact that Boltzmann's constant is $k = R/N_A$, where R is the universal gas constant and N_A is Avogadro's number, and realizing that $N/N_A = n$, we find that

Monatomic ideal gas $U = \frac{3}{2}nRT$ (14.7)

Thus, the internal energy depends on the number of moles and Kelvin temperature of the gas. In fact, it can be shown that the internal energy is proportional to the Kelvin temperature for *any type* of ideal gas (e.g., monatomic, diatomic, etc.). For example, when hot-air balloonists turn on the burner, they're increasing the temperature, and hence the internal energy, of the air inside the balloon (see Figure 14.12).

Figure 14.12 Since air behaves approximately as an ideal gas, the internal energy inside a hot-air balloon increases as the temperature rises.

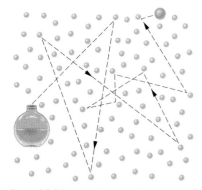

Figure 14.13 A drop of ink placed in water eventually becomes completely dispersed because of diffusion.

Initially   Later

*14.4   DIFFUSION

You can smell the fragrance of a perfume at some distance from an open bottle because perfume molecules evaporate from the liquid, where they are relatively concentrated, and spread out into the air, where they are less concentrated. During their journey, they collide with other molecules, so their paths resemble the zigzag paths characteristic of Brownian motion. The process in which molecules move from a region of higher concentration to one of lower concentration is called *diffusion.* Diffusion also occurs in liquids and solids, and Figure 14.13 illustrates ink diffusing through water. However, compared to the rate of diffusion in gases, the rate is generally smaller in liquids and even smaller in solids. The host medium, such as the air or water in the examples above, is referred to as the *solvent,* while the diffusing substance, like the perfume molecules or the ink in Figure 14.13, is known as the *solute.* Relatively speaking, diffusion is a slow process, even in a gas. Conceptual Example 7 illustrates why.

CONCEPTUAL EXAMPLE 7   Why Diffusion Is Relatively Slow

We have seen in Example 6 that a gas molecule has a translational rms speed of hundreds of meters per second at room temperature. At such a speed, a molecule could travel across an ordinary room in just a fraction of a second. Yet, it often takes several seconds, and sometimes minutes, for the fragrance of a perfume to reach the other side of a room. Why does it take so long?

Reasoning and Solution When a perfume molecule diffuses through air, it makes millions of collisions each second with air molecules. As Figure 14.14 illustrates, the speed and direction of motion change abruptly as a result of each collision. Between collisions, the perfume molecule moves in a straight line at a constant speed. Although a perfume molecule does move very fast between collisions, it wanders only slowly away from the bottle because of the zigzag path resulting from the collisions. It would take a long time for a molecule to diffuse in this manner across a room. Usually, however, convection currents are present and carry the fragrance across the room in a matter of seconds or minutes.

Related Homework Material: *Problems 42, 43, and 44*

Figure 14.14 A perfume molecule collides with millions of air molecules during its journey, so the path has a zigzag shape. Although the air molecules are shown as stationary, they are also moving.

The diffusion process can be described in terms of the arrangement in Figure 14.15*a*. A hollow channel of length L and cross-sectional area A is filled with a fluid. The left end of the channel is connected to a container in which the solute concentration C_2 is relatively high, while the right end is connected to a container in which the solute concentration C_1 is lower. These concentrations are defined as the total mass of the solute molecules divided by the volume of the solution (e.g., 0.1 kg/m³). Because of the difference in concentration between the ends of the channel, $\Delta C = C_2 - C_1$, there is a net diffusion of the solute from the left end to the right end.

Figure 14.15*a* is similar to Figure 13.8 for the conduction of heat along a bar, which, for convenience, is reproduced in Figure 14.15*b*. When the ends of the bar are maintained at different temperatures, T_2 and T_1, the heat Q conducted along the bar in a time t is

$$Q = \frac{(kA\,\Delta T)t}{L} \tag{13.1}$$

where $\Delta T = T_2 - T_1$, and k is the thermal conductivity. Whereas conduction is the flow of heat from a region of higher temperature to a region of lower temperature, diffusion is the mass flow of solute from a region of higher concentration to a re-

Figure 14.15 (*a*) Solute diffuses through the channel from the region of higher concentration to the region of lower concentration. (*b*) Heat is conducted along a bar whose ends are maintained at different temperatures.

gion of lower concentration. By analogy with Equation 13.1, it is possible to write an equation for diffusion: (1) replace Q by the mass m of solute that is diffusing through the channel, (2) replace $\Delta T = T_2 - T_1$ by the difference in concentrations $\Delta C = C_2 - C_1$, and (3) replace k by a constant known as the diffusion constant D. The resulting equation, first formulated by the German physiologist Adolf Fick (1829–1901), is referred to as **Fick's law of diffusion.**

● **FICK'S LAW OF DIFFUSION**

The mass m of solute that diffuses in a time t through a solvent contained in a channel of length L and cross-sectional area A is

$$m = \frac{(DA\,\Delta C)t}{L} \qquad (14.8)$$

where ΔC is the concentration difference between the ends of the channel and D is the diffusion constant.

SI Unit for the Diffusion Constant: m^2/s

It can be verified from Equation 14.8 that the diffusion constant has units of m^2/s, the exact value depending on the nature of the solute and the solvent. For example, the diffusion constant for ink in water is different from that for ink in benzene. Example 8 illustrates an important application of Fick's law.

EXAMPLE 8 **Water Given Off by Plant Leaves**

Large amounts of water can be given off by plants. It has been estimated, for instance, that a single sunflower plant can lose up to a pint of water a day during the growing season. Figure 14.16 shows a cross-sectional view of a leaf. Inside the leaf,

THE PHYSICS OF...

water loss from plant leaves.

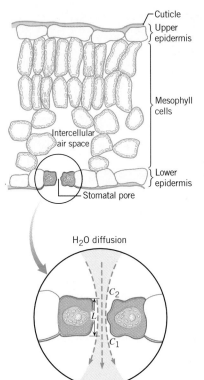

Figure 14.16 A cross-sectional view of a leaf. Water vapor diffuses out of the leaf through a stomatal pore.

water passes from the liquid phase to the vapor phase at the walls of the mesophyll cells. The water vapor then diffuses through the intercellular air spaces and eventually exits the leaf through small openings, called stomatal pores. The diffusion constant for water vapor in air is $D = 2.4 \times 10^{-5}$ m²/s. A stomatal pore has a cross-sectional area of about $A = 8.0 \times 10^{-11}$ m² and a length of about $L = 2.5 \times 10^{-5}$ m. The concentration of water vapor on the interior side of a pore is roughly $C_2 = 0.022$ kg/m³, while that on the outside is approximately $C_1 = 0.011$ kg/m³. Determine the mass of water vapor that passes through a stomatal pore in one hour.

Reasoning and Solution Fick's law of diffusion shows that

$$m = \frac{(DA\,\Delta C)t}{L} \tag{14.8}$$

$$m = \frac{(2.4 \times 10^{-5}\ \text{m}^2/\text{s})(8.0 \times 10^{-11}\ \text{m}^2)(0.022\ \text{kg/m}^3 - 0.011\ \text{kg/m}^3)(3600\ \text{s})}{2.5 \times 10^{-5}\ \text{m}}$$

$$= \boxed{3.0 \times 10^{-9}\ \text{kg}}$$

This amount of water may not seem significant. However, a single leaf may have a million or so stomatal pores, so the water lost by an entire plant can be substantial.

14.5 CONCEPTS & CALCULATIONS

This chapter introduces the ideal gas law, which is a relation between the pressure, volume, temperature, and the number of moles in an ideal gas. In Section 10.1, we examined how the compression of a spring depends on the force applied to it. Example 9 reviews how a gas produces a force and why an ideal gas at different temperatures causes a spring to compress by different amounts.

CONCEPTS & CALCULATIONS EXAMPLE 9   The Ideal Gas Law and Springs

Figure 14.17 shows three identical chambers containing a piston and a spring whose spring constant is $k = 5.8 \times 10^4$ N/m. The chamber in part a is completely evacuated, and the piston just touches its left end. In this position, the spring is unstrained. In part b of the drawing, 0.75 mol of an ideal gas, gas 1, is introduced into the chamber, and the spring compresses by $x_1 = 15$ cm. In part c, 0.75 mol of another ideal gas, gas 2, is introduced into the chamber, and the spring compresses by $x_2 = 24$ cm. Find the temperature of each gas.

Concept Questions and Answers Which gas exerts the greater force on the piston?

> *Answer* We know that a greater force is required to compress a spring by a greater amount. Therefore, gas 2 exerts the greater force.

How is the force required to compress a spring related to the displacement of the spring from its unstrained position?

> *Answer* According to Equation 10.1, the force F required to compress a spring is directly proportional to the displacement x of the spring from its unstrained position; $F = kx$, where k is the spring constant of the spring.

Figure 14.17 Two ideal gases at different temperatures compress the spring by different amounts.

Which gas exerts the greater pressure on the piston?

Answer   According to Equation 11.3, pressure is defined as the magnitude F of the force acting perpendicular to the surface of the piston divided by the area A of the piston; $P = F/A$. Since gas 2 exerts the greater force, and the area of the piston is the same for both gases, gas 2 exerts the greater pressure.

Which gas has the greater temperature?

Answer   According to the ideal gas law, $T = PV/(nR)$, gas 2 has the greater temperature. Both gases contain the same number n of moles. However, gas 2 has both a greater pressure P and volume V. Thus, it has the greater temperature.

Solution We can use the ideal gas law in the form $T = PV/(nR)$ to determine the temperature T of each gas. First, however, we must find the pressure. According to Equation 11.3, the pressure is the magnitude F of the force that the gas exerts on the piston divided by the area A of the piston, so $P = F/A$. The force applied to the piston is related to the displacement x of the spring by Equation 10.1, $F = kx$. Thus, the pressure is $P = kx/A$. Using this expression for the pressure in the ideal gas law gives

$$T = \frac{PV}{nR} = \frac{\left(\dfrac{kx}{A}\right)V}{nR}$$

However, the cylindrical volume V of the gas is equal to the product of the displacement x and the area A, so $V = xA$. With this substitution, the temperature of the gas becomes

$$T = \frac{\left(\dfrac{kx}{A}\right)V}{nR} = \frac{\left(\dfrac{kx}{A}\right)(xA)}{nR} = \frac{kx^2}{nR}$$

The temperatures of the gases are

Gas 1 $T_1 = \dfrac{kx_1^2}{nR} = \dfrac{(5.8 \times 10^4 \text{ N/m})(15 \times 10^{-2} \text{ m})^2}{(0.75 \text{ mol})[8.31 \text{ J/(mol} \cdot \text{K)}]} = \boxed{210 \text{ K}}$

Gas 2 $T_2 = \dfrac{kx_2^2}{nR} = \dfrac{(5.8 \times 10^4 \text{ N/m})(24 \times 10^{-2} \text{ m})^2}{(0.75 \text{ mol})[8.31 \text{ J/(mol} \cdot \text{K)}]} = \boxed{540 \text{ K}}$

As anticipated, gas 2, which compresses the spring more, has the higher temperature.

The kinetic theory of gases is important because it allows us to understand the relation between the macroscopic properties of a gas, such as pressure and temperature, and the microscopic properties of its particles, such as speed and mass. The following example reviews the essential features of this theory.

CONCEPTS & CALCULATIONS EXAMPLE 10   Hydrogen Atoms in Outer Space

In outer space the density of matter is extremely low, about one atom per cm^3. The matter is mainly hydrogen atoms ($m = 1.67 \times 10^{-27}$ kg) whose rms speed is 260 m/s. A cubical box, 2.0 m on a side, is placed in outer space, and the hydrogen atoms are allowed to enter. (a) What is the magnitude of the force that the atoms exert on one wall of the box? (b) Determine the pressure that the atoms exert. (c) Does outer space have a temperature, and, if so, what is it?

Concept Questions and Answers Why do hydrogen atoms exert a force on the walls of the box?

Answer   Every time an atom collides with a wall and rebounds, the atom exerts a force on the wall. Imagine opening your hand so it is flat and having someone throwing a ball straight at it. Your hand is like the wall, and as the ball rebounds, you can feel the force. Intuitively, you would expect the force to become greater as the speed and mass of the ball become greater. This is indeed the case.

Do the atoms generate a pressure on the walls of the box?

Answer Yes. Pressure is defined as the magnitude of the force exerted perpendicularly on a wall divided by the area of the wall. Since the atoms exert a force, they also produce a pressure.

Do hydrogen atoms in outer space have a temperature? If so, how is the temperature related to the microscopic properties of the atoms?

Answer Yes. The Kelvin temperature is proportional to the average kinetic energy of an atom. The average kinetic energy, in turn, is proportional to the mass of an atom and the square of the rms speed. Since we know both these quantities, we can determine the temperature of the gas.

Solution

(a) The magnitude F of the force exerted on a wall is given by (see Section 14.3)

$$F = \left(\frac{N}{3}\right)\left(\frac{mv_{rms}^2}{L}\right)$$

where N is the number of atoms in the box, m is the mass of a single atom, v_{rms} is the rms speed of the atoms, and L is the length of one side of the box. The volume of the cubical box is $(2.0 \times 10^2 \text{ cm})^3 = 8.0 \times 10^6 \text{ cm}^3$. The number N of atoms is equal to the number of atoms per cubic centimeter times the volume of the box in cubic centimeters:

$$N = \left(\frac{1 \text{ atom}}{\text{cm}^3}\right)(8.0 \times 10^6 \text{ cm}^3) = 8.0 \times 10^6$$

The magnitude of the force acting on one wall is

$$F = \left(\frac{N}{3}\right)\left(\frac{mv_{rms}^2}{L}\right) = \left(\frac{8.0 \times 10^6}{3}\right)\left[\frac{(1.67 \times 10^{-27} \text{ kg})(260 \text{ m/s})^2}{2.0 \text{ m}}\right]$$

$$= \boxed{1.5 \times 10^{-16} \text{ N}}$$

(b) The pressure is the magnitude of the force divided by the area A of a wall:

$$P = \frac{F}{A} = \frac{1.5 \times 10^{-16} \text{ N}}{(2.0 \text{ m})^2} = \boxed{3.8 \times 10^{-17} \text{ Pa}}$$

(c) According to Equation 14.6, the Kelvin temperature T of the hydrogen atoms is related to the average kinetic energy of an atom by $\frac{3}{2}kT = \frac{1}{2}mv_{rms}^2$, where k is Boltzmann's constant. Solving this equation for the temperature gives

$$T = \frac{mv_{rms}^2}{3k} = \frac{(1.67 \times 10^{-27} \text{ kg})(260 \text{ m/s})^2}{3(1.38 \times 10^{-23} \text{ J/K})} = \boxed{2.7 \text{ K}}$$

This is a frigid 2.76 kelvins above absolute zero.

At the end of the problem set for this chapter, you will find homework problems that contain both conceptual and quantitative parts. These problems are grouped under the heading *Concepts & Calculations, Group Learning Problems*. They are designed for use by students working alone or in small learning groups. The conceptual part of each problem provides a convenient focus for group discussions.

SUMMARY

Each element in the periodic table is assigned an **atomic mass**. One **atomic mass unit** (u) is exactly one-twelfth the mass of an atom of carbon-12. The **molecular mass** of a molecule is the sum of the atomic masses of its atoms. The number of moles n contained in a sample is equal to the number of particles N (atoms or molecules) in the sample divided by the number of particles per mole N_A, $n = N/N_A$, where N_A is **Avogadro's number** and has a value of $N_A = 6.022 \times 10^{23}$ particles per mole. The number of moles is also equal to the mass of the sample (expressed in grams) divided by the mass per mole (expressed in grams per mole). The mass per mole of a substance has the same numerical value as the atomic or molecular mass of one of its particles.

The **ideal gas law** relates the absolute pressure P, the volume V, the number of moles n, and the Kelvin temperature T of an ideal gas according to $PV = nRT$, where $R = 8.31$ J/(mol·K) is the universal gas constant. An alternative form of the ideal gas law is $PV = NkT$, where N is the number of particles and $k = R/N_A$ is Boltzmann's constant. A real gas behaves as an ideal gas when the density of the real gas is low enough that its particles do not interact, except via elastic collisions.

The distribution of particle speeds in an ideal gas at constant temperature is the **Maxwell speed distribution.** According to the **kinetic theory of gases,** an ideal gas consists of a large number of particles (atoms or molecules) that are in constant random motion. The particles are far apart compared to their dimensions, so they do not interact except when elastic collisions occur. The pressure on the walls of a container is produced by the impacts of the particles with the walls. According to the kinetic theory of gases, the Kelvin temperature T of an ideal gas is a measure of the average translational kinetic energy $\overline{\text{KE}}$ per particle through the relation $\overline{\text{KE}} = \frac{3}{2}kT$. The **internal energy** U of n moles of a monatomic ideal gas is $U = \frac{3}{2}nRT$. The internal energy of any type of ideal gas is proportional to its Kelvin temperature.

Diffusion is the process whereby solute molecules move through a solvent from a region of higher concentration to a region of lower concentration. **Fick's law of diffusion** states that the mass m of solute that diffuses in a time t through a solvent contained in a channel of length L and cross-sectional area A is given by $m = (DA \Delta C)t/L$, where ΔC is the concentration difference between the ends of the channel and D is the diffusion constant.

CONCEPTUAL QUESTIONS

1. (a) Which, if either, contains a greater number of molecules, a mole of hydrogen (H_2) or a mole of oxygen (O_2)? (b) Which one has more mass? Give reasons for your answers.

2. Suppose two different substances, A and B, have the same mass densities. (a) In general, does one mole of substance A have the same mass as one mole of substance B? (b) Does 1 m^3 of substance A have the same mass as 1 m^3 of substance B? Justify each answer.

3. A tightly sealed house has a large ceiling fan that blows air out of the house and into the attic. This fan is turned on, and the owners forget to open any windows or doors. What happens to the air pressure in the house after the fan has been on for a while, and does it become easier or harder for the fan to do its job? Explain.

4. Above the liquid in a can of hair spray is a gas at a relatively high pressure. The label on the can includes the warning "Do not store at high temperatures." Use the ideal gas law and explain why the warning is given.

5. Assuming that air behaves like an ideal gas, explain what happens to the pressure in a tightly sealed house when the electric furnace turns on for a while.

6. ✍ When you climb a mountain, your eardrums "pop" outward as the air pressure decreases. When you come down, they "pop" inward as the pressure increases. At the sea coast, there is a cave that can be entered only by swimming through a completely submerged passage and emerging into a pocket of air within the cave. The cave is not vented to the external atmosphere. As the tide comes in, the water level in the cave rises, and your eardrums "pop." Is this "popping" analogous to what happens as you climb up or down a mountain? Give your reasoning.

7. Atmospheric pressure decreases with increasing altitude. With this fact in mind, explain why helium-filled weather balloons are underinflated when they are launched from earth.

8. A slippery cork is being pressed into a very full (but not 100% full) bottle of wine. When released, the cork slowly slides back out. However, if some wine is removed from the bottle before the cork is inserted, the cork does not slide out. Account for these observations in terms of the ideal gas law.

9. A commonly used packing material consists of "bubbles" of air trapped between bonded layers of plastic, as the photograph shows. Using the ideal gas law, explain why this packing material offers less protection on cold days than on warm days.

10. The kinetic theory of gases assumes that a gas molecule rebounds with the same speed after colliding with the wall of a container. Suppose the speed after the collision is less than that before the collision. Would the pressure of the gas be greater than, equal to, or less than that predicted by kinetic theory? Account for your answer.

11. If the translational speed of each molecule in an ideal gas were tripled, would the Kelvin temperature also triple? If not, by what factor would the Kelvin temperature increase? Account for your answer.

12. If the temperature of an ideal gas is doubled from 50 to 100 °C, does the average translational kinetic energy of each particle double? If not, why not?

13. It is possible for both the pressure and volume of a monatomic ideal gas to change *without* causing the internal energy of the gas to change. Explain how this could occur.

14. Suppose that the atoms in a container of helium (He) have the same translational rms speed as the molecules in a container of argon (Ar). Treating each gas as an ideal gas, explain which, if either, has the greater temperature.

15. The ionosphere is the uppermost part of the earth's atmosphere. The temperature of the ionized gas in the ionosphere is about 1000 K. However, the density of the gas is extremely low, being on the order of 10^{11} molecules/m^3. Although the temperature is very high, an astronaut in the ionosphere would not burn up. Why?

16. 🔬 In the lungs, oxygen in very small sacs called alveoli diffuses into the blood. The diffusion occurs directly through the walls of the sacs. The walls are very thin, so the oxygen diffuses over a distance L that is quite small. Because there are so many alveoli, the effective area A across which diffusion occurs is very large. Use this information, together with Fick's law of diffusion, to explain why the mass of oxygen per second that diffuses into the blood is large.

PROBLEMS

Note: The pressures referred to in these problems are absolute pressures, unless indicated otherwise.

---

ssm Solution is in the Student Solutions Manual **www** Solution is available on the World Wide Web at http://www.wiley.com/college/cutnell
🔬 **This icon represents a biomedical application.**

Section 14.1 The Mole, Avogadro's Number, and Molecular Mass

1. ssm Hemoglobin has a molecular mass of 64 500 u. Find the mass (in kg) of one molecule of hemoglobin.

2. The chlorophyll-*a* molecule, $C_{55}H_{72}MgN_4O_5$, is important in photosynthesis. (a) Determine its molecular mass (in atomic mass units). (b) What is the mass (in grams) of 3.00 moles of chlorophyll-*a* molecules?

3. The artificial sweetener NutraSweet is a chemical called aspartame ($C_{14}H_{18}N_2O_5$). What is (a) its molecular mass (in atomic mass units) and (b) the mass (in kg) of an aspartame molecule?

4. 🔬 Manufacturers of headache remedies routinely claim that their own brands are more potent pain relievers than the competing brands. The best way of making the comparison is to compare the number of molecules in the standard dosage. Tylenol uses 325 mg of acetaminophen ($C_8H_9NO_2$) as the standard dose, while Advil uses 2.00×10^2 mg of ibuprofen ($C_{13}H_{18}O_2$). Find the number of molecules of pain reliever in the standard doses of (a) Tylenol and (b) Advil.

5. ssm A mass of 135 g of an element is known to contain 30.1×10^{23} atoms. What is the element?

*6. A cylindrical glass of water (H_2O) has a radius of 4.50 cm and a height of 12.0 cm. The density of water is 1.00 g/cm^3. How many moles of water molecules are contained in the glass?

*7. ssm www At the normal boiling point of a material, the liquid phase has a density of 958 kg/m^3, and the vapor phase has a

density of 0.598 kg/m^3. Determine the ratio of the distance between neighboring molecules in the gas phase to that in the liquid phase. (*Hint: Assume that the volume of each phase is filled with many cubes, with one molecule at the center of each cube.*)

*8. A sample of ethyl alcohol (C_2H_5OH) has a density of 806 kg/m^3 and a volume of 2.00×10^{-3} m^3. (a) Determine the mass (in kg) of a molecule of ethyl alcohol, and (b) find the number of molecules in the liquid.

Section 14.2 The Ideal Gas Law

9. A Goodyear blimp typically contains 5400 m^3 of helium (He) at an absolute pressure of 1.1×10^5 Pa. The temperature of the helium is 280 K. What is the mass (in kg) of the helium in the blimp?

10. At the start of a trip, a driver adjusts the absolute pressure in her tires to be 2.81×10^5 Pa when the outdoor temperature is 284 K. At the end of the trip she measures the pressure to be 3.01×10^5 Pa. Ignoring the expansion of the tires, find the air temperature inside the tires at the end of the trip.

11. ssm In a diesel engine, the piston compresses air at 305 K to a volume that is one-sixteenth of the original volume and a pressure that is 48.5 times the original pressure. What is the temperature of the air after the compression?

12. Two rooms, A and B, have the same dimensions and are connected by an open hallway. The temperature of room A is −8 °C and that of B is 19 °C. Find the ratio N_A/N_B of the number of air molecules in room A to that in B.

13. In a portable oxygen system, the oxygen (O_2) is contained in a cylinder whose volume is 0.0028 m³. A full cylinder has an absolute pressure of 1.5×10^7 Pa when the temperature is 296 K. Find the mass (in kg) of oxygen in the cylinder.

14. ♨ Oxygen for hospital patients is kept in special tanks, where the oxygen has a pressure of 65.0 atmospheres and a temperature of 288 K. The tanks are stored in a separate room, and the oxygen is pumped to the patient's room, where it is administered at a pressure of 1.00 atmosphere and a temperature of 297 K. What volume does 1.00 m³ of oxygen in the tanks occupy at the conditions in the patient's room?

15. ssm A young male adult takes in about 5.0×10^{-4} m³ of fresh air during a normal breath. Fresh air contains approximately 21% oxygen. Assuming that the pressure in the lungs is 1.0×10^5 Pa and air is an ideal gas at a temperature of 310 K, find the number of oxygen molecules in a normal breath.

16. A closed cylindrical tank has a height of 0.80 m. The tank is initially filled with air at a pressure of 2.0 atm. Water is pumped into the tank until the air pressure reaches 6.0 atm. The temperature is constant. Determine the height of the compressed air.

17. Two ideal gases have the same mass density and the same absolute pressure. One of the gases is helium (He), and its temperature is 175 K. The other gas is neon (Ne). What is the temperature of the neon?

18. On the sunlit surface of Venus, the atmospheric pressure is 9.0×10^6 Pa, and the temperature is 740 K. On the earth's surface the atmospheric pressure is 1.0×10^5 Pa, while the surface temperature can reach 320 K. These data imply that Venus has a "thicker" atmosphere at its surface than does the earth, which means that the number of molecules per unit volume (N/V) is greater on the surface of Venus than on the earth. Find the ratio $(N/V)_{\text{Venus}}/(N/V)_{\text{Earth}}$.

*19. ssm The relative humidity is 55% on a day when the temperature is 30.0 °C. Using the graph that accompanies problem 71 in Chapter 12, determine the number of moles of water vapor per cubic meter of air.

*20. A tank contains 11.0 g of chlorine gas (Cl_2) at a temperature of 82 °C and an absolute pressure of 5.60×10^5 Pa. The mass per mole of Cl_2 gas is 70.9 g/mol. (a) Determine the volume of the tank. (b) Later, the temperature of the tank has dropped to 31 °C and, due to a leak, the pressure has dropped to 3.80×10^5 Pa. How many grams of chlorine gas have leaked out of the tank?

*21. Assume that the pressure in a room remains constant at 1.01×10^5 Pa and the air is composed only of nitrogen (N_2). The volume of the room is 60.0 m³. When the temperature increases from 289 to 302 K, what mass of air (in kg) escapes from the room?

*22. Review Conceptual Example 3 before starting this problem. A bubble, located 0.200 m beneath the surface of the beer, rises to the top. The air pressure at the top is 1.01×10^5 Pa. Assume that the density of beer is the same as that of fresh water. If the temperature and number of moles of CO_2 remain constant as the bubble rises, find the ratio of its volume at the top to that at the bottom.

*23. ssm www A primitive diving bell consists of a cylindrical tank with one end open and one end closed. The tank is lowered into a freshwater lake, open end downward. Water rises into the tank, compressing the trapped air, whose temperature remains constant during the descent. The tank is brought to a halt when the distance between the surface of the water in the tank and the surface of the lake is 40.0 m. Atmospheric pressure at the surface of the lake is 1.01×10^5 Pa. Find the fraction of the tank's volume that is filled with air.

*24. The drawing shows two thermally insulated tanks. They are connected by a valve that is initially closed. Each tank contains neon gas at the pressure, temperature, and volume indicated in the drawing.

$P_1 = 5.0 \times 10^5$ Pa
$T_1 = 220$ K
$V_1 = 2.0$ m³

$P_2 = 2.0 \times 10^5$ Pa
$T_2 = 580$ K
$V_2 = 5.8$ m³

When the valve is opened, the contents of the two tanks mix, and the pressure becomes constant throughout. (a) What is the final temperature? Ignore any change in temperature of the tanks themselves. (*Hint: The heat gained by the gas in one tank is equal to that lost by the other.*) (b) What is the final pressure?

**25. The mass of a hot-air balloon and its occupants is 320 kg (excluding the hot air inside the balloon). The air outside the balloon has a pressure of 1.01×10^5 Pa and a density of 1.29 kg/m³. To lift off, the air inside the balloon is heated. The volume of the heated balloon is 650 m³. The pressure of the heated air remains the same as that of the outside air. To what temperature (in kelvins) must the air be heated so that the balloon just lifts off? The molecular mass of air is 29 u.

**26. A gas fills the right portion of a horizontal cylinder whose radius is 5.00 cm. The initial pressure of the gas is 1.01×10^5 Pa. A frictionless movable piston separates the gas from the left portion of the cylinder that is evacuated and contains an ideal spring, as the drawing shows. The piston is initially held in place by a pin. The spring is initially unstrained, and the length of the gas-filled portion is 20.0 cm. When the pin is removed and the gas is allowed to expand, the length of the gas-filled chamber doubles. The initial and final temperatures are equal. Determine the spring constant of the spring.

**27. ssm A cylindrical glass beaker of height 1.520 m rests on a table. The bottom half of the beaker is filled with a gas, and the top half is filled with liquid mercury that is exposed to the atmosphere. The gas and mercury do not mix because they are separated by a frictionless, movable piston of negligible mass and thickness. The initial temperature is 273 K. The temperature is increased until a value is reached when one-half of the mercury has spilled out. Ignore the thermal expansion of the glass and mercury, and find this temperature.

Section 14.3 Kinetic Theory of Gases

28. Two moles of nitrogen (N_2) gas are placed in a container whose volume is 8.5×10^{-3} m³. The pressure of the gas is 4.5×10^5 Pa. What is the average translational kinetic energy of a nitrogen molecule?

29. The surface of the sun has a temperature of about 6.0×10^3 K. This hot gas contains hydrogen atoms ($m = 1.67 \times 10^{-27}$ kg). Find the rms speed of these atoms.

30. Near the surface of Venus, the rms speed of carbon dioxide molecules (CO_2) is 650 m/s. What is the temperature (in kelvins) of the atmosphere at that point?

31. ssm The average value of the squared speed $\overline{v^2}$ does not equal the square of the average speed $(\overline{v})^2$. To verify this fact, consider three particles with the following speeds: $v_1 = 3.0$ m/s, $v_2 = 7.0$ m/s, and $v_3 = 9.0$ m/s. Calculate (a) $\overline{v^2} = \frac{1}{3}(v_1^2 + v_2^2 + v_3^2)$ and (b) $(\overline{v})^2 = [\frac{1}{3}(v_1 + v_2 + v_3)]^2$.

32. There are three moles of oxygen molecules (O_2) in a container whose volume is 0.0247 m³. The rms speed of the molecules is 483 m/s, and each has a mass of 5.31×10^{-26} kg. What is the pressure exerted on a wall of the container?

33. ssm Suppose that a tank contains 680 m³ of neon at an absolute pressure of 1.01×10^5 Pa. The temperature is changed from 293.2 to 294.3 K. What is the increase in the internal energy of the neon?

34. Initially, the translational rms speed of a molecule of an ideal gas is 463 m/s. The pressure and volume of this gas are kept constant, while the number of molecules is doubled. What is the final translational rms speed of the molecules?

*__**35.**__ The temperature near the surface of the earth is 295 K. An argon atom (atomic mass = 39.948 u) has a kinetic energy equal to the average translational kinetic energy and is moving straight up. If the atom does not collide with any other atoms or molecules, how high up would it go before coming to rest? Assume that the acceleration due to gravity is constant throughout the ascent.

*__**36.**__ Compressed air can be pumped underground into huge caverns as a form of energy storage. The volume of a cavern is 5.6×10^5 m³, and the pressure of the air in it is 7.7×10^6 Pa. Assume that air is a diatomic ideal gas whose internal energy U is given by $U = \frac{5}{2}nRT$. If one home uses 30.0 kW·h of energy per day, how many homes could this internal energy serve for one day?

__**__37. ssm www In a TV, electrons with a speed of 8.4×10^7 m/s strike the screen from behind, causing it to glow. The electrons come to a halt after striking the screen. Each electron has a mass of 9.11×10^{-31} kg, and there are 6.2×10^{16} electrons per second hitting the screen over an area of 1.2×10^{-7} m². What is the pressure that the electrons exert on the screen?

__**__38. When perspiration on the human body absorbs heat, some of the perspiration turns into water vapor. The latent heat of vaporization at body temperature (37 °C) is 2.42×10^6 J/kg. The heat absorbed is approximately equal to the average energy E given to

a single water molecule (H_2O) times the number of water molecules that are vaporized. What is E?

Section 14.4 Diffusion

39. A tube has a length of 0.015 m and a cross-sectional area of 7.0×10^{-4} m². The tube is filled with a solution of sucrose in water. The diffusion constant of sucrose in water is 5.0×10^{-10} m²/s. A difference in concentration of 3.0×10^{-3} kg/m³ is maintained between the ends of the tube. How much time is required for 8.0×10^{-13} kg of sucrose to be transported through the tube?

40. It is found that the amino acid glycine diffuses through water at a mass rate of $m/t = 6.00 \times 10^{-14}$ kg/s. The diffusion constant is 10.6×10^{-10} m²/s. A tube of water has a radius of 1.40 cm. What difference in concentration per unit length of the tube $\Delta C/L$ must be maintained to give this flow?

41. ssm The diffusion constant of ethanol in water is 12.4×10^{-10} m²/s. A cylinder has a cross-sectional area of 4.00 cm² and a length of 2.00 cm. A difference in ethanol concentration of 1.50 kg/m³ is maintained between the ends of the cylinder. In one hour, what mass of ethanol diffuses through the cylinder?

42. Review Conceptual Example 7 before starting this problem. Ammonia, which has a strong smell, is diffusing through air contained in a tube of length L and cross-sectional area of 4.0×10^{-4} m². The diffusion constant for ammonia in air is 4.2×10^{-5} m²/s. In a certain time, 8.4×10^{-8} kg of ammonia diffuses through the air when the difference in ammonia concentration between the ends of the tube is 3.5×10^{-2} kg/m³. Find the speed at which ammonia diffuses through the air, that is, the length of the tube divided by the time to travel this length.

43. Review Conceptual Example 7 before working this problem. For water vapor in air at 293 K, the diffusion constant is $D = 2.4 \times 10^{-5}$ m²/s. As outlined in problem 44(a), the time required for the first solute molecules to traverse a channel of length L is $t = L^2/(2D)$, according to Fick's law. Find the time t for $L = 0.010$ m. (b) For comparison, how long would a water molecule take to travel $L = 0.010$ m at the translational rms speed of water molecules (assumed to be an ideal gas) at a temperature of 293 K? (c) Explain why the answer to part (a) is so much longer than the answer to part (b).

*__**44.**__ Review Conceptual Example 7 as background for this problem. It is possible to convert Fick's law into a form that is useful when the concentration is zero at one end of the diffusion channel ($C_1 = 0$ in Figure 14.15a). (a) Noting that AL is the volume V of the channel and that m/V is the average concentration of solute in the channel, show that Fick's law becomes $t = L^2/(2D)$. This form of Fick's law can be used to estimate the time required for the first solute molecules to traverse the channel. (b) A bottle of perfume is opened in a room where convection currents are absent. Assuming that the diffusion constant for perfume in air is 1.0×10^{-5} m²/s, estimate the minimum time required for the perfume to be smelled 2.5 cm away.

__**__45. The drawing shows a container that is partially filled with

2.0 grams of water. The temperature is maintained at a constant 20 °C. The space above the liquid contains air that is completely saturated with water vapor. A tube of length 0.15 m and cross-sectional area 3.0×10^{-4} m^2 connects the water vapor at one end to air that remains completely dry at the other end. The diffusion constant for water vapor in air is 2.4×10^{-5} m^2/s. How long does it take for the water in the container to evaporate completely? *(Hint: Refer to problem 71 in Chapter 12 to find the pressure of the water vapor.)*

ADDITIONAL PROBLEMS

46. An ultrahigh vacuum pump can reduce the absolute pressure to 1.2×10^{-7} Pa. In a volume of 2.0 m^3 and at a temperature of 27 °C, how many molecules of gas are present at this pressure?

47. ssm A bicycle tire whose volume is 4.1×10^{-4} m^3 has a temperature of 296 K and an absolute pressure of 4.8×10^5 Pa. A cyclist brings the pressure up to 6.2×10^5 Pa without changing the temperature or volume. How many moles of air must be pumped into the tire?

48. An ideal gas at 15.5 °C and a pressure of 1.72×10^5 Pa occupies a volume of 2.81 m^3. (a) How many moles of gas are present? (b) If the volume is raised to 4.16 m^3 and the temperature raised to 28.2 °C, what will be the pressure of the gas?

49. ssm www Carbon tetrachloride (CCl$_4$) is diffusing through benzene (C$_6$H$_6$), as the drawing illustrates. The concentration of CCl$_4$ at the left end of the tube is maintained at 1.00×10^{-2} kg/m^3, and the diffusion constant is 20.0×10^{-10} m^2/s. The CCl$_4$ enters the tube at a mass rate of 5.00×10^{-13} kg/s. Using these data and those shown in the drawing, find (a) the mass of CCl$_4$ per second that passes point A and (b) the concentration of CCl$_4$ at point A.

50. At a temperature of 0 °C, the rms speed of the atoms in a monatomic gas is 1300 m/s. (a) Determine the mass of an atom in atomic mass units. (b) Using the periodic table, identify the element.

51. If the translational rms speed of the water vapor molecules (H$_2$O) in air is 648 m/s, what is the translational rms speed of the carbon dioxide molecules (CO$_2$) in the same air? Both gases are at the same temperature.

52. A frictionless gas-filled cylinder is fitted with a movable piston, as the drawing shows. The block resting on the top of the piston determines the constant pressure that the gas has. The height h is 0.120 m when the temperature is 273 K and increases as the temperature increases. What is the value of h when the temperature reaches 318 K?

*53. ssm A drop of water has a radius of 9.00×10^{-4} m. How many water molecules are in the drop?

*54. Helium (He), a monatomic gas, fills a 0.010-m^3 container. The pressure of the gas is 6.2×10^5 Pa. How long would a 0.25-hp engine have to run (1 hp = 746 W) to produce an amount of energy equal to the internal energy of this gas?

**55. ssm Estimate the spacing between the centers of neighboring atoms in a piece of solid aluminum, based on a knowledge of the density (2700 kg/m^3) and atomic mass (26.9815 u) of aluminum. *(Hint: Assume that the volume of the solid is filled with many small cubes, with one atom at the center of each.)*

**56. In 10.0 s, 200 bullets strike and embed themselves in a wall. The bullets strike the wall perpendicularly. Each bullet has a mass of 5.0×10^{-3} kg and a speed of 1200 m/s. (a) What is the average change in momentum per second for the bullets? (b) Determine the average force exerted on the wall. (c) Assuming the bullets are spread out over an area of 3.0×10^{-4} m^2, obtain the average pressure they exert on this region of the wall.

**57. A spherical balloon is made from a material whose mass is 3.00 kg. The thickness of the material is negligible compared to the 1.50-m radius of the balloon. The balloon is filled with helium (He) at a temperature of 305 K and just floats in air, neither rising nor falling. The density of the surrounding air is 1.19 kg/m^3. Find the absolute pressure of the helium gas.

CONCEPTS & CALCULATIONS
GROUP LEARNING PROBLEMS

Note: Each of these problems consists of Concept Questions followed by a related quantitative Problem. They are designed for use by students working alone or in small learning groups. The Concept Questions involve little or no mathematics and are intended to stimulate group discussions. They focus on the concepts with which the problems deal. Recognizing the concepts is the essential initial step in any problem-solving technique.

58. Concept Questions Suppose you know the mass per mole (in units of g/mol) of a substance. (a) Without doing any calculations, how can you determine the mass of one of its atoms (expressed in atomic mass units)? (b) What information about the sample is obtained when you divide its mass by the mass per mole?

Problem Gold has a mass per mole of 196.967 g/mol. What is the mass of a single gold atom in (a) atomic mass units and (b) kilograms? (c) How many moles of gold atoms are in a 285-g sample?

59. Concept Questions The temperature of an ideal gas is doubled while the volume is kept constant. Does the absolute pressure of the gas double when the temperature that doubles is (a) the Kelvin temperature and (b) the Celsius temperature? Explain.

Problem Determine the ratio P_2/P_1 of the final pressure to the initial pressure when the temperature rises (a) from 35.0 to 70.0 K and (b) from 35.0 to 70.0 °C. Check to see that your answers are consistent with your answers to the Concept Questions.

60. Concept Question Four closed tanks, A, B, C, and D, each contain identical numbers of moles of an ideal gas. The table gives the absolute pressure and volume of the gas in each tank. Which tanks (if any) have the same temperature? Account for your answer.

	A	B	C	D
Absolute pressure (Pa)	25.0	30.0	20.0	2.0
Volume (m³)	4.0	5.0	5.0	75

Problem Each of the tanks contains 0.10 mol of gas. Using this number and the data in the table, compute the temperature of each gas. Verify that your answers are consistent with your answer to the Concept Questions.

61. Concept Question Four containers are filled with monatomic ideal gases. For each container, the mass of an individual atom and the rms speed of the atoms are expressed in terms of m and v_{rms}, respectively (see the table). Rank the gases according to their temperatures, highest first. Justify your ranking.

	A	B	C	D
Mass	m	m	$2m$	$2m$
Rms speed	v_{rms}	$2v_{rms}$	v_{rms}	$2v_{rms}$

Problem Suppose that $m = 3.32 \times 10^{-26}$ kg, and $v_{rms} = 1223$ m/s. Find the temperature of the gas in each tank. Be sure that your answers are consistent with your answer to the Concept Question.

62. Concept Questions Imagine a gas, such as helium or neon, diffusing through air in an arrangement like that in Figure 14.15a. The diffusion rate is the number of gas atoms per second diffusing from the left end to the right end of the channel. The faster the atoms move, the greater the diffusion rate. Therefore, the diffusion rate is proportional to the rms speed of the atoms. Recall, from the kinetic theory of gases, how the average kinetic energy of an atom is related to the temperature of the gas. (a) As the temperature of the gas increases, would you expect the diffusion rate to increase, decrease, or remain the same? Provide a reason for your answer. (b) For a given temperature, which atom, helium or neon, would you expect to have the greater diffusion rate? Why?

Problem The mass of a helium atom is 4.002 60 u and a neon atom is 20.179 u. (a) Suppose the temperature of the helium gas is raised to 580 from 290 K. Find the ratio of the diffusion rate at the higher temperature to that at the lower temperature. (b) When the temperature is 290 K, determine the ratio of the diffusion rate of helium to that of neon. Check your answers for consistency with respect to your answers to the Concept Questions.

The lava from a volcanic eruption is red-hot because of heat absorbed deep within the earth. The laws that govern heat and work form the basis of thermodynamics, the subject of this chapter.

CHAPTER 15

THERMODYNAMICS

Figure 15.1 The hot air in a balloon is one example of a thermodynamic system.

Figure 15.2 (*a*) Systems A and B are surrounded by adiabatic walls and register the same temperature on the thermometer. (*b*) When A is put into thermal contact with B through diathermal walls, no net flow of heat occurs between the systems.

15.1 THERMODYNAMIC SYSTEMS AND THEIR SURROUNDINGS

We have studied heat and work as separate topics. Often, however, they occur simultaneously. In an automobile engine, for instance, fuel is burned at a relatively high temperature, some of its internal energy is used for doing the work of driving the pistons up and down, and the excess heat is removed by the cooling system to prevent overheating. ***Thermodynamics*** is the branch of physics that is built upon the fundamental laws that heat and work obey.

In thermodynamics the collection of objects upon which attention is being focused is called the ***system,*** while everything else in the environment is called the ***surroundings.*** For example, the system in an automobile engine could be the burning gasoline, while the surroundings would then include the pistons, the exhaust system, the radiator, and the outside air. The system and its surroundings are separated by walls of some kind. Walls that permit heat to flow through them, such as those of the engine block, are called ***diathermal walls.*** Perfectly insulating walls that do not permit heat to flow between the system and its surroundings are called ***adiabatic walls.***

To understand what the laws of thermodynamics have to say about the relationship between heat and work, it is necessary to describe the physical condition or ***state of a system.*** We might be interested, for instance, in the hot air within the balloon in Figure 15.1. The hot air itself would be the system, and the skin of the balloon provides the walls that separate this system from the surrounding cooler air. The state of the system would be specified by giving values for the pressure, volume, temperature, and mass of the hot air.

As this chapter discusses, there are four laws of thermodynamics. We begin with the one known as the zeroth law and then consider the remaining three.

15.2 THE ZEROTH LAW OF THERMODYNAMICS

The zeroth law of thermodynamics deals with the concept of ***thermal equilibrium.*** Two systems are said to be in thermal equilibrium if there is no net flow of heat between them when they are brought into thermal contact. For instance, you are *not* in thermal equilibrium with the water in Lake Michigan in January. Just dive into it, and you will find out how quickly your body loses heat to the frigid water. To help explain the central idea of the zeroth law of thermodynamics, Figure 15.2a shows two systems labeled A and B. Each is within a container whose adiabatic walls are made from insulation that prevents the flow of heat, and each has the same temperature, as indicated by the thermometers. In part *b*, one wall of each container is replaced by a thin silver sheet, and the two sheets are touched together. Silver has a large thermal conductivity, so heat flows through it readily and the silver sheets behave as diathermal walls. Even though the diathermal walls would permit it, no net flow of heat occurs in part *b*, indicating that the two systems are in thermal equilibrium. There is no net flow of heat because the two systems have the same temperature. We see, then, that ***temperature is the indicator of thermal equilibrium in the sense that there is no net flow of heat between two systems in thermal contact that have the same temperature.***

In Figure 15.2 the thermometer plays an important role. System A is in equilibrium with the thermometer, and so is system B. In each case, the thermometer registers the same temperature, thereby indicating that the two systems are equally hot. Consequently, systems A and B are found to be in thermal equilibrium with each other. In effect, the thermometer is a third system. The fact that system A and system B are each in thermal equilibrium with this third system at the same temperature means that they are in thermal equilibrium with each other. This finding is an example of the *zeroth law of thermodynamics.*

● **THE ZEROTH LAW OF THERMODYNAMICS**

Two systems individually in thermal equilibrium with a third system* are in thermal equilibrium with each other.

The zeroth law establishes temperature as the indicator of thermal equilibrium and implies that all parts of a system must be in thermal equilibrium if the system is to have a definable single temperature. In other words, there can be no flow of heat within a system in thermal equilibrium.

15.3 THE FIRST LAW OF THERMODYNAMICS

● *Concepts at a Glance*   In Chapter 6 we saw that forces can do work and that work can change the kinetic and potential energy of an object. For example, the atoms and molecules of a substance exert forces on one another. As a result, they have kinetic and potential energy. These and other kinds of molecular energy constitute the internal energy of a substance. When a substance participates in a process involving energy in the form of work and heat, the internal energy of the substance can change. The relationship between work, heat, and changes in the internal energy is known as the first law of thermodynamics and is illustrated in the Concepts-at-a-Glance chart in Figure 15.3. We will now see that the first law of thermodynamics is an expression of the conservation of energy. ●

*The state of the third system is the same when it is in thermal equilibrium with either of the two systems. In Figure 15.2, for example, the mercury is at the same position in the thermometer in either system.

Figure 15.3 *Concepts at a Glance* The concepts of work, heat, and the internal energy of a system are related by the first law of thermodynamics, which is an expression of the law of conservation of energy. The body of the athlete in the photograph contains internal energy. Part of this energy is used to do the work of running up the stadium stairs, and part is lost in the form of heat carried off by evaporating sweat.

Concepts at a Glance

Work (Sections 6.1 and 6.9)

Internal Energy (Section 12.6)

Heat (Section 12.6)

First Law of Thermodynamics: The Conservation of Energy

Suppose that a system gains heat Q and that this is the only effect occurring. Consistent with the law of conservation of energy, the internal energy of the system increases from an initial value of U_i to a final value of U_f, the change being $\Delta U = U_f - U_i = Q$. In writing this equation, we use the convention that *heat Q is positive when the system gains heat and negative when the system loses heat.* The internal energy of a system can also change because of work. If a system does work W on its surroundings and there is no heat flow, energy conservation indicates that the internal energy of the system decreases from U_i to U_f, the change now being $\Delta U = U_f - U_i = -W$. The minus sign is included because we follow the convention that *work is positive when it is done by the system and negative when it is done on the system.* A system can gain or lose energy simultaneously in the form of heat Q and work W. The change in internal energy due to both factors is given by Equation 15.1. Thus, the *first law of thermodynamics* is just the conservation of energy principle applied to heat, work, and the change in the internal energy.

● THE FIRST LAW OF THERMODYNAMICS

The internal energy of a system changes from an initial value U_i to a final value of U_f due to heat Q and work W:

$$\Delta U = U_f - U_i = Q - W \qquad (15.1)$$

Q is positive when the system gains heat and negative when it loses heat. W is positive when work is done by the system and negative when work is done on the system.

Example 1 illustrates the use of Equation 15.1 and the sign conventions for Q and W.

EXAMPLE 1   Positive and Negative Work

Figure 15.4 illustrates a system and its surroundings. In part a, the system gains 1500 J of heat from its surroundings, and 2200 J of work is done *by* the system on the surroundings. In part b, the system also gains 1500 J of heat, but 2200 J of work is done *on* the system by the surroundings. In each case, determine the change in the internal energy of the system.

Reasoning In Figure 15.4a the system loses more energy in doing work than it gains in the form of heat, so the internal energy of the system decreases. Thus, we expect the change in the internal energy, $\Delta U = U_f - U_i$, to be negative. In part b of the drawing, the system gains energy both in the form of heat and work. The internal energy of the system increases, and we expect ΔU to be positive.

Solution

(a) The heat is positive, $Q = +1500$ J, since it is gained by the system. The work is positive, $W = +2200$ J, since it is done *by* the system. According to the first law of thermodynamics

$$\Delta U = Q - W = (+1500 \text{ J}) - (+2200 \text{ J}) = \boxed{-700 \text{ J}} \qquad (15.1)$$

The minus sign for ΔU indicates that the internal energy has decreased, as expected.

(a)

(b)

Figure 15.4 (a) The system gains energy in the form of heat but loses energy because work is done by the system. (b) The system gains energy in the form of heat and also gains energy because work is done on the system.

(b) The heat is positive, $Q = +1500$ J, since it is gained by the system. But the work is negative, $W = -2200$ J, since it is done *on* the system. Thus,

$$\Delta U = Q - W = (+1500 \text{ J}) - (-2200 \text{ J}) = \boxed{+3700 \text{ J}} \qquad (15.1)$$

The plus sign for ΔU indicates that the internal energy has increased, as expected.

In the first law of thermodynamics, the internal energy U, heat Q, and work W are energy quantities, and each is expressed in energy units such as joules. However, there is a fundamental difference between U, on the one hand, and Q and W on the other. The next example sets the stage for explaining this difference.

EXAMPLE 2   An Ideal Gas

The temperature of three moles of a monatomic ideal gas is reduced from $T_i = 540$ K to $T_f = 350$ K by two different methods. In the first method 5500 J of heat flows into the gas, while in the second, 1500 J of heat flows into it. In each case find (a) the change in the internal energy and (b) the work done by the gas.

Reasoning Since the internal energy of a monatomic ideal gas is $U = \frac{3}{2}nRT$ (Equation 14.7) and since the number of moles n is fixed, only a change in temperature T can alter the internal energy. Since the change in T is the same in both methods, the change in U is also the same. From the given temperatures, the change ΔU in internal energy can be determined. Then, the first law of thermodynamics can be used with ΔU and the given heat values to calculate the work.

Solution

(a) Using Equation 14.7 for the internal energy of a monatomic ideal gas, we find for each method of adding heat that

$$\Delta U = \tfrac{3}{2}nR(T_f - T_i) = \tfrac{3}{2}(3.0 \text{ mol})[8.31 \text{ J/(mol·K)}](350 \text{ K} - 540 \text{ K}) = \boxed{-7100 \text{ J}}$$

(b) Since ΔU is now known and the heat is given in each method, Equation 15.1 can be used to determine the work:

1st method $W = Q - \Delta U = 5500 \text{ J} - (-7100 \text{ J}) = \boxed{12\,600 \text{ J}}$

2nd method $W = Q - \Delta U = 1500 \text{ J} - (-7100 \text{ J}) = \boxed{8600 \text{ J}}$

In each method the gas does work, but it does more in the first method.

To understand the difference between U and either Q or W, consider the value for ΔU in Example 2. In both methods ΔU is the same. Its value is determined once the initial and final temperatures are specified because the internal energy of an ideal gas depends only on the temperature. Temperature is one of the variables (along with pressure and volume) that define the state of a system. *The internal energy depends only on the state of a system, not on the method by which the system arrives at a given state.* In recognition of this characteristic, internal energy is referred to as a ***function of state.**** In contrast, heat and work are not functions of state because they have different values for each different method used to make the system change from one state to another, as in Example 2.

---

* The fact that an ideal gas is used in Example 2 does not restrict our conclusion. Had a real (nonideal) gas or other material been used, the only difference would have been that the expression for the internal energy would have been more complicated. It might have involved the volume V, as well as the temperature T, for instance.

Figure 15.5 The substance in the chamber is expanding isobarically because the pressure is held constant by the external atmosphere and the weight of the piston and the block.

15.4 THERMAL PROCESSES

A system can interact with its surroundings in many ways, and the heat and work that come into play always obey the first law of thermodynamics. This section introduces four common thermal processes. In each case, the process is assumed to be *quasi-static*, which means that it occurs slowly enough that a uniform pressure and temperature exist throughout all regions of the system at all times.

An isobaric process is one that occurs at constant pressure. For instance, Figure 15.5 shows a substance (solid, liquid, or gas) contained in a chamber fitted with a frictionless piston. The pressure P experienced by the substance is always the same and is determined by the external atmosphere and the weight of the piston and the block resting on it. Heating the substance makes it expand and do work W in lifting the piston and block through the displacement **s**. The work can be calculated from $W = Fs$ (Equation 6.1), where F is the magnitude of the force and s is the magnitude of the displacement. The force is generated by the pressure P acting on the bottom surface of the piston (area $= A$), according to $F = PA$ (Equation 10.19). With this substitution for F, the work becomes $W = (PA)s$. But the product $A \cdot s$ is the change in volume of the material, $\Delta V = V_f - V_i$, where V_f and V_i are the final and initial volumes, respectively. Thus, the expression for the work is

Isobaric process
$$W = P\Delta V = P(V_f - V_i) \qquad (15.2)$$

Consistent with our sign convention, this result predicts a positive value for the work done *by a system* when it expands isobarically (V_f exceeds V_i). Equation 15.2 also applies to an isobaric compression (V_f less than V_i). Then, the work is negative, since work must be done *on the system* to compress it. Example 3 emphasizes that $W = P\Delta V$ applies to any system, solid, liquid, or gas, as long as the pressure remains constant while the volume changes.

EXAMPLE 3   Isobaric Expansion of Water

One gram of water is placed in the cylinder in Figure 15.5, and the pressure is maintained at 2.0×10^5 Pa. The temperature of the water is raised by 31 C°. In one case, the water is in the liquid phase and expands by the small amount of 1.0×10^{-8} m³. In another case, the water is in the gas phase and expands by the much greater amount of 7.1×10^{-5} m³. For the water in each case, find (a) the work done and (b) the change in the internal energy.

Reasoning In both cases the material expands, so work is done by the system (water) on the surroundings and is, therefore, positive. The work is done at a constant pressure, so it can be found from $W = P\Delta V$. Once the work is known, the first law of thermodynamics, $\Delta U = Q - W$, can be used to find the change in internal energy, provided a value for the heat Q can be found. The heat needed to raise the temperature can be obtained from $Q = cm\Delta T$ (Equation 12.4), where the specific heat capacity of liquid water is $c = 4186$ J/(kg·C°) (see Table 12.2) and the specific heat capacity of water vapor at constant pressure is $c_P = 2020$ J/(kg·C°).

Solution

(a) In both cases the process is isobaric, so the work is given by Equation 15.2:

$$W_{\text{liquid}} = P\Delta V = (2.0 \times 10^5 \text{ Pa})(1.0 \times 10^{-8} \text{ m}^3) = \boxed{0.0020 \text{ J}}$$

$$W_{\text{gas}} = P\Delta V = (2.0 \times 10^5 \text{ Pa})(7.1 \times 10^{-5} \text{ m}^3) = \boxed{14 \text{ J}}$$

Under comparable conditions, liquids (and solids) change volume much less than gases do, which leads to a much smaller work of expansion or compression for liquids than for gases.

(b) Using $\Delta U = Q - W$ (Equation 15.1) and $Q = cm\,\Delta T$ (Equation 12.4), we find that $\Delta U = cm\,\Delta T - W$. Therefore,

$$\Delta U_{\text{liquid}} = [4186 \text{ J/(kg} \cdot \text{C}^\circ)](0.0010 \text{ kg})(31 \text{ C}^\circ) - 0.0020 \text{ J}$$
$$= 130 \text{ J} - 0.0020 \text{ J} = \boxed{130 \text{ J}}$$

$$\Delta U_{\text{gas}} = [2020 \text{ J/(kg} \cdot \text{C}^\circ)](0.0010 \text{ kg})(31 \text{ C}^\circ) - 14 \text{ J}$$
$$= 63 \text{ J} - 14 \text{ J} = \boxed{49 \text{ J}}$$

For the liquid, virtually all the 130 J of heat serves to change the internal energy, since the volume change and the corresponding work of expansion are so small. In contrast, a significant fraction of the 63 J of heat added to the vapor causes work of expansion to be done, so that only 49 J is left to change the internal energy.

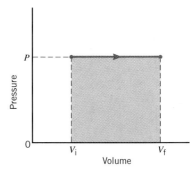

Figure 15.6 For an isobaric process, a pressure-versus-volume plot is a horizontal straight line, and the work done $[W = P(V_f - V_i)]$ is the colored rectangular area under the graph.

It is often convenient to display thermal processes graphically. For instance, Figure 15.6 shows a plot of pressure versus volume for an isobaric expansion. Since the pressure is constant, the graph is a horizontal straight line, beginning at the initial volume V_i and ending at the final volume V_f. In terms of such a plot, the work $W = P(V_f - V_i)$ is the area under the graph, which is the shaded rectangle of height P and width $V_f - V_i$.

Another common thermal process is an ***isochoric process, one that occurs at constant volume.*** Figure 15.7*a* illustrates an isochoric process in which a substance (solid, liquid, or gas) is heated. The substance would expand if it could, but the rigid container keeps the volume constant, so the pressure–volume plot shown in Figure 15.7*b* is a vertical straight line. Because the volume is constant, the pressure inside rises, and the substance exerts more and more force on the walls. While enormous forces can be generated in the closed container, no work is done, since the walls do not move. Consistent with zero work being done, the area under the vertical straight line in Figure 15.7*b* is zero. Since no work is done, the first law of thermodynamics indicates that the heat in an isochoric process serves only to change the internal energy: $\Delta U = Q - W = Q$.

A third important thermal process is an ***isothermal process, one that takes place at constant temperature.*** The next section illustrates the important features of an isothermal process when the system is an ideal gas.

Last, there is the ***adiabatic process, one that occurs without the transfer of heat.*** Since there is no heat transfer, Q equals zero, and the first law indicates that $\Delta U = Q - W = -W$. Thus, when work is done by a system adiabatically, the internal energy of the system decreases by exactly the amount of the work done. When work is done on a system adiabatically, the internal energy increases correspondingly. The next section discusses an adiabatic process for an ideal gas.

A process may be complex enough that it is not recognizable as one of the four just discussed. For instance, Figure 15.8 shows a process for a gas in which the pressure, volume, and temperature are changed along the straight line from X to Y. With the aid of integral calculus, it can be shown that ***the area under a pressure–volume graph is the work for any kind of process,*** so the area representing the work has been colored in the drawing. The volume increases, and work is done by the gas. This work is positive by convention, as is the area. In contrast, if a process reduces the volume, work is done on the gas, and this work is negative by

Figure 15.7 (*a*) The substance in the chamber is being heated isochorically because the rigid chamber keeps the volume constant. (*b*) The pressure–volume plot for an isochoric process is a vertical straight line. The area under the graph is zero, indicating that no work is done.

Figure 15.8 The colored area gives the work done by the gas for the process from X to Y.

convention. Correspondingly, the area under the pressure–volume graph would be assigned a negative value. In Example 4, we determine the work for the case shown in Figure 15.8.

EXAMPLE 4 Work and the Area Under a Pressure–Volume Graph

Determine the work for the process in which the pressure, volume, and temperature of a gas are changed along the straight line from X to Y in Figure 15.8.

Reasoning The work is given by the area (in color) under the straight line between X and Y. Since the volume increases, work is done by the gas on the surroundings, so the work is positive. The area can be found by counting squares in Figure 15.8 and multiplying by the area per square.

Solution We estimate that there are 8.9 colored squares in the drawing. The area of one square is $(2.0 \times 10^5 \text{ Pa})(1.0 \times 10^{-4} \text{ m}^3) = 2.0 \times 10^1$ J, so the work is

$$W = +(8.9 \text{ squares})(2.0 \times 10^1 \text{ J/square}) = \boxed{+180 \text{ J}}$$

15.5 THERMAL PROCESSES THAT UTILIZE AN IDEAL GAS

Isothermal Expansion or Compression

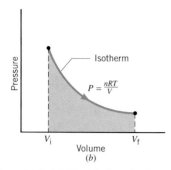

Figure 15.9 (a) The ideal gas in the cylinder is expanding isothermally at temperature T. The force holding the piston in place is reduced slowly, so the expansion occurs quasi-statically. (b) The work done by the gas is given by the colored area.

When a system performs work isothermally, the temperature remains constant. In Figure 15.9a, for instance, a metal cylinder contains n moles of an ideal gas, and the large mass of hot water maintains the cylinder and gas at a constant Kelvin temperature T. The piston is held in place initially so the volume of the gas is V_i. As the external force applied to the piston is reduced quasi-statically, the gas expands to the final volume V_f. Figure 15.9b gives a plot of pressure ($P = nRT/V$) versus volume for the process. The solid red line in the graph is called an isotherm (meaning "constant temperature") because it represents the relation between pressure and volume when the temperature is held constant. The work W done by the gas is *not* given by $W = P \Delta V = P(V_f - V_i)$ because the pressure is not constant. Nevertheless, the work is equal to the area under the graph. The techniques of integral calculus lead to the following result* for W:

Isothermal expansion or compression of an ideal gas

$$W = nRT \ln\left(\frac{V_f}{V_i}\right) \qquad (15.3)$$

Where does the energy for this work originate? Since the internal energy of any ideal gas is proportional to the Kelvin temperature ($U = \frac{3}{2}nRT$ for a monatomic ideal gas, for example), the internal energy remains constant throughout an isothermal process, and the change in internal energy is zero. The first law of thermodynamics becomes $\Delta U = 0 = Q - W$. In other words, $Q = W$, and the energy for the work originates in the hot water. Heat flows into the gas from the water, as Figure 15.9a illustrates. If the gas is compressed isothermally, Equation 15.3 still applies, and heat flows out of the gas into the water. The following example deals with the isothermal expansion of an ideal gas.

---

* In this result, "ln" denotes the natural logarithm to the base $e = 2.71828$. The natural logarithm is related to the common logarithm to the base ten by $\ln(V_f/V_i) = 2.303 \log(V_f/V_i)$.

EXAMPLE 5   Isothermal Expansion of an Ideal Gas

Two moles of the monatomic gas argon expand isothermally at 298 K, from an initial volume of $V_i = 0.025$ m^3 to a final volume of $V_f = 0.050$ m^3. Assuming that argon is an ideal gas, find (a) the work done by the gas, (b) the change in the internal energy of the gas, and (c) the heat supplied to the gas.

Reasoning and Solution

(a) The work done by the gas can be found from Equation 15.3:

$$W = nRT \ln\left(\frac{V_f}{V_i}\right)$$

$$= (2.0 \text{ mol})[8.31 \text{ J}/(\text{mol} \cdot \text{K})](298 \text{ K}) \ln\left(\frac{0.050 \text{ m}^3}{0.025 \text{ m}^3}\right)$$

$$= \boxed{+3400 \text{ J}}$$

(b) The internal energy of a monatomic ideal gas is $U = \frac{3}{2}nRT$ and does not change when the temperature is constant. Therefore, $\boxed{\Delta U = 0}$.

(c) The heat Q supplied can be determined from the first law of thermodynamics:

$$Q = \Delta U + W = 0 + 3400 \text{ J} = \boxed{+3400 \text{ J}} \qquad (15.1)$$

INTERACTIVE LEARNINGWARE #30

A bubble from the tank of a scuba diver contains 3.5×10^{-4} mol of gas. The bubble expands as it rises to the surface from a freshwater depth of 10.3 m. Assuming that the gas is an ideal gas and the temperature remains constant at 291 K, find the amount of heat that flows into the bubble.

Here is a common situation in which the first law of thermodynamics is at play. This Interactive LearningWare example is interesting because it gives you the opportunity to review Section 11.3, which discusses how pressure in a fluid depends on depth. It also provides the chance to review ideal gas behavior, which Chapter 14 discusses.

Go to our Web site at

http://www.wiley.com/college/cutnell

Adiabatic Expansion or Compression

When a system performs work adiabatically, no heat flows into or out of the system. Figure 15.10a shows an arrangement in which n moles of an ideal gas do work under adiabatic conditions, expanding quasi-statically from an initial volume V_i to a final volume V_f. The arrangement is similar to that in Figure 15.9 for isothermal expansion. However, a different amount of work is done here, because the cylinder is now surrounded by insulating material that prevents the flow of heat, so $Q = 0$. According to the first law of thermodynamics, the change in internal energy is $\Delta U = Q - W = -W$. Since the internal energy of an ideal monatomic gas is $U = \frac{3}{2}nRT$, it follows that $\Delta U = U_f - U_i = \frac{3}{2}nR(T_f - T_i)$, where T_i and T_f are the initial and final Kelvin temperatures. With this substitution, the relation $\Delta U = -W$ becomes

Adiabatic expansion or compression of a monatomic ideal gas

$$W = \frac{3}{2}nR(T_i - T_f) \qquad (15.4)$$

When an ideal gas expands adiabatically, it does positive work, so W is positive in Equation 15.4. Therefore, the term $T_i - T_f$ is also positive, and the final temperature of the gas must be less than the initial temperature. The internal energy of the gas is reduced to provide the necessary energy to do the work, and because the internal energy is proportional to the Kelvin temperature, the temperature decreases. Figure 15.10b shows a plot of pressure versus volume for the adiabatic process. The adiabatic curve (red) intersects the isotherms (blue) at the higher initial temperature

Metal cylinder

Insulating material
(a)

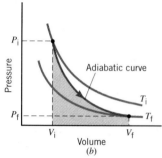

(b)

Figure 15.10 *(a)* The ideal gas in the cylinder is expanding adiabatically. The force holding the piston in place is reduced slowly, so the expansion occurs quasi-statically. *(b)* A plot of pressure versus volume yields the adiabatic curve shown in red, which intersects the isotherms (blue) at the initial temperature T_i and the final temperature T_f. The work done by the gas is given by the colored area.

$[T_i = P_i V_i/(nR)]$ and the lower final temperature $[T_f = P_f V_f/(nR)]$. The colored area under the adiabatic curve represents the work done.

The reverse of an adiabatic expansion is an adiabatic compression (W is negative), and Equation 15.4 indicates that the final temperature exceeds the initial temperature. The energy provided by the agent doing the work increases the internal energy of the gas. As a result, the gas becomes hotter.

The equation that gives the adiabatic curve (red) between the initial pressure and volume (P_i, V_i) and the final pressure and volume (P_f, V_f) in Figure 15.10*b* can be derived using integral calculus. The result is

Adiabatic expansion or compression of an ideal gas

$$P_i V_i^\gamma = P_f V_f^\gamma \tag{15.5}$$

where the exponent γ is the ratio of the specific heat capacities at constant pressure and constant volume, $\gamma = c_P/c_V$. Equation 15.5 applies in conjunction with the ideal gas law, for *each point* on the adiabatic curve satisfies the relation $PV = nRT$.

15.6 SPECIFIC HEAT CAPACITIES AND THE FIRST LAW OF THERMODYNAMICS

In this section the first law of thermodynamics is used to gain an understanding of the factors that determine the specific heat capacity of a material. Remember, when the temperature of a substance changes as a result of heat flow, the change in temperature ΔT and the amount of heat Q are related according to $Q = cm\,\Delta T$ (Equation 12.4). In this expression c denotes the specific heat capacity in units of J/(kg·C°), and m is the mass in kilograms. It is more convenient now to express the amount of material as the number of moles n, rather than the number of kilograms. Therefore, we replace the expression $Q = cm\,\Delta T$ with the following analogous expression:

$$Q = Cn\,\Delta T \tag{15.6}$$

where the capital letter C (as opposed to the lowercase c) refers to the ***molar specific heat capacity*** in units of J/(mol·K). In addition, the unit for measuring the temperature change ΔT is the Kelvin (K) rather than the Celsius degree (C°). For gases it is necessary to distinguish between the molar specific heat capacities C_P and C_V, which apply, respectively, to conditions of constant pressure and constant volume. With the help of the first law of thermodynamics and an ideal gas as an example, it is possible to see why C_P and C_V differ.

To determine the molar specific heat capacities, we must first calculate the heat Q needed to raise the temperature of an ideal gas from T_i to T_f. According to the first law, $Q = \Delta U + W$. We also know that the internal energy of a monatomic ideal gas is $U = \tfrac{3}{2}nRT$ (Equation 14.7). As a result, $\Delta U = U_f - U_i = \tfrac{3}{2}nR(T_f - T_i)$. When the heating process occurs at constant pressure, the work done is given by Equation 15.2: $W = P\,\Delta V = P(V_f - V_i)$. For an ideal gas $PV = nRT$, so the work becomes $W = nR(T_f - T_i)$. On the other hand, when the volume is constant, $\Delta V = 0$, and the work done is zero. The calculation of the heat is summarized below:

$$Q = \Delta U + W$$

$$Q_{\text{constant pressure}} = \tfrac{3}{2}nR(T_f - T_i) + nR(T_f - T_i)$$

$$Q_{\text{constant volume}} = \tfrac{3}{2}nR(T_f - T_i) + 0$$

The molar specific heat capacities can now be determined, since Equation 15.6 indicates that $C = Q/[n(T_f - T_i)]$:

**Constant pressure
for a monatomic
ideal gas**

$$C_P = \tfrac{3}{2}R + R = \tfrac{5}{2}R \qquad (15.7)$$

**Constant volume
for a monatomic
ideal gas**

$$C_V = \tfrac{3}{2}R \qquad (15.8)$$

The ratio γ of the specific heats is

**Monatomic
ideal gas**

$$\gamma = \frac{C_P}{C_V} = \frac{\tfrac{5}{2}R}{\tfrac{3}{2}R} = \frac{5}{3} \qquad (15.9)$$

For real monatomic gases near room temperature, experimental values of C_P and C_V give ratios very close to the theoretical value of $\tfrac{5}{3}$.

The difference between C_P and C_V arises because work is done when the gas expands in response to the addition of heat under conditions of constant pressure, whereas no work is done under conditions of constant volume. For a monatomic ideal gas, C_P exceeds C_V by an amount equal to R, the ideal gas constant:

$$C_P - C_V = R \qquad (15.10)$$

In fact, it can be shown that Equation 15.10 applies to any kind of ideal gas — monatomic, diatomic, etc.

15.7 THE SECOND LAW OF THERMODYNAMICS

Ice cream melts when left out on a warm day. A cold can of soda warms up on a hot day at a picnic. Ice cream and soda never become colder when left in a hot environment, for heat always flows spontaneously from hot to cold, and never from cold to hot. The spontaneous flow of heat is the focus of one of the most profound laws in all of science, the *second law of thermodynamics.*

● **THE SECOND LAW OF THERMODYNAMICS:
THE HEAT FLOW STATEMENT**

Heat flows spontaneously from a substance at a higher temperature to a substance at a lower temperature and does not flow spontaneously in the reverse direction.

● *Concepts at a Glance*   It is important to realize that the second law of thermodynamics deals with a different aspect of nature than does the first law of thermodynamics. The second law is a statement about the natural tendency of heat to flow from hot to cold, whereas the first law deals with energy conservation and focuses on both heat and work. A number of important devices depend on heat and work in their operation, and to understand such devices both laws are needed. For instance, an automobile engine is a type of heat engine because it uses heat to produce work. In discussing heat engines Sections 15.8 and 15.9 will bring together the first and second laws to analyze engine efficiency, as the Concepts-at-a-Glance chart in Figure 15.11 indicates. Then, in Section 15.10 we will see that refrigera-

Concepts at a Glance

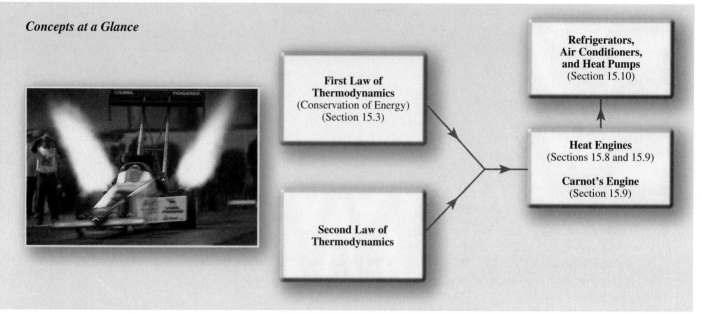

Figure 15.11 *Concepts at a Glance*
The first and second laws of thermodynamics are used to evaluate the performance of heat engines, as well as refrigerators, air conditioners, and heat
pumps. The photograph shows a top
fuel dragster, which uses a heat engine
to propel itself forward.

tors, air conditioners, and heat pumps also utilize heat and work and are closely related to heat engines. The way in which these three appliances operate also depends
on both the first and second laws of thermodynamics.

15.8 HEAT ENGINES

A *heat engine* is any device that uses heat to perform work. It has three essential
features:

1. Heat is supplied to the engine at a relatively high temperature from a place
   called the *hot reservoir.*
2. Part of the input heat is used to perform work by the *working substance* of
   the engine, which is the material within the engine that actually does the
   work (e.g., the gasoline–air mixture in an automobile engine).
3. The remainder of the input heat is rejected at a temperature lower than the
   input temperature to a place called the *cold reservoir.*

Figure 15.12 emphasizes these features in a schematic fashion. The symbol Q_H
refers to the magnitude of the input heat, and the subscript H indicates the hot reservoir. The symbol Q_C refers to the magnitude of the rejected heat, and the subscript
C denotes the cold reservoir. The symbol W stands for the magnitude of the work
done. These three symbols refer to magnitudes only, without reference to algebraic
signs. Therefore, when these symbols appear in an equation, they do not have negative values assigned to them.

To be highly efficient, a heat engine must produce a relatively large amount of
work from as little input heat as possible. Thus, the *efficiency e* of a heat engine is
defined as the ratio of the work W done by the engine to the input heat Q_H:

Figure 15.12 This schematic representation of a heat engine shows the input
heat (magnitude $= Q_H$) that originates
from the hot reservoir, the work (magnitude $= W$) that the engine does, and
the heat (magnitude $= Q_C$) that the engine rejects to the cold reservoir.

$$e = \frac{\text{Work done}}{\text{Input heat}} = \frac{W}{Q_H} \qquad (15.11)$$

If the input heat were converted entirely into work, the engine would have an efficiency of 1.00, since $W = Q_H$; such an engine would be 100% efficient. **Efficiencies are often quoted as percentages obtained by multiplying the ratio W/Q_H by a factor of 100.** Thus, an efficiency of 68% would mean that a value of 0.68 is used for the value of the efficiency in Equation 15.11.

An engine, like any device, must obey the principle of conservation of energy. Some of the engine's input heat Q_H is converted into work W, and the remainder Q_C is rejected to the cold reservoir. If there are no other losses in the engine, the principle of energy conservation requires that

$$Q_H = W + Q_C \qquad (15.12)$$

Solving this equation for W and substituting the result into Equation 15.11 leads to the following alternative expression for the efficiency e of a heat engine:

$$e = \frac{Q_H - Q_C}{Q_H} = 1 - \frac{Q_C}{Q_H} \qquad (15.13)$$

Example 6 illustrates how the concepts of efficiency and energy conservation are applied to a heat engine.

THE PHYSICS OF...

a heat engine.

EXAMPLE 6   An Automobile Engine

An automobile engine has an efficiency of 22.0% and produces 2510 J of work. How much heat is rejected by the engine?

Reasoning Energy conservation indicates that the amount of heat rejected to the cold reservoir is the part of the input heat that is not converted into work. The amount Q_C rejected is $Q_C = Q_H - W$, according to Equation 15.12. To use this equation, however, we need a value for the input heat Q_H, so we begin our solution by finding this value.

Solution From Equation 15.11 for the efficiency e, we find that $Q_H = W/e$. Substituting this result into Equation 15.12, we see that the rejected heat is

$$Q_C = Q_H - W = \frac{W}{e} - W = (2510\ \text{J})\left(\frac{1}{0.220} - 1\right) = \boxed{8900\ \text{J}}$$

● **PROBLEM SOLVING INSIGHT**
When efficiency is stated as a percentage (e.g., 22.0%), it must be converted to a decimal fraction (e.g., 0.220) before being used in an equation.

In Example 6, less than one-quarter of the input heat is converted into work because the efficiency of the automobile engine is only 22.0%. If the engine were 100% efficient, all the input heat would be converted into work. Unfortunately, nature does not permit 100% efficient heat engines to exist, as the next section discusses.

15.9   CARNOT'S PRINCIPLE AND THE CARNOT ENGINE

What is it that allows a heat engine to operate with maximum efficiency? The French engineer Sadi Carnot (1796–1832) proposed that a heat engine has maximum efficiency when the processes within the engine are reversible. **A reversible process is one in which both the system and its environment can be returned to exactly the states they were in before the process occurred.**

In a reversible process, *both* the system and its environment can be returned to their initial states. Therefore, a process that involves an energy-dissipating mechanism, such as friction, cannot be reversible because the energy wasted due to friction would alter the system or the environment or both. There are also reasons other than friction why a process may not be reversible. For instance, the spontaneous flow of heat from a hot substance to a cold substance is irreversible, even though friction is not present. For heat to flow in the reverse direction, work must be done, as we will see in Section 15.10. The agent doing such work must be located in the environment of the hot and cold substances, and, therefore, the environment must change while the heat is moved back from cold to hot. Since the system and the environment cannot *both* be returned to their initial states, the process of spontaneous heat flow is irreversible. In fact, all spontaneous processes are irreversible, such as the explosion of an unstable chemical or the bursting of a bubble. When the word "reversible" is used in connection with engines, it does not just mean a gear that allows the engine to operate a device in reverse. All cars have a reverse gear, for instance, but no automobile engine is thermodynamically reversible, since friction exists no matter which way the car moves.

Today, the idea that the efficiency of a heat engine is a maximum when the engine operates reversibly is referred to as **Carnot's principle.**

● CARNOT'S PRINCIPLE: AN ALTERNATIVE STATEMENT OF THE SECOND LAW OF THERMODYNAMICS

No irreversible engine operating between two reservoirs at constant temperatures can have a greater efficiency than a reversible engine operating between the same temperatures. Furthermore, all reversible engines operating between the same temperatures have the same efficiency.

Carnot's principle is quite remarkable, for no mention is made of the working substance of the engine. It does not matter if the working substance is a gas, a liquid, or a solid. As long as the process is reversible, the efficiency of the engine is a maximum. Carnot's principle does *not* state, or even imply, that a reversible engine has an efficiency of 100%.

It can be shown that if Carnot's principle were not valid, it would be possible for heat to flow spontaneously from a cold substance to a hot substance, in violation of the second law of thermodynamics. In effect, then, Carnot's principle is another way of expressing the second law.

No real engine operates reversibly. Nonetheless, the idea of a reversible engine provides a useful standard for judging the performance of real engines. Figure 15.13 shows a reversible engine, called a **Carnot engine,** that is particularly useful as an idealized model. An important feature of the Carnot engine is that all input heat Q_H originates from a hot reservoir *at a single temperature T_H* and all rejected heat Q_C goes into a cold reservoir *at a single temperature T_C*.

Carnot's principle implies that the efficiency of a reversible engine is independent of the working substance of the engine, and therefore can depend only on the temperatures of the hot and cold reservoirs. Since Efficiency $= 1 - Q_C/Q_H$ according to Equation 15.13, the ratio Q_C/Q_H can depend only on the reservoir temperatures. This observation led Lord Kelvin to propose a **thermodynamic temperature scale.** He proposed that the thermodynamic temperatures of the cold and hot reservoirs be defined such that their ratio is equal to Q_C/Q_H. Thus, the thermodynamic temperature scale is related to the heats absorbed and rejected by a Carnot engine, and is independent of the working substance. If a reference temperature is properly chosen, it can be shown that the scale is identical to the Kelvin scale intro-

Figure 15.13 A Carnot engine is a reversible engine in which all input heat Q_H originates from a hot reservoir at a single temperature T_H, and all rejected heat Q_C goes into a cold reservoir at a single temperature T_C. The work done by the engine is W.

duced in Section 12.2 and used in the ideal gas law. As a result, the ratio of the rejected heat Q_C to the input heat Q_H is

$$\frac{Q_C}{Q_H} = \frac{T_C}{T_H} \qquad (15.14)$$

where the temperatures T_C and T_H *must be expressed in kelvins.*

The efficiency of a Carnot engine can be written in a particularly useful way by substituting Equation 15.14 into Equation 15.13 for the efficiency, $e = 1 - Q_C/Q_H$:

$$\text{Efficiency of a Carnot engine} = 1 - \frac{T_C}{T_H} \qquad (15.15)$$

This relation gives the maximum possible efficiency for a heat engine operating between two Kelvin temperatures T_C and T_H, and the next example illustrates its application.

EXAMPLE 7   A Tropical Ocean as a Heat Engine

Water near the surface of a tropical ocean has a temperature of 298.2 K (25.0 °C), while water 700 m beneath the surface has a temperature of 280.2 K (7.0 °C). It has been proposed that the warm water be used as the hot reservoir and the cool water as the cold reservoir of a heat engine. Find the maximum possible efficiency for such an engine.

THE PHYSICS OF...

extracting work from a warm ocean.

Reasoning The maximum possible efficiency is the efficiency that a Carnot engine would have (Equation 15.15) operating between temperatures of $T_H = 298.2$ K and $T_C = 280.2$ K.

Solution Using $T_H = 298.2$ K and $T_C = 280.2$ K in Equation 15.15, we find that

$$\text{Efficiency of a Carnot engine} = 1 - \frac{T_C}{T_H} = 1 - \frac{280.2 \text{ K}}{298.2 \text{ K}} = \boxed{0.060 \ (6.0\%)}$$

• PROBLEM SOLVING INSIGHT
When determining the efficiency of a Carnot engine, be sure the temperatures T_C and T_H of the cold and hot reservoirs are expressed in kelvins; degrees Celsius or degrees Fahrenheit will not do.

In Example 7 the maximum possible efficiency is only 6.0%. The small efficiency arises because the Kelvin temperatures of the hot and cold reservoirs are so close. A greater efficiency is possible only when there is a greater difference between the reservoir temperatures. However, there are limits on how large the efficiency of a heat engine can be, as Conceptual Example 8 discusses.

CONCEPTUAL EXAMPLE 8   Natural Limits on the Efficiency of a Heat Engine

Consider a hypothetical engine that receives 1000 J of heat as input from a hot reservoir and delivers 1000 J of work, rejecting no heat to a cold reservoir whose temperature is above 0 K. Decide whether this engine violates the first or the second law of thermodynamics, or both.

Reasoning and Solution The first law of thermodynamics is an expression of energy conservation. From the point of view of energy conservation, nothing is wrong with an engine that converts 1000 J of heat into 1000 J of work. Energy has been neither created nor destroyed; it has only been transformed from one form (heat) to another (work). This engine does, however, violate the second law of thermodynamics. Since all of the input heat is converted into work, the efficiency of the engine is 1, or 100%. But Equation 15.15, which is based on the second law, indicates that the maximum possible efficiency is $1 - T_C/T_H$, where T_C and T_H are the temperatures of the cold and hot reservoirs, respectively. Since we know that T_C is above 0 K, it is clear that the ratio T_C/T_H is greater than zero, so the maximum possible efficiency is less than 1, or less than 100%. It is important to understand that the first and second laws of thermodynamics address different aspects of nature. *It is the second law, not the first law, that limits the efficiencies of heat engines to values less than 100%.*

Example 8 has emphasized that *even a perfect heat engine has an efficiency that is less than 1.0 or 100%.* In this regard, we note that the maximum possible efficiency, as given by Equation 15.15, approaches 1.0 when T_C approaches absolute zero (0 K). However, experiments have shown that it is not possible to cool a substance to absolute zero (see Section 15.12), so nature does not permit the existence of a 100% efficient heat engine. As a result, there will always be heat rejected to a cold reservoir whenever a heat engine is used to do work, even if friction and other irreversible processes are eliminated completely. This rejected heat is a form of thermal pollution. Thus, the second law of thermodynamics requires that at least some thermal pollution be generated whenever heat engines are used to perform work. This kind of thermal pollution can be reduced only if society reduces its dependence on heat engines to do work.

15.10 REFRIGERATORS, AIR CONDITIONERS, AND HEAT PUMPS

The natural tendency of heat is to flow from hot to cold, as indicated by the second law of thermodynamics. However, if work is used, heat can be *made* to flow from cold to hot, against its natural tendency. Refrigerators, air conditioners, and heat pumps are, in fact, devices that do just that. As Figure 15.14 illustrates, these devices use work W to extract an amount of heat Q_C from the cold reservoir and deposit an amount of heat Q_H into the hot reservoir. Generally speaking, such a process is called a ***refrigeration process.*** A comparison of this drawing with Figure 15.13 shows that the directions of the arrows symbolizing heat and work in a refrigeration process are opposite to those in an engine process. Nonetheless, energy is conserved during a refrigeration process, just as it is in an engine process, so $Q_H = W + Q_C$. Moreover, if the process occurs reversibly, we have ideal devices that are called Carnot refrigerators, Carnot air conditioners, and Carnot heat pumps. For these ideal devices, the relation $Q_C/Q_H = T_C/T_H$ applies, just as it does for the Carnot engine.

In a ***refrigerator,*** the interior of the unit is the cold reservoir, while the warmer exterior is the hot reservoir. As Figure 15.15 illustrates, the refrigerator takes heat from the food inside and deposits it into the kitchen, along with the energy needed to do the work of making the heat flow from cold to hot. For this reason, the outside surfaces (usually the sides and back) of most refrigerators are warm to the touch while the units operate. Thus, a refrigerator warms the kitchen.

An ***air conditioner*** is like a refrigerator, except that the room itself is the cold reservoir and the outdoors is the hot reservoir. Figure 15.16 shows a window unit, which cools a room by removing heat and depositing it outside, along with the work used to make the heat flow from cold to hot. Conceptual Example 9 considers a common misconception about refrigerators and air conditioners.

Figure 15.14 In a refrigeration process, work W is used to remove heat Q_C from the cold reservoir and deposit heat Q_H into the hot reservoir.

Figure 15.15 A refrigerator.

CONCEPTUAL EXAMPLE 9
You Can't Beat the Second Law of Thermodynamics

Is it possible to cool your kitchen by leaving the refrigerator door open or cool your bedroom by putting a window air conditioner on the floor by the bed?

Reasoning and Solution Whatever heat Q_C is removed from the air directly in front of the open refrigerator is deposited directly back into the kitchen at the rear of the unit. Moreover, according to the second law, work W is needed to move that heat from cold to hot, and the energy from this work is also deposited into the kitchen as additional heat. Thus, the open refrigerator puts into the kitchen an amount of heat $Q_H = Q_C + W$, which is more than it removes. ***Rather than cooling the kitchen, the open refrigerator warms it up.*** Putting a window air conditioner on the floor to cool your bedroom is similarly a no-win game. The heat pumped out the back of the air conditioner and into the bedroom is greater than the heat pulled into the front of the unit. Consequently, ***the air conditioner actually warms the bedroom.***

Related Homework Material: *Problem 66*

INTERACTIVE LEARNINGWARE #31

START
Visualization
Reasoning
Knowns and Unknowns
Modeling the Problem
Numerical Solution
FINISH

The water in a glass on a table has a mass of 0.25 kg (about eight liquid ounces). It has the same temperature as the room, which is 27 °C. A Carnot refrigerator is operating in this room. The water is put in the refrigerator, the inside of which is maintained at 3 °C. How much work is done by the refrigerator in cooling the water from 27 to 3 °C?

This Interactive LearningWare example deals with the kind of job that a refrigerator does routinely. The example assumes a Carnot or perfect refrigerator, so the value of the work calculated is the minimum amount required to do the job. A real refrigerator would have to do more work than that calculated. The example also gives you the opportunity to review Section 12.7, which discusses the concept of specific heat capacity.

Go to our Web site at

http://www.wiley.com/college/cutnell

The quality of a refrigerator or air conditioner is rated according to its coefficient of performance. Such appliances perform well when they remove a relatively large amount of heat Q_C from the cold reservoir with as little work W as possible. Therefore, the coefficient of performance is defined as the ratio of Q_C to W, and the greater this ratio is, the better the performance is:

$$\text{\textit{Refrigerator or air conditioner}} \qquad \text{Coefficient of performance} = \frac{Q_C}{W} \qquad (15.16)$$

Commercially available refrigerators and air conditioners have coefficients of performance in the range 2–6, depending on the temperatures involved. The coefficients of performance for these real devices are less than those for ideal, or Carnot, refrigerators and air conditioners.

In a sense, refrigerators and air conditioners operate like pumps. They pump heat "uphill" from a lower temperature to a higher temperature, just as a water pump forces water uphill from a lower elevation to a higher elevation. It would be appropriate to call them heat pumps. However, the name "heat pump" is reserved for the device illustrated in Figure 15.17, which is a home heating appliance. The ***heat pump*** uses work W to make heat Q_C from the wintry outdoors (the cold reservoir) flow up the temperature "hill" into a warm house (the hot reservoir). According to the conservation of energy, the heat pump deposits inside the house an amount of heat $Q_H = Q_C + W$. The air conditioner and the heat pump do closely related jobs. The air conditioner refrigerates the inside of the house and heats up the outdoors, while the heat pump refrigerates the outdoors and heats up the inside. These jobs are so closely related that most heat pump systems serve in a dual capacity, being equipped with a switch that converts them from heaters in the winter into air conditioners in the summer.

Heat pumps are popular for home heating in today's energy-conscious world, and it is easy to understand why. Suppose 1000 J of energy is available to use for home heating. Figure 15.18 shows that a conventional electric heating system uses this 1000 J to heat a coil of wire, just as in a toaster. A fan blows air across the hot coil, and forced convection carries the 1000 J of heat into the house. In contrast, the heat pump in Fig-

Figure 15.16 A window air conditioner removes heat from a room, which is the cold reservoir, and deposits heat outdoors, which is the hot reservoir.

THE PHYSICS OF...

a heat pump.

Figure 15.17 In a heat pump the cold reservoir is the wintry outdoors, and the hot reservoir is the inside of the house.

● **PROBLEM SOLVING INSIGHT**
When applying Equation 15.14 ($Q_C/Q_H = T_C/T_H$) to heat pumps, refrigerators, or air conditioners, be sure the temperatures T_C and T_H are expressed in kelvins; degrees Celsius or degrees Fahrenheit will not do.

Figure 15.18 This conventional electric heating system is delivering 1000 J of heat to the living room.

ure 15.17 does not use the 1000 J directly as heat. Instead, it uses the 1000 J to do the work W of pumping heat Q_C from the cooler outdoors into the warmer house. The heat pump delivers to the house an amount of energy $Q_H = Q_C + W$. With $W = 1000$ J, this becomes $Q_H = Q_C + 1000$ J, so that the heat pump delivers more than 1000 J of heat into the house, whereas the conventional electric heating system delivers only 1000 J. The next example shows how the basic relations $Q_H = W + Q_C$ and $Q_C/Q_H = T_C/T_H$ are used with heat pumps.

EXAMPLE 10   A Heat Pump

An ideal or Carnot heat pump is used to heat a house to a temperature of $T_H = 294$ K (21 °C). How much work must be done by the pump to deliver $Q_H = 3350$ J of heat into the house when the outdoor temperature T_C is (a) 273 K (0 °C) and (b) 252 K (−21 °C)?

Reasoning The conservation of energy ($Q_H = W + Q_C$) applies to the heat pump. Thus, the work can be determined from $W = Q_H − Q_C$, provided we can obtain a value for Q_C, the heat taken by the pump from the outside. To determine Q_C, we use the fact that the pump is a Carnot heat pump and operates reversibly. Therefore, the relation $Q_C/Q_H = T_C/T_H$ (Equation 15.14) applies. Solving it for Q_C, we obtain $Q_C = Q_H(T_C/T_H)$. Using this result, we find that

$$W = Q_H − Q_C = Q_H − Q_H \left(\frac{T_C}{T_H} \right) = Q_H \left(1 − \frac{T_C}{T_H} \right)$$

Solution

(a) At an indoor temperature of $T_H = 294$ K and an outdoor temperature of $T_C = 273$ K, the work needed is

$$W = Q_H \left(1 − \frac{T_C}{T_H} \right) = (3350 \text{ J}) \left(1 − \frac{273 \text{ K}}{294 \text{ K}} \right) = \boxed{240 \text{ J}}$$

(b) This solution is identical to that in part (a), except that it is now cooler outside, so $T_C = 252$ K. The necessary work is $W = \boxed{479 \text{ J}}$, which is more than in part (a). More work must be done because the heat is pumped up a greater temperature "hill" when the outside is colder than when it is warmer.

It is also possible to specify a coefficient of performance for heat pumps. However, unlike refrigerators and air conditioners, the job of a heat pump is to heat, not to cool. As a result, the coefficient of performance of a heat pump is the ratio of the heat Q_H delivered into the house to the work W required to deliver it:

Heat pump $$\begin{array}{c} \text{Coefficient} \\ \text{of} \\ \text{performance} \end{array} = \frac{Q_H}{W} \qquad (15.17)$$

The coefficient of performance depends on the indoor and outdoor temperatures. Commercial units have coefficients of about 3−4 under favorable conditions.

15.11   ENTROPY AND THE SECOND LAW OF THERMODYNAMICS

A Carnot engine has the maximum possible efficiency for its operating conditions because the processes occurring within it are reversible. Irreversible processes, such as friction, cause real engines to operate at less than maximum efficiency, for they reduce our ability to use heat to perform work. As an extreme example, imagine that a hot object is placed in thermal contact with a cold object, so heat flows spontaneously, and hence irreversibly, from hot to cold. Eventually both objects reach the same temperature, and $T_C = T_H$. A Carnot engine using these two objects as heat reservoirs is unable to do work, because the efficiency of the engine is zero [Efficiency $= 1 - (T_C/T_H) = 0$]. In general, irreversible processes cause us to lose some, but not necessarily all, of the ability to perform work. This partial loss can be expressed in terms of a concept called ***entropy.***

To introduce the idea of entropy we recall the relation $Q_C/Q_H = T_C/T_H$ that applies to a Carnot engine. This equation can be rearranged as $Q_C/T_C = Q_H/T_H$, which focuses attention on the heat Q divided by the Kelvin temperature T. The quantity Q/T is called the change in the entropy ΔS:

$$\Delta S = \left(\frac{Q}{T}\right)_R \qquad (15.18)$$

In this expression the temperature T must be in kelvins, and the subscript R refers to the word "reversible." It can be shown that Equation 15.18 applies to any process in which heat Q enters or leaves a system reversibly at a constant temperature. Such is the case for the heat that flows into and out of the reservoirs of a Carnot engine. Equation 15.18 indicates that the SI unit for entropy is a joule per kelvin (J/K).

Entropy, like internal energy, is a function of the state or condition of the system. Only the state of a system determines the entropy S that a system has. Therefore, the change in entropy ΔS is equal to the entropy of the final state of the system minus the entropy of the initial state.

We can now describe what happens to the entropy of a Carnot engine. As the engine operates, the entropy of the hot reservoir decreases, since heat Q_H departs at a Kelvin temperature T_H. The change in the entropy of the hot reservoir is $\Delta S_H = -Q_H/T_H$, where the minus sign is needed to indicate a decrease in entropy, since the symbol Q_H denotes only the magnitude of the heat. In contrast, the entropy of the cold reservoir increases by an amount $\Delta S_C = +Q_C/T_C$, for the rejected heat enters the cold reservoir at a Kelvin temperature T_C. The total change in entropy is

$$\Delta S_C + \Delta S_H = \frac{Q_C}{T_C} - \frac{Q_H}{T_H} = 0$$

because $Q_C/T_C = Q_H/T_H$ according to Equation 15.14.

The fact that the total change in entropy is zero for a Carnot engine is a specific illustration of a general result. It can be proved that when *any* reversible process occurs, the change in the entropy of the universe is zero; $\Delta S_{universe} = 0$ for a reversible process. The word "universe" means that $\Delta S_{universe}$ takes into account the entropy changes of all parts of the system and all parts of the environment. ***Reversible processes do not alter the total entropy of the universe.*** To be sure, the entropy of one part of the universe may change because of a reversible process, but if so, the entropy of another part changes in the opposite way by the same amount.

To understand what happens to the entropy of the universe when an *irreversible* process occurs is more difficult, for the expression $\Delta S = (Q/T)_R$ does not apply directly. However, if a system changes irreversibly from an initial state to a final state, this expression can be used to calculate ΔS indirectly, as Figure 15.19 indicates. We imagine a hypothetical reversible process that causes the system to change between *the same initial and final states* and then find ΔS for this reversible process. The value obtained for ΔS also applies to the irreversible process that actually occurs, since only the nature of the initial and final states, and not the path between them,

Figure 15.19 Although the relation $\Delta S = (Q/T)_R$ applies to reversible processes, it can be used as part of an indirect procedure to find the entropy change for an irreversible process. This drawing illustrates the procedure discussed in the text.

Hot reservoir
$T_H = 650$ K

Copper
rod
1200 J

Cold reservoir
$T_C = 350$ K

Figure 15.20 Heat flows spontaneously from a hot reservoir to a cold reservoir.

Hot reservoir
$T_H = 650$ K

$Q_H = 1200$ J

Carnot
engine $W = 920$ J

$Q_C = 280$ J

Cold reservoir
$T_C = 150$ K

(a)

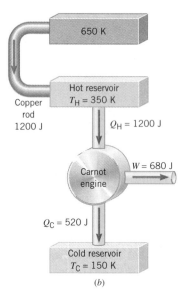

650 K

Copper
rod
1200 J

Hot reservoir
$T_H = 350$ K $Q_H = 1200$ J

Carnot
engine $W = 680$ J

$Q_C = 520$ J

Cold reservoir
$T_C = 150$ K

(b)

Figure 15.21 Heat in the amount of 1200 J is used as input for an engine under two different conditions in parts *a* and *b*.

determines ΔS. Example 11 illustrates this indirect method and shows that spontaneous (irreversible) processes increase the entropy of the universe.

EXAMPLE 11   The Entropy of the Universe Increases

Figure 15.20 shows 1200 J of heat flowing spontaneously through a copper rod from a hot reservoir at 650 K to a cold reservoir at 350 K. Determine the amount by which this irreversible process changes the entropy of the universe, assuming that no other changes occur.

Reasoning The hot-to-cold heat flow is irreversible, so the relation $\Delta S = (Q/T)_R$ is applied to a hypothetical process whereby the 1200 J of heat is taken reversibly from the hot reservoir and added reversibly to the cold reservoir.

Solution The total entropy change of the universe is the algebraic sum of the entropy changes for each reservoir:

$$\Delta S_{universe} = -\underbrace{\frac{1200 \text{ J}}{650 \text{ K}}}_{\substack{\text{Entropy lost} \\ \text{by the hot} \\ \text{reservoir}}} + \underbrace{\frac{1200 \text{ J}}{350 \text{ K}}}_{\substack{\text{Entropy gained} \\ \text{by the cold} \\ \text{reservoir}}} = \boxed{+1.6 \text{ J/K}}$$

The irreversible process causes the entropy of the universe to increase by 1.6 J/K.

Example 11 is a specific illustration of a general result: **any irreversible process increases the entropy of the universe.** In other words, $\Delta S_{universe} > 0$ for an irreversible process. Reversible processes do not alter the entropy of the universe, whereas irreversible processes cause the entropy to increase. Therefore, the entropy of the universe continually increases, like time itself, and entropy is sometimes called "time's arrow." It can be shown that this behavior of the entropy of the universe provides a completely general statement of the second law of thermodynamics, which applies not only to heat flow but also to all kinds of other processes.

● THE SECOND LAW OF THERMODYNAMICS STATED IN TERMS OF ENTROPY

The total entropy of the universe does not change when a reversible process occurs ($\Delta S_{universe} = 0$) and increases when an irreversible process occurs ($\Delta S_{universe} > 0$).

When an irreversible process occurs and the entropy of the universe increases, the energy available for doing work decreases, as the next example illustrates.

EXAMPLE 12   Energy Unavailable for Doing Work

Suppose that 1200 J of heat is used as input for an engine under two different conditions. In Figure 15.21a the heat is supplied by a hot reservoir whose temperature is 650 K. In part *b* of the drawing, the heat flows irreversibly through a copper rod into a second reservoir whose temperature is 350 K and then enters the engine. In either case, a 150-K reservoir is used as the cold reservoir. For each case, determine the maximum amount of work that can be obtained from the 1200 J of heat.

Reasoning According to Equation 15.11, the work obtained from the engine is the product of its efficiency and the input heat: $W = (\text{Efficiency})Q_H = (\text{Efficiency}) \times (1200 \text{ J})$.

The maximum amount of work is obtained when the efficiency is a maximum, that is, when the engine is a Carnot engine. The efficiency of a Carnot engine is given by Equation 15.15 as Efficiency $= 1 - T_C/T_H$. Therefore, the efficiency may be determined from the Kelvin temperatures of the hot and cold reservoirs.

Solution

Before
irreversible
heat flow
$$\text{Efficiency} = 1 - \frac{T_C}{T_H} = 1 - \frac{150 \text{ K}}{650 \text{ K}} = 0.77$$
$$W = (\text{Efficiency})(1200 \text{ J}) = (0.77)(1200 \text{ J}) = \boxed{920 \text{ J}}$$

After
irreversible
heat flow
$$\text{Efficiency} = 1 - \frac{T_C}{T_H} = 1 - \frac{150 \text{ K}}{350 \text{ K}} = 0.57$$
$$W = (\text{Efficiency})(1200 \text{ J}) = (0.57)(1200 \text{ J}) = \boxed{680 \text{ J}}$$

When the 1200 J of input heat is taken from the 350-K reservoir instead of the 650-K reservoir, the efficiency of the Carnot engine is smaller. As a result, less work (680 J versus 920 J) can be extracted from the input heat.

Example 12 shows that 240 J less work (920 J − 680 J) can be performed when the input heat is obtained from the reservoir with the smaller temperature. In other words, the irreversible process of heat flow through the copper rod causes energy to become unavailable for doing work in the amount of $W_{\text{unavailable}} = 240$ J. Example 11 shows that this irreversible process also causes the entropy of the universe to increase by an amount $\Delta S_{\text{universe}} = +1.6$ J/K. These values for $W_{\text{unavailable}}$ and $\Delta S_{\text{universe}}$ are in fact related. If you multiply $\Delta S_{\text{universe}}$ by 150 K, which is the lowest Kelvin temperature in Example 12, you obtain $W_{\text{unavailable}} = (150 \text{ K}) \times (1.6 \text{ J/K}) = 240$ J. This is one illustration of the following general result:

$$W_{\text{unavailable}} = T_0 \Delta S_{\text{universe}} \qquad (15.19)$$

where T_0 is the Kelvin temperature of the coldest heat reservoir. Since irreversible processes cause the entropy of the universe to increase, they cause energy to be degraded, in the sense that part of the energy becomes unavailable for the performance of work. In contrast, there is no penalty when reversible processes occur, because for them $\Delta S_{\text{universe}} = 0$, and there is no loss of work.

Entropy can also be interpreted in terms of order and disorder. As an example, consider a block of ice (Figure 15.22) with each of its H_2O molecules fixed rigidly in place in a highly structured and ordered arrangement. In comparison, the puddle of water into which the ice melts is disordered and unorganized, for the molecules in a liquid are free to move from place to place. Heat is required to melt the ice and produce the disorder. Moreover, heat flow into a system increases the entropy of the

Figure 15.22 A block of ice is an example of an ordered system relative to a puddle of water.

system, according to $\Delta S = (Q/T)_R$. We associate an increase in entropy, then, with an increase in disorder. Conversely, we associate a decrease in entropy with a decrease in disorder or a greater degree of order. Example 13 illustrates an order-to-disorder change and the increase of entropy that accompanies it.

EXAMPLE 13   Order to Disorder

Find the change in entropy that results when a 2.3-kg block of ice melts slowly (reversibly) at 273 K (0 °C).

Reasoning Since the phase change occurs reversibly at a constant temperature, the change in entropy can be found by using Equation 15.18, $\Delta S = (Q/T)_R$, where Q is the heat absorbed by the melting ice. This heat can be determined by using the relation $Q = mL_f$ (Equation 12.5), where m is the mass and $L_f = 3.35 \times 10^5$ J/kg is the latent heat of fusion of water.

Solution Using Equation 15.18 and Equation 12.5, we find that the change in entropy is

$$\Delta S = \left(\frac{Q}{T}\right)_R = \frac{mL_f}{T} = \frac{(2.3 \text{ kg})(3.35 \times 10^5 \text{ J/kg})}{273 \text{ K}} = \boxed{+2.8 \times 10^3 \text{ J/K}}$$

a result that is positive, since the ice absorbs heat as it melts.

Figure 15.23 With the aid of 100 pounds of dynamite, demolition experts caused this hotel-casino in Las Vegas to go from an ordered state (lower entropy) to a disordered state (higher entropy).

Figure 15.23 shows another order-to-disorder change that can be described in terms of entropy.

15.12 THE THIRD LAW OF THERMODYNAMICS

To the zeroth, first, and second laws of thermodynamics we add the third (and last) law. The ***third law of thermodynamics*** indicates that it is impossible to reach a temperature of absolute zero.

● **THE THIRD LAW OF THERMODYNAMICS**

It is not possible to lower the temperature of any system to absolute zero in a finite number of steps.

This law, like the second law, can be expressed in a number of ways, but a discussion of them is beyond the scope of this text. The third law is needed to explain a number of experimental observations that cannot be explained by the other laws of thermodynamics.

15.13 CONCEPTS   &   CALCULATIONS

The first law of thermodynamics is basically a restatement of the conservation-of-energy principle in terms of heat and work. Example 14 emphasizes this important fact by showing that the conservation principle and the first law provide the same approach to a problem. In addition, the example reviews the concept of latent heat of sublimation (see Section 12.8) and the ideal gas law (see Section 14.2).

CONCEPTS & CALCULATIONS EXAMPLE 14   The Sublimation of Zinc

The sublimation of zinc (mass per mole = 0.0654 kg/mol) occurs at a temperature of 6.00×10^2 K, and the latent heat of sublimation is 1.99×10^6 J/kg. The pressure remains constant during the sublimation. Assume that the zinc vapor can be treated as a monatomic ideal gas and that the volume of solid zinc is negligible compared to the corresponding vapor. What is the change in the internal energy of the zinc when 1.50 kg of zinc sublimates?

Concept Questions and Answers What is sublimation and what is the latent heat of sublimation?

Answer Sublimation is the process whereby a solid phase changes directly into a gas phase in response to the input of heat. The heat per kilogram needed to cause the phase change is called the latent heat of sublimation L_s. The heat Q needed to bring about the sublimation of a mass m of solid material is given by Equation 12.5 as $Q = mL_s$.

When a solid phase changes to a gas phase, does the volume of the material increase or decrease, and by how much?

Answer Gases, in general, have greater volumes than solids do, so the volume of the material increases. The increase in volume is $\Delta V = V_{gas} - V_{solid}$. Since the volume of the solid V_{solid} is negligibly small compared to the volume of the gas V_{gas}, we have $\Delta V = V_{gas}$. Using the ideal gas law as given in Equation 14.1, it follows that $V_{gas} = nRT/P$, so that $\Delta V = nRT/P$. In this result, n is the number of moles of material, R is the universal gas constant, and T is the Kelvin temperature.

As the material changes from a solid to a gas, does it do work on the environment or does the environment do work on it? How much work is involved?

Answer To make room for itself, the expanding material must push against the environment and, in so doing, does work on the environment. Since the pressure remains constant, the work done by the material is given by Equation 15.2 as $W = P \Delta V$. Since $\Delta V = nRT/P$, the work becomes $W = P(nRT/P) = nRT$.

In this problem we begin with heat Q, and it is used for two purposes: First, it makes the solid change into a gas, which entails a change ΔU in the internal energy of the material, $\Delta U = U_{solid} - U_{gas}$. Second, it allows the expanding material to do work W on the environment. According to the conservation-of-energy principle, how is Q related to ΔU and W?

Answer According to the conservation-of-energy principle, energy can neither be created nor destroyed, but can only be converted from one form to another (see Section 6.8). Therefore, part of the heat Q is used for ΔU and part for W, with the result that $Q = \Delta U + W$.

According to the first law of thermodynamics, how is Q related to ΔU and W?

Answer As indicated in Equation 15.1, the first law of thermodynamics is $\Delta U = Q - W$. Rearranging this equation gives $Q = \Delta U + W$, which is identical to the result obtained from the conservation-of-energy principle.

Solution Using the fact that $Q = \Delta U + W$, that $Q = mL_s$, and that $W = nRT$, we have that

$$Q = \Delta U + W \quad \text{or} \quad mL_s = \Delta U + nRT$$

Solving for ΔU gives

$$\Delta U = mL_s - nRT$$

In this result, n is the number of moles of the ideal gas. According to the discussion in Section 14.1, the number of moles of gaseous zinc is the mass m of the sample divided by the mass per mole of zinc or $n = m/(0.0654 \text{ kg/mol})$. Therefore, we find

$$\Delta U = mL_s - nRT$$

$$= (1.50 \text{ kg})\left(1.99 \times 10^6 \frac{\text{J}}{\text{kg}}\right) - \left(\frac{1.50 \text{ kg}}{0.0654 \text{ kg/mol}}\right)\left(8.31 \frac{\text{J}}{\text{mol} \cdot \text{K}}\right)(6.00 \times 10^2 \text{ K})$$

$$= \boxed{2.87 \times 10^6 \text{ J}}$$

Heat engines can be used to perform work, as we have seen in this chapter. The concept of work, however, was first introduced in Chapter 6, along with the idea of kinetic energy and the work–energy theorem. The next example reviews some of the main features of heat engines, as well as kinetic energy and the work–energy theorem.

CONCEPTS & CALCULATIONS EXAMPLE 15   The Work–Energy Theorem

Each of two Carnot engines uses the same cold reservoir at a temperature of 275 K for its exhaust heat. Each engine receives 1450 J of input heat. The work from either of these engines is used to drive a pulley arrangement that uses a rope to accelerate a 125-kg crate from rest along a horizontal frictionless surface, as Figure 15.24 suggests. With engine 1 the crate attains a speed of 2.00 m/s, while with engine 2 it attains a speed of 3.00 m/s. Find the temperature of the hot reservoir for each engine.

Concept Questions and Answers With which engine is the change in the crate's kinetic energy greater?

Answer   The change is greater with engine 2. Kinetic energy is $\text{KE} = \frac{1}{2}mv^2$, according to Equation 6.2, where m is the mass of the crate and v is its speed. The change in the kinetic energy is the final minus the initial value, or $\text{KE}_f - \text{KE}_0$. Since the crate starts from rest, it has zero initial kinetic energy. Thus, the change is equal to the final kinetic energy. Since engine 2 gives the crate the greater final speed, it causes the greater change in kinetic energy.

Which engine does more work?

Answer   The work–energy theorem, as stated in Equation 6.3, indicates that the net work done on an object equals the change in the object's kinetic energy, or $W = \text{KE}_f - \text{KE}_0$. The net work is the work done by the net force. In Figure 15.24 the surface is horizontal, and the crate does not leave it. Therefore, the upward normal force that the surface applies to the crate must balance the downward weight of the crate. Furthermore, the surface is frictionless, so there is no friction force. The net force acting on the crate, then, consists of the single force due to the tension in the rope, which arises from the action of the engine. Thus, the work done by the engine is, in fact, the net work done on the crate. But we know that engine 2 causes the crate's kinetic energy to change by the greater amount, so that engine must do more work.

For which engine is the temperature of the hot reservoir greater?

Answer   The temperature of the hot reservoir for engine 2 is greater. We know that engine 2 does more work, but each engine receives the same 1450 J of input heat. Therefore, engine 2 derives more work from the input heat. In other words, it is more efficient. But the efficiency of a Carnot engine depends only on the Kelvin temperatures of its hot and cold reservoirs. Since both engines use the same cold reservoir whose temperature is 275 K, only the temperatures of the hot reservoirs are different. Higher temperatures for the hot reservoir are associated with greater efficiencies, so the temperature of the hot reservoir for engine 2 is greater.

Figure 15.24 With the aid of pulleys and a rope, a Carnot engine provides the work that is used to accelerate the crate from rest along a horizontal frictionless surface. See Example 15.

Solution According to Equation 15.11, the efficiency e of a heat engine is the work W divided by the input heat Q_H, or $e = W/Q_H$. According to Equation 15.15, the efficiency of a Carnot engine is Efficiency $= 1 - T_C/T_H$, where T_C and T_H are, respectively, the Kelvin temperatures of the cold and hot reservoirs. Combining these two equations, we have

$$e = 1 - \frac{T_C}{T_H} = \frac{W}{Q_H}$$

But W is the net work done on the crate, and it equals the change in the crate's kinetic energy, or $W = KE_f - KE_0 = \frac{1}{2}mv^2$, according to Equations 6.2 and 6.3. With this substitution the efficiency expression becomes

$$1 - \frac{T_C}{T_H} = \frac{\frac{1}{2}mv^2}{Q_H}$$

Solving for the temperature T_H, we find

$$T_H = \frac{T_C}{1 - \dfrac{mv^2}{2Q_H}}$$

As expected, the value for T_H for engine 2 is greater:

Engine 1 $\quad T_H = \dfrac{275 \text{ K}}{1 - \dfrac{(125 \text{ kg})(2.00 \text{ m/s})^2}{2(1450 \text{ J})}} = \boxed{332 \text{ K}}$

Engine 2 $\quad T_H = \dfrac{275 \text{ K}}{1 - \dfrac{(125 \text{ kg})(3.00 \text{ m/s})^2}{2(1450 \text{ J})}} = \boxed{449 \text{ K}}$

At the end of the problem set for this chapter, you will find homework problems that contain both conceptual and quantitative parts. These problems are grouped under the heading *Concepts & Calculations, Group Learning Problems.* They are designed for use by students working alone or in small learning groups. The conceptual part of each problem provides a convenient focus for group discussions.

S U M M A R Y

Thermodynamics is the branch of physics built upon the laws obeyed by energy in the form of work and heat. A thermodynamic **system** is the collection of objects on which attention is being focused, and the **surroundings** are everything else. The **state of a system** is the physical condition of the system, as described by values for physical parameters, often pressure, volume, and temperature.

Two systems are in **thermal equilibrium** if there is no net flow of heat between them when they are brought into thermal contact. The **zeroth law of thermodynamics** states that two systems individually in thermal equilibrium with a third system are in thermal equilibrium with each other. **Temperature** is the indicator of thermal equilibrium in the sense that there is no net flow of heat between two systems in thermal contact that have the same temperature.

The **first law of thermodynamics** states that due to heat Q and work W, the internal energy of a system changes from its initial value of U_i to a final value of U_f, according to $\Delta U = U_f - U_i = Q - W$. The first law is the conservation-of-energy principle applied to heat, work, and change in the internal energy. The internal energy is called a **function of state** because it depends only on the state of the system and not on the method by which the system came to be in a given state. Heat and work are not functions of state because they depend on how the system is changed from one state to another.

Thermal processes are **quasi-static** when they occur slowly enough that a uniform pressure and temperature exist throughout the system. An **isobaric process** is one that occurs at constant pressure. The work W done when a system changes at a constant pressure P from an initial volume V_i to a final volume V_f is $W = P(V_f - V_i)$. An **isochoric process** is one that takes place at constant volume, and no work is done in such a process. An **isothermal process** is one that occurs at constant temperature. An **adiabatic process** is one that takes place without the transfer of heat. The work done in a quasi-static process is given by the area under the corresponding pressure-versus-volume graph.

When n moles of an ideal gas change quasi-statically from an initial volume V_i to a final volume V_f at a constant Kelvin temperature T, the work done is $W = nRT \ln(V_f/V_i)$. When n moles of a monatomic ideal gas change quasi-statically and adiabatically from an initial temperature T_i to a final temperature T_f, the work done is $W = \frac{3}{2}nR(T_i - T_f)$. Along with the ideal gas law, an ideal gas also obeys the relation $P_iV_i^{\gamma} = P_fV_f^{\gamma}$ in an adiabatic process, where $\gamma = C_P/C_V$ is the ratio of the specific heat capacities at constant pressure and constant volume.

The **molar specific heat capacity** C of a substance determines how much heat Q is added or removed when the temperature of n moles of the substance changes by an amount ΔT: $Q = Cn\Delta T$. For a monatomic ideal gas, the molar specific heat capacities at constant pressure and constant volume are, respectively, $C_P = \frac{5}{2}R$ and $C_V = \frac{3}{2}R$, where R is the ideal gas constant. For any type of ideal gas $C_P - C_V = R$.

A **reversible process** is one in which both the system and its environment can be returned to exactly the initial states they were in before the process occurred. All spontaneous processes (such as the conduction of heat) and any process involving friction are irreversible.

The **second law of thermodynamics** can be stated in a number of equivalent forms. In terms of heat flow, the second law declares that heat flows spontaneously from a substance at a higher temperature to a substance at a lower temperature and does not flow spontaneously in the reverse direction. In the form known as **Carnot's principle,** the second law states that no irreversible engine operating between two reservoirs at constant temperatures can have a greater efficiency than a reversible engine operating between the same temperatures. Furthermore, all reversible engines operating between the same temperatures have the same efficiency. In terms of **entropy,** the second law states that the total entropy of the universe does not change when a reversible process occurs and increases when an irreversible process occurs.

A **heat engine** produces work W from input heat Q_H that is extracted from a heat reservoir at a relatively high temperature. The engine rejects heat Q_C into a reservoir at a relatively low temperature. The **efficiency** e of a heat engine is defined as $e = W/Q_H$. In addition, the principle of the conservation of energy requires that $Q_H = W + Q_C$.

A **Carnot engine** is a reversible engine in which all input heat Q_H originates from a hot reservoir at a single Kelvin temperature T_H and all rejected heat Q_C goes into a cold reservoir at a single Kelvin temperature T_C. For a Carnot engine, $Q_C/Q_H = T_C/T_H$. The **efficiency of a Carnot engine** is the maximum efficiency that an engine operating between two fixed temperatures can have: Efficiency of a Carnot engine $= 1 - T_C/T_H$.

Refrigerators, air conditioners, and **heat pumps** are devices that utilize work W to make heat Q_C flow from a lower Kelvin temperature T_C to a higher Kelvin temperature T_H. In the process (the refrigeration process) they deposit an amount of heat Q_H at the higher temperature. The principle of the conservation of energy requires that $Q_H = W + Q_C$. If the refrigeration process is perfect, in the sense that it occurs reversibly, the devices are called Carnot devices and the relation $Q_C/Q_H = T_C/T_H$ holds. The **coefficient of performance of a refrigerator or an air conditioner** is Q_C/W, while the **coefficient of performance of a heat pump** is Q_H/W.

The **change in entropy** ΔS for a process in which heat Q enters or leaves a system reversibly at a constant Kelvin temperature T is $\Delta S = (Q/T)_R$, where R stands for "reversible." Irreversible processes cause energy to be degraded in the sense that part of the energy becomes unavailable for the performance of work. The **energy that is unavailable for doing work** because of an irreversible process is $W_{unavailable} = T_0\Delta S_{universe}$, where $\Delta S_{universe}$ is the total entropy change of the universe and T_0 is the Kelvin temperature of the coldest reservoir into which heat is rejected. Increased entropy is associated with a greater degree of disorder and decreased entropy with a lesser degree of disorder (more order).

The **third law of thermodynamics** states that it is not possible to lower the temperature of any system to absolute zero in a finite number of steps.

CONCEPTUAL QUESTIONS

1. Ignore friction and assume that air behaves as an ideal gas. The plunger of a bicycle tire pump is pushed down rapidly with the end of the pump sealed so that no air escapes and there is little time for heat to flow through the cylinder wall. Explain why the cylinder of the pump becomes warm to the touch.

2. One hundred joules of heat is added to a gas, and the gas expands at constant pressure. Is it possible that the internal energy increases by 200 J? Account for your answer with the aid of the first law of thermodynamics.

3. A gas is compressed isothermally, and its internal energy increases. Is the gas an ideal gas? Justify your answer.

4. A material undergoes an isochoric process that is also adiabatic. Is the internal energy of the material at the end of the process greater than, less than, or the same as it was at the start? Justify your answer.

5. (a) Is it possible for the temperature of a substance to rise without heat flowing into it? (b) Does the temperature of a substance necessarily have to change because heat flows into or out of it? In each case, give your reasoning and use the example of an ideal gas.

6. The drawing shows a pressure–volume graph in which a gas expands at constant pressure from A to B, and then goes from B to C at constant volume. Complete the table below by deciding whether each of the remaining four entries should be positive ($+$), negative ($-$), or zero (0). Give a reason for each answer that you choose.

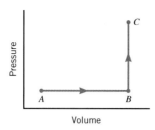

	ΔU	Q	W
$A \rightarrow B$	$+$		
$B \rightarrow C$		$+$	

7. The drawing shows an arrangement for an adiabatic free expansion or "throttling" process. The process is adiabatic because the entire arrangement is contained within perfectly insulating walls. The gas in chamber A rushes suddenly into chamber B through a hole in the partition. Chamber B is evacuated, so the gas expands there under zero external pressure and the work $W = P \, \Delta V$ is zero. Assume that the gas is an ideal gas and explain how the final temperature of the gas after expansion compares to its initial temperature.

8. Suppose a material contracts when it is heated. Follow the same line of reasoning used in the text to reach Equations 15.7 and 15.8 and deduce which specific heat capacity for the material is larger, C_P or C_V.

9. When a solid melts at constant pressure, the volume of the resulting liquid does not differ much from the volume of the solid. Using what you know about the first law of thermodynamics and the latent heat of fusion, describe how the internal energy of the liquid compares to the internal energy of the solid.

10. Suppose that you want to heat a gas so that its temperature will be as high as possible. Would you heat it under conditions of constant pressure or constant volume? Why?

11. Consider a hypothetical device that takes 10 000 J of heat from a hot reservoir and 5000 J of heat from a cold reservoir and produces 15 000 J of work. (a) Does this device violate the first law of thermodynamics? Explain. (b) Does the device violate the second law of thermodynamics? Explain.

12. If you saw an advertisement for an automobile that claimed the same gas mileage with and without the air conditioner operating, would you be suspicious? Explain, using what the second law of thermodynamics has to say about work and the direction of heat flow in an air conditioner.

13. The second law of thermodynamics, in the form of Carnot's principle, indicates that the most efficient heat engine operating between two temperatures is a reversible one. Does this mean that a reversible engine operating between the temperatures of 600 and 400 K must be more efficient than an *irreversible* engine

operating between 700 and 300 K? Provide a reason for your answer.

14. Three reversible engines, A, B, and C, use the same cold reservoir for their exhaust heats. However, they use different hot reservoirs that have the following temperatures: (A) 1000 K, (B) 1100 K, and (C) 900 K. Rank these engines in order of increasing efficiency (smallest efficiency first). Account for your answer.

15. Suppose you wish to improve the efficiency of a Carnot engine. Compare the improvement to be realized via each of the following alternatives: (a) lower the Kelvin temperature of the cold reservoir by a factor of four, (b) raise the Kelvin temperature of the hot reservoir by a factor of four, (c) cut the Kelvin temperature of the cold reservoir in half and double the Kelvin temperature of the hot reservoir. Give your reasoning.

16. A refrigerator is kept in a garage that is not heated in the cold winter or air-conditioned in the hot summer. Does it cost more for this refrigerator to make a kilogram of ice cubes in the winter or in the summer? Give your reasoning.

17. Is it possible for a Carnot heat pump to have a coefficient of performance that is less than one? Justify your answer.

18. Air conditioners and refrigerators both remove heat from a cold reservoir and deposit it in a hot reservoir. Why, then, does an air conditioner cool the inside of a house while a refrigerator warms the house?

19. It has been said that heat pumps can't possibly deliver more energy into your house than they consume in operating. Can they or can't they? Explain.

20. A refrigerator is advertised as being easier to "live with" during the summer because it puts into your kitchen only the heat that it removes from the food. Does this advertising claim violate the second law of thermodynamics? Account for your answer.

21. On a summer day a window air conditioner cycles on and off, according to how the temperature within the room changes. Are you more likely to be able to fry an egg on the outside part of the unit when the unit is on or when it is off? Explain.

22. An event happens somewhere in the universe and, as a result, the entropy of an object changes by -5 J/K. Which one (or more) of the following is a possible value for the entropy change for the rest of the universe: -5 J/K, 0 J/K, $+5$ J/K, $+10$ J/K? Account for your choice(s) in terms of the second law of thermodynamics.

23. When water freezes from a less-ordered liquid to a more-ordered solid, its entropy decreases. Why doesn't this decrease in entropy violate the second law of thermodynamics?

24. In each of the following cases, which has the greater entropy: (a) a handful of popcorn kernels or the popcorn that results from them, (b) a salad before or after it has been tossed, (c) a messy apartment or a neat apartment? Why?

25. A glass of water contains a teaspoon of dissolved sugar. After a while, the water evaporates, leaving behind sugar crystals. The entropy of the sugar crystals is less than the entropy of the dissolved sugar because the sugar crystals are in a more-ordered state. Why doesn't this process violate the second law of thermodynamics?

26. A builder uses lumber to construct a building, which is unfortunately destroyed in a fire. Thus, the lumber existed at one time or another in three different states: (1) as unused building material, (2) as a building, and (3) as a burned-out shell of a building. Rank these three states in order of decreasing entropy (largest first). Provide a reason for the ranking.

PROBLEMS

Section 15.3 The First Law of Thermodynamics

1. ssm When one gallon of gasoline is burned in a car engine, 1.19×10^8 J of internal energy is released. Suppose that 1.00×10^8 J of this energy flows directly into the surroundings (engine block and exhaust system) in the form of heat. If 6.0×10^5 J of work is required to make the car go one mile, how many miles can the car travel on one gallon of gas?

2. In moving out of a dormitory at the end of the semester, a student does 1.6×10^4 J of work. In the process, his internal energy decreases by 4.2×10^4 J. Determine each of the following quantities (including the algebraic sign): (a) W, (b) ΔU, and (c) Q.

3. The work done to compress one mole of a monatomic ideal gas is 6200 J. The temperature of the gas changes from 350 to 550 K. (a) How much heat flows between the gas and its surroundings? (b) Determine whether the heat flows into or out of the gas.

4. A system does 164 J of work on its environment and gains 77 J of heat in the process. Find the change in the internal energy of (a) the system and (b) the environment.

5. ssm The internal energy of a system changes because the system gains 165 J of heat and performs 312 J of work. In returning to its initial state, the system loses 114 J of heat. During this return process, (a) how much work is involved, and (b) is work done by the system or is work done on the system?

*6. ⚕ In exercising, a weight lifter loses 0.150 kg of water through evaporation, the heat required to evaporate the water coming from the weight lifter's body. The work done in lifting weights is 1.40×10^5 J. (a) Assuming that the latent heat of vaporization of perspiration is 2.42×10^6 J/kg, find the change in the internal energy of the weight lifter. (b) Determine the minimum number of nutritional calories of food (1 nutritional calorie = 4186 J) that must be consumed to replace the loss of internal energy.

Section 15.4 Thermal Processes

7. A gas is compressed under isobaric conditions, and its volume changes from 7.0×10^{-3} to 2.0×10^{-3} m^3. The pressure of the gas is 1.5×10^5 Pa. (a) Determine the work done (including the algebraic sign). (b) How much work is done (including the algebraic sign) if the gas then expands from 2.0×10^{-3} to 8.0×10^{-3} m^3 under isobaric conditions?

8. A system undergoes a two-step process. In the first step, the internal energy of the system increases by 228 J when 166 J of work is done on the system. In the second step, the internal energy of the system increases by 115 J when 177 J of work is done

on the system. For the overall process, find the heat. What type of process is the overall process? Explain.

9. ssm The volume of a gas is changed along the curved line between A and B in the drawing. Do not assume that the curved line is an isotherm or that the gas is ideal. (a) Find the magnitude of the work for the process, and (b) determine whether the work is positive or negative.

10. (a) Using the data presented in the accompanying pressure-versus-volume graph, estimate the magnitude of the work done when the system changes from A to B to C along the path shown. (b) Determine whether the work is done by the system or on the system and, hence, whether the work is positive or negative.

11. A gas is contained in a chamber such as that in Figure 15.5. Suppose the region outside the chamber is evacuated and the total mass of the block and the movable piston is 135 kg. When 2050 J of heat flows into the gas, the internal energy of the gas increases by 1730 J. What is the distance s through which the piston rises?

12. The pressure and volume of a gas are changed along the path $ABCA$. Determine the work done (including the algebraic sign) in each segment of the path: (a) A to B, (b) B to C, and (c) C to A.

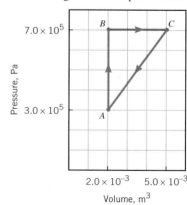

Problem 12

13. ssm When a .22-caliber rifle is fired, the expanding gas from the burning gunpowder creates a pressure behind the bullet. This pressure causes the force that pushes the bullet through the barrel. The barrel has a length of 0.61 m and an opening whose radius is 2.8 $\times 10^{-3}$ m. A bullet (mass = 2.6×10^{-3} kg) has a speed of 370 m/s after passing through this barrel. Ignore friction and determine the average pressure of the expanding gas.

*14. A piece of aluminum has a volume of 1.4×10^{-3} m³. The coefficient of volume expansion for aluminum is $\beta = 69 \times 10^{-6}$ (C°)$^{-1}$. The temperature of this object is raised from 20 to 320 °C. How much work is done by the expanding aluminum if the air pressure is 1.01×10^5 Pa?

*15. ssm www A monatomic ideal gas expands isobarically. Using the first law of thermodynamics, prove that the heat Q is positive, so that it is impossible for heat to flow out of the gas.

*16. Refer to the drawing that accompanies problem 10. When a system changes from A to B along the path shown on the pressure-versus-volume graph, it gains 2700 J of heat. What is the change in the internal energy of the system?

**17. Water is heated in an open pan where the air pressure is one atmosphere. The water remains a liquid, which expands by a small amount as it is heated. Determine the ratio of the work done by the water to the heat absorbed by the water.

Section 15.5 Thermal Processes That Utilize an Ideal Gas

18. The temperature of three moles of an ideal gas is 373 K. How much work does the gas do in expanding isothermally to four times its initial volume?

19. ssm One-half mole of a monatomic ideal gas absorbs 1200 J of heat while 2500 J of work is done by the gas. (a) What is the temperature change of the gas? (b) Is the change an increase or a decrease?

20. Six grams of helium (molecular mass = 4.0 u) expands isothermally at 370 K and does 9600 J of work. Assuming that helium is an ideal gas, determine the ratio of the final volume of the gas to the initial volume.

21. A monatomic ideal gas ($\gamma = \frac{5}{3}$) is compressed adiabatically, and its volume is reduced by a factor of two. Determine the factor by which its pressure increases.

22. One-half a mole of a monatomic ideal gas expands adiabatically and does 610 J of work. By how many kelvins does its temperature change? Specify whether the change is an increase or a decrease.

23. ssm A monatomic ideal gas has an initial temperature of 405 K. This gas expands and does the same amount of work whether the expansion is adiabatic or isothermal. When the expansion is adiabatic, the final temperature of the gas is 245 K. What is the ratio of the final to the initial volume when the expansion is isothermal?

*24. The drawing refers to one mole of a monatomic ideal gas and shows a process that has four steps, two isobaric (A to B, C to D) and two isochoric (B to C, D to A). Complete the following table by calculating ΔU, W, and Q (including the algebraic signs) for each of the four steps.

	ΔU	W	Q
A to B			
B to C			
C to D			
D to A			

*25. A diesel engine does not use spark plugs to ignite the fuel and air in the cylinders. Instead, the temperature required to ignite the fuel occurs because the pistons compress the air in the cylinders. Suppose air at an initial temperature of 27 °C is compressed adiabatically to a temperature of 681 °C. Assume the air to be an ideal gas for which $\gamma = \frac{7}{5}$. Find the compression ratio, which is the ratio of the initial volume to the final volume.

*26. A monatomic ideal gas ($\gamma = \frac{5}{3}$) expands adiabatically, and the ratio of its initial to its final Kelvin temperature is $T_i/T_f = 2.00$. Find the ratio of its initial to its final pressure P_i/P_f.

*27. The pressure and volume of an ideal monatomic gas change from A to B to C, as the drawing shows. The curved line between A and C is an isotherm. (a) Determine the total heat for the process and (b) state whether the flow of heat is into or out of the gas.

**28. The drawing shows an adiabatically isolated cylinder that is divided initially into two identical parts by an adiabatic partition. Both sides contain one mole of a monatomic ideal gas ($\gamma = \frac{5}{3}$), with the initial temperature being 525 K on the left and 275 K on the right. The partition is then allowed to move slowly (i.e., quasi-statically) to the right, until the pressures on each side of the partition are the same. Find the final temperatures on the (a) left and (b) right.

**29. ssm The work done by one mole of a monatomic ideal gas ($\gamma = \frac{5}{3}$) in expanding adiabatically is 825 J. The initial temperature and volume of the gas are 393 K and 0.100 m³. Obtain (a) the final temperature and (b) final volume of the gas.

15.6 Specific Heat Capacities and the First Law of Thermodynamics

30. Argon is a monatomic gas whose atomic mass is 39.9 u. The temperature of eight grams of argon is raised by 75 K under conditions of constant pressure. Assuming that argon is an ideal gas, how much heat is required?

31. ssm How much heat is required to change the temperature of 1.5 mol of a monatomic ideal gas by 77 K if the pressure is held constant?

32. Heat is added to a monatomic ideal gas. Twice as much heat is added at constant pressure as at constant volume. The change in temperature of the gas at constant volume is 35 K. What is the change in temperature at constant pressure?

33. ssm The temperature of 2.5 mol of a monatomic ideal gas is 350 K. The internal energy of this gas is doubled by the addition of heat. How much heat is needed when it is added at (a) constant volume and (b) constant pressure?

34. Three moles of a monatomic ideal gas are heated at a constant volume of 1.50 m³. The amount of heat added is 5.24×10^3 J. (a) What is the change in the temperature of the gas? (b) Find the change in its internal energy. (c) Determine the change in pressure.

35. Heat Q is added to a monatomic ideal gas at constant pressure. As a result, the gas does work W. Find the ratio Q/W.

*36. A monatomic ideal gas is heated at a constant volume of 1.00×10^{-3} m³, using a ten-watt heater. The pressure of the gas increases by 5.0×10^4 Pa. How long was the heater on?

*37. ssm A monatomic ideal gas expands at constant pressure. (a) What percentage of the heat being supplied to the gas is used to increase the internal energy of the gas? (b) What percentage is used for doing the work of expansion?

*38. Suppose that 31.4 J of heat is added to an ideal gas. The gas expands at a constant pressure of 1.40×10^4 Pa while changing its volume from 3.00×10^{-4} to 8.00×10^{-4} m³. The gas is not monatomic, so the relation $C_P = \frac{5}{2}R$ does not apply. (a) Determine the change in the internal energy of the gas. (b) Calculate its molar specific heat capacity C_P.

**39. One mole of neon, a monatomic gas, starts out at conditions of standard temperature and pressure. The gas is heated at constant volume until its pressure is tripled, then further heated at constant pressure until its volume is doubled. Assume that neon behaves as an ideal gas. For the entire process, find the heat added to the gas.

15.8 Heat Engines

40. Due to a tune-up, the efficiency of an automobile engine increases by 5.0%. For an input heat of 1300 J, how much more work does the engine produce after the tune-up than before?

41. ssm In doing 16 600 J of work, an engine rejects 9700 J of heat. What is the efficiency of the engine?

42. The input heat for an engine is 2.41×10^4 J, and the rejected heat is 5.86×10^3 J. Find the work done by the engine.

43. An engine has an efficiency of 64% and produces 5500 J of work. Determine (a) the input heat and (b) the rejected heat.

*44. The amount of heat that an engine rejects to the cold reservoir is three times the amount of work that it does. What is the efficiency of the engine?

**45. ssm www Engine A receives three times more input

heat, produces five times more work, and rejects two times more heat than engine B. Find the efficiency of (a) engine A and (b) engine B.

Section 15.9 Carnot's Principle and the Carnot Engine

46. An engine has a hot reservoir temperature of 950 K and a cold reservoir temperature of 620 K. The engine operates at three-fifths maximum efficiency. What is the efficiency of the engine?

47. Five thousand joules of heat is put into a Carnot engine whose hot and cold reservoirs have temperatures of 500 and 200 K, respectively. How much heat is converted into work?

48. A power-generating plant produces work (energy) with an efficiency of 33%. The temperature of its hot reservoir is 521 K. What is the highest possible value for the temperature of its cold reservoir?

49. ssm A Carnot engine has an efficiency of 0.700, and the temperature of its cold reservoir is 378 K. (a) Determine the temperature of its hot reservoir. (b) If 5230 J of heat is rejected to the cold reservoir, what amount of heat is put into the engine?

50. A Carnot engine produces three times more work than discarded heat. The temperature of its cold reservoir is 292 K. What is the temperature of its hot reservoir?

51. A Carnot engine uses a hot reservoir consisting of a large amount of boiling water and a cold reservoir consisting of a large tub of ice and water. When 6800 J of heat is put into the engine and the engine produces work, how many kilograms of ice in the tub are melted due to the heat delivered to the cold reservoir?

*52. From a hot reservoir at a temperature of T_1, Carnot engine A takes an input heat of 5550 J, delivers 1750 J of work, and rejects heat to a cold reservoir that has a temperature of 503 K. This cold reservoir at 503 K also serves as the hot reservoir for Carnot engine B, which uses the rejected heat of the first engine as input heat. Engine B also delivers 1750 J of work, while rejecting heat to an even colder reservoir that has a temperature of T_2. Find the temperatures (a) T_1 and (b) T_2.

*53. ssm A power plant taps steam superheated by geothermal energy to 505 K (the temperature of the hot reservoir) and uses the steam to do work in turning the turbine of an electric generator. The steam is then converted back into water in a condenser at 323 K (the temperature of the cold reservoir), after which the water is pumped back down into the earth where it is heated again. The output power (work per unit time) of the plant is 84 000 kilowatts. Determine (a) the maximum efficiency at which this plant can operate and (b) the minimum amount of rejected heat that must be removed from the condenser every twenty-four hours.

*54. The hot reservoir for a Carnot engine has a temperature of 890 K, while the cold reservoir has a temperature of 670 K. The heat input for this engine is 4800 J. The 670-K reservoir also serves as the hot reservoir for a second Carnot engine. This second engine uses the rejected heat of the first engine as input and extracts additional work from it. The rejected heat from the second engine goes into a reservoir that has a temperature of 420 K. Find the total work delivered by the two engines.

**55. ssm A nuclear-fueled electric power plant utilizes a so-called "boiling water reactor." In this type of reactor, nuclear energy causes water under pressure to boil at 285 °C (the temperature of the hot reservoir). After the steam does the work of turning the turbine of an electric generator, the steam is converted back into water in a condenser at 40 °C (the temperature of the cold reservoir). To keep the condenser at 40 °C, the rejected heat must be carried away by some means, for example, by water from a river. The plant operates at three-fourths of its Carnot efficiency, and the electrical output power of the plant is 1.2×10^9 watts. A river with a water flow rate of 1.0×10^5 kg/s is available to remove the rejected heat from the plant. Find the number of Celsius degrees by which the temperature of the river rises.

**56. The hot and cold reservoirs of a Carnot engine have temperatures of 845 and 395 K, respectively. The engine does the work of lifting a 15.0-kg block straight up from rest, so that at a height of 5.00 m the block has a speed of 8.50 m/s. How much heat must be put into the engine?

Section 15.10 Refrigerators, Air Conditioners, and Heat Pumps

57. The water in a deep underground well is used as the cold reservoir of a Carnot heat pump that maintains the temperature of a house at 301 K. To deposit 14 200 J of heat in the house, the heat pump requires 800 J of work. Determine the temperature of the well water.

58. The inside of a Carnot refrigerator is maintained at a temperature of 277 K, while the temperature in the kitchen is 299 K. Using 2500 J of work, how much heat can this refrigerator remove from its inside compartment?

59. ssm www The temperatures indoors and outdoors are 299 and 312 K, respectively. A Carnot air conditioner deposits 6.12×10^5 J of heat outdoors. How much heat is removed from the house?

60. The coefficient of performance of a refrigerator is 4.6. How much electrical energy is used in removing 4100 J of heat from the food inside?

61. A heat pump removes 2090 J of heat from the outdoors and delivers 3140 J of heat to the inside of a house. (a) How much work does the heat pump need? (b) What is the coefficient of performance of the heat pump?

62. A Carnot air conditioner maintains the temperature in a house at 297 K on a day when the temperature outside is 311 K. What is the coefficient of performance of the air conditioner?

63. A Carnot engine has an efficiency of 0.70. If this engine were run backward as a heat pump, what would be the coefficient of performance?

*64. Two kilograms of liquid water at 0 °C is put into the freezer compartment of a Carnot refrigerator. The temperature of the compartment is −15 °C, and the temperature of the kitchen is 27 °C. If the cost of electrical energy is ten cents per kilowatt·hour, how much does it cost to make two kilograms of ice at 0 °C?

*65. ssm www A Carnot refrigerator transfers heat from its inside (6.0 °C) to the room air outside (20.0 °C). (a) Find the coefficient of performance of the refrigerator. (b) Determine the magnitude of the minimum work needed to cool 5.00 kg of water from 20.0 to 6.0 °C when it is placed in the refrigerator.

*66. Review Conceptual Example 9 before attempting this problem. A window air conditioner has an average coefficient of performance of 2.0. This unit has been placed on the floor by the bed, in a futile attempt to cool the bedroom. During this attempt 7.6×10^4 J of heat is pulled in the front of the unit. The room is sealed and contains 3800 mol of air. Assuming that the molar specific heat capacity of the air is $C_V = \frac{5}{2}R$, determine the rise in temperature caused by operating the air conditioner in this manner.

**67. ssm A Carnot engine uses hot and cold reservoirs that have temperatures of 1684 and 842 K, respectively. The input heat for this engine is Q_H. The work delivered by the engine is used to operate a Carnot heat pump. The pump removes heat from the 842-K reservoir and puts it into a hot reservoir at a temperature T'. The amount of heat removed from the 842-K reservoir is also Q_H. Find the temperature T'.

Section 15.11 Entropy and the Second Law of Thermodynamics

68. Four kilograms of carbon dioxide sublimes from solid "dry ice" to a gas at a pressure of one atmosphere and a temperature of 194.7 K. The latent heat of sublimation is 5.77×10^5 J/kg. Find the change in entropy of the carbon dioxide.

69. The inside of a house is at 20.0 °C, and the outside is at −15 °C. The house loses 1.30×10^5 J of heat. Find $\Delta S_{universe}$, the change in entropy of the universe.

70. Heat Q flows spontaneously from a reservoir at 394 K into a reservoir that has a lower temperature T. Because of the spontaneous flow, thirty percent of Q is rendered unavailable for work when a Carnot engine operates between the reservoir at temperature T and a reservoir at 248 K. Find the temperature T.

71. ssm A process occurs in which the entropy of a system increases by 125 J/K. During the process, the energy that becomes unavailable for doing work is zero. (a) Is this process reversible or irreversible? Give your reasoning. (b) Determine the change in the entropy of the surroundings.

*72. (a) Find the equilibrium temperature that results when one kilogram of liquid water at 373 K is added to two kilograms of liquid water at 283 K in a perfectly insulated container. (b) When heat is added to or removed from a solid or liquid of mass m and specific heat capacity c, the change in entropy can be shown to be $\Delta S = mc \; \ln(T_f/T_i)$, where T_i and T_f are the initial and final Kelvin temperatures. Use this equation to calculate the entropy change for each amount of water. Then combine the two entropy changes algebraically to obtain the total entropy change of the universe. Note that the process is irreversible, so the total entropy change of the universe is greater than zero. (c) Assuming that the coldest reservoir at hand has a temperature of 273 K, determine the amount of energy that becomes unavailable for doing work because of the irreversible process.

*73. (a) Five kilograms of water at 80.0 °C is mixed in a perfect thermos with 2.00 kg of ice at 0.0 °C, and the mixture is allowed to reach equilibrium. Using the expression $\Delta S = mc \ln(T_f/T_i)$ [see problem 72] and the change in entropy for melting, find the change in entropy that occurs. (b) Should the entropy of the universe increase or decrease as a result of the mixing process? Give your reasoning and state whether your answer in part (a) is consistent with your answer here.

ADDITIONAL PROBLEMS

74. Five moles of a monatomic ideal gas expand adiabatically, and its temperature decreases from 370 to 290 K. Determine (a) the work done (including the algebraic sign) by the gas, and (b) the change in its internal energy.

75. ssm Three moles of an ideal gas are compressed from 5.5×10^{-2} to 2.5×10^{-2} m^3. During the compression, 6.1×10^3 J of work is done on the gas, and heat is removed to keep the temperature of the gas constant at all times. Find (a) ΔU, (b) Q, and (c) the temperature of the gas.

76. Engine A discards 72% of its input heat into a cold reservoir. Engine B has twice the efficiency as engine A. What percentage of its input heat does engine B discard?

77. The specific heat capacity of a material is 1100 J/(kg·C°). The temperature of 2.0 kg of this solid material is raised by 6.0 C°. Ignoring the work that corresponds to the small change in the volume of the material, determine the change in the internal energy of the material.

78. A Carnot refrigerator maintains the food inside it at 276 K, while the temperature of the kitchen is 298 K. The refrigerator removes 3.00×10^4 J of heat from the food. How much heat is delivered to the kitchen?

79. ssm Find the change in entropy of the H$_2$O molecules when (a) three kilograms of ice melts into water at 273 K and (b) three kilograms of water changes into steam at 373 K. (c) On the basis of the answers to parts (a) and (b), discuss which change creates more disorder in the collection of H$_2$O molecules.

80. The temperature of 2.5 mol of helium (a monatomic gas) is lowered by 35 K under conditions of constant volume. Assuming that helium behaves as an ideal gas, how much heat is removed from the gas?

81. A Carnot heat pump operates between an outdoor temperature of 265 K and an indoor temperature of 298 K. Find its coefficient of performance.

82. A Carnot engine operates between temperatures of 650 and 350 K. To improve the efficiency of the engine, it is decided either to raise the temperature of the hot reservoir by 40 K or to lower the temperature of the cold reservoir by 40 K. Which change gives the greatest improvement? Justify your answer by calculating the efficiency in each case.

83. ssm A system gains 1500 J of heat, while the internal energy of the system increases by 4500 J and the volume decreases by 0.010 m^3. Assume the pressure is constant and find its value.

*84. Refer to the drawing in problem 9, where the curve between A and B is now an isotherm. An ideal gas begins at A and is changed along the horizontal line from A to C and then along the vertical line from C to B. (a) Find the heat for the process ACB and (b) determine whether it flows into or out of the gas.

*85. How long would a 3.00-kW space heater have to run to put into a kitchen the same amount of heat as a refrigerator (coefficient of performance = 3.00) does when it freezes 1.50 kg of water at 20.0 °C into ice at 0.0 °C?

*86. When a monatomic ideal gas expands at a constant pressure of 2.6×10^5 Pa, the volume of the gas increases by 6.2×10^{-3} m^3. (a) Determine the heat that flows into or out of the gas. (b) Specify the direction of the flow.

*87. ssm A monatomic ideal gas ($\gamma = \frac{5}{3}$) is contained within a perfectly insulated cylinder that is fitted with a movable piston. The initial pressure of the gas is 1.50×10^5 Pa. The piston is pushed so as to compress the gas, with the result that the Kelvin temperature doubles. What is the final pressure of the gas?

*88. Suppose a monatomic ideal gas is contained within a vertical cylinder that is fitted with a movable piston. The piston is frictionless and has a negligible mass. The area of the piston is 3.14×10^{-2} m^2, and the pressure outside the cylinder is 1.01×10^5 Pa. Heat (2093 J) is removed from the gas. Through what distance does the piston drop?

*89. ⚖ Even at rest, the human body generates heat. The heat arises because of the body's metabolism, that is, the chemical reactions that are always occurring in the body to generate energy. In rooms designed for use by large groups, adequate ventilation or air conditioning must be provided to remove this heat. Consider a classroom containing 200 students. Assume that the metabolic rate of generating heat is 130 W for each student and that the heat accumulates during a fifty-minute lecture. In addition, assume that the air has a molar specific heat of $C_V = \frac{5}{2}R$ and that the room (volume = 1200 m^3, initial pressure = 1.01×10^5 Pa, and initial temperature = 21 °C) is sealed shut. If all the heat generated by the students were absorbed by the air, by how much would the air temperature rise during a lecture?

*90. Suppose the gasoline in a car engine burns at 631 °C, while the exhaust temperature (the temperature of the cold reservoir) is 139 °C and the outdoor temperature is 27 °C. Assume that the engine can be treated as a Carnot engine (a gross oversimplification). In an attempt to increase mileage performance, an inventor builds a second engine that functions between the exhaust and outdoor temperatures and uses the exhaust heat to produce additional work. Assume that the inventor's engine can also be treated as a Carnot engine. Determine the ratio of the total work produced by both engines to that produced by the first engine alone.

**91. ssm An engine has an efficiency e_1. The engine takes input heat Q_H from a hot reservoir and delivers work W_1. The heat rejected by this engine is used as input heat for a second engine, which has an efficiency e_2 and delivers work W_2. The overall efficiency of this two-engine device is the total work delivered $(W_1 + W_2)$ divided by the input heat Q_H. Find an expression for the overall efficiency e in terms of e_1 and e_2.

**92. One mole of a monatomic ideal gas has an initial pressure, volume, and temperature of P_0, V_0, and 438 K, respectively. It undergoes an isothermal expansion that triples the volume of the gas. Then, the gas undergoes an isobaric compression back to its original volume. Finally, the gas undergoes an isochoric increase in pressure, so that the final pressure, volume, and temperature are P_0, V_0, and 438 K, respectively. Find the total heat for this three-step process, and state whether it is absorbed by or given off by the gas.

CONCEPTS CALCULATIONS

GROUP LEARNING PROBLEMS

Note: Each of these problems consists of Concept Questions followed by a related quantitative Problem. They are designed for use by students working alone or in small learning groups. The Concept Questions involve little or no mathematics and are intended to stimulate group discussions. They focus on the concepts with which the problems deal. Recognizing the concepts is the essential initial step in any problem-solving technique.

93. Concept Questions A system does 4.8×10^4 J of work, and 7.6×10^4 J of heat flows into the system in the process. (a) Considered by itself, does the work increase or decrease the internal energy of the system? (b) Considered by itself, does the heat increase or decrease the internal energy? (c) Considering the work and heat together, does the internal energy of the system increase, decrease, or remain the same? Explain.

Problem Find the change in the internal energy of the system. Make sure that your answer is consistent with your answers to the Concept Questions.

94. Concept Questions A system gains a certain amount of energy in the form of heat at constant pressure, and the internal energy of the system increases by an even greater amount. (a) Is any work done and, if so, is it done on or by the system? (b) If there is work, is it positive or negative, according to our convention? (c) Does the volume of the system increase, decrease, or remain the same? Give your reasoning.

Problem A system gains 2780 J of heat at a constant pressure of 1.26×10^5 Pa, and its internal energy increases by 3990 J. What is the change in volume of the system, and is it an increase or a decrease? Verify that your answer is consistent with your answers to the Concept Questions.

95. Concept Questions An ideal gas expands isothermally, doing work, while heat flows into the gas. (a) Does the internal energy of the gas increase, decrease, or remain the same? (b) Is the work done by the gas greater than, less than, or equal to the heat that flows into the gas? Account for your answers.

Problem Three moles of neon expand isothermally from 0.100 to 0.250 m^3. Into the gas flows 4.75×10^3 J of heat. Assuming that neon is an ideal gas, find its temperature.

96. Concept Questions A mountain climber, starting from rest, does work in climbing upward. At the top, she is again at rest. In the process, her body generates 4.1×10^6 J of energy via metabolic processes. In fact, her body acts like a heat engine, the efficiency of which is given by Equation 15.11 as $e = W/Q_H$, where W is the work and Q_H is the input heat. (a) Is the 4.1×10^6 J of energy equal to W or Q_H? (b) How is the work done in climbing upward related to the vertical height of the climb? Explain.

Problem The vertical height of the climb is 730 m. The climber has a mass of 52 kg. Find her efficiency as a heat engine.

97. Concept Questions Two Carnot engines, A and B, utilize the same hot reservoir, but engine A is less efficient than engine B. (a) Which engine produces more work for a given heat input? (b) Which engine has the lower cold-reservoir temperature? Give your reasoning.

Problem Carnot engine A has an efficiency of 0.60, while Carnot engine B has an efficiency of 0.80. Both engines utilize the same hot reservoir, which has a temperature of 650 K and delivers 1200 J of heat to each engine. Find the work produced by each engine and the temperatures of the cold reservoirs that they use. Check to see that your answers are consistent with your answers to the Concept Questions.

98. Concept Questions Two Carnot air conditioners, A and B, are removing heat from different rooms. The outside temperature is the same for both, but the room temperatures are different. The room serviced by unit A is kept colder than the room serviced by unit B. The heat removed from both rooms is the same. (a) Which unit requires the greater amount of work? (b) Which unit deposits the greater amount of heat outside? Explain.

Problem The outside temperature is 309.0 K. The room serviced by unit A is kept at a temperature of 294.0 K, while the room serviced by unit B is kept at 301.0 K. The heat removed from either room is 4330 J. For both units, find the work required and find the heat deposited outside. Verify that your answers are consistent with your answers to the Concept Questions.

Having relatively large ears enhances the ability of the bat-eared fox to hear weak sounds.

CHAPTER 16

WAVES AND SOUND

16.1 THE NATURE OF WAVES

Water waves have two features common to all waves:

1. A wave is a traveling disturbance.
2. A wave carries energy from place to place.

In Figure 16.1 the wave created by the motorboat travels across the lake and disturbs the fisherman. However, *there is no bulk flow of water* outward from the motorboat. The wave is not a bulk movement of water such as a river, but, rather, a disturbance traveling on the surface of the lake. Part of the wave's energy in Figure 16.1 is transferred to the fisherman and his boat.

We will consider two basic types of waves, transverse and longitudinal. Figure 16.2 illustrates how a transverse wave can be generated using a Slinky, a remarkable toy in the form of a long, loosely coiled spring. If one end of the Slinky is jerked up and down, as in part *a*, an upward pulse is sent traveling toward the right. If the end is then jerked down and up, as in part *b*, a downward pulse is generated and also moves to the right. If the end is continually moved up and down in simple harmonic motion, an entire wave is produced. As part *c* illustrates, the wave consists of a series of alternating upward and downward sections that propagate to the right, disturbing the vertical position of the Slinky in the process. To focus attention on the disturbance, a colored dot is attached to the Slinky in part *c* of the drawing. As the wave advances, the dot is displaced up and down in simple harmonic motion. The motion of the dot occurs perpendicular, or transverse, to the direction in which the wave travels. Thus, *a transverse wave is one in which the disturbance is perpendicular to the direction of travel of the wave.* Radio waves, light waves, and microwaves are transverse waves. Transverse waves also travel on the strings of instruments such as guitars and banjos.

A longitudinal wave can also be generated with a Slinky, and Figure 16.3 demonstrates how. When one end of the Slinky is pushed forward along its length (i.e., longitudinally) and then returned to its starting point, as in part *a*, a region where the coils are squeezed together or compressed is sent traveling to the right. If the end is pulled backward and then returned to its starting point, as in part *b*, a region where the coils are pulled apart or stretched is formed and also moves to the right. If the end is continually moved back and forth in simple harmonic motion, an entire wave is created. As part *c* shows, the wave consists of a series of alternating

Figure 16.1 The wave created by the motorboat travels across the lake and disturbs the fisherman.

Figure 16.2 (*a*) An upward pulse moves to the right, followed by (*b*) a downward pulse. (*c*) When the end of the Slinky is moved up and down continuously, a transverse wave is produced.

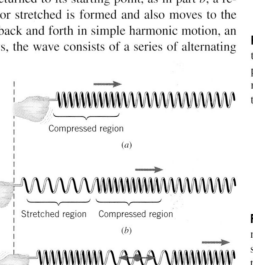

Compressed region

(*a*)

Stretched region   Compressed region

(*b*)

(*c*)

Figure 16.3 (*a*) A compressed region moves to the right, followed by (*b*) a stretched region. (*c*) When the end of the Slinky is moved back and forth continuously, a longitudinal wave is produced.

Figure 16.4 A water wave is neither transverse nor longitudinal, since water particles at the surface move clockwise on nearly circular paths as the wave moves from left to right.

Figure 16.5 *Concepts at a Glance* For periodic waves, the terms *cycle, amplitude, period,* and *frequency* have the same meaning as they do in simple harmonic motion. Sound, such as that produced by this clarinet, is a periodic wave and is described using this terminology.

compressed and stretched regions that travel to the right and disturb the separation between adjacent coils. A colored dot is once again attached to the Slinky to emphasize the vibratory nature of the disturbance. In response to the wave, the dot moves back and forth in simple harmonic motion along the line of travel of the wave. Thus, *a longitudinal wave is one in which the disturbance is parallel to the line of travel of the wave.* A sound wave is a longitudinal wave.

Some waves are neither transverse nor longitudinal. For instance, in a water wave the motion of the water particles is not strictly perpendicular or strictly parallel to the line along which the wave travels. Instead, the motion includes both transverse and longitudinal components, since the water particles at the surface move on nearly circular paths, as Figure 16.4 indicates.

16.2 PERIODIC WAVES

● *Concepts at a Glance* The transverse and longitudinal waves that we have been discussing are called *periodic waves* because they consist of cycles or patterns that are produced over and over again by the source. In Figures 16.2 and 16.3 the repetitive patterns occur as a result of the simple harmonic motion of the left end of the Slinky. Every segment of the Slinky vibrates in simple harmonic motion. Sections 10.1 and 10.2 discuss the simple harmonic motion of an object on a spring and introduce the concepts of cycle, amplitude, period, and frequency. As the Concepts-at-a-Glance chart in Figure 16.5 illustrates, this same terminology is used to describe periodic waves, such as the sound waves we hear and the light waves we see (discussed in Chapter 24). ●

Figure 16.6 uses a graphical representation of a transverse wave to review this terminology. One *cycle* of a wave is shaded in color in both parts of the drawing. A wave is a series of many cycles. In part *a* the vertical position of the Slinky is plotted on the vertical axis, while the corresponding distance along the length of the Slinky is plotted on the horizontal axis. Such a graph is equivalent to a photograph of the wave taken at one instant in time and shows the disturbance that exists at each point along the Slinky's length. As marked on this graph, the **amplitude** A is the maximum excursion of a particle of the medium from the particle's undisturbed

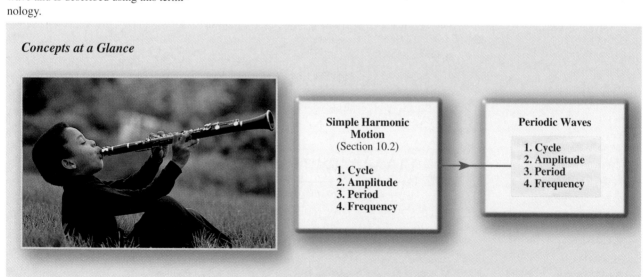

Concepts at a Glance

Simple Harmonic Motion (Section 10.2)	Periodic Waves
1. Cycle	1. Cycle
2. Amplitude	2. Amplitude
3. Period	3. Period
4. Frequency	4. Frequency

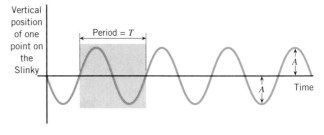

Vertical position of the Slinky

Wavelength = λ

Distance

Undisturbed position

A

(a) At a particular time

Vertical position of one point on the Slinky

Period = T

Time

A

(b) At a particular location

Figure 16.6 One cycle of the wave is shaded in color, and the amplitude of the wave is denoted as A.

position. The amplitude is the distance between a crest, or highest point on the wave pattern, and the undisturbed position; it is also the distance between a trough, or lowest point on the wave pattern, and the undisturbed position. The **wavelength** λ is the horizontal length of one cycle of the wave, as shown in Figure 16.6a. The wavelength is also the horizontal distance between two successive crests, two successive troughs, or any two successive equivalent points on the wave.

Part b of Figure 16.6 shows a graph in which time, rather than distance, is plotted on the horizontal axis. This graph is obtained by observing a single point on the Slinky. As the wave passes, the point under observation oscillates up and down in simple harmonic motion. As indicated on the graph, the **period** T is the time required for one complete up/down cycle, just as it is for an object vibrating on a spring. Equivalently, the period is the time required for the wave to travel a distance of one wavelength. The period T is related to the **frequency** f, just as it is for any example of simple harmonic motion:

$$f = \frac{1}{T} \tag{10.5}$$

The period is commonly measured in seconds, and frequency is measured in cycles per second or hertz (Hz). If, for instance, one cycle of a wave takes one-tenth of a second to pass an observer, then ten cycles pass per second, as Equation 10.5 indicates [$f = 1/(0.1 \text{ s}) = 10 \text{ cycles/s} = 10 \text{ Hz}$].

A simple relation exists between the period, the wavelength, and the speed of a wave, a relation that Figure 16.7 helps to introduce. Imagine waiting at a railroad crossing, while a freight train moves by at a constant speed v. The train consists of a long line of identical boxcars, each of which has a length λ and requires a time T to pass, so the speed is $v = \lambda/T$. This same equation applies for a wave and relates the speed of the wave to the wavelength λ and the period T. Since the frequency of a wave is $f = 1/T$, the expression for the speed is

$$v = \frac{\lambda}{T} = f\lambda \tag{16.1}$$

The time for one car to pass is the period T

Velocity = $\mathbf{v}$

Wavelength = λ

Figure 16.7 A train moving at a constant speed serves as an analogy for a traveling wave.

The terminology just discussed and the fundamental relations $f = 1/T$ and $v = f\lambda$ apply to longitudinal as well as to transverse waves. Example 1 illustrates how the wavelength of a wave is determined by the wave speed and the frequency established by the source.

EXAMPLE 1 The Wavelengths of Radio Waves

AM and FM radio waves are transverse waves that consist of electric and magnetic disturbances. These waves travel at a speed of 3.00×10^8 m/s. A station broadcasts an AM radio wave whose frequency is 1230×10^3 Hz (1230 kHz on the dial) and an

FM radio wave whose frequency is 91.9×10^6 Hz (91.9 MHz on the dial). Find the distance between adjacent crests in each wave.

Reasoning The distance between adjacent crests is the wavelength λ. Since the speed of each wave is $v = 3.00 \times 10^8$ m/s and the frequencies are known, the relation $v = f\lambda$ can be used to determine the wavelengths.

Solution

• **PROBLEM SOLVING INSIGHT**
The equation $v = f\lambda$ applies to any kind of periodic wave.

AM
$$\lambda = \frac{v}{f} = \frac{3.00 \times 10^8 \text{ m/s}}{1230 \times 10^3 \text{ Hz}} = \boxed{244 \text{ m}}$$

FM
$$\lambda = \frac{v}{f} = \frac{3.00 \times 10^8 \text{ m/s}}{91.9 \times 10^6 \text{ Hz}} = \boxed{3.26 \text{ m}}$$

Notice that the wavelength of an AM radio wave is longer than two and one-half football fields!

16.3   THE SPEED OF A WAVE ON A STRING

The properties of the material* or medium through which a wave travels determine the speed of the wave. For example, Figure 16.8 shows a transverse wave on a string and draws attention to four string particles that have been drawn as colored dots. As the wave moves to the right, each particle is displaced, one after the other, from its undisturbed position. In the drawing, particles 1 and 2 have already been displaced upward, while particles 3 and 4 are not yet affected by the wave. Particle 3 will be next to move because the section of string immediately to its left (i.e., particle 2) will pull it upward.

Figure 16.8 leads us to conclude that the speed with which the wave moves to the right depends on how quickly one particle of the string is accelerated upward in response to the net pulling force exerted by its adjacent neighbors. In accord with Newton's second law, a stronger net force results in a greater acceleration, and, thus, a faster-moving wave. The ability of one particle to pull on its neighbors depends on how tightly the string is stretched, that is, on the tension (see Section 4.10 for a review of tension). The greater the tension, the greater the pulling force the particles exert on each other, and the faster the wave travels, other things being equal. Along with the tension, a second factor influences the wave speed. According to Newton's second law, the inertia or mass of particle 3 in Figure 16.8 also affects how quickly it responds to the upward pull of particle 2. For a given net pulling force, a smaller mass has a greater acceleration than a larger mass. Therefore, other things being equal, a wave travels faster on a string whose particles have a small mass, or, as it turns out, on a string that has a small mass per unit length. The mass per unit length is called the *linear density* of the string. It is the mass m of the string divided by its length L, or m/L. The effects of the tension F and the mass per unit

Figure 16.8 As a transverse wave moves to the right with speed v, each string particle is displaced, one after the other, from its undisturbed position.

---

* Electromagnetic waves (discussed in Chapter 24) can move through a vacuum, as well as through materials such as glass and water.

length are evident in the following expression for the speed v of a small-amplitude wave on a string:

$$v = \sqrt{\frac{F}{m/L}} \tag{16.2}$$

The motion of transverse waves along a string is important in the operation of musical instruments, such as the guitar, the violin, and the piano. In these instruments, the strings are either plucked, bowed, or struck to produce transverse waves. Example 2 discusses the speed of the waves on the strings of a guitar.

EXAMPLE 2   Waves Traveling on Guitar Strings

Transverse waves travel on the strings of an electric guitar after the strings are plucked (see Figure 16.9). The length of each string between its two fixed ends is 0.628 m, and the mass is 0.208 g for the highest pitched E string and 3.32 g for the lowest pitched E string. Each string is under a tension of 226 N. Find the speeds of the waves on the two strings.

Reasoning The speed of a wave on a guitar string, as expressed by Equation 16.2, depends on the tension F in the string and its linear density m/L. Since the tension is the same for both strings, and smaller linear densities give rise to greater speeds, we expect the wave speed to be greatest on the string with the smallest linear density.

Solution The speeds of the waves are given by Equation 16.2 as

High-pitched E $v = \sqrt{\dfrac{F}{m/L}} = \sqrt{\dfrac{226\ \text{N}}{(0.208 \times 10^{-3}\ \text{kg})/(0.628\ \text{m})}} = \boxed{826\ \text{m/s}}$

Low-pitched E $v = \sqrt{\dfrac{F}{m/L}} = \sqrt{\dfrac{226\ \text{N}}{(3.32 \times 10^{-3}\ \text{kg})/(0.628\ \text{m})}} = \boxed{207\ \text{m/s}}$

Notice how fast the waves move; the speeds correspond to 1850 and 463 mi/h.

Conceptual Example 3 offers additional insight into the nature of a wave as a traveling disturbance.

CONCEPTUAL EXAMPLE 3   Wave Speed Versus Particle Speed

Is the speed of a transverse wave on a string the same as the speed at which a particle on the string moves (see Figure 16.10)?

Reasoning and Solution The speed v_{wave} of a transverse wave on a string specifies how fast the disturbance travels along the string. According to Equation 16.2, this speed is determined by the tension F and the linear density m/L of the string. Thus, the speed of the wave depends solely on the *properties of the string*, and, if these properties do not change, the speed remains constant.

The particle speed v_{particle} specifies how fast the particle is moving as it oscillates up and down, and it is different from the wave speed. If the source of the wave (e.g., the hand in Figure 16.2c) vibrates in simple harmonic motion, each string particle vibrates in a like manner, with the same amplitude and frequency as the source. Moreover, the particle speed, unlike the wave speed, is not constant. As for any object in simple harmonic motion, the particle speed is greatest when the particle is passing through the undisturbed position of the string and zero when the particle is at its maximum displacement. According to Equation 10.7, the particle speed depends on the amplitude A, the angular frequency ω, and the time t through the relation $v_{\text{particle}} = A\omega \sin \omega t$, where the minus sign has been omitted because we are interested only in the speed, which is the magnitude of the particle's velocity. Thus, the

THE PHYSICS OF...

waves on guitar strings.

Transverse vibration of the string

Figure 16.9 Transverse waves are generated on a guitar string by plucking it.

v_{particle}

String particle

v_{wave}

Undisturbed position of string

Figure 16.10 A transverse wave on a string is moving to the right with a constant speed v_{wave}. A string particle moves up and down in simple harmonic motion about the undisturbed position of the string. The speed of the particle v_{particle} changes from moment to moment as the wave passes.

speed of a string particle is determined by the *properties of the source* creating the wave and not on the properties of the string itself. In contrast, the speed of the wave is determined by the properties of the string. We see, then, that **the two speeds, v_{wave} and $v_{particle}$, are not the same.**

Related Homework Material: *Question 4, Problems 18 and 98*

*16.4   THE MATHEMATICAL DESCRIPTION OF A WAVE

When a wave travels through a medium, it displaces the particles of the medium from their undisturbed positions. Suppose a particle is located at a distance x from a coordinate origin. We would like to know the displacement y of this particle from its undisturbed position at any time t as the wave passes. For periodic waves that result from simple harmonic motion of the source, the expression for the displacement involves a sine or cosine, a fact that is not surprising. After all, in Chapter 10 simple harmonic motion is described using sinusoidal equations, and the graphs for a wave in Figure 16.6 look like a plot of displacement versus time for an object oscillating on a spring (see Figure 10.6).

Our tack will be to present the expression for the displacement and then show graphically that it gives a correct description. Equation 16.3 represents the displacement of a particle caused by a wave traveling in the $+x$ direction (to the right), with an amplitude A, frequency f, and wavelength λ. Equation 16.4 applies to a wave moving in the $-x$ direction (to the left).

Wave motion toward $+x$ $$y = A \sin\left(2\pi ft - \frac{2\pi x}{\lambda}\right)$$ (16.3)

Wave motion toward $-x$ $$y = A \sin\left(2\pi ft + \frac{2\pi x}{\lambda}\right)$$ (16.4)

These equations apply to transverse or longitudinal waves and assume that $y = 0$ m when $x = 0$ m and $t = 0$ s.

Consider a transverse wave moving in the $+x$ direction along a string. The term $(2\pi ft - 2\pi x/\lambda)$ in Equation 16.3 is called the *phase angle* of the wave. A string particle located at the origin ($x = 0$ m) exhibits simple harmonic motion with a phase angle of $2\pi ft$; that is, its displacement as a function of time is $y = A \sin(2\pi ft)$. A particle located at a distance x also exhibits simple harmonic motion, but its phase angle is

$$2\pi ft - \frac{2\pi x}{\lambda} = 2\pi f\left(t - \frac{x}{f\lambda}\right) = 2\pi f\left(t - \frac{x}{v}\right)$$

The quantity x/v is the time needed for the wave to travel the distance x. In other words, the simple harmonic motion that occurs at x is delayed by the time interval x/v compared to the motion at the origin.

Figure 16.11 shows the displacement y plotted as a function of position x along the string at a series of time intervals separated by one-fourth of the period $T(t = 0 \text{ s}, \frac{1}{4}T, \frac{2}{4}T, \frac{3}{4}T, T)$. These graphs are constructed by substituting the corresponding value for t into Equation 16.3, remembering that $f = 1/T$, and then calculating y at a series of values for x. The graphs are like photographs taken at various times as the wave moves to the right. For reference, the colored square on each graph marks the place on the wave that is located at $x = 0$ m when $t = 0$ s. As time

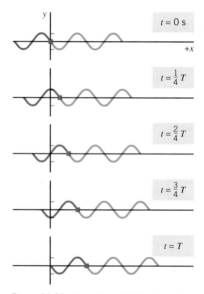

Figure 16.11 Equation 16.3 is plotted here at a series of times separated by one-fourth of the period T. The colored square in each of the graphs marks the place on the wave that is located at $x = 0$ m when $t = 0$ s. As time passes, the wave moves to the right.

passes, the colored square moves to the right, along with the wave. Note that the phase angle $(2\pi ft - 2\pi x/\lambda)$ is measured in *radians,* not degrees. **When a calculator is used to evaluate the function sin $(2\pi ft - 2\pi x/\lambda)$, it must be set to its radian mode.** In a similar manner, it can be shown that Equation 16.4 represents a wave moving in the $-x$ direction.

16.5 THE NATURE OF SOUND

Longitudinal Sound Waves

Sound is a longitudinal wave that is created by a vibrating object, such as a guitar string, the human vocal cords, or the diaphragm of a loudspeaker. Moreover, sound can be created or transmitted only in a medium, such as a gas, liquid, or solid. As we will see, the particles of the medium must be present for the disturbance of the wave to move from place to place. Sound cannot exist in a vacuum.

To see how sound waves are produced and why they are longitudinal, consider the vibrating diaphragm of a loudspeaker. When the diaphragm moves outward, it compresses the air directly in front of it, as in Figure 16.12a. This compression causes the air pressure to rise slightly. The region of increased pressure is called a *condensation,* and it travels away from the speaker at the speed of sound. The condensation is analogous to the compressed region of coils in a longitudinal wave on a Slinky, which is included in Figure 16.12a for comparison. After producing a condensation, the diaphragm reverses its motion and moves inward, as in part *b* of the drawing. The inward motion produces a region known as a *rarefaction,* where the air pressure is slightly less than normal. The rarefaction is similar to the stretched region of coils in a longitudinal Slinky wave. Following immediately behind the condensation, the rarefaction also travels away from the speaker at the speed of sound. Figure 16.13 further emphasizes the similarity between a sound wave and a longitudinal Slinky wave. As the wave passes, the colored dots attached both to the Slinky and to an air molecule execute simple harmonic motion about their undisturbed positions. The colored arrows on either side of the dots indicate that the simple harmonic motion occurs parallel to the line of travel. The drawing also shows that the wavelength λ is the distance between the centers of two successive condensations; λ is also the distance between the centers of two successive rarefactions.

THE PHYSICS OF...

how a loudspeaker diaphragm produces sound.

(a)

(b)

Figure 16.12 (*a*) When the speaker diaphragm moves outward, it creates a condensation. (*b*) When the diaphragm moves inward, it creates a rarefaction. The condensation and rarefaction on the Slinky are included for comparison. In reality, the velocity of the wave on the Slinky v_{Slinky} is much smaller than the velocity of sound in air **v**. For simplicity, the two waves are shown here to have the same velocity.

Figure 16.13 Both the wave on the Slinky and the sound wave are longitudinal waves. The colored dots attached to the Slinky and to an air molecule vibrate back and forth parallel to the line of travel of the wave.

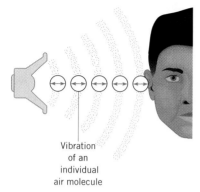

Vibration
of an
individual
air molecule

Figure 16.14 Although the condensations and rarefactions travel from the speaker to the listener, the individual air molecules do not move with the wave. A given molecule vibrates back and forth about a fixed location.

Figure 16.14 illustrates a sound wave spreading out in space after being produced by a loudspeaker. When the condensations and rarefactions arrive at the ear, they force the eardrum to vibrate at the same frequency as the speaker diaphragm. The vibratory motion of the eardrum is interpreted by the brain as sound. It should be emphasized that sound is not a mass movement of air, like the wind. As the condensations and rarefactions of the sound wave travel outward from the vibrating diaphragm in Figure 16.14, the individual air molecules are not carried along with the wave. Rather, each molecule executes simple harmonic motion about a fixed location. In doing so, one molecule collides with its neighbor and passes the condensations and rarefactions forward. The neighbor, in turn, repeats the process.

The Frequency of a Sound Wave

Each cycle of a sound wave includes one condensation and one rarefaction, and the *frequency* is the number of cycles per second that passes by a given location. For example, if the diaphragm of a speaker vibrates back and forth in simple harmonic motion at a frequency of 1000 Hz, then 1000 condensations, each followed by a rarefaction, are generated every second, thus forming a sound wave whose frequency is also 1000 Hz. A sound with a single frequency is called a *pure tone.* Experiments have shown that a healthy young person hears all sound frequencies from approximately 20 to 20 000 Hz (20 kHz). The ability to hear the high frequencies decreases with age, however, and a normal middle-aged adult hears frequencies only up to 12–14 kHz.

Pure tones are used in push-button telephones, such as that shown in Figure 16.15. These phones simultaneously produce two pure tones when each button is pressed, a different pair of tones for each different button. The tones are transmitted electronically to the central telephone office, where they activate switching circuits that complete the call. For example, the drawing indicates that pressing the "5" button produces pure tones of 770 and 1336 Hz simultaneously, while the "9" button generates tones of 852 and 1477 Hz.

Sound can be generated whose frequency lies below 20 Hz or above 20 kHz, although humans normally do not hear it. Sound waves with frequencies below 20 Hz are said to be *infrasonic,* while those with frequencies above 20 kHz are referred to as *ultrasonic.* Rhinoceroses use infrasonic frequencies as low as 5 Hz to call one another (Figure 16.16), while bats use ultrasonic frequencies up to 100 kHz for locating their food sources and navigating (Figure 16.17).

THE PHYSICS OF...

push-button telephones.

Figure 16.15 A push-button telephone and a schematic showing the two pure tones produced when each button is pressed.

Figure 16.16 Rhinoceroses call to one another using infrasonic sound waves. Photo courtesy of San Diego Zoo.

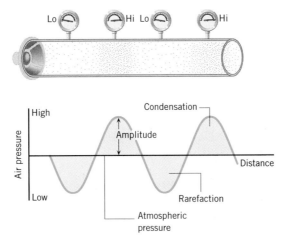

Figure 16.17 Bats use ultrasonic sound waves for navigating and locating food sources.

Frequency is an objective property of a sound wave because frequency can be measured with an electronic frequency counter. A listener's perception of frequency, however, is subjective. The brain interprets the frequency detected by the ear primarily in terms of the subjective quality called *pitch.* A pure tone with a large (high) frequency is interpreted as a high-pitched sound, while a pure tone with a small (low) frequency is interpreted as a low-pitched sound. A piccolo produces high-pitched sounds, and a tuba produces low-pitched sounds.

The Pressure Amplitude of a Sound Wave

Figure 16.18 illustrates a pure-tone sound wave traveling in a tube. Attached to the tube is a series of gauges that indicate the pressure variations along the wave. The graph shows that the air pressure varies sinusoidally along the length of the tube. Although this graph has the appearance of a transverse wave, remember that the sound itself is a longitudinal wave. The graph also shows the *pressure amplitude* of the wave, which is the magnitude of the maximum change in pressure, measured relative to the undisturbed or atmospheric pressure. The pressure fluctuations in a

Figure 16.18 A sound wave is a series of alternating condensations and rarefactions. The graph shows that the condensations are regions of higher-than-normal air pressure, and the rarefactions are regions of lower-than-normal air pressure.

TABLE 16.1   Speed of Sound in Gases, Liquids, and Solids

Substance	Speed (m/s)
Gases	
Air (0 °C)	331
Air (20 °C)	343
Carbon	
dioxide (0 °C)	259
Oxygen (0 °C)	316
Helium (0 °C)	965
Liquids	
Chloroform (20 °C)	1004
Ethanol (20 °C)	1162
Mercury (20 °C)	1450
Fresh water (20 °C)	1482
Seawater (20 °C)	1522
Solids	
Copper	5010
Glass (Pyrex)	5640
Lead	1960
Steel	5960

Figure 16.19   An ultrasonic ruler uses sound with a frequency greater than 20 kHz to measure the distance x to the wall. The blue arcs and blue arrow denote the outgoing sound wave, while the red arcs and red arrow denote the wave reflected from the wall.

sound wave are normally very small. For instance, in a typical conversation between two people the pressure amplitude is about 3×10^{-2} Pa, certainly a small amount compared with the atmospheric pressure of $1.01 \times 10^{+5}$ Pa. The ear is remarkable in being able to detect such small changes.

Loudness is an attribute of sound that depends primarily on the amplitude of the wave: the larger the amplitude, the louder the sound. The pressure amplitude is an objective property of a sound wave, since it can be measured. Loudness, on the other hand, is subjective. Each individual determines what is loud, depending on the acuteness of his or her hearing.

16.6   THE SPEED OF SOUND

Gases

Sound travels through gases, liquids, and solids at considerably different speeds, as Table 16.1 reveals. Near room temperature, the speed of sound in air is 343 m/s (767 mi/h) and is markedly greater in liquids and solids. For example, sound travels more than four times faster in water and more than seventeen times faster in steel than it does in air. In general, sound travels slowest in gases, faster in liquids, and fastest in solids.

Like the speed of a wave on a guitar string, the speed of sound depends on the properties of the medium. In a gas, it is only when molecules collide that the condensations and rarefactions of a sound wave can move from place to place. It is reasonable, then, to expect the speed of sound in a gas to have the same order of magnitude as the average molecular speed between collisions. For an ideal gas this average speed is the translational rms-speed given by Equation 14.6: $v_{\text{rms}} = \sqrt{3kT/m}$, where T is the Kelvin temperature, m is the mass of a molecule, and k is Boltzmann's constant. Although the expression for v_{rms} overestimates the speed of sound, it does give the correct dependence on Kelvin temperature and particle mass. Careful analysis shows that the speed of sound in an ideal gas is given by

Ideal gas
$$v = \sqrt{\frac{\gamma kT}{m}} \qquad (16.5)$$

where $\gamma = c_P/c_V$ is the ratio of the specific heat capacity at constant pressure c_P to the specific heat capacity at constant volume c_V.

The factor γ is introduced in Section 15.5, where the adiabatic compression and expansion of an ideal gas is discussed. It appears in Equation 16.5 because the condensations and rarefactions of a sound wave are formed by adiabatic compressions and expansions of the gas. The regions that are compressed (the condensations) become slightly warmed, and the regions that are expanded (the rarefactions) become slightly cooled. However, no appreciable heat flows from a condensation to an adjacent rarefaction because the distance between the two (half a wavelength) is relatively large for most audible sound waves and a gas is a poor thermal conductor. Thus, the compression and expansion process is adiabatic. Example 4 illustrates the use of Equation 16.5.

EXAMPLE 4   An Ultrasonic Ruler

Figure 16.19 shows an ultrasonic ruler that is used to measure the distance between itself and a target, such as a wall. To initiate the measurement, the ruler generates a pulse of ultrasonic sound that travels to the wall and, like an echo, reflects from it. The reflected pulse returns to the ruler, which measures the time it takes for the round-trip. Using a preset value for the speed of sound, the unit determines the distance to the wall and displays it on a digital readout. Suppose the round-trip travel time is 20.0 ms on a day when the air temperature is 23 °C. Assuming that air is an ideal gas for which $\gamma = 1.40$ and that the average mass of a molecule in the air is 28.9 u, find the distance x to the wall.

Reasoning The distance between the ruler and the wall is $x = vt$, where v is the speed of sound and t is the time for the sound pulse to reach the wall. The time t is one-half the round-trip time, so $t = 10.0$ ms. The speed of sound in air can be obtained directly from Equation 16.5, provided the temperature and mass are expressed in the SI units of kelvins and kilograms, respectively.

Solution To convert the air temperature of 23 °C to the Kelvin temperature scale, we add 23 to 273.15 (see Equation 12.1); $T = 23 + 273.15 = 296$ K. The molecular mass (in kilograms) can be obtained from the conversion relation between atomic mass units and kilograms (see Section 14.1), 1 u = 1.6605×10^{-27} kg;

$$m = (28.9\text{ u})\left(\frac{1.6605 \times 10^{-27}\text{ kg}}{1\text{ u}}\right) = 4.80 \times 10^{-26}\text{ kg}$$

For the speed of sound, we find

$$v = \sqrt{\frac{\gamma kT}{m}} = \sqrt{\frac{(1.40)(1.38 \times 10^{-23}\text{ J/K})(296\text{ K})}{4.80 \times 10^{-26}\text{ kg}}} = 345\text{ m/s} \quad (16.5)$$

The distance to the wall is

$$x = vt = (345\text{ m/s})(10.0 \times 10^{-3}\text{ s}) = \boxed{3.45\text{ m}}$$

● **PROBLEM SOLVING INSIGHT**
When using equation $v = \sqrt{\gamma kT/m}$ to calculate the speed of sound in an ideal gas, be sure to express the temperature T in kelvins and not in degrees Celsius or Fahrenheit.

Sonar (**so**und **na**vigation **r**anging) is a technique for determining water depth and locating underwater objects, such as reefs, submarines, and schools of fish. The core of a sonar unit consists of an ultrasonic transmitter and receiver mounted on the bottom of a ship. The transmitter emits a short pulse of ultrasonic sound, and at a later time the reflected pulse returns and is detected by the receiver. The water depth is determined from the electronically measured round-trip time of the pulse and a knowledge of the speed of sound in water; the depth registers automatically on an appropriate meter. Such a depth measurement is similar to the distance measurement discussed for the ultrasonic ruler in Example 4.

Conceptual Example 5 illustrates how the speed of sound in air can be used to estimate the distance to a thunderstorm, using a handy rule of thumb.

CONCEPTUAL EXAMPLE 5   Lightning, Thunder, and a Rule of Thumb

There is a rule of thumb for estimating how far away a thunderstorm is. After you see a flash of lightning, count off the seconds until the thunder is heard. Divide the number of seconds by five. The result gives the approximate distance (in miles) to the thunderstorm. Why does this rule work?

Reasoning and Solution Figure 16.20 shows a lightning bolt from a thunderstorm and a person who is standing one mile (1.6×10^3 m) away. When the lightning occurs, light and sound (thunder) are produced very nearly at the same instant. Light travels so rapidly ($v_{\text{light}} = 3.0 \times 10^8$ m/s) that it reaches the observer almost instantaneously. Its travel time is only $(1.6 \times 10^3\text{ m})/(3.0 \times 10^8\text{ m/s}) = 5.3 \times 10^{-6}$ s. In comparison, sound travels very slowly ($v_{\text{sound}} = 343$ m/s). The time for the thunder to reach the person is $(1.6 \times 10^3\text{ m})/(343\text{ m/s}) = 5$ s. Thus, the time interval be-

Figure 16.20 A lightning bolt from a thunderstorm generates a flash of light and sound (thunder). The speed of light is much greater than the speed of sound. Therefore, the light reaches the person first, followed about 5 seconds later by the sound.

INTERACTIVE LEARNINGWARE #32

START

Visualization

Reasoning

Knowns and Unknowns

Modeling the Problem

Numerical Solution

FINISH

A rhinoceros is calling to her mate using infrasonic sound whose frequency is 5.0 Hz. Her mate is 480 m away. The air temperature is 35 °C. Assume that air is an ideal gas for which $\gamma = 1.40$ and that the average mass of a molecule in the air is 4.80×10^{-26} kg. How many cycles of the sound wave are between the two animals?

This Interactive LearningWare problem brings together two concepts that are important in understanding the properties of a sound wave. First, the fundamental relation between the speed, frequency, and wavelength of a wave is reviewed. Second, the problem examines how the speed of sound depends on the temperature of the air and the mass of an air molecule.

Go to our Web site at

http://www.wiley.com/college/cutnell

tween seeing the flash and hearing the thunder is about 5 seconds for every mile of travel. ***This rule of thumb works because the speed of light is so much greater than the speed of sound*** that the time needed for the light to reach the observer is negligible compared to the time needed for the sound.

Related Homework Material: *Problem 38*

Liquids

In a liquid, the speed of sound depends on the density ρ and the *adiabatic* bulk modulus B_{ad} of the liquid:

Liquid $$v = \sqrt{\frac{B_{ad}}{\rho}} \tag{16.6}$$

The bulk modulus is introduced in Section 10.7 in a discussion of the volume deformation of liquids and solids. There it is tacitly assumed that the temperature remains constant while the volume of the material changes; i.e., the compression or expansion is isothermal. However, the condensations and rarefactions in a sound wave occur under *adiabatic* rather than isothermal conditions. Thus, the adiabatic bulk modulus B_{ad} must be used when calculating the speed of sound in liquids. Values of B_{ad} will be provided as needed in this text.

Solid Bars

When sound travels through a long slender solid bar, the speed of the sound depends on the properties of the medium according to

Long slender solid bar $$v = \sqrt{\frac{Y}{\rho}} \tag{16.7}$$

where Y is Young's modulus (defined in Section 10.7) and ρ is the density.

16.7 SOUND INTENSITY

Sound waves carry energy that can be used to do work, like forcing the eardrum to vibrate. Or, in an extreme case such as a sonic boom, the energy can be sufficient to cause damage to windows and buildings. The amount of energy transported per second by a sound wave is called the ***power*** of the wave and is measured in SI units of joules per second (J/s) or watts (W).

● *Concepts at a Glance* When a sound wave leaves a source, such as the loudspeaker in Figure 16.21, the power spreads out and passes through imaginary surfaces that have increasingly larger areas. For instance, the same sound power passes through the surfaces labeled 1 and 2 in the drawing. However, the power is spread out over a greater area in surface 2. The Concepts-at-a-Glance chart in Figure 16.22 indicates how we take this spreading-out effect into account. We will bring together the ideas of sound power and the area through which the power passes and, in the

Figure 16.21 The power carried by a sound wave spreads out after leaving a source, such as a loudspeaker. Thus, the power passes perpendicularly through surface 1 and then through surface 2, which has the larger area.

Concepts at a Glance

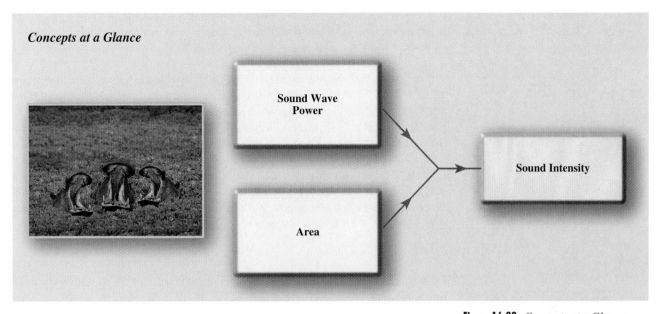

Figure 16.22 *Concepts at a Glance* The concept of sound intensity takes into account both the sound power and the area through which the power passes. A single bellowing bull hippopotamus can produce a sound intensity nearly as large as that at the loudest rock concert.

process, formulate the concept of sound intensity. The idea of wave intensity is not confined to sound waves. It will recur, for example, in Chapter 24 when we discuss another important type of waves, electromagnetic waves. ●

The **sound intensity** I is defined as the sound power P that passes perpendicularly through a surface divided by the area A of that surface:

$$I = \frac{P}{A} \qquad (16.8)$$

The unit of sound intensity is power per unit area, or W/m^2. The next example illustrates how the sound intensity changes as the distance from a loudspeaker changes.

EXAMPLE 6   Sound Intensities

In Figure 16.21, 12×10^{-5} W of sound power passes perpendicularly through the surfaces labeled 1 and 2. These surfaces have areas of $A_1 = 4.0 \text{ m}^2$ and $A_2 = 12 \text{ m}^2$. Determine the sound intensity at each surface and discuss why listener 2 hears a quieter sound than listener 1.

Reasoning The sound intensity I is the sound power P passing perpendicularly through a surface divided by the area A of that surface. Since the same sound power passes through both surfaces, but surface 2 has the greater area, the sound intensity is less at surface 2.

Solution The sound intensity at each surface follows from Equation 16.8:

Surface 1 $I_1 = \dfrac{P}{A_1} = \dfrac{12 \times 10^{-5} \text{ W}}{4.0 \text{ m}^2} = \boxed{3.0 \times 10^{-5} \text{ W/m}^2}$

Surface 2 $I_2 = \dfrac{P}{A_2} = \dfrac{12 \times 10^{-5} \text{ W}}{12 \text{ m}^2} = \boxed{1.0 \times 10^{-5} \text{ W/m}^2}$

● **PROBLEM SOLVING INSIGHT**
Sound intensity I and sound power P are different concepts. They are related, however, since intensity equals power per unit area.

The sound intensity is less at the more-distant surface, where the same power passes through a threefold greater area. The ear of a listener, with its fixed area, intercepts

Sound source at
center of sphere

Figure 16.23 The sound source at the center of the sphere emits sound uniformly in all directions. In this drawing, only a hemisphere is shown for clarity.

less power where the intensity, or power per unit area, is smaller. Thus, listener 2 intercepts less of the sound power than listener 1. With less power striking the ear, the sound is quieter.

For a 1000-Hz tone, the smallest sound intensity that the human ear can detect is about 1×10^{-12} W/m²; this intensity is called the ***threshold of hearing.*** On the other extreme, continuous exposure to intensities greater than 1 W/m² can be painful and result in permanent hearing damage. The human ear is remarkable for the wide range of intensities to which it is sensitive.

If a source emits sound *uniformly in all directions,* the intensity depends on distance in a simple way. Figure 16.23 shows such a source at the center of an imaginary sphere (for clarity only a hemisphere is shown). The radius of the sphere is r. Since all the radiated sound power P passes through the spherical surface of area $A = 4\pi r^2$, the intensity at a distance r is

$$\textbf{Spherically uniform radiation} \qquad I = \frac{P}{4\pi r^2} \qquad (16.9)$$

From this we see that the intensity of a source that radiates sound uniformly in all directions varies as $1/r^2$. For example, if the distance increases by a factor of two, the sound intensity decreases by a factor of $2^2 = 4$. Example 7 illustrates the effect of the $1/r^2$ dependence of intensity on distance.

EXAMPLE 7   Fireworks

During a fireworks display, a rocket explodes high in the air, as Figure 16.24 illustrates. Assume that the sound spreads out uniformly in all directions and that reflections from the ground can be ignored. When the sound reaches listener 2, who is $r_2 = 640$ m away from the explosion, the sound has an intensity of $I_2 = 0.10$ W/m². What is the sound intensity detected by listener 1, who is $r_1 = 160$ m away from the explosion?

Reasoning Listener 1 is four times closer to the explosion than listener 2. Therefore, the sound intensity detected by listener 1 is $4^2 = 16$ times greater than that detected by listener 2.

Solution The ratio of the sound intensities can be found using Equation 16.9:

$$\frac{I_1}{I_2} = \frac{\dfrac{P}{4\pi r_1^2}}{\dfrac{P}{4\pi r_2^2}} = \frac{r_2^2}{r_1^2} = \frac{(640 \text{ m})^2}{(160 \text{ m})^2} = 16$$

As a result, $I_1 = (16)I_2 = (16)(0.10 \text{ W/m}^2) = \boxed{1.6 \text{ W/m}^2}$.

● **PROBLEM SOLVING INSIGHT**
Equation 16.9 can be used only when the sound spreads out uniformly in all directions and there are no reflections of the sound waves.

Figure 16.24 If an explosion in a fireworks display radiates sound uniformly in all directions, the intensity at any distance r is $I = P/(4\pi r^2)$, where P is the sound power of the explosion.

Equation 16.9 is valid only when no walls, ceilings, floors, etc. are present to reflect the sound and cause it to pass through the same surface more than once. Conceptual Example 8 demonstrates why this is so.

CONCEPTUAL EXAMPLE 8   Reflected Sound and Sound Intensity

Suppose the person singing in the shower in Figure 16.25 produces a sound power P. Sound reflects from the surrounding shower stall. At a distance r in front of the person, does Equation 16.9, $I = P/(4\pi r^2)$, underestimate, overestimate, or give the correct sound intensity?

Reasoning and Solution In arriving at Equation 16.9, it was assumed that the sound spreads out uniformly from the source and passes only once through the imaginary surface that surrounds it (see Figure 16.23). Only part of this imaginary surface is shown in Figure 16.25. The drawing illustrates three paths by which the sound passes through the surface. The "direct" sound travels directly along a path from its source to the surface. It is the intensity of the direct sound that is given by $I = P/(4\pi r^2)$. The remaining paths are two of the many that characterize the sound reflected from the shower stall. The total sound power that passes through the surface is the sum of the direct and reflected powers. Thus, the total sound intensity at a distance r from the source is greater than that of the direct sound alone, and **the relation $I = P/(4\pi r^2)$ underestimates the sound intensity from the singing because it does not take into account the reflected sound.** People like to sing in the shower because their voices sound so much louder due to the enhanced intensity caused by the reflected sound.

Related Homework Material: *Problems 52 and 99*

Figure 16.25 When someone sings in the shower, the sound power passing through part of an imaginary spherical surface (shown in pink) is the sum of the direct sound power and the reflected sound power.

16.8 DECIBELS

The *decibel* (dB) is a measurement unit used when comparing two sound intensities. The simplest method of comparison would be to compute the ratio of the intensities. For instance, we could compare $I = 8 \times 10^{-12}$ W/m^2 to $I_0 = 1 \times 10^{-12}$ W/m^2 by computing $I/I_0 = 8$ and stating that I is eight times greater than I_0. However, because of the way in which the human hearing mechanism responds to intensity, it is more appropriate to use a logarithmic scale for the comparison. For this purpose, the *intensity level β* (expressed in decibels) is defined as follows:

$$\beta = (10 \text{ dB}) \log \left(\frac{I}{I_0} \right) \qquad (16.10)$$

where "log" denotes the logarithm to the base ten. I_0 is the intensity of the reference level to which I is being compared and is often the threshold of hearing, $I_0 = 1.00 \times 10^{-12}$ W/m^2. With the aid of a calculator, the intensity level can be evaluated for the values of I and I_0 given above:

$$\beta = (10 \text{ dB}) \log \left(\frac{8 \times 10^{-12} \text{ W/m}^2}{1 \times 10^{-12} \text{ W/m}^2} \right) = (10 \text{ dB}) \log 8 = (10 \text{ dB})(0.9) = 9 \text{ dB}$$

This result indicates that I is 9 decibels greater than I_0. Although β is called the "intensity level," it is *not* an intensity and does *not* have intensity units of W/m^2. In fact, the decibel, like the radian, is dimensionless.

TABLE 16.2   Typical Sound Intensities and Intensity Levels Relative to the Threshold of Hearing

	Intensity I (W/m^2)	Intensity Level β (dB)
Threshold of hearing	1.0×10^{-12}	0
Rustling leaves	1.0×10^{-11}	10
Whisper	1.0×10^{-10}	20
Normal conversation (1 meter)	3.2×10^{-6}	65
Inside car in city traffic	1.0×10^{-4}	80
Car without muffler	1.0×10^{-2}	100
Live rock concert	1.0	120
Threshold of pain	10	130

Notice that if both I and I_0 are at the threshold of hearing, then $I = I_0$, and the intensity level is 0 dB according to Equation 16.10:

$$\beta = (10 \text{ dB}) \log \left(\frac{I_0}{I_0} \right) = (10 \text{ dB}) \log 1 = 0$$

since $\log 1 = 0$. Thus, *an intensity level of zero decibels does not mean that the sound intensity I is zero; it means that I = I$_0$.*

Table 16.2 lists the intensities I and the associated intensity levels β for some common sounds, using the threshold of hearing as the reference level. Intensity levels can be measured with a sound level meter, such as the one in Figure 16.26. The intensity level β is displayed on its scale, assuming that the threshold of hearing is 0 dB.

When a sound wave reaches a listener's ear, the sound is interpreted by the brain as loud or soft, depending on the intensity of the wave. Greater intensities give rise to louder sounds. However, the relation between intensity and loudness is not a simple proportionality, for doubling the intensity does *not* double the loudness, as we will now see.

Suppose you are sitting in front of a stereo system that is producing an intensity level of 90 dB. If the volume control on the amplifier is turned up slightly to produce a 91 dB level, you would just barely notice the change in loudness. *Hearing tests have revealed that a one-decibel (1-dB) change in the intensity level corresponds to approximately the smallest change in loudness that an average listener can detect.* Since 1 dB is the smallest perceivable increment in loudness, a change of 3 dB, say, from 90 to 93 dB, is still a rather small change in loudness. Example 9 determines the factor by which the sound intensity must be increased to achieve this small change in loudness.

Figure 16.26 A sound level meter and a close-up view of its decibel scale.

EXAMPLE 9   Comparing Sound Intensities

Audio system 1 produces an intensity level of $\beta_1 = 90.0$ dB, while system 2 produces an intensity level of $\beta_2 = 93.0$ dB. The corresponding intensities (in W/m^2) are I_1 and I_2. Determine the ratio I_2/I_1.

Reasoning Intensity levels are related to intensities by logarithms (see Equation 16.10), and it is a property of logarithms (see Appendix D) that $\log A - \log B = \log (A/B)$. Subtracting the two intensity levels and using this property, we find that

$$\beta_2 - \beta_1 = (10 \text{ dB}) \log\left(\frac{I_2}{I_0}\right) - (10 \text{ dB}) \log\left(\frac{I_1}{I_0}\right) = (10 \text{ dB}) \log\left(\frac{I_2/I_0}{I_1/I_0}\right)$$

$$= (10 \text{ dB}) \log\left(\frac{I_2}{I_1}\right)$$

Solution Using the result just obtained, we find

$$93.0 \text{ dB} - 90.0 \text{ dB} = (10 \text{ dB}) \log\left(\frac{I_2}{I_1}\right)$$

$$0.30 = \log\left(\frac{I_2}{I_1}\right) \quad \text{or} \quad \frac{I_2}{I_1} = 10^{0.30} = \boxed{2.0}$$

Since doubling the intensity changes the loudness by only a small amount (3 dB) and does not double it, there is no simple proportionality between intensity and loudness.

Figure 16.27 In spite of its tenfold greater power, the 200-watt audio system has only about double the loudness of the 20-watt system, when both are set for maximum volume.

To double the loudness of a sound, the intensity must be increased by more than a factor of two. *__Experiment shows that if the intensity level increases by 10 dB, the new sound seems approximately twice as loud as the original sound.__* For instance, a 70-dB intensity level sounds about twice as loud as a 60-dB level, and an 80-dB intensity level sounds about twice as loud as a 70-dB level. The factor by which the sound intensity must be increased to double the loudness can be determined by the method used in Example 9:

$$\beta_2 - \beta_1 = 10.0 \text{ dB} = (10 \text{ dB})\left[\log\left(\frac{I_2}{I_0}\right) - \log\left(\frac{I_1}{I_0}\right)\right]$$

Solving this equation reveals that $I_2/I_1 = 10.0$. Thus, increasing the sound intensity by a factor of ten will double the perceived loudness. Consequently, with both audio systems in Figure 16.27 set at maximum volume, the 200-watt system will sound only twice as loud as the much cheaper 20-watt system.

16.9 THE DOPPLER EFFECT

Have you ever heard an approaching fire truck and noticed the distinct change in the sound of the siren as the truck passes? The effect is similar to what you get when you put together the two syllables "eee" and "yow" to produce "eee-yow." While the truck approaches, the pitch of the siren is relatively high ("eee"), but as the truck passes and moves away, the pitch suddenly drops ("yow"). Something similar, but less familiar, occurs when an observer moves toward or away from a stationary source of sound. Such phenomena were first identified in 1842 by the Austrian physicist Christian Doppler (1803–1853) and are collectively referred to as the Doppler effect.

　●　*Concepts at a Glance*   To explain why the Doppler effect occurs, we will bring together concepts that we have discussed previously, namely, the velocity of an object and the wavelength and frequency of a sound wave (Section 16.5). As the Concepts-at-a-Glance chart in Figure 16.28 indicates, we will incorporate the effect of the velocities of the source and observer of the sound into the definitions of wavelength and frequency. In so doing, we will learn that the *__Doppler effect__* is the change in frequency or pitch of the sound detected by an observer because the

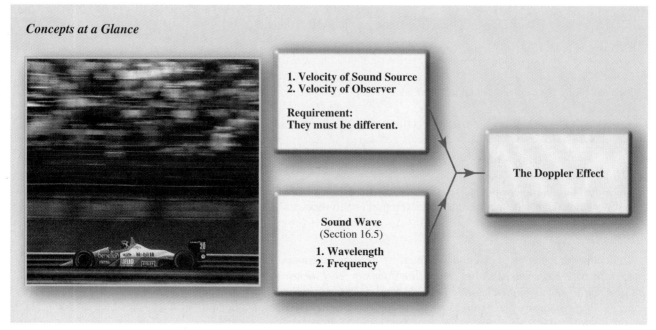

Concepts at a Glance

1. Velocity of Sound Source
2. Velocity of Observer

Requirement:
They must be different.

Sound Wave
(Section 16.5)

1. Wavelength
2. Frequency

The Doppler Effect

Figure 16.28 *Concepts at a Glance*
The Doppler effect arises when the source and the observer of the sound wave have different velocities with respect to the medium through which the sound travels. At an Indy-car racing event, the Doppler effect creates the characteristic sound that you hear when the cars pass by. It is similar to that heard when a train blowing its horn passes you at a high speed.

sound source and the observer have different velocities with respect to the medium of sound propagation. ●

Moving Source

To see how the Doppler effect arises, consider the sound emitted by a siren on the stationary fire truck in Figure 16.29a. Like the truck, the air is assumed to be stationary with respect to the earth. Each solid blue arc in the drawing represents a condensation of the sound wave. Since the sound pattern is symmetrical, listeners standing in front of or behind the truck detect the same number of condensations per second and, consequently, hear the same frequency. Once the truck begins to move, the situation changes, as part *b* of the picture illustrates. Ahead of the truck, the condensations are now closer together, resulting in a decrease in the wavelength of the sound. This "bunching-up" occurs because the moving truck "gains ground" on a previously emitted condensation before emitting the next one. Since the condensations are closer together, the observer standing in front of the truck senses more of them arriving per second than she does when the truck is stationary. The

Figure 16.29 (*a*) When the truck is stationary, the wavelength of the sound is the same in front of and behind the truck. (*b*) When the truck is moving, the wavelength in front of the truck becomes smaller, while the wavelength behind the truck becomes larger.

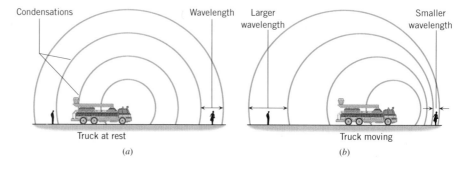

Condensations   Wavelength   Larger wavelength   Smaller wavelength

Truck at rest
(a)

Truck moving
(b)

increased rate of arrival corresponds to a greater sound frequency, which the observer hears as a higher pitch. Behind the moving truck, the condensations are farther apart than they are when the truck is stationary. This increase in the wavelength occurs because the truck pulls away from condensations emitted toward the rear. Consequently, fewer condensations per second arrive at the ear of an observer behind the truck, corresponding to a smaller sound frequency or lower pitch.

If the stationary siren in Figure 16.29a emits a condensation at the time $t = 0$ s, it will emit the next one at time T, where T is the period of the wave. The distance between these two condensations is the wavelength λ of the sound produced by the stationary source, as Figure 16.30a indicates. When the truck is moving with a speed v_s (the subscript "s" stands for the "source" of sound) toward a stationary observer, the siren also emits condensations at $t = 0$ s and at time T. However, prior to emitting the second condensation, the truck moves closer to the observer by a distance $v_s T$, as Figure 16.30b shows. As a result, the distance between successive condensations is no longer the wavelength λ created by the stationary siren, but, rather, a wavelength λ' that is shortened by the amount $v_s T$:

$$\lambda' = \lambda - v_s T$$

Let's denote the frequency perceived by the stationary observer as f_o, where the subscript "o" stands for "observer." According to Equation 16.1, f_o is equal to the speed of sound v divided by the shortened wavelength λ':

$$f' = \frac{v}{\lambda'} = \frac{v}{\lambda - v_s T}$$

But for the stationary siren, $\lambda = v/f_s$ and $T = 1/f_s$, where f_s is the frequency of the sound emitted by the source (not the frequency f_o perceived by the observer). With the aid of these substitutions for λ and T, the expression for f_o can be arranged to give the following result:

Source moving toward stationary observer
$$f_o = f_s \left(\frac{1}{1 - \dfrac{v_s}{v}} \right) \qquad (16.11)$$

Since the term $1 - v_s/v$ is in the denominator in Equation 16.11 and is less than one, the frequency f_o heard by the observer is *greater* than the frequency f_s emitted by the source. The difference between these two frequencies, $f_o - f_s$, is called the ***Doppler shift,*** and its magnitude depends on the ratio of the speed of the source v_s to the speed of sound v.

When the siren moves away from, rather than toward, the observer, the wavelength λ' becomes *greater* than λ according to

$$\lambda' = \lambda + v_s T$$

Notice the presence of the "+" sign in this equation, in contrast to the "−" sign that appeared earlier. The same reasoning that led to Equation 16.11 can be used to obtain an expression for the observed frequency f_o:

Source moving away from stationary observer
$$f_o = f_s \left(\frac{1}{1 + \dfrac{v_s}{v}} \right) \qquad (16.12)$$

The denominator $1 + v_s/v$ in Equation 16.12 is greater than one, so the frequency f_o heard by the observer is *less* than the frequency f_s emitted by the source. The next example illustrates how large the Doppler shift is in a familiar situation.

Figure 16.30 (a) When the fire truck is stationary, the distance between successive condensations is one wavelength λ. (b) When the truck moves with a speed v_s, the wavelength of the sound in front of the truck is shortened to λ'.

EXAMPLE 10   The Sound of a Passing Train

A high-speed train is traveling at a speed of 44.7 m/s (100 mi/h) when the engineer sounds the 415-Hz warning horn. The speed of sound is 343 m/s. What are the frequency and wavelength of the sound, as perceived by a person standing at a crossing, when the train is (a) approaching and (b) leaving the crossing?

Reasoning When the train approaches, the person at the crossing hears a sound whose frequency is greater than 415 Hz because of the Doppler effect. As the train moves away, the person hears a frequency that is less than 415 Hz. We may use Equations 16.11 and 16.12, respectively, to determine these frequencies. In either case, the observed wavelength can be obtained according to Equation 16.1 as the speed of sound divided by the observed frequency.

Solution

(a) When the train approaches, the observed frequency is

$$f_o = f_s \left(\frac{1}{1 - \dfrac{v_s}{v}} \right) = (415\ \text{Hz}) \left(\frac{1}{1 - \dfrac{44.7\ \text{m/s}}{343\ \text{m/s}}} \right) = \boxed{477\ \text{Hz}} \quad (16.11)$$

The observed wavelength is

$$\lambda' = \frac{v}{f_o} = \frac{343\ \text{m/s}}{477\ \text{Hz}} = \boxed{0.719\ \text{m}} \quad (16.1)$$

(b) When the train leaves the crossing, the observed frequency is

$$f_o = f_s \left(\frac{1}{1 + \dfrac{v_s}{v}} \right) = (415\ \text{Hz}) \left(\frac{1}{1 + \dfrac{44.7\ \text{m/s}}{343\ \text{m/s}}} \right) = \boxed{367\ \text{Hz}} \quad (16.12)$$

In this case, the observed wavelength is

$$\lambda' = \frac{v}{f_o} = \frac{343\ \text{m/s}}{367\ \text{Hz}} = \boxed{0.935\ \text{m}}$$

Moving Observer

Figure 16.31 shows how the Doppler effect arises when the sound source is stationary and the observer moves, again assuming the air is stationary. The observer moves with a speed v_o ("o" stands for "observer") toward the stationary source and covers a distance $v_o t$ in a time t. During this time, the moving observer encounters all the condensations that he would if he were stationary, *plus an additional number.* The additional number of condensations encountered is the distance $v_o t$ divided by the distance λ between successive condensations, or $v_o t/\lambda$. Thus, the additional number of condensations encountered per second is v_o/λ. Since a stationary observer would hear a frequency f_s emitted by the source, the moving observer hears a higher frequency f_o given by

$$f_o = f_s + \frac{v_o}{\lambda} = f_s \left(1 + \frac{v_o}{f_s \lambda} \right)$$

Using the fact that $v = f_s \lambda$, we find that

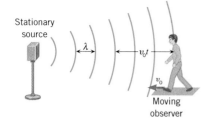

Figure 16.31 An observer moving with a speed v_o toward the stationary source intercepts more wave condensations per unit of time than does a stationary observer.

Observer moving toward stationary source

$$f_o = f_s \left(1 + \frac{v_o}{v} \right) \quad (16.13)$$

An observer moving *away from* a stationary source moves in the same direction as the sound wave and, as a result, intercepts *fewer* condensations per second than a stationary observer does. In this case, the moving observer hears a smaller frequency f_o that is given by

Observer moving away
from stationary source
$$f_o = f_s \left(1 - \frac{v_o}{v} \right) \qquad (16.14)$$

The physical mechanism producing the Doppler effect in the case of the moving observer is different from that in the case of the moving source. When the source moves and the observer is stationary, the wavelength of the sound changes, giving rise to the frequency f_o heard by the observer. On the other hand, when the observer moves and the source is stationary, the *wavelength of the sound does not change.* Instead, a moving observer intercepts a different number of wave condensations per second than does a stationary observer and, therefore, detects a different frequency f_o.

General Case

It is possible for both the sound source and the observer to move with respect to the medium of sound propagation. If the medium is stationary, Equations 16.11–16.14 may be combined to give the observed frequency f_o as

Source and observer
both moving
$$f_o = f_s \left(\frac{1 \pm \dfrac{v_o}{v}}{1 \mp \dfrac{v_s}{v}} \right) \qquad (16.15)$$

In the numerator, the plus sign applies when the observer moves toward the source, and the minus sign applies when the observer moves away from the source. In the denominator, the minus sign is used when the source moves toward the observer, and the plus sign is used when the source moves away from the observer. The symbols v_o, v_s, and v denote numbers without an algebraic sign because the direction of travel has been taken into account by the plus and minus signs that appear directly in this equation.

Nexrad

NEXRAD stands for **Nex**t Generation Weather **Rad**ar and is a new nationwide system used by the National Weather Service to provide dramatically improved early warning of severe storms, such as the tornado in Figure 16.32. The system is based on radar waves, which are a type of electromagnetic wave (see Chapter 24) and, like sound waves, can exhibit the Doppler effect. The Doppler effect is at the heart of NEXRAD. As the drawing illustrates, a tornado is a swirling mass of air and water droplets. Radar pulses are sent out by a NEXRAD unit, whose protective covering is shaped like a soccer ball. The waves reflect from the water droplets and return to the unit, where the frequency is observed and compared to the outgoing

THE PHYSICS OF...

Next Generation Weather Radar.

INTERACTIVE LEARNINGWARE #33

START

Visualization

Reasoning

Knowns and Unknowns

Modeling the Problem

Numerical Solution

FINISH

A hawk flies directly toward a bird-watcher at a speed of 11.0 m/s. The frequency heard by the stationary bird-watcher is 894 Hz when the speed of sound is 343 m/s. What is the frequency heard by the bird-watcher when the hawk flies away from the watcher at the same speed?

This Interactive LearningWare problem illustrates the Doppler effect in two different situations. As the hawk approaches, the bird-watcher hears a higher frequency because the wavelength of the sound is smaller. As the hawk moves away, the bird-watcher hears a lower frequency because the wavelength is now larger.

Go to our Web site at

http://www.wiley.com/college/cutnell

(a)

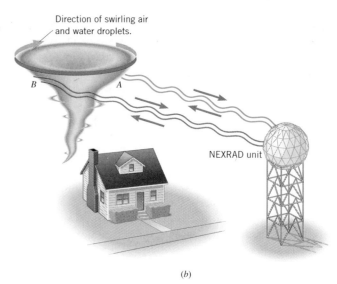

Direction of swirling air
and water droplets.

B   A

NEXRAD unit

(b)

Figure 16.32   (a) A tornado is one of nature's most dangerous storms. (b) The National Weather Service uses the NEXRAD system, which is based on Doppler-shifted radar, to identify the storms that are likely to spawn tornadoes.

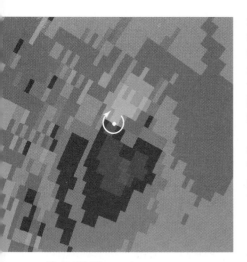

Figure 16.33   In this color-enhanced NEXRAD view of a tornado, shades of green denote winds moving toward the NEXRAD station, and shades of red denote winds moving away from the station, which is located below and to the right of the figure. The white dot marks the center of the storm, and the arrow indicates how the winds are circulating.

THE PHYSICS OF...

ultrasonic imaging.

frequency. For instance, droplets at point *A* in the drawing are moving toward the unit, and the radar waves reflected from them have their frequency Doppler-shifted to higher values. Droplets at point *B*, however, are moving away from the unit. The frequency of the waves reflected from these droplets is Doppler-shifted to lower values. Computer processing of the Doppler frequency shifts leads to color-enhanced views on display screens (see Figure 16.33). These views reveal the direction and magnitude of the wind velocity and can identify, from distances up to 140 mi, the swirling air masses that are likely to spawn tornadoes. The equations that specify the Doppler frequency shifts are different from those given for sound waves by Equations 16.11–16.15. The reason for the difference is that radar waves propagate from one place to another by a different mechanism than that of sound waves (see Section 24.5).

16.10   APPLICATIONS OF SOUND IN MEDICINE

When ultrasonic waves are used in medicine for diagnostic purposes, high-frequency sound pulses are produced by a transmitter and directed into the body. As in sonar, reflections occur. They occur each time a pulse encounters a boundary between two tissues that have different densities or a boundary between a tissue and the adjacent fluid. By scanning ultrasonic waves across the body and detecting the echoes generated from various internal locations, it is possible to obtain an image or sonogram of the inner anatomy. Ultrasonic waves are employed extensively in obstetrics to examine the developing fetus (Figure 16.34). The fetus, surrounded by the amniotic sac, can be distinguished from other anatomical features so that fetal size, position, and possible abnormalities can be detected.

Ultrasound is also used in other medically related areas. For instance, malignancies in the liver, kidney, brain, and pancreas can be detected with ultrasound. Yet

Figure 16.34 An ultrasonic scanner can be used to produce an image of the fetus as it develops in the uterus.

another application involves monitoring the real-time movement of pulsating structures, such as heart valves ("echocardiography") and large blood vessels.

When ultrasound is used to locate internal anatomical features or foreign objects in the body, the wavelength of the sound wave must be about the same size, or smaller, than the object to be located. Therefore, high frequencies in the range from 1 to 15 MHz (1 MHz = 1 megahertz = 1×10^6 Hz) are the norm. For instance, the wavelength of 5-MHz ultrasound is $\lambda = v/f = 0.3$ mm, if a value of 1540 m/s is used for the speed of sound through tissue. A sound wave with a frequency higher than 5 MHz and a correspondingly shorter wavelength is required for locating objects smaller than 0.3 mm.

Ultrasound also has applications other than locating objects in the body. Neurosurgeons use a device called a **c**avitron **u**ltrasonic **s**urgical **a**spirator (CUSA) to remove brain tumors once thought to be inoperable. Ultrasonic sound waves cause the slender tip of the CUSA probe (see Figure 16.35) to vibrate at approximately 23 kHz. The probe shatters any section of the tumor that it touches, and the fragments are flushed out of the brain with a saline solution. Because the tip of the probe is small, the surgeon can selectively remove small bits of malignant tissue without damaging the surrounding healthy tissue.

The Doppler flow meter is a particularly interesting medical application of the Doppler effect. This device measures the speed of blood flow, using transmitting and receiving elements that are placed directly on the skin, as in Figure 16.36. The transmitter emits a continuous sound whose frequency is typically about 5 MHz. When the sound is reflected from the red blood cells, its frequency is changed in a kind of Doppler effect because the cells are moving. The receiving element detects the reflected sound, and an electronic counter measures its frequency, which is Doppler-shifted relative to the transmitter frequency. From the change in frequency the speed of the blood flow can be determined. Typically, the change in frequency is around 600 Hz for flow speeds of about 0.1 m/s. The Doppler flow meter can be used to locate regions where blood vessels have narrowed, since greater flow speeds occur in the narrowed regions, according to the equation of continuity (see Section 11.8). In addition, the Doppler flow meter can be used to detect the motion of a fetal heart as early as 8–10 weeks after conception.

THE PHYSICS OF...

the cavitron ultrasonic surgical aspirator. (See below.)

THE PHYSICS OF...

the Doppler flow meter. (See below.)

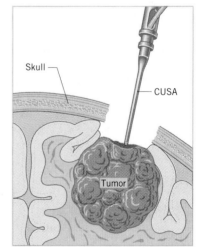

Figure 16.35 Neurosurgeons use a cavitron ultrasonic surgical aspirator (CUSA) to "cut out" brain tumors without adversely affecting the surrounding healthy tissue.

hearing.

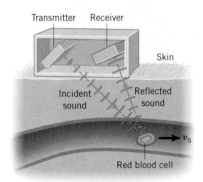

Figure 16.36   A Doppler flow meter measures the speed of red blood cells.

Figure 16.37   Each curve represents the intensity levels at which sounds of various frequencies have the same loudness. The curves are labeled by their intensity levels at 1000 Hz. These curves are known as the Fletcher–Munson curves.

*16.11   THE SENSITIVITY OF THE HUMAN EAR

Although the ear is capable of detecting sound intensities as small as 1×10^{-12} W/m^2, it is *not* equally sensitive to all frequencies, as Figure 16.37 shows. This figure displays a series of graphs known as the **Fletcher–Munson curves,** after H. Fletcher and M. Munson, who first determined them in 1933. In these graphs the audible sound frequencies are plotted on the horizontal axis, while the sound intensity levels (in decibels) are plotted on the vertical axis. Each curve is a *constant loudness* curve because it shows the sound intensity level needed at each frequency to make the sound appear to have the same loudness. For example, the lowest (red) curve represents the threshold of hearing. It shows the intensity levels at which sounds of different frequencies just become audible. The graph indicates that the intensity level of a 100-Hz sound must be about 37 dB greater than the intensity level of a 1000-Hz sound to be at the threshold of hearing. Therefore, the ear is *less sensitive* to a 100-Hz sound than it is to a 1000-Hz sound. In general, Figure 16.37 reveals that the ear is most sensitive in the range from about 1 to 5 kHz, and becomes progressively less sensitive at higher and lower frequencies.

Each curve in Figure 16.37 represents a different loudness, and each is labeled according to its intensity level at 1000 Hz. For instance, the curve labeled "60" represents all sounds that have the same loudness as that of a 1000-Hz sound whose intensity level is 60 dB. These constant-loudness curves become "flatter" as the loudness increases, the relative flatness indicating that the ear is nearly equally sensitive to all frequencies when the sound is loud. Thus, when you listen to loud sounds, you hear the low frequencies, the middle frequencies, and the high frequencies about equally well. However, when you listen to quiet sounds, the high and low frequencies seem to "disappear," for the ear is relatively insensitive to these frequencies under such conditions.

16.12   CONCEPTS & CALCULATIONS

One of the important concepts that we encountered in this chapter is a transverse wave. For instance, transverse waves travel along a guitar string when it is plucked or along a violin string when it is bowed. The next example reviews how the travel speed depends on the properties of the string and on the tension in it.

CONCEPTS & CALCULATIONS EXAMPLE 11
What Determines the Speed of a Wave on a String?

Figure 16.38 shows waves traveling on two strings. The loose ends of the strings are attached to a box whose weight is 28.0 N. String 1 has a mass of 8.00 g and a length of 4.00 cm, while string 2 has a mass of 12.0 g and a length of 8.00 cm. Determine the speed of the wave on each string.

Concept Questions and Answers   Is the tension the same in each string?

Answer   Yes, the tension is the same. The two strings support the box, so the tension in each string is one-half the weight of the box. The fact that the strings have different masses and lengths does not affect the tension, which is determined only by the weight of the hanging box.

Is the speed of each wave the same?

Answer Not necessarily. The speed of a wave on a string depends on both the tension and the linear density, as Equation 16.2 indicates. The tension is the same in both strings, but if the linear densities of the strings are different, the speeds are different.

String 1 has a smaller mass and, hence, less inertia than string 2. Does this mean that the speed of the wave on string 1 is greater than that on string 2?

Answer Maybe yes, maybe no. The speed of a wave depends on the linear density of the string, which is its mass divided by its length. Depending on the lengths of the strings, string 1 could have a larger linear density and, hence, smaller speed, than string 2. The solution below illustrates this point.

Solution The speed of a wave on a string is given by Equation 16.2 as $v = \sqrt{F/(m/L)}$, where F is the tension and m/L is the mass per unit length or linear density. Since both strings support the box, the tension in each is one-half the weight of the box, or $F = \frac{1}{2}(28.0 \text{ N}) = 14.0 \text{ N}$. The linear densities of the strings are

$$\frac{m_1}{L_1} = \frac{8.00 \text{ g}}{4.00 \text{ cm}} = 2.00 \text{ g/cm} = 0.200 \text{ kg/m}$$

$$\frac{m_2}{L_2} = \frac{12.0 \text{ g}}{8.00 \text{ cm}} = 1.50 \text{ g/cm} = 0.150 \text{ kg/m}$$

The speed of each wave is

$$v_1 = \sqrt{\frac{F}{m_1/L_1}} = \sqrt{\frac{14.0 \text{ N}}{0.200 \text{ kg/m}}} = \boxed{8.37 \text{ m/s}}$$

$$v_2 = \sqrt{\frac{F}{m_2/L_2}} = \sqrt{\frac{14.0 \text{ N}}{0.150 \text{ kg/m}}} = \boxed{9.66 \text{ m/s}}$$

Figure 16.38 A wave travels on each of the two strings. The strings have different masses and lengths and together support the 28.0-N box. Which is the faster wave?

The next example illustrates how the Doppler effect arises when an observer is moving away from or toward a stationary source of sound. In fact, we will see that it's possible for both situations to occur at the same time.

CONCEPTS & CALCULATIONS EXAMPLE 12
The Doppler Effect for a Moving Observer

A siren, mounted on a tower, emits a sound whose frequency is 2140 Hz. A person is driving a car away from the tower at a speed of 27.0 m/s. As Figure 16.39 illustrates, the sound reaches the person by two paths: the sound reflected from a building in front of the car, and the sound coming directly from the siren. The speed of sound is 343 m/s. What frequency does the person hear for the (a) reflected and (b) direct sounds?

Concept Questions and Answers One way that the Doppler effect can arise is that the wavelength of the sound changes. For either the direct or the reflected sound, does the wavelength change?

Figure 16.39 The sound from the siren reaches the car by a reflected path and a direct path. Because of the Doppler effect, the driver hears a different frequency for each sound.

Answer   No. The wavelength changes only when the source of the sound is moving, as illustrated in Figure 16.29. The siren is stationary, so the wavelength does not change.

Why does the driver hear a frequency for the reflected sound that is different than 2140 Hz, and is it greater than or smaller than 2140 Hz?

Answer   The car and the reflected sound are traveling in opposite directions, the car to the right and the reflected sound to the left. The driver intercepts more wave cycles per second than if the car were stationary. Consequently, the driver hears a frequency greater than 2140 Hz.

Why does the driver hear a frequency for the direct sound that is different than 2140 Hz, and is it greater than or smaller than 2140 Hz?

Answer   The car and the direct sound are traveling in the same direction. As the sound passes the car, the number of wave cycles per second intercepted by the driver is now less than if the car were stationary. Thus, the driver hears a frequency that is less than 2140 Hz.

Solution

(a) The frequency f_o that the driver (the "observer") hears is equal to the frequency f_s of the waves emitted by the siren *plus* an additional number of cycles per second because the car and the reflected sound are moving in opposite directions. The additional number of cycles per second is v_o/λ, where v_o is the speed of the car and λ is the wavelength of the sound (see the subsection "Moving Observer" in Section 16.9). According to Equation 16.1, the wavelength is equal to the speed of sound v divided by the frequency of the siren, $\lambda = v/f_s$. Thus, the frequency heard by the driver can be written as

$$f_o = f_s + \underbrace{\frac{v_o}{\lambda}}_{\substack{\text{Additional}\\\text{number of}\\\text{cycles intercepted}\\\text{per second}}} = f_s + \frac{v_o}{v/f_s}$$

$$= f_s\left(1 + \frac{v_o}{v}\right) = (2140 \text{ Hz})\left(1 + \frac{27.0 \text{ m/s}}{343 \text{ m/s}}\right) = \boxed{2310 \text{ Hz}}$$

(b) The frequency f_o that the driver hears for the direct sound is equal to the frequency f_s of the waves emitted by the siren *minus* v_o/λ, because the car and direct sound are moving in the same direction:

$$f_o = f_s - \frac{v_o}{\lambda} = f_s - \frac{v_o}{v/f_s}$$

$$= f_s\left(1 - \frac{v_o}{v}\right) = (2140 \text{ Hz})\left(1 - \frac{27.0 \text{ m/s}}{343 \text{ m/s}}\right) = \boxed{1970 \text{ Hz}}$$

As expected for the reflected wave, the driver hears a frequency greater than 2140 Hz, while for the direct sound he hears a frequency less than 2140 Hz.

At the end of the problem set for this chapter, you will find homework problems that contain both conceptual and quantitative parts. These problems are grouped under the heading *Concepts & Calculations, Group Learning Problems.* They are designed for use by students working alone or in small learning groups. The conceptual part of each problem provides a convenient focus for group discussions.

SUMMARY

A **wave** is a traveling disturbance and carries energy from place to place. In a **transverse wave**, the disturbance is perpendicular to the direction of travel of the wave. In a **longitudinal wave**, the disturbance is parallel to the line along which the wave travels.

In a **periodic wave**, the pattern of the disturbance is produced over and over again by the source of the wave. The **amplitude** of the wave is the maximum excursion of a particle of the medium from the particle's undisturbed position. The **wavelength** λ is the distance along the length of the wave between two successive equivalent points, such as two crests or two troughs. The **period** T is the time required for the wave to travel a distance of one wavelength. The **frequency** f (in hertz) is the number of wave cycles per second that pass an observer and is the reciprocal of the period (in seconds): $f = 1/T$. The **speed** v of a wave is related to the wavelength and the frequency according to $v = f\lambda$.

The **speed of a wave** depends on the properties of the medium in which the wave travels. For a transverse wave on a string that has a tension F and a mass per unit length m/L, the wave speed is $v = \sqrt{F/(m/L)}$.

Sound is a longitudinal wave that consists of alternating regions of greater-than-normal pressure (condensations) and less-than-normal pressure (rarefactions). Each cycle of a sound wave includes one condensation and one rarefaction. A sound wave with a large frequency is interpreted by the brain as a high-pitched sound, while one with a small frequency is interpreted as a low-pitched sound. The **pressure amplitude** of a sound wave is the magnitude of the maximum change in pressure, measured relative to the undisturbed pressure. The larger the pressure amplitude, the louder the sound.

The **speed of sound** v depends on the properties of the medium. In an ideal gas, the speed of sound is $v = \sqrt{\gamma kT/m}$, where $\gamma = c_P/c_V$ is the ratio of the specific heat capacities at constant pressure and constant volume, k is Boltz-mann's constant, T is the Kelvin temperature, and m is the mass of a molecule of the gas. In a liquid, the speed of sound is $v = \sqrt{B_{ad}/\rho}$, where B_{ad} is the adiabatic bulk modulus and ρ is the mass density. For a solid in the shape of a long slender bar, the expression for the speed of sound is $v = \sqrt{Y/\rho}$, where Y is Young's modulus.

The **intensity** I of a sound wave is the power P that passes perpendicularly through a surface divided by the area A of the surface: $I = P/A$. The SI unit of intensity is watts per square meter (W/m^2). The smallest sound intensity that humans can detect is known as the **threshold of hearing** and is about 1×10^{-12} W/m^2 for a 1-kHz sound. When a source emits sound uniformly in all directions and no reflections are present, the intensity of the sound is inversely proportional to the square of the distance from the source.

The **intensity level** β (in decibels) is used to compare a sound intensity I to the sound intensity I_0 of a reference level: $\beta = (10\ dB) \log(I/I_0)$.

The **Doppler effect** is the change in frequency detected by an observer because the sound source and the observer have different velocities with respect to the medium of sound propagation. If the observer and source move with speeds v_o and v_s, respectively, and if the medium is stationary, the frequency f_o detected by the observer is

$$f_o = f_s \left(\frac{1 \pm v_o/v}{1 \mp v_s/v} \right)$$

where f_s is the frequency of the sound emitted by the source and v is the speed of sound. In the numerator, the plus sign applies when the observer moves toward the source, and the minus sign applies when the observer moves away from the source. In the denominator, the minus sign is used when the source moves toward the observer, and the plus sign is used when the source moves away from the observer.

CONCEPTUAL QUESTIONS

1. Considering the nature of a water wave (see Figure 16.4), describe the motion of a fishing float on the surface of a lake when a wave passes beneath the float. Is it really correct to say that the float bobs straight "up and down"? Explain.

2. "Domino Toppling" is one entry in the *Guiness Book of World Records*. The event consists of lining up an incredible number of dominoes and then letting them topple, one after another. Is the disturbance that propagates along the line of dominoes transverse, longitudinal, or partly both? Explain.

3. Suppose that a longitudinal wave moves along a Slinky at a speed of 5 m/s. Does one coil of the Slinky move through a distance of 5 m in one second? Justify your answer.

4. Examine Conceptual Example 3 before addressing this question. A wave moves on a string with a constant velocity. Does this mean that the particles of the string always have zero acceleration? Justify your answer.

5. A wire is strung tightly between two immovable posts. Discuss how an increase in temperature affects the speed of a transverse wave on this wire. Give your reasoning, ignoring any change in the mass per unit length of the wire.

6. A rope of mass m is hanging down from the ceiling. Nothing is attached to the loose end of the rope. A transverse wave is traveling on the rope. As the wave travels up the rope, does the speed of the wave increase, decrease, or remain the same? Give a reason for your choice.

7. One end of each of two identical strings is attached to a wall. Each string is being pulled tight by someone at the other end. A transverse pulse is sent traveling along one of the strings. A bit later an identical pulse is sent traveling along the other string. What, if anything, can be done to make the second pulse catch up with and pass the first pulse? Account for your answer.

8. In Section 4.10 the concept of a "massless" rope is discussed. Would it take any time for a transverse wave to travel the length of a massless rope? Justify your answer.

9. In a traveling sound wave, are there any particles that are *always* at rest as the wave passes by? Justify your answer.

10. Do you expect an echo to return to you more quickly or less quickly on a hot day as compared to a cold day, other things being equal? Account for your answer.

11. A loudspeaker produces a sound wave. Does the wavelength of the sound increase, decrease, or remain the same, when the wave travels from air into water? Justify your answer. *(Hint: The frequency does not change as the sound enters the water.)*

12. JELL-O starts out as a liquid and then sets to a gel. What would you expect to happen to the speed of sound in this material as the JELL-O sets? Does it increase, decrease, or remain the same? Give your reasoning.

13. ♪ Some animals rely on an acute sense of hearing for survival, and the visible part of the ear on such animals is often relatively large. Explain how this anatomical feature helps to increase the sensitivity of the animal's hearing for low-intensity sounds.

14. A source is emitting sound uniformly in all directions. There are no reflections anywhere. A *flat* surface faces the source. Is the sound intensity the same at all points on the surface? Give your reasoning.

15. If two people talk simultaneously and each creates an intensity level of 65 dB at a certain point, does the total intensity level at this point equal 130 dB? Account for your answer.

16. Two cars, one behind the other, are traveling in the same direction at the same speed. Does either driver hear the other's horn at a frequency that is different from that heard when both cars are at rest? Justify your answer.

17. A source of sound produces the same frequency underwater as it does in air. This source has the same velocity in air as it does underwater. The observer of the sound is stationary, both in air and underwater. Is the Doppler effect greater in air or underwater when the source (a) approaches and (b) moves away from the observer? Explain.

18. A music fan at a swimming pool is listening to a radio on a diving platform. The radio is playing a constant frequency tone when this fellow, clutching his radio, jumps. Describe the Doppler effect heard by (a) a person left behind on the platform and (b) a person down below floating on a rubber raft. In each case, specify (1) whether the observed frequency is greater or smaller than the frequency produced by the radio, (2) whether the observed frequency is constant, and (3) how the observed frequency changes during the fall, if it does change. Give your reasoning.

19. When a car is at rest, its horn emits a frequency of 600 Hz. A person standing in the middle of the street hears the horn with a frequency of 580 Hz. Should the person jump out of the way? Account for your answer.

20. The text discusses how the Doppler effect arises when (1) the observer is stationary and the source moves and (2) the observer moves and the source is stationary. A car is speeding toward a large wall and sounds the horn. Is the Doppler effect present in the echo that the driver hears? If it is present, from which of the above situations does it arise, (1) or (2) or both? Explain.

PROBLEMS

ssm Solution is in the Student Solutions Manual. **www** Solution is available on the World Wide Web at http://www.wiley.com/college/cutnell
♪ This icon represents a biomedical application.

Section 16.1 The Nature of Waves,
Section 16.2 Periodic Waves

1. ssm A person standing in the ocean notices that after a wave crest passes by, ten more crests pass in a time of 120 s. What is the frequency of the wave?

2. The magnetic tape of a cassette deck moves with a speed of 0.048 m/s ($1\frac{7}{8}$ inches per second). The recording head records a 15-kHz tone on the tape. What is the wavelength λ of the magnetized regions?

3. ssm In Figure 16.2 the hand moves the end of the Slinky up and down through two complete cycles in one second. The wave moves along the Slinky at a speed of 0.50 m/s. Find the distance between two adjacent crests on the wave.

4. Consider the freight train in Figure 16.7. Suppose 15 boxcars pass by in a time of 12.0 s and each has a length of 14.0 m. (a) What is the frequency at which each boxcar passes? (b) What is the speed of the train?

5. A longitudinal wave with a frequency of 3.0 Hz takes 1.7 s to travel the length of a 2.5-m Slinky (see Figure 16.3). Determine the wavelength of the wave.

6. Tsunamis are fast-moving waves often generated by underwater earthquakes. In the deep ocean their amplitude is barely noticeable, but upon reaching shore, they can rise up to the astonishing height of a six-story building. One tsunami, generated off the Aleutian islands in Alaska, had a wavelength of 750 km and traveled a distance of 3700 km in 5.3 h. (a) What was the speed (in m/s) of the wave? For reference, the speed of a 747 jetliner is about 250 m/s. Find the wave's (b) frequency and (c) period.

7. ssm Suppose the amplitude and frequency of the transverse wave in Figure 16.2c are, respectively, 1.3 cm and 5.0 Hz. Find the *total vertical distance* (in cm) through which the colored dot moves in 3.0 s.

8. A person fishing from a pier observes that four wave crests pass by in 7.0 s and estimates the distance between two successive crests as 4.0 m. The timing starts with the first crest and ends with the fourth. What is the speed of the wave?

*9. The speed of a transverse wave on a string is 450 m/s, while the wavelength is 0.18 m. The amplitude of the wave is 2.0 mm. How much time is required for a particle of the string to move through a total distance of 1.0 km?

*10. In Figure 16.3c the colored "dot" exhibits simple harmonic motion as the longitudinal wave passes. The wave has an amplitude of 5.4×10^{-3} m and a frequency of 4.0 Hz. Find the maximum acceleration of the dot.

**11. A water-skier is moving at a speed of 12.0 m/s. When she skis in the same direction as a traveling wave, she springs upward every 0.600 s because of the wave crests. When she skis in the direction opposite to that in which the wave moves, she springs upward every 0.500 s in response to the crests. The speed of the skier is greater than the speed of the wave. Determine (a) the speed and (b) the wavelength of the wave.

Section 16.3 The Speed of a Wave on a String

12. A 0.75-m string is stretched so the tension is 2.3 N. A transverse wave with a frequency of 150 Hz and a wavelength of 0.40 m travels on the string. What is the mass of the string?

13. ssm The linear density of the A string on a violin is 7.8×10^{-4} kg/m. A wave on the string has a frequency of 440 Hz and a wavelength of 65 cm. What is the tension in the string?

14. A vibrator moves one end of a rope up and down to generate a wave. The tension in the rope is 58 N. The frequency is then doubled. To what value must the tension be adjusted, so the new wave has the same wavelength as the old one?

15. ssm www The middle C string on a piano is under a tension of 944 N. The period and wavelength of a wave on this string are 3.82 ms and 1.26 m, respectively. Find the linear density of the string.

16. Two wires are parallel, and one is directly above the other. Each has a length of 50.0 m and a mass per unit length of 0.020 kg/m. However, the tension in wire A is 6.00×10^2 N, while the tension in wire B is 3.00×10^2 N. Transverse wave pulses are generated simultaneously, one at the left end of wire A and one at the right end of wire B. The pulses travel toward each other. How much time does it take until the pulses pass each other?

17. To measure the acceleration due to gravity on a distant planet, an astronaut hangs a 0.085-kg ball from the end of a wire. The wire has a length of 1.5 m and a linear density of 3.1×10^{-4} kg/m. Using electronic equipment, the astronaut measures the time for a transverse pulse to travel the length of the wire and ob-

tains a value of 0.083 s. The mass of the wire is negligible compared to the mass of the ball. Determine the acceleration due to gravity.

*18. Review Conceptual Example 3 before starting this problem. The amplitude of a transverse wave on a string is 4.5 cm. The ratio of the maximum particle speed to the speed of the wave is 3.1. What is the wavelength (in cm) of the wave?

*19. ssm Two blocks are connected by a wire that has a mass per unit length of 8.50×10^{-4} kg/m. One block has a mass of 19.0 kg, while the other has a mass of 42.0 kg. These blocks are being pulled across a horizontal frictionless floor by a horizontal force **P** that is applied to the less massive block. A transverse wave travels on the wire between the blocks with a speed of 352 m/s (relative to the wire). The mass of the wire is negligible compared to the mass of the blocks. Find the magnitude of **P**.

**20. A copper wire, whose cross-sectional area is 1.1×10^{-6} m², has a linear density of 7.0×10^{-3} kg/m and is strung between two walls. At the ambient temperature, a transverse wave travels with a speed of 46 m/s on this wire. The coefficient of linear expansion for copper is 17×10^{-6} (C°)$^{-1}$, and Young's modulus for copper is 1.1×10^{11} N/m². What will be the speed of the wave when the temperature is lowered by 14 C°? Ignore any change in the linear density caused by the change in temperature.

**21. The drawing shows a 15.0-kg ball being whirled in a circular path on the end of a string. The motion occurs on a frictionless, horizontal table. The angular speed of the ball is $\omega = 12.0$ rad/s. The string has a mass of 0.0230 kg. How much time does it take for a wave on the string to travel from the center of the circle to the ball?

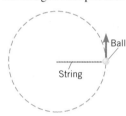

Section 16.4 The Mathematical Description of a Wave

(Note: The phase angles $(2\pi ft - 2\pi x/\lambda)$ and $(2\pi ft + 2\pi x/\lambda)$ are measured in radians, not degrees.)

22. A wave is moving in the $+x$ direction. Assuming that the wave has the following properties, write the equation of the wave (similar to Equation 16.3 or 16.4): speed = 7.1 m/s, amplitude = 0.15 m, wavelength = 0.28 m.

23. ssm A wave has the following properties: amplitude = 0.37 m, period = 0.77 s, wave speed = 12 m/s. The wave is traveling in the $-x$ direction. What is the mathematical expression (similar to Equation 16.3 or 16.4) for the wave?

24. The drawing shows two graphs that represent a transverse wave on a string. The wave is moving in the $+x$ direction. Using the information contained in these graphs, write the mathematical expression (similar to Equation 16.3 or 16.4) for the wave.

25. The displacement (in meters) of a wave is $y = 0.26 \sin(\pi t - 3.7\pi x)$, where t is in seconds and x is in meters. (a) Is the wave traveling in the $+x$ or $-x$ direction? (b) What is the displacement y when $t = 38$ s and $x = 13$ m?

****26.** The tension in a string is 15 N, and its linear density is 0.85 kg/m. A wave on the string travels toward the $-x$ direction; it has an amplitude of 3.6 cm and a frequency of 12 Hz. What are the (a) speed and (b) wavelength of the wave? (c) Write down a mathematical expression (like Equation 16.3 or 16.4) for the wave, substituting numbers for the variables A, f, and λ.

****27. ssm** A transverse wave is traveling on a string. The displacement y of a particle from its equilibrium position is given by $y = (0.021 \text{ m}) \sin(25t - 2.0x)$. Note that the phase angle $25t - 2.0x$ is in radians, t is in seconds, and x is in meters. The linear density of the string is 1.6×10^{-2} kg/m. What is the tension in the string?

*****28.** A transverse wave on a string has an amplitude of 0.20 m and a frequency of 175 Hz. Consider the particle of the string at $x = 0$ m. It begins with a displacement of $y = 0$ m when $t = 0$ s, according to Equation 16.3 or 16.4. How much time passes between the first two instants when this particle has a displacement of $y = 0.10$ m?

Section 16.5 The Nature of Sound,
Section 16.6 The Speed of Sound

29. ssm The speed of a sound in a container of hydrogen at 201 K is 1220 m/s. What would be the speed of sound if the temperature were raised to 405 K? Assume that hydrogen behaves like an ideal gas.

30. A sonar unit on a submarine sends out a pulse of sound into seawater. The pulse returns 1.30 s later. What is the distance to the object that reflects the pulse back to the submarine?

31. The right-most key on a piano produces a sound wave that has a frequency of 4185.6 Hz. Assuming that the speed of sound in air is 343 m/s, find the corresponding wavelength.

32. Have you ever listened for an approaching train by kneeling next to a railroad track and putting your "ear to the rail?" Young's modulus for steel is $Y = 2.0 \times 10^{11}$ N/m^2, and the density of steel is $\rho = 7860$ kg/m^3. On a day when the temperature is 20 °C, how many times greater is the speed of sound in the rail than in the air?

33. ssm Argon (molecular mass $= 39.9$ u) is a monatomic gas. Assuming that it behaves like an ideal gas at 298 K ($\gamma = 1.67$), find (a) the rms speed of argon atoms and (b) the speed of sound in argon.

34. A sound wave is incident on a pool of fresh water. The sound enters the water perpendicularly and travels a distance of 0.45 m before striking a 0.15-m-thick copper block lying on the bottom. The sound passes through the block, reflects from the bottom surface of the block, and returns to the top of the water along the same path. How much time elapses between when the sound enters and leaves the water?

35. An explosion occurs at the end of a pier. The sound reaches the other end of the pier by traveling through three media: air, fresh water, and a slender handrail of solid steel. The speeds of sound in air, water, and the handrail are 343, 1482, and 5040 m/s, respectively. The sound travels a distance of 125 m in each medium. (a) Through which medium does the sound arrive first, second, and third? (b) After the first sound arrives, how much later do the second and third sounds arrive?

36. As the drawing illustrates, a siren can be made by blowing a jet of air through 20 equally spaced holes in a rotating disk. The time it takes for successive holes to move past the air jet is the period of the sound. The siren is to produce a 2200-Hz tone. What must be the angular speed ω (in rad/s) of the disk?

37. At what temperature is the speed of sound in helium (atomic mass $= 4.003$ u) the same as the speed of sound in oxygen at 0 °C? Consider helium to be an ideal gas ($\gamma = 1.67$).

****38.** Review the rule of thumb in Conceptual Example 5. Suppose the speed of light were 961 m/s, instead of 3.0×10^8 m/s, but the speed of sound remained at 343 m/s. The new rule of thumb is stated as follows:

> After you see a flash of lightning, count off the seconds until the thunder is heard. Divide the number of seconds by the number X. The result gives the approximate distance (in miles) to the thunderstorm.

What is the number X?

****39. ssm** A long slender bar is made from an unknown material. The length of the bar is 0.83 m, its cross-sectional area is 1.3×10^{-4} m^2, and its mass is 2.1 kg. A sound wave travels from one end of the bar to the other end in 1.9×10^{-4} s. From which one of the materials listed in Table 10.1 is the bar most likely to be made?

****40.** When an earthquake occurs, two types of sound waves are generated and travel through the earth. The primary, or P, wave has a speed of about 8.0 km/s and the secondary, or S, wave has a speed of about 4.5 km/s. A seismograph, located some distance away, records the arrival of the P wave and then, 78 s later, records the arrival of the S wave. Assuming that the waves travel in a straight line, how far is the seismograph from the earthquake?

****41. ssm** A hunter is standing on flat ground between two vertical cliffs that are directly opposite one another. He is closer to one cliff than the other. He fires a gun and, after a while, hears three echoes. The second echo arrives 1.6 s after the first, and the third echo arrives 1.1 s after the second. Assuming that the speed of sound is 343 m/s and that there are no reflections of sound from the ground, find the distance between the cliffs.

****42.** At a height of ten meters above the surface of a lake, a sound pulse is generated. The echo from the bottom of the lake returns to the point of origin 0.140 s later. The air and water temperatures are 20 °C. How deep is the lake?

*43. A monatomic ideal gas ($\gamma = 1.67$) is contained within a box whose volume and pressure are 2.5 m^3 and 3.5 $\times$ 10^5 Pa, respectively. The total mass of the gas is 2.3 kg. Find the speed of sound in the gas.

*44. Both krypton (Kr) and neon (Ne) can be approximated as monatomic ideal gases. The atomic mass of krypton is 83.8 u, while that of neon is 20.2 u. A loudspeaker produces a sound whose wavelength in krypton is 1.25 m. If the loudspeaker were used to produce sound of the same frequency in neon at the same temperature, what would be the wavelength?

**45. In a mixture of argon (atomic mass = 39.9 u) and neon (atomic mass = 20.2 u), the speed of sound is 363 m/s at 3.00 $\times$ 10^2 K. Assume that both monatomic gases behave as ideal gases. Find the percentage of the atoms that are argon and the percentage that are neon.

**46. The sonar unit on a boat is designed to measure the depth of fresh water ($\rho = 1.00 \times 10^3$ kg/m^3, $B_{ad} = 2.20 \times 10^9$ Pa). When the boat moves into salt water ($\rho = 1025$ kg/m^3, $B_{ad} = 2.37 \times 10^9$ Pa), the sonar unit is no longer calibrated properly. In salt water, the sonar unit indicates the water depth to be 10.0 m. What is the actual depth of the water?

**47. ssm www As a prank, someone drops a water-filled balloon out of a window. The balloon is released from rest at a height of 10.0 m above the ears of a man who is the target. Because of a guilty conscience, however, the prankster shouts a warning after the balloon is released. The warning will do no good, however, if shouted after the balloon reaches a certain point, even if the man could react infinitely quickly. Assuming that the air temperature is 20 °C and ignoring the effect of air resistance on the balloon, determine how far above the man's ears this point is.

Section 16.7 Sound Intensity

48. A typical adult ear has a surface area of 2.1 $\times$ 10^{-3} m^2. The sound intensity during a normal conversation is about 3.2 $\times$ 10^{-6} W/m^2 at the listener's ear. Assume the sound strikes the surface of the ear perpendicularly. How much power is intercepted by the ear?

49. ssm www At a distance of 3.8 m from a siren, the sound intensity is 3.6 $\times$ 10^{-2} W/m^2. Assuming that the siren radiates sound uniformly in all directions, find the total power radiated.

50. The average sound intensity inside a busy restaurant is 3.2 $\times$ 10^{-5} W/m^2. How much energy goes into each ear (area = 2.1 $\times$ 10^{-3} m^2) during a one-hour meal?

51. A rocket in a fireworks display explodes high in the air. The sound spreads out uniformly in all directions. The intensity of the sound is 2.0 $\times$ 10^{-6} W/m^2 at a distance of 120 m from the explosion. Find the distance from the source at which the intensity is 0.80 $\times$ 10^{-6} W/m^2.

52. Suppose in Conceptual Example 8 (see Figure 16.25) that the person is producing 1.1 mW of sound power. Some of the sound is reflected from the floor and ceiling. The intensity of this reflected sound at a distance of 3.0 m from the source is 4.4 $\times$ 10^{-6} W/m^2. What is the total sound intensity due to both the direct and reflected sounds, at this point?

53. ssm A loudspeaker has a circular opening with a radius of 0.0950 m. The electrical power needed to operate the speaker is 25.0 W. The average sound intensity at the opening is 17.5 W/m^2. What percentage of the electrical power is converted by the speaker into sound power?

*54. ✍ Deep ultrasonic heating is used to promote healing of torn tendons. It is produced by applying ultrasonic sound to the body. The sound transducer (generator) is circular with a radius of 1.8 cm, and it produces a sound intensity of 5.9 $\times$ 10^3 W/m^2. How much time is required for the transducer to emit 4800 J of sound energy?

*55. A dish of lasagna is being heated in a microwave oven. The effective area of the lasagna that is exposed to the microwaves is 1.6 $\times$ 10^{-2} m^2. The mass of the lasagna is 0.25 kg, and its specific heat capacity is 3400 J/(kg·C°). The temperature rises by 80.0 C° in 7.0 minutes. What is the intensity of the microwaves in the oven?

*56. When a helicopter is hovering 1450 m directly overhead, an observer on the ground measures a sound intensity I. Assume that sound is radiated uniformly from the helicopter and that ground reflections are negligible. How far must the helicopter fly in a straight line parallel to the ground before the observer measures a sound intensity of $\frac{1}{4}I$?

**57. ssm A rocket, starting from rest, travels straight up with an acceleration of 58.0 m/s^2. When the rocket is at a height of 562 m, it produces sound that eventually reaches a ground-based monitoring station directly below. The sound is emitted uniformly in all directions. The monitoring station measures a sound intensity I. Later, the station measures an intensity $\frac{1}{3}I$. Assuming that the speed of sound is 343 m/s, find the time that has elapsed between the two measurements.

Section 16.8 Decibels

58. The volume control on a stereo amplifier is adjusted so the sound intensity level increases from 23 to 61 dB. What is the ratio of the final sound intensity to the original sound intensity?

59. The bellow of a territorial bull hippopotamus has been measured at 115 dB above the threshold of hearing. What is the sound intensity?

60. A recording engineer works in a soundproofed room that is 44.0 dB quieter than the outside. If the sound intensity in the room is 1.20 $\times$ 10^{-10} W/m^2, what is the intensity outside?

61. ssm When a person wears a hearing aid, the sound intensity level increases by 30.0 dB. By what factor does the sound intensity increase?

62. The equation $\beta = (10 \text{ dB}) \log (I/I_0)$, which defines the decibel, is sometimes written in terms of power P (in watts) rather than intensity I (in watts/meter^2). The form $\beta = (10 \text{ dB}) \log (P/P_0)$ can be used to compare two power levels in terms of decibels. Suppose that stereo amplifier A is rated at $P = 250$ watts per channel, while amplifier B has a rating of $P_0 = 45$

watts per channel. (a) Expressed in decibels, how much more powerful is A compared to B? (b) Will A sound more than twice as loud as B? Justify your answer.

63. ssm A listener doubles his distance from a source that emits sound uniformly in all directions. By how many decibels does the sound intensity level change?

64. For information, read problem 62 before working this problem. Stereo manufacturers express the power output of a stereo amplifier using the decibel, abbreviated as dBW, where the "W" indicates that a reference power level of $P_0 = 1.00$ W has been used. If an amplifier has a power rated at 17.5 dBW, how many watts of power can this amplifier deliver?

***65.** Sound is coming through an open window whose dimensions are 1.1 m × 0.75 m. The sound intensity level is 95 dB above the threshold of hearing. How much sound *energy* comes through the window in one hour?

***66.** When a single person shouts at a football game, the sound intensity level at the center of the field is 60.0 dB. When all the people shout together, the intensity level increases to 109 dB. Assuming that each person generates the same sound intensity at the center of the field, how many people are at the game?

***67.** A portable radio is sitting at the edge of a balcony 5.1 m above the ground. The unit is emitting sound uniformly in all directions. By accident, it falls from rest off the balcony and continues to play on the way down. A gardener is working in a flower bed directly below the falling unit. From the instant the unit begins to fall, how much time is required for the sound intensity level heard by the gardener to increase by 10.0 dB?

****68.** A source emits sound uniformly in all directions. A radial line is drawn from this source. On this line, determine the positions of two points, 1.00 m apart, such that the intensity level at one point is 2.00 dB greater than that at the other.

****69. ssm** Suppose that when a certain sound intensity level (in dB) triples, the sound intensity (in W/m²) also triples. Determine this sound intensity level.

Section 16.9 The Doppler Effect

70. A source is generating circular waves on the surface of a lake. The waves have a wavelength of 13.4 m and travel outward at a speed of 6.70 m/s. You are in a boat whose speed is 4.20 m/s and heading directly toward the source of the waves. What do you observe for the frequency of the waves?

71. The security alarm on a parked car goes off and produces a frequency of 960 Hz. The speed of sound is 343 m/s. As you drive toward this parked car, pass it, and drive away, you observe the frequency to change by 95 Hz. At what speed are you driving?

72. Suppose you are stopped for a traffic light, and an ambulance approaches you from behind with a speed of 18 m/s. The siren on the ambulance produces sound with a frequency of 955 Hz. The speed of sound in air is 343 m/s. What is the wavelength of the sound reaching your ears?

73. ssm From a vantage point very close to the track at a stock car race, you hear the sound emitted by a moving car. You detect a frequency that is 0.86 times smaller than that emitted by the car when it is stationary. The speed of sound is 343 m/s. What is the speed of the car?

***74.** A loudspeaker in a parked car is producing sound whose frequency is 20 510 Hz. The sound cannot be heard because the frequency is too high. When the car is moving, however, a person standing on the street hears the sound. (a) Is the car moving toward or away from the person? Why? (b) If the speed of sound is 343 m/s, what is the minimum speed of the car if the person is a student with normal hearing?

***75. ssm** An aircraft carrier has a speed of 13.0 m/s relative to the water. A jet is catapulted from the deck and has a speed of 67.0 m/s relative to the water. The engines produce a 1550-Hz whine, and the speed of sound is 343 m/s. What is the frequency of the sound heard by the crew on the ship?

***76.** A microphone is moving in air toward a stationary source of sound (speed of sound = 343 m/s). The detected frequency is 83.0 Hz greater than the emitted frequency. When the microphone moves at the same speed toward the same stationary source in a liquid, the detected frequency is only 23.0 Hz greater than the emitted frequency. What is the speed of sound in the liquid?

***77.** A bungee jumper jumps from rest and screams with a frequency of 589 Hz. The air temperature is 20 °C. What is the frequency heard by the people on the ground below when she has fallen a distance of 11.0 m? Assume that the bungee cord has not yet taken effect, so she is in free-fall.

***78.** Two submarines are underwater and approaching each other head-on. Sub A has a speed of 12 m/s and sub B has a speed of 8 m/s. Sub A sends out a 1550-Hz sonar wave that travels at a speed of 1522 m/s. (a) What is the frequency detected by sub B? (b) Part of the sonar wave is reflected from B and returns to A. What frequency does A detect for this reflected wave?

***79. ssm** A motorcycle starts from rest and accelerates along a straight line at 2.81 m/s². The speed of sound is 343 m/s. A siren at the starting point remains stationary. How far has the motorcycle gone when the driver hears the frequency of the siren at 90.0% of the value it has when the motorcycle is stationary?

****80.** A microphone is attached to a spring that is suspended from the ceiling, as the drawing indicates. Directly below on the floor is a stationary 440-Hz source of sound. The microphone vibrates up and down in simple harmonic motion with a period of 2.0 s. The difference between the maximum and minimum sound frequencies detected by the microphone is 2.1 Hz. Ignoring any reflections of sound in the room and using 343 m/s for the speed of sound, determine the amplitude of the simple harmonic motion.

Sound source

ADDITIONAL PROBLEMS

81. An amplified guitar has a sound intensity level that is 14 dB greater than the same unamplified sound. What is the ratio of the amplified intensity to the unamplified intensity?

82. To navigate, a porpoise emits a sound wave that has a wavelength of 2.5 cm. The speed at which sound travels in seawater is 1470 m/s. Find the period of the wave.

83. ssm At 20 °C the densities of fresh water and ethanol are, respectively, 998 and 789 kg/m³. Find the ratio of the adiabatic bulk modulus of fresh water to the adiabatic bulk modulus of ethanol at 20 °C.

84. A transverse wave is traveling with a speed of 300 m/s on a horizontal string. If the tension in the string is increased by a factor of four, what is the speed of the wave?

85. At a football game, a stationary spectator is watching the halftime show. A trumpet player in the band is playing a 784-Hz tone while marching directly toward the spectator at a speed of 0.83 m/s. On a day when the speed of sound is 343 m/s, what frequency does the spectator hear?

86. A person lying on an air mattress in the ocean rises and falls through one complete cycle every five seconds. The crests of the wave causing the motion are 20.0 m apart. Determine (a) the frequency and (b) the speed of the wave.

87. ssm Humans can detect a difference in sound intensity levels as small as 1.0 dB. What is the ratio of the sound intensities?

88. Suppose that sound is emitted uniformly in all directions by a public address system. The intensity at a location 22 m away from the sound source is 3.0×10^{-4} W/m². What is the intensity at a spot that is 78 m away?

89. The distance between a loudspeaker and the left ear of a listener is 2.70 m. (a) Calculate the time required for sound to travel this distance if the air temperature is 20 °C. (b) Assuming that the sound frequency is 523 Hz, how many wavelengths of sound are contained in this distance?

90. A wave has a displacement (in meters) of $y = (0.45) \sin (8.0\pi t + \pi x)$, where t and x are expressed in seconds and meters, respectively. (a) Find the amplitude, the frequency, the wavelength, and the speed of the wave. (b) Is this wave traveling in the $+x$ or $-x$ direction?

91. ssm A speeder looks in his rearview mirror. He notices that a police car has pulled behind him and is matching his speed of 38 m/s. The siren on the police car has a frequency of 860 Hz when the police car and the listener are stationary. The speed of sound is 343 m/s. What frequency does the speeder hear when the siren is turned on in the moving police car?

92. The wavelength of a sound wave in air is 2.74 m at 20 °C. What is the wavelength of this sound wave in fresh water at 20 °C? *(Hint: The frequency of the sound is the same in both media.)*

93. A rocket engine emits 2.0×10^5 J of sound energy every second. The sound is emitted uniformly in all directions. What is

the sound intensity level, measured relative to the threshold of hearing, at a distance of 85 m away from the engine?

* **94.** Two identical rifles are shot at the same time, and the sound intensity level is 80.0 dB. What would be the sound intensity level if only one rifle were shot? *(Hint: The answer is not 40.0 dB.)*

* **95. ssm www** The drawing shows a frictionless incline and pulley. The two blocks are connected by a wire (mass per unit length = 0.0250 kg/m) and remain stationary. A transverse wave on the wire has a speed of 75.0 m/s. Neglecting the weight of the wire relative to the tension in the wire, find the masses m_1 and m_2 of the blocks.

* **96.** A 3.49 rad/s ($33\frac{1}{3}$ rpm) record has a 5.00-kHz tone cut in the groove. If the groove is located 0.100 m from the center of the record (see drawing), what is the "wavelength" in the groove?

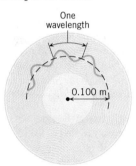

Problem 96

* **97.** A steel cable of cross-sectional area 2.83×10^{-3} m² is kept under a tension of 1.00×10^4 N. The density of steel is 7860 kg/m³ (this is *not* the linear density). At what speed does a transverse wave move along the cable?

* **98.** Review Conceptual Example 3 before starting this problem. A horizontal wire is under a tension of 315 N and has a mass per unit length of 6.50×10^{-3} kg/m. A transverse wave with an amplitude of 2.50 mm and a frequency of 585 Hz is traveling on this wire. As the wave passes, a particle of the wire moves up and down in simple harmonic motion. Obtain (a) the speed of the wave and (b) the maximum speed with which the particle moves up and down.

* **99. ssm** Review Conceptual Example 8 as background for this problem. A loudspeaker is generating sound in a room. At a certain point, the sound waves coming directly from the speaker (without reflecting from the walls) create an intensity level of 75.0 dB. The waves reflected from the walls create, by themselves, an intensity level of 72.0 dB at the same point. What is the total intensity level? *(Hint: The answer is not 147.0 dB.)*

* **100.** The sound intensity level of a person speaking normally is about 65 dB above the threshold of hearing. What is the minimum number of people speaking simultaneously, each with this intensity level, that is necessary to produce a sound intensity level at least 78 dB above the threshold of hearing?

** **101.** A jet is flying horizontally, as the drawing shows. When the plane is directly overhead at B, a person on the ground hears the sound coming from *A* in the drawing. The average temperature of the air is 20 °C. If the speed of the plane at *A* is 164 m/s, what is its speed at *B*, assuming that it has a constant acceleration?

102. Civil engineers use a transit theodolite when surveying. A modern version of this device determines distance by measuring the time required for an ultrasonic pulse to reach a target, reflect from it, and return. Effectively, such a theodolite is calibrated properly when it is programmed with the speed of sound appropriate for the ambient air temperature. (a) Suppose the round-trip time for the pulse is 0.580 s on a day when the air temperature is 293 K, the temperature for which the instrument is calibrated. How far is the target from the theodolite? (b) Assume that air behaves as an ideal gas. If the air temperature were 298 K, rather than the calibration temperature of 293 K, what percentage error would there be in the distance measured by the theodolite?

CONCEPTS & CALCULATIONS

GROUP LEARNING PROBLEMS

Note: Each of these problems consists of Concept Questions followed by a related quantitative Problem. They are designed for use by students working alone or in small learning groups. The Concept Questions involve little or no mathematics and are intended to stimulate group discussions. They focus on the concepts with which the problems deal. Recognizing the concepts is the essential initial step in any problem-solving technique.

103. Concept Questions The drawing shows a snapshot of two waves traveling to the right at the same speed. (a) Rank the waves according to their wavelengths, largest first. (b) Which wave, if either, has the higher frequency? (c) If a particle were attached to each wave, like that in Figure 16.10, which particle would have the greater maximum speed as it moves up and down? Justify your answers.

Problem (a) From the data in the drawing, determine the wavelength of each wave. (b) If the speed of the waves is 12 m/s, calculate the frequency of each one. (c) What is the maximum speed for a particle attached to each wave? Check that your answers are consistent with those for the Concept Questions.

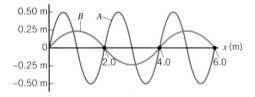

104. Concept Questions Example 4 in the text discusses an ultrasonic ruler that displays the distance between the ruler and an object, such as a wall. The ruler sends out a pulse of ultrasonic sound and measures the time it takes for the pulse to reflect from the object and return. The ruler uses this time, along with a preset value for the speed of sound in air, to determine the distance. Suppose you use this ruler underwater, rather than in air. (a) Is the speed of sound in water greater than, less than, or equal to the speed of sound in air? (b) Is the reading on the ruler greater than, less than, or equal to the actual distance? Provide reasons for your answers.

Problem The actual distance from the ultrasonic ruler to an object is 25.0 m. The adiabatic bulk modulus and density of seawater are $B = 2.31 \times 10^9$ Pa and $\rho = 1025$ kg/m^3, respectively. Assume that the ruler uses a preset value of 343 m/s for the speed of sound in air, and determine the distance reading on its display. Verify that your answer is consistent with your answer to the Concept Questions.

105. Concept Question Suppose you are part of a team that is trying to break the "sound barrier" with a jet-powered car, which means that it must travel faster than the speed of sound in air. Would you attempt this feat early in the morning when the temperature is cool, later in the afternoon when the temperature is warmer, or does it even matter what the temperature is?

Problem In the morning, the air temperature is 0 °C and the speed of sound is 331 m/s. What must be the speed of your car if it is to break the sound barrier when the temperature has risen to 43 °C in the afternoon? Assume that air behaves like an ideal gas.

106. Concept Questions A source of sound is located at the center of two concentric spheres, parts of which are shown in the drawing. The source emits sound uniformly in all directions. On the spheres are drawn three small patches that may, or may not, have equal areas. However, the same sound power passes through each patch. (a) Rank the sound intensity at each patch, greatest first. (b) Rank the area of each patch, largest first. Provide reasons for your answers.

Problem The source produces 2.3 W of sound power, and the radii of the concentric spheres are $r_A = 0.60$ m and $r_B = 0.80$ m.

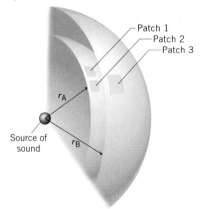

(a) Determine the sound intensity at each of the three patches. (b) The sound power that passes through each of the patches is 1.8×10^{-3} W. Find the area of each patch. Verify that your answers are consistent with those to the Concept Questions.

107. Concept Questions The threshold of hearing for an average young person is 0 dB. Two individuals who are not average have thresholds of hearing that are $\beta_1 = -8.00$ dB and $\beta_2 = +12.0$ dB. (a) Is the sound intensity at a level of -8.00 dB greater than, or less than, that at 0 dB? (b) Which individual has the better hearing? Why?

Problem How many times greater (or less) is the sound intensity when person 1 ($\beta_1 = -8.00$ dB) hears it at the threshold of hearing compared to when person 2 ($\beta_2 = +12.0$ dB) hears it at the threshold of hearing? Check to see that your answer is consistent with those to the Concept Questions.

108. Concept Questions The table shows three situations in which the Doppler effect may arise. The first two columns indicate the velocities of the sound source and the observer, where the length of each arrow is proportional to the speed. For each situation, fill in the empty columns by deciding whether the wavelength of the sound and the frequency heard by the observer increase, de-

crease, or remain the same compared to the case when there is no Doppler effect. Provide a reason for each answer.

	Velocity of Sound Source	Velocity of Observer	Wavelength	Frequency Heard by Observer
(a)	0 m/s	0 m/s		
(b)	$\longrightarrow$	0 m/s		
(c)	$\longrightarrow$	$\leftarrow$		

Problem The siren on an ambulance is emitting a sound whose frequency is 2450 Hz. The speed of sound is 343 m/s. (a) If the ambulance is stationary and you (the "observer") are sitting in a parked car, what is the wavelength of the sound and the frequency heard by you? (b) Suppose the ambulance is moving toward you at a speed of 26.8 m/s. Determine the wavelength of the sound and the frequency heard by you. (c) If the ambulance is moving toward you at a speed of 26.8 m/s and you are moving toward it at a speed of 14.0 m/s, find the wavelength of the sound and the frequency that you hear. Be sure that your answers are consistent with your answers to the Concept Questions.

Virtually all musical instruments, such as the bells in this photograph, produce their sound in a way that involves the principle of linear superposition. This chapter discusses the principle and a number of other topics that relate to it.

CHAPTER 17

THE PRINCIPLE OF LINEAR SUPERPOSITION AND INTERFERENCE PHENOMENA

17.1   THE PRINCIPLE OF LINEAR SUPERPOSITION

As we have seen in the last chapter, sound is a pressure wave. Often, two or more sound waves are present at the same place at the same time, such as when everyone is talking at a party or when music plays from the speakers of a stereo system. To illustrate what happens when several waves pass simultaneously through the same region, let's first consider Figures 17.1 and 17.2, which show two transverse pulses of equal heights moving toward each other along a Slinky. In Figure 17.1 both pulses are "up," while in Figure 17.2 one is "up" and the other is "down." Part *a* of each drawing shows the two pulses beginning to overlap. The pulses merge, and the Slinky assumes a shape that is *the sum of the shapes of the individual pulses.* Thus, when the two "up" pulses overlap completely, as in Figure 17.1*b*, the Slinky has a pulse height that is twice the height of an individual pulse. Likewise, when the "up" pulse and the "down" pulse overlap exactly, as in Figure 17.2*b*, they momentarily cancel, and the Slinky becomes straight. In either case, the two pulses move apart after overlapping, and the Slinky once again conforms to the shapes of the individual pulses.

The adding together of individual pulses to form a resultant pulse is an example of a more general concept called the ***principle of linear superposition.***

● THE PRINCIPLE OF LINEAR SUPERPOSITION

When two or more waves are present simultaneously at the same place, the resultant disturbance is the sum of the disturbances from the individual waves.

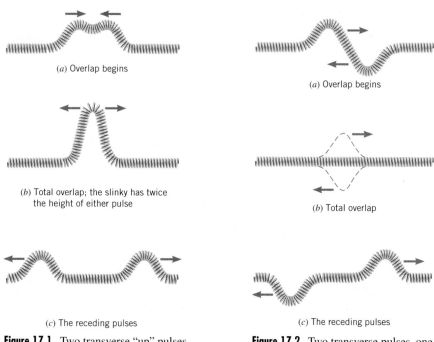

(a) Overlap begins

(b) Total overlap; the slinky has twice the height of either pulse

(c) The receding pulses

Figure 17.1   Two transverse "up" pulses passing through each other.

(a) Overlap begins

(b) Total overlap

(c) The receding pulses

Figure 17.2   Two transverse pulses, one "up" and one "down," passing through each other.

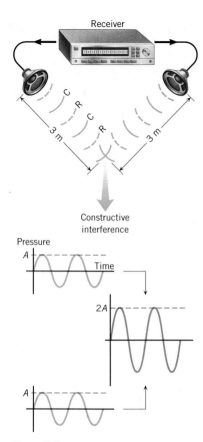

Figure 17.3 As a result of constructive interference between the two sound waves (amplitude = *A*), a loud sound (amplitude = 2*A*) is heard at an overlap point located equally distant from two in-phase speakers (C, condensation; R, rarefaction).

THE PHYSICS OF...

noise-canceling headphones.

This principle can be applied to all types of waves, including sound waves, water waves, and electromagnetic waves such as light. It embodies one of the most important concepts in all of physics, and the remainder of this chapter deals with examples related to it.

17.2 CONSTRUCTIVE AND DESTRUCTIVE INTERFERENCE OF SOUND WAVES

Suppose that the sounds from two speakers overlap in the middle of a listening area, as in Figure 17.3, and that each speaker produces a sound wave of the same amplitude and frequency. For convenience, the wavelength of the sound is chosen to be $\lambda = 1$ m. In addition, assume the diaphrams of the speakers vibrate in phase; that is, they move outward together and inward together. If the distance of each speaker from the overlap point is the same (3 m in the drawing), the condensations (C) of one wave always meet the condensations of the other when the waves come together; similarly, rarefactions (R) always meet rarefactions. According to the principle of linear superposition, the combined pattern is the sum of the individual patterns. As a result, the pressure fluctuations at the overlap point have twice the amplitude *A* that the individual waves have, and a listener at this spot hears a louder sound than that coming from either speaker alone. When two waves always meet condensation-to-condensation and rarefaction-to-rarefaction (or crest-to-crest and trough-to-trough), they are said to be *exactly in phase* and to exhibit *constructive interference.*

Now consider what happens if one of the speakers is moved. The result is surprising. In Figure 17.4, the left speaker is moved away* from the overlap point by a distance equal to one-half of the wavelength, or 0.5 m. Therefore, at the overlap point, a condensation arriving from the left meets a rarefaction arriving from the right. Likewise, a rarefaction arriving from the left meets a condensation arriving from the right. According to the principle of linear superposition, the net effect is a mutual cancellation of the two waves. The condensations from one wave offset the rarefactions from the other, leaving only a *constant air pressure.* A constant air pressure, devoid of condensations and rarefactions, means that a listener detects no sound. When two waves always meet condensation-to-rarefaction (or crest-to-trough), they are said to be *exactly out of phase* and to exhibit *destructive interference.*

When two waves meet, they interfere constructively if they always meet exactly in phase and destructively if they always meet exactly out of phase. In either case, the wave patterns do not shift relative to one another as time passes. Sources that produce waves in this fashion are called *coherent sources.*

Destructive interference is the basis of a useful technique for reducing the loudness of undesirable sounds. For instance, Figure 17.5 shows a pair of noise-canceling headphones. Small microphones are mounted inside the headphones and detect noise such as the engine noise that an airplane pilot would hear. The headphones also contain circuitry to process the electronic signals from the microphones and reproduce the noise in a form that is exactly out of phase compared to the original. This out-of-phase version is played back through the headphone speakers and,

---

* When the left speaker is moved back, its sound intensity and, hence, its pressure amplitude decrease at the overlap point. In this chapter assume that the power delivered to the left speaker by the receiver is increased slightly to keep the amplitudes equal at the overlap point.

because of destructive interference, combines with the original noise to produce a quieter background.

It should be apparent that if the left speaker in Figure 17.4 were moved away from the overlap point by *another* one-half wavelength ($3\frac{1}{2}$ m + $\frac{1}{2}$ m = 4 m), the two waves would again be in phase, and constructive interference would occur. The listener would hear a loud sound because the left wave travels one whole wavelength (λ = 1 m) farther than the right wave and, at the overlap point, condensation meets condensation and rarefaction meets rarefaction. In general, the important issue is the *difference* in the distances traveled by each wave in reaching the overlap point.

> *For two wave sources vibrating in phase, a difference in path lengths that is zero or an integer number (1, 2, 3, . . .) of wavelengths leads to constructive interference; a difference in path lengths that is a half-integer number ($\frac{1}{2}$, $1\frac{1}{2}$, $2\frac{1}{2}$, . . .) of wavelengths leads to destructive interference.*

Interference effects can also be detected if the two speakers are fixed in position and the listener moves about the room. Consider Figure 17.6, where the sound waves spread outward from each speaker, as indicated by the concentric circular arcs. Each solid arc represents the center of a condensation, while each dashed arc represents the center of a rarefaction. Where the two waves overlap, there are places of constructive interference and places of destructive interference. Constructive interference occurs at any spot where two condensations or two rarefactions intersect, and the drawing shows four such places as solid dots. A listener stationed at any one of these locations hears a loud sound. On the other hand, destructive interference occurs at any place where a condensation and a rarefaction intersect, such as the two open dots in the picture. A listener situated at a point of destructive interference hears no sound. At locations where neither constructive nor destructive interference occurs, the two waves partially reinforce or partially cancel, depending on the position relative to the speakers. Thus, it is possible for a listener to walk about the overlap region and hear marked variations in loudness.

The individual sound waves from the speakers in Figure 17.6 carry energy, and the energy delivered to the overlap region is the sum of the energies of the individual waves. This fact is consistent with the principle of conservation of energy, which we first encountered in Section 6.8. This principle states that energy can neither be created nor destroyed, but can only be converted from one form to another. One of the interesting consequences of interference is that the energy is redistributed, so there are places within the overlap region where the sound is loud and other places where there is no sound at all. Interference, so to speak, "robs Peter to

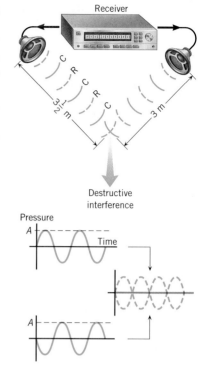

Figure 17.4 The speakers in this drawing vibrate in phase. However, the left speaker is one-half of a wavelength ($\frac{1}{2}$ m) farther from the overlap point than the right speaker. Because of destructive interference, no sound is heard at the overlap point (C, condensation; R, rarefaction).

Figure 17.5 Noise-canceling headphones utilize destructive interference.

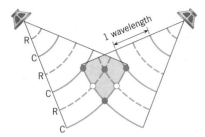

Figure 17.6 Two sound waves overlap in the shaded region. The solid lines denote the centers of condensations (C), while the dashed lines denote the centers of rarefactions (R). Constructive interference occurs at each solid dot (●) and destructive interference at each open dot (○).

● **PROBLEM SOLVING INSIGHT**
To decide whether two sources of sound produce constructive or destructive interference at a point, determine the difference in path lengths between each source and that point and compare it to the wavelength of the sound.

Figure 17.7 Example 1 assumes that the speakers are vibrating in phase and discusses whether this setup leads to constructive or destructive interference at point C for 214-Hz sound waves.

THE PHYSICS OF...

wiring stereo speakers.

pay Paul," but energy is always conserved in the process. Example 1 illustrates how to decide what a listener hears.

EXAMPLE 1   What Does a Listener Hear?

In Figure 17.7 two in-phase loudspeakers, A and B, are separated by 3.20 m. A listener is stationed at point C, which is 2.40 m in front of speaker B. The triangle ABC is a right triangle. Both speakers are playing identical 214-Hz tones, and the speed of sound is 343 m/s. Does the listener hear a loud sound or no sound?

Reasoning The listener will hear either a loud sound or no sound, depending upon whether the interference occurring at point C is constructive or destructive. To determine which it is, we need to find the difference in the distances traveled by the two sound waves that reach point C and see whether the difference is an integer or half-integer number of wavelengths. In either event, the wavelength can be found from Equation 16.1 ($\lambda = v/f$).

Solution Since the triangle ABC is a right triangle, the distance AC is given by the Pythagorean theorem as $\sqrt{(3.20 \text{ m})^2 + (2.40 \text{ m})^2} = 4.00$ m. The distance BC is given as 2.40 m. Thus, we find that the difference in the travel distances for the waves is 4.00 m $-$ 2.40 m $=$ 1.60 m. The wavelength of the sound is

$$\lambda = \frac{v}{f} = \frac{343 \text{ m/s}}{214 \text{ Hz}} = 1.60 \text{ m} \qquad (16.1)$$

Since the difference in the distances is one wavelength, constructive interference occurs at point C, and **the listener hears a loud sound.**

Up to this point, we have been assuming that the speaker diaphragms vibrate synchronously or in phase; that is, they move outward together and inward together. This may not be the case, however, and Conceptual Example 2 considers what happens then.

CONCEPTUAL EXAMPLE 2   Out-of-Phase Speakers

To make a speaker operate, two wires must be connected between the speaker and the receiver (amplifier), as in Figure 17.8. To ensure that the diaphragms of two speakers vibrate in phase, it is necessary to make these connections in exactly the same way. If the wires for one speaker are not connected just as they are for the other speaker, the two diaphragms will vibrate out of phase. Whenever one diaphragm moves outward, the other will move inward, and vice versa. Suppose that in Figures 17.3 and 17.4 the connections are made so that the speaker diaphragms vibrate out of phase, everything else remaining the same. In each case, what kind of interference would result at the overlap point?

Reasoning and Solution Since the diaphragms are vibrating out of phase, one of them is now moving exactly opposite to the way it was moving originally. Let us assume that it is the one on the left in the drawings. The effect of this change is that every condensation originating from the left speaker becomes a rarefaction and every rarefaction becomes a condensation. As a result, in Figure 17.3 a rarefaction from the left now meets a condensation from the right at the overlap point, and **destructive interference results.** In Figure 17.4, a condensation from the left now meets a condensation from the right at the overlap point, **leading to constructive interference.** These results are opposite to those shown in the figures.

Instructions for connecting stereo systems specifically warn users to avoid out-of-phase vibration of the speaker diaphragms. One way to check your system is to play some music that is rich in low-frequency bass tones. Use the monaural mode on your receiver so the same sound comes from each speaker, and slide the speakers toward each other. If the diaphragms are moving in phase, the bass sound will either remain the same or slightly improve as the speakers come together. If the diaphragms are moving out of phase, the bass sound will decrease noticeably because of destructive interference when the speakers are right next to each other. In this event, simply interchange the wires to the terminals on one (but not both) of the speakers.

Related Homework Material: *Problem 8*

The phenomena of constructive and destructive interference are exhibited by all types of waves, not just sound waves. We will encounter interference effects again in Chapter 27, in connection with light waves.

17.3 DIFFRACTION

● *Concepts at a Glance*   Section 16.5 discusses the fact that sound is a pressure wave created by a vibrating object, such as a loudspeaker. The previous two sections of this chapter have examined what happens when two sound waves are present simultaneously at the same place; according to the principle of linear superposition, a resultant disturbance is formed from the sum of the individual waves. This principle reveals that overlapping sound waves exhibit interference effects, whereby the sound energy is redistributed within the overlap region. As the Concepts-at-a-Glance chart in Figure 17.9 shows, we will now use the principle of linear superposition to explore another interference effect, that of diffraction. ●

When a wave encounters an obstacle or the edges of an opening, it bends around them. For instance, a sound wave produced by a stereo system bends around the edges of an open doorway, as Figure 17.10*a* illustrates. If such bending did not occur, sound could be heard outside the room only at locations directly in front of the doorway, as part *b* of the drawing suggests. (It is assumed that no sound is transmit-

Figure 17.8 A loudspeaker is connected to a receiver (amplifier) by two wires.

Figure 17.9 *Concepts at a Glance* By combining our knowledge of sound waves with the principle of linear superposition, we can understand the interference phenomenon of diffraction. In the drawing a water wave moves to the right and passes through an opening in a breakwater. To the right of the breakwater, the bending of the wave is an example of diffraction.

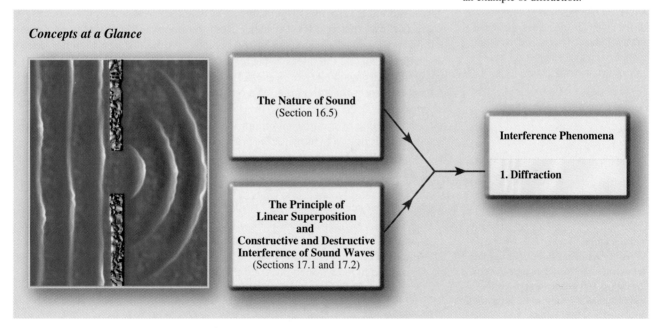

Concepts at a Glance

The Nature of Sound
(Section 16.5)

The Principle of
Linear Superposition
and
Constructive and Destructive
Interference of Sound Waves
(Sections 17.1 and 17.2)

Interference Phenomena

1. Diffraction

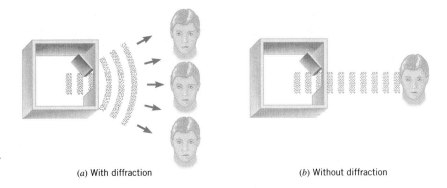

Figure 17.10 (*a*) The bending of a sound wave around the edges of the doorway is an example of diffraction. The source of the sound within the room is not shown. (*b*) If diffraction did not occur, the sound wave would not bend as it passes through the doorway.

(*a*) With diffraction   (*b*) Without diffraction

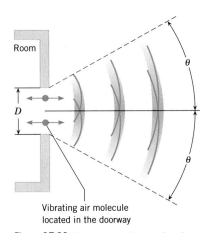

Figure 17.11 Each vibrating molecule of the air in the doorway generates a sound wave that expands outward and bends, or diffracts, around the edges of the doorway. Because of interference effects among the sound waves produced by all the molecules, the sound intensity is mostly confined to the region defined by the angle θ on either side of the doorway.

ted directly through the walls.) The bending of a wave around an obstacle or the edges of an opening is called ***diffraction.*** All kinds of waves exhibit diffraction.

To demonstrate how the bending of waves arises, Figure 17.11 shows an expanded view of Figure 17.10. When the sound wave reaches the doorway, the air in the doorway is set into longitudinal vibration. In effect, each molecule of the air in the doorway becomes a source of a sound wave in its own right, and, for purposes of illustration, the drawing shows two of the molecules. Each produces a sound wave that expands outward in three dimensions, much like a water wave does in two dimensions when a stone is dropped into a pond. The sound waves generated by all the molecules in the doorway must be added together to obtain the total sound intensity at any location outside the room, in accord with the principle of linear superposition. However, even considering only the waves from the two molecules in the picture, it is clear that the expanding wave patterns reach locations off to either side of the doorway. The net effect is a "bending" or diffraction of the sound around the edges of the opening. Further insight into the origin of diffraction can be obtained with the aid of Huygens' principle (see Section 27.5).

When the sound waves generated by every molecule in the doorway are added together, it is found that there are places where the intensity is a maximum and places where it is zero, in a fashion similar to that discussed in the previous section. Analysis shows that at a great distance from the doorway the intensity is a maximum directly opposite the center of the opening. As the distance to either side of the center increases, the intensity decreases and reaches zero, then rises again to a maximum, falls again to zero, rises back to a maximum, and so on. Only the maximum at the center is a strong one. The other maxima are weak and become progressively weaker at greater distances from the center. In Figure 17.11 the angle θ defines the location of the first minimum intensity point on either side of the center. Equation 17.1 gives θ in terms of the wavelength λ and the width D of the doorway and assumes that the doorway can be treated like a slit whose height is very large compared to its width:

Single slit—first minimum $\sin \theta = \dfrac{\lambda}{D}$ (17.1)

● **PROBLEM SOLVING INSIGHT**
When a wave passes through an opening, the extent of diffraction is greater when the ratio λ/D is greater, where λ is the wavelength of the wave and D is the width or diameter of the opening.

Waves also bend around the edges of openings other than single slits. Particularly important is the diffraction of sound by a circular opening. In this case, the angle θ is related to the wavelength λ and the diameter D of the opening by

*Circular opening
—first minimum* $\sin \theta = 1.22 \dfrac{\lambda}{D}$ (17.2)

An important point to remember about Equations 17.1 and 17.2 is that the extent of the diffraction depends on the ratio of the wavelength to the size of the opening. If the ratio λ/D is small, then θ is small. Little diffraction occurs, and the waves are beamed in the forward direction as they leave an opening, much like the light from a flashlight. Such sound waves are said to have "narrow dispersion." Since high-frequency sound has a relatively small wavelength, it tends to have a narrow dispersion. On the other hand, for larger values of the ratio λ/D, the angle θ is larger. The waves spread out over a larger region and are said to have a "wide dispersion." Low-frequency sound, with its relatively large wavelength, typically has a wide dispersion.

In a stereo loudspeaker, a wide dispersion of the sound is desirable. Example 3 illustrates, however, that there are limitations to the dispersion that can be achieved, depending on the loudspeaker design.

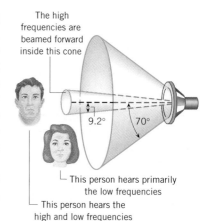

The high frequencies are beamed forward inside this cone

9.2°   70°

This person hears primarily the low frequencies

This person hears the high and low frequencies

Figure 17.12 Because the dispersion of high frequencies is less than that of low frequencies, you should be directly in front of the speaker to hear both the high and low frequencies equally well.

EXAMPLE 3 Designing a Loudspeaker for Wide Dispersion

A 1500-Hz sound and a 8500-Hz sound each emerge from a loudspeaker through a circular opening whose diameter is 0.30 m. Assuming that the speed of sound in air is 343 m/s, find the diffraction angle θ for each sound.

Reasoning The diffraction angle θ for each sound wave is given by $\sin\theta = 1.22(\lambda/D)$. However, it will first be necessary to calculate the wavelengths of the sounds from $\lambda = v/f$ (Equation 16.1).

Solution The wavelengths of the two sounds are

$$\lambda_{1500} = \frac{343 \text{ m/s}}{1500 \text{ Hz}} = 0.23 \text{ m} \quad \text{and} \quad \lambda_{8500} = \frac{343 \text{ m/s}}{8500 \text{ Hz}} = 0.040 \text{ m}$$

THE PHYSICS OF...

tweeter loudspeakers. (See below.)

The diffraction angles can now be determined:

1500-Hz sound $\sin\theta = 1.22\dfrac{\lambda_{1500}}{D} = 1.22\left(\dfrac{0.23 \text{ m}}{0.30 \text{ m}}\right) = 0.94$ (17.2)

$\theta = \sin^{-1} 0.94 = \boxed{70°}$

8500-Hz sound $\sin\theta = 1.22\dfrac{\lambda_{8500}}{D} = 1.22\left(\dfrac{0.040 \text{ m}}{0.30 \text{ m}}\right) = 0.16$ (17.2)

$\theta = \sin^{-1} 0.16 = \boxed{9.2°}$

Figure 17.12 illustrates these results. With a 0.30-m opening, the dispersion of the higher-frequency sound is limited to only 9.2°. To increase the dispersion, a smaller opening is needed. It is for this reason that loudspeaker designers use a small-diameter speaker called a *tweeter* for high-frequency sound, as Figure 17.13 indicates.

As we have seen, diffraction is an interference effect, one in which some of the wave's energy is directed into regions that would otherwise not be accessible. Energy, of course, is conserved during this process, for energy is only redistributed during diffraction; no energy is created or destroyed.

Tweeter

17.4 BEATS

● *Concepts at a Glance* In situations where waves with the *same frequency* overlap, we have seen how the principle of linear superposition leads to constructive and destructive interference and how it explains diffraction. We will see in this

Figure 17.13 Small-diameter speakers, called tweeters, are used to produce high-frequency sound. The small diameter helps to promote a wider dispersion of the sound.

Concepts at a Glance

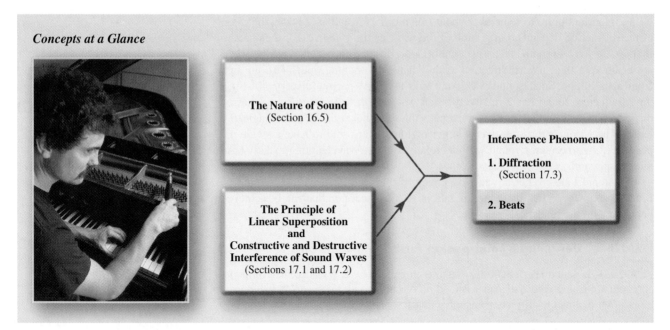

Figure 17.14 *Concepts at a Glance*
This figure, which is an extension of the concept chart in Figure 17.9, shows that the principle of linear superposition can be used to explain the phenomenon of beats. Piano tuners and other musicians routinely use beats in the process of tuning their instruments.

section that two overlapping waves with *slightly different frequencies* give rise to the phenomenon of beats. However, as the Concepts-at-a-Glance chart in Figure 17.14 illustrates, the principle of linear superposition again provides an explanation of what happens when the waves overlap. This chart is an expansion of that in Figure 17.9. ●

A tuning fork has the property of producing a single-frequency sound wave when struck with a sharp blow. Figure 17.15 shows sound waves coming from two tuning forks placed side by side. The tuning forks in the drawing are identical, and each is designed to produce a 440-Hz tone. However, a small piece of putty has been attached to one fork, whose frequency is lowered to 438 Hz because of the added mass. When the forks are sounded simultaneously, the loudness of the resulting sound rises and falls periodically—faint, then loud, then faint, then loud, and so on. The periodic variations in loudness are called *beats* and result from the interference between two sound waves with slightly different frequencies.

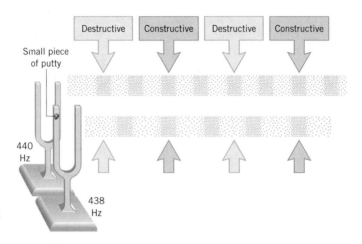

Figure 17.15 Two tuning forks have slightly different frequencies of 440 and 438 Hz. The phenomenon of beats occurs when the forks are sounded simultaneously. The sound waves are not drawn to scale.

For clarity, Figure 17.15 shows the condensations and rarefactions of the sound waves separately. In reality, however, the waves spread out and overlap. In accord with the principle of linear superposition, the ear detects the combined total of the two. Notice that there are places where the waves interfere constructively and places where they interfere destructively. When a region of constructive interference reaches the ear, a loud sound is heard. When a region of destructive interference arrives, the sound intensity drops to zero (assuming each of the waves has the same amplitude). The number of times per second that the loudness rises and falls is the *beat frequency* and is the *difference* between the two sound frequencies. Thus, in the situation illustrated in Figure 17.15, an observer hears the sound loudness rise and fall at the rate of 2 times per second (440 Hz − 438 Hz).

Figure 17.16 helps to explain why the beat frequency is the difference between the two frequencies. The drawing displays graphical representations of the pressure patterns of a 10-Hz wave and a 12-Hz wave, along with the pressure pattern that results when the two overlap. These frequencies have been chosen for convenience, even though they lie below the audio range and are inaudible. Audible sound waves behave exactly in the same way. The top two drawings in blue show the pressure variations in a one-second interval of each wave. The third drawing in red shows the

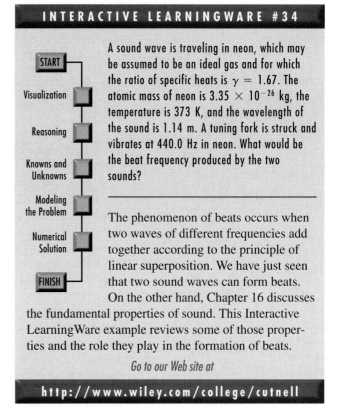

result of adding together the blue patterns according to the principle of linear superposition. Notice that the amplitude in the red drawing is not constant, as it is in the individual waves. Instead, the amplitude changes from a minimum to a maximum, back to a minimum, and so on. When such pressure variations reach the ear and occur in the audible frequency range, they produce a loud sound when the amplitude is a maximum and a faint sound when the amplitude is a minimum. Two loud-faint

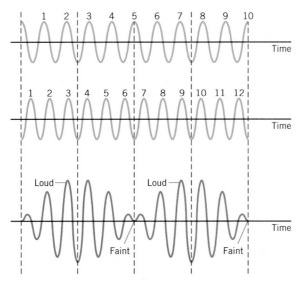

Figure 17.16 A 10-Hz sound wave and a 12-Hz sound wave, when added together, produce a wave with a beat frequency of 2 Hz. The drawings show the pressure patterns (in blue) of the individual waves and the pressure pattern (in red) that results when the two overlap. The time interval shown is one second.

cycles, or beats, occur in the one-second interval shown in the drawing, corresponding to a beat frequency of 2 Hz. Thus, the beat frequency is the difference between the frequencies of the individual waves, or 12 Hz − 10 Hz = 2 Hz.

Musicians often tune their instruments by listening to a beat frequency. For instance, a guitar player sounds an out-of-tune string along with a tone from a source known to have the correct frequency. The guitarist adjusts the tension in the string until the beats vanish, ensuring that the string is vibrating at the correct frequency.

THE PHYSICS OF...

tuning a musical instrument.

17.5 TRANSVERSE STANDING WAVES

● *Concepts at a Glance*   A standing wave is another interference effect that can occur when two waves overlap. Standing waves can arise with transverse waves, such as those on a guitar string, and also with longitudinal sound waves, such as those in a flute. In any case, the principle of linear superposition provides an explanation of the effect, just as it does for diffraction and beats. The Concepts-at-a-Glance chart in Figure 17.17 summarizes the role that this simple but powerful principle has played in this chapter. It is an expansion of the charts in Figures 17.9 and 17.14. ●

Figure 17.18 shows some of the essential features of transverse standing waves. In this figure the left end of each string is vibrated back and forth, while the right end is attached to a wall. Regions of the string move so fast that they appear only as a "blur" in the photographs. Each of the patterns shown is called a ***transverse standing wave pattern.*** Notice that the patterns include special places called nodes and antinodes. The ***nodes*** are places that do not vibrate at all, and the ***antinodes*** are places where maximum vibration occurs. To the right of each photograph is a series of superimposed drawings that help us to visualize the motion of the string as it vibrates in a standing wave pattern. These drawings freeze the shape of the string at various times and emphasize the maximum vibration that occurs at an antinode with the aid of a red "dot" attached to the string.

Figure 17.17 *Concepts at a Glance* This chart, which is a continuation of those in Figures 17.9 and 17.14, indicates that transverse and longitudinal standing waves are related to the principle of linear superposition. Guitars produce sound by utilizing the transverse standing waves that form on each plucked string.

Concepts at a Glance

The Nature of Sound
(Section 16.5)

The Principle of
Linear Superposition
and
Constructive and Destructive
Interference of Sound Waves
(Sections 17.1 and 17.2)

Interference Phenomena

1. Diffraction
   (Section 17.3)
2. Beats
   (Section 17.4)
3. Transverse and
   Longitudinal
   Standing Waves
   (Sections 17.5 and 17.6)

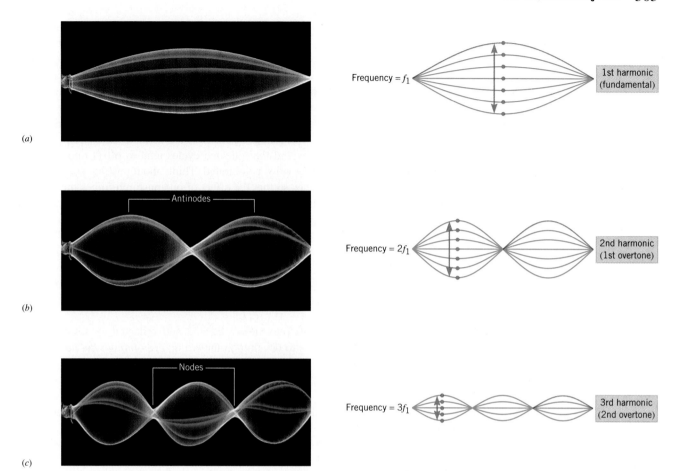

(a)

(b)

Antinodes

(c)

Nodes

Frequency = f_1 — 1st harmonic (fundamental)

Frequency = $2f_1$ — 2nd harmonic (1st overtone)

Frequency = $3f_1$ — 3rd harmonic (2nd overtone)

Figure 17.18 Vibrating a string at certain unique frequencies sets up transverse standing wave patterns, such as the three shown in the photographs on the left. Each drawing on the right shows the various shapes that the string assumes at various times as it vibrates. The red dots attached to the strings focus attention on the maximum vibration that occurs at an antinode. In each of the drawings, one-half of a wave cycle is outlined in red.

Each standing wave pattern is produced at a unique frequency of vibration. These frequencies form a series, the smallest frequency f_1 corresponding to the one-loop pattern and the larger frequencies being integer multiples of f_1, as Figure 17.18 indicates. Thus, if f_1 is 10 Hz, the frequency needed to establish the 2-loop pattern is $2f_1$ or 20 Hz, while that needed to create the 3-loop pattern is $3f_1$ or 30 Hz, and so on. The frequencies in this series (f_1, $2f_1$, $3f_1$, etc.) are called **harmonics.** The lowest frequency f_1 is called the first harmonic, and the higher frequencies are labeled as the second harmonic ($2f_1$), the third harmonic ($3f_1$), and so forth. The harmonic number (1st, 2nd, 3rd, etc.) corresponds to the number of loops in the standing wave pattern. The frequencies in this series are also referred to as the fundamental frequency, the first overtone, the second overtone, and so on. Thus, frequencies above the fundamental are **overtones** (see Figure 17.18).

Standing waves arise because identical waves travel on the string in *opposite directions* and combine in accord with the principle of linear superposition. A standing wave is said to be "standing," since it does not travel in one direction or the other, as do the individual waves that produce it. Figure 17.19 shows why there are waves traveling in both directions on the string. At the top of the picture, one-half of a wave cycle (the remainder of the wave is omitted for clarity) is moving toward the wall on the right. When the half-cycle reaches the wall, it causes the string to pull upward on the wall. Consistent with Newton's action–reaction law, the wall

Figure 17.19 In reflecting from the wall, a forward-traveling half-cycle becomes a backward-traveling half-cycle that is inverted.

• **PROBLEM SOLVING INSIGHT**
The distance between two successive nodes (or between two successive antinodes) of a standing wave is equal to one-half of a wavelength.

pulls downward on the string, and a downward-pointing half-cycle is sent back toward the left. Thus, the wave reflects from the wall. Upon arriving back at the point of origin, the wave reflects again, this time from the hand vibrating the string. For small vibration amplitudes, the hand is essentially fixed and behaves as the wall does in causing reflections. Repeated reflections at both ends of the string create a multitude of wave cycles traveling in both directions.

As each new cycle is formed by the vibrating hand, previous cycles that have reflected from the wall arrive and reflect again from the hand. Unless the timing is right, however, the new cycles and the reflected cycles tend to offset one another, and the formation of a standing wave is inhibited. Think about pushing someone on a swing and timing your pushes so that the effect of one push reinforces that of another. Such reinforcement in the case of the wave cycles leads to a large amplitude standing wave. Suppose the string has a length L and its left end is being vibrated at a frequency f_1. The time required to create a new wave cycle is the period T of the wave, where $T = 1/f_1$. On the other hand, the time needed for a cycle to travel from the hand to the wall and back, a distance of $2L$, is $2L/v$, where v is the wave speed. Reinforcement between new and reflected cycles occurs if these two times are equal; that is, if $1/f_1 = 2L/v$. Thus, a standing wave is established when the string is vibrated with a frequency of $f_1 = v/(2L)$.

Repeated reinforcement between newly created and reflected cycles causes a large amplitude standing wave to develop on the string, *even though the hand itself vibrates with only a small amplitude*. Thus, the motion of the string is a resonance effect, analogous to that discussed in Section 10.6 for an object attached to a spring. The frequency f_1 at which resonance occurs is sometimes called a ***natural frequency*** of the string, similar to the frequency at which an object oscillates on a spring.

There is a difference between the resonance of the string and the resonance of a spring system, however. An object on a spring has only a single natural frequency, whereas the string has a *series* of natural frequencies. The series arises because a reflected wave cycle need not return to its point of origin in time to reinforce *every* newly created cycle. Reinforcement can occur, for instance, on *every other* new cycle, as it does if the string is vibrated at twice the frequency f_1, or $f_2 = 2f_1$. Likewise, if the vibration frequency is $f_3 = 3f_1$, reinforcement occurs on *every third* new cycle. Similar arguments apply for any frequency $f_n = nf_1$, where n is an integer. As a result, the series of natural frequencies that lead to standing waves on a string fixed at both ends is given by

String fixed at both ends $\qquad f_n = n\left(\dfrac{v}{2L}\right) \qquad n = 1, 2, 3, 4, \ldots \qquad (17.3)$

It is also possible to obtain Equation 17.3 in another way. In Figure 17.18, one-half of a wave cycle is outlined in red for each of the harmonics, to show that each "loop" in a standing wave pattern corresponds to one-half a wavelength. Since the two fixed ends of the string are nodes, the length L of the string must contain an integer number n of half-wavelengths: $L = n(\frac{1}{2}\lambda_n)$ or $\lambda_n = 2L/n$. Using this result for the wavelength in the relation $f_n\lambda_n = v$ shows that $f_n(2L/n) = v$, which can be rearranged to give Equation 17.3.

Standing waves on a string play an important role in the way many musical instruments produce sound. For instance, a guitar string is stretched between two supports and, when plucked, vibrates according to the series of natural frequencies given by Equation 17.3. The next two examples deal with how this series of frequencies governs the playing and the design of a guitar.

EXAMPLE 4 Playing a Guitar

The heaviest string on an electric guitar has a linear density of $m/L = 5.28 \times 10^{-3}$ kg/m and is stretched with a tension of $F = 226$ N. This string produces the musical note E when vibrating along its entire length in a standing wave at the fundamental frequency of 164.8 Hz. (a) Find the length L of the string between its two fixed ends (see Figure 17.20a). (b) A guitar player wants the string to vibrate at a fundamental frequency of 2×164.8 Hz $= 329.6$ Hz, as it must if the musical note E is to be sounded one octave higher in pitch. To accomplish this, he presses the string against the proper fret and then plucks the string (see part b of the drawing). Find the distance L between the fret and the bridge of the guitar.

Reasoning The fundamental frequency f_1 is given by Equation 17.3 with $n = 1$: $f_1 = v/(2L)$. Since f_1 is known in both parts (a) and (b), the length L in each case can be calculated directly from this expression, once the speed v is known. The speed, in turn, is related to the tension F and the linear density m/L according to Equation 16.2.

Solution

(a) The speed is

$$v = \sqrt{\frac{F}{m/L}} = \sqrt{\frac{226 \text{ N}}{5.28 \times 10^{-3} \text{ kg/m}}} = 207 \text{ m/s} \qquad (16.2)$$

According to $f_1 = v/(2L)$, the length of the string is

$$L = \frac{v}{2f_1} = \frac{207 \text{ m/s}}{2(164.8 \text{ Hz})} = \boxed{0.628 \text{ m}}$$

(b) The distance L that locates the fret can be determined exactly as in part (a) by using the wave speed $v = 207$ m/s and noting that the frequency is now $f_1 = 329.6$ Hz: $\boxed{L = 0.314 \text{ m}}$. This length is exactly half that determined in part (a) because the frequencies have a ratio of $2:1$.

(a)

(b)

Figure 17.20 These drawings show the standing waves (in blue) that exist on a guitar string under different playing conditions.

CONCEPTUAL EXAMPLE 5 The Frets on a Guitar

Figure 17.21 shows the frets on the neck of a guitar. They allow the player to produce a complete sequence of musical notes using a single string. Starting with the fret at the top of the neck, each successive fret indicates where the player should press to get the next note in the sequence. Musicians call the sequence the chromatic scale, and every thirteenth note in it corresponds to one octave or a doubling of the

THE PHYSICS OF...

the frets on a guitar.

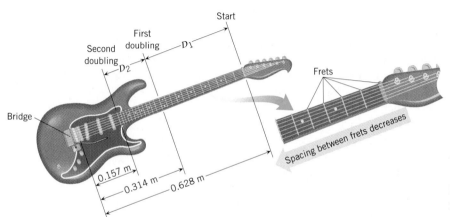

Figure 17.21 The spacing between the frets on the neck of a guitar decreases going down the neck toward the bridge.

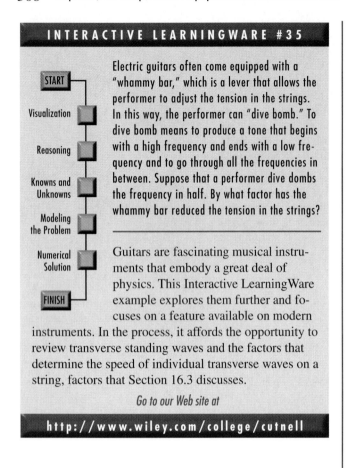

INTERACTIVE LEARNINGWARE #35

START
Visualization
Reasoning
Knowns and Unknowns
Modeling the Problem
Numerical Solution
FINISH

Electric guitars often come equipped with a "whammy bar," which is a lever that allows the performer to adjust the tension in the strings. In this way, the performer can "dive bomb." To dive bomb means to produce a tone that begins with a high frequency and ends with a low frequency and to go through all the frequencies in between. Suppose that a performer dive dombs the frequency in half. By what factor has the whammy bar reduced the tension in the strings?

Guitars are fascinating musical instruments that embody a great deal of physics. This Interactive LearningWare example explores them further and focuses on a feature available on modern instruments. In the process, it affords the opportunity to review transverse standing waves and the factors that determine the speed of individual transverse waves on a string, factors that Section 16.3 discusses.

Go to our Web site at

http://www.wiley.com/college/cutnell

sound frequency. The spacing between the frets is greatest at the top of the neck and decreases with each additional fret further on down. The spacing eventually becomes smaller than the width of a finger, limiting the number of frets that can be used. Why does the spacing between the frets decrease going down the neck?

Reasoning and Solution Our reasoning is based on Equation 17.3, with $n = 1$ [$f_1 = v/(2L)$]. The value of n is 1 because a string vibrates mainly at its fundamental frequency when plucked, as mentioned in Example 4. This equation indicates that the fundamental frequency f_1 is inversely proportional to the length L between a given fret and the bridge of the guitar. Thus, in Example 4 we found that the E-string had a length of 0.628 m corresponding to a frequency of $f_1 = 164.8$ Hz. We also found that the length between the bridge and the fret that must be pressed to double this frequency to 329.6 Hz is one-half of 0.628 m, or 0.314 m. To understand why the spacing between frets decreases going down the neck, consider the fret that must be pressed to double the frequency again, from 329.6 Hz to 659.2 Hz. The length between the bridge and this fret would be one-half of 0.314 m, or 0.157 m. The three lengths of 0.628 m, 0.314 m, and 0.157 m are shown in Figure 17.21. The distances D_1 and D_2 are also shown and give the spacings between the starting fret at the top of the neck, the fret for the first doubling, and the fret for the second doubling. Clearly, D_1 is greater than D_2, and the frets near the top of the neck have more space between them than those further on down.

Related Homework Material: *Problem 32*

17.6 LONGITUDINAL STANDING WAVES

Standing wave patterns can also be formed from longitudinal waves. For example, when sound reflects from a wall, the forward- and backward-going waves can produce a standing wave. Figure 17.22 illustrates the vibrational motion in a longitudinal standing wave on a Slinky. As in a transverse standing wave, there are nodes and antinodes. At the nodes the coils of the Slinky do not vibrate at all; that is, they have no displacement. At the antinodes the coils vibrate with maximum amplitude, as indicated by the red dots in the picture. At an antinode the coils have a maximum displacement. The vibration occurs along the line of travel of the individual waves, as is to be expected for longitudinal waves. In a standing wave of sound, the molecules or atoms of the medium behave as the red dots do.

Figure 17.22 A longitudinal standing wave on a Slinky showing the displacement nodes (N) and antinodes (A).

Musical instruments in the wind family depend on longitudinal standing waves in producing sound. Since wind instruments (trumpet, flute, clarinet, pipe organ, etc.) are modified tubes or columns of air, it is useful to examine the standing waves that can be set up in such tubes. Figure 17.23 shows two cylindrical columns of air that are open at both ends. Sound waves, originating from a tuning fork, travel up and down within each tube, since they reflect from the ends of the tubes, even though the ends are open. If the frequency f of the tuning fork matches one of the natural frequencies of the air column, the downward- and upward-traveling waves combine to form a standing wave, and the sound of the tuning fork becomes markedly louder. To emphasize the longitudinal nature of the standing wave patterns, the left side of the drawing replaces the air in the tubes with Slinkies, on which the nodes and antinodes are indicated with red dots. As an additional aid in visualizing the standing waves, the right side of the drawing shows blurred blue patterns within each tube. These patterns symbolize the amplitude of the vibrating air molecules at various locations. Wherever the pattern is widest, the amplitude of vibration is greatest (a displacement antinode), and wherever the pattern is narrowest there is no vibration (a displacement node).

To determine the natural frequencies of the air columns in Figure 17.23, notice that there is a displacement antinode at each end of the tube because the air molecules there are free to move.* As in a transverse standing wave, the distance between two successive antinodes is one-half of a wavelength, so the length L of the tube must be an integer number n of half-wavelengths: $L = n(\frac{1}{2}\lambda_n)$ or $\lambda_n = 2L/n$. Using this wavelength in the relation $f_n = v/\lambda_n$ shows that the natural frequencies f_n of the tube are

Tube open at both ends $$f_n = n\left(\frac{v}{2L}\right) \quad n = 1, 2, 3, 4, \ldots \quad (17.4)$$

At these frequencies, large amplitude standing waves develop within the tube due to resonance. Example 6 illustrates how Equation 17.4 is involved when a flute is played.

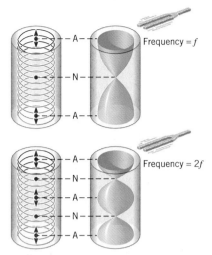

Figure 17.23 A pictorial representation of longitudinal standing waves on a Slinky (left side) and in a tube of air (right side) that is open at both ends (A, antinode; N, node).

EXAMPLE 6   Playing a Flute

When all the holes are closed on one type of flute, the lowest note it can sound is a middle C, whose fundamental frequency is 261.6 Hz. (a) The air temperature is 293 K, and the speed of sound is 343 m/s. Assuming the flute is a cylindrical tube open at both ends, determine the distance L in Figure 17.24, that is, the distance from the mouthpiece to the end of the tube. (This distance is only approximate, since the antinode does not occur exactly at the mouthpiece.) (b) A flautist can alter the length of the flute by adjusting the extent to which the head joint is inserted into the main stem of the instrument. If the air temperature rises to 305 K, to what length must the flute be adjusted to play a middle C?

Reasoning The fundamental frequency f_1 is given by Equation 17.4 with $n = 1$: $f_1 = v/(2L)$. This expression can be used to calculate the length as $L = v/(2f_1)$. When the speed of sound v changes, as it does when the temperature changes, the length of the flute must be changed. The effect of temperature on the speed of sound in air is given by $v = \sqrt{\gamma kT/m}$ (Equation 16.5), assuming air behaves as an ideal gas. Thus, the speed is proportional to the square root of the Kelvin temperature ($v \propto \sqrt{T}$), a fact that we can use to find the speed at the higher temperature.

THE PHYSICS OF...

a flute.

Figure 17.24 The length L of a flute between the mouthpiece and the end of the instrument determines the fundamental frequency of the lowest playable note.

---

* In reality, the antinode does not occur exactly at the open end. However, if the tube's diameter is small compared to its length, little error is made in assuming that the antinode is located right at the end.

Solution

(a) At a temperature of 293 K, when the speed of sound is $v = 343$ m/s, the length of the flute is

$$L = \frac{v}{2f_1} = \frac{343 \text{ m/s}}{2(261.6 \text{ Hz})} = \boxed{0.656 \text{ m}}$$

(b) Since $v \propto \sqrt{T}$, it follows that

$$\frac{v_{305 \text{ K}}}{v_{293 \text{ K}}} = \frac{\sqrt{305 \text{ K}}}{\sqrt{293 \text{ K}}} = 1.02$$

As a result, $v_{305 \text{ K}} = 1.02(v_{293 \text{ K}}) = 1.02(343 \text{ m/s}) = 3.50 \times 10^2$ m/s. The adjusted flute length is

$$L = \frac{v}{2f_1} = \frac{3.50 \times 10^2 \text{ m/s}}{2(261.6 \text{ Hz})} = \boxed{0.669 \text{ m}}$$

Thus, to play in tune at the higher temperature, a flautist must lengthen the flute by 0.013 m.

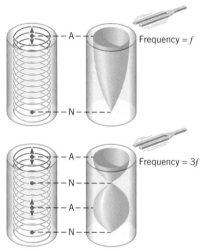

Figure 17.25 A pictorial representation of the longitudinal standing waves on a Slinky (left side) and in a tube of air (right side) that is open only at one end (A, antinode; N, node).

Standing waves can also exist in a tube with only one end open, as the patterns in Figure 17.25 indicate. Note the difference between these patterns and those in Figure 17.23. Here the standing waves have a displacement antinode at the open end and a displacement node at the closed end, where the air molecules are not free to move. Since the distance between a node and an adjacent antinode is one-fourth of a wavelength, the length L of the tube must be an odd number of quarter-wavelengths: $L = 1(\frac{1}{4}\lambda)$ and $L = 3(\frac{1}{4}\lambda)$ for the two standing wave patterns in Figure 17.25. In general, then, $L = n(\frac{1}{4}\lambda_n)$, where n is any odd integer ($n = 1, 3, 5, \ldots$). From this result it follows that $\lambda_n = 4L/n$, and the natural frequencies f_n can be obtained from the relation $f_n = v/\lambda_n$:

Tube open at only one end $$f_n = n\left(\frac{v}{4L}\right) \qquad n = 1, 3, 5, \ldots \qquad (17.5)$$

A tube open at only one end can develop standing waves only at the odd harmonic frequencies f_1, f_3, f_5, etc. In contrast, a tube open at both ends can develop standing waves at all harmonic frequencies f_1, f_2, f_3, etc. Moreover, the fundamental frequency f_1 of a tube open at only one end (Equation 17.5) is one-half that of a tube open at both ends (Equation 17.4). In other words, a tube open only at one end needs to be only one-half as long as a tube open at both ends in order to produce the *same* fundamental frequency.

Energy is also conserved when a standing wave is produced, either on a string or in a tube of air. The energy of the standing wave is the sum of the energies of the individual waves that comprise the standing wave. Once again, interference redistributes the energy of the individual waves to create locations of greatest energy (displacement antinodes) and locations of no energy (displacement nodes).

*17.7 COMPLEX SOUND WAVES

Musical instruments produce sound in a way that depends on standing waves. Examples 4 and 5 illustrate the role of transverse standing waves on the string of an electric guitar, while Example 6 stresses the role of longitudinal standing waves in the air column within a flute. In each example sound is produced at the fundamental frequency of the instrument.

In general, however, a musical instrument does not produce just the fundamental frequency when it plays a note, but generates a number of harmonics as well. Different instruments, such as a violin and a trumpet, generate harmonics to different extents, and the harmonics give the instruments their characteristic sound qualities or timbres. Suppose, for instance, that a violinist and a trumpet player both sound concert A, a note whose fundamental frequency is 440 Hz. Even though both instruments are playing the same note, most people can distinguish the sound of the violin from that of the trumpet. The instruments sound different because the relative amplitudes of the harmonics (880 Hz, 1320 Hz, etc.) that the instruments create are different.

The sound wave corresponding to a note produced by a musical instrument or a singer is called a **complex sound wave** because it consists of a mixture of the fundamental and harmonic frequencies. The pattern of pressure fluctuations in a complex wave can be obtained by using the principle of linear superposition, as Figure 17.26 indicates. This drawing shows a bar graph in which the heights of the bars give the relative amplitudes of the harmonics contained in a note such as a singer might produce. When the individual pressure patterns for each of the three harmonics are added together, they yield the complex pressure pattern shown at the top of the picture.*

In practice, a bar graph such as that in Figure 17.26 is determined with the aid of an electronic instrument known as a spectrum analyzer. When the note is produced, the complex sound wave is detected by a microphone that converts the wave into an electrical signal. The electrical signal, in turn, is fed into the spectrum analyzer, as Figure 17.27 illustrates. The spectrum analyzer then determines the amplitude and frequency of each harmonic present in the complex wave and displays the results on its screen.

THE PHYSICS OF...

a spectrum analyzer. (See below.)

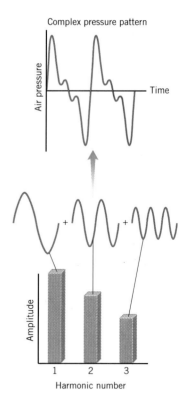

Figure 17.26 The graph on the top shows the pattern of pressure fluctuations such as a singer might produce. The pattern is the sum of the first three harmonics. The relative amplitudes of the harmonics correspond to the heights of the vertical bars in the bar graph.

17.8 CONCEPTS & CALCULATIONS

Diffraction is the bending of a traveling wave around an obstacle or the edges of an opening and is one of the consequences of the principle of linear superposition. As Equations 17.1 and 17.2 indicate, the extent of diffraction when a sound wave passes through an opening depends on the ratio of the wavelength of the sound to the width or diameter of the opening. Example 7 compares diffraction in two different media and reviews some of the fundamental properties of sound waves.

CONCEPTS & CALCULATIONS EXAMPLE 7   Diffraction in Two Different Media

A sound wave with a frequency of 15 kHz emerges through a circular opening that has a diameter of 0.20 m. Find the diffraction angle θ when the sound travels in air at a speed of 343 m/s and in water at a speed of 1482 m/s.

Concept Questions and Answers The diffraction angle for a circular opening is given by Equation 17.2 as $\sin \theta = 1.22\lambda/D$, where λ is the wavelength of the sound and D is the diameter of the opening. How is the wavelength related to the frequency of the sound?

Answer According to Equation 16.1, the wavelength is given by $\lambda = v/f$, where v is the speed of sound and f is the frequency.

---

* In carrying out the addition, we assume that each individual pattern begins at zero at the origin when the time equals zero.

Pressure pattern

Spectrum analyzer

Figure 17.27 A microphone detects a complex sound wave, and a spectrum analyzer determines the amplitude and frequency of each harmonic present in the wave.

Is the wavelength of the sound in air greater than, smaller than, or equal to the wavelength in water?

Answer According to Equation 16.1, the wavelength is proportional to the speed v for a given value of the frequency f. Since sound travels at a slower speed in air than in water, the wavelength in air is smaller than the wavelength in water.

Is the diffraction angle in air greater than, smaller than, or equal to the diffraction angle in water?

Answer The extent of diffraction is determined by λ/D, the ratio of the wavelength to the diameter of the opening. Smaller ratios lead to less diffraction or smaller diffraction angles. The wavelength in air is smaller than in water, and the diameter of the opening is the same in both cases. Therefore, the ratio λ/D is smaller in air than in water, and the diffraction angle in air is smaller than the diffraction angle in water.

Solution Using Equation 17.2 ($\sin \theta = 1.22\lambda/D$) and Equation 16.1 ($\lambda = v/f$), we have

$$\sin \theta = 1.22 \frac{\lambda}{D} = 1.22 \frac{v}{fD}$$

Applying this result in air and in water, we find

Air $\theta = \sin^{-1}\left(1.22 \frac{v}{fD}\right) = \sin^{-1}\left[1.22 \frac{(343 \text{ m/s})}{(15\ 000 \text{ Hz})(0.20 \text{ m})}\right] = \boxed{8.0°}$

Water $\theta = \sin^{-1}\left(1.22 \frac{v}{fD}\right) = \sin^{-1}\left[1.22 \frac{(1482 \text{ m/s})}{(15\ 000 \text{ Hz})(0.20 \text{ m})}\right] = \boxed{37°}$

As expected, the diffraction angle in air is smaller.

The next example deals with standing waves of sound in a gas. One of the factors that affects the formation of standing waves is the speed at which the individual waves travel. This example reviews how the speed of sound depends on the properties of the gas.

CONCEPTS & CALCULATIONS EXAMPLE 8   Standing Waves of Sound

Two tubes of gas are identical and are open at only one end. One tube contains neon (Ne) and the other krypton (Kr). Both are monatomic gases, have the same temperature, and may be assumed to be ideal gases. The fundamental frequency of the tube containing neon is 481 Hz. What is the fundamental frequency of the tube containing krypton?

Concept Questions and Answers For a gas-filled tube open at only one end, the fundamental frequency is given by Equation 17.5 as $f_1 = v/(4L)$, where v is the speed of sound and L is the length of the tube. How is the speed related to the properties of the gas?

Answer According to Equation 16.5, the speed is given by $v = \sqrt{\gamma kT/m}$, where γ is the ratio of the specific heat capacities at constant pressure and constant volume, k is Boltzmann's constant, T is the Kelvin temperature, and m is the mass of an atom of the gas.

All of the factors that affect the speed of sound are the same except the atomic mass. The periodic table located on the inside of the back cover gives the atomic masses of neon and krypton as 20.179 u and 83.80 u, respectively. Is the speed of sound in krypton greater than, smaller than, or equal to the speed of sound in neon?

Answer   According to Equation 16.5 the speed of sound is $v = \sqrt{\gamma kT/m}$, so the speed is inversely proportional to the square root of the mass m of an atom. Thus, the speed is smaller when the mass is greater. Since krypton has the greater mass, the speed of sound in krypton is smaller than in neon.

Is the fundamental frequency of the tube containing krypton greater than, smaller than, or equal to that of the tube containing neon?

Answer   The fundamental frequency is given by Equation 17.5 as $f_1 = v/(4L)$. Since the speed of sound in krypton is smaller than in neon, the fundamental frequency of the krypton-filled tube is smaller than that of the neon-filled tube.

Solution   Using Equation 17.5 [$f_1 = v/(4L)$] and Equation 16.5 ($v = \sqrt{\gamma kT/m}$), we have

$$f_1 = \frac{v}{4L} = \frac{1}{4L} \sqrt{\frac{\gamma kT}{m}}$$

Applying this result to both tubes and taking the ratio of the frequencies, we obtain

$$\frac{f_{1,\,\mathrm{Kr}}}{f_{1,\,\mathrm{Ne}}} = \frac{\dfrac{1}{4L} \sqrt{\gamma kT/m_{\mathrm{Kr}}}}{\dfrac{1}{4L} \sqrt{\gamma kT/m_{\mathrm{Ne}}}} = \sqrt{\frac{m_{\mathrm{Ne}}}{m_{\mathrm{Kr}}}}$$

Solving for $f_{1,\,\mathrm{Kr}}$ gives

$$f_{1,\,\mathrm{Kr}} = f_{1,\,\mathrm{Ne}} \sqrt{\frac{m_{\mathrm{Ne}}}{m_{\mathrm{Kr}}}} = (481 \text{ Hz}) \sqrt{\frac{20.179 \text{ u}}{83.80 \text{ u}}} = \boxed{236 \text{ Hz}}$$

As expected, the fundamental frequency for the krypton-filled tube is smaller than that for the neon-filled tube.

At the end of the problem set for this chapter, you will find homework problems that contain both conceptual and quantitative parts. These problems are grouped under the heading *Concepts & Calculations, Group Learning Problems.* They are designed for use by students working alone or in small learning groups. The conceptual part of each problem provides a convenient focus for group discussions.

S U M M A R Y

The **principle of linear superposition** states that when two or more waves are present simultaneously at the same place, the resultant disturbance is the sum of the disturbances from the individual waves. **Constructive interference** occurs at a point when two waves meet there crest-to-crest and trough-to-trough, thus reinforcing each other. **Destructive interference** occurs when the waves meet crest-to-trough and cancel each other. When the waves meet crest-to-crest and trough-to-trough, they are **exactly in phase.** When they meet crest-to-trough, they are **exactly out of phase.**

Diffraction is the bending of a wave around an obstacle or the edges of an opening. The angle through which the wave bends depends on the ratio of the wavelength λ of the wave to the width D of the opening; the greater the ratio λ/D, the greater the angle.

Beats are the periodic variations in amplitude that arise from the linear superposition of two waves that have slightly different frequencies. When the waves are sound waves, the variations in amplitude cause the loudness to vary at the **beat frequency,** which is the difference between the frequencies of the waves.

A transverse or longitudinal **standing wave** is the pattern of disturbance that results when oppositely traveling waves of the same frequency and amplitude pass through each other. A standing wave has places of minimum and maximum vibration called, respectively, **nodes** and **antinodes.** Under resonant conditions, standing waves can be established only at certain frequencies f_n, known as the **natural frequencies.** For a string that is fixed at both ends and has a length L, the natural frequencies are $f_n = nv/(2L)$, where v is the speed of the wave on the string and n is any positive integer, $n = 1, 2, 3, \ldots$. For a gas in a cylindrical tube open at both ends, the natural frequencies are given by the same expression, where v is the speed of sound in the gas and $n = 1, 2, 3, \ldots$. However, if the cylindrical tube is open only at one end, the natural frequencies are $f_n = nv/(4L)$, where $n = 1, 3, 5, \ldots$

A **complex sound wave** consists of a mixture of a fundamental frequency and overtone frequencies.

CONCEPTUAL QUESTIONS

1. Does the principle of linear superposition imply that two sound waves, passing through the same place at the same time, always create a louder sound than either wave alone? Explain.

2. Suppose you are sitting at the overlap point between the two speakers in Figure 17.4. Because of destructive interference, you hear no sound, even though both speakers are emitting identical sound waves. One of the two speakers is suddenly shut off. Describe what you would hear.

3. Review Figure 17.3. As you walk along a line that is perpendicular to the line between the speakers and passes through the overlap point, you do not observe the loudness to change from loud to faint to loud. However, as you walk along a line through the overlap point and parallel to the line between the speakers, you do observe the loudness to alternate between faint and loud. Explain why your observations are different in the two cases.

4. At an open-air rock concert the music is played through speakers that are on the stage. The speakers point straight forward and not toward the sides. You are wandering about and notice that the sounds of the female vocalists can be heard in front of the stage but not off to the sides. The rhythmic bass, however, can be heard both in front of the stage and off to the sides. How does the phenomenon of diffraction explain your observations?

5. Refer to Example 1 in Section 16.2. Which type of radio wave, AM or FM, diffracts more readily around a given obstacle? Give your reasoning.

6. A tuning fork has a frequency of 440 Hz. The string of a violin and this tuning fork, when sounded together, produce a beat frequency of 1 Hz. From these two pieces of information alone, is it possible to determine the exact frequency of the violin string? Explain.

7. When the regions of constructive and destructive interference in Figure 17.15 move past a listener's ear, a beat frequency of 2 Hz is heard. Suppose that the tuning forks in the drawing were sounded underwater and that the listener is also underwater. The forks vibrate at 438 and 440 Hz, just as they do in air. However, sound travels four times faster in water than in air. Explain why the listener hears a beat frequency of 2 Hz, just as he does in air, even though the regions of constructive and destructive interference are moving past his ear four times faster.

8. The tension in a guitar string is doubled. Does the frequency of oscillation also double? If not, by what factor does the frequency change? Specify whether the change is an increase or a decrease.

9. A string is attached to a wall and vibrates back and forth, as in Figure 17.18. The vibration frequency and length of the string are fixed. The tension in the string is changed, and it is observed that at certain values of the tension a standing wave pattern develops. Account for the fact that no standing waves are observed once the tension is increased beyond a certain value.

10. A string is vibrating back and forth as in Figure 17.18a. The tension in the string is decreased by a factor of four, with the frequency and the length of the string remaining the same. Draw the new standing wave pattern that develops on the string. Give your reasoning.

11. A rope is hanging vertically straight down. The top end is being vibrated back and forth. Standing waves can develop on the rope analogous to those on a horizontal rope. (a) There is a node at the top end. Is there a node or an antinode at the bottom end? Explain. (b) The separation between successive nodes is not the same everywhere on the rope, as it would be if the rope were horizontal. Is the separation greater near the top or near the bottom? Taking into account the mass of the rope, give your reasoning.

12. By blowing across the top of an empty soda bottle, a relatively loud sound can be created because a standing wave develops in the bottle. Explain why the pitch of the sound becomes higher when the bottle is partially filled with water.

13. In Figure 17.23 the tubes are filled with air. Suppose, instead, that the tube at the bottom of the drawing is filled with a gas in which the speed of sound is twice what it is in air. The frequency of the tuning fork remains unchanged. Draw the standing wave pattern that would exist in the tube. Justify your drawing.

14. Standing waves can ruin the acoustics of a concert hall if there is excessive reflection of the sound waves that the performers generate. For example, suppose a performer generates a 2093-Hz tone. If a large amplitude standing wave is present, it is possible for a listener to move a distance of only 4.1 cm and hear the loudness of the tone change from loud to faint. Account for this observation in terms of standing waves, pointing out why the distance is 4.1 cm.

15. The natural frequencies of a wind instrument change when it is brought inside from the cold outdoors. Why do the frequencies change, and is the change an increase or a decrease? Justify your answers.

16. The tones produced by a typical orchestra are complex sound waves, and most have fundamental frequencies less than 5000 Hz. However, a high-quality stereo system must be able to reproduce frequencies up to 20 000 Hz accurately. Explain why.

PROBLEMS

ssm Solution is in the Student Solutions Manual. **www** Solution is available on the World Wide Web at http://www.wiley.com/college/cutnell
⚕ This icon represents a biomedical application.

Section 17.1 The Principle of Linear Superposition,
Section 17.2 Constructive and Destructive Interference of Sound Waves

1. ssm The drawing shows a string on which two rectangular pulses are traveling at a constant speed of 1 cm/s at time $t = 0$. Using the principle of linear superposition, draw the shape of the string at $t = 1$ s, 2 s, 3 s, and 4 s.

2. Repeat problem 1, assuming that the pulse on the right is pointing downward rather than upward.

3. Two pulses are traveling toward each other, each having a speed of 1 cm/s. At $t = 0$, their positions are shown in the drawing. When $t = 1$ s, what is the height of the resultant pulse at (a) $x = 3$ cm and at (b) $x = 4$ cm?

4. The sound produced by the loudspeaker in the drawing has a frequency of 12 000 Hz and arrives at the microphone via two different paths. The sound travels through the left tube *LXM,* which has a fixed length. Simultaneously, the sound travels through the right tube *LYM,* the length of which can be changed by moving the sliding section. At *M,* the sound waves coming from the two paths interfere. As the length of the path *LYM* is changed, the sound loudness detected by the microphone changes. When the sliding section is pulled out by 0.020 m, the loudness changes from a maximum to a minimum. Find the speed at which sound travels through the gas in the tube.

5. ssm Review Example 1 in the text. Speaker A is moved further to the left, while ABC remains a right triangle. What is the separation between the speakers when constructive interference occurs again at point C?

6. In Figure 17.7, suppose that the separation between speakers A and B is 5.00 m and the speakers are vibrating in-phase. They are playing identical 125-Hz tones, and the speed of sound is 343 m/s. What is the largest possible distance between speaker B and the observer at C, such that he observes destructive interference?

7. ssm www The drawing shows a loudspeaker A and point C, where a listener is positioned. A second loudspeaker B is located somewhere to the right of A. Both speakers vibrate in phase and are playing a 68.6-Hz tone. The speed of sound is 343 m/s. What is the closest to speaker A that speaker B can be located, so that the listener hears no sound?

*8. Review Conceptual Example 2 in preparation for this problem. Assume that the two loudspeakers in Figure 17.7 are vibrating *out of phase* instead of in phase. The speed of sound is 343 m/s. What is the smallest frequency that will produce destructive interference at point C?

**9. Speakers A and B are vibrating in phase. They are directly facing each other, are 7.80 m apart, and are each playing a 73.0-Hz tone. The speed of sound is 343 m/s. On the line *between* the speakers there are three points where constructive interference occurs. What are the distances of these three points from speaker A?

Section 17.3 Diffraction

10. Sound emerges through a doorway, as in Figure 17.11. The width of the doorway is 77 cm, and the speed of sound is 343 m/s. Find the diffraction angle θ when the frequency of the sound is (a) 5.0 kHz and (b) 5.0×10^2 Hz.

11. ssm A speaker has a diameter of 0.30 m. (a) Assuming that the speed of sound is 343 m/s, find the diffraction angle θ for a 2.0-kHz tone. (b) What speaker diameter D should be used to generate a 6.0-kHz tone whose diffraction angle is as wide as that for the 2.0-kHz tone in part (a)?

12. In a diffraction horn loudspeaker, the sound exits through a rectangular opening, like a small doorway. The width D of a diffraction horn and the diameter of a circular speaker are equal.

The sound produced by the two speakers has the same diffraction angle θ. What is the ratio of the wavelength of the sound produced by the diffraction horn to that produced by the circular speaker?

13. Sound exits a diffraction horn loudspeaker through a rectangular opening, like a small doorway. A person is sitting at an angle θ off to the side of a diffraction horn that has a width D of 0.060 m. The speed of sound is 343 m/s. This individual does not hear a sound wave that has a frequency of 8100 Hz. When she is sitting at an angle $\theta/2$, there is a different frequency that she does not hear. What is it?

*14. A 3.00-kHz tone is being produced by a speaker with a diameter of 0.175 m. The air temperature changes from 0 to 29 °C. Assuming air to be an ideal gas, find the *change* in the diffraction angle θ.

*15. A row of seats is parallel to a stage at a distance of 8.7 m from it. At the center and front of the stage is a diffraction horn loudspeaker. This speaker sends out its sound through an opening that is like a small doorway with a width D of 7.5 cm. The speaker is playing a tone that has a frequency of 1.0×10^4 Hz. The speed of sound is 343 m/s. What is the separation between two seats, located on opposite sides of the center of the row, at which the tone cannot be heard?

Section 17.4 Beats

16. When a tuning fork is sounded together with a 492-Hz tone, a beat frequency of 2 Hz is heard. Then a small piece of putty is stuck to the tuning fork, and the tuning fork is again sounded along with the 492-Hz tone. The beat frequency decreases. What is the frequency of the tuning fork?

17. ssm Two ultrasonic sound waves combine and form a beat frequency that is in the range of human hearing. The frequency of one of the ultrasonic waves is 70 kHz. What is (a) the smallest possible and (b) the largest possible value for the frequency of the other ultrasonic wave?

18. Two out-of-tune flutes play the same note. One produces a tone that has a frequency of 262 Hz, while the other produces 266 Hz. When a tuning fork is sounded together with the 262-Hz tone, a beat frequency of 1 Hz is produced. When the same tuning fork is sounded together with the 266-Hz tone, a beat frequency of 3 Hz is produced. What is the frequency of the tuning fork?

19. ssm Two pure tones are sounded together. The drawing shows the pressure variations of the two sound waves, measured with respect to atmospheric pressure. What is the beat frequency?

20. When a guitar string is sounded along with a 440-Hz tuning fork, a beat frequency of 5 Hz is heard. When the same string is sounded along with a 436-Hz tuning fork, the beat frequency is 9 Hz. What is the frequency of the string?

*21. A sound wave is traveling in seawater, where the adiabatic bulk modulus and density are 2.31×10^9 Pa and 1025 kg/m^3, respectively. The wavelength of the sound is 3.35 m. A tuning fork is struck underwater and vibrates at 440.0 Hz. What would be the beat frequency heard by an underwater swimmer?

**22. Two tuning forks X and Y have different frequencies and produce an 8-Hz beat frequency when sounded together. When X is sounded along with a 392-Hz tone, a 3-Hz beat frequency is detected. When Y is sounded along with the 392-Hz tone, a 5-Hz beat frequency is heard. What are the frequencies f_X and f_Y when (a) f_X is greater than f_Y and (b) f_X is less than f_Y?

Section 17.5 Transverse Standing Waves

23. ssm The A string on a string bass is tuned to vibrate at a fundamental frequency of 55.0 Hz. If the tension in the string were increased by a factor of four, what would be the new fundamental frequency?

24. If the string in Figure 17.18 is vibrating at a frequency of 4.0 Hz and the distance between two successive nodes is 0.30 m, what is the speed of the waves on the string?

25. The G string on a guitar has a fundamental frequency of 196 Hz and a length of 0.62 m. This string is pressed against the proper fret to produce the note C, whose fundamental frequency is 262 Hz. What is the distance L between the fret and the end of the string at the bridge of the guitar (see Figure 17.20b)?

26. The fundamental frequency of a string fixed at both ends is 256 Hz. How long does it take for a wave to travel the length of this string?

27. ssm Ideally, the strings on a violin are stretched with the same tension. Each has the same length between its two fixed ends. The musical notes and corresponding fundamental frequencies of two of these strings are G (196.0 Hz) and E (659.3 Hz). The linear density of the E string is 3.47×10^{-4} kg/m. What is the linear density of the G string?

28. A 41-cm length of wire has a mass of 6.0 g. It is stretched between two fixed supports and is under a tension of 160 N. What is the fundamental frequency of this wire?

29. A string of length 2.50 m is fixed at both ends. When the string vibrates at a frequency of 85.0 Hz, a standing wave with five loops is formed. (a) What is the wavelength of the waves that travel on the string? (b) What is the speed of the waves? (c) What would be the fundamental frequency of this string?

*30. Two strings have different lengths and linear densities, as the drawing shows. They are joined together and stretched so that the tension in each string is 190.0 N. The free ends of the joined string are fixed in place. Find the lowest frequency that permits standing waves in both strings with a node at the junction. The standing wave pattern in each string may have a different number of loops.

*31. ssm The E string on an electric bass guitar has a length of 0.628 m and, when producing the note E, vibrates at a fundamental frequency of 41.2 Hz. Players sometimes add to their instruments a device called a "D-tuner." This device allows the E string to be used to produce the note D, which has a fundamental frequency of 36.7 Hz. The D-tuner works by extending the length of the string, keeping all other factors the same. By how much does a D-tuner extend the length of the E string?

**32. Review Conceptual Example 5 before attempting this problem. As the drawing shows, the length of a guitar string is 0.628 m. The frets are numbered for convenience. A performer can play a musical scale on a single string because the spacing *between the frets* is designed according to the following rule: When the string is pushed against any fret j, the fundamental frequency of the shortened string is larger by a factor of the twelfth root of two $\left(\sqrt[12]{2}\right)$ than it is when the string is pushed against the fret $j - 1$. Assuming that the tension in the string is the same for any note, find the spacing (a) between fret 1 and fret 0 and (b) between fret 7 and fret 6.

**33. ssm www The drawing shows an arrangement in which a block (mass = 15.0 kg) is held in position on a frictionless incline by a cord (length = 0.600 m). The mass per unit length of the cord is 1.20×10^{-2} kg/m, so the mass of the cord is negligible compared to the mass of the block. The cord is being vibrated at a frequency of 165 Hz (vibration source not shown in the drawing). What are the values of the angle θ between 15.0° and 90.0° at which a standing wave exists on the cord?

Section 17.6 Longitudinal Standing Waves,
Section 17.7 Complex Sound Waves

34. ◢ Sound enters the ear, travels through the auditory canal, and reaches the eardrum. The auditory canal is approximately a tube open at only one end. The other end is closed by the eardrum. A typical length for the auditory canal in an adult is about 2.9 cm. The speed of sound is 343 m/s. What is the fundamental frequency of the canal? (Interestingly, the fundamental frequency is in the frequency range where human hearing is most sensitive.)

35. A piccolo and a flute can be approximated as cylindrical tubes with both ends open. The lowest fundamental frequency produced by one kind of piccolo is 587.3 Hz, while that produced by a flute is 261.6 Hz. What is the ratio of the length of the piccolo to that of the flute?

36. The range of human hearing is roughly from twenty hertz to twenty kilohertz. Based on these limits and a value of 343 m/s for the speed of sound, what are the lengths of the longest and shortest pipes (open at both ends and producing sound at their fundamental frequencies) that you expect to find in a pipe organ?

37. ssm A tube of air is open at only one end and has a length of 1.5 m. This tube sustains a standing wave at its third harmonic. What is the distance between one node and the adjacent antinode?

38. A tube with a cap on one end, but open at the other end, produces a standing wave whose fundamental frequency is 130.8 Hz. The speed of sound is 343 m/s. (a) If the cap is removed, what is the new fundamental frequency? (b) How long is the tube?

39. ssm Both neon (Ne) and helium (He) are monatomic gases and can be assumed to be ideal gases. The fundamental frequency of a tube of neon is 268 Hz. What is the fundamental frequency of the tube if the tube is filled with helium, all other factors remaining the same?

*40. A tube, open at both ends, contains an unknown ideal gas for which $\gamma = 1.40$. At 293 K, the shortest tube in which a standing wave can be set up with a 294-Hz tuning fork has a length of 0.248 m. Find the mass of a gas molecule.

*41. Two loudspeakers face each other, vibrate in phase, and produce identical 440-Hz tones. A listener walks from one speaker toward the other at a constant speed and hears the loudness change (loud–soft–loud) at a frequency of 3.0 Hz. The speed of sound is 343 m/s. What is the walking speed?

*42. A person hums into the top of a well and finds that standing waves are established at frequencies of 42, 70.0, and 98 Hz. The frequency of 42 Hz is not necessarily the fundamental frequency. The speed of sound is 343 m/s. How deep is the well?

*43. ssm www A vertical tube is closed at one end and open to air at the other end. The air pressure is 1.01×10^5 Pa. The tube has a length of 0.75 m. Mercury (mass density =

13 600 kg/m³) is poured into it to shorten the effective length for standing waves. What is the absolute pressure at the bottom of the mercury column, when the fundamental frequency of the shortened, air-filled tube is equal to the third harmonic of the original tube?

**44. A tube, open at only one end, is cut into two shorter (non-equal) lengths. The piece open at both ends has a fundamental frequency of 425 Hz, while the piece open only at one end has a fundamental frequency of 675 Hz. What is the fundamental frequency of the original tube?

ADDITIONAL PROBLEMS

45. A stretched rubber band has a length of 0.10 m and a fundamental frequency of 440 Hz. What is the speed at which waves travel on the rubber band?

46. The fundamental frequency of a vibrating system is 400 Hz. For each of the following systems, give the three lowest frequencies (excluding the fundamental) at which standing waves can occur: (a) a string fixed at both ends, (b) a cylindrical pipe with both ends open, and (c) a cylindrical pipe with only one end open.

47. ssm Two loudspeakers are vibrating in phase. They are set up as in Figure 17.7, and point C is located as shown there. The speed of sound is 343 m/s. The speakers play the same tone. What is the smallest frequency that will produce destructive interference at point C?

48. A tuning fork vibrates at a frequency of 524 Hz. An out-of-tune piano string vibrates at 529 Hz. How much time separates successive beats?

49. The fundamental frequencies of two air columns are the same. Column A is open at both ends, while column B is open at only one end. The length of column A is 0.60 m. What is the length of column B?

50. Repeat problem 1, assuming that the pulses have the shape (half up and half down) shown in the drawing.

51. ssm On a cello, the string with the largest linear density $(1.56 \times 10^{-2} \text{ kg/m})$ is the C string. This string produces a fundamental frequency of 65.4 Hz and has a length of 0.800 m between the two fixed ends. Find the tension in the string.

*52. The graph shows a transverse standing wave on a string. (a) What is the wavelength of each wave that combines to form the standing wave? (b) If the velocity of one of the component waves is +12.0 cm/s, what is the velocity of the other? (c) What is the frequency of each component wave? (d) Suppose that a "dot" is attached to the string at $x = 2.0$ cm. Determine the maximum speed (in cm/s) of this dot as it vibrates up and down.

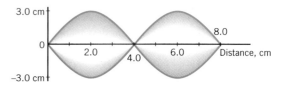

*53. A cylindrical pipe is *closed at both ends*. Derive an expression for the frequencies of the allowed standing waves, similar in form to Equations 17.4 and 17.5, in terms of the speed of sound v, the length of the pipe L, and the harmonic number n. State which integer values of n are allowed.

**54. Two loudspeakers are mounted on a merry-go-round whose radius is 9.01 m. When stationary, the speakers both play a tone whose frequency is 100.0 Hz. As the drawing illustrates, they are situated at opposite ends of a diameter. The speed of sound is 343.00 m/s, and the merry-go-round revolves once every 20.0 s. What is the beat frequency that is detected by the listener when the merry-go-round is near the position shown?

Merry-go-round
(top view)

Listener

**55. ssm www The note that is three octaves above middle C is supposed to have a fundamental frequency of 2093 Hz. On a certain piano the steel wire that produces this note has a cross-sectional area of $7.85 \times 10^{-7} \text{ m}^2$. The wire is stretched between two pegs. When the piano is tuned properly to produce the correct frequency at 25.0 °C, the wire is under a tension of 818.0 N. Suppose the temperature drops to 20.0 °C. In addition, as an approximation, assume that the wire is kept from contracting as the temperature drops. Consequently, the tension in the wire changes. What beat frequency is produced when this piano and another instrument (properly tuned) sound the note simultaneously?

CONCEPTS & CALCULATIONS

GROUP LEARNING PROBLEMS

Note: Each of these problems consists of Concept Questions followed by a related quantitative Problem. They are designed for use by students working alone or in small learning groups. The Concept Questions involve little or no mathematics and are intended to stimulate group discussions. They focus on the concepts with which the problems deal. Recognizing the concepts is the essential initial step in any problem-solving technique.

56. Concept Questions Both drawings show the same square, at one corner of which an observer *O* is stationed. Two loudspeakers are located at corners of the square, either as in drawing 1 or as in drawing 2. The speakers produce the same single-frequency tone in either drawing and are in phase. Constructive interference occurs in drawing 1, but destructive interference occurs in drawing 2. (a) Will only certain frequencies lead to the constructive interference in drawing 1, or will it occur for any frequency at all? (b) Will only certain frequencies lead to the destructive interference in drawing 2, or will it occur for any frequency at all? Justify your answers.

Problem One side of the square has a length of $L = 0.75$ m. The speed of sound is 343 m/s. Find the single smallest frequency that will produce both constructive interference in drawing 1 and destructive interference in drawing 2.

Drawing 1   Drawing 2

57. Concept Questions (a) When sound emerges from a loudspeaker, is the diffraction angle determined by the wavelength, the diameter of the speaker, or a combination of these two factors? (b) How is the wavelength of a sound related to its frequency? Explain your answers.

Problem The following two lists give diameters and sound frequencies for three loudspeakers. Pair each diameter with a frequency, so that the diffraction angle is the same for each of the speakers. The speed of sound is 343 m/s. Find the common diffraction angle.

Diameter, D	Frequency, f
0.050 m	6.0 kHz
0.10 m	4.0 kHz
0.15 m	12.0 kHz

58. Concept Questions Two cars have identical horns, each emitting a frequency f_s. One of the cars is moving toward a bystander waiting at a corner, and the other is parked. The two horns sound simultaneously. (a) From the moving horn, does the bystander hear a frequency that is greater than, less than, or equal to f_s? (b) From the stationary horn, does the bystander hear a frequency that is greater than, less than, or equal to f_s? (c) Does the bystander hear a beat frequency from the combined sound of the two horns? Account for your answers.

Problem The frequency that the horns emit is $f_s = 395$ Hz. The speed of the moving car is 12.0 m/s and the speed of sound is 343 m/s. What is the beat frequency heard by the bystander?

59. Concept Questions Two wires are stretched between two fixed supports and have the same length. On wire A there is a second-harmonic standing wave whose frequency is 660 Hz. However, the same frequency of 660 Hz is the third harmonic on wire B. (a) Is the fundamental frequency of wire A greater than, less than, or equal to the fundamental frequency of wire B? Explain. (b) How is the fundamental frequency related to the length L of the wire and the speed v at which individual waves travel back and forth on the wire? (c) Do the individual waves travel on wire A with a greater, smaller, or the same speed as on wire B? Give your reasoning.

Problem The common length of the wires is 1.2 m. Find the speed at which individual waves travel on each wire. Verify that your answer is consistent with your answers to the Concept Questions.

60. Concept Questions A copper block is suspended in air from a wire in Part 1 of the drawing. A container of mercury is then raised up around the block as in Part 2. (a) The fundamental frequency of the wire is given by Equation 17.3 with $n = 1$: $f_1 = v/(2L)$. How is the speed v at which individual waves travel on the wire related to the tension in the wire? (b) Is the tension in the wire in Part 2 less than, greater than, or equal to the tension in Part 1? (c) Is the fundamental frequency of the wire in Part 2 less than, greater than, or equal to the fundamental frequency in Part 1? Justify each of your answers.

Problem In Part 2 of the drawing one-half of the block's volume is submerged in the mercury. The density of copper is 8890 kg/m³, and the density of mercury is 13 600 kg/m³. Find the ratio of the fundamental frequency of the wire in Part 2 to the fundamental

Copper block

Part 1   Part 2

frequency in Part 1. Check to see that your answer is consistent with your answers to the Concept Questions.

61. Concept Questions One method for measuring the speed of sound uses standing waves. A cylindrical tube is open at both ends, and one end admits sound from a tuning fork. A movable plunger is inserted into the other end. The distance between the end of the tube where the tuning fork is and the plunger is L. For a fixed frequency, the plunger is moved until the smallest value of L is measured that allows a standing wave to be formed. (a) When a stand- ing wave is formed in the tube, is there a node or an antinode at the end of the tube where the tuning fork is? (b) When a standing wave is formed, is there a node or an antinode at the plunger? (c) How is the smallest value of L related to the wavelength of the sound? Explain your answers.

Problem The tuning fork produces a 485-Hz tone, and the smallest value observed for L is 0.264 m. What is the speed of the sound in the gas in the tube?

Lightning, which is the flow of electric charge in the atmosphere, is nature's most spectacular display of electricity.

CHAPTER 18

ELECTRIC FORCES AND ELECTRIC FIELDS

⊖ electron
⊕ proton
⊙ neutron

Figure 18.1 An atom contains a small, positively charged nucleus, about which the negatively charged electrons move. The closed-loop paths shown here are symbolic only. In reality, the electrons do not follow discrete paths, as Section 30.5 discusses.

18.1 THE ORIGIN OF ELECTRICITY

The electrical nature of matter is inherent in atomic structure. An atom consists of a small, relatively massive nucleus that contains particles called protons and neutrons. A proton has a mass of 1.673×10^{-27} kg, while a neutron has a slightly greater mass of 1.675×10^{-27} kg. Surrounding the nucleus is a diffuse cloud of orbiting particles called electrons, as Figure 18.1 suggests. An electron has a mass of 9.11×10^{-31} kg. Like mass, *electric charge* is an intrinsic property of protons and electrons, and only two types of charge have been discovered, positive and negative. A proton has a positive charge, and an electron has a negative charge. A neutron has no net electric charge.

Experiment reveals that the magnitude of the charge on the proton *exactly equals* the magnitude of the charge on the electron; the proton carries a charge $+e$, and the electron carries a charge $-e$. The SI unit for measuring the magnitude of an electric charge is the **coulomb* (C),** and e has been determined experimentally to have the value

$$e = 1.60 \times 10^{-19} \text{ C}$$

It should be noted that the symbol e represents only the magnitude of the charge on a proton or an electron and does not include the algebraic sign that indicates whether the charge is positive or negative. In nature, atoms are normally found with equal numbers of protons and electrons. Usually, then, an atom carries no net charge because the algebraic sum of the positive charge of the nucleus and the negative charge of the electrons is zero. When an atom, or any object, carries no net charge, the object is said to be *electrically neutral.* The neutrons in the nucleus are electrically neutral particles.

The charge on an electron or a proton is the *smallest* amount of free charge that has been discovered. Charges of larger magnitude are built up on an object by adding or removing electrons. Thus, any charge of magnitude q is an integer multiple of e; that is, $q = Ne$, where N is an integer. Because any electric charge q occurs in integer multiples of elementary, indivisible charges of magnitude e, electric charge is said to be *quantized.* Example 1 emphasizes the quantized nature of electrical charge.

EXAMPLE 1   A Lot of Electrons

How many electrons are there in one coulomb of negative charge?

Reasoning The negative charge is due to the presence of excess electrons, since they carry negative charge. Because an electron has a charge whose magnitude is $e = 1.60 \times 10^{-19}$ C, the number of electrons is equal to the charge q divided by the charge e on each electron.

Solution The number N of electrons is

$$N = \frac{q}{e} = \frac{1.00 \text{ C}}{1.60 \times 10^{-19} \text{ C}} = \boxed{6.25 \times 10^{18}}$$

This number is enormous.

---

* The definition of the coulomb depends on electric currents and magnetic fields, concepts that will be discussed later. Therefore, we postpone its definition until Section 21.7.

18.2 CHARGED OBJECTS AND THE ELECTRIC FORCE

Electricity has many useful applications, and they are related to the fact that it is possible to transfer electric charge from one object to another. Usually electrons are transferred, and the body that gains electrons acquires an excess of negative charge. The body that loses electrons has an excess of positive charge. Such separation of charge occurs often when two unlike materials are rubbed together. For example, when an ebonite (hard, black rubber) rod is rubbed against animal fur, some of the electrons from atoms of the fur are transferred to the rod. The ebonite becomes negatively charged, and the fur becomes positively charged, as Figure 18.2 indicates. Similarly, if a glass rod is rubbed with a silk cloth, some of the electrons are removed from the atoms of the glass and deposited on the silk, leaving the silk negatively charged and the glass positively charged. There are many familiar examples of charge separation, as when you walk across a nylon rug, vigorously run a comb through dry hair, or remove a pullover sweater. In each case, objects become "electrified" as surfaces rub against one another and a transfer of electrons occurs.

When an ebonite rod is rubbed with animal fur, the rubbing process serves only to separate electrons and protons already present in the materials. No electrons or protons are created or destroyed. Whenever an electron is transferred to the rod, a proton is left behind on the fur. Since the charges on the electron and proton have identical magnitudes but opposite signs, the algebraic sum of the two charges is zero, and the transfer does not change the net charge of the fur/rod system. If each material contains an equal number of protons and electrons to begin with, the net charge of the system is zero initially and remains zero at all times during the rubbing process.

Electric charges play a role in many situations other than rubbing two surfaces together. They are involved, for instance, in chemical reactions, electric circuits, and radioactive decay. A great number of experiments have verified that in any situation, the ***law of conservation of electric charge*** is obeyed.

● **LAW OF CONSERVATION OF ELECTRIC CHARGE**
During any process, the net electric charge of an isolated system remains constant (is conserved).

It is easy to demonstrate that two electrically charged objects exert a force on one another. Consider Figure 18.3a, which shows two small balls that have been *oppositely charged* and are light and free to move. The balls attract each other. On the other hand, balls with the *same* type of charge, either both positive or both negative, repel each other, as parts *b* and *c* of the drawing indicate. It is a fundamental characteristic of electric charges that ***like charges repel and unlike charges attract each other.***

　● *Concepts at a Glance*　Like other forces that we have encountered, the electric force (also called the electrostatic force) can alter the motion of an object. It can do so by contributing to the net external force $\Sigma\mathbf{F}$ that acts on the object. Newton's second law, $\Sigma\mathbf{F} = m\mathbf{a}$, specifies the acceleration $\mathbf{a}$ that arises because of the net external force. The Concepts-at-a-Glance chart in Figure 18.4 is an expanded version of the charts in Figures 4.9, 10.4, and 11.5, and emphasizes that any external electric force that acts on an object must be taken into account when determining the net external force to be used in the second law. ●

Figure 18.2 When an ebonite rod is rubbed against animal fur, electrons from atoms of the fur are transferred to the rod. This transfer gives the rod a negative charge (−) and leaves a positive charge (+) on the fur.

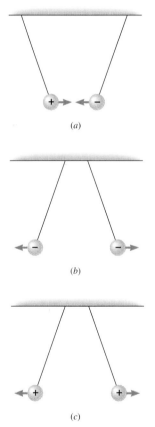

Figure 18.3 (*a*) A positive charge (+) and a negative charge (−) attract each other. (*b*) Two negative charges repel each other. (*c*) Two positive charges repel each other.

Concepts at a Glance

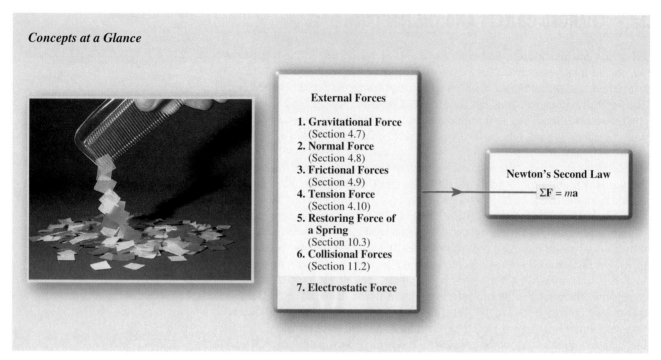

Figure 18.4 *Concepts at a Glance* Any external electrostatic forces that act on an object must be included when determining the net external force that is used in Newton's second law. An electrically charged comb, for example, exerts an attractive electrostatic force on the molecules that make up the paper squares.

THE PHYSICS OF...

electronic ink.

The sign in Figure 18.5 looks innocent enough, but the technology behind it, called electronic ink, may revolutionize the way books and other printed matter are made. It is not your ordinary ink. Letters and graphics on a page are formed by using the electric force and can be changed instantly, much like the symbols displayed on a computer monitor. Figure 18.5a illustrates the essential features of electronic ink. It consists of millions of clear microcapsules, each having the diameter of a human hair and filled with a dark inky liquid. Inside each microcapsule are several dozen extremely tiny white beads that carry a slightly negative charge. The microcapsules are sandwiched between two sheets, an opaque base layer and a transparent top layer, at which the reader looks. When a positive charge is applied to a small region of the base layer, as shown in part (b) of the drawing, the negatively charged white beads are drawn to it, leaving dark ink at the top layer. Thus, a viewer sees only the dark liquid. When a negative charge is applied to a region of the base layer, the negatively charged white beads are repelled from it and are forced to the top of the microcapsules; now a viewer sees a white area due to the beads. Thus, electronic ink is based on the principle that like charges repel and unlike charges attract each other; a positive charge causes one color to appear, and a negative charge causes another color to appear. Each small region, whether dark or light, is known as a pixel (short for "picture element"). Computer chips provide the instructions to produce the negative and positive charges on the base layer of each pixel. Letters and graphics are produced by the patterns generated with the two colors.

(a)

(b)

Figure 18.5 (*a*) Electronic ink consists of microcapsules filled with a dark, inky liquid and dozens of white beads. (*b*) Dark and light pixels are formed when positive and negative charges are placed in the base layer by electronic circuitry.

18.3 CONDUCTORS AND INSULATORS

Not only can electric charge exist *on an object,* but it can also move *through an object.* However, materials differ vastly in their abilities to allow electric charge to move or be conducted through them. To help illustrate such differences in conductivity, Figure 18.6*a* recalls the conduction of heat through a bar of material whose ends are maintained at different temperatures. As Section 13.2 discusses, metals

Figure 18.6 (*a*) Heat is conducted from the hotter end of the metal bar to the cooler end. (*b*) Electrons are conducted from the negatively charged end of the metal bar to the positively charged end.

(*a*)   (*b*)

conduct heat readily and, therefore, are known as thermal conductors. On the other hand, substances that conduct heat poorly are referred to as thermal insulators.

A situation analogous to the conduction of heat arises when a metal bar is placed between two charged objects, as in Figure 18.6*b*. Electrons are conducted through the bar from the negatively charged object toward the positively charged object. Substances that readily conduct electric charge are called ***electrical conductors.*** Although there are exceptions, good thermal conductors are generally good electrical conductors. Metals such as copper, aluminum, silver, and gold are excellent electrical conductors and, therefore, are used in electrical wiring. Materials that conduct electric charge *poorly* are known as ***electrical insulators.*** In many cases, thermal insulators are also electrical insulators. Common electrical insulators are rubber, many plastics, and wood. Insulators, such as the rubber or plastic that coats electrical wiring, prevent electric charge from going where it is not wanted.

The difference between electrical conductors and insulators is related to atomic structure. As electrons orbit the nucleus, those in the outer orbits experience a weaker force of attraction to the nucleus than do those in the inner orbits. Consequently, the outermost electrons (also called the valence electrons) can be dislodged more easily than the inner ones. In a good conductor, some valence electrons become detached from a parent atom and wander more or less freely throughout the material, belonging to no one atom in particular. The exact number of electrons detached from each atom depends on the nature of the material, but is usually between one and three. When one end of a conducting bar is placed in contact with a negatively charged object and the other end in contact with a positively charged object, as in Figure 18.6*b*, the "free" electrons are able to move readily away from the negative end and toward the positive end. The ready movement of electrons is the hallmark of a good conductor. In an insulator the situation is different, for there are very few electrons free to move throughout the material. Virtually every electron remains bound to its parent atom. Without the "free" electrons, there is very little flow of charge when the material is placed between two oppositely charged bodies, so the material is an electrical insulator.

18.4 CHARGING BY CONTACT AND BY INDUCTION

Figure 18.7 (*a*) Electrons are transferred by rubbing the negatively charged rod on the metal sphere. (*b*) When the rod is removed, the electrons distribute themselves over the surface of the sphere.

When a negatively charged ebonite rod is rubbed on a metal object, such as the sphere in Figure 18.7*a*, some of the excess electrons from the rod are transferred to the object. Once the electrons are on the metal sphere, where they can move readily, they repel one another and spread out over the sphere's surface. The insulated stand prevents them from flowing to the earth, where they could spread out even more. When the rod is removed, as in part *b* of the picture, the sphere is left with a negative charge distributed over its surface. In a similar manner, the sphere would be left with a positive charge after being rubbed with a positively charged rod. In this case, electrons from the sphere would be transferred to the rod. The process of giving one object a net electric charge by placing it in contact with another object that is already charged is known as ***charging by contact.***

Figure 18.8 (*a*) When a charged rod is brought near the metal sphere without touching it, some of the positive and negative charges in the sphere are separated. (*b*) Some of the electrons leave the sphere through the grounding wire, with the result (*c*) that the sphere acquires a positive net charge.

It is also possible to charge a conductor in a way that does not involve contact. In Figure 18.8, a negatively charged rod is brought close to, *but does not touch,* a metal sphere. In the sphere, the free electrons closest to the rod move to the other side, as part *a* of the drawing indicates. As a result, the part of the sphere nearest the rod becomes positively charged and the part farthest away becomes negatively charged. These positively and negatively charged regions have been "induced" or "persuaded" to form because of the repulsive force between the negative rod and the free electrons on the sphere. If the rod were removed, the free electrons would return to their original places, and the charged regions would disappear.

Under most conditions the earth is a good electrical conductor. So when a metal wire is attached between the sphere and the ground, as in Figure 18.8*b*, some of the free electrons leave the sphere and distribute themselves over the much larger earth. If the grounding wire is then removed, followed by the ebonite rod, the sphere is left with a positive net charge, as part *c* of the picture shows. The process of giving one object a net electric charge *without touching* the object to a second charged object is called ***charging by induction.*** The process could also be used to give the sphere a negative net charge, if a positively charged rod were used. Then, electrons would be drawn up from the ground through the grounding wire and onto the sphere.

If the sphere in Figure 18.8 were made from an insulating material like plastic, instead of metal, the method of producing a net charge by induction would not work, because very little charge would flow through the insulating material and down the grounding wire. However, the electric force of the charged rod would have some effect, as Figure 18.9 illustrates. The electric force would cause the positive and negative charges in the molecules of the insulating material to separate slightly, with the negative charges being "pushed" away from the negative rod. Although no net charge is created, the surface of the plastic does acquire a slight induced positive charge and is attracted to the negative rod. For a similar reason, one piece of cloth can stick to another in the phenomenon known as "static cling," which occurs when an article of clothing has acquired an electric charge while being tumbled about in a clothes drier.

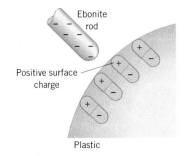

Figure 18.9 The negatively charged rod induces a slight positive surface charge on the plastic.

18.5 COULOMB'S LAW

The Force That Point Charges Exert on Each Other

The electrostatic force that stationary charged objects exert on each other depends on the amount of charge on the objects and the distance between them. Experiments reveal that the greater the charge and the closer together they are, the greater is the force. To set the stage for explaining these features in more detail, Figure 18.10 shows two charged bodies. These objects are so small, compared to the distance r between them, that they can be regarded as mathematical points. The "point

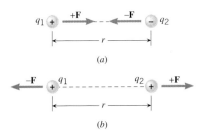

Figure 18.10 Each point charge exerts a force on the other. Regardless of whether the forces are (*a*) attractive or (*b*) repulsive, they are directed along the line between the charges and have equal magnitudes.

charges" have magnitudes q_1 and q_2. If the charges have *unlike* signs, as in part *a* of the picture, each object is *attracted* to the other by a force that is directed along the line between them; $+\mathbf{F}$ is the electric force exerted on object 1 by object 2 and $-\mathbf{F}$ is the electric force exerted on object 2 by object 1. If, as in part *b*, the charges have the *same* sign (both positive or both negative), each object is repelled from the other. The repulsive forces, like the attractive forces, act along the line between the charges. Whether attractive or repulsive, the two forces are equal in magnitude but opposite in direction. These forces always exist as a pair, each one acting on a different object, in accord with Newton's action–reaction law.

The French physicist Charles-Augustin Coulomb (1736–1806) carried out a number of experiments to determine how the electric force that one point charge applies to another depends on the amount of each charge and the separation between them. His result, now known as **Coulomb's law,** is stated below.

● COULOMB'S LAW

The magnitude F of the electrostatic force exerted by one point charge on another point charge is directly proportional to the magnitudes q_1 and q_2 of the charges and inversely proportional to the square of the distance r between them:

$$F = k\frac{q_1 q_2}{r^2} \tag{18.1}$$

where k is a proportionality constant whose value in SI units is $k = 8.99 \times 10^9$ N·m²/C². The electrostatic force. is directed along the line joining the charges, and it is attractive if the charges have unlike signs and repulsive if the charges have like signs.

It is common practice to express k in terms of another constant ϵ_0, by writing $k = 1/(4\pi\epsilon_0)$; ϵ_0 is called the **permittivity of free space** and has a value of $\epsilon_0 = 1/(4\pi k) = 8.85 \times 10^{-12}$ C²/(N·m²). Equation 18.1 gives the magnitude of the electrostatic force that each point charge exerts on the other. When using this equation, then, it is important to remember to substitute only the charge magnitudes (without algebraic signs) for q_1 and q_2, as Example 2 illustrates.

EXAMPLE 2   A Large Attractive Force

Two objects, whose charges are $+1.0$ and -1.0 C, are separated by 1.0 km. Compared to 1.0 km, the sizes of the objects are small. Find the magnitude of the attractive force that either charge exerts on the other.

Reasoning Considering that the sizes of the objects are small compared to the separation distance, we can treat the charges as point charges. Coulomb's law may then be used to find the magnitude of the attractive force, provided that only the *magnitudes of the charges* are used for the symbols q_1 and q_2 that appear in the law.

Solution The magnitude of the force is

$$F = k\frac{q_1 q_2}{r^2} = \frac{(8.99 \times 10^9 \text{ N·m}^2/\text{C}^2)(1.0 \text{ C})(1.0 \text{ C})}{(1.0 \times 10^3 \text{ m})^2} = \boxed{9.0 \times 10^3 \text{ N}} \tag{18.1}$$

The force calculated in Example 2 corresponds to about 2000 pounds and is so large because charges of ± 1.0 C are enormous. Such large charges are encountered only in the most severe conditions, as in a lightning bolt, where as much as 25 coulombs can be transferred between the cloud and the ground. The typical charges produced in the laboratory are much smaller and are measured conveniently in microcoulombs (1 microcoulomb = 1 μC = 10^{-6} C).

Coulomb's law has a form remarkably similar to Newton's law of gravitation ($F = Gm_1m_2/r^2$). The force in both laws depends on the inverse square ($1/r^2$) of the distance between the two objects and is directed along the line between them. In addition, the force is proportional to the product of an intrinsic property of each of the objects, the magnitudes of the charges q_1 and q_2 in Coulomb's law and the masses m_1 and m_2 in the gravitation law. But there is a major difference between the two laws. The electrostatic force can be either repulsive or attractive, depending on whether or not the charges have the same sign; in contrast, the gravitational force is always an attractive force.

Section 5.5 discusses how the gravitational attraction between the earth and a satellite provides the centripetal force that keeps a satellite in orbit. Example 3 illustrates that the electrostatic force of attraction plays a similar role in a famous model of the atom created by the Danish physicist Niels Bohr (1885–1962).

INTERACTIVE LEARNINGWARE #36

START
Visualization
Reasoning
Knowns and Unknowns
Modeling the Problem
Numerical Solution
FINISH

An object of mass of 215 kg is located just above the surface of the earth (radius = 6.38×10^6 m). Suppose that the object and the earth have an identical charge q. Therefore, two forces — the gravitational force and the electrostatic force — act on the object. Assuming that the earth's charge is located at the center of the earth, determine q (magnitude and sign) when the object is in equilibrium.

This problem reviews several important concepts. It involves the electric force due to the presence of the charges, as well as the gravitational force due to the masses. The solution uses these two forces in an equilibrium application, thus affording a nice review of Newton's second law of motion.

Go to our Web site at

http://www.wiley.com/college/cutnell

EXAMPLE 3   A Model of the Hydrogen Atom

In the Bohr model of the hydrogen atom, the electron ($-e$) is in orbit about the nuclear proton ($+e$) at a radius of $r = 5.29 \times 10^{-11}$ m, as Figure 18.11 shows. Determine the speed of the electron, assuming the orbit to be circular.

Reasoning Recall from Section 5.2 that any object moving with speed v on a circular path of radius r has a centripetal acceleration of $a_c = v^2/r$. This acceleration is directed toward the center of the circle. Newton's second law specifies that the net force ΣF needed to create this acceleration is $\Sigma F = ma_c = mv^2/r$, where m is the mass of the object. This equation can be solved for the speed: $v = \sqrt{(\Sigma F)r/m}$. Since the mass of the electron is $m = 9.11 \times 10^{-31}$ kg and the radius is given, we can calculate the speed, once a value for the net force is available. For the electron in the hydrogen atom, the net force is provided almost exclusively by the electrostatic force, as given by Coulomb's law. The electron is also pulled toward the proton by the gravitational force. However, the gravitational force is negligible in comparison to the electrostatic force.

Solution The electron experiences an electrostatic force of attraction because of the proton, and the magnitude of this force is

$$F = k\frac{q_1q_2}{r^2} = \frac{(8.99 \times 10^9 \text{ N}\cdot\text{m}^2/\text{C}^2)(1.60 \times 10^{-19} \text{ C})(1.60 \times 10^{-19} \text{ C})}{(5.29 \times 10^{-11} \text{ m})^2}$$

$$= 8.22 \times 10^{-8} \text{ N}$$

Using this value for the net force, we find

$$v = \sqrt{\frac{(\Sigma F)r}{m}} = \sqrt{\frac{(8.22 \times 10^{-8} \text{ N})(5.29 \times 10^{-11} \text{ m})}{9.11 \times 10^{-31} \text{ kg}}} = \boxed{2.18 \times 10^6 \text{ m/s}}$$

This orbital speed is almost five million miles per hour.

Figure 18.11 In the Bohr model of the hydrogen atom, the electron ($-e$) orbits the proton ($+e$) at a distance of $r = 5.29 \times 10^{-11}$ m. The velocity of the electron is **v**.

● **PROBLEM SOLVING INSIGHT**
When using Coulomb's law ($F = kq_1q_2/r^2$), remember that the symbols q_1 and q_2 stand for the charge magnitudes. Do not substitute negative numbers for these symbols.

THE PHYSICS OF...

adhesion.

Figure 18.12 After a strip of adhesive tape has been pulled off a metal surface, there are tiny pits (approximately one ten-millionth of a meter in diameter) in the sticky surface of the tape, as this image shows. It was obtained using an atomic-force microscope.

Since the electrostatic force depends on the inverse square of the distance between the charges, it becomes larger for smaller distances, such as those involved when a strip of adhesive tape is stuck to a smooth surface. Electrons shift over the small distances between the tape and the surface. As a result, the materials become oppositely charged. Since the distance between the charges is relatively small, the electrostatic force of attraction is large enough to contribute to the adhesive bond. Figure 18.12 shows an image of the sticky surface of a piece of tape after it has been pulled off a metal surface. The image was obtained using an atomic-force microscope and reveals the tiny pits left behind when microscopic portions of the adhesive remain stuck to the metal because of the strong adhesive bonding forces.

The Force on a Point Charge Due to Two or More Other Point Charges

Up to now, we have been discussing the electrostatic force on a point charge (magnitude q_1) due to another point charge (magnitude q_2). Suppose that a third point charge (magnitude q_3) is also present. What would be the net force on q_1 due to both q_2 and q_3? It is convenient to deal with such a problem in parts. First, find the magnitude and direction of the force exerted on q_1 by q_2 (ignoring q_3). Then, determine the force exerted on q_1 by q_3 (ignoring q_2). The *net force* on q_1 is the *vector sum* of these forces. Examples 4 and 5 illustrate this approach when the charges lie along a straight line and on a plane, respectively.

EXAMPLE 4 Three Charges on a Line

Figure 18.13*a* shows three point charges that lie along the *x* axis in a vacuum. Determine the magnitude and direction of the net electrostatic force on q_1.

Reasoning Part *b* of the drawing shows a free-body diagram of the forces that act on q_1. Since q_1 and q_2 have opposite signs, they attract one another. Thus, the force exerted on q_1 by q_2 is $\mathbf{F}_{12}$, and it points to the left. Similarly, the force exerted on q_1 by q_3 is $\mathbf{F}_{13}$ and is an attractive force. It points to the right in Figure 18.13*b*. The magnitudes of these forces can be obtained from Coulomb's law. The net force is the vector sum of $\mathbf{F}_{12}$ and $\mathbf{F}_{13}$.

Solution The magnitudes of the forces are

$$F_{12} = k \frac{q_1 q_2}{r_{12}^2} = \frac{(8.99 \times 10^9 \ \text{N} \cdot \text{m}^2/\text{C}^2)(3.0 \times 10^{-6} \ \text{C})(4.0 \times 10^{-6} \ \text{C})}{(0.20 \ \text{m})^2} = 2.7 \ \text{N}$$

$$F_{13} = k \frac{q_1 q_3}{r_{13}^2} = \frac{(8.99 \times 10^9 \ \text{N} \cdot \text{m}^2/\text{C}^2)(3.0 \times 10^{-6} \ \text{C})(7.0 \times 10^{-6} \ \text{C})}{(0.15 \ \text{m})^2} = 8.4 \ \text{N}$$

Since $\mathbf{F}_{12}$ points in the negative *x* direction, and $\mathbf{F}_{13}$ points in the positive *x* direction, the net force $\mathbf{F}$ is

$$\mathbf{F} = \mathbf{F}_{12} + \mathbf{F}_{13} = (-2.7 \ \text{N}) + (8.4 \ \text{N}) = \boxed{+5.7 \ \text{N}}$$

The plus sign in the answer indicates that the net force points to the right in the drawing.

● **PROBLEM SOLVING INSIGHT**
Often charge magnitudes are specified in microcoulombs (μC). When using Coulomb's law, be sure to convert microcoulombs into coulombs (1 μC = 10^{-6} C) before substituting for the charge magnitudes q_1 and q_2.

(a)

(b) Free-body diagram for q_1

Figure 18.13 (*a*) Three charges lying along the *x* axis. (*b*) The force exerted on q_1 by q_2 is $\mathbf{F}_{12}$, while the force exerted on q_1 by q_3 is $\mathbf{F}_{13}$.

EXAMPLE 5 Three Charges in a Plane

Figure 18.14*a* shows three point charges that lie in the *x*, *y* plane in a vacuum. Find the magnitude and direction of the net electrostatic force on q_1.

Reasoning The force exerted on q_1 by q_2 is $\mathbf{F}_{12}$ and is an attractive force because the two charges have opposite signs. It points along the line between the charges. The

Figure 18.14 (*a*) Three charges lying in a plane. (*b*) The net force acting on q_1 is $\mathbf{F} = \mathbf{F_{12}} + \mathbf{F_{13}}$. The angle that $\mathbf{F}$ makes with the $+x$ axis is θ.

force exerted on q_1 by q_3 is $\mathbf{F_{13}}$ and is also an attractive force. It points along the line between q_1 and q_3. Coulomb's law specifies the magnitudes of these forces. Since the forces point in different directions (see Figure 18.14*b*), we will use vector components to find the net force.

Solution The magnitudes of the forces are

$$F_{12} = k\frac{q_1 q_2}{r_{12}^2} = \frac{(8.99 \times 10^9 \text{ N} \cdot \text{m}^2/\text{C}^2)(4.0 \times 10^{-6} \text{ C})(6.0 \times 10^{-6} \text{ C})}{(0.15 \text{ m})^2} = 9.6 \text{ N}$$

$$F_{13} = k\frac{q_1 q_3}{r_{13}^2} = \frac{(8.99 \times 10^9 \text{ N} \cdot \text{m}^2/\text{C}^2)(4.0 \times 10^{-6} \text{ C})(5.0 \times 10^{-6} \text{ C})}{(0.10 \text{ m})^2} = 18 \text{ N}$$

The net force $\mathbf{F}$ is the vector sum of $\mathbf{F_{12}}$ and $\mathbf{F_{13}}$, as part *b* of the drawing shows. The components of $\mathbf{F}$ that lie in the x and y directions are $\mathbf{F_x}$ and $\mathbf{F_y}$, respectively. Our approach to finding $\mathbf{F}$ is the same as that used in Chapters 1 and 4. The forces $\mathbf{F_{12}}$ and $\mathbf{F_{13}}$ are resolved into x and y components. Then, the x components are combined to give $\mathbf{F_x}$, and the y components are combined to give $\mathbf{F_y}$. Once $\mathbf{F_x}$ and $\mathbf{F_y}$ are known, the magnitude and direction of $\mathbf{F}$ can be determined using trigonometry.

● **PROBLEM SOLVING INSIGHT**
The electrostatic force is a vector and has a direction as well as a magnitude. When adding electrostatic forces, take into account the directions of all forces, using vector components as needed.

Force	x component	y component
$\mathbf{F_{12}}$	$+(9.6 \text{ N}) \cos 73° = +2.8 \text{ N}$	$+(9.6 \text{ N}) \sin 73° = +9.2 \text{ N}$
$\mathbf{F_{13}}$	$+18 \text{ N}$	0 N
$\mathbf{F}$	$\mathbf{F_x} = +21 \text{ N}$	$\mathbf{F_y} = +9.2 \text{ N}$

The magnitude F and the angle θ of the net force are

$$F = \sqrt{F_x^2 + F_y^2} = \sqrt{(21 \text{ N})^2 + (9.2 \text{ N})^2} = \boxed{23 \text{ N}}$$

$$\theta = \tan^{-1}\left(\frac{F_y}{F_x}\right) = \tan^{-1}\left(\frac{9.2 \text{ N}}{21 \text{ N}}\right) = \boxed{24°}$$

18.6 THE ELECTRIC FIELD

Definition

As we know, a charge can experience an electrostatic force due to the presence of other charges. For instance, the positive charge q_0 in Figure 18.15 experiences a

Figure 18.15 A positive charge q_0 experiences an electrostatic force $\mathbf{F}$ due to the surrounding charges on the ebonite rod and the two spheres.

A positive charge is placed on the x axis ($q = +7.00$ μC, $x = 0.600$ m), and another positive charge is placed on the y axis ($q = +9.00$ μC, $y = 0.400$ m). A third charge ($q = -6.00$ μC, $m = 5.00 \times 10^{-8}$ kg) is placed at the coordinate origin. If the charge at the origin were free to move, what would be the magnitude of its acceleration?

This problem provides additional practice in evaluating the net electrostatic force that two charges exert on a third one. The example reviews Coulomb's law and the process of vector addition. It also reviews the all-important Newton's second law of motion and shows how the acceleration is obtained from the net force and the mass of the accelerating body.

Go to our Web site at

http://www.wiley.com/college/cutnell

force **F**, which is the vector sum of the forces exerted by the charges on the rod and the two spheres. It is useful to think of q_0 as a **test charge** for determining the extent to which the surrounding charges generate a force. However, in using a test charge, we must be careful to select one with a very small magnitude, so that it does not alter the locations of the other charges. The next example illustrates how the concept of a test charge is applied.

EXAMPLE 6   A Test Charge

The positive test charge in Figure 18.15 is $q_0 = +3.0 \times 10^{-8}$ C and experiences a force $F = 6.0 \times 10^{-8}$ N in the direction shown in the drawing. (a) Find the *force per coulomb* that the test charge experiences. (b) Using the result of part (a), predict the force that a charge of $+12 \times 10^{-8}$ C would experience if it replaced q_0.

Reasoning The charges in the environment apply a force **F** to the test charge q_0. The force per coulomb experienced by the test charge is $\mathbf{F}/q_0$. If q_0 is replaced by a new charge q, then the force on this new charge is the force per coulomb times q.

Solution

(a) The force per coulomb of charge is

$$\frac{F}{q_0} = \frac{6.0 \times 10^{-8}\ \text{N}}{3.0 \times 10^{-8}\ \text{C}} = \boxed{2.0\ \text{N/C}}$$

(b) The result from part (a) indicates that the surrounding charges can exert 2.0 newtons of force per coulomb of charge. Thus, a charge of $+12 \times 10^{-8}$ C would experience a force whose magnitude is

$$F = (2.0\ \text{N/C})(12 \times 10^{-8}\ \text{C}) = \boxed{24 \times 10^{-8}\ \text{N}}$$

The direction of this force would be the same as that experienced by the test charge, since both have the same positive sign.

● *Concepts at a Glance*   The electric force per coulomb, $\mathbf{F}/q_0$, calculated in Example 6(a) is one illustration of an idea that is very important in the study of electricity. The idea is called the *electric field.* As the Concepts-at-a-Glance chart in Figure 18.16 illustrates, the notion of an electric field emerges when we combine the electrostatic force with the idea of a test charge. In Chapter 21, we will combine the magnetic force and a moving test charge to define another important type of field, the magnetic field. Equation 18.2 presents the definition of the electric field. ●

● **DEFINITION OF THE ELECTRIC FIELD**

The electric field **E** that exists at a point is the electrostatic force **F** experienced by a small test charge q_0 placed at that point divided by the charge itself:

$$\mathbf{E} = \frac{\mathbf{F}}{q_0} \tag{18.2}$$

Concepts at a Glance

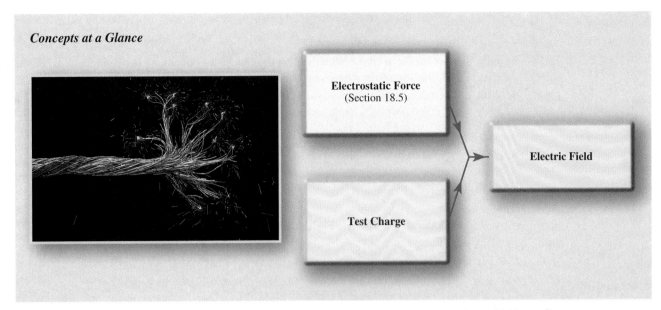

Figure 18.16 *Concepts at a Glance*
A test charge and the electrostatic force
that acts on it are brought together to
define the concept of an electric field.
The sparks shooting off from the ends
of some of the frayed wires are caused
by the electric field between two wires
that are nearly touching.

The electric field is a vector, and its direction is the same as the direction of the force **F** on a positive test charge.

SI Unit of Electric Field: newton per coulomb (N/C)

Equation 18.2 indicates that the unit for the electric field is that of force divided by charge, which is a newton/coulomb (N/C) in SI units. The definition also emphasizes that the electric field is a vector with the same direction as the force on a *positive* test charge.

It is the surrounding charges that create an electric field at a given point. Any positive or negative charge placed at the point interacts with the field and, as a result, experiences a force, as the next example indicates.

EXAMPLE 7   An Electric Field Leads to a Force

In Figure 18.17*a* the charges on the two metal spheres and the ebonite rod create an electric field **E** at the spot indicated. This field has a magnitude of 2.0 N/C and is directed as in the drawing. Determine the force on a charge placed at that spot, if the charge has a value of (a) $q_0 = +18 \times 10^{-8}$ C and (b) $q_0 = -24 \times 10^{-8}$ C.

Reasoning The electric field at a given spot can exert a variety of forces, depending on the magnitude and sign of the charge placed there. The charge is assumed to be small enough that it does not alter the locations of the surrounding charges that create the field.

Solution

(a) The magnitude of the force is the product of the magnitudes of q_0 and **E**:

$$F = q_0 E = (18 \times 10^{-8}\,\text{C})(2.0\,\text{N/C}) = \boxed{36 \times 10^{-8}\,\text{N}} \qquad (18.2)$$

Since q_0 is positive, the force points in the same direction as the electric field, as part *a* of the drawing indicates.

Figure 18.17 The electric field **E** that exists at a given spot can exert a variety of forces. The force exerted depends on the magnitude and sign of the charge placed at that spot. (*a*) The force on a positive charge points in the same direction as **E**, while (*b*) the force on a negative charge points opposite to **E**.

(b) In this case, the magnitude of the force is

$$F = q_0 E = (24 \times 10^{-8}\ \mathrm{C})(2.0\ \mathrm{N/C}) = \boxed{48 \times 10^{-8}\ \mathrm{N}} \qquad (18.2)$$

The force on the negative charge points in the direction *opposite* to the force on the positive charge, that is, opposite to the electric field (see part *b* of the drawing).

At a particular point in space, each of the surrounding charges contributes to the net electric field that exists there. To determine the net field, it is necessary to obtain the various contributions separately and then find the vector sum of them all. Such an approach is an illustration of the principle of linear superposition, as applied to electric fields. (This principle is introduced in Section 17.1, in connection with waves.) Example 8 emphasizes the vector nature of the electric field.

EXAMPLE 8   Electric Fields Add as Vectors Do

Figure 18.18 shows two charged objects, *A* and *B*. Each contributes as follows to the net electric field at point *P*: $E_A = 3.00\ \mathrm{N/C}$ directed to the right, and $E_B = 2.00\ \mathrm{N/C}$ directed downward. Thus, $\mathbf{E_A}$ and $\mathbf{E_B}$ are perpendicular. What is the net electric field at *P*?

Reasoning The net electric field **E** is the vector sum of $\mathbf{E_A}$ and $\mathbf{E_B}$: $\mathbf{E} = \mathbf{E_A} + \mathbf{E_B}$. As illustrated in Figure 18.18, $\mathbf{E_A}$ and $\mathbf{E_B}$ are perpendicular, so **E** is the diagonal of the rectangle shown in the drawing. Thus, we can use the Pythagorean theorem to find the magnitude of **E** and trigonometry to find the directional angle θ.

Solution The magnitude of the net electric field is

$$E = \sqrt{E_A^2 + E_B^2} = \sqrt{(3.00\ \mathrm{N/C})^2 + (2.00\ \mathrm{N/C})^2} = \boxed{3.61\ \mathrm{N/C}}$$

The direction of **E** is given by the angle θ in the drawing:

$$\theta = \tan^{-1}\left(\frac{E_B}{E_A}\right) = \tan^{-1}\left(\frac{2.00\ \mathrm{N/C}}{3.00\ \mathrm{N/C}}\right) = \boxed{33.7^\circ}$$

Point Charges

A more complete understanding of the electric field concept can be gained by considering the field created by a point charge, as in the following example.

EXAMPLE 9   The Electric Field of a Point Charge

There is an isolated point charge of $q = +15\ \mu\mathrm{C}$ in a vacuum at the left in Figure 18.19*a*. Using a test charge of $q_0 = +0.80\ \mu\mathrm{C}$, determine the electric field at point *P*, which is 0.20 m away.

Reasoning Following the definition of the electric field, we place the test charge q_0 at point *P*, determine the force acting on the test charge, and then divide the force by the test charge.

Solution Coulomb's law (Equation 18.1), gives the magnitude of the force:

$$F = \frac{k q_0 q}{r^2} = \frac{(8.99 \times 10^9\ \mathrm{N \cdot m^2/C^2})(0.80 \times 10^{-6}\ \mathrm{C})(15 \times 10^{-6}\ \mathrm{C})}{(0.20\ \mathrm{m})^2} = 2.7\ \mathrm{N}$$

Equation 18.2 gives the magnitude of the electric field:

$$E = \frac{F}{q_0} = \frac{2.7\ \mathrm{N}}{0.80 \times 10^{-6}\ \mathrm{C}} = \boxed{3.4 \times 10^6\ \mathrm{N/C}}$$

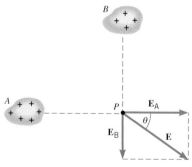

Figure 18.18 The electric field contributions $\mathbf{E_A}$ and $\mathbf{E_B}$, which come from the two charge distributions, are added vectorially to obtain the net field **E** at point *P*.

The electric field **E** points in the *same direction* as the force **F** on the positive test charge. Since the test charge experiences a force of repulsion directed to the right, the electric field vector also points to the right, as Figure 18.19b shows.

The electric field produced by a point charge q can be obtained in general terms from Coulomb's law. First, note that the magnitude of the force exerted by the charge q on a test charge q_0 is $F = kqq_0/r^2$. Then, divide this value by q_0 to obtain the magnitude of the field. Since q_0 is eliminated algebraically from the result, *the electric field does not depend on the test charge:*

Point charge q
$$E = \frac{kq}{r^2} \tag{18.3}$$

As in Coulomb's law, only the magnitude of q is used in Equation 18.3, without regard to whether q is positive or negative. If q is positive, then **E** is directed away from q, as in Figure 18.19b. On the other hand, if q is negative, then **E** is directed toward q, since a negative charge attracts a positive test charge. For instance, Figure 18.19c shows the electric field that would exist at P if there were a charge of $-q$ instead of $+q$ at the left of the drawing. Example 10 reemphasizes the fact that all the surrounding charges make a contribution to the electric field that exists at a given place.

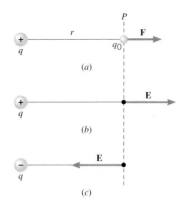

Figure 18.19 (a) At location P, a positive test charge q_0 experiences a repulsive force **F** due to the positive point charge q. (b) At P the electric field **E** is directed to the right. (c) If the charge q were negative rather than positive, the electric field would have the same magnitude as in (b) but point to the left.

EXAMPLE 10   The Electric Fields from Separate Charges May Cancel

Two positive point charges, $q_1 = +16 \ \mu C$ and $q_2 = +4.0 \ \mu C$, are separated in a vacuum by a distance of 3.0 m, as Figure 18.20 illustrates. Find the spot on the line between the charges where the net electric field is zero.

Reasoning Between the charges the two field contributions have opposite directions, and the net electric field is zero at the place where the magnitude of $\mathbf{E_1}$ equals that of $\mathbf{E_2}$. However, since q_2 is smaller than q_1, this location must be *closer* to q_2, in order that the field of the smaller charge can balance the field of the larger charge. In the drawing, the cancellation spot is labeled P, and its distance from q_1 is d.

Solution At P, $E_1 = E_2$, and using the expression $E = kq/r^2$, we have

$$\frac{k(16 \times 10^{-6} \ \text{C})}{d^2} = \frac{k(4.0 \times 10^{-6} \ \text{C})}{(3.0 \ \text{m} - d)^2}$$

Rearranging this expression shows that $4.0(3.0 \ \text{m} - d)^2 = d^2$. Taking the square root of each side of this equation reveals that

$$2.0(3.0 \ \text{m} - d) = \pm d$$

The plus and minus signs on the right occur because either the positive or negative root can be taken. Therefore, there are two possible values for d: $+2.0$ m and $+6.0$ m. The value $+6.0$ m corresponds to a location off to the right of both charges, where the magnitudes of $\mathbf{E_1}$ and $\mathbf{E_2}$ are equal, but where the directions are the same. Thus, $\mathbf{E_1}$ and $\mathbf{E_2}$ do not cancel at this spot. The other value for d corresponds to the location shown in the drawing and is the zero-field location: $\boxed{d = +2.0 \ \text{m}}$.

● **PROBLEM SOLVING INSIGHT**
Equation 18.3 gives only the magnitude of the electric field produced by a point charge. Therefore, do not use negative numbers for the symbol q in this equation.

When point charges are arranged in a symmetrical fashion, it is often possible to deduce useful information about the magnitude and direction of the electric field by taking advantage of the symmetry. Conceptual Example 11 illustrates the use of this technique.

Figure 18.20 The two point charges q_1 and q_2 create electric fields $\mathbf{E_1}$ and $\mathbf{E_2}$ that cancel at a location P on the line between the charges.

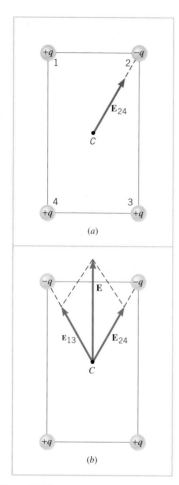

Figure 18.21 Charges of identical magnitude, but different signs, are placed at the corners of a rectangle. The charges give rise to different electric fields at the center *C* of the rectangle, depending on the signs of the charges.

CONCEPTUAL EXAMPLE 11   Symmetry and the Electric Field

Figure 18.21 shows point charges fixed to the corners of a rectangle in two different ways. The charges all have the same magnitudes, but they have different signs. Consider the net electric field at the center *C* of the rectangle in each case. Which field is larger?

Reasoning and Solution In Figure 18.21*a*, the charges at corners 1 and 3 are both $+q$. The positive charge at corner 1 produces an electric field at *C* that points toward corner 3. In contrast, the positive charge at corner 3 produces an electric field at *C* that points toward corner 1. Thus, the two fields have opposite directions. The magnitudes of the fields are identical because the charges have the same magnitude and are equally far from the center. Therefore, the fields from the two positive charges cancel.

Now, let's look at the electric field produced by the charges on corners 2 and 4 in Figure 18.21*a*. The electric field due to the negative charge at corner 2 points toward corner 2, and the field due to the positive charge at corner 4 points the same way. Furthermore, the magnitudes of these fields are equal because each charge has the same magnitude and is located at the same distance from the center of the rectangle. As a result, the two fields combine to give the net electric field $\mathbf{E}_{24}$ shown in the drawing.

In Figure 18.21*b*, the charges on corners 2 and 4 are identical to those in part *a* of the drawing, so this pair gives rise to the electric field labeled $\mathbf{E}_{24}$. The charges on corners 1 and 3 are identical to those on corners 2 and 4, so they give rise to an electric field labeled $\mathbf{E}_{13}$, which has the same magnitude as $\mathbf{E}_{24}$. The net electric field $\mathbf{E}$ is the vector sum of $\mathbf{E}_{24}$ and $\mathbf{E}_{13}$, and is also shown in the drawing. Clearly, this sum is greater than $\mathbf{E}_{24}$ alone. Therefore, ***the net field in part b is larger than that in part a.***

Related Homework Material: *Conceptual Question 12, Problem 36*

The Parallel Plate Capacitor

Equation 18.3, which gives the electric field of a point charge, is a very useful result. With the aid of integral calculus, this equation can be applied in a variety of situations where point charges are distributed over one or more surfaces. One such example that has considerable practical importance is the ***parallel plate capacitor***. As Figure 18.22 shows, this device consists of two parallel metal plates, each with area *A*. A charge $+q$ is spread uniformly over one plate, while a charge $-q$ is spread uniformly over the other plate. In the region between the plates and away from the edges, the electric field points from the positive plate toward the negative plate and is perpendicular to both. Using Gauss' law (see Section 18.9), it can be shown that the electric field has a magnitude of

Parallel plate capacitor
$$E = \frac{q}{\epsilon_0 A} = \frac{\sigma}{\epsilon_0} \qquad (18.4)$$

where ϵ_0 is the permittivity of free space. In this expression the Greek symbol sigma (σ) denotes the charge per unit area ($\sigma = q/A$) and is sometimes called the charge density. Except in the region near the edges, the field has the same value at all places between the plates. The field does *not* depend on the distance from the charges, in distinct contrast to the field created by an isolated point charge.

Figure 18.22 A parallel plate capacitor.

18.7 ELECTRIC FIELD LINES

As we have seen, electric charges create an electric field in the space surrounding them. It is useful to have a kind of "map" that gives the direction and indicates the strength of the field at various places. The great English physicist Michael Faraday (1791–1867) proposed an idea that provides such a "map," the idea of *electric field lines*. Since the electric field is the electric force per unit charge, the electric field lines are sometimes called *lines of force.*

To introduce the electric-field-line concept, Figure 18.23a shows a positive point charge $+q$. At the locations numbered 1–8, a positive test charge would experience a repulsive force, as the arrows in the drawing indicate. Therefore, the electric field created by the charge $+q$ is directly radially outward. The electric field lines are lines drawn to show this direction, as part b of the drawing illustrates. They begin on the charge $+q$ and point radially *outward*. Figure 18.24 shows the field lines in the vicinity of a negative charge $-q$. In this case they are directed radially *inward* because the force on a positive test charge is one of attraction, indicating that the electric field points inward. In general, *electric field lines are always directed away from positive charges and toward negative charges.*

The electric field lines in Figures 18.23 and 18.24 are drawn in only two dimensions, as a matter of convenience. Field lines radiate from the charges in three dimensions, and an infinite number of lines could be drawn. However, for clarity only a small number is ever included in pictures. The number is chosen to be proportional to the magnitude of the charge; thus, five times as many lines would emerge from a $+5q$ charge as from a $+q$ charge.

The pattern of electric field lines also provides information about the magnitude or strength of the field. Notice in Figures 18.23 and 18.24 that near the charges, where the electric field is stronger, the lines are closer together. At distances far from the charges, where the electric field is weaker, the lines are more spread out. It

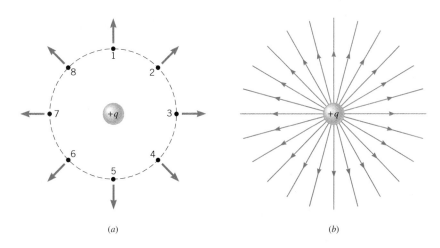

(a)   (b)

Figure 18.23 (a) At any of the eight marked spots around a positive point charge $+q$, a positive test charge would experience a repulsive force directed radially outward. (b) The electric field lines are directed radially outward from a positive point charge $+q$.

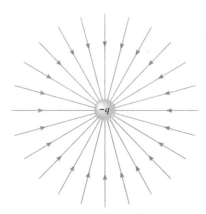

Figure 18.24 The electric field lines are directed radially inward toward a negative point charge $-q$.

Edge view

Figure 18.25 In the central region of a parallel plate capacitor, the electric field lines are parallel and evenly spaced, indicating that the electric field there has the same magnitude and direction at all points.

is true in general that the electric field is stronger in regions where the field lines are closer together. In fact, no matter how many charges are present, the number of lines per unit area passing perpendicularly through a surface is proportional to the magnitude of the electric field.

In regions where the electric field lines are equally spaced, there is the same number of lines per unit area everywhere, and the electric field has the same strength at all points. For example, Figure 18.25 shows that the field lines between the plates of a parallel plate capacitor are parallel and equally spaced, except near the edges where they bulge outward. The equally spaced, parallel lines indicate that the electric field has the same magnitude and direction at all points in the central region of the capacitor.

Often, electric field lines are curved, as in the case of an *electric dipole.* An electric dipole consists of two separated point charges that have the same magnitude but opposite signs. The electric field of a dipole is proportional to the product of the magnitude of one of the charges and the distance between the charges. This product is called the *dipole moment.* Many molecules, such as H_2O and HCl, have dipole moments. Figure 18.26 depicts the field lines in the vicinity of a dipole. For a curved field line, the electric field vector at a point is *tangent* to the line at that point (see points 1, 2, and 3 in the drawing). The pattern of the lines for the dipole indicates that the electric field is greatest in the region between and immediately surrounding the two charges, since the lines are closest together there.

Notice in Figure 18.26 that a given field line starts on the positive charge and ends on the negative charge. In general, *electric field lines always begin on a positive charge and end on a negative charge and do not start or stop in midspace. Furthermore, the number of lines leaving a positive charge or entering a negative charge is proportional to the magnitude of the charge*. This means, for example, that if 100 lines are drawn leaving a $+4$ μC charge, then 75 lines would have to end on a -3 μC charge and 25 lines on a -1 μC charge. Thus, 100 lines leave the charge of $+4$ μC and end on a total charge of -4 μC, so the lines begin and end on equal amounts of total charge.

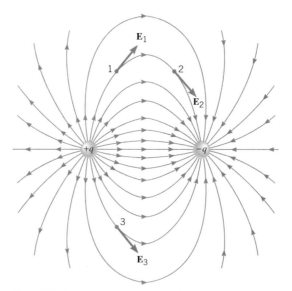

Figure 18.26 The electric field lines of an electric dipole are curved and extend from the positive to the negative charge. At any point, such as 1, 2, or 3, the field created by the dipole is tangent to the line through the point.

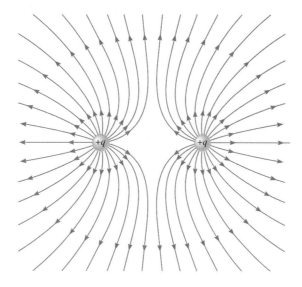

Figure 18.27 The electric field lines for two identical positive point charges. If the charges were both negative, the directions of the lines would be reversed.

The electric field lines are also curved in the vicinity of two identical charges. Figure 18.27 shows the pattern associated with two positive point charges and reveals that there is an absence of lines in the region between the charges. The absence of lines indicates that the electric field is relatively weak between the charges.

Some of the important properties of electric field lines are reexamined in Conceptual Example 12.

CONCEPTUAL EXAMPLE 12   Drawing Electric Field Lines

Figure 18.28a shows three negative point charges ($-q$, $-q$, and $-2q$) and one positive point charge ($+4q$), along with some electric field lines drawn between the charges. There are three things wrong with this drawing. What are they?

Reasoning and Solution One aspect of Figure 18.28a that is incorrect is that electric field lines cross at point P. Field lines can never cross, and here's why. An electric charge placed at P experiences a single net force due to the presence of the other charges in its environment. Therefore, there is only one value for the electric field (which is the force per unit charge) at that point. If two field lines intersected, there would be two electric fields at the point of intersection, one associated with each line. Since there can be only one value of the electric field at any point, there can be only one electric field line passing through that point, so electric field lines never cross each other.

Another mistake in Figure 18.28a is the number of electric field lines that end on the negative charges. Remember, the number of field lines leaving a positive charge or entering a negative charge is proportional to the magnitude of the charge. The $-2q$ charge has half the magnitude of the $+4q$ charge. Therefore, since 8 lines leave the $+4q$ charge, 4 of them (one-half of them) must enter the $-2q$ charge. Of the remaining 4 lines that leave the positive charge, 2 enter each of the $-q$ charges, according to a similar line of reasoning.

The third error in Figure 18.28a is the way in which the electric field lines are drawn between the $+4q$ charge and the $-q$ charge at the left of the drawing. As drawn, the lines are parallel and evenly spaced. This would indicate that the electric field everywhere in this region has a constant magnitude and direction, as is the case in the central region of a parallel plate capacitor. But the electric field between the $+4q$ and $-q$ charges is not constant everywhere. It certainly is stronger in places

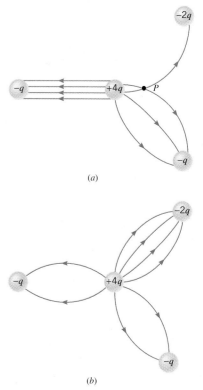

(a)

(b)

Figure 18.28 (*a*) Incorrectly and (*b*) correctly drawn electric field lines.

(a)

(b)

Figure 18.29 (a) Excess charge within a conductor (copper) moves quickly (b) to the surface.

close to the $+4q$ or $-q$ charge than it is midway between them. The field lines, therefore, should be drawn with a curved nature, similar (but not identical) to those that surround a dipole. Figure 18.28b shows more nearly correct representations of the field lines for the four charges.

Related Homework Material: *Problems 26 and 54*

18.8 THE ELECTRIC FIELD INSIDE A CONDUCTOR: SHIELDING

In conducting materials such as copper, electric charges move readily in response to the forces that electric fields exert. This characteristic property of conducting materials has a major effect on the electric field that can exist within and around them. Suppose that a piece of copper carries a number of excess electrons somewhere within it, as in Figure 18.29a. Each electron would experience a force of repulsion because of the electric field of its neighbors. And, since copper is a conductor, the excess electrons move readily in response to that force. In fact, as a consequence of the $1/r^2$ dependence on distance in Coulomb's law, they rush to the surface of the copper. Once static equilibrium is established with all of the excess charge on the surface, no further movement of charge occurs, as part b of the drawing indicates. Similarly, excess positive charge also moves to the surface of a conductor. In general, *at equilibrium under electrostatic conditions, any excess charge resides on the surface of a conductor.*

Now consider the interior of the copper in Figure 18.29b. The interior is electrically neutral, although there are still free electrons that can move under the influence of an electric field. The absence of a net movement of these free electrons indicates that there is no net electric field present within the conductor. In fact, the excess charges arrange themselves on the conductor surface precisely in the manner needed to make the electric field zero within the material. Thus, *at equilibrium under electrostatic conditions, the electric field is zero at any point within a conducting material.* This fact has some fascinating implications.

Figure 18.30a shows an uncharged, solid, cylindrical conductor at equilibrium in the central region of a parallel plate capacitor. Induced charges on the surface of the cylinder alter the electric field lines of the capacitor. Since an electric field cannot exist within the conductor under these conditions, the electric field lines do not penetrate the cylinder. Instead, they end or begin on the induced charges. Consequently, a test charge placed *inside* the conductor would feel no force due to the presence of

Figure 18.30 (a) A cylindrical conductor (shown in cross section) is placed between the oppositely charged plates of a capacitor. The electric field lines do not penetrate the conductor. The blowup shows that, just outside the conductor, the electric field lines are perpendicular to its surface. (b) The electric field is zero in a cavity within the conductor.

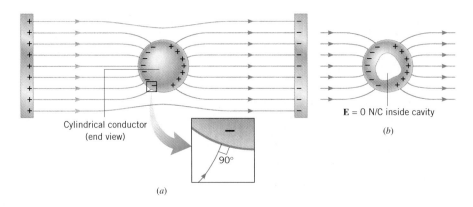

the charges on the capacitor. In other words, *the conductor shields any charge within it from electric fields created outside the conductor.* The shielding results from the induced charges on the conductor surface.

Since the electric field is zero inside the conductor, nothing is disturbed if a cavity is cut from the interior of the material, as in part *b* of the drawing. Thus, the interior of the cavity is also shielded from external electric fields, a fact that has important applications, particularly for shielding electronic circuits. "Stray" electric fields are produced by various electrical appliances (e.g., hair driers, blenders, and vacuum cleaners), and these fields can interfere with the operation of sensitive electronic circuits, such as those in stereo amplifiers, televisions, and computers. To eliminate such interference, circuits are often enclosed within metal boxes that provide shielding from external fields.

The blowup in Figure 18.30*a* shows another aspect of how conductors alter the electric field lines created by external charges. The lines are altered because *the electric field just outside the surface of a conductor is perpendicular to the surface at equilibrium under electrostatic conditions*. If the field were not perpendicular, there would be a component of the field parallel to the surface. Since the free electrons on the surface of the conductor can move, they would do so under the force exerted by that parallel component. But, in reality, no electron flow occurs at equilibrium. Therefore, there can be no parallel component, and the electric field is perpendicular to the surface.

The preceding discussion deals with features of the electric field within and around a conductor at equilibrium under electrostatic conditions. These features are related to the fact that conductors contain mobile free electrons and *do not apply to insulators,* which contain very few free electrons. Example 13 further explores the behavior of a conducting material in the presence of an electric field.

THE PHYSICS OF...

shielding electronic circuits.

CONCEPTUAL EXAMPLE 13   A Conductor in an Electric Field

A charge $+q$ is suspended at the center of a hollow, electrically neutral, spherical conductor, as Figure 18.31 illustrates. Show that this charge induces (a) a charge of $-q$ on the interior surface and (b) a charge of $+q$ on the exterior surface of the conductor.

Reasoning and Solution

(a) Electric field lines emanate from the positive charge $+q$. Since the electric field inside the metal conductor must be zero at equilibrium under electrostatic conditions, each field line ends when it reaches the conductor, as the picture shows. Since field lines terminate only on negative charges, there must be an induced *negative* charge on the interior surface of the conductor. Furthermore, the lines begin and end on equal amounts of charge, so the magnitude of the total induced charge is the same as the magnitude of the charge at the center. Thus, the total induced charge on the interior surface is $-q$.

(b) Before the charge $+q$ is introduced, the conductor is electrically neutral. Therefore, it carries no net charge. We have also seen that there can be no excess charge within the metal. Thus, since an induced charge of $-q$ appears on the interior surface, a charge of $+q$ must be induced on the outer surface. The positive charge on the outer surface generates field lines that radiate outward (see the drawing) as if they originated from the central charge and the conductor were absent. The conductor does not shield the outside from the field produced by the charge on the inside.

Related Homework Material: *Problem 56*

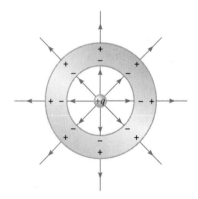

Figure 18.31 A positive charge $+q$ is suspended at the center of a hollow spherical conductor that is electrically neutral. Induced charges appear on the inner and outer surfaces of the conductor. The electric field within the conductor itself is zero.

Concepts at a Glance

Electric Field
(Section 18.6)

Surface Through Which Field Passes

Electric Flux

Gauss' Law

Figure 18.32 *Concepts at a Glance* The electric field and the surface through which it passes are brought together to define the concept of electric flux. It is in terms of electric flux that Gauss' law is formulated. When the electric field is produced by a distribution of charges, such as that on the silvery metal sphere in the photograph, Gauss' law relates the electric flux to the charge.

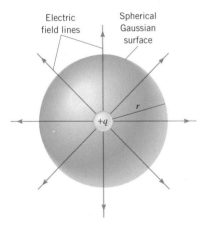

Figure 18.33 A positive point charge is located at the center of an imaginary spherical surface of radius r. Such a surface is one example of a Gaussian surface. Here the electric field is perpendicular to the surface and has the same magnitude everywhere on it.

18.9 GAUSS' LAW

Section 18.6 discusses how a point charge creates an electric field in the space around the charge. There are also many situations in which an electric field is produced by charges that are spread out over a region, rather than by a single point charge. Such an extended collection of charges is called a charge distribution. For example, the electric field within the parallel plate capacitor in Figure 18.22 is produced by positive charges spread uniformly over one plate and an equal number of negative charges spread over the other plate. As we will see, Gauss' law describes the relationship between a charge distribution and the electric field it produces.

Concepts at a Glance In presenting Gauss' law, it will be necessary to introduce a new idea called electric flux, and the Concepts-at-a-Glance chart in Figure 18.32 outlines the approach we will follow. This figure is an extended version of the concept chart in Figure 18.16 because the idea of flux involves both the electric field and the surface through which it passes. In Figure 18.23*b*, for instance, the electric field from the point charge can be viewed as passing through an imaginary spherical surface centered on the charge. By bringing together the electric field and the surface through which it passes, we will be able to define electric flux and then present Gauss' law. The notion of flux will come up again in Chapter 22, where we will encounter magnetic flux.

We begin by developing a version of Gauss' law that applies only to a point charge. The electric field lines for a positive point charge radiate outward in all directions from the charge, as Figure 18.23*b* indicates. The magnitude E of the electric field at a distance r from the charge is $E = kq/r^2$, according to Equation 18.3. As mentioned in Section 18.5, the constant k can be expressed as $k = 1/(4\pi\epsilon_0)$, where ϵ_0 is the permittivity of free space. With this substitution, the magnitude of the electric field becomes $E = q/(4\pi\epsilon_0 r^2)$. We now place this point charge at the center of an imaginary spherical surface of radius r, as Figure 18.33 shows. Such a hypothetical closed surface is called a *Gaussian surface,* although in general it need not be spherical. The surface area A of a sphere is $A = 4\pi r^2$, and the magnitude of the electric field can be written in terms of this area as $E = q/(A\epsilon_0)$, or

Gauss' law for a point charge

$$\underbrace{EA}_{\substack{\text{Electric} \\ \text{flux, } \Phi_E}} = \frac{q}{\epsilon_0} \qquad (18.5)$$

The left side of Equation 18.5 is the product of the magnitude E of the electric field at any point on the Gaussian surface and the area A of the surface. In Gauss' law this product is especially important and is called the *electric flux,* Φ_E: $\Phi_E = EA$. (It will be necessary to modify this definition of flux when we consider the general case of a Gaussian surface with an arbitrary shape.)

Equation 18.5 is the result we have been seeking, for it is the form of Gauss' law that applies to a point charge. This result indicates, aside from the constant ϵ_0, that the electric flux Φ_E depends only on the charge q within the Gaussian surface and is independent of the radius r of the surface. We will now see how to generalize Equation 18.5 to account for situations where the electric field is produced by a distribution of charges and the Gaussian surface has an arbitrary shape.

Figure 18.34 shows a charge distribution whose net charge is labeled Q. The charge distribution is surrounded by a Gaussian surface, i.e., an imaginary closed surface. The surface can have *any arbitrary shape* (it need not be spherical), but *it must be closed* (an open surface would be like that of half an eggshell). The direction of the electric field is not necessarily perpendicular to the Gaussian surface. Furthermore, the magnitude of the electric field need not be constant on the surface, but can vary from point to point.

To determine the electric flux through such a surface, we divide the surface up into many tiny sections with areas, ΔA_1, ΔA_2, and so on. Each section is so small that it is essentially flat and the electric field $\mathbf{E}$ is a constant (both in magnitude and direction) over it. For reference, a dashed line called the "normal" is drawn perpendicular to each section on the outside of the surface. To determine the electric flux for each of the sections, we use only the component of $\mathbf{E}$ that is perpendicular to the surface, that is, the component of the electric field that passes through the surface. From the drawing it can be seen that this component has a magnitude of $E \cos \phi$, where ϕ is the angle between the electric field and the normal. The electric flux through any individual section is then $(E \cos \phi)\Delta A$. The electric flux Φ_E that passes through the entire Gaussian surface is the sum of all of these individual fluxes: $\Phi_E = (E_1 \cos \phi_1)\Delta A_1 + (E_2 \cos \phi_2)\Delta A_2 + \cdots$, or

$$\Phi_E = \Sigma(E \cos \phi)\, \Delta A \qquad (18.6)$$

where, as usual, the symbol Σ means "the sum of." Gauss' law relates the electric flux Φ_E to the net charge Q enclosed by the arbitrarily shaped Gaussian surface.

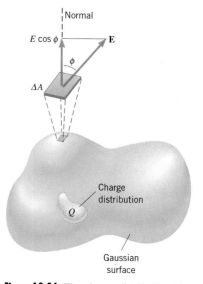

Figure 18.34 The charge distribution Q is surrounded by an arbitrarily shaped Gaussian surface. The electric flux Φ through any tiny segment of the surface is the product of $E \cos \phi$ and the area ΔA of the segment: $\Phi = (E \cos \phi)\, \Delta A$. The angle ϕ is the angle between the electric field and the normal to the surface.

● GAUSS' LAW

The electric flux Φ_E through a Gaussian surface is equal to the net charge Q enclosed by the surface divided by ϵ_0, the permittivity of free space:

$$\underbrace{\sum (E \cos \phi)\Delta A}_{\text{Electric flux, } \Phi_E} = \frac{Q}{\epsilon_0} \qquad (18.7)$$

SI Unit of Electric Flux: $\text{N} \cdot \text{m}^2/\text{C}$

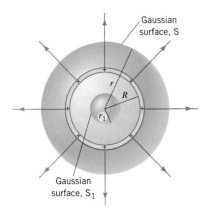

Figure 18.35 A uniform distribution of positive charge resides on a thin spherical shell of radius R. The spherical Gaussian surfaces S and S_1 are used in Example 14 to evaluate the electric flux outside and inside the shell, respectively.

Gauss' law is often used to find the magnitude of the electric field produced by a distribution of charges. The law is most useful when the distribution is uniform and symmetrical. In the next two examples we will see how to apply Gauss' law in such situations.

EXAMPLE 14   The Electric Field of a Charged Thin Spherical Shell

Figure 18.35 shows a thin spherical shell of radius R. A positive charge Q is spread uniformly over the shell. Find the magnitude of the electric field at any point (a) outside the shell and (b) inside the shell.

Reasoning Because the charge is distributed uniformly over the spherical shell, the electric field is symmetric. This means that the electric field is directed radially outward in all directions, and its magnitude is the same at all points that are equidistant from the shell. All such points lie on a sphere, so the symmetry is called *spherical symmetry*. With this symmetry in mind, we will use a spherical Gaussian surface to evaluate the electric flux Φ_E. We will then use Gauss' law to determine the magnitude of the electric field.

Solution

(a) To find the magnitude of the electric field outside the charged shell, we evaluate the electric flux $\Phi_E = \Sigma(E \cos \phi)\Delta A$ by using a spherical Gaussian surface of radius r $(r > R)$ that is concentric with the shell. See the surface labeled S in Figure 18.35. Since the electric field **E** is everywhere perpendicular to the Gaussian surface, $\phi = 0°$ and $\cos \phi = 1$. In addition, E has the same value at all points on the surface, since they are equidistant from the charged shell. Being constant over the surface, E can be factored outside the summation, with the result that $\Phi_E = \Sigma(E \cos 0°)\Delta A = E(\Sigma\Delta A)$. The term $\Sigma\Delta A$ is just the sum of the tiny areas that make up the Gaussian surface. Since the area of a spherical surface is $4\pi r^2$, we have $\Sigma\Delta A = 4\pi r^2$. The electric flux is then $\Phi_E = E(4\pi r^2)$. Setting the electric flux equal to Q/ϵ_0, as specified by Gauss' law, yields $E(4\pi r^2) = Q/\epsilon_0$. This equation can be solved for E to show that

$$E = \frac{Q}{4\pi\epsilon_0 r^2} \qquad \text{(for } r > R)$$

This is a surprising result, for it is exactly the same as that for a point charge (see Equation 18.3). Thus, the electric field outside a uniformly charged spherical shell is the same as if all the charge Q were concentrated as a point charge at the center of the shell.

(b) To find the magnitude of the electric field inside the charged shell, we select a spherical Gaussian surface that lies inside the shell and is concentric with it. See the surface labeled S_1 in Figure 18.35. Inside the charged shell, the electric field (if it exists) must also have spherical symmetry. Therefore, using reasoning like that in part (a), the electric flux through the Gaussian surface is $\Phi_E = \Sigma(E \cos \phi)\Delta A = E(4\pi r_1^2)$. In accord with Gauss' law, the electric flux must be equal to Q/ϵ_0, where Q is the net charge *inside* the Gaussian surface. But now $Q = 0$ C, since all the charge lies on the shell that is *outside* the surface S_1. Consequently, we have $E(4\pi r_1^2) = Q/\epsilon_0 = 0$, or

$$E = 0 \text{ N/C} \qquad \text{(for } r < R)$$

Gauss' law allows us to deduce that there is no electric field inside a uniform spherical shell of charge. An electric field exists only on the outside.

EXAMPLE 15   The Electric Field Inside a Parallel Plate Capacitor

According to Equation 18.4, the electric field inside a parallel plate capacitor, and away from the edges, is constant and has a magnitude of $E = \sigma/\epsilon_0$, where σ is the charge density (the charge per unit area) on a plate. Use Gauss' law to prove this result.

Reasoning Figure 18.36a shows the electric field inside a parallel plate capacitor. Because the positive and negative charges are distributed uniformly over the surfaces of the plates, symmetry requires that the electric field be perpendicular to the plates. We will take advantage of this symmetry by choosing our Gaussian surface to be a small cylinder whose axis is perpendicular to the plates (see part b of the figure). With this choice, we will be able to evaluate the electric flux and then, with the aid of Gauss' law, determine E.

Solution Figure 18.36b shows that we have placed our Gaussian cylinder so that its left end is inside the positive metal plate, and the right end is in the space between the plates. To determine the electric flux through this Gaussian surface, we evaluate the flux through each of the three parts—labeled 1, 2, and 3 in the drawing—that make up the total surface of the cylinder and then add up the fluxes.

Surface 1—the left, flat end of the cylinder—is embedded inside the positive metal plate. As discussed in Section 18.8, the electric field is zero everywhere inside a conductor that is in equilibrium under electrostatic conditions. Since $E = 0$ N/C, the electric flux through this surface is also zero: $\Phi_1 = \Sigma(E \cos \phi)\Delta A = 0$.

Surface 2—the curved wall of the cylinder—is everywhere parallel to the electric field between the plates, so that $\cos \phi = \cos 90° = 0$. Therefore, the electric flux through this surface is also zero: $\Phi_2 = \Sigma(E \cos 90°)\Delta A = 0$.

Surface 3—the right, flat end of the cylinder—is perpendicular to the electric field between the plates, so $\cos \phi = \cos 0° = 1$. The electric field is constant over this surface, so E can be taken outside the summation in Equation 18.6. As a result, the electric flux through this surface is $\Phi_3 = \Sigma(E \cos 0°)\Delta A = E(\Sigma\Delta A) = EA$, where $\Sigma\Delta A = A$ is the area of surface 3.

The electric flux through the entire Gaussian cylinder is the sum of the three fluxes determined above:

$$\Phi_E = \Phi_1 + \Phi_2 + \Phi_3 = 0 + 0 + EA = EA$$

According to Gauss' law, we set the electric flux equal to Q/ϵ_0, where Q is the net charge *inside* the Gaussian cylinder: $EA = Q/\epsilon_0$. But Q/A is the charge per unit area, σ, on the plate. Therefore, we arrive at the value of the electric field inside a parallel plate capacitor: $\boxed{E = \sigma/\epsilon_0}$. Notice that the distance of the right end of the Gaussian cylinder from the positive plate does not appear in this result, indicating that the electric field has the same value everywhere between the plates.

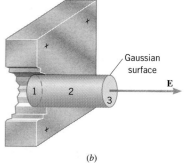

Figure 18.36 (a) A side view of a parallel plate capacitor, showing some of the electric field lines. (b) The Gaussian surface is a cylinder oriented so its axis is perpendicular to the positive plate and its left end is inside the plate.

*18.10   COPIERS AND COMPUTER PRINTERS

The electrostatic force that charged particles exert on one another plays the central role in an office copier. The copying process is called *xerography*, from the Greek *xeros* and *graphos*, meaning "dry writing." The heart of a copier is the xerographic drum, an aluminum cylinder coated with a layer of selenium (see Figure 18.37a). Aluminum is an excellent electrical conductor. Selenium, on the other hand, is a photoconductor; it is an insulator in the dark but becomes a conductor when exposed to light. Consequently, a positive charge deposited on the selenium surface will remain there, provided the selenium is kept in the dark. When the drum is exposed to light, however, electrons from the aluminum pass through the conducting selenium and neutralize the positive charge.

THE PHYSICS OF...

xerography.

Figure 18.37 (*a*) This cutaway view shows the essential elements of a copying machine. (*b*) The five steps in the xerographic process.

The photoconductive property of selenium is critical to the xerographic process, as Figure 18.37*b* illustrates. First, an electrode called a *corotron* gives the entire selenium surface a positive charge in the dark. Second, a series of lenses and mirrors focuses an image of a document onto the revolving drum. The dark and light areas of the document produce corresponding areas on the drum. The dark areas retain their positive charge, but the light areas become conducting and lose their positive charge, ending up neutralized. Thus, a positive-charge image of the document remains on the selenium surface. In the third step, a special dry, black powder, called the *toner,* is given a negative charge and then spread onto the drum, where it adheres selectively to the positively charged areas. The fourth step involves transferring the toner onto a blank piece of paper. However, the attraction of the positive-charge image holds the toner to the drum. To transfer the toner, the paper is given a *greater positive charge* than that of the image, with the aid of another corotron. Lastly, the paper and adhering toner pass through heated pressure rollers, which melt the toner into the fibers of the paper and produces the finished copy.

THE PHYSICS OF...

a laser printer.

A laser printer is used with computers to provide high-quality copies of text and graphics. It is similar in operation to the xerographic machine, except the information to be reproduced is not on paper. Instead, the information is transferred from the computer's memory to the printer, and laser light is used to copy it onto the selenium–aluminum drum. A laser beam, focused to a fine point, is scanned rapidly from side to side across the rotating drum, as Figure 18.38 indicates. While the light remains on, the positive charge on the drum is neutralized. As the laser beam moves, the computer turns the beam off at the right moments during each scan to produce the desired positive-charge image, which is the letter "A" in the picture.

An inkjet printer is another type of printer that uses electric charges in its operation. While shuttling back and forth across the paper, the inkjet printhead ejects a thin stream of ink. Figure 18.39 illustrates the elements of one type of printhead. The ink is forced out of a small nozzle and breaks up into extremely small droplets, with diameters less than 1×10^{-4} m. About 150 000 droplets leave the nozzle each second and travel with a speed of approximately 18 m/s toward the paper. During their flight, the droplets pass through two electrical components, an *electrode* and the *deflection plates* (a parallel plate capacitor). When the printhead moves over regions of the paper that are not to be inked, the charging control is turned on and an electric field is established between the printhead and the electrode. As the drops pass through the electric field, they acquire a net charge by the process of induction. The deflection plates divert the charged droplets into a gutter and thus prevent them from reaching the paper. Whenever ink is to be placed on the paper, the charging control, responding to instructions from the computer, turns off the electric field. The uncharged droplets fly straight through the deflection plates and strike the paper.

Figure 18.38 As the laser beam scans back and forth across the surface of the xerographic drum, a positive-charge image of the letter "A" is created.

THE PHYSICS OF...

an inkjet printer.

18.11 CONCEPTS & CALCULATIONS

In this chapter we have studied electric forces and electric fields. We conclude now by presenting some examples that review important features of these concepts. The three-part format of the examples stresses the role of conceptual understanding in problem solving. First, the problem statement is given. Then, there is a concept question-and-answer section, followed by the solution section. The purpose of the concept question-and-answer section is to provide help in understanding the solution and to illustrate how a review of the concepts can help in anticipating some of the characteristics of the numerical answers.

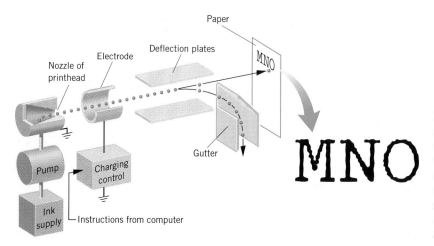

Figure 18.39 An inkjet printhead ejects a steady flow of ink droplets. Charged droplets are deflected into a gutter by the deflection plates, while uncharged droplets fly straight onto the paper. Letters formed by an inkjet printer look normal, except when greatly enlarged and the patterns from the drops become apparent.

(a)

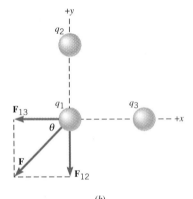

(b)

Figure 18.40 (*a*) Three equal charges lie on the *x* and *y* axes. (*b*) The net force exerted on q_1 by the other two charges is **F**.

CONCEPTS & CALCULATIONS EXAMPLE 16
The Vector Nature of Electric Forces

The charges on three identical metal spheres are -12 μC, $+4.0$ μC, and $+2.0$ μC. The spheres are brought together so they simultaneously touch each other. They are then separated and placed on the *x* and *y* axes, as in Figure 18.40*a*. What is the net force (magnitude and direction) exerted on the sphere at the origin? Treat the spheres as if they were particles.

Concept Questions and Answers Is the net charge on the three spheres the same before and after touching?

Answer   Yes. The conservation of electric charge states that, during any process, the net electric charge of an isolated system remains constant (is conserved). Therefore, the net charge on the three spheres before they touch (-12.0 μC $+ 4.0$ μC $+ 2.0$ μC $= -6.0$ μC) is the same as the net charge after they touch.

After the spheres touch and are separated, do they have identical charges?

Answer   Yes. Since the spheres are identical, the net charge (-6.0 μC) distributes itself equally over the three spheres. After they are separated, each has one-third of the net charge: $q_1 = q_2 = q_3 = \frac{1}{3}(-6.0 \ \mu C) = -2.0 \ \mu C$.

Do q_2 and q_3 exert forces of equal magnitude on q_1?

Answer   Yes. The charges q_2 and q_3 have equal magnitudes and are the same distance from q_1. According to Coulomb's law, then, they exert forces of equal magnitude on q_1.

Is the magnitude of the net force exerted on q_1 equal to $2F$, where F is the magnitude of the force that either q_2 or q_3 exerts on q_1?

Answer   No. Although the two forces that act on q_1 have equal magnitudes, they have different directions. The forces are repulsive forces, since all of the charges in part *a* of the drawing are identical. Figure 18.40*b* shows the force $\mathbf{F}_{12}$ exerted on q_1 by q_2 and the force $\mathbf{F}_{13}$ exerted on q_1 by q_3. To obtain the net force $\mathbf{F}$, we must take these directions into account by using vector addition.

Solution

The magnitude F_{12} of the force exerted on q_1 by q_2 is given by Coulomb's law, Equation 18.1, as

$$F_{12} = \frac{k q_1 q_2}{r_{12}^2} = \frac{(8.99 \times 10^9 \ \text{N} \cdot \text{m}^2/\text{C}^2)(2.0 \times 10^{-6} \ \text{C})(2.0 \times 10^{-6} \ \text{C})}{(3.5 \times 10^{-3} \ \text{m})^2}$$

$$= 2.9 \times 10^3 \ \text{N}$$

Note that we have used the magnitudes of q_1 and q_2 in Coulomb's law. As mentioned previously, the magnitude of the force $\mathbf{F}_{13}$ exerted on q_1 by q_3 has the same value as F_{12}, so $F_{13} = 2.9 \times 10^3$ N. Since the forces $\mathbf{F}_{12}$ and $\mathbf{F}_{13}$ are perpendicular to each other, we may use the Pythagorean theorem to find the magnitude F of the net force:

$$F = \sqrt{F_{12}^2 + F_{13}^2} = \sqrt{(2.9 \times 10^3 \ \text{N})^2 + (2.9 \times 10^3 \ \text{N})^2} = \boxed{4.1 \times 10^3 \ \text{N}}$$

The angle θ that the net force makes with the $-x$ axis (see part *b* of the drawing) is

$$\theta = \tan^{-1}\left(\frac{F_{12}}{F_{13}}\right) = \tan^{-1}\left(\frac{2.9 \times 10^3 \ \text{N}}{2.9 \times 10^3 \ \text{N}}\right) = \boxed{45°}$$

CONCEPTS & CALCULATIONS EXAMPLE 17
Becoming Familiar with Electric Fields

Two point charges are lying on the y axis in Figure 18.41a: $q_1 = -4.00 \ \mu C$ and $q_2 = +4.00 \ \mu C$. They are equidistant from the point P, which lies on the x axis. (a) What is the net electric field at P? (b) A small object of charge $q_0 = +8.00 \ \mu C$ and mass $m = 1.20$ g is placed at P. When it is released, what is its acceleration?

Concept Questions and Answers There is no charge at P in part (a). Is there an electric field at P?

   Answer Yes. An electric field is produced by the charges q_1 and q_2, and it exists throughout the entire region that surrounds them. If a test charge were placed at this point, it would experience a force due to the electric field. The force would be the product of the charge and the electric field.

The charge q_1 produces an electric field at the point P. What is the direction of this field?

   Answer The electric field created by a charge always points away from a positive charge and toward a negative charge. Since q_1 is negative, the electric field $\mathbf{E}_1$ points toward it (see Figure 18.41b).

What is the direction of the electric field produced by q_2 at P?

   Answer Since q_2 is positive, the electric field $\mathbf{E}_2$ it produces points away from q_2, as shown in the drawing.

Is the magnitude of the net electric field equal to $E_1 + E_2$, where E_1 and E_2 are the magnitudes of the electric fields produced by q_1 and q_2?

   Answer No, because the electric fields have different directions. We must add the individual fields as vectors to obtain the net electric field. Only then can we determine its magnitude.

Solution

(a) The magnitude of the electric fields that q_1 and q_2 produce at P are given by Equation 18.3, where the distances are specified in the drawing:

$$E_1 = \frac{kq_1}{r_1^2} = \frac{(8.99 \times 10^9 \ \text{N} \cdot \text{m}^2/\text{C}^2)(4.00 \times 10^{-6} \ \text{C})}{(0.700 \ \text{m})^2} = 7.34 \times 10^4 \ \text{N/C}$$

$$E_2 = \frac{kq_2}{r_2^2} = \frac{(8.99 \times 10^9 \ \text{N} \cdot \text{m}^2/\text{C}^2)(4.00 \times 10^{-6} \ \text{C})}{(0.700 \ \text{m})^2} = 7.34 \times 10^4 \ \text{N/C}$$

The x and y components of these fields and the total field $\mathbf{E}$ are given in the following table.

Electric field	x component	y component
$\mathbf{E}_1$	$-E_1 \cos 31.0° = -6.29 \times 10^4 \ \text{N/C}$	$+E_1 \sin 31.0° = +3.78 \times 10^4 \ \text{N/C}$
$\mathbf{E}_2$	$+E_2 \cos 31.0° = +6.29 \times 10^4 \ \text{N/C}$	$+E_2 \sin 31.0° = +3.78 \times 10^4 \ \text{N/C}$
$\mathbf{E}$	$E_x = 0 \ \text{N/C}$	$E_y = +7.56 \times 10^4 \ \text{N/C}$

The net electric field $\mathbf{E}$ has only a component along the $+y$ axis. So,

$$\boxed{\mathbf{E} = 7.56 \times 10^4 \ \text{N/C, directed along the } +y \text{ axis.}}$$

(b) According to Newton's second law, Equation 4.2, the acceleration $\mathbf{a}$ of an object placed at this point is equal to the net force acting on it divided by its mass. The net

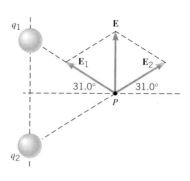

Figure 18.41 (a) Two charges q_1 and q_2 produce an electric field at the point P. (b) The electric fields $\mathbf{E}_1$ and $\mathbf{E}_2$ add to give the net electric field $\mathbf{E}$.

force $\mathbf{F}$ is the product of the charge and the net electric field, $\mathbf{F} = q_0\mathbf{E}$, as indicated by Equation 18.2. Thus, the acceleration is

$$\mathbf{a} = \frac{\mathbf{F}}{m} = \frac{q_0\mathbf{E}}{m}$$

$$= \frac{(8.00 \times 10^{-6}\ \text{C})(7.56 \times 10^4\ \text{N/C})}{1.20 \times 10^{-3}\ \text{kg}} = \boxed{5.04 \times 10^2\ \text{m/s}^2,\ \text{along the} +y\ \text{axis}}$$

At the end of the problem set for this chapter, you will find homework problems that contain both conceptual and quantitative parts. These problems are grouped under the heading *Concepts & Calculations, Group Learning Problems*. They are designed for use by students working alone or in small learning groups. The conceptual part of each problem provides a convenient focus for group discussions.

SUMMARY

There are two kinds of **electric charge:** positive and negative; the SI unit of electric charge is the **coulomb** (C). The magnitude of the charge on an electron or a proton is $e = 1.60 \times 10^{-19}$ C. The electron carries a charge of $-e$, while the proton carries a charge of $+e$. The charge on any object, whether positive or negative, is **quantized,** in the sense that the charge consists of an integer number of protons or electrons. **The law of conservation of electric charge** states that the net electric charge of an isolated system remains constant during any process.

An **electrical conductor** is a material, such as copper, that conducts electric charge readily. An **electrical insulator** is a material, such as rubber, that conducts electric charge poorly. **Charging by contact** is the process of giving one object a net electric charge by placing it in contact with an object that is already charged. **Charging by induction** is the process of giving an object a net electric charge without touching it to a charged object.

One charge exerts an electric force on another charge. **For like charges the force is a repulsion, while for unlike charges the force is an attraction.** Coulomb's law gives the magnitude F of the electric force that two point charges exert on each other as $F = kq_1q_2/r^2$, where q_1 and q_2 are the magnitudes of the charges, r is the distance between the charges, and $k = 8.99 \times 10^9$ N$\cdot$m^2/C^2. The force acts along the line between the two point charges. The permittivity of free space ϵ_0 is defined by the relation $k = 1/(4\pi\epsilon_0)$.

The **electric field E** at a given spot is a vector and is the electrostatic force $\mathbf{F}$ experienced by a small test charge q_0 placed at

that spot divided by the charge itself: $\mathbf{E} = \mathbf{F}/q_0$. The direction of the electric field is the same as the direction of the force on a positive test charge. The SI unit for the electric field is the newton per coulomb (N/C). The source of the electric field at any spot is the charged objects surrounding that spot. The magnitude of the electric field created by a point charge q is $E = kq/r^2$, where r is the distance from the charge. The magnitude of the electric field between the plates of a parallel plate capacitor is $E = \sigma/\epsilon_0$, where σ is the charge per unit area on either plate.

Electric field lines are lines that can be thought of as a "map" providing information about the direction and strength of the electric field. Electric field lines are directed away from positive charges and toward negative charges. The direction of the lines gives the direction of the electric field, since the electric field vector at a point is tangent to the line at that point. The electric field is strongest in regions where the number of lines per unit area passing perpendicularly through a surface is the greatest, that is, where the lines are packed together most tightly.

Excess negative or positive charge resides on the surface of a conductor at equilibrium under electrostatic conditions. In such a situation, the electric field at any point within the conducting material is zero, and the electric field just outside the surface of the conductor is perpendicular to the surface.

Gauss' law states that the electric flux Φ_E (see Equation 18.6) through a closed surface (a Gaussian surface) is equal to the net charge Q enclosed by the surface divided by ϵ_0, the permittivity of free space: $\Phi_E = Q/\epsilon_0$.

CONCEPTUAL QUESTIONS

1. In Figure 18.8 the grounding wire is removed first, followed by the rod, and the sphere is left with a positive charge. If the rod were removed first, followed by the grounding wire, would the sphere be left with a charge? Account for your answer.

2. A metallic object is given a positive charge by the process of induction, as illustrated in Figure 18.8. (a) Does the mass of the object increase, decrease, or remain the same? Why? (b) What happens to the mass of the object if it is given a negative charge by induction? Explain.

3. A rod made from insulating material carries a net charge, while a copper sphere is neutral. The rod and the sphere do not touch. Is it possible for the rod and the sphere to (a) attract one another and (b) repel one another? Explain.

4. On a dry day, just after washing your hair to remove natural oils and drying it thoroughly, run a plastic comb through it. Small bits of paper will be attracted to the comb. Explain why.

5. Blow up a balloon and rub it against your shirt a number of times. In so doing you give the balloon a net electric charge. Now touch the balloon to the ceiling. Upon being released, the balloon will remain stuck to the ceiling. Why?

6. A proton and an electron are held in place on the x axis. The proton is at $x = -d$, while the electron is at $x = +d$. They are released simultaneously, and the only force that affects their motions is the electrostatic force of attraction that each applies to the other. Which particle reaches the origin first? Give your reasoning.

7. A particle is attached to a spring and is pushed so that the spring is compressed more and more. As a result, the spring exerts a greater and greater force on the particle. Similarly, a charged particle experiences a greater and greater force when pushed closer and closer to another particle that is fixed in position and has a charge of the same polarity. In spite of the similarity, the charged particle will *not* exhibit simple harmonic motion upon being released, as will the particle on the spring. Explain why not.

8. Identical point charges are fixed to opposite corners of a square. Where does a third point charge experience the greater force, at one of the empty corners or at the center of the square? Account for your answer.

9. On a thin, nonconducting rod, positive charges are spread evenly, so that there is the same amount of charge per unit length at every point. On another identical rod, positive charges are spread evenly over only the left half, and the same amount of negative charges are spread evenly over the right half. For each rod, deduce the *direction* of the electric field at a point that is located directly above the midpoint of the rod. Give your reasoning.

10. There is an electric field at point P. A very small charge is placed at this point and experiences a force. Another very small charge is then placed at this point and experiences a force that differs in both magnitude and direction from that experienced by the first charge. How can these two different forces result from the single electric field that exists at point P?

11. Three point charges are fixed to the corners of a square, one to a corner, in such a way that the net electric field at the empty corner is zero. Do these charges all have (a) the same sign and (b) the same magnitude (but, possibly, different signs)? Justify your answers.

12. Review Conceptual Example 11 as an aid in answering this question. Suppose in Figure 18.21 that charges $+q$ are placed on corners 1 and 3 of the rectangle, and charges $-q$ are placed on corners 2 and 4. What is the net electric field at the center C of the rectangle?

13. In Figure 18.26 there is no place on the line through the charges where the electric field is zero, neither to the left of the positive charge, nor between the charges, nor to the right of the

negative charge. Now, suppose the magnitude of the negative charge were greater than the magnitude of the positive charge. Is there any place on the line through the charges where the electric field is zero? Justify your answer.

14. Drawings I and II show two examples of electric field lines. Decide which of the following statements are true and which are false, defending your choice in each case. (a) In both I and II the electric field is the same everywhere. (b) As you move from left to right in each case, the electric field becomes stronger. (c) The electric field in I is the same everywhere but becomes stronger in II as you move from left to right. (d) The electric fields in both I and II could be created by negative charges located somewhere on the left and positive charges somewhere on the right. (e) Both I and II arise from a single positive point charge located somewhere on the left.

15. A positively charged particle is moving horizontally when it enters the region between the plates of a capacitor, as the drawing illustrates. (a) Draw the trajectory that the particle follows in moving through the capacitor. (b) When the particle is within the capacitor, which of the following four vectors, if any, are *parallel* to the electric field inside the capacitor: the particle's displacement, its velocity, its linear momentum, its acceleration? For each vector explain why the vector is, or is not, parallel to the electric field of the capacitor.

16. Refer to Figure 18.27. Imagine a plane that is perpendicular to the line between the charges, midway between them, and is half into and half out of the paper. The electric flux through this plane is zero. Explain why.

17. Two charges, $+q$ and $-q$, are inside a Gaussian surface. Since the net charge inside the Gaussian surface is zero, Gauss' law states that the electric flux through the surface is also zero, i.e., $\Phi_E = 0$. Does the fact that $\Phi_E = 0$ imply that the electric field E at any point on the Gaussian surface is also zero? Justify your answer. (*Hint: Imagine a Gaussian surface that encloses the two charges in Figure 18.26.*)

18. The drawing shows three charges, labeled q_1, q_2, and q_3. A Gaussian surface is drawn around q_1 and q_2. (a) Which charges determine the electric flux through the Gaussian surface? (b) Which charges produce the electric field at the point P? Justify your answers.

19. A charge $+q$ is placed inside a spherical Gaussian surface. The charge is *not* located at the center of the sphere. (a) Can Gauss' law tell us exactly where the charge is located inside the sphere? Justify your answer. (b) Can Gauss' law tell us about the magnitude of the electric flux through the Gaussian surface? Why?

PROBLEMS

Problems that are not marked with a star are considered the easiest to solve. Problems that are marked with a single star (*) are more difficult, while those marked with a double star (**) are the most difficult. Note: All charges are point charges, unless specified otherwise.

---

ssm Solution is in the Student Solutions Manual. **www** Solution is available on the World Wide Web at http://www.wiley.com/college/cutnell
🔊 This icon represents a biomedical application.

Section 18.1 The Origin of Electricity, Section 18.2 Charged Objects and the Electric Force, Section 18.3 Conductors and Insulators, Section 18.4 Charging by Contact and by Induction

1. ssm How many electrons must be removed from an electrically neutral silver dollar to give it a charge of $+2.4\ \mu C$?

2. A rod has a charge of $-2.0\ \mu C$. How many electrons must be removed so that the charge becomes $+3.0\ \mu C$?

3. A metal sphere has a charge of $+8.0\ \mu C$. What is the net charge after 6.0×10^{13} electrons have been placed on it?

4. Object A is metallic and electrically neutral. It is charged by induction so that it acquires a charge of -3.0×10^{-6} C. Object B is identical to object A and is also electrically neutral. It is charged by induction so that it acquires a charge of $+3.0 \times 10^{-6}$ C. Find the *difference* in mass between the charged objects and state which has the greater mass.

5. ssm Consider three identical metal spheres, A, B, and C. Sphere A carries a charge of $+5q$. Sphere B carries a charge of $-q$. Sphere C carries no net charge. Spheres A and B are touched together and then separated. Sphere C is then touched to sphere A and separated from it. Lastly, sphere C is touched to sphere B and separated from it. (a) How much charge ends up on sphere C? What is the total charge on the three spheres (b) before they are allowed to touch each other and (c) after they have touched?

*6. Water has a mass per mole of 18.0 g/mol, and each water molecule (H_2O) has 10 electrons. (a) How many electrons are there in one liter (1.00×10^{-3} m³) of water? (b) What is the net charge of all these electrons?

Section 18.5 Coulomb's Law

7. ssm www Two very small spheres are initially neutral and separated by a distance of 0.50 m. Suppose that 3.0×10^{13} electrons are removed from one sphere and placed on the other. (a) What is the magnitude of the electrostatic force that acts on each sphere? (b) Is the force attractive or repulsive? Why?

8. The nucleus of the helium atom contains two protons that are separated by about 3.0×10^{-15} m. Find the magnitude of the electrostatic force that each proton exerts on the other. (The protons remain together in the nucleus because the repulsive electrostatic force is balanced by an attractive force called the strong nuclear force.)

9. The force of repulsion that two like charges exert on each other is 3.5 N. What will the force be if the distance between the charges is increased to five times its original value?

10. In a vacuum, two particles have charges of q_1 and q_2, where $q_1 = +3.5\ \mu C$. They are separated by a distance of 0.26 m, and particle 1 experiences an attractive force of 3.4 N. What is q_2 (magnitude and sign)?

11. ssm Three charges are fixed to an xy coordinate system. A charge of $+18\ \mu C$ is on the y axis at $y = +3.0$ m. A charge of $-12\ \mu C$ is at the origin. Lastly, a charge of $+45\ \mu C$ is on the x axis at $x = +3.0$ m. Determine the magnitude and direction of the net electrostatic force on the charge at $x = +3.0$ m. Specify the direction relative to the $-x$ axis.

12. A charge of $-3.00\ \mu C$ is fixed at the center of a compass. Two additional charges are fixed on the circle of the compass (radius $= 0.100$ m). The charges on the circle are $-4.00\ \mu C$ at the position due north and $+5.00\ \mu C$ at the position due east. What is the magnitude and direction of the net electrostatic force acting on the charge at the center? Specify the direction relative to due east.

13. An equilateral triangle has sides of 0.15 m. Charges of -9.0, $+8.0$, and $+2.0\ \mu C$ are located at the corners of the triangle. Find the magnitude of the net electrostatic force exerted on the 2.0-μC charge.

14. Two tiny conducting spheres are identical and carry charges of $-20.0\ \mu C$ and $+50.0\ \mu C$. They are separated by a distance of 2.50 cm. (a) What is the magnitude of the force that each sphere experiences, and is the force attractive or repulsive? (b) The spheres are brought into contact and then separated to a distance of 2.50 cm. Determine the magnitude of the force that each sphere now experiences, and state whether the force is attractive or repulsive.

15. ssm www Two particles, with identical positive charges and a separation of 2.60×10^{-2} m, are released from rest. Immediately after the release, particle 1 has an acceleration $\mathbf{a_1}$ whose magnitude is 4.60×10^3 m/s², while particle 2 has an acceleration $\mathbf{a_2}$ whose magnitude is 8.50×10^3 m/s². Particle 1 has a mass of 6.00×10^{-6} kg. Find (a) the charge on each particle and (b) the mass of particle 2.

*16. A point charge of $-0.70\ \mu C$ is fixed to one corner of a square. An identical charge is fixed to the diagonally opposite corner. A point charge q is fixed to each of the remaining corners. The net force acting on either of the charges q is zero. Find the magnitude and algebraic sign of q.

*17. Two small objects, A and B, are fixed in place and separated by 2.00 cm in a vacuum. Object A has a charge of $+1.00\ \mu C$, and object B has a charge of $-1.00\ \mu C$. How many electrons must be removed from A and put onto B to make the electrostatic force that acts on each object an attractive force whose magnitude is 45.0 N?

*18. The drawing shows three point charges fixed in place. The charge at the coordinate origin has a value of $q_1 = +8.00 \mu C$; the other two have identical magnitudes, but opposite signs: $q_2 = -5.00 \mu C$ and $q_3 = +5.00 \mu C$.

(a) Determine the net force (magnitude and direction) exerted on q_1 by the other two charges. (b) If q_1 had a mass of 1.50 g and it were free to move, what would be its acceleration?

*19. ssm Two spheres are mounted on identical horizontal springs and rest on a friction-less table, as in the

drawing. When the spheres are uncharged, the spacing between them is 0.0500 m, and the springs are unstrained. When each sphere has a charge of $+1.60 \mu C$, the spacing doubles. Assuming that the spheres have a negligible diameter, determine the spring constant of the springs.

*20. Two positive charges, when combined, give a total charge of $+9.00 \mu C$. When the charges are separated by 3.00 m, the force exerted by one charge on the other has a magnitude of 8.00×10^{-3} N. Find the amount of each charge.

*21. An electrically neutral model airplane is flying in a horizontal circle on a 3.0-m guideline, which is nearly parallel to the ground. The line breaks when the kinetic energy of the plane is 50.0 J. Reconsider the same situation, except that now there is a point charge of $+q$ on the plane and a point charge of $-q$ at the other end of the guideline. In this case, the line breaks when the kinetic energy of the plane is 51.8 J. Find the magnitude of the charges.

**22. Two objects are identical and small enough that their sizes can be ignored relative to the distance between them, which is 0.200 m. In a vacuum, each object carries a different charge, and they attract each other with a force of 1.20 N. The objects are brought into contact, so the net charge is shared equally, and then they are returned to their initial positions. Now it is found that the objects repel one another with a force whose magnitude is equal to that of the initial attractive force. What is the initial charge on each object? Note that there are two answers.

**23. ssm A small spherical insulator of mass 8.00×10^{-2} kg and charge $+0.600 \mu C$ is hung by a thin wire of negligible mass. A charge of $-0.900 \mu C$ is held 0.150 m away from the sphere and directly to the right of it, so the wire makes an an-

gle θ with the vertical (see the drawing). Find (a) the angle θ and (b) the tension in the wire.

**24. Two identical, small insulating balls are suspended by separate 0.25-m threads that are attached to a common point on the ceiling. Each ball has a mass of 8.0×10^{-4} kg. Initially the balls

are uncharged and hang straight down. They are then given identical positive charges and, as a result, spread apart with an angle of 36° between the threads. Determine (a) the charge on each ball and (b) the tension in the threads.

Section 18.6 The Electric Field, Section 18.7 Electric Field Lines, Section 18.8 The Electric Field Inside a Conductor: Shielding

25. An electric field of 260 000 N/C points due west at a certain spot. What are the magnitude and direction of the force that acts on a charge of $-7.0 \mu C$ at this spot?

26. Review Conceptual Example 12 as an aid in working this problem. Charges of $-4q$ are fixed to opposite corners of a square. A charge of $+5q$ is fixed to one of the remaining corners, and a charge of $+3q$ is fixed to the last corner. Assuming that ten electric field lines emerge from the $+5q$ charge, sketch the field lines in the vicinity of the four charges.

27. ssm A charge of $+3.0 \times 10^{-5}$ C is located at a place where there is an electric field that points due east and has a magnitude of 15 000 N/C. What are the magnitude and direction of the force acting on the charge?

28. Two charges are placed on the x axis. One charge ($q_1 = +8.5 \mu C$) is at $x_1 = +3.0$ cm and the other ($q_2 = -21 \mu C$) is at $x_1 = +9.0$ cm. Find the net electric field (magnitude and direction) at (a) $x = 0$ cm and (b) $x = +6.0$ cm.

29. Two charges, -16 and $+4.0 \mu C$, are fixed in place and separated by 3.0 m. (a) At what spot along a line through the charges is the net electric field zero? Locate this spot relative to the positive charge. (Hint: The spot does not necessarily lie between the two charges.) (b) What would be the force on a charge of $+14 \mu C$ placed at this spot?

30. An electric field with a magnitude of 160 N/C exists at a spot that is 0.15 m away from a charge. At a place that is 0.45 m away from this charge, what is the electric field strength?

31. ssm A 3.0-μC point charge is placed in an external uniform electric field of 1.6×10^4 N/C. At what distance from the charge is the net electric field zero?

32. A charge of $q = +7.50 \mu C$ is located in an electric field. The x and y components of the electric field are $E_x = 6.00 \times 10^3$ N/C and $E_y = 8.00 \times 10^3$ N/C, respectively. (a) What is the magnitude of the force on the charge? (b) Determine the angle that the force makes with the $+x$ axis.

33. The magnitude of the electric field between the plates of a parallel plate capacitor is 2.4×10^5 N/C. Each plate carries a charge whose magnitude is 0.15 μC. What is the area of each plate?

34. Two charges are located on the x axis: $q_1 = +6.0 \mu C$ at $x_1 = +4.0$ cm, and $q_2 = +6.0 \mu C$ at $x_2 = -4.0$ cm. Two other charges are located on the y axis: $q_3 = +3.0 \mu C$ at $y_3 = +5.0$ cm, and $q_4 = -8.0 \mu C$ at $y_4 = +7.0$ cm. Find the net electric field (magnitude and direction) at the origin.

35. ssm A small drop of water is suspended motionless in air by a uniform electric field that is directed upward and has a

magnitude of 8480 N/C. The mass of the water drop is 3.50×10^{-9} kg. (a) Is the excess charge on the water drop positive or negative? Why? (b) How many excess electrons or protons reside on the drop?

36. Review Conceptual Example 11 before attempting this problem. The magnitude of each of the charges in Figure 18.21 is 8.60×10^{-12} C. The lengths of the sides of the rectangles are 3.00 cm and 5.00 cm. Find the magnitude of the electric field at the center of the rectangle in Figure 18.21a and b.

*37. A proton is moving parallel to a uniform electric field. The electric field accelerates the proton and increases its linear momentum to 5.0×10^{-23} kg·m/s from 1.5×10^{-23} kg·m/s in a time of 6.3×10^{-6} s. What is the magnitude of the electric field?

*38. An electron is released from rest at the negative plate of a parallel plate capacitor. The charge per unit area on each plate is $\sigma = 1.8 \times 10^{-7}$ C/m², and the plates are separated by a distance of 1.5×10^{-2} m. How fast is the electron moving just before it reaches the positive plate?

*39. **ssm www** A rectangle has a length of $2d$ and a height of d. Each of the following three charges is located at a corner of the rectangle: $+q_1$ (upper left corner), $+q_2$ (lower right corner), and $-q$ (lower left corner). The net electric field at the (empty) upper right corner is zero. Find the magnitudes of q_1 and q_2. Express your answers in terms of q.

*40. A small object, which has a charge $q = 7.5$ μC and mass $m = 9.0 \times 10^{-5}$ kg, is placed in a constant electric field. Starting from rest, the object accelerates to a speed of 2.0×10^3 m/s in a time of 0.96 s. Determine the magnitude of the electric field.

*41. The drawing shows two positive charges q_1 and q_2 fixed to a circle. At the center of the circle they produce a net electric field that is directed upward along the vertical axis. Determine the ratio q_2/q_1.

Problem 41

**42. The drawing shows an electron entering the lower left side of a parallel plate capacitor and exiting at the upper right side. The initial speed of the electron is 7.00×10^6 m/s. The capacitor is 2.00 cm long, and its plates are separated by 0.150 cm. Assume that the electric field between the plates is uniform everywhere and find its magnitude.

43. **ssm Two point charges of the same magnitude but opposite signs are fixed to either end of the base of an isosceles triangle, as the drawing shows. The electric field at the midpoint M between the charges has a magnitude E_M. The field directly above the

midpoint at point P has a magnitude E_P. The ratio of these two field magnitudes is $E_M/E_P = 9.0$. Find the angle α in the drawing.

**44. The magnitude of the electric field between the plates of a parallel plate capacitor is 480 N/C. A silver dollar is placed between the plates and oriented parallel to the plates. (a) Ignoring the edges of the coin, find the induced charge density σ on each face of the coin. (b) Assuming the coin has a radius of 1.9 cm, find the magnitude of the total charge on each face of the coin.

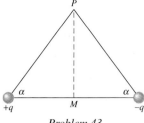

Problem 43

**45. A small plastic ball of mass 6.50×10^{-3} kg and charge $+0.150$ μC is suspended from an insulating thread and hangs between the plates of a capacitor (see the drawing). The ball is in equilibrium, with the thread making an angle of 30.0° with respect to the vertical. The area of each plate is 0.0150 m². What is the magnitude of the charge on each plate?

Section 18.9 Gauss' Law

46. A spherical surface completely surrounds a collection of charges. Find the electric flux through the surface if the collection consists of (a) a single $+3.5 \times 10^{-6}$ C charge, (b) a single -2.3×10^{-6} C charge, and (c) both of the charges in (a) and (b).

47. **ssm The drawing shows an edge-on view of two planar surfaces that intersect and are mutually perpendicular. Surface 1 has an area of 1.7 m², while surface 2 has an area of 3.2 m². The electric field $\mathbf{E}$ in the drawing is uniform and has a magnitude of 250 N/C. Find the electric flux through (a) surface 1 and (b) surface 2.

48. A rectangular surface (0.16 m × 0.38 m) is oriented in a uniform electric field of 580 N/C. What is the maximum possible electric flux through the surface?

49. A surface completely surrounds a $+2.0 \times 10^{-6}$ C charge. Find the electric flux through this surface when the surface is (a) a sphere with a radius of 0.50 m, (b) a sphere with a radius of 0.25 m, and (c) a cube with edges that are 0.25 m long.

50. A vertical wall (5.9 m × 2.5 m) in a house faces due east. A uniform electric field has a magnitude of 150 N/C. This field is parallel to the ground and points 35° north of east. What is the electric flux through the wall?

*51. **ssm** A cube is located with one corner at the origin of an x, y, z, coordinate system. One of the cube's faces lies in the x, y plane, another in the y, z plane, and another in the x, z plane. In other words, the cube is in the first octant of the coordinate system. The edges of the cube are 0.20 m long. A uniform electric field is parallel to the x, y plane and points in the direction of the $+y$ axis. The magnitude of the field is 1500 N/C. (a) Find the electric flux through each of the six faces of the cube. (b) Add the six values obtained in part (a) to show that the electric flux through the cubical surface is zero, as Gauss' law predicts, since there is no net charge within the cube.

*52. Two spherical shells have a common center. A -1.6×10^{-6}-C charge is spread uniformly over the inner shell, which has a radius of 0.050 m. A $+5.1 \times 10^{-6}$-C charge is spread uni-

formly over the outer shell, which has a radius of 0.15 m. Find the magnitude and direction of the electric field at a distance (measured from the common center) of (a) 0.20 m, (b) 0.10 m, and (c) 0.025 m.

**53. A long, thin, straight wire of length L has a positive charge Q distributed uniformly along it. Use Gauss' law to show that the electric field created by this wire at a radial distance r has a magnitude of $E = \lambda/(2\pi\epsilon_0 r)$, where $\lambda = Q/L$. (Hint: For a Gaussian surface, use a cylinder aligned with its axis along the wire and note that the cylinder has a flat surface at either end, as well as a curved surface.)

ADDITIONAL PROBLEMS

54. Review the important features of electric field lines discussed in Conceptual Example 12. Three point charges ($+q$, $+2q$, and $-3q$) are at the corners of an equilateral triangle. Sketch in six electric field lines between the three charges.

55. ssm Two charges attract each other with a force of 1.5 N. What will be the force if the distance between them is reduced to one-ninth of its original value?

56. Conceptual Example 13 in the text deals with the hollow spherical conductor in Figure 18.31. The conductor is initially electrically neutral, and then a charge $+q$ is placed at the center of the hollow space. Suppose the conductor initially has a net charge of $+2q$ instead of being neutral. What is the total charge on the interior and on the exterior surface when the $+q$ charge is placed at the center?

57. A charge $+q$ is located at the origin, while an identical charge is located on the x axis at $x = +0.50$ m. A third charge of $+2\,q$ is located on the x axis at such a place that the net electrostatic force on the charge at the origin doubles, its direction remaining unchanged. Where should the third charge be located?

58. Three charges are located on the $+x$ axis as follows: $q_1 = +25\ \mu C$ at $x = 0$ m, $q_2 = +11\ \mu C$ at $x = +2.0$ m, and $q_3 = +45\ \mu C$ at $x = +3.5$ m. (a) Find the electrostatic force (magnitude and direction) acting on q_2. (b) Suppose q_2 were $-11\ \mu C$, rather than $+11\ \mu C$. Without performing any further detailed calculations, specify the magnitude and direction of the force exerted on q_2. Give your reasoning.

59. ssm www A tiny ball (mass $= 0.012$ kg) carries a charge of $-18\ \mu C$. What electric field (magnitude and direction) is needed to cause the ball to float above the ground?

60. A long, thin rod (length $= 4.0$ m) lies along the x axis, with its midpoint at the origin. In a vacuum, a $+8.0\ \mu C$ point charge is fixed to one end of the rod, while a $-8.0\ \mu C$ point charge is fixed to the other end. Everywhere in the x, y plane there is a constant external electric field (magnitude $= 5.0 \times 10^3$ N/C) that is perpendicular to the rod. With respect to the z axis, find the magnitude of the net torque applied to the rod.

61. ssm Two tiny spheres have the same mass and carry charges of the same magnitude. The mass of each sphere is $2.0 \times$

10^{-6} kg. The gravitational force that each sphere exerts on the other is balanced by the electric force. (a) What algebraic signs can the charges have? (b) Determine the charge magnitude.

*62. Two charges are placed between the plates of a parallel plate capacitor. One charge is $+q_1$ and the other is $q_2 = +5.00\ \mu C$. The charge per unit area on each plate has a magnitude of $\sigma = 1.30 \times 10^{-4}$ C/m^2. The force on q_1 due to q_2 equals the force on q_1 due to the electric field of the parallel plate capacitor. What is the distance r between the two charges?

*63. A small object has a mass of 2.0×10^{-3} kg and a charge of $-25\ \mu C$. It is placed at a certain spot where there is an electric field. When released, the object experiences an acceleration of 3.5×10^3 m/s^2 in the direction of the $+x$ axis. Determine the magnitude and direction of the electric field.

*64. Two small charged objects are attached to a horizontal spring, one at each end. The magnitudes of the charges are equal, and the spring constant is 220 N/m. The spring is observed to be stretched by 0.020 m relative to its unstrained length of 0.32 m. Determine (a) the possible algebraic signs and (b) the magnitude of the charges.

*65. ssm In the rectangle in the drawing, a charge is to be placed at the empty corner to make the net force on the charge at corner A point along the vertical direction. What charge (magnitude and algebraic sign) must be placed at the empty corner?

*66. A uniform electric field has a magnitude of 2.3×10^3 N/C. In a vacuum, a proton begins with a speed of 2.5×10^4 m/s and moves in the direction of this field. Find the speed of the proton after it has moved a distance of 2.0 mm.

**67. There are four charges, each with a magnitude of $2.0\ \mu C$. Two are positive and two are negative. The charges are fixed to the corners of a 0.30-m square, one to a corner, in such a way that the net force on any charge is directed toward the center of the square. Find the magnitude of the net electrostatic force experienced by any charge.

CONCEPTS & CALCULATIONS

GROUP LEARNING PROBLEMS

Note: Each of these problems consists of Concept Questions followed by a related quantitative Problem. They are designed for use by students working alone or in small learning groups. The Concept Questions involve little or no mathematics and are intended to stimulate group discussions. They focus on the concepts with which the problems deal. Recognizing the concepts is the essential initial step in any problem-solving technique.

68. Concept Questions Two identical metal spheres have charges of q_1 and q_2. They are brought together so they touch, and then they are separated. (a) How is the net charge on the two spheres before they touch related to the net charge after they touch? (b) After they touch and are separated, is the charge on each sphere the same? Why?

Problem Four identical metal spheres have charges of $q_A = -8.0 \ \mu C$, $q_B = -2.0 \ \mu C$, $q_C = +5.0 \ \mu C$, and $q_D = +12.0 \ \mu C$. (a) Two of the spheres are brought together so they touch and then they are separated. Which spheres are they, if the final charge on each one is $+5.0 \ \mu C$? (b) In a similar manner, which three spheres are brought together and then separated, if the final charge on each one is $+3.0 \ \mu C$? (c) How many electrons would have to be added to one of the spheres in part (b) to make it electrically neutral?

69. Concept Questions The drawings show three charges that have the same magnitude, but different signs. In all cases the distance between charges 1 and 2 and between 2 and 3 is the same. (a) Draw the electrical force that each charge exerts on charge 2. Each force should be drawn in the correct direction, and its magnitude should be correct relative to that of the other force. (b) Rank the magnitudes of the net electrical force on charge 2, largest first. Explain.

Problem The magnitude of the charges is $q = 8.6 \ \mu C$, and the distance between them is 3.8 mm. Determine the magnitude of the net force on charge 2 for each of the three drawings. Verify that your answers are consistent with your answers to the Concept Questions.

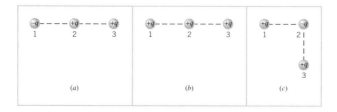

70. Concept Questions Suppose you want to neutralize the gravitational attraction between the earth and the moon by placing equal amounts of charge on each. (a) Should the charges be both positive, both negative, or one positive and the other negative? Why? (b) Do you need to know the distance between the earth and the moon to find the magnitude of the charge? Why or why not?

Problem The masses of the earth and moon are 5.98×10^{24} and 7.35×10^{22} kg, respectively. Identical amounts of charge are placed on each body, such that the net force (gravitational plus electrical) on each is zero. What is the magnitude of the charge?

71. Concept Questions Suppose you want to determine the electric field in a certain region of space. You have a small object of known charge and an instrument that measures the magnitude and direction of the force exerted on the object by the electric field. How would you determine the magnitude and direction of the electric field if the object were (a) positively charged and (b) negatively charged?

Problem The object has a charge of $+20.0 \ \mu C$ and the instrument indicates that the electric force exerted on it is 40.0 μN, due east. What is the magnitude and direction of the electric field? (b) What is the magnitude and direction of the electric field if the object has a charge of $-10.0 \ \mu C$ and the instrument indicates that the force is 20.0 μN due west?

72. Concept Question The drawing shows two situations in which charges are placed on the x and y axes. They are all located at the same distance from the origin. Without doing any calculations, does the electric field at the origin in part (a) have a magnitude that is greater than, less than, or equal to the magnitude of the field at the origin in part (b)? Justify your answer.

Problem The distance between each of the charges and the origin is 6.1 cm. For each of the situations shown in the drawing, determine the magnitude of the electric field at the origin. Check to see that your results are consistent with your answer to the Concept Question.

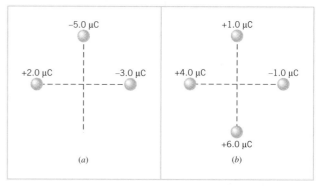

73. Concept Questions A proton and an electron are moving due east in a constant electric field that also points due east. (a) Does each experience an electric force of the same magnitude and direction? (b) What is the direction of the proton's acceleration and the direction of the electron's acceleration? (c) Is the magnitude of the proton's acceleration greater than, less than, or the same as that of the electron's acceleration? Explain your answers.

Problem The electric field points due east and has a magnitude of 8.0×10^4 N/C. Determine the magnitude of the acceleration of the proton and the electron. Check that your answers are consistent with part c of the Concept Questions.

Towers such as these carry the cables that distribute electricity throughout the country. The electricity in the cables has a high voltage, and voltage is related to the concept of electric potential energy, as discussed in this chapter.

CHAPTER 19

ELECTRIC POTENTIAL ENERGY AND THE ELECTRIC POTENTIAL

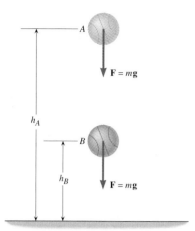

Figure 19.1 Gravity exerts a force, $\mathbf{F} = m\mathbf{g}$, on the basketball of mass m. Work is done by the gravitational force as the ball falls from A to B.

19.1 POTENTIAL ENERGY

In the last chapter we discussed the electrostatic force that two point charges exert on each other, $F = kq_1q_2/r^2$. The form of this equation is similar to that for the gravitational force that two particles exert on each other, which is $F = Gm_1m_2/r^2$, according to Newton's law of universal gravitation. Both of these forces are conservative and, as Section 6.4 explains, a potential energy can be associated with a conservative force. Thus, an electric potential energy exists that is analogous to the gravitational potential energy. To set the stage for a discussion of the electric potential energy, let's review some of the important aspects of the gravitational counterpart.

Figure 19.1, which is essentially Figure 6.10, shows a basketball of mass m falling from point A to point B. The gravitational force, $m\mathbf{g}$, is the only force acting on the ball, where g is the magnitude of the acceleration due to gravity. As Section 6.3 discusses, the work W_{AB} done by the gravitational force when the ball falls from a height of h_A to a height of h_B is

$$W_{AB} = \underbrace{mgh_A}_{\substack{\text{Initial} \\ \text{gravitational} \\ \text{potential energy,} \\ \text{GPE}_A}} - \underbrace{mgh_B}_{\substack{\text{Final} \\ \text{gravitational} \\ \text{potential energy,} \\ \text{GPE}_B}} = \text{GPE}_A - \text{GPE}_B \qquad (6.4)$$

Recall that the quantity mgh is the gravitational potential energy* of the ball, $\text{GPE} = mgh$ (Equation 6.5), and represents the energy that the ball has by virtue of its position relative to the surface of the earth. Thus, the work done by the gravitational force equals the initial gravitational potential energy minus the final gravitational potential energy.

Figure 19.2 clarifies the analogy between electric and gravitational potential energies. In this drawing a positive test charge $+q_0$ is situated at point A between two oppositely charged plates. Because of the charges on the plates, an electric field $\mathbf{E}$ exists in the region between them. Consequently, the test charge experiences an electric force, $\mathbf{F} = q_0\mathbf{E}$, that is directed downward, toward the lower plate. (The gravitational force is being neglected here.) As the charge moves from A to B, work is done by this force, in a fashion analogous to the work done by the gravitational force in Figure 19.1. The work W_{AB} done by the electric force equals the difference between the electric potential energy EPE at A and that at B:

$$W_{AB} = \text{EPE}_A - \text{EPE}_B \qquad (19.1)$$

This expression is similar to Equation 6.4. Because the electric force is a conservative force, the path along which the test charge moves from A to B is of no consequence, for the work W_{AB} is the same for all paths.

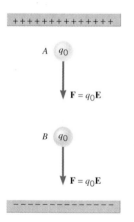

Figure 19.2 Because of the electric field $\mathbf{E}$, an electric force, $\mathbf{F} = q_0\mathbf{E}$, is exerted on a positive test charge $+q_0$. Work is done by the force as the charge moves from A to B.

19.2 THE ELECTRIC POTENTIAL DIFFERENCE

● *Concepts at a Glance* Since the electric force is $\mathbf{F} = q_0\mathbf{E}$, the work that it does as the charge moves from A to B in Figure 19.2 depends on the charge q_0. It is useful, therefore, to express this work on a per-unit-charge basis, by dividing both sides of Equation 19.1 by the charge:

$$\frac{W_{AB}}{q_0} = \frac{\text{EPE}_A}{q_0} - \frac{\text{EPE}_B}{q_0} \qquad (19.2)$$

---

* The gravitational potential energy is now being denoted by GPE to distinguish it from the electric potential energy EPE.

Concepts at a Glance

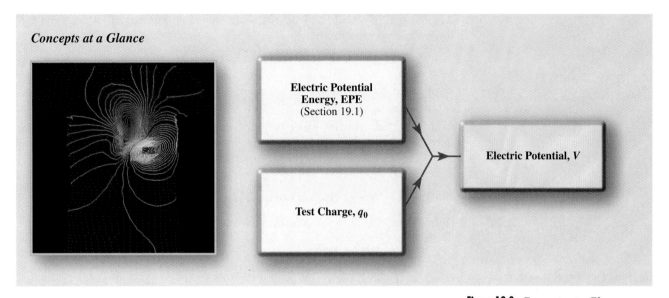

Figure 19.3 *Concepts at a Glance* The electric potential energy and the notion of a test charge are combined to produce the idea of an electric potential. Lines of constant electric potential surround the human heart in the photograph. As we will see later in the chapter, the pattern of the lines suggests that the heart behaves like an electric dipole.

Notice that the right-hand side of this equation is the difference between two terms, each of which is an electric potential energy divided by the test charge, EPE/q_0. The quantity EPE/q_0 is the electric potential energy per unit charge and is an important concept in electricity. It is called the *electric potential* or, simply, the *potential* and is referred to with the symbol V, as in Equation 19.3. The Concepts-at-a-Glance chart in Figure 19.3 illustrates the fact that the electric potential is a concept that incorporates both the electric potential energy and the idea of a test charge.

● **DEFINITION OF ELECTRIC POTENTIAL**

The electric potential V at a given point is the electric potential energy EPE of a small test charge q_0 situated at that point divided by the charge itself:

$$V = \frac{\text{EPE}}{q_0} \qquad (19.3)$$

SI Unit of Electric Potential: joule/coulomb = volt (V)

The SI unit of electric potential is a joule per coulomb, a quantity known as a *volt.* The name honors Alessandro Volta (1745–1827), who invented the voltaic pile, the forerunner of the battery. In spite of the similarity in names, the electric potential energy EPE and the electric potential V are *not* the same. The electric potential energy, as its name implies, is an *energy* and, therefore, is measured in joules. In contrast, the electric potential is an *energy per unit charge* and is measured in joules per coulomb, or volts.

According to Equations 19.2 and 19.3, the electric potential difference, $V_B - V_A$, between two points A and B is related to the work per unit charge in the following manner:

$$V_B - V_A = \frac{\text{EPE}_B}{q_0} - \frac{\text{EPE}_A}{q_0} = \frac{-W_{AB}}{q_0} \qquad (19.4)$$

Often the "delta" notation is used to express the difference (final value minus initial value) in potentials and in potential energies: $\Delta V = V_B - V_A$ and

The energy used to operate this electric vehicle comes from the charging station and is stored in the vehicle's battery pack. The voltage of the pack and the magnitude of the charge passing through it are the main factors that determine how much energy is used to propel the vehicle.

$\Delta(\text{EPE}) = \text{EPE}_B - \text{EPE}_A$. In terms of this notation, Equation 19.4 takes the following more compact form:

$$\Delta V = \frac{\Delta(\text{EPE})}{q_0} = \frac{-W_{AB}}{q_0} \qquad (19.4)$$

Neither the potential V nor the potential energy EPE can be determined in an absolute sense, for only the *differences* ΔV and $\Delta(\text{EPE})$ are measurable in terms of the work W_{AB}. The gravitational potential energy has this same characteristic, since only the value at one height relative to that at some reference height has any significance. Example 1 emphasizes the relative nature of the electric potential.

EXAMPLE 1   Work, Electric Potential Energy, and Electric Potential

In Figure 19.2, the work done by the electric force as the test charge ($q_0 = +2.0 \times 10^{-6}$ C) moves from A to B is $W_{AB} = +5.0 \times 10^{-5}$ J. (a) Find the difference, $\Delta(\text{EPE}) = \text{EPE}_B - \text{EPE}_A$, in the electric potential energies of the charge between these points. (b) Determine the potential difference, $\Delta V = V_B - V_A$, between the points.

Reasoning The work done by the electric force when the charge moves from A to B is $W_{AB} = \text{EPE}_A - \text{EPE}_B$, according to Equation 19.1. Therefore, the difference in the electric potential energies (final value minus initial value) is $\Delta(\text{EPE}) = \text{EPE}_B - \text{EPE}_A = -W_{AB}$. The potential difference, $\Delta V = V_B - V_A$, is the difference in the electric potential energies divided by the charge q_0, according to Equation 19.4.

Solution

(a) The difference in the electric potential energies of the charge at points A and B is

$$\underbrace{\text{EPE}_B - \text{EPE}_A}_{= \Delta(\text{EPE})} = -W_{AB} = \boxed{-5.0 \times 10^{-5} \text{ J}} \qquad (19.1)$$

Thus the charge has a higher electric potential energy at A than at B.

(b) The potential difference ΔV between A and B is

$$\underbrace{V_B - V_A}_{= \Delta V} = \frac{\text{EPE}_B - \text{EPE}_A}{q_0} = \frac{-5.0 \times 10^{-5} \text{ J}}{2.0 \times 10^{-6} \text{ C}} = \boxed{-25 \text{ V}} \qquad (19.4)$$

We can see, then, that the electric potential is higher at A than at B.

In Figure 19.1 the speed of the basketball increases as it falls from A to B. Since point A has a greater gravitational potential energy than point B, we see that an object of mass m accelerates when it moves from a region of higher potential energy toward a region of lower potential energy. Likewise, the positive charge in Figure 19.2 accelerates as it moves from A to B because of the electric repulsion from the upper plate and the attraction to the lower plate. Since point A has a higher electric potential energy than point B, we conclude that *a positive charge accelerates from a region of higher electric potential energy (or higher potential) toward a region of lower electric potential energy (or lower potential).* On the other hand, a negative charge placed between the plates in Figure 19.2 behaves in the opposite fashion, since the electric force acting on the negative charge is directed opposite to that on the positive charge. *A negative charge accelerates from a region of lower potential toward a region of higher potential.* The next example illustrates the way positive and negative charges behave.

CONCEPTUAL EXAMPLE 2
The Accelerations of Positive and Negative Charges

Three points, A, B, and C, are located along a horizontal line, as Figure 19.4 illustrates. A positive test charge is released from rest at A and accelerates toward B. Upon reaching B, the test charge continues to accelerate toward C. Assuming that only motion along the line is possible, what will a negative test charge do when it is released from rest at B?

Reasoning and Solution A negative charge will accelerate from a region of lower potential toward a region of higher potential. Therefore, before we can decide what the negative test charge will do, it is necessary to know how the electric potentials compare at points A, B, and C. This information can be deduced from the behavior of the positive test charge, which accelerates from a region of higher toward a region of lower potential. Since the positive test charge accelerates from A to B, the potential at A must exceed that at B. And since the positive test charge accelerates from B to C, the potential at B must exceed that at C. The potential at point B, then, must lie between that at points A and C, as Figure 19.4 illustrates. When the negative test charge is released from rest at B, it will accelerate toward the region of higher potential. In other words, *it will begin moving toward A.*

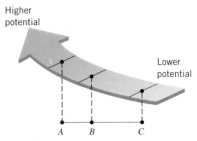

Figure 19.4 The electric potentials at points A, B, and C are different. Under the influence of these potentials, positive and negative charges accelerate in opposite directions.

As a familiar application of electric potential energy and electric potential, Figure 19.5 shows a 12-V automobile battery with a headlight connected between its terminals. The positive terminal, point A, has a potential that is 12 V higher than the potential at the negative terminal, point B; in other words, $V_A - V_B = 12$ V. Positive charges are repelled from the positive terminal and travel through the wires and headlight toward the negative terminal.* As the charges pass through the headlight, virtually all their potential energy is converted into heat, which causes the filament to glow "white hot" and emit light. When the charges reach the negative terminal, they no longer have any potential energy. The battery then gives the charges an additional "shot" of potential energy by moving them to the higher-potential positive terminal, and the cycle is repeated. In raising the potential energy of the charges, the battery does work on them and draws from its reserve of chemical energy to do so. Example 3 illustrates the concepts of electric potential energy and electric potential as applied to a battery.

EXAMPLE 3   Operating a Headlight

Determine the number of particles, each carrying a charge of 1.60×10^{-19} C (the magnitude of the charge on an electron), that pass between the terminals of a 12-V car battery when a 60.0-W headlight burns for one hour.

Reasoning To obtain the number of charged particles, we determine the total charge needed to provide the energy consumed by the headlight in one hour. Dividing the total charge by the charge on each particle gives the number of particles.

Solution Using energy at a rate of 60.0 joules per second (60.0 watts) for one hour, the headlight uses a total energy of

$$\text{Energy} = \text{Power} \times \text{Time} = (60.0 \text{ W})(3600 \text{ s}) = 2.2 \times 10^5 \text{ J} \qquad (6.10b)$$

Figure 19.5 A headlight connected to a 12-V battery.

---

* Historically, it was believed that positive charges flow in the wires of an electric circuit. Today, it is known that negative charges flow in wires from the negative toward the positive terminal. Here, however, we follow the customary practice of describing the flow of negative charges by specifying the opposite but equivalent flow of positive charges. This hypothetical flow of positive charges is called the conventional electric current, as we will see in Section 20.1.

Since this energy comes from the battery, it represents the difference between the electrical potential energy of the charges at point A and that at point B: 2.2×10^5 J $=$ EPE$_A -$ EPE$_B$. According to Equation 19.4, we have

$$\frac{\text{EPE}_A - \text{EPE}_B}{q_0} = V_A - V_B$$

$$q_0 = \frac{\text{EPE}_A - \text{EPE}_B}{V_A - V_B} = \frac{2.2 \times 10^5 \text{ J}}{12 \text{ V}} = 1.8 \times 10^4 \text{ C}$$

The number of particles whose individual charges combine to provide this total charge is $(1.8 \times 10^4 \text{ C})/(1.60 \times 10^{-19} \text{ C}) = \boxed{1.1 \times 10^{23}}$.

As used in connection with batteries, the volt is a familiar unit for measuring electric potential difference. The word "volt" also appears in another context, as part of a unit that is used to measure energy, particularly the energy of an atomic particle, such as an electron or a proton. This energy unit is called the *electron volt* (eV). ***One electron volt is the amount by which the potential energy of an electron ($q_0 = 1.60 \times 10^{-19}$ C) changes when the electron moves through a potential difference of one volt.*** Since the change in potential energy equals $q_0 \, \Delta V$, one electron volt is equal to $(1.60 \times 10^{-19}$ C$) \times (1.00$ V$) = 1.60 \times 10^{-19}$ J; thus,

$$1 \text{ eV} = 1.60 \times 10^{-19} \text{ J}$$

One million (10^{+6}) electron volts of energy is referred to as one MeV, and one billion (10^{+9}) electron volts of energy is one GeV, where the "G" stands for the prefix "giga" (pronounced "jig′a").

● *Concepts at a Glance*   The Concepts-at-a-Glance chart in Figure 19.6 emphasizes that the total energy of an object, which is the sum of its kinetic and potential energies, is an important concept. Its significance lies in the fact that the total energy remains the same (is conserved) during the object's motion, provided that nonconservative forces, such as friction, are either absent or do no net work. While the sum of the energies at each instant remains constant, energy may be converted from one form to another, e.g., gravitational potential energy is converted into kinetic energy as a ball falls. As Figure 19.6 illustrates, we now include the electric potential energy EPE as part of the total energy that an object can have:

$$\underset{\substack{\text{Total} \\ \text{energy}}}{E} = \underset{\substack{\text{Translational} \\ \text{kinetic energy}}}{\tfrac{1}{2}mv^2} + \underset{\substack{\text{Rotational} \\ \text{kinetic energy}}}{\tfrac{1}{2}I\omega^2} + \underset{\substack{\text{Gravitational} \\ \text{potential} \\ \text{energy}}}{mgh} + \underset{\substack{\text{Elastic} \\ \text{potential} \\ \text{energy}}}{\tfrac{1}{2}kx^2} + \underset{\substack{\text{Electric} \\ \text{potential} \\ \text{energy}}}{\text{EPE}}$$

If the total energy is conserved as the object moves, then its final energy E_f is equal to its initial energy E_0, or $E_f = E_0$. Example 4 illustrates how the conservation of energy is applied to a charge moving in an electric field. ●

EXAMPLE 4   The Conservation of Energy

● **PROBLEM SOLVING INSIGHT**
A positive charge accelerates from a region of higher potential toward a region of lower potential. In contrast, a negative charge accelerates from a region of lower potential toward a region of higher potential.

(a) A particle has a mass of $m = 1.8 \times 10^{-5}$ kg and a positive charge of $q_0 = +3.0 \times 10^{-5}$ C. It is released from rest at point A and accelerates horizontally until it reaches point B, as Figure 19.7a shows. The only force acting on the particle is an electric force (not shown in the drawing), and the electric potential at A is 25 V greater than that at B. In other words, $V_A - V_B = 25$ V. What is the speed v_B of the particle when it reaches point B? (b) If the same particle had a *negative*

Concepts at a Glance

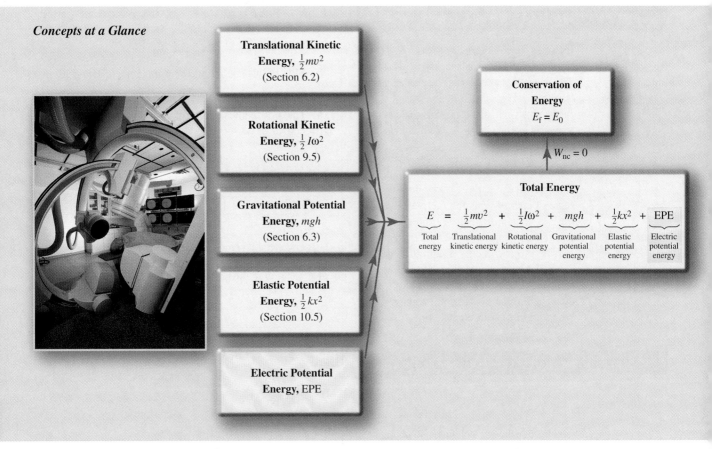

Figure 19.6 *Concepts at a Glance* The electric potential energy is another type of energy. The total energy is conserved when nonconservative forces do no net work ($W_{nc} = 0$). This chart is a continuation of those in Figures 6.16, 9.22, and 10.18. The photograph shows an X-ray imaging device. To produce X-rays, electrons are accelerated from rest and allowed to collide with a metal target. As each electron is accelerated, electric potential energy is converted into kinetic energy; the total energy is conserved during the process.

charge and were released from rest at B (see Figure 19.7b), what would be its speed v_A at A?

Reasoning Since the only force acting on the particle is the conservative electric force, the total energy of the particle is conserved. This means that the total energy is the same at points A and B:

$$\frac{1}{2}mv_B^2 + \underbrace{\frac{1}{2}I\omega_B^2}_{\substack{= \,0\text{ J, since} \\ \omega_B = 0\text{ rad/s}}} + mgh_B + \underbrace{\frac{1}{2}kx_B^2}_{\substack{= \,0\text{ J, since no} \\ \text{elastic forces} \\ \text{are present}}} + \text{EPE}_B$$

$$= \frac{1}{2}mv_A^2 + \underbrace{\frac{1}{2}I\omega_A^2}_{\substack{= \,0\text{ J, since} \\ \omega_A = 0\text{ rad/s}}} + mgh_A + \underbrace{\frac{1}{2}kx_A^2}_{\substack{= \,0\text{ J, since no} \\ \text{elastic forces} \\ \text{are present}}} + \text{EPE}_A$$

In this expression the angular speeds ω_A and ω_B are zero because a particle cannot rotate. Moreover, we set $h_A = h_B$ (the particle moves horizontally) and note that

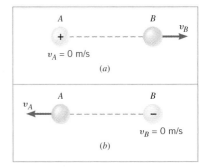

Figure 19.7 (*a*) A positive charge starts from rest at point A and accelerates toward point B. (*b*) A negative charge starts from rest at B and accelerates toward A.

$\text{EPE}_A - \text{EPE}_B = q_0(V_A - V_B)$; the conservation of energy equation then reduces to

$$\tfrac{1}{2}mv_B^2 = \tfrac{1}{2}mv_A^2 + q_0(V_A - V_B)$$

We will use this equation to determine the final speeds.

Solution

(a) The positively charged particle starts from rest at A ($v_A = 0$ m/s). Its speed v_B at B can be found from the equation above:

$$v_B = \sqrt{\frac{2q_0(V_A - V_B)}{m}}$$

$$= \sqrt{\frac{2(+3.0 \times 10^{-5}\ \text{C})(25\ \text{V})}{1.8 \times 10^{-5}\ \text{kg}}} = \boxed{9.1\ \text{m/s}}$$

(b) The negatively charged particle starts from rest at B ($v_B = 0$ m/s). Its speed v_A at A can also be found from the last equation in the reasoning section:

$$v_A = \sqrt{\frac{-2q_0(V_A - V_B)}{m}}$$

$$= \sqrt{\frac{-2(-3.0 \times 10^{-5}\ \text{C})(25\ \text{V})}{1.8 \times 10^{-5}\ \text{kg}}} = \boxed{9.1\ \text{m/s}}$$

19.3 THE ELECTRIC POTENTIAL DIFFERENCE CREATED BY POINT CHARGES

A positive point charge $+q$ creates an electric potential in a fashion that Figure 19.8 helps to explain. This picture shows two locations A and B, at distances r_A and r_B from the charge. At any position between A and B an electrostatic force of repulsion **F** acts on a positive test charge $+q_0$. The magnitude of the force is given by Coulomb's law as $F = kq_0q/r^2$. When the test charge moves from A to B, work is done by this force. Since r varies between r_A and r_B, the force F also varies, and the work is not simply the product of the force and the distance between the points. (Recall from Section 6.1 that work equals force times distance only if the force is constant.) However, the work W_{AB} can be found by using the methods of integral calculus, and the result is

$$W_{AB} = \frac{kqq_0}{r_A} - \frac{kqq_0}{r_B}$$

The potential difference, $V_B - V_A$, between A and B can now be determined by substituting this expression for W_{AB} into Equation 19.4:

$$V_B - V_A = \frac{-W_{AB}}{q_0} = \frac{kq}{r_B} - \frac{kq}{r_A} \tag{19.5}$$

As point B is located farther and farther from the charge q, r_B becomes larger and larger. In the limit that r_B is infinitely large, the term kq/r_B becomes zero, and it is customary to set V_B equal to zero also. In this limit, Equation 19.5 becomes

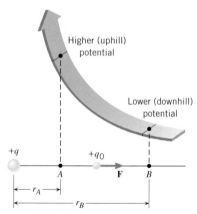

Figure 19.8 The positive test charge $+q_0$ experiences a repulsive force **F** due to the positive point charge $+q$. As a result, work is done by this force when the test charge moves from A to B. Consequently, the electric potential is higher (uphill) at A and lower (downhill) at B.

$V_A = kq/r_A$, and it is standard convention to omit the subscripts and write the potential in the following form:

Potential of a point charge
$$V = \frac{kq}{r} \qquad (19.6)$$

The symbol V in this equation does not refer to the potential in any absolute sense. Rather, $V = kq/r$ stands for the amount by which the potential at a distance r from a point charge differs from the potential at an infinite distance away. In other words, V refers to a potential difference with the arbitrary assumption that the potential at infinity is zero.

With the aid of Equation 19.6, we can describe the effect that a point charge q has on the surrounding space. When q is positive, the value of $V = kq/r$ is also positive, indicating that the positive charge has everywhere raised the potential above the zero reference value. Conversely, when q is negative, the potential V is also negative, indicating that the negative charge has everywhere decreased the potential below the zero reference value. The next example deals with these effects quantitatively.

EXAMPLE 5   The Potential of a Point Charge

Using a zero reference potential at infinity, determine the amount by which a point charge of 4.0×10^{-8} C alters the electric potential at a spot 1.2 m away when the charge is (a) positive and (b) negative.

Reasoning A point charge q alters the potential at every location in the surrounding space. In the expression $V = kq/r$, the effect of the charge in increasing or decreasing the potential is conveyed by the algebraic sign for the value of q.

Solution

(a) Figure 19.9a shows the potential when the charge is positive:

$$V = \frac{kq}{r} = \frac{(8.99 \times 10^9 \ \text{N·m}^2\text{/C}^2)(+4.0 \times 10^{-8} \ \text{C})}{1.2 \ \text{m}} = \boxed{+300 \ \text{V}} \qquad (19.6)$$

(b) Part b of the drawing illustrates the results when the charge is negative. A calculation similar to that in part (a) shows the potential is now negative: $\boxed{-300 \ \text{V}}$.

A single point charge raises or lowers the potential at a given location, depending on whether the charge is positive or negative. ***When two or more charges are present, the potential due to all the charges is obtained by adding together the individual potentials,*** as Examples 6 and 7 show.

EXAMPLE 6   The Total Electric Potential

At locations A and B in Figure 19.10, find the total electric potential due to the two point charges.

Reasoning At each location, each charge contributes to the total electric potential. We obtain the individual contributions by using $V = kq/r$ and find the total potential by adding the individual contributions algebraically. The two charges have the same magnitude. Thus, at A the total potential is positive because this spot is closer to the positive charge, whose effect dominates over that of the more distant negative charge. At B, midway between the charges, the total potential is zero, since the potential of one charge exactly offsets that of the other.

Figure 19.9 A point charge of 4.0×10^{-8} C alters the potential at a spot 1.2 m away. The potential is (a) increased by 300 V when the charge is positive and (b) decreased by 300 V when the charge is negative, relative to a zero reference potential at infinity.

Figure 19.10 Both the positive and negative charges affect the electric potential at locations A and B.

Solution

Location	Contribution from + Charge
A	$\dfrac{(8.99 \times 10^9 \text{ N·m}^2/\text{C}^2)(+8.0 \times 10^{-9} \text{ C})}{0.20 \text{ m}}$
B	$\dfrac{(8.99 \times 10^9 \text{ N·m}^2/\text{C}^2)(+8.0 \times 10^{-9} \text{ C})}{0.40 \text{ m}}$

Contribution from − Charge	Total Potential
$+\dfrac{(8.99 \times 10^9 \text{ N·m}^2/\text{C}^2)(-8.0 \times 10^{-9} \text{ C})}{0.60 \text{ m}}$	$= \boxed{+240 \text{ V}}$
$+\dfrac{(8.99 \times 10^9 \text{ N·m}^2/\text{C}^2)(-8.0 \times 10^{-9} \text{ C})}{0.40 \text{ m}}$	$= \boxed{0 \text{ V}}$

CONCEPTUAL EXAMPLE 7   Where Is the Potential Zero?

Two point charges are fixed in place, as in Figure 19.11. The positive charge is $+2q$ and has twice the magnitude of the negative charge, which is $-q$. On the line that passes through the charges, how many places are there at which the total potential is zero?

Reasoning and Solution   The total potential is the algebraic sum of the individual potentials created by each charge. It will be zero if the potential due to the positive charge is exactly offset by the potential due to the negative charge. The potential of a point charge is directly proportional to the charge and inversely proportional to the distance from the charge. Thus, the individual potentials cannot exactly offset one another anywhere to the left of the positive charge. The positive charge has the larger magnitude and is closer to any place in this region than the negative charge is. As a result, the potential of the positive charge here always dominates over that of the negative charge.

Between the charges there is a location where the individual potentials do balance. We saw a similar situation in Example 6, where the balance occurred at the midpoint between two charges that had equal magnitudes. Now the charges have unequal magnitudes, so the balance does not occur at the midpoint. Instead, it occurs at a location that is closer to the charge with the smaller magnitude, namely, the negative charge. Then, since the potential of a point charge is inversely proportional to the distance from the charge, the effect of the smaller charge will be able to offset that of the more distant larger charge.

To the right of the negative charge, there is another location at which the individual potentials exactly offset one another. All places on this section of the line are closer to the negative than to the positive charge. In this region, therefore, a location exists at which the potential of the smaller negative charge again exactly offsets that of the more distant and larger positive charge.

Related Homework Material: *Problem 18*

• **PROBLEM SOLVING INSIGHT**
At any location, the total electric potential is the algebraic sum of the individual potentials created by each point charge that is present.

The total potential is zero at these points

Figure 19.11 Two point charges, one positive and one negative. The positive charge, $+2q$, has twice the magnitude of the negative charge, $-q$.

In Example 6 we determined the total potential at a spot due to several point charges. In Example 8 we now extend this technique to find the total potential energy of three charges.

EXAMPLE 8   The Potential Energy of a Group of Charges

Figure 19.12 shows three point charges; initially, they are infinitely far apart. They are then brought together and placed at the corners of an equilateral triangle. Each side of the triangle has a length of 0.50 m. Determine the electric potential energy of the triangular group. In other words, determine the amount by which the electric potential energy of the group differs from that of the three charges in their initial, infinitely separated locations.

Reasoning We will proceed in steps by adding charges to the triangle, one at a time, and then determining the electric potential energy at each step. According to Equation 19.3, $EPE = q_0 V$, the electric potential energy is the product of the charge and the electric potential at the spot where the charge is placed. The total electric potential energy of the triangular group is the sum of the energies of each step in assembling the group.

Solution The order in which the charges are put on the triangle does not matter; we begin with the charge of $+5.0\ \mu$C. When this charge is placed at a corner of the triangle, it has no electric potential energy, according to $EPE = q_0 V$. This is because the total potential V produced by the other two charges is zero at this corner, since they are infinitely far away. Once the charge is in place, the potential it creates at either empty corner ($r = 0.50$ m) is

$$V = \frac{kq}{r} = \frac{(8.99 \times 10^9\ \text{N}\cdot\text{m}^2/\text{C}^2)(+5.0 \times 10^{-6}\ \text{C})}{0.50\ \text{m}} = +9.0 \times 10^4\ \text{V} \quad (19.6)$$

Therefore, when the $+6.0$-μC charge is placed at the second corner of the triangle, its electric potential energy is

$$EPE = qV = (+6.0 \times 10^{-6}\ \text{C})(+9.0 \times 10^4\ \text{V}) = +0.54\ \text{J} \quad (19.3)$$

The electric potential at the remaining empty corner is the sum of the potentials due to the two charges that are already in place:

$$V = \frac{(8.99 \times 10^9\ \text{N}\cdot\text{m}^2/\text{C}^2)(+5.0 \times 10^{-6}\ \text{C})}{0.50\ \text{m}}$$
$$+ \frac{(8.99 \times 10^9\ \text{N}\cdot\text{m}^2/\text{C}^2)(+6.0 \times 10^{-6}\ \text{C})}{0.50\ \text{m}} = +2.0 \times 10^5\ \text{V}$$

When the third charge, $-2.0\ \mu$C, is placed at the remaining empty corner, its electric potential energy is

$$EPE = qV = (-2.0 \times 10^{-6}\ \text{C})(+2.0 \times 10^5\ \text{V}) = -0.40\ \text{J} \quad (19.3)$$

The total potential energy of the triangular group differs from that of the widely separated charges by an amount that is the sum of the potential energies calculated above:

$$\text{Total potential energy} = 0\ \text{J} + 0.54\ \text{J} - 0.40\ \text{J} = \boxed{+0.14\ \text{J}}$$

This energy originates in the work done to bring the charges together.

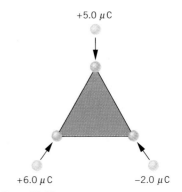

Figure 19.12 Three point charges are placed on the corners of an equilateral triangle. Example 8 illustrates how to determine the total electric potential energy of this group of charges.

● **PROBLEM SOLVING INSIGHT**
Be careful to distinguish between the concepts of potential V and electric potential energy EPE. Potential is electric potential energy per unit charge: $V = EPE/q$.

19.4 EQUIPOTENTIAL SURFACES AND THEIR RELATION TO THE ELECTRIC FIELD

An *equipotential surface* is a surface on which the electric potential is the same everywhere. The easiest equipotential surfaces to visualize are those that surround an isolated point charge. According to Equation 19.6, the potential at a distance r

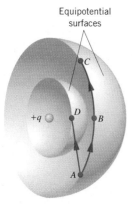

Equipotential surfaces

Figure 19.13 The equipotential surfaces that surround the point charge $+q$ are spherical. The electric force does no work as a charge moves on a path that lies on an equipotential surface, such as the path ABC. However, work is done by the electric force when a charge moves between two equipotential surfaces, as along the path AD.

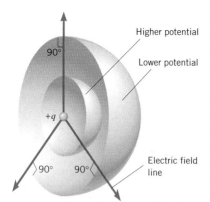

Higher potential

Lower potential

Electric field line

Figure 19.14 The radially directed electric field of a point charge is perpendicular to the spherical equipotential surfaces that surround the charge. The electric field points in the direction of *decreasing* potential.

Figure 19.15 A cross-sectional view of the equipotential surfaces (in blue) of an electric dipole. The surfaces are drawn so that at every point they are perpendicular to the electric field lines (in red) of the dipole.

from a point charge q is $V = kq/r$. Thus, wherever r is the same, the potential is the same, and the equipotential surfaces are spherical surfaces centered on the charge. There are an infinite number of such surfaces, one for every value of r, and Figure 19.13 illustrates two of them. The larger the distance r, the smaller is the potential of the equipotential surface.

The net electric force does no work as a charge moves on an equipotential surface. This important characteristic arises because when an electric force does work W_{AB} as a charge moves from A to B, the potential changes according to Equation 19.4, $V_B - V_A = -W_{AB}/q_0$. Since the potential remains the same on an equipotential surface, $V_A = V_B$, and we see that $W_{AB} = 0$ J. In Figure 19.13, for instance, the electric force does no work as a test charge moves along the circular arc ABC, which lies on an equipotential surface. In contrast, the electric force does work when a charge moves *between* equipotential surfaces, as from A to D in the picture.

The spherical equipotential surfaces that surround an isolated point charge illustrate another characteristic of such surfaces. Figure 19.14 shows two of the surfaces around a positive point charge, along with some electric field lines. The electric field lines give the direction of the electric field, and for a positive point charge, the electric field is directed radially outward. Therefore, at each location on an equipotential sphere the electric field is perpendicular to the surface and points outward in the direction of decreasing potential, as the drawing emphasizes. This perpendicular relation is valid whether or not the equipotential surfaces result from a positive charge or have a spherical shape; *the electric field created by any group of charges is everywhere perpendicular to the associated equipotential surfaces and points in the direction of decreasing potential.* For example, Figure 19.15 shows the electric field lines around an electric dipole, along with some equipotential surfaces (shown in cross section). Since the field lines are not simply radial, the equipotential surfaces are no longer spherical but, instead, have the shape necessary to be everywhere perpendicular to the field lines.

To see why an equipotential surface must be perpendicular to the electric field, consider Figure 19.16, which shows a hypothetical situation in which the perpendicular relation does *not* hold. If **E** were not perpendicular to the equipotential surface, there would be a component of **E** parallel to the surface. This field component would exert an electric force on a test charge placed on the surface. As the charge moves along the surface, work would be done by this electric force. The work, according to Equation 19.4, would cause the potential to change, and, thus, the surface could not be an equipotential surface as assumed. The only way out of the

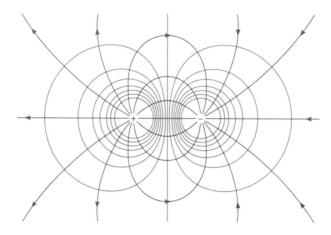

dilemma is for the electric field to be perpendicular to the surface, so there is no component of the field parallel to the surface.

We have already encountered one equipotential surface in the last chapter. In Section 18.8, we found that the direction of the electric field just outside an electrical conductor is perpendicular to the conductor's surface, when the conductor is at equilibrium under electrostatic conditions. Thus, the surface of any conductor is an equipotential surface under such conditions. In fact, since the electric field is zero everywhere inside a conductor whose charges are in equilibrium, the entire conductor can be regarded as an equipotential volume.

There is a quantitative relation between the electric field and the equipotential surfaces. One example that illustrates this relation is the parallel plate capacitor in Figure 19.17. As Section 18.6 discusses, the electric field $\mathbf{E}$ between the metal plates is perpendicular to them and is the same everywhere, ignoring fringe fields at the edges. To be perpendicular to the electric field, the equipotential surfaces must be planes that are parallel to the plates, which themselves are equipotential surfaces. The potential difference between the plates is given by Equation 19.4 as $\Delta V = V_B - V_A = -W_{AB}/q_0$, where A is a point on the positive plate and B is a point on the negative plate. The work done by the electric force as a positive test charge q_0 moves from A to B is $W_{AB} = F\Delta s$, where F refers to the electric force and Δs to the displacement along a line perpendicular to the plates. The force equals the product of the charge and the electric field E ($F = q_0 E$), so the work becomes $W_{AB} = F\Delta s = q_0 E\Delta s$. Therefore, the potential difference between the capacitor plates can be written in terms of the electric field as $\Delta V = -W_{AB}/q_0 = -q_0 E\Delta s/q_0$, or

$$E = -\frac{\Delta V}{\Delta s} \qquad (19.7)*$$

The quantity $\Delta V/\Delta s$ is referred to as the *potential gradient* and has units of volts per meter. The next example deals further with the equipotential surfaces between the plates of a capacitor.

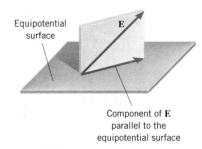

Figure 19.16 In this hypothetical situation, the electric field $\mathbf{E}$ is not perpendicular to the equipotential surface. As a result, there is a component of $\mathbf{E}$ parallel to the surface.

EXAMPLE 9   The Electric Field and Potential Are Related

The plates of the capacitor in Figure 19.17 are separated by a distance of 0.032 m, and the potential difference between them is $\Delta V = V_B - V_A = -64$ V. Between the two equipotential surfaces shown in color there is a potential difference of -3.0 V. Find the spacing between the two colored surfaces.

Reasoning The electric field is $E = -\Delta V/\Delta s$. To find the spacing between the two colored equipotential surfaces, we solve this equation for Δs, with $\Delta V = -3.0$ V and E equal to the electric field between the plates of the capacitor. A value for E can be obtained by using the values given for the distance and potential difference between the plates.

Solution The electric field between the capacitor plates is

$$E = -\frac{\Delta V}{\Delta s} = -\frac{-64 \text{ V}}{0.032 \text{ m}} = 2.0 \times 10^3 \text{ V/m} \qquad (19.7)$$

The spacing between the colored equipotential surfaces can now be determined

$$\Delta s = -\frac{\Delta V}{E} = -\frac{-3.0 \text{ V}}{2.0 \times 10^3 \text{ V/m}} = \boxed{1.5 \times 10^{-3} \text{ m}}$$

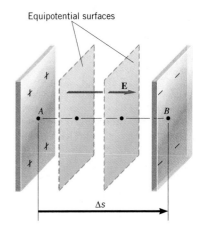

Figure 19.17 The metal plates of a parallel plate capacitor are equipotential surfaces. Two additional equipotential surfaces are shown between the plates. These two equipotential surfaces are parallel to the plates and are perpendicular to the electric field $\mathbf{E}$ between the plates.

---

* In general, the relation $E = -\Delta V/\Delta s$ gives only the component of the electric field along the displacement Δs; it does not give the perpendicular component.

Figure 19.18 A parallel plate capacitor consists of two metal plates, one carrying a charge $+q$ and the other a charge $-q$. The potential of the positive plate exceeds that of the negative plate by an amount V. The region between the plates is filled with a dielectric.

19.5   CAPACITORS AND DIELECTRICS

The Capacitance of a Capacitor

In Section 18.6 we saw that a parallel plate capacitor consists of two parallel metal plates placed near one another but not touching. This type of capacitor is only one among many. In general, a **capacitor** consists of two conductors of any shape placed near one another without touching. For a reason that will become clear later on, it is common practice to fill the region between the conductors or plates with an electrically insulating material called a **dielectric,** as Figure 19.18 illustrates.

A capacitor stores electric charge. Each capacitor plate carries a charge of the *same magnitude,* one positive and the other negative. Because of the charges, the electric potential of the positive plate exceeds that of the negative plate by an amount V, as Figure 19.18 indicates. Experiment shows that when the magnitude q of the charge on each plate is doubled, the electric potential difference V is also doubled, so q is proportional to V: $q \propto V$. Equation 19.8 expresses this proportionality with the aid of a proportionality constant C, which is the **capacitance** of the capacitor.

> ● **THE RELATION BETWEEN CHARGE AND POTENTIAL DIFFERENCE FOR A CAPACITOR**
>
> The magnitude q of the charge on each plate of a capacitor is directly proportional to the magnitude V of the potential difference between the plates:
>
> $$q = CV \qquad (19.8)$$
>
> where C is the capacitance.
>
> **SI Unit of Capacitance:** coulomb/volt $=$ farad (F)

Equation 19.8 shows that the SI unit of capacitance is the coulomb per volt (C/V). This unit is called the *farad* (F), named after the English scientist Michael Faraday (1791–1867). One farad is an enormous capacitance. Usually smaller amounts, such as a microfarad (1 μF $= 10^{-6}$ F) or a picofarad (1 pF $= 10^{-12}$ F), are used in electric circuits. The capacitance reflects the ability of the capacitor to store charge, in the sense that a larger capacitance C allows more charge q to be put onto the plates for a given value of the potential difference V.

THE PHYSICS OF...

random-access memory (RAM) chips.

The ability of a capacitor to store charge lies at the heart of the random-access memory (RAM) chips used in computers, where information is stored in the form of the "ones" and "zeros" that comprise binary numbers. Figure 19.19 illustrates the role of a capacitor in a RAM chip. The capacitor is connected to a transistor switch, to which two lines are connected, an address line and a data line. A single RAM chip often contains millions of such transistor–capacitor units. The address line is used by the computer to locate a particular transistor–capacitor combination, and the data line carries the data to be stored. A pulse on the address line turns on the transistor switch. With the switch turned on, a pulse coming in on the data line can cause the capacitor to charge. A charged capacitor means that a "one" has been stored, while an uncharged capacitor means that a "zero" has been stored.

Figure 19.19 A transistor–capacitor combination is part of a RAM chip used in computer memories.

The Dielectric Constant

If a dielectric is inserted between the plates of a capacitor, the capacitance can increase markedly because of the way in which the dielectric alters the electric field between the plates. Figure 19.20 shows how this effect comes about. In part *a*, the region between the charged plates is empty. The field lines point from the positive toward the negative plate. In part *b*, a dielectric has been inserted between the plates. Since the capacitor is not connected to anything, the charge on the plates remains constant as the dielectric is inserted. In many materials (e.g., water) the molecules possess permanent dipole moments, even though the molecules are electrically neutral. The dipole moment exists, because one end of a molecule has a slight excess of negative charge while the other end has a slight excess of positive charge. When such molecules are placed between the charged plates of the capacitor, the negative ends are attracted to the positive plate and the positive ends are attracted to the negative plate. As a result, the dipolar molecules tend to orient themselves end to end, as in part *b*. Whether or not a molecule has a permanent dipole moment, the electric field can cause the electrons to shift position within a molecule, making one end slightly negative and the opposite end slightly positive. Because of the end-to-end orientation, the left surface of the dielectric becomes positively charged, and the right surface becomes negatively charged. The surface charges are shown in red in the picture.

Because of the surface charges on the dielectric, not all the electric field lines generated by the charges on the plates pass through the dielectric. As Figure 19.20*c* shows, some of the field lines end on the negative surface charges and begin again on the positive surface charges. Thus, the electric field inside the dielectric is less than the electric field inside the empty capacitor, assuming the charge on the plates remains constant. This reduction in the electric field is described by the ***dielectric constant*** κ, which is the ratio of the field magnitude E_0 without the dielectric to the field magnitude E inside the dielectric;

$$\kappa = \frac{E_0}{E} \tag{19.9}$$

Being a ratio of two field strengths, the dielectric constant is a number without units. Moreover, since the field $\mathbf{E}_0$ without the dielectric is greater than the field $\mathbf{E}$ inside the dielectric, the dielectric constant is greater than unity. The value of κ depends on the nature of the dielectric material, as Table 19.1 indicates.

The Capacitance of a Parallel Plate Capacitor

The capacitance of a capacitor is affected by the geometry of the plates and the dielectric constant of the material between them. For example, Figure 19.18 shows a parallel plate capacitor in which the area of each plate is A and the separation between the plates is d. The magnitude of the electric field inside the dielectric is given by Equation 19.7 (without the minus sign) as $E = V/d$, where V is the magnitude of the potential difference between the plates. If the charge on each plate is kept fixed, the electric field inside the dielectric is related to the electric field in the absence of the dielectric via Equation 19.9. Therefore,

$$E = \frac{E_0}{\kappa} = \frac{V}{d}$$

Since the electric field within an empty capacitor is $E_0 = q/(\epsilon_0 A)$ (see Equation 18.4), it follows that $q/(\kappa\epsilon_0 A) = V/d$, which can be solved for q to give

$$q = \left(\frac{\kappa\epsilon_0 A}{d}\right)V$$

(a)

(b)

(c)

Figure 19.20 (*a*) The electric field lines inside an empty capacitor. (*b*) The electric field produced by the charges on the plates creates an end-to-end alignment of the molecular dipoles within the dielectric. (*c*) The resulting positive and negative surface charges on the dielectric cause a reduction in the electric field. The space between the dielectric and the plates is added for clarity. In reality, the dielectric fills the region between the plates.

TABLE 19.1   Dielectric Constants of Some Common Substances[a]

Substance	Dielectric Constant, κ
Vacuum	1
Air	1.00054
Teflon	2.1
Benzene	2.28
Paper (royal gray)	3.3
Ruby mica	5.4
Neoprene rubber	6.7
Methyl alcohol	33.6
Water	80.4

[a] Near room temperature.

A comparison of this expression with $q = CV$ (Equation 19.8) reveals that the capacitance C is

Parallel plate capacitor filled with a dielectric
$$C = \frac{\kappa \epsilon_0 A}{d} \qquad (19.10)$$

Notice that only the geometry of the plates (A and d) and the dielectric constant κ affect the capacitance. With C_0 representing the capacitance of the empty capacitor ($\kappa = 1$), Equation 19.10 shows that $C = \kappa C_0$. In other words, the capacitance with the dielectric present is increased by a factor of κ over the capacitance without the dielectric. It can be shown that the relation $C = \kappa C_0$ applies to any capacitor, not just to a parallel plate capacitor. One reason, then, that capacitors are filled with dielectric materials is to increase the capacitance. Example 10 illustrates the effect that increasing the capacitance has on the charge stored by a capacitor.

EXAMPLE 10   Storing Electric Charge

The capacitance of an empty capacitor is 1.2 μF. The capacitor is connected to a 12-V battery and charged up. With the capacitor connected to the battery, a slab of dielectric material is inserted between the plates. As a result, 2.6×10^{-5} C of *additional* charge flows from one plate, through the battery, and on to the other plate. What is the dielectric constant κ of the material?

Reasoning The charge stored by a capacitor is $q = CV$, according to Equation 19.8. The battery maintains a constant potential difference of $V = 12$ volts between the plates of the capacitor while the dielectric is inserted. Inserting the dielectric causes the capacitance C to increase, so that with V held constant, the charge q must increase. Thus, additional charge flows onto the plates. To find the dielectric constant, we apply Equation 19.8 to the empty capacitor and then to the capacitor filled with the dielectric material.

Solution The empty capacitor has a capacitance $C_0 = 1.2$ μF and stores an amount of charge $q_0 = C_0 V$. With the dielectric material in place, the capacitor has a capacitance $C = \kappa C_0$ and stores an amount of charge $q = (\kappa C_0)V$. The additional charge that the battery supplies is $q - q_0 = (\kappa C_0)V - C_0 V = 2.6 \times 10^{-5}$ C. Solving for the dielectric constant, we find that

$$\kappa = \frac{2.6 \times 10^{-5} \text{ C}}{C_0 V} + 1 = \frac{2.6 \times 10^{-5} \text{ C}}{(1.2 \times 10^{-6} \text{ F})(12 \text{ V})} + 1 = \boxed{2.8}$$

In Example 10 the capacitor remains connected to the battery while the dielectric is inserted between the plates. The next example discusses what happens if the capacitor is disconnected from the battery before the dielectric is inserted.

CONCEPTUAL EXAMPLE 11
The Effect of a Dielectric When a Capacitor Has a Constant Charge

An empty capacitor is connected to a battery and charged up. The capacitor is then disconnected from the battery, and a slab of dielectric material is inserted between the plates. Does the voltage across the plates increase, remain the same, or decrease?

Reasoning and Solution Our reasoning is guided by the following fact: once the capacitor is disconnected from the battery, the charge on its plates remains constant, for there is no longer any way for charge to be added or removed. The charge q stored by the capacitor is $q = CV$. Inserting the dielectric causes the capacitance C to increase. Therefore, *the voltage V across the plates must decrease in order for q to re-*

main unchanged. The amount by which the voltage decreases from the value initially established by the battery depends on the dielectric constant of the slab.

Related Homework Material: *Problem 46*

Capacitors are used often in electronic devices, and Example 12 deals with one familiar application.

EXAMPLE 12 A Computer Keyboard

THE PHYSICS OF...

a computer keyboard.

One common kind of computer keyboard is based on the idea of capacitance. Each key is mounted on one end of a plunger, the other end being attached to a movable metal plate (see Figure 19.21). The movable plate is separated from a fixed plate, the two plates forming a capacitor. When the key is pressed, the movable plate is pushed closer to the fixed plate, and the capacitance increases. Electronic circuitry enables the computer to detect the *change* in capacitance, thereby recognizing which key has been pressed. The separation of the plates is normally 5.00×10^{-3} m but decreases to 0.150×10^{-3} m when a key is pressed. The plate area is 9.50×10^{-5} m^2, and the capacitor is filled with a material whose dielectric constant is 3.50. Determine the change in capacitance that is detected by the computer.

Reasoning We can use Equation 19.10 directly to find the capacitance of the key, since the dielectric constant κ, the plate area A, and the plate separation d are known. We will use this relation twice, once to find the capacitance when the key is pressed and once when it is not pressed. The change in capacitance will be the difference between these two values.

Solution When the key is pressed, the capacitance is

$$C = \frac{\kappa \epsilon_0 A}{d} = \frac{(3.50)[8.85 \times 10^{-12} \text{ C}^2/(\text{N} \cdot \text{m}^2)](9.50 \times 10^{-5} \text{ m}^2)}{0.150 \times 10^{-3} \text{ m}}$$

$$= 19.6 \times 10^{-12} \text{ F} \quad (19.6 \text{ pF}) \tag{19.10}$$

A similar calculation reveals that when the key is *not* pressed, the capacitance has a value of 0.589×10^{-12} F (0.589 pF). The *change* in capacitance is $\boxed{19.0 \times 10^{-12} \text{ F} \ (19.0 \text{ pF})}$. The presence of the dielectric increases the *change* in the capacitance, which makes it easier for the circuitry within the computer to detect it.

Key
Plunger
Movable metal plate
Dielectric
Fixed metal plate

Figure 19.21 In one kind of computer keyboard, each key, when pressed, changes the separation between the plates of a capacitor.

Energy Storage in a Capacitor

When a capacitor stores charge, it also stores energy. In charging up a capacitor, for example, a battery does work in transferring an increment of charge from one plate of the capacitor to the other plate. The work done is equal to the product of the charge increment and the potential difference between the plates. However, as each increment of charge is moved, the potential difference increases slightly, and a larger amount of work is needed to move the next increment. The total work W done in completely charging the capacitor is the product of the total charge q transferred and the average potential difference $\bar{V}$; $W = q\bar{V}$. Since the average potential difference is one-half the final potential V, or $\bar{V} = \frac{1}{2}V$, the total work done by the battery is $W = \frac{1}{2}qV$. This work does not disappear but is stored as electric potential energy in the capacitor, so that EPE $= \frac{1}{2}qV$. Since $q = CV$, the energy stored becomes

$$\text{Energy} = \tfrac{1}{2}(CV)V = \tfrac{1}{2}CV^2 \tag{19.11}$$

It is also possible to regard the energy as being stored in the electric field between the plates. The relation between energy and field strength can be obtained

INTERACTIVE LEARNINGWARE #40

START
Visualization
Reasoning
Knowns and Unknowns
Modeling the Problem
Numerical Solution
FINISH

The bulb in a flashlight uses a power of 1.2 W. A 550-μF capacitor stores the energy that this bulb uses in ten minutes of operation. (a) What is the voltage across the plates of this capacitor and (b) what is the charge on one of those plates?

This Interactive LearningWare deals with the energy and charge stored in a capacitor. It also provides a review of the concept of power, which, in this problem, is the rate at which energy is used (see Section 6.7).

Go to our Web site at

http://www.wiley.com/college/cutnell

for a parallel plate capacitor by substituting $V = Ed$ and $C = \kappa\epsilon_0 A/d$ into Equation 19.11:

$$\text{Energy} = \frac{1}{2}\left(\frac{\kappa\epsilon_0 A}{d}\right)(Ed)^2$$

Since the area A times the separation d is the volume between the plates, the energy per unit volume or **energy density** is

$$\text{Energy density} = \frac{\text{Energy}}{\text{Volume}} = \tfrac{1}{2}\kappa\epsilon_0 E^2 \qquad (19.12)$$

It can be shown that this expression is valid for any electric field strength, not just that between the plates of a capacitor.

The energy-storing capability of a capacitor is often put to good use in electronic circuits. For example, in an electronic flash attachment for a camera, energy from the battery pack is stored in capacitors. The capacitors are then discharged between the electrodes of the flash tube, which converts the energy into light. Flash duration times range from 1/200 to 1/1 000 000 second or less, with the shortest flashes being used in high-speed photography (see Figure 19.22). Some flash attachments automatically control the flash duration by monitoring the light reflected from the photographic subject and quickly stopping or quenching the capacitor discharge when the reflected light reaches a predetermined level.

During a heart attack, the heart produces a rapid, unregulated pattern of beats, a condition known as cardiac fibrillation. Cardiac fibrillation can often be stopped by sending a very fast discharge of electrical energy through the heart. Emergency medical personnel use defibrillators, such as the one shown in Figure 19.23. A paddle is connected to each plate of a large capacitor, and the paddles are placed on the chest near the heart. The capacitor is charged to a potential difference of about a thousand volts. The capacitor is then discharged in a few thousandths of a second; the discharge current passes through a paddle, the heart, and the other paddle. Within a few seconds, the heart often returns to its normal beating pattern.

THE PHYSICS OF...

an electronic flash attachment for a camera.

THE PHYSICS OF...

a defibrillator.

*19.6 BIOMEDICAL APPLICATIONS OF ELECTRIC POTENTIAL DIFFERENCES

Conduction of Electrical Signals in Neurons

The human nervous system is remarkable for its ability to transmit information in the form of electrical signals. These signals are carried by the nerves, and the concept of electric potential difference plays an important role in the process. For example, sensory information from our eyes and ears is carried to the brain by the optic nerve and auditory nerves, respectively. Other nerves transmit signals from the brain or spinal column to muscles, causing them to contract. Still other nerves carry signals within the brain.

A nerve consists of a bundle of *axons,* and each axon is one part of a nerve cell, or *neuron.* As Figure 19.24 illustrates, a typical neuron consists of a cell body with numerous extensions, called *dendrites,* and a single axon. The dendrites convert

Figure 19.22 This time-lapse photo was obtained using an electronic flash attachment with the camera. The energy for each "flash" comes from the electrical energy stored in a capacitor.

stimuli, such as pressure or heat, into electrical signals that travel through the neuron. The axon sends the signal to the nerve endings, which transmit the signal across a gap (called a *synapse*) to the next neuron or to a muscle.

The fluid inside a cell, the intracellular fluid, is quite different from that outside the cell, the extracellular fluid. Both fluids contain concentrations of positive and negative ions. However, the extracellular fluid is rich in sodium (Na^+) and chlorine (Cl^-) ions, and the intracellular fluid is rich in potassium (K^+) ions and negatively charged proteins. These concentration differences between the fluids are extremely important to the life of the cell. If the cell membrane were freely permeable, the ions would diffuse across it until the concentrations on both sides were equal. (See Section 14.4 for a review of diffusion.) This does not happen because a living cell has a selectively permeable membrane. Ions can enter or leave the cell only through membrane channels, and the permeability of the channels varies markedly from one ion to another. For example, it is much easier for K^+ ions to diffuse out of the cell than it is for Na^+ to enter the cell. As a result of selective membrane permeability, there is a small buildup of negative charges just inside the membrane and an equal amount of positive charges on the outside (see Figure 19.25). The buildup of charge occurs very close to the membrane, so the membrane acts like a capacitor (see problems 39 and 53). Elsewhere in the intracellular and extracellular fluids, there are equal numbers of positive and negative ions, so the fluids are overall electrically neutral. Such a separation of positive and negative charges gives rise to an electric potential difference across the membrane, called the *resting membrane potential*. In neurons, the resting membrane potential ranges from −40 to −90 mV, with a typical value being −70 mV. The minus sign indicates that the inside of the membrane is negative relative to the outside.

A "resting" neuron is one that is not conducting an electrical signal. The *change* in the resting membrane potential is the key factor in the initiation and conduction of a signal. When a sufficiently strong stimulus is applied to a given point on the neuron, "gates" in the membrane open and sodium ions flood into the cell, as Figure 19.26 illustrates. The sodium ions are driven into the cell by attraction to the negative ions on the inside of the membrane as well as by the relatively high concentration of sodium ions outside the cell. The large influx of Na^+ ions first neutralizes the negative ions on the interior of the membrane and then causes it to become positively charged. As a result, the membrane potential in this localized region goes from −70 mV, the resting potential, to about +30 mV in a very short time (see Figure 19.27). The sodium gates then close, and the cell membrane quickly returns to its normal resting potential. This change in potential, from −70 mV to +30 mV

Figure 19.23 A defibrillator uses the electrical energy stored in a capacitor to deliver a controlled electric current that can restore normal heart rhythm in a heart attack victim. Airplane personnel are now being trained to use portable units such as the one shown here to save lives in the air.

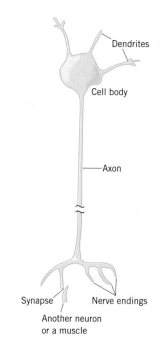

Figure 19.24 The anatomy of a typical neuron.

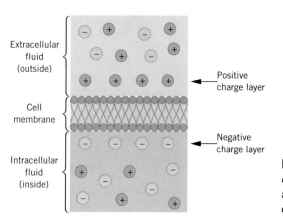

Figure 19.25 Positive and negative charge layers form on the outside and inside surfaces of a membrane during its resting state.

Figure 19.26 When a stimulus is applied to the cell, positive sodium ions (Na^+) rush into the cell, causing the interior surface of the membrane to become momentarily positive.

THE PHYSICS OF...

an action potential.

and back to −70 mV, is known as the *action potential.* The action potential lasts for a few milliseconds, and it is the electrical signal that propagates down the axon, typically at a speed of about 50 m/s, to the next neuron or to a muscle cell.

Medical Diagnostic Techniques

Several important medical diagnostic techniques depend on the fact that the surface of the human body is *not* an equipotential surface. Between various points on the body there are small potential differences (approximately 30–500 μV), which provide the basis for electrocardiography, electroencephalography, and electroretinography. The potential differences can be traced to the electrical characteristics of muscle cells and nerve cells. In carrying out their biological functions, these cells utilize positively charged sodium and potassium ions and negatively charged chlorine ions that exist within the cells and in the intercellular fluid. As a result of such charged particles, electric fields are generated that extend to the surface of the body and lead to small potential differences.

Figure 19.28 shows some locations on the body where electrodes are placed to measure potential differences in electrocardiography. The potential difference between two points changes as the heart beats and forms a repetitive pattern. The recorded pattern of potential difference versus time is called an electrocardiogram (ECG or EKG), and its shape depends on which pair of points in the picture (*A* and *B*, *B* and *C*, etc.) is used to locate the electrodes. The figure also shows some EKGs and indicates the regions (*P, Q, R, S,* and *T*) that can be associated with specific parts of the heart's beating cycle. The distinct differences between the EKGs of healthy and damaged hearts provide physicians with a valuable diagnostic tool.

In electroencephalography the electrodes are placed at specific locations on the head, as Figure 19.29 indicates, and they record the potential differences that characterize brain behavior. The graph of potential difference versus time is known as an electroencephalogram (EEG). The various parts of the patterns in an EEG are often referred to as "waves" or "rhythms." The drawing shows an example of the

Figure 19.27 The action potential is caused by the rush of positive sodium ions into the cell and a subsequent return of the cell to its resting potential.

THE PHYSICS OF...

electrocardiography.

THE PHYSICS OF...

electroencephalography.

Figure 19.28 The potential differences generated by heart muscle activity provide the basis for electrocardiography. The normal and abnormal EKG patterns correspond to one heartbeat.

Figure 19.29 In electroencephalography the potential differences created by the electrical activity of the brain are used for diagnosing abnormal behavior.

Electronic flash

Figure 19.30 The electrical activity of the retina of the eye generates the potential differences used in electroretinography.

main resting rhythm of the brain, the so-called alpha rhythm, and also illustrates the distinct differences that are found between the EEGs generated by healthy and diseased (abnormal) tissue.

The electrical characteristics of the retina of the eye lead to the potential differences measured in electroretinography. Figure 19.30 shows a typical electrode placement used to record the pattern of potential difference versus time that occurs when the eye is stimulated by a flash of light. One electrode is mounted on a contact lens, while the other is often placed on the forehead. The recorded pattern is called an electroretinogram (ERG), parts of the pattern being referred to as the "*A* wave" and the "*B* wave." As the graphs show, the ERGs of normal and diseased eyes can differ markedly.

THE PHYSICS OF...

electroretinography.

19.7 CONCEPTS & CALCULATIONS

The conservation of energy (Chapter 6) and the conservation of linear momentum (Chapter 7) are two of the most broadly applicable principles in all of science. In this chapter, we have seen that electrically charged particles obey the conservation-of-energy principle, provided that the electric potential energy is taken into account. The behavior of electrically charged particles, however, must also be consistent with the conservation-of-momentum principle, as the first example in this section emphasizes.

CONCEPTS & CALCULATIONS EXAMPLE 13   Conservation Principles

One particle has a mass of $m_1 = 3.6 \times 10^{-6}$ kg, while another has a mass of $m_2 = 6.2 \times 10^{-6}$ kg. Each has the same electrical charge. These particles are initially held at rest, and the two-particle system has an initial electric potential energy of 0.150 J. Suddenly, the particles are released and fly apart because of the repulsive electrical force that acts on each one. The effects of the gravitational force are negligible, and no other forces act on the particles. At one instant following the release, the speed of the less massive particle is measured to be $v_1 = 170$ m/s. What is the electric potential energy of the two-particle system at this instant?

Concept Questions and Answers What type of energy does the two-particle system have initially?

Answer Initially, the particles are at rest, so they have no kinetic energy. However, they do have electric potential energy. They also have gravitational potential energy, but it is negligible.

What type of energy does the two-particle system have at the instant that is being asked about?

Answer Since each particle is moving, each has kinetic energy. The two-particle system also has electric potential energy at this instant. Gravitational potential energy remains negligible.

Does the principle of conservation of energy apply?

Answer Yes. Since the gravitational force is negligible, the only force acting here is the conservative electric force. Nonconservative forces are absent. Thus, the principle of conservation of mechanical energy applies.

Does the principle of conservation of linear momentum apply to the two particles as they fly apart?

Answer Yes. This principle states that the total linear momentum of an isolated system remains constant. An isolated system is one for which the vector sum of the external forces acting on the system is zero. There are no external forces acting here. The only appreciable force acting on each of the two particles is the force of electric repulsion, which is an internal force.

Solution The conservation-of-energy principle indicates that the total energy of the two-particle system is the same at the later instant as it is initially. Considering that only kinetic energy and electric potential energy are present, the principle can be stated as follows:

$$\underbrace{KE_f + EPE_f}_{\text{Final total energy}} = \underbrace{KE_0 + EPE_0}_{\text{Initial total energy}} \quad \text{or} \quad EPE_f = EPE_0 - KE_f$$

We have used the fact that KE_0 is zero, since the particles are initially at rest. The final kinetic energy of the system is $KE_f = \frac{1}{2}m_1v_{f1}^2 + \frac{1}{2}m_2v_{f2}^2$. In this expression, we have values for both masses and the speed v_{f1}. We can find the speed v_{f2} by using the principle of conservation of momentum:

$$\underbrace{m_1v_{f1} + m_2v_{f2}}_{\substack{\text{Final total} \\ \text{momentum}}} = \underbrace{m_1v_{01} + m_2v_{02}}_{\substack{\text{Initial total} \\ \text{momentum}}} \quad \text{or} \quad v_{f2} = -\frac{m_1}{m_2}v_{f1}$$

We have again used the fact that the particles are initially at rest, so that $v_{01} = v_{02} = 0$ m/s. Using this result for v_{f2}, we can now obtain the final electric potential energy from the conservation-of-energy equation:

$$EPE_f = EPE_0 - \left[\frac{1}{2}m_1v_{f1}^2 + \frac{1}{2}m_2\left(-\frac{m_1}{m_2}v_{f1}\right)^2\right]$$

$$= EPE_0 - \frac{1}{2}m_1\left(1 + \frac{m_1}{m_2}\right)v_{f1}^2$$

$$= 0.150 \text{ J} - \frac{1}{2}(3.6 \times 10^{-6} \text{ kg})\left(1 + \frac{3.6 \times 10^{-6} \text{ kg}}{6.2 \times 10^{-6} \text{ kg}}\right)(170 \text{ m/s})^2$$

$$= \boxed{0.068 \text{ J}}$$

Chapter 18 introduces the electric field, and the present chapter introduces the electric potential. These two concepts are central to the study of electricity, and it is important to distinguish between them, as the next example emphasizes.

CONCEPTS & CALCULATIONS EXAMPLE 14   Electric Field and Electric Potential

Two identical point charges $(+2.4 \times 10^{-9}\text{ C})$ are fixed in place, separated by 0.50 m. Find the electric field and the electric potential at the midpoint of the line between the charges q_A and q_B.

Concept Questions and Answers The electric field is a vector and has a direction. At the midpoint, what are the directions of the individual electric-field contributions from q_A and q_B?

Answer Since both charges are positive, the individual electric-field contributions from q_A and q_B point away from each charge. Thus, at the midpoint they point in opposite directions.

Is the magnitude of the net electric field at the midpoint greater than, less than, or equal to zero?

Answer The magnitude of the net electric field is zero. This is because the electric field from q_A cancels the electric field from q_B. These charges have the same magnitude q, and the midpoint is the same distance r from each of them. According to Equation 18.3 $(E = kq/r^2)$, then, each charge produces a field of the same strength. Since the individual electric-field contributions from q_A and q_B point in opposite directions at the midpoint, their vector sum is zero.

Is the total electric potential at the midpoint positive, negative, or zero?

Answer The total electric potential at the midpoint is the algebraic sum of the individual contributions from each charge. According to Equation 19.6 $(V = kq/r)$, each contribution is positive, since each charge is positive. Thus, the total electric potential is also positive.

Does the total electric potential have a direction associated with it?

Answer No. The electric potential is a scalar quantity, not a vector quantity. Therefore, it has no direction associated with it.

Solution As discussed in the second concept question, the total electric field at the midpoint is $\boxed{E = 0\text{ N/C}}$. Using Equation 19.6 $(V = kq/r)$, we find that the total potential at the midpoint is

$$V_{\text{Total}} = \frac{kq_A}{r_A} + \frac{kq_B}{r_B}$$

Since $q_A = q_B = +2.4 \times 10^{-9}\text{ C}$ and $r_A = r_B = 0.25\text{ m}$, we have

$$V = \frac{2kq_A}{r_A} = \frac{2(8.99 \times 10^9\text{ N} \cdot \text{m}^2/\text{C}^2)(+2.4 \times 10^{-9}\text{ C})}{0.25\text{ m}} = \boxed{170\text{ V}}$$

At the end of the problem set for this chapter, you will find homework problems that contain both conceptual and quantitative parts. These problems are grouped under the heading *Concepts & Calculations, Group Learning Problems*. They are designed for use by students working alone or in small learning groups. The conceptual part of each problem provides a convenient focus for group discussions.

SUMMARY

When work W_{AB} is done by an electric force as a positive test charge $+q_0$ moves from point A to point B, the work equals the **electric potential energy** EPE at A minus that at B: $W_{AB} = \text{EPE}_A - \text{EPE}_B$. The **electric potential** V is the electric potential energy per unit charge, $V = \text{EPE}/q_0$. The electric potential difference between two points is $V_B - V_A = (\text{EPE}_B - \text{EPE}_A)/q_0 = -W_{AB}/q_0$. A positive charge accelerates from a region of higher potential toward a region of lower potential. Conversely, a negative charge accelerates from a region of lower potential toward a region of higher potential. The electric potential at a distance r from a **point charge** q is $V = kq/r$. This expression for V assumes the potential is zero at an infinite distance away from the charge.

The **electron volt** (eV) is a unit of energy. One electron volt corresponds to 1.60×10^{-19} J and is the amount by which the potential energy of an electron changes when the electron moves through a potential difference of one volt.

An **equipotential surface** is a surface on which the electric potential is the same everywhere. The electric force does no work as a charge moves on an equipotential surface. The electric field created by any group of charges is always perpendicular to the associated equipotential surfaces and points

in the direction of decreasing potential. The electric field is given by $E = -\Delta V/\Delta s$, where ΔV is the potential difference and Δs is the displacement perpendicular to the equipotential surfaces.

A **capacitor** is a device that can store charge and consists of two conductors or plates that are near one another, but not touching. The magnitude q of the charge on each plate is given by $q = CV$, where V is the magnitude of the potential difference between the plates and C is the **capacitance.** The SI unit for capacitance is the farad (F), and one farad is one coulomb per volt (C/V). The insulating material included between the plates is called a **dielectric.** The dielectric constant κ of the material is $\kappa = E_0/E$, where E_0 and E are, respectively, the magnitudes of the electric fields between the plates without and with a dielectric, assuming the charge on the plates is kept fixed. The capacitance of a parallel plate capacitor is $C = \kappa\epsilon_0 A/d$, where $\epsilon_0 = 8.85 \times 10^{-12}$ $\text{C}^2/(\text{N}\cdot\text{m}^2)$ is the permittivity of free space, A is the area of each plate, and d is the distance between the plates. The **electric potential energy** stored in a capacitor is $\frac{1}{2}CV^2$. The **energy density** or energy stored per unit volume is $\frac{1}{2}\kappa\epsilon_0 E^2$.

CONCEPTUAL QUESTIONS

1. The drawing shows three possibilities for the potentials at two points, A and B. In each case, the same positive charge is moved from A to B. In which case, if any, is the most work done on the positive charge by the electric force? Account for your answer.

A	B	A	B	A	B
•	•	•	•	•	•
150 V	100 V	25 V	−25 V	−10 V	−60 V
Case 1		Case 2		Case 3	

2. A positive point charge and a negative point charge have equal magnitudes. One charge is fixed to one corner of a square, and the other is fixed to another corner. On which corners should the charges be placed, so that the same potential exists at the empty corners? Give your reasoning.

3. Three point charges have identical magnitudes, but two of the charges are positive and one is negative. These charges are fixed to the corners of a square, one to a corner. No matter how the charges are arranged, the potential at the empty corner is positive. Explain why.

4. What point charges, all having the same magnitude, would you place at the corners of a square (one charge per corner), so that both the electric field and the electric potential (assuming a zero reference value at infinity) are zero at the center of the square? Account for the fact that the charge distribution gives rise to *both* a zero field and a zero potential.

5. Positive charge is spread uniformly around a circular ring, as the drawing illustrates. Equation 19.6 gives the correct potential at points along the line perpendicular to the plane of the ring at its center. However, the equation does not give the correct potential at points that do not lie on this line. In the equation, q represents the total charge on the ring. Why does Equation 19.6 apply for points on the line, but not for points off the line?

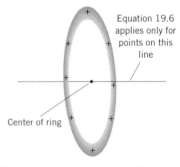

Equation 19.6 applies only for points on this line

Center of ring

6. The electric field at a single location is zero. Does this fact necessarily mean that the electric potential at the same place is zero? Use a spot on the line between two identical point charges as an example to support your reasoning.

7. An electric potential energy exists when two protons are separated by a certain distance. Does the electric potential energy increase, decrease, or remain the same when (a) both protons are replaced by electrons, and (b) only one of the protons is replaced by an electron? Justify your answers.

8. A proton is fixed in place. An electron is released from rest and allowed to collide with the proton. Then the roles of the proton and electron are interchanged, and the same experiment is repeated. Which is traveling faster when the collision occurs, the proton or the electron? Justify your answer.

9. The potential is constant throughout a given region of space. Is the electric field zero or nonzero in this region? Explain.

10. In a region of space where the electric field is constant everywhere, as it is inside a parallel plate capacitor, is the potential constant everywhere? Account for your answer.

11. A positive test charge is placed in an electric field. In what direction should the charge be moved relative to the field, such that the charge experiences a constant electric potential? Explain.

12. The location marked P in the drawing lies midway between the point charges $+q$ and $-q$. The blue lines labeled A, B, and C are edge-on views of three planes. Which one of these planes is an equipotential surface? Why?

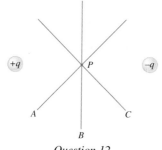

Question 12

13. Imagine that you are moving a positive test charge along the line between two identical point charges. With regard to

the electric potential, is the midpoint on the line analogous to the top of a mountain or the bottom of a valley when the two point charges are (a) positive and (b) negative? In each case, explain your answer.

14. Repeat question 13, assuming that you are moving a negative instead of a positive test charge.

15. The potential at a point in space has a certain value, which is not zero. Is the electric potential energy the same for every charge that is placed at that point? Give your reasoning.

16. A proton and an electron are released from rest at the midpoint between the plates of a charged parallel plate capacitor. Except for these particles, nothing else is between the plates. Ignore the attraction between the proton and the electron, and decide which particle strikes a capacitor plate first. Why?

17. A parallel plate capacitor is charged up by a battery. The battery is then disconnected, but the charge remains on the plates. The plates are then pulled apart. Explain whether each of the following quantities increases, decreases, or remains the same as the distance between the plates increases: (a) the capacitance of the capacitor, (b) the potential difference between the plates, (c) the electric field between the plates, and (d) the electric potential energy stored by the capacitor. Give reasons for your answers.

PROBLEMS

Note: All charges are assumed to be point charges unless specified otherwise.

**Section 19.1 Potential Energy,
Section 19.2 The Electric Potential Difference**

1. ssm ⚕ Suppose that the electric potential outside a living cell is higher than that inside the cell by 0.070 V. How much work is done by the electric force when a sodium ion (charge = $+e$) moves from the outside to the inside?

2. A particle has a charge of $+1.5$ μC and moves from point A to point B, a distance of 0.20 m. The particle experiences a constant electric force, and its motion is along the line of action of the force. The difference between the particle's electric potential energy at A and B is $EPE_A - EPE_B = +9.0 \times 10^{-4}$ J. (a) Find the magnitude and direction of the electric force that acts on the particle. (b) Find the magnitude and direction of the electric field that the particle experiences.

3. Just as you touch a metal door knob, a spark of electricity (electrons) jumps from your hand to the knob. The electric potential of the knob is 2.0×10^4 V greater than that of your hand. The work done by the electric force on the electrons is 1.5×10^{-7} J. How many electrons jump from your hand to the knob?

4. The anode (positive terminal) of an X-ray tube is at a potential of $+125\,000$ V with respect to the cathode (negative terminal). (a) How much work (in joules) is done by the electric force when

an electron is accelerated from the cathode to the anode? (b) If the electron is initially at rest, what kinetic energy does the electron have when it arrives at the anode?

5. ssm In a television picture tube, electrons strike the screen after being accelerated from rest through a potential difference of 25 000 V. The speeds of the electrons are quite large, and for accurate calculations of the speeds, the effects of special relativity must be taken into account. Ignoring such effects, find the electron speed just before the electron strikes the screen.

6. A particle with a charge of -1.5 μC and a mass of 2.5×10^{-6} kg is released from rest at point A and accelerates toward point B, arriving there with a speed of 42 m/s. (a) What is the potential difference $V_B - V_A$ between A and B? (b) Which point is at the higher potential? Give your reasoning.

7. An electric car accelerates for 8.0 s by drawing energy from its 320-V battery pack. During this time, 1300 C of charge pass through the battery pack. Find the minimum horsepower rating of the car.

*8. A typical 12-V car battery can deliver about 7.5×10^5 C of charge before dying. This is not very much. To get a feel for this, calculate the maximum number of kilograms of water (100 °C) that could be boiled into steam (100 °C) using energy from this battery.

*9. ssm www The potential at location A is 452 V. A positively charged particle is released there from rest and arrives at location B with a speed v_B. The potential at location C is 791 V, and when released from rest from this spot, the particle arrives at B with twice the speed it previously had, or $2v_B$. Find the potential at B.

**10. A particle is uncharged and is thrown vertically upward from ground level with a speed of 25.0 m/s. As a result, it attains a maximum height h. The particle is then given a positive charge $+q$ and reaches the same maximum height h when thrown vertically upward with a speed of 30.0 m/s. The electric potential at the height h exceeds the electric potential at ground level. Finally, the particle is given a negative charge $-q$. Ignoring air resistance, determine the speed with which the negatively charged particle must be thrown vertically upward, so that it attains exactly the maximum height h. In all three situations, be sure to include the effect of gravity.

Section 19.3 The Electric Potential Difference Created by Point Charges

11. ssm There is an electric potential of $+130$ V at a spot that is 0.25 m away from a charge. Find the magnitude and sign of the charge.

12. Two charges A and B are fixed in place, at different distances from a certain spot. At this spot the potentials due to the two charges are equal. Charge A is 0.18 m from the spot, while charge B is 0.43 m from it. Find the ratio q_B/q_A of the charges.

13. An electron and a proton are initially very far apart (effectively an infinite distance apart). They are then brought together to form a hydrogen atom, in which the electron orbits the proton at an average distance of 5.29×10^{-11} m. What is $\text{EPE}_{\text{final}} - \text{EPE}_{\text{initial}}$, which is the change in the electric potential energy?

14. Location A is 2.00 m from a charge of -3.00×10^{-8} C, while location B is 3.00 m from the charge. Find the potential difference $V_B - V_A$ between the two points, and state which point is at the higher potential.

15. ssm www Two identical point charges are fixed to diagonally opposite corners of a square that is 0.500 m on a side. Each charge is $+3.0 \times 10^{-6}$ C. How much work is done by the electric force as one of the charges moves to an empty corner?

16. Two positive point charges are held in place, 0.74 m apart. They are then moved so that their electric potential energy doubles. What is the new separation between the charges?

17. A charge of $+9q$ is fixed to one corner of a square, while a charge of $-8q$ is fixed to the opposite corner. Expressed in terms of q, what charge should be fixed to the center of the square, so the potential is zero at each of the two empty corners?

18. Review Conceptual Example 7 as background for this problem. Two charges are fixed in place with a separation d. One charge is positive and has twice the magnitude of the other charge, which is negative. The positive charge lies to the left of the negative charge, as in Figure 19.11. Relative to the negative charge, locate the two spots on the line through the charges where the total potential is zero.

*19. ssm A charge of -3.00 μC is fixed in place. From a horizontal distance of 0.0450 m, a particle of mass 7.20×10^{-3} kg and charge -8.00 μC is fired with an initial speed of 65.0 m/s directly toward the fixed charge. How far does the particle travel before its speed is zero?

*20. Four identical charges ($+2.0$ μC each) are brought from infinity and fixed to a straight line. The charges are located 0.40 m apart. Determine the electric potential energy of this group.

*21. Two protons are moving directly toward one another. When they are very far apart, their initial speeds are 1.5×10^6 m/s. What is the distance of closest approach?

*22. Identical point charges of $+1.7$ μC are fixed to diagonally opposite corners of a square. A third charge is then fixed at the center of the square, such that it causes the potentials at the empty corners to change signs without changing magnitudes. Find the sign and magnitude of the third charge.

*23. ssm Determine the electric potential energy for the array of three charges shown in the drawing, relative to its value when the charges are infinitely far away.

Problem 23

**24. A positive charge $+q_1$ is located to the left of a negative charge $-q_2$. On a line passing through the two charges, there are two places where the total potential is zero. The first place is between the charges and is 4.00 cm to the left of the negative charge. The second place is 7.00 cm to the right of the negative charge. (a) What is the distance between the charges? (b) Find q_1/q_2, the ratio of the magnitudes of the charges.

**25. Charges q_1 and q_2 are fixed in place, q_2 being located at a distance d to the right of q_1. A third charge q_3 is then fixed to the line joining q_1 and q_2 at a distance d to the right of q_2. The third charge is chosen so the potential energy of the group is zero; that is, the potential energy has the same value as that of the three charges when they are widely separated. Determine q_3, assuming that (a) $q_1 = q_2 = q$ and (b) $q_1 = q$ and $q_2 = -q$. Express your answers in terms of q.

**26. Two particles each have a mass of 6.0×10^{-3} kg. One has a charge of $+5.0 \times 10^{-6}$ C, and the other has a charge of -5.0×10^{-6} C. They are initially held at rest at a distance of 0.80 m apart. Both are then released and accelerate toward each other. How fast is each particle moving when the separation between them is one-half its initial value?

Section 19.4 Equipotential Surfaces and Their Relation to the Electric Field

27. ssm An equipotential surface that surrounds a $+3.0 \times 10^{-7}$ C point charge has a radius of 0.15 m. What is the potential of this surface?

28. What point charge is at the center of a $+36$-V equipotential surface that has a radius of 0.18 m?

29. A spark plug in an automobile engine consists of two metal conductors that are separated by a distance of 0.75 mm. When an electric spark jumps between them, the magnitude of the electric field is 4.7×10^7 V/m. What is the magnitude of the potential difference ΔV between the conductors?

30. ⌁ The inner and outer surfaces of a cell membrane carry a negative and positive charge, respectively. Because of these charges, a potential difference of about 0.070 V exists across the membrane. The thickness of the membrane is 8.0×10^{-9} m. What is the magnitude of the electric field in the membrane?

31. ssm Two points, A and B, are separated by 0.016 m. The potential at A is +95 V, and that at B is +28 V. Determine the magnitude and direction of the electric field between the two points.

*32. At a distance of 1.60 m from a point charge of $+2.00 \times 10^{-6}$ C, there is an equipotential surface. At greater distances there are additional equipotential surfaces. The potential difference between any two successive surfaces is 1.00×10^3 V. Starting at a distance of 1.60 m and moving radially outward, how many of the additional equipotential surfaces are crossed by the time the electric field has shrunk to one-half its initial value? Do not include the starting surface.

*33. The electric field has a constant value of 3.0×10^3 V/m and is directed downward. The field is the same everywhere. The potential at a point P within this region is 135 V. Find the potential at the following points: (a) 8.0×10^{-3} m directly above P, (b) 3.3×10^{-3} m directly below P, (c) 5.0×10^{-3} m directly to the right of P.

*34. The drawing shows the potential at five points on a set of axes. Each of the four outer points is 6.0×10^{-3} m from the point at the origin. From the data shown, find the magnitude and direction of the electric field in the vicinity of the origin.

Problem 34

*35. ssm Equipotential surface A has a potential of 5650 V, while equipotential surface B has a potential of 7850 V. A particle has a mass of 5.00×10^{-2} kg and a charge of $+4.00 \times 10^{-5}$ C. The particle has a speed of 2.00 m/s on surface A. An outside force is applied to the particle, and it moves to surface B, arriving there with a speed of 3.00 m/s. How much work is done by the outside force in moving the particle from A to B?

Section 19.5 Capacitors and Dielectrics

36. What voltage is required to store 7.2×10^{-5} C of charge on the plates of a 6.0-μF capacitor?

37. ssm ⌁ The electric potential energy stored in the capacitor of a defibrillator is 73 J, and the capacitance is 120 μF. What is the potential difference across the capacitor plates?

38. A capacitor stores 5.3×10^{-5} C of charge when connected to a 6.0-V battery. How much charge does the capacitor store when connected to a 9.0-V battery?

39. ssm ⌁ The membrane that surrounds a certain type of living cell has a surface area of 5.0×10^{-9} m^2 and a thickness of 1.0×10^{-8} m. Assume that the membrane behaves like a parallel plate capacitor and has a dielectric constant of 5.0. (a) The potential on the outer surface of the membrane is +60.0 mV greater than that on the inside surface. How much charge resides on the outer surface? (b) If the charge in part (a) is due to K^+ ions (charge $+e$), how many such ions are present on the outer surface?

40. A capacitor has a capacitance of 2.5×10^{-8} F. In the charging process, electrons are removed from one plate and placed on the other plate. When the potential difference between the plates is 450 V, how many electrons have been transferred?

41. The electronic flash attachment for a camera contains a capacitor for storing the energy used to produce the flash. In one such unit, the potential difference between the plates of a 750-μF capacitor is 330 V. (a) Determine the energy that is used to produce the flash in this unit. (b) Assuming that the flash lasts for 5.0×10^{-3} s, find the effective power or "wattage" of the flash.

42. Two identical capacitors store different amounts of energy: capacitor A stores 310 J and capacitor B stores 34 J. The voltage across the plates of capacitor B is 12 V. Find the voltage across the plates of capacitor A.

*43. ssm What is the potential difference between the plates of a 3.3-F capacitor that stores sufficient energy to operate a 75-W light bulb for one minute?

*44. Two hollow metal spheres are concentric with each other. The inner sphere has a radius of 0.1500 m and a potential of 85.0 V. The radius of the outer sphere is 0.1520 m and its potential is 82.0 V. If the region between the spheres is filled with Teflon, find the electric energy contained in this space.

*45. The dielectric strength of an insulating material is the maximum electric field strength to which the material can be subjected without electrical breakdown occurring. Suppose a parallel plate capacitor is filled with a material whose dielectric constant is 3.5 and whose dielectric strength is 1.4×10^7 N/C. If this capacitor is to store 1.7×10^{-7} C of charge on each plate without suffering breakdown, what must be the radius of its circular plates?

*46. Review Conceptual Example 11 before attempting this problem. An empty capacitor is connected to a 12.0-V battery and charged up. The capacitor is then disconnected from the battery, and a slab of dielectric material ($\kappa = 2.8$) is inserted between the plates. Find the amount by which the potential difference across the plates changes. Specify whether the change is an increase or a decrease.

**47. ssm www The potential difference between the plates of a capacitor is 175 V. Midway between the plates, a proton and an electron are released. The electron is released from rest. The proton is projected perpendicularly toward the negative plate with an initial speed. The proton strikes the negative plate at the same instant the electron strikes the positive plate. Ignore the attraction between the two particles, and find the initial speed of the proton.

**48. The plate separation of a charged capacitor is 0.0800 m. A proton and an electron are released from rest at the midpoint between the plates. Ignore the attraction between the two particles, and determine how far the proton has traveled by the time the electron strikes the positive plate.

ADDITIONAL PROBLEMS

49. At a distance of 0.20 m from a charge, the electric potential is 164 V. What is the potential at a distance of 0.80 m?

50. An electric force moves a charge of $+1.80 \times 10^{-4}$ C from point A to point B and performs 5.80×10^{-3} J of work on the charge. (a) What is the difference ($\text{EPE}_A - \text{EPE}_B$) between the electric potential energies of the charge at the two points? (b) Determine the potential difference ($V_A - V_B$) between the two points. (c) State which point is at the higher potential.

51. ssm A parallel plate capacitor has a capacitance of 7.0 μF when filled with a dielectric. The area of each plate is 1.5 m^2 and the separation between the plates is 1.0×10^{-5} m. What is the dielectric constant of the dielectric?

52. Point A is at a potential of $+250$ V, and point B is at a potential of -150 V. An α-particle is a helium nucleus that contains two protons and two neutrons; the neutrons are electrically neutral. An α-particle starts from rest at A and accelerates toward B. When the α-particle arrives at B, what kinetic energy (in electron volts) does it have?

53. ⚖ An axon is the relatively long tail-like part of a neuron, or nerve cell. The outer surface of the axon membrane (dielectric constant = 5, thickness = 1×10^{-8} m) is charged positively, and the inner portion is charged negatively. Thus, the membrane is a kind of capacitor. Assuming that an axon can be treated like a parallel plate capacitor with a plate area of 5×10^{-6} m^2, what is its capacitance?

54. A charge of $+125$ μC is fixed at the center of a square that is 0.64 m on a side. How much work is done by the electric force as a charge of $+7.0$ μC moves from one corner of the square to any other empty corner? Explain.

55. ssm www When you walk across a rug on a dry day, your body can become electrified, and its electric potential can change. When the potential becomes large enough, a spark of negative charges can jump between your hand and a metal surface. A spark occurs when the electric field strength created by the charges on your body reaches the dielectric strength of the air. The dielectric strength of the air is 3.0×10^6 N/C and is the electric field strength at which the air suffers electrical breakdown. Suppose a spark 3.0 mm long jumps between your hand and a metal doorknob. Assuming that the electric field is uniform, find the potential difference ($V_{knob} - V_{hand}$) between your hand and the doorknob.

*****56.** The drawing shows the electric potential as a function of distance along the x axis. Determine the magnitude of the electric field in the region (a) A to B, (b) B to C, and (c) C to D.

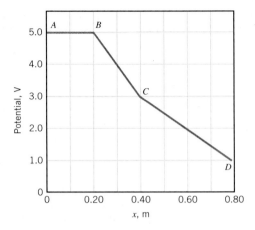

*****57.** An empty capacitor has a capacitance of 2.7 μF and is connected to a 12-V battery. A dielectric material ($\kappa = 4.0$) is inserted between the plates of this capacitor. What is the magnitude of the surface charge on the dielectric that is adjacent to either plate of the capacitor? *(Hint: The surface charge is equal to the difference in the charge on the plates with and without the dielectric.)*

****58.** One particle has a mass of 3.00×10^{-3} kg and a charge of $+8.00$ μC. A second particle has a mass of 6.00×10^{-3} kg and the same charge. The two particles are initially held in place and then released. The particles fly apart, and when the separation between them is 0.100 m, the speed of the 3.00×10^{-3}-kg particle is 125 m/s. Find the initial separation between the particles.

****59. ssm** The drawing shows a parallel plate capacitor. One-half of the region between the plates is filled with a material that has a dielectric constant κ_1. The other half is filled with a material that has a dielectric constant κ_2. The area of each plate is A, and the plate separation is d. The potential difference across the plates is V. Note especially that the charge stored by the capacitor is $q_1 + q_2 = CV$, where q_1 and q_2 are the charges on the area of the plates in contact with materials 1 and 2, respectively. Show that $C = \epsilon_0 A(\kappa_1 + \kappa_2)/(2d)$.

****60.** A positive charge of $+q_1$ is located 3.00 m to the left of a negative charge $-q_2$. The charges have different magnitudes. On the line through the charges, the net *electric field* is zero at a spot 1.00 m to the right of the negative charge. On this line there are also two spots where the potential is zero. Locate these two spots relative to the negative charge.

CONCEPTS & CALCULATIONS

GROUP LEARNING PROBLEMS

Note: Each of these problems consists of Concept Questions followed by a related quantitative Problem. They are designed for use by students working alone or in small learning groups. The Concept Questions involve little or no mathematics and are intended to stimulate group discussions. They focus on the concepts with which the problems deal. Recognizing the concepts is the essential initial step in any problem-solving technique.

61. Concept Questions The drawing shows a square, on two corners of which are fixed different positive charges. A third charge that is negative is brought to one of the empty corners. (a) At which of the empty corners, A or B, is the potential greater? (b) Is the electric potential energy of the third charge positive or negative? (c) For which location of the third charge, corner A or B, is the magnitude of the electric potential energy greater? Explain your answers.

Problem The length of a side of the square is $L = 0.25$ m. Find the electric potential energy of a charge $q_3 = -6.0 \times 10^{-9}$ C placed at corner A and then at corner B. Compare your answers for consistency with your answers to the Concept Questions.

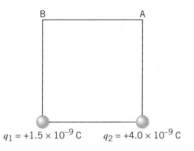

$q_1 = +1.5 \times 10^{-9}$ C $q_2 = +4.0 \times 10^{-9}$ C

62. Concept Questions Charges of $-q$ and $+2q$ are fixed in place, with a distance d between them. A dashed line is drawn through the negative charge, perpendicular to the line between the charges. On the dashed line, at a distance L from the negative charge, there is at least one spot where the total potential is zero. (a) At this spot, is the magnitude of the potential from the positive charge greater than, less than, or equal to the magnitude of the potential from the negative charge? (b) Is the distance from the positive charge to the zero-potential spot greater than, less than, or equal to L? (c) How many spots on the dashed line are there where the total potential is zero? Account for your answers.

Problem The distance between the charges is $d = 2.00$ m. Find L.

63. Concept Questions An electron and a proton, starting from rest, are accelerated through an electric potential difference of the same magnitude. In the process, the electron acquires a speed v_e, while the proton acquires a speed v_p. (a) As each particle accelerates from rest, it gains kinetic energy. Does it gain or lose electric

potential energy? (b) Does the electron gain more, less, or the same amount of kinetic energy as the proton does? (c) Is v_e greater than, less than, or equal to v_p? Justify your answers.

Problem Find the ratio v_e/v_p. Verify that your answer is consistent with your answers to the Concept Questions.

64. Concept Questions A positive point charge is surrounded by an equipotential surface A, which has a radius of r_A. A positive test charge moves from surface A to another equipotential surface B, which has a radius of r_B. In the process, the electric force does negative work. (a) Does the electric force acting on the test charge have the same or opposite direction as the displacement of the test charge? (b) Is r_B greater than or less than r_A? Explain your answers.

Problem The positive point charge is $q = +7.2 \times 10^{-8}$ C, and the test charge is $q_0 = +4.5 \times 10^{-11}$ C. The work done as the test charge moves from surface A to surface B is $W_{AB} = -8.1 \times 10^{-9}$ J. The radius of surface A is $r_A = 1.8$ m. Find r_B. Check to see that your answer is consistent with your answers to the Concept Questions.

65. Concept Questions Two capacitors have the same plate separation. However, one has square plates, while the other has circular plates. The square plates are a length L on each side, and the diameter of the circular plates is L. (a) If the same dielectric material were between the plates in each capacitor, which one would have the greater capacitance? (b) By putting different dielectric materials between the capacitor plates, the two capacitors can be made to have the same capacitance. Which capacitor should contain the dielectric material with the greater dielectric constant? Give your reasoning in each case.

Problem The capacitors have the same capacitance because they contain different dielectric materials. The dielectric constant of the material between the square plates has a value of $\kappa_{square} = 3.00$. What is the dielectric constant κ_{circle} of the material between the circular plates? Be sure that your answer is consistent with your answers to the Concept Questions.

66. Concept Questions Capacitor A and capacitor B each have the same voltage across their plates. However, the energy of capacitor A can melt m kilograms of ice at 0 °C, while the energy of capacitor B can boil away the same amount of water at 100 °C. (a) Which requires more energy, melting the ice or boiling the water? (b) Which capacitor has the greater capacitance? Explain your answers.

Problem The capacitance of capacitor A is 9.3 μF. What is the capacitance of capacitor B? Be sure that your answer is consistent with your answers to the Concept Questions.

The lights in this night view of the San Francisco skyline symbolize the dependence of all modern cities on electricity.

CHAPTER 20
ELECTRIC CIRCUITS

20.1 ELECTROMOTIVE FORCE AND CURRENT

Look around you. Chances are that there is an electrical device nearby—a radio, a hair drier, a computer—something that needs electrical energy to operate. The energy needed to run a portable cassette player, for instance, comes from batteries, as Figure 20.1 illustrates. The transfer of energy takes place via an electric circuit, in which the energy source (the battery pack) and the energy-consuming device (the cassette player) are connected by conducting wires, through which electric charges move.

Within a battery, a chemical reaction occurs that transfers electrons from one terminal (leaving it positively charged) to another terminal (leaving it negatively charged). Figure 20.2 shows the two terminals of a car battery and a flashlight battery. The drawing also illustrates the symbol $\left(\underset{+}{\vdash}\!\!\vdash\right)$ used to represent a battery in circuit drawings. Because of the positive and negative charges on the battery terminals, an electric potential difference exists between them. The maximum potential difference is called the ***electromotive force* (emf)*** of the battery, for which the symbol $\mathscr{E}$ is used. In a typical car battery, the chemical reaction maintains the potential of the positive terminal at a maximum of 12 volts (12 joules/coulomb) higher than the potential of the negative terminal, so the emf is $\mathscr{E} = 12$ V. Thus, one coulomb of charge emerging from the battery and entering a circuit has at most 12 joules of energy. In a typical flashlight battery the emf is 1.5 V. In reality, the potential difference between the terminals of a battery is somewhat less than the maximum value indicated by the emf, for reasons that Section 20.9 discusses.

● ***Concepts at a Glance*** In a circuit such as that in Figure 20.1, the battery creates an electric field within and parallel to the wire, directed from the positive toward the negative terminal. The electric field exerts a force on the free electrons in the wire, and they respond by moving. Figure 20.3 shows charges moving inside a wire and crossing an imaginary surface that is perpendicular to their motion. This flow of charge is known as an ***electric current***. The Concepts-at-a-Glance chart in Figure 20.4 illustrates that we will formulate this new concept of electric current by bringing together the amount of charge Δq that passes through the surface and the time interval Δt during which it does so. ●

The electric current I is defined as the amount of charge per unit time that crosses the imaginary surface in Figure 20.3, in much the same sense that a river current is the amount of water flowing per unit of time. If the rate is constant, the current is

$$I = \frac{\Delta q}{\Delta t} \tag{20.1}$$

Figure 20.1 In an electric circuit, energy is transferred from a source (the battery pack) to a device (the cassette player) by charges that move through a conducting wire.

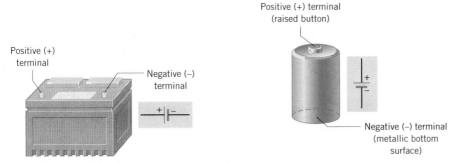

Figure 20.2 Typical batteries and the symbol $\left(\underset{+}{\vdash}\!\!\vdash\right)$ used to represent them in electric circuits.

Figure 20.3 The electric current is the amount of charge per unit time that passes through a surface that is perpendicular to the motion of the charges.

---

* The word "force" appears in this context for historical reasons, even though it is incorrect; electric potential is energy per unit charge, which is not force.

Concepts at a Glance

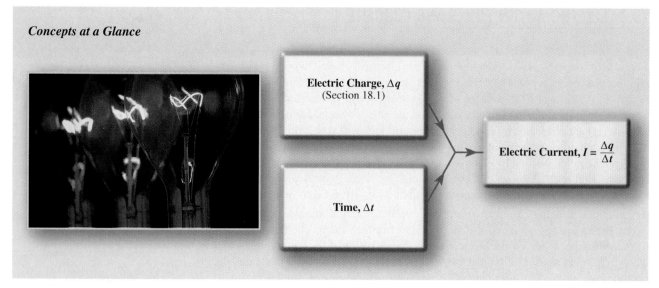

Figure 20.4 *Concepts at a Glance* The concept of electric current is formulated from the concept of electric charge and the time during which it flows in a circuit. The electric current in these incandescent light bulbs causes the filaments to glow.

If the rate of flow is not constant, then Equation 20.1 gives the average current. Since the units for charge and time are the coulomb (C) and the second (s), the SI unit for current is a coulomb per second (C/s). One coulomb per second is referred to as an ***ampere*** (A), after the French mathematician André-Marie Ampère (1775–1836).

If the charges move around a circuit in the same direction at all times, the current is said to be ***direct current (dc),*** which is the kind produced by batteries. In contrast, the current is said to be ***alternating current (ac)*** when the charges move first one way and then the opposite way, changing direction from moment to moment. Many energy sources produce alternating current, e.g., generators at power companies and microphones. Example 1 deals with direct current.

EXAMPLE 1   A Pocket Calculator

The current from the 3.0-V battery pack of a pocket calculator is 0.17 mA. In one hour of operation, (a) how much charge flows in the circuit and (b) how much energy does the battery deliver to the calculator circuit?

Reasoning Since current is defined as charge per unit time, the charge that flows in one hour is the product of the current and the time (3600 s). The charge that leaves the 3.0-V battery pack has 3.0 joules of energy per coulomb of charge. Thus, the total energy delivered to the calculator circuit is the charge (in coulombs) times the energy per unit charge (in volts or joules/coulomb).

Solution

(a) The charge that flows in one hour can be determined from Equation 20.1:

$$\Delta q = I(\Delta t) = (0.17 \times 10^{-3}\ \text{A})(3600\ \text{s}) = \boxed{0.61\ \text{C}}$$

(b) The energy delivered to the calculator circuit is

$$\text{Energy} = \text{Charge} \times \underbrace{\frac{\text{Energy}}{\text{Charge}}}_{\text{Battery emf}} = (0.61\ \text{C})(3.0\ \text{V}) = \boxed{1.8\ \text{J}}$$

Today, it is known that electrons flow in metal wires. Figure 20.5 shows the negative electrons emerging from the negative terminal of the battery and moving around the circuit toward the positive terminal. It is customary, however, *not* to use the flow of electrons when discussing circuits. Instead, a so-called **conventional current** is used, for reasons that date back to the time when it was believed that positive charges moved through metal wires. Conventional current is the hypothetical flow of positive charges that would have the same effect in the circuit as the movement of negative charges that actually does occur. In Figure 20.5, negative electrons leave the negative terminal of the battery, pass through the device, and arrive at the positive terminal. The same effect would have been achieved if an equivalent amount of positive charge had left the positive terminal, passed through the device, and arrived at the negative terminal. Therefore, the drawing shows the conventional current originating from the positive terminal. A conventional current of hypothetical positive charges is consistent with our earlier use of a positive test charge for defining electric fields and potentials. The direction of conventional current is always from a point of higher potential toward a point of lower potential, that is, from the positive toward the negative terminal. In this text, the symbol I stands for conventional current.

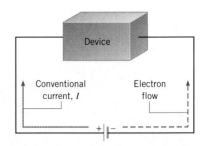

Figure 20.5 In a circuit, electrons actually flow through the metal wires. However, it is customary to use a conventional current I to describe the flow of charges.

20.2 OHM'S LAW

The current that a battery can push through a wire is analogous to the water flow that a pump can push through a pipe. Greater pump pressures lead to larger water flow rates, and, similarly, greater battery voltages* lead to larger electric currents. In the simplest case, the current I is directly proportional to the voltage V; that is, $I \propto V$. Thus, a 12-V battery leads to twice as much current as a 6-V battery, when each is connected to the same circuit.

   Concepts at a Glance In a water pipe, the flow rate is not only determined by the pump pressure but is also affected by the length and diameter of the pipe. Longer and narrower pipes offer higher resistance to the moving water and lead to smaller flow rates for a given pump pressure. A similar situation exists in electrical circuits, and to deal with it we introduce the concept of electrical resistance. As the Concepts-at-a-Glance chart in Figure 20.6 indicates, electrical resistance is defined in terms of two ideas that have already been discussed—the electric potential difference or voltage (see Section 19.2) and the electric current (see Section 20.1).

   The *resistance (R)* is defined as the ratio of the voltage V applied across a piece of material to the current I through the material, or $R = V/I$. When only a small current results from a large voltage, there is a high resistance to the moving charge. For many materials (e.g., metals), the ratio V/I is the same for a given piece of material over a wide range of voltages and currents. In such a case, the resistance is a constant. Then, as Figure 20.6 indicates, the relation $R = V/I$ is referred to as *Ohm's law*, after the German physicist Georg Simon Ohm (1789–1854), who discovered it.

---

* The potential difference between two points, such as the terminals of a battery, is commonly called the voltage between the points.

Concepts at a Glance

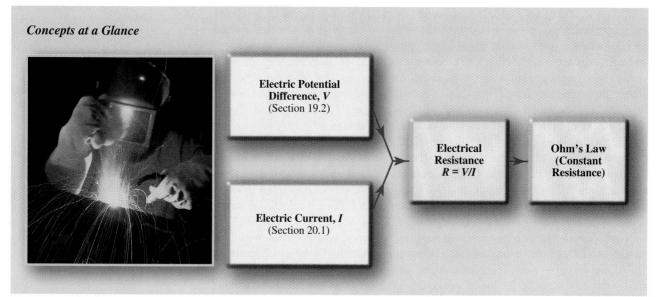

Figure 20.6 *Concepts at a Glance*
The concepts of electric potential difference or voltage *V* and the electric current *I* are brought together to define the electrical resistance *R*. Ohm's law specifies that the resistance remains constant as the voltage and current change. Electric arc welding sends sparks when a current is sent through the metals being welded together. Because of the resistance of the metals, the current generates sufficient heat to melt the metals and form the joints.

● OHM'S LAW

The ratio *V/I* is a constant, where *V* is the voltage applied across a piece of material (such as a wire) and *I* is the current through the material:

$$\frac{V}{I} = R = \text{constant} \quad \text{or} \quad V = IR \qquad (20.2)$$

R is the resistance of the piece of material.

SI Unit of Resistance: volt/ampere (V/A) = ohm (Ω)

The SI unit for resistance is a volt per ampere, which is called an *ohm* and is represented by the Greek capital letter omega (Ω). Ohm's law is not a fundamental law of nature like Newton's laws of motion. It is only a statement of the way certain materials behave in electric circuits.

To the extent that a wire or an electrical device offers resistance to the flow of charges, it is called a **resistor.** The resistance can have a wide range of values. The copper wires in a television set, for instance, have a very small resistance. On the other hand, commercial resistors can have resistances up to many kilohms (1 kΩ = 10^3 Ω) or megohms (1 MΩ = 10^6 Ω). Such resistors play an important role in electric circuits, where they are used to limit the amount of current and establish proper voltage levels.

In drawing electric circuits we follow the usual conventions: (1) a zigzag line (–⌇⌇⌇–) represents a resistor and (2) a straight line (———) represents an ideal conducting wire, or one with a negligible resistance. Example 2 illustrates an application of Ohm's law to the circuit in a flashlight.

EXAMPLE 2   A Flashlight

The filament in a light bulb is a resistor in the form of a thin piece of wire. The wire becomes hot enough to emit light because of the current in it. Figure 20.7 shows a flashlight that uses two 1.5-V batteries (effectively a single 3.0-V battery) to provide a current of 0.40 A in the filament. Determine the resistance of the glowing filament.

Reasoning The filament resistance is assumed to be the only resistance in the circuit. The potential difference applied across the filament is that of the 3.0-V battery. The resistance, given by Equation 20.2, is equal to this potential difference divided by the current.

Solution The resistance of the filament is

$$R = \frac{V}{I} = \frac{3.0 \text{ V}}{0.40 \text{ A}} = \boxed{7.5 \ \Omega} \qquad (20.2)$$

20.3 RESISTANCE AND RESISTIVITY

In a water pipe, the length and cross-sectional area of the pipe determine the resistance that the pipe offers to the flow of water. Longer pipes with smaller cross-sectional areas offer greater resistance. Analogous effects are found in the electrical case. For a wide range of materials, the resistance of a piece of material of length L and cross-sectional area A is

$$R = \rho \frac{L}{A} \qquad (20.3)$$

where ρ is a proportionality constant known as the ***resistivity*** of the material. It can be seen from Equation 20.3 that the unit for resistivity is the ohm·meter ($\Omega \cdot$m), and Table 20.1 lists values for various materials. All the conductors in the table are metals and have small resistivities. Insulators such as rubber have large resistivities. Materials like germanium and silicon have intermediate resistivity values and are, accordingly, called *semiconductors*.

Resistivity is an inherent property of a material, inherent in the same sense that density is an inherent property. Resistance, on the other hand, depends on both the resistivity and the geometry of the material. Thus, two wires can be made from copper, which has a resistivity of $1.72 \times 10^{-8} \ \Omega \cdot$m, but Equation 20.3 indicates that a short wire with a large cross-sectional area has a smaller resistance than does a long, thin wire. Wires that carry large currents, such as main power cables, are thick rather than thin so that the resistance of the wires is kept as small as possible. Simi-

Figure 20.7 The circuit in this flashlight consists of a resistor (the filament of the light bulb) connected to a 3.0-V battery.

TABLE 20.1 Resistivities[a] of Various Materials

Material	Resistivity ρ ($\Omega \cdot$m)	Material	Resistivity ρ ($\Omega \cdot$m)
Conductors		***Semiconductors***	
Aluminum	2.82×10^{-8}	Carbon	3.5×10^{-5}
Copper	1.72×10^{-8}	Germanium	0.5^{b}
Gold	2.44×10^{-8}	Silicon	$20-2300^{b}$
Iron	9.7×10^{-8}	***Insulators***	
Mercury	95.8×10^{-8}	Mica	$10^{11}-10^{15}$
Nichrome (alloy)	100×10^{-8}	Rubber (hard)	$10^{13}-10^{16}$
Silver	1.59×10^{-8}	Teflon	10^{16}
Tungsten	5.6×10^{-8}	Wood (maple)	3×10^{10}

[a] The values pertain to temperatures near 20 °C.
[b] Depending on purity.

larly, electric tools that are to be used far away from wall sockets require thicker extension cords, as Example 3 illustrates.

EXAMPLE 3   Longer Extension Cords

THE PHYSICS OF...

electrical extension cords.

The instructions for an electric lawn mower suggest that a 20-gauge extension cord can be used for distances up to 35 m, but a thicker 16-gauge cord should be used for longer distances, to keep the resistance of the wire as small as possible. The cross-sectional area of 20-gauge wire is $5.2 \times 10^{-7} \text{ m}^2$, while that of 16-gauge wire is $13 \times 10^{-7} \text{ m}^2$. Determine the resistance of (a) 35 m of 20-gauge copper wire and (b) 75 m of 16-gauge copper wire.

Reasoning According to Equation 20.3, the resistance of a copper wire depends on the resistivity of copper and the length and cross-sectional area of the wire. The resistivity can be obtained from Table 20.1, while the length and cross-sectional area are given in the problem statement.

Solution From Table 20.1 the resistivity of copper is $1.72 \times 10^{-8} \ \Omega \cdot \text{m}$. The resistance of the wires can be found using Equation 20.3:

20-gauge wire $\quad R = \dfrac{\rho L}{A} = \dfrac{(1.72 \times 10^{-8} \ \Omega \cdot \text{m})(35 \text{ m})}{5.2 \times 10^{-7} \text{ m}^2} = \boxed{1.2 \ \Omega}$

16-gauge wire $\quad R = \dfrac{\rho L}{A} = \dfrac{(1.72 \times 10^{-8} \ \Omega \cdot \text{m})(75 \text{ m})}{13 \times 10^{-7} \text{ m}^2} = \boxed{0.99 \ \Omega}$

Even though it is more than twice as long, the thicker 16-gauge wire has less resistance than the thinner 20-gauge wire. It is necessary to keep the resistance as low as possible in order to minimize heating of the wire, thereby reducing the possibility of a fire, as Conceptual Example 7 in Section 20.5 emphasizes.

THE PHYSICS OF...

impedance plethysmography.

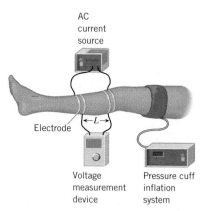

Figure 20.8 In the technique of impedance plethysmography, the electrical resistance of the calf is measured to diagnose deep venous thrombosis (blood clotting in the veins).

Equation 20.3 provides the basis for an important medical diagnostic technique known as impedance (or resistance) plethysmography. Figure 20.8 shows how the technique is applied to diagnose blood clotting in the veins (deep venous thrombosis) near the knee. A pressure cuff, like that used in blood pressure measurements, is placed around the midthigh, while electrodes are attached around the calf. The two outer electrodes are connected to a source of a small amount of ac current. The two inner electrodes are separated by a distance L, and the voltage between them is measured. The voltage divided by the current gives the resistance. The key to this technique is the fact that resistance can be related to the volume between the inner electrodes. The volume is the product of the length L and the cross-sectional area A of the calf, or $Volume = LA$. Solving for A and substituting in Equation 20.3 shows that

$$R = \rho \frac{L}{A} = \rho \frac{L}{Volume/L} = \rho \frac{L^2}{Volume}$$

Thus, resistance is inversely proportional to volume, a fact that is exploited in diagnosing deep venous thrombosis. Blood flows from the heart into the calf through arteries in the leg and returns through the system of veins. The pressure cuff in Figure 20.8 is inflated to the point where it cuts off the venous flow, but not the arterial flow. As a result, more blood enters than leaves the calf, the volume of the calf increases, and the electrical resistance decreases. When the cuff pressure is removed suddenly, the volume returns to a normal value, and so does the electrical resistance. With healthy (unclotted) veins, there is a rapid return to normal values. A slow return, however, reveals the presence of clotting.

The resistivity of a material depends on temperature. In metals, the resistivity increases with increasing temperature, whereas in semiconductors the reverse is true. For many materials and limited temperature ranges, it is possible to express the temperature dependence of the resistivity as follows:

$$\rho = \rho_0[1 + \alpha(T - T_0)] \qquad (20.4)$$

In this expression ρ and ρ_0 are the resistivities at temperatures T and T_0, respectively. The term α has the unit of reciprocal temperature and is the **temperature coefficient of resistivity**. When the resistivity increases with increasing temperature, α is positive, as it is for metals. When the resistivity decreases with increasing temperature, α is negative, as it is for the semiconductors carbon, germanium, and silicon. Since resistance is given by $R = \rho L/A$, both sides of Equation 20.4 can be multiplied by L/A to show that resistance depends on temperature according to

$$R = R_0[1 + \alpha(T - T_0)] \qquad (20.5)$$

The next example illustrates the role of the resistivity and its temperature coefficient in determining the electrical resistance of a piece of material.

EXAMPLE 4   The Heating Element of an Electric Stove

Figure 20.9*a* shows a cherry-red heating element on an electric stove. The element contains a wire (length = 1.1 m, cross-sectional area = 3.1×10^{-6} m^2) through which electric charge flows. As Figure 20.9*b* shows, this wire is embedded within an electrically insulating material that is contained within a metal casing. The wire becomes hot in response to the flowing charge and heats the casing. The material of the wire has a resistivity of $\rho_0 = 6.8 \times 10^{-5}\ \Omega \cdot$m at $T_0 = 320\ °$C and a temperature coefficient of resistivity of $\alpha = 2.0 \times 10^{-3}\ (C°)^{-1}$. Determine the resistance of the heater wire at an operating temperature of 420 °C.

Reasoning Equation 20.3 ($R = \rho L/A$) can be used to find the resistance of the wire at 420 °C, once the resistivity ρ is determined at this temperature. Since the resistivity at 320 °C is given, Equation 20.4 can be employed to find the resistivity at 420 °C.

Solution At the operating temperature of 420 °C, the material of the wire has a resistivity of

$$\rho = \rho_0[1 + \alpha(T - T_0)]$$
$$\rho = (6.8 \times 10^{-5}\ \Omega \cdot \text{m})[1 + (2.0 \times 10^{-3}\ (C°)^{-1})(420\ °\text{C} - 320\ °\text{C})]$$
$$= 8.2 \times 10^{-5}\ \Omega \cdot \text{m}$$

This value of the resistivity can be used along with the given length and cross-sectional area to find the resistance of the heater wire:

$$R = \frac{\rho L}{A} = \frac{(8.2 \times 10^{-5}\ \Omega \cdot \text{m})(1.1\ \text{m})}{3.1 \times 10^{-6}\ \text{m}^2} = \boxed{29\ \Omega} \qquad (20.3)$$

There is an important class of materials whose resistivity suddenly goes to zero below a certain temperature T_c called the *critical temperature*, commonly a few degrees above absolute zero. Below this temperature, such materials are called *superconductors*. The name derives from the fact that with zero resistivity, these materials offer no resistance to electric current and are, therefore, perfect conductors. One of the remarkable properties of zero resistivity is that once a current is established in a superconducting ring, it continues indefinitely without the need of an emf. Currents have persisted in superconductors for many years without measurable decay.

THE PHYSICS OF...

a heating element on an electric stove.

(a)

Heater wire
$(A = 3.1 \times 10^{-6}\ \text{m}^2)$

Metal casing

$L = 1.1$ m

(b)

Figure 20.9 A heating element from an electric stove.

In contrast, the current in a nonsuperconducting material drops to zero almost immediately after the emf is removed.

Many metals become superconductors only at very low temperatures, such as aluminum ($T_c = 1.18$ K), tin ($T_c = 3.72$ K), lead ($T_c = 7.20$ K), and niobium ($T_c = 9.25$ K). Materials involving copper oxide complexes have been made that undergo the transition to the superconducting state at 175 K. Certain polymers, with structural properties similar to DNA, have been fabricated whose transition temperature is in excess of 500 K, well above room temperature. Superconductors have many technological applications, including magnetic resonance imaging (Section 21.7), magnetic levitation of trains (Section 21.9), cheaper transmission of electrical power, powerful (yet small) electric motors, and faster computer chips.

20.4 ELECTRIC POWER

Suppose that an amount of charge Δq emerges from a battery in a time Δt and that the potential difference between the battery terminals is V. According to the definition of electric potential given in Equation 19.3, the energy of this charge is the product of the charge and the potential difference, or $(\Delta q)V$. Since the change in energy per unit time is the power P (see Equation 6.10b), the electrical power provided to the circuit is

$$P = \frac{(\Delta q)V}{\Delta t} = \underbrace{\frac{\Delta q}{\Delta t}}_{\text{Current } I} V$$

The term $\Delta q / \Delta t$ is the charge per unit time or the current I in the circuit. Thus, the electric power is the product of current and voltage.

> ● **ELECTRIC POWER**
>
> When there is a current I in a circuit as a result of a voltage V, the electric power P delivered to the circuit is
>
> $$P = IV \qquad (20.6)$$
>
> *SI Unit of Power:* watt (W)

Power is measured in watts, and Equation 20.6 indicates that the product of an ampere and a volt is equal to a watt.

Many electrical devices are essentially resistors that become hot when provided with sufficient electric power: toasters, irons, space heaters, heating elements on electric stoves, and incandescent light bulbs, to name a few. In such cases, it is possible to obtain two additional, but equivalent, expressions for the power. These two expressions follow directly upon substituting $V = IR$, or equivalently $I = V/R$, into the relation $P = IV$:

$$P = IV \qquad (20.6\text{a})$$

$$P = I(IR) = I^2R \qquad (20.6\text{b})$$

$$P = \left(\frac{V}{R}\right)V = \frac{V^2}{R} \qquad (20.6\text{c})$$

Example 5 deals with the electric power delivered to the bulb of a flashlight.

EXAMPLE 5   The Power and Energy Used in a Flashlight

In the flashlight in Figure 20.7, the current is 0.40 A, and the voltage is 3.0 V. Find (a) the power delivered to the bulb and (b) the energy dissipated in the bulb in 5.5 minutes of operation.

Reasoning The electrical power delivered to the bulb is the product of the current and voltage. Since power is energy per unit time, the energy delivered to the bulb is the product of the power and time.

Solution

(a) The power is
$$P = IV = (0.40 \text{ A})(3.0 \text{ V}) = \boxed{1.2 \text{ W}} \qquad (20.6a)$$

The "wattage" rating of this bulb would, therefore, be 1.2 W.

(b) The energy consumed in 5.5 minutes (330 s) follows from the definition of power as energy per unit time:
$$\text{Energy} = P \, \Delta t = (1.2 \text{ W})(330 \text{ s}) = \boxed{4.0 \times 10^2 \text{ J}}$$

Monthly electric bills specify the cost for the energy consumed during the month. Energy is the product of power and time, and electric companies compute your energy consumption by expressing power in kilowatts and time in hours. Therefore, a commonly used unit for energy is the *kilowatt-hour* (kWh). For instance, if you used an average power of 1440 watts (1.44 kW) for 30 days (720 h), your energy consumption would be (1.44 kW)(720 h) = 1040 kWh. At a cost of $0.10 per kWh, your monthly bill would be $104. As shown in Example 10 in Chapter 12, 1 kWh = 3.60×10^6 J of energy.

20.5   ALTERNATING CURRENT

Many electric circuits use batteries and involve direct current (dc). However, there are considerably more circuits that operate with alternating current (ac), in which the direction of charge flow reverses periodically. The common generators that create ac electricity depend on magnetic forces for their operation and are discussed in Chapter 22. In an ac circuit, these generators serve the same purpose as a battery serves in a dc circuit; that is, they give energy to the moving electric charges. This section deals with ac circuits that contain only resistance.

Since the electrical outlets in a house provide alternating current, we all use ac circuits routinely. Figure 20.10 shows the ac circuit that is formed when a toaster is plugged into a wall socket. The heating element of a toaster is essentially a thin wire of resistance R and becomes red hot when electrical energy is dissipated in it. The circuit schematic in the picture introduces the symbol $\ominus$ that is used to represent the generator. In this case, the generator is located at the electric power company.

Figure 20.11 shows a graph that records the voltage V produced between the terminals of the ac generator in Figure 20.10 at each moment of time. This is the most common type of ac voltage. It fluctuates sinusoidally between positive and negative values as a function of time t:
$$V = V_0 \sin 2\pi f t \qquad (20.7)$$

where V_0 is the maximum or peak value of the voltage, and f is the frequency (in cycles/s or Hz) at which the voltage oscillates. The angle $2\pi f t$ in Equation 20.7 is expressed in radians, so a calculator must be set to its radian mode before the sine

Figure 20.10 This circuit consists of a toaster (resistance = R) and an ac generator at the electric power company.

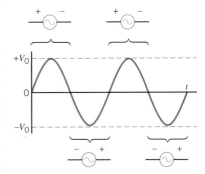

Figure 20.11 The voltage V produced between the terminals of an ac generator fluctuates sinusoidally in time, in the most common case. The relative polarity of the generator terminals during the positive and negative parts of the sinusoidal graph is indicated.

of this angle can be evaluated. In the United States, the voltage at most home wall outlets has a *peak value* of approximately $V_0 = 170$ volts and oscillates with a frequency of $f = 60$ Hz. Thus, the period of each cycle is $\frac{1}{60}$ s, and the polarity of the generator terminals reverses twice during each cycle, as Figure 20.11 indicates.

The current in an ac circuit also oscillates. In circuits that contain only resistance, the current reverses direction each time the polarity of the generator terminals reverses. Thus, the current in a circuit like that in Figure 20.10 would have a frequency of 60 Hz and would change direction twice during each cycle. Substituting $V = V_0 \sin 2\pi ft$ into $V = IR$ shows that the current can be represented as

$$I = \frac{V_0}{R} \sin 2\pi ft = I_0 \sin 2\pi ft \qquad (20.8)$$

The peak current is given by $I_0 = V_0/R$, so it can be determined if the peak voltage and the resistance are known.

The power delivered to an ac circuit by the generator is given by $P = IV$, just as it is in a dc circuit. However, since both I and V depend on time, the power fluctuates as time passes. Substituting Equations 20.7 and 20.8 for V and I into $P = IV$ gives

$$P = I_0 V_0 \sin^2 2\pi ft \qquad (20.9)$$

This expression is plotted in Figure 20.12.

Since the power fluctuates in an ac circuit, it is customary to consider the average power $\overline{P}$ which is one-half the peak power, as Figure 20.12 indicates:

$$\overline{P} = \tfrac{1}{2} I_0 V_0 \qquad (20.10)$$

On the basis of this expression, a kind of average current and average voltage can be introduced that are very useful when discussing ac circuits. A slight rearrangement of Equation 20.10 reveals that

$$\overline{P} = \left(\frac{I_0}{\sqrt{2}}\right)\left(\frac{V_0}{\sqrt{2}}\right) = I_{rms} V_{rms} \qquad (20.11)$$

I_{rms} and V_{rms} are called the **root mean square (rms)** current and voltage, respectively, and may be calculated from the peak values by dividing them by $\sqrt{2}*$:

$$I_{rms} = \frac{I_0}{\sqrt{2}} \qquad (20.12)$$

$$V_{rms} = \frac{V_0}{\sqrt{2}} \qquad (20.13)$$

Since the maximum ac voltage at a home wall socket in the United States is typically $V_0 = 170$ volts, the corresponding rms voltage is $V_{rms} = 170$ volts$/\sqrt{2} = 120$ volts. Thus, when the instructions for an electric appliance specify 120 V, it is an rms voltage that is meant. Similarly, when we specify an ac voltage or current in this text, it is an rms value, unless indicated otherwise. Likewise, when we specify ac power, it is an average power, unless stated otherwise.

Except for dealing with average quantities, the relation $\overline{P} = I_{rms} V_{rms}$ has the same form as Equation 20.6a ($P = IV$). Moreover, Ohm's law can be written conveniently in terms of rms quantities:

$$V_{rms} = I_{rms} R \qquad (20.14)$$

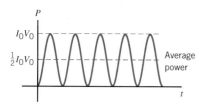

Figure 20.12 In an ac circuit, the power P delivered to a resistor oscillates between zero and a peak value of $I_0 V_0$, where I_0 and V_0 are the peak current and voltage, respectively.

---

* Equations 20.12 and 20.13 apply only for sinusoidal current and voltage.

Substituting Equation 20.14 into $\overline{P} = I_{rms}V_{rms}$ shows that the average power can be expressed in the following ways:

$$\overline{P} = I_{rms}V_{rms} \qquad (20.15a)$$

$$\overline{P} = I_{rms}^2 R \qquad (20.15b)$$

$$\overline{P} = \frac{V_{rms}^2}{R} \qquad (20.15c)$$

These expressions are completely analogous to $P = IV = I^2R = V^2/R$ for dc circuits. Example 6 deals with the average power in one familiar ac circuit.

EXAMPLE 6   Electrical Power Sent to a Loudspeaker

A stereo receiver applies a peak ac voltage of 34 V to a speaker. The speaker behaves approximately* as if it has a resistance of 8.0 Ω, as the circuit in Figure 20.13 indicates. Determine (a) the rms voltage, (b) the rms current, and (c) the average power for this circuit.

Reasoning The rms voltage is, by definition, equal to the peak voltage divided by $\sqrt{2}$. We can find the rms current in the circuit from Ohm's law, as the rms voltage divided by the resistance of the speaker. The average power for the circuit is the product of the rms current and the rms voltage.

Solution

(a) The peak value of the voltage is $V_0 = 34$ V, so the corresponding rms value is

$$V_{rms} = \frac{V_0}{\sqrt{2}} = \frac{34 \text{ V}}{\sqrt{2}} = \boxed{24 \text{ V}} \qquad (20.13)$$

(b) The rms current can be obtained from Ohm's law:

$$I_{rms} = \frac{V_{rms}}{R} = \frac{24 \text{ V}}{8.0 \text{ }\Omega} = \boxed{3.0 \text{ A}} \qquad (20.14)$$

(c) The average power is

$$\overline{P} = I_{rms}V_{rms} = (3.0 \text{ A})(24 \text{ V}) = \boxed{72 \text{ W}} \qquad (20.15a)$$

8.0-Ω speaker

Receiver

$V_0 = 34$ V $R = 8.0$ Ω

Figure 20.13 A receiver applies an ac voltage (peak value = 34 V) to an 8.0-Ω speaker.

● **PROBLEM SOLVING INSIGHT**
The rms values of the ac voltage and the ac current, V_{rms} and I_{rms}, respectively, are not the same as the peak values V_0 and I_0. The rms values are always smaller than the peak values by a factor of $\sqrt{2}$.

The electrical power dissipated in a resistor causes it to heat up. Excessive power can lead to a potential fire hazard, as Conceptual Example 7 discusses.

CONCEPTUAL EXAMPLE 7   Extension Cords and a Potential Fire Hazard

During the winter, many people use portable electric space heaters to keep warm. Sometimes, however, the heater must be located far from a 120-V wall receptacle, so an extension cord must be used (see Figure 20.14). However, manufacturers often warn against using an extension cord. Or, if one must be used, they recommend a certain wire gauge, or smaller. Why the warning, and why are smaller-gauge wires better than larger-gauge wires?

Reasoning and Solution An electric space heater contains a heater element that is a piece of wire of resistance R, which is heated to a high temperature. The heating occurs because of the power $I_{rms}^2 R$ dissipated in the heater element. A typical heater uses a relatively large current I_{rms} of about 12 A. On its way to the heater, this current

Figure 20.14 When an extension cord is used with a space heater, the cord must have a resistance that is sufficiently small to prevent overheating of the cord.

---

* Other factors besides resistance can affect the current and voltage in ac circuits; they are discussed in Chapter 23.

passes through the wires of any extension cord being used. Since these additional wires offer resistance to the current, the extension cord can also heat up, just as the heater element does. The extent of the heating depends on the resistance of the extension cord. If the heating becomes excessive, the insulation around the wire can melt, and a fire can result. Thus, *manufacturers of space heaters warn about extension cords in order to prevent a fire hazard.* To keep the heating of the extension cord to a safe level, the resistance of the wire must be kept small. Recall from Section 20.3 that the resistance of a wire depends inversely on its cross-sectional area, so a large cross-sectional area gives rise to a small resistance. A wire with a large cross-sectional area is one that has a small gauge number, as discussed in Example 3. *Smaller-gauge wires are better because they offer less resistance than larger-gauge wires.*

Related Homework Material: *Problem 34*

20.6 SERIES WIRING

Thus far, we have dealt with circuits that include only a single device, such as a light bulb or a loudspeaker. There are, however, many circuits in which more than one device is connected to a voltage source. This section introduces one method by which such connections may be made, namely, series wiring. *Series wiring means that the devices are connected in such a way that there is the same electric current through each device.* Figure 20.15 shows a circuit in which two different devices, represented by resistors R_1 and R_2, are connected in series with a battery. Note that if the current in one resistor is interrupted, the current in the other is too. This could occur, for example, if two light bulbs were connected in series and the filament of one bulb broke. Because of the series wiring, the voltage V supplied by the battery is divided between the two resistors. The drawing indicates that the portion of the voltage across R_1 is V_1, while the portion across R_2 is V_2, so $V = V_1 + V_2$. For each resistance, the definition of resistance is $R = V/I$, which may be solved for the corresponding voltage $V = IR$. Therefore, we have

$$V = V_1 + V_2 = IR_1 + IR_2 = I(R_1 + R_2) = IR_S$$

where R_S is called the *equivalent resistance* of the series circuit. Thus, two resistors in series are equivalent to a single resistor whose resistance is $R_S = R_1 + R_2$, in the sense that there is the same current through R_S as there is through the series combination of R_1 and R_2. This line of reasoning can be extended to any number of resistors in series, with the result that

Series resistors $R_S = R_1 + R_2 + R_3 + \cdots$ (20.16)

Example 8 illustrates the concept of equivalent resistance in a series circuit.

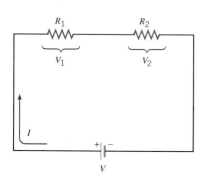

Figure 20.15   When two resistors are connected in series, the same current I is in both of them.

EXAMPLE 8   Resistors in a Series Circuit

A 6.00-Ω resistor and a 3.00-Ω resistor are connected in series with a 12.0-V battery, as Figure 20.16 indicates. Assuming that the battery contributes no resistance to the circuit, find (a) the current, (b) the power dissipated in each resistor, and (c) the total power delivered to the resistors by the battery.

Reasoning The current I can be determined from Ohm's law as $I = V/R_S$, where $R_S = R_1 + R_2$ is the equivalent resistance of the two resistors. The power delivered to each resistor is given by Equation 20.6b as $P = I^2 R$, where R is the resistance of

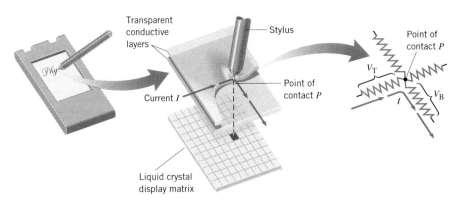

Figure 20.16 A 6.00-Ω and a 3.00-Ω resistor connected in series are equivalent to a single 9.00-Ω resistor.

the resistor being considered and I is the current through it. The total power delivered by the battery is the power delivered to the 6.00-Ω resistor plus the power delivered to the 3.00-Ω resistor.

Solution

(a) The equivalent resistance is

$$R_S = 6.00 \ \Omega + 3.00 \ \Omega = 9.00 \ \Omega \qquad (20.16)$$

Applying Ohm's law yields the current as

$$I = \frac{V}{R_S} = \frac{12.0 \ \text{V}}{9.00 \ \Omega} = \boxed{1.33 \ \text{A}} \qquad (20.2)$$

(b) Now that the current is known, the power dissipated in each resistor can be obtained from $P = I^2R$:

6.00-Ω resistor $P = I^2R = (1.33 \ \text{A})^2(6.00 \ \Omega) = \boxed{10.6 \ \text{W}}$ (20.6b)

3.00-Ω resistor $P = I^2R = (1.33 \ \text{A})^2(3.00 \ \Omega) = \boxed{5.31 \ \text{W}}$

(c) The total power delivered by the battery is the sum of the contributions in part (b): $P = 10.6 \ \text{W} + 5.31 \ \text{W} = 15.9 \ \text{W}$. Alternatively, the total power can be obtained directly by using the equivalent resistance $R_S = 9.00 \ \Omega$ and the current from part (a):

$$P = I^2R_S = (1.33 \ \text{A})^2(9.00 \ \Omega) = \boxed{15.9 \ \text{W}}$$

In general, the total power delivered to any number of resistors in series is equal to the power delivered to the equivalent resistor.

Pressure-sensitive pads form the heart of computer input devices that function as personal "electronic or digital assistants," and they offer an interesting application of series resistors. These devices are simple to use. You write directly on the pad with a plastic stylus that itself contains no electronics (see Figure 20.17). The writ-

THE PHYSICS OF...

personal electronic assistants.

Figure 20.17 The pressure pad on which the user writes in a personal "electronic or digital assistant" is based on the use of resistances that are connected in series.

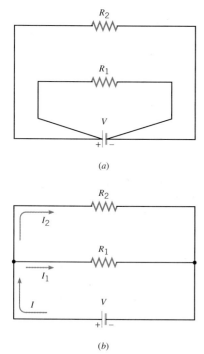

Figure 20.18 (*a*) When two resistors are connected in parallel, the same voltage *V* is applied across each resistor. (*b*) This circuit drawing is equivalent to that in part *a*. I_1 and I_2 are, respectively, the currents in R_1 and R_2.

Figure 20.19 This drawing shows some of the parallel connections found in a typical home. Each wall socket provides 120 V to the appliance connected to it. In addition, 120 V is applied to the light bulb when the switch is turned on.

ing appears as the stylus is moved, and recognition software interprets it as input for the built-in computer. The pad utilizes two transparent conductive layers that are separated by a small distance, except where pressure from the stylus brings them into contact (see point *P* in the drawing). Current *I* enters the positive side of the top layer, flows into the bottom layer through point *P*, and exits that layer through its negative side. Each layer provides resistance to the current, the amounts depending on where the point *P* is located. As the right side of the drawing indicates, the resistances from the layers are in series, since the same current flows through both of them. The voltage across the top-layer resistance is V_T, while the voltage across the bottom-layer resistance is V_B. In each case, the voltage is the current times the resistance. These two voltages are used to locate the point *P* and to activate (darken) one of the elements or pixels in a liquid crystal display matrix that lies beneath the transparent layers (see Section 24.6 for a discussion of liquid crystal displays). As the stylus writes, the writing becomes visible as one element after another in the display matrix is activated.

20.7 PARALLEL WIRING

Parallel wiring is another method of connecting electrical devices. *Parallel wiring means that the devices are connected in such a way that the same voltage is applied across each device.* Figure 20.18 shows two resistors connected in parallel between the terminals of a battery. Part *a* of the picture is drawn so as to emphasize that the entire voltage of the battery is applied across each resistor. Actually, parallel connections are rarely drawn in this manner; instead they are drawn as in part *b*, where the dots indicate the points where the wires for the two branches are joined together. Parts *a* and *b* are equivalent representations of the same circuit.

Parallel wiring is very common. For example, when an electrical appliance is plugged into a wall socket, the appliance is connected in parallel with other appliances, as in Figure 20.19, where the entire voltage of 120 V is applied across the television, the stereo, and the light bulb. The presence of the unused socket or other devices that are turned off does not affect the operation of those devices that are turned on. Moreover, if the current in one device is interrupted (perhaps by an opened switch or a broken wire), the current in the other devices is not interrupted. In contrast, if household appliances were connected in series, there would be no current through any appliance if the current in the circuit were halted at any point.

When two resistors R_1 and R_2 are connected as in Figure 20.18, each receives current from the battery as if the other were not present. Therefore, R_1 and R_2 together draw more current from the battery than does either resistor alone. According to the definition of resistance, $R = V/I$, a larger current implies a smaller resistance. Thus, the two parallel resistors behave as a single equivalent resistance that is *smaller* than either R_1 or R_2. Figure 20.20 returns to the water flow analogy to provide additional insight into this important feature of parallel wiring. In part *a*, two sections of pipe that have the same length are connected in parallel with a pump. In part *b* these two sections have been replaced with a single pipe of the same length, whose cross-sectional area equals the combined cross-sectional areas of section 1 and section 2. The pump can push more water per second through the wider pipe in part *b* than it can through *either* of the narrower pipes in part *a*. In effect, the wider pipe offers less resistance to the flow than either of the narrower pipes offers individually.

As in a series circuit, it is possible to replace a parallel combination of resistors with an equivalent resistor that results in the same total current and power for a

Figure 20.20 (*a*) Two equally long pipe sections, with cross-sectional areas A_1 and A_2, are connected in parallel to a water pump. (*b*) The two parallel pipe sections in part *a* are equivalent to a single pipe of the same length whose cross-sectional area is $A_1 + A_2$.

given voltage as the original combination. To determine the equivalent resistance for the two resistors in Figure 20.18, note that the total current I from the battery is the sum of I_1 and I_2, where I_1 is the current in resistor R_1 and I_2 is the current in resistor R_2: $I = I_1 + I_2$. Since the same voltage V is applied across each resistor, the definition of resistance indicates that $I_1 = V/R_1$ and $I_2 = V/R_2$. Therefore,

$$I = I_1 + I_2 = \frac{V}{R_1} + \frac{V}{R_2} = V\left(\frac{1}{R_1} + \frac{1}{R_2}\right) = V\left(\frac{1}{R_P}\right)$$

where R_P is the equivalent resistance. Hence, when two resistors are connected in parallel, they are equivalent to a single resistor whose resistance R_P can be obtained from $1/R_P = 1/R_1 + 1/R_2$. For any number of resistors in parallel, a similar line of reasoning reveals that

Parallel resistors
$$\frac{1}{R_P} = \frac{1}{R_1} + \frac{1}{R_2} + \frac{1}{R_3} + \cdots \tag{20.17}$$

The next example deals with a parallel combination of resistors that occurs in a stereo system.

EXAMPLE 9   Main and Remote Stereo Speakers

Most receivers allow the user to connect "remote" speakers (to play music in another room, for instance) in addition to the main speakers. Figure 20.21 shows that the remote speaker and the main speaker for the right stereo channel are connected to the receiver in parallel (for clarity, the speakers for the left channel are not shown). At the instant represented in the picture, the ac voltage across the speakers is 6.00 V. The main-speaker resistance is 8.00 Ω, and the remote-speaker resistance is 4.00 Ω.* Determine (a) the equivalent resistance of the two speakers, (b) the total current supplied by the receiver, (c) the current in each speaker, (d) the power dissipated in each speaker, and (e) the total power delivered by the receiver.

Reasoning The total current supplied to the two speakers by the receiver can be obtained as $I_{rms} = V_{rms}/R_P$, where R_P is the equivalent resistance of the two speakers in parallel and can be obtained from $1/R_P = 1/R_1 + 1/R_2$. The current in each speaker is different, however, since the speakers have different resistances. The average

THE PHYSICS OF...

main and remote stereo speakers.

---

* In reality, frequency-dependent characteristics of the speaker (see Chapter 23) play a role in the operation of a loudspeaker. We assume here, however, that the frequency of the sound is low enough that the speakers behave as pure resistances.

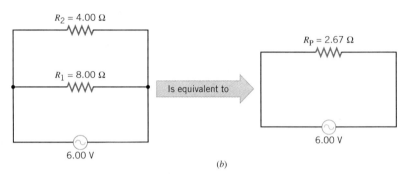

Figure 20.21   (a) The main and remote speakers in a stereo system are connected in parallel to the receiver. (b) The circuit schematic shows the situation when the ac voltage across the speakers is 6.00 V.

● **PROBLEM SOLVING INSIGHT**

The equivalent resistance R_P of a number of resistors in parallel has a reciprocal given by $R_P^{-1} = R_1^{-1} + R_2^{-1} + R_3^{-1} + \cdots$, where R_1, R_2, and R_3 are the individual resistances. After adding together the reciprocals R_1^{-1}, R_2^{-1}, and R_3^{-1}, do not forget to take the reciprocal of the result to find R_P.

power delivered to a given speaker is the product of its current and voltage. In the parallel connection the same voltage is applied to each speaker.

Solution

(a) According to Equation 20.17, the equivalent resistance of the two speakers is given by

$$\frac{1}{R_P} = \frac{1}{8.00\ \Omega} + \frac{1}{4.00\ \Omega} = \frac{3}{8.00\ \Omega} \quad \text{or} \quad R_P = \frac{8.00\ \Omega}{3} = \boxed{2.67\ \Omega}$$

This result is illustrated in part b of the drawing.

(b) Using the equivalent resistance in Ohm's law shows that the total current is

$$I_{\text{rms}} = \frac{V_{\text{rms}}}{R_P} = \frac{6.00\ \text{V}}{2.67\ \Omega} = \boxed{2.25\ \text{A}} \tag{20.14}$$

(c) Applying Ohm's law to each speaker gives the individual speaker currents:

8.00-Ω speaker $\quad I_{\text{rms}} = \dfrac{V_{\text{rms}}}{R} = \dfrac{6.00\ \text{V}}{8.00\ \Omega} = \boxed{0.750\ \text{A}}$

4.00-Ω speaker $\quad I_{\text{rms}} = \dfrac{V_{\text{rms}}}{R} = \dfrac{6.00\ \text{V}}{4.00\ \Omega} = \boxed{1.50\ \text{A}}$

The sum of these currents is equal to the total current obtained in part (b).

(d) The average power dissipated in each speaker can be calculated using $\overline{P} = I_{\text{rms}} V_{\text{rms}}$ with the individual currents obtained in part (c):

8.00-Ω speaker $\quad \overline{P} = (0.750\ \text{A})(6.00\ \text{V}) = \boxed{4.50\ \text{W}} \tag{20.15a}$

4.00-Ω speaker $\quad \overline{P} = (1.50\ \text{A})(6.00\ \text{V}) = \boxed{9.00\ \text{W}} \tag{20.15a}$

(e) The total power delivered by the receiver is the sum of the individual values found in part (d), $\overline{P} = 4.50\ \text{W} + 9.00\ \text{W} = 13.5\ \text{W}$. Alternatively, the total power can be obtained from the equivalent resistance $R_P = 2.67\ \Omega$ and the total current in part (b):

$$\overline{P} = I_{\text{rms}}^2 R_P = (2.25\ \text{A})^2 (2.67\ \Omega) = \boxed{13.5\ \text{W}} \tag{20.15b}$$

In a parallel combination of resistances, it is the *smallest* resistance that has the largest impact in determining the equivalent resistance. In fact, if one resistance approaches zero, then according to Equation 20.17, the equivalent resistance also approaches zero. In such a case, the near-zero resistance is said to *short out* the other resistances by providing a near-zero resistance path for the current to follow as a shortcut around the other resistances.

An interesting application of parallel wiring occurs in a 3-way light bulb, as Conceptual Example 10 discusses.

CONCEPTUAL EXAMPLE 10   A 3-Way Light Bulb and Parallel Wiring

Three-way light bulbs are popular because they can provide three levels of illumination (e.g., 50 W, 100 W, and 150 W) using a 120-V socket. The socket, however, must be equipped with a special 3-way switch that enables one to select the illumination level. This switch, however, does not select different voltages, for a 3-way bulb uses a single voltage of 120 V. Within the bulb are two separate filaments. When one of these burns out (i.e., vaporizes), the bulb can produce only one level of illumination (either its lowest or its intermediate level) but not the highest level. How are the two filaments connected, in series or in parallel? And how can two filaments be used to produce three different illumination levels?

Reasoning and Solution If the filaments were wired in series and one of them burned out, no current would pass through the bulb and none of the illumination levels would be available, contrary to what is observed. Therefore, ***the filaments must be connected in parallel,*** as in Figure 20.22.

As to how two filaments can be used to produce three illumination levels, we note that the power dissipated in a resistance R is $\overline{P} = V_{rms}^2/R$, according to Equation 20.15c. With a single value of 120 V for the voltage V_{rms}, three different wattage ratings for the bulb can be obtained only if three different values for the resistance R can occur. In a 50 W/100 W/150 W three-way bulb, for example, one resistance R_{50} is provided by the 50-W filament, and the second resistance R_{100} comes from the 100-W filament. The third resistance R_{150} is the parallel combination of the other two and can be obtained from $1/R_{150} = 1/R_{50} + 1/R_{100}$. Figure 20.22 illustrates a simplified version of how the 3-way switch operates to provide these three alternatives. In the first position of the switch, contact A is closed and contact B is open, so current passes only through the 50-W filament. In the second position, contact A is open and contact B is closed, so current passes only through the 100-W filament. In the third position, both contacts A and B are closed, and both filaments light up, giving 150-W of illumination.

Related Homework Material: *Problem 50*

THE PHYSICS OF...

a 3-way light bulb.

100-W filament
50-W filament

B
A

Simplified version of 3-way switch in lamp socket

Figure 20.22   A 3-way light bulb uses two filaments connected in parallel. The filaments can be turned on one at a time, or both together.

20.8 CIRCUITS WIRED PARTIALLY IN SERIES AND PARTIALLY IN PARALLEL

Often an electric circuit is wired partially in series and partially in parallel. The key to determining the current, voltage, and power in such a case is to deal with the circuit in parts, with the resistances in each part being either in series or in parallel with each other. Example 11 shows how this analysis is carried out.

EXAMPLE 11   A Four-Resistor Circuit

Figure 20.23 shows a circuit composed of a 24-V battery and four resistors, whose resistances are 110, 180, 220, and 250 Ω. Find (a) the total current supplied by the battery and (b) the voltage between points A and B in the circuit.

Figure 20.23   The four circuits shown in this picture are equivalent.

Reasoning   The total current supplied by the battery can be obtained from Ohm's law, $I = V/R$, where R is the equivalent resistance of the four resistors. The equivalent resistance can be calculated by dealing with the circuit in parts. The voltage V_{AB} between the points A and B is also given by Ohm's law, $V_{AB} = IR_{AB}$, where I is the current and R_{AB} is the equivalent resistance between the two points.

Solution

(a) The 220-Ω resistor and the 250-Ω resistor are in series, so they are equivalent to a single resistor whose resistance is 220 Ω + 250 Ω = 470 Ω (see Figure 20.23). The 470-Ω resistor is in parallel with the 180-Ω resistor. Their equivalent resistance can be obtained from Equation 20.17:

$$\frac{1}{R_{AB}} = \frac{1}{470 \ \Omega} + \frac{1}{180 \ \Omega} = 0.0077 \ \Omega^{-1} \quad \text{or} \quad R_{AB} = \frac{1}{0.0077 \ \Omega^{-1}} = 130 \ \Omega$$

The circuit is now equivalent to a circuit containing a 110-Ω resistor in series with a 130-Ω resistor (see the third circuit from the top in the drawing). This combination acts like a single resistor whose resistance is R = 110 Ω + 130 Ω = 240 Ω. The total current from the battery is, then,

$$I = \frac{V}{R} = \frac{24 \ \text{V}}{240 \ \Omega} = \boxed{0.10 \ \text{A}}$$

(b) The current I = 0.10 A passes through the resistance between points A and B. Therefore, Ohm's law indicates that the voltage across the 130-Ω resistor between points A and B is

$$V_{AB} = IR_{AB} = (0.10 \ \text{A})(130 \ \Omega) = \boxed{13 \ \text{V}}$$

20.9   INTERNAL RESISTANCE

So far, the circuits we have considered include batteries or generators that contribute only their emfs to a circuit. In reality, however, such devices also add some resistance. This resistance is called the *internal resistance* of the battery or generator because it is located inside the device. In a battery, the internal resistance is due to the chemicals within the battery. In a generator, the internal resistance is the resistance of wires and other components within the generator.

Figure 20.24 shows a schematic representation of the internal resistance r of a battery. The drawing emphasizes that when an external resistance R is connected to the battery, the resistance is connected *in series* with the internal resistance. The internal resistance of a functioning battery is typically small (several thousandths of

Figure 20.24   When an external resistance R is connected between the terminals of a battery, the resistance is connected in series with the internal resistance r of the battery.

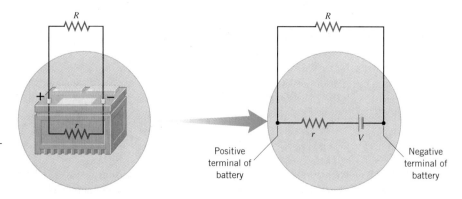

Positive terminal of battery

Negative terminal of battery

an ohm for a new car battery). Nevertheless, the effect of the internal resistance may not be negligible. Example 12 illustrates that when current is drawn from a battery, the internal resistance causes the voltage between the terminals to drop below the maximum value specified by the battery's emf. The actual voltage between the terminals of a battery is known as the *terminal voltage.*

EXAMPLE 12   The Terminal Voltage of a Battery

Figure 20.25 shows a car battery whose emf is 12.0 V and whose internal resistance is 0.010 Ω. This resistance is relatively large because the battery is old and the terminals are corroded. What is the terminal voltage when the current I drawn from the battery is (a) 10.0 A and (b) 100.0 A?

Reasoning The voltage between the terminals is not the entire 12.0-V emf, because part of the emf is needed to make the current go through the internal resistance. The amount of voltage needed can be determined from Ohm's law as the current I through the battery times the internal resistance r. For larger currents, a larger amount of voltage is needed, leaving less of the emf between the terminals.

Solution

(a) The voltage needed to make a current of $I = 10.0$ A go through an internal resistance of $r = 0.010$ Ω is $V = Ir = (10.0$ A$)(0.010$ $\Omega) = 0.10$ V. To find the terminal voltage, remember that the direction of conventional current is always from a higher toward a lower potential. To emphasize this fact in the drawing, plus and minus signs have been included at the right and left ends, respectively, of the resistance r. The terminal voltage can be calculated by starting at the negative terminal and keeping track of how the voltage increases and decreases as we move toward the positive terminal. The voltage rises by 12.0 V due to the battery's emf. However, the voltage drops by 0.10 V because of the potential difference across the internal resistance. Therefore, the terminal voltage is 12.0 V − 0.10 V = $\boxed{11.9 \text{ V}}$.

(b) When the current through the battery is 100.0 A, the voltage needed to make the current go through the internal resistance is $V = (100.0$ A$)(0.010$ $\Omega) = 1.0$ V. The terminal voltage now decreases to 12.0 V − 1.0 V = $\boxed{11.0 \text{ V}}$.

To car's electrical system
(ignition, lights, radio, etc.)

Positive   Negative
terminal   terminal

$r = 0.010$ Ω 12 V

Figure 20.25 A car battery whose emf is 12 V and whose internal resistance is r.

Example 12 indicates that the terminal voltage of a battery is smaller when the current drawn from the battery is larger, an effect that any car owner can demonstrate. Turn the headlights on before starting your car, so the current through the battery is about 10 A, as in part (a) of Example 12. Then start the car. The starter motor draws a large amount of additional current from the battery, momentarily increasing the total current by an appreciable amount. Consequently, the terminal voltage of the battery decreases, causing the headlights to dim.

THE PHYSICS OF...

automobile batteries.

20.10   KIRCHHOFF'S RULES

Electric circuits that contain a number of resistors can often be analyzed by combining individual groups of resistors in series and parallel, as Section 20.8 discusses. However, there are many circuits in which no two resistors are in series or in parallel. To deal with such circuits it is necessary to employ methods other than the series–parallel method. One alternative is to take advantage of Kirchhoff's rules, named after their developer Gustav Kirchhoff (1824–1887).

 ● *Concepts at a Glance* There are two rules, the *junction rule* and the *loop rule.* The Concepts-at-a-Glance chart in Figure 20.26 emphasizes that these rules

Concepts at a Glance

Figure 20.26 *Concepts at a Glance* Kirchhoff's rules are applications to electric circuits of the law of conservation of electric charge and the principle of conservation of energy. The currents and voltages in complex electrical circuits, such as this computer circuit, can be analyzed using Kirchhoff's rules.

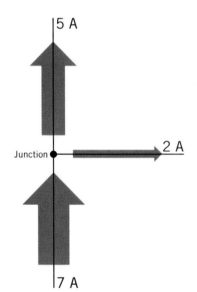

Figure 20.27 A junction is a point in a circuit where a number of wires are connected together. If 7 A of current is directed into the junction, then a total of 7 A (5 A + 2 A) of current must be directed out of the junction.

arise from principles and ideas that we have encountered earlier. The junction rule is an application of the law of conservation of electric charge (see Section 18.2) to the electric current in a circuit. The loop rule is an application of the principle of conservation of energy (see Section 6.8) to the electric potential (see Section 19.2) that exists at various places in a circuit. ●

Figure 20.27 illustrates in greater detail the basic idea behind Kirchhoff's junction rule. The picture shows a junction where several wires are connected together. As Section 18.2 discusses, electric charge is conserved. Therefore, since there is no continual accumulation of charges at the junction itself, the total charge per second flowing into the junction must equal the total charge per second flowing out of it. In other words, *the junction rule states that the total current directed into a junction must equal the total current directed out of the junction,* or 7 A = 5 A + 2 A for the specific case shown in the picture.

To help explain Kirchhoff's loop rule, Figure 20.28 shows a circuit in which a 12-V battery is connected to a series combination of a 5-Ω and a 1-Ω resistor. The plus and minus signs associated with each resistor remind us that, outside a battery, conventional current is directed from a higher toward a lower potential. From left to right, there is a potential drop of 10 V across the first resistor and another drop of 2 V across the second resistor. Keeping in mind that potential is the electric potential energy per unit charge, let us follow a positive test charge clockwise* around the circuit. Starting at the negative terminal of the battery, we see that the test charge gains

* The choice of the clockwise direction is arbitrary.

potential energy because of the 12-V rise in potential due to the battery. The test charge then loses potential energy because of the 10-V and 2-V drops in potential across the resistors, ultimately arriving back at the negative terminal. In traversing the closed circuit loop, the test charge is like a skier gaining gravitational potential energy in going up a hill on a chair lift and then losing it to friction in coming down and stopping. When the skier returns to the starting point, the gain equals the loss, so there is no net change in potential energy. Similarly, when the test charge arrives back at its starting point, there is no net change in electric potential energy, the gains matching the losses. *The loop rule expresses this example of energy conservation in terms of the electric potential and states that for a closed circuit loop, the total of all the potential rises (12 V) is the same as the total of all the potential drops (10 V + 2 V).*

Kirchhoff's rules can be applied to any circuit, even when the resistors are not in series or in parallel. The two rules are summarized below, and Examples 13 and 14 illustrate how to use them.

Figure 20.28 Following a positive test charge clockwise around the circuit, we see that the total voltage drop of 10 V + 2 V across the two resistors equals the voltage rise of 12 V provided by the battery. The plus and minus signs marking the ends of the resistors emphasize that, outside a battery, the conventional current of 2 A is directed from a higher potential (+) toward a lower potential (−).

● KIRCHHOFF'S RULES

Junction rule. The sum of the magnitudes of the currents directed into a junction equals the sum of the magnitudes of the currents directed out of the junction.

Loop rule. Around any closed circuit loop, the sum of the potential drops equals the sum of the potential rises.

EXAMPLE 13   Using Kirchhoff's Loop Rule

Figure 20.29 shows a circuit that contains two batteries and two resistors. Determine the current I in the circuit.

Reasoning The first step is to draw the current, which we have chosen to be clockwise around the circuit. This direction is **arbitrary**, and if it is incorrect, I will turn out to be negative.

The second step is to mark the resistors with plus and minus signs, which serve as an aid in identifying the potential drops and rises for Kirchhoff's loop rule. *Remember that, outside a battery, conventional current is always directed from a higher potential (+) toward a lower potential (−).* Thus, we *must* mark the resistors as indicated in Figure 20.29, to be consistent with the clockwise direction chosen for the current.

We may now apply Kirchhoff's loop rule to the circuit, starting at point A, proceeding clockwise around the loop, and identifying the potential drops and rises as we go. The potential across each resistor is given by Ohm's law as $V = IR$. The clockwise direction is arbitrary, and the same result is obtained with a counterclockwise path.

Figure 20.29 A single-loop circuit that contains two batteries and two resistors.

Solution Starting at corner A and moving clockwise around the loop, there is

1. A potential drop (+ to −) of $IR = I(12\ \Omega)$ across the 12-Ω resistor
2. A potential drop (+ to −) of 6.0 V across the 6.0-V battery
3. A potential drop (+ to −) of $IR = I(8.0\ \Omega)$ across the 8.0-Ω resistor
4. A potential rise (− to +) of 24 V across the 24-V battery

Setting the sum of the potential drops equal to the sum of the potential rises, as required by Kirchhoff's loop rule, gives

$$\underbrace{I(12\ \Omega) + 6.0\ \text{V} + I(8.0\ \Omega)}_{\text{Potential drops}} = \underbrace{24\ \text{V}}_{\text{Potential rises}}$$

INTERACTIVE LEARNINGWARE #41

START

Visualization

Reasoning

Knowns and
Unknowns

Modeling
the Problem

Numerical
Solution

FINISH

The circuit in the drawing contains three
resistors and two batteries. Find the currents in
the 3.0-Ω resistor and the 4.0-Ω resistor.

This Interactive LearningWare example
further reviews all the important features
in using Kirchhoff's rules. It shows how
to apply the junction and loop rules, how
to mark the resistors with the appropriate + and − symbols, and illustrates what happens when the current is initially chosen to be in the wrong direction. Finally, the example emphasizes that it does not matter how one moves around a circuit, clockwise or counterclockwise, when evaluating the potential rises and drops for the loop rule.

Go to our Web site at

http://www.wiley.com/college/cutnell

THE PHYSICS OF...

an automobile electrical system.

• **PROBLEM SOLVING INSIGHT**
In problems involving Kirchhoff's rules, it is always helpful to mark the resistors with plus and minus signs to keep track of the potential rises and drops in the circuit.

Solving this equation for the current yields $\boxed{I = 0.90 \text{ A}}$. The current is a positive number, indicating that our initial choice for the direction (clockwise) of the current was correct.

EXAMPLE 14   The Electrical System of a Car

In a car, the headlights are connected to the battery and would discharge the battery if it were not for the alternator, which is run by the engine. Figure 20.30 indicates how the car battery, headlights, and alternator are connected. The circuit includes an internal resistance of 0.0100 Ω for the car battery and its leads and a resistance of 1.20 Ω for the headlights. For the sake of simplicity, the alternator is approximated as an additional 14.00-V battery with an internal resistance of 0.100 Ω. Determine the currents through the car battery (I_{B}), the headlights (I_{H}), and the alternator (I_{A}).

Reasoning The drawing shows the directions chosen arbitrarily for the currents I_{B}, I_{H}, and I_{A}. If any direction is incorrect, the analysis will reveal a negative value for the corresponding current.

Next, we mark the resistors with the plus and minus signs that serve as an aid in identifying the potential drops and rises for the loop rule, recalling that, outside a battery, conventional current is always directed from a higher potential (+) toward a lower potential (−). Thus, given the directions selected for I_{B}, I_{H}, and I_{A}, the plus and minus signs *must* be those indicated in Figure 20.30.

Kirchhoff's junction and loop rules can now be used.

Solution The junction rule can be applied to junction *B* or junction *E*. In either case, the same equation results:

*Junction rule
applied at B*

$$\underbrace{I_{\text{A}} + I_{\text{B}}}_{\substack{\text{Into} \\ \text{junction}}} = \underbrace{I_{\text{H}}}_{\substack{\text{Out of} \\ \text{junction}}}$$

In applying the loop rule to the lower loop *BEFA*, we start at point *B*, move clockwise around the loop, and identify the potential drops and rises. There is a potential

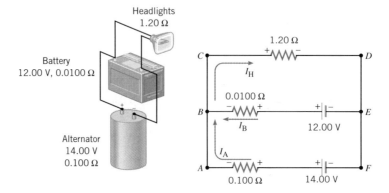

Figure 20.30   A circuit showing the headlight(s), battery, and alternator of a car.

rise ($-$ to $+$) of I_B (0.0100 Ω) across the 0.0100-Ω resistor and then a drop ($+$ to $-$) of 12.00 V due to the car battery. Continuing around the loop, we find a 14.00-V rise ($-$ to $+$) across the alternator, followed by a potential drop ($+$ to $-$) of I_A (0.100 Ω) across the 0.100-Ω resistor. Setting the sum of the potential drops equal to the sum of the potential rises gives the following result:

Loop rule: BEFA $\underbrace{I_A(0.100\ \Omega)\ +\ 12.00\ \text{V}}_{\text{Potential drops}}\ =\ \underbrace{I_B(0.0100\ \Omega)\ +\ 14.00\ \text{V}}_{\text{Potential rises}}$

Since there are three unknown variables in this problem, I_B, I_H, and I_A, a third equation is needed for a solution. To obtain the third equation, we apply the loop rule to the upper loop *CDEB*, choosing a clockwise path for convenience. The result is

Loop rule: CDEB $\underbrace{I_B(0.0100\ \Omega)\ +\ I_H(1.20\ \Omega)}_{\substack{\text{Potential} \\ \text{drops}}}\ =\ \underbrace{12.00\ \text{V}}_{\substack{\text{Potential} \\ \text{rises}}}$

These three equations can be solved simultaneously to show that

$$\boxed{I_B\ =\ -9.0\ \text{A},\ I_H\ =\ 10.1\ \text{A},\ I_A\ =\ 19.1\ \text{A}}.$$

The negative answer for I_B indicates that the current through the battery is not directed from right to left, as drawn in Figure 20.30. Instead, the 9.0-A current is directed from left to right, opposite to the way current would be directed if the alternator were not connected. It is the left-to-right current created by the alternator that keeps the battery charged.

20.11 THE MEASUREMENT OF CURRENT AND VOLTAGE

Current and voltage can be measured with devices known, respectively, as ammeters and voltmeters. There are two types of such devices: those that use digital electronics and those that do not. The essential feature of nondigital devices is the dc *galvanometer*. As Figure 20.31*a* illustrates, a galvanometer consists of a magnet, a coil of wire, a spring, a pointer, and a calibrated scale. The coil is mounted in such a way that it can rotate, which causes the pointer to move in relation to the scale. As Section 21.6 will discuss, the coil rotates in response to the torque applied by the magnet when there is a current in the coil. The coil stops rotating when this torque is balanced by the torque of the spring.

A galvanometer has two important characteristics that must be considered when it is used as part of a measurement device. First, the amount of dc current that causes a full-scale deflection of the pointer indicates the sensitivity of the galvanometer. For instance, Figure 20.31*a* shows an instrument that deflects full scale when the current in the coil is 0.10 mA. The second important characteristic is the resistance R_C of the wire in the coil. Figure 20.31*b* shows how a galvanometer with a coil resistance of R_C is represented in a circuit diagram.

Since an **ammeter** is an instrument that measures current, it must be inserted in the circuit so the current passes directly through it, as Figure 20.32 shows. An ammeter includes a galvanometer and one or more *shunt resistors*, which are connected in parallel with the galvanometer and provide a bypass for current in excess of the galvanometer's full-scale limit. The bypass allows the ammeter to be used to measure a current exceeding the full-scale limit. In Figure 20.33, for instance, a current of 60.0 mA enters terminal *A* of an ammeter, which uses a galvanometer with a full-scale current of 0.100 mA. The shunt resistor *R* can be selected so that

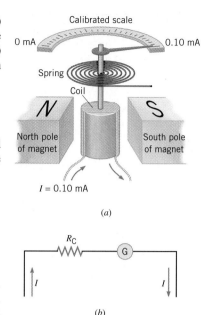

(*a*)

(*b*)

Figure 20.31 (*a*) The essential parts of a dc galvanometer. The coil of wire and pointer rotate when there is a current in the wire. (*b*) This picture shows how a galvanometer with a coil resistance of R_C is represented in a circuit diagram.

THE PHYSICS OF...

an ammeter.

Figure 20.32 An ammeter must be inserted into a circuit so that the current passes directly through it.

Ammeter

Figure 20.33 If a galvanometer with a full-scale limit of 0.100 mA is to be used to measure a current of 60.0 mA, a shunt resistance R must be used, so the excess current of 59.9 mA can detour around the galvanometer coil.

THE PHYSICS OF...

a voltmeter.

0.100 mA of current enters the galvanometer, while 59.9 mA bypasses it. In such a case, the measurement scale on the ammeter would be labeled 0 to 60.0 mA. To determine the shunt resistance, it is necessary to know the coil resistance R_C (see problem 80).

When an ammeter is inserted into a circuit, the equivalent resistance of the coil and the shunt resistor adds to the circuit resistance. Any increase in circuit resistance causes a reduction in current, and this is a problem, for an ammeter should only measure the current, not change it. Therefore, an *ideal* ammeter would have zero resistance. In practice, a good ammeter is designed with a sufficiently small equivalent resistance, so there is only a negligible reduction of the current in the circuit when the ammeter is inserted.

A ***voltmeter*** is an instrument that measures the voltage between two points, A and B, in a circuit. Figure 20.34 shows that the voltmeter must be connected between the points and is *not* inserted into the circuit as an ammeter is. A voltmeter includes a galvanometer whose scale is calibrated in volts. Suppose, for instance, that the galvanometer in Figure 20.35 has a full-scale current of 0.1 mA and a coil resistance of 50 Ω. Under full-scale conditions, the voltage across the coil would be $V = IR_C = (0.1 \times 10^{-3}\text{ A})(50\ \Omega) = 0.005$ V. Thus, this galvanometer could be used to register voltages in the range $0 - 0.005$ V. A voltmeter, then, is a galvanometer used in this fashion, along with some provision for adjusting the range of voltages to be measured. To adjust the range, an additional resistance R can be connected in series with the coil resistance R_C. The value of R needed to extend the measurement range to a full-scale voltage V can be determined from the full-scale current I by applying Ohm's law in the form $V = I(R + R_C)$ (see problem 81).

Ideally, the voltage registered by a voltmeter should be the same as the voltage that exists when the voltmeter is not connected. However, a voltmeter takes some current from a circuit and, thus, alters the circuit voltage to some extent. An *ideal* voltmeter would have infinite resistance and draw away only an infinitesimal amount of current. In reality, a good voltmeter is designed with a resistance that is large enough so the unit does not appreciably alter the voltage in the circuit to which it is connected.

Figure 20.34 To measure the voltage between two points, A and B, in a circuit, a voltmeter is connected between the points.

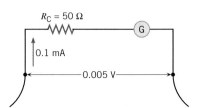

Figure 20.35 The galvanometer shown has a full-scale deflection of 0.1 mA and a coil resistance of 50 Ω.

20.12 CAPACITORS IN SERIES AND PARALLEL

Figure 20.36 shows two different capacitors connected in parallel to a battery. Since the capacitors are in parallel, they have the same voltage V across their plates. However, the capacitors *contain different amounts of charge*. The charge stored by a capacitor is $q = CV$ (Equation 19.8), so $q_1 = C_1V$ and $q_2 = C_2V$.

As with resistors, it is always possible to replace a parallel combination of capacitors with an *equivalent capacitor* that stores the same charge and energy for a given voltage as the combination does. To determine the equivalent capacitance C_P, note that the total charge q stored by the two capacitors is

$$q = q_1 + q_2 = C_1V + C_2V = (C_1 + C_2)V = C_PV$$

This result indicates that two capacitors in parallel can be replaced by an equivalent capacitor whose capacitance is $C_P = C_1 + C_2$. For any number of capacitors in parallel, the equivalent capacitance is

Parallel capacitors $C_P = C_1 + C_2 + C_3 + \cdots$ (20.18)

Capacitances in parallel simply add together to give an equivalent capacitance. This behavior contrasts with that of resistors in parallel, which combine as reciprocals, according to Equation 20.17.

The equivalent capacitor not only stores the same amount of charge as the parallel combination of capacitors, but also stores the same amount of energy. For instance, the energy stored in a single capacitor is $\frac{1}{2}CV^2$ (Equation 19.11), so the total energy stored by two capacitors in parallel is

$$\text{Total energy} = \tfrac{1}{2}C_1V^2 + \tfrac{1}{2}C_2V^2 = \tfrac{1}{2}(C_1 + C_2)V^2 = \tfrac{1}{2}C_PV^2$$

which is equal to the energy stored in the equivalent capacitor C_P.

When capacitors are connected in series, the equivalent capacitance is different than when they are in parallel. As an example, Figure 20.37 shows two capacitors in series and reveals the following important fact: **All capacitors in series, regardless of their capacitances, contain charges of the same magnitude, $+q$ and $-q$, on their plates.** The battery places a charge of $+q$ on plate a of capacitor C_1, and this charge induces a charge of $+q$ to depart from the opposite plate a', leaving behind a charge $-q$. The $+q$ charge that leaves plate a' is deposited on plate b of ca-

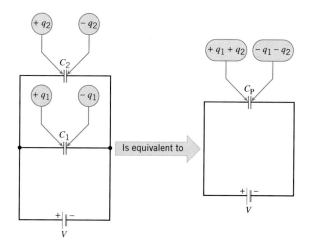

Figure 20.36 In a parallel combination of capacitances C_1 and C_2, the voltage V across each capacitor is the same, but the charges q_1 and q_2 on each capacitor are different.

Figure 20.37   In a series combination of capacitances C_1 and C_2, the same amount of charge q is on the plates of each capacitor, but the voltages V_1 and V_2 across each capacitor are different.

pacitor C_2 (since these two plates are connected by a wire), where it induces a $+q$ charge to move away from the opposite plate b', leaving behind a charge of $-q$. Thus, all capacitors in series contain charges of the same magnitude on their plates. Note the difference between charging capacitors in parallel and in series. When charging parallel capacitors, the battery moves a charge q that is the sum of the charges moved for each of the capacitors: $q = q_1 + q_2 + q_3 + \cdots$. In contrast, when charging a series combination of n capacitors, the battery only moves a charge q, not nq, because the charge q passes by induction from one capacitor directly to the next one in line.

The equivalent capacitance C_S for the series connection in Figure 20.37 can be determined by observing that the battery voltage V is shared by the two capacitors. The drawing indicates that the voltages across C_1 and C_2 are V_1 and V_2, so that $V = V_1 + V_2$. Since the voltages across the capacitors are $V_1 = q/C_1$ and $V_2 = q/C_2$, we find that

$$V = V_1 + V_2 = \frac{q}{C_1} + \frac{q}{C_2} = q\left(\frac{1}{C_1} + \frac{1}{C_2}\right) = q\left(\frac{1}{C_S}\right)$$

Thus, two capacitors in series can be replaced by an equivalent capacitor whose capacitance C_S can be obtained from $1/C_S = 1/C_1 + 1/C_2$. For any number of capacitors connected in series the equivalent capacitance is given by

Series capacitors
$$\frac{1}{C_S} = \frac{1}{C_1} + \frac{1}{C_2} + \frac{1}{C_3} + \cdots \tag{20.19}$$

Equation 20.19 indicates that capacitances in series combine as reciprocals and do not simply add together as resistors in series do. It is left as an exercise (problem 91) to show that the equivalent series capacitance stores the same electrostatic energy as the sum of the energies of the individual capacitors.

It is possible to simplify circuits containing a number of capacitors in the same general fashion as that outlined for resistors in Example 11 and Figure 20.23. The capacitors in a parallel grouping can be combined according to Equation 20.18, and those in a series grouping can be combined according to Equation 20.19.

20.13   RC CIRCUITS

Many electric circuits contain both resistors and capacitors. Figure 20.38 illustrates an example of a resistor–capacitor or *RC* circuit. Part *a* of the drawing shows the circuit at a time *t* after the switch has been closed and the battery

Figure 20.38   Charging a capacitor.

has begun to charge up the capacitor plates. The charge on the plates builds up gradually to its equilibrium value of $q_0 = CV_0$, where V_0 is the voltage of the battery. Assuming that the capacitor is uncharged at time $t = 0$ s when the switch is closed, it can be shown that the magnitude q of the charge on the plates at time t is

Capacitor charging $q = q_0[1 - e^{-t/(RC)}]$ (20.20)

where the exponential e has the value of 2.718. . . . Part *b* of the drawing shows a graph of this expression, which indicates that the charge is $q = 0$ C when $t = 0$ s and increases gradually toward the equilibrium value of $q_0 = CV_0$. The voltage V across the capacitor at any time can be obtained from Equation 20.20 by dividing the charges q and q_0 by the capacitance C, since $V = q/C$ and $V_0 = q_0/C$.

The term RC in the exponent in Equation 20.20 is called the **time constant** τ of the circuit:

$$\tau = RC \qquad (20.21)$$

The time constant is measured in seconds; verification of the fact that an ohm times a farad is equivalent to a second is left as an exercise (see conceptual question 16). The time constant is the amount of time required for the capacitor to accumulate 63.2% of its equilibrium charge, as can be seen by substituting $t = \tau = RC$ in Equation 20.20; $q_0(1 - e^{-1}) = q_0(0.632)$. The charge approaches its equilibrium value rapidly when the time constant is small and slowly when the time constant is large.

Figure 20.39 shows a circuit at a time t after a switch is closed to allow a charged capacitor to begin discharging. There is no battery in this circuit, so the charge $+q$ on the left plate of the capacitor can flow counterclockwise through the resistor and neutralize the charge $-q$ on the right plate. Assuming that the capacitor has a charge q_0 at time $t = 0$ s when the switch is closed, it can be shown that

Capacitor discharging $q = q_0 e^{-t/(RC)}$ (20.22)

where q is the amount of charge remaining on either plate at time t. The graph of this expression in part *b* of the drawing shows that the charge begins at q_0 when $t = 0$ s and decreases gradually toward zero. Smaller values of the time constant RC lead to a more rapid discharge. Equation 20.22 indicates that when $t = \tau = RC$, the magnitude of the charge remaining on each plate is $q_0 e^{-1} = q_0(0.368)$. Therefore, the time constant is also the amount of time required for a charged capacitor to *lose* 63.2% of its charge.

The charging/discharging of a capacitor has many applications. Heart pacemakers, for instance, incorporate RC circuits to control the timing of voltage pulses that are delivered to a malfunctioning heart to regulate its beating cycle. The pulses are delivered by electrodes attached externally to the chest or located internally near the heart when the pacemaker is implanted surgically (see Figure 20.40). A voltage pulse is delivered when the capacitor discharges to a preset level, following which the capacitor is recharged rapidly and the cycle repeats. The value of the time constant RC controls the pulsing rate, which is about once per second.

The charging/discharging of a capacitor is also used in automobiles that have windshield wipers equipped for intermittent operation during a light drizzle. In this mode of operation, the wipers remain off for a while and then turn on briefly. The timing of the on–off cycle is determined by the time constant of a resistor–capacitor combination.

(a)

(b)

Figure 20.39   Discharging a capacitor.

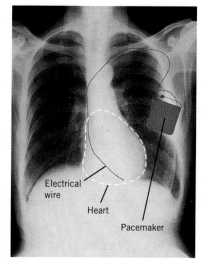

Figure 20.40   This X-ray photograph shows a heart pacemaker that has been implanted surgically.

THE PHYSICS OF...

heart pacemakers.

THE PHYSICS OF...

windshield wipers.

Figure 20.41 (*a*) A normally operating clothes drier that is connected to a wall socket via a two-prong plug. (*b*) The drier malfunctions, because an internal wire accidentally touches the metal casing. A person who touches the casing can receive an electrical shock.

20.14 SAFETY AND THE PHYSIOLOGICAL EFFECTS OF CURRENT

Electric circuits, while very useful, can also be hazardous. To reduce the danger inherent in using circuits, proper *electrical grounding* is necessary. The next two figures help to illustrate what electrical grounding means and how it is achieved.

Figure 20.41*a* shows a clothes drier connected to a wall socket via a two-prong plug. The drier is operating normally; that is, the wires inside are insulated from the metal casing of the drier, so no charge flows through the casing itself. Notice that one terminal of the ac generator is customarily connected to ground (⏚) by the electric power company. Part *b* of the drawing shows the hazardous result that occurs if a wire comes loose and contacts the metal casing. A person touching it receives a shock, since electric charge flows through the casing, the person's body, and the ground on the way back to the generator.

Figure 20.42 shows the same appliance connected to a wall socket via a three-prong plug that provides safe electrical grounding. The third prong connects the metal casing directly to a copper rod driven into the ground or to a copper water pipe that is in the ground. This arrangement protects against electrical shock in the event that a broken wire touches the metal casing, for charge would flow through the casing, through the third prong, and into the ground, returning eventually to the generator. No charge would flow through the person's body, since the copper rod provides much less electrical resistance than does the body.

Serious and sometimes fatal injuries can result from electric shock. The severity of the injury depends on the magnitude of the current and the parts of the body through which the moving charges pass. The amount of current that causes a mild tingling sensation is about 0.001 A. Currents on the order of 0.01 – 0.02 A can lead to muscle spasms, in which a person "can't let go" of the object causing the shock. Currents of approximately 0.2 A are potentially fatal because they can make the heart fibrillate, or beat in an uncontrolled manner. Substantially larger currents stop the heart completely. However, since the heart often begins beating normally again

THE PHYSICS OF...

safe electrical grounding.

THE PHYSICS OF...

the physiological effects of current.

Figure 20.42 When the drier malfunctions, a person touching it receives no shock, since electric charge flows through the third prong and into the ground via a copper rod, rather than through the person's body.

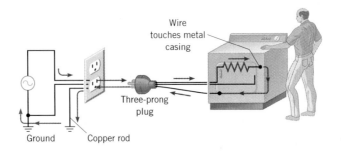

after the current ceases, the larger currents can be less dangerous than the smaller currents that cause fibrillation.

20.15 CONCEPTS & CALCULATIONS

Series and parallel wiring are two common ways in which devices, such as light bulbs, can be connected to a circuit. The next example reviews the concepts of voltage, current, resistance, and power in the context of these two types of circuits.

CONCEPTS & CALCULATIONS EXAMPLE 15
The Brightness of Light Bulbs Wired in Series and in Parallel

A circuit contains a 48-V battery and a single light bulb whose resistance is 240 Ω. A second, identical, light bulb can be wired either in series or parallel with the first one (see Figure 20.43). Determine the power in a single bulb when the circuit contains (a) only one bulb, (b) two bulbs in series, and (c) two bulbs in parallel. Assume that the battery has no internal resistance.

Concept Questions and Answers How is the power P consumed by a light bulb related to its resistance R and the voltage V across it?

Answer   Equation 20.6c gives the answer as $P = V^2/R$.

When there is only one light bulb in the circuit, what is the voltage V?

Answer   Since there is only one bulb in the circuit, V is the voltage of the battery, $V = 48$ V.

The more power dissipated in a light bulb, the brighter it is. When two bulbs are wired in series, does the brightness of each bulb increase, decrease, or remain the same relative to the brightness when only a single bulb is in the circuit?

Answer   Since the bulbs are identical and are wired in series, each receives one-half the battery voltage V. The power consumed by each bulb, then, is

$$P = \frac{(\frac{1}{2}V)^2}{R} = \frac{1}{4}\left(\frac{V^2}{R}\right)$$

This result shows that the power dissipated in each bulb in the series circuit is only one-fourth the power dissipated in the single-bulb circuit. Thus, the brightness of each bulb decreases.

When two bulbs are wired in parallel, does the brightness of each bulb increase, decrease, or remain the same relative to the brightness when only a single bulb is in the circuit?

Answer   Since the bulbs are wired in parallel, each receives the full battery voltage V. Thus, the power consumed by each bulb remains the same as when only one bulb is in the circuit, so the brightness of each bulb does not change.

Solution

(a) When only one bulb is in the circuit, the power it consumes is

$$P = \frac{V^2}{R} = \frac{(48 \text{ V})^2}{240 \ \Omega} = \boxed{96 \text{ W}} \qquad (20.6c)$$

(b) When the two bulbs are wired in series, the current in the circuit is given by Equation 20.2 as $I = V/R_S$, where $R_S = R + R$ is the equivalent resistance of the

Figure 20.43   Two identical light bulbs are wired in series and in parallel.

series circuit. The power consumed by one of the light bulbs can be expressed using Equation 20.6b:

$$P = I^2R = \left(\frac{V}{R+R}\right)^2 R = \frac{V^2}{4R} = \frac{(48\ \text{V})^2}{4(240\ \Omega)} = \boxed{24\ \text{W}}$$

As expected, the power dissipated in each bulb is only one-fourth that when there is only one bulb in the circuit.

(c) When the two bulbs are wired in parallel, the voltage across each is the same as the voltage of the battery. Therefore, the power dissipated in each bulb is given by

$$P = \frac{V^2}{R} = \frac{(48\ \text{V})^2}{240\ \Omega} = \boxed{96\ \text{W}} \qquad (20.6\text{c})$$

As expected, the power dissipated, and hence the brightness, does not change relative to that in the single-bulb circuit.

Kirchhoff's junction rule and loop rule are important tools for analyzing the currents and voltages in complex circuits. The rules are easy to use, once some of the subtleties are understood. This example explores these subtleties in a two-loop circuit.

CONCEPTS & CALCULATIONS EXAMPLE 16   Using Kirchhoff's Rules

For the circuit in Figure 20.44, use Kirchhoff's junction rule and loop rule to find the currents through the three resistors.

Concept Questions and Answers Notice that there are two loops, labeled 1 and 2, in this circuit. Does it matter that there is no battery in loop 1, but only two resistors?

> *Answer* No, it doesn't matter. A loop can have any number of batteries, including none at all.

The currents through the three resistors are labeled as I_1, I_2 and I_3. Does it matter which direction, left-to-right or right-to-left, has been chosen for each current?

> *Answer* No. If we initially select the wrong direction for a current, there is no problem. The value obtained for that particular current will turn out to be a negative number, indicating that the actual current is in the opposite direction.

When we place the $+$ and $-$ signs at the ends of each resistor, does it matter which end is $+$ and which is $-$?

> *Answer* Yes it does. Once the direction of the current has been selected, the $+$ and $-$ signs must be chosen so that the current goes from the $+$ end toward the $-$ end of the resistor. Notice that this is the case for the three resistors in Figure 20.44.

When we evaluate the potential drops and rises around a closed loop, does it matter which direction, clockwise or counterclockwise, is chosen for the evaluation?

> *Answer* No, the direction is arbitrary. If we choose a clockwise direction, for example, we will have a certain number of potential drops and rises. If we choose a counterclockwise direction, all the drops become rises, and vice versa. Since we always set the potential drops equal to the potential rises, it does not matter which direction is picked for evaluating them. In the example that follows, we will use a counterclockwise direction for loop 1 and a clockwise direction for loop 2.

Figure 20.44 The three currents in this circuit are evaluated using Kirchhoff's loop and junction rules.

Solution Let's begin by expressing the current I_3 in terms of I_1 and I_2. Notice that I_1 and I_2 flow into junction c and I_3 flows out of it. Kirchhoff's junction rule states that

$$\underbrace{I_1 + I_2}_{\substack{\text{Currents} \\ \text{into junction}}} = \underbrace{I_3}_{\substack{\text{Currents} \\ \text{out of junction}}} \qquad (20.23)$$

We now apply the loop rule to loop 1, starting at point a and proceeding counterclockwise around the closed circuit ($abcdea$):

$$\underbrace{I_2(6.0\ \Omega)}_{\text{Potential drops}} = \underbrace{I_1(3.0\ \Omega)}_{\text{Potential rises}} \qquad (20.24)$$

The two equations above contain three unknown variables, so we now apply the loop rule to loop 2 to obtain another equation that contains these variables. Starting at point a, we will move clockwise around loop 2 ($abcfgha$):

$$\underbrace{I_2(6.0\ \Omega) + I_3(4.0\ \Omega)}_{\text{Potential drops}} = \underbrace{6.0\ \text{V}}_{\text{Potential rises}} \qquad (20.25)$$

Substituting Equation 20.23 for I_3 into Equation 20.25 gives

$$I_2(6.0\ \Omega) + (I_1 + I_2)(4.0\ \Omega) = 6.0\ \text{V} \qquad (20.26)$$

Solving Equation 20.24 for I_2 in terms of I_1, substituting the result into Equation 20.26, and solving the resulting equation for I_1 yields $I_1 = \boxed{\frac{2}{3}\ \text{A}}$. Substituting this result back into Equation 20.24 and solving for I_2 gives $I_2 = \boxed{\frac{1}{3}\ \text{A}}$. Equation 20.23 indicates, then, that $I_3 = I_1 + I_2 = \boxed{1.0\ \text{A}}$.

At the end of the problem set for this chapter, you will find homework problems that contain both conceptual and quantitative parts. These problems are grouped under the heading *Concepts & Calculations, Group Learning Problems*. They are designed for use by students working alone or in small learning groups. The conceptual part of each problem provides a convenient focus for group discussions.

SUMMARY

There must be at least one source or generator of electrical energy in an electric circuit. The **electromotive force (emf)** of a generator, such as a battery, is the maximum potential difference (in volts) that exists between the terminals of the generator.

The rate of flow of charge is called the **electric current.** If the rate is constant, the current I is given by $I = (\Delta q)/(\Delta t)$, where Δq is the magnitude of the charge crossing a surface in a time Δt, the surface being perpendicular to the motion of the charge. The SI unit for current is the coulomb per second (C/s), which is referred to as an ampere (A). When the charges flow only in one direction around a circuit, the current is called **direct current (dc).** When the direction of charge flow changes from moment to moment, the current is known as **alternating current (ac).**

The definition of **electrical resistance** is $R = V/I$, where V is the voltage applied across a piece of material and I is the current through the material. If the ratio V/I is constant for all values of V and I, the relation $R = V/I$ or $V = IR$ is referred to as **Ohm's law.** Resistance is measured in volts per ampere, a unit called an ohm (Ω).

The resistance of a piece of material of length L and cross-sectional area A is $R = \rho L/A$, where ρ is the **resistivity** of the material. The resistivity of a material depends on the temperature. For many materials and limited temperature ranges, the temperature dependence is given by $\rho = \rho_0[1 + \alpha(T - T_0)]$, where ρ and ρ_0 are the resistivities at temperatures T and T_0, respectively, and α is the **temperature coefficient of resistivity.**

In a circuit in which a current I results from a voltage V, the **electric power** delivered to the circuit is $P = IV$. Since $V = IR$, the power dissipated in a resistance R is also given by $P = I^2R$ or $P = V^2/R$.

The **alternating voltage between the terminals of an ac generator** can be represented by $V = V_0 \sin 2\pi ft$, where V_0 is the peak value of the voltage, t is the time, and f is the frequency at which the voltage oscillates. Correspondingly, in a circuit containing only resistance, the **ac current** is $I = I_0 \sin 2\pi ft$, where I_0 is the peak value of the current and is related to the peak voltage via $I_0 = V_0/R$. The **root mean square (rms)** voltage and current are related to the peak values according to $V_{rms} = V_0/\sqrt{2}$ and $I_{rms} = I_0/\sqrt{2}$. The power in an ac circuit is the product of the current and the voltage and oscillates in time. The **average power** is $\overline{P} = I_{rms}V_{rms}$. Since $V_{rms} = I_{rms}R$, the average power can also be written as $\overline{P} = I_{rms}^2 R$ or $\overline{P} = V_{rms}^2/R$.

When devices are connected **in series**, there is the same electric current through each device. The **equivalent resistance R_S of a series combination of resistances** (R_1, R_2, R_3, etc.) is $R_S = R_1 + R_2 + R_3 + \cdots$. The equivalent resistance dissipates the same total power as the series combination.

When devices are connected **in parallel,** the same voltage is applied across each device. In general, devices wired in parallel carry different currents. The **reciprocal of the equivalent resistance R_P of a parallel combination of resistances** is $1/R_P = 1/R_1 + 1/R_2 + 1/R_3 + \cdots$. The equivalent resistance dissipates the same total power as the parallel combination.

The **internal resistance** of a battery or generator is the resistance within the battery or generator. The **terminal voltage** is the voltage between the terminals of a battery or generator and is equal to the emf only when there is no current through the device. When there is a current I, the internal resistance r causes the terminal voltage to be less than the emf by an amount Ir.

Kirchhoff's junction rule states that the sum of the magnitudes of the currents directed into a junction equals the sum of the magnitudes of the currents directed out of the junction. **Kirchhoff's loop rule** states that, around any closed circuit loop, the sum of the potential drops equals the sum of the potential rises.

A **galvanometer** is a device that responds to electric current and is used in nondigital ammeters and voltmeters. An **ammeter** is an instrument that measures current and must be inserted into a circuit so the current passes directly through the ammeter. A **voltmeter** is an instrument for measuring the voltage between two points in a circuit. A voltmeter must be connected between the two points and is not inserted into a circuit as an ammeter is.

The **equivalent capacitance C_P for a parallel combination of capacitances** (C_1, C_2, C_3, etc.) is $C_P = C_1 + C_2 + C_3 + \cdots$. In general, each capacitor in a parallel combination carries a different amount of charge. The equivalent capacitor carries the same total charge and stores the same total energy as the parallel combination.

The **reciprocal of the equivalent capacitance C_S for a series combination of capacitances** is $1/C_S = 1/C_1 + 1/C_2 + 1/C_3 + \cdots$. Each capacitor in the combination carries the same amount of charge. The equivalent capacitor carries the same amount of charge as *any one* of the capacitors in the combination and stores the same total energy as the entire combination.

The **charging or discharging of a capacitor** in a dc series circuit (resistance R, capacitance C) does not occur instantaneously. Charge builds up gradually according to the relation $q = q_0[1 - e^{-t/(RC)}]$, where q is the charge on the capacitor at time t, q_0 is the equilibrium value of the charge, and RC is the **time constant** of the circuit. The discharging of a capacitor through a resistor is described by $q = q_0 e^{-t/(RC)}$, where q_0 is the charge on the capacitor at time $t = 0$ s.

CONCEPTUAL QUESTIONS

1. The drawing shows a circuit in which a light bulb is connected to the household ac voltage via two switches S_1 and S_2. This is the kind of wiring, for example, that allows you to turn a carport light on and off from either inside the house or out in the carport. Explain which position A or B of S_2 turns the light on when S_1 is set to (a) position A and (b) position B.

2. When an incandescent light bulb is turned on, the tungsten filament becomes white hot. The temperature coefficient of resistivity for tungsten is a positive number. What happens to the power delivered to the bulb as the filament heats up? Does the power increase, remain the same, or decrease? Justify your answer.

3. Two materials have different resistivities. Two wires of the same length are made, one from each of the materials. Is it possible for each wire to have the same resistance? Explain.

4. Does the resistance of a copper wire increase or decrease when both the length and the diameter of the wire are doubled? Justify your answer.

5. One electrical appliance operates with a voltage of 120 V, while another operates with 240 V. Based on this information alone, is it correct to say that the second appliance uses more power than the first? Give your reasoning.

6. Two light bulbs are designed for use at 120 V and are rated at 75 W and 150 W. Which light bulb has the greater filament resistance? Why?

7. Often, the instructions for an electrical appliance do not state how many watts of power the appliance uses. Instead, a statement such as "10 A, 120 V" is given. Explain why this statement is equivalent to telling you the power consumption.

8. The drawing shows a circuit that includes a bimetallic strip (made from brass and steel, see Section 12.4) with a resistance heater wire wrapped around it.

When the switch is initially closed, a current appears in the circuit, because charges flow through the heater wire (which becomes hot), the strip itself, the contact point, and the light bulb. The bulb glows in response. As long as the switch remains closed, does the bulb continue to glow, eventually turn off permanently, or flash on and off? Account for your answer.

9. The power rating of a 1000-W heater specifies the power consumed when the heater is connected to an ac voltage of 120 V. Explain why the power consumed by two of these heaters connected in series with a voltage of 120 V is not 2000 W.

10. A number of light bulbs are to be connected to a single electrical outlet. Will the bulbs provide more brightness if they are connected in series or in parallel? Why?

11. A car has two headlights. The filament of one burns out. However, the other headlight stays on. Draw a circuit diagram that shows how the lights are connected to the battery. Give your reasoning.

12. In one of the circuits in the drawing, none of the resistors is in series or in parallel. Which is it? Explain.

(a)   (b)   (c)

13. You have four identical resistors, each with a resistance of R. You are asked to connect these four together so that the equivalent resistance of the resulting combination is R. How many ways can you do it? There is more than one way. Justify your answers.

14. Compare the resistance of an ideal ammeter with the resistance of an ideal voltmeter and explain why the resistances are so different.

15. Describe what would happen to the current in a circuit if a voltmeter, inadvertently mistaken for an ammeter, were inserted into the circuit.

16. The time constant of a series RC circuit is $\tau = RC$. Verify that an ohm times a farad is equivalent to a second.

PROBLEMS

Note: For problems that involve ac conditions, the current and voltage are rms values and the power is an average value, unless indicated otherwise.

Section 20.1 Electromotive Force and Current,
Section 20.2 Ohm's Law

1. ssm A portable compact disc player is designed to play for 2.0 h on a fully charged battery pack. If the battery pack provides a total of 180 C of charge, how much current does the player use in operating?

2. A defibrillator is used during a heart attack to restore the heart to its normal beating pattern (see Section 19.5). A defibrillator passes 18 A of current through the torso of a person in 2.0 ms. (a) How much charge moves during this time? (b) How many electrons pass through the wires connected to the patient?

3. The filament of a light bulb has a resistance of 580 Ω. A voltage of 120 V is connected across the filament. How much current is in the filament?

4. A toaster has a resistance of 14 Ω and is plugged into a 120-V outlet. What is the current in the toaster?

5. ssm The heating element of a clothes drier has a resistance of 11 Ω and is connected across a 240-V electrical outlet. What is the current in the heating element?

6. A battery charger is connected to a dead battery and delivers a current of 6.0 A for 5.0 hours, keeping the voltage across the battery terminals at 12 V in the process. How much energy is delivered to the battery?

*__7.__ The resistance of a bagel toaster is 14 Ω. To prepare a bagel, the toaster is operated for one minute from a 120-V outlet. How much energy is delivered to the toaster?

*__8.__ A car battery has a rating of 220 ampere · hours (A · h). This rating is one indication of the *total charge* that the battery can provide to a circuit before failing. (a) What is the total charge (in coulombs) that this battery can provide? (b) Determine the maximum current that the battery can provide for 38 minutes.

**__9. ssm__ A beam of protons is moving toward a target in a particle accelerator. This beam constitutes a current whose value is

0.50 μA. (a) How many protons strike the target in 15 seconds? (b) Each proton has a kinetic energy of 4.9×10^{-12} J. Suppose the target is a 15-gram block of aluminum, and all the kinetic energy of the protons goes into heating it up. What is the change in temperature of the block at the end of 15 s?

Section 20.3 Resistance and Resistivity

10. High-voltage power lines are a familiar sight throughout the country. The aluminum wire used for some of these lines has a cross-sectional area of 4.9×10^{-4} m². What is the resistance of ten kilometers of this wire?

11. ssm www Two wires have the same length and the same resistance. One is made from aluminum and the other from copper. Obtain the ratio of the cross-sectional area of the aluminum wire to that of the copper wire.

12. A coil of wire has a resistance of 38.0 Ω at 25 °C and 43.7 Ω at 55 °C. What is the temperature coefficient of resistivity?

13. A cylindrical copper cable carries a current of 1200 A. There is a potential difference of 1.6×10^{-2} V between two points on the cable that are 0.24 m apart. What is the radius of the cable?

14. A cylindrical wire of length 2.80 m and radius 1.03 mm carries a current of 1.35 A. The voltage across the ends of the wire is 0.0320 V. From what material in Table 20.1 is the wire made?

15. A wire of unknown composition has a resistance of $R_0 = 35.0 \ \Omega$ when immersed in water at 20.0 °C. When the wire is placed in boiling water, its resistance rises to 47.6 Ω. What is the temperature of a hot summer day when the wire has a resistance of 37.8 Ω?

* **16.** A tungsten wire has a radius of 0.075 mm and is heated from 20.0 to 1320 °C. The temperature coefficient of resistivity is $\alpha = 4.5 \times 10^{-3}$ (C°)$^{-1}$. When 120 V is applied across the ends of the hot wire, a current of 1.5 A is produced. How long is the wire? Neglect any effects due to thermal expansion.

* **17. ssm www** A wire has a resistance of 21.0 Ω. It is melted down, and from the same volume of metal a new wire is made that is three times longer than the original wire. What is the resistance of the new wire?

* **18.** A toaster uses a Nichrome heating wire. When the toaster is turned on at 20 °C, the initial current is 1.50 A. A few seconds later, the toaster warms up and the current has a value of 1.30 A. The average temperature coefficient of resistivity for Nichrome wire is 4.5×10^{-4} (C°)$^{-1}$. What is the temperature of the heating wire?

* **19. ssm www** Two wires have the same cross-sectional area and are joined end to end to form a single wire. One is tungsten, which has a temperature coefficient of resistivity of $\alpha = 0.0045$ (C°)$^{-1}$. The other is carbon, for which $\alpha = -0.0005$ (C°)$^{-1}$. The total resistance of the composite wire is the sum of the resistances of the pieces. The total resistance of the composite does *not change with temperature*. What is the ratio of the lengths of the tungsten and carbon sections? Ignore any changes in length due to thermal expansion.

** **20.** An aluminum wire is hung between two towers and has a length of 175 m. A current of 125 A exists in the wire, and the

potential difference between the ends of the wire is 0.300 V. The density of aluminum is 2700 kg/m³. Find the mass of the wire.

Section 20.4 Electric Power

21. An automobile battery is being charged at a voltage of 12.0 V and a current of 19.0 A. How much power is being produced by the charger?

22. A cigarette lighter in a car is a resistor that, when activated, is connected across the 12-V battery. Suppose a lighter dissipates 33 W of power. Find (a) the resistance of the lighter and (b) the current that the battery delivers to the lighter.

23. ssm In doing a load of clothes, a clothes drier uses 16 A of current at 240 V for 45 min. A personal computer, in contrast, uses 2.7 A of current at 120 V. With the energy used by the clothes drier, how long (in hours) could you use this computer to "surf" the internet?

24. There are approximately 110 million TVs in the United States. Each uses, on average, 75 W of power and is turned on for 6.0 hours a day. If electrical energy costs $0.10 per KWh, how much money is spent every day in keeping the TVs turned on?

25. An electric blanket is connected to a 120-V outlet and consumes 140 W of power. What is the current in the wire in the blanket?

26. A blow-drier and a vacuum cleaner each operate with an ac voltage of 120 V. The current rating of the blow-drier is 11 A, while that of the vacuum cleaner is 4.0 A. Determine the power consumed by (a) the blow-drier and (b) the vacuum cleaner. (c) Determine the ratio of the energy used by the blow-drier in 15 minutes to the energy used by the vacuum cleaner in one-half an hour.

* **27. ssm** Tungsten has a temperature coefficient of resistivity of 0.0045 (C°)$^{-1}$. A tungsten wire is connected to a source of constant voltage via a switch. At the instant the switch is closed, the temperature of the wire is 28 °C, and the initial power dissipated in the wire is P_0. At what wire temperature has the power dissipated in the wire decreased to $\frac{1}{2}P_0$?

* **28.** A piece of Nichrome wire has a radius of 6.5×10^{-4} m. It is used in a laboratory to make a heater that dissipates 4.00×10^2 W of power when connected to a voltage source of 120 V. Ignoring the effect of temperature on resistance, estimate the necessary length of wire.

** **29.** An iron wire has a resistance of 12 Ω at 20.0 °C and a mass of 1.3×10^{-3} kg. A current of 0.10 A is sent through the wire for one minute and causes the wire to become hot. Assuming that all the electrical energy is dissipated in the wire and remains there, find the final temperature of the wire. (*Hint: Use the average resistance of the wire during the heating process, and see Table 12.2 for the specific heat capacity of iron. Note $\alpha = 0.0050$ (C°)$^{-1}$.*)

Section 20.5 Alternating Current

30. The average power dissipated in a stereo speaker is 55 W. Assuming that the speaker can be treated as a 4.0-Ω resistance, find the peak value of the ac voltage applied to the speaker.

31. The current in a circuit is ac and has a peak value of 2.50 A. Determine the rms current.

32. A light bulb is connected to a 120.0-V wall socket. The current in the bulb depends on the time t according to the relation $I = (0.707 \text{ A})\sin [(314 \text{ Hz})t]$, where the peak current is 0.707 A. (a) What is the frequency of the alternating current? (b) Determine the resistance of the bulb's filament. (c) What is the average power consumed by the light bulb?

33. ssm The heating element in an iron has a resistance of 16 Ω and is connected to a 120-V wall socket. (a) What is the average power consumed by the iron, and (b) the peak power?

34. Review Conceptual Example 7 as an aid in solving this problem. A portable electric heater uses 18 A of current. The manufacturer recommends that an extension cord attached to the heater dissipates no more than 2.0 W of power per meter of length. What is the smallest radius of copper wire that can be used in the extension cord? *(Note: An extension cord contains two wires.)*

35. An electric furnace runs nine hours a day to heat a house during January (31 days). The heating element has a resistance of 5.3 Ω and carries a current of 25 A. The cost of electricity is $0.10/kWh. Find the monthly cost of running the furnace.

*36. On its highest setting, a heating element on an electric stove (see Figure 20.9) is connected to an ac voltage of 240 V. This element has a resistance of 29 Ω. (a) Find the power dissipated in the element. (b) Assuming that three-fourths of the heat produced by the element is used to heat a pot of water (the rest being wasted), find the time required to bring 1.9 kg of water (half a gallon) at 15 °C to a boil.

*37. ssm The *recovery time* of a hot water heater is the time required to heat all the water in the unit to the desired temperature. Suppose that a 52-gal (1.00 gal = 3.79×10^{-3} m^3) unit starts with cold water at 11 °C and delivers hot water at 53 °C. The unit is electric and utilizes a resistance heater (120 V ac, 3.0 Ω) to heat the water. Assuming that no heat is lost to the environment, determine the recovery time (in hours) of the unit.

**38. To save on heating costs, the owner of a green house keeps 660 kg of water around in barrels. During a winter day, the water is heated by the sun to 10.0 °C. During the night the water freezes into ice at 0.0 °C in nine hours. What is the minimum ampere rating of an electric heating system (240 V) that would provide the same heating effect as the water does?

Section 20.6 Series Wiring

39. ssm Three resistors, 25, 45, and 75 Ω, are connected in series, and a 0.51-A current passes through them. What is (a) the equivalent resistance and (b) the potential difference across the three resistors?

40. The current in a 47-Ω resistor is 0.12 A. This resistor is in series with a 28-Ω resistor, and the series combination is connected across a battery. What is the battery voltage?

41. A 36.0-Ω resistor and an 18.0-Ω resistor are connected in series across a 15.0-V battery. What is the voltage across (a) the 36.0-Ω resistor and (b) the 18.0-Ω resistor?

42. A 60.0-W lamp is placed in series with a resistor and a 120.0-V source. If the voltage across the lamp is 25 V, what is the resistance R of the resistor?

43. ssm The current in a series circuit is 15.0 A. When an additional 8.00-Ω resistor is inserted in series, the current drops to 12.0 A. What is the resistance in the original circuit?

44. Three resistors, 9.0, 5.0, and 1.0 Ω, are connected in series across a 24-V battery. Find (a) the current in, (b) the voltage across, and (c) the power dissipated in each resistor.

*45. Three resistors are connected in series across a battery. The value of each resistance and its maximum power rating are as follows: 5.0 Ω and 20.0 W, 30.0 Ω and 10.0 W, and 15.0 Ω and 10.0 W. (a) What is the greatest voltage that the battery can have without one of the resistors burning up? (b) How much power does the battery deliver to the circuit in (a)?

*46. One heater uses 340 W of power when connected by itself to a battery. Another heater uses 240 W of power when connected by itself to the same battery. How much total power do the heaters use when they are both connected in series across the battery?

**47. Two resistances, R_1 and R_2, are connected in series across a 12-V battery. The current increases by 0.20 A when R_2 is removed, leaving R_1 connected across the battery. However, the current increases by just 0.10 A when R_1 is removed, leaving R_2 connected across the battery. Find (a) R_1 and (b) R_2.

Section 20.7 Parallel Wiring

48. What resistance must be placed in parallel with a 155-Ω resistor to make the equivalent resistance 115 Ω?

49. ssm A 16-Ω loudspeaker and an 8.0-Ω loudspeaker are connected in parallel across the terminals of an amplifier. Assuming the speakers behave as resistors, determine the equivalent resistance of the two speakers.

50. For the 3-way bulb (50 W, 100 W, 150 W) discussed in Conceptual Example 10, find the resistance of each of the two filaments. Assume that the wattage ratings are not limited by significant figures and ignore any heating effects on the resistances.

51. ssm Two resistors, 42.0 and 64.0 Ω, are connected in parallel. The current through the 64.0-Ω resistor is 3.00 A. (a) Determine the current in the other resistor. (b) What is the total power consumed by the two resistors?

52. A wire whose resistance is R is cut into three equally long pieces, which are then connected in parallel. In terms of R, what is the resistance of the parallel combination?

53. A coffee cup heater and a lamp are connected in parallel to the same 120-V outlet. Together, they use a total of 84 W of power. The resistance of the heater is 6.0×10^2 Ω. Find the resistance of the lamp.

54. The drawing shows an electric heater ($R = 9.60$ Ω) and a toaster ($R = 17.0$ Ω). (a) What is the voltage across the heater? (b) Determine the current in the toaster. (c) What is the total power supplied to the heater and the toaster?

*55. ssm The total current delivered to a number of devices connected in parallel is the sum of the individual currents in each device. Circuit breakers are resettable automatic switches that protect against a dangerously

Heater

9.60 Ω

Toaster

17.0 Ω

120.0 V

Problem 54

large total current by "open-ing" to stop the current at a specified safe value. A 1650-W toaster, a 1090-W iron, and a 1250-W microwave oven are turned on in a kitchen. As the drawing shows, they are all connected through a 20-A circuit breaker to an ac voltage of 120 V. (a) Find the equivalent resistance of the three devices. (b) Obtain the total current delivered by the source and determine whether the breaker will "open" to prevent an accident.

*56. A resistor (resistance = R) is connected first in parallel and then in series with a 2.00-Ω resistor. A battery delivers five times as much current to the parallel combination than it does to the series combination. Determine the two possible values for R.

**57. The rear window defogger of a car consists of thirteen thin wires (resistivity = 88.0×10^{-8} Ω·m) embedded in the glass. The wires are connected in parallel to the 12.0-V battery, and each has a length of 1.30 m. The defogger can melt 2.10×10^{-2} kg of ice at 0 °C into water at 0 °C in two minutes. Assume that all the power dissipated in the wires is used immediately to melt the ice. Find the cross-sectional area of each wire.

Section 20.8 Circuits Wired Partially in Series and Partially in Parallel

58. A 14-Ω coffee maker and a 16-Ω frying pan are connected in series across a 120-V source of voltage. A 23-Ω bread maker is also connected across the 120-V source and is in parallel with the series combination. Find the total current supplied by the source of voltage.

59. **ssm** For the combination of resistors shown in the drawing, determine the equivalent resistance between points A and B.

60. Find the equivalent resistance between points A and B in the drawing.

61. **ssm** Determine the equivalent resistance between the points A and B for the group of resistors in the drawing.

62. A 60.0-Ω resistor is connected in parallel with a 120.0-Ω resistor. This parallel group is connected in series with a 20.0-Ω resistor. The total combination is connected across a 15.0-V battery. Find (a) the current and (b) the power dissipated in the 120.0-Ω resistor.

*63. Determine the power dissipated in the 5.0-Ω resistor in the circuit shown in the drawing.

*64. The circuit in the drawing contains five identical resistors. The 45-V battery delivers 58 W of power to the circuit. What is the resistance R of each resistor?

*65. **ssm www** The current in the 8.00-Ω resistor in the drawing is 0.500 A. Find the current in (a) the 20.0-Ω resistor and in (b) the 9.00-Ω resistor.

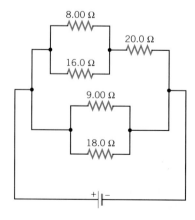

**66. Three identical resistors are connected in parallel. The equivalent resistance increases by 700 Ω when one resistor is removed and connected in series with the remaining two, which are still in parallel. Find the resistance of each resistor.

Section 20.9 Internal Resistance

67. A 1.40-Ω resistor is connected across a 9.00-V battery. The voltage between the terminals of the battery is observed to be only 8.30 V. Find the internal resistance of the battery.

68. A new "D" battery has an emf of 1.5 V. When a wire of negligible resistance is connected between the terminals of the battery, a current of 28 A is produced. Find the internal resistance of the battery.

69. ssm A battery has an internal resistance of 0.50 Ω. A number of identical light bulbs, each with a resistance of 15 Ω, are connected in parallel across the battery terminals. The terminal voltage of the battery is observed to be one-half the emf of the battery. How many bulbs are connected?

70. A battery delivering a current of 55.0 A to a circuit has a terminal voltage of 23.4 V. The electric power being dissipated by the internal resistance of the battery is 34.0 W. Find the emf of the battery.

71. A battery has an emf of 12.0 V and an internal resistance of 0.15 Ω. What is the terminal voltage when the battery is connected to a 1.50-Ω resistor?

*72. A 75.0-Ω and a 45.0-Ω resistor are connected in parallel. When this combination is connected across a battery, the current delivered by the battery is 0.294 A. When the 45.0-Ω resistor is disconnected, the current from the battery drops to 0.116 A. Determine (a) the emf and (b) the internal resistance of the battery.

Section 20.10 Kirchhoff's Rules

73. ssm Consider the circuit in the drawing. Determine (a) the magnitude of the current in the circuit and (b) the magnitude of the voltage between the points labeled A and B. (c) State which point, A or B, is at the higher potential.

74. A current of 2.0 A exists in the partial circuit shown in the drawing. What is the magnitude of the potential difference between the points (a) A and B, and (b) A and C?

75. Two batteries, each with an internal resistance of 0.015 Ω, are connected as in the drawing. In effect, the 9.0-V battery is being used to charge the 8.0-V battery. What is the current in the circuit?

76. For the circuit shown in the drawing, find the current I through the 2.00-Ω resistor and the voltage V of the battery to the left of this resistor.

*77. ssm Determine the voltage across the 5.0-Ω resistor in the drawing. Which end of the resistor is at the higher potential?

*78. For the circuit in the drawing, find the current in the 10.0-Ω resistor. Specify the direction of the current.

**79. The circuit in the drawing is known as a Wheatstone bridge circuit. Find the voltage between points B and D, and state which point is at the higher potential.

Section 20.11 The Measurement of Current and Voltage

80. A galvanometer has a full-scale current of 0.100 mA and a coil resistance of 50.0 Ω. This instrument is used with a shunt resistor to form an ammeter that will register full scale for a current of 60.0 mA. Determine the resistance of the shunt resistor.

81. ssm A voltmeter utilizes a galvanometer that has a 180-Ω coil resistance and a full-scale current of 8.30 mA. The voltmeter measures voltages up to 30.0 V. Determine the resistance that is connected in series with the galvanometer.

82. The coil of a galvanometer has a resistance of 20.0 Ω, and its meter deflects full scale when a current of 6.20 mA passes through it. To make the galvanometer into an ammeter, a 24.8-mΩ shunt resistor is added to it. What is the maximum current that this ammeter can read?

83. The equivalent resistance of a voltmeter is 140 000 Ω. The voltmeter uses a galvanometer that has a full-scale current of 180 μA. What is the maximum voltage that can be measured by the voltmeter?

*__84.__ Two scales on a voltmeter measure voltages up to 20.0 and 30.0 V, respectively. The resistance connected in series with the galvanometer is 1680 Ω for the 20.0-V scale and 2930 Ω for the 30.0-V scale. Determine the coil resistance and the full-scale current of the galvanometer that is used in the voltmeter.

**__85. ssm__ In measuring a voltage, a voltmeter uses some current from the circuit. Consequently, the voltage measured is only an approximation to the voltage present when the voltmeter is not connected. Consider a circuit consisting of two 1550-Ω resistors connected in series across a 60.0-V battery. (a) Find the voltage across one of the resistors. (b) A voltmeter has a full-scale voltage of 60.0 V and uses a galvanometer with a full-scale deflection of 5.00 mA. Determine the voltage that this voltmeter registers when it is connected across the resistor used in part (a).

Section 20.12 Capacitors in Series and Parallel

86. A 4.0-μF and an 8.0-μF capacitor are connected in parallel across a 25-V battery. Find (a) the equivalent capacitance and (b) the total charge stored on the two capacitors.

87. Determine the equivalent capacitance between A and B for the group of capacitors in the drawing.

88.
Three capacitors (4.0, 6.0, and 12.0 μF) are connected in series across a 50.0-V battery. Find the voltage across the 4.0-μF capacitor.

89. ssm A 3.0-μF capacitor and a 4.0-μF capacitor are connected in series across a 40.0-V battery. A 10.0-μF capacitor is also connected directly across the battery terminals. Find the total charge that the battery delivers to the capacitors.

90. Three capacitors have identical geometries. One is filled with a material whose dielectric constant is 2.50. Another is filled with a material whose dielectric constant is 4.00. The third capacitor is filled with a material whose dielectric constant κ is such that this single capacitor has the same capacitance as the series combination of the other two. Determine κ.

91. Suppose two capacitors (C_1 and C_2) are connected in series. Show that the sum of the energies stored in these capacitors is equal to the energy stored in the equivalent capacitor. [Hint: The energy stored in a capacitor can be expressed as $q^2/(2C)$.]

*__92.__ A 3.00-μF and a 5.00-μF capacitor are connected in series across a 30.0-V battery. A 7.00-μF capacitor is then connected in parallel across the 3.00-μF capacitor. Determine the voltage across the 7.00-μF capacitor.

*__93. ssm__ A 7.0-μF and a 3.0-μF capacitor are connected in series across a 24-V battery. What voltage is required to charge a parallel combination of the two capacitors to the same total energy?

**__94.__ The drawing shows two fully charged capacitors (C_1 = 2.00 μF, q_1 = 6.00 μC; C_2 = 8.00 μF, q_2 = 12.0 μC). The switch is closed, and charge flows until equilibrium is re-established (i.e., until both capacitors have the same voltage across their plates). Find the resulting voltage across either capacitor.

Section 20.13 *RC* Circuits

95. ssm In a heart pacemaker, a pulse is delivered to the heart 81 times per minute. The capacitor that controls this pulsing rate discharges through a resistance of 1.8×10^6 Ω. One pulse is delivered every time the fully charged capacitor loses 63.2% of its original charge. What is the capacitance of the capacitor?

96. An electronic flash attachment for a camera produces a flash by using the energy stored in a 750-μF capacitor. Between flashes, the capacitor recharges through a resistor whose resistance is chosen so the capacitor recharges with a time constant of 3.0 s. Determine the value of the resistance.

97. The circuit in the drawing contains two resistors and two capacitors that are connected to a battery via a switch. When the switch is closed, the capacitors begin to charge up. What is the time constant for the charging process?

*98. Three identical capacitors are connected with a resistor in two different ways. When they are connected as in part *a* of the drawing, the time constant to charge up this circuit is 0.34 s. What is the time constant when they are connected with the same resistor as in part *b*?

(a)   (b)

**99. How many time constants must elapse before a capacitor in a series *RC* circuit is charged to within 0.10% of its equilibrium charge?

ADDITIONAL PROBLEMS

100. The heating element in an iron has a resistance of 24 Ω. The iron is plugged into a 120-V outlet. What is the power dissipated by the iron?

101. ssm Three capacitors (3.0, 7.0, and 9.0 μF) are connected in series. What is their equivalent capacitance?

102. A lightning bolt delivers a charge of 35 C to the ground in a time of 1.0×10^{-3} s. What is the current?

103. ssm In the arctic, electric socks are useful. A pair of socks uses a 9.0-V battery pack for each sock. A current of 0.11 A is drawn from each battery pack by wire woven into the socks. Find the resistance of the wire in one sock.

104. Find the magnitude and direction of the current in the 2.0-Ω resistor in the drawing.

105. In Section 12.3 it was mentioned that temperatures are often measured with electrical resistance thermometers made of platinum wire. Suppose that the resistance of a platinum resistance thermometer is 125 Ω when its temperature is 20.0 °C. The wire is then immersed in boiling chlorine, and the resistance drops to 99.6 Ω. The temperature coefficient of resistivity of platinum is $\alpha = 3.72 \times 10^{-3}$ (C°)$^{-1}$. What is the temperature of the boiling chlorine?

106. The circuit in the drawing shows two resistors, a capacitor, and a battery. When the capacitor is fully charged, what is the magnitude q of the charge on one of its plates?

107. A galvanometer with a coil resistance of 12.0 Ω and a full-scale current of 0.150 mA is used with a shunt resistor to make an ammeter. The ammeter registers a maximum current of 4.00 mA. Find the equivalent resistance of the ammeter.

*108. Two cylindrical rods, one copper and the other iron, are identical in lengths and cross-sectional areas. They are joined, end-to-end, to form one long rod. A 12-V battery is connected across the free ends of the copper–iron rod. What is the voltage between the ends of the copper rod?

*109. ssm Eight different values of resistance can be obtained by connecting together three resistors (1.00, 2.00, and 3.00 Ω) in all possible ways. What are they?

*110. A cylindrical aluminum pipe of length 1.50 m has an inner radius of 2.00×10^{-3} m and an outer radius of 3.00×10^{-3} m. The interior of the pipe is completely filled with copper. What is the resistance of this unit? (*Hint: Imagine that the pipe is connected between the terminals of a battery and decide whether the aluminum and copper parts of the pipe are in series or in parallel.*)

*111. ssm An extension cord is used with an electric weed trimmer that has a resistance of 15.0 Ω. The extension cord is made of copper wire that has a cross-sectional area of 1.3×10^{-6} m^2.

The combined length of the two wires in the extension cord is 92 m. (a) Determine the resistance of the extension cord. (b) The extension cord is plugged into a 120-V socket. What voltage is applied to the trimmer itself?

*112. A resistor has a resistance R, and a battery has an internal resistance r. When the resistor is connected across the battery, ten percent less power is dissipated in R than there would be if the battery had no internal resistance. Find the ratio r/R.

*113. An iron wire has a resistance of 5.90 Ω at 20.0 °C, and a gold wire has a resistance of 6.70 Ω at the same temperature. The temperature coefficient of resistivity for iron is 0.0050 (C°)$^{-1}$, while for gold it is 0.0034 (C°)$^{-1}$. At what temperature do the wires have the same resistance?

*114. A 47-Ω resistor can dissipate up to 0.25 W of power without burning up. What is the smallest number of such resistors that

can be connected in series across a 9.0-V battery without any one of them burning up?

**115. ssm A sheet of gold foil (negligible thickness) is placed between the plates of a capacitor and has the same area as each of the plates. The foil is parallel to the plates, at a position one-third of the way from one to the other. Before the foil is inserted, the capacitance is C_0. What is the capacitance after the foil is in place? Express your answer in terms of C_0.

**116. 🔊 A digital thermometer uses a thermistor as the temperature-sensing element. A thermistor is a kind of semiconductor and has a large negative temperature coefficient of resistivity α. Suppose $\alpha = -0.060$ (C°)$^{-1}$ for the thermistor in a digital thermometer used to measure the temperature of a patient. The resistance of the thermistor decreases to 85% of its value at the normal body temperature of 37.0 °C. What is the patient's temperature?

CONCEPTS CALCULATIONS

G R O U P   L E A R N I N G   P R O B L E M S

Note: Each of these problems consists of Concept Questions followed by a related quantitative Problem. They are designed for use by students working alone or in small learning groups. The Concept Questions involve little or no mathematics and are intended to stimulate group discussions. They focus on the concepts with which the problems deal. Recognizing the concepts is the essential initial step in any problem-solving technique.

117. Concept Questions The resistance and the magnitude of the current depend on the path that the current takes. The drawing shows three situations in which the current takes different paths through a piece of material. Rank them according to (a) resistance and (b) current, largest first. Give your reasoning.

Problem Each of the rectangular pieces is made from a material whose resistivity is $\rho = 1.50 \times 10^{-2}$ $\Omega \cdot$m, and the unit of length in the drawing is $L_0 = 5.00$ cm. If the material is connected to a 3.00-V battery, find (a) the resistance and (b) the current in each case. Verify that your answers are consistent with your answers to the Concept Questions.

(a)   (b)   (c)

118. Concept Questions Each of the four circuits in the drawing consists of a single resistor whose resistance is either R or $2R$, and a single battery whose voltage is either V or $2V$. Rank the circuits according to (a) the power and (b) the current delivered to the resistor, largest to smallest. Explain your answers.

Problem The unit of voltage in each circuit is $V = 12.0$ V and the unit of resistance is $R = 6.00$ Ω. Determine (a) the power dissipated in each resistor and (b) the current delivered to each resistor. Check to see that your answers are consistent with your answers to the Concept Questions.

(a)   (b)   (c)   (d)

119. Concept Questions The drawing shows three different resistors in two different circuits. The resistances are such that $R_1 > R_2 > R_3$. (a) For the circuit on the left, rank the current through each resistor and the voltage across each one, largest first. (b) Repeat part (a) for the circuit on the right. Justify your answers.

Problem The battery has a voltage of $V = 24.0$ V, and the resistors have values of $R_1 = 50.0$ Ω, $R_2 = 25.0$ Ω, and $R_3 = 10.0$ Ω. (a) For the circuit on the left, determine the current through and the voltage across each resistor. (b) Repeat part (a) for the circuit on the right. Be sure your answers are consistent with your answers to the Concept Questions.

(a)   (b)

120. Concept Question The circuit in the drawing contains three identical resistors. Consider the equivalent resistance between the two points *a* and *b*, *b* and *c*, and *a* and *c*. Rank the equivalent resistances in decreasing order. Explain your reasoning.

Problem Each resistor has a value of 10.0 Ω. Determine the equivalent resistance between the points *a* and *b*, *b* and *c*, and *a* and *c*. Check to see that your answers are consistent with your answer to the Concept Question.

121. Concept Question Parts *a* and *b* of the drawing contain a resistor and a battery. Using the direction of the current as a guide, label the ends of each resistor with + and − signs. In both cases, proceed from point *A* to point *B* and determine the potential drops and potential rises. Express the drops and rises in terms of the current *I*, the resistance *R*, and the battery voltage *V*.

	Potential drops	Potential rises
Part *a*		
Part *b*		

Problem Using Kirchhoff's loop rule, find the value of the current *I* in part *c* of the drawing.

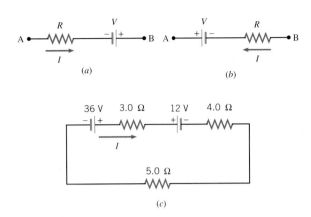

(a)

(b)

(c)

122. Concept Question Two capacitors, C_1 and C_2 are connected to a battery whose voltage is *V*. Recall from Section 19.5 that the electrical energy stored by each capacitor is $\frac{1}{2}C_1V_1^2$ and $\frac{1}{2}C_2V_2^2$, where V_1 and V_2 are, respectively, the voltages across C_1 and C_2. If the capacitors are connected in series, is the total energy stored by them greater than, less than, or equal to the total energy stored when they are connected in parallel? Justify your answer.

Problem The battery voltage is $V = 60.0$ V, and the capacitances are $C_1 = 2.00\ \mu$F and $C_2 = 4.00\ \mu$F. Determine the total energy stored by the two capacitors when they are wired (a) in parallel and (b) in series. Check to make sure that your answer is consistent with your answer to the Concept Question.

Magnetic resonance imaging (MRI) uses a magnetic field in a noninvasive way to provide doctors with a powerful diagnostic tool. Three-dimensional whole body images, such as the one shown here, are now available.

CHAPTER 21

MAGNETIC FORCES AND MAGNETIC FIELDS

21.1 MAGNETIC FIELDS

Permanent magnets have long been used in navigational compasses. As Figure 21.1 illustrates, the compass needle is a permanent magnet supported so it can rotate freely in a plane. When the compass is placed on a horizontal surface, the needle rotates until one end points approximately to the north. The end of the needle that points north is labeled the ***north magnetic pole;*** the opposite end is the ***south magnetic pole.***

Magnets can exert forces on each other. Figure 21.2 shows that the magnetic forces between north and south poles have the property that ***like poles repel each other, and unlike poles attract.*** This behavior is similar to that of like and unlike electric charges. However, there is a significant difference between magnetic poles and electric charges. It is possible to separate positive from negative electric charges and produce isolated charges of either kind. In contrast, no one has found a magnetic monopole (an isolated north or south pole). Any attempt to separate north and south poles by cutting a bar magnet in half fails, because each piece becomes a smaller magnet with its own north and south poles.

Surrounding a magnet, there is a ***magnetic field.*** The magnetic field is analogous to the electric field that exists in the space around electric charges. Like the electric field, the magnetic field has both a magnitude and a direction. We postpone a discussion of the magnitude until Section 21.2, concentrating our attention here only on the direction of the field. ***The direction of the magnetic field at any point in space is the direction indicated by the north pole of a small compass needle placed at that point.*** In Figure 21.3 the compass needle is symbolized by an arrow, the head of the arrow being the north pole. The drawing shows how compasses can be used to map out the magnetic field in the space surrounding a bar magnet. Since like poles repel and unlike poles attract, the needle of each compass becomes aligned relative to the bar magnet in the manner shown in the picture. The compass needles provide a visual picture of the magnetic field that the bar magnet creates in the surrounding space.

As an aid in visualizing the electric field, we introduced the notion of electric field lines in Section 18.7. In a similar fashion, it is possible to draw magnetic field lines in the vicinity of a magnet. Figure 21.4a illustrates some of the lines around a bar magnet. The lines appear to originate from the north pole and to end on the south pole; they do not start or stop in midspace. A visual image of the magnetic field lines in a plane can be created by sprinkling finely ground iron filings on a piece of paper that covers the magnet. Iron filings in a magnetic field behave like tiny compasses and align themselves along the field lines, as the photo in part *b* of the figure shows.

As is the case with electric field lines, the magnetic field at any point is tangent to the magnetic field line at that point. Furthermore, the strength of the field is proportional to the number of lines per unit area that passes through a surface oriented perpendicular to the lines. Thus, the magnetic field is stronger in regions where the field lines are relatively close together and weaker where they are relatively far apart. For instance, in Figure 21.4a the lines are closest together near the north and south poles, reflecting the fact that the strength of the field is greatest in these regions. Away from the poles, the magnetic field becomes weaker. Notice in part *c* of the drawing that the field lines in the gap between the poles of the horseshoe magnet are nearly parallel and equally spaced, indicating that the magnetic field there is approximately constant.

Figure 21.1 The needle of a compass is a permanent magnet that has a north magnetic pole (N) at one end and a south magnetic pole (S) at the other.

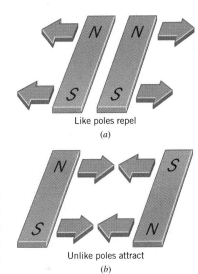

Like poles repel
(a)

Unlike poles attract
(b)

Figure 21.2 Bar magnets have a north magnetic pole at one end and a south magnetic pole at the other end. (a) Like poles repel each other, and (b) unlike poles attract.

Figure 21.3 At any location in the vicinity of a magnet, the north pole (the arrowhead in this drawing) of a small compass needle points in the direction of the magnetic field at that location.

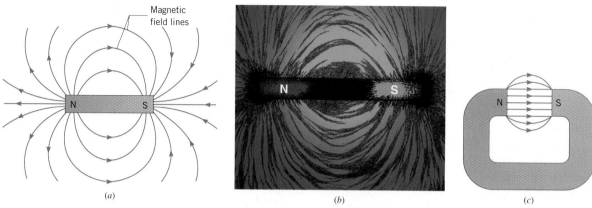

Figure 21.4 (*a*) The magnetic field lines and (*b*) the pattern of iron filings in the vicinity of a bar magnet. (*c*) The magnetic field lines in the gap of a horseshoe magnet.

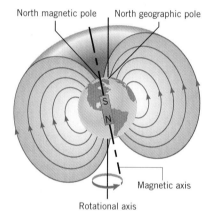

Figure 21.5 The earth behaves magnetically almost as if a bar magnet were located near its center. The axis of this fictitious bar magnet does not coincide with the earth's rotational axis; the two axes are currently about 11.5° apart.

Although the north pole of a compass needle points northward, it does not point exactly at the north geographic pole. The north geographic pole is that point where the earth's axis of rotation crosses the surface in the northern hemisphere (see Figure 21.5). Measurements of the magnetic field surrounding the earth show that the earth behaves magnetically almost as if it were a bar magnet.* As the drawing illustrates, the orientation of this fictitious bar magnet defines a magnetic axis for the earth. The location where the magnetic axis crosses the surface in the northern hemisphere is known as the north magnetic pole. The north magnetic pole is so named because it is the location toward which the north end of a compass needle points. Since unlike poles attract, the south pole of the earth's fictitious bar magnet lies beneath the north magnetic pole, as Figure 21.5 indicates.

The north magnetic pole does not coincide with the north geographic pole but, instead, lies in Hudson Bay, Canada, some 1300 km to the south. It is interesting that the position of the north magnetic pole is not fixed but moves over the years. For example, its current location is about 770 km northwest of its position in 1904. Pointing as it does at the north magnetic pole, a compass needle deviates from the north geographic pole. The angle that a compass needle deviates is called the *angle of declination*. For New York City, the present angle of declination is about 12° west, meaning that a compass needle points 12° west of geographic north.

Figure 21.5 shows that the earth's magnetic field lines are not parallel to the surface at all points. For instance, near the north magnetic pole the field lines are almost perpendicular to the surface of the earth. The angle that the magnetic field makes with respect to the surface at any point is known as the *angle of dip*.

21.2 THE FORCE THAT A MAGNETIC FIELD EXERTS ON A MOVING CHARGE

⬤ *Concepts at a Glance*   When a charge is placed in an electric field, it experiences an electric force, as Section 18.6 discusses. When a charge is placed in a magnetic field, it also experiences a force, provided that certain conditions are met, as we will see. The ***magnetic force,*** like all the forces we have studied (e.g., the

---

* At present it is not known with certainty what causes the earth's magnetic field. The magnetic field seems to arise from electric currents that circulate within the liquid outer region of the earth's core. Section 21.7 discusses how a current produces a magnetic field.

Concepts at a Glance

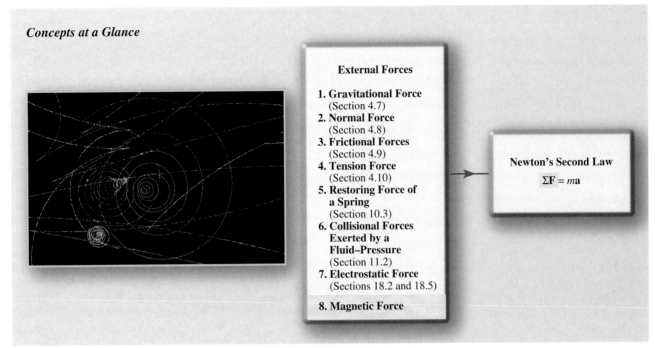

External Forces

1. **Gravitational Force**
   (Section 4.7)
2. **Normal Force**
   (Section 4.8)
3. **Frictional Forces**
   (Section 4.9)
4. **Tension Force**
   (Section 4.10)
5. **Restoring Force of**
   a Spring
   (Section 10.3)
6. **Collisional Forces**
   Exerted by a
   Fluid–Pressure
   (Section 11.2)
7. **Electrostatic Force**
   (Sections 18.2 and 18.5)

8. **Magnetic Force**

Newton's Second Law

$$\Sigma\mathbf{F} = m\mathbf{a}$$

gravitational, elastic, and electric forces), may contribute to the net force that causes an object to accelerate. Thus, when present, the magnetic force must be included in Newton's second law, as the Concepts-at-a-Glance chart in Figure 21.6 illustrates. This chart is an extension of the charts in Figures 4.9, 10.4, 11.5, and 18.4 and emphasizes that the second law is remarkable for taking into account the forces that arise in such a wide variety of situations. ●

The following two conditions must be met for a charge to experience a magnetic force when placed in a magnetic field:

1. The charge must be moving, for no magnetic force acts on a stationary charge.

2. The velocity of the moving charge must have a component that is perpendicular to the direction of the magnetic field.

To examine the second condition more closely, consider Figure 21.7, which shows a positive test charge $+q_0$ moving with a velocity **v** through a magnetic field. In the drawing, the magnetic field vector is labeled by the symbol **B**. The field is produced by an arrangement of magnets not shown in the drawing and is assumed to be constant in both magnitude and direction. If the charge moves *parallel or antiparallel* to the field, as in part *a* of the drawing, the charge experiences *no magnetic force*. If, on the other hand, the charge moves *perpendicular* to the field, as in part *b*, the charge experiences the *maximum possible force* $\mathbf{F}_{\text{max}}$. In general, if a charge moves at an angle θ^* with respect to the field (see part *c* of the drawing), only the velocity component $v \sin \theta$, which is perpendicular to the field, gives rise to a magnetic force. This force **F** is smaller than the maximum possible force. The component of the velocity that is parallel to the magnetic field yields no force.

Figure 21.6 *Concepts at a Glance* The magnetic force, like the other forces we have encountered, may contribute to the net force $\Sigma\mathbf{F}$ that acts on an object. According to Newton's second law, the acceleration **a** is directly proportional to the net force. A bubble chamber is one device that can be used to detect charged particles. In a bubble chamber, magnetic forces cause such particles to follow curved paths or tracks, as the photograph illustrates.

---

* The angle θ between the velocity of the charge and the magnetic field is chosen so that it lies in the range $0 \leq \theta \leq 180°$.

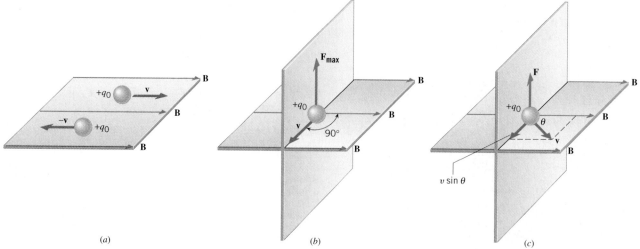

(a)   (b)   (c)

$v \sin \theta$

Figure 21.7 (a) No magnetic force acts on a charge moving with a velocity **v** that is parallel or antiparallel to a magnetic field **B**. (b) The charge experiences a maximum force $\mathbf{F}_{max}$ when the charge moves perpendicular to the field. (c) If the charge travels at an angle θ with respect to **B**, only the velocity component perpendicular to **B** gives rise to a magnetic force **F**, which is smaller than $\mathbf{F}_{max}$. This component is $v \sin \theta$.

Figure 21.7 shows that the direction of the magnetic force **F** is perpendicular to both the velocity **v** and the magnetic field **B**; in other words, **F** is perpendicular to the plane defined by **v** and **B**. As an aid in remembering the direction of the force, it is convenient to use *Right-Hand Rule No. 1 (RHR-1)*, as Figure 21.8 illustrates:

> ***Right-Hand Rule No. 1.*** Extend the right hand so the fingers point along the direction of the magnetic field **B** and the thumb points along the velocity **v** of the charge. The palm of the hand then faces in the direction of the magnetic force **F** that acts on a positive charge.

It is as if the open palm of the right hand pushes on the positive charge in the direction of the magnetic force. If the moving charge is *negative* instead of positive, the direction of the magnetic force is *opposite* to that predicted by RHR-1. Thus, there is an easy method for finding the force on a moving negative charge. First, assume that the charge is positive and use RHR-1 to find the direction of the force. Then, reverse this direction to find the direction of the force acting on the negative charge.

◉ *Concepts at a Glance* We will now use what we know about the magnetic force to define the magnetic field, in a procedure that is analogous to that used in Section 18.6 to define the electric field. The Concepts-at-a-Glance chart in Figure 18.16 illustrates that a test charge and the electrostatic force acting on it are brought together to define the electric field. The Concepts-at-a-Glance chart in Figure 21.9 illustrates that a similar procedure is used in the magnetic case. Recall that the electric field at any point in space is the force per unit charge that acts on a test charge q_0 placed at that point. In other words, to determine the electric field **E**, we divide the electrostatic force **F** by the charge q_0: $\mathbf{E} = \mathbf{F}/q_0$. In the magnetic case, however, the test charge is moving, and the force depends not only on the charge q_0, but also on the velocity component $v \sin \theta$ that is perpendicular to the magnetic field. Therefore, to determine the magnitude of the magnetic field, we divide the magnitude of the magnetic force not only by q_0, but also by $v \sin \theta$, according to the following definition. ◉

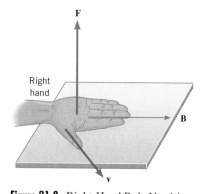

Figure 21.8 Right-Hand Rule No. 1 is illustrated. When the right hand is oriented so the fingers point along the magnetic field **B** and the thumb points along the velocity **v** of a positively charged particle, the palm faces in the direction of the magnetic force **F** applied to the particle.

● **DEFINITION OF THE MAGNETIC FIELD**

The magnitude B of the magnetic field at any point in space is defined as

$$B = \frac{F}{q_0(v \sin \theta)} \tag{21.1}$$

Concepts at a Glance

Figure 21.9 *Concepts at a Glance*
The magnetic field is defined in terms of the magnetic force and a (moving) test charge. This chart is analogous to the chart in Figure 18.16, which illustrates how the electric field is defined. The earth's magnetic field plays a role in the way honeybees navigate between their hive and the flowers they visit.

where F is the magnitude of the magnetic force on a positive test charge q_0 and $\mathbf{v}$ is the velocity of the charge and makes an angle θ ($0 \le \theta \le 180°$) with the direction of the magnetic field. The magnetic field $\mathbf{B}$ is a vector, and its direction can be determined by using a small compass needle.

SI Unit of Magnetic Field: $\dfrac{\text{newton} \cdot \text{second}}{\text{coulomb} \cdot \text{meter}} = 1$ tesla (T)

The unit of magnetic field strength that follows from Equation 21.1 is the $\text{N} \cdot \text{s}/(\text{C} \cdot \text{m})$. This unit is called the *tesla* (T), a tribute to the Croatian-born, American engineer Nikola Tesla (1856–1943). Thus, one tesla is the strength of the magnetic field in which a unit test charge, traveling perpendicular to the magnetic field with a speed of one meter per second, experiences a force of one newton. Because a coulomb per second is an ampere (1 C/s = 1 A), the tesla is often written as $1\,\text{T} = 1\,\text{N}/(\text{A} \cdot \text{m})$.

In many situations the magnetic field has a value that is considerably less than one tesla. For example, the strength of the magnetic field near the earth's surface is approximately 10^{-4} T. In such circumstances, a magnetic field unit called the *gauss* (G) is sometimes used. Although not an SI unit, the gauss is a convenient size for many applications involving magnetic fields. The relation between the gauss and the tesla is

$$1 \text{ gauss} = 10^{-4} \text{ tesla}$$

Example 1 deals with the magnetic force exerted on a moving proton and on a moving electron.

EXAMPLE 1 Magnetic Forces on Charged Particles

A proton in a particle accelerator has a speed of 5.0×10^6 m/s. The proton encounters a magnetic field whose magnitude is 0.40 T and whose direction makes an angle of $\theta = 30.0°$ with respect to the proton's velocity (see Figure 21.7c). Find (a) the magnitude and direction of the magnetic force on the proton and (b) the acceleration of the proton. (c) What would be the force and acceleration if the particle were an electron instead of a proton?

Reasoning For both the proton and the electron, the magnitude of the magnetic force is given by Equation 21.1. The magnetic forces that act on these particles have opposite directions, however, because the charges have opposite signs. In either case, the acceleration is given by Newton's second law, which applies to the magnetic force just as it does to any force. In using the second law, we must take into account the fact that the masses of a proton and an electron are different.

Solution

(a) The positive charge on a proton is 1.60×10^{-19} C, and according to Equation 21.1, the magnitude of the magnetic force is $F = q_0 v B \sin \theta$. Therefore,

$$F = (1.60 \times 10^{-19} \text{ C})(5.0 \times 10^6 \text{ m/s})(0.40 \text{ T})(\sin 30.0°) = \boxed{1.6 \times 10^{-13} \text{ N}}$$

The direction of the magnetic force is given by RHR-1 and is directed **upward** in Figure 21.7c, with the magnetic field pointing to the right.

(b) The magnitude a of the proton's acceleration follows directly from Newton's second law as the magnitude of the net force divided by the mass m_p of the proton. Since the only force acting on the proton is the magnetic force F, it is the net force. Thus,

$$a = \frac{F}{m_p} = \frac{1.6 \times 10^{-13} \text{ N}}{1.67 \times 10^{-27} \text{ kg}} = \boxed{9.6 \times 10^{13} \text{ m/s}^2} \tag{4.1}$$

The direction of the acceleration is the same as the direction of the net force (the magnetic force).

(c) The magnitude of the magnetic force on the electron is the same as that on the proton, since both have the same speed and charge magnitude. However, the direction of the force on the electron is opposite to that on the proton, or **downward** in Figure 21.7c, since the electron charge is negative. Furthermore, the electron has a smaller mass m_e and, therefore, experiences a significantly greater acceleration:

$$a = \frac{F}{m_e} = \frac{1.6 \times 10^{-13} \text{ N}}{9.11 \times 10^{-31} \text{ kg}} = \boxed{1.8 \times 10^{17} \text{ m/s}^2}$$

The direction of this acceleration is downward in Figure 21.7c.

21.3 THE MOTION OF A CHARGED PARTICLE IN A MAGNETIC FIELD

Comparing Particle Motion in Electric and Magnetic Fields

The motion of a charged particle in an electric field is noticeably different from the motion in a magnetic field. For example, Figure 21.10a shows a positive charge moving between the plates of a parallel plate capacitor. Initially, the charge is moving perpendicular to the direction of the electric field. Since the direction of the electric force on a positive charge is in the same direction as the electric field, the particle is deflected sideways. Part b of the drawing shows the same particle traveling initially at right angles to a magnetic field. An application of RHR-1 shows that when the charge enters the field, the charge is deflected upward (not sideways) by the magnetic force. As the charge moves upward, the direction of the magnetic force changes, always remaining perpendicular to both the magnetic field and the velocity. Conceptual Example 2 focuses on the difference in how electric and magnetic fields apply forces to a moving charge.

(a)

(b)

Figure 21.10 (a) The electric force **F** that acts on a positive charge is parallel to the electric field **E** and causes the particle's trajectory to bend in a horizontal plane. (b) The magnetic force **F** is perpendicular to both the magnetic field **B** and the velocity **v** and causes the particle's trajectory to bend in a vertical plane.

CONCEPTUAL EXAMPLE 2   A Velocity Selector

A velocity selector is a device for measuring the velocity of a charged particle. The device operates by applying electric and magnetic forces to the particle in such a way that these forces balance. Figure 21.11a shows a particle with a positive charge $+q$ and a velocity **v**, which is perpendicular to a constant magnetic field* **B**. How should an electric field **E** be directed so that the force it applies to the particle can balance the magnetic force produced by **B**?

Reasoning and Solution If the electric and magnetic forces are to balance, they must have opposite directions. By applying RHR-1, we find that the magnetic force acting on the positively charged particle in Figure 21.11a is directed upward. Therefore, the electric force must be directed downward. But the force applied to a positive charge by an electric field has the same direction as the field itself. Therefore, ***the electric field must point downward*** if the electric force is to balance the magnetic force. In other words, the electric and magnetic fields are perpendicular. Figure 21.11b shows a velocity selector that uses perpendicular electric and magnetic fields. The device is a cylindrical tube located within the magnetic field **B**. Inside the tube is a parallel plate capacitor that produces the electric field **E**. The charged particle enters the left end of the tube perpendicular to both fields. If the strengths of the fields **E** and **B** are adjusted properly, the electric and magnetic forces acting on the particle will cancel each other. With no net force acting on the particle, the velocity remains unchanged, according to Newton's first law. As a result, the particle moves in a straight line at a constant speed and exits at the right end of the tube. The magnitude of the velocity "selected" can be determined from a knowledge of the strengths of the electric and magnetic fields. Particles with velocities different from the one "selected" are deflected and do not exit at the right end of the tube.

Related Homework Material: *Problems 16, 20, and 23*

THE PHYSICS OF...

a velocity selector.

We have seen that a charged particle traveling in a magnetic field experiences a magnetic force that is always perpendicular to the field. In contrast, the force applied by an electric field is always parallel (or antiparallel) to the field direction. Because of the difference in the way that electric and magnetic fields exert forces, the work done on a charged particle by each field is different, as we now discuss.

The Work Done on a Charged Particle Moving through Electric and Magnetic Fields

In Figure 21.10a an electric field applies a force to a positively charged particle, and, consequently, the path of the particle bends in the direction of the force. Because there is a component of the particle's displacement in the direction of the electric force, the force does work on the particle, according to Equation 6.1. This work increases the kinetic energy and, hence, the speed of the particle, in accord

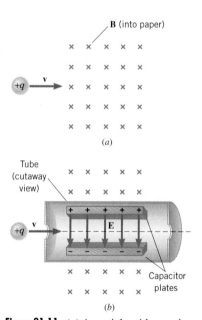

Figure 21.11 (*a*) A particle with a positive charge q and velocity **v** moves perpendicularly into a magnetic field **B**. (*b*) A velocity selector consists of a tube in which an electric field **E** is oriented perpendicular to a magnetic field, and the field magnitudes are adjusted so that the electric and magnetic forces acting on the particle balance.

---

* In many instances it is convenient to orient the magnetic field **B** so its direction is perpendicular to the page. In these cases it is customary to use a dot to symbolize the magnetic field pointing out of the page (toward the reader); this dot symbolizes the tip of the arrow representing the **B** vector. A region where a magnetic field is directed *into the page* is drawn as a series of crosses that indicate the tail feathers of the arrows representing the **B** vectors. Therefore, regions where a magnetic field is directed out of the page or into the page are drawn as shown below:

· · · ·   × × × ×
· · · ·   × × × ×
· · · ·   × × × ×

Out of page   Into page

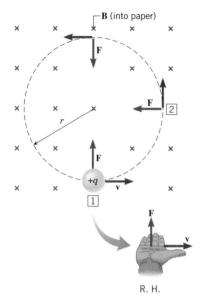

Figure 21.12 A positively charged particle is moving perpendicular to a constant magnetic field. The magnetic force **F** causes the particle to move on a circular path (R.H. = right hand).

with the work–energy theorem (see Section 6.2). In contrast, the magnetic force in Figure 21.10b always acts in a direction that is perpendicular to the motion of the charge. Consequently, the displacement of the moving charge never has a component in the direction of the magnetic force. As a result, *the magnetic force cannot do work and change the kinetic energy of the charged particle* in Figure 21.10b, although the force does alter the direction of the motion.

The Circular Trajectory

To describe the motion of a charged particle in a constant magnetic field more completely, let's discuss the special case in which the velocity of the particle is perpendicular to a uniform magnetic field. As Figure 21.12 illustrates, the magnetic force serves to move the particle in a circular path. To understand why the path is circular, consider two points on the circumference labeled 1 and 2. When the positively charged particle is at point 1, the magnetic force **F** is perpendicular to the velocity **v** and points directly upward in the drawing. This force causes the trajectory to bend upward. When the particle reaches point 2, the magnetic force still remains perpendicular to the velocity but is now directed to the left in the drawing. *The magnetic force always remains perpendicular to the velocity and is directed toward the center of the circular path.*

To find the radius of the path in Figure 21.12, we use the concept of centripetal force from Section 5.3. The centripetal force is the net force, directed toward the center of the circle, that is needed to keep a particle moving along a circular path. The magnitude F_c of this force depends on the speed v and mass m of the particle, as well as the radius r of the circle:

$$F_c = \frac{mv^2}{r} \qquad (5.3)$$

In the present situation, the magnetic force furnishes the centripetal force. Being perpendicular to the velocity, the magnetic force does no work in keeping the charge $+q$ on the circular path. According to Equation 21.1, the magnitude of the magnetic force is $qvB \sin 90°$, so $qvB = mv^2/r$ or

$$r = \frac{mv}{qB} \qquad (21.2)$$

This result shows that the radius of the circle is inversely proportional to the magnitude of the magnetic field, with stronger fields producing "tighter" circular paths. Example 3 illustrates an application of Equation 21.2.

EXAMPLE 3   The Motion of a Proton

A proton is released from rest at point A, which is located next to the positive plate of a parallel plate capacitor (see Figure 21.13). The proton then accelerates toward the negative plate, leaving the capacitor at point B through a small hole in the plate. The electric potential of the positive plate is 2100 V greater than that of the negative plate, so $V_A - V_B = 2100$ V. Once outside the capacitor, the proton travels at a constant velocity until it enters a region of constant magnetic field of magnitude 0.10 T. The velocity is perpendicular to the magnetic field, which is directed out of the page in Figure 21.13. Find (a) the speed v_B of the proton when it leaves the negative plate of the capacitor, and (b) the radius r of the circular path on which the proton moves in the magnetic field.

Figure 21.13 The proton, starting from rest at the positive plate of the capacitor, accelerates toward the negative plate. After leaving the capacitor, the proton enters a magnetic field, where it moves on a circular path of radius r.

Reasoning The only force that acts on the proton (charge $= +e$) while it is between the capacitor plates is the conservative electric force. Thus, we can use the conservation of energy to find the speed of the proton when it leaves the negative plate. The total energy of the proton is the sum of its kinetic energy, $\frac{1}{2}mv^2$, and electric potential energy, EPE. Following Example 4 in Chapter 19, we set the total energy at point B equal to the total energy at point A:

$$\underbrace{\tfrac{1}{2}mv_B^2 + \text{EPE}_B}_{\text{Total energy at } B} = \underbrace{\tfrac{1}{2}mv_A^2 + \text{EPE}_A}_{\text{Total energy at } A}$$

We note that $v_A = 0$ m/s, since the proton starts from rest, and use Equation 19.3 to set $\text{EPE}_A - \text{EPE}_B = e(V_A - V_B)$. Then the conservation of energy reduces to $\frac{1}{2}mv_B^2 = e(V_A - V_B)$. Solving for v_B gives $v_B = \sqrt{2e(V_A - V_B)/m}$. The proton enters the magnetic field with this speed and moves on a circular path with a radius that is given by Equation 21.2.

Solution

(a) The speed of the proton is

$$v_B = \sqrt{\frac{2e(V_A - V_B)}{m}} = \sqrt{\frac{2(1.60 \times 10^{-19}\text{ C})(2100\text{ V})}{1.67 \times 10^{-27}\text{ kg}}} = \boxed{6.3 \times 10^5\text{ m/s}}$$

(b) When the proton moves in the magnetic field, the radius of the circular path is

$$r = \frac{mv_B}{eB} = \frac{(1.67 \times 10^{-27}\text{ kg})(6.3 \times 10^5\text{ m/s})}{(1.60 \times 10^{-19}\text{ C})(0.10\text{ T})} = \boxed{6.6 \times 10^{-2}\text{ m}} \qquad (21.2)$$

One of the exciting areas in physics today is the study of elementary particles, which are the basic building blocks from which all matter is constructed. Important information about an elementary particle can be obtained from its motion in a magnetic field, with the aid of a device known as a bubble chamber. A bubble chamber contains a superheated liquid such as hydrogen, which will boil and form bubbles readily. When an electrically charged particle passes through the chamber, a thin track of bubbles is left in its wake. This track can be photographed to show how a magnetic field affects the particle motion. Conceptual Example 4 illustrates how physicists deduce information from such photographs.

CONCEPTUAL EXAMPLE 4   Particle Tracks in a Bubble Chamber

Figure 21.14*a* shows the bubble-chamber tracks resulting from an event that begins at point A. At this point a gamma ray, which is emitted by certain radioactive substances and travels in from the left, spontaneously transforms into two charged particles. There is no track from the gamma ray itself. These particles move away from point A, producing the two spiral tracks. A third charged particle is knocked out of a hydrogen atom and moves forward, producing the long track with the slight upward curvature. Each of the three particles has the same mass and carries a charge of the same magnitude. A uniform magnetic field is directed out of the paper toward you. Guided by RHR-1 and Equation 21.2, deduce the sign of the charge carried by each particle, identify which particle is moving most rapidly, and account for the two spiral paths.

Reasoning and Solution To help in our reasoning. Figure 21.14*b* shows a positively charged particle traveling with a velocity **v** that is perpendicular to a magnetic field. The field is directed out of the paper. RHR-1 indicates that the magnetic force points downward. Therefore, downward-curving tracks in the photograph indicate a positive charge, while upward-curving tracks indicate a negative charge. ***Particles 1 and***

(*a*)

(*b*)

Figure 21.14 (*a*) A photograph of tracks in a bubble chamber. A magnetic field is directed out of the paper. At point *A* a gamma ray (not visible) spontaneously transforms into an electron (e⁻) and a positron (e⁺), which produce the spirals. In addition, an electron is knocked forward out of a hydrogen atom in the chamber. (*b*) In accord with RHR-1, the magnetic field applies a downward force to a positively charged particle that moves to the right.

INTERACTIVE LEARNINGWARE #42

START
Visualization
Reasoning
Knowns and Unknowns
Modeling the Problem
Numerical Solution
FINISH

A particle has a charge of 3.3×10^{-7} C. In a 0.20-T magnetic field, it travels on a circular path that has a radius of 0.18 m. The centripetal acceleration of the particle is 2.0×10^{10} m/s². What is the magnitude of the centripetal force that acts on the particle?

We have seen that a charged particle follows a circular path when it moves perpendicular to a magnetic field. When any object moves on a circular path, it has a centripetal acceleration due to a centripetal force, as Chapter 5 discusses. This Interactive LearningWare example reviews the ideas of centripetal acceleration and centripetal force as they are applied to the motion of a charged particle in a magnetic field.

Go to our Web site at

http://www.wiley.com/college/cutnell

3, then, must carry a negative charge. They are, in fact, electrons (e⁻). In contrast, **particle 2 must have a positive charge.** It is called a positron (e⁺), an elementary particle that has the same mass as an electron but an opposite charge.

In Equation 21.2 the mass m, the charge magnitude q, and the magnetic field strength B are the same for each particle. Therefore, the radius r is proportional to the speed v, and a greater radius means a greater speed. The track for particle 3 has the greatest radius, so **particle 3 has the greatest speed.**

Each spiral indicates that the radius is decreasing as the particle moves. Since the radius is proportional to the speed, the speeds of particles 1 and 2 must be decreasing as these particles move. Correspondingly, the kinetic energies of these particles must be decreasing. **Particles 1 and 2 are losing energy each time they collide with a hydrogen atom in the bubble chamber.**

Related Homework Material: *Questions 5 and 6, Problem 22*

21.4 THE MASS SPECTROMETER

Physicists use mass spectrometers for determining the relative masses and abundances of isotopes.* Chemists use these instruments to help identify unknown molecules produced in chemical reactions. Mass spectrometers are also used during surgery, where they give the anesthesiologist information on the gases, including the anesthetic, in the patient's lungs.

In the type of mass spectrometer illustrated in Figure 21.15, the atoms or molecules are first vaporized and then ionized by the ion source. The ionization process removes one electron from the particle, leaving it with a net positive charge of $+e$. The positive ions are then accelerated through the potential difference V, which is applied between the ion source and the metal plate. With a speed v, the ions pass through a hole in the plate and enter a region of constant magnetic field **B**, where they are deflected in semicircular paths. Only those ions following a path with the proper radius r strike the detector, which records the number of ions arriving per second.

The mass m of the detected ions can be expressed in terms of r, B, and v by recalling that the radius of the path followed by a particle of charge $+e$ is $r = mv/(eB)$ (Equation 21.2). In addition, the Reasoning section in Example 3 shows that the ion speed v can be expressed in terms of the potential difference V as $v = \sqrt{2eV/m}$. This expression for v is the same at that used in Example 3, except that, for convenience, we have replaced the potential difference, $V_A - V_B$, by the symbol V. Eliminating v from these two equations algebraically and solving for the mass gives

$$m = \left(\frac{er^2}{2V} \right) B^2$$

B (out of paper)

m m_1

r

V

Metal plate

Detector

Ion source

Figure 21.15 The basic features of a mass spectrometer. The dashed lines are the paths traveled by ions of different masses. Ions with mass m follow the path of radius r and enter the detector. Ions with the larger mass m_1 follow the outer path and miss the detector.

---

* Isotopes are atoms that have the same atomic number but different atomic masses due to the presence of different numbers of neutrons in the nucleus. They are discussed in Section 31.1.

This result shows that the mass of each ion reaching the detector is proportional to B^2. By experimentally changing the value of B and keeping the term in the parentheses constant, ions of different masses are allowed to enter the detector. A plot of the detector output as a function of B^2 then gives an indication of what masses are present and the abundance of each mass.

Figure 21.16 shows a record obtained by a mass spectrometer for naturally occurring neon gas. The results show that the element neon has three isotopes whose atomic mass numbers are 20, 21, and 22. These isotopes occur because neon atoms exist with different numbers of neutrons in the nucleus. Notice that the isotopes have different abundances, with neon-20 being the most abundant.

Figure 21.16 The mass spectrum (not to scale) of naturally occurring neon, showing three isotopes whose atomic mass numbers are 20, 21, and 22. The larger the peak, the more abundant the isotope.

21.5 THE FORCE ON A CURRENT IN A MAGNETIC FIELD

As we have seen, a charge moving through a magnetic field can experience a magnetic force. Since an electric current is a collection of moving charges, a current in the presence of a magnetic field can also experience a magnetic force. In Figure 21.17, for instance, a current-carrying wire is placed between the poles of a magnet. When the direction of the current I is as shown, the moving charges experience a magnetic force that pushes the wire to the right in the drawing. The direction of the force is determined in the usual manner by using RHR-1, with the minor modification that the direction of the velocity of a positive charge is replaced by the direction of the conventional current I. If the current in the drawing were reversed by switching the leads to the battery, the direction of the force would be reversed, and the wire would be pushed to the left.

When a charge moves through a magnetic field, the magnitude of the force that acts on the charge is $F = qvB \sin \theta$. With the aid of Figure 21.18, this expression can be put into a form that is more suitable for use with an electric current. The drawing shows a wire of length L that carries a current I. The wire is oriented at an angle θ with respect to a magnetic field $\mathbf{B}$. This picture is similar to Figure 21.7c, except that now the charges move in a wire. The magnetic force exerted on this length of wire is the net force acting on the total amount of charge moving in the wire. Suppose that an amount of charge Δq travels the length of the wire in a time interval Δt. The magnitude of the magnetic force on this amount of charge is given by Equation 21.1 as $F = (\Delta q)vB \sin \theta$. Multiplying and dividing the right side of this equation by Δt, we find that

$$F = \underbrace{\left(\frac{\Delta q}{\Delta t}\right)}_{I} \underbrace{(v\,\Delta t)}_{L}\, B \sin \theta$$

According to Equation 20.1, the term $\Delta q/\Delta t$ is the current I in the wire, and the term $v\,\Delta t$ is the length L of the wire. With these two substitutions, the expression for the magnetic force exerted on a current-carrying wire becomes

Magnetic force on a current-carrying wire of length L

$$F = ILB \sin \theta \qquad (21.3)$$

As in the case of a single charge traveling in a magnetic field, the magnetic force on a current-carrying wire is a maximum when the wire is oriented perpendicular to

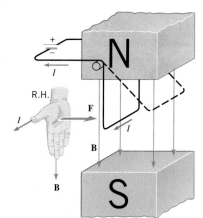

Figure 21.17 The wire carries a current I, and the bottom segment of the wire is oriented perpendicular to a magnetic field $\mathbf{B}$. A magnetic force deflects the wire to the right.

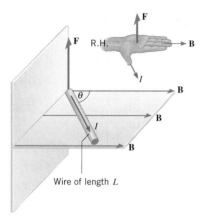

Figure 21.18 The current *I* in the wire, oriented at an angle θ with respect to a magnetic field **B**, is acted upon by a magnetic force **F**.

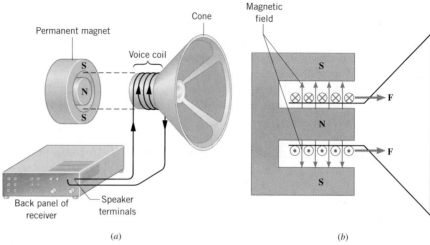

(a)   *(b)*

Figure 21.19 (*a*) An "exploded" view of one type of speaker design, which shows a cone, a voice coil, and a permanent magnet. (*b*) Because of the current in the voice coil (shown as ⊗ and ⊙), the magnetic field causes a force **F** to be exerted on the voice coil and cone.

the field ($\theta = 90°$) and vanishes when the current is parallel or antiparallel to the field ($\theta = 0°$ or $180°$). The direction of the magnetic force is given by RHR-1, as Figure 21.18 indicates.

Most loudspeakers operate on the principle that a magnetic field exerts a force on a current-carrying wire. Figure 21.19*a* shows a speaker design that consists of three parts: a cone, a voice coil, and a permanent magnet. The cone is mounted so it can vibrate back and forth. When vibrating, it pushes and pulls on the air in front of it, thereby creating sound waves. Attached to the apex of the cone is the voice coil, which is a hollow cylinder, around which coils of wire are wound. The voice coil is slipped over one of the poles of the stationary permanent magnet (the north pole in the drawing) and can move freely. The two ends of the voice-coil wire are connected to the speaker terminals on the back panel of a receiver.

THE PHYSICS OF...

how a loudspeaker produces sound.

The receiver acts as an ac generator, sending an alternating current to the voice coil. The alternating current interacts with the magnetic field to generate an alternating force that pushes and pulls on the voice coil and the attached cone. To see how the magnetic force arises, consider Figure 21.19*b*, which is a cross-sectional view of the voice coil and the magnet. In the cross-sectional view, the current is directed into the page in the upper half of the voice coil (⊗⊗⊗) and out of the page in the lower half (⊙⊙⊙). In both cases the magnetic field is perpendicular to the current, so the maximum possible force is exerted on the wire. An application of RHR-1 to both the upper and lower halves of the voice coil shows that the magnetic force **F** in the drawing is directed to the right, causing the cone to accelerate in that direction. One-half of a cycle later when the current is reversed, the direction of the magnetic force is also reversed, and the cone accelerates to the left. If, for example, the alternating current from the receiver has a frequency of 1000 Hz, the alternating magnetic force causes the cone to vibrate back and forth at the same frequency, and a 1000-Hz sound wave is produced. Thus, it is the magnetic force on a current-carrying wire that is responsible for converting an electrical signal into a sound wave. In Example 5 a typical force and acceleration in a loudspeaker are determined.

● **PROBLEM SOLVING INSIGHT**
Whenever the current in a wire reverses direction, the force exerted on the wire by a given magnetic field also reverses direction.

EXAMPLE 5
The Force and Acceleration in a Loudspeaker

The voice coil of a speaker has a diameter of $d = 0.025$ m, contains 55 turns of wire, and is placed in a 0.10-T magnetic field. The current in the voice coil is 2.0 A. (a) Determine the magnetic force that acts on the coil and cone. (b) The voice coil and cone have a combined mass of 0.020 kg. Find their acceleration.

Reasoning The magnetic force that acts on the current-carrying voice coil is given by Equation 21.3 as $F = ILB \sin \theta$. The effective length L of the wire in the voice coil is very nearly the number of turns N times the circumference (πd) of one turn: $L = N\pi d$. The acceleration of the voice coil and cone is given by Newton's second law as the magnetic force divided by the combined mass.

Solution

(a) Since the magnetic field acts perpendicular to all parts of the wire, $\theta = 90°$ and the force on the voice coil is

$$F = ILB \sin \theta = (2.0 \text{ A})[55\pi(0.025 \text{ m})](0.10 \text{ T})\sin 90°$$
$$= \boxed{0.86 \text{ N}} \quad (21.3)$$

(b) According to Newton's second law, the acceleration of the voice coil and cone is

$$a = \frac{F}{m} = \frac{0.86 \text{ N}}{0.020 \text{ kg}} = \boxed{43 \text{ m/s}^2} \quad (4.1)$$

This acceleration is more than four times the acceleration due to gravity.

INTERACTIVE LEARNINGWARE #43

START
Visualization
Reasoning
Knowns and Unknowns
Modeling the Problem
Numerical Solution
FINISH

Two conducting rails are 1.6 m apart and are parallel to the ground at the same height. As the drawing shows, a 0.20-kg aluminum rod is lying on top of the rails, and a 0.050-T magnetic field points upward, perpendicular to the ground. There is a current I in the rod, directed as in the drawing. The coefficient of static friction between the rod and each rail is $\mu_s = 0.45$. How much current is needed to make the rod begin moving, and in which direction will it move?

Conducting rails
B B B
90°
I
1.6 m

When an object carries an electric current, a magnetic field can apply a force to it. But other forces may also act on the object. What happens to the object depends on how the various forces compare with one another or, more specifically, on the net force that they jointly create. This Interactive LearningWare example deals with such a case and gives you the opportunity to review what we learned about the static frictional force in Section 4.9.

Go to our Web site at

http://www.wiley.com/college/cutnell

Magnetohydrodynamic (MHD) propulsion is a revolutionary type of propulsion system that does not use propellers to power ships and submarines, but, instead, uses a magnetic force on a current. The method eliminates motors, drive shafts, gears, as well as propellers, so it promises to be a low-noise system with great reliability at relatively low cost. Figure 21.20a shows the first ship to use MHD technology, the *Yamato 1*. In part *b* the schematic side view shows one of the two MHD propulsion units that are mounted underneath the vessel. Seawater enters the front of the unit and is expelled from the rear. A jet engine uses air in a similar fashion, taking it in the front and pushing it out the back to propel the plane forward.

Figure 21.20c presents an enlarged view of a propulsion unit. An electromagnet (see Section 21.7) within the vessel uses superconducting wire to produce a strong magnetic field. Electrodes (metal plates) mounted on either side of the unit are attached to a dc electrical generator. The generator sends an electric current through the seawater, from one electrode to the other, perpendicular to the magnetic field. Consistent with RHR-1, the field applies a force **F** to the current, as the drawing shows. The magnetic force pushes the seawater, similar to the way in which it pushes the wire in Figure 21.17. As a result, a water jet is expelled from the tube. Since the MHD unit exerts a magnetic force on the water, the water exerts a force on the unit and the vessel attached to it. According to Newton's third law, this "reaction" force is equal in magnitude, but opposite in direction to the magnetic force, and it is the reaction force that provides the thrust to drive the vessel.

THE PHYSICS OF...

magnetohydrodynamic propulsion.

Figure 21.20 (*a*) The *Yamato 1* is the first ship to use magnetohydrodynamic (MHD) propulsion. (*b*) A schematic side view of the *Yamato 1*. (*c*) In the MHD unit, the magnetic force exerted on the current forces water out the back.

21.6 THE TORQUE ON A CURRENT-CARRYING COIL

We have seen that a current-carrying wire can experience a force when placed in a magnetic field. If a loop of wire is suspended properly in a magnetic field, the magnetic force produces a torque that tends to rotate the loop. This torque is responsible for the operation of a widely used type of electric motor.

Figure 21.21*a* shows a rectangular loop of wire attached to a vertical shaft. The shaft is mounted such that it is free to rotate in a uniform magnetic field. When there is a current in the loop, the loop rotates because magnetic forces act on the two vertical sides, labeled 1 and 2 in the drawing. Part *b* shows a top view of the loop and the magnetic forces **F** and −**F** on the two sides. These two forces have the same magnitude, but an application of RHR-1 shows that they point in opposite directions, so the loop experiences no net force. The loop does, however, experience a net torque that tends to rotate it in a clockwise fashion about the vertical shaft. Figure 21.22*a* shows that the torque is maximum when the normal to the plane of the loop is perpendicular

Figure 21.21 (*a*) A current-carrying loop of wire, which can rotate about a vertical shaft, is situated in a magnetic field. (*b*) A top view of the loop. The current in side 1 is directed out of the page (⊙), while that in side 2 is directed into the page (⊗). The current in side 1 experiences a force **F** that is opposite to the force exerted on side 2. The two forces produce a clockwise torque about the shaft.

to the field. In contrast, part *b* shows that the torque is zero when the normal is parallel to the field. ***When a current-carrying loop is placed in a magnetic field, the loop tends to rotate such that its normal becomes aligned with the magnetic field.*** In this respect, a current loop behaves like a magnet (e.g., a compass needle) suspended in a magnetic field, since a magnet also rotates to align itself with the magnetic field.

It is possible to determine the magnitude of the torque on the loop. From Equation 21.3 the magnetic force on each vertical side has a magnitude of $F = ILB \sin 90°$, where L is the length of side 1 or side 2, and $\theta = 90°$ because the current I always remains perpendicular to the magnetic field as the loop rotates. As Section 9.1 discusses, the torque produced by a force is the product of the magnitude of the force and the lever arm. In Figure 21.21*b* the lever arm is the perpendicular distance from the line of action of the force to the shaft. This distance is given by $(w/2) \sin \phi$, where w is the width of the loop, and ϕ is the angle between the normal to the plane of the loop and the direction of the magnetic field. The net torque is the sum of the torques on the two sides, so

$$\text{Net torque} = \tau = ILB \left(\tfrac{1}{2}w \sin \phi\right) + ILB \left(\tfrac{1}{2}w \sin \phi\right) = IAB \sin \phi$$

In this result the product Lw has been replaced by the area A of the loop. If the wire is wrapped so as to form a coil containing N loops, each of area A, the force on each side is N times larger, and the torque becomes proportionally greater:

$$\tau = NIAB \sin \phi \qquad (21.4)$$

Equation 21.4 has been derived for a rectangular coil, but it is valid for any shape of flat coil, such as a circular coil. The torque depends on the geometric properties of the coil and the current in it through the quantity NIA. This quantity is known as the ***magnetic moment*** of the coil, and its units are ampere · meter^2. The greater the magnetic moment of a current-carrying coil, the greater the torque that the coil experiences when placed in a magnetic field. Example 6 discusses the torque that a magnetic field applies to such a coil.

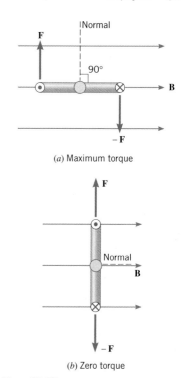

(a) Maximum torque

(b) Zero torque

Figure 21.22 (*a*) Maximum torque occurs when the normal to the plane of the loop is perpendicular to the magnetic field. (*b*) The torque is zero when the normal is parallel to the field.

EXAMPLE 6   The Torque Exerted on a Current-Carrying Coil

A coil of wire has an area of $2.0 \times 10^{-4} \text{ m}^2$, consists of 100 loops or turns, and contains a current of 0.045 A. The coil is placed in a uniform magnetic field of magnitude 0.15 T. (a) Determine the magnetic moment of the coil. (b) Find the maximum torque that the magnetic field can exert on the coil.

Reasoning and Solution

(a) The magnetic moment of the coil is

$$\text{Magnetic moment} = NIA = (100)(0.045 \text{ A})(2.0 \times 10^{-4} \text{ m}^2)$$
$$= \boxed{9.0 \times 10^{-4} \text{ A} \cdot \text{m}^2}$$

(b) According to Equation 21.4, the torque is the product of the magnetic moment NIA and $B \sin \phi$. However, the maximum torque occurs when $\phi = 90°$, so

$$\tau = \underbrace{(NIA)}_{\substack{\text{Magnetic} \\ \text{moment}}} (B \sin 90°) = (9.0 \times 10^{-4} \text{ A} \cdot \text{m}^2)(0.15 \text{ T}) = \boxed{1.4 \times 10^{-4} \text{ N} \cdot \text{m}}$$

The electric motor is found in many devices, such as CD players, tape decks, automobiles, washing machines, and air conditioners. Figure 21.23 shows that a direct-current (dc) motor consists of a coil of wire placed in a magnetic field and free to rotate about a vertical shaft. The coil of wire contains many turns and is wrapped around an iron cylinder that rotates with the coil, although these features have been omitted to

Figure 21.23 The basic components of a dc motor. The platter of a turntable is shown as it might be attached to the motor.

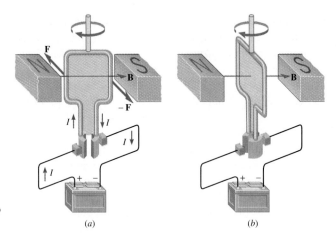

Figure 21.24 (*a*) When a current exists in the coil, the coil experiences a torque. (*b*) Because of its inertia, the coil continues to rotate when there is no current.

THE PHYSICS OF...

a direct-current electric motor.

simplify the drawing. The coil and iron cylinder assembly is known as the armature. Each end of the wire coil is attached to a metallic half-ring. Rubbing against each of the half-rings is a graphite contact called a brush. While the half-rings rotate with the coil, the graphite brushes remain stationary. The two half-rings and the associated brushes are referred to as a split-ring commutator, the purpose of which will be explained shortly.

The operation of a motor can be understood by considering Figure 21.24. In part *a* the current from the battery enters the coil through the left brush and half-ring, goes around the coil, and then leaves through the right half-ring and brush. According to RHR-1, the directions of the forces **F** and −**F** on the two sides of the coil are as shown in the drawing. These forces produce the torque that turns the coil. Eventually the coil reaches the position shown in part *b* of the drawing. In this position the half-rings momentarily lose electrical contact with the brushes, so that there is no current in the coil and no applied torque. However, like any moving object, the rotating coil does not stop immediately, for its inertia carries it onward. When the half-rings reestablish contact with the brushes, there again is a current in the coil, and a magnetic torque again rotates the coil in the same direction. The split-ring commutator ensures that the current is always in the proper direction to yield a torque that produces a continuous rotation of the coil.

21.7 MAGNETIC FIELDS PRODUCED BY CURRENTS

We have seen that a current-carrying wire can experience a magnetic force when placed in a magnetic field that is produced by an external source, such as a permanent magnet. *A current-carrying wire also produces a magnetic field of its own,* as we will see in this section. Hans Christian Oersted (1777–1851) first discovered this effect in 1820 when he observed that a current-carrying wire influenced the orientation of a nearby compass needle. The compass needle aligns itself with the net magnetic field produced by the current and the magnetic field of the earth. Oersted's discovery, which linked the motion of electric charges with the creation of a magnetic field, marked the beginning of an important discipline called *electromagnetism.*

A Long, Straight Wire

Figure 21.25*a* illustrates Oersted's discovery with a very long, straight wire. When a current is present, the compass needles point in a circular pattern about the wire. The pattern indicates that the magnetic field lines produced by the current are circles centered on the wire. If the direction of the current is reversed, the needles also reverse their directions, indicating that the direction of the magnetic field has reversed. The direction of the field can be obtained by using *Right-Hand Rule No. 2 (RHR-2)*, as part *b* of the drawing indicates:

Right-Hand Rule No. 2. Curl the fingers of the right hand into the shape of a half-circle. Point the thumb in the direction of the conventional current *I*, and the tips of the fingers will point in the direction of the magnetic field **B**.

Experimentally, it is found that the magnitude *B* of the magnetic field produced by a long, straight wire is directly proportional to the current *I* and inversely proportional to the radial distance *r* from the wire: $B \propto I/r$. As usual, this proportionality is converted into an equation by introducing a proportionality constant, which, in this instance, is written as $\mu_0/2\pi$. Thus, the magnitude of the magnetic field is

Long, straight wire
$$B = \frac{\mu_0 I}{2\pi r} \qquad (21.5)$$

The constant μ_0 is known as the **permeability of free space,** and its value is $\mu_0 = 4\pi \times 10^{-7}$ T·m/A. The magnetic field becomes stronger nearer the wire, where *r* is smaller. Therefore, the field lines near the wire are closer together than those located farther away, where the field is weaker. Figure 21.26 shows the pattern of field lines.

The magnetic field that surrounds a current-carrying wire can exert a force on a moving charge, as the next example illustrates.

EXAMPLE 7   A Current Exerts a Magnetic Force on a Moving Charge

Figure 21.27 shows a long, straight wire carrying a current of $I = 3.0$ A. A particle of charge $q_0 = +6.5 \times 10^{-6}$ C is moving parallel to the wire at a distance of $r = 0.050$ m; the speed of the particle is $v = 280$ m/s. Determine the magnitude and direction of the magnetic force exerted on the moving charge by the current in the wire.

Reasoning The current generates a magnetic field in the space around the wire. A charge moving through this field experiences a magnetic force **F** whose magnitude is given by Equation 21.1 as $F = q_0 v B \sin \theta$, where θ is the angle between the magnetic field and the velocity of the charge. The magnitude of the magnetic field follows from Equation 21.5 as $B = \mu_0 I/(2\pi r)$. Thus, the magnitude of the magnetic force can be expressed as

$$F = q_0 v B \sin \theta = q_0 v \left(\frac{\mu_0 I}{2\pi r} \right) \sin \theta$$

The direction of the magnetic force is predicted by RHR-1 (see Section 21.2).

Solution The drawing shows that the magnetic field **B** lies in the plane that is perpendicular to both the wire and velocity **v** of the particle. Thus, the angle between **B** and **v** is $\theta = 90°$, and the magnitude of the magnetic force is

$$F = q_0 v \left(\frac{\mu_0 I}{2\pi r} \right) \sin 90°$$

$$F = (6.5 \times 10^{-6}\text{ C})(280\text{ m/s}) \left[\frac{(4\pi \times 10^{-7}\text{ T·m/A})(3.0\text{ A})}{2\pi(0.050\text{ m})} \right]$$

$$= \boxed{2.2 \times 10^{-8}\text{ N}}$$

The direction of the magnetic force is predicted by RHR-1 and, as the drawing shows, is radially inward toward the wire.

Figure 21.25 (*a*) A long, straight, current-carrying wire produces magnetic field lines that are circular about the wire. One such circular line is indicated by the compass needles. (*b*) If the thumb of the right hand (R.H.) is pointed in the direction of the current *I*, the curled fingers point in the direction of the magnetic field, according to RHR-2.

Figure 21.26 The magnetic field becomes stronger as the radial distance *r* decreases, so the field lines are closer together near the wire.

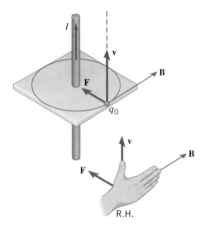

Figure 21.27 The moving positive charge q_0 experiences a magnetic force **F** because of the magnetic field **B** produced by the current in the wire.

We have now seen that an electric current can create a magnetic field of its own. We have also seen earlier that an electric current can experience a force created by another magnetic field. Therefore, the magnetic field that one current creates can exert a force on another nearby current. Examples 8 and 9 deal with this magnetic interaction between currents.

EXAMPLE 8
Two Current-Carrying Wires Exert Magnetic Forces on One Another*

Figure 21.28 shows two parallel straight wires. The wires are separated by a distance of $r = 0.065$ m and carry currents of $I_1 = 15$ A and $I_2 = 7.0$ A. Find the magnitude and direction of the force that the magnetic field of wire 1 applies to a 1.5-m length of wire 2 when the currents are (a) in opposite directions and (b) in the same direction.

Reasoning The current I_2 in wire 2 is situated in the magnetic field produced by the current in wire 1. The magnitude F of the magnetic force experienced by a length L of wire 2 is given by Equation 21.3 as $F = I_2 LB \sin \theta$. Here B is the magnitude of the magnetic field produced by wire 1 and is given by Equation 21.5 as $B = \mu_0 I_1/(2\pi r)$. The direction of the magnetic force can be determined by using RHR-1.

Solution

(a) At wire 2, the magnitude of the magnetic field created by wire 1 is

$$B = \frac{\mu_0 I_1}{2\pi r} = \frac{(4\pi \times 10^{-7}\,\text{T·m/A})(15\,\text{A})}{2\pi(0.065\,\text{m})} = 4.6 \times 10^{-5}\,\text{T} \qquad (21.5)$$

The direction of this field is upward at the location of wire 2, as part *a* of the figure shows. The direction can be obtained using RHR-2 (thumb of right hand along I_1, curled fingers point upward at wire 2 and indicate the direction of **B**). The magnetic field is perpendicular to wire 2 ($\theta = 90°$), so the magnitude of the force on a 1.5-m section of its length is

$$F = I_2 LB \sin \theta = (7.0\,\text{A})(1.5\,\text{m})(4.6 \times 10^{-5}\,\text{T}) \sin 90°$$
$$= \boxed{4.8 \times 10^{-4}\,\text{N}} \qquad (21.3)$$

---

* In this example the currents in the two wires and the distance between them are known; therefore, the magnetic force that one wire exerts on the other can be calculated. If, instead, the force and the distance were known and the wires carried the same current, that current could be calculated. This is, in fact, the procedure used to define the ampere, which is the unit for electric current. This procedure is chosen because force and distance are quantities that can be measured with a high degree of precision. One ampere is the amount of electric current in each of two long, parallel wires that gives rise to a magnetic force per unit length of 2×10^{-7} N/m on each wire when the wires are separated by one meter. With the ampere defined in terms of force and distance, the coulomb is defined as the quantity of electrical charge that passes a given point in one second when the current is one ampere or one coulomb per second.

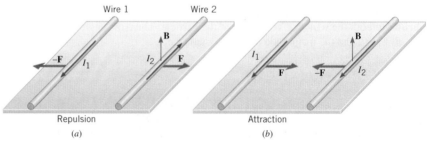

Figure 21.28 (*a*) Two long, parallel wires carrying currents I_1 and I_2 in opposite directions repel each other. (*b*) The wires attract each other when the currents are in the same direction.

The direction of the magnetic force on wire 2 is away from wire 1, as part *a* of the drawing indicates; the force direction is found by using RHR-1 (fingers of the right hand extended upward along **B**, thumb points along I_2, palm pushes in the direction of the force **F**).

In a like manner, the current in wire 2 also creates a magnetic field that produces a force on wire 1. Reasoning similar to that above shows that a 1.5-m length of wire 1 is repelled from wire 2 with a force that also has a magnitude of 4.8×10^{-4} N. Thus, each wire generates a force on the other, and, if the currents are in *opposite* directions, the wires *repel* each other. The fact that the two wires exert equal but oppositely directed forces on each other is consistent with Newton's third law, the action–reaction law.

(b) If the current in wire 2 is reversed, as part *b* of the drawing indicates, wire 2 is attracted to wire 1 because the direction of the magnetic force is reversed. However, the magnitude of the force is the same as that calculated in part *a* above. Likewise, wire 1 is attracted to wire 2. Two parallel wires carrying currents in the *same* direction *attract* each other.

Figure 21.29 A straight wire carries a current I_1, while a rectangular coil carries a current I_2. The dashed line is parallel to the wire and locates a small segment on each short side of the coil.

CONCEPTUAL EXAMPLE 9 **The Net Force That a Current-Carrying Wire Exerts on a Current-Carrying Coil**

Figure 21.29 shows a straight wire carrying a current I_1 and a rectangular coil carrying a current I_2. The wire and the coil lie in the same plane, with the wire parallel to the long sides of the rectangle. Is the coil attracted to or repelled from the wire?

Reasoning and Solution The current in the straight wire exerts a force on each of the four sides of the coil. The net force acting on the coil is the vector sum of these four forces. To decide whether the net force is attractive or repulsive, we need to consider the directions and magnitudes of the individual forces. In the long side of the coil near the wire, the current I_2 has the same direction as the current I_1, and we have just seen in Example 8 that two such currents attract each other. In the long side of the coil farthest from the wire, I_2 has a direction opposite to that of I_1, and according to Example 8, they repel one another. However, the attractive force is stronger than the repulsive force because the magnetic field produced by the current I_1 is stronger at shorter distances than at greater distances. Consequently, we reach the preliminary conclusion that the coil is attracted to the wire.

But what about the forces that act on the two short sides? Consider a small segment of each of the short sides, located at the same distance from the straight wire, as indicated by the dashed line in Figure 21.29. Each of these segments experiences the same magnetic field from the current I_1. RHR-2 shows that this field is directed downward into the plane of the paper, so that it is perpendicular to the current I_2 in each segment. But the directions of I_2 in the segments are opposite. As a result, RHR-1 reveals that the magnetic field from the straight wire applies a force to one segment that is opposite to that applied to the other segment. Thus, the forces acting on the two short sides of the coil cancel, and our preliminary conclusion is valid; *the coil is attracted to the wire.*

Related Homework Material: *Problem 54*

Loop of Wire

If a current-carrying wire is bent into a circular loop, the magnetic field lines around the loop have the pattern shown in Figure 21.30*a*. At the *center* of a loop of radius *R*, the magnetic field is perpendicular to the plane of the loop and has the value $B = \mu_0 I/(2R)$, where *I* is the current in the loop. Often, the loop consists of *N*

Figure 21.30 (*a*) The magnetic field lines in the vicinity of a current-carrying circular loop. (*b*) The direction of the magnetic field at the center of the loop is given by RHR-2.

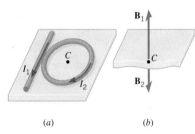

(a)   (b)

Figure 21.31 (a) A long, straight wire carrying a current I_1 lies next to a circular loop that carries a current I_2. (b) The magnetic fields at the center C of the loop produced by the long, straight wire ($\mathbf{B}_1$) and the circular loop ($\mathbf{B}_2$).

turns of wire that are wound sufficiently close together that they form a flat coil with a single radius. In this case, the magnetic fields of the individual turns add together to give a net field that is N times greater than that of a single loop. For such a coil the magnetic field at the center is

Center of a circular loop $$B = N\frac{\mu_0 I}{2R}$$ (21.6)

The direction of the magnetic field at the center of the loop can be determined with the help of RHR-2. If the thumb of the right hand is pointed in the direction of the current and the curled fingers are placed at the center of the loop, as in Figure 21.30b, the fingers indicate that the magnetic field points from right to left.

Example 10 shows how the magnetic fields produced by the current in a loop of wire and the current in a long, straight wire combine to form a net magnetic field.

EXAMPLE 10   Finding the Net Magnetic Field

A long, straight wire carries a current of $I_1 = 8.0$ A. As Figure 21.31a illustrates, a circular loop of wire lies immediately to the right of the straight wire. The loop has a radius of $R = 0.030$ m and carries a current of $I_2 = 2.0$ A. Assuming that the thickness of the wires is negligible, find the magnitude and direction of the net magnetic field at the center C of the loop.

Reasoning The net magnetic field at the point C is the sum of two contributions: (1) the field $\mathbf{B}_1$ produced by the long, straight wire, and (2) the field $\mathbf{B}_2$ produced by the circular loop. An application of RHR-2 shows that at point C the field $\mathbf{B}_1$ is directed upward, perpendicular to the plane containing the straight wire and the loop (see part b of the drawing). Similarly, RHR-2 shows that the magnetic field $\mathbf{B}_2$ is directed downward, opposite to the direction of $\mathbf{B}_1$.

Solution If we choose the upward direction in Figure 21.31b as positive, the net magnetic field at the point C is

$$B = \underbrace{\frac{\mu_0 I_1}{2\pi r}}_{\substack{\text{Long, straight}\\\text{wire}}} - \underbrace{\frac{\mu_0 I_2}{2R}}_{\substack{\text{Center of a}\\\text{circular loop}}} = \frac{\mu_0}{2}\left(\frac{I_1}{\pi r} - \frac{I_2}{R}\right)$$

$$B = \frac{(4\pi \times 10^{-7}\ \text{T}\cdot\text{m/A})}{2}\left[\frac{8.0\ \text{A}}{\pi(0.030\ \text{m})} - \frac{2.0\ \text{A}}{0.030\ \text{m}}\right] = \boxed{1.1 \times 10^{-5}\ \text{T}}$$

The net field is positive, so it is directed upward, perpendicular to the plane.

● **PROBLEM SOLVING INSIGHT**
Do not confuse the formula for the magnetic field produced at the center of a circular loop with that of a long, straight wire. The formulas are similar, differing only by a factor of π in the denominator.

A comparison of the magnetic field lines around the current loop in Figure 21.30a with those in the vicinity of the short bar magnet in Figure 21.32a shows that the two patterns are quite similar. Not only are the patterns similar, but the loop itself behaves as a bar magnet with a "north pole" on one side and a "south pole" on the other side. To emphasize that the loop may be imagined to be a bar magnet, Figure 21.32b includes a "phantom" bar magnet at the center of the loop. The side of the loop that acts like a north pole can be determined with the aid of RHR-2; curl the fingers of the right hand into the shape of a half-circle, point the thumb along the current I, and place the curled fingers at the center of the loop. The fingers not only point in the direction of $\mathbf{B}$, but they also point toward the north pole.

Because a current-carrying loop acts like a bar magnet, two adjacent loops can be either attracted to or repelled from each other, depending on the relative directions of the currents. Figure 21.33 includes a "phantom" magnet for each loop and

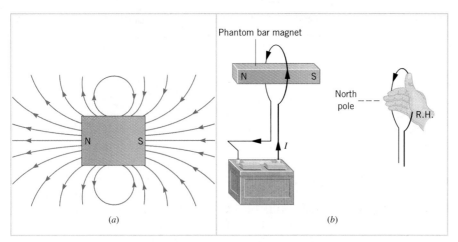

Figure 21.32 (*a*) The field lines around the bar magnet resemble those around the loop in Figure 21.30*a*. (*b*) The current loop can be imagined to be a phantom bar magnet with a north pole and a south pole.

shows that the loops are attracted to each other when the currents are in the same direction and repelled from each other when the currents are in opposite directions. This behavior is analogous to that of the two long, straight wires discussed in Example 8 (see Figure 21.28).

A Solenoid

A solenoid is a long coil of wire in the shape of a helix (see Figure 21.34). If the wire is wound so the turns are packed close to each other and the solenoid is long compared to its diameter, the magnetic field lines have the appearance shown in the drawing. Notice that the field inside the solenoid and away from its ends is nearly constant in magnitude and directed parallel to the axis. The direction of the field inside the solenoid is given by RHR-2, just as it is for a circular current loop. The magnitude of the magnetic field in the interior of a long solenoid is

Interior of a long solenoid

$$B = \mu_0 n I \tag{21.7}$$

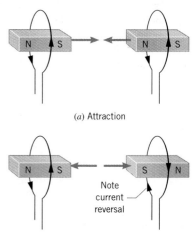

(*a*) Attraction

(*b*) Repulsion

Figure 21.33 (*a*) The two current loops attract each other if the directions of the currents are the same and (*b*) repel each other if the directions are opposite. The "phantom" magnets help explain the attraction and repulsion.

Figure 21.34 A solenoid and a cross-sectional view of it, showing the magnetic field lines and the north and south poles.

Figure 21.35 A magnetic resonance imaging machine. The patient is inserted into a circular opening at one end of a solenoid.

THE PHYSICS OF...

magnetic resonance imaging (MRI).

where n is the number of turns per unit length of the solenoid and I is the current. If, for example, the solenoid contains 100 turns and has a length of 0.05 m, the number of turns per unit length is $n = (100 \text{ turns})/(0.05 \text{ m}) = 2000 \text{ turns/m}$. The magnetic field outside the solenoid is not constant and is much weaker than the interior field. In fact, the magnetic field outside is nearly zero if the length of the solenoid is much greater than its diameter.

As with a single loop of wire, a solenoid can also be imagined to be a bar magnet, for the solenoid is just an array of connected current loops. And, as with a circular current loop, the location of the north pole can be determined with RHR-2. Figure 21.34 shows that the left end of the solenoid acts as a north pole, and the right end behaves as a south pole. Solenoids are often referred to as *electromagnets*, and they have several advantages over permanent magnets. For one thing, the strength of the magnetic field can be altered by changing the current and/or the number of turns per unit length. Furthermore, the north and south poles of an electromagnet can be readily switched by reversing the current.

Applications of the magnetic field produced by a current-carrying solenoid are widespread. An exciting medical application is in the technique of magnetic resonance imaging (MRI). With this technique, detailed pictures of the internal parts of the body can be obtained in a noninvasive way that involves none of the risks inherent in the use of X-rays. Figure 21.35 shows a patient about to be placed inside a magnetic resonance imaging machine. The circular opening into which the patient is inserted is one end of a solenoid, which is typically made from superconducting wire. The superconducting wire facilitates the use of large currents to produce a strong magnetic field. In the presence of this field, the nuclei of certain atoms can be made to behave as tiny radio transmitters and emit radio waves similar to those used by FM stations. The hydrogen atom, which is so prevalent in the human body, can be made to behave in this fashion. The strength of the magnetic field determines where a given collection of hydrogen atoms will broadcast on an imaginary "FM dial." With a magnetic field that has a slightly different strength at different places, it is possible to associate the location on this imaginary FM dial with a physical location within the body. Computer processing of these locations produces the magnetic resonance image. When hydrogen atoms are used in this way, the image is essentially a map showing their distribution within the body. Remarkably detailed magnetic resonance images can now be obtained, such as those shown in Figure 21.36. They pro-

Figure 21.36 Magnetic resonance imaging provides one way to diagnose brain disorders. The MRI of the diseased brain shown here reveals the presence of a tumor (large red area). The colors are computer generated as an aid in distinguishing between different types of tissue.

Healthy brain.   Diseased brain.

vide doctors with a powerful diagnostic tool that complements those available from X-ray and other techniques. Surgeons can now perform operations more accurately by stepping inside specially designed MRI scanners along with the patient and seeing live MRI images of the area into which they are cutting.

Television sets and computer display monitors use electromagnets (solenoids) to produce images by exerting magnetic forces on moving electrons. An evacuated glass tube, called a cathode-ray tube (CRT), contains an electron gun that sends a narrow beam of high-speed electrons toward the screen of the tube, as Figure 21.37a illustrates. The inner surface of the screen is covered with a phosphor coating, and when the electrons strike it, they generate a spot of visible light. This spot is called a pixel (a contraction of "picture element").

To create a black-and-white picture, the electron beam is scanned rapidly from left to right across the screen. As the beam makes each horizontal scan, the number of electrons per second striking the screen is changed by electronics controlling the electron gun, making the scan line brighter in some places and darker in others. When the beam reaches the right side of the screen, it is turned off and returned to the left side slightly below where it started (see part b of the figure). The beam is then scanned across the next line, and so on. In current TV sets, a complete picture consists of 525 scan lines (or 625 in Europe) and is formed in $\frac{1}{30}$ of a second. High-definition TV sets are being developed that use a greater number of scan lines.

The electron beam is deflected by a pair of electromagnets placed around the neck of the tube, between the electron gun and the screen. One electromagnet is responsible for producing the horizontal deflection of the beam and the other for the vertical deflection. For clarity, Figure 21.37a shows the net magnetic field gener-

---

INTERACTIVE LEARNINGWARE #44

START
Visualization
Reasoning
Knowns and Unknowns
Modeling the Problem
Numerical Solution
FINISH

A long, straight wire is oriented in the north–south direction, and the current in the wire is directed to the north. The horizontal component of the earth's magnetic field is 4.5×10^{-5} T and points due north. A small horizontal compass needle is located directly below the wire, 1.9 cm from it. The compass needle points 35° north of west. What is the current in the wire?

The magnetic field is a vector and can arise in a number of ways. Current-carrying wires produce magnetic fields, and the earth produces a magnetic field, for instance. The net field is the vector sum of all of the contributions that are present. The addition of vectors has appeared throughout this text, and Sections 1.6 and 1.8 discuss the general methods for adding vectors. This Interactive LearningWare example deals with both magnetic fields and vector addition.

Go to our Web site at

http://www.wiley.com/college/cutnell

Figure 21.37 (a) A cathode-ray tube contains an electron gun, a magnetic field for deflecting the electron beam, and a phosphor-coated screen. (b) The image is formed by scanning the electron beam across the screen. (c) The red, green, and blue phosphors of a color TV.

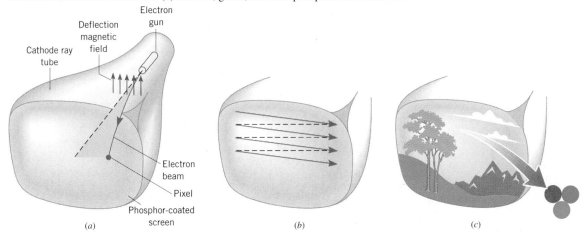

(a) (b) (c)

THE PHYSICS OF...

television screens and computer display monitors.

ated by the electromagnets at one instant, and not the electromagnets themselves. The electric current in the electromagnets produces a net magnetic field that exerts a force on the moving electrons, causing their trajectories to bend and reach different points on the screen. Changing the current changes the field, so the electrons can be deflected to any point on the screen.

A color TV operates with three electron guns instead of one. And the single phosphor of a black-and-white TV is replaced by a large number of three-dot clusters of phosphors that glow red, green, and blue when struck by an electron beam, as indicated in Figure 21.37c. Each red, green, and blue color in a cluster is produced when electrons from one of the three guns strike the corresponding phosphor dot. The three dots are so close together that, from a normal viewing distance, they cannot be separately distinguished. Red, green, and blue are primary colors, so all other colors can be created by varying the intensities of the three beams focused on a cluster.

21.8 AMPERE'S LAW

Concepts at a Glance We have seen that an electric current creates a magnetic field. However, the magnitude and direction of the field at any point in space depends on the specific geometry of the wire carrying the current. For instance, distinctly different magnetic fields surround a long, straight wire, a circular loop of wire, and a solenoid. Although different, each of these fields can be obtained from a general law known as ***Ampere's law,*** which is valid for a wire of any geometrical shape. The Concepts-at-a-Glance chart in Figure 21.38, which is an extension of the chart in Figure 21.9, illustrates that Ampere's law specifies the relationship between a current and its associated magnetic field.

To see how Ampere's law is stated, consider Figure 21.39, which shows two wires carrying currents I_1 and I_2. In general, there may be any number of currents. Around the wires we construct an arbitrarily shaped but closed path. Note that this

Figure 21.38 *Concepts at a Glance* As the concept chart in Figure 21.9 illustrates, the magnetic field is defined in terms of the magnetic force that acts on a moving test charge; this earlier chart is reproduced here in the three upper-left panels. Electric currents produce magnetic fields, and the relation between the two is expressed by Ampere's law. This law was discovered by Andre Marie Ampere (1775–1836), a gifted physicist and mathematician.

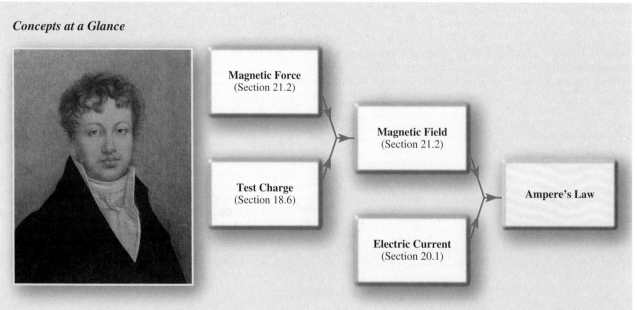

Concepts at a Glance

Magnetic Force (Section 21.2)

Test Charge (Section 18.6)

Magnetic Field (Section 21.2)

Electric Current (Section 20.1)

Ampere's Law

path encloses a surface. The path is constructed from a large number of short segments, each of length $\Delta\ell$. Ampere's law deals with the product of $\Delta\ell$ and $B_\parallel$ for each segment, where $B_\parallel$ is the component of the magnetic field that is *parallel* to $\Delta\ell$ (see the blow-up view in the drawing). For magnetic fields that do not change as time passes, the law states that the sum of all the $B_\parallel \Delta\ell$ terms is proportional to the net current $I = I_1 + I_2$ passing through the surface enclosed by the path. Ampere's law is stated in equation form as follows:

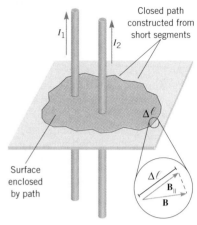

Figure 21.39 This setup is used in the text to explain Ampere's law.

● **AMPERE'S LAW FOR STATIC MAGNETIC FIELDS**

For any current geometry that produces a magnetic field that does not change in time,

$$\Sigma B_\parallel \Delta\ell = \mu_0 I \qquad (21.8)$$

where $\Delta\ell$ is a small segment of length along a closed path of arbitrary shape around the current, $B_\parallel$ is the component of the magnetic field parallel to $\Delta\ell$, I is the net current passing through the surface bounded by the path, and μ_0 is the permeability of free space. The symbol Σ indicates that the sum of all $B_\parallel \Delta\ell$ terms must be taken around the closed path.

To illustrate the use of Ampere's law, we apply it in Example 11 to the special case of the current in a long, straight wire and show that it leads to the proper expression for the magnetic field.

EXAMPLE 11   A Long, Straight Current-Carrying Wire

Use Ampere's law to obtain the magnetic field produced by the current in a long, straight wire.

Reasoning Figure 21.25*a* shows that compass needles point in a circular pattern around the current-carrying wire, so we know that the magnetic field lines are circular. Therefore, it is convenient to use a circular path of radius r when applying Ampere's law, as Figure 21.40 indicates.

Solution Along the circular path in Figure 21.40, the magnetic field is everywhere parallel to $\Delta\ell$ and has a constant magnitude, since each point is at the same distance from the wire. Thus, $B_\parallel = B$, and according to Ampere's law, we have

$$\Sigma B_\parallel \Delta\ell = B(\Sigma \Delta\ell) = \mu_0 I$$

But $\Sigma \Delta\ell$ is just the circumference $2\pi r$ of the circle, so Ampere's law reduces to

$$B(\Sigma \Delta\ell) = B(2\pi r) = \mu_0 I$$

Dividing both sides by $2\pi r$ shows that $\boxed{B = \mu_0 I/(2\pi r)}$, as given earlier in Equation 21.5.

Figure 21.40 Example 11 uses Ampere's law to find the magnetic field in the vicinity of this long, straight, current-carrying wire.

21.9 MAGNETIC MATERIALS

Ferromagnetism

The similarity between the magnetic field lines in the neighborhood of a bar magnet and those around a current loop suggests that the magnetism in each case arises from a common cause. The field that surrounds the loop is created by the charges moving in the wire. The magnetic field around a bar magnet is also due to the mo-

Figure 21.41 (*a*) Each magnetic domain is a highly magnetized region that behaves like a small magnet (represented by an arrow whose head indicates a north pole). An unmagnetized piece of iron consists of many domains that are randomly aligned. The size of each domain is exaggerated for clarity. (*b*) The external magnetic field of the permanent magnet causes those domains that are parallel or nearly parallel to the field to grow in size (shown in gold).

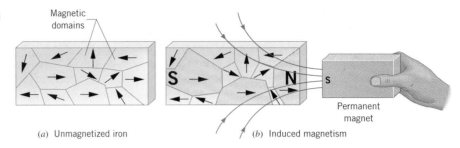

(*a*) Unmagnetized iron   (*b*) Induced magnetism

tion of charges, but the motion is not that of a bulk current through the magnetic material. Instead, the motion responsible for the magnetism is that of the electrons within the atoms of the material.

The magnetism produced by electrons within an atom can arise from two motions. First, each electron orbiting the nucleus behaves like an atomic-sized loop of current that generates a small magnetic field, similar to the field created by the current loop in Figure 21.30*a*. Second, each electron possesses a spin that also gives rise to a magnetic field. The net magnetic field created by the electrons within an atom is due to the combined fields created by their orbital and spin motions.

In most substances the magnetism produced at the atomic level tends to cancel out, with the result that the substance is nonmagnetic overall. However, there are some materials, known as ***ferromagnetic materials,*** in which the cancellation does not occur for groups of approximately $10^{16} - 10^{19}$ neighboring atoms, because they have electron spins that are naturally aligned parallel to each other. This alignment results from a special type of quantum mechanical* interaction between the spins. The result of the interaction is a small but highly magnetized region of about 0.01 to 0.1 mm in size, depending on the nature of the material; this region is called a ***magnetic domain.*** Each domain behaves as a small magnet with its own north and south poles. Common ferromagnetic materials are iron, nickel, cobalt, chromium dioxide, and alnico (an *al*uminum–*ni*ckel–*co*balt alloy).

Induced Magnetism

Often the magnetic domains in a ferromagnetic material are arranged randomly, as Figure 21.41*a* illustrates for a piece of iron. In such a situation, the magnetic fields of the domains cancel each other, so the iron displays little, if any, overall magnetism. However, an unmagnetized piece of iron can be magnetized by placing it in an external magnetic field provided by a permanent magnet or an electromagnet. The external magnetic field penetrates the unmagnetized iron and *induces* (or "brings about") a state of magnetism in the iron by causing two effects. Those domains whose magnetism is parallel or nearly parallel to the external magnetic field grow in size at the expense of other domains that are not so oriented. Part *b* of the drawing shows the growing domains in gold. In addition, the magnetic alignment of some domains may rotate and become more oriented in the direction of the external field. The resulting preferred alignment of the domains gives the iron an overall magnetism, so the iron behaves like a magnet with associated north and south poles. In some types of ferromagnetic materials, such as the chromium dioxide used in cassette tapes, the domains remain aligned for the most part when the external magnetic field is removed, and the material thus becomes permanently magnetized.

Figure 21.42 Because of induced magnetism, powerful magnetic grippers enable this worker to climb the metal bulkhead of a ship.

---

* The branch of physics called quantum mechanics is mentioned in Section 29.5, although a detailed discussion of it is beyond the scope of this book.

The magnetism induced in a ferromagnetic material can be surprisingly large, even in the presence of a weak external field. For instance, it is not unusual for the induced magnetic field to be a hundred to a thousand times stronger than the external field that causes the alignment. For this reason, high-field electromagnets are constructed by wrapping the current-carrying wire around a solid core made from iron or other ferromagnetic material.

Induced magnetism explains why a permanent magnet sticks to a refrigerator door and why an electromagnet can pick up scrap iron at a junkyard. The photo in Figure 21.42 illustrates yet another example of induced magnetism. Notice in Figure 21.41*b* that there is a north pole at the end of the iron that is closest to the south pole of the permanent magnet. The net result is that the two opposite poles give rise to an attraction between the iron and the permanent magnet. Conversely, the north pole of the permanent magnet would also attract the piece of iron by inducing a south pole in the nearest side of the iron. In nonferromagnetic materials, such as aluminum and copper, the formation of magnetic domains does not occur, so magnetism cannot be induced into these substances. Consequently, magnets do not stick to aluminum cans or to copper pennies.

Magnetic Tape Recording

The process of magnetic tape recording uses induced magnetism, as Figure 21.43 illustrates. The weak electrical signal from a microphone is routed to an amplifier where it is amplified. The current from the output of the amplifier is then sent to the recording head, which is a coil of wire wrapped around an iron core. The iron core has the approximate shape of a horseshoe with a small gap between the two ends. The ferromagnetic iron substantially enhances the magnetic field produced by the current in the wire.

When there is a current in the coil, the recording head becomes an electromagnet with a north pole at one end and a south pole at the other end. The magnetic field lines pass through the iron core and cross the gap. Within the gap, the lines are directed from the north pole to the south pole. Some of the field lines in the gap "bow outward," as Figure 21.43 indicates, the bowed region of magnetic field being called the *fringe field*. The fringe field penetrates the magnetic coating on the tape and induces magnetism in the coating. This induced magnetism is retained when the tape leaves the vicinity of the recording head and, thus, provides a means for storing audio information. Audio information is stored, because at any instant in time the way in which the tape is magnetized depends on the amount and direction of current in the recording head. The current, in turn, depends on the sound picked up by the microphone, so that changes in the sound that occur from moment to moment are preserved as changes in the tape's induced magnetism.

Maglev Trains

A magnetically levitated train—maglev, for short—uses forces that arise from induced magnetism to levitate or float above a guideway. Since it rides a few centimeters above the guideway, a maglev does not need wheels. Freed from friction with the guideway, the train can achieve significantly greater speeds than do conventional trains. For example, the Transrapid maglev in Figure 21.44 has achieved speeds of 110 m/s (250 mph).

Figure 21.44*a* shows that the Transrapid maglev achieves levitation with electromagnets mounted on arms that extend around and under the guideway. When a current is sent to an electromagnet, the resulting magnetic field creates induced mag-

THE PHYSICS OF...

magnetic tape recording.

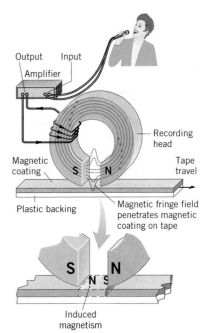

Figure 21.43 The magnetic fringe field of the recording head penetrates the magnetic coating on the tape and causes the coating to become magnetized.

Figure 21.44 (*a*) The Transrapid maglev (a German train) has achieved speeds of 110 m/s (250 mph). The levitation electromagnets are drawn up toward the rail in the guideway, levitating the train. (*b*) The magnetic propulsion system.

Guideway   Rail

Arm

Levitation electromagnet

Guideway

(*a*)

(*b*)

netism in a rail mounted in the guideway. The upward attractive force from the induced magnetism is balanced by the weight of the train, so the train moves without touching the rail or the guideway.

Magnetic levitation only lifts the train and does not move it forward. Figure 21.44*b* illustrates how magnetic propulsion is achieved. In addition to the levitation electromagnets, propulsion electromagnets are also placed along the guideway. By controlling the direction of the currents in the train and guideway electromagnets, it is possible to create an unlike pole in the guideway just ahead of each electromagnet on the train and a like pole just behind. Each electromagnet on the train is thus both pulled and pushed forward by electromagnets in the guideway. By adjusting the timing of the like and unlike poles in the guideway, the speed of the train can be adjusted. Reversing the poles in the guideway electromagnets serves to brake the train.

THE PHYSICS OF...

a magnetically levitated train.

21.10 CONCEPTS & CALCULATIONS

Both magnetic fields and electric fields can apply forces to an electric charge. However, there are distinct differences in the way the two types of fields apply their forces. This chapter discusses the magnetic field, and Chapter 18 deals with the electric field. Example 13 serves to review how the two types of fields behave.

CONCEPTS & CALCULATIONS EXAMPLE 12   Magnetic and Electric Fields

Figure 21.45 The net force acting on the moving charge in this drawing is calculated in Example 12.

Figure 21.45 shows a negatively charged particle whose charge magnitude is $q = 2.80 \times 10^{-6}$ C. It is moving along the $+y$ axis at a speed of $v = 4.80 \times 10^{6}$ m/s. A magnetic field of magnitude 3.35×10^{-5} T is directed along the $+z$ axis, and an electric field of magnitude 123 N/C points along the $-x$ axis. Determine the magnitude and direction of the net force that acts on the particle.

Concept Questions and Answers What is the net force?

Answer   The net force is the vector sum of all the forces acting on the particle, which in this case are the magnetic force and the electric force.

How do you determine the direction of the magnetic force acting on the negative charge?

Answer   First, remember that the direction for either a positive or negative charge is perpendicular to the plane formed by the magnetic field vector **B** and the velocity vector **v**. Then, apply RHR-1 to find the direction of the force as if the charge were positive. Finally, reverse the direction indicated by RHR-1 to determine the direction of the force on the negative charge.

How do you determine the direction of the electric force acting on the negative charge?

Answer   The direction for either a positive or negative charge is along the line of the electric field. The electric force on a positive charge is in the same direction as the electric field, while the force on a negative charge is in the direction opposite to the electric field.

Does the fact that the charge is moving affect the values of the magnetic and electric forces?

Answer   The fact that the charge is moving affects the value of the magnetic force, for without motion, there is no magnetic force. The electric force, in contrast, has the same value whether or not the charge is moving.

> • **PROBLEM SOLVING INSIGHT**
> The direction of the magnetic force exerted on a negative charge is opposite to that exerted on a positive charge, assuming both charges are moving in the same direction in the same magnetic field.

Solution An application of RHR-1 (see Figure 21.8) reveals that if the charge were positive, the magnetic force would point along the $+x$ axis. But the charge is negative, so the magnetic force points along the $-x$ axis. According to Equation 21.1, the magnitude of the magnetic force is $F_{\text{magnetic}} = qvB \sin \theta$, where q is the magnitude of the charge. Since the velocity and the magnetic field are perpendicular, $\theta = 90°$, and we have $F_{\text{magnetic}} = qvB$.

The electric field points along the $-x$ axis and would apply a force in that same direction if the charge were positive. But the charge is negative, so the electric force points in the opposite direction, or along the $+x$ axis. According to Equation 18.2, the magnitude of the electric force is $F_{\text{electric}} = qE$.

Using plus and minus signs to take into account the different directions of the magnetic and electric forces, we find that the net force is

$$\Sigma F = +F_{\text{electric}} - F_{\text{magnetic}} = +qE - qvB$$
$$= +(2.80 \times 10^{-6} \text{ C})(123 \text{ N/C}) - (2.80 \times 10^{-6} \text{ C})(4.80 \times 10^6 \text{ m/s})$$
$$\times (3.35 \times 10^{-5} \text{ T})$$
$$= \boxed{-1.06 \times 10^{-4} \text{ N}}$$

The fact that the answer is negative means that the net force points along the $-x$ axis.

A number of forces can act simultaneously on a rigid object, some of them producing torques, as Chapter 9 discusses. If, as a result of the forces and torques, the object has no acceleration of any kind, it is in equilibrium. Section 9.2 discusses the equilibrium of rigid objects. The next example illustrates that the magnetic force can be one of the forces keeping an object in equilibrium.

CONCEPTS & CALCULATIONS EXAMPLE 13   Equilibrium

A 125-turn rectangular coil of wire is hung from one arm of a balance, as Figure 21.46 shows. With the magnetic field turned off, a mass M is added to the pan on the other arm to balance the weight of the coil. When a constant 0.200-T magnetic field

Figure 21.46 This setup is the focus of Example 13.

is turned on and there is a current of 8.50 A in the coil, how much additional mass m must be added to regain the balance?

Concept Questions and Answers In a balanced or equilibrium condition the device has no angular acceleration. What does this imply about the net torque acting on the device?

Answer According to Newton's second law as applied to rotational motion (Equation 9.7), when the angular acceleration is zero, the net torque is zero. Counterclockwise torques must balance clockwise torques for an object that is in equilibrium.

What is a torque?

Answer According to Equation 9.1, the torque produced by a force is the magnitude of the force times the lever arm of the force. The lever arm is the perpendicular distance between the line of action of the force and the axis of rotation. Counterclockwise torques are positive, and clockwise torques are negative.

In calculating the torques acting on an object in equilibrium, where do you locate the axis of rotation?

Answer The location of the axis is arbitrary because, if there is no angular acceleration, the torques must add up to zero with respect to any axis whatsoever. Choose an axis that simplifies your calculations.

Solution In a condition of equilibrium the sum of the torques acting on the device is zero. For the purpose of calculating torques, we use an axis that is perpendicular to the page at the knife-edge supporting the balance arms (see the drawing). The forces that can produce torques are the weights of the pans, the masses in the left pan, and the coil hanging from the right pan, and the magnetic force on the coil. The weight of each pan is W_{pan}, and the weight of the coil is W_{coil}. The weights of the masses M and m are Mg and mg, where g is the magnitude of the acceleration due to gravity. The lever arm for each of these forces is ℓ (see the drawing). When the magnetic field is turned off and the mass m is not present, the sum of the torques is

Field off $\qquad \Sigma\tau = \underbrace{(+W_{pan}\ell + Mg\ell)}_{\text{Left pan}} + \underbrace{(-W_{pan}\ell - W_{coil}\ell)}_{\text{Right pan}} = Mg\ell - W_{coil}\ell = 0$

When the field is turned on, a magnetic force is applied to the 0.0150-m segment of the coil that is in the field. Since the magnetic field points into the page and the current is from right to left, an application of RHR-1 reveals that the magnetic force points downward. According to Equation 21.3, the magnitude of the magnetic force is $F = ILB \sin \theta$, where I is the current and L is the length of the wire in the field. Since the current and the field are perpendicular, $\theta = 90°$, and we have $F = ILB$. The sum of the torques with the field turned on and the mass m added to the left pan is

Field on $\qquad \Sigma\tau = \underbrace{(+W_{pan}\ell + Mg\ell + mg\ell)}_{\text{Left pan}} + \underbrace{(-W_{pan}\ell - W_{coil}\ell - ILB\ell)}_{\text{Right pan}}$

$\qquad\qquad = Mg\ell + mg\ell - W_{coil}\ell - ILB\ell = 0$

From the field-off equation, it follows that $Mg\ell - W_{coil}\ell = 0$. With this substitution, the field-on equation simplifies to the following:

$$mg\ell - ILB\ell = 0$$

Solving this expression for the additional mass m gives

$$m = \frac{ILB}{g} = \frac{(8.50 \text{ A})[125(0.0150 \text{ m})](0.200 \text{ T})}{9.80 \text{ m/s}^2} = \boxed{0.325 \text{ kg}}$$

In this calculation we have used the fact that the coil has 125 turns, so that the length of the wire in the field is 125 (0.0150 m).

At the end of the problem set for this chapter, you will find homework problems that contain both conceptual and quantitative parts. These problems are grouped under the heading *Concepts & Calculations, Group Learning Problems.* They are designed for use by students working alone or in small learning groups. The conceptual part of each problem provides a convenient focus for group discussions.

SUMMARY

A magnet has a north pole and a south pole. The north pole is the end that points toward the north magnetic pole of the earth when the magnet is freely suspended. **Like poles repel each other, and unlike poles attract each other.**

A **magnetic field** exists in the space around a magnet. The magnetic field is a vector whose direction at any point is the direction indicated by the north pole of a small compass needle placed at that point. The magnitude B of the magnetic field at any point in space is defined as $B = F/(q_0 v \sin \theta)$, where F is the magnitude of the magnetic force that acts on a charge q_0 whose velocity $\mathbf{v}$ makes an angle θ with respect to the magnetic field. The SI unit for the magnetic field is the tesla (T). The direction of the magnetic force is perpendicular to both $\mathbf{v}$ and $\mathbf{B}$, and for a positive charge the direction can be determined with the aid of Right-Hand Rule No. 1 (RHR-1, see Section 21.2). The magnetic force on a moving negative charge is opposite to the force on a moving positive charge.

If a particle of charge q and mass m moves with speed v perpendicular to a uniform magnetic field $\mathbf{B}$, the magnetic force causes the charge to move on a circular path of radius $r = mv/(qB)$. A constant magnetic force does no work on a charged particle because the direction of the force is always perpendicular to the motion of the particle. Being unable to do work, the magnetic force cannot change the kinetic energy of the particle; however, the magnetic force does change the direction in which the particle moves.

An electric current, being composed of moving charges, can experience a magnetic force when placed in a magnetic field. For a straight wire that has a length L and carries a current I, the magnetic force has a magnitude of $F = ILB \sin \theta$, where θ is the angle between the directions of I and $\mathbf{B}$. The direction of the force is perpendicular to both I and $\mathbf{B}$ and is given by RHR-1.

Magnetic forces can exert a **torque** on a current-carrying loop of wire and thus cause the loop to rotate. If a current I exists in a coil of wire with N turns, each of area A, in the presence of a magnetic field $\mathbf{B}$, the coil experiences a torque of magnitude $\tau = NIAB \sin \phi$, where ϕ is the angle between the direction of the magnetic field and the normal to the plane of the coil.

An electric current produces a magnetic field, with different current geometries giving rise to different field patterns. For a **long, straight wire,** the magnetic field lines are circles centered on the wire, and their direction is given by Right-Hand Rule No. 2 (RHR-2, see Section 21.7). The magnitude of the field at a radial distance r from the wire is $B = \mu_0 I/(2\pi r)$, where I is the current and μ_0 is the permeability of free space ($\mu_0 = 4\pi \times 10^{-7}$ T·m/A). The magnetic field at the center of **a flat circular coil** consisting of N turns, each of radius R, is $B = N\mu_0 I/(2R)$. The coil has associated with it a north pole on one side and a south pole on the other side. The side of the coil that behaves like a north pole can be predicted by using RHR-2. A **solenoid** is a coil of wire wound in the shape of a helix. Inside a long solenoid the magnetic field is nearly constant and has the value $B = \mu_0 n I$, where n is the number of turns per unit length of the solenoid. One end of a solenoid behaves like a north pole, and the other end like a south pole, as can be predicted by using RHR-2.

Ampere's law, as presented in Equation 21.8, specifies the relationship between a current and its associated magnetic field.

Ferromagnetic materials, such as iron, are made up of tiny regions called domains, each of which behaves as a small magnet. In an unmagnetized ferromagnetic material, the domains are randomly aligned. In a permanent magnet, many of the domains are aligned, and a high degree of magnetism results. An unmagnetized ferromagnetic material can be induced into becoming magnetized by placing it in an external magnetic field.

CONCEPTUAL QUESTIONS

1. Magnetic field lines, like electric field lines, never intersect. Suppose it were possible for two magnetic field lines to intersect at a point in space. Discuss what this would imply about the force(s) that act on a charge moving through such a point, thereby ruling out the possibility of field lines crossing.

2. Suppose you accidentally use your left hand, instead of your right hand, to determine the direction of the magnetic force on a

positive charge moving in a magnetic field. Do you get the correct answer? If not, what direction do you get?

3. A charged particle, passing through a certain region of space, has a velocity whose magnitude and direction remain constant. (a) If it is known that the external magnetic field is zero everywhere in this region, can you conclude that the external electric field is also zero? Explain. (b) If it is known that the external electric field is zero everywhere, can you conclude that the external magnetic field is also zero? Explain.

4. Suppose that the positive charge in Figure 21.10*a* were launched from the negative plate toward the positive plate, in a direction opposite to the electric field. A sufficiently strong electric field would prevent the charge from striking the positive plate. Suppose the positive charge in Figure 21.10*b* were launched from the south pole toward the north pole, in a direction opposite to the magnetic field. Would a sufficiently strong magnetic field prevent the charge from reaching the north pole? Account for your answer.

5. Review Conceptual Example 4 as background for this question. Three particles move through a constant magnetic field and follow the paths shown in the drawing. Determine whether each particle is positively charged, negatively charged, or neutral. Give a reason for each answer.

6. Review Conceptual Example 4 as background for this question. Three particles have identical charges and masses. They enter a constant magnetic field and follow the paths shown in the picture. Which particle is moving the fastest, and which is moving the slowest? Justify your answers.

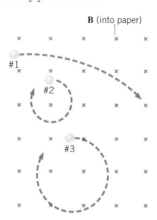

7. Refer to Figure 21.12. Assume that the particle in the picture is a proton. If an electron is projected at point 1 with the same velocity v, it will not follow the same path as the proton, unless the magnetic field is adjusted. Explain how the magnitude and direction of the field must be changed.

8. The drawing shows a top view of four interconnected chambers. A negative charge is fired into chamber 1. By turning on separate magnetic fields in each chamber, the charge can be made to exit from chamber 4, as shown. (a) Describe how the magnetic field in each chamber should be directed. (b) If the speed of the charge is v when it enters chamber 1, what is the speed of the charge when it exits chamber 4? Why?

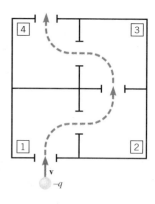

9. A positive charge moves along a circular path under the influence of a magnetic field. The magnetic field is perpendicular to the plane of the circle, as in Figure 21.12. If the velocity of the particle is reversed at some point along the path, will the particle retrace its path? If not, draw the new path. Explain.

10. When one end of a bar magnet is placed near a TV screen, the picture becomes distorted. Why?

11. The drawing shows a particle carrying a positive charge $+q$ at the coordinate origin, as well as a target located in the third quadrant. A uniform magnetic field is directed perpendicularly into the plane of the paper. The charge can be projected in the plane of the paper only, along the positive or negative x or y axis. Thus, there are four possible directions for the initial velocity of the particle. The particle can be made to hit the target for only two of the four directions. Which two are they? Give your reasoning, and draw the two paths that the particle can follow on its way to the target.

12. Refer to Figure 21.17. (a) What happens to the direction of the magnetic force if the current is reversed? (b) What happens to the direction of the force if *both* the current and the magnetic poles are reversed? Explain your answers.

13. The drawing shows a conducting wire wound into a helical shape. The helix acts like a spring and expands back toward its original shape after its coils are squeezed together and released. The bottom end of the wire just barely touches the mercury (a good electrical conductor) in the cup. After the switch is closed, current in the circuit causes the light bulb to glow. Does the bulb glow continually, glow briefly and then go out, or repeatedly turn on and off like a turn signal on a car? Explain.

14. In Figure 21.28, assume that the current I_1 is larger than the current I_2. In parts a and b, decide whether there are places where the total magnetic field is zero. State whether they are located to the left of both wires, between the wires, or to the right of both wires. Give your reasoning.

15. The drawing shows an end-on view of three parallel wires that are perpendicular to the plane of the paper. In two of the wires the current is directed into the paper, while in the remaining wire the current is directed out of the paper. The two outermost wires are held rigidly in place. Which way will the middle wire move? Explain.

16. For each electromagnet at the left of the drawing, explain whether it will be attracted to or repelled from the permanent magnet at the right.

17. For each electromagnet at the left of the drawing, explain whether it will be attracted to or repelled from the adjacent electromagnet at the right.

18. Refer to Figure 21.5. If the earth's magnetism is assumed to originate from a large circular loop of current within the earth, how is the plane of this current loop oriented relative to the magnetic axis, and what is the direction of the current around the loop?

19. There are four wires viewed end-on in the drawing. They are long, straight, and perpendicular to the plane of the paper. Their cross sections lie at the corners of a square. Currents of the same magnitude are in each of these wires. Choose the direction of the current for each wire, so that when any single current is turned off, the total magnetic field at point P (the center of the square) is directed toward a corner of the square. Account for your answer.

20. Suppose you have two bars, one of which is a permanent magnet and the other of which is not a magnet, but is made from a ferromagnetic material like iron. The two bars look exactly alike. (a) Using a third bar, which is known to be a magnet, how can you determine which of the look-alike bars is the permanent magnet and which is not? (b) Can you determine the identities of the look-alike bars with the aid of a third bar that is not a magnet, but is made from a ferromagnetic material? Give reasons for your answers.

21. In a TV commercial that advertises a soda pop, a strong electromagnet picks up a delivery truck carrying cans of the soft drink. The picture switches to the interior of the truck, where cans are seen to fly upward and stick to the roof just beneath the electromagnet. Are these cans made entirely of aluminum? Justify your answer.

PROBLEMS

ssm Solution is in the Student Solutions Manual.   **www** Solution is available on the World Wide Web at http://www.wiley.com/college/cutnell
⚕ **This icon represents a biomedical application.**

Section 21.1 Magnetic Fields, Section 21.2 The Force That a Magnetic Field Exerts on a Moving Charge

1. ssm A charge of 12 μC, traveling with a speed of 9.0×10^6 m/s in a direction perpendicular to a magnetic field, experiences a magnetic force of 8.7×10^{-3} N. What is the magnitude of the field?

2. A particle with a charge of $+8.4$ μC and a speed of 45 m/s enters a uniform magnetic field whose magnitude is 0.30 T. For each of the cases in the drawing, find the magnitude and direction of the magnetic force on the particle.

3. A charge $q_1 = 25.0$ μC moves with a speed of 4.50×10^3 m/s perpendicular to a uniform magnetic field. The charge experiences a magnetic force of 7.31×10^{-3} N. A second charge

$q_2 = 5.00$ μC travels at an angle of 40.0° with respect to the same magnetic field and experiences a 1.90×10^{-3}-N force. Determine (a) the magnitude of the magnetic field and (b) the speed of q_2.

4. Two charged particles move in the same direction with respect to the same magnetic field. Particle 1 travels three times faster than particle 2. However, each particle experiences a magnetic force of the same magnitude. Find the ratio q_1/q_2 of the magnitudes of the charges.

5. ssm At a certain location, the horizontal component of the earth's magnetic field is 2.5×10^{-5} T, due north. A proton moves eastward with just the right speed, so the magnetic force on it balances its weight. Find the speed of the proton.

6. When a charged particle moves at an angle of 25° with respect to a magnetic field, it experiences a magnetic force of magnitude F. At what angle (less than 90°) with respect to this field will this particle, moving at the same speed, experience a magnetic force of magnitude $2F$?

7. An electron is moving through a magnetic field whose magnitude is 8.70×10^{-4} T. The electron experiences only a magnetic force and has an acceleration of magnitude 3.50×10^{14} m/s^2. At a certain instant, it has a speed of 6.80×10^6 m/s. Determine the angle θ (less than 90°) between the electron's velocity and the magnetic field.

*__8.__ There is a 0.200-T magnetic field directed along the $+x$ axis and a field of unknown magnitude along the $+y$ axis. A particle carrying a charge of 6.50×10^{-5} C experiences a maximum force of 0.455 N when traveling at a speed of 2.00×10^4 m/s through the region where the fields are. Find the magnitude of the unknown field.

*__9. ssm www__ The electrons in the beam of a television tube have a kinetic energy of 2.40×10^{-15} J. Initially, the electrons move horizontally from west to east. The vertical component of the earth's magnetic field points down, toward the surface of the earth, and has a magnitude of 2.00×10^{-5} T. (a) In what direction are the electrons deflected by this field component? (b) What is the acceleration of an electron in part (a)?

Section 21.3 The Motion of a Charged Particle in a Magnetic Field, Section 21.4 The Mass Spectrometer

10. An electron moves at a speed of 6.0×10^6 m/s perpendicular to a constant magnetic field. The path is a circle of radius 1.3×10^{-3} m. (a) Draw a sketch showing the magnetic field and the electron's path. (b) What is the magnitude of the field? (c) Find the magnitude of the electron's acceleration.

11. An ionized helium atom has a mass of 6.6×10^{-27} kg and a speed of 4.4×10^5 m/s. The atom moves perpendicular to a 0.75-T magnetic field on a circular path of radius 0.012 m. Determine whether the charge of the ionized helium atom is $+e$ or $+2e$.

12. A magnetic field has a magnitude of 1.2×10^{-3} T, and an electric field has a magnitude of 4.6×10^3 N/C. Both fields point in the same direction. A positive 1.8-μC charge moves at a speed of 3.1×10^6 m/s in a direction that is perpendicular to both fields. Determine the magnitude of the net force that acts on the charge.

13. ssm A beam of protons moves in a circle of radius 0.25 m. The protons move perpendicular to a 0.30-T magnetic field. (a) What is the speed of each proton? (b) Determine the magnitude of the centripetal force that acts on each proton.

14. The solar wind is a thin, hot gas given off by the sun. Charged particles in this gas enter the magnetic field of the earth and can experience a magnetic force. Suppose a charged particle traveling with a speed of 9.0×10^6 m/s encounters the earth's magnetic field at an altitude where the field has a magnitude of 1.2×10^{-7} T. Assuming that the particle's velocity is perpendicular to the magnetic field, find the radius of the circular path on which the particle would move if it were (a) an electron and (b) a proton.

15. ssm www A charged particle with a charge-to-mass ratio of $q/m = 5.7 \times 10^8$ C/kg travels on a circular path that is perpendicular to a magnetic field whose magnitude is 0.72 T. How much time does it take for the particle to complete one revolution?

16. Review Conceptual Example 2 before attempting this problem. Derive an expression for the magnitude v of the velocity "selected" by the velocity selector. This expression should give v in terms of the strengths E and B of the electric and magnetic fields, respectively.

17. Suppose that an ion source in a mass spectrometer produces *doubly* ionized gold ions (Au^{2+}), each with a mass of 3.27×10^{-25} kg. The ions are accelerated from rest through a potential difference of 1.00 kV. Then, a 0.500-T magnetic field causes the ions to follow a circular path. Determine the radius of the path.

*__18.__ The ion source in a mass spectrometer produces both singly and doubly ionized species, X^+ and X^{2+}. The difference in mass between these species is too small to be detected. Both species are accelerated through the same electric potential difference, and both experience the same magnetic field, which causes them to move on circular paths. The radius of the path for the species X^+ is r_1, while the radius for species X^{2+} is r_2. Find the ratio r_1/r_2 of the radii.

*__19. ssm__ A particle of charge $+7.3$ μC and mass 3.8×10^{-8} kg is traveling perpendicular to a 1.6-T magnetic field, as the drawing shows. The speed of the particle is 44 m/s. (a) What is the value of the angle θ, such that the particle's subsequent path will intersect the y axis at the greatest possible value of y? (b) Determine this value of y.

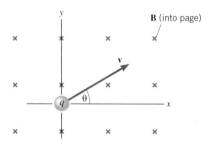

*20. Review Conceptual Example 2 as background for this problem. A charged particle moves through a velocity selector at a constant speed in a straight line. The electric field of the velocity selector is 3.80×10^3 N/C, while the magnetic field is 0.360 T. When the electric field is turned off, the charged particle travels on a circular path whose radius is 4.30 cm. Find the charge-to-mass ratio of the particle.

*21. A proton with a speed of 2.2×10^6 m/s is shot into a region between two plates that are separated by a distance of 0.18 m. As the drawing shows, a magnetic field exists between the plates, and it is perpendicular to the velocity of the proton. What must be the magnitude of the magnetic field, so the proton just misses colliding with the opposite plate?

*22. As preparation for this problem, review Conceptual Example 4. The radius of the track for particle 3 (kinetic energy = KE_3) is exactly 22 times larger than the initial radius of the track for particle 1 (initial kinetic energy = KE_1). Determine the ratio KE_3/KE_1.

*23. **ssm** Review Conceptual Example 2 as an aid in understanding this problem. The drawing shows a positively charged particle entering a 0.52-T magnetic field. The particle has a speed of 270 m/s and moves perpendicular to the magnetic field. Just as the particle enters the magnetic field, an electric field is turned on. What must be the magnitude and direction of the electric field such that the *net* force on the particle is twice the magnetic force?

*24. An α-particle is the nucleus of a helium atom; the orbiting electrons are missing. The α-particle contains two protons and two neutrons, and has a mass of 6.64×10^{-27} kg. Suppose an α-particle is accelerated from rest through a potential difference and then enters a region where its velocity is perpendicular to a 0.0210-T magnetic field. With what angular speed ω does the α-particle move on its circular path?

**25. Refer to question 11 (not problem 11) before starting this problem. Suppose that the target discussed there is located at the coordinates $x = -0.10$ m and $y = -0.10$ m. In addition, suppose that the particle is a proton and the magnetic field has a magnitude of 0.010 T. The speed at which the particle is projected is the same for either of the two paths leading to the target. Find the speed.

Section 21.5 The Force on a Current in a Magnetic Field

26. A wire carries a current of 0.66 A. This wire makes an angle of 58° with respect to a magnetic field of magnitude 4.7×10^{-5} T. The wire experiences a magnetic force of magnitude 7.1×10^{-5} N. What is the length of the wire?

27. **ssm** An electric power line carries a current of 1400 A in a location where the earth's magnetic field is 5.0×10^{-5} T. The line makes an angle of 75° with respect to the field. Deter-

mine the magnitude of the magnetic force on a 120-m length of line.

28. A straight wire in a magnetic field experiences a force of 0.030 N when the current in the wire is 2.7 A. What is the current in the wire when it experiences a force of 0.047 N?

29. A square coil of wire containing a single turn is placed in a uniform 0.25-T magnetic field, as the drawing shows. Each side has a length of 0.32 m, and the current in the coil is 12 A. Determine the magnitude of the magnetic force on each of the four sides.

30. The triangular loop of wire shown in the drawing carries a current of $I = 4.70$ A. A uniform magnetic field is directed parallel to side AB of the triangle and has a magnitude of 1.80 T. (a) Find the magnitude and direction of the magnetic force exerted on each side of the triangle. (b) Determine the magnitude of the net force exerted on the triangle.

31. **ssm** A wire of length 0.655 m carries a current of 21.0 A. In the presence of a 0.470-T magnetic field, the wire experiences a force of 5.46 N. What is the angle (less than 90°) between the wire and the magnetic field?

32. The x, y, and z components of a magnetic field are $B_x = 0.10$ T, $B_y = 0.15$ T, and $B_z = 0.17$ T. A 25-cm wire is oriented along the z axis and carries a current of 4.3 A. What is the magnitude of the magnetic force that acts on this wire?

*33. **ssm** A copper rod of length 0.85 m is lying on a frictionless table (see the drawing). Each end of the rod is attached to a fixed wire by an unstretched spring whose spring constant is $k = 75$ N/m. A magnetic field with a strength of 0.16 T is oriented perpendicular to the surface of the table. (a) What must be the direction of the current in the copper rod that causes the springs to stretch? (b) If the current is 12 A, by how much does each spring stretch?

Table
(top view)

*34. The drawing shows a thin, uniform rod, which has a length of 0.40 m and a mass of 0.080 kg. This rod lies in the plane of the paper and is attached to the floor by a hinge at point P. A uniform magnetic field of 0.31 T is directed perpendicular into the plane of the paper. There is a current $I = 3.8$ A in the rod, which does not rotate clockwise or counterclockwise. Find the angle θ. (*Hint: The magnetic force may be taken to act at the center of gravity.*)

**35. The two conducting rails in the drawing are tilted upward so they each make an angle of 30.0° with respect to the ground. The vertical magnetic field has a magnitude of 0.050 T. The 0.20-kg aluminum rod (length = 1.6 m) slides *without friction* down the rails at a constant velocity. How much current flows through the bar?

Conducting rails

1.6 m

30°

Section 21.6 The Torque on a Current-Carrying Coil

36. A 0.50-m length of wire is formed into a single-turn, square loop in which there is a current of 12 A. The loop is placed in a magnetic field of 0.12 T, as in Figure 21.22a. What is the maximum torque that the loop can experience?

37. **ssm** A circular coil of wire has a radius of 0.10 m. The coil has 50 turns and a current of 15 A, and is placed in a magnetic field whose magnitude is 0.20 T. (a) Determine the magnetic moment of the coil. (b) What is the maximum torque the coil can experience in this field?

38. Two coils have the same number of circular turns and carry the same current. Each rotates in a magnetic field as in Figure 21.21. Coil 1 has a radius of 5.0 cm and rotates in a 0.18-T field. Coil 2 rotates in a 0.42-T field. Each coil experiences the same maximum torque. What is the radius (in cm) of coil 2?

39. Suppose the current-carrying triangle in problem 30 is free to rotate about an axis that is attached along side AC. Using the data in that problem, find (a) the magnetic moment of the triangle and (b) the magnitude of the net torque exerted on it by the magnetic field.

40. A coil carries a current and experiences a torque due to a magnetic field. The value of the torque is 80.0% of the maximum possible torque. (a) What is the smallest angle between the magnetic field and the normal to the plane of the coil? (b) Make a drawing, showing how this coil would be oriented relative to the magnetic field. Be sure to include the angle in the drawing.

41. **ssm www** The rectangular loop in the drawing consists of 75 turns and carries a current of $I = 4.4$ A. A 1.8-T magnetic field is directed along the $+y$ axis. The loop is free to rotate about the z axis. (a) Determine the magnitude of the net torque exerted on the loop and (b) state whether the 35° angle will increase or decrease.

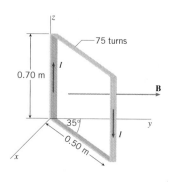

42. The maximum torque experienced by a coil in a 0.75-T magnetic field is 8.4×10^{-4} N·m. The coil is circular and consists of only one turn. The current in the coil is 3.7 A. What is the length of the wire from which the coil is made?

*43. The coil in Figure 21.22a contains 380 turns and has an area per turn of 2.5×10^{-3} m². The magnetic field is 0.12 T, and the current in the coil is 0.16 A. A brake shoe is pressed perpendicularly against the shaft to keep the coil from turning. The coefficient of static friction between the shaft and the brake shoe is 0.70. The radius of the shaft is 0.010 m. What is the magnitude of the minimum normal force that the brake shoe exerts on the shaft?

*44. A square coil and a rectangular coil are each made from the same length of wire. Each contains a single turn. The long sides of the rectangle are twice as long as the short sides. Find the ratio $\tau_{square}/\tau_{rectangle}$ of the maximum torques that these coils experience in the same magnetic field when they contain the same current.

45. **ssm In the model of the hydrogen atom due to Niels Bohr, the electron moves around the proton at a speed of 2.2×10^6 m/s in a circle of radius 5.3×10^{-11} m. Considering the orbiting electron to be a small current loop, determine the magnetic moment associated with this motion. (*Hint: The electron travels around the circle in a time equal to the period of the motion.*)

Section 21.7 Magnetic Fields Produced by Currents

46. A long, straight wire carries a current of 48 A. The magnetic field produced by this current at a certain point is 8.0×10^{-5} T. How far is the point from the wire?

47. In a lightning bolt, 15 C of charge flows in a time of 1.5×10^{-3} s. Assuming that the lightning bolt can be represented as a long, straight line of current, what is the magnitude of the magnetic field at a distance of 25 m from the bolt?

48. What must be the radius of a circular loop of wire so the magnetic field at its center is 1.8×10^{-4} T when the loop carries a current of 12 A?

49. **ssm** A long solenoid consists of 1400 turns of wire and has a length of 0.65 m. There is a current of 4.7 A in the wire. What is the magnitude of the magnetic field within the solenoid?

50. Suppose in Figure 21.28a that $I_1 = I_2 = 25$ A and that the separation between the wires is 0.016 m. By applying an external

magnetic field (created by a source other than the wires) it is possible to cancel the mutual repulsion of the wires. This external field must point along the vertical direction. (a) Does the external field point up or down? Explain. (b) What is the magnitude of the external field?

51. ssm Two rigid rods are oriented parallel to each other and to the ground. The rods carry the same current in the same direction. The length of each rod is 0.85 m, while the mass of each is 0.073 kg. One rod is held in place above the ground, and the other floats beneath it at a distance of 8.2×10^{-3} m. Determine the current in the rods.

52. Two circular loops of wire, each containing a single turn, have the same radius of 4.0 cm and a common center. The planes of the loops are perpendicular. Each carries a current of 1.7 A. What is the magnitude of the net magnetic field at the common center?

53. Two long, straight wires are separated by 0.120 m. The wires carry currents of 8.0 A in opposite directions, as the drawing indicates. Find the magnitude of the net magnetic field at the points labeled (a) A and (b) B.

*54. As background for this problem, review Conceptual Example 9. A rectangular current loop is located near a long, straight wire that carries a current of 12 A (see the drawing). The current in the loop is 25 A. Determine the magnitude of the net magnetic force that acts on the loop.

*55. Two circular coils are concentric and lie in the same plane. The inner coil contains 120 turns of wire, has a radius of 0.012 m, and carries a current of 6.0 A. The outer coil contains 150 turns and has a radius of 0.017 m. What must be the magnitude and direction (relative to the current in the inner coil) of the current in the outer coil, such that the net magnetic field at the common center of the two coils is zero?

*56. Two parallel rods are each 0.50 m in length. They are attached at their centers to a spring that is initially neither stretched nor compressed. The spring has a spring constant of 150 N/m. When 950 A of current is in each rod in the same direction, the spring is observed to be compressed by 2.0 cm. Treat the rods as

long, straight wires and find the separation between them when the current is present.

*57. ssm A piece of copper wire has a resistance per unit length of 5.90×10^{-3} Ω/m. The wire is wound into a thin, flat coil of many turns that has a radius of 0.140 m. The ends of the wire are connected to a 12.0-V battery. Find the magnetic field strength at the center of the coil.

**58. The drawing shows two long, straight wires that are suspended from a ceiling. The mass per unit length of each wire is 0.050 kg/m. Each of the four strings suspending the wires has a length of 1.2 m. When the wires carry identical currents in opposite directions, the angle between the strings holding the two wires is 15°. What is the current in each wire?

**59. The drawing shows an end-on view of three wires. They are long, straight, and perpendicular to the plane of the paper. Their cross sections lie at the corners of a square. The currents in wires 1 and 2 are $I_1 = I_2 = I$ and are directed into the paper. What is the direction of the current in wire 3, and what is the ratio I_3/I, such that the net magnetic field at the empty corner is zero?

Section 21.8 Ampere's Law

60. Suppose a uniform magnetic field is everywhere perpendicular to this page. The field points directly upward toward you. A circular path is drawn on the page. Use Ampere's law to show that there can be no net current passing through the circular surface.

61. ssm The wire in Figure 21.40 carries a current of 12 A. Suppose that a second long, straight wire is placed right next to this wire. The current in the second wire is 28 A. Use Ampere's law to find the magnitude of the magnetic field at a distance of $r = 0.72$ m from the wires when the currents are (a) in the same direction and (b) in opposite directions.

*62. A very long, hollow cylinder is formed by rolling up a thin sheet of copper. Electric charges flow along the copper sheet parallel to the axis of the cylinder. The arrangement is, in effect, a hollow tube of current I. Use Ampere's law to show that the magnetic field (a) is $\mu_0 I/(2\pi r)$ outside the cylinder at a distance r from the axis and (b) is zero at any point within the hollow interior of the cylinder. (*Hint: For closed paths, use* *circles perpendicular to and centered on the axis of the cylinder.)*

** 63. A long, cylindrical conductor is solid throughout and has a radius R. Electric charges flow parallel to the axis of the cylinder and pass uniformly through the entire cross section. The arrangement is, in effect, a solid tube of current I_0. The current per unit cross-sectional area (i.e., the current density) is $I_0/(\pi R^2)$. Use Ampere's law to show that the magnetic field inside the conductor at a distance r from the axis is $\mu_0 I_0 r/(2\pi R^2)$. *(Hint: For a closed path, use a circle of radius r perpendicular to and centered on the axis. Note that the current through any surface is the area of the surface times the current density.)*

ADDITIONAL PROBLEMS

64. In a television set, electrons are accelerated from rest through a potential difference of 19 kV. The electrons then pass through a 0.28-T magnetic field that deflects them to the appropriate spot on the screen. Find the magnitude of the maximum magnetic force that an electron can experience.

65. ssm A long solenoid has 1400 turns per meter of length, and it carries a current of 3.5 A. A small circular coil of wire is placed inside the solenoid with the normal to the coil oriented at an angle of 90.0° with respect to the axis of the solenoid. The coil consists of 50 turns, has an area of 1.2×10^{-3} m^2, and carries a current of 0.50 A. Find the torque exerted on the coil.

66. A proton, traveling with a velocity of 4.5×10^6 m/s due east, experiences a maximum magnetic force of 8.0×10^{-14} N due south. (a) What are the magnitude and direction of the magnetic field? (b) Answer part (a), assuming the proton is replaced by an electron.

67. The proton has an intrinsic magnetic moment of 1.4×10^{-26} A·m^2. If the magnetic moment makes an angle of $\phi = 64°$ with respect to a 0.65-T magnetic field, what is the torque exerted on the proton?

68. A long, straight wire carrying a current of 305 A is placed in a uniform magnetic field whose magnitude is 7.00×10^{-3} T. The wire is perpendicular to the field. Find a point in space where the net magnetic field is zero. Locate this point by specifying its perpendicular distance from the wire.

69. ssm Due to friction with the air, an airplane has acquired a net charge of 1.70×10^{-5} C. The plane moves with a speed of 2.80×10^2 m/s at an angle θ with respect to the earth's magnetic field, the magnitude of which is 5.00×10^{-5} T. The magnetic force on the airplane has a magnitude of 2.30×10^{-7} N. Find the angle θ. (There are two possible angles.)

70. Near the equator in South America the earth's magnetic field has a strength of 3.2×10^{-5} T; the field is parallel to the surface of the earth and points due north. A straight wire, 46 m in length, has an east-west orientation and experiences a magnetic force of 0.058 N, directed vertically down (toward the earth). What is the magnitude and direction of the current in the wire?

71. An ion source in a mass spectrometer produces deuterons. (A deuteron is a particle that has approximately twice the mass of a proton but the same charge.) Each deuteron is accelerated from rest through a potential difference of 2.00×10^3 V, after which it enters a 0.600-T magnetic field. Find the radius of its circular path.

72. A circular loop of wire and a long, straight wire carry currents of I_1 and I_2 (see the drawing), where $I_2 = 6.6I_1$. The loop and the straight wire lie in the same plane. The net magnetic field at the center of the loop is zero. Find the distance H, expressing your answer in terms of R, the radius of the loop.

*73. ssm www A particle of mass 6.0×10^{-8} kg and charge $+7.2$ μC is traveling due east. It enters perpendicularly a magnetic field whose magnitude is 3.0 T. After entering the field, the particle completes one-half of a circle and exits the field traveling due west. How much time does the particle spend in the magnetic field?

*74. Two long, straight, parallel wires A and B are separated by a distance of one meter. They carry currents in opposite directions, and the current in wire A is one-third of that in wire B. On a line drawn perpendicular to the wires, find the point where the net magnetic field is zero. Determine this point relative to wire A.

*75. The drawing shows a parallel plate capacitor that is moving with a speed of 32 m/s through a 3.6-T magnetic field. The velocity **v** is perpendicular to the magnetic field. The electric field within the capacitor has a value of 170 N/C, and each plate has an area of 7.5×10^{-4} m^2. What is the magnetic force (magnitude and direction) exerted on the positive plate of the capacitor?

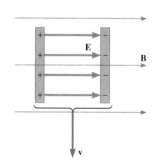

** 76. The drawing shows two wires that carry the same current of $I = 85.0$ A and are oriented perpendicular to the plane of the paper. The current in one wire is directed out of the paper, while the current in the other is directed into the paper. Find the magnitude and direction of the net magnetic field at point P.

** 77. ssm A charge of 4.0×10^{-6} C is placed on a small conducting sphere that is located at the end of a thin insulating rod whose length is 0.20 m. The rod rotates with an angular speed of $\omega = 150$ rad/s about an axis that passes perpendicularly through its other end. Find the magnetic moment of the rotating charge. (*Hint: The charge travels around a circle in a time equal to the period of the motion.*)

CONCEPTS *&* CALCULATIONS

G R O U P   L E A R N I N G   P R O B L E M S

Note: Each of these problems consists of Concept Questions followed by a related quantitative Problem. They are designed for use by students working alone or in small learning groups. The Concept Questions involve little or no mathematics and are intended to stimulate group discussions. They focus on the concepts with which the problems deal. Recognizing the concepts is the essential initial step in any problem-solving technique.

78. Concept Questions (a) A charge moves along the $+x$ axis and experiences no magnetic force, although there is a magnetic field. What can you conclude about the direction of the magnetic field? (b) A moving charge experiences the maximum possible magnetic force when moving in a magnetic field. What can you conclude about the angle θ that the charge's velocity makes with respect to the magnetic field? Explain your answers.

Problem A particle that has a 8.2-μC charge moves with a velocity of magnitude 5.0×10^5 m/s. When the velocity points along the $+x$ axis, the particle experiences no magnetic force, although there is a magnetic field present. The maximum possible magnetic force that the charge could experience has a magnitude of 0.48 N. Find the magnitude and direction of the magnetic field. Note that there are two possible answers for the direction of the field. Make sure that your answers are consistent with your answers to the Concept Questions.

79. Concept Questions (a) A proton is projected perpendicularly into a magnetic field with a certain velocity and follows a circular path. Then an electron is projected perpendicularly into the same magnetic field with the same velocity. Does the electron follow the exact same circular path that the proton followed? (b) To make the electron follow the exact same circular path, should the field direction be kept the same or reversed, and (c) should the field magnitude be increased, reduced, or kept the same? Account for your answers.

Problem A proton is projected perpendicularly into a magnetic field that has a magnitude of 0.50 T. The field is then adjusted so that an electron will follow the exact same circular path when it is projected perpendicularly into the field with the same velocity that the proton had. What is the magnitude of the field used for the electron? Verify that your answer is consistent with your answers to the Concept Questions.

80. Concept Questions Particle 1 and particle 2 carry the same charge q, but particle 1 has a smaller mass than particle 2. These two particles accelerate from rest through the same electric potential difference V and enter the same magnetic field, which has a magnitude B. The particles travel perpendicular to the field on circular paths. Upon entering the field region, which particle, if either, has the greater (a) kinetic energy and (b) speed? Give your reasoning.

Problem The masses of the particles are $m_1 = 2.3 \times 10^{-8}$ kg and $m_2 = 5.9 \times 10^{-8}$ kg. The radius of the circular path for particle 1 is $r_1 = 12$ cm. What is the radius of the circular path for particle 2?

81. Concept Question The drawing shows a wire comprised of three segments, AB, BC, and CD. There is a current I in the wire. There is also a magnetic field $\mathbf{B}$ that is the same everywhere and points in the direction of the $+z$ axis. Rank the wire segments according to the magnetic force (largest first) that they experience. Justify your ranking.

Problem The lengths of the wire segments are $L_{AB} = 1.1$ m, $L_{BC} = 0.55$ m, and $L_{CD} = 0.55$ m. The current is $I = 2.8$ A, and the magnitude of the magnetic field is $B = 0.26$ T. Find the magnitude of the force that acts on each segment. Be sure your answers are consistent with your answer to the Concept Question.

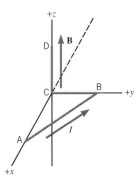

82. Concept Question You have a wire of length L from which to make the square coil of a dc motor. In a given magnetic field a coil of N turns each with area A produces more torque when its total effective area of NA is greater rather than smaller. This follows directly from Equation 21.4. Is more torque obtained by using the length of wire to make a single-turn coil or a two-turn coil? Explain.

Problem The length of the wire is $L = 1.00$ m. The current in the coil is $I = 1.7$ A, and the magnetic field of the motor is 0.34 T. Find the maximum torque when the wire is used to make a single-turn square coil and a two-turn square coil. Verify that your answers are consistent with your answer to the Concept Question.

83. Concept Questions The drawing shows two perpendicular, long, straight wires that lie in the plane of the paper. Each wire carries the same current I. What is the direction of the net magnetic field at (a) point A and (b) point B? (c) Is the magnitude of the net field at point A greater than, less than, or equal to the magnitude of the net field at point B?

Problem The current in each of the wires is $I = 5.6$ A. Find the magnitudes of the net fields at points A and B. Verify that your answers are consistent with your answers to the Concept Questions.

Lauryn Hill, like most singers, uses a microphone when she performs. Many microphones operate using magnetism and a phenomenon known as electromagnetic induction.

CHAPTER 22

ELECTROMAGNETIC INDUCTION

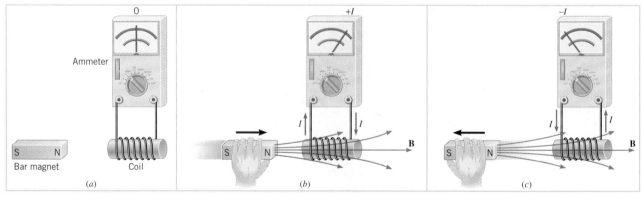

Figure 22.1 (a) When there is no relative motion between the coil of wire and the bar magnet, there is no current in the coil. (b) A current is created in the coil when the magnet moves toward the coil. (c) A current also exists when the magnet moves away from the coil, but the direction of the current is opposite to that in (b).

22.1 INDUCED EMF AND INDUCED CURRENT

There are a number of ways a magnetic field can be used to generate an electric current, and Figure 22.1 illustrates one of them. This drawing shows a bar magnet and a helical coil of wire to which an ammeter is connected. When there is no *relative* motion between the magnet and the coil, as in part *a* of the drawing, the ammeter reads zero, indicating that no current exists. However, when the magnet moves toward the coil, as in part *b*, a current *I* appears. As the magnet approaches, the magnetic field that it creates at the location of the coil becomes stronger and stronger, and it is this *changing* field that produces the current. When the magnet moves away from the coil, as in part *c*, a current is also produced, but with a reversed direction. Now the magnetic field at the coil becomes weaker as the magnet moves away, and once again it is the *changing* field that generates the current.

A current would also be created in Figure 22.1 if the magnet were held stationary and the coil were moved, because the magnetic field at the coil would be changing as the coil approached or receded from the magnet. Only relative motion between the magnet and the coil is needed to generate a current; it does not matter which one moves.

The current in the coil is called an ***induced current*** because it is brought about (or "induced") by a changing magnetic field. Since a source of emf is always

Figure 22.2 Induced emf lies at the heart of an automobile cruise control device. The emf is induced in a sensing coil by magnets attached to the rotating drive shaft.

needed to produce a current, the coil itself behaves as if it were a source of emf. This emf is known as an ***induced emf.*** Thus, a changing magnetic field induces an emf in the coil, and the emf leads to an induced current.

Induced emf and induced current are frequently used in the cruise control devices found in many cars. Figure 22.2 illustrates how a cruise control device operates. Usually two magnets are mounted on opposite sides of the vehicle's drive shaft, with a stationary sensing coil positioned nearby. As the shaft turns, the magnets pass by the coil and cause an induced emf and current to appear in it. A microprocessor (the "brain" of a computer) counts the pulses of current and, with the aid of its internal clock and a knowledge of the shaft's radius, determines the rotational speed of the drive shaft. The rotational speed, in turn, is related to the car's speed. Thus, once the driver sets the desired cruising speed with the speed control switch (mounted near the steering wheel), the microprocessor can compare it with the measured speed. To the extent that the selected cruising speed and the measured speed differ, a signal is sent to a servo or control mechanism, which causes the throttle/fuel injector to send more or less fuel to the engine. The car speeds up or slows down, accordingly, until the desired cruising speed is reached.

Figure 22.3 shows another way to induce an emf and a current in a coil. An emf can be induced by *changing the area* of a coil in a constant magnetic field. Here the shape of the coil is being distorted so as to reduce the area. As long as the area is changing, an induced emf and current exist; they vanish when the area is no longer changing. If the distorted coil is returned to its original shape, thereby increasing the area, an oppositely directed current is generated while the area is changing.

In each of the previous examples, both an emf and a current are induced in the coil because the coil is part of a complete, or closed, circuit. If the circuit were open—perhaps because of an open switch—there would be no induced current. However, an emf would still be induced in the coil, whether the current exists or not.

Changing a magnetic field and changing the area of a coil are methods that can be used to create an induced emf. The phenomenon of producing an induced emf with the aid of a magnetic field is called ***electromagnetic induction.*** The next section discusses yet another method by which an induced emf can be created.

22.2 MOTIONAL EMF

The Emf Induced in a Moving Conductor

When a conducting rod moves through a constant magnetic field, an emf is induced in the rod. This special case of electromagnetic induction arises as a result of the magnetic force (see Section 21.2) that acts on a moving charge. Consider the metal rod of length L moving to the right in Figure 22.4a. The velocity **v** of the rod is constant and is perpendicular to a uniform magnetic field **B**. Each charge q within the rod also moves with a velocity **v** and experiences a magnetic force of magnitude $F = qvB$, according to Equation 21.1. By using RHR-1, it can be seen that the mobile, free electrons are driven to the bottom of the rod, leaving behind an equal amount of positive charge at the top. (Remember to reverse the direction of the force that RHR-1 predicts, since the electrons have a negative charge. See Section 21.2.) The positive and negative charges accumulate until the attractive electric force that they exert on each other becomes equal in magnitude to the magnetic force. When the two forces balance, equilibrium is reached and no further charge separation occurs.

THE PHYSICS OF...

an automobile cruise control device.

Figure 22.3 While the area of the coil is changing, an induced emf and current are generated.

(a)

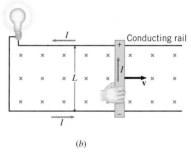

(b)

Figure 22.4 (a) When a conducting rod moves at right angles to a constant magnetic field, the magnetic force causes opposite charges to appear at the ends of the rod, giving rise to an induced emf. (b) The induced emf causes an induced current I to appear in the circuit.

The separated charges on the ends of the moving conductor give rise to an induced emf, called a **motional emf,** because it originates from the motion of charges through a magnetic field. The emf exists as long as the rod moves. If the rod is brought to a halt, the magnetic force vanishes, with the result that the attractive electric force reunites the positive and negative charges and the emf disappears. The emf of the moving rod is analogous to that between the terminals of a battery. However, the emf of a battery is produced by chemical reactions, whereas the motional emf is created by the agent that moves the rod through the magnetic field (like the hand in Figure 22.4*b*.)

The fact that the electric and magnetic forces balance at equilibrium in Figure 22.4*a* can be used to determine the magnitude of the motional emf $\mathscr{E}$. Acccording to Equation 18.2, the magnitude of the electric force acting on the positive charge q at the top of the rod is Eq, where E is the magnitude of the electric field due to the separated charges. And according to Equation 19.7 (without the minus sign), the electric field magnitude is given by the voltage between the ends of the rod (the emf $\mathscr{E}$) divided by the length L of the rod. Thus, the electric force is $Eq = (\mathscr{E}/L)q$. The magnetic force is qvB, according to Equation 21.1, because the charge q moves perpendicular to the magnetic field. Since these two forces balance, it follows that $(\mathscr{E}/L)q = qvB$. The emf, then, is

Motional emf when **v, B,**
and L are mutually $$\mathscr{E} = vBL \qquad\qquad (22.1)$$
perpendicular

As expected, $\mathscr{E} = 0$ V when $v = 0$ m/s, for no motional emf is developed in a stationary rod. Greater speeds and stronger magnetic fields lead to greater emfs for a given length L. As with batteries, $\mathscr{E}$ is expressed in volts. In Figure 22.4*b* the rod is sliding on conducting rails that form part of a closed circuit, and L is the length of the rod between the rails. Due to the emf, electrons flow in a clockwise direction around the circuit. Positive charge would flow in the direction opposite to the electron flow, so the conventional current I is drawn counterclockwise in the picture. Example 1 illustrates how to determine the electrical energy that the motional emf delivers to a device such as the light bulb in the drawing.

EXAMPLE 1   Operating a Light Bulb with Motional Emf

Suppose the rod in Figure 22.4*b* is moving at a speed of 5.0 m/s in a direction perpendicular to a 0.80-T magnetic field. The rod has a length of 1.6 m and a negligible electrical resistance. The rails also have negligible resistance. The light bulb, however, has a resistance of 96 Ω. Find (a) the emf produced by the rod, (b) the induced current in the circuit, (c) the electrical power delivered to the bulb, and (d) the energy used by the bulb in 60.0 s.

Reasoning The moving rod acts like an imaginary battery and supplies a motional emf of vBL to the circuit. The induced current can be determined from Ohm's law as the motional emf divided by the resistance of the bulb. The electric power delivered to the bulb is the product of the induced current and the potential difference across the bulb (which, in this case, is the motional emf). The energy used is the product of the power and the time.

Solution

(a) The motional emf is given by Equation 22.1 as

$$\mathscr{E} = vBL = (5.0 \text{ m/s})(0.80 \text{ T})(1.6 \text{ m}) = \boxed{6.4 \text{ V}}$$

(b) According to Ohm's law, the induced current is equal to the motional emf divided by the resistance of the circuit:

$$I = \frac{\mathcal{E}}{R} = \frac{6.4 \text{ V}}{96 \ \Omega} = \boxed{0.067 \text{ A}} \tag{20.2}$$

(c) The electrical power P delivered to the light bulb is the product of the current I and the potential difference across the bulb:

$$P = I\mathcal{E} = (0.067 \text{ A})(6.4 \text{ V}) = \boxed{0.43 \text{ W}} \tag{20.6a}$$

(d) Since power is energy per unit time, the energy E used in 60.0 s is the product of the power and the time:

$$E = Pt = (0.43 \text{ W})(60.0 \text{ s}) = \boxed{26 \text{ J}} \tag{6.10b}$$

Motional Emf and Electrical Energy

Motional emf arises because a magnetic force acts on the charges in a conductor that is moving through a magnetic field. Whenever this emf causes a current, a second magnetic force enters the picture. In Figure 22.4b, for instance, the second force arises because the current I in the rod is perpendicular to the magnetic field. The current, and hence the rod, experiences a magnetic force $\mathbf{F}$ whose magnitude is given by Equation 21.3 as $F = ILB \sin 90°$. Using the values of I, L, and B from Example 1, we see that $F = (0.067 \text{ A})(1.6 \text{ m})(0.80 \text{ T}) = 0.086 \text{ N}$. The direction of $\mathbf{F}$ is specified by RHR-1 and is *opposite* to the velocity $\mathbf{v}$ of the rod, as Figure 22.5 shows. By itself, $\mathbf{F}$ would *slow down* the rod, and here lies the crux of the matter. To keep the rod moving to the right with a constant velocity, a counterbalancing force must be applied to the rod by an external agent, such as the hand in the picture. The counterbalancing force must have a magnitude of 0.086 N and must be directed opposite to the magnetic force $\mathbf{F}$. If the counterbalancing force were removed, the rod would decelerate under the influence of $\mathbf{F}$ and eventually come to rest. During the deceleration, the motional emf would decrease and the light bulb would eventually go out.

We can now answer an important question—Who or what provides the 26 J of electrical energy that the light bulb in Example 1 uses in sixty seconds? The provider is the external agent that applies the 0.086-N counterbalancing force needed to keep the rod moving. This agent does work, and Example 2 shows that the work done is equal to the electrical energy used by the bulb.

Figure 22.5 A magnetic force $\mathbf{F}$ is exerted on the current I in the moving rod and is opposite to the velocity $\mathbf{v}$. The rod will slow down unless a counterbalancing force is applied by the hand.

EXAMPLE 2   The Work Needed to Keep the Light Bulb Burning

In Example 1, an external agent (the hand in Figure 22.5) supplies a 0.086-N force that keeps the rod moving at a constant speed of 5.0 m/s. Determine the work done in 60.0 s by the external agent.

Reasoning The work W done by the hand in Figure 22.5 is given by Equation 6.1 as $W = (F \cos \theta)x$. In this equation F is the magnitude of the external force, θ is the angle between the external force and the displacement of the rod ($\theta = 0°$), and x is the distance that the rod moves in a time t. The distance is the product of the speed of the rod and the time, $x = vt$, so the work can be expressed as $W = (F \cos 0°)vt$.

Solution The work done by the external agent is

$$W = F \, vt = (0.086 \text{ N})(5.0 \text{ m/s})(60.0 \text{ s}) = \boxed{26 \text{ J}}$$

Figure 22.6 The current cannot be directed clockwise in this circuit, because the magnetic force **F** exerted on the rod would then be in the same direction as the velocity **v**. The rod would accelerate to the right and create energy on its own, violating the principle of conservation of energy.

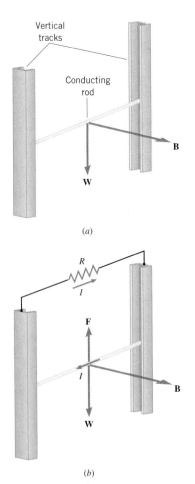

(a)

(b)

Figure 22.7 (*a*) Because there is no kinetic friction between the falling rod and the tracks, the only force acting on the rod is its weight **W**. (*b*) When an induced current *I* exists in the circuit, a magnetic force **F** also acts on the rod.

The 26 J of work done on the rod by the external agent is the same as the 26 J of energy used by the bulb. Hence, the moving rod and the magnetic force convert mechanical work into electrical energy, much as a battery converts chemical energy into electrical energy.

It is important to realize that the direction of the current in Figure 22.5 is consistent with the principle of conservation of energy. Consider what would happen if the direction of the current were reversed, as in Figure 22.6. With the direction of the current reversed, the direction of the magnetic force **F** would also be reversed and point in the direction of the velocity **v** of the rod. As a result, the force would cause the rod to accelerate rather than decelerate. The rod would accelerate without the need for an external force (like that provided by the hand in Figure 22.5) and create a motional emf that supplies energy to the light bulb. Thus, this hypothetical generator would produce energy out of nothing, since there is no external agent. Such a device cannot exist because it violates the principle of conservation of energy, which states that energy cannot be created or destroyed, but can only be converted from one form to another. Therefore, the current cannot be directed clockwise around the circuit, as in Figure 22.6. In situations such as that in Examples 1 and 2, when a motional emf leads to an induced current, a magnetic force always appears that opposes the motion, in accord with the principle of conservation of energy. Conceptual Example 3 deals further with the important issue of energy conservation.

CONCEPTUAL EXAMPLE 3　Conservation of Energy

Figure 22.7*a* illustrates a conducting rod that is free to slide down between two vertical copper tracks. There is no kinetic friction between the rod and the tracks, although the rod maintains electrical contact with the tracks during its fall. A constant magnetic field **B** is directed perpendicular to the motion of the rod, as the drawing shows. Because there is no friction, the only force that acts on the rod is its weight **W**, so the rod falls with an acceleration equal to the acceleration due to gravity, $a = 9.8$ m/s². Suppose that a resistance R is connected between the tops of the tracks, as in part *b* of the drawing. (a) Does the rod now fall with the acceleration due to gravity? (b) How does energy conservation apply to what happens in part *b* of the drawing?

Reasoning and Solution

(a) As the rod falls perpendicular to the magnetic field, a motional emf is induced between its ends. This emf is induced whether or not the resistance R is attached between the tracks. However, when R is present, a complete circuit is formed, and the emf produces an induced current I that is perpendicular to the field. The direction of this current is such that the rod experiences an upward magnetic force **F**, opposite to the falling motion and the weight of the rod (see part *b* of the drawing and use RHR-1). The net force acting downward on the rod is **W** + **F**, which is *less* than the weight, since **F** points upward and the weight **W** points downward. In accord with Newton's second law of motion, the downward acceleration is proportional to the net force, so ***the rod falls with an acceleration that is less than the acceleration due to gravity.*** In fact, as the speed of the rod in Figure 22.7*b* increases during the descent, the magnetic force becomes larger and larger. There will be a time when the magnitude of the magnetic force becomes equal to the magnitude of the rod's weight. When this occurs, the net force on the rod will be zero, and the rod's acceleration will also be zero. From then on, the rod will fall at a constant velocity.

(b) For the freely falling rod in Figure 22.7*a*, gravitational potential energy (GPE) is converted into kinetic energy (KE) as the rod picks up speed on the way down. In Figure 22.7*b*, however, the rod always has a smaller speed than that of a freely falling rod at the same place. This is because the acceleration is less than the acceleration due to gravity. In terms of energy conservation, the smaller speed arises because only part of the GPE is converted into KE, with part also being dissipated as heat in the resistance R. In fact, when the rod reaches the point when it falls with a constant velocity, none of the GPE is converted into KE because all of it is dissipated as heat in the resistance.

Related Homework Material: *Problem 9*

22.3 MAGNETIC FLUX

Motional Emf and Magnetic Flux

⬤ *Concepts at a Glance* Motional emf, as well as any other type of induced emf, can be described in terms of a concept called magnetic flux. Magnetic flux is analogous to electric flux, which deals with the electric field and the surface through which it passes (see Section 18.9 and Figure 18.32). Magnetic flux is defined in a similar way. The Concepts-at-a-Glance chart in Figure 22.8 outlines the ideas involved. The upper three boxes in the middle of the chart are the same as those in Figure 21.9, and they indicate that a magnetic field is defined in terms of the magnetic force that acts on a moving test charge. By bringing together the magnetic field and the surface through which it passes, we will see how the concept of magnetic flux arises. ⬤

Now let us see how the motional emf given in Equation 22.1 ($\mathcal{E} = vBL$) can be written in terms of the magnetic flux. Figure 22.9a shows the rod used to derive Equation 22.1; it is moving through a magnetic field between a time $t = 0$ s and a later time t_0. During this interval, it has moved a distance x_0 to the right. At a later

Figure 22.8 *Concepts at a Glance* Magnetic flux incorporates both the idea of a magnetic field and the surface through which it passes.

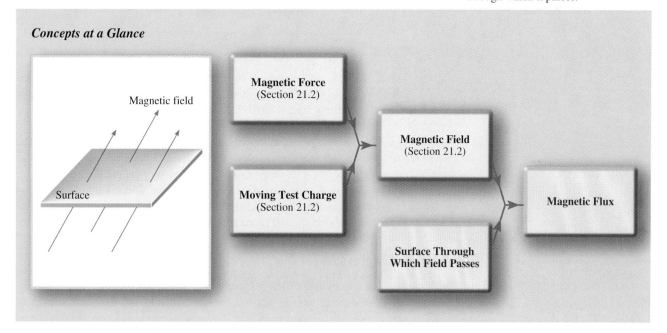

Concepts at a Glance

Magnetic field

Surface

Magnetic Force (Section 21.2)

Moving Test Charge (Section 21.2)

Magnetic Field (Section 21.2)

Surface Through Which Field Passes

Magnetic Flux

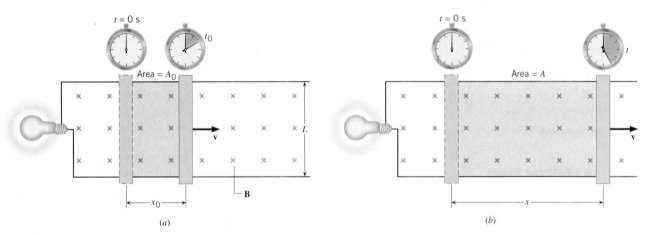

Figure 22.9 (a) In a time t_0, the moving rod sweeps out an area $A_0 = x_0 L$. (b) The area swept out in a time t is $A = xL$. In both parts of the figure the areas are shaded in color.

time t, the rod has moved an even greater distance x, as part b of the drawing indicates. The speed v of the rod is the distance traveled divided by the elapsed time: $v = (x - x_0)/(t - t_0)$. Substituting this expression for v into $\mathscr{E} = vBL$ gives

$$\mathscr{E} = \left(\frac{x - x_0}{t - t_0} \right) BL = \left(\frac{xL - x_0 L}{t - t_0} \right) B$$

As the drawing indicates, the term $x_0 L$ is the area A_0 swept out by the rod in moving a distance x_0, while xL is the area A swept out in moving a distance x. In terms of these areas, the emf is

$$\mathscr{E} = \left(\frac{A - A_0}{t - t_0} \right) B = \frac{(BA) - (BA)_0}{t - t_0}$$

Notice how the product BA of the magnetic field strength and the area appears in the numerator of this expression. The quantity BA is given the name **magnetic flux** and is represented by the symbol Φ (Greek capital letter *phi*); thus $\Phi = BA$. The magnitude of the induced emf is the *change* in flux $\Delta\Phi = \Phi - \Phi_0$ divided by the time interval $\Delta t = t - t_0$ during which the change occurs:

$$\mathscr{E} = \frac{\Phi - \Phi_0}{t - t_0} = \frac{\Delta\Phi}{\Delta t}$$

In other words, the induced emf equals the time rate of change of the magnetic flux.

You will almost always see the equation above written with a minus sign, namely, $\mathscr{E} = -\Delta\Phi/\Delta t$. The minus sign is introduced for the following reason: The direction of the current induced in the circuit is such that the magnetic force **F** acts on the rod to *oppose* its motion, thereby tending to slow down the rod (see Figure 22.5). The minus sign is a reminder that the polarity of the induced emf sends the induced current in the proper direction so as to give rise to this opposing magnetic force.

The advantage of writing the induced emf as $\mathscr{E} = -\Delta\Phi/\Delta t$ is that this relation is far more general than our present discussion suggests. In Section 22.4 we will see that $\mathscr{E} = -\Delta\Phi/\Delta t$ can be applied to *all possible ways of generating induced emfs.*

A General Expression for Magnetic Flux

In Figure 22.9 the direction of the magnetic field **B** is perpendicular to the surface swept out by the moving rod. In general, however, **B** may not be perpendicular to the surface. For instance, in Figure 22.10 the direction perpendicular to the surface is indicated by the normal to the surface, but the magnetic field is inclined at an angle ϕ with respect to this direction. In such a case the flux is computed using only the component of the field that is perpendicular to the surface, $B \cos \phi$. The general expression for magnetic flux is

$$\Phi = (B \cos \phi)A = BA \cos \phi \qquad (22.2)$$

If either the magnitude B of the magnetic field or the angle ϕ is not constant over the surface, an average value of the product $B \cos \phi$ must be used to compute the flux. Equation 22.2 shows that the unit of magnetic flux is the tesla·meter^2 (T·m^2). This unit is called a *weber* (Wb), after the German physicist Wilhelm Weber (1804–1891): 1 Wb = 1 T·m^2. Example 4 illustrates how to determine the magnetic flux for three different orientations of the surface of a coil relative to the magnetic field.

Figure 22.10 When computing the magnetic flux, the component of the magnetic field that is perpendicular to the surface must be used; this component is $B \cos \phi$.

EXAMPLE 4   Magnetic Flux

A rectangular coil of wire is situated in a constant magnetic field whose magnitude is 0.50 T. The coil has an area of 2.0 m^2. Determine the magnetic flux for the three orientations, $\phi = 0°$, 60.0°, and 90.0°, shown in Figure 22.11.

Reasoning The magnetic flux Φ is defined as $\Phi = BA \cos \phi$, where B is the magnitude of the magnetic field, A is the area of the surface through which the magnetic field passes, and ϕ is the angle between the magnetic field and the normal to the surface.

Solution The magnetic flux for the three cases is:

$\phi = 0°$	$\Phi = (0.50 \text{ T})(2.0 \text{ m}^2) \cos 0° = \boxed{1.0 \text{ Wb}}$
$\phi = 60.0°$	$\Phi = (0.50 \text{ T})(2.0 \text{ m}^2) \cos 60.0° = \boxed{0.50 \text{ Wb}}$
$\phi = 90.0°$	$\Phi = (0.50 \text{ T})(2.0 \text{ m}^2) \cos 90.0° = \boxed{0 \text{ Wb}}$

● **PROBLEM SOLVING INSIGHT**
The magnetic flux Φ is determined by more than just the magnitude B of the magnetic field and the area A. It also depends on the angle ϕ (see Figure 22.10 and Equation 22.2).

Graphical Interpretation of Magnetic Flux

It is possible to interpret the magnetic flux graphically because the magnitude of the magnetic field **B** is proportional to the number of field lines per unit area that passes through a surface perpendicular to the lines (see Section 21.1). For instance, the magnitude of **B** in Figure 22.12a is three times larger than it is in part b of the drawing, since the number of field lines drawn through the identical surfaces is in the ratio of 3:1. Because Φ is directly proportional to B for a given area, the flux in

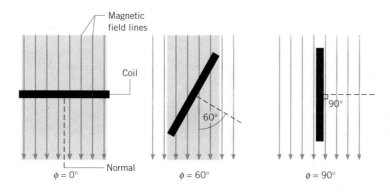

Figure 22.11 This picture shows three orientations of a rectangular coil (drawn as an edge view), relative to the magnetic field lines. The magnetic field lines that pass through the coil are those in the regions shaded in blue.

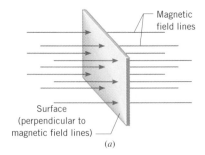

(a)

Magnetic field lines

Surface (perpendicular to magnetic field lines)

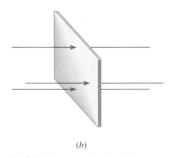

(b)

Figure 22.12 The magnitude of the magnetic field in (*a*) is three times as great as that in (*b*), because the number of magnetic field lines crossing the surfaces is in the ratio of 3:1.

part *a* is also three times larger than that in part *b*. Therefore, ***the magnetic flux is proportional to the number of field lines that passes through a surface.***

The graphical interpretation of flux also applies when the surface is oriented at an angle with respect to **B**. For example, as the coil in Figure 22.11 is rotated from $\phi = 0°$ to 60° to 90°, the number of magnetic field lines passing through the surface (see the field lines in the regions shaded in blue) changes in the ratio of 8:4:0 or 2:1:0. The results of Example 4 show that the flux in the three orientations changes by the same ratio. Because the magnetic flux is proportional to the number of field lines passing through a surface, one often encounters phrases like "the flux that passes through a surface bounded by a loop of wire."

22.4 FARADAY'S LAW OF ELECTROMAGNETIC INDUCTION

Two scientists are given credit for the discovery of electromagnetic induction: the Englishman Michael Faraday (1791–1867) and the American Joseph Henry (1797–1878). Although Henry was the first to observe electromagnetic induction, Faraday investigated it in more detail and published his findings first. Consequently, the law that describes the phenomenon bears his name.

● *Concepts at a Glance* Faraday discovered that whenever there is a *change in flux* through a loop of wire, an emf is induced in the loop. In this context, the word "change" refers to a change as time passes. A flux that is constant in time creates no emf. The Concepts-at-a-Glance chart in Figure 22.13, which extends the chart in Figure 22.8, emphasizes that *Faraday's law of electromagnetic induction* is expressed by bringing together the idea of magnetic flux and the time interval

Figure 22.13 *Concepts at a Glance* Faraday's law of electromagnetic induction specifies the emf created by a change in the magnetic flux that occurs as time passes. The change in the flux and the time interval over which it occurs both appear in the statement of the law. The pickups in Keith Richards' guitar use electromagnetic induction to convert the vibratory motion of the strings into an induced emf.

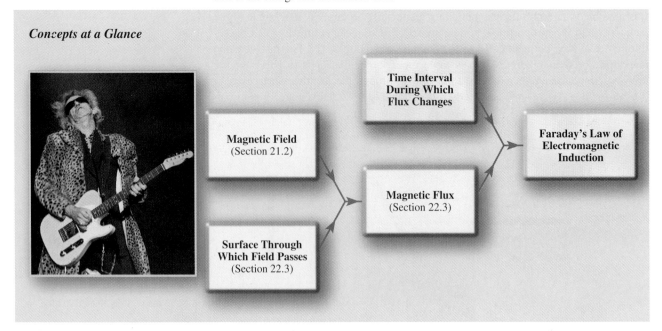

Concepts at a Glance

Magnetic Field (Section 21.2)

Surface Through Which Field Passes (Section 22.3)

Time Interval During Which Flux Changes

Magnetic Flux (Section 22.3)

Faraday's Law of Electromagnetic Induction

during which it changes. In fact, Faraday found that the magnitude of the induced emf is equal to the time rate of change of the magnetic flux. This is consistent with the relation we obtained in Section 22.3 for the specific case of motional emf: $\mathscr{E} = -\Delta\Phi/\Delta t$.

Often the magnetic flux passes through a coil of wire containing more than one loop (or turn). If the coil consists of N loops, and if the same flux passes through each loop, it is found experimentally that the total induced emf is N times that induced in a single loop. An analogous situation occurs in a flashlight when two 1.5-V batteries are stacked in series on top of one another to give a total emf of 3.0 volts. For the general case of N loops, the total induced emf is described by Faraday's law of electromagnetic induction in the following manner.

● **FARADAY'S LAW OF ELECTROMAGNETIC INDUCTION**

The average emf $\mathscr{E}$ induced in a coil of N loops is

$$\mathscr{E} = -N\left(\frac{\Phi - \Phi_0}{t - t_0}\right) = -N\frac{\Delta\Phi}{\Delta t} \qquad (22.3)$$

where $\Delta\Phi$ is the change in magnetic flux through one loop and Δt is the time interval during which the change occurs. The term $\Delta\Phi/\Delta t$ is the average time rate of change of the flux that passes through one loop.

SI Unit of Induced Emf: volt (V)

Faraday's law states that an emf is generated if the magnetic flux changes for any reason. Since the flux is given by Equation 22.2 as $\Phi = BA \cos \phi$, it depends on three factors, B, A, and ϕ, any of which may change. Example 5 considers a change in B.

EXAMPLE 5 The Emf Induced by a Changing Magnetic Field

A coil of wire consists of 20 turns, each of which has an area of 1.5×10^{-3} m^2. A magnetic field is perpendicular to the surface of each loop at all times, so that $\phi = \phi_0 = 0°$. At time $t_0 = 0$ s, the magnitude of the magnetic field at the location of the coil is $B_0 = 0.050$ T. At a later time $t = 0.10$ s, the magnitude of the field at the coil has increased to $B = 0.060$ T. (a) Find the average emf induced in the coil during this time. (b) What would be the value of the average induced emf if the magnitude of the magnetic field decreased from 0.060 T to 0.050 T in 0.10 s?

Reasoning To find the induced emf, we use Faraday's law of electromagnetic induction (Equation 22.3), combining it with the definition of magnetic flux from Equation 22.2. We note that only the magnitude of the magnetic field changes in time. All other factors remain constant.

Solution

(a) Since $\phi = \phi_0$, the induced emf is

$$\mathscr{E} = -N\left(\frac{\Phi - \Phi_0}{t - t_0}\right) = -N\left(\frac{BA \cos \phi - B_0 A \cos \phi_0}{t - t_0}\right)$$

$$= -NA \cos \phi \left(\frac{B - B_0}{t - t_0}\right)$$

$$\mathscr{E} = -(20)(1.5 \times 10^{-3} \text{ m}^2)(\cos 0°)\left(\frac{0.060 \text{ T} - 0.050 \text{ T}}{0.10 \text{ s} - 0 \text{ s}}\right) = \boxed{-3.0 \times 10^{-3} \text{ V}}$$

● **PROBLEM SOLVING INSIGHT**
The change in any quantity is the final value minus the initial value: e.g., the change in flux is $\Delta\Phi = \Phi - \Phi_0$ and the change in time is $\Delta t = t - t_0$.

INTERACTIVE LEARNINGWARE #46

START

Visualization

Reasoning

Knowns and Unknowns

Modeling the Problem

Numerical Solution

FINISH

A 75-turn conducting coil has an area of 8.5×10^{-3} m², and the normal to the coil is parallel to a magnetic field **B**. The coil has a resistance of 14 Ω. At what rate (in T/s) must the magnitude of **B** change for an induced current of 7.0 mA to exist in the coil?

Two important concepts are combined in this Interactive LearningWare example. The first involves the use of Faraday's law of electromagnetic induction when the magnitude of the magnetic field is changing in time. The second reviews the definition of electrical resistance which states that resistance is the ratio of the emf applied across a piece of material divided by the current through the material (see Section 20.2).

Go to our Web site at

http://www.wiley.com/college/cutnell

(b) The calculation here is similar to that in part (a), except the initial and final values of B are interchanged. This interchange reverses the sign of the emf, so $\boxed{\mathscr{E} = +3.0 \times 10^{-3} \text{ V}}$. Because the algebraic sign or polarity of the emf is reversed, the direction of the induced current would be opposite to that in part (a).

The next example demonstrates that an emf can be created when a coil is rotated in a magnetic field.

EXAMPLE 6 The Emf Induced in a Rotating Coil

A flat coil of wire has an area of 0.020 m² and consists of 50 turns. At $t_0 = 0$ s the coil is oriented so the normal to its surface is parallel ($\phi_0 = 0°$) to a constant magnetic field of magnitude 0.18 T. The coil is then rotated through an angle of $\phi = 30.0°$ in a time of 0.10 s (see Figure 22.11). (a) Determine the average induced emf. (b) What would be the induced emf if the coil were returned to its initial orientation in the same time of 0.10 s?

Reasoning As in Example 5 we can determine the induced emf by using Faraday's law of electromagnetic induction, along with the definition of magnetic flux. In the present case, however, only ϕ (the angle between the normal to the surface of the coil and the magnetic field) changes in time. All other factors remain constant.

Solution

(a) Faraday's law yields

$$\mathscr{E} = -N\left(\frac{\Phi - \Phi_0}{t - t_0}\right) = -N\left(\frac{BA \cos \phi - BA \cos \phi_0}{t - t_0}\right)$$

$$= -NBA\left(\frac{\cos \phi - \cos \phi_0}{t - t_0}\right)$$

$$\mathscr{E} = -(50)(0.18 \text{ T})(0.020 \text{ m}^2)\left(\frac{\cos 30.0° - \cos 0°}{0.10 \text{ s} - 0 \text{ s}}\right) = \boxed{+0.24 \text{ V}}$$

(b) When the coil is rotated back to its initial orientation in a time of 0.10 s, the initial and final values of ϕ are interchanged. As a result, the induced emf has the same magnitude, but opposite polarity, so $\boxed{\mathscr{E} = -0.24 \text{ V}}$.

THE PHYSICS OF...

a ground fault interrupter.

One application of Faraday's law that is found in the home is a safety device called a ground fault interrupter. This device protects against electric shock from an appliance, such as a clothes drier. It plugs directly into a wall socket, as in Figure 22.14, or in new home construction, replaces the socket entirely. The interrupter consists of a circuit breaker that can be triggered to stop the current to the drier, depending on whether an induced voltage appears across a sensing coil. This coil is wrapped around an iron ring, through which the current-carrying wires pass. In the drawing, the current going to the drier is shown in red, while the returning current is shown in green. Each of the currents creates a magnetic field that encircles the corresponding wire, according to RHR-2 (see Section 21.7). However, the field lines have opposite directions since the currents have opposite directions. As the

Figure 22.14 The clothes drier is connected to the wall socket through a ground fault interrupter. The drier is operating normally.

drawing shows, the iron ring guides the field lines through the sensing coil. Since the current is ac, the fields from the red and green current are changing, but the red and green field lines always have opposite directions and the opposing fields cancel at all times. As a result, the net flux through the coil remains zero, and no induced emf appears in the coil. Thus, when the drier operates normally, the circuit breaker is not triggered and does not shut down the current. The picture changes if the drier malfunctions, as when a wire inside the unit breaks and accidentally contacts the metal case. When someone touches the case, some of the current begins to pass through the person's body and into the ground, returning to the ac generator *without using the return wire that passes through the ground fault interrupter.* Under this condition, the net magnetic field through the sensing coil is no longer zero and changes with time, since the current is ac. The changing flux causes an induced voltage to appear in the sensing coil, which triggers the circuit breaker to stop the current. Ground fault interrupters work very fast (in less than a millisecond) and turn off the current before it reaches a dangerous level.

Conceptual Example 7 discusses another application of electromagnetic induction, namely, how a stove can cook food without getting hot.

Figure 22.15 The water in the metal pot is boiling. Yet, the water in the glass pot is not boiling, and the stove top is cool to the touch. The stove operates in this way by using electromagnetic induction.

CONCEPTUAL EXAMPLE 7   An Induction Stove

Figure 22.15 shows two pots of water that were placed on an induction stove at the same time. There are two interesting features in this drawing. First, the stove itself is cool to the touch. Second, the water in the metal pot is boiling while that in the glass pot is not. How can such a "cool" stove boil water, and why isn't the water in the glass pot boiling?

Reasoning and Solution The key to this puzzle must be related to the fact that one pot is metal and one is glass. We know that metals are good conductors, while glass is an insulator. Perhaps the stove causes electricity to flow directly in the metal pot. This is exactly what happens. The stove is called an *induction stove* because it operates by using electromagnetic induction. Just beneath the cooking surface is a metal coil that carries an ac current. This alternating current produces an alternating magnetic field that extends outward to the location of the metal pan. As the changing magnetic field crosses the bottom surface of the pan, an emf is induced in it. Because the pan is

THE PHYSICS OF...

an induction stove.

metallic, an induced current is generated by the induced emf. The pan has a finite resistance to the induced current and heats up as energy is dissipated in this resistance. This heat is conducted to the water and eventually raises its temperature to the boiling point. An emf is also induced in the glass pot and the cooking surface of the stove. However, these materials are insulators, so that very little induced current exists within them. Thus, they do not heat up very much and remain cool to the touch.

Related Homework Material: *Question 8*

22.5   LENZ'S LAW

An induced emf drives current around a circuit just as the emf of a battery does. With a battery, conventional current is directed out of the positive terminal, through the attached device, and into the negative terminal. The same is true for an induced emf, although the location of the positive and negative terminals is generally not as obvious. Therefore, a method is needed for determining the polarity or algebraic sign of the induced emf, so the terminals can be identified. As we discuss this method, it will be helpful to keep in mind that the net magnetic field penetrating a coil of wire results from two contributions. One is the original magnetic field that produces the changing flux that leads to the induced emf. The other arises because of the induced current, which, like any current, creates its own magnetic field. The field created by the induced current is called the *induced magnetic field.*

To determine the polarity of the induced emf, we will use a method based on a discovery made by the Russian physicist Heinrich Lenz (1804–1865). This discovery is known as *Lenz's law.*

● **LENZ'S LAW**

The induced emf resulting from a changing magnetic flux has a polarity that leads to an induced current whose direction is such that the induced magnetic field opposes the original flux change.

Lenz's law is best illustrated with examples. Each will be worked out according to the following reasoning strategy:

REASONING STRATEGY

Determining the Polarity of the Induced Emf

1. Determine whether the magnetic flux that penetrates a coil is increasing or decreasing.

2. Find what the direction of the induced magnetic field must be so that it can *oppose the change in flux* by adding to or subtracting from the original field.

3. Having found the direction of the induced magnetic field, use RHR-2 (see Section 21.7) to determine the direction of the induced current. Then the polarity of the induced emf can be assigned because conventional current is directed out of the positive terminal, through the external circuit, and into the negative terminal.

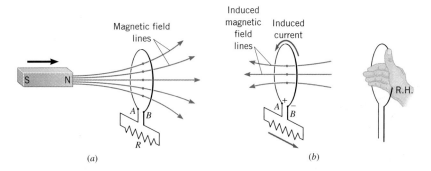

Figure 22.16 (*a*) As the magnet moves to the right, the magnetic flux through the loop increases. The external circuit attached to the loop has a resistance *R*. (*b*) The polarity of the induced emf is indicated by the + and − symbols.

CONCEPTUAL EXAMPLE 8   The Emf Produced by a Moving Magnet

Figure 22.16*a* shows a permanent magnet approaching a loop of wire. The external circuit attached to the loop consists of the resistance *R*, which could be the resistance of the filament in a light bulb, for instance. Find the direction of the induced current and the polarity of the induced emf.

Reasoning and Solution We apply Lenz's law, the essence of which is that the change in magnetic flux must be opposed by the induced magnetic field. The magnetic flux through the loop is increasing, since the magnitude of the magnetic field at the loop is increasing as the magnet approaches. To oppose the increase in the flux, the direction of the induced magnetic field must be opposite to the field of the bar magnet. Since the field of the bar magnet passes through the loop from left to right in part *a* of the drawing, the induced field must pass through the loop from right to left, as in part *b*. To create such an induced field, *the induced current must be directed counterclockwise around the loop, when viewed from the side nearest the magnet.* (See the application of RHR-2 in the drawing.) The loop behaves as a source of emf, just like a battery. Since conventional current is directed into the external circuit from the positive terminal, *point A in Figure 22.16b must be the positive terminal, while point B must be the negative terminal.*

In Conceptual Example 8 the direction of the induced magnetic field is opposite to the direction of the external field of the bar magnet. However, the induced field is not always opposite to the external field, for Lenz's law requires only that it must oppose the *change* in the flux that generates the emf. Conceptual Example 9 illustrates this point.

CONCEPTUAL EXAMPLE 9   The Emf Produced by a Moving Copper Ring

In Figure 22.17 there is a constant magnetic field in a rectangular region of space. This field is directed perpendicularly into the plane of the paper. Outside this region there is no magnetic field. A copper ring slides through the region, from position 1 to position 5. For each of the five positions, determine if an induced current exists in the ring and, if so, find its direction.

Reasoning and Solution Lenz's law requires that the induced magnetic field must oppose the change in flux. Sometimes this means that the induced field will be opposite to the external field, as in Example 8, but not always. We will see that the induced field must sometimes have the same direction as the external field in order to oppose the flux change.

Position 1: No flux passes through the ring because the magnetic field is zero outside the rectangular region. Consequently, *there is no change in flux and no induced emf or current.*

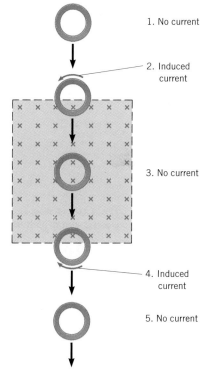

Figure 22.17 A copper ring passes through a rectangular region where a constant magnetic field is directed into the page. The picture shows that a current is induced in the ring at locations 2 and 4.

Position 2: As the ring moves into the field region, the flux increases and there is an induced emf and current. To determine the direction of the current, we require that the induced field point out of the plane of the paper, opposite to the external field. Only then can the induced field oppose the increase in the flux, in accord with Lenz's law. To produce such an induced field, RHR-2 indicates that ***the direction of the induced current must be counterclockwise,*** as the drawing shows.

Position 3: Even though a flux passes through the moving ring, ***there is no induced emf or current*** because the flux remains constant within the rectangular region. To induce an emf, it is not sufficient just to have a flux; the flux must change as time passes.

Position 4: As the ring leaves the field region, the flux decreases, and the induced field must oppose this change. Since the change is a decrease in flux, the induced field must point in the *same* direction as the external field, namely, into the paper. With this orientation, the induced field will increase the net magnetic field through the ring and thereby increase the flux. To produce an induced field pointing into the paper, RHR-2 indicates that ***the induced current must be clockwise,*** opposite to what it is in position 2. By comparing the results for positions 2 and 4, it should be clear that the induced magnetic field is not always opposite to the external field.

Position 5: As in position 1, ***there is no induced current,*** since the magnetic field is everywhere zero.

Related Homework Material: *Problem 31*

Lenz's law should not be thought of as an independent law, for it is a consequence of the law of conservation of energy. The connection between energy conservation and induced emf has already been discussed in Section 22.2 for the specific case of motional emf. However, the connection is valid for any type of induced emf. In fact, the polarity of the induced emf, as specified by Lenz's law, ensures that energy is conserved.

22.6 APPLICATIONS OF ELECTROMAGNETIC INDUCTION TO THE REPRODUCTION OF SOUND

THE PHYSICS OF...

the electric guitar pickup.

Electromagnetic induction plays an important role in the technology used for the reproduction of sound. Virtually all electric guitars, for example, use electromagnetic pickups in which an induced emf is generated in a coil of wire by a vibrating string. Most guitars have at least two pickups located below each string and some, as Figure 22.18 illustrates, have three. Each pickup is sensitive to different harmonics produced by the vibrating string. The string is made from a magnetizable

Figure 22.18 When the string of an electric guitar vibrates, an emf is induced in the coil of the pickup. The two ends of the coil are connected to the input of an amplifier.

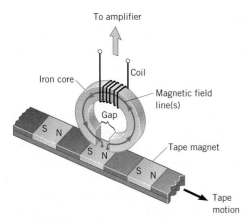

Figure 22.19 The magnetic playback head of a tape deck. As each tape magnet passes by the gap in the iron core, some of the magnetic field lines are routed through the core and the coil. The change in flux through the coil creates an induced emf. The width of the gap has been exaggerated.

Figure 22.20 A moving coil microphone.

metal, while the pickup consists of a coil of wire within which a permanent magnet is located. The permanent magnet produces a magnetic field that penetrates the guitar string, causing it to become magnetized with north and south poles. When the magnetized string is plucked, it oscillates, thereby changing the magnetic flux that passes through the coil. The changing flux induces an emf in the coil, and the polarity of this emf alternates with the vibratory motion of the string. A string vibrating at 440 Hz, for example, induces a 440-Hz ac emf in the coil. This signal, after being amplified, is sent to the speakers, which produce a 440-Hz sound wave (concert A).

The playback head of a cassette deck uses a moving tape to generate an emf in a coil of wire. Figure 22.19 shows a section of magnetized tape in which a series of "tape magnets" have been created in the magnetic layer of the tape during the recording process (see Section 21.9). The tape moves beneath the playback head, which consists of a coil of wire wrapped around an iron core. The iron core has the approximate shape of a horseshoe with a small gap between the two ends. The drawing shows an instant when a tape magnet is under the gap. Some of the field lines of the tape magnet are routed through the highly magnetizable iron core, and hence through the coil, as they proceed from the north pole to the south pole. Consequently, the flux through the coil changes as the tape moves past the gap. The change in flux leads to an ac emf, which is amplified and sent to the speakers, where the original sound is reproduced.

There are a number of types of microphones, and Figure 22.20 illustrates the one known as a moving coil microphone. When a sound wave strikes the diaphragm of the microphone, the diaphragm vibrates back and forth, and the attached coil moves along with it. Nearby is a stationary magnet. As the coil alternately approaches and recedes from the magnet, the flux through the coil changes. Consequently, an ac emf is induced in the coil. This electrical signal is sent to an amplifier and then to the speakers. In the type of microphone called a moving magnet microphone, the magnet is attached to the diaphragm and moves relative to a stationary coil.

THE PHYSICS OF...

the playback head of a tape deck.

THE PHYSICS OF...

a moving coil and a moving magnet microphone.

22.7   THE ELECTRIC GENERATOR

How a Generator Produces an Emf

THE PHYSICS OF...

the electric generator.

Figure 22.21 Electric generators such as these supply electrical power by producing an induced emf according to Faraday's law of electromagnetic induction.

Electric generators, such as those in Figure 22.21, produce virtually all of the world's electrical energy. A generator produces electrical energy from mechanical work, just the opposite of what a motor does. In a motor, an *input* electric current causes a coil to rotate, thereby doing mechanical work on any object attached to the shaft of the motor. In a generator, the shaft is rotated by some mechanical means, such as an engine or a turbine, and an emf is induced in a coil. If the generator is connected to an external circuit, an electric current is the *output* of the generator.

In its simplest form, an ac generator consists of a coil of wire that is rotated in a uniform magnetic field, as Figure 22.22a indicates. Although not shown in the picture, the wire is usually wound around an iron core. As in an electric motor, the coil/core combination is called the armature. Each end of the wire forming the coil is connected to the external circuit by means of a metal ring that rotates with the coil. Each ring slides against a stationary carbon brush, to which the external circuit (the lamp in the drawing) is connected.

To see how current is produced by the generator, consider the two vertical sides of the coil in Figure 22.22b. Since each is moving in a magnetic field **B**, the magnetic force exerted on the charges in the wire causes them to flow, thus creating a current. With the aid of RHR-1 (fingers of extended right hand point along **B**, thumb along the velocity **v**, palm pushes in the direction of the force on a positive charge), it can be seen that the direction of the current is from bottom to top in the left side and from top to bottom in the right side. Thus, charge flows around the loop. The upper and lower segments of the loop are also moving. However, these segments can be ignored because the magnetic force on the charges within them points toward the sides of the wire and not along the length.

Figure 22.22 (*a*) This electric generator consists of a coil (only one loop is shown) of wire that is rotated in a magnetic field **B** by some mechanical means. (*b*) The current *I* arises because of the magnetic force exerted on the charges in the moving wire. (*c*) The dimensions of the coil.

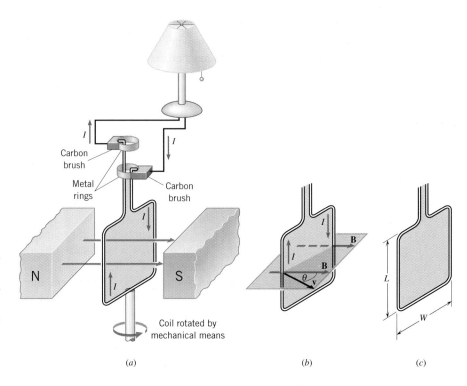

The magnitude of the motional emf developed in a conductor moving through a magnetic field is given by Equation 22.1. To apply this expression to the left side of the coil, whose length is L (see Figure 22.22c), we need to use the velocity component $v_\perp$ that is perpendicular to $\mathbf{B}$. Letting θ be the angle between $\mathbf{v}$ and $\mathbf{B}$, it follows that $v_\perp = v \sin\theta$, and the emf can be written as

$$\mathscr{E} = BLv_\perp = BLv \sin\theta$$

The emf induced in the right side has the same magnitude as that for the left side. Since the emfs from both sides drive current in the same direction around the loop, the emf for the complete loop is $\mathscr{E} = 2BLv \sin\theta$. If the coil consists of N loops, the net emf is N times as great as that of one loop, so

$$\mathscr{E} = N(2BLv \sin\theta)$$

It is convenient to express the variables v and θ in terms of the angular speed ω at which the coil rotates. Equation 8.2 shows that the angle θ is the product of the angular speed and the time, $\theta = \omega t$, if it is assumed that $\theta = 0$ rad when $t = 0$ s. Furthermore, any point on each vertical side moves on a circular path of radius $r = W/2$, where W is the width of the coil (see Figure 22.22c). Thus, the tangential speed v of each side is related to the angular speed ω via Equation 8.9 as $v = r\omega = (W/2)\omega$. Substituting these expressions for θ and v in the previous equation for $\mathscr{E}$, and recognizing that the product LW is the area A of the coil, we can write the induced emf as

Emf induced in a rotating planar coil $$\mathscr{E} = NAB\omega \sin\omega t = \mathscr{E}_0 \sin\omega t \qquad \text{where } \omega = 2\pi f \qquad (22.4)$$

In this result, the angular speed ω is in radians per second and is related to the frequency f (in cycles per second or hertz [H$_z$]) according to $\omega = 2\pi f$ (Equation 10.6).

Although Equation 22.4 was derived for a rectangular coil, the result is valid for any planar shape of area A (e.g., circular) and shows that the emf varies sinusoidally with time. The peak, or maximum, emf $\mathscr{E}_0$ occurs when $\sin\omega t = 1$ and has the value $\mathscr{E}_0 = NAB\omega$. Figure 22.23 shows a plot of Equation 22.4 and reveals that the emf changes polarity as the coil rotates. This changing polarity is exactly the same as that discussed for an ac voltage in Section 20.5 and illustrated in Figure 20.11. If the external circuit connected to the generator is a closed circuit, an alternating current results that changes direction at the same frequency f as the emf changes polarity. Therefore, this electric generator is also called an *alternating current (ac) generator*. The next two examples show how Equation 22.4 is applied.

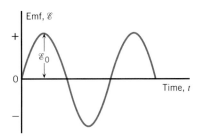

Figure 22.23 An ac generator produces this alternating emf $\mathscr{E}$ according to $\mathscr{E} = \mathscr{E}_0 \sin\omega t$.

EXAMPLE 10   An AC Generator

In Figure 22.22 the coil of the ac generator rotates at a frequency of $f = 60.0$ Hz and develops an emf of 120 V(rms). The coil has an area of $A = 3.0 \times 10^{-3}$ m^2 and consists of $N = 500$ turns. Find the magnitude of the magnetic field in which the coil rotates.

Reasoning The magnetic field can be found from the relation $\mathscr{E}_0 = NAB\omega$. However, in using this equation we must remember that $\mathscr{E}_0$ is the peak emf, while the given value of 120 V is not a peak value but an rms value. The peak emf is related to the rms emf by $\mathscr{E}_0 = \sqrt{2}\mathscr{E}_{rms}$, according to Equation 20.13.

• **PROBLEM SOLVING INSIGHT**
When using the equation $\mathscr{E}_0 = NAB\omega$, remember that the angular frequency ω must be in rad/s and is related to the frequency f (in Hz) according to $\omega = 2\pi f$ (Equation 10.6).

Solution The peak emf is $\mathscr{E}_0 = \sqrt{2}\,\mathscr{E}_{\text{rms}} = \sqrt{2}(120\ \text{V}) = 170\ \text{V}$. Solving $\mathscr{E}_0 = NAB\omega$ for B and using the fact that $\omega = 2\pi f$, we find that the magnitude of the magnetic field is

$$B = \frac{\mathscr{E}_0}{NA\omega} = \frac{\mathscr{E}_0}{NA2\pi f} = \frac{170\ \text{V}}{(500)(3.0 \times 10^{-3}\ \text{m}^2)2\pi(60.0\ \text{Hz})} = \boxed{0.30\ \text{T}}$$

EXAMPLE 11   A Bike Generator

A generator is mounted on a bicycle to power a headlight. A small wheel on the shaft of the generator is pressed against the bike tire and turns the armature 44 times for each revolution of the tire. The tire has a radius of 0.33 m. The armature has 75 turns, each with an area of $2.6 \times 10^{-3}\ \text{m}^2$, and rotates in a 0.10-T magnetic field. When the peak emf being generated is 6.0 V, what is the linear speed of the bike?

Reasoning The relation $\mathscr{E}_0 = NAB\omega$ provides a solution to this problem, if we take advantage of two additional facts. The first is that the angular speed ω of the armature is 44 times larger than the angular speed ω_{tire} of the bike tire. The second is that the tire is rolling, so that ω_{tire} is related to the linear speed v of the bike according to $\omega_{\text{tire}} = v/r$ (Equation 8.12), where r is the radius of the tire.

Solution Using $\omega = 44\omega_{\text{tire}} = 44(v/r)$, we find that the peak emf is $\mathscr{E}_0 = NAB \times 44(v/r)$. Solving for the linear speed v gives

$$v = \frac{\mathscr{E}_0 r}{44NAB} = \frac{(6.0\ \text{V})(0.33\ \text{m})}{44(75)(2.6 \times 10^{-3}\ \text{m}^2)(0.10\ \text{T})} = \boxed{2.3\ \text{m/s}}$$

The Electrical Energy Delivered by a Generator and the Countertorque

Some power-generating stations burn fossil fuel (coal, gas, or oil) to heat water and produce pressurized steam for turning the blades of a turbine, whose shaft is linked to the generator. Others use nuclear fuel or falling water as a source of energy. As the turbine rotates, the generator coil also rotates and mechanical work is transformed into electrical energy.

The devices to which the generator supplies electricity are known collectively as the "load," because they place a burden or load on the generator by taking electrical energy from it. If all the devices are switched off, the generator runs under a no-load condition, because there is no current in the external circuit and the generator does not supply electrical energy. Then, work needs to be done on the turbine only to overcome friction and other mechanical losses within the generator itself, and fuel consumption is at a minimum.

Figure 22.24 illustrates a situation in which a load is connected to a generator. Because there is now a current $I = I_1 + I_2$ in the coil, and the coil is situated in a magnetic field, the current experiences a magnetic force **F**. Figure 22.25 shows the magnetic force acting on the left side of the coil, the direction of **F** being given by RHR-1. A force of equal magnitude but opposite direction acts on the right side of the coil, although this force is not shown in the drawing. The magnetic force **F** gives rise to a *countertorque* that opposes the rotational motion. The greater the current drawn from the generator, the greater the countertorque, and the harder it is for the turbine to turn the coil. To compensate for this countertorque and keep the coil rotating at a constant angular speed, work must be done by the turbine, which means more fuel must be burned. This is another example of the law of conservation of energy, since the electrical energy consumed by the load must ultimately come from the energy source used to drive the turbine.

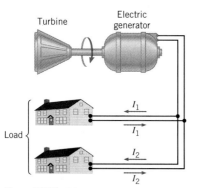

Figure 22.24 The generator supplies a total current of $I = I_1 + I_2$, to the load.

The Back Emf Generated by an Electric Motor

A generator converts mechanical work into electrical energy; in contrast, an electric motor converts electrical energy into mechanical work. Both devices are similar and consist of a coil of wire that rotates in a magnetic field. In fact, as the armature of a motor rotates, the magnetic flux passing through the coil changes and an emf is induced in the coil. Thus, when a motor is operating, two sources of emf are present: (1) the applied emf V that provides current to drive the motor (e.g., from a 120-V outlet), and (2) the emf $\mathscr{E}$ induced by the generator-like action of the rotating coil. The circuit diagram in Figure 22.26 shows these two emfs.

Consistent with Lenz's law, the induced emf $\mathscr{E}$ acts to oppose the applied emf V and is called the **back emf** or the **counter emf** of the motor. The greater the speed of the motor, the greater the flux change through the coil, and the greater is the back emf. Because V and $\mathscr{E}$ have opposite polarities, the net emf in the circuit is $V - \mathscr{E}$. In Figure 22.26, R is the resistance of the wire in the coil, and the current I drawn by the motor is determined from Ohm's law as the net emf divided by the resistance:

$$I = \frac{V - \mathscr{E}}{R} \tag{22.5}$$

The next example uses this result to illustrate that the current in a motor depends on both the applied emf V and the back emf $\mathscr{E}$.

EXAMPLE 12   Operating a Motor

The coil of an ac motor has a resistance of $R = 4.1\ \Omega$. The motor is plugged into an outlet where $V = 120.0$ volts (rms), and the coil develops a back emf of $\mathscr{E} = 118.0$ volts (rms) when rotating at normal speed. The motor is turning a wheel. Find (a) the current when the motor first starts up and (b) the current when the motor is operating at normal speed.

Reasoning Once normal operating speed is attained, the motor need only work to compensate for frictional losses. But in bringing the wheel up to speed from rest, the motor must also do work to increase the wheel's rotational kinetic energy. Thus, bringing the wheel up to speed requires more work, and hence more current, than maintaining the normal operating speed. We expect our answers to parts (a) and (b) to reflect this fact.

Solution

(a) When the motor just starts up, the coil is not rotating, so there is no back emf induced in the coil and $\mathscr{E} = 0$ V. The start-up current drawn by the motor is

$$I = \frac{V - \mathscr{E}}{R} = \frac{120\ \text{V} - 0\ \text{V}}{4.1\ \Omega} = \boxed{29\ \text{A}} \tag{22.5}$$

Shaft

(a)

Top view of coil

F

Shaft

Current I (out of paper)

B

(b)

Figure 22.25 (a) A current I exists in the rotating coil of the generator. (b) A top view of the coil, showing the magnetic force **F** exerted on the left side of the coil.

● **PROBLEM SOLVING INSIGHT**
The current in an electric motor depends on both the applied emf V and any back emf $\mathscr{E}$ developed because the coil of the motor is rotating.

Figure 22.26 The applied emf V supplies the current I to drive the motor. The circuit on the right shows V along with the electrical equivalent of the motor, including the resistance R of its coil and the back emf $\mathscr{E}$.

Ac motor

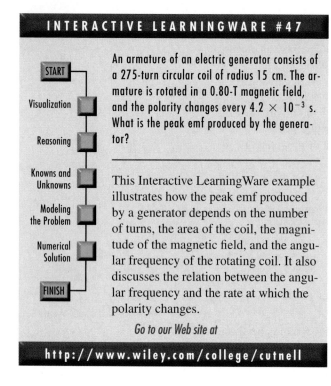

INTERACTIVE LEARNINGWARE #47

START

Visualization

Reasoning

Knowns and
Unknowns

Modeling
the Problem

Numerical
Solution

FINISH

An armature of an electric generator consists of a 275-turn circular coil of radius 15 cm. The armature is rotated in a 0.80-T magnetic field, and the polarity changes every 4.2×10^{-3} s. What is the peak emf produced by the generator?

This Interactive LearningWare example illustrates how the peak emf produced by a generator depends on the number of turns, the area of the coil, the magnitude of the magnetic field, and the angular frequency of the rotating coil. It also discusses the relation between the angular frequency and the rate at which the polarity changes.

Go to our Web site at

http://www.wiley.com/college/cutnell

(b) At normal speed, the motor develops a back emf of $\mathcal{E} = 118.0$ volts, so the current is

$$I = \frac{V - \mathcal{E}}{R} = \frac{120.0 \text{ V} - 118.0 \text{ V}}{4.1 \ \Omega} = \boxed{0.49 \text{ A}}$$

Example 12 illustrates that when a motor is just starting, there is little back emf, and, consequently, a relatively large current exists in the coil. As the motor speeds up, the back emf increases until it reaches a maximum value when the motor is rotating at normal speed. The back emf becomes almost equal to the applied emf, and the current is reduced to a relatively small value, which is sufficient to provide the torque on the coil needed to overcome frictional and other losses in the motor and to drive the load (e.g., a fan).

22.8 MUTUAL INDUCTANCE AND SELF-INDUCTANCE

Mutual Inductance

We have seen that an emf can be induced in a coil by keeping the coil stationary and moving a magnet nearby, or by moving the coil near a stationary magnet. Figure 22.27 illustrates another important method of inducing an emf. Here, two coils of wire are placed close to each other, the *primary coil* and the *secondary coil*. The primary coil is the one connected to an ac generator, which sends an alternating current I_p through it. The secondary coil is not attached to a generator, although a voltmeter is connected across it to register any induced emf.

The current-carrying primary coil is an electromagnet and creates a magnetic field in the surrounding region. If the two coils are close to each other, a significant fraction of this magnetic field penetrates the secondary coil and produces a magnetic flux. The flux is changing, since the current in the primary coil and its associated magnetic field are changing. Because of the change in flux, an emf is induced in the secondary coil.

The effect in which a changing current in one circuit induces an emf in another circuit is called **mutual induction.** According to Faraday's law of electromagnetic induction, the emf $\mathcal{E}_s$ induced in the secondary coil is proportional to the change in flux $\Delta\Phi_s$ passing through it. However, $\Delta\Phi_s$ is produced by the change in current ΔI_p in the primary coil. Therefore, it is convenient to recast Faraday's law into a form that relates $\mathcal{E}_s$ to ΔI_p. To see how this recasting is accomplished, note that the net magnetic flux passing through the secondary coil is $N_s\Phi_s$, where N_s is the number of loops in the secondary coil and Φ_s is the flux through one loop (assumed to be the same for all loops). The net flux is proportional to the magnetic field, which, in turn, is proportional to the current I_p in the primary. Thus, we can write $N_s\Phi_s \propto I_p$. This proportionality can be converted into an equation in the usual manner by introducing a proportionality constant M, known as the **mutual inductance:**

$$N_s\Phi_s = MI_p \quad \text{or} \quad M = \frac{N_s\Phi_s}{I_p} \tag{22.6}$$

Figure 22.27 An alternating current I_p in the primary coil creates an alternating magnetic field. This changing field induces an emf in the secondary coil.

Substituting this equation into Faraday's law, we find that

$$\mathscr{E}_s = -N_s \frac{\Delta \Phi_s}{\Delta t} = -\frac{\Delta(N_s \Phi_s)}{\Delta t} = -\frac{\Delta(MI_p)}{\Delta t} = -M \frac{\Delta I_p}{\Delta t}$$

Emf due to
mutual induction
$$\mathscr{E}_s = -M \frac{\Delta I_p}{\Delta t} \qquad (22.7)$$

Writing Faraday's law in this manner makes it clear that the emf $\mathscr{E}_s$ induced in the secondary coil is due to the change in the current ΔI_p in the primary coil.

Equation 22.7 shows that the measurement unit for the mutual inductance M is $V \cdot s/A$, which is called a henry (H) in honor of Joseph Henry: $1\ V \cdot s/A = 1\ H$. The mutual inductance depends on the geometry of the coils and the nature of any ferromagnetic core material that is present. Although M can be calculated for some highly symmetrical arrangements, it is usually measured experimentally. In most situations, values of M are less than 1 H and are often on the order of millihenries ($1\ mH = 1 \times 10^{-3}\ H$) or microhenries ($1\ \mu H = 1 \times 10^{-6}\ H$).

Self-Inductance

In all the examples of induced emfs presented so far, the magnetic field has been produced by an external source, such as a permanent magnet or an electromagnet. However, the magnetic field need not arise from an external source. An emf can be induced in a current-carrying coil by a change in the magnetic field that the current itself produces. For instance, Figure 22.28 shows a coil connected to an ac generator. The alternating current creates an alternating magnetic field that, in turn, creates a changing flux through the coil. The change in flux induces an emf in the coil, in accord with Faraday's law. The effect in which a changing current in a circuit induces an emf in the same circuit is referred to as *self-induction.*

When dealing with self-induction, as with mutual induction, it is customary to recast Faraday's law into a form in which the induced emf is proportional to the change in current in the coil rather than to the change in flux. If Φ is the magnetic flux that passes through one turn of the coil, then $N\Phi$ is the net flux through a coil of N turns. Since Φ is proportional to the magnetic field, and the magnetic field is proportional to the current I, it follows that $N\Phi \propto I$. By inserting a constant L, called the *self-inductance* or simply the *inductance* of the coil, we can convert this proportionality into Equation 22.8:

$$N\Phi = LI \quad \text{or} \quad L = \frac{N\Phi}{I} \qquad (22.8)$$

Faraday's law of induction now gives the induced emf as

$$\mathscr{E} = -N \frac{\Delta \Phi}{\Delta t} = -\frac{\Delta(N\Phi)}{\Delta t} = -\frac{\Delta(LI)}{\Delta t} = -L \frac{\Delta I}{\Delta t}$$

Emf due to
self-induction
$$\mathscr{E} = -L \frac{\Delta I}{\Delta t} \qquad (22.9)$$

Like mutual inductance, L is measured in henries. The magnitude of L depends on the geometry of the coil and on the core material. By wrapping the coil around a ferromagnetic (iron) core, the magnetic flux—and therefore the inductance—can be increased substantially relative to that for an air core. Because of their self-inductance, coils are known as *inductors* and are widely used in electronics. Inductors come in all sizes, typically in the range between millihenries and microhenries. Example 13 shows how to determine the inductance of a solenoid.

Figure 22.28 The alternating current in the coil generates an alternating magnetic field that induces an emf in the coil.

EXAMPLE 13 The Self-Inductance of a Long Solenoid

A long solenoid of length $\ell = 8.0 \times 10^{-2}$ m and cross-sectional area $A = 5.0 \times 10^{-5}$ m^2 contains $n = 6500$ turns per meter. (a) Find the self-inductance of the solenoid, assuming the core is air. (b) Determine the emf induced in the solenoid when the current increases from 0 to 1.5 A in a time of 0.20 s.

Reasoning

(a) The self-inductance can be found by using Equation 22.8 ($L = N\Phi/I$) provided the flux Φ can be determined. The flux is given by Equation 22.2 as $\Phi = BA \cos \phi$. In the case of a solenoid, the interior magnetic field is directed perpendicular to the plane of the loops (see Sections 21.7 and 21.8), so $\phi = 0°$ and $\Phi = BA$. The magnetic field inside a long solenoid has the value $B = \mu_0 nI$, according to Equation 21.7, where n is the number of turns per unit length.

(b) The emf induced in the solenoid can be obtained from the relation $\mathscr{E} = -L(\Delta I/\Delta t)$, since L is determined in part (a) and ΔI and Δt are given.

Solution

(a) The self-inductance of the solenoid is

$$L = \frac{N\Phi}{I} = \frac{N(BA)}{I} = \frac{N(\mu_0 nI)A}{I} = \mu_0 nNA = \mu_0 n^2 A\ell$$

where we have replaced N by $n\ell$. Substituting the given values into this result yields

$$L = \mu_0 n^2 A\ell = (4\pi \times 10^{-7}\,\text{T}\cdot\text{m/A})(6500\,\text{turns/m})^2$$
$$\times (5.0 \times 10^{-5}\,\text{m}^2)(8.0 \times 10^{-2}\,\text{m}) = \boxed{2.1 \times 10^{-4}\,\text{H}}$$

(b) The induced emf that results from the increasing current is

$$\mathscr{E} = -L\frac{\Delta I}{\Delta t} = -(2.1 \times 10^{-4}\,\text{H})\left(\frac{1.5\,\text{A} - 0\,\text{A}}{0.20\,\text{s}}\right) = \boxed{-1.6 \times 10^{-3}\,\text{V}} \quad (22.9)$$

The negative sign reminds us that the induced emf opposes the increasing current.

The Energy Stored in an Inductor

An inductor, like a capacitor, can store energy. This stored energy arises because a generator does work to establish a current in an inductor. Suppose an inductor is connected to a generator whose terminal voltage can be varied continuously from zero to some final value. As the voltage is increased, the current I in the circuit rises continuously from zero to its final value. While the current is rising, an induced emf $\mathscr{E} = -L(\Delta I/\Delta t)$ appears across the inductor. Conforming to Lenz's law, the polarity of the induced emf $\mathscr{E}$ is opposite to that of the generator voltage, so as to oppose the increase in the current. Thus, the generator must do work to push the charges through the inductor against this induced emf. The increment of work ΔW done by the generator in moving a small amount of charge ΔQ through the inductor is $\Delta W = (\Delta Q)\mathscr{E} = (\Delta Q)L(\Delta I/\Delta t)$, according to Equation 19.4. To ensure that the work done is positive, as it must be since the generator is driving charge against an opposing emf, the minus sign in front of the $L(\Delta I/\Delta t)$ term has been removed. Since $\Delta Q/\Delta t$ is the current I, the work done is

$$\Delta W = LI(\Delta I)$$

In this expression ΔW represents the work done by the generator to increase the current in the inductor by an amount ΔI. To determine the total work W done while the current is changed from zero to its final value, all the small increments ΔW must be added together. This summation is left as an exercise (see problem 52). The result is $W = \frac{1}{2}LI^2$, where I represents the final current in the inductor. This work is stored as energy in the inductor, so that

Energy stored in an inductor

$$\text{Energy} = \tfrac{1}{2}LI^2 \qquad (22.10)$$

It is possible to regard the energy in an inductor as being stored in its magnetic field. For the special case of a long solenoid, Example 13 shows that the self-inductance is $L = \mu_0 n^2 A\ell$, where n is the number of turns per unit length, A is the cross-sectional area, and ℓ is the length of the solenoid. As a result, the energy stored in a long solenoid is

$$\text{Energy} = \tfrac{1}{2}LI^2 = \tfrac{1}{2}\mu_0 n^2 A\ell I^2$$

Since $B = \mu_0 nI$ at the interior of a long solenoid (Equation 21.7), this energy can be expressed as

$$\text{Energy} = \frac{1}{2\mu_0} B^2 A\ell$$

The term $A\ell$ is the volume inside the solenoid where the magnetic field exists, so the energy per unit volume or **energy density** is

$$\text{Energy density} = \frac{\text{Energy}}{\text{Volume}} = \frac{1}{2\mu_0} B^2 \qquad (22.11)$$

Although this result was obtained for the special case of a long solenoid, it is quite general and is valid for any point where a magnetic field exists in air or vacuum or in a nonmagnetic material. Thus, energy can be stored in a magnetic field, just as it can in an electric field.

22.9 TRANSFORMERS

One of the most important applications of mutual induction and self-induction takes place in a transformer. A **transformer** is a device for increasing or decreasing an ac voltage. For instance, whenever cordless appliances (e.g., a hand-held vacuum cleaner) are plugged into a wall receptacle to recharge the batteries, a transformer plays a role in reducing the 120-V ac voltage to a much smaller value. Typically, between 3 and 9 V are needed to energize batteries. In another example, a picture tube in a television set needs about 15 000 V to accelerate the electron beam, and a transformer is used to obtain this high voltage from the 120 V at a wall socket.

Figure 22.29 shows a drawing of a transformer. The transformer consists of an iron core on which two coils are wound: a primary coil with N_p turns, and a sec-

Figure 22.29 A transformer consists of a primary coil and a secondary coil, both wound on an iron core. The changing magnetic flux produced by the current in the primary coil induces an emf in the secondary coil. At the far right is the symbol for a transformer.

ondary coil with N_s turns. The primary coil is the one connected to the ac generator. For the moment, suppose the switch in the secondary circuit is open, so there is no current in this circuit.

The alternating current in the primary coil establishes a changing magnetic field in the iron core. Because iron is easily magnetized, it greatly enhances the magnetic field relative to that in an air core and guides the field lines to the secondary coil. In a well-designed core, nearly all the magnetic flux Φ that passes through each turn of the primary also goes through each turn of the secondary. Since the magnetic field is changing, the flux through the primary and secondary coils is also changing, and consequently an emf is induced in both coils. In the secondary coil the induced emf $\mathscr{E}_s$ arises from mutual induction and is given by Faraday's law as

$$\mathscr{E}_s = -N_s \frac{\Delta \Phi}{\Delta t}$$

In the primary coil the induced emf $\mathscr{E}_p$ is due to self-induction and is specified by Faraday's law as

$$\mathscr{E}_p = -N_p \frac{\Delta \Phi}{\Delta t}$$

The term $\Delta \Phi / \Delta t$ is the same in both of these equations, since the same flux penetrates each turn of both coils. Dividing the two equations shows that

$$\frac{\mathscr{E}_s}{\mathscr{E}_p} = \frac{N_s}{N_p}$$

In a high-quality transformer the resistances of the coils are negligible, so the magnitudes of the emfs, $\mathscr{E}_s$ and $\mathscr{E}_p$, are nearly equal to the terminal voltages, V_s and V_p, across the coils (see Section 20.9 for a discussion of terminal voltage). The relation $\mathscr{E}_s / \mathscr{E}_p = N_s / N_p$ is called the ***transformer equation*** and is usually written in terms of the terminal voltages:

***Transformer
equation***
$$\frac{V_s}{V_p} = \frac{N_s}{N_p} \qquad (22.12)$$

According to the transformer equation, if N_s is greater than N_p, the secondary (output) voltage is greater than the primary (input) voltage. In this case we have a *step-up* transformer. On the other hand, if N_s is less than N_p, the secondary voltage is less than the primary voltage, and we have a *step-down* transformer. The ratio N_s / N_p is referred to as the *turns ratio* of the transformer. A turns ratio of 8/1 (often written as 8 : 1) means, for example, that the secondary coil has eight times more turns than does the primary coil. Conversely, a turns ratio of 1 : 8 implies that the secondary coil has one-eighth as many turns as the primary coil.

A transformer operates with ac electricity and not with dc. A steady direct current in the primary coil produces a flux that does not change in time, and thus no emf is induced in the secondary coil. The ease with which transformers can change voltages from one value to another is a principal reason why ac is preferred over dc.

If the switch in the secondary circuit of Figure 22.29 is closed, a current I_s exists in the circuit and electrical energy is fed to the TV tube. This energy comes from the ac generator connected to the primary coil. Although the secondary voltage V_s may be larger or smaller than the primary voltage V_p, energy is not being created or destroyed by the transformer. Energy conservation requires that the energy delivered to the secondary coil must be the same as the energy delivered to the primary coil, provided no energy is dissipated in heating these coils or is otherwise lost. In a

Power distribution stations use high-voltage transformers similar to this one to step-up or step-down voltages.

well-designed transformer, less than 1% of the input energy is lost in the form of heat. Noting that power is energy per unit time, and assuming 100% energy transfer, the average power $\overline{P}_p$ delivered to the primary coil is equal to the average power $\overline{P}_s$ delivered to the secondary coil: $\overline{P}_p = \overline{P}_s$. But $P = IV$ (Equation 20.15a), so $I_p V_p = I_s V_s$, or

$$\frac{I_s}{I_p} = \frac{V_p}{V_s} = \frac{N_p}{N_s} \qquad (22.13)$$

Observe that V_s/V_p is equal to the turns ratio N_s/N_p, while I_s/I_p is equal to the inverse turns ratio N_p/N_s. Consequently, *a transformer that steps up the voltage simultaneously steps down the current, and a transformer that steps down the voltage steps up the current.* However, the power is neither stepped up nor stepped down, since $\overline{P}_p = \overline{P}_s$. Example 14 emphasizes this fact.

EXAMPLE 14 A Step-Down Transformer

A step-down transformer inside a stereo receiver has 330 turns in the primary coil and 25 turns in the secondary coil. The plug connects the primary coil to a 120-V wall socket, and there is a current of 0.83 A in the primary coil while the receiver is turned on. Connected to the secondary coil are the transistor circuits of the receiver. Find (a) the voltage across the secondary coil, (b) the current in the secondary coil, and (c) the average electrical power delivered to the transistor circuits.

Reasoning The transformer equation, Equation 22.12, states that the secondary voltage V_s is equal to the product of the primary voltage V_p and the turns ratio N_s/N_p. On the other hand, the secondary current I_s is equal to the product of the primary current I_p and the inverse turns ratio N_p/N_s. The average power delivered to the transistor circuits is the product of the secondary current and the secondary voltage.

Solution

(a) The voltage across the secondary coil can be found from the transformer equation:

$$V_s = V_p \frac{N_s}{N_p} = (120 \text{ V}) \left(\frac{25}{330} \right) = \boxed{9.1 \text{ V}} \qquad (22.12)$$

(b) The current in the secondary coil follows from Equation 22.13 as

$$I_s = I_p \frac{N_p}{N_s} = (0.83 \text{ A}) \left(\frac{330}{25} \right) = \boxed{11 \text{ A}}$$

(c) The average power $\overline{P}_s$ delivered to the secondary is the product of I_s and V_s:

$$\overline{P}_s = I_s V_s = (11 \text{ A})(9.1 \text{ V}) = \boxed{1.0 \times 10^2 \text{ W}} \qquad (20.15a)$$

As a check on our calculation, we verify that the power delivered to the secondary coil is the same as that sent to the primary coil from the wall receptacle: $\overline{P}_p = I_p V_p = (0.83 \text{ A})(120 \text{ V}) = 1.0 \times 10^2 \text{ W}$.

Transformers play an important role in the transmission of power between electrical generating plants and the communities they serve. Whenever electricity is transmitted, there is always some loss of power in the transmission lines themselves due to resistive heating. Since the resistance of the wires is proportional to their length, the longer the wires the greater is the power loss. Power companies reduce this loss by using transformers that step up the voltage to high levels, while reducing the current. A smaller current means less power loss, since $P =$

Figure 22.30 Transformers play a key role in the transmission of electric power.

I^2R, where R is the resistance of the transmission wires (see problem 59). Figure 22.30 shows one possible way of transmitting power. The power plant produces a voltage of 12 000 V. This voltage is then raised to 240 000 V by a 20:1 step-up transformer. The high-voltage power is sent over the long-distance transmission line. Upon arrival at the city, the voltage is reduced to about 8000 V at a substation using a 1:30 step-down transformer. However, before any domestic use, the voltage is further reduced to 240 V (or possibly 120 V) by another step-down transformer that is often mounted on a utility pole. The power is then distributed to consumers.

22.10 CONCEPTS & CALCULATIONS

In this chapter we have seen that there are three ways to create an induced emf in a coil: by changing the magnitude of a magnetic field, by changing the direction of the field relative to the coil, and by changing the area of the coil. We have already explored the first two methods, and Example 15 now illustrates the third method. In the process, it provides a review of Faraday's law of electromagnetic induction.

CONCEPTS & CALCULATIONS EXAMPLE 15
The Emf Produced by a Changing Area

A circular coil of radius 0.11 m contains a single turn and is located in a constant magnetic field of magnitude 0.27 T. The magnetic field is parallel to the normal to the plane of the coil. The radius increases to 0.30 m in a time of 0.080 s. (a) Determine the emf induced in the coil. (b) The coil has a resistance of 0.70 Ω. Find the induced current.

Concept Questions and Answers Why is there an emf induced in the coil?

Answer According to Faraday's law of electromagnetic induction, Equation 22.3, an emf is induced whenever the magnetic flux through the coil changes. The magnetic flux, as expressed by Equation 22.2, depends on the area of the coil, which, in turn, depends on the radius. If the radius changes, the area changes, causing the flux to change and an induced emf to appear.

Does the magnitude of the induced emf depend on whether the area is increasing or decreasing?

Answer No. The magnitude of the induced emf depends only on the magnitude of the rate $\Delta\Phi/\Delta t$ at which the flux changes. It does not matter whether the flux is increasing or decreasing.

What determines the amount of current induced in the coil?

Answer According to Ohm's law (Equation 20.2), the current is equal to the emf induced in the coil divided by the resistance of the coil. Therefore, the larger the induced emf and the smaller the resistance, the larger will be the induced current.

If the coil is cut so it is no longer one continuous piece, is there an induced emf and induced current?

Answer An induced emf is generated in the coil regardless of whether the coil is whole or cut. However, an induced current exists only if the coil is continuous. Cutting the coil is like opening a switch. There is no longer a continuous path for the current to follow, so the current stops.

Solution

(a) The average induced emf is given by Faraday's law of electromagnetic induction.

$$\mathscr{E} = -N\frac{\Delta \Phi}{\Delta t} = -N\left(\frac{\Phi - \Phi_0}{t - t_0}\right) \tag{22.3}$$

where Φ and Φ_0 are the final and initial fluxes, respectively. The definition of magnetic flux is $\Phi = BA \cos \phi$, according to Equation 22.2, where $\phi = 0°$ because the field is parallel to the normal. The area of a circle is $A = \pi r^2$. With these expressions for Φ and A, Faraday's law becomes

$$\mathscr{E} = -N\left(\frac{BA \cos \phi - BA_0 \cos \phi}{t - t_0}\right) = -NB(\cos \phi)\left(\frac{\pi r^2 - \pi r_0^2}{t - t_0}\right)$$

$$= -(1)(0.27 \text{ T})(\cos 0°)\left[\frac{\pi(0.30 \text{ m})^2 - \pi(0.11 \text{ m})^2}{0.080 \text{ s}}\right] = \boxed{0.83 \text{ V}}$$

(b) The induced current I is equal to the induced emf divided by the resistance R of the coil:

$$I = \frac{\mathscr{E}}{R} = \frac{0.83 \text{ V}}{0.70 \text{ }\Omega} = \boxed{1.2 \text{ A}} \tag{20.2}$$

One of the most important applications of Faraday's law of electromagnetic induction is the electric generator, for it is the source of virtually all the electrical energy that we use. A generator produces an emf as the rotating coil changes its orientation relative to a fixed magnetic field. Example 16 reviews the characteristics of this type of induced emf.

CONCEPTS & CALCULATIONS EXAMPLE 16   The Emf Produced by Generator

Figure 22.31 shows the emf produced by a generator as a function of time. The coil of the generator has an area of $A = 0.15 \text{ m}^2$ and consists of $N = 10$ turns. The coil rotates in a magnetic field of magnitude 0.27 T. (a) Determine the period of the motion. (b) What is the angular frequency of the rotating coil? (c) Find the value of the emf when $t = \frac{1}{4}T$. (d) What is the emf when $t = 0.025$ s?

Concept Questions and Answers Can the period of the rotating coil be determined from the graph?

Answer Yes. The period is the time for the coil to rotate through one revolution, or cycle. During this time, the emf is positive for one half of the cycle and negative for the other half of the cycle.

The emf produced by a generator depends on its angular frequency. How is the angular frequency of the rotating coil related to the period?

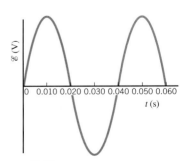

Figure 22.31 A plot of the emf produced by a generator as a function of time.

Answer According to Equation 10.6, the angular frequency ω is related to the period T by $\omega = 2\pi/T$.

Starting at $t = 0$ s, how much time is required for the generator to produce its peak emf? Express the answer in terms of the period T of the motion (e.g., $t = \frac{1}{10}T$.)

Answer An examination of the graph shows that the emf reaches its maximum or peak value one quarter of the way through a cycle. Since the time to complete one cycle is the period T, the time to reach the peak emf is $t = \frac{1}{4}T$.

How often does the polarity of the emf change in one cycle?

Answer The graph shows that in every cycle the emf has an interval of positive polarity followed by an interval of negative polarity. In other words, the polarity changes from positive to negative, and then from negative to positive during each cycle. Thus, the polarity changes twice during each cycle.

Solution

(a) The period is the time for the generator to make one complete cycle. From the graph it can be seen that this time is $T = \boxed{0.040 \text{ s}}$.

(b) The angular frequency ω is related to the period by $\omega = 2\pi/T$, so

$$\omega = \frac{2\pi}{T} = \frac{2\pi}{0.040 \text{ s}} = \boxed{160 \text{ rad/s}}$$

(c) When $t = \frac{1}{4}T$, the emf has reached its peak value. According to Equation 22.4, the peak emf is

$$\mathcal{E}_0 = NAB\omega = (10)(0.15 \text{ m}^2)(0.27 \text{ T})(160 \text{ rad/s}) = \boxed{65 \text{ V}}$$

(d) The emf produced by the generator as a function of time is given by Equation 22.4

$$\mathcal{E} = \mathcal{E}_0 \sin \omega t = (65 \text{ V})\sin[(160 \text{ rad/s})(0.025 \text{ s})] = \boxed{-49 \text{ V}}$$

● **PROBLEM SOLVING INSIGHT**
In evaluating sin ωt a calculator must be set to the radian mode (not the degree mode) because ω is expressed in units of rad/s.

At the end of the problem set for this chapter, you will find homework problems that contain both conceptual and quantitative parts. These problems are grouped under the heading *Concepts & Calculations, Group Learning Problems*. They are designed for use by students working alone or in small learning groups. The conceptual part of each problem provides a convenient focus for group discussions.

SUMMARY

The **magnetic flux Φ** that passes through a surface is $\Phi = BA \cos \phi$, where A is the area of the surface, B is the magnitude of the magnetic field at the surface, and ϕ is the angle between **B** and the normal to the surface.

Electromagnetic induction is the phenomenon in which an emf is induced in a coil of wire by a change in the magnetic flux that passes through the coil.

Faraday's law of electromagnetic induction states that the average emf $\mathcal{E}$ induced in a coil of N loops is

$$\mathcal{E} = -N\left(\frac{\Phi - \Phi_0}{t - t_0}\right) = -N\frac{\Delta\Phi}{\Delta t}$$

where $\Delta\Phi$ is the change in magnetic flux through one loop and Δt is the time interval during which the change occurs.

For the special case of a conductor of length L moving with speed v perpendicular to a magnetic field **B**, the induced emf is called motional emf and its value is given by $\mathcal{E} = vBL$.

Lenz's law provides a way to determine the polarity of an induced emf. Lenz's law is stated as follows: The induced emf resulting from a changing magnetic flux has a polarity that leads to an induced current whose direction is such that the induced magnetic field opposes the original flux change. This statement is a consequence of the law of conservation of energy.

In its simplest form, an **electric generator** consists of a coil of N loops that rotates in a uniform magnetic field **B**. The emf produced by this generator is $\mathcal{E} = NAB\omega \sin \omega t = \mathcal{E}_0 \sin \omega t$, where A is the area of the coil, ω is the angular speed (in rad/s) of the coil, and $\mathscr{E}_0$ is the peak emf. The angular speed in rad/s is related to the frequency f in cycles/s or Hz according to $\omega = 2\pi f$.

When an electric motor is running, it exhibits a generator-like behavior by producing an induced emf, called a **back emf.** The current I needed to keep the motor running at a constant speed is $I = (V - \mathscr{E})/R$, where V is the emf applied to the motor by an external source, $\mathscr{E}$ is the back emf, and R is the resistance of the motor coil.

Mutual induction is the effect in which a changing current in the primary coil induces an emf in the secondary coil. The emf $\mathscr{E}_s$ induced in the secondary coil by a change in current ΔI_p in the primary coil is $\mathscr{E}_s = -M(\Delta I_p/\Delta t)$, where Δt is the time interval during which the change occurs. The constant M is the **mutual inductance** between the two coils and is measured in henries (H).

Self-induction is the effect in which a change in current ΔI in a coil induces an emf $\mathscr{E} = -L(\Delta I/\Delta t)$ in the same coil.

The constant L is the **self-inductance** or **inductance** of the coil and is measured in henries. To establish a current I in an inductor, work must be done by an external agent. This work is stored as energy in the inductor, the amount being Energy $= \frac{1}{2}LI^2$. The energy stored in an inductor can be regarded as being stored in its magnetic field. At any point in air or vacuum or in a nonmagnetic material where a magnetic field $\mathbf{B}$ exists, the **energy density,** or the energy stored per unit volume, is Energy density $= B^2/(2\mu_0)$.

A **transformer** consists of a primary coil of N_p turns and a secondary coil of N_s turns. If the resistances of the coils are negligible, the voltage V_p across the primary and the voltage V_s across the secondary are related according to the **transformer equation** $V_s/V_p = N_s/N_p$. A transformer functions with ac electricity, not with dc. If the transformer is 100% efficient in transferring power from the primary coil to the secondary coil, the ratio of the secondary current I_s to the primary current I_p is $I_s/I_p = N_p/N_s$.

CONCEPTUAL QUESTIONS

1. Suppose the coil and the magnet in Figure 22.1a were each moving with the same velocity relative to the earth. Would there be an induced current in the coil? Explain.

2. In the discussion concerning Figure 22.5, we saw that a force of 0.086 N from an external agent was required to keep the rod moving at a constant speed. Suppose the light bulb in the figure is unscrewed from its socket. How much force would now be needed to keep the rod moving at a constant speed? Justify your answer.

3. Eddy currents are electric currents that can arise in a piece of metal when it moves through a region where the magnetic field is not the same everywhere. The picture shows, for example, a metal sheet moving to the right at a velocity **v** and a magnetic field **B** that is directed perpendicular to the sheet. At the instant represented, the magnetic field only extends over the left half of the sheet. An emf is induced that leads to the eddy current shown. Explain why this current causes the metal sheet to slow down. This action of eddy currents is used in various devices as a brake to damp out unwanted motion.

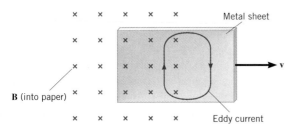

4. A magnetic field is necessary if there is to be a magnetic flux passing through a coil of wire. Yet, just because there is a magnetic field does not mean that a magnetic flux will pass through a coil. Account for this observation.

5. Suppose the magnetic flux through a 1-m² flat surface is known to be 2 Wb. From this data alone, it is possible to determine certain information about the average magnetic field at the surface, but not the magnitude and direction of the total field. What exactly can be ascertained about the field? Explain.

6. A conducting rod is free to slide along a pair of conducting rails, in a region where a uniform and constant (in time) magnetic field is directed into the plane of the paper, as the drawing illustrates. Initially the rod is at rest. Describe the rod's motion after the switch is closed. Be sure to account for the effect of any motional emf that develops.

7. Explain how a bolt of lightning can produce a current in the circuit of an electrical appliance, even when the lightning does not directly strike the appliance.

8. Review Conceptual Example 7 before answering this question. A solenoid is connected to an ac source. A copper ring is placed inside the solenoid, with the normal to the ring being parallel to the axis of the solenoid. The copper ring gets hot, yet nothing touches it. Why?

9. A robot is designed to move parallel to a cable hidden under the floor. The cable carries a steady direct current I. A sensor mounted on the robot consists of a coil of wire. The coil is near the floor and parallel to it. As long as the robot moves parallel to the cable, with the coil directly over it, no emf is induced in the coil, since the magnetic flux through the coil does not change. But when the robot deviates from the parallel path, an induced emf appears in the coil. The emf is sent to electronic circuits that bring the robot back to the path. Explain why an emf would be induced in the sensor coil.

10. In a car, the generator-like action of the alternator occurs while the engine is running and keeps the battery fully charged. The headlights would discharge an old and failing battery quickly if it were not for the alternator. Explain why the engine of a parked car runs more "quietly" with the headlights off than with them on when the battery is in bad shape.

11. In Figure 22.3 a coil of wire is being stretched. (a) Using Lenz's law, verify that the induced current in the coil has the direction shown in the drawing. (b) Deduce the direction of the induced current if the direction of the external magnetic field in the figure were reversed. Explain.

12. (a) When the switch in the circuit in the drawing is closed, a current is established in the coil and the metal ring "jumps" upward. Explain this be-

Iron core — Metal ring
Switch
Coil
+ | |− Battery

havior. (b) Describe what would happen to the ring if the battery polarity were reversed.

13. The string of an electric guitar vibrates in a standing wave pattern that consists of nodes and antinodes. (Section 17.5 discusses standing waves.) Where should an electromagnetic pickup be located in the standing wave pattern to produce a maximum emf, at a node or an antinode? Why?

14. An electric motor in a hair drier is running at normal speed and, thus, is drawing a relatively small current, as in part (b) of Example 12. What happens to the current drawn by the motor if the shaft is prevented from turning, so the back emf is suddenly reduced to zero? Remembering that the wire in the coil of the motor has some resistance, what happens to the temperature of the coil? Justify your answers.

15. One transformer is a step-up device, while another is step-down. These two units have the same voltage across and the same current in their primary coils. Does either one deliver more power to the circuit attached to the secondary coil? If so, which one? Ignore any heat loss within the transformers and account for your answer.

PROBLEMS

Section 22.2 Motional Emf

1. ssm A spark can jump between two nontouching conductors if the potential difference between them is sufficiently large. A potential difference of approximately 940 V is required to produce a spark in an air gap of 1.0×10^{-4} m. Suppose the light bulb in Figure 22.4b is replaced by such a gap. How fast would a 1.3-m rod have to be moving in a magnetic field of 4.8 T to cause a spark to jump across the gap?

2. Near San Francisco, where the vertically downward component of the earth's magnetic field is 4.8×10^{-5} T, a car is traveling forward at 25 m/s. An emf of 2.4×10^{-3} V is induced between the sides of the car. (a) Which side of the car is positive, the driver's side or the passenger's side? (b) What is the width of the car?

3. The wingspan (tip-to-tip) of a Boeing 747 jetliner is 59 m. The plane is flying horizontally at a speed of 220 m/s. The vertical component of the earth's magnetic field is 5.0×10^{-6} T. Find the emf induced between the wing tips.

4. In 1996 NASA performed an experiment, called the Tethered Satellite experiment. In this experiment a 2.0×10^4-m length of wire was let out by the space shuttle Atlantis to generate a motional emf. The shuttle had an orbital speed of 7.6×10^3 m/s, and the magnitude of the earth's magnetic field at the location of the wire was 5.1×10^{-5} T. If the wire had moved perpendicular to the earth's magnetic field, what would have been the motional emf generated between the ends of the wire?

5. ssm www The drawing shows three identical rods (A, B, and C) moving in different planes. A constant magnetic field of magnitude 0.45 T is directed along the $+y$ axis. The length of each rod is $L = 1.3$ m,

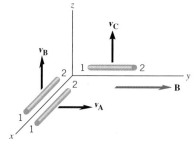

and the speeds are the same, $v_A = v_B = v_C = 2.7$ m/s. For each rod, find the magnitude of the motional emf, and indicate which end (1 or 2) of the rod is positive.

***6.** Suppose the light bulb in Figure 22.4b is replaced with a short wire of zero resistance, and the resistance of the rails is negligible. The only resistance is from the moving rod, which is iron (resistivity = 9.7×10^{-8} $\Omega \cdot$m). The rod has a cross-sectional area of 3.1×10^{-6} m^2 and moves with a speed of 2.0 m/s. The magnetic field has a magnitude of 0.050 T. What is the current in the rod?

***7. ssm** Suppose the light bulb in Figure 22.4b is replaced by a 6.0-Ω electric heater that consumes 15 W of power. The conducting bar moves to the right at a constant speed, the field strength is 2.4 T, and the length of the bar between the rails is 1.2 m. (a) How fast is the bar moving? (b) What force must be applied to the bar to keep it moving to the right at the constant speed found in part (a)?

*8. Refer to the drawing that accompanies conceptual question 6 (not problem 6). Suppose that the voltage of the battery in the circuit is 3.0 V, the magnitude of the magnetic field (directed perpendicularly into the plane of the paper) is 0.60 T, and the length of the rod between the rails is 0.20 m. Assuming that the rails are very long and have negligible resistance, find the maximum speed attained by the rod after the switch is closed.

**9. Review Conceptual Example 3 and Figure 22.7b as an aid in solving this problem. A conducting rod slides down between two frictionless vertical copper tracks at a constant speed of 4.0 m/s perpendicular to a 0.50-T magnetic field. The resistance of the rod and tracks is negligible. The rod maintains electrical contact with the tracks at all times and has a length of 1.3 m. A 0.75-Ω resistor is attached between the tops of the tracks. (a) What is the mass of the rod? (b) Find the change in the gravitational potential energy that occurs in a time of 0.20 s. (c) Find the electrical energy dissipated in the resistor in 0.20 s.

Section 22.3 Magnetic Flux

For problems in this set, assume that the magnetic flux is a positive quantity.

10. A hand is held flat and placed in a uniform magnetic field of magnitude 0.35 T. The hand has an area of 0.0160 m^2 and negligible thickness. Determine the magnetic flux that passes through the hand when the normal to the hand is (a) parallel and (b) perpendicular to the magnetic field.

11. **ssm** A standard door into a house rotates about a vertical axis through one side, as defined by the door's hinges. A uniform magnetic field is parallel to the ground and perpendicular to this axis. Through what angle must the door rotate so that the magnetic flux that passes through it decreases from its maximum value to one-third of its maximum value?

12. A loop of wire has the shape shown in the drawing. The top part of the wire is bent into a semicircle of radius $r = 0.20$ m. The normal to the plane of the loop is parallel to a constant magnetic field of magnitude 0.75 T. What is the change $\Delta\Phi$ in the magnetic flux that passes through the loop when, starting with the position shown in the drawing, the semicircle is rotated through half a revolution?

B (into paper)

13. The drawing shows three square surfaces, one lying in the xy plane, one in the xz plane, and one in the yz plane. The sides of each square have lengths of 2.0×10^{-2} m. A uniform magnetic field exists in this region, and its components are: $B_x = 0.50$ T, $B_y = 0.80$ T, and $B_z = 0.30$ T. Determine the magnetic flux that passes through the surface that lies in (a) the xy plane, (b) the xz plane, and (c) the yz plane.

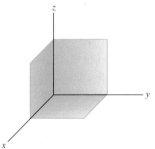

14. A house has a floor area of 112 m^2 and an outside wall that has an area of 28 m^2. The earth's magnetic field here has a horizontal component of 2.6×10^{-5} T that points due north and a vertical component of 4.2×10^{-5} T that points straight down, toward the earth. Determine the magnetic flux through the wall if the wall faces (a) north and (b) east. (c) Calculate the magnetic flux that passes through the floor.

*15. **ssm www** A five-sided object, whose dimensions are shown in the drawing, is placed in a uniform magnetic field. The magnetic field has a magnitude of 0.25 T and points along the positive y direction. Determine the magnetic flux through each of the five sides.

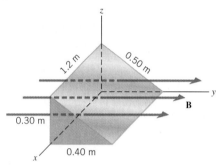

*16. A rectangular loop of wire is moving toward the bottom of the page with a speed of 0.020 m/s (see the drawing). The loop is leaving a region in which a 2.4-T magnetic field exists; the magnetic field outside this region is zero. During a time of 2.0 s, what is the magnitude of the *change* in the magnetic flux?

Section 22.4 Faraday's Law of Electromagnetic Induction

17. A 300-turn rectangular loop of wire has an area per turn of 5.0×10^{-3} m^2. At $t_0 = 0$ s a magnetic field is turned on, and its magnitude increases to 0.40 T when $t = 0.80$ s. The field is directed at an angle of $\phi = 30.0°$ with respect to the normal of the loop. (a) Find the magnitude of the average emf induced in the loop. (b) If the loop is a closed circuit whose resistance is 6.0 Ω, determine the average induced current.

18. The drawing shows a straight wire, a part of which is bent into the shape of a circle. The radius of the circle is 2.0 cm. A constant magnetic field of magnitude 0.55 T is directed perpendicular to the plane of the paper. Someone grabs the ends of the wire and pulls it taut, so the radius of the circle shrinks to zero in a time of 0.25 s. Find the magnitude of the average induced emf between the ends of the wire.

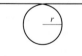

19. **ssm** A circular coil (950 turns, radius = 0.060 m) is rotating in a uniform magnetic field. At $t = 0$ s, the normal to the coil is perpendicular to the magnetic field. At $t = 0.010$ s, the normal

makes an angle of $\phi = 45°$ with the field because the coil has made one-eighth of a revolution. An average emf of magnitude 0.065 V is induced in the coil. Find the magnitude of the magnetic field at the location of the coil.

20. A planar coil of wire has a single turn. The normal to this coil is parallel to a uniform and constant (in time) magnetic field of 1.7 T. An emf that has a magnitude of 2.6 V is induced in this coil because the coil's area A is shrinking. What is the magnitude of $\Delta A/\Delta t$, which is the rate (in m^2/s) at which the area changes?

21. ⚕ Magnetic resonance imaging (MRI) is a medical technique for producing "pictures" of the interior of the body. The patient is placed within a strong magnetic field. One safety concern is what would happen to the positively and negatively charged particles in the body fluids if an equipment failure caused the magnetic field to be shut off suddenly. An induced emf could cause these particles to flow, producing an electric current within the body. Suppose the largest surface of the body through which flux passes has an area of 0.032 m^2 and a normal that is parallel to a magnetic field of 1.5 T. Determine the smallest time period during which the field can be allowed to vanish if the magnitude of the average induced emf is to be kept less than 0.010 V.

22. A constant magnetic field passes through a single rectangular loop whose dimensions are 0.35 m $\times$ 0.55 m. The magnetic field has a magnitude of 2.1 T and is inclined at an angle of 65° with respect to the normal to the plane of the loop. (a) If the magnetic field decreases to zero in a time of 0.45 s, what is the magnitude of the average emf induced in the loop? (b) If the magnetic field remains constant at its initial value of 2.1 T, what is the magnitude of the rate $\Delta A/\Delta t$ at which the area should change so that the average emf has the same magnitude?

23. A 1.5-m-long aluminum rod is rotating about an axis that is perpendicular to one end. A 0.16-T magnetic field is directed parallel to the axis. The rod rotates through one-fourth of a circle in 0.66 s. What is the magnitude of the average emf generated between the ends of the rod during this time?

*24. A piece of copper wire is formed into a single circular loop of radius 12 cm. A magnetic field is oriented parallel to the normal to the loop, and it increases from 0 to 0.60 T in a time of 0.45 s. The wire has a resistance per unit length of 3.3×10^{-2} Ω/m. What is the average electrical energy dissipated in the resistance of the wire?

*25. ssm www A conducting coil of 1850 turns is connected to a galvanometer, and the total resistance of the circuit is 45.0 Ω. The area of each turn is 4.70×10^{-4} m^2. This coil is moved from a region where the magnetic field is zero into a region where it is nonzero, the normal to the coil being kept parallel to the magnetic field. The amount of charge that is induced to flow around the circuit is measured to be 8.87×10^{-3} C. Find the magnitude of the magnetic field. (Such a device can be used to measure magnetic field strength and is called a *flux meter.*)

*26. The drawing shows a copper wire (negligible resistance) bent into a circular shape with a radius of 0.50 m. The radial section BC is fixed in place, while the copper bar AC sweeps around at

an angular speed of 15 rad/s. The bar makes electrical contact with the wire at all times. The wire and the bar have negligible resistance. A uniform magnetic field exists everywhere, is perpendicular to the plane of the circle, and has a magnitude of 3.8×10^{-3} T. Find the magnitude of the current induced in the loop ABC.

**27. Two 0.68-m-long conducting rods are rotating at the same speed in opposite directions, and both are perpendicular to a 4.7-T magnetic field. As the drawing shows, the ends of these rods come to within 1.0 mm of each other as they rotate. Moreover, the fixed ends about which the rods are rotating are connected by a wire, so these ends are at the same electric potential. If a potential difference of 4.5×10^3 V is required to cause a 1.0-mm spark in air, what is the angular speed (in rad/s) of the rods when a spark jumps across the gap?

Section 22.5 Lenz's Law

28. The drawing shows that a uniform magnetic field is directed perpendicularly out of the plane of the paper and fills the entire region to the left of the y axis. There is no magnetic field to the right of the y axis. A rigid right triangle ABC is made of copper wire. The triangle rotates counterclockwise about the origin at point C. What is the direction (clockwise or counterclockwise) of the induced current when the triangle is crossing (a) the $+y$ axis, (b) the $-x$ axis, (c) the $-y$ axis, and (d) the $+x$ axis? For each case, justify your answer.

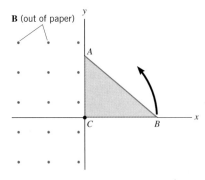

29. ssm In Figure 22.1, suppose the north and south poles of the magnet were interchanged. Determine the direction of the current through the ammeter in parts b and c of the picture (left-to-right or right-to-left). Give your rationale.

30. Review the drawing that accompanies Problem 12. The semicircular piece of wire rotates through half a revolution in the direction shown, starting from the position indicated in the drawing. Which end of the resistor, the left or the right end, is positive? Explain your reasoning.

31. ssm Review Conceptual Example 9 as an aid in understanding this problem. A long, straight wire lies on a table and carries a current I. As the drawing shows, a small circular loop of wire is pushed across the top of the table from position 1 to position 2. Determine the direction of the induced current, clockwise or counterclockwise, as the loop moves past (a) position 1 and (b) position 2. Justify your answers.

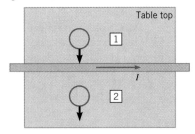
Table top

*32. The drawing shows a bar magnet falling through a metal ring. In part a the ring is solid all the way around, but in part b it has been cut through. (a) Explain why the motion of the magnet in part a is retarded when the magnet is above the

(a)   (b)

ring and below the ring as well. Draw any induced currents that appear in the ring. (b) Explain why the motion of the magnet is unaffected by the ring in part b.

*33. A circular loop of wire rests on a table. A long, straight wire lies on this loop, directly over its center, as the drawing illustrates. The current I in the straight wire is increasing. In what direction is the induced current, if any, in the loop? Give your reasoning.

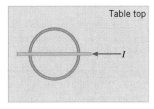
Table top

**34. A wire loop is suspended from a string that is attached to point P in the drawing. When released, the loop swings downward, from left to right, through a uniform magnetic field, with the plane of the loop remaining perpendicular to the plane of the paper at all times. (a) Determine the direction of the current induced in the loop as it swings past the locations labeled I and II. Specify the direction of the current in terms of the points x, y, and z on the loop (e.g., $x \rightarrow y \rightarrow z$ or $z \rightarrow y \rightarrow x$). The points x, y, and z lie behind the plane of the paper. (b) What is the direction of the induced current at the locations II and I when the loop swings back, from right to left? Provide reasons for your answers.

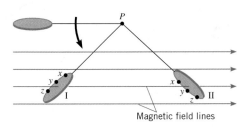
Magnetic field lines

Section 22.7 The Electric Generator

35. ssm www The drawing shows a plot of the output emf of a generator as a function of time t. The coil of this device has a cross-sectional area per turn of 0.020 m² and contains 150 turns. Find (a) the frequency f of the generator in hertz, (b) the angular speed ω in rad/s, and (c) the magnitude of the magnetic field.

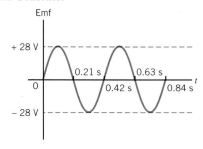

36. One generator uses a magnetic field of 0.10 T and has a coil area per turn of 0.045 m². A second generator has a coil area per turn of 0.015 m². The generator coils have the same number of turns and rotate at the same angular speed. What magnetic field should be used in the second generator, so that its peak emf is the same as that of the first generator?

37. You are requested to design a 60.0-Hz ac generator whose maximum emf is to be 5500 V. The generator is to contain a 150-turn coil whose area per turn is 0.85 m². What should be the magnitude of the magnetic field in which the coil rotates?

38. A vacuum cleaner is plugged into a 120.0-V socket and uses 3.0 A of current in normal operation when the back emf generated by the electric motor is 72.0 V. Find the coil resistance of the motor.

39. ssm A generator has a square coil consisting of 248 turns. The coil rotates at 79.1 rad/s in a 0.170-T magnetic field. The peak output of the generator is 75.0 V. What is the length of one side of the coil?

40. A 120.0-volt motor draws a current of 7.00 A when running at normal speed. The resistance of the armature wire is 0.720 Ω. (a) Determine the back emf generated by the motor. (b) What is the current at the instant when the motor is just turned on and has not begun to rotate? (c) What series resistance must be added to limit the starting current to 15.0 A?

*41. ssm At its normal operating speed, an electric fan motor draws only 15.0% of the current it draws when it just begins to turn the fan blade. The fan is plugged into a 120.0-V socket. What back emf does the motor generate at its normal operating speed?

*42. The coil of a generator has a radius of 0.14 m. When this coil is unwound, the wire from which it is made has a length of 5.7 m. The magnetic field of the generator is 0.20 T, and the coil rotates at an angular speed of 25 rad/s. What is the peak emf of this generator?

**43. A motor is designed to operate on 117 V and draws a current of 12.2 A when it first starts up. At its normal operating speed, the motor draws a current of 2.30 A. Obtain (a) the resistance of the armature coil, (b) the back emf developed at normal speed, and (c) the current drawn by the motor at one-third normal speed.

Section 22.8 Mutual Inductance and Self-Inductance

44. The average emf induced in the secondary coil is 0.12 V when the current in the primary coil changes from 3.4 to 1.6 A in 0.14 s. What is the mutual inductance of the coils?

45. ssm Suppose you wish to make a solenoid whose self-inductance is 1.4 mH. The inductor is to have a cross-sectional area of 1.2×10^{-3} m^2 and a length of 0.052 m. How many turns of wire are needed?

46. The current through a 3.2-mH inductor varies with time according to the graph shown in the drawing. What is the average induced emf during the time intervals (a) 0–2.0 ms, (b) 2.0–5.0 ms, and (c) 5.0–9.0 ms?

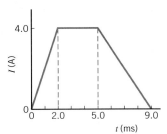

47. Mutual induction can be used as the basis for a metal detector. A typical setup uses two large coils that are parallel to each other and have a common axis. Because of mutual induction, the ac generator connected to the primary coil causes an emf of 0.46 V to be induced in the secondary coil. When someone without metal objects walks through the coils, the mutual inductance and, thus, the induced emf do not change much. But when a person carrying a handgun walks through, the mutual inductance increases. If the mutual inductance increases by a factor of three, find the new value of the induced emf. The change in emf can be used to trigger an alarm.

48. Two coils have a mutual inductance of 2.5 mH. In the primary coil the current changes by 3.0 A in 0.040 s. The circuit of the secondary coil has a resistance of 2.0 Ω. Find the magnitude of the average current induced in the secondary coil.

49. ssm The earth's magnetic field, like any magnetic field, stores energy. The maximum strength of the earth's field is about 7.0×10^{-5} T. Find the maximum magnetic energy stored in the space above a city if the space occupies an area of 5.0×10^8 m^2 and has a height of 1500 m.

*50. A magnetic field has a magnitude of 12 T. What is the magnitude of an electric field that stores the same energy per unit volume as this magnetic field?

*51. A long, current-carrying solenoid with an air core has 1750 turns per meter of length and a radius of 0.0180 m. A coil of 125 turns is wrapped tightly around the outside of the solenoid. What is the mutual inductance of this system?

*52. The purpose of this problem is to show that the work W needed to establish a final current I_f in an inductor is $W = \frac{1}{2}LI_f^2$ (Equation 22.10). In Section 22.8 we saw that the amount of work ΔW needed to change the current through an inductor by an amount ΔI is $\Delta W =$ $LI(\Delta I)$, where L is the inductance. The drawing shows a graph of LI versus I. Notice that $LI(\Delta I)$ is the area of the shaded vertical rectangle whose height is LI and whose width is ΔI. Use this fact

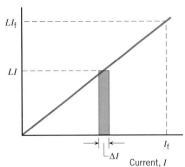

to show that the total work W needed to establish a current I_f is $W = \frac{1}{2}LI_f^2$.

**53. ssm A solenoid has a cross-sectional area of 6.0×10^{-4} m^2, consists of 400 turns per meter, and carries a current of 0.40 A. A 10-turn coil is wrapped tightly around the circumference of the solenoid. The ends of the coil are connected to a 1.5-Ω resistor. Suddenly, a switch is opened, and the current in the solenoid dies to zero in a time of 0.050 s. Find the average current induced in the coil.

Section 22.9 Transformers

54. A neon sign requires 12 000 V for its operation. It operates from a 220-V receptacle. (a) What type of transformer, step-up or step-down, is needed? (b) What must be the turns ratio N_s/N_p of the transformer?

55. ssm Electric doorbells found in many homes require 10.0 V to operate. To obtain this voltage from the standard 120-V supply, a transformer is used. Is a step-up or a step-down transformer needed, and what is its turns ratio N_s/N_p?

56. The batteries in a portable CD player are recharged by a unit that plugs into a wall socket. Inside the unit is a step-down transformer with a turns ratio of 1:13. The wall socket provides 120 V. What voltage does the secondary coil of the transformer provide?

57. The secondary coil of a step-up transformer provides the voltage that operates an electrostatic air filter. The turns ratio of the transformer is 43:1. The primary coil is plugged into a standard 120-V outlet. The current in the secondary coil is 1.5×10^{-3} A. Find the power consumed by the air filter.

58. In some places, insect "zappers," with their blue lights, are a familiar sight on a summer's night. These devices use a high voltage to electrocute insects. One such device uses an ac voltage of 4320 V, which is obtained from a standard 120.0-V outlet by means of a transformer. If the primary coil has 21 turns, how many turns are in the secondary coil?

*59. ssm A generating station is producing 1.2×10^6 W of power that is to be sent to a small town located 7.0 km away. Each of the two wires that comprise the transmission line has a resistance per kilometer of length of 5.0×10^{-2} Ω/km. (a) Find the power lost in heating the wires if the power is transmitted at 1200 V. (b) A 100:1 step-up transformer is used to raise the voltage before the power is transmitted. How much power is now lost in heating the wires?

*60. In a television set the power needed to operate the picture tube is 95 W and is derived from the secondary coil of a transformer. There is a current of 5.3 mA in the secondary coil. The primary coil is connected to a 120-V receptacle. Find the turns ratio N_s/N_p of the transformer.

**61. A generator is connected across the primary coil (N_p turns) of a transformer, while a resistance R_2 is connected across the secondary coil (N_s turns). This circuit is equivalent to a circuit in which a single resistance R_1 is connected directly across the generator, without the transformer. Show that $R_1 = (N_p/N_s)^2 R_2$, by starting with Ohm's law as applied to the secondary coil.

ADDITIONAL PROBLEMS

62. A magnetic field is perpendicular to a 0.040-m × 0.060-m rectangular coil of wire that has one hundred turns. In a time of 0.050 s, an average emf of magnitude 1.5 V is induced in the coil. What is the magnitude of the change in the magnetic field?

63. The maximum strength of the earth's magnetic field is about 7.0×10^{-5} T near the south magnetic pole. In principle, this field could be used with a rotating coil to generate 60.0-Hz ac electricity. What is the minimum number of turns (area per turn = 0.016 m^2) that the coil must have to produce an rms voltage of 120 V?

64. Parts a and b of the drawing show the same uniform and constant (in time) magnetic field **B** directed perpendicularly into the paper over a rectangular region. Outside this region, there is no field. Also shown is a rectangular coil (one turn), which lies in the plane of the paper. In part a the long side of the coil (length = L) is just at the edge of the field region, while in part b the short side (width = W) is just at the edge. It is known that $L/W = 3.0$. In both parts of the drawing the coil is pushed into the field with the same velocity **v** until it is completely within the field region. The magnitude of the average emf induced in the coil in part a is 0.15 V. What is its magnitude in part b?

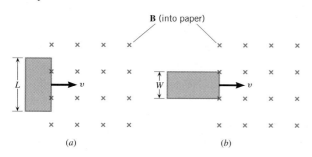

(a)   (b)

65. ssm Suppose in Figure 22.1 that the bar magnet is held stationary, but the coil of wire is free to move. Which way will current be directed through the ammeter, left-to-right or right-to-left, when the coil is moved (a) to the left and (b) to the right? Explain.

66. The resistances of the primary and secondary coils of a transformer are 56 and 14 Ω, respectively. Both coils are made from lengths of the same copper wire. The circular turns of each coil have the same diameter. Find the turns ratio N_s/N_p.

67. ssm The coil of an ac generator has an area per turn of 1.2×10^{-2} m^2 and consists of 500 turns. The coil is situated in a 0.13-T magnetic field and is rotating at an angular speed of 34 rad/s. What is the emf induced in the coil at the instant when the normal to the loop makes an angle of 27° with respect to the direction of the magnetic field?

***68.** A 3.0-μF capacitor has a voltage of 35 V between its plates. What must be the current in a 5.0-mH inductor, such that the energy stored in the inductor equals the energy stored in the capacitor?

***69.** A copper rod is sliding on two conducting rails that make an angle of 15° with respect to each other, as in the drawing. The rod is moving to the right with a constant speed of 0.40 m/s. A 0.42-T uniform magnetic field is perpendicular to the plane of the paper. Determine the magnitude of the average emf induced in the triangle ABC during the 5.0-s period after the rod has passed point A.

***70.** Indicate the direction of the electric field between the plates of the parallel plate capacitor shown in the drawing if the magnetic field is decreasing in time. Give your reasoning.

Problem 70

***71. ssm** A magnetic field is passing through a loop of wire whose area is 0.018 m^2. The direction of the magnetic field is parallel to the normal to the loop, and the magnitude of the field is increasing at the rate of 0.20 T/s. (a) Determine the magnitude of the emf induced in the loop. (b) Suppose the area of the loop can be enlarged or shrunk. If the magnetic field is increasing as in part (a), at what rate (in m^2/s) should the area be changed at the instant when $B = 1.8$ T if the induced emf is to be zero? Explain whether the area is to be enlarged or shrunk.

***72.** A long solenoid (cross-sectional area = 1.0×10^{-6} m^2, number of turns per unit length = 2400 turns/m) is bent into a circular shape so it looks like a doughnut. This wire-wound doughnut is called a toroid. Assume that the diameter of the solenoid is small compared to the radius of the toroid, which is 0.050 m. With this assumption, use the results of Example 13 to determine the self-inductance of the toroid.

****73.** The armature of an electric drill motor has a resistance of 15.0 Ω. When connected to a 120.0-V outlet, the motor rotates at its normal speed and develops a back emf of 108 V. (a) What is the current through the motor? (b) If the armature "freezes up" due to a lack of lubrication in the bearings and can no longer rotate, what is the current in the stationary armature? (c) What is the current when the motor runs at only half speed?

****74.** Coil 1 is a flat circular coil that has N_1 turns and a radius R_1. At its center is a much smaller flat, circular coil that has N_2 turns and radius R_2. The planes of the coils are parallel. Assume that coil 2 is so small that the magnetic field due to coil 1 has nearly the same value at all points covered by the area of coil 2. Determine an expression for the mutual inductance between these two coils in terms of μ_0, N_1, R_1, N_2, and R_2.

CONCEPTS & CALCULATIONS

GROUP LEARNING PROBLEMS

Note: Each of these problems consists of Concept Questions followed by a related quantitative Problem. They are designed for use by students working alone or in small learning groups. The Concept Questions involve little or no mathematics and are intended to stimulate group discussions. They focus on the concepts with which the problems deal. Recognizing the concepts is the essential initial step in any problem-solving techinque.

75. Concept Questions Two circuits contain an emf produced by a moving metal rod, like that in Figure 22.4*b*. The speed of the rod is the same in each circuit, but the bulb in circuit 1 has one-half the resistance of the bulb in circuit 2. The circuits are otherwise identical. Is (a) the motional emf and (b) the current in circuit 1 greater than, the same as, or less than, that in circuit 2? (c) If the speed of the rod in circuit 1 were doubled, how would the power delivered to the light bulb compare to that in circuit 2? Provide a reason for each of your answers.

Problem The resistance of the light bulb in circuit 1 is 55 Ω, while that in circuit 2 is 110 Ω. Determine (a) the ratio $\mathscr{E}_1/\mathscr{E}_2$ of the emfs and (b) the ratio I_1/I_2 of the currents. (c) If the speed of the rod in circuit 1 is twice that in circuit 2, what is the ratio P_1/P_2 of the powers? Check to see that your answers are consistent with your answers to the Concept Questions.

76. Concept Questions A magnetic field passes through a stationary wire loop, and its magnitude changes in time according to the graph in the drawing. The direction of the field remains constant, however. There are three equal time intervals indicated in the graph: 0–3.0 s, 3.0–6.0 s, and 6.0–9.0 s. (a) Is the induced emf equal to zero during any of the intervals? (b) During which interval is the magnitude of the induced emf the largest? (c) If the direction of the current induced during the first interval is clockwise, what is the direction during the third interval? In all cases, provide a reason for your answer.

Problem The loop consists of 50 turns of wire and has an area of 0.15 m². The magnetic field is oriented parallel to the normal to the loop. (a) For each interval, determine the magnitude of the induced emf. (b) The wire has a resistance of 0.50 Ω. Determine the induced current for the first and third intervals. Make sure your answers are consistent with your answers to the Concept Questions.

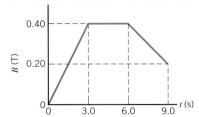

77. Concept Questions The drawing shows a straight wire carrying a current *I*. Above the wire is a rectangular loop that contains a resistor *R*. (a) Does the magnetic field produced by the current *I* penetrate the loop and generate a magnetic flux? (b) When is there an induced current in the loop, if the current *I* is constant or if it is decreasing in time? (c) When there is an induced magnetic field produced by the loop, does it always have a direction that is opposite to the direction of the magnetic field produced by the current *I*? Provide a reason for each answer.

Problem If the current *I* is decreasing in time, what is the direction of the induced current through the resistor *R*—left-to-right or right-to-left? Give your reasoning.

78. Concept Questions A flat coil of wire has an area *A*, *N* turns, and a resistance *R*. It is situated in a magnetic field, such that the normal to the coil is parallel to the magnetic field. The coil is then rotated through an angle of 90°, so that the normal becomes perpendicular to the magnetic field. (a) Why is an emf induced in the coil? (b) What determines the amount of induced current in the coil? (c) How is the amount of charge that flows related to the induced current?

Problem The coil has an area of 1.5×10^{-3} m², 50 turns, and a resistance of 140 Ω. During the time it is rotating, a charge of 8.5×10^{-5} C flows in the coil. What is the magnitude of the magnetic field?

79. Concept Questions A constant current *I* exists in a solenoid whose inductance is *L*. The current is then reduced to zero in a certain amount of time. (a) If the wire from which the solenoid is made has no resistance, is there a voltage across the solenoid during the time when the current is constant? (b) If the wire from which the solenoid is made has no resistance, is there an emf across the solenoid during the time that the current is being reduced to zero? (c) Does the solenoid store electrical energy when the current is constant? If so, express this energy in terms of the current and the inductance. (d) When the current is reduced from its constant value to zero, what is the rate at which energy is removed from the solenoid? Express your answer in terms of the initial current, the inductance, and the time during which the current goes to zero.

Problem A solenoid has an inductance of $L = 3.1$ H and carries a current of $I = 15$ A. (a) If the current goes from 15 to 0 A in a time of 75 ms, what is the emf induced in the solenoid? (b) How much electrical energy is stored in the solenoid? (c) At what rate must the electrical energy be removed from the solenoid when the current is reduced to zero in 75 ms?

80. Concept Questions The rechargeable batteries for a laptop computer need a much smaller voltage than what a wall socket provides. Therefore, a transformer is plugged into the wall socket and produces the necessary voltage for charging the batteries. (a) Is the transformer a step-up or a step-down transformer? (b) Is the current that goes through the batteries greater than, equal to, or smaller than the current coming from the wall socket? (c) If the transformer has a negligible resistance, is the electrical power delivered to the batteries greater than, equal to, or less than the power coming from the wall socket? In all cases, provide a reason for your answer.

Problem The batteries of a laptop computer are rated at 9.0 V, and a current of 225 mA is used to charge them. The wall socket provides a voltage of 120 V. (a) Determine the turns ratio of the transformer. (b) What is the current coming from the wall socket? (c) Find the power delivered by the wall socket and the power sent to the batteries. Be sure your answers are consistent with your answers to the Concept Questions.

These French high speed trains derive their operating power from ac electricity (25 000 V, 50 Hz) and can achieve speeds as high as 300 km/h (186 mph).

CHAPTER 23

ALTERNATING CURRENT CIRCUITS

23.1 CAPACITORS AND CAPACITIVE REACTANCE

Our experience with capacitors so far has been in dc circuits. As we have seen in Section 20.13, charge flows in a dc circuit only for the brief period after the battery voltage is applied across the capacitor. In other words, charge flows only while the capacitor is charging up. After the capacitor becomes fully charged, no more charge leaves the battery. However, suppose the battery connections to the fully charged capacitor were suddenly reversed. Then charge would flow again, but in the reverse direction, until the battery recharges the capacitor according to the new connections. What happens in an ac circuit is similar. The polarity of the voltage applied to the capacitor continually switches back and forth, and, in response, charges flow first one way around the circuit and then the other way. This flow of charge, surging back and forth, constitutes an alternating current. Thus, charge flows continuously in an ac circuit containing a capacitor.

To help set the stage for the present discussion, recall that, for a purely resistive ac circuit, the rms voltage V_{rms} across the resistor is related to the rms current I_{rms} through it by Equation 20.14 ($V_{rms} = I_{rms}R$). The resistance R has the same value for any frequency of the ac voltage or current. Figure 23.1 emphasizes this fact by showing that a graph of resistance versus frequency is a horizontal straight line.

For the rms voltage across a capacitor the following expression applies, which is analogous to $V_{rms} = I_{rms}R$:

$$V_{rms} = I_{rms}X_C \qquad (23.1)$$

The term X_C appears in place of the resistance R and is called the **capacitive reactance.** The capacitive reactance, like resistance, is measured in *ohms* and determines how much rms current exists in a capacitor in response to a given rms voltage across the capacitor. It is found experimentally that the capacitive reactance X_C is inversely proportional to both the frequency f and the capacitance C, according to the following equation:

$$X_C = \frac{1}{2\pi fC} \qquad (23.2)$$

For a fixed value of the capacitance C, Figure 23.2 gives a plot of X_C versus frequency, according to Equation 23.2. A comparison of this drawing with Figure 23.1 reveals that a capacitor and a resistor behave differently. As the frequency becomes very large, Figure 23.2 shows that X_C approaches zero, signifying that a capacitor offers only a negligibly small opposition to the alternating current. In contrast, in the limit of zero frequency (i.e., direct current), X_C becomes infinitely large, and a capacitor provides so much opposition to the motion of charges that there is no current. Example 1 illustrates how frequency and capacitance determine the amount of current in an ac circuit.

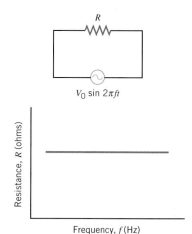

Figure 23.1 The resistance in a purely resistive circuit has the same value at all frequencies. The maximum emf of the generator is V_0.

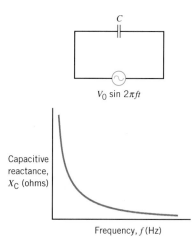

Figure 23.2 The capacitive reactance X_C is inversely proportional to the frequency f according to $X_C = 1/(2\pi fC)$.

EXAMPLE 1   A Capacitor in an AC Circuit

For the circuit in Figure 23.2, the capacitance of the capacitor is 1.50 μF, and the rms voltage of the generator is 25.0 V. What is the rms current in the circuit when the frequency of the generator is (a) 1.00×10^2 Hz and (b) 5.00×10^3 Hz?

Reasoning The current can be found from $I_{rms} = V_{rms}/X_C$, once the capacitive reactance is determined. The values for the capacitive reactance will reflect the fact that the capacitor provides more opposition to the current when the frequency is smaller.

Solution

(a) At a frequency of 1.00×10^2 Hz, we find

$$X_C = \frac{1}{2\pi fC} = \frac{1}{2\pi(1.00 \times 10^2 \text{ Hz})(1.50 \times 10^{-6} \text{ F})} = 1060 \ \Omega \quad (23.2)$$

$$I_{rms} = \frac{V_{rms}}{X_C} = \frac{25.0 \text{ V}}{1060 \ \Omega} = \boxed{0.0236 \text{ A}} \quad (23.1)$$

(b) When the frequency is 5.00×10^3 Hz, the calculations are similar:

$$X_C = \frac{1}{2\pi fC} = \frac{1}{2\pi(5.00 \times 10^3 \text{ Hz})(1.50 \times 10^{-6} \text{ F})} = 21.2 \ \Omega \quad (23.2)$$

$$I_{rms} = \frac{V_{rms}}{X_C} = \frac{25.0 \text{ V}}{21.2 \ \Omega} = \boxed{1.18 \text{ A}} \quad (23.1)$$

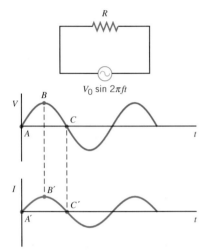

Figure 23.3 The instantaneous voltage V and current I in a purely resistive circuit are *in phase,* which means that they increase and decrease in step with one another.

We now consider the behavior of the instantaneous (not rms) voltage and current. For comparison, Figure 23.3 shows graphs of voltage and current versus time in a resistive circuit. These graphs indicate that, when only resistance is present, the voltage and current are proportional to each other at every moment. For example, when the voltage increases from A to B on the graph, the current follows along in step, increasing from A' to B' during the same time. Likewise, when the voltage decreases from B to C, the current decreases from B' to C'. For this reason, the current in a resistance R is said to be ***in phase*** with the voltage across the resistance.

For a capacitor, this in-phase relation between instantaneous voltage and current does *not* exist. Figure 23.4 shows graphs of the ac voltage and current versus time for a circuit that contains only a capacitor. As the voltage increases from A to B, the charge on the capacitor increases and reaches its full value at B. The current, however, is not the same thing as the charge. The current is the rate of flow of charge and has a maximum positive value at the start of the charging process at A'. It is a maximum because there is no charge on the capacitor at the start and hence no capacitor voltage to oppose the generator voltage. When the capacitor is fully charged at B, the capacitor voltage has a magnitude equal to that of the generator and completely opposes the generator voltage. The result is that the current decreases to zero at B'. While the capacitor voltage decreases from B to C, the charges flow out of the capacitor in a direction opposite to that of the charging current, as indicated by the negative current from B' to C'. Thus, voltage and current are not in phase but are, in fact, one-quarter wave cycle out of step, or out of phase. More specifically, assuming that the voltage fluctuates as $V_0 \sin (2\pi ft)$, the current varies as $I_0 \sin (2\pi ft + \pi/2) = I_0 \cos (2\pi ft)$. Since $\pi/2$ radians correspond to 90° and since the current reaches its maximum value *before* the voltage does, it is said that the current in a capacitor *leads* the voltage across the capacitor by a phase angle of 90°.

The fact that the current and voltage for a capacitor are 90° out of phase has an important consequence from the point of view of electric power, since power is the product of current and voltage. For the time interval between points A and B (or A' and B') in Figure 23.4, both current and voltage are positive. Therefore, the instantaneous power is also positive, meaning that the generator is delivering energy to the capacitor. However, during the period between B and C (or B' and C'), the current is negative while the voltage remains positive, and the power, being the product of the two, is negative. During this period, the capacitor is returning energy to the generator. Thus, the power alternates between positive and negative values for equal periods of time. In other words, the capacitor alternately absorbs and releases en-

Figure 23.4 In a circuit containing only a capacitor, the instantaneous voltage and current are not in phase. Instead, the current *leads* the voltage by one-quarter of a cycle or by a phase angle of 90°.

ergy. Consequently, *on the average, the power is zero and a capacitor uses no energy in an ac circuit.*

To simplify the analysis of ac circuits later on, it will prove useful to use a model for the voltage and current in such circuits. In this model, voltage and current are represented by rotating arrows, often called *phasors,* whose lengths correspond to the maximum voltage V_0 and maximum current I_0, as Figure 23.5 indicates. These phasors rotate counterclockwise at a frequency f. For a resistor, the phasors are co-linear as they rotate (see part *a* of the drawing) because voltage and current are in phase. For a capacitor (see part *b*), the phasors remain perpendicular while rotating because the phase angle between the current and the voltage is 90°. Since current leads voltage for a capacitor, the current phasor is ahead of the voltage phasor in the direction of rotation.

Note from the two circuit drawings in Figure 23.5 that the instantaneous voltage across the resistor or the capacitor is $V_0 \sin (2\pi ft)$. We can find this instantaneous value of the voltage directly from the phasor diagram. Imagine that the voltage phasor V_0 in this diagram represents the hypotenuse of a right triangle. Then, the *vertical component* of the phasor would be $V_0 \sin (2\pi ft)$. In a similar manner, the instantaneous current can be found as the vertical component of the current phasor.

23.2 INDUCTORS AND INDUCTIVE REACTANCE

As Section 22.8 discusses, an inductor is usually a coil of wire, and the basis of its operation is Faraday's law of electromagnetic induction. According to Faraday's law, an inductor develops a voltage that opposes a change in the current. This voltage V is given by $V = -L(\Delta I/\Delta t)$ (see Equation 22.9*), where $\Delta I/\Delta t$ is the rate at which the current changes and L is the inductance of the inductor. In an ac circuit the current is always changing, and Faraday's law can be used to show that the rms voltage across an inductor is

$$V_{rms} = I_{rms}X_L \qquad (23.3)$$

Equation 23.3 is analogous to $V_{rms} = I_{rms}R$, with the term X_L appearing in place of the resistance R and being called the *inductive reactance.* The inductive reactance is measured in ohms and determines how much rms current exists in an inductor for a given rms voltage across the inductor. It is found experimentally that the inductive reactance X_L is directly proportional to the frequency f and the inductance L, as indicated by the following equation:

$$X_L = 2\pi fL \qquad (23.4)$$

This relation indicates that the larger the inductance, the larger the inductive reactance. Note that the inductive reactance is directly proportional to the frequency ($X_L \propto f$). In contrast, the capacitive reactance is inversely proportional to the frequency ($X_C \propto 1/f$).

Figure 23.6 shows a graph of the inductive reactance versus frequency for a fixed value of the inductance, according to Equation 23.4. As the frequency becomes very large, X_L also becomes very large. In such a situation, an inductor provides a large opposition to the alternating current. In the limit of zero frequency

---

* When an inductor is used in a circuit, the notation is simplified if we designate the potential difference across the inductor as the voltage V, rather than the emf $\mathscr{E}$.

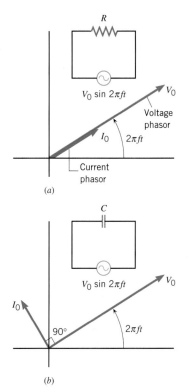

Figure 23.5 These rotating-arrow models represent the voltage and the current in ac circuits that contain (*a*) only a resistor and (*b*) only a capacitor.

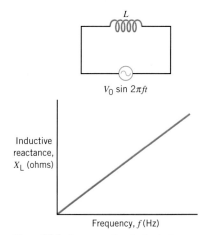

Figure 23.6 In an ac circuit the inductive reactance X_L is directly proportional to the frequency f, according to $X_L = 2\pi fL$.

Figure 23.7 The instantaneous voltage and current in a circuit containing only an inductor are not in phase. The current *lags behind* the voltage by one-quarter of a cycle or by a phase angle of 90°.

• **PROBLEM SOLVING INSIGHT**
The inductive reactance X_L is directly proportional to the frequency f of the voltage. If the frequency increases by a factor of 50, for example, the inductive reactance increases by a factor of 50.

(i.e., direct current), X_L becomes zero, indicating that an inductor does not oppose direct current at all. The next example demonstrates the effect of inductive reactance on the current in an ac circuit.

EXAMPLE 2   An Inductor in an AC Circuit

The circuit in Figure 23.6 contains a 3.60-mH inductor. The rms voltage of the generator is 25.0 V. Find the rms current in the circuit when the generator frequency is (a) 1.00×10^2 Hz and (b) 5.00×10^3 Hz.

Reasoning The current can be calculated from $I_{rms} = V_{rms}/X_L$, provided the inductive reactance is obtained first. The inductor offers more opposition to the changing current when the frequency is larger, and the values for the inductive reactance will reflect this fact.

Solution

(a) At a frequency of 1.00×10^2 Hz, we find

$$X_L = 2\pi f L = 2\pi(1.00 \times 10^2 \text{ Hz})(3.60 \times 10^{-3} \text{ H}) = 2.26 \ \Omega \quad (23.4)$$

$$I_{rms} = \frac{V_{rms}}{X_L} = \frac{25.0 \text{ V}}{2.26 \ \Omega} = \boxed{11.1 \text{ A}} \quad (23.3)$$

(b) The calculation is similar when the frequency is 5.00×10^3 Hz:

$$X_L = 2\pi f L = 2\pi(5.00 \times 10^3 \text{ Hz})(3.60 \times 10^{-3} \text{ H}) = 113 \ \Omega \quad (23.4)$$

$$I_{rms} = \frac{V_{rms}}{X_L} = \frac{25.0 \text{ V}}{113 \ \Omega} = \boxed{0.221 \text{ A}} \quad (23.3)$$

By virtue of its inductive reactance, an inductor affects the amount of current in an ac circuit. The inductor also influences the current in another way, as Figure 23.7 shows. This figure displays graphs of voltage and current versus time for a circuit containing only an inductor. At a maximum or minimum on the current graph, the current does not change much with time, so the voltage generated by the inductor to oppose a change in the current is zero. At the points on the current graph where the current is zero, the graph is at its steepest, and the current has the largest rate of increase or decrease. Correspondingly, the voltage generated by the inductor to oppose a change in the current has the largest positive or negative value. Thus, current and voltage are not in phase but are one-quarter of a wave cycle out of phase. If the voltage varies as $V_0 \sin (2\pi f t)$, the current fluctuates as $I_0 \sin (2\pi f t - \pi/2) = -I_0 \cos (2\pi f t)$. The current reaches its maximum *after* the voltage does, and it is said that the current *lags behind* the voltage by a phase angle of 90° ($\pi/2$ radians). In a purely capacitive circuit, in contrast, the current leads the voltage by 90°.

In an inductor the 90° phase difference between current and voltage leads to the same result for average power that it does in a capacitor. An inductor alternately absorbs and releases energy for equal periods of time, so **on the average, the power is zero and an inductor uses no energy in an ac circuit.**

As an alternative to the graphs in Figure 23.7, Figure 23.8 uses phasors to describe the instantaneous voltage and current in a circuit containing only an inductor. The voltage and current phasors remain perpendicular as they rotate, for there is a 90° phase angle between them. The current phasor lags behind the voltage phasor, relative to the direction of rotation, in contrast to the equivalent picture in Figure 23.5*b* for a capacitor. Once again, the instantaneous values for voltage and current are given by the vertical components of the phasors.

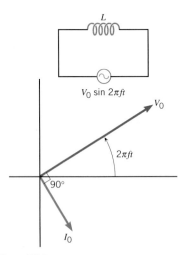

Figure 23.8 This phasor model represents the voltage and current in a circuit that contains only an inductor.

23.3 CIRCUITS CONTAINING RESISTANCE, CAPACITANCE, AND INDUCTANCE

Figure 23.9 A series RCL circuit contains a resistor, a capacitor, and an inductor.

Capacitors and inductors can be combined along with resistors in a single circuit. The simplest combination contains a resistor, a capacitor, and an inductor in series, as Figure 23.9 shows. In a series RCL circuit the total opposition to the flow of charge is called the *impedance* of the circuit and comes partially from (1) the resistance R, (2) the capacitive reactance X_C, and (3) the inductive reactance X_L. It is tempting to follow the analogy of a series combination of resistors and calculate the impedance by simply adding together R, X_C, and X_L. However, such a procedure is not correct. Instead, the phasors shown in Figure 23.10 must be used. The lengths of the voltage phasors in this drawing represent the maximum voltages V_R, V_C, and V_L across the resistor, the capacitor, and the inductor, respectively. The current is the same for each device, since the circuit is wired in series. The length of the current phasor represents the maximum current I_0. Notice that the drawing shows the current phasor to be (1) in phase with the voltage phasor for the resistor, (2) ahead of the voltage phasor for the capacitor by 90°, and (3) behind the voltage phasor for the inductor by 90°. These three facts are consistent with our earlier discussion in Sections 23.1 and 23.2.

The basis for dealing with the voltage phasors in Figure 23.10 is Kirchhoff's loop rule. In an ac circuit this rule applies to the *instantaneous* voltages across each circuit component and the generator. Therefore, it is necessary to take into account the fact that these voltages do not have the same phase; that is, the phasors V_R, V_C, and V_L point in different directions in the drawing. Kirchhoff's loop rule indicates that the phasors add together to give the total voltage V_0 that is supplied to the circuit by the generator. The addition, however, must be like a vector addition, to take into account the different directions. Since V_L and V_C point in opposite directions, they combine to give a resultant phasor of $V_L - V_C$, as Figure 23.11 shows. In this drawing the resultant $V_L - V_C$ is perpendicular to V_R and may be combined with it to give the total voltage V_0. Using the Pythagorean theorem, we find

$$V_0{}^2 = V_R{}^2 + (V_L - V_C)^2$$

In this equation each of the symbols stands for a maximum voltage and when divided by $\sqrt{2}$ gives the corresponding rms voltage. Therefore, it is possible to divide both sides of the equation by $(\sqrt{2})^2$ and obtain a result for $V_{rms} = V_0/\sqrt{2}$. This result has exactly the same form as that above, but involves the rms voltages $V_{R\text{-rms}}$, $V_{C\text{-rms}}$, and $V_{L\text{-rms}}$. However, to avoid such awkward symbols, we simply interpret V_R, V_C, and V_L as rms quantities in the following expression:

$$V_{rms}^2 = V_R{}^2 + (V_L - V_C)^2 \qquad (23.5)$$

The last step in determining the impedance of the circuit is to remember that $V_R = I_{rms}R$, $V_C = I_{rms}X_C$, and $V_L = I_{rms}X_L$. With these substitutions Equation 23.5 can be written as

$$V_{rms} = I_{rms}\sqrt{R^2 + (X_L - X_C)^2}$$

Therefore, for the entire RCL circuit, it follows that

$$V_{rms} = I_{rms}Z \qquad (23.6)$$

where the impedance Z of the circuit is defined as

Series RCL combination

$$Z = \sqrt{R^2 + (X_L - X_C)^2} \qquad (23.7)$$

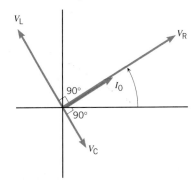

Figure 23.10 The three voltage phasors (V_R, V_C, and V_L) and the current phasor (I_0) for a series RCL circuit.

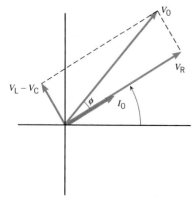

Figure 23.11 This simplified version of Figure 23.10 results when the phasors V_L and V_C, which point in opposite directions, are combined to give a resultant of $V_L - V_C$.

The impedance of the circuit, like R, X_C, and X_L, is measured in ohms. In Equation 23.7, $X_L = 2\pi fL$ and $X_C = 1/(2\pi fC)$.

The phase angle between the current in and the voltage across a series RCL combination is the angle ϕ between the current phasor I_0 and the voltage phasor V_0 in Figure 23.11. According to the drawing, the tangent of this angle is

$$\tan \phi = \frac{V_L - V_C}{V_R} = \frac{I_{rms}X_L - I_{rms}X_C}{I_{rms}R}$$

Series RCL combination

$$\tan \phi = \frac{X_L - X_C}{R} \tag{23.8}$$

The phase angle ϕ is important because it has a major effect on the average power $\overline{P}$ dissipated by the circuit. Remember, on the average, only the resistance consumes power; that is, $\overline{P} = I_{rms}^2 R$ (Equation 20.15b). According to Figure 23.11, $\cos \phi = V_R/V_0 = (I_{rms}R)/(I_{rms}Z) = R/Z$, so that $R = Z \cos \phi$. Therefore,

$$\overline{P} = I_{rms}^2 Z \cos \phi = I_{rms}(I_{rms}Z) \cos \phi$$
$$\overline{P} = I_{rms}V_{rms} \cos \phi \tag{23.9}$$

where V_{rms} is the rms voltage of the generator. The term $\cos \phi$ is called the *power factor* of the circuit. As a check on the validity of Equation 23.9, note that if no resistance is present, $R = 0$, and $\cos \phi = R/Z = 0$. Consequently, $\overline{P} = I_{rms}V_{rms} \cos \phi = 0$, a result that is expected since neither a capacitor nor an inductor consumes energy on the average. Conversely, if only resistance is present, $Z = \sqrt{R^2 + (X_L - X_C)^2} = R$, and $\cos \phi = R/Z = 1$. In this case, $\overline{P} = I_{rms}V_{rms} \cos \phi = I_{rms}V_{rms}$, which is the expression for the average power dissipated in a resistor. Example 3 deals with the current, voltages, and power for a series RCL circuit.

EXAMPLE 3 Current, Voltages, and Power in a Series RCL Circuit

A series RCL circuit contains a 148-Ω resistor, a 1.50-μF capacitor, and a 35.7-mH inductor. The generator has a frequency of 512 Hz and an rms voltage of 35.0 V. Obtain (a) the rms voltage across each circuit element and (b) the average electric power consumed by the circuit.

Reasoning The rms voltages across each circuit element can be determined from $V_R = I_{rms}R$, $V_C = I_{rms}X_C$, and $V_L = I_{rms}X_L$, as soon as the rms current and the reactances X_C and X_L are known. Since the rms current can be found from $I_{rms} = V_{rms}/Z$ and $V_{rms} = 35.0$ V, the first step in the solution is to find the impedance Z from the individual reactances. The average power consumed is given by $\overline{P} = I_{rms}V_{rms} \cos \phi$, where the phase angle ϕ can be obtained from $\tan \phi = (X_L - X_C)/R$.

Solution

(a) The individual reactances are

$$X_C = \frac{1}{2\pi fC} = \frac{1}{2\pi(512 \text{ Hz})(1.50 \times 10^{-6} \text{ F})} = 207 \ \Omega \tag{23.2}$$

$$X_L = 2\pi fL = 2\pi(512 \text{ Hz})(35.7 \times 10^{-3} \text{ H}) = 115 \ \Omega \tag{23.4}$$

The impedance of the circuit is

$$Z = \sqrt{R^2 + (X_L - X_C)^2} = \sqrt{(148 \ \Omega)^2 + (115 \ \Omega - 207 \ \Omega)^2} = 174 \ \Omega \tag{23.7}$$

The rms current in each circuit element is

$$I_{rms} = \frac{V_{rms}}{Z} = \frac{35.0 \text{ V}}{174 \ \Omega} = 0.201 \text{ A} \tag{23.6}$$

● **PROBLEM SOLVING INSIGHT**
In an RCL series circuit, the rms voltages across the resistor, capacitor, and inductor do not add up to equal the rms voltage across the generator.

The rms voltage across each circuit element now follows immediately:

$$V_R = I_{rms}R = (0.201 \text{ A})(148 \ \Omega) = \boxed{29.7 \text{ V}} \qquad (20.14)$$

$$V_C = I_{rms}X_C = (0.201 \text{ A})(207 \ \Omega) = \boxed{41.6 \text{ V}} \qquad (23.1)$$

$$V_L = I_{rms}X_L = (0.201 \text{ A})(115 \ \Omega) = \boxed{23.1 \text{ V}} \qquad (23.3)$$

Observe that these three rms voltages do not add up to give the generator's rms voltage, which is 35.0 V. Instead, the rms voltages satisfy Equation 23.5. It is the sum of the *instantaneous* voltages across R, C, and L, rather than the sum of the rms voltages, that equals the generator's *instantaneous* voltage, according to Kirchhoff's loop rule.

(b) The average power consumed by the circuit is $\overline{P} = I_{rms}V_{rms}\cos\phi$. Therefore, a value for the phase angle ϕ is needed and can be obtained from $\tan\phi = (X_L - X_C)/R$ as follows:

$$\phi = \tan^{-1}\left(\frac{X_L - X_C}{R}\right) = \tan^{-1}\left(\frac{115 \ \Omega - 207 \ \Omega}{148 \ \Omega}\right) = -32° \qquad (23.8)$$

The phase angle is negative since the circuit is more capacitive than inductive (X_C is greater than X_L), and the current leads the voltage. The average power consumed is

$$\overline{P} = I_{rms}V_{rms}\cos\phi = (0.201 \text{ A})(35.0 \text{ V})\cos(-32°) = \boxed{6.0 \text{ W}} \qquad (23.9)$$

In addition to the series RCL circuit, there are many different ways to connect resistors, capacitors, and inductors. In analyzing these additional possibilities, it helps to keep in mind the behavior of capacitors and inductors at the extreme limits of the frequency. When the frequency approaches zero (i.e., dc conditions), the reactance of a capacitor becomes so large that no charge can flow through the capacitor. It is as if the capacitor were cut out of the circuit, leaving an open gap in the connecting wire. In the limit of zero frequency the reactance of an inductor is vanishingly small. The inductor offers no opposition to a dc current. It is as if the inductor were replaced with a wire of zero resistance. In the limit of very large frequency, the behaviors of a capacitor and an inductor are reversed. The capacitor has a very small reactance and offers little opposition to the current, as if it were replaced by a wire with zero resistance. In contrast, the inductor has a very large reactance when the frequency is very large. The inductor offers so much opposition to the current that it might as well be cut out of the circuit, leaving an open gap in the connecting wire. Conceptual Example 4 illustrates how to gain insight into more complicated circuits using the limiting behaviors of capacitors and inductors.

CONCEPTUAL EXAMPLE 4
The Limiting Behavior of Capacitors and Inductors

Figure 23.12a shows two circuits. The rms voltage of the generator is the same in each case. The values of the resistance R, the capacitance C, and the inductance L in these circuits are the same. The frequency of the ac generator is very near zero. In which circuit does the generator supply more rms current?

Reasoning and Solution According to Equation 23.6, the rms current is given by $I_{rms} = V_{rms}/Z$. However, the impedance Z is not given by Equation 23.7, since the circuits in Figure 23.12a are not series RCL circuits. Since V_{rms} is the same in each case, the greater current is delivered to the circuit with the smaller impedance Z. In the limit of very small frequency, the capacitors behave as if they were cut out of the circuit, leaving gaps in the connecting wires. In this limit, however, the inductors behave as

Circuit I

Circuit II

(a)

Circuit I
(Low-frequency limit)

Circuit II
(Low-frequency limit)

(b)

Figure 23.12 (a) These circuits are discussed in the limit of very small or low frequency in Conceptual Example 4. (b) For a frequency very near zero, the circuits in part a behave as if they were as shown here.

if they were replaced by wires with zero resistance. Figure 23.12*b* shows the circuits as they would appear according to these changes. Circuit I behaves as if it contained only two identical resistors in series, with a total impedance of $Z = 2R$. In contrast, circuit II behaves as if it contained two identical resistors in parallel, in which case the total impedance is given by $1/Z = 1/R + 1/R$, or $Z = R/2$. At a frequency very near zero, then, ***circuit II has the smaller impedance and its generator supplies the greater rms current.***

Related Homework Material: *Questions 6 and 7, Problems 22, 24, and 38*

23.4   RESONANCE IN ELECTRIC CIRCUITS

Figure 23.13 *Concepts at a Glance*
The phenomenon of resonance can occur when an object vibrates on a spring or when standing waves are established on a string or in a tube of air. Resonance can also occur in a series RCL circuit. The antenna on the cellular phone in the photograph is connected to a resonant circuit and enables the phone to detect the radio waves that carry the conversation. The resonant frequency of this circuit is matched to the frequency of the radio waves.

⬤ *Concepts at a Glance* The behavior of current and voltage in a series RCL circuit can give rise to a condition of ***resonance.*** Resonance occurs when the frequency of a vibrating force exactly matches a natural (resonant) frequency of the object to which the force is applied. When resonance occurs, the force can transmit a large amount of energy to the object, leading to a large-amplitude vibratory motion. As the Concepts-at-a-Glance chart in Figure 23.13 illustrates, we have already encountered several examples of resonance. First, resonance can occur when a vibrating force is applied to an object of mass m that is attached to a spring whose spring constant is k (Section 10.6). In this case there is one natural frequency f_0, whose value is $f_0 = (1/2\pi)\sqrt{k/m}$. Second, resonance occurs when standing waves are set up on a string (Section 17.5) or in a tube of air (Section 17.6). As the right-hand side of the concept chart shows, the string and tube of air have many natural frequencies, one for each allowed standing wave. As we will now see, a condition of resonance can also be established in a series RCL circuit. In this case there is

Concepts at a Glance

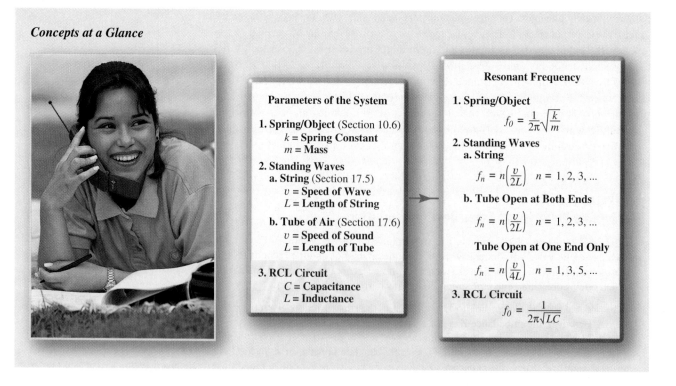

Parameters of the System

1. **Spring/Object** (Section 10.6)
   k = Spring Constant
   m = Mass

2. **Standing Waves**
   a. **String** (Section 17.5)
      v = Speed of Wave
      L = Length of String
   b. **Tube of Air** (Section 17.6)
      v = Speed of Sound
      L = Length of Tube

3. **RCL Circuit**
   C = Capacitance
   L = Inductance

Resonant Frequency

1. **Spring/Object**
   $$f_0 = \frac{1}{2\pi}\sqrt{\frac{k}{m}}$$

2. **Standing Waves**
   a. **String**
      $$f_n = n\left(\frac{v}{2L}\right) \quad n = 1, 2, 3, \dots$$
   b. **Tube Open at Both Ends**
      $$f_n = n\left(\frac{v}{2L}\right) \quad n = 1, 2, 3, \dots$$

   Tube Open at One End Only
      $$f_n = n\left(\frac{v}{4L}\right) \quad n = 1, 3, 5, \dots$$

3. **RCL Circuit**
   $$f_0 = \frac{1}{2\pi\sqrt{LC}}$$

Figure 23.14 The oscillation of an object on a spring is analogous to the oscillation of the electric and magnetic fields that occur, respectively, in a capacitor and in an inductor.

only one natural frequency, and the vibrating force is provided by the oscillating electric field that is related to the voltage of the generator.

Figure 23.14 helps us to understand why there is a resonant frequency for an ac circuit. This drawing presents an analogy between the electrical case (ignoring resistance) and the mechanical case of an object attached to a horizontal spring (ignoring friction). Part *a* shows a fully stretched spring that has just been released, with the initial speed *v* of the object being zero. All the energy is stored in the form of elastic potential energy. When the object begins to move, it gradually loses potential energy and picks up kinetic energy. In part *b*, the object moves with maximum kinetic energy through the position where the spring is unstretched (zero potential energy). Because of its inertia, the moving object coasts through this position and eventually comes to a halt in part *c* when the spring is fully compressed and all kinetic energy has been converted back into elastic potential energy. Part *d* of the picture is like part *b*, except the direction of motion is reversed. The resonant frequency f_0 of the object on the spring is the natural frequency at which the object vibrates and is given as $f_0 = (1/2\pi)\sqrt{k/m}$ according to Equations 10.6 and 10.11. In this expression, *m* is the mass of the object, and *k* is the spring constant.

In the electrical case, Figure 23.14*a* begins with a fully charged capacitor that has just been connected to an inductor. At this instant the energy is stored in the electric field between the capacitor plates. As the capacitor discharges, the electric field between the plates decreases, while a magnetic field builds up around the inductor because of the increasing current in the circuit. The maximum current and the maximum magnetic field exist at the instant when the capacitor is completely discharged, as in part *b* of the figure. Energy is now stored entirely in the magnetic field of the inductor. The voltage induced in the inductor keeps the charges flowing until the capacitor again becomes fully charged, but now with reverse polarity, as in part *c*. Once again, the energy is stored in the electric field between the plates, and no energy resides in the magnetic field of the inductor. Part *d* of the cycle repeats part *b*, but with reversed directions of current and magnetic field. Thus, an ac circuit can have a resonant frequency because there is a natural tendency for energy to shuttle back and forth between the electric field of the capacitor and the magnetic field of the inductor.

To determine the resonant frequency at which energy shuttles back and forth between the capacitor and the inductor, we note that the current in a series RCL circuit is $I_{rms} = V_{rms}/Z$. In this expression Z is the impedance of the circuit and is given

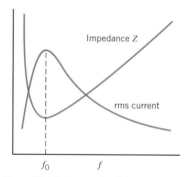

Figure 23.15 In a series RCL circuit the impedance is a minimum, and the current is a maximum, when the frequency f equals the resonant frequency f_0 of the circuit.

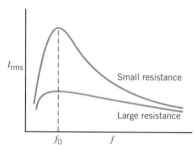

Figure 23.16 The effect of resistance on the current in a series RCL circuit.

Figure 23.17 A heterodyne metal detector can be used to locate buried metal objects.

by $Z = \sqrt{R^2 + (X_L - X_C)^2}$. As Figure 23.15 illustrates, the rms current is a maximum when the impedance is a minimum, assuming a given generator voltage. The minimum impedance of $Z = R$ occurs when the frequency is f_0, such that $X_L = X_C$ or $2\pi f_0 L = 1/(2\pi f_0 C)$. This result can be solved for f_0, which is the resonant frequency:

$$f_0 = \frac{1}{2\pi\sqrt{LC}} \tag{23.10}$$

The resonant frequency is determined by the inductance and the capacitance, but not the resistance.

The effect of resistance on electrical resonance is to make the "sharpness" of the circuit response less pronounced, as Figure 23.16 indicates. When the resistance is small, the current-versus-frequency graph falls off suddenly on either side of the maximum current. When the resistance is large, the falloff is more gradual, and there is less current at the maximum.

The following example deals with one application of resonance in electrical circuits. In this example the focus is on the oscillation of energy between a capacitor and an inductor. Once a capacitor/inductor combination is energized, the energy will oscillate indefinitely as in Figure 23.14, provided there is some provision to replace any dissipative losses that occur because of resistance. Circuits that include this type of provision are called oscillator circuits.

EXAMPLE 5   A Heterodyne Metal Detector

Figure 23.17 shows a heterodyne metal detector being used. As Figure 23.18 illustrates, this device utilizes two capacitor/inductor oscillator circuits, A and B. Each circuit produces its own resonant frequency, $f_{0A} = 1/(2\pi\sqrt{L_A C})$ and $f_{0B} = 1/(2\pi\sqrt{L_B C})$. Any difference between these two frequencies is detected through earphones as a beat frequency $f_{0B} - f_{0A}$, similar to the beat frequency that two musical tones produce. In the absence of any nearby metal object, the inductances L_A and L_B are the same, and f_{0A} and f_{0B} are identical. There is no beat frequency. When inductor B (the search coil) comes near a piece of metal, the inductance L_B decreases, the corresponding oscillator frequency f_{0B} increases, and a beat frequency is heard. Suppose that initially each inductor is adjusted so $L_B = L_A$, and each oscillator has a resonant frequency of 855.5 kHz. Assuming that the inductance of search coil B decreases by 1.00% due to a nearby piece of metal, determine the beat frequency heard through the earphones.

Reasoning To find the beat frequency $f_{0B} - f_{0A}$, we need to determine the amount by which the resonant frequency f_{0B} changes because of a 1.00% decrease in the inductance L_B.

Solution We begin by obtaining the ratio of f_{0B} to f_{0A}:

$$\frac{f_{0B}}{f_{0A}} = \frac{\dfrac{1}{2\pi\sqrt{L_B C}}}{\dfrac{1}{2\pi\sqrt{L_A C}}} = \sqrt{\frac{L_A}{L_B}}$$

Due to the 1.00% decrease in the inductance of coil B, $L_B = 0.9900 L_A$, so that

$$\frac{f_{0B}}{f_{0A}} = \sqrt{\frac{L_A}{0.9900 L_A}} = 1.005$$

Therefore, the new value for f_{0B} is $f_{0B} = 1.005 f_{0A} = 1.005 \times (855.5\ \text{kHz}) = 859.8\ \text{kHz}$. As a result, the detected beat frequency is

$$f_{0B} - f_{0A} = 859.8\ \text{kHz} - 855.5\ \text{kHz} = \boxed{4.3\ \text{kHz}}$$

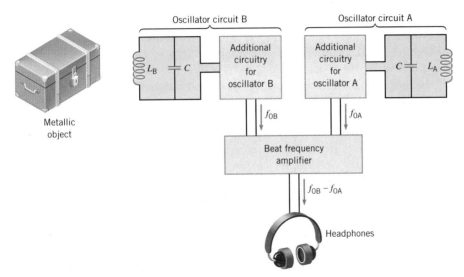

Figure 23.18 A heterodyne metal detector uses two electrical oscillators, *A* and *B,* in its operation. When the resonant frequency of oscillator *B* is changed due to the proximity of a metallic object, a beat frequency, whose value is $f_{0B} - f_{0A}$, is heard in the headphones.

23.5 SEMICONDUCTOR DEVICES

Semiconductor devices such as diodes and transistors are widely used in modern electronics, and Figure 23.19 illustrates one application. The drawing shows an audio system in which small ac voltages (originating in a compact disc player, an FM tuner, or a cassette deck) are amplified so they can drive the speaker(s). The electric circuits that accomplish the amplification do so with the aid of a dc voltage provided by the power supply. In portable units the power supply is simply a battery. In nonportable units, however, the power supply is a separate electric circuit containing diodes, along with other elements. As we will see, the diodes convert the 60-Hz ac voltage present at a wall outlet into the dc voltage needed by the amplifier, which, in turn, performs its job of amplification with the aid of transistors.

n-Type and *p*-Type Semiconductors

The materials used in diodes and transistors are semiconductors, such as silicon and germanium. However, they are not pure materials because small amounts of "impurity" atoms (about one part in a million) have been added to them to change their

Figure 23.19 In a typical audio system, diodes are used in the power supply to create a dc voltage from the ac voltage present at the wall socket. This dc voltage is necessary so the transistors in the amplifier can perform their task of enlarging the small ac voltages originating in the compact disc player, etc.

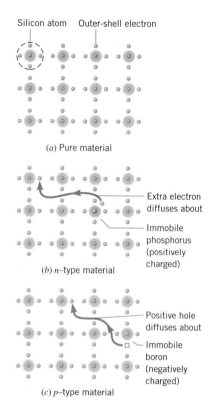

Silicon atom   Outer-shell electron

(a) Pure material

Extra electron diffuses about

Immobile phosphorus (positively charged)

(b) n–type material

Positive hole diffuses about

Immobile boron (negatively charged)

(c) p–type material

Figure 23.20 A silicon crystal that is (a) undoped or pure, (b) doped with phosphorus to produce an n-type material, and (c) doped with boron to produce a p-type material.

conductive properties. For instance, Figure 23.20a shows an array of atoms that symbolizes the crystal structure in pure silicon. Each silicon atom has four outer-shell* electrons, and each electron participates with electrons from neighboring atoms in forming the bonds that hold the crystal together. Since they participate in forming bonds, these electrons generally do not move throughout the crystal. Consequently, pure silicon and germanium are not good conductors of electricity. It is possible, however, to increase their conductivities by adding tiny amounts of impurity atoms, such as phosphorus or arsenic, whose atoms have five outer-shell electrons. For example, when a phosphorus atom replaces a silicon atom in the crystal, only four of the five outer-shell electrons of phosphorus fit into the crystal structure. The extra fifth electron does not fit in and is relatively free to diffuse throughout the crystal, as part b of the drawing suggests. A semiconductor containing small amounts of phosphorus can, therefore, be envisioned as containing immobile, positively charged phosphorus atoms and a pool of electrons that are free to wander throughout the material. These mobile electrons allow the semiconductor to conduct electricity.

The process of adding impurity atoms is called *doping.* A semiconductor doped with an impurity that contributes mobile electrons is called an **n-type semiconductor,** since the mobile charge carriers have a **n**egative charge. Note that an n-type semiconductor is overall electrically neutral, since it contains equal numbers of positive and negative charges.

It is also possible to dope a silicon crystal with an impurity whose atoms have only three outer-shell electrons (e.g., boron or gallium). Because of the missing fourth electron, there is a "hole" in the lattice structure at the boron atom, as Figure 23.20c illustrates. An electron from a neighboring silicon atom can move into this hole, in which event the region around the boron atom, having acquired the electron, becomes negatively charged. Of course, when a nearby electron does move, it leaves behind a hole. This hole is positively charged, since it results from the removal of an electron from the vicinity of a neutral silicon atom. The vast majority of atoms in the lattice are silicon, so the hole is almost always next to another silicon atom. Consequently, an electron from one of these adjacent atoms can move into the hole, with the result that the hole moves to yet another location. In this fashion, a positively charged hole can wander through the crystal. This type of semiconductor can, therefore, be viewed as containing immobile, negatively charged boron atoms and an equal number of positively charged, mobile holes. Because of the mobile holes, the semiconductor can conduct electricity. In this case the charge carriers are positive. A semiconductor doped with an impurity that introduces mobile **p**ositive holes is called a **p-type semiconductor.**

The Semiconductor Diode

A **p-n junction diode** is a device formed from a p-type semiconductor and an n-type semiconductor. The p-n junction between the two materials is of fundamental importance to the operation of diodes and transistors. Figure 23.21 shows separate p-type and n-type semiconductors, each electrically neutral. Figure 23.22a shows them joined together to form a diode. Mobile electrons from the n-type semiconductor and mobile holes from the p-type semiconductor flow across the junction and combine. This process leaves the n-type material with a positive charge layer and the p-type material with a negative charge layer, as part b of the drawing indicates. The positive and negative charge layers on the two sides of the junction set up an electric field **E**, much like that in a parallel plate capacitor. This electric field

* Section 30.6 discusses the electronic structure of the atom in terms of "shells."

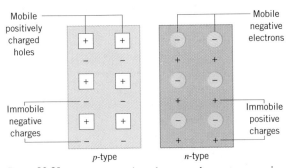

Figure 23.21 A *p*-type semiconductor and an *n*-type semiconductor.

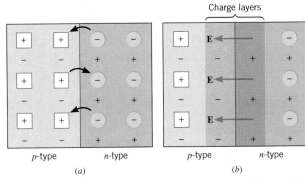

Figure 23.22 (*a*) At the junction between *n* and *p* materials, mobile electrons and holes combine and (*b*) create positive and negative charge layers. The electric field produced by the charge layers is **E**.

tends to prevent any further movement of charge across the junction, and all charge flow quickly stops.

Suppose now that a battery is connected across the *p-n* junction, as in Figure 23.23*a*, where the negative terminal of the battery is attached to the *n*-material, and the positive terminal is attached to the *p*-material. In this situation the junction is said to be in a condition of ***forward bias,*** and, as a result, there is a current in the circuit. The negative terminal of the battery repels the mobile electrons in the *n*-type material, and they move toward the junction. Likewise, the positive terminal repels the positive holes in the *p*-type material, and they also move toward the junction. At the junction the electrons fill the holes. In the meantime, the negative terminal provides a fresh supply of electrons to the *n*-material, and the positive terminal pulls off electrons from the *p*-material, forming new holes in the process. Consequently, a continual flow of charge, and hence a current, is maintained.

In Figure 23.23*b* the battery polarity has been reversed, and the *p-n* junction is in a condition known as ***reverse bias.*** The battery forces electrons in the *n*-material and holes in the *p*-material away from the junction. As a result, the potential across the junction builds up until it opposes the battery potential, and very little current can be sustained through the diode. The diode, then, is a unidirectional device, for it allows current to pass only in one direction.

The graph in Figure 23.24 shows the dependence of the current on the magnitude and polarity of the voltage applied across a *p-n* junction diode. The exact val-

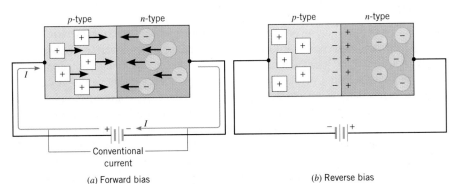

(*a*) Forward bias

(*b*) Reverse bias

Figure 23.23 (*a*) There is an appreciable current through the diode when the diode is forward biased. (*b*) Under a reverse bias condition, there is almost no current through the diode.

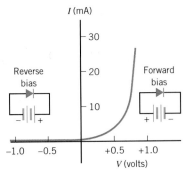

Figure 23.24 The current-versus-voltage characteristics of a typical *p-n* junction diode.

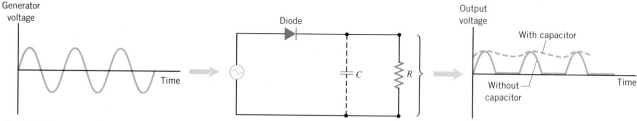

Figure 23.25 A half-wave rectifier circuit, together with a capacitor, constitutes a dc power supply because the rectifier converts an ac voltage into a dc voltage.

ues of the current depend on the nature of the semiconductor and the extent of the doping. Also shown in the drawing is the symbol used for a diode (⟶⊢) . The direction of the arrowhead in the symbol indicates the direction of the conventional current in the diode under a forward bias condition. In a forward bias condition, the side of the symbol that contains the arrowhead has a positive potential relative to the other side.

Because diodes are unidirectional devices, they are commonly used in *rectifier circuits,* which convert an ac voltage into a dc voltage. For instance, Figure 23.25 shows a circuit in which charges flow through the resistance R only while the ac generator biases the diode in the forward direction. Since current occurs only during one-half of every generator voltage cycle, the circuit is called a half-wave rectifier. A plot of the output voltage across the resistor reveals that only the positive halves of each cycle are present. If a capacitor is added in parallel with the resistor, as indicated in the drawing, the capacitor charges up and keeps the voltage from dropping to zero between each positive half-cycle. It is also possible to construct full-wave rectifier circuits, in which both halves of every cycle of the generator voltage drive current through the load resistor in the same direction (see conceptual question 11).

When a circuit such as that in Figure 23.25 includes a capacitor and also a transformer to establish the desired voltage level, the circuit is called a power supply. In the audio system in Figure 23.19, the power supply receives the 60-Hz ac voltage from a wall socket and produces a dc output voltage that is used for the transistors within the amplifier. Power supplies using diodes are also found in virtually all electronic appliances, such as televisions and microwave ovens.

Solar Cells

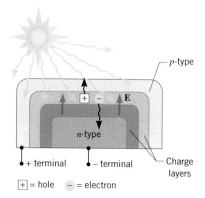

Figure 23.26 A solar cell formed from a *p-n* junction. When sunlight strikes the solar cell it acts like a battery, with + and − terminals.

Solar cells use *p-n* junctions to convert sunlight directly into electricity, as Figure 23.26 illustrates. The solar cell in this drawing consists of a *p*-type semiconductor surrounding an *n*-type semiconductor. As discussed earlier, charge layers form at the junction between the two types of semiconductors, leading to an electric field **E** pointing from the *n*-type toward the *p*-type layer. The outer covering of *p*-type material is so thin that sunlight penetrates into the charge layers and ionizes some of the atoms there. In the process of ionization, the energy of the sunlight causes a negative electron to be ejected from the atom, leaving behind a positive hole. As the drawing indicates, the electric field in the charge layers causes the electron and the hole to move away from the junction. The electron moves into the *n*-type material, and the hole moves into the *p*-type material. As a result, the sunlight causes the solar cell to develop negative and positive terminals, much like the terminals of a battery. The current that a single solar cell can provide is small, so applications of solar cells often use many of them mounted to form large panels, as Figure 23.27 illustrates.

THE PHYSICS OF...

solar cells.

Transistors

A number of different kinds of transistors are in use today. One type is the *bipolar junction transistor,* which consists of two *p-n* junctions formed by three layers of doped semiconductors. As Figure 23.28 indicates, there are *pnp* and *npn* transistors. In either case, the middle region is made very thin compared to the outer regions.

A transistor is useful because it can be used in circuits that amplify a smaller voltage into a larger one. A transistor plays the same kind of role in an amplifier circuit that a valve does when it controls the flow of water through a pipe. A small change in the valve setting produces a large change in the amount of water per second that flows through the pipe. In other words, a small change in the voltage input to a transistor produces a large change in the output from the transistor.

Figure 23.29 shows a *pnp* transistor connected to two batteries, labeled V_E and V_C. The voltages V_E and V_C are applied in such a way that the *p-n* junction on the left has a forward bias, while the *p-n* junction on the right has a reverse bias. Moreover, the voltage V_C is usually much larger than V_E for a reason to be discussed shortly. The drawing also shows the standard symbol and nomenclature for the three sections of the transistor, namely, the *emitter,* the *base,* and the *collector.* The arrowhead in the symbol points in the direction of the conventional current through the emitter.

The positive terminal of V_E pushes the mobile positive holes in the *p*-type material of the emitter toward the emitter/base junction. And since this junction has a forward bias, the holes enter the base region readily. Once in the base region, the holes come under the strong influence of V_C and are attracted to its negative terminal. Since the base is so thin (about 10^{-6} m or so), approximately 98% of the holes are drawn through the base and into the collector. The remaining 2% of the holes combine with free electrons in the base region, thereby giving rise to a small base current I_B. As the drawing shows, the moving holes in the emitter and collector constitute currents that are labeled I_E and I_C, respectively. From Kirchhoff's junction rule it follows that $I_C = I_E - I_B$.

Because the base current I_B is small, the collector current is determined primarily by current from the emitter ($I_C = I_E - I_B \approx I_E$). This means that a change in I_E

Figure 23.27 In regions where commercially available electricity is limited, solar-powered public telephone booths are especially convenient. The solar cells are located on the panel on the top of the booth.

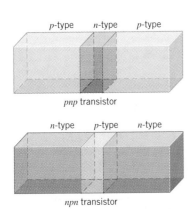

Figure 23.28 There are two kinds of bipolar junction transistors, *pnp* and *npn*.

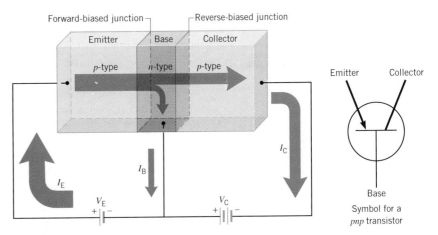

Figure 23.29 A *pnp* transistor, along with its bias voltages V_E and V_C. On the symbol for the *pnp* transistor, the emitter is marked with an arrowhead that denotes the direction of conventional current through the emitter.

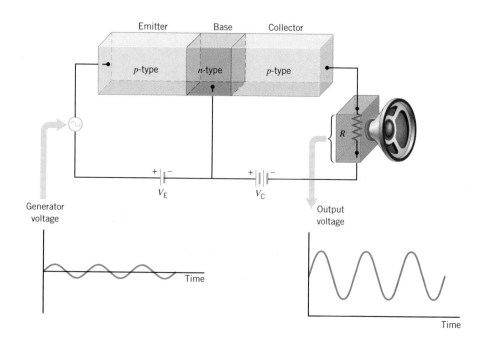

Figure 23.30 The basic *pnp* transistor amplifier in this drawing amplifies a small generator voltage to produce an enlarged voltage across the resistance *R*.

THE PHYSICS OF...

transistors.

will cause a change in I_C of nearly the same amount. Furthermore, a substantial change in I_E can be caused by only a small change in the forward bias voltage V_E. To see that this is the case, look back at Figure 23.24 and notice how steep the current-versus-voltage curve is for a *p-n* junction; small changes in the forward bias voltage give rise to large changes in the current.

With the help of Figure 23.30 we can now appreciate what was meant by the earlier statement that a small change in the voltage input to a transistor leads to a large change in the output. This picture shows an ac generator connected in series with the battery V_E and a resistance *R* connected in series with the collector. The generator voltage could originate from many sources, such as an electric guitar pickup or a compact disc player, while the resistance *R* could represent a loud-speaker. The generator introduces small voltage changes in the forward bias across the emitter/base junction and, thus, causes large corresponding changes in the current I_C leaving the collector and passing through the resistance *R*. As a result, the output voltage across *R* is an enlarged or amplified version of the input voltage of the generator. The operation of an *npn* transistor is similar to that of a *pnp* transistor. The main difference is that the bias voltages and current directions are reversed.

It is important to realize that the increased power available at the output of a transistor amplifier does *not* come from the transistor itself. Rather, it comes from the power provided by the voltage source V_C. The transistor, acting like an automatic valve, merely allows the small, weak signals from the input generator to control the power taken from the source V_C and delivered to the resistance *R*.

Today it is possible to combine arrays of tens of thousands of transistors, diodes, resistors, and capacitors on a tiny "chip" of silicon that usually measures less than a centimeter on a side. These arrays are called integrated circuits (ICs) and can be designed to perform almost any desired electronic function. Integrated circuits, such as the type in Figure 23.31, have revolutionized the electronics industry and lie at the heart of computers, cellular phones, digital watches, and programmable appliances.

Figure 23.31 Integrated circuit (IC) chips are manufactured on wafers of semiconductor material. Shown here is one wafer containing many chips. Some of the so-called smart cards in which the chips are used are also shown.

23.6 CONCEPTS & CALCULATIONS

A capacitor is one of the important elements found in ac circuits. As we have seen in this chapter, the capacitance of a capacitor influences the amount of current in a circuit. The capacitance, in turn, is determined by the geometry of the capacitor and the material that fills the space between its plates, as Section 19.5 discusses. When capacitors are connected together, the equivalent capacitance depends on the nature of the connection, for example, whether it is a series or a parallel connection, as Section 20.12 discusses. The next example provides a review of these issues concerning capacitors.

CONCEPTS & CALCULATIONS EXAMPLE 6   **Capacitors in AC Circuits**

Two parallel plate capacitors are filled with the same dielectric material and have the same plate area. However, capacitor 1 has a plate separation that is twice that of capacitor 2. When capacitor 1 is connected across the terminals of an ac generator, the generator delivers an rms current of 0.60 A. What is the current delivered by the generator when both capacitors are connected in parallel across its terminals? In both cases the generator produces the same frequency and voltage.

Concept Questions and Answers Which of the two capacitors has the greater capacitance?

Answer The capacitance of a capacitor is given by Equation 19.10 as $C = \kappa \epsilon_0 A/d$, where κ is the dielectric constant of the material between the plates, ϵ_0 is the permittivity of free space, A is the area of each plate, and d is the separation between the plates. Since the dielectric constant and the plate area are the same for each capacitor, the capacitance is inversely proportional to the plate separation. Therefore, capacitor 2, with the smaller plate separation, has the greater capacitance.

Is the equivalent capacitance of the parallel combination greater or smaller than the capacitance of capacitor 1?

Answer According to Equation 20.18, the equivalent capacitance of the two capacitors in parallel is $C_P = C_1 + C_2$, where C_1 and C_2 are the individual capacitances. Therefore, the value for C_P is greater than the value for C_1.

Is the capacitive reactance for C_P greater or smaller than for C_1?

Answer The capacitive reactance is given by $X_C = 1/(2\pi f C)$, according to Equation 23.2, where f is the frequency. For a given frequency, the reactance is inversely proportional to the capacitance. Since the capacitance C_P is greater than C_1, the corresponding reactance is smaller.

When both capacitors are connected in parallel across the terminals of the generator, is the current from the generator greater or smaller than when capacitor 1 is connected alone?

Answer According to Equation 23.1, the current is given by $I_{rms} = V_{rms}/X_C$, where V_{rms} is the rms voltage of the generator and X_C is the capacitive reactance. For a given voltage, the current is inversely proportional to the reactance. Since the reactance in the parallel case is smaller than for C_1 alone, the current in the parallel case is greater.

Solution

Using Equation 23.1 to express the current as $I_{rms} = V_{rms}/X_C$ and Equation 23.2 to express the reactance as $X_C = 1/(2\pi f C)$, we find that the current is

$$I_{rms} = \frac{V_{rms}}{X_C} = \frac{V_{rms}}{1/(2\pi f C)} = V_{rms}2\pi f C$$

Applying this result to the case where C_1 is connected alone to the generator and to the case where the two capacitors are connected in parallel, we obtain

$$\underbrace{I_{1,\text{rms}} = V_{\text{rms}}2\pi f C_1}_{C_1 \text{ alone}} \quad \text{and} \quad \underbrace{I_{P,\text{rms}} = V_{\text{rms}}2\pi f C_P}_{C_1 \text{ and } C_2 \text{ in parallel}}$$

Dividing the two expressions gives

$$\frac{I_{P,\text{rms}}}{I_{1,\text{rms}}} = \frac{V_{\text{rms}}2\pi f C_P}{V_{\text{rms}}2\pi f C_1} = \frac{C_P}{C_1}$$

According to Equation 20.18, the effective capacitance of the two capacitors in parallel is $C_P = C_1 + C_2$, so that the result for the current ratio becomes

$$\frac{I_{P,\text{rms}}}{I_{1,\text{rms}}} = \frac{C_1 + C_2}{C_1} = 1 + \frac{C_2}{C_1}$$

Since the capacitance of a capacitor is given by Equation 19.10 as $C = \kappa\epsilon_0 A/d$, we find that

$$\frac{I_{P,\text{rms}}}{I_{1,\text{rms}}} = 1 + \frac{\kappa\epsilon_0 A/d_2}{\kappa\epsilon_0 A/d_1} = 1 + \frac{d_1}{d_2}$$

We know that $d_1 = 2d_2$, so that the current in the parallel case is

$$I_{P,\text{rms}} = I_{1,\text{rms}}\left(1 + \frac{d_1}{d_2}\right) = (0.60 \text{ A})\left(1 + \frac{2d_2}{d_2}\right) = \boxed{1.8 \text{ A}}$$

As expected, the current in the parallel case is greater.

In ac circuits that contain capacitance, inductance, and resistance, it is only the resistance that, on average, dissipates power. The average power dissipated in a capacitor or an inductor is zero. However, the presence of a capacitor and/or an inductor does influence the rms current in the circuit. When the current changes for any reason, the power dissipated in a resistor also changes, as Example 7 illustrates.

CONCEPTS & CALCULATIONS EXAMPLE 7   Only Resistance Dissipates Power

An ac generator has a frequency of 1200 Hz and a constant rms voltage. When a 470-Ω resistor is connected between the terminals of the generator, an average power of 0.25 W is dissipated in the resistor. Then, a 0.080-H inductor is connected in series with the resistor, and the combination is connected between the generator terminals. What is the average power dissipated in the series combination?

Concept Questions and Answers In which case does the generator deliver a greater rms current?

Answer When only the resistor is present, the current is given by Equation 20.14 as $I_{\text{rms}} = V_{\text{rms}}/R$, where V_{rms} is the rms voltage of the generator. When the resistor and the inductor are connected in series, the current is given by $I_{\text{rms}} = V_{\text{rms}}/Z$, according to Equation 23.6. In this expression Z is the impedance and is given by Equation 23.7 as $Z = \sqrt{R^2 + X_L^2}$, where X_L is the inductive reactance and is given by Equation 23.4 as $X_L = 2\pi f L$. Since Z is greater than R, the current is greater when only the resistance is present.

In which case is a greater average power dissipated in the circuit?

Answer Only the resistor in an ac circuit dissipates power on average, the amount of the average power being $\overline{P} = I_{\text{rms}}^2 R$, according to Equation 20.15b. Since the resistance R is the same in both cases, a greater average power is dissipated when the current I_{rms} is greater. Since the current is greater when only the resistor is present, more power is dissipated in that case.

Solution When only the resistor is present, the average power is $\overline{P} = I^2_{rms}\,R$, according to Equation 20.15b, and the current is given by $I_{rms} = V_{rms}/R$, according to Equation 20.14. Therefore, we find in this case that

Resistor Only
$$\overline{P} = I^2_{rms}\,R = \left(\frac{V_{rms}}{R}\right)^2 R = \frac{V^2_{rms}}{R}$$

When the inductor is also present, the average power is still $\overline{P} = I^2_{rms}\,R$, but the current is now given by $I_{rms} = V_{rms}/Z$, according to Equation 23.6, in which Z is the impedance. In this case, then, we have

Resistor and Inductor
$$\overline{P} = I^2_{rms}\,R = \left(\frac{V_{rms}}{Z}\right)^2 R = \frac{V^2_{rms}\,R}{Z^2}$$

Dividing this result by the analogous result for the resistor-only case, we obtain

$$\frac{\overline{P}_{Resistor\ and\ Inductor}}{\overline{P}_{Resistor\ Only}} = \frac{V^2_{rms}\,R/Z^2}{V^2_{rms}/R} = \frac{R^2}{Z^2}$$

In this expression Z is the impedance and is given by Equation 23.7 as $Z = \sqrt{R^2 + X^2_L}$, where X_L is the inductive reactance and is given by Equation 23.4 as $X_L = 2\pi fL$. With these substitutions, the ratio of the powers becomes

$$\frac{\overline{P}_{Resistor\ and\ Inductor}}{\overline{P}_{Resistor\ Only}} = \frac{R^2}{R^2 + (2\pi fL)^2}$$

The power dissipated in the series combination, then, is

$$\overline{P}_{Resistor\ and\ Inductor} = \overline{P}_{Resistor\ Only}\left[\frac{R^2}{R^2 + (2\pi fL)^2}\right]$$
$$= (0.25\ \text{W})\,\frac{(470\ \Omega)^2}{(470\ \Omega)^2 + [2\pi(1200\ \text{Hz})(0.080\ \text{H})]^2} = \boxed{0.094\ \text{W}}$$

As expected, more power is dissipated in the resistor-only case.

At the end of the problem set for this chapter, you will find homework problems that contain both conceptual and quantitative parts. These problems are grouped under the heading *Concepts & Calculations, Group Learning Problems*. They are designed for use by students working alone or in small learning groups. The conceptual part of each problem provides a convenient focus for group discussions.

SUMMARY

In an ac circuit the rms voltage across a capacitor is related to the rms current according to $V_{rms} = I_{rms}X_C$, where X_C is the **capacitive reactance.** The capacitive reactance is measured in ohms and is given by $X_C = 1/(2\pi fC)$ for a capacitance C and a frequency f. The current in a capacitor leads the voltage across the capacitor by a phase angle of $90°$, and, as a result, a capacitor consumes no power, on average.

For an inductor the rms voltage and the rms current are related by $V_{rms} = I_{rms}X_L$, where X_L is the **inductive reactance.**

For an inductance L and frequency f the inductive reactance is given in ohms as $X_L = 2\pi fL$. The ac current in an inductor lags behind the voltage by a phase angle of $90°$. Consequently, an inductor, like a capacitor, consumes no power, on average.

When a resistor, a capacitor, and an inductor are connected in series, the rms voltage across the combination is related to the rms current according to $V_{rms} = I_{rms}Z$, where Z is the **impedance** of the combination. The impedance (in ohms) for the series combination is $Z = \sqrt{R^2 + (X_L - X_C)^2}$. The phase angle ϕ between current and voltage for a series RCL combination is given by $\tan \phi = (X_L - X_C)/R$. Only the resistor in the combination dissipates power, on average, according to the relation $\overline{P} = I_{rms}V_{rms} \cos \phi$. The term $\cos \phi$ is the **power factor** of the circuit.

A series RCL circuit has a **resonant frequency** f_0 that is $f_0 = 1/(2\pi\sqrt{LC})$. At resonance the impedance of the circuit has a minimum value equal to the resistance R, and the rms current has a maximum value.

In an **n-type semiconductor,** mobile negative electrons carry the current. An n-type material is produced by doping a semiconductor such as silicon with a small amount of impurity such as phosphorus. In a **p-type semiconductor,** mobile positive holes in the crystal structure carry the current. A p-type material is made by doping a semiconductor with an impurity such as boron. These two types of semiconductors are used in the **p-n junction diode** and in **pnp** and **npn** bipolar junction transistors.

CONCEPTUAL QUESTIONS

1. A light bulb and a parallel plate capacitor (including a dielectric material between the plates) are connected in series to the 60-Hz ac voltage present at a wall outlet. Describe what happens to the brightness of the bulb when the dielectric material is removed from the space between the plates. Explain.

2. The ends of a long, straight wire are connected to the terminals of an ac generator, and the current is measured. The wire is then disconnected, wound into the shape of a multiple-turn coil, and reconnected to the generator. In which case, if either, does the generator deliver a larger current? Account for your answer.

3. An air-core inductor is connected in series with a light bulb, and this circuit is plugged into an electrical outlet. What happens to the brightness of the bulb when a piece of iron is inserted inside the inductor? Give a reason for your answer.

4. Consider the circuit in Figure 23.9. With the capacitor and the inductor present, a certain amount of current is in the circuit. Then the capacitor and the inductor are removed, so that only the resistor remains connected to the generator. Is it possible that the current in the simplified circuit has the same rms value as that in the original circuit? Give your reasoning.

5. Can an ac circuit, such as the one in Figure 23.9, have the same impedance at two different generator frequencies? Justify your answer.

6. Review Conceptual Example 4 as an aid in understanding this question. An inductor and a capacitor are connected in parallel across the terminals of a generator. What happens to the current from the generator as the frequency becomes (a) very large and (b) very small? Give your reasoning.

7. Review Conceptual Example 4. For which of the two circuits discussed there does the generator deliver more current when the frequency is very large? Account for your answer.

8. In a series RCL circuit at resonance, does the current lead or lag behind the voltage across the generator or is it in phase with the voltage? Explain.

9. Is it possible for two series RCL circuits to have the same resonant frequencies and yet have (a) different R values and (b) different C and L values? Justify your answers.

10. Suppose the generator connected to a series RCL circuit has a frequency that is greater than the resonant frequency of the circuit. Suppose, in addition, that it is necessary to match the resonant frequency of the circuit to the frequency of the generator. To accomplish this, should you add a second capacitor in series or in parallel with the one already present? Explain your choice.

11. The drawing shows a full-wave rectifier circuit, in which the direction of the current through the load resistor R is the same for both positive and negative halves of the generator's voltage cycle. Draw the current through the circuit when

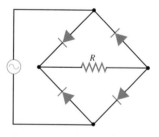

(a) the top of the generator is positive and the bottom is negative and (b) the top of the generator is negative and the bottom is positive.

PROBLEMS

Note: For problems in this set, the ac current and voltage are rms values, and the power is an average value, unless indicated otherwise.

ssm Solution is in the Student Solutions Manual. **www** Solution is available on the World Wide Web at http://www.wiley.com/college/cutnell
⚕ This icon represents a biomedical application.

Section 23.1 Capacitors and Capacitive Reactance

1. ssm At what frequency does a 7.50-μF capacitor have a reactance of 168 Ω?

2. What voltage is needed to create a current of 35 mA in a circuit containing only a 0.86-μF capacitor, when the frequency is 3.4 kHz?

3. A capacitor in an ac circuit has a capacitive reactance of 36 Ω when the frequency is 170 Hz. What is the capacitive reactance when the frequency is 510 Hz?

4. A capacitor is connected across the terminals of an ac generator that has a frequency of 440 Hz and supplies a voltage of 24 V. When a second capacitor is connected in parallel with the first one, the current from the generator increases by 0.18 A. Find the capacitance of the second capacitor.

5. ssm A circuit consists of a 3.00-μF and a 6.00-μF capacitor connected in series across the terminals of a 510-Hz generator. The voltage of the generator is 120 V. (a) Determine the equivalent capacitance of the two capacitors. (b) Find the current in the circuit.

*6. A capacitor is connected across an ac generator whose frequency is 750 Hz and whose *peak* output voltage is 140 V. The rms current in the circuit is 3.0 A. (a) What is the capacitance of the capacitor? (b) What is the magnitude of the *maximum* charge on one plate of the capacitor?

**7. A capacitor (capacitance C_1) is connected across the terminals of an ac generator. Without changing the voltage or frequency of the generator, a second capacitor (capacitance C_2) is added in series with the first one. As a result, the current delivered by the generator decreases by a factor of three. Suppose the second capacitor had been added in parallel with the first one, instead of in series. By what factor would the current delivered by the generator have increased?

Section 23.2 Inductors and Inductive Reactance

8. A 0.047-H inductor is wired across the terminals of a generator that has a voltage of 2.1 V and supplies a current of 0.023 A. Find the frequency of the generator.

9. ssm At what frequency (in Hz) are the reactances of a 52-mH inductor and a 76-μF capacitor equal?

10. The reactance of an inductor is 480 Ω when the frequency is 1350 Hz. What is the reactance when the frequency is 450 Hz?

11. ssm www An 8.2-mH inductor is connected to an ac generator (10.0 V rms, 620 Hz). Determine the *peak value* of the current supplied by the generator.

12. Two ac generators supply the same voltage. However, the first generator has a frequency of 1.5 kHz, while the second has a frequency of 6.0 kHz. When an inductor is connected across the terminals of the first generator, the current delivered is 0.30 A. How much current is delivered when this inductor is connected across the terminals of the second generator?

*13. The rms current in a solenoid is 0.036 A when the solenoid is connected to an 18-kHz generator. The solenoid has a cross-sectional area of 3.1×10^{-5} m^2 and a length of 2.5 cm. The solenoid has 135 turns. Determine the *peak voltage* of the generator.

**14. Two inductors are connected in parallel across the terminals of a generator. One has an inductance of $L_1 = 0.030$ H, while the other has an inductance of $L_2 = 0.060$ H. A single inductor, with an inductance L, is connected across the terminals of a second generator that has the same frequency and voltage as the first one. The current delivered by the second generator is equal to the *total* current delivered by the first generator. Find L.

Section 23.3 Circuits Containing Resistance, Capacitance, and Inductance

15. ssm A series RCL circuit includes a resistance of 275 Ω, an inductive reactance of 648 Ω, and a capacitive reactance of 415 Ω. The current in the circuit is 0.233 A. What is the voltage of the generator?

16. The elements in a series RCL circuit are a 106-Ω resistor, a 3.30-μF capacitor, and a 0.0310-H inductor. The frequency is 609 Hz. What is (a) the impedance of the circuit and (b) the phase angle between the current and the voltage of the generator?

17. A series RCL circuit contains a 47.0-Ω resistor, a 2.00-μF capacitor, and a 4.00-mH inductor. When the frequency is 2550 Hz, what is the power factor of the circuit?

18. An ac generator has a frequency of 4.80 kHz and produces a current of 0.0400 A in a series circuit that contains only a 232-Ω resistor and a 0.250-μF capacitor. Obtain (a) the voltage of the generator and (b) the phase angle between the current and the voltage across the resistor/capacitor combination.

19. ssm A circuit consists of a 215-Ω resistor and a 0.200-H inductor. These two elements are connected in series across a generator that has a frequency of 106 Hz and a voltage of 234 V. (a) What is the current in the circuit? (b) Determine the phase angle between the current and the voltage of the generator.

20. A circuit consists of a resistor in series with an inductor and an ac generator that supplies a voltage of 115 V. The inductive reactance is 52.0 Ω, and the current in the circuit is 1.75 A. Find the average power dissipated in the circuit.

*21. ssm www A circuit consists of an 85-Ω resistor in series with a 4.0-μF capacitor, the two being connected between the terminals of an ac generator. The voltage of the generator is fixed. At what frequency is the current in the circuit one-half the value that exists when the frequency is very large?

*22. Refer to Conceptual Example 4 as background for this problem. For the circuit shown in the drawing, find the current provided by the generator when the frequency is (a) very large and (b) very small.

$V_{rms} = 75$ V

Problem 22

*23. A series RCL circuit contains only a capacitor ($C = 8.80$ μF), an inductor ($L = 4.30$ mH), and a generator (*peak* voltage = 64.0 V, frequency = 1.00×10^3 Hz). When

$t = 0$ s, the instantaneous value of the voltage is zero, and it rises to a maximum one-quarter of a period later. (a) Find the *instantaneous* value of the voltage across the capacitor/inductor combination when $t = 1.50 \times 10^{-4}$ s. (b) What is the *instantaneous* value of the current when $t = 1.50 \times 10^{-4}$ s? *(Hint: The instantaneous values of the voltage and current are, respectively, the vertical components of the voltage and current phasors.)*

*24. Review Conceptual Example 4 before attempting this problem. Then refer to the drawing, in which the generator delivers four times as much current at very low frequencies than it does at very high frequencies. Find the ratio R_2/R_1 of the resistances.

Problem 24

**25. When a resistor is connected by itself to an ac generator, the average power dissipated in the resistor is 1.000 W. When a capacitor is added in series with the resistor, the power dissipated is 0.500 W. When an inductor is added in series with the resistor (without the capacitor), the power dissipated is 0.250 W. Determine the power dissipated when both the capacitor and the inductor are added in series with the resistor.

Section 23.4 Resonance in Electric Circuits

26. A series RCL circuit has a resonant frequency of 740 kHz. If the value of the inductance is 15 μH, what is the value of the capacitance?

27. ssm A 10.0-Ω resistor, a 12.0-μF capacitor, and a 17.0-mH inductor are connected in series with a 155-V generator. (a) At what frequency is the current a maximum? (b) What is the maximum value of the rms current?

28. The power dissipated in a series RCL circuit is 65.0 W, and the current is 0.530 A. The circuit is at resonance. Determine the voltage of the generator.

29. ssm The resonant frequency of a series RCL circuit is 9.3 kHz. The inductance and capacitance of the circuit are each tripled. What is the new resonant frequency?

30. The resonant frequency of a RCL circuit is 1.3 kHz, and the value of the inductance is 7.0 mH. What is the resonant frequency (in kHz) when the value of the inductance is 1.5 mH?

31. A series RCL circuit has a resonant frequency of 1500 Hz. When operating at a frequency other than 1500 Hz, the circuit has a capacitive reactance of 5.0 Ω and an inductive reactance of 30.0 Ω. What are the values of (a) L and (b) C?

*32. The ratio of the inductive reactance to the capacitive reactance is observed to be 5.36 in a series RCL circuit. The resonant frequency of the circuit is 225 Hz. What is the frequency (nonresonant) of the generator that is connected to the circuit?

*33. ssm Suppose you have a number of capacitors. Each of these capacitors is identical to the capacitor that is already in a series RCL circuit. How many of these additional capacitors must be inserted in series in the circuit, so the resonant frequency triples?

**34. In a series RCL circuit the dissipated power drops by a factor of two when the frequency of the generator is changed from the resonant frequency to a nonresonant frequency. The peak voltage is held constant while this change is made. Determine the power factor of the circuit at the nonresonant frequency.

**35. When the frequency is twice the resonant frequency, the impedance of a series RCL circuit is twice the value of the impedance at resonance. Obtain the ratios of the inductive and capacitive reactances to the resistance; that is, obtain (a) X_L/R and (b) X_C/R when the frequency is twice the resonant frequency.

ADDITIONAL PROBLEMS

36. Suppose that the inductance is zero ($L = 0$ H) in the series RCL circuit shown in Figure 23.10. The rms voltages across the generator and the resistor are 45 and 24 V, respectively. What is the rms voltage across the capacitor?

37. ssm A capacitor is attached to a 5.00-Hz generator. The instantaneous current is observed to reach a maximum value at a certain time. What is the least amount of time that passes before the instantaneous voltage across the capacitor reaches its maximum value?

38. Review Conceptual Example 4 and Figure 23.12. Find the ratio of the current in circuit I to that in circuit II in the *high-frequency limit* for the same generator voltage.

39. A light bulb has a resistance of 240 Ω. It is connected to a standard wall socket (120 V, 60.0 Hz). (a) Determine the current in the bulb. (b) Determine the current in the bulb after a 10.0-μF capacitor is added in series in the circuit. (c) It is possible to re-

turn the current in the bulb to the value calculated in part (a) by adding an inductor in series with the bulb and the capacitor. What is the value of the inductance of this inductor?

40. The current in an inductor is 0.20 A, and the frequency is 750 Hz. If the inductance is 0.080 H, what is the voltage across the inductor?

41. ssm www A 40.0-μF capacitor is connected across a 60.0-Hz generator. An inductor is then connected in parallel with the capacitor. What is the value of the inductance if the rms currents in the inductor and capacitor are equal?

42. In a series circuit, a generator (1350 Hz, 15.0 V) is connected to a 16.0-Ω resistor, a 4.10-μF capacitor, and a 5.30-mH inductor. Find the voltage across each circuit element.

*43. ssm www A series RCL circuit contains a 5.10-μF capacitor and a generator whose voltage is 11.0 V. At a resonant frequency of 1.30 kHz the power dissipated in the circuit is 25.0 W.

Find the values of (a) the inductance and (b) the resistance. (c) Calculate the power factor when the generator frequency is 2.31 kHz.

*44. In reality, there is some resistance R in the wire from which an inductor is made. Therefore, an actual inductor should be represented as a resistor in series with an ideal (resistanceless) inductor. With this in mind, suppose the current in an actual 4.5-mH inductor is I_0 when the inductor is connected to a battery. However, when the battery is replaced with a

1200-Hz generator whose rms voltage is the same as the battery voltage, the current is $I_0/4$. What is the resistance R of the wire?

**45. A 108-Ω resistor, a 0.200-μF capacitor, and a 5.42-mH inductor are connected in series to a generator whose voltage is 26.0 V. The current in the circuit is 0.141 A. Because of the shape of the current-versus-frequency graph (see Figure 23.15), there are two possible values for the frequency that correspond to this current. Obtain these two values.

CONCEPTS CALCULATIONS
GROUP LEARNING PROBLEMS

Note: Each of these problems consists of Concept Questions followed by a related quantitative Problem. They are designed for use by students working alone or in small learning groups. The Concept Questions involve little or no mathematics and are intended to stimulate group discussions. They focus on the concepts with which the problems deal. Recognizing the concepts is the essential initial step in any problem-solving technique.

46. Concept Questions (a) Does the capacitance of a parallel plate capacitor increase or decrease when a dielectric material is inserted between the plates? (b) Is the equivalent capacitance of two capacitors in parallel greater or smaller than the capacitance of either capacitor alone? (c) A capacitor is connected between the terminals of an ac generator. When a second capacitor is connected in parallel with the first, does the current supplied by the generator increase or decrease? Explain your answers.

Problem Two parallel plate capacitors are identical, except that one of them is empty and the other contains a material with a dielectric constant of 4.2 in the space between the plates. The empty capacitor is connected between the terminals of an ac generator that has a fixed frequency and rms voltage. The generator delivers a current of 0.22 A. What current does the generator deliver after the other capacitor is connected in parallel with the first one? Verify that your answer is consistent with your answers to the Concept Questions.

47. Concept Questions (a) An inductance L_1 is connected across the terminals of an ac generator, which delivers a current to it. Then a second inductance L_2 is connected in parallel with L_1. Does the presence of L_2 alter the current in L_1? (b) The generator delivers a current to L_2 as well as L_1. Would the removal of L_1 alter the current in L_2? (c) Is the current delivered to the parallel combination greater or smaller than the current to either inductance alone? Give your reasoning.

Problem An ac generator has a frequency of 2.2 kHz and a voltage of 240 V. An inductance $L_1 = 6.0$ mH is connected across its terminals. Then a second inductance $L_2 = 9.0$ mH is connected in parallel with L_1. Find the current that the generator delivers to L_1 and to the parallel combination. Check to see that your answers are consistent with your answers to the Concept Questions.

48. Concept Questions (a) An ac circuit contains only a resistor and a capacitor in series. Is the phase angle ϕ between the current and the voltage of the generator positive or negative, and how is the impedance Z of the circuit related to the resistance R and the capacitive reactance X_C? (b) An ac circuit contains only a resistor and an inductor in series. Is the phase angle ϕ positive or negative, and how is the impedance Z of the circuit related to the resistance R and the inductive reactance X_L? Account for your answers.

Problem A series circuit has an impedance of 192 Ω, and the phase angle is $\phi = -75.0°$. The circuit contains a resistor and either a capacitor or an inductor. Find the resistance R and the capacitive reactance X_C or the inductive reactance X_L, whichever is appropriate.

49. Concept Questions (a) A series RCL circuit contains three elements—a resistor, a capacitor, and an inductor. Which of the elements, on average, dissipate(s) power? (b) How is the current in the circuit related to the generator voltage and the impedance of the circuit? (c) What is the impedance of the circuit at resonance? (d) At resonance, how is the average dissipated power related to the generator voltage and the resistance?

Problem The resistor in a series RCL circuit has a resistance of 92 Ω, while the voltage of the generator is 3.0 V. At resonance, what is the average power dissipated in the circuit?

50. Concept Questions The capacitance in a series RCL circuit is C_1. The generator in the circuit has a fixed frequency that is less than the resonant frequency of the circuit. By adding a second capacitance C_2 to the circuit, it is desired to reduce the resonant frequency so that it matches the generator frequency. (a) To reduce the resonant frequency, should the circuit capacitance be increased or decreased? (b) Should C_2 be added in parallel or in series with C_1? Give your reasoning.

Problem The initial circuit capacitance is $C_1 = 2.60$ μF, and the corresponding resonant frequency is $f_{01} = 7.30$ kHz. The generator frequency is 5.60 kHz. What is the value of the capacitance C_2 that should be added so that the circuit will have a resonant frequency that matches the generator frequency?

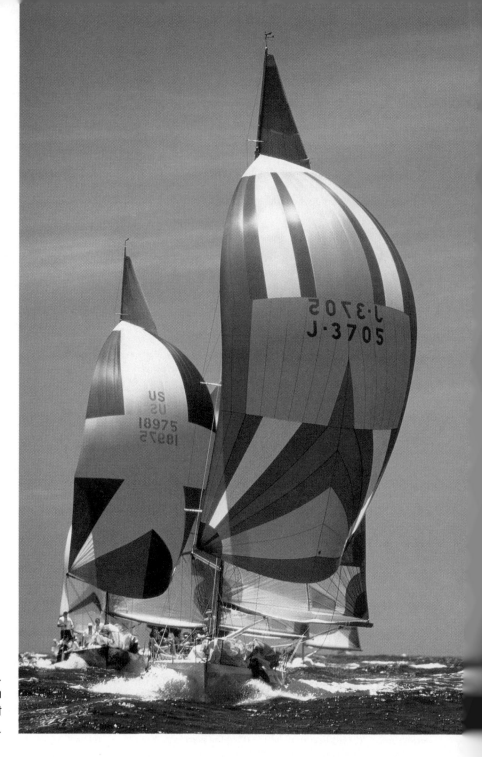

These racing yachts have brilliantly colored sails.
Each color corresponds to a different wavelength
in the mixture of electromagnetic waves that
composes the sunlight striking the sails.

CHAPTER 24

ELECTROMAGNETIC WAVES

24.1 THE NATURE OF ELECTROMAGNETIC WAVES

● *Concepts at a Glance*   In Section 13.3 we saw that energy is transported to us from the sun via a class of waves known as electromagnetic waves. This class includes the familiar visible, ultraviolet, and infrared waves. In Sections 18.6, 21.1, and 21.2 we studied the concepts of electric and magnetic fields. It was the great Scottish physicist James Clerk Maxwell (1831–1879) who showed that these two fields fluctuating together can form a propagating *electromagnetic wave.* The Concepts-at-a-Glance chart in Figure 24.1 emphasizes that we will now bring together our knowledge of electric and magnetic fields in order to understand this important type of wave. ●

Figure 24.2 illustrates one way to create an electromagnetic wave. The setup consists of two straight metal wires that are connected to the terminals of an ac generator and serve as an antenna. The potential difference between the terminals changes sinusoidally with time t and has a period T. Part a shows the instant $t = 0$ s, when there is no charge at the ends of either wire. Since there is no charge, there is no electric field at the point P just to the right of the antenna. As time passes, the top wire becomes positively charged and the bottom wire negatively charged. One-quarter of a cycle later ($t = \frac{1}{4}T$), the charges have attained their maximum values, as part b of the drawing indicates. The corresponding electric field $\mathbf{E}$ at point P is represented by the red arrow and has increased to its maximum strength in the downward direction.* Part b also shows that the electric field created at earlier times (see the black arrow in the picture) has not disappeared but has moved to the right. Here lies the crux of the matter: At distant points, the electric field of the charges is not felt immediately. Instead, the field is created first near the wires and then, like the effect of a pebble dropped into a pond, moves outward as a wave in all directions. Only the field moving to the right is shown in the picture for the sake of clarity.

Figure 24.1 *Concepts at a Glance* Electric and magnetic fields fluctuating together form an electromagnetic wave. Radio telescopes, such as those in the photograph, have dish-shaped surfaces that are designed to reflect the incoming electromagnetic waves to a common point so that as much energy as possible is collected.

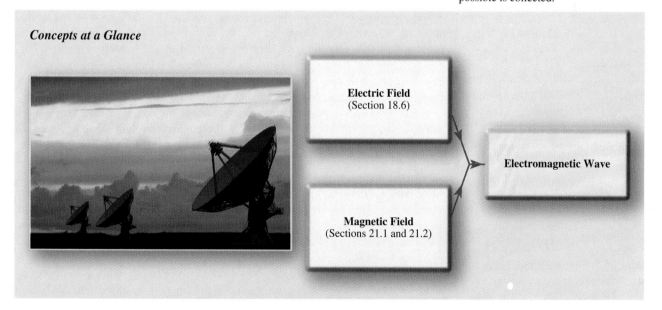

Concepts at a Glance

Electric Field
(Section 18.6)

Magnetic Field
(Sections 21.1 and 21.2)

Electromagnetic Wave

---

* The direction of the electric field can be obtained by imagining a positive test charge at P and determining the direction in which it would be pushed because of the charges on the wires.

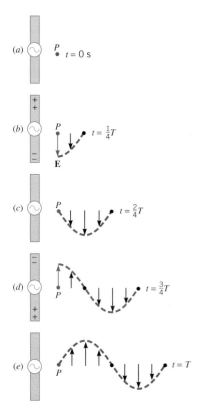

Figure 24.2 In each part of the drawing, the red arrow represents the electric field **E** produced at point P by the oscillating charges on the antenna at the indicated time. The black arrows represent the electric fields created at earlier times. All fields propagate to the right. Fields also propagate in other directions, but these are omitted for simplicity.

Figure 24.3 The oscillating current I in the antenna wires creates a magnetic field **B** at point P that is tangent to a circle centered on the wires. The field is directed into the page when the current is upward and out of the page when the current is downward.

Parts $c-e$ of Figure 24.2 show the creation of the electric field at point P (red arrow) at later times during the generator cycle. In each part, the fields produced earlier in the sequence (black arrows) continue propagating toward the right. Part d shows the charges on the wires when the polarity of the generator has reversed, so the top wire is negative and the bottom wire is positive. As a result, the electric field at P has reversed its direction and points upward. In part e of the sequence, a complete sine wave has been drawn through the tips of the electric field vectors to emphasize that the field changes sinusoidally.

Along with the electric field in Figure 24.2, a magnetic field **B** is also created, because the charges flowing in the antenna constitute an electric current, which produces a magnetic field. Figure 24.3 illustrates the field direction at point P at the instant when the current in the antenna wire is upward. With the aid of Right-Hand Rule No. 2 (thumb of right hand points along the current I, fingers curl in the direction of **B**), the magnetic field at P can be seen to point into the page. As the oscillating current changes, the magnetic field changes accordingly. The magnetic fields created at earlier times propagate outward as a wave, just as the electric fields do.

Notice that the magnetic field in Figure 24.3 is perpendicular to the page, whereas the electric field in Figure 24.2 lies in the plane of the page. Thus, the electric and magnetic fields created by the antenna are mutually perpendicular and remain so as they move outward. Moreover, both fields are perpendicular to the direction of travel. These perpendicular electric and magnetic fields, moving together, constitute an electromagnetic wave.

The electric and magnetic fields in Figures 24.2 and 24.3 decrease to zero rapidly with increasing distance from the antenna. Therefore, they exist mainly near the antenna and together are called the ***near field.*** Electric and magnetic fields do form a wave at large distances from the antenna, however. These fields arise from an effect that is different from that which produces the near field and are referred to as the ***radiation field.*** Faraday's law of induction provides part of the basis for the radiation field. As Section 22.4 discusses, this law describes the emf or potential difference produced by a changing magnetic field. And, as Section 19.4 explains, a potential difference can be related to an electric field. Thus, a changing magnetic field produces an electric field. Maxwell predicted that the reverse effect also occurs, namely, that a changing electric field produces a magnetic field. The radiation field arises because the changing magnetic field creates an electric field that fluctuates in time and the changing electric field creates the magnetic field.

Figure 24.4 shows the electromagnetic wave of the radiation field far from the antenna. The picture shows only the part of the wave traveling along the $+x$ axis. The parts traveling in the other directions have been omitted for clarity. It should be clear from the drawing that ***an electromagnetic wave is a transverse wave*** because the electric and magnetic fields are both perpendicular to the direction in which the wave travels. Moreover, an electromagnetic wave, unlike a wave on a string or a sound wave, does not require a medium in which to propagate. ***Electromagnetic waves can travel through a vacuum or a material substance,*** since electric and magnetic fields can exist in either one.

Electromagnetic waves can be produced in situations that do not involve a wire antenna. In general, any electric charge that is accelerating emits an electromagnetic wave, whether the charge is inside a wire or not. In an alternating current, an electron oscillates in simple harmonic motion along the length of the wire and is one example of an accelerating charge.

All electromagnetic waves move through a vacuum at the same speed, and the symbol c is used to denote its value. This speed is called the ***speed of light in a vac-***

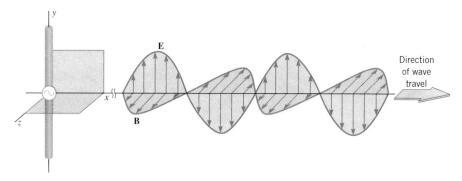

Figure 24.4 This picture shows the wave of the radiation field far from the antenna. Observe that **E** and **B** are perpendicular to each other, and both are perpendicular to the direction of travel.

uum and is $c = 3.00 \times 10^8$ m/s. In air, electromagnetic waves travel at nearly the same speed as they do in a vacuum, but, in general, they move through a substance such as glass at a speed that is substantially less than c.

The frequency of an electromagnetic wave is determined by the oscillation frequency of the electric charges at the source of the wave. In Figures 24.2–24.4 the wave frequency would equal the frequency of the ac generator. Suppose, for example, that the antenna is broadcasting electromagnetic waves known as radio waves. The frequencies of AM radio waves lie between 545 and 1605 kHz, these numbers corresponding to the limits of the AM broadcast band on the dial. The frequencies of FM radio waves lie between 88 and 108 MHz on the dial. Television channels 2–6, on the other hand, utilize electromagnetic waves with frequencies between 54 and 88 MHz, while channels 7–13 use frequencies between 174 and 216 MHz.

Radio and television reception involves a process that is the reverse of that outlined earlier for the creation of electromagnetic waves. When broadcasted waves reach a receiving antenna, they interact with the electric charges in the antenna wires. Either the electric field or the magnetic field of the waves can be used. To take full advantage of the electric field, the wires of the receiving antenna must be parallel to the electric field, as Figure 24.5 indicates. The electric field acts on the electrons in the wire, forcing them to oscillate back and forth along the length of the wire. Consequently, an ac current exists in the antenna and the circuit connected to it. The variable-capacitor C ($\dashv\!\!\!\nearrow\!\!\!\vdash$) and the inductor L in the circuit provide one way to select the frequency of the desired electromagnetic wave. By adjusting the value of the capacitance, it is possible to adjust the corresponding resonant frequency f_0 of the circuit [$f_0 = 1/(2\pi\sqrt{LC})$, Equation 23.10] to match the frequency of the wave. Under the condition of resonance there will be a maximum oscillating current in the inductor. Because of mutual inductance, this current creates a maximum voltage in the second coil in the drawing, and this voltage can then be amplified and processed by the remaining radio or television circuitry.

THE PHYSICS OF...

radio and television reception.

Figure 24.5 A radio wave can be detected with a receiving antenna wire that is parallel to the electric field of the wave. The magnetic field of the radio wave has been omitted for simplicity.

Figure 24.6 With a receiving antenna in the form of a loop, the magnetic field of a broadcasted radio wave can be detected. The normal to the plane of the loop should be parallel to the magnetic field for best reception. For clarity, the electric field of the radio wave has been omitted.

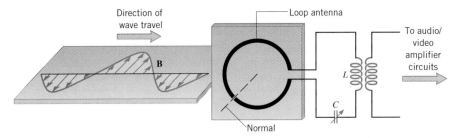

To detect the magnetic field of a broadcasted radio wave, a receiving antenna in the form of a loop can be used, as Figure 24.6 shows. For best reception, the normal to the plane of the wire loop is oriented parallel to the magnetic field. Then, as the wave sweeps by, the magnetic field penetrates the loop, and the changing magnetic flux induces a voltage and a current in the loop, in accord with Faraday's law. Once again, the resonant frequency of a capacitor/inductor combination can be adjusted to match the frequency of the desired electromagnetic wave. Both straight wire and loop antennas can be seen on the ship in Figure 24.7.

Cochlear implants use the broadcasting and receiving of radio waves to provide assistance to hearing-impaired people who have auditory nerves that are at least partially intact. These implants utilize radio waves to bypass the damaged part of the hearing mechanism and access the auditory nerve directly, as Figure 24.8 illustrates. An external microphone (often set into an ear mold) detects sound waves and sends a corresponding electrical signal to a speech processor small enough to be carried in a pocket. The speech processor encodes these signals into a radio wave, which is broadcast from an external transmitter coil placed over the site of a miniature receiver (and its receiving antenna) that has been surgically inserted beneath the skin. The receiver acts much like a radio. It detects the broadcasted wave and from the encoded audio information produces electrical signals that represent the sound wave. These signals are sent along a wire to electrodes that are implanted in the cochlea of the inner ear. The electrodes stimulate

Figure 24.7 This cruise ship uses both straight wire and loop antennas to communicate with other vessels and on-shore stations.

THE PHYSICS OF...

cochlear implants.

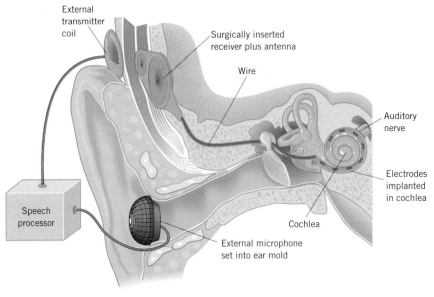

Figure 24.8 Hearing-impaired people can sometimes recover part of their hearing with the help of a cochlear implant. The broadcasting and receiving of electromagnetic waves lies at the heart of this device, as the text discusses.

the auditory nerves that feed directly between structures within the cochlea and the brain. To the extent that the nerves are intact, a person can learn to recognize sounds.

Radio waves are only one part of the broad spectrum of electromagnetic waves that has been discovered. The next section discusses the entire spectrum.

24.2 THE ELECTROMAGNETIC SPECTRUM

An electromagnetic wave, like any periodic wave, has a frequency f and a wavelength λ that are related to the speed v of the wave by $v = f\lambda$ (Equation 16.1). For electromagnetic waves traveling through a vacuum or, to a good approximation, through air, the speed is $v = c$, so $c = f\lambda$.

As Figure 24.9 shows, electromagnetic waves exist with an enormous range of frequencies, from values less than 10^4 Hz to greater than 10^{22} Hz. Since all these waves travel through a vacuum at the same speed of $c = 3.00 \times 10^8$ m/s, Equation 16.1 can be used to find the correspondingly wide range of wavelengths that the picture also displays. The ordered series of electromagnetic wave frequencies or wavelengths in Figure 24.9 is called the ***electromagnetic spectrum.*** Historically, regions of the spectrum have been given names such as radio waves and infrared waves. Although the boundary between adjacent regions is shown as a sharp line in the drawing, the boundary is not so well defined in practice, and the regions often overlap.

Beginning on the left in Figure 24.9, we find radio waves. Lower-frequency radio waves are generally produced by electric oscillator circuits, while higher-frequency radio waves (called microwaves) are usually generated using electron tubes called klystrons. Infrared radiation, sometimes loosely called heat waves, originates with the vibration and rotation of molecules within a material. Visible light is emitted by hot objects, such as the sun, a burning log, or the filament of an incandescent light bulb, when the temperature is high enough to excite the electrons within an atom. Ultraviolet frequencies can be produced from the discharge of an electric arc. X-rays are produced by the sudden deceleration of high-speed electrons. And, finally, gamma rays are radiation from nuclear decay.

Two views of our galaxy (the Milky Way) as observed in two different regions of the electromagnetic spectrum: (*a*) the visible region and (*b*) the infrared region. (The infrared region is invisible to the eye, so the colors in this picture have been computer generated.)

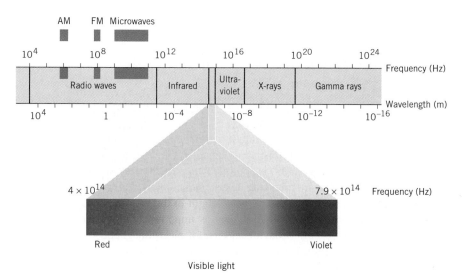

Figure 24.9 The electromagnetic spectrum.

a pyroelectric ear thermometer.

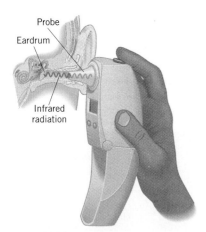

Probe

Eardrum

Infrared
radiation

Figure 24.10 A pyroelectric thermometer measures body temperature by determining the amount of infrared radiation emitted by the eardrum and surrounding tissue.

The human body, like any object, radiates infrared radiation, and the amount emitted depends on the temperature of the body. Although infrared radiation cannot be seen by the human eye, it can be detected by sensors. An ear thermometer, like the pyroelectric thermometer shown in Figure 24.10, determines the body's temperature by measuring the amount of infrared radiation that emanates from the ear drum and surrounding tissue. The ear is one of the best places to measure body temperature because it is close to the hypothalamus, an area at the bottom of the brain that controls body temperature. The ear is also not cooled or warmed by eating, drinking, or breathing. When the probe of the thermometer is inserted into the ear canal, infrared radiation travels down the barrel of the probe and strikes the sensor. The absorption of infrared radiation warms the sensor, and, as a result, its electrical conductivity changes. The change in electrical conductivity is measured by an electronic circuit. The output from the circuit is sent to a microprocessor, which calculates the body temperature and displays the result on a digital readout.

Of all the frequency ranges in the electromagnetic spectrum, the most familiar is that of visible light, although it is the smallest in Figure 24.9. Only waves with frequencies between about 4.0×10^{14} Hz and 7.9×10^{14} Hz are perceived by the human eye as visible light. Usually visible light is discussed in terms of wavelengths (in vacuum) rather than frequencies. As Example 1 indicates, the wavelengths of visible light are extremely small and, therefore, are normally expressed in *nanometers* (nm); 1 nm = 10^{-9} m. An obsolete (non-SI) unit occasionally used for wavelengths is the *angstrom* (Å); 1 Å = 10^{-10} m.

EXAMPLE 1   The Wavelengths of Visible Light

Find the range in wavelengths (in vacuum) for visible light in the frequency range between 4.0×10^{14} Hz (red light) and 7.9×10^{14} Hz (violet light). Express the answers in nanometers.

Reasoning According to Equation 16.1, the wavelength (in vacuum) λ of a light wave is equal to the speed of light c in a vacuum divided by the frequency f of the wave, $\lambda = c/f$.

Solution The wavelength corresponding to a frequency of 4.0×10^{14} Hz is

$$\lambda = \frac{c}{f} = \frac{3.00 \times 10^8 \text{ m/s}}{4.0 \times 10^{14} \text{ Hz}} = 7.5 \times 10^{-7} \text{ m}$$

Since 1 nm = 10^{-9} m, it follows that $\lambda = (7.5 \times 10^{-7}$ m$)[(1$ nm$)/(10^{-9}$ m$)] = \boxed{750 \text{ nm}}$.

The calculation for a frequency of 7.9×10^{14} Hz is similar:

$$\lambda = \frac{c}{f} = \frac{3.00 \times 10^8 \text{ m/s}}{7.9 \times 10^{14} \text{ Hz}} = 3.8 \times 10^{-7} \text{ m} \quad \text{or} \quad \boxed{\lambda = 380 \text{ nm}}$$

The eye/brain recognizes light of different wavelengths as different colors. A wavelength of 750 nm (in vacuum) is approximately the longest wavelength of red light, whereas 380 nm (in vacuum) is approximately the shortest wavelength of violet light. Between these limits are found the other familiar colors, as Figure 24.9 indicates.

The association between color and wavelength in the visible part of the electromagnetic spectrum is well known. But, just as important, the wavelength plays a central role in governing the behavior and use of electromagnetic waves in all regions of the spectrum. For instance, Conceptual Example 2 considers the influence of the wavelength on diffraction.

CONCEPTUAL EXAMPLE 2   The Diffraction of AM and FM Radio Waves

As we have seen in Section 17.3, diffraction is the ability of a wave to bend around an obstacle or the edges of an opening. Based on the discussion in that chapter, would you expect AM or FM radio waves to bend more readily around an obstacle such as a building?

Reasoning and Solution Section 17.3 points out that, other things being equal, sound waves exhibit diffraction to a greater extent when the wavelength is longer than when it is shorter. Based on this information, we expect that longer-wavelength electromagnetic waves will bend more readily around obstacles than will shorter-wavelength waves. Figure 24.9 shows that AM radio waves have considerably longer wavelengths than FM waves. Therefore, *AM waves have a greater ability to bend around buildings than do FM waves.* The inability of FM waves to diffract around obstacles is a major reason why FM stations broadcast their signals essentially in a "line-of-sight" fashion.

The picture of light as a wave is supported by experiments that will be discussed in Chapter 27. However, there are also experiments indicating that light can behave as if it were composed of discrete particles rather than waves. These experiments will be discussed in Chapter 29. Wave theories and particle theories of light have been around for hundreds of years, and it is now widely accepted that light, as well as other electromagnetic radiation, exhibits a dual nature. Either wave-like or particle-like behavior can be observed, depending on the kind of experiment being performed.

24.3 THE SPEED OF LIGHT

At a speed of 3.00×10^8 m/s, light travels from the earth to the moon in a little over a second, so the time required for light to travel between two places on earth is very short. Therefore, the earliest attempts at measuring the speed of light were met with limited success. One of the first accurate measurements employed a rotating mirror, and Figure 24.11 shows a simplified version of the setup. It was used first by the French scientist Jean Foucault (1819–1868) and later in a more refined version by the American physicist Albert Michelson (1852–1931). If the angular speed of the rotating eight-sided mirror in Figure 24.11 is adjusted correctly, light reflected from one side travels to the fixed mirror, reflects, and can be detected after reflecting from another side that has rotated into place at just the right time. The minimum angular speed must be such that one side of the mirror rotates one-eighth of a revolution during the time it takes for the light to make the round trip between the mirrors. For one of his experiments, Michelson placed mirrors on Mt. San Antonio and Mt. Wilson in California, a distance of 35 km apart. From a value of the minimum angular speed in such experiments, he obtained the value of $c = (2.997\ 96 \pm 0.000\ 04) \times 10^8$ m/s in 1926.

Today, the speed of light has been determined with such high accuracy that it is used to define the meter. As discussed in Section 1.2, the speed of light is now *defined* to be

Speed of light in a vacuum

$$c = 299\ 792\ 458 \text{ m/s}$$

(although a value of 3.00×10^8 m/s is adequate for most calculations). The second is defined in terms of a cesium clock, and the meter is then defined as the distance light travels in a vacuum during a time of $1/(299\ 792\ 458)$ s. Although the speed of

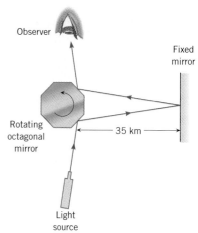

Figure 24.11 Between 1878 and 1931 Michelson used a rotating eight-sided mirror to measure the speed of light. This is a simplified version of the setup.

(a)

(b)

Figure 24.12 A view of the sky (a) before and (b) a few hours after the 1987 supernova.

light in a vacuum is large, it is finite, so it takes a finite amount of time for light to travel from one place to another. The travel time is especially long for light traveling between astronomical objects, as Conceptual Example 3 discusses.

CONCEPTUAL EXAMPLE 3   Looking Back in Time

A supernova is a violent explosion that occurs at the death of certain stars. For a few days after the explosion, the intensity of the emitted light can become a billion times greater than that of our own sun. But after several years, the intensity usually returns to zero. Supernovae are relatively rare events in the universe, for only six have been observed in our galaxy within the past 400 years. One of them was recorded in 1987. It occurred in a neighboring galaxy, approximately 1.66×10^{21} m away. Figure 24.12 shows a photograph of the sky (a) before the explosion, and (b) a few hours after the explosion. Why do astronomers say that viewing an event like the supernova is like "looking back in time?"

Reasoning and Solution The light from the supernova traveled to earth at a speed of $c = 3.00 \times 10^8$ m/s. Even at such a speed, the travel time is large because the distance $d = 1.66 \times 10^{21}$ m is enormous. The time t is $t = d/c = (1.66 \times 10^{21}$ m$)/(3.00 \times 10^8$ m/s$) = 5.53 \times 10^{12}$ s. This corresponds to 175 000 years. So *when astronomers saw the explosion in 1987, they were actually seeing the light that left the supernova 175 000 years earlier;* in other words, they were "looking back in time." In fact, whenever we view any celestial object, such as a star, we are seeing it as it was a long time ago. The farther the object is from earth, the longer it takes for the light to reach us, and the further back in time we are looking.

Related Homework Material: *Problem 16*

In 1865 Maxwell determined theoretically that electromagnetic waves propagate through a vacuum at a speed given by

$$c = \frac{1}{\sqrt{\epsilon_0 \mu_0}} \tag{24.1}$$

where $\epsilon_0 = 8.85 \times 10^{-12}$ C^2/(N·m^2) is the (electric) permittivity of free space and $\mu_0 = 4\pi \times 10^{-7}$ T·m/A is the (magnetic) permeability of free space. Originally ϵ_0 was introduced in Section 18.5 as an alternative way of writing the proportionality constant k in Coulomb's law [$k = 1/(4\pi\epsilon_0)$] and, hence, plays a basic role in determining the strengths of the electric fields created by point charges. The role of μ_0 is similar for magnetic fields; it was introduced in Section 21.7 as part of a proportionality constant in the expression for the magnetic field created by the current in a long straight wire. Substituting the values for ϵ_0 and μ_0 into Equation 24.1 shows that

$$c = \frac{1}{\sqrt{[8.85 \times 10^{-12} \text{ C}^2/(\text{N·m}^2)][4\pi \times 10^{-7} \text{ T·m/A}]}} = 3.00 \times 10^8 \text{ m/s}$$

The experimental and theoretical values for c agree. Maxwell's success in predicting c provided a basis for inferring that light behaves as a wave consisting of oscillating electric and magnetic fields.

24.4 THE ENERGY CARRIED BY ELECTROMAGNETIC WAVES

THE PHYSICS OF...

a microwave oven.

Electromagnetic waves, like water waves or sound waves, carry energy. The energy is carried by the electric and magnetic fields that comprise the wave. In a microwave oven, for example, microwaves penetrate food and deliver their energy to it, as Figure

24.13 illustrates. The electric field of the microwaves is largely responsible for delivering the energy, and water molecules in the food absorb it. The absorption occurs because each water molecule has a permanent dipole moment; that is, one end of a molecule has a slight positive charge, and the other end a negative charge of equal magnitude. As a result, the positive and negative ends of different molecules can form a bond. However, the electric field of the microwaves exerts forces on the positive and negative ends of a molecule, causing it to spin. Because the field is oscillating rapidly—about 2.4×10^9 times a second—the water molecules are kept spinning at a high rate. In the process, the energy of the microwaves is used to break bonds between neighboring water molecules and ultimately is converted into internal energy. As the internal energy increases, the temperature of the water increases, and the food cooks.

A measure of the energy stored in the electric field **E** of an electromagnetic wave, such as a microwave, is provided by the electric energy density. As we saw in Section 19.5, this density is the electric energy per unit volume of space in which the electric field exists:

$$\frac{\text{Electric energy}}{\text{density}} = \frac{\text{Electric energy}}{\text{Volume}} = \frac{1}{2}\kappa\epsilon_0 E^2 = \frac{1}{2}\epsilon_0 E^2 \quad (19.12)$$

In this equation, the dielectric constant κ has been set equal to unity, since we are dealing with an electric field in a vacuum (or air). From Section 22.8, the analogous expression for the magnetic energy density is

$$\frac{\text{Magnetic energy}}{\text{density}} = \frac{\text{Magnetic energy}}{\text{Volume}} = \frac{1}{2\mu_0}B^2 \quad (22.11)$$

The *total energy density* u of an electromagnetic wave in a vacuum is the sum of these two energy densities:

$$u = \frac{\text{Total energy}}{\text{Volume}} = \frac{1}{2}\epsilon_0 E^2 + \frac{1}{2\mu_0}B^2 \quad (24.2)$$

In an electromagnetic wave propagating through a vacuum or air, the electric field and the magnetic field carry equal amounts of energy per unit volume of space. Since $\frac{1}{2}\epsilon_0 E^2 = \frac{1}{2}(B^2/\mu_0)$, it is possible to rewrite Equation 24.2 for the total energy density in two additional, but equivalent, forms:

$$u = \frac{1}{2}\epsilon_0 E^2 + \frac{1}{2\mu_0}B^2 \quad (24.2a)$$

$$u = \epsilon_0 E^2 \quad (24.2b)$$

$$u = \frac{1}{\mu_0}B^2 \quad (24.2c)$$

The fact that the two energy densities are equal implies that the electric and magnetic fields are related. To see how, we set the electric energy density equal to the magnetic energy density and obtain

$$\frac{1}{2}\epsilon_0 E^2 = \frac{1}{2\mu_0}B^2 \quad \text{or} \quad E^2 = \frac{1}{\epsilon_0\mu_0}B^2$$

But according to Equation 24.1, $c = 1/\sqrt{\epsilon_0\mu_0}$, so it follows that $E^2 = c^2 B^2$. Taking the square root of both sides of this result shows that the relation between the electric and magnetic fields in an electromagnetic wave is

$$E = cB \quad (24.3)$$

Figure 24.13 A microwave oven. The rotating fan blades reflect the microwaves to all parts of the oven.

In an electromagnetic wave, the electric and magnetic fields fluctuate sinus-oidally in time, so Equations 24.2 give the energy density of the wave at any instant in time. If an average value $\bar{u}$ for the total energy density is desired, average values are needed for E^2 and B^2. In Section 20.5 we faced a similar situation for alternating currents and voltages and introduced rms quantities. Using an analogous procedure here, we find that the rms values for the electric and magnetic fields, E_{rms} and B_{rms}, are related to the maximum values of these fields, E_0 and B_0, by

$$E_{rms} = \frac{1}{\sqrt{2}} E_0 \quad \text{and} \quad B_{rms} = \frac{1}{\sqrt{2}} B_0$$

Equations 24.2a–c can now be interpreted as giving the average energy density $\bar{u}$, provided the symbols E and B are interpreted to mean the rms values given above. The average density of the sunlight reaching the earth is determined in the next example.

EXAMPLE 4   The Average Energy Density of Sunlight

Sunlight enters the top of the earth's atmosphere with an electric field whose rms value is $E_{rms} = 720$ N/C. Find (a) the average total energy density of this electro-magnetic wave and (b) the rms value of the sunlight's magnetic field.

Reasoning The average total density $\bar{u}$ of the sunlight can be obtained from Equation 24.2b, provided the rms value is used for the electric field. Since the magnitudes of the magnetic and electric fields are related according to Equation 24.3, the rms value of the magnetic field is $B_{rms} = E_{rms}/c$.

Solution

(a) According to Equation 24.2b, the average total energy density is

$$\bar{u} = \epsilon_0 E_{rms}^2 = [8.85 \times 10^{-12} \text{ C}^2/(\text{N} \cdot \text{m}^2)](720 \text{ N/C})^2 = \boxed{4.6 \times 10^{-6} \text{ J/m}^3}$$

(b) Using Equation 24.3, we find that the rms magnetic field is

$$B_{rms} = \frac{E_{rms}}{c} = \frac{720 \text{ N/C}}{3.0 \times 10^8 \text{ m/s}} = \boxed{2.4 \times 10^{-6} \text{ T}}$$

● *Concepts at a Glance*   As an electromagnetic wave moves through space, it carries energy from one region to another. This energy transport is characterized by the *intensity* of the wave. We have encountered the concept of intensity before, in connection with sound waves in Section 16.7. The Concepts-at-a-Glance chart in Figure 16.22 illustrates that the intensity of a sound wave is formulated by bringing together the power of the wave and the area through which it passes. The intensity of an electromagnetic wave is defined in an analogous fashion, as the Concepts-at-a-Glance chart in Figure 24.14 illustrates. For a sound wave, the intensity is the sound power that passes perpendicularly through a surface divided by the area of the surface. For an electromagnetic wave, the intensity is the electromagnetic power per unit area. ●

Using this definition of intensity, we can show that the electromagnetic intensity S is related to the energy density u. According to Equation 16.8 the intensity is the power P that passes perpendicularly through a surface divided by the area A of that surface, or $S = P/A$. Furthermore, the power is equal to the total energy passing through the surface divided by the elapsed time t (Equation 6.10b), so that $P = $ (Total energy)/t. Combining these two relations gives

$$S = \frac{P}{A} = \frac{\text{Total energy}}{tA}$$

Concepts at a Glance

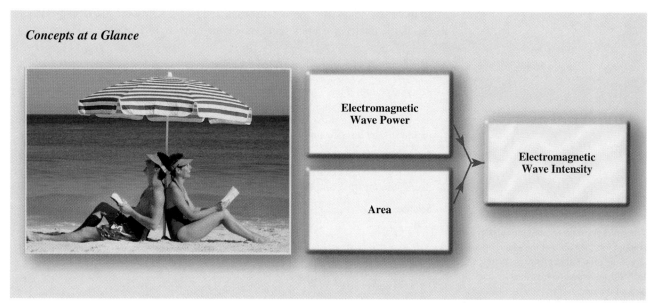

Now, consider Figure 24.15, which shows an electromagnetic wave traveling in a vacuum along the x axis. In a time t the wave travels the distance ct, passing through the surface of area A. Consequently, the volume of space through which the wave passes is ctA. The total (electric and magnetic) energy in this volume is

Total energy = (Total energy density) × Volume = $u(ctA)$

Using this result in the expression for the intensity, we obtain

$$S = \frac{\text{Total energy}}{tA} = \frac{uctA}{tA} = cu \qquad (24.4)$$

Thus, the intensity and the energy density are related by the speed of light, c. Substituting Equations 24.2a–c, one at a time, into Equation 24.4 shows that the intensity of an electromagnetic wave depends on the electric and magnetic fields according to the following equivalent relations:

$$S = cu = \frac{1}{2}\,c\epsilon_0 E^2 + \frac{c}{2\mu_0}\,B^2 \qquad (24.5a)$$

$$S = c\epsilon_0 E^2 \qquad (24.5b)$$

$$S = \frac{c}{\mu_0}\,B^2 \qquad (24.5c)$$

Figure 24.14 *Concepts at a Glance* The intensity of an electromagnetic wave is formulated just as it is for a sound wave, by bringing together the power of the wave and the area through which the power passes. The ocean lovers in this photograph are using an umbrella to protect themselves from the intensity of the sun's electromagnetic waves.

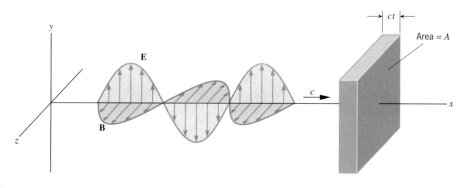

Figure 24.15 In a time t, an electromagnetic wave moves a distance ct along the x axis and passes through a surface of area A.

If the rms values for the electric and magnetic fields are used in Equations 24.5a–c, the intensity becomes an average intensity, $\overline{S}$, as Example 5 illustrates.

EXAMPLE 5   A Neodymium-Glass Laser

A neodymium-glass laser emits short pulses of high-intensity electromagnetic waves. The electric field has an rms value of $E_{\text{rms}} = 2.0 \times 10^9$ N/C. Find the average power of each pulse that passes through a 1.6×10^{-5}-m^2 surface that is perpendicular to the laser beam.

Reasoning Since the intensity of a wave is the power per unit area that passes perpendicularly through a surface, the average power $\overline{P}$ of the wave is the product of the average intensity $\overline{S}$ and the area A.

Solution The average power is $\overline{P} = \overline{S}A$. But from Equation 24.5b, $\overline{S} = c\epsilon_0 E_{\text{rms}}^2$ so that

$$\overline{P} = c\epsilon_0 E_{\text{rms}}^2 A$$
$$\overline{P} = (3.0 \times 10^8 \text{ m/s})[8.85 \times 10^{-12} \text{ C}^2/(\text{N}\cdot\text{m}^2)](2.0 \times 10^9 \text{ N/C})^2(1.6 \times 10^{-5} \text{ m}^2)$$
$$= \boxed{1.7 \times 10^{11} \text{ W}}$$

● **PROBLEM SOLVING INSIGHT**
The concepts of power and intensity are similar, but they are not the same. Intensity is the power that passes perpendicularly through a surface divided by the area of the surface.

24.5 THE DOPPLER EFFECT AND ELECTROMAGNETIC WAVES

Section 16.9 presents a discussion of the Doppler effect that sound waves exhibit when either the source of a sound wave, the observer of the wave, or both are moving with respect to the medium of propagation (e.g., air). This effect is one in which the observed sound frequency is greater or smaller than the frequency emitted by the source. A different Doppler effect arises when the source moves than when the observer moves.

Electromagnetic waves also can exhibit a Doppler effect, but it differs from that for sound waves for two reasons. First, sound waves require a medium such as air in which to propagate. In the Doppler effect for sound, it is the motion (of the source, the observer, and the waves themselves) relative to this medium that is important. In the Doppler effect for electromagnetic waves, motion relative to a medium plays no role, because the waves do not require a medium in which to propagate. They can travel in a vacuum. Second, in the equations for the Doppler effect in Section 16.9, the speed of sound plays an important role, and it depends on the reference frame relative to which it is measured. For example, the speed of sound with respect to moving air is different than it is with respect to stationary air. As we will see in Section 28.2, electromagnetic waves behave in a different way. The speed at which they travel has the same value, whether it is measured relative to a stationary observer or relative to one moving at a constant velocity. For these two reasons, the same Doppler effect arises for electromagnetic waves when either the source or the observer of the waves moves; only the relative motion of the source and the observer with respect to one another is important.

When electromagnetic waves and the source and the observer of the waves all travel along the same line in a vacuum (or in air, to a good degree of approximation), the single equation that specifies the Doppler effect is

$$f_{\text{o}} = f_{\text{s}} \left(1 \pm \frac{v_{\text{rel}}}{c} \right) \qquad \text{if } v_{\text{rel}} \ll c \qquad (24.6)$$

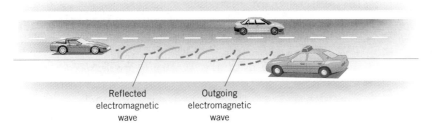

Reflected
electromagnetic
wave

Outgoing
electromagnetic
wave

Figure 24.16 The radar gun used by police emits an electromagnetic wave in the radio frequency region of the spectrum. The Doppler effect exhibited by the wave after it reflects from a moving vehicle is used to determine the vehicle's speed (see Example 6).

In this expression, f_o is the observed frequency, while f_s is the frequency emitted by the source. The symbol v_{rel} stands for the speed of the source and the observer relative to one another, and c is the speed of light in a vacuum. Equation 24.6 applies only if $v_{rel} \ll c$, that is, only if v_{rel} is very small compared to c. The plus sign in this equation applies when the source and observer are moving toward one another, while the minus sign applies when they are moving apart.

It is essential to realize that v_{rel} is the *relative* speed of the source and the observer. Thus, if the source is moving due east at a speed of 28 m/s with respect to the earth, while the observer is moving due east at a speed of 22 m/s, the value for v_{rel} is 28 m/s − 22 m/s = 6 m/s. Because v_{rel} is the relative *speed,* it has no algebraic sign. The direction of the relative motion is taken into account by choosing the plus or minus sign in Equation 24.6. Example 6 illustrates one familiar use of the Doppler effect for electromagnetic waves.

EXAMPLE 6   Radar Guns and Speed Traps

Police use radar guns and the Doppler effect to catch speeders. One such gun emits an electromagnetic wave with a frequency of $f_s = 8.0 \times 10^9$ Hz, as Figure 24.16 illustrates. In this picture, a car is approaching a police car parked on the side of the road. The direction of approach is essentially head-on. The wave from the radar gun reflects from the speeding car and returns to the police car, where on-board equipment measures its frequency to be greater than that of the emitted wave by 2100 Hz. Find the speed of the car with respect to the highway.

Reasoning The Doppler effect depends on the relative speed v_{rel} between the speeding car and the police car. We will find this speed and then relate it to the speed of the moving car with respect to the highway by using the fact that the police car is at rest. There are two Doppler frequency changes in this situation. First, the speeder's car observes the wave frequency coming from the radar gun to have a frequency f_o that is different from the emitted frequency f_s. According to Equation 24.6 (with the plus sign, since the two cars are coming together), $f_o - f_s = f_s(v_{rel}/c)$. Then, the wave reflects and returns to the police car, where it is observed to have a frequency f_o' that is different than its frequency f_o at the instant of reflection. Again using Equation 24.6, we find that $f_o' - f_o = f_o(v_{rel}/c)$. Adding the two previous equations gives the following result for the total Doppler change in frequency:

$$(f_o' - f_o) + (f_o - f_s) = f_o' - f_s = f_o\left(\frac{v_{rel}}{c}\right) + f_s\left(\frac{v_{rel}}{c}\right) \approx 2f_s\left(\frac{v_{rel}}{c}\right)$$

where we have assumed that f_o and f_s differ by only a negligibly small amount, since v_{rel} is small compared to the speed of light c. Solving for the relative speed v_{rel} gives

$$v_{rel} \approx \left(\frac{f_o' - f_s}{2f_s}\right)c$$

THE PHYSICS OF...

radar speed traps.

● PROBLEM SOLVING INSIGHT
The Doppler effect for electromagnetic waves depends on the speed v_{rel} of the observer and the source of the waves relative to one another. In general, do not use the speed of the observer or of the source with respect to the ground in Equation 24.6.

Solution The speed of the moving car relative to the police car is

$$v_{\text{rel}} \approx \left(\frac{f_\text{o}' - f_\text{s}}{2f_\text{s}}\right) c = \left[\frac{2100 \text{ Hz}}{2(8.0 \times 10^9 \text{ Hz})}\right] (3.0 \times 10^8 \text{ m/s}) = 39 \text{ m/s}$$

Since the police car is at rest, its speed is $v_\text{P} = 0$ m/s, and the speeder has a speed of $v_{\text{rel}} = v - v_\text{P} = v = \boxed{39 \text{ m/s}}$.

The Doppler effect of electromagnetic waves provides a powerful tool for astronomers. For instance, Example 10 in Chapter 5 discusses how astronomers have identified a supermassive black hole at the center of galaxy M87 by using the Hubble space telescope. They focused the telescope on regions to either side of the center of the galaxy (see Figure 5.15). From the light emitted by these two regions, they were able to use the Doppler effect to determine that one side is moving away from the earth, while the other side is moving toward the earth. In other words, the galaxy is rotating. The speeds of recession and approach enabled astronomers to determine the rotational speed of the galaxy, and Example 10 in Chapter 5 shows how the value for this speed leads to the identification of the black hole. Astronomers routinely study the Doppler effect of the light that reaches the earth from distant parts of the universe. From such studies, they have determined the speeds at which distant light emitting objects are receding from the earth.

24.6 POLARIZATION

Polarized Electromagnetic Waves

Direction of rope vibrations

Direction of wave travel

(a)

(b)

Figure 24.17 A transverse wave is linearly polarized when its vibrations always occur along one direction. (a) A linearly polarized wave on a rope can pass through a slit that is parallel to the direction of the rope vibrations, but (b) cannot pass through a slit that is perpendicular to the vibrations.

One of the essential features of electromagnetic waves is that they are transverse waves, and because of this feature they can be polarized. Figure 24.17 illustrates the idea of polarization by showing a transverse wave as it travels along a rope toward a slit. The wave is said to be *linearly polarized,* which means that its vibrations always occur along one direction. This direction is called the direction of polarization. In part *a* of the picture, the direction of polarization is vertical, parallel to the slit. Consequently, the wave passes through easily. However, when the slit is turned perpendicular to the direction of polarization, as in part *b*, the wave cannot pass, because the slit prevents the rope from oscillating. For longitudinal waves, such as sound waves, the notion of polarization has no meaning. In a longitudinal wave the direction of vibration is along the direction of travel, and the orientation of the slit would have no effect on the wave.

In an electromagnetic wave such as that in Figure 24.4, the electric field oscillates along the *y* axis. Similarly, the magnetic field oscillates along the *z* axis. Therefore, the wave is linearly polarized, with the direction of polarization taken arbitrarily to be that along which the electric field oscillates. If the wave is a radio wave generated by a straight-wire antenna, the direction of polarization is determined by the orientation of the antenna. In comparison, the electromagnetic waves given off as light by an incandescent light bulb are completely unpolarized. In this case the light waves are emitted by a large number of atoms in the hot filament of the bulb. When an electron in an atom oscillates, the atom behaves as a miniature antenna that broadcasts light for brief periods of time, about 10^{-8} seconds. However, the directions of these atomic antennas change randomly as a result of collisions. Unpolarized light, then, consists of many individual waves, emitted in short bursts by many "atomic antennas," each with its own direction of polarization. Figure 24.18 compares polarized and unpolarized light. In the unpolarized case, the arrows shown around the direction of wave travel symbolize the random directions of polarization of the individual waves that comprise the light.

Linearly polarized light can be produced from unpolarized light with the aid of certain materials. One commercially available material goes under the name of Polaroid. Such materials allow only the component of the electric field along one direction to pass through, while absorbing the field component perpendicular to this direction. As Figure 24.19 indicates, the direction of polarization that a polarizing material allows through is called the *transmission axis.* No matter how the transmission axis is oriented, the intensity of the transmitted polarized light is one-half that of the incident unpolarized light. The reason for this is that the unpolarized light contains all polarization directions to an equal extent. Moreover, the electric field for each direction can be resolved into components perpendicular and parallel to the transmission axis, with the result that the average components perpendicular and parallel to the transmission axis are equal. As a result, the polarizing material absorbs as much of the electric (and magnetic) field strength as it transmits.

Malus' Law

Once polarized light has been produced with a piece of polarizing material, it is possible to use a second piece to change the polarization direction and to adjust the intensity of the light. Figure 24.20 shows how. In this picture the first piece of polarizing material is called the *polarizer,* while the second piece is referred to as the *analyzer.* The transmission axis of the analyzer is oriented at an angle θ relative to the transmission axis of the polarizer. If the electric field strength of the polarized light incident on the analyzer is E, the field strength passing through is the component parallel to the transmission axis, or $E \cos \theta$. According to Equation 24.5b, the intensity is proportional to the square of the electric field strength. Consequently, the average intensity of polarized light passing through the analyzer is proportional to $\cos^2 \theta$. Thus, both the polarization direction and the intensity of the light can be adjusted by rotating the transmission axis of the analyzer relative to that of the po-

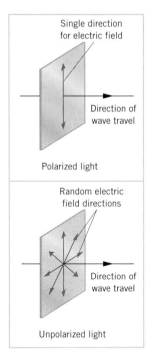

Figure 24.18 Polarized light consists of an electromagnetic wave in which the electric field fluctuates along a single direction. Unpolarized light consists of short bursts of electromagnetic waves emitted by many different atoms. The electric field directions of these bursts are perpendicular to the direction of wave travel but are distributed randomly about it.

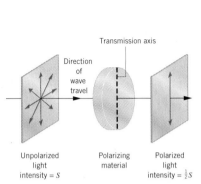

Figure 24.19 With the aid of a piece of polarizing material, polarized light may be produced from unpolarized light. The transmission axis of the material is the direction of polarization of the light that passes through the material.

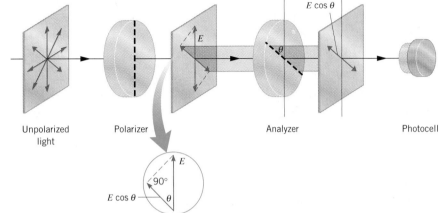

Figure 24.20 Two sheets of polarizing material, called the polarizer and the analyzer, may be used to adjust the polarization direction and intensity of the light reaching the photocell. This can be done by changing the angle θ between the transmission axes of the polarizer and analyzer.

larizer. The average intensity $\overline{S}$ of the light leaving the analyzer, then, is

Malus' law $$\overline{S} = \overline{S}_0 \cos^2 \theta \qquad (24.7)$$

where $\overline{S}_0$ is the average intensity of the light entering the analyzer. Equation 24.7 is sometimes called **Malus' law,** for it was discovered by the French engineer Etienne-Louis Malus (1775–1812). Example 7 illustrates the use of Malus' law.

EXAMPLE 7   Using Polarizers and Analyzers

What value of θ should be used in Figure 24.20, so the average intensity of the polarized light reaching the photocell is one-tenth the average intensity of the unpolarized light?

Reasoning Both the polarizer and the analyzer reduce the intensity of the light. The polarizer reduces the intensity by a factor of one-half, as discussed earlier. Therefore, if the average intensity of the unpolarized light is $\overline{I}$, the average intensity of the polarized light leaving the polarizer and striking the analyzer is $\overline{S}_0 = \overline{I}/2$. The angle θ must now be selected so the average intensity of the light leaving the analyzer is $\overline{S} = \overline{I}/10$. The solution follows directly from Malus' law.

Solution Using $\overline{S}_0 = \overline{I}/2$ and $\overline{S} = \overline{I}/10$ in Malus' law, we find that $\overline{I}/10 = (\overline{I}/2) \cos^2 \theta$. Solving this relation for $\cos \theta$ yields

$$\cos \theta = \sqrt{\frac{1}{5}} \quad \text{and} \quad \theta = \cos^{-1}\left(\frac{1}{\sqrt{5}}\right) = \boxed{63.4°}$$

When $\theta = 90°$ in Figure 24.20, the polarizer and analyzer are said to be **crossed,** and no light is transmitted by the polarizer/analyzer combination. As an illustration of this effect, Figure 24.21 shows two pairs of Polaroid sunglasses in uncrossed and

● **PROBLEM SOLVING INSIGHT**
Remember that when unpolarized light strikes a polarizer, only one half of the incident light is transmitted, the other half being absorbed by the polarizer.

Figure 24.21 When Polaroid sunglasses are uncrossed (top photograph), the transmitted light is dimmed due to the extra thickness of tinted plastic. However, when they are crossed (bottom photograph), the intensity of the transmitted light is reduced to zero because of the effects of polarization.

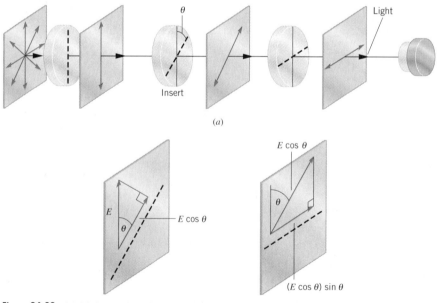

Figure 24.22 (a) Light reaches the photocell when a piece of polarizing material is inserted between a crossed polarizer and analyzer. (b) The component of the electric field parallel to the transmission axis of the insert is $E \cos \theta$. (c) Light incident on the analyzer has a component $(E \cos \theta) \sin \theta$ parallel to its transmission axis.

crossed configurations. Conceptual Example 8 illustrates an interesting result that occurs when a piece of polarizing material is inserted between a crossed polarizer and analyzer.

CONCEPTUAL EXAMPLE 8
How Can a Crossed Polarizer and Analyzer Transmit Light?

As explained earlier in the text, no light reaches the photocell when the polarizer and analyzer are crossed. Suppose now, that a third piece of polarizing material is inserted between the polarizer and analyzer, as in Figure 24.22a. Does light now reach the photocell?

Reasoning and Solution Surprisingly, the answer is yes, even though the polarizer and analyzer are crossed. To show why, we note that if any light is to pass through the analyzer, it must have an electric field component parallel to the transmission axis of the analyzer. Without the insert, there is no such component. But with the insert, there is, as parts *b* and *c* of the drawing illustrate. Part *b* shows that the electric field E of the light leaving the polarizer makes an angle θ with respect to the transmission axis of the insert. The component of the field that is parallel to the insert's transmission axis is $E \cos \theta$, and this component passes through the insert. Part *c* of the drawing illustrates that the field ($E \cos \theta$) incident on the analyzer has a component parallel to the transmission axis of the analyzer, namely ($E \cos \theta$) $\sin \theta$. This component passes through the analyzer, so that light now reaches the photocell.

Light will reach the photocell as long as the angle θ is between 0 and 90°. If the angle is 0 or 90°, however, no light reaches the photocell. Can you explain, without using any equations, why no light reaches the photocell under either of these conditions?

Related Homework Material: *Problems 38 and 40*

INTERACTIVE LEARNINGWARE #48

START
Visualization
Reasoning
Knowns and Unknowns
Modeling the Problem
Numerical Solution
FINISH

The drawing shows three sheets of polarizing material. The orientation of the transmission axis for each is labeled relative to the vertical direction. A beam of unpolarized light, whose intensity is 1550 W/m², is incident on the first sheet. What is the intensity of the beam transmitted through the three sheets when $\theta_1 = 30.0°$, $\theta_2 = 45.0°$, and $\theta_3 = 70.0°$?

Vertical
Transmitted beam
Incident beam
θ_1 θ_2 θ_3

This Interactive LearningWare problem illustrates two important features of problems that involve polarizers. First, it reviews the amount by which a polarizer reduces the intensity of unpolarized light. Second, it shows that the angle θ in Malus' law is the angle with respect to the polarization direction of the incident light, and not necessarily the angle with respect to the vertical.

Go to our Web site at

http://www.wiley.com/college/cutnell

THE PHYSICS OF...

a liquid crystal display. (See below.)

An application of a crossed polarizer/analyzer combination occurs in one kind of liquid crystal display (LCD). LCDs are widely used in pocket calculators and digital watches. The display usually consists of blackened numbers and letters set against a light gray background. As Figure 24.23 indicates, each number or letter is formed from a combination of liquid crystal segments that have been "turned on" and appear black. The liquid crystal part of an LCD segment consists of the liquid crystal material sandwiched between two transparent electrodes, as in Figure 24.24. When a voltage is applied between the electrodes, the liquid crystal is said to be "on." Part *a* of the picture shows that linearly polarized incident light passes through the "on" material without having its direction of polarization affected. When the voltage is removed, as in part *b*, the liquid crystal is said to be "off" and now rotates the direction of polarization by 90°. A complete LCD segment also includes a crossed polarizer/analyzer combination, as Figure 24.25 illustrates. The polarizer, analyzer, electrodes, and liquid crystal material are packaged as a single unit. The polarizer produces polarized light from incident unpolarized light. With the display segment turned on, as in Figure 24.25, the polarized light emerges from

Segments turned on

Figure 24.23 Liquid crystal displays use liquid crystal "segments" to form the numbers.

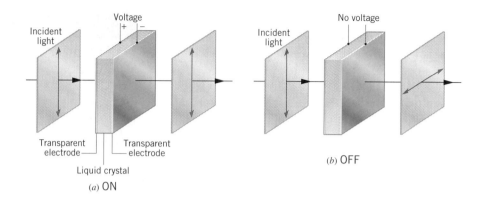

Figure 24.24 A liquid crystal in its (*a*) "on" state and (*b*) "off" state.

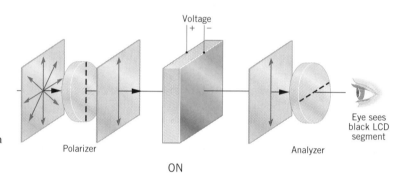

Figure 24.25 A liquid crystal display (LCD) incorporates a crossed polarizer/analyzer combination. When the LCD segment is turned on (voltage applied), no light is transmitted through the analyzer, and the observer sees a black segment.

Figure 24.26 A color LCD monitor is thin and lightweight.

the liquid crystal only to be absorbed by the analyzer, since the light is polarized perpendicular to the transmission axis of the analyzer. Since no light emerges from the analyzer, an observer sees a black segment against a light gray background, as in Figure 24.23. On the other hand, the segment is turned off when the voltage is removed, in which case the liquid crystal rotates the direction of polarization by 90° to coincide with the axis of the analyzer. The light now passes through the analyzer and enters the eye of the observer. However, the light coming from the segment has been designed to have the same color and shade (light gray) as the background of the display, so the segment becomes indistinguishable from the background.

Color LCD monitors are becoming very popular because they typically occupy only a third of the space of traditional monitors and weigh a quarter as much. An LCD monitor, such as that in Figure 24.26, uses thousands of LCD segments arranged like the squares on graph paper. To produce color, three segments are grouped together to form a tiny picture element (or "pixel"). Color filters are used to enable one segment in the pixel to produce red light, one to produce green, and one to produce blue. The eye blends the colors from each pixel into a composite color. By varying the intensity of the red, green, and blue colors, the pixel can generate an entire spectrum of colors.

The Occurrence of Polarized Light in Nature

THE PHYSICS OF...

Polaroid sunglasses.

Polaroid is a familiar material because of its widespread use in sunglasses. Such sunglasses are designed so that the transmission axis of the Polaroid is oriented vertically when the glasses are worn in the usual fashion. Thus, the glasses prevent any light that is polarized horizontally from reaching the eye. Light from the sun is unpolarized, but a considerable amount of horizontally polarized sunlight originates

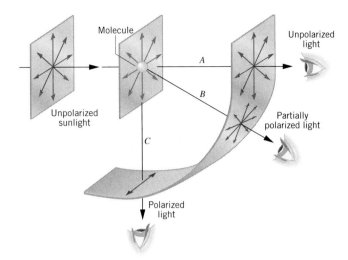

Figure 24.27 In the process of being scattered from atmospheric molecules, unpolarized light from the sun becomes partially polarized.

by reflection from horizontal surfaces such as that of a lake. Section 26.4 discusses this effect. Polaroid sunglasses reduce glare by preventing the horizontally polarized reflected light from reaching the eyes.

Polarized sunlight also originates from the scattering of light by molecules in the atmosphere. Figure 24.27 shows light being scattered by a single atmospheric molecule. The electric fields in the unpolarized sunlight cause the electrons in the molecule to vibrate perpendicular to the direction in which the light is traveling. The electrons, in turn, reradiate the electromagnetic waves in different directions, as the drawing illustrates. The light radiated straight ahead in direction A is unpolarized, just like the incident light. But light radiated perpendicular to the incident light in direction C is polarized. Light radiated in the intermediate direction B is partially polarized. There is experimental evidence that some bird species use such polarized light as a navigational aid.

24.7 CONCEPTS & CALCULATIONS

One of the central ideas of this chapter is that electromagnetic waves carry energy. Two concepts are used to describe this energy, the wave's intensity and its energy density. The next example reviews these important ideas.

CONCEPTS & CALCULATIONS EXAMPLE 9
Intensity and Energy Density of a Wave

Figure 24.28 shows the popular dish antenna that receives digital TV signals from a satellite. The average intensity of the electromagnetic wave that carries a particular TV program is $\overline{S} = 7.5 \times 10^{-14}$ W/m^2, and the circular aperture of the antenna has a radius of $r = 15$ cm. (a) Determine the electromagnetic energy delivered to the dish during a one-hour program. (b) What is the average energy density of the electromagnetic wave?

Concept Questions and Answers How is the average power passing through the circular aperture of the antenna related to the average intensity of the TV signal?

Answer According to Equation 16.8, the average intensity $\overline{S}$ of a wave is equal to the average power $\overline{P}$ that passes perpendicularly through a surface divided by the area A of the surface. Thus, the average power is $\overline{P} = \overline{S}A$, where $A = \pi r^2$.

Figure 24.28 A dish antenna picks up the TV signal from a satellite. The signal extends over the dish's entire aperture (radius r).

How much energy does the antenna receive in a time t?

Answer Since the average power is the energy per unit time according to Equation 6.10b, the energy received by the antenna is the product of the average power and the time, or Energy $= \bar{P}t$.

What is the average energy density, or average energy per unit volume, of the electromagnetic wave?

Answer The energy delivered to the dish is carried by the electromagnetic wave. The wave passes perpendicularly through an imaginary surface that has an area A that matches the circular area of the dish's aperture. In a time t, all the energy that passes through the surface is contained, therefore, in a cylinder of length x (see Figure 24.28). Since the energy is being carried by an electromagnetic wave, it travels at the speed of light c. Thus, the length of the cylinder is $x = ct$. The energy density u of the wave, or energy per unit volume, is the energy contained within the cylinder divided by its volume.

Solution

(a) According to Equation 6.10b, the energy received by the antenna during a time t is equal to the average power multiplied by the time, Energy $= \bar{P}t$. On the other hand, Equation 16.8 gives the average power as the product of the average intensity and the area, $\bar{P} = \bar{S}A$, so that

$$\text{Energy} = \underbrace{\bar{S}A}_{\substack{\text{Average} \\ \text{power}}}\, t = \bar{S}(\pi r^2)t$$

$$= (7.5 \times 10^{-14} \text{ W/m}^2)\pi(0.15 \text{ m})^2(3600 \text{ s}) = \boxed{1.9 \times 10^{-11} \text{ J}}$$

(b) The energy density is equal to the energy divided by the volume of the cylinder in Figure 24.28. The volume of the cylinder is equal to the cross-sectional area πr^2 times its length ct.

$$u = \frac{\text{Energy}}{\text{Volume}} = \frac{\text{Energy}}{(\pi r^2)(ct)}$$

$$= \frac{1.9 \times 10^{-11} \text{ J}}{\pi(0.15 \text{ m})^2(3.00 \times 10^8 \text{ m/s})(3600 \text{ s})} = \boxed{2.5 \times 10^{-22} \text{ J/m}^3}$$

This calculation is equivalent to using Equation 24.4, which gives the energy density as

$$u = \frac{\text{Energy}}{\text{Volume}} = \frac{\bar{S}}{c} = \frac{7.5 \times 10^{-14} \text{ W/m}^2}{3.00 \times 10^8 \text{ m/s}} = 2.5 \times 10^{-22} \text{ J/m}^3$$

We have seen how the intensities of completely polarized or completely unpolarized light beams can change as they pass through a polarizer. But what about light that is partially polarized and partially unpolarized? Can the concepts that we discussed in Section 24.6 be applied to such light? The answer is "yes," and Example 10 illustrates how.

CONCEPTS & CALCULATIONS EXAMPLE 10
Partially Polarized and Partially Unpolarized Light

The light beam in Figure 24.29 passes through a polarizer whose transmission axis makes an angle ϕ with the vertical. The beam is partially polarized and partially unpolarized, and the average intensity $\bar{S}_0$ of the incident light is the sum of the average intensity $\bar{S}_{0,\text{ polar}}$ of the polarized light and the average intensity $\bar{S}_{0,\text{ unpolar}}$ of the unpolarized light; $\bar{S}_0 = \bar{S}_{0,\text{ polar}} + \bar{S}_{0,\text{ unpolar}}$. The intensity $\bar{S}$ of the transmitted light is also the sum of two parts: $\overline{S} = \overline{S}_{polar} + \overline{S}_{unpolar}$. As the polarizer is rotated clockwise, the intensity of the transmitted light has a minimum value of $\overline{S} = 2.0 \text{ W/m}^2$ when $\phi = 20.0°$ and has a maximum value of $\overline{S} = 8.0 \text{ W/m}^2$ when the angle is $\phi = \phi_{max}$. (a) What is the intensity $\overline{S}_{0, \text{unpolar}}$ of the unpolarized incident light? (b) What is the intensity $\overline{S}_{0, \text{polar}}$ of the polarized incident light?

Figure 24.29 Light that is partially polarized and partially unpolarized is incident on a sheet of polarizing material.

Concept Questions and Answers How is $\overline{S}_{unpolar}$ related to $\overline{S}_{0, \text{unpolar}}$?

Answer When unpolarized light strikes a polarizer, one half is absorbed and the other half is transmitted, so that $\overline{S}_{unpolar} = \frac{1}{2}\overline{S}_{0, \text{unpolar}}$. This is true for any angle ϕ of the transmission axis.

How is $\overline{S}_{polar}$ related to $\overline{S}_{0, \text{polar}}$?

Answer When polarized light strikes a polarizer, the transmitted intensity is given by Malus' law (Equation 24.7), $\overline{S}_{polar} = \overline{S}_{0, \text{polar}} \cos^2 \theta$, where θ is the angle between the direction of polarization of the incident light and the transmission axis. Note that θ is not equal to ϕ because the direction of polarization of the incident beam is not given and may not be in the vertical direction.

The minimum transmitted intensity is 2.0 W/m². Why isn't it 0 W/m²?

Answer When the polarized part of the incident beam passes through the polarizer, the intensity of its transmitted portion changes as the angle ϕ changes, going from some maximum value to 0 W/m², in accord with Malus' law. However, when the unpolarized part of the beam passes through the polarizer, the intensity of its transmitted portion does not change as the angle ϕ changes. Thus, the minimum intensity of the transmitted beam is 2.0 W/m², rather than 0 W/m², because of the presence of the unpolarized light in the incident beam.

What is the angle ϕ_{max} at which the intensity of the light leaving the polarizing material is a maximum?

Answer When the polarized part of the beam passes through the polarizer, the intensity of its transmitted portion is zero when $\phi = 20.0°$. The intensity rises to a maximum when the polarizer is rotated by 90.0° relative to this position. Thus, the intensity is a maximum when $\phi = \phi_{max} = 20.0° + 90.0° = 110.0°$.

Solution

(a) The average intensity $\overline{S}$ of the transmitted light is (see Figure 24.29) $\overline{S} = \overline{S}_{polar} + \overline{S}_{unpolar}$. We are given that the minimum intensity is $\overline{S} = 2.0 \text{ W/m}^2$ when $\phi = 20.0°$, and we know that $\overline{S}_{polar} = 0 \text{ W/m}^2$ for this angle. Thus, $\overline{S}_{unpolar} = \overline{S} - \overline{S}_{polar} = 2.0 \text{ W/m}^2$. The intensity $\overline{S}_{0, \text{unpolar}}$ of the unpolarized incident light is twice this amount because the polarizer absorbs one-half the incident light:

$$\overline{S}_{0, \text{unpolar}} = 2\overline{S}_{unpolar} = 2(2.0 \text{ W/m}^2) = \boxed{4.0 \text{ W/m}^2}$$

(b) When $\phi = \phi_{max} = 110°$, the intensity of the transmitted beam is at its maximum value of $\overline{S} = 8.0 \text{ W/m}^2$. Writing the intensity of the transmitted beam as the sum of two parts, we have

$$\overline{S} = 8.0 \text{ W/m}^2 = \underbrace{\overline{S}_{0, \text{polar}} \cos^2 \theta}_{\overline{S}_{polar}} + \underbrace{2.0 \text{ W/m}^2}_{\overline{S}_{unpolar}}$$

The angle θ is the angle between the direction of polarization of the incident polarized light and the transmission axis of the polarizer. Since $\overline{S}_{polar}$ is a maximum, we know that $\theta = 0°$. Solving this equation for $\overline{S}_{0, \text{polar}}$ yields

$$\overline{S}_{0, \text{polar}} = 8.0 \text{ W/m}^2 - 2.0 \text{ W/m}^2 = \boxed{6.0 \text{ W/m}^2}$$

At the end of the problem set for this chapter, you will find homework problems that contain both conceptual and quantitative parts. These problems are grouped under the heading *Concepts & Calculations, Group Learning Problems*. They are designed for use by students working alone or in small learning groups. The conceptual part of each problem provides a convenient focus for group discussions.

SUMMARY

An **electromagnetic wave** in a vacuum consists of mutually perpendicular and oscillating electric and magnetic fields. The wave is a transverse wave, since the fields are perpendicular to the direction of wave propagation. All electromagnetic waves travel through a vacuum at the speed of light c ($c = 3.00 \times 10^8$ m/s).

The frequency f and wavelength λ of an electromagnetic wave in a vacuum are related to its speed through the relation $c = f\lambda$. The series of electromagnetic waves, arranged in order of their frequencies or wavelengths, is called the **electromagnetic spectrum.** The spectrum includes radio waves, infrared radiation, visible light, ultraviolet radiation, X-rays, and gamma rays. **Visible light** has frequencies between about 4.0×10^{14} Hz and 7.9×10^{14} Hz. The human eye and brain perceive different frequencies or wavelengths as different colors.

Maxwell showed that the speed of light in a vacuum is $c = 1/\sqrt{\epsilon_0\mu_0}$, where ϵ_0 is the (electric) permittivity of free space and μ_0 is the (magnetic) permeability of free space.

The **total energy density** u of an electromagnetic wave is the total energy per unit volume of the wave and, in a vacuum, is given by $u = \frac{1}{2}\epsilon_0 E^2 + B^2/(2\mu_0)$, where E and B are the magnitudes of the electric and magnetic fields. In a vacuum, E and B are related by $E = cB$, and the electric and magnetic parts of the total energy density are equal.

The **intensity** of an electromagnetic wave is the power that the wave carries perpendicularly through a surface divided by the area of the surface. In a vacuum, the intensity S is related to the total energy density u according to $S = cu$.

When electromagnetic waves and the source and observer of the waves all travel along the same line in a vacuum, the **Doppler effect** is given by $f_o = f_s(1 \pm v_{rel}/c)$, where f_o and f_s are, respectively, the observed and emitted wave frequencies. In this expression, v_{rel} is the relative speed of the source and observer and is assumed to be very small with respect to the speed of light c. The plus sign is used when the source and observer approach each other, and the minus sign when they move apart.

A **linearly polarized** electromagnetic wave is one in which all oscillations of the electric field occur along one direction, which is taken to be the direction of polarization. In **unpolarized light** the direction of polarization does not remain fixed, but fluctuates randomly in time.

Polarizing materials allow only the component of the wave's electric field along one direction to pass through them. The preferred transmission direction for the electric field is called the **transmission axis** of the material. When unpolarized light is incident on a piece of polarizing material, the transmitted polarized light has an intensity that is one-half that of the incident light. When two pieces of polarizing material are used one after the other, the first is called the polarizer, while the second is referred to as the analyzer. If the average intensity of polarized light falling on an analyzer is $\bar{S}_0$, the average intensity $\bar{S}$ of the light leaving the analyzer is given by **Malus' law** as $\bar{S} = \bar{S}_0 \cos^2 \theta$, where θ is the angle between the transmission axes of the polarizer and analyzer. When $\theta = 90°$, the polarizer and the analyzer are said to be "crossed," and no light passes through the analyzer.

CONCEPTUAL QUESTIONS

1. Which of the following concepts applies to both sound waves and electromagnetic waves: (a) intensity or (b) polarization? Account for your answer.

2. Refer to Figure 24.2. Between the times indicated in parts c and d in the drawing, what is the direction of the magnetic field at the point P for the electromagnetic wave being generated? Is it directed into or out of the plane of the paper? Justify your answer.

3. A transmitting antenna is located at the origin of an x, y, z axis system and broadcasts an electromagnetic wave whose electric field oscillates along the y axis. The wave travels along the $+x$ axis. Three possible wire loops can be used with an LC-tuned cir-

cuit to detect this wave: One loop lies in the xy plane, another in the xz plane, and the third in the yz plane. Which of the loops will detect the wave? Why?

4. Why does the peak value of the emf induced in a loop antenna (see Figure 24.6) depend on the frequency of the electromagnetic wave, while the peak value of the emf induced in a straight-wire antenna (see Figure 24.5) does not?

5. Suppose that the electric field of an electromagnetic wave decreases in magnitude. Does the magnetic field increase, decrease, or remain the same? Account for your answer.

6. An astronomer measures the Doppler change in frequency for the light reaching the earth from a distant star. From this measurement, can the astronomer tell whether the star is moving away from the earth or whether the earth is moving away from the star? Explain.

7. Is there any real difference between a polarizer and an analyzer? In other words, can a polarizer be used as an analyzer, and vice versa?

8. Malus' law applies to the setup in Figure 24.20, which shows the analyzer rotated through an angle θ and the polarizer held fixed. Does Malus' law apply when the analyzer is held fixed and the polarizer is rotated? Give your reasoning.

9. In Example 7, we saw that, when the angle between the polarizer and analyzer is 63.4°, the intensity of the transmitted light drops to one-tenth of that of the incident unpolarized light. What happens to the light intensity that is not transmitted?

10. Light is incident from the left on two pieces of polarizing material, 1 and 2. As part *a* of the drawing illustrates, the transmission axis of material 1 is along the vertical direction, while that of material 2 makes an angle of θ with respect to the vertical. In part *b* of the drawing the two polarizing materials are interchanged. (a) Assume that the incident light is unpolarized and determine whether the intensity of the transmitted light in part *a* is greater than, equal to, or less than that in part *b*. (b) Repeat part (a), assuming that the incident light is linearly polarized along the vertical direction. Justify your answers to both parts (a) and (b).

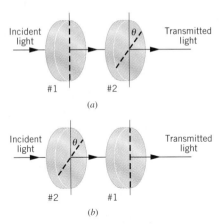

11. You are sitting upright on the beach near a lake on a sunny day, wearing Polaroid sunglasses. When you lie down on your side, facing the lake, the sunglasses don't work as well as they did while you were sitting upright. Why not?

P R O B L E M S

ssm Solution is in the Student Solutions Manual. **www** Solution is available on the World Wide Web at http://www.wiley.com/college/cutnell
⚕ This icon represents a biomedical application.

Section 24.1 The Nature of Electromagnetic Waves

1. ssm In astronomy, distances are often expressed in light-years. One light-year is the distance traveled by light in one year. The distance to Alpha Centauri, the closest star other than our own sun that can be seen by the naked eye, is 4.3 light-years. Express this distance in meters.

2. (a) Neil A. Armstrong was the first person to walk on the moon. The distance between the earth and the moon is 3.85×10^8 m. Find the time it took for his voice to reach earth via radio waves. (b) Determine the communication time for the first person who will some day walk on Mars, which is 5.6×10^{10} m from earth at the point of closest approach.

3. For an FM radio station broadcasting at a frequency of 88.0 MHz, the capacitance in Figure 24.5 must be adjusted to a value of 33.0×10^{-12} F. Assuming the inductance does not change, determine the value of the capacitance for an FM station broadcasting at 108.0 MHz.

4. An AM station is broadcasting a radio wave whose frequency is 1400 kHz. The value of the capacitance in Figure 24.5 is 8.4×10^{-11} F. What must be the value of the inductance in order that this station can be tuned in by the radio?

*5. ssm Equation 16.3, $y = A \sin (2\pi ft - 2\pi x/\lambda)$, gives the mathematical representation of a wave oscillating in the y direction and traveling in the positive x direction. Let y in this equation equal the electric field of an electromagnetic wave traveling in a vacuum. The maximum electric field is $A = 156$ N/C, and the frequency is $f = 1.50 \times 10^8$ Hz. Plot a graph of the electric field strength versus position, using for x the following values: 0, 0.50, 1.00, 1.50, and 2.00 m. Plot this graph for (a) a time $t = 0$ s and (b) a time t that is one-fourth of the wave's period.

**6. A flat coil of wire is used with an LC-tuned circuit as a receiving antenna. The coil has a radius of 0.25 m and consists of 450 turns. The transmitted radio wave has a frequency of 1.2 MHz. The magnetic field of the wave is parallel to the normal to the coil and has a maximum value of 2.0×10^{-13} T. Using Faraday's law of electromagnetic induction and the fact that the magnetic field changes from zero to its maximum value in one-quarter of a wave period, find the magnitude of the average emf induced in the antenna during this time.

Section 24.2 The Electromagnetic Spectrum

7. A truck driver is broadcasting at a frequency of 26.965 MHz with a CB (citizen's band) radio. Determine the wavelength of the electromagnetic wave being used.

8. Determine the range of wavelengths for FM radio waves with frequencies between 88.0 and 108.0 MHz.

9. ssm www Television sets sometimes use a "rabbit-ears" antenna. A rabbit-ears antenna consists of a pair of metal rods. The length of each rod can be adjusted to be one-quarter of a wavelength of an electromagnetic wave whose frequency is 60.0 MHz. How long is each rod?

10. ⚖ Magnetic resonance imaging or MRI (see Section 21.7) and positron emission tomography or PET scanning (see Section 32.6) are two medical diagnostic techniques. Both employ electromagnetic waves. For these waves, find the ratio of the MRI wavelength (frequency = 6.38×10^7 Hz) to the PET scanning wavelength (frequency = 1.23×10^{20} Hz).

11. ssm The human eye is most sensitive to light having a frequency of about 5.5×10^{14} Hz, which is in the yellow-green region of the electromagnetic spectrum. How many wavelengths of this light can fit across the width of your thumb, a distance of about 2.0 cm?

12. TV channel 3 (VHF) broadcasts at a frequency of 63.0 MHz. TV channel 23 (UHF) broadcasts at a frequency of 527 MHz. Find the ratio (VHF/UHF) of the wavelengths for these channels.

*13. Section 17.5 deals with transverse standing waves on a string. Electromagnetic waves also can form standing waves. In a standing wave pattern formed from microwaves, the distance between a node and an adjacent antinode is 0.50 cm. What is the microwave frequency?

Section 24.3 The Speed of Light

14. Ghost images are formed in a TV picture when the electromagnetic wave from the broadcasting antenna reflects from a building or other large object and arrives at the TV set shortly after the wave coming directly from the broadcasting antenna. If the reflected wave arrives 5.0×10^{-7} s after the direct wave, what is the difference in the distances traveled by the two waves?

15. ssm Two astronauts are 1.5 m apart in their spaceship. One speaks to the other. The conversation is transmitted to earth via electromagnetic waves. The time it takes for sound waves to travel at 343 m/s through the air between the astronauts equals the time it takes for the electromagnetic waves to travel to the earth. How far away from the earth is the spaceship?

16. Review Conceptual Example 3 before attempting this problem. The brightest star in the night sky is Sirius, which is at a distance of 8.3×10^{16} m. When we look at this star, how far back in time are we seeing it? Express your answer in years. (There are $365\frac{1}{4}$ days in one year.)

17. Figure 24.11 illustrates Michelson's setup for measuring the speed of light with the mirrors placed on Mt. San Antonio and Mt. Wilson in California, which are 35 km apart. Using a value of 3.00×10^8 m/s for the speed of light, find the minimum angular speed (in rev/s) for the rotating mirror.

18. A communications satellite is in a synchronous orbit that is 3.6×10^7 m directly above the equator. The satellite is located midway between Quito, Equador, and Belém, Brazil, two cities almost on the equator that are separated by a distance of 3.5×10^6 m. Find the time it takes for a telephone call to go by way of satellite between these cities. Ignore the curvature of the earth.

*19. ssm A mirror faces a cliff, located some distance away. Mounted on the cliff is a second mirror, directly opposite the first mirror and facing toward it. A gun is fired very close to the first mirror. The speed of sound is 343 m/s. How many times does the flash of the gunshot travel the round-trip distance between the mirrors before the echo of the gunshot is heard?

*20. A celebrity holds a press conference, which is televised live. A television viewer hears the sound picked up by a microphone directly in front of the celebrity. This viewer is seated 2.3 m from the television set. A reporter at the press conference is located 4.1 m from the microphone and hears the words directly *at the very same instant* that the television viewer hears them. Using a value of 343 m/s for the speed of sound, determine the maximum distance between the television viewer and the celebrity.

Section 24.4 The Energy Carried by Electromagnetic Waves

21. An industrial laser is used to burn a hole through a piece of metal. The average intensity of the light is $\bar{S} = 1.23 \times 10^9$ W/m². What is the rms value of (a) the electric field and (b) the magnetic field in the electromagnetic wave emitted by the laser?

22. Suppose the electric field in an electromagnetic wave has a maximum strength of 1470 N/C. What is the maximum strength of the magnetic field of the wave?

23. ssm The microwave radiation left over from the Big Bang explosion of the universe has an average energy density of 4×10^{-14} J/m³. What is the rms value of the electric field of this radiation?

24. The electromagnetic wave that delivers a cellular phone call to a car has a magnetic field with an rms value of 1.5×10^{-10} T. The wave passes perpendicularly through an open window, the area of which is 0.20 m². How much energy does this wave carry through the window during a 45-s phone call?

25. ssm A future space station in orbit about the earth is being powered by an electromagnetic beam from the earth. The beam has a cross-sectional area of 135 m² and transmits an average power of 1.20×10^4 W. What are the rms values of the (a) electric and (b) magnetic fields?

26. The average intensity of sunlight at the top of the earth's atmosphere is 1390 W/m². What is the maximum energy that a 25-m × 45-m solar panel could collect in one hour in this sunlight?

*27. The drawing shows an edge-on view of the solar panels on a communications satellite. The dashed line specifies the normal to the panels. Sunlight strikes the panels at an angle θ with respect to the normal. If the solar power impinging on the panels is 3200 W when $\theta = 55°$, what is it when $\theta = 35°$?

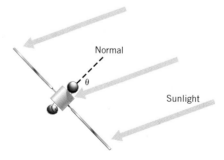

*28. ⚡ A heat lamp emits infrared radiation whose rms electric field is $E_{rms} = 2800$ N/C. (a) What is the average intensity of the radiation? (b) The radiation is focused on a person's leg over a circular area of radius 4.0 cm. What is the average power delivered to the leg? (c) The portion of the leg being radiated has a mass of 0.28 kg and a specific heat capacity of 3500 J/(kg·C°). How long does it take to raise its temperature by 2.0 C°? Assume that there is no other heat transfer into or out of the portion of the leg being heated.

*29. ssm www The power radiated by the sun is 3.9×10^{26} W. The earth orbits the sun in a nearly circular orbit of radius 1.5×10^{11} m. The earth's axis of rotation is tilted by 27° relative to the plane of the orbit (see the drawing), so sunlight does not strike the equator perpendicularly. What power strikes a 0.75-m² patch of flat land at the equator at point Q?

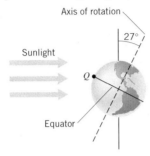

**30. The average intensity of sunlight reaching the earth is 1390 W/m². A charge of 2.6×10^{-8} C is placed in the path of this electromagnetic wave. (a) What is the magnitude of the maximum electric force that the charge experiences? (b) If the charge is moving at a speed of 3.7×10^4 m/s, what is the magnitude of the maximum magnetic force that the charge could experience?

Section 24.5 The Doppler Effect and Electromagnetic Waves

31. A distant galaxy emits light that has a wavelength of 500.7 nm. On earth, the wavelength of this light is measured to be 503.7 nm. (a) Decide whether this galaxy is approaching or receding from the earth. Give your reasoning. (b) Find the speed of the galaxy relative to the earth.

32. A speeder is pulling directly away and increasing his distance from a police car that is moving at 25 m/s with respect to the ground. The radar gun in the police car emits an electromagnetic wave with a frequency of 7.0×10^9 Hz. The wave reflects from the speeder's car and returns to the police car, where its frequency is measured to be 320 Hz less than the emitted frequency. Find the speeder's speed with respect to the ground.

*33. ssm A distant galaxy is simultaneously rotating and receding from the earth. As the drawing shows, the galactic center is receding from the earth at a relative speed of $u_G = 1.6 \times 10^6$ m/s. Relative to the center, the tangential speed is $v_T = 0.4 \times 10^6$ m/s for locations A and B, which are equidistant from the center. When the frequencies of the light coming from regions A and B are measured on earth, they are not the same and each is different from the emitted frequency of 6.200×10^{14} Hz. Find the measured frequency for the light from (a) region A and (b) region B.

Section 24.6 Polarization

34. Unpolarized light whose intensity is 1.10 W/m² is incident on the polarizer in Figure 24.20. (a) What is the intensity of the light leaving the polarizer? (b) If the analyzer is set at an angle of $\theta = 75°$ with respect to the polarizer, what is the intensity of the light that reaches the photocell?

35. ssm Linearly polarized light is incident on a piece of polarizing material. What is the ratio of the transmitted light intensity to the incident light intensity when the angle between the transmission axis and the incident electric field is (a) 25° and (b) 65°?

36. For each of the three sheets of polarizing material shown in the drawing, the orientation of the transmission axis is labeled relative to the vertical. The incident beam of light is unpolarized and has an intensity of 1260.0 W/m². What is the intensity of the beam transmitted through the three sheets when $\theta_1 = 19.0°$, $\theta_2 = 55.0°$, and $\theta_3 = 100.0°$?

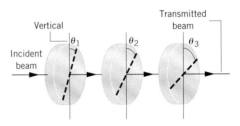

37. Polarized light strikes a piece of polarizing material. The incident light is polarized at an angle of 38° relative to the transmission axis of the material. What percentage of the light intensity is transmitted?

38. Review Conceptual Example 8 before solving this problem. Suppose unpolarized light of intensity 150 W/m² falls on the polarizer in Figure 24.22a, and the angle θ in the drawing is 30.0°. What is the light intensity reaching the photocell?

*39. ssm www More than one analyzer can be used in a setup like that in Figure 24.20, each analyzer following the previous one. Suppose that the transmission axis of the first analyzer is rotated 27° relative to the transmission axis of the polarizer, and that the transmission axis of each additional analyzer is rotated 27° relative to the transmission axis of the previous one. What is the minimum number of analyzers needed, so the light reaching the photocell has an intensity that is reduced by at least a factor of one hundred relative to that striking the first analyzer?

*40. Before attempting this problem, review Conceptual Example 8. The intensity of the light that reaches the photocell in Figure 24.22a is 110 W/m², when $\theta = 23°$. What would be the intensity reaching the photocell if the *analyzer* were removed from the setup, everything else remaining the same?

ADDITIONAL PROBLEMS

41. Some of the X-rays produced in an X-ray machine have a wavelength of 2.1 nm. What is the frequency of these electromagnetic waves?

42. A laser emits a narrow beam of light. The radius of the beam is 1.0×10^{-3} m, and the power is 1.2×10^{-3} W. What is the intensity of the laser beam?

43. ssm www The distance between earth and the moon can be determined from the time it takes for a laser beam to travel from earth to a reflector on the moon and back. If the round-trip time can be measured to an accuracy of one-tenth of a nanosecond (1 ns = 10^{-9} s), what is the corresponding error in the earth–moon distance?

44. On a cloudless day, the sunlight that reaches the surface of the earth has an average intensity of about 1.0×10^3 W/m². What is the average electromagnetic energy contained in 5.5 m³ of space just above the earth's surface?

45. The intensity of sunlight at the top of the earth's atmosphere is about 1390 W/m². The distance between the sun and earth is 1.50×10^{11} m, while that between the sun and Mars is 2.28×10^{11} m. What is the intensity of sunlight at the surface of Mars?

46. A radio station broadcasts a radio wave whose wavelength is 274 m. (a) What is the frequency of the wave? (b) Is this radio wave AM or FM? (See Figure 24.9.)

47. ssm In the polarizer/analyzer combination in Figure 24.20, 90.0% of the light intensity falling on the analyzer is absorbed. Determine the angle between the transmission axes of the polarizer and the analyzer.

48. Suppose that the police car in Example 6 is moving to the right at 27 m/s, while the speeder is coming up from behind at a speed of 39 m/s, both speeds being with respect to the ground. Assume that the electromagnetic wave emitted by the radar gun has a frequency of 8.0×10^9 Hz. (a) Find the magnitude of the difference between the frequency of the emitted wave and the wave that returns to the police car after reflecting from the speeder's car. (b) Which wave has the greater frequency? Why?

*49. In a traveling electromagnetic wave, the electric field is represented mathematically as

$$E = E_0 \sin [(1.5 \times 10^{10} \text{ s}^{-1})\, t - (5.0 \times 10^1 \text{ m}^{-1})\, x]$$

where E_0 is the maximum field strength. (a) What is the frequency of the wave? (b) This wave and the wave that results from its reflection can form a standing wave, in a way similar to that in which standing waves can arise on a string (see Section 17.5). What is the separation between adjacent nodes in the standing wave?

*50. A beam of polarized light has an average intensity of 15 W/m² and is sent through a polarizer. The transmission axis makes an angle of 25° with respect to the direction of polarization. Determine the rms value of the electric field of the transmitted beam.

*51. ssm The mean distance between earth and the sun is 1.50×10^{11} m. The average intensity of solar radiation incident on the upper atmosphere of the earth is 1390 W/m². Assuming the sun emits radiation uniformly in all directions, determine the total power radiated by the sun.

*52. An argon-ion laser produces a cylindrical beam of light whose average power is 0.750 W. How much energy is contained in a 2.50-m length of the beam?

*53. A tiny source of light emits light uniformly in all directions. The average power emitted is 60.0 W. For a point located 8.00 m away from this source, determine the rms (a) electric and (b) magnetic field strengths in the light waves.

**54. Suppose that the light falling on the polarizer in Figure 24.20 is partially polarized (average intensity = $\overline{S}_P$) and partially unpolarized (average intensity = $\overline{S}_U$). The total incident intensity is $\overline{S}_P + \overline{S}_U$, and the percentage polarization is $100\overline{S}_P/(\overline{S}_P + \overline{S}_U)$. When the polarizer is rotated in such a situation, the intensity reaching the photocell varies between a minimum value of $\overline{S}_{min}$ and a maximum value of $\overline{S}_{max}$. Show that the percentage polarization can be expressed as $100(\overline{S}_{max} - \overline{S}_{min})/(\overline{S}_{max} + \overline{S}_{min})$.

CONCEPTS CALCULATIONS

GROUP LEARNING PROBLEMS

Note: Each of these problems consists of Concept Questions followed by a related quantitative Problem. They are designed for use by students working alone or in small learning groups. The Concept Questions involve little or no mathematics and are intended to stimulate group discussions. They focus on the concepts with which the problems deal. Recognizing the concepts is the essential initial step in any problem-solving technique.

55. Concept Questions A certain type of laser puts out light of known frequency. The light, however, occurs as a series of short pulses, each lasting for a time t_0. (a) How is the wavelength of

the light related to its frequency? (b) How is the length (in meters) of each pulse related to the time t_0?

Problem A laser puts out a pulse of light that lasts for 2.7×10^{-11} s. The frequency of the light is 5.2×10^{14} Hz. (a) How many wavelengths are there in one pulse? (b) The light enters a pool of water. Its frequency remains the same, but the light slows down to a speed of 2.3×10^8 m/s. How many wavelengths are there now in one pulse?

56. Concept Questions (a) Suppose that the magnitude E of the electric field in an electromagnetic wave triples. By what factor does the intensity S of the wave change? (b) The magnitude B of the magnetic field is much smaller than E, because, according to Equation 24.3, $B = E/c$, where c is the speed of light in a vacuum. If B triples, by what factor does the intensity change? Account for your answers.

Problem The magnitude of the electric field of an electromagnetic field increases from 315 to 945 N/C. (a) Determine the intensities for the two values of the electric field. (b) What is the magnitude of the magnetic field associated with each electric field? (c) Determine the intensity for each value of the magnetic field. Make sure your answers are consistent with your answers to the Concept Questions.

57. Concept Questions A source is radiating light waves uniformly in all directions. At a certain distance r from the source a person measures the average intensity of the waves. (a) Does the average intensity increase, decrease, or remain the same as r increases? (b) If the magnitude of the electric field is determined from the average intensity, is the electric field the rms value or the peak value? In both cases, justify your answers.

Problem A light bulb emits light uniformly in all directions. The emitted power is 150.0 W. At a distance of 5.00 m from the bulb, what are (a) the average intensity and the magnitudes of the (b) rms and (c) peak electric fields?

58. Concept Questions An electric charge is placed in a laser beam. Does a stationary charge experience a force due to (a) the electric field and (b) the magnetic field of the electromagnetic wave? Now suppose that the charge is moving perpendicular to the magnetic field of the beam. Does it experience (c) an electric force and (d) a magnetic force? Account for your answers.

Problem A stationary particle of charge $q = 2.6 \times 10^{-8}$ C is placed in a laser beam whose intensity is 2.5×10^3 W/m². Determine the magnitude of the (a) electric and (b) magnetic forces exerted on the charge. If the charge is moving perpendicular to the magnetic field with a speed of 3.7×10^4 m/s, find the magnitudes of the (c) electric and (d) magnetic forces exerted on it. Verify that your answers are consistent with your answers to the Concept Questions.

59. Concept Questions The drawing shows light incident on a polarizer whose transmission axis is parallel to the z axis. The polarizer is rotated clockwise through an angle α between 0 and 90°. While the polarizer is being rotated, does the intensity of the transmitted light increase, decrease, or remain the same if the incident light is (a) unpolarized, (b) polarized parallel to the z axis, and (c) polarized parallel to the y axis? Provide a reason for each of your answers.

Problem The intensity of the incident light is 7.0 W/m². Determine the intensity of the transmitted light for each of the six cases shown in the table.

	Intensity of Transmitted Light	
Incident Light	$\alpha = 0°$	$\alpha = 35°$
Unpolarized		
Polarized parallel to z axis		
Polarized parallel to y axis		

Be sure that your answers are consistent with your answers to the Concept Questions.

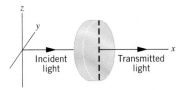

60. Concept Question The drawing shows three polarizer/analyzer pairs. The incident light on each pair is unpolarized and has the same intensity. Rank the pairs according to the intensity of the transmitted light, largest first. Provide reasons for your answers.

Problem The intensity of the unpolarized incident beam is 48 W/m². Find the intensity of the transmitted beams for each of the three cases shown in the drawing. Be sure your answers are consistent with your answer to the Concept Question.

With a wary eye, this colorful lizard sizes up its image reflected in a flat or plane mirror. This chapter discusses the images formed by the reflection of light from both plane and spherical mirrors.

CHAPTER 25

THE REFLECTION OF LIGHT: MIRRORS

25.1 WAVE FRONTS AND RAYS

Mirrors are usually close at hand. It is difficult, for example, to put on makeup, shave, or drive a car without them. We see images in mirrors because some of the light that strikes them is reflected into our eyes. To discuss reflection, it is necessary to introduce the concepts of a wave front and a ray of light. And we can do so by taking advantage of the familiar topic of sound waves (see Chapter 16). Both sound and light are kinds of waves. Sound is a pressure wave, while light is electromagnetic in nature. However, the wave properties of each are similar. In particular, the ideas of a wave front and a ray apply to both.

Consider a small spherical object whose surface is pulsating in simple harmonic motion. A sound wave is emitted that moves spherically outward from the object at a constant speed. To represent this wave, we draw surfaces through all points of the wave that are in the same phase of motion. These surfaces of constant phase are called *wave fronts.* Figure 25.1 shows a hemispherical view of the wave fronts. In this view they appear as concentric spherical shells about the vibrating object. If the wave fronts are drawn through the condensations, or crests, of the sound wave, as they are in the picture, the distance between adjacent wave fronts equals the wavelength λ. The radial lines pointing outward from the source and perpendicular to the wave fronts are called *rays.* The rays point in the direction of the velocity of the wave.

Figure 25.2a shows small sections of two adjacent spherical wave fronts. At large distances from the source, the wave fronts become less and less curved and approach the shape of flat surfaces, as part b of the drawing shows. Waves whose wave fronts are flat surfaces (i.e., planes) are known as *plane waves* and are important in understanding the properties of mirrors and lenses. Since rays are perpendicular to the wave fronts, the rays for a plane wave are parallel to each other.

The concepts of wave fronts and rays can also be applied to light waves. For light waves, the ray concept is particularly convenient for showing the path taken by the light. We will make frequent use of light rays, and they can be regarded essentially as narrow beams of light much like those that lasers produce.

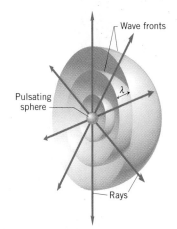

Figure 25.1 A hemispherical view of a sound wave emitted by a pulsating sphere. The wave fronts are drawn through the condensations of the wave, so the distance between two successive wave fronts is the wavelength λ. The rays are perpendicular to the wave fronts and point in the direction of the velocity of the wave.

25.2 THE REFLECTION OF LIGHT

Most objects reflect a certain portion of the light falling on them. Suppose a ray of light is incident on a flat, shiny surface, such as the mirror in Figure 25.3. As the drawing shows, the *angle of incidence* θ_i is the angle that the incident ray makes with respect to the normal, which is a line drawn perpendicular to the surface at the point of incidence. The *angle of reflection* θ_r is the angle that the reflected ray

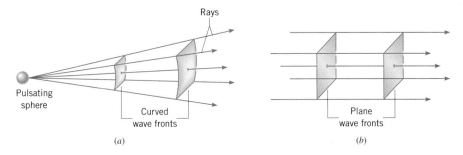

(a)   (b)

Figure 25.2 (a) Portions of two spherical wave fronts are shown. The rays are perpendicular to the wave fronts and diverge. (b) For a plane wave, the wave fronts are flat surfaces, and the rays are parallel to each other.

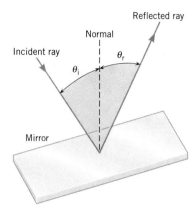

Figure 25.3 The angle of reflection θ_r equals the angle of incidence θ_i. These angles are measured with respect to the normal, which is a line drawn perpendicular to the surface of the mirror at the point of incidence.

makes with the normal. The **_law of reflection_** describes the behavior of the incident and reflected rays.

● LAW OF REFLECTION

The incident ray, the reflected ray, and the normal to the surface all lie in the same plane, and the angle of reflection θ_r equals the angle of incidence θ_i:

$$\theta_r = \theta_i$$

When parallel light rays strike a smooth, plane surface, such as that in Figure 25.4a, the reflected rays are parallel to each other. This type of reflection is one example of what is known as *specular reflection* and is important in determining the properties of mirrors. Most surfaces, however, are not perfectly smooth, for they contain irregularities the sizes of which are equal to or greater than the wavelength of light. The irregular surface reflects the light rays in various directions, as part *b* of the drawing suggests. This type of reflection is known as *diffuse reflection*. Common surfaces that give rise to diffuse reflection are most papers, wood, nonpolished metals, and walls covered with a "flat" (nongloss) paint.

A revolution of digital technology is poised to occur in the movie industry. Digital film techniques have already produced films such as *Star Wars, The Phantom Menace*. Until recently, however, films have been viewed primarily using projectors that shine light directly through a strip of film containing the images. Now, digital projectors are available that allow the movie to be viewed completely without film by using a digital representation (zeros and ones) of the images. These projectors depend on the law of reflection and tiny mirrors called micromirrors, each about the size of one-fourth the diameter of a human hair. Each micromirror creates a tiny portion of an individual movie frame on the screen and serves as a pixel, like one of the glowing spots that comprise the picture on a TV screen or computer monitor. This pixel action is possible because a micromirror pivots in one direction or the reverse in response to the "zero" or "one" in the digital representation of the frame. One of the directions puts a portion of the light from a powerful xenon lamp on the screen, and the other does not. The pivoting action can occur as fast as 1000 times per second, leading to a series of light pulses for each pixel that the eye and the brain combine and interpret as a continuously changing image. Present-generation digital projectors use about 800 000 micromirrors to reproduce each of the three primary colors (red, green, and blue) that comprise a color image.

THE PHYSICS OF...

digital movie projectors and micromirrors.

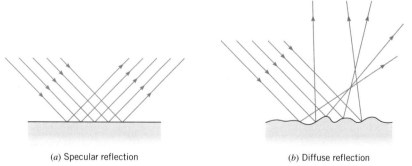

(a) Specular reflection   (b) Diffuse reflection

Figure 25.4 (a) The drawing shows specular reflection from a polished plane surface, such as a mirror. The reflected rays are parallel to each other. (b) A rough surface reflects the light rays in all directions; this type of reflection is known as diffuse reflection.

25.3 THE FORMATION OF IMAGES BY A PLANE MIRROR

When you look into a plane (flat) mirror, you see an image of yourself that has three properties:

1. The image is upright.

2. The image is the same size as you are.

3. The image is located as far behind the mirror as you are in front of it.

As Figure 25.5*a* illustrates, the image of yourself in the mirror is also reversed left to right. If you wave your *right* hand, it is the *left* hand of the image that waves back. Similarly, letters and words held up to a mirror are reversed. Ambulances and other emergency vehicles are often lettered in reverse, as in Figure 25.5*b*, so that the letters will appear normal when seen in the rearview mirror of a car.

To illustrate why an image appears to originate from behind a plane mirror, Figure 25.6*a* shows a light ray leaving the top of an object. This ray reflects from the mirror (angle of reflection equals angle of incidence) and enters the eye. To the eye, it appears that the ray originates from behind the mirror, somewhere back along the dashed line. Actually, rays going in all directions leave each point on the object. But only a small bundle of such rays is intercepted by the eye. Part *b* of the figure shows a bundle of two rays leaving the top of the object. All the rays that leave a given point on the object, no matter what angle θ they have when they strike the mirror, appear to originate from a corresponding point on the image behind the mirror (see the dashed lines in part *b*). For each point on the object, there is a single corresponding point on the image, and it is this fact that makes the image in a plane mirror a sharp and undistorted one.

Although rays of light *seem* to come from the image, it is evident from Figure 25.6*b* that they do not originate from behind the plane mirror where the image appears to be. Because all of the rays of light do not actually emanate from the image, it is called a **virtual image.** In this text the parts of the light rays that appear to come from a virtual image are represented by dashed lines. *Curved* mirrors, on the other hand, can produce images from which all the light rays actually do emanate. Such images are known as **real images** and are discussed in later sections.

With the aid of the law of reflection, it is possible to show that the image is located as far behind a plane mirror as the object is in front of it. In Figure 25.7 the object distance is d_o and the image distance is d_i. A ray of light leaves the base of the object, strikes the mirror at an angle of incidence θ, and is reflected at the same angle. To the eye, this ray appears to come from the base of the image. For the angles β_1 and β_2 in the drawing it follows that $\theta + \beta_1 = 90°$ and $\alpha + \beta_2 = 90°$. But the angle α is equal to the angle of reflection θ, since the two are opposite angles

Figure 25.5 (*a*) The person's right hand becomes the image's left hand when viewed in a plane mirror. (*b*) Many emergency vehicles are reverse-lettered so the lettering appears normal when viewed through the rearview mirror of a car.

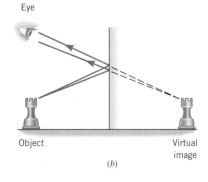

Figure 25.6 (*a*) A ray of light from the top of the chess piece reflects from the mirror. To the eye, the ray seems to come from behind the mirror. (*b*) The bundle of rays from the top of the object appears to originate from the image behind the mirror.

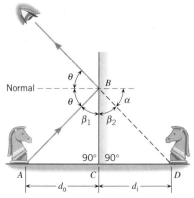

Figure 25.7 This drawing illustrates the geometry used with a plane mirror to show that the image distance d_i equals the object distance d_o.

formed by intersecting lines. Therefore, $\beta_1 = \beta_2$. As a result, triangles *ABC* and *DBC* are identical (congruent) because they share a common side *BC* and have equal angles ($\beta_1 = \beta_2$) at the top and equal angles ($90°$) at the base. Thus, the object distance d_o equals the image distance d_i.

By starting with a light ray from the top of the object, rather than the bottom, we can extend the line of reasoning given above to show also that the height of the image equals the height of the object.

Conceptual Examples 1 and 2 discuss some interesting features of plane mirrors.

CONCEPTUAL EXAMPLE 1   Full-Length Versus Half-Length Mirrors

In Figure 25.8 a woman is standing in front of a plane mirror. What is the minimum mirror height necessary for her to see her full image?

Reasoning and Solution The mirror is labeled *ABCD* in the drawing and is the same height as the woman. Light emanating from her body is reflected by the mirror, and some of this light enters her eyes. Consider a ray of light from her foot *F*. This ray strikes the mirror at *B* and enters her eyes at *E*. According to the law of reflection, the angles of incidence and reflection are both θ. Any light from her foot that strikes the mirror below *B* is reflected toward a point on her body that is below her eyes. Since light striking the mirror below *B* does not enter her eyes, the part of the mirror between *B* and *A* may be removed. The section *BC* of the mirror that produces the image is one-half the woman's height between *F* and *E*. This follows because the right triangles *FBM* and *EBM* are identical. They are identical because they share a common side *BM* and have two angles, θ and $90°$, that are the same.

The blowup in Figure 25.8 illustrates a similar line of reasoning, starting with a ray from the woman's head at *H*. This ray is reflected from the mirror at *P* and enters her eyes. The top mirror section *PD* can be removed without disturbing this reflection. The necessary section *CP* is one-half the woman's height between her head at *H* and her eyes at *E*. We find, then, that only the sections *BC* and *CP* are needed for the woman to see her full height. The height of section *BC* plus section *CP* is exactly one-half the woman's height. Thus, ***to view one's full length in a mirror, only a half-length mirror is needed.*** The conclusions here are valid regardless of how far the person stands from the mirror.

Related Homework Material: *Problems 2, 32*

CONCEPTUAL EXAMPLE 2   Multiple Reflections

A person is sitting in front of two mirrors that intersect at a $90°$ angle. As Figure 25.9*a* illustrates, the person sees three images of himself. (The person himself is not shown; only the images are present.) Why are there three images rather than two?

Reasoning and Solution Figure 25.9*b* shows a top view of the person in front of the mirrors. It is a straightforward matter to understand two of the images that he sees. These are the images that are normally seen when one stands in front of a mirror. Standing in front of mirror 1, he sees image 1, which is located as far behind that mirror as he is in front of it. He also sees image 2 behind mirror 2, at a distance that matches his distance in front of that mirror. Each of these images arises from light emanating from his body and reflecting from a single mirror. However, it is also possible for light to undergo two reflections in sequence, first from one mirror and then from the other. ***When a double reflection occurs, an additional image becomes possible.*** Figure 25.9*b* shows two rays of light that strike mirror 1. Each one, according to the law of reflection, has an angle of reflection that equals the angle of incidence. The rays then strike mirror 2, where they again are reflected according to the law of

Figure 25.8 For the woman to see her full-sized image, only a half-sized mirror is needed.

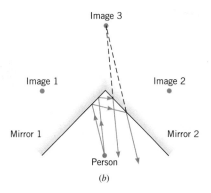

Figure 25.9 (*a*) These two perpendicular plane mirrors produce three images of the person sitting in front of them. (*b*) A "double" reflection, one from each mirror, gives rise to image 3.

reflection. When the outgoing rays are extended backward (see the dashed lines in the drawing), they intersect and appear to originate from image 3.

Related Homework Material: *Problem 4*

25.4 SPHERICAL MIRRORS

The most common type of curved mirror is a spherical mirror. As Figure 25.10 shows, a spherical mirror has the shape of a section from the surface of a sphere. If the inside surface of the mirror is polished, it is a ***concave mirror.*** If the outside surface is polished, it is a ***convex mirror.*** The drawing shows both types of mirrors, with a light ray reflecting from the polished surface. The law of reflection applies, just as it does for a plane mirror. For either type of spherical mirror, the normal is drawn perpendicular to the mirror at the point of incidence. For each type, the center of curvature is located at point C, and the radius of curvature is R. The ***principal axis*** of the mirror is a straight line drawn through C and the midpoint of the mirror.

Figure 25.11 shows a tree in front of a concave mirror. A point on this tree lies on the principal axis of the mirror and is beyond the center of curvature C. Light rays emanate from this point and reflect from the mirror, consistent with the law of reflection. If the rays are near the principal axis, they cross it at a common point after reflection. This point is called the image point. The rays continue to diverge from the image point as if there were an object there. Since light rays actually come from the image point, the image is a real image.

If the tree in Figure 25.11 is infinitely far from the mirror, the rays are parallel to each other and to the principal axis as they approach the mirror. Figure 25.12 shows rays near and parallel to the principal axis, as they reflect from the mirror and pass through an image point. In this special case the image point is referred to as the ***focal point F*** of the mirror. Therefore, an object infinitely far away on the principal

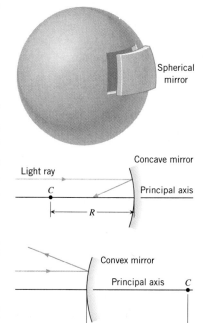

Figure 25.10 A spherical mirror has the shape of a segment of a spherical surface. The center of curvature is point C and the radius is R. For a concave mirror, the reflecting surface is the inner one, while for a convex mirror it is the outer one.

Figure 25.11 A point on the tree lies on the principal axis of the concave mirror. Rays from this point that are near the principal axis are reflected from the mirror and cross the axis at the image point.

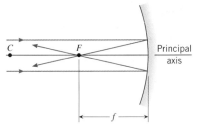

Figure 25.12 Light rays near and parallel to the principal axis are reflected from a concave mirror and converge at the focal point *F*. The focal length *f* is the distance between *F* and the mirror.

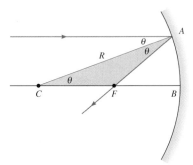

Figure 25.13 This drawing is used to show that the focal point *F* of a concave mirror is halfway between the center of curvature *C* and the mirror at point *B*.

THE PHYSICS OF...

capturing solar energy with mirrors; automobile headlights.

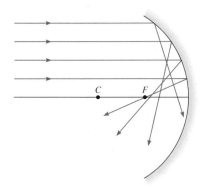

Figure 25.14 Rays that are farthest from the principal axis have the greatest angle of incidence and miss the focal point *F* after reflection from the mirror.

axis gives rise to an image at the focal point of the mirror. The distance between the focal point and the middle of the mirror is the ***focal length f*** of the mirror.

We can show that the focal point *F* lies halfway between the center of curvature *C* and the middle of a concave mirror. In Figure 25.13, a light ray parallel to the principal axis strikes the mirror at point *A*. The line *CA* is the radius of the mirror and, therefore, is the normal to the spherical surface at the point of incidence. The ray reflects from the mirror such that the angle of reflection θ equals the angle of incidence. Furthermore, the angle *ACF* is also θ because the radial line *CA* is a transversal of two parallel lines. Since two of its angles are equal, the colored triangle *CAF* is an isosceles triangle; thus, sides *CF* and *FA* are equal. But when the incoming ray lies close to the principal axis, the angle of incidence θ is small, and the distance *FA* does not differ appreciably from the distance *FB*. Therefore, in the limit that θ is small, $CF = FA = FB$, and so the focal point *F* lies halfway between the center of curvature and the mirror. In other words, the focal length *f* is one-half of the radius *R*:

Focal length of a concave mirror

$$f = \tfrac{1}{2}R \qquad (25.1)$$

Rays that lie close to the principal axis are known as ***paraxial rays,**** and Equation 25.1 is valid only for such rays. Rays that are far from the principal axis do not converge to a single point after reflection from the mirror, as Figure 25.14 shows. The result is a blurred image. The fact that a spherical mirror does not bring all rays parallel to the axis to a single image point is known as ***spherical aberration.*** Spherical aberration can be minimized by using a mirror whose height is small compared to the radius of curvature.

A sharp image point can be obtained with a large mirror, if the mirror is parabolic in shape instead of spherical. The shape of a parabolic mirror is such that all light rays parallel to the principal axis, regardless of their distance from the axis, are reflected through a single image point. However, parabolic mirrors are costly to manufacture and are used where the sharpest images are required, as in research-quality telescopes. Parabolic mirrors are also used in one method of capturing solar energy for commercial purposes. Figure 25.15 shows a long row of concave parabolic mirrors that reflect the sun's rays to the focal point. Located at the focal point and running the length of the row is an oil-filled pipe. The focused rays of the sun heat the oil. In a solar-thermal electric plant, the heat from many such rows is used to generate steam. The steam, in turn, drives a turbine connected to an electric generator. A parabolic mirror is also used in an automobile headlight. Here, however, the situation is reversed from the operation of a solar collector. In a headlight, a high-intensity light bulb is placed at the focal point of the mirror, and light emerges parallel to the principal axis.

A convex mirror also has a focal point, and Figure 25.16 illustrates its meaning. In this picture, parallel rays are incident on a convex mirror. Clearly, the rays diverge after being reflected. If the incident parallel rays are paraxial, the reflected rays seem to come from a single point *F* behind the mirror. This point is the focal point of the convex mirror, and its distance from the midpoint of the mirror is the focal length *f*. The focal length of a convex mirror is also one-half of the radius of curvature, just as it is for a concave mirror. However, we assign the focal length of a convex mirror a negative value because it will be convenient later on:

Focal length of a convex mirror

$$f = -\tfrac{1}{2}R \qquad (25.2)$$

---

* Paraxial rays are close to the principal axis but not necessarily parallel to it.

Figure 25.15 This long row of parabolic mirrors focuses the sun's rays to heat an oil-filled pipe located at the focal point of each mirror. It is one of many that are used by a solar-thermal electric plant in the Mojave Desert.

25.5 THE FORMATION OF IMAGES BY SPHERICAL MIRRORS

Concave Mirrors

● *Concepts at a Glance*   As we have seen, some of the light rays emitted from an object in front of a mirror strike the mirror, reflect from it, and form an image. We can analyze the image produced by either concave or convex mirrors by using a graphical method called ***ray tracing.*** As the Concepts-at-a-Glance chart in Figure 25.17 shows, this method is based on the law of reflection and the notion that a

Figure 25.16 When paraxial light rays that are parallel to the principal axis strike a convex mirror, the reflected rays appear to originate from the focal point *F*. The radius of curvature is *R* and the focal length is *f*.

Figure 25.17 *Concepts at a Glance*   The method of ray tracing is used to predict the location and size of an image produced by a spherical mirror. Ray tracing is based on the law of reflection and two points associated with the mirror, its center of curvature *C* and its focal point *F*. The red and the blue rays in the photograph, after reflecting from the mirror, pass through the focal point. Consistent with the law of reflection, the angle of reflection for each ray equals the angle of incidence.

Concepts at a Glance

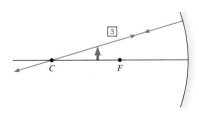

Figure 25.18 The rays labeled 1, 2, and 3 are useful in locating the image of an object that stands in front of a concave spherical mirror. The object is represented as a vertical arrow.

spherical mirror has a center of curvature C and a focal point F. Ray tracing enables us to find the location of the image, as well as its size, by taking advantage of the following fact: paraxial rays leave a point on the object and intersect at a corresponding point on the image after reflection. ●

Three specific paraxial rays are especially convenient to use. Figure 25.18 shows an object in front of a concave mirror, and these three rays leave a point on the top of the object. They are labeled 1, 2, and 3, and when tracing their paths, we use the following reasoning strategy.

REASONING STRATEGY

Ray Tracing for a Concave Mirror

Ray 1. This ray is initially parallel to the principal axis and, therefore, passes through the focal point F after reflection from the mirror.

Ray 2. This ray passes through the focal point F and is reflected parallel to the principal axis. Ray 2 is analogous to ray 1, except that the reflected, rather than the incident, ray is parallel to the principal axis.

Ray 3. This ray travels along a line that passes through the center of curvature C and follows a radius of the spherical mirror; as a result, the ray strikes the mirror perpendicularly and reflects back on itself.

If rays 1, 2, and 3 are superimposed on a scale drawing, they converge at a point on the top of the image, as can be seen in Figure 25.19a.* Although three rays have been used here to locate the image, only two are really needed; the third ray is usually drawn to serve as a check. In a similar fashion, rays from all other points on the object locate corresponding points on the image, and the mirror forms a complete

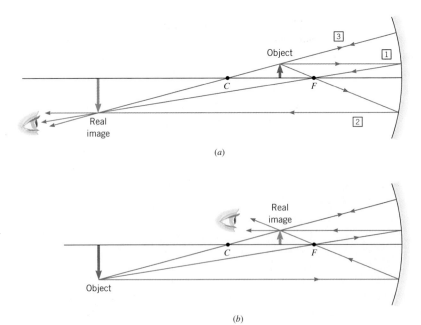

(a)

(b)

Figure 25.19 (a) When an object is placed between the focal point F and the center of curvature C of a concave mirror, a real image is formed. The image is enlarged and inverted relative to the object. (b) When the object is located beyond the center of curvature C, a real image is created that is reduced in size and inverted relative to the object.

---

* In the drawings that follow, we assume that the rays are paraxial, although the distance between the rays and the principal axis is often exaggerated for clarity.

(a)

(b)

Figure 25.20 (a) When an object is located between a concave mirror and its focal point F, an enlarged, upright, and virtual image is produced. (b) A makeup mirror is concave and forms a virtual image as shown in part a.

image of the object. If you place your eye as shown in the drawing, you will see an image that is *larger* and *inverted* relative to the object. The image is *real* because the light rays actually pass through the image.

If the locations of the object and image in Figure 25.19a are interchanged, the situation in part b of the drawing results. The three rays in part b are the same as those in part a, except the directions are reversed. These drawings illustrate the **principle of reversibility,** which states that **if the direction of a light ray is reversed, the light retraces its original path.** This principle is quite general and is not restricted to reflection from mirrors. The image is *real,* and it is *smaller* and *inverted* relative to the object.

When the object is placed between the focal point F and a concave mirror, as in Figure 25.20a, three rays can again be drawn to find the image. But now ray 2 does not go through the focal point on its way to the mirror, since the object is inside the focal point. However, when projected backward, ray 2 appears to come from the focal point. Therefore, after reflection, ray 2 is directed parallel to the principal axis. In this case the three reflected rays diverge from each other and do not converge to a common point. However, when projected behind the mirror, the three rays appear to come from a common point; thus, a *virtual* image is formed. This virtual image is *larger* than the object and *upright.* Makeup and shaving mirrors are concave mirrors. When you place your face between the mirror and its focal point, you see an enlarged virtual image of yourself, as part b of the drawing shows.

Concave mirrors are also used in a new method for displaying the speed of a car. The method presents a digital readout (e.g., 55 mph) that the driver sees when looking directly through the windshield, as in Figure 25.21a. The advantage of the method, which is called a head-up display (HUD), is that the driver does not need to take his or her eyes off the road to monitor the speed. Figure 25.21b shows how a HUD works. Located below the windshield is a readout device that displays the speed in digital form. This device is located in front of a concave mirror but within its focal point. The arrangement is similar to that in Figure 25.20a and produces a virtual, upright, and enlarged image of the speed readout (see virtual image 1 in Figure 25.21b). Light rays that appear to come from this image strike the windshield at a place where a so-called combiner is located. The purpose of the com-

THE PHYSICS OF...

makeup and shaving mirrors.

THE PHYSICS OF...

a head-up display for automobiles.

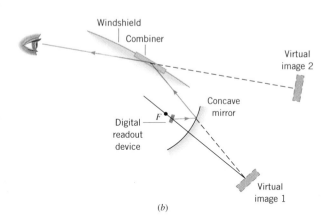

(a)

(b)

Figure 25.21 *(a)* A head-up display (HUD) presents the driver with a digital readout of the car's speed in the field of view seen through the windshield. *(b)* One version of a HUD uses a concave mirror.

biner is to combine the digital readout information with the field of view that the driver sees through the windshield. The combiner is virtually undetectable by the driver because it allows all colors except one to pass through it unaffected. The one exception is the color produced by the digital readout device. For this color, the combiner behaves as a plane mirror and reflects the light that appears to originate from image 1. Thus, the combiner produces image 2, which is what the driver sees. The location of image 2 is near the front bumper. The driver can then read the speed with his eyes focused just as they are to see the road.

Convex Mirrors

The procedure for determining the location and size of an image in a convex mirror is similar to that for a concave mirror. The same three rays are used. However, the focal point and center of curvature of a convex mirror lie behind the mirror, not in front of it. Figure 25.22a shows the rays. When tracing their paths, we use the following reasoning strategy, which takes into account the different locations of the focal point and center of curvature.

REASONING STRATEGY

Ray Tracing for a Convex Mirror

Ray 1. This ray is initially parallel to the principal axis and, therefore, appears to originate from the focal point *F* after reflection from the mirror.

Ray 2. This ray heads toward *F*, emerging parallel to the principal axis after reflection. Ray 2 is analogous to ray 1, except that the reflected, rather than the incident, ray is parallel to the principal axis.

Ray 3. This ray travels toward the center of curvature *C*; as a result, the ray strikes the mirror perpendicularly and reflects back on itself.

(a)

(b)

Figure 25.22 (a) An object placed in front of a convex mirror produces a virtual image behind the mirror. The virtual image is reduced in size and upright. (b) The sun shield in this pilot's helmet acts as a convex mirror and reflects an image of his plane.

The three rays in Figure 25.22a appear to come from a point on a *virtual* image that is behind the mirror. The virtual image is *diminished in size* and *upright,* relative to the object. A convex mirror *always* forms a virtual image of the object, no matter where in front of the mirror the object is placed. Figure 25.22b shows an example of such an image. Because of its shape a convex mirror gives a wider field of view than do other types of mirrors. Therefore, they are often used in stores for security purposes. A mirror with a wide field of view is also needed to give a driver a good rear view. Thus, the outside mirror on the passenger side is often a convex mirror. Printed on such a mirror is usually the warning "VEHICLES IN MIRROR ARE CLOSER THAN THEY APPEAR." The reason for the warning is that the virtual image in Figure 25.22a is reduced in size and therefore looks smaller, just as a distant object would appear in a plane mirror. An unwary driver, thinking that the side-view mirror is a plane mirror, might incorrectly deduce from the small size of the image that the car behind is far enough away to ignore.

THE PHYSICS OF...

passenger-side automobile mirrors.

25.6 THE MIRROR EQUATION AND THE MAGNIFICATION EQUATION

Concave Mirrors

● *Concepts at a Glance* Ray diagrams drawn to scale are useful for determining the location and size of the image formed by a mirror. However, for an accurate description of the image, a more analytical technique is needed. We will now derive two equations, known as the ***mirror equation*** and the ***magnification equation,*** that will provide a complete description of the image. The Concepts-at-a-Glance chart in Figure 25.23, which is an extension of the chart in Figure 25.17, emphasizes that these equations are based on the law of reflection and provide relationships between the following quantities:

Concepts at a Glance

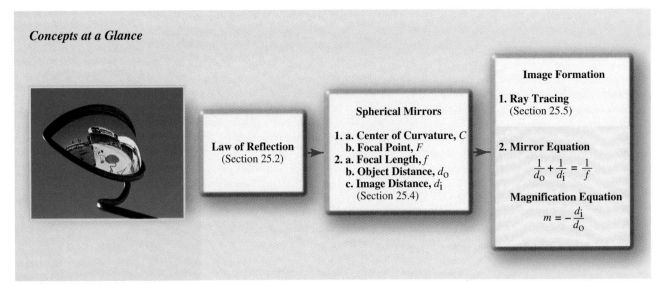

Figure 25.23 *Concepts at a Glance* The mirror equation and the magnification equation describe how spherical mirrors form images and are based on the law of reflection. These equations give the location and relative size of the image. The photograph shows the image of a school bus in a trucker's convex side-view mirror.

f = the focal length of the mirror
d_o = the object distance, which is the distance between the object and the mirror
d_i = the image distance, which is the distance between the image and the mirror
m = the magnification of the mirror, which is the ratio of the height of the image to the height of the object.

We begin our derivation of the mirror equation by referring to Figure 25.24a, which shows a ray leaving the top of the object and striking a concave mirror at the point where the principal axis intersects the mirror. Since the principal axis is perpendicular to the mirror, it is also the normal at this point of incidence. Therefore, the ray reflects at an equal angle and passes through the image. The two colored triangles are similar triangles because they have equal angles, so

$$\frac{h_o}{-h_i} = \frac{d_o}{d_i}$$

where h_o is the height of the object and h_i is the height of the image. The minus sign appears on the left in this equation because the image is inverted in Figure 25.24a. In part *b* another ray leaves the top of the object, this one passing through the focal point F, reflecting parallel to the principal axis, and then passing through the image. Provided the ray remains close to the axis, the two colored areas can be considered to be similar triangles, with the result that

$$\frac{h_o}{-h_i} = \frac{d_o - f}{f}$$

Setting the two equations above equal to each other yields $d_o/d_i = (d_o - f)/f$. Rearranging this result gives the *mirror equation:*

Mirror equation

$$\frac{1}{d_o} + \frac{1}{d_i} = \frac{1}{f} \qquad (25.3)$$

We have derived this equation for a real image formed in front of a concave mirror. In this case, the image distance is a positive quantity, as are the object distance and the focal length. However, we have seen in the last section that a concave mirror can also form a virtual image, if the object is located between the focal point and the mirror. Equation 25.3 can also be applied to such a situation, provided that we adopt the convention that d_i is negative for an image behind the mirror, as it is for a virtual image.

In deriving the magnification equation, we remember that the **magnification** m of a mirror is the ratio of the image height to the object height: $m = h_i/h_o$. If the image height is less than the object height, the magnitude of m is less than one. Conversely, if the image is larger than the object, the magnitude of m is greater than one. We have already shown that $h_o/(-h_i) = d_o/d_i$, so it follows that

Magnification equation
$$m = \frac{\text{Image height, } h_i}{\text{Object height, } h_o} = -\frac{d_i}{d_o} \qquad (25.4)$$

As Examples 3 and 4 show, the value of m is positive if the image is upright and negative if the image is inverted.

Figure 25.24 These diagrams are used to derive the mirror equation and the magnification equation. (*a*) The two colored triangles are similar triangles. (*b*) If the ray is close to the principal axis, the two colored regions are almost similar triangles.

EXAMPLE 3 A Real Image Formed by a Concave Mirror

A 2.0-cm-high object is placed 7.10 cm from a concave mirror whose radius of curvature is 10.20 cm. Find (a) the location of the image and (b) its size.

Reasoning Since $f = \frac{1}{2}R = (10.20 \text{ cm})/2 = 5.10 \text{ cm}$, the object is located between the focal point F and the center of curvature C of the mirror, as in Figure 25.19a. Based on this figure, we expect that the image is real and that, relative to the object, it is farther away from the mirror, inverted, and larger.

Solution

(a) With $d_o = 7.10 \text{ cm}$ and $f = 5.10 \text{ cm}$, the mirror equation (Equation 25.3) can be used to find the image distance:

$$\frac{1}{d_i} = \frac{1}{f} - \frac{1}{d_o} = \frac{1}{5.10 \text{ cm}} - \frac{1}{7.10 \text{ cm}} = 0.055 \text{ cm}^{-1} \quad \text{or} \quad \boxed{d_i = 18 \text{ cm}}$$

In this calculation, f and d_o are positive numbers, indicating that the focal point and the object are in front of the mirror. The positive answer for d_i means that the image is also in front of the mirror, and the reflected rays actually pass through the image, as Figure 25.19a shows. In other words, the positive value for d_i indicates that the image is a real image.

(b) The height of the image can be determined once the magnification m of the mirror is known. The magnification equation (Equation 25.4) can be used to find m:

$$m = -\frac{d_i}{d_o} = -\frac{18 \text{ cm}}{7.10 \text{ cm}} = -2.5$$

The image height is $h_i = mh_o = (-2.5)(2.0 \text{ cm}) = \boxed{-5.0 \text{ cm}}$. The image is 2.5 times larger than the object, the negative values for m and h_i indicating that the image is inverted with respect to the object, as in Figure 25.19a.

● **PROBLEM SOLVING INSIGHT**
According to the mirror equation, the image distance d_i has a reciprocal given by $d_i^{-1} = f^{-1} - d_o^{-1}$. After combining the reciprocals f^{-1} and d_o^{-1}, do not forget to take the reciprocal of the result to find d_i.

EXAMPLE 4 A Virtual Image Formed by a Concave Mirror

An object is placed 6.00 cm in front of a concave mirror that has a 10.0-cm focal length. (a) Determine the location of the image. (b) The object is 1.2 cm high. Find the image height.

Reasoning The object is located between the focal point and the mirror, as in Figure 25.20a. The setup is analogous to a person using a makeup or shaving mirror. Therefore, we expect that the image is virtual and that, relative to the object, it is upright and larger.

INTERACTIVE LEARNINGWARE #49

START

Visualization

Reasoning

Knowns and Unknowns

Modeling the Problem

Numerical Solution

FINISH

A gemstone is placed 20.0 cm in front of a concave mirror and is within the focal point. When the concave mirror is replaced by a plane mirror, the image moves 15.0 cm toward the mirror. Find the focal length of the concave mirror.

Both plane mirrors and spherical mirrors operate in a way that is based on the law of reflection, as we have seen in this chapter. This Interactive Learning-Ware example deals with both types of mirrors and, therefore, is an especially good review of this material.

Go to our Web site at

http://www.wiley.com/college/cutnell

Solution

(a) Using the mirror equation with $d_o = 6.00$ cm and $f = 10.0$ cm, we have

$$\frac{1}{d_i} = \frac{1}{f} - \frac{1}{d_o} = \frac{1}{10.0 \text{ cm}} - \frac{1}{6.00 \text{ cm}}$$

$$= -0.067 \text{ cm}^{-1} \quad \text{or} \quad \boxed{d_i = -15 \text{ cm}}$$

The answer for d_i is negative, indicating that the image is *behind* the mirror. Thus, as expected, the image is virtual.

(b) The image height h_i can be found from the magnification m and the object height h_o:

$$m = -\frac{d_i}{d_o} = -\frac{(-15 \text{ cm})}{6.00 \text{ cm}} = 2.5$$

The image height is $h_i = mh_o = (2.5)(1.2 \text{ cm}) = \boxed{3.0 \text{ cm}}$. The image is larger than the object, and the positive values for m and h_i indicate that the image is upright (see Figure 25.20a).

Convex Mirrors

The mirror equation and the magnification equation can also be used with convex mirrors, provided the focal length f is taken to be a *negative number,* as indicated earlier in Equation 25.2. One way to remember this is to recall that the focal point of a convex mirror lies *behind* the mirror. Example 5 deals with a convex mirror.

EXAMPLE 5   A Virtual Image Formed by a Convex Mirror

A convex mirror is used to reflect light from an object placed 66 cm in front of the mirror. The focal length of the mirror is $f = -46$ cm (note the minus sign). Find (a) the location of the image and (b) the magnification.

Reasoning We have seen that a convex mirror always forms a virtual image, as in Figure 25.22a, where the image is upright and smaller than the object. These characteristics should also be indicated by the results of our analysis here.

Solution

(a) With $d_o = 66$ cm and $f = -46$ cm, the mirror equation gives

$$\frac{1}{d_i} = \frac{1}{f} - \frac{1}{d_o} = \frac{1}{-46 \text{ cm}} - \frac{1}{66 \text{ cm}} = -0.037 \text{ cm}^{-1} \text{ or } \boxed{d_i = -27 \text{ cm}}$$

The negative sign for d_i indicates that the image is behind the mirror and, therefore, is a virtual image.

(b) According to the magnification equation, the magnification is

$$m = -\frac{d_i}{d_o} = -\frac{(-27 \text{ cm})}{66 \text{ cm}} = \boxed{0.41}$$

The image is smaller (m is less than one) and upright (m is positive) with respect to the object.

● **PROBLEM SOLVING INSIGHT**
When using the mirror equation, it is useful to construct a ray diagram to guide your thinking and to check your calculation.

Convex mirrors, like plane (flat) mirrors, always produce virtual images behind the mirror. However, the virtual image in a convex mirror is closer to the mirror than it would be if the mirror were planar, as Example 6 illustrates.

EXAMPLE 6   A Convex Versus a Plane Mirror

An object is placed 9.00 cm in front of a mirror. The image is 3.00 cm closer to the mirror when the mirror is convex than when it is planar. Find the focal length of the convex mirror.

Reasoning For a plane mirror, the image and the object are the same distance on either side of the mirror. Thus, the image would be 9.00 cm behind a plane mirror. If the image in a convex mirror is 3.00 cm closer than this, the image must be located 6.00 cm behind the convex mirror. In other words, when the object distance is $d_o = 9.00$ cm, the image distance for the convex mirror is $d_i = -6.00$ cm (negative because the image is virtual). The mirror equation can be used to find the focal length of the mirror.

Solution According to the mirror equation, the reciprocal of the focal length is

$$\frac{1}{f} = \frac{1}{d_o} + \frac{1}{d_i} = \frac{1}{9.00 \text{ cm}} + \frac{1}{-6.00 \text{ cm}} = -0.056 \text{ cm}^{-1} \quad \text{or} \quad \boxed{f = -18 \text{ cm}}$$

The Reasoning Strategy below summarizes the sign conventions that are used with the mirror equation and the magnification equation. These conventions apply to both concave and convex mirrors.

REASONING STRATEGY

Summary of Sign Conventions for Spherical Mirrors

Focal length

   f is + for a concave mirror.

   f is − for a convex mirror.

Object distance

   d_o is + if the object is in front of the mirror (real object).

   d_o is − if the object is behind the mirror (virtual object).*

Image distance

   d_i is + if the image is in front of the mirror (real image).

   d_i is − if the image is behind the mirror (virtual image).

Magnification

   m is + for an image that is upright with respect to the object.

   m is − for an image that is inverted with respect to the object.

---

* Sometimes optical systems use two (or more) mirrors, and the image formed by the first mirror serves as the object for the second mirror. Occasionally, such an object falls *behind* the second mirror. In this case the object distance is negative, and the object is said to be a virtual object.

25.7 CONCEPTS & CALCULATIONS

The reflections of the boy and girl in this fun-house mirror are distorted because the mirror is curved and not flat. It is not a spherical mirror, however.

Relative to the object in front of a spherical mirror, the image can differ in a number of respects. The image can be real (in front of the mirror) or virtual (behind the mirror). It can be larger or smaller than the object. And it can be upright or inverted. As you solve problems dealing with spherical mirrors, keep these image characteristics in mind. They can help guide you to the correct answer, as Examples 7 and 8 illustrate.

CONCEPTS & CALCULATIONS EXAMPLE 7   Finding the Focal Length

An object is located 7.0 cm in front of a mirror. The virtual image is located 4.5 cm away from the mirror and is smaller than the object. Find the focal length of the mirror.

Concept Questions and Answers Based solely on the fact that the image is virtual, is the mirror concave or convex or is either type possible?

Answer Either type is possible. A concave mirror can form a virtual image if the object is within the focal point of the mirror, as Figure 25.20a illustrates. A convex mirror always forms a virtual image, as Figure 25.22a shows.

The image is smaller than the object, as well as virtual. Do these characteristics together indicate a concave or convex mirror, or do they indicate either type?

Answer They indicate a convex mirror. A concave mirror can produce an image that is smaller than the object if the object is located beyond the center of curvature of the mirror, as in Figure 25.19b. However, the image in Figure 25.19b is real, not virtual. A concave mirror, in contrast, always produces an image that is virtual and smaller than the object, as Figure 25.22a illustrates.

Is the focal length positive or negative?

Answer The focal length is negative because the mirror is convex. A concave mirror has a positive focal length.

Solution The virtual image is located behind the mirror and, therefore, has a negative image distance, $d_i = -4.5$ cm. Using this value together with the object distance of $d_o = 7.0$ cm, we can apply the mirror equation to find the focal length:

$$\frac{1}{f} = \frac{1}{d_o} + \frac{1}{d_i} = \frac{1}{7.0 \text{ cm}} + \frac{1}{-4.5 \text{ cm}} = -0.079 \text{ cm}^{-1} \quad \text{or} \quad \boxed{f = -13 \text{ cm}}$$

As expected, the focal length is negative.

CONCEPTS & CALCULATIONS EXAMPLE 8   Two Choices

The radius of curvature of a mirror is 12 cm. A diamond ring is placed in front of this mirror. The image is twice the size of the ring. Find the object distance of the ring.

Concept Questions and Answers Is the mirror concave or convex or is either type possible?

Answer A convex mirror always forms an image that is smaller than the object, as Figure 25.22a shows. Therefore, the mirror must be concave.

How many places are there in front of a concave mirror where the ring can be placed and produce an image that is twice the size of the object?

Answer There are two places. Figure 25.19*a* illustrates that one of the places is between the center of curvature and the focal point. The enlarged image is real and inverted. Figure 25.20*a* shows that another possibility is between the focal point and the mirror, in which case the enlarged image is virtual and upright.

What are the possible values for the magnification of the image of the ring?

Answer Since the image is inverted in Figure 25.19*a*, the magnification for this possibility is $m = -2$. In Figure 25.20*a*, however, the image is upright, so the magnification is $m = +2$ for this option. In either case, the image is twice the size of the ring.

Solution According to the mirror equation and the magnification equation, we have

$$\underbrace{\frac{1}{d_o} + \frac{1}{d_i} = \frac{1}{f}}_{\text{Mirror equation}} \quad \text{and} \quad \underbrace{m = -\frac{d_i}{d_o}}_{\text{Magnification equation}}$$

We can solve the magnification equation for the image distance and obtain $d_i = -m d_o$. Substituting this expression for d_i into the mirror equation gives

$$\frac{1}{d_o} + \frac{1}{(-m d_o)} = \frac{1}{f} \quad \text{or} \quad d_o = \frac{f(m-1)}{m}$$

Applying this result with the two magnifications, we obtain

$m = -2 \qquad d_o = \frac{f(m-1)}{m} = \frac{(12 \text{ cm})(-2-1)}{-2} = \boxed{18 \text{ cm}}$

$m = +2 \qquad d_o = \frac{f(m-1)}{m} = \frac{(12 \text{ cm})(+2-1)}{+2} = \boxed{6.0 \text{ cm}}$

At the end of the problem set for this chapter, you will find homework problems that contain both conceptual and quantitative parts. These problems are grouped under the heading *Concepts & Calculations, Group Learning Problems*. They are designed for use by students working alone or in small learning groups. The conceptual part of each problem provides a convenient focus for group discussions.

SUMMARY

Wave fronts are surfaces on which all points of a wave are in the same phase of motion. If the wave fronts are flat surfaces, the wave is called a **plane wave. Rays** are lines that are perpendicular to the wave fronts and point in the direction of the velocity of the wave.

When light reflects from a smooth surface, the reflected light obeys the **law of reflection,** which states that (a) the incident ray, the reflected ray, and the normal to the surface all lie in the same plane, and (b) the angle of reflection equals the angle of incidence. The law of reflection explains how mirrors form images. A **virtual image** is one from which all the rays of light do not actually come but only appear to do so. A **real image** is one from which all the rays of light actually do emanate. A **plane mirror** forms an upright, virtual image that is located as far behind the mirror as the object is in front of it. In addition, the heights of the image and the object are equal.

A **spherical mirror** has the shape of a section from the surface of a sphere. The **principal axis** of a mirror is a straight line drawn through the center of curvature and the middle of the mirror's surface. Rays that lie close to the principal axis are known as **paraxial rays. The radius of curvature** R of the mirror is the distance from the center of curvature to the mirror. The **focal point** of a concave spherical mirror is a point on the principal axis, in front of the mirror. Incident paraxial rays that are parallel to the principal axis converge to the focal point after being reflected from the concave mirror. The focal point of a convex spherical mirror is a point on the principal axis behind the mirror. For a convex mirror, paraxial rays that are parallel to the principal axis seem to diverge from the focal point after reflecting from the mirror. The **focal length** f of a mirror is the distance along the principal axis between the focal point and the mirror. The focal length and the radius of curvature are related by $f = \frac{1}{2}R$ for a concave mirror and $f = -\frac{1}{2}R$ for a convex mirror.

The image produced by a concave mirror can be located by a graphical method known as **ray tracing,** using the three rays shown in Figure 25.18. Similarly, the image produced by a convex mirror can be found using the three rays shown in Figure 25.22.

The **mirror equation** can be used with either concave or convex mirrors and specifies the relation between the image distance d_i, the object distance d_o, and the focal length f of the mirror: $1/d_o + 1/d_i = 1/f$. The **magnification** m of a mirror is the ratio of the image height h_i to the object height h_o. The magnification is also related to d_i and d_o by the **magnification equation:** $m = h_i/h_o = -d_i/d_o$. The algebraic sign conventions for the variables appearing in these equations are summarized in the Reasoning Strategy at the end of Section 25.6.

CONCEPTUAL QUESTIONS

1. A sign painted on a store window is reversed when viewed from inside the store. If a person inside the store views the reversed sign in a plane mirror, does the sign appear as it would when viewed from outside the store? (Try it by writing some letters on a transparent sheet of paper and then holding the back side of the paper up to a mirror.) Explain.

2. If a clock is held in front of a mirror, its image is reversed left to right. From the point of view of a person looking into the mirror, does the image of the second hand rotate in the reverse (counterclockwise) direction? Justify your answer.

3. (a) Which kind of spherical mirror, concave or convex, can be used to start a fire with sunlight? (b) For the best results, how far from the mirror should the paper to be ignited be placed? Explain your answers.

4. The photograph shows an experimental device at Sandia National Laboratories in New Mexico. This device is a mirror that focuses sunlight to heat sodium to a boil, which then heats helium gas in an engine. The engine does the work of driving a generator to produce electricity. The sodium unit and the engine are labeled in the

Sodium unit and engine

photo. (a) What kind of mirror is being used, and (b) where is the sodium unit located relative to the mirror? Express your answer in terms of the focal length of the mirror. Give your reasoning.

5. Refer to Figure 25.14 and the related discussion about spherical aberration. To bring the top ray closer to the focal point F after reflection, describe how you would change the shape of the mirror. Would you open it up to produce a more gently curving shape or bring the top and bottom edges closer to the principal axis? Account for your answer using the law of reflection.

6. (a) Can the image formed by a concave mirror ever be projected directly onto a screen, without the help of other mirrors or lenses? If so, specify where the object should be placed relative

to the mirror. (b) Repeat part (a) assuming that the mirror is convex.

7. (a) When you look at the back side of a shiny teaspoon, held at arm's length, you see yourself upright. Why? (b) When you look at the other side of the spoon, you see yourself upside down. Why?

8. If you stand between two parallel plane mirrors, you see an infinite number of images of yourself. This occurs because an image in one mirror is reflected in the other mirror to produce another image, which is then re-reflected, and so forth. The multiple images are equally spaced. Suppose that you are facing a convex mirror, with a plane mirror behind you. Describe what you would see and comment about the spacing between any multiple images. Explain your reasoning.

9. Sometimes news personnel covering an event use a microphone arrangement that is designed to increase the ability of the mike to pick up weak sounds. The drawing shows that the arrange-

Microphone

ment consists of a "hollowed-out" shell behind the mike. The shell acts like a mirror for sound waves. Explain how this arrangement enables the mike to detect weak sounds.

10. When you see the image of yourself formed by a mirror, it is because (1) light rays actually coming from a real image enter your eyes or (2) light rays appearing to come from a virtual image enter your eyes. If light rays from the image do not enter your eyes, you do not see yourself. Are there any places on the principal axis where you cannot see yourself when you are in front of a mirror that is (a) convex and (b) concave? If so, where are these places?

11. Plane mirrors and convex mirrors form virtual images. With a plane mirror, the image may be infinitely far behind the mirror, depending on where the object is located in front of the mirror. For an object in front of a single convex mirror, what is the greatest distance behind the mirror at which the image can be found? Justify your answer.

12. Is it possible to use a convex mirror to produce an image that is larger than the object? Provide a reason for your answer.

13. Suppose you stand in front of a spherical mirror (concave or convex). Is it possible for your image to be (a) real and upright or (b) virtual and inverted? Justify your answers.

PROBLEMS

Section 25.2 The Reflection of Light,
Section 25.3 The Formation of Images by a Plane Mirror

1. ssm Two plane mirrors are separated by 120°, as the drawing illustrates. If a ray strikes mirror M_1 at a 65° angle of incidence, at what angle θ does it leave mirror M_2?

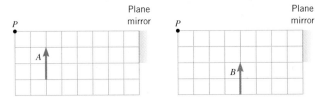

Problem 1

2. Review Conceptual Example 1 as an aid in understanding this problem. The drawings show two arrows, A and B, that are located in front of a plane mirror. A person at point P is viewing the image of each arrow. Which images can be seen in their entirety? Determine your answers by drawing a ray from the head and foot of each arrow that reflects from the mirror according to the law of reflection and reaches point P. Only if *both* rays reach point P after reflection can the image of that arrow be seen in its entirety.

3. A person stands 3.6 m in front of a wall that is covered floor-to-ceiling with a plane mirror. His eyes are 1.8 m above the floor. He holds a flashlight between his feet and manages to point it at the mirror. At what angle of incidence must the light strike the mirror so the light will reach his eyes?

4. Review Conceptual Example 2. Suppose that in Figure 25.9*b* the two perpendicular plane mirrors are represented by the $-x$ and $-y$ axes of an x, y coordinate system. An object is in front of these mirrors at a point whose coordinates are $x = -2.0$ m and $y = -1.0$ m. Find the coordinates that locate each of the three images.

5. ssm www Two diverging light rays, originating from the same point, have an angle of 10° between them. After the rays reflect from a plane mirror, what is the angle between them? Construct one possible ray diagram that supports your answer.

6. On the $+y$ axis a laser is located at $y = +3.0$ cm. The coordinates of a target are $x = +9.0$ cm and $y = +6.0$ cm. The $+x$ axis represents the edge-on view of a mirror. At what point on the $+x$ axis should the laser be aimed in order to hit the target after reflection?

*7. A ray of light strikes a plane mirror at a 45° angle of incidence. The mirror is then rotated by 15° into the position shown in red in the drawing, while the incident ray is kept fixed. (a) Through what angle ϕ does the reflected ray rotate? (b) What is the answer to part (a) if the angle of incidence is 60° instead of 45°?

*8. The drawing shows a top view of a square room. One wall is missing, and the other three are each mirrors. From point P in the center of the open side, a laser is fired, with the intent of hitting a small target located at the center of one wall. Identify five directions in which the laser can be fired and score a hit, assuming that the light does not strike any mirror more than once. Draw the rays to confirm your choices.

9. ssm A lamp is twice as far in front of a plane mirror as a person is. Light from the lamp reaches the person via two paths. It strikes the mirror at a 30.0° angle of incidence and reflects from it before reaching the person. It also travels directly to the person without reflecting. Find the ratio of the travel time along the reflected path to the travel time along the direct path.

Section 25.4 Spherical Mirrors,
Section 25.5 The Formation of Images by Spherical Mirrors

10. A 2.0-cm-high object is situated 15.0 cm in front of a concave mirror that has a radius of curvature of 10.0 cm. Using a ray diagram drawn to scale, measure (a) the location and (b) the height of the image. The mirror must be drawn to scale.

11. Repeat problem 10 for a concave mirror with a focal length of 20.0 cm, an object distance of 12.0 cm, and a 2.0-cm-high object.

12. A convex mirror has a focal length of -40.0 cm. A 12.0-cm-tall object is located 40.0 cm in front of this mirror. Using a ray diagram drawn to scale, determine the (a) location and (b) size of the image. Note that the mirror must be drawn to scale.

13. ssm Repeat problem 10 for a convex mirror with a radius of curvature of 1.00×10^2 cm, an object distance of 25.0 cm, and a 10.0-cm-high object.

14. Repeat problem 10 for a concave mirror with a focal length of 7.50 cm, an object distance of 11.0 cm, and a 1.0-cm-high object.

* **15.** A plane mirror and a concave mirror ($f = 8.0$ cm) are facing each other and are separated by a distance of 20.0 cm. An object is placed 10.0 cm in front of the plane mirror. Consider the light from the object that reflects first from the plane mirror and then from the concave mirror. Using a ray diagram drawn to scale, find the location of the image that this light produces in the concave mirror. Specify this distance relative to the concave mirror.

Section 25.6 The Mirror Equation and the Magnification Equation

16. A coin is placed 8.0 cm in front of a concave mirror. The mirror produces a real image that has a diameter 4.0 times larger than that of the coin. What is the image distance?

17. ssm A concave mirror has a focal length of 42 cm. The image formed by this mirror is 97 cm in front of the mirror. What is the object distance?

18. A clown is using a concave makeup mirror to get ready for a show and is 27 cm in front of the mirror. The image is 65 cm *behind* the mirror. Find (a) the focal length of the mirror and (b) the magnification.

19. A concave mirror ($R = 64.0$ cm) is used to project a transparent slide onto a wall. The slide is located at a distance of 38.0 cm from the mirror, and a small flashlight shines light through the slide and onto the mirror. The setup is similar to that in Figure 25.19a. (a) How far from the wall should the mirror be located? (b) The height of the object on the slide is 1.20 cm. What is the height of the image? (c) How should the slide be oriented, so that the picture on the wall looks normal?

20. An object that is 25 cm in front of a convex mirror has an image located 17 cm behind the mirror. How far behind the mirror is the image located when the object is 19 cm in front of the mirror?

21. ssm The image behind a convex mirror (radius of curvature = 68 cm) is located 22 cm from the mirror. (a) Where is the object located and (b) what is the magnification of the mirror? Determine whether the image is (c) upright or inverted and (d) larger or smaller than the object.

22. A convex mirror produces an image that is half the size of an object that is placed 13 cm in front of it. What is the focal length of the mirror?

23. ssm www A small postage stamp is placed in front of a concave mirror (radius = R), such that the image distance equals the object distance. (a) In terms of R, what is the object distance? (b) What is the magnification of the mirror? (c) State whether the image is upright or inverted relative to the object. Draw a ray diagram to guide your thinking.

* **24.** A dentist's mirror is placed 2.0 cm from a tooth. The *enlarged* image is located 5.6 cm behind the mirror. (a) What kind of mirror (plane, concave, or convex) is being used? (b) Determine the focal length of the mirror. (c) What is the magnification? (d) How is the image oriented relative to the object?

* **25.** An object is placed in front of a convex mirror, and the size of the image is one-third that of the object. What is the ratio d_o/f of the object distance to the focal length of the mirror?

* **26.** The same object is located at the same distance from two spherical mirrors, A and B. The magnifications produced by the mirrors are $m_A = 4.0$ and $m_B = 2.0$. Find the ratio f_A/f_B of the focal lengths of the mirrors.

* **27. ssm www** An image formed by a convex mirror ($f = -24.0$ cm) has a magnification of 0.150. Which way and by how much should the object be moved to double the size of the image?

** **28.** A concave mirror has a focal length of 30.0 cm. The distance between an object and its image is 45.0 cm. Find the object and image distances assuming that (a) the object lies beyond the center of curvature and (b) the object lies within the focal point.

** **29.** Using the mirror equation and the magnification equation, show that for a convex mirror the image is always (a) virtual (i.e., d_i is always negative) and (b) upright and smaller, relative to the object (i.e., m is positive and less than one).

ADDITIONAL PROBLEMS

30. The image of a very distant car is located 12 cm behind a convex mirror. (a) What is the radius of curvature of the mirror? (b) Draw a ray diagram to scale showing this situation.

31. ssm www When viewed in a spherical mirror, the image of a setting sun is a virtual image. The image lies 12.0 cm behind the mirror. (a) Is the mirror concave or convex? Why? (b) What is the radius of curvature of the mirror?

32. Review Conceptual Example 1 before attempting this problem. A person whose eyes are 1.70 m above the floor stands in front of a plane mirror. The top of her head is 0.12 m above her eyes. (a) What is the height of the shortest mirror in which she can see her entire image? (b) How far above the floor should the bottom edge of the mirror be placed?

33. The focal length of a concave mirror is 17 cm. An object is located 38 cm in front of this mirror. Where is the image located?

34. Convex mirrors are being used to monitor the aisles in a store. The mirrors have a radius of curvature of 4.0 m. (a) What is the image distance if a customer is 15 m in front of the mirror? (b) Is the image real or virtual? (c) If a customer is 1.6 m tall, how tall is the image?

35. ssm The image produced by a concave mirror is located 26 cm in front of the mirror. The focal length of the mirror is 12 cm. How far in front of the mirror is the object located?

* **36.** The drawing shows two plane mirrors that intersect at an angle of 50°. An incident light ray reflects from one mirror and then the other. What is the angle θ between the incident and outgoing rays?

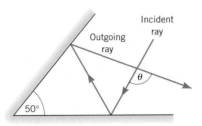

*37. In Figure 25.21*b* the head-up display is designed so that the distance between the digital readout device and virtual image 1 is 2.00 m. The magnification of virtual image 1 is 4.00. Find the focal length of the concave mirror. *(Hint: Remember that the image distance for virtual image 1 is a negative quantity.)*

*38. A candle is placed 15.0 cm in front of a convex mirror. When the convex mirror is replaced with a plane mirror, the image moves 7.0 cm farther away from the mirror. Find the focal length of the convex mirror.

*39. **ssm** An object is located 14.0 cm in front of a convex mirror, the image being 7.00 cm behind the mirror. A second object, twice as tall as the first one, is placed in front of the mirror, but at a different location. The image of this second object has the same height as the other image. How far in front of the mirror is the second object located?

**40. A spherical mirror is polished on both sides. When used as a convex mirror, the magnification is +1/4. What is the magnification when used as a concave mirror, the object remaining the same distance from the mirror?

**41. In the drawing for problem 8, a laser is fired from point *P* in the center of the open side of the square room. The laser is pointed at the mirrored wall on the right. At what angle of incidence must the light strike the right-hand wall, so that after being reflected, the light hits the left corner of the back wall?

CONCEPTS & CALCULATIONS
GROUP LEARNING PROBLEMS

Note: Each of these problems consists of Concept Questions followed by a related quantitative Problem. They are designed for use by students working alone or in small learning groups. The Concept Questions involve little or no mathematics and are intended to stimulate group discussions. They focus on the concepts with which the problems deal. Recognizing the concepts is the essential initial step in any problem-solving technique.

42. Concept Questions (a) Suppose that you are walking toward a stationary plane mirror. Following the method discussed in Section 3.4, express your image's velocity $\mathbf{v}_{IY}$ relative to you in terms of the image's velocity $\mathbf{v}_{IM}$ relative to the mirror and the mirror's velocity $\mathbf{v}_{MY}$ relative to you. (b) How is the mirror's velocity $\mathbf{v}_{MY}$ relative to you related to your velocity $\mathbf{v}_{YM}$ relative to the mirror? Explain. (c) Consider both velocities $\mathbf{v}_{YM}$ and $\mathbf{v}_{IM}$. Do they have the same magnitudes and the same directions? Explain.

Problem When you walk perpendicularly toward a stationary plane mirror with a velocity of +0.90 m/s, what is the velocity of your image relative to you? The direction in which you walk is the positive direction.

43. Concept Questions (a) Suppose that you are walking toward a plane mirror as in the drawing. The view is from above. Following the method discussed in Section 3.4, express your image's velocity $\mathbf{v}_{IY}$ relative to you in terms of the image's velocity $\mathbf{v}_{IM}$ relative to the mirror and the mirror's velocity $\mathbf{v}_{MY}$ relative to you. (b) How is the mirror's velocity $\mathbf{v}_{MY}$ relative to you related to your velocity $\mathbf{v}_{YM}$ relative to the mirror? Explain. (c) Consider both velocities $\mathbf{v}_{YM}$ and $\mathbf{v}_{IM}$. Do they have the same *x* and *y* components? Explain.

Problem You walk at an angle of $\theta = 50.0°$ toward a plane mirror, as in the drawing. Your walking velocity has a magnitude of 0.90 m/s. What is the velocity of your image relative to you (magnitude and direction)?

Problem 43

44. Concept Questions (a) For an image that is in front of a mirror, is the image distance positive or negative? (b) Given the image distance, what additional information is needed to determine the focal length? Explain. (c) For an inverted image is the image height positive or negative? (d) Given the object and image heights and a statement as to whether the image is upright or inverted, what additional information is needed to determine the object distance?

Problem A small statue has a height of 3.5 cm and is placed in front of a concave mirror. The image of the statue is inverted, 1.5 cm tall, and is located 13 cm in front of the mirror. Find the focal length of the mirror.

45. Concept Questions These questions refer to Figure 25.22*a*. (a) As the object distance increases, does reflected ray 1 change? (b) As the object distance increases, does reflected ray 3 make a greater or smaller angle with respect to the principal axis? (c) Extending the reflected rays 1 and 3 behind the mirror allows us to locate the top of the image. As the object distance increases, does the image height increase or decrease?

Problem A convex mirror has a focal length of −27.0 cm. Find the magnification produced by the mirror when the object distance is 9.0 cm and 18.0 cm. Verify that your answers are consistent with your answers to the Concept Questions.

This spectacular intrauterine photograph of a fetus was taken with the aid of optical fibers, which use the refraction of light in a way that is discussed in this chapter.

CHAPTER 26

THE REFRACTION OF LIGHT: LENSES AND OPTICAL INSTRUMENTS

26.1 THE INDEX OF REFRACTION

As Section 24.3 discusses, light travels through a vacuum at a speed $c = 3.00 \times 10^8$ m/s. It can also travel through many materials, such as air, water, and glass. Atoms in the material absorb, reemit, and scatter the light, however. Therefore, light travels through the material at a speed that is less than c, the actual speed depending on the nature of the material. In general, we will see that the change in speed as a ray of light goes from one material to another causes the ray to deviate from its incident direction. This change in direction is called **refraction,** and it is governed by Snell's law of refraction, which will be discussed in the next section.

    ● *Concepts at a Glance*   To describe the extent to which the speed of light in a material medium differs from that in a vacuum, we use a parameter called the ***index of refraction*** (or ***refractive index***). As the Concepts-at-a-Glance chart in Figure 26.1 illustrates, the index of refraction incorporates both the speed of light c in a vacuum and the speed v in the material. The index of refraction is an important parameter because it appears in Snell's law of refraction, the cornerstone of all the phenomena discussed in this chapter. ●

● DEFINITION OF THE INDEX OF REFRACTION

The index of refraction n of a material is the ratio of the speed c of light in a vacuum to the speed v of light in the material:

$$n = \frac{\text{Speed of light in a vacuum}}{\text{Speed of light in the material}} = \frac{c}{v} \qquad (26.1)$$

Table 26.1 lists the refractive indices for some common substances. The values of n are greater than unity because the speed of light in a material medium is less than it is in a vacuum. For example, the index of refraction for diamond is $n = 2.419$, so the speed of light in diamond is $v = c/n = (3.00 \times 10^8$ m/s)/2.419 =

Figure 26.1 *Concepts at a Glance* The index of refraction of a material incorporates both the speed of light in a vacuum and the speed in the material. The index of refraction appears in Snell's law of refraction, which describes how light changes its direction of travel in passing from one material to another. This change in direction of travel is evident in the photograph, which shows light traveling through air and being bent sharply downward upon entering a tank of water.

Concepts at a Glance

Speed c of Light in a Vacuum (Section 24.3)

Speed v of Light in a Material Medium

Index of Refraction

Snell's Law of Refraction (Section 26.2)

TABLE 26.1 Index of Refraction[a] for Various Substances

Substance	Index of Refraction, n
Solids at 20 °C	
Diamond	2.419
Glass, crown	1.523
Ice (0 °C)	1.309
Sodium chloride	1.544
Quartz	
Crystalline	1.544
Fused	1.458
Liquids at 20 °C	
Benzene	1.501
Carbon disulfide	1.632
Carbon tetrachloride	1.461
Ethyl alcohol	1.362
Water	1.333
Gases at 0 °C, 1 atm	
Air	1.000 293
Carbon dioxide	1.000 45
Oxygen, O_2	1.000 271
Hydrogen, H_2	1.000 139

[a] Measured with light whose wavelength in a vacuum is 589 nm.

1.24×10^8 m/s. In contrast, the index of refraction for air (and also for other gases) is so close to unity that $n_{air} = 1$ for most purposes. The index of refraction depends slightly on the wavelength of the light, and the values in Table 26.1 correspond to a wavelength of $\lambda = 589$ nm in a vacuum.

26.2 SNELL'S LAW AND THE REFRACTION OF LIGHT

Snell's Law

When light strikes the interface between two transparent materials, such as air and water, the light generally divides into two parts, as Figure 26.2a illustrates. Part of the light is reflected, with the angle of reflection equaling the angle of incidence. The remainder is transmitted across the interface. If the incident ray does not strike the interface at normal incidence, the transmitted ray has a different direction than the incident ray. The ray that enters the second material is said to be refracted.

In Figure 26.2a the light travels from a medium where the refractive index is smaller (air) into a medium where it is larger (water), and the refracted ray is bent *toward* the normal. Both the incident and refracted rays obey the principle of reversibility, so their directions can be reversed to give a situation like that in part b of the drawing. Here light travels from a material with a greater refractive index (water) into one with a smaller refractive index (air), and the refracted ray is bent *away* from the normal. In this case the reflected ray lies in the water rather than in the air. In both parts of the drawing the angles of incidence, refraction, and reflection are measured relative to the normal. Note that the index of refraction of air is labeled n_1 in part a, while it is labeled n_2 in part b, because *we label all variables associated with the incident (and reflected) ray with a subscript 1 and all variables associated with the refracted ray with a subscript 2.*

The angle of refraction θ_2 depends on the angle of incidence θ_1 and on the indices of refraction, n_2 and n_1, of the two media. The relation between these quantities is known as *Snell's law of refraction,* after the Dutch mathematician Willebrord Snell (1591–1626) who discovered it experimentally. A proof of Snell's law is presented at the end of this section.

● **SNELL'S LAW OF REFRACTION**

When light travels from a material with refractive index n_1 into a material with refractive index n_2, the refracted ray, the incident ray, and the normal to the interface between the materials all lie in the same plane. The angle of refraction θ_2 is related to the angle of incidence θ_1 by

$$n_1 \sin \theta_1 = n_2 \sin \theta_2 \qquad (26.2)$$

Example 1 illustrates the use of Snell's law.

● **PROBLEM SOLVING INSIGHT**
The angle of incidence θ_1 and the angle of refraction θ_2 that appear in Snell's law are measured with respect to the normal to the surface, and not with respect to the surface itself.

EXAMPLE 1   Determining the Angle of Refraction

A light ray strikes an air/water surface at an angle of 46° with respect to the normal. The refractive index for water is 1.33. Find the angle of refraction when the direction of the ray is (a) from air to water and (b) from water to air.

Reasoning Snell's law of refraction applies to both part (a) and part (b). However, in part (a) the incident ray is in air, while in part (b) it is in water. We keep track of this difference by always labeling the variables associated with the incident ray with a subscript 1 and the variables associated with the refracted ray with a subscript 2.

Solution

(a) The incident ray is in air, so $\theta_1 = 46°$ and $n_1 = 1.00$. The refracted ray is in water, so $n_2 = 1.33$. Snell's law can be used to find the angle of refraction θ_2:

$$\sin \theta_2 = \frac{n_1 \sin \theta_1}{n_2} = \frac{(1.00) \sin 46°}{1.33} = 0.54 \qquad (26.2)$$

$$\theta_2 = \sin^{-1}(0.54) = \boxed{33°}$$

Since θ_2 is less than θ_1, the refracted ray is bent *toward* the normal, as Figure 26.2a shows.

(b) With the incident ray in water, we find that

$$\sin \theta_2 = \frac{n_1 \sin \theta_1}{n_2} = \frac{(1.33) \sin 46°}{1.00} = 0.96$$

$$\theta_2 = \sin^{-1}(0.96) = \boxed{74°}$$

Since θ_2 is greater than θ_1, the refracted ray is bent *away* from the normal, as Figure 26.2b indicates.

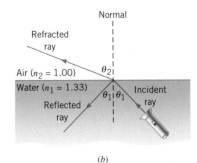

Figure 26.2 (a) When a ray of light is directed from air into water, part of the light is reflected at the interface and the remainder is refracted into the water. The refracted ray is bent *toward* the normal ($\theta_2 < \theta_1$). (b) When a ray of light is directed from water into air, the refracted ray in air is bent *away* from the normal ($\theta_2 > \theta_1$).

We have seen that reflection and refraction of light waves occur simultaneously at the interface between two transparent materials. It is important to keep in mind that light waves are composed of electric and magnetic fields, which carry energy. The principle of conservation of energy (see Chapter 6) indicates that the energy reflected plus the energy refracted must add up to equal the energy carried by the incident light, provided that none of the energy is absorbed by the materials. The percentage of incident energy that appears as reflected versus refracted light depends on the angle of incidence and the refractive indices of the materials on either side of the interface. For instance, when light travels from air toward water at perpendicular incidence, most of its energy is refracted and little is reflected. But when the angle of incidence is nearly 90° and the light barely grazes the water surface, most of its energy is reflected, with only a small amount refracted into the water. On a rainy night, you probably have experienced the annoying glare that results when light from an oncoming car just grazes the wet road. Under such conditions, most of the light energy reflects into your eyes.

The simultaneous reflection and refraction of light have applications in a number of devices. For instance, interior rearview mirrors in cars often have adjustment levers. One position of the lever is for day driving, while the other is for night driving and reduces glare from the headlights of the car behind. As Figure 26.3a indicates, this kind of mirror is a glass wedge, the back side of which is silvered and highly reflecting. Part *b* of the picture shows the day setting. Light from the car behind follows path *ABCD* in reaching the driver's eye. At points *A* and *C*, where the light strikes the air–glass surface, there are both reflected and refracted rays. The reflected rays are drawn as thin lines, the thinness denoting that only a small percentage (about 10%) of the light during the day is reflected at *A* and *C*. The weak reflected rays at *A* and *C* do not reach the driver's eye. In contrast, almost all the light reaching the silvered back surface at *B* is reflected toward the driver. Since most of the light follows path *ABCD*, the driver sees a bright image of the car behind. During the night, the adjustment lever can be used to rotate the mirror

THE PHYSICS OF...

rearview mirrors that have a day–night adjustment.

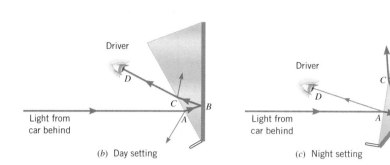

Figure 26.3 An interior rearview mirror with a day–night adjustment lever.

(a)

(b) Day setting

(c) Night setting

clockwise (see part *c* of the drawing), away from the driver. Now, most of the light from the headlights behind follows path *ABC* and does not reach the driver. Only the light that is weakly reflected from the front surface along path *AD* is seen. As a result, there is significantly less glare.

Apparent Depth

One interesting consequence of refraction is that an object lying under water appears to be closer to the surface than it actually is. Example 2 sets the stage for explaining why, by showing what must be done to shine a light on such an object.

Figure 26.4 The beam from the searchlight is refracted when it enters the water.

● **PROBLEM SOLVING INSIGHT**
Remember that the refractive indices are written as n_1 for the medium in which the incident light travels and n_2 for the medium in which the refracted light travels.

EXAMPLE 2 Finding a Sunken Chest

A searchlight on a yacht is being used at night to illuminate a sunken chest, as in Figure 26.4. At what angle of incidence θ_1 should the light be aimed?

Reasoning The angle of incidence θ_1 must be such that, after refraction, the light strikes the chest. The angle of incidence can be obtained from Snell's law, once the angle of refraction θ_2 is determined. This angle can be found using the data in Figure 26.4 and trigonometry. The light travels from a region of lower into a region of higher refractive index, so the light is bent toward the normal and we expect θ_1 to be greater than θ_2.

Solution From the data in the drawing it follows that $\tan \theta_2 = (2.0 \text{ m})/(3.3 \text{ m})$, so $\theta_2 = 31°$. With $n_1 = 1.00$ for air and $n_2 = 1.33$ for water, Snell's law gives

$$\sin \theta_1 = \frac{n_2 \sin \theta_2}{n_1} = \frac{(1.33) \sin 31°}{1.00} = 0.69$$
$$\theta_1 = \sin^{-1}(0.69) = \boxed{44°}$$

As expected, θ_1 is greater than θ_2.

When the sunken chest in Example 2 is viewed from the boat (Figure 26.5*a*), light rays from the chest pass upward through the water, refract away from the normal when they enter the air, and then travel to the observer. This picture is similar to Figure 26.4, except the direction of the rays is reversed and the searchlight is replaced by an observer. When the rays entering the air are extended back into the water (see the dashed lines), they indicate that the observer sees a virtual image of the chest at an *apparent depth* that is less than the actual depth. The image is virtual because light rays do not actually pass through it. For the situation shown in Figure 26.5*a*, it is difficult to determine the apparent depth. A case that is much simpler is shown in part *b* of the drawing, where the observer is *directly above* the submerged object, and the apparent depth d' is related to the actual depth d by

Apparent depth, observer directly above object

$$d' = d\left(\frac{n_2}{n_1}\right) \qquad (26.3)$$

In this result, n_1 is the refractive index of the medium associated with the incident ray (the medium in which the object is located), while n_2 refers to the medium associated with the refracted ray (the medium in which the observer is situated). The proof of Equation 26.3 is the focus of problem 17 at the end of the chapter. Example 3 illustrates that the effect of apparent depth is quite noticeable in water.

EXAMPLE 3   The Apparent Depth of a Swimming Pool

A swimmer is treading water (with her head above the water) at the surface of a 3.00-m-deep pool. She sees a coin on the bottom directly below. How deep does the coin appear to be?

Reasoning Equation 26.3 may be used to find the apparent depth, provided we remember that the light rays travel from the coin to the swimmer. Therefore, the incident ray is coming from the coin under the water ($n_1 = 1.33$), while the refracted ray is in the air ($n_2 = 1.00$).

Solution The apparent depth d' of the coin is

$$d' = d\left(\frac{n_2}{n_1}\right) = (3.00\text{ m})\left(\frac{1.00}{1.33}\right) = \boxed{2.26\text{ m}} \qquad (26.3)$$

In Example 3, a person sees a coin on the bottom of a pool at an apparent depth that is less than the actual depth. Conceptual Example 4 considers the reverse situation, namely, a person looking from under the water at a coin in the air.

CONCEPTUAL EXAMPLE 4   On the Inside Looking Out

A swimmer is under water and looking up at the surface. Someone holds a coin in the air, directly above the swimmer's eyes. To the swimmer, the coin appears to be at a certain height above the water. Is the apparent height of the coin greater than, less than, or the same as its actual height?

Reasoning and Solution Figure 26.6 shows that the apparent height of the coin above the water surface is greater than the actual height. Consider two rays of light leaving a point P on the coin. When the rays enter the water, they are refracted toward the normal because water has a larger index of refraction than air has. By extending the refracted rays backward (see the dashed lines in the drawing), we find that the rays appear to the swimmer as if they originate from a point P' that is farther from the water than point P is. Thus, the swimmer sees the coin at an apparent height d' that is *greater* than the actual height d. Equation 26.3 [$d' = d\,(n_2/n_1)$] reveals the same result, for n_1 represents the medium (air) associated with the incident ray and n_2 represents the medium (water) associated with the refracted ray. Since n_2 for water is greater than n_1 for air, the ratio n_2/n_1 is greater than one and d' is larger than d. This situation is the opposite of that in Figure 26.5*b*, where an object beneath the water appears to be closer to the surface than it actually is.

Related Homework Material: *Problems 12 and 18*

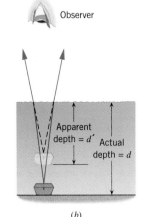

Figure 26.5 (*a*) Because light from the chest is refracted away from the normal when the light enters the air, the apparent depth of the image is less than the actual depth. (*b*) The observer is viewing the submerged object from directly overhead.

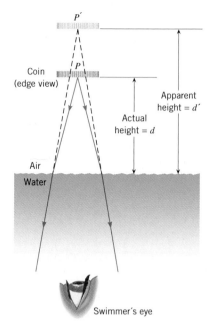

Figure 26.6 Rays from point P on a coin in the air above the water refract toward the normal as they enter the water. An underwater swimmer perceives the rays as originating from a point P' that is farther above the surface than the actual point P.

The Displacement of Light by a Transparent Slab of Material

A window pane is an example of a transparent slab of material. It consists of a plate of glass with parallel surfaces. When a ray of light passes through the glass, the emergent ray is parallel to the incident ray, but displaced from it, as Figure 26.7 shows. This result can be verified by applying Snell's law to each of the two glass surfaces, with the result that $n_1 \sin \theta_1 = n_2 \sin \theta_2 = n_3 \sin \theta_3$. Since air surrounds the glass, $n_1 = n_3$, and it follows that $\sin \theta_1 = \sin \theta_3$. Therefore, $\theta_1 = \theta_3$, and the emergent and incident rays are parallel. However, the drawing shows that the emergent ray is displaced laterally relative to the incident ray. The extent of the displacement depends on the angle of incidence, the thickness of the slab, and its refractive index.

Derivation of Snell's Law

Snell's law can be derived by considering what happens to the wave fronts when the light passes from one medium into another. Figure 26.8a shows light propagating from medium 1, where the speed is relatively large, into medium 2, where the speed is smaller; therefore, n_1 is less than n_2. The plane wave fronts in this picture are drawn perpendicular to the incident and refracted rays. Since the part of each wave front that penetrates medium 2 slows down first, the wave fronts in medium 2 are rotated clockwise relative to those in medium 1. Correspondingly, the refracted ray in medium 2 is bent toward the normal, as the drawing shows.

Although the incident and refracted waves have different speeds, *they have the same frequency f*. The fact that the frequency does not change can be understood in terms of the atomic mechanism underlying the generation of the refracted wave. When the electromagnetic wave strikes the surface, the oscillating electric field forces the electrons in the molecules of medium 2 to oscillate at the same frequency as the wave. The accelerating electrons behave like atomic antennas that radiate "extra" electromagnetic waves, which combine with the original wave. The net electromagnetic wave within medium 2 is a superposition of the original wave plus the extra radiated waves, and it is this superposition that constitutes the refracted wave. Since the extra waves are radiated at the same frequency as the original wave, the refracted wave also has the same frequency as the original wave.

The distance between successive wave fronts in Figure 26.8a has been chosen to be the wavelength λ. Since the frequencies are the same in both media, but the speeds are different, it follows from Equation 16.1 that the wavelengths are different: $\lambda_1 = v_1/f$ and $\lambda_2 = v_2/f$. Since v_1 is assumed to be larger than v_2, λ_1 is larger than λ_2, and the wave fronts are farther apart in medium 1.

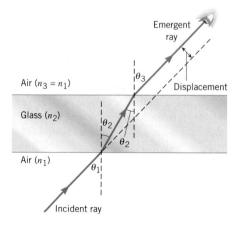

Figure 26.7 When a ray of light passes through a pane of glass that has parallel surfaces and is surrounded by air, the emergent ray is parallel to the incident ray ($\theta_3 = \theta_1$), but is displaced from it.

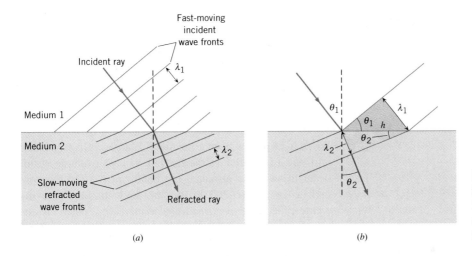

Fast-moving
incident
wave fronts

Incident ray

λ_1

Medium 1

Medium 2

Slow-moving
refracted
wave fronts

Refracted ray

λ_2

(a)

θ_1

θ_1 h

λ_1

λ_2

θ_2

θ_2

(b)

Figure 26.8 (*a*) The wave fronts are refracted as the light passes from medium 1 into medium 2. (*b*) An enlarged view of the incident and refracted wave fronts at the surface.

Figure 26.8*b* shows an enlarged view of the incident and refracted wave fronts at the surface. The angles θ_1 and θ_2 within the colored right triangles are, respectively, the angles of incidence and refraction. In addition, the triangles share the same hypotenuse h. Therefore,

$$\sin \theta_1 = \frac{\lambda_1}{h} = \frac{v_1/f}{h} = \frac{v_1}{hf} \quad \text{and} \quad \sin \theta_2 = \frac{\lambda_2}{h} = \frac{v_2/f}{h} = \frac{v_2}{hf}$$

Combining these two equations into a single equation by eliminating the common term hf gives

$$\frac{\sin \theta_1}{v_1} = \frac{\sin \theta_2}{v_2}$$

By multiplying each side of this result by c, the speed of light in a vacuum, and recognizing that the ratio c/v is the index of refraction n, we arrive at Snell's law of refraction: $n_1 \sin \theta_1 = n_2 \sin \theta_2$.

● *Concepts at a Glance* A number of important phenomena are related to refraction and Snell's law. The Concepts-at-a-Glance chart in Figure 26.9 is an expan-

Figure 26.9 *Concepts at a Glance* Four of the phenomena that we will discuss in this chapter are listed at the right of this chart. In the photograph, total internal reflection plays a role in allowing you to see around the bend in the curved plastic tube. Light rays emanating from the bee's body enter the tube and, in part, are guided around the 90° turn via total internal reflection, eventually exiting through the front surface and allowing you to see the bee.

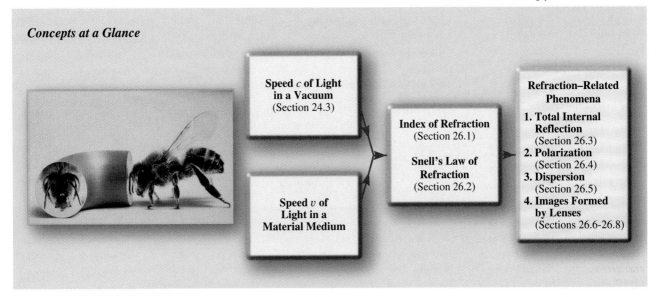

Concepts at a Glance

Speed *c* of Light in a Vacuum (Section 24.3)

Speed *v* of Light in a Material Medium

Index of Refraction (Section 26.1)

Snell's Law of Refraction (Section 26.2)

Refraction–Related Phenomena

1. **Total Internal Reflection** (Section 26.3)
2. **Polarization** (Section 26.4)
3. **Dispersion** (Section 26.5)
4. **Images Formed by Lenses** (Sections 26.6-26.8)

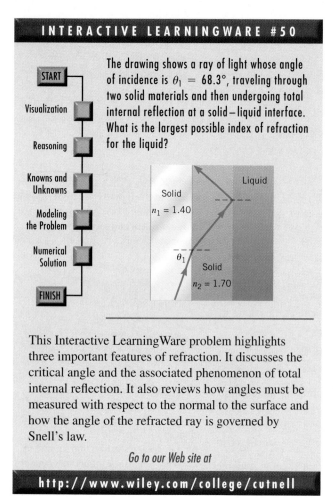

The drawing shows a ray of light whose angle of incidence is $\theta_1 = 68.3°$, traveling through two solid materials and then undergoing total internal reflection at a solid–liquid interface. What is the largest possible index of refraction for the liquid?

This Interactive LearningWare problem highlights three important features of refraction. It discusses the critical angle and the associated phenomenon of total internal reflection. It also reviews how angles must be measured with respect to the normal to the surface and how the angle of the refracted ray is governed by Snell's law.

sion of that in Figure 26.1 and lists four of these phenomena, each of which will be discussed in the remainder of this chapter. In the next section, we turn our attention to the first of them, total internal reflection. ●

26.3 TOTAL INTERNAL REFLECTION

When light passes from a medium of larger refractive index into one of smaller refractive index—for example, from water to air—the refracted ray bends *away* from the normal, as in Figure 26.10a. As the angle of incidence increases, the angle of refraction also increases. When the angle of incidence reaches a certain value, called the **critical angle θ_c**, the angle of refraction is 90°. Then the refracted ray points along the surface; part *b* illustrates what happens at the critical angle. When the angle of incidence exceeds the critical angle, as in part *c* of the drawing, there is no refracted light. All the incident light is reflected back into the medium from which it came, a phenomenon called **total internal reflection.** Total internal reflection occurs only when light travels from a higher-index medium toward a lower-index medium. It does not occur when light propagates in the reverse direction—for example, from air to water.

An expression for the critical angle θ_c can be obtained from Snell's law by setting $\theta_1 = \theta_c$ and $\theta_2 = 90°$ (see Figure 26.10b):

$$\textbf{\textit{Critical angle}} \quad \sin\theta_c = \frac{n_2 \sin 90°}{n_1} = \frac{n_2}{n_1} \quad (n_1 > n_2)$$

$$(26.4)$$

For instance, the critical angle for light traveling from water ($n_1 = 1.33$) to air ($n_2 = 1.00$) is $\theta_c = \sin^{-1}(1.00/1.33) = 48.8°$. For incident angles greater than 48.8°, Snell's law predicts that $\sin\theta_2$ is greater than unity, a value that is not possible. Thus, light rays with incident angles exceeding 48.8° yield no refracted light, and the light is totally reflected back into the water, as Figure 26.10c indicates. The next example illustrates how the critical angle changes when the indices of refraction change.

EXAMPLE 5 **Total Internal Reflection**

A beam of light is propagating through diamond ($n_1 = 2.42$) and strikes a diamond–air interface at an angle of incidence of 28°. (a) Will part of the beam enter

Figure 26.10 (a) When light travels from a higher-index medium (water) into a lower-index medium (air), the refracted ray is bent away from the normal. (b) When the angle of incidence is equal to the critical angle θ_c, the angle of refraction is 90°. (c) If θ_1 is greater than θ_c, there is no refracted ray, and total internal reflection occurs.

(a)

(b)   (c)

the air ($n_2 = 1.00$) or will the beam be totally reflected at the interface? (b) Repeat part (a), assuming that the diamond is surrounded by water ($n_2 = 1.33$).

Reasoning Total internal reflection occurs only when the beam of light has an angle of incidence that is greater than the critical angle θ_c. The critical angle is different in parts (a) and (b), since it depends on the ratio n_2/n_1 of the refractive indices of the incident (n_1) and refracting (n_2) media.

Solution

(a) The critical angle θ_c for total internal reflection at the diamond–air interface is given by Equation 26.4 as

$$\theta_c = \sin^{-1}\left(\frac{n_2}{n_1}\right) = \sin^{-1}\left(\frac{1.00}{2.42}\right) = 24.4°$$

Because the angle of incidence of 28° is greater than the critical angle, there is no refraction, and the light is totally reflected back into the diamond.

(b) If water, rather than air, surrounds the diamond, the critical angle for total internal reflection becomes larger:

$$\theta_c = \sin^{-1}\left(\frac{n_2}{n_1}\right) = \sin^{-1}\left(\frac{1.33}{2.42}\right) = 33.3°$$

Now a beam of light that has an angle of incidence of 28° (less than the critical angle of 33.3°) at the diamond–water interface is refracted into the water.

The critical angle plays an important role in why a diamond sparkles, as Conceptual Example 6 discusses.

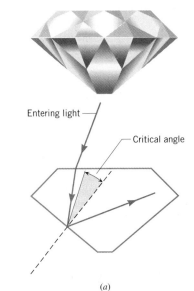

Entering light

Critical angle

(a)

CONCEPTUAL EXAMPLE 6   The Sparkle of a Diamond

A diamond gemstone is famous for its sparkle because the light coming from it glitters as it is moved about. Why does a diamond exhibit such brilliance? And why does a diamond lose much of its brilliance when placed under water?

Reasoning and Solution A diamond sparkles because when it is held a certain way, the intensity of the light coming from it is greatly enhanced. Figure 26.11 helps to explain that this enhancement is related to total internal reflection. Part a of the drawing shows a ray of light striking a bottom facet of the diamond at an angle of incidence that is greater than the critical angle. This ray, and all other rays whose angles of incidence exceed the critical angle, are totally reflected back into the diamond, eventually exiting the top surface to give the diamond its sparkle. Many of the rays striking a bottom facet behave in this fashion because diamond has a relatively small critical angle in air. As part (a) of Example 5 shows, the critical angle is 24.4°. The value is so small because the index of refraction of diamond ($n = 2.42$) is large compared to that of air ($n = 1.00$).

Now consider what happens to the same ray of light within the diamond when the diamond is placed in water. Because water has a larger index of refraction than air, the critical angle increases to 33.3°, as part (b) of Example 5 shows. Therefore, this particular ray is no longer totally internally reflected. As Figure 26.11b illustrates, only some of the light is reflected back into the diamond, and the remainder escapes into the water. Consequently, less light exits from the top of the diamond, causing it to lose much of its brilliance.

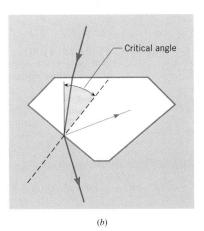

Critical angle

(b)

Figure 26.11 (a) Near the bottom of the diamond, light is totally internally reflected, because the incident angle exceeds the critical angle for diamond and air. (b) When the diamond is in water, the same light is partially reflected and partially refracted, since the incident angle is less than the critical angle for diamond and water.

THE PHYSICS OF...

why a diamond sparkles.

Many optical instruments, such as binoculars, periscopes, and telescopes, use glass prisms and total internal reflection to turn a beam of light through 90° or 180°. Figure 26.12a shows a light ray entering a 45°–45°–90° glass prism ($n_1 = 1.5$) and striking the hypotenuse of the prism at an angle of incidence of $\theta_1 = 45°$.

(a)   (b)   (c)

Figure 26.12 Total internal reflection at a glass–air interface can be used to turn a ray of light through an angle of (a) 90° or (b) 180°. (c) Two prisms, each reflecting the light twice by total internal reflection, are sometimes used in binoculars to produce a lateral displacement of a light ray.

The critical angle for a glass–air interface is $\theta_c = \sin^{-1}(n_2/n_1) = \sin^{-1}(1.0/1.5) = 42°$. Since the angle of incidence is greater than the critical angle, the light is totally reflected at the hypotenuse and is directed vertically upward in the drawing, having been turned through an angle of 90°. Part *b* of the picture shows how the same prism can turn the beam through 180° when total internal reflection occurs twice. Prisms can also be used in tandem to produce a lateral displacement of a light ray, while leaving its initial direction unaltered. Figure 26.12*c* illustrates such an application in binoculars.

An important application of total internal reflection occurs in fiber optics, where hair-thin threads of glass or plastic, called optical fibers, "pipe" light from one place to another. Figure 26.13*a* shows that an optical fiber consists of a cylindrical inner *core* that carries the light and an outer concentric shell, the *cladding*. The core is made from transparent glass or plastic that has a relatively high index of refraction. The cladding is also made of glass, but of a type that has a relatively low index of refraction. Light enters one end of the core, strikes the core/cladding interface at an angle of incidence greater than the critical angle, and, therefore, is reflected back into the core. Light, then, travels inside the optical fiber along a zigzag path. In a well-designed fiber, little light is lost as a result of absorption by the core, so light can travel many kilometers before its intensity diminishes appreciably. Fibers are often bundled together to produce cables. Because the fibers themselves are so thin, the cables are relatively small and flexible and can fit into places inaccessible to larger metal wires.

Optical fiber cables are the medium of choice for high-quality telecommunications because the cables are relatively immune to external electrical interference and because a light beam can carry information through an optical fiber just as electricity carries information through copper wires. The information-carrying capacity

THE PHYSICS OF...

fiber optics.

Figure 26.13 (a) Light can travel with little loss in a curved optical fiber because the light is totally reflected whenever it strikes the core–cladding interface and because the absorption of light by the core itself is small. (b) Light is being transmitted by a bundle of optical fibers.

Light ray   Cladding   Core

(a)

(b)

Figure 26.14 A bronchoscope is being used to look for signs of pulmonary disease.

Figure 26.15 A colonoscope, shown in use in this X-ray photograph, is used to examine the interior of the colon for signs of cancer.

Figure 26.16 Optical fibers have made arthroscopic surgery possible, such as the repair of a damaged knee shown in this photograph.

of light, however, is thousands of times greater than that of electricity. A laser beam traveling through a single optical fiber can carry tens of thousands of telephone conversations and several TV programs simultaneously.

In the field of medicine, optical fiber cables have had extraordinary impact. In the practice of endoscopy, for instance, a device called an endoscope is used to peer inside the body. Figure 26.14 shows a bronchoscope being used, which is a kind of endoscope that is inserted through the nose or mouth, down the bronchial tubes, and into the lungs. It consists of two optical fiber cables. One provides light to illuminate interior body parts, while the other sends back an image for viewing. A bronchoscope simplifies greatly the diagnosis of pulmonary disease. Tissue samples can even be collected with some bronchoscopes. A colonoscope is another kind of endoscope, and its design is similar to that of the bronchoscope. It is inserted through the rectum and used to examine the interior of the colon (see Figure 26.15). The colonoscope currently offers the best hope for diagnosing colon cancer in its early stages, when it can be treated.

The use of optical fibers has also revolutionized surgical techniques. In arthroscopic surgery, a small surgical instrument, several millimeters in diameter, is mounted at the end of an optical fiber cable. The surgeon can insert the instrument and cable into a joint, such as the knee, with only a tiny incision and minimal damage to the surrounding tissue (see Figure 26.16). Consequently, recovery from the procedure is relatively rapid compared to traditional surgical techniques.

THE PHYSICS OF...

endoscopy.

THE PHYSICS OF...

arthroscopic surgery.

26.4 POLARIZATION AND THE REFLECTION AND REFRACTION OF LIGHT

For incident angles other than 0°, unpolarized light becomes partially polarized in reflecting from a nonmetallic surface, such as water. To demonstrate this fact, rotate a pair of Polaroid sunglasses in the sunlight reflected from a lake. You will see that the light intensity transmitted through the glasses is a minimum when the glasses

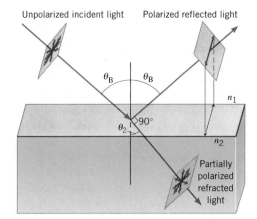

Figure 26.17 When unpolarized light is incident on a nonmetallic surface at the Brewster angle θ_B, the reflected light is 100% polarized in a direction parallel to the surface. The angle between the reflected and refracted rays is 90°.

are oriented as they are normally worn. Since the transmission axis of the glasses is aligned vertically, it follows that the light reflected from the lake is partially polarized in the horizontal direction.

There is one special angle of incidence at which the reflected light is completely polarized parallel to the surface, the refracted ray being only partially polarized. This angle is called the ***Brewster angle*** θ_B. Figure 26.17 summarizes what happens when unpolarized light strikes a nonmetallic surface at the Brewster angle. The value of θ_B is given by ***Brewster's law,*** in which n_1 and n_2 are, respectively, the refractive indices of the materials in which the incident and refracted rays propagate:

Brewster's law
$$\tan \theta_B = \frac{n_2}{n_1} \qquad (26.5)$$

This relation is named after the Scotsman David Brewster (1781–1868), who discovered it. Figure 26.17 also indicates that the reflected and refracted rays are perpendicular to each other when light strikes the surface at the Brewster angle (see problem 37 at the end of the chapter).

26.5 THE DISPERSION OF LIGHT: PRISMS AND RAINBOWS

Figure 26.18a shows a ray of monochromatic light passing through a glass prism surrounded by air. When the light enters the prism at the left face, the refracted ray is bent toward the normal, for the refractive index of glass is greater than that of air.

Figure 26.18 (*a*) A ray of light is refracted as it passes through the prism. The prism is surrounded by air. (*b*) Two different colors are refracted by different amounts. For clarity, the amount of refraction has been exaggerated. (*c*) Sunlight is dispersed into its color components by this prism.

When the light leaves the prism at the right face, it is refracted away from the normal. Thus, the net effect of the prism is to change the direction of the ray, causing it to bend downward upon entering the prism, and downward again upon leaving. Because the refractive index of the glass depends on wavelength (see Table 26.2), rays corresponding to different colors are bent by different amounts by the prism and depart traveling in different directions. The greater the index of refraction for a given color, the greater the bending, and part *b* of the drawing shows the refractions for the colors red and violet, which are at opposite ends of the visible spectrum. If a beam of sunlight, which contains all colors, is sent through the prism, the sunlight is separated into a spectrum of colors, as part *c* shows. The spreading of light into its color components is called ***dispersion.***

In Figure 26.18*a* the ray of light is refracted twice by a glass prism surrounded by air. Conceptual Example 7 explores what happens to the light when the prism is surrounded by materials other than air.

TABLE 26.2 Indices of Refraction *n* of Crown Glass at Various Wavelengths

Approximate Color	Wavelength in Vacuum (nm)	Index of Refraction, *n*
Red	660	1.520
Orange	610	1.522
Yellow	580	1.523
Green	550	1.526
Blue	470	1.531
Violet	410	1.538

CONCEPTUAL EXAMPLE 7
The Refraction of Light Depends on Two Refractive Indices

In Figure 26.18*a* the glass prism is surrounded by air and bends the ray of light downward. It is also possible for the prism to bend the ray upward, as in Figure 26.19*a*, or to not bend the ray at all, as in part *b* of the drawing. How can the situations illustrated in Figure 26.19 arise?

Reasoning and Solution Snell's law of refraction includes the refractive indices of *both* materials on either side of an interface. With this in mind, we note that the ray bends upward, or away from the normal, as it enters the prism in Figure 26.19*a*. A ray bends away from the normal when it travels from a medium with a larger refractive index into a medium with a smaller refractive index. When the ray leaves the prism, it again bends upward, which is toward the normal at the point of exit. A ray bends toward the normal when traveling from a smaller toward a larger refractive index. Thus, *the situation in Figure 26.19a could arise if the prism were immersed in a fluid, such as carbon disulfide, which has a larger refractive index than does glass* (see Table 26.1).

We have seen in Figures 26.18*a* and 26.19*a* that a glass prism can bend a ray of light either downward or upward, depending on whether the surrounding fluid has a smaller or larger index of refraction than the glass. It seems logical to conclude, then, that *the ray will not bend at all, either down or up, if the surrounding fluid has the same index of refraction as the glass* — a condition known as *index matching*. This is exactly what is happening in Figure 26.19*b*, for the ray proceeds straight through the prism as if it were not even there. If the index of refraction of the surrounding fluid equals that of the glass prism, then $n_1 = n_2$, and Snell's law ($n_1 \sin \theta_1 = n_2 \sin \theta_2$) reduces to $\sin \theta_1 = \sin \theta_2$. Therefore, the angle of refraction equals the angle of incidence, and no bending of the light occurs.

Related Homework Material: *Questions 18 and 21, Problems 10 and 42*

(*a*)

(*b*)

Figure 26.19 A ray of light passes through identical prisms, each surrounded by a different fluid. The ray of light is (*a*) refracted upward and (*b*) not refracted at all.

THE PHYSICS OF...

rainbows.

Another example of dispersion occurs in rainbows, in which refraction by water droplets gives rise to the colors. You can often see a rainbow just as a storm is leaving, if you look at the departing rain with the sun at your back. When light from the sun enters a spherical raindrop, as in Figure 26.20, light of each color is refracted or bent by an amount that depends on the refractive index of water for that wavelength. After reflection from the back surface of the droplet, the different colors are again refracted as they reenter the air. Although each droplet disperses the light into its full spectrum of colors, the observer in Figure 26.21*a* sees only one color of

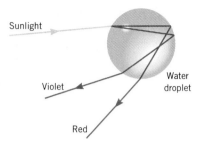

Figure 26.20 When sunlight emerges from a water droplet, the light is dispersed into its constituent colors, of which only two are shown.

light coming from any given droplet, since only one color travels in the right direction to reach the observer's eyes. However, all colors are visible in a rainbow (see Figure 26.21b) because each color originates from different droplets at different angles of elevation.

26.6 LENSES

The lenses used in optical instruments, such as eyeglasses, cameras, and telescopes, are made from transparent materials that refract light. They refract the light in such a way that an image of the source of the light is formed. Figure 26.22a shows a crude lens formed from two glass prisms. Suppose an object, centered on the principal axis, is infinitely far from the lens so the rays from the object are parallel to the principal axis. In passing through the prisms, these rays are bent toward the axis

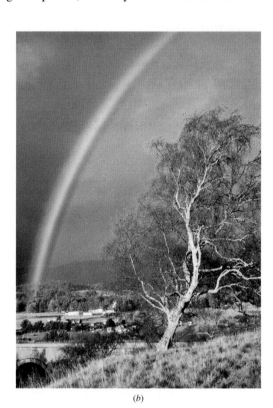

Figure 26.21 The different colors seen in a rainbow originate from water droplets at different angles of elevation.

Figure 26.22 (a) These two prisms cause rays of light that are parallel to the principal axis to change direction and cross the axis at different points. (b) With a converging lens, paraxial rays that are parallel to the principal axis converge to the focal point F after passing through the lens.

(a)   (b)

Figure 26.23 (*a*) These two prisms cause parallel rays to diverge. (*b*) With a diverging lens, paraxial rays that are parallel to the principal axis appear to originate from the focal point *F* after passing through the lens.

because of refraction. Unfortunately, the rays do not all cross the axis at the same place, and, therefore, such a crude lens gives rise to a "blurred" image of the object.

A better lens can be constructed from a single piece of transparent material with properly curved surfaces, often spherical, as in part *b* of the drawing. With this improved lens, rays that are near the principal axis (paraxial rays) and parallel to it converge to a single point on the axis after emerging from the lens. This point is called the *focal point F* of the lens. Thus, an object located infinitely far away on the principal axis leads to an image at the focal point of the lens. The distance between the focal point and the lens is the *focal length f.* In what follows, we assume the lens is so thin compared to *f* that it makes no difference whether *f* is measured between the focal point and either surface of the lens or the center of the lens. The type of lens in Figure 26.22*b* is known as a *converging lens* because it causes incident parallel rays to converge at the focal point.

Another type of lens found in optical instruments is a *diverging lens,* which causes incident parallel rays to diverge after exiting the lens. Two prisms can also be used to form a crude diverging lens, as in Figure 26.23*a*. In a properly designed diverging lens, such as that in part *b* of the picture, paraxial rays that are parallel to the principal axis appear to originate from a single point on the axis after passing through the lens. This point is the focal point *F*, and its distance *f* from the lens is the focal length. Again, we assume that the lens is thin compared to the focal length.

Converging and diverging lenses come in a variety of shapes, as Figure 26.24 illustrates. Observe that converging lenses are thicker at the center than at the edges, whereas diverging lenses are thinner at the center.

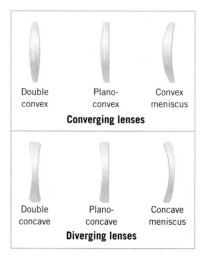

Figure 26.24 Converging and diverging lenses come in a variety of shapes.

26.7 THE FORMATION OF IMAGES BY LENSES

Ray Diagrams

Each point on an object emits light rays in all directions, and when some of these rays pass through a lens, they form an image. As with mirrors, ray diagrams can be drawn to determine the location and size of the image. Lenses differ from mirrors, however, in that light can pass through a lens from left to right or from right to left. Therefore, when constructing ray diagrams, begin by locating a focal point *F* on *each side of the lens;* each point lies on the principal axis at the same distance *f* from the lens. The lens is assumed to be thin, in that its thickness is small compared with the focal length and the distances of the object and the image from the lens. For convenience, it is also assumed that the object is located to the left of the lens and is oriented perpendicular to the principal axis. There are three paraxial rays that leave a point on the top of the object and are especially helpful in drawing ray diagrams. They are labeled 1, 2, and 3 in Figure 26.25. When tracing their paths, we use the following reasoning strategy.

REASONING STRATEGY

Ray Tracing for Converging and Diverging Lenses

Converging Lens	Diverging Lens

Ray 1

This ray initially travels parallel to the principal axis. In passing through a converging lens, the ray is refracted toward the axis and travels through the focal point on the right side of the lens, as Figure 26.25a shows.

This ray initially travels parallel to the principal axis. In passing through a diverging lens, the ray is refracted away from the axis, and *appears* to have originated from the focal point on the left of the lens. The dashed line in Figure 26.25d represents the apparent path of the ray.

Ray 2

This ray first passes through the focal point on the left and then is refracted by the lens in such a way that it leaves traveling parallel to the axis, as in part b.

This ray leaves the object and moves toward the focal point on the right of the lens. Before reaching the focal point, however, the ray is refracted by the lens so as to exit parallel to the axis. See part e, where the dashed line indicates the ray's path in the absence of the lens.

Ray 3*

This ray travels directly through the center of the thin lens without any appreciable bending, as in part c.

This ray travels directly through the center of the thin lens without any appreciable bending, as in part f.

* Ray 3 does not bend as it proceeds through the lens because the left and right surfaces of each type of lens are nearly parallel at the center. Thus, in either case, the lens behaves as a transparent slab. As Figure 26.7 shows, the rays incident on and exiting from a slab travel in the same direction with only a lateral displacement. If the lens is sufficiently thin, the displacement is negligibly small.

Image Formation by a Converging Lens

Figure 26.26a illustrates the formation of a real image by a converging lens. Here the object is located at a distance from the lens that is greater than twice the focal length (beyond the point labeled 2F). To locate the image, any two of the three special rays can be drawn from the tip of the object, although all three are shown in the

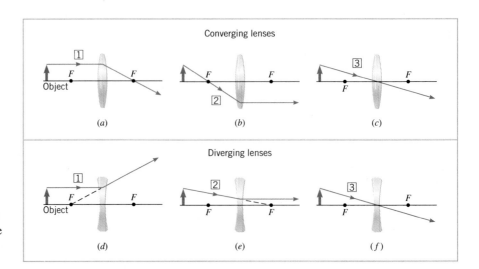

Figure 26.25 The rays shown here are useful in determining the nature of the images formed by converging and diverging lenses.

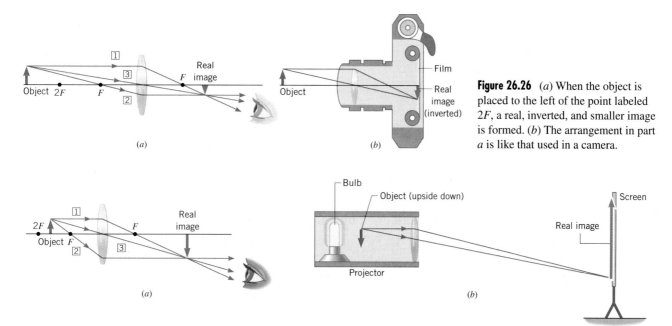

Figure 26.26 (*a*) When the object is placed to the left of the point labeled 2*F*, a real, inverted, and smaller image is formed. (*b*) The arrangement in part *a* is like that used in a camera.

Figure 26.27 (*a*) When the object is placed between 2*F* and *F*, the image is real, inverted, and larger than the object. (*b*) This arrangement is found in projectors.

drawing. The point on the right side of the lens where these rays intersect locates the tip of the image. The ray diagram indicates that the image is real, inverted, and smaller than the object. This optical arrangement is similar to that used in a camera, where a piece of film records the image (see part *b* of the drawing).

When the object is placed between 2*F* and *F*, as in Figure 26.27*a*, the image is still real and inverted; however, the image is now larger than the object. This optical system is used in a slide or film projector in which a small piece of film is the object and the enlarged image falls on a screen. However, to obtain an image that is right-side up, the film must be placed in the projector upside down.

When the object is located between the focal point and the lens, as in Figure 26.28, the rays diverge after leaving the lens. To a person viewing the diverging rays, they appear to come from an image behind (to the left of) the lens. Because all the rays do not actually come from the image, it is a virtual image. The ray diagram shows that the virtual image is upright and enlarged. A magnifying glass uses this arrangement, as can be seen in part *b* of the drawing.

THE PHYSICS OF...

a camera.

THE PHYSICS OF...

a slide or film projector.

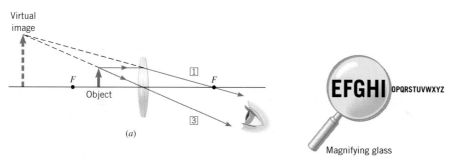

Figure 26.28 (*a*) When an object is placed within the focal point *F* of a converging lens, an upright, enlarged, and virtual image is created. (*b*) Such an image is seen when looking through a magnifying glass.

Figure 26.29 (*a*) A diverging lens always forms a virtual image of a real object. The image is upright and smaller relative to the object. (*b*) The image seen through a diverging lens.

(*a*)

ABC ABC DEF DEF

Diverging lens
(*b*)

Image Formation by a Diverging Lens

Light rays diverge upon leaving a diverging lens, as Figure 26.29 shows, and the ray diagram indicates that a virtual image is formed on the left side of the lens. In fact, regardless of the position of a real object, a diverging lens always forms a virtual image that is upright and smaller relative to the object.

26.8 THE THIN-LENS EQUATION AND THE MAGNIFICATION EQUATION

Figure 26.30 *Concepts at a Glance* The equations that describe how thin lenses form images are based on Snell's law of refraction. The thin-lens equation and the magnification equation are used to design lenses that correct vision, such as the contact lens that the woman in the photograph is putting in her eye.

● *Concepts at a Glance*   When an object is placed in front of a spherical mirror, we can determine the location, size, and nature of its image by using the technique of ray tracing or the mirror and magnification equations. Both options are based on the law of reflection. The mirror and magnification equations relate the distances d_o and d_i of the object and image from the mirror to the focal length f and magnification m. For an object placed in front of a lens, Snell's law of refraction leads to the technique of ray tracing and to equations that are identical to the mirror and magnification equations. The similarity between mirrors and lenses can be seen by comparing the Concepts-at-a-Glance chart in Figure 26.30 with that for mirrors in Figure 25.23. ●

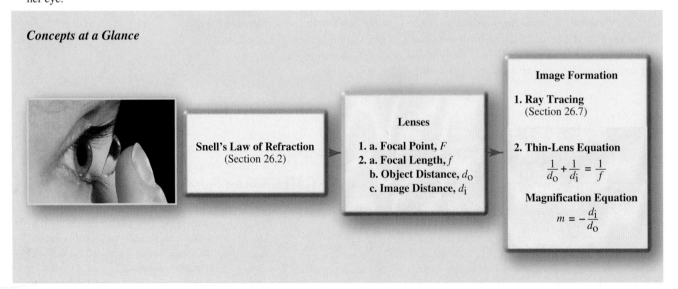

Concepts at a Glance

Snell's Law of Refraction
(Section 26.2)

Lenses

1. a. Focal Point, F
2. a. Focal Length, f
   b. Object Distance, d_o
   c. Image Distance, d_i

Image Formation

1. **Ray Tracing**
   (Section 26.7)

2. **Thin-Lens Equation**
$$\frac{1}{d_o} + \frac{1}{d_i} = \frac{1}{f}$$

Magnification Equation
$$m = -\frac{d_i}{d_o}$$

The chart in Figure 26.30 shows that Snell's law leads to equations that are referred to as the thin-lens equation and the magnification equation:

Thin-lens equation
$$\frac{1}{d_o} + \frac{1}{d_i} = \frac{1}{f}$$
(26.6)

Magnification equation $m = \dfrac{\text{Image height}}{\text{Object height}} = \dfrac{h_i}{h_o} = -\dfrac{d_i}{d_o}$ (26.7)

Figure 26.31 defines the symbols in these expressions with the aid of a thin converging lens, but the expressions also apply to a diverging lens, if it is thin. The derivations of these equations are presented at the end of this section.

Certain sign conventions accompany the use of the thin-lens and magnification equations, and the conventions are similar to those used with mirrors in Section 25.6. The issue of real-versus-virtual images, however, is slightly different with lenses than with mirrors. With a mirror, a real image is formed on the *same side* of the mirror as the object, in which case the image distance d_i is a positive number (see Figure 25.19). With a lens, a positive value for d_i also means the image is real. But, starting with an actual object, a real image is formed on the *opposite side* of the lens as the object (see Figure 26.31). The **sign conventions** listed in the following Reasoning Strategy apply to light rays traveling from left to right from a real object.

REASONING STRATEGY

Summary of Sign Conventions for Lenses

Focal length

f is + for a converging lens.

f is − for a diverging lens.

Object distance

d_o is + if the object is to the left of the lens (real object), as is usual.

d_o is − if the object is to the right of the lens (virtual object).*

Image distance

d_i is + for an image (real) formed to the right of the lens by a real object.

d_i is − for an image (virtual) formed to the left of the lens by a real object.

Magnification

m is + for an image that is upright with respect to the object.

m is − for an image that is inverted with respect to the object.

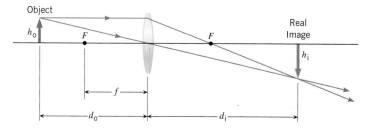

Figure 26.31 The drawing shows the focal length f, the object distance d_o, and the image distance d_i for a converging lens. The object and image heights are, respectively, h_o and h_i.

---

* This situation arises in systems containing more than one lens, where the image formed by the first lens becomes the object for the second lens. In such a case, the object of the second lens may lie to the right of that lens, in which event d_o is assigned a negative value and the object is called a virtual object.

Examples 8 and 9 illustrate the use of the thin-lens and magnification equations.

EXAMPLE 8   The Real Image Formed by a Camera Lens

A 1.70-m-tall person is standing 2.50 m in front of a camera. The camera uses a converging lens whose focal length is 0.0500 m. (a) Find the image distance (the distance between the lens and the film) and determine whether the image is real or virtual. (b) Find the magnification and the height of the image on the film.

Reasoning This optical arrangement is similar to that in Figure 26.26a, where the object distance is greater than twice the focal length of the lens. Therefore, we expect the image to be real, inverted, and smaller than the object.

Solution

• **PROBLEM SOLVING INSIGHT**
In the thin-lens equation, the reciprocal of the image distance d_i is given by $d_i^{-1} = f^{-1} - d_o^{-1}$, where f is the focal length and d_o is the object distance. After combining the reciprocals f^{-1} and d_o^{-1}, do not forget to take the reciprocal of the result to find d_i.

(a) To find the image distance d_i we use the thin-lens equation with $d_o = 2.50$ m and $f = 0.0500$ m:

$$\frac{1}{d_i} = \frac{1}{f} - \frac{1}{d_o} = \frac{1}{0.0500 \text{ m}} - \frac{1}{2.50 \text{ m}} = 19.6 \text{ m}^{-1} \quad \text{or} \quad \boxed{d_i = 0.0510 \text{ m}}$$

Since the image distance is a positive number, a $\boxed{\text{real image}}$ is formed on the film.

(b) The magnification follows from the magnification equation:

$$m = -\frac{d_i}{d_o} = -\frac{0.0510 \text{ m}}{2.50 \text{ m}} = \boxed{-0.0204}$$

The image is 0.0204 times as large as the object, and it is inverted since m is negative. Since the object height is $h_o = 1.70$ m, the image height is

$$h_i = mh_o = (-0.0204)(1.70 \text{ m}) = \boxed{-0.0347 \text{ m}}$$

EXAMPLE 9   The Virtual Image Formed by a Diverging Lens

An object is placed 7.10 cm to the left of a diverging lens whose focal length is $f = -5.08$ cm (a diverging lens has a negative focal length). (a) Find the image distance and determine whether the image is real or virtual. (b) Obtain the magnification.

Reasoning This situation is similar to that in Figure 26.29a. The ray diagram shows that the image is virtual, erect, and smaller than the object.

Solution

(a) The thin-lens equation can be used to find the image distance d_i:

$$\frac{1}{d_i} = \frac{1}{f} - \frac{1}{d_o} = \frac{1}{-5.08 \text{ cm}} - \frac{1}{7.10 \text{ cm}} = -0.338 \text{ cm}^{-1} \quad \text{or} \quad \boxed{d_i = -2.96 \text{ cm}}$$

The image distance is negative, indicating that the image is $\boxed{\text{virtual}}$ and located to the left of the lens.

(b) Since d_i and d_o are known, the magnification equation shows that

$$m = -\frac{d_i}{d_o} = -\frac{-2.96 \text{ cm}}{7.10 \text{ cm}} = \boxed{0.417}$$

The image is upright (m is +) and smaller ($m < 1$) than the object.

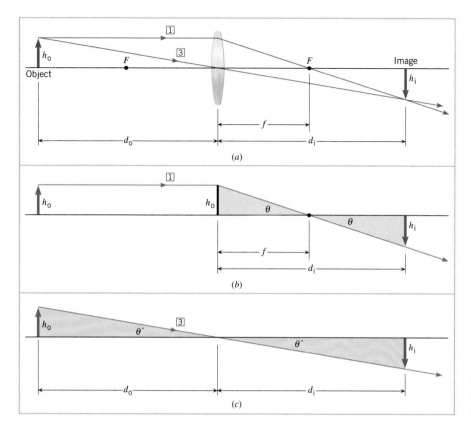

Figure 26.32 These ray diagrams are used for deriving the thin-lens and magnification equations.

The thin-lens and magnification equations can be derived by considering rays 1 and 3 in Figure 26.32a. Ray 1 is shown separately in part b of the drawing, where the angle θ is the same in each of the two colored triangles. Thus, $\tan \theta$ is the same for each triangle:

$$\tan \theta = \frac{h_o}{f} = \frac{-h_i}{d_i - f}$$

A minus sign has been inserted in the numerator of the ratio $h_i/(d_i - f)$ for the following reason. The angle θ in Figure 26.32b is assumed to be positive. Since the image is inverted relative to the object, the image height h_i is a negative number. The insertion of the minus sign ensures that the term $-h_i/(d_i - f)$, and hence $\tan \theta$, is a positive quantity.

Ray 3 is shown separately in part c of the drawing, where the angles θ' are the same. Therefore,

$$\tan \theta' = \frac{h_o}{d_o} = \frac{-h_i}{d_i}$$

A minus sign has been inserted in the numerator of the term h_i/d_i for the same reason that a minus sign was inserted earlier, namely, to ensure that $\tan \theta'$ is a positive quantity. The first equation gives $h_i/h_o = -(d_i - f)/f$, while the second equation yields $h_i/h_o = -d_i/d_o$. Equating these two expressions for h_i/h_o and rearranging the result produce the thin-lens equation, $1/d_o + 1/d_i = 1/f$. The magnification equation follows directly from the equation $h_i/h_o = -d_i/d_o$, if we recognize that h_i/h_o is the magnification m of the lens.

26.9 LENSES IN COMBINATION

Many optical instruments, such as microscopes and telescopes, use a number of lenses together to produce an image. Among other things, a multiple-lens system can produce an image that is magnified more than is possible with a single lens. For instance, Figure 26.33a shows a two-lens system used in a microscope. The first lens, the lens closest to the object, is referred to as the *objective.* The second lens is known as the *eyepiece* (or *ocular*). The object is placed just outside the focal point F_o of the objective. The image formed by the objective—called the "first image" in the drawing—is real, inverted, and enlarged compared to the object. This first image then serves as the object for the eyepiece. Since the first image falls between the eyepiece and its focal point F_e, the eyepiece forms an enlarged, virtual, final image, which is what the observer sees.

The location of the final image in a multiple-lens system can be determined by applying the thin-lens equation to each lens separately. The key point to remember in such situations is that ***the image produced by one lens serves as the object for the next lens,*** as the next example illustrates.

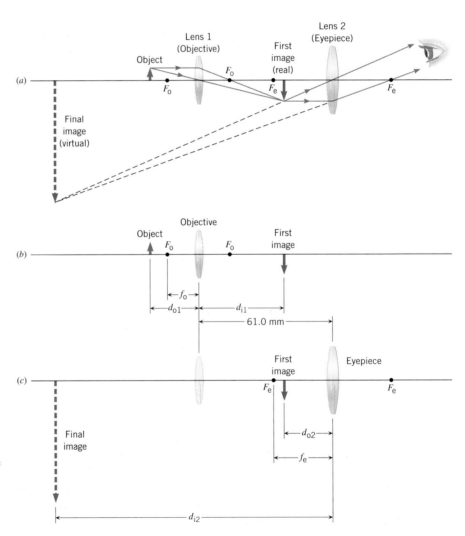

Figure 26.33 (*a*) This two-lens system can be used as a compound microscope to produce a virtual, enlarged, and inverted final image. (*b*) The objective forms the first image and (*c*) the eyepiece forms the final image.

EXAMPLE 10
A Microscope—Two Lenses in Combination

The objective and eyepiece of the compound microscope in Figure 26.33 are both converging lenses and have focal lengths of $f_o = 15.0$ mm and $f_e = 25.5$ mm. A distance of 61.0 mm separates the lenses. The microscope is being used to examine an object placed $d_{o1} = 24.1$ mm in front of the objective. Find the final image distance.

Reasoning We begin by using the thin-lens equation to find the location of the image produced by the first lens, the objective. This image then becomes the object for the second lens, the eyepiece. The thin-lens equation can be used again to locate the final image produced by the eyepiece.

Solution The "first image" distance d_{i1} can be determined using the thin-lens equation with $d_{o1} = 24.1$ mm and $f_o = 15.0$ mm (see Figure 26.33b):

$$\frac{1}{d_{i1}} = \frac{1}{f_o} - \frac{1}{d_{o1}} = \frac{1}{15.0 \text{ mm}} - \frac{1}{24.1 \text{ mm}}$$
$$= 0.0252 \text{ mm}^{-1} \quad \text{or} \quad d_{i1} = 39.7 \text{ mm}$$

The first image now becomes the object for the eyepiece (see part c of the drawing). Since the distance between the lenses is 61.0 mm, the object distance for the eyepiece is $d_{o2} = 61.0$ mm $- d_{i1} = 61.0$ mm $- 39.7$ mm $= 21.3$ mm. Noting that the focal length of the eyepiece is $f_e = 25.5$ mm, we can determine the final image distance with the aid of the thin-lens equation:

$$\frac{1}{d_{i2}} = \frac{1}{f_e} - \frac{1}{d_{o2}} = \frac{1}{25.5 \text{ mm}} - \frac{1}{21.3 \text{ mm}}$$
$$= -0.0077 \text{ mm}^{-1} \quad \text{or} \quad \boxed{d_{i2} = -130 \text{ mm}}$$

The fact that d_{i2} is negative indicates that the final image is virtual. It lies 130 mm to the left of the eyepiece, as the drawing shows.

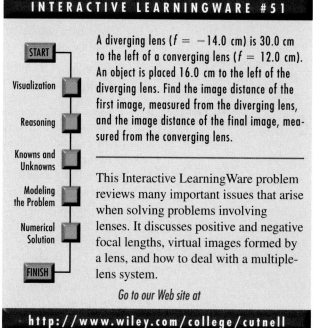

INTERACTIVE LEARNINGWARE #51

START
Visualization
Reasoning
Knowns and Unknowns
Modeling the Problem
Numerical Solution
FINISH

A diverging lens ($f = -14.0$ cm) is 30.0 cm to the left of a converging lens ($f = 12.0$ cm). An object is placed 16.0 cm to the left of the diverging lens. Find the image distance of the first image, measured from the diverging lens, and the image distance of the final image, measured from the converging lens.

This Interactive LearningWare problem reviews many important issues that arise when solving problems involving lenses. It discusses positive and negative focal lengths, virtual images formed by a lens, and how to deal with a multiple-lens system.

Go to our Web site at

http://www.wiley.com/college/cutnell

26.10 THE HUMAN EYE

Anatomy

Without doubt, the human eye is the most remarkable of all optical devices. Figure 26.34 shows some of its main anatomical features. The eyeball is approximately spherical with a diameter of about 25 mm. Light enters the eye through a transparent membrane (the *cornea*). This membrane covers a clear liquid region (the *aqueous humor*), behind which is a diaphragm (the *iris*), the *lens,* a region filled with a jelly-like substance (the *vitreous humor*), and, finally, the *retina.* The retina is the light-sensitive part of the eye, consisting of millions of structures called *rods* and *cones.* When stimulated by light, these structures send electrical impulses via the *optic nerve* to the brain, which interprets the image on the retina.

The iris is the colored portion of the eye and controls the amount of light reaching the retina. The iris acts as a controller because it is a muscular diaphragm with a variable opening at its center, through which the light passes. The opening is called the *pupil.* The diameter of the pupil varies from about 2 to 7 mm, decreasing in bright light and increasing (dilating) in dim light.

Figure 26.34 A cross-sectional view of the human eye.

Of prime importance to the operation of the eye is the fact that the lens is flexible, and its shape can be altered by the action of the *ciliary muscle.* The lens is connected to the ciliary muscle by the *suspensory ligaments* (see the drawing). We will see shortly how the shape-changing ability of the lens affects the focusing ability of the eye.

Optics

THE PHYSICS OF...

the human eye.

Optically, the eye and the camera are similar; both have a lens system and a diaphragm with a variable opening or aperture at its center. Moreover, the retina of the eye and the film in the camera serve similar functions, for both record the image formed by the lens system. In the eye, the image formed on the retina is real, inverted, and smaller than the object, just as it is in a camera. Although the image on the retina is inverted, it is interpreted by the brain as being right-side up.

For clear vision, the eye must refract the incoming light rays, so as to form a sharp image on the retina. In reaching the retina, the light travels through five different media, each with a different index of refraction n: air ($n = 1.00$), the cornea ($n = 1.38$), the aqueous humor ($n = 1.33$), the lens ($n = 1.40$, on the average), and the vitreous humor ($n = 1.34$). Each time light passes from one medium into another, it is refracted at the boundary. The greatest amount of refraction, about 70% or so, occurs at the air/cornea boundary. According to Snell's law, the large refraction at this interface occurs primarily because the refractive index of air ($n = 1.00$) is so different from that of the cornea ($n = 1.38$). The refraction at all the other boundaries is relatively small because the indices of refraction on either side of these boundaries are nearly equal. The lens itself contributes only about 20–25% of the total refraction, since the surrounding aqueous and vitreous humors have indices of refraction that are nearly the same as that of the lens.

Even though the lens contributes only a quarter of the total refraction or less, its function is an important one. The eye has a fixed image distance; that is, the distance between the lens and the retina is constant. Therefore, the only way for images to be produced on the retina for objects located at different distances is for the focal length of the lens to be adjustable. And it is the ciliary muscle that adjusts the focal length. When the eye looks at a very distant object, the ciliary muscle is not tensed. The lens has its least curvature and, consequently, its longest focal length. Under this condition the eye is said to be "fully relaxed," and the rays form a sharp image on the retina, as in Figure 26.35*a.* When the object moves closer to the eye, the ciliary muscle tenses automatically, thereby increasing the curvature of the lens, shortening the focal length, and permitting a sharp image to form again on the retina (Figure 26.35*b*). When a sharp image of an object is formed on the retina, we say the eye is "focused" on the object. The process in which the lens changes its focal length to focus on objects at different distances is called ***accommodation*** and occurs so swiftly that we are usually unaware of it.

When you hold a book too close, the print is blurred because the lens cannot adjust enough to bring the book into focus. The point nearest the eye at which an object can be placed and still produce a sharp image on the retina is called the ***near point*** of the eye. The ciliary muscle is fully tensed when an object is placed at the near point. For people in their early twenties with normal vision, the near point is located about 25 cm from the eye. It increases to about 50 cm at age 40 and to roughly 500 cm at age 60. Since most reading material is held at a distance of 25–45 cm from the eye, older adults typically need eyeglasses to overcome the loss of accommodation. The ***far point*** of the eye is the location of the farthest object on which the fully relaxed eye can focus. A person with normal eyesight can see objects very far away, such as the planets and stars, and thus has a far point located nearly at infinity.

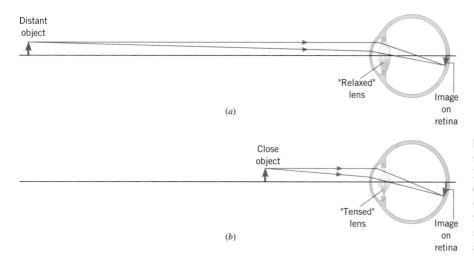

Figure 26.35 (*a*) When fully relaxed, the lens of the eye has its longest focal length, and an image of a very distant object is formed on the retina.
(*b*) When the ciliary muscle is tensed, the lens has a shorter focal length. Consequently, an image of a closer object is also formed on the retina.

Nearsightedness

A person who is **nearsighted (myopic)** can focus on nearby objects but cannot clearly see objects far away. For such a person, the far point of the eye is not at infinity and may even be as close to the eye as three or four meters. When a nearsighted eye tries to focus on a distant object, the eye is fully relaxed, like a normal eye. However, the nearsighted eye has a focal length that is shorter than it should be, so rays from the distant object form a sharp image in front of the retina, as Figure 26.36*a* shows, and blurred vision results.

THE PHYSICS OF...

nearsightedness.

Figure 26.36 (*a*) When a nearsighted person views a distant object, the image is formed in front of the retina. The result is blurred vision. (*b*) With a diverging lens in front of the eye, the image is moved onto the retina and clear vision results. (*c*) The diverging lens is designed to form a virtual image at the far point of the nearsighted eye.

The nearsighted eye can be corrected with glasses or contacts that use *diverging* lenses, as Figure 26.36b suggests. The rays from the object diverge after leaving the eyeglass lens. Therefore, when they are subsequently refracted toward the principal axis by the eye, a sharp image is formed farther back and falls on the retina. Since the relaxed (but nearsighted) eye can focus on an object at the eye's far point—but not on objects farther away—the diverging lens is designed to transform a very distant object into an image located at the far point. Part c of the drawing shows this transformation, and the next example illustrates how to determine the focal length of the diverging lens that accomplishes it.

EXAMPLE 11   Eyeglasses for the Nearsighted Person

A nearsighted person has a far point located only 521 cm from the eye. Assuming that eyeglasses are to be worn 2 cm in front of the eye, find the focal length needed for the diverging lenses of the glasses so the person can see distant objects.

Reasoning In Figure 26.36c the far point is 521 cm away from the eye. Since the glasses are worn 2 cm from the eye, the far point is 519 cm to the left of the diverging lens. The image distance, then, is −519 cm, the negative sign indicating that the image is a virtual image formed to the left of the lens. The object is assumed to be infinitely far from the diverging lens. The thin-lens equation can be used to find the focal length of the eyeglasses. We expect the focal length to be negative, since the lens is a diverging lens.

Solution With $d_i = -519$ cm and $d_o = \infty$, the focal length can be found as follows:

$$\frac{1}{f} = \frac{1}{d_o} + \frac{1}{d_i} = \frac{1}{\infty} + \frac{1}{-519 \text{ cm}} \quad \text{or} \quad \boxed{f = -519 \text{ cm}} \qquad (26.6)$$

The value for f is negative, as expected, for a diverging lens.

● **PROBLEM SOLVING INSIGHT**
Eyeglasses are worn about 2 cm from the eyes. Be sure, if necessary, to take this 2 cm into account when determining the object and image distances (d_o and d_i) that are used in the thin-lens equation.

Farsightedness

THE PHYSICS OF...

farsightedness.

A *farsighted (hyperopic)* person can usually see distant objects clearly, but cannot focus on those nearby. Whereas the near point of a young and normal eye is located about 25 cm from the eye, the near point of a farsighted eye may be considerably farther away than that, perhaps as far as several hundred centimeters. When a farsighted eye tries to focus on a book held closer than the near point, it accommodates and shortens its focal length as much as it can. However, even at its shortest, the focal length is longer than it should be. Therefore, the light rays from the book would form a sharp image behind the retina if they could do so, as Figure 26.37a suggests. In reality, no light passes through the retina, but a blurred image does form on it.

Figure 26.37b shows that farsightedness can be corrected by placing a *converging* lens in front of the eye. The lens refracts the light rays more toward the principal axis before they enter the eye. Consequently, when the rays are refracted even more by the eye, they converge to form an image on the retina. Part c of the figure illustrates what the eye sees when it looks through the converging lens. The lens is designed so that the eye perceives the light to be coming from a virtual image located at the near point. Example 12 shows how the focal length of the converging lens is determined to correct for farsightedness.

EXAMPLE 12   Contact Lenses for the Farsighted Person

A farsighted person has a near point located 210 cm from the eyes. Obtain the focal length of the converging lenses in a pair of contacts that can be used to read a book held 25.0 cm from the eyes.

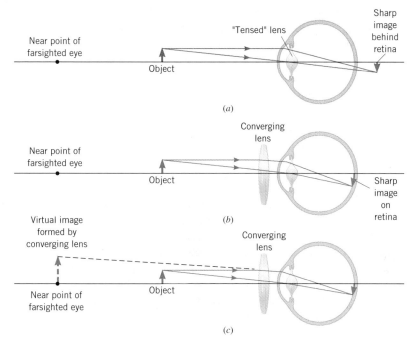

Figure 26.37 (*a*) When a farsighted person views an object located inside the near point, a sharp image would be formed behind the retina if light could pass through it. Only a blurred image forms on the retina. (*b*) With a converging lens in front of the eye, the sharp image is moved onto the retina and clear vision results. (*c*) The converging lens is designed to form a virtual image at the near point of the farsighted eye.

Reasoning A contact lens is placed directly against the eye. Thus, the object distance, which is the distance from the book to the lens, is 25.0 cm. The lens forms an image of the book at the near point of the eye, so the image distance is −210 cm. The minus sign indicates that the image is a virtual image formed to the left of the lens, as in Figure 26.37*c*. The focal length can be obtained from the thin-lens equation.

Solution With d_o = 25.0 cm and d_i = −210 cm, the focal length can be determined from the thin-lens equation as follows:

$$\frac{1}{f} = \frac{1}{d_o} + \frac{1}{d_i} = \frac{1}{25.0 \text{ cm}} + \frac{1}{-210 \text{ cm}} = 0.0352 \text{ cm}^{-1} \quad \text{or} \quad \boxed{f = 28.4 \text{ cm}}$$

The Refractive Power of a Lens—The Diopter

The extent to which rays of light are refracted by a lens depends on its focal length. However, the optometrists who prescribe correctional lenses and the opticians who make the lenses do not specify the focal length directly in prescriptions. Instead, they use the concept of ***refractive power*** to describe the extent to which a lens refracts light:

$$\begin{array}{c} \text{Refractive power} \\ \text{of a lens} \\ \text{(in diopters)} \end{array} = \frac{1}{f \text{ (in meters)}} \qquad (26.8)$$

The refractive power is measured in units of *diopters*. One diopter is 1 m^{-1}.

Equation 26.8 shows that a converging lens has a refractive power of 1 diopter if it focuses parallel light rays to a focal point 1 m beyond the lens. If a lens refracts parallel rays even more and converges them to a focal point only 0.25 m beyond the lens, the lens has four times more refractive power, or 4 diopters. Since a converging lens has a positive focal length and a diverging lens has a negative focal length, the refractive power of a converging lens is positive while that of a diverging lens is negative. For instance, the eyeglasses in Example 11 would be described in a prescription from an optometrist in the following way: Refractive power = $1/(-5.19 \text{ m}) = -0.193$ diopters. The contact lenses in Example 12 would be described in a similar fashion: Refractive power = $1/(0.284 \text{ m}) = 3.52$ diopters.

Figure 26.38 The angle θ is the angular size of both the image and the object.

26.11 ANGULAR MAGNIFICATION AND THE MAGNIFYING GLASS

If you hold a penny at arm's length, the penny looks larger than the moon. The reason is that the penny, being so close, forms a larger image on the retina of the eye than does the more distant moon. The brain interprets the larger image of the penny as arising from a larger object. The size of the image on the retina determines how large an object appears to be. However, the size of the image on the retina is difficult to measure. Alternatively, the angle θ subtended by the image can be used as an indication of the image size. Figure 26.38 shows this alternative, which has the advantage that θ is also the angle subtended by the object and, hence, can be measured more easily. The angle θ is called the **angular size** of both the image and the object. The larger the angular size, the larger the image on the retina, and the larger the object appears to be.

According to Equation 8.1, the angle θ (measured in radians) is the length of the circular arc that is subtended by the angle divided by the radius of the arc, as Figure 26.39a indicates. Part b of the drawing shows the situation for an object of height h_o viewed at a distance d_o from the eye. When θ is small, h_o is approximately equal to the arc length and d_o is nearly equal to the radius, so that

$$\theta \text{ (in radians)} = \text{Angular size} \approx \frac{h_o}{d_o}$$

This approximation is good to within one percent for angles of $9°$ or smaller. In the next example the angular size of a penny is compared with that of the moon.

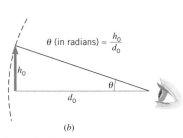

Figure 26.39 (a) The angle θ, measured in radians, is the arc length divided by the radius. (b) For small angles (less than $9°$), θ is approximately equal to h_o/d_o, where h_o and d_o are the object height and distance.

EXAMPLE 13   A Penny and the Moon

Compare the angular size of a penny (diameter = $h_o = 1.9$ cm) held at arm's length ($d_o = 71$ cm) with that of the moon (diameter = $h_o = 3.5 \times 10^6$ m, and $d_o = 3.9 \times 10^8$ m).

Reasoning The angular size θ of an object is given approximately by its height h_o divided by its distance d_o from the eye, $\theta \approx h_o/d_o$, provided that the angle involved is less than roughly $9°$. As we shall see, this approximation applies here. In this example, the "heights" of the penny and the moon are their diameters.

Solution The angular sizes of the penny and moon are

Penny $\theta \approx \dfrac{h_o}{d_o} = \dfrac{1.9 \text{ cm}}{71 \text{ cm}} = \boxed{0.027 \text{ rad } (1.5°)}$

Moon $\theta \approx \dfrac{h_o}{d_o} = \dfrac{3.5 \times 10^6 \text{ m}}{3.9 \times 10^8 \text{ m}} = \boxed{0.0090 \text{ rad } (0.52°)}$

The penny thus appears to be about three times as large as the moon.

An optical instrument, such as a magnifying glass, allows us to view small or distant objects because it produces a larger image on the retina than would be possible otherwise. In other words, an optical instrument magnifies the angular size of the object. The **angular magnification** (or **magnifying power**) M is the angular size θ' of the final image produced by the instrument divided by a reference angular size θ. The reference angular size is the angular size of the object when seen without the instrument.

$$
\begin{array}{c}
\textit{Angular} \\
\textit{magnification}
\end{array}
\qquad
M = \dfrac{\begin{array}{c}\text{Angular size of} \\ \text{final image produced} \\ \text{by optical instrument}\end{array}}{\begin{array}{c}\text{Reference angular size} \\ \text{of object seen without} \\ \text{optical instrument}\end{array}} = \dfrac{\theta'}{\theta}
\qquad (26.9)
$$

A magnifying glass is the simplest device that provides angular magnification. In this case, the reference angular size θ is chosen to be the angular size of the object when placed at the near point of the eye and seen without the magnifying glass. Since an object cannot be brought closer than the near point and still produce a sharp image on the retina, θ represents the largest angular size obtainable without the magnifying glass. Figure 26.40a indicates that the reference angular size is $\theta \approx h_o/N$, where N is the distance from the eye to the near point. To compute θ', recall from Section 26.7 and Figure 26.28 that a magnifying glass is usually a single converging lens, with the object located inside the focal point. In this situation, Figure 26.40b indicates that the lens produces a virtual image that is enlarged and upright with respect to the object. Assuming the eye is next to the magnifying glass, the angular size θ' seen by the eye is $\theta' \approx h_o/d_o$, where d_o is the object distance. The angular magnification is

$$
M = \frac{\theta'}{\theta} \approx \frac{h_o/d_o}{h_o/N} = \frac{N}{d_o}
$$

According to the thin-lens equation, d_o is related to the image distance d_i and the focal length f of the lens by

$$
\frac{1}{d_o} = \frac{1}{f} - \frac{1}{d_i}
$$

Substituting this expression for $1/d_o$ into the previous expression for M leads to the following result:

$$
\begin{array}{c}
\textit{Angular magnification} \\
\textit{of a magnifying glass}
\end{array}
\qquad
M = \frac{\theta'}{\theta} \approx \left(\frac{1}{f} - \frac{1}{d_i} \right) N
\qquad (26.10)
$$

Two special cases of this result are of interest, depending on whether the image is located as close to the eye as possible or as far away as possible. To be seen clearly, the closest the image can be relative to the eye is at the near point, or $d_i = -N$. The minus sign indicates that the image lies to the left of the lens and is virtual. In this event, Equation 26.10 becomes $M \approx N/f + 1$. The farthest the image can be from the eye is at infinity ($d_i = -\infty$); this occurs when the object is placed at the focal point of the lens. When the image is at infinity, Equation 26.10 simplifies to $M \approx N/f$. Clearly, the angular magnification is greater when the image is at the near point of the eye rather than at infinity. In either case, however, the greatest magnification is achieved by using a magnifying glass with the shortest possible focal length. Example 14 illustrates how to determine the angular magnification of a magnifying glass that is used in these two ways.

THE PHYSICS OF...

a magnifying glass.

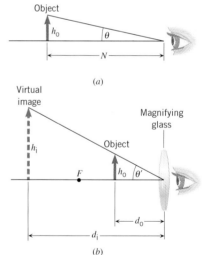

Figure 26.40 (*a*) Without a magnifying glass, the largest angular size θ occurs when the object is placed at the near point, a distance N from the eye. (*b*) A magnifying glass produces an enlarged, virtual image of an object placed inside the focal point F of the lens. The angular size of both the image and the object is θ'.

This curator at the Smithsonian Institution, aided by a magnifying glass, is examining an ancient sword.

EXAMPLE 14   Examining a Diamond with a Magnifying Glass

A jeweler, whose near point is 40.0 cm from his eye and whose far point is at infinity, is using a small magnifying glass (called a loupe) to examine a diamond. The lens of the magnifying glass has a focal length of 5.00 cm, and the image of the gem is −185 cm from the lens. The image distance is negative because the image is virtual and is formed on the same side of the lens as the object. (a) Determine the angular magnification of the magnifying glass. (b) Where should the image be located so the jeweler's eye is fully relaxed and has the least strain? What is the angular magnification under this "least strain" condition?

Reasoning The angular magnification of the magnifying glass can be determined from Equation 26.10. In part (a) the image distance is −185 cm. In part (b) the ciliary muscle of the jeweler's eye is fully relaxed, so the image must be located infinitely far from the eye, at its far point, as Section 26.10 discusses.

Solution

(a) With $f = 5.00$ cm, $d_i = -185$ cm, and $N = 40.0$ cm, the angular magnification is

$$M = \left(\frac{1}{f} - \frac{1}{d_i}\right) N = \left(\frac{1}{5.00 \text{ cm}} - \frac{1}{-185 \text{ cm}}\right)(40.0 \text{ cm}) = \boxed{8.22}$$

(b) With $f = 5.00$ cm, $d_i = -\infty$, and $N = 40.0$ cm, the angular magnification is

$$M = \left(\frac{1}{f} - \frac{1}{d_i}\right) N = \left(\frac{1}{5.00 \text{ cm}} - \frac{1}{-\infty}\right)(40.0 \text{ cm}) = \boxed{8.00}$$

Jewelers often prefer to minimize eye strain when viewing objects, even though it means a slight reduction in angular magnification.

THE PHYSICS OF...

the compound microscope.

Figure 26.41 A compound microscope.

26.12 THE COMPOUND MICROSCOPE

To increase the angular magnification beyond that possible with a magnifying glass, an additional converging lens can be included to "premagnify" the object before the magnifying glass comes into play. The result is an optical instrument known as the **compound microscope** (Figure 26.41). As discussed in Section 26.9, the magnifying glass is called the eyepiece, and the additional lens is called the objective.

The angular magnification of the compound microscope is $M = \theta'/\theta$ (Equation 26.9), where θ' is the angular size of the final image and θ is the reference angular size. As with the magnifying glass in Figure 26.40, the reference angular size is determined by the height h_o of the object when the object is located at the near point of the unaided eye: $\theta \approx h_o/N$, where N is the distance between the eye and the near point. Assuming that the object is placed just outside the focal point F_o of the objective (see Figure 26.33a) and that the final image is very far from the eyepiece (i.e., near infinity, see Figure 26.33c), it can be shown that

Angular magnification of a compound microscope $$M \approx -\frac{(L - f_e)N}{f_o f_e} \qquad (26.11)$$

In Equation 26.11, f_o and f_e are, respectively, the focal lengths of the objective and the eyepiece. The angular magnification is greatest when f_o and f_e are as small as possible (since they are in the denominator in Equation 26.11) and when the distance L between the lenses is as large as possible. Furthermore, L must be greater than the sum of f_o and f_e for this equation to be valid. Example 15 deals with the angular magnification of a compound microscope.

EXAMPLE 15   The Angular Magnification of a Compound Microscope

The focal length of the objective of a compound microscope is $f_o = 0.40$ cm, while that of the eyepiece is $f_e = 3.0$ cm. The two lenses are separated by a distance of $L = 20.0$ cm. A person with a near point distance of $N = 25$ cm is using the microscope. (a) Determine the angular magnification of the microscope. (b) Compare the answer in part (a) with the largest angular magnification obtainable by using the eyepiece alone as a magnifying glass.

Reasoning The angular magnification of the compound microscope can be obtained directly from Equation 26.11, since all the variables are known. When the eyepiece is used alone as a magnifying glass, as in Figure 26.40b, the largest angular magnification occurs when the image seen through the eyepiece is as close as possible to the eye. The image in this case is at the near point, and according to Equation 26.10, the angular magnification is $M \approx (N/f_e) + 1$.

Solution

(a) The angular magnification of the compound microscope is

$$M \approx -\frac{(L - f_e)N}{f_o f_e} = -\frac{(20.0\ \text{cm} - 3.0\ \text{cm})(25\ \text{cm})}{(0.40\ \text{cm})(3.0\ \text{cm})} = \boxed{-350}$$

The minus sign indicates that the final image is inverted relative to the initial object.

(b) The maximum angular magnification of the eyepiece by itself is

$$M \approx \frac{N}{f_e} + 1 = \frac{25\ \text{cm}}{3.0\ \text{cm}} + 1 = \boxed{9.3}$$

The effect of the objective is to increase the angular magnification of the compound microscope by a factor of $350/9.3 = 38$ compared to that of a magnifying glass.

26.13 THE TELESCOPE

A telescope is an instrument for magnifying distant objects, such as stars and planets. Like a microscope, a telescope consists of an objective and an eyepiece (also called the ocular). Since the object is usually far away, the light rays striking the telescope are nearly parallel, and the "first image" is formed just beyond the focal point F_o of the objective, as Figure 26.42a illustrates. The first image is real and inverted. Unlike that in the compound microscope, however, this image is *smaller* than the object. If, as in part b of the drawing, the telescope is constructed so the first image lies just inside the focal point F_e of the eyepiece, the eyepiece acts like a magnifying glass. It forms a final image that is greatly enlarged, vir-

THE PHYSICS OF...

the telescope.

Figure 26.42 (*a*) An astronomical telescope is used to view distant objects. (Note the "break" in the principal axis, between the object and the objective.) The objective produces a real, inverted, first image. (*b*) The eyepiece magnifies the first image to produce the final image near infinity.

(a)

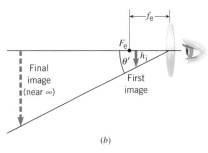

(b)

tual, and located near infinity. This final image can then be viewed with a fully relaxed eye.

The angular magnification M of a telescope, like that of a magnifying glass or a microscope, is the angular size θ' subtended by the final image of the telescope divided by the reference angular size θ of the object. For an astronomical object, such as a planet, it is convenient to use as a reference the angular size of the object seen in the sky with the unaided eye. Since the object is far away, the angular size seen by the unaided eye is nearly the same as the angle θ subtended at the objective of the telescope in Figure 26.42a. Moreover, θ is also the angle subtended by the first image, so $\theta \approx -h_i/f_o$, where h_i is the height of the first image and f_o is the focal length of the objective. A minus sign has been inserted into this equation because the first image is inverted relative to the object and the image height h_i is a negative number. The insertion of the minus sign ensures that the term $-h_i/f_o$, and hence θ, is a positive quantity. To obtain an expression for θ', refer to Figure 26.42b and note that the first image is located very near the focal point F_e of the eyepiece, which has a focal length f_e. Therefore, $\theta' \approx h_i/f_e$. The angular magnification of the telescope is, then,

Angular magnification of an astronomical telescope
$$M = \frac{\theta'}{\theta} \approx \frac{h_i/f_e}{-h_i/f_o} = -\frac{f_o}{f_e} \qquad (26.12)$$

The angular magnification is determined by the ratio of the focal length of the objective to the focal length of the eyepiece. For large angular magnifications, the objective should have a long focal length and the eyepiece a short one. Some of the design features of a telescope are the topic of the next example.

EXAMPLE 16   The Angular Magnification of an Astronomical Telescope

The telescope shown in Figure 26.43 has the following specifications: $f_o = 985$ mm and $f_e = 5.00$ mm. From these data, find (a) the angular magnification of the telescope and (b) the approximate length of the telescope.

Reasoning The angular magnification of the telescope follows directly from Equation 26.12, since the focal lengths of the objective and eyepiece are known. We can find the length of the telescope by noting that it is approximately equal to the distance L between the objective and eyepiece. Figure 26.42 shows that the first image is located just beyond the focal point F_o of the objective and just inside the focal point F_e of the eyepiece. These two focal points are, therefore, very close together, so the distance L is approximately the sum of the two focal lengths: $L \approx f_o + f_e$.

Solution

(a) The angular magnification is

$$M = -\frac{f_o}{f_e} = -\frac{985 \text{ mm}}{5.00 \text{ mm}} = \boxed{-197} \qquad (26.12)$$

(b) The approximate length of the telescope is

$$L \approx f_o + f_e = 985 \text{ mm} + 5.00 \text{ mm} = \boxed{990 \text{ mm}}$$

Figure 26.43 An astronomical telescope. The viewfinder is a separate, small telescope with low magnification and serves as an aid in locating the object. Once the object has been found, the viewer looks through the eyepiece to obtain the full magnification of the telescope.

26.14   LENS ABERRATIONS

Rather than forming a sharp image, a single lens typically forms an image that is slightly out of focus. This lack of sharpness arises because the rays originating from a single point on the object are not focused to a single point on the image. As a result, each point on the image becomes a small "blur." The lack of point-to-point correspondence between object and image is called an aberration.

One common type of aberration is ***spherical aberration,*** and it occurs with converging and diverging lenses made with spherical surfaces. Figure 26.44*a* shows how spherical aberration arises with a converging lens. Ideally, all rays traveling parallel to the principal axis are refracted so they cross the axis at the same point after passing through the lens. However, rays far from the principal axis are refracted more by the lens than those closer in. Consequently, the outer rays cross the axis closer to the lens than do the inner rays, so a lens with spherical aberration does not have a unique focal point. Instead, as the drawing suggests, there is a location along the principal axis where the light converges to the smallest cross-sectional area. This area is circular and is known as the ***circle of least confusion.*** The circle of least confusion is where the most satisfactory image can be formed by the lens.

Spherical aberration can be reduced substantially by using a variable-aperture diaphragm to allow only those rays close to the principal axis to pass through the lens. Figure 26.44*b* indicates that a reasonably sharp focal point can be achieved by this method, although less light now passes through the lens. Lenses with parabolic surfaces are also used to reduce this type of aberration, but they are difficult and expensive to make.

Chromatic aberration also causes blurred images. It arises because the index of refraction of the material from which the lens is made varies with wavelength. Section 26.5 discusses how this variation leads to the phenomenon of dispersion, in which different colors refract by different amounts. Figure 26.45*a* shows sunlight incident on a converging lens, in which the light spreads into its color spectrum because of dispersion. For clarity, however, the picture shows only the colors at the opposite ends of the visible spectrum—red and violet. Violet is refracted more than red, so the violet ray crosses the principal axis closer to the lens than does the red ray. Thus, the focal length of the lens is shorter for violet than for red, with intermediate values of the focal length corresponding to the colors in between. As a result of chromatic aberration, an undesirable color fringe surrounds the image.

Chromatic aberration can be greatly reduced by using a compound lens, such as the combination of a converging lens and a diverging lens shown in Figure 26.45*b*. Each lens is made from a different type of glass. With this lens combination the red and violet rays almost come to a common focus and, thus, chromatic aberration is reduced. A lens combination designed to reduce chromatic aberration is called an *achromatic lens* (from the Greek "achromatos," meaning "without color"). All high-quality cameras use achromatic lenses.

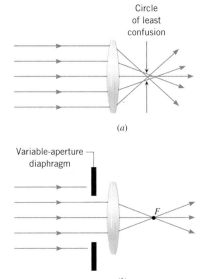

Figure 26.44 (*a*) In a converging lens, spherical aberration prevents light rays parallel to the principal axis from converging to a common point. (*b*) Spherical aberration can be reduced by allowing only rays near the principal axis to pass through the lens. The refracted rays now converge more nearly to a single focal point F.

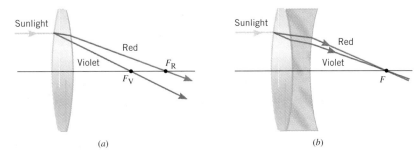

Figure 26.45 (*a*) Chromatic aberration arises when different colors are focused at different points along the principal axis: F_V = focal point for violet light, F_R = focal point for red light. (*b*) A converging and a diverging lens in tandem can be designed to bring different colors more nearly to the same focal point F.

26.15 CONCEPTS & CALCULATIONS

One important phenomenon discussed in this chapter is how a ray of light is refracted when it goes from one medium into another. Example 17 reviews some of the important aspects of refraction, including Snell's law, the concept of a critical angle, and the notion of index matching.

CONCEPTS & CALCULATIONS EXAMPLE 17   Refraction

A ray of light is incident on a glass/water interface at the critical angle θ_c, as Figure 26.46 illustrates. The reflected light then passes through a liquid (immiscible with water) and into air. The indices of refraction for the four substances are given in the drawing. Determine the angle of refraction θ_5 for the ray as it passes into the air.

Concept Questions and Answers What determines the critical angle when the ray strikes the glass/water interface?

Answer According to Equation 26.4, the critical angle θ_c is determined by the indices of refraction n_{glass} and n_{water}; $\theta_c = \sin^{-1}(n_{water}/n_{glass})$.

When light is incident at the glass/water interface at the critical angle, what is the angle of refraction, and how is the angle of reflection θ_1 related to the critical angle?

Answer When light is incident at the critical angle, the angle of refraction is $90°$, so the refracted ray skims along the surface. (For simplicity, this ray is not shown in the drawing.) We know from our work with mirrors that the angle of reflection is equal to the angle of incidence. Therefore, the angle of reflection is equal to the critical angle, so $\theta_1 = \theta_c$.

When the reflected ray strikes the glass/liquid interface, how is the angle of refraction θ_3 related to the angle of incidence θ_2? Note that the two materials have the same indices of refraction.

Answer According to Snell's law, Equation 26.2, the angle of refraction θ_3 is related to the angle of incidence θ_2 by $\sin \theta_3 = (n_{glass} \sin \theta_2)/n_{liquid}$. Since the two indices of refraction are equal, a condition known as "index matching," $\sin \theta_3 = \sin \theta_2$, and the two angles are equal. Therefore, the ray crosses the glass/liquid interface without being refracted.

When the ray passes from the liquid into the air, is the ray refracted?

Answer Yes. Because the two materials have different indices of refraction, the angle of refraction θ_5 is different from the angle of incidence θ_4. The angle of refraction can be found from Snell's law.

Solution Let us follow the ray of light as it progresses from left to right in Figure 26.46. It strikes the glass/water interface at the critical angle of

$$\theta_c = \sin^{-1}\left(\frac{n_{water}}{n_{glass}}\right) = \sin^{-1}\left(\frac{1.33}{1.52}\right) = 61.0°$$

The angle of reflection θ_1 from the glass/water interface is the same as the critical angle, so $\theta_1 = 61.0°$.

The ray then strikes the glass/liquid interface with an angle of incidence of $\theta_2 = 90.0° - 61.0° = 29.0°$. Note, as usual, that the angle of incidence is measured relative to the normal. At the glass/liquid interface, the angle of refraction θ_3 is the same as the angle of incidence θ_2, since the indices of refraction of the two materials are the same ($n_{glass} = n_{liquid} = 1.52$). Thus, $\theta_3 = \theta_2 = 29.0°$.

At the liquid/air interface we use Snell's law, Equation 26.2, to determine the angle θ_5 at which the refracted ray enters the air: $n_{liquid} \sin \theta_4 = n_{air} \sin \theta_5$. Note from

Figure 26.46 A ray of light strikes the glass/water interface at the critical angle θ_c. The reflected ray passes through a liquid and then into air.

the drawing that the angle of incidence θ_4 is the same as θ_3, so $\theta_4 = 29.0°$. The angle of refraction is

$$\theta_5 = \sin^{-1}\left(\frac{1.52 \sin 29.0°}{1.00}\right) = \boxed{47.5°}$$

One of the most important uses of refraction is in lenses, the behavior of which is governed by the thin-lens and magnification equations. Example 18 discusses how these equations are applied to a two-lens system and reviews the all-important conventions that must be followed (see the Reasoning Strategy in Section 26.8). In particular, the example shows how to account for a virtual object, in which the object lies to the right (rather than to the left) of a lens.

CONCEPTS & CALCULATIONS EXAMPLE 18   A Two-Lens System

Figure 26.47 illustrates a coverging lens ($f_1 = +20.0$ cm) and a diverging lens ($f_2 = -15.0$ cm) that are separated by a distance of 10.0 cm. An object with a height of $h_{o1} = 5.00$ mm is placed 45.0 cm to the left of the first (converging) lens. What is (a) the image distance d_{i1} and (b) the height h_{i1} of the image produced by the first lens? (c) What is the object distance for the second (diverging) lens? Find (d) the image distance d_{i2} and (e) the height h_{i2} of the image produced by the second lens.

Figure 26.47 The image produced by the first lens, the "first image," falls to the right of the second lens, so the image becomes a virtual object for the second lens.

Concept Questions and Answers Is the image produced by the first (converging) lens real or virtual?

Answer A converging lens can form either a real or a virtual image, depending on where the object is located relative to the focal point. If the object is to the left of the focal point, as it is in this example, the image is real and falls to the right of the lens. If, on the other hand, the object had been located between the focal point and the lens, the image would have been a virtual image located to the left of the converging lens.

As far as the second lens is concerned, what role does the image produced by the first lens play?

Answer The image produced by the first lens acts as the object for the second lens.

Note in Figure 26.47 that the image produced by the first lens is called the "first image," and it falls to the right of the second lens. This image acts as the object for the second lens. Normally, however, an object would lie to the left of a lens. How do we take into account that this object lies to the right of the diverging lens?

Answer According to the Reasoning Strategy in Section 26.8, an object that is located to the right of a lens is called a virtual object and is assigned a negative object distance.

How do we find the location of the image produced by the second lens when its object is a virtual object?

Answer Once the object distance for the second lens has been assigned a negative number, we can use the thin-lens equation in the usual manner to find the image distance.

Solution

(a) The distance d_{i1} of the image from the first lens can be found from the thin-lens equation, Equation 26.6. The focal length, $f_1 = +20.0$ cm, is positive because the lens is a converging lens, and the object distance, $d_{o1} = +45.0$ cm, is positive because it lies to the left of the lens:

$$\frac{1}{d_{i1}} = \frac{1}{f_1} - \frac{1}{d_{o1}} = \frac{1}{20.0 \text{ cm}} - \frac{1}{45.0 \text{ cm}} = 0.0278 \text{ cm}^{-1} \quad \text{or} \quad d_{i1} = \boxed{36.0 \text{ cm}}$$

(b) The height h_{i1} of the image produced by the first lens can be obtained from the magnification equation, Equation 26.7:

$$h_{i1} = h_{o1}\left(-\frac{d_{i1}}{d_{o1}}\right) = (5.00 \text{ mm})\left(-\frac{36.0 \text{ cm}}{45.0 \text{ cm}}\right) = \boxed{-4.00 \text{ mm}}$$

The minus sign indicates that the first image is inverted with respect to the object (see Figure 26.47).

(c) The first image falls 36.0 cm to the right of the first lens. However, the second lens is located 10.0 cm to the right of the first lens, so the first image is located 36.0 cm $-$ 10.0 cm = 26.0 cm to the right of the second lens. This image acts as the object for the second lens. Since the object for the second lens lies to the right of it, the object is a virtual object and assigned a negative number; $d_{o2} = \boxed{-26.0 \text{ cm}}$.

(d) The distance d_{i2} of the image from the second lens can be found from the thin-lens equation. The focal length, $f_2 = -15.0$ cm, is negative because the lens is a diverging lens, and the object distance, $d_{o2} = -26.0$ cm, is negative because it lies to the right of the lens:

$$\frac{1}{d_{i2}} = \frac{1}{f_2} - \frac{1}{d_{o2}} = \frac{1}{-15.0 \text{ cm}} - \frac{1}{-26.0 \text{ cm}} = -0.0282 \text{ cm}^{-1}$$

or

$$d_{i2} = \boxed{-35.5 \text{ cm}}$$

The negative sign for d_{i2} means that the image is formed to the left of the diverging lens and, hence, is a virtual image.

(e) The image height h_{i1} produced by the first lens becomes the object height h_{o2} for the second lens; $h_{o2} = h_{i1} = -4.00$ mm. The height h_{i2} of the image produced by the second lens follows from the magnification equation:

$$h_{i2} = h_{o2}\left(-\frac{d_{i2}}{d_{o2}}\right) = (-4.00 \text{ mm})\left(-\frac{-35.5 \text{ cm}}{-26.0 \text{ cm}}\right) = \boxed{5.46 \text{ mm}}$$

The positive value for h_{i2} means that the final image has the same orientation as the first image, that is, inverted.

At the end of the problem set for this chapter, you will find homework problems that contain both conceptual and quantitative parts. These problems are grouped under the heading *Concepts & Calculations, Group Learning Problems.* They are designed for use by students working alone or in small learning groups. The conceptual part of each problem provides a convenient focus for group discussions.

SUMMARY

When light strikes the interface between two media, part of the light is reflected and the remainder is transmitted across the interface. The change in the direction of travel as light passes from one medium into another is called **refraction.** The **index of refraction** n of a material is the ratio of the speed of light c in a vacuum to the speed of light v in the material: $n = c/v$. **Snell's law of refraction** states that (1) the refracted ray, the incident ray, and the normal to the interface all lie in the same plane, and (2) the angle of refraction θ_2 is related to the angle of incidence θ_1 by $n_1 \sin \theta_1 = n_2 \sin \theta_2$, where n_1 and n_2 are the indices of refraction of the incident and refracting media, respectively. The angles are measured relative to the normal.

Because of refraction, a submerged object has an **apparent depth** that is different from its actual depth. If the observer is directly above (below) the object, the apparent depth (height) d' is related to the actual depth (height) d by $d' = d(n_2/n_1)$, where n_1 and n_2 are the refractive indices of the media in which the object and the observer, respectively, are located.

When light passes from a medium of larger refractive index n_1 into one of smaller refractive index, n_2, the refracted ray is bent away from the normal. If the incident ray is at the **critical angle** θ_c, the angle of refraction is 90°. The critical angle is determined from Snell's law and is given by $\sin \theta_c = n_2/n_1$. When the angle of incidence exceeds the critical angle, all the incident light is reflected back into the medium from which it came, a phenomenon known as **total internal reflection.**

When light is incident on a nonmetallic surface at the **Brewster angle** θ_B, the reflected light is completely polarized parallel to the surface. The Brewster angle is given by $\tan \theta_B = n_2/n_1$, where n_1 and n_2 are the refractive indices of the incident and refracting media, respectively. When light is incident at the Brewster angle, the reflected and refracted rays are perpendicular to each other.

A glass prism can spread a beam of sunlight into a spectrum of colors because the index of refraction of the glass depends on the wavelength of the light. The spreading of light into its color components is known as **dispersion.**

Converging lenses and **diverging lenses** depend on the phenomenon of refraction in forming an image. With a converging lens, paraxial rays that are parallel to the principal axis are focused to a point on the axis by the lens. This point is called the **focal point** of the lens, and its distance from the lens is the **focal length** f. Paraxial light rays that are parallel to the principal axis of a diverging lens appear to originate from its focal point after passing through the lens. The image produced by a converging or a diverging lens can be located with the help of a **ray diagram,** which can be constructed using the rays shown in Figure 26.25.

The **thin-lens equation** can be used with either converging or diverging lenses that are thin, and it relates the object distance d_o, the image distance d_i, and the focal length f of the lens: $1/d_o + 1/d_i = 1/f$. The magnification m of a lens is the ratio of the image height h_i to the object height h_o. The magnification is also related to d_i and d_o by the **magnification equation:** $m = -(d_i/d_o)$. The algebraic sign conventions for the variables appearing in the thin-lens and magnification equations are summarized in Section 26.8. When two or more lenses are used in combination, the image produced by one lens serves as the object for the next lens.

In the **human eye,** a real, inverted image is formed on a light-sensitive surface, called the retina. **Accommodation** is the process by which the focal length of the eye is automatically adjusted, so that objects at different distances produce sharp images on the retina. The **near point** of the eye is the point nearest the eye at which an object can be placed and still have a sharp image produced on the retina. The **far point** of the eye is the location of the farthest object on which the fully relaxed eye can focus. For a young and normal eye, the near point is located 25 cm from the eye, and the far point is located at infinity.

A **nearsighted (myopic)** eye is one that can focus on nearby objects, but not on distant ones. Nearsightedness can be corrected by wearing eyeglasses or contacts made from diverging lenses. A **farsighted (hyperopic)** eye can see distant objects clearly, but not those close up. Farsightedness can be corrected by using converging lenses.

The **refractive power** of a lens is measured in diopters and is given by $1/f$, where f is the focal length of the lens in meters. A converging lens has a positive refractive power, while a diverging lens has a negative refractive power.

The **angular size** of an object is the angle that it subtends at the eye of the viewer. For small angles, the angular size in radians is $\theta \approx h_o/d_o$, where h_o is the height of the object and d_o is the object distance. The **angular magnification** M of an optical instrument is the angular size θ' of the final image produced by the instrument divided by the reference angular size θ of the object, which is that seen without the instrument: $M = \theta'/\theta$.

A **magnifying glass** is usually a single converging lens that forms an enlarged, upright, and virtual image of an object placed at or inside the focal point of the lens. For a magnifying glass held close to the eye, the angular magnification is approximately $M \approx (1/f - 1/d_i)N$, where f is the focal length of the lens, d_i is the image distance, and N is the distance of the viewer's near point from the eye.

A **compound microscope** usually consists of two lenses, an objective and an eyepiece. The final image is enlarged, inverted, and virtual. The angular magnification of such a microscope is $M \approx -(L - f_e)N/(f_o f_e)$, where f_o and f_e are, respectively, the focal lengths of the objective and eyepiece, L is the distance between the two lenses, and N is the distance of the viewer's near point from the eye.

An **astronomical telescope** magnifies distant objects with the aid of an objective and eyepiece, and it produces a final image that is inverted and virtual. The angular magnification of a telescope is $M \approx -f_o/f_e$, where f_o and f_e are the focal lengths of the objective and eyepiece.

Lens aberrations limit the formation of perfectly focused or sharp images by optical instruments. **Spherical aberration** occurs because rays that pass through the outer edge of a lens with spherical surfaces are not focused at the same point as those that pass through the center of the lens. **Chromatic aberration** arises because a lens focuses different colors at different points.

CONCEPTUAL QUESTIONS

1. In Figure 26.2a, suppose that a layer of oil were added on top of the water. The angle θ_1 at which the incident light travels through the air remains the same. Assuming that light still enters the water, does the angle of refraction at which it does so change because of the presence of the oil? Explain.

2. Two slabs with parallel faces are made from different types of glass. A ray of light travels through air and enters each slab at the same angle of incidence, as the drawing shows. Which slab has the greater index of refraction? Why?

3. Two blocks, made from the same transparent material, are immersed in different liquids. A ray of light strikes each block at the same angle of incidence. From the drawing, determine which liquid has the greater index of refraction. Justify your answer.

Question 2

4. When an observer peers over the edge of a deep empty bowl, he does not see the entire bottom surface, so a small object lying on the bottom is hidden from view. However, when the bowl is filled with water, the object can be seen. Explain this effect.

5. Two identical containers, one filled with water ($n = 1.33$) and the other filled with ethyl alcohol ($n = 1.36$), are viewed from directly above. Which container (if either) appears to have a greater depth of fluid? Why?

6. When you look through an aquarium window at a fish, is the fish as close as it appears? Explain.

7. At night, when it's dark outside and you are standing in a brightly lit room, it is easy to see your reflection in a window. During the day it is not so easy. Account for these facts.

8. A man is fishing from a dock. (a) If he is using a bow and arrow, should he aim above the fish, at the fish, or below the fish, to strike it? (b) How would he aim if he were using a laser gun? Give your reasoning.

9. Two rays of light converge to a point on a screen. A plane-parallel plate of glass is placed in the path of this converging light, and the glass plate is parallel to the screen. Will the point of convergence remain on the screen? If not, will the point move toward the glass or away from it? Justify your answer by drawing a diagram and showing how the rays are affected by the glass.

10. A person sitting at the beach is wearing a pair of Polaroid sunglasses and notices little discomfort due to the glare from the water on a bright sunny day. When she lies on her side, however, she notices that the glare increases. Why?

11. You are sitting by the shore of a lake on a sunny and windless day. Your Polaroid sunglasses are not equally effective at all times of the day in reducing the glare of the sunlight reflected from the lake. Account for this observation.

12. The dispersion of sunlight by a prism is discussed in connection with Figure 26.18. Is it possible to direct a mixture of colors into a prism and have sunlight emerge from it? If so, explain how this could be done.

13. Refer to Figure 26.7. Note that the ray within the glass slab is traveling from a medium with a larger refractive index toward a medium with a smaller refractive index. Is it possible, for θ_1 less than $90°$, that the ray within the glass will experience total internal reflection at the glass/air interface? Account for your answer.

14. Suppose you want to make a rainbow by spraying water from a garden hose into the air. (a) Where must you stand relative to the water and the sun to see the rainbow? (b) Why can't you ever walk under the rainbow?

15. A person is floating on an air mattress in the middle of a swimming pool. His friend is sitting on the side of the pool. The person on the air mattress claims that there is a light shining up from the bottom of the pool directly beneath him. His friend insists, however, that she cannot see any light from where she sits on the side. Can both individuals be correct? Give your reasoning.

16. A beam of blue light is propagating in glass. When the light reaches the boundary between the glass and the surrounding air, the beam is totally reflected back into the glass. However, red light with the same angle of incidence is not totally reflected, and some of the light is refracted into the air. Why do these two colors behave differently?

17. A beacon in a lighthouse is to produce a parallel beam of light. The beacon consists of a bulb and a converging lens. Should the bulb be placed outside the focal point, at the focal point, or inside the focal point of the lens? State your reason.

18. Review Conceptual Example 7 as an aid in answering this question. Is it possible that a converging lens (in air) behaves as a diverging lens when surrounded by another medium? Give a reason for your answer.

19. A spherical mirror and a lens are immersed in water. Compared to the way they work in air, which one do you expect will be more affected by the water? Why?

20. A converging lens is used to project a real image onto a screen, as in Figure 26.27b. A piece of black tape is then placed over the upper half of the lens. Will only the lower half of the image be visible on the screen? Justify your answer by drawing rays from various points on the object to the corresponding points on the image.

21. Review Conceptual Example 7 as an aid in answering this question. A converging lens is made from glass whose index of refraction is n. The lens is surrounded by a fluid whose index of refraction is also n. Can this lens still form an image, either real or virtual, of an object? Why?

22. In a TV mystery program, a photographic negative is introduced as evidence in a court trial. The negative shows an image of a house (now burned down) that was the scene of the crime. At the trial the defendant's acquittal depends on knowing exactly how far above the ground a window was. An expert called by the defense claims that this height can be calculated from only two pieces of information: (1) the measured height on the film, and (2) the focal length of the camera lens. Explain whether the expert is making sense, using the thin-lens and magnification equations to guide your thinking.

23. Two people who wear glasses are camping. One is nearsighted and the other is farsighted. Whose glasses may be useful in starting a fire with the sun's rays? Give your reasoning.

24. Suppose that a 21-year-old with normal vision (near point = 25 cm) is standing in front of a plane mirror. How close can he stand to the mirror and still see himself in focus? Explain.

25. ⚕ If we read for a long time, our eyes become "tired." When this happens, it helps to stop reading and look at a distant object. From the point of view of the ciliary muscle, why does this refresh the eyes?

26. To a swimmer under water, objects look blurred and out of focus. However, when the swimmer wears goggles that keep the water away from the eyes, the objects appear sharp and in focus. Why do goggles improve a swimmer's underwater vision?

27. ⚕ The refractive power of the lens of the eye is 15 diopters when surrounded by the aqueous and vitreous humors. If this lens is removed from the eye and surrounded by air, its refractive power increases to about 150 diopters. Why is the refractive power of the lens so much greater outside the eye?

28. The light shining through a full glass of wine forms an irregularly shaped bright spot on the table, but does not do so when the glass is empty. Explain.

29. Jupiter is the largest planet in our solar system. Yet, to the naked eye, it looks smaller than Venus. Why?

30. By means of a ray diagram, show that the eyes of a person wearing glasses appear to be (a) smaller when the glasses use diverging lenses to correct for nearsightedness and (b) larger when the glasses use converging lenses to correct for farsightedness.

31. Can a diverging lens be used as a magnifying glass? Justify your answer with a ray diagram.

32. Who benefits more from using a magnifying glass, a person whose near point is located 25 cm away from the eyes or a person whose near point is located 75 cm away from the eyes? Provide a reason for your answer.

33. Two lenses, whose focal lengths are 3.0 and 45 cm, are used to build a telescope. Which lens should be the objective? Why?

34. Two refracting telescopes have identical eyepieces, although one telescope is twice as long as the other. Which has the greater angular magnification? Provide a reason for your answer.

35. Suppose a well-designed optical instrument is composed of two converging lenses separated by 14 cm. The focal lengths of the lenses are 0.60 and 4.5 cm. Is the instrument a microscope or a telescope? Why?

36. It is often thought that virtual images are somehow less important than real images. To show that this is not true, identify which of the following instruments normally produce final images that are virtual: (a) a projector, (b) a camera, (c) a magnifying glass, (d) eyeglasses, (e) a compound microscope, and (f) an astronomical telescope.

37. Why does chromatic aberration occur in lenses, but not in mirrors?

PROBLEMS

Unless specified otherwise, use the values given in Table 26.1 for the refractive indices.

ssm Solution is in the Student Solutions Manual.   **www** Solution is available on the World Wide Web at http://www.wiley.com/college/cutnell
⚕ This icon represents a biomedical application.

Section 26.1 The Index of Refraction

1. ssm What is the speed of light in benzene?

2. Find the ratio of the speed of light in diamond to the speed of light in ice.

3. The frequency of a light wave is the same when the light travels in ethyl alcohol as it is when it travels in carbon disulfide. Find the ratio of the wavelength of the light in ethyl alcohol to that in carbon disulfide.

4. Light has a wavelength of 340.0 nm and a frequency of 5.403×10^{14} Hz when traveling through a certain substance. What substance from Table 26.1 could this be?

5. ssm www A glass window ($n = 1.5$) has a thickness of 4.0×10^{-3} m. How long does it take light to pass perpendicularly through the plate?

6. The speed of light is 1.25 times as large in material A than in material B. Determine the ratio n_A/n_B of the refractive indices of these materials.

*7. In a certain time, light travels 3.50 km in a vacuum. During the same time, light travels only 2.50 km in a liquid. What is the refractive index of the liquid?

*8. A flat sheet of ice has a thickness of 2.0 cm. It is on top of a flat sheet of crystalline quartz that has a thickness of 1.1 cm. Light strikes the ice perpendicularly and travels through it and then through the quartz. In the time it takes the light to travel through the two sheets, how far (in cm) would it have traveled in a vacuum?

Section 26.2 Snell's Law and the Refraction of Light

9. ssm A beam of light is traveling in air and strikes a material. The angles of incidence and refraction are 63.0° and 47.0°, respectively. Obtain the speed of light in the material.

10. Refer to Conceptual Example 7 as an aid in understanding this problem. The drawing shows a ray of light traveling through a gas ($n = 1.00$), a solid ($n = 1.55$), and a liquid ($n = 1.55$). At what angle θ does the light enter the liquid?

11. ssm The drawing shows a coin resting on the bottom of a beaker filled with an unknown liquid. A ray of light from the coin travels to the surface of the liquid and is refracted as it enters into the air. A person sees the ray as it skims just above the surface of the liquid. How fast is the light traveling in the liquid?

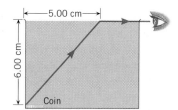

12. As an aid in understanding this problem, refer to Conceptual Example 4. A swimmer, who is looking up from under the water, sees a diving board directly above at an apparent height of 4.0 m above the water. What is the actual height of the diving board?

13. A layer of oil ($n = 1.45$) floats on an unknown liquid. A ray of light shines from the oil into the unknown liquid. The angles of incidence and refraction are, respectively, $65.0°$ and $53.0°$. What is the index of refraction of the unknown liquid?

14. Light in a vacuum is incident on a transparent glass slab. The angle of incidence is $35.0°$. The slab is then immersed in a pool of liquid. When the angle of incidence for the light striking the slab is $20.3°$, the angle of refraction for the light entering the slab is the same as when the slab was in a vacuum. What is the index of refraction of the liquid?

* **15. ssm www** In Figure 26.7, suppose that the angle of incidence is $\theta_1 = 30.0°$, the thickness of the glass pane is 6.00 mm, and the refractive index of the glass is $n_2 = 1.52$. Find the amount (in mm) by which the emergent ray is displaced relative to the incident ray.

16. A ray of sunlight hits a frozen lake at a $45°$ angle of incidence. At what angle of refraction does the ray penetrate (a) the ice and (b) the water beneath the ice?

* **17.** Refer to Figure 26.5a and assume the observer is nearly above the submerged object. For this situation, derive the expression for the apparent depth: $d' = d(n_2/n_1)$, Equation 26.3. (Hint: Use Snell's law of refraction and the fact that the angles of incidence and refraction are small, so $\tan \theta \approx \sin \theta$.)

* **18.** Review Conceptual Example 4 as background for this problem. A man in a boat is looking straight down at a fish in the water directly beneath him. The fish is looking straight up at the man. They are equidistant from the air/water interface. To the man, the fish appears to be 2.0 m beneath his eyes. To the fish, how far above its eyes does the man appear to be?

* **19.** A silver medallion is sealed within a transparent block of plastic. An observer in air, viewing the medallion from directly above, sees the medallion at an apparent depth of 1.6 cm beneath the top surface of the block. How far below the top surface would the medallion appear if the observer (not wearing goggles) and the block were under water?

** **20.** A beaker has a height of 30.0 cm. The lower half of the beaker is filled with water, and the upper half is filled with oil ($n = 1.48$). To a person looking down into the beaker from above, what is the apparent depth of the bottom?

** **21. ssm** A small logo is embedded in a thick block of crown glass ($n = 1.52$), 3.20 cm beneath the top surface of the glass. The block is put under water, so there is 1.50 cm of water above

the top surface of the block. The logo is viewed from directly above by an observer in air. How far beneath the top surface of the water does the logo appear to be?

** **22.** The back wall of a home aquarium is a mirror that is 30.0 cm away from the front wall. The walls of the tank are negligibly thin. A fish is swimming midway between the front and back walls. (a) How far from the front wall does the fish seem to be located? (b) An image of the fish appears behind the mirror. How far does this image appear to be from the front wall of the aquarium? (c) Would the refractive index of the liquid have to be larger or smaller in order for the image of the fish to appear in *front* of the mirror, rather than behind it? Why?

Section 26.3 Total Internal Reflection

23. ssm One method of determining the refractive index of a transparent solid is to measure the critical angle when the solid is in air. If θ_c is found to be $40.5°$, what is the index of refraction of the solid?

24. What is the critical angle for light emerging from carbon disulfide into air?

25. A ray of light is traveling in glass and strikes a glass/liquid interface. The angle of incidence is $58.0°$, and the index of refraction of glass is $n = 1.50$. (a) What must be the index of refraction of the liquid such that the direction of the light entering the liquid is not changed? (b) What is the largest index of refraction that the liquid can have, such that none of the light is transmitted into the liquid and all of it is reflected back into the glass?

26. A point source of light is submerged 2.2 m below the surface of a lake and emits rays in all directions. On the surface of the lake, directly above the source, the area illuminated is a circle. What is the maximum radius that this circle could have?

27. ssm A person is sitting in a small boat in the ocean. A shark is swimming under water at a depth of 4.5 m. When the shark is beyond a certain distance (measured horizontally) from the boat, the shark cannot be seen. Assume that the person's eyes are very near the surface of the water and find that distance.

28. The drawing shows a crown glass slab with a rectangular cross section. As illustrated, a laser beam strikes the upper surface at an angle of $60.0°$. After reflecting from the upper surface, the beam reflects from the side and bottom surfaces. (a) If the glass is surrounded by air, determine where part of the beam first exits the glass, at point A, B, or C. (b) Repeat part (a), assuming that the glass is surrounded by water.

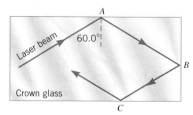

* **29.** The drawing shows a crystalline quartz slab with a rectangular cross section. A ray of light strikes the slab at an incident angle of $\theta_1 =$

34°, enters the quartz, and travels to point P. This slab is surrounded by a fluid with a refractive index n. What is the maximum value of n such that total internal reflection occurs at point P?

*30. The drawing shows an optical fiber that consists of a core made of flint glass ($n_{flint} = 1.667$) surrounded by a cladding made of crown glass ($n_{crown} = 1.523$).

A beam of light enters the fiber from air at an angle θ_1 with respect to the normal. What is θ_1 if the light strikes the core-cladding interface at the critical angle θ_c?

Section 26.4 Polarization and the Reflection and Refraction of Light

31. ssm Light is reflected from a glass coffee table. When the angle of incidence is 56.7°, the reflected light is completely polarized parallel to the surface of the glass. What is the index of refraction of the glass?

32. For light that originates within a liquid and strikes the liquid/air interface, the critical angle is 39°. What is Brewster's angle for this light?

33. At what angle of incidence is sunlight completely polarized upon being reflected from the surface of a lake (a) in the summer and (b) in the winter when the water is frozen?

34. Light is incident from air onto a beaker of carbon tetrachloride. If the reflected light is 100% polarized, what is the angle of refraction of the light that penetrates into the carbon tetrachloride?

35. ssm www When light strikes the surface between two materials from above, the Brewster angle is 65.0°. What is the Brewster angle when the light encounters the same surface from below?

*36. When red light in a vacuum is incident at the Brewster angle on a certain type of glass, the angle of refraction is 29.9°. What are (a) the Brewster angle and (b) the index of refraction of the glass?

*37. In Section 26.4 it is mentioned that the reflected and refracted rays are perpendicular to each other when light strikes the surface at the Brewster angle. This is equivalent to saying that the angle of reflection plus the angle of refraction is 90°. Using Snell's law and Brewster's law, prove that the angle of reflection plus the angle of refraction is 90°.

Section 26.5 The Dispersion of Light: Prisms and Rainbows

38. Yellow light ($n = 2.417$) strikes a diamond at a 45.0° angle of incidence and is refracted when it enters the diamond. Blue light ($n = 1.684$) strikes a piece of flint glass and has the same angle of refraction as does the yellow light in the diamond. What is the angle of incidence of the blue light?

39. ssm A beam of sunlight encounters a plate of crown glass at a 45.00° angle of incidence. Using the data in Table 26.2, find the angle between the violet ray and the red ray in the glass.

40. A ray of sunlight is passing from diamond into crown glass; the angle of incidence is 35.00°. The indices of refraction for the blue and red components of the ray are: blue ($n_{diamond} = 2.444$, $n_{crown\ glass} = 1.531$), and red ($n_{diamond} = 2.410$, $n_{crown\ glass} = 1.520$). Determine the angle between the refracted blue and red rays in the crown glass.

41. Horizontal rays of red light ($\lambda = 660$ nm, in vacuum) and violet light ($\lambda = 410$ nm, in vacuum) are incident on the flint-glass prism shown in the drawing. The indices of refraction for the red and violet light are 1.662 and 1.698, respectively. What is the angle of refraction for each ray as it emerges from the prism?

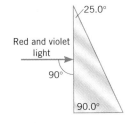

*42. Refer to Conceptual Example 7 as background material for this problem. The drawing shows a horizontal beam of light that is incident on an ice prism. The base of the prism is also horizontal. The prism

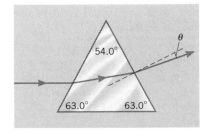

($n = 1.31$) is surrounded by oil whose index of refraction is 1.48. Determine the angle θ that the exiting light makes with the normal to the right face of the prism.

*43. ssm This problem relates to Figure 26.18 which illustrates the dispersion of light by a prism. The prism is made from glass, and its cross section is an equilateral triangle. The indices of refraction for the red and violet light are 1.662 and 1.698, respectively. The angle of incidence for both the red and violet light is 60.0°. Find the angles of refraction at which the red and violet rays emerge into the air from the prism.

Section 26.6 Lenses, Section 26.7 The Formation of Images by Lenses, Section 26.8 The Thin-Lens Equation and the Magnification Equation

(Note: When drawing ray diagrams, be sure that the object height h_o is much smaller than the focal length f of the lens or mirror. This ensures that the rays are paraxial rays.)

44. When a diverging lens is held 13 cm above a line of print, as in Figure 26.29, the image is 5.0 cm beneath the lens. What is the focal length of the lens?

45. ssm A macroscopic (or macro) lens for a camera is usually a converging lens of normal focal length built into a lens barrel that can be adjusted to provide the additional lens-to-film distance needed when focusing at very close range. Suppose that a macro lens ($f = 50.0$ mm) has a maximum lens-to-film distance of 275 mm. How close can the object be located in front of the lens?

46. To focus a camera on objects at different distances, the converging lens is moved toward or away from the film, so a sharp

image always falls on the film. A camera with a telephoto lens ($f = 200.0$ mm) is to be focused on an object located first at a distance of 3.5 m and then at 50.0 m. Over what distance must the lens be movable?

47. A diverging lens has a focal length of -25 cm. (a) Find the image distance when an object is placed 38 cm from the lens. (b) Is the image real or virtual?

48. A movie camera has a converging lens with a focal length of 85.0 mm. It takes a picture of a 145-cm tall person standing 16.0 m away. What is the height of the image on the film? Is the image upright or inverted relative to the object? Give your reasoning.

49. ssm An object is located 30.0 cm to the left of a converging lens whose focal length is 50.0 cm. (a) Draw a ray diagram to scale and from it determine the image distance and the magnification. (b) Use the thin-lens and magnification equations to verify your answers to part (a).

50. A slide projector has a converging lens whose focal length is 105.00 mm. (a) How far (in meters) from the lens must the screen be located if a slide is placed 108.00 mm from the lens? (b) If the slide measures 24.0 mm $\times$ 36.0 mm, what are the dimensions (in mm) of its image?

51. A camera is supplied with two interchangeable lenses, whose focal lengths are 35.0 and 150.0 mm. A woman whose height is 1.80 m stands 8.00 m in front of the camera. What is the height (including sign) of her image on the film, as produced by (a) the 35.0-mm lens and (b) the 150.0-mm lens?

52. The distance between an object and its image formed by a diverging lens is 6.0 cm. The focal length of the lens is -3.0 cm. Find (a) the image distance and (b) the object distance.

* **53. ssm** An object is 18 cm in front of a diverging lens that has a focal length of -12 cm. How far in front of the lens should the object be placed so that the size of its image is reduced by a factor of 2.0?

* **54.** When a converging lens is used in a camera (as in Figure 26.26b), the film must be placed at a distance of 0.210 m from the lens to record an image of an object that is 4.00 m from the lens. The same lens is then used in a projector (see Figure 26.27b), with the screen 0.500 m from the lens. How far from the projector lens should the film be placed?

* **55. ssm www** The moon's diameter is 3.48×10^6 m, and its mean distance from the earth is 3.85×10^8 m. The moon is being photographed by a camera whose lens has a focal length of 50.0 mm. (a) Find the diameter of the moon's image on the slide film. (b) When the slide is projected onto a screen that is 15.0 m from the lens of the projector ($f = 110.0$ mm), what is the diameter of the moon's image on the screen?

* **56.** From a distance of 72 m, a photographer uses a telephoto lens ($f = 300.0$ mm) to take a picture of a charging rhinoceros. How far from the rhinoceros would the photographer have to be to record an image of the same size using a lens whose focal length is 50.0 mm?

** **57.** A converging lens ($f = 25.0$ cm) is used to project an image of an object onto a screen. The object and the screen are 125 cm apart, and between them the lens can be placed at either of two locations. Find the two object distances.

** **58.** An object is 20.0 cm from a converging lens, and the image falls on a screen. When the object is moved 4.00 cm closer to the lens, the screen must be moved 2.70 cm farther away from the lens to register a sharp image. Determine the focal length of the lens.

Section 26.9 Lenses in Combination

59. ssm A converging lens ($f = 12.0$ cm) is located 30.0 cm to the left of a diverging lens ($f = -6.00$ cm). A postage stamp is placed 36.0 cm to the left of the converging lens. (a) Locate the final image of the stamp relative to the diverging lens. (b) Find the overall magnification. (c) Is the final image real or virtual? With respect to the original object, is the final image (d) upright or inverted, and is it (e) larger or smaller?

60. Two identical diverging lenses are separated by 16 cm. The focal length of each lens is -8.0 cm. An object is located 4.0 cm to the left of the lens that is on the left. Determine the final image distance relative to the lens on the right.

61. A converging lens has a focal length of 0.080 m. An object is located 0.040 m to the left of this lens. A second converging lens has the same focal length as the first one and is located 0.120 m to the right of it. Relative to the second lens, where is the final image located?

62. A converging lens ($f_1 = 24.0$ cm) is located 56.0 cm to the left of a diverging lens ($f_2 = -28.0$ cm). An object is placed to the left of the converging lens, and the final image produced by the two-lens combination lies 20.7 cm to the left of the diverging lens. How far is the object from the converging lens?

63. An object, 0.75 cm tall, is placed 12.0 cm to the left of a diverging lens ($f = -8.00$ cm). A converging lens is placed 8.00 cm to the right of the diverging lens. The final image is virtual and is 29.0 cm to the left of the diverging lens. Determine (a) the focal length of the converging lens and (b) the height of the final image.

* **64.** A coin is located 20.0 cm to the left of a converging lens ($f = 16.0$ cm). A second, identical lens is placed to the right of the first lens, such that the image formed by the combination has the same size and orientation as the original coin. Find the separation between the lenses.

* **65. ssm** An object is placed 20.0 cm to the left of a diverging lens ($f = -8.00$ cm). A concave mirror ($f = 12.0$ cm) is placed 30.0 cm to the right of the lens. (a) Find the final image distance, measured relative to the mirror. (b) Is the final image real or virtual? (c) Is the final image upright or inverted with respect to the original object?

** **66.** Two converging lenses ($f_1 = 9.00$ cm and $f_2 = 6.00$ cm) are separated by 18.0 cm. The lens on the left has the longer focal length. An object stands 12.0 cm to the left of the left-hand lens in the combination. (a) Locate the final image relative to the lens on the right. (b) Obtain the overall magnification. (c) Is the final image real or virtual? With respect to the original object, is the final image (d) upright or inverted and is it (e) larger or smaller?

Section 26.10 The Human Eye

67. ⚞ A farsighted person has a near point that is 48.0 cm from her eyes. She wears eyeglasses that are designed to enable her to read a newspaper held at a distance of 25.0 cm from her eyes. Find the focal length of the eyeglasses, assuming that they are worn (a) 2.0 cm from the eyes and (b) 3.0 cm from the eyes.

68. ⚞ A woman can read the large print in a newspaper only when it is at a distance of 65 cm or more from her eyes. (a) Is she myopic or hyperopic? (b) What should be the refractive power of her glasses (worn 2.0 cm from the eyes), so she can read the newspaper at a distance of 25 cm from her eyes?

69. ssm ⚞ A person has far points of 5.0 m from the right eye and 6.5 m from the left eye. Write a prescription for the refractive power of each corrective contact lens.

70. ⚞ A student is reading a lecture written on a blackboard. The lenses in her eyes have a refractive power of 57.50 diopters, and the lens-to-retina distance is 1.750 cm. (a) How far (in meters) is the blackboard from her eyes? (b) If the writing on the blackboard is 5.00 cm high, what is the size of the image on her retina?

71. ⚞ A nearsighted person cannot read a sign that is more than 5.2 m from his eyes. To deal with this problem, he wears contact lenses that do not correct his vision completely, but do allow him to read signs located up to distances of 12.0 m from his eyes. What is the focal length of the contacts?

72. ⚞ A person holds a book 25 cm in front of the effective lens of her eye; the print in the book is 2.0 mm high. If the effective lens of the eye is located 1.7 cm from the retina, what is the size (including the sign) of the print image on the retina?

*73. ⚞ A nearsighted person wears contacts to correct for a far point that is only 3.62 m from his eyes. The near point of his unaided eyes is 25.0 cm from his eyes. If he does not remove the lenses when reading, how close can he hold a book and see it clearly?

**74. ⚞ Bill is farsighted and has a near point located 125 cm from his eyes. Anne is also farsighted, but her near point is 75.0 cm from her eyes. Both have glasses that correct their vision to a normal near point (25.0 cm from the eyes), and both wear the glasses 2.0 cm from the eyes. Relative to the eyes, what is the closest object that can be seen clearly (a) by Anne when she wears Bill's glasses and (b) by Bill when he wears Anne's glasses?

**75. ⚞ The far point of a nearsighted person is 6.0 m from her eyes, and she wears contacts that enable her to see distant objects clearly. A tree is 18.0 m away and 2.0 m high. (a) When she looks through the contacts at the tree, what is its image distance? (b) How high is the image formed by the contacts?

Section 26.11 Angular Magnification and the Magnifying Glass

76. A spectator, seated in the left field stands, is watching a 1.9-m-tall baseball player who is 75 m away. On a TV screen, located 3.0 m from a person watching the game at home, the same player has a 0.12-m image. Find the angular size of the player as seen by (a) the spectator watching the game live and (b) the TV viewer. (c) To whom does the player appear to be larger?

77. ssm A quarter (diameter = 2.4 cm) is held at arms length (70.0 cm). The sun has a diameter of 1.39×10^9 m and is 1.50×10^{11} m from the earth. What is the ratio of the angular size of the quarter to that of the sun?

78. An object has an angular size of 0.0150 rad when placed at the near point (21.0 cm) of an eye. When the eye views this object using a magnifying glass, the largest possible angular size of the image is 0.0380 rad. What is the focal length of the magnifying glass?

79. A magnifying glass is held next to the eye, above a magazine. The image formed by the magnifying glass is located at the near point of the eye. The near point is 0.30 m away from the eye, and the angular magnification is 3.4. Find the focal length of the magnifying glass.

80. The angular magnification of a magnifying glass (f = 22 cm) used as in Figure 26.40b is 2.5. The person using it has a near point that is 36 cm from his eyes. What is the image distance for the object being examined?

*81. ssm A person using a magnifying glass as in Figure 26.40b observes that for clear vision its maximum angular magnification is 1.25 times as large as its minimum angular magnification. Assuming that the person has a near point located 25 cm from her eye, what is the focal length of the magnifying glass?

**82. A farsighted person can read printing as close as 25.0 cm when she wears contacts that have a focal length of 45.4 cm. One day, however, she forgets her contacts and uses a magnifying glass, as in Figure 26.40b. It has a maximum angular magnification of 7.50 for a young person with a normal near point of 25.0 cm. What is the maximum angular magnification that the magnifying glass can provide for her?

Section 26.12 The Compound Microscope

83. A microscope for viewing blood cells has an objective with a focal length of 0.50 cm and an eyepiece with a focal length of 2.5 cm. The distance between the objective and eyepiece is 14.0 cm. If a blood cell subtends an angle of 2.1×10^{-5} rad when viewed with the naked eye at a near point of 25.0 cm, what angle (magnitude only) does it subtend when viewed through the microscope?

84. The distance between the lenses in a microscope is 18 cm. The focal length of the objective is 1.5 cm. If the microscope is to provide an angular magnification of −83 when used by a person with a normal near point (25 cm from the eye), what must be the focal length of the eyepiece?

85. ssm A compound microscope has a barrel whose length is 16.0 cm and an eyepiece whose focal length is 1.4 cm. The viewer has a near point located 25 cm from his eyes. What focal length must the objective have so the angular magnification of the microscope is −320?

86. An anatomist is viewing heart muscle cells with a microscope that has two selectable objectives with refracting powers of

100 and 300 diopters. When she uses the 100-diopter objective, the image of a cell subtends an angle of 3×10^{-3} rad with the eye. What angle is subtended when she uses the 300-diopter objective?

*87. The maximum angular magnification of a magnifying glass is 12.0 when a person uses it who has a near point that is 25.0 cm from his eyes. The same person finds that a microscope, using this magnifying glass as the eyepiece, has an angular magnification of −525. The separation between the eyepiece and the objective of the microscope is 23.0 cm. Obtain the focal length of the objective.

*88. In a compound microscope, the focal length of the objective is 3.50 cm and that of the eyepiece is 6.50 cm. The distance between the lenses is 26.0 cm. (a) What is the angular magnification of the microscope if the person using it has a near point of 35.0 cm? (b) If, as normal, the first image lies just inside the focal point of the eyepiece (see Figure 26.33), how far is the object from the objective? (c) What is the magnification (not the angular magnification) of the objective?

Section 26.13 The Telescope

89. ssm An astronomical telescope has an angular magnification of −184 and uses an objective with a focal length of 48.0 cm. What is the focal length of the eyepiece?

90. An astronomical telescope for hobbyists has an angular magnification of −155. The eyepiece has a focal length of 5.00 mm. (a) Determine the focal length of the objective. (b) About how long is the telescope?

91. Mars subtends an angle of 8.0×10^{-5} rad at the unaided eye. An astronomical telescope has an eyepiece with a focal length of 0.032 m. When Mars is viewed using this telescope, it subtends an angle of 2.8×10^{-3} rad. Find the focal length of the telescope's objective lens.

92. A telescope has an objective and an eyepiece that have refractive powers of 1.25 diopters and 250 diopters, respectively. Find the angular magnification of the telescope.

93. ssm An amateur astronomer decides to build a telescope from a discarded pair of eyeglasses. One of the lenses has a refractive power of 11 diopters, while the other has a refractive power of 1.3 diopters. (a) Which lens should be the objective? (b) How far apart should the lenses be separated? (c) What is the angular magnification of the telescope?

*94. The objective and eyepiece of an astronomical telescope are 1.25 m apart, and the eyepiece has a focal length of 5.0 cm. What is the angular magnification of the telescope?

*95. The telescope at Yerkes Observatory in Wisconsin has an objective whose focal length is 19.4 m. Its eyepiece has a focal length of 10.0 cm. (a) What is the angular magnification of the telescope? (b) If the telescope is used to look at a lunar crater (diameter = 1500 m), what is the size of the first image, assuming the surface of the moon is 3.77×10^8 m from the surface of the earth? (c) How close does the crater appear to be when seen through the telescope?

**96. An astronomical telescope is being used to examine a relatively close object that is only 114.00 m away from the objective of the telescope. The objective and eyepiece have focal lengths of 1.500 and 0.070 m, respectively. Noting that the expression $M \approx -f_o/f_e$ is no longer applicable because the object is so close, use the thin-lens and magnification equations to find the angular magnification of this telescope. (*Hint: See Figure 26.42 and note that the focal points F_o and F_e are so close together that the distance between them may be ignored.*)

ADDITIONAL PROBLEMS

97. ssm An object is located 9.0 cm in front of a converging lens ($f = 6.0$ cm). Using an accurately drawn ray diagram, determine where the image is located.

98. An insect subtends an angle of only 4.0×10^{-3} rad at the unaided eye when placed at the near point. What is the angular size (magnitude only) when the insect is viewed through a microscope whose angular magnification has a magnitude of 160?

99. A glass block ($n = 1.60$) is immersed in a liquid. A ray of light within the glass hits a glass–liquid surface at a 65.0° angle of incidence. Some of the light enters the liquid. What is the smallest possible refractive index for the liquid?

100. A spotlight on a boat is 2.5 m above the water, and the light strikes the water at a point that is 8.0 m horizontally displaced from the spotlight (see the drawing). The depth of the water is 4.0 m. Determine the distance d, which locates the point where the light strikes the bottom.

101. ssm A nearsighted person has a far point located only 220 cm from his eyes. Determine the focal length of contact lenses that will enable him to see distant objects clearly.

102. Amber ($n = 1.546$) is a transparent brown-yellow fossil resin. An insect, trapped and preserved within the amber, appears to be 2.5 cm beneath the surface, when viewed directly from above. How far below the surface is the insect actually located?

103. A tourist takes a picture of a mountain 14 km away using a camera that has a lens with a focal length of 50 mm. She then takes a second picture when the mountain is only 5.0 km away. What is the ratio of the height of the mountain's image on the film for the second picture to its height on the film for the first picture?

104. A diverging lens has a focal length of −32 cm. An object is placed 19 cm in front of this lens. Calculate (a) the image distance and (b) the magnification. Is the image (c) real or virtual, (d) upright or inverted, and (e) enlarged or reduced in size?

105. ssm A light ray in air is incident on a water surface at a 43° angle of incidence. Find (a) the angle of reflection and (b) the angle of refraction.

106. A jeweler whose near point is 72 cm from his eye uses a magnifying glass as in Figure 26.40*b* to examine a watch. The watch is held 4.0 cm from the magnifying glass. Find the angular magnification of the magnifying glass.

107. ssm ⚎ An optometrist prescribes contact lenses that have a focal length of 55.0 cm. (a) Are the lenses converging or diverging, and (b) is the person who wears them nearsighted or farsighted? (c) Where is the unaided near point of the person located, if the lenses are designed so that objects no closer than 35.0 cm can be seen clearly?

108. A ray of light traveling in material *A* strikes the interface between materials *A* and *B* at an angle of incidence of 72°. The angle of refraction is 56°. Find the ratio n_A/n_B of the refractive indices of the two materials.

109. A camper is trying to start a fire by focusing sunlight onto a piece of paper. The diameter of the sun is 1.39×10^9 m, and its mean distance from the earth is 1.50×10^{11} m. The camper is using a converging lens whose focal length is 10.0 cm. (a) What is the area of the sun's image on the paper? (b) If 0.530 W of sunlight pass through the lens, what is the intensity of the sunlight at the paper?

110. (a) For a diverging lens ($f = -20.0$ cm), construct a ray diagram to scale and find the image distance for an object that is 20.0 cm from the lens. (b) Determine the magnification of the lens from the diagram.

*111. ssm An office copier uses a lens to place an image of a document onto a rotating drum. The copy is made from this image. (a) What kind of lens is used? If the document and its copy are to have the same size, but are inverted with respect to one another, (b) how far from the document is the lens located and (c) how far from the lens is the image located? Express your answers in terms of the focal length *f* of the lens.

*112. A stamp collector is viewing a stamp with a magnifying glass held next to her eye. Her near point is 25 cm from her eye. (a) What is the refractive power of a magnifying glass that has an angular magnification of 6.0 when the image of the stamp is located at the near point? (b) What is the angular magnification when the image of the stamp is 45 cm from the eye?

*113. ssm ⚎ At age forty, a man requires contact lenses ($f = 65.0$ cm) to read a book held 25.0 cm from his eyes. At age forty-five, he finds that while wearing these contacts he must now hold a book 29.0 cm from his eyes. (a) By what distance has his near point *changed?* (b) What focal length lenses does he require at age forty-five to read a book at 25.0 cm?

*114. An object is in front of a converging lens ($f = 0.30$ m). The magnification of the lens is $m = 4.0$. (a) Relative to the lens, in what direction should the object be moved so that the magnification changes to $m = -4.0$? (b) Through what distance should the object be moved?

**115. ssm The angular magnification of a telescope is 32 800 times as large when you look through the correct end of the telescope than when you look through the wrong end. What is the angular magnification of the telescope?

**116. The equation

$$\frac{1}{d_o} + \frac{1}{d_i} = \frac{1}{f}$$

is called the *Gaussian* form of the thin-lens equation. The drawing shows the variables d_o, d_i, and *f*. The drawing also shows the distances *x* and *x'*, which are, respectively, the distance from the object to the focal point on the left of the lens and the distance from the focal point on the right of the lens to the image. An equivalent form of the thin-lens equation, involving *x*, *x'*, and *f*, is called the *Newtonian* form. Show that the Newtonian form of the thin-lens equation can be written as $xx' = f^2$.

**117. ⚎ The contacts worn by a farsighted person allow her to see objects clearly that are as close as 25.0 cm, even though her uncorrected near point is 79.0 cm from her eyes. When she is looking at a poster, the contacts form an image of the poster at a distance of 217 cm from her eyes. (a) How far away is the poster actually located? (b) If the poster is 0.350 m tall, how tall is the image formed by the contacts?

CONCEPTS & CALCULATIONS
GROUP LEARNING PROBLEMS

Note: Each of these problems consists of Concept Questions followed by a related quantitative Problem. They are designed for use by students working alone or in small learning groups. The Concept Questions involve little or no mathematics and are intended to stimulate group discussions. They focus on the concepts with which the problems deal. Recognizing the concepts is the essential initial step in any problem-solving technique.

118. Concept Questions The drawing shows four different situations in which a light ray is traveling from one medium into another.

Without doing any calculations, but taking note of the relative sizes of the angles of incidence and refraction, decide which situations (if any) show a refraction that is physically possible. Provide a reason as to why the refraction is possible or impossible.

Problem For the first three cases, the angle of incidence is 55°; for the fourth case, the angle of incidence is 0°. For each case, determine the actual angle of refraction. Check to be sure that your answers are consistent with your answers to the Concept Questions.

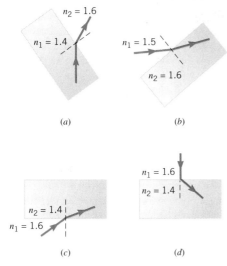

(a)   (b)

(c)   (d)

119. Concept Question The drawing shows a ray of light traveling through three materials whose surfaces are parallel to each other. The refracted rays (but not the reflected rays) are shown as the light passes through each material. Taking into account the relative sizes of the angles of incidence and refraction, rank the materials according to their indices of refraction, greatest first. Provide reasons for your ranking.

Problem A ray of light strikes the a–b interface at a 50.0° angle of incidence. The index of refraction of material a is $n_a = 1.20$. The angles of refraction in materials b and c are, respectively, 45.0° and 56.7°. Find the indices of refraction in these two media. Verify that your answers are consistent with your answers to the Concept Question.

120. Concept Question The drawing shows three materials, a, b, and c. A ray of light strikes the a–b interface at its critical angle. The reflected ray then strikes the a–c interface at its critical angle. Rank the three materials according to their indices of refraction, largest first.

Problem A ray of light is incident at the a–b interface with an angle of incidence equal to the critical angle, $\theta_c = 40.0°$. The index of refraction of material a is $n_a = 1.80$. Find the indices of refraction for the two other materials. Be sure your ranking of the indices is consistent with that determined in the Concept Question.

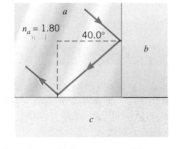

121. Concept Question The drawing shows three layers of different materials, with air above and below the layers. The interfaces between the layers are parallel. The index of refraction of each layer is given in the drawing. Identical rays of light are sent into the layers, and each ray zigzags through the layer, reflecting from the top and bottom surfaces. Fill in the table below, specifying a "yes" or "no" as to whether total internal reflection is possible from the top and bottom surfaces of each layer. Provide a reason for each of your answers.

Layer	Is total internal reflection possible?	
	Top surface of layer	Bottom surface of layer
a		
b		
c		

Problem For each layer, the ray of light has an angle of incidence of 75.0°. For the cases where total internal reflection is possible from either the top or bottom surface of a layer, determine the amount by which the angle of incidence exceeds the critical angle.

122. Concept Questions An object is placed to the left of a lens, and a real image is formed to the right of the lens. The image is inverted relative to the object and is one-half the size of the object. (a) What kind of lens, converging or diverging, is used to produce this image? (b) How is the height h_i of the image related to the height h_o of the object? Don't forget to take into account the fact that the image is inverted relative to the object. (c) What is the ratio d_i/d_o of the image distance to the object distance?

Problem For the situation described in the Concept Questions, the distance between the object and the image is 90.0 cm. (a) How far from the lens is the object? (b) What is the focal length of the lens?

123. Concept Question Two systems are formed from a converging lens and a diverging lens, as shown in parts a and b of the drawing. (The point labeled "F" is the focal point of the converging lens.) An object is placed inside the focal point of lens 1. Without doing any calculations, determine for each system whether the final image lies to the left or to the right of lens 2. Provide a reason for each answer.

Problem The focal lengths of the converging and diverging lenses are 15.00 and −20.0 cm, respectively. The distance between the lenses is 50.0 cm, and an object is placed 10.00 cm to the left of lens 1. Determine the final image distance for each system, measured with respect to lens 2. Check to be sure your answers are consistent with your answers to the Concept Question.

(a)   (b)

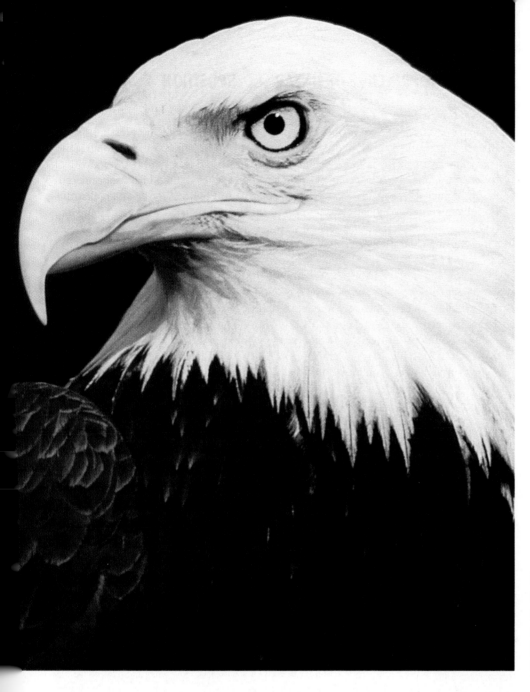

Eagles have eyes that can distinguish between two distant objects that are close together. This ability is related, in part, to the large pupils of the eyes and the wave nature of light.

CHAPTER 27

INTERFERENCE AND THE WAVE NATURE OF LIGHT

27.1 THE PRINCIPLE OF LINEAR SUPERPOSITION

● *Concepts at a Glance* Chapter 17 examines what happens when several sound waves are present at the same place at the same time. The pressure disturbance that results is governed by the principle of linear superposition, which states that the resultant disturbance is the sum of the disturbances from the individual waves. Light is also a wave, an electromagnetic wave, and it too obeys the superposition principle. The Concepts-at-a-Glance chart in Figure 27.1 illustrates that this principle is used to explain interference phenomena associated with light, including Young's double-slit interference, thin-film interference, and the interference effects that occur in the Michelson interferometer. This chart is analogous to the one in Figure 17.17 for the interference of sound waves. When two or more light waves pass through a given point, their electric fields combine according to the principle of linear superposition and produce a resultant electric field. According to Equation 24.5b, the square of the electric field strength is proportional to the intensity of the light, which, in turn, is related to its brightness. Thus, interference can and does alter the brightness of light, just as it affects the loudness of sound. ●

Figure 27.2 illustrates what happens when two identical waves (same wavelength λ and same amplitude) arrive at the point P in phase, that is, crest to crest and trough to trough. According to the principle of linear superposition, the waves reinforce each other and ***constructive interference*** occurs. The resulting total wave at P has an amplitude that is twice the amplitude of either individual wave, and in the case of light waves, the brightness at P is greater than that due to either wave alone. The waves start out in phase and are in phase at P because the distances, ℓ_1 and ℓ_2, between this spot and the sources of the waves differ by one wavelength. In Figure 27.2, these distances are $\ell_1 = 2\frac{1}{4}$ wavelengths and $\ell_2 = 3\frac{1}{4}$ wavelengths. In general, when the waves start out in phase, constructive interference will result at P whenever the distances are the same or differ by any integer number of wavelengths; in other words, assuming ℓ_2 is the larger distance, whenever $\ell_2 - \ell_1 = m\lambda$, where $m = 0, 1, 2, 3, \ldots .$

Figure 27.1 *Concepts at a Glance* The principle of linear superposition explains various interference phenomena associated with light waves. The beautiful iridescent colors on the wings of this tropical butterfly are due to the thin-film interference of light. A transparent cuticlelike material on the upper wing surface acts like a thin film.

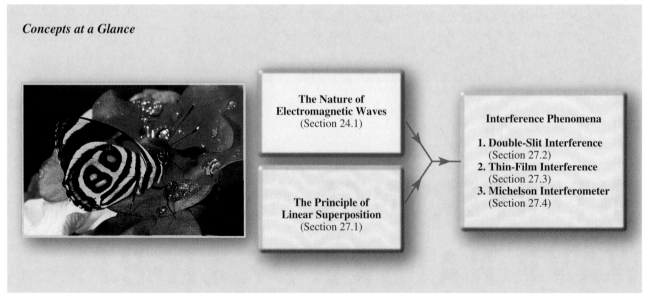

Concepts at a Glance

The Nature of
Electromagnetic Waves
(Section 24.1)

The Principle of
Linear Superposition
(Section 27.1)

Interference Phenomena

1. **Double-Slit Interference**
(Section 27.2)
2. **Thin-Film Interference**
(Section 27.3)
3. **Michelson Interferometer**
(Section 27.4)

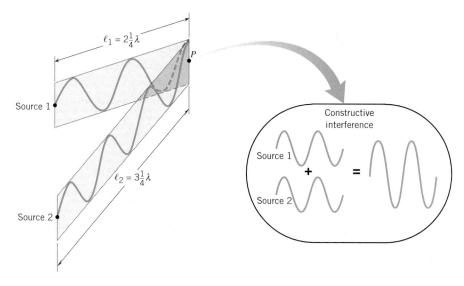

Figure 27.2 The waves emitted by source 1 and source 2 start out in phase and arrive at point P in phase, leading to constructive interference at that point.

Figure 27.3 shows what occurs when two identical waves arrive at the point P out of phase with one another, or crest to trough. Now the waves mutually cancel, according to the principle of linear superposition, and ***destructive interference*** results. With light waves this would mean that there is no brightness. The waves begin with the same phase but are out of phase at P because the distances through which they travel in reaching this spot differ by one-half of a wavelength ($\ell_1 = 2\frac{3}{4}\lambda$ and $\ell_2 = 3\frac{1}{4}\lambda$ in the drawing). For waves that start out in phase, destructive interference will take place at P whenever the distances differ by any odd integer number of half-wavelengths, that is, whenever $\ell_2 - \ell_1 = \frac{1}{2}\lambda, \frac{3}{2}\lambda, \frac{5}{2}\lambda \ldots$, where ℓ_2 is the larger distance. This is equivalent to $\ell_2 - \ell_1 = (m + \frac{1}{2})\lambda$, where $m = 0, 1, 2, 3, \ldots$.

If constructive or destructive interference is to continue occurring at a point, the sources of the waves must be ***coherent sources.*** Two sources are coherent if the waves they emit maintain a constant phase relation. Effectively, this means

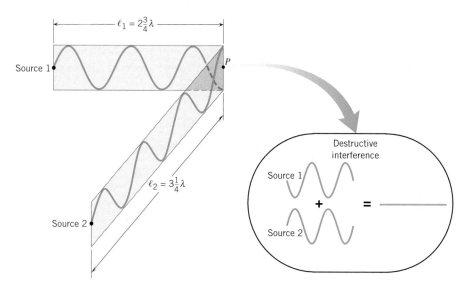

Figure 27.3 The waves emitted by the two sources have the same phase to begin with, but they arrive at point P out of phase. As a result, destructive interference occurs at P.

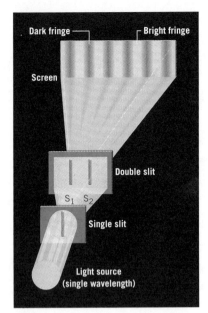

Figure 27.4 In Young's double-slit experiment, two slits S_1 and S_2 act as coherent sources of light. Light waves from these slits interfere constructively and destructively on the screen to produce, respectively, the bright and dark fringes. The slit widths and the distance between the slits have been exaggerated for clarity.

Figure 27.5 The waves from slits S_1 and S_2 interfere constructively (parts *a* and *b*) or destructively (part *c*) on the screen, depending on the difference in distances between the slits and the screen. The slit widths and the distance between the slits have been exaggerated for clarity.

that the waves do not shift relative to one another as time passes. For instance, suppose that the wave pattern of source 1 in Figure 27.3 shifted forward or backward by random amounts at random moments. Then, on average, neither constructive nor destructive interference would be observed at point *P* because there would be no stable relation between the two wave patterns. Lasers are coherent sources of light, while incandescent light bulbs and fluorescent lamps are incoherent sources.

27.2 YOUNG'S DOUBLE-SLIT EXPERIMENT

In 1801 the English scientist Thomas Young (1773–1829) performed a historic experiment that demonstrated the wave nature of light by showing that two overlapping light waves interfered with each other. His experiment was particularly important because he was also able to determine the wavelength of the light from his measurements, the first such determination of this important property. Figure 27.4 shows one arrangement of Young's experiment, in which light of a single wavelength (monochromatic light) passes through a single narrow slit and falls on two closely spaced, narrow slits S_1 and S_2. These two slits act as coherent sources of light waves that interfere constructively and destructively at different points on the screen to produce a pattern of alternating bright and dark fringes. The purpose of the single slit is to ensure that only light from one direction falls on the double slit. Without it, light coming from different points on the light source would strike the double slit from different directions and cause the pattern on the screen to be washed out. The slits S_1 and S_2 act as coherent sources of light waves because the light from each originates from the same primary source, namely, the single slit.

To help explain the origin of the bright and dark fringes, Figure 27.5 presents three top views of the double slit and the screen. Part *a* illustrates how a bright fringe arises directly opposite the midpoint between the two slits. In this part of the drawing the waves (identical) from each slit travel to the midpoint on the screen. At this location, the distances ℓ_1 and ℓ_2 to the slits are equal, each containing the same number of wavelengths. Therefore, constructive interference results, leading to the bright fringe. Part *b* indicates that constructive interference produces another bright fringe on one side of the midpoint when the distance ℓ_2 is larger than ℓ_1 by exactly one wavelength. A bright fringe also occurs symmetrically on the other side of the midpoint when the distance ℓ_1 exceeds ℓ_2 by one wavelength; for clarity, however, this bright fringe is not shown. Constructive interference produces additional bright

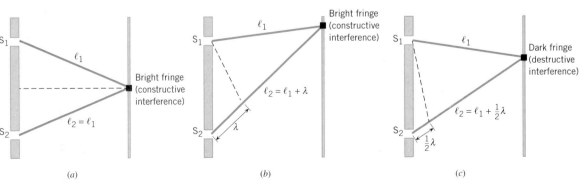

(a)   (b)   (c)

fringes on both sides of the middle wherever the difference between ℓ_1 and ℓ_2 is an integer number of wavelengths: λ, 2λ, 3λ, etc. Part *c* shows how the first dark fringe arises. Here the distance ℓ_2 is larger than ℓ_1 by exactly one-half a wavelength, so the waves interfere destructively, giving rise to the dark fringe. Destructive interference creates additional dark fringes on both sides of the center wherever the difference between ℓ_1 and ℓ_2 equals an odd integer number of half-wavelengths: $1(\frac{\lambda}{2})$, $3(\frac{\lambda}{2})$, $5(\frac{\lambda}{2})$, etc.

The brightness of the fringes in Young's experiment varies, as the photograph in Figure 27.6 shows. Below the photograph is a graph to suggest the way in which the intensity varies for the fringe pattern. The central fringe is labeled with a zero, while the other bright fringes are numbered in ascending order on either side of the center. It can be seen that the central fringe has the greatest intensity. To either side of the center, the intensities of the other fringes decrease symmetrically in a way that depends on how small the slit widths are relative to the wavelength of the light.

The position of the fringes observed on the screen in Young's experiment can be calculated with the aid of Figure 27.7. If the screen is located far away compared with the separation *d* of the slits, then the lines labeled ℓ_1 and ℓ_2 in part *a* are nearly parallel. Being nearly parallel, these lines make approximately equal angles θ with the horizontal. The distances ℓ_1 and ℓ_2 differ by an amount $\Delta\ell$, which is the length of the short side of the colored triangle in part *b* of the drawing. Since the triangle is a right triangle, it follows that $\Delta\ell = d \sin \theta$. Constructive interference occurs when the distances differ by an integer number *m* of wavelengths λ, or $\Delta\ell = d \sin \theta = m\lambda$. Therefore, the angle θ for the interference maxima can be determined from the following expression:

Bright fringes of a double slit

$$\sin \theta = m \frac{\lambda}{d} \qquad m = 0, 1, 2, 3, \ldots \qquad (27.1)$$

The value of *m* specifies the *order* of the fringe. Thus, $m = 2$ identifies the "second-order" bright fringe. Part *c* of the drawing stresses that the angle θ given by Equation 27.1 locates bright fringes on either side of the midpoint between the slits. A similar line of reasoning leads to the conclusion that the dark fringes, which lie between the bright fringes, are located according to

Dark fringes of a double slit

$$\sin \theta = \left(m + \frac{1}{2} \right) \frac{\lambda}{d} \qquad m = 0, 1, 2, 3, \ldots \qquad (27.2)$$

Example 1 illustrates how to determine the distance of a higher-order bright fringe from the central bright fringe with the aid of these expressions.

Figure 27.6 The results of Young's double-slit experiment, showing a photograph of the bright and dark fringes formed on the screen and a graph of the light intensity. The central or zeroth fringe is the brightest fringe (greatest intensity).

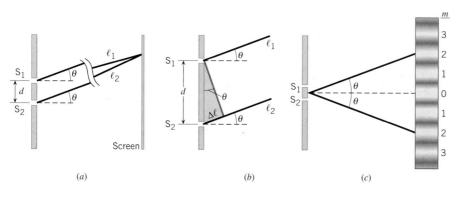

Figure 27.7 (*a*) Rays from slits S_1 and S_2, which make approximately the same angle θ with the horizontal, strike a distant screen at the same spot. (*b*) The difference in the path lengths of the two rays is $\Delta\ell = d \sin \theta$. (*c*) The angle θ is the angle at which a bright fringe ($m = 2$, here) occurs on either side of the central bright fringe ($m = 0$).

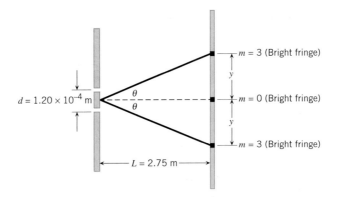

Figure 27.8 The third-order bright fringe ($m = 3$) is observed on the screen at a distance y from the central bright fringe ($m = 0$).

EXAMPLE 1   Young's Double-Slit Experiment

Red light ($\lambda = 664$ nm in vacuum) is used in Young's experiment with the slits separated by a distance $d = 1.20 \times 10^{-4}$ m. The screen in Figure 27.8 is located at a distance from the slits given by $L = 2.75$ m. Find the distance y on the screen between the central bright fringe and the third-order bright fringe.

Reasoning This problem can be solved by first using Equation 27.1 to determine the value of θ that locates the third-order bright fringe. Then trigonometry can be used to obtain the distance y.

Solution According to Equation 27.1, we find

$$\theta = \sin^{-1}\left(\frac{m\lambda}{d}\right) = \sin^{-1}\left[\frac{3(6.64 \times 10^{-9}\ \text{m})}{1.20 \times 10^{-4}\ \text{m}}\right] = 0.951°$$

According to Figure 27.8, the distance y can be calculated from $\tan \theta = y/L$:

$$y = L \tan \theta = (2.75\ \text{m}) \tan 0.951° = \boxed{0.0456\ \text{m}}$$

In the version of Young's experiment that we have discussed, monochromatic light has been used. It is possible, however, to use light that contains a mixture of wavelengths, and Conceptual Example 2 deals with some of the interesting features of the resulting interference pattern.

CONCEPTUAL EXAMPLE 2   White Light and Young's Experiment

Figure 27.9 shows a photograph that illustrates the kind of interference fringes that can result when white light, which is a mixture of all colors, is used in Young's experiment. Except for the central fringe, which is white, the bright fringes are a rainbow of colors. Why does Young's experiment separate white light into its constituent colors? In any group of colored fringes, such as the two singled out in Figure 27.9, why is red farther out from the central fringe than green is? And finally, why is the central fringe white rather than colored?

Reasoning and Solution To understand how the color separation arises, we need to remember that each color corresponds to a different wavelength λ and that constructive and destructive interference depend on the wavelength. According to Equation 27.1 ($\sin \theta = m\lambda/d$), there is a different angle that locates a bright fringe for each value of λ, or for each color. These different angles lead to the separation of colors on the observation screen. In fact, on either side of the central fringe, there is one group of colored fringes for $m = 1$ and another for each value of m.

Now, consider what it means that, within any single group of colored fringes, red is farther out from the central fringe than green is. It means that, in the equation

Figure 27.9 This photograph shows the results observed on the screen in one version of Young's experiment in which white light (a mixture of all colors) is used.

$\sin \theta = m\lambda/d$, red light has a larger angle θ than green light does. Does this make sense? Yes, because red has the larger wavelength (see Table 26.2, where $\lambda_{\text{red}} = 660$ nm and $\lambda_{\text{green}} = 550$ nm).

In Figure 27.9, the central fringe is distinguished from all the other colored fringes by being white. In Equation 27.1, the central fringe is different from the other fringes because it is the only one for which $m = 0$. In Equation 27.1, a value of $m = 0$ means that $\sin \theta = m\lambda/d = 0$, which reveals that $\theta = 0°$, no matter what the wavelength λ is. In other words, all wavelengths have a zeroth-order bright fringe located at the same place on the screen, so that all colors strike the screen there and mix together to produce the white central fringe.

Related Homework Material: *Problem 6*

Historically, Young's experiment provided strong evidence that light has a wavelike character. If light behaved only as a stream of "tiny particles," as others believed at the time,* then the two slits would deliver the light energy into only two bright fringes located directly opposite the slits on the screen. Instead, Young's experiment shows that wave interference redistributes the energy from the two slits into many bright fringes.

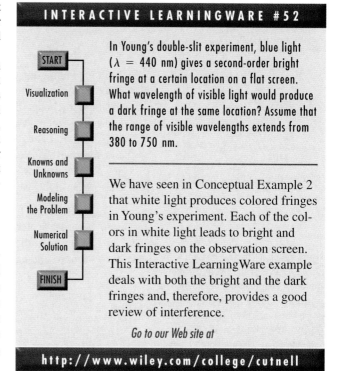

INTERACTIVE LEARNINGWARE #52

START
Visualization
Reasoning
Knowns and Unknowns
Modeling the Problem
Numerical Solution
FINISH

In Young's double-slit experiment, blue light ($\lambda = 440$ nm) gives a second-order bright fringe at a certain location on a flat screen. What wavelength of visible light would produce a dark fringe at the same location? Assume that the range of visible wavelengths extends from 380 to 750 nm.

We have seen in Conceptual Example 2 that white light produces colored fringes in Young's experiment. Each of the colors in white light leads to bright and dark fringes on the observation screen. This Interactive LearningWare example deals with both the bright and the dark fringes and, therefore, provides a good review of interference.

Go to our Web site at

http://www.wiley.com/college/cutnell

27.3 THIN-FILM INTERFERENCE

Young's double-slit experiment is one example of interference between light waves. Interference also occurs in more common circumstances. For instance, Figure 27.10 shows a thin film, such as gasoline floating on water. To begin with, let us assume that the film has a constant thickness. Consider what happens when monochromatic light (a single wavelength) strikes the film nearly perpendicularly. At the top surface of the film reflection occurs and produces the light wave represented by ray 1. However, refraction also occurs, and some light enters the film. Part of this light reflects from the bottom surface and passes back up through the film, eventually reentering the air. Thus, a second light wave, which is represented by ray 2, also exists. Moreover, this wave, having traversed the film twice, has traveled farther than wave 1. Because of the extra travel distance, there can be interference between the two waves. If constructive interference occurs, an observer, whose eyes detect the superposition of waves 1 and 2, would see a uniformly bright film. If destructive interference occurs, an observer would see a uniformly dark film. The controlling factor is whether the extra distance for wave 2 is an integer number of whole wavelengths or an odd integer number of half-wavelengths.

In Figure 27.10 the difference in path lengths between waves 1 and 2 occurs inside the thin film. Therefore, *the wavelength that is important for thin-film interference is the wavelength within the film,* not the wavelength in vacuum. The wavelength within the film can be calculated from the wavelength in vacuum by using the index of refraction n for the film. According to Equations 26.1 and 16.1,

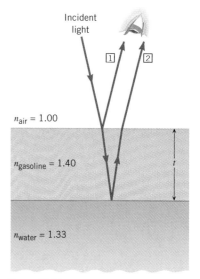

Incident light

$n_{\text{air}} = 1.00$

$n_{\text{gasoline}} = 1.40$

$n_{\text{water}} = 1.33$

Figure 27.10 Because of reflection and refraction, two light waves, represented by rays 1 and 2, enter the eye when light shines on a thin film of gasoline floating on a thick layer of water.

---

* It is now known that the particle or corpuscular theory of light, which Isaac Newton promoted, does indeed explain some experiments that the wave theory cannot explain. Today, light is regarded as having both particle and wave characteristics. Chapter 29 discusses this dual nature of light.

(a)

(b)

Figure 27.11 When a wave on a string reflects from a wall, the wave undergoes a phase change. Thus, an upward-pointing half-cycle of the wave becomes, after reflection, a downward-pointing half-cycle, and vice versa, as the numbered labels in the drawing indicate.

● **PROBLEM SOLVING INSIGHT**
When analyzing thin-film interference effects, remember to use the wavelength of the light in the film (λ_{film}) instead of the wavelength in a vacuum (λ_{vacuum}).

$n = c/v = (c/f)/(v/f) = \lambda_{\text{vacuum}}/\lambda_{\text{film}}$. In other words,

$$\lambda_{\text{film}} = \frac{\lambda_{\text{vacuum}}}{n} \tag{27.3}$$

In explaining the interference that can occur in Figure 27.10, we need to add one other important part to the story. Whenever waves reflect at a boundary, it is possible for them to change phase. Figure 27.11, for example, shows that a wave on a string is inverted when it reflects from the end that is tied to a wall (see also Figure 17.19). This inversion is equivalent to a half-cycle of the wave, as if the wave had traveled an additional distance of one-half of a wavelength. In contrast, a phase change does not occur when a wave on a string reflects from the end of a string that is hanging free. When light waves undergo reflection, similar phase changes occur as follows:

1. When light travels through a material with a smaller refractive index toward a material with a larger refractive index (e.g., air to gasoline), reflection at the boundary occurs along with a phase change that is equivalent to one-half of a wavelength.

2. When light travels from a larger toward a smaller refractive index, there is no phase change upon reflection at the boundary.

The next example indicates how the phase change that can accompany reflection is taken into account when dealing with thin-film interference.

EXAMPLE 3   Colored Thin Films of Gasoline

A thin film of gasoline floats on a puddle of water. Sunlight falls almost perpendicularly on the film and reflects into your eyes. Although sunlight is white, since it contains all colors, the film has a yellow hue because destructive interference eliminates the color of blue ($\lambda_{\text{vacuum}} = 469$ nm) from the reflected light. The refractive indices of the blue light in gasoline and water are 1.40 and 1.33, respectively. Determine the minimum nonzero thickness t of the film.

Reasoning To solve this problem, we must express the condition for destructive interference in terms of the film thickness t and the wavelength in the gasoline film λ_{film}. We must also take into account any phase changes that occur upon reflection.

Solution Equation 27.3, with $n = 1.40$, gives the wavelength of blue light in the film as $\lambda_{\text{film}} = (469 \text{ nm})/1.40 = 335$ nm. In Figure 27.10, the phase change for wave 1 is equivalent to one-half of a wavelength, since this light travels from a smaller refractive index ($n_{\text{air}} = 1.00$) toward a larger refractive index ($n_{\text{gasoline}} = 1.40$). In contrast, there is no phase change when wave 2 reflects from the bottom surface of the film, since this light travels from a larger refractive index ($n_{\text{gasoline}} = 1.40$) toward a smaller one ($n_{\text{water}} = 1.33$). The net phase change between waves 1 and 2 due to reflection is, thus, equivalent to one-half of a wavelength, $\frac{1}{2}\lambda_{\text{film}}$. This half-wavelength must be combined with the extra travel distance for wave 2, to determine the condition for destructive interference. For destructive interference, the combined total must be an odd integer number of half-wavelengths. Since wave 2 travels back and forth through the film and since light strikes the film nearly perpendicularly, the extra travel distance is twice the film thickness, or $2t$. Thus, the condition for destructive interference is

$$\underbrace{2t}_{\substack{\text{Extra distance} \\ \text{traveled by} \\ \text{wave 2}}} + \underbrace{\frac{1}{2}\lambda_{\text{film}}}_{\substack{\text{Half-wavelength} \\ \text{net phase change} \\ \text{due to reflection}}} = \underbrace{\frac{1}{2}\lambda_{\text{film}}, \frac{3}{2}\lambda_{\text{film}}, \frac{5}{2}\lambda_{\text{film}}, \cdots}_{\substack{\text{Condition for} \\ \text{destructive interference}}}$$

After subtracting the term $\frac{1}{2}\lambda_{\text{film}}$ from each side of this equation, we can solve for the thickness t of the film that yields destructive interference:

$$t = \frac{m\lambda_{\text{film}}}{2} \qquad m = 0, 1, 2, 3, \ldots$$

With $m = 1$, the expression above gives the minimum nonzero film thickness for which the blue color is missing in the reflected light:

$$t = \tfrac{1}{2}\lambda_{\text{film}} = \tfrac{1}{2}(335 \text{ nm}) = \boxed{168 \text{ nm}}$$

Example 3 deals with a thin film that has the same yellow color everywhere. In nature, such a uniformly colored thin film would be unusual, and the next example deals with a more realistic situation.

CONCEPTUAL EXAMPLE 4   Multicolored Thin Films

Under natural conditions, thin films, like the soap bubble in Figure 27.12 or like gasoline on water, have a multicolored appearance that often changes while you are watching them. Why are such films multicolored, and what can be inferred from the fact that the colors change in time?

Reasoning and Solution In Example 3 we have seen that a thin film can appear yellow if destructive interference removes blue light from the reflected sunlight. The thickness of the film is the key. If the thickness were different, so that destructive interference removed green light from the reflected sunlight, the film would appear magenta. Constructive interference can also cause certain colors to appear brighter than others in the reflected light and give the film a colored appearance. The colors that are enhanced by constructive interference, like those removed by destructive interference, depend on the thickness of the film. Thus, we conclude that ***the different colors in a soap bubble or a thin film of gasoline on water arise because the thickness is different in different places on the film. Moreover, the fact that the colors change as you watch them indicates that the thickness is changing.*** A number of factors can cause the thickness to change, including air currents, temperature fluctuations, and the pull of gravity, which tends to make a vertical film sag, leading to thicker regions at the bottom than at the top.

Related Homework Material: *Problems 12, 14, and 56*

Figure 27.12 A soap bubble is multicolored when viewed in sunlight because of the effects of thin-film interference.

The colors that you see when sunlight is reflected from a thin film also depend on your viewing angle. At an oblique angle, the light corresponding to ray 2 in Figure 27.10 would travel a greater distance within the film than it does at nearly perpendicular incidence. The greater distance would lead to destructive interference for a different wavelength.

Thin-film interference can be beneficial in optical instruments. For example, many cameras contain up to six or more lenses. Reflections from all the lens surfaces can reduce considerably the amount of light directly reaching the film. In addition, multiple reflections from the lenses often reach the film indirectly and degrade the quality of the image. To minimize such unwanted reflections, high-quality lenses are often covered with a thin nonreflective coating of magnesium fluoride ($n = 1.38$). The thickness of the coating is usually chosen to ensure that destructive interference eliminates the reflection of green light, which is in the middle of the visible spectrum. It should be pointed out that the absence of any reflected light does not mean that it has been destroyed by the nonreflective coating. Rather, the "missing" light has been transmitted into the coating and the lens.

THE PHYSICS OF...

nonreflecting lens coatings.

Another interesting illustration of thin-film interference is the air wedge. As Figure 27.13a shows, an air wedge is formed when two flat plates of glass are separated along one side, perhaps by a thin sheet of paper. The thickness of this film of air varies between zero, where the plates touch, and the thickness of the paper. When monochromatic light reflects from this arrangement, alternate bright and dark fringes are formed by constructive and destructive interference, as the picture indicates and Example 5 discusses.

EXAMPLE 5   An Air Wedge

(a) Assuming that green light ($\lambda_{vacuum} = 552$ nm) strikes the glass plates nearly perpendicularly in Figure 27.13, determine the number of bright fringes that occurs between the place where the plates touch and the edge of the sheet of paper (thickness $= 4.10 \times 10^{-5}$ m). (b) Explain why there is a dark fringe where the plates touch.

Reasoning A bright fringe occurs wherever there is constructive interference, as determined by any phase changes due to reflection and the effects of the thickness of the air wedge. We examine the phase changes and the effects of the thickness separately. There is no phase change upon reflection for wave 1, since this light travels from a larger (glass) toward a smaller (air) refractive index. In contrast, there is a half-wavelength phase change for wave 2, since the ordering of the refractive indices is reversed at the lower air/glass boundary where reflection occurs. The net phase change due to reflection for waves 1 and 2, then, is equivalent to a half wavelength. Now we combine any extra distance traveled by ray 2 with this half-wavelength and determine the condition for the constructive interference that creates the bright fringes. Constructive interference occurs whenever the *combination* yields an integer number of wavelengths. At nearly perpendicular incidence, the extra travel distance for wave 2 is approximately twice the thickness t of the wedge at any point, so the condition for constructive interference is

$$\underbrace{2t}_{\substack{\text{Extra distance} \\ \text{traveled by} \\ \text{wave 2}}} + \underbrace{\tfrac{1}{2}\lambda_{film}}_{\substack{\text{Half-wavelength} \\ \text{net phase change} \\ \text{due to reflection}}} = \underbrace{\lambda_{film}, 2\lambda_{film}, 3\lambda_{film}, \dots}_{\substack{\text{Condition for} \\ \text{constructive interference}}}$$

Figure 27.13 (*a*) The wedge of air formed between two flat glass plates causes an interference pattern of alternating dark and bright fringes to appear in reflected light. (*b*) A side view of the glass plates and the air wedge.

Light source

Sheet of paper

Glass plates

(*a*)

Incident light

$n_{glass} = 1.52$

$n_{air} = 1.00$

$n_{glass} = 1.52$

(*b*)

Subtracting the term $\frac{1}{2}\lambda_{\text{film}}$ from the left-hand side of this equation and from each term on the right-hand side yields

$$2t = \underbrace{\tfrac{1}{2}\lambda_{\text{film}}, \tfrac{3}{2}\lambda_{\text{film}}, \tfrac{5}{2}\lambda_{\text{film}}, \ldots}_{\substack{(m + \frac{1}{2})\,\lambda_{\text{film}} \\ m = 0, 1, 2, 3, \ldots}}$$

Therefore,

$$t = \frac{(m + \frac{1}{2})\lambda_{\text{film}}}{2} \qquad m = 0, 1, 2, 3, \ldots$$

In this expression, note that the "film" is a film of air. Since the refractive index of air is nearly one, λ_{film} is virtually the same as that in a vacuum, so $\lambda_{\text{film}} = 552$ nm.

Solution

(a) When t equals the thickness of the paper holding the plates apart, the corresponding value of m can be obtained from the equation above:

$$m = \frac{2t}{\lambda_{\text{film}}} - \frac{1}{2} = \frac{2(4.10 \times 10^{-5}\ \text{m})}{552 \times 10^{-9}\ \text{m}} - \frac{1}{2} = 148$$

Since the first bright fringe occurs when $m = 0$, the number of bright fringes is $m + 1 = \boxed{149}$.

(b) Where the plates touch, there is a dark fringe because of destructive interference between the light waves represented by rays 1 and 2. Destructive interference occurs, since the thickness of the wedge is zero here and the only difference between the rays is the half-wavelength phase change due to reflection from the lower plate.

Another type of air wedge can also be used to determine the degree to which the surface of a lens or mirror is spherical. When an accurate spherical surface is put in contact with an optically flat plate, as in Figure 27.14a, the circular interference fringes shown in part b of the figure can be observed. The circular fringes are called *Newton's rings*. They arise in the same way that the straight fringes arise in Figure 27.13a.

27.4 THE MICHELSON INTERFEROMETER

An interferometer is an apparatus that can be used to measure the wavelength of light by utilizing interference between two light waves. One particularly famous interferometer is that developed by Albert A. Michelson (1852–1931). The Michelson interferometer uses reflection to set up conditions where two light waves interfere. Figure 27.15 presents a schematic drawing of the instrument. Waves emitted by the monochromatic light source strike a *beam splitter*, so called because it splits the beam of light into two parts. The beam splitter is a glass plate, the far side of which is coated with a thin layer of silver that reflects part of the beam upward as wave A in the drawing. The coating is so thin, however, that it also allows the remainder of the beam to pass directly through as wave F. Wave A strikes an adjustable mirror and reflects back on itself. It again crosses the beam splitter and then enters the viewing telescope. Wave F strikes a fixed mirror and returns, to be partly reflected into the viewing telescope by the beam splitter. Note that wave A passes through the glass plate of the beam splitter three times in reaching the viewing scope, while wave F passes through it only once. The compensating plate in the path of wave F has the same thickness as the beam splitter plate and ensures that wave F also passes three times through the same thickness of glass on the way to

• **PROBLEM SOLVING INSIGHT**
Reflected light will experience a phase change only if the light travels from a material with a smaller refractive index toward a material with a larger refractive index. Be sure to take such a phase change into account when analyzing thin-film interference.

(a)

(b)

Figure 27.14 (a) The air wedge between a convex spherical glass surface and an optically flat plate leads to (b) a pattern of circular interference fringes that is known as Newton's rings.

THE PHYSICS OF...

the Michelson interferometer.

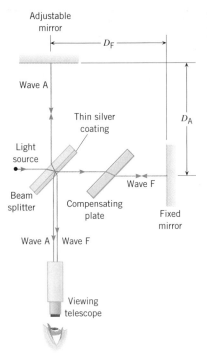

Figure 27.15 A schematic drawing of a Michelson interferometer.

the viewing scope. Thus, an observer who views the superposition of waves A and F through the telescope sees constructive or destructive interference, depending only on the difference in path lengths D_A and D_F traveled by the two waves.

Now suppose the mirrors are perpendicular to each other, the beam splitter makes a 45° angle with each, and the distances D_A and D_F are equal. Waves A and F travel the same distance, and the field of view in the telescope is uniformly bright due to constructive interference. However, if the adjustable mirror is moved away from the telescope by a distance of $\frac{1}{4}\lambda$, wave A travels back and forth by an amount that is twice this value, leading to an extra distance of $\frac{1}{2}\lambda$. Then, the waves are out of phase when they reach the viewing scope, destructive inteference occurs, and the viewer sees a dark field. If the adjustable mirror is moved farther, full brightness returns as soon as the waves are in phase and interfere constructively. The in-phase condition occurs when wave A travels a total extra distance of λ relative to wave F. Thus, as the mirror is continuously moved, the viewer sees the field of view change from bright to dark, then back to bright, and so on. The amount by which D_A has been changed can be measured and related to the wavelength of the light, since a bright field changes into a dark field and back again each time D_A is changed by a half wavelength. (The back-and-forth change in distance is λ.) If a sufficiently large number of wavelengths are counted in this manner, the Michelson interferometer can be used to obtain a very accurate value for the wavelength from the measured changes in D_A.

27.5 DIFFRACTION

⬤ *Concepts at a Glance*   In the previous sections of this chapter we have used the principle of linear superposition to investigate some interference phenomena associated with light waves. As the Concepts-at-a-Glance chart in Figure 27.16 shows, we will now use the principle of linear superposition to explore three more topics, diffraction, resolving power, and the diffraction grating, all of which involve interference phenomena. This chart is an extension of the earlier version in Figure 27.1. ⬤

Figure 27.16 The principle of linear superposition can be used to describe the topics highlighted in gold. The grooved surface of a CD behaves like a diffraction grating and separates white light into a spectrum of colors.

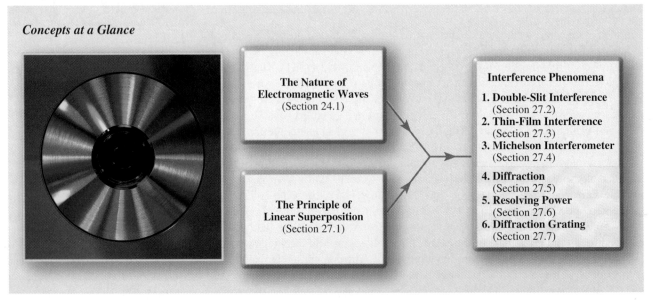

Concepts at a Glance

The Nature of Electromagnetic Waves (Section 24.1)	Interference Phenomena
	1. Double-Slit Interference (Section 27.2)
	2. Thin-Film Interference (Section 27.3)
	3. Michelson Interferometer (Section 27.4)
The Principle of Linear Superposition (Section 27.1)	**4. Diffraction** (Section 27.5)
	5. Resolving Power (Section 27.6)
	6. Diffraction Grating (Section 27.7)

As we have seen in Section 17.3, **diffraction** is the bending of waves around obstacles or the edges of an opening. In Figure 27.17, which is similar to Figure 17.11, sound waves are leaving a room through an open doorway. Because the exiting sound waves bend, or diffract, around the edges of the opening, a listener outside the room can hear the sound even when standing around the corner from the doorway.

Diffraction is an interference effect, and the Dutch scientist Christian Huygens (1629–1695) developed a principle that is useful in explaining why diffraction arises. **Huygens' principle** describes how a wave front that exists at one instant gives rise to the wave front that exists later on. This principle states that:

> *Every point on a wave front acts as a source of tiny wavelets that move forward with the same speed as the wave; the wave front at a later instant is the surface that is tangent to the wavelets.*

We begin by using Huygens' principle to explain the diffraction of sound waves in Figure 27.17. The drawing shows the top view of a plane wave front of sound approaching a doorway and identifies five points on the wave front just as it is leaving the opening. According to Huygens' principle, each of these points acts as a source of wavelets, which are shown as red circular arcs at some moment after they are emitted. The tangent to the wavelets from points 2, 3, and 4 indicates that in front of the doorway the wave front is flat and moving straight ahead. But at the edges, points 1 and 5 are the last points that produce wavelets. Huygens' principle suggests that in conforming to the curved shape of the wavelets near the edges, the new wave front moves into regions that it would not reach otherwise. The sound wave, then, bends or diffracts around the edges of the doorway.

Huygens' principle applies not only to sound waves, but to all kinds of waves. For instance, light has a wavelike nature and, consequently, exhibits diffraction. Therefore, you may ask, "Since I can hear around the edges of a doorway, why can't I also see around them?" As a matter of fact, light waves do bend around the edges of a doorway. However, the degree of bending is extremely small, so the diffraction of light is not enough to allow you to see around the corner.

As we will learn, the extent to which a wave bends around the edges of an opening is determined by the ratio λ/W, where λ is the wavelength of the wave and W is the width of the opening. The photographs in Figure 27.18 illustrate the effect of

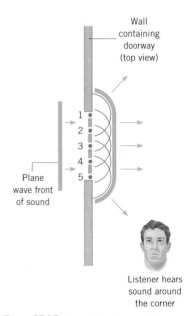

Figure 27.17 Sound bends or diffracts around the edges of a doorway, so even a person who is not standing directly in front of the opening can hear the sound. The five red points within the doorway act as sources and emit the five Huygens wavelets shown in red.

(a) Smaller value for λ/W, less diffraction.

(b) Larger value for λ/W, more diffraction.

Figure 27.18 These photographs show water waves (horizontal lines) approaching an opening whose width W is greater in (a) than in (b). In addition, the wavelength λ of the waves is smaller in (a) than in (b). Therefore, the ratio λ/W increases from (a) to (b) and so does the extent of the diffraction, as the red arrows indicate.

(a) Without diffraction (b) With diffraction

Figure 27.19 (a) If light were to pass through a very narrow slit *without* being diffracted, only the region on the screen directly opposite the slit would be illuminated. (b) Diffraction causes the light to bend around the edges of the slit into regions it would not otherwise reach, forming a pattern of alternating bright and dark fringes on the screen. The slit width has been exaggerated for clarity.

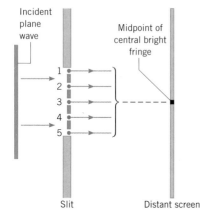

Figure 27.20 A plane wave front is incident on a single slit. This top view of the slit shows five sources of Huygens wavelets. The wavelets travel toward the midpoint of the central bright fringe on the screen, as the red rays indicate. The screen is very far from the slit.

this ratio on the diffraction of water waves. The degree to which the waves are diffracted or bent is indicated by the two red arrows in each photograph. In part a, the ratio λ/W is small because the wavelength (as indicated by the distance between the wave fronts) is small relative to the width of the opening. The wave fronts move through the opening with little bending or diffraction into the regions around the edges. In part b, the wavelength is larger and the width of the opening is smaller. As a result, the ratio λ/W is larger, and the degree of bending becomes more pronounced, with the wave fronts penetrating more into the regions around the edges of the opening.

Based on the pictures in Figure 27.18, we might expect that light waves of wavelength λ will bend or diffract appreciably when they pass through an opening whose width W is small enough to make the ratio λ/W sufficiently large. This is indeed the case, as Figure 27.19 illustrates. In this picture, it is assumed that parallel rays (or plane wave fronts) of light fall on a very narrow slit and illuminate a viewing screen that is located far from the slit. Part a of the drawing shows what would happen if light were *not* diffracted; it would pass through the slit without bending around the edges and produce an image of the slit on the screen. Part b shows what actually happens. The light diffracts around the edges of the slit and brightens regions on the screen that are not directly opposite the slit. The diffraction pattern on the screen consists of a bright central band, accompanied by a series of narrower faint fringes that are parallel to the slit itself.

To help explain how the pattern of diffraction fringes arises, Figure 27.20 shows a top view of a plane wave front approaching the slit and singles out five sources of Huygens wavelets. Consider how the light from these five sources reaches the midpoint on the screen. To simplify things, the screen is assumed to be so far from the slit that the rays from each Huygens source are nearly parallel.* Then, all the wavelets travel virtually the same distance to the midpoint, arriving there in phase. As a result, constructive interference creates a bright central fringe on the screen, directly opposite the slit.

---

* When the rays are parallel, the diffraction is called Fraunhofer diffraction in tribute to the German optician Josef Fraunhofer (1787–1826). When the rays are not parallel, the diffraction is referred to as Fresnel diffraction, named for the French physicist Augustin Jean Fresnel (1788–1827).

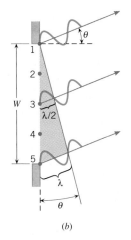

(a)

(b)

Figure 27.21 These drawings pertain to single-slit diffraction and show how destructive interference leads to the first dark fringe on either side of the central bright fringe. For clarity, only one of the dark fringes is shown. The screen is very far from the slit.

The wavelets emitted by the Huygens sources in the slit can also interfere destructively on the screen, as Figure 27.21 illustrates. Part *a* shows light rays directed from each source toward the first dark fringe. The angle θ gives the position of this dark fringe relative to the line between the midpoint of the slit and the midpoint of the central bright fringe. Since the screen is very far from the slit, the rays from each Huygens source are nearly parallel and oriented at nearly the same angle θ, as in part *b* of the drawing. The wavelet from source 1 travels the shortest distance to the screen, while that from source 5 travels the farthest. Destructive interference creates the first dark fringe when the extra distance traveled by the wavelet from source 5 is exactly one wavelength, as the colored right triangle in the drawing indicates. Under this condition, the extra distance traveled by the wavelet from source 3 at the center of the slit is exactly one-half of a wavelength. Therefore, wavelets from sources 1 and 3 in Figure 27.21*b* are exactly out of phase and interfere destructively when they reach the screen. Similarly, a wavelet that originates slightly below source 1 cancels a wavelet that originates the same distance below source 3. Thus, each wavelet from the upper half of the slit cancels a corresponding wavelet from the lower half, and no light reaches the screen. As can be seen from the colored right triangle, the angle θ locating the first dark fringe is given by $\sin \theta = \lambda/W$, where W is the width of the slit.

Figure 27.22 shows the condition that leads to destructive interference at the second dark fringe on either side of the midpoint on the screen. In reaching the screen, the light from source 5 now travels a distance of two wavelengths farther than the light from source 1. Under this condition, the wavelet from source 5 travels one wavelength farther than the wavelet from source 3, and the wavelet from source 3 travels one wavelength farther than the wavelet from source 1. Therefore, each half of the slit can be treated as the entire slit was in the previous paragraph; all the wavelets from the top half interfere destructively with each other, and all the wavelets from the bottom half do likewise. As a result, no light from either half reaches the screen, and another dark fringe occurs. The colored triangle in the drawing shows that this second dark fringe occurs when $\sin \theta = 2\lambda/W$. Similar arguments hold for the third- and higher-order dark fringes, with the general result being

Dark fringes for single-slit diffraction
$$\sin \theta = m\frac{\lambda}{W} \qquad m = 1, 2, 3, \ldots \qquad (27.4)$$

Between each pair of dark fringes there is a bright fringe due to constructive interference. The brightness of the fringes is related to the light intensity, just as loud-

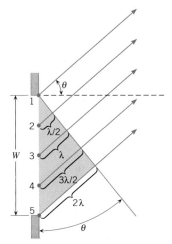

Figure 27.22 In a single-slit diffraction pattern, multiple dark fringes occur on either side of the central bright fringe. This drawing shows how destructive interference creates the second dark fringe on a very distant screen.

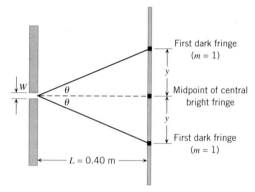

Figure 27.23 The photograph shows a single-slit diffraction pattern, with a bright and wide central fringe. The higher-order bright fringes are much less intense than the central fringe, as the graph indicates.

Figure 27.24 The distance $2y$ is the width of the central bright fringe.

ness is related to sound intensity. The intensity of the light at any location on the screen is the amount of light energy per second per unit area that strikes the screen there. Figure 27.23 gives a graph of the light intensity along with a photograph of a single-slit diffraction pattern. The central bright fringe, which is approximately twice as wide as the other bright fringes, has by far the greatest intensity. The width of the central fringe provides one indication of the extent of diffraction, as Example 6 illustrates.

EXAMPLE 6   Single-Slit Diffraction

Light passes through a slit and shines on a flat screen that is located $L = 0.40$ m away (see Figure 27.24). The width of the slit is $W = 4.0 \times 10^{-6}$ m. The distance between the middle of the central bright fringe and the first dark fringe is y. Determine the width $2y$ of the central bright fringe when the wavelength of the light in a vacuum is (a) $\lambda = 690$ nm (red) and (b) $\lambda = 410$ nm (violet).

Reasoning The width of the central bright fringe is determined by two factors. One is the angle θ that locates the first dark fringe on either side of the midpoint. The other is the distance L between the screen and the slit. Larger values for θ and L lead to a wider central bright fringe. Larger values of θ mean greater diffraction and occur when the ratio λ/W is larger. Thus, we expect the width of the central bright fringe to be greater when λ is greater.

Solution

(a) The angle θ in Equation 27.4 locates the first dark fringe when $m = 1$; $\sin \theta = (1)\lambda/W$. Therefore,

$$\theta = \sin^{-1}\left(\frac{\lambda}{W}\right) = \sin^{-1}\left(\frac{690 \times 10^{-9}\ \text{m}}{4.0 \times 10^{-6}\ \text{m}}\right) = 9.9°$$

According to Figure 27.24, $\tan \theta = y/L$, so

$$y = L \tan \theta = (0.40\ \text{m}) \tan 9.9° = 0.070\ \text{m}$$

The width of the central fringe, then, is $\boxed{2y = 0.14\ \text{m}}$.

(b) Repeating the calculation above with $\lambda = 410$ nm shows that $\theta = 5.9°$ and $\boxed{2y = 0.083 \text{ m}}$. As expected, the width $2y$ is greater in part (a), where the wavelength λ is greater.

In the production of computer chips it is important to minimize the effects of diffraction. As Figure 23.31 illustrates, such chips are very small and yet contain enormous numbers of electronic components. Such miniaturization is achieved using the techniques of photolithography. The patterns on the chip are created first on a "mask," which is similar to a photographic slide. Light is then directed through the mask onto silicon wafers that have been coated with a photosensitive material. The light-activated parts of the coating can be removed chemically, to leave the ultra-thin lines that correspond to the miniature patterns on the chip. As the light passes through the narrow slit-like patterns on the mask, the light spreads out due to diffraction. If excessive diffraction occurs, the light spreads out so much that sharp patterns are not formed on the photosensitive material coating the silicon wafer. Ultraminiaturization of the patterns requires the absolute minimum of diffraction, and currently this is achieved by using ultraviolet light, which has a wavelength shorter than that of visible light. The shorter the wavelength λ, the smaller the ratio λ/W, and the less the diffraction, as illustrated in Example 6. Intensive research programs are under way to develop the technique of X-ray lithography. The wavelengths of X-rays are much shorter than those of ultraviolet light and, thus, will reduce diffraction even more, allowing further miniaturization.

Another example of diffraction can be seen when light from a point source falls on an opaque disk, such as a coin (Figure 27.25). The effects of diffraction modify the dark shadow cast by the disk in several ways. First, the light waves diffracted around the circular edge of the disk interfere constructively at the center of the shadow to produce a small bright spot. There are also circular bright fringes in the shadow area. In addition, the boundary between the circular shadow and the lighted screen is not sharply defined but consists of concentric bright and dark fringes. The various fringes are analogous to those produced by a single slit and are due to interference between Huygens wavelets that originate from different points near the edge of the disk.

27.6 RESOLVING POWER

Figure 27.26 shows three photographs of an automobile's headlights taken at progressively greater distances from the camera. In parts *a* and *b*, the two separate headlights can be seen clearly. In part *c*, however, the car is so far away that the headlights are barely distinguishable and appear almost as a single light. The ***resolving power*** of an optical instrument, such as a camera, is its ability to distinguish between two closely spaced objects. If a camera with a higher resolving power had taken these pictures, the photograph in part *c* would have shown two distinct and separate headlights. Any instrument used for viewing objects that are close together must have a high resolving power. This is true, for example, for a telescope used to view distant stars or for a microscope used to view tiny organisms. We will now see that diffraction occurs when

THE PHYSICS OF...

producing computer chips using photolithography.

Opaque disk

Light

Figure 27.25 The diffraction pattern formed by an opaque disk consists of a small bright spot in the center of the dark shadow, circular bright fringes within the shadow, and concentric bright and dark fringes surrounding the shadow.

(a)   (b)   (c)

Figure 27.26 These automobile headlights were photographed at various distances from the camera, closest in part (*a*) and farthest in part (*c*). In part (*c*), the headlights are so far away that they are barely distinguishable.

light passes through the circular, or nearly circular, openings that admit light into cameras, telescopes, microscopes, and human eyes. The resulting diffraction pattern places a natural limit on the resolving power of these instruments.

Figure 27.27 shows the diffraction pattern created by a small circular opening when the viewing screen is far from the opening. The pattern consists of a central bright circular region, surrounded by alternating bright and dark circular fringes. These fringes are analogous to the rectangular fringes that a single slit produces. The angle θ in the picture locates the first circular dark fringe relative to the central bright region and is given by

$$\sin \theta = 1.22 \frac{\lambda}{D} \tag{27.5}$$

where λ is the wavelength of the light and D is the diameter of the opening. This expression is similar to Equation 27.4 for a slit ($\sin \theta = \lambda/W$, when $m = 1$) and is valid when the distance to the screen is much larger than the diameter D.

An optical instrument with the ability to resolve two closely spaced objects can produce images of them that can be identified separately. Think about the images on the film when light from two widely separated point objects passes through the circular aperture of a camera. As Figure 27.28 illustrates, each image is a circular

Figure 27.27 When light passes through a small circular opening, a circular diffraction pattern is formed on a screen. The angle θ locates the first dark fringe relative to the central bright region. The intensities of the bright fringes and the diameter of the opening have been exaggerated for clarity.

Figure 27.28 When light from two point objects passes through the circular aperture of a camera, two circular diffraction patterns are formed as images on the film. The images here are completely separated or resolved because the objects are widely separated.

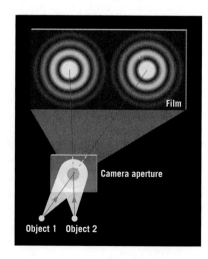

Film

Camera aperture

Object 1   Object 2

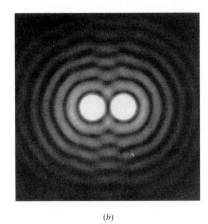

Figure 27.29 (*a*) According to the Rayleigh criterion, two point objects are just resolved when the first dark fringe (zero intensity) of one of the images falls on the central bright fringe (maximum intensity) of the other image. (*b*) This photograph shows two overlapping but still resolvable diffraction patterns.

diffraction pattern, but the two patterns do not overlap and are completely resolved. On the other hand, if the objects are sufficiently close together, the intensity patterns created by the diffraction overlap, as Figure 27.29*a* suggests. In fact, if the overlap is extensive, it may no longer be possible to distinguish the patterns separately. In such a case, the picture from a camera would show a single blurred object instead of two separate objects. In Figure 27.29*b* the diffraction patterns overlap, but not enough to prevent us from seeing that two objects are present. Ultimately, then, diffraction limits the ability of an optical instrument to produce distinguishable images of objects that are close together.

It is useful to have a criterion for judging whether two closely spaced objects are resolved by an optical instrument. Figure 27.29*a* presents the ***Rayleigh criterion*** for resolution, first proposed by Lord Rayleigh (1842–1919):

> ***Two point objects are just resolved when the first dark fringe in the diffraction pattern of one falls directly on the central bright fringe in the diffraction pattern of the other.***

The minimum angle θ_{min} between the two objects in the drawing is that given by Equation 27.5. If θ_{min} is small and is expressed in radians, $\sin \theta_{min} \approx \theta_{min}$. Then, Equation 27.5 can be rewritten as

$$\theta_{min} \approx 1.22 \frac{\lambda}{D} \qquad (\theta_{min} \text{ in radians}) \qquad (27.6)$$

For a given wavelength λ and aperture diameter D, this result specifies the smallest angle that two point objects can subtend at the aperture and still be resolved. According to Equation 27.6, optical instruments designed to resolve closely spaced objects (small values of θ_{min}) must utilize the smallest possible wavelength and the largest possible aperture diameter. For example, when short-wavelength ultraviolet light is collected by its large 2.4-m-diameter mirror, the Hubble Space Telescope is capable of resolving two closely spaced stars that have an angular separation of about $\theta_{min} = 1 \times 10^{-7}$ rad. This angle is equivalent to resolving two objects only 1 cm apart when they are 1×10^5 m (about 62 miles) from the telescope. Example 7 deals with the resolving power of the human eye.

THE PHYSICS OF...

comparing human eyes and eagle eyes.

• **PROBLEM SOLVING INSIGHT**
The minimum angle θ_{min} between two objects that are just resolved must be expressed in radians, not degrees, when using Equation 27.6 ($\theta_{min} \approx 1.22\lambda/D$).

θ_{min} $H = 120$ m

s Object 1   Object 2

Figure 27.30 The Rayleigh criterion can be used to estimate the smallest distance s that can separate two objects on the ground, if a person on a hang glider is to have any hope of distinguishing one object from the other.

EXAMPLE 7   The Human Eye Versus the Eagle's Eye

(a) A hang glider is flying at an altitude of $H = 120$ m. Green light (wavelength = 555 nm in vacuum) enters the pilot's eye through a pupil that has a diameter $D = 2.5$ mm. Determine how far apart two point objects must be on the ground if the pilot is to have any hope of distinguishing between them (see Figure 27.30). (b) An eagle's eye has a pupil with a diameter of $D = 6.2$ mm. Repeat part (a) for an eagle flying at the same altitude as the glider.

Reasoning According to the Rayleigh criterion, the two objects must be separated by a distance s sufficient to subtend an angle $\theta_{min} \approx 1.22\lambda/D$ at the pupil of the pilot's eye.* Remembering to use radians for the angle, we will relate θ_{min} to the altitude H and the desired distance s.

Solution

(a) Using the Rayleigh criterion, we find

$$\theta_{min} \approx 1.22\frac{\lambda}{D} = 1.22\left(\frac{555 \times 10^{-9}\text{ m}}{2.5 \times 10^{-3}\text{ m}}\right) = 2.7 \times 10^{-4}\text{ rad} \qquad (27.6)$$

According to Equation 8.1, θ_{min} in radians is $\theta_{min} \approx s/H$, so

$$s \approx \theta_{min}\,H = (2.7 \times 10^{-4}\text{ rad})(120\text{ m}) = \boxed{0.032\text{ m}}$$

(b) Since the pupil of an eagle's eye is larger than that of a human eye, diffraction creates less of a limitation for the eagle. A calculation like that above, using $D = 6.2$ mm, reveals that the diffraction limit for the eagle is $\boxed{s = 0.013\text{ m}}$.

Many optical instruments have a resolving power that exceeds that of the human eye. The typical camera does, for instance. Conceptual Example 8 compares the abilities of the human eye and a camera to resolve two closely spaced objects.

CONCEPTUAL EXAMPLE 8   What You See Is Not What You Get

The French postimpressionist artist Georges Seurat developed a technique of painting in which dots of color are placed close together on the canvas. From sufficiently far away the individual dots are not distinguishable, and the images in the picture take on a more normal appearance. Figure 27.31 shows a person in an art museum, looking at one of Seurat's paintings. Suppose that this person stands close to the painting, then backs up until the dots just become indistinguishable to his eyes and takes a picture from this position. When he gets home and has the film developed, however, he can see the individual dots in the enlarged photograph. Why does the camera resolve the individual dots, while his eyes do not? Assume that light enters his eyes through pupils that have diameters of 2.5 mm and enters the camera through an aperture or opening with a diameter of 25 mm.

Reasoning and Solution To answer this question, we turn to the Rayleigh criterion for resolving two point objects, namely, $\theta_{min} \approx 1.22\lambda/D$. According to the Rayleigh criterion, a larger value of the diameter D of the opening means that the value for θ_{min} is smaller, so that an optical instrument can be farther away and still resolve the objects. In other words, a larger value of D means a greater resolving power. For the eye and

---

* In applying the Rayleigh criterion, we use the given vacuum wavelength because it is nearly identical to the wavelength in air. We use the wavelength in air or vacuum, even though the diffraction occurs within the eye, where the index of refraction is about $n = 1.36$ and the wavelength is $\lambda_{eye} = \lambda_{vacuum}/n$, according to Equation 27.3. The reason is that, in entering the eye, the light is refracted according to Snell's law, which also includes an effect due to the index of refraction. If the angle of incidence is small, the effect of n in Snell's law cancels the effect of n in Equation 27.3, to a good degree of approximation.

Figure 27.31 This person is about to take a photograph of a famous painting by Georges Seurat, who developed the technique of using tiny dots of color to construct his images. Conceptual Example 8 discusses what the person sees when the film is developed.

the camera the diameters are $D_{eye} = 2.5$ mm and $D_{camera} = 25$ mm, so the diameter for the camera is ten times larger than that for the eye. Thus, at the distance at which the eye loses its ability to resolve the individual dots in the painting, the camera can easily resolve the dots. As discussed in the footnote to Example 7, we can ignore the effect on the wavelength of the index of refraction of the material from which the eye is made. We conclude, then, that *the camera succeeds in resolving the dots in the painting while the eye fails because the camera gathers light through a much larger opening.*

Related Homework Material: *Question 14, Problem 33*

27.7 THE DIFFRACTION GRATING

Diffraction patterns of bright and dark fringes occur when monochromatic light passes through a single or double slit. Fringe patterns also result when light passes through more than two slits, and an arrangement consisting of a large number of parallel, closely spaced slits is called a ***diffraction grating.*** Gratings with as many as 40 000 slits per centimeter can be made, depending on the production method. In one method a diamond-tipped cutting tool is used to inscribe closely spaced parallel lines on a glass plate, the spaces between the lines serving as the slits. In fact, the number of slits per centimeter is often quoted as the number of lines per centimeter.

Figure 27.32 illustrates how light travels to a distant viewing screen from each of five slits in a grating and forms the central bright fringe and the first-order bright

THE PHYSICS OF...

the diffraction grating.

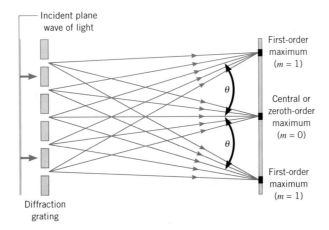

Figure 27.32 When light passes through a diffraction grating, a central bright fringe ($m = 0$) and higher-order bright fringes ($m = 1, 2, \ldots$) form when the light falls on a distant viewing screen.

First-order maximum

Second-order maximum

Figure 27.33 The conditions shown here lead to the first- and second-order intensity maxima in the diffraction pattern produced by the diffraction grating on the right.

fringes on either side. Higher-order bright fringes are also formed but are not shown in the drawing. Each bright fringe is located by an angle θ relative to the central fringe. These bright fringes are sometimes called the *principal fringes* or *principal maxima,* since they are places where the light intensity is a maximum. The term "principal" distinguishes them from other, much less bright fringes that are referred to as secondary fringes or secondary maxima.

Constructive interference creates the principal fringes. To show how, we assume the screen is far from the grating, so that the rays remain nearly parallel while the light travels toward the screen, as in Figure 27.33. In reaching the place on the screen where the first-order maximum is located, light from slit 2 travels a distance of one wavelength farther than light from slit 1. Similarly, light from slit 3 travels one wavelength farther than light from slit 2, and so forth, as emphasized by the four colored right triangles on the right-hand side of the drawing. For the first-order maximum, the enlarged view of one of these right triangles shows that constructive interference occurs when $\sin \theta = \lambda/d$, where d is the separation between the slits. The second-order maximum forms when the extra distance traveled by light from adjacent slits is two wavelengths, so that $\sin \theta = 2\lambda/d$. The general result is

Grating (5 slits)

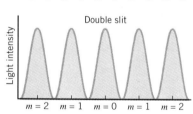

Double slit

Figure 27.34 The bright fringes produced by a diffraction grating are much narrower than those produced by a double slit. Note the three small secondary bright fringes between the principal bright fringes of the grating. For a large number of slits, these secondary fringes become very small.

Principal maxima of a diffraction grating

$$\sin \theta = m\frac{\lambda}{d} \qquad m = 0, 1, 2, 3, \ldots \qquad (27.7)$$

The separation d between the slits can be calculated from the number of slits per centimeter of grating; for instance, a grating with 2500 slits per centimeter has a slit separation of $d = (1/2500)$ cm $= 4.0 \times 10^{-4}$ cm. Equation 27.7 is identical to Equation 27.1 for the double slit. A grating, however, produces bright fringes that are much *narrower* or *sharper* than those from a double slit, as the intensity patterns in Figure 27.34 reveal. Note also from the drawing that between the principal maxima of a diffraction grating there are secondary maxima with much smaller intensities.

The next example illustrates the ability of a grating to separate the components in a mixture of colors.

EXAMPLE 9   Separating Colors with a Diffraction Grating

A mixture of violet light ($\lambda = 410$ nm in vacuum) and red light ($\lambda = 660$ nm in vacuum) falls on a grating that contains 1.0×10^4 lines/cm. For each wavelength, find the angle θ that locates the first-order maxima.

Reasoning Before Equation 27.7 can be used here, a value for the separation d between the slits is needed: $d = 1/(1.0 \times 10^4 \text{ lines/cm}) = 1.0 \times 10^{-4}$ cm, or 1.0×10^{-6} m. For violet light, the angle θ_{violet} for the first-order maxima ($m = 1$) is given by $\sin \theta_{\text{violet}} = m\lambda_{\text{violet}}/d$, with an analogous equation applying for the red light.

Solution For violet light, the angle locating the first-order maxima is

$$\theta_{\text{violet}} = \sin^{-1} \frac{\lambda_{\text{violet}}}{d} = \sin^{-1} \left(\frac{410 \times 10^{-9} \text{ m}}{1.0 \times 10^{-6} \text{ m}} \right) = \boxed{24°}$$

For red light, a similar calculation with $\lambda_{\text{red}} = 660 \times 10^{-9}$ m shows that $\boxed{\theta_{\text{red}} = 41°}$. Because θ_{violet} and θ_{red} are different, separate first-order bright fringes are seen for violet and red light on a viewing screen.

If the light in Example 9 had been sunlight, the angles for the first-order maxima would cover all values in the range between 24° and 41°, since sunlight contains all colors or wavelengths between violet and red. Consequently, a rainbow-like dispersion of the colors would be observed to either side of the central fringe on a screen, as can be seen in Figure 27.35. This drawing shows that the spectrum of colors associated with the $m = 2$ order is completely separate from that of the $m = 1$ order. For higher orders, however, the spectra from adjacent orders may overlap (see problems 43 and 55). Note also that the central maximum ($m = 0$) is white because all the colors overlap there.

An instrument designed to measure the angles at which the principal maxima of a grating occur is called a grating spectroscope. With a measured value of the angle, calculations such as those in Example 9 can be turned around to provide the corresponding value of the wavelength. As we will point out in Chapter 30, the atoms in a hot gas emit discrete wavelengths, and determining the values of these wavelengths is one important technique used to identify the atoms. Figure 27.36 contains a sketch of a grating spectroscope. The slit that admits light from the source (e.g., a hot gas) is located at the focal point of the collimating lens, so the light rays striking the grating are parallel. The telescope is used to detect the bright fringes and, hence, to measure the angle θ.

27.8 COMPACT DISCS, DIGITAL VIDEO DISCS, AND THE USE OF INTERFERENCE

The compact disc (CD) and the digital video disc (DVD) have revolutionized how text, graphics, music, and movies are stored for use in computers, stereo sound systems, and televisions. The operation of these discs uses interference effects in some interesting ways. Each contains a spiral track that holds the information and is detected using a laser beam that reflects from the bottom of the disc, as Figure 27.37 illustrates. The information is encoded in the form of raised areas on the bottom of the disc. These raised areas appear as "pits" when viewed from the *top* or labeled side of the disc. They are separated by flat areas called "land." The pits and land are covered with a transparent plastic coating, which has been omitted from the drawing for simplicity.

As the disc rotates, the laser beam reflects off the bottom and into a detector. The reflected light intensity fluctuates as the pits and land areas pass by, and the fluctuations convey the information as a series of binary numbers (zeros and ones). To

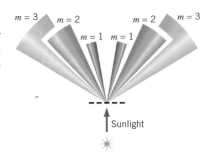

Figure 27.35 When sunlight falls on a diffraction grating, a rainbow of colors is produced at each principal maximum ($m = 1, 2, \ldots$). The central maximum ($m = 0$), however, is white but is not shown in the drawing.

THE PHYSICS OF...

a grating spectroscope.

Figure 27.36 A grating spectroscope.

THE PHYSICS OF...

retrieving information from compact discs and digital video discs.

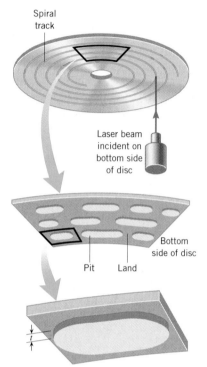

Figure 27.37 The bottom surface of a compact disc (CD) or digital video disc (DVD) carries information in the form of raised areas ("pits") and land along a spiral track. A CD or DVD is played by using a laser beam that strikes the bottom surface and reflects from it.

THE PHYSICS OF...

the three-beam tracking method for compact discs.

make the fluctuations easier to detect, the pit thickness t (see Figure 27.37) is chosen with destructive interference in mind. As the laser beam overlaps the edges of a pit, part of the beam is reflected from the raised pit surface and part from the land. The part that reflects from the land travels an additional distance of $2t$. The thickness of the pits is chosen so that $2t$ is one-half of a wavelength of the laser beam in the plastic coating. With this choice, destructive interference occurs when the two parts of the reflected beam combine. As a result, there is markedly less reflected intensity when the laser beam passes over a pit edge than when it passes over other places on the surface. Thus, the fluctuations in reflected light that occur while the disc rotates are large enough to detect because of the effects of destructive interference. Example 10 determines the theoretical thickness of the pits on a compact disc. In reality, a value slightly less than that obtained in the example is used for technical reasons that are not pertinent here.

EXAMPLE 10   Pit Thickness on a Compact Disc

The laser in a CD player has a wavelength of 780 nm in a vacuum. The plastic coating over the pits has an index of refraction of $n = 1.5$. Find the thickness of the pits on a CD.

Reasoning As we have discussed, the thickness t is chosen so that $2t = \frac{1}{2}\lambda_{coating}$ in order to achieve destructive interference. Equation 27.3 gives the wavelength in the plastic coating as $\lambda_{coating} = \lambda_{vacuum}/n$.

Solution The thickness of the pits is

$$t = \frac{\lambda_{coating}}{4} = \frac{\lambda_{vacuum}}{4n} = \frac{780 \times 10^{-9} \text{ m}}{4(1.5)} = \boxed{1.3 \times 10^{-7} \text{ m}}$$

The pit thickness calculated in Example 10 applies only to a CD. The pit thickness (and other dimensions as well) are smaller on a DVD because the lasers used for DVDs have smaller wavelengths (635 nm, for example). The fact that the pit dimensions are smaller is one of the reasons that a DVD has more information storage capacity than a CD, from seven to twenty-six times more, depending on the type of DVD.

As a CD or DVD rotates, the laser beam must accurately follow the pits and land along the spiral track. One type of CD player utilizes a three-beam tracking method to ensure that the laser beam follows the spiral properly. Figure 27.38 shows that a diffraction grating is the key element in this method. Before the laser beam strikes the CD, the beam passes through a grating that produces a central maximum and two first-order maxima, one on either side. As the picture indicates, the central maximum beam falls directly on the spiral track. This beam reflects into a detector, and the reflected light intensity fluctuates as the pits and land areas pass by, the fluctuations conveying the information. The two first-order maxima beams are called tracking beams. They hit the CD between the arms of the spiral and also reflect into detectors of their own. Under perfect conditions, the intensities of the two reflected tracking beams do not fluctuate, since they originate from the smooth surface between the arms of the spiral where there are no pits. As a result, each tracking-beam detector puts out the same constant electrical signal. However, if the tracking drifts to either side, the reflected intensity of each tracking beam changes because of the pits. In response, the tracking-beam detectors produce different electrical signals. The difference between the signals is used in a "feedback" circuit to correct for the drift and put the three beams back into their proper positions.

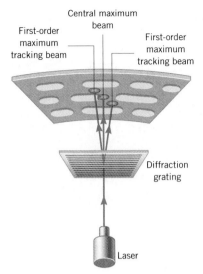

Figure 27.38 A three-beam tracking method is sometimes used in CD players to ensure that the laser follows the spiral track correctly. The three beams are derived from a single laser beam by using a diffraction grating.

Figure 27.39 In this drawing of the crystalline structure of sodium chloride, the small red spheres represent positive sodium ions, while the large blue spheres represent negative chloride ions.

27.9 X-RAY DIFFRACTION

Not all diffraction gratings are commercially made. Nature also creates diffraction gratings, although these gratings do not look like an array of closely spaced slits. Instead, nature's gratings are the arrays of regularly spaced atoms that exist in crystalline solids. For example, Figure 27.39 shows a crystal of ordinary salt (NaCl). Typically, the atoms in a crystalline solid are separated by distances of about 1.0×10^{-10} m, so we might expect a crystalline array of atoms to act like a grating with roughly this "slit" spacing for electromagnetic waves of the appropriate wavelength. Assuming that $\sin \theta = 0.5$ and that $m = 1$ in Equation 27.7, it follows that $0.5 = \lambda/d$. A value of $d = 1.0 \times 10^{-10}$ m in this equation gives a wavelength of $\lambda = 0.5 \times 10^{-10}$ m. This wavelength is much shorter than that of visible light and falls in the X-ray region of the electromagnetic spectrum. (See Figure 24.9.)

A diffraction pattern does indeed result when X-rays are directed onto a crystalline material, as Figure 27.40a illustrates for a crystal of NaCl. The pattern consists of a complicated arrangement of spots because a crystal has a complex three-dimensional structure. It is from patterns such as these that the spacing between atoms and the nature of the crystal structure can be determined. X-ray diffraction has also been applied with great success toward understanding the structure of biologically important molecules, such as proteins and nucleic acids. One of the most famous results was the discovery in 1953 by J. D. Watson and F. H. C. Crick that the structure of the nucleic acid DNA is a double helix. X-ray diffraction patterns such as that in Figure 27.40b played the pivotal role in their research.

THE PHYSICS OF...

X-ray diffraction.

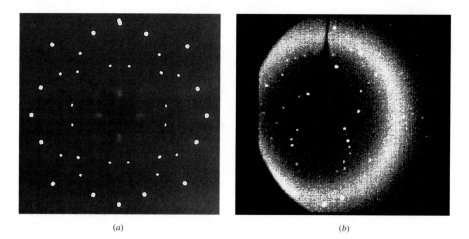

Figure 27.40 The X-ray diffraction patterns from (a) crystalline NaCl and (b) crystalline DNA.

(a)   (b)

27.10 CONCEPTS & CALCULATIONS

The ability to exhibit interference effects is a fundamental characteristic of any kind of wave. Our understanding of these effects depends on the principle of linear superposition, which we first encountered in Chapter 17. Only by means of this principle can we understand the constructive and destructive interference of light waves that lie at the heart of every topic in this chapter. Therefore, it is fitting that we review the essence of the principle in the next example.

CONCEPTS & CALCULATIONS EXAMPLE 11
The Principle of Linear Superposition

A square is 3.5 m on a side, and point A is the midpoint of one of its sides. On the side opposite this spot, two in-phase loudspeakers are located at adjacent corners, as Figure 27.41 indicates. Standing at point A, you hear a loud sound because constructive interference occurs between the identical sound waves coming from the speakers. As you walk along the side of the square toward either empty corner, the loudness diminishes gradually but does not entirely disappear until you reach either empty corner, where you hear no sound at all. Thus, at each empty corner destructive interference occurs. Find the wavelength of the sound waves.

Concept Questions and Answers Why does constructive interference occur at point A?

Answer The loudspeakers are in-phase sources of identical sound waves. The waves from one speaker travel a distance ℓ_1 in reaching point A and the waves from the second speaker travel a distance ℓ_2. The condition that leads to constructive interference is $\ell_2 - \ell_1 = m\lambda$, where λ is the wavelength of the waves and $m = 0, 1, 2, 3. \ldots$. Thus, the two distances are the same or differ by an integer number of wavelengths. Point A is the midpoint of a side of the square, so that ℓ_1 and ℓ_2 are the same, and constructive interference occurs.

What is the general condition that leads to destructive interference?

Answer The waves start out in phase, so the general condition that leads to destructive interference is $\ell_2 - \ell_1 = (m + \frac{1}{2})\lambda$, where $m = 0, 1, 2, 3. \ldots$.

The general condition that leads to destructive interference entails a number of possibilities. Which one of them, if any, applies at either empty corner of the square?

Answer   The general condition that leads to destructive interference is $\ell_2 - \ell_1 = (m + \frac{1}{2})\lambda$, where $m = 0, 1, 2, 3. \ldots$. There are many possible values for m, and we are being asked what the specific value is. As you walk from point A toward either empty corner, one of the distances decreases and the other increases; the loudness diminishes gradually and disappears at either empty corner. Therefore, destructive interference occurs for the first time at the corner, where the difference between the distances has attained the smallest possible value of $\ell_2 - \ell_1 = \frac{1}{2}\lambda$. This means that we are dealing with the specific case in which $m = 0$.

Solution   Consider the destructive interference that occurs at either empty corner. Using L to denote the length of a side of the square and taking advantage of the Pythagorean theorem, we have

$$\ell_1 = L \quad \text{and} \quad \ell_2 = \sqrt{L^2 + L^2} = \sqrt{2}\,L$$

The specific condition for the destructive interference at the empty corner is

$$\ell_2 - \ell_1 = \sqrt{2}\,L - L = \frac{1}{2}\lambda$$

Solving for the wavelength of the waves gives

$$\lambda = 2L\,(\sqrt{2} - 1) = 2(3.5 \text{ m})\,(\sqrt{2} - 1) = \boxed{2.9 \text{ m}}$$

Figure 27.41   Two loudspeakers are located at the corners of a square and produce identical sound waves. As Example 11 discusses, constructive interference occurs at point A, which is at the midpoint of the side of the square, and destructive interference occurs at either empty corner.

The next example deals with thin-film interference and serves as a review of the factors that must be considered in such cases.

CONCEPTS & CALCULATIONS EXAMPLE 12   A Soap Film

A soap film ($n = 1.33$) is 375 nm thick and is surrounded on both sides by air. Sunlight, whose wavelengths (in vacuum) extend from 380 to 750 nm, strikes the film nearly perpendicularly. For which wavelength(s) in this range does constructive interference cause the film to look bright in reflected light?

Concept Questions and Answers   What, if any, phase change occurs when light, traveling in air, reflects from the air–film interface?

Answer   In air the index of refraction is nearly $n = 1$, while in the film it is $n = 1.33$. A phase change occurs whenever light travels through a material with a smaller refractive index toward a material with a larger refractive index and reflects from the boundary between the two. The phase change is equivalent to $\frac{1}{2}\lambda_{\text{film}}$, where λ_{film} is the wavelength in the film.

What, if any, phase change occurs when light, traveling in the film, reflects from the film–air interface?

Answer   In the film the index of refraction is $n = 1.33$, while in air it is nearly $n = 1$. A phase change does *not* occur when light travels through a material with a larger refractive index toward a material with a smaller refractive index and reflects from the boundary between the two.

Is the wavelength of the light in the film greater than, smaller than, or equal to the wavelength in a vacuum?

Answer   The wavelength of the light in the film (refractive index $= n$) is given by Equation 27.3 as $\lambda_{\text{film}} = \lambda_{\text{vacuum}}/n$. Since the refractive index of the film is $n = 1.33$, the wavelength in the film is less than the wavelength in a vacuum.

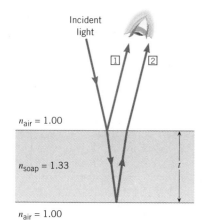

Figure 27.42 The constructive interference that occurs with this thin soap film is analyzed in Example 12.

Solution Figure 27.42 shows the soap film and the two rays of light that represent the interfering light waves. At nearly perpendicular incidence, ray 2 travels a distance of $2t$ farther than ray 1, where t is the thickness of the film. In addition, as discussed in the Concept Questions, ray 2 experiences no phase change upon reflection at the bottom film surface, while ray 1 does experience a phase shift of $\frac{1}{2}\lambda_{film}$ at the upper film surface. Therefore, the net phase change for the two reflected rays is $\frac{1}{2}\lambda_{film}$. We must combine this amount with the extra travel distance to determine the condition for constructive interference. For constructive interference, the combined total must be an integer number of wavelengths in the film:

$$\underbrace{2t}_{\substack{\text{Extra distance} \\ \text{traveled by} \\ \text{ray 2}}} + \underbrace{\tfrac{1}{2}\lambda_{film}}_{\substack{\text{Half-wavelength} \\ \text{net phase change} \\ \text{due to reflection}}} = \underbrace{\lambda_{film}, 2\lambda_{film}, 3\lambda_{film}, \ldots}_{\substack{\text{Condition for} \\ \text{constructive interference}}}$$

Subtracting the term $\frac{1}{2}\lambda_{film}$ from the left side of this equation and from each term on the right side, we obtain

$$2t = \underbrace{\tfrac{1}{2}\lambda_{film}, \tfrac{3}{2}\lambda_{film}, \tfrac{5}{2}\lambda_{film}, \ldots}_{\substack{(m + \frac{1}{2})\,\lambda_{film} \\ m = 0, 1, 2, 3, \ldots}}$$

Solving for the wavelength gives

$$\lambda_{film} = \frac{2t}{m + \frac{1}{2}} \qquad m = 0, 1, 2, 3, \ldots$$

According to Equation 27.3, we have $\lambda_{film} = \lambda_{vacuum}/n$. Using this substitution in our result for λ_{film}, we obtain

$$\lambda_{vacuum} = \frac{2nt}{m + \frac{1}{2}} \qquad m = 0, 1, 2, 3, \ldots$$

For the first four values of m and the given values for n and t, we find

m = 0 $\qquad \lambda_{vacuum} = \dfrac{2nt}{m + \frac{1}{2}} = \dfrac{2(1.33)(375\text{ nm})}{0 + \frac{1}{2}} = 2000\text{ nm}$

m = 1 $\qquad \lambda_{vacuum} = 665\text{ nm}$

m = 2 $\qquad \lambda_{vacuum} = 399\text{ nm}$

m = 3 $\qquad \lambda_{vacuum} = 285\text{ nm}$

The range of visible wavelengths (in vacuum) extends from 380 to 750 nm. Therefore, the visible wavelengths in which the film appears bright due to constructive interference are $\boxed{399 \text{ and } 665 \text{ nm}}$.

At the end of the problem set for this chapter, you will find homework problems that contain both conceptual and quantitative parts. These problems are grouped under the heading *Concepts & Calculations, Group Learning Problems*. They are designed for use by students working alone or in small learning groups. The conceptual part of each problem provides a convenient focus for group discussions.

SUMMARY

The **principle of linear superposition** states that when two or more waves are present simultaneously in the same region of space, the resultant disturbance is the sum of the distur-

bances from the individual waves. According to this principle, two or more light waves can interfere constructively or destructively when they exist at the same place at the same time, provided they originate from **coherent sources.** Two sources are coherent if they emit waves that have a constant phase relationship.

In **Young's double-slit experiment,** light passes through a pair of closely spaced narrow slits and produces a pattern of alternating bright and dark fringes on a viewing screen. The fringes arise because of constructive and destructive interference. The angle θ for the mth higher-order bright fringe is given by $\sin \theta = m\lambda/d$, where d is the spacing between the narrow slits, λ is the wavelength of the light, and $m = 0, 1, 2, 3, \ldots$. Similarly, the angle for the dark fringes is given by $\sin \theta = (m + \frac{1}{2}) \lambda/d$.

Constructive and destructive interference of light waves can occur with **thin films** of transparent materials. The interference occurs between light waves that reflect from the top and bottom surfaces of the film. One important factor in thin-film interference is the thickness of the film relative to the wavelength of the light within the film. The wavelength within the film is $\lambda_{film} = \lambda_{vacuum}/n$, where n is the refractive index of the film. A second important factor is the phase change that can occur when light undergoes reflection at each surface of the film; the exact nature of the phase change is discussed in Section 27.3.

Diffraction is a bending of waves around obstacles or the edges of an opening. Diffraction is an interference effect that can be explained with the aid of **Huygens' principle.** This principle states that every point on a wave front acts as a source of tiny wavelets that move forward with the same speed as the wave; the wave front at a later instant is the surface that is tangent to the wavelets. When light passes through a single narrow slit and falls on a viewing screen, a pattern of bright and dark fringes is formed because of the superposition of such wavelets. The angle θ for the mth dark fringe on either side of the central bright fringe is given by $\sin \theta = m\lambda/W$, where W is the slit width and $m = 1, 2, 3, \ldots$.

The **resolving power** of an optical instrument is the ability of the instrument to distinguish between two closely spaced objects. Resolving power is limited by the diffraction that occurs when light waves enter an instrument, often through a circular opening. Consideration of the diffraction fringes for a circular opening leads to the **Rayleigh criterion** for resolution. This criterion specifies that two point objects are just resolved when the first dark fringe in the diffraction pattern of one falls directly on the central bright fringe in the diffraction pattern of the other. According to this specification, the minimum angle (in radians) that two point objects can subtend at an aperture of diameter D and still be resolved as separate objects is $\theta_{min} \approx 1.22\lambda/D$, where λ is the wavelength of the light.

A **diffraction grating** consists of a large number of parallel, closely spaced slits. When light passes through a diffraction grating and falls on a viewing screen, the light forms a pattern of bright and dark fringes. The bright fringes are referred to as principal maxima and are found at an angle θ, such that $\sin \theta = m\lambda/d$, where d is the separation between two successive slits and $m = 0, 1, 2, 3, \ldots$.

CONCEPTUAL QUESTIONS

1. Suppose that a radio station broadcasts simultaneously from two transmitting antennas at two different locations. Is it clear that your radio will have better reception with two transmitting antennas rather than one? Justify your answer.

2. Two sources of waves are in phase and produce identical waves. These sources are mounted at the corners of a square. At the center of the square waves from the sources produce constructive interference, no matter which two corners of the square are occupied by the sources. Explain why.

3. (a) How would the pattern of bright and dark fringes produced in a Young's double-slit experiment change if the light waves coming from *both* slits had their phases shifted by an amount equivalent to a half wavelength? (b) How would the pattern change if the light coming from *only one* of the slits had its phase shifted by an amount equivalent to a half wavelength?

4. Replace slits S_1 and S_2 in Figure 27.4 with identical in-phase loudspeakers and use the same ac electrical signal to drive them. The two sound waves produced will then be identical, and you will have the audio equivalent of Young's double-slit experiment. In terms of loudness and softness, describe what you would hear as you walked along the screen, starting from the center and going to either end.

5. In Young's double-slit experiment, is it possible to see interference fringes when the wavelength of the light is greater than the distance between the slits? Provide a reason for your answer.

6. A camera lens is covered with a nonreflective coating that eliminates the reflection of perpendicularly incident green light. Recalling Snell's law of refraction (see Section 26.2), would you expect the reflected green light to be eliminated if it were incident on the nonreflective coating at an angle of 45° rather than perpendicularly? Justify your answer.

7. When sunlight reflects from a thin film of soapy water, the film appears multicolored, in part because destructive interference removes different wavelengths from the light reflected at different places, depending on the thickness of the film. As the film becomes thinner and thinner, it looks darker and darker in reflected light, appearing black just before it breaks. The blackness means that destructive interference removes *all* wavelengths from the reflected light when the film is very thin. Explain why.

8. Two pieces of the same glass are covered with thin films of different materials. The thickness of each film is the same. In reflected sunlight, however, the films have different colors. Give a reason that accounts for the difference in colors.

9. In Figure 27.14*b* there is a dark spot at the center of the pattern of Newton's rings. By considering the phase changes that occur when light reflects from the upper curved surface and the lower flat surface, account for the dark spot.

10. A thin film of a material is floating on water ($n = 1.33$). When the material has a refractive index of $n = 1.20$, the film looks bright in reflected light as its thickness approaches zero. But when the material has a refractive index of $n = 1.45$, the film looks black in reflected light as its thickness approaches zero. Explain these observations in terms of constructive and destructive interference and the phase changes that occur when light waves undergo reflection.

11. A transparent coating is deposited on a glass plate and has a refractive index that is *larger than that of the glass,* not smaller, as it is for a typical nonreflective coating. For a certain wavelength within the coating, the thickness of the coating is a quarter wavelength. The coating *enhances* the reflection of the light corresponding to this wavelength. Explain why, referring to Example 3 in the text to guide your thinking.

12. On most cameras one can select the *f*-number setting, or *f*-stop. The *f*-number gives the ratio of the focal length of the camera lens to the diameter of the aperture through which light enters the camera. If one wishes to resolve two closely spaced objects in a picture, should a small or a large *f*-number setting be used? Account for your answer.

13. Explain why a sound wave diffracts much more than a light wave does when the two pass through the same doorway.

14. Review Conceptual Example 8 before answering this question. A person is viewing one of Seurat's paintings that consists of dots of color. She is so close to the painting that the dots are distinguishable. Without moving, however, she can squint, which makes the painting take on a more normal appearance. In terms of the Rayleigh criterion, why does squinting make the painting look more normal?

15. Four light bulbs are arranged at the corners of a rectangle that is three times longer than it is wide. You look at this arrangement perpendicular to the plane of the rectangle. From very far away, your eyes cannot resolve the individual bulbs and you see a single "smear" of light. From close in, you see the individual bulbs. Between these two extremes, what do you see? Draw two pictures to illustrate the possibilities that exist, depending on how far away you are. Explain your drawings.

16. ⚕ Suppose the pupil of your eye were elliptical instead of circular in shape, with the long axis of the ellipse oriented in the vertical direction. (a) Would the resolving power of your eye be the same in the horizontal and vertical directions? (b) In which direction would the resolving power be greatest? Justify your answers by discussing how the diffraction of light waves would differ in the two directions.

17. ⚕ Suppose you were designing an eye and could select the size of the pupil and the wavelength of the electromagnetic waves to which the eye is sensitive. As far as the limitation created by diffraction is concerned, rank the following design choices in order of decreasing resolving power (greatest first): (a) large pupil and ultraviolet wavelengths, (b) small pupil and infrared wavelengths, and (c) small pupil and ultraviolet wavelengths. Justify your answer.

18. In our discussion of single-slit diffraction, we considered the ratio of the wavelength λ to the width W of the slit. We ignored the height of the slit, in effect assuming that the height was much larger than the width. Suppose the height and width were the same size, so that diffraction in both dimensions occurred. How would the diffraction pattern in Figure 27.19*b* be altered? Give your reasoning.

19. What would happen to the distance between the bright fringes produced by a diffraction grating if the entire interference apparatus (light source, grating, and screen) were immersed in water? Why?

PROBLEMS

Section 27.1 The Principle of Linear Superposition, Section 27.2 Young's Double-Slit Experiment

1. ssm The transmitting antenna for a radio station is 7.00 km from your house. The frequency of the electromagnetic wave broadcast by this station is 536 kHz. The station builds a second transmitting antenna that broadcasts an identical electromagnetic wave in phase with the original one. The new antenna is 8.12 km from your house. Does constructive or destructive interference occur at the receiving antenna of your radio? Show your calculations.

2. A Young's double-slit experiment is performed using light that has a wavelength of 630 nm. The separation between the slits is 5.3×10^{-5} m. Find the angles that locate the (a) first-, (b) second-, and (c) third-order bright fringes on the screen.

3. In a Young's double-slit experiment, the angle that locates the second dark fringe on either side of the central bright fringe is 5.4°. Find the ratio of the slit separation d to the wavelength λ of the light.

4. Two in-phase sources of waves are separated by a distance of 4.00 m. These sources produce identical waves that have a wavelength of 5.00 m. On the line between them, there are two places at which the same type of interference occurs. (a) Is it constructive or destructive interference, and (b) where are the places located?

5. ssm In a Young's double-slit experiment, the angle that locates the second-order bright fringe is 2.0°. The slit separation is 3.8×10^{-5} m. What is the wavelength of the light?

*6. Review Conceptual Example 2 before attempting this problem. Two slits are 0.158 mm apart. A mixture of red light (wavelength = 665 nm) and yellow-green light (wavelength = 565 nm) falls on the slits. A flat observation screen is located 2.24 m away. What is the distance on the screen between the third-order red fringe and the third-order yellow-green fringe?

*7. In a Young's double-slit experiment the separation y between the first-order bright fringe and the central bright fringe on a flat screen is 0.0240 m, when the light has a wavelength of 475 nm. Assume that the angles that locate the fringes on the screen are small enough so that $\sin\theta \approx \tan\theta$. Find the separation y when the light has a wavelength of 611 nm.

**8. In Young's experiment a mixture of orange light (611 nm) and blue light (471 nm) shines on the double slit. The centers of the first-order bright blue fringes lie at the outer edges of a screen that is located 0.500 m away from the slits. However, the first-order bright orange fringes fall off the screen. By how much and in which direction (toward or away from the slits) should the screen be moved, so that the centers of the first-order bright orange fringes just appear on the screen? It may be assumed that θ is small, so that $\sin\theta \approx \tan\theta$.

**9. ssm www A sheet of plastic ($n = 1.60$) covers *one slit* of a double slit (see the drawing). When the double slit is illuminated by monochromatic light ($\lambda_{\text{vacuum}} = 586$ nm), the center of the screen appears dark rather than bright. What is the minimum thickness of the plastic?

Section 27.3 Thin-Film Interference

10. A layer of transparent plastic ($n = 1.61$) on glass ($n = 1.52$) looks dark when viewed in reflected light whose wavelength is 589 nm in vacuum. Find the two smallest possible nonzero values for the thickness of the layer.

Problem 9

11. ssm A nonreflective coating of magnesium fluoride ($n = 1.38$) covers the glass ($n = 1.52$) of a camera lens. Assuming that the coating prevents reflection of yellow-green light (wavelength in vacuum = 565 nm), determine the minimum nonzero thickness that the coating can have.

12. For background on this problem, review Conceptual Example 4. A mixture of yellow light (wavelength = 580 nm in vacuum) and violet light (wavelength = 410 nm in vacuum) falls perpendicularly on a film of gasoline that is floating on a puddle of water. For both wavelengths, the refractive index of gasoline is $n = 1.40$ and that of water is $n = 1.33$. What is the minimum nonzero thickness of the film in a spot that looks (a) yellow and (b) violet because of destructive interference?

13. Example 5(a) in the text deals with the air wedge formed between two plates of glass ($n = 1.52$). Repeat this example, assuming that the wedge of air is replaced by water ($n = 1.33$).

14. Review Conceptual Example 4 before beginning this problem. A soap film with different thicknesses at different places has

an unknown refractive index n and air on both sides. In reflected light it looks multicolored. One region looks yellow because destructive interference has removed blue ($\lambda_{\text{vacuum}} = 469$ nm) from the reflected light, while another looks magenta because destructive interference has removed green ($\lambda_{\text{vacuum}} = 555$ nm). In these regions the film has the minimum nonzero thickness t required for the destructive interference to occur. Find the ratio $t_{\text{magenta}}/t_{\text{yellow}}$.

*15. ssm www Orange light ($\lambda_{\text{vacuum}} = 611$ nm) shines on a soap film ($n = 1.33$) that has air on either side of it. The light strikes the film perpendicularly. What is the minimum thickness of the film for which constructive interference causes it to look bright in reflected light?

*16. The drawing shows a cross section of a plano-concave lens resting on a flat glass plate. (A plano-concave lens has one surface that is a plane and the other that is concave spherical.) The thickness t is 1.37×10^{-5} m. The lens is illuminated with monochromatic light ($\lambda_{\text{vacuum}} = 550$ nm), and a series of concentric bright and dark rings is formed, much like Newton's rings. How many bright rings are there?

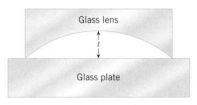

*17. A film of oil lies on wet pavement. The refractive index of the oil exceeds that of the water. The film has the minimum nonzero thickness such that it appears dark due to destructive interference when viewed in red light (wavelength = 660 nm in vacuum). Assuming that the visible spectrum extends from 380 to 750 nm, what are the visible wavelength(s) (in vacuum) for which the film will appear bright due to constructive interference?

**18. A uniform layer of water ($n = 1.33$) lies on a glass plate ($n = 1.52$). Light shines perpendicularly on the layer. Because of constructive interference, the layer looks maximally bright when the wavelength of the light is 432 nm in vacuum and *also* when it is 648 nm in vacuum. (a) Obtain the minimum thickness of the film. (b) Assuming that the film has the minimum thickness and that the visible spectrum extends from 380 to 750 nm, determine the visible wavelength(s) (in vacuum) for which the film appears completely dark.

Section 27.5 Diffraction

19. ssm A diffraction pattern forms when light passes through a single slit. The wavelength of the light is 675 nm. Determine the angle that locates the first dark fringe when the width of the slit is (a) 1.8×10^{-4} m and (b) 1.8×10^{-6} m.

20. A doorway is 0.91 m wide. (a) Obtain the angle that locates the first dark fringe in the Fraunhofer diffraction pattern formed when red light (wavelength = 660 nm) passes through the doorway. (b) Repeat part (a) for a 440-Hz sound wave (concert A), assuming that the speed of sound is 343 m/s.

21. A single slit has a width of 2.1×10^{-6} m and is used to form a diffraction pattern. Find the angle that locates the second dark

fringe when the wavelength of the light is (a) 430 nm and (b) 660 nm.

22. The first dark fringe in the diffraction pattern of a single slit is located at an angle of $\theta_A = 34°$. With the same light, the first dark fringe formed with another single slit is at an angle of $\theta_B = 56°$. Find the ratio W_A/W_B of the widths of the two slits.

23. ssm www Light shines through a single slit whose width is 5.6×10^{-4} m. A diffraction pattern is formed on a flat screen located 4.0 m away. The distance between the middle of the central bright fringe and the first dark fringe is 3.5 mm. What is the wavelength of the light?

*24. The width of a slit is 2.0×10^{-5} m. Light with a wavelength of 480 nm passes through this slit and falls on a screen that is located 0.50 m away. In the diffraction pattern, find the width of the bright fringe that is next to the central bright fringe.

*25. ssm The central bright fringe in a single-slit diffraction pattern has a width that equals the distance between the screen and the slit. Find the ratio λ/W of the wavelength of the light to the width of the slit.

*26. In a single-slit diffraction pattern on a flat screen, the central bright fringe is 1.2 cm wide when the slit width is 3.2×10^{-5} m. When the slit is replaced by a second slit, the wavelength of the light and the distance to the screen remaining unchanged, the central bright fringe broadens to a width of 1.9 cm. What is the width of the second slit? It may be assumed that θ is so small that $\sin \theta \approx \tan \theta$.

**27. In a single-slit diffraction pattern, the central fringe is 450 times wider than the slit. The screen is 18 000 times farther from the slit than the slit is wide. What is the ratio λ/W, where λ is the wavelength of the light shining through the slit and W is the width of the slit? Assume that the angle that locates a dark fringe on the screen is small, so that $\sin \theta \approx \tan \theta$.

Section 27.6 Resolving Power

28. You are looking down at the earth from inside a jetliner flying at an altitude of 8690 m. The pupil of your eye has a diameter of 2.00 mm. Determine how far apart two cars must be on the ground if you are to have any hope of distinguishing between them in (a) red light (wavelength = 665 nm in vacuum) and (b) violet light (wavelength = 405 nm in vacuum).

29. It is claimed that some professional baseball players can see which way the ball is spinning as it travels toward home plate. One way to judge this claim is to estimate the distance at which a batter can first hope to resolve two points on opposite sides of a baseball, which has a diameter of 0.0738 m. (a) Estimate this distance, assuming that the pupil of the eye has a diameter of 2.0 mm and the wavelength of the light is 550 nm in vacuum. (b) Considering that the distance between the pitcher's mound and home plate is 18.4 m, can you rule out the claim based on your answer to part (a)?

30. Two asteroids are traveling close to each other through the solar system at a distance of 2.0×10^{10} m from earth. With light of wavelength 550 nm, they are just resolved by the Hubble

Space Telescope, whose aperture has a diameter of 2.4 m. How far apart are the asteroids?

31. ssm The largest refracting telescope in the world is at the Yerkes Observatory in Williams Bay, Wisconsin. The objective of the telescope has a diameter of 1.02 m. Two objects are 3.75×10^4 m from the telescope. With light of wavelength 565 nm, how close can the objects be to each other so that they are just resolved by the telescope?

32. Astronomers have discovered a planetary system orbiting the star Upsilon Andromedae, which is at a distance of 4.2×10^{17} m from the earth. One planet is believed to be located at a distance of 1.2×10^{11} m from the star. Using visible light with a vacuum wavelength of 550 nm, what is the minimum necessary aperture diameter that a telescope must have so that it can resolve the planet and the star?

33. Review Conceptual Example 8 as background for this problem. In addition to the data given there, assume that the dots in the painting are separated by 1.5 mm and that the wavelength of the light is $\lambda_{vacuum} = 550$ nm. Find the distance at which the dots can just be resolved by (a) the eye and (b) the camera.

*34. The pupil of an eagle's eye has a diameter of 6.0 mm. Two field mice are separated by 0.010 m. From a distance of 176 m, the eagle sees them as one unresolved object and dives toward them at a speed of 17 m/s. Assume that the eagle's eye detects light that has a wavelength of 550 nm in a vacuum. How much time passes until the eagle sees the mice as separate objects?

*35. ssm www In an experiment, red light (wavelength = 694.3 nm) is sent to the moon. At the surface of the moon, which is 3.77×10^8 m away, the light strikes a reflector left there by astronauts. The reflected light returns to the earth, where it is detected. When it leaves the spotlight, the circular beam of light has a diameter of about 0.20 m, and diffraction causes the beam to spread as the light travels to the moon. In effect, the first circular dark fringe in the diffraction pattern defines the size of the central bright spot on the moon. Determine the diameter (not the radius) of the central bright spot on the moon.

**36. Two concentric circles of light emit light whose wavelength is 555 nm. The larger circle has a radius of 4.0 cm, while the smaller circle has a radius of 1.0 cm. When taking a picture of these lighted circles, a camera admits light through an aperture whose diameter is 12.5 mm. What is the maximum distance at which the camera can (a) distinguish one circle from the other and (b) reveal that the inner circle is a circle of light rather than a solid disk of light?

Section 27.7 The Diffraction Grating,
Section 27.8 Compact Discs, Digital Video Discs,
and the Use of Interference

37. ssm The diffraction gratings discussed in the text are transmission gratings because light *passes through* them. There are also gratings in which the light *reflects from* the grating to form a pattern of fringes. Equation 27.7 also applies to a reflection grating with straight parallel lines when the incident light shines perpendicularly on the grating. The surface of a compact disc (CD) has a multicolored appearance because it acts like a reflection grating and spreads sunlight into its colors. The arms of the spiral track on the CD are separated by 1.1×10^{-6} m. Using Equation 27.7, estimate the angle that corresponds to the first-order maximum for a wavelength of (a) 660 nm (red) and (b) 410 nm (violet).

38. For a wavelength of 420 nm, a diffraction grating produces a bright fringe at an angle of 26°. For an unknown wavelength, the same grating produces a bright fringe at an angle of 41°. In both cases the bright fringes are of the same order m. What is the unknown wavelength?

39. A diffraction grating produces a first-order bright fringe that is 0.0894 m away from the central bright fringe on a flat screen. The separation between the slits of the grating is 4.17×10^{-6} m, and the distance between the grating and the screen is 0.625 m. What is the wavelength of the light shining on the grating?

40. When a grating is used with light that has a wavelength of 621 nm, a third-order maximum is formed at an angle of 18.0°. How many lines per centimeter does this grating have?

41. ssm The wavelength of the laser beam used in a compact disc player is 780 nm. Suppose that a diffraction grating produces first-order tracking beams that are 1.2 mm apart at a distance of 3.0 mm from the grating. Estimate the spacing between the slits of the grating.

*42. The first-order maximum produced by a grating is located at an angle of $\theta = 16.0°$. What is the angle for the third-order maximum with the same light?

*43. ssm Violet light (wavelength = 410 nm) and red light (wavelength = 660 nm) lie at opposite ends of the visible spectrum. (a) For each wavelength, find the angle θ that locates the first-order maximum produced by a grating with 3300 lines/cm. This grating converts a mixture of all colors between violet and red into a rainbow-like dispersion between the two angles. Repeat the calculation above for (b) the second-order maximum and (c) the third-order maximum. (d) From your results, decide whether there is an overlap between any of the "rainbows" and, if so, specify which orders overlap.

*44. Three, and only three, bright fringes can be seen on either side of the central maximum when a grating is illuminated with light ($\lambda = 510$ nm). What is the maximum number of lines/cm for the grating?

**45. Two gratings A and B have slit separations d_A and d_B, respectively. They are used with the same light and the same observation screen. When grating A is replaced with grating B, it is observed that the first-order maximum of A is exactly replaced by the second-order maximum of B. (a) Determine the ratio d_B/d_A of the spacings between the slits of the gratings. (b) Find the next two principal maxima of grating A and the principal maxima of B that exactly replace them when the gratings are switched. Identify these maxima by their order numbers.

ADDITIONAL PROBLEMS

46. A rock concert is being held in an open field. Two loudspeakers are separated by 7.00 m. As an aid in arranging the seating, a test is conducted in which both speakers vibrate in phase and produce an 80.0-Hz bass tone simultaneously. The speed of sound is 343 m/s. A reference line is marked out in front of the speakers, perpendicular to the midpoint of the line between the speakers. Relative to either side of this reference line, what is the smallest angle that locates the places where destructive interference occurs? People seated in these places would have trouble hearing the 80.0-Hz bass tone.

47. ssm A flat observation screen is placed at a distance of 4.5 m from a pair of slits. The separation on the screen between the central bright fringe and the first-order bright fringe is 0.037 m. The light illuminating the slits has a wavelength of 490 nm. Determine the slit separation.

48. A slit whose width is 4.30×10^{-5} m is located 1.32 m from a flat screen. Light shines through the slit and falls on the screen. Find the width of the central fringe of the diffraction pattern when the wavelength of the light is 635 nm.

49. A mixture of red light ($\lambda_{vacuum} = 661$ nm) and green light ($\lambda_{vacuum} = 551$ nm) shines perpendicularly on a soap film ($n = 1.33$) that has air on either side. What is the minimum nonzero thickness of the film, so that destructive interference causes it to look red in reflected light?

50. Two parallel slits are illuminated by light composed of two wavelengths, one of which is 645 nm. On a viewing screen, the light whose wavelength is known produces its third dark fringe at the same place where the light whose wavelength is unknown produces its fourth-order bright fringe. The fringes are counted relative to the central or zeroth-order bright fringe. What is the unknown wavelength?

51. ssm Late one night on a highway, a car speeds by you and fades into the distance. Under these conditions the pupils of your eyes have diameters of about 7.0 mm. The taillights of this car are separated by a distance of 1.2 m and emit red light (wavelength = 660 nm in vacuum). How far away from you is this car when its taillights appear to merge into a single spot of light because of the effects of diffraction?

*52. Monochromatic light shines on a diffraction grating. When the light source and the grating are in air, the first-order maximum occurs at an angle of 33°. At what angle does the first-order maximum occur when the source and the grating are immersed in water ($n = 1.33$)?

*53. You are using a microscope to examine a blood sample. Recall from Section 26.12 that the sample should be placed just outside the focal point of the objective lens of the microscope. (a) The specimen is being illuminated with light of wavelength λ and the diameter of the objective equals its focal length. Determine the closest distance between two blood cells that can just be re-solved. Express your answer in terms of λ. (b) Based on your answer to (a), should you use light with a longer wavelength or a shorter wavelength if you wish to resolve two blood cells that are even closer together?

*54. The separation between the slits of a grating is 2.2×10^{-6} m. This grating is used with light that contains all wavelengths between 410 and 660 nm. Rainbow-like spectra form on a screen 3.2 m away. How wide (in meters) is (a) the first-order spectrum and (b) the second-order spectrum?

*55. ssm The same diffraction grating is used with two different wavelengths of light, λ_A and λ_B. The fourth-order principal maximum of light A exactly overlaps the third-order principal maximum of light B. Find the ratio λ_A/λ_B.

*56. Review Conceptual Example 4 before attempting this problem. A film of gasoline ($n = 1.40$) floats on water ($n = 1.33$). Yellow light (wavelength = 580 nm in vacuum) shines perpen-

dicularly on this film. (a) Determine the minimum nonzero thickness of the film, such that the film appears bright yellow due to constructive interference. (b) Repeat part (a), assuming that the gasoline film is on glass ($n = 1.52$) instead of water.

**57. There are 5620 lines per centimeter in a grating that is used with light whose wavelength is 471 nm. A flat observation screen is located at a distance of 0.750 m from the grating. What is the minimum width that the screen must have so the *centers* of all the principal maxima formed on either side of the central maximum fall on the screen?

**58. A piece of curved glass has a radius of curvature of 10.0 m and is used to form Newton's rings, as in Figure 27.14. Not counting the dark spot at the center of the pattern, there are one hundred dark fringes, the last one being at the outer edge of the curved piece of glass. The light being used has a wavelength of 654 nm in vacuum. What is the radius of the outermost dark ring in the pattern?

CONCEPTS & CALCULATIONS

GROUP LEARNING PROBLEMS

Note: Each of these problems consists of Concept Questions followed by a related quantitative Problem. They are designed for use by students working alone or in small learning groups. The Concept Questions involve little or no mathematics and are intended to stimulate group discussions. They focus on the concepts with which the problems deal. Recognizing the concepts is the essential initial step in any problem-solving technique.

59. Concept Questions (a) Point A is the midpoint of one of the sides of a square. On the side opposite this spot, two in-phase loudspeakers are located at adjacent corners, as in Figure 27.41. Standing at point A, you hear a loud sound because constructive interference occurs between the identical sound waves coming from the speakers. Why does constructive interference occur at point A? (b) As you walk along the side of the square toward either empty corner, the loudness diminishes gradually to nothing and then increases again until you hear a maximally loud sound at the corner. Explain these observations in terms of destructive and constructive interference. (c) The general condition that leads to constructive interference entails a number of possibilities for the variable m, which designates the order of the interference maxima. Which one of those possibilities, if any, applies at either empty corner of the square? Explain.

Problem The square referred to in the Concept Questions is 4.6 m on a side. Find the wavelength of the sound waves.

60. Concept Questions (a) The screen in Figure 27.8 has a fixed width and is centered on the midpoint between the two slits. Should the screen be moved to the left or to the right to ensure that the third-order bright fringe does *not* lie on the screen? (b) Do fewer fringes lie on the screen when the distance L between the double-slit and the screen is smaller or larger? Account for your answers.

Problem In a setup like that in Figure 27.8, a wavelength of 625 nm is used in a Young's double-slit experiment. The separation between the slits is $d = 1.4 \times 10^{-5}$ m. The total width of the screen is 0.20 m. In one version of the setup, the separation between the double slit and the screen is $L_A = 0.35$ m, while in another version it is $L_B = 0.50$ m. On one side of the central bright fringe, how many bright fringes lie on the screen in the two versions of the setup? Do not include the central bright fringe in your counting. Verify that your answer is consistent with your answers to the Concept Questions.

61. Concept Questions (a) What, if any, phase change occurs when light, traveling in air, reflects from the interface between the air and a soap film ($n = 1.33$)? (b) What, if any, phase change occurs when light, traveling in a soap film, reflects from the interface between the soap film and a glass plate ($n = 1.52$)? (c) Is the wavelength of the light in a soap film greater than, smaller than, or equal to the wavelength in a vacuum?

Problem A soap film ($n = 1.33$) is 465 nm thick and lies on a glass plate ($n = 1.52$). Sunlight, whose wavelengths (in vacuum) extend from 380 to 750 nm, travels through the air and strikes the film perpendicularly. For which wavelength(s) in this range does destructive interference cause the film to look dark in reflected light?

62. Concept Questions (a) In a single-slit diffraction pattern the width of the central bright fringe is defined by the location of the first dark fringe that lies on either side of it. For a given slit width, does the width of the central bright fringe increase, decrease, or remain the same as the wavelength of the light increases? (b) For a given wavelength, does the width of the central bright fringe increase, decrease, or remain the same as the slit

width increases? (c) When both the wavelength and the slit width change, it is possible for the width of the central bright fringe to remain the same. What condition must be satisfied for this to happen? In each case, give your reasoning.

Problem A slit has a width of $W_1 = 2.3 \times 10^{-6}$ m. When light with a wavelength of $\lambda_1 = 510$ nm passes through this slit, the width of the central bright fringe on a flat observation screen has a certain value. Keeping the screen in the same place, this slit is replaced with a second slit (width W_2) and a wavelength of $\lambda_2 = 740$ nm is used. The width of the central bright fringe on the screen is observed to be unchanged. Find W_2.

63. ✍ **Concept Questions** An inkjet color printer uses tiny dots of red, green, and blue ink to produce an image. At normal viewing distances, the eye does not resolve the individual dots, so that the image has a normal look. The angle θ_{min} is the minimum angle that two dots can subtend at the eye and still be resolved separately. (a) For which color does θ_{min} have the largest value? (b) For which color does θ_{min} have the smallest value? (c) Corresponding to each value of θ_{min}, there is a value for the separation distance s between the dots. How is θ_{min} related to s and the viewing distance L? (d) Assume that the dot separation on the printed page is the same for all colors and is chosen so that none of the colored dots can be seen as separate objects. Should the maxi-

mum allowable dot separation be s_{red}, s_{green}, or s_{blue}? For each answer, give your reasoning.

Problem The wavelengths for red, green, and blue are $\lambda_{red} = 660$ nm, $\lambda_{green} = 550$ nm, and $\lambda_{blue} = 470$ nm. The diameter of the pupil through which light enters the eye is 2.0 mm. For a viewing distance of 0.40 m, find the maximum separation distance that the dots can have and not be resolved separately. Check to see that your answer is consistent with your answers to the Concept Questions.

64. **Concept Questions** (a) Two diffraction gratings are located at the same distance from observation screens. Light with the same wavelength λ is used for each. The principal maxima of grating A are observed to be closer together on the screen than the principal maxima of grating B. Which grating diffracts the light to a greater extent? (b) Which grating has the smaller slit separation d? (c) Which grating has the greater number of lines per meter? Justify each of your answers.

Problem The separation between adjacent principal maxima for grating A is 2.7 cm, while for grating B it is 3.2 cm. Grating A has 2000 lines per meter. How many lines per meter does grating B have? The diffraction angles are small enough that $\sin \theta \approx \tan \theta$. Be sure that your answer is consistent with your answers to the Concept Questions.

Albert Einstein (1879–1955), the author of the theory of special relativity, is one of the most famous scientists of the twentieth century. This photograph shows the Einstein Memorial Sculpture by Robert Berks, which is located in Washington, DC, at the National Academy of Sciences.

CHAPTER 28
SPECIAL RELATIVITY

28.1 EVENTS AND INERTIAL REFERENCE FRAMES

In the theory of special relativity, an *event,* such as the launching of the space shuttle in Figure 28.1, is a physical "happening" that occurs at a certain place and time. In this drawing two observers are watching the lift-off, one standing on the earth and one in an airplane that is flying at a constant velocity relative to the earth. To record the event, each observer uses a *reference frame* that consists of a set of *x, y, z* axes (called a *coordinate system*) and a clock. The coordinate systems are used to establish where the event occurs, and the clocks to specify when. Each observer is at rest relative to his own reference frame. However, the earth-based observer and the airborne observer are moving relative to each other and so are their respective reference frames.

The theory of special relativity deals with a "special" kind of reference frame, called an *inertial reference frame.* As Section 4.2 discusses, an inertial reference frame is one in which Newton's law of inertia is valid. That is, if the net force acting on a body is zero, the body either remains at rest or moves at a constant velocity. In other words, the acceleration of such a body is zero when measured in an inertial reference frame. Rotating and otherwise accelerating reference frames are not inertial reference frames. The earth-based reference frame in Figure 28.1 is not quite an inertial frame because it is subjected to centripetal accelerations as the earth spins on its axis and revolves around the sun. In most situations, however, the effects of these accelerations are small, and we can neglect them. To the extent that the earth-based reference frame is an inertial frame, so is the plane-based reference frame, for the plane moves at a constant velocity relative to the earth. The next section discusses why inertial reference frames are important in relativity.

28.2 THE POSTULATES OF SPECIAL RELATIVITY

Einstein built his theory of special relativity on two fundamental assumptions or postulates about the way nature behaves.

Figure 28.1 Using an earth-based reference frame, an observer standing on the earth records the location and time of an event (the space shuttle lift-off). Likewise, an observer in the airplane uses a plane-based reference frame to describe the event.

- **THE POSTULATES OF SPECIAL RELATIVITY**
  1. *The Relativity Postulate.* The laws of physics are the same in every inertial reference frame.
  2. *The Speed of Light Postulate.* The speed of light in a vacuum, measured in any inertial reference frame, always has the same value of *c,* no matter how fast the source of light and the observer are moving relative to each other.

It is not too difficult to accept the relativity postulate. For instance, in Figure 28.1 each observer, using his own inertial reference frame, can make measurements on the motion of the space shuttle. The relativity postulate asserts that both observers find their data to be consistent with Newton's laws of motion. Similarly, both observers find that the behavior of the electronics on the space shuttle is described by the laws of electromagnetism. According to the relativity postulate, *any inertial reference frame is as good as any other for expressing the laws of physics.* As far as inertial reference frames are concerned, nature does not play favorites.

Since the laws of physics are the same in all inertial reference frames, there is no experiment that can distinguish between an inertial frame that is at rest and one that is moving at a constant velocity. When you are seated on the aircraft in Figure 28.1,

Figure 28.2 Both the person on the truck and the observer on the earth measure the speed of the light to be c, regardless of the speed of the truck.

for instance, it is just as valid to say that you are at rest and the earth is moving as it is to say the reverse. It is not possible to single out one particular inertial reference frame as being at "absolute rest." Consequently, it is meaningless to talk about the "absolute velocity" of an object—that is, its velocity measured relative to a reference frame at "absolute rest." Thus, the earth moves relative to the sun, which itself moves relative to the center of our galaxy. And the galaxy moves relative to other galaxies, and so on. According to Einstein, only the relative velocity between objects, not their absolute velocities, can be measured and is physically meaningful.

While the relativity postulate seems plausible, the speed of light postulate defies common sense. For instance, Figure 28.2 illustrates a person standing on the bed of a truck that is moving at a constant speed of 15 m/s relative to the ground. Now, suppose you are standing on the ground and the person on the truck shines a flashlight at you. The person on the truck observes the speed of light to be c. What do you measure for the speed of light? You might guess that the speed of light would be $c + 15$ m/s. However, this guess is inconsistent with the speed of light postulate, which states that all observers in inertial reference frames measure the speed of light to be c—nothing more, nothing less. Therefore, you must also measure the speed of light to be c, the same as that measured by the person on the truck. According to the speed of light postulate, the fact that the flashlight is moving has no influence whatsoever on the speed of the light approaching you. This property of light, although surprising, has been verified many times by experiment.

Since waves, such as water waves and sound waves, require a medium through which to propagate, it was natural for scientists before Einstein to assume that light did too. This hypothetical medium was called the *luminiferous ether* and was assumed to fill all of space. Furthermore, it was believed that light traveled at the speed c only when measured with respect to the ether. According to this view, an observer moving relative to the ether would measure a speed for light that was slower or faster than c, depending on whether the observer moved with or against the light. During the years 1883–1887, however, the American scientists A. A. Michelson and E. W. Morley carried out a series of famous experiments whose results were not consistent with the ether theory. Their results indicated that the speed of light is indeed the same in all inertial reference frames and does not depend on the motion of the observer. These experiments, and others, led eventually to the demise of the ether theory and the acceptance of the theory of special relativity.

⬤ *Concepts at a Glance*   The remainder of this chapter reexamines, from the viewpoint of special relativity, a number of fundamental concepts that have been discussed in earlier chapters from the viewpoint of classical physics. These concepts are time, length, momentum, kinetic energy, and the addition of velocities. We will see that each is modified by special relativity in a way that depends on the speed v of a moving object relative to the speed c of light in a vacuum. The middle of the Concepts-at-a-Glance chart in Figure 28.3 illustrates that when the object moves slowly (v is much smaller than c [$v \ll c$]), the modification is negligibly small, and the classical view of each concept provides an accurate description of re-

Concepts at a Glance

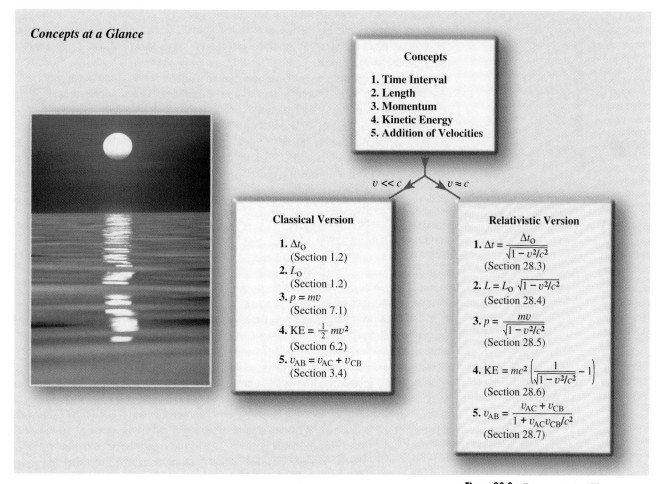

Concepts

1. Time Interval
2. Length
3. Momentum
4. Kinetic Energy
5. Addition of Velocities

$v \ll c$ $v \approx c$

Classical Version

1. Δt_O
   (Section 1.2)
2. L_O
   (Section 1.2)
3. $p = mv$
   (Section 7.1)
4. $KE = \frac{1}{2} mv^2$
   (Section 6.2)
5. $v_{AB} = v_{AC} + v_{CB}$
   (Section 3.4)

Relativistic Version

1. $\Delta t = \dfrac{\Delta t_O}{\sqrt{1 - v^2/c^2}}$
   (Section 28.3)
2. $L = L_O \sqrt{1 - v^2/c^2}$
   (Section 28.4)
3. $p = \dfrac{mv}{\sqrt{1 - v^2/c^2}}$
   (Section 28.5)
4. $KE = mc^2 \left(\dfrac{1}{\sqrt{1 - v^2/c^2}} - 1 \right)$
   (Section 28.6)
5. $v_{AB} = \dfrac{v_{AC} + v_{CB}}{1 + v_{AC}v_{CB}/c^2}$
   (Section 28.7)

Figure 28.3 *Concepts at a Glance* This chart emphasizes that it is the speed v of a moving object, as compared to the speed c of light in a vacuum, that determines whether the effects of special relativity are measurably important. One astonishing result of special relativity is that mass and energy are equivalent. Our sun, for instance, loses mass as it radiates the electromagnetic energy upon which life on earth depends.

ality. However, when the object moves so rapidly that v is an appreciable fraction of the speed of light, the effects of special relativity must be considered. In this case, the lower right-hand side of Figure 28.3 lists the various equations that convey the modifications imposed by special relativity. Each of these equations will be discussed in later sections of this chapter.

It is important to realize that the modifications imposed by special relativity do not imply that the classical concepts of time, length, momentum, kinetic energy, and the addition of velocities, as developed by Newton and others, are wrong. They are just limited to speeds that are very small compared to the speed of light. In contrast, the relativistic view of the concepts applies to all speeds between zero and the speed of light.

28.3 THE RELATIVITY OF TIME: TIME DILATION

Time Dilation

Common experience indicates that time passes just as quickly for a person standing on the ground as it does for an astronaut in a spacecraft. In contrast, special relativity reveals that the person on the ground measures time passing more slowly for the astronaut than for himself. We can see how this curious effect arises with the help of the clock illustrated in Figure 28.4, which uses a pulse of light to mark time. A

Figure 28.4 A light clock.

short pulse of light is emitted by a light source, reflects from a mirror, and then strikes a detector that is situated next to the source. Each time a pulse reaches the detector, a "tick" registers on the chart recorder, another short pulse of light is emitted, and the cycle repeats. Thus, the time interval between successive "ticks" is marked by a beginning event (the firing of the light source) and an ending event (the pulse striking the detector). The source and detector are so close to each other that the two events can be considered to occur at the same location.

Suppose two identical clocks are built. One is kept on earth, and the other is placed aboard a spacecraft that travels at a constant velocity relative to the earth. The astronaut is at rest with respect to the clock on the spacecraft and, therefore, sees the light pulse move along the up/down path shown in Figure 28.5a. According to the astronaut, the time interval Δt_0 required for the light to follow this path is the distance $2D$ divided by the speed of light c; $\Delta t_0 = 2D/c$. To the astronaut, Δt_0 is the time interval between the "ticks" of the spacecraft clock, that is, the time interval between the beginning and ending events of the clock. An earth-based observer, however, does *not* measure Δt_0 as the time interval between these two events. Since the spacecraft is moving, the earth-based observer sees the light pulse follow the diagonal path shown in red in part *b* of the drawing. This path is longer than the up/down path seen by the astronaut. But light travels at the *same speed c* for both observers, in accord with the speed of light postulate. Therefore, the earth-based observer measures a time interval Δt between the two events that is *greater* than the time interval Δt_0 measured by the astronaut. In other words, the earth-based observer, using his own earth-based clock to measure the performance of the astronaut's clock, finds that the astronaut's clock runs slowly. This result of special relativity is known as **time dilation.** (To *dilate* means to expand, and the time interval Δt is "expanded" relative to Δt_0.)

The time interval Δt that the earth-based observer measures in Figure 28.5b can be determined as follows. While the light pulse travels from the source to the detector, the spacecraft moves a distance $2L = v\,\Delta t$ to the right, where v is the speed of the spacecraft relative to the earth. From the drawing it can be seen that the light pulse travels a total diagonal distance of $2s$ during the time interval Δt. Applying the Pythagorean theorem, we find that

$$2s = 2\sqrt{D^2 + L^2} = 2\sqrt{D^2 + \left(\frac{v\,\Delta t}{2}\right)^2}$$

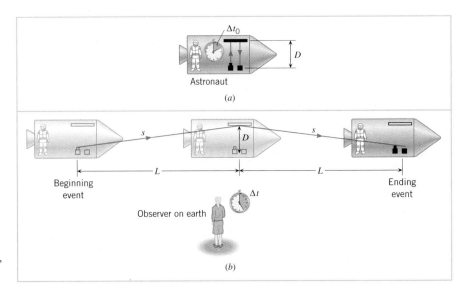

Figure 28.5 (*a*) The astronaut measures the time interval Δt_0 between successive "ticks" of his light clock. (*b*) An observer on earth watches the astronaut's clock and sees the light pulse travel a greater distance between "ticks" than it does in part *a*. Consequently, the earth-based observer measures a time interval Δt between "ticks" that is greater than Δt_0.

But the distance $2s$ is also equal to the speed of light times the time interval Δt, so $2s = c\,\Delta t$. Therefore,

$$c\,\Delta t = 2\sqrt{D^2 + \left(\frac{v\,\Delta t}{2}\right)^2}$$

Squaring this result and solving for Δt gives

$$\Delta t = \frac{2D}{c}\frac{1}{\sqrt{1 - \dfrac{v^2}{c^2}}}$$

But $2D/c = \Delta t_0$, the time interval between successive "ticks" of the spacecraft's clock as measured by the astronaut. With this substitution, the equation for Δt can be expressed as

Time dilation
$$\Delta t = \frac{\Delta t_0}{\sqrt{1 - \dfrac{v^2}{c^2}}} \tag{28.1}$$

The symbols in this formula are summarized as follows:

Δt_0 = proper time interval between two events, as measured by an observer who is at rest with respect to the events and who views them as occurring *at the same place*

Δt = time interval measured by an observer who is in motion with respect to the events and who views them as occurring at *different places*

v = relative speed between the two observers

c = speed of light in a vacuum

For a speed v that is less than c, the term $\sqrt{1 - v^2/c^2}$ in Equation 28.1 is less than 1, and the dilated time interval Δt is greater than Δt_0. Example 1 illustrates this time dilation effect.

EXAMPLE 1 **Time Dilation**

The spacecraft in Figure 28.5 is moving past the earth at a constant speed v that is 0.92 times the speed of light. Thus, $v = (0.92)(3.0 \times 10^8$ m/s), which is often written as $v = 0.92c$. The astronaut measures the time interval between successive "ticks" of the spacecraft clock to be $\Delta t_0 = 1.0$ s. What is the time interval Δt that an earth observer measures between "ticks" of the astronaut's clock?

Reasoning Since the clock on the spacecraft is moving relative to the earth observer, the earth observer measures a greater time interval Δt between "ticks" than does the astronaut, who is at rest relative to the clock. The dilated time interval Δt can be determined from the time dilation relation, Equation 28.1.

Solution The dilated time interval is

$$\Delta t = \frac{\Delta t_0}{\sqrt{1 - \dfrac{v^2}{c^2}}} = \frac{1.0 \text{ s}}{\sqrt{1 - \left(\dfrac{0.92c}{c}\right)^2}} = \boxed{2.6 \text{ s}}$$

From the point of view of the earth-based observer, the astronaut is using a clock that is running slowly, for the earth-based observer measures a time between "ticks" that is longer (2.6 s) than what the astronaut measures (1.0 s).

Astronaut Jerry M. Linenger "floating" inside the *Mir* spacecraft.

Proper Time Interval

In Figure 28.5 both the astronaut and the person standing on the earth are measuring the time interval between a beginning event (the firing of the light source) and an ending event (the light pulse striking the detector). For the astronaut, who is at rest with respect to the light clock, the two events occur at the same location. (Remember, we are assuming that the light source and detector are so close together that they are considered to be at the same place.) Being at rest with respect to a clock is the usual or "proper" situation, so the time interval Δt_0 measured by the astronaut is called the ***proper time interval***. In general, the proper time interval Δt_0 between two events is the time interval measured by an observer who is at rest relative to the events and sees them at the *same location* in space. On the other hand, the earth-based observer does not see the two events occurring at the same location in space, since the spacecraft is in motion. The time interval Δt that the earth-based observer measures is, therefore, not a proper time interval in the sense that we have defined it.

To understand situations involving time dilation, it is essential to distinguish between Δt_0 and Δt. In such situations it is helpful if one first identifies the two events that define the time interval. These events may be something other than the firing of a light source and the light pulse striking a detector. Then determine the reference frame in which the two events occur at the same place. An observer at rest in this reference frame measures the proper time interval Δt_0.

Space Travel

THE PHYSICS OF...

space travel and special relativity.

One of the intriguing aspects of time dilation occurs in conjunction with space travel. Since enormous distances are involved, travel to even the closest star outside our solar system would take a long time. However, as the following example shows, the travel time can be considerably less for the passengers than one might guess.

EXAMPLE 2   Space Travel

Alpha Centauri, a nearby star in our galaxy, is 4.3 light-years away. This means that, as measured by a person on earth, it would take light 4.3 years to reach this star. If a rocket leaves for Alpha Centauri at a speed of $v = 0.95c$ relative to the earth, by how much will the passengers have aged, according to their own clock, when they reach their destination? Assume that the earth and Alpha Centauri are stationary with respect to one another.

Reasoning The two events in this problem are the departure from earth and the arrival at Alpha Centauri. At departure, earth is just outside the spaceship. Upon arrival at the destination, Alpha Centauri is just outside. Therefore, relative to the passengers, the two events occur at the same place, namely, just outside the spaceship. Thus, the passengers measure the proper time interval Δt_0 on their clock, and it is this interval that we must find. For a person left behind on earth, the events occur at *different places,* so such a person measures the dilated time interval Δt rather than the proper time interval. To find Δt we note that the time to travel a given distance is inversely proportional to the speed. Since it takes 4.3 years to traverse the distance between earth and Alpha Centauri at the speed of light, it would take even longer at the slower speed of $v = 0.95c$. Thus, a person on earth measures the dilated time interval to be $\Delta t = (4.3 \text{ years})/0.95 = 4.5$ years. This value can be used with the time-dilation equation to find the proper time interval Δt_0.

Solution Using the time-dilation equation, we find that the proper time interval by which the passengers judge their own aging is

$$\Delta t_0 = \Delta t \sqrt{1 - \frac{v^2}{c^2}} = (4.5 \text{ years}) \sqrt{1 - \left(\frac{0.95c}{c}\right)^2} = \boxed{1.4 \text{ years}}$$

● **PROBLEM SOLVING INSIGHT**
In dealing with time dilation, decide which interval is the proper time interval as follows: (1) Identify the two events that define the interval. (2) Determine the reference frame in which the events occur at the same place; an observer at rest in this frame measures the proper time interval Δt_0.

Thus, the people aboard the rocket have aged by only 1.4 years when they reach Alpha Centauri, and not the 4.5 years an earthbound observer has calculated.

Verification of Time Dilation

A striking confirmation of time dilation was achieved in 1971 by an experiment carried out by J. C. Hafele and R. E. Keating.* They transported very precise cesium-beam atomic clocks around the world on commercial jets. Since the speed of a jet plane is considerably less than c, the time-dilation effect is extremely small. However, the atomic clocks were accurate to about $\div$ 10^{-9} s, so the effect could be measured. The clocks were in the air for 45 hours, and their times were compared to reference atomic clocks kept on earth. The experimental results revealed that, within experimental error, the readings on the clocks on the planes were different from those on earth by an amount that agreed with the prediction of relativity.

The behavior of subatomic particles called *muons* provides additional confirmation of time dilation. These particles are created high in the atmosphere, at altitudes of about 10 000 m. When at rest, muons exist only for about 2.2×10^{-6} s before disintegrating. With such a short lifetime, these particles could never make it down to the earth's surface, even traveling at nearly the speed of light. However, *a large number of muons do reach the earth.* The only way they can do so is to live longer because of time dilation, as Example 3 illustrates.

EXAMPLE 3   The Lifetime of a Muon

The average lifetime of a muon at rest is 2.2×10^{-6} s. A muon created in the upper atmosphere, thousands of meters above sea level, travels toward the earth at a speed of $v = 0.998c$. Find, on the average, (a) how long a muon lives according to an observer on earth, and (b) how far the muon travels before disintegrating.

Reasoning The two events of interest are the generation and subsequent disintegration of the muon. When the muon is at rest, these events occur at the same place, so the muon's average (at rest) lifetime of 2.2×10^{-6} s is a proper time interval Δt_0. When the muon moves at a speed $v = 0.998c$ relative to the earth, an observer on the earth measures a dilated lifetime Δt that is given by Equation 28.1. The average distance x traveled by a muon, as measured by an earth observer, is equal to the muon's speed times the dilated time interval.

Solution

(a) The observer on earth measures a dilated lifetime given by

$$\Delta t = \frac{\Delta t_0}{\sqrt{1 - \dfrac{v^2}{c^2}}} = \frac{2.2 \times 10^{-6}\ \text{s}}{\sqrt{1 - \left(\dfrac{0.998c}{c}\right)^2}} = \boxed{35 \times 10^{-6}\ \text{s}} \quad (28.1)$$

• **PROBLEM SOLVING INSIGHT**
The proper time interval Δt_0 is always shorter than the dilated time interval Δt.

(b) The distance traveled by the muon before it disintegrates is

$$x = v\ \Delta t = (0.998)(3.00 \times 10^8\ \text{m/s})(35 \times 10^{-6}\ \text{s}) = \boxed{1.0 \times 10^4\ \text{m}}$$

Thus, the dilated, or extended, lifetime provides sufficient time for the muon to reach the surface of the earth. If its lifetime were only 2.2×10^{-6} s, a muon would travel only 660 m before disintegrating and could never reach the earth.

---

* J. C. Hafele and R. E. Keating, Around the world atomic clocks: relativistic time gains observed. *Science,* Vol. 168, 14 July 1972.

28.4 THE RELATIVITY OF LENGTH: LENGTH CONTRACTION

Because of time dilation, observers moving at a constant velocity relative to each other measure different time intervals between two events. For instance, Example 2 in the previous section illustrates that a trip from earth to Alpha Centauri at a speed of $v = 0.95c$ takes 4.5 years according to a clock on earth, but only 1.4 years according to a clock in the rocket. These two times differ by the factor $\sqrt{1 - v^2/c^2}$. Since the times for the trip are different, one might ask if the observers measure different distances between earth and Alpha Centauri. The answer, according to special relativity, is yes. After all, both the earth-based observer and the rocket passenger agree that the relative speed between the rocket and earth is $v = 0.95c$. Since speed is distance divided by time and the time is different for the two observers, it follows that the distances must also be different, if the relative speed is to be the same for both individuals. Thus, the earth observer determines the distance to Alpha Centauri to be $L_0 = v \Delta t = (0.95c)(4.5 \text{ years}) = 4.3$ light-years. On the other hand, a passenger aboard the rocket finds the distance is only $L = v \Delta t_0 = (0.95c)(1.4 \text{ years}) = 1.3$ light-years. The passenger, measuring the shorter time, also measures the shorter distance. This shortening of the distance between two points is one example of a phenomenon known as **length contraction.**

The relation between the distances measured by two observers in relative motion at a constant velocity can be obtained with the aid of Figure 28.6. Part *a* of the drawing shows the situation from the point of view of the earth-based observer. This person measures the time of the trip to be Δt, the distance to be L_0, and the relative speed of the rocket to be $v = L_0/\Delta t$. Part *b* of the drawing presents the point of view of the passenger, for whom the rocket is at rest, and the earth and Alpha Centauri appear to move by at a speed v. The passenger determines the distance of the trip to be L, the time to be Δt_0, and the relative speed to be $v = L/\Delta t_0$. Since the relative speed computed by the passenger equals that computed by the earth-based observer, it follows that $v = L/\Delta t_0 = L_0/\Delta t$. Using this result and the time-dilation equation, Equation 28.1, we obtain the following relation between L and L_0:

● **PROBLEM SOLVING INSIGHT**
The proper length L_0 is always larger than the contracted length L.

Figure 28.6 (*a*) As measured by an observer on the earth, the distance to Alpha Centauri is L_0, and the time required to make the trip is Δt. (*b*) According to the passenger on the spacecraft, the earth and Alpha Centauri move with speed v relative to the craft. The passenger measures the distance and time of the trip to be L and Δt_0, respectively, both quantities less than those in part *a*.

Length contraction

$$L = L_0 \sqrt{1 - \frac{v^2}{c^2}} \qquad (28.2)$$

The length L_0 is called the **proper length;** it is the length (or distance) between two points *as measured by an observer at rest with respect to them.* Since v is less than c, the term $\sqrt{1 - v^2/c^2}$ is less than 1, and L is less than L_0. It is important to note that this length contraction occurs only along the direction of the motion. Those dimensions that are perpendicular to the motion are not shortened, as the next example discusses.

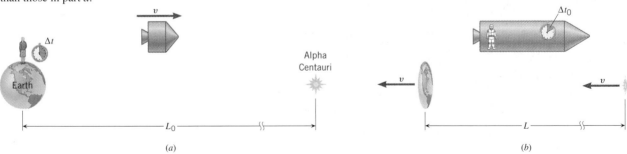

(*a*)   (*b*)

EXAMPLE 4 The Contraction of a Spacecraft

An astronaut, using a meter stick that is at rest relative to a cylindrical spacecraft, measures the length and diameter of the spacecraft to be 82 and 21 m, respectively. The spacecraft moves with a constant speed of $v = 0.95c$ relative to the earth, as in Figure 28.6. What are the dimensions of the spacecraft, as measured by an observer on earth?

Reasoning The length of 82 m is a proper length L_0, since it is measured using a meter stick that is at rest relative to the spacecraft. The length L measured by the observer on earth can be determined from the length-contraction formula, Equation 28.2. On the other hand, the diameter of the spacecraft is perpendicular to the motion, so the earth observer does not measure any change in the diameter.

Solution The length L of the spacecraft, as measured by the observer on earth, is

$$L = L_0 \sqrt{1 - \frac{v^2}{c^2}} = (82 \text{ m}) \sqrt{1 - \left(\frac{0.95c}{c}\right)^2} = \boxed{26 \text{ m}}$$

Both the astronaut and the observer on earth measure the same value for the diameter of the spacecraft: $\boxed{\text{Diameter} = 21 \text{ m}}$. Figure 28.6a shows the size of the spacecraft as measured by the earth observer, while part b shows the size measured by the astronaut.

When dealing with relativistic effects we need to distinguish carefully between the criteria for the proper time interval and the proper length. The proper time interval Δt_0 between two events is the time interval measured by an observer who is at rest relative to the events and sees them occurring at the *same place*. All other moving inertial observers will measure a larger value for this time interval. The proper length L_0 of an object is the length measured by an observer who is *at rest* with respect to the object. All other moving inertial observers will measure a shorter value for this length. The observer who measures the proper time interval may not be the same one who measures the proper length. For instance, Figure 28.6 shows that the astronaut measures the proper time interval Δt_0 for the trip between earth and Alpha Centauri, while the earth-based observer measures the proper length (or distance) L_0 for the trip.

It should be emphasized that the word "proper" in the phrases proper time and proper length does *not* mean that these quantities are the correct or preferred quantities in any absolute sense. If this were so, the observer measuring these quantities would be using a preferred reference frame for making the measurement, a situation that is prohibited by the relativity postulate. According to this postulate, there is no preferred inertial reference frame. When two observers are moving relative to each other at a constant velocity, each measures the other person's clock to run more slowly than his own, and each measures the other person's length, along that person's motion, to be contracted.

28.5 RELATIVISTIC MOMENTUM

Thus far we have discussed how time intervals and distances between two events are measured by observers moving at a constant velocity relative to each other. Special relativity also alters our ideas about momentum and energy.

Recall from Chapter 7 that when two or more objects interact, the principle of conservation of linear momentum applies if the system of objects is isolated. This principle states that the total linear momentum of an isolated system remains constant at all times. (An isolated system is one in which the sum of the external forces acting on the objects is zero.) The conservation of linear momentum is a law of physics and, in accord with the relativity postulate, is valid in all inertial reference frames. That is, when the total linear momentum is conserved in one inertial reference frame, it is conserved in all inertial reference frames.

As an example of momentum conservation, suppose several people are watching two billiard balls collide on a frictionless pool table. One person is standing next to the pool table, and the other is moving past the table with a constant velocity. Since the two balls constitute an isolated system, the relativity postulate requires that both observers must find the total linear momentum of the two-ball system to be the same before, during, and after the collision. For this kind of situation, Section 7.1 defines the classical linear momentum **p** of an object to be the product of its mass m and velocity **v**. As a result, the magnitude of the classical momentum is $p = mv$. As long as the speed of an object is considerably smaller than the speed of light, this definition is adequate. However, when the speed approaches the speed of light, an analysis of the collision shows that the total linear momentum is not conserved in all inertial reference frames if one defines linear momentum simply as the product of mass and velocity. In order to preserve the conservation of linear momentum, it is necessary to modify this definition. The theory of special relativity reveals that the magnitude of the **relativistic momentum** must be defined as in Equation 28.3:

Magnitude of the relativistic momentum

$$p = \frac{mv}{\sqrt{1 - \dfrac{v^2}{c^2}}} \qquad (28.3)$$

The total relativistic momentum of an isolated system is conserved in all inertial reference frames.

From Equation 28.3, we can see that the magnitudes of the relativistic and non-relativistic momenta differ by the same factor of $\sqrt{1 - v^2/c^2}$ that occurs in the time-dilation and length-contraction equations. Since this factor is always less than 1 and occurs in the denominator in Equation 28.3, the relativistic momentum is always larger than the nonrelativistic momentum. To illustrate how the two quantities differ as the speed v increases, Figure 28.7 shows a plot of the ratio of the momentum magnitudes (relativistic/nonrelativistic) as a function of v. According to Equation 28.3, this ratio is just $1/\sqrt{1 - v^2/c^2}$. The graph shows that for speeds attained by ordinary objects, such as cars and planes, the relativistic and nonrelativistic momenta are almost equal because their ratio is nearly 1. Thus, at speeds much less than the speed of light, either the nonrelativistic momentum or the relativistic momentum can be used to describe collisions. On the other hand, when the speed of the object becomes comparable to the speed of light, the relativistic momentum becomes significantly greater than the nonrelativistic momentum and must be used. Example 5 deals with the relativistic momentum of an electron traveling close to the speed of light.

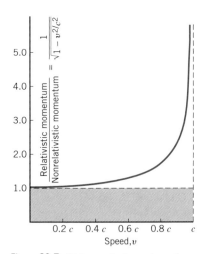

Figure 28.7 This graph shows how the ratio of the magnitude of the relativistic momentum to the magnitude of the nonrelativistic momentum increases as the speed of an object approaches the speed of light.

EXAMPLE 5   The Relativistic Momentum of a High-Speed Electron

The particle accelerator at Stanford University (Figure 28.8) is three kilometers long and accelerates electrons to a speed of 0.999 999 999 7c, which is very nearly equal to the speed of light. Find the magnitude of the relativistic momentum of an electron that emerges from the accelerator, and compare it with the nonrelativistic value.

Reasoning and Solution The magnitude of the electron's relativistic momentum can be obtained from Equation 28.3 if we recall that the mass of an electron is $m = 9.11 \times 10^{-31}$ kg:

$$p = \frac{mv}{\sqrt{1 - \dfrac{v^2}{c^2}}} = \frac{(9.11 \times 10^{-31}\text{ kg})(0.999\,999\,999\,7c)}{\sqrt{1 - \dfrac{(0.999\,999\,999\,7c)^2}{c^2}}} = \boxed{1 \times 10^{-17}\text{ kg} \cdot \text{m/s}}$$

This value for the magnitude of the momentum agrees with that measured experimentally. The relativistic momentum is greater than the nonrelativistic momentum by a factor of

$$\frac{1}{\sqrt{1 - \dfrac{v^2}{c^2}}} = \frac{1}{\sqrt{1 - \dfrac{(0.999\,999\,999\,7c)^2}{c^2}}} = \boxed{4 \times 10^4}$$

Figure 28.8 The Stanford three-kilometer linear accelerator accelerates electrons to nearly the speed of light.

28.6 THE EQUIVALENCE OF MASS AND ENERGY

The Total Energy of an Object

One of the most astonishing results of special relativity is that mass and energy are equivalent, in the sense that a gain or loss of mass can be regarded equally well as a gain or loss of energy. Consider, for example, an object of mass m traveling at a speed v. Einstein showed that the **total energy** E of the moving object is related to its mass and speed by the following relation:

Total energy of an object

$$E = \frac{mc^2}{\sqrt{1 - \dfrac{v^2}{c^2}}} \qquad (28.4)$$

To gain some understanding of Equation 28.4, consider the special case when the object is at rest. When $v = 0$ m/s, the total energy is called the *rest energy E_0*, and Equation 28.4 reduces to Einstein's now-famous equation:

Rest energy of an object

$$E_0 = mc^2 \qquad (28.5)$$

The rest energy represents the energy equivalent of the mass of an object at rest. As Example 6 shows, even a small mass is equivalent to an enormous amount of energy.

EXAMPLE 6 The Energy Equivalent of a Golf Ball

A 0.046-kg golf ball is lying on the green. (a) Find the rest energy of the golf ball. (b) If this rest energy were used to operate a 75-W light bulb, for how many years could the bulb stay on?

Reasoning The rest energy E_0 that is equivalent to the mass m of the golf ball is found from the relation $E_0 = mc^2$. The power used by the bulb is 75 W, which means that it consumes 75 J of energy per second. If the entire rest energy of the ball were available for use, the bulb could stay on for a time equal to the rest energy divided by the power.

Solution

(a) The rest energy of the ball is

$$E_0 = mc^2 = (0.046\text{ kg})(3.0 \times 10^8\text{ m/s})^2 = \boxed{4.1 \times 10^{15}\text{ J}} \qquad (28.5)$$

(b) This rest energy can keep the bulb burning for a time t given by

$$t = \frac{\text{Rest energy}}{\text{Power}} = \frac{4.1 \times 10^{15} \text{ J}}{75 \text{ W}} = 5.5 \times 10^{13} \text{ s} \qquad (6.10\text{b})$$

Since one year contains 3.2×10^7 s, we find $\boxed{t = 1.7 \times 10^6 \text{ yr}}$, or 1.7 million years!

When an object is accelerated from rest to a speed v, the object acquires kinetic energy in addition to its rest energy. The total energy E is the sum of the rest energy E_0 and the kinetic energy KE, or $E = E_0 + \text{KE}$. Therefore, the kinetic energy is the difference between the object's total energy and its rest energy. Using Equations 28.4 and 28.5, we can write the kinetic energy as

$$\text{KE} = E - E_0 = mc^2 \left(\frac{1}{\sqrt{1 - \dfrac{v^2}{c^2}}} - 1 \right) \qquad (28.6)$$

This equation is the relativistically correct expression for the kinetic energy of an object of mass m moving at speed v.

Equation 28.6 looks nothing like the kinetic energy expression introduced in Chapter 6, namely, $\text{KE} = \frac{1}{2}mv^2$. However, for speeds much less than the speed of light (v " c), the relativistic equation for the kinetic energy reduces to $\text{KE} = \frac{1}{2}mv^2$, as can be seen by using the binomial expansion* to represent the square root term in Equation 28.6:

$$\frac{1}{\sqrt{1 - \dfrac{v^2}{c^2}}} = 1 + \frac{1}{2}\left(\frac{v^2}{c^2}\right) + \frac{3}{8}\left(\frac{v^2}{c^2}\right)^2 + \cdots$$

Suppose v is much smaller than c, say $v = 0.01c$. The second term in the expansion has the value $\frac{1}{2}(v^2/c^2) = 5.0 \times 10^{-5}$, while the third term has the much smaller value $\frac{3}{8}(v^2/c^2)^2 = 3.8 \times 10^{-9}$. The additional terms are smaller still, so if v " c, we can neglect the third and additional terms in comparison with the first and second terms. Substituting the first two terms into Equation 28.6 gives

$$\text{KE} \approx mc^2 \left(1 + \frac{1}{2}\frac{v^2}{c^2} - 1 \right) = \tfrac{1}{2}mv^2$$

which is the familiar form for the kinetic energy. However, Equation 28.6 gives the correct kinetic energy for all speeds and must be used for speeds near the speed of light, as in Example 7.

EXAMPLE 7   A High-Speed Electron

An electron ($m = 9.109 \times 10^{-31}$ kg) is accelerated from rest to a speed of $v = 0.9995c$ in a particle accelerator. Determine the electron's (a) rest energy, (b) total energy, and (c) kinetic energy in millions of electron volts or MeV.

Reasoning and Solution

(a) The electron's rest energy is

$$E_0 = mc^2 = (9.109 \times 10^{-31} \text{ kg})(2.998 \times 10^8 \text{ m/s})^2 = 8.187 \times 10^{-14} \text{ J} \qquad (28.5)$$

---

* The binomial expansion states that $(1 - x)^n = 1 - nx + n(n - 1)x^2/2 + \cdots$. In our case, $x = v^2/c^2$ and $n = -1/2$.

Since 1 eV = 1.602×10^{-19} J, the electron's rest energy is

$$(8.187 \times 10^{-14} \text{ J}) \left(\frac{1 \text{ eV}}{1.602 \times 10^{-19} \text{ J}} \right) = \boxed{5.11 \times 10^5 \text{ eV} \quad \text{or} \quad 0.511 \text{ MeV}}$$

(b) The total energy of an electron traveling at a speed of $v = 0.9995c$ is

$$E = \frac{mc^2}{\sqrt{1 - \dfrac{v^2}{c^2}}} = \frac{(9.109 \times 10^{-31} \text{ kg})(2.998 \times 10^8 \text{ m/s})^2}{\sqrt{1 - \left(\dfrac{0.9995c}{c} \right)^2}}$$

$$= \boxed{2.59 \times 10^{-12} \text{ J} \quad \text{or} \quad 16.2 \text{ MeV}} \tag{28.4}$$

(c) The kinetic energy is the difference between the total energy and the rest energy:

$$\text{KE} = E - E_0 = 2.59 \times 10^{-12} \text{ J} - 8.2 \times 10^{-14} \text{ J}$$

$$= \boxed{2.51 \times 10^{-12} \text{ J} \quad \text{or} \quad 15.7 \text{ MeV}} \tag{28.6}$$

For comparison, if the kinetic energy of the electron had been calculated from $\frac{1}{2}mv^2$, a value of only 0.26 MeV would have been obtained.

Since mass and energy are equivalent, any change in one is accompanied by a corresponding change in the other. For instance, life on earth is dependent on electromagnetic energy (light) from the sun. Because this energy is leaving the sun (see Figure 28.9), there is a decrease in the sun's mass. Example 8 illustrates how to determine this decrease.

EXAMPLE 8   The Sun Is Losing Mass

The sun radiates electromagnetic energy at the rate of 3.92×10^{26} W. (a) What is the change in the sun's mass during each second that it is radiating energy? (b) The mass of the sun is 1.99×10^{30} kg. What fraction of the sun's mass is lost during a human lifetime of 75 years?

Reasoning Since 1 W = 1 J/s, the amount of electromagnetic energy radiated during each second is 3.92×10^{26} J. Thus, during each second, the sun's rest energy decreases by this amount. The change ΔE_0 in the sun's rest energy is related to the change Δm in its mass by $\Delta E_0 = (\Delta m)c^2$, according to Equation 28.5.

Solution

(a) For each second that the sun radiates energy, the change in its mass is

$$\Delta m = \frac{\Delta E_0}{c^2} = \frac{3.92 \times 10^{26} \text{ J}}{(3.00 \times 10^8 \text{ m/s})^2} = \boxed{4.36 \times 10^9 \text{ kg}}$$

Over 4 billion kilograms of mass are lost by the sun during each second.

(b) The amount of mass lost by the sun in 75 years is

$$\Delta m = (4.36 \times 10^9 \text{ kg/s}) \left(\frac{3.16 \times 10^7 \text{ s}}{1 \text{ year}} \right) (75 \text{ years}) = 1.0 \times 10^{19} \text{ kg}$$

While this is an enormous amount of mass, it represents only a tiny fraction of the sun's total mass:

$$\frac{\Delta m}{m_{\text{sun}}} = \frac{1.0 \times 10^{19} \text{ kg}}{1.99 \times 10^{30} \text{ kg}} = \boxed{5.0 \times 10^{-12}}$$

Any change in the energy of a system causes a change in the mass of the system according to $\Delta E_0 = (\Delta m)c^2$. It does not matter whether the change in energy is due

Visible light image.

X-ray image.

Figure 28.9 The sun emits electromagnetic energy over a broad portion of the electromagnetic spectrum. These photographs were obtained using that energy in the indicated regions of the spectrum.

to a change in electromagnetic energy, potential energy, thermal energy, or so on. While any change in energy gives rise to a change in mass, in most instances the change in mass is too small to be detected. For instance, when 4186 J of heat is used to raise the temperature of 1 kg of water by 1 C°, the mass changes by only $\Delta m = \Delta E_0/c^2 = (4186 \text{ J})/(3.00 \times 10^8 \text{ m/s})^2 = 4.7 \times 10^{-14}$ kg. Conceptual Example 9 illustrates further how a change in the energy of an object leads to an equivalent change in its mass.

CONCEPTUAL EXAMPLE 9   When Is a Massless Spring Not Massless?

Figure 28.10a shows a top view of a spring lying on a horizontal table. The spring is initially unstrained and assumed to be massless. Suppose that the spring is either stretched or compressed by an amount x from its unstrained length, as part b of the drawing shows. Is the mass of the spring still zero, or has it changed? And, if the mass has changed, is the change greater, smaller, or the same when the spring is stretched rather than compressed?

Reasoning and Solution Whenever a spring is stretched or compressed, its elastic potential energy changes. In Section 10.3 we showed that the elastic potential energy of an ideal spring is equal to $\frac{1}{2}kx^2$, where k is the spring constant and x is the amount of stretch or compression. Consistent with the theory of special relativity, any change in the total energy of a system, including a change in the elastic potential energy, is equivalent to a change in the mass of the system. Thus, the mass of a strained spring is greater than that of an unstrained (and massless) spring. Furthermore, since the elastic potential energy depends on x^2, the increase in mass of the spring is the same whether it is compressed or stretched, provided the magnitude of x is the same in both cases.

The increase is exceedingly small because it is equal to the elastic potential energy divided by c^2 and c^2 is so large.

Related Homework Material: *Questions 8, 9, and 10*

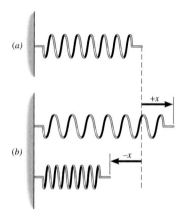

Figure 28.10 (a) This spring is unstrained and assumed to have no mass. (b) When the spring is either stretched or compressed by an amount x, it gains elastic potential energy and, hence, mass.

It is also possible to transform matter itself into other forms of energy. For example, the positron (see Section 31.4) has the same mass as an electron but an opposite electrical charge. If these two particles of matter collide, they are completely annihilated, and a burst of high-energy electromagnetic waves is produced. Thus, matter is transformed into electromagnetic waves, the energy of the electromagnetic waves being equal to the total energies of the two colliding particles.

The transformation of electromagnetic waves into matter also happens. In one experiment, an extremely high-energy electromagnetic wave, called a gamma ray (see Section 31.4), passes close to the nucleus of an atom. If the gamma ray has sufficient energy, it can create an electron and a positron. The gamma ray disappears, and the two particles of matter appear in its place. Except for picking up some momentum, the nearby nucleus remains unchanged. The process in which the gamma ray is transformed into the two particles is known as *pair production*.

The Speed of Light Is the Ultimate Speed

One of the important consequences of the theory of special relativity is that objects with mass cannot reach the speed of light in a vacuum. Thus, the speed of light represents the ultimate speed. To see that this speed limitation is a consequence of special relativity, consider Equation 28.6, which gives the kinetic energy of a moving object. As v approaches the speed of light c, the $\sqrt{1 - v^2/c^2}$ term in the denominator approaches zero. Hence, the kinetic energy becomes infinitely large. However, the work–energy theorem (Chapter 6) tells us that an infinite amount of work would have to be done to give the object an infinite kinetic energy. Since an infinite amount of work is not available, we are left with the conclusion that objects with mass cannot attain the speed of light c.

28.7 THE RELATIVISTIC ADDITION OF VELOCITIES

The velocity of an object relative to an observer plays a central role in special relativity, and to determine this velocity, it is sometimes necessary to add two or more velocities together. We first encountered relative velocity in Section 3.4, so we will begin by reviewing some of the ideas presented there. Figure 28.11 illustrates a truck moving at a constant velocity of $v_{TG} = +15$ m/s relative to an observer standing on the ground, where the plus sign denotes a direction to the right. Suppose someone on the truck throws a baseball toward the observer at a velocity of $v_{BT} = +8.0$ m/s relative to the truck. We might conclude that the observer on the ground sees the ball approaching at a velocity of $v_{BG} = v_{BT} + v_{TG} = 8.0$ m/s $+$ 15 m/s $= +23$ m/s. These symbols are similar to those used in Section 3.4 and have the following meaning:

$v_{\boxed{BG}}$ = velocity of the $\boxed{\text{Baseball}}$ relative to the $\boxed{\text{Ground}}$ = $+23$ m/s

$v_{\boxed{BT}}$ = velocity of the $\boxed{\text{Baseball}}$ relative to the $\boxed{\text{Truck}}$ = $+8.0$ m/s

$v_{\boxed{TG}}$ = velocity of the $\boxed{\text{Truck}}$ relative to the $\boxed{\text{Ground}}$ = $+15.0$ m/s

Although the result that $v_{BG} = +23$ m/s seems reasonable, careful measurements would show that it is not quite right. According to special relativity, the equation $v_{BG} = v_{BT} + v_{TG}$ is not valid for the following reason. If the velocity of the truck had a magnitude sufficiently close to the speed of light, the equation would predict that the observer on the earth could see the baseball moving faster than the speed of light. This is not possible, since no object with a finite mass can move faster than the speed of light.

For the case where the truck and ball are moving along the same straight line, the theory of special relativity reveals that the velocities are related according to

$$v_{BG} = \frac{v_{BT} + v_{TG}}{1 + \dfrac{v_{BT} v_{TG}}{c^2}}$$

The subscripts in this equation have been chosen for the specific situation shown in Figure 28.11. For the general situation, the relative velocities are related by the **velocity-addition formula:**

Velocity addition
$$v_{AB} = \frac{v_{AC} + v_{CB}}{1 + \dfrac{v_{AC} v_{CB}}{c^2}} \tag{28.7}$$

$v_{BT} = +8.0$ m/s

$v_{TG} = +15$ m/s

Ground-based observer

Figure 28.11 The truck is approaching the ground-based observer at a relative velocity of $v_{TG} = +15$ m/s. The velocity of the baseball relative to the truck is $v_{BT} = +8.0$ m/s.

where all the velocities are assumed to be constant and the symbols have the following meanings:

$v_{\boxed{AB}}$ = velocity of $\boxed{\text{object A}}$ relative to $\boxed{\text{object B}}$

$v_{\boxed{AC}}$ = velocity of $\boxed{\text{object A}}$ relative to $\boxed{\text{object C}}$

$v_{\boxed{CB}}$ = velocity of $\boxed{\text{object C}}$ relative to $\boxed{\text{object B}}$

The ordering of the subscripts in Equation 28.7 follows the discussion in Section 3.4. For motion along a straight line, the velocities can have either positive or negative values, depending on whether they are directed along the positive or negative direction. Furthermore, switching the order of the subscripts changes the sign of the velocity, so, for example, $v_{BA} = -v_{AB}$ (see Example 11 in Chapter 3).

Equation 28.7 differs from the nonrelativistic formula ($v_{AB} = v_{AC} + v_{CB}$) by the presence of the $v_{AC}v_{CB}/c^2$ term in the denominator. This term arises because of the effects of time dilation and length contraction that occur in special relativity. When v_{AC} and v_{CB} are small compared to c, the $v_{AC}v_{CB}/c^2$ term is small compared to 1, so the velocity-addition formula reduces to $v_{AB} \approx v_{AC} + v_{CB}$. However, when either v_{AC} or v_{CB} is comparable to c, the results can be quite different, as Example 10 illustrates.

EXAMPLE 10   The Relativistic Addition of Velocities

Imagine a hypothetical situation in which the truck in Figure 28.11 is moving relative to the ground with a velocity of $v_{TG} = +0.8c$. A person riding on the truck throws a baseball at a velocity relative to the truck of $v_{BT} = +0.5c$. What is the velocity v_{BG} of the baseball relative to a person standing on the ground?

Reasoning The observer on the ground does *not* see the baseball approaching at $v_{BG} = 0.5c + 0.8c = 1.3c$. This cannot be because the speed of the ball would then exceed the speed of light. The velocity-addition formula gives the correct velocity, which has a magnitude less than the speed of light.

Solution The ground-based observer sees the ball approaching with a velocity of

$$v_{BG} = \frac{v_{BT} + v_{TG}}{1 + \dfrac{v_{BT}v_{TG}}{c^2}} = \frac{0.5c + 0.8c}{1 + \dfrac{(0.5c)(0.8c)}{c^2}} = \boxed{0.93c}$$

Example 10 discusses how the speed of a baseball is viewed by observers in different inertial reference frames. The next example deals with a similar situation, except that the baseball is replaced by the light of a laser beam.

CONCEPTUAL EXAMPLE 11   The Speed of a Laser Beam

Figure 28.12 shows an intergalactic cruiser approaching a hostile spacecraft. The velocity of the cruiser relative to the spacecraft is $v_{CS} = +0.7c$. Both vehicles are moving at a constant velocity. The cruiser fires a beam of laser light at the enemy. The velocity of the laser beam relative to the cruiser is $v_{LC} = +c$. (a) What is the velocity of the laser beam v_{LS} relative to the renegades aboard the spacecraft? (b) At what velocity do the renegades aboard the spacecraft see the laser beam move away from the cruiser?

Reasoning and Solution

(a) Since both vehicles move at a constant velocity, each constitutes an inertial reference frame. According to the speed of light postulate, *all* observers in inertial ref-

Figure 28.12 An intergalactic cruiser, closing in on a hostile spacecraft, fires a beam of laser light.

erence frames measure the speed of light in a vacuum to be c. Thus, the renegades aboard the hostile spacecraft see the laser beam travel toward them at the speed of light, even though the beam is emitted from the cruiser, which itself is moving at seven-tenths the speed of light.

(b) The renegades aboard the spacecraft see the cruiser approach them at a relative velocity of $v_{CS} = +0.7c$, and they also see the laser beam approach them at a relative velocity of $v_{LS} = +c$. Both these velocities are measured relative to the *same* inertial reference frame, namely, that of the spacecraft. Therefore, the renegades aboard the spacecraft see the laser beam move away from the cruiser at a velocity that is the difference between these two velocities, or $+c - (+0.7c) = +0.3c$. The velocity-addition formula, Equation 28.7, is not applicable here because both velocities are measured relative to the *same* inertial reference frame (the spacecraft's reference frame). The velocity-addition formula can be used only when the velocities are measured relative to different inertial reference frames.

Related Homework Material: *Question 12, Problem 40*

It is a straightforward matter to show that the velocity-addition formula is consistent with the speed of light postulate. Consider Figure 28.13, which shows a person riding on a truck and holding a flashlight. The velocity of the light, relative to the person on the truck, is $v_{LT} = +c$. The velocity v_{LG} of the light relative to the observer standing on the ground is given by the velocity-addition formula as

$$v_{LG} = \frac{v_{LT} + v_{TG}}{1 + \frac{v_{LT}v_{TG}}{c^2}} = \frac{c + v_{TG}}{1 + \frac{cv_{TG}}{c^2}} = \frac{(c + v_{TG})c}{(c + v_{TG})} = c$$

Thus, the velocity-addition formula indicates that the observer on the ground and the person on the truck both measure the speed of light to be c, independent of the relative velocity v_{TG} between them. This is exactly what the speed of light postulate states.

28.8 CONCEPTS & CALCULATIONS

There are many astonishing consequences of special relativity, two of which are time dilation and length contraction. Example 12 reviews these important concepts

Figure 28.13 The speed of the light emitted by the flashlight is c relative to both the truck and the observer on the ground.

Ground-based observer

in the context of a golf game in a world where the speed of light is a little faster than that of a golf cart.

CONCEPTS & CALCULATIONS EXAMPLE 12   Golf and Special Relativity

Imagine playing golf in a world where the speed of light is only $c = 3.40$ m/s. Golfer A drives a ball down a flat horizontal fairway for a distance that he measures as 75.0 m. Golfer B, riding in a cart, happens to pass by just as the ball is hit. Golfer A stands at the tee and watches while golfer B moves down the fairway toward the ball at a constant speed of 2.80 m/s. (a) How far is the ball hit according to a measurement made by golfer B? (b) According to each golfer, how much time does it take for golfer B to reach the ball?

Concept Questions and Answers Who measures the proper length of the drive, and who measures the contracted length?

Answer Consider two coordinate systems, one attached to the earth and the other to the golf cart. The proper length L_0 is the distance between two points as measured by an observer *who is at rest with respect to them.* Since golfer A is standing at the tee and is at rest relative to the earth, golfer A measures the proper length of the drive, which is $L_0 = 75.0$ m. Golfer B is not at rest with respect to the two points, however, and measures a contracted length for the drive that is less than 75.0 m.

Who measures the proper time interval, and who measures the dilated time interval?

Answer The proper time interval Δt_0 is the time interval measured by an observer who is at rest in his or her coordinate system and *who views the beginning and ending events as occurring at the same place.* The beginning event is when the ball is hit, and the ending event is when golfer B arrives at the ball. When the ball is hit, it is along side the origin of golfer B's coordinate system. When golfer B arrives at the ball down the fairway, it is again along side the origin of his coordinate system. Thus, golfer B measures the proper time interval. Golfer A, who does not see the beginning and ending events occurring at the same place, measures a dilated or longer time interval.

Solution

(a) Golfer A, being at rest with respect to the beginning and ending points, measures the proper length of the drive, so $L_0 = 75.0$ m. Golfer B, who is moving, measures a contracted length L given by Equation 28.2:

$$L = L_0 \sqrt{1 - \frac{v^2}{c^2}} = (75.0 \text{ m}) \sqrt{1 - \frac{(2.80 \text{ m/s})^2}{(3.40 \text{ m/s})^2}} = \boxed{42.5 \text{ m}}$$

Thus, the moving golfer measures the length of the drive to be a shortened 42.5 m rather than the 75.0 m measured by the stationary golfer.

(b) According to golfer B, the time interval Δt_0 it takes to reach the ball is equal to the contracted length L that he measures divided by the speed of the ground with respect to him. The speed of the ground with respect to golfer B is the same as the speed v of the cart with respect to the ground. Thus, we find

$$\Delta t_0 = \frac{L}{v} = \frac{42.5 \text{ m}}{2.80 \text{ m/s}} = \boxed{15.2 \text{ s}}$$

Golfer A, standing at the tee, measures a dilated time interval Δt, which is related to the proper time interval by Equation 28.1:

$$\Delta t = \frac{\Delta t_0}{\sqrt{1 - \frac{v^2}{c^2}}} = \frac{15.2 \text{ s}}{\sqrt{1 - \frac{(2.80 \text{ m/s})^2}{(3.40 \text{ m/s})^2}}} = \boxed{26.8 \text{ s}}$$

In summary, golfer A measures the proper length (75.0 m) and a dilated time interval (26.8 s), while golfer B measures a shortened length (42.5 m) and a proper time interval (15.2 s).

Other important consequences of special relativity are the equivalence of mass and energy, and how kinetic energy depends on the total and rest energies. Example 13 compares these properties for three different particles.

CONCEPTS & CALCULATIONS EXAMPLE 13 Mass and Energy

The rest energy E_0 and the total energy E of three particles, expressed in terms of a basic amount of energy $E' = 5.98 \times 10^{-10}$ J, are listed in the table below. The speeds of these particles are large, in some cases approaching the speed of light. For each particle, determine its mass and kinetic energy.

Particle	Rest Energy	Total Energy
a	E'	$2E'$
b	E'	$4E'$
c	$5E'$	$6E'$

Concept Questions and Answers Given the rest energies specified in the table, what is the ranking (largest first) of the masses of the particles?

Answer The rest energy is the energy that an object has when its speed is zero. According to special relativity, the rest energy E_0 and the mass m are equivalent. The relation between the two is given by Equation 28.5 as $E_0 = mc^2$, where c is the speed of light in a vacuum. Thus, the rest energy is directly proportional to the mass. From the table it can be seen that particles a and b have identical rest energies, so they have identical masses. Particle c has the greatest rest energy, so it has the greatest mass. The ranking of the masses, largest first, is c, then a and b (a tie).

Is the kinetic energy KE given by the expression $KE = \frac{1}{2}mv^2$, and what is the ranking (largest first) of the kinetic energies of the particles?

Answer No. The expression $KE = \frac{1}{2}mv^2$ is approximately valid only when the speed of the object is much, much less than the speed of light. According to special relativity, the kinetic energy is the difference between the total energy E and the rest energy E_0, so $KE = E - E_0$. Therefore, we can examine the table and determine the kinetic energy of each particle in terms of E'. The kinetic energies of particles a, b, and c are, respectively, $2E' - E' = E'$, $4E' - E' = 3E'$, and $6E' - 5E' = E'$. The ranking of the kinetic energies, largest first, is b, then a and c (a tie).

Solution

(a) The mass of particle a can be found from its rest energy $E_0 = mc^2$. Since $E_0 = E'$ (see the table), its mass is

$$m_a = \frac{E'}{c^2} = \frac{5.98 \times 10^{-10} \text{ J}}{(3.00 \times 10^8 \text{ m/s})^2} = \boxed{6.64 \times 10^{-27} \text{ kg}}$$

In a similar manner, we find that the masses of particles b and c are

$$\boxed{m_b = 6.64 \times 10^{-27} \text{ kg}} \quad \text{and} \quad \boxed{m_c = 33.2 \times 10^{-27} \text{ kg}}$$

As expected, the ranking is $m_c > m_a = m_b$.

(b) According to Equation 28.6, the kinetic energy KE of a particle is equal to its total energy E minus its rest energy E_0; $KE = E - E_0$. For particle a, its total energy is $E = 2E'$ and its rest energy is $E_0 = E'$, so its kinetic energy is

$$KE_a = 2E' - E' = E' = \boxed{5.98 \times 10^{-10} \text{ J}}$$

The kinetic energies of particles b and c can be determined in a similar fashion:

$$\boxed{KE_b = 17.9 \times 10^{-10} \text{ J}} \quad \text{and} \quad \boxed{KE_c = 5.98 \times 10^{-10} \text{ J}}$$

As anticipated, the ranking is $KE_b > KE_a = KE_c$.

At the end of the problem set for this chapter, you will find homework problems that contain both conceptual and quantitative parts. These problems are grouped under the heading *Concepts & Calculations, Group Learning Problems*. They are designed for use by students working alone or in small learning groups. The conceptual part of each problem provides a convenient focus for group discussions.

SUMMARY

The special theory of relativity is based on two postulates. The **relativity postulate** states that the laws of physics are the same in every inertial reference frame. The **speed of light postulate** says that the speed of light in a vacuum, measured in any inertial reference frame, always has the same value of c, no matter how fast the source of light and the observer are moving relative to each other.

The **proper time interval** Δt_0 between two events is the time interval measured by an observer who is at rest relative to the events and views them occurring at the same place. An observer who is in motion with respect to the events and who views them as occurring at different places measures a dilated time interval Δt. The dilated time interval is greater than the proper time interval, according to the **time-dilation equation**: $\Delta t = \Delta t_0 / \sqrt{1 - v^2/c^2}$. In this expression, v is the relative speed between the observer who measures Δt_0 and the observer who measures Δt.

The **proper length** L_0 between two points is the length measured by an observer who is at rest relative to the points.

An observer moving with a relative speed v parallel to the line between the two points does not measure the proper length. Instead, such an observer measures a contracted length L given by the **length-contraction formula**: $L = L_0 \sqrt{1 - v^2/c^2}$.

An object of mass m, moving with speed v, has a **relativistic momentum** whose magnitude is given by $p = mv / \sqrt{1 - v^2/c^2}$.

Energy and mass are equivalent. The total energy E of an object of mass m, moving at speed v, is $E = mc^2 / \sqrt{1 - v^2/c^2}$. The total energy of an object is the sum of its rest energy, $E_0 = mc^2$, and its kinetic energy KE: $E = E_0 + KE$. The kinetic energy is, therefore, $KE = E - E_0$. Objects with mass cannot attain the speed of light, which is the **ultimate speed** for such objects.

When an object is moving with respect to a reference frame that itself is moving relative to an observer, the **velocity-addition formula** (Equation 28.7) gives the velocity of the object as measured by the observer.

CONCEPTUAL QUESTIONS

1. The speed of light in water is c/n, where $n = 1.33$ is the index of refraction of water. Thus, the speed of light in water is less than c. Why doesn't this violate the speed of light postulate?

2. A baseball player at home plate hits a pop fly straight up (the beginning event) that is caught by the catcher at home plate (the ending event). Which of the following observers record the proper time interval between the two events: (a) a spectator sitting in the stands, (b) a spectator sitting on the couch and watching the game on TV, and (c) the third baseman running in to cover the play? Explain your answers.

3. The earth spins on its axis once each day. To a person viewing the earth from an inertial reference frame in space, which clock runs slower, a clock at the north pole or one at the equator? Why? (Ignore the orbital motion of the earth about the sun.)

4. Suppose you are standing at a railroad crossing, watching a train go by. (a) Both you and a passenger in the train are looking at a clock on the train. Which of you measures the proper time interval? (b) Who measures the proper length of the train car? (c) Who measures the proper distance between the railroad ties under the track? Justify your answers.

5. There are tables that list data for the various particles of matter that physicists have discovered. Often, such tables list the masses of the particles in units of energy, such as in MeV (million electron volts), rather than in kilograms. Why is this possible?

6. Why is it easier to accelerate an electron to a speed that is close to the speed of light, compared to accelerating a proton to the same speed?

7. Light travels in water at a speed of 2.26×10^8 m/s. It is possible, in principle, for a particle that has mass to travel through water at a faster speed than this. In what sense, then, does the speed of light represent the ultimate speed attainable by objects that have mass?

8. Review Conceptual Example 9 for background pertinent to this question. Do two positive electric charges separated by a finite distance have more mass than when they are infinitely far apart (assume that the charges remain stationary)? Provide a reason for your answer.

9. Review Conceptual Example 9 for background relevant to this question. One system consists of two stationary electrons, separated by a distance r. Another consists of a positron and an electron, both stationary and separated by the same distance r. A positron has the same mass as an electron, but it has a positive electric charge $+e$. Which system of charges, if either, has the greater mass? Provide a reason for your answer.

10. Review Conceptual Example 9 for background pertinent to this question. Suppose that a parallel plate capacitor is initially uncharged. The capacitor is charged up by removing electrons from one plate and placing them on the other plate. Does the uncharged or the charged capacitor, if either, have the greater mass? Why?

11. The speed limit on many interstate highways is 65 miles per hour. If the speed of light were 65 miles per hour, would you be able to drive at the speed limit? Give your reasoning.

12. Review Conceptual Example 11 as an aid in answering this question. A person is approaching you in a truck that is traveling very close to the speed of light. This person throws a baseball toward you. Relative to the truck, the ball is thrown with a speed nearly equal to the speed of light, so the person on the truck sees the baseball move away from the truck at a very high speed. Yet you see the baseball move away from the truck very slowly. Why? Use the velocity-addition formula to guide your thinking.

13. Which of the following quantities will two observers always measure to be the *same*, regardless of the relative velocity between the observers: (a) the time interval between two events; (b) the length of an object; (c) the speed of light in a vacuum; (d) the relative speed between the observers? In each case, give a reason for your answer.

14. If the speed of light were infinitely large instead of 3.0×10^8 m/s, would the effects of time dilation and length contraction be observable? Explain, using the equations presented in the text to support your reasoning.

PROBLEMS

Before doing any calculations involving time dilation or length contraction, it is useful to identify which observer measures the proper time interval Δt_0 *or the proper length* L_0.

ssm Solution is in the Student Solutions Manual.   **www** Solution is available on the World Wide Web at http://www.wiley.com/college/cutnell
⚕ This icon represents a biomedical application.

Section 28.3 The Relativity of Time: Time Dilation

1. ssm A law enforcement officer in an intergalactic "police car" turns on a red flashing light and sees it generate a flash every 1.5 s. A person on earth measures that the time between flashes is 2.5 s. How fast is the "police car" moving relative to the earth?

2. Suppose you are planning a trip in which a spacecraft is to travel at a constant velocity for exactly six months, as measured by a clock on board the spacecraft, and then return home at the same speed. Upon return, the people on earth will have advanced exactly one hundred years into the future. According to special relativity, how fast must you travel? Express your answer to 5 significant figures as a multiple of c, e.g., $0.955\,85c$.

3. ssm A radar antenna is rotating at an angular speed of 0.25 rad/s, as measured on earth. To an observer moving past the antenna at a speed of $0.80c$, what is its angular speed?

4. Suppose that you are traveling on board a spacecraft that is moving with respect to the earth at a speed of $0.975c$. You are breathing at a rate of 8.0 breaths per minute. As monitored on earth, what is your breathing rate?

*5. A 5.00-kg object oscillates back and forth at the end of a spring whose spring constant is 49.3 N/m. An observer is traveling at a speed of 2.80×10^8 m/s relative to the fixed end of the spring. What does this observer measure for the period of oscillation?

*6. An astronaut travels at a speed of 7800 m/s relative to the earth, a speed that is very small compared to c. According to a clock on the earth, the trip lasts 15 days. Determine the *difference* (in seconds) between the time recorded by the earth clock and the astronaut's clock. [*Hint: When* $v \ll c$, *the following approximation is valid:* $\sqrt{1 - v^2/c^2} \approx 1 - \frac{1}{2}(v^2/c^2)$.]

**7. As observed on earth, a certain type of bacteria is known to double in number every 24.0 hours. Two cultures of these bacteria are prepared, each consisting initially of one bacterium. One culture is left on earth and the other placed on a rocket that travels at a speed of $0.866c$ relative to the earth. At a time when the earthbound culture has grown to 256 bacteria, how many bacteria are in the culture on the rocket, according to an earth-based observer?

Section 28.4 The Relativity of Length: Length Contraction

8. Suppose the straight-line distance between New York and San Francisco is 4.2×10^6 m (neglecting the curvature of the earth). A UFO is flying between these two cities at a speed of $0.70c$ relative to the earth. What do the voyagers aboard the UFO measure for this distance?

9. ssm A UFO streaks across the sky at a speed of $0.90c$ relative to the earth. A person on earth determines the length of the UFO to be 230 m along the direction of its motion. What length does the person measure for the UFO when it lands?

10. The distance from earth to the center of our galaxy is about 23 000 ly (1 ly = 1 light year = 9.47×10^{15} m), as measured by an earth-based observer. A spaceship is to make this journey at a speed of $0.9990c$. According to a clock on board the spaceship, how long will it take to make the trip? Express your answer in years (1 yr = 3.16×10^{7} s).

11. ssm www Two spaceships A and B are exploring a new planet. Relative to this planet, spaceship A has a speed of $0.60c$, while spaceship B has a speed of $0.80c$. What is the ratio D_A/D_B of the values for the planet's diameter that each spaceship measures in a direction that is parallel to its motion?

12. Suppose you are traveling in space and pass a rectangular landing pad on a planet. Your spacecraft has a speed of $0.85c$ relative to the planet and moves in a direction parallel to the length of the pad. While moving, you measure the length to be 1800 m and the width to be 1500 m. What are the dimensions of the landing pad according to the engineer who built it?

13. A space traveler moving at a speed of $0.60c$ with respect to the earth makes a trip to a distant star that is stationary relative to the earth. He measures the length of this trip to be 8.0 light-years. What would be the length of this same trip (in light-years) as measured by a traveler moving at a speed of $0.80c$ with respect to the earth?

*14. As the drawing shows, a carpenter on a space station has constructed a 30.0° ramp. A rocket moves past the space station with a relative speed of $0.730c$ in a direction parallel to side x. What does a person aboard the rocket measure for the angle of the ramp?

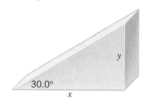

**15. ssm www A rectangle has the dimensions of 3.0 m × 2.0 m when viewed by someone at rest with respect to it. When you move past the rectangle along one of its sides, the rectangle looks like a square. What dimensions do you observe when you move at the same speed along the adjacent side of the rectangle?

**16. Twins who are 19.0 years of age leave the earth and travel to a distant planet 12.0 light-years away. Assume that the planet and earth are at rest with respect to each other. The twins depart at the same time on different spaceships. One twin travels at a speed of $0.900c$, while the other twin travels at $0.500c$. (a) According to the theory of special relativity, what is the difference between their ages when they meet again at the earliest possible time? (b) Which twin is older?

Section 28.5 Relativistic Momentum

17. At what speed is the magnitude of the relativistic momentum of a particle three times the magnitude of the nonrelativistic momentum?

18. A car, whose mass is 1550 kg, is traveling at 15.0 m/s. If the speed of light were 25.0 m/s, what would be the magnitude of the momentum of the car as measured by a person standing on the ground?

19. ssm A rocket of mass 1.40×10^5 kg has a relativistic momentum the magnitude of which is 3.15×10^{13} kg·m/s. How fast is the rocket traveling?

20. A woman is 1.6 m tall and has a mass of 55 kg. She moves past an observer with the direction of the motion parallel to her height. The observer measures her relativistic momentum to have a magnitude of 2.0×10^{10} kg·m/s. What does the observer measure for her height?

*21. Starting from rest, two skaters "push off" against each other on smooth level ice, where friction is negligible. One is a woman and one is a man. The woman moves away with a velocity of +2.5 m/s relative to the ice. The mass of the woman is 54 kg, and the mass of the man is 88 kg. Assuming that the speed of light is 3.0 m/s, so that the relativistic momentum must be used, find the recoil velocity of the man relative to the ice. (Hint: This problem is similar to Example 6 in Chapter 7.)

Section 28.6 The Equivalence of Mass and Energy

22. Radium is a radioactive element whose nucleus emits an α particle (a helium nucleus) that has a kinetic energy of about 7.8×10^{-13} J (4.9 MeV). To what amount of mass is this energy equivalent?

23. ssm How much work must be done on an electron to accelerate it from rest to a speed of $0.990c$?

24. Suppose one gallon of gasoline produces 1.1×10^8 J of energy, and this energy is sufficient to operate a car for twenty miles. An aspirin tablet has a mass of 325 mg. If the aspirin could be converted completely into thermal energy, how many miles could the car go on a single tablet?

25. A nuclear power reactor generates 3.0×10^9 W of power. In one year, what is the change in the mass of the nuclear fuel due to the energy being taken from the reactor?

26. Four kilograms of water are heated from 20.0 °C to 60.0 °C. (a) How much heat is required to produce this change in temperature? [The specific heat capacity of water is 4186 J/(kg·C°).] (b) By how much does the mass of the water increase?

27. ssm Determine the ratio of the relativistic kinetic energy to the nonrelativistic kinetic energy ($\frac{1}{2}mv^2$) when a particle has a speed of (a) $1.00 \times 10^{-3}c$ and (b) $0.970c$.

*28. An electron is accelerated from rest through a potential difference of 2.40×10^7 V. (a) What is the relativistic kinetic energy (in joules) of the electron? (b) What is the speed of the electron?

*29. Using Equations 28.3 and 28.4, prove that the total energy of an object is related to its relativistic momentum p according to $E^2 = p^2 c^2 + (mc^2)^2$.

Section 28.7 The Relativistic Addition of Velocities

30. Galaxy A is moving away from us with a speed of $0.75c$ relative to the earth. Galaxy B is moving away from us in the opposite direction with a relative speed of $0.55c$. Assume that the earth and the galaxies are moving at constant velocities, so they are inertial reference frames. How fast is galaxy A moving according to an observer in galaxy B?

31. ssm A spacecraft approaching the earth launches an exploration vehicle. After the launch, an observer on earth sees the spacecraft approaching at a speed of $0.50c$ and the exploration vehicle approaching at a speed of $0.70c$. What is the speed of the exploration vehicle relative to the spaceship?

32. It has been proposed that spaceships of the future will be powered by ion propulsion engines in which ions are ejected from the back of the ship to drive it forward. In one such engine the ions are to be ejected with a speed of $0.80c$ relative to the engine. If the ship were traveling away from the earth with a velocity of $0.70c$, what would be the velocity of the ions relative to the earth? (Be sure to assign the correct plus or minus signs to the velocities, assuming that the direction away from earth is positive.)

*__33. ssm www__ The crew of a rocket that is moving away from the earth launches an escape pod, which they measure to be 45 m long. The pod is launched toward the earth with a speed of $0.55c$

relative to the rocket. After the launch, the rocket's speed relative to the earth is $0.75c$. What is the length of the escape pod as determined by an observer on earth?

*__34.__ A person on earth notices a rocket approaching from the right at a speed of $0.75c$ and another rocket approaching from the left at $0.65c$. What is the relative speed between the two rockets, as measured by a passenger on one of them?

**__35.__ Two atomic particles approach each other in a head-on collision. Each particle has a mass of 2.16×10^{-25} kg. The speed of each particle is 2.10×10^8 m/s when measured by an observer standing in the laboratory. (a) What is the speed of one particle as seen by the other particle? (b) Determine the relativistic momentum of one particle, as would be observed by the other.

ADDITIONAL PROBLEMS

36. An electron and a positron each have a mass of 9.11×10^{-31} kg. They collide and both vanish, with only electromagnetic radiation appearing after the collision. If each particle is moving at a speed of $0.20c$ relative to the laboratory before the collision, determine the energy of the electromagnetic radiation.

37. ssm www How fast must a meter stick be moving if its length is observed to shrink to one-half a meter?

38. A Klingon spacecraft has a speed of $0.75c$ with respect to the earth. The Klingons measure 37.0 h for the time interval between two events on the earth. What value for the time interval would they measure if their ship had a speed of $0.94c$ with respect to the earth?

39. A spaceship is approaching the earth at a relative speed of $0.85c$. The mass of the ship is 2.0×10^7 kg. Find the magnitude of (a) the classical momentum and (b) the relativistic momentum of the ship.

*__40.__ Refer to Conceptual Example 11 as an aid in solving this problem. An intergalactic cruiser has two types of guns: a photon cannon that fires a beam of laser light, and an ion gun that shoots ions at a velocity of $0.950c$ relative to the cruiser. The cruiser

closes in on an alien spacecraft at a velocity of $0.800c$ relative to this spacecraft. The captain fires both types of guns. At what velocity do the aliens see (a) the laser light and (b) the ions approach them? At what velocity do the aliens see (c) the laser light and (d) the ions move away from the cruiser?

*__41. ssm__ An unstable particle is at rest and suddenly breaks up into two fragments. No external forces act on the particle or its fragments. One of the fragments has a velocity of $+0.800c$ and a mass of 1.67×10^{-27} kg, while the other has a mass of 5.01×10^{-27} kg. What is the velocity of the more massive fragment? *(Hint: This problem is similar to Example 6 in Chapter 7.)*

*__42.__ How close would two stationary electrons have to be positioned so that their total mass is twice what it is when the electrons are very far apart?

**__43.__ A woman and a man are on separate rockets, which are flying parallel to each other and have a relative speed of $0.940c$. The woman measures the same value for the length of her own rocket and for the length of the man's rocket. What is the ratio of the value that the man measures for the length of his own rocket to the value he measures for the length of the woman's rocket?

CONCEPTS & CALCULATIONS
GROUP LEARNING PROBLEMS

Note: Each of these problems consists of Concept Questions followed by a related quantitative Problem. They are designed for use by students working alone or in small learning groups. The Concept Questions involve little or no mathematics and are intended to stimulate group discussions. They focus on the concepts with which the problems deal. Recognizing the concepts is the essential initial step in any problem-solving technique.

44. __Concept Questions__ A Martian leaves Mars in a spaceship that is heading to Venus. On the way, the spaceship passes earth with a speed v relative to it. Assume that the three planets do not move

relative to each other during the trip. (a) Who, if anyone, measures the proper time interval Δt_0 for the trip—the Martian or a person on earth? (b) Who, if anyone, measures the proper length between Mars and Venus? (c) As measured by the Martian, how is the time of the trip related to the length of the trip and the speed v of the spaceship?

__Problem__ The spaceship has a speed of $0.80c$ relative to earth, and the distance between Mars and Venus is 1.20×10^{11} m, as measured by a person on earth. (a) What does the Martian measure for the distance between Mars and Venus? (b) What is the time of the trip (in seconds) as measured by the Martian?

45. Concept Question Examine Equation 28.3 for the magnitude p of the relativistic momentum. Decide whether doubling the speed v causes p to double or more than double. Decide whether halving the mass m causes p to decrease by a factor of two or more than a factor of two. With this in mind, consider the three particles in the table and rank them according to the magnitude of their relativistic momenta, largest first. Provide a reason for your ranking.

Particle	Mass	Speed
a	m	v
b	$\frac{1}{2}m$	$2v$
c	$\frac{1}{4}m$	$4v$

Problem The mass and speed in the table are $m = 1.20 \times 10^{-8}$ kg and $v = 0.200c$. Determine the momentum for each particle according to special relativity. Check to see that your answers are consistent with your answer to the Concept Question.

46. Concept Questions (a) Suppose that a certain mass of ice at 0 °C is melted into liquid water at 0 °C. According to special relativity, which has the greater mass, the ice or the water? (b) The same mass of liquid water at 100 °C is boiled into steam at 100 °C. According to special relativity, which has the greater mass, the water or the steam? (c) Is the change in mass in question (b) greater than, less than, or the same as the change in mass in question (a)? Explain your answers.

Problem Two kilograms of water are changed (a) from ice at 0 °C into liquid water at 0 °C and (b) from liquid water at 100 °C into steam at 100 °C. For each situation, determine the change in mass of the water. Verify that your answers are consistent with your answers to the Concept Questions.

47. Concept Questions Suppose you're driving down the highway at a speed of 25 m/s. A truck in the next lane is traveling toward you with a speed of 35 m/s relative to the ground. Relative to you, the truck is approaching at a speed of 25 m/s + 35 m/s = 60 m/s. (a) Since the speeds are much less than the speed of light, would the relative speed be the same if you were going 5.0 m/s relative to the ground and the truck were moving at 55 m/s relative to the ground? (b) Suppose the speed of light were $c = 65$ m/s. Would the relative speeds be the same for the two situations? Give a reason for each of your answers.

Problem Using the data presented in the Concept Questions and assuming that the speed of light is $c = 65$ m/s, determine the speed of the truck relative to you for the two situations when your speed and the truck's speed are, respectively, (a) 25 m/s and 35 m/s, and (b) 5.0 m/s and 55 m/s. Check to see that your answers are consistent with your answer to Concept Question (b).

Newly born star
sitting on surface
of EGG

This is the central portion of the Eagle Nebula located about 7000 light-years from earth, as seen by the Hubble Space Telescope. Among the topics in this chapter are particlelike entities called photons. As we will see, they play the central role in the process of photoevaporation, which allows astronomers to peer into the dense star-forming regions of the nebula.

CHAPTER 29
PARTICLES AND WAVES

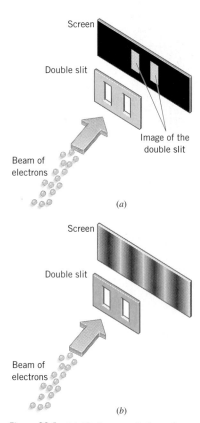

Figure 29.1 (*a*) If electrons behaved as discrete particles with no wave properties, they would pass through one or the other of the two slits and strike the screen, causing it to glow and produce exact images of the slits. (*b*) In reality, the screen reveals a pattern of bright and dark fringes, similar to the pattern produced when a beam of light is used and interference occurs between the light waves coming from each slit.

29.1 THE WAVE–PARTICLE DUALITY

The ability to exhibit interference effects is an essential characteristic of waves. For instance, Section 27.2 discusses Young's famous experiment in which light passes through two closely spaced slits and produces a pattern of bright and dark fringes on a screen (see Figure 27.4). The fringe pattern is a direct indication that interference is occurring between the light waves coming from each slit.

One of the most incredible discoveries of twentieth-century physics is that particles can also behave like waves and exhibit interference effects. For instance, Figure 29.1 shows a version of Young's experiment performed by directing *a beam of electrons* onto a double slit. In this experiment, the screen is like a television screen and glows wherever an electron strikes it. Part *a* of the picture indicates the pattern that would be seen on the screen if each electron, behaving strictly as a particle, were to pass through one slit or the other and strike the screen. The pattern would consist of an image of each slit. Part *b* shows the pattern actually observed, which consists of bright and dark fringes, reminiscent of that obtained when light waves pass through a double slit. The fringe pattern indicates that the electrons are exhibiting interference effects that are associated with waves.

But how can electrons behave like waves in the experiment shown in Figure 29.1*b*? And what kind of waves are they? The answers to these profound questions will be discussed later in this chapter. For the moment, we intend only to emphasize that the picture of an electron as a tiny discrete particle of matter does not account for the fact that the electron can behave as a wave in some circumstances. In other words, the electron exhibits a dual nature, with both particle-like characteristics and wave-like characteristics.

Here is another interesting question: If a particle can exhibit wave-like properties, can waves exhibit particle-like behavior? As the next three sections reveal, the answer is yes. In fact, experiments that demonstrated the particle-like behavior of waves were performed near the beginning of the twentieth century, before the experiments that demonstrated the wave-like properties of the electrons. Scientists now accept the ***wave–particle duality*** as an essential part of nature:

Waves can exhibit particlelike characteristics, and particles can exhibit wavelike characteristics.

Section 29.2 begins the remarkable story of the wave–particle duality by discussing the electromagnetic waves that are radiated by a perfect blackbody. It is appropriate to begin with blackbody radiation, because it provided the first link in the chain of experimental evidence leading to our present understanding of the wave–particle duality.

29.2 BLACKBODY RADIATION AND PLANCK'S CONSTANT

All bodies, no matter how hot or cold, continuously radiate electromagnetic waves. For instance, we see the glow of very hot objects because they emit electromagnetic waves in the visible region of the spectrum. Our sun, which has a surface temperature of about 6000 K, appears yellow, while the cooler star Betelgeuse has a red-orange appearance due to its lower surface temperature of 2900 K. However, at relatively low temperatures, cooler objects emit visible light waves only weakly and, as a result, do not appear to be glowing. Certainly the human body, at only 310 K, does not emit enough visible light to be seen in the dark with the unaided eye. But the body does emit electromagnetic waves in the infrared region of the spectrum, and these can be detected with infrared sensitive devices.

At a given temperature, the intensities of the electromagnetic waves emitted by an object vary from wavelength to wavelength throughout the visible, infrared, and other regions of the spectrum. Figure 29.2 illustrates how the intensity per unit wavelength depends on wavelength for a perfect blackbody emitter. As Section 13.3 discusses, a perfect blackbody at a constant temperature absorbs and reemits all the electromagnetic radiation that falls on it. The two curves in the drawing show that at a higher temperature the maximum emitted intensity per unit wavelength increases and shifts to shorter wavelengths, toward the visible region of the spectrum. In accounting for the shape of these curves, the German physicist Max Planck (1858–1947) took the first step toward our present understanding of the wave–particle duality.

In 1900 Planck calculated the blackbody radiation curves, using a model that represents a blackbody as a large number of atomic oscillators, each of which emits and absorbs electromagnetic waves. To obtain agreement between the theoretical and experimental curves, Planck assumed that the energy E of an atomic oscillator could have only the discrete values of $E = 0$, hf, $2hf$, $3hf$, and so on. In other words, he assumed that

$$E = nhf \qquad n = 0, \ 1, \ 2, \ 3, \ \ldots \qquad (29.1)$$

where n is a positive integer, f is the frequency of vibration (in hertz), and h is a constant now called **Planck's constant.*** Experiment has shown that Planck's constant has a value of

$$h = 6.626\,068\,91 \times 10^{-34} \text{ J} \cdot \text{s}$$

The radical feature of Planck's assumption was that the energy of an atomic oscillator could have only discrete values (hf, $2hf$, $3hf$, etc.), with energies in between these values being forbidden. Whenever the energy of a system can have only certain definite values, and nothing in between, the energy is said to be *quantized*. This quantization of the energy was unexpected on the basis of the traditional physics of the time. However, it was soon realized that energy quantization had wide-ranging implications.

Conservation of energy requires that the energy carried off by the radiated electromagnetic waves must equal the energy lost by the atomic oscillators in Planck's model. Suppose, for example, that an oscillator with an energy of $3hf$ emits an electromagnetic wave. According to Equation 29.1, the next smallest allowed value for the energy of the oscillator is $2hf$. In such a case, the energy carried off by the electromagnetic wave would have the value of hf, equaling the amount of energy lost by the oscillator. Thus, Planck's model for blackbody radiation sets the stage for the idea that electromagnetic energy occurs as a collection of discrete amounts or packets of energy, the energy of a packet being equal to hf. As the next section discusses, it was Einstein who made the specific proposal that light consists of such energy packets.

Figure 29.2 The electromagnetic radiation emitted by a perfect blackbody has an intensity per unit wavelength that varies from wavelength to wavelength, as each curve indicates. At the higher temperature, the intensity per unit wavelength is greater, and the maximum occurs at a shorter wavelength.

29.3 PHOTONS AND THE PHOTOELECTRIC EFFECT

● *Concepts at a Glance*   The total energy E and the linear momentum **p** are fundamental concepts in physics. We have seen in Chapters 6 and 7 how they apply to moving particles, such as electrons and protons. The total energy of a (nonrela-

---

* It is now known that the energy of a harmonic oscillator is $E = (n + \frac{1}{2})hf$, the extra term of $\frac{1}{2}$ being unimportant to the present discussion.

Concepts at a Glance

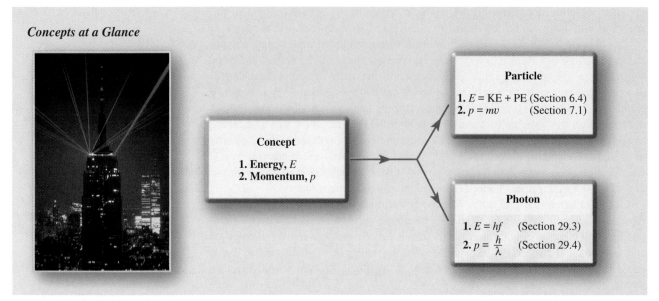

Figure 29.3 *Concepts at a Glance*
A moving particle has energy E and momentum p. An electromagnetic wave is composed of particlelike entities called photons, each of which also has energy and momentum. Although the beams of laser light in this photograph look continuous, each is composed of discrete photons.

Figure 29.4 In the photoelectric effect, light with a sufficiently high frequency ejects electrons from a metal surface. These photoelectrons, as they are called, are drawn to the positive collector, thus producing a current.

tivistic) particle is the sum of its kinetic (KE) and potential (PE) energies, or $E = $ KE + PE. The magnitude of the particle's momentum is the product of its mass m and speed v, or $p = mv$. These particle concepts are listed in the upper-right portion of the Concepts-at-a-Glance chart in Figure 29.3. We will now discuss the fact that electromagnetic waves are composed of particlelike entities called *photons,* and the lower-right portion of the chart shows that the ideas of energy and momentum also apply to them. However, as we will see, the equations defining photon energy ($E = hf$) and momentum ($p = h/\lambda$) are different from those for a particle, as the chart indicates.

Experimental evidence that light consists of photons comes from a phenomenon called the *photoelectric effect,* in which electrons are emitted from a metal surface when light shines on it. Figure 29.4 illustrates the effect. The electrons are emitted if the light being used has a sufficiently high frequency. The ejected electrons move toward a positive electrode called the collector and cause a current to register on the ammeter. Because the electrons are ejected with the aid of light, they are called *photoelectrons.* As will be discussed shortly, a number of features of the photoelectric effect could not be explained solely with the ideas of classical physics.

In 1905 Einstein presented an explanation of the photoelectric effect that took advantage of Planck's work concerning blackbody radiation. It was primarily for his theory of the photoelectric effect that he was awarded the Nobel prize in physics in 1921. In his photoelectric theory, Einstein proposed that light of frequency f could be regarded as a collection of discrete packets of energy (photons), each packet containing an amount of energy E given by

Energy of a photon
$$E = hf \tag{29.2}$$

where h is Planck's constant. The light energy given off by a light bulb, for instance, is carried by photons. The brighter the light shining on a given area, the greater is the number of photons per second that strike the area. Example 1 estimates the number of photons emitted per second by a typical light bulb.

EXAMPLE 1   Photons from a Light Bulb

In converting electrical energy into light energy, a sixty-watt incandescent light bulb operates at about 2.1% efficiency. Assuming that all the light is green light (vacuum wavelength = 555 nm), determine the number of photons per second given off by the bulb.

Reasoning The number of photons emitted per second can be found by dividing the amount of light energy emitted per second by the energy E of one photon. The energy of a single photon is $E = hf$, according to Equation 29.2. The frequency f of the photon is related to its wavelength λ by Equation 16.1 as $f = c/\lambda$.

Solution At an efficiency of 2.1%, the light energy emitted per second by a sixty-watt bulb is $(0.021)(60.0 \text{ J/s}) = 1.3$ J/s. The energy of a single photon is

$$E = hf = \frac{hc}{\lambda} = \frac{(6.63 \times 10^{-34} \text{ J} \cdot \text{s})(3.00 \times 10^8 \text{ m/s})}{555 \times 10^{-9} \text{ m}}$$
$$= 3.58 \times 10^{-19} \text{ J}$$

Therefore,

$$\begin{array}{l}\text{Number of}\\ \text{photons emitted}\\ \text{per second}\end{array} = \frac{1.3 \text{ J/s}}{3.58 \times 10^{-19} \text{ J/photon}}$$
$$= \boxed{3.6 \times 10^{18} \text{ photons/s}}$$

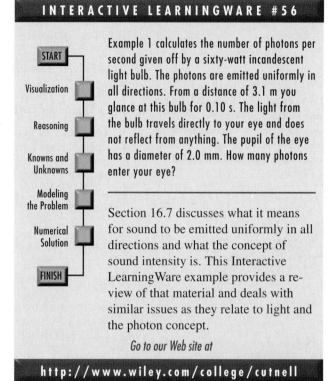

INTERACTIVE LEARNINGWARE #56

START
Visualization
Reasoning
Knowns and Unknowns
Modeling the Problem
Numerical Solution
FINISH

Example 1 calculates the number of photons per second given off by a sixty-watt incandescent light bulb. The photons are emitted uniformly in all directions. From a distance of 3.1 m you glance at this bulb for 0.10 s. The light from the bulb travels directly to your eye and does not reflect from anything. The pupil of the eye has a diameter of 2.0 mm. How many photons enter your eye?

Section 16.7 discusses what it means for sound to be emitted uniformly in all directions and what the concept of sound intensity is. This Interactive LearningWare example provides a review of that material and deals with similar issues as they relate to light and the photon concept.

Go to our Web site at

http://www.wiley.com/college/cutnell

According to Einstein, when light shines on a metal, a photon can give up its energy to an electron in the metal. If the photon has enough energy to do the work of removing the electron from the metal, the electron can be ejected. The work required depends on how strongly the electron is held. For the *least strongly* held electrons, the necessary work has a minimum value W_0 and is called the *work function* of the metal. If a photon has energy in excess of the work needed to remove an electron, the excess appears as kinetic energy of the ejected electron. Thus, the least strongly held electrons are ejected with the maximum kinetic energy KE_{max}. Einstein applied the conservation of energy principle and proposed the following relation to describe the photoelectric effect:

$$\underbrace{hf}_{\substack{\text{Photon}\\\text{energy}}} = \underbrace{\text{KE}_{\text{max}}}_{\substack{\text{Maximum}\\\text{kinetic energy}\\\text{of ejected}\\\text{electron}}} + \underbrace{W_0}_{\substack{\text{Minimum}\\\text{work needed to}\\\text{eject electron}}} \qquad (29.3)$$

According to this equation, $\text{KE}_{\text{max}} = hf - W_0$, which is plotted in Figure 29.5, with KE_{max} along the ordinate and f along the abscissa. The graph is a straight line that crosses the abscissa at $f = f_0$. At this frequency, the electron departs from the metal with no kinetic energy ($\text{KE}_{\text{max}} = 0$ J). According to Equation 29.3, when $\text{KE}_{\text{max}} = 0$ J the energy hf_0 of the incident photon is equal to the work function W_0 of the metal: $hf_0 = W_0$.

The photon picture provides an explanation for a number of features of the photoelectric experiment that are difficult to explain without photons. It is observed, for instance, that only light with a frequency above a certain minimum value f_0 will

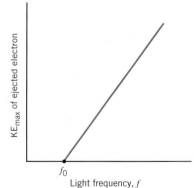

Figure 29.5 Photons can eject electrons from a metal when the light frequency is above a minimum value f_0. For frequencies above this value, ejected electrons have a maximum kinetic energy KE_{max} that is linearly related to the frequency, as the graph shows.

eject electrons. If the frequency of the light is below this value, no electrons are ejected, regardless of how intense the light is. The next example determines the minimum frequency value for a silver surface.

EXAMPLE 2   The Photoelectric Effect for a Silver Surface

The work function for a silver surface is $W_0 = 4.73$ eV. Find the minimum frequency that light must have to eject electrons from this surface.

Reasoning The minimum frequency f_0 is that frequency at which the photon energy equals the work function W_0 of the metal, so the electron is ejected with zero kinetic energy. Since 1 eV $= 1.60 \times 10^{-19}$ J, the work function expressed in joules is $W_0 = (4.73 \text{ eV})(1.60 \times 10^{-19} \text{ J/1 eV}) = 7.57 \times 10^{-19}$ J. Using Equation 29.3, we find

$$hf_0 = \underbrace{\text{KE}_{max}}_{= 0 \text{ J}} + W_0 \quad \text{or} \quad f_0 = \frac{W_0}{h}$$

Solution The minimum frequency f_0 is

$$f_0 = \frac{W_0}{h} = \frac{7.57 \times 10^{-19} \text{ J}}{6.63 \times 10^{-34} \text{ J} \cdot \text{s}} = \boxed{1.14 \times 10^{15} \text{ Hz}}$$

Photons with frequencies less than f_0 do not have enough energy to eject electrons from a silver surface. Since $\lambda_0 = c/f_0$, the wavelength of this light is $\lambda_0 = 263$ nm, which is in the ultraviolet region of the electromagnetic spectrum.

Another significant feature of the photoelectric effect is that the maximum kinetic energy of the ejected electrons remains the same when the intensity of the light increases, provided the light frequency remains the same. As the light intensity increases, more photons per second strike the metal, and consequently more electrons per second are ejected. However, since the frequency is the same for each photon, the energy of each photon is also the same. Thus, the ejected electrons always have the same maximum kinetic energy.

Whereas the photon model of light explains the photoelectric effect satisfactorily, the electromagnetic wave picture of light does not. Certainly, it is possible to imagine that the electric field of an electromagnetic wave would cause electrons in the metal to oscillate and tear free from the surface when the amplitude of oscillation becomes large enough. However, were this the case, higher intensity light would eject electrons with a greater maximum kinetic energy, a fact that experiment does not confirm. Moreover, in the electromagnetic wave picture, a relatively long time would be required with low-intensity light before the electrons would build up a sufficiently large oscillation amplitude to tear free. Instead, experiment shows that even the weakest light intensity causes electrons to be ejected almost instantaneously, provided the frequency of the light is above the minimum value f_0. The failure of the electromagnetic wave picture to explain the photoelectric effect does not mean that the wave model should be abandoned. But we must recognize that the wave picture does not account for all the characteristics of light. The photon model also makes an important contribution to our understanding of the way light behaves when it interacts with matter.

Because a photon has energy, the photon can eject an electron from a metal surface when it interacts with the electron. However, a photon is different from a normal particle. A normal particle has a mass and can travel at speeds up to, but not equal to, the speed of light. A photon, on the other hand, travels at the speed of light in a vacuum and does not exist as an object at rest. The energy of a photon is en-

● **PROBLEM SOLVING INSIGHT**
The work function of a metal is the minimum energy needed to eject an electron from the metal. An electron that has received this minimum energy has no kinetic energy once outside the metal.

(a)

(b)

Figure 29.6 (a) Digital cameras like this one use an array of charge-coupled devices instead of film to capture a picture. (b) If a sufficient number of charge-coupled devices are used, this type of camera produces pictures as good as those obtained with many film cameras. This photograph was taken with a digital camera.

tirely kinetic in nature, for it has no rest energy and no mass. To show that a photon has no mass, we rewrite Equation 28.4 for the total energy E as

$$E \sqrt{1 - \frac{v^2}{c^2}} = mc^2$$

The term $\sqrt{1 - (v^2/c^2)}$ is zero because a photon travels at the speed of light, $v = c$. Since the energy E of the photon is finite, the left side of the equation above is zero. Thus, the right side must also be zero, so $m = 0$ kg and the photon has no mass.

One of the most exciting and useful applications of the photoelectric effect is the charge-coupled device (CCD). An array of these devices is used instead of film in digital cameras (see Figure 29.6) to capture pictures in the form of many small groups of electrons. CCD arrays are also used in digital camcorders and electronic scanners, and they provide the method of choice with which astronomers capture those spectacular images of the planets and the stars. For use with visible light, a CCD array consists of a sandwich of semiconducting silicon, insulating silicon dioxide, and a number of electrodes, as Figure 29.7 shows. The array is divided into many small sections or pixels, sixteen of which are shown in the drawing. Each pixel captures a small part of a picture. (Digital cameras that use 1.3 million pixels are now available for less than $500.) The blow-up in Figure 29.7 shows a single pixel. Incident photons of visible light strike the silicon and generate electrons via the photoelectric effect. The range of energies of the visible photons is such that one electron is released when a photon interacts with a silicon atom. The electrons are trapped within a pixel because of a positive voltage applied to the electrodes beneath the insulating layer. Thus, the number of electrons that are released and trapped is proportional to the number of photons striking the pixel. In this fashion, each pixel in the CCD array accumulates an accurate representation of the light intensity at that point on the picture. Color information is provided using red, green, or blue filters or a system of prisms to separate the colors. Astronomers use CCD arrays not only in the visible region of the electromagnetic spectrum but in other regions as well.

In addition to trapping the photoelectrons, the electrodes beneath the pixels are used to read out the electron representation of the picture. By changing the positive voltages applied to the electrodes, it is possible to cause all of the electrons trapped in one row of pixels to be transferred to the adjacent row. In this fashion, for instance, row 1 in Figure 29.7 is transferred into row 2, row 2 into row 3, and row 3 into the bottom row, which serves a special purpose. The bottom row functions as a horizontal shift register, from which the contents of each pixel can be shifted to the right, one at a time, and read into an analog signal processor. This processor senses the varying number of electrons in each pixel in the shift register as a kind of wave that has a fluctuating amplitude. After another shift in rows, the information in the next row is read out, and so forth, and so on. The output of the analog signal processor is sent to an analog-to-digital converter, which produces a digital representation of the picture in terms of the zeros and ones that a computer can recognize.

Another application of the photoelectric effect depends on the fact that the moving photoelectrons in Figure 29.4 constitute a current, a current that changes as the intensity of the light changes. All automatic garage door openers have a safety feature that prevents the door from closing when it encounters an obstruction (person, vehicle, etc.). As Figure 29.8 illustrates, a sending unit transmits an invisible (infrared) beam across the opening of the door. The beam is detected by a receiving unit that contains a photodiode. A photodiode is a type of *pn* junction diode (see Section 23.5). When infrared photons strike the photodiode, electrons bound to the atoms absorb the pho-

THE PHYSICS OF...

charge-coupled devices and digital cameras.

Figure 29.7 A CCD array can be used to capture photographic images using the photoelectric effect.

THE PHYSICS OF...

a safety feature of garage door openers.

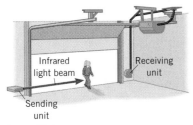

Figure 29.8 When an obstruction prevents the infrared light beam from reaching the receiving unit, the current in the receiving unit drops. This drop in current is detected by an electronic circuit that stops the downward movement of the door and then causes it to open.

THE PHYSICS OF...

photoevaporation and star formation.

tons and become liberated. These liberated, mobile electrons cause the current in the photodiode to increase. When a person walks through the beam, the light is momentarily blocked from reaching the receiving unit, and the current in the photodiode decreases. The change in current is sensed by electronic circuitry that immediately stops the downward motion of the door and then causes it to raise up.

Look back at the spectacular photograph at the beginning of this chapter. It shows the central portion of the Eagle Nebula, a giant star-forming region some 7000 light-years from earth. The photo was taken by the Hubble Space Telescope and reveals towering clouds of molecular gas and dust, in which there is dramatic evidence of the energy carried by photons. These clouds extend more than a light-year from base to tip and are the birthplace of stars. A star begins to form within a cloud when the gravitational force pulls together sufficient gas to create a high-density "ball." When the gaseous ball becomes sufficiently dense, thermonuclear fusion (see Section 32.5) occurs at its core, and the star begins to shine. The newly born stars are buried within the cloud and cannot be seen from earth. However, the process of photoevaporation allows astronomers to see many of the high-density regions where stars are being formed. Photoevaporation is the process in which high-energy, ultraviolet (UV) photons from hot stars outside the cloud heat it up, much like microwave photons heat food in a microwave oven. Figure 29.9 includes a reproduction of the upper left portion of the chapter-opening photo and shows streamers of gas photoevaporating from the cloud as it is illuminated by stars located beyond the picture's upper edge. As photoevaporation proceeds, globules of gas that are denser than their surroundings are exposed. The globules are known as evaporating gaseous globules (EGGs), and they are slightly larger than our solar system. The drawing in part *a* of Figure 29.9 shows that the EGGs shade the gas and dust behind them from the UV photons, creating the many finger-like protrusions seen on the surface of the cloud. Astronomers believe that some of these EGGs contain young stars within them. In some cases, so much gas has boiled off that a newborn star can be seen on the surface of an EGG (see the circled feature in the chapter-opening photograph).

29.4 THE MOMENTUM OF A PHOTON AND THE COMPTON EFFECT

Although Einstein presented his photon model for the photoelectric effect in 1905, it was not until 1923 that the photon picture began to achieve widespread acceptance. It was then that the American physicist Arthur H. Compton (1892–1962) used the photon model to explain his research on the scattering of X-rays by the electrons in graphite. X-rays are high-frequency electromagnetic waves and, like light, they are composed of photons.

Figure 29.10 illustrates what happens when an X-ray photon strikes an electron in a piece of graphite. Like two billiard balls colliding on a pool table, the X-ray photon scatters in one direction, and the electron recoils in another direction after the collision. Compton observed that the scattered photon has a frequency f' that is smaller than the frequency f of the incident photon, indicating that the photon loses energy during the collision. In addition, he found that the difference between the two frequencies depends on the angle θ at which the scattered photon leaves the collision. The phenomenon in which an X-ray photon is scattered from an electron, the scattered photon having a smaller frequency than the incident photon, is called the *Compton effect.*

Cloud of gas

UV photons

EGG

(a)

Streaming gas

"Fingers" containing EGGs at their tips

(b)

Figure 29.9 (*a*) This drawing illustrates the photoevaporation that is occurring in the photograph in part (*b*). (*b*) This is the upper left portion of the chapter-opening photograph. Photoevaporation produces fingerlike projections on the surface of the gas clouds in the Eagle Nebula. At the fingertips are high-density evaporating gaseous globules (EGGs).

In Section 7.3 the collision between two objects is analyzed using the fact that the total kinetic energy and the total linear momentum of the objects are the same before and after the collision. Similar analysis can be applied to the collision between a photon and an electron. The electron is assumed to be initially at rest and essentially free, that is, not bound to the atoms of the material. According to the principle of conservation of energy,

$$\underbrace{hf}_{\substack{\text{Energy of}\\\text{incident}\\\text{photon}}} = \underbrace{hf'}_{\substack{\text{Energy of}\\\text{scattered}\\\text{photon}}} + \underbrace{\text{KE}}_{\substack{\text{Kinetic energy}\\\text{of recoil}\\\text{electron}}} \qquad (29.4)$$

where the relation $E = hf$ has been used for the photon energies. It follows, then, that $hf' = hf - \text{KE}$, which shows that the energy and corresponding frequency f' of the scattered photon are less than the energy and frequency of the incident photon, just as Compton observed. Since $\lambda' = c/f'$, the wavelength of the scattered X-rays is larger than that of the incident X-rays.

For an initially stationary electron, conservation of total linear momentum requires that

$$\begin{array}{c}\text{Momentum of}\\\text{incident photon}\end{array} = \begin{array}{c}\text{Momentum of}\\\text{scattered photon}\end{array} + \begin{array}{c}\text{Momentum of}\\\text{recoil electron}\end{array} \qquad (29.5)$$

To find an expression for the magnitude p of the photon's momentum, we use Equations 28.3 and 28.4. According to these equations, the momentum of any particle is $p = mv/\sqrt{1 - (v^2/c^2)}$ and its total energy is $E = mc^2/\sqrt{1 - (v^2/c^2)}$. Dividing these two equations, we find that $p/E = v/c^2$. Since a photon travels at the speed of light, $v = c$ and $p/E = 1/c$. Therefore, the momentum of a photon is $p = E/c$. But the energy of a photon is $E = hf$, while the wavelength is $\lambda = c/f$. Therefore, the magnitude of the momentum is

$$p = \frac{hf}{c} = \frac{h}{\lambda} \qquad (29.6)$$

Using Equations 29.4, 29.5, and 29.6, Compton showed that the difference between the wavelength λ' of the scattered photon and the wavelength λ of the incident photon is related to the scattering angle θ by

$$\lambda' - \lambda = \frac{h}{mc}(1 - \cos\theta) \qquad (29.7)$$

In this equation m is the mass of the electron. The quantity $h/(mc)$ is called the **Compton wavelength of the electron** and has the value $h/(mc) = 2.43 \times 10^{-12}$ m.

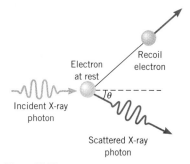

Electron at rest

Recoil electron

Incident X-ray photon

θ

Scattered X-ray photon

Figure 29.10 In an experiment performed by Arthur H. Compton, an X-ray photon collides with a stationary electron. The scattered photon and the recoil electron depart the collision in different directions.

Since $\cos \theta$ varies between $+1$ and -1, the shift $\lambda' - \lambda$ in the wavelength can vary between zero and $2h/(mc)$, depending on the value of θ, a fact observed by Compton.

The photoelectric effect and the Compton effect provide compelling evidence that light can exhibit particle-like characteristics attributable to energy packets called photons. But what about the interference phenomena discussed in Chapter 27, such as Young's double-slit experiment and single slit diffraction, which demonstrate that light behaves as a wave? Does light have two distinct personalities, in which it behaves like a stream of particles in some experiments and like a wave in others? The answer is yes, for physicists now believe that this wave–particle duality is an inherent property of light. Light is a far more interesting (and complex) phenomenon than just a stream of particles or an electromagnetic wave.

In the Compton effect the electron recoils because it gains some of the photon's momentum. In principle, then, the momentum that photons have can be used to make other objects move. Conceptual Example 3 considers a propulsion system for an interstellar spaceship that is based on the momentum of a photon.

CONCEPTUAL EXAMPLE 3   Solar Sails and Interstellar Spaceships

One propulsion method for interstellar spaceships found in science fiction novels uses a large sail. The intent is that sunlight striking the sail creates a force that pushes the ship away from the sun (Figure 29.11), much as the wind propels a sailboat. Does such a design have any hope of working and, if so, should the surface of the sail facing the sun be shiny like a mirror or black, in order to produce the greatest possible force?

Reasoning and Solution There is certainly reason to believe that the design might work. In Conceptual Example 3 in Chapter 7, we found that hailstones striking the roof of a car exert a force on it because they have a momentum. Photons also have momentum, so, like the hailstones, they can apply a force to the sail.

As in Chapter 7, we will be guided by the impulse–momentum theorem in assessing the force. This theorem (Equation 7.4) states that when a net force acts on an object, the impulse of the net force is equal to the change in momentum of the object. Greater impulses lead to greater forces for a given time interval. Thus, when a photon collides with the sail, the photon's momentum changes because of the force applied by the sail to the photon. Newton's action–reaction law indicates that the photon simultaneously applies a force of equal magnitude, but opposite direction to the sail. It is this reaction force that propels the spaceship, and it will be greater when the momentum change experienced by the photon is greater. The surface of the sail facing the sun, then, should be such that it causes the largest possible momentum change for the impinging photons.

As we found in Section 13.3, radiation reflects from a shiny mirror-like surface and is absorbed by a black surface. Now, consider a photon that strikes the sail perpendicularly. In reflecting from a mirror-like surface, its momentum changes from its value in the forward direction to a value of the same magnitude in the reverse direction. This is a greater change than what occurs when the photon is absorbed by a black surface. Then, the momentum changes only from its value in the forward direction to a value of zero. Consequently, the surface of the sail facing the sun should be shiny in order to produce the greatest possible propulsion force. A shiny surface causes the photons to bounce like the hailstones on the roof of the car and, in so doing, apply a greater force to the sail.

Related Homework Material: *Question 10, Problem 45*

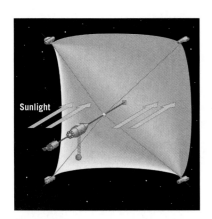

Figure 29.11 A solar sail provides the propulsion for this interstellar spaceship.

Concepts at a Glance

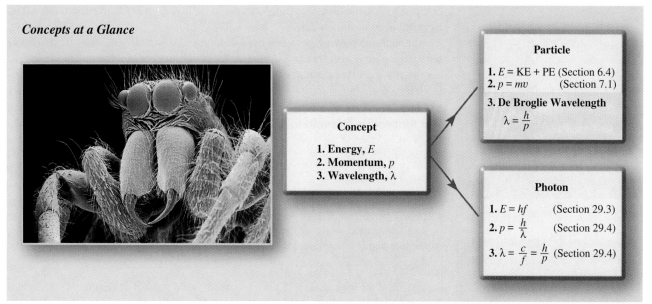

Concept

1. **Energy,** E
2. **Momentum,** p
3. **Wavelength,** λ

Particle

1. $E = \mathrm{KE} + \mathrm{PE}$ (Section 6.4)
2. $p = mv$ (Section 7.1)
3. **De Broglie Wavelength**

$\lambda = \dfrac{h}{p}$

Photon

1. $E = hf$ (Section 29.3)
2. $p = \dfrac{h}{\lambda}$ (Section 29.4)
3. $\lambda = \dfrac{c}{f} = \dfrac{h}{p}$ (Section 29.4)

Figure 29.12 *Concepts at a Glance* Moving particles and waves both possess energy, momentum, and a wavelength. The wavelength of a particle is known as its de Broglie wavelength. This photograph shows a zebra jumping spider, made with a scanning electron microscope. This microscope uses electrons instead of light. The resolution of fine detail is exceptional because the wavelength of an electron can be made much smaller than that of light.

29.5 THE DE BROGLIE WAVELENGTH AND THE WAVE NATURE OF MATTER

● *Concepts at a Glance* As a graduate student in 1923, Louis de Broglie (1892–1987) made the astounding suggestion that since light waves could exhibit particle-like behavior, particles of matter should exhibit wave-like behavior. De Broglie proposed that all moving matter has a wavelength associated with it, just as a wave does. The Concept-at-a-Glance chart in Figure 29.12, which is a continuation of the chart in Figure 29.3, shows that the notions of energy, momentum, and wavelength are applicable to particles as well as to waves. ●

De Broglie made the explicit proposal that the wavelength λ of a particle is given by the same relation (Equation 29.6) that applies to a photon:

De Broglie wavelength
$$\lambda = \frac{h}{p} \tag{29.8}$$

where h is Planck's constant and p is the magnitude of the relativistic momentum of the particle. Today, λ is known as the ***de Broglie wavelength*** of the particle.

Confirmation of de Broglie's suggestion came in 1927 from the experiments of the American physicists Clinton J. Davisson (1881–1958) and Lester H. Germer (1896–1971) and, independently, the English physicist George P. Thomson (1882–1975). Davisson and Germer directed a beam of electrons onto a crystal of nickel and observed that the electrons exhibited a diffraction behavior, analogous to that seen when X-rays are diffracted by a crystal (see Section 27.9 for a discussion of X-ray diffraction). The wavelength of the electrons revealed by the diffraction pattern matched that predicted by de Broglie's hypothesis, $\lambda = h/p$. More recently, Young's double-slit experiment has been performed with electrons and reveals the effects of wave interference illustrated in Figure 29.1.

Particles other than electrons can also exhibit wave-like properties. For instance, neutrons are sometimes used in diffraction studies of crystal structure. Figure 29.13

(a)

(b)

Figure 29.13 (a) The neutron diffraction pattern and (b) the X-ray diffraction pattern for a crystal of sodium chloride (NaCl).

compares the neutron diffraction pattern and the X-ray diffraction pattern caused by a crystal of rock salt (NaCl).

Although all moving particles have a de Broglie wavelength, the effects of this wavelength are observable only for particles whose masses are very small, on the order of the mass of an electron or a neutron, for instance. Example 4 illustrates why.

EXAMPLE 4
The de Broglie Wavelength of an Electron and a Baseball

Determine the de Broglie wavelength for (a) an electron (mass $= 9.1 \times 10^{-31}$ kg) moving at a speed of 6.0×10^6 m/s and (b) a baseball (mass $= 0.15$ kg) moving at a speed of 13 m/s.

Reasoning In each case, the de Broglie wavelength is given by Equation 29.8 as Planck's constant divided by the magnitude of the momentum. Since the speeds are small compared to the speed of light, we can ignore relativistic effects and express the magnitude of the momentum as the product of the mass and the speed.

Solution

(a) Since the magnitude p of the momentum is the product of the mass m of the particle and its speed v, we have $p = mv$. Using this expression in Equation 29.8 for the de Broglie wavelength, we obtain

$$\lambda = \frac{h}{p} = \frac{h}{mv} = \frac{6.63 \times 10^{-34} \text{ J} \cdot \text{s}}{(9.1 \times 10^{-31} \text{ kg})(6.0 \times 10^6 \text{ m/s})} = \boxed{1.2 \times 10^{-10} \text{ m}}$$

A de Broglie wavelength of 1.2×10^{-10} m is about the size of the interatomic spacing in a solid, such as the nickel crystal used by Davisson and Germer, and, therefore, leads to the observed diffraction effects.

(b) A calculation similar to that in part (a) shows that the de Broglie wavelength of the baseball is $\boxed{\lambda = 3.3 \times 10^{-34} \text{ m}}$. This wavelength is incredibly small, even by comparison with the size of an atom (10^{-10} m) or a nucleus (10^{-14} m). Thus, the ratio λ/W of this wavelength to the width W of an ordinary opening, such as a window, is so small that the diffraction of a baseball passing through the window cannot be observed.

The de Broglie equation for particle wavelength provides no hint as to what kind of wave is associated with a particle of matter. To gain some insight into the nature of this wave, we turn our attention to Figure 29.14. Part *a* shows the fringe pattern on the screen when electrons are used in a version of Young's double-slit experiment. The bright fringes occur in places where particle waves coming from each slit interfere constructively, while the dark fringes occur in places where the waves interfere destructively.

When an electron passes through the double-slit arrangement and strikes a spot on the screen, the screen glows at that spot, and parts *b*, *c*, and *d* of Figure 29.14 illustrate how the spots accumulate in time. As more and more electrons strike the screen, the spots eventually form the fringe pattern that is evident in part *d*. Bright fringes occur where there is a high probability of electrons striking the screen, and dark fringes occur where there is a low probability. Here lies the key to understanding particle waves. ***Particle waves are waves of probability,*** waves whose magnitude at a point in space gives an indication of the probability that the particle will be found at that point. At the place where the screen is located, the pattern of probabilities conveyed by the particle waves causes the fringe pattern to emerge. The fact that no fringe pattern is apparent in part *b* of the picture does not mean that there are no probability waves present; it just means that too few electrons have struck the screen for the pattern to be recognizable.

The pattern of probabilities that leads to the fringes in Figure 29.14 is analogous to the pattern of light intensities that is responsible for the fringes in Young's original experiment with light waves (see Figure 27.4). Section 24.4 discusses the fact that the intensity of the light is proportional to either the square of the electric field strength or the square of the magnetic field strength of the wave. In an analogous fashion in the case of particle waves, the probability is proportional to the square of the magnitude Ψ (Greek letter psi) of the wave. Ψ is referred to as the ***wave function*** of the particle.

In 1925 the Austrian physicist Erwin Schrödinger (1887–1961) and the German physicist Werner Heisenberg (1901–1976) independently developed theoretical frameworks for determining the wave function. In so doing, they established a new branch of physics called ***quantum mechanics.*** The word "quantum" refers to the fact that in the world of the atom, where particle waves must be considered, the particle energy is quantized, so only certain energies are allowed. To understand the structure of the atom and the phenomena related to it, quantum mechanics is essential, and the Schrödinger equation for calculating the wave function is now widely used. In the next chapter, we will explore the structure of the atom based on the ideas of quantum mechanics.

29.6 THE HEISENBERG UNCERTAINTY PRINCIPLE

As the previous section discusses, the bright fringes in Figure 29.14 indicate the places where there is a high probability of an electron striking the screen. And since there are a number of bright fringes, there is more than one place where each electron has some probability of hitting. As a result, it is not possible to specify in advance exactly where on the screen an individual electron will hit. All we can do is speak of the probability that the electron may end up in a number of different places. No longer is it possible to say, as Newton's laws would suggest, that a single electron, fired through the double slit, will travel directly forward in a straight line and strike the screen. This simple picture just does not apply when a particle as small as an electron passes through a pair of closely spaced narrow slits. Because the wave nature of particles is important in such circumstances, we lose the ability to predict with 100% certainty the path that a single particle will follow. Instead, only the average behavior of large numbers of particles is predictable, and the behavior of any individual particle is uncertain.

To see more clearly into the nature of the uncertainty, consider electrons passing through a single slit, as in Figure 29.15. After a sufficient number of electrons strike the screen, a diffraction pattern emerges. The electron diffraction pattern consists of alternating bright and dark fringes and is analogous to that for light waves shown in Figure 27.23. Figure 29.15 shows the slit and locates the first dark fringe on either side of the central bright fringe. The central fringe is bright because electrons strike the screen over the entire region between the dark fringes. If the electrons striking the screen outside the central bright fringe are neglected, the extent to which the electrons are diffracted is given by the angle θ in the drawing. To reach locations within the central bright fringe, some electrons must have acquired momentum in the y direction, despite the fact that they enter the slit traveling along the x direction and have no momentum in the y direction to start with. The figure illustrates that the y component of the momentum may be as large as Δp_y. The notation Δp_y indicates the difference between the maximum value of the y component of the momentum after the electron passes through the slit and its value of zero before the elec-

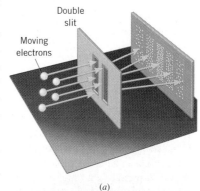

Double slit

Moving electrons

(a)

(b)   After 100 electrons

(c)   After 3000 electrons

(d)   After 70 000 electrons

Figure 29.14 In this electron version of Young's double-slit experiment, the characteristic fringe pattern becomes recognizable only after a sufficient number of electrons have struck the screen. (*Source:* A. Tonomura, J. Endo, T. Matsuda, and T. Kawasaki, *Am. J. Phys.* 57(2): 117, Feb. 1989.)

Figure 29.15 When a sufficient number of electrons pass through a single slit and strike the screen, a diffraction pattern of bright and dark fringes emerges. (Only the central bright fringe is shown.) This pattern is due to the wave nature of the electrons and is analogous to that produced by light waves.

tron passes through the slit. Δp_y represents the *uncertainty* in the y component of the momentum in that the y component may have any value from zero to Δp_y.

It is possible to relate Δp_y to the width W of the slit. To do this, we assume that Equation 27.4, which applies to light waves, also applies to particle waves whose de Broglie wavelength is λ. This equation, $\sin \theta = \lambda/W$, specifies the angle θ that locates the first dark fringe. If θ is small, $\sin \theta \approx \tan \theta$. Moreover, Figure 29.15 indicates that $\tan \theta = \Delta p_y/p_x$, where p_x is the x component of the momentum of the electron. Therefore, $\Delta p_y/p_x \approx \lambda/W$. But $p_x = h/\lambda$ according to de Broglie's equation, so that

$$\frac{\Delta p_y}{p_x} = \frac{\Delta p_y}{h/\lambda} \approx \frac{\lambda}{W}$$

As a result,

$$\Delta p_y \approx \frac{h}{W} \tag{29.9}$$

which indicates that a smaller slit width leads to a larger uncertainty in the y component of the electron's momentum.

It was Heisenberg who first suggested that the uncertainty Δp_y in the y component of the momentum is related to the uncertainty in the y position of the electron as the electron passes through the slit. Since the electron can pass through anywhere over the width W, the uncertainty in the y position of the electron is $\Delta y = W$. Substituting Δy for W in Equation 29.9 shows that $\Delta p_y \approx h/\Delta y$ or $(\Delta p_y)(\Delta y) \approx h$. The result of Heisenberg's more complete analysis is given below in Equation 29.10 and is known as the **Heisenberg uncertainty principle.** Note that the Heisenberg principle is a general principle with wide applicability. It does not just apply to the case of single slit diffraction, which we have used here for the sake of convenience.

● **THE HEISENBERG UNCERTAINTY PRINCIPLE**

Momentum and position

$$(\Delta p_y)(\Delta y) \geq \frac{h}{2\pi} \tag{29.10}$$

Δy = uncertainty in a particle's position along the y direction
Δp_y = uncertainty in the y component of the linear momentum of the particle

Energy and time

$$(\Delta E)(\Delta t) \geq \frac{h}{2\pi} \qquad (29.11)$$

ΔE = uncertainty in the energy of a particle when the particle is in a certain state

Δt = time interval during which the particle is in the state

The Heisenberg uncertainty principle places limits on the accuracy with which the momentum and position of a particle can be specified simultaneously. These limits are not just limits due to faulty measuring techniques. They are fundamental limits imposed by nature, and there is no way to circumvent them. Equation 29.10 indicates that Δp_y and Δy cannot both be arbitrarily small at the same time. If one is small, then the other must be large, so that their product equals or exceeds ($\geq$) Planck's constant divided by 2π. For example, if the position of a particle is known exactly, so that Δy is zero, then Δp_y is an infinitely large number, and the momentum of the particle is completely uncertain. Conversely, if we assume that Δp_y is zero, then Δy is an infinitely large number, and the position of the particle is completely uncertain. In other words, the Heisenberg uncertainty principle states that it is impossible to specify precisely both the momentum and position of a particle at the same time.

There is also an uncertainty principle that deals with energy and time, as expressed by Equation 29.11. The product of the uncertainty ΔE in the energy of a particle and the time interval Δt during which the particle remains in a given energy state is greater than or equal to Planck's constant divided by 2π. Therefore, the shorter the lifetime of a particle in a given state, the greater is the uncertainty in the energy of that state.

Example 5 shows that the uncertainty principle has significant consequences for the motion of tiny particles such as electrons but has little effect on the motion of macroscopic objects, even those with as little mass as a Ping-Pong ball.

EXAMPLE 5 The Heisenberg Uncertainty Principle

Assume that the position of an object is known so precisely that the uncertainty in the position is only $\Delta y = 1.5 \times 10^{-11}$ m. (a) Determine the minimum uncertainty in the momentum of the object. Find the corresponding minimum uncertainty in the speed of the object, if the object is (b) an electron (mass = 9.1×10^{-31} kg) and (c) a Ping-Pong ball (mass = 2.2×10^{-3} kg).

Reasoning The minimum uncertainty Δp_y in the y component of the momentum is given by the Heisenberg uncertainty principle as $\Delta p_y = h/(2\pi \Delta y)$, where Δy is the uncertainty in the position of the object. Both the electron and the Ping-Pong ball have the same uncertainty in their momenta because they have the same uncertainty in their positions. However, these objects have very different masses. As a result, we will find that the uncertainty in the speeds of these objects is very different.

Solution

(a) The minimum uncertainty in the y component of the momentum is

$$\Delta p_y = \frac{h}{2\pi \Delta y} = \frac{6.63 \times 10^{-34}\ \text{J} \cdot \text{s}}{2\pi(1.5 \times 10^{-11}\ \text{m})} = \boxed{7.0 \times 10^{-24}\ \text{kg} \cdot \text{m/s}} \qquad (29.10)$$

(b) Since $\Delta p_y = m\,\Delta v_y$, the minimum uncertainty in the speed of the electron is

$$\Delta v_y = \frac{\Delta p_y}{m} = \frac{7.0 \times 10^{-24}\ \text{kg} \cdot \text{m/s}}{9.1 \times 10^{-31}\ \text{kg}} = \boxed{7.7 \times 10^{6}\ \text{m/s}}$$

• **PROBLEM SOLVING INSIGHT**
The Heisenberg uncertainty principle states that the product of Δp_y and Δy is greater than or equal to $h/2\pi$. The minimum uncertainty occurs when the product is equal to $h/2\pi$.

Thus, the small uncertainty in the y position of the electron gives rise to a large uncertainty in the speed of the electron.

(c) The uncertainty in the speed of the Ping-Pong ball is

$$\Delta v_y = \frac{\Delta p_y}{m} = \frac{7.0 \times 10^{-24}\ \text{kg} \cdot \text{m/s}}{2.2 \times 10^{-3}\ \text{kg}} = \boxed{3.2 \times 10^{-21}\ \text{m/s}}$$

Because the mass of the Ping-Pong ball is relatively large, the small uncertainty in its y position gives rise to an uncertainty in its speed that is much smaller than that for the electron. Thus, in contrast to the electron, we can know simultaneously where the ball is and how fast it is moving, to a very high degree of certainty.

Example 5 emphasizes how the uncertainty principle imposes different uncertainties on the speeds of an electron (small mass) and a Ping-Pong ball (large mass). For objects like the ball, which have relatively large masses, the uncertainties in position and speed are so small that they have no effect on our ability to determine simultaneously where such objects are and how fast they are moving. The uncertainties calculated in Example 5 depend on more than just the mass, however. They also depend on Planck's constant, which is a very small number. It is interesting to speculate about what life would be like if Planck's constant were much larger than 6.63×10^{-34} J $\cdot$ s. Conceptual Example 6 deals with just such speculation.

CONCEPTUAL EXAMPLE 6 What If Planck's Constant Were Large?

A bullet leaving the barrel of a gun is analogous to an electron passing through the single slit in Figure 29.15. With this analogy in mind, what would hunting be like if Planck's constant had a relatively large value instead of its small value of 6.63×10^{-34} J $\cdot$ s?

Reasoning and Solution It takes two to hunt, the hunter and the hunted. Consider the hunter first and what it would mean to take aim if Planck's constant were large. The bullet moving down the barrel of the gun has a de Broglie wavelength given by Equation 29.8 as $\lambda = h/p$. This wavelength is large, since Planck's constant h is now large. (We are assuming that the magnitude p of the bullet's momentum is not abnormally large.) Remember from Section 27.5 that the diffraction of light waves through an opening is greater when the wavelength is greater, other things being equal. We expect, therefore, that with λ being comparable to the opening in the barrel, the hunter will have to contend with appreciable diffraction when the bullet leaves the barrel. This means that, in spite of being aimed directly at the target, the bullet may come nowhere near it. In Figure 29.15, for comparison, diffraction allows the electron to strike the screen on either side of the maximum intensity point of the central bright fringe. Under such circumstances, the hunter might as well not bother aiming. He'll have just as much success by scattering his shots around randomly, hoping to hit something.

Now consider the target and what it means that the hunter "aimed directly at the target." This must mean that the target's location is known, at least approximately. In other words, the uncertainty Δy in the target's position is reasonably small. But the uncertainty principle indicates that the minimum uncertainty in the target's momentum is $\Delta p = h/(2\pi \Delta y)$. Since Planck's constant is now large, Δp is also large. The target, therefore, may have a large momentum and be moving rapidly. Even if the bullet happens to reach the place where the hunter aimed, the target may no longer be there. Hunting would be vastly different indeed, if Planck's constant had a very large value.

Related Homework Material: *Problem 34*

29.7 CONCEPTS & CALCULATIONS

In the photoelectric effect electrons can be emitted from a metal surface when light shines on it, as we have seen. Einstein explained this phenomenon in terms of photons and the conservation of energy. The same basic explanation applies when the light has a single wavelength or when it consists of a mixture of wavelengths, as sunlight does. Our earlier discussion in Section 29.3 deals with a single wavelength. Example 7, in contrast, considers a mixture of wavelengths and reviews the basic concepts that come into play in this important phenomenon.

CONCEPTS & CALCULATIONS EXAMPLE 7
Sunlight and the Photoelectric Effect

Sunlight, whose visible wavelengths range from 380 to 750 nm, is incident on a sodium surface. The work function is $W_0 = 2.28$ eV. Find the maximum kinetic energy KE_{max} (in joules) of the photoelectrons emitted from the surface and the range of wavelengths that will cause photoelectrons to be emitted.

Concept Questions and Answers Will electrons with the greatest value of KE_{max} be emitted when the incident photons have a relatively greater or a relatively smaller amount of energy?

Answer When a photon ejects an electron from the sodium, the photon energy is used for two purposes. Some of it is used to do the work of ejecting the electron, and what remains is carried away as kinetic energy of the emitted electron. The minimum work to eject an electron is equal to the work function W_0, and in this case the electron carries away a maximum amount KE_{max} of kinetic energy. According to the principle of conservation of energy, the energy of an incident photon is equal to the sum of W_0 and KE_{max}. Thus, for a given value of W_0, greater values of KE_{max} will occur when the incident photon has a greater energy to begin with.

In the range of visible wavelengths, which wavelength corresponds to incident photons that carry the greatest energy?

Answer According to Equation 29.2, the energy of a photon is related to the frequency f and Planck's constant h according to $E = hf$. But the frequency is given by Equation 16.1 as $f = c/\lambda$, where c is the speed of light and λ is the wavelength. Therefore, we have $E = hc/\lambda$, and the smallest value of the wavelength corresponds to the greatest energy.

What is the smallest value of KE_{max} with which an electron can be ejected from the sodium?

Answer The smallest value of KE_{max} with which an electron can be ejected from the sodium is 0 J. This occurs when an incident photon carries an energy that exactly equals the work function W_0, with no energy left over to be carried away by the electron as kinetic energy.

Solution The value given for the work function is in electron volts, so we first convert it into joules by using the fact that 1 eV = 1.60×10^{-19} J:

$$W_0 = (2.28 \text{ eV}) \frac{1.60 \times 10^{-19} \text{ J}}{1 \text{ eV}} = 3.65 \times 10^{-19} \text{ J}$$

The incident photons that have the greatest energy are those with the smallest wavelength or $\lambda = 380$ nm. Then, according to Equation 29.3 and Equation 16.1 ($f = c/\lambda$), the greatest value for KE_{max} is

$$KE_{max} = hf - W_0 = \frac{hc}{\lambda} - W_0$$

$$= \frac{(6.63 \times 10^{-34} \text{ J·s})(3.00 \times 10^8 \text{ m/s})}{380 \times 10^{-9} \text{ m}} - 3.65 \times 10^{-19} \text{ J} = \boxed{1.58 \times 10^{-19} \text{ J}}$$

As the wavelength increases, the energy of the incident photons decreases until it equals the work function, at which point electrons are ejected with zero kinetic energy. Thus, we have

$$KE_{max} = 0 \text{ J} = \frac{hc}{\lambda} - W_0$$

$$\lambda = \frac{hc}{W_0} = \frac{(6.63 \times 10^{-34} \text{ J·s})(3.00 \times 10^8 \text{ m/s})}{3.65 \times 10^{-19} \text{ J}} = 5.45 \times 10^{-7} \text{ m} = 545 \text{ nm}$$

The range of wavelengths over which sunlight ejects electrons from the sodium surface is, then, $\boxed{380 - 545 \text{ nm}}$.

Kinetic energy and momentum, respectively, are discussed in Chapters 6 and 7. In the present chapter we have seen that moving particles of matter not only have kinetic energy and momentum but are also characterized by a wavelength, which is known as the de Broglie wavelength. We conclude with an example that reviews kinetic energy and momentum and focuses on the relation between these fundamental physical quantities and the de Broglie wavelength.

CONCEPTS & CALCULATIONS EXAMPLE 8   The de Broglie Wavelength

An electron and a proton have the same kinetic energy and are moving at speeds much less than the speed of light. Determine the ratio of the de Broglie wavelength of the electron to that of the proton.

Concept Questions and Answers How is the de Broglie wavelength λ related to the magnitude p of the momentum?

Answer   According to Equation 29.8, the de Broglie wavelength is given by $\lambda = h/p$, where h is Planck's constant. The greater the momentum, the smaller the de Broglie wavelength, and vice versa.

How is the magnitude of the momentum related to the kinetic energy of a particle of mass m that is moving at a speed that is much less than the speed of light?

Answer   According to Equation 7.2 the magnitude of the momentum is $p = mv$, where v is the speed. Equation 6.2 gives the kinetic energy as $KE = \frac{1}{2}mv^2$, which can be solved to show that $v = \sqrt{2(KE)/m}$. Substituting this result into Equation 7.2 gives $p = m\sqrt{2(KE)/m} = \sqrt{2m(KE)}$.

Which has the greater wavelength, the electron or the proton?

Answer   According to $\lambda = h/p$, the particle with the greater wavelength is the one with the smaller momentum. But $p = \sqrt{2m(KE)}$ indicates that, for a given kinetic energy, the particle with the smaller momentum is the one with the smaller mass. The masses of the electron and proton are $m_{electron} = 9.11 \times 10^{-31}$ kg and $m_{proton} = 1.67 \times 10^{-27}$ kg. Thus, the electron, with its smaller mass, has the greater de Broglie wavelength.

Solution Using Equation 29.8 for the de Broglie wavelength and the fact that the magnitude of the momentum is related to the kinetic energy by $p = \sqrt{2m(KE)}$, we have

$$\lambda = \frac{h}{p} = \frac{h}{\sqrt{2m(KE)}}$$

Applying this result to the electron and the proton gives

$$\frac{\lambda_{\text{electron}}}{\lambda_{\text{proton}}} = \frac{h/\sqrt{2m_{\text{electron}}\,(\text{KE})}}{h/\sqrt{2m_{\text{proton}}\,(\text{KE})}} = \sqrt{\frac{m_{\text{proton}}}{m_{\text{electron}}}} = \sqrt{\frac{1.67 \times 10^{-27}\ \text{kg}}{9.11 \times 10^{-31}\ \text{kg}}} = \boxed{42.8}$$

As expected, the wavelength for the electron is greater than that for the proton.

At the end of the problem set for this chapter, you will find homework problems that contain both conceptual and quantitative parts. These problems are grouped under the heading *Concepts & Calculations, Group Learning Problems.* They are designed for use by students working alone or in small learning groups. The conceptual part of each problem provides a convenient focus for group discussions.

SUMMARY

The **wave–particle duality** refers to the fact that a wave can exhibit particlelike characteristics and a particle can exhibit wavelike characteristics.

At a constant temperature, a perfect blackbody absorbs and reemits all the electromagnetic radiation that falls on it. Max Planck calculated the emitted radiation intensity per unit wavelength as a function of wavelength. In his theory, Planck assumed that a blackbody consists of atomic oscillators that can have only quantized energies. Planck's quantized energies are given by $E = nhf$, where $n = 0, 1, 2, 3, \ldots$, h is **Planck's constant** (6.63×10^{-34} J·s), and f is the vibration frequency.

All electromagnetic radiation consists of **photons**, which are packets of energy. The energy of a photon is $E = hf$, where h is Planck's constant and f is the frequency of the light. A photon in a vacuum always travels at the speed of light c and has no mass. The **photoelectric effect** is the phenomenon in which light shining on a metal surface causes electrons to be ejected from the surface. The **work function** W_0 of a metal is the minimum work that must be done to eject an electron from the metal. In accordance with the conservation of energy, the electrons ejected from a metal have a maximum kinetic energy KE_{max} that is related to the energy hf of the incident photon by $hf = \text{KE}_{\text{max}} + W_0$.

The **Compton effect** is the scattering of a photon by an electron in a material, the scattered photon having a smaller frequency than the incident photon. The magnitude p of the photon's momentum is $p = h/\lambda$, and in the Compton effect, part of it is transferred to the recoiling electron. The difference between the wavelength λ' of the scattered photon and the wavelength λ of the incident photon is related to the scattering angle θ by $\lambda' - \lambda = [h/(mc)](1 - \cos\theta)$, where m is the mass of the electron and the quantity $h/(mc)$ is known as the Compton wavelength of the electron.

The **de Broglie wavelength** of a particle is $\lambda = h/p$, where p is the magnitude of the relativistic momentum of the particle. Because of its de Broglie wavelength, a particle can exhibit wave-like characteristics. The wave associated with a particle is a wave of probability.

The **Heisenberg uncertainty principle** places limits on our knowledge about the behavior of a particle. The uncertainty principle indicates that $(\Delta p_y)(\Delta y) \geq h/(2\pi)$, where Δy and Δp_y are, respectively, the uncertainties in the position and momentum of the particle. The uncertainty principle also states that $(\Delta E)(\Delta t) \geq h/(2\pi)$, where ΔE is the uncertainty in the energy of a particle when the particle is in a certain state and Δt is the time interval during which the particle is in the state.

CONCEPTUAL QUESTIONS

1. The photons emitted by a source of light do *not* all have the same energy. Is the source monochromatic? Give your reasoning.

2. Which colored light bulb, red, orange, yellow, green, or blue, emits photons with (a) the least energy and (b) the greatest energy? Account for your answers.

3. Does a photon emitted by a higher-wattage red light bulb have more energy than a photon emitted by a lower-wattage red bulb? Justify your answer.

4. When a sufficient number of visible light photons strike a piece of photographic film, the film becomes exposed. An X-ray photon is more energetic than a visible light photon. Yet, most photographic films are not exposed by the X-ray machines used at airport security checkpoints. Explain what these observations imply about the number of photons emitted by the X-ray machines.

5. Radiation of a given wavelength causes electrons to be emitted from the surface of one metal but not from the surface of another metal. Explain why this could be.

6. In the photoelectric effect, suppose that the intensity of the light is increased, while the frequency is kept constant. The frequency is greater than the minimum frequency f_0. State whether each of the following will increase, decrease, or remain constant, and explain your choice: (a) the current in the phototube, (b) the number of electrons emitted per second from the metal surface, (c) the maximum kinetic energy that an electron could have, (d) the maximum momentum that an electron could have, and (e) the minimum de Broglie wavelength that an electron could have.

7. In a Compton scattering experiment, an electron is accelerated straight ahead in the same direction as that of the incident X-ray photon. Which way does the scattered photon move? Explain your reasoning, using the principle of conservation of momentum.

8. A photon can undergo Compton scattering from a molecule such as nitrogen, just as it does from an electron. However, the change in photon wavelength is much less than when the scattering is from an electron. Explain why, using Equation 29.7 for a nitrogen molecule instead of an electron.

9. When the speed of a particle doubles, its momentum doubles, and its kinetic energy becomes four times greater. When the momentum of a photon doubles, does its energy become four times greater? Provide a reason for your answer.

10. Review Conceptual Example 3 as background for this question. The photograph shows a device called a radiometer. The four rectangular panels are black on one side and shiny like a mirror on the other side. In bright light the panel arrangement spins around, in a direction from the black side of a panel toward the shiny side. Do photon collisions with both sides of the panels (collisions such as those discussed in Conceptual Example 3) cause the observed spinning? Give your reasoning.

11. A stone is dropped from the top of a building. As the stone falls, does its de Broglie wavelength increase, decrease, or remain the same? Provide a reason for your answer.

12. An electron and a neutron have different masses. Is it possible, according to Equation 29.8, that they can have the same de Broglie wavelength? Account for your answer.

13. In Figure 29.1 replace the electrons with protons that have the same speed. With the aid of Equation 27.1 for the bright fringes in Young's double-slit experiment and Equation 29.8, decide whether the angular separation between the fringes would increase, decrease, or remain the same when compared to that produced by the electrons.

PROBLEMS

In working these problems, ignore relativistic effects.

ssm Solution is in the Student Solutions Manual.   **www** Solution is available on the World Wide Web at http://www.wiley.com/college/cutnell

⚕ This icon represents a biomedical application.

Section 29.3 Photons and the Photoelectric Effect

1. ssm Ultraviolet light is responsible for sun tanning. Find the wavelength (in nm) of an ultraviolet photon whose energy is 6.4×10^{-19} J.

2. Two photons have energies of 3.3×10^{-16} J and 1.3×10^{-20} J. Using Figure 24.9, identify the appropriate region in the electromagnetic spectrum for each of these photons.

3. An FM radio station broadcasts at a frequency of 98.1 MHz. The power radiated from the antenna is 5.0×10^4 W. How many photons per second does the antenna emit?

4. The maximum wavelength that an electromagnetic wave can have and still eject electrons from a metal surface is 485 nm. What is the work function W_0 of this metal? Express your answer in electron volts.

5. ssm Ultraviolet light with a frequency of 3.00×10^{15} Hz strikes a metal surface and ejects electrons that have a maximum kinetic energy of 6.1 eV. What is the work function (in eV) of the metal?

6. Radiation of a certain wavelength causes electrons with a maximum kinetic energy of 0.68 eV to be ejected from a metal whose work function is 2.75 eV. What will be the maximum kinetic energy (in eV) with which this same radiation ejects electrons from another metal whose work function is 2.17 eV?

7. An AM radio station broadcasts an electromagnetic wave at a frequency of 665 kHz, while an FM station broadcasts at 91.9 MHz. How many AM photons are needed to have a total energy equal to that of one FM photon?

*8. Radiation with a wavelength of 238 nm shines on a metal surface and ejects electrons that have a maximum speed of 3.75×10^5 m/s. Which one of the following metals is it, the values in parentheses being the work functions: potassium (2.24 eV), calcium (2.71 eV), uranium (3.63 eV), aluminum (4.08 eV), or gold (4.82 eV)?

*9. ssm An owl has good night vision because its eyes can detect a light intensity as small as 5.0×10^{-13} W/m². What is the minimum number of photons per second that an owl eye can detect if its pupil has a diameter of 8.5 mm and the light has a wavelength of 510 nm?

*10. The maximum wavelength for which an electromagnetic wave can eject electrons from a platinum surface is 196 nm. When radiation with a wavelength of 141 nm shines on the surface, what is the maximum speed of the ejected electrons?

**11. ssm www (a) How many photons (wavelength = 620 nm) must be absorbed to melt a 2.0-kg block of ice at 0 °C into water at 0 °C? (b) On the average, how many H_2O molecules does one photon convert from the ice phase to the water phase?

**12. A laser emits 1.30×10^{18} photons per second in a beam of light that has a diameter of 2.00 mm and a wavelength of 514.5 nm. Determine (a) the average electric field strength and (b) the average magnetic field strength for the electromagnetic wave that constitutes the beam.

Section 29.4 The Momentum of a Photon and the Compton Effect

13. The microwaves used in a microwave oven have a wavelength of about 0.13 m. What is the momentum of a microwave photon?

14. A photon has the same momentum as an electron moving with a speed of 2.0×10^5 m/s. What is the wavelength of the photon?

15. ssm Incident X-rays have a wavelength of 0.3120 nm and are scattered by the "free" electrons in graphite. The scattering angle in Figure 29.10 is $\theta = 135.0°$. What is the magnitude of the momentum of (a) the incident photon and (b) the scattered photon? (For accuracy, use $h = 6.626 \times 10^{-34}$ J·s and $c = 2.998 \times 10^8$ m/s.)

16. Figure 29.10 shows the setup for measuring the Compton effect. With a fixed incident wavelength, a wavelength of λ_1' is measured for a scattering angle of $\theta_1 = 30.0°$, while a wavelength of λ_2' is measured for a scattering angle of $\theta_2 = 70.0°$. Find the difference in wavelengths, $\lambda_2' - \lambda_1'$.

17. An incident X-ray photon of wavelength 0.2800 nm is scattered from an electron that is initially at rest. The photon is scattered at an angle of $\theta = 180.0°$ in Figure 29.10 and has a wavelength of 0.2849 nm. Use the conservation of linear momentum to find the momentum gained by the electron.

*18. The X-rays detected at a scattering angle of $\theta = 163°$ in Figure 29.10 have a wavelength of 0.1867 nm. Find (a) the wavelength of an incident photon, (b) the energy of an incident photon, (c) the energy of a scattered photon, and (d) the kinetic energy of the recoil electron. (For accuracy, use $h = 6.626 \times 10^{-34}$ J·s and $c = 2.998 \times 10^8$ m/s.)

*19. ssm www What is the maximum amount by which the wavelength of an incident photon could change when it undergoes Compton scattering from a nitrogen molecule (N_2)?

**20. An X-ray photon is scattered at an angle of $\theta = 180.0°$ from an electron that is initially at rest. After scattering, the electron has a speed of 4.67×10^6 m/s. Find the wavelength of the incident X-ray photon.

Section 29.5 The de Broglie Wavelength and the Wave Nature of Matter

21. A particle has a speed of 1.2×10^6 m/s. Its de Broglie wavelength is 8.4×10^{-14} m. What is the mass of the particle?

22. The interatomic spacing in a crystal of table salt is 0.282 nm. This crystal is being studied in a neutron diffraction experiment, similar to the one that produced the photograph in Figure 29.13a. How fast must a neutron (mass = 1.67×10^{-27} kg) be moving to have a de Broglie wavelength of 0.282 nm?

23. ssm As Section 27.5 discusses, sound waves diffract or bend around the edges of a doorway. Larger wavelengths diffract more than smaller wavelengths. (a) The speed of sound is 343 m/s. With what speed would a 55.0-kg person have to move through a doorway to diffract to the same extent as a 128-Hz bass tone? (b) At the speed calculated in part (a), how long (in years) would it take the person to move a distance of one meter?

24. What is (a) the wavelength of a 5.0-eV photon and (b) the de Broglie wavelength of a 5.0-eV electron?

25. Recall from Section 14.3 that the average kinetic energy of an atom in a monatomic ideal gas is given by $\overline{KE} = \frac{3}{2}kT$, where $k = 1.38 \times 10^{-23}$ J/K and T is the Kelvin temperature of the gas. Determine the de Broglie wavelength of a helium atom (mass = 6.65×10^{-27} kg) that has the average kinetic energy at room temperature (293 K).

26. Neutrons ($m = 1.67 \times 10^{-27}$ kg) are being emitted by two objects. The neutrons are moving with a speed of 2.80×10^3 m/s and pass through a circular aperture whose diameter is 0.100 mm. This situation is analogous to that shown in Figure 27.29a for light. According to the Rayleigh criterion, what is the minimum angle θ_{min} (in radians) between the two objects, such that they are just resolved using neutrons?

*27. ssm From a cliff 9.5 m above a lake, a young woman (mass = 41 kg) jumps from rest, straight down into the water. At the instant she strikes the water, what is her de Broglie wavelength?

*28. Particle A is at rest. Particle B collides head-on with it. Since the collision is completely inelastic, the two particles stick together and move off with a common velocity after the collision. The masses of the particles are different, and no external forces act on them. The de Broglie wavelength of particle B before the collision is 2.0×10^{-34} m. What is the de Broglie wavelength of the object that moves off after the collision?

*29. In a television picture tube the electrons are accelerated from rest through a potential difference V. Just before an electron strikes the screen, its de Broglie wavelength is 1.0×10^{-11} m. What is the potential difference?

**30. The kinetic energy of a particle is equal to the energy of a photon. The particle moves at 5.0% of the speed of light. Find the ratio of the photon wavelength to the de Broglie wavelength of the particle.

Section 29.6 The Heisenberg Uncertainty Principle

31. ssm www In the lungs there are tiny sacs of air, which are called alveoli. The average diameter of one of these sacs

is 0.25 mm. Consider an oxygen molecule (mass = 5.3×10^{-26} kg) trapped within a sac. What is the minimum uncertainty in the velocity of this oxygen molecule?

32. Consider a line that is 2.5 m long. A moving object is somewhere along this line, but its position is not known. (a) Find the minimum uncertainty in the momentum of the object. Find the minimum uncertainty in the object's velocity, assuming that the object is (b) a golf ball (mass = 0.045 kg) and (c) an electron.

33. A proton is confined to a nucleus whose diameter is 5.5×10^{-15} m. If this distance is considered to be the uncertainty in the position of the proton, what is the minimum uncertainty in its momentum?

34. Review Conceptual Example 6 as background for this problem. When electrons pass through a single slit, as in Figure 29.15, they form a diffraction pattern. As Section 29.6 discusses, the central bright fringe extends to either side of the midpoint, according to an angle θ given by $\sin \theta = \lambda/W$, where λ is the de Broglie wavelength of the electron and W is the width of the slit. When λ is the same size as W, $\theta = 90°$, and the central fringe fills the entire observation screen. In this case, an electron passing through the slit has roughly the same probability of hitting the screen either straight ahead or anywhere off to one side or the other. Now, imagine yourself in a world where Planck's constant is large enough so you exhibit similar effects when you walk through a 0.90-m-wide doorway. Your mass is 82 kg and you walk at a speed of 0.50 m/s. How large would Planck's constant have to be in this hypothetical world?

35. ssm www Particles pass through a single slit of width 0.200 mm (see Figure 29.15). The de Broglie wavelength of each particle is 633 nm. After the particles pass through the slit, they spread out over a range of angles. Use the Heisenberg uncertainty principle to determine the minimum range of angles.

*__36.__ Suppose the minimum uncertainty in the position of a particle is equal to its de Broglie wavelength. If the particle has an average speed of 4.5×10^5 m/s, what is the minimum uncertainty in its speed?

ADDITIONAL PROBLEMS

37. A honeybee (mass = 1.3×10^{-4} kg) is crawling at a speed of 0.020 m/s. What is the de Broglie wavelength of the bee?

38. The dissociation energy of a molecule is the energy required to break apart the molecule into its separate atoms. The dissociation energy for the cyanogen molecule is 1.22×10^{-18} J. Suppose that this energy is provided by a single photon. Determine the (a) wavelength and (b) frequency of the photon. (c) In what region of the electromagnetic spectrum does this photon lie (see Figure 24.9)?

39. ssm The de Broglie wavelength of a proton in a particle accelerator is 1.30×10^{-14} m. Determine the kinetic energy (in joules) of the proton.

40. An electron and a proton have the same speed. What is the ratio $\lambda_{electron}/\lambda_{proton}$ of their de Broglie wavelengths?

41. A photon of red light has a wavelength of 730 nm, while a photon of violet light has a wavelength of 380 nm. Find the ratio p_{violet}/p_{red} of the photon momenta.

42. A magnesium surface has a work function of 3.68 eV. Electromagnetic waves with a wavelength of 215 nm strike the surface and eject electrons. Find the maximum kinetic energy of the ejected electrons. Express your answer in electron volts.

*__43.__ **ssm** The width of the central bright fringe in a diffraction pattern on a screen is identical when either electrons or red light (vacuum wavelength = 661 nm) pass through a single slit. The distance between the screen and the slit is the same in each case and is large compared to the slit width. How fast are the electrons moving?

*__44.__ A proton is located at a distance of 0.420 m from a point charge of $+8.30\ \mu C$. The repulsive electric force moves the proton until it is at a distance of 1.58 m from the charge. Suppose that the electric potential energy lost by the system is carried off by a photon that is emitted during the process. What is its wavelength?

**__45.__ Review Conceptual Example 3 before attempting this problem. A beam of visible light has a wavelength of 395 nm and shines perpendicularly on a surface. As a result, there are 3.0×10^{18} photons per second striking the surface. By using the impulse–momentum theorem (Section 7.1), obtain the average force that this beam applies to the surface when (a) the surface is a mirror, so the momentum of each photon is reversed after reflection, and (b) the surface is black, so each photon is absorbed and there are no reflected photons.

CONCEPTS & CALCULATIONS

GROUP LEARNING PROBLEMS

Note: Each of these problems consists of Concept Questions followed by a related quantitative Problem. They are designed for use by students working alone or in small learning groups. The Concept Questions involve little or no mathematics and are intended to stimulate group discussions. They focus on the concepts with which the problems deal. Recognizing the concepts is the essential initial step in any problem-solving technique.

46. Concept Questions The maximum speed of the electrons emitted from a metal surface is v_A when the wavelength of the light used is λ_A. However, a smaller maximum speed v_B is observed when a wavelength λ_B is used. (a) In which case is there a greater maximum kinetic energy KE_{max}? (b) In which case is the frequency of the light greater? (c) Is wavelength λ_A greater or smaller than λ_B? Explain your answers.

Problem The work function of a metal surface is 4.80×10^{-19} J. The maximum speed of the electrons emitted from the surface is $v_A = 7.30 \times 10^5$ m/s when the wavelength of the light used is λ_A. However, a speed of $v_B = 5.00 \times 10^5$ m/s is observed when a wavelength λ_B is used. Find the wavelengths λ_A and λ_B. Be sure your answers are consistent with your answers to the Concept Questions.

47. Concept Questions Heat is a form of energy. One way to supply the heat needed to warm an object is to irradiate the object with electromagnetic waves. (a) How is the heat Q needed to raise the temperature of an object by ΔT degrees related to its specific heat capacity $c_{specific\ heat}$ and mass m? (b) Is the energy carried by an infrared photon greater than, smaller than, or the same as the energy carried by a visible photon of light? Explain. (c) Assuming that all the photons are absorbed by the object, is the number of infrared photons needed to supply a given amount of heat greater or smaller than the number of visible photons? Explain.

Problem A glass plate has a mass of 0.50 kg and a specific heat capacity of 840 J/(kg·C°). The wavelength of infrared light is 6.0×10^{-5} m, while the wavelength of blue visible light is 4.7×10^{-7} m. Find the number of infrared photons and the number of visible photons needed to raise the temperature of the glass plate by 2.0 C°, assuming that all photons are absorbed by the glass. Verify that your answers are consistent with your answers to the Concept Questions.

48. Concept Questions In the Compton effect, momentum conservation applies, so the total momentum of the photon and the electron is the same before and after the scattering occurs. Suppose in Figure 29.10 that the direction of the incident photon is in the $+x$ direction and that the scattered photon emerges at an angle of $\theta = 90.0°$, which is the $-y$ direction. (a) Before the scattering, the total momentum has an x component. Does it have a y component? (b) After the scattering, the total momentum has an x component. Does it have a y component? (c) Does the momentum of the scattered electron have an x component? (d) Considering your answer to Question (b), and considering the fact that the scattered photon is moving along the $-y$ axis, does the momentum of the scattered electron have a y component? In each case justify your answer.

Problem In Figure 29.10 the incident photon has a wavelength of 9.00×10^{-12} m. The scattered photon emerges at $\theta = 90.0°$. Find the x and y components of the momentum of the scattered electron. Check that your answers are consistent with your answers to the Concept Questions.

49. Concept Questions Consider a Young's double-slit experiment that uses electrons. In versions A and B of the experiment the first-order bright fringes are located at angles θ_A and θ_B, where θ_B is greater than θ_A. In these two versions of the experiment, the separation between the slits is the same but the electrons have different momenta. (a) In which case is the wavelength of an electron greater? (b) In which case is the momentum of an electron greater? Explain your answers.

Problem In a Young's double-slit experiment the angle that locates the first-order bright fringes is $\theta_A = 1.6 \times 10^{-4}$ degrees when the magnitude of the electron momentum is $p_A = 1.2 \times 10^{-22}$ kg·m/s. With the same double slit, what momentum magnitude p_B is necessary, so that an angle of $\theta_B = 4.0 \times 10^{-4}$ degrees locates the first-order bright fringes?

The laser is used extensively in medicine and communications. Here, though, it has an artistic application, as Tim Walsh plays the laser harp.

CHAPTER 30

THE NATURE OF THE ATOM

30.1 RUTHERFORD SCATTERING AND THE NUCLEAR ATOM

An atom contains a small, positively charged nucleus (radius $\approx 10^{-15}$ m), which is surrounded at relatively large distances (radius $\approx 10^{-10}$ m) by a number of electrons, as Figure 30.1 illustrates. In the natural state, an atom is electrically neutral because the nucleus contains a number of protons (each with a charge of $+e$) that equals the number of electrons (each with a charge of $-e$). This model of the atom is universally accepted now and is referred to as the "nuclear atom."

The nuclear atom is a relatively recent idea. In the early part of the twentieth century a widely accepted model, due to the English physicist Joseph J. Thomson (1856–1940), pictured the atom very differently. In Thomson's view there was no nucleus at the center of an atom. Instead, the positive charge was assumed to be spread throughout the atom, forming a kind of "paste" or "pudding," in which the negative electrons were suspended like "plums."

The "plum-pudding" model was discredited in 1911 when the New Zealand physicist Ernest Rutherford (1871–1937) published experimental results that the model could not explain. As Figure 30.2 indicates, Rutherford and his co-workers directed a beam of alpha particles (α particles) at a thin metal foil made of gold. Alpha particles are positively charged particles (the nuclei of helium atoms, although this was not recognized at the time) emitted by some radioactive materials. If the "plum-pudding" model were correct, the α particles would be expected to pass nearly straight through the foil. After all, there is nothing in this model to deflect the relatively massive α particles, since the electrons have a comparatively small mass and the positive charge is spread out in a diluted "pudding." Using a zinc sulfide screen, which flashed briefly when struck by an α particle, Rutherford and co-workers were able to determine that not all the α particles passed straight through the foil. Instead, some were deflected at large angles, even backward. Rutherford himself said, "It was almost as incredible as if you had fired a fifteen-inch shell at a piece of tissue and it came back and hit you." Rutherford concluded that the positive charge, instead of being distributed thinly and uniformly throughout the atom, was concentrated in a small region called the nucleus.

But how could the electrons in a nuclear atom remain separated from the positively charged nucleus? If the electrons were stationary, they would be pulled inward by the attractive electric force of the nuclear charge. Therefore, the electrons must be moving around the nucleus in some fashion, like planets revolving around the sun. In fact, the nuclear model of the atom is sometimes referred to as the planetary model. The dimensions of the atom, however, are such that it contains a larger fraction of empty space than our solar system does, as Conceptual Example 1 discusses.

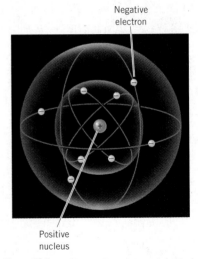

Figure 30.1 In the nuclear atom a small positively charged nucleus is surrounded at relatively large distances by a number of electrons.

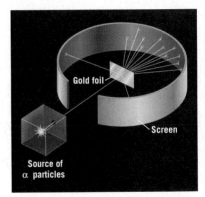

Figure 30.2 A Rutherford scattering experiment in which α particles are scattered by a thin gold foil. The entire apparatus is placed in a vacuum chamber (not shown).

CONCEPTUAL EXAMPLE 1   Atoms Are Mostly Empty Space

In the planetary model of the atom, the nucleus (radius $\approx 1 \times 10^{-15}$ m) is analogous to the sun (radius $\approx 7 \times 10^8$ m). Electrons orbit (radius $\approx 1 \times 10^{-10}$ m) the nucleus like the earth orbits (radius $\approx 1.5 \times 10^{11}$ m) the sun. If the dimensions of the solar system had the same proportions as those of the atom, would the earth be closer to or farther away from the sun than it actually is?

Reasoning and Solution The radius of an electron orbit is one hundred thousand times larger than the radius of the nucleus: $(1 \times 10^{-10}$ m$)/(1 \times 10^{-15}$ m$) = 10^5$. If the radius of the earth's orbit were one hundred thousand times larger than the radius of the sun, the earth's orbit would have a radius of $10^5 \times (7 \times 10^8$ m) $= 7 \times 10^{13}$ m. This is more than four hundred times greater than the actual orbital radius of 1.5×10^{11} m, so **the earth would be much farther away from the sun.** In fact, it would be more than ten times farther from the sun than Pluto is, which is our most distant planet and has an orbital radius of only about 6×10^{12} m. An atom, then, contains a much greater fraction of empty space than our solar system does.

Related Homework Material: *Problem 2*

While the planetary model of the atom is easy to visualize, it too is fraught with difficulties. For instance, an electron moving on a curved path has a centripetal acceleration, as Section 5.2 discusses. And when an electron is accelerating, it radiates electromagnetic waves, which carry away energy. With their energy constantly being depleted, the electrons would spiral inward and eventually collapse into the nucleus. Since matter is stable, such a collapse does not occur. Thus, the planetary model, while providing a more realistic picture of the atom than the "plum-pudding" model, must be telling only part of the story. The full story of atomic structure is fascinating, and the next section describes another aspect of it.

30.2 LINE SPECTRA

We have seen in Sections 13.3 and 29.2 that all objects emit electromagnetic waves, and we will see in Section 30.3 how this radiation arises. For a solid object, such as the hot filament of a light bulb, these waves have a continuous range of wavelengths, some of which are in the visible region of the spectrum. The continuous range of wavelengths is characteristic of the entire collection of atoms that make up the solid. In contrast, individual atoms, free of the strong interactions that are present in a solid, emit only certain specific wavelengths rather than a continuous range. These wavelengths are characteristic of the atom and provide important clues about its structure. To study the behavior of individual atoms, low-pressure gases are used in which the atoms are relatively far apart.

A low-pressure gas in a sealed tube can be made to emit electromagnetic waves by applying a sufficiently large potential difference between two electrodes located within the tube. With a grating spectroscope like that in Figure 27.36, the individual wavelengths emitted by the gas can be separated and identified as a series of bright fringes. The series of fringes is called a *line spectrum* because each bright fringe appears as a thin rectangle (a "line") resulting from the large number of parallel, closely spaced slits in the grating. Figure 30.3 shows the visible parts of the line spectra for neon and mercury. The specific visible wavelengths emitted by neon and mercury give neon signs and mercury vapor street lamps their characteristic colors.

THE PHYSICS OF...

neon signs and mercury vapor street lamps.

The simplest line spectrum is that of atomic hydrogen, and much effort has been devoted to understanding the pattern of wavelengths that it contains. Figure 30.4 illustrates in schematic for some of the groups or series of lines in the spectrum of atomic hydrogen. The group of lines in the visible region is known as the **Balmer series,** in recognition of Johann J. Balmer (1825–1898), a Swiss schoolteacher who found an empirical equation that gave the values for the observed wavelengths. This equation is given next, along with similar equations that apply to the **Lyman series**

Neon (Ne)

Mercury (Hg)

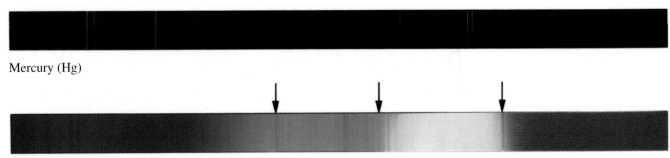

Solar absorption spectrum (Fraunhofer lines)

Figure 30.3 The line spectra for neon and mercury, along with the continuous spectrum of the sun. The dark lines in the sun's spectrum are called Fraunhofer lines, three of which are marked by arrows.

at shorter wavelengths and the ***Paschen series*** at longer wavelengths, which are also shown in the drawing:

Lyman series $\dfrac{1}{\lambda} = R\left(\dfrac{1}{1^2} - \dfrac{1}{n^2}\right)$ $n = 2, 3, 4, \ldots$ (30.1)

Balmer series $\dfrac{1}{\lambda} = R\left(\dfrac{1}{2^2} - \dfrac{1}{n^2}\right)$ $n = 3, 4, 5, \ldots$ (30.2)

Paschen series $\dfrac{1}{\lambda} = R\left(\dfrac{1}{3^2} - \dfrac{1}{n^2}\right)$ $n = 4, 5, 6, \ldots$ (30.3)

In these equations, the constant term R has the value of $R = 1.097 \times 10^7 \text{ m}^{-1}$ and is called the *Rydberg constant*. An essential feature of each group of lines is that there is a long and a short wavelength limit, with the lines being increasingly crowded toward the short wavelength limit. Figure 30.4 also gives the wavelength limits for each series, and Example 2 determines them for the Balmer series.

EXAMPLE 2   The Balmer Series

Find (a) the longest and (b) the shortest wavelengths of the Balmer series.

Reasoning Each wavelength in the series corresponds to one value for the integer n in Equation 30.2. Longer wavelengths are associated with smaller values of n. The longest wavelength occurs when n has its smallest value of $n = 3$. The shortest wavelength arises when n has a very large value, so that $1/n^2$ is essentially zero.

Figure 30.4 Line spectrum of atomic hydrogen. Only the Balmer series lies in the visible region of the electromagnetic spectrum.

Solution

(a) With $n = 3$, Equation 30.2 reveals that for the longest wavelength

$$\frac{1}{\lambda} = R\left(\frac{1}{2^2} - \frac{1}{n^2}\right) = (1.097 \times 10^7 \text{ m}^{-1})\left(\frac{1}{2^2} - \frac{1}{3^2}\right)$$

$$= 1.524 \times 10^6 \text{ m}^{-1} \quad \text{or} \quad \boxed{\lambda = 656 \text{ nm}}$$

(b) With $1/n^2 = 0$, Equation 30.2 reveals that for the shortest wavelength

$$\frac{1}{\lambda} = (1.097 \times 10^7 \text{ m}^{-1})\left(\frac{1}{2^2} - 0\right) = 2.743 \times 10^6 \text{ m}^{-1} \quad \text{or} \quad \boxed{\lambda = 365 \text{ nm}}$$

Equations 30.1–30.3 are useful because they reproduce the wavelengths that hydrogen atoms radiate. However, these equations are empirical, and they provide no insight as to *why* certain wavelengths are radiated and others are not. It was the great Danish physicist, Niels Bohr (1885–1962), who provided the first model of the atom that predicted the discrete wavelengths emitted by atomic hydrogen. Bohr's model started us on the way toward understanding how the structure of the atom restricts the radiated wavelengths to certain values. In 1922 Bohr received the Nobel Prize in Physics for his accomplishment.

30.3 THE BOHR MODEL OF THE HYDROGEN ATOM

The Model

In 1913 Bohr presented a model that led to equations such as Balmer's for predicting the specific wavelengths that the hydrogen atom radiates. Bohr's theory begins with Rutherford's picture of an atom as a nucleus surrounded by electrons moving in circular orbits. In analyzing this picture, Bohr made a number of assumptions in order to combine the new quantum ideas of Planck and Einstein with the traditional description of a particle in uniform circular motion.

Adopting Planck's idea of quantized energy levels (see Section 29.2), Bohr hypothesized that in a hydrogen atom there can be only certain values of the total energy (electron kinetic energy plus potential energy). These allowed energy levels correspond to different orbits for the electron as it moves around the nucleus, the larger orbits being associated with larger total energies. Figure 30.5 illustrates two of the orbits. In addition, Bohr assumed that an electron in one of these orbits *does not* radiate electromagnetic waves. For this reason, the orbits are called **stationary orbits** or **stationary states.** Bohr recognized that radiationless orbits violated the laws of physics, as they were then known. But the assumption of such orbits was necessary because the traditional laws indicated that an electron radiates electromagnetic waves as it accelerates around a circular path, and the loss of the energy carried by the waves would lead to the collapse of the orbit.

To incorporate Einstein's photon concept, Bohr theorized that a photon is emitted only when the electron *changes* orbits from a larger one with a higher energy to a smaller one with a lower energy, as Figure 30.5 indicates. But how do electrons get into the higher-energy orbits in the first place? They get there by picking up energy when atoms collide, which happens more often when a gas is heated, or by acquiring energy when a high voltage is applied to a gas.

When an electron in an initial orbit with a larger energy E_i changes to a final orbit with a smaller energy E_f, the emitted photon has an energy of $E_i - E_f$, consistent with the law of conservation of energy. But according to Einstein, the energy of a photon is hf, where f is its frequency and h is Planck's constant. As a result, we

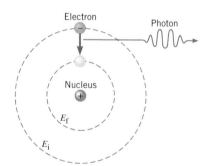

Figure 30.5 In the Bohr model, a photon is emitted when the electron drops from a larger, higher energy orbit (energy = E_i) to a smaller, lower energy orbit (energy = E_f).

Concepts at a Glance

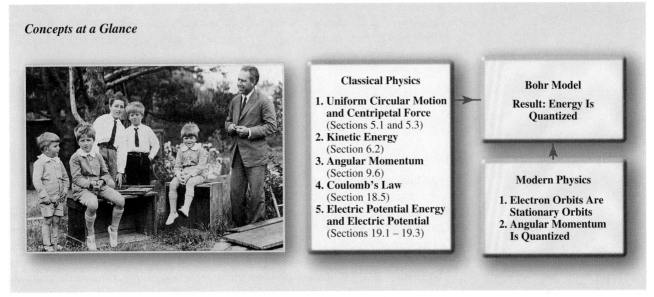

Classical Physics

1. **Uniform Circular Motion and Centripetal Force** (Sections 5.1 and 5.3)
2. **Kinetic Energy** (Section 6.2)
3. **Angular Momentum** (Section 9.6)
4. **Coulomb's Law** (Section 18.5)
5. **Electric Potential Energy and Electric Potential** (Sections 19.1 – 19.3)

Bohr Model

Result: Energy Is Quantized

Modern Physics

1. **Electron Orbits Are Stationary Orbits**
2. **Angular Momentum Is Quantized**

Figure 30.6 *Concepts at a Glance* By combining ideas from classical and modern physics, Niels Bohr was able to obtain values for the quantized energy levels of the electron in a hydrogen atom. He is shown here with his five sons.

find that

$$E_i - E_f = hf \tag{30.4}$$

Since the frequency of an electromagnetic wave is related to the wavelength by $f = c/\lambda$, Bohr could use Equation 30.4 to determine the wavelengths radiated by a hydrogen atom. First, however, he had to derive expressions for the energies E_i and E_f.

The Energies and Radii of the Bohr Orbits

● *Concepts at a Glance* To derive an expression for the quantized energy levels, Bohr brought together concepts from the older classical physics and the newer modern physics that emerged at the beginning of the twentieth century. The Concepts-at-a-Glance chart in Figure 30.6 shows the classical and modern physics that he merged to obtain the correct expression for the electron energy levels in atomic hydrogen. ●

Let's now bring together the concepts outlined in Figure 30.6 to derive Bohr's result. For an electron of mass m and speed v in an orbit of radius r, the total energy is the kinetic energy ($\mathrm{KE} = \frac{1}{2}mv^2$) of the electron plus the electric potential energy EPE. The potential energy is the product of the charge ($-e$) on the electron and the electric potential produced by the positive nuclear charge, in accord with Equation 19.3. We assume that the nucleus contains Z protons,* for a total nuclear charge of $+Ze$. The electric potential at a distance r from a point charge of $+Ze$ is given as $+kZe/r$ by Equation 19.6, where $k = 8.988 \times 10^9 \ \mathrm{N \cdot m^2/C^2}$. The electric potential energy is, then, $\mathrm{EPE} = (-e)(+kZe/r)$. Consequently, the total energy E of the atom is

$$E = \mathrm{KE} + \mathrm{EPE} = \frac{1}{2}mv^2 - \frac{kZe^2}{r} \tag{30.5}$$

But a centripetal force of magnitude mv^2/r (Equation 5.3) acts on a particle in uniform circular motion. As Figure 30.7 indicates, the centripetal force is

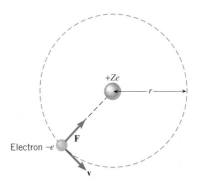

Figure 30.7 In the Bohr model, the electron is in uniform circular motion around the nucleus. The centripetal force **F** is the electrostatic force of attraction that the positive nuclear charge exerts on the electron.

---

* For hydrogen, Z = 1, but we also wish to consider situations in which Z is greater than 1.

provided by the electrostatic force of attraction **F** that the protons in the nucleus exert on the electron. According to Coulomb's law (Equation 18.1), the magnitude of the electrostatic force is $F = kZe^2/r^2$. Therefore, $mv^2/r = kZe^2/r^2$, or

$$mv^2 = \frac{kZe^2}{r} \tag{30.6}$$

We can use this relation to eliminate the term mv^2 from Equation 30.5, with the result that

$$E = \frac{1}{2}\left(\frac{kZe^2}{r}\right) - \frac{kZe^2}{r} = -\frac{kZe^2}{2r} \tag{30.7}$$

The total energy of the atom is negative because the negative electric potential energy is larger in magnitude than the positive kinetic energy.

A value for the radius r is needed, if Equation 30.7 is to be useful. To determine r, Bohr made an assumption about the angular momentum of the electron. The angular momentum L is given by Equation 9.10 as $L = I\omega$, where $I = mr^2$ is the moment of inertia of the electron moving on its circular path and $\omega = v/r$ is the angular speed of the electron in radians per second. Thus, the angular momentum is $L = (mr^2)(v/r) = mvr$. Bohr conjectured that the angular momentum can assume only certain discrete values; in other words, L is quantized. He postulated that the allowed values are integer multiples of Planck's constant divided by 2π:

$$L_n = mv_n r_n = n\frac{h}{2\pi} \qquad n = 1, 2, 3, \ldots \tag{30.8}$$

Solving this equation for v_n and substituting the result into Equation 30.6 lead to the following expression for the radius r_n of the nth Bohr orbit:

$$r_n = \left(\frac{h^2}{4\pi^2 mke^2}\right)\frac{n^2}{Z} \qquad n = 1, 2, 3, \ldots \tag{30.9}$$

With $h = 6.626 \times 10^{-34}$ J·s, $m = 9.109 \times 10^{-31}$ kg, $k = 8.988 \times 10^9$ N·m²/C², and $e = 1.602 \times 10^{-19}$ C, this expression reveals that

Radii for Bohr orbits (in meters) $\qquad r_n = (5.29 \times 10^{-11} \text{ m})\frac{n^2}{Z} \qquad n = 1, 2, 3, \ldots \tag{30.10}$

Therefore, in the hydrogen atom ($Z = 1$) the smallest Bohr orbit ($n = 1$) has a radius of $r_1 = 5.29 \times 10^{-11}$ m. This particular value is called the **Bohr radius.** Figure 30.8 shows the first three Bohr orbits for the hydrogen atom.

The expression for the radius of a Bohr orbit can be substituted into Equation 30.7 to show that the corresponding total energy for the nth orbit is

$$E_n = -\left(\frac{2\pi^2 mk^2 e^4}{h^2}\right)\frac{Z^2}{n^2} \qquad n = 1, 2, 3, \ldots \tag{30.11}$$

Substituting values for h, m, k, and e into this expression yields

Bohr energy levels in joules $\qquad E_n = -(2.18 \times 10^{-18} \text{ J})\frac{Z^2}{n^2} \qquad n = 1, 2, 3, \ldots \tag{30.12}$

Often, atomic energies are expressed in electron volts rather than joules. Since 1.60×10^{-19} J = 1 eV, Equation 30.12 can be rewritten as

Bohr energy levels in electron volts $\qquad E_n = -(13.6 \text{ eV})\frac{Z^2}{n^2} \qquad n = 1, 2, 3, \ldots \tag{30.13}$

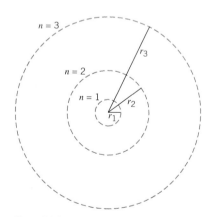

Figure 30.8 The first Bohr orbit in the hydrogen atom has a radius $r_1 = 5.29 \times 10^{-11}$ m. The second and third Bohr orbits have radii $r_2 = 4r_1$ and $r_3 = 9r_1$, respectively.

Energy Level Diagrams

It is useful to represent the energy values given by Equation 30.13 on an *energy level diagram,* as in Figure 30.9. In this diagram, which applies to the hydrogen atom ($Z = 1$), the highest energy level corresponds to $n = \infty$ in Equation 30.13 and has an energy of 0 eV. This is the energy of the atom when the electron is completely removed ($r = \infty$) from the nucleus and is at rest. In contrast, the lowest energy level corresponds to $n = 1$ and has a value of -13.6 eV. The lowest energy level is called the **ground state,** to distinguish it from the higher levels, which are called **excited states.** Observe how the energies of the excited states come closer and closer together as n increases.

The electron in a hydrogen atom at room temperature spends most of its time in the ground state. To raise the electron from the ground state ($n = 1$) to the highest possible excited state ($n = \infty$), 13.6 eV of energy must be supplied. Supplying this amount of energy removes the electron from the atom, producing the positive hydrogen ion H^+. The energy needed to remove the electron is called the **ionization energy.** Thus, the Bohr model predicts that the ionization energy of atomic hydrogen is 13.6 eV, in excellent agreement with the experimental value. In Example 3 the Bohr model is applied to doubly ionized lithium.

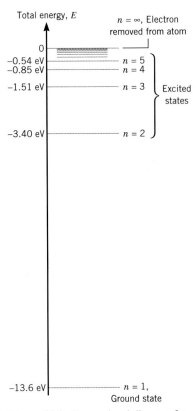

Figure 30.9 Energy level diagram for the hydrogen atom.

EXAMPLE 3   The Ionization Energy of Li^{2+}

The Bohr model does not apply when more than one electron orbits the nucleus because it does not account for the electrostatic force that one electron exerts on another. For instance, an electrically neutral lithium atom (Li) contains three electrons in orbit around a nucleus that includes three protons ($Z = 3$), and Bohr's analysis is not applicable. However, the Bohr model can be used for the doubly charged positive ion of lithium (Li^{2+}) that results when two electrons are removed from the neutral atom, leaving only one electron to orbit the nucleus. Obtain the ionization energy that is needed to remove the remaining electron from Li^{2+}.

Reasoning The lithium ion Li^{2+} contains three times the positive nuclear charge as that of the hydrogen atom. Therefore, the orbiting electron is attracted more strongly to the nucleus in Li^{2+} than in the hydrogen atom. As a result, we expect that more energy is required to ionize Li^{2+} than the 13.6 eV required for atomic hydrogen.

Solution The Bohr energy levels for Li^{2+} are given by Equation 30.13 with $Z = 3$; $E_n = -(13.6 \text{ eV})(3^2/n^2)$. Therefore, the ground state ($n = 1$) energy is

$$E_1 = -(13.6 \text{ eV}) \frac{3^2}{1^2} = -122 \text{ eV}$$

To remove the electron from Li^{2+}, 122 eV of energy must be supplied: $\boxed{\text{Ionization energy} = 122 \text{ eV}}$. This value for the ionization energy agrees well with the experimental value of 122.4 eV and, as expected, is greater than the 13.6 eV required for atomic hydrogen.

The Line Spectra of the Hydrogen Atom

● *Concepts at a Glance* To predict the wavelengths in the line spectrum of the hydrogen atom, Bohr combined his ideas about atoms (electron orbits are stationary orbits and the angular momentum of an electron is quantized) with Einstein's idea of the photon. This synthesis is illustrated in the Concepts-at-a-Glance chart in Figure 30.10, which is an extension of the one in Figure 30.6. ●

Concepts at a Glance

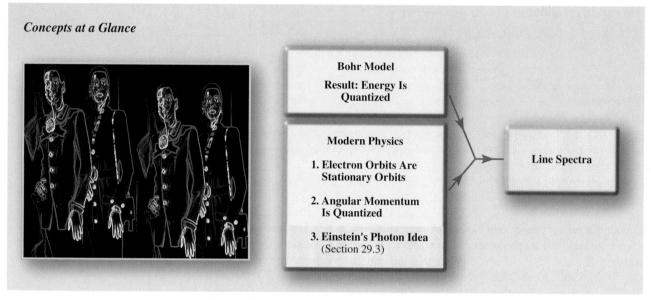

Figure 30.10 *Concepts at a Glance* This chart extends the chart in Figure 30.6 and shows that Bohr was able to predict the line spectrum of the hydrogen atom by combining his quantized energy level expression with Einstein's photon concept. The visible wavelengths in the line spectrum of atoms are used to achieve special effects, such as the red hair and blue colors in this sign.

Figure 30.11 The Lyman and Balmer series of lines in the hydrogen atom spectrum correspond to transitions that the electron makes between higher and lower energy levels, as indicated here.

As applied by Bohr, the photon concept is inherent in Equation 30.4, $E_i - E_f = hf$, which states that the frequency f of the photon is proportional to the difference between two energy levels of the hydrogen atom. If we substitute Equation 30.11 for the total energies E_i and E_f into Equation 30.4 and recall from Equation 16.1 that $f = c/\lambda$, we obtain the following result:

$$\frac{1}{\lambda} = \frac{2\pi^2 m k^2 e^4}{h^3 c}(Z^2)\left(\frac{1}{n_f^2} - \frac{1}{n_i^2}\right) \tag{30.14}$$

$$n_i, n_f = 1, 2, 3, \ldots \quad \text{and} \quad n_i > n_f$$

With the known values for h, m, k, e, and c, it can be seen that $2\pi^2 m k^2 e^4/(h^3 c) = 1.097 \times 10^7$ m^{-1}, in agreement with the Rydberg constant R that appears in Equations 30.1–30.3. The agreement between the theoretical and experimental values of the Rydberg constant was a major accomplishment of Bohr's theory.

With $Z = 1$ and $n_f = 1$, Equation 30.14 reproduces Equation 30.1 for the Lyman series. Thus, Bohr's model shows that the Lyman series of lines occurs when electrons make transitions from higher energy levels with $n_i = 2, 3, 4, \ldots$ to the first energy level where $n_f = 1$. Figure 30.11 shows these transitions. Notice that when an electron makes a transition from $n_i = 2$ to $n_f = 1$, the longest wavelength photon in the Lyman series is emitted, since the energy change is the smallest possible. When an electron makes a transition from the highest level where $n_i = \infty$ to the lowest level where $n_f = 1$, the shortest wavelength is emitted, since the energy change is the largest possible. Since the higher energy levels are increasingly close together, the lines in the series become more and more crowded toward the short wavelength limit, as can be seen in Figure 30.4. Figure 30.11 also shows the energy level transitions for the Balmer series, where $n_i = 3, 4, 5, \ldots$, while $n_f = 2$. In the Paschen series (see Figure 30.4) $n_i =$

4, 5, 6, . . . , while $n_f = 3$. The next example deals further with the line spectrum of the hydrogen atom.

EXAMPLE 4
The Brackett Series for Atomic Hydrogen

In the line spectrum of atomic hydrogen there is also a group of lines known as the Brackett series. These lines are produced when electrons, excited to high energy levels, make transitions to the $n = 4$ level. Determine (a) the longest wavelength in this series and (b) the wavelength that corresponds to the transition from $n_i = 6$ to $n_f = 4$. (c) Refer to Figure 24.9 and identify the spectral region in which these lines are found.

Reasoning The longest wavelength corresponds to the transition that has the smallest energy change, which is between the $n_i = 5$ and $n_f = 4$ levels in Figure 30.9. The wavelength for this transition, as well as that for the transition from $n_i = 6$ to $n_f = 4$, can be obtained from Equation 30.14.

Solution

(a) Using Equation 30.14 with $Z = 1$, $n_i = 5$, and $n_f = 4$, we find that

$$\frac{1}{\lambda} = (1.097 \times 10^7 \text{ m}^{-1})(1^2)\left(\frac{1}{4^2} - \frac{1}{5^2}\right)$$

$$= 2.468 \times 10^5 \text{ m}^{-1} \quad \text{or} \quad \boxed{\lambda = 4051 \text{ nm}}$$

(b) The calculation here is similar to that in part (a):

$$\frac{1}{\lambda} = (1.097 \times 10^7 \text{ m}^{-1})(1^2)\left(\frac{1}{4^2} - \frac{1}{6^2}\right)$$

$$= 3.809 \times 10^5 \text{ m}^{-1} \quad \text{or} \quad \boxed{\lambda = 2625 \text{ nm}}$$

(c) According to Figure 24.9, these lines lie in the infrared region of the spectrum.

INTERACTIVE LEARNINGWARE #57

START — Visualization — Reasoning — Knowns and Unknowns — Modeling the Problem — Numerical Solution — FINISH

A stationary hydrogen atom ($m = 1.67 \times 10^{-27}$ kg) emits a photon by making a transition from the $n_i = 3$ to the $n_f = 1$ state. (a) What is the momentum of the emitted photon? (b) Determine the recoil speed of the hydrogen atom.

This is a great Interactive Learning-Ware problem because it reviews three concepts that we have studied: the emission of a photon when an electron in a hydrogen atom goes from a higher to a lower energy level, the relation between the energy and momentum of a photon (Section 29.4), and the conservation of linear momentum (Section 7.2). These ideas are brought together to find the speed of the hydrogen atom just after it emits the photon.

Go to our Web site at

http://www.wiley.com/college/cutnell

• **PROBLEM SOLVING INSIGHT**
In the line spectrum of atomic hydrogen, all lines in a given series (e.g., the Lyman series) are identified by a single value of the quantum number n_f for the *lower energy level into which an electron falls*. Each line in a given series, however, corresponds to a different value of the quantum number n_i for the *higher energy level where an electron originates*.

THE PHYSICS OF...

absorption lines in the sun's spectrum.

The various lines in the hydrogen atom spectrum are produced when electrons change from higher to lower energy levels and photons are emitted. Consequently, the spectral lines are called *emission lines.* Electrons can also make transitions in the reverse direction, from lower to higher levels, in a process known as absorption. In this case, an atom absorbs a photon that has precisely the energy needed to produce the transition. Thus, if photons with a continuous range of wavelengths pass through a gas and then are analyzed with a grating spectroscope, a series of dark *absorption lines* appear in the continuous spectrum. The dark lines indicate the wavelengths that have been removed by the absorption process. Such absorption lines can be seen in Figure 30.3 in the spectrum of the sun, where they are called Fraunhofer lines, after their discoverer. They are due to atoms, located in the outer and cooler layers of the sun, that absorb radiation coming from the interior. The interior portion of the sun emits a continuous spectrum of wavelengths, since it is too hot for individual atoms to retain their structures.

The Bohr model provides a great deal of insight into atomic structure. However, this model is now known to be oversimplified and has been superseded by a more detailed picture provided by quantum mechanics and the Schrödinger equation (see Section 30.5).

30.4 DE BROGLIE'S EXPLANATION OF BOHR'S ASSUMPTION ABOUT ANGULAR MOMENTUM

Of all the assumptions Bohr made in his model of the hydrogen atom, perhaps the most puzzling is the one about the angular momentum of the electron ($L_n = mv_nr_n = nh/2\pi$; $n = 1, 2, 3, \ldots$). Why should the angular momentum have only those values that are integer multiples of Planck's constant divided by 2π? In 1923, ten years after Bohr's work, de Broglie pointed out that his own theory for the wavelength of a moving particle could provide an answer to this question.

In de Broglie's way of thinking, the electron in its circular Bohr orbit must be pictured as a particle wave. And like waves traveling on a string, particle waves can lead to standing waves under resonant conditions. Section 17.5 discusses these conditions for a string. Standing waves form when the total distance traveled by a wave down the string and back is one wavelength, two wavelengths, or any integer number of wavelengths. The total distance around a Bohr orbit of radius r is the circumference of the orbit or $2\pi r$. By the same reasoning, then, the condition for standing particle waves for the electron in a Bohr orbit would be

$$2\pi r = n\lambda \qquad n = 1, 2, 3, \ldots$$

where n is the number of whole wavelengths that fit into the circumference of the circle. But according to Equation 29.8 the de Broglie wavelength of the electron is $\lambda = h/p$, where p is the magnitude of the electron's momentum. If the speed of the electron is much less than the speed of light, the momentum is $p = mv$, and the condition for standing particle waves becomes $2\pi r = nh/(mv)$. A rearrangement of this result gives

$$mvr = n \frac{h}{2\pi} \qquad n = 1, 2, 3, \ldots$$

which is just what Bohr assumed for the angular momentum of the electron. As an example, Figure 30.12 illustrates the standing particle wave on a Bohr orbit for which $2\pi r = 4\lambda$.

De Broglie's explanation of Bohr's assumption about angular momentum emphasizes an important fact, namely, that particle waves play a central role in the structure of the atom. Moreover, the theoretical framework of quantum mechanics provides the basis for determining the wave function Ψ (Greek letter *psi*) that represents a particle wave. The next section deals with the picture that quantum mechanics gives for atomic structure, a picture that supersedes the Bohr model. In any case, the Bohr expression for the energy levels (Equation 30.11) can be applied when a single electron orbits the nucleus, while the theoretical framework of quantum mechanics can be applied, in principle, to atoms that contain an arbitrary number of electrons.

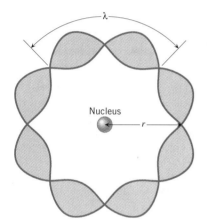

Figure 30.12 De Broglie suggested standing particle waves as an explanation for Bohr's angular momentum assumption. Here, a standing particle wave is illustrated on a Bohr orbit where four de Broglie wavelengths fit into the circumference of the orbit.

30.5 THE QUANTUM MECHANICAL PICTURE OF THE HYDROGEN ATOM

The picture of the hydrogen atom that quantum mechanics and the Schrödinger equation provide differs in a number of ways from the Bohr model. The Bohr model uses a single integer number n to identify the various electron orbits and the associated energies. Because this number can have only discrete values, rather than a continuous range of values, n is called a ***quantum number.*** In contrast, quantum mechanics reveals that four different quantum numbers are needed to describe each state of the hydrogen atom. These four are described below.

1. **The principal quantum number n.** As in the Bohr model, this number determines the total energy of the atom and can have only integer values: $n = 1, 2, 3, \ldots$. In fact, the Schrödinger equation predicts* that the energy of the hydrogen atom is identical to that obtained from the Bohr model: $E_n = -(13.6 \text{ eV}) Z^2/n^2$.

2. **The orbital quantum number ℓ.** This number determines the angular momentum of the electron due to its orbital motion. The values that ℓ can have depend on the value of n, and only the following integers are allowed:
$$\ell = 0, 1, 2, \ldots , (n - 1)$$

   For instance, if $n = 1$, the orbital quantum number can have only the value $\ell = 0$, but if $n = 4$, the values $\ell = 0, 1, 2$, and 3 are possible. The magnitude L of the angular momentum of the electron is

$$L = \sqrt{\ell(\ell + 1)} \, \frac{h}{2\pi} \qquad (30.15)$$

3. **The magnetic quantum number m_ℓ.** The word "magnetic" is used here because an externally applied magnetic field influences the energy of the atom, and this quantum number is used in describing the effect. Since the effect was discovered by the Dutch physicist Pieter Zeeman (1865–1943), it is known as the *Zeeman effect.* When there is no external magnetic field, m_ℓ plays no role in determining the energy. In either event, the magnetic quantum number determines the component of the angular momentum along a specific direction, which is called the z direction by convention. The values that m_ℓ can have depend on the value of ℓ, with only the following positive and negative integers being permitted:
$$m_\ell = -\ell, \ldots , -2, -1, 0, +1, +2, \ldots , +\ell$$

   For example, if the orbital quantum number is $\ell = 2$, then the magnetic quantum number can have the values $m_\ell = -2, -1, 0, +1$, and $+2$. The component L_z of the angular momentum in the z direction is

$$L_z = m_\ell \, \frac{h}{2\pi} \qquad (30.16)$$

4. **The spin quantum number m_s.** This number is needed because the electron has an intrinsic property called spin angular momentum. Loosely speaking, we can view the electron as spinning while it orbits the nucleus, analogous to the way the earth spins as it orbits the sun. There are two possible values for the spin quantum number of the electron:
$$m_s = +\tfrac{1}{2} \quad \text{or} \quad m_s = -\tfrac{1}{2}$$

   Sometimes the phrases "spin up" and "spin down" are used to refer to the directions of the spin angular momentum associated with the values for m_s.

---

* This prediction requires that small relativistic effects and small interactions within the atom be ignored, and assumes that the hydrogen atom is not located in an external magnetic field.

Table 30.1 summarizes the four quantum numbers that are needed to describe each state of the hydrogen atom. One set of values for n, ℓ, m_ℓ, and m_s corresponds to one state. As the principal quantum number n increases, the number of possible combinations of the four quantum numbers rises rapidly, as Example 5 illustrates.

EXAMPLE 5   Quantum Mechanical States of the Hydrogen Atom

Determine the number of possible states for the hydrogen atom when the principal quantum number is (a) $n = 1$ and (b) $n = 2$.

Reasoning Each different combination of the four quantum numbers summarized in Table 30.1 corresponds to a different state. We begin with the value for n and find the allowed values for ℓ. Then, for each ℓ value we find the possibilities for m_ℓ. Finally, m_s may be $+\frac{1}{2}$ or $-\frac{1}{2}$ for each group of values for n, ℓ, and m_ℓ.

Solution

(a) The diagram below shows the possibilities for ℓ, m_ℓ, and m_s when $n = 1$:

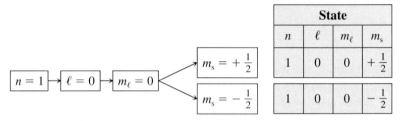

Thus, there are two different states for the hydrogen atom. These two states have the same energy, since they have the same value of n.

(b) When $n = 2$, there are eight possible combinations for the values of n, ℓ, m_ℓ, and m_s, as the diagram below indicates:

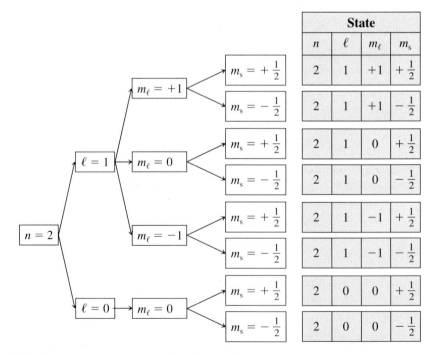

With the same value of $n = 2$, all eight states have the same energy.

TABLE 30.1   Quantum Numbers for the Hydrogen Atom

Name	Symbol	Allowed Values
Principal quantum number	n	1, 2, 3, . . .
Orbital quantum number	ℓ	0, 1, 2, . . . , $(n-1)$
Magnetic quantum number	m_ℓ	$-\ell, \ldots, -2, -1, 0, +1, +2, \ldots, +\ell$
Spin quantum number	m_s	$-\frac{1}{2}, +\frac{1}{2}$

Quantum mechanics provides a more accurate picture of atomic structure than does the Bohr model. It is important to realize that the two pictures differ substantially, as Conceptual Example 6 illustrates.

CONCEPTUAL EXAMPLE 6   The Bohr Model Versus Quantum Mechanics

Consider two hydrogen atoms. There are no external magnetic fields present, and the electron in each atom has the same energy. Is it possible for the electrons in these atoms (a) to have zero orbital angular momentum and (b) to have different orbital angular momenta? Answer these questions according to the Bohr model and to quantum mechanics.

Reasoning and Solution

(a) In both the Bohr model and quantum mechanics, the energy is proportional to $1/n^2$, according to Equation 30.13, where n is the principal quantum number. Moreover, the value of n may be $n = 1, 2, 3, \ldots$, and may not be zero. *In the Bohr model, the fact that n may not be zero means that it is not possible for the orbital angular momentum to be zero* because the angular momentum is proportional to n, according to Equation 30.8. In the quantum mechanical picture the magnitude of the orbital angular momentum is proportional to $\sqrt{\ell(\ell+1)}$, as given by Equation 30.15. Here, ℓ is the orbital quantum number and may take on the values $\ell = 0, 1, 2, \ldots , (n-1)$. We note that ℓ [and therefore $\sqrt{\ell(\ell+1)}$] may be zero, no matter what the value for n. Consequently, *the orbital angular momentum may be zero according to quantum mechanics,* in contrast to the case for the Bohr model.

(b) If the electrons have the same energy, they have the same value for the principal quantum number n. *In the Bohr model, this means that they cannot have different values for the orbital angular momentum* L_n, since $L_n = nh/(2\pi)$, according to Equation 30.8. In quantum mechanics, the energy is also determined by n when external magnetic fields are absent, but the orbital angular momentum is determined by ℓ. Since $\ell = 0, 1, 2, \ldots , (n-1)$, different values of ℓ are compatible with the same value of n. For instance, if $n = 2$ for both electrons, one of them could have $\ell = 0$, while the other could have $\ell = 1$. *According to quantum mechanics, then, the electrons could have different orbital angular momenta, even though they have the same energy.*

The following table summarizes the discussion from parts (a) and (b):

	Bohr Model	Quantum Mechanics
(a) For a given n, can the angular momentum ever be zero?	No	Yes
(b) For a given n, can the angular momentum have different values?	No	Yes

Related Homework Material: *Question 5, Problem 26*

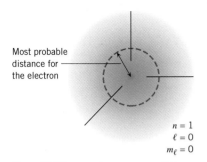

Most probable distance for the electron

$n = 1$
$\ell = 0$
$m_\ell = 0$

Figure 30.13 The electron probability cloud for the ground state ($n = 1$, $\ell = 0$, $m_\ell = 0$) of the hydrogen atom.

According to the Bohr model, the nth orbit is a circle of radius r_n, and every time the position of the electron in this orbit is measured, the electron is found exactly at a distance r_n away from the nucleus. This simplistic picture is now known to be incorrect, and the quantum mechanical picture of the atom has replaced it. Suppose the electron is in a quantum mechanical state for which $n = 1$, and we imagine making a number of measurements of the electron's position with respect to the nucleus. We would find that its position is uncertain, in the sense that there is a probability of finding the electron sometimes very near the nucleus, sometimes very far from the nucleus, and sometimes at intermediate locations. The probability is determined by the wave function Ψ, as Section 29.5 discusses. We can make a three-dimensional picture of our findings by marking a dot at each location where the electron is found. More dots occur at places where the probability of finding the electron is higher, and after a sufficient number of measurements, a picture of the quantum mechanical state emerges. Figure 30.13 shows the spatial distribution for an electron in a state for which $n = 1$, $\ell = 0$, and $m_\ell = 0$. This picture is constructed from so many measurements that the individual dots are no longer visible but have merged to form a kind of probability "cloud" whose density changes gradually from place to place. The dense regions indicate places where the probability of finding the electron is higher, while the less dense regions indicate places where the probability is lower. Also indicated in Figure 30.13 is the radius where quantum mechanics predicts the greatest probability per unit radial distance of finding the electron in the $n = 1$ state. This radius matches exactly the radius of 5.29×10^{-11} m found for the first Bohr orbit.

For a principal quantum number of $n = 2$, the probability clouds are different than for $n = 1$. In fact, more than one cloud shape is possible because with $n = 2$ the orbital quantum number can be either $\ell = 0$ or $\ell = 1$. While the value of ℓ does not affect the energy of the hydrogen atom, the value does have a significant effect on the shape of the probability clouds. Figure 30.14a shows the cloud for $n = 2$, $\ell = 0$, and $m_\ell = 0$. Part b of the drawing shows that when $n = 2$, $\ell = 1$, and $m_\ell = 0$, the cloud has a two-lobe shape with the nucleus at the center between the lobes. For larger values of n, the probability clouds become increasingly complex and are spread out over larger volumes of space.

The probability cloud picture of the electron in a hydrogen atom is very different from the well-defined orbit of the Bohr model. The fundamental reason for this difference is to be found in the Heisenberg uncertainty principle, as Conceptual Example 7 discusses.

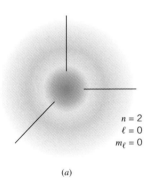

$n = 2$
$\ell = 0$
$m_\ell = 0$

(a)

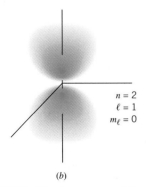

$n = 2$
$\ell = 1$
$m_\ell = 0$

(b)

Figure 30.14 The electron probability clouds for the hydrogen atom when (a) $n = 2$, $\ell = 0$, $m_\ell = 0$ and (b) $n = 2$, $\ell = 1$, $m_\ell = 0$.

CONCEPTUAL EXAMPLE 7
The Uncertainty Principle and the Hydrogen Atom

In the Bohr model of the hydrogen atom, the electron in the ground state ($n = 1$) is in an orbit that has a radius of exactly 5.29×10^{-11} m. Furthermore, as calculated in Example 3 in Chapter 18, the speed of the electron in this orbit is exactly 2.18×10^6 m/s. Considering the Heisenberg uncertainty principle, is this a realistic picture of atomic structure?

Reasoning and Solution The Bohr model indicates that the electron is located exactly at a radius of 5.29×10^{-11} m, so the uncertainty Δy in its radial position is zero. We will now show that an uncertainty of $\Delta y = 0$ m is not consistent with the Heisenberg principle, so that the Bohr picture is not realistic. What does the Heisenberg principle [Equation 29.10, $(\Delta p_y)(\Delta y) \geq h/(2\pi)$] say about Δy? According to the principle, the minimum uncertainty in the radial position of the electron is $\Delta y = h/(2\pi \, \Delta p_y)$, where Δp_y is the uncertainty in the momentum. The magnitude of the electron's momentum is the product of its mass m and speed v, so that $\Delta p_y = $

$\Delta(mv) = m\,\Delta v$. It seems reasonable to assume that the uncertainty Δv in the speed should be less than 2.18×10^6 m/s. This means that the speed is somewhere between zero and twice the value known to apply to the ground state in the Bohr model. A much larger uncertainty would mean that the electron could have so much energy that it would not likely remain in orbit. With $\Delta v = 2.18 \times 10^6$ m/s and $m = 9.11 \times 10^{-31}$ kg for the mass of the electron, we find that the minimum uncertainty in the radial position of the electron is $\Delta y = h/(2\pi m\,\Delta v) = 5.3 \times 10^{-11}$ m. This uncertainty is quite large, for it is virtually equal to the Bohr radius. According to the uncertainty principle, then, the radial position of the electron can be anywhere from approximately zero to twice the Bohr radius. **The single-radius orbit of the Bohr model does not correctly represent this aspect of reality at the atomic level.** Quantum mechanics, however, does correctly represent it in terms of a probability cloud picture of atomic structure.

30.6 THE PAULI EXCLUSION PRINCIPLE AND THE PERIODIC TABLE OF THE ELEMENTS

Except for hydrogen, all electrically neutral atoms contain more than one electron, the number being given by the atomic number Z of the element. In addition to being attracted by the nucleus, the electrons repel each other. This repulsion contributes to the total energy of a multiple-electron atom. As a result, the one-electron energy expression for hydrogen, $E_n = -(13.6 \text{ eV})\,Z^2/n^2$, does not apply to other neutral atoms. However, the simplest approach for dealing with a multiple-electron atom still uses the four quantum numbers n, ℓ, m_ℓ, and m_s.

Detailed quantum mechanical calculations reveal that the energy level of each state of a multiple-electron atom depends on both the principal quantum number n and the orbital quantum number ℓ. Figure 30.15 illustrates that the energy generally increases as n increases, but there are exceptions, as the drawing indicates. Furthermore, for a given n, the energy also increases as ℓ increases.

In a multiple-electron atom, all electrons with the same value of n are said to be in the same **shell.** Electrons with $n = 1$ are in a single shell (sometimes called the K shell), electrons with $n = 2$ are in another shell (the L shell), those with $n = 3$ are in a third shell (the M shell), and so on. Those electrons with the same values for both n and ℓ are often referred to as being in the same **subshell.** The $n = 1$ shell consists of a single $\ell = 0$ subshell. The $n = 2$ shell has two subshells, one with $\ell = 0$ and one with $\ell = 1$. Similarly, the $n = 3$ shell has three subshells, one with $\ell = 0$, one with $\ell = 1$, and one with $\ell = 2$.

In the hydrogen atom near room temperature, the electron spends most of its time in the lowest energy level or ground state, namely, in the $n = 1$ shell. Similarly, when an atom contains more than one electron and is near room temperature, the electrons spend most of their time in the lowest energy levels possible. The lowest energy state for an atom is called the **ground state.** However, when a multiple-electron atom is in its ground state, not every electron is in the $n = 1$ shell in general because the electrons obey a principle discovered by the Austrian physicist Wolfgang Pauli (1900–1958).

● THE PAULI EXCLUSION PRINCIPLE

No two electrons in an atom can have the same set of values for the four quantum numbers n, ℓ, m_ℓ, and m_s.

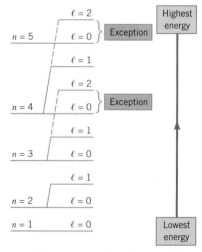

Figure 30.15 When there is more than one electron in an atom, the total energy of a given state depends on the principal quantum number n and the orbital quantum number ℓ. Generally, the energy increases with increasing n and, for a fixed n, with increasing ℓ. There are exceptions to the general rule, however, as indicated here. For clarity, levels for $n = 6$ and higher are not shown.

For instance, suppose two electrons in an atom have three quantum numbers that are identical: $n = 3$, $m_\ell = 1$, and $m_s = -\frac{1}{2}$. According to the exclusion principle, it is not possible for each to have $\ell = 2$, for example, since each would then have the same four quantum numbers. Each electron must have a different value for ℓ($\ell = 1$ and $\ell = 2$, for instance) and, consequently, would be in a different subshell. With the aid of the Pauli exclusion principle, we can determine which energy levels are occupied by the electrons in an atom in its ground state, as the next example demonstrates.

EXAMPLE 8   Ground States of Atoms

Determine which of the energy levels in Figure 30.15 are occupied by the electrons in the ground state of hydrogen (1 electron), helium (2 electrons), lithium (3 electrons), beryllium (4 electrons), and boron (5 electrons).

Reasoning In the ground state of an atom the electrons are in the lowest available energy levels. Consistent with the Pauli exclusion principle, they fill those levels "from the bottom up," that is, from the lowest to the highest energy.

Solution As the colored dot (●) in Figure 30.16 indicates, the electron in the hydrogen atom (H) is in the $n = 1$, $\ell = 0$ subshell, which has the lowest possible energy. A second electron is present in the helium atom (He), and both electrons can have the quantum numbers $n = 1$, $\ell = 0$, and $m_\ell = 0$. However, in accord with the Pauli exclusion principle, each electron must have a different spin quantum number, $m_s = +\frac{1}{2}$ for one electron and $m_s = -\frac{1}{2}$ for the other. Thus, the drawing shows both electrons in the lowest energy level.

The third electron that is present in the lithium atom (Li) would violate the exclusion principle if it were also in the $n = 1$, $\ell = 0$ subshell, no matter what the value for m_s. Thus, the $n = 1$, $\ell = 0$ subshell is filled when occupied by two electrons. With this level filled, the $n = 2$, $\ell = 0$ subshell becomes the next lowest energy level available and is where the third electron of lithium is found (see Figure 30.16). In the beryllium atom (Be), the fourth electron is in the $n = 2$, $\ell = 0$ subshell, along with the third electron. This is possible, since the third and fourth electrons can have different values for m_s.

With the first four electrons in place as just discussed, the fifth electron in the boron atom (B) cannot fit into the $n = 1$, $\ell = 0$ or the $n = 2$, $\ell = 0$ subshell without violating the exclusion principle. Therefore, the fifth electron is found in the $n = 2$, $\ell = 1$ subshell, which is the next available energy level with the lowest energy, as Figure 30.16 indicates. For this electron, m_ℓ can be -1, 0, or $+1$, and m_s can be $+\frac{1}{2}$ or $-\frac{1}{2}$ in each case. However, in the absence of an external magnetic field, all six of these possibilities correspond to the same energy.

Because of the Pauli exclusion principle, there is a maximum number of electrons that can fit into an energy level or subshell. Example 8 shows that the $n = 1$, $\ell = 0$ subshell can hold at most two electrons. The $n = 2$, $\ell = 1$ subshell, however, can hold six electrons because with $\ell = 1$, there are three possibilities for m_ℓ

Figure 30.16 The electrons (●) in the ground state of an atom fill the available energy levels "from the bottom up," that is, from the lowest to the highest energy, consistent with the Pauli exclusion principle. The ranking of the energy levels in this figure is meant to apply for a given atom, not between one atom and another.

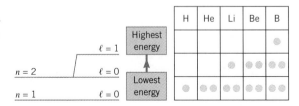

(-1, 0, and $+1$), and for each of these, the value of m_s can be $+\frac{1}{2}$ or $-\frac{1}{2}$. In general, m_ℓ can have the values $0, \pm 1, \pm 2, \ldots, \pm \ell$, for $2\ell + 1$ possibilities. Since each of these can be combined with two possibilities for m_s, the total number of different combinations for m_ℓ and m_s is $2(2\ell + 1)$. This, then, is the maximum number of electrons the ℓth subshell can hold, as Figure 30.17 summarizes.

For historical reasons, there is a widely used convention in which each subshell of an atom is referred to by a letter rather than by the value of its orbital quantum number ℓ. For instance, an $\ell = 0$ subshell is called an s subshell. An $\ell = 1$ subshell and an $\ell = 2$ subshell are known as p and d subshells, respectively. The higher values of $\ell = 3$, 4, etc., are referred to as f, g, etc., in alphabetical sequence, as Table 30.2 indicates.

This convention of letters is used in a shorthand notation that is convenient for indicating simultaneously the principal quantum number n, the orbital quantum number ℓ, and the number of electrons in the n, ℓ subshell. The notation is as follows:

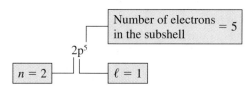

With this notation, the arrangement or configuration of the electrons in an atom can be specified efficiently. For instance, in Example 8, we found that the electron configuration for boron has two electrons in the $n = 1$, $\ell = 0$ subshell, two in the $n = 2$, $\ell = 0$ subshell, and one in the $n = 2$, $\ell = 1$ subshell. In shorthand notation this arrangement is expressed as $1s^2\, 2s^2\, 2p^1$. Table 30.3 gives the ground state electron configurations written in this fashion for elements containing up to thirteen electrons. The first five entries are those worked out in Example 8.

Each entry in the periodic table of the elements often includes the ground state electronic configuration, as Figure 30.18 illustrates for argon. To save space, only

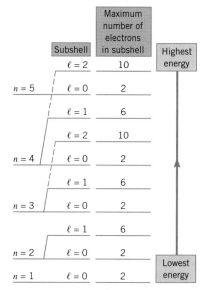

Figure 30.17 The maximum number of electrons that the ℓth subshell can hold is $2(2\ell + 1)$.

TABLE 30.2   The Convention of Letters Used to Refer to the Orbital Quantum Number

Orbital Quantum Number ℓ	Letter
0	s
1	p
2	d
3	f
4	g
5	h

TABLE 30.3   Ground State Electronic Configurations of Atoms

Element	Number of Electrons	Configuration of the Electrons
Hydrogen (H)	1	$1s^1$
Helium (He)	2	$1s^2$
Lithium (Li)	3	$1s^2\, 2s^1$
Beryllium (Be)	4	$1s^2\, 2s^2$
Boron (B)	5	$1s^2\, 2s^2\, 2p^1$
Carbon (C)	6	$1s^2\, 2s^2\, 2p^2$
Nitrogen (N)	7	$1s^2\, 2s^2\, 2p^3$
Oxygen (O)	8	$1s^2\, 2s^2\, 2p^4$
Fluorine (F)	9	$1s^2\, 2s^2\, 2p^5$
Neon (Ne)	10	$1s^2\, 2s^2\, 2p^6$
Sodium (Na)	11	$1s^2\, 2s^2\, 2p^6\, 3s^1$
Magnesium (Mg)	12	$1s^2\, 2s^2\, 2p^6\, 3s^2$
Aluminum (Al)	13	$1s^2\, 2s^2\, 2p^6\, 3s^2\, 3p^1$

Figure 30.18 The entries in the periodic table of the elements often include the ground state configuration of the outermost electrons.

Figure 30.19 In an X-ray tube, electrons are emitted by a heated filament, accelerate through a large potential difference V, and strike a metal target. The X-rays originate when the electrons interact with the metal target.

THE PHYSICS OF...

X-rays.

Figure 30.20 When a molybdenum target is bombarded with electrons that have been accelerated from rest through a potential difference of 45 000 V, this X-ray spectrum is produced. The vertical axis is not to scale.

the configuration of the outermost electrons and unfilled subshells is specified using the shorthand notation just discussed. Originally the periodic table was developed by the Russian chemist Dmitri Mendeleev (1834–1907) on the basis that certain groups of elements exhibit similar chemical properties. There are eight of these groups, plus the transition elements in the middle of the table, the lanthanide series, and the actinide series. The similar chemical properties within a group can be explained on the basis of the configurations of the outer electrons of the elements in the group. Thus, quantum mechanics and the Pauli exclusion principle offer an explanation for the chemical behavior of the atoms. The full periodic table can be found on the inside of the back cover.

30.7 X-RAYS

X-rays were discovered by Wilhelm K. Roentgen (1845–1923), a Dutch physicist who performed much of his work in Germany. X-rays can be produced when electrons, accelerated through a large potential difference, collide with a metal target made from molybdenum or platinum, for example. The target is contained within an evacuated glass tube, as Figure 30.19 shows. A plot of X-ray intensity per unit wavelength versus the wavelength looks similar to Figure 30.20 and consists of sharp peaks or lines superimposed on a broad continuous spectrum. The sharp peaks are called characteristic lines or ***characteristic X-rays*** because they are characteristic of the target material. The broad continuous spectrum is referred to as ***Bremsstrahlung*** (German for "braking radiation") and is emitted when the electrons decelerate or "brake" upon hitting the target.

In Figure 30.20 the characteristic lines are marked K_α and K_β because they involve the $n = 1$ or K shell of a metal atom. If an electron with enough energy strikes the target, one of the K-shell electrons can be knocked out. An electron in one of the outer shells can then fall into the K shell, and an X-ray photon is emitted in the process. Example 9 shows that a large potential difference is needed to operate an X-ray tube, so the electrons impinging on the metal target have sufficient energy to generate the characteristic X-rays.

EXAMPLE 9   Operating an X-Ray Tube

Strictly speaking, the Bohr model does not apply to multiple-electron atoms, but it can be used to make estimates. Use the Bohr model to estimate the minimum energy that an incoming electron must have to knock a K-shell electron entirely out of an atom in a platinum ($Z = 78$) target in an X-ray tube.

Reasoning According to the Bohr model, the energy of a K-shell electron is given by Equation 30.13 with $n = 1$: $E_n = -(13.6 \text{ eV}) Z^2/n^2$. When striking a platinum target, an incoming electron must have at least enough energy to raise the K-shell electron from this low energy level up to the 0-eV level that corresponds to a very large distance from the nucleus. Only then will the incoming electron knock the K-shell electron entirely out of a target atom.

Solution The energy of the Bohr $n = 1$ level is

$$E_1 = -(13.6 \text{ eV}) \frac{Z^2}{n^2} = -(13.6 \text{ eV}) \frac{77^2}{1^2} = -8.1 \times 10^4 \text{ eV}$$

In this calculation we have used 77 rather than 78 for the value of Z. In so doing, we account approximately for the fact that each of the two K-shell electrons applies a repulsive force to the other. This repulsive force tends to balance the attractive force of one nuclear proton. In effect, one electron shields the other from the force of that proton. Therefore, to raise the K-shell electron up to the 0-eV level, the minimum energy for an incoming electron is $\boxed{8.1 \times 10^4 \text{ eV}}$. One electron volt is the kinetic energy acquired when an electron accelerates from rest through a potential difference of one volt. Thus, a potential difference of 81 000 V must be applied to the X-ray tube.

The K_α line in Figure 30.20 arises when an electron in the $n = 2$ level falls into the vacancy that the impinging electron has created in the $n = 1$ level. Similarly, the K_β line arises when an electron in the $n = 3$ level falls to the $n = 1$ level. Example 10 determines an estimate for the K_α wavelength of platinum.

EXAMPLE 10
The K_α Characteristic X-Ray for Platinum

Use the Bohr model to estimate the wavelength of the K_α line in the X-ray spectrum of platinum.

Reasoning This example is very similar to Example 4, which deals with the emission line spectrum of the hydrogen atom. As in that example, we use Equation 30.14, this time with the initial value of n being $n_i = 2$ and the final value being $n_f = 1$. As in Example 9, a value of 77 rather than 78 is used for Z to account approximately for the shielding effect of the single K-shell electron in canceling out the attraction of one nuclear proton.

Solution Using Equation 30.14, we find that

$$\frac{1}{\lambda} = (1.097 \times 10^7 \text{ m}^{-1})(77^2)\left(\frac{1}{1^2} - \frac{1}{2^2}\right)$$
$$= 4.9 \times 10^{10} \text{ m}^{-1} \quad \text{or} \quad \boxed{\lambda = 2.0 \times 10^{-11} \text{ m}}$$

This answer is close to an experimental value of 1.9×10^{-11} m.

Another interesting feature of the X-ray spectrum in Figure 30.20 is the sharp cutoff that occurs at a wavelength of λ_0 on the short wavelength side of the Bremsstrahlung. This cutoff wavelength is independent of the target material but depends on the energy of the impinging electrons. An impinging electron cannot give up any more than all of its kinetic energy when decelerated by the metal target in an X-ray tube. Thus, at most, an emitted X-ray photon can have an energy equal to the kinetic energy KE of the electron and a frequency given by Equation 29.2 as $f = (\text{KE})/h$, where h is Planck's constant. But the kinetic energy acquired by an electron in accelerating from rest through a potential difference V is eV, according to earlier discussions in Section 19.2; V is the potential difference applied across the X-ray tube. Thus, the maximum photon frequency is $f_0 = (eV)/h$. Since $f_0 = c/\lambda_0$, a maximum frequency corresponds to a minimum wavelength, which is the cutoff wavelength λ_0:

$$\lambda_0 = \frac{hc}{eV} \qquad (30.17)$$

INTERACTIVE LEARNINGWARE #58

START
Visualization
Reasoning
Knowns and Unknowns
Modeling the Problem
Numerical Solution
FINISH

High-energy electrons impinge on a metal target in an X-ray tube. The target is a mixture of two metals, one of which is cobalt. Both metals emit K_α X-ray photons, and the wavelength that originates from the cobalt is 1.247 times as great as that produced by the unknown metal. What is the unknown metal?

A K_α X-ray arises when an electron jumps from a higher to a lower energy level in an atom, and this Interactive LearningWare problem reviews the essential features of the process. It also illustrates how one can identify an unknown element by measuring the wavelength of its K_α photon relative to that produced by cobalt.

Go to our Web site at

http://www.wiley.com/college/cutnell

An X-ray technician is preparing this patient for a diagnostic procedure.

THE PHYSICS OF...

CAT scanning.

Figure 30.20, for instance, assumes a potential difference of 45 000 V, which corresponds to a cutoff wavelength of

$$\lambda_0 = \frac{(6.63 \times 10^{-34} \text{ J} \cdot \text{s})(3.00 \times 10^8 \text{ m/s})}{(1.60 \times 10^{-19} \text{ C})(45\,000 \text{ V})} = 2.8 \times 10^{-11} \text{ m}$$

The medical profession began using X-rays for diagnostic purposes almost immediately after their discovery. When a conventional X-ray is obtained, the patient is typically positioned in front of a piece of photographic film, and a single burst of radiation is directed through the patient and onto the film. Since the dense structure of bone absorbs X-rays much more than soft tissue does, a shadow-like picture is recorded on the film. As useful as such pictures are, they have an inherent limitation. The image on the film is a superposition of all the "shadows" that result as the radiation passes through one layer of body material after another. Interpreting which part of a conventional X-ray corresponds to which layer of body material is very difficult.

The technique known as CAT scanning or CT scanning has greatly extended the ability of X-rays to provide images from specific locations within the body. The acronym CAT stands for computerized axial tomography or computer-assisted tomography, while the shorter version CT stands for computerized tomography. In this technique a series of X-ray images are obtained as indicated in Figure 30.21. A number of X-ray beams form a "fanned out" array of radiation and pass simultaneously through the patient. Each of the beams is detected on the other side by a detector, which records the beam intensity. The various intensities are different, depending on the nature of the body material through which the beams have passed. The feature of CAT scanning that leads to dramatic improvements over the conventional technique is that the X-ray source can be rotated to different orientations, so that the "fanned-out" array of beams can be sent through the patient from various directions. Figure 30.21*a* singles out two directions for illustration. In reality many different orientations are used, and the intensity of each beam in the array is recorded as a function of orientation. The way in which the intensity of a beam changes from one orientation to another is used as input to a computer. The computer then constructs a highly resolved image of the cross-sectional slice of body material through which the "fan" of radiation has passed. In effect, the CAT scanning technique makes it possible to take an X-ray picture of a cross-sectional "slice" that is perpendicular to the body's long axis. In fact, the word "axial" in the phrase "computerized axial tomography" refers to the body's long axis. Figure 30.22*a* shows a two-dimensional CAT scan of the type discussed here, while part *b* of the figure shows a three-dimensional version.

Figure 30.21 (*a*) In CAT scanning, a "fanned-out" array of X-ray beams is sent through the patient from different orientations. (*b*) A patient being positioned in a CAT scanner.

(*a*)   (*b*)

(a)   (b)

Figure 30.22 (*a*) This two-dimensional CAT scan of a brain reveals a large intracranial tumor (colored purple). (*b*) This three-dimensional CAT scan reveals an arachnoid cyst (colored yellow) within a skull. In both photographs the colors are artificial, having been computer generated to aid in distinguishing anatomical features.

30.8 THE LASER

The laser is one of the most useful inventions of the twentieth century. Today, there are many types of lasers, and most of them work in a way that depends directly on the quantum mechanical structure of the atom.

When an electron makes a transition from a higher energy state to a lower energy state, a photon is emitted. The emission process can be one of two types, spontaneous emission or stimulated emission. In ***spontaneous emission*** (see Figure 30.23*a*), the photon is emitted spontaneously, in a random direction, without external provocation. In ***stimulated emission*** (see Figure 30.23*b*), an incoming photon induces or stimulates the electron to change energy levels. To produce stimulated emission, however, the incoming photon must have an energy that exactly matches the difference between the energies of the two levels, namely, $E_i - E_f$. Stimulated emission is similar to a resonant process, in which the incoming photon "jiggles" the electron at just the frequency to which it is particularly sensitive and causes the change between energy levels. This frequency is given by Equation 30.4 as $f = (E_i - E_f)/h$. The operation of lasers depends on stimulated emission.

Stimulated emission has three important features. First, one photon goes in and two photons come out (see Figure 30.23*b*). In this sense, the process amplifies the number of photons. In fact, this is the origin of the word "laser," which is an acronym for **l**ight **a**mplification by the **s**timulated **e**mission of **r**adiation. Second, the emitted photon travels in the same direction as the incoming photon. And third, the emitted photon is exactly in step with or has the same phase as the incoming photon. In other words, the two electromagnetic waves that these two photons represent are coherent (see Section 17.2) and are locked in step with one another. In contrast, two photons emitted by the filament of an incandescent light bulb are emitted independently. They are not coherent, since one does not stimulate the emission of the other.

While stimulated emission plays a pivotal role in a laser, other factors are also important. For instance, an external source of energy must be available to excite electrons into higher energy levels. The energy can be provided in a number of ways, including intense flashes of ordinary light and high-voltage discharges. If sufficient energy is delivered to the atoms, more electrons will be excited to a higher energy level than remain in a lower level, a condition known as a ***population inversion.*** Figure 30.24 compares a normal energy level population with a population inversion. The population inversions used in lasers involve a higher energy state that is ***metastable,*** in the sense that electrons remain in it for a much longer period of time than they do in an ordinary excited state (10^{-3} s versus 10^{-8} s, for example).

THE PHYSICS OF...

the laser.

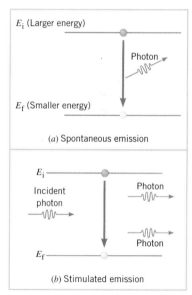

Figure 30.23 (*a*) Spontaneous emission of a photon occurs when the electron (●) makes an unprovoked transition from a higher to a lower energy level, the photon departing in a random direction. (*b*) Stimulated emission of a photon occurs when an incoming photon with the correct energy induces an electron to change energy levels, the emitted photon traveling in the same direction as the incoming photon.

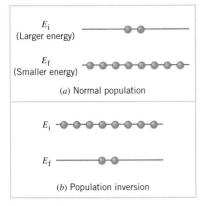

Figure 30.24 (*a*) In a normal situation at room temperature, most of the electrons in atoms are found in a lower or ground state energy level. (*b*) If an external energy source is provided to excite electrons into a higher energy level, a population inversion can be created in which more electrons are in the higher level than in the lower level.

The requirement of a metastable higher energy state is essential, so that there is more time to enhance the population inversion.

Figure 30.25 shows the widely used helium/neon laser. To sustain the necessary population inversion, a high voltage is discharged across a low-pressure mixture of 15% helium and 85% neon contained in a glass tube. The laser process begins when an atom, via spontaneous emission, emits a photon parallel to the axis of the tube. This photon, via stimulated emission, causes another atom to emit two photons parallel to the tube axis. These two photons, in turn, stimulate two more atoms, yielding four photons. Four yield eight, and so on, in a kind of avalanche. To ensure that more and more photons are created by stimulated emission, both ends of the tube are silvered to form mirrors that reflect the photons back and forth through the helium/neon mixture. One end is only partially silvered, however, so that some of the photons can escape from the tube to form the laser beam. When the stimulated emission in a laser involves only a single pair of energy levels, the output beam has a single frequency or wavelength and is said to be monochromatic.

A laser beam is also exceptionally narrow. The width is determined by the size of the opening through which the beam exits, and very little spreading-out occurs, except that due to diffraction around the edges of the opening. A laser beam does not spread much because any photons emitted at an angle with respect to the tube axis are quickly reflected out the sides of the tube by the silvered ends (see Figure 30.25). These ends are carefully arranged to be perpendicular to the tube axis. Since all the power in a laser beam can be confined to a narrow region, the intensity, or power per unit area, can be quite large.

Figure 30.26 shows the pertinent energy levels for a helium/neon laser. By coincidence, helium and neon have nearly identical metastable higher energy states, respectively located 20.61 and 20.66 eV above the ground state. The high-voltage discharge across the gaseous mixture excites electrons in helium atoms to the 20.61-eV state. Then, when an excited helium atom collides inelastically with a neon atom, the 20.61 eV of energy is given to an electron in the neon atom, along with 0.05 eV of kinetic energy from the moving atoms. As a result, the electron in the neon atom is raised to the 20.66-eV state. In this fashion, a population inversion

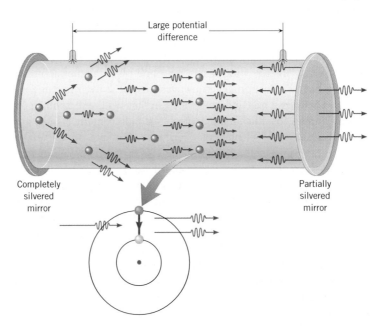

Figure 30.25 A schematic drawing of a helium/neon laser.

is sustained in the neon, relative to an energy level that is 18.70 eV above the ground state. In producing the laser beam, stimulated emission causes electrons in neon to drop from the 20.66-eV level to the 18.70-eV level. The energy change of 1.96 eV corresponds to a wavelength of 633 nm, which is in the red region of the visible spectrum.

The helium/neon laser is not the only kind of laser. There are many different types, including the ruby laser, the argon-ion laser, the carbon dioxide laser, the gallium arsenide solid-state laser, and chemical dye lasers. Depending on the type and whether the laser operates continuously or in pulses, the available beam power ranges from milliwatts to megawatts. Since lasers provide coherent monochromatic electromagnetic radiation that can be confined to an intense narrow beam, they are useful in a wide variety of situations. Today they are used to reproduce music in compact disc players, to weld parts of automobile frames together, to measure distances accurately in surveying, to transmit telephone conversations and other forms of communication over long distances, and to study molecular structure. Many other uses have been found since the laser was invented in 1960, and the next section discusses some of the interesting ones in the field of medicine.

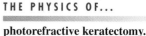

Figure 30.26 These energy levels are involved in the operation of a helium/neon laser.

*30.9 MEDICAL APPLICATIONS OF THE LASER

One of the medical areas in which the laser has had a substantial impact is in ophthalmology, which deals with the structure, function, and diseases of the eye. Section 26.10 discusses the human eye and the use of contact lenses and eyeglasses to correct nearsightedness and farsightedness. In these conditions, the eye cannot refract light properly and produces blurred images on the retina.

A laser-based procedure known as photorefractive keratectomy (PRK) offers an alternative treatment for nearsightedness and farsightedness that does not rely on lenses. It involves the use of a laser to remove small amounts of tissue from the cornea of the eye (see Figure 26.34) and thereby change its curvature. As Section 26.10 points out, light enters the eye through the cornea, and it is at the air/cornea boundary that most of the refraction of the light occurs. Therefore, changing the curvature of that boundary can correct deficiencies in the way the eye refracts light, thus causing the image to be focused onto the retina where it belongs. Ideally, the cornea is dome-shaped. If the dome is too steep, however, the rays of light are focused in front of the retina and nearsightedness results. As Figure 30.27a shows, the laser light removes tissue from the center of the cornea, thereby flattening it and increasing its focal length. On the other hand, if the shape of the cornea is too flat, light rays would come to a focus behind the retina if they could, and farsightedness occurs. As part b of the drawing illustrates, the center of the cornea is now masked and the laser is used to remove peripheral tissue. This steepens the shape of the cornea, thereby shortening its focal point and allowing rays to be focused on the retina.

The laser light in the PRK technique is pulsed and comes from an ultraviolet excimer laser that produces a wavelength of 193 nm. The cornea absorbs this wavelength extremely well, so that weak pulses can be used, leading to highly precise and controllable removal of corneal tissue. Typically, 0.1 to 0.5 μm of tissue is removed by each pulse without damaging adjacent layers. During the procedure, eye movement presents a major problem. It can be overcome, however, with the aid of a mechanical device that stabilizes the eye, or, more recently, with the aid of systems capable of tracking eye movements. These movements occur roughly every 15 ms, and the tracking system follows them and retargets the laser accordingly.

THE PHYSICS OF...

photorefractive keratectomy.

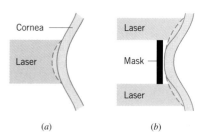

Figure 30.27 (*a*) To correct for myopia (nearsightedness), the laser vaporizes tissue (dashed line) on the center of the cornea, thereby flattening it. (*b*) To correct for hyperopia (farsightedness), the laser vaporizes tissue on the peripheral region of the cornea, thereby steepening its contour.

(a)   (b)

Figure 30.28 A patient with a port-wine stain (a) before and (b) after treatment using a pulsed dye laser (wavelength = 585 nm). From Tan, E. and Vinciullo, C. Pulsed dye laser treatment of port-wine stains: a review of patients treated in Western Australia. MJA 1996; 6; 333–336. ©Copyright 1996. *The Medical Journal of Australia.* Reproduced with permission.

THE PHYSICS OF...

removing port-wine stains.

Another medical application of the laser is in the treatment of congenital capillary malformations or port-wine stains, which affect 0.3% of children at birth. These birthmarks are usually found on the head and neck, as Figure 30.28a illustrates. Preferred treatment for port-wine stains now utilizes a pulsed dye laser. Figure 30.28b shows an example of an excellent result after irradiation with laser light of wavelength 585 nm, in the form of pulses lasting 0.45 ms and occurring every 3 s. The laser beam was focused to a 5-mm-diameter spot. The light is absorbed by oxyhemoglobin in the malformed capillaries, which are destroyed in the process without damage to adjacent normal tissue. Eventually the destroyed capillaries are replaced by normal blood vessels, which causes the port-wine stain to fade.

THE PHYSICS OF...

photodynamic therapy for cancer.

In the treatment of cancer, the laser is being used along with light-activated drugs in photodynamic therapy. The procedure involves administering the drug intravenously, so that the tumor can absorb it from the bloodstream, the advantage being that the drug is then located right near the cancer cells. When the drug is activated by laser light, a chemical reaction ensues that disintegrates the cancer cells and the small blood vessels that feed them. In Figure 30.29 a patient is being treated for cancer of the esophagus. An endoscope that uses optical fibers is inserted down the patient's throat to guide the red laser light to the tumor site and activate the drug. Photodynamic therapy works best with small tumors in their early stages.

Figure 30.29 Photodynamic therapy to treat cancer of the esophagus is being administered to this patient. Red laser light is routed to the tumor site with an endoscope that incorporates optical fibers.

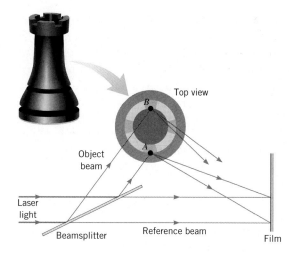

Figure 30.30 An arrangement used to produce a hologram.

*30.10 HOLOGRAPHY

One of the most familiar applications of lasers is in holography, which is a process for producing three-dimensional images. The information used to produce a holographic image is captured on photographic film, which is referred to as a hologram. Figure 30.30 illustrates how a hologram is made. Laser light strikes a half-silvered mirror, or beamsplitter, which reflects part and transmits part of the light. In the drawing, the reflected part is called the object beam because it illuminates the object (a chess piece). The transmitted part is called the reference beam. The object beam reflects from the chess piece at points such as A and B and, together with the reference beam, falls on the film. One of the main characteristics of laser light is that it is coherent. Thus, the light from the two beams has a stable phase relationship, like the light from the two slits in Young's double slit experiment (see Section 27.2). Because of the stable phase relationship and because the two beams travel different distances, an interference pattern is formed on the film. This pattern is the hologram and, although much more complex, is analogous to the pattern of bright and dark fringes formed in the double-slit experiment.

Figure 30.31 shows in greater detail how a holographic interference pattern arises. This drawing considers only the reference beam and the light (wavelength = λ) coming from point A on the chess piece. As we know from Section 27.1, constructive interference between the two light waves leads to a bright fringe; it occurs when the waves, in reaching the film, travel distances that differ by an integer number m of wavelengths. In the drawing, ℓ_m is the distance between point A and the place on the film where the mth order bright fringe occurs. ℓ_0 is the perpendicular distance that the reference beam would travel from point A to the $m = 0$ bright fringe. In addition, r_m is the distance along the film that locates the bright fringe. In terms of these distances, we know that

$$\ell_m - \ell_0 = m\lambda \qquad \text{(condition for constructive interference)}$$
$$\ell_0{}^2 + r_m{}^2 = \ell_m{}^2 \qquad \text{(Pythagorean theorem)}$$

The first of these equations indicates that $\ell_m = m\lambda + \ell_0$, which can be substituted into the second equation. The result can be rearranged to show that

$$r_m{}^2 = m\lambda(m\lambda + 2\ell_0)$$

THE PHYSICS OF...

holography.

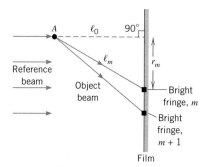

Figure 30.31 This drawing helps to explain how the interference pattern arises on the film when light from point A (see Figure 30.30) and light from the reference beam combine there.

Since ℓ_0 is typically much larger than λ ($\ell_0 \approx 10^{-1}$ m and $\lambda \approx 10^{-6}$ m), it follows that $r_m \approx \sqrt{m\lambda 2\ell_0}$. In other words, r_m is roughly proportional to $\sqrt{m}$. Therefore, the fringes are farther apart near the top of the film than they are near the bottom. For example, for the $m = 1$ and $m = 2$ fringes, we have $r_2 - r_1 \propto \sqrt{2} - \sqrt{1} = 0.41$, while for the $m = 2$ and $m = 3$ fringes, we have $r_3 - r_2 \propto \sqrt{3} - \sqrt{2} = 0.32$.

In addition to the fringe pattern just discussed, the total interference pattern on the hologram includes interference effects that are related to light coming from point B and other locations on the object in Figure 30.30. The total pattern is very complicated. Nevertheless, the fringe pattern for point A alone is sufficient to illustrate the fact that a hologram can be used to produce both a virtual image and a real image of the object, as we will now see.

To produce the holographic images, the laser light is directed through the interference pattern on the film, as in Figure 30.32. The pattern can be thought of as a kind of diffraction grating, with the bright fringes analogous to the space between the slits of the grating. Section 27.7 discusses how light passing through a grating is split into higher-order bright fringes that are symmetrically located on either side of a central bright fringe. Figure 30.32 shows the three rays corresponding to the central and first-order bright fringes, as they originate from a spot near the top and a spot near the bottom of the film. The angle θ, which locates the first-order fringes relative to the central fringe, is given by Equation 27.7 as $\sin \theta = \lambda/d$, where d is the separation between the slits of the grating. When the slit separation is greater, as it is near the top of the film, the angle is smaller. When the slit separation is smaller, as it is near the bottom, the angle is larger. Thus, Figure 30.32 has been drawn with θ_{TOP} smaller than θ_{BOTTOM}. Of the three rays emerging from the film at the top and the three at the bottom, we use the uppermost one in each case to locate the real image of point A on the chess piece. The real image is located where these two rays intersect, when extended to the right. To locate the virtual image, we use the lower ray in each of the three-ray bundles at the top and bottom of the film. When projected to the left, they appear to be originating from the spot where the projections intersect, that is, from the virtual image.

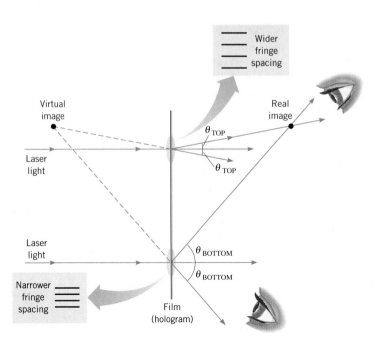

Figure 30.32 When the laser light used to produce a hologram is shone through it, both a real and a virtual image of the object are produced.

(a)

(b)

Figure 30.33 A hologram of a galleon spaceship approaching earth. Two views of the same hologram are shown: (*a*) looking at the hologram perpendicularly and (*b*) looking at an angle to the right of the perpendicular.

A holographic image differs greatly from a photographic image. The most obvious difference is that a hologram provides a three-dimensional image, while photographs are two dimensional. The reason that holographic images are three dimensional is inherent in the interference pattern formed on the film. In Figure 30.30 part of this pattern arises because the light emitted by point *A* travels different distances in reaching different spots on the film than does the light in the reference beam. The same can be said about the light emitted from point *B* and other places as well. As a result, the total interference pattern contains information about how much farther from the film the various points on the object are, and because of this information holographic images are three dimensional. Furthermore, as Figure 30.33 illustrates, it is possible to "walk around" a holographic image and view it from different angles, as you would the original object.

There is also a vast difference between the methods used to produce holograms and photographs. As Section 26.7 discusses, a converging lens is used in a camera to produce a photograph. The lens focuses the light rays originating from a point on the object to a corresponding point on the film. To produce a hologram, lenses are not used in such a fashion, and a point on the object *does not* correspond to a single point on the film. In Figure 30.30, light from point *A* diverges on its way to the film, and there is no lens to make it converge to a single corresponding point. The light falls over the entire exposed region of the film and contributes everywhere to the interference pattern that is formed. Therefore, a hologram may be cut into smaller pieces, and each piece will contain some information about the light originating from point *A*. For this reason, each smaller piece can be used to produce a three-dimensional image of the object. In contrast, it is not possible to reconstruct the entire image in a photograph from only a small piece of the original film.

The holograms discussed here are typically viewed with the aid of the laser light used to produce them. There are also other kinds of holograms. Credit cards, for example, sometimes use holograms for identification purposes. This kind of hologram is called a rainbow hologram and is designed to be viewed in white light that is reflected from it. Other applications of holography include head-up displays for the instrument panels in high-performance fighter planes, laser scanners at supermarket checkout counters, computerized data storage and retrieval systems, and methods for high-precision biomedical measurements.

30.11 CONCEPTS & CALCULATIONS

The Bohr model of the hydrogen atom introduces a number of important features that characterize the quantum picture of atomic structure. Among them are the concepts of quantized energy levels and the photon emission that occurs when an electron makes a transition from a higher to a lower energy state. Example 11 deals with these ideas.

CONCEPTS & CALCULATIONS EXAMPLE 11   Hydrogen Atom Energy Levels

A hydrogen atom ($Z = 1$) is in the third excited state. It makes a transition to a different state, and a photon is either absorbed or emitted. Determine the quantum number n_f of the final state and the energy of the photon when the photon is (a) emitted with the shortest possible wavelength, (b) emitted with the longest possible wavelength, and (c) absorbed with the longest possible wavelength.

Concept Questions and Answers What is the quantum number of the third excited state?

Answer   The lowest-energy state is the ground state, and its quantum number is $n = 1$. The first excited state corresponds to $n = 2$, the second excited state to $n = 3$, and the third excited state to $n = 4$.

When an atom emits a photon, is the final quantum number n_f of the atom greater than or less than the initial quantum number n_i?

Answer   Since the photon carries energy away from the atom, the final energy of the atom is less than its initial energy. Lower energies correspond to lower quantum numbers (see Figure 30.9). Therefore, the final quantum number is less than the initial quantum number.

When an atom absorbs a photon, is the final quantum number n_f of the atom greater than or less than the initial quantum number n_i?

Answer   When an atom absorbs a photon, the atom gains the energy of the photon, so the final energy of the atom is greater than its initial energy. Greater energies correspond to higher quantum numbers, so the final quantum number is greater than the initial quantum number.

How is the wavelength of a photon related to its energy?

Answer   The energy E of a photon is given by Equation 29.2 as $E = hf$, where h is Planck's constant and f is its frequency. But the frequency and wavelength λ are related by $f = c/\lambda$, according to Equation 16.1, where c is the speed of light. Combining these relations gives $E = hc/\lambda$, or $\lambda = hc/E$. Thus, we see that the wavelength is inversely proportional to the energy.

Solution

(a) When a photon is emitted with the shortest possible wavelength, it has the largest possible energy since the wavelength is inversely proportional to the energy. The largest possible energy arises when the electron jumps from the initial state ($n_i = 4$) to the ground state ($n_f = 1$), as shown by transition A in Figure 30.34. Therefore, the quantum number of the final state is $\boxed{n_f = 1}$. The energy E of the photon is the difference between the energies of the two states, so $E = E_4 - E_1$. The energy of the nth state is given by Equation 30.13 as $E_n = -(13.6 \text{ eV})Z^2/n^2$, so the energy of the photon is

$$E = (-13.6 \text{ eV})(1)^2\left(\frac{1}{4^2} - \frac{1}{1^2}\right) = \boxed{12.8 \text{ eV}}$$

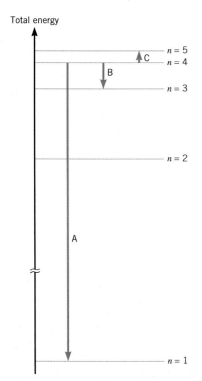

Figure 30.34 A photon is emitted when an electron in the $n = 4$ state jumps to either the $n = 1$ or the $n = 3$ state. A photon is absorbed when an electron makes a transition from the $n = 4$ state to the $n = 5$ state.

(b) When a photon is emitted with the longest possible wavelength, it has the smallest possible energy. The smallest possible energy arises when the electron jumps from the initial state ($n_i = 4$) to the next lower state ($n_f = 3$), as shown by transition B in Figure 30.34. Therefore, the quantum number of the final state is $\boxed{n_f = 3}$. The energy E of the photon is the difference between the energies of the two states, so $E = E_4 - E_3$:

$$E = (-13.6 \text{ eV})(1)^2 \left(\frac{1}{4^2} - \frac{1}{3^2} \right) = \boxed{0.661 \text{ eV}}$$

(c) When a photon is absorbed by the hydrogen atom, the electron jumps to a higher energy state. The photon has the longest possible wavelength when its energy is the smallest. The smallest possible energy change in the hydrogen atom arises when the electron jumps from the initial state ($n_i = 4$) to the next higher state ($n_f = 5$), as shown by transition C in the drawing. Therefore, the quantum number of the final state is $\boxed{n_f = 5}$. The energy E of the photon is the difference between the energies of the two states, so $E = E_5 - E_4$:

$$E = (-13.6 \text{ eV})(1)^2 \left(\frac{1}{5^2} - \frac{1}{4^2} \right) = \boxed{0.306 \text{ eV}}$$

The next example reviews the physics of how a K_α X-ray is produced, how its energy is related to the ionization energies of the target atom, and how to determine the minimum voltage needed to produce it in an X-ray tube.

CONCEPTS & CALCULATIONS EXAMPLE 12   The Production of K_α X-Rays

The K-shell and L-shell ionization energies of copper are 8979 eV and 951 eV, respectively. (a) What must be the minimum voltage across an X-ray tube with a copper target to produce K_α X-ray photons? (b) Determine the wavelength of a K_α photon.

Concept Questions and Answers How is the K_α photon produced and how much energy does it have?

Answer As transition A in Figure 30.35 shows, a K_α photon is produced when an electron in a copper atom jumps from the higher energy $n = 2$ state to the lower energy $n = 1$ state. The energy E of the photon is the difference between the energies of these two states: $E = E_2 - E_1$.

What must be the minimum voltage across the X-ray tube to produce a K_α photon?

Answer It takes energy to produce a K_α photon, and this energy comes from the electrons striking the target in the X-ray tube. According to Equation 19.3, the energy possessed by each electron when it arrives at the target is eV, where e is the magnitude of the electron's charge and V is the potential difference across the tube. Since the energy of the photon is $E_2 - E_1$, the minimum voltage across the tube must be such that $eV_{\min} = E_2 - E_1$, or $V_{\min} = (E_2 - E_1)/e$.

What is meant by the phrases "K-shell ionization energy" and "L-shell ionization energy?"

Answer The K-shell ionization energy is the energy needed to completely remove an electron in the K shell ($n_i = 1$) from the atom (see transition B in Figure 30.35). The removed electron is assumed to have no kinetic energy and is infinitely far away ($n_f = \infty$) so that it has no electric potential energy. Thus, the total energy of the removed electron is zero. Similarly, the L-shell ionization energy is the energy needed to completely remove an electron in the L shell ($n_i = 2$) from the atom (see transition C in Figure 30.35).

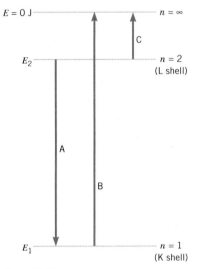

Figure 30.35 A K_α X-ray photon is emitted when an electron jumps from the $n = 2$ state to the $n = 1$ state (transition A). The K-shell ionization energy is the energy required to raise an electron from the $n = 1$ to the $n = \infty$ state, where it has zero total energy (transition B). Similarly, the L-shell ionization energy is the energy needed for the transition from $n = 2$ to $n = \infty$ (transition C). For clarity, the energy levels are not drawn to scale.

What does the difference between the K-shell and L-shell ionization energies represent?

Answer   As transitions B and C in Figure 30.35 suggest, the difference between the two ionization energies is equal to the difference between the total energy (not the ionization energy) E_2 of the L shell and the total energy E_1 of the K shell.

Solution

(a) The minimum voltage across the X-ray tube is $V_{min} = (E_2 - E_1)/e$. But the difference $E_2 - E_1$ in energies between the $n = 2$ and $n = 1$ states is equal to the difference in energies between the K-shell and L-shell ionization energies, or 8979 eV − 951 eV = 8028 eV. Thus, the minimum voltage is

$$V_{min} = \frac{E_2 - E_1}{e} = \frac{(8028 \text{ eV})\left(\dfrac{1.602 \times 10^{-19} \text{ J}}{1 \text{ eV}}\right)}{1.602 \times 10^{-19} \text{ C}} = \boxed{8028 \text{ V}}$$

(b) The wavelength of the K_α photon is $\lambda = c/f$, where f is its frequency. The frequency is related to the energy $E_2 - E_1$ of the photon by Equation 29.2, $f = (E_2 - E_1)/h$, where h is Planck's constant. Combining these relations gives $\lambda = hc/(E_2 - E_1)$, so the wavelength of the emitted photon is

$$\lambda = \frac{hc}{E_2 - E_1} = \frac{(6.626 \times 10^{-34} \text{ J}\cdot\text{s})(2.998 \times 10^8 \text{ m/s})}{(8028 \text{ eV})\left(\dfrac{1.602 \times 10^{-19} \text{ J}}{1 \text{ eV}}\right)} = \boxed{1.545 \times 10^{-10} \text{ m}}$$

At the end of the problem set for this chapter, you will find homework problems that contain both conceptual and quantitative parts. These problems are grouped under the heading *Concepts & Calculations, Group Learning Problems*. They are designed for use by students working alone or in small learning groups. The conceptual part of each problem provides a convenient focus for group discussions.

S U M M A R Y

The idea of a **nuclear atom** originated in 1911 as a result of experiments by Ernest Rutherford in which α particles were scattered by a thin metal foil. The phrase "nuclear atom" refers to the fact that an atom consists of a small, positively charged nucleus surrounded at relatively large distances by a number of electrons, whose total negative charge equals the positive nuclear charge when the atom is electrically neutral.

A **line spectrum** is a series of discrete electromagnetic wavelengths emitted by the atoms of a low-pressure gas that is subjected to a sufficiently high potential difference. Certain groups of discrete wavelengths are referred to as "series." The line spectrum of atomic hydrogen includes, among others, the **Lyman series,** the **Balmer series,** and the **Paschen series** of wavelengths.

The Bohr model applies to atoms or ions that have only a single electron orbiting a nucleus containing Z protons. This model assumes that the electron exists in circular orbits that are called **stationary orbits** because the electron does not radiate electromagnetic waves while in them. According to this model, a photon is emitted only when an electron changes from a higher energy orbit to a lower energy orbit. The model also assumes that the orbital angular momentum L_n of the electron can only have values that are integer multiples of Planck's constant divided by 2π: $L_n = nh/(2\pi)$; $n = 1, 2, 3, \ldots$. With the assumptions above, it can be shown that the nth Bohr orbit has a radius of $r_n = (5.29 \times 10^{-11} \text{ m}) \times (n^2/Z)$ and that the total energy associated with this orbit is $E_n = -(13.6 \text{ eV})(Z^2/n^2)$. The **ionization energy** is the energy needed to remove an electron completely from an atom. The Bohr model predicts that the wavelengths comprising the line spectrum emitted by a hydrogen atom are given by Equation 30.14.

Quantum mechanics describes the hydrogen atom in terms of four quantum numbers: (1) **the principal quantum number** n, which can have the integer values $n = 1, 2, 3, \ldots$; (2) **the orbital quantum number** ℓ, which can have the integer values $\ell = 0, 1, 2, \ldots, (n - 1)$; (3) **the magnetic quantum number** m_ℓ, which can have the positive and negative integer values $m_\ell = -\ell, \ldots, -2, -1, 0, +1, +2, \ldots, +\ell$, and (4) **the spin quantum number** m_s, which, for an electron, can be either $m_s = +\frac{1}{2}$ or $m_s = -\frac{1}{2}$. According to quantum mechanics, an electron does not reside in a circular orbit but, rather, has some probability of being found at various distances from the nucleus.

The Pauli exclusion principle states that no two electrons in an atom can have the same set of values for the four quantum numbers n, ℓ, m_ℓ, and m_s. This principle determines the way in which the electrons in multiple-electron atoms are distributed into shells (defined by the value of n) and subshells (defined by the values of n and ℓ). Table 30.2 summarizes the conventional notation used to refer to the orbital quantum number. The arrangement of the periodic table of the elements is related to the exclusion principle.

X-rays are electromagnetic waves emitted when high-energy electrons strike a metal target contained within an evacuated glass tube. The emitted X-ray spectrum of wavelengths consists of sharp "peaks" or "lines," called **characteristic X-rays,** superimposed on a broad continuous range of wavelengths called **Bremsstrahlung.** The minimum wavelength, or cutoff wavelength, of the Bremsstrahlung is determined by the kinetic energy of the electrons striking the target in the X-ray tube, according to Equation 30.17.

A **laser** is a device that generates electromagnetic waves via a process known as **stimulated emission.** In this process, one photon stimulates the production of another photon by causing an electron in an atom to fall from a higher energy level to a lower energy level. Because of this mechanism of photon production, the electromagnetic waves generated by a laser are coherent and may be confined to a very narrow beam.

CONCEPTUAL QUESTIONS

1. At room temperature, most of the atoms of atomic hydrogen contain electrons that are in the ground state or $n = 1$ energy level. A tube is filled with atomic hydrogen. Electromagnetic radiation with a continuous spectrum of wavelengths, including those in the Lyman, Balmer, and Paschen series, enters one end of this tube and leaves the other end. The exiting radiation is found to contain absorption lines. To which one (or more) of the series do the wavelengths of these absorption lines correspond? Assume that once an electron absorbs a photon and jumps to a higher energy level, it does not absorb yet another photon and jump to an even higher energy level. Explain your answer.

2. Refer to Question 1. Suppose the electrons in the atoms are mostly in excited states. Do you expect there to be a greater or lesser number of absorption lines in the exiting radiation, compared to the situation when the electrons are in the ground state? Account for your answer.

3. When the outermost electron in an atom is in an excited state, the atom is more easily ionized than when the outermost electron is in the ground state. Why?

4. In the Bohr model for the hydrogen atom, the closer the electron is to the nucleus, the smaller is the total energy of the atom. Is this also true in the quantum mechanical picture of the hydrogen atom? Justify your answer.

5. Conceptual Example 6 provides background for this question. Consider two different hydrogen atoms. The electron in each atom is in an excited state. Is it possible for the electrons to have different energies but the same orbital angular momentum L, according to (a) the Bohr model and (b) quantum mechanics? In each case, give a reason for your answer.

6. In the quantum mechanical picture of the hydrogen atom, the orbital angular momentum of the electron *may* be zero in any of the possible energy states. For which energy state *must* the orbital angular momentum be zero? Give your reasoning.

7. Can a 5g subshell contain (a) 22 electrons and (b) 17 electrons? Why?

8. An electronic configuration for manganese ($Z = 25$) is written as $1s^2\,2s^2\,2p^6\,3s^2\,3p^6\,4s^2\,3d^4\,4p^1$. Does this configuration of the electrons represent the ground state or an excited state? Consult Figure 30.17 for the order in which the subshells fill, and account for your answer.

9. The drawing shows the X-ray spectra produced by an X-ray tube when the tube is operated at two different potential differences. Explain why the characteristic lines occur at the same wavelengths in the two spectra, while the cutoff wavelength λ_0 shifts to the right when a smaller voltage is used to operate the tube.

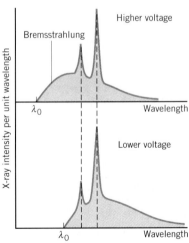

10. In the production of X-rays, it is possible to create Bremsstrahlung X-rays without producing the characteristic X-rays. Explain how this can be accomplished by adjusting the electric potential difference used to operate the X-ray tube.

11. The short wavelength side of X-ray spectra ends abruptly at a cutoff wavelength λ_0 (see Figure 30.20). Does this cutoff wavelength depend on the target material used in the X-ray tube? Give your reasoning.

12. Explain why a laser beam focused to a small spot can cut through a piece of metal.

13. Suppose that a laser produces green light instead of the red light produced by a helium/neon laser. Are the photons emitted by this laser more energetic, less energetic, or of equal energy as compared to those emitted by the helium/neon laser? Justify your answer.

PROBLEMS

In working these problems, ignore relativistic effects.

ssm Solution is in the Student Solutions Manual. **www** Solution is available on the World Wide Web at http://www.wiley.com/college/cutnell
⌁ This icon represents a biomedical application.

Section 30.1 Rutherford Scattering and the Nuclear Atom

1. ssm The nucleus of the hydrogen atom has a radius of about 1×10^{-15} m. The electron is normally at a distance of about 5.3×10^{-11} m from the nucleus. Assuming the hydrogen atom is a sphere with a radius of 5.3×10^{-11} m, find (a) the volume of the atom, (b) the volume of the nucleus, and (c) the percentage of the volume of the atom that is occupied by the nucleus.

2. Review Conceptual Example 1 and use the information therein as an aid in working this problem. Suppose you're building a scale model of the hydrogen atom, and the nucleus is represented by a ball of radius 3.2 cm (somewhat smaller than a baseball). How many miles away (1 mi = 1.61×10^5 cm) should the electron be placed?

3. The nucleus of a hydrogen atom is a single proton, which has a radius of about 1.0×10^{-15} m. The single electron in a hydrogen atom normally orbits the nucleus at a distance of 5.3×10^{-11} m. What is the ratio of the density of the hydrogen nucleus to the density of the complete hydrogen atom?

4. The mass of an α particle is 6.64×10^{-27} kg. An α particle used in a scattering experiment has a kinetic energy of 7.00×10^{-13} J. What is the de Broglie wavelength of the particle?

∗5. ssm The nucleus of a copper atom contains 29 protons and has a radius of 4.8×10^{-15} m. How much work (in electron volts) is done by the electric force as a proton is brought from infinity, where it is at rest, to the "surface" of a copper nucleus?

∗6. There are Z protons in the nucleus of an atom, where Z is the atomic number of the element. An α particle carries a charge of $+2e$. In a scattering experiment, an α particle, heading directly toward a nucleus in a metal foil, will come to a halt when all the particle's kinetic energy is converted to electric potential energy. In such a situation, how close will an α particle with a kinetic energy of 5.0×10^{-13} J come to a gold nucleus ($Z = 79$)?

Section 30.2 Line Spectra, Section 30.3 The Bohr Model of the Hydrogen Atom

7. ssm www What is the radius for the $n = 5$ Bohr orbit in a doubly ionized lithium atom Li^{2+} ($Z = 3$)?

8. The electron in a hydrogen atom is in the first excited state, when the electron acquires an additional 2.86 eV of energy. What

is the quantum number n of the state into which the electron moves?

9. In the line spectrum of atomic hydrogen there is also a group of lines known as the Pfund series. These lines are produced when electrons, excited to high energy levels, make transitions to the $n = 5$ level. Determine (a) the longest wavelength and (b) the shortest wavelength in this series. (c) Refer to Figure 24.9 and identify the region of the electromagnetic spectrum in which these lines are found.

10. A singly ionized helium atom (He^+) has only one electron in orbit about the nucleus. What is the radius of the ion when it is in the second excited state?

11. ssm Find the energy (in joules) of the photon that is emitted when the electron in a hydrogen atom undergoes a transition from the $n = 7$ energy level to produce a line in the Paschen series.

12. A hydrogen atom is in the ground state. It absorbs energy and makes a transition to the $n = 3$ excited state. The atom returns to the ground state by emitting two photons. What are their wavelengths?

13. Consider the Bohr energy expression (Equation 30.13) as it applies to singly ionized helium He^+ ($Z = 2$) and doubly ionized lithium Li^{2+} ($Z = 3$). This expression predicts equal electron energies for these two species for certain values of the quantum number n (the quantum number is different for each species). For quantum numbers less than or equal to 9, what are the lowest three energies (in electron volts) for which the helium energy level is equal to the lithium energy level?

14. In the hydrogen atom, what is the total energy (in electron volts) of an electron that is in an orbit with a radius of 4.761×10^{-10} m?

∗15. ssm For atomic hydrogen, the Paschen series of lines occurs when $n_f = 3$, while the Brackett series occurs when $n_f = 4$ in Equation 30.14. Using this equation, show that the ranges of wavelengths in these two series overlap.

∗16. In the Bohr model, Equation 30.13 gives the total energy (kinetic plus potential). In a certain Bohr orbit, the total energy is -4.90 eV. For this orbit, determine the kinetic energy and the electric potential energy of the electron.

*17. **ssm** The Bohr model can be applied to singly ionized helium He$^+$ ($Z = 2$). Using this model, consider the series of lines that is produced when the electron makes a transition from higher energy levels into the $n_f = 4$ level. Some of the lines in this series lie in the visible region of the spectrum (380–750 nm). What are the values of n_i for the energy levels from which the electron makes the transitions corresponding to these lines?

*18. Doubly ionized lithium Li^{2+} ($Z = 3$) and triply ionized beryllium Be^{3+} ($Z = 4$) each emit a line spectrum. For a certain series of lines in the lithium spectrum, the shortest wavelength is 40.5 nm. For the same series of lines in the beryllium spectrum, what is the shortest wavelength?

**19. A diffraction grating is used in the first order to separate the wavelengths in the Balmer series of atomic hydrogen. (Section 27.7 discusses diffraction gratings.) The grating and an observation screen are separated by a distance of 81.0 cm, as Figure 27.32 illustrates. You may assume that θ is small, so $\sin \theta \approx \theta$ when radian measure is used for θ. How many lines per centimeter should the grating have, so the longest and the next-to-the-longest wavelengths in the series are separated by 3.00 cm on the screen?

**20. A certain species of ionized atoms produces an emission line spectrum according to the Bohr model, but the number of protons Z in the nucleus is unknown. A group of lines in the spectrum forms a series in which the shortest wavelength is 22.79 nm and the longest wavelength is 41.02 nm. Find the next-to-the-longest wavelength in the series of lines.

Section 30.5 The Quantum Mechanical Picture of the Hydrogen Atom

21. The orbital quantum number for the electron in a hydrogen atom is $\ell = 5$. What is the smallest possible value (algebraically) for the total energy of this electron? Give your answer in electron volts.

22. Suppose the value of the principal quantum number is $n = 4$. What are the possible values for the magnetic quantum number m_ℓ?

23. **ssm www** The principal quantum number for an electron in an atom is $n = 6$, while the magnetic quantum number is $m_\ell = 2$. What possible values for the orbital quantum number ℓ could this electron have?

24. A hydrogen atom is in its second excited state. Determine, according to quantum mechanics, (a) the total energy (in eV) of the atom, (b) the magnitude of the maximum angular momentum the electron can have in this state, and (c) the maximum value that the z component L_z of the angular momentum can have.

*25. The total orbital angular momentum of the electron in a hydrogen atom has a magnitude of $L = 3.66 \times 10^{-34}$ J·s. In the quantum mechanical picture of the atom, what values can the angular momentum component L_z have?

*26. Review Conceptual Example 6 as background for this problem. For the hydrogen atom, the Bohr model and quantum mechanics both give the same value for the energy of the nth state.

However, they do not give the same value for the orbital angular momentum L. (a) For $n = 1$, determine the values of L (in units of $h/2\pi$) predicted by the Bohr model and quantum mechanics. (b) Repeat part (a) for $n = 3$, noting that quantum mechanics permits more than one value of ℓ when the electron is in the $n = 3$ state.

27. **ssm www An electron is in the $n = 5$ state. What is the smallest possible value for the angle between the z component of the orbital angular momentum and the orbital angular momentum?

Section 30.6 The Pauli Exclusion Principle and the Periodic Table of the Elements

28. Following the style used in Table 30.3, determine the electronic configuration of the ground state for yttrium Y($Z = 39$). Refer to Figure 30.17 to see the order in which the subshells fill.

29. In the style shown in Table 30.3, write down the ground state electronic configuration for arsenic As ($Z = 33$). Refer to Figure 30.17 for the order in which the subshells fill.

30. Two of the three electrons in a lithium atom have quantum numbers of $n = 1$, $\ell = 0$, $m_\ell = 0$, $m_s = +\frac{1}{2}$ and $n = 1$, $\ell = 0$, $m_\ell = 0$, $m_s = -\frac{1}{2}$. What quantum numbers can the third electron have if the atom is in (a) its ground state and (b) its first excited state?

31. **ssm** When an electron makes a transition between energy levels of an atom, there are no restrictions on the initial and final values of the principal quantum number n. According to quantum mechanics, however, there is a rule that restricts the initial and final values of the orbital quantum number ℓ. This rule is called a *selection rule* and states that $\Delta \ell = \pm 1$. In other words, when an electron makes a transition between energy levels, the value of ℓ can only increase or decrease by one. The value of ℓ may not remain the same or increase or decrease by more than one. According to this rule, which of the following energy level transitions are allowed: (a) 2s → 1s, (b) 2p → 1s, (c) 4p → 2p, (d) 4s → 2p, and (e) 3d → 3s?

*32. What is the atom with the smallest atomic number that contains the same number of electrons in its s subshells as it does in its d subshell? Refer to Figure 30.17 for the order in which the subshells fill.

Section 30.7 X-Rays

33. **ssm** Molybdenum has an atomic number of $Z = 42$. Using the Bohr model, estimate the wavelength of the K_α X-ray.

34. By using the Bohr model decide which element is likely to emit a K_α X-ray with a wavelength of 4.5×10^{-9} m.

35. An X-ray tube is being operated at a potential difference of 45.0 kV. What is the Bremsstrahlung wavelength that corresponds to 25.0% of the kinetic energy with which an electron collides with the metal target in the tube?

36. What is the minimum potential difference that must be applied to an X-ray tube to knock a K-shell electron completely out of an atom in a copper ($Z = 29$) target? Use the Bohr model as needed.

*37. **ssm** An X-ray tube contains a silver ($Z = 47$) target. The high voltage in this tube is increased from zero. Using the Bohr model, find the value of the voltage at which the K_α X-ray just appears in the X-ray spectrum.

*38. The highest-energy X-rays produced by an X-ray tube have a wavelength of 1.20×10^{-10} m. What is the speed of the electrons in Figure 30.19 just before they strike the metal target?

Section 30.8 The Laser

39. The ultraviolet excimer laser used in photorefractive keratectomy (see Section 30.9) has a wavelength of 193 nm. What is the energy (in joules) of a photon that this laser produces?

40. A pulsed laser emits light in a series of short pulses, each having a duration of 25.0 ms. The average power of each pulse is 5.00 mW, and the wavelength of the light is 633 nm. Find (a) the energy of each pulse and (b) the number of photons in each pulse.

41. **ssm www** A laser is used in eye surgery to weld a detached retina back into place. The wavelength of the laser beam is 514 nm, while the power is 1.5 W. During surgery, the laser beam is turned on for 0.050 s. During this time, how many photons are emitted by the laser?

42. The dye laser used in the treatment of the port-wine stain in Figure 30.28 (see Section 30.9) has a wavelength of 585 nm. A carbon dioxide laser produces a wavelength of 1.06×10^{-5} m. What is the minimum number of photons that the carbon dioxide laser must produce to deliver at least as much or more energy to a target as does a single photon from the dye laser?

*43. Fusion is the process by which the sun produces energy. One experimental technique for creating controlled fusion utilizes a solid-state laser that emits a wavelength of 1060 nm and can produce a power of 1.0×10^{14} W for a pulse duration of 1.1×10^{-11} s. In contrast, the helium/neon laser used at the checkout counter in a bar-code scanner emits a wavelength of 633 nm and produces a power of about 1.0×10^{-3} W. How long (in days) would the helium/neon laser have to operate to produce the same number of photons that the solid-state laser produces in 1.1×10^{-11} s?

ADDITIONAL PROBLEMS

44. In the style indicated in Table 30.3, write down the ground state electronic configuration of manganese Mn ($Z = 25$). Refer to Figure 30.17 for the order in which the subshells fill.

45. **ssm** It is possible to use electromagnetic radiation to ionize atoms. To do so, the atoms must absorb the radiation, the photons of which must have enough energy to remove an electron from an atom. What is the longest radiation wavelength that can be used to ionize the ground state hydrogen atom?

46. The atomic number of lead is $Z = 82$. According to the Bohr model, what is the energy (in joules) of a K_α X-ray photon?

47. Write down the fourteen sets of the four quantum numbers that correspond to the electrons in a completely filled 4f subshell.

48. The K_β characteristic X-ray line for tungsten has a wavelength of 1.84×10^{-11} m. What is the difference in energy between the two energy levels that give rise to this line? Express the answer in (a) joules and (b) electron volts.

49. **ssm** For a doubly ionized lithium atom Li^{2+} ($Z = 3$), what is the principal quantum number of the state in which the electron has the same total energy as a ground state electron has in the hydrogen atom?

*50. The energy of the $n = 2$ Bohr orbit is -30.6 eV for an unidentified ionized atom in which only one electron moves about the nucleus. What is the radius of the $n = 5$ orbit for this species?

*51. For an electron in a hydrogen atom, the z component of the angular momentum has a *maximum* value of $L_z = 2.11 \times 10^{-34}$ J·s. Find the three smallest possible values (algebraically) for the total energy (in electron volts) that this atom could have.

*52. Consider a particle of mass m that can exist only between $x = 0$ m and $x = +L$ on the x axis. We could say that this particle is confined to a "box" of length L. In this situation, imagine the standing de Broglie waves that can fit into the box. For example, the drawing shows the first three possibilities. Note in this picture that there are either one, two, or three half-wavelengths that fit into the distance L. Use Equation 29.8 for the de Broglie wavelength of a particle and derive an expression for the allowed energies (only kinetic energy) that the particle can have. This expression involves m, L, Planck's constant, and a quantum number n that can have only the values 1, 2, 3, ...

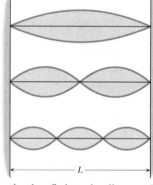

*53. **ssm** A wavelength of 410.2 nm is emitted by the hydrogen atoms in a high-voltage discharge tube. What are the initial and final values of the quantum number n for the energy level transition that produces this wavelength?

**54. (a) Derive an expression for the speed of the electron in the nth Bohr orbit, in terms of Z, n, and the constants k, e, and h. For the hydrogen atom, determine the speed in (b) the $n = 1$ orbit and (c) the $n = 2$ orbit. (d) Generally, when speeds are less than one-tenth the speed of light, the effects of special relativity can be ignored. Are the speeds found in (b) and (c) consistent with ignoring relativistic effects in the Bohr model?

CONCEPTS & CALCULATIONS

GROUP LEARNING PROBLEMS

Note: Each of these problems consists of Concept Questions followed by a related quantitative Problem. They are designed for use by students working alone or in small learning groups. The Concept Questions involve little or no mathematics and are intended to stimulate group discussions. They focus on the concepts with which the problems deal. Recognizing the concepts is the essential initial step in any problem-solving technique.

55. Concept Questions A hydrogen atom is in an excited state. (a) What is meant by the phrase "ionization energy of the excited state"? (b) Is the ionization energy of an excited state greater than, less than, or the same as the ionization energy for the ground state? Why?

Problem (a) What is the ionization energy of a hydrogen atom that is in the $n = 4$ state? (b) Determine the ratio of the ionization energy for $n = 4$ to that of the ground state. Check to make sure that your answer is consistent with your answers to the Concept Questions.

56. Concept Questions Answer the following questions using the quantum mechanical model of the atom. (a) Can the orbital quantum number ℓ be greater than the principal quantum number n? (b) Can the orbital quantum number have negative values? (c) Is it possible for the magnetic quantum number m_ℓ to be greater than the orbital quantum number? (d) Can the magnetic quantum number have negative values?

Problem The table lists the quantum numbers for five states of the hydrogen atom. Which (if any) of them are not possible?

	n	ℓ	m_ℓ
(a)	3	3	0
(b)	2	1	−1
(c)	4	2	3
(d)	5	−3	2
(e)	4	0	0

57. Concept Questions Answer the following questions using the quantum mechanical model of the atom. (a) For a given principal quantum number n, what values of the angular momentum quantum number ℓ are possible? (b) For a given angular momentum, what values of the magnetic quantum number m_ℓ are possible?

Problem Which of the following subshell configurations are not allowed? For those that are not allowed, give the reason why. (a) $3s^1$, (b) $2d^2$, (c) $3s^4$, (d) $4p^8$, $5f^{12}$.

58. Concept Questions Suppose that the molybdenum ($Z = 42$) target in an X-ray tube is replaced by a silver ($Z = 47$) target. Do (a) the cutoff wavelength λ_0 and (b) the wavelength of the K_α X-ray photon increase, decrease, or remain the same? Assume that the voltage across the tube is constant and is sufficient to produce characteristic X-rays from both targets. Provide a reason for each answer.

Problem The voltage across the X-ray tube is 35.0 kV. Determine (a) the cutoff wavelength λ_0 and (b) the wavelengths of the K_α X-ray photons emitted by the molybdenum and silver targets. Verify that your answers are consistent with your answers to the Concept Questions.

59. Concept Questions The drawing shows three energy levels of a laser that are involved in the lasing action. These levels are analogous to the levels in the Ne atoms of a He-Ne laser. The E_2 level is a metastable level, and the E_0 level is the ground state. (a) What two levels are involved when an external source of energy initially causes the electrons to jump to a higher energy state? (b) Between what two levels does the population inversion occur? (c) Between what two levels does the lasing action occur? Provide a reason for each answer.

Problem The difference between the energy levels of the laser is shown in the drawing. (a) What energy (in eV per electron) must an external source provide to start the lasing action? (b) What is the wavelength of the laser light? (c) In what region of the electromagnetic spectrum does the laser light lie (see Figure 24.9)? Check your answers for consistency with respect to your answers to the Concept Questions.

Paleontologists use the disintegration of radioactive nuclei to date the fossil remains of extinct species, such as this dinosaur (*Carcharodontosaurus*), whose skull is slightly larger than that of the famous *T. Rex*.

CHAPTER 31

NUCLEAR PHYSICS AND RADIOACTIVITY

31.1 NUCLEAR STRUCTURE

Atoms consist of electrons in orbit about a central nucleus. As we have seen in Chapter 30, the electron orbits are quantum mechanical in nature and have interesting characteristics. Little has been said about the nucleus, however. Since the nucleus is interesting in its own right, we now consider it in greater detail.

The nucleus of an atom consists of neutrons and protons, collectively referred to as **nucleons**. The **neutron**, discovered in 1932 by the English physicist James Chadwick (1891–1974), carries no electrical charge and has a mass slightly larger than that of a proton (see Table 31.1).

The number of protons in the nucleus is different in different elements and is given by the **atomic number Z**. In an electrically neutral atom, the number of nuclear protons equals the number of electrons in orbit around the nucleus. The number of neutrons in the nucleus is N. The total number of protons and neutrons is referred to as the **atomic mass number A** because the total nuclear mass is *approximately* equal to A times the mass of a single nucleon:

$$\underbrace{A}_{\substack{\text{Number of protons} \\ \text{and neutrons} \\ \text{(atomic mass number} \\ \text{or nucleon number)}}} = \underbrace{Z}_{\substack{\text{Number of} \\ \text{protons} \\ \text{(atomic number)}}} + \underbrace{N}_{\substack{\text{Number of} \\ \text{neutrons}}} \tag{31.1}$$

Sometimes A is also called the **nucleon number.** A shorthand notation is often used to specify Z and A along with the chemical symbol for the element. For instance, the nuclei of all naturally occurring aluminum atoms have $A = 27$, and the atomic number for aluminum is $Z = 13$. In shorthand notation, then, the aluminum nucleus is specified as $^{27}_{13}\text{Al}$. The number of neutrons in an aluminum nucleus is $N = A - Z = 14$. In general, for an element whose chemical symbol is X, the symbol for the nucleus is

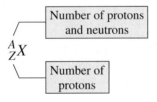

For a proton the symbol is ^1_1H, since the proton is the nucleus of a hydrogen atom. A neutron is denoted by ^1_0n. In the case of an electron we use $^0_{-1}\text{e}$, where $A = 0$ because an electron is not composed of protons or neutrons and $Z = -1$ because the electron has a negative charge.

TABLE 31.1   Properties of Particles in the Atom

Particle	Electric Charge (C)	Mass	
		Kilograms (kg)	Atomic Mass Units (u)
Electron	-1.60×10^{-19}	$9.109\,390 \times 10^{-31}$	$5.485\,799 \times 10^{-4}$
Proton	$+1.60 \times 10^{-19}$	$1.672\,623 \times 10^{-27}$	$1.007\,276$
Neutron	0	$1.674\,929 \times 10^{-27}$	$1.008\,665$

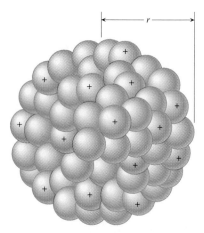

Figure 31.1 The nucleus is approximately spherical (radius = r) and contains protons ($\oplus$) clustered closely together with neutrons ($\bullet$).

Nuclei that contain the same number of protons, but a different number of neutrons, are known as *isotopes*. Carbon, for example, occurs in nature in two stable forms. In most carbon atoms (98.90%), the nucleus is the $^{12}_{6}C$ isotope and consists of six protons and six neutrons. A small fraction (1.10%), however, contains nuclei that have six protons and seven neutrons, namely, the $^{13}_{6}C$ isotope. The percentages given above are the natural abundances of the isotopes. The atomic masses in the periodic table are average atomic masses, taking into account the abundances of the various isotopes.

The protons and neutrons in the nucleus are clustered together to form an approximately spherical region, as Figure 31.1 illustrates. Experiment shows that the radius r of the nucleus depends on the atomic mass number A and is given approximately in meters by

$$r \approx (1.2 \times 10^{-15} \text{ m})A^{1/3} \tag{31.2}$$

The radius of the aluminum nucleus ($A = 27$), for example, is $r \approx (1.2 \times 10^{-15} \text{ m})27^{1/3} = 3.6 \times 10^{-15}$ m. Equation 31.2 leads to an important conclusion concerning the nuclear density of different atoms, as Conceptual Example 1 discusses.

CONCEPTUAL EXAMPLE 1   Nuclear Density

It is well known that lead and oxygen contain different atoms and that the density of solid lead is much greater than that of gaseous oxygen. Using Equation 31.2, decide whether the density of the *nucleus* in a lead atom is greater than, approximately equal to, or less than that in an oxygen atom.

Reasoning and Solution We know from Equation 11.1 that density is mass divided by volume. The total mass M of a nucleus is approximately equal to the number of nucleons A times the mass m of a single nucleon, since the masses of a proton and a neutron are nearly the same: $M \approx Am$. The values of A are different for lead and oxygen atoms, however. The volume V of the nucleus is approximately spherical and has a radius r, so that $V = \frac{4}{3}\pi r^3$. But r^3 is proportional to the number of nucleons A, as we can see by cubing each side of Equation 31.2. Therefore, the volume V is also proportional to A. Thus, when the total mass M is divided by the volume V, the factor of A appears in both the numerator and denominator and is eliminated algebraically from the result, no matter what the value of A is. We find, then, that **the density of the nucleus in a lead atom is approximately the same as it is in an oxygen atom.** In general, Equation 31.2 implies that the nuclear density has nearly the same value in all atoms. The difference in densities between solid lead and gaseous oxygen arises mainly because of the difference in how closely the atoms themselves are packed together in the solid and gaseous phases.

Related Homework Material: *Problems 8, 9*

31.2 THE STRONG NUCLEAR FORCE AND THE STABILITY OF THE NUCLEUS

Two positive charges that are as close together as they are in a nucleus repel one another with a very strong electrostatic force. What, then, keeps the nucleus from flying apart? Clearly, some kind of attractive force must hold the nucleus together, since many kinds of naturally occurring atoms contain stable nuclei. The gravitational force of attraction between nucleons is too weak to counteract the repulsive electric force, so a different type of force must hold the nucleus together. This force is the ***strong nuclear force*** and is one of only three fundamental forces that have been discovered, fundamental in the sense that all forces in nature can be explained in terms of these three. The gravitational force is also one of these forces, as is the electroweak force (see Section 31.5).

Many features of the strong nuclear force are well known. For example, it is almost independent of electric charge. At a given separation distance, nearly the same nuclear force of attraction exists between two protons, two neutrons, or between a proton and a neutron. The range of action of the strong nuclear force is extremely short, with the force of attraction being very strong when two nucleons are as close as 10^{-15} m and essentially zero at larger distances. In contrast, the electric force between two protons decreases to zero only gradually as the separation distance increases to large values and, therefore, has a relatively long range of action.

The limited range of action of the strong nuclear force plays an important role in the stability of the nucleus. For a nucleus to be stable, the electrostatic repulsion between the protons must be balanced by the attraction between the nucleons due to the strong nuclear force. But one proton repels all other protons within the nucleus, since the electrostatic force has such a long range of action. In contrast, a proton or a neutron attracts only its nearest neighbors via the strong nuclear force. As the number Z of protons in the nucleus increases under these conditions, the number N of neutrons has to increase even more, if stability is to be maintained. Figure 31.2 shows a plot of N versus Z for naturally occurring elements that have stable nuclei. For reference, the plot also includes the straight line that represents the condition $N = Z$. With few exceptions, the points representing stable nuclei fall above this reference line, reflecting the fact that the number of neutrons becomes greater than the number of protons as the atomic number Z increases.

As more and more protons occur in a nucleus, there comes a point when a balance of repulsive and attractive forces cannot be achieved by an increased number of neutrons. Eventually, the limited range of action of the strong nuclear force prevents extra neutrons from balancing the long-range electric repulsion of extra protons. The stable nucleus with the largest number of protons ($Z = 83$) is that of bismuth, $^{209}_{83}\text{Bi}$, which contains 126 neutrons. All nuclei with more than 83 protons (e.g., uranium, $Z = 92$) are unstable and spontaneously break apart or rearrange their internal structures as time passes. This spontaneous disintegration or rearrangement of internal structure is called ***radioactivity,*** first discovered in 1896 by the French physicist Henri Becquerel (1852–1908). Section 31.4 discusses radioactivity in greater detail.

31.3 THE MASS DEFECT OF THE NUCLEUS AND NUCLEAR BINDING ENERGY

Because of the strong nuclear force, the nucleons in a stable nucleus are held tightly together. Therefore, energy is required to separate a stable nucleus into its constituent protons and neutrons, as Figure 31.3 illustrates. The more stable the nucleus is, the greater the amount of energy needed to break it apart. The required energy is called the ***binding energy*** of the nucleus.

Figure 31.2 With few exceptions, the naturally occurring stable nuclei have a number N of neutrons that equals or exceeds the number Z of protons. Each dot in this plot represents a stable nucleus.

Figure 31.3 Energy, called the binding energy, must be supplied to break the nucleus apart into its constituent protons and neutrons. Each of the separated nucleons is at rest and out of the range of the forces of the other nucleons.

Concepts at a Glance

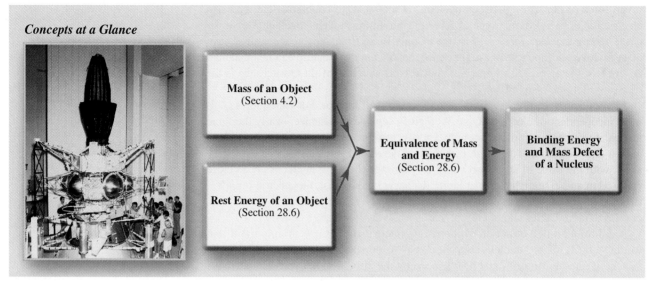

Figure 31.4 *Concepts at a Glance*
The mass and rest energy of an object are equivalent, in the sense that if one increases (or decreases), the other does too. The binding energy of a nucleus is the mass of the separated nucleons minus the mass of the intact nucleus, expressed in units of energy. The photograph shows a test model of NASA's *Galileo* spacecraft, which operates by using a process called fission to provide energy. This process depends on the fact that different nuclei have different binding energies and is discussed in Chapter 32.

● *Concepts at a Glance*   As the Concepts-at-a-Glance chart in Figure 31.4 illustrates, two ideas that we have studied previously come into play as we discuss the binding energy of a nucleus. These are mass (Section 4.2) and the rest energy of an object (Section 28.6). In Einstein's theory of special relativity, mass and energy are equivalent. A change Δm in the mass of a system is equivalent to a change ΔE_0 in the rest energy of the system by an amount $\Delta E_0 = (\Delta m)c^2$, where c is the speed of light in a vacuum. Thus, in Figure 31.3, the binding energy used to disassemble the nucleus appears as extra mass of the separated nucleons. In other words, the sum of the individual masses of the separated protons and neutrons is greater by an amount Δm than the mass of the stable nucleus. The difference in mass Δm is known as the ***mass defect*** of the nucleus. ●

As Example 2 shows, the binding energy of a nucleus can be determined from the mass defect according to Equation 31.3:

$$\text{Binding energy} = (\text{Mass defect})c^2 = (\Delta m)c^2 \qquad (31.3)$$

EXAMPLE 2   The Binding Energy of the Helium Nucleus

The most abundant isotope of helium has a $^4_2$H nucleus whose mass is 6.6447×10^{-27} kg. For this nucleus, find (a) the mass defect and (b) the binding energy.

Reasoning The symbol $^4_2$He indicates that the helium nucleus contains $Z = 2$ protons and $N = 4 - 2 = 2$ neutrons. To obtain the mass defect Δm, we first determine the sum of the individual masses of the separated protons and neutrons. Then we subtract from this sum the mass of the $^4_2$H nucleus. Finally, we use Equation 31.3 to calculate the binding energy from the value for Δm.

Solution

(a) Using data from Table 31.1, we find that the sum of the individual masses of the nucleons is

$$\underbrace{2(1.6726 \times 10^{-27}\ \text{kg})}_{\text{Two protons}} + \underbrace{2(1.6749 \times 10^{-27}\ \text{kg})}_{\text{Two neutrons}} = 6.6950 \times 10^{-27}\ \text{kg}$$

This value is greater than the mass of the intact $^4_2$He nucleus, and the mass defect is

$$\Delta m = 6.6950 \times 10^{-27}\ \text{kg} - 6.6447 \times 10^{-27}\ \text{kg} = \boxed{0.0503 \times 10^{-27}\ \text{kg}}$$

(b) According to Equation 31.3, the binding energy is

$$\frac{\text{Binding}}{\text{energy}} = (\Delta m)c^2 = (0.0503 \times 10^{-27}\ \text{kg})(3.00 \times 10^8\ \text{m/s})^2 = 4.53 \times 10^{-12}\ \text{J}$$

Usually, binding energies are expressed in energy units of electron volts instead of joules (1 eV = 1.60×10^{-19} J):

$$\frac{\text{Binding}}{\text{energy}} = (4.53 \times 10^{-12}\ \text{J})\left(\frac{1\ \text{eV}}{1.60 \times 10^{-19}\ \text{J}}\right) = 2.83 \times 10^7\ \text{eV} = \boxed{28.3\ \text{MeV}}$$

In this result, one million electron volts is denoted by the unit MeV. The value of 28.3 MeV is more than two million times greater than the energy required to remove an orbital electron from an atom.

In calculations such as that in Example 2, it is customary to use the ***atomic mass unit*** (u) instead of the kilogram. As introduced in Section 14.1, the atomic mass unit is one-twelfth of the mass of a $^{12}_6$C atom of carbon. In terms of this unit, the mass of a $^{12}_6$C atom is exactly 12 u. Table 31.1 also gives the masses of the electron, the proton, and the neutron in atomic mass units. For future use, the energy equivalent of one atomic mass unit can be calculated by observing that the mass of a proton is 1.6726×10^{-27} kg or 1.0073 u, so that

$$1\ \text{u} = (1\ \text{u})\left(\frac{1.6726 \times 10^{-27}\ \text{kg}}{1.0073\ \text{u}}\right) = 1.6605 \times 10^{-27}\ \text{kg}$$

and

$$\Delta E_0 = (\Delta m)c^2 = (1.6605 \times 10^{-27}\ \text{kg})(2.9979 \times 10^8\ \text{m/s})^2 = 1.4924 \times 10^{-10}\ \text{J}$$

In electron volts, therefore, one atomic mass unit is equivalent to

$$1\ \text{u} = (1.4924 \times 10^{-10}\ \text{J})\left(\frac{1\ \text{eV}}{1.6022 \times 10^{-19}\ \text{J}}\right) = 9.315 \times 10^8\ \text{eV} = 931.5\ \text{MeV}$$

Data tables for isotopes give masses in atomic mass units. Typically, however, the given masses are not nuclear masses. They are *atomic masses*, that is, the masses of neutral atoms, including the mass of the orbital electrons. Example 3 deals again with the $^4_2$He nucleus and shows how to take into account the effect of the orbital electrons when using such data to determine binding energies.

EXAMPLE 3   The Binding Energy of the Helium Nucleus, Revisited

The atomic mass of $^4_2$He is 4.0026 u, and the atomic mass of $^1_1$H is 1.0078 u. Using atomic mass units instead of kilograms, obtain the binding energy of the $^4_2$He nucleus.

Reasoning To determine the binding energy, we calculate the mass defect in atomic mass units and then use the fact that one atomic mass unit is equivalent to 931.5 MeV of energy. The mass of 4.0026 u for $^4_2$He *includes the mass of the two electrons in the neutral helium atom*. To calculate the mass defect, we must subtract 4.0026 u from the sum of the individual masses of the nucleons, including the mass

Figure 31.5 Data tables usually give the mass of the neutral atom (including the orbital electrons) rather than the mass of the nucleus. When using data from such tables to determine the mass defect of a nucleus, the mass of the orbital electrons must be taken into account, as this drawing illustrates for the $^4_2$He isotope of helium. See Example 3.

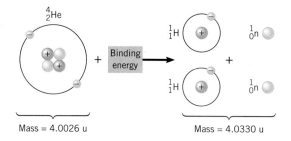

Mass = 4.0026 u   Mass = 4.0330 u

of the electrons. As Figure 31.5 illustrates, the electron mass will be included if the masses of two hydrogen atoms are used in the calculation instead of the masses of two protons. The mass of a $^1_1$H hydrogen atom is given as 1.0078 u, and the mass of a neutron is given in Table 31.1 as 1.0087 u.

Solution The sum of the individual masses is

$$\underbrace{2(1.0078\ u)}_{\substack{\text{Two hydrogen}\\\text{atoms}}} + \underbrace{2(1.0087\ u)}_{\text{Two neutrons}} = 4.0330\ u$$

The mass defect is $\Delta m = 4.0330\ u - 4.0026\ u = 0.0304\ u$. Since 1 u is equivalent to 931.5 MeV, the binding energy is $\boxed{\text{Binding energy} = 28.3\ \text{MeV}}$, which matches that obtained in Example 2.

To see how the nuclear binding energy varies from nucleus to nucleus, it is necessary to compare the binding energy for each nucleus on a per-nucleon basis. Figure 31.6 shows a graph in which the binding energy divided by the nucleon number A is plotted against the nucleon number itself. In the graph, the peak for the $^4_2$He isotope of helium indicates that the $^4_2$He nucleus is particularly stable. The binding energy per nucleon increases rapidly for nuclei with small masses and reaches a maximum of approximately 8.7 MeV/nucleon for a nucleon number of about $A = 60$. For greater nucleon numbers, the binding energy per nucleon decreases gradually. Eventually, the binding energy per nucleon decreases enough so there is insufficient binding energy to hold the nucleus together. Nuclei more massive than the $^{209}_{83}$Bi nucleus of bismuth are unstable and hence radioactive.

Figure 31.6 A plot of binding energy per nucleon versus the nucleon number A.

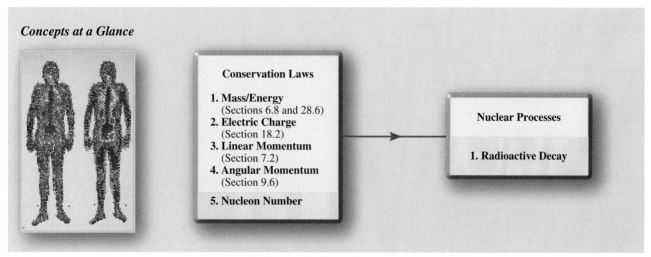

Concepts at a Glance

Conservation Laws

1. **Mass/Energy**
   (Sections 6.8 and 28.6)
2. **Electric Charge**
   (Section 18.2)
3. **Linear Momentum**
   (Section 7.2)
4. **Angular Momentum**
   (Section 9.6)
5. **Nucleon Number**

Nuclear Processes

1. **Radioactive Decay**

Figure 31.7 *Concepts at a Glance*
The conservation laws listed at the left side of this chart are obeyed when a nucleus undergoes radioactive decay. The three types of naturally occurring decay are α decay, β decay, and γ decay. Nuclear medicine uses radioactive decay to produce scans of living organisms. This photograph shows scans of the front (left) and back views of a healthy man injected with a radioactive isotope that tends to concentrate in the bones, especially the spine, ribs, and pelvis.

31.4 RADIOACTIVITY

When an unstable or radioactive nucleus disintegrates spontaneously, certain kinds of particles and/or high-energy photons are released. These particles and photons are collectively called "rays." Three kinds of rays are produced by naturally occurring radioactivity: **α rays, β rays,** and **γ rays.** They are named according to the first three letters of the Greek alphabet, alpha (α), beta (β), and gamma (γ), to indicate the extent of their ability to penetrate matter. α rays are the least penetrating, being blocked by a thin (≈ 0.01 mm) sheet of lead, while β rays penetrate lead to a much greater distance (≈ 0.1 mm). γ rays are the most penetrating and can pass through an appreciable thickness (≈ 100 mm) of lead.

　Concepts at a Glance The nuclear disintegration process that produces α, β, and γ rays must obey the laws of physics that we have studied in previous chapters. As the Concepts-at-a-Glance chart in Figure 31.7 reminds us, these laws are called conservation laws because each of them deals with a property (such as mass/energy, electric charge, linear momentum, and angular momentum) that is conserved or does not change during a process. To the first four conservation laws in Figure 31.7, we now add a fifth, the conservation of nucleon number. In all radioactive decay processes it has been observed that the number of nucleons present before the decay is equal to the number of nucleons after the decay. Therefore, the number of nucleons is conserved during a nuclear disintegration. As applied to the disintegration of a nucleus, the conservation laws require that the energy, electric charge, linear momentum, angular momentum, and nucleon number that a nucleus possesses must remain unchanged when it disintegrates into nuclear fragments and accompanying α, β, or γ rays. ●

The three types of radioactivity that occur naturally can be observed in a relatively simple experiment. A piece of radioactive material is placed at the bottom of a narrow hole in a lead cylinder. The cylinder is located within an evacuated chamber, as Figure 31.8 illustrates. A magnetic field is directed perpendicular to the plane of the paper, and a photographic plate is positioned to the right of the hole. Three spots appear on the developed plate, which are associated with the radioactivity of the nuclei in the material. Since moving particles are deflected by a magnetic field only when they are electrically charged, this experiment reveals that two

Figure 31.8 α and β rays are deflected by a magnetic field and, therefore, consist of moving charged particles. γ rays are not deflected by a magnetic field and, consequently, must be uncharged.

types of radioactivity (α and β rays, as it turns out) consist of charged particles, while the third type (γ rays) does not.

α Decay

When a nucleus disintegrates and produces α rays, it is said to undergo **α decay**. Experimental evidence shows that α rays consist of positively charged particles, each one being the ^4_2He nucleus of helium. Thus, an α particle has a charge of $+2e$ and a nucleon number of $A = 4$. Since the grouping of 2 protons and 2 neutrons in a ^4_2He nucleus is particularly stable, as we have seen in connection with Figure 31.6, it is not surprising that an α particle can be ejected as a unit from a more massive unstable nucleus.

Figure 31.9 shows the disintegration process for one example of α decay:

$$^{238}_{92}\text{U} \longrightarrow {}^{234}_{90}\text{Th} + {}^4_2\text{He}$$

| Parent nucleus (uranium) | Daughter nucleus (thorium) | α particle (helium nucleus) |

The original nucleus is referred to as the parent nucleus (P), and the nucleus remaining after disintegration is called the daughter nucleus (D). Upon emission of an α particle, the uranium $^{238}_{92}\text{U}$ parent is converted into the $^{234}_{90}\text{Th}$ daughter, which is an isotope of thorium. The parent and daughter nuclei are different, so α decay converts one element into another, a process known as **transmutation.**

Electric charge is conserved during α decay. In Figure 31.9, for instance, 90 of the 92 protons in the uranium nucleus end up in the thorium nucleus, and the remaining 2 protons are carried off by the α particle. The total number of 92, however, is the same before and after disintegration. α decay also conserves the number of nucleons, for the number is the same before (238) and after ($234 + 4$) disintegration. Consistent with the conservation of electric charge and nucleon number, the general form for α decay is

α **decay** $$^A_Z\text{P} \longrightarrow {}^{A-4}_{Z-2}\text{D} + {}^4_2\text{He}$$

| Parent nucleus | Daughter nucleus | α particle (helium nucleus) |

When a nucleus releases an α particle, the nucleus also releases energy. In fact, the energy released by radioactive decay is responsible, in part, for keeping the interior of the earth hot and, in some places, even molten. The following example

Figure 31.9 α decay occurs when an unstable parent nucleus emits an α particle and in the process is converted into a different or daughter nucleus.

shows how the conservation of mass/energy can be used to determine the amount of energy released in α decay.

EXAMPLE 4 α Decay and the Release of Energy

The atomic mass of uranium $^{238}_{92}$U is 238.0508 u, while that of thorium $^{234}_{90}$Th is 234.0436 u, and that of an alpha particle $^{4}_{2}$He is 4.0026 u. Determine the energy released when α decay converts $^{238}_{92}$U into $^{234}_{90}$Th.

Reasoning Since energy is released during the decay, the combined mass of the $^{234}_{90}$Th daughter nucleus and the α particle is less than the mass of the $^{238}_{92}$U parent nucleus. The difference in mass is equivalent to the energy released. We will determine the difference in mass in atomic mass units and then use the fact that 1 u is equivalent to 931.5 MeV.

Solution The decay and the masses are shown below:

$$^{238}_{92}\text{U} \longrightarrow \underbrace{^{234}_{90}\text{Th} \;+\; ^{4}_{2}\text{He}}_{238.0462\ \text{u}}$$

238.0508 u   234.0436 u   4.0026 u

The decrease in mass is 238.0508 u − 238.0462 u = 0.0046 u. As usual, the masses are atomic masses and include the mass of the orbital electrons. But this causes no error here because the same total number of electrons is included for $^{238}_{92}$U, on the one hand, and for $^{234}_{90}$Th plus $^{4}_{2}$He, on the other. Since 1 u is equivalent to 931.5 MeV, the released energy is $\boxed{4.3 \text{ MeV}}$.

When α decay occurs as in Example 4, the energy released appears as kinetic energy of the recoiling $^{234}_{90}$Th nucleus and the α particle, except for a small portion carried away as a γ ray. Conceptual Example 5 discusses how the $^{234}_{90}$Th nucleus and the α particle share in the released energy.

CONCEPTUAL EXAMPLE 5
How Energy Is Shared During the α Decay of $^{238}_{92}$U

In Example 4, the energy released by the α decay of $^{238}_{92}$U is found to be 4.3 MeV. Since this energy is carried away as kinetic energy of the recoiling $^{234}_{90}$Th nucleus and the α particle, it follows that $KE_{Th} + KE_\alpha = 4.3$ MeV. However, KE_{Th} and KE_α are not equal. Which particle carries away more kinetic energy, the $^{234}_{90}$Th nucleus or the α particle?

Reasoning and Solution Kinetic energy depends on the mass m and speed v of a particle, since $KE = \frac{1}{2}mv^2$. The $^{234}_{90}$Th nucleus has a much greater mass than the α particle, and since the kinetic energy is proportional to the mass, it is tempting to conclude that the $^{234}_{90}$Th nucleus has the greater kinetic energy. This conclusion is not correct, however, since it does not take into account the fact that the $^{234}_{90}$Th nucleus and the α particle have different speeds after the decay. In fact, we expect the thorium nucleus to recoil with the smaller speed precisely *because* it has the greater mass. The decaying $^{238}_{92}$U is like a father and his young daughter on ice skates, pushing off against one another. The more massive father recoils with much less speed than the daughter. We can use the principle of conservation of linear momentum to verify our expectation.

As Section 7.2 discusses, the conservation principle states that the total linear momentum of an isolated system remains constant. An isolated system is one for which the vector sum of the external forces acting on the system is zero, and the

Figure 31.10 A smoke detector.

decaying $^{238}_{92}U$ nucleus fits this description. It is stationary initially, and since momentum is mass times velocity, its initial momentum is zero. In its final form, the system consists of the $^{234}_{90}Th$ nucleus and the α particle and has a final total momentum of $m_{Th}v_{Th} + m_\alpha v_\alpha$. According to momentum conservation, the initial and final values of the total momentum of the system must be the same, so that $m_{Th}v_{Th} + m_\alpha v_\alpha = 0$. Solving this equation for the velocity of the thorium nucleus, we find that $v_{Th} = -m_\alpha v_\alpha / m_{Th}$. Since m_{Th} is much greater than m_α, we can see that the speed of the thorium nucleus is less than the speed of the α particle. Moreover, the kinetic energy depends on the square of the speed and only the first power of the mass. As a result, of its much greater speed, *the α particle has the greater kinetic energy.*

Related Homework Material: *Problem 54*

One widely used application of α decay is in smoke detectors. Figure 31.10 illustrates how a smoke detector operates. Two small and parallel metal plates are separated by a distance of about one centimeter. A tiny amount of radioactive material at the center of one of the plates emits α particles, which collide with air molecules. During the collisions, the air molecules are ionized to form positive and negative ions. The voltage from a battery causes one plate to be positive and the other negative, so that each plate attracts ions of opposite charge. As a result there is a current in the circuit attached to the plates. The presence of smoke particles between the plates reduces the current, since the ions that collide with a smoke particle are usually neutralized. The drop in current that smoke particles cause is used to trigger an alarm.

THE PHYSICS OF...

radioactivity and smoke detectors.

β Decay

The β rays in Figure 31.8 are deflected by the magnetic field in a direction opposite to that of the positively charged α rays. Consequently, these β rays, which are the most common kind, consist of negatively charged particles or β^- particles. Experiment shows that β^- particles are electrons. As an illustration of β^- decay, consider the thorium $^{234}_{90}Th$ nucleus, which decays by emitting a β^- particle, as in Figure 31.11:

$$^{234}_{90}Th \longrightarrow ^{234}_{91}Pa + ^{0}_{-1}e$$

| Parent nucleus (thorium) | Daughter nucleus (protactinium) | β^- particle (electron) |

Figure 31.11 β decay occurs when a neutron in an unstable parent nucleus decays into a proton and an electron, the electron being emitted as the β^- particle. In the process, the parent nucleus is transformed into the daughter nucleus.

β^- decay, like α decay, causes a transmutation of one element into another. In this case, thorium $^{234}_{90}Th$ is converted into protactinium $^{234}_{91}Pa$. The law of conservation of charge is obeyed, since the net number of positive charges is the same before (90) and after (91 − 1) the β^- emission. The law of conservation of nucleon number is obeyed, since the nucleon number remains at $A = 234$. The general form for β^- decay is

β^- decay $$^{A}_{Z}P \longrightarrow ^{A}_{Z+1}D + ^{0}_{-1}e$$

| Parent nucleus | Daughter nucleus | β^- particle (electron) |

The electron emitted in β^- decay does *not* actually exist within the parent nucleus and is *not* one of the orbital electrons. Instead, the electron is created when a neutron decays into a proton and an electron; when this occurs, the proton number of the parent nucleus increases from Z to $Z + 1$ and the nucleon number remains unchanged. The electron is usually fast-moving and escapes from the atom, leaving behind a positively charged atom.

Example 6 illustrates that energy is released during β^- decay, just as it is during α decay, and that the conservation of mass/energy applies.

EXAMPLE 6 β^- Decay and the Release of Energy

The atomic mass of thorium $^{234}_{90}\text{Th}$ is 234.043 59 u, while that of protactinium $^{234}_{91}\text{Pa}$ is 234.043 30 u. Find the energy released when β^- decay changes $^{234}_{90}\text{Th}$ into $^{234}_{91}\text{Pa}$.

Reasoning and Solution To find the energy released, we follow the usual procedure of determining how much the mass has decreased because of the decay and then calculating the equivalent energy. The decay and the masses are shown below:

$$^{234}_{90}\text{Th} \longrightarrow \underbrace{^{234}_{91}\text{Pa} + {}^{0}_{-1}\text{e}}_{234.043\ 30\ \text{u}}$$
$$234.043\ 59\ \text{u}$$

When the $^{234}_{90}\text{Th}$ nucleus of a thorium atom is converted into a $^{234}_{91}\text{Pa}$ nucleus, the number of orbital electrons remains the same, so the resulting protactinium atom is missing one orbital electron. However, the given mass includes all 91 electrons of a neutral protactinium atom. In effect, then, the value of 234.043 30 u for $^{234}_{91}\text{Pa}$ already includes the mass of the β^- particle. The mass decrease that accompanies the β^- decay is 234.043 59 u − 234.043 30 u = 0.000 29 u. The equivalent energy (1 u = 931.5 MeV) is $\boxed{0.27\ \text{MeV}}$. This is the maximum kinetic energy that the emitted electron can have.

● **PROBLEM SOLVING INSIGHT**
In β^- decay, be careful not to include the mass of the electron (${}^{0}_{-1}\text{e}$) twice. As discussed here for the daughter atom ($^{234}_{91}\text{Pa}$), the atomic mass already includes the mass of the emitted electron.

A second kind of β decay sometimes occurs.† In this process the particle emitted by the nucleus is a ***positron*** rather than an electron. A positron, also called a β^+ particle, has the same mass as an electron but carries a charge of $+e$ instead of $-e$. The disintegration process for β^+ decay is

β^+ decay $^{A}_{Z}\text{P} \longrightarrow {}^{A}_{Z-1}\text{D} + {}^{0}_{1}\text{e}$

| Parent | Daughter | β^+ particle |
| nucleus | nucleus | (positron) |

The emitted positron does *not* exist within the nucleus but, rather, is created when a nuclear proton is transformed into a neutron. In the process, the proton number of the parent nucleus decreases from Z to $Z - 1$, and the nucleon number remains the same. As with β^- decay, the laws of conservation of charge and nucleon number are obeyed, and there is a transmutation of one element into another.

γ Decay

The nucleus, like the orbital electrons, exists only in discrete energy states or levels. When a nucleus changes from an excited energy state (denoted by an asterisk *) to a lower energy state, a photon is emitted. The process is similar to the one dis-

---

† A third kind of β decay also occurs in which a nucleus pulls in or captures one of the orbital electrons from outside the nucleus. The process is called ***electron capture***, or **K capture**, since the electron normally comes from the innermost or K shell.

● **PROBLEM SOLVING INSIGHT**
The energy ΔE of a γ-ray photon, like that of photons in other regions of the electromagnetic spectrum (visible, infrared, microwave, etc.), is equal to the product of Planck's constant h and the frequency f of the photon: $\Delta E = hf$.

cussed in Section 30.3 for the photon emission that leads to the hydrogen atom line spectrum. With nuclear energy levels, however, the photon has a much greater energy and is called a γ ray. The γ decay process is written as follows:

$$\gamma \ decay \qquad ^A_Z P^* \longrightarrow ^A_Z P + \gamma$$

$$\underset{\substack{\text{Excited} \\ \text{energy state}}}{} \qquad \underset{\substack{\text{Lower} \\ \text{energy state}}}{} \qquad \underset{\gamma \text{ ray}}{}$$

γ decay does *not* cause a transmutation of one element into another. In the next example the wavelength of one particular γ ray photon is determined.

EXAMPLE 7
The Wavelength of a Photon Emitted During γ Decay

What is the wavelength of the 0.186 MeV γ ray photon emitted by radium $^{226}_{88}$Ra?

Reasoning The photon energy is the difference between two nuclear energy levels. Equation 30.4 gives the relation between the energy level separation ΔE and the frequency f of the photon as $\Delta E = hf$. Since $f\lambda = c$, the wavelength of the photon is $\lambda = hc/\Delta E$.

Solution First we must convert the photon energy into joules:

$$\Delta E = (0.186 \times 10^6 \text{ eV})\left(\frac{1.60 \times 10^{-19} \text{ J}}{1 \text{ eV}}\right) = 2.98 \times 10^{-14} \text{ J}$$

The wavelength of the photon is

$$\lambda = \frac{hc}{\Delta E} = \frac{(6.63 \times 10^{-34} \text{ J} \cdot \text{s})(3.00 \times 10^8 \text{ m/s})}{2.98 \times 10^{-14} \text{ J}} = \boxed{6.67 \times 10^{-12} \text{ m}}$$

Gamma Knife radiosurgery is becoming a very promising medical procedure for treating certain problems of the brain, including benign and cancerous tumors, as well as blood vessel malformations. The procedure, which involves no knife at all, uses powerful, highly focused beams of γ rays aimed at the tumor or malformation. The γ rays are emitted by a radioactive cobalt-60 source. As Figure 31.12a illus-

Figure 31.12 (*a*) In Gamma Knife radiosurgery, a protective metal helmet containing 201 small holes is placed over the patient's head. (*b*) The holes focus the beams of γ rays to a tiny target within the brain.

(*a*)

(*b*)

trates, the patient wears a protective metal helmet that is perforated with 201 small holes. Part *b* of the figure shows that the holes focus the γ rays to a single tiny target within the brain. The target tissue thus receives a very intense dose of radiation and is destroyed, while the surrounding healthy tissue is undamaged. Gamma Knife surgery is a noninvasive, painless, and bloodless procedure that is often performed under local anesthesia. Hospital stays are 70 to 90 percent shorter than with conventional surgery, and patients often return to work within a few days.

31.5 THE NEUTRINO

When a β particle is emitted by a radioactive nucleus, energy is simultaneously released, as Example 6 illustrates. Experimentally, however, it is found that most β particles do not have enough kinetic energy to account for all the energy released. If a β particle carries away only part of the energy, where does the remainder go? The question puzzled physicists until 1930, when Wolfgang Pauli proposed that part of the energy is carried away by another particle that is emitted along with the β particle. This additional particle is called the ***neutrino***, and its existence was verified experimentally in 1956. The Greek letter nu (ν) is used to symbolize the neutrino. For instance, the β^- decay of thorium $^{234}_{90}\text{Th}$ (see Section 31.4) is more correctly written as

$$^{234}_{90}\text{Th} \longrightarrow \, ^{234}_{91}\text{Pa} + \, ^{0}_{-1}\text{e} + \bar{\nu}$$

The bar above the ν is included because the neutrino emitted in this particular decay process is an antimatter neutrino or antineutrino. A normal neutrino (ν without the bar) is emitted when β^+ decay occurs.

The neutrino has zero electrical charge and is extremely difficult to detect because it interacts very weakly with matter. For example, the average neutrino can penetrate one light-year of lead (about 9.5×10^{15} m) without interacting with it. Thus, even though trillions of neutrinos pass through our bodies every second, they have no effect. Although difficult, it is possible to detect neutrinos. Figure 31.13 shows the Super Kamiokande neutrino detector in Japan. It is located 915 m underground and consists of a steel cylindrical tank, ten stories tall, whose inner wall is lined with 11 000 photomultiplier tubes. The tank is filled with 12.5 million gallons of ultrapure water. Neutrinos colliding with the water molecules produce light patterns that the photomultiplier tubes detect.

One of the major scientific questions of our time is whether neutrinos have mass. The question is important because neutrinos are so plentiful in the universe. Even a very small mass could account for a significant portion of the mass in the universe and, possibly, have an effect on the formation of galaxies. In 1998 the Super Kamiokande detector yielded the first strong, but indirect, evidence that neutrinos do indeed have mass, in an amount that appears to be a tiny fraction of the mass of an electron. This finding implies that neutrinos travel at less than the speed of light. If the neutrino's mass were zero, like that of a photon, it would travel at the speed of light.

The emission of neutrinos and β particles involves a force called the weak nuclear force because it is much weaker than the strong nuclear force. It is now known that the weak nuclear force and the electromagnetic force are two different manifestations of a single, more fundamental force, the ***electroweak force.*** The theory for the electroweak force was developed by Sheldon Glashow (1932–), Abdus Salam (1926–), and Steven Weinberg (1933–), who shared a Nobel

Figure 31.13 The Super Kamiokande neutrino detector in Japan consists of a steel cylindrical tank containing 12.5 million gallons of ultrapure water. Its inner wall is lined with 11 000 photomultiplier tubes.

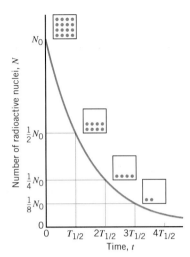

Figure 31.14 The half-life $T_{1/2}$ of a radioactive decay is the time in which one-half of the radioactive nuclei disintegrate.

prize for their achievement in 1979. The electroweak force, the gravitational force, and the strong nuclear force are the three fundamental forces in nature.

31.6 RADIOACTIVE DECAY AND ACTIVITY

The question of which radioactive nucleus in a group of nuclei disintegrates at a given instant is decided like the drawing of numbers in a state lottery; individual disintegrations occur randomly. As time passes, the number N of parent nuclei decreases, as Figure 31.14 shows. This graph of N versus time indicates that the decrease occurs in a smooth fashion, with N approaching zero after enough time has passed. To help describe the graph, it is useful to define the **half-life** $T_{1/2}$ of a radioactive isotope as the time required for one-half of the nuclei present to disintegrate. For example, radium $^{226}_{88}\text{Ra}$ has a half-life of 1600 years, for it takes this amount of time for one-half of a given quantity of this isotope to disintegrate. In another 1600 years, one-half of the remaining radium atoms will disintegrate, leaving only one-fourth of the original number intact. In Figure 31.14, the number of nuclei present at time $t = 0$ s is $N = N_0$, while the number present at $t = T_{1/2}$ is $N = \frac{1}{2}N_0$. The number present at $t = 2T_{1/2}$ is $N = \frac{1}{4}N_0$, and so forth. The value of the half-life depends on the nature of the radioactive nucleus. Values ranging from a fraction of a second to billions of years have been found, as Table 31.2 indicates.

Radon $^{222}_{86}\text{Rn}$ is a naturally occurring radioactive gas produced when radium $^{226}_{88}\text{Ra}$ undergoes α decay. There is a nationwide concern about radon as a health hazard because radon in the soil is gaseous and can enter the basement of homes through cracks in the foundation. Once inside, the concentration of radon can rise markedly, depending on the type of housing construction and the concentration of radon in the surrounding soil. Radon gas decays into daughter nuclei that are also radioactive. The radioactive nuclei can attach to dust and smoke particles that can be inhaled, and they remain in the lungs to release tissue-damaging radiation. Prolonged exposure to high levels of radon can lead to lung cancer. Since radon gas concentrations can be measured with inexpensive monitoring devices, it is recommended that all homes be tested for radon. Example 8 deals with the half-life of radon $^{222}_{86}\text{Rn}$.

EXAMPLE 8   The Radioactive Decay of Radon Gas

THE PHYSICS OF...

radioactive radon gas in houses.

Suppose 3.0×10^7 radon atoms are trapped in a basement at the time the basement is sealed against further entry of the gas. The half-life of radon is 3.83 days. How many radon atoms remain after 31 days?

Reasoning During each half-life, the number of radon atoms is reduced by a factor of two. Thus, we determine the number of half-lives there are in a period of 31 days and reduce the number of radon atoms by a factor of two for each one.

Solution In a period of 31 days there are 31 days/3.83 days = 8.1 half-lives. In 8 half-lives the number of radon atoms is reduced by a factor of $2^8 = 256$. Ignoring the difference between 8 and 8.1 half-lives, we find that the number of atoms remaining is $3.0 \times 10^7/256 = \boxed{1.2 \times 10^5}$.

The **activity** of a radioactive sample is the number of disintegrations per second that occur. Each time a disintegration occurs, the number N of radioactive nuclei decreases. As a result, the activity can be obtained by dividing ΔN, the change in the number of nuclei, by Δt, the time interval during which the change takes place; the average activity over the time interval Δt is the magnitude of $\Delta N/\Delta t$. Since the decay of any individual nucleus is completely random, the number of disintegrations per second that occurs in a sample is proportional to the number of radioactive nuclei present, so that

$$\frac{\Delta N}{\Delta t} = -\lambda N \qquad (31.4)$$

where λ is a proportionality constant referred to as the *decay constant*. The minus sign is present in this equation because each disintegration decreases the number N of nuclei originally present.

The SI unit for activity is the *becquerel* (Bq); one becquerel equals one disintegration per second. Activity is also measured in terms of a unit called the *curie* (Ci), in honor of Marie (1867–1934) and Pierre (1859–1906) Curie, the discoverers of radium and polonium. Historically, the curie was chosen as a unit because it is roughly the activity of one gram of pure radium. In terms of becquerels,

$$1 \text{ Ci} = 3.70 \times 10^{10} \text{ Bq}$$

The activity of the radium put into the dial of a watch to make it glow in the dark is about 4×10^4 Bq, and the activity used in radiation therapy for cancer treatment is approximately a billion times greater, or 4×10^{13} Bq.

The mathematical expression for the graph of N versus t shown in Figure 31.14 can be obtained from Equation 31.4 with the aid of calculus. The result for the number N of radioactive nuclei present at time t is

$$N = N_0 e^{-\lambda t} \qquad (31.5)$$

assuming that the number at $t = 0$ s is N_0. The exponential e has the value $e = 2.718 \ldots$, and many calculators provide the value of e^x. We can relate the half-life $T_{1/2}$ of a radioactive nucleus to its decay constant λ in the following manner. By substituting $N = \frac{1}{2}N_0$ and $t = T_{1/2}$ into Equation 31.5, we find that $\frac{1}{2} = e^{-\lambda T_{1/2}}$. Taking the natural logarithm of both sides of this equation reveals that $\ln 2 = \lambda T_{1/2}$ or

$$T_{1/2} = \frac{\ln 2}{\lambda} = \frac{0.693}{\lambda} \qquad (31.6)$$

The following example illustrates the use of Equations 31.5 and 31.6.

TABLE 31.2   Some Half-lives for Radioactive Decay

Isotope		Half-life
Polonium	$^{214}_{84}\text{Po}$	1.64×10^{-4} s
Krypton	$^{89}_{36}\text{Kr}$	3.16 min
Radon	$^{222}_{86}\text{Rn}$	3.83 da
Strontium	$^{90}_{38}\text{Sr}$	28.5 yr
Radium	$^{226}_{88}\text{Ra}$	1.6×10^3 yr
Carbon	$^{14}_{6}\text{C}$	5.73×10^3 yr
Uranium	$^{238}_{92}\text{U}$	4.47×10^9 yr
Indium	$^{115}_{49}\text{In}$	4.41×10^{14} yr

EXAMPLE 9   The Activity of Radon $^{222}_{86}$Rn

As in Example 8, suppose there are 3.0×10^7 radon atoms ($T_{1/2} = 3.83$ days or 3.31×10^5 s) trapped in a basement. (a) How many radon atoms remain after 31 days? Find the activity (b) just after the basement is sealed against further entry of radon and (c) 31 days later.

Reasoning The number N of radon atoms remaining after a time t is given by $N = N_0 e^{-\lambda t}$, where $N_0 = 3.0 \times 10^7$ is the original number of atoms when $t = 0$ s and λ is the decay constant. The decay constant is related to the half-life $T_{1/2}$ of the radon atoms by $\lambda = 0.693/T_{1/2}$. The activity can be obtained from Equation 31.4, $\Delta N/\Delta t = -\lambda N$.

Solution

(a) The decay constant is

$$\lambda = \frac{0.693}{T_{1/2}} = \frac{0.693}{3.83 \text{ days}} = 0.181 \text{ days}^{-1} \qquad (31.6)$$

and the number N of radon atoms remaining after 31 days is

$$N = N_0 e^{-\lambda t} = (3.0 \times 10^7)e^{-(0.181 \text{ days}^{-1})(31 \text{ days})} = \boxed{1.1 \times 10^5} \qquad (31.5)$$

This value is slightly different from that found in Example 8 because there we ignored the difference between 8.0 and 8.1 half-lives.

(b) The activity can be obtained from Equation 31.4, provided the decay constant is expressed in reciprocal seconds: $\lambda = 0.693/(3.31 \times 10^5 \text{ s}) = 2.09 \times 10^{-6} \text{ s}^{-1}$. Thus, the number of disintegrations per second is

$$\frac{\Delta N}{\Delta t} = -\lambda N = -(2.09 \times 10^{-6} \text{ s}^{-1})(3.0 \times 10^7) = -63 \text{ disintegrations/s}$$

The activity is the magnitude of $\Delta N/\Delta t$, so initially $\boxed{\text{Activity} = 63 \text{ Bq}}$.

(c) From part (a), the number of radioactive nuclei remaining at the end of 31 days is $N = 1.1 \times 10^5$, and reasoning similar to that in part (b) reveals that $\boxed{\text{Activity} = 0.23 \text{ Bq}}$.

31.7 RADIOACTIVE DATING

One important application of radioactivity is the determination of the age of archeological or geological samples. If an object contains radioactive nuclei when it is formed, then the decay of these nuclei marks the passage of time like a clock, half of the nuclei disintegrating during each half-life. If the half-life is known, a measurement of the number of nuclei present today relative to the number present initially can give the age of the sample. According to Equation 31.4, the activity of a sample is proportional to the number of radioactive nuclei, so one way to obtain the age is to compare present activity with initial activity. A more accurate way is to determine the present number of radioactive nuclei with the aid of a mass spectrometer.

The present activity of a sample can be measured, but how is it possible to know what the original activity was, perhaps thousands of years ago? Radioactive dating methods entail certain assumptions that make it possible to estimate the original activity. For instance, the radiocarbon technique utilizes the $^{14}_{6}C$ isotope of carbon, which undergoes β^- decay with a half-life of 5730 yr. This isotope is present in the earth's atmosphere at an equilibrium concentration of about one atom for every 8.3×10^{11} atoms of normal carbon $^{12}_{6}C$. It is often assumed* that this value has remained constant over the years because $^{14}_{6}C$ is created when cosmic rays interact with the earth's upper atmosphere, a production method that offsets the loss via β^- decay. Moreover, nearly all living organisms ingest the equilibrium concentration of $^{14}_{6}C$. However, once an organism dies, metabolism no longer sustains the input of $^{14}_{6}C$, and β^- decay causes half of the $^{14}_{6}C$ nuclei to disintegrate every 5730 yr. Example 10 illustrates how to determine the $^{14}_{6}C$ activity of one gram of carbon in a living organism.

This mummified Incan child was one of three recently discovered on the top of a northern Argentine mountain. Radioactive dating is one of the techniques used to determine the age of such remains.

---

* The assumption that the $^{14}_{6}C$ concentration has always been at its present equilibrium value has been evaluated by comparing $^{14}_{6}C$ ages with ages determined by counting tree rings. More recently, ages determined using the radioactive decay of uranium $^{238}_{92}U$ have been used for comparison. These comparisons indicate that the equilibrium value of the $^{14}_{6}C$ concentration has indeed remained constant for the past 1000 years. However, from there back about 30 000 years, it appears that the $^{14}_{6}C$ concentration in the atmosphere was larger than its present value by up to 40%. As a first approximation we ignore such discrepancies.

EXAMPLE 10 $^{14}_{6}C$ Activity Per Gram of Carbon in a Living Organism

(a) Determine the number of carbon $^{14}_{6}C$ atoms present for every gram of carbon $^{12}_{6}C$ in a living organism. Find (b) the decay constant and (c) the activity of this sample.

Reasoning The total number of carbon $^{12}_{6}C$ atoms in one gram of carbon $^{12}_{6}C$ is equal to the corresponding number of moles times Avogadro's number (see Section 14.1). Since there is only one $^{14}_{6}C$ atom for every 8.3×10^{11} atoms of $^{12}_{6}C$, the number of $^{14}_{6}C$ atoms is equal to the total number of $^{12}_{6}C$ atoms divided by 8.3×10^{11}. The decay constant λ for $^{14}_{6}C$ is $\lambda = 0.693/T_{1/2}$, where $T_{1/2}$ is the half-life. The activity is equal to the magnitude of $\Delta N/\Delta t$, which is the decay constant times the number of $^{14}_{6}C$ atoms present, according to Equation 31.4.

Solution

(a) One gram of carbon $^{12}_{6}C$ (atomic mass = 12 u) is equivalent to 1.0/1.2 mol. Since Avogadro's number is 6.02×10^{23} atoms/mol and since there is one $^{14}_{6}C$ atom for every 8.3×10^{11} atoms of $^{12}_{6}C$, the number of $^{14}_{6}C$ atoms is

$$\text{Number of } ^{14}_{6}C \text{ atoms for every 1.0 gram of carbon } ^{12}_{6}C = \left(\frac{1.0}{12} \text{ mol}\right)\left(6.02 \times 10^{23} \frac{\text{atoms}}{\text{mol}}\right)\left(\frac{1}{8.3 \times 10^{11}}\right)$$

$$= \boxed{6.0 \times 10^{10} \text{ atoms}}$$

(b) Since the half-life of $^{14}_{6}C$ is 5730 yr (1.81×10^{11} s), the decay constant is

$$\lambda = \frac{0.693}{T_{1/2}} = \frac{0.693}{1.81 \times 10^{11} \text{ s}} = \boxed{3.83 \times 10^{-12} \text{ s}^{-1}} \qquad (31.6)$$

(c) Equation 31.4 indicates that $\Delta N/\Delta t = -\lambda N$. The activity is the magnitude of $\Delta N/\Delta t$ or λN. Thus, we find

$$\text{Activity of } ^{14}_{6}C \text{ for every 1.0 gram of carbon } ^{12}_{6}C \text{ in a living organism} = \lambda N = (3.83 \times 10^{-12} \text{ s}^{-1})(6.0 \times 10^{10} \text{ atoms}) = \boxed{0.23 \text{ Bq}}$$

An organism that lived thousands of years ago presumably had an activity of about 0.23 Bq per gram of carbon. When the organism died, the activity began decreasing. From a sample of the remains, the current activity per gram of carbon can be measured and compared to the value of 0.23 Bq to determine the time that has transpired since death. This procedure is illustrated in Example 11.

EXAMPLE 11   The Iceman

On 19 September 1991, a German tourist made a startling discovery during a walking trip in the Italian Alps. He found the Iceman, a Stone Age traveler whose body had become trapped in the ice of a glacier. Figure 31.15 shows the remarkably well-preserved remains, which were dated using the radiocarbon method. Material found with the body had a $^{14}_{6}C$ activity of about 0.121 Bq per gram of carbon. Determine the age of the Iceman's remains.

Reasoning According to Equation 31.5, the number of nuclei remaining at time t is $N = N_0 e^{-\lambda t}$. Multiplying both sides of this expression by the decay constant λ and recognizing that the product of λ and N is the activity A, we find that $A = A_0 e^{-\lambda t}$, where $A_0 = 0.23$ Bq is the activity at time $t = 0$ for one gram of carbon. The decay constant λ can be determined from the value of 5730 yr for the half-life of $^{14}_{6}C$,

Figure 31.15 These remains of the Iceman were discovered in the ice of a glacier in the Italian Alps in 1991. Radiocarbon dating reveals his age (see Example 11).

In the radioactive dating technique that uses the carbon $^{14}_{6}$C isotope, the relatively few $^{14}_{6}$C atoms can be detected by measuring their activity, as we have seen. It is also possible to detect these atoms more accurately using an accelerator mass spectrometer. This photograph shows the linear accelerator used in the procedure.

using Equation 31.6. With known values for A_0 and λ, the given activity of $A = 0.121$ Bq per gram of carbon can be used to find t, the Iceman's age.

Solution For $^{14}_{6}$C, the decay constant is $\lambda = 0.693/T_{1/2} = 0.693/(5730 \text{ yr}) = 1.21 \times 10^{-4} \text{ yr}^{-1}$. Since $A = 0.121$ Bq and $A_0 = 0.23$ Bq, the age can be determined from

$$A = 0.121 \text{ Bq} = (0.23 \text{ Bq})e^{-(1.21 \times 10^{-4} \text{ yr}^{-1})t}$$

Taking the natural logarithm of both sides of this result gives

$$\ln\left(\frac{0.121 \text{ Bq}}{0.23 \text{ Bq}}\right) = -(1.21 \times 10^{-4} \text{ yr}^{-1})t$$

which gives an age for the sample of $\boxed{t = 5300 \text{ yr}}$.

Radiocarbon dating is not the only radioactive dating method. For example, other methods utilize uranium $^{238}_{92}$U, potassium $^{40}_{19}$K, and lead $^{210}_{82}$Pb. For such methods to be useful, the half-life of the radioactive species must be neither too short nor too long relative to the age of the sample to be dated, as Conceptual Example 12 discusses.

CONCEPTUAL EXAMPLE 12   Dating a Bottle of Wine

A bottle of red wine is thought to have been sealed about 5 years ago. The wine contains a number of different kinds of atoms, including carbon, oxygen, and hydrogen. Each of these has a radioactive isotope. The radioactive isotope of carbon is the familiar $^{14}_{6}$C, with a half-life of 5730 yr. The radioactive isotope of oxygen is $^{15}_{8}$O and has a half-life of 122.2 s. The radioactive isotope of hydrogen is $^{3}_{1}$H and is called tritium; its half-life is 12.33 yr. The activity of each of these isotopes is known at the time the bottle was sealed. However, only one of the isotopes is useful for determining the age of the wine accurately. Which is it?

Reasoning and Solution In a dating method that measures the activity of a radioactive isotope, the age of the sample is related to the change in the activity during the time period in question. Here the expected age is about 5 years. This period is only a small fraction of the 5730-yr half-life of $^{14}_{6}$C. As a result, relatively few of the $^{14}_{6}$C nuclei would decay during the wine's life, and the measured activity would change little from its initial value. To obtain an accurate age from such a small change would require prohibitively precise measurements. Nor is the $^{15}_{8}$O isotope very useful. The difficulty is its relatively short half-life of 122.2 s. During a 5-year period, so many half-lives of 122.2 s would occur that the activity would decrease to a vanishingly small level. It would not be even possible to measure it. The only remaining option is *the tritium isotope of hydrogen.* The expected age of 5 yr is long enough relative to the half-life of 12.33 yr that a measurable change in activity will occur, but not so long that the activity will have completely vanished for all practical purposes.

Related Homework Material: *Question 14, Problem 42*

31.8 RADIOACTIVE DECAY SERIES

When an unstable parent nucleus decays, the resulting daughter nucleus is sometimes also unstable. If so, the daughter then decays and produces its own daughter, and so on, until a completely stable nucleus is produced. This sequential decay of

Figure 31.16 The radioactive decay series that begins with uranium $^{238}_{92}$U and ends with lead $^{206}_{82}$Pb. Half-lives are given in seconds (s), minutes (m), hours (h), days (d), or years (y). The insert in the upper left identifies the type of decay that each nucleus undergoes.

one nucleus after another is called a ***radioactive decay series.*** Examples 4–6 discuss the first two steps of a series that begins with uranium $^{238}_{92}$U:

$$\begin{array}{cc} \text{Uranium} & \text{Thorium} \\ ^{238}_{92}\text{U} & \longrightarrow\ ^{234}_{90}\text{Th} + ^{4}_{2}\text{He} \end{array}$$

$$\longrightarrow\ ^{234}_{91}\text{Pa} + ^{0}_{-1}\text{e}$$

Protactinium

Furthermore, Example 8 deals with radon $^{222}_{86}$Rn, which is formed down the line in the $^{238}_{92}$U radioactive decay series. Figure 31.16 shows the entire series. At several points branches occur because more than one kind of decay is possible for an intermediate species. Ultimately, however, the series ends with lead $^{206}_{82}$Pb, which is stable.

The $^{238}_{92}$U series and other such series are the only sources of some of the radioactive elements found in nature. Radium $^{226}_{88}$Ra, for instance, has a half-life of 1600 yr, which is short enough that all the $^{226}_{88}$Ra created when the earth was formed billions of years ago has now disappeared. The $^{238}_{92}$U series provides a continuing supply of $^{226}_{88}$Ra, however.

31.9 RADIATION DETECTORS

There are a number of devices that can be used to detect the particles and photons (γ rays) emitted when a radioactive nucleus decays. Such devices detect the ionization that these particles and photons cause as they pass through matter.

THE PHYSICS OF...

radiation detectors.

Figure 31.17   A Geiger counter.

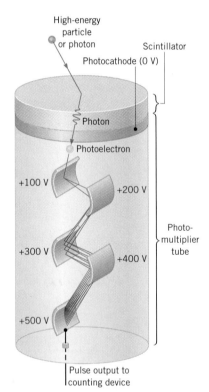

Figure 31.18   A scintillation counter.

The most familiar detector is the ***Geiger counter,*** which Figure 31.17 illustrates. The Geiger counter consists of a gas-filled metal cylinder. The α, β, or γ rays enter the cylinder through a thin window at one end. γ rays can also penetrate directly through the metal. A wire electrode runs along the center of the tube and is kept at a high positive voltage (1000–3000 V) relative to the outer cylinder. When a high-energy particle or photon enters the cylinder, it collides with and ionizes a gas molecule. The electron produced from the gas molecule accelerates toward the positive wire, ionizing other molecules in its path. Additional electrons are formed, and an avalanche of electrons rushes toward the wire, leading to a pulse of current through the resistor R. This pulse can be counted or made to produce a "click" in a loudspeaker. The number of counts or clicks is related to the number of disintegrations that produced the particles or photons.

The ***scintillation counter*** is another important radiation detector. As Figure 31.18 indicates, this device consists of a scintillator mounted on a photomultiplier tube. Often the scintillator is a crystal (e.g., cesium iodide) containing a small amount of impurity (thallium), but plastic, liquid, and gaseous scintillators are also used. In response to ionizing radiation, the scintillator emits a flash of visible light. The photons of the flash then strike the photocathode of the photomultiplier tube. The photocathode is made of a material that emits electrons because of the photoelectric effect. These photoelectrons are then attracted to a special electrode kept at a voltage of about +100 V relative to the photocathode. The electrode is coated with a substance that emits several additional electrons for every electron striking it. The additional electrons are attracted to a second similar electrode (voltage = +200 V) where they generate even more electrons. Commercial photomultiplier tubes contain as many as 15 of these special electrodes, so photoelectrons resulting from the light flash of the scintillator lead to a cascade of electrons and a pulse of current. As in a Geiger tube, the current pulses can be counted.

Ionizing radiation can also be detected with several types of ***semiconductor detectors.*** Such devices utilize *n*- and *p*-type materials (see Section 23.5), and their operation depends on the electrons and holes formed in the materials as a result of the radiation. One of the main advantages of semiconductor detectors is their ability to discriminate between two particles with only slightly different energies.

A number of instruments provide a pictorial representation of the path that high-energy particles follow after they are emitted from unstable nuclei. In a ***cloud chamber,*** a gas is cooled just to the point where it will condense into droplets, provided nucleating agents are available on which the droplets can form. When a high-energy particle, such as an α particle or a β particle, passes through the gas, the ions it leaves behind serve as nucleating agents, and droplets form along the path of the particle. A ***bubble chamber*** works in a similar fashion, except it contains a liquid that is just at the point of boiling. Tiny bubbles form along the trail of a high-energy particle passing through the liquid. The paths revealed in a cloud or bubble chamber can be photographed to provide a permanent record of the event. Figure 31.19 shows a photograph of the tracks in a bubble chamber. A ***photographic emulsion*** also can be used directly to produce a record of the path taken by a particle of ionizing radiation. The ions formed as the particle passes through the emulsion cause silver to be deposited along the track when the emulsion is developed.

31.10 CONCEPTS & CALCULATIONS

Radioactive decay obeys the conservation laws of physics, and we have studied five of these laws, as the Concepts-at-a-Glance chart in Figure 31.7 illustrates. Two of them are particularly important in understanding the types of radioactivity that occur and the nuclear changes that accompany them. These are the conservation of electric charge and the conservation of nucleon number. Example 13 emphasizes their importance and reviews how they are applied.

Figure 31.19 A photograph of particle tracks in a bubble chamber.

CONCEPTS & CALCULATIONS EXAMPLE 13
Electric Charge and Nucleon Number

Thorium $^{228}_{90}$Th produces a daughter nucleus that is radioactive. The daughter, in turn, produces its own radioactive daughter, and so on. This process continues until bismuth $^{212}_{83}$Bi is reached. What are the total number N_α of α particles and the total number N_β of β^- particles that are generated in this series of radioactive decays?

Concept Questions and Answers How many of the 90 protons in the thorium nucleus are carried off by the N_α alpha particles?

Answer Each α particle is a helium $^4_2$He nucleus and carries off two protons. Therefore, the total number of protons carried off by the α particles is $N_\alpha(2)$.

How many protons are left behind when the N_β beta-minus particles are emitted?

Answer Each β^- particle is an electron $^{\,0}_{-1}$e and is emitted when a neutron in the nucleus decays into a proton and an electron. Therefore, the total number of protons left behind by the β^- particles is N_β.

How many of the 228 nucleons in the thorium nucleus are carried off by the N_α alpha particles?

Answer Each α particle is a helium $^4_2$He nucleus and carries off four nucleons. Therefore, the total number of nucleons carried off by the α particles is $N_\alpha(4)$.

Does the departure of a β^- particle alter the number of nucleons?

Answer No. Each β^- particle is an electron $^{\,0}_{-1}$e and is emitted when a neutron in the nucleus decays into a proton and an electron. In effect, a neutron is replaced by a proton. But since each is a nucleon, the number of nucleons is not changed.

Solution The overall decay process can be written as

$$^{228}_{90}\text{Th} \longrightarrow {}^{212}_{83}\text{Bi} + N_\alpha({}^4_2\text{He}) + N_\beta({}^{\,0}_{-1}\text{e})$$

Since electric charge must be conserved, we know that the 90 protons on the left must equal the total number of protons on the right, the result being

$$90 = 83 + N_\alpha(2) + N_\beta(-1)$$

Since the nucleon number must be conserved also, we know that the 228 nucleons on the left must equal the total number of nucleons on the right, so that

$$228 = 212 + N_\alpha(4) + N_\beta(0)$$

Solving this result for the number of α particles gives $\boxed{N_\alpha = 4}$. Substituting $N_\alpha = 4$ into the conservation-of-charge equation, we find that the number of β^- particles is $\boxed{N_\beta = 1}$.

Energy in the form of heat is needed to raise the temperature of an object, as Section 12.7 discusses. One source of the heat can be radioactive decay, as the next example discusses.

CONCEPTS & CALCULATIONS EXAMPLE 14
The Energy from Radioactive Decay

A one-gram sample of thorium $^{228}_{90}$Th contains 2.64×10^{21} atoms and undergoes α decay with a half-life of 1.913 yr (1.677×10^4 h). Each disintegration releases an energy of 5.52 MeV (8.83×10^{-13} J). Assuming that all of the energy is used to heat a 3.8-kg sample of water, find the change in temperature of the water that occurs in one hour.

Concept Questions and Answers How much heat Q is needed to raise the temperature of a mass m of water by ΔT degrees?

Answer According to Equation 12.4, the heat needed is $Q = cm\,\Delta T$, where $c = 4186$ J/(kg·C°) is the specific heat capacity of water.

The energy released by each disintegration is E. What is the total energy E_{Total} released by a number n of disintegrations?

Answer The total energy released is just the number of disintegrations times the energy for each one, or $E_{\text{Total}} = nE$.

What is the number n of disintegrations that occur during a time t?

Answer The number of radioactive nuclei remaining after a time t is given by Equations 31.5 and 31.6 as

$$N = N_0 e^{-0.693t/T_{1/2}}$$

where N_0 is the number of radioactive nuclei present at $t = 0$ s and $T_{1/2}$ is the half-life for the decay. Therefore, the number n of disintegrations that occur during the time t is $N_0 = N$ or

$$n = N_0 - N = N_0(1 - e^{-0.693t/T_{1/2}})$$

Solution According to Equation 12.4, the heat needed to change the temperature of the water is $Q = cm\,\Delta T$. This heat is provided by the total energy released in one hour of radioactive decay, or $E_{\text{Total}} = nE$. Setting $Q = E_{\text{Total}}$, we obtain $nE = cm\,\Delta T$, which can be solved for the change in temperature ΔT:

$$\Delta T = \frac{nE}{cm}$$

Using the expression that we obtained in the Concept Questions for the number of disintegrations n that occur in the time t, we obtain

$$\Delta T = \frac{N_0(1 - e^{-0.693t/T_{1/2}})E}{cm}$$

$$= \frac{(2.64 \times 10^{21})[1 - e^{-0.693(1.000\,\text{h})/(1.677\times 10^4\,\text{h})}](8.83 \times 10^{-13}\,\text{J})}{[4186\,\text{J/(kg·C°)}](3.8\,\text{kg})} = \boxed{6.1\,\text{C°}}$$

At the end of the problem set for this chapter, you will find homework problems that contain both conceptual and quantitative parts. These problems are grouped under the heading *Concepts & Calculations, Group Learning Problems.* They are designed for use by students working alone or in small learning groups. The conceptual part of each problem provides a convenient focus for group discussions.

SUMMARY

The nucleus of an atom consists of protons and neutrons, which are collectively referred to as **nucleons.** A **neutron** is an electrically neutral particle whose mass is slightly larger than that of a proton. The **atomic number** Z is the number of protons in the nucleus. The **atomic mass number** or **nucleon number** A is the total number of protons and neutrons in the nucleus: $A = Z + N$, where N is the number of neutrons. Nuclei that contain the same number of protons, but a different number of neutrons, are called **isotopes.** The approximate radius r (in meters) of a nucleus is given by $r \approx (1.2 \times 10^{-15} \text{ m})A^{1/3}$.

The **strong nuclear force** is the force of attraction between nucleons and is one of the three fundamental forces of nature. This force balances the electrostatic force of repulsion between protons and holds the nucleus together. The strong nuclear force has a very short range of action and is almost independent of electric charge.

The **binding energy** of a nucleus is the energy required to separate the nucleus into its constituent protons and neutrons. The binding energy is equal to $(\Delta m)c^2$, where c is the speed of light in a vacuum and Δm is the mass defect of the nucleus. The **mass defect** is the amount by which the sum of the individual masses of the protons and neutrons exceeds the mass of the parent nucleus.

When specifying nuclear masses, it is customary to use the **atomic mass unit** (u), which is one-twelfth of the mass of a $^{12}_{6}$C atom. One atomic mass unit has a mass of 1.6605×10^{-27} kg and is equivalent to an energy of 931.5 MeV.

Unstable nuclei spontaneously decay by breaking apart or rearranging their internal structures in a process called **radioactivity.** Naturally occurring radioactivity produces α rays, β rays, and γ rays. **α rays** consist of positively charged particles, each particle being the $^{4}_{2}$He nucleus of helium. The most common kind of **β ray** consists of negatively charged particles, or β^- particles, which are electrons. Another kind of β ray consists of positively charged particles, or β^+ particles. A β^+ particle, also called a **positron,** has the same mass as an electron but carries a charge of $+e$ instead of $-e$. **γ rays** are high-energy photons. If a radioactive parent nucleus disintegrates into a daughter nucleus that has a different atomic number, one element has been converted into another element, the conversion being referred to as a **transmutation.**

The **neutrino** is an electrically neutral particle that is emitted along with β particles and has a mass that appears to be much smaller than the mass of an electron.

The **half-life** $T_{1/2}$ of a radioactive isotope is the time required for one-half of the nuclei present to disintegrate or decay. The **activity** is the number of disintegrations per second that occur. In other words, the activity is the magnitude of $\Delta N/\Delta t$, where ΔN is the change in the number N of radioactive nuclei and Δt is the time interval during which the change occurs. The SI unit for activity is the becquerel (Bq), one becquerel being one disintegration per second. Radioactive decay obeys the following relation: $\Delta N/\Delta t = -\lambda N$, where λ is the **decay constant.** This equation can be solved to show that $N = N_0 e^{-\lambda t}$, where N_0 is the number of nuclei present initially. The decay constant is related to the half-life $T_{1/2}$ according to $T_{1/2} = 0.693/\lambda$.

The sequential decay of one nucleus after another is called a **radioactive decay series,** and Figure 31.16 illustrates one such series.

CONCEPTUAL QUESTIONS

1. A material is known to be an isotope of lead, although the particular isotope is not known. From such limited information, which of the following quantities can you specify: (a) its atomic number, (b) its neutron number, and (c) its atomic mass number? Explain.

2. Two nuclei have different nucleon numbers A_1 and A_2. Are the two nuclei necessarily isotopes of the same element? Give your reasoning.

3. Two nuclei have the same radius, even though they contain different numbers of protons and different numbers of neutrons. Explain how this is possible.

4. Using Figure 31.6, rank the following nuclei in ascending order (smallest first) according to the binding energy per nucleon: phosphorus $^{31}_{15}$P, cobalt $^{59}_{27}$Co, tungsten $^{184}_{74}$W, and thorium $^{232}_{90}$Th.

5. Describe qualitatively how the radius of the daughter nucleus compares to that of the parent nucleus for (a) α decay and (b) β decay. Justify your answers in terms of Equation 31.2.

6. Why do α and β decay produce new elements, but γ decay does not?

7. Uranium $^{238}_{92}$U decays into thorium $^{234}_{90}$Th by means of α decay, as Example 4 in the text discusses. A reasonable question to ask is, "Why doesn't the $^{238}_{92}$U nucleus just emit a single proton instead of an α particle?" This hypothetical decay scheme is shown below, along with the pertinent atomic masses:

$$^{238}_{92}\text{U} \longrightarrow {}^{237}_{91}\text{Pa} + {}^{1}_{1}\text{H}$$

| Uranium | Protactinium | Proton |
| 238.050 78 u | 237.051 14 u | 1.007 83 u |

For a decay to be possible, it must bring the parent nucleus toward a more stable state by allowing the release of energy. Compare the total mass of the products of this hypothetical decay with the mass of $^{238}_{92}$U and decide whether the emission of a single proton is possible for $^{238}_{92}$U. Explain.

8. Explain why unstable nuclei with short half-lives typically have only a small or zero natural abundance.

9. The half-life of indium $^{115}_{49}$In is 4.41×10^{14} yr. Thus, one-half of the nuclei in a sample of this isotope will decay in this time, which is very long. Is it possible for any single nucleus in the sample to decay after only one second? Justify your answer.

10. (a) Is it possible for two different samples that contain different radioactive elements to have the same activity? (b) Is it possible for two different samples of the same radioactive element to have different activities? In each case, defend your answer.

11. To which of the following objects, each about 1000 yr old, can the radiocarbon dating technique *not* be applied: a wooden box, a gold statue, and some plant seeds? Explain.

12. Two isotopes have half-lives that are short, relative to the known age of the earth. Today, one of these isotopes is found in nature and one is not. How is this possible, considering that both should have long since decayed to zero? Explain.

13. Suppose there were a greater number of carbon $^{14}_{6}$C atoms in a plant living 5000 yr ago than is currently believed. When the

seeds of this plant are tested today using radiocarbon dating, is the age obtained too small or too large? Give your reasoning.

14. Review Conceptual Example 12 as an aid in answering this question. Tritium is an isotope of hydrogen and undergoes β^- decay with a half-life of 12.33 yr. Like carbon $^{14}_{6}$C, tritium is produced in the atmosphere because of cosmic rays and can be used in a radioactive dating technique. Can tritium dating be used to determine a reliable date for a sample that is about 700 yr old? Account for your answer.

15. Because of radioactive decay, one element can be transmuted into another. Thus, a container of uranium $^{238}_{92}$U ultimately becomes a container of lead $^{206}_{82}$Pb, as Figure 31.16 indicates. How long does the transmutation of $^{238}_{92}$U entirely into $^{206}_{82}$Pb require—several decades, several centuries, thousands of years, millions of years, or billions of years? State your reasoning.

16. In Figure 31.8 the trajectory of the β^- particles has a greater curvature than that of the α particles. Assuming that the speeds of the two types of particles are the same, explain why the curvatures are different. Consider the effects of both mass and charge.

PROBLEMS

The data given for atomic masses in these problems include the mass of the electrons orbiting the nucleus of the electrically neutral atom.

ssm Solution is in the Student Solutions Manual. **www** Solution is available on the World Wide Web at http://www.wiley.com/college/cutnell
⚕ This icon represents a biomedical application.

Section 31.1 Nuclear Structure, Section 31.2 The Strong Nuclear Force and the Stability of the Nucleus

1. ssm In each of the following cases, what element does the symbol X represent and how many neutrons are in the nucleus: (a) $^{195}_{78}$X, (b) $^{32}_{16}$X, (c) $^{63}_{29}$X, (d) $^{11}_{5}$X, and (e) $^{239}_{94}$X? Use the periodic table on the inside of the back cover as needed.

2. How many protons and neutrons are there in the nucleus of (a) oxygen $^{18}_{8}$O and (b) tin $^{120}_{50}$Sn?

3. By what factor does the nucleon number of a nucleus have to increase in order for the nuclear radius to double?

4. For $^{208}_{82}$Pb find (a) the net electrical charge of the nucleus, (b) the number of neutrons, (c) the number of nucleons, (d) the approximate radius of the nucleus, and (e) the nuclear density.

5. ssm www The largest stable nucleus has a nucleon number of 209, while the smallest has a nucleon number of 1. If each nucleus is assumed to be a sphere, what is the ratio (largest/smallest) of the surface areas of these spheres?

*6. An unknown nucleus contains 70 neutrons and has twice the volume of the nickel $^{60}_{28}$Ni nucleus. Identify the unknown nucleus in the form $^{A}_{Z}$X. Use the periodic table on the inside of the back cover as needed.

*7. One isotope (X) contains an equal number of protons and neutrons. Another isotope (Y) of the same element has twice the number of neutrons as the first isotope does. Determine the ratio r_Y/r_X of the nuclear radii of the isotopes.

**8. Conceptual Example 1 provides some useful background for this problem. (a) Determine an approximate value for the density (in kg/m^3) of the nucleus. (b) If a BB (radius = 2.3 mm) from an air rifle had a density equal to the nuclear density, what mass would the BB have? (c) Assuming the mass of a supertanker is about 1.5×10^8 kg, how many "supertankers" of mass would this hypothetical BB have?

**9. ssm www Refer to Conceptual Example 1 for a discussion of nuclear densities. A neutron star is composed of neutrons and has a density that is approximately the same as that of a nucleus. What is the radius of a neutron star whose mass is 0.40 times the mass of the sun?

Section 31.3 The Mass Defect of the Nucleus and Nuclear Binding Energy (*The atomic mass for hydrogen $^{1}_{1}$H is 1.007 825 u, including the mass of one electron.*)

10. Determine the mass defect of the nucleus for cobalt $^{59}_{27}$Co, which has an atomic mass of 58.933 198 u. Express your answer in (a) atomic mass units and (b) kilograms.

11. ssm Use the binding energy per nucleon curve in Figure 31.6 to determine the total binding energy of $^{16}_{8}$O.

12. Find the binding energy (in MeV) for aluminum $^{27}_{13}$Al (atomic mass = 26.981 539 u).

13. The earth revolves around the sun, and the two represent a bound system that has a binding energy of 2.6×10^{33} J. Suppose the earth and sun were completely separated, so that they were infinitely far apart and at rest. What would be the difference between the mass of the separated system and that of the bound system?

14. For lead $^{206}_{82}Pb$ (atomic mass = 205.974 440 u) obtain (a) the mass defect in atomic mass units, (b) the binding energy (in MeV), and (c) the binding energy per nucleon (in MeV).

*__15. ssm__ Two isotopes of a certain element have binding energies that differ by 5.03 MeV. The isotope with the larger binding energy contains one more neutron than the other isotope. Find the difference in atomic mass between the two isotopes.

*__16.__ (a) Energy is required to separate a nucleus into its constituent nucleons, as Figure 31.3 indicates; this energy is the *total* binding energy of the nucleus. In a similar way one can speak of the energy that binds a single nucleon to the remainder of the nucleus. For example, separating nitrogen $^{14}_{7}N$ into nitrogen $^{13}_{7}N$ and a neutron takes energy equal to the binding energy of the neutron, as shown below:

$$^{14}_{7}N + Energy \longrightarrow ^{13}_{7}N + ^{1}_{0}n$$

Find the energy (in MeV) that binds the neutron to the $^{14}_{7}N$ nucleus by considering the mass of $^{13}_{7}N$ (atomic mass = 13.005 738 u) and the mass of $^{1}_{0}n$ (atomic mass = 1.008 665 u), as compared to the mass of $^{14}_{7}N$ (atomic mass = 14.003 074 u). (b) Similarly, one can speak of the energy that binds a single proton to the $^{14}_{7}N$ nucleus:

$$^{14}_{7}N + Energy \longrightarrow ^{13}_{6}C + ^{1}_{1}H$$

Following the procedure outlined in part (a), determine the energy (in MeV) that binds the proton (atomic mass = 1.007 825 u) to the $^{14}_{7}N$ nucleus. (c) Which nucleon is more tightly bound, the neutron or the proton?

Section 31.4 Radioactivity

17. α decay occurs for each of the nuclei given below. Write the decay process for each, including the chemical symbols and values for Z and A for the daughter nuclei: (a) $^{212}_{84}Po$ and (b) $^{232}_{92}U$.

18. In the form $^{A}_{Z}X$, identify the daughter nucleus that results when (a) plutonium $^{242}_{94}Pu$ undergoes α decay, (b) sodium $^{24}_{11}Na$ undergoes β^- decay, and (c) nitrogen $^{13}_{7}N$ undergoes β^+ decay.

19. For the following nuclei, each undergoing β^- decay, write the decay process, identifying each daughter nucleus with its chemical symbol and values for Z and A: (a) $^{14}_{6}C$ and (b) $^{212}_{82}Pb$.

20. Carbon $^{14}_{6}C$ (atomic mass = 14.003 241 u) is converted into nitrogen $^{14}_{7}N$ (atomic mass = 14.003 074 u) via β^- decay. (a) Write this process in symbolic form, giving Z and A for the parent and daughter nuclei and the β^- particle. (b) Determine the energy (in MeV) released.

21. ssm Write the β^- decay process for $^{35}_{16}S$, including the chemical symbol and values for Z and A.

22. Find the energy (in MeV) released when α decay converts radium $^{226}_{88}Ra$ (atomic mass = 226.025 40 u) into radon $^{222}_{86}Rn$ (atomic mass = 222.017 57 u). The atomic mass of an α particle is 4.002 603 u.

23. ssm www Write the β^+ decay process for each of the following nuclei, being careful to include Z and A and the proper chemical symbol for each daughter nucleus: (a) $^{18}_{9}F$ and (b) $^{15}_{8}O$.

*__24.__ Radon $^{220}_{86}Rn$ produces a daughter nucleus that is radioactive. The daughter, in turn, produces its own radioactive daughter, and so on. This process continues until lead $^{208}_{82}Pb$ is reached. What are the total number N_α of α particles and the total number N_β of β^- particles that are generated in this series of radioactive decays?

*__25.__ Plutonium $^{239}_{94}Pu$ (atomic mass = 239.052 16 u) undergoes α decay. Assuming that all the released energy is in the form of kinetic energy of the α particle (atomic mass = 4.002 603 u) and ignoring the recoil of the daughter nucleus, find the speed of the α particle. Ignore relativistic effects.

*__26.__ Determine the symbol $^{A}_{Z}X$ for the parent nucleus whose α decay produces the same daughter as the β^- decay of thallium $^{208}_{81}Tl$.

**__27. ssm__ Find the energy (in MeV) released when β^+ decay converts sodium $^{22}_{11}Na$ (atomic mass = 21.994 434 u) into neon $^{22}_{10}Ne$ (atomic mass = 21.991 383 u). Notice that the atomic mass for $^{22}_{11}Na$ includes the mass of 11 electrons, whereas the atomic mass for $^{22}_{10}Ne$ includes the mass of only 10 electrons.

**__28.__ An isotope of beryllium (atomic mass = 7.017 u) emits a gamma ray and recoils with a speed of 2.19×10^4 m/s. Assuming that the beryllium nucleus is stationary to begin with, find the wavelength of the gamma ray.

Section 31.6 Radioactive Decay and Activity

29. In 9.0 days the number of radioactive nuclei decreases to one-eighth the number present initially. What is the half-life (in days) of the material?

30. The $^{32}_{15}P$ isotope of phosphorus has a half-life of 14.28 da. What is its decay constant in units of s^{-1}?

31. ssm The number of radioactive nuclei present at the start of an experiment is 4.60×10^{15}. The number present twenty days later is 8.14×10^{14}. What is the half-life (in days) of the nuclei?

32. ⚗ Iodine $^{131}_{53}I$ is used in diagnostic and therapeutic techniques in the treatment of thyroid disorders. This isotope has a half-life of 8.04 days. What percentage of an initial sample of $^{131}_{53}I$ remains after 30.0 days?

33. ⚗ Strontium $^{90}_{38}Sr$ has a half-life of 28.5 yr. It is chemically similar to calcium, enters the body through the food chain, and collects in the bones. Consequently, $^{90}_{38}Sr$ is a particularly serious health hazard. How long (in years) will it take for 99.9900% of the $^{90}_{38}Sr$ released in a nuclear reactor accident to disappear?

34. ⚗ A device used in radiation therapy for cancer contains 0.50 g of cobalt $^{60}_{27}Co$ (59.933 819 u). The half-life of $^{60}_{27}Co$ is 5.27 yr. Determine the activity of the radioactive material.

35. ssm If the activity of a radioactive substance is initially 398 disintegrations/min and two days later it is 285 disintegrations/min, what is the activity four days later still, or six days after the start? Give your answer in disintegrations/min.

*__36.__ To make the dial of a watch glow in the dark, 1.000×10^{-9} kg of radium $^{226}_{88}Ra$ is used. The half-life of this isotope is 1.60×10^3 yr. How many kilograms of radium *disappear* while the watch is in use for fifty years?

*37. To see why one curie of activity was chosen to be 3.7×10^{10} Bq, determine the activity (in disintegrations per second) of one gram of radium $^{226}_{88}$Ra $(T_{1/2} = 1.6 \times 10^3$ yr).

*38. What mass (in grams) of krypton $^{92}_{36}$Kr $(T_{1/2} = 1.840$ s, mass per mole = 91.9 g/mol) has the same activity as 2.00 g of xenon $^{140}_{54}$Xe $(T_{1/2} = 13.6$ s, mass per mole = 139.9 g/mol)?

*39. **ssm** Two radioactive nuclei A and B are present in equal numbers to begin with. Three days later, there are three times as many A nuclei as there are B nuclei. The half-life of species B is 1.50 days. Find the half-life of species A.

Section 31.7 Radioactive Dating

40. Bones of the woolly mammoth have been found in North America. The youngest of these bones has a $^{14}_{6}$C activity per gram of carbon that is about 21% of what was present in the live animal. How long ago (in years) did this animal disappear from North America?

41. ssm The practical limit to ages that can be determined by radiocarbon dating is about 41 000 yr. In a 41 000-yr-old sample, what percentage of the original $^{14}_{6}$C atoms remains?

42. Review Conceptual Example 12 before starting to solve this problem. The number of unstable nuclei remaining after a time $t = 5.00$ yr is N, and the number present initially is N_0. Find the ratio N/N_0 for (a) $^{14}_{6}$C (half-life = 5730 yr), (b) $^{15}_{8}$O (half-life = 122.2 s; use $t = 1.00$ h, since otherwise the answer is out of the range of your calculator), and (c) $^{3}_{1}$H (half-life = 12.33 yr). Verify that your answers are consistent with the reasoning in Conceptual Example 12.

43. A sample has a $^{14}_{6}$C activity of 0.0061 Bq per gram of carbon. (a) Find the age of the sample, assuming that the activity per gram of carbon in a living organism has been constant at a value of 0.23 Bq. (b) Evidence suggests that the value of 0.23 Bq might have been as much as 40% larger. Repeat part (a), taking into account this 40% increase.

44. The shroud of Turin is a religious artifact known since the Middle Ages. In 1988 its age was measured using the radiocarbon dating technique, which revealed that the shroud could not have been made before 1200 AD. Of the $^{14}_{6}$C nuclei that were present in the living matter from which the shroud was made, what percentage remained in 1988?

*45. **ssm** When any radioactive dating method is used, experimental error in the measurement of the sample's activity leads to error in the estimated age. In an application of the radiocarbon dating technique to certain fossils, an activity of 0.10 Bq per gram of carbon is measured to within an accuracy of ± ten percent. Find the age of the fossils and the maximum error (in years) in the value obtained. Assume that there is no error in the 5730-year half-life of $^{14}_{6}$C nor in the value of 0.23 Bq per gram of carbon in a living organism.

46. (a) A sample is being dated by the radiocarbon technique. If the sample were uncontaminated, its activity would be 0.011 Bq per gram of carbon. Find the true age (in years) of the sample. (b) Suppose the sample is contaminated, so that only 98.0% of its carbon is ancient carbon. The remaining 2.0% is fresh carbon, in the sense that the $^{14}_{6}$C it contains has not had any time to decay. Assuming that the lab technician is unaware of the contamination, what apparent age (in years) would be determined for the sample?

ADDITIONAL PROBLEMS

47. Mercury $^{202}_{80}$Hg has an atomic mass of 201.970 617 u. Obtain the binding energy *per nucleon* (in MeV/nucleon).

48. What is the radius of a nucleus of titanium $^{48}_{22}$Ti?

49. ssm How many half-lives are required for the number of radioactive nuclei to decrease to one-millionth of the initial number?

50. Tritium $^{3}_{1}$H (atomic mass = 3.016 050 u) undergoes β^- decay and produces $^{3}_{2}$He (atomic mass = 3.016 030 u). Determine the energy (in MeV) released.

51. Complete the following decay processes by stating what the symbol "X" represents $(X = \alpha, \beta^-, \beta^+,$ or $\gamma)$: (a) $^{211}_{82}$Pb $\rightarrow$ $^{211}_{83}$Bi + X, (b) $^{11}_{6}$C $\rightarrow$ $^{11}_{5}$B + X, (c) $^{231}_{90}$Th* $\rightarrow$ $^{231}_{90}$Th + X, and (d) $^{210}_{84}$Po $\rightarrow$ $^{206}_{82}$Pb + X.

52. An archeological specimen containing 9.2 g of carbon has an activity of 1.6 Bq. How old (in years) is the specimen?

*53. **ssm www** The photomultiplier tube in a commercial scintillation counter contains 15 of the special electrodes or dynodes. Each dynode produces 3 electrons for every electron that strikes it. One photoelectron strikes the first dynode. What is the maximum number of electrons that strike the 15th dynode?

*54. Review Conceptual Example 5 as background for this problem. The α decay of uranium $^{238}_{92}$U produces thorium $^{234}_{90}$Th (atomic mass = 234.0436 u). In Example 4, the energy released in this decay is determined to be 4.3 MeV. Determine how much of this energy is carried away by the recoiling $^{234}_{90}$Th daughter nucleus and how much by the α particle (atomic mass = 4.002 603 u). Assume that the energy of each particle is kinetic energy, and ignore the small amount of energy carried away by the γ ray that is also emitted. In addition, ignore relativistic effects.

*55. Two isotopes of the same element have the same binding energy. One isotope contains two more neutrons than the other. What is the difference between the atomic masses (in atomic mass units) of these isotopes?

56. ☀ Both gold $^{198}_{79}$Au $(T_{1/2} = 2.69$ days) and iodine $^{131}_{53}$I $(T_{1/2} = 8.04$ days) are used in diagnostic medicine related to the liver. At the time laboratory supplies are monitored, the activity of the gold is observed to be five times greater than the activity of the iodine. How many days later will the two activities be equal?

CONCEPTS & CALCULATIONS

G R O U P   L E A R N I N G   P R O B L E M S

Note: Each of these problems consists of Concept Questions followed by a related quantitative Problem. They are designed for use by students working alone or in small learning groups. The Concept Questions involve little or no mathematics and are intended to stimulate group discussions. They focus on the concepts with which the problems deal. Recognizing the concepts is the essential initial step in any problem-solving technique.

57. Concept Questions (a) In a nucleus, each proton experiences a repulsive electrostatic force from each of the other protons. Write an expression for the magnitude of the force that one proton (charge $= +e$) applies to another proton that is located a distance r away. (b) The force that acts on either of two particular protons in the nucleus has the smallest possible magnitude. Relative to one another, where in the nucleus must these two protons be located? Explain.

Problem In a nucleus of gold $^{197}_{79}$Au what is the magnitude of the least possible electrostatic force of repulsion that one proton can exert on another?

58. Concept Questions (a) How is the total released energy related to the decrease in mass that accompanies β^- decay? (b) When a parent nucleus undergoes β^- decay, a daughter nucleus, a β^- particle, and an antineutrino $\overline{\gamma}$ are produced. The released energy is shared among these three particles. Assume that the kinetic energy of the β^- particle is known and that the antineutrino carries away the maximum possible energy. What then can you say about the kinetic energy of the recoiling daughter nucleus?

Problem The β^- decay of phosphorus $^{32}_{15}$P atomic mass = 31.973 907 u) produces a daughter nucleus that is sulfur $^{32}_{16}$S (atomic mass = 31.972 070 u), a β^- particle, and an antineutrino $\overline{\gamma}$. The kinetic energy of the β^- particle is 0.90 MeV. Find the maximum possible energy that the antineutrino could carry away.

59. Concept Questions (a) Two radioactive nuclei A and B have half-lives of $T_{1/2, A}$ and $T_{1/2, B}$, where $T_{1/2, A}$ is greater than $T_{1/2, B}$. During the same time period, is the fraction of nuclei A that decay greater than, smaller than, or the same as the fraction of nuclei B that decay? (b) The numbers of these nuclei present initially are $N_{0, A}$ and $N_{0, B}$, the ratio of the two being $N_{0, A}/N_{0, B}$. Is the ratio N_A/N_B of the number of nuclei present at a later time greater than, smaller than, or the same as $N_{0, A}/N_{0, B}$? Justify your answers.

Problem Two waste products from nuclear reactors are strontium $^{90}_{38}$Sr ($T_{1/2}$ = 28.5 yr) and cesium $^{134}_{55}$Cs ($T_{1/2}$ = 2.06 yr). These two species are present initially in a ratio of $N_{0, Sr}/N_{0, Cs}$ = 7.80 × 10^{-3}. What is the ratio N_{Sr}/N_{Cs} fifteen years later? Verify that your answer is consistent with your answers to the Concept Questions.

60. Concept Questions (a) Write an expression for the heat Q needed to melt a mass m of water. (b) In α decay how is the released energy related to the decrease in mass that accompanies the decay? (c) During α decay, the energy released by each disintegration is E. What is the total energy E_{Total} released by a number n of disintegrations? (d) If the number of radioactive nuclei present initially is N_0, what is the number of disintegrations that occur in one half-life?

Problem A one-gram sample of radium $^{224}_{88}$Ra (atomic mass = 224.020 186 u, $T_{1/2}$ = 3.66 da) contains 2.69 × 10^{21} nuclei and undergoes α decay to produce radon $^{220}_{86}$Rn (atomic mass = 220.011 368 u). The atomic mass of an alpha particle is 4.002 603 u. With the energy released in 3.66 da, how many kilograms of ice could be melted at 0 °C?

One method, called the "Z pinch," being tried to produce controlled nuclear fusion uses a burst of electric current from giant capacitors to create an extremely hot plasma (ionized gas). During the burst, electric discharges illuminate the surface of the water that surrounds the fusion process.

CHAPTER 32

IONIZING RADIATION, NUCLEAR ENERGY, AND ELEMENTARY PARTICLES

32.1 BIOLOGICAL EFFECTS OF IONIZING RADIATION

Ionizing radiation consists of photons and/or moving particles that have sufficient energy to knock an electron out of an atom or molecule, thus forming an ion. The photons usually lie in the ultraviolet, X-ray, or γ-ray regions of the electromagnetic spectrum (see Figure 24.9), while the moving particles can be the α and β particles emitted during radioactive decay. An energy of roughly 1 to 35 eV is needed to ionize an atom or molecule, and the particles and γ rays emitted during nuclear disintegration often have energies of several million eV. Therefore, a single α particle, β particle, or γ ray can ionize thousands of molecules.

Nuclear radiation is potentially harmful to humans because the ionization it produces can alter significantly the structure of molecules within a living cell. The alterations can lead to the death of the cell and even the organism itself. Despite the potential hazards, ionizing radiation is used in medicine for diagnostic and therapeutic purposes, such as locating bone fractures and treating cancer. The hazards can be minimized only if the fundamentals of radiation exposure, including dose units and the biological effects of radiation, are understood.

Exposure is a measure of the ionization produced in air by X-rays or γ rays, and it is defined in the following manner. A beam of X-rays or γ rays is sent through a mass *m* of dry air at standard temperature and pressure (STP: 0 °C, 1 atm pressure). In passing through the air, the beam produces positive ions whose total charge is *q*. Exposure is defined as the total charge per unit mass of air: exposure = q/m. The SI unit for exposure is coulombs per kilogram (C/kg). However, the first radiation unit to be defined was the *roentgen* (R), and it is still used today. With *q* expressed in coulombs (C) and *m* in kilograms (kg), the exposure in roentgens is given by

$$\text{Exposure (in roentgens)} = \left(\frac{1}{2.58 \times 10^{-4}}\right)\frac{q}{m} \tag{32.1}$$

Thus, when X-rays or γ rays produce an exposure of one roentgen, $q = 2.58 \times 10^{-4}$ C of positive charge are produced in $m = 1$ kg of dry air:

$$1\,\text{R} = 2.58 \times 10^{-4}\,\text{C/kg} \qquad \text{(dry air, at STP)}$$

Since the concept of exposure is defined in terms of the ionizing abilities of X-rays and γ rays in air, it does not specify the effect of radiation on living tissue. For biological purposes, the *absorbed dose* is a more suitable quantity because it is the energy absorbed from the radiation per unit mass of absorbing material:

$$\text{Absorbed dose} = \frac{\text{Energy absorbed}}{\text{Mass of absorbing material}} \tag{32.2}$$

The SI unit of absorbed dose is the *gray* (Gy), which is the unit of energy divided by the unit of mass: 1 Gy = 1 J/kg. Equation 32.2 is applicable to all types of radiation and absorbing media. Another unit is often used for absorbed dose, namely, the *rad* (rd). The word rad is an acronym for **r**adiation **a**bsorbed **d**ose. The rad and the gray are related by

$$1\,\text{rad} = 0.01\,\text{gray}$$

The amount of biological damage produced by ionizing radiation is different for different kinds of radiation. For instance, a 1-rad dose of neutrons is far more likely to produce eye cataracts than a 1-rad dose of X-rays. To compare the damage caused by different types of radiation, the *relative biological effectiveness* (RBE) is

THE PHYSICS OF...

the biological effects of ionizing radiation.

used.* The relative biological effectiveness of a particular type of radiation compares the dose of that radiation needed to produce a certain biological effect to the dose of 200-keV X-rays needed to produce the same biological effect:

$$\text{Relative biological effectiveness (RBE)} = \frac{\text{The dose of 200-keV X-rays that produces a certain biological effect}}{\text{The dose of radiation that produces the same biological effect}} \quad (32.3)$$

The RBE depends on the nature of the ionizing radiation and its energy, as well as the type of tissue being irradiated. Table 32.1 lists some typical RBE values for different kinds of radiation, assuming that an "average" biological tissue is being irradiated. A value of RBE = 1 for γ rays and β^- particles indicates that they produce the same biological damage as do 200-keV X-rays. The larger RBE values for protons, α particles, and fast neutrons indicate that they cause substantially more damage. The RBE is often used in conjunction with the absorbed dose to reflect the damage-producing character of the radiation. The product of the absorbed dose in rads (not in grays) and the RBE is the *biologically equivalent dose*:

$$\begin{array}{c}\text{Biologically equivalent dose} \\ \text{(in rem)}\end{array} = \begin{array}{c}\text{Absorbed dose} \\ \text{(in rad)}\end{array} \times \text{RBE} \quad (32.4)$$

The unit for the biologically equivalent dose is the *rem*, short for **r**oentgen **e**quivalent, **m**an. Example 1 illustrates the use of the biologically equivalent dose.

TABLE 32.1 Relative Biological Effectiveness (RBE) for Various Types of Radiation

Type of Radiation	RBE
200-keV X-rays	1
γ rays	1
β^- particles (electrons)	1
Protons	10
α particles	10–20
Neutrons	
Slow	2
Fast	10

EXAMPLE 1   Comparing Absorbed Doses of γ Rays and Neutrons

A biological tissue is irradiated with γ rays that have an RBE of 0.70. The absorbed dose of γ rays is 850 rd. The tissue is then exposed to neutrons whose RBE is 3.5. The biologically equivalent dose of the neutrons is the same as that of the γ rays. What is the absorbed dose of neutrons?

Reasoning The biologically equivalent doses of the neutrons and the γ rays are the same. Therefore, the tissue damage produced in each case is the same. However, the RBE of the neutrons is larger than the RBE of the γ rays by a factor of 3.5/0.70 = 5.0. Consequently, we will find that the absorbed dose of the neutrons is only one-fifth as great as that of the γ rays.

Solution The biologically equivalent dose of the γ rays is the product of the absorbed dose (in rad) and the RBE:

$$\begin{array}{c}\text{Biologically equivalent} \\ \text{dose of } \gamma \text{ rays}\end{array} = (850 \text{ rad})(0.70) = 6.0 \times 10^2 \text{ rem} \quad (32.4)$$

For the neutrons (RBE = 3.5), the biologically equivalent dose is the same. Therefore, 6.0×10^2 rem = (Absorbed dose of neutrons)(3.5) and

$$\begin{array}{c}\text{Absorbed dose} \\ \text{of neutrons}\end{array} = \frac{6.0 \times 10^2 \text{ rem}}{3.5} = \boxed{170 \text{ rd}}$$

TABLE 32.2 Average Biologically Equivalent Doses of Radiation Received by a U. S. Resident [a]

Source of Radiation	Biologically Equivalent Dose (mrem/yr) [b]
Natural background radiation	
Cosmic rays	28
Radioactive earth and air	28
Internal radioactive nuclei	39
Inhaled radon	≈200
Man-made radiation	
Consumer products	10
Medical/dental diagnostics	39
Nuclear medicine	14
Rounded total:	360

[a] National Council on Radiation Protection and Measurement, Report No. 93, "Ionizing Radiation Exposure of the Population of the United States," 1987.

[b] 1 mrem = 10^{-3} rem.

Everyone is continually exposed to background radiation from natural sources, such as cosmic rays (high-energy particles that come from outside the solar system), radioactive materials in the environment, radioactive nuclei (primarily carbon $^{14}_{6}$C and potassium $^{40}_{19}$K) within our own bodies, and radon. Table 32.2 lists the average biologically equivalent doses received from these sources by a person in the

* The RBE is sometimes called the *quality factor* (QF).

United States. According to this table, radon is a major contributor to the natural background radiation. Radon is an odorless radioactive gas and poses a health hazard because, when inhaled, it can damage the lungs and cause cancer. Radon is found in soil and rocks and enters houses through cracks and crevices in the foundation. The amount of radon in the soil varies greatly throughout the country, with some localities having significant amounts and others having virtually none. Accordingly, the dose that any individual receives can vary widely from the average value of 200 mrem/yr given in Table 32.2 (1 mrem $= 10^{-3}$ rem). In many houses, the entry of radon can be reduced significantly by sealing the foundation against entry of the gas and providing good ventilation so it does not accumulate.

To the natural background of radiation, a significant amount of man-made radiation has been added, mostly from medical/dental diagnostic X-rays. Table 32.2 indicates an average total dose of 360 mrem/yr from all sources.

The effects of radiation on humans can be grouped into two categories, according to the time span between initial exposure and the appearance of physiological symptoms: (1) short-term or acute effects that appear within a matter of minutes, days, or weeks, and (2) long-term or latent effects that appear years, decades, or even generations later.

Radiation sickness is the general term applied to the acute effects of radiation. Depending on the severity of the dose, a person with radiation sickness can exhibit nausea, vomiting, fever, diarrhea, and loss of hair. Ultimately, death can occur. The severity of radiation sickness is related to the dose received, and in the following discussion the biologically equivalent doses quoted are whole-body, single doses. A dose less than 50 rem causes no short-term ill effects. A dose between 50 and 300 rem brings on radiation sickness, the severity increasing with increasing dosage. A whole-body dose in the range of 400–500 rem is classified as an LD_{50} dose, meaning that it is a lethal dose (LD) for about 50% of the people so exposed; death occurs within a few months. Whole-body doses greater than 600 rem result in death for almost all individuals.

Long-term or latent effects of radiation may appear as a result of high-level brief exposure or low-level exposure over a long period of time. Some long-term effects are hair loss, eye cataracts, and various kinds of cancer. In addition, genetic defects caused by mutated genes may be passed on from one generation to the next.

Because of the hazards of radiation, the federal government has established dose limits. The permissible dose for an individual is defined as the dose, accumulated over a long period of time or resulting from a single exposure, that carries negligible probability of a severe health hazard. Current federal standards (1991) state that an individual in the general population should not receive more than 500 mrem of man-made radiation each year, *exclusive* of medical sources. A person exposed to radiation in the workplace (e.g., a radiation therapist) should not receive more than 5 rem per year from work-related sources.

32.2 INDUCED NUCLEAR REACTIONS

Section 31.4 discusses how a radioactive parent nucleus disintegrates spontaneously into a daughter nucleus. It is also possible to bring about or "induce" the disintegration of a stable nucleus by striking it with another nucleus, an atomic or subatomic particle, or a γ-ray photon. A **nuclear reaction** is said to occur whenever the incident nucleus, particle, or photon causes a change to occur in a target nucleus.

Concepts at a Glance

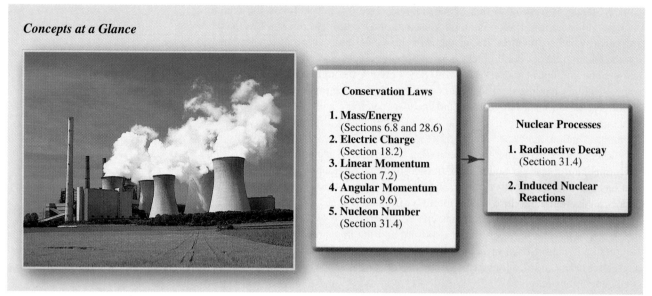

Figure 32.1 *Concepts at a Glance*
In any nuclear process, such as radioactive decay or an induced nuclear reaction, the conservation laws of physics are obeyed. Such processes provide the basis for the commercial generation of power, as in the German nuclear power plant shown in the photograph.

In 1919 Ernest Rutherford observed that when an α particle strikes a nitrogen nucleus, an oxygen nucleus and a proton are produced. This nuclear reaction is written as

$$\underbrace{{}^{4}_{2}\text{He}}_{\substack{\text{Incident} \\ \alpha \text{ particle}}} + \underbrace{{}^{14}_{7}\text{N}}_{\substack{\text{Nitrogen} \\ \text{(target)}}} \longrightarrow \underbrace{{}^{17}_{8}\text{O}}_{\text{Oxygen}} + \underbrace{{}^{1}_{1}\text{H}}_{\text{Proton, } p}$$

Because the incident α particle induces the transmutation of nitrogen into oxygen, this reaction is an example of an ***induced nuclear transmutation***.

Nuclear reactions are often written in a shorthand form. For example, the reaction above is designated by ${}^{14}_{7}\text{N}\,(\alpha, p)\,{}^{17}_{8}\text{O}$. The first and last symbols represent the initial and final nuclei, respectively. The symbols within the parentheses denote the incident α particle (on the left) and the small emitted particle or proton (on the right). Some other induced nuclear transmutations are listed below, together with the corresponding shorthand notations:

Nuclear Reaction	Notation
${}^{1}_{0}\text{n} + {}^{10}_{5}\text{B} \rightarrow {}^{7}_{3}\text{Li} + {}^{4}_{2}\text{He}$	${}^{10}_{5}\text{B}\,(n, \alpha)\,{}^{7}_{3}\text{Li}$
$\gamma + {}^{25}_{12}\text{Mg} \rightarrow {}^{24}_{11}\text{Na} + {}^{1}_{1}\text{H}$	${}^{25}_{12}\text{Mg}\,(\gamma, p)\,{}^{24}_{11}\text{Na}$
${}^{1}_{1}\text{H} + {}^{13}_{6}\text{C} \rightarrow {}^{14}_{7}\text{N} + \gamma$	${}^{13}_{6}\text{C}\,(p, \gamma)\,{}^{14}_{7}\text{N}$

⬤ *Concepts at a Glance*   Induced nuclear reactions, like the radioactive decay process discussed in Section 31.4, obey the conservation laws of physics. We have discussed these laws in previous chapters, as the Concepts-at-a-Glance chart in Figure 32.1 indicates. This chart is similar to that shown earlier in Figure 31.7 and emphasizes the important role that the conservation laws play in all nuclear processes. In particular, both the total electric charge of the nucleons and the total number of nucleons are conserved during an induced nuclear reaction. The fact that these quantities are conserved makes it possible to identify the nucleus produced in a reaction, as the next example illustrates. ⬤

EXAMPLE 2   An Induced Nuclear Transmutation

An α particle strikes an aluminum $^{27}_{13}\text{Al}$ nucleus. As a result, an unknown nucleus $^{A}_{Z}\text{X}$ and a neutron $^{1}_{0}\text{n}$ are produced:

$$^{4}_{2}\text{He} + {}^{27}_{13}\text{Al} \longrightarrow {}^{A}_{Z}\text{X} + {}^{1}_{0}\text{n}$$

Identify the nucleus produced, including its atomic number Z (the number of protons) and its atomic mass number A (the number of nucleons).

Reasoning The total electric charge of the nucleons is conserved, so that we can set the total number of protons before the reaction equal to the total number after the reaction. The total number of nucleons is also conserved, so that we can set the total number before the reaction equal to the total number after the reaction. These two conserved quantities will allow us to identify the nucleus $^{A}_{Z}\text{X}$.

Solution The conservation of total electric charge and total number of nucleons leads to the equations listed below:

Conserved Quantity	Before Reaction	After Reaction
Total electric charge (number of protons)	$2 + 13$ =	$Z + 0$
Total number of nucleons	$4 + 27$ =	$A + 1$

Solving these equations for Z and A gives $Z = 15$ and $A = 30$. Since $Z = 15$ identifies the element as phosphorus, the nucleus produced is $\boxed{^{30}_{15}\text{P}}$.

Induced nuclear transmutations can be used to produce isotopes that are not found naturally. In 1934, Enrico Fermi suggested a method for producing elements with a higher atomic number than uranium ($Z = 92$). These elements—neptunium ($Z = 93$), plutonium ($Z = 94$), americium ($Z = 95$), and so on—are known as *transuranium elements*, and none occurs naturally. They are created in a nuclear reaction between a suitably chosen lighter element and a small incident particle, usually a neutron or an α particle. For example, Figure 32.2 shows a reaction that produces plutonium from uranium. A neutron is captured by a uranium $^{238}_{92}\text{U}$ nucleus, producing $^{239}_{92}\text{U}$ and a γ ray. The $^{239}_{92}\text{U}$ nucleus is radioactive and decays with a half-life of 23.5 min into neptunium $^{239}_{93}\text{Np}$. Neptunium is also radioactive and disintegrates with a half-life of 2.4 days into plutonium $^{239}_{94}\text{Pu}$. Plutonium is the final product and has a half-life of 24 100 yr.

The neutrons that participate in nuclear reactions can have kinetic energies that cover a wide range. In particular, those that have a kinetic energy of about 0.04 eV or less are called **thermal neutrons.** The name derives from the fact that such a relatively small kinetic energy is comparable to the average translational kinetic energy of a molecule at room temperature.

Figure 32.2 An induced nuclear reaction is shown in which $^{238}_{92}\text{U}$ is transmuted into the transuranium element plutonium $^{239}_{94}\text{Pu}$.

32.3 NUCLEAR FISSION

In 1939 four German scientists, Otto Hahn, Lise Meitner, Fritz Strassmann, and Otto Frisch, made an important discovery that ushered in the atomic age. They found that a uranium nucleus, after absorbing a neutron, splits into two fragments, each with a smaller mass than the original nucleus. The splitting of a massive nucleus into two less-massive fragments is known as **nuclear fission.**

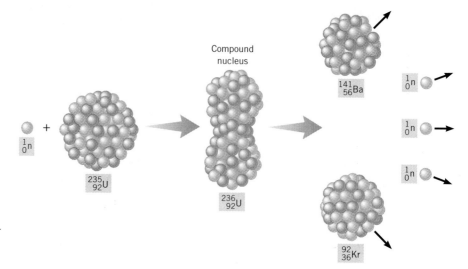

Figure 32.3 A slowly moving neutron causes the uranium nucleus $^{235}_{92}U$ to fission into barium $^{141}_{56}Ba$, krypton $^{92}_{36}Kr$, and three neutrons.

Figure 32.3 shows a fission reaction in which a uranium $^{235}_{92}U$ nucleus is split into barium $^{141}_{56}Ba$ and krypton $^{92}_{36}Kr$ nuclei. The reaction begins when $^{235}_{92}U$ absorbs a slowly moving neutron, creating a "compound nucleus," $^{236}_{92}U$. The compound nucleus disintegrates quickly into $^{141}_{56}Ba$, $^{92}_{36}Kr$, and three neutrons according to the following reaction:

$$\underset{\substack{\text{Compound} \\ \text{nucleus} \\ \text{(unstable)}}}{^{1}_{0}n + {}^{235}_{92}U \longrightarrow {}^{236}_{92}U} \longrightarrow \underset{\text{Barium}}{^{141}_{56}Ba} + \underset{\text{Krypton}}{^{92}_{36}Kr} + \underset{\text{3 neutrons}}{3{}^{1}_{0}n}$$

This reaction is only one of the many possible reactions that can occur when uranium fissions. For example, another reaction is

$$\underset{\substack{\text{Compound} \\ \text{nucleus} \\ \text{(unstable)}}}{^{1}_{0}n + {}^{235}_{92}U \longrightarrow {}^{236}_{92}U} \longrightarrow \underset{\text{Xenon}}{^{140}_{54}Xe} + \underset{\text{Strontium}}{^{94}_{38}Sr} + \underset{\text{2 neutrons}}{2{}^{1}_{0}n}$$

Some reactions produce as many as 5 neutrons; however, the average number produced per fission is 2.5.

When a neutron collides with and is absorbed by a uranium nucleus, the uranium nucleus begins to vibrate and becomes distorted. The vibration continues until the distortion becomes so severe that the attractive strong nuclear force can no longer balance the electrostatic repulsion between the nuclear protons. At this point, the nucleus bursts apart into fragments, which carry off energy, primarily in the form of kinetic energy. The energy carried off by the fragments is enormous and was stored in the original nucleus primarily in the form of electric potential energy. An average of roughly 200 MeV of energy is released per fission. This energy is approximately 10^8 times greater than the energy released per molecule in an ordinary chemical reaction, such as the combustion of gasoline or coal. Example 3 demonstrates how to estimate the energy released during the fission of a nucleus.

EXAMPLE 3
The Energy Released During Nuclear Fission

Estimate the amount of energy released when a massive nucleus ($A = 240$) fissions.

Reasoning Figure 31.6 shows that the binding energy of a nucleus with $A = 240$ is about 7.6 MeV per nucleon. We assume that this nucleus fissions into two fragments, each with $A \approx 120$. According to Figure 31.6, the binding energy of the fragments increases to about 8.5 MeV per nucleon. Consequently, when a massive nucleus fissions, there is a release of about 8.5 MeV − 7.6 MeV = 0.9 MeV of energy per nucleon.

Solution Since there are 240 nucleons involved in the fission process, the total energy released per fission is approximately (0.9 MeV/nucleon)(240 nucleons) $\approx$ 200 MeV.

INTERACTIVE LEARNINGWARE #60

START
Visualization
Reasoning
Knowns and Unknowns
Modeling the Problem
Numerical Solution
FINISH

Consider the following neutron-induced fission reaction:

$$\underset{1.008\,665\,u}{{}^{1}_{0}n} + \underset{239.052\,157\,u}{{}^{239}_{94}Pu} \longrightarrow$$

$$\underset{95.921\,750\,u}{{}^{96}_{38}Sr} + \underset{139.910\,581\,u}{{}^{A}_{Z}X} + \underset{4(1.008\,665\,u)}{4\,{}^{1}_{0}n}$$

(a) Determine the atomic mass number A and the atomic number Z for the unknown nucleus ${}^{A}_{Z}X$. What element is it? (b) Find the energy (in MeV) released by this reaction.

This Interactive LearningWare problem illustrates several important concepts involved in nuclear reactions, including the conservation of nucleon number and the conservation of electric charge. The problem also shows how to determine the energy released by this fission reaction.

Go to our Web site at

http://www.wiley.com/college/cutnell

Virtually all naturally occurring uranium is composed of two isotopes. These isotopes and their natural abundances are ${}^{238}_{92}U$ (99.275%) and ${}^{235}_{92}U$ (0.720%). Although ${}^{238}_{92}U$ is by far the most abundant isotope, the probability that it will capture a neutron and fission is very small. For this reason, ${}^{238}_{92}U$ is not the isotope of choice for generating nuclear energy. In contrast, the isotope ${}^{235}_{92}U$ readily captures a neutron and fissions, *provided the neutron is a thermal neutron* (kinetic energy $\approx$ 0.04 eV or less). The probability of a thermal neutron causing ${}^{235}_{92}U$ to fission is about five hundred times greater than a neutron whose energy is relatively high, say 1 MeV. Thermal neutrons can also be used to fission other nuclei, such as plutonium ${}^{239}_{94}Pu$. Conceptual Example 4 deals with one of the reasons why thermal neutrons are useful for inducing nuclear fission.

CONCEPTUAL EXAMPLE 4
Thermal Neutrons Versus Thermal Protons or Alpha Particles

Why is it possible for a thermal neutron (i.e., one with a relatively small kinetic energy) to penetrate a nucleus, whereas a proton or an α particle would need a much larger amount of energy to penetrate the same nucleus?

Reasoning and Solution To penetrate a nucleus, a particle such as a neutron, a proton, or an α particle must have enough kinetic energy to do the work of overcoming any repulsive force that it encounters along the way. A repulsive force can arise because protons in the nucleus are electrically charged. *Since a neutron is electrically neutral, however, it encounters no electrostatic force of repulsion as it approaches the nuclear protons and, hence, needs relatively little energy to reach the nucleus.* In contrast, a proton and an α particle both carry positive charge, and each encounters an electrostatic force of repulsion as it approaches the nuclear protons. It is true that each also experiences the attractive strong nuclear force from the nuclear protons and neutrons. But the strong nuclear force has an extremely short range of action and, therefore, comes into play only when an impinging particle reaches the nucleus. In comparison, the electrostatic force of repulsion has a long range of action and is encountered by an incoming charged particle throughout the entire journey to the target

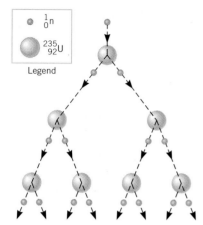

Figure 32.4 A chain reaction. For clarity, it is assumed that each fission generates two neutrons (2.5 neutrons are actually liberated on the average). The fission fragments are not shown.

nucleus. Consequently, ***an impinging proton or α particle needs a relatively large amount of kinetic energy to do the work of overcoming the repulsive electrostatic force.***

Related Homework Material: *Problem 40*

The fact that the uranium fission reaction releases 2.5 neutrons, on the average, makes it possible for a self-sustaining series of fissions to occur. As Figure 32.4 illustrates, each neutron released can initiate another fission event, resulting in the emission of still more neutrons, followed by more fissions, and so on. A ***chain reaction*** is a series of nuclear fissions whereby some of the neutrons produced by each fission cause additional fissions. During an uncontrolled chain reaction, it would not be unusual for the number of fissions to increase a thousandfold within a few millionths of a second. With an average energy of about 200 MeV being released per fission, an uncontrolled chain reaction can generate an incredible amount of energy in a very short time, as happens in an atomic bomb (which is actually a *nuclear* bomb).

By limiting the number of neutrons in the environment of the fissile nuclei, it is possible to establish a condition whereby each fission event contributes, on average, only *one neutron* that fissions another nucleus (see Figure 32.5). In this manner, the chain reaction and the rate of energy production are *controlled*. The controlled fission chain reaction is the principle behind the nuclear reactors used in the commercial generation of electric power.

32.4 NUCLEAR REACTORS

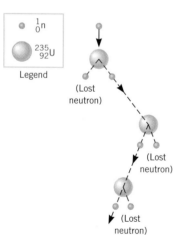

Figure 32.5 In a controlled chain reaction, only one neutron, on average, from each fission event causes another nucleus to fission. As a result, energy is released at a steady or controlled rate.

A nuclear reactor is a type of furnace in which energy is generated by a controlled fission chain reaction. The first nuclear reactor was built by Enrico Fermi in 1942, on the floor of a squash court under the west stands of Stagg Field at the University of Chicago. Today, there are many kinds and sizes of reactors, but they all have three basic components: fuel elements, a neutron moderator, and control rods. Figure 32.6 illustrates these components.

The ***fuel elements*** contain the fissile fuel and, for example, may be in the shape of thin rods about 1 cm in diameter. In a large power reactor there may be thousands of fuel elements placed close together, the entire region of fuel elements being known as the ***reactor core.***

Uranium $^{235}_{92}U$ is a common reactor fuel. Since the natural abundance of this isotope is only about 0.7%, there are special uranium-enrichment plants to increase the percentage. Most commercial reactors use uranium in which the amount of $^{235}_{92}U$ has been enriched to about 3%.

While neutrons with energies of about 0.04 eV (or less) readily fission $^{235}_{92}U$, the neutrons released during the fission process have significantly greater energies of several MeV or so. Consequently, a nuclear reactor must contain some type of material that will decrease or moderate the speed of such energetic neutrons so they can readily fission additional $^{235}_{92}U$ nuclei. The material that slows down the neutrons is called a ***moderator.*** One commonly used moderator is water. When an energetic neutron leaves a fuel element, the neutron enters the surrounding water and collides with water molecules. With each collision, the neutron loses an appreciable fraction of its energy and slows down. Once slowed down to thermal energy by the moderator, a process that takes less than 10^{-3} s, the neutron is capable of initiating a fission event upon reentering a fuel element.

If the output power from a reactor is to remain constant, only one neutron from each fission event must trigger a new fission, as Figure 32.5 suggests. When each fission leads to one additional fission—no more or no less—the reactor is said to be *critical*. A reactor normally operates in a critical condition, for then it produces a steady output of energy. The reactor is *subcritical* when, on average, the neutrons from each fission trigger *less than one* subsequent fission. In a subcritical reactor, the chain reaction is not self-sustaining and eventually dies out. When the neutrons from each fission trigger *more than one* additional fission, the reactor is *supercritical*. During a supercritical condition, the energy released by a reactor increases. If left unchecked, the increasing energy can lead to a partial or total meltdown of the reactor core, with the possible release of radioactive material into the environment.

Clearly, a control mechanism is needed to keep the reactor in its normal or critical state. This control is accomplished by a number of **control rods** that can be moved into and out of the reactor core (see Figure 32.6). The control rods contain an element, such as boron or cadmium, that readily absorbs neutrons without fissioning. If the reactor becomes supercritical, the control rods are automatically moved farther into the core to absorb the excess neutrons causing the condition. In response, the reactor returns to its critical state. Conversely, if the reactor becomes subcritical, the control rods are partially withdrawn from the core, so fewer neutrons are absorbed. Thus, more neutrons are available for fission, and the reactor returns to its critical state.

Figure 32.7 illustrates a pressurized water reactor. In such a reactor, the heat generated within the fuel rods is carried away by water that surrounds the rods. To remove as much heat as possible, the water temperature is allowed to rise to a high value (about 300 °C). To prevent boiling, which occurs at 100 °C at 1 atmosphere of pressure, the water is pressurized in excess of 150 atmospheres. The hot water is pumped through a heat exchanger, where heat is transferred to water flowing in a second, closed system. The heat transferred to the second system produces steam that drives a turbine. The turbine is coupled to an electric generator, whose output electrical power is delivered to consumers via high-voltage transmission lines. After exiting the turbine, the steam is condensed back into water that is returned to the heat exchanger.

Figure 32.6 A nuclear reactor consists of fuel elements, control rods, and a moderator (in this case, water).

Figure 32.7 Diagram of a nuclear power plant that uses a pressurized water reactor.

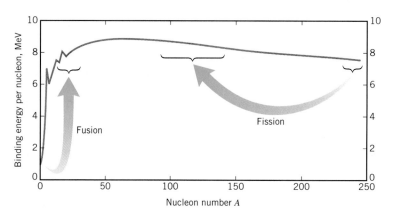

Figure 32.8 When fission occurs, a massive nucleus divides into two fragments whose binding energy per nucleon is greater than that of the original nucleus. When fusion occurs, two low-mass nuclei combine to form a more massive nucleus whose binding energy per nucleon is greater than that of the original nuclei.

32.5 NUCLEAR FUSION

In Example 3 in Section 32.3, the binding-energy-per-nucleon curve is used to estimate the amount of energy released in the fission process. As summarized in Figure 32.8, the massive nuclei at the right end of the curve have a binding energy of about 7.6 MeV per nucleon. The less-massive fission fragments are near the center of the curve and have a binding energy of approximately 8.5 MeV per nucleon. The energy released per nucleon by fission is the difference between these two values, or about 0.9 MeV per nucleon.

A glance at the far left end of the diagram in Figure 32.8 suggests another means of generating energy. Two very-low-mass nuclei with relatively small binding energies per nucleon could be combined or "fused" into a single, more massive nucleus that has a greater binding energy per nucleon. This process is called ***nuclear fusion.*** A substantial amount of energy can be released during a fusion reaction, as Example 5 shows.

EXAMPLE 5   The Energy Released During Nuclear Fusion

Two isotopes of hydrogen, $^2_1$H (deuterium, D) and $^3_1$H (tritium, T), fuse to form $^4_2$He and a neutron according to the following reaction:

$$^2_1\text{H} + ^3_1\text{H} \longrightarrow ^4_2\text{He} + ^1_0\text{n}$$

Determine the energy released by this fusion reaction, which is illustrated in Figure 32.9.

Reasoning Energy is released, so the total mass of the final nuclei is less than the total mass of the initial nuclei. To determine the energy released, we find the amount (in atomic mass units u) by which the total mass has decreased. Then, we use the fact that 1 u is equivalent to 931.5 MeV of energy, as determined in Section 31.3. This approach is the same as that used in Section 31.4 for radioactive decay.

Solution The masses of the initial and final nuclei in this reaction, as well as that of the neutron, are

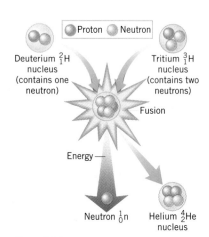

Figure 32.9 Deuterium and tritium are fused together to form a helium nucleus ($^4_2$He). The result is the release of an enormous amount of energy, mainly carried by a single high-energy neutron ($^1_0$n).

	Initial Masses		Final Masses
$^2_1$H	2.014 u	$^4_2$He	4.003 u
$^3_1$H	3.016 u	$^1_0$n	1.009 u
Total:	5.030 u	Total:	5.012 u

The decrease in mass, or the mass defect, is $\Delta m = (5.030 \text{ u} - 5.012 \text{ u}) = 0.018$ u. Since 1 u is equivalent to 931.5 MeV, the energy released is $\boxed{17 \text{ MeV}}$.

The deuterium nucleus contains 2 nucleons, while the tritium nucleus contains 3. Thus, there are 5 nucleons that participate in the fusion, so the energy released per nucleon is about 3.4 MeV. This energy per nucleon is greater than that released in a fission process (≈ 0.9 MeV per nucleon). Thus, for a given mass of fuel, a fusion reaction yields more energy than a fission reaction.

Because fusion reactions release so much energy, there is considerable interest in fusion reactors, although to date no commercial units have been constructed. The difficulties in building a fusion reactor arise mainly because the two low-mass nuclei must be brought sufficiently near each other so that the short-range strong nuclear force can pull them together, leading to fusion. But each nucleus has a positive charge and repels the other electrically. For the nuclei to get sufficiently close in the presence of the repulsive electric force, they must have large kinetic energies, and hence large temperatures, to start with. For example, a temperature of around a hundred million °C is needed to start the deuterium–tritium reaction discussed in Example 5.

Reactions that require such extremely high temperatures are called ***thermonuclear reactions.*** The most important thermonuclear reactions occur in stars, such as our own sun. The energy radiated by the sun comes from deep within its core, where the temperature is high enough to initiate the fusion process. One group of reactions thought to occur in the sun is the *proton–proton* cycle, which is a series of reactions whereby six protons form a helium nucleus, two positrons, two γ rays, two protons, and two neutrinos. The energy released by the proton–proton cycle is about 25 MeV.

Man-made fusion reactions have been carried out in a fusion-type nuclear bomb—commonly called a hydrogen bomb. In a hydrogen bomb, the fusion reaction is ignited by a fission bomb using uranium or plutonium. The temperature produced by the fission bomb is sufficiently high to initiate a thermonuclear reaction where, for example, hydrogen isotopes are fused into helium, releasing even more energy. For fusion to be useful as a commercial energy source, the energy must be released in a steady, controlled manner—unlike that in a bomb. To date, scientists have not succeeded in constructing a fusion device that produces more energy on a continual basis than is expended in operating the device. A fusion device uses a high temperature to start a reaction, and under such a condition, all the atoms are completely ionized to form a *plasma* (a gas composed of charged particles, like $^2_1\text{H}^+$ and e^-). The problem is to confine the hot plasma for a long enough time so that collisions among the ions can lead to fusion.

One ingenious method of confining the plasma is called *magnetic confinement* because it uses a magnetic field to contain and compress the charges in the plasma. Charges moving in the magnetic field are subject to magnetic forces. As the forces increase, the associated pressure builds, and the temperature rises. The gas becomes a superheated plasma, ultimately fusing when the pressure and temperature are high enough.

Another type of confinement scheme, known as *inertial confinement*, is also being developed. Tiny, solid pellets of fuel are dropped into a container. As each pellet reaches the center of the container, a number of high-intensity laser beams strike the pellet simultaneously. The heating causes the exterior of the pellet to vaporize almost instantaneously. However, the inertia of the vaporized atoms keeps them from expanding outward as fast as the vapor is being formed. As a result, high pressures,

THE PHYSICS OF...

nuclear fusion using magnetic confinement.

THE PHYSICS OF...

nuclear fusion using inertial confinement.

high densities, and high temperatures are achieved at the center of the pellet, thus causing fusion. A variant of inertial confinement fusion, called "Z pinch," is under development at Sandia National Laboratories in New Mexico (see the chapter-opening photo). This device would also implode tiny fuel pellets, but without using lasers. Instead, scientists are using a cylindrical array of fine tungsten wires that are connected to a gigantic capacitor. When the capacitor is discharged, a huge current is sent through the wires. The heated wires vaporize almost instantly, generating a hot gas of ions known as a plasma. The plasma is driven inward upon itself by the huge magnetic field produced by the current. The compressed plasma becomes super hot and generates a gigantic X-ray pulse. This pulse, it is hoped, would implode the solid fuel pellets to temperatures and pressures where fusion would occur.

When compared to fission, fusion has some attractive features as an energy source. As we have seen in Example 5, fusion yields more energy than fission, for a given mass of fuel. Moreover, one type of fuel, ^2_1H (deuterium), is found in the waters of the oceans and is plentiful, cheap, and relatively easy to separate from the common ^1_1H isotope of hydrogen. Fissile materials like naturally occurring uranium $^{235}_{92}\text{U}$ are much less available, and supplies could be depleted within a century or two. However, the commercial use of fusion to provide cheap energy remains in the future.

32.6 ELEMENTARY PARTICLES

Setting the Stage

By 1932 the electron, the proton, and the neutron had been discovered and were thought to be nature's three *elementary particles,* in the sense that they were the basic building blocks from which all matter is constructed. Experimental evidence obtained since then, however, shows that several hundred additional particles exist, and scientists no longer believe that the proton and the neutron are elementary particles.

Most of these new particles have masses greater than the electron's mass, and many are more massive than protons or neutrons. Virtually all the new particles are unstable and decay with times between about 10^{-6} and 10^{-23} s.

Often, new particles are produced by accelerating protons or electrons to high energies and letting them collide with a target nucleus. For example, Figure 32.10 shows a collision between an energetic proton and a stationary proton. If the incoming proton has sufficient energy, the collision produces an entirely new particle, the *neutral pion* (π^0). The π^0 particle lives for only about 0.8×10^{-16} s before it decays into two γ-ray photons. Since the pion did not exist before the collision, it was created from part of the incident proton's energy. Because a new particle such as the neutral pion is often created from energy, it is customary to report the mass of the particle in terms of its equivalent *rest energy* (see Equation 28.5). Often energy units of MeV are used. For instance, detailed analyses of experiments reveal that the mass of the π^0 particle is equivalent to a rest energy of 135.0 MeV. For comparison, the more massive proton has a rest energy of 938.3 MeV. Analyses of experiments also provide the electric charge and other properties of particles created in high-energy collisions. In the limited space available here, it is not possible to describe all the new particles that have been found. However, we will highlight some of the more significant discoveries.

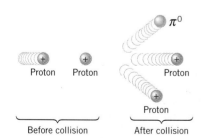

Figure 32.10 When an energetic proton collides with a stationary proton, a neutral pion (π^0) is produced. Part of the energy of the incident proton goes into creating the pion.

Neutrinos

In 1930 Wolfgang Pauli suggested that a particle called the *neutrino* (now known as the electron neutrino) should accompany the β decay of a radioactive nucleus. As

Section 31.5 discusses, the neutrino has no electric charge, has a very small mass (a tiny fraction of the mass of an electron), and travels at speeds approaching (but less than) the speed of light. Neutrinos were finally discovered in 1956. Today, neutrinos are created in abundance in nuclear reactors and particle accelerators and are thought to be plentiful in the universe.

Positrons and Antiparticles

The year 1932 saw the discovery of the *positron* (a contraction for "positive electron"). The positron has the same mass as the electron but carries an opposite charge of $+e$. A collision between a positron and an electron is likely to annihilate both particles, converting them into electromagnetic energy in the form of γ rays. For this reason, positrons never coexist with ordinary matter for any appreciable length of time. The mutual annihilation of a positron and an electron lies at the heart of an important medical diagnostic technique, as Conceptual Example 6 discusses.

CONCEPTUAL EXAMPLE 6   Positron Emission Tomography

Positron emission tomography, or PET scanning, as it is known, utilizes positrons in the following way. Certain radioactive isotopes decay by positron emission, for example, oxygen $^{15}_{8}O$. Such isotopes are injected into the body, where they collect at specific sites. The positron ($^{0}_{1}e$) emitted during the decay of the isotope encounters an electron ($_{-1}^{0}e$) in the body tissue almost at once. The resulting mutual annihilation produces two γ-ray photons ($^{0}_{1}e + _{-1}^{0}e \rightarrow \gamma + \gamma$), which are detected by devices mounted on a ring around the patient, as Figure 32.11a shows. The two photons strike oppositely positioned detectors and, in doing so, reveal the line on which the annihilation occurred. Such information leads to a computer-generated image that can be use-

THE PHYSICS OF...

PET scanning.

(a)   (b)

Figure 32.11 (a) In positron emission tomography, or PET scanning, a radioactive isotope is injected into the body. The isotope decays by emitting a positron, which annihilates an electron in the body tissue, producing two γ-ray photons. These photons strike detectors mounted on opposite sides of a ring that surrounds the patient. (b) PET scanning is being used to evaluate a brain disease.

Healthy   Diseased
brain   brain

Figure 32.12 Positron emission tomography provides an important medical diagnostic technique. The PET scan of the diseased brain shown here reveals Alzheimer's disease. The colors are computer generated.

ful in diagnosing abnormalities at the site where the radioactive isotope collected (see Figure 32.12). How does the principle of conservation of linear momentum explain the fact that the photons strike detectors located opposite one another?

Reasoning and Solution The momentum conservation principle states that the total linear momentum of an isolated system remains constant (see Section 7.2). Therefore, the total momentum of the γ-ray photons will be equal to the total momentum of the positron and electron before annihilation, provided that the system is isolated. An isolated system is one on which no net external force acts. The positron and the electron do exert electrostatic forces on one another, however, since they carry electric charges. But these are internal, not external, forces and cannot change the total linear momentum of the two-particle system. The total photon momentum, then, must equal the total momentum of the positron and the electron, which is nearly zero, to the extent that the two particles have much less momentum than the photons do. With a total linear momentum of zero, the momentum vector of one photon must point opposite to the momentum vector of the other photon. Thus, the two photons depart from the annihilation site traveling in opposite directions and strike oppositely located detectors.

Related Homework Material: *Problem 36*

The positron is an example of an antiparticle, and after its discovery, scientists came to realize that for every type of particle there is a corresponding type of antiparticle. The antiparticle is a form of matter that has the same mass as the particle but carries an opposite electric charge (e.g., the electron–positron pair) or a magnetic moment that is oriented in an opposite direction relative to the spin (e.g., the neutrino–antineutrino pair). A few electrically neutral particles, like the photon and the neutral pion (π^0), are their own antiparticles.

Muons and Pions

In 1937 the American physicists S. H. Neddermeyer and C. D. Anderson discovered a new charged particle whose mass was about 207 times greater than the mass of the electron. The particle is designated by the Greek letter μ (mu) and is known as a *muon*. There are two muons that have the same mass but opposite charge: the particle μ^- and its antiparticle μ^+. The μ^- muon has the same charge as the electron, while the μ^+ muon has the same charge as the positron. Both muons are unstable, with a lifetime of 2.2×10^{-6} s. The μ^- muon decays into an electron (β^-), a muon neutrino (ν_μ), and an electron antineutrino ($\bar{\nu}_e$), according to the following reaction:

$$\mu^- \longrightarrow \beta^- + \nu_\mu + \bar{\nu}_e$$

The μ^+ muon decays into a positron (β^+), a muon antineutrino ($\bar{\nu}_\mu$), and an electron neutrino (ν_e):

$$\mu^+ \longrightarrow \beta^+ + \bar{\nu}_\mu + \nu_e$$

Muons interact with protons and neutrons via the weak nuclear force (see Section 31.5).

The Japanese physicist Hidekei Yukawa (1907–1981) predicted in 1935 that *pions* exist, but they were not discovered until 1947. Pions come in three varieties: one that is positively charged, the negatively charged antiparticle with the same mass, and the neutral pion, mentioned earlier, which is its own antiparticle. The symbols for these pions are, respectively, π^+, π^-, and π^0. The charged pions are unstable and have a lifetime of 2.6×10^{-8} s. The decay of a charged pion almost always produces a muon:

$$\pi^- \longrightarrow \mu^- + \bar{\nu}_\mu$$
$$\pi^+ \longrightarrow \mu^+ + \nu_\mu$$

As mentioned earlier, the neutral pion π^0 is also unstable and decays into two γ-ray photons, the lifetime being 0.8×10^{-16} s. The pions are of great interest because, unlike the muons, the pions interact with protons and neutrons via the strong nuclear force.

Classification of Particles

It is useful to group the known particles into three families, the photons, the leptons, and the hadrons, as Table 32.3 summarizes. This grouping is made according to the nature of the force by which a particle interacts with other particles. The **photon family,** for instance, has only one member, the photon. The photon interacts only with charged particles, and the interaction is only via the *electromagnetic force.* No other particle behaves in this manner.

The **lepton family** consists of particles that interact by means of the *weak nuclear force.* Leptons can also exert gravitational and electromagnetic (if the leptons are charged) forces on other particles. The four better-known leptons are the electron, the muon, the electron neutrino ν_e, and the muon neutrino ν_μ. Table 32.3 lists these particles together with their antiparticles. Recently, two other leptons have been discovered, the tau particle (τ) and its neutrino (ν_τ), bringing the number of particles in the lepton family to six.

The **hadron family** contains the particles that interact by means of the *strong nuclear force and the weak nuclear force.* Hadrons can also interact by gravitational

TABLE 32.3   Some Particles and Their Properties

Family	Particle	Particle Symbol	Antiparticle Symbol	Rest Energy (MeV)	Lifetime (s)
Photon	Photon	γ	Self[a]	0	Stable
Lepton	Electron	e^- or β^-	e^+ or β^+	0.511	Stable
	Muon	μ^-	μ^+	105.7	2.2×10^{-6}
	Tau	τ^-	τ^+	1784	10^{-13}
	Electron neutrino	ν_e	$\overline{\nu}_e$	≈ 0	Stable
	Muon neutrino	ν_μ	$\overline{\nu}_\mu$	≈ 0	Stable
	Tau neutrino	ν_τ	$\overline{\nu}_\tau$	≈ 0	Stable
Hadron					
Mesons					
	Pion	π^+	π^-	139.6	2.6×10^{-8}
		π^0	Self[a]	135.0	0.8×10^{-16}
	Kaon	K^+	K^-	493.7	1.2×10^{-8}
		K^0_S	$\overline{K}^0_S$	497.7	0.9×10^{-10}
		K^0_L	$\overline{K}^0_L$	497.7	5.2×10^{-8}
	Eta	η^0	Self[a]	548.8	$<10^{-18}$
Baryons					
	Proton	p	$\overline{p}$	938.3	Stable
	Neutron	n	$\overline{n}$	939.6	900
	Lambda	Λ^0	$\overline{\Lambda}^0$	1116	2.6×10^{-10}
	Sigma	Σ^+	$\overline{\Sigma}^-$	1189	0.8×10^{-10}
		Σ^0	$\overline{\Sigma}^0$	1192	6×10^{-20}
		Σ^-	$\overline{\Sigma}^+$	1197	1.5×10^{-10}
	Omega	Ω^-	Ω^+	1672	0.8×10^{-10}

[a] The particle is its own antiparticle.

and electromagnetic forces, but at short distances ($\leq 10^{-15}$ m) the strong nuclear force dominates. Among the hadrons are the proton, the neutron, and the pions. As Table 32.3 indicates, most hadrons are short-lived. The hadrons are subdivided into two groups, the **mesons** and the **baryons,** for a reason that will be discussed in connection with the idea of quarks.

Quarks

As more and more hadrons were discovered, it became clear that they were not all elementary particles. The suggestion was made that the hadrons are made up of smaller, more elementary particles called **quarks.** In 1963 a quark theory was advanced independently by M. Gell–Mann (1929–) and G. Zweig (1937–). The theory proposed that there are three quarks and three corresponding antiquarks, and that hadrons are constructed from combinations of these. Thus, the quarks are elevated to the status of elementary particles for the hadron family. The particles in the photon and lepton families are considered to be elementary, and as such they are not composed of quarks.

The three quarks were named *up* (*u*), *down* (*d*), and *strange* (*s*), and were assumed to have, respectively, fractional charges of $+\frac{2}{3}e$, $-\frac{1}{3}e$, and $-\frac{1}{3}e$. In other words, a quark possesses a charge magnitude smaller than that of an electron, which has a charge of $-e$. Table 32.4 lists the symbols and electric charges of these quarks and the corresponding antiquarks. Experimentally, quarks should be recognizable by their fractional charges, but in spite of an extensive search for them, free quarks have never been found.

According to the original quark theory, the mesons are different from the baryons, for each meson consists of only two quarks—a quark and an antiquark—while a baryon contains three quarks. For instance, the π^- pion (a meson) is composed of a *d* quark and a $\bar{u}$ antiquark, $\pi^- = d + \bar{u}$, as Figure 32.13 shows. These two quarks combine to give the π^- pion a net charge of $-e$. Similarly, the π^+ pion is a combination of the $\bar{d}$ and *u* quarks, $\pi^+ = \bar{d} + u$. In contrast, protons and neutrons, being baryons, consist of three quarks. A proton contains the combination $d + u + u$, while a neutron contains the combination $d + d + u$ (see Figure 32.13). These groups of three quarks give the correct charges for the proton and neutron.

The original quark model was extremely successful in predicting not only the correct charges for the hadrons, but other properties as well. However, in 1974 a new particle, the J/ψ meson, was discovered. This meson has a rest energy of 3100 MeV, much larger than the rest energies of other known mesons. The existence of the J/ψ meson could be explained only if a new quark–antiquark pair existed; this new quark was called *charmed* (*c*). With the discovery of more and more particles, it has been necessary to postulate a fifth and a sixth quark; their names are

Figure 32.13 According to the original quark model of hadrons, all mesons consist of a quark and an antiquark, while baryons contain three quarks.

TABLE 32.4   Quarks and Antiquarks

| Name | Quarks | | Antiquarks | |
	Symbol	Charge	Symbol	Charge
Up	u	$+\frac{2}{3}e$	$\bar{u}$	$-\frac{2}{3}e$
Down	d	$-\frac{1}{3}e$	$\bar{d}$	$+\frac{1}{3}e$
Strange	s	$-\frac{1}{3}e$	$\bar{s}$	$+\frac{1}{3}e$
Charmed	c	$+\frac{2}{3}e$	$\bar{c}$	$-\frac{2}{3}e$
Top	t	$+\frac{2}{3}e$	$\bar{t}$	$-\frac{2}{3}e$
Bottom	b	$-\frac{1}{3}e$	$\bar{b}$	$+\frac{1}{3}e$

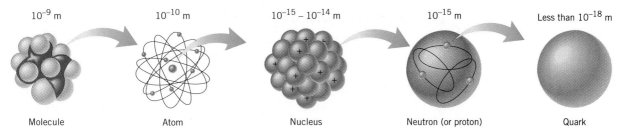

| 10^{-9} m | 10^{-10} m | 10^{-15} – 10^{-14} m | 10^{-15} m | Less than 10^{-18} m |
| Molecule | Atom | Nucleus | Neutron (or proton) | Quark |

Figure 32.14 The current view of how matter is composed of basic units, starting with a molecule and ending with a quark. The approximate sizes of each unit are also listed.

top (*t*) and *bottom* (*b*), although some scientists prefer to call these quarks *truth* and *beauty*. Today, there is firm evidence for all six quarks, each with its corresponding antiquark. All of the hundreds of the known hadrons can be accounted for in terms of these six quarks and their antiquarks.

In addition to their electric charges, quarks also have other properties. For example, each quark is believed to possess a characteristic called **color**, for which there are three possibilities: blue, green, or red. The corresponding possibilities for the antiquarks are antiblue, antigreen, and antired. The use of the term "color" and the specific choices of blue, green, and red is arbitrary and pure whimsy, for the visible colors of the electromagnetic spectrum have nothing to do with quark properties. The quark property called color, however, is an important one, for it brings the quark model into agreement with the Pauli exclusion principle and enables the model to account for experimental observations that are otherwise difficult to explain.

The Standard Model

The various elementary particles that have been discovered can interact via one or more of the following four forces: the gravitational force, the strong nuclear force, the weak nuclear force, and the electromagnetic force. In particle physics, the phrase "**the standard model**" refers to the currently accepted explanation for the strong nuclear force, the weak nuclear force, and the electromagnetic force. In this model, the strong nuclear force between quarks is described in terms of the concept of color, the theory being referred to as quantum chromodynamics. And according to the standard model, the weak nuclear force and the electromagnetic force are separate manifestations of a single even more fundamental interaction, referred to as the electroweak interaction, as we have seen in Section 31.5.

In the standard model, our understanding of the building blocks of matter follows the hierarchical pattern illustrated in Figure 32.14. Molecules, such as water (H_2O) and glucose ($C_6H_{12}O_6$), are composed of atoms. Each atom consists of a nucleus that is surrounded by a cloud of electrons. The nucleus, in turn, is made up of protons and neutrons, which are composed of quarks.

32.7 COSMOLOGY

Cosmology is the study of the structure and evolution of the universe. In this study, both the very large and the very small aspects of the universe are important. Astronomers, for example, study stars located at enormous distances from the earth, up to billions of light-years. In contrast, particle physicists focus their efforts on the very small elementary particles (10^{-18} m or smaller) that comprise matter. The synergy between the work of astronomers and particle physicists has led to significant advances in our understanding of the universe. Central to that understanding is the belief that the universe is expanding, and we begin by discussing the evidence that justifies this belief.

The Expanding Universe and the Big Bang

The idea that the universe is expanding originated with the astronomer Edwin P. Hubble (1889–1953). He found that light reaching the earth from distant galaxies is Doppler-shifted toward greater wavelengths, that is, toward the red end of the visible spectrum. As Section 24.5 discusses, this type of Doppler shift results when the observer and the source of the light are moving away from each other. The speed at which a galaxy is receding from the earth can be determined from the measured Doppler shift in wavelength. Hubble found that a galaxy located at a distance d from the earth recedes from the earth at a speed v given by

Hubble's law $$v = Hd \qquad (32.5)$$

where H is a constant known as the Hubble parameter. In other words, the recession speed is proportional to the distance d, so that more distant galaxies are moving away from the earth at greater speeds. Equation 32.5 is referred to as Hubble's law.

Hubble's picture of an expanding universe does not mean that the earth is at the center of the expansion. In fact, there is no literal center. Imagine a loaf of raisin bread expanding as it bakes. Each raisin moves away from every other raisin, without any single one acting as a center for the expansion. Galaxies in the universe behave in a similar fashion. Observers in other galaxies would see distant galaxies moving away, just as we do.

Experimental measurements by astronomers indicate that an approximate value for the Hubble parameter is

$$H = 0.022 \, \frac{\text{m}}{\text{s} \cdot \text{light-year}}$$

The value for the Hubble parameter is believed to be accurate within 10%. Scientists are very interested in obtaining an accurate value for H because it can be related to an age for the universe, as the next example illustrates.

EXAMPLE 7   An Age for the Universe

Determine an estimate of the age of the universe using Hubble's law.

Reasoning Consider a galaxy currently located at a distance d from the earth. According to Hubble's law, this galaxy is moving away from us at a speed of $v = Hd$. At an earlier time, therefore, this galaxy must have been closer. We can imagine, in fact, that in the remote past the separation distance was relatively small and that the universe originated at such a time. To estimate the age of the universe, we calculate the time it has taken the galaxy to recede to its present position. For this purpose, time is simply distance divided by speed, or $t = d/v$.

Solution Using Hubble's law and the fact that a distance of 1 light-year is 9.46×10^{15} m, we estimate the age of the universe to be

$$t = \frac{d}{v} = \frac{d}{Hd} = \frac{1}{H}$$

$$t = \frac{1}{0.022 \, \dfrac{\text{m}}{\text{s} \cdot \text{light-year}}} = \frac{1}{\left(0.022 \, \dfrac{\text{m}}{\text{s} \cdot \text{light-year}}\right)\left(\dfrac{1 \text{ light-year}}{9.46 \times 10^{15} \text{ m}}\right)}$$

$$= 4.3 \times 10^{17} \text{ s} \quad \text{or} \quad \boxed{1.4 \times 10^{10} \text{ yr}}$$

The idea presented in Example 7 that our galaxy and other galaxies in the universe were very close together at some earlier instant in time lies at the heart of the **Big Bang theory.** This theory postulates that the universe had a definite beginning in a cataclysmic event, sometimes called the primeval fireball. Dramatic additional evidence supporting the theory was discovered in 1965 by Arno A. Penzias (1926–) and Robert W. Wilson (1936–). Using a radio telescope, they discovered that the earth is being bathed in weak electromagnetic waves in the microwave region of the spectrum (wavelength = 7.35 cm, see Figure 24.9). They observed that the intensity of these waves is the same, no matter where in the sky they pointed their telescope, and concluded that the waves originated outside of our galaxy. This microwave background radiation, as it is called, represents the radiation left over from the Big Bang and is a kind of cosmic afterglow. Subsequent measurements have confirmed the research of Penzias and Wilson and shown that the microwave radiation is consistent with a perfect blackbody (see Sections 13.3 and 29.2) radiating at a temperature of 2.7 K, in agreement with the theoretical analysis of the Big Bang. In 1978, Penzias and Wilson received a Nobel prize for their discovery.

The Standard Model for the Evolution of the Universe

Based on the recent experimental and theoretical research in particle physics, scientists have proposed an evolutionary sequence of events following the Big Bang. This sequence is known as the **standard cosmological model** and is illustrated in Figure 32.15.

Immediately following the Big Bang, the temperature of the universe was incredibly high, about 10^{32} K. During this initial period, the three fundamental forces (the gravitational force, the strong nuclear force, and the electroweak force) all behaved as a single unified force. Very quickly, in about 10^{-43} s, the gravitational force took on a separate identity all its own, as Figure 32.15 indicates. Meanwhile, the strong nuclear force and the electroweak force continued to act as a single force, which is sometimes referred to as the GUT force. GUT stands for the Grand Unified Theory that presumably would explain such a force. Slightly later, at about 10^{-35} s after the Big Bang, the GUT force separated into the strong nuclear force and the electroweak force (see Figure 32.15), the universe expanding and cooling somewhat to a temperature of roughly 10^{28} K. From this point on, the strong nuclear force behaved as we know it today, while the electroweak force maintained its

Figure 32.15 According to the standard cosmological model, the universe has evolved as illustrated here. In this model, the universe is presumed to have originated with a cataclysmic event known as the Big Bang. The times shown are those following this event.

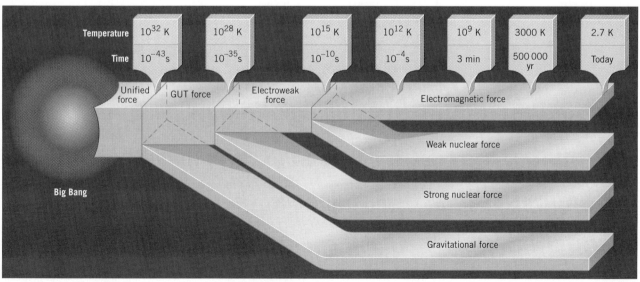

| Temperature | 10^{32} K | 10^{28} K | 10^{15} K | 10^{12} K | 10^{9} K | 3000 K | 2.7 K |
| Time | 10^{-43} s | 10^{-35} s | 10^{-10} s | 10^{-4} s | 3 min | 500 000 yr | Today |

identity. In this scenario, note that the weak nuclear force and the electromagnetic force have not yet manifested themselves as separate entities. The disappearance of the electroweak force and the appearance of the weak nuclear force and the electromagnetic force eventually occurred at approximately 10^{-10} s after the Big Bang, when the temperature of the expanding universe had cooled to about 10^{15} K.

From the Big Bang up until the strong nuclear force separated from the GUT force at a time of 10^{-35} s, all particles of matter were similar, and there was no distinction between quarks and leptons. After this time, quarks and leptons became distinguishable. Eventually the quarks and antiquarks formed hadrons, such as protons and neutrons and their antiparticles. By a time of 10^{-4} s after the Big Bang, however, the temperature had cooled to approximately 10^{12} K, and the hadrons had mostly disappeared. Protons and neutrons survived only as a very small fraction of the total number of particles, the majority of which were leptons such as electrons, positrons, and neutrinos. Like most of the hadrons before them, most of the electrons and positrons eventually disappeared. However, they did leave behind a relatively small number of electrons to join the small number of protons and neutrons at a time of about 3 min following the Big Bang. At this time the temperature of the expanding universe had decreased to about 10^9 K, and small nuclei such as that of helium began forming. Later, when the universe was about 500 000 years old and the temperature had dropped to near 3000 K, hydrogen and helium atoms began forming. As the temperature decreased further, stars and galaxies formed, and today we find a temperature of 2.7 K characterizing the cosmic background radiation of the universe.

32.8 CONCEPTS & CALCULATIONS

When considering the biological effects of ionizing radiation, the concept of the biologically equivalent dose is especially important. Its importance lies in the fact that the biologically equivalent dose incorporates both the amount of energy per unit mass that is absorbed and the effectiveness of a particular type of radiation in producing a certain biological effect. Example 8 examines this concept, as well as reviewing the notions of power (Section 6.7) and intensity (Section 16.7) of a wave.

CONCEPTS & CALCULATIONS EXAMPLE 8   The Biologically Equivalent Dose

A patient is being given a chest X-ray. The X-ray beam is turned on for 0.20 s, and its intensity is 0.40 W/m^2. The area of the chest being exposed is 0.072 m^2, and the radiation is absorbed by 3.6 kg of tissue. If the relative biological effectiveness (RBE) of the X-rays for this tissue is 1.1, determine the biologically equivalent dose received by the patient.

Concept Questions and Answers How is the power of the beam related to the beam intensity?

Answer   The power P is equal to the intensity I times the cross-sectional area A of the beam, as indicated by Equation 16.8 ($P = IA$).

How is the energy absorbed by the tissue related to the power of the beam?

Answer   The energy E is equal to the product of the power P delivered by the beam and the time t of exposure, according to Equation 6.10b ($E = Pt$).

What is the absorbed dose?

Answer   The absorbed dose is defined by Equation 32.2 as the energy E absorbed by the tissue divided by its mass m, or Absorbed dose = E/m. The unit of the absorbed dose is the gray (Gy).

How is the biologically equivalent dose related to the absorbed dose?

Answer   The biologically equivalent dose is equal to the product of the absorbed dose and the relative biological effectiveness (RBE), as indicated by Equation 32.4. However, the absorbed dose must be expressed in rad units, not in gray units. The relation between the two units is 1 rad = 0.01 Gy.

Solution   The biologically equivalent dose is given by Equation 32.4 as

$$\text{Biologically equivalent dose} = [\text{Absorbed dose (in rad)}](\text{RBE})$$

At this point we need to convert the absorbed dose from rad units to Gy units:

$$\text{Absorbed dose (in rad)} = [\text{Absorbed dose (in Gy)}]\left(\frac{1\ \text{rad}}{0.01\ \text{Gy}}\right)$$

The biologically equivalent dose then becomes

$$\text{Biologically equivalent dose} = [\text{Absorbed dose (in Gy)}]\left(\frac{1\ \text{rad}}{0.01\ \text{Gy}}\right)(\text{RBE})$$

Equation 32.2 indicates that the Absorbed dose (in Gy) is equal to the energy E divided by the mass m of the tissue, so that we have

$$\text{Biologically equivalent dose} = \left(\frac{E}{m}\right)\left(\frac{1\ \text{rad}}{0.01\ \text{Gy}}\right)(\text{RBE})$$

From the answers to the first and second Concept Questions, we can conclude that the absorbed energy E is related to the intensity I, the cross-sectional area A, and the time duration t of the beam via $E = Pt = IAt$. Substituting this result into the equation for the biologically equivalent dose gives

$$\begin{aligned}
\text{Biologically equivalent dose} &= \left(\frac{IAt}{m}\right)\left(\frac{1\ \text{rad}}{0.01\ \text{Gy}}\right)(\text{RBE}) \\
&= \frac{(0.40\ \text{W/m}^2)(0.072\ \text{m}^2)(0.20\ \text{s})}{3.6\ \text{kg}}\left(\frac{1\ \text{rad}}{0.01\ \text{Gy}}\right)(1.1) = \boxed{0.18\ \text{rem}}
\end{aligned}$$

The next example illustrates the decay of a particle into two photons, so matter is completely converted into electromagnetic waves. The example discusses how to find the energy, frequency, and wavelength of each photon. It also provides an opportunity to review the principle of conservation of energy and the principle of conservation of linear momentum.

CONCEPTS & CALCULATIONS EXAMPLE 9   The Decay of the π^0 Meson

The π^0 meson is a particle that has a rest energy of 135.0 MeV (see Table 32.3). It lives for a very short time and then decays into two γ-ray photons in the following manner: $\pi^0 \rightarrow \gamma + \gamma$. Suppose that one of the γ-ray photons travels along the $+x$ axis. If the π^0 is at rest when it decays, find (a) the energy (in MeV), (b) the frequency and wavelength, and (c) the momentum of each γ-ray photon.

Concept Questions and Answers   How is the energy E of each γ-ray photon related to the rest energy E_0 of the π^0 particle?

Answer   According to the principle of conservation of energy, the sum of the energies of the two γ-ray photons must equal the rest energy E_0 of the π^0 particle, since it is at rest. Therefore, the energy E of each photon is $E = \frac{1}{2}E_0$.

How can the frequency and wavelength of the photon be determined from its energy?

Answer   A photon is a packet of electromagnetic energy, and each contains an amount of energy given by Equation 29.2 as $E = hf$, where h is Planck's constant and f is the frequency. Equation 16.1 indicates that the wavelength λ of the photon is inversely proportional to its frequency, $\lambda = c/f$, where c is the speed of light.

How is the total linear momentum of the photons related to the momentum of the π^0 particle, and what is the momentum of each photon?

Answer   The principle of conservation of linear momentum states that the total linear momentum of an isolated system is constant (see Section 7.2). Since the π^0 particle is at rest, the total linear momentum before the decay is zero. Since momentum is conserved, the total linear momentum of the two γ-ray photons must also be zero. The magnitude p of the momentum of a photon is given by Equation 29.6 as $p = h/\lambda$. Thus, if one photon travels along the $+x$ axis with a momentum of $+h/\lambda$, the other must travel along the $-x$ axis with a momentum of $-h/\lambda$, ensuring that the total momentum is zero.

Solution

(a) Since the sum of the energies of the two γ-ray photons must be equal to the rest energy E_0 of the π^0 particle, the energy of each photon is

$$E = \tfrac{1}{2}E_0 = \tfrac{1}{2}(135.0 \text{ MeV}) = \boxed{67.5 \text{ MeV}}$$

(b) The frequency of each photon is given by Equation 29.2 as its energy (in units of joules) divided by Planck's constant. Using the fact that $1 \text{ eV} = 1.60 \times 10^{-19} \text{ J}$, we find that the frequency of the photon is

$$f = \frac{E}{h} = \frac{(67.5 \times 10^6 \text{ eV})\left(\dfrac{1.60 \times 10^{-19} \text{ J}}{1 \text{ eV}}\right)}{6.63 \times 10^{-34} \text{ J} \cdot \text{s}} = \boxed{1.63 \times 10^{22} \text{ Hz}}$$

The wavelength of the photon can be found directly from Equation 16.1:

$$\lambda = \frac{c}{f} = \frac{3.00 \times 10^8 \text{ m/s}}{1.63 \times 10^{22} \text{ Hz}} = \boxed{1.84 \times 10^{-14} \text{ m}}$$

(c) The magnitude of a photon's momentum is related to its wavelength by Equation 29.6:

$$p = \frac{h}{\lambda} = \frac{6.63 \times 10^{-34} \text{ J} \cdot \text{s}}{1.84 \times 10^{-14} \text{ m}} = 3.60 \times 10^{-20} \text{ kg} \cdot \text{m/s}$$

Since the photons move along the x axis in opposite directions, their momenta are $\boxed{+3.60 \times 10^{-20} \text{ kg} \cdot \text{m/s}}$ and $\boxed{-3.60 \times 10^{-20} \text{ kg} \cdot \text{m/s}}$.

At the end of the problem set for this chapter, you will find homework problems that contain both conceptual and quantitative parts. These problems are grouped under the heading *Concepts & Calculations, Group Learning Problems*. They are designed for use by students working alone or in small learning groups. The conceptual part of each problem provides a convenient focus for group discussions.

SUMMARY

Ionizing radiation consists of photons and/or moving particles that have enough energy to ionize an atom or molecule. **Exposure** is a measure of the ionization produced in air by X-rays or γ rays. An exposure of one roentgen (R) produces 2.58×10^{-4} coulombs of positive charge in 1 kg of dry air at STP conditions.

The **absorbed dose** is the amount of energy absorbed from the radiation per unit mass of absorbing material. The SI unit of absorbed dose is the gray (Gy); $1 \text{ Gy} = 1 \text{ J/kg}$. However, the rad (rd) is another unit that is often used; $1 \text{ rd} = 0.01 \text{ Gy}$.

The amount of biological damage produced by ionizing radiation is different for different types of radiation. The **rela-**

tive biological effectiveness (RBE) is the absorbed dose of 200-keV X-rays required to produce a certain biological effect divided by the dose of a given type of radiation that produces the same biological effect. The **biologically equivalent dose** is the product of the absorbed dose (in rads) and the RBE. The unit for the biologically equivalent dose is the rem.

An **induced nuclear transmutation** is the process whereby an incident particle or photon strikes a nucleus and causes the production of a new element.

Nuclear fission occurs when a massive nucleus splits into two less-massive fragments. Fission can be induced by the absorption of a thermal neutron. When a massive nucleus fissions, energy is released because the binding energy per nucleon is greater for the fragments than for the original nucleus. Neutrons are also released during nuclear fission. These neutrons can, in turn, induce other nuclei to fission and lead to a process known as a **chain reaction**. A **nuclear reactor** is a device that generates energy by a controlled chain reaction.

In a **fusion** process, two nuclei with smaller masses combine to form a single nucleus with a larger mass. Energy is re-leased by fusion when the binding energy per nucleon is greater for the larger nucleus than for the smaller nuclei.

Subatomic particles are divided into three families: the **photon** family, the **lepton** family, and the **hadron** family. **Elementary particles** are the basic building blocks of matter. All members of the photon and lepton families are elementary particles. The quark theory proposes that the hadrons are not elementary particles, but are composed of elementary particles called **quarks.** Currently, the hundreds of hadron particles can be accounted for in terms of six quarks and their antiquarks. The **standard model** consists of two parts: (1) the currently accepted explanation for the strong nuclear force in terms of the quark concept of color and (2) the theory of the electroweak interaction.

Cosmology is the study of the structure and evolution of the universe. Our universe is expanding. The speed v at which a distant galaxy recedes from the earth is given by **Hubble's law** as $v = Hd$, where d is the distance of the galaxy from the earth, and $H = 0.022$ m/(s·light-year) is called the Hubble parameter. The **standard cosmological model** for the evolution of the universe is discussed in Section 32.7.

CONCEPTUAL QUESTIONS

1. Two different types of radiation have the same RBE. Does this mean that each type delivers the same amount of energy to the tissue that it irradiates? Justify your answer.

2. ⚕ When a dentist X-rays your teeth, a lead apron is placed over your chest and lower body. What is the purpose of this apron?

3. State whether the two quantities in each of the following cases are related and, if so, give the relation between them: (a) rads and grays, and (b) rads and roentgens.

4. Explain why the following reactions are *not* allowed:
(a) $^{60}_{28}\text{Ni}$ (α, p) $^{62}_{29}\text{Cu}$, (b) $^{27}_{13}\text{Al}$ (n, n) $^{28}_{13}\text{Al}$, (c) $^{39}_{19}\text{K}$ (p, α) $^{36}_{17}\text{Cl}$.

5. Thermal neutrons and thermal protons have nearly the same speed. But thermal neutrons and thermal electrons have very different speeds. Account for this observation, and explain whether the neutrons or the electrons have the greater speed.

6. Would a release of energy accompany the fission of a nucleus of nucleon number 25 into two fragments of about equal mass? Using the curve in Figure 32.8, account for your answer.

7. In the fission of $^{235}_{92}\text{U}$ there are, on the average, 2.5 neutrons released per fission. Suppose a *different* element is being fissioned and, on the average, only 1.0 neutron is released per fission. If a small fraction of the thermal neutrons absorbed by the nuclei does *not* produce a fission, can a self-sustaining chain reaction be produced using this element? Explain.

8. The mass of coal consumed in a coal-burning electric power plant is about two million times greater than the mass of $^{235}_{92}\text{U}$ used to fuel a comparable nuclear power plant. Why?

9. Refer to Figure 32.8 and decide if the fusion of two nuclei, each with a nucleon number of 60, would release energy. Give your reasoning.

10. Explain the difference between fission and fusion and why each process produces energy.

PROBLEMS

Section 32.1 Biological Effects of Ionizing Radiation

1. ssm A beam of γ rays passes through 4.0×10^{-3} kg of dry air and generates 1.7×10^{12} ions, each with a charge of $+e$. What is the exposure (in roentgens)?

2. What absorbed dose (in rads) of α particles (RBE = 20) causes as much biological damage as a 60-rad dose of protons (RBE = 10)?

3. ⚕ A film badge worn by a radiologist indicates that she has received an absorbed dose of 2.5×10^{-5} Gy. The mass of the radiologist is 65 kg. How much energy has she absorbed?

4. A person receives a single whole-body dose of α particles. The absorbed dose is 38 rad, and the RBE of the α particles is 12. (a) Determine the biologically equivalent dose received by this person. (b) With this dose, which one of the following would you expect to happen: no short-term ill effects, the onset of radiation sickness, a 50% chance of dying, almost certain death?

5. ssm www A 75-kg person is exposed to 45 mrem of α particles (RBE = 12). How much energy has this person absorbed?

6. $\triangleleft$ A beam of particles is directed at a 0.015-kg tumor. There are 1.6×10^{10} particles per second reaching the tumor, and the energy of each particle is 4.0 MeV. The RBE for the radiation is 14. Find the biologically equivalent dose given to the tumor in 25 s.

*7. ssm What absorbed dose (in rad) of γ rays is required to change a block of ice at 0.0 °C into steam at 100.0 °C?

*8. A beam of nuclei is used for cancer therapy. Each nucleus has an energy of 130 MeV, and the relative biological effectiveness (RBE) of this type of radiation is 16. The beam is directed onto a 0.17-kg tumor, which receives a biologically equivalent dose of 180 rem. How many nuclei are in the beam?

*9. $\triangleleft$ A 2.0-kg tumor is being irradiated by a radioactive source. The tumor receives an absorbed dose of 12 Gy in a time of 850 s. Each disintegration of the radioactive source produces a particle that enters the tumor and delivers an energy of 0.40 MeV. What is the activity $\Delta N/\Delta t$ (see Section 31.6) of the radioactive source?

Section 32.2 Induced Nuclear Reactions

10. What is the atomic number Z and the element X in the reaction $^{10}_{5}\text{B}\,(\alpha, p)\,^{13}_{Z}\text{X}$?

11. ssm A nitrogen $^{14}_{7}\text{N}$ nucleus absorbs a deuterium $^{2}_{1}\text{H}$ nucleus during a nuclear reaction. What is the name, atomic number, and nucleon number of the compound nucleus?

12. What is the nucleon number A in the reaction $^{27}_{13}\text{Al}\,(\alpha, n)\,^{A}_{15}\text{P}$?

13. Complete the following nuclear reactions, assuming that the unknown quantity signified by the question mark is a single entity:

    (a) $^{43}_{20}\text{Ca}\,(\alpha, ?)\,^{46}_{21}\text{Sc}$ **(d)** $?\,(\alpha, p)\,^{17}_{8}\text{O}$

    (b) $^{9}_{4}\text{Be}\,(?, n)\,^{12}_{6}\text{C}$ **(e)** $^{55}_{25}\text{Mn}\,(n, \gamma)\,?$

    (c) $^{9}_{4}\text{Be}\,(p, \alpha)\,?$

14. Write the reactions below in the shorthand form discussed in the text.

    (a) $^{1}_{0}\text{n} + ^{14}_{7}\text{N} \rightarrow ^{14}_{6}\text{C} + ^{1}_{1}\text{H}$

    (b) $^{1}_{0}\text{n} + ^{238}_{92}\text{U} \rightarrow ^{239}_{92}\text{U} + \gamma$

    (c) $^{1}_{0}\text{n} + ^{24}_{12}\text{Mg} \rightarrow ^{23}_{11}\text{Na} + ^{2}_{1}\text{H}$

*15. ssm Consider the induced nuclear reaction $^{2}_{1}\text{H} + ^{14}_{7}\text{N} \rightarrow ^{12}_{6}\text{C} + ^{4}_{2}\text{He}$. The atomic masses are $^{2}_{1}\text{H}$ (2.014 102 u), $^{14}_{7}\text{N}$ (14.003 074 u), $^{12}_{6}\text{C}$ (12.000 000 u), and $^{4}_{2}\text{He}$ (4.002 603 u). Determine the energy (in MeV) released when the $^{12}_{6}\text{C}$ and $^{4}_{2}\text{He}$ nuclei are formed in this manner.

*16. During a nuclear reaction, an unknown particle is absorbed by a copper $^{63}_{29}\text{Cu}$ nucleus, and the reaction products are $^{62}_{29}\text{Cu}$, a neutron, and a proton. What is the name, atomic number, and nucleon number of the *compound nucleus?*

Section 32.3 Nuclear Fission, Section 32.4 Nuclear Reactors

17. $^{235}_{92}\text{U}$ absorbs a thermal neutron and fissions into rubidium $^{93}_{37}\text{Rb}$ and cesium $^{141}_{55}\text{Cs}$. What nucleons are produced by the fission, and how many are there?

18. Determine the number of neutrons released during the following fission reaction: $^{1}_{0}\text{n} + ^{235}_{92}\text{U} \rightarrow ^{133}_{51}\text{Sb} + ^{99}_{41}\text{Nb} + $ neutrons.

19. ssm When a $^{235}_{92}\text{U}$ (235.043 924 u) nucleus fissions, about 200 MeV of energy is released. What is the ratio of this energy to the rest energy of the uranium nucleus?

20. Uranium $^{235}_{92}\text{U}$ fissions into two fragments plus three neutrons: $^{1}_{0}\text{n} + ^{235}_{92}\text{U} \rightarrow (2 \text{ fragments}) + 3^{1}_{0}\text{n}$. The mass of a neutron is 1.008 665 u and that of $^{235}_{92}\text{U}$ is 235.043 924 u. If 225.0 MeV of energy is released during the fission, what is the combined mass of the two fragments?

21. Neutrons released by a fission reaction must be slowed by collisions with the moderator nuclei before the neutrons can cause further fissions. Suppose a 1.5-MeV neutron leaves each collision with 65% of its incident energy. How many collisions are required to reduce the neutron's energy to at least 0.040 eV, which is the energy of a thermal neutron?

22. A neutron causes $^{232}_{90}\text{Th}$ to change according to the reaction

$$^{1}_{0}\text{n} + ^{232}_{90}\text{Th} \rightarrow ^{A}_{Z}\text{X} + \gamma$$

(a) Identify the unknown nucleus $^{A}_{Z}\text{X}$, giving its atomic mass number A, its atomic number Z, and the symbol X for the element. (b) The $^{A}_{Z}\text{X}$ nucleus subsequently undergoes β^{-} decay, and its daughter does too. Identify the final nucleus, giving its atomic mass number, atomic number, and name.

*23. ssm When 1.0 kg of coal is burned, about 3.0×10^{7} J of energy is released. If the energy released per $^{235}_{92}\text{U}$ fission is 2.0×10^{2} MeV, how many kilograms of coal must be burned to produce the same energy as 1.0 kg of $^{235}_{92}\text{U}$?

*24. The energy consumed in one year in the United States is about 9.3×10^{19} J. When each $^{235}_{92}\text{U}$ nucleus fissions, about 2.0×10^{2} MeV of energy is released. How many kilograms of $^{235}_{92}\text{U}$ would be needed to generate this energy if all the nuclei fissioned?

*25. Suppose the $^{239}_{94}\text{Pu}$ nucleus fissions into two fragments whose mass ratio is 0.32:0.68. With the aid of Figure 32.8, estimate the energy (in MeV) released during this fission.

**26. A 20.0 kiloton atomic bomb releases as much energy as 20.0 kilotons of TNT (1.0 kiloton of TNT releases about 5.0×10^{12} J of energy). Recall that about 2.0×10^{2} MeV of energy is released when each $^{235}_{92}\text{U}$ nucleus fissions. (a) How many $^{235}_{92}\text{U}$ nuclei are fissioned to produce the bomb's energy? (b) How many grams of uranium are fissioned? (c) What is the equivalent mass (in grams) of the bomb's energy?

**27. ssm www A nuclear power plant is 25% efficient, meaning that 25% of the power it generates goes into producing electricity. The remaining 75% is wasted as heat. The plant generates 8.0×10^{8} watts of electric power. If each fission releases 2.0×10^{2} MeV of energy, how many kilograms of $^{235}_{92}\text{U}$ are fissioned per year?

Section 32.5 Nuclear Fusion

28. Tritium ($^3_1$H) is a rare isotope of hydrogen that can be produced by the fusion reaction

$$\underbrace{^A_Z X}_{1.0087 \text{ u}} + \underbrace{^A_1 Y}_{2.0141 \text{ u}} \longrightarrow \underbrace{^3_1 H}_{3.0161 \text{ u}} + \gamma$$

(a) Determine the atomic mass number A, the atomic number Z, and the names X and Y of the unknown particles. (b) Using the masses given in the reaction, determine how much energy (in MeV) is released by this reaction.

29. The fusion of two deuterium nuclei ($^2_1$H, mass $= 2.0141$ u) can yield a tritium nucleus ($^3_1$H, mass $= 3.0161$ u) and a proton ($^1_1$H, mass $= 1.0078$ u). What is the energy (in MeV) released in this reaction?

30. Two deuterium ($^2_1$H) nuclei can fuse and form $^3_2$He and a neutron. The atomic masses are $^2_1$H (2.0141 u), $^3_2$He (3.0160 u), and $^1_0$n (1.0087 u). Find the energy (in MeV) released.

*31. ssm One proposed fusion reaction combines lithium $^6_3$Li (6.015 u) with deuterium $^2_1$H (2.014 u) to give helium $^4_2$He (4.003 u): $^2_1$H $+ ^6_3$Li $\rightarrow 2^4_2$He. How many kilograms of lithium would be needed to supply the energy needs of one household for a year, estimated to be 3.8×10^{10} J?

*32. Deuterium($^2_1$H) is an attractive fuel for fusion reactions because it is abundant in the waters of the oceans. In the oceans, about 0.015% of the hydrogen atoms in the water (H_2O) are deuterium atoms. (a) How many deuterium atoms are there in one kilogram of water? (b) If each deuterium nucleus produces about 7.2 MeV in a fusion reaction, how many kilograms of ocean water would be needed to supply the energy needs of the United States for one year, estimated to be 9.3×10^{19} J?

**33. The proton–proton cycle thought to occur in the sun consists of the following sequence of reactions:

(1) $^1_1$H $+ ^1_1$H $\longrightarrow ^2_1$H $+ ^0_1$e $+ \nu$
(2) $^1_1$H $+ ^2_1$H $\longrightarrow ^3_2$He $+ \gamma$
(3) $^3_2$He $+ ^3_2$He $\longrightarrow ^4_2$He $+ ^1_1$H $+ ^1_1$H

In these reactions $^0_1$e is a positron (mass $= 0.000\ 549$ u), ν is a neutrino (mass ≈ 0 u), and γ is a gamma ray photon (mass $= 0$ u). Note that reaction (3) uses two $^3_2$He nuclei, which are

formed by *two* reactions of type (1) and *two* reactions of type (2). Verify that the proton–proton cycle generates about 25 MeV of energy. The atomic masses are $^1_1$H (1.007 825 u), $^2_1$H (2.014 102 u), $^3_2$He (3.016 030 u), and $^4_2$He (4.002 603 u). Be sure to account for the fact that there are two electrons in two hydrogen atoms, while there is one electron in a single deuterium ($^2_1$H) atom. The mass of one electron is 0.000 549 u.

Section 32.6 Elementary Particles

34. The main decay mode for the negative pion is $\pi^- \rightarrow \mu^- + \bar{\nu}_\mu$. Find the energy (in MeV) released in this decay. Consult Table 32.3 as needed.

35. ssm The lambda particle Λ^0 has an electric charge of zero. It is a baryon and, hence, is composed of three quarks. They are all different. One of these quarks is the up quark u, and there are no antiquarks present. Make a list of the three possibilities for the quarks contained in Λ^0. (Other information is needed to decide which one of these possibilities is the Λ^0 particle.)

36. Review Conceptual Example 6 as background for this problem. An electron and its antiparticle annihilate each other, producing two γ-ray photons. The kinetic energies of the particles are negligible. For each photon, determine (a) its energy (in MeV), (b) its wavelength, and (c) the magnitude of its momentum.

37. What are the quarks from which an antiproton is constructed?

38. A high-energy proton collides with a statioary proton, and the reaction $p + p \rightarrow n + p + \pi^+$ occurs. The rest energy of the π^+ pion is 139.6 MeV. Ignore momentum conservation and find the minimum energy (in MeV) the incident proton must have.

*39. ssm www Suppose a neutrino is created and has an energy of 35 MeV. (a) If the neutrino, like the photon, has no mass and travels at the speed of light, find the momentum of the neutrino. (b) Determine the de Broglie wavelength of the neutrino.

*40. Review Conceptual Example 4 as background for this problem. An energetic proton is fired at a stationary proton. For the reaction to produce new particles, the two protons must approach each other to within a distance of about 8.0×10^{-15} m. The moving proton must have a sufficient speed to overcome the repulsive Coulomb force. What must be the minimum initial kinetic energy (in MeV) of the proton?

ADDITIONAL PROBLEMS

41. During an X-ray examination, a person is exposed to radiation at a rate of 3.1×10^{-5} grays per second. The exposure time is 0.10 s, and the mass of the exposed tissue is 1.2 kg. Determine the energy absorbed.

42. A collision between two protons produces three new particles: $p + p \rightarrow p + \pi^+ + \Lambda^0 + K^0$. The rest energies of the new particles are π^+ (139.6 MeV), Λ^0 (1116 MeV), and K^0 (497.7 MeV). Note that one proton disappears during the reaction. How much of the protons' incident energy (in MeV) is transformed into matter during this reaction?

43. ssm www Write the equation for the reaction $^{17}_8$O $(\gamma, \alpha n) \ ^{12}_6$C. The notation "$\alpha n$" means that an α particle and a neutron are produced by the reaction.

44. What energy (in MeV) is liberated by the following fission reaction?

$$\underbrace{^1_0 n}_{1.009 \text{ u}} + \underbrace{^{235}_{92} U}_{235.044 \text{ u}} \longrightarrow \underbrace{^{141}_{56} Ba}_{140.914 \text{ u}} + \underbrace{^{92}_{36} Kr}_{91.926 \text{ u}} + \underbrace{3^1_0 n}_{3(1.009 \text{ u})}$$

*45. ssm Imagine your car is powered by a fusion engine in which the following reaction occurs: 3^2_1H $\rightarrow ^4_2$He $+ ^1_1$H $+ ^1_0$n.

The masses are $_1^2$H (2.0141 u), $_2^4$He (4.0026 u), $_1^1$H (1.0078 u), and $_0^1$n (1.0087 u). The engine uses 6.1×10^{-6} kg of deuterium $_1^2$H fuel. If one gallon of gasoline produces 2.1×10^9 J of energy, how many gallons of gasoline would have to be burned to equal the energy released by all the deuterium fuel?

*46. A water sample receives a 750-rad dose of radiation. Find the rise in the water temperature.

*47. The water that cools a reactor core enters the reactor at 216 °C and leaves at 287 °C. (The water is pressurized, so it does not turn to steam.) The core is generating 5.6×10^9 W of power. Assume that the specific heat capacity of water is 4420 J/(kg · C°) over the temperature range stated above, and find the mass of water that passes through the core each second.

*48. (a) If each fission of a $_{92}^{235}$U nucleus releases about 2.0×10^2 MeV of energy, determine the energy (in joules) released by the complete fissioning of 1.0 gram of $_{92}^{235}$U. (b) How many grams of $_{92}^{235}$U are consumed in one year, in order to supply the energy needs of a household that uses 30.0 kWh of energy per day, on the average?

**49. One kilogram of dry air at STP conditions is exposed to 1.0 R of X-rays. One roentgen is defined by Equation 32.1. An equivalent definition can be based on the fact that an exposure of one roentgen deposits 8.3×10^{-3} J of energy per kilogram of dry air. Using the two definitions and assuming that all ions produced are singly charged, determine the average energy (in eV) needed to produce one ion in air.

CONCEPTS & CALCULATIONS

GROUP LEARNING PROBLEMS

Note: Each of these problems consists of Concept Questions followed by a related quantitative Problem. They are designed for use by students working alone or in small learning groups. The Concept Questions involve little or no mathematics and are intended to stimulate group discussions. They focus on the concepts with which the problems deal. Recognizing the concepts is the essential initial step in any problem-solving technique.

50. Concept Questions Identical body tissues are being irradiated by two types of radiation (e.g., X-rays and protons). (a) If the absorbed doses are the same, which type of radiation, the one with the larger RBE or the smaller RBE, produces the greater biological effect on the tissue? (b) If the two types of radiation have the same RBE, which type, the one that generates the greater absorbed dose or the smaller absorbed dose, produces the greater biological effect? Provide a reason for each answer.

Problem Someone stands near a radioactive source and receives doses of the following types of radiation: γ rays (20 mrad, RBE = 1), electrons (30 mrad, RBE = 1), protons (5 mrad, RBE = 10), and slow neutrons (5 mrad, RBE = 2). Rank the types of radiation, highest first, as to which produces the largest biologically equivalent dose. Verify that your answers are consistent with your answers to the Concept Questions.

51. Concept Question The definition of the biologically equivalent dose (Equation 32.4) uses the concept of the absorbed dose measured in a unit called the rad. The definition of the absorbed dose measured in a unit called the gray is given by Equation 32.2 as the energy absorbed divided by the mass of the absorbing material. How does one convert from the absorbed dose (in rad) to the absorbed dose (in Gy)?

Problem The biologically equivalent dose for a typical chest X-ray is 2.5×10^{-2} rem. The mass of the exposed tissue is 21 kg, and it absorbs 6.2×10^{-3} J of energy. What is the relative biological effectiveness (RBE) for the radiation on this particular type of tissue?

52. Concept Questions (a) Which of the following are nucleons: protons, electrons, neutrons, γ-ray photons? (b) In a nuclear reaction, what is meant by the phrase "the total number of nucleons is conserved?" (c) Which of the following have electric charge: protons, electrons, neutrons, γ-ray photons? (d) In a nuclear reaction, what is meant by the phrase "the total electric charge is conserved?"

Problem For each of the nuclear reactions listed below, determine the unknown particle $_Z^A$X:

(a) $_Z^A$X + $_7^{14}$N → $_1^1$H + $_8^{17}$O
(b) $_7^{15}$N + $_Z^A$X → $_6^{12}$C + $_2^4$He
(c) $_1^1$H + $_{13}^{27}$Al → $_Z^A$X + $_0^1$n
(d) $_3^7$Li + $_1^1$H → $_2^4$He + $_Z^A$X

53. Concept Questions During a nuclear reaction, whether it is a fission reaction or a fusion reaction, energy is released. (a) Is the total mass of the particles after the reaction greater than or less than the total mass of the particles before the reaction? (b) How is this difference in total mass related to the energy released by the reaction?

Problem In one type of fusion reaction a proton fuses with a neutron to form a deuterium nucleus: $_1^1$H + $_0^1$n → $_1^2$H. The masses are $_1^1$H (1.0078 u), $_0^1$n (1.0087 u), and $_1^2$H (2.0141 u). How much energy (in MeV) is released by this reaction?

APPENDIX A POWERS OF TEN AND SCIENTIFIC NOTATION

In science, very large and very small decimal numbers are conveniently expressed in terms of powers of ten, some of which are listed below:

$$10^3 = 10 \times 10 \times 10 = 1000 \qquad 10^{-3} = \frac{1}{10 \times 10 \times 10}$$
$$= 0.001$$

$$10^2 = 10 \times 10 = 100 \qquad 10^{-2} = \frac{1}{10 \times 10} = 0.01$$

$$10^1 = 10 \qquad 10^{-1} = \frac{1}{10} = 0.1$$

$$10^0 = 1$$

Using powers of ten, we can write the radius of the earth in the following way, for example:

$$\text{Earth radius} = 6\,380\,000 \text{ m} = 6.38 \times 10^6 \text{ m}$$

The factor of ten raised to the sixth power is ten multiplied by itself six times, or one million, so the earth's radius is 6.38 million meters. Alternatively, the factor of ten raised to the sixth power indicates that the decimal point in the term 6.38 is to be moved six places *to the right* to obtain the radius as a number without powers of ten.

For numbers less than one, negative powers of ten are used. For instance, the Bohr radius of the hydrogen atom is

$$\text{Bohr radius} = 0.000\,000\,000\,0529 \text{ m} = 5.29 \times 10^{-11} \text{ m}$$

The factor of ten raised to the minus eleventh power indicates that the decimal point in the term 5.29 is to be moved eleven places *to the left* to obtain the radius as a number without powers of ten. Numbers expressed with the aid of powers of ten are said to be in ***scientific notation.***

Calculations that involve the multiplication and division of powers of ten are carried out as in the following examples:

$$(2.0 \times 10^6)(3.5 \times 10^3) = (2.0 \times 3.5) \times 10^{6+3} = 7.0 \times 10^9$$

$$\frac{9.0 \times 10^7}{2.0 \times 10^4} = \left(\frac{9.0}{2.0}\right) \times 10^7 \times 10^{-4}$$

$$= \left(\frac{9.0}{2.0}\right) \times 10^{7-4} = 4.5 \times 10^3$$

The general rules for such calculations are

$$\frac{1}{10^n} = 10^{-n} \tag{A-1}$$

$$10^n \times 10^m = 10^{n+m} \qquad \text{(Exponents added)} \tag{A-2}$$

$$\frac{10^n}{10^m} = 10^{n-m} \qquad \text{(Exponents subtracted)} \tag{A-3}$$

where n and m are any positive or negative number.

Scientific notation is convenient because of the ease with which it can be used in calculations. Moreover, scientific notation provides a convenient way to express the significant figures in a number, as Appendix B discusses.

APPENDIX B SIGNIFICANT FIGURES

The number of ***significant figures*** in a number is the number of digits whose values are known with certainty. For instance, a person's height is measured to be 1.78 m, with the measurement error being in the third decimal place. All three digits are known with certainty, so that the number contains three significant figures. If a zero is given as the last digit to the right of the decimal point, the zero is presumed to be significant. Thus, the number 1.780 m contains four significant figures. As another example, consider a distance of 1500 m. This number contains only two significant figures, the one and the five. The zeros immediately to the left of the unexpressed decimal point are not counted as significant figures. However, zeros located between significant figures are significant, so a distance of 1502 m contains four significant figures.

Scientific notation is particularly convenient from the point of view of significant figures. Suppose it is known that a certain distance is fifteen hundred meters, to four significant figures. Writing the number as 1500 m presents a problem because it implies that only two significant figures are known. In contrast, the scientific notation of 1.500×10^3 m has the advantage of indicating that the distance is known to four significant figures.

When two or more numbers are used in a calculation, the number of significant figures in the answer is limited by the number of significant figures in the original data. For instance, a rectangular garden with sides of 9.8 m and 17.1 m has an area of (9.8 m)(17.1 m). A calculator gives 167.58 m² for this product. However, one of the original lengths is known only to two significant figures, so the final answer is limited to only two significant figures and should be rounded off to 170 m². In general, **when numbers are multiplied or divided, the number of significant figures in the final answer equals the smallest number of significant figures in any of the original factors.**

The number of significant figures in the answer to an addition or a subtraction is also limited by the original data. Consider the total distance along a biker's trail that consists of three segments with the distances shown as follows:

$$
\begin{array}{rl}
 & 2.5 \ \ \text{km} \\
 & 11 \ \ \ \ \text{km} \\
 & \underline{5.26} \ \text{km} \\
\text{Total} & 18.76 \ \text{km}
\end{array}
$$

The distance of 11 km contains no significant figures to the right of the decimal point. Therefore, neither does the sum of the three distances, and the total distance should not be reported as 18.76 km. Instead, the answer is rounded off to 19 km. In general, **when numbers are added or subtracted, the last significant figure in the answer occurs in the last column (counting from left to right) containing a number that results from a combination of digits that are all significant.** In the answer of 18.76 km, the eight is the sum of $2 + 1 + 5$, each digit being significant. However, the seven is the sum of $5 + 0 + 2$, and the zero is not significant, since it comes from the 11-km distance, which contains no significant figures to the right of the decimal point.

APPENDIX C ALGEBRA

C.1 PROPORTIONS AND EQUATIONS

Physics deals with physical variables and the relations between them. Typically, variables are represented by the letters of the English and Greek alphabets. Sometimes, the relation between variables is expressed as a proportion or inverse proportion. Other times, however, it is more convenient or necessary to express the relation by means of an equation, which is governed by the rules of algebra.

If two variables are **directly proportional** and one of them doubles, then the other variable also doubles. Similarly, if one variable is reduced to one-half its original value, then the other is also reduced to one-half its original value. In general, if x is directly proportional to y, then increasing or decreasing one variable by a given factor causes the other variable to change in the same way by the same factor. This kind of relation is expressed as $x \propto y$, where the symbol $\propto$ means "is proportional to."

Since the proportional variables x and y always increase and decrease by the same factor, the ratio of x to y must have a constant value, or $x/y = k$, where k is a constant, independent of the values for x and y. Consequently, a proportionality such as $x \propto y$ can also be expressed in the form of an equation: $x = ky$. The constant k is referred to as a **proportionality constant.**

If two variables are **inversely proportional** and one of them increases by a given factor, then the other decreases by the same factor. An inverse proportion is written as $x \propto 1/y$. This kind of proportionality is equivalent to the following equation: $xy = k$, where k is a proportionality constant, independent of x and y.

C.2 SOLVING EQUATIONS

Some of the variables in an equation typically have known values, and some do not. It is often necessary to solve the equation so that a variable whose value is unknown is expressed in terms of the known quantities. **In the process of solving an equation, it is permissible to manipulate the equation in any way, as long as a change made on one side of the equals sign is also made on the other side.** For example, consider the equation $v = v_0 + at$. Suppose values for v, v_0, and a are available, and the value of t is required. To solve the equation for t, we begin by subtracting v_0 from *both* sides:

$$
\begin{array}{rcl}
v & = & v_0 + at \\
\underline{-v_0} & & \underline{-v_0} \\
v - v_0 & = & at
\end{array}
$$

Next, we divide both sides of $v - v_0 = at$ by the quantity a:

$$
\frac{v - v_0}{a} = \frac{at}{a} = (1)t
$$

On the right side, the a in the numerator divided by the a in the denominator equals one, so that

$$t = \frac{v - v_0}{a}$$

It is always possible to check the correctness of the algebraic manipulations performed in solving an equation by substituting the answer back into the original equation. In the previous example, we substitute the answer for t into $v = v_0 + at$:

$$v = v_0 + a\left(\frac{v - v_0}{a}\right) = v_0 + (v - v_0) = v$$

The result $v = v$ implies that our algebraic manipulations were done correctly.

Algebraic manipulations other than addition, subtraction, multiplication, and division may play a role in solving an equation. The same basic rule applies, however: Whatever is done to the left side of an equation must also be done to the right side. As another example, suppose it is necessary to express v_0 in terms of v, a, and x, where $v^2 = v_0^2 + 2ax$. By subtracting $2ax$ from both sides, we isolate v_0^2 on the right:

$$\begin{array}{rcl} v^2 &=& v_0^2 + 2ax \\ -2ax &=& -2ax \\ \hline v^2 - 2ax &=& v_0^2 \end{array}$$

To solve for v_0, we take the positive and negative square root of *both* sides of $v^2 - 2ax = v_0^2$:

$$v_0 = \pm \sqrt{v^2 - 2ax}$$

C.3 SIMULTANEOUS EQUATIONS

When more than one variable in a single equation is unknown, additional equations are needed if solutions are to be found for all of the unknown quantities. Thus, the equation $3x + 2y = 7$ cannot be solved by itself to give unique values for both x and y. However, if x and y simultaneously obey the equation $x - 3y = 6$, then both unknowns can be found.

There are a number of methods by which such simultaneous equations can be solved. One method is to solve one equation for x in terms of y and substitute the result into the other equation to obtain an expression containing only the single unknown variable y. The equation $x - 3y = 6$, for instance, can be solved for x by adding $3y$ to each side, with the result that $x = 6 + 3y$. The substitution of this expression for x into the equation $3x + 2y = 7$ is shown below:

$$3x + 2y = 7$$
$$3(6 + 3y) + 2y = 7$$
$$18 + 9y + 2y = 7$$

We find, then, that $18 + 11y = 7$, a result that can be solved for y:

$$\begin{array}{rcl} 18 + 11y &=& 7 \\ -18 & & -18 \\ \hline 11y &=& -11 \end{array}$$

Dividing both sides of this result by 11 shows that $y = -1$. The value of $y = -1$ can be substituted in either of the original equations to obtain a value for x:

$$\begin{array}{rcl} x - 3y &=& 6 \\ x - 3(-1) &=& 6 \\ x + 3 &=& 6 \\ -3 & & -3 \\ \hline x &=& 3 \end{array}$$

C.4 THE QUADRATIC FORMULA

Equations occur in physics that include the square of a variable. Such equations are said to be *quadratic* in that variable and often can be put into the following form:

$$ax^2 + bx + c = 0 \qquad \text{(C-1)}$$

where a, b, and c are constants independent of x. This equation can be solved to give the **quadratic formula,** which is

$$x = \frac{-b \pm \sqrt{b^2 - 4ac}}{2a} \qquad \text{(C-2)}$$

The $\pm$ in the quadratic formula indicates that there are two solutions. For instance, if $2x^2 - 5x + 3 = 0$, then $a = 2$, $b = -5$, and $c = 3$. The quadratic formula gives the two solutions as follows:

Solution 1:
Plus sign

$$x = \frac{-b + \sqrt{b^2 - 4ac}}{2a}$$
$$= \frac{-(-5) + \sqrt{(-5)^2 - 4(2)(3)}}{2(2)}$$
$$= \frac{+5 + \sqrt{1}}{4} = \frac{3}{2}$$

Solution 2:
Minus sign

$$x = \frac{-b - \sqrt{b^2 - 4ac}}{2a}$$
$$= \frac{-(-5) - \sqrt{(-5)^2 - 4(2)(3)}}{2(2)}$$
$$= \frac{+5 - \sqrt{1}}{4} = 1$$

APPENDIX D EXPONENTS AND LOGARITHMS

Appendix A discusses powers of ten, such as 10^3, which means ten multiplied by itself three times, or $10 \times 10 \times 10$. The three is referred to as an ***exponent.*** The use of exponents extends beyond powers of ten. In general, the term y^n means the factor y is multiplied by itself n times. For example, y^2, or y squared, is familiar and means $y \times y$. Similarly, y^5 means $y \times y \times y \times y \times y$.

The rules that govern algebraic manipulations of exponents are the same as those given in Appendix A (see Equations A-1, A-2, and A-3) for powers of ten:

$$\frac{1}{y^n} = y^{-n} \tag{D-1}$$

$$y^n y^m = y^{n+m} \qquad \text{(Exponents added)} \tag{D-2}$$

$$\frac{y^n}{y^m} = y^{n-m} \qquad \text{(Exponents subtracted)} \tag{D-3}$$

To the three rules above we add two more that are useful. One of these is

$$y^n z^n = (yz)^n \tag{D-4}$$

The following example helps to clarify the reasoning behind this rule:

$$3^2 5^2 = (3 \times 3)(5 \times 5) = (3 \times 5)(3 \times 5) = (3 \times 5)^2$$

The other additional rule is

$$(y^n)^m = y^{nm} \qquad \text{(Exponents multiplied)} \tag{D-5}$$

To see why this rule applies, consider the following example:

$$(5^2)^3 = (5^2)(5^2)(5^2) = 5^{2+2+2} = 5^{2 \times 3}$$

Roots, such as a square root or a cube root, can be represented with fractional exponents. For instance,

$$\sqrt{y} = y^{1/2} \quad \text{and} \quad \sqrt[3]{y} = y^{1/3}$$

In general, the nth root of y is given by

$$\sqrt[n]{y} = y^{1/n} \tag{D-6}$$

The rationale for Equation D-6 can be explained using the fact that $(y^n)^m = y^{nm}$. For instance, the fifth root of y is the number that, when multiplied by itself five times, gives back

y. As shown below, the term $y^{1/5}$ satisfies this definition:

$$(y^{1/5})(y^{1/5})(y^{1/5})(y^{1/5})(y^{1/5}) = (y^{1/5})^5 = y^{(1/5) \times 5} = y$$

Logarithms are closely related to exponents. To see the connection between the two, note that it is possible to express any number y as another number B raised to the exponent x. In other words,

$$y = B^x \tag{D-7}$$

The exponent x is called the ***logarithm*** of the number y. The number B is called the ***base number.*** One of two choices for the base number is usually used. If $B = 10$, the logarithm is known as the *common logarithm,* for which the notation "log" applies:

Common logarithm $y = 10^x \quad \text{or} \quad x = \log y$ (D-8)

If $B = e = 2.718 \ldots$, the logarithm is referred to as the *natural logarithm,* and the notation "ln" is used:

Natural logarithm $y = e^z \quad \text{or} \quad z = \ln y$ (D-9)

The two kinds of logarithms are related by

$$\ln y = 2.3026 \log y \tag{D-10}$$

Both kinds of logarithms are often given on calculators.

The logarithm of the product or quotient of two numbers A and C can be obtained from the logarithms of the individual numbers according to the rules below. These rules are illustrated here for natural logarithms, but they are the same for any kind of logarithm.

$$\ln (AC) = \ln A + \ln C \tag{D-11}$$

$$\ln \left(\frac{A}{C} \right) = \ln A - \ln C \tag{D-12}$$

Thus, the logarithm of the product of two numbers is the sum of the individual logarithms, and the logarithm of the quotient of two numbers is the difference between the individual logarithms. Another useful rule concerns the logarithm of a number A raised to an exponent n:

$$\ln A^n = n \ln A \tag{D-13}$$

Rules D-11, D-12, and D-13 can be derived from the definition of the logarithm and the rules governing exponents.

APPENDIX E GEOMETRY AND TRIGONOMETRY

E.1 Geometry

Angles

Two angles are equal if

1. They are vertical angles (see Figure E1).
2. Their sides are parallel (see Figure E2).

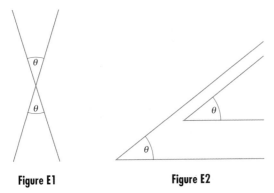

Figure E1   **Figure E2**

3. Their sides are mutually perpendicular (see Figure E3).

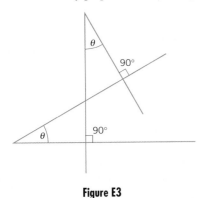

Figure E3

Triangles

1. The *sum of the angles* of any triangle is 180° (see Figure E4).

$\alpha + \beta + \gamma = 180°$

Figure E4

2. A *right triangle* has one angle that is 90°.
3. An *isosceles triangle* has two sides that are equal.
4. An *equilateral triangle* has three sides that are equal. Each angle of an equilateral triangle is 60°.
5. Two triangles are **similar** if two of their angles are equal (see Figure E5). The corresponding sides of similar triangles are proportional to each other:

$$\frac{a_1}{a_2} = \frac{b_1}{b_2} = \frac{c_1}{c_2}$$

Figure E5

6. Two similar triangles are **congruent** if they can be placed on top of one another to make an exact fit.

Circumferences, Areas, and Volumes of Some Common Shapes

1. Triangle of base b and altitude h (see Figure E6):

$$\text{Area} = \tfrac{1}{2} bh$$

Figure E6

2. Circle of radius r:

$$\text{Circumference} = 2\pi r$$
$$\text{Area} = \pi r^2$$

3. Sphere of radius r:

$$\text{Surface area} = 4\pi r^2$$
$$\text{Volume} = \tfrac{4}{3}\pi r^3$$

4. Right circular cylinder of radius r and height h (see Figure E7):

$$\text{Surface area} = 2\pi r^2 + 2\pi rh$$
$$\text{Volume} = \pi r^2 h$$

Figure E7

E.2 Trigonometry

Basic Trigonometric Functions

1. For a right triangle, the sine, cosine, and tangent of an angle θ are defined as follows (see Figure E8):

$$\sin\theta = \frac{\text{Side opposite }\theta}{\text{Hypotenuse}} = \frac{h_o}{h}$$

$$\cos\theta = \frac{\text{Side adjacent to }\theta}{\text{Hypotenuse}} = \frac{h_a}{h}$$

$$\tan\theta = \frac{\text{Side opposite }\theta}{\text{Side adjacent to }\theta} = \frac{h_o}{h_a}$$

Figure E8

2. The secant ($\sec\theta$), cosecant ($\csc\theta$), and cotangent ($\cot\theta$) of an angle θ are defined as follows:

$$\sec\theta = \frac{1}{\cos\theta} \qquad \csc\theta = \frac{1}{\sin\theta} \qquad \cot\theta = \frac{1}{\tan\theta}$$

Triangles and Trigonometry

1. The **Pythagorean theorem** states that the square of the hypotenuse of a right triangle is equal to the sum of the squares of the other two sides (see Figure E8):

$$h^2 = h_o{}^2 + h_a{}^2$$

2. The **law of cosines** and the **law of sines** apply to any triangle, not just a right triangle, and they relate the angles and the lengths of the sides (see Figure E9):

Figure E9

Law of cosines $c^2 = a^2 + b^2 - 2ab\cos\gamma$

Law of sines $\dfrac{a}{\sin\alpha} = \dfrac{b}{\sin\beta} = \dfrac{c}{\sin\gamma}$

Other Trigonometric Identities

1. $\sin(-\theta) = -\sin\theta$
2. $\cos(-\theta) = \cos\theta$
3. $\tan(-\theta) = -\tan\theta$
4. $(\sin\theta)/(\cos\theta) = \tan\theta$
5. $\sin^2\theta + \cos^2\theta = 1$
6. $\sin(\alpha + \beta) = \sin\alpha\cos\beta + \cos\alpha\sin\beta$

$$\text{If } \alpha = 90°, \sin(90° + \beta) = \cos\beta$$
$$\text{If } \alpha = \beta, \sin 2\beta = 2\sin\beta\cos\beta$$

7. $\cos(\alpha + \beta) = \cos\alpha\cos\beta \mp \sin\alpha\sin\beta$

$$\text{If } \alpha = 90°, \cos(90° + \beta) = \mp\sin\beta$$
$$\text{If } \alpha = \beta, \cos 2\beta = \cos^2\beta - \sin^2\beta = 1 - 2\sin^2\beta$$

ANSWERS TO ODD-NUMBERED PROBLEMS

CHAPTER 1
1. 1.37 lb
3. (a) 5700 s (b) 86 400 s
5. (a) 5×10^{-3} g (b) 5 mg
   (c) 5×10^3 μg
7. (a) correct (b) not correct
   (c) not correct (d) correct
   (e) correct
9. 3.13×10^8 m^3
11. 80.1 km, 25.9° south of west
13. 713 m
15. 3.73 m
17. 0.487 nm
19. 99° opposite the 190-cm side
    3.0×10^1 degrees opposite the 95-cm side
    51° opposite the 150-cm side
21. 200 N due east or 600 N due west
23. (a) 5.31 km, south
    (b) 5.31 km, north
25. 0.90 km, 56° north of west
27. (a) 5600 newtons
    (b) along the dashed line
29. (a) 4746 m
    (b) 21.9° south of east
31. (a) C (b) B
33. (a) 147 km (b) 47.9 km
35. 222 m, 55.8° below the $-x$ axis
37. (a) $A_x = 650$ units, $A_y = 380$ units
    (b) $A_{x'} = 570$ units, $A_{y'} = -480$ units
39. (a) 322 newtons (b) 209 newtons
    (c) 279 newtons
41. 7.1 m, 9.9° north of east
43. 30.2 m, 10.2°
45. 0.532
47. (a) 10.4 units (b) 12.0 units
49. 6.88 km, 26.9°
51. 13.5 m
53. (a) 15.8 m/s
    (b) 6.37 m/s
55. 3.00 m, 42.8° above the $-x$ axis
57. (a) 5.70×10^2 newtons
    (b) 33.6° south of west
59. (a) 6.43 m, 70.0° above the $-x$ axis
    (b) 7.66 m, 20.0° below the $-x$ axis
61. (a) 142 N, $\theta = 67°$ south of east
    (b) 142 N, $\theta = 67°$ north of west
63. 288 units due west and 156 units due north

CHAPTER 2
1. (a) 12.4 km
   (b) 8.8 km, due east
3. 0.80 s
5. 28 m
7. 52 m
9. 7.2×10^3 m
11. 2.1 s
13. (a) 1.5 m/s^2 (b) 1.5 m/s^2
    (c) Yes, 76 m
15. 3.44 m/s, due west
17. +26.0 m/s
19. (a) 11.4 s (b) +44.7 m/s^2
21. 4.5 m
23. 3.1 m/s^2, opposite to the direction of the motion
25. 0.17 s
27. 0.74 m/s
29. 107 m
31. 1.50 km (acceleration phase), 0.50 km (deceleration phase)
33. (a) 11 s (b) 3.0×10^1 m
35. 14 s
37. 44.1 m/s
39. $d_1 = 0.018$ m, $d_2 = 0.071$ m, $d_3 = 0.16$ m
41. 1.1 s
43. (a) -7.9 m/s (b) 3.2 m
45. 1.7 s
47. 12 m
49. 13.2 m/s
51. 0.40 s
53. 10.6 m
55. 16 m and 41 m
57. The answer is in graphical form.
59. -2.0×10^1 km/h (segment A), 1.0×10^1 km/h (segment B), 40 km/h (segment C)
61. -8.3 km/h^2
63. (a) 6.6 s (b) 5.3 m/s
65. 91.5 m/s
67. (a) 4.0 s (b) 4.0 s
69. 9.1
71. 43.7 m/s, downward
73. (a) 2.67×10^4 m
    (b) 6.74 m/s, due north
75. 2.0×10^1 m
77. 15 s
79. (a) 13 m/s (b) 0.93 m/s^2

CHAPTER 3
1. 242 m/s
3. 8600 m
5. 8.8×10^2 m
7. (a) 0.33 m/s^2 (b) 0.16 m/s^2
   (c) -0.29 m/s^2
9. 27.0°
11. (a) 2.99×10^4 m/s
    (b) 2.69×10^4 m/s
13. 4.42 s
15. 14.6 s
17. 14.1 m/s
19. (a) 9.0 m (b) 1.1 s
21. 30.0 m
23. 39 m/s
25. (a) 239 m/s, 57.1° with respect to the horizontal
    (b) 239 m/s, 57.1° with respect to the horizontal
27. 24 buses
29. 1.7 s
31. 5.2 m
33. 14.9 m
35. 17.8 m/s
37. 11 m/s
39. 4.00
41. 42°
43. $D = 850$ m, $H = 31$ m
45. $v_{0B} = 8.79$ m/s, $\theta_B = 81.5°$
47. (a) 41 m/s, due east
    (b) 41 m/s, due west
49. 31 s
51. (a) 2.0×10^3 s (b) 1.8×10^3 m
53. 6.3 m/s, 18° north of east
55. 2.3 m/s
57. 5.2 m/s, 52° west of south
59. 3.05 m/s, 14.8° north of west
61. 2.40 m
63. (a) 1.78 s (b) 20.8 m/s
65. (a) 1.1 s (b) 1.3 s
67. 15.7°
69. 63.9 m/s, 85° west of south
71. 0.141° and 89.860°
73. 4.2 m/s

CHAPTER 4
1. 93 N
3. 37 N
5. 130 N

7. 3.5×10^4 N

9. (a) $F_A = 3.6$ N   (b) $F_B = 0.40$ N

11. 18.0 m/s^2, 56.3° above the $+x$ axis

13. 30.9 m/s^2, 27.2° above the $+x$ axis

15. -470 N

17. 18.4 N, 68° north of east

19. (a) $m = 115$ kg, $W = 1.13 \times 10^3$ N
    (b) $m = 115$ kg, $W = 0$ N

21. 1.8×10^{-7} N

23. (a) 10.5 m/s^2 (b) 1.07

25. 1.76×10^{24} kg

27. 0.223 m/s^2

29. 0.0050

31. 4.7 kg

33. $0.414L$ or $-2.414L$

35. 7.3×10^2 N

37. The block will move., 3.72 m/s^2

39. (a) 447 N   (b) 241 N

41. (a) 21.2 N   (b) 23.8 N
    (c) 18.6 N

43. (a) 0.980 m/s^2, opposite to the
        skater's motion
    (b) 29.5 m

45. 68°

47. 58.8 N

49. 9.70 N

51. 39 N

53. 1.00×10^2 N, 53.1° south of east

55. 13 kg/s

57. 1.9×10^2 N

59. 40.0 N, 59.8° above the horizontal

61. 0.665

63. 220 N, 64° north of east

65. 1730 N, due west

67. (a) 2.99 m/s^2 (b) 129 N

69. 6.6 m/s

71. 4290 N

73. 29 400 N

75. 8.7 s

77. 3.9 m/s^2

79. 0.265 m

81. (a) 5.2 m/s^2 (b) 11 m/s^2

83. 1.2 s

85. (a) 0.60 m/s^2
    (b) 104 N (left string), 230 N
        (right string)

87. (a) 3.75 m/s^2 (b) 2.4×10^2 N

89. 929 N

91. 0.444

93. (a) 914 N   (b) 822 N

95. (a) 1.3 N   (b) 6.5 N

97. (a) 0.0640 m/s^2 (b) 1.58×10^4 km

99. (a) 3.56 m/s^2 (b) 281 N

101. 0.200

103. 0.14 m/s^2

105. (a) $\Delta T_A = 0$ N, $\Delta T_B = -4.7$ N,
        $\Delta T_C = 0$ N
    (b) $\Delta T_A = 0$ N, $\Delta T_B = 0$ N,
        $\Delta T_C = +4.7$ N

107. 16.3 N

109. (a) 13.7 N   (b) 1.37 m/s^2

111. 406 N

CHAPTER 5

1. 160 s

3. 0.016 s

5. 0.71

7. 61°

9. 2.2

11. (a) 1.2×10^4 N   (b) 1.7×10^4 N

13. 12 m/s

15. (a) 0.189 N   (b) 4.00

17. 3500 N

19. 28°

21. 2.0×10^1 m/s

23. (a) 19 m/s   (b) 23 m/s

25. 2.12×10^6 N

27. 1.33×10^4 m/s

29. 4.20×10^4 m/s

31. (a) 2.99×10^4 m/s
    (b) 2.01×10^{30} kg

33. 2.45×10^4 N

35. (a) 912 m   (b) 228 m
    (c) 2.50 m/s^2

37. 4.72×10^3 m

39. (a) 1.70×10^3 N   (b) 1.66×10^3 N

41. 14.0 m/s

43. 0.68 m/s

45. (a) 5.0×10^1 m/s^2
    (b) zero
    (c) 2.0×10^1 m/s^2

47. 332 m

49. 2.9×10^4 N

51. 2.0×10^2 m/s^2

53. 23 N at 19.0 m/s and 77 N at 38.0 m/s

55. 0.250

CHAPTER 6

1. 2.2×10^3 J

3. (a) 2980 J   (b) 3290 J

5. 42.8°

7. (a) 54.9 N   (b) 1060 J
    (c) -1060 J   (d) 0 J

9. 45 N

11. 2.07×10^3 N

13. (a) 3.1×10^3 J   (b) 2.2×10^2 J

15. 6.4×10^5 J

17. (a) 38 J   (b) 3.8×10^3 N

19. 18%

21. 10.9 m/s

23. 1.4×10^{11} J

25. 2.39×10^5 J

27. (a) -3.0×10^4 J   (b) no

29. 5.24×10^5 J

31. 2.3×10^4 J

33. 3.9 m/s

35. At $h = 20.0$ m, KE $= 0$ J, PE $=$
    392 J, and $E = 392$ J. At $h = 10.0$ m,
    KE $= 196$ J, PE $= 196$ J, and $E =$
    392 J. At $h = 0$ m, KE $= 392$ J, PE $=$
    0 J, and $E = 392$ J.

37. 4.8 m/s

39. 1.7 m/s

41. 6.33 m

43. 40.8 kg

45. -4.51×10^4 J

47. (a) -1086 J
    (b) 2.01 m below the starting point

49. 45.9 m/s

51. 21.5°

53. 4.17 m/s

55. 2.1×10^3 W

57. 3.6×10^6 J

59. (a) 1.0×10^4 W   (b) 13 hp

61. 6.7×10^2 N

63. (a) bow 1   (b) 25 J

65. (a) 93 J   (b) 0 J
    (c) 2.3 m/s

67. (a) 1.50×10^2 J   (b) 7.07 m/s

69. 4.13 m

71. -2.6×10^6 J

73. Work done: 1.0×10^3 J (force **P**),
    -940 J (kinetic frictional force), 0 J
    (normal force), 0 J (weight of the box)

75. -1.21×10^6 J

77. 4.43 m/s

79. 13.5 m

CHAPTER 7

1. 11 N·s, in the same direction as the
    average force

3. 165 N, upward

5. -8.7 kg·m/s

7. $+69$ N

9. 3.7 N·s

11. 9.00×10^2 N

13. 4.28 N·s, upward

15. -1.5×10^{-4} m/s

17. $+9.3$ m/s

19. (a) -0.14 m/s
    (b) -7.1×10^{-3} m/s

21. 1.5×10^{-10} m/s

23. 0.097 m

25. 84 kg

27. 3.00 m

29. +9.09 m/s

31. (a) −0.400 m/s (5.00-kg ball),
+1.60 m/s (7.50-kg ball)
(b) +0.800 m/s

33. 0.19 s

35. (a) 73.0° (b) 4.28 m/s

37. (a) +3.17 m/s (b) 0.0171

39. (a) 5.56 m/s
(b) −2.83 m/s (1.50-kg ball),
+2.73 m/s (4.60-kg ball)
(c) 0.409 m (1.50-kg ball),
0.380 m (4.60-kg ball)

41. 4.67×10^6 m

43. 6.46×10^{-11} m

45. (a) −1.5 m/s (b) +1.1 m/s

47. (a) Bonzo, since he has the recoil
velocity with the smaller
magnitude.
(b) 1.7

49. 3.5×10^2 m/s

51. +4500 m/s, in the same direction as
the rocket before the explosion.

53. $m_1 = 1.00$ kg, $m_2 = 1.00$ kg

55. (a) +9.0 m/s
(b) -3.7×10^4 N·s (c) 5.2 m

57. +547 m/s

CHAPTER 8

1. 63.7 grad

3. 13 rad/s

5. 6.4×10^{-3} rad/s²

7. 492 rad/s

9. 825 m

11. 336 m/s

13. 6.05 m

15. 25 rev

17. (a) 4.00×10^1 rad
(b) 15.0 rad/s

19. (a) 1.2×10^4 rad (b) 1.1×10^2 s

21. (a) 4.60×10^3 rad
(b) 2.00×10^2 rad/s²

23. 12.5 s

25. 3.37 s

27. 7.37 s

29. (a) 0.332 rad/s (b) 128 m

31. 157 m/s

33. 0.34 m

35. (a) 1.25 m/s (b) 7.98 rev/s

37. 1.43×10^{-1} m/s

39. 380 m/s²

41. (a) 2.5 m/s² (b) 3.1 m/s²

43. $1/\sqrt{3}$

45. 1.00 rad

47. 8.71 rad/s²

49. (a) 7.50 rad/s

(b) -1.73×10^{-3} rad/s², The angular
velocity is decreasing.

51. 0.300 m/s

53. (a) 8.33 rad/s, counterclockwise
(b) 14.7 rad/s, clockwise

55. 20.6 rad

57. (a) 9.00 m/s²
(b) Radially inward, toward the center
of the track

59. 157.3 rad/s

61. (a) 54.0 rad/s (b) 486 rad

63. 28.0 rad/s

65. 7.45 s

67. 2 revolutions

CHAPTER 9

1. 0.25 m

3. 1.70×10^3 N·m

5. 1.3

7. 23.3 N·m, counterclockwise

9. 0.667 m

11. (a) 3.40×10^3 N (b) 2.28×10^3 N

13. 0.591

15. (a) 27 N, to the left
(b) 27 N, to the right
(c) 27 N, to the right
(d) 143 N, downward and to the left
(79° below the horizontal)

17. 1.0×10^3 N, to the left

19. (a) 1.60×10^5 N (b) 4.20×10^5 N

21. (a) 7.40×10^2 N (b) 0.851 N

23. (a) 1.21×10^3 N
(b) 1.01×10^3 N, downward

25. 69 N

27. 17.5 N

29. 1.25 kg·m²

31. (a) −11 N·m (b) −9.2 rad/s²

33. 0.060 kg·m²

35. -3.32×10^{-2} N·m

37. (a) 2.67 kg·m² (b) 1.16 m

39. 0.78 N

41. 2.12 s

43. 6.1×10^5 rev/min

45. (a) $v_{T1} = 12.0$ m/s, $v_{T2} = 9.00$ m/s,
$v_{T3} = 18.0$ m/s
(b) 1.08×10^3 J (c) 60.0 kg·m²
(d) 1.08×10^3 J

47. 1.18

49. 3/4

51. 2.0 m

53. 4.4 kg·m²

55. (a) 0.14 rad/s
(b) A net external torque must be
applied in a direction opposite to
the angular deceleration.

57. 8% increase

59. 0.573 m

61. 4.2 N·m

63. (a) 487 N (b) 209 N

65. (a) 0.131 kg·m²
(b) 3.6×10^{-4} kg·m²
(c) 0.149 kg·m²

67. 37.6°

69. 14 rad/s

71. 51.4 N

CHAPTER 10

1. 237 N

3. (a) 7.44 N (b) 7.44 N

5. 0.012 m

7. 6.3×10^{-2} m

9. 0.236 m

11. 2.29×10^{-3} m

13. (a) 1.00×10^3 N/m
(b) 0.340

15. 3.5×10^4 N/m

17. (a) 0.080 m (b) 1.6 rad/s
(c) 2.0 N/m (d) 0 m/s
(e) 0.20 m/s²

19. (a) 140 m/s
(b) 1.7×10^{15} m/s²

21. 4.3 kg

23. 0.806

25. 7.18×10^{-2} m

27. 6.55 m/s

29.

h (meters)	KE	PE (gravity)	PE (elastic)	E
0	0	0	8.76 J	8.76 J
0.200	1.00 J	3.92 J	3.84 J	8.76 J
0.400	0	7.84 J	0.92 J	8.76 J

31. 4.0 m/s

33. (a) 9.0×10^{-2} m (b) 2.1 m/s

35. 3.00×10^1 rad/s

37. 2.36×10^3 N/m

39. 21 m

41. 6.0 m/s²

43. 0.816

45. $L = (7/5)R$

47. 3.7×10^{-5} m

49. 7.7×10^{-5} m

51. 260 m

53. 1.4×10^{-6}

55. 6.6×10^4 N, directed perpendicular to
the sheet

57. -2.8×10^{-4}

59. (a) 6.3×10^{-2} m (b) 7.3×10^{-2} m

61. 4.6×10^{-4}

63. 1.0×10^{-3} m

65. -4.4×10^{-5}

67. (a) 5.08×10^{-2} m
(b) 5.00×10^{-2} s
(c) 6.40 m/s
69. 0.99 m
71. 1.6×10^5 N
73. 0.091°
75. 0.240 m
77. 2.1×10^{-5} m
79. 4.0×10^{-5} m
81. 66 Hz
83. (a) 0.25 s   **(b)** 0.75 s

CHAPTER 11

1. 6.6×10^6 kg
3. 6.8×10^9 N
5. 317 m²
7. 0.13 m
9. 63%
11. (a) 4.50×10^2 N   **(b)** 53.2 N
13. 4.33×10^6 Pa
15. 29 bricks
17. 0.750 m
19. 1.2×10^5 Pa
21. (a) 3.5×10^6 N   **(b)** 1.2×10^6 N
23. (a) 1.26×10^5 Pa   **(b)** 19.4 m
25. 46.2 mm Hg
27. 31.3 rad/s
29. 1.46 m
31. 3.8×10^5 N
33. 2.1×10^4 N
35. 8.50×10^5 N·m
37. 108 N
39. 59 N
41. 2.2×10^3 kg
43. 2.04×10^{-3} m³
45. 0.20 m
47. 7.6×10^{-2} m
49. 60.3%
51. 4.5×10^{-5} kg/s
53. (a) 7.0×10^{-5} m³/s
(b) 2.5×10^{-4} m/s
55. (a) 1.6×10^{-4} m³/s
(b) 2.0×10^1 m/s
57. (a) 150 Pa
(b) The pressure inside the roof is greater than that outside the roof, so there is a net outward force.
59. 1.92×10^5 N
61. (a) 2.48×10^5 Pa   **(b)** 1.01×10^5 Pa
(c) 0.342 m³/s
63. 38 m/s
65. (a) 14 m/s   **(b)** 0.98 m³/s
67. 21 m/s
69. 7.78 m/s
71. (a) 1.01×10^5 Pa   **(b)** 1.19×10^5 Pa
73. 4.4 Pa

75. (a) 2.8×10^{-5} N   **(b)** 1.0×10^1 m/s
77. 20.6 m
79. 4.89 m
81. 3400 N
83. 1.91 m/s
85. (a) 0.055 m   **(b)** 5.4 J
87. (a) 2.45×10^5 Pa   **(b)** 1.73×10^5 Pa
89. 6.3×10^{-3} kg
91. 1.19×10^5 Pa
93. (a) 32.8 m/s   **(b)** 54.9 m
95. 1.41×10^5 N, straight down

CHAPTER 12

1. 0.2 C°
3. (a) 10.0 °C and 40.0 °C
(b) 283.15 K and 313.15 K
5. −459.67 °F
7. −164 °C
9. 0.084 m
11. 1500 m
13. 8.0×10^{-4}
15. -2.82×10^{-4} m
17. 49 °C
19. 2.0027 s
21. 8.3×10^{-5} m²
23. 0.6
25. 3.1×10^{-3} m³
27. 18
29. 8.9×10^{-8} m³
31. one penny
33. 45 atm
35. 13 200 kg/m³
37. 6.9
39. 690 J/(kg·C°)
41. 940 °C
43. 27.05 °C
45. $230
47. 4.4×10^3 N
49. 6.7×10^2 W
51. 3.9×10^5 J
53. 9.3 kg
55. 3.9×10^{-3} kg
57. 1.85×10^5 J
59. (a) 1.49×10^{14} J   **(b)** 993 homes
61. 0.223
63. 1.0×10^{-3} kg
65. 3.50×10^2 m/s
67. 0 °C
69. 120 °C
71. 2400 Pa
73. 33%
75. 28%
77. 0.10 m
79. 9.49×10^{-3} kg
81. 110 C°
83. 0.33 gal

85. 10 °C
87. 1.9×10^4 J/kg
89. 930 W
91. 0.016 C°
93. 1.3×10^5 J

CHAPTER 13

1. 12 J
3. 1.47×10^{-2} J/(s·m·C°)
5. 2.0×10^{-3} m
7. 17
9. 3.0×10^{-2} kg
11. (a) 130 °C   **(b)** 830 J
(c) 237 °C
13. 107.2 °C
15. 0.74
17. 0.3
19. 14.5 da
21. 532 K
23. 0.70
25. 12
27. 8.0×10^2 J/s
29. 380 K
31. 320 K
33. 732 K
35. (a) 2.0   **(b)** 0.61

CHAPTER 14

1. 1.07×10^{22} kg
3. (a) 294.307 u
(b) 4.887×10^{-25} kg
5. aluminum
7. 11.7
9. 1.0×10^3 kg
11. 925 K
13. 0.550 kg
15. 2.5×10^{21} molecules
17. 882 K
19. 0.93 mol/m³
21. 3.0 kg
23. 0.205
25. 440 K
27. 308 K
29. 1.2×10^4 m/s
31. (a) 46.3 m²/s²   **(b)** 40.1 m²/s²
33. 3.9×10^5 J
35. 9400 m
37. 4.0×10^1 Pa
39. 11 s
41. 1.34×10^{-7} kg
43. (a) 2.1 s   **(b)** 1.6×10^{-5} s
45. 2.3×10^6 s
47. 2.3×10^{-2} mol
49. (a) 5.00×10^{-13} kg/s
(b) 5.8×10^{-3} kg/m³
51. 414 m/s

53. 1.02×10^{20} molecules

55. 2.6×10^{-10} m

57. 6.19×10^5 Pa

CHAPTER 15

1. 32 miles

3. (a) -3700 J
(b) Heat flows out of the gas.

5. (a) -261 J
(b) Work is done on the system.

7. (a) -7.5×10^2 J   (b) $+9.0 \times 10^2$ J

9. (a) 3100 J   (b) Negative

11. 0.24 m

13. 1.2×10^7 Pa

15. The answer is a proof.

17. 4.99×10^{-6}

19. (a) -2.1×10^2 K (b) Decrease

21. 3.17

23. 1.81

25. 18.0

27. (a) -8.00×10^4 J
(b) Heat flows out of the gas.

29. (a) 327 K   (b) 0.132 m^3

31. 2400 J

33. (a) 1.1×10^4 J   (b) 1.8×10^4 J

35. 5/2

37. (a) 60.0%   (b) 40.0%

39. 2.38×10^4 J

41. 0.631

43. (a) 8600 J   (b) 3100 J

45. (a) 5/9   (b) 1/3

47. 3000 J

49. (a) 1260 K   (b) 1.74×10^4 J

51. 0.015 kg

53. (a) 0.360   (b) 1.3×10^{13} J

55. 5.7 C°

57. 284 K

59. 5.86×10^5 J

61. (a) 1050 J   (b) 2.99

63. 1.4

65. (a) 2.0×10^1 (b) 1.5×10^4 J

67. 1.26×10^3 K

69. 6.0×10^1 J/K

71. (a) Reversible   (b) -125 J/K

73. (a) $+541$ J/K
(b) The entropy of the universe
   increases.

75. (a) 0 J   (b) -6.1×10^3 J
(c) 310 K

77. 13 000 J

79. (a) $+3.68 \times 10^3$ J/K
(b) $+1.82 \times 10^4$ J/K
(c) The vaporization process creates
   more disorder.

81. 9.03

83. 3.0×10^5 Pa

85. 279 s

87. 8.49×10^5 Pa

89. 75 K

91. $e = e_1 + e_2 - e_1e_2$

CHAPTER 16

1. 0.083 Hz

3. 0.25 m

5. 0.49 m

7. 78 cm

9. 5.0×10^1 s

11. (a) 1.1 m/s   (b) 6.55 m

13. 64 N

15. 8.68×10^{-3} kg/m

17. 1.2 m/s^2

19. 153 N

21. 3.26×10^{-3} s

23. $y = (0.37 \text{ m}) \sin (2.6\, \pi t + 0.22\, \pi x)$,
   where x and y are in meters and t is
   in seconds.

25. (a) $+x$ direction (b) -0.080 m

27. 2.5 N

29. 1730 m/s

31. 8.19×10^{-2} m

33. (a) 431 m/s   (b) 322 m/s

35. (a) steel, water, air
(b) 0.059 s, 0.339 s

37. 28.8 K

39. tungsten

41. 650 m

43. 8.0×10^2 m/s

45. 57% argon, 43% neon

47. 0.404 m

49. 6.5 W

51. 190 m

53. 1.98%

55. 1.0×10^4 W/m^2

57. 2.6 s

59. 0.316 W/m^2

61. 1000

63. -6.0 dB

65. 9.5 J

67. 0.84 s

69. 2.39 dB

71. 17 m/s

73. 56 m/s

75. 1350 Hz

77. 615 Hz

79. 209 m

81. 25

83. 2.06

85. 786 Hz

87. 1.3

89. (a) 7.87×10^{-3} s
(b) 4.12 wavelengths

91. 860 Hz

93. 120 dB

95. $m_1 = 28.7$ kg, $m_2 = 14.3$ kg

97. 21.2 m/s

99. 76.8 dB

101. 239 m/s

CHAPTER 17

1. The answer is a series of drawings.

3. (a) 2 cm   (b) 1 cm

5. 5.06 m

7. 3.89 m

9. 3.90 m, 1.55 m, 6.25 m

11. (a) 44°   (b) 0.10 m

13. 1.5×10^4 Hz

15. 8.9 m

17. (a) 50 kHz
(b) 90 kHz

19. 8 Hz

21. 8 Hz

23. 1.10×10^2 Hz

25. 0.46 m

27. 3.93×10^{-3} kg/m

29. (a) 1.00 m   (b) 85.0 m/s
(c) 17.0 Hz

31. 0.077 m

33. 20.8° and 53.1°

35. 0.445

37. 0.50 m

39. 602 Hz

41. 1.2 m/s

43. 1.68×10^5 Pa

45. 88 m/s

47. 107 Hz

49. 0.30 m

51. 171 N

53. $f_n = n \left(\dfrac{v}{2L} \right)$, where $n = 1, 2, 3, \ldots$

55. 12 Hz

CHAPTER 18

1. 1.5×10^{13} electrons

3. -1.6 μC

5. (a) $+1.5\, q$ (b) $+4\, q$
(c) $+4\, q$

7. (a) 0.83 N
(b) Attractive, because the spheres
   carry charges of opposite sign.

9. 0.14 N

11. 0.38 N, 49° below the $-x$ axis.

13. 6.8 N

15. (a) 4.56×10^{-8} C
(b) 3.25×10^{-6} kg

17. 2.6×10^{12} electrons

19. 92.0 N/m

21. 3.5×10^{-5} C

23. (a) 15.4°   (b) 0.813 N

25. 1.8 N, due east

27. 0.45 N, due east

29. (a) 3.0 m   (b) 0 N

31. 1.3 m

33. 7.1×10^{-2} m^2

35. (a) Positive, so that the electrostatic force points upward

    (b) 2.53×10^7 protons

37. 35 N/C

39. $q_1 = 0.716\ q$, $q_2 = 0.0895\ q$

41. 0.364

43. 61°

45. 3.25×10^{-8} C

47. (a) 350 N·m^2/C   (b) 460 N·m^2/C

49. (a) 2.3×10^5 N·m^2/C

    (b) 2.3×10^5 N·m^2/C

    (c) 2.3×10^5 N·m^2/C

51. (a) The flux through the face in the x, z plane at $y = 0$ m is -6.0×10^1 N·m^2/C. The flux through the face parallel to the x, z plane at $y = 0.20$ m is $+6.0 \times 10^1$ N·m^2/C. The flux through each of the remaining four faces is zero.

    (b) 0 N·m^2/C

53. The answer is a proof.

55. 120 N

57. $x = +0.71$ m

59. 6.5×10^3 N/C, downward

61. (a) both positive or both negative

    (b) 1.7×10^{-16} C

63. 2.8×10^5 N/C, $-x$ direction

65. -3.3×10^{-6} C

67. 0.37 N

CHAPTER 19

1. 1.1×10^{-20} J

3. 4.7×10^7

5. 9.4×10^7 m/s

7. 7.0×10^1 hp

9. 339 V

11. $+3.6 \times 10^{-9}$ C

13. -4.35×10^{-18} J

15. -4.7×10^{-2} J

17. $-q/\sqrt{2}$

19. 0.0342 m

21. 6.1×10^{-14} m

23. -0.746 J

25. (a) $-2q/3$ (b) $-2q$

27. 18 000 V

29. 3.5×10^4 V

31. 4.2×10^3 V/m, directed from A to B

33. (a) 159 V   (b) 125 V

    (c) 135 V

35. 0.213 J

37. 1.1×10^3 V

39. (a) 1.3×10^{-12} C (b) 8.1×10^6

41. (a) 41 J   (b) 8200 W

43. 52 V

45. 1.1×10^{-2} m

47. 2.77×10^6 m/s

49. 41 V

51. 5.3

53. 2×10^{-8} F

55. $+9.0 \times 10^3$ V

57. 1.0×10^{-4} C

59. The answer is a proof.

CHAPTER 20

1. 0.025 A

3. 0.21 A

5. 22 A

7. 6.2×10^4 J

9. (a) 4.7×10^{13} protons (b) 17 C°

11. 1.64

13. 9.9×10^{-3} m

15. 37.8 °C

17. 189 Ω

19. 70

21. 228 W

23. 8.9 h

25. 1.2 A

27. 250 °C

29. 33 °C

31. 1.77 A

33. (a) 9.0×10^2 W   (b) 1.8×10^3 W

35. $92

37. 2.0 h

39. (a) 145 Ω   (b) 74 V

41. (a) 10.0 V   (b) 5.00 V

43. 32 Ω

45. (a) 28.9 V   (b) 16.7 W

47. (a) 35 Ω   (b) 5.0×10^1 Ω

49. 5.3 Ω

51. (a) 4.57 A   (b) 1450 W

53. 240 Ω

55. (a) 3.6 Ω

    (b) 33 A, breaker will open

57. 3.58×10^{-8} m^2

59. 1.0×10^2 Ω

61. 4.6 Ω

63. 2.2 W

65. (a) 0.750 A   (b) 2.11 A

67. 0.12 Ω

69. 30 bulbs

71. 10.9 V

73. (a) 0.38 A   (b) 2.0×10^1 V

    (c) Point B

75. 33 A

77. 0.75 V, the left end

79. 0.94 V, point D

81. 3.43×10^3 Ω

83. 25 V

85. (a) 30.0 V   (b) 28.1 V

87. 2.0 µF

89. 4.69×10^{-4} C

91. The answer is a proof.

93. 11 V

95. 4.1×10^{-7} F

97. 1.2×10^{-2} s

99. 6.9

101. 1.7 µF

103. 82 Ω

105. -34.6 °C

107. 0.450 Ω

109. 6.00 Ω, 0.545 Ω, 3.67 Ω, 2.75 Ω, 2.20 Ω, 1.50 Ω, 1.33 Ω, 0.833 Ω

111. (a) 1.2 Ω   (b) 110 V

113. 140 °C

115. C_0

CHAPTER 21

1. 8.1×10^{-5} T

3. (a) 6.50×10^{-2} T

    (b) 9.10×10^3 m/s

5. 4.1×10^{-3} m/s

7. 19.7°

9. (a) Due south

    (b) 2.55×10^{14} m/s^2

11. $+2e$

13. (a) 7.2×10^6 m/s (b) 3.5×10^{-13} N

15. 1.5×10^{-8} s

17. 0.0904 m

19. (a) $\theta = 0°$ (b) 0.29 m

21. 0.13 T

23. 140 V/m, directed toward the bottom of the page

25. 9.6×10^4 m/s

27. 8.1 N

29. 0.96 N (top and bottom sides), 0 N (left and right sides)

31. 57.6°

33. (a) Left-to-right   (b) 1.1×10^{-2} m

35. 14 A

37. (a) 24 A·m^2 (b) 4.8 N·m

39. (a) 13.4 A·m^2 (b) 24.1 N·m

41. (a) 170 N·m   (b) Increase

43. 2.6 N

45. 9.3×10^{-24} A·m^2

47. 8.0×10^{-5} T

49. 1.3×10^{-2} T

51. 190 A

53. (a) 4.3×10^{-5} T (b) 5.3×10^{-5} T

55. 6.8 A, opposite to the direction of the current in the inner coil

57. 1.04×10^{-2} T

59. I_3 is directed out of the paper, $I_3/I = 2$

61. (a) 1.1×10^{-5} T   (b) 4.4×10^{-6} T

63. The answer is a proof.
65. 1.9×10^{-4} N·m
67. 8.2×10^{-27} N·m
69. 75.1° and 105°
71. 0.0152 m
73. 8.7×10^{-3} s
75. 1.3×10^{-10} N, directed out of the page
77. 1.2×10^{-5} A·m²

CHAPTER 22

1. 150 m/s
3. 0.065 V
5. (Rod *A*) emf = 0 V, (Rod *B*) emf = 1.6 V and end 2 is positive, (Rod *C*) emf = 0 V
7. (a) 3.3 m/s   (b) 4.6 N
9. (a) 0.23 kg   (b) −1.8 J
    (c) 1.8 J
11. 70.5°
13. (a) 1.2×10^{-4} Wb
    (b) 3.2×10^{-4} Wb
    (c) 2.0×10^{-4} Wb
15. (Two triangular ends) 0 Wb, (Bottom surface) 0 Wb, (1.2 m × 0.30 m surface) 0.090 Wb, (1.2 m × 0.50 m surface) 0.090 Wb
17. (a) 0.65 V   (b) 0.11 A
19. 8.6×10^{-5} T
21. 4.8 s
23. 0.43 V
25. 0.459 T
27. 2100 rad/s
29. (Figure 22.1*b*) right to left, (Figure 22.1*c*) left to right
31. (a) clockwise   (b) clockwise
33. no induced current
35. (a) 2.4 Hz   (b) 15 rad/s
    (c) 0.62 T
37. 0.11 T
39. 0.150 m
41. 102 V
43. (a) 9.59 Ω   (b) 95 V
    (c) 8.9 A
45. 220 turns
47. 1.4 V
49. 1.5×10^9 J
51. 2.80×10^{-4} H
53. 1.6×10^{-5} A
55. 1/12, step down
57. 7.7 W
59. (a) 7.0×10^5 W   (b) 7.0×10^1 W
61. The answer is a proof.
63. 4.0×10^5 turns
65. (a) left to right   (b) right to left
67. 12 V
69. 0.045 V

71. (a) 3.6×10^{-3} V
    (b) 2.0×10^{-3} m²/s, shrunk
73. (a) 0.80 A   (b) 8.00 A
    (c) 4.40 A

CHAPTER 23

1. 126 Hz
3. 12 Ω
5. (a) 2.00 μF   (b) 0.77 A
7. 3/2
9. 8.0×10^1 Hz
11. 0.44 A
13. 0.17 V
15. 83.9 V
17. 0.819
19. (a) 0.925 A   (b) 31.8°
21. 270 Hz
23. (a) 51.8 V   (b) −4.21 A
25. 0.651 W
27. (a) 352 Hz   (b) 15.5 A
29. 3.1 kHz
31. (a) 1.3×10^{-3} H   (b) 8.7×10^{-6} F
33. 8
35. (a) $4/\sqrt{3}$ (b) $1/\sqrt{3}$
37. 5.00×10^{-2} s
39. (a) 0.50 A   (b) 0.34 A
    (c) yes, 0.704 H
41. 176 mH
43. (a) 2.94×10^{-3} H
    (b) 4.84 Ω
    (c) 0.163
45. 3.11×10^3 Hz and 7.50×10^3 Hz

CHAPTER 24

1. 4.1×10^{16} m
3. 2.19×10^{-11} F
5. The answers are in graphical form.
7. 11.118 m
9. 1.25 m
11. 3.7×10^4 wavelengths
13. 1.5×10^{10} Hz
15. 1.3×10^6 m
17. 540 rev/s
19. 8.75×10^5 times
21. (a) 6.81×10^5 N/C
    (b) 2.27×10^{-3} T
23. 0.07 N/C
25. (a) 183 N/C   (b) 6.10×10^{-7} T
27. 4600 W
29. 920 W
31. (a) receding   (b) 1.8×10^6 m/s
33. (a) 6.175×10^{14} Hz
    (b) 6.159×10^{14} Hz
35. (a) 0.82   (b) 0.18
37. 62%
39. 20 analyzers

41. 1.4×10^{17} Hz
43. 0.015 m
45. 602 W/m²
47. 71.6°
49. (a) 2.4×10^9 Hz   (b) 0.063 m
51. 3.93×10^{26} W
53. (a) 5.30 N/C   (b) 1.77×10^{-8} T

CHAPTER 25

1. 55°
3. 14°
5. 10°
7. (a) 30°   (b) 30°
9. 1.73
11. (a) Image distance is 3.0×10^1 cm behind the mirror.
    (b) Image height is 5.0 cm.
13. (a) Image distance is 16.7 cm behind the mirror.
    (b) Image height is 6.67 cm.
15. 10.9 cm
17. +74 cm
19. (a) 2.0×10^2 cm   (b) −6.3 cm
    (c) Upside down
21. (a) +62 cm   (b) +0.35
    (c) Upright   (d) Smaller
23. (a) *R*   (b) −1
    (c) Inverted
25. −2
27. 80.0 cm, toward the mirror
29. (a) The answer is a proof.
    (b) The answer is a proof.
31. (a) Convex   (b) 24.0 cm
33. d_i = +31 cm
35. d_o = +22 cm
37. 0.533 m
39. +42.0 cm
41. 33.7°

CHAPTER 26

1. 2.00×10^8 m/s
3. 1.198
5. 2.0×10^{-11} s
7. 1.40
9. 2.46×10^8 m/s
11. 1.92×10^8 m/s
13. 1.64
15. 1.19 mm
17. The answer is a derivation.
19. 2.1 cm
21. 3.23 cm
23. 1.54
25. (a) 1.50   (b) 1.27
27. 5.1 m
29. 1.35
31. 1.52

33. (a) 53.12°   **(b)** 52.62°
35. 25.0°
37. The answer is a proof.
39. 0.35°
41. (Red) 44.6°, (Violet) 45.9°
43. (Red) 52.7°, (Violet) 56.2°
45. 61.1 mm
47. (a) −15 cm   **(b)** virtual
49. (a) $d_i = -75$ cm, $m = 2.5$
    (b) $d_i = -75.0$ cm, $m = 2.50$
51. (a) -7.90×10^{-3} m
    (b) -3.44×10^{-2} m
53. 48 cm
55. (a) 4.52×10^{-4} m
    (b) 6.12×10^{-2} m
57. +35 cm and +90.5 cm
59. (a) 4.00 cm to the left of the
        diverging lens
    (b) −0.167   **(c)** virtual
    (d) inverted   **(e)** smaller
61. 0.13 m to the right of the second lens
63. (a) 19.6 cm   **(b)** 0.87 cm
65. (a) 18.1 cm   **(b)** real
    (c) inverted
67. (a) 46.0 cm   **(b)** 43.0 cm
69. (Right eye) −0.20 diopters,
    (Left eye) −0.15 diopters
71. −9.2 cm
73. 26.9 cm
75. (a) −4.5 m   **(b)** 0.50 m
77. 3.7
79. 0.13 m
81. 6.3 cm
83. 4.8×10^{-3} rad
85. 0.81 cm
87. 0.435 cm
89. 0.261 cm
91. 1.1 m
93. (a) the 1.3-diopter lens
    (b) 0.86 m
    (c) −8.5
95. (a) −194
    (b) -7.8×10^{-5} m
    (c) 1.94×10^{6} m
97. $d_i = 18$ cm
99. 1.45
101. −220 cm
103. 2.8
105. (a) 43°   **(b)** 31°
107. (a) converging   **(b)** farsighted
    (c) 96.3 cm from the eyes
109. (a) 6.74×10^{-7} m²
    (b) 7.86×10^{5} W/m²
111. (a) converging   **(b)** $d_o = 2f$
    (c) $d_i = 2f$
113. (a) 11.8 cm   **(b)** 47.8 cm
115. −181
117. (a) 31.3 cm   **(b)** 2.43 m

CHAPTER 27

1. Constructive
3. 16
5. 660 nm
7. 0.0309 m
9. 487 nm
11. 102 nm
13. 198
15. 115 nm
17. 440 nm
19. (a) 0.21°   **(b)** 22°
21. (a) 24°   **(b)** 39°
23. 490 nm
25. 0.447
27. 0.013
29. (a) 220 m   **(b)** No
31. 0.0254 m
33. (a) 5.6 m   **(b)** 56 m
35. 3.2×10^{3} m
37. (a) 37°   **(b)** 22°
39. 5.90×10^{-7} m
41. 4.0×10^{-6} m
43. (a) 7.9° (violet), 13° (red)
    (b) 16° (violet), 26° (red)
    (c) 24° (violet), 41° (red)
    (d) The second and third orders overlap.
45. (a) 2
    (b) $m_B = 4$ and $m_A = 2$; $m_B = 6$ and
        $m_A = 3$
47. 6.0×10^{-5} m
49. 207 nm
51. 1.0×10^{4} m
53. (a) 1.22λ   **(b)** Shorter
55. 3/4
57. 1.95 m

CHAPTER 28

1. 2.4×10^{8} m/s
3. 0.15 rad/s
5. 5.57 s
7. 16
9. 530 m
11. 1.3
13. 6.0 light years
15. 3.0 m × 1.3 m
17. 2.83×10^{8} m/s
19. 1.80×10^{8} m/s
21. −2.0 m/s
23. 5.0×10^{-13} J
25. 1.1 kg
27. (a) 1.0   **(b)** 6.6
29. The answer is a proof.
31. 0.31c

33. 42 m
35. (a) 0.940c
    (b) 1.8×10^{-16} kg·m/s
37. 2.60×10^{8} m/s
39. (a) 5.1×10^{15} kg·m/s
    (b) 9.7×10^{15} kg·m/s
41. −0.406c
43. 8.59

CHAPTER 29

1. 310 nm
3. 7.7×10^{29} photons/s
5. 6.3 eV
7. 138
9. 73 photons/s
11. (a) 2.1×10^{24} photons
    (b) 32 molecules/photon
13. 5.1×10^{-33} kg·m/s
15. (a) 2.124×10^{-24} kg·m/s
    (b) 2.096×10^{-24} kg·m/s
17. 4.692×10^{-24} kg·m/s
19. 9.50×10^{-17} m
21. 6.6×10^{-27} kg
23. (a) 4.50×10^{-36} m/s
    (b) 7.05×10^{27} years
25. 7.38×10^{-11} m
27. 1.2×10^{-36} m
29. 1.5×10^{4} V
31. 8.0×10^{-6} m/s
33. 1.9×10^{-20} kg·m/s
35. $-0.0289° \leq \theta \leq +0.0289°$
37. 2.6×10^{-28} m
39. 7.77×10^{-13} J
41. 1.9
43. 1.10×10^{3} m/s
45. (a) 1.0×10^{-8} N   **(b)** 5.0×10^{-9} N

CHAPTER 30

1. (a) 6.2×10^{-31} m³
    (b) 4×10^{-45} m³
    (c) 7×10^{-13} %
3. 1.5×10^{14}
5. -8.7×10^{6} eV
7. 4.41×10^{-10} m
9. (a) 7458 nm   **(b)** 2279 nm
    (c) infrared region
11. 1.98×10^{-19} J
13. −13.6 eV, −3.40 eV, −1.51 eV
15. The answer is a proof.
17. $6 \leq n_i \leq 19$
19. 2180 lines/cm
21. −0.378 eV
23. 2, 3, 4, 5
25. $\pm 3.16 \times 10^{-34}$ J·s,
    $\pm 2.11 \times 10^{-34}$ J·s,
    $\pm 1.05 \times 10^{-34}$ J·s, 0 J·s

27. $26.6°$

29. $1s^2\, 2s^2\, 2p^6\, 3s^2\, 3p^6\, 4s^2\, 3d^{10}\, 4p^3$

31. (a) not permitted   (b) permitted
    (c) not permitted   (d) permitted
    (e) not permitted

33. 7.230×10^{-11} m

35. 1.11×10^{-10} m

37. 21 600 V

39. 1.03×10^{-18} J

41. 1.9×10^{17} photons

43. 21 days

45. 91.2 nm

47.

n	ℓ	m_ℓ	m_s
4	3	3	1/2
4	3	3	−1/2
4	3	2	1/2
4	3	2	−1/2
4	3	1	1/2
4	3	1	−1/2
4	3	0	1/2
4	3	0	−1/2
4	3	−1	1/2
4	3	−1	−1/2
4	3	−2	1/2
4	3	−2	−1/2
4	3	−3	1/2
4	3	−3	−1/2

49. 3

51. -1.51 eV, -0.850 eV, -0.544 eV

53. $n_i = 6$ and $n_f = 2$

CHAPTER 31

1. (a) X = Pt, 117 neutrons
    (b) X = S, 16 neutrons

(c) X = Cu, 34 neutrons

(d) X = B, 6 neutrons

(e) X = Pu, 145 neutrons

3. 8

5. 35.2

7. 1.14

9. 9.4×10^3 m

11. 128 MeV

13. 2.9×10^{16} kg

15. 1.003 27 u

17. (a) $^{212}_{84}\text{Po} \rightarrow\; ^{208}_{82}\text{Pb} +\; ^{4}_{2}\text{He}$
    (b) $^{232}_{92}\text{U} \rightarrow\; ^{228}_{90}\text{Th} +\; ^{4}_{2}\text{He}$

19. (a) $^{14}_{6}\text{C} \rightarrow\; ^{14}_{7}\text{N} +\; ^{0}_{-1}\text{e}$
    (b) $^{212}_{82}\text{Pb} \rightarrow\; ^{212}_{83}\text{Bi} +\; ^{0}_{-1}\text{e}$

21. $^{35}_{16}\text{S} \rightarrow\; ^{35}_{17}\text{Cl} +\; ^{0}_{-1}\text{e}$

23. (a) $^{18}_{9}\text{F} \rightarrow\; ^{18}_{8}\text{O} +\; ^{0}_{+1}\text{e}$
    (b) $^{15}_{8}\text{O} \rightarrow\; ^{15}_{7}\text{N} +\; ^{0}_{+1}\text{e}$

25. 1.59×10^7 m/s

27. 1.82 MeV

29. 3.0 days

31. 8.00 days

33. 379 yr

35. 146 disintegrations/min

37. 3.7×10^{10} Bq

39. 7.23 days

41. 0.70%

43. (a) 3.0×10^4 yr
    (b) 3.3×10^4 yr

45. 6900 yr, maximum error is 900 yr

47. 7.90 MeV per nucleon

49. 19.9

51. (a) β^-
    (b) β^+
    (c) γ
    (d) α

53. 4 782 969 electrons

55. 2.017 330 u

CHAPTER 32

1. 0.26 roentgens

3. 1.6×10^{-3} J

5. 2.8×10^{-3} J

7. 3.01×10^8 rd

9. 4.4×10^{11} s^{-1}

11. oxygen $^{16}_{8}\text{O}$, $Z = 8$, $A = 16$

13. (a) ? = proton $^{1}_{1}\text{H}$
    (b) ? = alpha particle $^{4}_{2}\text{He}$
    (c) ? = lithium $^{6}_{3}\text{Li}$
    (d) ? = nitrogen $^{14}_{7}\text{N}$
    (e) ? = manganese $^{56}_{25}\text{Mn}$

15. 13.6 MeV

17. 2 neutrons

19. 9.0×10^{-4}

21. 41 collisions

23. 2.7×10^6 kg

25. 160 MeV

27. 1200 kg

29. 4.0 MeV

31. 1.1×10^{-4} kg

33. released energy = 24.7 Mev

35. possibility #1 = u, d, s
    possibility #2 = u, d, b
    possibility #3 = u, s, b

37. $\bar{d}, \bar{u},$ and $\bar{u}$

39. (a) 1.9×10^{-20} kg·m/s
    (b) 3.5×10^{-14} m

41. 3.7×10^{-6} J

43. $\gamma +\; ^{17}_{8}\text{O} \rightarrow\; ^{12}_{6}\text{C} +\; ^{4}_{2}\text{He} +\; ^{1}_{0}\text{n}$

45. 1.0 gal

47. 1.8×10^4 kg/s

49. 32 eV

PHOTO CREDITS

CHAPTER 1
Opener: Everett Collection, Inc. *Page 3:* Courtesy Bureau International des Poids et Mesures, France. *Page 4:* Courtesy National Institute of Standards and Measures. *Page 5:* James Marshall/The Stock Market. *Page 6:* David Joel/Tony Stone Images/New York, Inc. *Page 10:* Dennis O'Clair/Tony Stone Images/New York, Inc.

CHAPTER 2
Opener: Phillip Wallick/The Stock Market. *Page 29* (top): Grafton Marshall Smith/The Stock Market. *Page 29* (bottom): M.L. Sinibaldi/The Stock Market. *Page 31:* Terje Rakke/The Image Bank. *Page 32:* Lori Adamski Peek/Tony Stone Images/New York, Inc. *Page 34:* ©Geoff Stunkard. *Page 35:* John Gichigi/Allsport. *Page 36:* John Nicholson/Allsport. *Page 37:* Lou Jones/The Image Bank. *Page 40:* Manfred Danegger/Tony Stone Images/New York, Inc.

CHAPTER 3
Opener: Tim Davis/Tony Stone Images/New York, Inc. *Page 62:* Courtesy NASA. *Page 64:* Michel Hans/Vandystadt/Allsport. *Page 72:* B. Graham/FPG International.

CHAPTER 4
Opener: Warrem Bolster/Tony Stone Images/New York, Inc. *Page 85:* (top): Todd Warshaw/Allsport. *Page 85:* (center): Jean-François Causse/Tony Stone Images/New York, Inc. *Page 85:* (bottom): Joe McBride/Tony Stone Images/New York, Inc. *Page 86:* Brian Bahr/Allsport. *Page 92:* Tim Davis/Tony Stone Images/New York, Inc. *Page 93:* Arthur Tilley/FPG International. *Page 97:* World Perspectives/Tony Stone Images/New York, Inc. *Page 99:* Lester Sloan/Woodfin Camp & Associates. *Page 101:* Courtesy Goodyear Tire & Rubber Co. *Page 103* (top): Brian Bailey/Adventure Photo. *Page 103* (bottom): Frans Lanting/Minden

Pictures, Inc. *Page 106:* Tom Hanson/Gamma Liaison.

CHAPTER 5
Opener: Courtesy Joe Schwartz. *Page 135:* Gray Mortimore/Allsport. *Page 136:* Courtesy Ringling Bros. and Barnum & Bailey Combined Shows, Inc. *Pages 140 and 142:* Courtesy NASA. *Page 141:* Courtesy NASA and Space Telescope Science Institute.

CHAPTER 6
Opener: Joseph McBride/Tony Stone Images/New York, Inc. *Page 155:* David Madison/Duomo Photography, Inc. *Page 157:* Ben Radford/Allsport. Page 163: James D. Wilson/Gamma Liaison. *Page 167:* Gail Shumway/FPG International. *Page 168:* Grafton M. Smith/The Stock Market. *Page 171:* Photo by Dan Feicht. Courtesy Cedar Point. *Page 175:* Ron Chapple/FPG International.

CHAPTER 7
Opener: ©Telegraph Colour Library/FPG International. *Page 188:* ©AP/Wide World Photos. *Page 189* (top left): Ed Bock/The Stock Market. *Page 189* (top right): Mark Gamba/The Stock Market. *Page 189* (bottom): W. Sallaz/The Image Bank. *Page 190:* Terje Rakke/The Image Bank. *Page 194:* David Leah/Tony Stone Images/New York, Inc.

CHAPTER 8
Opener: Globus Brothers/The Stock Market. *Page 215:* ©Roger Ressmeyer/CORBIS. *Page 216:* Alan Thornton/Tony Stone Images/New York, Inc. *Page 217:* Gary Brettnacher/Tony Stone Images/New York, Inc. *Page 219:* Peter Gridley/FPG International.

CHAPTER 9
Opener: Paul Grebliunas/Tony Stone Images/New York, Inc. *Page 241:* Patrick Ingrand/Tony Stone Images/New York, Inc. *Page 247:* ©AP/Wide World Photos. *Page*

252 (top): ©Tony Stone Images/New York, Inc. *Page 252* (bottom): Amwell/Tony Stone Images/New York, Inc. *Page 257:* Dean Conger/Corbis Images. *Page 258:* Glyn Kirk/Tony Stone Images/New York, Inc. *Page 264:* ©Paul Miller.

CHAPTER 10
Opener: Adam Jones/Natural Selection. *Page 276:* Ellis Herwig/Stock, Boston/PNI. *Page 283:* Terje Rakke/The Image Bank. *Page 287:* Robert Mathena/Fundamental Photographs. *Page 289:* Everett Johnson/Leo de Wys, Inc. *Page 290:* Courtesy Ringling Bros. and Barnum & Bailey Combined Shows, Inc.

CHAPTER 11
Opener: Darryl Torckler/Tony Stone Images/New York, Inc. *Page 310:* Daniel J. Cox/Tony Stone Images/New York, Inc. *Page 311:* Jeff Hunter/The Image Bank. *Page 312:* Jim Richardson/Westlight/ Corbis Images. *Page 313:* Richard Megna/Fundamental Photographs. *Page 317:* Phil Prosen/The Image Bank. *Page 320:* Rick Rickman/Duomo Photography, Inc. *Page 321:* Kevin Daly/The Stock Market. *Page 322:* Nicholas Pinturas/Tony Stone Images/New York, Inc. *Page 325:* Reuters/RTV/Archive Photos. *Page 329:* Gary Gladstone/The Image Bank. *Page 332:* Paul McCormick/The Image Bank.

CHAPTER 12
Opener: Ian Shaw/Tony Stone Images/New York, Inc. *Page 349* (top): Courtesy Omega Engineering, Inc. *Page 349* (bottom): Science Photo Library/Photo Researchers. *Page 350* (top): Courtesy Dr. Sumio Uematsu, Johns Hopkins Hospital. *Page 350* (bottom): Courtesy NOAA. *Page 352:* Richard Choy/Peter Arnold, Inc. Page 359: G. Brad Lewis/Tony Stone Images/New York, Inc. *Page 360:* Frank Oberle/Tony Stone Images/New York, Inc. *Page 363:* ©Telegraph Colour Library/FPG International. *Page 367:* Gregg Adams/Tony Stone Images/New York, Inc.

CHAPTER 13

Opener: Johny Johnson/Tony Stone Images/New York, Inc. *Page 382:* Michael Simpson/FPG International. *Page 383:* Doris De Witt/Tony Stone Images/New York, Inc. *Page 384:* Robert Frerck/Tony Stone Images/New York, Inc. *Page 389:* ©Pierre Du Charme, *The Ledger,* Lakeland, Florida. *Page 390:* Timothy O'Keefe/Index Stock. *Page 391* (top): Wolfgang Kaehler/Corbis Images. *Page 391* (bottom): Nigel Dennis-Wildlife Pictures/Peter Arnold, Inc. *Page 394:* Courtesy NASA.

CHAPTER 14

Opener: James H. Carmichael/The Image Bank. *Page 404:* Courtesy Smithsonian Institution. *Page 406:* Andy Washnik. *Page 407:* Courtesy Richard Zare, Stanford University. *Page 408:* David Feetham/Visuals Unlimited. *Page 413:* John Gillmoure/The Stock Market. *Page 419:* Andy Washnik.

CHAPTER 15

Opener: Glen Allison/Tony Stone Images/New York, Inc. *Page 426:* Mort Pechter/The Stock Market. *Page 427:* Gregg Adam/Tony Stone Images/New York, Inc. *Page 436:* Jamie Squire/Allsport. *Page 446:* Reuters/Aaron Mayes/Archive Photos.

CHAPTER 16

Opener: Art Wolfe/Tony Stone Images/New York, Inc. *Page 460:* Bryan F. Peterson/The Stock Market. *Page 466:* Photo by Ron Garrison, courtesy Zoological Society of San Diego. *Page 467:* Merlin D. Tuttle/Bat Conservation International/Photo Researchers. *Page 471:* Renee Lynn/Tony Stone Images/New York, Inc. *Page 476:* Simon Bruty/Tony Stone Images/New York, Inc. *Page 480* (top): Warren Faidley/International Stock Photo. *Page 480* (bottom): Courtesy Kurt Hondl, National Severe Storms Laboratory, Norman, OK. *Page 481* (left): G. Marche/FPG International. *Page 481* (right): ©Howard Sochurek.

CHAPTER 17

Opener: Owen Franken/Corbis Images. *Page 499:* Andy Washnik. *Page 501:*

Courtesy SONY Corporation. *Page 502:* Ellis Herwig/Stock, Boston/ PNI. *Page 504:* DiMaggio/Kalish/The Stock Market. *Page 505:* Richard Megna/Fundamental Photographs.

CHAPTER 18

Opener: Ralph Wetmore/Tony Stone Images/New York, Inc. *Page 524:* ©1968 FP/Fundamental Photographs. *Page 530:* Courtesy Louis Scudiero and J. Thomas Dickinson, Washington University. *Page 533:* David A. Wagner/Phototake. *Page 542:* Paul Souders/Corbis Images.

CHAPTER 19

Opener: Benjamin Rondel/The Stock Market. *Page 559:* Courtesy Christopher Johnson and Robert MacLeod, University of Utah. Page 560: ©AP/Wide World Photos. *Page 563:* Mason Morfit/FPG International. *Page 574:* Kim Taylor/Bruce Coleman, Inc. *Page 575:* Courtesy Hewlett Packard Co.

CHAPTER 20

Opener: Chad Ehlers/Tony Stone Images/New York, Inc. *Page 588:* Michael Orton/Tony Stone Images/New York, Inc. *Page 590:* Dick Reed/The Stock Market. *Page 593:* Courtesy General Electric Appliances. *Page 606:* Mitch Caesar/Tony Stone Images/New York, Inc. *Page 613:* RNHRD NHS Trust/Tony Stone Images/New York, Inc.

CHAPTER 21

Opener: GJLP/CNRI/Phototake. *Page 630:* Yoav Levy/Phototake. *Page 631:* Rob Atkins/The Image Bank. *Page 633:* Christoph Burki/Tony Stone Images/New York, Inc. *Page 637:* Lawrence Berkeley Laboratory/Photo Researchers. *Page 642:* Dennis Budd Gray. *Page 650* (top): Charles Thatcher/Tony Stone Images/New York, Inc. *Page 650* (bottom): Howard Sochurek/The Stock Market. *Page 652:* Jean-Loup Charmet/Science Photo Library/Photo Researchers. *Page 654:* ©Trident Technologies. Reproduced with permission. *Page 656:* Courtesy Deutsche Bundesbahn/Transrapid.

CHAPTER 22

Opener and *Page 678:* ©AP/Wide World Photos. *Page 686:* Michael Melford/The

Image Bank. *Page 694:* Charles Thatcher/Tony Stone Images/New York, Inc.

CHAPTER 23

Opener: Chris Kapolka/Tony Stone Images/New York, Inc. *Page 716:* Roy Morsch/The Stock Market. *Page 718:* Sandy Felsenthal/Corbis Images. *Page 723:* ©Telegraph Colour Library/FPG International. *Page 724:* Courtesy ORGA Card Systems, Inc.

CHAPTER 24

Opener: James Andrew Bareham/Tony Stone Images/New York, Inc. *Page 733:* Lester Lefkowitz/Tony Stone Images/New York, Inc. *Page 736:* Harvey Lloyd/The Stock Market. *Page 737:* Courtesy Astronomical Society of the Pacific. *Page 740:* Courtesy Anglo-Australian Telescope Board. *Page 743:* Tony Stone Images/New York, Inc. *Page 748:* Diane Schiumo/Fundamental Photographs. *Page 750:* Courtesy ViewSonic Corporation.

CHAPTER 25

Opener: Art Wolfe/Tony Stone Images/New York, Inc. *Page 763:* Yoav Levy/Phototake. *Page 765:* Dan McCoy/Rainbow. *Page 767* (top): ©Mastrorillo/The Stock Market. *Page 767* (bottom): Richard Megna/Fundamental Photographs. *Page 769:* Ken Karp. *Page 770:* ©GM Corporation. All rights reserved. *Page 771:* Chad Slattery/Tony Stone Images/New York, Inc. *Page 772:* Alan Schein/The Stock Market. *Page 776:* Jeremy Horner/Tony Stone Images/New York, Inc. *Page 778:* Courtesy Sandia National Laboratories.

CHAPTER 26

Opener: ©Lennart Nilsson, from *A Child Is Born. Page 783:* Richard Megna/Fundamental Photographs. *Page 789:* Courtesy Schott Fiber Optics, Inc., Southbridge, MA. *Page 792:* Masashiro Sano/The Stock Market. *Page 793* (left): Leonard Lessin/Peter Arnold, Inc. *Page 793* (center): Courtesy Johns Hopkins University School of Medicine. *Page 793* (right): Margaret Rose Orthopaedic Hospital/Photo Researchers. *Page 794:* Pete Saloutos/The Stock Market. *Page 796:* David Woodfall/Tony Stone

Images/New York, Inc. *Page 800:* Med. Illus. SBHA/Tony Stone Images/New York, Inc. *Page 801:* Richard T. Nowitz/Phototake. *Page 814:* Courtesy Tasco.

CHAPTER 27

Opener: Frans Lanting/Minden Pictures, Inc. *Page 830:* James Carmichael/The Image Bank. *Page 833:* From *Atlas of Optical Phenomena,* Michael Cagnet, Springer-Verlag, Berlin. *Page 834:* Andy Washnik. *Page 837:* AP/Wide World Photos. *Page 839:* Courtesy Bausch & Lomb. *Page 840:* ©Bruno/The Stock Market. *Page 841:* Courtesy Education Development Center. *Page 844:* From *Atlas of Optical Phenomena,* Michael Cagnet, Springer-Verlag, Berlin. *Pages 846* and *847:* ©Truax/The Image Finders. *Page 849: A Sunday on La Grande Jatte* by Georges Seurat; ©1996. All rights reserved. Reproduced with permission and courtesy of The Art Institute of Chicago. *Page 854* (left): Courtesy Edwin Jones, University of South Carolina. *Page 854* (right): James King-Holmes/OCMS/Science Photo Library/Photo Researchers.

CHAPTER 28

Opener: The Einstein Memorial Sculpture (©Robert Berks, 1978) at the National Academy of Sciences. Photograph by Diana H. Walker, courtesy National Academy of Sciences. *Page 867:* John Darling/Tony Stone Images/New York, Inc. *Page 870:* Courtesy NASA. *Page 875:* Bill Marsh/Photo Researchers. *Page 877* (top): NASA/Mark Marten/Photo Researchers. *Page 877* (bottom): Dr. Leon Golub/Photo Researchers.

CHAPTER 29

Opener: Courtesy NASA. *Page 892:* Peter B. Kaplan/Photo Researchers. *Page 895* (left): Courtesy Olympus America, Inc. *Page 895* (right): EOS D2000 digital image supplied courtesy of Canon U.S.A., Inc. *Page 897:* Courtesy NASA. *Page 899:* Andrew Syred/Tony Stone Images/New York, Inc. *Page 900* (top): From *Physics Review* 73:527 (1948), by Wollan, Shull and Marney. *Page 900* (bottom): Courtesy Edwin Jones, University of South Carolina. *Page 901:* Courtesy Akira Tonomura, J. Endo, T. Matsuda and T. Kawasaki, *Am. J. Phys.* 57(2): 117, February 1989. *Page 908:* Courtesy Central Scientific Co.

CHAPTER 30

Opener: Photo by Allen Kitchen, ©1999 Laser Spectacles, Inc. *Page 915:* Courtesy Bausch & Lomb. *Page 917:* Courtesy A.I.P. Niels Bohr Library, Margrethe Bohr Collection. *Page 920:* Jacques Lieberman/Index Stock. *Page 932* (top):

Randy Duchaine/The Stock Market. *Page 932* (bottom): Hank Morgan /Photo Researchers. *Page 933* (left): Mehau Kulyk/Photo Researchers. *Page 933* (right): ©Photo Researchers. *Page 936* (top): From Tan, E. and Vinciullo, C. Pulsed dye laser treatment of port-wine stains: a review of patients treated in Western Australia. MJA 1996; 6; 333–336. ©Copyright 1996. *The Medical Journal of Australia.* Reproduced with permission. *Page 936* (bottom): Fritz Hoffmann/The Image Works. *Page 939:* ©Holographics North, Inc., and Sprint. Reproduced with permission of Sprint Corporation.

CHAPTER 31

Opener: HCE Larsson and PC Sereno. *Page 952:* Courtesy NASA. *Page 955:* Elscint/Photo Researchers. *Page 960:* Courtesy Elekta Instruments, Inc. *Page 961:* ©Kyodo News International. *Page 964:* Reuters/Rickey Rogers/Archive Photos. *Page 965:* ©Corbis Sygma. *Page 966:* James King-Holmes/Photo Researchers. *Page 969:* ©Omikron/S.S./Photo Researchers. *Page 976:* Ray Nelson/Phototake.

CHAPTER 32

Opener: Norbert Schafer/The Stock Market. *Page 989:* Kay Chernush/The Image Bank. *Page 990:* National Institutes of Health/Photo Researchers.

INDEX

SI UNITS

Quantity	Name of Unit	Symbol	Expression in Terms of Other SI Units	Quantity	Name of Unit	Symbol	Expression in Terms of Other SI Units
Length	meter	m	Base unit	Pressure, stress	pascal	Pa	N/m^2
Mass	kilogram	kg	Base unit	Viscosity	—	—	$Pa \cdot s$
Time	second	s	Base unit	Electric charge	coulomb	C	$A \cdot s$
Electric current	ampere	A	Base unit	Electric field	—	—	N/C
Temperature	kelvin	K	Base unit	Electric potential	volt	V	J/C
Amount of substance	mole	mol	Base unit	Resistance	ohm	Ω	V/A
Velocity	—	—	m/s	Capacitance	farad	F	C/V
Acceleration	—	—	m/s^2	Inductance	henry	H	$V \cdot s/A$
Force	newton	N	$kg \cdot m/s^2$	Magnetic field	tesla	T	$N \cdot s/(C \cdot m)$
Work, energy	joule	J	$N \cdot m$	Magnetic flux	weber	Wb	$T \cdot m^2$
Power	watt	W	J/s	Specific heat capacity	—	—	$J/(kg \cdot K)$ or $J/(kg \cdot C°)$
Impulse, momentum	—	—	$kg \cdot m/s$	Thermal conductivity	—	—	$J/(s \cdot m \cdot K)$ or $J/(s \cdot m \cdot C°)$
Plane angle	radian	rad	m/m	Entropy	—	—	J/K
Angular velocity	—	—	rad/s	Radioactive activity	becquerel	Bq	s^{-1}
Angular acceleration	—	—	rad/s^2	Absorbed dose	gray	Gy	J/kg
Torque	—	—	$N \cdot m$	Exposure	—	—	C/kg
Frequency	hertz	Hz	s^{-1}				
Density	—	—	kg/m^3				

THE GREEK ALPHABET

Alpha	A	α	Iota	I	ι	Rho	P	ρ
Beta	B	β	Kappa	K	κ	Sigma	Σ	σ
Gamma	Γ	γ	Lambda	Λ	λ	Tau	T	τ
Delta	Δ	δ	Mu	M	μ	Upsilon	Υ	υ
Epsilon	E	ϵ	Nu	N	ν	Phi	Φ	ϕ
Zeta	Z	ζ	Xi	Ξ	ξ	Chi	X	χ
Eta	H	η	Omicron	O	o	Psi	Ψ	ψ
Theta	Θ	θ	Pi	Π	π	Omega	Ω	ω